How to Use This Book Effectively

This book has several features that facilitate study and contribute further to understanding:

Examples

- ▶ Numerous annotated solved examples are provided that feature the *solution methodology* presented in Sec. 1.9 and illustrated in Example 1.1. We encourage you to study these examples, including the accompanying comments.
- ▶ Each solved example concludes with a list of the *Skills Developed* in solving the example and a *Quick Quiz* that allows an immediate check of understanding.
- ▶ Less formal examples are given throughout the text. They open with ➡ **FOR EXAMPLE...** and close with ⬅. These examples also should be studied.

Exercises

- ▶ Each chapter has a set of discussion questions under the heading *Exercises: Things engineers think about* that may be done on an individual or small-group basis. They are intended to allow you to gain a deeper understanding of the text material, think critically, and test yourself.
- ▶ A large number of end-of-chapter problems also are provided under the heading *Problems: Developing engineering skills.* The problems are sequenced to coordinate with the subject matter and are listed in increasing order of difficulty. The problems are also classified under headings to expedite the process of selecting review problems to solve. Answers to selected problems are provided on the student companion site that accompanies this book: *www.wiley.com/go/global/moran.*
- ▶ Because one purpose of this book is to help you prepare to use thermodynamics in engineering practice, design considerations related to thermodynamics are included. Every chapter has a set of problems under the heading *Design & open ended problems: exploring engineering practice* that provide brief design experiences to help you develop creativity and engineering judgment. They also provide opportunities to develop your communication skills.

Further Study Aids

- ▶ Each chapter opens with an introduction giving the *engineering context,* stating the *chapter objective*, and listing the *learning objectives.*
- ▶ Each chapter concludes with a *chapter summary and study guide* that provides a point of departure to study for examinations.
- ▶ For easy reference, each chapter also concludes with lists of *Key Engineering Concepts* and *Key Equations.*
- ▶ Key words are listed in the margins and coordinated with the text material at those locations.
- ▶ Key equations are set off by a double horizontal bar, as, for example, Eq. 1.8.
- ▶ *Take Note...* in the margin identifies where we refine our problem-solving methodology, as on p. 9, or introduce conventions such as rounding the temperature 273.15 K to 273 K, as on p. 18.
- ▶ The icon 💻 denotes end-of-chapter problems where the use of appropriate computer software is recommended.
- ▶ For quick reference, conversion factors and important constants are provided on the next page.
- ▶ A list of symbols is provided on the inside back cover and facing page.

Conversion Factors

Mass and Density

1 kg = 2.2046 lb
1 g/cm^3 = 10^3 kg/m^3
1 g/cm^3 = 62.428 lb/ft^3
1 lb = 0.4536 kg
1 lb/ft^3 = 0.016018 g/cm^3
1 lb/ft^3 = 16.018 kg/m^3

Length

1 cm = 0.3937 in.
1 m = 3.2808 ft
1 in. = 2.54 cm
1 ft = 0.3048 m

Velocity

1 km/h = 0.62137 mile/h
1 mile/h = 1.6093 km/h

Volume

1 cm^3 = 0.061024 in.^{3}
1 m^3 = 35.315 ft^3
1 L = 10^{-3} m^3
1 L = 0.0353 ft^3
1 in.^{3} = 16.387 cm^3
1 ft^3 = 0.028317 m^3
1 gal = 0.13368 ft^3
1 gal = 3.7854×10^{-3} m^3

Force

1 N = 1 kg · m/s^2
1 N = 0.22481 lbf
1 lbf = 32.174 lb · ft/s^2
1 lbf = 4.4482 N

Pressure

1 Pa = 1 N/m^2
= 1.4504×10^{-4} lbf/in.^{2}
1 bar = 10^5 N/m^2
1 atm = 1.01325 bar
1 lbf/in.^{2} = 6894.8 Pa
1 lbf/in.^{2} = 144 lbf/ft^2
1 atm = 14.696 lbf/in.^{2}

Energy and Specific Energy

1 J = 1 N · m = 0.73756 ft · lbf
1 kJ = 737.56 ft · lbf
1 kJ = 0.9478 Btu
1 kJ/kg = 0.42992 Btu/lb
1 ft · lbf = 1.35582 J
1 Btu = 778.17 ft · lbf
1 Btu = 1.0551 kJ
1 Btu/lb = 2.326 kJ/kg
1 kcal = 4.1868 kJ

Energy Transfer Rate

1 W = 1 J/s = 3.413 Btu/h
1 kW = 1.341 hp
1 Btu/h = 0.293 W
1 hp = 2545 Btu/h
1 hp = 550 ft · lbf/s
1 hp = 0.7457 kW

Specific Heat

1 kJ/kg · K = 0.238846 Btu/lb · °R
1 kcal/kg · K = 1 Btu/lb · °R
1 Btu/lb · °R = 4.1868 kJ/kg · K

Constants

Universal Gas Constant

$$\bar{R} = \begin{cases} 8.314 \text{ kJ/kmol} \cdot \text{K} \\ 1545 \text{ ft} \cdot \text{lbf/lbmol} \cdot {}^\circ\text{R} \\ 1.986 \text{ Btu/lbmol} \cdot {}^\circ\text{R} \end{cases}$$

Standard Acceleration of Gravity

$$g = \begin{cases} 9.80665 \text{ m/s}^2 \\ 32.174 \text{ ft/s}^2 \end{cases}$$

Standard Atmospheric Pressure

$$1 \text{ atm} = \begin{cases} 1.01325 \text{ bar} \\ 14.696 \text{ lbf/in.}^2 \\ 760 \text{ mm Hg} = 29.92 \text{ in. Hg} \end{cases}$$

Temperature Relations

$T(°R) = 1.8\, T(K)$
$T(°C) = T(K) - 273.15$
$T(°F) = T(°R) - 459.67$

Principles of Engineering Thermodynamics

Preface

Thermodynamics became a formal area of study in the nineteenth century through consideration of the capacity of hot bodies to produce work. Throughout the twentieth century engineering applications of thermodynamics helped pave the way for increased human well being with advances in major areas such as transportation, power generation, and heating/cooling of buildings. In the twenty-first century, thermodynamics will continue to provide concepts and methods essential for addressing critical societal issues.

Twenty-first century issues where thermodynamics will contribute significantly include using fossil fuels more effectively, fostering renewable energy technologies, and developing more fuel-efficient transportation systems. Other critical areas where thermodynamics will play a role are in mitigating global warming, air pollution, and water pollution. Applications in bioengineering, biomedical systems, and nanotechnology are also emerging. This book provides the tools needed by specialists working in all such fields. For non-specialists, the book provides background for making decisions about technology related to thermodynamics—on the job and as informed citizens.

The seventh edition considers many of these new applications while retaining the basic organization and level of the previous editions. Several enhancements to improve student learning have been introduced. Included are new text elements and interior design features that help students to better understand and apply the subject matter. With this edition we continue our effective pedagogy, clear and concise presentations at a level appropriate for college sophomores and juniors, sound developments of the fundamentals, and state-of-the-art engineering applications.

Core Text Features

This text continues to provide the core features that have made the text the global leader in engineering thermodynamics education.

- ▶ Develops an intuitive understanding of thermodynamics by focusing on physical explanations. See, for example, the introduction to the control volume energy balance on p. 155 and the Carnot Corollary discussion on p. 226.
- ▶ Develops basic principles in an engaging, readable way using applications of contemporary interest. See, as examples, *Bio...connections* on pp. 4, 155, 172, *Energy & Environment* on pp. 33, 66, 229, and *Horizons* on pp. 37, 120, 160.
- ▶ Teaches students to model systems and solve engineering problems involving thermodynamics. Our methodology also teaches students to think systematically and helps them reduce errors. See Sec. 1.9 and all solved examples.
- ▶ Provides a substantial selection of problems in every chapter organized to help students develop engineering skills in three modes:
  - ▶ **Conceptual.** See *Exercises: things engineers think about.*
  - ▶ **Skill Building.** See *Problems: developing engineering skills.*
  - ▶ **Design.** See *Design & open ended problems: exploring engineering practice.*

  Many of the conceptual and design problems build on the engaging *Bio...connections, Energy & Environment*, and *Horizons* discussions placed throughout the text.
- ▶ Prepares students to use thermodynamics in engineering practice through
  - ▶ **Sound developments of the application areas.** Included in Chaps. 8–14 are comprehensive developments of power and refrigeration cycles, psychrometrics,

and combustion from which instructors can choose various levels of coverage ranging from short introductions to in-depth studies.

- **Emphasis on engineering design and analysis.** Specific text material on the design process is included in Sec. 1.8: *Engineering Design and Analysis* and Sec. 7.7: *Thermoeconomics.* Each chapter also provides carefully crafted *Design and open ended problems* that allow students to develop an appreciation of engineering practice and to enhance a variety of skills such as creativity, formulating and solving problems, making engineering judgments, and communicating ideas.

- Employs an effective development of the second law of thermodynamics. The text features the *entropy balance* (Chap. 6), recognized globally as the most effective way for students to learn how to apply the second law. Also, the presentation of *exergy analysis* (Chaps. 7 and 13) has become the state-of-the-art model for learning that subject matter.
- Uses software to enhance problem solving for deeper learning. In every chapter, end-of-chapter problems are identified by a special icon to signal the use of computer software. All of these problems can be solved using *Interactive Thermodynamics (IT)*, the interactive software program available with the text. Tutorials on the use of *IT* are given in several places throughout the text, beginning in Sec. 3.7. As in previous editions, the text presentation is carefully structured so instructors wanting to omit the use of software can do so seamlessly.
- Encourages flexibility in units. The text allows an SI or mixed SI/English presentation. Careful use of units and systematic application of unit conversion factors is emphasized throughout the text. A handy set of conversion factors is inserted inside the front cover of the book for easy access.
- Provides many text features that are proven to enhance student learning. We recognize that the first course in thermodynamics covers a lot of material in a short period of time. To help students learn and study the material more effectively, we have included a variety of features to assist them. For easy reference, these features are inserted inside the front cover under the heading "How to Use This Book Effectively".

New in the Seventh Edition

- Each chapter begins with a set of **Learning Objectives.**
- Each solved example concludes with a list of the **Skills Developed** in solving the example that reinforce specific skills related to the learning objectives.
- Each solved example has a follow-up **Quick Quiz** that gives the student an immediate check of understanding of an important skill.
- A set of **Key Equations** has been added at the close of each chapter for easy reference and to assist in studying.
- Engaging new features called **Bio...connections, Energy & Environment,** and **Horizons** have been introduced throughout. They tie the subject matter to contemporary applications in biomedicine, bioengineering, energy resource use, environmental engineering, and emerging technologies.
- **End-of-Chapter problems** in each of three modes: conceptual, skill building, and design have been **substantially refreshed.** Some build on the engaging *Bio...connections, Energy & Environment,* and *Horizons* discussions introduced in this edition. The skill-building problems are classified under headings to assist instructors and students alike in problem selection. These problems range from confidence-building exercises illustrating basic skills to more challenging ones that may

involve several components and require high-order thinking. As in previous editions, a generous collection of problems is provided.

- **New and revised class-tested content** contributes to student learning and instructor effectiveness:
  - The discussion of pressure measurement has been expanded with new content on manometers and barometers in Sec. 1.6.1.
  - A discussion of *net* radiant exchange between two surfaces has been added in Sec. 2.4.2.
  - Example 3.6 considers the measurement of the calorie value of cooking oil.
  - More detailed modeling considerations are provided in Secs. 4.6–4.10 for nozzles and diffusers, turbines, compressors and pumps, heat exchangers, and throttling devices, respectively.
  - Example 4.6 considers the analysis of a pumping system.
  - An *entropy statement of the second law* has been introduced in Sec. 5.2.3. This allows instructors to elect an axiomatic approach to the second law, bypass much of the Chap. 5 detail, and go quickly to the entropy and entropy balance developments of Chap. 6 if they choose.
  - The discussion of entropy and probability has been expanded with new content in Secs. 6.1.3 and 6.8.2 to provide a brief introduction that relates to technologies involving microscopic analysis.
  - By focusing on physical explanations and stressing the relationship of exergy to fossil fuel use, the introduction to exergy analysis in Secs. 7.1–7.5 has been placed in a more digestible, abbreviated form.
  - Ideal gas model *reviews* have been added at key locations in Sec. 6.5 (Table 6.1) and in Sec. 9.1 (Table 9.1).
- In response to users and to student needs, **class-tested changes** have been introduced. Many contribute to a more *just-in-time* presentation.
  - To emphasize the central role of effective modeling in engineering analysis, we introduce the more descriptive term *Engineering Model* in our solution methodology (Sec. 1.9) in lieu of the term *Assumptions.*
  - The introduction to *quasiequilibrium processes* has moved from Chap. 1 to Sec. 2.2.5.
  - The introduction to *thermodynamic cycles* has moved from Chap. 1 to Sec. 2.6.
  - The discussion of *phase* and *pure substance* has moved from Chap. 1 to Sec. 3.1.1.
  - The discussions of the *constant-volume gas thermometer* and *gas scale of temperature* has moved from Chap. 1 to Sec. 5.8.2.
  - The development of the *Clausius integral* has moved from Chap. 6 to Sec. 5.11. This brings all introductory cycle-related second law content into Chap. 5 and allows Chap. 6 to deal exclusively with entropy and the entropy balance.
- The software accompanying the text, *Interactive Thermodynamics: IT,* has been significantly enhanced with an up-to-date graphical user interface and text-editing features, an equation editor for easier entry of mathematical notation, and *Excel* input of *IT* subroutines. Brief tutorials of *IT* are included within the text and the use of *IT* is illustrated within some of the solved examples. For instructors preferring to use *Engineering Equation Solver (EES)*, detailed *EES* solutions to the computer problems are provided on the instructor's companion website: *www.wiley.com/go/global/moran* (available on request), making it easy to choose that software as an alternative.

Supplements

The following supplements are available on request only to instructors:

- ▶ Figures from the text suitable for creating presentations and/or exams.
- ▶ PowerPoint lecture slides.
- ▶ Electronic files of solutions to all computer problems in both *Interactive Thermodynamics (IT)* and *Engineering Equation Solver (EES)* formats.

To request, please send a mail to acadmktg@wiley.com.

▶ *Interactive Thermodynamics (IT) software*. Available online from the Book Companion Site at *www.wiley.com/go/global/moran*.

Ways to Meet Different Course Needs

In recognition of the evolving nature of engineering curricula, and in particular of the diverse ways engineering thermodynamics is presented, the text is structured to meet a variety of course needs. The following table illustrates several possible uses of the text assuming a semester basis (3 credits). Coverage would be adjusted somewhat for courses on a quarter basis depending on credit value. Courses could be taught in the second or third year to engineering students with appropriate background.

Type of course	Intended audience	Chapter coverage
Surveys	Nonmajors	• Principles. Chaps. 1–6. • Applications. Selected topics from Chaps. 8–10 (omit compressible flow in Chap. 9).
	Majors	• Principles. Chaps. 1–6. • Applications. Same as above plus selected topics from Chaps. 12 and 13.
Two-course sequences	Majors	• First course. Chaps. 1–7. (Chap. 7 may be deferred to second course or omitted.) • Second course. Selected topics from Chaps. 8–14 to meet particular course needs.

Acknowledgments

We thank the many users of our previous editions, located at more than 200 universities and colleges in the United States and Canada, and over the globe, who contributed to this revision through their comments and constructive criticism. Special thanks are owed to Stephen McClain, The University of Alabama at Birmingham, and Margaret Drake, The Ohio State University, for contributing problems in this edition.

Others who have contributed to the development of this edition are listed below. We greatly appreciate these contributions.

John Abraham, *Purdue University*
Suresh Aggarwal, *University of Illinois, Chicago*
Gregory A. Buck, *South Dakota School of Mines and Technology*
Seth Butler, *Ohio State University*
Sahraoui Chaieb, *University of Illinois, Urbana Champaign*
Ashley Emery, *University of Washington*
David Erickson, *Cornell University*
Andrei G. Fedorov, *Georgia Institute of Technology*
Donald L. Fenton, *Kansas State University*
Mohamed A. Gaballa, *University of Arizona*
Afshin Ghajar, *Oklahoma State University*
S. Mostafa Ghiaasiaan, *Georgia Institute of Technology*
Sarah Gong, *University of Wisconsin, Milwaukee*
Matthew R. Jones, *Brigham Young University*
A. Murty Kanury, *Oregon State University*
Amir Karimi, *University of Texas, San Antonio*
Andrew Kean, *Cal Poly San Luis Obispo*
Joseph Kmec, *Purdue University*
Wayne B. Krause, *South Dakota School of Mines and Technology*
John R. Lloyd, *Michigan State University*
Edward Lumsdaine, *Michigan Technological University*
Adam Mahaffey, *Ohio State University*
Sandip Mazumder, *Ohio State University*
Donna J. Michalek, *Michigan Technological University*
Sohail Murad, *University of Illinois, Chicago*
Keiko Nomura, *University of California, San Diego*
Laurent Pilon, *University of California, Los Angeles*
R. P. Roy, *Arizona State University, Tempe*
Harry J. Sauer, *University of Missouri, Rolla*
Brian Savilonis, *Worcester Polytechnic Institute*
Enrico Sciubba, *University of Rome (Italy)*
Ahmad K. Sleiti, *University of Central Florida*
Paul Sojka, *Purdue University*
Charles Stanier, *University of Iowa*
George Tsatsaronis, *Technical University of Berlin (Germany)*
Etim Ubong, *Kettering University*
Antonio Valero, *University of Zargoza (Spain)*
Jon Van Gerpen, *University of Idaho*
R. J. Watkins, *United States Naval Academy*
Ebtisam S. Wilkins, *University of New Mexico, Albuquerque*
Nathan Williams, *United States Naval Academy*
Nicholas Zabaras, *Cornell University*

We also recognize the efforts of many individuals in the John Wiley and Sons, Inc., organization who have contributed their talents and energy to this edition. We applaud their professionalism and commitment.

We are extremely gratified by the reception this book has enjoyed, and have worked to make it even more effective in this seventh edition. As always, we welcome your comments, criticisms, and suggestions.

Michael J. Moran
moran.4@osu.edu

Howard N. Shapiro
hshapiro@wayne.edu

Daisie D. Boettner
boettnerd@aol.com

Margaret B. Bailey
margaret.bailey@rit.edu

Contents

Answers to Selected Problems: Visit the student companion site at *www.wiley.com/go/global/moran*.

1

Getting Started

Introductory Concepts and Definitions

Engineering Context

Although aspects of thermodynamics have been studied since ancient times, the formal study of thermodynamics began in the early nineteenth century through consideration of the capacity of hot objects to produce work. Today the scope is much larger. Thermodynamics now provides essential concepts and methods for addressing critical twenty-first-century issues, such as using fossil fuels more effectively, fostering renewable energy technologies, and developing more fuel-efficient means of transportation. Also critical are the related issues of greenhouse gas emissions and air and water pollution.

Thermodynamics is both a branch of science and an engineering specialty. The scientist is normally interested in gaining a fundamental understanding of the physical and chemical behavior of fixed quantities of matter at rest, and uses the principles of thermodynamics to relate the properties of matter. Engineers are generally interested in studying *systems* and how they interact with their *surroundings*. To facilitate this, thermodynamics has been extended to the study of systems through which matter flows, including bioengineering and biomedical systems.

The **objective** of this chapter is to introduce you to some of the fundamental concepts and definitions that are used in our study of engineering thermodynamics. In most instances the introduction is brief, and further elaboration is provided in subsequent chapters.

Learning Objectives

When you complete your study of this chapter you will be able to...

- ✔ demonstrate understanding of several fundamental concepts used throughout the book . . . including closed system, control volume, boundary and surroundings, property, state, process, the distinction between extensive and intensive properties, and equilibrium.
- ✔ apply SI Engineering units, including units for specific volume, pressure, and temperature.
- ✔ work with the Kelvin and Celsius temperature scales.
- ✔ apply the problem-solving methodology used in this book.

1.1 Using Thermodynamics

Engineers use principles drawn from thermodynamics and other engineering sciences, such as fluid mechanics and heat and mass transfer, to analyze and design things intended to meet human needs. The wide realm of application of these principles is suggested by Table 1.1, which lists a few of the areas where engineering thermodynamics is important.

Engineers seek to achieve improved designs and better performance, as measured by factors such as an increase in the output of some desired product, a reduced input of a scarce resource, a reduction in total costs, or a lesser environmental impact. The principles of engineering thermodynamics play an important part in achieving these goals.

1.2 Defining Systems

An important step in any engineering analysis is to describe precisely what is being studied. In mechanics, if the motion of a body is to be determined, normally the first step is to define a *free body* and identify all the forces exerted on it by other bodies. Newton's second law of motion is then applied. In thermodynamics the term *system* is used to identify the subject of the analysis. Once the system is defined and the relevant interactions with other systems are identified, one or more physical laws or relations are applied.

system

The ***system*** is whatever we want to study. It may be as simple as a free body or as complex as an entire chemical refinery. We may want to study a quantity of matter contained within a closed, rigid-walled tank, or we may want to consider something such as a pipeline through which natural gas flows. The composition of the matter inside the system may be fixed or may be changing through chemical or nuclear reactions. The shape or volume of the system being analyzed is not necessarily constant, as when a gas in a cylinder is compressed by a piston or a balloon is inflated.

surroundings

boundary

Everything external to the system is considered to be part of the system's ***surroundings.*** The system is distinguished from its surroundings by a specified ***boundary,*** which may be at rest or in motion. You will see that the interactions between a system and its surroundings, which take place across the boundary, play an important part in engineering thermodynamics.

Two basic kinds of systems are distinguished in this book. These are referred to, respectively, as *closed systems* and *control volumes*. A closed system refers to a fixed quantity of matter, whereas a control volume is a region of space through which mass may flow. The term *control mass* is sometimes used in place of closed system, and the term *open system* is used interchangeably with control volume. When the terms control mass and control volume are used, the system boundary is often referred to as a *control surface*.

1.2.1 Closed Systems

closed system

isolated system

A ***closed system*** is defined when a particular quantity of matter is under study. A closed system always contains the same matter. There can be no transfer of mass across its boundary. A special type of closed system that does not interact in any way with its surroundings is called an ***isolated system.***

Figure 1.1 shows a gas in a piston–cylinder assembly. When the valves are closed, we can consider the gas to be a closed system. The boundary lies just inside the piston and cylinder walls, as shown by the dashed lines on the figure. Since the portion of the boundary between the gas and the piston moves with the piston, the system volume varies. No mass would cross this or any other part of the boundary. If combustion

Table 1.1

Selected Areas of Application of Engineering Thermodynamics

- Aircraft and rocket propulsion
- Alternative energy systems
  - Fuel cells
  - Geothermal systems
  - Magnetohydrodynamic (MHD) converters
  - Ocean thermal, wave, and tidal power generation
  - Solar-activated heating, cooling, and power generation
  - Thermoelectric and thermionic devices
  - Wind turbines
- Automobile engines
- Bioengineering applications
- Biomedical applications
- Combustion systems
- Compressors, pumps
- Cooling of electronic equipment
- Cryogenic systems, gas separation, and liquefaction
- Fossil and nuclear-fueled power stations
- Heating, ventilating, and air-conditioning systems
  - Absorption refrigeration and heat pumps
  - Vapor-compression refrigeration and heat pumps
- Steam and Gas Turbines
  - Power production
  - Propulsion

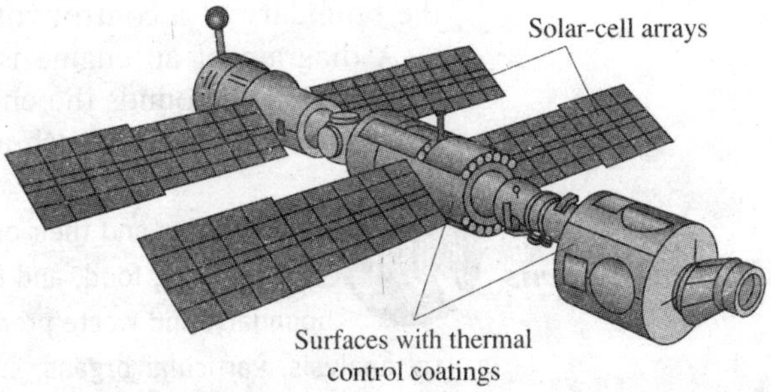

International Space Station

Refrigerator

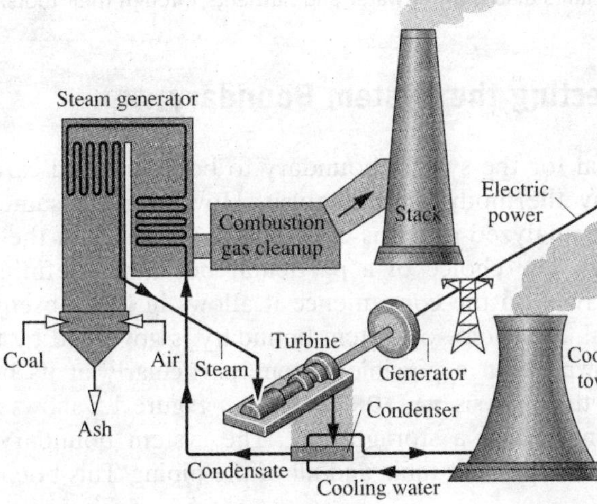

Electrical power plant

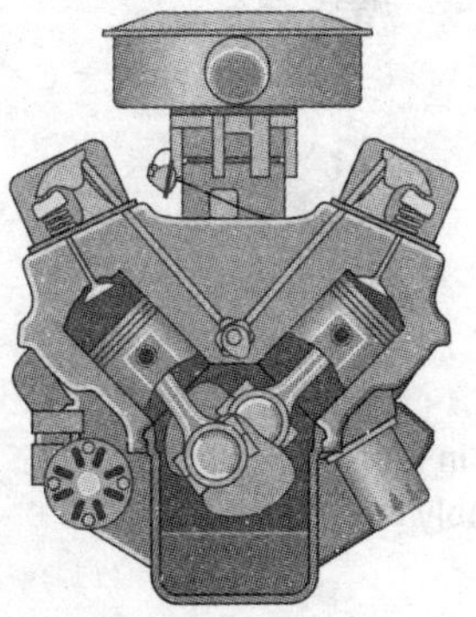
Automobile engine

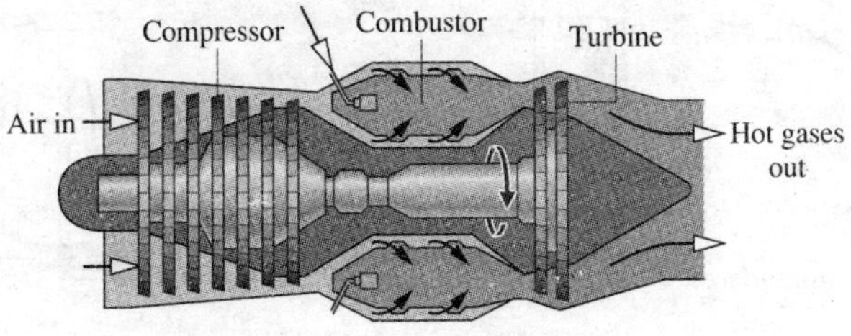

Turbojet engine

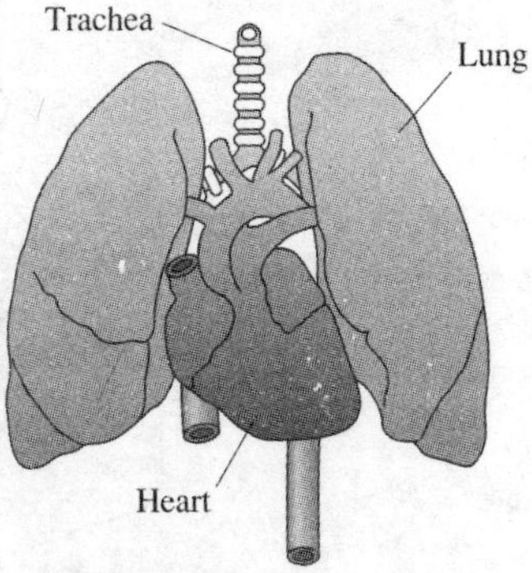

Biomedical applications

occurs, the composition of the system changes as the initial combustible mixture becomes products of combustion.

1.2.2 Control Volumes

In subsequent sections of this book, thermodynamic analyses are made of devices such as turbines and pumps through which mass flows. These analyses can be conducted in principle by studying a particular quantity of matter, a closed system, as it passes through the device. In most cases it is simpler to think instead in terms of a given region of space through which mass flows. With this approach, a *region* within a prescribed boundary is studied. The region is called a ***control volume.*** Mass may cross the boundary of a control volume.

control volume

A diagram of an engine is shown in Fig. 1.2*a*. The dashed line defines a control volume that surrounds the engine. Observe that air, fuel, and exhaust gases cross the boundary. A schematic such as in Fig. 1.2*b* often suffices for engineering analysis.

Bio... connections

Living things and their organs can be studied as control volumes. For the pet shown in Fig. 1.3*a*, air, food, and drink essential to sustain life and for activity enter across the boundary, and waste products exit. A schematic such as Fig. 1.3*b* can suffice for biological analysis. Particular organs, such as the heart, also can be studied as control volumes. As shown in Fig. 1.4, plants can be studied from a control volume viewpoint. Intercepted solar radiation is used in the production of essential chemical substances within plants by *photosynthesis.* During photosynthesis, plants take in carbon dioxide from the atmosphere and discharge oxygen to the atmosphere. Plants also draw in water and nutrients through their roots.

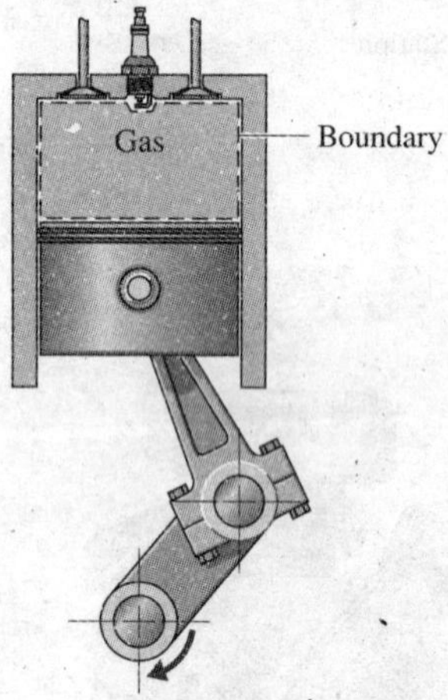

Fig. 1.1 Closed system: A gas in a piston–cylinder assembly.

1.2.3 Selecting the System Boundary

It is essential for the system boundary to be delineated carefully before proceeding with any thermodynamic analysis. However, the same physical phenomena often can be analyzed in terms of alternative choices of the system, boundary, and surroundings. The choice of a particular boundary defining a particular system depends heavily on the convenience it allows in the subsequent analysis.

In general, the choice of system boundary is governed by two considerations: (1) what is known about a possible system, particularly at its boundaries, and (2) the objective of the analysis. ➡ **FOR EXAMPLE...** Figure 1.5 shows a sketch of an air compressor connected to a storage tank. The system boundary shown on the figure encloses the compressor, tank, and all of the piping. This boundary might be selected

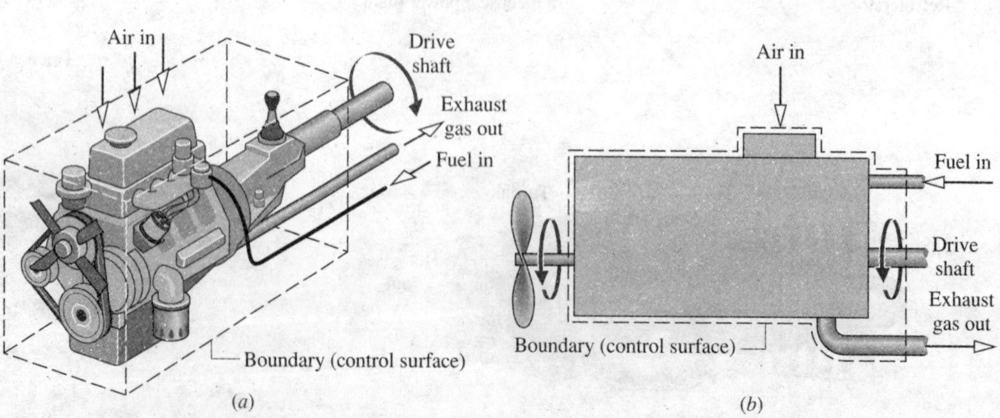

Fig. 1.2 Example of a control volume (open system). An automobile engine.

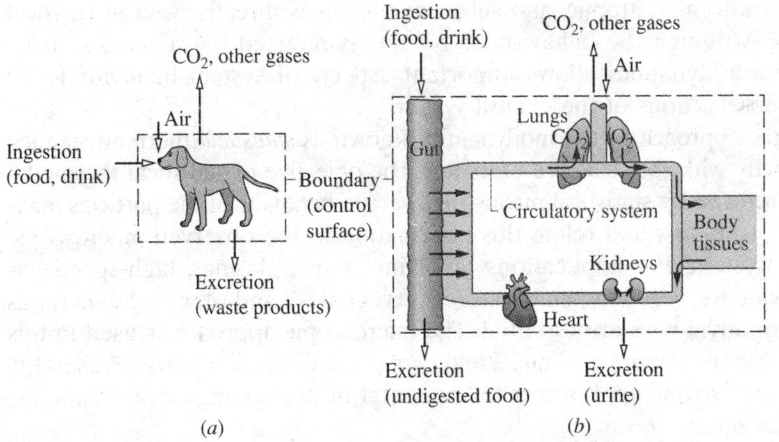

Fig. 1.3 Example of a control volume (open system) in biology.

if the electrical power input is known, and the objective of the analysis is to determine how long the compressor must operate for the pressure in the tank to rise to a specified value. Since mass crosses the boundary, the system would be a control volume. A control volume enclosing only the compressor might be chosen if the condition of the air entering and exiting the compressor is known, and the objective is to determine the electric power input. ◀

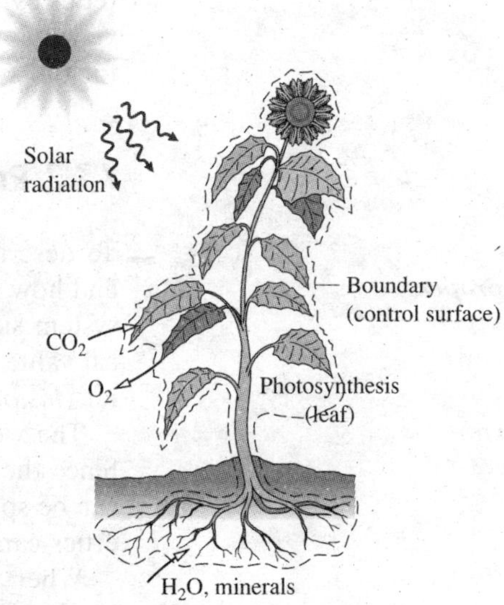

Fig. 1.4 Example of a control volume (open system) in botany.

1.3 Describing Systems and Their Behavior

Engineers are interested in studying systems and how they interact with their surroundings. In this section, we introduce several terms and concepts used to describe systems and how they behave.

1.3.1 Macroscopic and Microscopic Views of Thermodynamics

Systems can be studied from a macroscopic or a microscopic point of view. The macroscopic approach to thermodynamics is concerned with the gross or overall behavior. This is sometimes called *classical* thermodynamics. No model of the structure

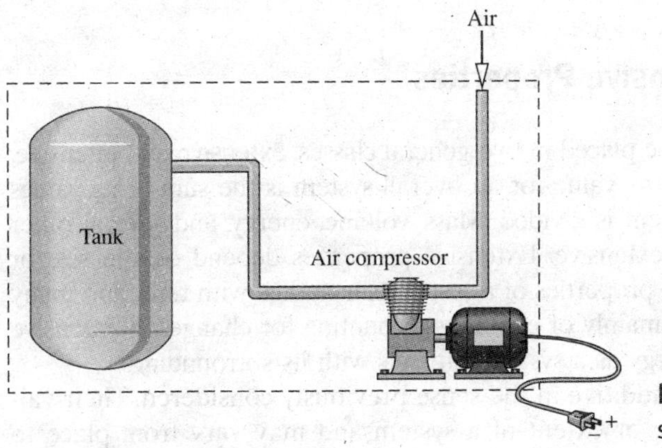

Fig. 1.5 Air compressor and storage tank.

of matter at the molecular, atomic, and subatomic levels is directly used in classical thermodynamics. Although the behavior of systems is affected by molecular structure, classical thermodynamics allows important aspects of system behavior to be evaluated from observations of the overall system.

The microscopic approach to thermodynamics, known as *statistical* thermodynamics, is concerned directly with the structure of matter. The objective of statistical thermodynamics is to characterize by statistical means the average behavior of the particles making up a system of interest and relate this information to the observed macroscopic behavior of the system. For applications involving lasers, plasmas, high-speed gas flows, chemical kinetics, very low temperatures (cryogenics), and others, the methods of statistical thermodynamics are essential. The microscopic approach is used in this text to interpret *internal energy* in Chap. 2 and *entropy* in Chap. 6. Moreover, as noted in Chap. 3, the microscopic approach is instrumental in developing certain data, for example, *ideal gas specific heats*.

For a wide range of engineering applications, classical thermodynamics not only provides a considerably more direct approach for analysis and design but also requires far fewer mathematical complications. For these reasons, the macroscopic viewpoint is the one adopted in this book. Finally, relativity effects are not significant for the systems under consideration in this book.

1.3.2 Property, State, and Process

property

To describe a system and predict its behavior requires knowledge of its properties and how those properties are related. A ***property*** is a macroscopic characteristic of a system such as mass, volume, energy, pressure, and temperature to which a numerical value can be assigned at a given time without knowledge of the previous behavior (*history*) of the system.

state

The word ***state*** refers to the condition of a system as described by its properties. Since there are normally relations among the properties of a system, the state often can be specified by providing the values of a subset of the properties. All other properties can be determined in terms of these few.

process

steady state

When any of the properties of a system change, the state changes and the system is said to have undergone a ***process.*** A process is a transformation from one state to another. However, if a system exhibits the same values of its properties at two different times, it is in the same state at these times. A system is said to be at ***steady state*** if none of its properties changes with time.

Many properties are considered during the course of our study of engineering thermodynamics. Thermodynamics also deals with quantities that are not properties, such as mass flow rates and energy transfers by work and heat. Additional examples of quantities that are not properties are provided in subsequent chapters. For a way to distinguish properties from *non*properties, see the box.

1.3.3 Extensive and Intensive Properties

extensive property

Thermodynamic properties can be placed in two general classes: extensive and intensive. A property is called ***extensive*** if its value for an overall system is the sum of its values for the parts into which the system is divided. Mass, volume, energy, and several other properties introduced later are extensive. Extensive properties depend on the size or extent of a system. The extensive properties of a system can change with time, and many thermodynamic analyses consist mainly of carefully accounting for changes in extensive properties such as mass and energy as a system interacts with its surroundings.

intensive property

Intensive properties are not additive in the sense previously considered. Their values are independent of the size or extent of a system and may vary from place to place within the system at any moment. Thus, intensive properties may be functions

Distinguishing Properties from Nonproperties

At a given state each property has a definite value that can be assigned without knowledge of how the system arrived at that state. Therefore, the change in value of a property as the system is altered from one state to another is determined solely by the two end states and is independent of the particular way the change of state occurred. That is, the change is independent of the details of the process. Conversely, if the value of a quantity is independent of the process between two states, then that quantity is the change in a property. This provides a test for determining whether a quantity is a property: ***A quantity is a property if, and only if, its change in value between two states is independent of the process.*** It follows that if the value of a particular quantity depends on the details of the process, and not solely on the end states, that quantity cannot be a property.

of both position and time, whereas extensive properties can vary only with time. Specific volume (Sec. 1.5), pressure, and temperature are important intensive properties; several other intensive properties are introduced in subsequent chapters.

➡ **FOR EXAMPLE...** to illustrate the difference between extensive and intensive properties, consider an amount of matter that is uniform in temperature, and imagine that it is composed of several parts, as illustrated in Fig. 1.6. The mass of the whole is the sum of the masses of the parts, and the overall volume is the sum of the volumes of the parts. However, the temperature of the whole is not the sum of the temperatures of the parts; it is the same for each part. Mass and volume are extensive, but temperature is intensive. ⬅

1.3.4 Equilibrium

Classical thermodynamics places primary emphasis on equilibrium states and changes from one equilibrium state to another. Thus, the concept of ***equilibrium*** is fundamental. In mechanics, equilibrium means a condition of balance maintained by an equality of opposing forces. In thermodynamics, the concept is more far-reaching, including not only a balance of forces but also a balance of other influences. Each kind of influence refers to a particular aspect of thermodynamic, or complete, equilibrium. Accordingly, several types of equilibrium must exist individually to fulfill the condition of complete equilibrium; among these are mechanical, thermal, phase, and chemical equilibrium.

equilibrium

Criteria for these four types of equilibrium are considered in subsequent discussions. For the present, we may think of testing to see if a system is in thermodynamic equilibrium by the following procedure: Isolate the system from its surroundings and watch for changes in its observable properties. If there are no changes, we conclude that the system was in equilibrium at the moment it was isolated. The system can be said to be at an ***equilibrium state.***

equilibrium state

When a system is isolated, it does not interact with its surroundings; however, its state can change as a consequence of spontaneous events occurring internally as its intensive properties, such as temperature and pressure, tend toward uniform values.

(a) (b)

Fig. 1.6 Figure used to discuss the extensive and intensive property concepts.

When all such changes cease, the system is in equilibrium. At equilibrium, temperature is uniform throughout the system. Also, pressure can be regarded as uniform throughout as long as the effect of gravity is not significant; otherwise a pressure variation can exist, as in a vertical column of liquid.

There is no requirement that a system undergoing a process be in equilibrium *during* the process. Some or all of the intervening states may be nonequilibrium states. For many such processes we are limited to knowing the state before the process occurs and the state after the process is completed.

1.4 Measuring Mass, Length, Time, and Force

When engineering calculations are performed, it is necessary to be concerned with the *units* of the physical quantities involved. A unit is any specified amount of a quantity by comparison with which any other quantity of the same kind is measured. For example, meters, centimeters, and kilometers are all *units of length.* Seconds, minutes, and hours are alternative *time units.*

Because physical quantities are related by definitions and laws, a relatively small number of physical quantities suffice to conceive of and measure all others. These are called *primary dimensions.* The others are measured in terms of the primary dimensions and are called *secondary*. For example, if length and time were regarded as primary, velocity and area would be secondary.

A set of primary dimensions that suffice for applications in *mechanics* are mass, length, and time. Additional primary dimensions are required when additional physical phenomena come under consideration. Temperature is included for thermodynamics, and electric current is introduced for applications involving electricity.

base unit

Once a set of primary dimensions is adopted, a ***base unit*** for each primary dimension is specified. Units for all other quantities are then derived in terms of the base units. Let us illustrate these ideas by considering briefly the system of units: SI units.

1.4.1 SI Units

SI base units

In the present discussion we consider the system of units called SI that takes mass, length, and time as primary dimensions and regards force as secondary. SI is the abbreviation for Système International d'Unités (International System of Units), which is the legally accepted system in most countries. The conventions of the SI are published and controlled by an international treaty organization. The ***SI base units*** for mass, length, and time are listed in Table 1.2 and discussed in the following paragraphs. The SI base unit for temperature is the kelvin, K.

The SI base unit of mass is the kilogram, kg. It is equal to the mass of a particular cylinder of platinum–iridium alloy kept by the International Bureau of Weights

Table 1.2
Units for Mass, Length, Time, and Force

Quantity	SI Unit	SI Symbol
mass	kilogram	kg
length	meter	m
time	second	s
force	newton (= 1 kg · m/s^2)	N

and Measures near Paris. The mass standard for the United States is maintained by the National Institute of Standards and Technology. The kilogram is the only base unit still defined relative to a fabricated object.

The SI base unit of length is the meter (metre), m, defined as the length of the path traveled by light in a vacuum during a specified time interval. The base unit of time is the second, s. The second is defined as the duration of 9,192,631,770 cycles of the radiation associated with a specified transition of the cesium atom.

The SI unit of force, called the newton, is a secondary unit, defined in terms of the base units for mass, length, and time. Newton's second law of motion states that the net force acting on a body is proportional to the product of the mass and the acceleration, written $F \propto ma$. The newton is defined so that the proportionality constant in the expression is equal to unity. That is, Newton's second law is expressed as the equality

$$F = ma \tag{1.1}$$

The newton, N, is the force required to accelerate a mass of 1 kilogram at the rate of 1 meter per second per second. With Eq. 1.1

$$1 \text{ N} = (1 \text{ kg})(1 \text{ m/s}^2) = 1 \text{ kg} \cdot \text{m/s}^2 \tag{1.2}$$

FOR EXAMPLE... to illustrate the use of the SI units introduced thus far, let us determine the weight in newtons of an object whose mass is 1000 kg, at a place on the earth's surface where the acceleration due to gravity equals a *standard* value defined as 9.80665 m/s^2. Recalling that the weight of an object refers to the force of gravity, and is calculated using the mass of the object, m, and the local acceleration of gravity, g, with Eq. 1.1 we get

$$\begin{aligned} F &= mg \\ &= (1000 \text{ kg})(9.80665 \text{ m/s}^2) = 9806.65 \text{ kg} \cdot \text{m/s}^2 \end{aligned}$$

This force can be expressed in terms of the newton by using Eq. 1.2 as a *unit conversion factor*. That is,

$$F = \left(9806.65 \frac{\text{kg} \cdot \text{m}}{\text{s}^2}\right) \left| \frac{1 \text{ N}}{1 \text{ kg} \cdot \text{m/s}^2} \right| = 9806.65 \text{ N}$$

Take Note...

Observe that in the calculation of force in newtons, the unit conversion factor is set off by a pair of vertical lines. This device is used throughout the text to identify unit conversions.

Since weight is calculated in terms of the mass and the local acceleration due to gravity, the weight of an object can vary because of the variation of the acceleration of gravity with location, but its mass remains constant. **FOR EXAMPLE...** if the object considered previously were on the surface of a planet at a point where the acceleration of gravity is one-tenth of the value used in the above calculation, the mass would remain the same but the weight would be one-tenth of the calculated value.

SI units for other physical quantities are also derived in terms of the SI base units. Some of the derived units occur so frequently that they are given special names and symbols, such as the newton. SI units for quantities pertinent to thermodynamics are given as they are introduced in the text. Since it is frequently necessary to work with extremely large or small values when using the SI unit system, a set of standard prefixes is provided in Table 1.3 to simplify matters. For example, km denotes kilometer, that is, 10^3 m.

Table 1.3

SI Unit Prefixes

Factor	Prefix	Symbol
10^{12}	tera	T
10^{9}	giga	G
10^{6}	mega	M
10^{3}	kilo	k
10^{2}	hecto	h
10^{-2}	centi	c
10^{-3}	milli	m
10^{-6}	micro	μ
10^{-9}	nano	n
10^{-12}	pico	p

1.4.2 English Engineering Units

Although SI units are the worldwide standard, at the present time many segments of the engineering community in the United States regularly use some other units. A large portion of America's stock of tools and industrial machines and much valuable engineering data utilize units other than SI units. For many years to come, engineers in the United States will have to be conversant with a variety of units.

English base units

In this section we consider a system of units that is commonly used in the United States, called the English Engineering system. The ***English base units*** for mass, length, and time are listed in Table 1.2 and discussed in the following paragraphs. English units for other quantities pertinent to thermodynamics are given as they are introduced in the text.

The base unit for length is the foot, ft, defined in terms of the meter as

$$1 \text{ ft} = 0.3048 \text{ m} \tag{1.3}$$

The inch, in., is defined in terms of the foot

$$12 \text{ in.} = 1 \text{ ft}$$

One inch equals 2.54 cm. Although units such as the minute and the hour are often used in engineering, it is convenient to select the second as the English Engineering base unit for time.

The English Engineering base unit of mass is the pound mass, lb, defined in terms of the kilogram as

$$1 \text{ lb} = 0.45359237 \text{ kg} \tag{1.4}$$

The symbol lbm also may be used to denote the pound mass.

Once base units have been specified for mass, length, and time in the English Engineering system of units, a force unit can be defined, as for the newton, using Newton's second law written as Eq. 1.1. From this viewpoint, the English unit of force, the pound force, lbf, is the force required to accelerate one pound mass at 32.1740 ft/s^2, which is the standard acceleration of gravity. Substituting values into Eq. 1.1

$$1 \text{ lbf} = (1 \text{ lb})(32.1740 \text{ ft/s}^2) = 32.1740 \text{ lb} \cdot \text{ft/s}^2 \tag{1.5}$$

With this approach force is regarded as *secondary.*

The pound force, lbf, is not equal to the pound mass, lb, introduced previously. Force and mass are fundamentally different, as are their units. The double use of the word "pound" can be confusing, however, and care must be taken to avoid error. ➡ **FOR EXAMPLE...** to show the use of these units in a single calculation, let us determine the weight of an object whose mass is 1000 lb at a location where the local acceleration of gravity is 32.0 ft/s^2. By inserting values into Eq. 1.1 and using Eq. 1.5 as a unit conversion factor, we get

$$F = mg = (1000 \text{ lb})\left(32.0\frac{\text{ft}}{\text{s}^2}\right)\left|\frac{1 \text{ lbf}}{32.1740 \text{ lb} \cdot \text{ft/s}^2}\right| = 994.59 \text{ lbf}$$

This calculation illustrates that the pound force is a unit of force distinct from the pound mass, a unit of mass. ⬅

1.5 Specific Volume

Three measurable intensive properties that are particularly important in engineering thermodynamics are specific volume, pressure, and temperature. Specific volume is considered in this section. Pressure and temperature are considered in Secs. 1.6 and 1.7, respectively.

From the macroscopic perspective, the description of matter is simplified by considering it to be distributed continuously throughout a region. The correctness of this idealization, known as the *continuum* hypothesis, is inferred from the fact that for an extremely large class of phenomena of engineering interest the resulting description of the behavior of matter is in agreement with measured data.

When substances can be treated as continua, it is possible to speak of their intensive thermodynamic properties "at a point." Thus, at any instant the density ρ at a point is defined as

$$\rho = \lim_{V \to V'} \left(\frac{m}{V} \right) \tag{1.6}$$

where V' is the smallest volume for which a definite value of the ratio exists. The volume V' contains enough particles for statistical averages to be significant. It is the smallest volume for which the matter can be considered a continuum and is normally small enough that it can be considered a "point." With density defined by Eq. 1.6, density can be described mathematically as a continuous function of position and time.

The density, or local mass per unit volume, is an intensive property that may vary from point to point within a system. Thus, the mass associated with a particular volume V is determined in principle by integration

$$m = \int_V \rho \, dV \tag{1.7}$$

and *not* simply as the product of density and volume.

specific volume

The ***specific volume*** v is defined as the reciprocal of the density, $v = 1/\rho$. It is the volume per unit mass. Like density, specific volume is an intensive property and may vary from point to point. SI units for density and specific volume are kg/m^3 and m^3/kg, respectively. However, they are also often expressed, respectively, as g/cm^3 and cm^3/g.

molar basis

In certain applications it is convenient to express properties such as specific volume on a molar basis rather than on a mass basis. A mole is an amount of a given substance numerically equal to its molecular weight. In this book we express the amount of substance on a ***molar basis*** in terms of the kilomole (kmol) as appropriate. In each case we use

$$n = \frac{m}{M} \tag{1.8}$$

The number of kilomoles of a substance, n, is obtained by dividing the mass, m, in kilograms by the molecular weight, M, in kg/kmol. When m is in grams, Eq. 1.8 gives n in gram moles, or *mol* for short. Recall from chemistry that the number of molecules in a gram mole, called Avogadro's number, is 6.022×10^{23}. Appendix Table A-1 provides molecular weights for several substances.

To signal that a property is on a molar basis, a bar is used over its symbol. Thus, $\bar{v}$ signifies the volume per kmol, as appropriate. In this text, the unit used for $\bar{v}$ is $m^3/kmol$. With Eq. 1.8, the relationship between $\bar{v}$ and v is

$$\bar{v} = Mv \tag{1.9}$$

where M is the molecular weight in kg/kmol, as appropriate.

1.6 Pressure

Next, we introduce the concept of pressure from the continuum viewpoint. Let us begin by considering a small area A passing through a point in a fluid at rest. The fluid on one side of the area exerts a compressive force on it that is normal to the

Horizons

Big Hopes For Nanotechnology

Nanoscience is the study of molecules and molecular structures, called nanostructures, having one or more dimensions less than about 100 nanometers. One nanometer is one billionth of a meter: $1 \text{ nm} = 10^{-9}$ m. To grasp this level of smallness, a stack of 10 hydrogen atoms would have a height of 1 nm, while a human hair has a diameter about 50,000 nm. *Nanotechnology* is the engineering of nanostructures into useful products. At the nanotechnology scale, behavior may differ from our macroscopic expectations. For example, the *averaging* used to assign property values *at a point* in the continuum model may no longer apply owing to the interactions among the atoms under consideration. Also at these scales, the nature of physical phenomena such as current flow may depend explicitly on the physical size of devices. After a decade of research, nanotechnology is now poised to provide new products with a broad range of uses, including implantable chemotherapy devices, biosensors for glucose detection in diabetics, novel electronic devices, new energy conversion technologies, and 'smart materials', as for example fabrics that allow water vapor to escape while keeping liquid water out.

area, F_{normal}. An equal but oppositely directed force is exerted on the area by the fluid on the other side. For a fluid at rest, no other forces than these act on the area. The ***pressure*** p at the specified point is defined as the limit

pressure

$$p = \lim_{A \to A'} \left(\frac{F_{\text{normal}}}{A} \right) \tag{1.10}$$

where A′ is the area at the "point" in the same limiting sense as used in the definition of density.

If the area A′ was given new orientations by rotating it around the given point, and the pressure determined for each new orientation, it would be found that the pressure at the point is the same in all directions *as long as the fluid is at rest.* This is a consequence of the equilibrium of forces acting on an element of volume surrounding the point. However, the pressure can vary from point to point within a fluid at rest; examples are the variation of atmospheric pressure with elevation and the pressure variation with depth in oceans, lakes, and other bodies of water.

Consider next a fluid in motion. In this case the force exerted on an area passing through a point in the fluid may be resolved into three mutually perpendicular components: one normal to the area and two in the plane of the area. When expressed on a unit area basis, the component normal to the area is called the *normal stress,* and the two components in the plane of the area are termed *shear stresses*. The magnitudes of the stresses generally vary with the orientation of the area. The state of stress in a fluid in motion is a topic that is normally treated thoroughly in *fluid mechanics.* The deviation of a normal stress from the pressure, the normal stress that would exist were the fluid at rest, is typically very small. In this book we assume that the normal stress at a point is equal to the pressure at that point. This assumption yields results of acceptable accuracy for the applications considered. Also, the term pressure, unless stated otherwise, refers to ***absolute pressure:*** pressure with respect to the zero pressure of a complete vacuum.

absolute pressure

1.6.1 Pressure Measurement

Manometers and barometers measure pressure in terms of the length of a column of liquid such as mercury, water, or oil. The manometer shown in Fig. 1.7 has one end open to the atmosphere and the other attached to a tank containing a gas at a uniform pressure. Since pressures at equal elevations in a *continuous* mass of a liquid or gas *at rest* are equal, the pressures at points *a* and *b* of Fig. 1.7 are equal. Applying an elementary force balance, the gas pressure is

$$p = p_{atm} + \rho g L \tag{1.11}$$

where p_{atm} is the local atmospheric pressure, ρ is the density of the manometer liquid, g is the acceleration of gravity, and L is the difference in the liquid levels.

The barometer shown in Fig. 1.8 is formed by a closed tube filled with liquid mercury and a small amount of mercury vapor inverted in an open container of liquid mercury. Since the pressures at points a and b are equal, a force balance gives the atmospheric pressure as

$$p_{atm} = p_{vapor} + \rho_m g L \tag{1.12}$$

where ρ_m is the density of liquid mercury. Because the pressure of the mercury vapor is much less than that of the atmosphere, Eq. 1.12 can be approximated closely as $p_{atm} = \rho_m g L$. For short columns of liquid, ρ and g in Eqs. 1.11 and 1.12 may be taken as constant.

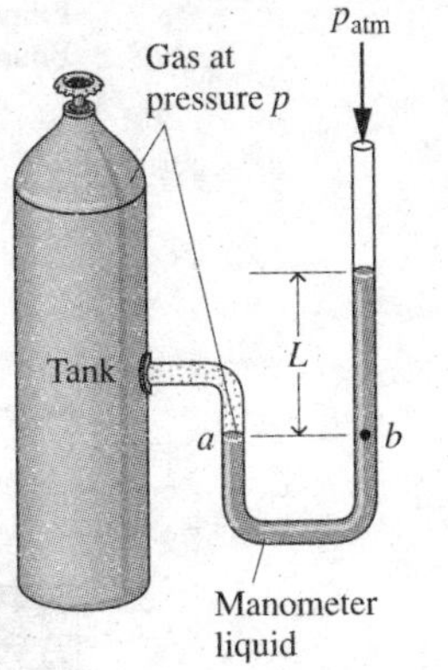

Fig. 1.7 Manometer.

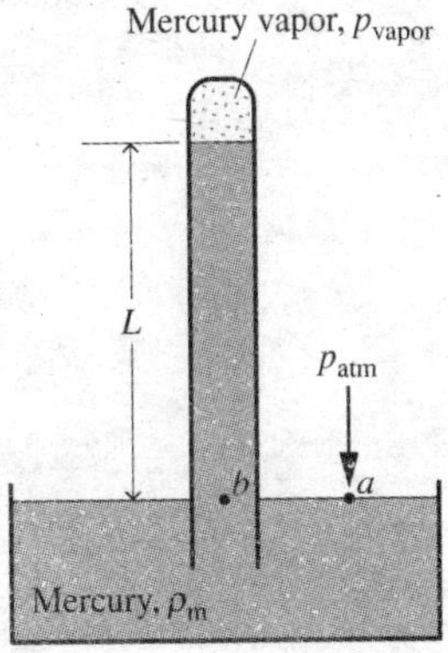

Fig. 1.8 Barometer.

Pressures measured with manometers and barometers are frequently expressed in terms of the length L in millimeters of mercury (mmHg), inches of mercury (inHg), inches of water (inH_2O), and so on. ➡ **FOR EXAMPLE...** a barometer reads 750 mmHg. If $\rho_m = 13.59 \text{ g/cm}^3$ and $g = 9.81 \text{ m/s}^2$, the atmospheric pressure, in N/m^2, is calculated as follows:

$$\begin{aligned} p_{atm} &= \rho_m g L \\ &= \left[\left(13.59 \frac{\text{g}}{\text{cm}^3}\right)\left|\frac{1 \text{ kg}}{10^3 \text{ g}}\right|\left|\frac{10^2 \text{ cm}}{1 \text{ m}}\right|^3\right]\left[9.81 \frac{\text{m}}{\text{s}^2}\right]\left[(750 \text{ mmHg})\left|\frac{1 \text{ m}}{10^3 \text{ mm}}\right|\right]\left|\frac{1 \text{ N}}{1 \text{ kg}\cdot\text{m/s}^2}\right| \\ &= 10^5 \text{ N/m}^2 \end{aligned}$$ ⬅

A Bourdon tube gage is shown in Fig. 1.9. The figure shows a curved tube having an elliptical cross section with one end attached to the pressure to be measured and the other end connected to a pointer by a mechanism. When fluid under pressure fills the tube, the elliptical section tends to become circular, and the tube straightens. This motion is transmitted by the mechanism to the pointer. By calibrating the deflection of the pointer for known pressures, a graduated scale can be determined from which any applied pressure can be read in suitable units. Because of its construction, the Bourdon tube measures the pressure relative to the pressure of the surroundings existing at the instrument. Accordingly, the dial reads zero when the inside and outside of the tube are at the same pressure.

Pressure can be measured by other means as well. An important class of sensors utilize the *piezoelectric* effect: A charge is generated within certain solid materials when they are deformed. This mechanical input/electrical output provides the basis for pressure measurement as well as displacement and force measurements. Another important type of sensor employs a diaphragm that deflects when a force is applied, altering an inductance, resistance, or capacitance. Figure 1.10 shows a piezoelectric pressure sensor together with an automatic data acquisition system.

1.6.2 Pressure Units

The SI unit of pressure and stress is the pascal.

$$1 \text{ pascal} = 1 \text{ N/m}^2$$

However, multiples of the pascal: the kPa, the bar, and the MPa are frequently used.

$$\begin{aligned} 1 \text{ kPa} &= 10^3 \text{ N/m}^2 \\ 1 \text{ bar} &= 10^5 \text{ N/m}^2 \\ 1 \text{ MPa} &= 10^6 \text{ N/m}^2 \end{aligned}$$

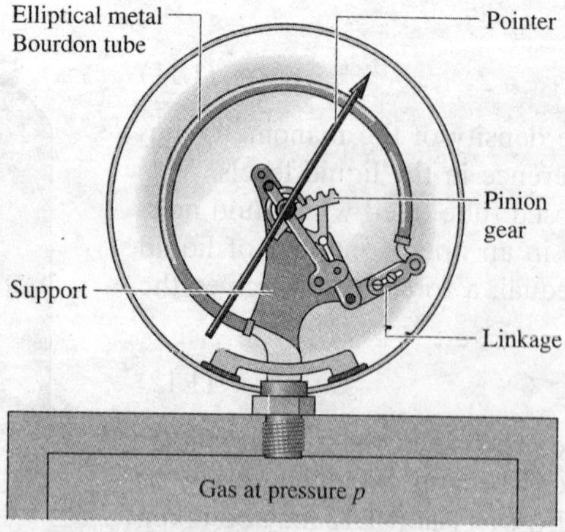

Fig. 1.9 Pressure measurement by a Bourdon tube gage.

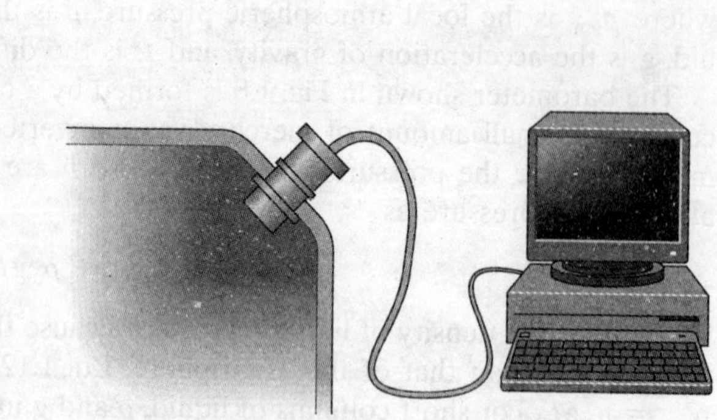

Fig. 1.10 Pressure sensor with automatic data acquisition.

Although atmospheric pressure varies with location on the earth, a standard reference value can be defined and used to express other pressures.

$$1 \text{ standard atmosphere (atm)} = \begin{cases} 1.01325 \times 10^5 \text{ N/m}^2 \\ 760 \text{ mmHg} = 29.92 \text{ inHg} \end{cases} \tag{1.13}$$

Since 1 bar (10^5 N/m^2) closely equals one standard atmosphere, it is a convenient pressure unit despite not being a standard SI unit. When working in SI, the bar, MPa, and kPa are all used in this text.

Although absolute pressures must be used in thermodynamic relations, pressure-measuring devices often indicate the *difference* between the absolute pressure of a system and the absolute pressure of the atmosphere existing outside the measuring device. The magnitude of the difference is called a ***gage pressure*** or a ***vacuum pressure.*** The term gage pressure is applied when the pressure of the system is greater than the local atmospheric pressure, p_{atm}.

gage pressure

vacuum pressure

$$p(\text{gage}) = p(\text{absolute}) - p_{\text{atm}}(\text{absolute}) \tag{1.14}$$

When the local atmospheric pressure is greater than the pressure of the system, the term vacuum pressure is used.

$$p(\text{vacuum}) = p_{\text{atm}}(\text{absolute}) - p(\text{absolute}) \tag{1.15}$$

Engineers in the United States frequently use the letters a and g to distinguish between absolute and gage pressures. The relationship among the various ways of expressing pressure measurements is shown in Fig. 1.11.

Bio... connections

One in three Americans is said to have high blood pressure. Since this can lead to heart disease, strokes, and other serious medical complications, medical practitioners recommend regular blood pressure checks for everyone. Blood pressure measurement aims to determine the maximum pressure (systolic pressure) in an artery when the heart is pumping blood and the minimum pressure (diastolic pressure) when the heart is resting, each pressure expressed in millimeters of mercury, mmHg. The systolic and diastolic pressures of healthy persons should be less than about 120 mmHg and 80 mmHg, respectively.

The standard blood pressure measurement apparatus in use for decades involving an inflatable cuff, mercury manometer, and stethoscope is gradually being replaced because of concerns over mercury toxicity and in response to special requirements, including monitoring during clinical exercise and during anesthesia. Also, for home use and self-monitoring, many patients prefer easy-to-use automated devices that provide digital displays of blood pressure data. This has prompted biomedical engineers to rethink blood pressure measurement and develop new mercury-free and stethoscope-free approaches. One of these uses a highly sensitive pressure transducer to detect pressure oscillations within an inflated cuff placed around the patient's arm. The monitor's software uses these data to calculate the systolic and diastolic pressures, which are displayed digitally.

1.7 Temperature

In this section the intensive property temperature is considered along with means for measuring it. A concept of temperature, like our concept of force, originates with our sense perceptions. Temperature is rooted in the notion of the "hotness" or "coldness" of objects. We use our sense of touch to distinguish hot objects from cold objects and to arrange objects in their order of "hotness," deciding that 1 is hotter than 2, 2 hotter than 3, and so on. But however sensitive human touch may be, we are unable to gauge this quality precisely.

A definition of temperature in terms of concepts that are independently defined or accepted as primitive is difficult to give. However, it is possible to arrive at an objective understanding of *equality* of temperature by using the fact that when the temperature of an object changes, other properties also change.

To illustrate this, consider two copper blocks, and suppose that our senses tell us that one is warmer than the other. If the blocks were brought into contact and isolated from their surroundings, they would interact in a way that can be described as a ***thermal (heat) interaction.*** During this interaction, it would be observed that the volume of the warmer block decreases somewhat with time, while the volume of the colder block

thermal (heat) interaction

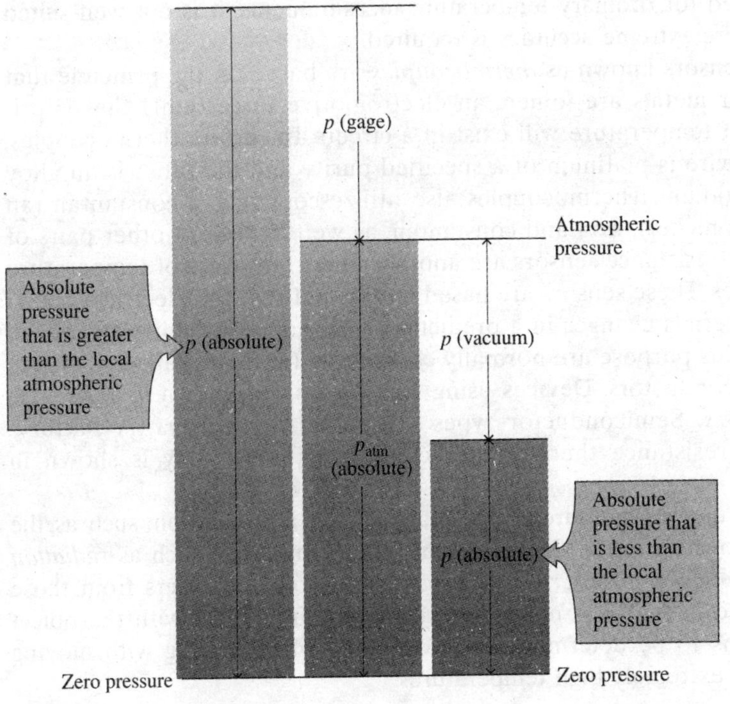

Fig. 1.11 Relationships among the absolute, atmospheric, gage, and vacuum pressures.

increases with time. Eventually, no further changes in volume would be observed, and the blocks would feel equally warm. Similarly, we would be able to observe that the electrical resistance of the warmer block decreases with time, and that of the colder block increases with time; eventually the electrical resistances would become constant also. When all changes in such observable properties cease, the interaction is at an end. The two blocks are then in ***thermal equilibrium.*** Considerations such as these lead us to infer that the blocks have a physical property that determines whether they will be in thermal equilibrium. This property is called ***temperature,*** and we postulate that when the two blocks are in thermal equilibrium, their temperatures are equal.

thermal equilibrium

temperature

It is a matter of experience that when two objects are in thermal equilibrium with a third object, they are in thermal equilibrium with one another. This statement, which is sometimes called the ***zeroth law of thermodynamics,*** is tacitly assumed in every measurement of temperature. Thus, if we want to know if two objects are at the same temperature, it is not necessary to bring them into contact and see whether their observable properties change with time, as described previously. It is necessary only to see if they are individually in thermal equilibrium with a third object. The third object is usually a *thermometer.*

zeroth law of thermodynamics

1.7.1 Thermometers

Any object with at least one measurable property that changes as its temperature changes can be used as a thermometer. Such a property is called a ***thermometric property.*** The particular substance that exhibits changes in the thermometric property is known as a *thermometric* substance.

thermometric property

A familiar device for temperature measurement is the liquid-in-glass thermometer pictured in Fig. 1.12*a*, which consists of a glass capillary tube connected to a bulb filled with a liquid such as alcohol and sealed at the other end. The space above the liquid is occupied by the vapor of the liquid or an inert gas. As temperature increases, the liquid expands in volume and rises in the capillary. The length L of the liquid in the capillary depends on the temperature. Accordingly, the liquid is the thermometric substance and L is the thermometric property. Although this type of thermometer is commonly used for ordinary temperature measurements, it is not well suited for applications where extreme accuracy is required.

More accurate sensors known as *thermocouples* are based on the principle that when two dissimilar metals are joined, an electromotive force (emf) that is primarily a function of temperature will exist in a circuit. In certain thermocouples, one thermocouple wire is platinum of a specified purity and the other is an alloy of platinum and rhodium. Thermocouples also utilize copper and constantan (an alloy of copper and nickel), iron and constantan, as well as several other pairs of materials. Electrical-resistance sensors are another important class of temperature measurement devices. These sensors are based on the fact that the electrical resistance of various materials changes in a predictable manner with temperature. The materials used for this purpose are normally conductors (such as platinum, nickel, or copper) or semiconductors. Devices using conductors are known as *resistance temperature detectors*. Semiconductor types are called *thermistors*. A battery-powered electrical-resistance thermometer commonly used today is shown in Fig. 1.12(*b*).

A variety of instruments measure temperature by sensing radiation, such as the ear thermometer shown in Fig. 1.12(*c*). They are known by terms such as *radiation thermometers* and *optical pyrometers*. This type of thermometer differs from those previously considered in that it is not required to come in contact with the object whose temperature is to be determined, an advantage when dealing with moving objects or objects at extremely high temperatures.

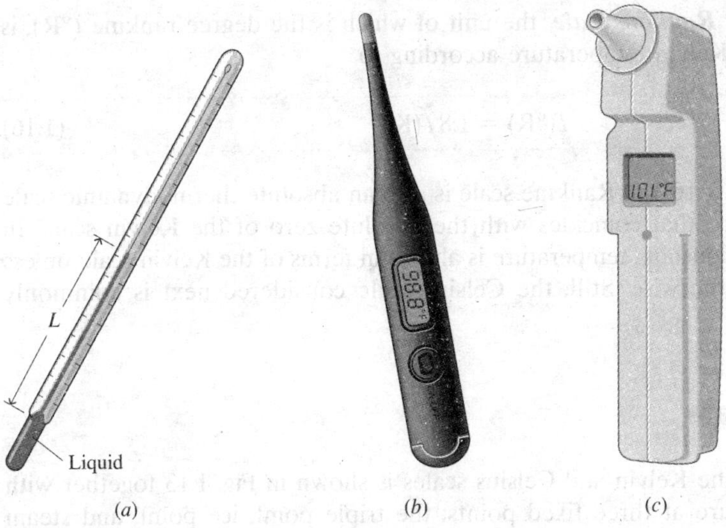

Fig. 1.12 Thermometers. (*a*) Liquid-in-glass. (*b*) Electrical-resistance. (*c*) Infrared-sensing ear thermometer.

Energy & Environment

The mercury-in-glass fever thermometers, once found in nearly every medicine cabinet, are a thing of the past. The *American Academy of Pediatrics* has designated mercury as too toxic to be present in the home. Families are turning to safer alternatives and disposing of mercury thermometers. Proper disposal is an issue, experts say.

The safe disposal of millions of obsolete mercury-filled thermometers has emerged in its own right as an environmental issue. For proper disposal, thermometers must be taken to hazardous-waste collection stations rather than simply thrown in the trash where they can be easily broken, releasing mercury. Loose fragments of broken thermometers and anything that contacted mercury should be transported in closed containers to appropriate disposal sites.

The present generation of liquid-in-glass fever thermometers for home use contains patented liquid mixtures that are nontoxic, safe alternatives to mercury. Other types of thermometers also are used in the home, including battery-powered electrical-resistance thermometers.

1.7.2 Kelvin Temperature Scale

Empirical means of measuring temperature, such as those considered in Sec. 1.7.1, have inherent limitations. ➡ **FOR EXAMPLE...** the tendency of the liquid in a liquid-in-glass thermometer to freeze at low temperatures imposes a lower limit on the range of temperatures that can be measured. At high temperatures liquids vaporize, and therefore these temperatures also cannot be determined by a liquid-in-glass thermometer. Accordingly, several *different* thermometers might be required to cover a wide temperature interval. ⬅

In view of the limitations of empirical means for measuring temperature, it is desirable to have a procedure for assigning temperature values that does not depend on the properties of any particular substance or class of substances. Such a scale is called a *thermodynamic* temperature scale. The ***Kelvin scale*** is an absolute thermodynamic temperature scale that provides a continuous definition of temperature, valid over all ranges of temperature. The unit of temperature on the Kelvin scale is the kelvin (K). The kelvin is the SI base unit for temperature.

Kelvin scale

To develop the Kelvin scale, it is necessary to use the conservation of energy principle and the second law of thermodynamics; therefore, further discussion is deferred to Sec. 5.8 after these principles have been introduced. However, we note here that the Kelvin scale has a zero of 0 K, and lower temperatures than this are not defined.

Rankine scale

By definition, the ***Rankine scale***, the unit of which is the degree rankine (°R), is proportional to the Kelvin temperature according to

$$T(°\text{R}) = 1.8T(\text{K}) \tag{1.16}$$

As evidenced by Eq. 1.16, the Rankine scale is also an absolute thermodynamic scale with an absolute zero that coincides with the absolute zero of the Kelvin scale. In thermodynamic relationships, temperature is always in terms of the Kelvin scale unless specifically stated otherwise. Still, the Celsius scale considered next is commonly encountered.

1.7.3 Celsius Scale

The relationship of the Kelvin and Celsius scales is shown in Fig. 1.13 together with values for temperature at three fixed points: the triple point, ice point, and steam point.

triple point

By international agreement, temperature scales are defined by the numerical value assigned to the easily reproducible ***triple point*** of water: the state of equilibrium between steam, ice, and liquid water (Sec. 3.2). As a matter of convenience, the temperature at this standard fixed point is defined as 273.16 kelvins, abbreviated as 273.16 K. This makes the temperature interval from the *ice point*[1] (273.15 K) to the *steam point*[2] equal to 100 K and thus in agreement over the interval with the Celsius scale, which assigns 100 Celsius degrees to it.

Celsius scale

The ***Celsius temperature scale*** uses the unit degree Celsius (°C), which has the same magnitude as the kelvin. Thus, temperature *differences* are identical on both scales. However, the zero point on the Celsius scale is shifted to 273.15 K, as shown by the following relationship between the Celsius temperature and the Kelvin temperature

$$T(°\text{C}) = T(\text{K}) - 273.15 \tag{1.17}$$

From this it can be concluded that on the Celsius scale the triple point of water is 0.01°C, and that 0 K corresponds to −273.15°C. These values are shown in Fig. 1.13.

Fahrenheit scale

A degree of the same size as that on the Rankine scale is used in the ***Fahrenheit scale***, but the zero point is shifted according to the relation

$$T(°\text{F}) = T(°\text{R}) - 459.67 \tag{1.18}$$

Substituting Eqs. 1.17 and 1.18 into Eq. 1.16, it follows that

$$T(°\text{F}) = 1.8T(°\text{C}) + 32 \tag{1.19}$$

This equation shows that the Fahrenheit temperature of the ice point (0°C) is 32°F and of the steam point (100°C) is 212°F. The 100 Celsius or Kelvin degrees between the ice point and steam point correspond to 180 Fahrenheit or Rankine degrees, as shown in Fig. 1.13.

Take Note...

When making engineering calculations, it's usually okay to round off the last numbers in Eqs. 1.17 and 1.18 to 273 and 460, respectively. This is frequently done in this book.

[1]The state of equilibrium between ice and air-saturated water at a pressure of 1 atm.

[2]The state of equilibrium between steam and liquid water at a pressure of 1 atm.

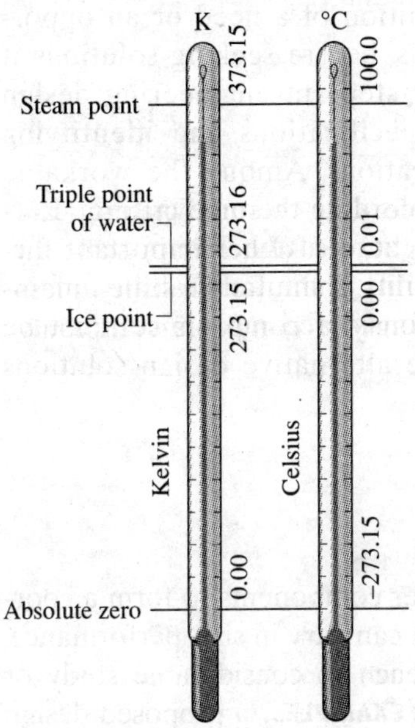

Fig. 1.13 Comparison of temperature scales.

Bio... connections

Cryobiology, the science of life at low temperatures, comprises the study of biological materials and systems (proteins, cells, tissues, and organs) at temperatures ranging from the cryogenic (below about 120 K) to the hypothermic (low body temperature). Applications include freeze-drying pharmaceuticals, cryosurgery for removing unhealthy tissue, study of cold-adaptation of animals and plants, and long-term storage of cells and tissues (called *cryopreservation*).

Cryobiology has challenging engineering aspects owing to the need for refrigerators capable of achieving the low temperatures required by researchers. Freezers to support research requiring cryogenic temperatures in the low-gravity environment of the International Space Station, shown in Table 1.1, are illustrative. Such freezers must be extremely compact and miserly in power use. Further, they must pose no hazards. On-board research requiring a freezer might include the growth of near-perfect protein crystals, important for understanding the structure and function of proteins and, ultimately, in the design of new drugs.

1.8 Engineering Design and Analysis

An important engineering function is to design and analyze things intended to meet human needs. Design and analysis are considered in this section.

1.8.1 Design

Engineering design is a decision-making process in which principles drawn from engineering and other fields such as economics and statistics are applied, usually iteratively, to devise a system, system component, or process. Fundamental elements of design include the establishment of objectives, synthesis, analysis, construction, testing, and evaluation. Designs typically are subject to a variety of ***constraints*** related to economics, safety, environmental impact, and so on.

design constraints

Design projects usually originate from the recognition of a need or an opportunity that is only partially understood initially. Thus, before seeking solutions it is important to define the design objectives. Early steps in engineering design include pinning down quantitative performance specifications and identifying alternative *workable* designs that meet the specifications. Among the workable designs are generally one or more that are "best" according to some criteria: lowest cost, highest efficiency, smallest size, lightest weight, etc. Other important factors in the selection of a final design include reliability, manufacturability, maintainability, and marketplace considerations. Accordingly, a compromise must be sought among competing criteria, and there may be alternative design solutions that are feasible.[3]

1.8.2 Analysis

Design requires synthesis: selecting and putting together components to form a coordinated whole. However, as each individual component can vary in size, performance, cost, and so on, it is generally necessary to subject each to considerable study or analysis before a final selection can be made. ➡ **FOR EXAMPLE...** a proposed design for a fire-protection system might entail an overhead piping network together with numerous sprinkler heads. Once an overall configuration has been determined, detailed engineering analysis would be necessary to specify the number and type of the spray heads, the piping material, and the pipe diameters of the various branches of the network. The analysis must also aim to ensure that all components form a smoothly working whole while meeting relevant cost constraints and applicable codes and standards. ⬅

Engineers frequently do analysis, whether explicitly as part of a design process or for some other purpose. Analyses involving systems of the kind considered in this book use, directly or indirectly, one or more of three basic laws. These laws, which are independent of the particular substance or substances under consideration, are

1. the conservation of mass principle
2. the conservation of energy principle
3. the second law of thermodynamics

In addition, relationships among the properties of the particular substance or substances considered are usually necessary (Chaps. 3, 6, 11–14). Newton's second law of motion (Chaps. 1, 2, 9), relations such as Fourier's conduction model (Chap. 2), and principles of engineering economics (Chap. 7) may also play a part.

The first steps in a thermodynamic analysis are definition of the system and identification of the relevant interactions with the surroundings. Attention then turns to the pertinent physical laws and relationships that allow the behavior of the system to be described in terms of an ***engineering model.*** The objective in modeling is to obtain a simplified representation of system behavior that is sufficiently faithful for the purpose of the analysis, even if many aspects exhibited by the actual system are ignored. For example, idealizations often used in mechanics to simplify an analysis and arrive at a manageable model include the assumptions of point masses, frictionless pulleys, and rigid beams. Satisfactory modeling takes experience and is a part of the *art* of engineering.

engineering model

Engineering analysis is most effective when it is done systematically. This is considered next.

[3]For further discussion, see A. Bejan, G. Tsatsaronis, and M. J. Moran, *Thermal Design and Optimization,* John Wiley & Sons, New York, 1996, Chap. 1.

1.9 Methodology for Solving Thermodynamics Problems

A major goal of this textbook is to help you learn how to solve engineering problems that involve thermodynamic principles. To this end, numerous solved examples and end-of-chapter problems are provided. It is extremely important for you to study the examples *and* solve problems, for mastery of the fundamentals comes only through practice.

To maximize the results of your efforts, it is necessary to develop a systematic approach. You must think carefully about your solutions and avoid the temptation of starting problems *in the middle* by selecting some seemingly appropriate equation, substituting in numbers, and quickly "punching up" a result on your calculator. Such a haphazard problem-solving approach can lead to difficulties as problems become more complicated. Accordingly, we strongly recommend that problem solutions be organized using the *five steps* in the box below, which are employed in the solved examples of this text.

❶ **Known:** State briefly in your own words what is known. This requires that you read the problem carefully *and* think about it.

❷ **Find:** State concisely in your own words what is to be determined.

❸ **Schematic and Given Data:** Draw a sketch of the system to be considered. Decide whether a closed system or control volume is appropriate for the analysis, and then carefully identify the boundary. Label the diagram with relevant information from the problem statement.

Record all property values you are given or anticipate may be required for subsequent calculations. Sketch appropriate property diagrams (see Sec. 3.2), locating key state points and indicating, if possible, the processes executed by the system.

The importance of good sketches of the system and property diagrams cannot be overemphasized. They are often instrumental in enabling you to think clearly about the problem.

❹ **Engineering Model:** To form a record of how you *model* the problem, list all *simplifying assumptions* and *idealizations* made to reduce it to one that is manageable. Sometimes this information also can be noted on the sketches of the previous step. The development of an appropriate model is a key aspect of successful problem solving.

❺ **Analysis:** Using your assumptions and idealizations, reduce the appropriate governing equations and relationships to forms that will produce the desired results.

It is advisable to work with equations as long as possible before substituting numerical data. When the equations are reduced to final forms, consider them to determine what additional data may be required. Identify the tables, charts, or property equations that provide the required values. Additional property diagram sketches may be helpful at this point to clarify states and processes.

When all equations and data are in hand, substitute numerical values into the equations. Carefully check that a consistent and appropriate set of units is being employed. Then perform the needed calculations.

Finally, consider whether the magnitudes of the numerical values are reasonable and the algebraic signs associated with the numerical values are correct.

The problem solution format used in this text is intended to *guide* your thinking, not substitute for it. Accordingly, you are cautioned to avoid the rote application of these five steps, for this alone would provide few benefits. Indeed, as a particular solution evolves you may have to return to an earlier step and revise it in light of a better understanding of the problem. For example, it might be necessary to add or delete an assumption, revise a sketch, determine additional property data, and so on.

The solved examples provided in the book are frequently annotated with various comments intended to assist learning, including commenting on what was learned, identifying key aspects of the solution, and discussing how better results might be obtained by relaxing certain assumptions. Such comments are optional in your solutions.

In some of the earlier examples and end-of-chapter problems, the solution format may seem unnecessary or unwieldy. However, as the problems become more complicated you will see that it reduces errors, saves time, and provides a deeper understanding of the problem at hand.

The example to follow illustrates the use of this solution methodology together with important system concepts introduced previously, including identification of interactions occurring at the boundary.

Example 1.1 USING THE SOLUTION METHODOLOGY AND SYSTEM CONCEPTS

A wind turbine–electric generator is mounted atop a tower. As wind blows steadily across the turbine blades, electricity is generated. The electrical output of the generator is fed to a storage battery.

(a) Considering only the wind turbine–electric generator as the system, identify locations on the system boundary where the system interacts with the surroundings. Describe changes occurring within the system with time.

(b) Repeat for a system that includes only the storage battery.

Solution

Known: A wind turbine–electric generator provides electricity to a storage battery.

Find: For a system consisting of (a) the wind turbine–electric generator and (b) the storage battery, identify locations where the system interacts with its surroundings, and describe changes occurring within the system with time.

Schematic and Given Data:

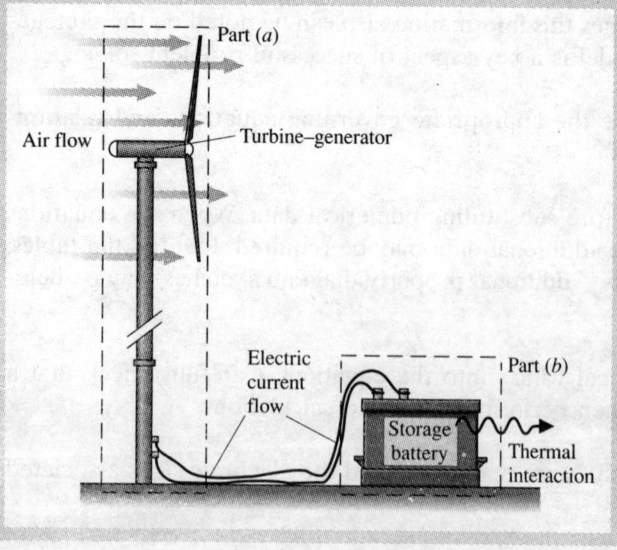

Engineering Model:

1. In part (a), the system is the control volume shown by the dashed line on the figure.
2. In part (b), the system is the closed system shown by the dashed line on the figure.
3. The wind is steady.

Fig. E1.1

Analysis:

(a) In this case, the wind turbine is studied as a control volume with air flowing across the boundary. Another principal interaction between the system and surroundings is the electric current passing through the wires. From the macroscopic perspective, however, such an interaction is not considered a mass transfer. With a steady wind, the turbine–generator is likely to reach steady-state operation, where the rotational speed of the blades is constant and a steady electric current is generated.

❶ **(b)** In this case, the battery is studied as a closed system. The principal interaction between the system and its surroundings is the electric current passing into the battery through the wires. As noted in part (a), this interaction is not considered a mass transfer. As the battery is charged and chemical reactions occur within it, the temperature of the battery surface may become somewhat elevated and a thermal interaction might occur between the battery and its surroundings. This interaction is likely to be of secondary importance. Also, as the battery is charged, the state within changes with time. The battery is not at steady state.

---

❶ Using terms familiar from a previous physics course, the system of part (a) involves the *conversion* of kinetic energy to electricity, whereas the system of part (b) involves energy *storage* within the battery.

Skills Developed: Ability to...

✓ apply the problem-solving methodology used in this book.

✓ define a control volume and identify interactions on its boundary.

✓ define a closed system and identify interactions on its boundary.

✓ distinguish steady-state operation from nonsteady operation.

Quick Quiz

May an *overall* system consisting of the turbine-generator and battery be considered as operating at steady state? Explain.

Ans. No. A system is at steady state only if *none* of its properties change with time.

✓ Chapter Summary and Study Guide

In this chapter, we have introduced some of the fundamental concepts and definitions used in the study of thermodynamics. The principles of thermodynamics are applied by engineers to analyze and design a wide variety of devices intended to meet human needs.

An important aspect of thermodynamic analysis is to identify systems and to describe system behavior in terms of properties and processes. Three important properties discussed in this chapter are specific volume, pressure, and temperature.

In thermodynamics, we consider systems at equilibrium states and systems undergoing processes (changes of state). We study processes during which the intervening states are not equilibrium states and processes during which the departure from equilibrium is negligible.

In this chapter, we have introduced SI units for mass, length, time, force, and temperature.

Chapter 1 concludes with discussions of how thermodynamics is used in engineering design and how to solve thermodynamics problems systematically.

This book has several features that facilitate study and contribute to understanding. For an overview, see *How To Use This Book Effectively* inside the front cover of the book.

The following checklist provides a study guide for this chapter. When your study of the text and the end-of-chapter exercises has been completed, you should be able to

✓ write out the meanings of the terms listed in the margin throughout the chapter and understand each of the related concepts. The subset of key concepts listed below is particularly important in subsequent chapters.

✓ use Si units for mass, length, time, force, and temperature and apply appropriately Newton's second law and Eqs. 1.16–1.19.

✓ work on a molar basis using Eq. 1.8.

✓ identify an appropriate system boundary and describe the interactions between the system and its surroundings.

✓ apply the methodology for problem solving discussed in Sec. 1.9.

Key Engineering Concepts

surroundings *p. 2*
boundary *p. 2*
closed system *p. 2*
control volume *p. 4*
property *p. 6*
state *p. 6*
process *p. 6*
extensive property *p. 6*
intensive property *p. 6*
equilibrium *p. 7*
specific volume *p. 11*
pressure *p. 12*
temperature *p. 16*
Kelvin scale *p. 17*

Key Equations

$n = m/M$	(1.8)	Relation between amounts of matter on a mass basis, m, and on a molar basis, n.
$T(°R) = 1.8\,T(K)$	(1.16)	Relation between the Rankine and Kelvin temperatures.
$T(°C) = T(K) - 273.15$	(1.17)	Relation between the Celsius and Kelvin temperatures.
$T(°F) = T(°R) - 459.67$	(1.18)	Relation between the Fahrenheit and Rankine temperatures.
$T(°F) = 1.8T\,(°C) + 32$	(1.19)	Relation between the Fahrenheit and Celsius temperatures.

Exercises: things engineers think about

1. Consider a leaf blower driven by a gasoline engine as the system. Would it be best analyzed as a closed system or a control volume? Are there any environmental impacts associated with such a leaf blower? Repeat if it were an *electrically-driven* leaf blower.

2. What components would be required for a high school laboratory project aimed at determining the electricity generated by a hamster running on its exercise wheel?

3. Based on the macroscopic view, a quantity of air at 100 kPa, 20°C is in equilibrium. Yet the atoms and molecules of the air are in constant motion. How do you reconcile this apparent contradiction?

4. Air at 1 atm, 20°C in a closed tank adheres to the continuum hypothesis. Yet when sufficient air has been drawn from the tank, the hypothesis no longer applies to the remaining air. Why?

5. What is a *nanotube*?

6. Can the value of an intensive property such as pressure or temperature be uniform with position throughout a system? Be constant with time? Both?

7. You may have used the mass unit *kg* in previous engineering or physics courses. What is the relation between the kg and gram mass? Is the kg a convenient mass unit?

8. Laura takes an elevator from the tenth floor of her office building to the lobby. Should she expect the air pressure on the two levels to differ much?

9. Are the *systolic* and *diastolic* pressures reported in blood pressure measurements absolute, gage, or vacuum pressures?

10. A data sheet indicates that the pressure at the inlet to a pump is −15 kPa. What does the negative sign denote?

11. When buildings have large exhaust fans, exterior doors can be difficult to open due to a pressure difference between the inside and outside. Do you think you could open a 1.2 m by 2.1 m door if the inside pressure were 0.28 kPa (vacuum)?

12. How do dermatologists remove pre-cancerous skin blemishes *cryosurgically*?

13. When the instrument panel of a car provides the outside air temperature, where is the temperature sensor located?

Problems: developing engineering skills

Exploring System Concepts

1.1 Using the Internet, obtain information about the operation of an application listed or shown in Table 1.1. Obtain sufficient information to provide a full description of the application, together with relevant thermodynamic aspects. Present your findings in a memorandum.

1.2 As illustrated in Fig. P1.2, water circulates through a piping system, servicing various household needs. Considering the water heater as a system, identify locations on the system boundary where the system interacts with its surroundings and describe events within the system. Repeat for the dishwasher and for the shower. Present your findings in a memorandum.

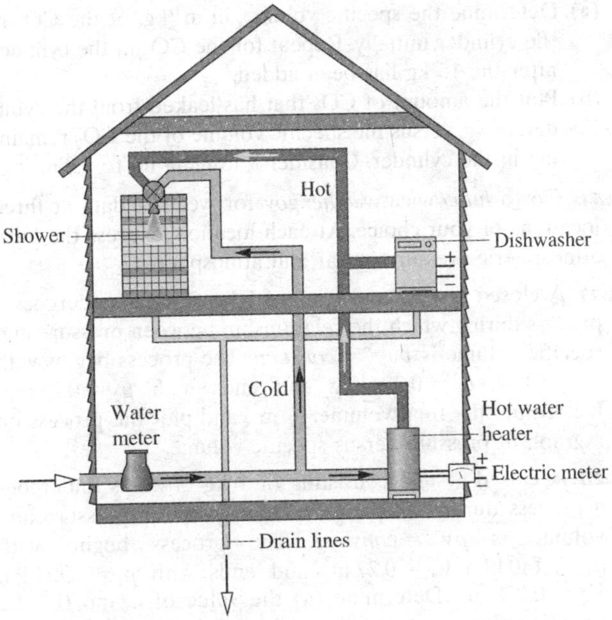

Fig. P1.2

1.3 Reef aquariums such as shown in Fig. P1.3 are popular attractions. Such facilities employ a variety of devices, including heaters, pumps, filters, and controllers, to create a healthy environment for the living things residing in the aquarium, which typically include species of fish, together with corals, clams, and anemone. Considering a reef aquarium as a system, identify locations on the system boundary where the system interacts with its surroundings. Using the Internet, describe events within the system, and comment on measures for the health and safety of the aquatic life. Present your findings in a memorandum.

Fig. P1.3

Working with Force and Mass

1.4 An object has a mass of 4.5 kg. Determine its weight, in lbf, at a location where the acceleration of gravity is 9.81 m/s^2.

1.5 An object weighs 25 kN at a location where the acceleration of gravity is 9.8 m/s^2. Determine its mass, in kg.

1.6 If Superman has a mass of 100 kg on his birth planet Krypton, where the acceleration of gravity is 25 m/s^2, determine (a) his weight on Krypton, in N, and (b) his mass, in kg, and weight, in N, on Earth where g = 9.81 m/s^2.

1.7 A person whose mass is 68 kg weighs 640.5 N. Determine (a) the *local* acceleration of gravity, in m/s^2, and (b) the person's mass, in kg and weight, in N, if g = 9.81 m/s^2.

1.8 A gas occupying a volume of 0.68 m^3 weighs 15.6 N on the moon, where the acceleration of gravity is 1.67 m/s^2. Determine its weight, in N, and density, in kg/m^3, on Mars, where g = 3.92 m/s^2.

1.9 In severe head-on automobile accidents, a deceleration of 60 g's or more (1 g = 9.81 m/s^2) often results in a fatality. What force, in N, acts on a child whose mass is 22.7 kg, when subjected to a deceleration of 60 g's?

1.10 Atomic and molecular weights of some common substances are listed in Appendix Table A-1. Using data from the appropriate table, determine

(a) the mass, in kg, of 10 kmol of each of the following: air, H_2O, Cu, SO_2.

(b) the number of kmol in 10 kg of each of the following: Ar, H_2, N_2, C.

1.11 When an object of mass 5 kg is suspended from a spring, the spring is observed to stretch by 8 cm. The deflection of the spring is related linearly to the weight of the suspended mass. What is the proportionality constant, in newtons per cm, if g = 9.81 m/s^2?

1.12 A spring compresses in length by 3 mm, for every 4.5 N of applied force. Determine the deflection, in inches, of the spring caused by the weight of an object whose mass is 6.8 kg. The local acceleration of gravity is g = 9.8 m/s^2.

1.13 A simple instrument for measuring the acceleration of gravity employs a *linear* spring from which a mass is suspended. At a location on Earth where the acceleration of gravity is 9.81 m/s^2, the spring extends 7.391 mm. If the spring extends 2.946 mm when the instrument is on Mars, what is the Martian acceleration of gravity? How much would the spring extend on the moon, where g = 1.668 m/s^2?

1.14 Estimate the magnitude of the force, in N, exerted by a seat belt on a 100 kg driver during a frontal collision that decelerates a car from 1.5 km/h to rest in 0.1 s. Express the car's deceleration in multiples of the standard acceleration of gravity, or g's.

1.15 An object whose mass is 2 kg is subjected to an applied upward force. The only other force acting on the object is the force of gravity. The net acceleration of the object is upward with a magnitude of 5 m/s^2. The acceleration of gravity is 9.81 m/s^2. Determine the magnitude of the applied upward force, in N.

1.16 An object whose mass is 16 kg is subjected to an applied upward force of 66.7 N. The only other force acting on the object is the force of gravity. Determine the net acceleration of the object, in m/s^2, assuming the acceleration of gravity is constant, g = 9.81 m/s^2. Is the net acceleration upward or downward?

1.17 An astronaut weighs 700 N on Earth where g = 9.81 m/s^2. What is the astronaut's weight, in N, on an orbiting space station where the acceleration of gravity is 6 m/s²? Express each weight in N.

1.18 If the variation of the acceleration of gravity, in m/s², with elevation z, in m, above sea level is $g = 9.81 - (3.3 \times 10^{-6})z$, determine the percent change in weight of an airliner landing from a cruising altitude of 10 km on a runway at sea level.

1.19 The storage tank of a water tower is nearly spherical in shape with a radius of 9 m. If the density of the water is 1000 kg/m³, what is the mass of water stored in the tower, in lb, when the tank is full? What is the weight, in N, of the water if the local acceleration of gravity is 9.8 m/s²?

1.20 As shown in Fig. P1.20, a cylinder of compacted scrap metal measuring 2 m in length and 0.5 m in diameter is suspended from a spring scale at a location where the acceleration of gravity is 9.78 m/s². If the scrap metal density, in kg/m³, varies with position z, in m, according to $\rho = 7800 - 360(z/L)^2$, determine the reading of the scale, in N.

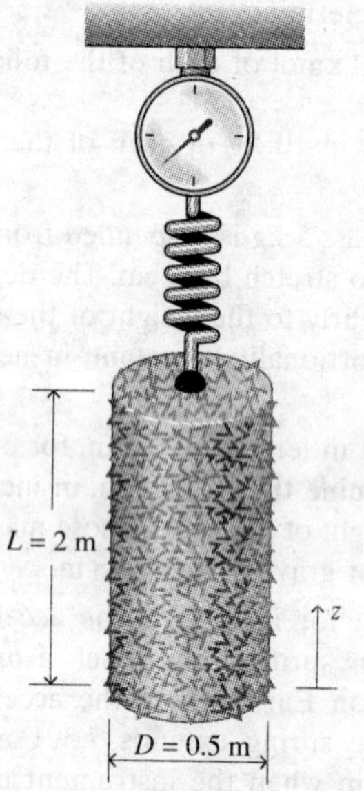

Fig. P1.20

Using Specific Volume and Pressure

1.21 A closed system consists of 0.5 kmol of ammonia occupying a volume of 6 m³. Determine (a) the weight of the system, in N, and (b) the specific volume, in m³/kmol and m³/kg. Let $g = 9.81$ m/s².

1.22 A spherical balloon holding 16 kg of air has a diameter of 3 m. For the air, determine (a) the specific volume, in m³/kg and m³/kmol, and (b) the weight, in N. Let $g = 9.8$ m/s².

1.23 A closed vessel having a volume of 1 liter holds 2.5×10^{22} molecules of ammonia vapor. For the ammonia, determine (a) the amount present, in kg and kmol, and (b) the specific volume, in m³/kg and m³/kmol.

1.24 The specific volume of water vapor at 0.3 MPa, 160°C is 0.651 m³/kg. If the water vapor occupies a volume of 2 m³, determine (a) the amount present, in kg and kmol, and (b) the number of molecules.

1.25 Fifteen kg of carbon dioxide (CO_2) gas is fed to a cylinder having a volume of 20 m³ and initially containing 15 kg of CO_2 at a pressure of 10 bar. Later a pinhole develops and the gas slowly leaks from the cylinder.

(a) Determine the specific volume, in m³/kg, of the CO_2 in the cylinder initially. Repeat for the CO_2 in the cylinder after the 15 kg has been added.

(b) Plot the amount of CO_2 that has leaked from the cylinder, in kg, versus the specific volume of the CO_2 remaining in the cylinder. Consider v ranging up to 1.0 m³/kg.

1.26 Go to *http://www.weather.gov* for weather data at three locations of your choice. At each location, express the local atmospheric pressure in bar and atmospheres.

1.27 A closed system consisting of 5 kg of a gas undergoes a process during which the relationship between pressure and specific volume is $pv^{1.3} = constant$. The process begins with $p_1 = 1$ bar, $v_1 = 0.2$ m³/kg and ends with $p_2 = 0.25$ bar. Determine the final volume, in m³, and plot the process on a graph of pressure versus specific volume.

1.28 A closed system consisting of 1 kg of a gas undergoes a process during which the relation between pressure and volume is $pV^n = constant$. The process begins with $p_1 = 140$ kPa, $V_1 = 0.27$ m³ and ends with $p_2 = 690$ kPa, $V_2 = 0.078$ m³. Determine (a) the value of n and (b) the specific volume at states 1 and 2, each in m³/kg. (c) Sketch the process on pressure–volume coordinates.

1.29 A system consists of nitrogen (N_2) in a piston–cylinder assembly, initially at $p_1 = 140$ kPa, and occupying a volume of 0.068 m³. The nitrogen is compressed to $p_2 = 690$ kPa and a final volume of 0.041 m³. During the process, the relation between pressure and volume is linear. Determine the pressure, in kPa, at an intermediate state where the volume is 0.057 m³, and sketch the process on a graph of pressure versus volume.

1.30 A gas initially at $p_1 = 1$ bar and occupying a volume of 1 liter is compressed within a piston–cylinder assembly to a final pressure $p_2 = 4$ bar.

(a) If the relationship between pressure and volume during the compression is $pV = constant$, determine the volume, in liters, at a pressure of 3 bar. Also plot the overall process on a graph of pressure versus volume.

(b) Repeat for a linear pressure–volume relationship between the same end states.

1.31 A gas contained within a piston–cylinder assembly undergoes three processes in series:

Process 1–2: Compression with $pV = constant$ from $p_1 = 1$ bar, $V_1 = 1.0$ m³ to $V_2 = 0.2$ m³

Process 2–3: Constant-pressure expansion to $V_3 = 1.0$ m³

Process 3–1: Constant volume

Sketch the processes in series on a p–V diagram labeled with pressure and volume values at each numbered state.

1.32 As shown in Fig. 1.7, a manometer is attached to a tank of gas in which the pressure is 104.0 kPa. The manometer liquid is mercury, with a density of 13.59 g/cm³. If $g = 9.81$ m/s² and the atmospheric pressure is 101.33 kPa, calculate

(a) the difference in mercury levels in the manometer, in cm.

(b) the gage pressure of the gas, in kPa.

1.33 A vacuum gage at the intake duct to a fan gives a reading of 105 mm of manometer fluid. The surrounding atmospheric pressure is 100 kPa. Determine the absolute pressure inside the duct, in kPa. The density of the manometer fluid is 800 kg/m^3, and the acceleration of gravity is 9.81 m/s^2.

1.34 The absolute pressure inside a tank is 0.4 bar, and the surrounding atmospheric pressure is 98 kPa. What reading would a Bourdon gage mounted in the tank wall give, in kPa? Is this a *gage* or *vacuum* reading?

1.35 The barometer shown in Fig. P1.35 contains mercury (ρ = 13.59 g/cm^3). If the local atmospheric pressure is 100 kPa and g = 9.81 m/s^2, determine the height of the mercury column, L, in mmHg and inHg.

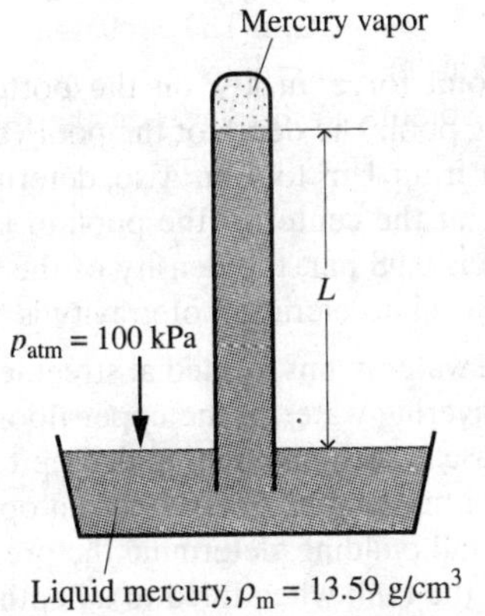

Fig. P1.35

1.36 Water flows through a *Venturi meter,* as shown in Fig. P1.36. The pressure of the water in the pipe supports columns of water that differ in height by 250 mm. Determine the difference in pressure between points a and b, in kPa. Does the pressure increase or decrease in the direction of flow? The atmospheric pressure is 101.3 kPa, the specific volume of water is 0.001 m^3/kg, and the acceleration of gravity is g = 9.8 m/s^2.

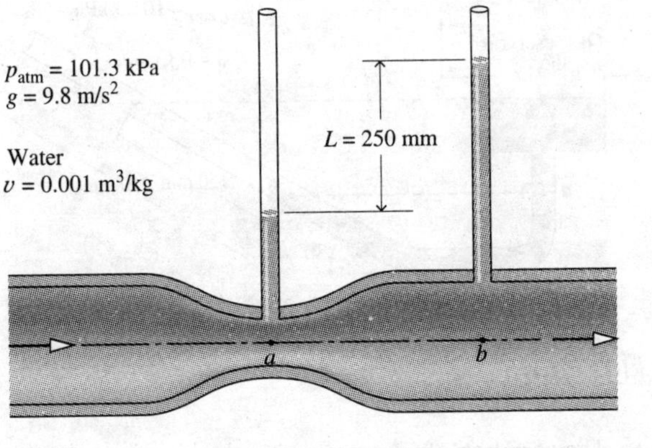

Fig. P1.36

1.37 Figure P1.37 shows a tank within a tank, each containing air. The absolute pressure in tank A is 267.7 kPa. Pressure gage A is located inside tank B and reads 140 kPa. The U-tube manometer connected to tank B contains mercury. Using data on the diagram, determine the absolute pressure inside tank B, in kPa, and the column length L, in cm. The atmospheric pressure surrounding tank B is 101 kPa. The acceleration of gravity is g = 9.81 m/s^2.

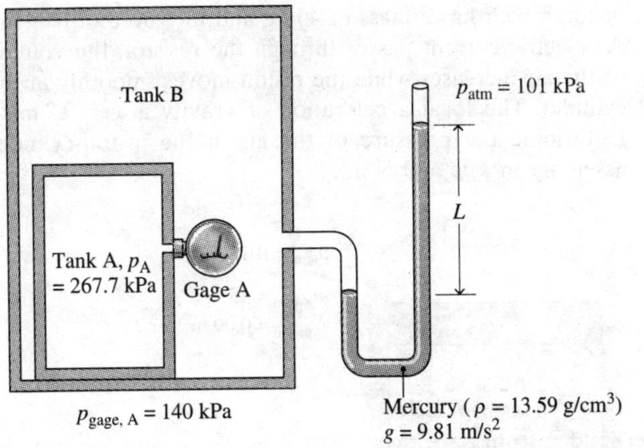

Fig. P1.37

1.38 As shown in Fig. P1.38, an underwater exploration vehicle submerges to a depth of 300 m. If the atmospheric pressure at the surface is 101.3 kPa, the water density is 1000 kg/m^3, and g = 9.8 m/s^2, determine the pressure on the vehicle, in atm.

Fig. P1.38

1.39 A vacuum gage indicates that the pressure of carbon dioxide in a closed tank is −10 kPa. A mercury barometer gives the local atmospheric pressure as 750 mmHg. Determine the absolute pressure of the carbon dioxide, in kPa. The density of mercury is 13.59 g/cm^3 and g = 9.81 m/s^2.

1.40 Refrigerant 22 vapor enters the compressor of a refrigeration system at an absolute pressure of 140 kPa. A pressure gage at the compressor exit indicates a pressure of 1930 kPa (gage). The atmospheric pressure is 101.3 kPa. Determine the change in absolute pressure from inlet to exit, in kPa, and the ratio of exit to inlet pressure.

1.41 As shown in Fig. P1.41, air is contained in a vertical piston-cylinder assembly fitted with an electrical resistor. The atmosphere exerts a pressure of 101.3 kPa on the top of the piston, which has a mass of 45 kg and face area of 0.09 m^2. As electric current passes through the resistor, the volume of the air increases while the piston moves smoothly in the cylinder. The local acceleration of gravity is $g = 9.8\ m/s^2$. Determine the pressure of the air in the piston-cylinder assembly, in kPa and N/m^2.

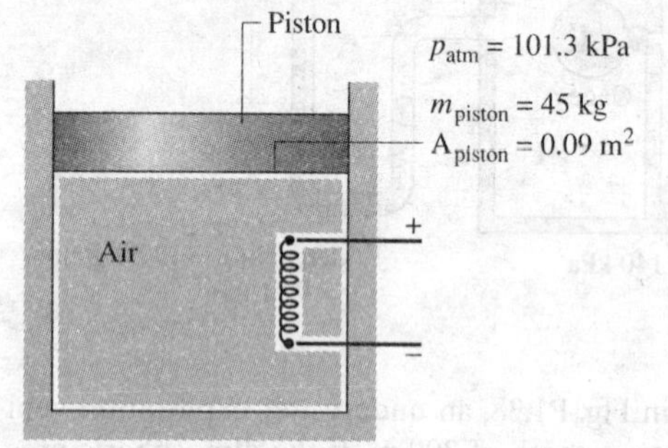

Fig. P1.41

1.42 Warm air is contained in a piston-cylinder assembly oriented horizontally as shown in Fig. P1.42. The air cools slowly from an initial volume of 0.003 m^3 to a final volume of 0.002 m^3. During the process, the spring exerts a force that varies linearly from an initial value of 900 N to a final value of zero. The atmospheric pressure is 100 kPa, and the area of the piston face is 0.018 m^2. Friction between the piston and the cylinder wall can be neglected. For the air in the piston-cylinder assembly, determine the initial and final pressures, each in kPa and atm.

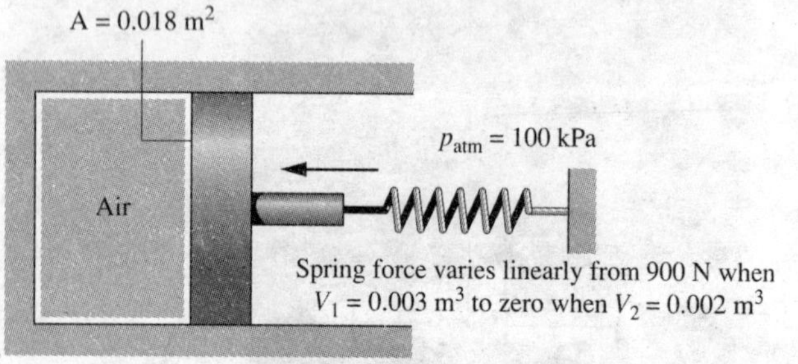

Fig. P1.42

1.43 Air contained within a vertical piston–cylinder assembly is shown in Fig. P1.43. On its top, the 10-kg piston is attached to a spring and exposed to an atmospheric pressure of 1 bar. Initially, the bottom of the piston is at $x = 0$, and the spring exerts a negligible force on the piston. The valve is opened and air enters the cylinder from the supply line, causing the volume of the air within the cylinder to increase by $3.9 \times 10^{-4}\ m^3$. The force exerted by the spring as the air expands within the cylinder varies linearly with x according to

$$F_{spring} = kx$$

where k = 10,000 N/m. The piston face area is $7.8 \times 10^{-3} m^2$. Ignoring friction between the piston and the cylinder wall, determine the pressure of the air within the cylinder, in bar, when the piston is in its initial position. Repeat when the piston is in its final position. The local acceleration of gravity is 9.81 m/s^2.

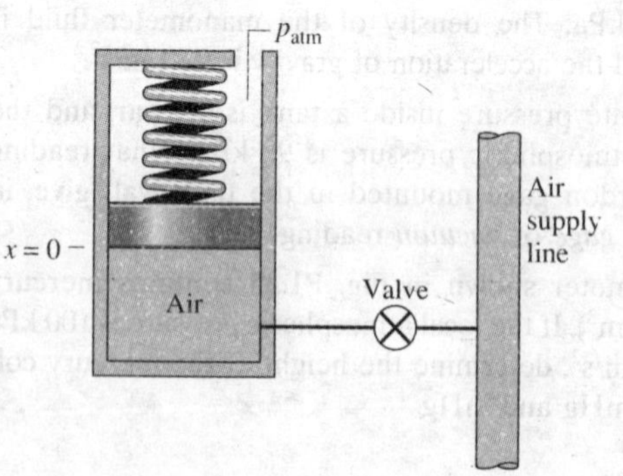

Fig. P1.43

1.44 Determine the total force, in kN, on the bottom of a 100 × 50 m swimming pool. The depth of the pool varies linearly along its length from 1 m to 4 m. Also, determine the pressure on the floor at the center of the pool, in kPa. The atmospheric pressure is 0.98 bar, the density of the water is 998.2 kg/m^3, and the local acceleration of gravity is 9.8 m/s^2.

1.45 The pressure from water mains located at street level may be insufficient for delivering water to the upper floors of tall buildings. In such a case, water may be pumped up to a tank that feeds water to the building by gravity. For an open storage tank atop a 90 m tall building, determine the pressure, in kPa, at the bottom of the tank when filled to a depth of 6 m. The density of water is 1000 kg/m^3, $g = 9.8\ m/s^2$, and the local atmospheric pressure is 101.3 kPa.

1.46 As shown in Fig. P1.46, an *inclined* manometer is used to measure the pressure of the gas within the reservoir. (a) Using data on the figure, determine the gas pressure, in kPa. (b) Express the pressure as a gage or a vacuum pressure, as appropriate, in kPa. (c) What advantage does an inclined manometer have over the U-tube manometer shown in Fig. 1.7?

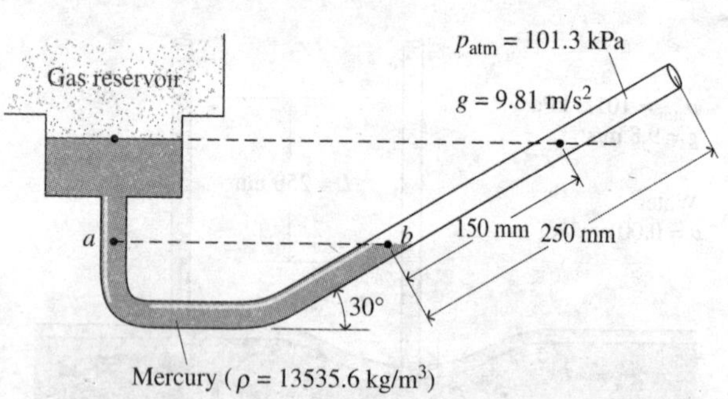

Fig. P1.46

1.47 The variation of pressure within the biosphere affects not only living things but also systems such as aircraft and undersea exploration vehicles.

(a) Plot the variation of atmospheric pressure, in atm, versus elevation z above sea level, in km, ranging from 0 to 10 km. Assume that the specific volume of the atmosphere, in m^3/kg, varies with the local pressure p, in kPa, according to $v = 72.435/p$.

(b) Plot the variation of pressure, in atm, versus depth z below sea level, in km, ranging from 0 to 2 km. Assume that the specific volume of seawater is constant, $v = 0.956 \times 10^{-3}$ m³/kg.

In each case, $g = 9.81$ m/s² and the pressure at sea level is 1 atm.

1.48 One thousand kg of natural gas at 100 bar and 255 K is stored in a tank. If the pressure, p, specific volume, v, and temperature, T, of the gas are related by the following expression

$$p = [(5.18 \times 10^{-3})T/(v - 0.002668)] - (8.91 \times 10^{-3})/v^2$$

where v is in m³/kg, T is in K, and p is in bar, determine the volume of the tank, in m³. Also, plot pressure versus specific volume for the *isotherms* T = 250, 500, and 1000 K.

1.49 A 2.2 m³ tank contains water vapor at 10340 kPa and 633 K. If the pressure, p, specific volume, v, and temperature, T, of water vapor are related by the expression

$$p = [(7183.04)T/(v - 0.0169)] - (25.02 \times 10^3)/v^2$$

where v is in m³/kg, T is in K, and p is in kPa, determine the mass of water in the tank. Also, plot pressure versus specific volume for the *isotherms* T = 667, 778, and 889 K.

Exploring Temperature

1.50 Convert the following temperatures from °C to K: (a) 21°C, (b) −40°C, (c) 500°C, (d) 0°C, (e) 100°C, (f) −273.15°C.

1.51 Convert the following temperatures from K to °C: (a) 293.15 K, (b) 233.15 K, (c) 533.15 K, (d) 255.4 K, (e) 373.15 K, (f) 0 K.

1.52 Natural gas is burned with air to produce gaseous products at 1985°C. Express this temperature in K.

1.53 The temperature of a child ill with a fever is measured as 40°C. The child's normal temperature is 37°C. Express both temperatures in K.

1.54 Does the Celsius degree represent a larger or smaller temperature unit than the Kelvin degree? Explain.

1.55 As shown in Fig. P1.55, a small-diameter water pipe passes through the 150 mm thick exterior wall of a dwelling. Assuming that temperature varies linearly with position x through the wall from 20°C to −7°C, would the water in the pipe freeze? Explain.

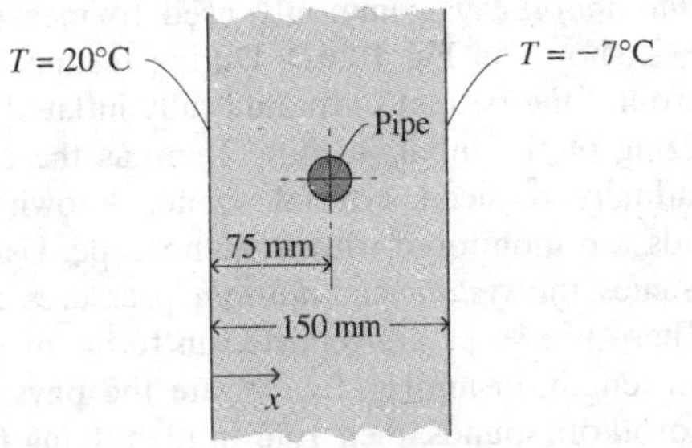

Fig. P1.55

1.56 What is (a) the lowest *naturally* occurring temperature recorded on Earth, (b) the lowest temperature recorded in a laboratory on Earth, (c) the lowest temperature recorded in the Earth's solar system, and (d) the temperature of deep space, each in K?

1.57 What is the maximum increase and maximum decrease in body temperature, each in °C, from a normal body temperature of 37°C that humans can experience before serious medical complications result?

1.58 For liquid-in-glass thermometers, the *thermometric* property is the change in length of the thermometer liquid with temperature. However, other effects are present that can affect the temperature reading of such thermometers. What are some of these?

Design & open ended problems: Exploring engineering practice

1.1D The issue of *global warming* is receiving considerable attention these days. Write a technical report including at least three references on the subject of global warming. Explain what is meant by the term global warming and discuss objectively the scientific evidence that is cited as the basis for the argument that global warming is occurring.

1.2D Barometers and liquid-in-glass thermometers have customarily used mercury, which is now recognized as a biohazard. Investigate medical complications of mercury exposure. Write a report including at least three references.

1.3D List several aspects of engineering economics relevant to design. What are the important contributors to *cost* that should be considered in engineering design? Discuss what is meant by *annualized costs.* Present your findings in a memorandum.

1.4D Determine the respective contributions to the electric power provided to customers by the electric utility serving your locale attributable to coal, natural gas, oil, biomass, nuclear power, hydropower, wind power, and solar power. Summarize your findings in a *pie chart.* For each contribution providing 1%, or more, of the total, identify associated air emissions, solid waste produced, including radioactive waste, and wildlife impacts. Write a report including at least three references.

1.5D Magnetic resonance imaging (MRI) employs a strong magnetic field to produce detailed pictures of internal organs and tissues. As shown in Fig. P1.5D, the patient reclines on a table that slides into the cylindrical opening where the field is created. Considering a MRI scanner as a system, identify locations on the system boundary where the system interacts with its surroundings. Also describe events occurring within the system and the measures taken for patient comfort and safety. Write a report including at least three references.

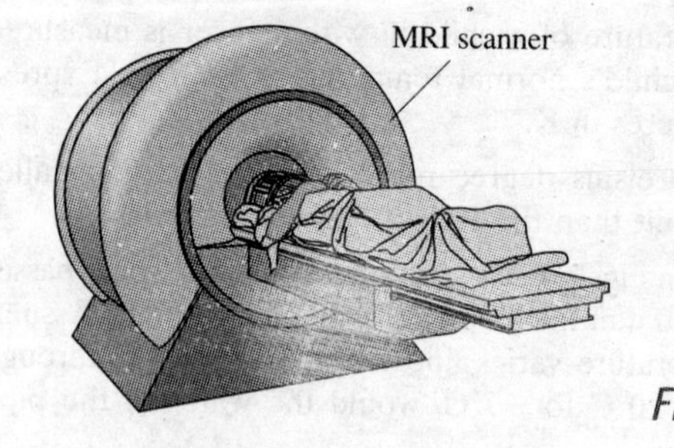

Fig. P1.5D

1.6D The *sphygmomanometer,* commonly used to measure blood pressure, is shown in Fig. P1.6D. During testing, the cuff is placed around the patient's arm and fully inflated by repeated squeezing of the inflation bulb. Then, as the cuff pressure is gradually reduced, arterial sounds known as *Korotkoff* sounds are monitored with a stethoscope. Using these sounds as cues, the *systolic* and *diastolic* pressures can be identified. These pressures are reported in terms of the mercury column length, in mmHg. Investigate the physical basis for the Korotkoff sounds, their role in identifying the systolic and diastolic pressures, and why these pressures are significant in medical practice. Write a report including at least three references.

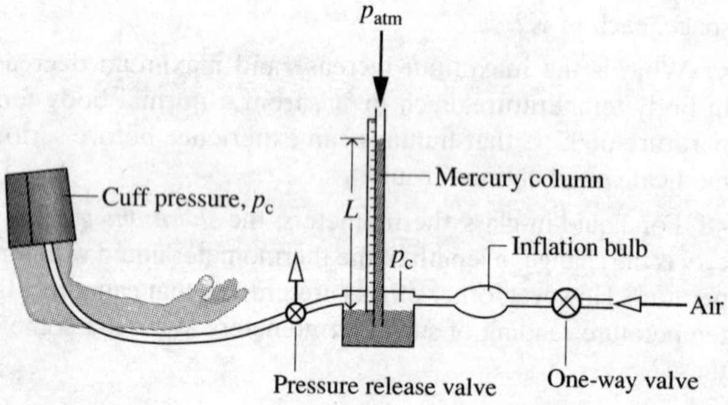

Fig. P1.6D

1.7D Design a low-cost, compact, light-weight, hand-held, human-powered air pump capable of directing a stream of air for cleaning computer keyboards, circuit boards, and hard-to-reach locations in electronic devices. The pump cannot use electricity, including batteries, nor employ any chemical propellants. All materials must be recyclable. Owing to existing patent protections, the pump must be a *distinct alternative* to the familiar tube and plunger bicycle pump and to existing products aimed at accomplishing the specified computer and electronic cleaning tasks.

1.8D In Bangladesh, unsafe levels of arsenic, which is a tasteless, odorless, and colorless poison, are present in underground wells providing drinking water to millions of people living in rural areas. The task is to identify affordable, easy-to-use treatment technologies for removing arsenic from their drinking water. Technologies considered should include, but not be limited to, applications of *smart materials* and other nanotechnology approaches. Write a report including at least three references.

1.9D Conduct a term-length design project in the realm of bioengineering done on either an independent or a small-group basis. The project might involve a device or technique for minimally invasive surgery, an implantable drug-delivery device, a biosensor, artificial blood, or something of special interest to you or your design group. Take several days to research your project idea and then prepare a brief written proposal, including several references, that provides a general statement of the core concept plus a list of objectives. During the project, observe good design practices such as discussed in Sec. 1.3 of *Thermal Design and Optimization,* John Wiley & Sons Inc., New York, 1996, by A. Bejan, G. Tsatsaronis, and M.J. Moran. Provide a well-documented final report, including several references.

1.10D Conduct a term-length design project involving the International Space Station pictured in Table 1.1, done on either an independent or a small-group basis. The project might involve an experiment that is best conducted in a low-gravity environment, a device for the comfort or use of the astronauts, or something of special interest to you or your design group. Take several days to research your project idea and then prepare a brief written proposal, including several references, that provides a general statement of the core concept plus a list of objectives. During the project, observe good design practices such as discussed in Sec. 1.3 of *Thermal Design and Optimization,* John Wiley & Sons Inc., New York, 1996, by A. Bejan, G. Tsatsaronis, and M. J. Moran. Provide a well-documented final report, including several references.

2

Energy and the First Law of Thermodynamics

Engineering Context

Energy is a fundamental concept of thermodynamics and one of the most significant aspects of engineering analysis. In this chapter we discuss energy and develop equations for applying the principle of conservation of energy. The current presentation is limited to closed systems. In Chap. 4 the discussion is extended to control volumes.

Energy is a familiar notion, and you already know a great deal about it. In the present chapter several important aspects of the energy concept are developed. Some of these you have encountered before. A basic idea is that energy can be *stored* within systems in various forms. Energy also can be *converted* from one form to another and *transferred* between systems. For closed systems, energy can be transferred by *work* and *heat transfer*. The total amount of energy is *conserved* in all conversions and transfers.

The **objective** of this chapter is to organize these ideas about energy into forms suitable for engineering analysis. The presentation begins with a review of energy concepts from mechanics. The thermodynamic concept of energy is then introduced as an extension of the concept of energy in mechanics.

Learning Objectives

When you complete your study of this chapter you will be able to...

- ✔ demonstrate understanding of key concepts related to energy and the first law of thermodynamics . . . including internal, kinetic, and potential energy, work and power, heat transfer and heat transfer modes, heat transfer rate, power cycle, refrigeration cycle, and heat pump cycle.
- ✔ apply closed system energy balances, appropriately modeling the case at hand, and correctly observing sign conventions for work and heat transfer.
- ✔ conduct energy analyses of systems undergoing thermodynamic cycles, evaluating as appropriate thermal efficiencies of power cycles and coefficients of performance of refrigeration and heat pump cycles.

2.1 Reviewing Mechanical Concepts of Energy

Building on the contributions of Galileo and others, Newton formulated a general description of the motions of objects under the influence of applied forces. Newton's laws of motion, which provide the basis for classical mechanics, led to the concepts of *work, kinetic energy,* and *potential energy,* and these led eventually to a broadened concept of energy. The present discussion begins with an application of Newton's second law of motion.

2.1.1 Work and Kinetic Energy

The curved line in Fig. 2.1 represents the path of a body of mass m (a closed system) moving relative to the x–y coordinate frame shown. The velocity of the center of mass of the body is denoted by $\mathbf{V}$.[1] The body is acted on by a resultant force $\mathbf{F}$, which may vary in magnitude from location to location along the path. The resultant force is resolved into a component $\mathbf{F}_s$ along the path and a component $\mathbf{F}_n$ normal to the path. The effect of the component $\mathbf{F}_s$ is to change the magnitude of the velocity, whereas the effect of the component $\mathbf{F}_n$ is to change the direction of the velocity. As shown in Fig. 2.1, s is the instantaneous position of the body measured along the path from some fixed point denoted by 0. Since the magnitude of $\mathbf{F}$ can vary from location to location along the path, the magnitudes of $\mathbf{F}_s$ and $\mathbf{F}_n$ are, in general, functions of s.

Let us consider the body as it moves from $s = s_1$, where the magnitude of its velocity is V_1, to $s = s_2$, where its velocity is V_2. Assume for the present discussion that the only interaction between the body and its surroundings involves the force $\mathbf{F}$. By Newton's second law of motion, the magnitude of the component $\mathbf{F}_s$ is related to the change in the magnitude of $\mathbf{V}$ by

$$F_s = m\frac{dV}{dt} \tag{2.1}$$

Using the chain rule, this can be written as

$$F_s = m\frac{dV}{ds}\frac{ds}{dt} = mV\frac{dV}{ds} \tag{2.2}$$

where $V = ds/dt$. Rearranging Eq. 2.2 and integrating from s_1 to s_2 gives

$$\int_{V_1}^{V_2} mV\,dV = \int_{s_1}^{s_2} F_s\,ds \tag{2.3}$$

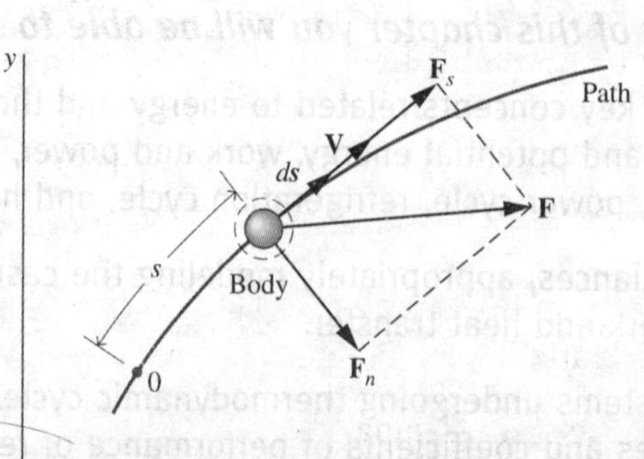

Fig. 2.1 Forces acting on a moving system.

[1]Boldface symbols denote vectors. Vector magnitudes are shown in lightface type.

The integral on the left of Eq. 2.3 is evaluated as follows

$$\int_{V_1}^{V_2} mV\,dV = \frac{1}{2}mV^2\Big]_{V_1}^{V_2} = \frac{1}{2}m(V_2^2 - V_1^2) \tag{2.4}$$

kinetic energy

The quantity $\frac{1}{2}mV^2$ is the ***kinetic energy,*** KE, of the body. Kinetic energy is a scalar quantity. The *change* in kinetic energy, ΔKE, of the body is[2]

$$\Delta KE = KE_2 - KE_1 = \frac{1}{2}m(V_2^2 - V_1^2) \tag{2.5}$$

The integral on the right of Eq. 2.3 is the *work* of the force F_s as the body moves from s_1 to s_2 along the path. Work is also a scalar quantity.

With Eq. 2.4, Eq. 2.3 becomes

$$\frac{1}{2}m(V_2^2 - V_1^2) = \int_{s_1}^{s_2} \mathbf{F} \cdot d\mathbf{s} \tag{2.6}$$

where the expression for work has been written in terms of the scalar product (dot product) of the force vector **F** and the displacement vector $d\mathbf{s}$. Equation 2.6 states that the work of the resultant force on the body equals the change in its kinetic energy. When the body is accelerated by the resultant force, the work done on the body can be considered a *transfer* of energy *to* the body, where it is *stored* as kinetic energy.

Kinetic energy can be assigned a value knowing only the mass of the body and the magnitude of its instantaneous velocity relative to a specified coordinate frame, without regard for how this velocity was attained. Hence, *kinetic energy is a property* of the body. Since kinetic energy is associated with the body as a whole, it is an *extensive* property.

Energy & Environment

Did you ever wonder what happens to the kinetic energy when you step on the brakes of your moving car? Automotive engineers have, and the result is the *hybrid electric* vehicle combining an electric motor with a small conventional engine.

When a hybrid is braked, some of its kinetic energy is harvested and stored in batteries. The electric motor calls on the stored energy to help the car start up again. A specially designed transmission provides the proper split between the engine and the electric motor to minimize fuel use. Because stored energy assists the engine, these cars get better fuel economy than comparably sized conventional vehicles. Tomorrow's hybrids may get even better fuel economy if plans develop for a *plug-in* version whereby the storage batteries are charged from an electrical outlet when the vehicle is idle, thereby allowing drivers of plug-in hybrids to get most of the energy they need for transportation from the electricity grid—not by way of the gas pump.

Better fuel economy not only stretches increasingly scarce and costly oil supplies but also reduces the release of CO_2 into the atmosphere, which has been linked to *global warming.* Each liter of gasoline burned by a vehicle produces about 2.4 kg of CO_2. Annually, a conventional vehicle produces several tons of CO_2. Fuel-thrifty hybrids produce much less.

2.1.2 Potential Energy

Equation 2.6 is a principal result of the previous section. Derived from Newton's second law, the equation gives a relationship between two *defined* concepts: kinetic energy and work. In this section it is used as a point of departure to extend the concept of

[2]The symbol Δ always means "final value minus initial value."

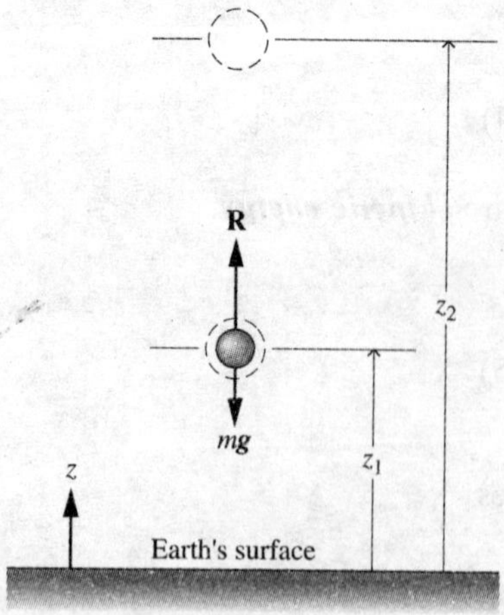

Fig. 2.2 Illustration used to introduce the potential energy concept.

energy. To begin, refer to Fig. 2.2, which shows a body of mass m that moves vertically from an elevation z_1 to an elevation z_2 relative to the surface of the earth. Two forces are shown acting on the system: a downward force due to gravity with magnitude mg and a vertical force with magnitude R representing the resultant of all *other* forces acting on the system.

The work of each force acting on the body shown in Fig. 2.2 can be determined by using the definition previously given. The total work is the algebraic sum of these individual values. In accordance with Eq. 2.6, the total work equals the change in kinetic energy. That is

$$\frac{1}{2}m(\mathrm{V}_2^2 - \mathrm{V}_1^2) = \int_{z_1}^{z_2} R\,dz - \int_{z_1}^{z_2} mg\,dz \tag{2.7}$$

A minus sign is introduced before the second term on the right because the gravitational force is directed downward and z is taken as positive upward.

The first integral on the right of Eq. 2.7 represents the work done by the force **R** on the body as it moves vertically from z_1 to z_2. The second integral can be evaluated as follows:

$$\int_{z_1}^{z_2} mg\,dz = mg(z_2 - z_1) \tag{2.8}$$

where the acceleration of gravity has been assumed to be constant with elevation. By incorporating Eq. 2.8 into Eq. 2.7 and rearranging

$$\frac{1}{2}m(\mathrm{V}_2^2 - \mathrm{V}_1^2) + mg(z_2 - z_1) = \int_{z_1}^{z_2} R\,dz \tag{2.9}$$

gravitational potential energy

The quantity mgz is the ***gravitational potential energy,*** PE. The *change* in gravitational potential energy, ΔPE, is

$$\Delta\mathrm{PE} = \mathrm{PE}_2 - \mathrm{PE}_1 = mg(z_2 - z_1) \tag{2.10}$$

Potential energy is associated with the force of gravity and is, therefore, an attribute of a system consisting of the body and the earth together. However, evaluating the force of gravity as mg enables the gravitational potential energy to be determined for a specified value of g knowing only the mass of the body and its elevation. With this view, potential energy is regarded as an *extensive property* of the body. Throughout this book, it is assumed that elevation differences are small enough that the gravitational force can be considered constant. The concept of gravitational potential energy can be formulated to account for the variation of the gravitational force with elevation, however.

To assign a value to the kinetic energy or the potential energy of a system, it is necessary to assume a datum and specify a value for the quantity at the datum. Values of kinetic and potential energy are then determined relative to this arbitrary choice of datum and reference value. However, since only *changes* in kinetic and potential energy between two states are required, these arbitrary reference specifications cancel.

2.1.3 Units for Energy

Work has units of force times distance. The units of kinetic energy and potential energy are the same as for work. In SI, the energy unit is the newton-meter, N · m, called the joule, J. In this book, it is convenient to use the kilojoule, kJ.

When a system undergoes a process where there are changes in kinetic and potential energy, special care is required to obtain a consistent set of units.

FOR EXAMPLE... to illustrate the proper use of units in the calculation of such terms, consider a system having a mass of 1 kg whose velocity increases from 15 m/s to 30 m/s while its elevation decreases by 10 m at a location where $g = 9.7 \text{ m/s}^2$. Then

$$\Delta KE = \frac{1}{2}m(V_2^2 - V_1^2)$$

$$= \frac{1}{2}(1 \text{ kg})\left[\left(30\frac{\text{m}}{\text{s}}\right)^2 - \left(15\frac{\text{m}}{\text{s}}\right)^2\right]\left|\frac{1 \text{ N}}{1 \text{ kg} \cdot \text{m/s}^2}\right|\left|\frac{1 \text{ kJ}}{10^3 \text{ N} \cdot \text{m}}\right|$$

$$= 0.34 \text{ kJ}$$

$$\Delta PE = mg(z_2 - z_1)$$

$$= (1 \text{ kg})\left(9.7\frac{\text{m}}{\text{s}^2}\right)(-10 \text{ m})\left|\frac{1 \text{ N}}{1 \text{ kg} \cdot \text{m/s}^2}\right|\left|\frac{1 \text{ kJ}}{10^3 \text{ N} \cdot \text{m}}\right|$$

$$= -0.10 \text{ kJ}$$

For a system having a mass of 1 lb whose velocity increases from 50 ft/s to 100 ft/s while its elevation decreases by 40 ft at a location where $g = 32.0 \text{ ft/s}^2$, we have

$$\Delta KE = \frac{1}{2}(1 \text{ lb})\left[\left(100\frac{\text{ft}}{\text{s}}\right)^2 - \left(50\frac{\text{ft}}{\text{s}}\right)^2\right]\left|\frac{1 \text{ lbf}}{32.2 \text{ lb} \cdot \text{ft/s}^2}\right|\left|\frac{1 \text{ Btu}}{778 \text{ ft} \cdot \text{lbf}}\right|$$

$$= 0.15 \text{ Btu}$$

$$\Delta PE = (1 \text{ lb})\left(32.0\frac{\text{ft}}{\text{s}^2}\right)(-40 \text{ ft})\left|\frac{1 \text{ lbf}}{32.2 \text{ lb} \cdot \text{ft/s}^2}\right|\left|\frac{1 \text{ Btu}}{778 \text{ ft} \cdot \text{lbf}}\right|$$

$$= -0.05 \text{ Btu}$$

2.1.4 Conservation of Energy in Mechanics

Equation 2.9 states that the total work of all forces acting on the body from the surroundings, with the exception of the gravitational force, equals the sum of the changes in the kinetic and potential energies of the body. When the resultant force causes the elevation to be increased, the body to be accelerated, or both, the work done by the force can be considered a *transfer* of energy *to* the body, where it is stored as gravitational potential energy and/or kinetic energy. The notion that *energy is conserved* underlies this interpretation.

The interpretation of Eq. 2.9 as an expression of the conservation of energy principle can be reinforced by considering the special case of a body on which the only force acting is that due to gravity, for then the right side of the equation vanishes and the equation reduces to

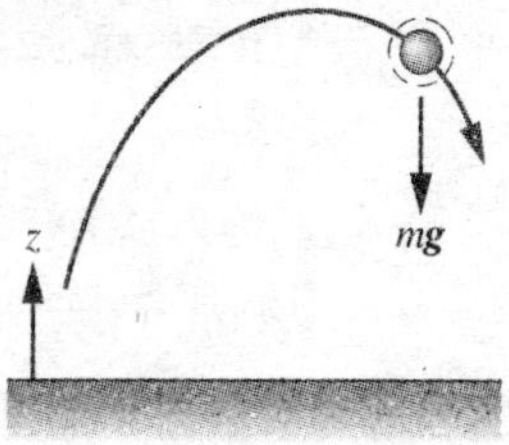

$$\frac{1}{2}m(V_2^2 - V_1^2) + mg(z_2 - z_1) = 0$$

or

$$\frac{1}{2}mV_2^2 + mgz_2 = \frac{1}{2}mV_1^2 + mgz_1 \tag{2.11}$$

Under these conditions, the *sum* of the kinetic and gravitational potential energies *remains constant.* Equation 2.11 also illustrates that energy can be *converted* from one form to another: For an object falling under the influence of gravity *only,* the potential energy would decrease as the kinetic energy increases by an equal amount.

2.1.5 Closing Comment

The presentation thus far has centered on systems for which applied forces affect only their overall velocity and position. However, systems of engineering interest normally interact with their surroundings in more complicated ways, with changes in other properties as well. To analyze such systems, the concepts of kinetic and potential energy alone do not suffice, nor does the rudimentary conservation of energy principle introduced in this section. In thermodynamics the concept of energy is broadened to account for other observed changes, and the principle of *conservation of energy* is extended to include a wide variety of ways in which systems interact with their surroundings. The basis for such generalizations is experimental evidence. These extensions of the concept of energy are developed in the remainder of the chapter, beginning in the next section with a fuller discussion of work.

2.2 Broadening Our Understanding of Work

The work W done by, or on, a system evaluated in terms of macroscopically observable forces and displacements is

$$W = \int_{s_1}^{s_2} \mathbf{F} \cdot d\mathbf{s} \tag{2.12}$$

This relationship is important in thermodynamics, and is used later in the present section to evaluate the work done in the compression or expansion of gas (or liquid), the extension of a solid bar, and the stretching of a liquid film. However, thermodynamics also deals with phenomena not included within the scope of mechanics, so it is necessary to adopt a broader interpretation of work, as follows.

thermodynamic definition of work

A particular interaction is categorized as a work interaction if it satisfies the following criterion, which can be considered the ***thermodynamic definition of work:*** *Work is done by a system on its surroundings if the sole effect on everything external to the system could have been the raising of a weight.* Notice that the raising of a weight is, in effect, a force acting through a distance, so the concept of work in thermodynamics is a natural extension of the concept of work in mechanics. However, the test of whether a work interaction has taken place is not that the elevation of a weight has actually taken place, or that a force has actually acted through a distance, but that the sole effect *could have been* an increase in the elevation of a weight.

FOR EXAMPLE... consider Fig. 2.3 showing two systems labeled A and B. In system A, a gas is stirred by a paddle wheel: the paddle wheel does work on the gas. In principle, the work could be evaluated in terms of the forces and motions at the boundary

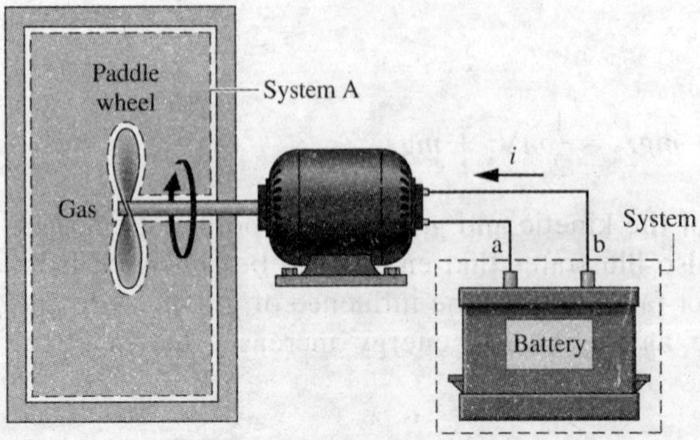

Fig. 2.3 Two examples of work.

Horizons

Nanoscale Machines on the Move

Engineers working in the field of nanotechnology, the engineering of molecular-sized devices, look forward to the time when practical nanoscale machines can be fabricated that are capable of movement, sensing and responding to stimuli such as light and sound, delivering medication within the body, performing computations, and numerous other functions that promote human well being. For inspiration, engineers study biological nanoscale *machines* in living things that perform functions such as creating and repairing cells, circulating oxygen, and digesting food. These studies have yielded positive results. Molecules mimicking the function of mechanical devices have been fabricated, including gears, rotors, ratchets, brakes, switches, and abacus-like structures. A particular success is the development of molecular motors that convert light to rotary or linear motion. Although devices produced thus far are rudimentary, they do demonstrate the feasibility of constructing nanomachines, researchers say.

between the paddle wheel and the gas. Such an evaluation of work is consistent with Eq. 2.12, where work is the product of force and displacement. By contrast, consider system B, which includes only the battery. At the boundary of system B, forces and motions are not evident. Rather, there is an electric current i driven by an electrical potential difference existing across the terminals a and b. That this type of interaction at the boundary can be classified as work follows from the thermodynamic definition of work given previously: We can imagine the current is supplied to a *hypothetical* electric motor that lifts a weight in the surroundings. ◀

Work is a means for transferring energy. Accordingly, the term work does not refer to what is being transferred between systems or to what is stored within systems. Energy is transferred and stored when work is done.

2.2.1 Sign Convention and Notation

Engineering thermodynamics is frequently concerned with devices such as internal combustion engines and turbines whose purpose is to do work. Hence, in contrast to the approach generally taken in mechanics, it is often convenient to consider such work as positive. That is,

$W > 0$: work done *by* the system
$W < 0$: work done *on* the system

This ***sign convention*** is used throughout the book. In certain instances, however, it is convenient to regard the work done *on* the system to be positive, as has been done in the discussion of Sec. 2.1. To reduce the possibility of misunderstanding in any such case, the direction of energy transfer is shown by an arrow on a sketch of the system, and work is regarded as positive in the direction of the arrow. ***sign convention for work***

To evaluate the integral in Eq. 2.12, it is necessary to know how the force varies with the displacement. This brings out an important idea about work: The value of W depends on the details of the interactions taking place between the system and surroundings during a process and not just the initial and final states of the system. It follows that ***work is not a property*** of the system or the surroundings. In addition, the limits on the integral of Eq. 2.12 mean "from state 1 to state 2" and cannot be interpreted as the *values* of work at these states. The notion of work at a state *has no meaning,* so the value of this integral should never be indicated as $W_2 - W_1$. ***work is not a property***

The differential of work, δW, is said to be *inexact* because, in general, the following integral cannot be evaluated without specifying the details of the process

$$\int_1^2 \delta W = W$$

On the other hand, the differential of a property is said to be *exact* because the change in a property between two particular states depends in no way on the details of the process linking the two states. For example, the change in volume between two states can be determined by integrating the differential dV, without regard for the details of the process, as follows

$$\int_{V_1}^{V_2} dV = V_2 - V_1$$

where V_1 is the volume *at* state 1 and V_2 is the volume *at* state 2. The differential of every property is exact. Exact differentials are written, as above, using the symbol d. To stress the difference between exact and inexact differentials, the differential of work is written as δW. The symbol δ is also used to identify other inexact differentials encountered later.

2.2.2 Power

power

Many thermodynamic analyses are concerned with the time rate at which energy transfer occurs. The rate of energy transfer by work is called ***power*** and is denoted by $\dot{W}$. When a work interaction involves a macroscopically observable force, the rate of energy transfer by work is equal to the product of the force and the velocity at the point of application of the force

$$\dot{W} = \mathbf{F} \cdot \mathbf{V} \tag{2.13}$$

A dot appearing over a symbol, as in $\dot{W}$, is used throughout this book to indicate a time rate. In principle, Eq. 2.13 can be integrated from time t_1 to time t_2 to get the total work done during the time interval

$$W = \int_{t_1}^{t_2} \dot{W}\, dt = \int_{t_1}^{t_2} \mathbf{F} \cdot \mathbf{V}\, dt \tag{2.14}$$

units for power

The same sign convention applies for $\dot{W}$ as for W. Since power is a time rate of doing work, it can be expressed in terms of any units for energy and time. In SI, the ***unit for power*** is J/s, called the watt. In this book the kilowatt, kW, is generally used.

FOR EXAMPLE... to illustrate the use of Eq. 2.13, let us evaluate the power required for a bicyclist traveling at 32 km per hour to overcome the drag force imposed by the surrounding air. This *aerodynamic drag* force is given by

$$F_d = \tfrac{1}{2} C_d A \rho V^2$$

where C_d is a constant called the *drag coefficient*, A is the frontal area of the bicycle and rider, and ρ is the air density. By Eq. 2.13 the required power is $\mathbf{F}_d \cdot \mathbf{V}$ or

$$\dot{W} = (\tfrac{1}{2} C_d A \rho V^2) V = \tfrac{1}{2} C_d A \rho V^3$$

Using typical values: $C_d = 0.88$, $A = 0.362\ m^2$, and $\rho = 1.2\ kg/m^3$, together with $V = 8.94$ m/s, and also converting units to horsepower, the power required is

$$\dot{W} = \frac{1}{2}(0.88)(0.362\ m^2)(1.2\ kg/m^3)(8.94\ m/s)^3 = 136.6\ W$$

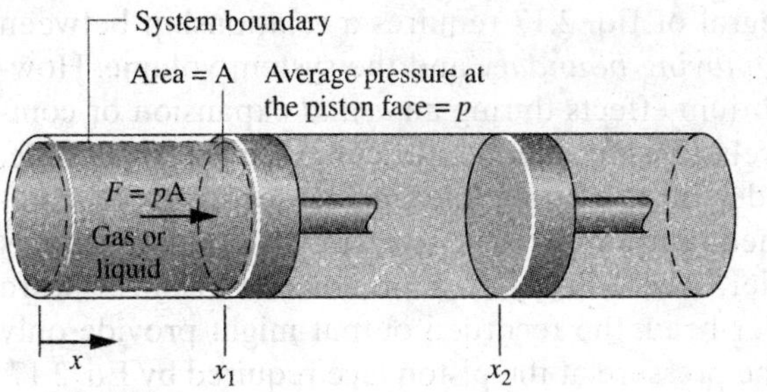

Fig. 2.4 Expansion or compression of a gas or liquid.

2.2.3 Modeling Expansion or Compression Work

There are many ways in which work can be done by or on a system. The remainder of this section is devoted to considering several examples, beginning with the important case of the work done when the volume of a quantity of a gas (or liquid) changes by expansion or compression.

Let us evaluate the work done by the closed system shown in Fig. 2.4 consisting of a gas (or liquid) contained in a piston–cylinder assembly as the gas expands. During the process the gas pressure exerts a normal force on the piston. Let p denote the pressure acting at the interface between the gas and the piston. The force exerted by the gas on the piston is simply the product pA, where A is the area of the piston face. The work done by the system as the piston is displaced a distance dx is

$$\delta W = pA\, dx \tag{2.15}$$

The product A dx in Eq. 2.15 equals the change in volume of the system, dV. Thus, the work expression can be written as

$$\delta W = p\, dV \tag{2.16}$$

Since dV is positive when volume increases, the work at the moving boundary is positive when the gas expands. For a compression, dV is negative, and so is work found from Eq. 2.16. These signs are in agreement with the previously stated sign convention for work.

For a change in volume from V_1 to V_2, the work is obtained by integrating Eq. 2.16

$$W = \int_{V_1}^{V_2} p\, dV \tag{2.17}$$

Although Eq. 2.17 is derived for the case of a gas (or liquid) in a piston–cylinder assembly, it is applicable to systems of *any* shape provided the pressure is uniform with position over the moving boundary.

2.2.4 Expansion or Compression Work in Actual Processes

There is no requirement that a system undergoing a process be in equilibrium *during* the process. Some or all of the intervening states may be nonequilibrium states. For many such processes we are limited to knowing the state before the process occurs and the state after the process is completed. Typically, at a nonequilibrium state intensive properties vary with position at a given time. Also, at a specified position intensive properties may vary with time, sometimes chaotically. In certain cases, spatial and temporal variations in properties such as temperature, pressure, and velocity can be measured, or obtained by solving appropriate governing equations, which are generally differential equations.

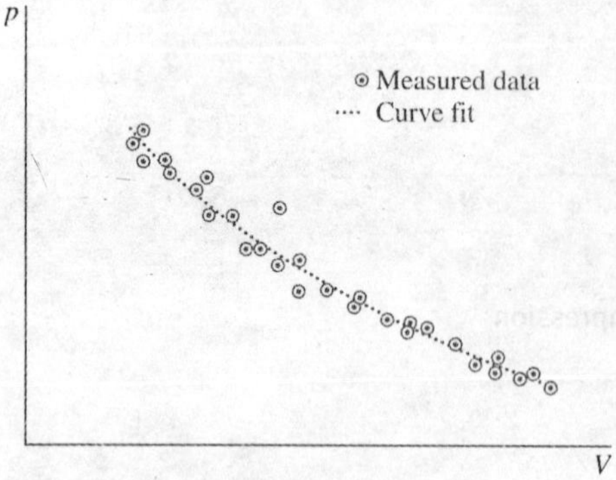

Fig. 2.5 Pressure–volume data.

To perform the integral of Eq. 2.17 requires a relationship between the gas pressure *at the moving boundary* and the system volume. However, due to nonequilibrium effects during an *actual* expansion or compression process, this relationship may be difficult, or even impossible, to obtain. In the cylinder of an automobile engine, for example, combustion and other nonequilibrium effects give rise to nonuniformities throughout the cylinder. Accordingly, if a pressure transducer were mounted on the cylinder head, the recorded output might provide only an approximation for the pressure at the piston face required by Eq. 2.17. Moreover, even when the measured pressure is essentially equal to that at the piston face, scatter might exist in the pressure–volume data, as illustrated in Fig. 2.5. Still, performing the integral of Eq. 2.17 based on a curve fitted to the data could give a *plausible estimate* of the work. We will see later that in some cases where lack of the required pressure–volume relationship keeps us from evaluating the work from Eq. 2.17, the work can be determined alternatively from an *energy balance* (Sec. 2.5).

2.2.5 Expansion or Compression Work in Quasiequilibrium Processes

quasiequilibrium process

Processes are sometime modeled as an idealized type of process called a ***quasiequilibrium (or quasistatic) process.*** A quasiequilibrium process is one in which the departure from thermodynamic equilibrium is at most infinitesimal. All states through which the system passes in a quasiequilibrium process may be considered equilibrium states. Because nonequilibrium effects are inevitably present during actual processes, systems of engineering interest can at best approach, but never realize, a quasiequilibrium process. Still the quasiequilibrium process plays a role in our study of engineering thermodynamics. For details, see the box.

To consider how a gas (or liquid) might be expanded or compressed in a quasiequilibrium fashion, refer to Fig. 2.6, which shows a system consisting of a gas initially at an equilibrium state. As shown in the figure, the gas pressure is maintained uniformly throughout by a number of small masses resting on the freely moving piston. Imagine that one of the masses is removed, allowing the piston to move upward as the gas expands slightly. During such an expansion, the state of the gas would depart only slightly from equilibrium. The system would eventually come to a new equilibrium state, where the pressure and all other intensive properties would again be uniform in value. Moreover, were the mass replaced, the gas would be restored to its initial state, while again the departure from equilibrium would be slight. If several

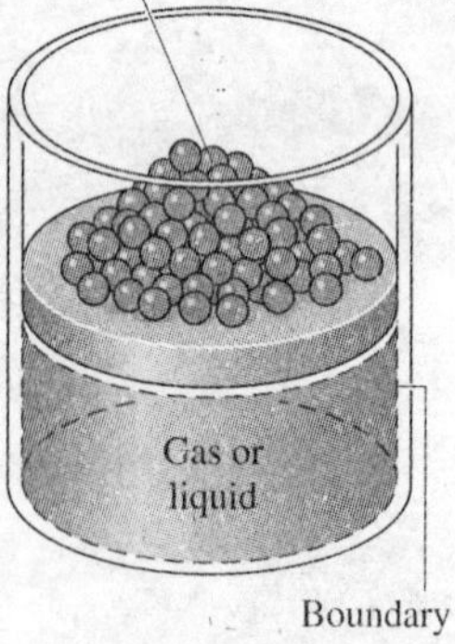

Fig. 2.6 Illustration of a quasiequilibrium expansion or compression.

Using the Quasiequilibrium Process Concept

Our interest in the quasiequilibrium process concept stems mainly from two considerations:

- Simple thermodynamic models giving at least *qualitative* information about the behavior of actual systems of interest often can be developed using the quasiequilibrium process concept. This is akin to the use of idealizations such as the point mass or the frictionless pulley in mechanics for the purpose of simplifying an analysis.
- The quasiequilibrium process concept is instrumental in deducing relationships that exist among the properties of systems at equilibrium (Chaps. 3, 6, and 11).

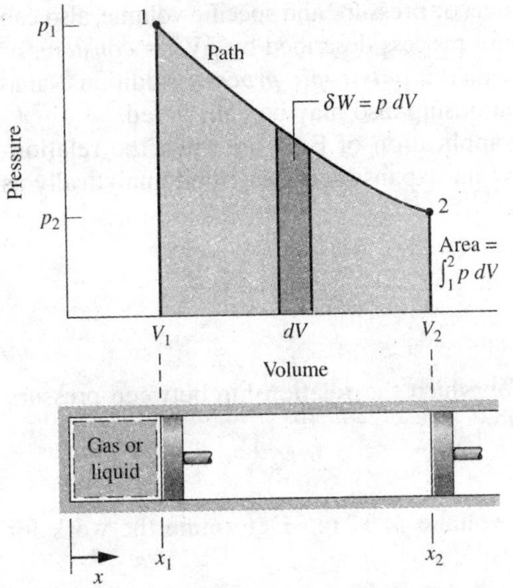

Fig. 2.7 Work of a quasiequilibrium expansion or compression process.

of the masses were removed one after another, the gas would pass through a sequence of equilibrium states without ever being far from equilibrium. In the limit as the increments of mass are made vanishingly small, the gas would undergo a quasiequilibrium expansion process. A quasiequilibrium compression can be visualized with similar considerations.

Equation 2.17 can be applied to evaluate the work in quasiequilibrium expansion or compression processes. For such idealized processes the pressure p in the equation is the pressure of the entire quantity of gas (or liquid) undergoing the process, and not just the pressure at the moving boundary. The relationship between the pressure and volume may be graphical or analytical. Let us first consider a graphical relationship.

A graphical relationship is shown in the pressure–volume diagram (p–V diagram) of Fig. 2.7. Initially, the piston face is at position x_1, and the gas pressure is p_1; at the conclusion of a quasiequilibrium expansion process the piston face is at position x_2, and the pressure is reduced to p_2. At *each* intervening piston position, the uniform pressure throughout the gas is shown as a point on the diagram. The curve, or *path*, connecting states 1 and 2 on the diagram represents the equilibrium states through which the system has passed during the process. The work done by the gas on the piston during the expansion is given by $\int p\,dV$, which can be interpreted as the area under the curve of pressure versus volume. Thus, the shaded area on Fig. 2.7 is equal to the work for the process. Had the gas been *compressed* from 2 to 1 along the same path on the p–V diagram, the *magnitude* of the work would be the same, but the sign would be negative, indicating that for the compression the energy transfer was from the piston to the gas.

The area interpretation of work in a quasiequilibrium expansion or compression process allows a simple demonstration of the idea that work depends on the process. This can be brought out by referring to Fig. 2.8. Suppose the gas in a piston–cylinder assembly goes from an initial equilibrium state 1 to a final equilibrium state 2 along two different paths, labeled A and B on Fig. 2.8. Since the area beneath each path represents the work for that process, the work depends on the details of the process as defined by the particular curve and not just on the end states. Using the test for a property given in Sec. 1.3, we can conclude again (Sec. 2.2.1) that *work is not a property.* The value of work depends on the nature of the process between the end states.

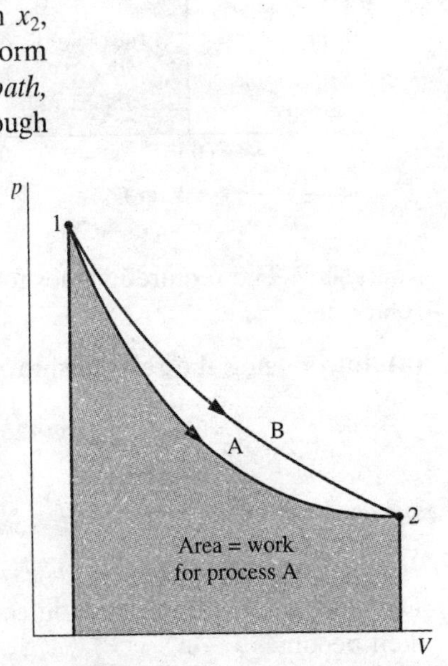

Fig. 2.8 Illustration that work depends on the process.

polytropic process

The relation between pressure and volume, or pressure and specific volume, also can be described analytically. A quasiequilibrium process described by $pV^n = constant$, or $pv^n = constant$, where n is a constant, is called a ***polytropic process.*** Additional analytical forms for the pressure–volume relationship also may be considered.

The example to follow illustrates the application of Eq. 2.17 when the relationship between pressure and volume during an expansion is described analytically as $pV^n = constant$.

Example 2.1 EVALUATING EXPANSION WORK

A gas in a piston–cylinder assembly undergoes an expansion process for which the relationship between pressure and volume is given by

$$pV^n = constant$$

The initial pressure is 3 bar, the initial volume is 0.1 m³, and the final volume is 0.2 m³. Determine the work for the process, in kJ, if (a) $n = 1.5$, (b) $n = 1.0$, and (c) $n = 0$.

Solution

Known: A gas in a piston–cylinder assembly undergoes an expansion for which $pV^n = constant$.

Find: Evaluate the work if (a) $n = 1.5$, (b) $n = 1.0$, (c) $n = 0$.

Schematic and Given Data: The given p–V relationship and the given data for pressure and volume can be used to construct the accompanying pressure–volume diagram of the process.

1

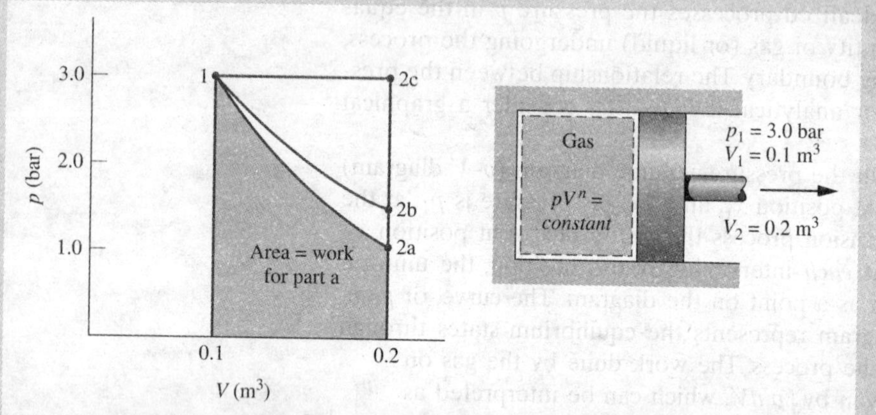

Engineering Model:

1. The gas is a closed system.
2. The moving boundary is the only work mode.
3. The expansion is a polytropic process. 2

Fig. E2.1

Analysis: The required values for the work are obtained by integration of Eq. 2.17 using the given pressure–volume relation.

(a) Introducing the relationship $p = constant/V^n$ into Eq. 2.17 and performing the integration

$$W = \int_{V_1}^{V_2} p\,dV = \int_{V_1}^{V_2} \frac{constant}{V^n}\,dV$$
$$= \frac{(constant)\,V_2^{1-n} - (constant)\,V_1^{1-n}}{1-n}$$

The constant in this expression can be evaluated at either end state: $constant = p_1V_1^n = p_2V_2^n$. The work expression then becomes

$$W = \frac{(p_2V_2^n)V_2^{1-n} - (p_1V_1^n)\,V_1^{1-n}}{1-n} = \frac{p_2V_2 - p_1V_1}{1-n} \quad (1)$$

This expression is valid for all values of n except $n = 1.0$. The case $n = 1.0$ is taken up in part (b).

To evaluate W, the pressure at state 2 is required. This can be found by using $p_1V_1^n = p_2V_2^n$, which on rearrangement yields

$$p_2 = p_1\left(\frac{V_1}{V_2}\right)^n = (3 \text{ bar})\left(\frac{0.1}{0.2}\right)^{1.5} = 1.06 \text{ bar}$$

Accordingly,

3 $$W = \left(\frac{(1.06 \text{ bar})(0.2 \text{ m}^3) - (3)(0.1)}{1 - 1.5}\right)\left|\frac{10^5 \text{ N/m}^2}{1 \text{ bar}}\right|\left|\frac{1 \text{ kJ}}{10^3 \text{ N} \cdot \text{m}}\right|$$

$$= +17.6 \text{ kJ}$$

(b) For $n = 1.0$, the pressure–volume relationship is $pV = constant$ or $p = constant/V$. The work is

$$W = constant \int_{V_1}^{V_2} \frac{dV}{V} = (constant) \ln \frac{V_2}{V_1} = (p_1V_1) \ln \frac{V_2}{V_1} \qquad (2)$$

Substituting values

$$W = (3 \text{ bar})(0.1 \text{ m}^3)\left|\frac{10^5 \text{ N/m}^2}{1 \text{ bar}}\right|\left|\frac{1 \text{ kJ}}{10^3 \text{ N} \cdot \text{m}}\right| \ln\left(\frac{0.2}{0.1}\right) = +20.79 \text{ kJ}$$

4 **(c)** For $n = 0$, the pressure–volume relation reduces to $p = constant$, and the integral becomes $W = p(V_2 - V_1)$, which is a special case of the expression found in part (a). Substituting values and converting units as above, $W = +30$ kJ.

---

1 In each case, the work for the process can be interpreted as the area under the curve representing the process on the accompanying p–V diagram. Note that the relative areas are in agreement with the numerical results.

2 The assumption of a polytropic process is significant. If the given pressure–volume relationship were obtained as a fit to experimental pressure–volume data, the value of $\int p\,dV$ would provide a plausible estimate of the work only when the measured pressure is essentially equal to that exerted at the piston face.

3 Observe the use of unit conversion factors here and in part (b).

4 It is not necessary to identify the gas (or liquid) contained within the piston–cylinder assembly. The calculated values for W are determined by the process path and the end states. However, if it is desired to evaluate a property such as temperature, both the nature and amount of the substance must be provided because appropriate relations among the properties of the particular substance would then be required.

Skills Developed: *Ability to...*
- ✓ apply the problem-solving methodology.
- ✓ define a closed system and identify interactions on its boundary.
- ✓ evaluate work using Eq. 2.17.
- ✓ apply the pressure–volume relation $pV^n = constant$.

Quick Quiz

Evaluate the work, in kJ, for a two-step process consisting of an expansion with $n = 1.0$ from $p_1 = 3$ bar, $V_1 = 0.1$ m^3 to $V = 0.15$ m^3, followed by an expansion with $n = 0$ from $V = 0.15$ m^3 to $V_2 = 0.2$ m^3.

Ans. 22.16 kJ.

2.2.6 Further Examples of Work

To broaden our understanding of the work concept, we now briefly consider several other examples of work.

Extension of a Solid Bar

Consider a system consisting of a solid bar under tension, as shown in Fig. 2.9. The bar is fixed at $x = 0$, and a force F is applied at the other end. Let the force be represented as $F = \sigma \mathrm{A}$, where A is the cross-sectional area of the bar and σ the *normal stress acting at the end* of the bar. The work done as the end of the bar moves a distance dx is given by $\delta W = -\sigma \mathrm{A}\, dx$. The minus sign is required because work is done *on* the bar when dx is positive. The work for a change in length from x_1 to x_2 is found by integration

$$W = -\int_{x_1}^{x_2} \sigma \mathrm{A}\, dx \tag{2.18}$$

Equation 2.18 for a solid is the counterpart of Eq. 2.17 for a gas undergoing an expansion or compression.

Stretching of a Liquid Film

Figure 2.10 shows a system consisting of a liquid film suspended on a wire frame. The two surfaces of the film support the thin liquid layer inside by the effect of *surface tension,* owing to microscopic forces between molecules near the liquid–air interfaces. These forces give rise to a macroscopically measurable force perpendicular to any line in the surface. The force per unit length across such a line is the surface tension. Denoting the surface tension *acting at the movable wire* by τ, the force F indicated on the figure can be expressed as $F = 2l\tau$, where the factor 2 is introduced because two film surfaces act at the wire. If the movable wire is displaced by dx, the work is given by $\delta W = -2l\tau\, dx$. The minus sign is required because work is done *on* the system when dx is positive. Corresponding to a displacement dx is a change in the total area of the surfaces in contact with the wire of $dA = 2l\, dx$, so the expression for work can be written alternatively as $\delta W = -\tau\, d\mathrm{A}$. The work for an increase in surface area from A_1 to A_2 is found by integrating this expression

$$W = -\int_{\mathrm{A}_1}^{\mathrm{A}_2} \tau\, d\mathrm{A} \tag{2.19}$$

Power Transmitted by a Shaft

A rotating shaft is a commonly encountered machine element. Consider a shaft rotating with angular velocity ω and exerting a torque $\mathcal{T}$ on its surroundings. Let the torque

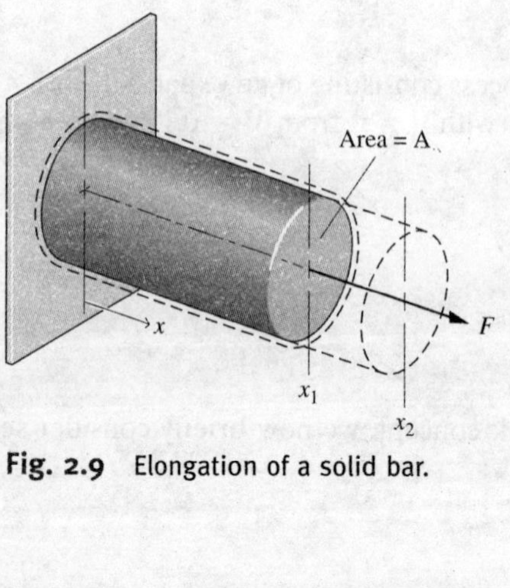

Fig. 2.9 Elongation of a solid bar.

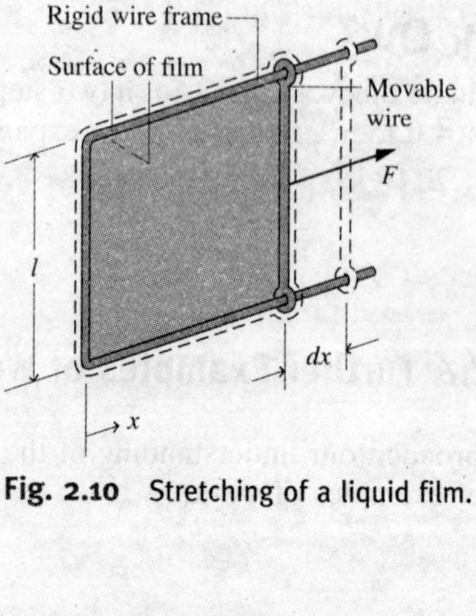

Fig. 2.10 Stretching of a liquid film.

be expressed in terms of a tangential force F_t and radius R: $\mathcal{T} = F_t R$. The velocity at the point of application of the force is $V = R\omega$, where ω is in radians per unit time. Using these relations with Eq. 2.13, we obtain an expression for the *power* transmitted from the shaft to the surroundings

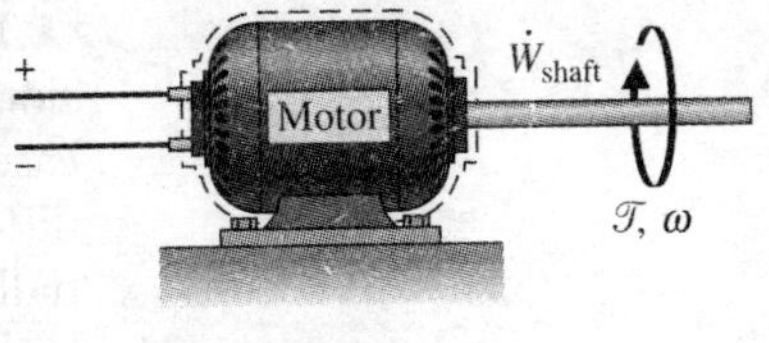

$$\dot{W} = F_t V = (\mathcal{T}/R)(R\omega) = \mathcal{T}\omega \tag{2.20}$$

A related case involving a gas stirred by a paddle wheel is considered in the discussion of Fig. 2.3.

Electric Power

Shown in Fig. 2.11 is a system consisting of an electrolytic cell. The cell is connected to an external circuit through which an electric current, i, is flowing. The current is driven by the electrical potential difference $\mathcal{E}$ existing across the terminals labeled a and b. That this type of interaction can be classed as work is considered in the discussion of Fig. 2.3.

The rate of energy transfer by work, or the power, is

$$\dot{W} = -\mathcal{E}i \tag{2.21}$$

Since the current i equals dZ/dt, the work can be expressed in differential form as

$$\delta W = -\mathcal{E}\,dZ \tag{2.22}$$

where dZ is the amount of electrical charge that flows into the system. The minus signs are required to be in accord with our previously stated sign convention for work. When the power is evaluated in terms of the watt, and the unit of current is the ampere (an SI base unit), the unit of electric potential is the volt, defined as 1 watt per ampere.

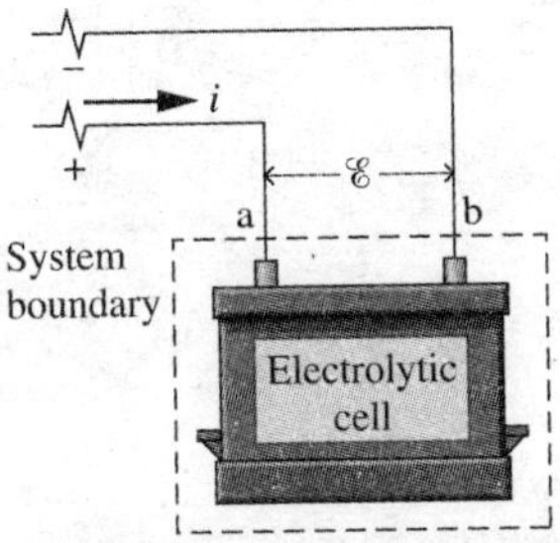

Fig. 2.11 Electrolytic cell used to discuss electric power.

Work Due to Polarization or Magnetization

Let us next refer briefly to the types of work that can be done on systems residing in electric or magnetic fields, known as the work of polarization and magnetization, respectively. From the microscopic viewpoint, electrical dipoles within dielectrics resist turning, so work is done when they are aligned by an electric field. Similarly, magnetic dipoles resist turning, so work is done on certain other materials when their magnetization is changed. Polarization and magnetization give rise to *macroscopically* detectable changes in the total dipole moment as the particles making up the material are given new alignments. In these cases the work is associated with forces imposed on the overall system by fields in the surroundings. Forces acting on the material in the system interior are called *body forces*. For such forces the appropriate displacement in evaluating work is the displacement of the matter on which the body force acts.

2.2.7 Further Examples of Work in Quasiequilibrium Processes

Systems other than a gas or liquid in a piston–cylinder assembly also can be envisioned as undergoing processes in a quasiequilibrium fashion. To apply the quasiequilibrium process concept in any such case, it is necessary to conceive of an *ideal situation* in which the external forces acting on the system can be varied so slightly that the resulting imbalance is infinitesimal. As a consequence, the system undergoes a process without ever departing significantly from thermodynamic equilibrium.

The extension of a solid bar and the stretching of a liquid surface can readily be envisioned to occur in a quasiequilibrium manner by direct analogy to the piston–cylinder case. For the bar in Fig. 2.9 the external force can be applied in such a way that it differs only slightly from the opposing force within. The normal stress is then essentially uniform throughout and can be determined as a function of the instantaneous length: $\sigma = \sigma(x)$. Similarly, for the liquid film shown in Fig. 2.10 the external force can be applied to the movable wire in such a way that the force differs only slightly from the opposing force within the film. During such a process, the surface tension is essentially uniform throughout the film and is functionally related to the instantaneous area: $\tau = \tau(\mathrm{A})$. In each of these cases, once the required functional relationship is known, the work can be evaluated using Eq. 2.18 or 2.19, respectively, in terms of properties of the system as a whole as it passes through equilibrium states.

Other systems also can be imagined as undergoing quasiequilibrium processes. For example, it is possible to envision an electrolytic cell being charged or discharged in a quasiequilibrium manner by adjusting the potential difference across the terminals to be slightly greater, or slightly less, than an ideal potential called the cell *electromotive force* (emf). The energy transfer by work for passage of a differential quantity of charge *to* the cell, dZ, is given by the relation

$$\delta W = -\mathscr{E}\, dZ \tag{2.23}$$

In this equation $\mathscr{E}$ denotes the cell emf, an intensive property of the cell, and not just the potential difference across the terminals as in Eq. 2.22.

Consider next a dielectric material residing in a *uniform electric field*. The energy transferred by work from the field when the polarization is increased slightly is

$$\delta W = -\mathbf{E} \cdot d(V\mathbf{P}) \tag{2.24}$$

where the vector $\mathbf{E}$ is the electric field strength within the system, the vector $\mathbf{P}$ is the electric dipole moment per unit volume, and V is the volume of the system. A similar equation for energy transfer by work from a *uniform magnetic field* when the magnetization is increased slightly is

$$\delta W = -\mu_0 \mathbf{H} \cdot d(V\mathbf{M}) \tag{2.25}$$

where the vector $\mathbf{H}$ is the magnetic field strength within the system, the vector $\mathbf{M}$ is the magnetic dipole moment per unit volume, and μ_0 is a constant, the permeability of free space. The minus signs appearing in the last three equations are in accord with our previously stated sign convention for work: W takes on a negative value when the energy transfer is *into* the system.

2.2.8 Generalized Forces and Displacements

The similarity between the expressions for work in the quasiequilibrium processes considered thus far should be noted. In each case, the work expression is written in the form of an intensive property and the differential of an extensive property. This is brought out by the following expression, which allows for one or more of these work modes to be involved in a process

$$\delta W = p\, dV - \sigma d(\mathrm{A}x) - \tau\, d\mathrm{A} - \mathscr{E}\, dZ - \mathbf{E} \cdot d(V\mathbf{P}) - \mu_0 \mathbf{H} \cdot d(V\mathbf{M}) + \cdots \tag{2.26}$$

where the last three dots represent other products of an intensive property and the differential of a related extensive property that account for work. Because of the notion of work being a product of force and displacement, the intensive property in these relations is sometimes referred to as a "generalized" force and the extensive property as a "generalized" displacement, even though the quantities making up the work expressions may not bring to mind actual forces and displacements.

Owing to the underlying quasiequilibrium restriction, Eq. 2.26 does not represent every type of work of practical interest. An example is provided by a paddle wheel that stirs a gas or liquid taken as the system. Whenever any shearing action takes place, the system necessarily passes through nonequilibrium states. To appreciate more fully the implications of the quasiequilibrium process concept requires consideration of the second law of thermodynamics, so this concept is discussed again in Chap. 5 after the second law has been introduced.

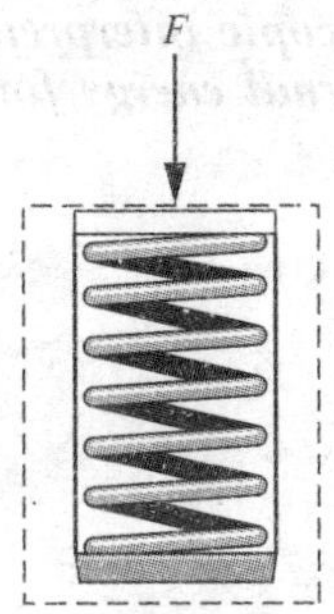

2.3 Broadening Our Understanding of Energy

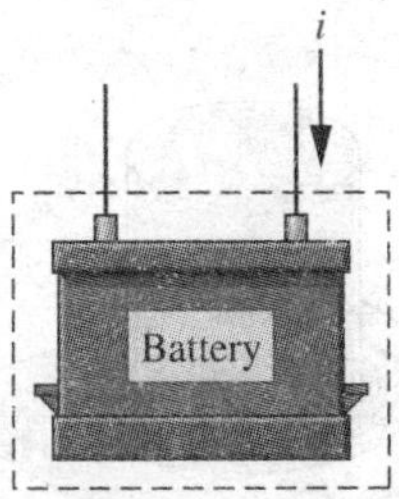

The objective in this section is to use our deeper understanding of work developed in Sec. 2.2 to broaden our understanding of the energy of a system. In particular, we consider the *total* energy of a system, which includes kinetic energy, gravitational potential energy, and other forms of energy. The examples to follow illustrate some of these forms of energy. Many other examples could be provided that enlarge on the same idea.

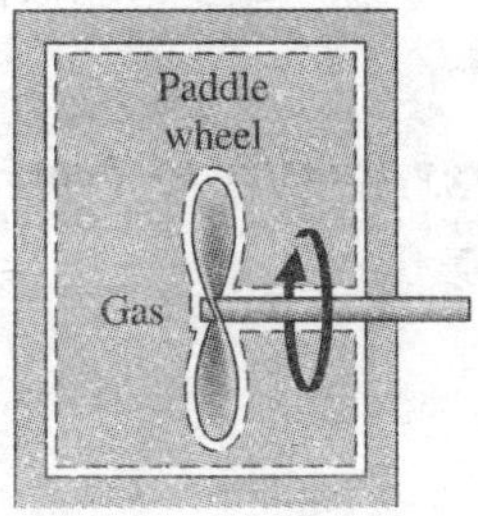

When work is done to compress a spring, energy is stored within the spring. When a battery is charged, the energy stored within it is increased. And when a gas (or liquid) initially at an equilibrium state in a closed, insulated vessel is stirred vigorously and allowed to come to a final equilibrium state, the energy of the gas is increased in the process. In each of these examples the change in system energy cannot be attributed to changes in the system's *overall* kinetic or gravitational potential energy as given by Eqs. 2.5 and 2.10, respectively. The change in energy can be accounted for in terms of *internal energy,* as considered next.

In engineering thermodynamics the change in the total energy of a system is considered to be made up of three *macroscopic* contributions. One is the change in kinetic energy, associated with the motion of the system *as a whole* relative to an external coordinate frame. Another is the change in gravitational potential energy, associated with the position of the system *as a whole* in the earth's gravitational field. All other energy changes are lumped together in the ***internal energy*** of the system. Like kinetic energy and gravitational potential energy, *internal energy is an extensive property* of the system, as is the total energy.

internal energy

Internal energy is represented by the symbol U, and the change in internal energy in a process is $U_2 - U_1$. The specific internal energy is symbolized by u or $\bar{u}$, respectively, depending on whether it is expressed on a unit mass or per mole basis.

The change in the total energy of a system is

$$E_2 - E_1 = (U_2 - U_1) + (\mathrm{KE}_2 - \mathrm{KE}_1) + (\mathrm{PE}_2 - \mathrm{PE}_1)$$

or

$$\Delta E = \Delta U + \Delta \mathrm{KE} + \Delta \mathrm{PE} \tag{2.27}$$

All quantities in Eq. 2.27 are expressed in terms of the energy units previously introduced.

The identification of internal energy as a macroscopic form of energy is a significant step in the present development, for it sets the concept of energy in thermodynamics apart from that of mechanics. In Chap. 3 we will learn how to evaluate changes in internal energy for practically important cases involving gases, liquids, and solids by using empirical data.

To further our understanding of internal energy, consider a system we will often encounter in subsequent sections of the book, a system consisting of a gas contained

microscopic interpretation of internal energy for a gas

in a tank. Let us develop a ***microscopic interpretation of internal energy*** by thinking of the energy attributed to the motions and configurations of the individual molecules, atoms, and subatomic particles making up the matter in the system. Gas molecules move about, encountering other molecules or the walls of the container. Part of the internal energy of the gas is the *translational* kinetic energy of the molecules. Other contributions to the internal energy include the kinetic energy due to *rotation* of the molecules relative to their centers of mass and the kinetic energy associated with *vibrational* motions within the molecules. In addition, energy is stored in the chemical bonds between the atoms that make up the molecules. Energy storage on the atomic level includes energy associated with electron orbital states, nuclear spin, and binding forces in the nucleus. In dense gases, liquids, and solids, intermolecular forces play an important role in affecting the internal energy.

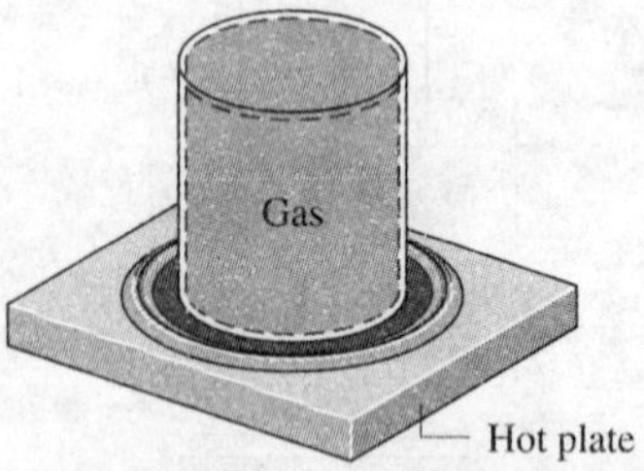

2.4 Energy Transfer by Heat

Thus far, we have considered quantitatively only those interactions between a system and its surroundings that can be classed as work. However, closed systems also can interact with their surroundings in a way that cannot be categorized as work. ➡ **FOR EXAMPLE...** when a gas in a rigid container interacts with a hot plate, the energy of the gas is increased even though no work is done. ⬅ This type of interaction is called an ***energy transfer by heat.***

energy transfer by heat

On the basis of experiment, beginning with the work of Joule in the early part of the nineteenth century, we know that energy transfers by heat are induced only as a result of a temperature difference between the system and its surroundings and occur only in the direction of decreasing temperature. Because the underlying concept is so important in thermodynamics, this section is devoted to a further consideration of energy transfer by heat.

2.4.1 Sign Convention, Notation, and Heat Transfer Rate

The symbol Q denotes an amount of energy transferred across the boundary of a system in a heat interaction with the system's surroundings. Heat transfer *into* a system is taken to be *positive,* and heat transfer *from* a system is taken as *negative.*

sign convention for heat transfer

$Q > 0$: heat transfer *to* the system
$Q < 0$: heat transfer *from* the system

This ***sign convention*** is used throughout the book. However, as was indicated for work, it is sometimes convenient to show the direction of energy transfer by an arrow on a sketch of the system. Then the heat transfer is regarded as positive in the direction of the arrow.

The sign convention for heat transfer is just the *reverse* of the one adopted for work, where a positive value for W signifies an energy transfer *from* the system to the surroundings. These signs for heat and work are a legacy from engineers and scientists who were concerned mainly with steam engines and other devices that develop a work output from an energy input by heat transfer. For such applications, it was convenient to regard both the work developed and the energy input by heat transfer as positive quantities.

heat is not a property

The value of a heat transfer depends on the details of a process and not just the end states. Thus, like work, ***heat is not a property,*** and its differential is written as δQ. The amount of energy transfer by heat for a process is given by the integral

$$Q = \int_1^2 \delta Q \tag{2.28}$$

where the limits mean "from state 1 to state 2" and do not refer to the values of heat at those states. As for work, the notion of "heat" at a state has no meaning, and the integral should *never* be evaluated as $Q_2 - Q_1$.

rate of heat transfer

The net ***rate of heat transfer*** is denoted by $\dot{Q}$. In principle, the amount of energy transfer by heat during a period of time can be found by integrating from time t_1 to time t_2

$$Q = \int_{t_1}^{t_2} \dot{Q}\, dt \tag{2.29}$$

To perform the integration, it is necessary to know how the rate of heat transfer varies with time.

In some cases it is convenient to use the *heat flux*, $\dot{q}$, which is the heat transfer rate per unit of system surface area. The net rate of heat transfer, $\dot{Q}$, is related to the heat flux $\dot{q}$ by the integral

$$\dot{Q} = \int_{A} \dot{q}\, dA \tag{2.30}$$

where A represents the area on the boundary of the system where heat transfer occurs.

The units for heat transfer Q and heat transfer rate $\dot{Q}$ are the same as those introduced previously for W and $\dot{W}$, respectively. The units for the heat flux are those of the heat transfer rate per unit area: kW/m^2.

adiabatic

The word ***adiabatic*** means *without heat transfer.* Thus, if a system undergoes a process involving no heat transfer with its surroundings, that process is called an *adiabatic process.*

Bio... connections

Medical researchers have found that by gradually increasing the temperature of cancerous tissue to 41–45°C the effectiveness of chemotherapy and radiation therapy is enhanced for some patients. Different approaches can be used, including raising the temperature of the entire body with heating devices and, more selectively, by beaming microwaves or ultrasound onto the tumor or affected organ. Speculation about why a temperature increase may be beneficial varies. Some say it helps chemotherapy penetrate certain tumors more readily by dilating blood vessels. Others think it helps radiation therapy by increasing the amount of oxygen in tumor cells, making them more receptive to radiation. Researchers report that further study is needed before the efficacy of this approach is established and the mechanisms whereby it achieves positive results are known.

2.4.2 Heat Transfer Modes

Methods based on experiment are available for evaluating energy transfer by heat. These methods recognize two basic transfer mechanisms: *conduction* and *thermal radiation*. In addition, empirical relationships are available for evaluating energy transfer involving a *combined* mode called *convection*. A brief description of each of these is given next. A detailed consideration is left to a course in engineering heat transfer, where these topics are studied in depth.

Conduction

Energy transfer by *conduction* can take place in solids, liquids, and gases. Conduction can be thought of as the transfer of energy from the more energetic particles of a substance to adjacent particles that are less energetic due to interactions between particles. The time rate of energy transfer by conduction is quantified macroscopically by *Fourier's law*. As an elementary application, consider Fig. 2.12 showing a

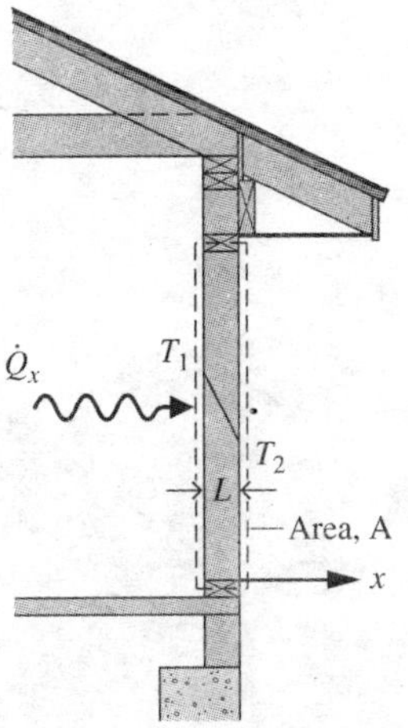

Fig. 2.12 Illustration of Fourier's conduction law.

plane wall of thickness L at steady state, where the temperature $T(x)$ varies linearly with position x. By ***Fourier's law***, the rate of heat transfer across any plane normal to the x direction, $\dot{Q}_x$, is proportional to the wall area, A, and the temperature gradient in the x direction, dT/dx

Fourier's law

$$\dot{Q}_x = -\kappa A \frac{dT}{dx} \tag{2.31}$$

where the proportionality constant κ is a property called the *thermal conductivity*. The minus sign is a consequence of energy transfer in the direction of *decreasing* temperature. ➡ **FOR EXAMPLE...** in the case of Fig. 2.12 the temperature varies linearly; thus, the temperature gradient is

$$\frac{dT}{dx} = \frac{T_2 - T_1}{L} (< 0)$$

and the rate of heat transfer in the x direction is then

$$\dot{Q}_x = -\kappa A \left[\frac{T_2 - T_1}{L}\right]$$ ⬅

Values of thermal conductivity are given in Table A-19 for common materials. Substances with large values of thermal conductivity such as copper are good conductors, and those with small conductivities (cork and polystyrene foam) are good insulators.

Radiation

Thermal radiation is emitted by matter as a result of changes in the electronic configurations of the atoms or molecules within it. The energy is transported by electromagnetic waves (or photons). Unlike conduction, thermal radiation requires no intervening medium to propagate and can even take place in a vacuum. Solid surfaces, gases, and liquids all emit, absorb, and transmit thermal radiation to varying degrees. The rate at which energy is emitted, $\dot{Q}_e$, *from* a surface of area A is quantified macroscopically by a modified form of the ***Stefan–Boltzmann law***

Stefan–Boltzmann law

$$\dot{Q}_e = \varepsilon \sigma A T_b^4 \tag{2.32}$$

which shows that thermal radiation is associated with the fourth power of the absolute temperature of the surface, T_b. The emissivity, ε, is a property of the surface that indicates how effectively the surface radiates $(0 \leq \varepsilon \leq 1.0)$, and σ is the Stefan–Boltzmann constant:

$$\sigma = 5.67 \times 10^{-8} \text{ W/m}^2 \cdot \text{K}^4$$

In general, the *net* rate of energy transfer by thermal radiation between two surfaces involves relationships among the properties of the surfaces, their orientations with respect to each other, the extent to which the intervening medium scatters, emits, and absorbs thermal radiation, and other factors. A special case that occurs frequently is radiation exchange between a surface at temperature T_b and a much larger surrounding surface at T_s, as shown in Fig. 2.13. The *net* rate of radiant exchange between the smaller surface, whose area is A and emissivity is ε, and the larger surroundings is

$$\dot{Q}_e = \varepsilon \sigma A [T_b^4 - T_s^4] \tag{2.33}$$

Convection

Energy transfer between a solid surface at a temperature T_b and an adjacent gas or liquid at another temperature T_f plays a prominent role in the performance of

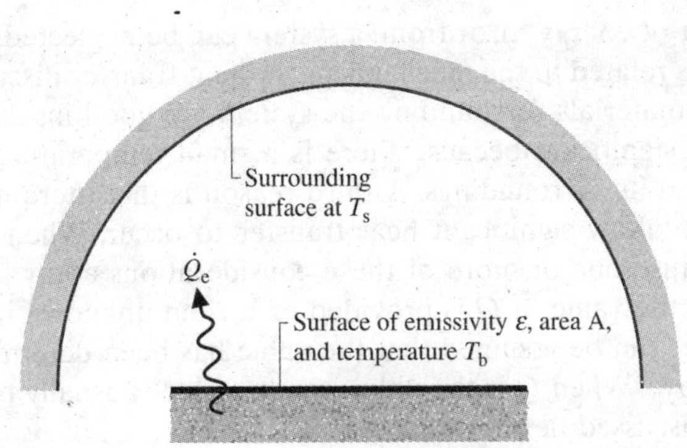

Fig. 2.13 Net radiation exchange.

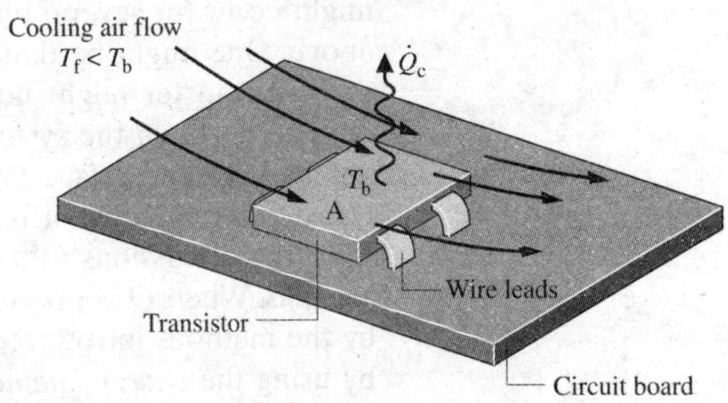

Fig. 2.14 Illustration of Newton's law of cooling.

many devices of practical interest. This is commonly referred to as *convection*. As an illustration, consider Fig. 2.14, where $T_b > T_f$. In this case, energy is transferred *in the direction indicated by the arrow* due to the *combined* effects of conduction within the air and the bulk motion of the air. The rate of energy transfer *from* the surface *to* the air can be quantified by the following *empirical* expression:

$$\dot{Q}_c = hA(T_b - T_f) \tag{2.34}$$

known as ***Newton's law of cooling.*** In Eq. 2.34, A is the surface area and the proportionality factor, h, is called the *heat transfer coefficient.* In subsequent applications of Eq. 2.34, a minus sign may be introduced on the right side to conform to the sign convention for heat transfer introduced in Sec. 2.4.1.

Newton's law of cooling

The heat transfer coefficient is *not* a thermodynamic property. It is an empirical parameter that incorporates into the heat transfer relationship the nature of the flow pattern near the surface, the fluid properties, and the geometry. When fans or pumps cause the fluid to move, the value of the heat transfer coefficient is generally greater than when relatively slow buoyancy-induced motions occur. These two general categories are called *forced* and *free* (or natural) convection, respectively. Table 2.1 provides typical values of the convection heat transfer coefficient for forced and free convection.

2.4.3 Closing Comments

The first step in a thermodynamic analysis is to define the system. It is only after the system boundary has been specified that possible heat interactions with the surroundings are considered, for these are *always* evaluated at the system boundary. In ordinary conversation, the term *heat* is often used when the word *energy* would be more correct thermodynamically. For example, one might hear, "Please close the door or 'heat' will be lost." In *thermodynamics,* heat refers only to a particular means whereby energy is transferred. It does not refer to what is being transferred between systems or to what is stored within systems. Energy is transferred and stored, not heat.

Table 2.1

Typical Values of the Convection Heat Transfer Coefficient

Applications	h ($W/m^2 \cdot K$)
Free convection	
Gases	2–25
Liquids	50–1,000
Forced convection	
Gases	25–250
Liquids	50–20,000

Sometimes the heat transfer of energy to, or from, a system can be neglected. This might occur for several reasons related to the mechanisms for heat transfer discussed above. One might be that the materials surrounding the system are good insulators, or heat transfer might not be significant because there is a small temperature difference between the system and its surroundings. A third reason is that there might not be enough surface area to allow significant heat transfer to occur. When heat transfer is neglected, it is because one or more of these considerations apply.

In the discussions to follow, the value of Q is provided or it is an unknown in the analysis. When Q is provided, it can be assumed that the value has been determined by the methods introduced above. When Q is the unknown, its value is usually found by using the *energy balance,* discussed next.

2.5 Energy Accounting: Energy Balance for Closed Systems

As our previous discussions indicate, the *only ways* the energy of a closed system can be changed are through transfer of energy by work or by heat. Further, based on the experiments of Joule and others, a fundamental aspect of the energy concept is that *energy is conserved;* we call this the ***first law of thermodynamics.*** For further discussion of the first law, see the box.

first law of thermodynamics

These considerations are summarized in words as follows:

$$\begin{bmatrix}\textit{change}\text{ in the amount}\\ \text{of energy contained}\\ \text{within a system}\\ \text{during some time}\\ \text{interval}\end{bmatrix} = \begin{bmatrix}\textit{net}\text{ amount of energy}\\ \text{transferred }\textit{in}\text{ across}\\ \text{the system boundary by}\\ \textit{heat}\text{ transfer during}\\ \text{the time interval}\end{bmatrix} - \begin{bmatrix}\textit{net}\text{ amount of energy}\\ \text{transferred }\textit{out}\text{ across}\\ \text{the system boundary}\\ \text{by }\textit{work}\text{ during the}\\ \text{time interval}\end{bmatrix}$$

This word statement is just an accounting balance for energy, an energy balance. It requires that in any process of a closed system the energy of the system increases or decreases by an amount equal to the net amount of energy transferred across its boundary.

The phrase *net amount* used in the word statement of the energy balance must be carefully interpreted, for there may be heat or work transfers of energy at many different places on the boundary of a system. At some locations the energy transfers may be into the system, whereas at others they are out of the system. The two terms on the right side account for the *net* results of all the energy transfers by heat and work, respectively, taking place during the time interval under consideration.

The ***energy balance*** can be expressed in symbols as

$$E_2 - E_1 = Q - W \tag{2.35a}$$

energy balance

Introducing Eq. 2.27 an alternative form is

$$\Delta\text{KE} + \Delta\text{PE} + \Delta U = Q - W \tag{2.35b}$$

which shows that an energy transfer across the system boundary results in a change in one or more of the macroscopic energy forms: kinetic energy, gravitational potential energy, and internal energy. All previous references to energy as a conserved quantity are included as special cases of Eqs. 2.35.

Note that the algebraic signs before the heat and work terms of Eqs. 2.35 are different. This follows from the sign conventions previously adopted. A minus sign

Joule's Experiments and the First Law

In classic experiments conducted in the early part of the nineteenth century, Joule studied processes by which a closed system can be taken from one equilibrium state to another. In particular, he considered processes that involve work interactions but no thermal interactions between the system and its surroundings. Any such process is an *adiabatic process,* in keeping with the discussion of Sec. 2.4.1.

Based on his experiments Joule deduced that the value of the net work is the same for *all* adiabatic processes between two equilibrium states. In other words, the value of the net work done by or on a closed system undergoing an adiabatic process between two given states *depends solely on the end states* and not on the details of the adiabatic process.

If the net work is the same for all adiabatic processes of a closed system between a given pair of end states, it follows from the definition of property (Sec. 1.3) that the net work for such a process is the change in some property of the system. This property is called *energy.*

Following Joule's reasoning, the *change in energy* between the two states is *defined* by

$$E_2 - E_1 = -W_{ad} \tag{a}$$

where the symbol E denotes the energy of a system and W_{ad} represents the net work for *any* adiabatic process between the two states. The minus sign before the work term is in accord with the previously stated sign convention for work. Finally, note that since any arbitrary value E_1 can be assigned to the energy of a system at a given state 1, no particular significance can be attached to the value of the energy at state 1 or at *any* other state. Only *changes* in the energy of a system have significance.

The foregoing discussion is based on experimental evidence beginning with the experiments of Joule. Because of inevitable experimental uncertainties, it is not possible to prove by measurements that the net work is *exactly* the same for *all* adiabatic processes between the same end states. However, the preponderance of experimental findings supports this conclusion, so it is adopted as a fundamental principle that the work actually is the same. This principle is an alternative formulation of the *first law,* and has been used by subsequent scientists and engineers as a springboard for developing the *conservation of energy* concept and the *energy balance* as we know them today.

appears before W because energy transfer by work *from* the system *to* the surroundings is taken to be positive. A plus sign appears before Q because it is regarded to be positive when the heat transfer of energy is *into* the system *from* the surroundings.

2.5.1 Important Aspects of the Energy Balance

Various special forms of the energy balance can be written. For example, the energy balance in differential form is

$$dE = \delta Q - \delta W \tag{2.36}$$

where dE is the differential of energy, a property. Since Q and W are not properties, their differentials are written as δQ and δW, respectively.

The instantaneous ***time rate form of the energy balance*** is

time rate form of the energy balance

$$\frac{dE}{dt} = \dot{Q} - \dot{W} \tag{2.37}$$

The rate form of the energy balance expressed in words is

$$\begin{bmatrix}\text{time } \textit{rate of change} \\ \text{of the energy} \\ \text{contained within} \\ \text{the system } \textit{at} \\ \textit{time } t\end{bmatrix} = \begin{bmatrix}\text{net } \textit{rate} \text{ at which} \\ \text{energy is being} \\ \text{transferred in} \\ \text{by heat transfer} \\ \textit{at time } t\end{bmatrix} - \begin{bmatrix}\text{net } \textit{rate} \text{ at which} \\ \text{energy is being} \\ \text{transferred out} \\ \text{by work } \textit{at} \\ \textit{time } t\end{bmatrix}$$

Since the time rate of change of energy is given by

$$\frac{dE}{dt} = \frac{d\text{KE}}{dt} + \frac{d\text{PE}}{dt} + \frac{dU}{dt}$$

Equation 2.37 can be expressed alternatively as

$$\frac{d\text{KE}}{dt} + \frac{d\text{PE}}{dt} + \frac{dU}{dt} = \dot{Q} - \dot{W} \tag{2.38}$$

Equations 2.35 through 2.38 provide alternative forms of the energy balance that may be convenient starting points when applying the principle of conservation of energy to closed systems. In Chap. 4 the conservation of energy principle is expressed in forms suitable for the analysis of control volumes. When applying the energy balance in *any* of its forms, it is important to be careful about signs and units and to distinguish carefully between rates and amounts. In addition, it is important to recognize that the location of the system boundary can be relevant in determining whether a particular energy transfer is regarded as heat or work.

➡ **FOR EXAMPLE...** consider Fig. 2.15, in which three alternative systems are shown that include a quantity of a gas (or liquid) in a rigid, well-insulated container. In Fig. 2.15*a*, the gas itself is the system. As current flows through the copper plate, there is an energy transfer from the copper plate to the gas. Since this energy transfer occurs as a result of the temperature difference between the plate and the gas, it is classified as a heat transfer. Next, refer to Fig. 2.15*b*, where the boundary is drawn to include the copper plate. It follows from the thermodynamic definition of work that the energy transfer that occurs as current crosses the boundary of this system must be regarded as work. Finally, in Fig. 2.15*c*, the boundary is located so that no energy is transferred across it by heat or work. ⬅

Thus far, we have been careful to emphasize that the quantities symbolized by W and Q in the foregoing equations account for transfers of *energy* and not transfers of work and heat, respectively. The terms work and heat denote different *means* whereby energy is transferred and not *what* is transferred. However, to achieve economy of expression in subsequent discussions, W and Q are often referred to simply as work and heat transfer, respectively. This less formal manner of speaking is commonly used in engineering practice.

The examples provided in Secs. 2.5.2.–2.5.4 bring out important ideas about energy and the energy balance. They should be studied carefully, and similar approaches should be used when solving the end-of-chapter problems. In this text, most applications of the energy balance will not involve significant kinetic or potential energy changes. Thus, to expedite the solutions of many subsequent examples and end-of-chapter problems, we indicate in the problem statement that such changes can be neglected. If this is not made explicit in a problem statement, you should decide on the basis of the problem at hand how best to handle the kinetic and potential energy terms of the energy balance.

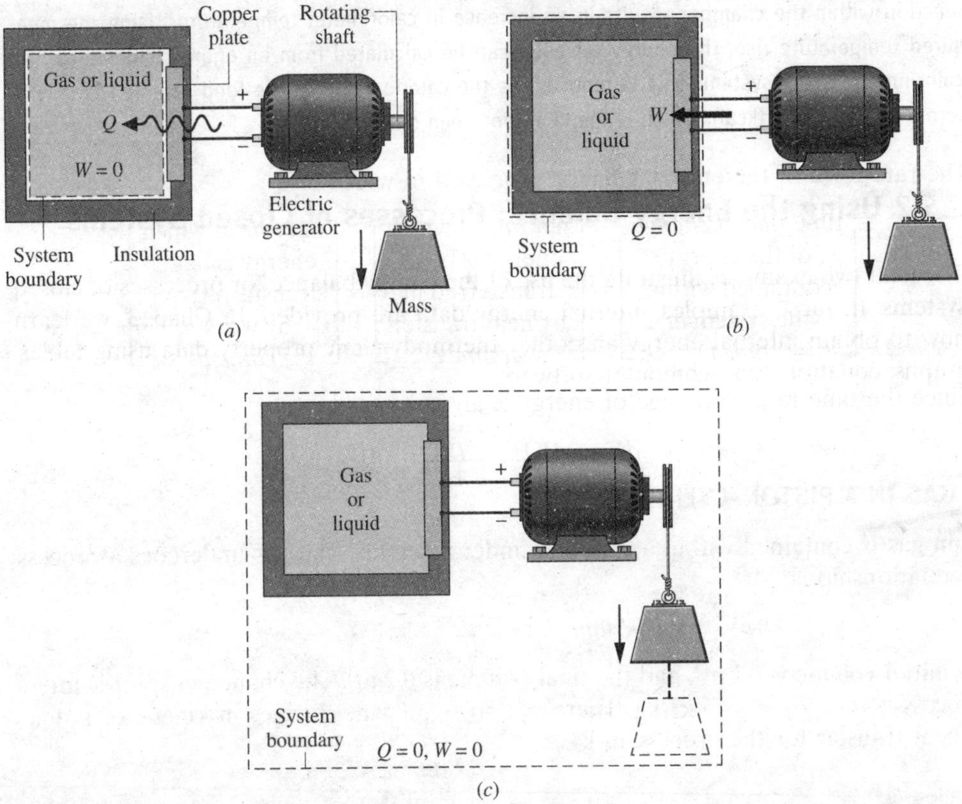

Fig. 2.15 Alternative choices for system boundaries.

Bio... connections

The energy required by animals to sustain life is derived from oxidation of ingested food. We often speak of food being *burned* in the body. This is an appropriate expression because experiments show that when food is burned with oxygen in a chamber, approximately the same energy is released as when the food is oxidized in the body. Such a chamber is the well-insulated, constant-volume *calorimeter* shown in Fig. 2.16.

A carefully weighed food sample is placed in the chamber of the calorimeter together with oxygen (O_2). The entire chamber is submerged in the calorimeter's water bath. The chamber contents are then electrically ignited, fully oxidizing the food sample. The energy released during the

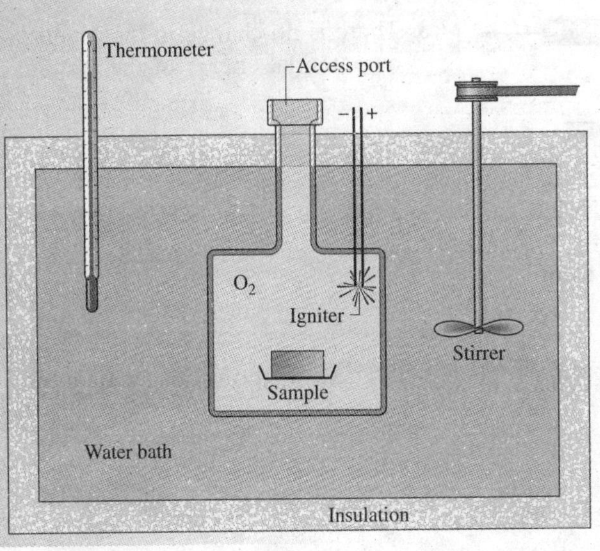

Fig. 2.16 Constant-volume calorimeter.

reaction within the chamber results in an increase in calorimeter temperature. Using the measured temperature rise, the energy released can be calculated from an energy balance for the calorimeter as the system. This is reported as the calorie value of the food sample, usually in terms of kilocalorie (kcal), which is the "Calorie" seen on food labels.

2.5.2 Using the Energy Balance: Processes of Closed Systems

The next two examples illustrate the use of the energy balance for processes of closed systems. In these examples, internal energy data are provided. In Chap. 3, we learn how to obtain internal energy and other thermodynamic property data using tables, graphs, equations, and computer software.

Example 2.2 COOLING A GAS IN A PISTON–CYLINDER

Four-tenths kilogram of a certain gas is contained within a piston–cylinder assembly. The gas undergoes a process for which the pressure–volume relationship is

$$pV^{1.5} = constant$$

The initial pressure is 3 bar, the initial volume is 0.1 m^3, and the final volume is 0.2 m^3. The change in specific internal energy of the gas in the process is $u_2 - u_1 = -55$ kJ/kg. There are no significant changes in kinetic or potential energy. Determine the net heat transfer for the process, in kJ.

Solution

Known: A gas within a piston–cylinder assembly undergoes an expansion process for which the pressure–volume relation and the change in specific internal energy are specified.

Find: Determine the net heat transfer for the process.

Schematic and Given Data:

❶

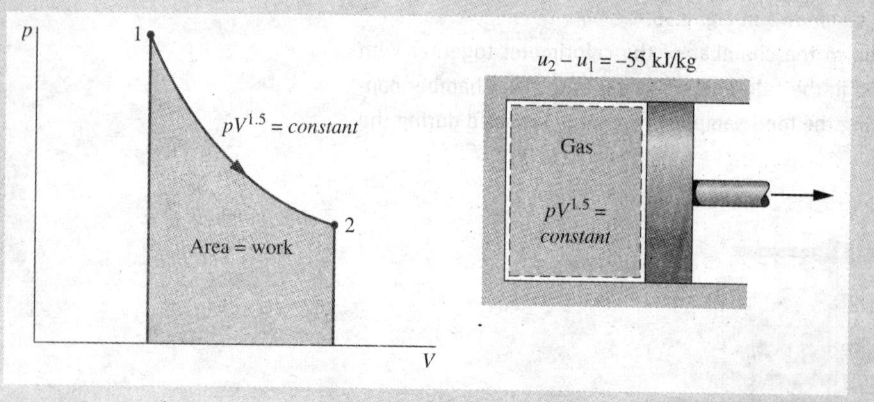

Engineering Model:

1. The gas is a closed system.
2. The process is described by $pV^{1.5} = constant$.
3. There is no change in the kinetic or potential energy of the system.

Fig. E2.2

Analysis: An energy balance for the closed system takes the form

$$\cancel{\Delta KE}^{0} + \cancel{\Delta PE}^{0} + \Delta U = Q - W$$

where the kinetic and potential energy terms drop out by assumption 3. Then, writing ΔU in terms of specific internal energies, the energy balance becomes

$$m(u_2 - u_1) = Q - W$$

where m is the system mass. Solving for Q

$$Q = m(u_2 - u_1) + W$$

The value of the work for this process is determined in the solution to part (a) of Example 2.1: $W = +17.6$ kJ. The change in internal energy is obtained using given data as

$$m(u_2 - u_1) = 0.4\ \text{kg}\left(-55\frac{\text{kJ}}{\text{kg}}\right) = -22\ \text{kJ}$$

Substituting values

❷ $$Q = -22 + 17.6 = -4.4\ \text{kJ}$$

---

❶ The given relationship between pressure and volume allows the process to be represented by the path shown on the accompanying diagram. The area under the curve represents the work. Since they are not properties, the values of the work and heat transfer depend on the details of the process and cannot be determined from the end states only.

❷ The minus sign for the value of Q means that a net amount of energy has been transferred from the system to its surroundings by heat transfer.

Skills Developed: *Ability to...*

✓ define a closed system and identify interactions on its boundary.

✓ apply the closed-system energy balance.

Quick Quiz

If the gas undergoes a process for which $pV = constant$ and $\Delta u = 0$, determine the heat transfer, in kJ, keeping the initial pressure and given volumes fixed.

Ans. 20.79 kJ.

In the next example, we follow up the discussion of Fig. 2.15 by considering two alternative systems. This example highlights the need to account correctly for the heat and work interactions occurring on the boundary as well as the energy change.

Example 2.3 CONSIDERING ALTERNATIVE SYSTEMS

Air is contained in a vertical piston–cylinder assembly fitted with an electrical resistor. The atmosphere exerts a pressure of 1 bar on the top of the piston, which has a mass of 45 kg and a face area of 0.09 m^2. Electric current passes through the resistor, and the volume of the air slowly increases by 0.045 m^3 while its pressure remains constant. The mass of the air is 0.27 kg, and its specific internal energy increases by 42 kJ/kg. The air and piston are at rest initially and finally. The piston–cylinder material is a ceramic composite and thus a good insulator. Friction between the piston and cylinder wall can be ignored, and the local acceleration of gravity is $g = 9.81$ m/s^2. Determine the heat transfer from the resistor to the air, in kJ, for a system consisting of **(a)** the air alone, **(b)** the air and the piston.

Solution

Known: Data are provided for air contained in a vertical piston–cylinder fitted with an electrical resistor.

Find: Considering each of two alternative systems, determine the heat transfer from the resistor to the air.

Schematic and Given Data:

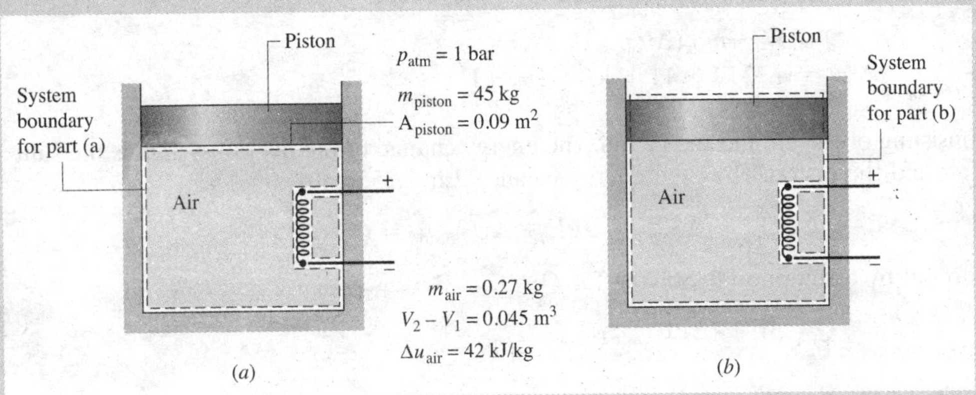

Fig. E2.3

Engineering Model:

1. Two closed systems are under consideration, as shown in the schematic.
2. The only significant heat transfer is from the resistor to the air, during which the air expands slowly and its pressure remains constant.
3. There is no net change in kinetic energy; the change in potential energy of the air is negligible; and since the piston material is a good insulator, the internal energy of the piston is not affected by the heat transfer.
4. Friction between the piston and cylinder wall is negligible.
5. The acceleration of gravity is constant; $g = 9.81\ \text{m/s}^2$.

Analysis: **(a)** Taking the air as the system, the energy balance, Eq. 2.35, reduces with assumption 3 to

$$(\cancel{\Delta KE}^{0} + \cancel{\Delta PE}^{0} + \Delta U)_{\text{air}} = Q - W$$

Or, solving for Q

$$Q = W + \Delta U_{\text{air}}$$

For this system, work is done by the force of the pressure p acting on the *bottom* of the piston as the air expands. With Eq. 2.17 and the assumption of constant pressure

$$W = \int_{V_1}^{V_2} p\,dV = p(V_2 - V_1)$$

To determine the pressure p, we use a force balance on the slowly moving, frictionless piston. The upward force exerted by the air on the *bottom* of the piston equals the weight of the piston plus the downward force of the atmosphere acting on the *top* of the piston. In symbols

$$p\text{A}_{\text{piston}} = m_{\text{piston}}\,g + p_{\text{atm}}\text{A}_{\text{piston}}$$

Solving for p and inserting values

$$\begin{aligned} p &= \frac{m_{\text{piston}}\,g}{\text{A}_{\text{piston}}} + p_{\text{atm}} \\ &= \frac{(45\ \text{kg})(9.81\ \text{m/s}^2)}{0.09\ \text{m}^2}\left|\frac{1\ \text{bar}}{10^5\ \text{N/m}^2}\right| + 1\ \text{bar} = 1.049\ \text{bar} \end{aligned}$$

Thus, the work is

$$\begin{aligned} W &= p(V_2 - V_1) \\ &= (1.049\ \text{bar})\,(0.045\ \text{m}^2)\left|\frac{10^5\ \text{N/m}^2}{1\ \text{bar}}\right|\left|\frac{1\ \text{kJ}}{10^3\ \text{N}\cdot\text{m}}\right| = 4.72\ \text{kJ} \end{aligned}$$

With $\Delta U_{\text{air}} = m_{\text{air}}(\Delta u_{\text{air}})$, the heat transfer is

$$\begin{aligned} Q &= W + m_{\text{air}}(\Delta u_{\text{air}}) \\ &= 4.72\ \text{kJ} + 11.07\ \text{kJ} = 15.8\ \text{kJ} \end{aligned}$$

(b) Consider next a system consisting of the air and the piston. The energy change of the overall system is the sum of the energy changes of the air and the piston. Thus, the energy balance, Eq. 2.35, reads

$$(\cancel{\Delta KE}^{0} + \cancel{\Delta PE}^{0} + \Delta U)_{\text{air}} + (\cancel{\Delta KE}^{0} + \Delta PE + \cancel{\Delta U}^{0})_{\text{piston}} = Q - W$$

where the indicated terms drop out by assumption 3. Solving for Q

$$Q = W + (\Delta PE)_{\text{piston}} + (\Delta U)_{\text{air}}$$

For this system, work is done at the *top* of the piston as it pushes aside the surrounding atmosphere. Applying Eq. 2.17

$$W = \int_{V_1}^{V_2} p\, dV = p_{\text{atm}}(V_2 - V_1)$$

$$= (1 \text{ bar})(0.045 \text{ m}^2) \left| \frac{10^5 \text{ N/m}^2}{1 \text{ bar}} \right| \left| \frac{1 \text{ kJ}}{10^3 \text{ N} \cdot \text{m}} \right| = 4.5 \text{ kJ}$$

The elevation change, Δz, required to evaluate the potential energy change of the piston can be found from the volume change of the air and the area of the piston face as

$$\Delta z = \frac{V_2 - V_1}{A_{\text{piston}}} = \frac{0.045 \text{ m}^3}{0.09 \text{ m}^2} = 0.5 \text{ m}$$

Thus, the potential energy change of the piston is

$$(\Delta \text{PE})_{\text{piston}} = m_{\text{piston}}\, g \Delta z$$
$$= (45 \text{ kg})(9.81 \text{ m/s}^2)(0.5 \text{ m}) = 0.22 \text{ kJ}$$

Finally,

$$Q = W + (\Delta \text{PE})_{\text{piston}} + m_{\text{air}} \Delta u_{\text{air}}$$
$$= 4.5 \text{ kJ} + 0.22 \text{ kJ} + 11.07 \text{ kJ} = 15.8 \text{ kJ}$$

❶ which agrees with the result of part (a).

---

❶ Although the value of Q is the same for each system, observe that the values for W differ. Also, observe that the energy changes differ, depending on whether the air alone or the air and the piston is the system.

Skills Developed: *Ability to...*
- ✓ define alternative closed systems and identify interactions on their boundaries.
- ✓ evaluate work using Eq. 2.17.
- ✓ apply the closed-system energy balance.

Quick Quiz

What is the change in potential energy of the air, in kJ?

Ans. $\approx 1.055 \times 10^{-3}$ kJ

2.5.3 Using the Energy Rate Balance: Steady-State Operation

A system is at steady state if none of its properties change with time (Sec. 1.3). Many devices operate at steady state or nearly at steady state, meaning that property variations with time are small enough to ignore. The two examples to follow illustrate the application of the energy rate equation to closed systems at steady state.

Example 2.4 GEARBOX AT STEADY STATE

During steady-state operation, a gearbox receives 60 kW through the input shaft and delivers power through the output shaft. For the gearbox as the system, the rate of energy transfer by convection is

$$\dot{Q} = -\text{hA}(T_b - T_f)$$

where h = 0.171 kW/m^2 · K is the heat transfer coefficient, A = 1.0 m^2 is the outer surface area of the gearbox, T_b = 300 K (27°C) is the temperature at the outer surface, and T_f = 293 K (20°C) is the temperature of the surrounding air away from the immediate vicinity of the gearbox. For the gearbox, evaluate the heat transfer rate and the power delivered through the output shaft, each in kW.

Solution

Known: A gearbox operates at steady state with a known power input. An expression for the heat transfer rate from the outer surface is also known.

Find: Determine the heat transfer rate and the power delivered through the output shaft, each in kW.

Schematic and Given Data:

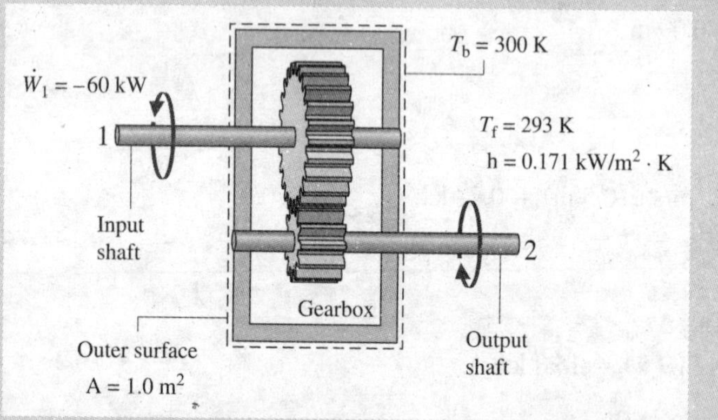

Engineering Model:

1. The gearbox is a closed system at steady state.
2. For the gearbox, convection is the dominant heat transfer mode.

Fig. E2.4

Analysis: Using the given expression for $\dot{Q}$ together with known data, the rate of energy transfer by heat is

❶
$$\begin{aligned}\dot{Q} &= -\text{hA}(T_b - T_f)\\ &= -\left(0.171\frac{\text{kW}}{\text{m}^2 \cdot \text{K}}\right)(1.0\ \text{m}^2)(300 - 293)\ \text{K}\\ &= -1.2\ \text{kW}\end{aligned}$$

The minus sign for $\dot{Q}$ signals that energy is carried *out* of the gearbox by heat transfer.

The energy rate balance, Eq. 2.37, reduces at steady state to

❷
$$\frac{dE}{dt}^{0} = \dot{Q} - \dot{W} \quad \text{or} \quad \dot{W} = \dot{Q}$$

The symbol $\dot{W}$ represents the *net* power from the system. The net power is the sum of $\dot{W}_1$ and the output power $\dot{W}_2$

$$\dot{W} = \dot{W}_1 + \dot{W}_2$$

With this expression for $\dot{W}$, the energy rate balance becomes

$$\dot{W}_1 + \dot{W}_2 = \dot{Q}$$

Solving for $\dot{W}_2$, inserting $\dot{Q} = -1.2$ kW, and $\dot{W}_1 = -60$ kW, where the minus sign is required because the input shaft brings energy *into* the system, we have

❸
$$\begin{aligned}\dot{W}_2 &= \dot{Q} - \dot{W}_1\\ &= (-1.2\ \text{kW}) - (-60\ \text{kW})\\ &= +58.8\ \text{kW}\end{aligned}$$

❹ The positive sign for $\dot{W}_2$ indicates that energy is transferred from the system through the output shaft, as expected.

---

❶ In accord with the sign convention for the heat transfer rate in the energy rate balance (Eq. 2.37), Eq. 2.34 is written with a minus sign: $\dot{Q}$ is negative since T_b is greater than T_f.

❷ Properties of a system at steady state do not change with time. Energy E is a property, but heat transfer and work are not properties.

❸ For this system, energy transfer by work occurs at two different locations, and the signs associated with their values differ.

❹ At steady state, the rate of heat transfer from the gear box accounts for the difference between the input and output power. This can be summarized by the following energy rate "balance sheet" in terms of *magnitudes:*

	Input	**Output**
	60 kW (input shaft)	58.8 kW (output shaft)
		1.2 kW (heat transfer)
Total:	60 kW	60 kW

Skills Developed: *Ability to...*

- ✓ define a closed system and identify interactions on its boundary.
- ✓ evaluate the rate of energy transfer by convection.
- ✓ apply the energy rate balance for *steady-state* operation.
- ✓ develop an *energy rate balance sheet.*

Quick Quiz

For an emissivity of 0.8 and taking $T_s = T_f$, use Eq. 2.33 to determine the net rate at which energy is radiated from the outer surface of the gearbox, in kW.

Ans. 0.03 kW.

Example 2.5 SILICON CHIP AT STEADY STATE

A silicon chip measuring 5 mm on a side and 1 mm in thickness is embedded in a ceramic substrate. At steady state, the chip has an electrical power input of 0.225 W. The top surface of the chip is exposed to a coolant whose temperature is 20°C. The heat transfer coefficient for convection between the chip and the coolant is 150 W/m^2 · K. If heat transfer by conduction between the chip and the substrate is negligible, determine the surface temperature of the chip, in °C.

Solution

Known: A silicon chip of known dimensions is exposed on its top surface to a coolant. The electrical power input and convective heat transfer coefficient are known.

Find: Determine the surface temperature of the chip at steady state.

Schematic and Given Data:

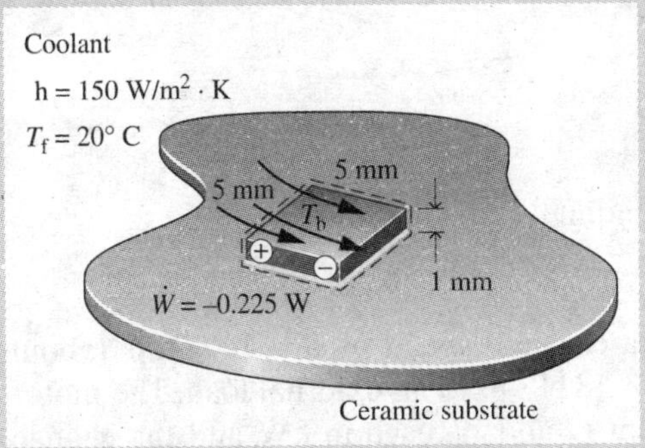

Fig. E2.5

Engineering Model:

1. The chip is a closed system at steady state.
2. There is no heat transfer between the chip and the substrate.

Analysis: The surface temperature of the chip, T_b, can be determined using the energy rate balance, Eq. 2.37, which at steady state reduces as follows

❶
$$\frac{dE}{dt}^{\,0} = \dot{Q} - \dot{W}$$

With assumption 2, the only heat transfer is by convection to the coolant. In this application, Newton's law of cooling, Eq. 2.34, takes the form

❷
$$\dot{Q} = -\text{hA}(T_b - T_f)$$

Collecting results

$$0 = -\text{hA}(T_b - T_f) - \dot{W}$$

Solving for T_b

$$T_b = \frac{-\dot{W}}{\text{hA}} + T_f$$

In this expression, $\dot{W} = -0.225$ W, A $= 25 \times 10^{-6}$ m^2, h $= 150$ W/m$^2 \cdot$ K, and $T_f = 293$ K, giving

$$T_b = \frac{-(-0.225\ \text{W})}{(150\ \text{W/m}^2 \cdot \text{K})(25 \times 10^{-6}\ \text{m}^2)} + 293\ \text{K}$$
$$= 353\ \text{K}\ (80°\text{C})$$

❶ Properties of a system at steady state do not change with time. Energy E is a property, but heat transfer and work are not properties.

❷ In accord with the sign convention for heat transfer in the energy rate balance (Eq. 2.37), Eq. 2.34 is written with a minus sign: $\dot{Q}$ is negative since T_b is greater than T_f.

Skills Developed: *Ability to...*
- ✓ define a closed system and identify interactions on its boundary.
- ✓ evaluate the rate of energy transfer by convection.
- ✓ apply the energy rate balance for *steady-state* operation.

Quick Quiz

If the surface temperature of the chip must be no greater than 60°C, what is the corresponding range of values required for the convective heat transfer coefficient, assuming all other quantities remain unchanged? **Ans.** h ≥ 225 W/m$^2 \cdot$ K.

2.5.4 Using the Energy Rate Balance: Transient Operation

Many devices undergo periods of transient operation where the state changes with time. This is observed during startup and shutdown periods. The next example illustrates the application of the energy rate balance to an electric motor during startup. The example also involves both electrical work and power transmitted by a shaft.

[redacted] TRANSIENT OPERATION OF A MOTOR

The rate of heat transfer between a certain electric motor and its surroundings varies with time as

$$\dot{Q} = -0.2[1 - e^{(-0.05t)}]$$

where t is in seconds and $\dot{Q}$ is in kW. The shaft of the motor rotates at a constant speed of $\omega = 100$ rad/s (about 955 revolutions per minute, or RPM) and applies a constant torque of $\mathcal{T} = 18$ N $\cdot$ m to an external load. The motor draws a constant electric power input equal to 2.0 kW. For the motor, plot $\dot{Q}$ and $\dot{W}$, each in kW, and the change in energy ΔE, in kJ, as functions of time from $t = 0$ to $t = 120$ s. Discuss.

Solution

Known: A motor operates with constant electric power input, shaft speed, and applied torque. The time-varying rate of heat transfer between the motor and its surroundings is given.

Find: Plot $\dot{Q}$, $\dot{W}$, and ΔE versus time. Discuss.

Schematic and Given Data:

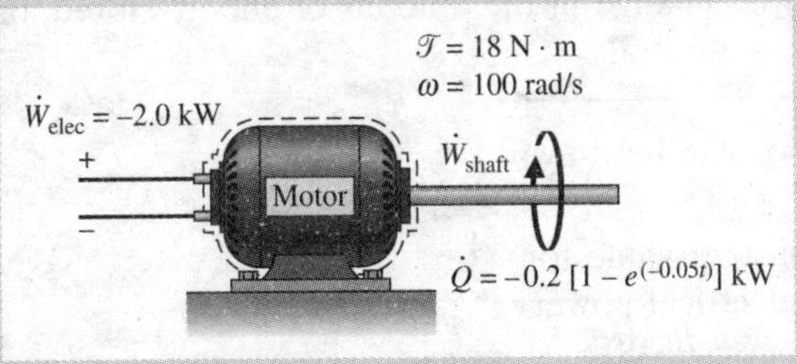

Engineering Model: The system shown in the accompanying sketch is a closed system.

Fig. E2.6a

Analysis: The time rate of change of system energy is

$$\frac{dE}{dt} = \dot{Q} - \dot{W}$$

$\dot{W}$ represents the *net* power *from* the system: the sum of the power associated with the rotating shaft, $\dot{W}_{shaft}$, and the power associated with the electricity flow, $\dot{W}_{elec}$

$$\dot{W} = \dot{W}_{shaft} + \dot{W}_{elec}$$

The rate $\dot{W}_{elec}$ is known from the problem statement: $\dot{W}_{elec} = -2.0$ kW, where the negative sign is required because energy is carried into the system by electrical work. The term $\dot{W}_{shaft}$ can be evaluated with Eq. 2.20 as

$$\dot{W}_{shaft} = \mathcal{T}\omega = (18 \text{ N} \cdot \text{m})(100 \text{ rad/s}) = 1800 \text{ W} = +1.8 \text{ kW}$$

Because energy exits the system along the rotating shaft, this energy transfer rate is positive.

In summary,

$$\dot{W} = \dot{W}_{elec} + \dot{W}_{shaft} = (-2.0 \text{ kW}) + (+1.8 \text{ kW}) = -0.2 \text{ kW}$$

where the minus sign means that the electrical power input is greater than the power transferred out along the shaft.

With the foregoing result for $\dot{W}$ and the given expression for $\dot{Q}$, the energy rate balance becomes

$$\frac{dE}{dt} = -0.2[1 - e^{(-0.05t)}] - (-0.2) = 0.2e^{(-0.05t)}$$

Integrating

$$\Delta E = \int_0^t 0.2e^{(-0.05t)}\, dt$$

$$= \frac{0.2}{(-0.05)} e^{(-0.05t)} \Big]_0^t = 4[1 - e^{(-0.05t)}]$$

1 The accompanying plots, Figs. E2.6*b*, *c*, are developed using the given expression for $\dot{Q}$ and the expressions for $\dot{W}$ and ΔE obtained in the analysis. Because of our sign conventions for heat and work, the values of $\dot{Q}$ and $\dot{W}$ are

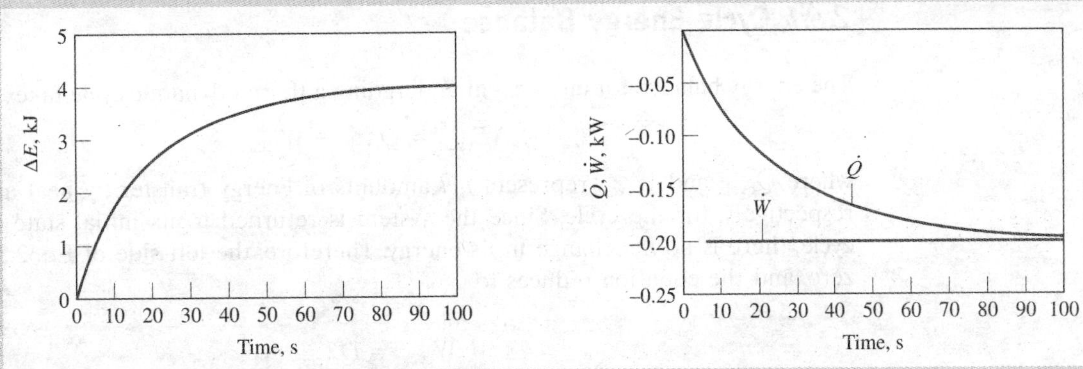

Fig. E2.6b, c

negative. In the first few seconds, the *net* rate that energy is carried in by work greatly exceeds the rate that energy is carried out by heat transfer. Consequently, the energy stored in the motor increases rapidly as the motor "warms up." As time elapses, the value of $\dot{Q}$ approaches $\dot{W}$, and the rate of energy storage diminishes. After about 100 s, this *transient* operating mode is nearly over, and there is little further change in the amount of energy stored, or ❷ in any other property. We may say that the motor is then at steady state.

❶ Figures E.2.6*b*, *c* can be developed using appropriate software or can be drawn by hand.

❷ At steady state, the value of $\dot{Q}$ is constant at −0.2 kW. This constant value for the heat transfer rate can be thought of as the portion of the electrical power input that is not obtained as a mechanical power output because of effects within the motor such as electrical resistance and friction.

Skills Developed: *Ability to...*

- ✓ define a closed system and identify interactions on its boundary.
- ✓ apply the energy rate balance for *transient* operation.
- ✓ develop and interpret graphical data.

Quick Quiz

If the dominant mode of heat transfer from the outer surface of the motor is convection, determine *at steady state* the temperature T_b on the outer surface, in K, for h = 0.17 kW/m² · K, A = 0.3 m², and T_f = 293 K.

Ans. 297 K.

2.6 Energy Analysis of Cycles

thermodynamic cycle

In this section the energy concepts developed thus far are illustrated further by application to systems undergoing thermodynamic cycles. A ***thermodynamic cycle*** is a sequence of processes that begins and ends at the same state. At the conclusion of a cycle all properties have the same values they had at the beginning. Consequently, over the cycle the system experiences no *net* change of state. Cycles that are repeated periodically play prominent roles in many areas of application. For example, steam circulating through an electrical power plant executes a cycle.

The study of systems undergoing cycles has played an important role in the development of the subject of engineering thermodynamics. Both the first and second laws of thermodynamics have roots in the study of cycles. Additionally, there are many important practical applications involving power generation, vehicle propulsion, and refrigeration for which an understanding of thermodynamic cycles is essential. In this section, cycles are considered from the perspective of the conservation of energy principle. Cycles are studied in greater detail in subsequent chapters, using both the conservation of energy principle and the second law of thermodynamics.

2.6.1 Cycle Energy Balance

The energy balance for any system undergoing a thermodynamic cycle takes the form

$$\Delta E_{\text{cycle}} = Q_{\text{cycle}} - W_{\text{cycle}} \tag{2.39}$$

where Q_{cycle} and W_{cycle} represent *net* amounts of energy transfer by heat and work, respectively, for the cycle. Since the system is returned to its initial state after the cycle, there is no *net* change in its energy. Therefore, the left side of Eq. 2.39 equals zero, and the equation reduces to

$$W_{\text{cycle}} = Q_{\text{cycle}} \tag{2.40}$$

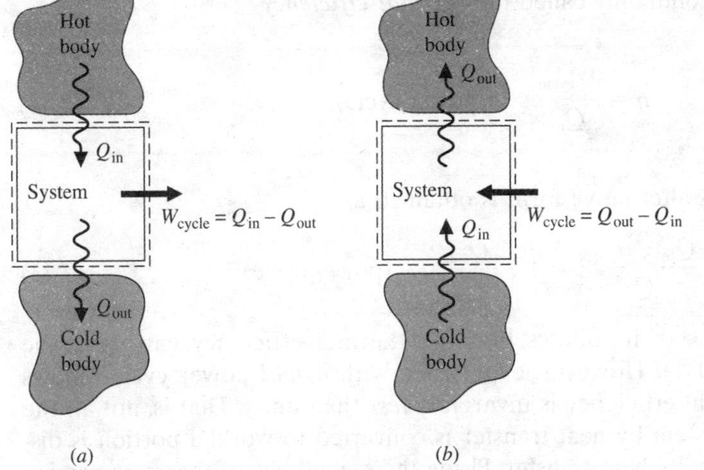

Fig. 2.17 Schematic diagrams of two important classes of cycles. (*a*) Power cycles. (*b*) Refrigeration and heat pump cycles.

Equation 2.40 is an expression of the conservation of energy principle that must be satisfied by *every* thermodynamic cycle, regardless of the sequence of processes followed by the system undergoing the cycle or the nature of the substances making up the system.

Figure 2.17 provides simplified schematics of two general classes of cycles considered in this book: power cycles and refrigeration and heat pump cycles. In each case pictured, a system undergoes a cycle while communicating thermally with two bodies, one hot and the other cold. These bodies are systems located in the surroundings of the system undergoing the cycle. During each cycle there is also a net amount of energy exchanged with the surroundings by work. Carefully observe that in using the symbols Q_{in} and Q_{out} on Fig. 2.17 we have departed from the previously stated sign convention for heat transfer. In this section it is advantageous to regard Q_{in} and Q_{out} as transfers of energy in the *directions indicated by the arrows.* The direction of the net work of the cycle, W_{cycle}, is *also indicated by an arrow.* Finally, note that the directions of the energy transfers shown in Fig. 2.17*b* are opposite to those of Fig. 2.17*a*.

Take Note...

When analyzing cycles, we normally take energy transfers as positive in the directions of arrows on a sketch of the system and write the energy balance accordingly.

2.6.2 Power Cycles

Systems undergoing cycles of the type shown in Fig. 2.17*a* deliver a net work transfer of energy to their surroundings during each cycle. Any such cycle is called a ***power cycle.*** From Eq. 2.40, the net work output equals the net heat transfer to the cycle, or

power cycle

$$W_{cycle} = Q_{in} - Q_{out} \quad \text{(power cycle)} \qquad (2.41)$$

where Q_{in} represents the heat transfer of energy *into* the system from the hot body, and Q_{out} represents heat transfer *out* of the system to the cold body. From Eq. 2.41 it is clear that Q_{in} must be greater than Q_{out} for a *power* cycle. The energy supplied by heat transfer to a system undergoing a power cycle is normally derived from the combustion of fuel or a moderated nuclear reaction; it can also be obtained from solar radiation. The energy Q_{out} is generally discharged to the surrounding atmosphere or a nearby body of water.

The performance of a system undergoing a *power cycle* can be described in terms of the extent to which the energy added by heat, Q_{in}, is *converted* to a net work output, W_{cycle}. The extent of the energy conversion from heat to work is expressed by the following ratio, commonly called the ***thermal efficiency***

thermal efficiency

$$\eta = \frac{W_{cycle}}{Q_{in}} \quad \text{(power cycle)} \tag{2.42}$$

Introducing Eq. 2.41, an alternative form is obtained as

$$\eta = \frac{Q_{in} - Q_{out}}{Q_{in}} = 1 - \frac{Q_{out}}{Q_{in}} \quad \text{(power cycle)} \tag{2.43}$$

Since energy is conserved, it follows that the thermal efficiency can never be greater than unity (100%). However, experience with *actual* power cycles shows that the value of thermal efficiency is invariably *less* than unity. That is, not all the energy added to the system by heat transfer is converted to work; a portion is discharged to the cold body by heat transfer. Using the second law of thermodynamics, we will show in Chap. 5 that the conversion from heat to work cannot be fully accomplished by any power cycle. The thermal efficiency of *every* power cycle must be less than unity: $\eta < 1$.

Energy & Environment

Today fossil-fueled power plants can have thermal efficiencies of 40%, or more. This means that up to 60% of the energy added by heat transfer during the power plant cycle is discharged from the plant other than by work, principally by heat transfer. One way power plant cooling is achieved is to use water drawn from a nearby river or lake. The water is eventually returned to the river or lake but at a higher temperature, which is a practice having several possible environmental consequences.

The return of large quantities of warm water to a river or lake can affect its ability to hold dissolved gases, including the oxygen required for aquatic life. If the return water temperature is greater than about 35°C, the dissolved oxygen may be too low to support some species of fish. If the return water temperature is too great, some species also can be stressed. As rivers and lakes become warmer, non-native species that thrive in the warmth can take over. Warmer water also fosters bacterial populations and algae growth.

Regulatory agencies have acted to limit warm water discharges from power plants, which has made cooling towers (Sec. 12.9) adjacent to power plants a common sight.

2.6.3 Refrigeration and Heat Pump Cycles

refrigeration and heat pump cycles

Next, consider the ***refrigeration and heat pump cycles*** shown in Fig. 2.17*b*. For cycles of this type, Q_{in} is the energy transferred by heat *into* the system undergoing the cycle *from* the cold body, and Q_{out} is the energy discharged by heat transfer *from* the system *to* the hot body. To accomplish these energy transfers requires a net work *input*, W_{cycle}. The quantities Q_{in}, Q_{out}, and W_{cycle} are related by the energy balance, which for refrigeration and heat pump cycles takes the form

$$W_{cycle} = Q_{out} - Q_{in} \quad \text{(refrigeration and heat pump cycles)} \tag{2.44}$$

Since W_{cycle} is positive in this equation, it follows that Q_{out} is greater than Q_{in}.

Although we have treated them as the same to this point, refrigeration and heat pump cycles actually have different objectives. The objective of a refrigeration cycle is to cool a refrigerated space or to maintain the temperature within a dwelling or other building *below* that of the surroundings. The objective of a heat pump is to maintain the temperature within a dwelling or other building *above* that of the surroundings or to provide heating for certain industrial processes that occur at elevated temperatures.

Since refrigeration and heat pump cycles have different objectives, their performance parameters, called *coefficients of performance,* are defined differently. These coefficients of performance are considered next.

Refrigeration Cycles

The performance of *refrigeration cycles* can be described as the ratio of the amount of energy received by the system undergoing the cycle from the cold body, Q_{in}, to the net work into the system to accomplish this effect, W_{cycle}. Thus, the ***coefficient of performance,*** β, is

$$\beta = \frac{Q_{in}}{W_{cycle}} \quad \text{(refrigeration cycle)} \tag{2.45}$$

coefficient of performance: refrigeration

Introducing Eq. 2.44, an alternative expression for β is obtained as

$$\beta = \frac{Q_{in}}{Q_{out} - Q_{in}} \quad \text{(refrigeration cycle)} \tag{2.46}$$

For a household refrigerator, Q_{out} is discharged to the space in which the refrigerator is located. W_{cycle} is usually provided in the form of electricity to run the motor that drives the refrigerator.

FOR EXAMPLE... in a refrigerator the inside compartment acts as the cold body and the ambient air surrounding the refrigerator is the hot body. Energy Q_{in} passes to the circulating refrigerant *from* the food and other contents of the inside compartment. For this heat transfer to occur, the refrigerant temperature is necessarily below that of the refrigerator contents. Energy Q_{out} passes *from* the refrigerant *to* the surrounding air. For this heat transfer to occur, the temperature of the circulating refrigerant must necessarily be above that of the surrounding air. To achieve these effects, a work *input* is required. For a refrigerator, W_{cycle} is provided in the form of electricity.

Heat Pump Cycles

The performance of *heat pumps* can be described as the ratio of the amount of energy discharged from the system undergoing the cycle to the hot body, Q_{out}, to the net work into the system to accomplish this effect, W_{cycle}. Thus, the ***coefficient of performance,*** γ, is

$$\gamma = \frac{Q_{out}}{W_{cycle}} \quad \text{(heat pump cycle)} \tag{2.47}$$

coefficient of performance: heat pump

Introducing Eq. 2.44, an alternative expression for this coefficient of performance is obtained as

$$\gamma = \frac{Q_{out}}{Q_{out} - Q_{in}} \quad \text{(heat pump cycle)} \tag{2.48}$$

From this equation it can be seen that the value of γ is never less than unity. For residential heat pumps, the energy quantity Q_{in} is normally drawn from the surrounding atmosphere, the ground, or a nearby body of water. W_{cycle} is usually provided by electricity.

The coefficients of performance β and γ are defined as ratios of the desired heat transfer effect to the cost in terms of work to accomplish that effect. Based on the definitions, it is desirable thermodynamically that these coefficients have values that are as large as possible. However, as discussed in Chap. 5, coefficients of performance must satisfy restrictions imposed by the second law of thermodynamics.

✓ Chapter Summary and Study Guide

In this chapter, we have considered the concept of energy from an engineering perspective and have introduced energy balances for applying the conservation of energy principle to closed systems. A basic idea is that energy can be stored within systems in three macroscopic forms: internal energy, kinetic energy, and gravitational potential energy. Energy also can be transferred to and from systems.

Energy can be transferred to and from closed systems by two means only: work and heat transfer. Work and heat transfer are identified at the system boundary and are not properties. In mechanics, work is energy transfer associated with macroscopic forces and displacements at the system boundary. The thermodynamic definition of work introduced in this chapter extends the notion of work from mechanics to include other types of work. Energy transfer by heat to or from a system is due to a temperature difference between the system and its surroundings, and occurs in the direction of decreasing temperature. Heat transfer modes include conduction, radiation, and convection. These sign conventions are used for work and heat transfer:

✓ $W, \dot{W}$ $\begin{cases} > 0\text{: work done by the system} \\ < 0\text{: work done on the system} \end{cases}$

✓ $Q, \dot{Q}$ $\begin{cases} > 0\text{: heat transfer to the system} \\ < 0\text{: heat transfer from the system} \end{cases}$

Energy is an extensive property of a system. Only changes in the energy of a system have significance. Energy changes are accounted for by the energy balance. The energy balance for a process of a closed system is Eq. 2.35 and an accompanying time rate form is Eq. 2.37. Equation 2.40 is a special form of the energy balance for a system undergoing a thermodynamic cycle.

The following checklist provides a study guide for this chapter. When your study of the text and end-of-chapter exercises has been completed, you should be able to

- ✓ write out the meanings of the terms listed in the margins throughout the chapter and understand each of the related concepts. The subset of key concepts listed below is particularly important in subsequent chapters.
- ✓ evaluate these energy quantities
  - –kinetic and potential energy changes using Eqs. 2.5 and 2.10, respectively.
  - –work and power using Eqs. 2.12 and 2.13, respectively.
  - –expansion or compression work using Eq. 2.17.
- ✓ apply closed system energy balances in each of several alternative forms, appropriately modeling the case at hand, correctly observing sign conventions for work and heat transfer, and carefully applying SI units.
- ✓ conduct energy analyses for systems undergoing thermodynamic cycles using Eq. 2.40, and evaluating, as appropriate, the thermal efficiencies of power cycles and coefficients of performance of refrigeration and heat pump cycles.

Key Engineering Concepts

kinetic energy p. 33
potential energy p. 34
work p. 36
power p. 38
internal energy p. 47
heat transfer p. 48
first law of thermodynamics p. 52
energy balance p. 52
thermodynamic cycle p. 64
power cycle p. 65
refrigeration cycle p. 66
heat pump cycle p. 66

Key Equations

$\Delta E = \Delta U + \Delta \text{KE} + \Delta \text{PE}$	(2.27)	Change in total energy of a system.
$\Delta \text{KE} = \text{KE}_2 - \text{KE}_1 = \frac{1}{2}m(\text{V}_2^2 - \text{V}_1^2)$	(2.5)	Change in kinetic energy of a mass m.
$\Delta \text{PE} = \text{PE}_2 - \text{PE}_1 = mg(z_2 - z_1)$	(2.10)	Change in gravitational potential energy of a mass m at constant g.

$E_2 - E_1 = Q - W$	(2.35a)	Energy balance for closed systems.
$\frac{dE}{dt} = \dot{Q} - \dot{W}$	(2.37)	Energy rate balance for closed systems.
$W = \int_{s_1}^{s_2} \mathbf{F} \cdot ds$	(2.12)	Work due to action of a force **F**.
$\dot{W} = \mathbf{F} \cdot \mathbf{V}$	(2.13)	Power due to action of a force **F**.
$W = \int_{V_1}^{V_2} p\, dV$	(2.17)	Expansion or compression work related to fluid pressure. See Fig. 2.4.
Thermodynamic Cycles		
$W_{cycle} = Q_{in} - Q_{out}$	(2.41)	Energy balance for a *power cycle.* As in Fig. 2.17a, all quantities are regarded as positive.
$\eta = \frac{W_{cycle}}{Q_{in}}$	(2.42)	Thermal efficiency of a power cycle.
$W_{cycle} = Q_{out} - Q_{in}$	(2.44)	Energy balance for a *refrigeration* or *heat pump cycle.* As in Fig. 2.17b, all quantities are regarded as positive.
$\beta = \frac{Q_{in}}{W_{cycle}}$	(2.45)	Coefficient of performance of a refrigeration cycle.
$\gamma = \frac{Q_{out}}{W_{cycle}}$	(2.47)	Coefficient of performance of a heat pump cycle.

Exercises: things engineers think about

1. What forces act on the bicycle and rider considered in Sec. 2.2.2? Sketch a free-body diagram.

2. Why is it incorrect to say that a system *contains* heat?

3. What examples of heat transfer by conduction, radiation, and convection do you encounter when using a charcoal grill?

4. When a meteor impacts the earth and comes to rest, what happens to its kinetic and potential energies?

5. After running 8 km on a treadmill at her campus rec center, Ashley observes that the treadmill belt is warm to the touch. Why is the belt warm?

6. When cooking ingredients are mixed in a blender, what happens to the energy transferred to the ingredients? Is the energy transfer by work, by heat transfer, or by work and heat transfer?

7. When microwaves are beamed onto a tumor during cancer therapy to increase the tumor's temperature, this interaction is considered work and not heat transfer. Why?

8. Experimental molecular motors are reported to exhibit movement upon the absorption of light, thereby achieving a conversion of electromagnetic radiation into motion. Should the incident light be considered work or heat transfer?

9. Why are the symbols ΔU, ΔKE, and ΔPE used to denote the energy change during a process, but the work and heat transfer for the process are represented, respectively, simply as W and Q?

10. If the change in energy of a closed system is known for a process between two end states, can you determine if the energy change was due to work, to heat transfer, or to some combination of work and heat transfer?

11. Referring to Fig. 2.8, which process, A or B, has the greater heat transfer?

12. What form does the energy balance take for an *isolated* system?

13. How would you define an *efficiency* for the motor of Example 2.6?

14. How many tons of CO_2 are produced annually by a conventional automobile?

Problems: developing engineering skills

Exploring Energy Concepts

2.1 A baseball has a mass of 0.14 kg. What is the kinetic energy relative to home plate of a 151 km/hr fastball, in kJ?

Fig. P2.1

2.2 An object whose mass is 400 kg is located at an elevation of 25 m above the surface of the earth. For $g = 9.78$ m/s^2, determine the gravitational potential energy of the object, in kJ, relative to the surface of the earth.

2.3 An object whose mass is 45 kg experiences a decrease in kinetic energy of 0.68 kJ and an increase in potential energy of 2 kJ. The initial velocity and elevation of the object, each relative to the surface of the earth, are 12 m/s and 9 m, respectively. If $g = 9.8$ m/s^2, determine

(a) the final velocity, in m/s
(b) the final elevation, in m

2.4 A 63.5 × 88 × 152 mm brick whose density is 1,922 kg/m^3 slips off the top of a building under construction and falls 21 m. For $g = 9.81$ m/s^2, determine the change in gravitational potential energy of the brick, in N · m.

2.5 What is the overall change in potential energy, in N · m and kJ, of an automobile having a mass of 1,134 kg in a drive from San Diego, CA to Santa Fe, NM? Take g constant.

2.6 An object of mass 1,000 kg, initially having a velocity of 100 m/s, decelerates to a final velocity of 20 m/s. What is the change in kinetic energy of the object, in kJ?

2.7 A 30-seat turboprop airliner whose mass is 14,000 kg takes off from an airport and eventually achieves its cruising speed of 620 km/h at an altitude of 10,000 m. For $g = 9.78$ m/s^2, determine the change in kinetic energy and the change in gravitational potential energy of the airliner, each in kJ.

2.8 An object whose mass is 0.5 kg has a velocity of 30 m/s. Determine

(a) the final velocity, in m/s, if the kinetic energy of the object decreases by 130 J.
(b) the change in elevation, in m, associated with a 130 J change in potential energy. Let $g = 9.81$ m/s^2.

2.9 An object whose mass is 2 kg is accelerated from a velocity of 200 m/s to a final velocity of 500 m/s by the action of a resultant force **F**. Determine the work done by the resultant force, in kJ, if there are no other interactions between the object and its surroundings.

2.10 An object whose mass is 136 kg experiences changes in its kinetic and potential energies owing to the action of a resultant force **R**. The work done on the object by the resultant force is 148 kJ. There are no other interactions between the object and its surroundings. If the object's elevation increases by 30.5 m and its final velocity is 61 m/s, what is its initial velocity, in m/s? Let $g = 9.81$ m/s^2.

2.11 A disk-shaped flywheel, of uniform density ρ, outer radius R, and thickness w, rotates with an angular velocity ω, in rad/s.

(a) Show that the moment of inertia, $I = \int_{vol} \rho r^2 \, dV$, can be expressed as $I = \pi \rho w R^4/2$, and the kinetic energy can be expressed as $KE = I\omega^2/2$.
(b) For a steel flywheel rotating at 3,000 RPM, determine the kinetic energy, in N · m, and the mass, in kg, if $R = 0.38$ m and $w = 0.025$ m.
(c) Determine the radius, in m, and the mass, in kg, of an aluminum flywheel having the same width, angular velocity, and kinetic energy as in part (b).

2.12 Using $KE = I\omega^2/2$ from Problem 2.11a, how fast would a flywheel whose moment of inertia is 8.4 kg · m^2 have to spin, in RPM, to store an amount of kinetic energy equivalent to the potential energy of a 45 kg mass raised to an elevation of 9 m above the surface of the earth? Let $g = 9.81$ m/s^2.

2.13 Two objects having different masses fall freely under the influence of gravity from rest and the same initial elevation. Ignoring the effect of air resistance, show that the magnitudes of the velocities of the objects are equal at the moment just before they strike the earth.

2.14 An object whose mass is 25 kg is projected upward from the surface of the earth with an initial velocity of 60 m/s. The only force acting on the object is the force of gravity. Plot the velocity of the object versus elevation. Determine the elevation of the object, in m, when its velocity reaches zero. The acceleration of gravity is $g = 9.8$ m/s^2.

2.15 A block of mass 10 kg moves along a surface inclined 30° relative to the horizontal. The center of gravity of the block is elevated by 3.0 m and the kinetic energy of the block *decreases* by 50 J. The block is acted upon by a constant force **R** parallel to the incline, and by the force of gravity. Assume frictionless surfaces and let $g = 9.81$ m/s^2. Determine the magnitude and direction of the constant force **R**, in N.

2.16 Beginning from rest, an object of mass 200 kg slides down a 10-m-long ramp. The ramp is inclined at an angle of 40° from the horizontal. If air resistance and friction between the object and the ramp are negligible, determine the velocity of the object, in m/s, at the bottom of the ramp. Let $g = 9.81$ m/s^2.

2.17 Jack, whose mass is 68 kg, runs 8 km in 43 minutes on a treadmill set at a one-degree incline. The treadmill display shows he has *burned* 620 kcal. For Jack to break even caloriewise, how much vanilla ice cream, in cups, may he have after his workout?

Fig. P2.17

Evaluating Work

2.18 A system with a mass of 5 kg, initially moving horizontally with a velocity of 40 m/s, experiences a constant horizontal *deceleration* of 2 m/s^2 due to the action of a resultant force. As a result, the system comes to rest. Determine the length of time, in s, the force is applied and the amount of energy transfer by work, in kJ.

2.19 An object of mass 36 kg, initially at rest, experiences a constant horizontal acceleration of 3.7 m/s^2 due to the action of a resultant force applied for 6.5 s. Determine the work of the resultant force, in N · m, and in kJ.

2.20 The drag force, F_d, imposed by the surrounding air on a vehicle moving with velocity V is given by

$$F_d = C_d A \tfrac{1}{2} \rho V^2$$

where C_d is a constant called the drag coefficient, A is the projected frontal area of the vehicle, and ρ is the air density. Determine the power, in kW, required to overcome aerodynamic drag for a truck moving at 110 km/h, if $C_d = 0.65$, A = 10 m^2, and ρ = 1.1 kg/m^3.

2.21 A major force opposing the motion of a vehicle is the rolling resistance of the tires, F_r, given by

$$F_r = f \mathcal{W}$$

where f is a constant called the rolling resistance coefficient and $\mathcal{W}$ is the vehicle weight. Determine the power, in kW, required to overcome rolling resistance for a truck weighing 322.5 kN that is moving at 110 km/h. Let $f = 0.0069$.

2.22 The two major forces opposing the motion of a vehicle moving on a level road are the rolling resistance of the tires, F_r, and the aerodynamic drag force of the air flowing around the vehicle, F_d, given respectively by

$$F_r = fmg, \qquad F_d = C_d A \tfrac{1}{2} \rho V^2$$

where f and C_d are constants known as the rolling resistance coefficient and drag coefficient, respectively, m and A are the vehicle mass and projected frontal area, respectively, V is the vehicle velocity, and ρ is the air density. For a passenger car with m = 1,610 kg, A = 2.2 m^2, and $C_d = 0.34$, and when $f = 0.02$ and ρ = 1.28 kg/m^3

(a) determine the power required, in kW, to overcome rolling resistance and aerodynamic drag when V is 88.5 km/hr.

(b) plot versus vehicle velocity ranging from 0 to 120.7 km/hr (i) the power to overcome rolling resistance, (ii) the power to overcome aerodynamic drag, and (iii) the total power, all in kW.

What implication for vehicle fuel economy can be deduced from the results of part (b)?

2.23 Measured data for pressure versus volume during the compression of a refrigerant within the cylinder of a refrigeration compressor are given in the table below. Using data from the table, complete the following:

(a) Determine a value of n such that the data are fit by an equation of the form $pV^n = constant$.
(b) Evaluate analytically the work done on the refrigerant, in kJ, using Eq. 2.17 along with the result of part (a).
(c) Using graphical or numerical integration of the data, evaluate the work done on the refrigerant, in kJ.
(d) Compare the different methods for estimating the work used in parts (b) and (c). Why are they estimates?

Data Point	p (bar)	V (cm^3)
1	7.7	203.13
2	9.0	171.88
3	10.8	140.63
4	13.6	109.38
5	18.6	78.13
6	29.2	46.88

2.24 Measured data for pressure versus volume during the expansion of gases within the cylinder of an internal combustion engine are given in the table below. Using data from the table, complete the following:

(a) Determine a value of n such that the data are fit by an equation of the form, $pV^n = constant$.
(b) Evaluate analytically the work done by the gases, in kJ, using Eq. 2.17 along with the result of part (a).
(c) Using graphical or numerical integration of the data, evaluate the work done by the gases, in kJ.
(d) Compare the different methods for estimating the work used in parts (b) and (c). Why are they estimates?

Data Point	p (bar)	V (cm^3)
1	15	300
2	12	361
3	9	459
4	6	644
5	4	903
6	2	1,608

2.25 One-fourth kg of a gas contained within a piston–cylinder assembly undergoes a constant-pressure process at 5 bar beginning at v_1 = 0.20 m^3/kg. For the gas as the system, the work is −15 kJ. Determine the final volume of the gas, in m^3.

2.26 Carbon dioxide (CO_2) gas within a piston–cylinder assembly undergoes an expansion from a state where p_1 = 138 kPa, V_1 = 0.014 m^3 to a state where p_2 = 34.5 kPa, V_2 = 0.07 m^3. The relationship between pressure and volume during the process is $p = A + BV$, where A and B are constants. (a) For the CO_2, evaluate the work, in N · m and kJ. (b) Evaluate A, in kPa, and B, in (kPa)/m^3.

2.27 A gas is compressed from $V_1 = 0.3\ \text{m}^3$, $p_1 = 1$ bar to $V_2 = 0.1\ \text{m}^3$, $p_2 = 3$ bar. Pressure and volume are related linearly during the process. For the gas, find the work, in kJ.

2.28 Nitrogen (N_2) gas within a piston–cylinder assembly undergoes a compression from $p_1 = 0.2$ MPa, $V_1 = 2.75\ \text{m}^3$ to a state where $p_2 = 2$ MPa. The relationship between pressure and volume during the process is $pV^{1.35} = constant$. For the N_2, determine (a) the volume at state 2, in m^3, and (b) the work, in kJ.

2.29 Oxygen (O_2) gas within a piston–cylinder assembly undergoes an expansion from a volume $V_1 = 0.01\ \text{m}^3$ to a volume $V_2 = 0.03\ \text{m}^3$. The relationship between pressure and volume during the process is $p = AV^{-1} + B$, where $A = 0.06\ \text{bar} \cdot \text{m}^3$ and $B = 3.0$ bar. For the O_2, determine (a) the initial and final pressures, each in bar, and (b) the work, in kJ.

2.30 A closed system consisting of 0.09 kg of air undergoes a polytropic process from $p_1 = 138$ kPa, $v_1 = 0.72\ \text{m}^3/\text{kg}$ to a final state where $p_2 = 552$ kPa, $v_2 = 0.25\ \text{m}^3/\text{kg}$. Determine the amount of energy transfer by work, in kJ, for the process.

2.31 Warm air is contained in a piston–cylinder assembly oriented horizontally as shown in Fig. P2.31. The air cools slowly from an initial volume of $0.003\ \text{m}^3$ to a final volume of $0.002\ \text{m}^3$. During the process, the spring exerts a force that varies linearly from an initial value of 900 N to a final value of zero. The atmospheric pressure is 100 kPa, and the area of the piston face is $0.018\ \text{m}^2$. Friction between the piston and the cylinder wall can be neglected. For the air, determine the initial and final pressures, in kPa, and the work, in kJ.

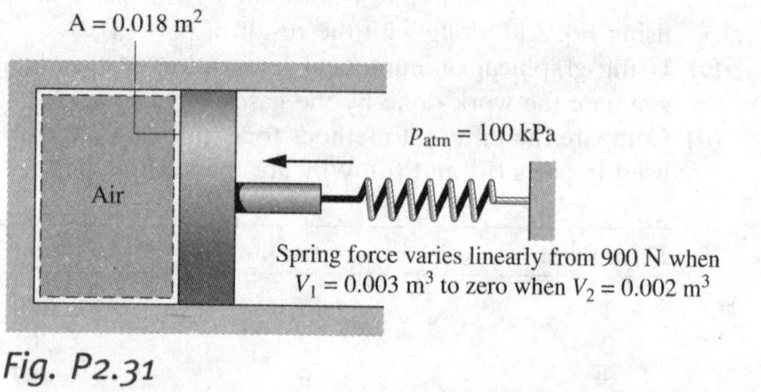

Fig. P2.31

2.32 Air contained within a piston–cylinder assembly is slowly heated. As shown in Fig. P2.32, during this process the pressure first varies linearly with volume and then remains constant. Determine the total work, in kJ.

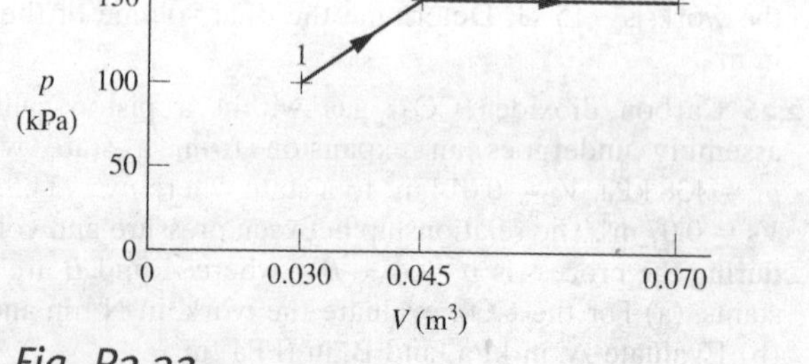

Fig. P2.32

2.33 Carbon monoxide gas (CO) contained within a piston–cylinder assembly undergoes three processes in series:

Process 1–2: Expansion from $p_1 = 5$ bar, $V_1 = 0.2\ \text{m}^3$ to $V_2 = 1\ \text{m}^3$, during which the pressure–volume relationship is $pV = constant$.

Process 2–3: Constant-volume heating from state 2 to state 3, where $p_3 = 5$ bar.

Process 3–1: Constant-pressure compression to the initial state.

Sketch the processes in series on p-V coordinates and evaluate the work for each process, in kJ.

2.34 Air contained within a piston–cylinder assembly undergoes three processes in series:

Process 1–2: Compression at constant pressure from $p_1 = 69$ kPa, $V_1 = 0.11\ \text{m}^3$ to state 2.

Process 2–3: Constant-volume heating to state 3, where $p_3 = 345$ kPa.

Process 3–1: Expansion to the initial state, during which the pressure–volume relationship is $pV = constant$.

Sketch the processes in series on p-V coordinates. Evaluate (a) the volume at state 2, in m^3, and (b) the work for each process, in kJ.

2.35 The belt sander shown in Fig. P2.35 has a belt speed of 7.6 m/s. The coefficient of friction between the sander and a plywood surface being finished is 0.2. If the downward (normal) force on the sander is 66.7 N, determine (a) the power transmitted by the belt, in kJ/s and kW, and (b) the work done in one minute of sanding, in kJ.

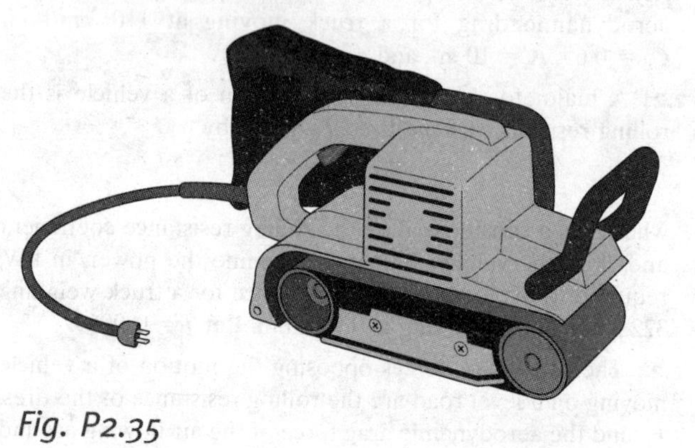

Fig. P2.35

2.36 The driveshaft of a building's air-handling fan is turned at 300 RPM by a belt running on a 0.3-m-diameter pulley. The net force applied by the belt on the pulley is 2,000 N. Determine the torque applied by the belt on the pulley, in N · m, and the power transmitted, in kW.

2.37 A 10-V battery supplies a constant current of 0.5 amp to a resistance for 30 min. (a) Determine the resistance, in ohms. (b) For the battery, determine the amount of energy transfer by work, in kJ.

2.38 A 12-V automotive storage battery is charged with a constant current of 2 amp for 24 h. If electricity costs $0.08 per kW · h, determine the cost of recharging the battery.

2.39 As shown in Fig. P2.39, a steel wire suspended vertically having a cross-section area A and an initial length x_0 is stretched by a downward force F applied to the end of the wire. The normal stress in the wire varies linearly according to $\sigma = C\varepsilon$, where ε is the *strain*, given by $\varepsilon = (x - x_0)/x_0$, and x is the stretched length of the wire. C is a material constant (Young's modulus). Assuming the cross-sectional area remains constant

(a) obtain an expression for the work done on the wire.
(b) evaluate the work done on the wire, in N · m, and the magnitude of the downward force, in N, if $x_0 = 3.048$ m, $x = 3.051$ m, $A = 64.5\ \text{mm}^2$, and $C = 17.2 \times 10^7$ kPa.

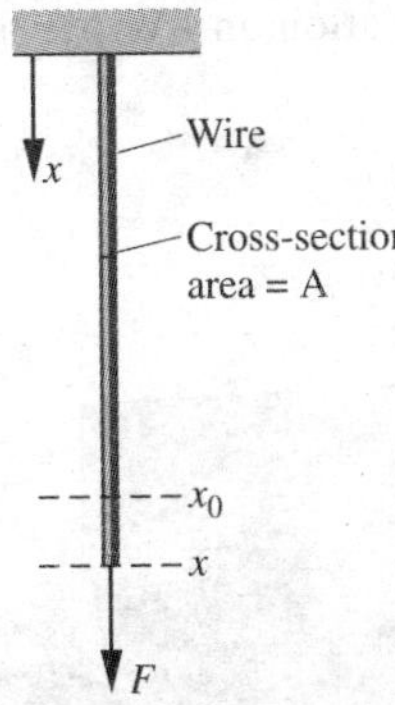

Fig. P2.39

2.40 A soap film is suspended on a wire frame, as shown in Fig. 2.10. The movable wire is displaced by an applied force, F. If the surface tension remains constant

(a) obtain an expression for the work done in stretching the film in terms of the surface tension τ, length ℓ, and displacement Δx.
(b) evaluate the work done, in J, if $\ell = 5$ cm, $\Delta x = 0.5$ cm, and $\tau = 25 \times 10^{-5}$ N/cm.

2.41 As shown in Fig. P2.41, a spring having an initial unstretched length of ℓ_0 is stretched by a force F applied at its end. The stretched length is ℓ. By *Hooke's law,* the force is linearly related to the spring extension by $F = k(\ell - \ell_0)$, where k is the *stiffness.* If stiffness is constant

(a) obtain an expression for the work done in changing the spring's length from ℓ_1 to ℓ_2.
(b) evaluate the work done, in J, if $\ell_0 = 3$ cm, $\ell_1 = 6$ cm, $\ell_2 = 10$ cm, and the stiffness is $k = 10^4$ N/m.

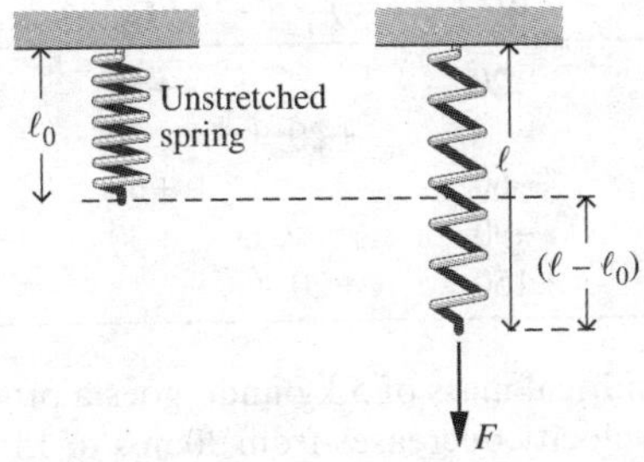

Fig. P2.41

Evaluating Heat Transfer

2.42 A flat surface having an area of $2\ \text{m}^2$ and a temperature of 350 K is cooled convectively by a gas at 300 K. Using data from Table 2.1, determine the largest and smallest heat transfer rates, in kW, that might be encountered for (a) free convection, (b) forced convection.

2.43 A 0.2-m-thick plane wall is constructed of concrete. At steady state, the energy transfer rate by conduction through a 1-m^2 area of the wall is 0.15 kW. If the temperature distribution is linear through the wall, what is the temperature difference across the wall, in K?

2.44 As shown in Fig. P2.44, the 152 mm-thick exterior wall of a building has an average thermal conductivity of 0.55×10^{-3} kW/m · K. At steady state, the temperature of the wall decreases linearly from $T_1 = 21.1°\text{C}$ on the inner surface to T_2 on the outer surface. The outside ambient air temperature is $T_0 = -3.9°\text{C}$ and the convective heat transfer coefficient is $28.97 \times 10^{-3}\ \text{kW/m}^2 \cdot \text{K}$. Determine (a) the temperature T_2, in °C, and (b) the rate of heat transfer through the wall, in kW/m^2 per of surface area.

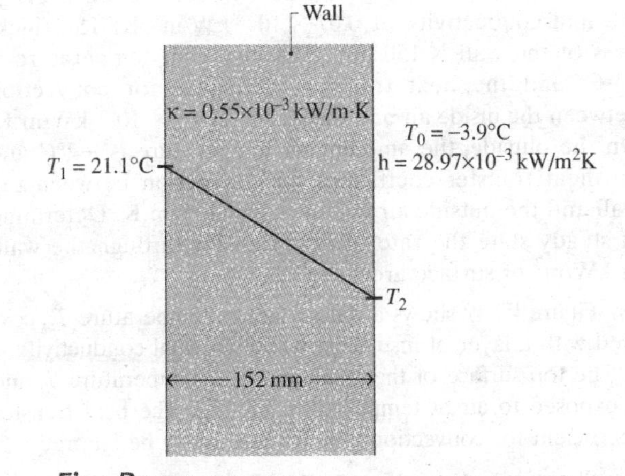

Fig. P2.44

2.45 As shown in Fig. P2.45, an oven wall consists of a 6.35 mm-thick layer of steel ($\kappa_s = 15.05 \times 10^{-3}$ kW/m · K) and a layer of brick ($\kappa_b = 0.73 \times 10^{-3}$ kW/m · K). At steady state, a temperature decrease of 0.67°C occurs over the steel layer. The inner temperature of the steel layer is 282°C. If the temperature of the outer surface of the brick must be no greater than 40°C, determine the minimum thickness of brick, in in., that ensures this limit is met.

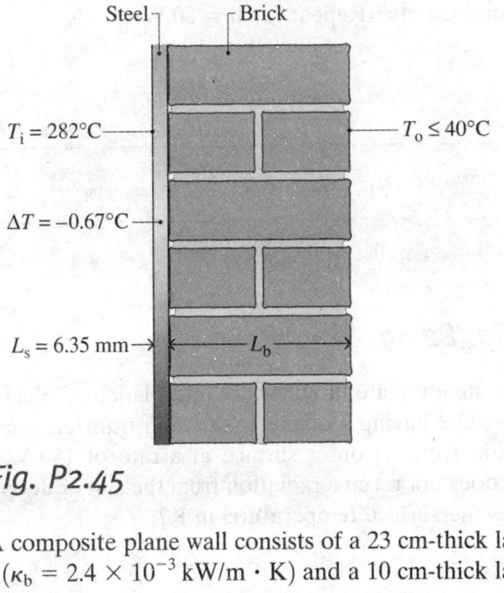

Fig. P2.45

2.46 A composite plane wall consists of a 23 cm-thick layer of brick ($\kappa_b = 2.4 \times 10^{-3}$ kW/m · K) and a 10 cm-thick layer of insulation ($\kappa_i = 0.09 \times 10^{-3}$ kW/m · K). The outer surface temperatures of the brick and insulation are 700 K and 310 K, respectively, and there is perfect contact at the interface between the two layers. Determine at steady state the instantaneous rate of heat transfer, in kW/m^2 of surface area, and the temperature, in K, at the interface between the brick and the insulation.

2.47 A composite plane wall consists of a 75-mm-thick layer of insulation ($\kappa_i = 0.05$ W/m · K) and a 25-mm-thick layer of siding ($\kappa = 0.10$ W/m · K). The inner temperature of the insulation is 20°C. The outer temperature of the siding is −13°C. Determine at steady state (a) the temperature at the interface of the two layers, in °C, and (b) the rate of heat transfer through the wall, in W per m^2 of surface area.

2.48 An insulated frame wall of a house has an average thermal conductivity of 0.07×10^{-3} kW/m · K. The thickness of the wall is 150 mm. The inside air temperature is 20°C, and the heat transfer coefficient for convection between the inside air and the wall is 11.36×10^{-3} kW/m^{2}K. On the outside, the ambient air temperature is −2°C and the heat transfer coefficient for convection between the wall and the outside air is 28.4×10^{-3} kW/m^{2}K. Determine at steady state the rate of heat transfer through the wall, in kW/m^2 of surface area.

2.49 Figure P2.49 shows a flat surface at temperature T_s, covered with a layer of insulation whose thermal conductivity is κ. The top surface of the insulation is at temperature T_i and is exposed to air at temperature $T_0(<T_s)$. The heat transfer coefficient for convection is h. Radiation can be ignored.

(a) Show that at steady state, the temperature T_i is given by

$$T_i = \frac{BT_0 + T_s}{B + 1}$$

where $B = hL/\kappa$.

(b) For fixed T_s and T_0, sketch the variation of T_i versus B. Discuss.

(c) For $T_s = 300$°C and $T_0 = 30$°C, determine the minimum value of B such that T_i is no greater than 60°C. If h = 4 W/m^2 · K and $L = 0.06$ m, what restriction does this limit on B place on the allowed values for thermal conductivity? Repeat for h = 20 W/m^2 · K.

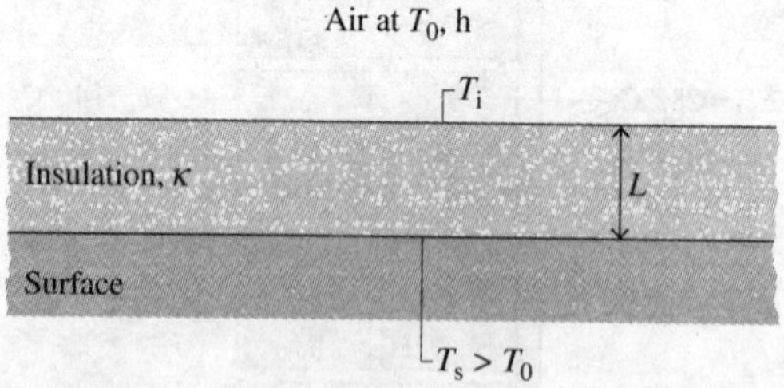

Fig. P2.49

2.50 At steady state, a spherical interplanetary electronics-laden probe having a diameter of 0.5 m transfers energy by radiation from its outer surface at a rate of 150 W. If the probe does not receive radiation from the sun or deep space, what is the surface temperature, in K?

2.51 A body whose surface area is 0.5 m^2, emissivity is 0.8, and temperature is 150°C is placed in a large, evacuated chamber whose walls are at 25°C. What is the rate at which radiation is *emitted* by the surface, in W? What is the *net* rate at which radiation is *exchanged* between the surface and the chamber walls, in W?

2.52 The outer surface of the grill hood shown in Fig. P2.52 is at 47°C and the emissivity is 0.93. The heat transfer coefficient for convection between the hood and the surroundings at 27°C is 10 W/m^2 · K. Determine the net rate of heat transfer between the grill hood and the surroundings by convection and radiation, in kW per m^2 of surface area.

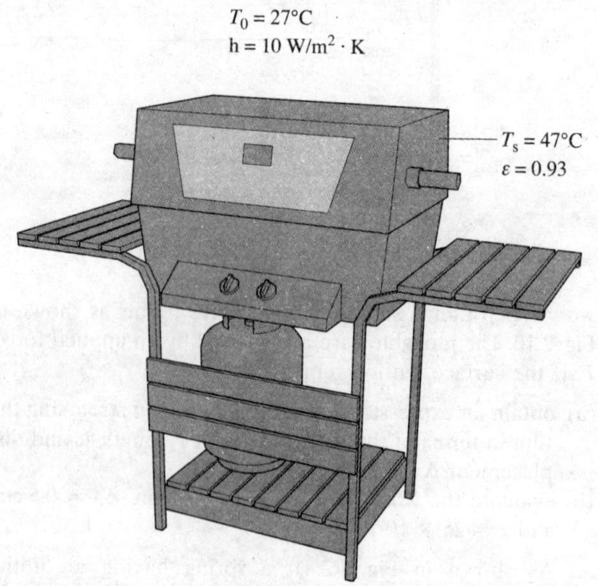

Fig. P2.52

Using the Energy Balance

2.53 Each line in the following table gives information about a process of a closed system. Each entry has the same energy units. Fill in the blank spaces in the table.

Process	Q	W	E_1	E_2	ΔE
a		−20		+50	+70
b	+50		+20	+50	
c		−60		+60	+20
d		−90		+50	0
e	+50	+150	+20		

2.54 A vertical cylindrical mass of 5 kg undergoes a process during which the velocity decreases from 30 m/s to 15 m/s, while the elevation remains unchanged. The initial specific internal energy of the mass is 1.2 kJ/kg and the final specific internal energy is 1.9 kJ/kg. During the process, the mass receives 2 kJ of energy by heat transfer through its bottom surface and loses 1 kJ of energy by heat transfer through its top surface. The lateral surface experiences no heat transfer. For this process, evaluate (a) the change in kinetic energy of the mass in kJ, and (b) the work in kJ.

2.55 A mass of 10 kg undergoes a process during which there is heat transfer from the mass at a rate of 5 kJ per kg, an elevation decrease of 50 m, and an increase in velocity from 15 m/s to 30 m/s. The specific internal energy decreases by 5 kJ/kg and the acceleration of gravity is constant at 9.7 m/s^2. Determine the work for the process, in kJ.

2.56 As shown in Fig. P2.56, 5 kg of steam contained within a piston–cylinder assembly undergoes an expansion from state 1, where the specific internal energy is u_1 = 2,709.9 kJ/kg, to state 2, where u_2 = 2,659.6 kJ/kg. During the process, there is heat transfer *to* the steam with a magnitude of 80 kJ. Also, a paddle wheel transfers energy *to* the steam by work in the amount of 18.5 kJ. There is no significant change in the kinetic or potential energy of the steam. Determine the energy transfer by work from the steam to the piston during the process, in kJ.

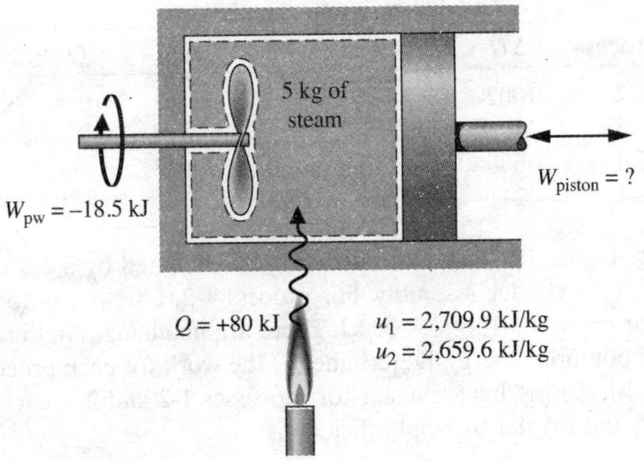

Fig. P2.56

2.57 A gas contained within a piston–cylinder assembly undergoes two processes, A and B, between the *same end states*, 1 and 2, where p_1 = 10 bar, V_1 = 0.1 m^3, U_1 = 400 kJ and p_2 = 1 bar, V_2 = 1.0 m^3, U_2 = 200 kJ:

Process A: Process from 1 to 2 during which the pressure-volume relation is pV = *constant.*

Process B: Constant-volume process from state 1 to a pressure of 2 bar, followed by a linear pressure-volume process to state 2.

Kinetic and potential energy effects can be ignored. For each of the processes A and B, (a) sketch the process on p-V coordinates, (b) evaluate the work, in kJ, and (c) evaluate the heat transfer, in kJ.

2.58 An electric generator coupled to a windmill produces an average electric power output of 15 kW. The power is used to charge a storage battery. Heat transfer from the battery to the surroundings occurs at a constant rate of 1.8 kW. For 8 h of operation, determine the total amount of energy stored in the battery, in kJ.

2.59 An electric motor draws a current of 10 amp with a voltage of 110 V. The output shaft develops a torque of 10.2 N · m and a rotational speed of 1,000 RPM. For operation at steady state, determine for the motor, each in kW

(a) the electric power required.
(b) the power developed by the output shaft.
(c) the rate of heat transfer.

2.60 As shown in Fig. P2.60, the outer surface of a transistor is cooled convectively by a fan-induced flow of air at a temperature of 25°C and a pressure of 1 atm. The transistor's outer surface area is 5 × 10^{-4} m^2. At steady state, the electrical power to the transistor is 3 W. Negligible heat transfer occurs through the base of the transistor. The convective heat transfer coefficient is 100 W/m^2 · K. Determine (a) the rate of heat transfer between the transistor and the air, in W, and (b) the temperature at the transistor's outer surface, in °C.

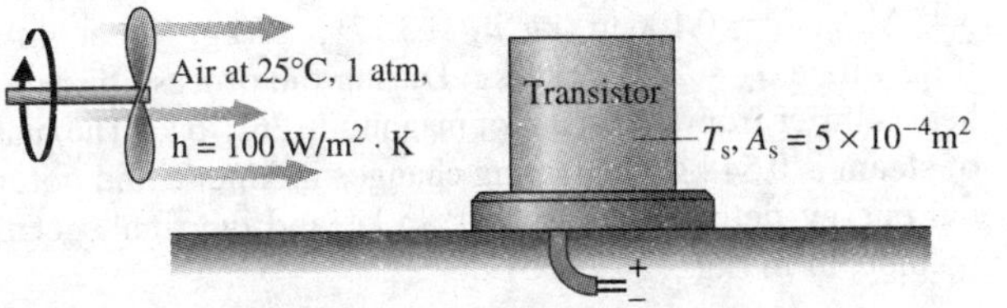

Fig. P2.60

2.61 One kg of Refrigerant 22, initially at p_1 = 0.9 MPa, u_1 = 232.92 kJ/kg, is contained within a rigid closed tank. The tank is fitted with a paddle wheel that transfers energy *to* the refrigerant at a constant rate of 0.1 kW. Heat transfer *from* the refrigerant *to* its surroundings occurs at a rate Kt, in kW, where K is a constant, in kW per minute, and t is time, in minutes. After 20 minutes of stirring, the refrigerant is at p_2 = 1.2 MPa, u_2 = 276.67 kJ/kg. No overall changes in kinetic or potential energy occur. (a) For the refrigerant, determine the work and heat transfer, each in kJ. (b) Determine value of the constant K appearing in the given heat transfer relation, in kW/min.

2.62 An electric generator supplies electricity to a storage battery at a rate of 15 kW for a period of 12 hours. During this 12-h period there also is heat transfer from the battery to the surroundings at a rate of 1.5 kW. Then, during the next 12-h period the battery discharges electricity to an external load at a rate of 5 kW while heat transfer from the battery to the surroundings occurs at a rate of 0.5 kW.

(a) For the first 12-h period, determine, in kW, the time rate of change of energy stored within the battery.
(b) For the second 12-h period, determine, in kW, the time rate of change of energy stored in the battery.
(c) For the *full* 24-h period, determine, in kJ, the *overall* change in energy stored in the battery.

2.63 A gas expands in a piston–cylinder assembly from p_1 = 8 bar, V_1 = 0.02 m^3 to p_2 = 2 bar in a process during which the relation between pressure and volume is $pV^{1.2}$ = *constant*. The mass of the gas is 0.25 kg. If the specific internal energy of the gas *decreases* by 55 kJ/kg during the process, determine the heat transfer, in kJ. Kinetic and potential energy effects are negligible.

2.64 Two kilograms of air is contained in a rigid well-insulated tank with a volume of 0.6 m^3. The tank is fitted with a paddle wheel that transfers energy to the air at a constant rate of 10 W for 1 h. If no changes in kinetic or potential energy occur, determine

(a) the specific volume at the final state, in m^3/kg.
(b) the energy transfer by work, in kJ.
(c) the change in specific internal energy of the air, in kJ/kg.

2.65 Helium gas is contained in a closed rigid tank. An electric resistor in the tank transfers energy *to* the gas at a constant rate of 1 kW. Heat transfer *from* the gas to its surroundings occurs at a rate of $5t$ watts, where t is time, in minutes. Plot the change in energy of the helium, in kJ, for $t \geq 0$ and comment.

2.66 Steam in a piston–cylinder assembly undergoes a polytropic process, with $n = 2$, from an initial state where $p_1 =$ 3.45 MPa, $v_1 = 0.106\ m^3/kg$, $u_1 = 3{,}171.1$ kJ/kg, to a final state where $u_2 = 2{,}303.9$ kJ/kg. During the process, there is a heat transfer from the steam of magnitude 361.76 kJ. The mass of steam is 0.54 kg. Neglecting changes in kinetic and potential energy, determine the work, in kJ, and the final specific volume, in m^3/kg.

2.67 A gas undergoes a process from state 1, where $p_1 =$ 415 kPa, $v_1 = 0.37\ m^3/kg$, to state 2 where $p_2 = 138$ kPa, according to $pv^{1.3} = constant$. The relationship between pressure, specific volume, and internal energy is

$$u = (1.443)pv - 221.557$$

where p is in kPa, v is in m^3/kg, and u is in kJ/kg. The mass of gas is 4.5 kg. Neglecting kinetic and potential energy effects, determine the heat transfer, in kJ.

2.68 Air is contained in a vertical piston–cylinder assembly by a piston of mass 50 kg and having a face area of 0.01 m^2. The mass of the air is 5 g, and initially the air occupies a volume of 5 liters. The atmosphere exerts a pressure of 100 kPa on the top of the piston. The volume of the air slowly decreases to 0.002 m^3 as the specific internal energy of the air decreases by 260 kJ/kg. Neglecting friction between the piston and the cylinder wall, determine the heat transfer to the air, in kJ.

2.69 Gaseous CO_2 is contained in a vertical piston–cylinder assembly by a piston of mass 50 kg and having a face area of 0.01 m^2. The mass of the CO_2 is 4 g. The CO_2 initially occupies a volume of 0.005 m^3 and has a specific internal energy of 657 kJ/kg. The atmosphere exerts a pressure of 100 kPa on the top of the piston. Heat transfer in the amount of 1.95 kJ occurs slowly from the CO_2 to the surroundings, and the volume of the CO_2 decreases to 0.0025 m^3. Friction between the piston and the cylinder wall can be neglected. The local acceleration of gravity is 9.81 m/s^2. For the CO_2, determine (a) the pressure, in kPa, and (b) the final specific internal energy, in kJ/kg.

2.70 A gas is contained in a vertical piston–cylinder assembly by a piston weighing 4,450 N and having a face area of 75 cm^2 The atmosphere exerts a pressure of 101.3 kPa on the top of the piston. An electrical resistor transfers energy to the gas in the amount of 5.3 kJ as the elevation of the piston increases by 0.6 m. The piston and cyclinder are poor thermal conductors and friction can be neglected. Determine the change in internal energy of the gas, in kJ, assuming it is the only significant internal energy change of any component present.

Analyzing Thermodynamic Cycles

2.71 The following table gives data, in kJ, for a system undergoing a thermodynamic cycle consisting of four processes in series. Determine

(a) the missing table entries, each in kJ.
(b) whether the cycle is a power cycle or a refrigeration cycle.

Process	ΔU	ΔKE	ΔPE	ΔE	Q	W
1–2	76	−5.3	−6.3		0	
2–3	67.5	0			95	
3–4	102		0			97
4–1		0	3.2			0

2.72 The following table gives data, in kJ, for a system undergoing a power cycle consisting of four processes in series. Determine (a) the missing table entries, each in kJ, and (b) the thermal efficiency.

Process	ΔU	ΔKE	ΔPE	ΔE	Q	W
1–2	1002	53	0		1055	
2–3		0	53	−475		475
3–4	−686		0	−633		0
4–1	211	−105.5	−53		0	

2.73 Figure P2.73 shows a power cycle executed by a gas in a piston–cylinder assembly. For process 1-2, $U_2 - U_1 = 15$ kJ. For process 3-1, $Q_{31} = 10$ kJ. There are no changes in kinetic or potential energy. Determine (a) the work for each process, in kJ, (b) the heat transfer for processes 1-2 and 2-3, each in kJ, and (c) the thermal efficiency.

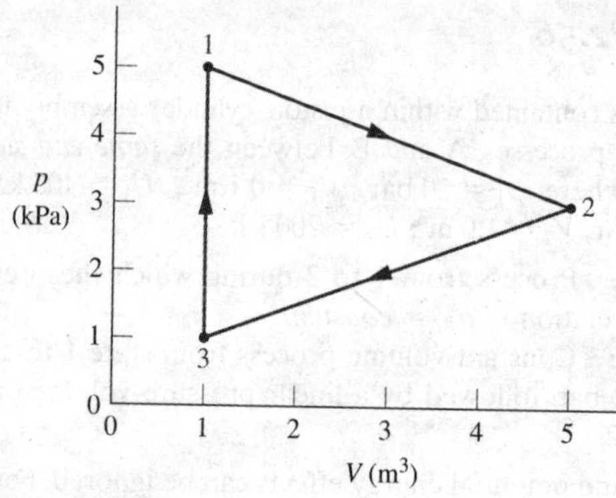

Fig. P2.73

2.74 A gas within a piston–cylinder assembly undergoes a thermodynamic cycle consisting of three processes in series, beginning at state 1 where $p_1 = 1$ bar, $V_1 = 1.5\ m^3$, as follows:

Process 1–2: Compression with $pV = constant$, $W_{12} = -104$ kJ, $U_1 = 512$ kJ, $U_2 = 690$ kJ.

Process 2–3: $W_{23} = 0$, $Q_{23} = -150$ kJ.

Process 3–1: $W_{31} = +50$ kJ.

There are no changes in kinetic or potential energy. (a) Determine Q_{12}, Q_{31}, and U_3, each in kJ. (b) Can this cycle be a power cycle? Explain.

2.75 A gas within a piston–cylinder assembly undergoes a thermodynamic cycle consisting of three processes:

Process 1–2: Compression with $pV = constant$, from $p_1 = 1$ bar, $V_1 = 1.6$ m^3 to $V_2 = 0.2$ m^3, $U_2 - U_1 = 0$.

Process 2–3: Constant pressure to $V_3 = V_1$.

Process 3–1: Constant volume, $U_1 - U_3 = -3{,}549$ kJ.

There are no significant changes in kinetic or potential energy. Determine the heat transfer and work for Process 2–3, in kJ. Is this a power cycle or a refrigeration cycle?

2.76 A gas within a piston–cylinder assembly undergoes a thermodynamic cycle consisting of three processes:

Process 1–2: Constant volume, $V = 0.028$ m^3, $U_2 - U_1 = 26.4$ kJ.

Process 2–3: Expansion with $pV = constant$, $U_3 = U_2$.

Process 3–1: Constant pressure, $p = 1.4$ bar, $W_{31} = -10.5$ kJ.

There are no significant changes in kinetic or potential energy.

(a) Sketch the cycle on a p–V diagram.
(b) Calculate the net work for the cycle, in kJ.
(c) Calculate the heat transfer for process 2–3, in kJ.
(d) Calculate the heat transfer for process 3–1, in kJ.

Is this a power cycle or a refrigeration cycle?

2.77 As shown in Fig. P2.77, a gas within a piston–cylinder assembly undergoes a thermodynamic cycle consisting of three processes in series:

Process 1–2: Compression with $U_2 = U_1$.

Process 2–3: Constant-volume cooling to $p_3 = 140$ kPa, $V_3 = 0.028$ m^3.

Process 3–1: Constant-pressure expansion with $W_{31} = 10.5$ kJ.

For the cycle, $W_{cycle} = -8.3$ kJ. There are no changes in kinetic or potential energy. Determine (a) the volume at state 1, in m^3, (b) the work and heat transfer for process 1–2, each in kJ. (c) Can this be a power cycle? A refrigeration cycle? Explain.

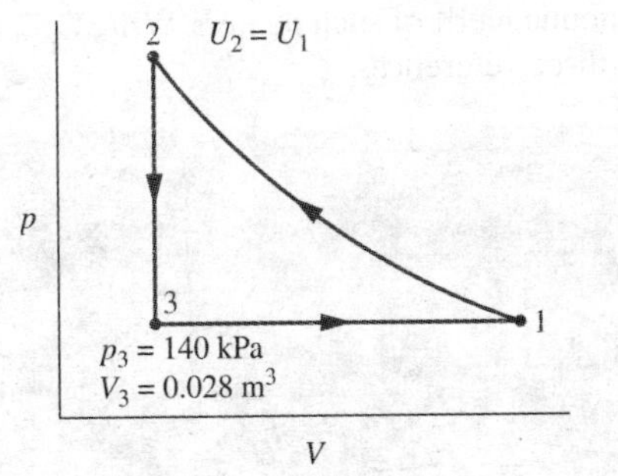

Fig. P2.77

2.78 For a power cycle operating as in Fig. 2.17*a*, the heat transfers are $Q_{in} = 50$ kJ and $Q_{out} = 35$ kJ. Determine the net work, in kJ, and the thermal efficiency.

2.79 The thermal efficiency of a power cycle operating as shown in Fig. 2.17*a* is 35%, and $Q_{out} = 40$ MJ. Determine the net work developed and the heat transfer Q_{in}, each in MJ.

2.80 For a power cycle operating as in Fig. 2.17a, $Q_{in} = 2{,}743$ kJ and $Q_{out} = 1{,}900$ kJ. What is the net work developed, in kJ, and the thermal efficiency?

2.81 For a power cycle operating as in Fig. 2.17*a*, $W_{cycle} = 844$ kJ and $Q_{out} = 1{,}900$ kJ. What is the thermal efficiency?

2.82 A power cycle receives energy by heat transfer from the combustion of fuel and develops power at a net rate of 150 MW. The thermal efficiency of the cycle is 40%.

(a) Determine the net rate at which the cycle receives energy by heat transfer, in MW.
(b) For 8,000 hours of operation annually, determine the net work output, in kW · h per year.
(c) Evaluating the net work output at $0.08 per kW · h, determine the value of net work, in $ per year.

2.83 A power cycle has a thermal efficiency of 40% and generates electricity at a rate of 100 MW. The electricity is valued at $0.08 per kW · h. Based on the cost of fuel, the cost to supply $\dot{Q}_{in}$ is $4.50 per GJ. For 8,000 hours of operation annually, determine, in $,

(a) the value of electricity generated per year.
(b) the annual fuel cost.
(c) Does the difference between the results of parts (a) and (b) represent *profit*? Discuss.

2.84 Shown in Fig. P2.84 is a *cogeneration* power plant operating in a thermodynamic cycle at steady state. The plant provides electricity to a community at a rate of 80 MW. The energy discharged from the power plant by heat transfer is denoted on the figure by $\dot{Q}_{out}$. Of this, 70 MW is provided to the community for water heating and the remainder is discarded to the environment without use. The electricity is valued at $0.08 per kW · h. If the cycle thermal efficiency is 40%, determine the (a) rate energy is added by heat transfer, $\dot{Q}_{in}$, in MW, (b) rate energy is discarded to the environment, in MW, and (c) value of the electricity generated, in $ per year.

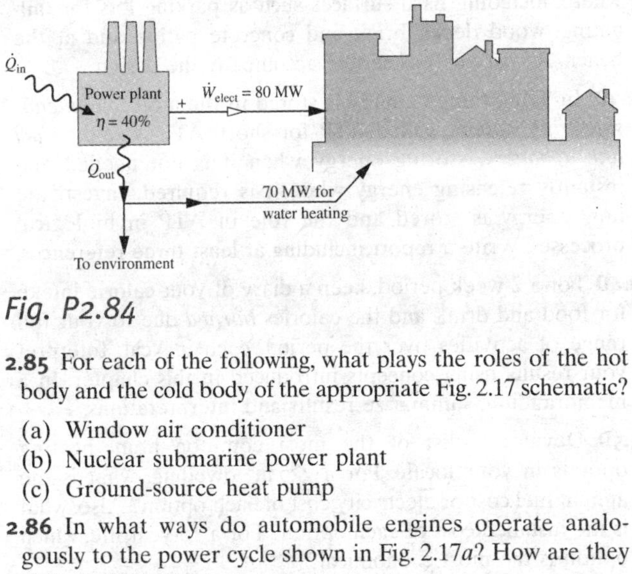

Fig. P2.84

2.85 For each of the following, what plays the roles of the hot body and the cold body of the appropriate Fig. 2.17 schematic?

(a) Window air conditioner
(b) Nuclear submarine power plant
(c) Ground-source heat pump

2.86 In what ways do automobile engines operate analogously to the power cycle shown in Fig. 2.17*a*? How are they different? Discuss.

2.87 A refrigeration cycle operating as shown in Fig. 2.17*b* has heat transfer $Q_{out} = 2{,}530$ kJ and net work of $W_{cycle} = 844$ kJ. Determine the coefficient of performance for the cycle.

2.88 A refrigeration cycle operates as shown in Fig. 2.17*b* with a coefficient of performance $\beta = 1.5$. For the cycle, $Q_{out} = 500$ kJ. Determine Q_{in} and W_{cycle}, each in kJ.

2.89 A household refrigerator operating steadily and with a coefficient of performance of 2.4 removes energy from a refrigerated space at a rate of 633 kJ/h. Evaluating electricity at \$0.08 per kW · h, determine the cost of electricity in a month when the refrigerator operates for 360 hours.

2.90 A heat pump cycle operating at steady state receives energy by heat transfer from well water at 10°C and discharges energy by heat transfer to a building at the rate of 1.2×10^5 kJ/h. Over a period of 14 days, an electric meter records that 1,490 kW · h of electricity is provided to the heat pump. These are the only energy transfers involved. Determine (a) the amount of energy that the heat pump receives over the 14-day period from the well water by heat transfer, in kJ, and (b) the heat pump's coefficient of performance.

2.91 A heat pump cycle whose coefficient of performance is 2.5 delivers energy by heat transfer to a dwelling at a rate of 20 kW.

(a) Determine the net power required to operate the heat pump, in kW.

(b) Evaluating electricity at \$0.08 per kW · h, determine the cost of electricity in a month when the heat pump operates for 200 hours.

2.92 A heat pump cycle delivers energy by heat transfer to a dwelling at a rate of 63,300 kJ/h. The power input to the cycle is 5.82 kW.

(a) Determine the coefficient of performance of the cycle.

(b) Evaluating electricity at \$0.08 per kW · h, determine the cost of electricity in a month when the heat pump operates for 200 hours.

2.93 An air-conditioning unit with a coefficient of performance of 2.93 provides 5,275 kJ/h of cooling while operating during the cooling season 8 hours per day for 125 days. If you pay 10 cents per kW · h for electricity, determine the cost, in dollars, for the cooling season.

Design & open ended problems: exploring engineering practice

2.1D Critically evaluate and compare the various hybrid electric automobiles on the market today. In a memorandum, summarize your findings and conclusions.

2.2D Design a go-anywhere, use-anywhere wind screen for outdoor recreational and casual-living activities, including sunbathing, reading, cooking, and picnicking. The wind screen must be lightweight, portable, easy to deploy, and low cost. A key constraint is that the wind screen can be set up anywhere, including hard surfaces such as parking lots for tailgating, wood decks, brick and concrete patios, and at the beach. A cost analysis should accompany the design.

2.3D In living things, energy is stored in the molecule *adenosine triphosphate,* called ATP for short. ATP is said to *act like a battery,* storing energy when it is not needed and instantly releasing energy when it is required. Investigate how energy is stored and the role of ATP in biological processes. Write a report including at least three references.

2.4D For a 2-week period, keep a diary of your calorie intake for food and drink and the calories *burned* due to your full range of activities over the period, each in kcal. Interpret your results using concepts introduced in this chapter. In a memorandum, summarize results and interpretations.

2.5D Develop a list of the most common home-heating options in your locale. For a 225 m^2 dwelling, what is the annual fuel cost or electricity cost of each option? Also, what is the installed cost of each option? For a 15-year life, which option is the most economical?

2.6D Homeowners concerned about interruption of electrical power from their local utility because of weather-related and utility system outages can acquire a standby generator that produces electricity using the home's existing natural gas or liquid propane (LP) fuel source. For a single-family dwelling of your choice, identify a standby generator that would provide electricity during an outage to a set of essential devices and appliances you specify. Summarize your findings in a memorandum including installed-cost and operating-cost data.

2.7D An advertisement describes a portable heater claimed to cut home-heating bills by up to 50%. The heater is said to be able to heat large rooms in minutes without having a high outer-surface temperature, reducing humidity and oxygen levels, or producing carbon monoxide. A typical deployment is shown in Fig. P2.7D. The heater is an enclosure containing electrically powered quartz infrared lamps that shine on copper tubes. Air drawn into the enclosure by a fan flows over the tubes and then is directed back into the living space. Heaters requiring 500 watts of power cost about \$400 while the 1,000-W model costs about \$500. Critically evaluate the technical and economic merit of such heaters. Write a report including at least three references.

Fig. P2.7D

2.8D Fossil-fuel power plants produce most of the electricity generated annually in the United States. The cost of electricity is determined by several factors, including the power plant thermal efficiency, the unit cost of the fuel, in \$ per kW · h, and the plant capital cost, in \$ per kW of power generated. Prepare a memorandum comparing typical ranges of these three factors for coal-fired steam power plants and natural gas-fired gas turbine power plants. Which type of plant is most prevalent in the United States?

2.9D According to the New York City Transit Authority, the operation of subways raise tunnel and station temperatures as much as 263–266 K above ambient temperature. Principal contributors to the temperature rise include train motor operation, lighting, and energy from the passengers themselves. Passenger discomfort can increase significantly in hot-weather periods if air conditioning is not provided. Still, because on-board air-conditioning units discharge energy by heat transfer to their surroundings, such units contribute to the overall tunnel and station energy management problem. Investigate the application to subways of *alternative* cooling strategies that provide substantial cooling with a minimal power requirement, including but not limited to thermal storage and nighttime ventilation. Write a report with at least three references.

2.10D An inventor proposes *borrowing* water from municipal water mains and storing it *temporarily* in a tank on the premises of a dwelling equipped with a heat pump. As shown in Fig. P2.10D, the stored water serves as the cold body for the heat pump and the dwelling itself serves as the hot body. Since the energy of the stored water decreases as energy is removed from it by the heat pump, water is drawn from the mains periodically to restore the energy level and an equal amount of lower-energy water is returned to the mains. As the invention requires no *net* water from the mains, the inventor maintains that nothing should be paid for water usage. The inventor also maintains that this approach not only gives a coefficient of performance superior to those of *air-source* heat pumps but also avoids the installation costs associated with *ground-source* heat pumps. In all, significant cost savings result, the inventor says. Critically evaluate the inventor's claims. Write a report including at least three references.

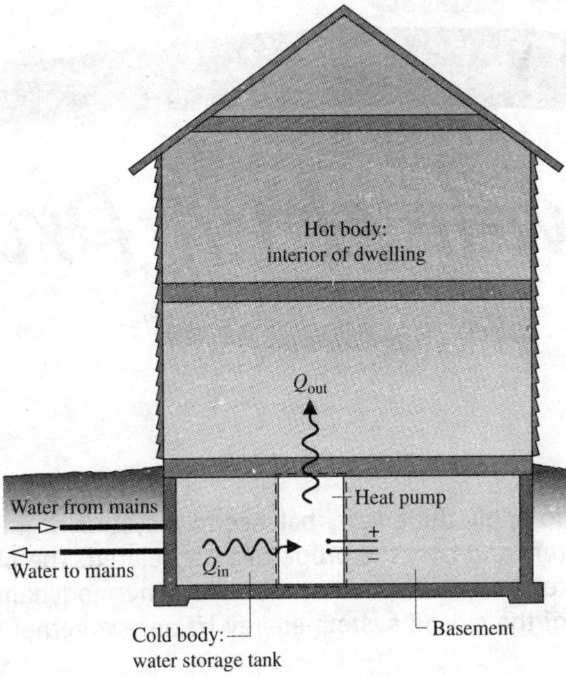

Fig. P2.10D

3

Evaluating Properties

Engineering Context

To apply the energy balance to a system of interest requires knowledge of the properties of the system and how the properties are related. The **objectives** of this chapter are to introduce property relations relevant to engineering thermodynamics and provide several examples illustrating the use of the closed system energy balance together with the property relations considered in this chapter.

Learning Objectives

When you complete your study of this chapter you will be able to...

- ✓ demonstrate understanding of key concepts... including phase and pure substance, state principle for simple compressible systems, p-v-T surface, saturation temperature and saturation pressure, two-phase liquid-vapor mixture, quality, enthalpy, and specific heats.
- ✓ apply the closed system energy balance with property data.
- ✓ sketch T-v, p-v, and phase diagrams, and locate states on these diagrams.
- ✓ retrieve property data from Tables A-1 through A-23.
- ✓ apply the ideal gas model for thermodynamic analysis, including determining when use of the model is warranted.

3.1 Getting Started

In this section, we introduce concepts that support our study of property relations, including phase, pure substance, and the state principle for simple systems.

3.1.1 Phase and Pure Substance

The term ***phase*** refers to a quantity of matter that is homogeneous throughout in both chemical composition and physical structure. Homogeneity in physical structure means that the matter is all *solid*, or all *liquid*, or all *vapor* (or equivalently all *gas*). A system can contain one or more phases.

phase

FOR EXAMPLE... a system of liquid water and water vapor (steam) contains *two* phases. A system of liquid water and ice, including the case of *slush*, also contains *two* phases. Gases, oxygen and nitrogen for example, can be mixed in any proportion to form a *single* gas phase. Certain liquids, such as alcohol and water, can be mixed to form a *single* liquid phase. But liquids such as oil and water, which are not miscible, form *two* liquid phases.

Two phases coexist during the *changes in phase* called *vaporization, melting*, and *sublimation.*

A ***pure substance*** is one that is uniform and invariable in chemical composition. A pure substance can exist in more than one phase, but its chemical composition must be the same in each phase.

pure substance

FOR EXAMPLE... if liquid water and water vapor form a system with two phases, the system can be regarded as a pure substance because each phase has the same composition. A uniform mixture of gases can be regarded as a pure substance provided it remains a gas and doesn't react chemically. Air can be regarded as a pure substance as long as it is a mixture of gases; but if a liquid phase should form on cooling, the liquid would have a different composition than the gas phase, and the system would no longer be considered a pure substance.

Changes in the composition due to chemical reaction are considered in Chap. 13.

3.1.2 Fixing the State

The state of a closed system at equilibrium is its condition as described by the values of its thermodynamic properties. From observation of many thermodynamic systems, it is known that not all of these properties are independent of one another, and the state can be uniquely determined by giving the values of the *independent* properties. Values for all other thermodynamic properties are determined once this independent subset is specified. A general rule known as the ***state principle*** has been developed as a guide in determining the number of independent properties required to fix the state of a system.

state principle

For most applications considered in this book, we are interested in what the state principle says about the *intensive* states of systems. Of particular interest are systems of commonly encountered pure substances, such as water or a uniform mixture of nonreacting gases. These systems are called ***simple compressible systems.*** Experience shows that the simple compressible systems occur in a wide range of engineering applications. For such systems, the state principle indicates that the number of independent intensive properties is *two*.

simple compressible systems

FOR EXAMPLE... in the case of a gas, temperature and another intensive property such as a specific volume might be selected as the two independent properties. The state principle then affirms that pressure, specific internal energy, and all other pertinent *intensive* properties could be determined as functions of T and v: $p = p(T, v)$, $u = u(T, v)$, and so on. The functional relations would be developed

> **State Principle for Simple Systems**
>
> Based on empirical evidence, there is one independent property for each way a system's energy can be varied independently. We saw in Chap. 2 that the energy of a closed system can be altered independently by heat or by work. Accordingly, an independent property can be associated with heat transfer as one way of varying the energy, and another independent property can be counted for each relevant way the energy can be changed through work. On the basis of experimental evidence, therefore, the *state principle* asserts that the number of independent properties is one plus the number of *relevant* work interactions. When counting the number of relevant work interactions, only those that would be significant in *quasiequilibrium* processes of the system need to be considered.
>
> The term *simple system* is applied when there is only *one* way the system energy can be significantly altered by work as the system undergoes quasiequilibrium processes. Therefore, counting one independent property for heat transfer and another for the single work mode gives a total of two independent properties needed to fix the state of a simple system. *This is the state principle for simple systems.* Although no system is ever truly simple, many systems can be modeled as simple systems for the purpose of thermodynamic analysis. The most important of these models for the applications considered in this book is the *simple compressible system.* Other types of simple systems are simple *elastic* systems and simple *magnetic* systems.

using experimental data and would depend explicitly on the particular chemical identity of the substances making up the system. The development of such functions is discussed in Chap. 11.

Intensive properties such as velocity and elevation that are assigned values relative to datums *outside* the system are excluded from present considerations. Also, as suggested by the name, changes in volume can have a significant influence on the energy of *simple compressible systems*. The only mode of energy transfer by work that can occur as a simple compressible system undergoes *quasiequilibrium* processes (Sec. 2.2) is associated with volume change, and is given by $\int p\, dV$. For further discussion of simple systems and the state principle, see the box.

Evaluating Properties: General Considerations

This part of the chapter is concerned generally with the thermodynamic properties of simple compressible systems consisting of *pure* substances. A pure substance is one of uniform and invariable chemical composition. Property relations for systems in which composition changes by chemical reaction are considered in Chap. 13. In the second part of this chapter, we consider property evaluation using the *ideal gas model.*

3.2 p–v–T Relation

We begin our study of the properties of pure, simple compressible substances and the relations among these properties with pressure, specific volume, and temperature. From experiment it is known that temperature and specific volume can be regarded

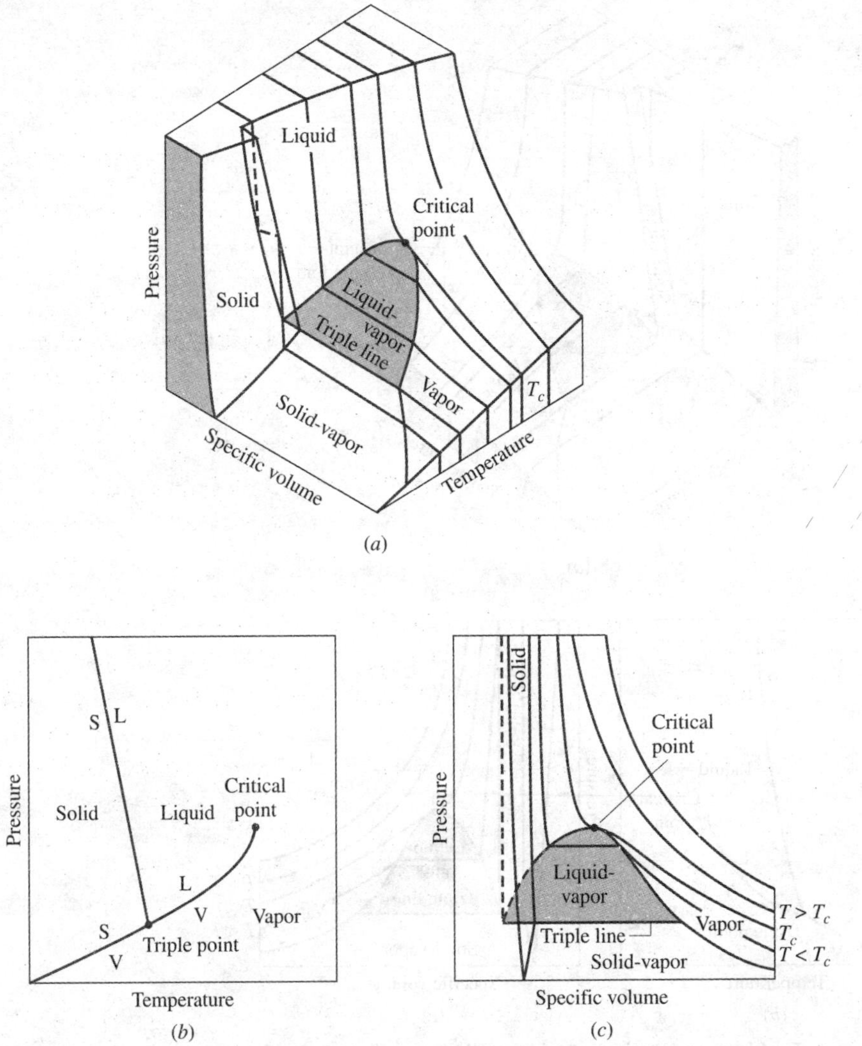

Fig. 3.1 *p–v–T* surface and projections for a substance that expands on freezing. (*a*) Three-dimensional view. (*b*) Phase diagram. (*c*) *p–v* diagram.

as independent and pressure determined as a function of these two: $p = p(T, v)$. The graph of such a function is a *surface,* the ***p–v–T surface.***

p–v–T surface

3.2.1 *p–v–T* Surface

Figure 3.1 is the *p–v–T* surface of a substance such as water that expands on freezing. Figure 3.2 is for a substance that contracts on freezing, and most substances exhibit this characteristic. The coordinates of a point on the *p–v–T* surfaces represent the values that pressure, specific volume, and temperature would assume when the substance is at equilibrium.

There are regions on the *p–v–T* surfaces of Figs. 3.1 and 3.2 labeled *solid, liquid,* and *vapor.* In these *single-phase* regions, the state is fixed by *any* two of the properties: pressure, specific volume, and temperature, since all of these are independent when there is a single phase present. Located between the single-phase regions are ***two-phase regions*** where two phases exist in equilibrium: liquid–vapor, solid–liquid, and solid–vapor. Two phases can coexist during changes in phase such as vaporization, melting, and sublimation. Within the two-phase regions pressure and temperature are not independent; one cannot be changed without changing the other. In these

two-phase regions

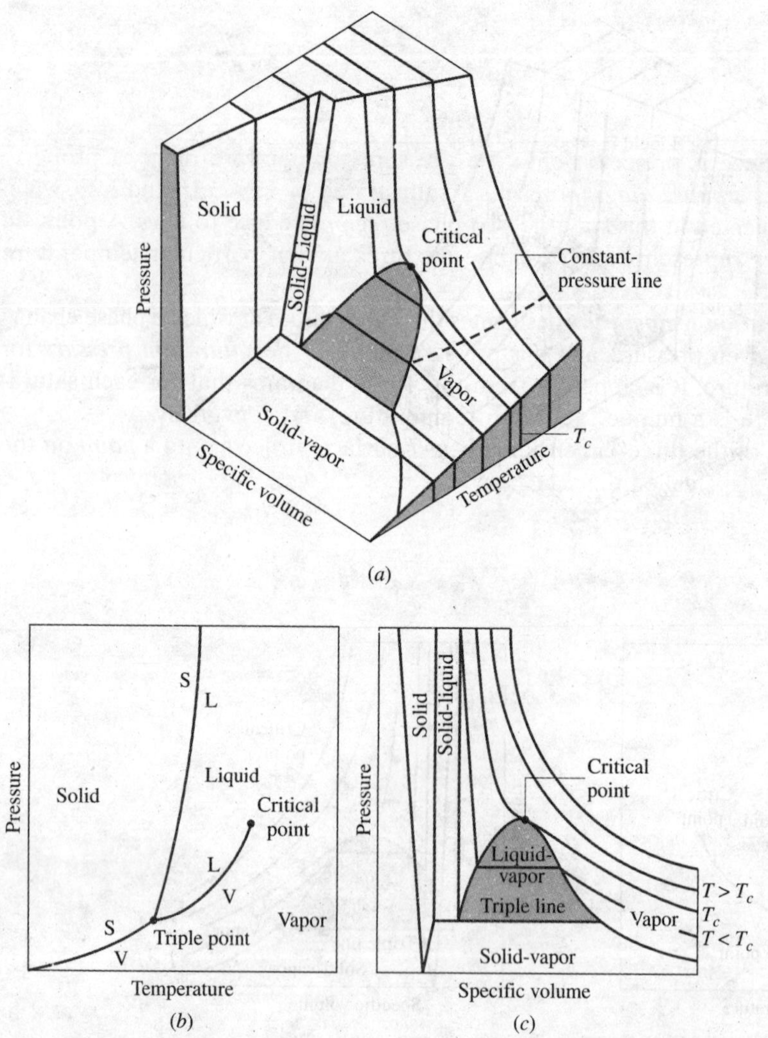

Fig. 3.2 *p–v–T* surface and projections for a substance that contracts on freezing. (*a*) Three-dimensional view. (*b*) Phase diagram. (*c*) *p–v* diagram.

regions the state cannot be fixed by temperature and pressure alone; however, the state can be fixed by specific volume and either pressure or temperature. Three phases can exist in equilibrium along the line labeled ***triple line.***

triple line

saturation state

A state at which a phase change begins or ends is called a ***saturation state.*** The dome-shaped region composed of the two-phase liquid–vapor states is called the ***vapor dome.*** The lines bordering the vapor dome are called saturated liquid and saturated vapor lines. At the top of the dome, where the saturated liquid and saturated vapor lines meet, is the ***critical point.*** The *critical temperature* T_c of a pure substance is the maximum temperature at which liquid and vapor phases can coexist in equilibrium. The pressure at the critical point is called the *critical pressure,* p_c. The specific volume at this state is the *critical specific volume*. Values of the critical point properties for a number of substances are given in Tables A-1 located in the Appendix.

vapor dome

critical point

The three-dimensional *p–v–T* surface is useful for bringing out the general relationships among the three phases of matter normally under consideration. However, it is often more convenient to work with two-dimensional projections of the surface. These projections are considered next.

3.2.2 Projections of the p–v–T Surface

The Phase Diagram

If the p–v–T surface is projected onto the pressure–temperature plane, a property diagram known as a ***phase diagram*** results. As illustrated by Figs. 3.1*b* and 3.2*b*, when the surface is projected in this way, the two-phase *regions* reduce to *lines*. A point on any of these lines represents all two-phase mixtures at that particular temperature and pressure.

phase diagram

The term ***saturation temperature*** designates the temperature at which a phase change takes place at a given pressure, and this pressure is called the ***saturation pressure*** for the given temperature. It is apparent from the phase diagrams that for each saturation pressure there is a unique saturation temperature, and conversely.

saturation temperature
saturation pressure

The triple *line* of the three-dimensional p–v–T surface projects onto a *point* on the phase diagram. This is called the ***triple point.*** Recall that the triple point of water is used as a reference in defining temperature scales (Sec. 1.7.3). By agreement, the temperature *assigned* to the triple point of water is 273.16 K. The *measured* pressure at the triple point of water is 0.6113 kPa.

triple point

The line representing the two-phase solid–liquid region on the phase diagram slopes to the left for substances that expand on freezing and to the right for those that contract. Although a single solid phase region is shown on the phase diagrams of Figs. 3.1 and 3.2, solids can exist in different solid phases. For example, seven different crystalline forms have been identified for water as a solid (ice).

p–v Diagram

Projecting the p–v–T surface onto the pressure–specific volume plane results in a ***p–v diagram,*** as shown by Figs. 3.1*c* and 3.2*c*. The figures are labeled with terms that have already been introduced.

p–v diagram

When solving problems, a sketch of the p–v diagram is frequently convenient. To facilitate the use of such a sketch, note the appearance of constant-temperature lines (isotherms). By inspection of Figs. 3.1*c* and 3.2*c*, it can be seen that for any specified temperature *less than* the critical temperature, pressure remains constant as the two-phase liquid–vapor region is traversed, but in the single-phase liquid and vapor regions the pressure decreases at fixed temperature as specific volume increases. For temperatures greater than or equal to the critical temperature, pressure decreases continuously at fixed temperature as specific volume increases. There is no passage across the two-phase liquid–vapor region. The critical isotherm passes through a point of inflection at the critical point and the slope is zero there.

T–v Diagram

Projecting the liquid, two-phase liquid–vapor, and vapor regions of the p–v–T surface onto the temperature–specific volume plane results in a ***T–v diagram*** as in Fig. 3.3. Since consistent patterns are revealed in the p–v–T behavior of all pure substances, Fig. 3.3 showing a T–v diagram for water can be regarded as representative.

T–v diagram

As for the p–v diagram, a sketch of the T–v diagram is often convenient for problem solving. To facilitate the use of such a sketch, note the appearance of constant-pressure lines (isobars). For pressures *less than* the critical pressure, such as the 10 MPa isobar on Fig. 3.3, the pressure remains constant with temperature as the two-phase region is traversed. In the single-phase liquid and vapor regions the temperature increases at fixed pressure as the specific volume increases. For pressures greater than or equal to the critical pressure, such as the one marked 30 MPa on Fig. 3.3, temperature increases continuously at fixed pressure as the specific volume increases. There is no passage across the two-phase liquid–vapor region.

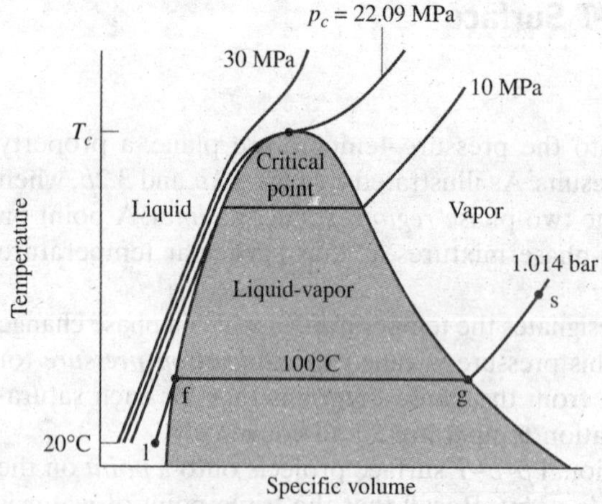

Fig. 3.3 Sketch of a temperature–specific volume diagram for water showing the liquid, two-phase liquid–vapor, and vapor regions (not to scale).

The projections of the p–v–T surface used in this book to illustrate processes are not generally drawn to scale. A similar comment applies to other property diagrams introduced later.

3.3 Studying Phase Change

It is instructive to study the events that occur as a pure substance undergoes a phase change. To begin, consider a closed system consisting of a unit mass (1 kg) of liquid water at 20°C contained within a piston–cylinder assembly, as illustrated in Fig. 3.4*a*. This state is represented by point l on Fig. 3.3. Suppose the water is slowly heated while its pressure is kept constant and uniform throughout at 1.014 bar.

Liquid States

As the system is heated at constant pressure, the temperature increases considerably while the specific volume increases slightly. Eventually, the system is brought to the state represented by f on Fig. 3.3. This is the saturated liquid state corresponding to the specified pressure. For water at 1.014 bar the saturation temperature is 100°C. The liquid states along the line segment l–f of Fig. 3.3 are sometimes referred to as ***subcooled liquid*** states because the temperature at these states is less than the saturation temperature at the given pressure. These states are also referred to as ***compressed liquid*** states because the pressure at each state is higher than the saturation pressure corresponding to the temperature at the state. The names liquid, subcooled liquid, and compressed liquid are used interchangeably.

subcooled liquid

compressed liquid

Two-Phase, Liquid–Vapor Mixture

When the system is at the saturated liquid state (state f of Fig. 3.3), additional heat transfer at fixed pressure results in the formation of vapor without any change in temperature but with a considerable increase in specific volume. As shown in Fig. 3.4*b*, the system would now consist of a two-phase liquid–vapor mixture. When a mixture of liquid and vapor exists in equilibrium, the liquid phase is a saturated liquid and the vapor phase is a saturated vapor. If the system is heated further until the last bit of liquid has vaporized, it is brought to point g on Fig. 3.3, the saturated vapor state.

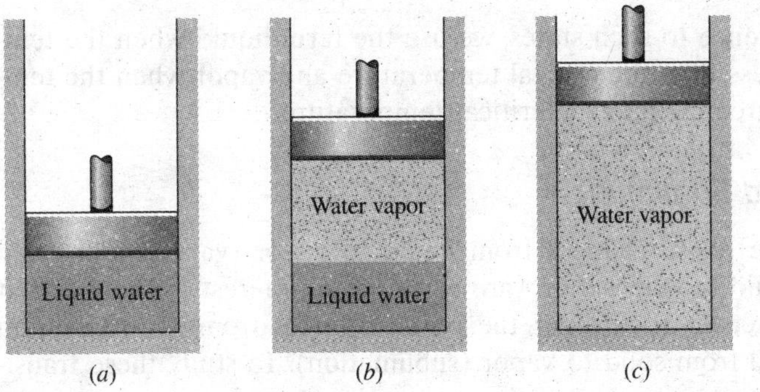

Fig. 3.4 Illustration of constant-pressure change from liquid to vapor for water.

The intervening ***two-phase liquid–vapor mixture*** states can be distinguished from one another by the *quality,* an intensive property.

two-phase liquid–vapor mixture

For a two-phase liquid–vapor mixture, the ratio of the mass of vapor present to the total mass of the mixture is its ***quality,*** *x*. In symbols,

quality

$$x = \frac{m_{\text{vapor}}}{m_{\text{liquid}} + m_{\text{vapor}}} \tag{3.1}$$

The value of the quality ranges from zero to unity: at saturated liquid states, $x = 0$, and at saturated vapor states, $x = 1.0$. Although defined as a ratio, the quality is frequently given as a percentage. Examples illustrating the use of quality are provided in Sec. 3.5. Similar parameters can be defined for two-phase solid–vapor and two-phase solid–liquid mixtures.

Vapor States

Let us return to a consideration of Figs. 3.3 and 3.4. When the system is at the saturated vapor state (state g on Fig. 3.3), further heating at fixed pressure results in increases in both temperature and specific volume. The condition of the system would now be as shown in Fig. 3.4*c*. The state labeled s on Fig. 3.3 is representative of the states that would be attained by further heating while keeping the pressure constant. A state such as s is often referred to as a ***superheated vapor*** state because the system would be at a temperature greater than the saturation temperature corresponding to the given pressure.

superheated vapor

Consider next the same thought experiment at the other constant pressures labeled on Fig. 3.3, 10 MPa, 22.09 MPa, and 30 MPa. The first of these pressures is less than the critical pressure of water, the second is the critical pressure, and the third is greater than the critical pressure. As before, let the system initially contain a liquid at 20°C. First, let us study the system if it were heated slowly at 10 MPa. At this pressure, vapor would form at a higher temperature than in the previous example, because the saturation pressure is higher (refer to Fig. 3.3). In addition, there would be somewhat less of an increase in specific volume from saturated liquid to vapor, as evidenced by the narrowing of the vapor dome. Apart from this, the general behavior would be the same as before.

Consider next the behavior of the system if it were heated at the critical pressure, or higher. As seen by following the critical isobar on Fig. 3.3, there would be no change in phase from liquid to vapor. At all states there would be only one phase. As shown by line *a-b-c* of the phase diagram sketched in Fig. 3.5, *vaporization* and the inverse process of *condensation* can occur only when the pressure is less than the critical pressure. Thus, at states where pressure is greater than the critical pressure, the terms liquid and vapor tend to lose their significance. Still, for

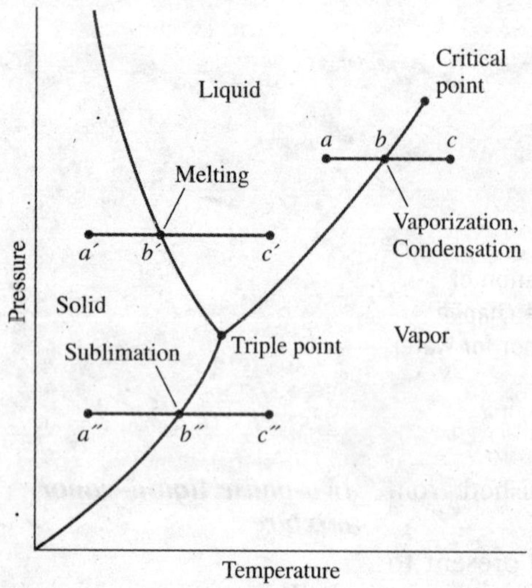

Fig. 3.5 Phase diagram for water (not to scale).

ease of reference to such states, we use the term liquid when the temperature is less than the critical temperature and vapor when the temperature is greater than the critical temperature.

Melting and Sublimation

Although the phase changes from liquid to vapor (vaporization) and vapor to liquid (condensation) are of principal interest in this book, it is also instructive to consider the phase changes from solid to liquid (melting) and from solid to vapor (sublimation). To study these transitions, consider a system consisting of a unit mass of ice at a temperature below the triple point temperature. Let us begin with the case where the pressure is greater than the triple point pressure and the system is at state a' of Fig. 3.5. Suppose the system is slowly heated while maintaining the pressure constant and uniform throughout. The temperature increases with heating until point b' on Fig. 3.5 is attained. At this state the ice is a saturated solid. Additional heat transfer at fixed pressure results in the formation of liquid without any change in temperature. As the system is heated further, the ice continues to melt until eventually the last bit melts, and the system contains only saturated liquid. During the melting process the temperature and pressure remain constant. For most substances, the specific volume increases during melting, but for water the specific volume of the liquid is less than the specific volume of the solid. Further heating at fixed pressure results in an increase in temperature as the system is brought to point c' on Fig. 3.5. Next, consider the case where the pressure is less than the triple point pressure and the system is at state a'' of Fig. 3.5. In this case, if the system is heated at constant pressure it passes through the two-phase solid–vapor region into the vapor region along the line a''–b''–c'' shown on Fig. 3.5. That is, sublimation occurs.

Bio... connections

While condensation of steam to a liquid and further cooling to a lower-temperature liquid state is easily imagined and even a part of our experience, liquefying gases such as nitrogen and oxygen may not be so familiar. Still, there are important medical and industrial applications for liquefied gases. One of these is the practice of *cryosurgery* by dermatologists, which is the localized freezing of skin tissue for the removal of unwanted lesions, including precancerous lesions. For this type of surgery, liquid nitrogen at about −195°C is applied as a spray or with a probe. Cryosurgery is quickly performed and generally without anesthetic. Dermatologists store liquid nitrogen required for up to several months in containers called *Dewar* flasks that are similar to vacuum bottles.

3.4 Retrieving Thermodynamic Properties

Thermodynamic property data can be retrieved in various ways, including tables, graphs, equations, and computer software. The emphasis of Secs. 3.5 and 3.6 to follow is on the use of *tables* of thermodynamic properties, which are commonly available for pure, simple compressible substances of engineering interest. The use of these tables is an important skill. The ability to locate states on property diagrams is an important related skill. The software available with this text, *Interactive Thermodynamics: IT*, is introduced in Sec. 3.7. *IT* is used selectively in examples and end-of-chapter problems throughout the book. Skillful use of tables and property diagrams is prerequisite for the effective use of software to retrieve thermodynamic property data.

Since tables for different substances are frequently set up in the same general format, the present discussion centers mainly on Tables A-2 through A-6 giving the properties of water; these are commonly referred to as the ***steam tables.*** Tables A-7 through A-9 for

steam tables

Refrigerant 22, Tables A-10 through A-12 for Refrigerant 134a, Tables A-13 through A-15 for ammonia, and Tables A-16 through A-18 for propane are used similarly, as are tables for other substances found in the engineering literature. Tables are provided in the Appendix in SI units.

The substances for which tabulated data are provided in this book have been selected because of their importance in current practice. Still, they are merely representative of a wide range of industrially important substances. To meet changing requirements and address special needs, new substances are frequently introduced while others become obsolete.

Energy & Environment

Naturally occurring refrigerants like ammonia, carbon dioxide, and certain hydrocarbons were used in refrigeration systems a century ago. These refrigerants were displaced by safer-to-use chlorine-containing refrigerants, known as chlorofluorocarbons (CFCs), paving the way for the refrigerators and air conditioners we enjoy today. Recently, because of concerns over the effects of chlorine on the earth's protective ozone layer, international agreements have been implemented to phase out use of CFCs. CFCs have been replaced by refrigerants such as Refrigerant 22 ($CHClF_2$) and Refrigerant 134a (CF_3CH_2F), having less potential to deplete the ozone layer. But studies now suggest that natural refrigerants may be preferable because of even lower overall impact on global warming.

Although decades of research and development have gone into today's refrigerators and air conditioners, and the use of natural refrigerants pose significant engineering challenges, naturals are again being actively considered. Ammonia, once widely used but dropped for domestic use owing to toxicity, is receiving renewed interest because it contains no chlorine. New energy-efficient refrigerators using propane are available on the global market, but are still of concern in the U.S. because of propane's flammability. Carbon dioxide is considered promising for small, lightweight systems, such as automotive and portable air-conditioning units. Although CO_2 is a global-warming gas, only a tiny amount would be required in a typical unit, and even this would be contained under proper maintenance and refrigeration unit disposal protocols, experts say.

3.5 Evaluating Pressure, Specific Volume, and Temperature

3.5.1 Vapor and Liquid Tables

The properties of water vapor are listed in Tables A-4 and of liquid water in Tables A-5. These are often referred to as the *superheated* vapor tables and *compressed* liquid tables, respectively. The sketch of the phase diagram shown in Fig. 3.6 brings out the structure of these tables. Since pressure and temperature are independent properties in the single-phase liquid and vapor regions, they can be used to fix the state in these regions. Accordingly, Tables A-4 and A-5 are set up to give values of several properties as functions of pressure and temperature. The first property listed is specific volume. The remaining properties are discussed in subsequent sections.

For each pressure listed, the values given in the superheated vapor table (Tables A-4) *begin* with the saturated vapor state and then proceed to higher temperatures. The data in the compressed liquid table (Tables A-5) *end* with saturated liquid states. That is, for a given pressure the property values are given as the temperature increases to the saturation temperature. In these tables, the value shown in parentheses after the pressure in the table heading is the corresponding saturation temperature. ➡ **FOR EXAMPLE...** in Tables A-4 and A-5, at a pressure of 10.0 MPa, the saturation temperature is listed as 311.06°C. ⬅

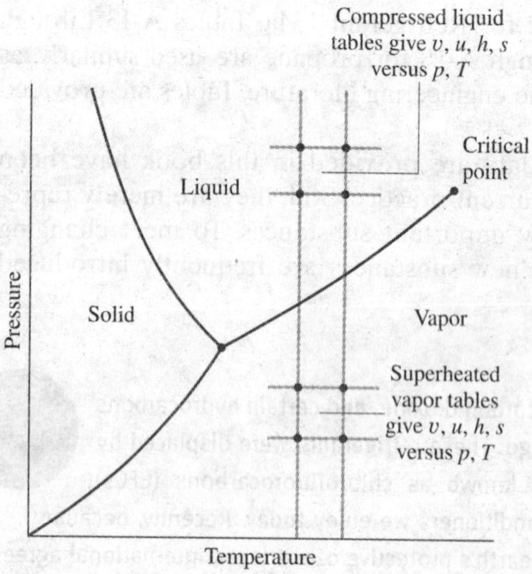

Fig. 3.6 Sketch of the phase diagram for water used to discuss the structure of the superheated vapor and compressed liquid tables (not to scale).

➡ **FOR EXAMPLE...** to gain more experience with Tables A-4 and A-5 verify the following: Table A-4 gives the specific volume of water vapor at 10.0 MPa and 600°C as 0.03837 m³/kg. At 10.0 MPa and 100°C, Table A-5 gives the specific volume of liquid water as 1.0385×10^{-3} m³/kg. ⬅

The states encountered when solving problems often do not fall exactly on the grid of values provided by property tables. *Interpolation* between adjacent table entries then becomes necessary. Care always must be exercised when interpolating table values. The tables provided in the Appendix are extracted from more extensive tables that are set up so that ***linear interpolation,*** illustrated in the following example, can be used with acceptable accuracy. Linear interpolation is assumed to remain valid when using the abridged tables of the text for the solved examples and end-of-chapter problems.

linear interpolation

➡ **FOR EXAMPLE...** let us determine the specific volume of water vapor at a state where p = 10 bar and T = 215°C. Shown in Fig. 3.7 is a sampling of data from Table A-4. At a pressure of 10 bar, the specified temperature of 215°C falls between the table values of 200 and 240°C, which are shown in boldface. The corresponding specific volume values are also shown in boldface. To determine the specific volume v corresponding to 215°C, we may think of the *slope* of a straight line joining the adjacent table entries, as follows

$$\textit{slope} = \frac{(0.2275 - 0.2060)\ \text{m}^3/\text{kg}}{(240 - 200)^\circ\text{C}} = \frac{(v - 0.2060)\ \text{m}^3/\text{kg}}{(215 - 200)^\circ\text{C}}$$

Solving for v, the result is v = 0.2141 m³/kg. ⬅

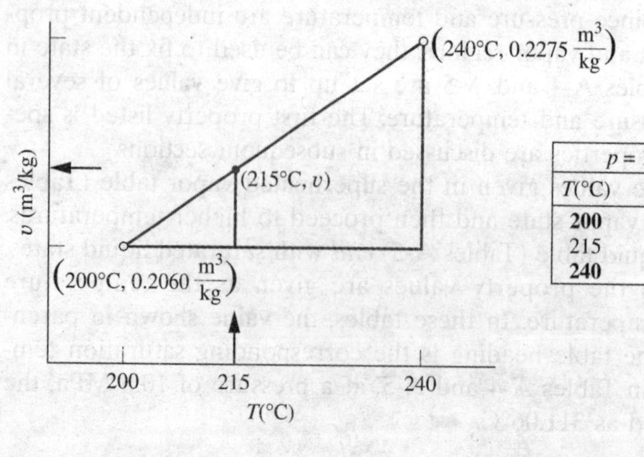

p = 10 bar	
T(°C)	v (m³/kg)
200	**0.2060**
215	v = ?
240	**0.2275**

Fig. 3.7 Illustration of linear interpolation.

The following example features the use of sketches of p–v and T–v diagrams in conjunction with tabular data to fix the end states of a process. In accord with the state principle, two independent intensive properties must be known to fix the states of the system under consideration.

Example 3.1 HEATING AMMONIA AT CONSTANT PRESSURE

A vertical piston–cylinder assembly containing 0.05 kg of ammonia, initially a saturated vapor, is placed on a hot plate. Due to the weight of the piston and the surrounding atmospheric pressure, the pressure of the ammonia is 1.5 bars. Heating occurs slowly, and the ammonia expands at constant pressure until the final temperature is 25°C. Show the initial and final states on T–v and p–v diagrams, and determine

(a) the volume occupied by the ammonia at each end state, in m³.

(b) the work for the process, in kJ.

Solution

Known: Ammonia is heated at constant pressure in a vertical piston–cylinder assembly from the saturated vapor state to a known final temperature.

Find: Show the initial and final states on T–v and p–v diagrams, and determine the volume at each end state and the work for the process.

Schematic and Given Data:

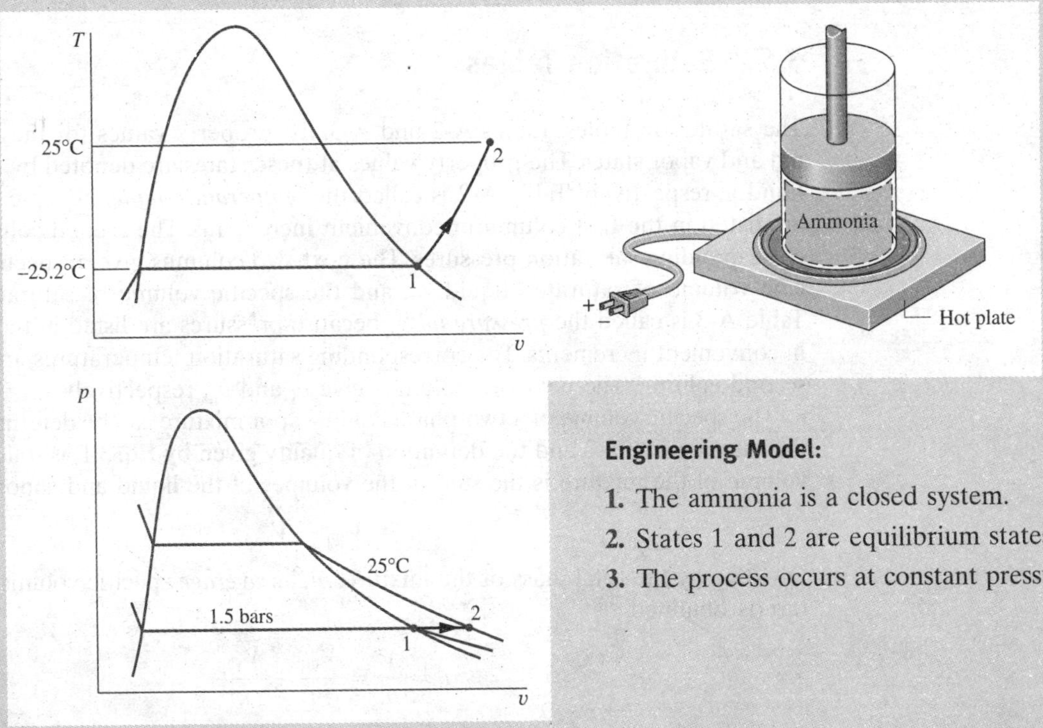

Engineering Model:

1. The ammonia is a closed system.
2. States 1 and 2 are equilibrium states.
3. The process occurs at constant pressure.

Fig. E3.1

Analysis: The initial state is a saturated vapor condition at 1.5 bars. Since the process occurs at constant pressure, the final state is in the superheated vapor region and is fixed by 1.5 bars. and T_2 = 25°C. The initial and final states are shown on the T–v and p–v diagrams above.

(a) The volumes occupied by the ammonia at states 1 and 2 are obtained using the given mass and the respective specific volumes. From Table A-14 at P_2 = 1.5 bars and corresponding to *Sat.* in the temperature column, we get $v_1 = v_g = 0.7787\ \text{m}^3/\text{kg}$. Thus

$$\begin{aligned} V_1 = mv_1 &= (0.05\ \text{kg})(0.7787\ \text{m}^3/\text{kg}) \\ &= 0.0389\ \text{m}^3 \end{aligned}$$

Interpolating in Table A-14 at p_2 = 1.5 bar and T_2 = 25°C, we get $v_2 = 0.9553\ \text{m}^3/\text{kg}$. Thus

$$V_2 = mv_2 = (0.05\ \text{kg})(0.9553\ \text{m}^3/\text{kg}) = 0.0478\ \text{m}^3$$

(b) In this case, the work can be evaluated using Eq. 2.17. Since the pressure is constant

$$W = \int_{V_1}^{V_2} p\,dV = p(V_2 - V_1)$$

Inserting values

❶ $$\begin{aligned} W &= (1.5\ \text{bars})(0.0478 - 0.0389)\text{m}^3 \left|\frac{10^5\ \text{N/m}^2}{1\ \text{bar}}\right|\left|\frac{\text{kJ}}{10^3\ \text{N}\cdot\text{m}}\right| \\ &= 1.335\ \text{kJ} \end{aligned}$$

❶ Note the use of conversion factors in this calculation.

Skills Developed: *Ability to...*

- ✓ define a closed system and identify interactions on its boundary.
- ✓ sketch *T-v* and *p-v* diagrams and locate states on them.
- ✓ evaluate work using Eq. 2.17.
- ✓ retrieve property data for ammonia at vapor states.

Quick Quiz

If heating continues at 1.5 bars from T_2 = 25°C to T_3 = 32°C, determine the work for Process 2-3, in kJ. **Ans.** 0.16 kJ

3.5.2 Saturation Tables

The saturation tables, Tables A-2 and A-3, list property values for the saturated liquid and vapor states. The property values at these states are denoted by the subscripts f and g, respectively. Table A-2 is called the *temperature table,* because temperatures are listed in the first column in convenient increments. The second column gives the corresponding saturation pressures. The next two columns give, respectively, the specific volume of saturated liquid, v_f, and the specific volume of saturated vapor, v_g. Table A-3 is called the *pressure table,* because pressures are listed in the first column in convenient increments. The corresponding saturation temperatures are given in the second column. The next two columns give v_f and v_g, respectively.

The specific volume of a two-phase liquid–vapor mixture can be determined by using the saturation tables and the definition of quality given by Eq. 3.1 as follows. The total volume of the mixture is the sum of the volumes of the liquid and vapor phases

$$V = V_{liq} + V_{vap}$$

Dividing by the total mass of the mixture, m, an *average* specific volume for the mixture is obtained

$$v = \frac{V}{m} = \frac{V_{liq}}{m} + \frac{V_{vap}}{m}$$

Since the liquid phase is a saturated liquid and the vapor phase is a saturated vapor, $V_{liq} = m_{liq}v_f$ and $V_{vap} = m_{vap}v_g$, so

$$v = \left(\frac{m_{liq}}{m}\right)v_f + \left(\frac{m_{vap}}{m}\right)v_g$$

Introducing the definition of quality, $x = m_{vap}/m$, and noting that $m_{liq}/m = 1 - x$, the above expression becomes

$$v = (1 - x)v_f + xv_g = v_f + x(v_g - v_f) \tag{3.2}$$

The increase in specific volume on vaporization $(v_g - v_f)$ is also denoted by v_{fg}.

➡ **FOR EXAMPLE...** consider a system consisting of a two-phase liquid–vapor mixture of water at 100°C and a quality of 0.9. From Table A-2 at 100°C, $v_f = 1.0435 \times 10^{-3}$ m³/kg and $v_g = 1.673$ m³/kg. The specific volume of the mixture is

$$v = v_f + x(v_g - v_f) = 1.0435 \times 10^{-3} + (0.9)(1.673 - 1.0435 \times 10^{-3}) = 1.506 \text{ m}^3/\text{kg}$$

where the v_f and v_g values are obtained from Table A-2. ⬅

To facilitate locating states in the tables, it is often convenient to use values from the saturation tables together with a sketch of a T–v or p–v diagram. For example, if the specific volume v and temperature T are known, refer to the appropriate temperature table, Table A-2, and determine the values of v_f and v_g. A T–v diagram illustrating these data is given in Fig. 3.8. If the given specific volume falls between v_f and v_g, the system consists of a two-phase liquid–vapor mixture, and the pressure is the saturation pressure corresponding to the given temperature. The quality can be found by solving Eq. 3.2. If the given specific volume is greater than v_g, the state is in the superheated vapor region. Then, by interpolating in Table A-4, the pressure and other properties listed can be determined. If the given specific volume is less than v_f, Table A-5 would be used to determine the pressure and other properties.

➡ **FOR EXAMPLE...** let us determine the pressure of water at each of three states defined by a temperature of 100°C and specific volumes, respectively, of $v_1 = 2.434$ m³/kg, $v_2 = 1.0$ m³/kg, and $v_3 = 1.0423 \times 10^{-3}$ m³/kg. Using the known temperature, Table A-2 provides the values of v_f and v_g: $v_f = 1.0435 \times 10^{-3}$ m³/kg, $v_g = 1.673$ m³/kg. Since v_1 is greater than v_g, state 1 is in the vapor region. Table A-4 gives the pressure as 0.70 bar. Next, since v_2 falls between v_f and v_g, the pressure is the saturation pressure corresponding to 100°C, which is 1.014 bar. Finally, since v_3 is less than v_f, state 3 is in the liquid region. Table A-5 gives the pressure as 25 bar. ⬅

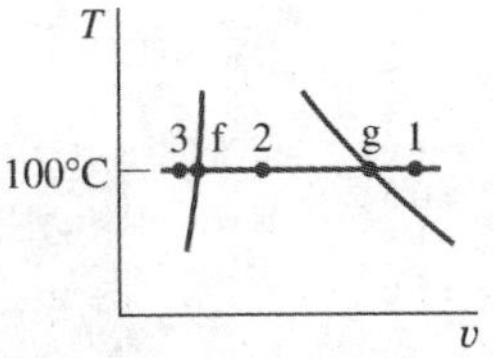

The following example features the use of a sketch of the T–v diagram in conjunction with tabular data to fix the end states of processes. In accord with the state principle, two independent intensive properties must be known to fix the states of the system under consideration.

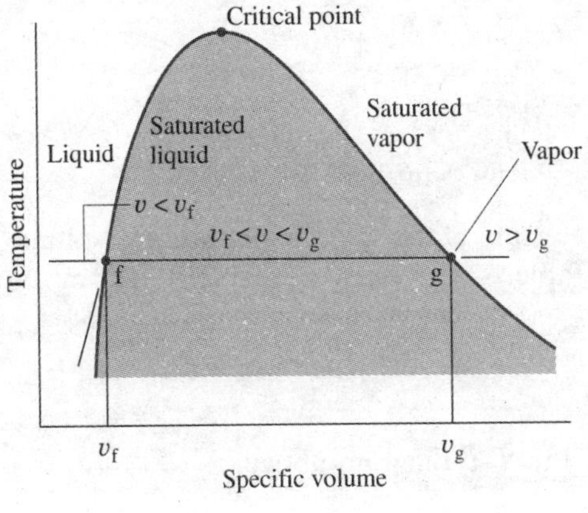

Fig. 3.8 Sketch of a T–v diagram for water used to discuss locating states in the tables.

Example 3.2 HEATING WATER AT CONSTANT VOLUME

A closed, rigid container of volume 0.5 m^3 is placed on a hot plate. Initially, the container holds a two-phase mixture of saturated liquid water and saturated water vapor at $p_1 = 1$ bar with a quality of 0.5. After heating, the pressure in the container is $p_2 = 1.5$ bar. Indicate the initial and final states on a T–v diagram, and determine

(a) the temperature, in °C, at states 1 and 2.

(b) the mass of vapor present at states 1 and 2, in kg.

(c) If heating continues, determine the pressure, in bar, when the container holds only saturated vapor.

Solution

Known: A two-phase liquid–vapor mixture of water in a closed, rigid container is heated on a hot plate. The initial pressure and quality and the final pressure are known.

Find: Indicate the initial and final states on a T–v diagram and determine at each state the temperature and the mass of water vapor present. Also, if heating continues, determine the pressure when the container holds only saturated vapor.

Schematic and Given Data:

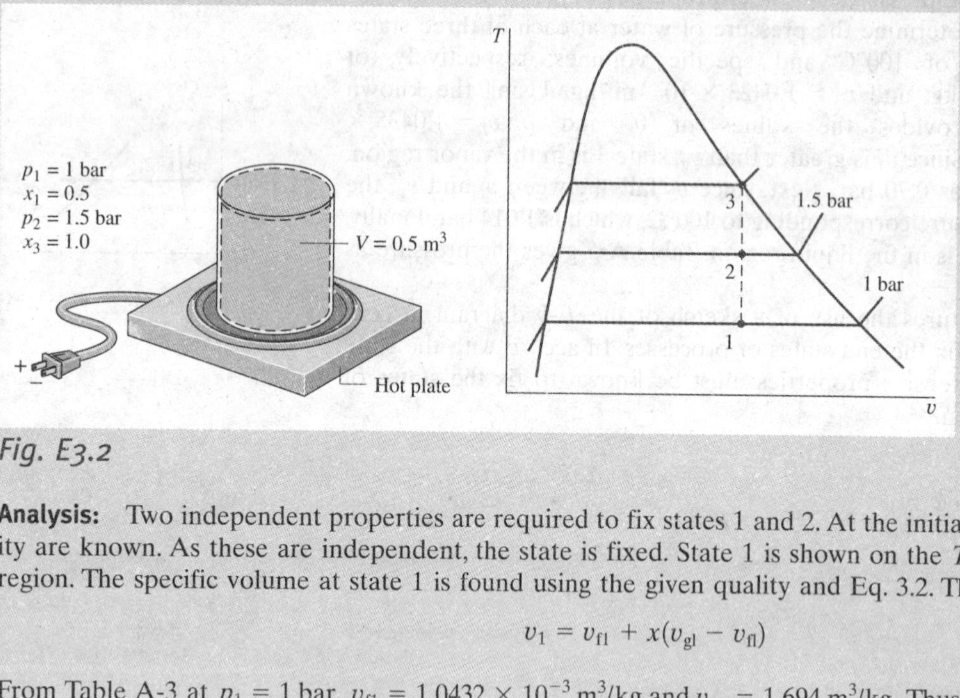

Engineering Model:

1. The water in the container is a closed system.
2. States 1, 2, and 3 are equilibrium states.
3. The volume of the container remains constant.

Fig. E3.2

Analysis: Two independent properties are required to fix states 1 and 2. At the initial state, the pressure and quality are known. As these are independent, the state is fixed. State 1 is shown on the T–v diagram in the two-phase region. The specific volume at state 1 is found using the given quality and Eq. 3.2. That is,

$$v_1 = v_{f1} + x(v_{g1} - v_{f1})$$

From Table A-3 at $p_1 = 1$ bar, $v_{f1} = 1.0432 \times 10^{-3}$ m^3/kg and $v_{g1} = 1.694$ m^3/kg. Thus,

$$v_1 = 1.0432 \times 10^{-3} + 0.5(1.694 - 1.0432 \times 10^{-3}) = 0.8475 \text{ m}^3\text{/kg}$$

At state 2, the pressure is known. The other property required to fix the state is the specific volume v_2. Volume and mass are each constant, so $v_2 = v_1 = 0.8475$ m^3/kg. For $p_2 = 1.5$ bar, Table A-3 gives $v_{f2} = 1.0582 \times 10^{-3}$ m^3/kg and $v_{g2} = 1.59$ m^3/kg. Since

❶ $$v_f < v_2 < v_{g2}$$

❷ state 2 must be in the two-phase region as well. State 2 is also shown on the T–v diagram above.

(a) Since states 1 and 2 are in the two-phase liquid–vapor region, the temperatures correspond to the saturation temperatures for the given pressures. Table A-3 gives

$$T_1 = 99.63°\text{C} \quad \text{and} \quad T_2 = 111.4°\text{C}$$

(b) To find the mass of water vapor present, we first use the volume and the specific volume to find the *total* mass, m. That is

$$m = \frac{V}{v} = \frac{0.5 \text{ m}^3}{0.8475 \text{ m}^3/\text{kg}} = 0.59 \text{ kg}$$

Then, with Eq. 3.1 and the given value of quality, the mass of vapor at state 1 is

$$m_{g1} = x_1 m = 0.5(0.59 \text{ kg}) = 0.295 \text{ kg}$$

The mass of vapor at state 2 is found similarly using the quality x_2. To determine x_2, solve Eq. 3.2 for quality and insert specific volume data from Table A-3 at a pressure of 1.5 bar, along with the known value of v, as follows

$$x_2 = \frac{v - v_{f2}}{v_{g2} - v_{f2}}$$
$$= \frac{0.8475 - 1.0528 \times 10^{-3}}{1.159 - 1.0528 \times 10^{-3}} = 0.731$$

Then, with Eq. 3.1

$$m_{g2} = 0.731(0.59 \text{ kg}) = 0.431 \text{ kg}$$

(c) If heating continued, state 3 would be on the saturated vapor line, as shown on the T–v diagram above. Thus, the pressure would be the corresponding saturation pressure. Interpolating in Table A-3 at $v_g = 0.8475 \text{ m}^3/\text{kg}$, we get $p_3 = 2.11$ bar.

Skills Developed: Ability to...
- ✓ define a closed system and identify interactions on its boundary.
- ✓ sketch a T-v diagram and locate states on it.
- ✓ retrieve property data for water at liquid-vapor states, using quality.

❶ The procedure for fixing state 2 is the same as illustrated in the discussion of Fig. 3.8.

❷ Since the process occurs at constant specific volume, the states lie along a vertical line.

Quick Quiz

If heating continues at constant specific volume from state 3 to a state where pressure is 3 bar, determine the temperature at that state, in °C.
Ans. 282°C

3.6 Evaluating Specific Internal Energy and Enthalpy

3.6.1 Introducing Enthalpy

In many thermodynamic analyses the sum of the internal energy U and the product of pressure p and volume V appears. Because the sum $U + pV$ occurs so frequently in subsequent discussions, it is convenient to give the combination a name, ***enthalpy,*** and a distinct symbol, H. By definition

enthalpy

$$H = U + pV \tag{3.3}$$

Since U, p, and V are all properties, this combination is also a property. Enthalpy can be expressed on a unit mass basis

$$h = u + pv \tag{3.4}$$

and per mole

$$\bar{h} = \bar{u} + p\bar{v} \tag{3.5}$$

Units for enthalpy are the same as those for internal energy.

3.6.2 Retrieving *u* and *h* Data

The property tables introduced in Sec. 3.5 giving pressure, specific volume, and temperature also provide values of specific internal energy u, enthalpy h, and entropy s. Use of these tables to evaluate u and h is described in the present section; the consideration of entropy is deferred until it is introduced in Chap. 6.

Data for specific internal energy u and enthalpy h are retrieved from the property tables in the same way as for specific volume. For saturation states, the values of u_f and u_g, as well as h_f and h_g, are tabulated versus both saturation pressure and saturation temperature. The specific internal energy for a two-phase liquid–vapor mixture is calculated for a given quality in the same way the specific volume is calculated

$$u = (1 - x)u_f + xu_g = u_f + x(u_g - u_f) \tag{3.6}$$

The increase in specific internal energy on vaporization $(u_g - u_f)$ is often denoted by u_{fg}. Similarly, the specific enthalpy for a two-phase liquid–vapor mixture is given in terms of the quality by

$$h = (1 - x)h_f + xh_g = h_f + x(h_g - h_f) \tag{3.7}$$

The increase in enthalpy during vaporization $(h_g - h_f)$ is often tabulated for convenience under the heading h_{fg}.

➡ **FOR EXAMPLE...** to illustrate the use of Eqs. 3.6 and 3.7, we determine the specific enthalpy of Refrigerant 22 when its temperature is 12°C and its specific internal energy is 144.58 kJ/kg. Referring to Table A-7, the given internal energy value falls between u_f and u_g at 12°C, so the state is a two-phase liquid–vapor mixture. The quality of the mixture is found by using Eq. 3.6 and data from Table A-7 as follows:

$$x = \frac{u - u_f}{u_g - u_f} = \frac{144.58 - 58.77}{230.38 - 58.77} = 0.5$$

Then, with the values from Table A-7, Eq. 3.7 gives

$$\begin{aligned} h &= (1 - x)h_f + xh_g \\ &= (1 - 0.5)(59.35) + 0.5(253.99) = 156.67 \text{ kJ/kg} \end{aligned}$$ ⬅

In the superheated vapor tables, u and h are tabulated along with v as functions of temperature and pressure. ➡ **FOR EXAMPLE...** let us evaluate T, v, and h for water at 0.10 MPa and a specific internal energy of 2,537.3 kJ/kg. Turning to Table A-3, note that the given value of u is greater than u_g at 0.1 MPa (u_g = 2,506.1 kJ/kg). This suggests that the state lies in the superheated vapor region. By inspection of Table A-4, we get T = 120°C, v = 1.793 m³/kg, and h = 2716.6 kJ/kg. Alternatively, h and u are related by the definition of h

$$\begin{aligned} h &= u + pv \\ &= 2537.3\frac{\text{kJ}}{\text{kg}} + \left(10^5\frac{\text{N}}{\text{m}^2}\right)\left(1.793\frac{\text{m}^3}{\text{kg}}\right)\left|\frac{1 \text{ kJ}}{10^3 \text{ N}\cdot\text{m}}\right| \\ &= 2537.3 + 179.3 = 2,716.6 \text{ kJ/kg} \end{aligned}$$

As another illustration, consider water at a state fixed by a pressure equal to 1.0 bar and a temperature of 120°C. From Table A-4, $v = 1.793\ \frac{\text{m}^3}{\text{kg}}$, $u = 2537.3\ \frac{\text{kJ}}{\text{kg}}$, and $h = 2716.6$ kJ/kg. As above, h may be calculated from u. Thus

$$h = u + pv$$

$$= 2537.3\frac{\text{kJ}}{\text{kg}} + (1.0\ \text{bar})\left(1.793\frac{\text{m}^3}{\text{kg}}\right)\left|\frac{100\ \text{kN/m}^2}{1\ \text{bar}}\right|$$

$$= (2537.3 + 179.3)\ \text{kJ/kg} = 2716.6\ \text{kJ/kg}$$

Specific internal energy and enthalpy data for liquid states of water are presented in Tables A-5. The format of these tables is the same as that of the superheated vapor tables considered previously. Accordingly, property values for liquid states are retrieved in the same manner as those of vapor states.

For water, Tables A-6 give the equilibrium properties of saturated solid and saturated vapor. The first column lists the temperature, and the second column gives the corresponding saturation pressure. These states are at pressures and temperatures *below* those at the triple point. The next two columns give the specific volume of saturated solid, v_i, and saturated vapor, v_g, respectively. The table also provides the specific internal energy, enthalpy, and entropy values for the saturated solid and the saturated vapor at each of the temperatures listed.

Bio... connections

What do first responders, military flight crews, costumed characters at theme parks, and athletes have in common? They share a need to avoid heat stress while performing their duty, job, and past-time, respectively. To meet this need, *wearable coolers* have been developed such as cooling vests and cooling collars. Wearable coolers may feature ice pack inserts, channels through which a cool liquid is circulated, encapsulated *phase-change materials*, or a combination. A familiar example of a phase-change material (PCM) is ice, which on melting at 0°C absorbs energy of about 334 kJ/kg.

When worn close to the body, PCM-laced apparel absorbs energy from persons working or exercising in hot environments, keeping them cool. When specifying a PCM for a wearable cooler, the material must change phase at the desired cooler operating temperature. Hydrocarbons known as *paraffins* are frequently used for such duty. Many coolers available today employ PCM beads with diameters as small as 0.5 microns, encapsulated in a durable polymer shell. Encapsulated phase-change materials also are found in other products.

3.6.3 Reference States and Reference Values

The values of u, h, and s given in the property tables are not obtained by direct measurement but are calculated from other data that can be more readily determined experimentally. The computational procedures require use of the second law of thermodynamics, so consideration of these procedures is deferred to Chap. 11 after the second law has been introduced. However, because u, h, and s are calculated, the matter of ***reference states*** and ***reference values*** becomes important and is considered briefly in the following paragraphs.

reference states
reference values

When applying the energy balance, it is *differences* in internal, kinetic, and potential energy between two states that are important, and *not* the values of these energy quantities at each of the two states. **FOR EXAMPLE...** consider the case of potential energy. The numerical value of potential energy determined relative to the surface of the earth is not the same as the value relative to the top of a tall building at the same location. However, the difference in potential energy between any two elevations is precisely the same regardless of the datum selected, because the datum cancels in the calculation.

Similarly, values can be assigned to specific internal energy and enthalpy relative to arbitrary reference values at arbitrary reference states. As for the case of potential energy considered above, the use of values of a particular property determined relative to an arbitrary reference is unambiguous as long as the calculations being performed involve only differences in that property, for then the reference value cancels. When chemical reactions take place among the substances under consideration, special attention must be given to the matter of reference states and values, however. A discussion of how property values are assigned when analyzing reactive systems is given in Chap. 13.

The tabular values of u and h for water, ammonia, propane, and Refrigerants 22 and 134a provided in the Appendix are relative to the following reference states and values. For water, the reference state is saturated liquid at 0.01°C. At this state, the specific internal energy is set to zero. Values of the specific enthalpy are calculated from $h = u + pv$, using the tabulated values for p, v, and u. For ammonia, propane, and the refrigerants, the reference state is saturated liquid at −40°C. At this reference state, the specific enthalpy is set to zero. Values of specific internal energy are calculated from $u = h - pv$ by using the tabulated values for p, v, and h. Notice in Table A-7 that this leads to a negative value for internal energy at the reference state, which emphasizes that it is not the numerical values assigned to u and h at a given state that are important but their *differences* between states. The values assigned to particular states change if the reference state or reference values change, but the differences remain the same.

3.7 Evaluating Properties Using Computer Software

The use of computer software for evaluating thermodynamic properties is becoming prevalent in engineering. Computer software falls into two general categories: those that provide data only at individual states and those that provide property data as part of a more general simulation package. The software available with this text, *Interactive Thermodynamics: IT,* is a tool that can be used not only for routine problem solving by providing data at individual state points, but also for simulation and analysis. Software other than *IT* also can be used for these purposes. See the box for discussion of software use in engineering thermodynamics.

IT provides data for substances represented in the Appendix tables. Generally, data are retrieved by simple call statements that are placed in the workspace of the program. **FOR EXAMPLE...** consider the two-phase, liquid–vapor mixture at state 1 of Example 3.2, for which $p = 1$ bar, $v = 0.8475$ m³/kg. The following illustrates how data for saturation temperature, quality, and specific internal energy are retrieved using *IT*. The functions for T, v, and u are obtained by selecting Water/Steam from the **Properties** menu. Choosing SI units from the **Units** menu, with p in bar, T in °C, and amount of substance in kg, the *IT* program is

```
p = 1//bar
v = 0.8475//m³/kg
T = Tsat_P("Water/Steam",p)
v = vsat_Px("Water/Steam",p,x)
u = usat_Px("Water/Steam",p,x)
```

Clicking the **Solve** button, the software returns values of $T = 99.63$°C, $x = 0.5$, and $u = 1{,}462$ kJ/kg. These values can be verified using data from Table A-3. Note that text inserted between the symbol // and a line return is treated as a comment.

The previous example illustrates an important feature of *IT.* Although the quality, x, is implicit in the list of arguments in the expression for specific volume, there is no

Using Software In Thermodynamics

The computer software tool ***Interactive Thermodynamics: IT*** is available for use with this text. Used properly, *IT* provides an important adjunct to learning engineering thermodynamics and solving engineering problems. The program is built around an equation solver enhanced with thermodynamic property data and other valuable features. With *IT* you can obtain a single numerical solution or vary parameters to investigate their effects. You also can obtain graphical output, and the Windows-based format allows you to use any Windows word-processing software or spreadsheet to generate reports. Additionally, functions in *IT* can be called from *Excel* through use of the *Excel Add-in Manager,* allowing you to use these thermodynamic functions while working within *Excel*. Other features of *IT* include:

- a guided series of help screens and a number of sample solved examples to help you learn how to use the program.
- drag-and-drop templates for many of the standard problem types, including a list of assumptions that you can customize to the problem at hand.
- predetermined scenarios for power plants and other important applications.
- thermodynamic property data for water, refrigerants 22 and 134a, ammonia, air–water vapor mixtures, and a number of ideal gases.
- the capability to input user-supplied data.
- the capability to interface with user-supplied routines.

Many features of *IT* are found in the popular *Engineering Equation Solver (EES)*. Readers already proficient with *EES* may prefer its use for solving problems in this text.

The use of computer software for engineering analysis is a powerful approach. Still, there are some rules to observe:

- Software *complements* and *extends* careful analysis, but does not substitute for it.
- Computer-generated values should be checked selectively against hand-calculated, or otherwise independently determined values.
- Computer-generated plots should be studied to see if the curves appear reasonable and exhibit expected trends.

need to solve the expression algebraically for x. Rather, the program can solve for x as long as the number of equations equals the number of unknowns.

IT also retrieves property values in the superheat region. **FOR EXAMPLE...** consider the superheated ammonia vapor at state 2 in Example 3.1, for which $p = 1.5$ bar and $T = 25°C$. Selecting Ammonia from the **Properties** menu and choosing SI units from the **Units** menu, data for specific volume, internal energy, and enthalpy are obtained from *IT* as follows:

```
p = 1.5 bar
T = 25°C
v = v_PT("Ammonia",p,T)
u = u_PT("Ammonia",p,T)
h = h_PT("Ammonia",p,T)
```

Clicking the **Solve** button, the software returns values of $v = 0.9553\ m^3/kg$, $u = 1{,}380.93$ kJ/kg and $h = 1{,}524.21$ kJ/kg, respectively. These values agree closely with the respective values obtained by interpolation in Table A-14.

3.8 Applying the Energy Balance Using Property Tables and Software

In Examples 3.3 and 3.4, closed systems undergoing processes are analyzed using the energy balance. In each case, sketches of p–v and/or T–v diagrams are used in conjunction with appropriate tables to obtain the required property data. Using property diagrams and table data introduces an additional level of complexity compared to similar problems in Chap. 2.

Example 3.3 STIRRING WATER AT CONSTANT VOLUME

A well-insulated rigid tank having a volume of 0.25 m^3 contains saturated water vapor at 100°C. The water is rapidly stirred until the pressure is 1.5 bars. Determine the temperature at the final state, in °C, and the work during the process, in kJ.

Solution

Known: By rapid stirring, water vapor in a well-insulated rigid tank is brought from the saturated vapor state at 100°C to a pressure of 1.5 bars.

Find: Determine the temperature at the final state and the work.

Schematic and Given Data:

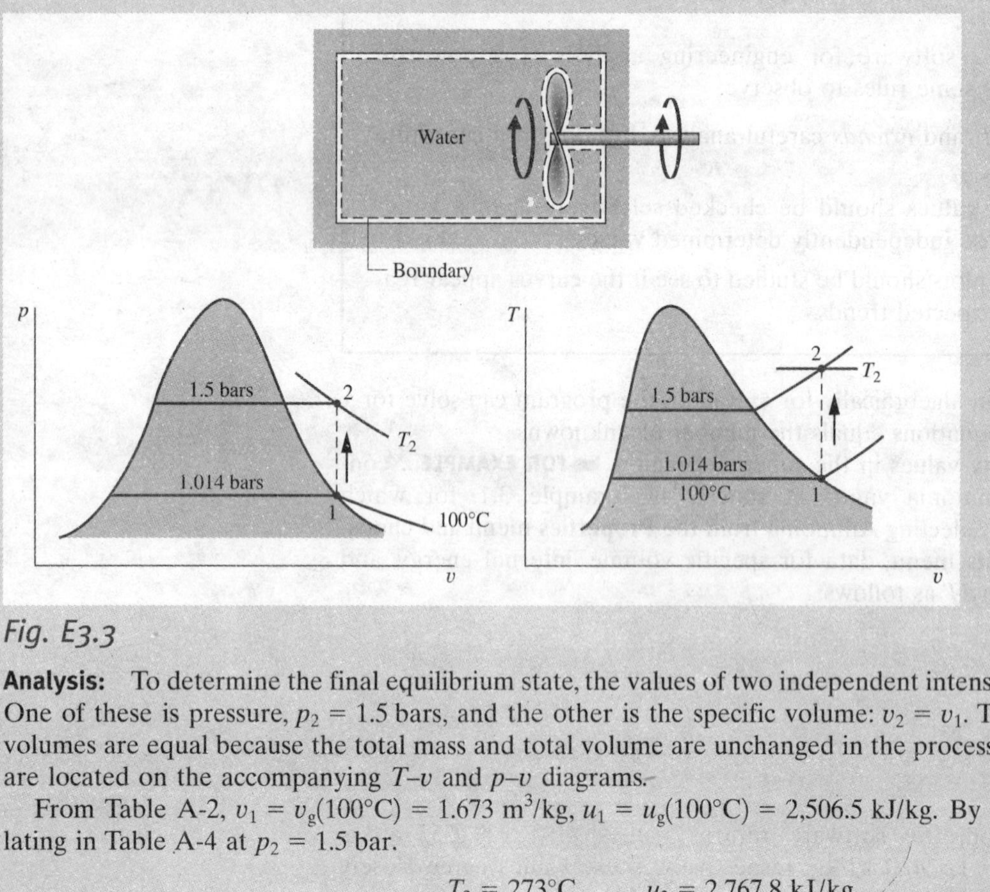

Engineering Model:

1. The water is a closed system.
2. The initial and final states are at equilibrium. There is no net change in kinetic or potential energy.
3. There is no heat transfer with the surroundings.
4. The tank volume remains constant.

❶

Fig. E3.3

Analysis: To determine the final equilibrium state, the values of two independent intensive properties are required. One of these is pressure, $p_2 = 1.5$ bars, and the other is the specific volume: $v_2 = v_1$. The initial and final specific volumes are equal because the total mass and total volume are unchanged in the process. The initial and final states are located on the accompanying T–v and p–v diagrams.

From Table A-2, $v_1 = v_g(100°\text{C}) = 1.673\ \text{m}^3/\text{kg}$, $u_1 = u_g(100°\text{C}) = 2{,}506.5\ \text{kJ/kg}$. By using $v_2 = v_1$ and interpolating in Table A-4 at $p_2 = 1.5$ bar.

$$T_2 = 273°\text{C}, \qquad u_2 = 2{,}767.8\ \text{kJ/kg}$$

Next, with assumptions 2 and 3 an energy balance for the system reduces to

$$\Delta U + \Delta \cancelto{0}{KE} + \Delta \cancelto{0}{PE} = \cancelto{0}{Q} - W$$

On rearrangement

$$W = -(U_2 - U_1) = -m(u_2 - u_1)$$

To evaluate W requires the system mass. This can be determined from the volume and specific volume

$$m = \frac{V}{v_1} = \left(\frac{0.25\ \text{m}^3}{1.673\ \text{m}^3/\text{kg}}\right) = 0.149\ \text{kg}$$

Finally, by inserting values into the expression for W

$$W = -(1.49\ \text{kg})(2{,}767.8 - 2{,}506.5)\ \text{kJ/kg} = -38.9\ \text{kJ}$$

where the minus sign signifies that the energy transfer by work is to the system.

---

❶ Although the initial and final states are equilibrium states, the intervening states are not at equilibrium. To emphasize this, the process has been indicated on the T–v and p–v diagrams by a dashed line. Solid lines on property diagrams are reserved for processes that pass through equilibrium states only (quasiequilibrium processes). The analysis illustrates the importance of carefully sketched property diagrams as an adjunct to problem solving.

Skills Developed: *Ability to...*

- ✓ define a closed system and identify interactions on its boundary.
- ✓ apply the energy balance with *steam table* data.
- ✓ sketch T-v and p-v diagrams and locate states on them.

Quick Quiz

If insulation were removed from the tank and the water cooled at constant volume from $T_2 = 229.4°\text{C}$ to $T_3 = 148.89°\text{C}$ while no stirring occurs, determine the heat transfer, in kJ.

Ans. −20.57 kJ.

Example 3.4 ANALYZING TWO PROCESSES IN SERIES

Water contained in a piston–cylinder assembly undergoes two processes in series from an initial state where the pressure is 10 bar and the temperature is 400°C.

Process 1–2: The water is cooled as it is compressed at a constant pressure of 10 bar to the saturated vapor state.

Process 2–3: The water is cooled at constant volume to 150°C.

(a) Sketch both processes on T–v and p–v diagrams.

(b) For the overall process, determine the work, in kJ/kg.

(c) For the overall process, determine the heat transfer, in kJ/kg.

Solution

Known: Water contained in a piston–cylinder assembly undergoes two processes: It is cooled and compressed while keeping the pressure constant, and then cooled at constant volume.

Find: Sketch both processes on T–v and p–v diagrams. Determine the net work and the net heat transfer for the overall process per unit of mass contained within the piston–cylinder assembly.

Schematic and Given Data:

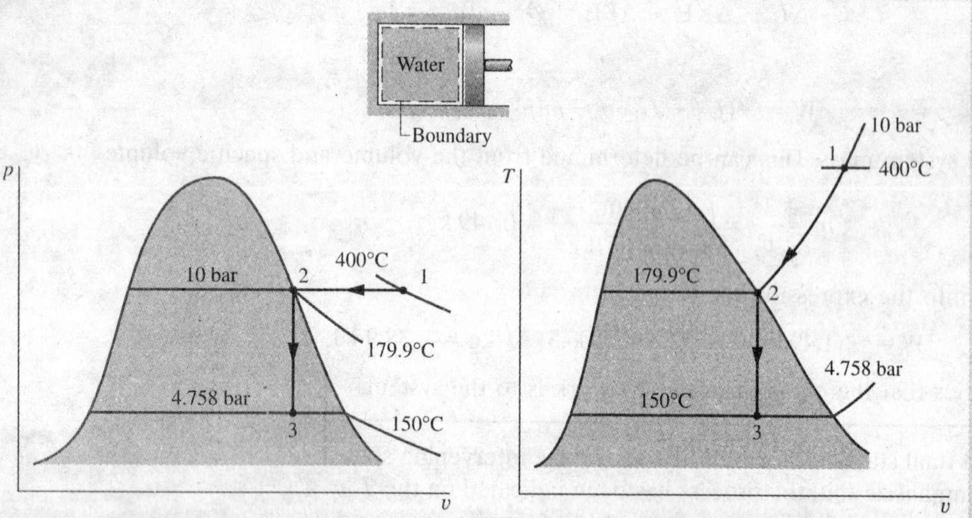

Engineering Model:

1. The water is a closed system.
2. The piston is the only work mode.
3. There are no changes in kinetic or potential energy.

Fig. E3.4

Analysis:

(a) The accompanying T–v and p–v diagrams show the two processes. Since the temperature at state 1, $T_1 = 400°C$, is greater than the saturation temperature corresponding to $p_1 = 10$ bar: 179.9°C, state 1 is located in the superheat region.

(b) Since the piston is the only work mechanism

$$W = \int_1^3 p\,dV = \int_1^2 p\,dV + \cancelto{0}{\int_2^3 p\,dV}$$

The second integral vanishes because the volume is constant in Process 2–3. Dividing by the mass and noting that the pressure is constant for Process 1–2

$$\frac{W}{m} = p(v_2 - v_1)$$

The specific volume at state 1 is found from Table A-4, using $p_1 = 10$ bar and $T_1 = 400°C$: $v_1 = 0.3066\ m^3/kg$. Also, $u_1 = 2{,}957.3$ kJ/kg. The specific volume at state 2 is the saturated vapor value at 10 bar: $v_2 = 0.1944\ m^3/kg$, from Table A-3. Hence

$$\frac{W}{m} = (10\text{ bar})(0.1944 - 0.3066)\left(\frac{m^3}{kg}\right)\left|\frac{10^5\ N/m^2}{1\text{ bar}}\right|\left|\frac{1\text{ kJ}}{10^3\ N\cdot m}\right|$$

$$= -112.2\text{ kJ/kg}$$

The minus sign indicates that work is done *on* the water vapor by the piston.

(c) An energy balance for the *overall* process reduces to

$$m(u_3 - u_1) = Q - W$$

By rearranging

$$\frac{Q}{m} = (u_3 - u_1) + \frac{W}{m}$$

To evaluate the heat transfer requires u_3, the specific internal energy at state 3. Since T_3 is given and $v_3 = v_2$, two independent intensive properties are known that together fix state 3. To find u_3, first solve for the quality

$$x_3 = \frac{v_3 - v_{f3}}{v_{g3} - v_{f3}} = \frac{0.1944 - 1.0905 \times 10^{-3}}{0.3928 - 1.0905 \times 10^{-3}} = 0.494$$

where v_{f3} and v_{g3} are from Table A-2 at 150°C. Then

$$u_3 = u_{f3} + x_3(u_{g3} - u_{f3}) = 631.68 + 0.494(2{,}559.5 - 631.98)$$
$$= 1{,}583.9 \text{ kJ/kg}$$

where u_{f3} and u_{g3} are from Table A-2 at 150°C.

Substituting values into the energy balance

$$\frac{Q}{m} = 1{,}583.9 - 2{,}957.3 + (-112.2) = -1{,}485.6 \text{ kJ/kg}$$

The minus sign shows that energy is transferred *out* by heat transfer.

Skills Developed: *Ability to...*

- ✓ define a closed system and identify interactions on its boundary.
- ✓ evaluate work using Eq. 2.17.
- ✓ apply the energy balance with *steam table* data.
- ✓ sketch *T-v* and *p-v* diagrams and locate states on them.

Quick Quiz

If the two specified processes were followed by Process 3-4, during which the water expands at a constant temperature of 150°C to saturated vapor, determine the work, in kJ/kg, for the overall process from 1 to 4.

Ans. $W/m = -17.8$ kJ/kg.

Example 3.5 illustrates the use of *Interactive Thermodynamics: IT* for solving problems. In this case, the software evaluates the property data, calculates the results, and displays the results graphically.

Example 3.5 PLOTTING THERMODYNAMIC DATA USING SOFTWARE

For the system of Example 3.2, plot the heat transfer, in kJ, and the mass of saturated vapor present, in kg, each versus pressure at state 2 ranging from 1 to 2 bar. Discuss the results.

Solution

Known: A two-phase liquid–vapor mixture of water in a closed, rigid container is heated on a hot plate. The initial pressure and quality are known. The pressure at the final state ranges from 1 to 2 bar.

Find: Plot the heat transfer and the mass of saturated vapor present, each versus pressure at the final state. Discuss.

Schematic and Given Data: See Figure E3.2.

Engineering Model:

1. There is no work.
2. Kinetic and potential energy effects are negligible.
3. See Example 3.2 for other assumptions.

Analysis: The heat transfer is obtained from the energy balance. With assumptions 1 and 2, the energy balance reduces to

$$\Delta U + \cancel{\Delta \text{KE}}^{0} + \cancel{\Delta \text{PE}}^{0} = Q - \cancel{W}^{0}$$

or

$$Q = m(u_2 - u_1)$$

Selecting Water/Steam from the **Properties** menu and choosing SI Units from the **Units** menu, the *IT* program for obtaining the required data and making the plots is

```
// Given data—State 1
p1 = 1//bar
x1 = 0.5
V = 0.5//m³
```

```
// Evaluate property data—State 1
   v1 = vsat_Px("Water/Steam", p1,x1)
   u1 = usat_Px("Water/Steam", p1,x1)
// Calculate the mass
   m = V/v1
// Fix state 2
   v2 = v1
   p2 = 1.5//bar
// Evaluate property data—State 2
   v2 = vsat_Px("Water/Steam", p2,x2)
   u2 = usat_Px("Water/Steam", p2,x2)
// Calculate the mass of saturated vapor present
   mg2 = x2 * m
// Determine the pressure for which the quality is unity
   v3 = v1
   v3 = vsat_Px( "Water/Steam" ,p3,1)
// Energy balance to determine the heat transfer
   m * (u2 – u1) = Q – W
   W = 0
```
❶

Click the **Solve** button to obtain a solution for p_2 = 1.5 bar. The program returns values of v_1 = 0.8475 m^3/kg and m = 0.59 kg. Also, at p_2 = 1.5 bar, the program gives m_{g2} = 0.4311 kg. These values agree with the values determined in Example 3.2.

Now that the computer program has been verified, use the **Explore** button to vary pressure from 1 to 2 bar in steps of 0.1 bar. Then, use the **Graph** button to construct the required plots. The results are:

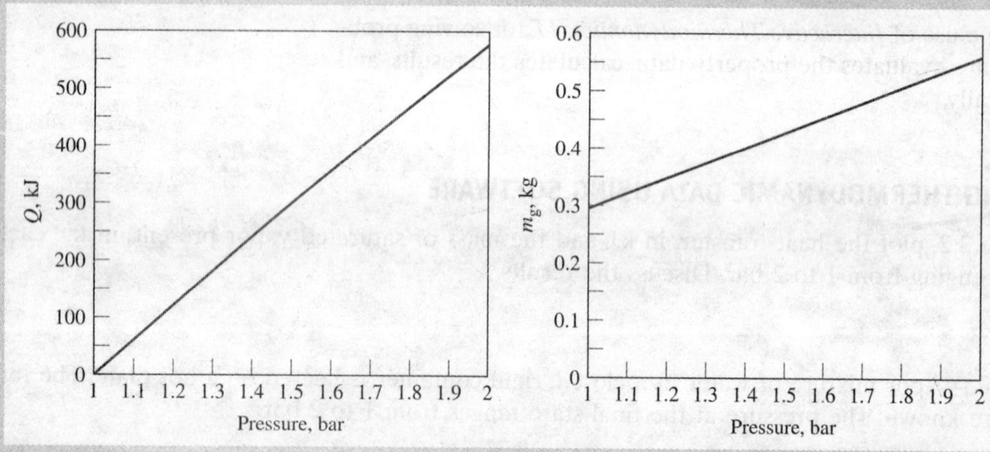

Fig. E3.5

We conclude from the first of these graphs that the heat transfer to the water varies directly with the pressure. The plot of m_g shows that the mass of saturated vapor present also increases as the pressure increases. Both of these results are in accord with expectations for the process.

---

❶ Using the **Browse** button, the computer solution indicates that the pressure for which the quality becomes unity is 2.096 bar. Thus, for pressures ranging from 1 to 2 bar, all of the states are in the two-phase liquid–vapor region.

Skills Developed: *Ability to...*
- ✓ apply the closed system energy balance.
- ✓ use *IT* to retrieve property data for water and plot calculated data.

Quick Quiz

If heating continues at constant specific volume to a state where the pressure is 3 bar, modify the *IT* program to give the temperature at that state, in °C.

Ans.
```
v4 = v1
p4 = 3//bar
v4 = v_PT ("Water/Steam", p4, T4)
T4 = 282.4°C
```

3.9 Introducing Specific Heats c_v and c_p

Several properties related to internal energy are important in thermodynamics. One of these is the property enthalpy introduced in Sec. 3.6.1. Two others, known as *specific heats,* are considered in this section. The specific heats are particularly useful for thermodynamic calculations involving the *ideal gas model* introduced in Sec. 3.12.

The intensive properties c_v and c_p are defined for pure, simple compressible substances as partial derivatives of the functions $u(T, v)$ and $h(T, p)$, respectively

$$c_v = \left(\frac{\partial u}{\partial T}\right)_v \tag{3.8}$$

$$c_p = \left(\frac{\partial h}{\partial T}\right)_p \tag{3.9}$$

where the subscripts v and p denote, respectively, the variables held fixed during differentiation. Values for c_v and c_p can be obtained via statistical mechanics using *spectroscopic* measurements. They also can be determined macroscopically through exacting property measurements. Since u and h can be expressed either on a unit mass basis or per mole, values of the specific heats can be similarly expressed. SI units are kJ/kg · K or kJ/kmol · K.

The property k, called the *specific heat ratio,* is simply the ratio

$$k = \frac{c_p}{c_v} \tag{3.10}$$

The properties c_v and c_v are referred to as ***specific heats*** (or *heat capacities*) because under certain *special conditions* they relate the temperature change of a system to the amount of energy added by heat transfer. However, it is generally preferable to think of c_v and c_p in terms of their definitions, Eqs. 3.8 and 3.9, and not with reference to this limited interpretation involving heat transfer. ***specific heats***

In general, c_v is a function of v and T (or p and T), and c_p depends on both p and T (or v and T). Figure 3.9 shows how c_p for water vapor varies as a function of temperature and pressure. The vapor phases of other substances exhibit similar behavior. Note that the figure gives the variation of c_p with temperature in the limit as pressure tends to zero. In this limit, c_p increases with increasing temperature, which is a characteristic exhibited by other gases as well. We will refer again to such *zero-pressure* values for c_v and c_p in Sec. 3.13.2.

Specific heat data are available for common gases, liquids, and solids. Data for gases are introduced in Sec. 3.13.2 as a part of the discussion of the ideal gas model. Specific heat values for some common liquids and solids are introduced in Sec. 3.10.2 as a part of the discussion of the incompressible substance model.

3.10 Evaluating Properties of Liquids and Solids

Special methods often can be used to evaluate properties of liquids and solids. These methods provide simple, yet accurate, approximations that do not require exact compilations like the compressed liquid tables for water, Tables A-5. Two such special methods are discussed next: approximations using saturated liquid data and the incompressible substance model.

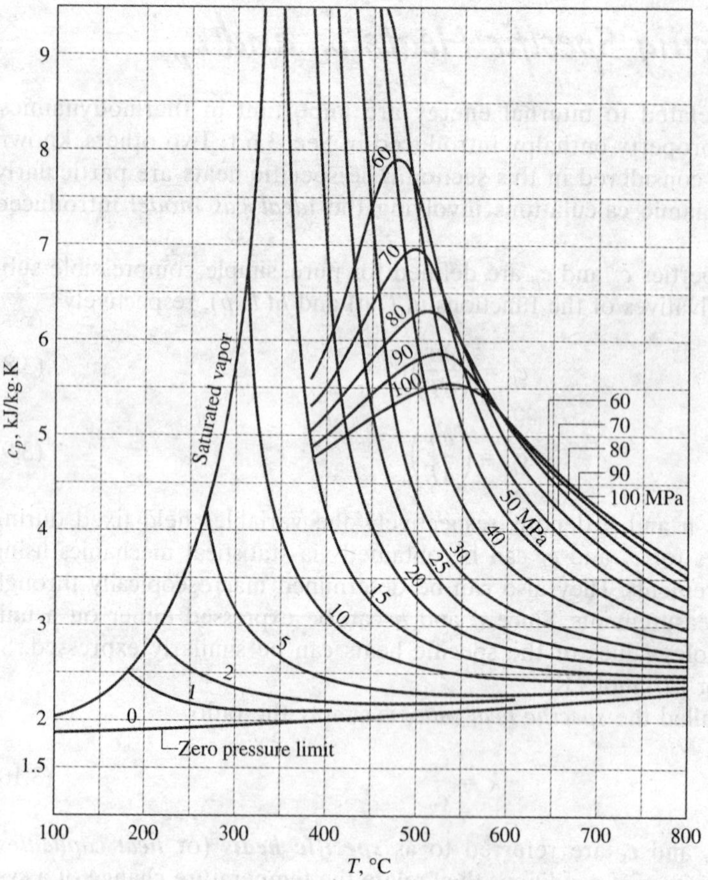

Fig. 3.9 c_p of water vapor as a function of temperature and pressure.

3.10.1 Approximations for Liquids Using Saturated Liquid Data

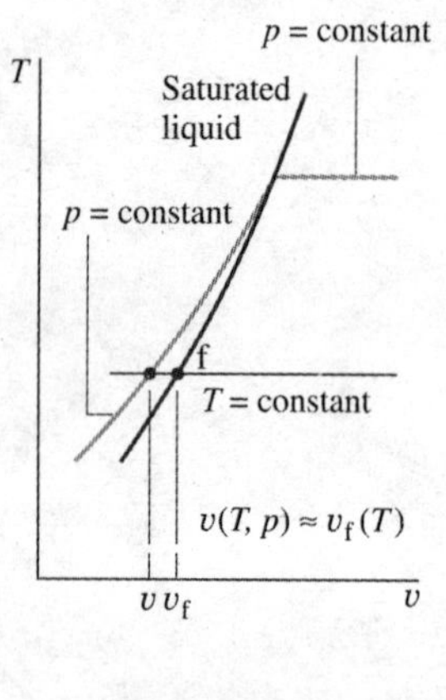

Approximate values for v, u, and h at liquid states can be obtained using saturated liquid data. To illustrate, refer to the compressed liquid tables, Tables A-5. These tables show that the specific volume and specific internal energy change very little with pressure *at a fixed temperature*. Because the values of v and u vary only gradually as pressure changes at fixed temperature, the following approximations are reasonable for many engineering calculations:

$$v(T, p) \approx v_f(T) \tag{3.11}$$

$$u(T, p) \approx u_f(T) \tag{3.12}$$

That is, for liquids v and u may be evaluated at the saturated liquid state corresponding to the temperature at the given state.

An approximate value of h at liquid states can be obtained by using Eqs. 3.11 and 3.12 in the definition $h = u + pv$; thus

$$h(T, p) \approx u_f(T) + pv_f(T)$$

This can be expressed alternatively as

$$h(T, p) \approx h_f(T) + v_f(T)[p - p_{sat}(T)] \tag{3.13}$$

where p_{sat} denotes the saturation pressure at the given temperature. The derivation is left as an exercise. When the contribution of the underlined term of Eq. 3.13 is small, the specific enthalpy can be approximated by the saturated liquid value, as for v and u. That is

$$h(T, p) \approx h_f(T) \tag{3.14}$$

Although the approximations given here have been presented with reference to liquid water, they also provide plausible approximations for other substances *when the only liquid data available are for saturated liquid states.* In this text, compressed liquid data are presented only for water (Tables A-5). Also note that *Interactive Thermodynamics: IT* does not provide compressed liquid data for *any* substance, but uses Eqs. 3.11, 3.12, and 3.14 to return liquid values for v, u, and h, respectively. When greater accuracy is required than provided by these approximations, other data sources should be consulted for more complete property compilations for the substance under consideration.

3.10.2 Incompressible Substance Model

As noted above, there are regions where the specific volume of liquid water varies little and the specific internal energy varies mainly with temperature. The same general behavior is exhibited by the liquid phases of other substances and by solids. The approximations of Eqs. 3.11–3.14 are based on these observations, as is the ***incompressible substance model*** under present consideration.

incompressible substance model

To simplify evaluations involving liquids or solids, the specific volume (density) is often assumed to be constant and the specific internal energy assumed to vary only with temperature. A substance idealized in this way is called *incompressible.*

Since the specific internal energy of a substance modeled as incompressible depends only on temperature, the specific heat c_v is also a function of temperature alone

$$c_v(T) = \frac{du}{dT} \quad \text{(incompressible)} \tag{3.15}$$

This is expressed as an ordinary derivative because u depends only on T.

Although the specific volume is constant and internal energy depends on temperature only, enthalpy varies with both pressure and temperature according to

$$h(T, p) = u(T) + pv \quad \text{(incompressible)} \tag{3.16}$$

For a substance modeled as incompressible, the specific heats c_v and c_p are equal. This is seen by differentiating Eq. 3.16 with respect to temperature while holding pressure fixed to obtain

$$\left(\frac{\partial h}{\partial T}\right)_p = \frac{du}{dT}$$

The left side of this expression is c_p by definition (Eq. 3.9), so using Eq. 3.15 on the right side gives

$$c_p = c_v \quad \text{(incompressible)} \tag{3.17}$$

Thus, for an incompressible substance it is unnecessary to distinguish between c_p and c_v, and both can be represented by the same symbol, c. Specific heats of some common liquids and solids are given in Tables A-19. Over limited temperature intervals the variation of c with temperature can be small. In such instances, the specific heat c can be treated as constant without a serious loss of accuracy.

Using Eqs. 3.15 and 3.16, the changes in specific internal energy and specific enthalpy between two states are given, respectively, by

$$u_2 - u_1 = \int_{T_1}^{T_2} c(T)\, dT \qquad \text{(incompressible)} \tag{3.18}$$

$$\begin{aligned} h_2 - h_1 &= u_2 - u_1 + v(p_2 - p_1) \\ &= \int_{T_1}^{T_2} c(T)\, dT + v(p_2 - p_1) \qquad \text{(incompressible)} \end{aligned} \tag{3.19}$$

If the specific heat c is taken as constant, Eqs. 3.18 and 3.19 become, respectively,

$$u_2 - u_1 = c(T_2 - T_1) \qquad \text{(incompressible, constant } c\text{)} \tag{3.20a}$$

$$h_2 - h_1 = c(T_2 - T_1) + \underline{v(p_2 - p_1)} \tag{3.20b}$$

In Eq. 3.20b, the underlined term is often small relative to the first term on the right side and then may be dropped.

The next example illustrates use of the incompressible substance model in an application involving the constant-volume calorimeter discussed in Sec. 2.5.1.

Example 3.6 MEASURING THE CALORIE VALUE OF COOKING OIL

One-tenth milliliter of cooking oil is placed in the chamber of a constant-volume calorimeter filled with sufficient oxygen for the oil to be completely burned. The chamber is immersed in a water bath. The mass of the water bath is 2.15 kg. For the purpose of this analysis, the metal parts of the apparatus are modeled as equivalent to an additional 0.5 kg of water. The calorimeter is well-insulated, and initially the temperature throughout is 25°C. The oil is ignited by a spark. When equilibrium is again attained, the temperature throughout is 25.3°C. Determine the change in internal energy of the chamber contents, in kcal per mL of cooking oil and in kcal per tablespoon of cooking oil.

Known: Data are provided for a constant-volume calorimeter testing cooking oil for caloric value.

Find: Determine the change in internal energy of the contents of the calorimeter chamber.

Schematic and Given Data:

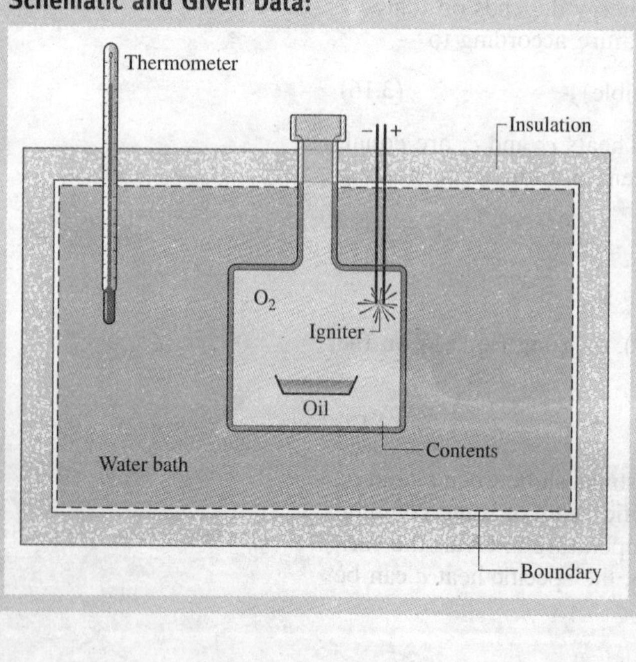

Fig. E3.6

Engineering Model:

1. The closed system is shown by the dashed line in the accompanying figure.
2. The total volume remains constant, including the chamber, water bath, and the amount of water modeling the metal parts.
3. Water is modeled as incompressible with constant specific heat c.
4. Heat transfer with the surroundings is negligible, and there is no change in kinetic or potential energy.

Analysis: With the assumptions listed, the closed system energy balance reads

$$\Delta U + \cancelto{0}{\Delta KE} + \cancelto{0}{\Delta PE} = \cancelto{0}{Q} - \cancelto{0}{W}$$

or

$$(\Delta U)_{contents} + (\Delta U)_{water} = 0$$

thus

$$(\Delta U)_{contents} = -(\Delta U)_{water} \tag{a}$$

The change in internal energy of the contents is equal and opposite to the change in internal energy of the water.

Since water is modeled as incompressible, Eq. 3.20a is used to evaluate the right side of Eq. (a), giving

❶ ❷ $$(\Delta U)_{contents} = -m_w c_w (T_2 - T_1) \tag{b}$$

With $m_w = 2.15\ \text{kg} + 0.5\ \text{kg} = 2.65\ \text{kg}$, $(T_2 - T_1) = 0.3\ \text{K}$, and $c_w = 4.18\ \text{kJ/kg} \cdot \text{K}$ from Table A-19, Eq. (b) gives

$$(\Delta U)_{contents} = -(2.65\ \text{kg})(4.18\ \text{kJ/kg} \cdot \text{K})(0.3\ \text{K}) = -3.32\ \text{kJ}$$

Converting to kcal, and expressing the result on a per milliliter of oil basis using the oil volume, 0.1 mL, we get

$$\frac{(\Delta U)_{contents}}{V_{oil}} = \frac{-3.32\ \text{kJ}}{0.1\ \text{mL}} \left| \frac{1\ \text{kcal}}{4.1868\ \text{kJ}} \right|$$

$$= -7.9\ \text{kcal/mL}$$

The calorie value of the cooking oil is the magnitude—that is, 7.9 kcal/mL. Labels on cooking oil containers usually give calorie value for a serving size of 1 tablespoon (15 mL). Using the calculated value, we get 119 kcal per tablespoon.

❶ The change in internal energy for water can be found alternatively using Eq. 3.12 together with saturated liquid internal energy data from Table A-2.

❷ The change in internal energy of the chamber contents cannot be evaluated using a specific heat because specific heats are defined (Sec. 3.9) only for *pure* substances—that is, substances that are unchanging in composition.

Skills Developed: *Ability to...*
- ✓ define a closed system and identify interactions within it and on its boundary.
- ✓ apply the energy balance using the incompressible substance model.

Quick Quiz

Using Eq. 3.12 together with saturated liquid internal energy data from Table A-2 find the change in internal energy of the water, in kJ, and compare with the value obtained assuming water is incompressible.

Ans. 3.32 kJ

3.11 *Generalized Compressibility Chart*

The object of the present section is to gain a better understanding of the relationship among pressure, specific volume, and temperature of gases. This is important not only as a basis for analyses involving gases but also for the discussions of the second part of the chapter, where the *ideal gas model* is introduced. The current presentation is conducted in terms of the *compressibility factor* and begins with the introduction of the *universal gas constant.*

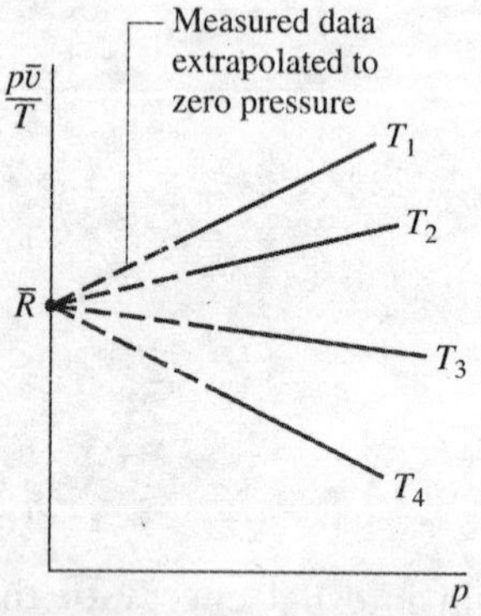

Fig. 3.10 Sketch of $p\bar{v}/T$ versus pressure for a gas at several specified values of temperature.

3.11.1 Universal Gas Constant, $\bar{R}$

Let a gas be confined in a cylinder by a piston and the entire assembly held at a constant temperature. The piston can be moved to various positions so that a series of equilibrium states at constant temperature can be visited. Suppose the pressure and specific volume are measured at each state and the value of the ratio $p\bar{v}/T$ ($\bar{v}$ is volume per mole) determined. These ratios can then be plotted versus pressure at constant temperature. The results for several temperatures are sketched in Fig. 3.10. When the ratios are extrapolated to zero pressure, *precisely the same limiting value is obtained* for each curve. That is,

$$\lim_{p \to 0} \frac{p\bar{v}}{T} = \bar{R} \tag{3.21}$$

universal gas constant

where $\bar{R}$ denotes the common limit for all temperatures. If this procedure were repeated for other gases, it would be found in every instance that the limit of the ratio $p\bar{v}/T$ as p tends to zero at fixed temperature is the same, namely $\bar{R}$. Since the same limiting value is exhibited by all gases, $\bar{R}$ is called the ***universal gas constant.*** Its value as determined experimentally is

$$\bar{R} = 8.314 \text{ kJ/kmol} \cdot K \tag{3.22}$$

Having introduced the universal gas constant, we turn next to the compressibility factor.

3.11.2 Compressibility Factor, Z

compressibility factor

The dimensionless ratio $p\bar{v}/\bar{R}T$ is called the ***compressibility factor*** and is denoted by Z. That is,

$$Z = \frac{p\bar{v}}{\bar{R}T} \tag{3.23}$$

As illustrated by subsequent calculations, when values for p, $\bar{v}$, $\bar{R}$, and T are used in consistent units, Z is unitless.

With $\bar{v} = Mv$ (Eq. 1.9), where M is the atomic or molecular weight, the compressibility factor can be expressed alternatively as

$$Z = \frac{pv}{RT} \tag{3.24}$$

where

$$R = \frac{\bar{R}}{M} \tag{3.25}$$

R is a constant for the particular gas whose molecular weight is M. Alternative units for R are kJ/kg · K. Table 3.1 provides a sampling of values for the gas constant R calculated from Eq. 3.25.

Equation 3.21 can be expressed in terms of the compressibility factor as

$$\lim_{p \to 0} Z = 1 \tag{3.26}$$

That is, the compressibility factor Z tends to unity as pressure tends to zero at fixed temperature. This can be illustrated by reference to Fig. 3.11, which shows Z for hydrogen plotted versus pressure at a number of different temperatures. In general, at states of a gas where pressure is small relative to the critical pressure, Z is approximately 1.

Table 3.1

Values of the gas constant *R* of Selected Elements and Compounds

Substance	Chemical Formula	*R* (kJ/kg · K)
Air	—	0.2870
Ammonia	NH_3	0.4882
Argon	Ar	0.2082
Carbon dioxide	CO_2	0.1889
Carbon monoxide	CO	0.2968
Helium	He	2.0769
Hydrogen	H_2	4.1240
Methane	CH_4	0.5183
Nitrogen	N_2	0.2968
Oxygen	O_2	0.2598
Water	H_2O	0.4614

Source: *R* values are calculated in terms of the universal gas constant $\overline{R} = 8.314$ kJ/kmol · K and the molecular weight *M* provided in Table A-1 using $R = \overline{R}/M$ (Eq. 3.25).

3.11.3 Generalized Compressibility Data, *Z* Chart

Figure 3.11 gives the compressibility factor for hydrogen versus pressure at specified values of temperature. Similar charts have been prepared for other gases. When these charts are studied, they are found to be *qualitatively* similar. Further study shows that when the coordinates are suitably modified, the curves for several different gases coincide closely when plotted together on the same coordinate axes, and so *quantitative* similarity also can be achieved. This is referred to as the *principle of corresponding states.* In one such approach, the compressibility factor *Z* is plotted versus a dimensionless ***reduced pressure*** p_R and ***reduced temperature*** T_R, defined as

reduced pressure and temperature

$$p_R = p/p_c \tag{3.27}$$

$$T_R = T/T_c \tag{3.28}$$

where p_c and T_c denote the critical pressure and temperature, respectively. This results in a ***generalized compressibility chart*** of the form $Z = f(p_R, T_R)$. Figure 3.12

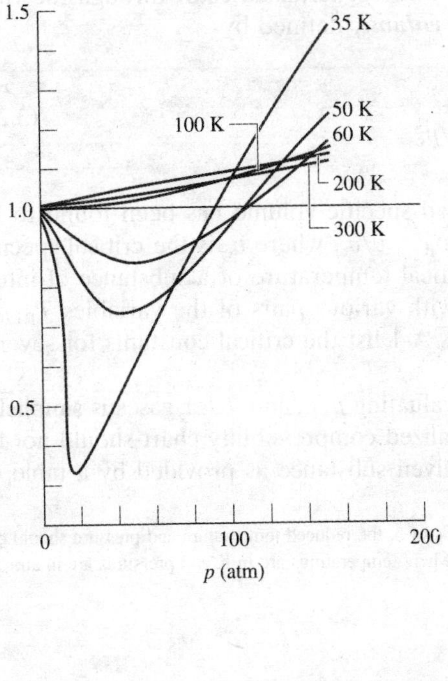

Fig. 3.11 Variation of the compressibility factor of hydrogen with pressure at constant temperature.

generalized compressibility chart

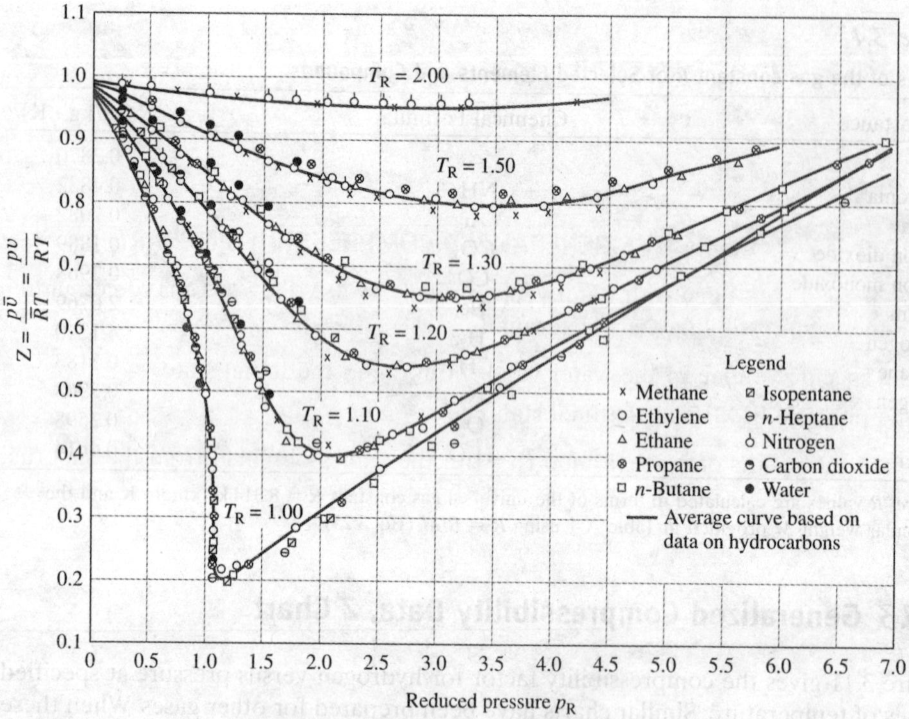

Fig. 3.12 Generalized compressibility chart for various gases.

shows experimental data for 10 different gases on a chart of this type. The solid lines denoting reduced isotherms represent the best curves fitted to the data.

A generalized chart more suitable for problem solving than Fig. 3.12 is given in the Appendix as Figs. A-1, A-2, and A-3. In Fig. A-1, p_R ranges from 0 to 1.0; in Fig. A-2, p_R ranges from 0 to 10.0; and in Fig. A-3, p_R ranges from 10.0 to 40.0. At any one temperature, the deviation of observed values from those of the generalized chart increases with pressure. However, for the 30 gases used in developing the chart, the deviation is *at most* on the order of 5% and for most ranges is much less.[1] From Figs. A-1 and A-2 it can be seen that the value of Z tends to unity for all temperatures as pressure tends to zero, in accord with Eq. 3.26. Figure A-3 shows that Z also approaches unity for all pressures at very high temperatures.

pseudoreduced specific volume

Values of specific volume are included on the generalized chart through the variable v'_R, called the ***pseudoreduced specific volume,*** defined by

$$v'_R = \frac{\bar{v}}{\bar{R}T_c/p_c} \tag{3.29}$$

For correlation purposes, the pseudoreduced specific volume has been found to be preferable to the *reduced* specific volume $v_R = \bar{v}/\bar{v}_c$, where $\bar{v}_c$ is the critical specific volume. Using the critical pressure and critical temperature of a substance of interest, the generalized chart can be entered with various pairs of the variables T_R, p_R, and v'_R: (T_R, p_R), (p_R, v'_R), or (T_R, v'_R). Tables A-1 list the critical constants for several substances.

The merit of the generalized chart for evaluating p, v, and T for gases is simplicity coupled with accuracy. However, the generalized compressibility chart should not be used as a substitute for p–v–T data for a given substance as provided by a table or

[1] To determine Z for hydrogen, helium, and neon above a T_R of 5, the reduced temperature and pressure should be calculated using $T_R = T/(T_c + 8)$ and $p_R = p/(p_c + 8)$, where temperatures are in K and pressures are in atm.

computer software. The chart is mainly useful for obtaining reasonable estimates in the absence of more accurate data.

The next example provides an illustration of the use of the generalized compressibility chart.

Example 3.7 USING THE GENERALIZED COMPRESSIBILITY CHART

A closed, rigid tank filled with water vapor, initially at 20 MPa, 520°C, is cooled until its temperature reaches 400°C. Using the compressibility chart, determine

(a) the specific volume of the water vapor in m^3/kg at the initial state.

(b) the pressure in MPa at the final state.

Compare the results of parts (a) and (b) with the values obtained from the superheated vapor table, Table A-4.

Solution

Known: Water vapor is cooled at constant volume from 20 MPa, 520°C to 400°C.

Find: Use the compressibility chart and the superheated vapor table to determine the specific volume and final pressure and compare the results.

Schematic and Given Data:

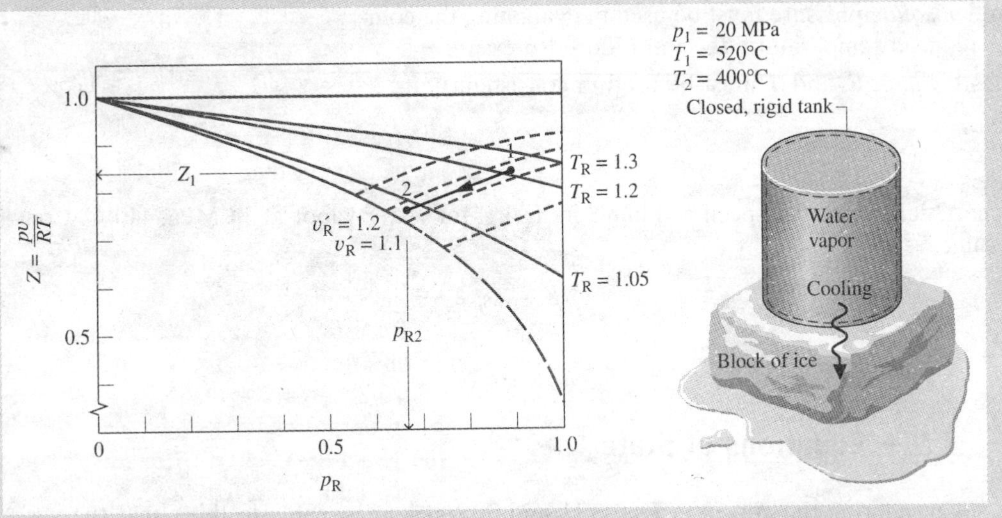

Fig. E3.7

Engineering Model:

1. The water vapor is a closed system.
2. The initial and final states are at equilibrium.
3. The volume is constant.

Analysis:

(a) From Table A-1, T_c = 647.3 K and p_c = 22.09 MPa for water. Thus

❶ $$T_{R1} = \frac{793}{647.3} = 1.23, \qquad p_{R1} = \frac{20}{22.09} = 0.91$$

With these values for the reduced temperature and reduced pressure, the value of Z obtained from Fig. A-1 is approximately 0.83. Since $Z = pv/RT$, the specific volume at state 1 can be determined as follows:

$$v_1 = Z_1 \frac{RT_1}{p_1} = 0.83 \frac{\bar{R}T_1}{Mp_1}$$

❷ $$= 0.83 \left(\frac{8314 \dfrac{\text{N}\cdot\text{m}}{\text{kmol}\cdot\text{K}}}{18.02 \dfrac{\text{kg}}{\text{kmol}}} \right) \left(\frac{793\ \text{K}}{20 \times 10^6 \dfrac{\text{N}}{\text{m}^2}} \right) = 0.0152\ \text{m}^3/\text{kg}$$

The molecular weight of water is from Table A-1.

Turning to Table A-4, the specific volume at the initial state is 0.01551 m^3/kg. This is in good agreement with the compressibility chart value, as expected.

(b) Since both mass and volume remain constant, the water vapor cools at constant specific volume, and thus at constant v'_R. Using the value for specific volume determined in part (a), the constant v'_R value is

$$v'_R = \frac{vp_c}{RT_c} = \frac{\left(0.0152\frac{m^3}{kg}\right)\left(22.09 \times 10^6\frac{N}{m^2}\right)}{\left(\frac{8314}{18.02}\frac{N \cdot m}{kg \cdot K}\right)(647.3\ K)} = 1.12$$

At state 2

$$T_{R2} = \frac{673}{647.3} = 1.04$$

Locating the point on the compressibility chart where $v'_R = 1.12$ and $T_R = 1.04$, the corresponding value for p_R is about 0.69. Accordingly

$$p_2 = p_c(p_{R2}) = (22.09\ \text{MPa})(0.69) = 15.24\ \text{MPa}$$

Interpolating in the superheated vapor tables gives $p_2 = 15.16$ MPa. As before, the compressibility chart value is in good agreement with the table value.

❶ *Absolute* temperature and *absolute* pressure must be used in evaluating the compressibility factor Z, the reduced temperature T_R, and reduced pressure p_R.

❷ Since Z is unitless, values for p, v, R, and T must be used in consistent units.

Skills Developed: *Ability to...*

✓ retrieve p-v-T data from the generalized compressibility chart.

✓ retrieve p-v-T data from the steam tables.

Quick Quiz

Using the compressibility chart, determine the specific volume, in m^3/kg, for water vapor at 14 MPa, 440°C. Compare with the steam table value.

Ans. 0.0195 m^3/kg

3.11.4 Equations of State

Considering the curves of Figs. 3.11 and 3.12, it is reasonable to think that the variation with pressure and temperature of the compressibility factor for gases might be expressible as an equation, at least for certain intervals of p and T. Two expressions can be written that enjoy a theoretical basis. One gives the compressibility factor as an infinite series expansion in pressure:

$$Z = 1 + \hat{B}(T)p + \hat{C}(T)p^2 + \hat{D}(T)p^3 + \cdots \tag{3.30}$$

where the coefficients $\hat{B}, \hat{C}, \hat{D}, \ldots$ depend on temperature only. The dots in Eq. 3.30 represent higher-order terms. The other is a series form entirely analogous to Eq. 3.30 but expressed in terms of $1/\bar{v}$ instead of p

$$Z = 1 + \frac{B(T)}{\bar{v}} + \frac{C(T)}{\bar{v}^2} + \frac{D(T)}{\bar{v}^3} + \cdots \tag{3.31}$$

virial equations of state

Equations 3.30 and 3.31 are known as ***virial equations of state,*** and the coefficients $\hat{B}, \hat{C}, \hat{D}, \ldots$ and $B, C, D, \ldots$ are called *virial coefficients*. The word *virial* stems from the Latin word for force. In the present usage it is force interactions among molecules that are intended.

The virial expansions can be derived by the methods of statistical mechanics, and physical significance can be attributed to the coefficients: $B/\overline{v}$ accounts for two-molecule interactions, $C/\overline{v}^2$ accounts for three-molecule interactions, etc. In principle, the virial coefficients can be calculated by using expressions from statistical mechanics derived from consideration of the force fields around the molecules of a gas. The virial coefficients also can be determined from experimental p–v–T data. The virial expansions are used in Sec. 11.1 as a point of departure for the further study of analytical representations of the p–v–T relationship of gases known generically as *equations of state*.

The virial expansions and the physical significance attributed to the terms making up the expansions can be used to clarify the nature of gas behavior in the limit as pressure tends to zero at fixed temperature. From Eq. 3.30, it is seen that if pressure decreases at fixed temperature, the terms $\hat{B}p$, $\hat{C}p^2$, etc., accounting for various molecular interactions tend to decrease, suggesting that the force interactions become weaker under these circumstances. In the limit as pressure approaches zero, these terms vanish, and the equation reduces to $Z = 1$ in accordance with Eq. 3.26. Similarly, since volume increases when the pressure decreases at fixed temperature, the terms $B/\overline{v}$, $C/\overline{v}^2$, etc., of Eq. 3.31 also vanish in the limit, giving $Z = 1$ when the force interactions between molecules are no longer significant.

Evaluating Properties Using the Ideal Gas Model

3.12 Introducing the Ideal Gas Model

In this section, the ideal gas model is introduced. The ideal gas model has many applications in engineering practice and is frequently used in subsequent sections of this text.

3.12.1 Ideal Gas Equation of State

As discussed in Sec. 3.11, study of a generalized compressibility chart such as Fig. A-2 shows that at states where the pressure p is small relative to the critical pressure p_c (low p_R) and/or the temperature T is large relative to the critical temperature T_c (high T_R), the compressibility factor, $Z = pv/RT$, is approximately 1. At such states, we can assume with reasonable accuracy that $Z = 1$, or

$$pv = RT \tag{3.32}$$

Known as the ***ideal gas equation of state,*** Eq. 3.32 underlies the second part of this chapter dealing with the ideal gas model.

ideal gas equation of state

Alternative forms of the same basic relationship among pressure, specific volume, and temperature are obtained as follows. With $v = V/m$, Eq. 3.32 can be expressed as

$$pV = mRT \tag{3.33}$$

In addition, using $v = \overline{v}/M$ and $R = \overline{R}/M$, which are Eqs. 1.9 and 3.25, respectively, where M is the molecular weight, Eq. 3.32 can be expressed as

$$p\overline{v} = \overline{R}T \tag{3.34}$$

or, with $\overline{v} = V/n$, as

$$pV = n\overline{R}T \tag{3.35}$$

3.12.2 Ideal Gas Model

For any gas whose equation of state is given *exactly* by $pv = RT$, the specific internal energy depends on temperature *only*. This conclusion is demonstrated formally in Sec. 11.4. It is also supported by experimental observations, beginning with the work of Joule, who showed in 1843 that the internal energy of air at low density depends primarily on temperature. Further motivation from the microscopic viewpoint is provided shortly. The specific enthalpy of a gas described by $pv = RT$ also depends on temperature only, as can be shown by combining the definition of enthalpy, $h = u + pv$, with $u = u(T)$ and the ideal gas equation of state to obtain $h = u(T) + RT$. Taken together, these specifications constitute the ***ideal gas model***, summarized as follows

ideal gas model

$$pv = RT \tag{3.32}$$

$$u = u(T) \tag{3.36}$$

$$h = h(T) = u(T) + RT \tag{3.37}$$

Take Note...

To expedite the solutions of many subsequent examples and end-of-chapter problems involving air, oxygen (O_2), nitrogen (N_2), carbon dioxide (CO_2), carbon monoxide (CO), hydrogen (H_2), and other common gases, we indicate in the problem statements that the ideal gas model should be used. If not indicated explicitly, the suitability of the ideal gas model should be checked using the Z chart or other data.

The specific internal energy and enthalpy of gases generally depend on two independent properties, not just temperature as presumed by the ideal gas model. Moreover, the ideal gas equation of state does not provide an acceptable approximation at all states. Accordingly, whether the ideal gas model is used depends on the error acceptable in a given calculation. Still, gases often do *approach* ideal gas behavior, and a particularly simplified description is obtained with the ideal gas model.

To verify that a gas can be modeled as an ideal gas, the states of interest can be located on a compressibility chart to determine how well $Z = 1$ is satisfied. As shown in subsequent discussions, other tabular or graphical property data can also be used to determine the suitability of the ideal gas model.

The next example illustrates the use of the ideal gas equation of state and reinforces the use of property diagrams to locate principal states during processes.

Example 3.8 AIR AS AN IDEAL GAS UNDERGOING A CYCLE

One pound of air in a piston-cylinder assembly undergoes a thermodynamic cycle consisting of three processes.

Process 1–2: Constant specific volume

Process 2–3: Constant-temperature expansion

Process 3–1: Constant-pressure compression

At state 1, the temperature is 300 K, and the pressure is 1 atm. At state 2, the pressure is 2 bars. Employing the ideal gas equation of state,

(a) sketch the cycle on p–v coordinates.

(b) determine the temperature at state 2, in K.

(c) determine the specific volume at state 3, in m^3/kg.

Solution

Known: Air executes a thermodynamic cycle consisting of three processes: Process 1–2, $v = constant$; Process 2–3, $T = constant$; Process 3–1, $p = constant$. Values are given for T_1, p_1, and p_2.

Find: Sketch the cycle on p–v coordinates and determine T_2 and v_3.

Schematic and Given Data:

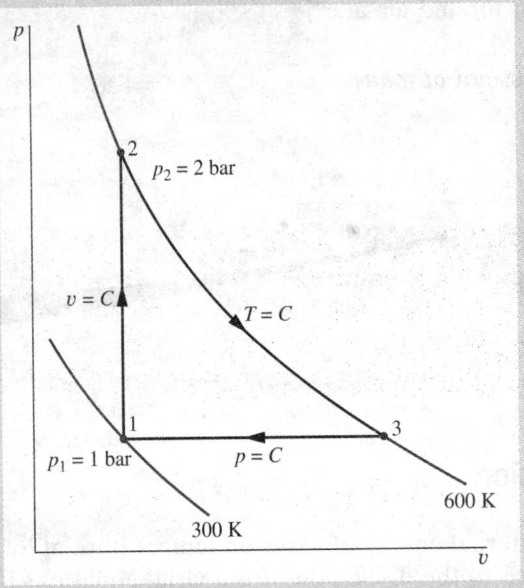

Fig. E3.8

Engineering Model:

1. The air is a closed system.
❶ 2. The air behaves as an ideal gas.
3. Volume change is the only work mode.

Analysis:

(a) The cycle is shown on p–v coordinates in the accompanying figure. Note that since $p = RT/v$ and temperature is constant, the variation of p with v for the process from 2 to 3 is nonlinear.

(b) Using $pv = RT$, the temperature at state 2 is

$$T_2 = p_2v_2/R$$

To obtain the specific volume v_2 required by this relationship, note that $v_2 = v_1$, so

❷
$$v_2 = RT_1/p_1$$

Combining these two results gives

$$T_2 = \frac{p_2}{p_1}T_1 = \left(\frac{2 \text{ bars}}{1 \text{ bar}}\right)(300 \text{ K}) = 600 \text{ K}$$

(c) Since $pv = RT$, the specific volume at state 3 is

$$v_3 = RT_3/p_3$$

Noting that $T_3 = T_2$, $p_3 = p_1$, and $R = \overline{R}/M$

$$v_3 = \frac{\overline{R}T_2}{Mp_1}$$

$$= \left(\frac{8.314 \dfrac{\text{kJ}}{\text{kmol} \cdot \text{J}}}{28.97 \dfrac{\text{kg}}{\text{mol}}}\right)\left(\frac{600 \text{ k}}{1 \text{ bar}}\right)\left(\frac{1 \text{ bar}}{10^5 \text{ N/m}^2}\right)\left(\frac{10^3 \text{N} \cdot \text{m}}{1 \text{ kJ}}\right)$$

$$= 1.72 \text{ m}^3/\text{kg}$$

where the molecular weight of air is from Table A-1.

❶ Table A-1 gives $p_c = 37.3$ bar, $T_c = 133$ K for air. Therefore, $p_{R2} = 0.054$, $T_{R2} = 4.52$. Referring to Fig. A-1, the value of the compressibility factor at this state is $Z \approx 1$. The same conclusion results when states 1 and 3 are checked. Accordingly, $pv = RT$ adequately describes the p–v–T relation for the air at these states.

❷ Carefully note that the equation of state $pv = RT$ requires the use of *absolute* temperature T and *absolute* pressure p.

Skills Developed: *Ability to...*
- ✓ evaluate p–v–T data using the ideal gas equation of state.
- ✓ sketch processes on a p–v diagram

Quick Quiz

Is the cycle sketched in Fig. E3.8 a power cycle or a refrigeration cycle? Explain.
Ans. A power cycle. As represented by enclosed area 1-2-3-1, the net work is positive.

3.12.3 Microscopic Interpretation

A picture of the dependence of the internal energy of gases on temperature at low density can be obtained with reference to the discussion of the virial equations in Sec. 3.11. As $p \to 0$ $(v \to \infty)$, the force interactions between molecules of a gas become weaker, and the virial expansions approach $Z = 1$ in the limit. The study of gases from the microscopic point of view shows that the dependence of the internal energy of a gas on pressure, or specific volume, at a specified temperature arises primarily because of molecular interactions. Accordingly, as the density of a gas decreases at fixed temperature, there comes a point where the effects of intermolecular forces are minimal. The internal energy is then determined principally by the temperature. From the microscopic point of view, the ideal gas model adheres to several idealizations: The gas consists of molecules that are in random motion and obey the laws of mechanics; the total number of molecules is large, but the volume of the molecules is a negligibly small fraction of the volume occupied by the gas; and no appreciable forces act on the molecules except during collisions. Further discussion of the ideal gas using the microscopic approach is provided in Sec. 3.13.2.

3.13 Internal Energy, Enthalpy, and Specific Heats of Ideal Gases

3.13.1 Δu, Δh, c_v, and c_p Relations

For a gas obeying the ideal gas model, specific internal energy depends only on temperature. Hence, the specific heat c_v, defined by Eq. 3.8, is also a function of temperature alone. That is,

$$c_v(T) = \frac{du}{dT} \quad \text{(ideal gas)} \tag{3.38}$$

This is expressed as an ordinary derivative because u depends only on T.

By separating variables in Eq. 3.38

$$du = c_v(T)\,dT \tag{3.39}$$

On integration, the change in specific internal energy is

$$u(T_2) - u(T_1) = \int_{T_1}^{T_2} c_v(T)\, dT \quad \text{(ideal gas)} \tag{3.40}$$

Similarly, for a gas obeying the ideal gas model, the specific enthalpy depends only on temperature, so the specific heat c_p, defined by Eq. 3.9, is also a function of temperature alone. That is

$$c_p(T) = \frac{dh}{dT} \quad \text{(ideal gas)} \tag{3.41}$$

Separating variables in Eq. 3.41

$$dh = c_p(T)\, dT \tag{3.42}$$

On integration, the change in specific enthalpy is

$$h(T_2) - h(T_1) = \int_{T_1}^{T_2} c_p(T)\, dT \quad \text{(ideal gas)} \tag{3.43}$$

An important relationship between the ideal gas specific heats can be developed by differentiating Eq. 3.37 with respect to temperature

$$\frac{dh}{dT} = \frac{du}{dT} + R$$

and introducing Eqs. 3.38 and 3.41 to obtain

$$c_p(T) = c_v(T) + R \quad \text{(ideal gas)} \tag{3.44}$$

On a molar basis, this is written as

$$\bar{c}_p(T) = \bar{c}_v(T) + \bar{R} \quad \text{(ideal gas)} \tag{3.45}$$

Although each of the two ideal gas specific heats is a function of temperature, Eqs. 3.44 and 3.45 show that the specific heats differ by just a constant: the gas constant. Knowledge of either specific heat for a particular gas allows the other to be calculated by using only the gas constant. The above equations also show that $c_p > c_v$ and $\bar{c}_p > \bar{c}_v$, respectively.

For an ideal gas, the specific heat ratio, k, is also a function of temperature only

$$k = \frac{c_p(T)}{c_v(T)} \quad \text{(ideal gas)} \tag{3.46}$$

Since $c_p > c_v$, it follows that $k > 1$. Combining Eqs. 3.44 and 3.46 results in

$$c_p(T) = \frac{kR}{k - 1} \tag{3.47a}$$

(ideal gas)

$$c_v(T) = \frac{R}{k - 1} \tag{3.47b}$$

Similar expressions can be written for the specific heats on a molar basis, with R being replaced by $\bar{R}$.

Horizons

What to Do About CO_2?

The concentration of carbon dioxide in the atmosphere is now over 30% greater than in pre-industrial times. Much of the increase is traceable to burning coal, oil, and natural gas. Evidence mounts that so much excess CO_2 in the atmosphere contributes to global warming and climate change. There is growing agreement that steps should be taken to reduce the CO_2 emitted to the atmosphere.

Reductions in CO_2 emissions can be achieved by using renewable energy such as wind and solar more, using fuel more efficiently, and switching to low-carbon fuels. Deep reductions will require additional measures, including the capture and storage (*sequestration*) of carbon dioxide. This would involve the removal of CO_2 from the exhaust gases of power plants, oil and gas refineries, and other industrial sources followed by storage of the captured CO_2. Some propose storage in depleted oil and gas reservoirs, unminable coal seams, deep salty aquifers, and other geological structures. Storage in the oceans, including injecting CO_2 to great depths from offshore pumping stations is another option, some say.

The deployment of CO_2 capture and storage technology faces major hurdles, including uncertainty over how long the injected gas will remain stored and possible collateral environmental impact when so much CO_2 is stored in nature. Another challenge is the development of effective technologies for removing CO_2 from gas streams. The expenditures of both fossil fuels and money required to capture CO_2, transport it to a storage site, and place it into storage are likely to be significant. Still, at some two-dozen test sites nationally, researchers are studying the viability of CO_2 capture and storage.

3.13.2 Using Specific Heat Functions

The foregoing expressions require the ideal gas specific heats as functions of temperature. These functions are available for gases of practical interest in various forms, including graphs, tables, and equations. Figure 3.13 illustrates the variation of $\bar{c}_p$ (molar basis) with temperature for a number of common gases. In the range of temperature shown, $\bar{c}_p$ increases with temperature for all gases, except for the

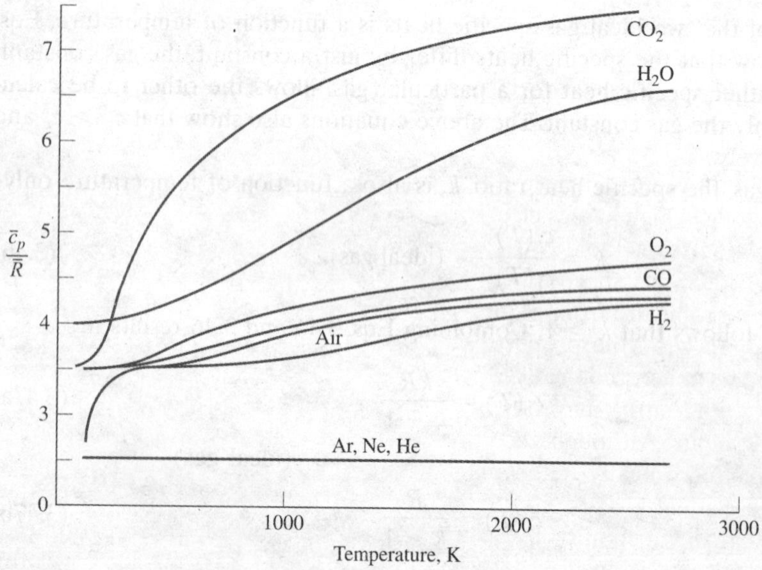

Fig. 3.13 Variation of $\bar{c}_p/\bar{R}$ with temperature for a number of gases modeled as ideal gases.

monatonic gases Ar, Ne, and He. For these, $\bar{c}_p$ is closely constant at the value predicted by kinetic theory: $\bar{c}_p = \frac{5}{2}\bar{R}$. Tabular specific heat data for selected gases are presented versus temperature in Tables A-20. Specific heats are also available in equation form. Several alternative forms of such equations are found in the engineering literature. An equation that is relatively easy to integrate is the polynomial form

$$\frac{\bar{c}_p}{\bar{R}} = \alpha + \beta T + \gamma T^2 + \delta T^3 + \varepsilon T^4 \tag{3.48}$$

Values of the constants α, β, γ, δ, and ε are listed in Tables A-21 for several gases in the temperature range 300 to 1000 K.

FOR EXAMPLE... to illustrate the use of Eq. 3.48, let us evaluate the change in specific enthalpy, in kJ/kg, of air modeled as an ideal gas from a state where $T_1 = 400$ K to a state where $T_2 = 900$ K. Inserting the expression for $\bar{c}_p(T)$ given by Eq. 3.48 into Eq. 3.43 and integrating with respect to temperature

$$h_2 - h_1 = \frac{\bar{R}}{M}\int_{T_1}^{T_2} (\alpha + \beta T + \gamma T^2 + \delta T^3 + \varepsilon T^4)\, dT$$

$$= \frac{\bar{R}}{M}\left[\alpha(T_2 - T_1) + \frac{\beta}{2}(T_2^2 - T_1^2) + \frac{\gamma}{3}(T_2^3 - T_1^3) + \frac{\delta}{4}(T_2^4 - T_1^4) + \frac{\varepsilon}{5}(T_2^5 - T_1^5)\right]$$

where the molecular weight M has been introduced to obtain the result on a unit mass basis. With values for the constants from Table A-21

$$h_2 - h_1 = \frac{8.314}{28.97}\left\{3.653(900 - 400) - \frac{1.337}{2(10)^3}[(900)^2 - (400)^2]\right.$$
$$+ \frac{3.294}{3(10)^6}[(900)^3 - (400)^3] - \frac{1.913}{4(10)^9}[(900)^4 - (400)^4]$$
$$\left. + \frac{0.2763}{5(10)^{12}}[(900)^5 - (400)^5]\right\} = 531.69 \text{ kJ/kg}$$

Specific heat functions $c_v(T)$ and $c_p(T)$ are also available in *IT: Interactive Thermodynamics* in the PROPERTIES menu. These functions can be integrated using the integral function of the program to calculate Δu and Δh, respectively. **FOR EXAMPLE...** let us repeat the immediately preceding example using *IT*. For air, the *IT* code is

```
cp = cp_T ("Air",T)
delh = Integral(cp, T)
```

Pushing SOLVE and sweeping T from 400 K to 900 K, the change in specific enthalpy is delh = 531.7 kJ/kg, which agrees closely with the value obtained by integrating the specific heat function from Table A-21, as illustrated above.

The source of ideal gas specific heat data is experiment. Specific heats can be determined macroscopically from painstaking property measurements. In the limit as pressure tends to zero, the properties of a gas tend to merge into those of its ideal gas model, so macroscopically determined specific heats of a gas extrapolated to very low pressures may be called either *zero-pressure* specific heats or *ideal gas* specific heats. Although zero-pressure specific heats can be obtained by extrapolating macroscopically determined experimental data, this is rarely done nowadays because ideal gas specific heats can readily be calculated with expressions from statistical mechanics by using *spectral* data, which can be obtained experimentally with precision. The determination of ideal gas specific heats is one of the important areas where the *microscopic approach* contributes significantly to the application of thermodynamics.

3.14 Applying the Energy Balance Using Ideal Gas Tables, Constant Specific Heats, and Software

Although changes in specific enthalpy and specific internal energy can be obtained by integrating specific heat expressions, as illustrated in Sec. 3.13, such evaluations are more easily conducted using the procedures considered in the present section. These procedures are illustrated in applications of the closed system energy balance.

3.14.1 Using Ideal Gas Tables

For a number of common gases, evaluations of specific internal energy and enthalpy changes are facilitated by the use of the *ideal gas tables*, Tables A-22 and A-23, which give u and h (or $\bar{u}$ and $\bar{h}$) versus temperature.

To obtain enthalpy versus temperature, write Eq. 3.43 as

$$h(T) = \int_{T_{ref}}^{T} c_p(T)\, dT + h(T_{ref})$$

where T_{ref} is an arbitrary reference temperature and $h(T_{ref})$ is an arbitrary value for enthalpy at the reference temperature. Tables A-22 and A-23 are based on the selection $h = 0$ at $T_{ref} = 0$ K. Accordingly, a tabulation of enthalpy versus temperature is developed through the integral[2]

$$h(T) = \int_{0}^{T} c_p(T)\, dT \tag{3.49}$$

Tabulations of internal energy versus temperature are obtained from the tabulated enthalpy values by using $u = h - RT$.

For air as an ideal gas, h and u are given in Table A-22 with units of kJ/kg. Values of molar specific enthalpy $\bar{h}$ and internal energy $\bar{u}$ for several other common gases modeled as ideal gases are given in Tables A-23 with units of kJ/kmol. Quantities other than specific internal energy and enthalpy appearing in these tables are introduced in Chap. 6 and should be ignored at present. Tables A-22 and A-23 are convenient for evaluations involving ideal gases, not only because the variation of the specific heats with temperature is accounted for automatically but also because the tables are easy to use.

The next example illustrates the use of the ideal gas tables, together with the closed system energy balance.

Example 3.9 USING THE ENERGY BALANCE AND IDEAL GAS TABLES

A piston–cylinder assembly contains 0.9 kg of air at a temperature of 300 K and a pressure of 1 bar. The air is compressed to a state where the temperature is 470 K and the pressure is 6 bars. During the compression, there is a heat transfer from the air to the surroundings equal to 20 kJ. Using the ideal gas model for air, determine the work during the process, in kJ.

Solution

Known: 0.9 kg of air are compressed between two specified states while there is heat transfer from the air of a known amount.

[2]The simple specific heat variation given by Eq. 3.48 is valid only for a limited temperature range, so tabular enthalpy values are calculated from Eq. 3.49 using other expressions that enable the integral to be evaluated accurately over wider ranges of temperature.

Find: Determine the work, in kJ.

Schematic and Given Data:

❶

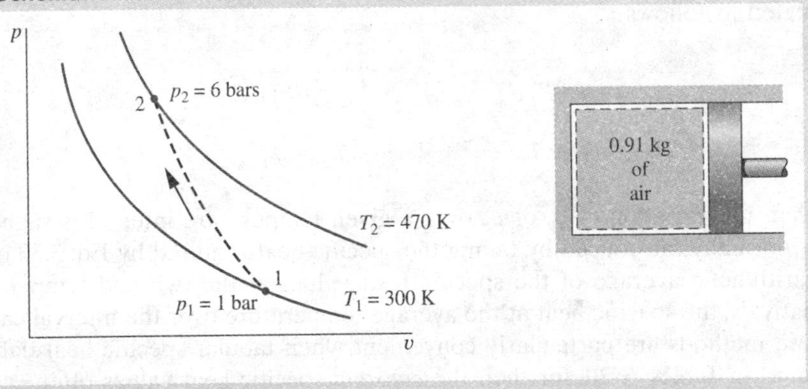

Engineering Model:

1. The air is a closed system.
2. The initial and final states are equilibrium states. There is no change in kinetic or potential energy.
3. The air is modeled as an ideal gas. ❷

Fig. E3.9

Analysis: An energy balance for the closed system is

$$\Delta \cancel{KE}^{\,0} + \Delta \cancel{PE}^{\,0} + \Delta U = Q - W$$

where the kinetic and potential energy terms vanish by assumption 2. Solving for W

❸ $$W = Q - \Delta U = Q - m(u_2 - u_1)$$

From the problem statement, $Q = -20$ kJ. Also, from Table A-22 at $T_1 = 300$ K, $u_1 = 214.07$ kJ/kg, and at $T_2 = 470$ K, $u_2 = 337.32$ kJ/kg. Accordingly

$$W = -20 \text{ kJ} - (0.9 \text{ kg})(337.32 - 214.07) = -130.9 \text{ kJ}$$

The minus sign indicates that work is done on the system in the process.

---

❶ Although the initial and final states are assumed to be equilibrium states, the intervening states are not necessarily equilibrium states, so the process has been indicated on the accompanying p–v diagram by a dashed line. This dashed line does not define a "path" for the process.

❷ Table A-1 gives $p_c = 37.7$ bars, $T_c = 133$ K for air. Therefore, at state 1, $p_{R1} = 0.03$, $T_{R1} = 2.26$, and at state 2, $p_{R2} = 0.16$, $T_{R2} = 3.51$. Referring to Fig. A-1, we conclude that at these states $Z \approx 1$, as assumed in the solution.

❸ In principle, the work could be evaluated through $\int p\, dV$, but because the variation of pressure at the piston face with volume is not known, the integration cannot be performed without more information.

Skills Developed: Ability to...
- ✓ define a closed system and identify interactions on its boundary.
- ✓ apply the energy balance using the ideal gas model.

Quick Quiz

Replacing air by carbon dioxide, but keeping all other problem statement details the same, evaluate work, in kJ. **Ans.** −131.99 kJ

3.14.2 Using Constant Specific Heats

When the specific heats are taken as constants, Eqs. 3.40 and 3.43 reduce, respectively, to

$$u(T_2) - u(T_1) = c_v(T_2 - T_1) \qquad (3.50)$$

$$h(T_2) - h(T_1) = c_p(T_2 - T_1) \qquad (3.51)$$

Equations 3.50 and 3.51 are often used for thermodynamic analyses involving ideal gases because they enable simple closed-form equations to be developed for many processes.

The constant values of c_v and c_p in Eqs. 3.50 and 3.51 are, strictly speaking, mean values calculated as follows

$$c_v = \frac{\int_{T_1}^{T_2} c_v(T)\, dT}{T_2 - T_1}, \qquad c_p = \frac{\int_{T_1}^{T_2} c_p(T)\, dT}{T_2 - T_1}$$

However, when the variation of c_v or c_p over a given temperature interval is slight, little error is normally introduced by taking the specific heat required by Eq. 3.50 or 3.51 as the arithmetic average of the specific heat values at the two end temperatures. Alternatively, the specific heat at the average temperature over the interval can be used. These methods are particularly convenient when tabular specific heat data are available, as in Tables A-20, for then the *constant* specific heat values often can be determined by inspection.

➡ **FOR EXAMPLE...** assuming the specific heat c_v is a constant and using Eq. 3.50, the expression for work in the solution of Example 3.9 reads

$$W = Q - mc_v(T_2 - T_1)$$

Evaluating c_v at the average temperature, 383 K (110°C), Table A-20 gives $c_v = 0.721$ kJ/kg · K. Inserting this value for c_v together with other data from Example 3.9

$$\begin{aligned} W &= -20 \text{ kJ} - (0.9 \text{ kg})\left(0.721 \frac{\text{kJ}}{\text{kg} \cdot \text{K}}\right)(470 - 300) \text{ K} \\ &= -130.31 \text{ kJ} \end{aligned}$$

which agrees closely with the answer obtained in Example 3.9 using Table A-22 data. ⬅

The following example illustrates the use of the closed system energy balance, together with the ideal gas model and the assumption of constant specific heats.

Example 3.10 USING THE ENERGY BALANCE AND CONSTANT SPECIFIC HEATS

Two tanks are connected by a valve. One tank contains 2 kg of carbon monoxide gas at 77°C and 0.7 bar. The other tank holds 8 kg of the same gas at 27°C and 1.2 bar. The valve is opened and the gases are allowed to mix while receiving energy by heat transfer from the surroundings. The final equilibrium temperature is 42°C. Using the ideal gas model with constant c_v, determine **(a)** the final equilibrium pressure, in bar **(b)** the heat transfer for the process, in kJ.

Solution

Known: Two tanks containing different amounts of carbon monoxide gas at initially different states are connected by a valve. The valve is opened and the gas allowed to mix while receiving a certain amount of energy by heat transfer. The final equilibrium temperature is known.

Find: Determine the final pressure and the heat transfer for the process.

Schematic and Given Data:

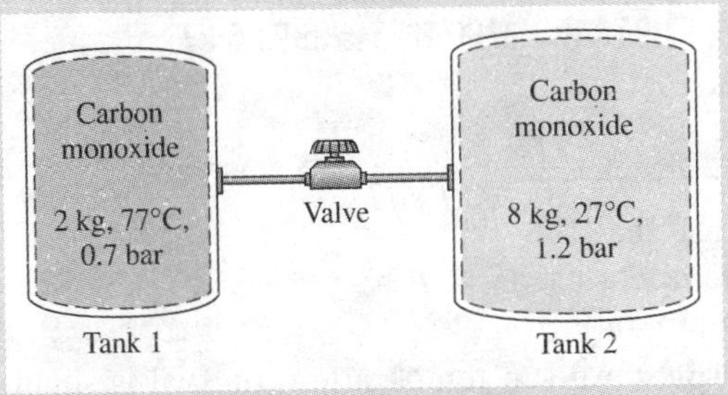

Fig. E3.10

Engineering Model:

1. The total amount of carbon monoxide gas is a closed system.
2. The gas is modeled as an ideal gas with constant c_v. ❶
3. The gas initially in each tank is in equilibrium. The final state is an equilibrium state.
4. No energy is transferred to, or from, the gas by work.
5. There is no change in kinetic or potential energy.

Analysis:

(a) The final equilibrium pressure p_f can be determined from the ideal gas equation of state

$$p_f = \frac{mRT_f}{V}$$

where m is the sum of the initial amounts of mass present in the two tanks, V is the total volume of the two tanks, and T_f is the final equilibrium temperature. Thus

$$p_f = \frac{(m_1 + m_2)RT_f}{V_1 + V_2}$$

Denoting the initial temperature and pressure in tank 1 as T_1 and p_1, respectively, $V_1 = m_1RT_1/p_1$. Similarly, if the initial temperature and pressure in tank 2 are T_2 and p_2, $V_2 = m_2RT_2/p_2$. Thus, the final pressure is

$$p_f = \frac{(m_1 + m_2)RT_f}{\left(\frac{m_1RT_1}{p_1}\right) + \left(\frac{m_2RT_2}{p_2}\right)} = \frac{(m_1 + m_2)T_f}{\left(\frac{m_1T_1}{p_1}\right) + \left(\frac{m_2T_2}{p_2}\right)}$$

Inserting values

$$p_f = \frac{(10 \text{ kg})(315 \text{ K})}{\frac{(2 \text{ kg})(350 \text{ K})}{0.7 \text{ bar}} + \frac{(8 \text{ kg})(300 \text{ K})}{1.2 \text{ bar}}} = 1.05 \text{ bar}$$

(b) The heat transfer can be found from an energy balance, which reduces with assumptions 4 and 5 to give

$$\Delta U = Q - \cancelto{0}{W}$$

or

$$Q = U_f - U_i$$

U_i is the initial internal energy, given by

$$U_i = m_1u(T_1) + m_2u(T_2)$$

where T_1 and T_2 are the initial temperatures of the CO in tanks 1 and 2, respectively. The final internal energy is U_f

$$U_f = (m_1 + m_2)u(T_f)$$

Introducing these expressions for internal energy, the energy balance becomes

$$Q = m_1[u(T_f) - u(T_1)] + m_2[u(T_f) - u(T_2)]$$

Since the specific heat c_v is constant (assumption 2)

$$Q = m_1c_v(T_f - T_1) + m_2c_v(T_f - T_2)$$

Evaluating c_v as the average of the values listed in Table A-20 at 300 K and 350 K, $c_v = 0.745$ kJ/kg · K. Hence

$$Q = (2\text{ kg})\left(0.745\frac{\text{kJ}}{\text{kg}\cdot\text{K}}\right)(315\text{ K} - 350\text{ K}) + (8\text{ kg})\left(0.745\frac{\text{kJ}}{\text{kg}\cdot\text{K}}\right)(315\text{ K} - 300\text{ K}) = +37.25\text{ kJ}$$

The plus sign indicates that the heat transfer is into the system.

❶ By referring to a generalized compressibility chart, it can be verified that the ideal gas equation of state is appropriate for CO in this range of temperature and pressure. Since the specific heat c_v of CO varies little over the temperature interval from 300 to 350 K (Table A-20), it can be treated as constant with acceptable accuracy.

Skills Developed: Ability to...

✓ define a closed system and identify interactions on its boundary.

✓ apply the energy balance using the ideal gas model when the specific heat c_v is constant.

Quick Quiz

Evaluate Q using specific internal energy values for CO from Table A-23. Compare with the result using constant c_v.
Ans. 36.99 kJ

3.14.3 Using Computer Software

Interactive Thermodynamics: IT also provides values of the specific internal energy and enthalpy for a wide range of gases modeled as ideal gases. Let us consider the use of *IT*, first for air, and then for other gases.

AIR. For air, *IT* uses the same reference state and reference value as in Table A-22, and the values computed by *IT* agree closely with table data. **FOR EXAMPLE...** let us use *IT* to evaluate the change in specific enthalpy of air from a state where $T_1 = 400$ K to a state where $T_2 = 900$ K. Selecting Air from the **Properties** menu, the following code would be used by *IT* to determine Δh (delh), in kJ/kg

```
h1 = h_T("Air",T1)
h2 = h_T("Air",T2)
T1 = 400//K
T2 = 900//K
delh = h2 – h1
```

Choosing K for the temperature unit and kg for the amount under the **Units** menu, the results returned by *IT* are $h_1 = 400.8$, $h_2 = 932.5$, and $\Delta h = 531.7$ kJ/kg, respectively. These values agree closely with those obtained from Table A-22: $h_1 = 400.98$, $h_2 = 932.93$, and $\Delta h = 531.95$ kJ/kg.

OTHER GASES. *IT* also provides data for each of the gases included in Table A-23. For these gases, the values of specific internal energy $\bar{u}$ and enthalpy $\bar{h}$ returned by *IT* are determined relative to a *standard reference state* that differs from that used in Table A-23. This equips *IT* for use in combustion applications; see Sec. 13.2.1 for further discussion. Consequently the values of $\bar{u}$ and $\bar{h}$ returned by *IT* for the gases of Table A-23 differ from those obtained directly from the table. Still, the property differences between two states remain the same, for datums cancel when differences are calculated.

FOR EXAMPLE... let us use *IT* to evaluate the change in specific enthalpy, in kJ/kmol, for carbon dioxide (CO_2) as an ideal gas from a state where $T_1 = 300$ K to a state where $T_2 = 500$ K. Selecting CO_2 from the **Properties** menu, the following code would be used by *IT*:

```
h1 = h_T("CO2",T1)
h2 = h_T("CO2",T2)
T1 = 300//K
T2 = 500//K
delh = h2 – h1
```

Choosing K for the temperature unit and moles for the amount under the **Units** menu, the results returned by *IT* are $\bar{h}_1 = -3.935 \times 10^5$, $\bar{h}_2 = -3.852 \times 10^5$, and $\Delta\bar{h} = 8{,}238$ kJ/kmol, respectively. The large negative values for $\bar{h}_1$ and $\bar{h}_2$ are a consequence of the reference state and reference value used by *IT* for CO_2. Although these values for specific enthalpy at states 1 and 2 differ from the corresponding values read from Table A-23: $\bar{h}_1 = 9{,}431$ and $\bar{h}_2 = 17{,}678$, which give $\Delta\bar{h} = 8{,}247$ kJ/kmol, the *difference* in specific enthalpy determined with each set of data agree closely.

The next example illustrates the use of software for problem solving with the ideal gas model. The results obtained are compared with those determined assuming the specific heat $\bar{c}_v$ is constant.

Example 3.11 USING THE ENERGY BALANCE AND SOFTWARE

One kmol of carbon dioxide gas (CO_2) in a piston–cylinder assembly undergoes a constant-pressure process at 1 bar from $T_1 = 300$ K to T_2. Plot the heat transfer to the gas, in kJ, versus T_2 ranging from 300 to 1,500 K. Assume the ideal gas model, and determine the specific internal energy change of the gas using

(a) $\bar{u}$ data from *IT*.

(b) a constant $\bar{c}_v$ evaluated at T_1 from *IT*.

Solution

Known: One kmol of CO_2 undergoes a constant-pressure process in a piston–cylinder assembly. The initial temperature, T_1, and the pressure are known.

Find: Plot the heat transfer versus the final temperature, T_2. Use the ideal gas model and evaluate $\Delta\bar{u}$ using (a) $\bar{u}$ data from *IT*, (b) constant $\bar{c}_v$ evaluated at T_1 from *IT*.

Schematic and Given Data:

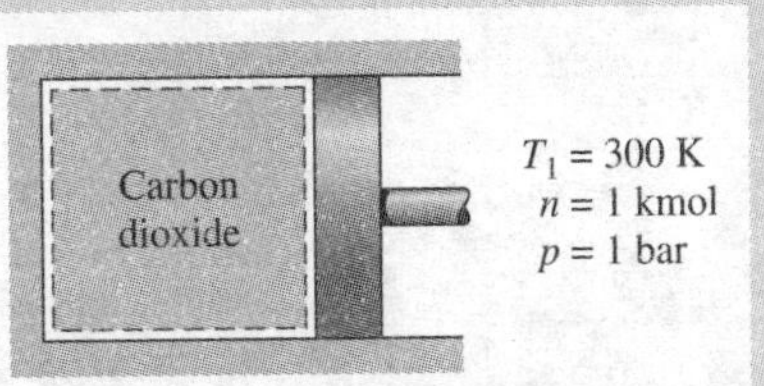

Engineering Model:

1. The carbon dioxide is a closed system.
2. The process occurs at constant pressure.
3. The carbon dioxide behaves as an ideal gas.
4. Kinetic and potential energy effects are negligible.

Fig. E3.11a

Analysis: The heat transfer is found using the closed system energy balance, which reduces to

$$U_2 - U_1 = Q - W$$

Using Eq. 2.17 at constant pressure (assumption 2)

$$W = p(V_2 - V_1) = pn(\bar{v}_2 - \bar{v}_1)$$

Then, with $\Delta U = n(\bar{u}_2 - \bar{u}_1)$, the energy balance becomes

$$n(\bar{u}_2 - \bar{u}_1) = Q - pn(\bar{v}_2 - \bar{v}_1)$$

Solving for Q

1 $$Q = n[(\bar{u}_2 - \bar{u}_1) + p(\bar{v}_2 - \bar{v}_1)]$$

With $p\bar{v} = \bar{R}T$, this becomes

$$Q = n[(\bar{u}_2 - \bar{u}_1) + \bar{R}(T_2 - T_1)]$$

The object is to plot Q versus T_2 for each of the following cases: **(a)** values for $\bar{u}_1$ and $\bar{u}_2$ at T_1 and T_2, respectively, are provided by *IT*, **(b)** Eq. 3.50 is used on a molar basis, namely

$$\bar{u}_2 - \bar{u}_1 = \bar{c}_v(T_2 - T_1)$$

where the value of $\bar{c}_v$ is evaluated at T_1 using *IT*.

The *IT* program follows, where Rbar denotes $\bar{R}$, cvb denotes $\bar{c}_v$, and ubar1 and ubar2 denote $\bar{u}_1$ and $\bar{u}_2$, respectively.

```
// Using the Units menu, select "mole" for the substance amount.
//Given Data
T1 = 300//K
T2 = 1500//K
n = 1//kmol
Rbar = 8.314//kJ/kmol · K

// (a) Obtain molar specific internal energy data using IT.
ubar1 = u_T ( "CO2", T1)
ubar2 = u_T ( "CO2", T2)
Qa = n*(ubar2 – ubar1) + n*Rbar*(T2 – T1)

// (b) Use Eq. 3.50 with cv evaluated at T1.
cvb = cv_T ( "CO2", T1)
Qb = n*cvb*(T2 – T1) + n*Rbar*(T2 – T1)
```

Use the **Solve** button to obtain the solution for the sample case of $T_2 = 1{,}500$ K. For part (a), the program returns $Q_a = 6.16 \times 10^4$ kJ. The solution can be checked using CO_2 data from Table A-23, as follows:

$$\begin{aligned} Q_a &= n[(\bar{u}_2 - \bar{u}_1) + \bar{R}(T_2 - T_1)] \\ &= (1\ \text{kmol})[(58{,}606 - 6{,}939)\,\text{kJ/kmol} + (8.314\ \text{kJ/kmol}\cdot\text{K})(1{,}500 - 300)\,\text{K}] \\ &= 61{,}644\ \text{kJ} \end{aligned}$$

Thus, the result obtained using CO_2 data from Table A-23 is in close agreement with the computer solution for the sample case. For part (b), *IT* returns $\bar{c}_v = 28.95$ kJ/kmol · K at T_1, giving $Q_b = 4.472 \times 10^4$ kJ when $T_2 = 1{,}500$ K. This value agrees with the result obtained using the specific heat c_v at 300 K from Table A-20, as can be verified.

Now that the computer program has been verified, use the **Explore** button to vary T_2 from 300 to 1,500 K in steps of 10. Construct the following graph using the **Graph** button:

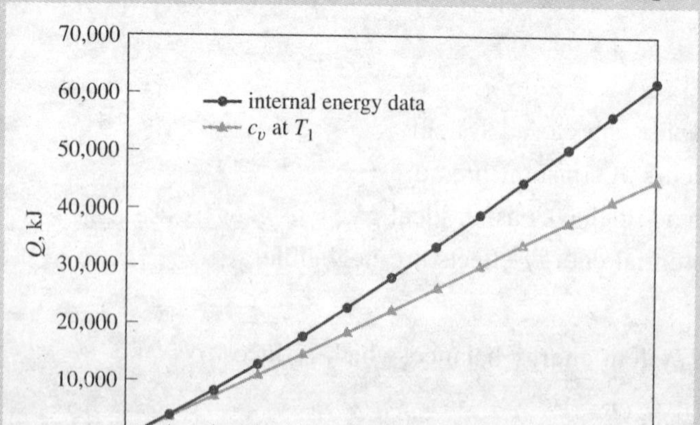

Fig. E3.11b

As expected, the heat transfer is seen to increase as the final temperature increases. From the plots, we also see that using constant $\bar{c}_v$ evaluated at T_1 for calculating $\Delta\bar{u}$, and hence Q, can lead to considerable error when compared to using $\bar{u}$ data. The two solutions compare favorably up to about 500 K, but differ by approximately 27% when heating to a temperature of 1500 K.

Skills Developed: *Ability to...*

- ✓ define a closed system and identify interactions on its boundary.
- ✓ apply the energy balance using the ideal gas model.
- ✓ use *IT* to retrieve property data for CO_2 as an ideal gas and plot calculated data.

❶ Alternatively, this expression for Q can be written as

$$Q = n[(\bar{u}_2 + p\bar{v}_2) - (\bar{u}_1 + p\bar{v}_1)]$$

Introducing $\bar{h} = \bar{u} + p\bar{v}$, the expression for Q becomes

$$Q = n(\bar{h}_2 - \bar{h}_1)$$

Quick Quiz

Repeat part (b) using $\bar{c}_v$ evaluated at $T_{average} = (T_1 + T_2)/2$. Which approach gives better agreement with the results of part (a): evaluating $\bar{c}_v$ at T_1 or at $T_{average}$?

Ans. At $T_{average}$.

3.15 Polytropic Process Relations

A *polytropic process* is a quasiequilibrium process (Sec. 2.2.5) described by

$$pV^n = constant \tag{3.52}$$

or, in terms of specific volumes, by $pv^n = constant$. In these expressions, n is a constant.

For a polytropic process between two states

$$p_1V_1^n = p_2V_2^n$$

or

$$\frac{p_2}{p_1} = \left(\frac{V_1}{V_2}\right)^n \tag{3.53}$$

The exponent n may take on any value from $-\infty$ to $+\infty$, depending on the particular process. When $n = 0$, the process is an isobaric (constant-pressure) process, and when $n = \pm\infty$, the process is an isometric (constant-volume) process.

For a polytropic process

$$\int_1^2 p\,dV = \frac{p_2V_2 - p_1V_1}{1 - n} \qquad (n \neq 1) \tag{3.54}$$

for any exponent n except $n = 1$. When $n = 1$,

$$\int_1^2 p\,dV = p_1V_1 \ln\frac{V_2}{V_1} \qquad (n = 1) \tag{3.55}$$

Example 2.1 provides the details of these integrations.

Equations 3.52 through 3.55 apply to *any* gas (or liquid) undergoing a polytropic process. When the *additional* idealization of ideal gas behavior is appropriate, further relations can be derived. Thus, when the ideal gas equation of state is introduced into Eqs. 3.53, 3.54, and 3.55, the following expressions are obtained, respectively

$$\frac{T_2}{T_1} = \left(\frac{p_2}{p_1}\right)^{(n-1)/n} = \left(\frac{V_1}{V_2}\right)^{n-1} \quad \text{(ideal gas)} \tag{3.56}$$

$$\int_1^2 p\, dV = \frac{mR(T_2 - T_1)}{1-n} \quad \text{(ideal gas, } n \neq 1\text{)} \tag{3.57}$$

$$\int_1^2 p\, dV = mRT \ln \frac{V_2}{V_1} \quad \text{(ideal gas, } n = 1\text{)} \tag{3.58}$$

For an ideal gas, the case $n = 1$ corresponds to an isothermal (constant-temperature) process, as can readily be verified.

Example 3.12 illustrates the use of the closed system energy balance for a system consisting of an ideal gas undergoing a polytropic process.

Example 3.12 POLYTROPIC PROCESS OF AIR AS AN IDEAL GAS

Air undergoes a polytropic compression in a piston–cylinder assembly from $p_1 = 1$ bar, $T_1 = 22°C$ to $p_2 = 5$ bars. Employing the ideal gas model with constant specific heat ratio k, determine the work and heat transfer per unit mass, in kJ/kg, if (a) $n = 1.3$, (b) $n = k$. Evaluate k at T_1.

Solution

Known: Air undergoes a polytropic compression process from a given initial state to a specified final pressure.

Find: Determine the work and heat transfer, each in kJ/kg.

Schematic and Given Data:

❶

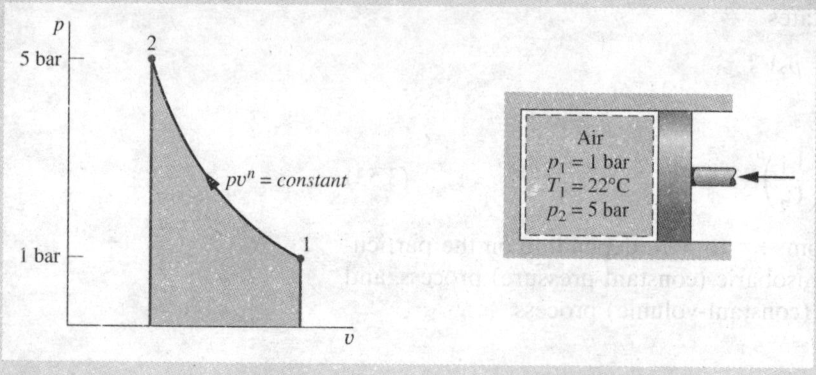

Engineering Model:

1. The air is a closed system.
2. The air behaves as an ideal gas with constant specific heat ratio k evaluated at the initial temperature.
3. The compression is polytropic.
4. There is no change in kinetic or potential energy.

Fig. E3.12

Analysis: The work can be evaluated in this case from the expression

$$W = \int_1^2 p\, dV$$

With Eq. 3.57

$$\frac{W}{m} = \frac{R(T_2 - T_1)}{1-n} \tag{a}$$

The heat transfer can be evaluated from an energy balance. Thus

$$\frac{Q}{m} = \frac{W}{m} + (u_2 - u_1)$$

Inspection of Eq. 3.47b shows that when the specific heat ratio k is constant, c_v is constant. Thus

$$\frac{Q}{m} = \frac{W}{m} + c_v(T_2 - T_1) \qquad \text{(b)}$$

(a) For $n = 1.3$, the temperature at the final state, T_2, can be evaluated from Eq. 3.56 as follows

$$T_2 = T_1\left(\frac{p_2}{p_1}\right)^{(n-1)/n} = 295 \text{ K}\left(\frac{5}{1}\right)^{(1.3-1)/1.3} = 427.68 \text{ K } (154.68°\text{C})$$

Using Eq. (a), the work is then

$$\frac{W}{m} = \frac{R(T_2 - T_1)}{1-n} = \left(287.21 \frac{\text{J}}{\text{kg} \cdot \text{K}}\right)\left(\frac{427.68 \text{ K} - 295 \text{ K}}{1 - 1.3}\right) = -127.02 \text{ kJ/kg}$$

At 22°C, Table A-20 gives $k = 1.401$ and $c_v = 0.716$ kJ/kg · K. Alternatively, c_v can be found using Eq. 3.47b, as follows:

$$c_v = \frac{R}{k-1} \qquad \text{(c)}$$

$$= \frac{287.21 \text{ J/kg} \cdot \text{K}}{(1.401 - 1)} = 0.716 \text{ kJ/kg} \cdot \text{K}$$

Substituting values into Eq. (b), we get

$$\frac{Q}{m} = -127.02 \frac{\text{kJ}}{\text{kg}} + \left(0.716 \frac{\text{kJ}}{\text{kg} \cdot \text{K}}\right)(427.68 - 295) \text{ K}$$

$$= -32.02 \frac{\text{kJ}}{\text{kg}}$$

(b) For $n = k$, substituting Eqs. (a) and (c) into Eq. (b) gives

$$\frac{Q}{m} = \frac{R(T_2 - T_1)}{1-k} + \frac{R(T_2 - T_1)}{k-1} = 0$$

That is, no heat transfer occurs in the polytropic process of an ideal gas for which $n = k$.

Skills Developed: *Ability to...*
- ✓ evaluate work using Eq. 2.17.
- ✓ apply the energy balance using the ideal gas model.
- ✓ apply the polytropic process concept.

❶ The states visited in a polytropic compression process are shown by the curve on the accompanying p–v diagram. The magnitude of the work per unit of mass is represented by the shaded area *below* the curve.

Quick Quiz

For $n = k$, evaluate the temperature at the final state, in K and °C.
Ans. 466.67 K(193.67°C)

✓ Chapter Summary and Study Guide

In this chapter, we have considered property relations for a broad range of substances in tabular, graphical, and equation form. Primary emphasis has been placed on the use of tabular data, but computer retrieval also has been considered.

A key aspect of thermodynamic analysis is fixing states. This is guided by the state principle for pure, simple compressible systems, which indicates that the intensive state is fixed by the values of *two* independent, intensive properties.

Another important aspect of thermodynamic analysis is locating principal states of processes on appropriate diagrams: p–v, T–v, and p–T diagrams. The skills of fixing states and using property diagrams are particularly important when solving problems involving the energy balance.

The ideal gas model is introduced in the second part of this chapter, using the compressibility factor as a point of departure. This arrangement emphasizes the limitations of the ideal gas model. When it is appropriate to use the ideal gas model, we stress that specific heats generally vary with temperature, and feature the use of the ideal gas tables in problem solving.

The following checklist provides a study guide for this chapter. When your study of the text and end-of-chapter exercises has been completed you should be able to

- ✓ write out the meanings of the terms listed in the margins throughout the chapter and understand each of the related concepts. The subset of key concepts listed below is particularly important in subsequent chapters.
- ✓ retrieve property data from Tables A-1 through A-23, using the state principle to fix states and linear interpolation when required.
- ✓ sketch T–v, p–v, and p–T diagrams, and locate principal states on such diagrams.
- ✓ apply the closed system energy balance with property data.
- ✓ evaluate the properties of two-phase, liquid–vapor mixtures using Eqs. 3.1, 3.2, 3.6, and 3.7.
- ✓ estimate the properties of liquids using Eqs. 3.11, 3.12, and 3.14.
- ✓ apply the incompressible substance model.
- ✓ use the generalized compressibility chart to relate p–v–T data of gases.
- ✓ apply the ideal gas model for thermodynamic analysis, including determining when use of the ideal gas model is warranted, and appropriately using ideal gas table data or constant specific heat data to determine Δu and Δh.

Key Engineering Concepts

phase *p. 81*
pure substance *p. 81*
state principle *p. 81*
simple compressible system *p. 81*
p–v–T surface *p. 83*
phase diagram *p. 85*
saturation temperature *p. 85*
saturation pressure *p. 85*
p–v diagram *p. 85*
T–v diagram *p. 85*
two-phase, liquid–vapor mixture *p. 87*
quality *p. 87*
superheated vapor *p. 87*
enthalpy *p. 95*
specific heats *p. 105*
ideal gas model *p. 116*

Key Equations

$x = \dfrac{m_{\text{vapor}}}{m_{\text{liquid}} + m_{\text{vapor}}}$	(3.1)	Quality, x, of a two-phase, liquid–vapor mixture.
$v = (1 - x)v_f + xv_g = v_f + x(v_g - v_f)$	(3.2)	Specific volume, internal energy and enthalpy of a two-phase, liquid–vapor mixture.
$u = (1 - x)u_f + xu_g = u_f + x(u_g - u_f)$	(3.6)	
$h = (1 - x)h_f + xh_g = h_f + x(h_g - h_f)$	(3.7)	
$v(T, p) \approx v_f(T)$	(3.11)	Specific volume, internal energy, and enthalpy of liquids, approximated by saturated liquid values, respectively.
$u(T, p) \approx u_f(T)$	(3.12)	
$h(T, p) \approx h_f(T)$	(3.14)	
Ideal Gas Model Relations		
$pv = RT$	(3.32)	Ideal gas model.
$u = u(T)$	(3.36)	
$h = h(T) = u(T) + RT$	(3.37)	
$u(T_2) - u(T_1) = \int_{T_1}^{T_2} c_v(T)\, dT$	(3.40)	Change in specific internal energy.
$u(T_2) - u(T_1) = c_v(T_2 - T_1)$	(3.50)	For constant c_v.
$h(T_2) - h(T_1) = \int_{T_1}^{T_2} c_p(T)\, dT$	(3.43)	Change in specific enthalpy.
$h(T_2) - h(T_1) = c_p(T_2 - T_1)$	(3.51)	For constant c_p.

Exercises: things engineers think about

1. If water contracted on freezing, what implications might this have for aquatic life?

2. At what temperature does water boil at the top of Mount McKinley, Alaska? In Death Valley, California?

3. Why do frozen water pipes tend to burst?

4. Why do many foods have high-altitude cooking instructions?

5. Can a substance that contracts on freezing exist as a solid when at a temperature greater than its triple-point temperature? Repeat for a substance that expands on freezing.

6. The specific internal energy is arbitrarily set to zero in Table A-2 for saturated liquid water at 0.01°C. If the reference value for u at this reference state were specified differently, would there be any significant effect on thermodynamic analyses using u and h?

7. For liquid water at 20°C and 1.0 MPa, what percent difference would there be if its specific enthalpy were evaluated using Eq. 3.14 instead of Eq. 3.13?

8. How do I measure the specific heat c_p of liquid water at atmospheric pressure and room temperature?

9. If a block of iron and a block of tin having equal volumes each received the same energy input by heat transfer, which block would experience the greater temperature increase?

10. How many minutes do I have to exercise to *burn* the calories in a helping of my favorite dessert?

11. What is my annual contribution to the emission of CO_2 into the atmosphere?

12. How do I estimate the mass of air contained in a bicycle tire?

13. Specific internal energy and enthalpy data for water vapor are provided in two tables: Tables A-4 and A-23. When would Table A-23 be used?

Problems: developing engineering skills

Exploring Concepts: Phase and Pure Substance

3.1 A system consists of liquid water in equilibrium with a gaseous mixture of air and water vapor. How many phases are present? Does the system consist of a pure substance? Explain. Repeat for a system consisting of ice and liquid water in equilibrium with a gaseous mixture of air and water vapor.

3.2 A system consists of liquid oxygen in equilibrium with oxygen vapor. How many phases are present? The system undergoes a process during which some of the liquid is vaporized. Can the system be viewed as being a pure substance during the process? Explain.

3.3 A system consisting of liquid water undergoes a process. At the end of the process, some of the liquid water has frozen, and the system contains liquid water and ice. Can the system be viewed as being a pure substance during the process? Explain.

3.4 A dish of liquid water is placed on a table in a room. After a while, all of the water evaporates. Taking the water and the air in the room to be a closed system, can the system be regarded as a pure substance *during* the process? *After* the process is completed? Discuss.

Using p–v–T Data

3.5 Determine the phase or phases in a system consisting of H_2O at the following conditions and sketch p–v and T–v diagrams showing the location of each state.

(a) $p = 5$ bar, $T = 151.9$°C.
(b) $p = 5$ bar, $T = 200$°C.
(c) $T = 200$°C, $p = 2.5$ MPa.
(d) $T = 160$°C, $p = 4.8$ bar.
(e) $T = -12$°C, $p = 1$ bar.

3.6 Determine the phase or phases in a system consisting of H_2O at the following conditions and sketch p–v and T–v diagrams showing the location of each state.

(a) $p = 5$ bar, $T = 151.9$°C.
(b) $p = 5$ bar, $T = 200$°C.
(c) $T = 200$°C, $p = 2.5$ MPa.
(d) $T = 160$°C, $p = 4.8$ bar.
(e) $T = -12$°C, $p = 1$ bar.

3.7 The following table lists temperatures and specific volumes of water vapor at two pressures:

$p = 1.0$ MPa		$p = 1.5$ MPa	
T(°C)	v (m^3/kg)	T (°C)	v (m^3/kg)
200	0.2060	200	0.1325
240	0.2275	240	0.1483
280	0.2480	280	0.1627

Data encountered in solving problems often do not fall exactly on the grid of values provided by property tables, and *linear interpolation* between adjacent table entries becomes necessary. Using the data provided here, estimate

(a) the specific volume at $T = 240°C$, $p = 1.25$ MPa, in m³/kg.
(b) the temperature at $p = 1.5$ MPa, $v = 0.1555$ m³/kg, in°C.
(c) the specific volume at $T = 220°C$, $p = 1.4$ MPa, in m³/kg.

3.8 The following table lists temperatures and specific volumes of ammonia vapor at two pressures:

$p = 345$ kPa		$p = 415$ kPa	
T (°C)	v (m³/kg)	T (°C)	v (m³/kg)
38	0.427	38	0.353
49	0.444	49	0.368
60	0.461	60	0.382

Data encountered in solving problems often do not fall exactly on the grid of values provided by property tables, and *linear interpolation* between adjacent table entries becomes necessary. Using the data provided here, estimate
(a) the specific volume at $T = 49°C$, $p = 372$ kPa, in m³/kg.
(b) the temperature at $p = 415$ kPa, $v = 0.373$ m³/kg, in °C.
(c) the specific volume at $T = 400$ kPa, $p = 43°C$, in m³/kg.

3.9 For H_2O, plot the following on a p–v diagram drawn to scale on log–log coordinates:

(a) the saturated liquid and saturated vapor lines from the triple point to the critical point, with pressure in MPa and specific volume in m³/kg.
(b) lines of constant temperature at 100 and 300°C.

3.10 For H_2O, determine the specified property at the indicated state. Locate the state on a sketch of the T–v diagram.

(a) $p = 300$ kPa, $v = 0.5$ m³/kg. Find T, in °C.
(b) $p = 28$ MPa, $T = 200°C$. Find v, in m³/kg.
(c) $p = 1$ MPa, $T = 405°C$. Find v, in m³/kg.
(d) $T = 100°C$, $x = 60\%$. Find v, in m³/kg.

3.11 For each case, determine the specific volume at the indicated state. Locate the state on a sketch of the T–v diagram.

(a) Water at $p = 101.3$ kPa, $T = 38°C$. Find v, in m³/kg.
(b) Ammonia at $T = -30°C$, $x = 50\%$. Find v, in m³/kg.
(c) Refrigerant 134a at $p = 1.5$ MPa, $T = 100°C$. Find v, in m³/kg.

3.12 For each case, determine the specified property at the indicated state. Locate the state on a sketch of the T–v diagram.

(a) Water at $v = 0.5$ m³/kg, $p = 3$ bar, determine T, in °C.
(b) Ammonia at $p = 0.76$ bar, $T = -28.8°C$, determine v, in m³/kg.
(c) Propane at $p = 1$ MPa, $T = 85°C$, determine v, in m³/kg.

3.13 Determine the mass, in kg, of 0.1 m³ of Refrigerant 134a at 4 bar, 100°C.

3.14 Determine the volume, in m³, occupied by 0.9 kg of H_2O at a pressure of 6.9 MPa and

(a) a temperature of 316°C.
(b) a quality of 80%.
(c) a temperature of 93°C.

3.15 A closed vessel with a volume of 0.054 m³ contains 2.27 kg of Refrigerant 134a. A pressure sensor in the tank wall reads 492.22 kPa (gage). If the atmospheric pressure is 99.3 kPa what is the temperature of the refrigerant, in °C?

3.16 Two kg of a two-phase, liquid–vapor mixture of carbon dioxide (CO_2) exists at −40°C in a 0.05 m³ tank. Determine the quality of the mixture, if the values of specific volume for saturated liquid and saturated vapor CO_2 at −40°C are $v_f = 0.896 \times 10^{-3}$ m³/kg and $v_g = 3.824 \times 10^{-2}$ m³/kg, respectively.

3.17 Determine the quality of a two-phase liquid–vapor mixture of

(a) H_2O at 20°C with a specific volume of 20 m³/kg.
(b) Propane at 15 bar with a specific volume of 0.02997 m³/kg.
(c) Refrigerant 134a at 60°C with a specific volume of 0.001 m³/kg.
(d) Ammonia at 1 MPa with a specific volume of 0.1 m³/kg.

3.18 Determine the quality of a two-phase liquid–vapor mixture of

(a) H_2O at 101.325 kPa with a specific volume of 1.56 m³/kg.
(b) Propane at 27°C with a specific volume of 0.0017 kg.
(c) Refrigerant 134a at 345 kPa with a specific volume of 0.03 m³/kg.
(d) Ammonia at 4°C with a specific volume of 125 m³/kg.

3.19 A two-phase liquid–vapor mixture of ammonia at 38°C has a specific volume of 0.06 m³/kg. Determine the quality of a two-phase liquid–vapor mixture at −17.8°C with the same specific volume.

3.20 A two-phase liquid–vapor mixture of a substance has a pressure of 150 bar and occupies a volume of 0.2 m³. The masses of saturated liquid and vapor present are 3.8 kg and 4.2 kg, respectively. Determine the specific volume of the mixture, in m³/kg.

3.21 As shown in Fig. P3.21, a closed, rigid cylinder contains different volumes of saturated liquid water and saturated water vapor at a temperature of 150°C. Determine the quality of the mixture, expressed as a percentage.

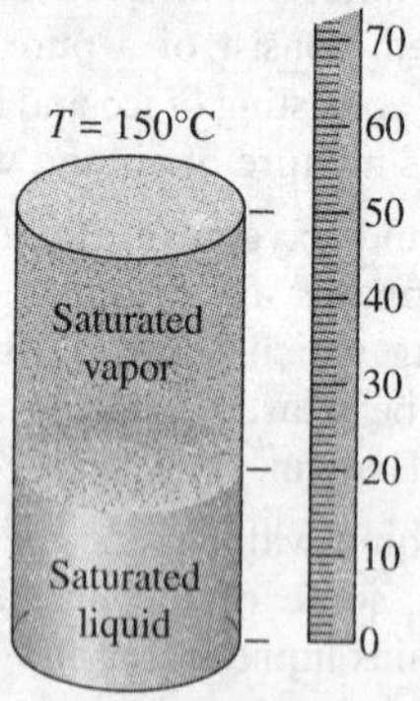

Fig. P3.21

3.22 A closed system consists of a two-phase liquid–vapor mixture of H_2O in equilibrium at 149°C. The quality of the mixture is 0.8 (80%) and the mass of saturated vapor present is 0.9 kg. Determine the mass of saturated liquid present, in kg, and the total volume of the system, in m³.

3.23 Ammonia, initially saturated vapor at −4°C, undergoes a constant-specific volume process to 200 kPa. At the final state, determine the temperature, in °C, and the quality. Locate each state on a sketch of the T–v diagram.

3.24 Water is contained in a closed, rigid, 0.2 m³ tank at an initial pressure of 5 bar and a quality of 50%. Heat transfer occurs until the tank contains only saturated vapor. Determine the final mass of vapor in the tank, in kg, and the final pressure, in bar.

3.25 A rigid tank contains 2.27 kg of a two-phase, liquid–vapor mixture of H_2O, initially at 127°C with a quality of 0.6. Heat transfer to the contents of the tank occurs until the temperature is 160°C. Show the process on a p–v diagram. Determine the mass of vapor, in kg, initially present in the tank and the final pressure, in kPa.

3.26 Two thousand kg of water, initially a saturated liquid at 150°C, is heated in a closed, rigid tank to a final state where the pressure is 2.5 MPa. Determine the final temperature, in °C, the volume of the tank, in m^3, and sketch the process on T–v and p–v diagrams.

3.27 Steam is contained in a closed rigid container with a volume of 1 m^3. Initially, the pressure and temperature of the steam are 7 bar and 500°C, respectively. The temperature drops as a result of heat transfer to the surroundings. Determine the temperature at which condensation first occurs, in °C, and the fraction of the total mass that has condensed when the pressure reaches 0.5 bar. What is the volume, in m^3, occupied by saturated liquid at the final state?

3.28 Water vapor is heated in a closed, rigid tank from saturated vapor at 160°C to a final temperature of 400°C. Determine the initial and final pressures, in bar, and sketch the process on T–v and p–v diagrams.

3.29 Ammonia undergoes an isothermal process from an initial state at $T_1 = 27°C$ and $v_1 = 0.6\ m^2/kg$ to saturated vapor. Determine the initial and final pressures, in kPa, and sketch the process on T–v and p–v diagrams.

3.30 One kilogram of water initially is at the critical point.

(a) If the water is cooled at constant-specific volume to a pressure of 30 bar, determine the quality at the final state.
(b) If the water undergoes a constant-temperature expansion to a pressure of 30 bar, determine the specific volume at the final state, in m^3/kg.

Show each process on a sketch of the T–v diagram.

3.31 1.36 kg of saturated water vapor, contained in a closed rigid tank whose volume is 0.36 m^3, is heated to a final temperature of 204°C. Sketch the process on a T–v diagram. Determine the pressures at the initial and final states, each in kPa.

3.32 0.9 kg of water vapor is compressed at a constant pressure of 1.7 MPa from a volume of 0.19 m^3 to a saturated vapor state. Determine the temperatures at the initial and final states, each in °C, and the work for the process, in kJ.

3.33 3.18 kg of propane in a piston-cylinder assembly, initially at $p_1 = 1.4$ MPa and $T_1 = 93°C$, undergoes a constant-pressure process to a final state. The work for the process is −93.73 kJ. At the final state, determine the temperature, in °C, if superheated, or the quality, if saturated.

3.34 Ammonia in a piston-cylinder assembly undergoes a constant-pressure process at 2.5 bar from $T_1 = 30°C$ to saturated vapor. Determine the work for the process, in kJ/kg of refrigerant.

3.35 Water vapor initially at 10 bar and 400°C is contained within a piston–cylinder assembly. The water is cooled at constant volume until its temperature is 150°C. The water is then condensed isothermally to saturated liquid. For the water as the system, evaluate the work, in kJ/kg.

3.36 Two kilograms of Refrigerant 22 undergo a process for which the pressure–volume relation is $pv^{1.05} = constant$. The initial state of the refrigerant is fixed by $p_1 = 2$ bar, $T_1 = -20°C$, and the final pressure is $p_2 = 10$ bar. Calculate the work for the process, in kJ.

3.37 Refrigerant 134a in a piston–cylinder assembly undergoes a process for which the pressure–volume relation is $pv^{1.058} = constant$. At the initial state, $p_1 = 200$ kPa, $T_1 = -10°C$. The final temperature is $T_2 = 50°C$. Determine the final pressure, in kPa, and the work for the process, in kJ/kg of refrigerant.

3.38 A piston–cylinder assembly contains 0.02 kg of Refrigerant 134a. The refrigerant is compressed from an initial state where $p_1 = 68.9$ kPa and $T_1 = -6.7°C$ to a final state where $p_2 = 1.1$ MPa. During the process, the pressure and specific volume are related by $pv = constant$. Determine the work, in kJ, for the refrigerant.

Using *u–h* Data

3.39 Determine the values of the specified properties at each of the following conditions.

(a) For Refrigerant 134a at $T = 60°C$ and $v = 0.072\ m^3/kg$, determine p in kPa and h in kJ/kg.
(b) For ammonia at $p = 8$ bar and $v = 0.005\ m^3/kg$, determine T in °C and u in kJ/kg.
(c) For Refrigerant 22 at $T = -10°C$ and $u = 200$ kJ/kg, determine p in bar and v in m^3/kg.

3.40 Determine the values of the specified properties at each of the following conditions.

(a) For Refrigerant 134a at $p = 965$ kPa and $h = 232.61$ kJ/kg, determine T in °C and v in m^3/kg.
(b) For ammonia at $T = -17.8°C$ and $v = 0.94\ m^3/kg$, determine p in kPa and h in kJ/kg.
(c) For Refrigerant 22 at $T = -1.1°C$ and $v = 0.07\ m^3/kg$, determine p in kPa and h in kJ/kg.

3.41 Using *IT*, determine the specified property data at the indicated states. Compare with results from the appropriate table.

(a) Cases (a), (b), and (c) of Problem 3.39.
(b) Cases (a), (b), and (c) of Problem 3.40.

3.42 Using the tables for water, determine the specified property data at the indicated states. In each case, locate the state by hand on sketches of the p–v and T–v diagrams.

(a) At $p = 3$ bar, $T = 240°C$, find v in m^3/kg and u in kJ/kg.
(b) At $p = 3$ bar, $v = 0.5\ m^3/kg$, find T in °C and u in kJ/kg.
(c) At $T = 400°C$, $p = 10$ bar, find v in m^3/kg and h in kJ/kg.
(d) At $T = 320°C$, $v = 0.03\ m^3/kg$, find p in MPa and u in kJ/kg.
(e) At $p = 28$ MPa, $T = 520°C$, find v in m^3/kg and h in kJ/kg.
(f) At $T = 100°C$, $x = 60\%$, find p in bar and v in m^3/kg.

(g) At $T = 10°C$, $v = 100\ m^3/kg$, find p in kPa and h in kJ/kg.
(h) At $p = 4$ MPa, $T = 160°C$, find v in m^3/kg and u in kJ/kg.

3.43 Using the tables for water, determine the specified property data at the indicated states. In each case, locate the state by hand on sketches of the p–v and T–v diagrams.

(a) At $p = 1.38$ bar, $T = 204°C$, find v in m^3/kg and u in kJ/kg.
(b) At $p = 1.38$ bar, $v = 1\ m^3/kg$, find T in °C and u in kJ/kg.
(c) At $T = 482°C$, $p = 11.7$ bar, find v in m^3/kg and h in kJ/kg.
(d) At $T = 315.6°C$, $v = 0.04\ m^3/kg$, find p in kPa and u in kJ/kg.
(e) At $p = 48$ bars, $T = 343°C$, find v in m^3/kg and h in kJ/kg.
(f) At $T = 204°C$, $x = 90\%$, find p in kPa and v in m^3/kg.
(g) At $T = 4°C$, $v = 121.7\ m^3/kg$, find p in kPa and h in kJ/kg.
(h) At $p = 41.37$ bars, $T = 160°C$, find v in m^3/kg and u in kJ/kg.

3.44 A quantity of water is at 15 MPa and 100°C. Evaluate the specific volume, in m^3/kg, and the specific enthalpy, in kJ/kg. using

(a) data from Table A-5.
(b) saturated liquid data from Table A-2.

3.45 Evaluate the specific volume, in m^3/kg, and the specific enthalpy, in kJ/kg, of water at 204°C and a pressure of 207 bars.

3.46 Evaluate the specific volume, in m^3/kg, and the specific enthalpy, in kJ/kg, of Refrigerant 134a at 35°C and 10.3 bar.

3.47 Evaluate the specific volume, in m^3/kg, and the specific enthalpy, in kJ/kg, of ammonia at 20°C and 1.0 MPa.

3.48 Evaluate the specific volume, in m^3/kg, and the specific enthalpy, in kJ/kg, of propane at 800 kPa and 0°C.

3.49 Plot versus pressure the percent changes in specific volume, specific internal energy, and specific enthalpy for water at 20°C from the saturated liquid state to the state where the pressure is 300 bar. Based on the resulting plots, discuss the implications regarding approximating compressed liquid properties using saturated liquid properties at 20°C, as discussed in Sec. 3.10.1.

Applying the Energy Balance

3.50 A piston-cylinder assembly contains 1 kg of water, initially occupying a volume of $0.5\ m^3$ at 1 bar. Energy transfer by heat to the water results in an expansion at constant temperature to a final volume of $1.694\ m^3$. Kinetic and potential energy effects are negligible. For the water, (a) show the process on a T–v diagram, (b) evaluate the work, in kJ, and (c) evaluate the heat transfer, in kJ.

3.51 Propane within a piston-cylinder assembly undergoes a constant-pressure process from saturated vapor at 400 kPa to a temperature of 40°C. Kinetic and potential energy effects are negligible. For the propane, (a) show the process on a p–v diagram, (b) evaluate the work, in kJ/kg, and (c) evaluate the heat transfer, in kJ/kg.

3.52 Refrigerant 134a is compressed with no heat transfer in a piston–cylinder assembly from 207 kPa, −6.7°C to 1.1 MPa. The mass of refrigerant is 0.02 kg. For the refrigerant as the system, $W = -0.59$ kJ. Kinetic and potential energy effects are negligible. Determine the final temperature, in °C.

3.53 Ammonia vapor in a piston-cylinder assembly undergoes a constant-pressure process from saturated vapor at 10 bar. The work is +16.5 kJ/kg. Changes in kinetic and potential energy are negligible. Determine (a) the final temperature of the ammonia, in °C, and (b) the heat transfer, in kJ/kg.

3.54 Water in a piston-cylinder assembly, initially at a temperature of 99.63°C and a quality of 65%, is heated at constant pressure to a temperature of 200°C. If the work during the process is +300 kJ, determine (a) the mass of water, in kg, and (b) the heat transfer, in kJ. Changes in kinetic and potential energy are negligible.

3.55 A piston-cylinder assembly containing water, initially a liquid at 10°C, undergoes a process at a constant pressure of 138 kPa to a final state where the water is a vapor at 149°C. Kinetic and potential energy effects are negligible. Determine the work and heat transfer, in kJ/kg, for each of three parts of the overall process: (a) from the initial liquid state to the saturated liquid state, (b) from saturated liquid to saturated vapor, and (c) from saturated vapor to the final vapor state, all at 138 kPa.

3.56 As shown in Fig. P3.56, 0.1 kg of propane is contained within a piston-cylinder assembly at a constant pressure of 0.2 MPa. Energy transfer by heat occurs slowly to the propane, and the volume of the propane increases from $0.0277\ m^3$ to $0.0307\ m^3$. Friction between the piston and cylinder is negligible. The local atmospheric pressure and acceleration of gravity are 100 kPa and 9.81 m/s^2, respectively. The propane experiences no significant kinetic and potential energy effects. For the propane, determine (a) the initial and final temperatures, in °C, (b) the work, in kJ, and (c) the heat transfer, in kJ.

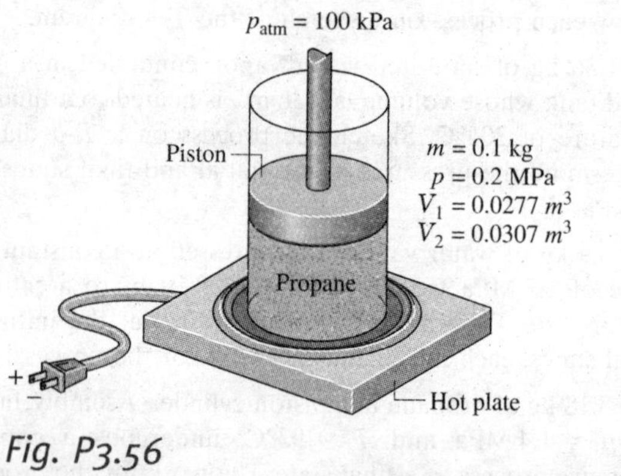

Fig. P3.56

3.57 One kg of Refrigerant 22, initially at 0.9 MPa, 20°C, is contained within a rigid, closed tank. The tank is fitted with a paddle wheel that transfers energy *to* the refrigerant at a constant rate of 0.1 kW. Heat transfer *from* the refrigerant *to* its surroundings occurs at a rate Kt, in kW, where the constant K is in kW per minute and t is time, in minutes. After

20 minutes of stirring, the refrigerant is at a pressure of 1.2 MPa. No overall changes in kinetic or potential energy occur. For the Refrigerant 22, determine

(a) the work, in kJ.
(b) the heat transfer, in kJ.
(c) Evaluate the constant K, in kW per minute.

3.58 A closed, rigid tank contains 2 kg of water initially at 80°C and a quality of 0.6. Heat transfer occurs until the tank contains only saturated vapor at a higher pressure. Kinetic and potential energy effects are negligible. For the water as the system, determine the amount of energy transfer by heat, in kJ.

3.59 As shown in Fig. P3.59, a rigid, closed tank having a volume of 0.54 m^3 and filled with 34.02 kg of Refrigerant 134a is exposed to the sun. At 9:00 AM, the refrigerant is at a pressure of 689.5 kPa. By 3:00 PM, owing to solar radiation, the refrigerant is a saturated vapor at a pressure greater than 689.5 kPa. For the refrigerant, the determine (a) the initial temperature, in °C, (b) the final pressure, in kPa, and (c) the heat transfer, in kJ.

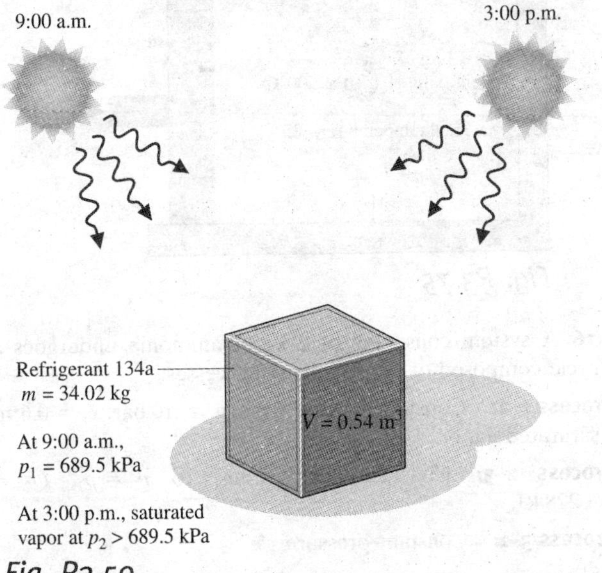

Fig. P3.59

3.60 A rigid, insulated tank fitted with a paddle wheel is filled with water, initially a two-phase liquid–vapor mixture at 138 kPa, consisting of 0.03 kg of saturated liquid and 0.03 kg of saturated vapor. The tank contents are stirred by the paddle wheel until all of the water is saturated vapor at a pressure greater than 138 kPa. Kinetic and potential energy effects are negligible. For the water, determine the

(a) volume occupied, in m^3.
(b) initial temperature, in °C.
(c) final pressure, in kPa.
(d) work, in kJ.

3.61 If the hot plate of Example 3.2 transfers energy at a rate of 0.1 kW to the two-phase mixture, determine the time required, in h, to bring the mixture from (a) state 1 to state 2, (b) state 1 to state 3.

3.62 A rigid tank is filled with 2.27 kg of water, initially at T_1 = 127°C and v_1 = 0.44 m^3/kg. The tank contents are heated until the temperature is T_2 = 177°C. Kinetic and potential energy effects are negligible. For the water, determine (a) the initial and final pressure, each in kPa, and (b) the heat transfer, in kJ.

3.63 Saturated water vapor is contained in a closed rigid tank. The water is cooled to a temperature of 110°C. At this temperature the masses of saturated vapor and liquid are each 1000 kg. Determine the heat transfer for the process, in kJ.

3.64 A rigid, well-insulated tank contains a two-phase mixture consisting of 0.00014 m^3 of saturated liquid water and 0.0324 m^3 of saturated water vapor, initially at 101.3 kPa. A paddle wheel stirs the mixture until only saturated vapor at a higher pressure remains in the tank. Kinetic and potential energy effects are negligible. For the water, determine the amount of energy transfer by work, in kJ.

3.65 A two-phase liquid–vapor mixture of H_2O, initially at 1.0 MPa with a quality of 90%, is contained in a rigid, well-insulated tank. The mass of H_2O is 2 kg. An electric resistance heater in the tank transfers energy to the water at a constant rate of 60 W for 1.95 h. Determine the final temperature of the water in the tank, in °C, and the final pressure, in bar.

3.66 Two kilograms of Refrigerant 134a, initially at 2 bar and occupying a volume of 0.12 m^3, undergoes a process at constant pressure until the volume has doubled. Kinetic and potential energy effects are negligible. Determine the work and heat transfer for the process, each in kJ.

3.67 0.9 kg of a two-phase liquid–vapor mixture of H_2O, initially at 689 kPa, are confined to one side of a rigid, well-insulated container by a partition. The other side of the container has a volume of 0.189 m^3 and is initially evacuated. The partition is removed and the water expands to fill the entire container. The pressure at the final equilibrium state is 275.8 kPa. Determine the quality of the mixture present initially and the overall volume of the container, in m^3.

3.68 One kilogram of saturated solid water at the triple point is heated to saturated liquid while the pressure is maintained constant. Determine the work and the heat transfer for the process, each in kJ. Show that the heat transfer equals the change in enthalpy of the water in this case.

3.69 A closed system initially contains a three-phase mixture consisting of 0.45 kg of saturated solid water (ice), 0.45 kg of saturated liquid water, and 0.1 kg of saturated water vapor at the triple point temperature and pressure. Heat transfer occurs while the pressure is maintained constant until only saturated vapor remains. Determine the amounts of energy transfer by work and heat, each in kJ.

3.70 A two-phase, liquid–vapor mixture of H_2O, initially at x = 30% and a pressure of 100 kPa, is contained in a piston-cylinder assembly, as shown in Fig P3.70. The mass of the piston is 10 kg, and its diameter is 15 cm. The pressure of the surroundings is 100 kPa. As the water is heated, the pressure inside the cylinder remains constant until the piston hits the stops. Heat transfer to the water continues at constant volume until the pressure is 150 kPa. Friction between the piston and the cylinder wall and kinetic and potential energy effects are negligible. For the overall process of the water, determine the work and heat transfer, each in kJ.

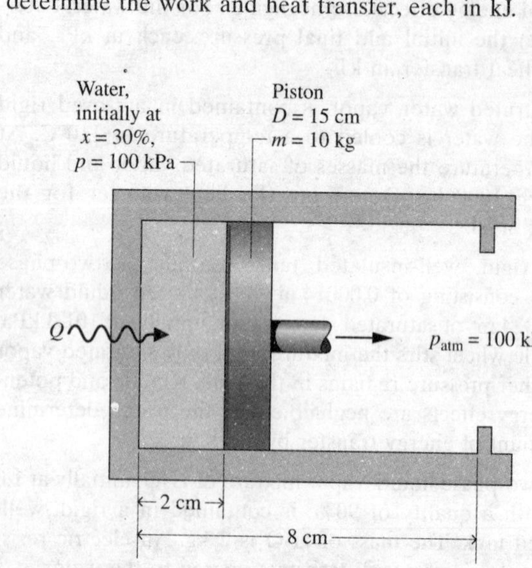

Fig. P3.70

3.71 A piston-cylinder assembly contains 0.9 kg of water, initially at 149°C. The water undergoes two processes in series: constant-volume heating followed by a constant-pressure process. At the end of the constant-volume process, the pressure is 690 kPa and the water is a two-phase, liquid–vapor mixture with a quality of 80%. At the end of the constant-pressure process, the temperature is 204°C. Neglect kinetic and potential energy effects.

(a) Sketch T–v and p–v diagrams showing the key states and the processes.

(b) Determine the work and heat transfer for each of the two processes, all in kJ.

3.72 A system consisting of 0.9 kg of water vapor, initially at 149°C and occupying a volume of 0.54 m³, is compressed isothermally to a volume of 0.24 m³. The system is then heated at constant volume to a final pressure of 827 kPa. During the isothermal compression there is energy transfer by work of magnitude 95.8 kJ *into* the system. Kinetic and potential energy effects are negligible. Determine the heat transfer, in kJ, for each process.

3.73 Ammonia in a piston-cylinder assembly undergoes two processes in series. At the initial state, $p_1 = 827$ kPa and the quality is 100%. Process 1–2 occurs at constant volume until the temperature is 38°C. The second process, from state 2 to state 3, occurs at constant temperature, with $Q_{23} = 104$ kJ, until the quality is again 100%. Kinetic and potential energy effects are negligible. For 1 kg of ammonia, determine **(a)** the heat transfer for Process 1–2 and **(b)** the work for Process 2–3, each in kJ.

3.74 1.4 kg of water is contained in a piston-cylinder assembly, initially occupying a volume $V_1 = 0.81$ m³ at $T_1 = 149$°C. The water undergoes two processes in series:

Process 1-2: Constant-temperature compression to $V_2 = 0.3$ m³, during which there is an energy transfer by heat *from* the water of 1345 kJ.

Process 2-3: Constant-volume heating to $p_3 = 827$ kPa.

Sketch the two processes in series on a T–v diagram. Neglecting kinetic and potential energy effects, determine the work in Process 1-2 and the heat transfer in Process 2-3, each in Btu.

3.75 As shown in Fig. P3.75, a piston-cylinder assembly fitted with stops contains 0.1 kg of water, initially at 1 MPa, 500°C. The water undergoes two processes in series:

Process 1-2: Constant-pressure cooling until the piston face rests against the stops. The volume occupied by the water is then one-half its initial volume.

Process 2-3: With the piston face resting against the stops, the water cools to 25°C.

Sketch the two processes in series on a p–v diagram. Neglecting kinetic and potential energy effects, evaluate for each process the work and heat transfer, each in kJ.

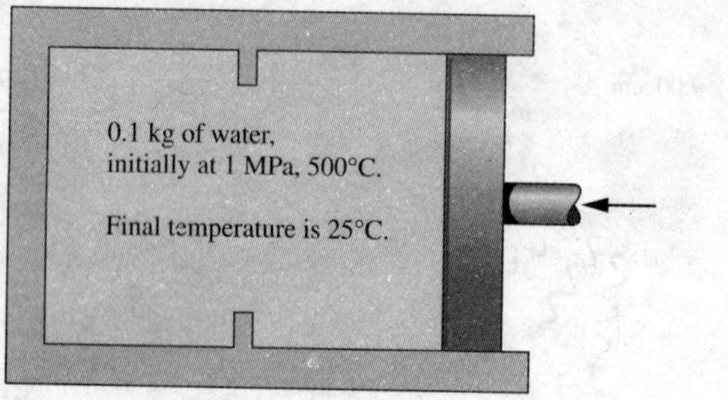

Fig. P3.75

3.76 A system consisting of 2 kg of ammonia undergoes a cycle composed of the following processes:

Process 1–2: Constant volume from $p_1 = 10$ bar, $x_1 = 0.6$ to saturated vapor.

Process 2–3: Constant temperature to $p_3 = p_1$, $Q_{23} = +228$ kJ.

Process 3–1: Constant pressure.

Sketch the cycle on p–v and T–v diagrams. Neglecting kinetic and potential energy effects, determine the net work for the cycle and the heat transfer for each process, all in kJ.

3.77 A system consisting of 1 kg of H_2O undergoes a power cycle composed of the following processes:

Process 1–2: Constant-pressure heating at 10 bar from saturated vapor.

Process 2–3: Constant-volume cooling to $p_3 = 5$ bar, $T_3 = 160$°C.

Process 3–4: Isothermal compression with $Q_{34} = -815.8$ kJ.

Process 4–1: Constant-volume heating.

Sketch the cycle on T–v and p–v diagrams. Neglecting kinetic and potential energy effects, determine the thermal efficiency.

3.78 A system consisting of 0.45 kg of Refrigerant 22 undergoes a cycle composed of the following processes:

Process 1–2: Constant pressure from $p_1 = 205$ kPa, $x_1 = 0.95$ to $T_2 = 4°C$.

Process 2–3: Constant temperature to saturated vapor with $W_{23} = -12.47$ kJ.

Process 3–1: Adiabatic expansion.

Sketch the cycle on p–v and T–v diagrams. Neglecting kinetic and potential energy effects, determine the net work for the cycle and the heat transfer for each process, all in kJ.

3.79 One-half kg of Refrigerant-22 is contained in a piston-cylinder assembly, initially saturated vapor at 5 bar. The refrigerant undergoes a process for which the pressure-specific volume relation is $pv = constant$ to a final pressure of 20 bar. Kinetic and potential energy effects can be neglected. Determine the work and heat transfer for the process, each in kJ.

3.80 Ten kilograms of Refrigerant 22 contained in a piston-cylinder assembly undergoes a process for which the pressure-specific volume relationship is $pv^n = constant$. The initial and final states of the refrigerant are fixed by $p_1 = 400$ kPa, $T_1 = -5°C$, and $p_2 = 2000$ kPa, $T_2 = 70°C$, respectively. Determine the work and heat transfer for the process, each in kJ.

3.81 A piston-cylinder assembly contains ammonia, initially at 0.8 bar and $-10°C$. The ammonia is compressed to a pressure of 5.5 bar. During the process, the pressure and specific volume are related by $pv = constant$. For 20 kg of ammonia, determine the work and heat transfer, each in kJ.

3.82 A piston-cylinder assembly contains ammonia, initially at $p_1 = 0.1$ MPa, $T_1 = 15°C$. The ammonia is compressed to a final state where $p_2 = 1$ MPa. During the process, the pressure and specific volume are related by $pv = constant$ and the work has a magnitude of 32 kJ. Sketch the process on a p–v diagram. Neglecting kinetic and potential energy effects, determine the heat transfer, in kJ.

3.83 Figure P3.83 shows a piston-cylinder assembly fitted with a spring. The cylinder contains water, initially at 538°C, and the spring is in a vacuum. The piston face, which has an area of 125 cm^2, is initially at $x_1 = 50$ cm. The water is cooled until the piston face is at $x_2 = 40$ cm. The force exerted by the spring varies linearly with x according to $F_{spring} = kx$, where k = 350 N/cm. Friction between the piston and cylinder is negligible. For the water, determine

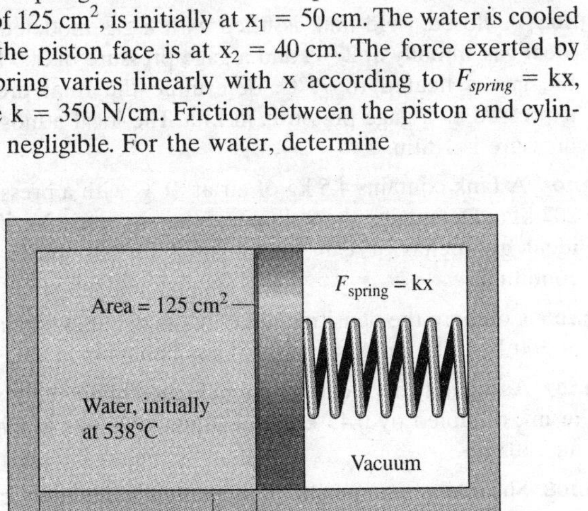

Fig. P3.83

(a) the initial and final pressures, each in kPa.
(b) the amount of water present, in kg.
(c) the work, in kJ.
(d) the heat transfer, in kJ.

3.84 As shown in Fig. P3.84, 0.5 kg of ammonia is contained in a piston-cylinder assembly, initially at $T_1 = -20°C$ and a quality of 25%. As the ammonia is slowly heated to a final state, where $T_2 = 20°C$, $p_2 = 0.6$ MPa, its pressure varies linearly with specific volume. There are no significant kinetic and potential energy effects. For the ammonia, (a) show the process on a sketch of the p–v diagram and (b) evaluate the work and heat transfer, each in kJ.

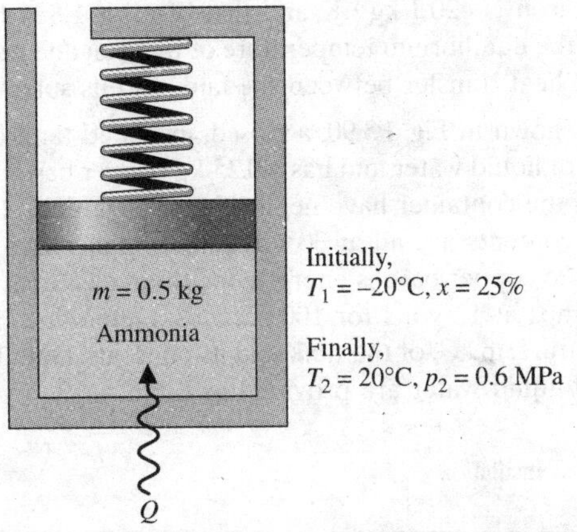

Fig. P3.84

3.85 Four liters of milk at 20°C is placed in a refrigerator. If energy is removed from the milk by heat transfer at a constant rate of 0.084 kJ/s, how long would it take, in minutes, for the milk to cool to 4°C? The specific heat and density of the milk are 4 kJ/kg · K and 1025 kg/m^3, respectively.

3.86 Shown in Fig. P3.86 is an insulated copper block that receives energy at a rate of 100 W from an embedded resistor. If the block has a volume of 10^{-3} m^3 and an initial temperature of 20°C, how long would it take, in minutes, for the temperature to reach 60°C? Data for copper are provided in Table A-19.

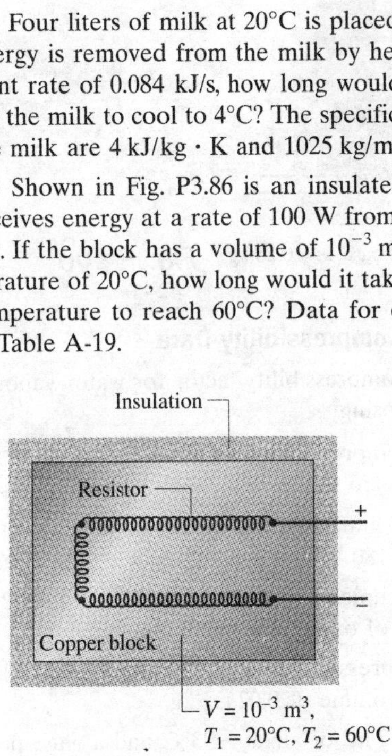

Fig. P3.86

3.87 A well-insulated copper tank of mass 13 kg contains 4 kg of liquid water. Initially, the temperature of the copper is 27°C and the temperature of the water is 50°C. An electrical resistor of neglible mass transfers 100 kJ of energy to the contents of the tank. The tank and its contents come to equilibrium. What is the final temperature, in °C?

3.88 A 5.4×10^{-3} m^3 aluminum bar, initially at 93°C, is placed in an open tank together with 0.108 m^3 of liquid water, initially at 21°C. For the water and the bar as the system, determine the final equilibrium temperature, in °C, ignoring heat transfer between the tank and its surroundings.

3.89 A 22.7 kg iron casting with an initial temperature of 370°C is quenched in a tank filled with oil. The tank is 0.9 m in diameter and 1.5 m tall. The oil can be considered incompressible with a density of 960 kg/m^3 and an initial temperature of 27°C. The mass of the tank is negligible. If the specific heat of iron is 420 J/kg · K and that of oil is 1,884 J/kg · K, what is the equilibrium temperature of the iron and oil, in °C? Neglect heat transfer between the tank and its surroundings.

3.90 As shown in Fig. P3.90, a closed, insulated tank contains 0.15 kg of liquid water and has a 0.25-kg copper base. The thin walls of the container have negligible mass. Initially, the tank and its contents are all at 30°C. A heating element embedded in the copper base is energized with an electrical current of 10 amps at 12 volts for 100 seconds. Determine the final temperature, in °C, of the tank and its contents. Data for copper and liquid water are provided in Table A-19.

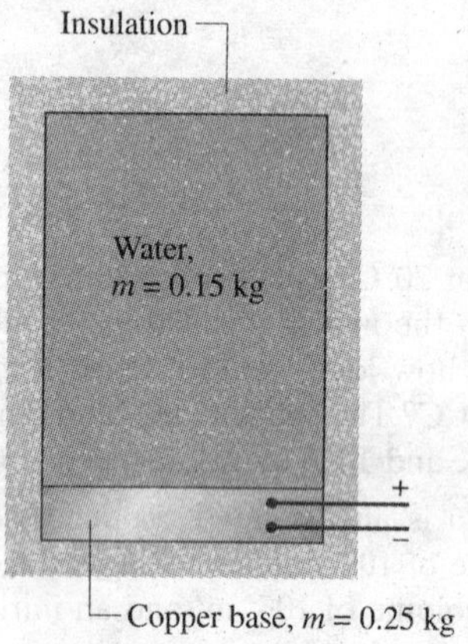

Fig. P3.90

Using Generalized Compressibility Data

3.91 Determine the compressibility factor for water vapor at 200 bar and 470°C, using

(a) data from the compressibility chart.
(b) data from the steam tables.

3.92 Determine the volume, in m^3, occupied by 40 kg of nitrogen (N_2) at 17 MPa, 180 K.

3.93 Oxygen (O_2) occupies a volume of 0.135 m^3 at 222 K, Determine the mass of oxygen, in kg.

3.94 Determine the pressure, in kPa, of ethane (C_2H_6) at 46°C and a specific volume of 0.02 m^3/kg.

3.95 A tank contains 2 m^3 of air at −93°C and a gage pressure of 1.4 MPa. Determine the mass of air, in kg. The local atmospheric pressure is 1 atm.

3.96 Butane (C_4H_{10}) in a piston-cylinder assembly undergoes an isothermal compression at 173°C from p_1 = 1.9 MPa to p_2 = 2.5 MPa. Determine the work, in kJ/kg.

3.97 Five kg of butane (C_4H_{10}) in a piston-cylinder assembly undergo a process from p_1 = 5 MPa, T_1 = 500 K to p_2 = 3 MPa, T_2 = 450 K during which the relationship between pressure and specific volume is pv^n = *constant*. Determine the work, in kJ.

3.98 2.27 kmol of carbon dioxide (CO_2), initially at 2.2 MPa, 367 K, is compressed at constant pressure in a piston–cylinder assembly. For the gas, W = −2,110 kJ. Determine the final temperature, in K.

3.99 For what ranges of pressure and temperature can air be considered an ideal gas? Explain your reasoning.

Working with the Ideal Gas Model

3.100 A tank contains 0.05 m^3 of nitrogen (N_2) at −21°C and 10 MPa. Determine the mass of nitrogen, in kg, using

(a) the ideal gas model.
(b) data from the compressibility chart.

Comment on the applicability of the ideal gas model for nitrogen at this state.

3.101 Determine the percent error in using the ideal gas model to determine the specific volume of

(a) water vapor at 14 MPa, 370°C.
(b) water vapor at 6.9 kPa, 93°C.
(c) ammonia at 415 kPa, 71°C.
(d) air at 1 bar, 1,111 K.
(e) Refrigerant 22 at 2.1 MPa, 60°C.

3.102 Check the applicability of the ideal gas model

(a) for water at 570°C and pressures of 11.03 MPa and 1.1 MPa.
(b) for carbon dioxide at 865 K and pressures of 75 bar and 3 bar.

3.103 Determine the temperature, in K, of oxygen (O_2) at 250 bar and a specific volume of 0.003 m^3/kg using generalized compressibility data and compare with the value obtained using the ideal gas model.

3.104 A closed, rigid tank is filled with a gas modeled as an ideal gas, initially at 27°C and a gage pressure of 300 kPa. If the gas is heated to 77°C, determine the final pressure, expressed as a gage pressure, in kPa. The local atmospheric pressure is 1 atm.

3.105 A tank contains 4.5 kg of air at 21°C with a pressure of 207 kPa. Determine the volume of the air, in m^3. Verify that ideal gas behavior can be assumed for air under these conditions.

3.106 Compare the densities, in kg/m^3, of helium and air, each at 300 K, 100 kPa. Assume ideal gas behavior.

3.107 Assuming the ideal gas model, determine the volume, in m^3, occupied by 0.45 kmol of argon (Ar) gas at 690 kPa and 305.6 K.

3.108 Show that the specific heat ratio of a monatomic ideal gas is equal to 5/3.

3.109 A balloon filled with helium, initially at 27°C, 1 bar, is released and rises in the atmosphere until the helium is at 17°C, 0.9 bar. Determine, as a percent, the change in volume of the helium from its initial volume.

3.110 By integrating $\bar{c}_p(T)$ obtained from Table A-21, determine the change in specific enthalpy, in kJ/kg, of methane (CH_4) from $T_1 = 320$ K, $p_1 = 2$ bar to $T_2 = 800$ K, $p_2 = 10$ bar. Check your result using *IT*.

3.111 Assuming ideal gas behavior for air, plot to scale the isotherms 300, 500, 1,000, and 2,000 K on a p–v diagram.

3.112 Assuming ideal gas behavior for air, plot to scale the isotherms 278 K, 555.6 K, 1,111 K, and 2,222 K on a p–v diagram.

Using the Energy Balance with the Ideal Gas Model

3.113 As shown in Fig. P3.113, a piston-cylinder assembly fitted with a paddle wheel contains air, initially at $p_1 = 207$ kPa, $T_1 = 282°C$, and $V_1 = 0.108\ m^3$. The air undergoes a process to a final state where $p_2 = 138$ kPa, $V_2 = 0.12\ m^3$. During the process, the paddle wheel transfers energy *to* the air by work in the amount 1.06 kJ, while the air transfers energy by work *to* the piston in the amount 3.5 kJ. Assuming ideal gas behavior, determine for the air (a) the temperature at state 2, in K and (b) the heat transfer, in kJ.

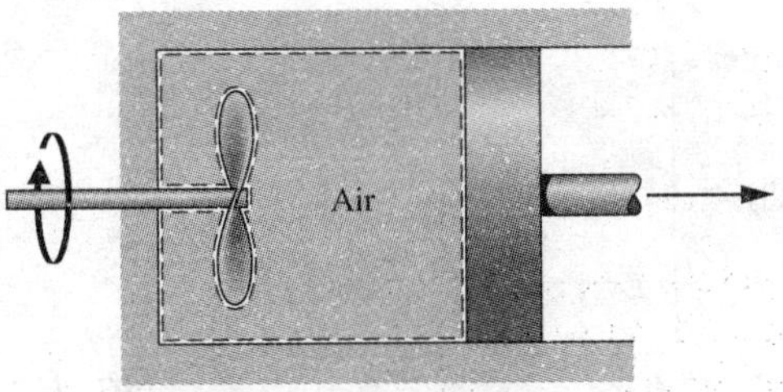

Initially, $p_1 = 207$ kPa, $T_1 = 282°C$, $V_1 = 0.108\ m^3$.
Finally, $p_2 = 138$ kPa, $V_2 = 0.12\ m^3$.

Fig. P3.113

3.114 Ammonia is contained in a rigid, well-insulated container. The initial pressure is 138 kPa, the mass is 0.05 kg, and the volume is $0.054\ m^3$. The gas is stirred by a paddle wheel, resulting in an energy transfer to the ammonia of magnitude 21 kJ. Assuming the ideal gas model, determine the final temperature of the ammonia, in K. Neglect kinetic and potential energy effects.

3.115 One kilogram of nitrogen fills the cylinder of a piston-cylinder assembly, as shown in Fig. P3.115. There is no friction between the piston and the cylinder walls, and the surroundings are at 1 atm. The initial volume and pressure in the cylinder are $1\ m^3$ and 1 atm, respectively. Heat transfer to the nitrogen occurs until the volume is doubled. Determine the heat transfer for the process, in kJ, assuming the specific heat ratio is constant, $k = 1.4$.

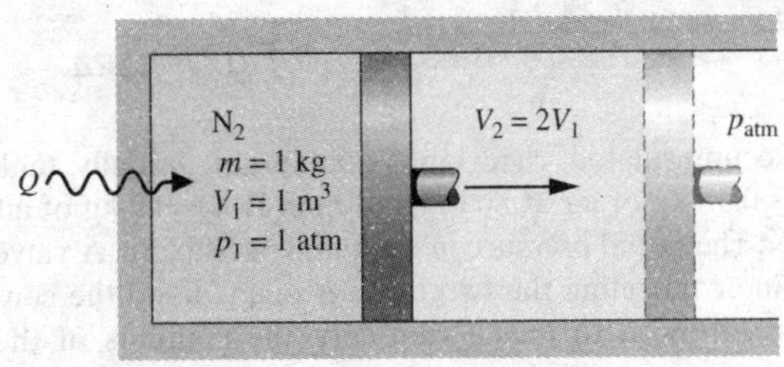

Fig. P3.115

3.116 A piston-cylinder assembly contains air at a pressure of 207 kPa and a volume of $0.02\ m^3$. The air is heated at constant pressure until its volume is doubled. Assuming the ideal gas model with constant specific heat ratio, $k = 1.4$, for the air, determine the work and heat transfer, each in kJ.

3.117 As shown in Fig. P3.117, a fan drawing electricity at a rate of 1.5 kW is located within a rigid enclosure, measuring 3 m × 4 m × 5 m. The enclosure is filled with air, initially at 27°C, 0.1 MPa. The fan operates steadily for 30 minutes. Assuming the ideal gas model, determine for the air (a) the mass, in kg, (b) the final temperature, in °C, and (c) the final pressure, in MPa. There is no heat transfer between the enclosure and the surroundings. Ignore the volume occupied by the fan itself and assume the fan stores no energy.

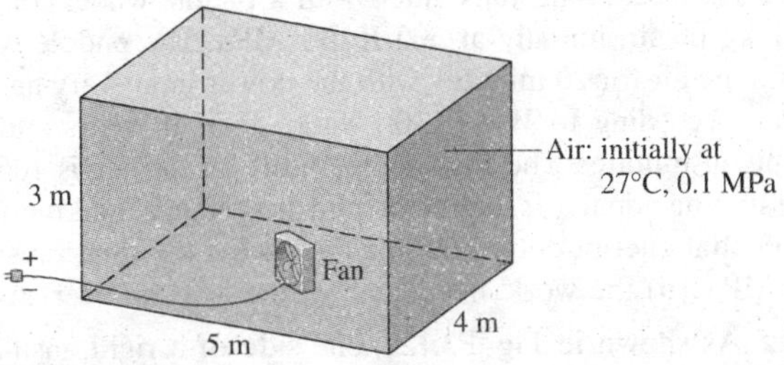

Fig. P3.117

3.118 A piston-cylinder assembly contains nitrogen (N_2), initially at 138 kPa, 27°C and occupying a volume of $0.054\ m^3$. The nitrogen is compressed to a final state where the pressure is 1.1 MPa and the temperature is 149°C. During compression, heat transfer of magnitude 1.69 kJ occurs *from* the nitrogen *to* its surroundings. Assuming ideal gas behavior, determine for the nitrogen, (a) the mass, in kg, and (b) the work, in kJ.

3.119 As shown in Fig. P3.119, a piston-cylinder assembly whose piston is resting on a set of stops contains 0.5 kg of helium gas, initially at 100 kPa and 25°C. The mass of the piston and the effect of the atmospheric pressure acting on the piston are such that a pressure of 500 kPa is required to raise it. How much energy must be transferred by heat to the helium, in kJ, before the piston starts rising? Assume ideal gas behavior for the helium.

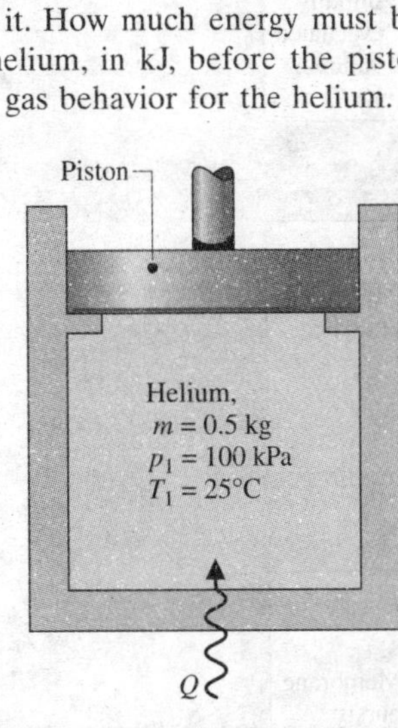

Fig. P3.119

3.120 As shown in Fig. P3.120, a tank fitted with an electrical resistor of negligible mass holds 2 kg of nitrogen (N_2), initially at 27°C, 0.1 MPa. Over a period of 10 minutes, electricity is provided to the resistor at a rate of 0.12 kW. During this same period, a heat transfer of magnitude 12.59 kJ occurs *from* the nitrogen *to* its surroundings. Assuming ideal gas behavior, determine the nitrogen's final temperature, in °C, and final pressure, in MPa.

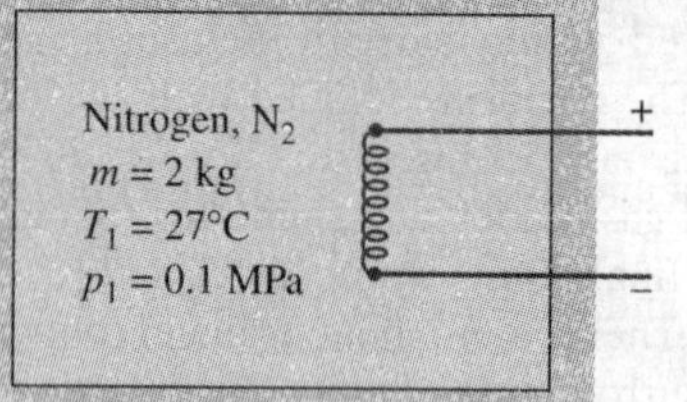

Fig. P3.120

3.121 A closed, rigid tank fitted with a paddle wheel contains 0.1 kg of air, initially at 300 K, 0.1 MPa. The paddle wheel stirs the air for 20 minutes, with the power input varying with time according to $\dot{W} = -10t$, where $\dot{W}$ is in watts and t is time, in minutes. The final temperature of the air is 1060 K. Assuming ideal gas behavior and no change in kinetic or potential energy, determine for the air (a) the final pressure, in MPa, (b) the work, in kJ, and (c) the heat transfer, in kJ.

3.122 As shown in Fig. P3.122, one side of a rigid, insulated container initially holds 2 m³ of air at 27°C, 0.3 MPa. The air is separated by a thin membrane from an evacuated volume of 3 m³. Owing to the pressure of the air, the membrane stretches and eventually bursts, allowing the air to occupy the full volume. Assuming the ideal gas model for the air, determine (a) the mass of the air, in kg, (b) the final temperature of the air, in K, and (c) the final pressure of the air, in MPa.

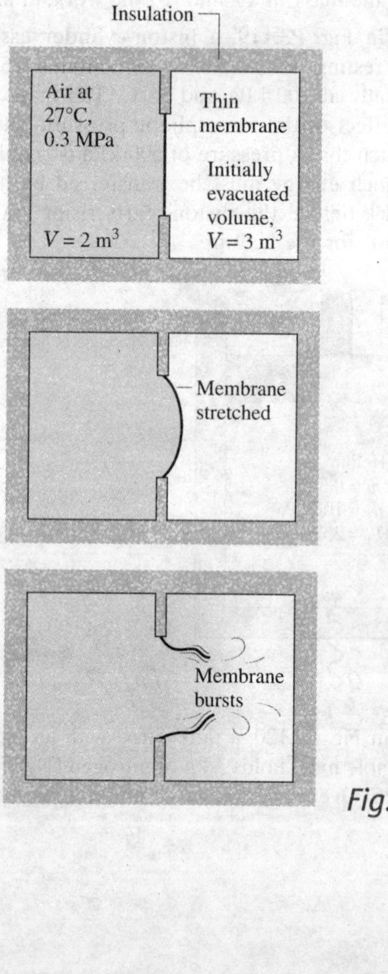

Fig. P3.122

3.123 Air is confined to one side of a rigid container divided by a partition, as shown in Fig. P3.123. The other side is initially evacuated. The air is initially at $p_1 = 5$ bar, $T_1 = 500$ K, and $V_1 = 0.2$ m³. When the partition is removed, the air expands to fill the entire chamber. Measurements show that $V_2 = 2\,V_1$ and $p_2 = p_1/4$. Assuming the air behaves as an ideal gas, determine (a) the final temperature, in K, and (b) the heat transfer, kJ.

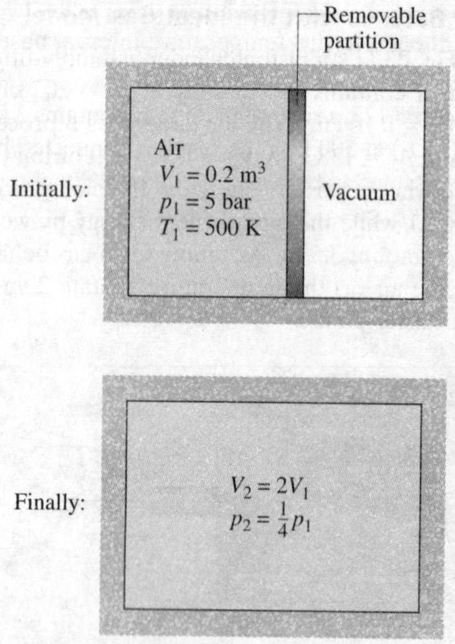

Fig. P3.123

3.124 Two kilograms of air, initially at 5 bar, 350 K and 4 kg of carbon monoxide (CO) initially at 2 bar, 450 K are confined to opposite sides of a rigid, well-insulated container by a partition, as shown in Fig. P3.124. The partition is free to move and allows conduction from one gas to the other without energy storage in the partition itself. The air and CO each behave as ideal gases with constant specific heat ratio, $k = 1.395$. Determine at equilibrium (a) the temperature, in K, (b) the pressure, in bar, and (c) the volume occupied by each gas, in m³.

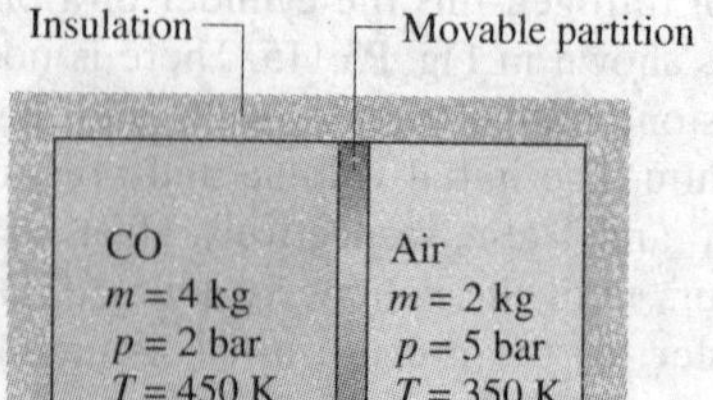

Fig. P3.124

3.125 Two uninsulated, rigid tanks contain air. Initially, tank A holds 0.45 kg of air at 800 K and tank B has 0.9 kg of air at 500 K. The initial pressure in each tank is 345 kPa. A valve in the line connecting the two tanks is opened and the contents are allowed to mix. Eventually, the contents of the tanks come to equilibrium at the temperature of the surroundings, 289 K. Assuming the ideal gas model, determine the amount of energy transfer by heat, in kJ, and the final pressure, in kPa.

3.126 Two kilograms of a gas with molecular weight 28 are contained in a closed, rigid tank fitted with an electric resistor. The resistor draws a constant current of 10 amps at a voltage of 12 V for 10 minutes. Measurements indicate that when equilibrium is reached, the temperature of the gas has increased by 40.3°C. Heat transfer to the surroundings is estimated to occur at a constant rate of 20 W. Assuming ideal gas behavior, determine an average value of the specific heat c_p, in kJ/kg · K, of the gas in this temperature interval based on the measured data.

3.127 As shown in Fig. P3.127, a rigid tank initially contains 3 kg of carbon dioxide (CO_2) at 500 kPa. The tank is connected by a valve to a piston-cylinder assembly located vertically above, initially containing 0.05 m³ of CO_2. Although the valve is closed, a slow leak allows CO_2 to flow into the cylinder from the tank until the tank pressure falls to 200 kPa. The weight of the piston and the pressure of the atmosphere maintain a constant pressure of 200 kPa in the cylinder. Owing to heat transfer, the temperature of the CO_2 throughout the tank and cylinder stays constant at 290 K. Assuming ideal gas behavior, determine for the CO_2 the work and heat transfer, each in kJ.

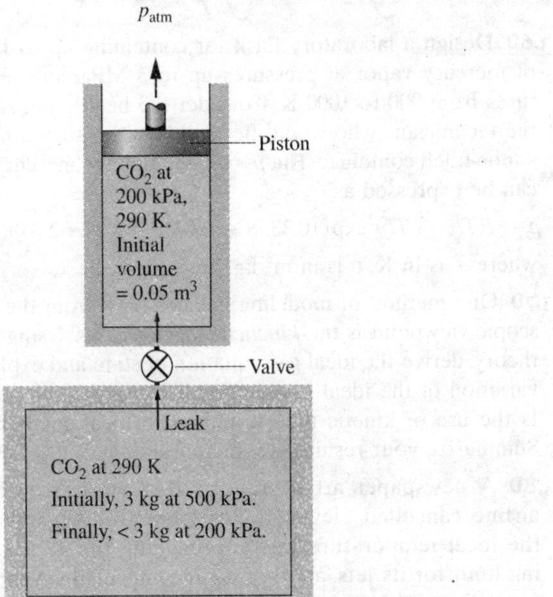

Fig. P3.127

3.128 A rigid tank, with a volume of 0.054 m³, contains air initially at 138 kPa, 278 K. If the air receives a heat transfer of magnitude 6.3 kJ, determine the final temperature, in K and the final pressure, in kPa. Assume ideal gas behavior, and use

(a) a constant specific heat value from Table A-20.
(b) a specific heat function from Table A-21.
(c) data from Table A-22.

3.129 Air is compressed adiabatically from $p_1 = 1$ bar, $T_1 = 300$ K to $p_2 = 15$ bar, $v_2 = 0.1227$ m³/kg. The air is then cooled at constant volume to $T_3 = 300$ K. Assuming ideal gas behavior, and ignoring kinetic and potential energy effects, calculate the work for the first process and the heat transfer for the second process, each in kJ per kg of air. Solve the problem each of two ways:

(a) using data from Table A-22.
(b) using a constant specific heat evaluated at 300 K.

3.130 Air is compressed in a piston-cylinder assembly from $p_1 = 68.94$ kPa, $T_1 = 278$ K to a final volume of $V_2 = 0.027$ m³ in a process described by $pv^{1.25} = constant$. The mass of air is 0.23 kg. Assuming ideal gas behavior and neglecting kinetic and potential energy effects, determine the work and the heat transfer, each in kJ, using (a) constant specific heats evaluated at 278 K, and (b) data from Table A-22. Compare the results and discuss.

3.131 Helium (He) gas initially at 2 bar, 200 K undergoes a polytropic process, with $n = k$, to a final pressure of 14 bar. Determine the work and heat transfer for the process, each in kJ per kg of helium. Assume ideal gas behavior.

3.132 A piston-cylinder assembly contains carbon monoxide modeled as an ideal gas with constant specific heat ratio, $k = 1.4$. The carbon monoxide undergoes a polytropic expansion with $n = k$ from an initial state, where $T_1 = 93$°C and $p_1 = 275.8$ kPa, to a final state, where the volume is twice the initial volume. Determine (a) the final temperature, in °C, and final pressure, in kPa, and (b) the work and heat transfer, each in kJ/kg.

3.133 Two-tenths kmol of nitrogen (N_2) in a piston-cylinder assembly undergoes two processes in series as follows:

Process 1-2: Constant pressure at 5 bar from $V_1 = 1.33$ m³ to $V_2 = 1$ m³.

Process 2-3: Constant volume to $p_3 = 4$ bar.

Assuming ideal gas behavior and neglecting kinetic and potential energy effects, determine the work and heat transfer for each process, in kJ.

3.134 One kilogram of air in a piston-cylinder assembly undergoes two processes in series from an initial state where $p_1 = 0.5$ MPa, $T_1 = 227$°C:

Process 1-2: Constant-temperature expansion until the volume is twice the initial volume.

Process 2-3: Constant-volume heating until the pressure is again 0.5 MPa.

Sketch the two processes in series on a p–v diagram. Assuming ideal gas behavior, determine (a) the pressure at state 2, in MPa, (b) the temperature at state 3, in °C, and for each of the processes (c) the work and heat transfer, each in kJ.

3.135 A piston-cylinder assembly contains air modeled as an ideal gas with a constant specific heat ratio, $k = 1.4$. The air undergoes a power cycle consisting of four processes in series:

Process 1-2: Constant-temperature expansion at 600 K from $p_1 = 0.5$ MPa to $p_2 = 0.4$ MPa.

Process 2-3: Polytropic expansion with $n = k$ to $p_3 = 0.3$ MPa.

Process 3-4: Constant-pressure compression to $V_4 = V_1$.

Process 4-1: Constant-volume heating.

Sketch the cycle on a p–v diagram. Determine (a) the work and heat transfer for each process, in kJ/kg, and (d) the thermal efficiency.

3.136 One kilogram of air undergoes a power cycle consisting of the following processes:

Process 1-2: Constant volume from $p_1 = 138$ kPa, $T_1 = 278$ K to $T_2 = 455.6$ K

Process 2-3: Adiabatic expansion to $v_3 = 1.4v_2$

Process 3-1: Constant-pressure compression

Sketch the cycle on a p–v diagram. Assuming ideal gas behavior, determine

(a) the pressure at state 2, in kPa.
(b) the temperature at state 3, in K.
(c) the thermal efficiency of the cycle.

3.137 A system consists of 2 kg of carbon dioxide gas initially at state 1, where $p_1 = 1$ bar, $T_1 = 300$ K. The system undergoes a power cycle consisting of the following processes:

Process 1-2: Constant volume to p_2, $p_2 > p_1$

Process 2-3: Expansion with $pv^{1.28} = constant$

Process 3-1: Constant-pressure compression

Assuming the ideal gas model and neglecting kinetic and potential energy effects,

(a) sketch the cycle on a p–v diagram.
(b) plot the thermal efficiency versus p_2/p_1 ranging from 1.05 to 4.

3.138 Air undergoes a polytropic process in a piston–cylinder assembly from $p_1 = 101.3$ kPa, $T_1 = 21°C$ to $p_2 = 690$ kPa. Using *IT*, plot the work and heat transfer, each in kJ per kg of air, for polytropic exponents ranging from 1.0 to 1.6. Investigate the error in the heat transfer introduced by assuming constant c_v evaluated at 21°C. Discuss.

3.139 Steam, initially at 5 MPa, 280°C undergoes a polytropic process in a piston–cylinder assembly to a final pressure of 20 MPa. Using *IT*, plot the heat transfer, in kJ per kg of steam, for polytropic exponents ranging from 1.0 to 1.6. Investigate the error in the heat transfer introduced by assuming ideal gas behavior for the steam. Discuss.

Design & open ended problems: Exploring engineering practice

3.1D Dermatologists remove skin blemishes from patients by applying sprays from canisters filled with liquid nitrogen (N_2). Investigate how liquid nitrogen is produced and delivered to physicians, and how physicians manage liquid nitrogen in their practices. Also investigate the advantages and disadvantages of this approach for removing blemishes compared to alternative approaches used today. Write a report including at least three references.

3.2D Although used for refrigerators and freezers elsewhere on the globe, hydrocarbon refrigerants have not taken hold in the United States thus far owing to concerns about liability if there is an accident. Research hydrocarbon refrigerant safety. Write a report including at least three references.

3.3D Methane-laden gas generated by the decomposition of landfill trash has economic value. Research literature on the possible uses of landfill gas. Contact the manager of a large landfill in your locale concerning the capture and use, if any, of landfill gas generated by the facility. Write a report including at least three references.

3.4D Metallurgists use phase diagrams to study *allotropic transformations,* which are phase transitions within the solid region. What features of the phase behavior of solids are important in the fields of metallurgy and materials processing? Write a report including at least three references.

3.5D Use of the *constant-volume* calorimeter to measure the calorie value of foods and other substances, as illustrated in Example 3.6, is one of three general types of biological calorimetry. Two others are *differential scanning calorimetry* and *isothermal titration calorimetry.* For these two, investigate the objectives and instrumentation of each. Present your findings in a memorandum, including sketches of the respective instrumentation together with summaries of how the instrumentation is used in the laboratory.

3.6D Design a laboratory flask for containing up to 10 kmol of mercury vapor at pressures up to 3 MPa, and temperatures from 900 to 1000 K. Consider the health and safety of the technicians who would be working with such a mercury vapor-filled container. The p–v–T relation for mercury vapor can be expressed as

$$p = RT/v - T/v^2 \exp(10.3338 - 0.0312095/T - 2.07950 \ln T)$$

where T is in K, v is in m³/kg, and p is in Pa.

3.7D One method of modeling gas behavior from the microscopic viewpoint is the *kinetic theory of gases.* Using kinetic theory, derive the ideal gas equation of state and explain the variation of the ideal gas specific heat c_v with temperature. Is the use of kinetic theory limited to ideal gas behavior? Summarize your results as a memorandum.

3.8D A newspaper article reports that on a day when an airline cancelled eleven flights from Las Vegas because the local temperature was approaching the 47°C operating limit for its jets, another airline canceled seven flights from Denver because the local temperature there was above the 40°C operating level for its propeller planes. Prepare a 30-min. presentation suitable for a middle school science class explaining the technical considerations behind these cancellations.

3.9D Following a rash of vehicle rollover accidents linked to under-inflated tires, the U.S. Congress enacted legislation mandating that new vehicles weighing less than 4500 kg be equipped with a tire-pressure warning system, and directed the U.S. National Highway Traffic Safety Administration to issue a rule. The rule is expected to require car manufacturers to install a four-tire pressure-monitoring system that can warn drivers when a tire is dangerously under-inflated. *Microelectromechanical* transducers are expected to play a role in such systems. Investigate the technical options for tire pressure-monitoring systems, including the merits of *battery-free* systems. Write a report including at least three references.

3.10D Carbon-dioxide capture and storage is under consideration today for lessening the effects on global warming and climate change traceable to burning fossil fuels. One frequently mentioned approach, shown in Fig. 3.10D, is to remove CO_2 from the exhaust gas of a power plant and inject it into an underground geological formation. Critically evaluate this approach for large-scale management of CO_2 generated by human activity, including technical aspects and related costs. Consider the leading methods for CO_2 capture from gas streams, issues related to injecting the CO_2 at great depths, and the consequences of CO_2 migration from storage. Formulate a position in favor of, or in opposition to, the deployment of such technology. Write a report including at least three references.

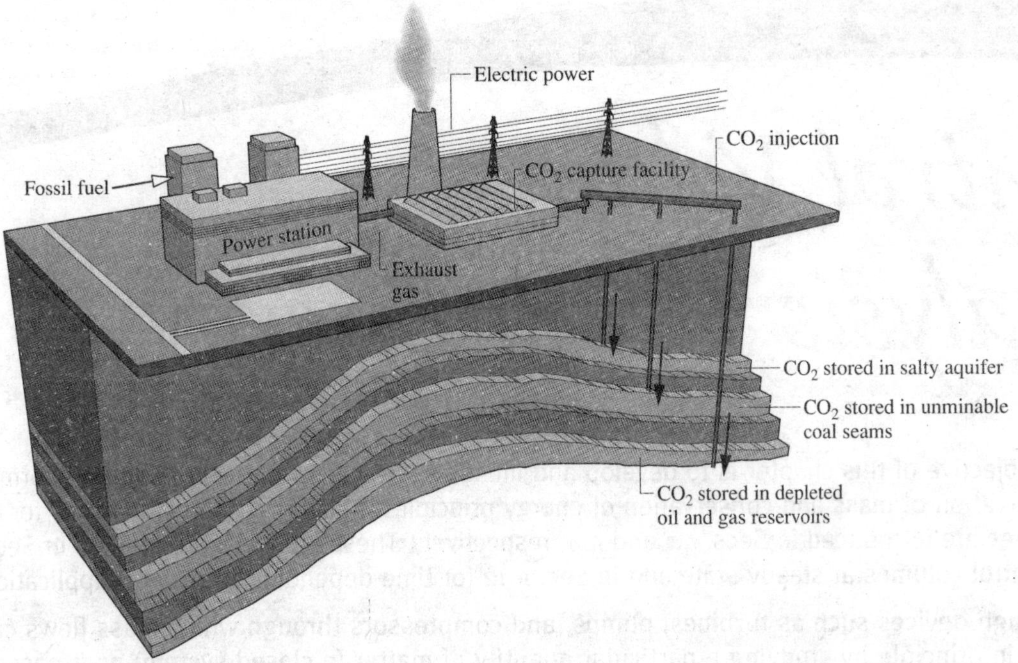

Fig. P3.10D

4

Control Volume Analysis Using Energy

Engineering Context

The **objective** of this chapter is to develop and illustrate the use of the control volume forms of the conservation of mass and conservation of energy principles. Mass and energy balances for control volumes are introduced in Secs. 4.1 and 4.4, respectively. These balances are applied in Secs. 4.5–4.11 to control volumes at steady state and in Sec. 4.12 for time-dependent (transient) applications.

Although devices such as turbines, pumps, and compressors through which mass flows can be analyzed in principle by studying a particular quantity of matter (a closed system) as it passes through the device, it is normally preferable to think of a region of space through which mass flows (a control volume). As in the case of a closed system, energy transfer across the boundary of a control volume can occur by means of work and heat. In addition, another type of energy transfer must be accounted for—the energy accompanying mass as it enters or exits.

Learning Objectives

When you complete your study of this chapter you will be able to...

- ✔ demonstrate understanding of key concepts related to control volume analysis, including distinguishing between steady-state and transient analysis, distinguishing between mass flow rate and volumetric flow rate, and the meanings of one-dimensional flow and flow work.
- ✔ apply mass and energy balances to control volumes.
- ✔ develop appropriate engineering models for control volumes, with particular attention to analyzing components commonly encountered in engineering practice such as nozzles, diffusers, turbines, compressors, heat exchangers, throttling devices, and integrated systems that incorporate two or more components.
- ✔ use property data in control volume analysis appropriately.

4.1 Conservation of Mass for a Control Volume

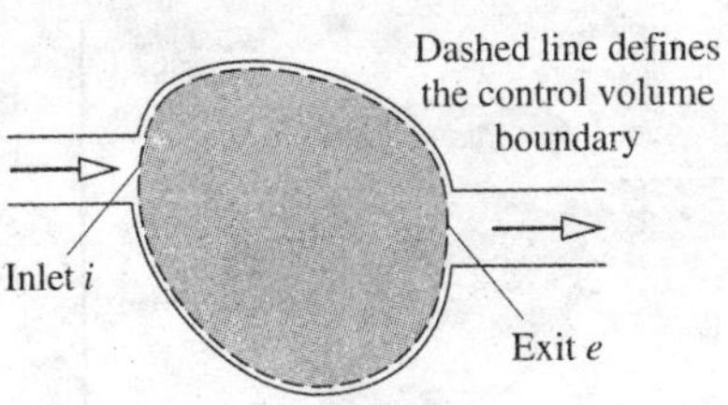

Fig. 4.1 One-inlet, one-exit control volume.

In this section an expression of the conservation of mass principle for control volumes is developed and illustrated. As a part of the presentation, the one-dimensional flow model is introduced.

4.1.1 Developing the Mass Rate Balance

The mass rate balance for control volumes is introduced by reference to Fig. 4.1, which shows a control volume with mass flowing in at i and flowing out at e, respectively. When applied to such a control volume, the ***conservation of mass*** principle states

conservation of mass

$$\begin{bmatrix} \text{time } \textit{rate of change} \text{ of} \\ \text{mass contained within} \\ \text{the control volume } \textit{at time } t \end{bmatrix} = \begin{bmatrix} \text{time } \textit{rate} \text{ of flow} \\ \text{of mass } \textit{in} \text{ across} \\ \text{inlet } i \textit{ at time } t \end{bmatrix} - \begin{bmatrix} \text{time } \textit{rate} \text{ of flow} \\ \text{of mass } \textit{out} \text{ across} \\ \text{exit } e \textit{ at time } t \end{bmatrix}$$

Denoting the mass contained within the control volume at time t by $m_{\text{cv}}(t)$, this statement of the conservation of mass principle can be expressed in symbols as

$$\frac{dm_{\text{cv}}}{dt} = \dot{m}_i - \dot{m}_e \tag{4.1}$$

where dm_{cv}/dt is the time rate of change of mass within the control volume, and $\dot{m}_i$ and $\dot{m}_e$ are the instantaneous ***mass flow rates*** at the inlet and exit, respectively. As for the symbols $\dot{W}$ and $\dot{Q}$, the dots in the quantities $\dot{m}_i$ and $\dot{m}_e$ denote time rates of transfer. In SI, all terms in Eq. 4.1 are expressed in kg/s. For a discussion of the development of Eq. 4.1, see the box.

mass flow rates

In general, there may be several locations on the boundary through which mass enters or exits. This can be accounted for by summing, as follows

$$\frac{dm_{\text{cv}}}{dt} = \sum_i \dot{m}_i - \sum_e \dot{m}_e \tag{4.2}$$

mass rate balance

Equation 4.2 is the ***mass rate balance*** for control volumes with several inlets and exits. It is a form of the conservation of mass principle commonly employed in engineering. Other forms of the mass rate balance are considered in discussions to follow.

Developing the Control Volume Mass Balance

For each of the extensive properties mass, energy, and entropy (Chap. 6), the control volume form of the property balance can be obtained by transforming the corresponding closed system form. Let us consider this for mass, recalling that the mass of a closed system is constant.

The figures in the margin at the top of the next page show a system consisting of a fixed quantity of matter m that occupies different regions at time t and a later time $t + \Delta t$. The mass under consideration is shown in color on the figures. At time t, the mass is the sum $m = m_{\text{cv}}(t) + m_i$, where $m_{\text{cv}}(t)$ is the mass contained within the control volume, and m_i is the mass within the small region labeled i adjacent to the control volume. Let us study the fixed quantity of matter m as time elapses.

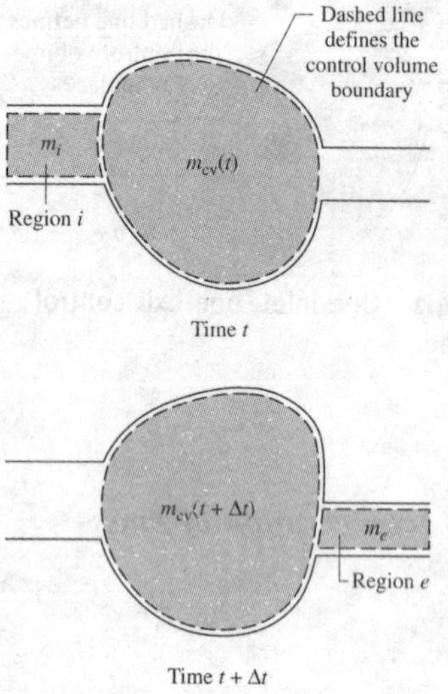

In a time interval Δt all the mass in region i crosses the control volume boundary, while some of the mass, call it m_e, initially contained within the control volume exits to fill the region labeled e adjacent to the control volume. Although the mass in regions i and e as well as in the control volume differ from time t to $t + \Delta t$, the *total* amount of mass is constant. Accordingly

$$m_{cv}(t) + m_i = m_{cv}(t + \Delta t) + m_e \tag{a}$$

or on rearrangement

$$m_{cv}(t + \Delta t) - m_{cv}(t) = m_i - m_e \tag{b}$$

Equation (b) is an *accounting* balance for mass. It states that the change in mass of the control volume during time interval Δt equals the amount of mass that enters less the amount of mass that exits.

Equation (b) can be expressed on a time rate basis. First, divide by Δt to obtain

$$\frac{m_{cv}(t + \Delta t) - m_{cv}(t)}{\Delta t} = \frac{m_i}{\Delta t} - \frac{m_e}{\Delta t} \tag{c}$$

Then, in the limit as Δt goes to zero, Eq. (c) becomes Eq. 4.1, the instantaneous *control volume rate equation* for mass

$$\frac{dm_{cv}}{dt} = \dot{m}_i - \dot{m}_e \tag{4.1}$$

where dm_{cv}/dt denotes the time rate of change of mass within the control volume, and $\dot{m}_i$ and $\dot{m}_e$ are the inlet and exit mass flow rates, respectively, all at time t.

4.1.2 Evaluating the Mass Flow Rate

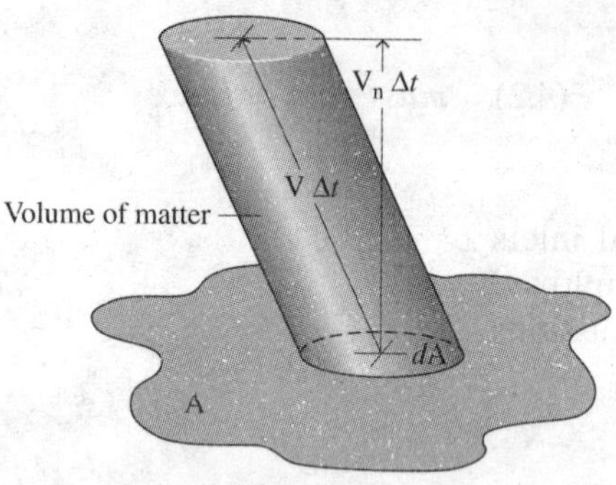

Fig. 4.2 Illustration used to develop an expression for mass flow rate in terms of local fluid properties.

An expression for the mass flow rate $\dot{m}$ of the matter entering or exiting a control volume can be obtained in terms of local properties by considering a small quantity of matter flowing with velocity V across an incremental area dA in a time interval Δt, as shown in Fig. 4.2. Since the portion of the control volume boundary through which mass flows is not necessarily at rest, the velocity shown in the figure is understood to be the velocity *relative* to the area dA. The velocity can be resolved into components normal and tangent to the plane containing dA. In the following development V_n denotes the component of the relative velocity normal to dA in the direction of flow.

The *volume* of the matter crossing dA during the time interval Δt shown in Fig. 4.2 is an oblique cylinder with a volume equal to the product of the area of its base dA and its altitude $V_n \Delta t$. Multiplying by the density ρ gives the amount of mass that crosses dA in time Δt

$$\begin{bmatrix} \text{amount of mass} \\ \text{crossing } dA \text{ during} \\ \text{the time interval } \Delta t \end{bmatrix} = \rho(V_n \Delta t)\, dA$$

Dividing both sides of this equation by Δt and taking the limit as Δt goes to zero, the instantaneous mass flow rate across incremental area dA is

$$\begin{bmatrix} \text{instantaneous rate} \\ \text{of mass flow} \\ \text{across } dA \end{bmatrix} = \rho V_n\, dA$$

When this is integrated over the area A through which mass passes, an expression for the mass flow rate is obtained

$$\dot{m} = \int_A \rho V_n \, dA \tag{4.3}$$

Equation 4.3 can be applied at the inlets and exits to account for the rates of mass flow into and out of the control volume.

4.2 Forms of the Mass Rate Balance

The mass rate balance, Eq. 4.2, is a form that is important for control volume analysis. In many cases, however, it is convenient to apply the mass balance in forms suited to particular objectives. Some alternative forms are considered in this section.

4.2.1 One-Dimensional Flow Form of the Mass Rate Balance

When a flowing stream of matter entering or exiting a control volume adheres to the following idealizations, the flow is said to be ***one-dimensional:***

one-dimensional flow

- ▶ The flow is normal to the boundary at locations where mass enters or exits the control volume.
- ▶ *All* intensive properties, including velocity and density, are *uniform with position* (bulk average values) over each inlet or exit area through which matter flows.

Take Note...
In subsequent control volume analyses, we routinely assume that the idealizations of one-dimensional flow are appropriate. Accordingly the assumption of one-dimensional flow is not listed explicitly in solved examples.

FOR EXAMPLE... Figure 4.3 illustrates the meaning of one-dimensional flow. The area through which mass flows is denoted by A. The symbol V denotes a single value that represents the velocity of the flowing air. Similarly T and v are single values that represent the temperature and specific volume, respectively, of the flowing air.

When the flow is one-dimensional, Eq. 4.3 for the mass flow rate becomes

$$\dot{m} = \rho A V \quad \text{(one-dimensional flow)} \tag{4.4a}$$

or in terms of specific volume

$$\dot{m} = \frac{AV}{v} \quad \text{(one-dimensional flow)} \tag{4.4b}$$

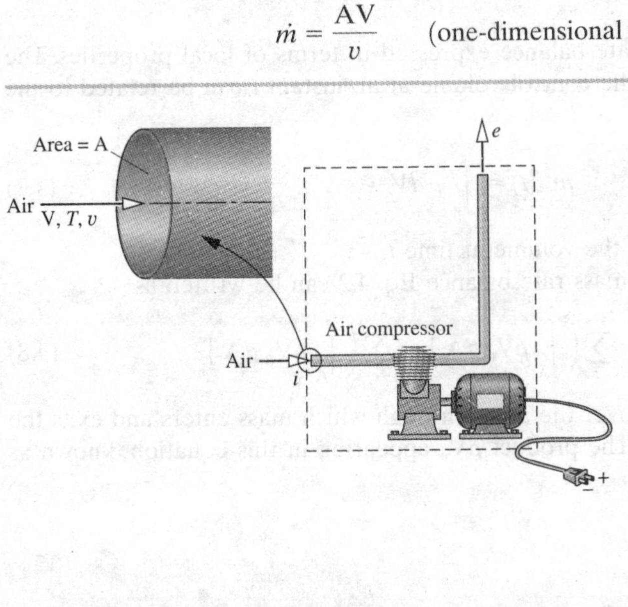

Fig. 4.3 Figure illustrating the one-dimensional flow model.

When area is in m^2, velocity is in m/s, and specific volume is in m^3/kg, the mass flow rate found from Eq. 4.4b is in kg/s, as can be verified. The product AV in Eqs. 4.4 is the ***volumetric flow rate.*** The volumetric flow rate is expressed in units of m^3/s.

volumetric flow rate

Substituting Eq. 4.4b into Eq. 4.2 results in an expression for the conservation of mass principle for control volumes limited to the case of one-dimensional flow at the inlets and exits

$$\frac{dm_{cv}}{dt} = \sum_i \frac{A_i V_i}{v_i} - \sum_e \frac{A_e V_e}{v_e} \quad \text{(one-dimensional flow)} \tag{4.5}$$

Note that Eq. 4.5 involves summations over the inlets and exits of the control volume. Each individual term in these sums applies to a particular inlet or exit. The area, velocity, and specific volume appearing in each term refers only to the corresponding inlet or exit.

4.2.2 Steady-State Form of the Mass Rate Balance

steady state

Many engineering systems can be idealized as being at ***steady state,*** meaning that *all* properties are unchanging in time. For a control volume at steady state, the identity of the matter within the control volume changes continuously, but the total amount present at any instant remains constant, so $dm_{cv}/dt = 0$ and Eq. 4.2 reduces to

$$\underset{\text{(mass rate in)}}{\sum_i \dot{m}_i} = \underset{\text{(mass rate out)}}{\sum_e \dot{m}_e} \tag{4.6}$$

That is, the total incoming and outgoing rates of mass flow are equal.

Note that equality of total incoming and outgoing rates of mass flow does not necessarily imply that a control volume is at steady state. Although the total amount of mass within the control volume at any instant would be constant, other properties such as temperature and pressure might be varying with time. When a control volume is at steady state, *every* property is independent of time. Also note that the steady-state assumption and the one-dimensional flow assumption are independent idealizations. One does not imply the other.

4.2.3 Integral Form of the Mass Rate Balance

We consider next the mass rate balance expressed in terms of local properties. The total mass contained within the control volume at an instant t can be related to the local density as follows

$$m_{cv}(t) = \int_V \rho\, dV \tag{4.7}$$

where the integration is over the volume at time t.

With Eqs. 4.3 and 4.7, the mass rate balance Eq. 4.2 can be written as

$$\frac{d}{dt}\int_V \rho\, dV = \sum_i \left(\int_A \rho V_n\, dA\right)_i - \sum_e \left(\int_A \rho V_n\, dA\right)_e \tag{4.8}$$

where the area integrals are over the areas through which mass enters and exits the control volume, respectively. The product ρV_n appearing in this equation, known as the ***mass flux,*** gives the time rate of mass flow per unit of area. To evaluate the terms of the right side of Eq. 4.8 requires information about the variation of the mass flux over the flow areas. The form of the conservation of mass principle given by Eq. 4.8 is usually considered in detail in fluid mechanics. ***mass flux***

4.3 Applications of the Mass Rate Balance

4.3.1 Steady-State Application

For a control volume at steady state, the conditions of the mass within the control volume and at the boundary do not vary with time. The mass flow rates also are constant with time.

Example 4.1 illustrates an application of the steady-state form of the mass rate balance to a control volume enclosing a mixing chamber called a *feedwater heater.* Feedwater heaters are components of the vapor power systems considered in Chap. 8.

Example 4.1 FEEDWATER HEATER AT STEADY STATE

A feedwater heater operating at steady state has two inlets and one exit. At inlet 1, water vapor enters at $p_1 = 7$ bar, $T_1 = 200°C$ with a mass flow rate of 40 kg/s. At inlet 2, liquid water at $p_2 = 7$ bar, $T_2 = 40°C$ enters through an area $A_2 = 25\ cm^2$. Saturated liquid at 7 bar exits at 3 with a volumetric flow rate of $0.06\ m^3/s$. Determine the mass flow rates at inlet 2 and at the exit, in kg/s, and the velocity at inlet 2, in m/s.

Solution

Known: A stream of water vapor mixes with a liquid water stream to produce a saturated liquid stream at the exit. The states at the inlets and exit are specified. Mass flow rate and volumetric flow rate data are given at one inlet and at the exit, respectively.

Find: Determine the mass flow rates at inlet 2 and at the exit, and the velocity V_2.

Schematic and Given Data:

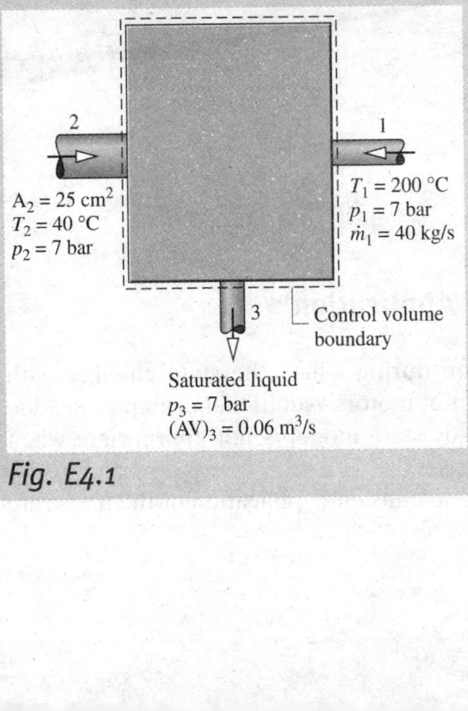

Fig. E4.1

Engineering Model: The control volume shown on the accompanying figure is at steady state.

Analysis: The principal relations to be employed are the mass rate balance (Eq. 4.2) and the expression $\dot{m} = AV/v$ (Eq. 4.4b). At steady state the mass rate balance becomes

❶
$$\frac{dm_{cv}}{dt}^{0} = \dot{m}_1 + \dot{m}_2 - \dot{m}_3$$

Solving for $\dot{m}_2$,

$$\dot{m}_2 = \dot{m}_3 - \dot{m}_1$$

The mass flow rate $\dot{m}_1$ is given. The mass flow rate at the exit can be evaluated from the given volumetric flow rate

$$\dot{m}_3 = \frac{(AV)_3}{v_3}$$

where v_3 is the specific volume at the exit. In writing this expression, one-dimensional flow is assumed. From Table A-3, $v_3 = 1.108 \times 10^{-3}$ m^3/kg. Hence,

$$\dot{m}_3 = \frac{0.06 \text{ m}^3/\text{s}}{(1.108 \times 10^{-3} \text{ m}^3/\text{kg})} = 54.15 \text{ kg/s}$$

The mass flow rate at inlet 2 is then

$$\dot{m}_2 = \dot{m}_3 - \dot{m}_1 = 54.15 - 40 = 14.15 \text{ kg/s}$$

For one-dimensional flow at 2, $\dot{m}_2 = A_2V_2/v_2$, so

$$V_2 = \dot{m}_2 v_2 / A_2$$

State 2 is a compressed liquid. The specific volume at this state can be approximated by $v_2 \approx v_f(T_2)$ (Eq. 3.11). From Table A-2 at 40°C, $v_2 = 1.0078 \times 10^{-3}$ m^3/kg. So,

$$V_2 = \frac{(14.15 \text{ kg/s})(1.0078 \times 10^{-3} \text{ m}^3/\text{kg})}{25 \text{ cm}^2}\left|\frac{10^4 \text{ cm}^2}{1 \text{ m}^2}\right| = 5.7 \text{ m/s}$$

❶ In accord with Eq. 4.6, the mass flow rate at the exit equals the sum of the mass flow rates at the inlets. It is left as an exercise to show that the volumetric flow rate at the exit *does not equal* the sum of the volumetric flow rates at the inlets.

Skills Developed: *Ability to...*
- ✓ apply the steady-state mass rate balance.
- ✓ apply the mass flow rate expression, Eq. 4.4b.
- ✓ retrieve property data for water.

Quick Quiz

Evaluate the volumetric flow rate, in m^3/s, at each inlet.
Ans. $(AV)_1 = 12$ m^3/s, $(AV)_2 = 0.01$ m^3/s

4.3.2 Time-Dependent (transient) Application

Many devices undergo periods of operation during which the state changes with time—for example, the startup and shutdown of motors. Additional examples include containers being filled or emptied. The steady-state model is not appropriate when analyzing such time-dependent (transient) cases.

Example 4.2 illustrates a time-dependent, or transient, application of the mass rate balance. In this case, a barrel is filled with water.

Example 4.2 FILLING A BARREL WITH WATER

Water flows into the top of an open barrel at a constant mass flow rate of 7 kg/s. Water exits through a pipe near the base with a mass flow rate proportional to the height of liquid inside: $\dot{m}_e = 1.4\ L$, where L is the instantaneous liquid height, in m. The area of the base is 0.2 m^2, and the density of water is 1000 kg/m^3. If the barrel is initially empty, plot the variation of liquid height with time and comment on the result.

Solution

Known: Water enters and exits an initially empty barrel. The mass flow rate at the inlet is constant. At the exit, the mass flow rate is proportional to the height of the liquid in the barrel.

Find: Plot the variation of liquid height with time and comment.

Schematic and Given Data:

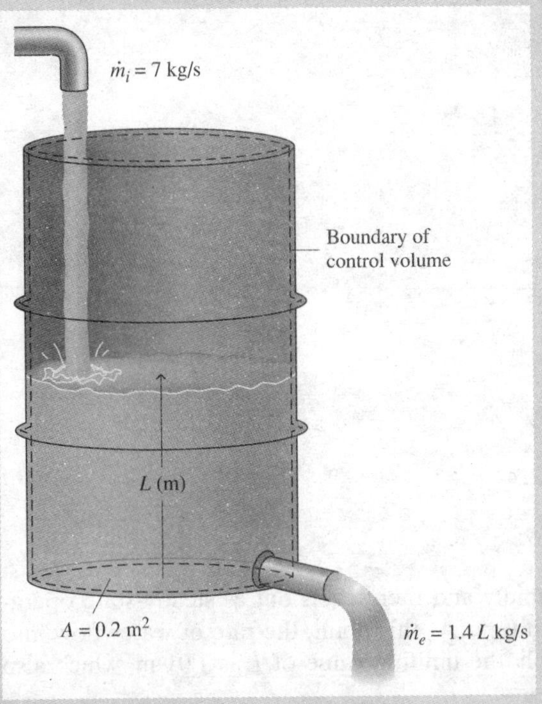

Fig. E4.2a

Engineering Model:

1. The control volume is defined by the dashed line on the accompanying diagram.
2. The water density is constant.

Analysis: For the one-inlet, one-exit control volume, Eq. 4.2 reduces to

$$\frac{dm_{cv}}{dt} = \dot{m}_i - \dot{m}_e$$

The mass of water contained within the barrel at time t is given by

$$m_{cv}(t) = \rho A L(t)$$

where ρ is density, A is the area of the base, and $L(t)$ is the instantaneous liquid height. Substituting this into the mass rate balance together with the given mass flow rates

$$\frac{d(\rho A L)}{dt} = 7 - 1.4\,L$$

Since density and area are constant, this equation can be written as

$$\frac{dL}{dt} + \left(\frac{1.4}{\rho A}\right)L = \frac{7}{\rho A}$$

which is a first-order, ordinary differential equation with constant coefficients. The solution is

1 $$L = 5 + C\exp\left(-\frac{-1.4t}{\rho A}\right)$$

where C is a constant of integration. The solution can be verified by substitution into the differential equation.

To evaluate C, use the initial condition: at $t = 0$, $L = 0$. Thus, $C = -5.0$, and the solution can be written as

$$L = 5.0[1 - \exp(-1.4t/\rho A)]$$

Substituting $\rho = 1000\ \text{kg/m}^3$ and $A = 0.2\ \text{m}^2$ results in

$$L = 5[1 - \exp(-0.007t)]$$

This relation can be plotted by hand or using appropriate software. The result is

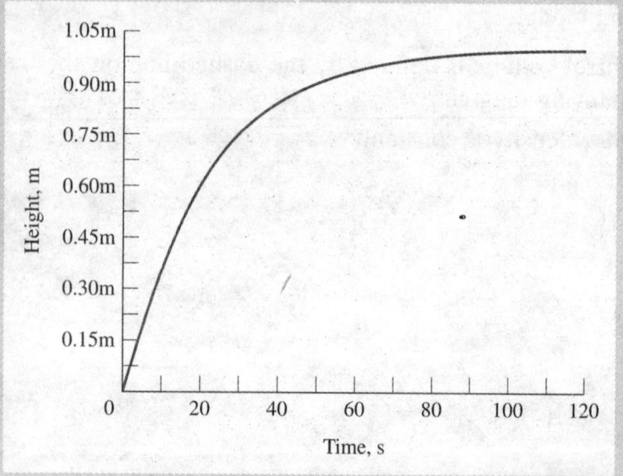

Fig. E4.2b

From the graph, we see that initially the liquid height increases rapidly and then levels out as steady-state operation is approached. After about 100 s, the height stays constant with time. At this point, the rate of water flow into the barrel equals the rate of flow out of the barrel. From the graph, the limiting value of L is 1.01 m, which also can be verified by taking the limit of the analytical solution as $t \to \infty$.

---

1 Alternatively, this differential equation can be solved using *Interactive Thermodynamics: IT*. The differential equation can be expressed as

```
der(L, t) + (1.4 * L)/(rho * A) = 7/(rho * A)
rho = 1000 // kg/m³
A = 0.2 // m²
```

where der(L,t) is dL/dt, rho is density ρ, and A is area. Using the **Explore** button, set the initial condition at $L = 0$, and sweep t from 0 to 200 in steps of 0.5. Then, the plot can be constructed using the **Graph** button.

Skills Developed: *Ability to...*
- ✓ apply the time-dependent mass rate balance.
- ✓ solve an ordinary differential equation and plot the solution.

Quick Quiz

If the mass flow rate of the water flowing into the barrel were 12.25 kg/s while all other data remains the same, what is the limiting value of the liquid height, L, in m?

Ans. 0.9 m

Bio... connections

The human heart provides a good example of how biological systems can be modeled as control volumes. Figure 4.4 shows the cross-section of a human heart. The flow is controlled by valves that intermittently allow blood to enter from veins and exit through arteries as the heart muscles pump. Work is done to increase the pressure of the blood leaving the heart to a level that will propel it through the cardiovascular system of the body. Observe that the boundary of the control volume enclosing the heart is not fixed but moves with time as the heart pulses.

Understanding the medical condition known as *arrhythmia* requires consideration of the time-dependent behavior of the heart. An arrhythmia is a change in the regular beat of the heart. This can take several forms. The heart may beat irregularly, skip a beat, or beat very fast or slow. An arrhythmia may be detectable by listening to the heart with a stethoscope, but an electrocardiogram offers a more precise approach. Although arrhythmia does occur in people without underlying heart disease, patients having serious symptoms may require treatment to keep their heartbeats regular. Many patients with arrhythmia may require no medical intervention at all.

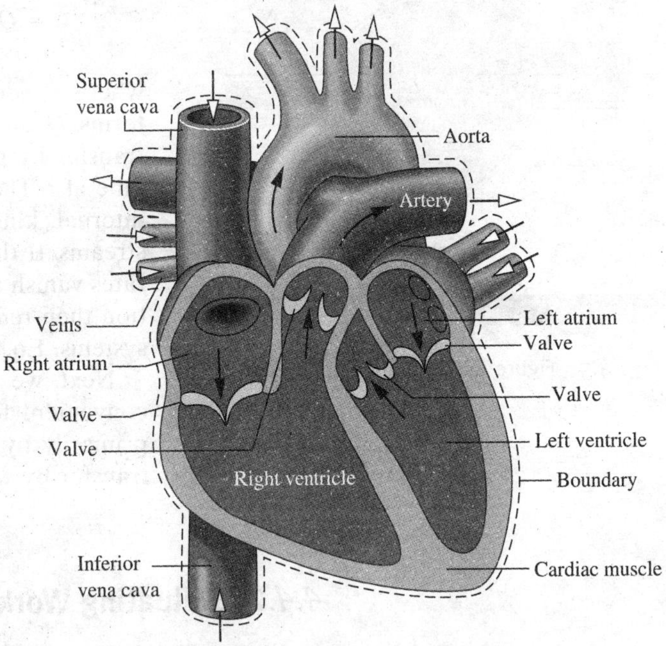

Fig. 4.4 Control volume enclosing the heart.

4.4 Conservation of Energy for a Control Volume

In this section, the rate form of the energy balance for control volumes is obtained. The energy rate balance plays an important role in subsequent sections of this book.

4.4.1 Developing the Energy Rate Balance for a Control Volume

We begin by noting that the control volume form of the energy rate balance can be derived by an approach closely paralleling that considered in the box of Sec. 4.1.1, where the control volume mass rate balance is obtained by transforming the closed system form. The present development proceeds less formally by arguing that, like mass, energy is an extensive property, so it too can be transferred into or out of a control volume as a result of mass crossing the boundary. Since this is the principal difference between the closed system and control volume forms, the control volume energy rate balance can be obtained by modifying the closed system energy rate balance to account for these energy transfers.

Accordingly, the *conservation of energy* principle applied to a control volume states:

$$\begin{bmatrix}\text{time } \textit{rate of change} \\ \text{of the energy} \\ \text{contained within} \\ \text{the control volume} \\ \textit{at time t}\end{bmatrix} = \begin{bmatrix}\text{net } \textit{rate} \text{ at which} \\ \text{energy is being} \\ \text{transferred in} \\ \text{by heat transfer} \\ \textit{at time t}\end{bmatrix} - \begin{bmatrix}\textit{net} \text{ rate at which} \\ \text{energy is being} \\ \text{transferred out} \\ \text{by work } \textit{at} \\ \textit{time t}\end{bmatrix} + \begin{bmatrix}\textit{net} \text{ rate of energy} \\ \text{transfer } \textit{into} \text{ the} \\ \text{control volume} \\ \text{accompanying} \\ \text{mass flow}\end{bmatrix}$$

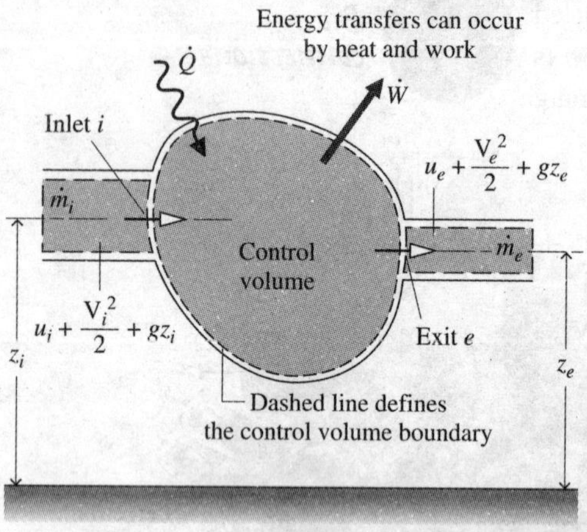

Fig. 4.5 Figure used to develop Eq. 4.9.

For the one-inlet one-exit control volume with one-dimensional flow shown in Fig. 4.5 the energy rate balance is

$$\frac{dE_{cv}}{dt} = \dot{Q} - \dot{W} + \underline{\dot{m}_i\left(u_i + \frac{V_i^2}{2} + gz_i\right)} - \underline{\dot{m}_e\left(u_e + \frac{V_e^2}{2} + gz_e\right)} \quad (4.9)$$

where E_{cv} denotes the energy of the control volume at time t. The terms $\dot{Q}$ and $\dot{W}$ account, respectively, for the net rate of energy transfer by heat and work across the boundary of the control volume at t. The underlined terms account for the rates of transfer of internal, kinetic, and potential energy of the entering and exiting streams. If there is no mass flow in or out, the respective mass flow rates vanish and the underlined terms of Eq. 4.9 drop out. The equation then reduces to the rate form of the energy balance for closed systems: Eq. 2.37.

Next, we will place Eq. 4.9 in an alternative form that is more convenient for subsequent applications. This will be accomplished primarily by recasting the work term $\dot{W}$, which represents the net rate of energy transfer by work across *all* portions of the boundary of the control volume.

4.4.2 Evaluating Work for a Control Volume

Because work is always done on or by a control volume where matter flows across the boundary, it is convenient to separate the work term $\dot{W}$ of Eq. 4.9 into *two contributions:* One contribution is the work associated with the fluid pressure as mass is introduced at inlets and removed at exits. The other contribution, denoted by $\dot{W}_{cv}$, includes *all other* work effects, such as those associated with rotating shafts, displacement of the boundary, and electrical effects.

Consider the work at an exit e associated with the pressure of the flowing matter. Recall from Eq. 2.13 that the rate of energy transfer by work can be expressed as the product of a force and the velocity at the point of application of the force. Accordingly, the *rate* at which work is done at the exit by the normal force (normal to the exit area in the direction of flow) due to pressure is the product of the normal force, $p_e A_e$, and the fluid velocity, V_e. That is

$$\begin{bmatrix}\text{time rate of energy transfer} \\ \text{by work } \textit{from} \text{ the control} \\ \text{volume at exit } e\end{bmatrix} = (p_e A_e)V_e \quad (4.10)$$

where p_e is the pressure, A_e is the area, and V_e is the velocity at exit e, respectively. A similar expression can be written for the rate of energy transfer by work into the control volume at inlet i.

With these considerations, the work term $\dot{W}$ of the energy rate equation, Eq. 4.9, can be written as

$$\dot{W} = \dot{W}_{cv} + (p_e A_e)V_e - (p_i A_i)V_i \quad (4.11)$$

where, in accordance with the sign convention for work, the term at the inlet has a negative sign because energy is transferred into the control volume there. A positive sign precedes the work term at the exit because energy is transferred out of the control volume there. With $AV = \dot{m}v$ from Eq. 4.4b, the above expression for work can be written as

$$\dot{W} = \dot{W}_{cv} + \dot{m}_e(p_e v_e) - \dot{m}_i(p_i v_i) \quad (4.12)$$

where $\dot{m}_i$ and $\dot{m}_e$ are the mass flow rates and v_i and v_e are the specific volumes evaluated at the inlet and exit, respectively. In Eq. 4.12, the terms $\dot{m}_i(p_i v_i)$ and $\dot{m}_e(p_e v_e)$ account for the work associated with the pressure at the inlet and exit, respectively. They are commonly referred to as ***flow work.*** The term $\dot{W}_{cv}$ accounts for *all other* energy transfers by work across the boundary of the control volume.

flow work

4.4.3 One-Dimensional Flow Form of the Control Volume Energy Rate Balance

Substituting Eq. 4.12 in Eq. 4.9 and collecting all terms referring to the inlet and the exit into separate expressions, the following form of the control volume energy rate balance results

$$\frac{dE_{cv}}{dt} = \dot{Q}_{cv} - \dot{W}_{cv} + \dot{m}_i\left(u_i + p_i v_i + \frac{V_i^2}{2} + gz_i\right) - \dot{m}_e\left(u_e + p_e v_e + \frac{V_e^2}{2} + gz_e\right) \quad (4.13)$$

The subscript "cv" has been added to $\dot{Q}$ to emphasize that this is the heat transfer rate over the boundary (control surface) of the *control volume.*

The last two terms of Eq. 4.13 can be rewritten using the specific enthalpy h introduced in Sec. 3.6.1. With $h = u + pv$, the energy rate balance becomes

$$\frac{dE_{cv}}{dt} = \dot{Q}_{cv} - \dot{W}_{cv} + \dot{m}_i\left(h_i + \frac{V_i^2}{2} + gz_i\right) - \dot{m}_e\left(h_e + \frac{V_e^2}{2} + gz_e\right) \quad (4.14)$$

The appearance of the sum $u + pv$ in the control volume energy equation is the principal reason for introducing enthalpy previously. It is brought in solely as a *convenience:* The algebraic form of the energy rate balance is simplified by the use of enthalpy and, as we have seen, enthalpy is normally tabulated along with other properties.

In practice there may be several locations on the boundary through which mass enters or exits. This can be accounted for by introducing summations as in the mass balance. Accordingly, the ***energy rate balance*** is

$$\frac{dE_{cv}}{dt} = \dot{Q}_{cv} - \dot{W}_{cv} + \sum_i \dot{m}_i\left(h_i + \frac{V_i^2}{2} + gz_i\right) - \sum_e \dot{m}_e\left(h_e + \frac{V_e^2}{2} + gz_e\right) \quad (4.15)$$

energy rate balance

In writing Eq. 4.15, the one-dimensional flow model is assumed where mass enters and exits the control volume.

Equation 4.15 is an *accounting* balance for the energy of the control volume. It states that the rate of energy increase or decrease within the control volume equals the difference between the rates of energy transfer in and out across the boundary. The mechanisms of energy transfer are heat and work, as for closed systems, and the energy that accompanies the mass entering and exiting.

Take Note...

Equation 4.15 is the most general form of the conservation of energy principle for control volumes used in this book. It serves as the starting point for applying the conservation of energy principle to control volumes in problem solving.

4.4.4 Integral Form of the Control Volume Energy Rate Balance

As for the case of the mass rate balance, the energy rate balance can be expressed in terms of local properties to obtain forms that are more generally applicable. Thus, the term $E_{cv}(t)$, representing the total energy associated with the control volume at time t, can be written as a volume integral

$$E_{cv}(t) = \int_V \rho e \, dV = \int_V \rho\left(u + \frac{V^2}{2} + gz\right) dV \quad (4.16)$$

Similarly, the terms accounting for the energy transfers accompanying mass flow and flow work at inlets and exits can be expressed as shown in the following form of the energy rate balance

$$\frac{d}{dt}\int_V \rho e\, dV = \dot{Q}_{cv} - \dot{W}_{cv} + \sum_i \left[\int_A \left(h + \frac{V^2}{2} + gz\right)\rho V_n\, dA\right]_i - \sum_e \left[\int_A \left(h + \frac{V^2}{2} + gz\right)\rho V_n\, dA\right]_e \qquad (4.17)$$

Additional forms of the energy rate balance can be obtained by expressing the heat transfer rate $\dot{Q}_{cv}$ as the integral of the *heat flux* over the boundary of the control volume, and the work $\dot{W}_{cv}$ in terms of normal and shear stresses at the moving portions of the boundary.

In principle, the change in the energy of a control volume over a time period can be obtained by integration of the energy rate balance with respect to time. Such integrations require information about the time dependences of the work and heat transfer rates, the various mass flow rates, and the states at which mass enters and leaves the control volume. Examples of this type of analysis are presented in Sec. 4.12.

4.5 Analyzing Control Volumes at Steady State

In this section we consider steady-state forms of the mass and energy rate balances. These balances are applied to a variety of devices of engineering interest in Secs. 4.6–4.11. The steady-state forms considered here do not apply to the transient startup or shutdown periods of operation of such devices, but only to periods of steady operation. This situation is commonly encountered in engineering.

4.5.1 Steady-State Forms of the Mass and Energy Rate Balances

For a control volume at steady state, the conditions of the mass within the control volume and at the boundary do not vary with time. The mass flow rates and the rates of energy transfer by heat and work are also constant with time. There can be no accumulation of mass within the control volume, so $dm_{cv}/dt = 0$ and the mass rate balance, Eq. 4.2, takes the form

$$\underset{\text{(mass rate in)}}{\sum_i \dot{m}_i} = \underset{\text{(mass rate out)}}{\sum_e \dot{m}_e} \qquad (4.6)$$

Furthermore, at steady state $dE_{cv}/dt = 0$, so Eq. 4.15 can be written as

$$0 = \dot{Q}_{cv} - \dot{W}_{cv} + \sum_i \dot{m}_i\left(h_i + \frac{V_i^2}{2} + gz_i\right) - \sum_e \dot{m}_e\left(h_e + \frac{V_e^2}{2} + gz_e\right) \qquad (4.18)$$

Alternatively

$$\underset{\text{(energy rate in)}}{\dot{Q}_{cv} + \sum_i \dot{m}_i\left(h_i + \frac{V_i^2}{2} + gz_i\right)} = \underset{\text{(energy rate out)}}{\dot{W}_{cv} + \sum_e \dot{m}_e\left(h_e + \frac{V_e^2}{2} + gz_e\right)} \qquad (4.19)$$

Equation 4.6 asserts that at steady state the total rate at which mass enters the control volume equals the total rate at which mass exits. Similarly, Eq. 4.19 asserts that the total rate at which energy is transferred into the control volume equals the total rate at which energy is transferred out.

Many important applications involve one-inlet, one-exit control volumes at steady state. It is instructive to apply the mass and energy rate balances to this special case. The mass rate balance reduces simply to $\dot{m}_1 = \dot{m}_2$. That is, the mass flow rate must be the same at the exit, 2, as it is at the inlet, 1. The common mass flow rate is designated simply by $\dot{m}$. Next, applying the energy rate balance and factoring the mass flow rate gives

$$0 = \dot{Q}_{cv} - \dot{W}_{cv} + \dot{m}\left[(h_1 - h_2) + \frac{(V_1^2 - V_2^2)}{2} + g(z_1 - z_2)\right] \tag{4.20a}$$

Or, dividing by the mass flow rate

$$0 = \frac{\dot{Q}_{cv}}{\dot{m}} - \frac{\dot{W}_{cv}}{\dot{m}} + (h_1 - h_2) + \frac{(V_1^2 - V_2^2)}{2} + g(z_1 - z_2) \tag{4.20b}$$

The enthalpy, kinetic energy, and potential energy terms all appear in Eqs. 4.20 as *differences* between their values at the inlet and exit. This illustrates that the datums used to assign values to specific enthalpy, velocity, and elevation cancel. In Eq. 4.20b, the ratios $\dot{Q}_{cv}/\dot{m}$ and $\dot{W}_{cv}/\dot{m}$ are rates of energy transfer *per unit mass flowing through the control volume.*

The foregoing steady-state forms of the energy rate balance relate only energy transfer quantities evaluated at the *boundary* of the control volume. No details concerning properties *within* the control volume are required by, or can be determined with, these equations. When applying the energy rate balance in any of its forms, it is necessary to use the same units for all terms in the equation. For instance, *every* term in Eq. 4.20b must have a unit such as kJ/kg. Appropriate unit conversions are emphasized in examples to follow.

4.5.2 Modeling Considerations for Control Volumes at Steady State

In this section, we provide the basis for subsequent applications by considering the modeling of control volumes *at steady state.* In particular, several applications are given in Secs. 4.6–4.11 showing the use of the principles of conservation of mass and energy, together with relationships among properties for the analysis of control volumes at steady state. The examples are drawn from applications of general interest to engineers and are chosen to illustrate points common to all such analyses. Before studying them, it is recommended that you review the methodology for problem solving outlined in Sec. 1.9. As problems become more complicated, the use of a systematic problem-solving approach becomes increasingly important.

When the mass and energy rate balances are applied to a control volume, simplifications are normally needed to make the analysis manageable. That is, the control volume of interest is *modeled* by making assumptions. The *careful* and *conscious* step of listing assumptions is necessary in every engineering analysis. Therefore, an important part of this section is devoted to considering various assumptions that are commonly made when applying the conservation principles to different types of devices. As you study the examples presented in Secs. 4.6–4.11, it is important to recognize the role played by careful assumption making in arriving at solutions. In each case considered, steady-state operation is assumed. The flow is regarded as one-dimensional at places where mass enters and exits the control volume. Also, at each of these locations, equilibrium property relations are assumed to apply.

Horizons

Smaller Can Be Better

Engineers are developing miniature systems for use where weight, portability, and/or compactness are critically important. Some of these applications involve tiny *micro systems* with dimensions in the micrometer to millimeter range. Other somewhat larger *meso-scale* systems can measure up to a few centimeters.

Microelectromechanical systems (MEMS) combining electrical and mechanical features are now widely used for sensing and control. Medical applications of MEMS include pressure sensors that monitor pressure within the balloon inserted into a blood vessel during angioplasty. Air bags are triggered in an automobile crash by tiny acceleration sensors. MEMS are also found in computer hard drives and printers.

Miniature versions of other technologies are being investigated. One study aims at developing an entire gas turbine power plant the size of a shirt button. Another involves micromotors with shafts the diameter of a human hair. Emergency workers wearing fire-, chemical-, or biological-protection suits might in the future be kept cool by tiny heat pumps imbedded in the suit material.

As designers aim at smaller sizes, frictional effects and heat transfers pose special challenges. Fabrication of miniature systems is also demanding. Taking a design from the concept stage to high-volume production can be both expensive and risky, industry representatives say.

In several of the examples to follow, the heat transfer term $\dot{Q}_{cv}$ is set to zero in the energy rate balance because it is small relative to other energy transfers across the boundary. This may be the result of one or more of the following factors:

- The outer surface of the control volume is well insulated.
- The outer surface area is too small for there to be effective heat transfer.
- The temperature difference between the control volume and its surroundings is so small that the heat transfer can be ignored.
- The gas or liquid passes through the control volume so quickly that there is not enough time for significant heat transfer to occur.

The work term $\dot{W}_{cv}$ drops out of the energy rate balance when there are no rotating shafts, displacements of the boundary, electrical effects, or other work mechanisms associated with the control volume being considered. The kinetic and potential energies of the matter entering and exiting the control volume are neglected when they are small relative to other energy transfers.

In practice, the properties of control volumes considered to be at steady state do vary with time. The steady-state assumption would still apply, however, when properties fluctuate only slightly about their averages, as for pressure in Fig. 4.6*a*. Steady state also might be assumed in cases where *periodic* time variations are observed, as

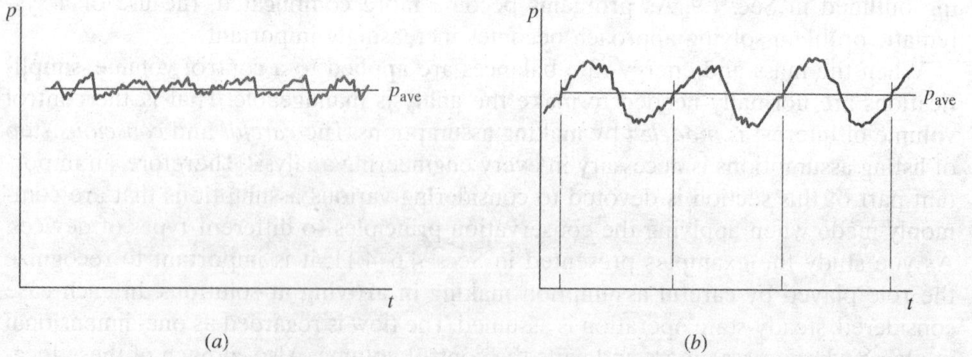

Fig. 4.6 Pressure variations about an average. (*a*) Fluctuation. (*b*) Periodic.

in Fig. 4.6*b*. For example, in reciprocating engines and compressors, the entering and exiting flows pulsate as valves open and close. Other parameters also might be time varying. However, the steady-state assumption can apply to control volumes enclosing these devices if the following are satisfied for each successive period of operation: (1) There is no *net* change in the total energy and the total mass within the control volume. (2) The *time-averaged* mass flow rates, heat transfer rates, work rates, and properties of the substances crossing the control surface all remain constant.

Next, we present brief discussions and examples illustrating the analysis of several devices of interest in engineering, including nozzles and diffusers, turbines, compressors and pumps, heat exchangers, and throttling devices. The discussions highlight some common applications of each device and the modeling typically used in thermodynamic analysis.

4.6 Nozzles and Diffusers

A ***nozzle*** is a flow passage of varying cross-sectional area in which the velocity of a gas or liquid increases in the direction of flow. In a ***diffuser,*** the gas or liquid decelerates in the direction of flow. Figure 4.7 shows a nozzle in which the cross-sectional area decreases in the direction of flow and a diffuser in which the walls of the flow passage diverge. Observe that as velocity increases, pressure decreases, and conversely.

nozzle
diffuser

For many readers, the most familiar application of a nozzle is its use with a garden hose. But nozzles and diffusers have several important engineering applications. In Fig. 4.8, a nozzle and diffuser are combined in a wind-tunnel test facility. Ducts with converging and diverging passages are commonly used in distributing cool and warm air in building air-conditioning systems. Nozzles and diffusers also are key components of turbojet engines (Sec. 9.9).

4.6.1 Nozzle and Diffuser Modeling Considerations

For a control volume enclosing a nozzle or diffuser, the only work is *flow work* at locations where mass enters and exits the control volume, so the term $\dot{W}_{cv}$ drops out of the energy rate balance. The change in potential energy from inlet to exit is negligible under most conditions. Thus, the underlined terms of Eq. 4.20a (repeated below) drop out, leaving the enthalpy, kinetic energy, and heat transfer terms, as shown by Eq. (a)

$$0 = \dot{Q}_{cv} - \underline{\dot{W}_{cv}} + \dot{m}\left[(h_1 - h_2) + \frac{(V_1^2 - V_2^2)}{2} + \underline{g(z_1 - z_2)}\right]$$

$$0 = \dot{Q}_{cv} + \dot{m}\left[(h_1 - h_2) + \frac{(V_1^2 - V_2^2)}{2}\right] \qquad \text{(a)}$$

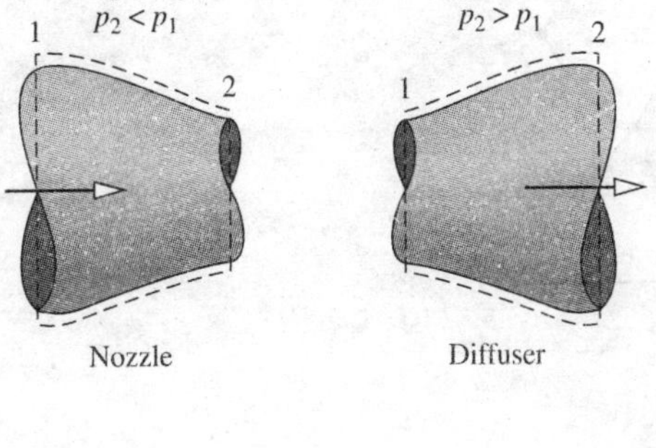

Fig. 4.7 Illustration of a nozzle and a diffuser.

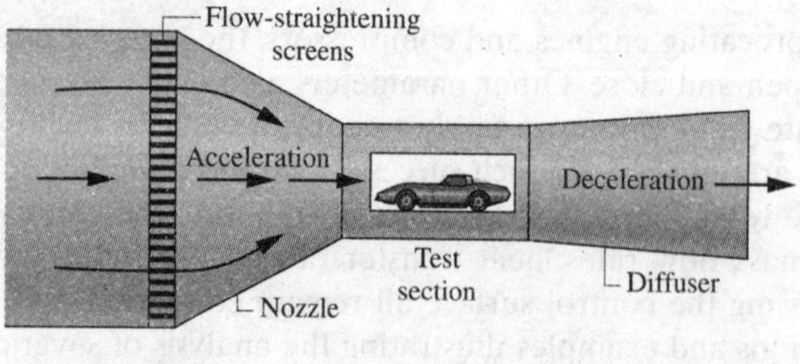

Fig. 4.8 Wind-tunnel test facility.

where $\dot{m}$ is the mass flow rate. The term $\dot{Q}_{cv}$ representing heat transfer with the surroundings normally would be unavoidable (or stray) heat transfer, and this is often small enough relative to the enthalpy and kinetic energy terms that it also can be neglected, giving simply

$$0 = (h_1 - h_2) + \left(\frac{V_1^2 - V_2^2}{2}\right) \tag{4.21}$$

4.6.2 Application to a Steam Nozzle

The modeling introduced in Sec. 4.6.1 is illustrated in the next example involving a steam nozzle. Particularly note the use of unit conversion factors in this application.

Example 4.3 CALCULATING EXIT AREA OF A STEAM NOZZLE

Steam enters a converging–diverging nozzle operating at steady state with $p_1 = 40$ bar, $T_1 = 400°C$, and a velocity of 10 m/s. The steam flows through the nozzle with negligible heat transfer and no significant change in potential energy. At the exit, $p_2 = 15$ bar, and the velocity is 665 m/s. The mass flow rate is 2 kg/s. Determine the exit area of the nozzle, in m^2.

Solution

Known: Steam flows through a nozzle at steady state with known properties at the inlet and exit, a known mass flow rate, and negligible effects of heat transfer and potential energy.

Find: Determine the exit area.

Schematic and Given Data:

❶

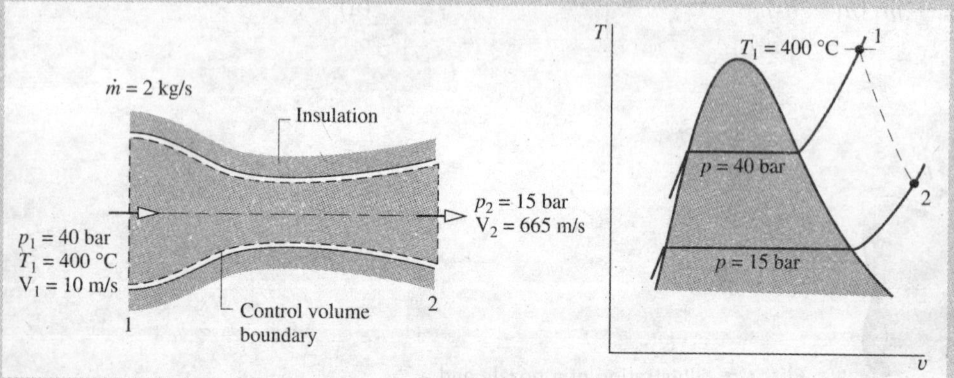

Fig. E4.3

Engineering Model:

1. The control volume shown on the accompanying figure is at steady state.
2. Heat transfer is negligible and $\dot{W}_{cv} = 0$.
3. The change in potential energy from inlet to exit can be neglected.

Analysis: The exit area can be determined from the mass flow rate $\dot{m}$ and Eq. 4.4b, which can be arranged to read

$$A_2 = \frac{\dot{m}v_2}{V_2}$$

To evaluate A_2 from this equation requires the specific volume v_2 at the exit, and this requires that the exit state be fixed.

The state at the exit is fixed by the values of two independent intensive properties. One is the pressure p_2, which is known. The other is the specific enthalpy h_2, determined from the steady-state energy rate balance Eq. 4.20a, as follows

$$0 = \overset{0}{\cancel{\dot{Q}_{cv}}} - \overset{0}{\cancel{\dot{W}_{cv}}} + \dot{m}\left[(h_1 - h_2) + \frac{(V_1^2 - V_2^2)}{2} + g(z_1 - z_2)\right]$$

The terms $\dot{Q}_{cv}$ and $\dot{W}_{cv}$ are deleted by assumption 2. The change in specific potential energy drops out in accordance with assumption 3 and $\dot{m}$ cancels, leaving

$$0 = (h_1 - h_2) + \left(\frac{V_1^2 - V_2^2}{2}\right)$$

Solving for h_2

$$h_2 = h_1 + \left(\frac{V_1^2 - V_2^2}{2}\right)$$

From Table A-4, $h_1 = 3213.6$ kJ/kg. The velocities V_1 and V_2 are given. Inserting values and converting the units of the kinetic energy terms to kJ/kg results in

❷
$$h_2 = 3213.6 \text{ kJ/kg} + \left[\frac{(10)^2 - (665)^2}{2}\right]\left(\frac{\text{m}^2}{\text{s}^2}\right)\left|\frac{1 \text{ N}}{1 \text{ kg}\cdot\text{m/s}^2}\right|\left|\frac{1 \text{ kJ}}{10^3 \text{ N}\cdot\text{m}}\right|$$
$$= 3213.6 - 221.1 = 2992.5 \text{ kJ/kg}$$

Finally, referring to Table A-4 at $p_2 = 15$ bar with $h_2 = 2992.5$ kJ/kg, the specific volume at the exit is $v_2 = 0.1627$ m³/kg. The exit area is then

$$A_2 = \frac{(2 \text{ kg/s})(0.1627 \text{ m}^3/\text{kg})}{665 \text{ m/s}} = 4.89 \times 10^{-4} \text{ m}^2$$

❶ Although equilibrium property relations apply at the inlet and exit of the control volume, the intervening states of the steam are not necessarily equilibrium states. Accordingly, the expansion through the nozzle is represented on the T–v diagram as a dashed line.

❷ Care must be taken in converting the units for specific kinetic energy to kJ/kg.

Skills Developed: *Ability to...*
- ✓ apply the steady-state energy rate balance to a control volume.
- ✓ apply the mass flow rate expression, Eq. 4.4b.
- ✓ develop an engineering model.
- ✓ retrieve property data for water.

Quick Quiz

Evaluate the nozzle inlet area, in m².

Ans. 1.47×10^{-2} m².

4.7 Turbines

turbine

A ***turbine*** is a device in which power is developed as a result of a gas or liquid passing through a set of blades attached to a shaft free to rotate. A schematic of an axial-flow steam or gas turbine is shown in Fig. 4.9. Such turbines are widely used for power generation in vapor power plants, gas turbine power plants, and aircraft engines (see Chaps. 8 and 9). In these applications, superheated steam or a gas enters the turbine and expands to a lower pressure as power is generated.

A *hydraulic* turbine installed in a dam is shown in Fig. 4.10. In this application, water falling through the propeller causes the shaft to rotate and power is generated. Hydroelectric power plants use hydraulic turbines, and like wind turbine-power plants require no fossil fuels to operate. They are considered *renewable power sources.*

Energy & Environment

One of the world's largest commercial wind turbines stands as tall as a 30-story building and produces electricity at a rate that would meet the needs of about 900 typical U.S. homes. The three-bladed rotor of this wind turbine has a diameter nearly the length of a football field and operates in winds up to 90 km per hour. This wind turbine features microprocessor control of all functions, ensuring that each blade is pitched at the correct angle for current wind conditions. Both the rotor and generator can vary their rotational speed during wind gusts, reducing fluctuations in the power provided to the electricity grid and the forces acting on the vital parts of the turbine.

Wind turbines are not without disadvantages. They are considered unsightly by some and noisy by others. Early wind turbines were criticized about excessive numbers of birds being struck and killed by rotor blades, but newer wind turbine installations have largely overcome this problem. Another disadvantage is that sufficient winds may not always be available when power is most needed. On the plus side, wind-generated electricity is quickly becoming competitive cost-wise with electricity from coal-fired power plants, and wind power is second only to hydroelectric power among the renewable power sources used by utilities today. Wind energy plants take less time to build than conventional plants and are modular, allowing additional units to be added as warranted. They also produce no global warming gases.

4.7.1 Steam and Gas Turbine Modeling Considerations

With a proper selection of the control volume enclosing a steam or gas turbine, the net kinetic energy of the matter flowing across the boundary is usually small enough to be neglected. The net potential energy of the flowing matter also is typically

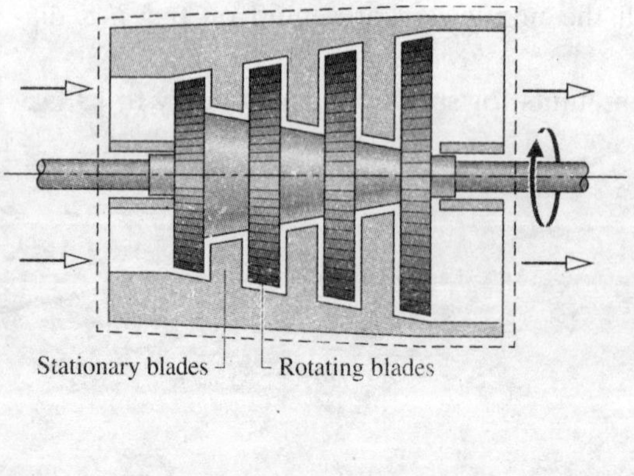

Fig. 4.9 Schematic of an axial-flow steam or gas turbine.

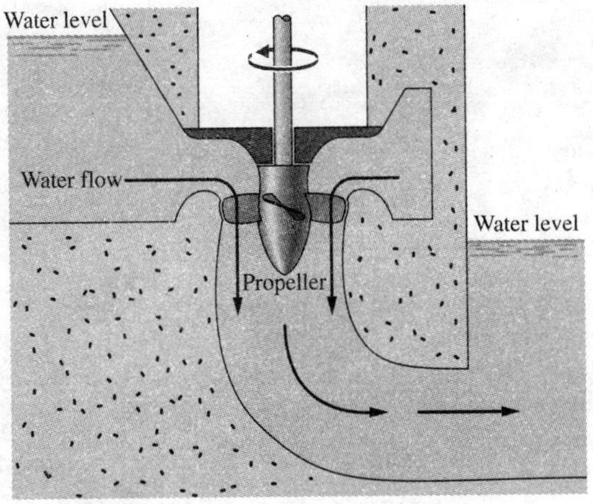

Fig. 4.10 Hydraulic turbine installed in a dam.

negligible. Thus, the underlined terms of Eq. 4.20a (repeated below) drop out, leaving the power, enthalpy, and heat transfer terms, as shown by Eq. (a)

$$0 = \dot{Q}_{cv} - \dot{W}_{cv} + \dot{m}\left[(h_1 - h_2) + \underline{\frac{(V_1^2 - V_2^2)}{2}} + \underline{g\,(z_1 - z_2)}\right]$$

$$0 = \dot{Q}_{cv} - \dot{W}_{cv} + \dot{m}\,(h_1 - h_2) \qquad \text{(a)}$$

where $\dot{m}$ is the mass flow rate. The only heat transfer between the turbine and surroundings normally would be unavoidable (or stray) heat transfer, and this is often small enough relative to the power and enthalpy terms that it also can be neglected, giving simply

$$\dot{W}_{cv} = \dot{m}(h_1 - h_2) \qquad \text{(b)}$$

4.7.2 Application to a Steam Turbine

In this section, modeling considerations for turbines are illustrated by application to a case involving the practically important steam turbine. Objectives in this example include assessing the significance of the heat transfer and kinetic energy terms of the energy balance and illustrating the appropriate use of unit conversion factors.

Example 4.4 CALCULATING HEAT TRANSFER FROM A STEAM TURBINE

Steam enters a turbine operating at steady state with a mass flow rate of 4600 kg/h. The turbine develops a power output of 1000 kW. At the inlet, the pressure is 60 bar, the temperature is 400°C, and the velocity is 10 m/s. At the exit, the pressure is 0.1 bar, the quality is 0.9 (90%), and the velocity is 30 m/s. Calculate the rate of heat transfer between the turbine and surroundings, in kW.

Solution

Known: A steam turbine operates at steady state. The mass flow rate, power output, and states of the steam at the inlet and exit are known.

Find: Calculate the rate of heat transfer.

Schematic and Given Data:

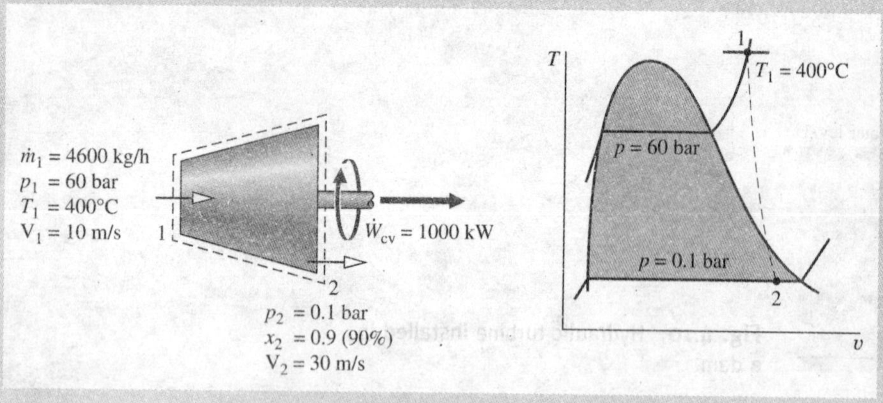

Fig. E4.4

Engineering Model:

1. The control volume shown on the accompanying figure is at steady state.
2. The change in potential energy from inlet to exit can be neglected.

Analysis: To calculate the heat transfer rate, begin with the one-inlet, one-exit form of the energy rate balance for a control volume at steady state, Eq. 4.20a. That is

$$0 = \dot{Q}_{cv} - \dot{W}_{cv} + \dot{m}\left[(h_1 - h_2) + \frac{(V_1^2 - V_2^2)}{2} + g(z_1 - z_2)\right]$$

where $\dot{m}$ is the mass flow rate. Solving for $\dot{Q}_{cv}$ and dropping the potential energy change from inlet to exit

$$\dot{Q}_{cv} = \dot{W}_{cv} + \dot{m}\left[(h_2 - h_1) + \left(\frac{V_2^2 - V_1^2}{2}\right)\right] \quad \text{(a)}$$

To compare the magnitudes of the enthalpy and kinetic energy terms, and stress the unit conversions needed, each of these terms is evaluated separately.

First, the specific *enthalpy difference* $h_2 - h_1$ is found. Using Table A-4, $h_1 = 3177.2$ kJ/kg. State 2 is a two-phase liquid–vapor mixture, so with data from Table A-3 and the given quality

$$h_2 = h_{f2} + x_2(h_{g2} - h_{f2})$$
$$= 191.83 + (0.9)(2392.8) = 2345.4 \text{ kJ/kg}$$

Hence

$$h_2 - h_1 = 2345.4 - 3177.2 = -831.8 \text{ kJ/kg}$$

Consider next the specific *kinetic energy difference*. Using the given values for the velocities

❶ $$\left(\frac{V_2^2 - V_1^2}{2}\right) = \left[\frac{(30)^2 - (10)^2}{2}\right]\left(\frac{\text{m}^2}{\text{s}^2}\right)\left|\frac{1 \text{ N}}{1 \text{ kg} \cdot \text{m/s}^2}\right|\left|\frac{1 \text{ kJ}}{10^3 \text{ N} \cdot \text{m}}\right|$$
$$= 0.4 \text{ kJ/kg}$$

Calculating $\dot{Q}_{cv}$ from Eq. (a)

❷ $$\dot{Q}_{cv} = (1000 \text{ kW}) + \left(4600\frac{\text{kg}}{\text{h}}\right)(-831.8 + 0.4)\left(\frac{\text{kJ}}{\text{kg}}\right)\left|\frac{1 \text{ h}}{3600 \text{ s}}\right|\left|\frac{1 \text{ kW}}{1 \text{ kJ/s}}\right|$$
$$= -62.3 \text{ kW}$$

❶ The magnitude of the change in specific kinetic energy from inlet to exit is much smaller than the specific enthalpy change. Note the use of unit conversion factors here and in the calculation of $\dot{Q}_{cv}$ to follow.

❷ The negative value of $\dot{Q}_{cv}$ means that there is heat transfer from the turbine to its surroundings, as would be expected. The magnitude of $\dot{Q}_{cv}$ is small relative to the power developed.

Skills Developed: *Ability to...*
- ✓ apply the steady-state energy rate balance to a control volume.
- ✓ develop an engineering model.
- ✓ retrieve property data for water.

Quick Quiz

If the change in kinetic energy from inlet to exit were neglected, evaluate the heat transfer rate, in kW, keeping all other data unchanged. Comment.

Ans. −62.9 kW

4.8 Compressors and Pumps

compressor
pump

Compressors and **pumps** are devices in which work is done on the substance flowing through them in order to change the state of the substance, typically to increase the pressure and/or elevation. The term *compressor* is used when the substance is a gas (vapor) and the term *pump* is used when the substance is a liquid. Various types of compressors are shown in Figs. 4.11 and 4.12.

The axial-flow compressor of Fig. 4.12a is a key component of turbojet engines (Sec. 9.9). Compressors also are essential components of refrigeration and heat pump systems (Chap. 10). In the study of Chap. 8, we find that pumps are important in vapor power systems. Pumps also are commonly used to fill water towers, remove water from flooded basements, and for numerous other domestic and industrial applications.

4.8.1 Compressor and Pump Modeling Considerations

For a control volume enclosing a compressor, the mass and energy rate balances reduce at steady state as for the case of turbines considered in Sec. 4.7.1. Thus, Eq. 4.20a reduces to read

$$0 = \dot{Q}_{cv} - \dot{W}_{cv} + \dot{m}(h_1 - h_2) \qquad \text{(a)}$$

Heat transfer with the surroundings is frequently a secondary effect that can be neglected, giving as for turbines

$$\dot{W}_{cv} = \dot{m}(h_1 - h_2) \qquad \text{(b)}$$

For pumps, heat transfer is generally a secondary effect, but the kinetic and potential energy terms of Eq. 4.20a may be significant depending on the application. Be

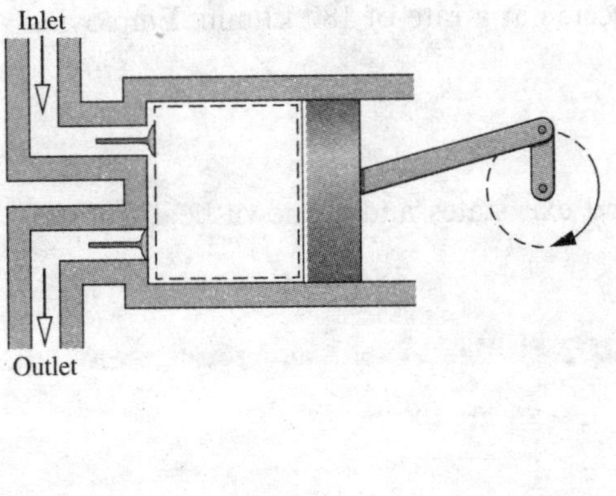

Fig. 4.11 Reciprocating compressor.

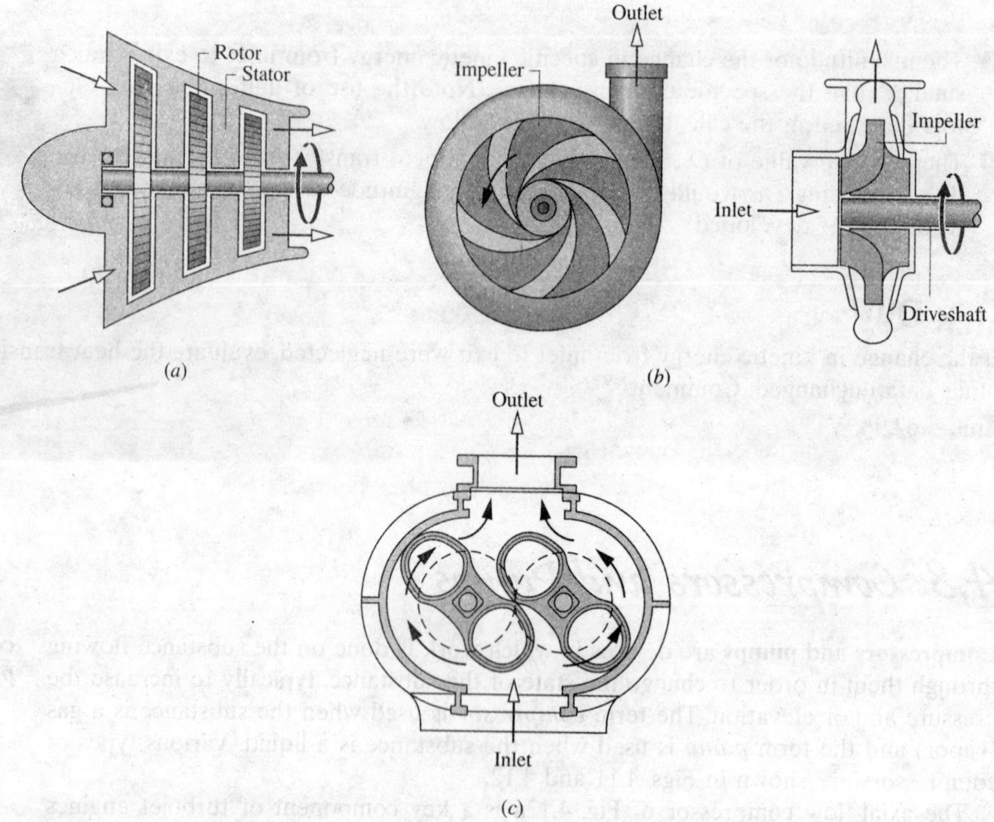

Fig. 4.12 Rotating compressors. (*a*) Axial flow. (*b*) Centrifugal. (*c*) Roots type.

sure to note that for compressors and pumps, the value of $\dot{W}_{cv}$ is *negative* because a power *input* is required.

4.8.2 Applications to an Air Compressor and a Pump System

In this section, modeling considerations for compressors and pumps are illustrated in Examples 4.5 and 4.6, respectively. In Example 4.5 the objectives include assessing the significance of the heat transfer and kinetic energy terms of the energy balance and illustrating the appropriate use of unit conversion factors.

Example 4.5 CALCULATING COMPRESSOR POWER

Air enters a compressor operating at steady state at a pressure of 1 bar, a temperature of 290 K, and a velocity of 6 m/s through an inlet with an area of 0.1 m^2. At the exit, the pressure is 7 bar, the temperature is 450 K, and the velocity is 2 m/s. Heat transfer from the compressor to its surroundings occurs at a rate of 180 kJ/min. Employing the ideal gas model, calculate the power input to the compressor, in kW.

Solution

Known: An air compressor operates at steady state with known inlet and exit states and a known heat transfer rate.

Find: Calculate the power required by the compressor.

Schematic and Given Data:

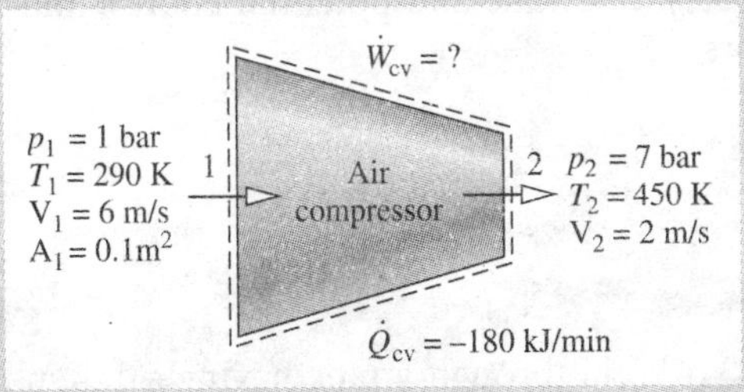

Fig. E4.5

Engineering Model:

1. The control volume shown on the accompanying figure is at steady state.
2. The change in potential energy from inlet to exit can be neglected.
❶ 3. The ideal gas model applies for the air.

Analysis: To calculate the power input to the compressor, begin with the one-inlet, one-exit form of the energy rate balance for a control volume at steady state, Eq. 4.20a. That is

$$0 = \dot{Q}_{cv} - \dot{W}_{cv} + \dot{m}\left[(h_1 - h_2) + \frac{(V_1^2 - V_2^2)}{2} + g(z_1 - z_2)\right]$$

Solving

$$\dot{W}_{cv} = \dot{Q}_{cv} + \dot{m}\left[(h_1 - h_2) + \left(\frac{V_1^2 - V_2^2}{2}\right)\right]$$

The change in potential energy from inlet to exit drops out by assumption 2.

The mass flow rate $\dot{m}$ can be evaluated with given data at the inlet and the ideal gas equation of state.

$$\dot{m} = \frac{A_1 V_1}{v_1} = \frac{A_1 V_1 p_1}{(\overline{R}/M)T_1} = \frac{(0.1\text{ m}^2)(6\text{ m/s})(10^5\text{ N/m}^2)}{\left(\dfrac{8314}{28.97}\dfrac{\text{N}\cdot\text{m}}{\text{kg}\cdot\text{K}}\right)(290\text{ K})} = 0.72\text{ kg/s}$$

The specific enthalpies h_1 and h_2 can be found from Table A-22. At 290 K, $h_1 = 290.16$ kJ/kg. At 450 K, $h_2 = 451.8$ kJ/kg. Substituting values into the expression for $\dot{W}_{cv}$, and applying appropriate unit conversion factors, we get

$$\dot{W}_{cv} = \left(-180\frac{\text{kJ}}{\text{min}}\right)\left|\frac{1\text{ min}}{60\text{ s}}\right| + 0.72\frac{\text{kg}}{\text{s}}\left[(290.16 - 451.8)\frac{\text{kJ}}{\text{kg}} + \left(\frac{(6)^2 - (2)^2}{2}\right)\left(\frac{\text{m}^2}{\text{s}^2}\right)\left|\frac{1\text{ N}}{1\text{ kg}\cdot\text{m/s}^2}\right|\left|\frac{1\text{ kJ}}{10^3\text{ N}\cdot\text{m}}\right|\right]$$

$$= -3\frac{\text{kJ}}{\text{s}} + 0.72\frac{\text{kg}}{\text{s}}(-161.64 + 0.02)\frac{\text{kJ}}{\text{kg}}$$

❷ $$= -119.4\frac{\text{kJ}}{\text{s}}\left|\frac{1\text{ kW}}{1\text{ kJ/s}}\right| = -119.4\text{ kW}$$

❶ The applicability of the ideal gas model can be checked by reference to the generalized compressibility chart.

❷ In this example $\dot{Q}_{cv}$ and $\dot{W}_{cv}$ have negative values, indicating that the direction of the heat transfer is *from* the compressor and work is done *on* the air passing through the compressor. The magnitude of the power *input* to the compressor is 119.4 kW.

Skills Developed: *Ability to…*
- ✓ apply the steady-state energy rate balance to a control volume.
- ✓ apply the mass flow rate expression, Eq. 4.4b.
- ✓ develop an engineering model.
- ✓ retrieve property data of air modeled as an ideal gas.

Quick Quiz

If the change in kinetic energy from inlet to exit were neglected, evaluate the compressor power, in kW, keeping all other data unchanged. Comment.

Ans. −119.4 kW.

In Example 4.6, a pump is a component of an overall system that delivers a high-velocity stream of water at an elevation greater than at the inlet. Note the modeling considerations in this case, particularly the roles of kinetic and potential energy, and the use of appropriate unit conversion factors.

Example 4.6 ANALYZING A PUMP SYSTEM

A pump steadily draws water from a pond at a volumetric flow rate of 0.83 m^3/min through a pipe having a 12-cm diameter inlet. The water is delivered through a hose terminated by a converging nozzle. The nozzle exit has a diameter of 3 cm and is located 10 m above the pipe inlet. Water enters at 20°C, 1 atm and exits with no significant change in temperature or pressure. The magnitude of the rate of heat transfer *from* the pump *to* the surroundings is 5% of the power input. The acceleration of gravity is 9.81 m/s^2. Determine (a) the velocity of the water at the inlet and exit, each in m/s, and (b) the power required by the pump, in kW.

Solution

Known: A pump system operates at steady state with known inlet and exit conditions. The rate of heat transfer from the pump is specified as a percentage of the power input.

Find: Determine the velocities of the water at the inlet and exit of the pump system and the power required.

Schematic and Given Data:

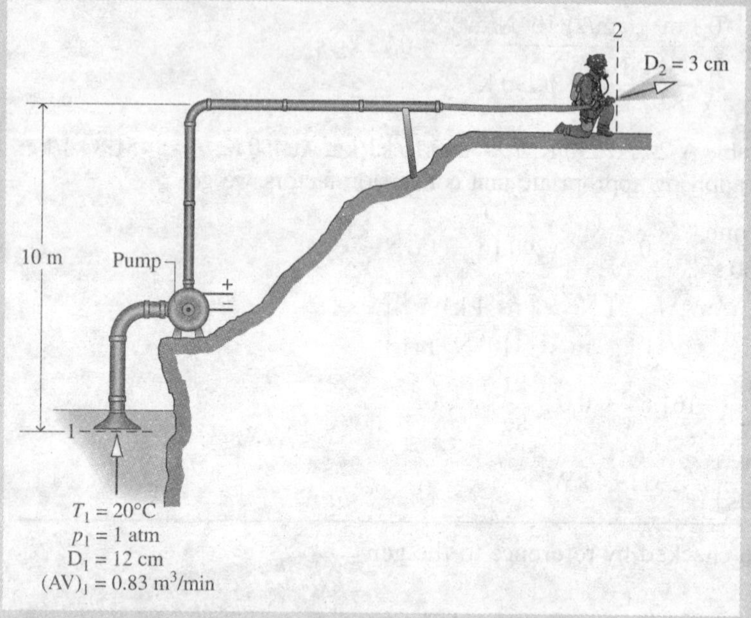

Fig. E4.6

Engineering Model:

1. A control volume encloses the pump, inlet pipe, and delivery hose.
2. The control volume is at steady state.
3. The magnitude of the heat transfer from the control volume is 5% of the power input.
4. For liquid water, $v \approx v_f(T)$, $h \approx h_f(T)$ (Eqs. 3.11 and 3.14).
5. $g = 9.81\ \text{m/s}^2$.

Analysis:

(a) A mass rate balance reduces at steady state to read $\dot{m}_2 = \dot{m}_1$. The common mass flow rate at the inlet and exit, $\dot{m}$, can be evaluated using Eq. 4.4b together with $v \approx v_f(20°\text{C}) = 1.0018 \times 10^{-3} \text{m}^3/\text{kg}$ from Table A-2. That is,

$$\dot{m} = \frac{\text{AV}}{v} = \left(\frac{0.83\ \text{m}^3/\text{min}}{1.0018 \times 10^{-3}\ \text{m}^3/\text{kg}}\right)\left|\frac{1\ \text{min}}{60\ \text{s}}\right|$$

$$= 13.8 \frac{\text{kg}}{\text{s}}$$

Thus, the inlet and exit velocities are, respectively,

❶
$$V_1 = \frac{\dot{m}v}{A_1} = \frac{(13.8 \text{ kg/s})(1.0018 \times 10^{-3} \text{ m}^3/\text{kg})}{\pi(0.12 \text{ m})^2/4} = 1.22 \text{ m/s}$$

$$V_2 = \frac{\dot{m}v}{A_2} = \frac{(13.8 \text{ kg/s})(1.0018 \times 10^{-3} \text{ m}^3/\text{kg})}{\pi(0.03 \text{ m})^2/4} = 19.56 \text{ m/s}$$

(b) To calculate the power input, begin with the one-inlet, one-exit form of the energy rate balance for a control volume at steady state, Eq. 4.20a. That is

$$0 = \dot{Q}_{cv} - \dot{W}_{cv} + \dot{m}\left[(h_1 - h_2) + \left(\frac{V_1^2 - V_2^2}{2}\right) + g(z_1 - z_2)\right]$$

❷ Introducing $\dot{Q}_{cv} = (0.05)\dot{W}_{cv}$, and solving for $\dot{W}_{cv}$

$$\dot{W}_{cv} = \frac{\dot{m}}{0.95}\left[(h_1 - h_2) + \left(\frac{V_1^2 - V_2^2}{2}\right) + g(z_1 - z_2)\right] \quad \text{(a)}$$

Since the temperature of the water does not change significantly and $h \approx h_f(T)$ (assumption 4), the enthalpy term drops out. Evaluating the kinetic energy term

$$\frac{V_1^2 - V_2^2}{2} = \frac{[(1.22)^2 - (19.56)^2]\left(\frac{\text{m}}{\text{s}}\right)^2}{2}\left|\frac{1 \text{ N}}{1 \text{ kg} \cdot \text{m/s}^2}\right|\left|\frac{1 \text{ kJ}}{10^3 \text{ N} \cdot \text{m}}\right| = -0.191 \text{ kJ/kg}$$

The potential energy term is

$$g(z_1 - z_2) = (9.81 \text{ m/s}^2)(0 - 10)\text{m}\left|\frac{1 \text{ N}}{1 \text{ kg} \cdot \text{m/s}^2}\right|\left|\frac{1 \text{ kJ}}{10^3 \text{ N} \cdot \text{m}}\right| = -0.098 \text{ kJ/kg}$$

Inserting values into Eq. (a)

$$\dot{W}_{cv} = \left(\frac{13.8 \text{ kg/s}}{0.95}\right)[0 - 0.191 - 0.098]\left(\frac{\text{kJ}}{\text{kg}}\right)\left|\frac{1 \text{ kW}}{1 \text{ kJ/s}}\right|$$
$$= -4.2 \text{ kW}$$

where the minus sign indicates that power is provided to the pump.

---

❶ Alternatively, V_1 can be evaluated from the volumetric flow rate at 1. This is left as an exercise.

❷ Since power is required to operate the pump, $\dot{W}_{cv}$ is negative in accord with our sign convention. The energy transfer by heat is from the control volume to the surroundings, and thus $\dot{Q}_{cv}$ is negative as well. Using the value of $\dot{W}_{cv}$ found below, $\dot{Q}_{cv} = (0.05)\dot{W}_{cv} = -0.21$ kW.

Skills Developed: *Ability to...*
- ✓ apply the steady-state energy rate balance to a control volume.
- ✓ apply the mass flow rate expression, Eq. 4.4b.
- ✓ develop an engineering model.
- ✓ retrieve properties of liquid water.

Quick Quiz

If the nozzle were removed and water exited directly from the hose, whose diameter is 5 cm, determine the velocity at the exit, in m/s, and the power required, in kW, keeping all other data unchanged.

Ans. 7.04 m/s, 1.77 kW.

4.9 Heat Exchangers

Heat exchangers have innumerable domestic and industrial applications, including use in home heating and cooling systems, automotive systems, electrical power generation, and chemical processing. Indeed, nearly every area of application listed in Table 1.1 (p. 3) involves heat exchangers.

heat exchanger

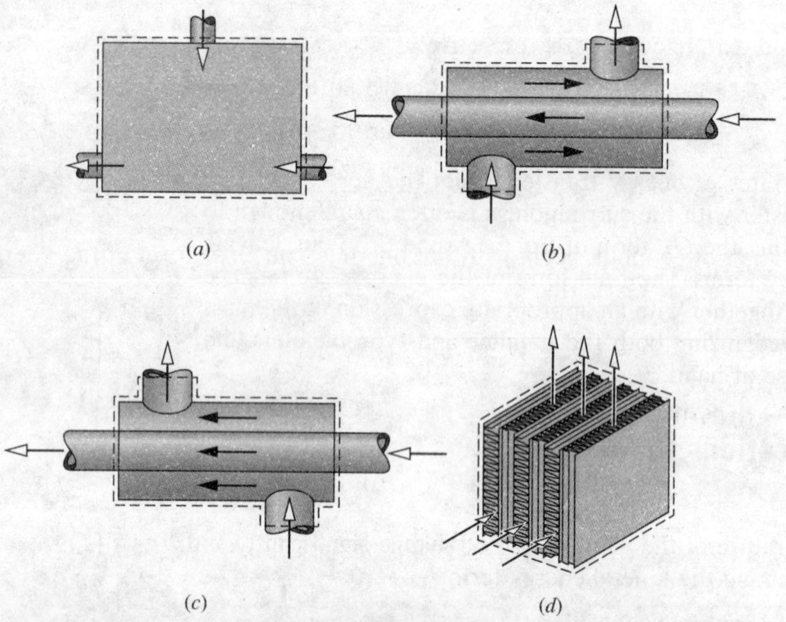

Fig. 4.13 Common heat exchanger types. (*a*) Direct contact heat exchanger. (*b*) Tube-within-a-tube counterflow heat exchanger. (*c*) Tube-within-a-tube parallel flow heat exchanger. (*d*) Cross-flow heat exchanger.

One common type of heat exchanger is a mixing chamber in which hot and cold streams are mixed directly as shown in Fig. 4.13*a*. The open feedwater heater, which is a component of the vapor power systems considered in Chap. 8, is an example of this type of device.

Another common type of exchanger is one in which a gas or liquid is *separated* from another gas or liquid by a wall through which energy is conducted. These heat exchangers, known as recuperators, take many different forms. Counterflow and parallel tube-within-a-tube configurations are shown in Figs. 4.13*b* and 4.13*c*, respectively. Other configurations include cross-flow, as in automobile radiators, and multiple-pass shell-and-tube condensers and evaporators. Figure 4.13*d* illustrates a cross-flow heat exchanger.

Bio... connections

Inflatable blankets such as shown in Fig. 4.14 are used to prevent subnormal body temperatures (hypothermia) during surgery. Typically, a heater and blower direct a stream of warm air into the blanket. Air exits the blanket through perforations in its surface. Such *thermal blankets* have been used safely and without incident in millions of surgical procedures. Still, there are obvious risks to patients if temperature controls fail and overheating occurs. Professional groups have called for reducing such hazards through improvements in the design of thermal blankets and better training of operating room personnel. This biomedical application is a good example of how engineers must bring into the design process their knowledge of heat exchanger principles, temperature sensing and control, and safety and reliability requirements.

4.9.1 Heat Exchanger Modeling Considerations

As shown by Fig. 4.13, heat exchangers can involve multiple inlets and exits. For a control volume enclosing a heat exchanger, the only work is flow work at the places where matter enters and exits, so the term $\dot{W}_{cv}$ drops out of the energy rate balance. In addition, the kinetic and potential energies of the flowing streams usually can be ignored at the inlets and exits. Thus, the underlined terms of Eq. 4.18 (repeated below) drop out, leaving the enthalpy and heat transfer terms, as shown by Eq. (a). That is,

$$0 = \dot{Q}_{cv} - \underline{\dot{W}_{cv}} + \sum_i \dot{m}_i \left(h_i + \underline{\frac{V_i^2}{2} + gz_i} \right) - \sum_e \dot{m}_e \left(h_e + \underline{\frac{V_e^2}{2} + gz_e} \right)$$

$$0 = \dot{Q}_{cv} + \sum_i \dot{m}_i h_i - \sum_e \dot{m}_e h_e \qquad \text{(a)}$$

Although high rates of energy transfer *within* the heat exchanger occur, heat transfer with the surroundings is often small enough to be neglected. Thus, the $\dot{Q}_{cv}$ term of Eq. (a) would drop out, leaving just the enthalpy terms. The final form of the energy rate balance must be solved together with an appropriate expression of the mass rate balance, recognizing both the number and type of inlets and exits for the case at hand.

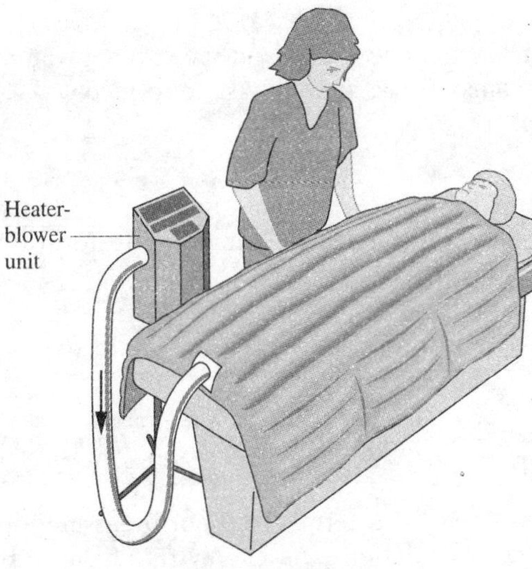

Fig. 4.14 Inflatable thermal blanket.

4.9.2 Applications to a Power Plant Condenser and Computer Cooling

The next example illustrates how the mass and energy rate balances are applied to a condenser at steady state. Condensers are commonly found in power plants and refrigeration systems.

Example 4.7 POWER PLANT CONDENSER

Steam enters the condenser of a vapor power plant at 0.1 bar with a quality of 0.95 and condensate exits at 0.1 bar and 45°C. Cooling water enters the condenser in a separate stream as a liquid at 20°C and exits as a liquid at 35°C with no change in pressure. Heat transfer from the outside of the condenser and changes in the kinetic and potential energies of the flowing streams can be ignored. For steady-state operation, determine .

(a) the ratio of the mass flow rate of the cooling water to the mass flow rate of the condensing steam.

(b) the rate of energy transfer from the condensing steam to the cooling water, in kJ per kg of steam passing through the condenser.

Solution

Known: Steam is condensed at steady state by interacting with a separate liquid water stream.

Find: Determine the ratio of the mass flow rate of the cooling water to the mass flow rate of the steam and the rate of energy transfer from the steam to the cooling water.

Schematic and Given Data:

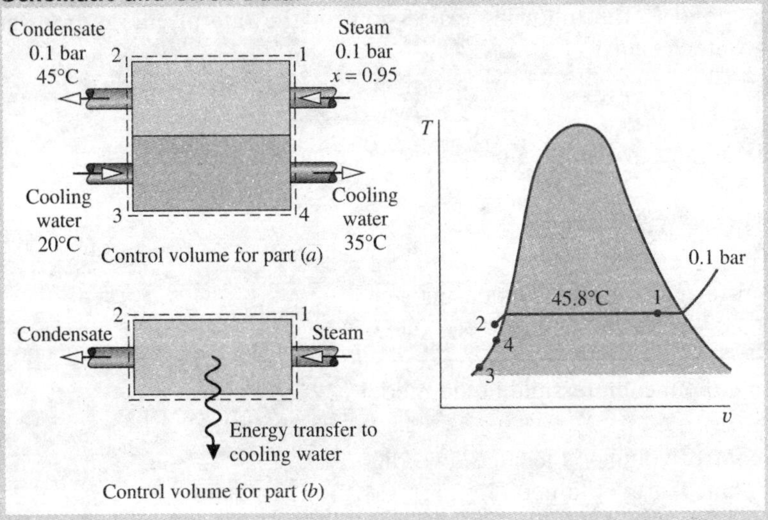

Fig. E4.7

Engineering Model:

1. Each of the two control volumes shown on the accompanying sketch is at steady-state.
2. There is no significant heat transfer between the overall condenser and its surroundings. $\dot{W}_{cv} = 0$.
3. Changes in the kinetic and potential energies of the flowing streams from inlet to exit can be ignored.
4. At states 2, 3, and 4, $h \approx h_f(T)$ (see Eq. 3.14).

Analysis: The steam and the cooling water streams do not mix. Thus, the mass rate balances for each of the two streams reduce at steady state to give

$$\dot{m}_1 = \dot{m}_2 \quad \text{and} \quad \dot{m}_3 = \dot{m}_4$$

(a) The ratio of the mass flow rate of the cooling water to the mass flow rate of the condensing steam, $\dot{m}_3/\dot{m}_1$, can be found from the steady-state form of the energy rate balance, Eq. 4.18, applied to the overall condenser as follows:

$$0 = \underline{\dot{Q}_{cv}} - \underline{\dot{W}_{cv}} + \dot{m}_1\left(h_1 + \underline{\frac{V_1^2}{2} + gz_1}\right) + \dot{m}_3\left(h_3 + \underline{\frac{V_3^2}{2} + gz_3}\right)$$
$$- \dot{m}_2\left(h_2 + \underline{\frac{V_2^2}{2} + gz_2}\right) - \dot{m}_4\left(h_4 + \underline{\frac{V_4^2}{2} + gz_4}\right)$$

The underlined terms drop out by assumptions 2 and 3. With these simplifications, together with the above mass flow rate relations, the energy rate balance becomes simply

$$0 = \dot{m}_1(h_1 - h_2) + \dot{m}_3(h_3 - h_4)$$

Solving, we get

$$\frac{\dot{m}_3}{\dot{m}_1} = \frac{h_1 - h_2}{h_4 - h_3}$$

The specific enthalpy h_1 can be determined using the given quality and data from Table A-3. From Table A-3 at 0.1 bar, $h_f = 191.83$ kJ/kg and $h_g = 2{,}584.7$ kJ/kg, so

$$h_1 = 191.83 + 0.95(2584.7 - 191.83) = 2{,}465.1 \text{ kJ/kg}$$

❶ Using assumption 4, the specific enthalpy at 2 is given by $h_2 \approx h_f(T_2) = 188.45$ kJ/kg. Similarly, $h_3 \approx h_f(T_3)$ and $h_4 \approx h_f(T_4)$, giving $h_4 - h_3 = 62.7$ kJ/kg. Thus,

$$\frac{\dot{m}_3}{\dot{m}_1} = \frac{2{,}465.1 - 188.45}{62.7} = 36.3$$

(b) For a control volume enclosing the steam side of the condenser only, begin with the steady-state form of energy rate balance, Eq. 4.20a.

❷
$$0 = \dot{Q}_{cv} - \underline{\dot{W}_{cv}} + \dot{m}_1\left[(h_1 - h_2) + \underline{\frac{(V_1^2 - V_2^2)}{2}} + \underline{g(z_1 - z_2)}\right]$$

The underlined terms drop out by assumptions 2 and 3. The following expression for the rate of energy transfer between the condensing steam and the cooling water results:

$$\dot{Q}_{cv} = \dot{m}_1(h_2 - h_1)$$

Dividing by the mass flow rate of the steam, $\dot{m}_1$, and inserting values

$$\frac{\dot{Q}_{cv}}{\dot{m}_1} = h_2 - h_1 = 188.45 - 2{,}465.1 = -2{,}276.7 \text{ kJ/kg}$$

where the minus sign signifies that energy is transferred *from* the condensing steam *to* the cooling water.

Skills Developed: *Ability to...*
- ✓ apply the steady-state mass and energy rate balances to a control volume.
- ✓ develop an engineering model.
- ✓ retrieve property data for water.

❶ Alternatively, $(h_4 - h_3)$ can be evaluated using the incompressible liquid model via Eq. 3.20b.

❷ Depending on where the boundary of the control volume is located, two different formulations of the energy rate balance are obtained. In part (a), both streams are included in the control volume. Energy transfer between them occurs internally and not across the boundary of the control volume, so the term $\dot{Q}_{cv}$ drops out of the energy rate balance. With the control volume of part (b), however, the term $\dot{Q}_{cv}$ must be included.

Quick Quiz

If the mass flow rate of the condensing steam is 125 kg/s, determine the mass flow rate of the cooling water, in kg/s. **Ans.** 4538 kg/s.

Excessive temperatures in electronic components are avoided by providing appropriate cooling. In the next example, we analyze the cooling of computer components, illustrating the use of the control volume form of energy rate balance together with property data for air.

Example 4.8 COOLING COMPUTER COMPONENTS

The electronic components of a computer are cooled by air flowing through a fan mounted at the inlet of the electronics enclosure. At steady state, air enters at 20°C, 1 atm. For noise control, the velocity of the entering air cannot exceed 1.3 m/s. For temperature control, the temperature of the air at the exit cannot exceed 32°C. The electronic components and fan receive, respectively, 80 W and 18 W of electric power. Determine the smallest fan inlet area, in cm^2, for which the limits on the entering air velocity and exit air temperature are met.

Solution

Known: The electronic components of a computer are cooled by air flowing through a fan mounted at the inlet of the electronics enclosure. Conditions are specified for the air at the inlet and exit. The power required by the electronics and the fan are also specified.

Find: Determine the smallest fan area for which the specified limits are met.

Schematic and Given Data:

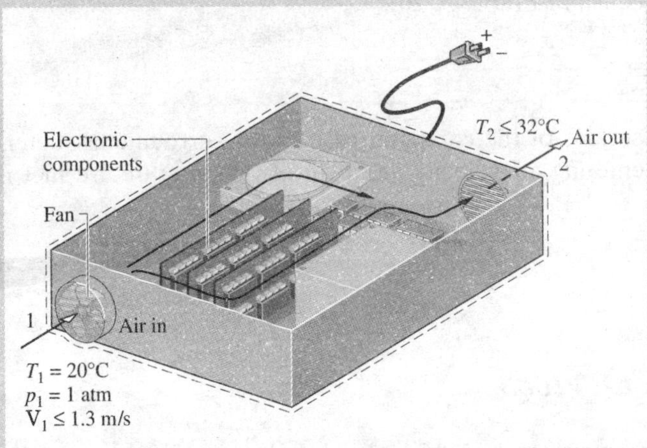

Fig. E4.8

Engineering Model:

1. The control volume shown on the accompanying figure is at steady state.
2. Heat transfer from the *outer* surface of the electronics enclosure to the surroundings is negligible. Thus, $\dot{Q}_{cv} = 0$.
3. ❶ Changes in kinetic and potential energies can be ignored.
4. ❷ Air is modeled as an ideal gas with $c_p = 1.005$ kJ/kg · K.

Analysis: The inlet area A_1 can be determined from the mass flow rate $\dot{m}$ and Eq. 4.4b, which can be rearranged to read

$$A_1 = \frac{\dot{m}v_1}{V_1} \tag{a}$$

The mass flow rate can be evaluated, in turn, from the steady-state energy rate balance, Eq. 4.20a.

$$0 = \underline{\dot{Q}_{cv}} - \dot{W}_{cv} + \dot{m}\left[(h_1 - h_2) + \underline{\left(\frac{V_1^2 - V_2^2}{2}\right)} + \underline{g(z_1 - z_2)}\right]$$

The underlined terms drop out by assumptions 2 and 3, leaving

$$0 = -\dot{W}_{cv} + \dot{m}(h_1 - h_2)$$

where $\dot{W}_{cv}$ accounts for the *total* electric power provided to the electronic components and the fan: $\dot{W}_{cv} = (-80 \text{ W}) + (-18 \text{ W}) = -98 \text{ W}$. Solving for $\dot{m}$, and using assumption 4 with Eq. 3.51 to evaluate $(h_1 - h_2)$

$$\dot{m} = \frac{(-\dot{W}_{cv})}{c_p(T_2 - T_1)}$$

Introducing this into the expression for A_1, Eq. (a), and using the ideal gas model to evaluate the specific volume v_1

$$A_1 = \frac{1}{V_1}\left[\frac{(-\dot{W}_{cv})}{c_p(T_2 - T_1)}\right]\left(\frac{RT_1}{p_1}\right)$$

From this expression we see that A_1 *increases* when V_1 and/or T_2 *decrease*. Accordingly, since $V_1 \leq 1.3$ m/s and $T_2 \leq 305$ K (32°C), the inlet area must satisfy

$$A_1 \geq \frac{1}{1.3 \text{ m/s}}\left[\frac{98 \text{ W}}{\left(1.005 \frac{\text{kJ}}{\text{kg}\cdot\text{K}}\right)(305 - 293)\text{K}}\left|\frac{1 \text{ kJ}}{10^3 \text{ J}}\right|\left|\frac{1 \text{ J/s}}{1 \text{ W}}\right|\right]\left(\frac{\left(\frac{8314 \text{ N}\cdot\text{m}}{28.97 \text{ kg}\cdot\text{K}}\right)293 \text{ K}}{1.01325 \times 10^5 \text{ N/m}^2}\right)\left|\frac{10^4 \text{ cm}^2}{1 \text{ m}^2}\right|$$

$$\geq 52 \text{ cm}^2$$

For the specified conditions, the smallest fan area is 52 cm².

❶ Cooling air typically enters and exits electronic enclosures at low velocities, and thus kinetic energy effects are insignificant.

❷ The applicability of the ideal gas model can be checked by reference to the generalized compressibility chart. Since the temperature of the air increases by no more than 12°C, the specific heat c_p is nearly constant (Table A-20).

Skills Developed: *Ability to...*

- ✓ apply the steady-state energy rate balance to a control volume.
- ✓ apply the mass flow rate expression, Eq. 4.4b.
- ✓ develop an engineering model.
- ✓ retrieve property data of air modeled as an ideal gas.

Quick Quiz

If heat transfer occurs at a rate of 11 W from the outer surface of the computer case to the surroundings, determine the smallest fan inlet area for which the limits on entering air velocity and exit air temperature are met if the total power input remains at 98 W.
Ans. 46 cm².

4.10 Throttling Devices

A significant reduction in pressure can be achieved simply by introducing a restriction into a line through which a gas or liquid flows. This is commonly done by means of a partially opened valve or a porous plug. These *throttling* devices are illustrated in Fig. 4.15.

An application of throttling occurs in vapor-compression refrigeration systems, where a valve is used to reduce the pressure of the refrigerant from the pressure at the exit of the *condenser* to the lower pressure existing in the *evaporator*. We consider this further in Chap. 10. Throttling also plays a role in the *Joule–Thomson*

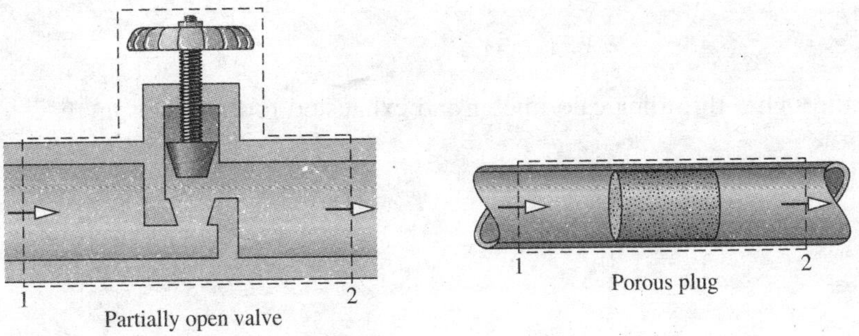

Fig. 4.15 Examples of throttling devices.

expansion considered in Chap. 11. Another application involves the ***throttling calorimeter,*** which is a device for determining the quality of a two-phase liquid–vapor mixture. The throttling calorimeter is considered in Example 4.9. ***throttling calorimeter***

4.10.1 Throttling Device Modeling Considerations

For a control volume enclosing a throttling device, the only work is flow work at locations where mass enters and exits the control volume, so the term $\dot{W}_{cv}$ drops out of the energy rate balance. There is usually no significant heat transfer with the surrounding, and the change in potential energy from inlet to exit is negligible. Thus, the underlined terms of Eq. 4.20a (repeated below) drop out, leaving the enthalpy and kinetic energy terms, as shown by Eq. (a). That is,

$$0 = \underline{\dot{Q}_{cv}} - \underline{\dot{W}_{cv}} + \dot{m}\left[(h_1 - h_2) + \frac{(V_1^2 - V_2^2)}{2} + \underline{g(z_1 - z_2)}\right]$$

$$0 = (h_1 - h_2) + \frac{V_1^2 - V_2^2}{2} \tag{a}$$

Although velocities may be relatively high in the vicinity of the restriction imposed by the throttling device on the flow through it, measurements made upstream and downstream of the reduced flow area show in most cases that the change in the specific kinetic energy of the flowing substance between these locations can be neglected. With this further simplification, Eq. (a) reduces to

$$h_2 = h_1 \qquad (p_2 < p_1) \tag{4.22}$$

When the flow through the valve or other restriction is idealized in this way, the process is called a ***throttling process.*** ***throttling process***

4.10.2 Application to a Throttling Calorimeter

Example 4.9 MEASURING STEAM QUALITY

A supply line carries a two-phase liquid–vapor mixture of steam at 20 bar. A small fraction of the flow in the line is diverted through a throttling calorimeter and exhausted to the atmosphere at 1 bar. The temperature of the exhaust steam is measured as 120°C. Determine the quality of the steam in the supply line.

Solution

Known: Steam is diverted from a supply line through a throttling calorimeter and exhausted to the atmosphere.

Find: Determine the quality of the steam in the supply line.

Schematic and Given Data:

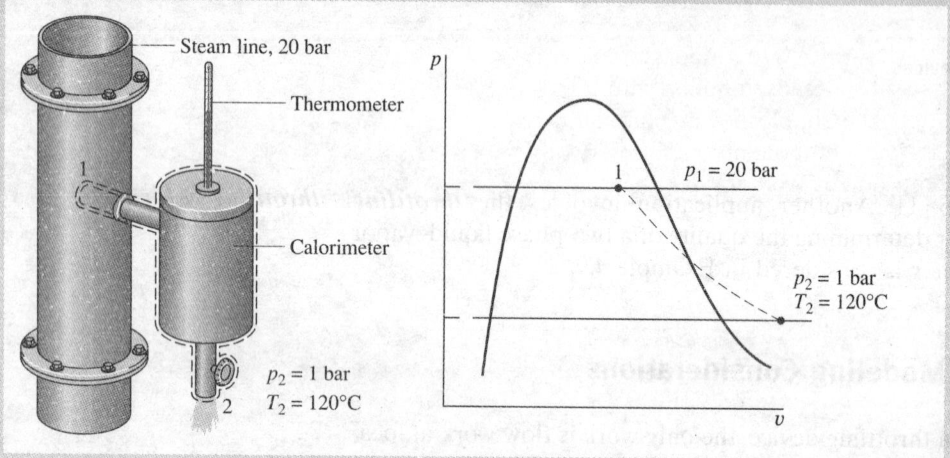

Fig. E4.9

Engineering Model:

1. The control volume shown on the accompanying figure is at steady state.
2. The diverted steam undergoes a throttling process.

Analysis: For a throttling process, the energy and mass balances reduce to give $h_2 = h_1$, which agrees with Eq. 4.22. Thus, with state 2 fixed, the specific enthalpy in the supply line is known, and state 1 is fixed by the known values of p_1 and h_1.

❶ As shown on the accompanying p–v diagram, state 1 is in the two-phase liquid–vapor region and state 2 is in the superheated vapor region. Thus,

$$h_2 = h_1 = h_{f1} + x_1(h_{g1} - h_{f1})$$

Solving for x_1,-

$$x_1 = \frac{h_2 - h_{f1}}{h_{g1} - h_{f1}}$$

From Table A-3 at 20 bar, h_{f1} = 908.79 kJ/kg and h_{g1} = 2,799.5 kJ/kg. At 1 bar and 120°C, h_2 = 2,766.6 kJ/kg from Table A-4. Inserting values into the above expression, the quality of the steam in the line is x_1 = 0.956 (95.6%).

❶ For throttling calorimeters exhausting to the atmosphere, the quality of the steam in the line must be greater than about 94% to ensure that the steam leaving the calorimeter is superheated.

Skills Developed: *Ability to...*
- ✔ apply Eq. 4.22 for a throttling process.
- ✔ retrieve property data for water.

Quick Quiz

If the supply line carried saturated vapor at 20.5 bar, determine the temperature at the calorimeter exit, in °C, for the same exit pressure, 1 bar.

Ans. 162.2°C.

4.11 System Integration

Thus far, we have studied several types of components selected from those commonly seen in practice. These components are usually encountered in combination, rather than individually. Engineers often must creatively combine components to achieve some overall objective, subject to constraints such as minimum total cost. This important engineering activity is called *system integration*.

In engineering practice and everyday life, integrated systems are regularly encountered. Many readers are already familiar with a particularly successful system integration: the simple power plant shown in Fig. 4.16. This system consists of four components in series: a turbine-generator, condenser, pump, and boiler. We consider such power plants in detail in subsequent sections of the book.

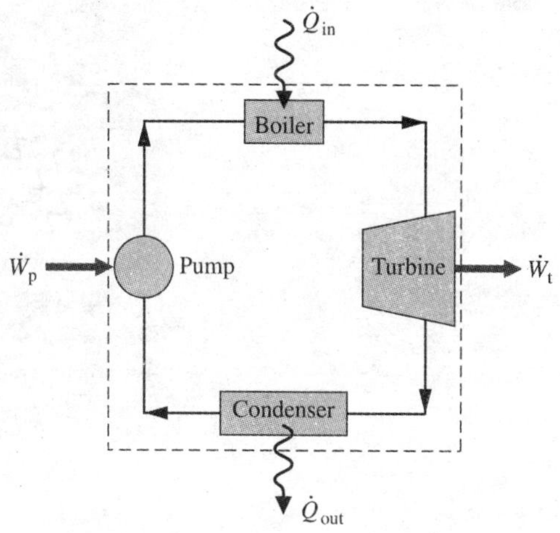

Fig. 4.16 Simple vapor power plant.

Energy & Environment

Healthy, comfortable homes that cut energy and water bills and protect the environment cost no more, builders say. The "I have a Dream House," a highly energy efficient and environmentally responsible house located close to the Atlanta boyhood home of Dr. Martin Luther King Jr., is a prime example.

The house, developed under U.S. Department of Energy auspices, can be heated and cooled for less than a dollar a day, and uses 57% less energy for heating and cooling than a conventional house. Still, construction costs are no more than for a conventional house.

Designers used a whole-house *integrated system* approach whereby components are carefully selected to be complementary in achieving an energy-thrifty, cost-effective outcome. The walls, roof, and floor of this 145-square-meter house are factory-built structural insulated panels incorporating foam insulation. This choice allowed designers to reduce the size of the heating and cooling equipment, thereby lowering costs. The house also features energy-efficient windows, tightly sealed ductwork, and a high-efficiency air conditioner that further contribute to energy savings.

Bio... connections

Living things also can be considered integrated systems. Figure 4.17 shows a control volume enclosing a tree receiving solar radiation. As indicated on the figure, a portion of the incident radiation is reflected to the surroundings. Of the net solar energy received by the tree, about 21% is returned to the surroundings by heat transfer, principally convection. Water management accounts for most of the remaining solar input.

Trees *sweat* as do people; this is called *evapotranspiration*. As shown in Fig. 4.17, about 78% of the net solar energy received by the tree is used to pump liquid water from the surroundings, primarily the ground, convert it to a vapor, and discharge it to the surroundings through tiny pores (called *stomata*) in the leaves. Nearly all the water taken up is lost in this manner and only a small fraction is used within the tree. Applying an energy balance to the control volume enclosing the tree, just 1% of the net solar energy received by the tree is left for use in the production of biomass (wood and leaves). Evapotranspiration benefits trees but also contributes significantly to water loss from watersheds, illustrating that in nature as in engineering there are *trade-offs*.

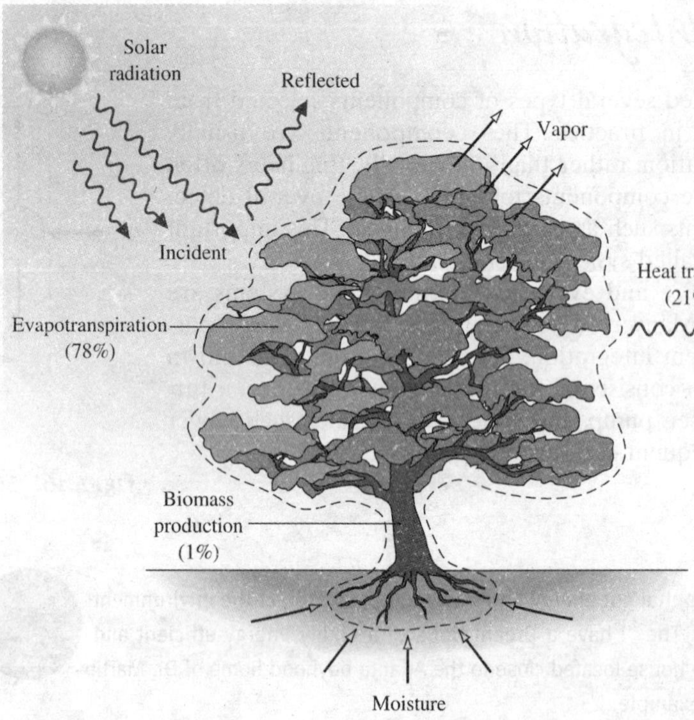

Fig. 4.17 Control volume enclosing a tree.

The example to follow provides another illustration of an integrated system. This case involves a *waste-heat* recovery system.

Example 4.10 WASTE HEAT RECOVERY SYSTEM

An industrial process discharges gaseous combustion products at 478°K, 1 bar with mass flow rate of 69.78 kg/s. As shown in Fig. E4.10, a proposed system for utilizing the combustion products combines a heat-recovery steam generator with a turbine. At steady state, combustion products exit the steam generator at 400°K, 1 bar and a separate stream of water enters at 0.275 MPa, 38.9°C with a mass flow rate of 2.079 kg/s. At the exit of the turbine, the pressure is 0.07 bar and the quality is 93%. Heat transfer from the outer surfaces of the steam generator and turbine can be ignored, as can the changes in kinetic and potential energies of the flowing streams. There is no significant pressure drop for the water flowing through the steam generator. The combustion products can be modeled as air as an ideal gas.

(a) Determine the power developed by the turbine, in kJ/s.

(b) Determine the turbine inlet temperature, in °C.

(c) Evaluating the power developed at \$0.08 per kW · h, which is a typical rate for electricity, determine the value of the power, in \$/year, for 8,000 hours of operation annually.

Solution

Known: Steady-state operating data are provided for a system consisting of a heat-recovery steam generator and a turbine.

Find: Determine the power developed by the turbine and the turbine inlet temperature. Evaluate the annual value of the power developed.

Schematic and Given Data:

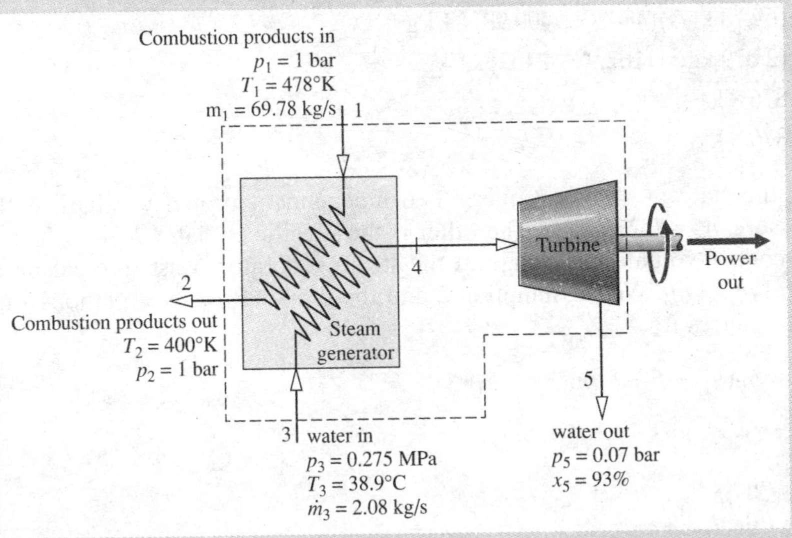

Fig. E4.10

Engineering Model:

1. The control volume shown on the accompanying figure is at steady state.
2. Heat transfer is negligible, and changes in kinetic and potential energy can be ignored.
3. There is no pressure drop for water flowing through the steam generator.
4. The combustion products are modeled as air as an ideal gas.

Analysis:

(a) The power developed by the turbine is determined from a control volume enclosing both the steam generator and the turbine. Since the gas and water streams do not mix, mass rate balances for each of the streams reduce, respectively, to give

$$\dot{m}_1 = \dot{m}_2, \qquad \dot{m}_3 = \dot{m}_5$$

For this control volume, the appropriate form of the steady-state energy rate balance is Eq. 4.18, which reads

$$0 = \underline{\dot{Q}_{cv}} - \dot{W}_{cv} + \dot{m}_1\left(h_1 + \underline{\frac{V_1^2}{2} + gz_1}\right) + \dot{m}_3\left(h_3 + \underline{\frac{V_3^2}{2} + gz_3}\right)$$

$$- \dot{m}_2\left(h_2 + \underline{\frac{V_2^2}{2} + gz_2}\right) - \dot{m}_5\left(h_5 + \underline{\frac{V_5^2}{2} + gz_5}\right)$$

The underlined terms drop out by assumption 2. With these simplifications, together with the above mass flow rate relations, the energy rate balance becomes

$$\dot{W}_{cv} = \dot{m}_1(h_1 - h_2) + \dot{m}_3(h_3 - h_5)$$

where $\dot{m}_1 = 69.78$ kg/s, $\dot{m}_3 = 2.08$ kg/s.

The specific enthalpies, h_1 and h_2, can be found from Table A-22: At 478 K, $h_1 = 480.35$ kJ/kg, and at 400 K, $h_2 = 400.98$ kJ/kg. At state 3, water is a liquid. Using Eq. 3.14 and saturated liquid data from Table A-2, $h_3 \approx h_f(T_3) = 162.9$ kJ/kg. State 5 is a two-phase liquid–vapor mixture. With data from Table A-3 and the given quality

$$h_5 = h_{f5} + x_5(h_{g5} - h_{f5})$$
$$= 161 + 0.93(2751.72 - 161) = 2403 \text{ kJ/kg}$$

Substituting values into the expression for $\dot{W}_{cv}$

$$\begin{aligned}\dot{W}_{cv} &= (69.78\ \text{kg/s})(480.3 - 400.98)\ \text{kJ/kg} \\ &\quad + (2.079\ \text{kg/s})\ (162.9 - 2403)\ \text{kJ/kg} \\ &= 876.8\ \text{kJ/s} \\ &= 876.8\ \text{kW}\end{aligned}$$

(b) To determine T_4, it is necessary to fix the state at 4. This requires two independent property values. With assumption 3, one of these properties is pressure, $p_4 = 0.275$ MPa. The other is the specific enthalpy h_4, which can be found from an energy rate balance for a control volume enclosing just the steam generator. Mass rate balances for each of the two streams give $\dot{m}_1 = \dot{m}_2$ and $\dot{m}_3 = \dot{m}_4$. With assumption 2 and these mass flow rate relations, the
❶ steady-state form of the energy rate balance reduces to

$$0 = \dot{m}_1(h_1 - h_2) + \dot{m}_3(h_3 - h_4)$$

Solving for h_4

$$\begin{aligned}h_4 &= h_3 + \frac{\dot{m}_1}{\dot{m}_3}(h_1 - h_2) \\ &= 162.9\frac{\text{kJ}}{\text{kg}} + \left(\frac{69.78}{2.079}\right)(480.3 - 400.98)\frac{\text{kJ}}{\text{kg}} \\ &= 2825\ \text{kJ/kg}\end{aligned}$$

Interpolating in Table A-4 at $p_4 = 0.275$ MPa with h_4, we get $T_4 = 180°\text{C}$.

(c) Using the result of part (a), together with the given economic data and appropriate conversion factors, the value of the power developed for 8000 hours of operation annually is

$$\begin{aligned}\text{Annual value} &= (876.8\ \text{KW})\left(8000\frac{\text{h}}{\text{year}}\right)\left(0.08\frac{\$}{\text{kW}\cdot\text{h}}\right) \\ &= 561{,}152\frac{\$}{\text{year}}\end{aligned}$$

❷

❶ Alternatively, to determine h_4 a control volume enclosing just the turbine can be considered.

❷ The decision about implementing this solution to the problem of utilizing the hot combustion products discharged from an industrial process would necessarily rest on the outcome of a detailed economic evaluation, including the cost of purchasing and operating the steam generator, turbine, and auxiliary equipment.

Skills Developed: Ability to...
- ✓ apply the steady-state mass and energy rate balances to a control volume.
- ✓ apply the mass flow rate expression, Eq. 4.4b.
- ✓ develop an engineering model.
- ✓ retrieve property data for water and for air modeled as an ideal gas.
- ✓ conduct an elementary economic evaluation.

Quick Quiz

Taking a control volume enclosing just the turbine, evaluate the turbine inlet temperature, in °C. **Ans.** 179°C.

4.12 Transient Analysis

transient

Many devices undergo periods of ***transient*** operation during which the state changes with time. Examples include the startup or shutdown of turbines, compressors, and motors. Additional examples include vessels being filled or emptied, as considered in Example 4.2 and in the discussion of Fig. 1.5. Because property values, work and heat transfer rates, and mass flow rates may vary with time during transient operation, the steady-state assumption is not appropriate when analyzing such cases. Special care must be exercised when applying the mass and energy rate balances, as discussed next.

4.12.1 The Mass Balance in Transient Analysis

First, we place the control volume mass balance in a form that is suitable for transient analysis. We begin by integrating the mass rate balance, Eq. 4.2, from time 0 to a final time t. That is

$$\int_0^t \left(\frac{dm_{\text{cv}}}{dt}\right)dt = \int_0^t \left(\sum_i \dot{m}_i\right)dt - \int_0^t \left(\sum_e \dot{m}_e\right)dt$$

This takes the form

$$m_{\text{cv}}(t) - m_{\text{cv}}(0) = \sum_i \left(\underline{\int_0^t \dot{m}_i\, dt}\right) - \sum_e \left(\underline{\int_0^t \dot{m}_e\, dt}\right)$$

Introducing the following symbols for the underlined terms

$$m_i = \int_0^t \dot{m}_i\, dt \quad \begin{cases} \text{amount of mass} \\ \text{entering the control} \\ \text{volume through inlet } i, \\ \text{from time 0 to } t \end{cases}$$

$$m_e = \int_0^t \dot{m}_e\, dt \quad \begin{cases} \text{amount of mass} \\ \text{exiting the control} \\ \text{volume through exit } e, \\ \text{from time 0 to } t \end{cases}$$

the mass balance becomes

$$m_{\text{cv}}(t) - m_{\text{cv}}(0) = \sum_i m_i - \sum_e m_e \tag{4.23}$$

In words, Eq. 4.23 states that the change in the amount of mass contained in the control volume equals the difference between the total incoming and outgoing amounts of mass.

4.12.2 The Energy Balance in Transient Analysis

Next, we integrate the energy rate balance, Eq. 4.15, ignoring the effects of kinetic and potential energy. The result is

$$U_{\text{cv}}(t) - U_{\text{cv}}(0) = Q_{\text{cv}} - W_{\text{cv}} + \sum_i \left(\underline{\int_0^t \dot{m}_i h_i\, dt}\right) - \sum_e \left(\underline{\int_0^t \dot{m}_e h_e\, dt}\right) \tag{4.24a}$$

where Q_{cv} accounts for the net amount of energy transferred by heat into the control volume and W_{cv} accounts for the net amount of energy transferred by work, except for flow work. The integrals shown underlined in Eq. 4.24a account for the energy carried in at the inlets and out at the exits.

For the *special case* where the states at the inlets and exits are *constant with time*, the respective specific enthalpies, h_i and h_e, would be constant, and the underlined terms of Eq. 4.24a become

$$\int_0^t \dot{m}_i h_i\, dt = h_i \int_0^t \dot{m}_i\, dt = h_i m_i$$

$$\int_0^t \dot{m}_e h_e\, dt = h_e \int_0^t \dot{m}_e\, dt = h_e m_e$$

Equation 4.24a then takes the following special form

$$U_{cv}(t) - U_{cv}(0) = Q_{cv} - W_{cv} + \sum_i m_i h_i - \sum_e m_e h_e \tag{4.24b}$$

Whether in the general form, Eq. 4.24a, or the special form, Eq. 4.24b, these equations account for the change in the amount of energy contained within the control volume as the difference between the total incoming and outgoing amounts of energy.

Another special case is when the intensive properties within the control volume are *uniform with position* at each instant. Accordingly, the specific volume and the specific internal energy are uniform throughout and can depend only on time, that is $v(t)$ and $u(t)$. Thus

$$\begin{aligned} m_{cv}(t) &= V_{cv}(t)/v(t) \\ U_{cv}(t) &= m_{cv}(t)u(t) \end{aligned} \tag{4.25}$$

When the control volume is comprised of different phases, the state of each phase would be assumed uniform throughout.

4.12.3 Transient Analysis Applications

The following examples provide illustrations of the transient analysis of control volumes using the conservation of mass and energy principles. In each case considered, we begin with the general forms of the mass and energy balances and reduce them to forms suited for the case at hand, invoking the idealizations discussed in this section when warranted.

The first example considers a vessel that is partially emptied as mass exits through a valve.

[redacted] WITHDRAWING STEAM FROM A TANK AT CONSTANT PRESSURE

A tank having a volume of 0.85 m³ initially contains water as a two-phase liquid–vapor mixture at 260°C and a quality of 0.7. Saturated water vapor at 260°C is slowly withdrawn through a pressure-regulating valve at the top of the tank as energy is transferred by heat to maintain the pressure constant in the tank. This continues until the tank is filled with saturated vapor at 260°C. Determine the amount of heat transfer, in kJ. Neglect all kinetic and potential energy effects.

Solution

Known: A tank initially holding a two-phase liquid–vapor mixture is heated while saturated water vapor is slowly removed. This continues at constant pressure until the tank is filled only with saturated vapor.

Find: Determine the amount of heat transfer.

Schematic and Given Data:

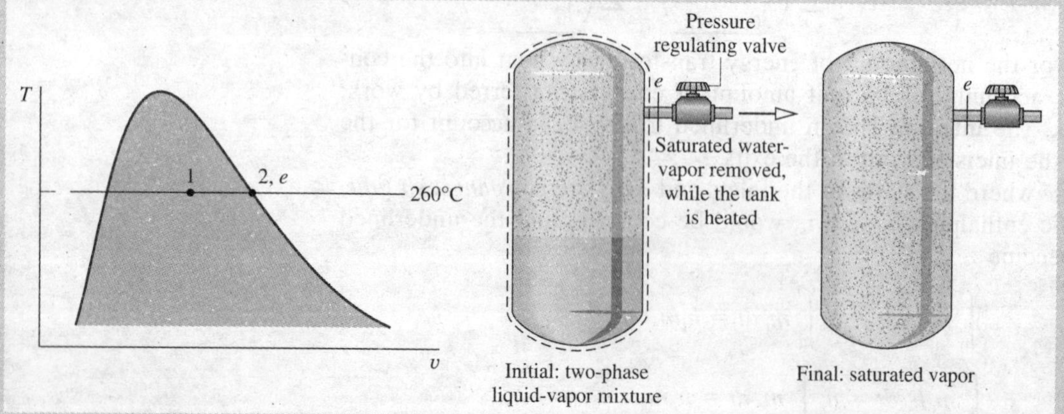

Fig. E4.11

Engineering Model:

1. The control volume is defined by the dashed line on the accompanying diagram.
2. For the control volume, $\dot{W}_{cv} = 0$ and kinetic and potential energy effects can be neglected.
3. At the exit the state remains constant.
4. ❶ The initial and final states of the mass within the vessel are equilibrium states.

Analysis: Since there is a single exit and no inlet, the mass rate balance Eq. 4.2 takes the form

$$\frac{dm_{cv}}{dt} = -\dot{m}_e$$

With assumption 2, the energy rate balance Eq. 4.15 reduces to

$$\frac{dU_{cv}}{dt} = \dot{Q}_{cv} - \dot{m}_e h_e$$

Combining the mass and energy rate balances results in

$$\frac{dU_{cv}}{dt} = \dot{Q}_{cv} + h_e \frac{dm_{cv}}{dt}$$

By assumption 3, the specific enthalpy at the exit is constant. Accordingly, integration of the last equation gives

$$\Delta U_{cv} = Q_{cv} + h_e \, \Delta m_{cv}$$

Solving for the heat transfer Q_{cv},

$$Q_{cv} = \Delta U_{cv} - h_e \, \Delta m_{cv}$$

or

❷
$$Q_{cv} = (m_2 u_2 - m_1 u_1) - h_e(m_2 - m_1)$$

where m_1 and m_2 denote, respectively, the initial and final amounts of mass within the tank.

The terms u_1 and m_1 of the foregoing equation can be evaluated with property values from Table A-2 at 260°C and the given value for quality. Thus

$$\begin{aligned} u_1 &= u_f + x_1(u_g - u_f) \\ &= 1128.4 + (0.7)(2599.0 - 1128.4) = 2157.8 \text{ kJ/kg} \end{aligned}$$

Also,

$$\begin{aligned} v_1 &= v_f + x_1(v_g - v_f) \\ &= 1.2755 \times 10^{-3} + (0.7)(0.04221 - 1.2755 \times 10^{-3}) = 29.93 \times 10^{-3} \text{ m}^3/\text{kg} \end{aligned}$$

Using the specific volume v_1, the mass initially contained in the tank is

$$m_1 = \frac{V}{v_1} = \frac{0.85 \text{ m}^3}{(29.93 \times 10^{-3} \text{ m}^3/\text{kg})} = 28.4 \text{ kg}$$

The final state of the mass in the tank is saturated vapor at 260°C, so Table A-2 gives

$$u_2 = u_g(260°\text{C}) = 2599.0 \text{ kJ/kg}, \qquad v_2 = v_g(260°\text{C}) = 42.21 \times 10^{-3} \text{ m}^3/\text{kg}$$

The mass contained within the tank at the end of the process is

$$m_2 = \frac{V}{v_2} = \frac{0.85 \text{ m}^3}{42.21 \times 10^{-3} \text{ m}^3/\text{kg}} = 20.14 \text{ kg}$$

Table A-2 also gives $h_e = h_g(260°\text{C}) = 2796.6$ kJ/kg.

Substituting values into the expression for the heat transfer yields

$$Q_{cv} = (20.14)(2599.0) - (28.4)(2157.8) - 2796.6(20.14 - 28.4)$$
$$= 14{,}162 \text{ kJ}$$

❶ In this case, idealizations are made about the state of the vapor exiting *and* the initial and final states of the mass contained within the tank.

❷ This expression for Q_{cv} could be obtained by applying Eq. 4.24b together with Eqs. 4.23 and 4.25.

Skills Developed: *Ability to...*
- ✓ apply the time-dependent mass and energy rate balances to a control volume.
- ✓ develop an engineering model.
- ✓ retrieve property data for water.

Quick Quiz

If the initial quality were 90%, determine the heat transfer, in kJ, keeping all other data unchanged.
Ans. 3707 kJ.

In the next two examples we consider cases where tanks are filled. In Example 4.12, an initially evacuated tank is filled with steam as power is developed. In Example 4.13, a compressor is used to store air in a tank.

Example 4.12 USING STEAM FOR EMERGENCY POWER GENERATION

Steam at a pressure of 15 bar and a temperature of 320°C is contained in a large vessel. Connected to the vessel through a valve is a turbine followed by a small initially evacuated tank with a volume of 0.6 m³. When emergency power is required, the valve is opened and the tank fills with steam until the pressure is 15 bar. The temperature in the tank is then 400°C. The filling process takes place adiabatically and kinetic and potential energy effects are negligible. Determine the amount of work developed by the turbine, in kJ.

Solution

Known: Steam contained in a large vessel at a known state flows from the vessel through a turbine into a small tank of known volume until a specified final condition is attained in the tank.

Find: Determine the work developed by the turbine.

Schematic and Given Data:

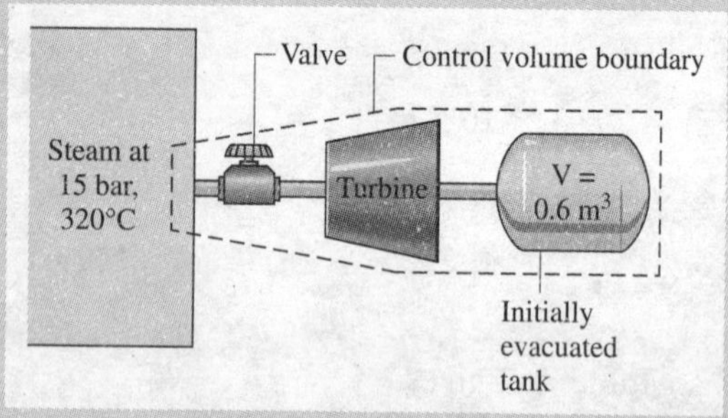

Fig. E4.12

Engineering Model:

1. The control volume is defined by the dashed line on the accompanying diagram.
2. For the control volume, $\dot{Q}_{cv} = 0$ and kinetic and potential energy effects are negligible.
3. ❶ The state of the steam within the large vessel remains constant. The final state of the steam in the smaller tank is an equilibrium state.
4. The amount of mass stored within the turbine and the interconnecting piping at the end of the filling process is negligible.

Analysis: Since the control volume has a single inlet and no exits, the mass rate balance, Eq. 4.2, reduces to

$$\frac{dm_{cv}}{dt} = \dot{m}_i$$

The energy rate balance, Eq. 4.15, reduces with assumption 2 to

$$\frac{dU_{cv}}{dt} = -\dot{W}_{cv} + \dot{m}_i h_i$$

Combining the mass and energy rate balances gives

$$\frac{dU_{cv}}{dt} = -\dot{W}_{cv} + h_i \frac{dm_{cv}}{dt}$$

Integrating

$$\Delta U_{cv} = -W_{cv} + h_i\, \Delta m_{cv}$$

❷ In accordance with assumption 3, the specific enthalpy of the steam entering the control volume is constant at the value corresponding to the state in the large vessel.

Solving for W_{cv}

$$W_{cv} = h_i\, \Delta m_{cv} - \Delta U_{cv}$$

ΔU_{cv} and Δm_{cv} denote, respectively, the changes in internal energy and mass of the control volume. With assumption 4, these terms can be identified with the small tank only.

Since the tank is initially evacuated, the terms ΔU_{cv} and Δm_{cv} reduce to the internal energy and mass within the tank at the end of the process. That is

$$\Delta U_{cv} = (m_2 u_2) - \cancelto{0}{(m_1 u_1)}, \qquad \Delta m_{cv} = m_2 - \cancelto{0}{m_1}$$

where 1 and 2 denote the initial and final states within the tank, respectively.

Collecting results yields

$$W_{cv} = m_2(h_i - u_2) \tag{a}$$

The mass within the tank at the end of the process can be evaluated from the known volume and the specific volume of steam at 15 bar and 400°C from Table A-4

$$m_2 = \frac{V}{v_2} = \frac{0.6\ \text{m}^3}{(0.203\ \text{m}^3/\text{kg})} = 2.96\ \text{kg}$$

The specific internal energy of steam at 15 bar and 400°C from Table A-4 is 2951.3 kJ/kg. Also, at 15 bar and 320°C, $h_1 = 3081.9$ kJ/kg.

Substituting values into Eq. (a)

$$W_{cv} = 2.96\ \text{kg}(3081.9 - 2951.3)\text{kJ/kg} = 386.6\ \text{kJ}$$

❶ In this case idealizations are made about the state of the steam entering the tank *and* the final state of the steam in the tank. These idealizations make the transient analysis manageable.

❷ A significant aspect of this example is the energy transfer into the control volume by flow work, incorporated in the *pv* term of the specific enthalpy at the inlet.

Skills Developed: *Ability to...*
- ✓ apply the time-dependent mass and energy rate balances to a control volume.
- ✓ develop an engineering model.
- ✓ retrieve property data for water.

Quick Quiz

If the turbine were removed, and the steam allowed to flow adiabatically into the small tank until the pressure in the tank is 15 bar, determine the final steam temperature in the tank, in °C.

Ans. 477°C.

Example 4.13 STORING COMPRESSED AIR IN A TANK

An air compressor rapidly fills a 0.28 m^3 tank, initially containing air at 21°C, 1 bar, with air drawn from the atmosphere at 21°C, 1 bar. During filling, the relationship between the pressure and specific volume of the air in the tank is $pv^{1.4} = constant$. The ideal gas model applies for the air, and kinetic and potential energy effects are negligible. Plot the pressure, in atm, and the temperature, in °C, of the air within the tank, each versus the ratio m/m_1, where m_1 is the initial mass in the tank and m is the mass in the tank at time $t > 0$. Also, plot the compressor work input, in kJ, versus m/m_1. Let m/m_1 vary from 1 to 3.

Solution

Known: An air compressor rapidly fills a tank having a known volume. The initial state of the air in the tank and the state of the entering air are known.

Find: Plot the pressure and temperature of the air within the tank, and plot the air compressor work input, each versus m/m_1 ranging from 1 to 3.

Schematic and Given Data:

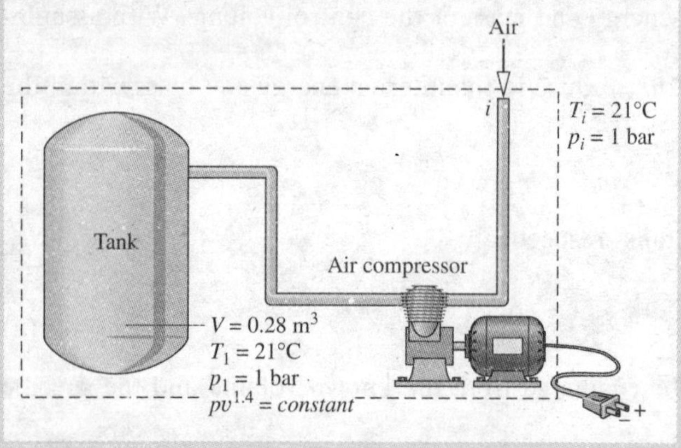

Fig. E4.13a

Engineering Model:

1. The control volume is defined by the dashed line on the accompanying diagram.
2. Because the tank is filled rapidly, $\dot{Q}_{cv}$ is ignored.
3. Kinetic and potential energy effects are negligible.
4. The state of the air entering the control volume remains constant.
5. The air stored within the air compressor and interconnecting pipes can be ignored.
6. ❶ The relationship between pressure and specific volume for the air in the tank is $pv^{1.4} = constant$.
7. The ideal gas model applies for the air.

Analysis: The required plots are developed using *Interactive Thermodynamics: IT*. The *IT* program is based on the following analysis. The pressure p in the tank at time $t > 0$ is determined from

$$pv^{1.4} = p_1 v_1^{1.4}$$

where the corresponding specific volume v is obtained using the known tank volume V and the mass m in the tank at that time. That is, $v = V/m$. The specific volume of the air in the tank initially, v_1, is calculated from the ideal gas equation of state and the known initial temperature, T_1, and pressure, p_1. That is

$$v_1 = \frac{RT_1}{p_1} = \frac{\left(\frac{8314\ \text{N}\cdot\text{m}}{28.97\ \text{kg}\cdot°\text{K}}\right)(294°\text{K})}{(1\ \text{bar})}\left|\frac{1\ \text{bar}}{10^5\ \text{N/m}^2}\right| = 0.8437\ \text{m}^3/\text{kg}$$

Once the pressure p is known, the corresponding temperature T can be found from the ideal gas equation of state, $T = pv/R$.

To determine the work, begin with the mass rate balance Eq. 4.2, which reduces for the single-inlet control volume to

$$\frac{dm_{cv}}{dt} = \dot{m}_i$$

Then, with assumptions 2 and 3, the energy rate balance Eq. 4.15 reduces to

$$\frac{dU_{cv}}{dt} = -\dot{W}_{cv} + \dot{m}_i h_i$$

Combining the mass and energy rate balances and integrating using assumption 4 gives

$$\Delta U_{cv} = -W_{cv} + h_i \Delta m_{cv}$$

Denoting the work *input* to the compressor by $W_{in} = -W_{cv}$ and using assumption 5, this becomes

$$W_{in} = mu - m_1 u_1 - (m - m_1)h_i \quad \text{(a)}$$

where m_1 is the initial amount of air in the tank, determined from

$$m_1 = \frac{V}{v_1} = \frac{0.28 \text{ m}^3}{0.8437 \text{ m}^3/\text{kg}} = 0.332 \text{ kg}$$

As a *sample* calculation to validate the *IT* program below, consider the case $m = 0.664$ kg, which corresponds to $m/m_1 = 2$. The specific volume of the air in the tank at that time is

$$v = \frac{V}{m} = \frac{0.28 \text{ m}^3}{0.664 \text{ kg}} = 0.422 \text{ m}^3/\text{kg}$$

The corresponding pressure of the air is

$$p = p_1\left(\frac{v_1}{v}\right)^{1.4} = (1 \text{ bar})\left(\frac{0.8437 \text{ m}^3/\text{kg}}{0.422 \text{ m}^3/\text{kg}}\right)^{1.4}$$
$$= 2.64 \text{ atm}$$

and the corresponding temperature of the air is

$$T = \frac{pv}{R} = \left(\frac{(2.64 \text{ bar})(0.422 \text{ m}^3/\text{kg})}{\left(\frac{8314}{28.97} \frac{\text{J}}{\text{kg} \cdot {}^\circ\text{K}}\right)}\right)\left|\frac{10^5 \text{ N/m}^2}{1 \text{ bar}}\right|$$
$$= 388^\circ\text{K } (114.9^\circ\text{C})$$

Evaluating u_1, u, and h_i at the appropriate temperatures from Table A-22, $u_1 = 209.8$ kJ/kg, $u = 27.75$ kJ/kg, $h_i = 294.2$ kJ/kg. Using Eq. (a), the required work input is

$$W_{in} = mu - m_1 u_1 - (m - m_1)h_i$$
$$= (0.664 \text{ kg})\left(277.5 \frac{\text{kJ}}{\text{kg}}\right) - (0.332 \text{ kg})\left(209.8 \frac{\text{kJ}}{\text{kg}}\right) - (0.332 \text{ kg})\left(294.2 \frac{\text{kJ}}{\text{kg}}\right)$$
$$= 16.9 \text{ kJ}$$

IT Program: Choosing SI units from the **Units** menu, and selecting Air from the **Properties** menu, the *IT* program for solving the problem is

```
//Given Data
p1 = 1//bar
T1 = 21//°C
Ti = 21//°C
V = 0.28//m³
n = 1.4

// Determine the pressure and temperature for t > o
v1 = v_TP("Air", T1, p1)
v = V/m
p * v ^n = p1 * v1 ^n
v = v_TP("Air", T, p)

// Specify the mass and mass ratio r
v1 = V/m1
r = m/m1
r = 2
```

```
// Calculate the work using Eq. (a)
Win = m * u – m1 * u1 – hi * (m – m1)
u1 = u_T("Air", T1)
u = u_T("Air", T)
hi = h_T("Air", Ti)
```

Using the **Solve** button, obtain a solution for the sample case $r = m/m_1 = 2$ considered above to validate the program. Good agreement is obtained, as can be verified. Once the program is validated, use the **Explore** button to vary the ratio m/m_1 from 1 to 3 in steps of 0.01. Then, use the **Graph** button to construct the required plots. The results are:

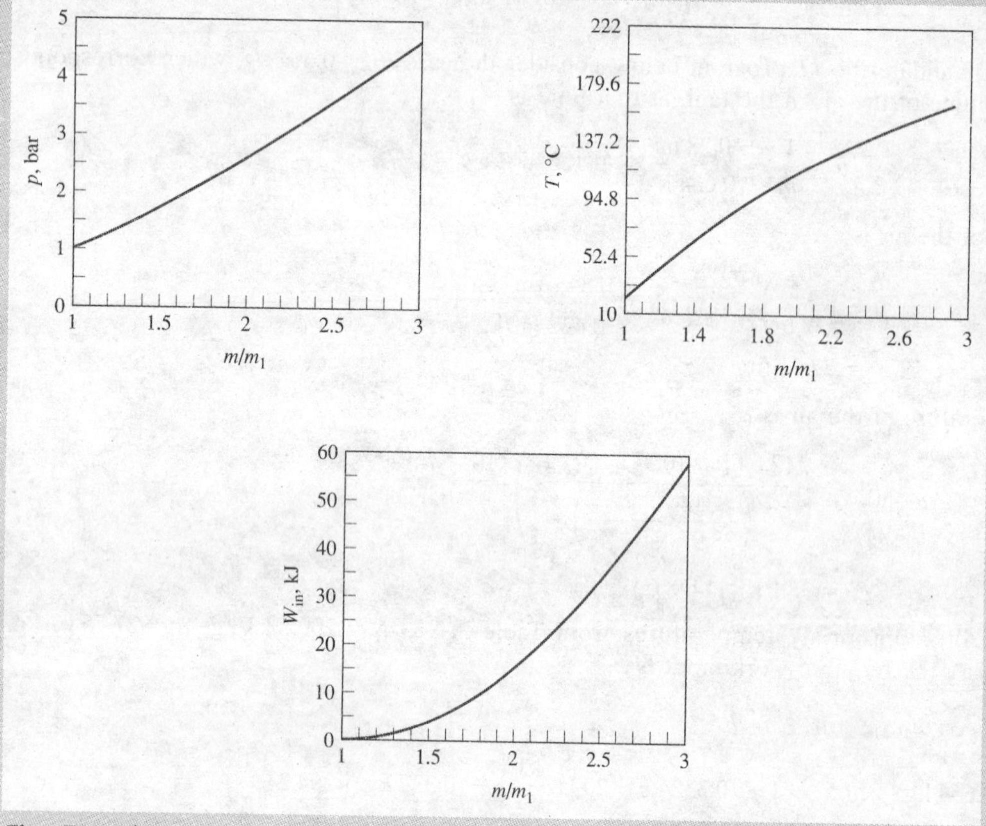

Fig. E4.13b

We conclude from the first two plots that the pressure and temperature each increase as the tank fills. The work required to fill the tank increases as well. These results are as expected.

---

❶ This pressure-specific volume relationship is in accord with what might be measured. The relationship is also consistent with the uniform state idealization, embodied by Eq. 4.25.

Skills Developed: *Ability to...*

- ✓ apply the time-dependent mass and energy rate balances to a control volume.
- ✓ develop an engineering model.
- ✓ retrieve property data for air modeled as an ideal gas.
- ✓ solve iteratively and plot the results using *IT*.

Quick Quiz

As a *sample* calculation, for the case $m = 1$ kg, evaluate p, in bar. Compare with the value read from the plot of Fig. E4.13 b.

Ans. 4.73 bar.

The final example of transient analysis is an application with a *well-stirred* tank. Such process equipment is commonly employed in the chemical and food processing industries.

Example 4.14 TEMPERATURE VARIATION IN A WELL-STIRRED TANK

A tank containing 45 kg of liquid water initially at 45°C has one inlet and one exit with equal mass flow rates. Liquid water enters at 45°C and a mass flow rate of 270 kg/h. A cooling coil immersed in the water removes energy at a rate of 7.6 kW. The water is well mixed by a paddle wheel so that the water temperature is uniform throughout. The power input to the water from the paddle wheel is 0.6 kW. The pressures at the inlet and exit are equal and all kinetic and potential energy effects can be ignored. Plot the variation of water temperature with time.

Solution

Known: Liquid water flows into and out of a well-stirred tank with equal mass flow rates as the water in the tank is cooled by a cooling coil.

Find: Plot the variation of water temperature with time.

Schematic and Given Data:

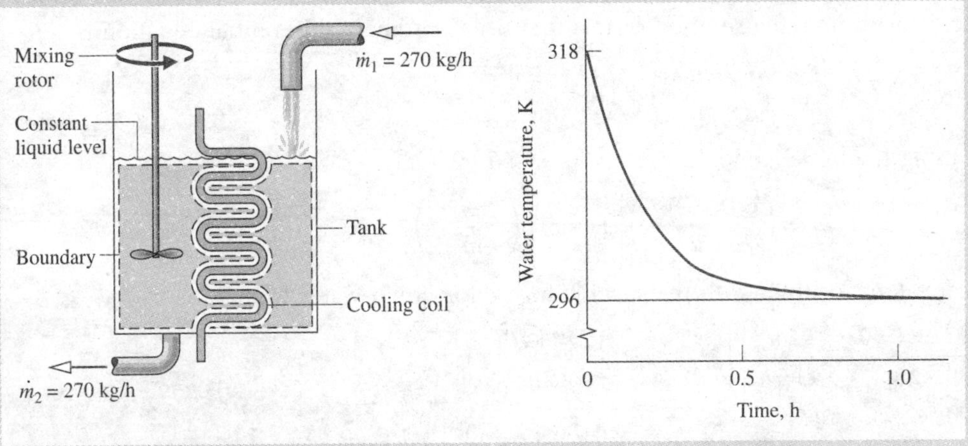

Fig. E4.14

Engineering Model:

1. The control volume is defined by the dashed line on the accompanying diagram.
2. For the control volume, the only significant heat transfer is with the cooling coil. Kinetic and potential energy effects can be neglected.
3. ❶ The water temperature is uniform with position throughout: $T = T(t)$.
4. The water in the tank is incompressible, and there is no change in pressure between inlet and exit.

Analysis: The energy rate balance, Eq. 4.15, reduces with assumption 2 to

$$\frac{dU_{cv}}{dt} = \dot{Q}_{cv} - \dot{W}_{cv} + \dot{m}(h_1 - h_2)$$

where $\dot{m}$ denotes the mass flow rate.

The mass contained within the control volume remains constant with time, so the term on the left side of the energy rate balance can be expressed as

$$\frac{dU_{cv}}{dt} = \frac{d(m_{cv}u)}{dt} = m_{cv}\frac{du}{dt}$$

Since the water is assumed incompressible, the specific internal energy depends on temperature only. Hence, the chain rule can be used to write

$$\frac{du}{dt} = \frac{du}{dT}\frac{dT}{dt} = c\frac{dT}{dt}$$

where c is the specific heat. Collecting results

$$\frac{dU_{cv}}{dt} = m_{cv}c\frac{dT}{dt}$$

With Eq. 3.20b the enthalpy term of the energy rate balance can be expressed as

$$h_1 - h_2 = c(T_1 - T_2) + v\overset{0}{\cancel{(p_1 - p_2)}}$$

where the pressure term is dropped by assumption 4. Since the water is well mixed, the temperature at the exit equals the temperature of the overall quantity of liquid in the tank, so

$$h_1 - h_2 = c(T_1 - T)$$

where T represents the uniform water temperature at time t.

With the foregoing considerations the energy rate balance becomes

$$m_{cv}c\frac{dT}{dt} = \dot{Q}_{cv} - \dot{W}_{cv} + \dot{m}c(T_1 - T)$$

As can be verified by direct substitution, the solution of this first-order, ordinary differential equation is

$$T = C_1 \exp\left(-\frac{\dot{m}}{m_{cv}}t\right) + \left(\frac{\dot{Q}_{cv} - \dot{W}_{cv}}{\dot{m}c}\right) + T_1$$

The constant C_1 is evaluated using the initial condition: at $t = 0$, $T = T_1$. Finally

$$T = T_1 + \left(\frac{\dot{Q}_{cv} - \dot{W}_{cv}}{\dot{m}c}\right)\left[1 - \exp\left(-\frac{\dot{m}}{m_{cv}}t\right)\right]$$

Substituting given numerical values together with the specific heat c for liquid water from Table A-19

$$T = 318\text{ K} + \left[\frac{[-7.6 - (-0.6)]\text{ kJ/s}}{\left(\frac{270}{3600}\frac{\text{kg}}{\text{s}}\right)\left(4.2\frac{\text{kJ}}{\text{kg}\cdot\text{K}}\right)}\right]\left[1 - \exp\left(-\frac{270\text{ kg/h}}{45\text{ kg}}t\right)\right]$$

$$= 318 - 22[1 - \exp(-6t)]$$

where t is in hours. Using this expression, we can construct the accompanying plot showing the variation of temperature with time.

---

❶ In this case idealizations are made about the state of the mass contained within the system and the states of the liquid entering and exiting. These idealizations make the transient analysis manageable.

Skills Developed: *Ability to...*

- ✓ apply the time-dependent mass and energy rate balances to a control volume.
- ✓ develop an engineering model.
- ✓ apply the incompressible substance model for water.
- ✓ solve an ordinary differential equation and plot the solution.

Quick Quiz

What is the water temperature, in °C, when *steady-state* is achieved?

Ans. 23°C.

✓ Chapter Summary and Study Guide

The conservation of mass and energy principles for control volumes are embodied in the mass and energy rate balances developed in this chapter. Although the primary emphasis is on cases in which one-dimensional flow is assumed, mass and energy balances are also presented in integral forms that provide a link to subsequent fluid mechanics and heat transfer courses. Control volumes at steady state are featured, but discussions of transient cases are also provided.

The use of mass and energy balances for control volumes at steady state is illustrated for nozzles and diffusers, turbines, compressors and pumps, heat exchangers, throttling devices, and integrated systems. An essential aspect of all such applications is the careful and explicit listing of appropriate assumptions. Such model-building skills are stressed throughout the chapter.

The following checklist provides a study guide for this chapter. When your study of the text and end-of-chapter exercises has been completed you should be able to

- ✓ write out the meanings of the terms listed in the margins throughout the chapter and understand each of the related concepts. The subset of key concepts listed below is particularly important in subsequent chapters.
- ✓ list the typical modeling assumptions for nozzles and diffusers, turbines, compressors and pumps, heat exchangers, and throttling devices.
- ✓ apply Eqs. 4.6 and 4.18–4.20 to control volumes at steady state, using appropriate assumptions and property data for the case at hand.
- ✓ apply mass and energy balances for the transient analysis of control volumes, using appropriate assumptions and property data for the case at hand.

Key Engineering Concepts

conservation of mass p. 147
mass flow rates p. 147
mass rate balance p. 147
one-dimensional flow p. 149
volumetric flow rate p. 150
steady state p. 150
flow work p. 157
energy rate balance p. 157
nozzle p. 161
diffuser p. 161
turbine p. 164
compressor p. 167
pump p. 167
heat exchanger p. 171
throttling process p. 177

Key Equations

$\dot{m} = \dfrac{\mathrm{AV}}{v}$	(4.4b)	Mass flow rate, one-dimensional flow (See Fig. 4.3.)
$\dfrac{dm_{\text{cv}}}{dt} = \sum_i \dot{m}_i - \sum_e \dot{m}_e$	(4.2)	Mass rate balance.
$\underset{\text{(mass rate in)}}{\sum_i \dot{m}_i} = \underset{\text{(mass rate out)}}{\sum_e \dot{m}_e}$	(4.6)	Mass rate balance at steady state.
$\dfrac{dE_{\text{cv}}}{dt} = \dot{Q}_{\text{cv}} - \dot{W}_{\text{cv}} + \sum_i \dot{m}_i\left(h_i + \dfrac{\mathrm{V}_i^2}{2} + gz_i\right) - \sum_e \dot{m}_e\left(h_e + \dfrac{\mathrm{V}_e^2}{2} + gz_e\right)$	(4.15)	Energy rate balance.
$0 = \dot{Q}_{\text{cv}} - \dot{W}_{\text{cv}} + \sum_i \dot{m}_i\left(h_i + \dfrac{\mathrm{V}_i^2}{2} + gz_i\right) - \sum_e \dot{m}_e\left(h_e + \dfrac{\mathrm{V}_e^2}{2} + gz_e\right)$	(4.18)	Energy rate balance at steady state.
$0 = \dot{Q}_{\text{cv}} - \dot{W}_{\text{cv}} + \dot{m}\left[(h_1 - h_2) + \dfrac{(\mathrm{V}_1^2 - \mathrm{V}_2^2)}{2} + g(z_1 - z_2)\right]$ $0 = \dfrac{\dot{Q}_{\text{cv}}}{\dot{m}} - \dfrac{\dot{W}_{\text{cv}}}{\dot{m}} + (h_1 - h_2) + \dfrac{(\mathrm{V}_1^2 - \mathrm{V}_2^2)}{2} + g(z_1 - z_2)$	(4.20a) (4.20b)	Energy rate balance for one-inlet, one-exit control volumes at steady state.
$h_2 = h_1 \qquad (p_2 < p_1)$	(4.22)	Throttling process. (See Fig. 4.15.)

Exercises: things engineers think about

1. Why does the relative velocity *normal* to the flow boundary, V_n, appear in Eqs. 4.3 and 4.8?

2. Why might a computer cooled by a *constant-speed* fan operate satisfactorily at sea level but overheat at high altitude?

3. By introducing enthalpy h to replace each of the $(u + pv)$ terms of Eq. 4.13, we get Eq. 4.14. An even simpler algebraic form would result by replacing each of the $(u + pv + V^2/2 + gz)$ terms by a single symbol, yet we have not done so. Why not?

4. When a slice of bread is placed in a toaster and the toaster is activated, is the toaster in steady-state operation, transient operation, both?

5. How is the work done by the heart measured?

6. Where do I encounter *microelectromechanical* systems in daily life?

7. Wind turbines and hydraulic turbines develop mechanical power from moving streams of air and water, respectively. In each case, what aspect of the stream is tapped for power?

8. For air flowing through a converging-diverging channel, sketch the variation of the air pressure as air accelerates in the converging section and decelerates in the diverging section.

9. Even though their outer surfaces would seem hot to the touch, large steam turbines in power plants might not be covered with much insulation. Why not?

10. One stream of a counterflow heat exchanger operating at steady state has saturated water vapor entering and saturated liquid exiting, each at 1 atm. The other stream has ambient air entering at 20°C and exiting at a higher temperature with no significant change in pressure. What is the variation of temperature with position of each stream as it passes through the heat exchanger? Sketch them.

11. Why does *evapotranspiration* in a tree require so much energy?

12. What are some examples of commonly encountered devices that undergo periods of transient operation? For each example, which type of system, closed system or control volume, would be most appropriate?

13. Why is it that when air at 1 atm is *throttled* to a pressure of 0.5 atm, its temperature at the valve exit is close to the temperature at the valve inlet, yet when air at 1 atm *leaks* into an insulated, rigid, initially evacuated tank until the tank pressure is 0.5 atm, the temperature of the air in the tank is greater than the air temperature outside the tank?

Problems: developing engineering skills

Applying Conservation of Mass

4.1 A 0.15 m³ tank, initially evacuated, develops a small hole, and air leaks in from the surroundings at a constant mass flow rate of 0.002 kg/s. Using the ideal gas model for air, determine the pressure, in kPa, in the tank after 30 s if the temperature is 20°C.

4.2 Liquid propane enters an initially empty cylindrical storage tank at a mass flow rate of 10 kg/s. The tank is 25-m long and has a 4-m diameter. The density of the liquid propane is 450 kg/m³. Determine the time, in h, to fill the tank.

4.3 A 0.5-m³ tank contains ammonia, initially at −40°C, 8 bar. A leak develops, and refrigerant flows out of the tank at a constant mass flow rate of 0.04 kg/s. The process occurs slowly enough that heat transfer from the surroundings maintains a constant temperature in the tank. Determine the time, in s, at which half of the mass has leaked out, and the pressure in the tank at that time, in bar.

4.4 Data are provided for the crude oil storage tank shown in Fig. P4.4. The tank initially contains 1,000 m³ of crude oil. Oil is pumped into the tank through a pipe at a rate of 2 m³/min and out of the tank at a velocity of 1.5 m/s through another pipe having a diameter of 0.15 m. The crude oil has a specific volume of 0.0015 m³/kg. Determine

(a) the mass of oil in the tank, in kg, after 24 hours, and
(b) the volume of oil in the tank, in m³, at that time.

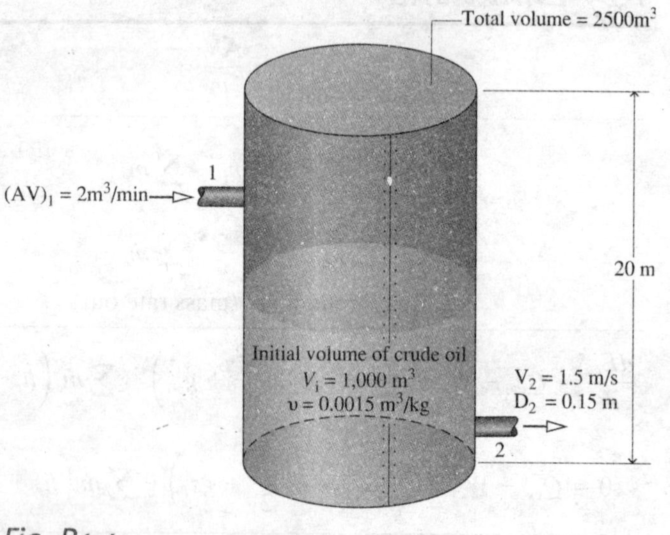

Fig. P4.4

4.5 Vegetable oil for cooking is dispensed from a cylindrical can fitted with a spray nozzle. According to the label, the can is able to deliver 560 sprays, each of duration 0.25 s and each having a mass of 0.25 g. Determine

(a) the mass flow rate of each spray, in g/s.
(b) the mass remaining in the can after 560 sprays, in g, if the initial mass in the can is 170 g.

4.6 Figure P4.6 shows a mixing tank initially containing 1,360 kg of liquid water. The tank is fitted with two inlet pipes, one delivering hot water at a mass flow rate of 0.4 kg/s and the other delivering cold water at a mass flow rate of 0.6 kg/s. Water exits through a single exit pipe at a mass flow rate of 1.12 kg/s. Determine the amount of water, in kg, in the tank after one hour.

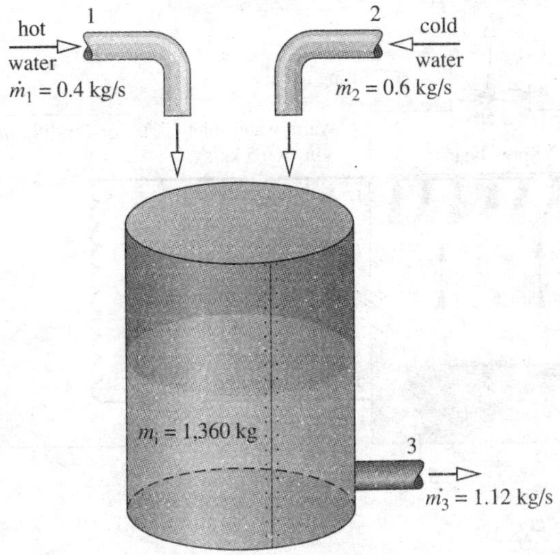

Fig. P4.6

4.7 A water storage tank initially contains 400 m^3 of water. The average daily usage is 40 m^3. If water is added to the tank at an average rate of $20[\exp(-t/20)]$ m^3 gallons per day, where t is time in days, for how many days will the tank contain water?

4.8 A carburetor in an internal combustion engine mixes air with fuel to achieve a combustible mixture in which the *air-to-fuel ratio* is 20 kg (air) per kg (fuel). For a fuel mass flow rate of 5×10^{-3} kg/s, determine the mass flow rate of the mixture, in kg/s.

4.9 Air enters a one-inlet, one-exit control volume at 8 bar, 600 K, and 40 m/s through a flow area of 20 cm^2. At the exit, the pressure is 2 bar, the temperature is 400 K, and the velocity is 350 m/s. The air behaves as an ideal gas. For steady-state operation, determine

(a) the mass flow rate, in kg/s.
(b) the exit flow area, in cm^2.

4.10 The small two-story office building shown in Fig. P4.10 has 1,020 m^3 of occupied space. Due to cracks around windows and outside doors, air leaks in on the windward side of the building and leaks out on the leeward side of the building. Outside air also enters the building when outer doors are opened. On a particular day, tests were conducted. The outdoor temperature was measured to be −9.5°C. The inside temperature was controlled at 20°C. Keeping the doors closed, the infiltration rate through the cracks was determined to be 0.03 m^3/s. The infiltration rate associated with door openings, averaged over the work day, was 0.02 m^3/s. The pressure difference was negligible between the inside and outside of the building. (a) Assuming ideal gas behavior, determine at steady state the volumetric flow rate of air exiting the building, in m^3/s. (b) When expressed in terms of the volume of the occupied space, determine the number of building air changes per hour.

4.11 Air enters a household electric furnace at 24°C, 1 bar, with a volumetric flow rate of 0.38 m^3/s. The furnace delivers air at 50°C, 1 bar to a duct system with three branches consisting of two 150 mm diameter ducts and a 300 mm duct. If the velocity in each 150 mm duct is 3 m/s, determine for steady-state operation

(a) the mass flow rate of air entering the furnace, in kg/s.
(b) the volumetric flow rate in each 150 mm duct, in m^3/s.
(c) the velocity in the 300 mm duct, in m/s.

4.12 Refrigerant 134a enters the condenser of a refrigeration system operating at steady state at 9 bar, 50°C, through a 2.5-cm-diameter pipe. At the exit, the pressure is 9 bar, the temperature is 30°C, and the velocity is 2.5 m/s. The

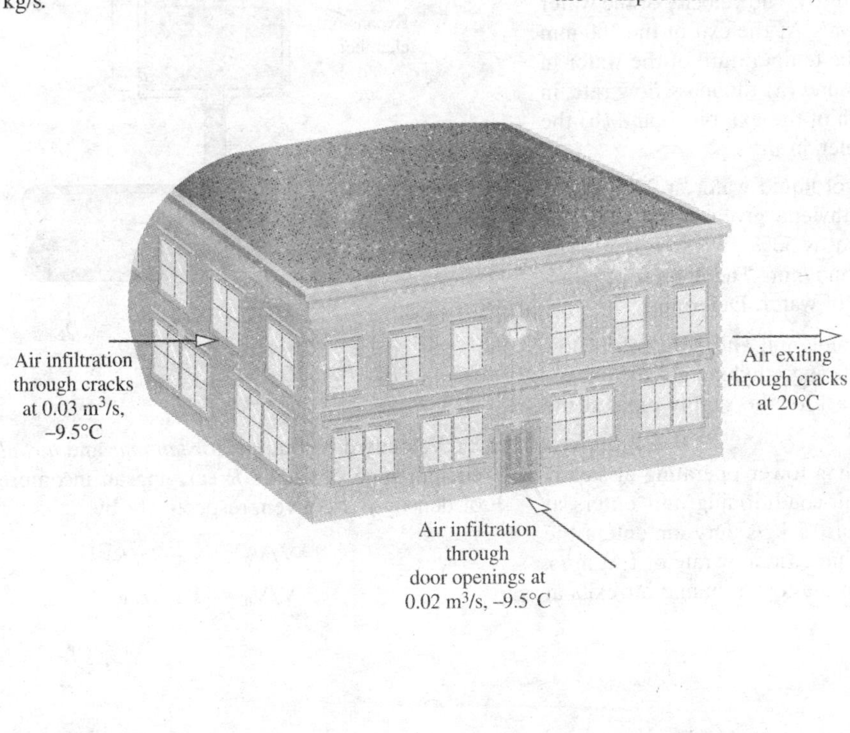

Fig. P4.10

mass flow rate of the entering refrigerant is 6 kg/min. Determine

(a) the velocity at the inlet, in m/s.
(b) the diameter of the exit pipe, in cm.

4.13 Steam at 160 bar, 480°C, enters a turbine operating at steady state with a volumetric flow rate of 800 m^3/min. Eighteen percent of the entering mass flow exits at 5 bar, 240°C, with a velocity of 25 m/s. The rest exits at another location with a pressure of 0.06 bar, a quality of 94%, and a velocity of 400 m/s. Determine the diameters of each exit duct, in m.

4.14 A substance flows through a 50 mm-diameter pipe with a velocity of 1 m/s at a particular location. Determine the mass flow rate, in kg/s, if the substance is

(a) water at 345 kPa, 25°C.
(b) nitrogen as an ideal gas at 345 kPa, 25°C.
(c) Refrigerant 22 at 345 kPa, 25°C.

4.15 Air enters a compressor operating at steady state with a pressure of 1 bar, a temperature of 20°C, and a volumetric flow rate of 0.25 m^3/s. The air velocity in the exit pipe is 210 m/s and the exit pressure is 1 MPa. If each unit mass of air passing from inlet to exit undergoes a process described by $pv^{1.34} = constant$, determine the exit temperature, in °C, and the diameter of the exit pipe, in mm.

4.16 Ammonia enters a control volume operating at steady state at $p_1 = 14$ bar, $T_1 = 28°C$, with a mass flow rate of 0.5 kg/s. Saturated vapor at 4 bar leaves through one exit, with a volumetric flow rate of 1.036 m^3/min, and saturated liquid at 4 bar leaves through a second exit. Determine

(a) the minimum diameter of the inlet pipe, in cm, so the ammonia velocity does not exceed 20 m/s.
(b) the volumetric flow rate of the second exit stream, in m^3/min.

4.17 Liquid water at 20°C enters a pump though an inlet pipe having a diameter of 152 mm. The pump operates at steady state and supplies water to two exit pipes having diameters of 76 mm and 100 mm, respectively. The velocity of the water exiting the 76 mm pipe is 0.4 m/s. At the exit of the 100 mm pipe the velocity is 0.2 m/s. The temperature of the water in each exit pipe is 22°C. Determine (a) the mass flow rate, in kg/s, in the inlet pipe and each of the exit pipes, and (b) the volumetric flow rate at the inlet, in m^3/s.

4.18 At steady state, a stream of liquid water at 20°C, 1 bar is mixed with a stream of ethylene glycol ($M = 62.07$) to form a refrigerant mixture that is 50% glycol by mass. The water molar flow rate is 4.2 kmol/min. The density of ethylene glycol is 1.115 times that of water. Determine

(a) the molar flow rate, in kmol/min, and volumetric flow rate, in m^3/min, of the entering ethylene glycol.
(b) the diameters, in cm, of each of the supply pipes if the velocity in each is 2.5 m/s.

4.19 Figure P4.19 shows a cooling tower operating at steady state. Warm water from an air-conditioning unit enters at 49°C with a mass flow rate of 0.5 kg/s. Dry air enters the tower at 21°C, 1 bar with a volumetric flow rate of 1.41 m^3/s. Because of evaporation within the tower, humid air exits at the top of the tower with a mass flow rate of 1.764 kg/s. Cooled liquid water is collected at the bottom of the tower for return to the air-conditioning unit together with makeup water. Determine the mass flow rate of the makeup water, in kg/s.

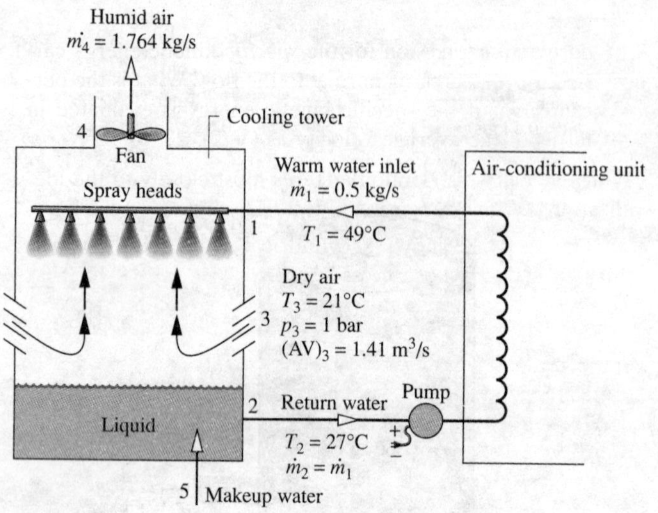

Fig. P4.19

4.20 A pipe carrying an incompressible liquid contains an expansion chamber as illustrated in Fig. P4.20.

(a) Develop an expression for the time rate of change of liquid level in the chamber, dL/dt, in terms of the diameters D_1, D_2, and D, and the velocities V_1 and V_2.
(b) Compare the relative magnitudes of the mass flow rates $\dot{m}_1$ and $\dot{m}_2$ when $dL/dt > 0$, $dL/dt = 0$, and $dL/dt < 0$, respectively.

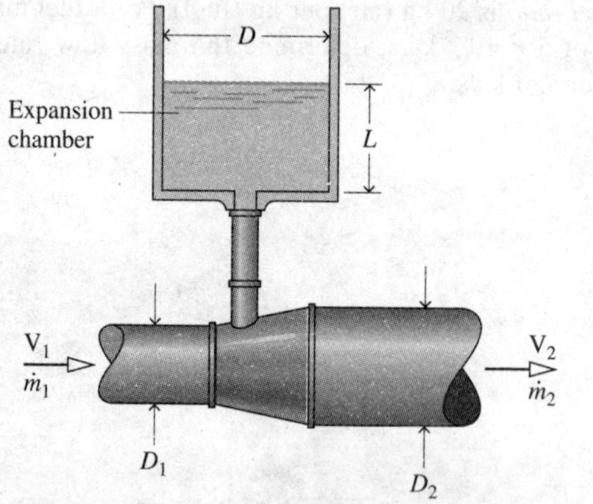

Fig. P4.20

4.21 Velocity distributions for *laminar* and *turbulent* flow in a circular pipe of radius R carrying an incompressible liquid of density ρ are given, respectively, by

$$V/V_0 = [1 - (r/R)^2]$$

$$V/V_0 = [1 - (r/R)]^{1/7}$$

where r is the radial distance from the pipe centerline and V_0 is the centerline velocity. For each velocity distribution

(a) plot V/V_0 versus r/R.
(b) derive expressions for the mass flow rate and the average velocity of the flow, V_{ave}, in terms of V_0, R, and ρ, as required.
(c) derive an expression for the *specific* kinetic energy carried through an area normal to the flow. What is the percent error if the specific kinetic energy is evaluated in terms of the average velocity as $(V_{ave})^2/2$?

Which velocity distribution adheres most closely to the idealizations of one-dimensional flow? Discuss.

4.22 Figure P4.22 shows a cylindrical tank being drained through a duct whose cross-sectional area is 3×10^{-4} m^2. The velocity of the water at the exit varies according to $(2gz)^{1/2}$, where z is the water level, in m, and g is the acceleration of gravity, 9.81 m/s^2. The tank initially contains 2500 kg of liquid water. Taking the density of the water as 10^3 kg/m^3, determine the time, in minutes, when the tank contains 900 kg of water.

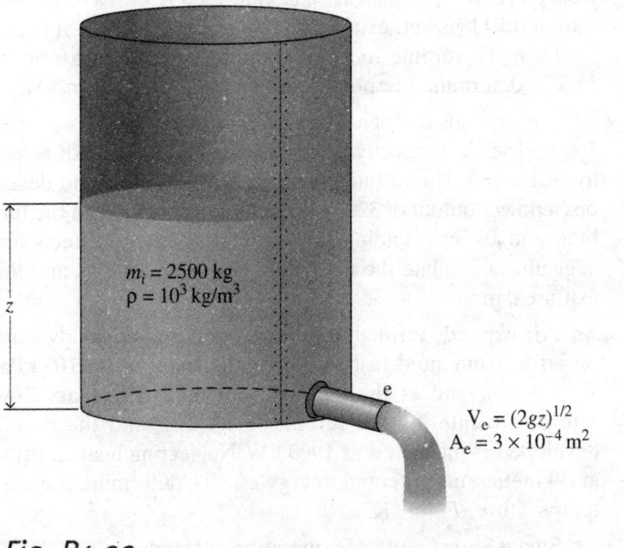

Fig. P4.22

Energy Analysis of Control Volumes at Steady State

4.23 Steam enters a horizontal pipe operating at steady state with a specific enthalpy of 3000 kJ/kg and a mass flow rate of 0.5 kg/s. At the exit, the specific enthalpy is 1700 kJ/kg. If there is no significant change in kinetic energy from inlet to exit, determine the rate of heat transfer between the pipe and its surroundings, in kW.

4.24 Refrigerant 134a enters a horizontal pipe operating at steady state at 40°C, 300 kPa and a velocity of 40 m/s. At the exit, the temperature is 50°C and the pressure is 240 kPa. The pipe diameter is 0.04 m. Determine (a) the mass flow rate of the refrigerant, in kg/s, (b) the velocity at the exit, in m/s, and (c) the rate of heat transfer between the pipe and its surroundings, in kW.

4.25 As shown in Fig. P4.25, air enters a pipe at 25°C, 100 kPa with a volumetric flow rate of 23 m^3/h. On the outer pipe surface is an electrical resistor covered with insulation. With a voltage of 120 V, the resistor draws a current of 4 amps. Assuming the ideal gas model with c_p = 1.005 kJ/kg · K for air and ignoring kinetic and potential energy effects, determine (a) the mass flow rate of the air, in kg/h, and (b) the temperature of the air at exit, in °C.

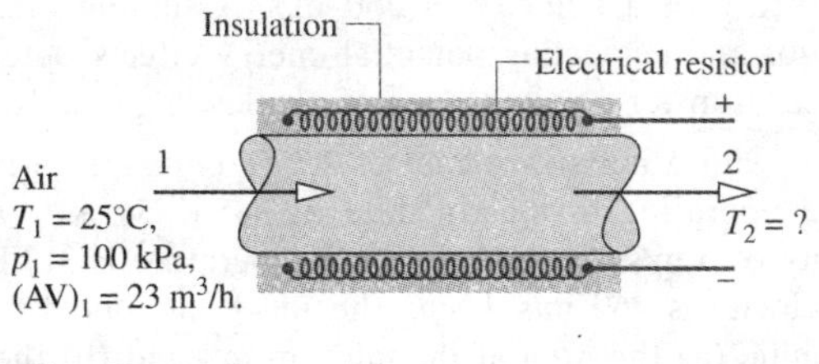

Fig. P4.25

4.26 Carbon dioxide gas is heated as it flows steadily through a 2.5-cm-diameter pipe. At the inlet, the pressure is 2 bar, the temperature is 300 K, and the velocity is 100 m/s. At the exit, the pressure and velocity are 0.9413 bar and 400 m/s, respectively. The gas can be treated as an ideal gas with constant specific heat c_p = 0.94 kJ/kg · K. Neglecting potential energy effects, determine the rate of heat transfer to the carbon dioxide, in kW.

4.27 Air at 600 kPa, 330 K enters a well-insulated, horizontal pipe having a diameter of 1.2 cm and exits at 120 kPa, 300 K. Applying the ideal gas model for air, determine at steady state (a) the inlet and exit velocities, each in m/s, and (b) the mass flow rate, in kg/s.

4.28 At steady state, air at 200 kPa, 52°C and a mass flow rate of 0.5 kg/s enters an insulated duct having differing inlet and exit cross-sectional areas. At the duct exit, the pressure of the air is 100 kPa, the velocity is 255 m/s, and the cross-sectional area is 2×10^{-3} m^2. Assuming the ideal gas model, determine

(a) the temperature of the air at the exit, in °C.
(b) the velocity of the air at the inlet, in m/s.
(c) the inlet cross-sectional area, in m^2.

4.29 Refrigerant 134a flows at steady state through a horizontal pipe having an inside diameter of 4 cm, entering as saturated vapor at −8°C with a mass flow rate of 17 kg/min. Refrigerant vapor exits at a pressure of 2 bar. If the heat transfer rate to the refrigerant is 3.41 kW, determine the exit temperature, in °C, and the velocities at the inlet and exit, each in m/s.

4.30 Steam enters a well-insulated, horizontal nozzle operating at steady state with a velocity of 10 m/s. If the specific enthalpy decreases by 45 kJ/kg from inlet to exit, determine the velocity at the exit, in m/s.

4.31 Steam enters a nozzle operating at steady state at 30 bar, 320°C, with a velocity of 100 m/s. The exit pressure and temperature are 10 bar and 200°C, respectively. The mass flow rate is 2 kg/s. Neglecting heat transfer and potential energy, determine

(a) the exit velocity, in m/s.
(b) the inlet and exit flow areas, in cm^2.

4.32 Steam enters a well-insulated nozzle at 2068 kPa, 316°C, with a velocity of 30 m/s and exits at 276 kPa with a velocity of 550 m/s. For steady-state operation, and neglecting potential energy effects, determine the exit temperature, in °C.

4.33 Air enters an uninsulated nozzle operating at steady state at 420°K with negligible velocity and exits the nozzle at 290°K with a velocity of 460 m/s. Assuming ideal gas behavior and neglecting potential energy effects, determine the heat transfer per unit mass of air flowing, in kJ/kg.

4.34 Air with a mass flow rate of 2 kg/s enters a horizontal nozzle operating at steady state at 445 K, 345 kPa and a velocity of 3 m/s. At the exit, the temperature is 317 K and the velocity is 460 m/s. Using the ideal gas model for air, determine (a) the area at the inlet, in m^2, and (b) the heat transfer between the nozzle and its surroundings, in kJ/kg of air flowing.

4.35 Helium gas flows through a well-insulated nozzle at steady state. The temperature and velocity at the inlet are 333 K and 53 m/s, respectively. At the exit, the temperature is 256 K and the pressure is 345 kPa. The mass flow rate is 0.5 kg/s. Using the ideal gas model, and neglecting potential energy effects, determine the exit area, in m^2.

4.36 Methane (CH_4) gas enters a horizontal, well-insulated nozzle operating at steady state at 80°C and a velocity of 10 m/s. Assuming ideal gas behavior for the methane, plot the temperature of the gas exiting the nozzle, in °C, versus the exit velocity ranging from 500 to 600 m/s.

4.37 As shown in Fig. P4.37, air enters the diffuser of a jet engine operating at steady state at 18 kPa, 216 K and a velocity of 265 m/s, all data corresponding to high-altitude flight. The air flows adiabatically through the diffuser and achieves a temperature of 250 K at the diffuser exit. Using the ideal gas model for air, determine the velocity of the air at the diffuser exit, in m/s.

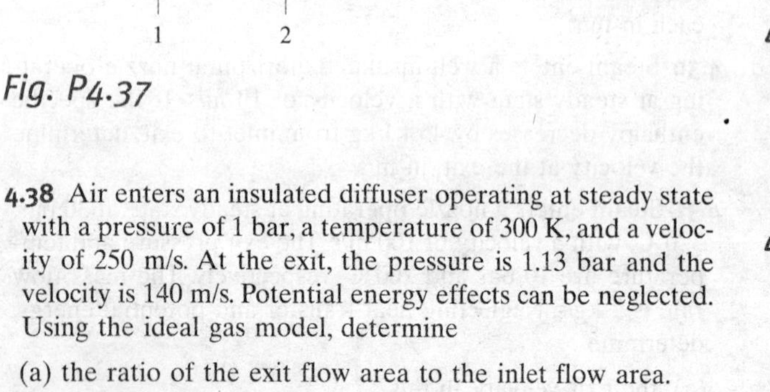

Fig. P4.37

4.38 Air enters an insulated diffuser operating at steady state with a pressure of 1 bar, a temperature of 300 K, and a velocity of 250 m/s. At the exit, the pressure is 1.13 bar and the velocity is 140 m/s. Potential energy effects can be neglected. Using the ideal gas model, determine

(a) the ratio of the exit flow area to the inlet flow area.
(b) the exit temperature, in K.

4.39 Refrigerant 134a enters an insulated diffuser as a saturated vapor at 7 bar with a velocity of 370 m/s. At the exit, the pressure is 16 bar and the velocity is negligible. The diffuser operates at steady state and potential energy effects can be neglected. Determine the exit temperature, in °C.

4.40 Carbon dioxide gas enters a well-insulated diffuser at 138 kPa, 278 K, with a velocity of 244 m/s through a flow area of 9 cm^2. At the exit, the flow area is 30 times the inlet area, and the velocity is 6 m/s. The potential energy change from inlet to exit is negligible. For steady-state operation, determine the exit temperature, in °K, the exit pressure, in kPa, and the mass flow rate, in kg/s.

4.41 Steam enters a well-insulated turbine operating at steady state at 4 MPa with a specific enthalpy of 3015.4 kJ/kg and a velocity of 10 m/s. The steam expands to the turbine exit where the pressure is 0.07 MPa, specific enthalpy is 2431.7 kJ/kg, and the velocity is 90 m/s. The mass flow rate is 11.95 kg/s. Neglecting potential energy effects, determine the power developed by the turbine, in kW.

4.42 Hot combustion gases, modeled as air behaving as an ideal gas, enter a turbine at 1000 kPa, 1500 K with a mass flow rate of 0.10 kg/s and exit at 200 kPa and 900 K. If heat transfer from the turbine to its surroundings occurs at a rate of 15 kW, determine the power output of the turbine, in kW.

4.43 Air expands through a turbine from 10 bar, 900 K to 1 bar, 500 K. The inlet velocity is small compared to the exit velocity of 100 m/s. The turbine operates at steady state and develops a power output of 3200 kW. Heat transfer between the turbine and its surroundings and potential energy effects are negligible. Calculate the mass flow rate of air, in kg/s, and the exit area, in m^2.

4.44 Air expands through a turbine operating at steady state on an instrumented test stand. At the inlet, p_1 = 1103 kPa, T_1 = 867 K, and at the exit, p_2 = 102 kPa. The mass flow rate of air entering the turbine is 4.8 kg/s, and the power developed is measured as 1900 kW. Neglecting heat transfer and kinetic and potential energy effects, determine the exit temperature, T_2, in °K.

4.45 Steam enters a turbine operating at steady state at 445 K and 3447 kPa and leaves at 5.5 kPa with a quality of 93%. The turbine develops 11,185 kW, and heat transfer from the turbine to the surroundings occurs at a rate of 733 kW. Neglecting kinetic and potential energy changes from inlet to exit, determine the volumetric flow rate of the steam at the inlet, in m^3/s.

4.46 A well-insulated turbine operating at steady state develops 23 MW of power for a steam flow rate of 40 kg/s. The steam enters at 360°C with a velocity of 35 m/s and exits as saturated vapor at 0.06 bar with a velocity of 120 m/s. Neglecting potential energy effects, determine the inlet pressure, in bar.

4.47 Steam enters a turbine operating at steady state with a mass flow of 10 kg/min, a specific enthalpy of 3100 kJ/kg, and a velocity of 30 m/s. At the exit, the specific enthalpy is 2300 kJ/kg and the velocity is 45 m/s. The elevation of the inlet is 3 m higher than at the exit. Heat transfer from the turbine to its surroundings occurs at a rate of 1.1 kJ per kg of steam flowing. Let $g = 9.81\ m/s^2$. Determine the power developed by the turbine, in kW.

4.48 Steam enters a turbine operating at steady state at 2 MPa, 360°C with a velocity of 100 m/s. Saturated vapor exits at 0.1 MPa and a velocity of 50 m/s. The elevation of the inlet is 3 m higher than at the exit. The mass flow rate of the steam is 15 kg/s, and the power developed is 7 MW. Let $g = 9.81\ m/s^2$. Determine (a) the area at the inlet, in m^2, and (b) the rate of heat transfer between the turbine and its surroundings, in kW.

4.49 The intake to a hydraulic turbine installed in a flood control dam is located at an elevation of 10 m above the turbine exit. Water enters at 20°C with negligible velocity and exits from the turbine at 10 m/s. The water passes through the turbine with no significant changes in temperature or pressure between the inlet and exit, and heat transfer is negligible. The acceleration of gravity is constant at $g = 9.81\ m/s^2$. If the power output at steady state is 500 kW, what is the mass flow rate of water, in kg/s?

4.50 Steam enters the first-stage turbine shown in Fig. P4.50 at 40 bar and 500°C with a volumetric flow rate of 90 m^3/min. Steam exits the turbine at 20 bar and 400°C. The steam is then reheated at constant pressure to 500°C before entering the second-stage turbine. Steam leaves the second stage as saturated vapor at 0.6 bar. For operation at steady state, and ignoring stray heat transfer and kinetic and potential energy effects, determine the

(a) mass flow rate of the steam, in kg/h.
(b) total power produced by the two stages of the turbine, in kW.
(c) rate of heat transfer to the steam flowing through the reheater, in kW.

4.51 Steam at 11032 kPa, 538°C, and a velocity of 0.6 m/s enters a turbine operating at steady state. As shown in Fig. P4.51, 22% of the entering mass flow is extracted at 1103 kPa, 232°C, with a velocity of 3 m/s. The rest of the steam exits as a two-phase liquid–vapor mixture at 7 kPa, with a quality of 85% and a velocity of 45 m/s. The turbine develops a power output of 2.6×10^5 kW. Neglecting potential energy effects and heat transfer between the turbine and its surroundings, determine

(a) the mass flow rate of the steam entering the turbine, in kg/h.
(b) the diameter of the extraction duct, in m.

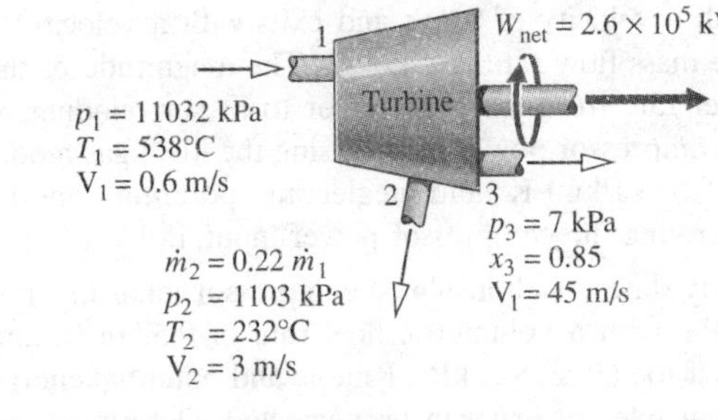

Fig. P4.51

4.52 Air enters a compressor operating at steady state at 1 atm with a specific enthalpy of 290 kJ/kg and exits at a higher pressure with a specific enthalpy of 1023 kJ/kg. The mass flow rate is 0.1 kg/s. If the compressor power input is 77 kW, determine the rate of heat transfer between the compressor and its surroundings, in kW. Neglect kinetic and potential energy effects.

4.53 Air enters a compressor operating at steady state at 1.05 bar, 300 K, with a volumetric flow rate of 12 m^3/min and exits at 12 bar, 400 K. Heat transfer occurs at a rate of 2 kW from the compressor to its surroundings. Assuming the ideal gas model for air and neglecting kinetic and potential energy effects, determine the power input, in kW.

4.54 Nitrogen is compressed in an axial-flow compressor operating at steady state from a pressure of 100 kPa and a temperature of 10°C to a pressure 400 kPa. The gas enters the compressor through a 150 mm-diameter duct with a velocity of 9 m/s and exits at 92°C with a velocity of 24 m/s. Using the ideal gas model, and neglecting stray heat transfer and potential energy effects, determine the compressor power input, in kW.

4.55 Refrigerant 134a enters a compressor operating at steady state as saturated vapor at 0.12 MPa and exits at 1.2 MPa and 70°C at a mass flow rate of 0.108 kg/s. As the refrigerant passes through the compressor, heat transfer to the surroundings occurs at a rate of 0.32 kJ/s. Determine at steady state the power input to the compressor, in kW.

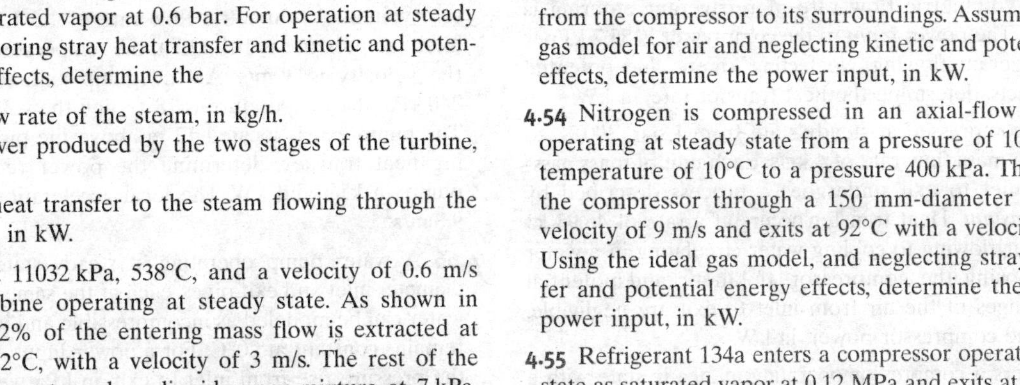

Fig. P4.50

4.56 Carbon dioxide gas is compressed at steady state from a pressure of 138 kPa and a temperature of 0°C to a pressure of 345 kPa and a temperature of 50°C. The gas enters the compressor with a velocity of 9 m/s and exits with a velocity of 24 m/s. The mass flow rate is 0.45 kg/s. The magnitude of the heat transfer rate from the compressor to its surroundings is 5% of the compressor power input. Using the ideal gas model with $c_p = 0.88$ kJ/kg · K and neglecting potential energy effects, determine the compressor power input, in kW.

4.57 At steady state, a well-insulated compressor takes in air at 15°C, 98 kPa, with a volumetric flow rate of 0.57 m^3/s, and compresses it to 260°C, 827 kPa. Kinetic and potential energy changes from inlet to exit can be neglected. Determine the compressor power, in kW, and the volumetric flow rate at the exit, in m^3/s.

4.58 Air enters a compressor operating at steady state with a pressure of 100 kPa, a temperature of 27°C, and a volumetric flow rate of 0.57 m^3/s. Air exits the compressor at 345 kPa. Heat transfer from the compressor to its surroundings occurs at a rate of 48 kJ/kg of air flowing. If the compressor power *input* is 78 kW, determine the exit temperature, in °C.

4.59 Refrigerant 134a enters an air conditioner compressor at 3.2 bar, 10°C, and is compressed at steady state to 10 bar, 70°C. The volumetric flow rate of refrigerant entering is 3.0 m^3/min. The power *input* to the compressor is 55.2 kJ per kg of refrigerant flowing. Neglecting kinetic and potential energy effects, determine the heat transfer rate, in kW.

4.60 Air is compressed at steady state from 1 bar, 300 K, to 6 bar with a mass flow rate of 4 kg/s. Each unit of mass passing from inlet to exit undergoes a process described by $pv^{1.27} = constant$. Heat transfer occurs at a rate of 46.95 kJ per kg of air flowing to cooling water circulating in a water jacket enclosing the compressor. If kinetic and potential energy changes of the air from inlet to exit are negligible, calculate the compressor power, in kW.

4.61 Air enters a compressor operating at steady state with a pressure of 101.4 kPa and a temperature of 21°C. The volumetric flow rate at the inlet is 0.5 m^3/s, and the flow area is 0.02 m^2. At the exit, the pressure is 240 kPa, the temperature is 138°C, and the velocity is 15 m/s. Heat transfer from the compressor to its surroundings occurs at a rate of 2.3 kJ/kg of air flowing. Potential energy effects are negligible, and the ideal gas model can be assumed for the air. Determine (a) the velocity of the air at the inlet, in m/s, (b) the mass flow rate, in kg/s, and (c) the compressor power, in kJ/s and kW.

4.62 Ammonia enters a refrigeration system compressor operating at steady state at −18°C, 138 kPa, and exits at 150°C, 1724 kPa. The magnitude of the power input to the compressor is 7.5 kW, and there is heat transfer from the compressor to the surroundings at a rate of 0.15 kW. Kinetic and potential energy effects are negligible. Determine the inlet volumetric flow rate, in m^3/s, first using data from Table A-15, and then assuming ideal gas behavior for the ammonia. Discuss.

4.63 Air enters a water-jacketed air compressor operating at steady state with a volumetric flow rate of 37 m^3/min at 136 kPa, 305 K and exits with a pressure of 680 kPa and a temperature of 400 K. The power input to the compressor is 155 kW. Energy transfer by heat from the compressed air to the cooling water circulating in the water jacket results in an increase in the temperature of the cooling water from inlet to exit with no change in pressure. Heat transfer from the outside of the jacket as well as all kinetic and potential energy effects can be neglected.

(a) Determine the temperature increase of the cooling water, in K, if the cooling water mass flow rate is 82 kg/min.
(b) Plot the temperature increase of the cooling water, in K, versus the cooling water mass flow rate ranging from 75 to 90 kg/min.

4.64 Air enters a compressor operating at steady state at 101.4 kPa and 15°C and is compressed to a pressure of 1034 kPa. As the air passes through the compressor, it is cooled at a rate of 23 kJ/kg of air flowing by water circulated through the compressor casing. The volumetric flow rate of the air at the inlet is 2 m^3/s, and the power input to the compressor is 520 kW. The air behaves as an ideal gas, there is no stray heat transfer, and kinetic and potential effects are negligible. Determine (a) the mass flow rate of the air, kg/s, and (b) the temperature of the air at the compressor exit, in °C.

4.65 A pump steadily draws water through a pipe from a reservoir at a volumetric flow rate of 1.3 L/s. At the pipe inlet, the pressure is 101.4 kPa, the temperature is 18°C, and the velocity is 3 m/s. At the pump exit, the pressure is 240 kPa, the temperature is 18°C, and the velocity is 12 m/s. The pump exit is located 12 m above the pipe inlet. Ignoring heat transfer, determine the power required by the pump, in kJ/s and kW. The local acceleration of gravity is 9.8 m/s^2.

4.66 A water pump operating at steady state has 76 mm-diameter inlet and exit pipes, each at the same elevation. The water can be modeled as incompressible and its temperature remains constant at 20°C. For a power input of 1.5 kW, plot the pressure rise from inlet to exit, in kPa, versus the volumetric flow rate ranging from 0.015 to 0.02 m^3/s.

4.67 A pump steadily delivers water through a hose terminated by a nozzle. The exit of the nozzle has a diameter of 2.5 cm and is located 4 m above the pump inlet pipe, which has a diameter of 5.0 cm. The pressure is equal to 1 bar at both the inlet and the exit, and the temperature is constant at 20°C. The magnitude of the power input required by the pump is 8.6 kW, and the acceleration of gravity is $g = 9.81$ m/s^2. Determine the mass flow rate delivered by the pump, in kg/s.

4.68 As shown in Fig. P4.68, a power washer used to clean the siding of a house has water entering through a hose at 20°C, 1 atm and a velocity of 0.2 m/s. A jet of water exits with a velocity of 20 m/s at an average elevation of 5 m with no significant change in temperature or pressure. At steady state, the magnitude of the heat transfer rate *from* the power washer to the surroundings is 10% of the electrical power *input.* Evaluating electricity at 8 cents per kW · h, determine the cost of the power required, in cents per liter of water delivered. Compare with the cost of water, assuming 0.05 cents per liter, and comment.

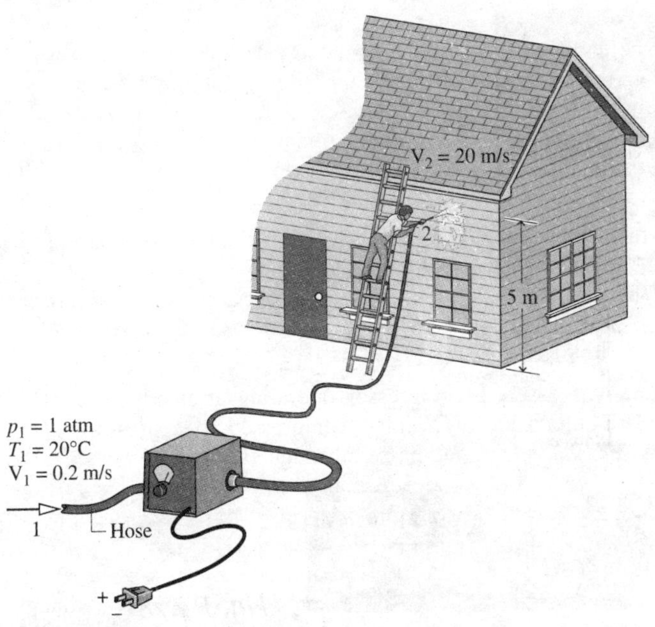

Fig. 4.68

4.69 An oil pump operating at steady state delivers oil at a rate of 5.5 kg/s through a 25 mm-diameter pipe. The oil, which can be modeled as incompressible, has a density of 1600 kg/m^3 and experiences a pressure rise from inlet to exit of 2.8 bar. There is no significant elevation difference between inlet and exit, and the inlet kinetic energy is negligible. Heat transfer between the pump and its surroundings is negligible, and there is no significant change in temperature as the oil passes through the pump. If pumps are available in 1/5 kW increments, determine the horsepower rating of the pump needed for this application.

4.70 Steam enters a heat exchanger operating at steady state at 0.07 MPa with a specific enthalpy of 2431.6 kJ/kg and exits at the same pressure as saturated liquid. The steam mass flow rate is 1.5 kg/min. A separate stream of air with a mass flow rate of 100 kg/min enters at 30°C and exits at 60°C. The ideal gas model with c_p = 1.005 kJ/kg · K can be assumed for air. Kinetic and potential energy effects are negligible. Determine (a) the quality of the entering steam and (b) the rate of heat transfer between the heat exchanger and its surroundings, in kW.

4.71 Refrigerant 134a enters a heat exchanger in a refrigeration system operating at steady state as saturated vapor at −17°C and exits at −7°C with no change in pressure. A separate liquid stream of Refrigerant 134a passes in counterflow to the vapor stream, entering at 41°C, 11 bar, and exiting at a lower temperature while experiencing no pressure drop. The outside of the heat exchanger is well insulated, and the streams have equal mass flow rates. Neglecting kinetic and potential energy effects, determine the exit temperature of the liquid stream, in °C.

4.72 Ammonia enters a heat exchanger operating at steady state as a superheated vapor at 14 bar, 60°C, where it is cooled and condensed to saturated liquid at 14 bar. The mass flow rate of the ammonia is 450 kg/h. A separate stream of air enters the heat exchanger at 17°C, 1 bar and exits at 42°C, 1 bar. Ignoring heat transfer from the outside of the heat exchanger and neglecting kinetic and potential energy effects, determine the mass flow rate of the air, in kg/min.

4.73 Refrigerant 134a enters a condenser operating at steady state at 4.1 bar and 60°C and is condensed to saturated liquid at 4.1 bar on the outside of tubes through which cooling water flows. In passing through the tubes, the cooling water increases in temperature by −8°C and experiences no significant pressure drop. The volumetric flow rate of cooling water is 132 L/m. Neglecting kinetic and potential energy effects and ignoring heat transfer from the outside of the condenser, determine

(a) the mass flow rate of refrigerant, in kg/h.
(b) the rate of energy transfer, in kJ/h, from the condensing refrigerant to the cooling water.

4.74 Steam at a pressure of 0.08 bar and a quality of 93.2% enters a shell-and-tube heat exchanger where it condenses on the outside of tubes through which cooling water flows, exiting as saturated liquid at 0.08 bar. The mass flow rate of the condensing steam is 3.4 × 10^5 kg/h. Cooling water enters the tubes at 15°C and exits at 35°C with negligible change in pressure. Neglecting stray heat transfer and ignoring kinetic and potential energy effects, determine the mass flow rate of the cooling water, in kg/h, for steady-state operation.

4.75 The cooling coil of an air-conditioning system is a heat exchanger in which air passes over tubes through which Refrigerant 22 flows. Air enters with a volumetric flow rate of 40 m^3/min at 27°C, 1.1 bar, and exits at 15°C, 1 bar. Refrigerant enters the tubes at 7 bar with a quality of 16% and exits at 7 bar, 15°C. Ignoring heat transfer from the outside of the heat exchanger and neglecting kinetic and potential energy effects, determine at steady state

(a) the mass flow rate of refrigerant, in kg/min.
(b) the rate of energy transfer, in kJ/min, from the air to the refrigerant.

4.76 Steam enters a heat exchanger operating at steady state at 250 kPa and a quality of 90% and exits as saturated liquid at the same pressure. A separate stream of oil with a mass flow rate of 29 kg/s enters at 20°C and exits at 100°C with no significant change in pressure. The specific heat of the oil is c = 2.0 kJ/kg · K. Kinetic and potential energy effects are negligible. If heat transfer from the heat exchanger to its surroundings is 10% of the energy required to increase the temperature of the oil, determine the steam mass flow rate, in kg/s.

4.77 Refrigerant 134a enters a heat exchanger at −12°C and a quality of 42% and exits as saturated vapor at the same temperature with a volumetric flow rate of 0.85 m^3/min. A separate stream of air enters at 22°C with a mass flow rate of 188 kg/min and exits at 17°C. Assuming the ideal gas model for air and ignoring kinetic and potential energy effects, determine (a) the mass flow rate of the Refrigerant 134a, in kg/min, and (b) the heat transfer between the heat exchanger and its surroundings, in kJ/min.

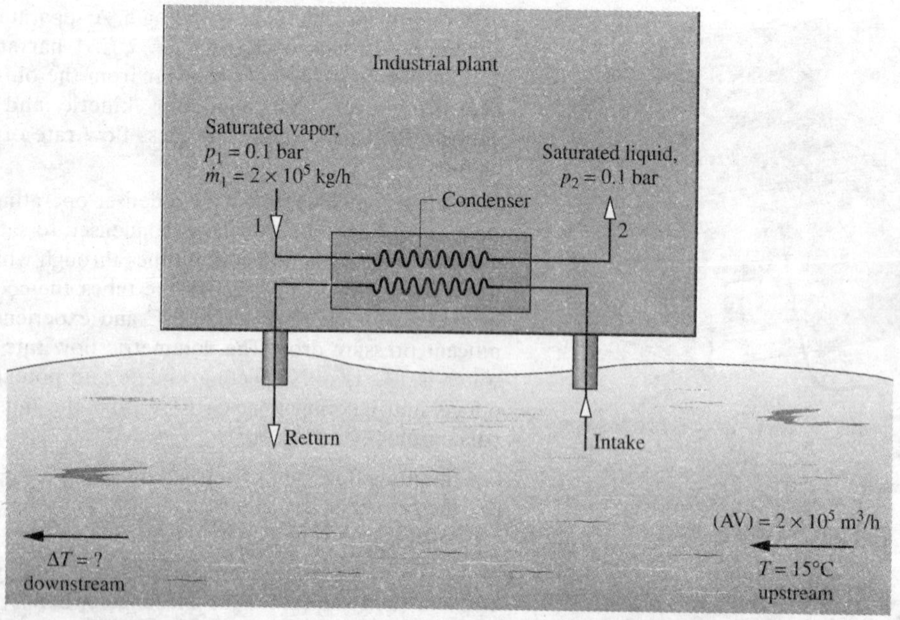

Fig. P4.78

4.78 As sketched in Fig. P4.78, a condenser using river water to condense steam with a mass flow rate of 2×10^5 kg/h from saturated vapor to saturated liquid at a pressure of 0.1 bar is proposed for an industrial plant. Measurements indicate that several hundred meters upstream of the plant, the river has a volumetric flow rate of 2×10^5 m³/h and a temperature of 15°C. For operation at steady state and ignoring changes in kinetic and potential energy, determine the river-water temperature rise, in °C, downstream of the plant traceable to use of such a condenser, and comment.

4.79 Figure P4.79 shows a solar collector panel embedded in a roof. The panel, which has a surface area of 2.16 m², receives energy from the sun at a rate of 2344 kJ/h per m² of collector surface. Twenty-five percent of the incoming energy is lost to the surroundings. The remaining energy is used to heat domestic hot water from 32°C to 49°C. The water passes through the solar collector with a negligible pressure drop. Neglecting kinetic and potential effects, determine at steady state how many gallons of water at 49°C the collector generates per hour.

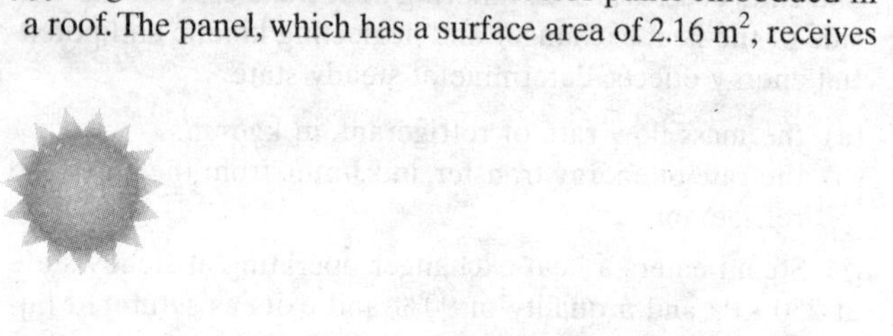

Fig. P4.79

4.80 A feedwater heater in a vapor power plant operates at steady state with liquid entering at inlet 1 with $T_1 = 45$°C and $p_1 = 3.0$ bar. Water vapor at $T_2 = 320$°C and $p_2 = 3.0$ bar enters at inlet 2. Saturated liquid water exits with a pressure of $p_3 = 3.0$ bar. Ignore heat transfer with the surroundings and all kinetic and potential energy effects. If the mass flow rate of the liquid entering at inlet 1 is $\dot{m}_1 = 3.2 \times 10^5$ kg/h, determine the mass flow rate at inlet 2, $\dot{m}_2$, in kg/h.

4.81 A feedwater heater operates at steady state with liquid water entering at inlet 1 at 7 bar, 42°C, and a mass flow rate of 70 kg/s. A separate stream of water enters at inlet 2 as a two-phase liquid–vapor mixture at 7 bar with a quality of 98%. Saturated liquid at 7 bar exits the feedwater heater at 3. Ignoring heat transfer with the surroundings and neglecting kinetic and potential energy effects, determine the mass flow rate, in kg/s, at inlet 2.

4.82 For the *desuperheater* shown in Fig. P4.82, liquid water at state 1 is injected into a stream of superheated vapor entering at state 2. As a result, saturated vapor exits at state 3. Data for steady state operation are shown on the figure. Ignoring stray heat transfer and kinetic and potential energy effects, determine the mass flow rate of the incoming superheated vapor, in kg/min.

4.83 As shown in Fig. P4.83, 15 kg/s of steam enters a *desuperheater* operating at steady state at 30 bar, 320°C, where it is mixed with liquid water at 25 bar and temperature T_2 to produce saturated vapor at 20 bar. Heat transfer between

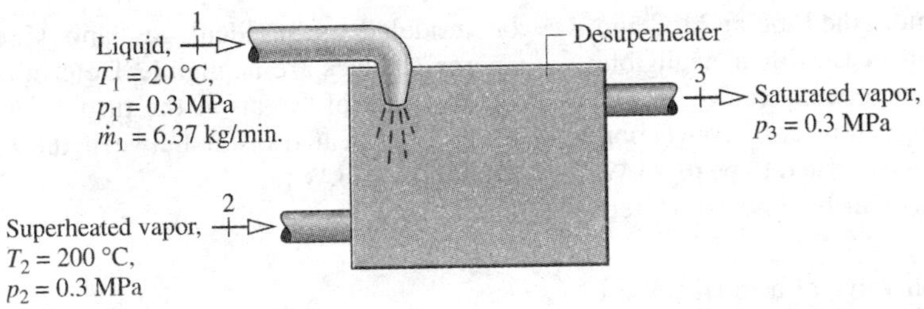

Fig. P4.82

the device and its surroundings and kinetic and potential energy effects can be neglected.

(a) If T_2 = 200°C, determine the mass flow rate of liquid, $\dot{m}_2$, in kg/s.

(b) Plot $\dot{m}_2$, in kg/s, versus T_2 ranging from 20 to 220°C.

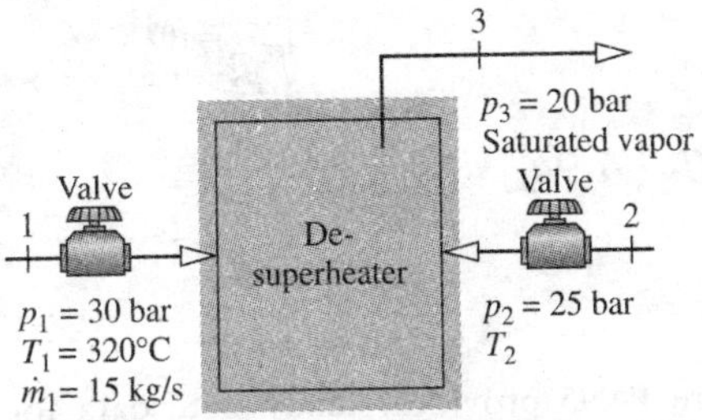

Fig. P4.83

4.84 Figure P4.84 provides steady-state data for the ducting ahead of the chiller coils in an air-conditioning system. Outside air at 32°C is mixed with return air at 24°C. Stray heat transfer is negligible, kinetic and potential energy effects can be ignored, and the pressure throughout is 1 bar. Modeling the air as an ideal gas with c_p = 1 kJ/kg · K, determine (a) the mixed-air temperature, in °C, and (b) the diameter of the mixed-air duct, in m.

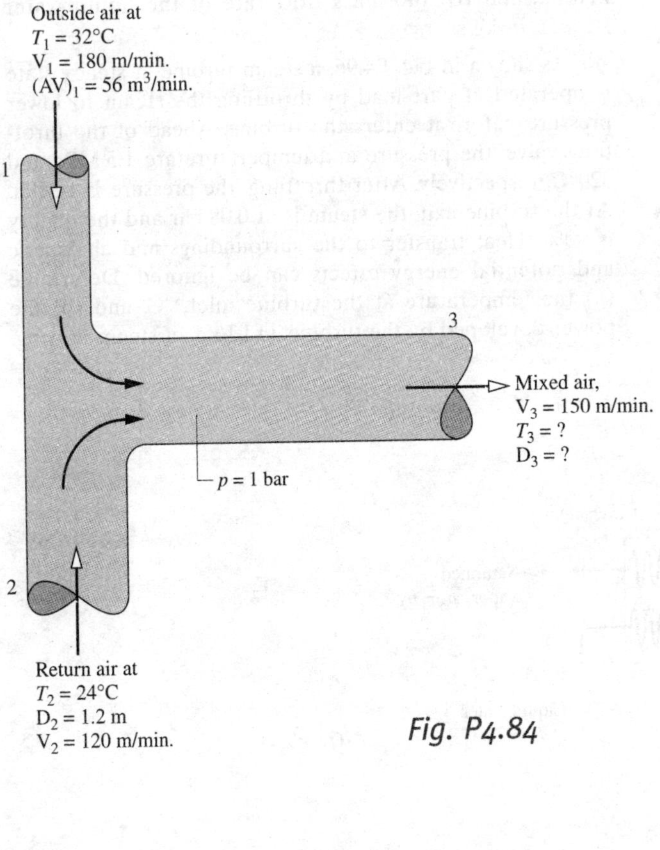

Fig. P4.84

4.85 The electronic components of Example 4.8 are cooled by air flowing through the electronics enclosure. The rate of energy transfer by forced convection from the electronic components to the air is $hA(T_s - T_a)$, where hA = 5 W/K, T_s denotes the average surface temperature of the components, and T_a denotes the average of the inlet and exit air temperatures. Referring to Example 4.8 as required, determine the largest value of T_s, in °C, for which the specified limits are met.

4.86 The electronic components of a computer consume 0.1 kW of electrical power. To prevent overheating, cooling air is supplied by a 25-W fan mounted at the inlet of the electronics enclosure. At steady state, air enters the fan at 20°C, 1 bar, and exits the electronics enclosure at 35°C. There is no significant energy transfer by heat from the outer surface of the enclosure to the surroundings, and the effects of kinetic and potential energy can be ignored. Determine the volumetric flow rate of the entering air, in m^3/s.

4.87 Ten kg/min of cooling water circulates through a water jacket enclosing a housing filled with electronic components. At steady state, water enters the water jacket at 22°C and exits with a negligible change in pressure at a temperature that cannot exceed 26°C. There is no significant energy transfer by heat from the outer surface of the water jacket to the surroundings, and kinetic and potential energy effects can be ignored. Determine the maximum electric power the electronic components can receive, in kW, for which the limit on the temperature of the exiting water is met.

4.88 As shown in Fig. P4.88, electronic components mounted on a flat plate are cooled by convection to the surroundings and by liquid water circulating through a U-tube bonded to

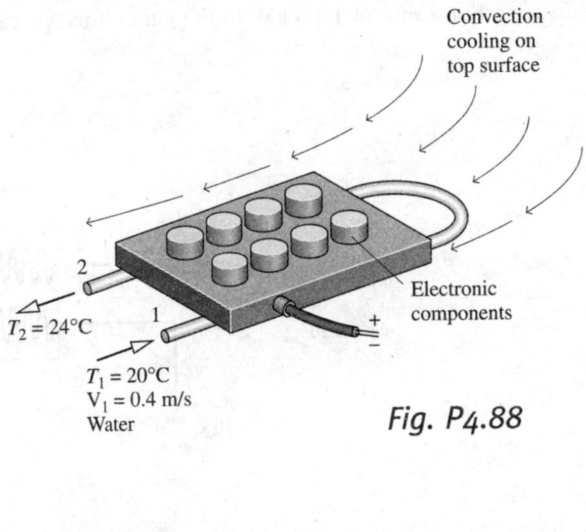

Fig. P4.88

the plate. At steady state, water enters the tube at 20°C and a velocity of 0.4 m/s and exits at 24°C with a negligible change in pressure. The electrical components receive 0.5 kW of electrical power. The rate of energy transfer by convection from the plate-mounted electronics is estimated to be 0.08 kW. Kinetic and potential energy effects can be ignored. Determine the tube diameter, in cm.

4.89 Ammonia enters the expansion valve of a refrigeration system at a pressure of 1.4 MPa and a temperature of 32°C and exits at 0.08 MPa. If the refrigerant undergoes a throttling process, what is the quality of the refrigerant exiting the expansion valve?

4.90 Propane vapor enters a valve at 1.6 MPa, 70°C, and leaves at 0.5 MPa. If the propane undergoes a throttling process, what is the temperature of the propane leaving the valve, in °C?

4.91 A large pipe carries steam as a two-phase liquid–vapor mixture at 1.0 MPa. A small quantity is withdrawn through a throttling calorimeter, where it undergoes a throttling process to an exit pressure of 0.1 MPa. For what range of exit temperatures, in °C, can the calorimeter be used to determine the quality of the steam in the pipe? What is the corresponding range of steam quality values?

4.92 Refrigerant 134a enters the expansion valve of an air-conditioning unit at 965 kPa, 27°C, and exits at 345 kPa. If the refrigerant undergoes a throttling process, what are the temperature, in °K, and the quality at the exit of the valve?

4.93 Steam at 3450 kPa, 260°C enters a well-insulated valve operating at steady state with a mass flow rate of 0.05 kg/s through a 25 mm-diameter pipe. The steam expands to 1380 kPa with no significant change in elevation.

(a) Determine the exit velocity, in m/s, and the exit temperature, in °C, if the ratio of inlet to exit pipe diameters, d_1/d_2, is 0.64.

(b) Plot the exit velocity, in m/s, the exit temperature, in °C, and the exit specific enthalpy, in kJ/kg, for d_1/d_2 ranging from 0.25 to 4.

4.94 Figure P4.94 shows a turbine operating at a steady state that provides power to an air compressor and an electric generator. Air enters the turbine with a mass flow rate of 5.4 kg/s at 527°C and exits the turbine at 107°C, 1 bar. The turbine provides power at a rate of 900 kW to the compressor and at a rate of 1400 kW to the generator. Air can be modeled as an ideal gas, and kinetic and potential energy changes are negligible. Determine (a) the volumetric flow rate of the air at the turbine exit, in m³/s, and (b) the rate of heat transfer between the turbine and its surroundings, in kW.

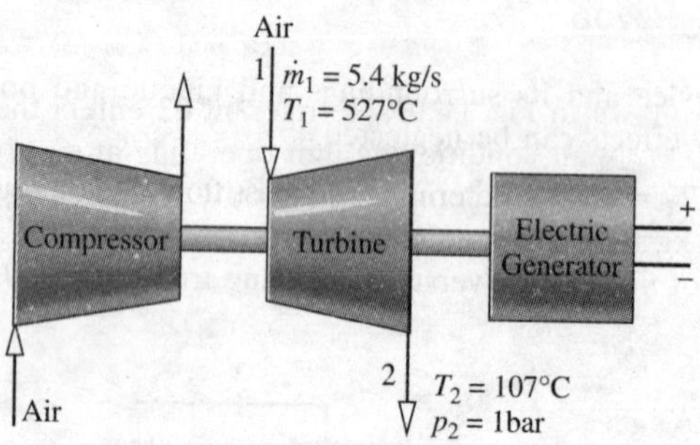

Fig. P4.94

4.95 Figure P4.95 provides steady-state data for a throttling valve in series with a heat exchanger. Saturated liquid Refrigerant 134a enters the valve at $T_1 = 36°C$ with a mass flow rate of 0.26 kg/s and is throttled to $T_2 = -8°C$. The refrigerant then enters the heat exchanger, exiting as saturated vapor with no significant decrease in pressure. In a separate stream, liquid water enters the heat exchanger at $T_4 = 20°C$ and exits as a liquid at $T_5 = 10°C$. Stray heat transfer and kinetic and potential energy effects can be ignored. Determine (a) the pressure at state 2, in kPa, and (b) the mass flow rate of the liquid water stream, in kg/s.

4.96 As shown in Fig. P4.96, a steam turbine at steady state is operated at part load by throttling the steam to lower pressure before it enters the turbine. Ahead of the throttling valve, the pressure and temperature are 1.5 MPa and 320°C, respectively. After throttling, the pressure is 1 MPa. At the turbine exit, the steam is at 0.08 bar and the quality is 90%. Heat transfer to the surroundings and all kinetic and potential energy effects can be ignored. Determine (a) the temperature at the turbine inlet, °C, and (b) the power developed by the turbine, in kJ/kg, of steam flowing.

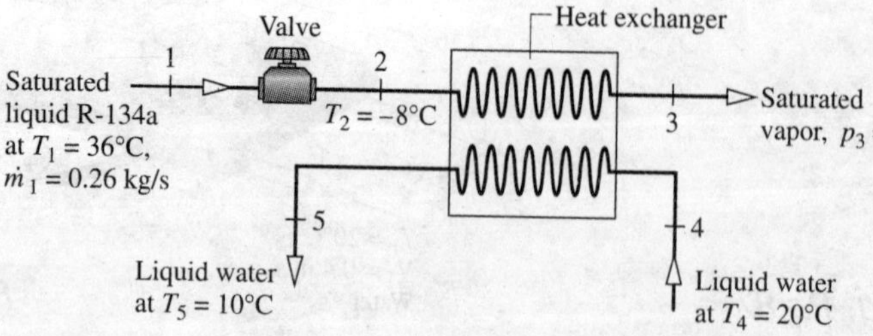

Fig. P4.95

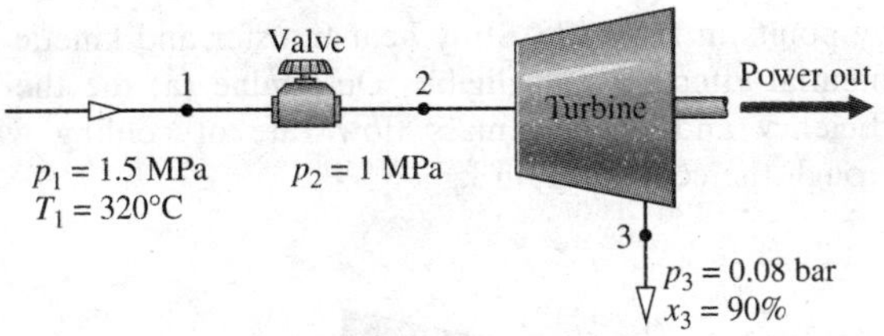

Fig. P4.96

4.97 As shown in Fig. P4.97, Refrigerant 22 enters the compressor of an air-conditioning unit operating at steady state at 4°C, 5.5 bar and is compressed to 60°C, 14 bar. The refrigerant exiting the compressor enters a condenser where energy transfer to air as a separate stream occurs and the refrigerant exits as a liquid at 14 bar, 32°C. Air enters the condenser at 27°C, 1 bar with a volumetric flow rate of 20.25 m³/min and exits at 43°C. Neglecting stray heat transfer and kinetic and potential energy effects, and assuming ideal gas behavior for the air, determine (a) the mass flow rate of refrigerant, in kg/min, and (b) the compressor power, in kilowatt.

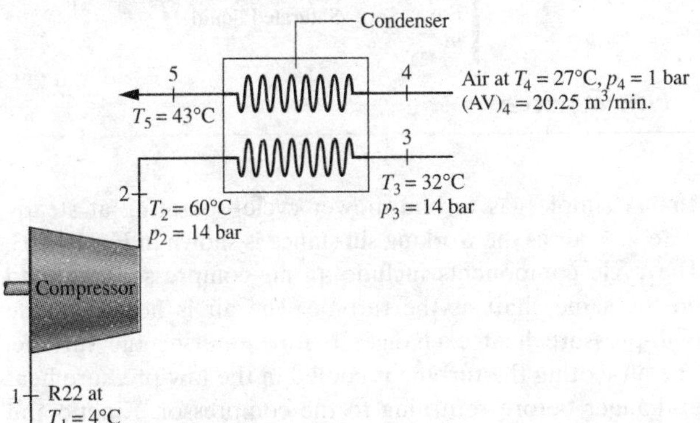

Fig. P4.97

4.98 Fig. P4.98 shows part of a refrigeration system consisting of a heat exchanger, an evaporator, a throttling valve, and associated piping. Data for steady-state operation with Refrigerant 134a are given in the figure. There is no significant heat transfer to or from the heat exchanger, valve, and piping. Kinetic and potential energy effects are negligible. Determine the rate of heat transfer between the evaporator and its surroundings, in kJ/h.

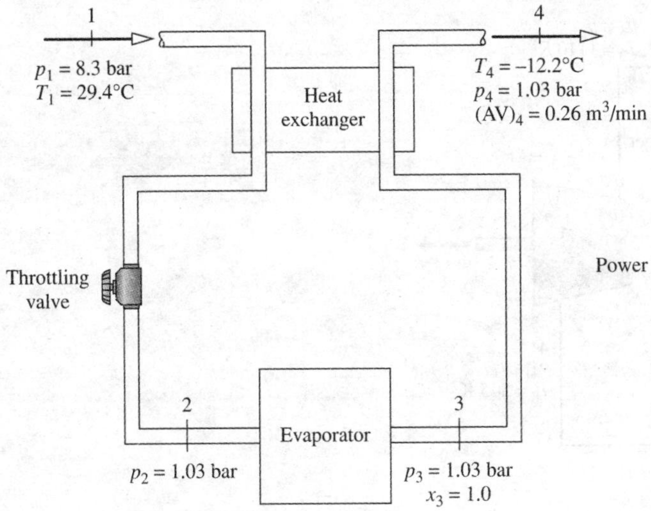

Fig. P4.98

4.99 Refrigerant 134a enters the flash chamber operating at steady state shown in Fig. P4.99 at 10 bar, 36°C, with a mass flow rate of 482 kg/h. Saturated liquid and saturated vapor exit as separate streams, each at pressure p. Heat transfer to the surroundings and kinetic and potential energy effects can be ignored.

(a) Determine the mass flow rates of the exiting streams, each in kg/h, if p = 4 bar.
(b) Plot the mass flow rates of the exiting streams, each in kg/h, versus p ranging from 1 to 9 bar.

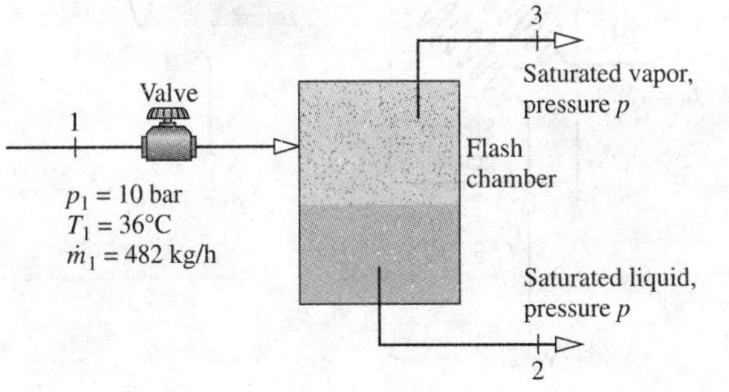

Fig. P4.99

4.100 Carbon dioxide (CO_2) modeled as an ideal gas flows through the compressor and heat exchanger shown in Fig. P4.100. The power input to the compressor is 100 kW. A separate liquid cooling water stream flows through the heat exchanger. All data are for operation at steady state. Stray heat transfer with the surroundings can be neglected, as can all kinetic and potential energy changes. Determine (a) the mass flow rate of the CO_2, in kg/s, and (b) the mass flow rate of the cooling water, in kg/s.

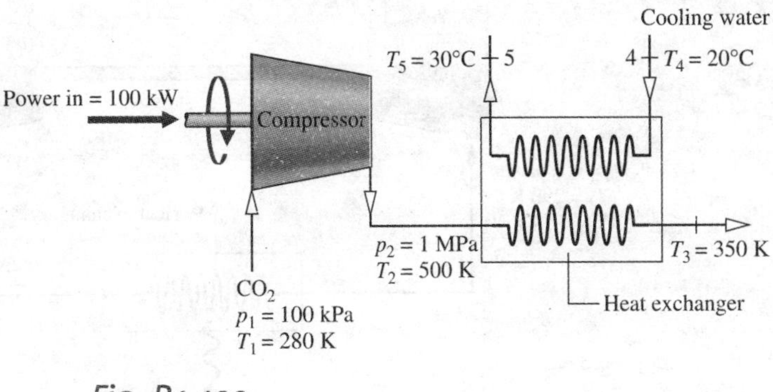

Fig. P4.100

4.101 At steady state, water enters the waste heat recovery-steam generator shown in Fig. P4.101 at 290 kPa, 104°C, and exits at 276 kPa, 160°C. The steam is then fed into a turbine from which it exits at 7 kPa and a quality of 90%. Air from an oven exhaust enters the steam generator at 182°C, 1 bar, with a volumetric flow rate of 2 m^3/s, and exits at 138°C, 1 bar. Ignore all stray heat transfer with the surroundings and all kinetic and potential energy effects. If the power developed is valued at 8 cents per kW · h, do you recommend implementation of this waste-heat recovery system? Provide supporting calculations.

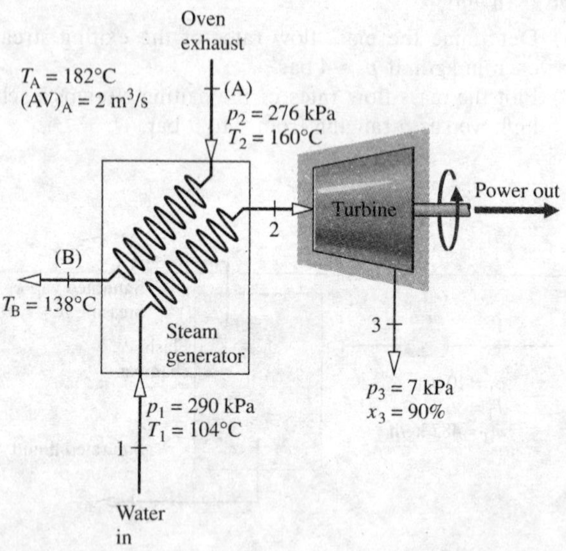

Fig. P4.101

4.102 A simple steam power plant operates at steady state with water circulating through the components with a mass flow rate of 60 kg/s. Figure P4.102 shows additional data at key points in the cycle. Stray heat transfer and kinetic and potential effects are negligible. Determine (a) the thermal efficiency and (b) the mass flow rate of cooling water through the condenser, in kg/s.

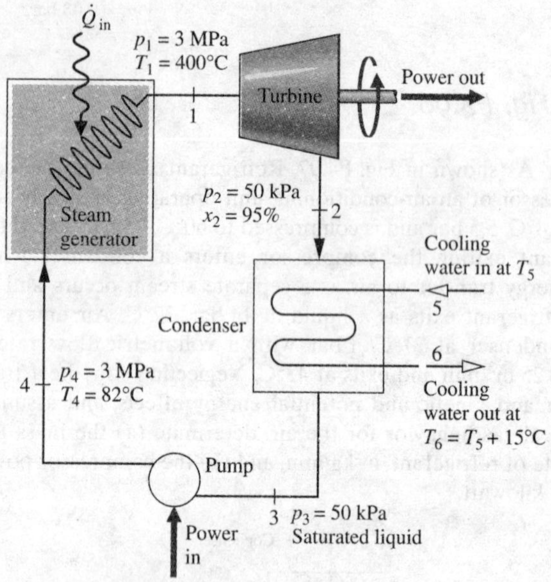

Fig. P4.102

4.103 A simple gas turbine power cycle operating at steady state with air as the working substance is shown in Fig. P4.103. The cycle components include an air compressor mounted on the same shaft as the turbine. The air is heated in the high-pressure heat exchanger before entering the turbine. The air exiting the turbine is cooled in the low-pressure heat exchanger before returning to the compressor. Kinetic and potential effects are negligible. The compressor and turbine

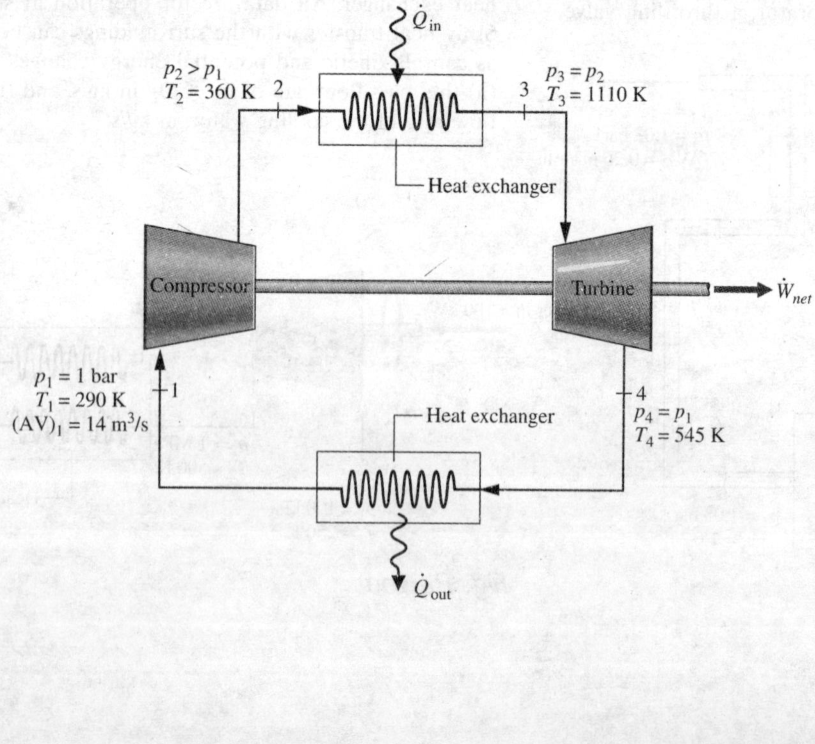

Fig. P4.103

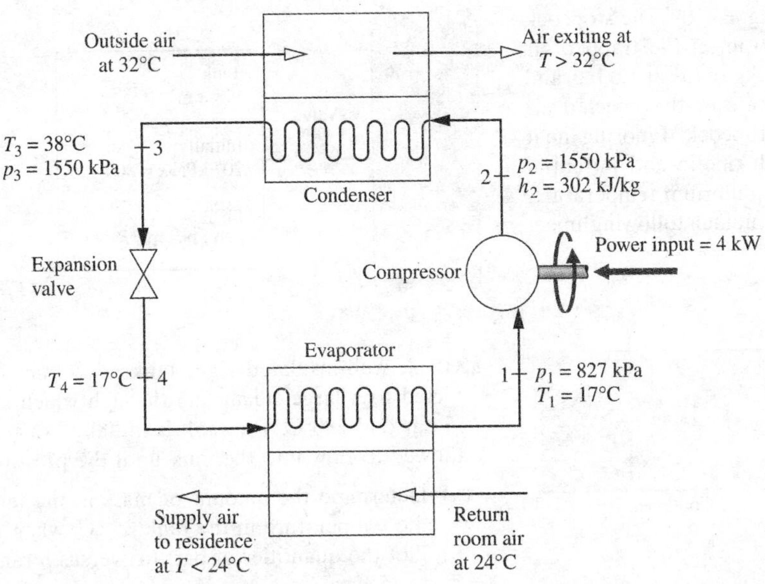

Fig. P4.104

are adiabatic. Using the ideal gas model for air, determine the (a) power required for the compressor, in kW, (b) power output of turbine, in kW, and (c) thermal efficiency of the cycle.

4.104 A residential air-conditioning system operates at steady state, as shown in Fig. P4.104. Refrigerant 22 circulates through the components of the system. Property data at key locations are given on the figure. If the evaporator removes energy by heat transfer from the room air at a rate of 10 kW, determine (a) the rate of heat transfer between the compressor and the surroundings, in kW, and (b) the coefficient of performance.

4.105 Separate streams of steam and air flow through the turbine and heat exchanger arrangement shown in Fig. P4.105. Steady-state operating data are provided on the figure. Heat transfer with the surroundings can be neglected, as can all kinetic and potential energy effects. Determine (a) T_3, in K, and (b) the power output of the second turbine, in kW.

Transient Analysis

4.106 A tiny hole develops in the wall of a rigid tank whose volume is 0.75 m^3, and air from the surroundings at 1 bar, 25°C leaks in. Eventually, the pressure in the tank reaches 1 bar. The process occurs slowly enough that heat transfer between the tank and the surroundings keeps the temperature of the air inside the tank constant at 25°C. Determine the amount of heat transfer, in kJ, if initially the tank

(a) is evacuated.
(b) contains air at 0.7 bar, 25°C.

4.107 A rigid tank of volume 0.75 m^3 is initially evacuated. A hole develops in the wall, and air from the surroundings at 1 bar, 25°C flows in until the pressure in the tank reaches 1 bar. Heat transfer between the contents of the tank and the surroundings is negligible. Determine the final temperature in the tank, in °C.

4.108 A rigid, well-insulated tank of volume 0.5 m^3 is initially evacuated. At time $t = 0$, air from the surroundings at 1 bar, 21°C begins to flow into the tank. An electric resistor transfers energy to the air in the tank at a constant rate of 100 W for 500 s, after which time the pressure in the tank is 1 bar. What is the temperature of the air in the tank, in °C, at the final time?

4.109 Two-tenths of a pound of air at 2 bar and 300 K is trapped within a syringe by a plunger at one end and a

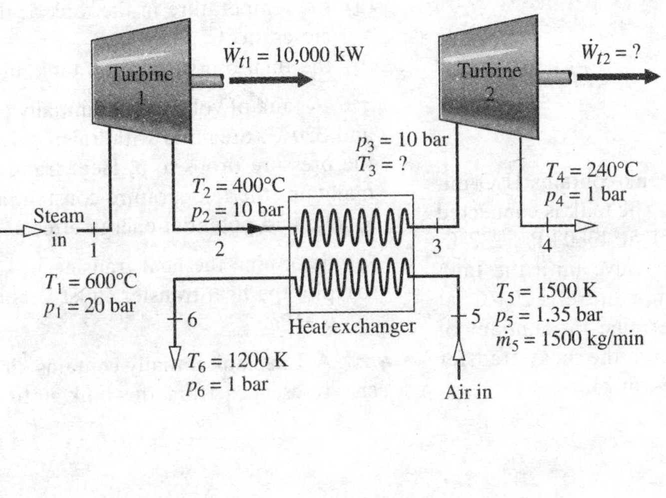

Fig. P4.105

stopcock at the other, as shown in Fig. P4.109. The stopcock is opened and the plunger moved to inject the trapped air into a container that initially holds 1 kg of air at 1.0 bar and 300 K. The plunger maintains the state of the injected air constant until all has passed the stopcock. Ignoring heat transfer with the surroundings, and kinetic and potential energy effects, determine the final equilibrium temperature, in K, and pressure, in kPa, in the container following injection of the air.

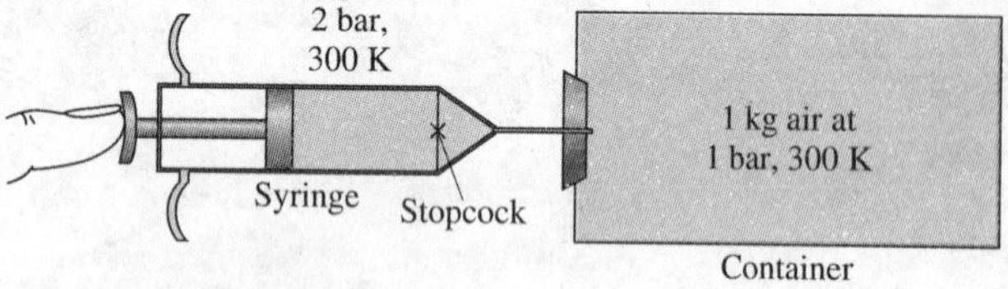

Fig. P4.109

4.110 The rigid tank illustrated in Fig. P4.110 has a volume of 0.06 m^3 and initially contains a two-phase liquid–vapor mixture of H_2O at a pressure of 15 bar and a quality of 20%. As the tank contents are heated, a pressure-regulating valve keeps the pressure constant in the tank by allowing saturated vapor to escape. Neglecting kinetic and potential energy effects

(a) determine the total mass in the tank, in kg, and the amount of heat transfer, in kJ, if heating continues until the final quality is $x = 0.5$.

(b) plot the total mass in the tank, in kg, and the amount of heat transfer, in kJ, versus the final quality x ranging from 0.2 to 1.0.

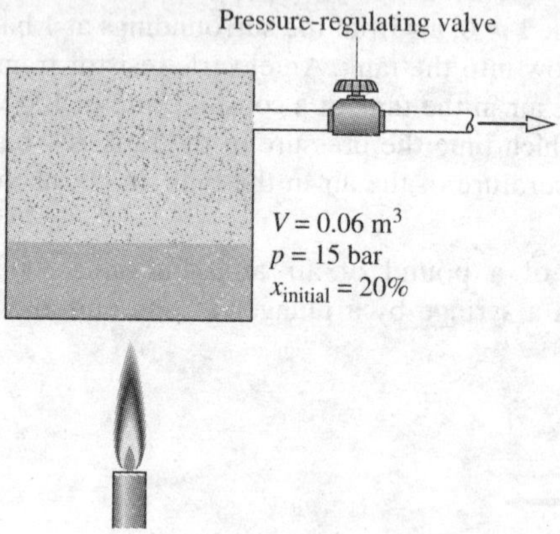

Fig. P4.110

4.111 As shown in Fig. P4.111, a 8 m^3 tank contains H_2O initially at 207 kPa and a quality of 80%. The tank is connected to a large steam line carrying steam at 1380 kPa, 232°C. Steam flows into the tank through a valve until the tank pressure reaches 690 kPa and the temperature is 204°C, at which time the valve is closed. Determine the amount of mass, in kg, that enters the tank and the heat transfer between the tank and its surroundings, in kJ.

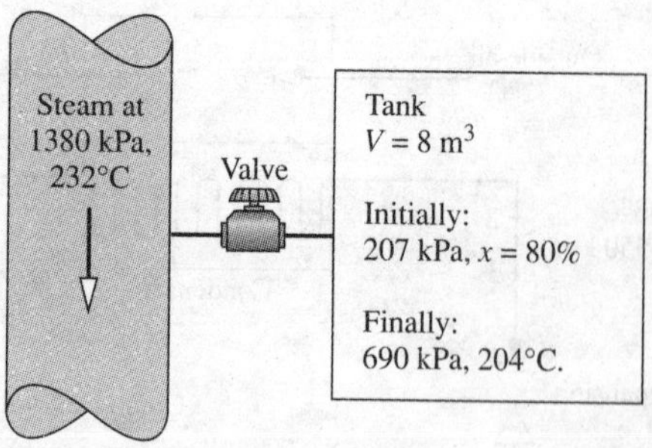

Fig. P4.111

4.112 A well-insulated rigid tank of volume 10 m^3 is connected to a large steam line through which steam flows at 15 bar and 280°C. The tank is initially evacuated. Steam is allowed to flow into the tank until the pressure inside is p.

(a) Determine the amount of mass in the tank, in kg, and the temperature in the tank, in °C, when $p = 15$ bar.

(b) Plot the quantities of part (a) versus p ranging from 0.1 to 15 bar.

4.113 A two-phase liquid–vapor mixture of Refrigerant 134a is contained in a 0.06 m^3, cylindrical storage tank at 670 kPa. Initially, saturated liquid occupies 0.04 m^3. The valve at the top of the tank develops a leak, allowing saturated vapor to escape slowly. Eventually, the volume of the liquid drops to 0.02 m^3. If the pressure in the tank remains constant, determine the mass of refrigerant that has escaped, in kg, and the heat transfer, in kJ.

4.114 A well-insulated rigid tank of volume 0.198 m^3 initially contains helium at 60°C and 200 kPa. A valve connected to the tank is opened, and helium is withdrawn slowly until the pressure within the tank drops to p. An electrical resistor inside the tank maintains the temperature at 70°C.

(a) Determine the mass of helium withdrawn, in kg, and the energy input to the resistor, in kJ, when $p = 124$ kPa.

(b) Plot the quantities of part (a) versus p ranging from 100 to 200 kPa.

4.115 A well-insulated rigid tank of volume 0.14 m^3 contains oxygen (O_2), initially at 275 kPa and 30°C. The tank is connected to a large supply line carrying oxygen at 690 kPa 38°C. A valve between the line and the tank is opened and gas flows into the tank until the pressure reaches 690 kPa, at which time the valve closes. The contents of the tank eventually cool back to 30°C. Determine

(a) the temperature in the tank at the time when the valve closes, in °C.

(b) the final pressure in the tank, in kPa.

4.116 A tank of volume 1 m^3 initially contains steam at 6 MPa and 320°C. Steam is withdrawn slowly from the tank until the pressure drops to p. Heat transfer to the tank contents maintains the temperature constant at 320°C. Neglecting all kinetic and potential energy effects

(a) determine the heat transfer, in kJ, if $p = 1.5$ MPa.

(b) plot the heat transfer, in kJ, versus p ranging from 0.5 to 6 MPa.

4.117 A 1 m^3 tank initially contains air at 300 kPa, 300 K. Air slowly escapes from the tank until the pressure drops to

100 kPa. The air that remains in the tank undergoes a process described by $pv^{1.2} = constant$. For a control volume enclosing the tank, determine the heat transfer, in kJ. Assume ideal gas behavior with constant specific heats.

4.118 A well-insulated tank contains 25 kg of Refrigerant 134a, initially at 300 kPa with a quality of 0.8 (80%). The pressure is maintained by nitrogen gas acting against a flexible bladder, as shown in Fig. P4.118. The valve is opened between the tank and a supply line carrying Refrigerant 134a at 1.0 MPa, 120°C. The pressure regulator allows the pressure in the tank to remain at 300 kPa as the bladder expands. The valve between the line and the tank is closed at the instant when all the liquid has vaporized. Determine the amount of refrigerant admitted to the tank, in kg.

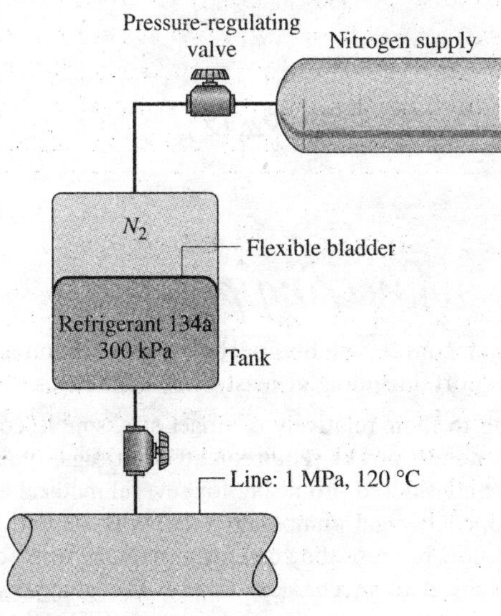

Fig. P4.118

4.119 A well-insulated piston–cylinder assembly is connected by a valve to an air supply line at 8 bar, as shown in Fig. P4.119. Initially, the air inside the cylinder is at 1 bar, 300 K, and the piston is located 0.5 m above the bottom of the cylinder. The atmospheric pressure is 1 bar, and the diameter of the piston face is 0.3 m. The valve is opened and air is admitted slowly until the volume of air inside the cylinder has doubled. The weight of the piston and the friction between the piston and the cylinder wall can be ignored. Using the ideal gas model, plot the final temperature, in K, and the final mass, in kg, of the air inside the cylinder for supply temperatures ranging from 300 to 500 K.

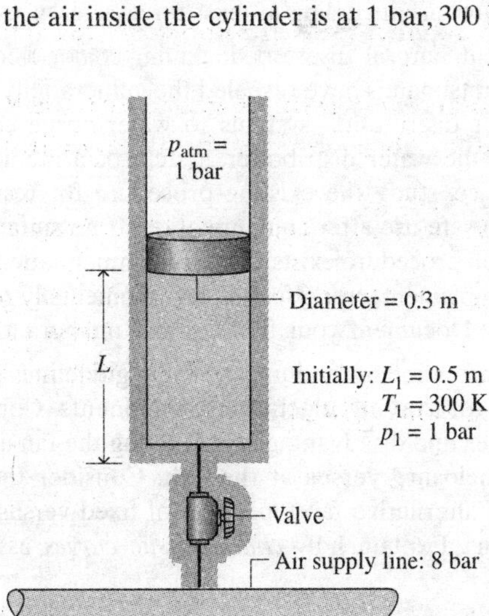

Fig. P4.119

4.120 A well-insulated piston–cylinder assembly is connected by a valve to an air supply at 6.9 bar, 300 K, as shown in Fig. P4.120. The air inside the cylinder is initially at 1 bar, 300 K, and occupies a volume of 0.0027 m^3. Initially, the piston face is located at x = 0 and the spring exerts no force on the piston. The atmospheric pressure is 1 bar, and the area of the piston face is 0.0204 m^2. The valve is opened, and air is admitted slowly until the volume of the air inside the cylinder is 0.0108 m^3. During the process, the spring exerts a force on the piston that varies according to F = kx. The ideal gas model applies for the air, and there is no friction between the piston and the cylinder wall. For the air within the cylinder, plot the final pressure, in bar, and the final temperature, in °C, versus k ranging from 9,500 to 10,950 N/m.

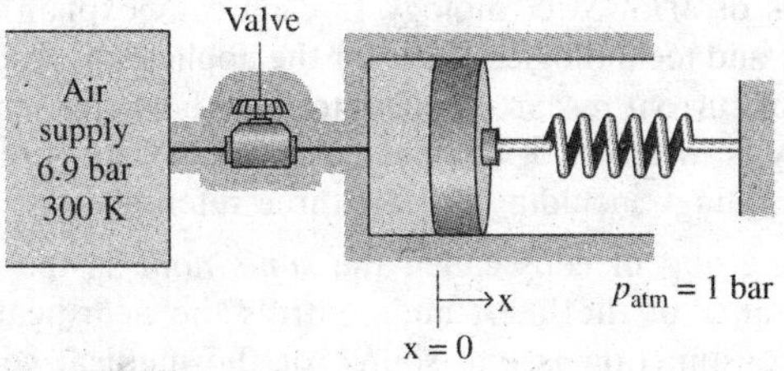

Fig. P4.120

4.121 Nitrogen gas is contained in a rigid 1-m tank, initially at 10 bar, 300 K. Heat transfer to the contents of the tank occurs until the temperature has increased to 400 K. During the process, a pressure-relief valve allows nitrogen to escape, maintaining constant pressure in the tank. Neglecting kinetic and potential energy effects, and using the ideal gas model with constant specific heats evaluated at 350 K, determine the mass of nitrogen that escapes, in kg, and the amount of energy transfer by heat, in kJ.

4.122 The air supply to a 56 m^3 office has been shut off overnight to conserve utilities, and the room temperature has dropped to 4°C. In the morning, a worker resets the thermostat to 21°C, and 6 m^3/min of air at 50°C begins to flow in through a supply duct. The air is well mixed within the room, and an equal mass flow of air at room temperature is withdrawn through a return duct. The air pressure is nearly 1 bar everywhere. Ignoring heat transfer with the surroundings and kinetic and potential energy effects, estimate how long it takes for the room temperature to reach 21°C. Plot the room temperature as a function of time.

4.123 A well-insulated chamber of volume 0.028 m^3 is shown in Fig. P4.123. Initially, the chamber contains air at 101.4 kPa and 38°C. Connected to the chamber are supply and discharge pipes equipped with valves that control the flow rates into and out of the chamber. The supply air is at 207 kPa, 93°C. Both valves are opened simultaneously, allowing air to flow with a mass flow rate $\dot{m}$ through each valve. The air within the chamber is well mixed, so the temperature and pressure at any time can be taken as uniform throughout. Neglecting kinetic and potential energy effects, and using the ideal gas model with constant specific heats for the air, plot the temperature, in °C, and the pressure, in kPa, of the air in the chamber versus time for $\dot{m}$ = 0.01, 0.02, and 0.04 kg/s.

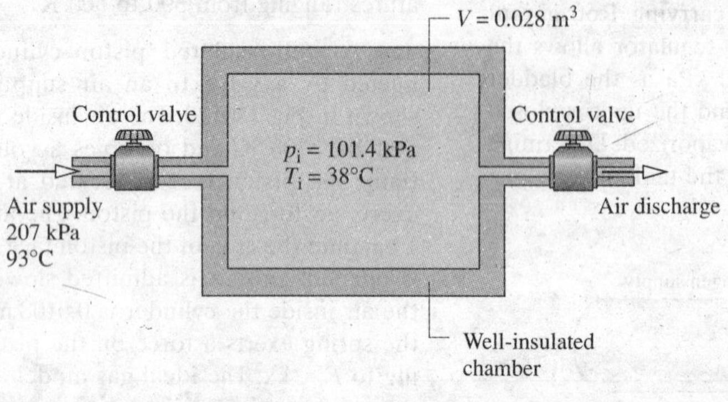

Fig. P4.123

Design & open ended problems: exploring engineering practice

4.1D Using the Internet, identify at least five medical applications of *MEMS* technology. In each case, explain the scientific and technological basis for the application, discuss the state of current research, and determine how close the technology is in terms of commercialization. Write a report of your findings, including at least three references.

4.2D A group of cells called the *sinus node* is the natural pacemaker of the heart and controls the heartbeat. Sinus node dysfunction is one source of the medical condition known as heart *arrhythmia:* irregular heartbeat. Significant arrhythmias are treated in several ways, including the use of an artificial pacemaker, which is an electrical device that sends the signals needed to make the heart beat properly. Research how both natural and artificial pacemakers operate to achieve their goal of maintaining a regular heartbeat. Place your findings in a memorandum that includes annotated sketches of each type of pacemaker.

4.3D Identify sites in your state where wind turbines for utility-scale electrical generation are feasible but do not yet exist. Prepare a memorandum to an appropriate governing or corporate entity with your recommendations as to whether wind-turbine electrical generation should be developed at the most promising sites. Consider engineering, economic, and societal aspects.

4.4D Generating power by harnessing ocean tides and waves is being studied across the globe. Underwater turbines develop power from tidal *currents.* Wave-power devices develop power from the *undulating motion* of ocean waves. Although tides and waves long have been used to meet modest power generation needs, investigators today are aiming at large-scale power generation systems. Some see the oceans as providing a potentially unlimited renewable source of power. Critically evaluate the viability of tidal and wave power, considering both technical and economic issues. Write a report, including at least three references.

4.5D Owing to their relatively compact size, simple construction, and modest power requirement, centrifugal-type blood pumps are considered promising for several medical applications. Still, centrifugal pumps have met with limited success for blood flow because they can cause damage to blood cells and are subject to mechanical failure. The goal of current development efforts is a device having sufficient long-term biocompatibility, performance, and reliability for widespread deployment. Investigate the status of centrifugal blood pump development, including identifying key technical challenges and prospects for overcoming them. Summarize your findings in a report, including at least three references.

4.6D Recent natural disasters, including major floods, hurricanes, and tsunamis have revealed the vulnerability of municipal water distribution systems to water-borne contamination. For the water distribution system of a municipality of your choice, study the existing procedure for restoring the system to safe use after contamination. If no suitable decontamination procedure exists, make recommendations. Suggest easy-to-implement, cost-effective, environmentally responsible measures. Document your findings in a memorandum.

4.7D Prepare a memorandum providing guidelines for selecting fans for cooling electronic components. Consider the advantages and disadvantages of locating the fan at the inlet of the enclosure versus at the exit. Consider the relative merits of alternative fan types and of fixed versus variable-speed fans. Explain how *characteristic curves* assist in fan selection.

4.8D Forced-air warming systems involving inflatable thermal blankets are commonly used to prevent subnormal body temperature (hypothermia) during and following surgery. A heater and blower provide a stream of warmed air to the blanket. While the air temperature leaving the heater/blower is monitored by a temperature sensor, the temperature of the air providing warming to patients can vary widely, causing in some instances overheating and localized burning of patients. The object of this project is to develop cost-effective modifications of existing thermal blankets that would control the air temperature and eliminate injurious "hot spots." The modifications must conform to standards governing the safety of systems involving heating in medical applications. Summarize your conclusions in a memorandum.

4.9D Residential *integrated systems* capable of generating electricity and providing space heating and water heating will reduce reliance on electricity supplied from central power plants. For a 230 m^2 dwelling in your locale, evaluate two alternative technologies for combined power and heating: a solar energy-based system and a natural gas fuel cell system. For each alternative, specify equipment and evaluate costs, including the initial system cost, installation cost, and operating cost. Compare total cost with that for conventional means for powering and heating the dwelling. Write a report summarizing your analysis and recommending either or both of the options if they are preferable to conventional means.

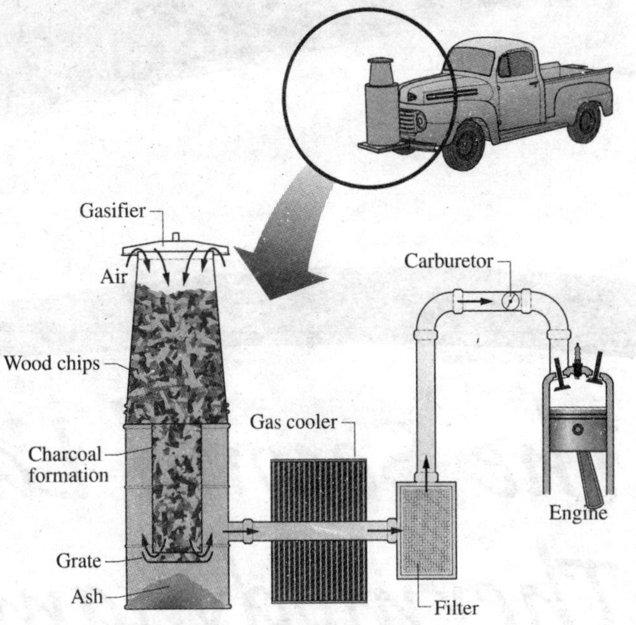

Fig. P4.10D

4.10D Figure P4.10D provides the schematic of a device for producing a combustible fuel gas from biomass. Owing to petroleum shortages, similar *gasifiers* were widely used in Europe during the 1940s war years, including use in tractors, trucks, fishing and ferry boats, and stationary engines. While several types of solid biomass can be employed in current gasifier designs, wood chips are commonly used. Wood chips are introduced at the top of the gasifier unit. Just below this level, the chips react with oxygen in the combustion air to produce charcoal. At the next depth, the charcoal reacts with hot combustion gases from the charcoal-formation stage to produce a fuel gas consisting mainly of hydrogen, carbon monoxide, and nitrogen from the combustion air. The fuel gas is then cooled, filtered, and ducted to the internal combustion engine served by the gasifier. Critically evaluate the suitability of this technology for use today in the event of a prolonged petroleum shortage in your locale. Document your conclusions in a memorandum.

5

The Second Law of Thermodynamics

Engineering Context

The presentation to this point has considered thermodynamic analysis using the conservation of mass and conservation of energy principles together with property relations. In Chaps. 2 through 4 these fundamentals are applied to increasingly complex situations. The conservation principles do not always suffice, however, and the second law of thermodynamics is also often required for thermodynamic analysis. The **objective** of this chapter is to introduce the second law of thermodynamics. A number of deductions that may be called corollaries of the second law are also considered, including performance limits for thermodynamic cycles. The current presentation provides the basis for subsequent developments involving the second law in Chaps. 6 and 7.

Learning Objectives

When you complete your study of this chapter, you will be able to...

- ✔ demonstrate understanding of key concepts related to the second law of thermodynamics, including alternative statements of the second law, the internally reversible process, and the Kelvin temperature scale.
- ✔ list several important irreversibilities.
- ✔ assess the performance of power cycles and refrigeration and heat pump cycles using, as appropriate, the corollaries of Secs. 5.6.2 and 5.7.2, together with Eqs. 5.9–5.11.
- ✔ describe the Carnot cycle.
- ✔ interpret the Clausius inequality as expressed by Eq. 5.13.

5.1 Introducing the Second Law

The objectives of the present section are to

1. motivate the need for and the usefulness of the second law.
2. introduce statements of the second law that serve as the point of departure for its application.

5.1.1 Motivating the Second Law

It is a matter of everyday experience that there is a definite direction for *spontaneous* processes. This can be brought out by considering the three systems pictured in Fig. 5.1.

► System a. An object at an elevated temperature T_i placed in contact with atmospheric air at temperature T_0 would eventually cool to the temperature of its much larger surroundings, as illustrated in Fig. 5.1*a*. In conformity with the conservation of energy principle, the decrease in internal energy of the body would appear as an increase in the internal energy of the surroundings. The *inverse* process would not take place *spontaneously,* even though energy could be conserved: The internal energy of the surroundings would not decrease spontaneously while the body warmed from T_0 to its initial temperature.

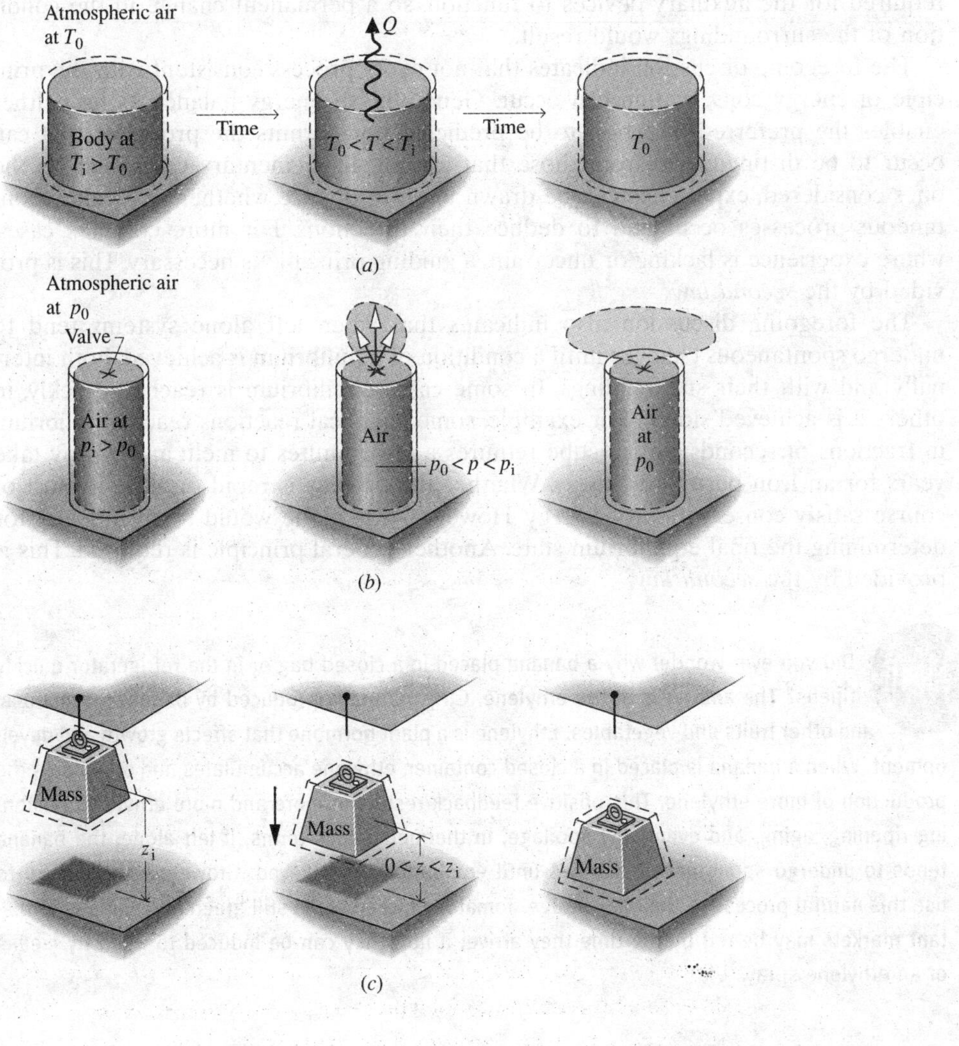

Fig. 5.1 Illustrations of spontaneous processes and the eventual attainment of equilibrium with the surroundings. (*a*) Spontaneous heat transfer. (*b*) Spontaneous expansion. (*c*) Falling mass.

- System b. Air held at a high pressure p_i in a closed tank would flow spontaneously to the lower pressure surroundings at p_0 if the interconnecting valve were opened, as illustrated in Fig. 5.1*b*. Eventually fluid motions would cease and all of the air would be at the same pressure as the surroundings. Drawing on experience, it should be clear that the *inverse* process would not take place *spontaneously,* even though energy could be conserved: Air would not flow spontaneously from the surroundings at p_0 into the tank, returning the pressure to its initial value.
- System c. A mass suspended by a cable at elevation z_i would fall when released, as illustrated in Fig. 5.1*c*. When it comes to rest, the potential energy of the mass in its initial condition would appear as an increase in the internal energy of the mass and its surroundings, in accordance with the conservation of energy principle. Eventually, the mass also would come to the temperature of its much larger surroundings. The *inverse* process would not take place *spontaneously,* even though energy could be conserved: The mass would not return spontaneously to its initial elevation while its internal energy or that of its surroundings decreased.

In each case considered, the initial condition of the system can be restored, but not in a spontaneous process. Some auxiliary devices would be required. By such auxiliary means the object could be reheated to its initial temperature, the air could be returned to the tank and restored to its initial pressure, and the mass could be lifted to its initial height. Also in each case, a fuel or electrical input normally would be required for the auxiliary devices to function, so a permanent change in the condition of the surroundings would result.

The foregoing discussion indicates that not every process consistent with the principle of energy conservation can occur. Generally, an energy balance alone neither enables the preferred direction to be predicted nor permits the processes that can occur to be distinguished from those that cannot. In elementary cases such as the ones considered, experience can be drawn upon to deduce whether particular spontaneous processes occur and to deduce their directions. For more complex cases, where experience is lacking or uncertain, a guiding principle is necessary. This is provided by the *second law.*

The foregoing discussion also indicates that when left alone systems tend to undergo spontaneous changes until a condition of equilibrium is achieved, both internally and with their surroundings. In some cases equilibrium is reached quickly, in others it is achieved slowly. For example, some chemical reactions reach equilibrium in fractions of seconds; an ice cube requires a few minutes to melt; and it may take years for an iron bar to rust away. Whether the process is rapid or slow, it must of course satisfy conservation of energy. However, that alone would be insufficient for determining the final equilibrium state. Another general principle is required. This is provided by the *second law.*

Bio... connections

Did you ever wonder why a banana placed in a closed bag or in the refrigerator quickly ripens? The answer is in the ethylene, C_2H_4, naturally produced by bananas, tomatoes, and other fruits and vegetables. Ethylene is a plant hormone that affects growth and development. When a banana is placed in a closed container, ethylene accumulates and stimulates the production of more ethylene. This positive feedback results in more and more ethylene, hastening ripening, aging, and eventually spoilage. In thermodynamic terms, if left alone, the banana tends to undergo spontaneous changes until equilibrium is achieved. Growers have learned to use this natural process to their advantage. Tomatoes picked while still green and shipped to distant markets may be red by the time they arrive; if not, they can be induced to ripen by means of an ethylene spray.

5.1.2 Opportunities for Developing Work

By exploiting the spontaneous processes shown in Fig. 5.1, it is possible, in principle, for work to be developed as equilibrium is attained. ➡ **FOR EXAMPLE...** instead of permitting the body of Fig. 5.1*a* to cool spontaneously with no other result, energy could be delivered by heat transfer to a system undergoing a power cycle that would develop a net amount of work (Sec. 2.6). Once the object attained equilibrium with the surroundings, the process would cease. Although there is an *opportunity* for developing work in this case, the opportunity would be wasted if the body were permitted to cool without developing any work. In the case of Fig. 5.1*b*, instead of permitting the air to expand aimlessly into the lower-pressure surroundings, the stream could be passed through a turbine and work could be developed. Accordingly, in this case there is also a possibility for developing work that would not be exploited in an uncontrolled process. In the case of Fig. 5.1*c*, instead of permitting the mass to fall in an uncontrolled way, it could be lowered gradually while turning a wheel, lifting another mass, and so on. ⬅

These considerations can be summarized by noting that when an imbalance exists between two systems, there is an opportunity for developing work that would be irrevocably lost if the systems were allowed to come into equilibrium in an uncontrolled way. Recognizing this possibility for work, we can pose two questions:

1. What is the theoretical maximum value for the work that could be obtained?
2. What are the factors that would preclude the realization of the maximum value?

That there should be a maximum value is fully in accord with experience, for if it were possible to develop unlimited work, few concerns would be voiced over our dwindling fossil fuel supplies. Also in accord with experience is the idea that even the best devices would be subject to factors such as friction that would preclude the attainment of the theoretical maximum work. The second law of thermodynamics provides the means for determining the theoretical maximum and evaluating quantitatively the factors that preclude attaining the maximum.

5.1.3 Aspects of the Second Law

The preceding discussions can be summarized by noting that the second law and deductions from it are useful because they provide means for:

1. predicting the direction of processes.
2. establishing conditions for equilibrium.
3. determining the best *theoretical* performance of cycles, engines, and other devices.
4. evaluating quantitatively the factors that preclude the attainment of the best theoretical performance level.

Additional uses of the second law include its roles in:

5. defining a temperature scale independent of the properties of any thermometric substance.
6. developing means for evaluating properties such as u and h in terms of properties that are more readily obtained experimentally.

Scientists and engineers have found many additional applications of the second law and deductions from it. It also has been used in economics, philosophy, and other areas far removed from engineering thermodynamics.

The six points listed can be thought of as aspects of the second law of thermodynamics and not as independent and unrelated ideas. Nonetheless, given the variety of these topic areas, it is easy to understand why there is no single statement of the second law that brings out each one clearly. There are several alternative, yet equivalent, formulations of the second law.

In the next section, three statements of the second law are introduced as *points of departure* for our study of the second law and its consequences. Although the exact relationship of these particular formulations to each of the second law aspects listed above may not be immediately apparent, all aspects listed can be obtained by deduction from these formulations or their corollaries. It is important to add that in every instance where a consequence of the second law has been tested directly or indirectly by experiment, it has been unfailingly verified. Accordingly, the basis of the second law of thermodynamics, like every other physical law, is experimental evidence.

5.2 Statements of the Second Law

Three alternative statements of the second law of thermodynamics are given in this section. They are the (1) Clausius, (2) Kelvin–Planck, and (3) Entropy statements. The Clausius and Kelvin–Planck statements are traditional formulations of the second law. You have likely encountered them before in an introductory physics course. Although the Clausius statement is more in accord with experience and thus easier to accept, the Kelvin–Planck statement provides a more effective means for bringing out important second law deductions related to systems undergoing thermodynamic cycles, which is the focus of the current chapter. The Kelvin–Planck statement also underlies the Entropy statement. The Entropy statement is the most effective form of the second law for an extremely wide range of engineering applications. The Entropy statement is the focus of Chapter 6.

5.2.1 Clausius Statement of the Second Law

Clausius statement

The ***Clausius statement*** of the second law asserts that:

It is impossible for any system to operate in such a way that the sole result would be an energy transfer by heat from a cooler to a hotter body.

The Clausius statement does not rule out the possibility of transferring energy by heat from a cooler body to a hotter body, for this is exactly what refrigerators and heat pumps accomplish. However, as the words "sole result" in the statement suggest, when a heat transfer from a cooler body to a hotter body occurs, there must be some *other effect* within the system accomplishing the heat transfer, its surroundings, or both. If the system operates in a thermodynamic cycle, its initial state is restored after each cycle, so the only place that must be examined for such *other* effects is its surroundings. ➡ **FOR EXAMPLE...** cooling of food is accomplished by refrigerators driven by electric motors requiring work from their surroundings to operate. The Clausius statement implies that it is impossible to construct a refrigeration cycle that operates without an input of work. ⬅

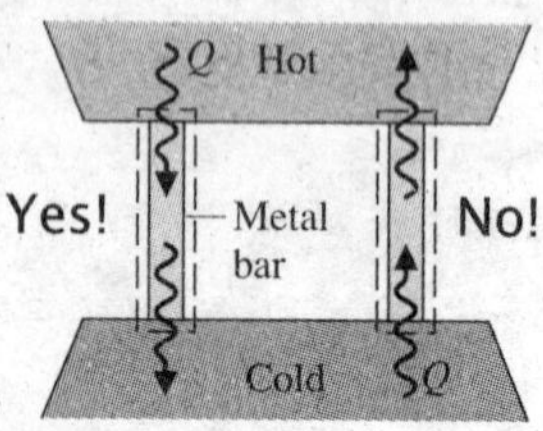

5.2.2 Kelvin–Planck Statement of the Second Law

thermal reservoir

Before giving the Kelvin–Planck statement of the second law, the concept of a ***thermal reservoir*** is introduced. A thermal reservoir, or simply a reservoir, is a special kind of system that always remains at constant temperature even though energy is added or removed by heat transfer. A reservoir is an idealization, of course, but such a system can be approximated in a number of ways—by the earth's atmosphere, large bodies of water (lakes, oceans), a large block of copper, and a system consisting of two phases (although the ratio of the masses of the two phases changes as the system is heated or cooled at constant pressure, the temperature remains constant as long as both phases coexist). Extensive properties of a thermal reservoir such as internal energy can change in interactions with other systems even though the reservoir temperature remains constant.

Having introduced the thermal reservoir concept, we give the ***Kelvin–Planck statement*** of the second law:

Kelvin–Planck statement

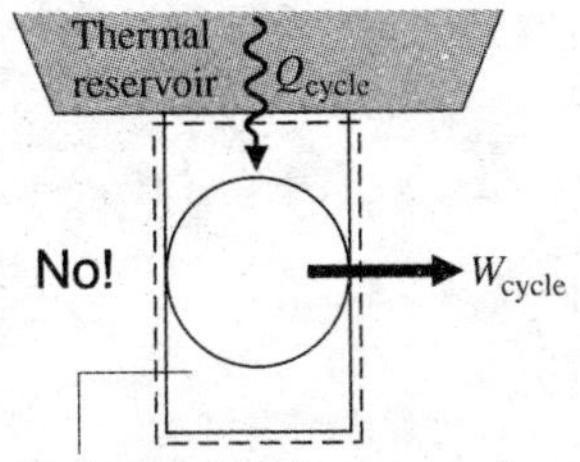

System undergoing a thermodynamic cycle

It is impossible for any system to operate in a thermodynamic cycle and deliver a net amount of energy by work to its surroundings while receiving energy by heat transfer from a single thermal reservoir.

The Kelvin–Planck statement does not rule out the possibility of a system developing a net amount of work from a heat transfer drawn from a single reservoir. It only denies this possibility if the system undergoes a thermodynamic cycle.

The Kelvin–Planck statement can be expressed analytically. To develop this, let us study a system undergoing a cycle while exchanging energy by heat transfer with a *single* reservoir. The first and second laws each impose constraints:

▶ A constraint is imposed by the first law on the net work and heat transfer between the system and its surroundings. According to the cycle energy balance (see Eq. 2.40 in Sec. 2.6),

$$W_{cycle} = Q_{cycle}$$

In words, the net work done by the system undergoing a cycle equals the net heat transfer to the system. Although the cycle energy balance allows the net work W_{cycle} to be positive or negative, the second law imposes a constraint on its direction, as considered next.

▶ According to the Kelvin–Planck statement, a system undergoing a cycle while communicating thermally with a single reservoir *cannot* deliver a net amount of work to its surroundings. That is, the net work of the cycle *cannot be positive.* However, the Kelvin–Planck statement does not rule out the possibility that there is a net work transfer of energy *to* the system during the cycle or that the net work is zero. Thus, the *analytical form of the Kelvin–Planck statement* is

$$W_{cycle} \leq 0 \quad \text{(single reservoir)} \tag{5.1}$$

where the words *single reservoir* are added to emphasize that the system communicates thermally only with a single reservoir as it executes the cycle. In Sec. 5.4, we associate the "less than" and "equal to" signs of Eq. 5.1 with the presence and absence of *internal irreversibilities,* respectively. The concept of irreversibilities is considered in Sec. 5.3.

The equivalence of the Clausius and Kelvin–Planck statements can be demonstrated by showing that the violation of each statement implies the violation of the other. For details, see the box.

Demonstrating the Equivalence of the Clausius and Kelvin–Planck Statements

The equivalence of the Clausius and Kelvin–Planck statements is demonstrated by showing that the violation of each statement implies the violation of the other.

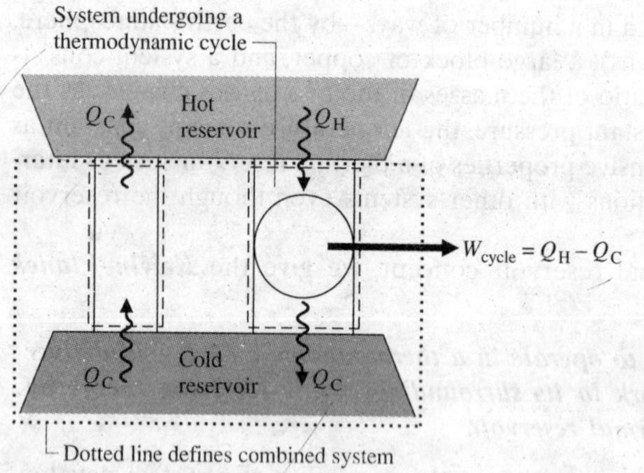

Fig. 5.2 Illustration used to demonstrate the equivalence of the Clausius and Kelvin–Planck statements of the second law.

That a violation of the Clausius statement implies a violation of the Kelvin–Planck statement is readily shown using Fig. 5.2, which pictures a hot reservoir, a cold reservoir, and two systems. The system on the left transfers energy Q_C from the cold reservoir to the hot reservoir by heat transfer without other effects occurring and thus *violates the Clausius statement.* The system on the right operates in a cycle while receiving Q_H (greater than Q_C) from the hot reservoir, rejecting Q_C to the cold reservoir, and delivering work W_{cycle} to the surroundings. The energy flows labeled on Fig. 5.2 are in the directions indicated by the arrows.

Consider the *combined* system shown by a dotted line on Fig. 5.2, which consists of the cold reservoir and the two devices. The combined system can be regarded as executing a cycle because one part undergoes a cycle and the other two parts experience no net change in their conditions. Moreover, the combined system receives energy ($Q_H - Q_C$) by heat transfer from a single reservoir, the hot reservoir, and produces an equivalent amount of work. Accordingly, the combined system violates the Kelvin–Planck statement. Thus, a violation of the Clausius statement implies a violation of the Kelvin–Planck statement. The equivalence of the two second-law statements is demonstrated completely when it is also shown that a violation of the Kelvin–Planck statement implies a violation of the Clausius statement. This is left as an exercise.

5.2.3 Entropy Statement of the Second Law

Mass and energy are familiar examples of extensive properties of systems. Entropy is another important extensive property. We show how the property entropy is evaluated in Secs. 6.2–6.5. Here we introduce several important aspects of entropy.

Just as mass and energy are *accounted for* by mass and energy balances, respectively, entropy is accounted for by an *entropy balance.* In words, the entropy balance states:

$$\begin{bmatrix}\textit{change}\text{ in the amount}\\ \text{of entropy contained}\\ \text{within the system}\\ \text{during some time}\\ \text{interval}\end{bmatrix} = \begin{bmatrix}\text{net amount of}\\ \text{entropy }\textit{transferred}\\ \textit{in}\text{ across the system}\\ \text{boundary during the}\\ \text{time interval}\end{bmatrix} + \begin{bmatrix}\text{amount of entropy}\\ \textit{produced within}\text{ the}\\ \text{system during the}\\ \text{time interval}\end{bmatrix} \tag{5.2}$$

Like mass and energy, *entropy can be transferred* across the system boundary. For closed systems, there is a single means for entropy transfer—namely, entropy transfer accompanying heat transfer. For control volumes entropy also is transferred in and out by streams of matter. These entropy transfers are considered further in Chap. 6.

Unlike mass and energy, which are conserved, *entropy is produced* (or *generated*) within systems whenever *nonidealities* (called *irreversibilities*) such as friction are present. The ***entropy statement of the second law*** states:

entropy statement of the second law

It is impossible for any system to operate in a way that entropy is destroyed.

It follows that the entropy production term of Eq. 5.2 may be positive or zero but *never* negative. Thus, entropy production is an indicator of whether a process is possible or impossible.

5.2.4 Second Law Summary

In the remainder of this chapter, we apply the Kelvin–Planck statement of the second law to draw conclusions about systems undergoing thermodynamic cycles. The chapter concludes with a discussion of the *Clausius inequality* (Sec. 5.11), which provides the basis for developing the entropy concept in Chap. 6. This is a traditional approach to the second law in engineering thermodynamics. However, the order can be reversed—namely, the Entropy statement can be adopted as the starting point for study of the second law aspects of systems.

5.3 Identifying Irreversibilities

One of the important uses of the second law of thermodynamics in engineering is to determine the best theoretical performance of systems. By comparing actual performance with the best theoretical performance, insights often can be gained into the potential for improvement. As might be surmised, the best performance is evaluated in terms of idealized processes. In this section such idealized processes are introduced and distinguished from actual processes involving *irreversibilities*.

5.3.1 Irreversible Processes

A process is called ***irreversible*** if the system and all parts of its surroundings cannot be exactly restored to their respective initial states after the process has occurred. A process is ***reversible*** if both the system and surroundings can be returned to their initial states. Irreversible processes are the subject of the present discussion. Reversible processes are considered again later in the section.

irreversible process

reversible processes

A system that has undergone an irreversible process is not necessarily precluded from being restored to its initial state. However, were the system restored to its initial state, it would not be possible also to return the surroundings to the state they were in initially. As illustrated below, the second law can be used to determine whether both the system and surroundings can be returned to their initial states after a process has occurred. That is, the second law can be used to determine whether a given process is reversible or irreversible.

It might be apparent from the discussion of the Clausius statement of the second law that any process involving a spontaneous heat transfer from a hotter body to a cooler body is irreversible. Otherwise, it would be possible to return this energy from the cooler body to the hotter body with no other effects within the two bodies or their surroundings. However, this possibility is contrary to our experience and is denied by the Clausius statement. Processes involving other kinds of spontaneous events are irreversible, such as an unrestrained expansion of a gas or liquid considered in Fig. 5.1. Friction, electrical resistance, hysteresis, and inelastic deformation are examples of effects whose presence during a process renders it irreversible.

In summary, irreversible processes normally include one or more of the following ***irreversibilities:***

irreversibilities

1. Heat transfer through a finite temperature difference
2. Unrestrained expansion of a gas or liquid to a lower pressure
3. Spontaneous chemical reaction
4. Spontaneous mixing of matter at different compositions or states
5. Friction—sliding friction as well as friction in the flow of fluids
6. Electric current flow through a resistance
7. Magnetization or polarization with hysteresis
8. Inelastic deformation

Although the foregoing list is not exhaustive, it does suggest that *all actual processes are irreversible*. That is, every process involves effects such as those listed, whether it is a naturally occurring process or one involving a device of our construction, from the simplest mechanism to the largest industrial plant. The term "irreversibility" is used to identify any of these effects. The list given previously comprises a few of the irreversibilities that are commonly encountered.

As a system undergoes a process, irreversibilities may be found within the system as well as within its surroundings, although in certain instances they may be located predominately in one place or the other. For many analyses it is convenient to divide the irreversibilities present into two classes. ***Internal irreversibilities*** are those that occur within the system. ***External irreversibilities*** are those that occur within the surroundings, often the immediate surroundings. As this distinction depends solely on the location of the boundary, there is some arbitrariness in the classification, for by extending the boundary to take in a portion of the surroundings, all irreversibilities become "internal." Nonetheless, as shown by subsequent developments, this distinction between irreversibilities is often useful.

internal and external irreversibilities

Engineers should be able to recognize irreversibilities, evaluate their influence, and develop practical means for reducing them. However, certain systems, such as brakes, rely on the effect of friction or other irreversibilities in their operation. The need to achieve profitable rates of production, high heat transfer rates, rapid accelerations, and so on invariably dictates the presence of significant irreversibilities. Furthermore, irreversibilities are tolerated to some degree in every type of system because the changes in design and operation required to reduce them would be too costly. Accordingly, although improved thermodynamic performance can accompany the reduction of irreversibilities, steps taken in this direction are constrained by a number of practical factors often related to costs.

FOR EXAMPLE... consider two bodies at different temperatures that are able to communicate thermally. With a *finite* temperature difference between them, a spontaneous heat transfer would take place and, as discussed previously, this would be a source of irreversibility. It might be expected that the importance of this irreversibility would diminish as the temperature difference approaches zero, and this is the case. From the study of heat transfer (Sec. 2.4), we know that the transfer of a finite amount of energy by heat between bodies whose temperatures differ only slightly would require a considerable amount of time, a larger (more costly) heat transfer surface area, or both. To eliminate this source of irreversibility, therefore, would require an infinite amount of time and/or an infinite surface area.

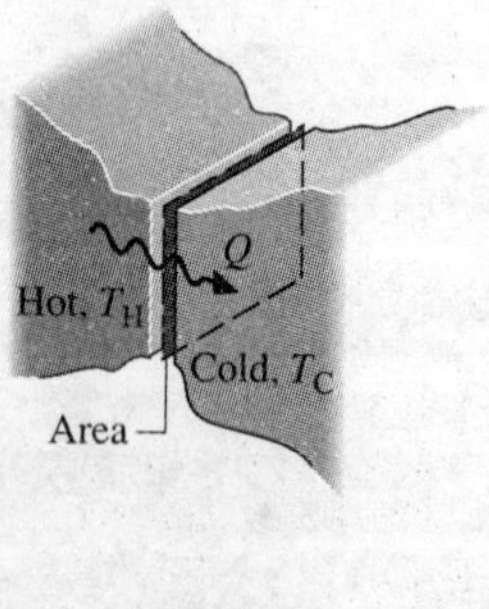

Whenever any irreversibility is present during a process, the process must necessarily be irreversible. However, the irreversibility of the process can be *demonstrated* using the Kelvin–Planck statement of the second law and the following procedure: (1) Assume there is a way to return the system and surroundings to their respective initial states. (2) Show that as a consequence of this assumption, it would be possible to devise a cycle that produces work while no effect occurs other than a heat transfer from a single reservoir. Since the existence of such a cycle is denied by the Kelvin–Planck statement, the initial assumption must be in error and it follows that the process is irreversible. This approach can be used to demonstrate that processes involving friction, heat transfer through a finite temperature difference, the unrestrained expansion of a gas or liquid to a lower pressure, and other effects from the list given previously are irreversible. A case involving friction is discussed in the box.

In most instances the use of the Kelvin–Planck statement to demonstrate the irreversibility of processes is cumbersome. It is normally easier to use the *entropy production* concept (Secs. 5.2.3, 6.7.1).

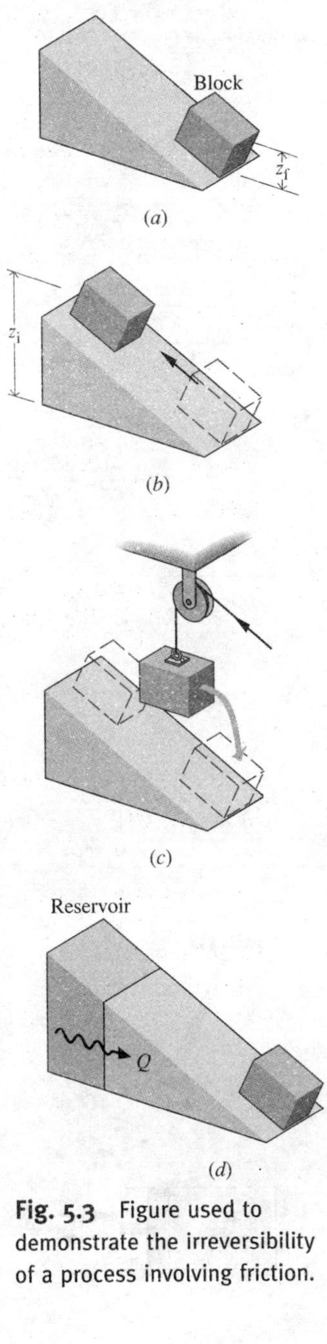

Fig. 5.3 Figure used to demonstrate the irreversibility of a process involving friction.

Demonstrating Irreversibility: Friction

Let us use the Kelvin–Planck statement to demonstrate the irreversibility of a process involving friction. Consider a system consisting of a block of mass m and an inclined plane. Initially the block is at rest at the top of the incline. The block then slides down the plane, eventually coming to rest at a lower elevation. There is no significant heat transfer between the system and its surroundings during the process.

Applying the closed system energy balance

$$(U_f - U_i) + mg(z_f - z_i) + \cancelto{0}{(KE_f - KE_i)} = \cancelto{0}{Q} - \cancelto{0}{W}$$

or

$$U_f - U_i = mg(z_i - z_f)$$

where U denotes the internal energy of the block–plane system and z is the elevation of the block. Thus, friction between the block and plane during the process acts to convert the potential energy decrease of the block to internal energy of the overall system. Since no work or heat interactions occur between the system and its surroundings, the condition of the surroundings remains unchanged during the process. This allows attention to be centered on the system only in demonstrating that the process is irreversible.

When the block is at rest after sliding down the plane, its elevation is z_f and the internal energy of the block–plane system is U_f. To demonstrate that the process is irreversible using the Kelvin–Planck statement, let us take this condition of the system, shown in Fig. 5.3*a*, as the initial state of a cycle consisting of three processes. We imagine that a pulley–cable arrangement and a thermal reservoir are available to assist in the demonstration.

Process 1: Assume that the inverse process can occur with no change in the surroundings. That is, as shown in Fig. 5.3*b*, assume that the block returns spontaneously to its initial elevation and the internal energy of the system decreases to its initial value, U_i. (This is the process we want to demonstrate is impossible.)

Process 2: As shown in Fig. 5.3*c*, use the pulley–cable arrangement provided to lower the block from z_i to z_f, allowing the decrease in potential energy to do work by lifting another mass located in the surroundings. The work done by the system equals the decrease in the potential energy of the block: $mg(z_i - z_f)$.

Process 3: The internal energy of the system can be increased from U_i to U_f by bringing it into communication with the reservoir, as shown in Fig. 5.3*d*. The heat transfer required is $Q = U_f - U_i$. Or, with the result of the energy balance on the system given above, $Q = mg(z_i - z_f)$. At the conclusion of this process the block is again at elevation z_f and the internal energy of the block–plane system is restored to U_f.

The net result of this cycle is to draw energy from a single reservoir by heat transfer and produce an equivalent amount of work. There are no other effects. However, such a cycle is denied by the Kelvin–Planck statement. Since both the heating of the system by the reservoir (Process 3) and the lowering of the mass by the pulley–cable while work is done (Process 2) are possible, it can be concluded that it is Process 1 that is impossible. Since Process 1 is the inverse of the original process where the block slides down the plane, it follows that the original process is irreversible.

5.3.2 Reversible Processes

A process of a system is *reversible* if the system and all parts of its surroundings can be exactly restored to their respective initial states after the process has taken place. It should be evident from the discussion of irreversible processes that reversible processes are purely hypothetical. Clearly, no process can be reversible that involves spontaneous heat transfer through a finite temperature difference, an unrestrained expansion of a gas or liquid, friction, or any of the other irreversibilities listed previously. In a strict sense of the word, a reversible process is one that is *perfectly executed.*

All actual processes are irreversible. Reversible processes do not occur. Even so, certain processes that do occur are approximately reversible. The passage of a gas through a properly designed nozzle or diffuser is an example (Sec. 6.12). Many other devices also can be made to approach reversible operation by taking measures to reduce the significance of irreversibilities, such as lubricating surfaces to reduce friction. A reversible process is the *limiting case* as irreversibilities, both internal and external, are reduced further and further.

Although reversible processes cannot actually occur, they can be imagined. Earlier in this section we considered how heat transfer would approach reversibility as the temperature difference approaches zero. Let us consider two additional examples:

- A particularly elementary example is a pendulum oscillating in an evacuated space. The pendulum motion approaches reversibility as friction at the pivot point is reduced. In the limit as friction is eliminated, the states of both the pendulum and its surroundings would be completely restored at the end of each period of motion. By definition, such a process is reversible.

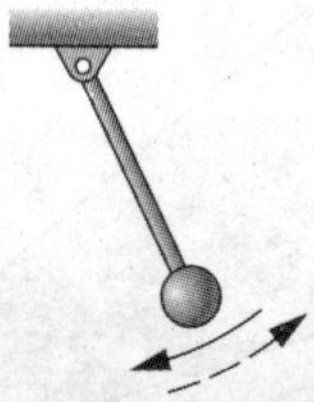

- A system consisting of a gas adiabatically compressed and expanded in a frictionless piston–cylinder assembly provides another example. With a very small increase in the external pressure, the piston would compress the gas slightly. At each intermediate volume during the compression, the intensive properties T, p, v, etc., would be uniform throughout: The gas would pass through a series of equilibrium states. With a small decrease in the external pressure, the piston would slowly move out as the gas expands. At each intermediate volume of the expansion, the intensive properties of the gas would be at the same uniform values they had at the corresponding step during the compression. When the gas volume returned to its initial value, all properties would be restored to their initial values as well. The work done *on* the gas during the compression would equal the work done *by* the gas during the expansion. If the work

Horizons

Second Law Takes Big Bite from Hydrogen

Hydrogen is not naturally occurring and thus must be produced. Hydrogen can be produced today from water by *electrolysis* and from natural gas by chemical processing called *reforming.* Hydrogen produced by these means and its subsequent utilization is burdened by the second law.

In electrolysis, an electrical input is employed to dissociate water to hydrogen according to $H_2O \rightarrow H_2 + \frac{1}{2}O_2$. When the hydrogen is subsequently used by a fuel cell to generate electricity, the cell reaction is $H_2 + \frac{1}{2}O_2 \rightarrow H_2O$. Although the cell reaction is the inverse of that occurring in electrolysis, the overall loop from electrical input–to hydrogen–to fuel cell-generated electricity is *not* reversible. Irreversibilities in the electrolyzer and the fuel cell conspire to ensure that the fuel cell-generated electricity is much less than the initial electrical input. Some say this is wasteful because the electricity provided for electrolysis could instead be *fully* directed to most applications envisioned for hydrogen, including transportation. Further, when fossil fuel is burned in a power plant to generate the electricity for electrolysis, the greenhouse gases that are formed can be associated with fuel cells by virtue of the hydrogen they consume. Although technical details differ, similar findings apply to the reforming of natural gas to hydrogen.

While hydrogen and fuel cells are expected to play a role in our energy future, second law barriers and other technical and economic issues stand in the way.

between the system and its surroundings were delivered to, and received from, a frictionless pulley–mass assembly, or the equivalent, there also would be no net change in the surroundings. This process would be reversible.

5.3.3 Internally Reversible Processes

In an irreversible process, irreversibilities are present within the system, its surroundings, or both. A reversible process is one in which there are no internal *or* external irreversibilities. An ***internally reversible process*** is one in which *there are no irreversibilities within the system.* Irreversibilities may be located within the surroundings, however, as when there is heat transfer between a portion of the boundary that is at one temperature and the surroundings at another.

internally reversible process

At every intermediate state of an internally reversible process of a closed system, all intensive properties are uniform throughout each phase present. That is, the temperature, pressure, specific volume, and other intensive properties do not vary with position. If there were a spatial variation in temperature, say, there would be a tendency for a spontaneous energy transfer by conduction to occur *within* the system in the direction of decreasing temperature. For reversibility, however, no spontaneous processes can be present. From these considerations it can be concluded that the internally reversible process consists of a series of equilibrium states: It is a quasiequilibrium process.

The use of the internally reversible process concept in thermodynamics is comparable to the idealizations made in mechanics: point masses, frictionless pulleys, rigid beams, and so on. In much the same way as these are used in mechanics to simplify an analysis and arrive at a manageable model, simple thermodynamic models of complex situations can be obtained through the use of internally reversible processes. Initial calculations based on internally reversible processes would be adjusted with efficiencies or correction factors to obtain reasonable estimates of actual performance under various operating conditions. Internally reversible processes are also useful in determining the best thermodynamic performance of systems.

Take Note...

The terms *internally reversible process* and *quasiequilibrium process* can be used interchangeably. However, to avoid having two terms that refer to the same thing, in subsequent sections we will refer to *any* such process as an internally reversible process.

Finally, using the internally reversible process concept, we refine the definition of the thermal reservoir introduced in Sec. 5.2.2 as follows: In subsequent discussions we assume that no internal irreversibilities are present within a thermal reservoir. That is, every process of a thermal reservoir is *internally reversible.*

5.4 Interpreting the Kelvin–Planck Statement

Let us briefly reconsider Eq. 5.1, the analytical form of the Kelvin–Planck statement of the second law, which is employed in subsequent sections to obtain a number of significant deductions. In each of these applications, the following idealizations are assumed: The thermal reservoir and the portion of the surroundings with which work interactions occur are free of irreversibilities. This allows the "less than" sign of Eq. 5.1 to be associated with irreversibilities *within* the system of interest. The "equal to" sign is employed only when no internal or external irreversibilities are present.

Accordingly, the ***analytical form of the Kelvin–Planck statement*** now takes the form

analytical form: Kelvin–Planck statement

$$W_{\text{cycle}} \leq 0 \begin{cases} < 0: & \text{Internal irreversibilities present.} \\ = 0: & \text{No internal irreversibilities.} \end{cases} \quad \text{(single reservoir)} \tag{5.3}$$

For details, see the box. Since Eq. 5.3 is more explicit than Eq. 5.1, it is used in subsequent discussions of this chapter as the preferred form of the Kelvin–Planck statement of the second law.

Associating Signs with the Kelvin–Planck Statement

Consider a system that undergoes a cycle while exchanging energy by heat transfer with a single reservoir, as shown in Fig. 5.4. Work is delivered to, or received from, the pulley–mass assembly located in the surroundings. A flywheel, spring, or some other device also can perform the same function. In subsequent applications of Eq. 5.3, the irreversibilities of primary interest are internal irreversibilities. To eliminate extraneous factors in such applications, therefore, assume that these are the only irreversibilities present. Hence, the pulley–mass assembly, flywheel, or other device to which work is delivered, or from which it is received, is idealized as free of irreversibilities. The thermal reservoir is also assumed free of irreversibilities.

To demonstrate the correspondence of the "equal to" sign of Eq. 5.1 with the absence of irreversibilities, consider a cycle operating as shown in Fig. 5.4 for which the equality applies. At the conclusion of one cycle,

- The system would necessarily be returned to its initial state.
- Since $W_{\text{cycle}} = 0$, there would be no *net* change in the elevation of the mass used to store energy in the surroundings.
- Since $W_{\text{cycle}} = Q_{\text{cycle}}$, it follows that $Q_{\text{cycle}} = 0$, so there also would be no *net* change in the condition of the reservoir.

Thus, the system and all elements of its surroundings would be exactly restored to their respective initial conditions. By definition, such a cycle is reversible. Accordingly, there can be no irreversibilities present within the system or its surroundings. It is left as an exercise to show the converse: If the cycle occurs reversibly, the equality applies. Since a cycle is either reversible or irreversible, it follows that the inequality sign implies the presence of irreversibilities, and the inequality applies whenever irreversibilities are present.

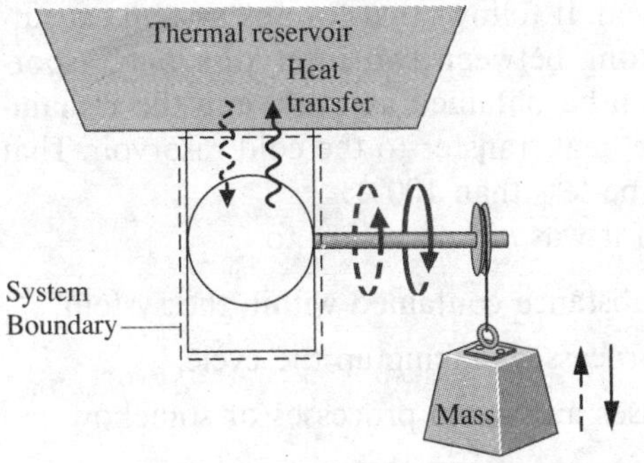

Fig. 5.4 System undergoing a cycle while exchanging energy by heat transfer with a single thermal reservoir.

5.5 Applying the Second Law to Thermodynamic Cycles

Several important applications of the second law related to power cycles and refrigeration and heat pump cycles are presented in Secs. 5.6, 5.7, 5.9, and 5.10. These applications further our understanding of the implications of the second law and provide the basis for important deductions from the second law introduced in Secs. 5.8 and 5.11 and in Chap. 6. Familiarity with thermodynamic cycles is required, and we recommend that you review Sec. 2.6, where cycles are considered from an energy, or first law, perspective and the thermal efficiency of power cycles and coefficients of performance for refrigeration and heat pump cycles are introduced. In particular, review Eqs. 2.41–2.48.

5.6 Second Law Aspects of Power Cycles Interacting with Two Reservoirs

5.6.1 Limit on Thermal Efficiency

A significant limitation on the performance of systems undergoing power cycles can be brought out using the Kelvin–Planck statement of the second law. Consider Fig. 5.5, which shows a system that executes a cycle while communicating thermally with *two* thermal reservoirs, a hot reservoir and a cold reservoir, and developing net work W_{cycle}. The thermal efficiency of the cycle is

$$\eta = \frac{W_{cycle}}{Q_H} = 1 - \frac{Q_C}{Q_H} \tag{5.4}$$

where Q_H is the amount of energy received by the system from the hot reservoir by heat transfer and Q_C is the amount of energy discharged from the system to the cold reservoir by heat transfer.

If the value of Q_C were zero, the system of Fig. 5.5 would withdraw energy Q_H from the hot reservoir and produce an equal amount of work, while undergoing a cycle. The thermal efficiency of such a cycle would have a value of unity (100%). However, this method of operation would violate the Kelvin–Planck

Take Note...
The energy transfers labeled on Fig 5.5 are positive in the directions indicated by the arrows.

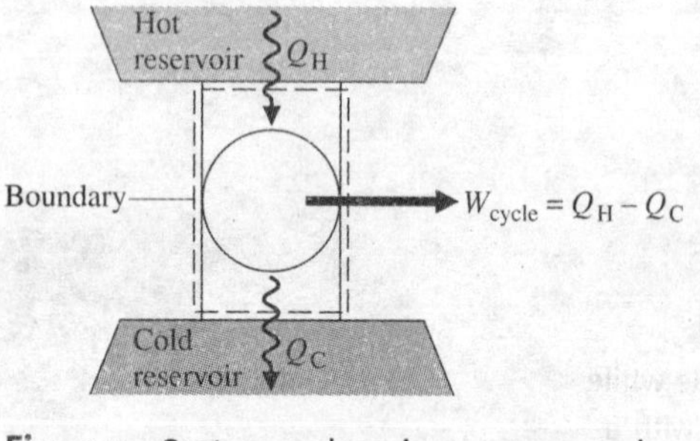

Fig. 5.5 System undergoing a power cycle while exchanging energy by heat transfer with two reservoirs.

statement and thus is not allowed. It follows that for any system executing a power cycle while operating between two reservoirs, only a portion of the heat transfer Q_H can be obtained as work, and the remainder, Q_C, must be discharged by heat transfer to the cold reservoir. That is, the thermal efficiency must be less than 100%.

In arriving at this conclusion it was *not* necessary to

- identify the nature of the substance contained within the system,
- specify the exact series of processes making up the cycle,
- indicate whether the processes are actual processes or somehow idealized.

The conclusion that the thermal efficiency must be less than 100% applies to *all* power cycles whatever their details of operation. This may be regarded as a corollary of the second law. Other corollaries follow.

5.6.2 Corollaries of the Second Law for Power Cycles

Carnot corollaries

Since no power cycle can have a thermal efficiency of 100%, it is of interest to investigate the maximum theoretical efficiency. The maximum theoretical efficiency for systems undergoing power cycles while communicating thermally with two thermal reservoirs at different temperatures is evaluated in Sec. 5.9 with reference to the following two corollaries of the second law, called the ***Carnot corollaries.***

1. The thermal efficiency of an irreversible power cycle is always less than the thermal efficiency of a reversible power cycle when each operates between the same two thermal reservoirs.
2. All reversible power cycles operating between the same two thermal reservoirs have the same thermal efficiency.

A cycle is considered *reversible* when there are no irreversibilities within the system as it undergoes the cycle and heat transfers between the system and reservoirs occur reversibly.

The idea underlying the first Carnot corollary is in agreement with expectations stemming from the discussion of the second law thus far. Namely, the presence of irreversibilities during the execution of a cycle is expected to exact a penalty. If two systems operating between the same reservoirs each receive the same amount of energy Q_H and one executes a reversible cycle while the other executes an irreversible cycle, it is in accord with intuition that the net work developed by the irreversible cycle will be less, and it will therefore have the smaller thermal efficiency.

The second Carnot corollary refers only to reversible cycles. All processes of a reversible cycle are perfectly executed. Accordingly, if two reversible cycles operating between the same reservoirs each receive the same amount of energy Q_H but one could produce more work than the other, it could only be as a result of more advantageous selections for the substance making up the system (it is conceivable that, say, air might be better than water vapor) *or* the series of processes making up the cycle (nonflow processes might be preferable to flow processes). This corollary denies both possibilities and indicates that the cycles must have the same efficiency whatever the choices for the working substance or the series of processes.

The two Carnot corollaries can be demonstrated using the Kelvin–Planck statement of the second law. For details, see the box.

Demonstrating the Carnot Corollaries

The first Carnot corollary can be demonstrated using the arrangement of Fig. 5.6. A reversible power cycle R and an irreversible power cycle I operate between the same two reservoirs and each receives the same amount of energy Q_H from the hot reservoir. The reversible cycle produces work W_R while the irreversible cycle produces work W_I. In accord with the conservation of energy principle, each cycle discharges energy to the cold reservoir equal to the difference between Q_H and the work produced. Let R now operate in the opposite direction as a refrigeration (or heat pump) cycle. Since R is reversible, the magnitudes of the energy transfers W_R, Q_H, and Q_C remain the same, but the energy transfers are oppositely directed, as shown by the dashed lines on Fig. 5.6. Moreover, with R operating in the opposite direction, the hot reservoir would experience *no net change* in its condition since it would receive Q_H *from* R while passing Q_H *to* I.

The demonstration of the first Carnot corollary is completed by considering the *combined system* shown by the dotted line on Fig. 5.6, which consists of the two cycles and the hot reservoir. Since its parts execute cycles or experience no net change, the combined system operates in a cycle. Moreover, the combined system exchanges energy by heat transfer with a single reservoir: the cold reservoir. Accordingly, the combined system must satisfy Eq. 5.3 expressed as

$$W_{cycle} < 0 \qquad \text{(single reservoir)}$$

where the inequality is used because the combined system is irreversible in its operation since irreversible cycle I is one of its parts. Evaluating W_{cycle} for the combined system in terms of the work amounts W_I and W_R, the above inequality becomes

$$W_I - W_R < 0$$

which shows that W_I must be less than W_R. Since each cycle receives the same energy input, Q_H, it follows that $\eta_I < \eta_R$ and this completes the demonstration.

The second Carnot corollary can be demonstrated in a parallel way by considering any two reversible cycles R_1 and R_2 operating between the same two reservoirs. Then, letting R_1 play the role of R and R_2 the role of I in the previous development, a combined system consisting of the two cycles and the hot reservoir may be formed that must obey Eq. 5.3. However, in applying Eq. 5.3 to this combined system, the equality is used because the system is reversible in operation. Thus, it can be concluded that $W_{R1} = W_{R2}$, and therefore, $\eta_{R1} = \eta_{R2}$. The details are left as an exercise.

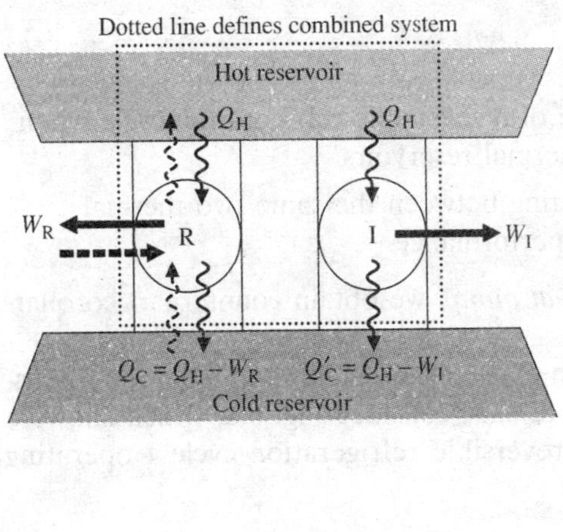

Fig. 5.6 Sketch for demonstrating that a reversible cycle R is more efficient than an irreversible cycle I when they operate between the same two reservoirs.

5.7 Second Law Aspects of Refrigeration and Heat Pump Cycles Interacting with Two Reservoirs

5.7.1 Limits on Coefficients of Performance

The second law of thermodynamics places limits on the performance of refrigeration and heat pump cycles as it does for power cycles. Consider Fig. 5.7, which shows a system undergoing a cycle while communicating thermally with two thermal reservoirs, a hot and a cold reservoir. The energy transfers labeled on the figure are in the directions indicated by the arrows. In accord with the conservation of energy principle, the cycle discharges energy Q_H by heat transfer to the hot reservoir equal to the sum of the energy Q_C received by heat transfer from the cold reservoir and the net work input. This cycle might be a refrigeration cycle or a heat pump cycle, depending on whether its function is to remove energy Q_C from the cold reservoir or deliver energy Q_H to the hot reservoir.

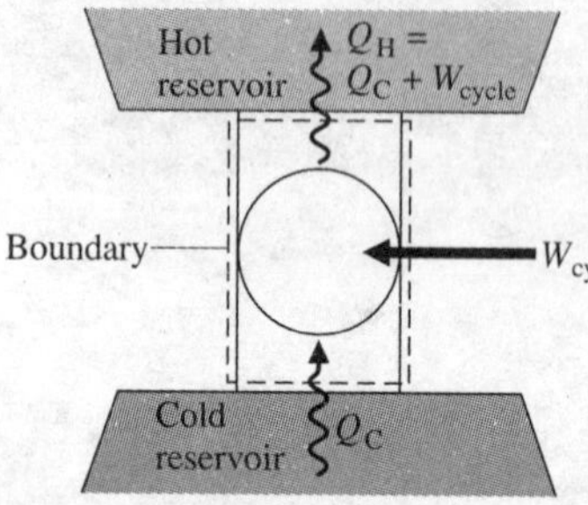

Fig. 5.7 System undergoing a refrigeration or heat pump cycle while exchanging energy by heat transfer with two reservoirs.

For a refrigeration cycle the coefficient of performance is

$$\beta = \frac{Q_C}{W_{cycle}} = \frac{Q_C}{Q_H - Q_C} \tag{5.5}$$

The coefficient of performance for a heat pump cycle is

$$\gamma = \frac{Q_H}{W_{cycle}} = \frac{Q_H}{Q_H - Q_C} \tag{5.6}$$

As the net work input to the cycle W_{cycle} tends to zero, the coefficients of performance given by Eqs. 5.5 and 5.6 approach a value of infinity. If W_{cycle} were identically zero, the system of Fig. 5.7 would withdraw energy Q_C from the cold reservoir and deliver energy Q_C to the hot reservoir, while undergoing a cycle. However, this method of operation would violate the Clausius statement of the second law and thus is not allowed. It follows that these coefficients of performance must invariably be finite in value. This may be regarded as another corollary of the second law. Further corollaries follow.

5.7.2 Corollaries of the Second Law for Refrigeration and Heat Pump Cycles

The maximum theoretical coefficients of performance for systems undergoing refrigeration and heat pump cycles while communicating thermally with two reservoirs at different temperatures are evaluated in Sec. 5.9 with reference to the following corollaries of the second law:

1. The coefficient of performance of an irreversible refrigeration cycle is always less than the coefficient of performance of a reversible refrigeration cycle when each operates between the same two thermal reservoirs.
2. All reversible refrigeration cycles operating between the same two thermal reservoirs have the same coefficient of performance.

By replacing the term *refrigeration* with *heat pump*, we obtain counterpart corollaries for heat pump cycles.

The first of these corollaries agrees with expectations stemming from the discussion of the second law thus far. To explore this, consider Fig. 5.8, which shows a reversible refrigeration cycle R and an irreversible refrigeration cycle I operating between the same two reservoirs. Each cycle removes the same energy Q_C from the cold reservoir. The net work input required to operate R is W_R, while the net work input for I is W_I. Each cycle discharges energy by heat transfer to the hot reservoir equal to the sum of Q_C and the net work input. The directions of the energy transfers are shown by arrows on Fig. 5.8. The presence of irreversibilities during the operation of a refrigeration cycle is expected to exact a penalty. If two refrigerators working between the same reservoirs each receive an identical energy transfer from the cold reservoir, Q_C, and one executes a reversible cycle while the other executes an irreversible cycle, we expect the irreversible cycle to require a greater net work input and thus to have a smaller coefficient of performance. By a simple extension it follows that all reversible refrigeration cycles operating between the same two reservoirs have the same coefficient of performance. Similar arguments apply to the counterpart heat pump cycle statements.

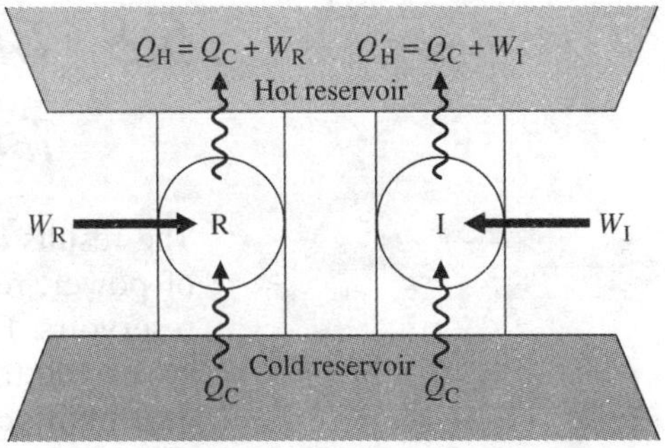

Fig. 5.8 Sketch for demonstrating that a reversible refrigeration cycle R has a greater coefficient of performance than an irreversible cycle I when they operate between the same two reservoirs.

These corollaries can be demonstrated formally using the Kelvin–Planck statement of the second law and a procedure similar to that employed for the Carnot corollaries. The details are left as an exercise.

Energy & Environment

Warm blankets of pollution-laden air surround major cities. Sunlight-absorbing rooftops and expanses of pavement, together with little greenery, conspire with other features of city living to raise urban temperatures several degrees above adjacent suburban areas. Figure 5.9 shows the variation of surface temperature in the vicinity of a city as measured by infrared measurements made from low-level flights over the area. Health-care professionals worry about the impact of these "heat islands," especially on the elderly. Paradoxically, the hot exhaust from the air conditioners city dwellers use to keep cool also make sweltering neighborhoods even hotter. Irreversibilities within air conditioners contribute to the warming effect. Air conditioners may account for as much as 20% of the urban temperature rise. Vehicles and commercial activity also are contributors. Urban planners are combating heat islands in many ways, including the use of highly reflective colored roofing products and the installation of roof-top gardens. The shrubs and trees of roof-top gardens absorb solar energy, leading to summer roof temperatures significantly below those of nearby buildings without roof-top gardens, reducing the need for air conditioning.

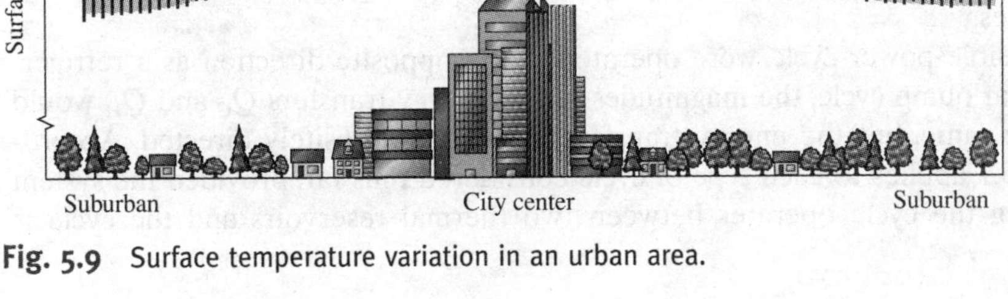

Fig. 5.9 Surface temperature variation in an urban area.

5.8 The Kelvin and International Temperature Scales

The results of Secs. 5.6 and 5.7 establish theoretical upper limits on the performance of power, refrigeration, and heat pump cycles communicating thermally with two reservoirs. Expressions for the *maximum* theoretical thermal efficiency of power cycles and the *maximum* theoretical coefficients of performance of refrigeration and heat pump cycles are developed in Sec. 5.9 using the Kelvin temperature scale considered next.

5.8.1 The Kelvin Scale

From the second Carnot corollary we know that all reversible power cycles operating between the same two thermal reservoirs have the same thermal efficiency, regardless of the nature of the substance making up the system executing the cycle or the series of processes. Since the efficiency is independent of these factors, its value can be related only to the nature of the reservoirs themselves. Noting that it is the difference in *temperature* between the two reservoirs that provides the impetus for heat transfer between them, and thereby for the production of work during the cycle, we reason that the efficiency depends *only* on the temperatures of the two reservoirs.

From Eq. 5.4 it also follows that for such reversible power cycles the ratio of the heat transfers Q_C/Q_H depends only on the reservoir temperatures. That is

$$\left(\frac{Q_C}{Q_H}\right)_{\substack{\text{rev}\\ \text{cycle}}} = \psi(\theta_C, \theta_H) \tag{a}$$

where θ_H and θ_C denote the temperatures of the reservoirs and the function ψ is for the present unspecified. Note that the words "rev cycle" are added to this expression to emphasize that it applies only to systems undergoing reversible cycles while operating between two thermal reservoirs.

Equation (a) provides a basis for defining a *thermodynamic* temperature scale: a scale independent of the properties of any substance. There are alternative choices for the function ψ that lead to this end. The ***Kelvin scale*** is obtained by making a particularly simple choice, namely, $\psi = T_C/T_H$, where T is the symbol used to denote temperatures on the Kelvin scale. With this, we get

Kelvin scale

$$\left(\frac{Q_C}{Q_H}\right)_{\substack{\text{rev}\\ \text{cycle}}} = \frac{T_C}{T_H} \tag{5.7}$$

Thus, two temperatures on the Kelvin scale are in the same ratio as the values of the heat transfers absorbed and rejected, respectively, by a system undergoing a reversible cycle while communicating thermally with reservoirs at these temperatures.

If a reversible power cycle were operated in the opposite direction as a refrigeration or heat pump cycle, the magnitudes of the energy transfers Q_C and Q_H would remain the same, but the energy transfers would be oppositely directed. Accordingly, Eq. 5.7 applies to each type of cycle considered thus far, provided the system undergoing the cycle operates between two thermal reservoirs and the cycle is reversible.

Take Note...

Some readers may prefer to proceed directly to Sec. 5.9 where Eq. 5.7 is applied.

More on the Kelvin Scale

Equation 5.7 gives only a ratio of temperatures. To complete the definition of the Kelvin scale, it is necessary to proceed as in Sec. 1.7 by assigning the value 273.16 K to the temperature at the triple point of water. Then, if a reversible cycle is operated between a reservoir at 273.16 K and another reservoir at temperature T, the two temperatures are related according to

$$T = 273.16\left(\frac{Q}{Q_{tp}}\right)_{\substack{rev\\cycle}} \tag{5.8}$$

where Q_{tp} and Q are the heat transfers between the cycle and reservoirs at 273.16 K and temperature T, respectively. In the present case, the heat transfer Q plays the role of the *thermometric property.* However, since the performance of a reversible cycle is independent of the makeup of the system executing the cycle, the definition of temperature given by Eq. 5.8 depends in no way on the properties of any substance or class of substances.

In Sec. 1.7 we noted that the Kelvin scale has a zero of 0 K, and lower temperatures than this are not defined. Let us take up these points by considering a reversible power cycle operating between reservoirs at 273.16 K and a lower temperature T. Referring to Eq. 5.8, we know that the energy rejected from the cycle by heat transfer Q would not be negative, so T must be nonnegative. Equation 5.8 also shows that the smaller the value of Q, the lower the value of T, and conversely. Accordingly, as Q approaches zero the temperature T approaches zero. It can be concluded that a temperature of zero on the Kelvin scale is the lowest conceivable temperature. This temperature is called the *absolute* zero, and the Kelvin scale is called an *absolute temperature scale.*

When numerical values of the thermodynamic temperature are to be determined, it is not possible to use reversible cycles, for these exist only in our imaginations. However, temperatures evaluated using the constant-volume gas thermometer discussed in Sec. 5.8.2 to follow are identical to those of the Kelvin scale in the range of temperatures where the gas thermometer can be used. Other empirical approaches can be employed for temperatures above and below the range accessible to gas thermometry. The Kelvin scale provides a continuous definition of temperature valid over all ranges and provides an essential connection between the several empirical measures of temperature.

5.8.2 The Gas Thermometer

The constant-volume gas thermometer shown in Fig. 5.10 is so exceptional in terms of precision and accuracy that it has been adopted internationally as the standard instrument for calibrating other thermometers. The *thermometric substance* is the gas (normally hydrogen or helium), and the *thermometric property* is the pressure exerted by the gas. As shown in the figure, the gas is contained in a bulb, and the pressure exerted by it is measured by an open-tube mercury manometer. As temperature increases, the gas expands, forcing mercury up in the open tube. The gas is kept at constant volume by raising or lowering the reservoir. The gas thermometer is used as a standard worldwide by bureaus of standards and research laboratories. However, because gas thermometers require elaborate apparatus and are large, slowly responding devices that demand painstaking experimental procedures, smaller, more rapidly responding thermometers are used for most temperature measurements and they are calibrated (directly or indirectly) against gas thermometers. For further discussion of gas thermometry, see the box.

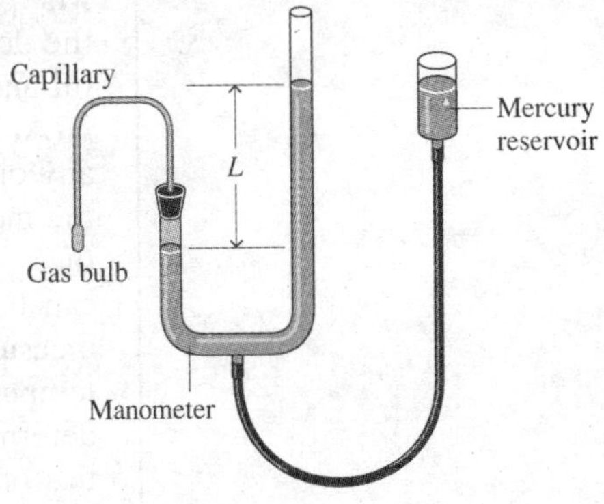

Fig. 5.10 Constant-volume gas thermometer.

Measuring Temperature with the Gas Thermometer—The Gas Scale

It is instructive to consider how numerical values are associated with levels of temperature by the gas thermometer shown in Fig. 5.10. Let p stand for the pressure in the bulb of a constant-volume gas thermometer in thermal equilibrium with a bath. A value can be assigned to the bath temperature by a linear relation

$$T = \alpha p \tag{a}$$

where α is an arbitrary constant.

The value of α is determined by inserting the thermometer into another bath maintained at the triple point of water and measuring the pressure, call it p_{tp}, of the confined gas at the triple point temperature, 273.16 K. Substituting values into Eq. (a) and solving for α

$$\alpha = \frac{273.16}{p_{tp}}$$

Inserting this in Eq. (a), the temperature of the original bath, at which the pressure of the confined gas is p, is then

$$T = 273.16\left(\frac{p}{p_{tp}}\right) \tag{b}$$

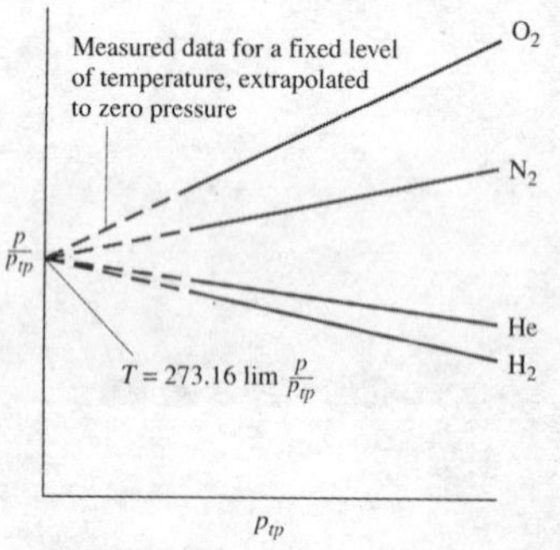

Fig. 5.11 Readings of constant-volume gas thermometers, when several gases are used.

However, since the values of both pressures, p and p_{tp}, depend *in part* on the amount of gas in the bulb, the value assigned by Eq. (b) to the bath temperature varies with the amount of gas in the thermometer. This difficulty is overcome in precision thermometry by repeating the measurements (in the original bath and the reference bath) several times with less gas in the bulb in each successive attempt. For each trial the ratio p/p_{tp} is calculated from Eq. (b) and plotted versus the corresponding reference pressure p_{tp} of the gas at the triple point temperature. When several such points have been plotted, the resulting curve is extrapolated to the ordinate where $p_{tp} = 0$. This is illustrated in Fig. 5.11 for constant-volume thermometers with a number of different gases.

Inspection of Fig. 5.11 shows that at each nonzero value of the reference pressure, the p/p_{tp} values differ with the gas employed in the thermometer. However, as pressure decreases, the p/p_{tp} values from thermometers with different gases approach one another, and in the limit as pressure tends to zero, *the same value for p/p_{tp} is obtained for each gas.* Based on these general results, the *gas temperature scale* is defined by the relationship

$$T = 273.16 \lim \frac{p}{p_{tp}} \tag{c}$$

where "lim" means that both p and p_{tp} tend to zero. It should be evident that the determination of temperatures by this means requires extraordinarily careful and elaborate experimental procedures.

Although the temperature scale of Eq. (c) is independent of the properties of any one gas, it still depends on the properties of gases in general. Accordingly, the measurement of low temperatures requires a gas that does not condense at these temperatures, and this imposes a limit on the range of temperatures that can be measured by a gas thermometer. The lowest temperature that can be measured with such an instrument is about 1 K, obtained with helium. At high temperatures gases dissociate, and therefore these temperatures also cannot be determined by a gas thermometer. Other empirical means, utilizing the properties of other substances, must be employed to measure temperature in ranges where the gas thermometer is inadequate. For further discussion see Sec. 5.8.3.

5.8.3 International Temperature Scale

To provide a standard for temperature measurement taking into account both theoretical and practical considerations, the International Temperature Scale (ITS) was adopted in 1927. This scale has been refined and extended in several revisions, most recently in 1990. *The International Temperature Scale of 1990 (ITS-90)* is defined in such a way that the temperature measured on it conforms with the thermodynamic temperature, the unit of which is the kelvin, to within the limits of accuracy of measurement obtainable in 1990. The ITS-90 is based on the assigned values of temperature of a number of reproducible *fixed points* (Table 5.1). Interpolation between the fixed-point temperatures is accomplished by formulas that give the relation between readings of standard instruments and values of the ITS. In the range from 0.65 to 5.0 K, ITS-90 is defined by equations giving the temperature as functions of the vapor pressures of particular helium isotopes. The range from 3.0 to 24.5561 K is based on measurements using a helium constant-volume gas thermometer. In the range from 13.8033 to 1234.93 K, ITS-90 is defined by means of certain platinum resistance thermometers. Above 1234.9 K the temperature is defined using *Planck's equation for blackbody radiation* and measurements of the intensity of visible-spectrum radiation.

Table 5.1

Defining Fixed Points of the International Temperature Scale of 1990

T (K)	Substance[a]	State[b]
3 to 5	He	Vapor pressure point
13.8033	e-H_2	Triple point
≈ 17	e-H_2	Vapor pressure point
≈ 20.3	e-H_2	Vapor pressure point
24.5561	Ne	Triple point
54.3584	O_2	Triple point
83.8058	Ar	Triple point
234.3156	Hg	Triple point
273.16	H_2O	Triple point
302.9146	Ga	Melting point
429.7485	In	Freezing point
505.078	Sn	Freezing point
692.677	Zn	Freezing point
933.473	Al	Freezing point
1234.93	Ag	Freezing point
1337.33	Au	Freezing point
1357.77	Cu	Freezing point

[a]He denotes ^3He or ^4He; e-H_2 is hydrogen at the equilibrium concentration of the ortho- and para-molecular forms.

[b]Triple point: temperature at which the solid, liquid, and vapor phases are in equilibrium. Melting point, freezing point: temperature, at a pressure of 101.325 kPa, at which the solid and liquid phases are in equilibrium.

Source: H. Preston-Thomas, "The International Temperature Scale of 1990 (ITS-90)," *Metrologia* 27, 3–10 (1990). See also www.ITS-90.com.

5.9 Maximum Performance Measures for Cycles Operating Between Two Reservoirs

The discussion continues in this section with the development of expressions for the maximum thermal efficiency of power cycles and the maximum coefficients of performance of refrigeration and heat pump cycles in terms of reservoir temperatures evaluated on the Kelvin scale. These expressions can be used as standards of comparison for actual power, refrigeration, and heat pump cycles.

5.9.1 Power Cycles

The use of Eq. 5.7 in Eq. 5.4 results in an expression for the thermal efficiency of a system undergoing a reversible *power cycle* while operating between thermal reservoirs at temperatures T_H and T_C. That is

Carnot efficiency

$$\eta_{max} = 1 - \frac{T_C}{T_H} \tag{5.9}$$

which is known as the ***Carnot efficiency.***

Recalling the two Carnot corollaries, it should be evident that the efficiency given by Eq. 5.9 is the thermal efficiency of *all* reversible power cycles operating between two reservoirs at temperatures T_H and T_C, and the *maximum* efficiency *any* power cycle can have while operating between the two reservoirs. By inspection, the value of the Carnot efficiency increases as T_H increases and/or T_C decreases.

Equation 5.9 is presented graphically in Fig. 5.12. The temperature T_C used in constructing the figure is 298 K in recognition that actual power cycles ultimately discharge energy by heat transfer at about the temperature of the local atmosphere or cooling water drawn from a nearby river or lake. Note that the possibility of increasing the thermal efficiency by reducing T_C below that of the environment is not practical, for maintaining T_C lower than the ambient temperature would require a refrigerator that would have to be supplied work to operate.

Figure 5.12 shows that the thermal efficiency increases with T_H. Referring to segment a–b of the curve, where T_H and η are relatively low, we see that η increases rapidly as T_H increases, showing that in this range even a small increase in T_H can have a large effect on efficiency. Though these conclusions, drawn as they are from Fig. 5.12, apply strictly only to systems undergoing reversible cycles, they are qualitatively correct for actual power cycles. The thermal efficiencies of actual cycles are observed to increase as the *average* temperature at which energy is added by heat transfer increases and/or the *average* temperature at which energy is discharged by

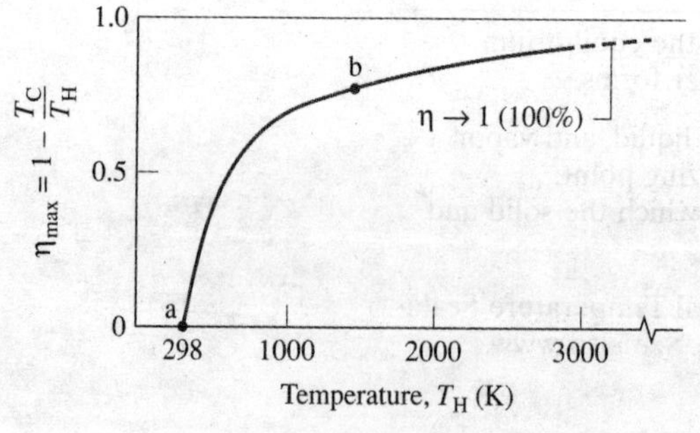

Fig. 5.12 Carnot efficiency versus T_H, for T_C = 298 K.

heat transfer is reduced. However, maximizing the thermal efficiency of a power cycle may not be the only objective. In practice, other considerations such as cost may be overriding.

Conventional power-producing cycles have thermal efficiencies ranging up to about 40%. This value may seem low, but the comparison should be made with an appropriate limiting value and not 100%. ➡ **FOR EXAMPLE...** consider a system executing a power cycle for which the average temperature of heat addition is 745 K and the average temperature at which heat is discharged is 298 K. For a reversible cycle receiving and discharging energy by heat transfer at these temperatures, the thermal efficiency given by Eq. 5.9 is 60%. When compared to this value, an actual thermal efficiency of 40% does not appear to be so low. The cycle would be operating at two-thirds of the theoretical maximum. ⬅

In the next example, we evaluate an inventor's claim about the performance of a power cycle, illustrating the use of the Carnot corollaries (Sec. 5.6.2) and the Carnot efficiency, Eq. 5.9.

Example 5.1 EVALUATING A POWER CYCLE PERFORMANCE CLAIM

An inventor claims to have developed a power cycle capable of delivering a net work output of 410 kJ for an energy input by heat transfer of 1000 kJ. The system undergoing the cycle receives the heat transfer from hot gases at a temperature of 500 K and discharges energy by heat transfer to the atmosphere at 300 K. Evaluate this claim.

Solution

Known: A system operates in a cycle and produces a net amount of work while receiving and discharging energy by heat transfer at fixed temperatures.

Find: Evaluate the claim that the cycle can develop 410 kJ of work for an energy input by heat of 1000 kJ.

Schematic and Given Data:

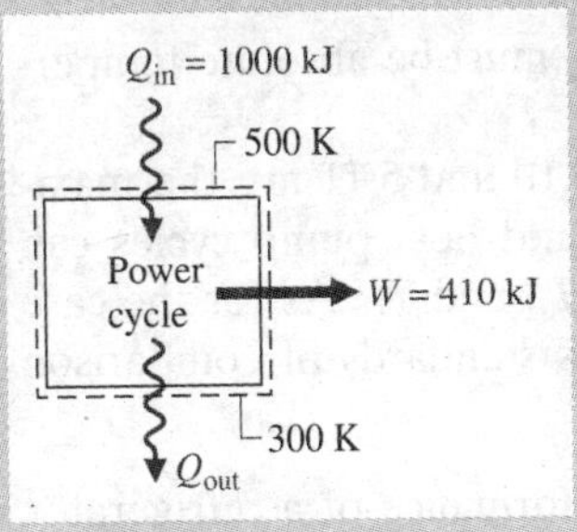

Fig. E5.1

Engineering Model:

1. The system shown on the accompanying figure executes a power cycle.
2. The hot gases and the atmosphere play the roles of hot and cold reservoirs, respectively.

Analysis: Inserting the values supplied by the inventor into Eq. 5.4, the cycle thermal efficiency is

$$\eta = \frac{410 \text{ kJ}}{1000 \text{ kJ}} = 0.41 (41\%)$$

The maximum thermal efficiency *any* power cycle can have while operating between reservoirs at $T_H = 500$ K and $T_C = 300$ K is given by Eq. 5.9.

❶
$$\eta_{max} = 1 - \frac{T_C}{T_H} = 1 - \frac{300 \text{ K}}{500 \text{ K}} = 0.40 (40\%)$$

The Carnot corollaries provide a basis for evaluating the claim: Since the thermal efficiency of the actual cycle exceeds the maximum theoretical value, the claim cannot be valid.

Skills Developed: *Ability to...*
✓ apply the Carnot corollaries, using Eqs. 5.4 and 5.9 appropriately.

❶ The temperatures T_C and T_H used in evaluating η_{max} *must* be in K.

Quick Quiz

If the cycle receives heat transfer from a hot gas at 600 K while all other data remain unchanged, evaluate the inventor's claim.

Ans. Claim is in accord with the second law.

5.9.2 Refrigeration and Heat Pump Cycles

Equation 5.7 is also applicable to reversible refrigeration and heat pump cycles operating between two thermal reservoirs, but for these Q_C represents the heat added to the cycle from the cold reservoir at temperature T_C on the Kelvin scale and Q_H is the heat discharged to the hot reservoir at temperature T_H. Introducing Eq. 5.7 in Eq. 5.5 results in the following expression for the coefficient of performance of any system undergoing a reversible refrigeration cycle while operating between the two reservoirs

$$\beta_{max} = \frac{T_C}{T_H - T_C} \tag{5.10}$$

Similarly, substituting Eq. 5.7 into Eq. 5.6 gives the following expression for the coefficient of performance of any system undergoing a reversible heat pump cycle while operating between the two reservoirs

$$\gamma_{max} = \frac{T_H}{T_H - T_C} \tag{5.11}$$

Note that the temperatures used to evaluate β_{max} and γ_{max} must be absolute temperatures on the Kelvin scale.

From the discussion of Sec. 5.7.2, it follows that Eqs. 5.10 and 5.11 are the maximum coefficients of performance that any refrigeration and heat pump cycles can have while operating between reservoirs at temperatures T_H and T_C. As for the case of the Carnot efficiency, these expressions can be used as standards of comparison for actual refrigerators and heat pumps.

In the next example, we evaluate the coefficient of performance of a refrigerator and compare it with the maximum theoretical value, illustrating the use of the second law corollaries of Sec. 5.7.2 together with Eq. 5.10.

Example 5.2 EVALUATING REFRIGERATOR PERFORMANCE

By steadily circulating a refrigerant at low temperature through passages in the walls of the freezer compartment, a refrigerator maintains the freezer compartment at −5°C when the air surrounding the refrigerator is at 22°C. The rate of heat transfer from the freezer compartment to the refrigerant is 8000 kJ/h and the power input required to operate the refrigerator is 3200 kJ/h. Determine the coefficient of performance of the refrigerator and compare with the coefficient of performance of a reversible refrigeration cycle operating between reservoirs at the same two temperatures.

Solution

Known: A refrigerator maintains a freezer compartment at a specified temperature. The rate of heat transfer from the refrigerated space, the power input to operate the refrigerator, and the ambient temperature are known.

Find: Determine the coefficient of performance and compare with that of a reversible refrigerator operating between reservoirs at the same two temperatures.

Schematic and Given Data:

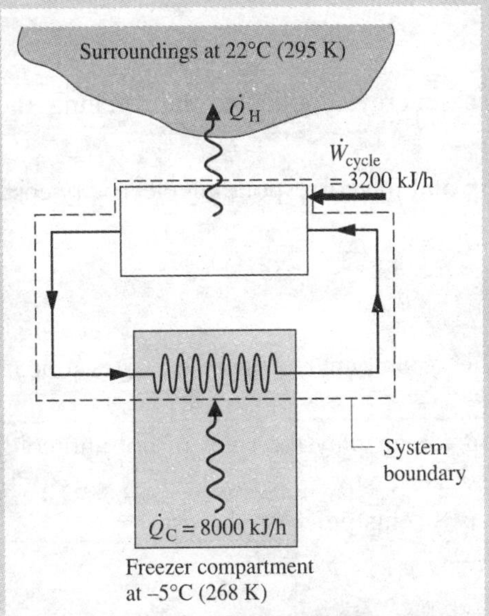

Fig. E5.2

Engineering Model:

1. The system shown on the accompanying figure is at steady state.
2. The freezer compartment and the surrounding air play the roles of cold and hot reservoirs, respectively.

Analysis: Inserting the given operating data into Eq. 5.5, the coefficient of performance of the refrigerator is

$$\beta = \frac{\dot{Q}_C}{\dot{W}_{cycle}} = \frac{8000 \text{ kJ/h}}{3200 \text{ kJ/h}} = 2.5$$

Substituting values into Eq. 5.10 gives the coefficient of performance of a reversible refrigeration cycle operating between reservoirs at $T_C = 268$ K and $T_H = 295$ K

❶ $$\beta_{max} = \frac{T_C}{T_H - T_C} = \frac{268 \text{ K}}{295 \text{ K} - 268 \text{ K}} = 9.9$$

In accord with the corollaries of Sec. 5.7.2, the coefficient of performance of the refrigerator is less than for a
❷ reversible refrigeration cycle operating between reservoirs at the same two temperatures.

---

❶ The temperatures T_C and T_H used in evaluating β_{max} *must* be in K or °R.

❷ The difference between the actual and maximum coefficients of performance suggests that there may be some potential for improving the thermodynamic performance. This objective should be approached judiciously, however, for improved performance may require increases in size, complexity, and cost.

Skills Developed: *Ability to...*
✓ apply the second law corollaries of Sec. 5.7.2, using Eqs. 5.5 and 5.10 appropriately.

Quick Quiz

An inventor claims the power required to operate the refrigerator can be reduced to 800 kJ/h while all other data remain the same. Evaluate this claim using the second law.

Ans. $\beta = 10$. Claim invalid.

In Example 5.3, we determine the minimum theoretical work input and cost for one day of operation of an electric heat pump, illustrating the use of the second law corollaries of Sec. 5.7.2 together with Eq. 5.11.

Example 5.3 EVALUATING HEAT PUMP PERFORMANCE

A dwelling requires 633 MJ per day to maintain its temperature at 21°C when the outside temperature is 0°C. (a) If an electric heat pump is used to supply this energy, determine the minimum theoretical work input for one day of operation, in kJ/day. (b) Evaluating electricity at 8 cents per kW · h, determine the minimum theoretical cost to operate the heat pump, in \$/day.

Solution

Known: A heat pump maintains a dwelling at a specified temperature. The energy supplied to the dwelling, the ambient temperature, and the unit cost of electricity are known.

Find: Determine the *minimum* theoretical work required by the heat pump and the corresponding electricity cost.

Schematic and Given Data:

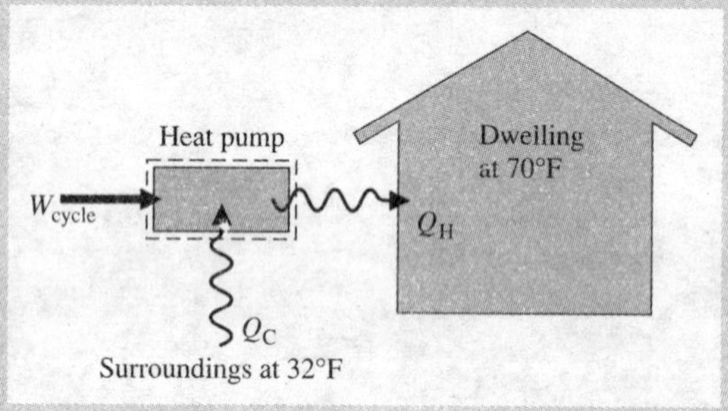

Fig. E5.3

Engineering Model:

1. The system shown on the accompanying figure executes a heat pump cycle.
2. The dwelling and the outside air play the roles of hot and cold reservoirs, respectively.
3. The value of electricity is 8 cents per kW · h.

Analysis:

(a) Using Eq. 5.6, the work for any heat pump cycle can be expressed as $W_{cycle} = Q_H/\gamma$. The coefficient of performance γ of an actual heat pump is less than, or equal to, the coefficient of performance γ_{max} of a reversible heat pump cycle when each operates between the same two thermal reservoirs: $\gamma \leq \gamma_{max}$. Accordingly, for a given value of Q_H, and using Eq. 5.11 to evaluate γ_{max}, we get

$$W_{cycle} \geq \frac{Q_H}{\gamma_{max}}$$

$$\geq \left(1 - \frac{T_C}{T_H}\right) Q_H$$

Inserting values

❶
$$W_{cycle} \geq \left(1 - \frac{273 \text{ K}}{294 \text{ K}}\right)\left(6.33 \times 10^5 \frac{\text{kJ}}{\text{day}}\right) = 4.5 \times 10^4 \frac{\text{kJ}}{\text{day}}$$

The *minimum* theoretical work input is 4.5×10^4 kJ/day.

(b) Using the result of part (a) together with the given cost data and an appropriate conversion factor

❷
$$\begin{bmatrix}\text{minimum} \\ \text{theoretical} \\ \text{cost per day}\end{bmatrix} = \left(4.5 \times 10^4 \frac{\text{kJ}}{\text{day}} \left|\frac{1 \text{ kW} \cdot \text{h}}{3600 \text{ kJ}}\right|\right)\left(0.08 \frac{\$}{\text{kW} \cdot \text{h}}\right) = 1 \frac{\$}{\text{day}}$$

❶ Note that the temperatures T_C and T_H *must* be in K.

❷ Because of irreversibilities, an actual heat pump must be supplied more work than the minimum to provide the same heating effect. The actual daily cost could be substantially greater than the minimum theoretical cost.

Skills Developed: *Ability to...*
✓ apply the second law corollaries of Sec. 5.7.2, using Eqs. 5.6 and 5.11 appropriately.
✓ conduct an elementary economic evaluation.

Quick Quiz

If the cost of electricity is 10 cents per kW · h, evaluate the minimum theoretical cost to operate the heat pump, in \$/day, keeping all other data the same.
Ans. \$1.26/day.

5.10 Carnot Cycle

The Carnot cycles introduced in this section provide specific examples of reversible cycles operating between two thermal reservoirs. Other examples are provided in Chap. 9: the Ericsson and Stirling cycles. In a ***Carnot cycle,*** the system executing the cycle undergoes a series of four internally reversible processes: two adiabatic processes alternated with two isothermal processes.

Carnot cycle

5.10.1 Carnot Power Cycle

Figure 5.13 shows the p–v diagram of a Carnot power cycle in which the system is a gas in a piston–cylinder assembly. Figure 5.14 provides details of how the cycle is executed. The piston and cylinder walls are nonconducting. The heat transfers are in the directions of the arrows. Also note that there are two reservoirs at temperatures T_H and T_C, respectively, and an insulating stand. Initially, the piston–cylinder assembly is on the insulating stand and the system is at state 1, where the temperature is T_C. The four processes of the cycle are

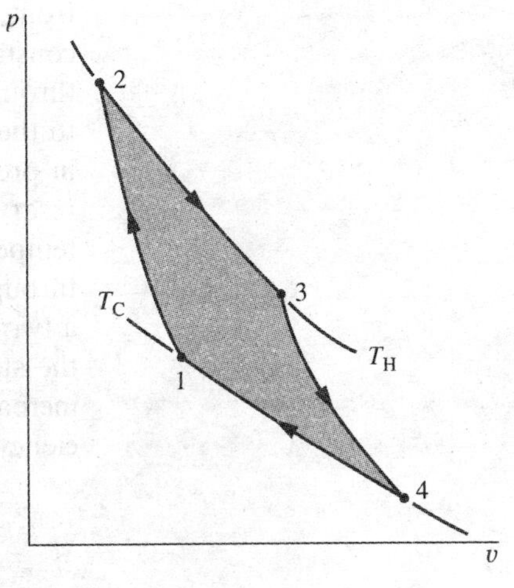

Fig. 5.13 p–v diagram for a Carnot gas power cycle.

Process 1–2: The gas is compressed *adiabatically* to state 2, where the temperature is T_H.

Process 2–3: The assembly is placed in contact with the reservoir at T_H. The gas expands *isothermally* while receiving energy Q_H from the hot reservoir by heat transfer.

Process 3–4: The assembly is again placed on the insulating stand and the gas is allowed to continue to expand *adiabatically* until the temperature drops to T_C.

Process 4–1: The assembly is placed in contact with the reservoir at T_C. The gas is compressed *isothermally* to its initial state while it discharges energy Q_C to the cold reservoir by heat transfer.

For the heat transfer during Process 2–3 to be reversible, the difference between the gas temperature and the temperature of the hot reservoir must be vanishingly small. Since the reservoir temperature remains constant, this implies that the temperature of the gas also remains constant during Process 2–3. The same can be concluded for the gas temperature during Process 4–1.

For each of the four internally reversible processes of the Carnot cycle, the work can be represented as an area on Fig. 5.13. The area under the adiabatic process line 1–2 represents the work done per unit of mass to compress the gas in this process. The areas under process lines 2–3 and 3–4 represent the work done per unit of mass by the gas as it expands in these processes. The area under process line 4–1 is the work done per unit of mass to compress the gas in this process. The enclosed area on the p–v diagram, shown shaded, is the net work developed by the cycle per unit of mass. The thermal efficiency of this cycle is given by Eq. 5.9.

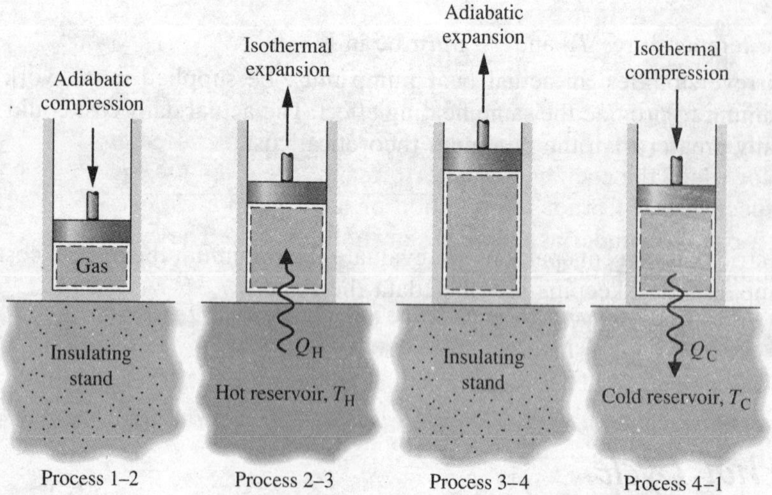

Fig. 5.14 Carnot power cycle executed by a gas in a piston–cylinder assembly.

The Carnot cycle is not limited to processes of a closed system taking place in a piston–cylinder assembly. Figure 5.15 shows the schematic and accompanying p–v diagram of a Carnot cycle executed by water steadily circulating through a series of four interconnected components that has features in common with the simple vapor power plant shown in Fig. 4.16. As the water flows through the boiler, a *change of phase* from liquid to vapor at constant temperature T_H occurs as a result of heat transfer from the hot reservoir. Since temperature remains constant, pressure also remains constant during the phase change. The steam exiting the boiler expands adiabatically through the turbine and work is developed. In this process the temperature decreases to the temperature of the cold reservoir, T_C, and there is an accompanying decrease in pressure. As the steam passes through the condenser, a heat transfer to the cold reservoir occurs and some of the vapor condenses at constant temperature T_C. Since temperature remains constant, pressure also remains constant as the water passes through the condenser. The fourth component is a pump, or compressor, that receives a two-phase liquid–vapor mixture from the condenser and returns it adiabatically to the state at the boiler entrance. During this process, which requires a work input to increase the pressure, the temperature increases from T_C to T_H. The thermal efficiency of this cycle also is given by Eq. 5.9.

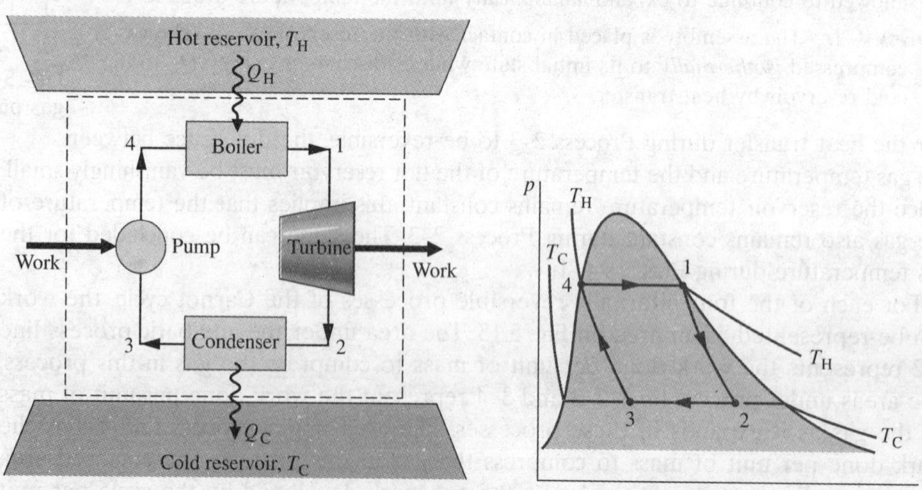

Fig. 5.15 Carnot vapor power cycle.

5.10.2 Carnot Refrigeration and Heat Pump Cycles

If a Carnot power cycle is operated in the opposite direction, the magnitudes of all energy transfers remain the same but the energy transfers are oppositely directed. Such a cycle may be regarded as a reversible refrigeration or heat pump cycle, for which the coefficients of performance are given by Eqs. 5.10 and 5.11, respectively. A Carnot refrigeration or heat pump cycle executed by a gas in a piston–cylinder assembly is shown in Fig. 5.16. The cycle consists of the following four processes in series:

Process 1–2: The gas expands *isothermally* at T_C while *receiving* energy Q_C from the cold reservoir by heat transfer.

Process 2–3: The gas is compressed *adiabatically* until its temperature is T_H.

Process 3–4: The gas is compressed *isothermally* at T_H while it *discharges* energy Q_H to the hot reservoir by heat transfer.

Process 4–1: The gas expands *adiabatically* until its temperature decreases to T_C.

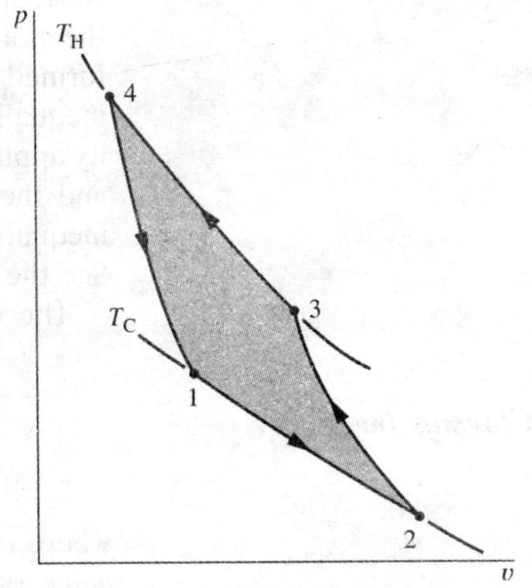

Fig. 5.16 *p–v* diagram for a Carnot gas refrigeration or heat pump cycle.

A refrigeration or heat pump effect can be accomplished in a cycle only if a net work input is supplied to the system executing the cycle. In the case of the cycle shown in Fig. 5.16, the shaded area represents the net work input per unit of mass.

5.10.3 Carnot Cycle Summary

In addition to the configurations discussed previously, Carnot cycles also can be devised that are composed of processes in which a capacitor is charged and discharged, a paramagnetic substance is magnetized and demagnetized, and so on. However, regardless of the type of device or the working substance used,

1. the Carnot cycle *always* has the same four internally reversible processes: two adiabatic processes alternated with two isothermal processes.
2. the thermal efficiency of the Carnot power cycle is *always* given by Eq. 5.9 in terms of the temperatures evaluated on the Kelvin scale.
3. the coefficients of performance of the Carnot refrigeration and heat pump cycles are *always* given by Eqs. 5.10 and 5.11, respectively, in terms of temperatures evaluated on the Kelvin scale.

5.11 Clausius Inequality

Corollaries of the second law developed thus far in this chapter are for systems undergoing cycles while communicating thermally with *one* or *two* thermal energy reservoirs. In the present section a corollary of the second law known as the *Clausius inequality* is introduced that is applicable to *any* cycle without regard for the body, or bodies, from which the cycle receives energy by heat transfer or to which the cycle rejects energy by heat transfer. The Clausius inequality provides the basis for further development in Chap. 6 of the entropy, entropy production, and entropy balance concepts introduced in Sec. 5.2.3.

The *Clausius inequality* states that for any thermodynamic cycle

$$\oint \left(\frac{\delta Q}{T}\right)_b \leq 0 \tag{5.12}$$

where δQ represents the heat transfer at a part of the system boundary during a portion of the cycle, and T is the absolute temperature at that part of the boundary. The subscript "b" serves as a reminder that the integrand is evaluated at the boundary of the system executing the cycle. The symbol $\oint$ indicates that the integral is to be performed over all parts of the boundary and over the entire cycle. The equality and inequality have the same interpretation as in the Kelvin–Planck statement: the equality applies when there are no internal irreversibilities as the system executes the cycle, and the inequality applies when internal irreversibilities are present. The Clausius inequality can be demonstrated using the Kelvin–Planck statement of the second law. See the box for details.

The ***Clausius inequality*** can be expressed equivalently as

Clausius inequality

$$\oint \left(\frac{\delta Q}{T}\right)_b = -\sigma_{cycle} \tag{5.13}$$

where σ_{cycle} can be interpreted as representing the "strength" of the inequality. The value of σ_{cycle} is positive when internal irreversibilities are present, zero when no internal irreversibilities are present, and can never be negative. In summary, the nature of a cycle executed by a system is indicated by the value for σ_{cycle} as follows:

$\sigma_{cycle} = 0$	no irreversibilities present within the system	
$\sigma_{cycle} > 0$	irreversibilities present within the system	(5.14)
$\sigma_{cycle} < 0$	impossible	

Accordingly, σ_{cycle} is a measure of the effect of the irreversibilities present within the system executing the cycle. This point is developed further in Sec. 6.7, where σ_{cycle} is identified as the *entropy produced* (or *generated*) by internal irreversibilities during the cycle.

Developing the Clausius Inequality

The Clausius inequality can be demonstrated using the arrangement of Fig. 5.17. A system receives energy δQ at a location on its boundary where the absolute temperature is T while the system develops work δW. In keeping with our sign convention for heat transfer, the phrase *receives energy* δQ includes the possibility of heat transfer *from* the system. The energy δQ is received from (or absorbed by) a thermal reservoir at T_{res}. To ensure that no irreversibility is introduced as a result of heat transfer between the reservoir and the system, let it be accomplished through an intermediary system that undergoes a cycle without irreversibilities of any kind. The cycle receives energy $\delta Q'$ from the reservoir and supplies δQ to the system while producing work $\delta W'$. From the definition of the Kelvin scale (Eq. 5.7), we have the following relationship between the heat transfers and temperatures:

$$\frac{\delta Q'}{T_{res}} = \left(\frac{\delta Q}{T}\right)_b \tag{a}$$

As temperature may vary, a multiplicity of such reversible cycles may be required.

Consider next the combined system shown by the dotted line on Fig. 5.17. An energy balance for the combined system is

$$dE_C = \delta Q' - \delta W_C$$

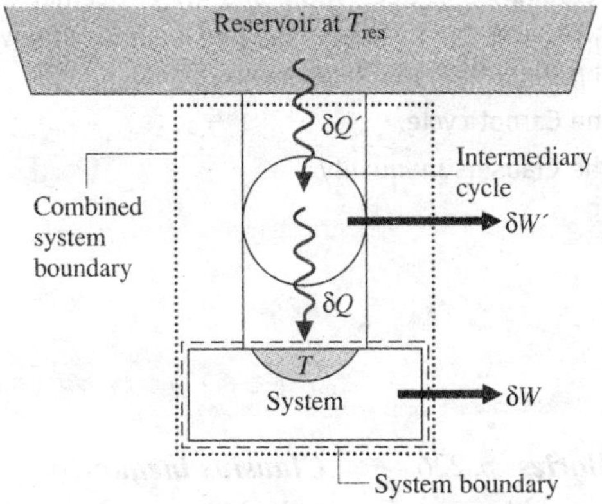

Fig. 5.17 Illustration used to develop the Clausius inequality.

where δW_C is the total work of the combined system, the sum of δW and $\delta W'$, and dE_C denotes the change in energy of the combined system. Solving the energy balance for δW_C and using Eq. (a) to eliminate $\delta Q'$ from the resulting expression yields

$$\delta W_C = T_{res}\left(\frac{\delta Q}{T}\right)_b - dE_C$$

Now, let the system undergo a single cycle while the intermediary system undergoes one or more cycles. The total work of the combined system is

$$W_C = \oint T_{res}\left(\frac{\delta Q}{T}\right)_b - \oint dE_C^{\,0} = T_{res}\oint\left(\frac{\delta Q}{T}\right)_b \qquad \text{(b)}$$

Since the reservoir temperature is constant, T_{res} can be brought outside the integral. The term involving the energy of the combined system vanishes because the energy change for any cycle is zero. The combined system operates in a cycle because its parts execute cycles. Since the combined system undergoes a cycle and exchanges energy by heat transfer with a single reservoir, Eq. 5.3 expressing the Kelvin–Planck statement of the second law must be satisfied. Using this, Eq. (b) reduces to give Eq. 5.12, where the equality applies when there are *no irreversibilities within the system* as it executes the cycle and the inequality applies when *internal irreversibilities are present.* This interpretation actually refers to the combination of system plus intermediary cycle. However, the intermediary cycle is regarded as free of irreversibilities, so the only possible site of irreversibilities is the system alone.

✓ Chapter Summary and Study Guide

In this chapter, we motivate the need for and usefulness of the second law of thermodynamics, and provide the basis for subsequent applications involving the second law in Chaps. 6 and 7. Three statements of the second law, the Clausius, Kelvin–Planck, and Entropy statements, are introduced together with several corollaries that establish the best theoretical performance for systems undergoing cycles while interacting with thermal reservoirs. The irreversibility concept is introduced and the related notions of irreversible, reversible, and internally reversible processes are discussed. The Kelvin temperature scale is defined and used to obtain expressions for the maximum performance measures of power, refrigeration, and heat pump cycles operating between two thermal reservoirs. The Carnot cycle is introduced to provide a specific example of a reversible cycle operating between two thermal reservoirs. Finally, the Clausius Inequality providing a bridge from Chap. 5 to Chap. 6 is presented and discussed.

The following checklist provides a study guide for this chapter. When your study of the text and end-of-chapter exercises has been completed you should be able to

- ✔ write out the meanings of the terms listed in the margins throughout the chapter and understand each of the related concepts. The subset of key concepts listed below is particularly important in subsequent chapters.
- ✔ give the Kelvin–Planck statement of the second law, correctly interpreting the "less than" and "equal to" signs in Eq. 5.3.
- ✔ list several important irreversibilities.
- ✔ apply the corollaries of Secs. 5.6.2 and 5.7.2 together with Eqs. 5.9, 5.10, and 5.11 to assess the performance of power cycles and refrigeration and heat pump cycles.
- ✔ describe the Carnot cycle.
- ✔ interpret the Clausius inequality.

Key Engineering Concepts

second law statements p. 216–219
irreversible process p. 219
irreversibilities p. 220
internal and external irreversibilities p. 220
internally reversible process p. 223
Carnot corollaries p. 226
Kelvin scale p. 230
Carnot efficiency p. 234
Clausius inequality p. 242

Key Equations

$W_{\text{cycle}} \le 0 \begin{cases} <0: & \text{Internal irreversibilities present.} \\ =0: & \text{No internal irreversibilities.} \end{cases}$ (single reservoir)	(5.3)	Analytical form of the Kelvin–Planck statement.
$\eta_{\max} = 1 - \dfrac{T_C}{T_H}$	(5.9)	Maximum thermal efficiency: power cycle operating between two reservoirs.
$\beta_{\max} = \dfrac{T_C}{T_H - T_C}$	(5.10)	Maximum coefficient of performance: refrigeration cycle operating between two reservoirs.
$\gamma_{\max} = \dfrac{T_H}{T_H - T_C}$	(5.11)	Maximum coefficient of performance: heat pump cycle operating between two reservoirs.
$\oint \left(\dfrac{\delta Q}{T} \right)_b = -\sigma_{\text{cycle}}$	(5.13)	Clausius inequality.

Exercises: things engineers think about

1. Extending the discussion of Sec. 5.1.2, how might work be developed when (a) T_i is less than T_0 in Fig. 5.1a, (b) p_i is less than p_0 in Fig. 5.1b?

2. A system consists of an ice cube in a cup of tap water. The ice cube melts and eventually equilibrium is attained. How might work be developed as the ice and water come to equilibrium?

3. Describe a process that would satisfy the conservation of energy principle, but does not actually occur in nature.

4. Are health risks associated with consuming tomatoes induced to ripen by an ethylene spray?

5. In years ahead, are we more likely to be driving plug-in hybrid electrical vehicles or fuel cell-powered vehicles operating on hydrogen?

6. Referring to Fig. 2.3, what are the internal irreversibilities associated with system A and system B, respectively?

7. For the gearbox of Example 2.4, what are the principal internal and external irreversibilities?

8. If a window air conditioner were placed on a table in a room and operated, would the room temperature increase, decrease, or remain the same? Explain.

9. When a power plant discharges cooling water to a river at a temperature higher than that of the river, what are the possible effects on the aquatic life of the river?

10. Referring to Eqs. 5.10 and 5.11, how might the coefficients of performance of refrigeration cycles and heat pumps be increased?

11. Is it possible for the coefficient of performance of a refrigeration cycle to be less than one? To be greater than one? Answer the same questions for a heat pump cycle.

12. A hot combustion gas enters a turbine operating at steady state and expands adiabatically to a lower pressure. Would you expect the power *output* to be greater in an internally reversible expansion or an actual expansion?

13. Refrigerant 22 enters a compressor operating at steady state and is compressed adiabatically to a higher pressure. Would you expect the power *input* to the compressor to be greater in an internally reversible compression or an actual compression?

Problems: developing engineering skills

Exploring the Second Law

5.1 Complete the demonstration of the equivalence of the Clausius and Kelvin–Planck statements of the second law given in Sec. 5.2 by showing that a violation of the Kelvin–Planck statement implies a violation of the Clausius statement.

5.2 An inventor claims to have developed a device that undergoes a thermodynamic cycle while communicating thermally with two reservoirs. The system receives energy Q_C from the cold reservoir and discharges energy Q_H to the hot reservoir while delivering a net amount of work to its surroundings. There are no other energy transfers between the device and its surroundings. Evaluate the inventor's claim using (a) the Clausius statement of the second law, and (b) the Kelvin–Planck statement of the second law.

5.3 Classify the following processes of a closed system as *possible, impossible,* or *indeterminate.*

	Entropy Change	Entropy Transfer	Entropy Production
(a)	>0	0	
(b)	<0		>0
(c)	0	>0	
(d)	>0	>0	
(e)	0	<0	
(f)	>0		<0
(g)	<0	<0	

5.4 As shown in Fig. P5.4, a hot thermal reservoir is separated from a cold thermal reservoir by a cylindrical rod insulated on its lateral surface. Energy transfer by conduction between the two reservoirs takes place through the rod, which remains at steady state. Using the Kelvin–Planck statement of the second law, demonstrate that such a process is irreversible.

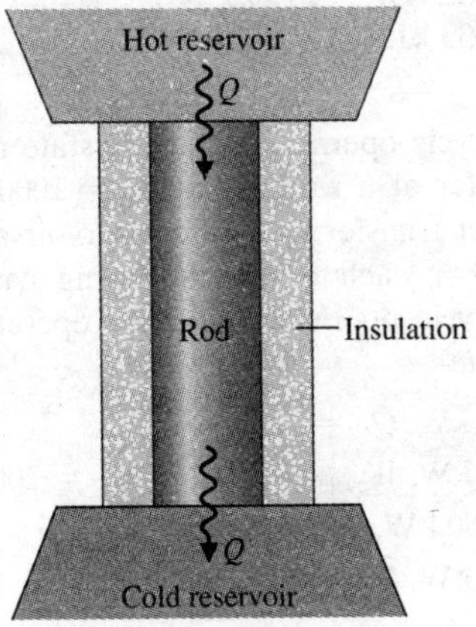

Fig. P5.4

5.5 As shown in Fig. P5.5, a rigid insulated tank is divided into halves by a partition. On one side of the partition is a gas. The other side is initially evacuated. A valve in the partition is opened and the gas expands to fill the entire volume. Using the Kelvin–Planck statement of the second law, demonstrate that this process is irreversible.

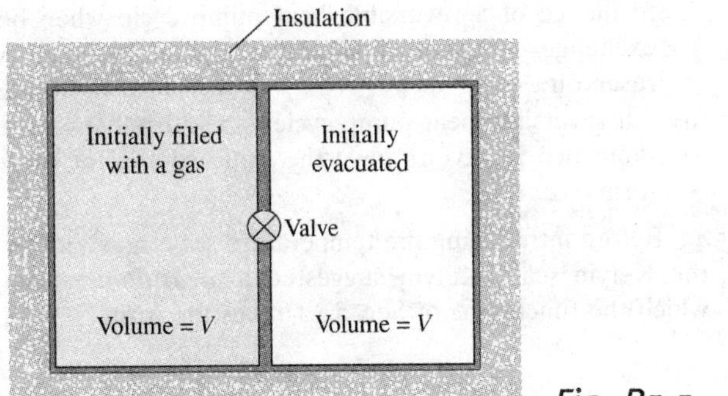

Fig. P5.5

5.6 Answer the following true or false.

(a) A process that violates the second law of thermodynamics violates the first law of thermodynamics.

(b) When a net amount of work is done on a closed system undergoing an *internally reversible* process, a net heat transfer of energy from the system also occurs.

(c) A closed system can experience an increase in entropy only when a net amount of entropy is transferred into the system.

(d) The change in entropy of a closed system is the same for every process between two specified end states.

5.7 Complete the discussion of the Kelvin–Planck statement of the second law in the box of Sec. 5.4 by showing that if a system undergoes a thermodynamic cycle reversibly while communicating thermally with a single reservoir, the equality in Eq. 5.3 applies.

5.8 A reversible power cycle R and an irreversible power cycle I operate between the same two reservoirs.

(a) If each cycle receives the same amount of energy Q_H from the hot reservoir, show that cycle I necessarily discharges more energy Q_C to the cold reservoir than cycle R. Discuss the implications of this for actual power cycles.

(b) If each cycle develops the same net work, show that cycle I necessarily receives more energy Q_H from the hot reservoir than cycle R. Discuss the implications of this for actual power cycles.

5.9 A power cycle I and a reversible power cycle R operate between the same two reservoirs, as shown in Fig. 5.6. Cycle I has a thermal efficiency equal to two-thirds of that for cycle R. Using the Kelvin–Planck statement of the second law, prove that cycle I must be irreversible.

5.10 Provide the details left to the reader in the demonstration of the second Carnot corollary given in the box of Sec. 5.6.2.

5.11 Using the Kelvin–Planck statement of the second law of thermodynamics, demonstrate the following corollaries:

(a) The coefficient of performance of an irreversible refrigeration cycle is always less than the coefficient of performance of a reversible refrigeration cycle when both exchange energy by heat transfer with the same two reservoirs.
(b) All reversible refrigeration cycles operating between the same two reservoirs have the same coefficient of performance.
(c) The coefficient of performance of an irreversible heat pump cycle is always less than the coefficient of performance of a reversible heat pump cycle when both exchange energy by heat transfer with the same two reservoirs.
(d) All reversible heat pump cycles operating between the same two reservoirs have the same coefficient of performance.

5.12 Before introducing the temperature scale now known as the Kelvin scale, Kelvin suggested a *logarithmic* scale in which the function ψ of Sec. 5.8.1 takes the form

$$\psi = \exp \theta_C / \exp \theta_H$$

where θ_H and θ_C denote, respectively, the temperatures of the hot and cold reservoirs on this scale.

(a) Show that the relation between the Kelvin temperature T and the temperature θ on the logarithmic scale is

$$\theta = \ln T + C$$

where C is a constant.
(b) On the Kelvin scale, temperatures vary from 0 to $+\infty$. Determine the range of temperature values on the logarithmic scale.
(c) Obtain an expression for the thermal efficiency of any system undergoing a reversible power cycle while operating between reservoirs at temperatures θ_H and θ_C on the logarithmic scale.

5.13 Demonstrate that the *gas temperature scale* (Sec. 5.8.2) is identical to the *Kelvin temperature scale* (Sec. 5.8.1).

5.14 The platinum resistance thermometer is said to be the most important of the three thermometers specified in ITS-90 because it covers the broad, practically significant interval from 13.8 K to 1234.93 K. What is the operating principle of resistance thermometry and why is platinum specified for use in ITS-90?

5.15 The relation between resistance R and temperature T for a *thermistor* closely follows

$$R = R_0 \exp\left[\beta\left(\frac{1}{T} - \frac{1}{T_0}\right)\right]$$

where R_0 is the resistance, in ohms (Ω), measured at temperature T_0 (K) and β is a material constant with units of K. For a particular thermistor $R_0 = 2.2\ \Omega$ at $T_0 = 310$ K. From a calibration test, it is found that $R = 0.31\ \Omega$ at $T = 422$ K. Determine the value of β for the thermistor and make a plot of resistance versus temperature.

5.16 Over a limited temperature range, the relation between electrical resistance R and temperature T for a *resistance temperature detector* is

$$R = R_0[1 + \alpha(T - T_0)]$$

where R_0 is the resistance, in ohms (Ω), measured at reference temperature T_0 (in °C) and α is a material constant with units of $(°C)^{-1}$. The following data are obtained for a particular resistance thermometer:

	T (°C)	R (Ω)
Test 1	0	51.39
Test 2	91	51.72

What temperature would correspond to a resistance of 51.47 Ω on this thermometer?

Power Cycle Applications

5.17 The data listed below are claimed for a power cycle operating between hot and cold reservoirs at 1000 K and 300 K, respectively. For each case, determine whether the cycle operates *reversibly*, operates *irreversibly*, or is *impossible*.

(a) $Q_H = 600$ kJ, $W_{cycle} = 300$ kJ, $Q_C = 300$ kJ
(b) $Q_H = 400$ kJ, $W_{cycle} = 280$ kJ, $Q_C = 120$ kJ
(c) $Q_H = 700$ kJ, $W_{cycle} = 300$ kJ, $Q_C = 500$ kJ
(d) $Q_H = 800$ kJ, $W_{cycle} = 600$ kJ, $Q_C = 200$ kJ

5.18 A power cycle receives energy Q_H by heat transfer from a hot reservoir at $T_H = 833$ K and rejects energy Q_C by heat transfer to a cold reservoir at $T_C = 278$ K. For each of the following cases, determine whether the cycle operates *reversibly*, operates *irreversibly*, or is *impossible*.

(a) $Q_H = 950$ kJ, $W_{cycle} = 475$ kJ
(b) $Q_H = 950$ kJ, $Q_C = 315$ kJ
(c) $W_{cycle} = 633$ kJ, $Q_C = 422$ kJ
(d) $\eta = 70\%$

5.19 A power cycle operating at steady state receives energy by heat transfer at a rate $\dot{Q}_H$ at $T_H = 1000$ K and rejects energy by heat transfer to a cold reservoir at a rate $\dot{Q}_C$ at $T_C = 300$ K. For each of the following cases, determine whether the cycle operates *reversibly*, operates *irreversibly*, or is *impossible*.

(a) $\dot{Q}_H = 500$ kW, $\dot{Q}_C = 100$ kW
(b) $\dot{Q}_H = 500$ kW, $\dot{W}_{cycle} = 250$ kW, $\dot{Q}_C = 200$ kW
(c) $\dot{W}_{cycle} = 350$ kW, $\dot{Q}_C = 150$ kW
(d) $\dot{Q}_H = 500$ kW, $\dot{Q}_C = 200$ kW

5.20 As shown in Fig. P5.20, a reversible power cycle receives energy Q_H by heat transfer from a hot reservoir at T_H and rejects energy Q_C by heat transfer to a cold reservoir at T_C.

(a) If $T_H = 1200$ K and $T_C = 300$ K, what is the thermal efficiency?

(b) If $T_H = 500°C$, $T_C = 20°C$, and $W_{cycle} = 1000$ kJ, what are Q_H and Q_C, each in kJ?

(c) If $\eta = 60\%$ and $T_C = 4°C$, what is T_H, in °C?

(d) If $\eta = 40\%$ and $T_H = 727°C$, what is T_C, in °C?

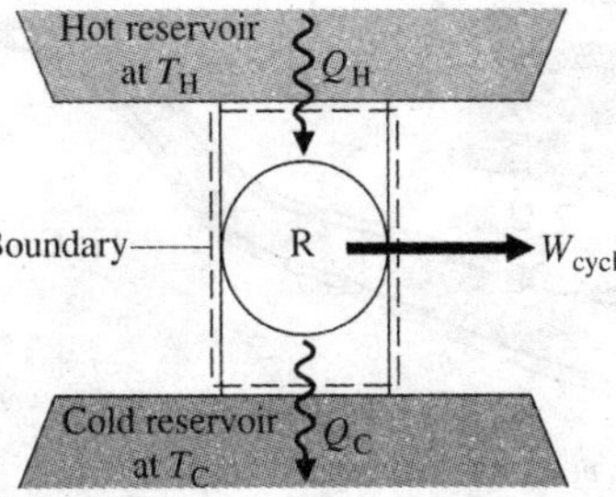

Fig. P5.20

5.21 A reversible power cycle receives 100 kJ by heat transfer from a hot reservoir at 327°C and rejects 40 kJ by heat transfer to a cold reservoir at T_C. Determine (a) the thermal efficiency and (b) the temperature T_C of the cold reservoir, in °C.

5.22 Determine the maximum theoretical thermal efficiency for *any* power cycle operating between hot and cold reservoirs at 602°C and 112°C, respectively.

5.23 A reversible power cycle operating as in Fig. 5.5 receives energy Q_H by heat transfer from a hot reservoir at T_H and rejects energy Q_C by heat transfer to a cold reservoir at 4°C. If $W_{cycle} = 3\,Q_C$, determine (a) the thermal efficiency and (b) T_H, in °C.

5.24 A reversible power cycle has the same thermal efficiency for hot and cold reservoirs at temperature T and 500 K, respectively, as for hot and cold reservoirs at 2000 and 1000 K, respectively. Determine T, in K.

5.25 As shown in Fig. P5.25, two reversible cycles arranged in series each produce the same net work, W_{cycle}. The first cycle receives energy Q_H by heat transfer from a hot reservoir at 555 K and rejects energy Q by heat transfer to a reservoir at an intermediate temperature, T. The second cycle receives energy Q by heat transfer from the reservoir at temperature T and rejects energy Q_C by heat transfer to a reservoir at 222 K. All energy transfers are positive in the directions of the arrows. Determine

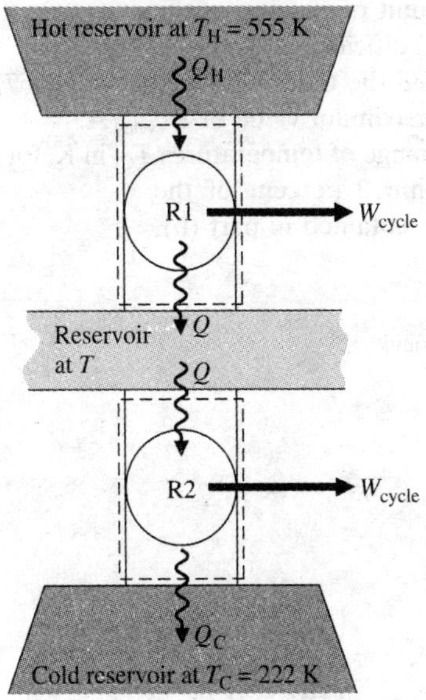

Fig. P5.25

(a) the intermediate temperature T, in K, and the thermal efficiency for each of the two power cycles.

(b) the thermal efficiency of a *single* reversible power cycle operating between hot and cold reservoirs at 555 K and 222 K, respectively. Also, determine the net work developed by the single cycle, expressed in terms of the net work developed by each of the two cycles, W_{cycle}.

5.26 A reversible power cycle operating between hot and cold reservoirs at 1000 K and 300 K, respectively, receives 100 kJ by heat transfer from the hot reservoir for each cycle of operation. Determine the net work developed in 10 cycles of operation, in kJ.

5.27 A reversible power cycle operating between hot and cold reservoirs at 560°C and 4°C, respectively, develops net work in the amount of 633 kJ for each cycle of operation. For three cycles of operation, determine the energy received by heat transfer from the hot reservoir, in kJ.

5.28 An inventor claims to have developed a power cycle having a thermal efficiency of 40%, while operating between hot and cold reservoirs at temperature T_H and $T_C = 300$ K, respectively, where T_H is (a) 600 K, (b) 500 K, (c) 400 K. Evaluate the claim for each case.

5.29 During January, at a location in Alaska winds at −30°C can be observed. Several meters below ground the temperature remains at 13°C, however. An inventor claims to have devised a power cycle exploiting this situation that has a thermal efficiency of 14%. Evaluate this claim.

5.30 An inventor claims to have developed a power cycle operating between hot and cold reservoirs at 1000 K and 250 K, respectively, that develops net work equal to a multiple of the amount of energy, Q_C, rejected to the cold reservoir—that is $W_{cycle} = NQ_C$, where all quantities are positive. What is the maximum theoretical value of the number N for any such cycle?

5.31 A power cycle operates between a reservoir at temperature T and a lower-temperature reservoir at 280 K. At steady state, the cycle develops 40 kW of power while rejecting 1000 kJ/min of energy by heat transfer to the cold reservoir. Determine the minimum theoretical value for T, in K.

5.32 At steady state, a new power cycle is claimed by its inventor to develop power at a rate of 74.6 kW for a heat addition rate of 5.3×10^5 kJ/h, while operating between hot and cold reservoirs at 1000 and 500 K, respectively. Evaluate this claim.

5.33 At steady state, a power cycle develops a power output of 10 kW while receiving energy by heat transfer at the rate of 10 kJ *per cycle of operation* from a source at temperature T. The cycle rejects energy by heat transfer to cooling water at a lower temperature of 300 K. If there are 100 cycles per minute, what is the minimum theoretical value for T, in K?

5.34 At steady state, a power cycle having a thermal efficiency of 38% generates 100 MW of electricity while discharging energy by heat transfer to cooling water at an average temperature of 21°C. The average temperature of the steam passing through the boiler is 480°C. Determine

(a) the rate at which energy is discharged to the cooling water, in kJ/h.

(b) the *minimum* theoretical rate at which energy could be discharged to the cooling water, in kJ/h. Compare with the actual rate and discuss.

5.35 *Ocean temperature energy conversion (OTEC)* power plants generate power by utilizing the naturally occurring decrease with depth of the temperature of ocean water. Near Florida, the ocean surface temperature is 27°C, while at a depth of 700 m the temperature is 7°C.

(a) Determine the maximum thermal efficiency for any power cycle operating between these temperatures.

(b) The thermal efficiency of existing OTEC plants is approximately 2 percent. Compare this with the result of part (a) and comment.

5.36 As shown in Fig. P5.36, a system undergoing a power cycle develops a net power output of 1 MW while receiving energy by heat transfer from steam condensing from saturated vapor to saturated liquid at a pressure of 100 kPa. Energy is discharged from the cycle by heat transfer to a nearby lake at 17°C. These are the only significant heat transfers. Kinetic and potential energy effects can be ignored. For operation at steady state, determine the minimum theoretical steam mass flow rate, in kg/s, required by any such cycle.

5.37 A power cycle operating at steady state receives energy by heat transfer from the combustion of fuel at an average temperature of 1000 K. Owing to environmental considerations, the cycle discharges energy by heat transfer to the atmosphere at 300 K at a rate no greater than 60 MW. Based on the cost of fuel, the cost to supply the heat transfer is \$4.50 per GJ. The power developed by the cycle is valued at \$0.08 per kW · h. For 8000 hours of operation annually, determine for any such cycle, in \$ per year, (a) the maximum value of the power generated and (b) the minimum fuel cost.

5.38 At steady state, a 750-MW power plant receives energy by heat transfer from the combustion of fuel at an average temperature of 317°C. As shown in Fig. P5.38, the plant discharges energy by heat transfer to a river whose mass flow rate is 1.65×10^5 kg/s. Upstream of the power plant the river is at 17°C. Determine the increase in the temperature of the river, ΔT, traceable to such heat transfer, in °C, if the thermal efficiency of the power plant is (a) the Carnot efficiency of a power cycle operating between hot and cold reservoirs at 317°C and 17°C, respectively, (b) two-thirds of the Carnot efficiency found in part (a). Comment.

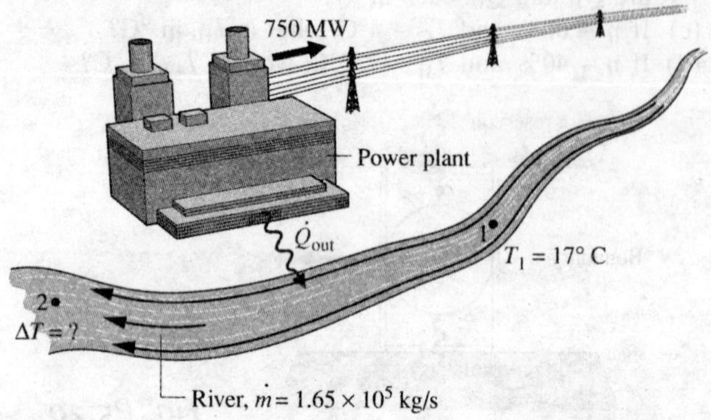

Fig. P5.38

5.39 At steady state, a power cycle receives energy by heat transfer at an average temperature of 463°C and discharges energy by heat transfer to a river. Upstream of the power plant the river has a volumetric flow rate of 67.82 m^3/s and a temperature of 20°C. From environmental considerations, the temperature of the river downstream of the plant can be no more than 22°C. Determine the maximum theoretical power that can be developed, in MW, subject to this constraint.

5.40 The preliminary design of a space station calls for a power cycle that at steady state receives energy by heat transfer at $T_H = 600$ K from a nuclear source and rejects energy to space by thermal radiation according to Eq. 2.32. For the radiative surface, the temperature is T_C, the emissivity is 0.6, and the surface *receives* no radiation from any source. The thermal efficiency of the power cycle is one-half that of a reversible power cycle operating between reservoirs at T_H and T_C.

(a) For $T_C = 400$ K, determine $\dot{W}_{cycle}/A$, the net power developed per unit of radiator surface area, in kW/m^2, and the thermal efficiency.

(b) Plot $\dot{W}_{cycle}/A$ and the thermal efficiency versus T_C, and determine the maximum value of $\dot{W}_{cycle}/A$.

(c) Determine the range of temperatures T_C, in K, for which $\dot{W}_{cycle}/A$ is within 2 percent of the maximum value obtained in part (b).

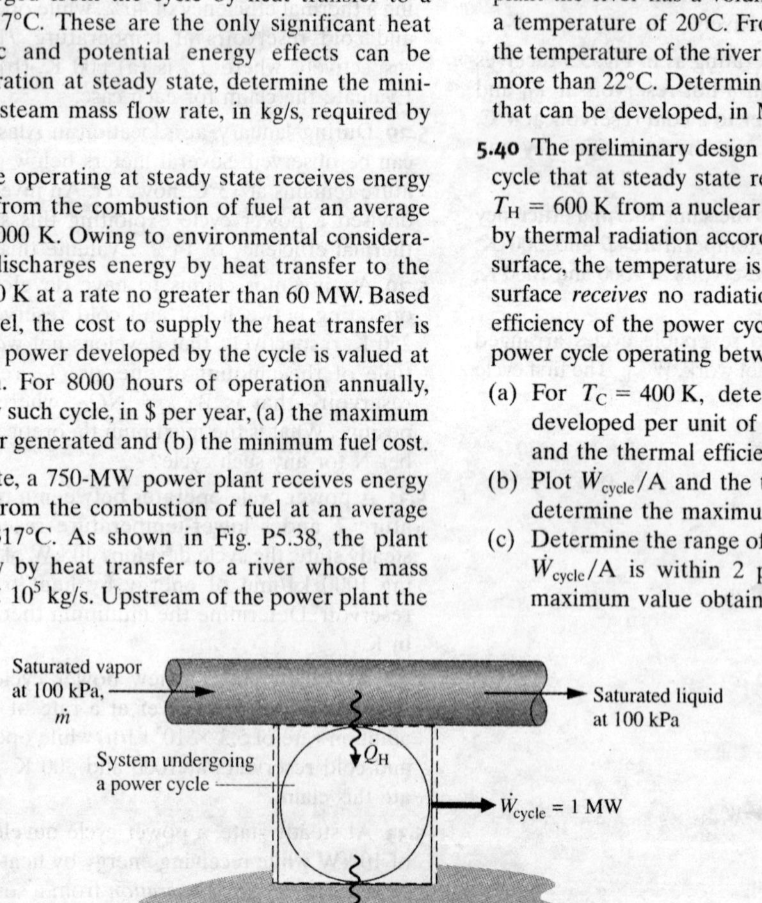

Fig. P5.36

5.41 To increase the thermal efficiency of a reversible power cycle operating between reservoirs at T_H and T_C, would you increase T_H while keeping T_C constant, or decrease T_C while keeping T_H constant? Are there any *natural* limits on the increase in thermal efficiency that might be achieved by such means?

5.42 Two reversible power cycles are arranged in series. The first cycle receives energy by heat transfer from a hot reservoir at temperature T_H and rejects energy by heat transfer to a reservoir at an intermediate temperature $T < T_H$. The second cycle receives energy by heat transfer from the reservoir at temperature T and rejects energy by heat transfer to a cold reservoir at temperature $T_C < T$.

(a) Obtain an expression for the thermal efficiency of a *single* reversible power cycle operating between hot and cold reservoirs at T_H and T_C, respectively, in terms of the thermal efficiencies of the two cycles.
(b) Obtain an expression for the intermediate temperature T in terms of T_H and T_C for the *special case* where the thermal efficiencies of the two cycles are equal.

Refrigeration and Heat Pump Cycle Applications

5.43 A refrigeration cycle operating between two reservoirs receives energy Q_C from a cold reservoir at $T_C = 280$ K and rejects energy Q_H to a hot reservoir at $T_H = 320$ K. For each of the following cases determine whether the cycle operates *reversibly,* operates *irreversibly,* or is *impossible*:

(a) $Q_C = 1500$ kJ, $W_{cycle} = 150$ kJ.
(b) $Q_C = 1400$ kJ, $Q_H = 1600$ kJ.
(c) $Q_H = 1600$ kJ, $W_{cycle} = 400$ kJ.
(d) $\beta = 5$.

5.44 A reversible refrigeration cycle operates between cold and hot reservoirs at temperatures T_C and T_H, respectively.

(a) If the coefficient of performance is 3.5 and $T_H = 27°C$, determine T_C, in °C.
(b) If $T_C = -30°C$ and $T_H = 30°C$, determine the coefficient of performance.
(c) If $Q_C = 528$ kJ, $Q_H = 844$ kJ, and $T_C = -7°C$, determine T_H, in °C.
(d) If $T_C = -1°C$ and $T_H = 38°C$, determine the coefficient of performance.
(e) If the coefficient of performance is 8.9 and $T_C = -5°C$, find T_H, in °C.

5.45 At steady state, a reversible heat pump cycle discharges energy at the rate $\dot{Q}_H$ to a hot reservoir at temperature T_H, while receiving energy at the rate $\dot{Q}_C$ from a cold reservoir at temperature T_C.

(a) If $T_H = 21°C$ and $T_C = 7°C$, determine the coefficient performance.
(b) If $\dot{Q}_H = 10.5$ kW, $\dot{Q}_C = 8.75$ kW, and $T_C = 0°C$, determine T_H, in °C.
(c) If the coefficient of performance is 10 and $T_H = 27°C$, determine T_C, in °C.

5.46 Two reversible cycles operate between hot and cold reservoirs at temperature T_H and T_C, respectively.

(a) If one is a power cycle and the other is a heat pump cycle, what is the relation between the coefficient of performance of the heat pump cycle and the thermal efficiency of the power cycle?
(b) If one is a refrigeration cycle and the other is a heat pump cycle, what is the relation between their coefficients of performance?

5.47 A refrigeration cycle rejects $Q_H = 528$ kJ per cycle to a hot reservoir at $T_H = 300$ K, while receiving $Q_C = 395$ kJ per cycle from a cold reservoir at temperature T_C. For 10 cycles of operation, determine (a) the net work input, in kJ, and (b) the minimum theoretical temperature T_C, in K.

5.48 A reversible heat pump cycle operates as in Fig. 5.7 between hot and cold reservoirs at $T_H = 27°C$ and $T_C = -3°C$, respectively. Determine the fraction of the heat transfer Q_H discharged at T_H provided by (a) the net work input, and (b) the heat transfer Q_C from the cold reservoir T_C.

5.49 A reversible power cycle and a reversible heat pump cycle operate between hot and cold reservoirs at temperature $T_H = 555$ K and T_C, respectively. If the thermal efficiency of the power cycle is 60%, determine (a) T_C, in K, and (b) the coefficient of performance of the heat pump.

5.50 An inventor has developed a refrigerator capable of maintaining its freezer compartment at −7°C while operating in a kitchen at 21°C, and claims the device has a coefficient of performance of (a) 10, (b) 9.5, (c) 4. Evaluate the claim in each of the three cases.

5.51 An inventor claims to have developed a refrigerator that at steady state requires a net power input of 0.52 kW to remove 12,660 kJ/h of energy by heat transfer from the freezer compartment at −17.8°C and discharge energy by heat transfer to a kitchen at 21°C. Evaluate this claim.

5.52 An inventor claims to have devised a refrigeration cycle operating between hot and cold reservoirs at 300 K and 250 K, respectively, that removes an amount of energy Q_C by heat transfer from the cold reservoir that is a multiple of the net work input—that is, $Q_C = NW_{cycle}$, where all quantities are positive. Determine the maximum theoretical value of the number N for any such cycle.

5.53 Data are provided for two reversible refrigeration cycles. One cycle operates between hot and cold reservoirs at 27°C and −8°C, respectively. The other cycle operates between the same hot reservoir at 27°C and a cold reservoir at −28°C. If each refrigerator removes the same amount of energy by heat transfer from its cold reservoir, determine the ratio of the net work input values of the two cycles.

5.54 By removing energy by heat transfer from its freezer compartment at a rate of 1.25 kW, a refrigerator maintains the freezer at −26°C on a day when the temperature of the surroundings is 22°C. Determine the minimum theoretical power, in kW, required by the refrigerator at steady state.

5.55 Determine the minimum theoretical power, in kJ/s, required at steady state by a refrigeration system to maintain a cryogenic sample at −126°C in a laboratory at 21°C, if energy leaks by heat transfer to the sample from its surroundings at a rate of 0.09 kJ/s.

5.56 For each kW of power input to an ice maker at steady state, determine the maximum rate that ice can be produced, in kg/h, from liquid water at 0°C. Assume that 333 kJ/kg of energy must be removed by heat transfer to freeze water at 0°C, and that the surroundings are at 20°C.

5.57 At steady state, a reversible refrigeration cycle operates between hot and cold reservoirs at 300 K and 270 K, respectively. Determine the minimum theoretical net power input required, in kW per kW of heat transfer from the cold reservoir.

5.58 At steady state, a refrigeration cycle operating between hot and cold reservoirs at 300 K and 275 K, respectively, removes energy by heat transfer from the cold reservoir at a rate of 600 kW.

(a) If the cycle's coefficient of performance is 4, determine the power input required, in kW.
(b) Determine the minimum theoretical power required, in kW, for *any* such cycle.

5.59 As shown in Fig P5.59, an air conditioner operating at steady state maintains a dwelling at 21°C on a day when the outside temperature is 32°C. If the rate of heat transfer into the dwelling through the walls and roof is 31,650 kJ/h, might a net power input to the air conditioner compressor of 2.2 kW be sufficient? If yes, determine the coefficient of performance. If no, determine the minimum theoretical power input, in kW.

5.60 A heat pump cycle is used to maintain the interior of a building at 20°C. At steady state, the heat pump receives energy by heat transfer from well water at 10°C and discharges energy by heat transfer to the building at a rate of 120,000 kJ/h. Over a period of 14 days, an electric meter records that 1490 kW · h of electricity is provided to the heat pump. Determine

(a) the amount of energy that the heat pump receives over the 14-day period from the well water by heat transfer, in kJ.
(b) the heat pump's coefficient of performance.
(c) the coefficient of performance of a reversible heat pump cycle operating between hot and cold reservoirs at 20°C and 10°C.

5.61 A refrigeration cycle has a coefficient of performance equal to 75% of the value for a reversible refrigeration cycle operating between cold and hot reservoirs at −5°C and 40°C, respectively. For operation at steady state, determine the net power input, in kW per kW of cooling, required by (a) the actual refrigeration cycle and (b) the reversible refrigeration cycle. Compare values.

5.62 By removing energy by heat transfer from a room, a window air conditioner maintains the room at 22°C on a day when the outside temperature is 32°C.

(a) Determine, in kW per kW of cooling, the *minimum* theoretical power required by the air conditioner.
(b) To achieve required rates of heat transfer with practical-sized units, air conditioners typically receive energy by heat transfer at a temperature *below* that of the room being cooled and discharge energy by heat transfer at a temperature *above* that of the surroundings. Consider the effect of this by determining the *minimum* theoretical power, in kW per kW of cooling, required when $T_C = 18°C$ and $T_H = 36°C$, and compare with the value found in part (a).

5.63 The refrigerator shown in Fig. P5.63 operates at steady state with a coefficient of performance of 4.5 and a power input of 0.8 kW. Energy is rejected from the refrigerator to the surroundings at 20°C by heat transfer from metal coils whose average surface temperature is 28°C. Determine

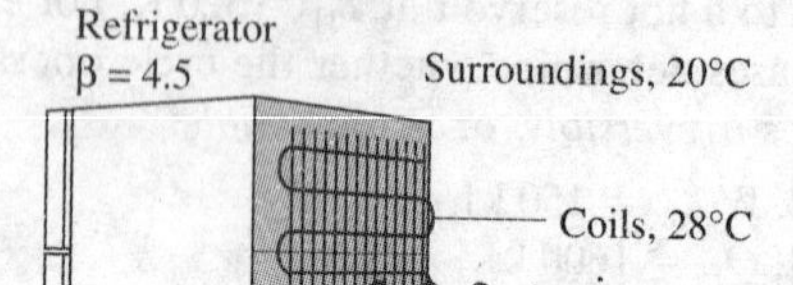

Fig. P5.63

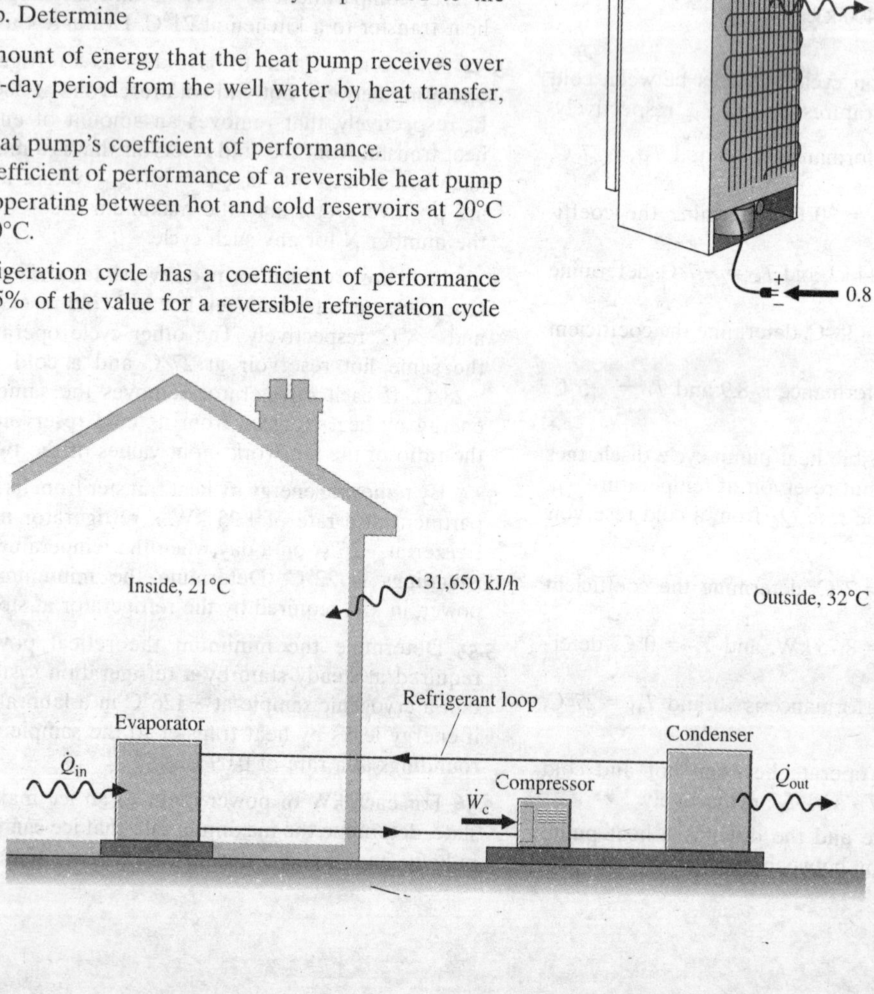

Fig. P5.59

(a) the rate energy is rejected, in kW.
(b) the lowest theoretical temperature *inside* the refrigerator, in K.
(c) the maximum theoretical power, in kW, that could be developed by a power cycle operating between the coils and the surroundings. Would you recommend making use of this opportunity for developing power?

5.64 At steady state, a heat pump provides 31,650 kJ/h to maintain a dwelling at 20°C on a day when the outside temperature is 1.7°C. The power input to the heat pump is 3.73 kW. If electricity costs 8 cents per kW · h, compare the actual operating cost with the minimum theoretical operating cost for each day of operation.

5.65 By supplying energy at an average rate of 21,100 kJ/h, a heat pump maintains the temperature of a dwelling at 21°C. If electricity costs 8 cents per kW · h, determine the minimum theoretical operating cost for each day of operation if the heat pump receives energy by heat transfer from

(a) the outdoor air at −5°C.
(b) well water at 8°C.

5.66 A heat pump with a coefficient of performance of 3.8 provides energy at an average rate of 75,000 kJ/h to maintain a building at 21°C on a day when the outside temperature is 0°C. If electricity costs 8 cents per kW · h

(a) determine the actual operating cost and the minimum theoretical operating cost, each in $/day.
(b) compare the results of part (a) with the cost of electrical-resistance heating.

5.67 A heat pump maintains a dwelling at temperature T when the outside temperature averages 5°C. The heat transfer rate through the walls and roof is 2000 kJ/h per degree of temperature difference between the inside and outside. If electricity costs 8 cents per kW · h

(a) determine the minimum theoretical operating cost for each day of operation when $T = 20°C$.
(b) plot the minimum theoretical operating cost for each day of operation as a function of T ranging from 18 to 23°C.

5.68 A heat pump maintains a dwelling at temperature T when the outside temperature is −7°C. The heat transfer rate through the walls and roof is 0.44 kJ/s per degree temperature difference between the inside and outside.

(a) If electricity costs 8 cents per kW · h, plot the minimum theoretical operating cost for each day of operation for T ranging from 20 to 24°C.
(b) If $T = 21°C$, plot the minimum theoretical operating cost for each day of operation for a cost of electricity ranging from 4 to 12 cents per kW · h.

5.69 Two reversible refrigeration cycles operate in series. The first cycle receives energy by heat transfer from a cold reservoir at 300 K and rejects energy by heat transfer to a reservoir at an intermediate temperature T greater than 300 K. The second cycle receives energy by heat transfer from the reservoir at temperature T and rejects energy by heat transfer to a higher-temperature reservoir at 883 K. If the refrigeration cycles have the same coefficient of performance, determine (a) T, in K, and (b) the value of each coefficient of performance.

5.70 Two reversible heat pump cycles operate in series. The first cycle receives energy by heat transfer from a cold reservoir at 250 K and rejects energy by heat transfer to a reservoir at an intermediate temperature T greater than 250 K. The second cycle receives energy by heat transfer from the reservoir at temperature T and rejects energy by heat transfer to a higher-temperature reservoir at 1440 K. If the heat pump cycles have the same coefficient of performance, determine (a) T, in K, and (b) the value of each coefficient of performance.

5.71 Two reversible refrigeration cycles are arranged in series. The first cycle receives energy by heat transfer from a cold reservoir at temperature T_C and rejects energy by heat transfer to a reservoir at an intermediate temperature T greater than T_C. The second cycle receives energy by heat transfer from the reservoir at temperature T and rejects energy by heat transfer to a higher-temperature reservoir at T_H. Obtain an expression for the coefficient of performance of a *single* reversible refrigeration cycle operating directly between cold and hot reservoirs at T_C and T_H, respectively, in terms of the coefficients of performance of the two cycles.

5.72 Repeat Problem 5.71 for the case of two reversible heat pump cycles.

5.73 A reversible power cycle receives Q_H from a hot reservoir at temperature T_H and rejects energy by heat transfer to the surroundings at temperature T_0. The work developed by the power cycle is used to drive a reversible refrigeration cycle that removes Q_C from a cold reservoir at temperature T_C and discharges energy by heat transfer to the same surroundings at T_0. Develop an expression for the ratio Q_C/Q_H in terms of the temperature ratios T_H/T_0 and T_C/T_0.

5.74 A reversible power cycle receives energy Q_H from a reservoir at temperature T_H and rejects Q_C to a reservoir at temperature T_C. The work developed by the power cycle is used to drive a reversible heat pump that removes energy Q'_C from a reservoir at temperature T'_C and rejects energy Q'_H to a reservoir at temperature T'_H.

(a) Develop an expression for the ratio Q'_H/Q_H in terms of the temperatures of the four reservoirs.
(b) What must be the relationship of the temperatures T_H, T_C, T'_C, and T'_H for Q'_H/Q_H to exceed a value of unity?

Carnot Cycle Applications

5.75 Two kilograms of water execute a Carnot power cycle. During the isothermal expansion, the water is heated until it is a saturated vapor from an initial state where the pressure is 40 bar and the quality is 15%. The vapor then expands adiabatically to a pressure of 1.5 bar while doing 491.5 kJ/kg of work.

(a) Sketch the cycle on p–v coordinates.
(b) Evaluate the heat and work for each process, in kJ.
(c) Evaluate the thermal efficiency.

5.76 One-half kg of water executes a Carnot power cycle. During the isothermal expansion, the water is heated at 315°C from a saturated liquid to a saturated vapor. The vapor then expands adiabatically to a temperature of 32°C and a quality of 64.3%.

(a) Sketch the cycle on p–v coordinates.
(b) Evaluate the heat and work for each process, in kJ.
(c) Evaluate the thermal efficiency.

5.77 One kilogram of air as an ideal gas executes a Carnot power cycle having a thermal efficiency of 50%. The heat transfer to the air during the isothermal expansion is 50 kJ. At the end of the isothermal expansion, the pressure is 574 kPa and the volume is 0.3 m^3. Determine

(a) the maximum and minimum temperatures for the cycle, in K.
(b) the pressure and volume at the beginning of the isothermal expansion in bar and m^3, respectively.
(c) the work and heat transfer for each of the four processes, in kJ.
(d) Sketch the cycle on p–v coordinates.

5.78 The pressure–volume diagram of a Carnot power cycle executed by an ideal gas with constant specific heat ratio k is shown in Fig. P5.78. Demonstrate that

(a) $V_4V_2 = V_1V_3$.
(b) $T_2/T_3 = (p_2/p_3)^{(k-1)/k}$.
(c) $T_2/T_3 = (V_3/V_2)^{k-1}$.

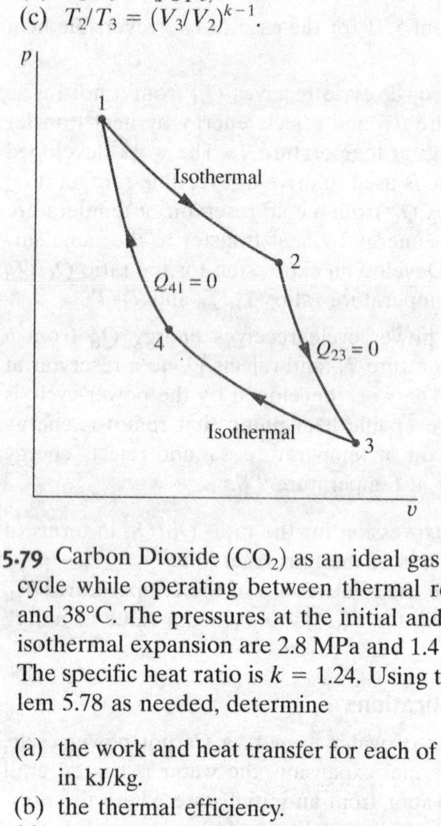

Fig. P5.78

5.79 Carbon Dioxide (CO_2) as an ideal gas executes a Carnot cycle while operating between thermal reservoirs at 232°C and 38°C. The pressures at the initial and final states of the isothermal expansion are 2.8 MPa and 1.4 MPa, respectively. The specific heat ratio is $k = 1.24$. Using the results of Problem 5.78 as needed, determine

(a) the work and heat transfer for each of the four processes, in kJ/kg.
(b) the thermal efficiency.
(c) the pressures at the initial and final states of the isothermal compression, in kPa.

5.80 One-tenth kilogram of air as an ideal gas with $k = 1.4$ executes a Carnot refrigeration cycle, as shown in Fig. 5.13. The isothermal expansion occurs at −23°C with a heat transfer to the air of 3.4 kJ. The isothermal compression occurs at 27°C to a final volume of 0.01 m^3. Using the results of Prob. 5.78 as needed, determine

(a) the pressure, in kPa, at each of the four principal states.
(b) the work, in kJ, for each of the four processes.
(c) the coefficient of performance.

Clausius Inequality Applications

5.81 A system executes a power cycle while receiving 1000 kJ by heat transfer at a temperature of 500 K and discharging energy by heat transfer at a temperature of 300 K. There are no other heat transfers. Applying Eq. 5.13, determine σ_{cycle} if the thermal efficiency is (a) 60%, (b) 40%, (c) 20%. Identify the cases (if any) that are internally reversible or impossible.

5.82 A system executes a power cycle while receiving 1690 kJ by heat transfer at a temperature of 1390 K and discharging 211 kJ by heat transfer at 278 K. A heat transfer from the system also occurs at a temperature of 833 K. There are no other heat transfers. If no internal irreversibilites are present, determine the thermal efficiency.

5.83 As shown in Fig. P5.83, a system executes a power cycle while receiving 750 kJ by heat transfer at a temperature of 1500 K and discharging 100 kJ by heat transfer at a temperature of 500 K. Another heat transfer from the system occurs at a temperature of 1000 K. Using Eq. 5.13, determine the thermal efficiency if σ_{cycle} is (a) 0.1 kJ/K, (b) 0.2 kJ/K, (c) 0.35 kJ/K.

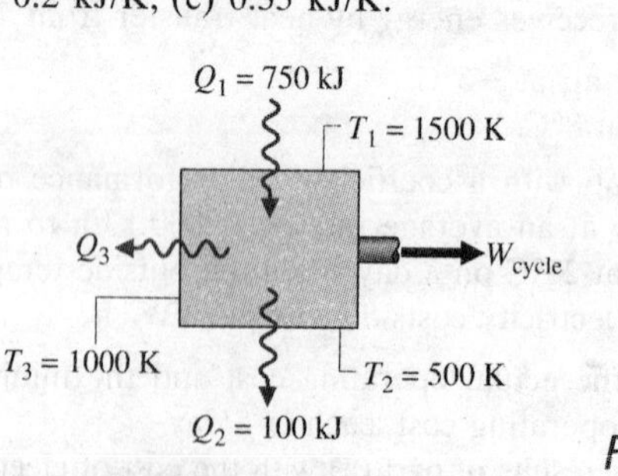

Fig. P5.83

5.84 A simple vapor power plant operating at steady state is shown in Fig. P5.84. In each of the following three cases, use Eq. 5.13 to determine whether the cycle is *possible, internally reversible,* or *impossible.* Processes 1–2 and 3–4 are adiabatic. Kinetic and potential energy changes can be ignored.

(a) Process 4–1: constant-pressure at 1 MPa from saturated liquid to saturated vapor. Process 2–3: constant-pressure at 20 kPa from $x_3 = 88\%$ to $x_4 = 18\%$.
(b) Process 4–1: constant-pressure at 8 MPa from saturated liquid to saturated vapor. Process 2–3: constant-pressure at 8 kPa from $x_2 = 67.5\%$ to $x_3 = 34.2\%$.
(c) Process 4–1: constant-pressure at 0.15 MPa from saturated liquid to saturated vapor. Process 2–3: constant-pressure at 20 kPa from $x_2 = 90\%$ to $x_3 = 10\%$.

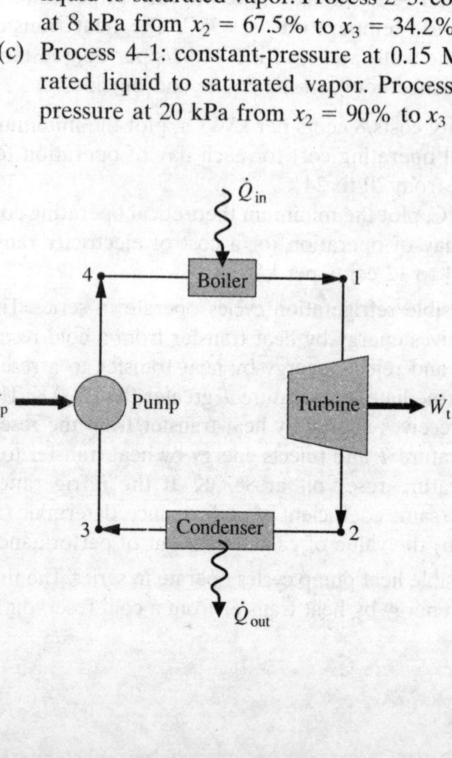

Fig. P5.84

5.85 A reversible power cycle R and an irreversible power cycle I operate between the same two reservoirs. Each receives Q_H from the hot reservoir. The reversible cycle develops work W_R, while the irreversible cycle develops work W_I. The reversible cycle discharges Q_C to the cold reservoir, while the irreversible cycle discharges Q'_C.

(a) Using Eq. 5.13, evaluate σ_{cycle} for cycle I in terms of W_I, W_R, and temperature T_C of the cold reservoir only.
(b) Demonstrate that $W_I < W_R$ and $Q'_C > Q_C$.

5.86 A reversible refrigeration cycle R and an irreversible refrigeration cycle I operate between the same two reservoirs and each removes Q_C from the cold reservoir. The net work input required by R is W_R, while the net work input for I is W_I. The reversible cycle discharges Q_H to the hot reservoir, while the irreversible cycle discharges Q'_H. Using Eq. 5.13, show that $W_I > W_R$ and $Q'_H > Q_H$.

5.87 Using Eq. 5.13, complete the following involving reversible and irreversible cycles:

(a) Reversible and irreversible *power cycles* each discharge energy Q_C to a cold reservoir at temperature T_C and receive energy Q_H from hot reservoirs at temperatures T_H and T'_H, respectively. There are no other heat transfers. Show that $T'_H > T_H$.
(b) Reversible and irreversible *refrigeration* cycles each discharge energy Q_H to a hot reservoir at temperature T_H and receive energy Q_C from cold reservoirs at temperatures T_C and T'_C, respectively. There are no other heat transfers. Show that $T'_C > T_C$.
(c) Reversible and irreversible *heat pump* cycles each receive energy Q_C from a cold reservoir at temperature T_C and discharge energy Q_H to hot reservoirs at temperatures T_H and T'_H, respectively. There are no other heat transfers. Show that $T'_H < T_H$.

5.88 Figure P5.88 shows a system consisting of a power cycle driving a heat pump. At steady state, the power cycle receives $\dot{Q}_s$ by heat transfer at T_s from the high-temperature source and delivers $\dot{Q}_1$ to a dwelling at T_d. The heat pump receives $\dot{Q}_0$ from the outdoors at T_0, and delivers $\dot{Q}_2$ to the dwelling. Using Eq. 5.13, obtain an expression for the maximum theoretical value of the performance parameter $(\dot{Q}_1 + \dot{Q}_2)/\dot{Q}_s$ in terms of the temperature ratios T_s/T_d and T_0/T_d.

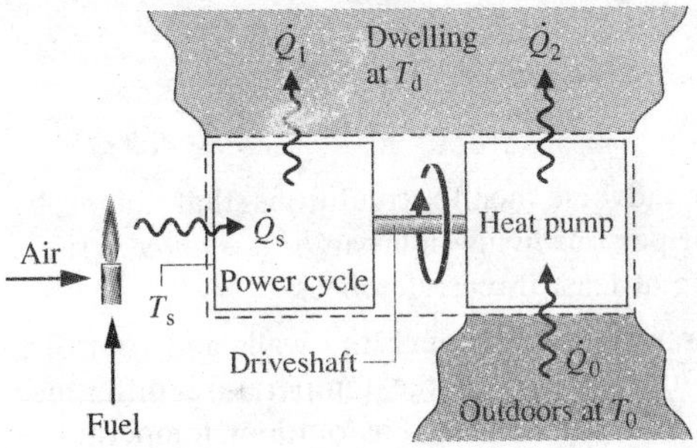

Fig. P5.88

Design & open ended problems: exploring engineering practice

5.1D Experiments by Australian researchers suggest that the second law can be violated at the micron scale over time intervals of up to 2 seconds. If validated, some say these findings could place a limit on the engineering of nanomachines because such devices may not behave simply like miniaturized versions of their larger counterparts. Investigate the implications, if any, for nanotechnology of these experimental findings. Write a report including at least three references.

5.2D The U.S. Food and Drug Administration (FDA) has long permitted the application of citric acid, ascorbic acid, and other substances to keep fresh meat looking red longer. In 2002, the FDA began allowing meat to be treated with carbon monoxide. Carbon monoxide reacts with *myoglobin* in the meat to produce a substance that resists the natural browning of meat, thereby giving meat a longer shelf life. Investigate the use of carbon monoxide for this purpose. Identify the nature of myoglobin and explain its role in the reactions that cause meat to brown or, when treated with carbon monoxide, allows the meat to appear red longer. Consider the hazards, if any, that may accompany this practice for consumers and for meat industry workers. Report your findings in a memorandum.

5.3D In an *Internet* paper titled, "How Lightning Strikes are Produced and Why the Second Law of Thermodynamics Is Invalid," the author claims that electricity can be produced economically by extracting "constant-temperature ambient heat from the atmosphere or the oceans and converting it fully into work." Critically evaluate this claim, and report your findings in a memorandum.

5.4D Figure P5.4D shows one of those bobbing toy birds that seemingly takes an endless series of sips from a cup filled with water. Prepare a 30-min presentation suitable for a middle school science class explaining the operating principles of this device and whether or not its behavior is at odds with the second law.

Fig. P5.4D

5.5D Figure P5.5D shows that the typical thermal efficiency of U.S. power plants increased rapidly in the early and mid-1900s, but has increased only gradually since then. Investigate the most important factors contributing to this plateauing of thermal efficiency and the most promising near-term and long-term technologies that might lead to appreciable thermal efficiency gains. Report your findings in a memorandum.

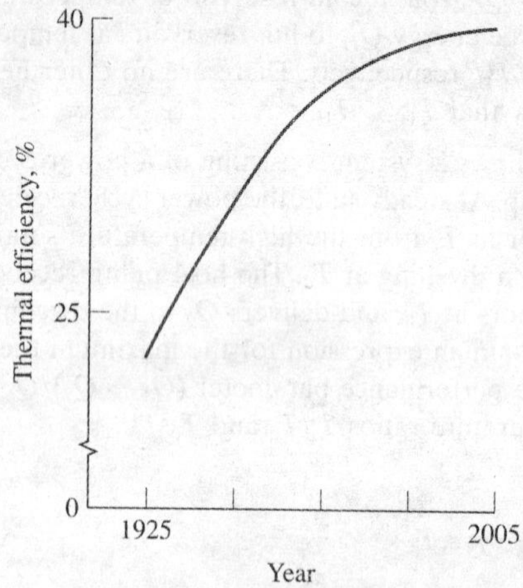

Fig. P5.5D

5.6D Investigate adverse health conditions that might be exacerbated for persons living in urban *heat islands*. Write a report including at least three references.

5.7D The heat transfer rate through the walls and roof of a building is 3570 kJ/h per degree temperature difference between the inside and outside. For outdoor temperatures ranging from 20 to −20°C, make a comparison of the daily cost of maintaining the building at 20°C by means of an electric heat pump, direct electric resistance heating, and a conventional gas-fired furnace. Report your findings in a memorandum.

5.8D Insulin and several other pharmaceuticals required daily by those suffering from diabetes and other medical conditions have relatively low thermal stability. Those living and traveling in hot climates are especially at risk by heat-induced loss of potency of their pharmaceuticals. Design a wearable, lightweight, and reliable cooler for transporting temperature-sensitive pharmaceuticals. The cooler also must be solely powered by human motion. While the long-term goal is a moderately priced consumer product, the final project report need only provide the costing of a single prototype.

5.9D The International Temperature Scale was first adopted by the International Committee on Weights and Measures in 1927 to provide a global standard for temperature measurement. This scale has been refined and extended in several revisions, most recently in 1990 (International Temperature Scale of 1990, ITS-90). What are some reasons for revising the scale? What are some principal changes that have been made since 1927? Summarize your findings in a memorandum.

5.10D A technical article considers hurricanes as an example of a *natural* Carnot engine: K. A. Emmanuel, "Toward a General Theory of Hurricanes," *American Scientist,* 76, 371–379, 1988. Also see *Physics Today,* 59, No. 8, 74–75, 2006, for a related discussion by this author. U.S. Patent (No. 4,885,913) is said to have been inspired by such an analysis. Does the concept have scientific merit? Engineering merit? Summarize your conclusions in a memorandum.

6

Using Entropy

Engineering Context

Up to this point, our study of the second law has been concerned primarily with what it says about systems undergoing thermodynamic cycles. In this chapter means are introduced for analyzing systems from the second law perspective as they undergo processes that are not necessarily cycles. The property *entropy* and the *entropy production* concept introduced in Chap. 5 play prominent roles in these considerations.

The **objective** of the present chapter is to develop an understanding of entropy concepts, including the use of entropy balances for closed systems and control volumes in forms effective for the analysis of engineering systems. The Clausius inequality developed in Sec. 5.11, expressed as Eq. 5.13, provides the basis.

Learning Objectives

When you complete your study of this chapter you will be able to...

- ✓ demonstrate understanding of key concepts related to entropy and the second law...including entropy transfer, entropy production, and the increase in entropy principle.
- ✓ evaluate entropy, evaluate entropy change between two states, and analyze isentropic processes, using appropriate property data.
- ✓ represent heat transfer in an internally reversible process as an area on a temperature-entropy diagram.
- ✓ apply entropy balances to closed systems and control volumes.
- ✓ evaluate isentropic efficiencies for turbines, nozzles, compressors, and pumps.

6.1 Entropy—A System Property

The word *energy* is so much a part of the language that you were undoubtedly familiar with the term before encountering it in early science courses. This familiarity probably facilitated the study of energy in these courses and in the current course in engineering thermodynamics. In the present chapter you will see that the analysis of systems from a second law perspective is effectively accomplished in terms of the property *entropy.* Energy and entropy are both abstract concepts. However, unlike energy, the word entropy is seldom heard in everyday conversation, and you may never have dealt with it quantitatively before. Energy and entropy play important roles in the remaining chapters of this book.

6.1.1 Defining Entropy Change

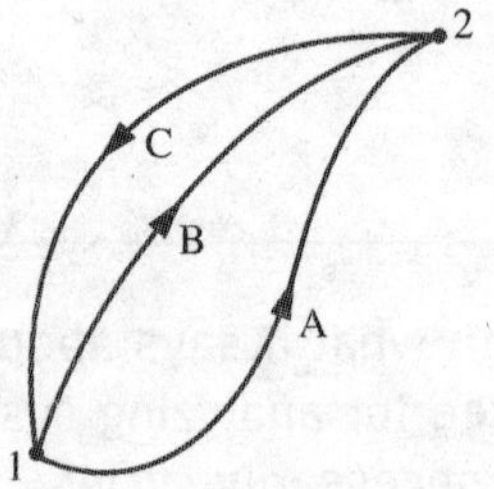

Fig. 6.1 Two internally reversible cycles.

A quantity is a property if, and only if, its change in value between two states is independent of the process (Sec. 1.3). This aspect of the property concept is used in the present section together with the Clausius inequality to introduce entropy change as follows:

Two cycles executed by a closed system are represented in Fig. 6.1. One cycle consists of an internally reversible process A from state 1 to state 2, followed by internally reversible process C from state 2 to state 1. The other cycle consists of an internally reversible process B from state 1 to state 2, followed by the same process C from state 2 to state 1 as in the first cycle. For the first cycle, Eq. 5.13 (the Clausius Inequality) takes the form

$$\left(\int_1^2 \frac{\delta Q}{T}\right)_A + \left(\int_2^1 \frac{\delta Q}{T}\right)_C = -\overset{0}{\cancel{\sigma}}_{\text{cycle}} \tag{6.1a}$$

$$\oint \left(\frac{\delta Q}{T}\right)_b = -\sigma_{\text{cycle}}$$
(Eq. 5.13)

For the second cycle, Eq. 5.13 takes the form

$$\left(\int_1^2 \frac{\delta Q}{T}\right)_B + \left(\int_2^1 \frac{\delta Q}{T}\right)_C = -\overset{0}{\cancel{\sigma}}_{\text{cycle}} \tag{6.1b}$$

In writing Eq. 6.1, the term σ_{cycle} has been set to zero since the cycles are composed of internally reversible processes.

When Eq. 6.1b is subtracted from Eq. 6.1a, we get

$$\left(\int_1^2 \frac{\delta Q}{T}\right)_A = \left(\int_1^2 \frac{\delta Q}{T}\right)_B$$

This shows that the integral of $\delta Q/T$ is the same for both processes. Since A and B are arbitrary, it follows that the integral of $\delta Q/T$ has the same value for *any* internally reversible process between the two states. In other words, the value of the integral depends on the end states only. It can be concluded, therefore, that the integral represents the change in some property of the system.

Selecting the symbol S to denote this property, which is called *entropy,* the ***change in entropy*** is given by

definition of entropy change

$$S_2 - S_1 = \left(\int_1^2 \frac{\delta Q}{T}\right)_{\substack{\text{int}\\\text{rev}}} \tag{6.2a}$$

where the subscript "int rev" is added as a reminder that the integration is carried out for any internally reversible process linking the two states. On a differential basis, the defining equation for entropy change takes the form

$$dS = \left(\frac{\delta Q}{T}\right)_{\substack{\text{int}\\\text{rev}}} \tag{6.2b}$$

Entropy is an extensive property.

The ***SI unit for entropy*** is J/K. However, in this book it is convenient to work in terms of kJ/K. Units in SI for *specific* entropy are kJ/kg · K for s and kJ/kmol · K for $\bar{s}$.

units for entropy

It should be clear that entropy is defined and evaluated in terms of a particular expression (Eq. 6.2a) for which *no accompanying physical picture is given.* We encountered this previously with the property enthalpy. Enthalpy is introduced without physical motivation in Sec. 3.6.1. Then, in Chap. 4, we learned how enthalpy is used for thermodynamic analysis of control volumes. As for the case of enthalpy, to gain an appreciation for entropy you need to understand *how* it is used and *what* it is used for. This is the aim of the rest of this chapter.

6.1.2 Evaluating Entropy

Since entropy is a property, the change in entropy of a system in going from one state to another is the same for *all* processes, both internally reversible and irreversible, between these two states. Thus, Eq. 6.2a allows the determination of the change in entropy, and once it has been evaluated, this is the magnitude of the entropy change for all processes of the system between the two states.

The defining equation for entropy change, Eq. 6.2a, serves as the basis for evaluating entropy relative to a reference value at a reference state. Both the reference value and the reference state can be selected arbitrarily. The value of entropy at any state y relative to the value at the reference state x is obtained in principle from

$$S_y = S_x + \left(\int_x^y \frac{\delta Q}{T}\right)_{\substack{\text{int}\\\text{rev}}} \tag{6.3}$$

where S_x is the reference value for entropy at the specified reference state.

The use of entropy values determined relative to an arbitrary reference state is satisfactory as long as they are used in calculations involving entropy differences, for then the reference value cancels. This approach suffices for applications where composition remains constant. When chemical reactions occur, it is necessary to work in terms of *absolute* values of entropy determined using the *third law of thermodynamics* (Chap. 13).

6.1.3 Entropy and Probability

The presentation of engineering thermodynamics provided in this book takes a *macroscopic* view as it deals mainly with the gross, or overall, behavior of matter. The macroscopic concepts of engineering thermodynamics introduced thus far, including energy and entropy, rest on operational definitions whose validity is shown directly or indirectly through experimentation. Still, insights concerning energy and entropy can result from considering the microstructure of matter. This brings in the use of *probability* and the notion of *disorder*. Further discussion of entropy, probability, and disorder is provided in Sec. 6.8.2.

6.2 Retrieving Entropy Data

In Chap. 3, we introduced means for retrieving property data, including tables, graphs, equations, and the software available with this text. The emphasis there is on evaluating the properties p, v, T, u, and h required for application of the conservation of mass and energy principles. For application of the second law, entropy values are usually required. In this section, means for retrieving entropy data for water and several refrigerants are considered.

Tables of thermodynamic data are introduced in Secs. 3.5 and 3.6 (Tables A-2 through A-18). Specific entropy is tabulated in the same way as considered there for the properties v, u, and h, and entropy values are retrieved similarly. The specific entropy values given in Tables A-2 through A-18 are relative to the following *reference states and values.* For water, the entropy of saturated liquid at 0.01°C is set to zero. For the refrigerants, the entropy of the saturated liquid at −40°C is assigned a value of zero.

6.2.1 Vapor Data

In the superheat regions of the tables for water and the refrigerants, specific entropy is tabulated along with v, u, and h versus temperature and pressure. ➡ **FOR EXAMPLE...** consider two states of water. At state 1 the pressure is 3 MPa and the temperature is 500°C. At state 2, the pressure is 0.3 MPa and the specific entropy is the same as at state 1, $s_2 = s_1$. The object is to determine the temperature at state 2. Using T_1 and p_1, we find the specific entropy at state 1 from Table A-4 as $s_1 = 7.2338$ kJ/kg · K. State 2 is fixed by the pressure, $p_2 = 0.3$ MPa, and the specific entropy, $s_2 = 7.2338$ kJ/kg · K. Returing to Table A-4 at 0.3 MPa and interpolating with s_2 between 160 and 200°C results in $T_2 = 183$°C. ⬅

6.2.2 Saturation Data

For saturation states, the values of s_f and s_g are tabulated as a function of either saturation pressure or saturation temperature. The specific entropy of a two-phase liquid–vapor mixture is calculated using the quality

$$\begin{aligned} s &= (1 - x)s_f + xs_g \\ &= s_f + x(s_g - s_f) \end{aligned} \tag{6.4}$$

These relations are identical in form to those for v, u, and h (Secs. 3.5 and 3.6). ➡ **FOR EXAMPLE...** let us determine the specific entropy of Refrigerant 134a at a state where the temperature is 0°C and the specific internal energy is 138.43 kJ/kg. Referring to Table A-10, we see that the given value for u falls between u_f and u_g at 0°C, so the system is a two-phase liquid–vapor mixture. The quality of the mixture can be determined from the known specific internal energy

$$x = \frac{u - u_f}{u_g - u_f} = \frac{138.43 - 49.79}{227.06 - 49.79} = 0.5$$

Then with values from Table A-10, Eq. 6.4 gives

$$\begin{aligned} s &= (1 - x)s_f + xs_g \\ &= (0.5)(0.1970) + (0.5)(0.9190) = 0.5580 \text{ kJ/kg} \cdot \text{K} \end{aligned}$$ ⬅

6.2.3 Liquid Data

Compressed liquid data are presented for water in Tables A-5. In these tables s, v, u, and h are tabulated versus temperature and pressure as in the superheat tables, and the tables are used similarly. In the absence of compressed liquid data, the value of the specific entropy can be estimated in the same way as estimates for v and u are obtained for liquid states (Sec. 3.10.1), by using the saturated liquid value at the given temperature

$$s(T, p) \approx s_f(T) \tag{6.5}$$

FOR EXAMPLE... suppose the value of specific entropy is required for water at 25 bar, 200°C. The specific entropy is obtained directly from Table A-5 as s = 2.3294 kJ/kg · K. Using the saturated liquid value for specific entropy at 200°C from Table A-2, the specific entropy is approximated with Eq. 6.5 as s = 2.3309 kJ/kg · K, which agrees closely with the previous value.

6.2.4 Computer Retrieval

The software available with this text, *Interactive Thermodynamics: IT,* provides data for the substances considered in this section. Entropy data are retrieved by simple call statements placed in the workspace of the program. **FOR EXAMPLE...** consider a two-phase liquid–vapor mixture of H_2O at p = 1 bar, v = 0.8475 m^3/kg. The following illustrates how specific entropy and quality x are obtained using *IT*

```
p = 1 // bar
v = 0.8475 // m³/kg
v = vsat_Px("Water/Steam",p,x)
s = ssat_Px("Water/Steam",p,x)
```

The software returns values of x = 0.5 and s = 4.331 kJ/kg · K, which can be checked using data from Table A-3. Note that quality x is implicit in the expression for specific volume, and it is not necessary to solve explicitly for x. As another example, consider superheated ammonia vapor at p = 1.5 bar, T = 8°C. Specific entropy is obtained from *IT* as follows:

```
p = 1.5 // bar
T = 8 // °C
s = s_PT("Ammonia",p,T)
```

The software returns s = 5.981 kJ/kg · K, which agrees closely with the value obtained by interpolation in Table A-15.

Take Note...

Note that *IT* does not provide compressed liquid data for *any* substance. *IT* returns liquid entropy data using the approximation of Eq. 6.5. Similarly, Eqs. 3.11, 3.12, and 3.14 are used to return liquid values for v, u, and h, respectively.

6.2.5 Using Graphical Entropy Data

The use of property diagrams as an adjunct to problem solving is emphasized throughout this book. When applying the second law, it is frequently helpful to locate states and plot processes on diagrams having entropy as a coordinate. Two commonly used figures having entropy as one of the coordinates are the temperature–entropy diagram and the enthalpy–entropy diagram.

Temperature–Entropy Diagram

T–s diagram

The main features of a ***temperature–entropy diagram*** are shown in Fig. 6.2. For detailed figures for water in SI units, see Fig. A-7. Observe that lines of constant enthalpy are shown on these figures. Also note that in the superheated vapor region, constant specific volume lines have a steeper slope than constant-pressure lines. Lines of constant quality are shown in the two-phase liquid–vapor region. On some figures, lines of constant quality are marked as *percent moisture* lines. The percent moisture is defined as the ratio of the mass of liquid to the total mass.

In the superheated vapor region of the T–s diagram, constant specific enthalpy lines become nearly horizontal as pressure is reduced. These states are shown as the shaded area on Fig. 6.2. For states in this region of the diagram, the enthalpy is determined primarily by the temperature: $h(T, p) \approx h(T)$. This is the region of the diagram where the ideal gas model provides a reasonable approximation. For superheated

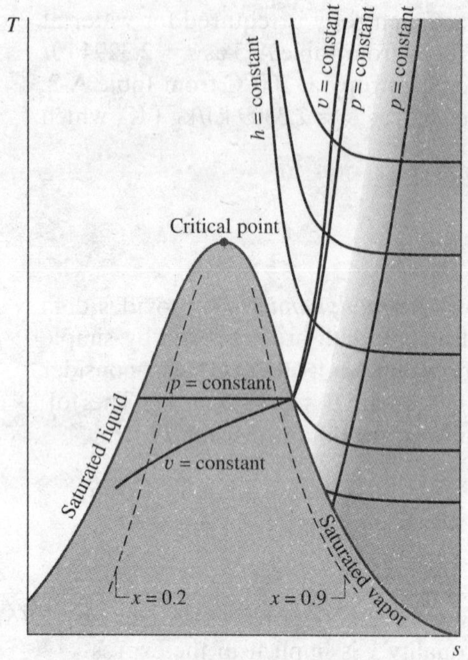

Fig. 6.2 Temperature–entropy diagram.

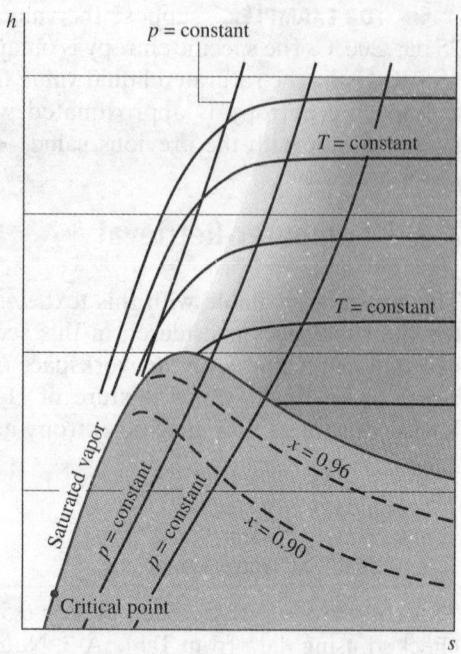

Fig. 6.3 Enthalpy–entropy diagram.

vapor states outside the shaded area, both temperature and pressure are required to evaluate enthalpy, and the ideal gas model is not suitable.

Enthalpy–Entropy Diagram

Mollier diagram

The essential features of an enthalpy–entropy diagram, commonly known as a ***Mollier diagram,*** are shown in Fig. 6.3. For detailed figures for water in SI, see Fig. A-8. Note the location of the critical point and the appearance of lines of constant temperature and constant pressure. Lines of constant quality are shown in the two-phase liquid–vapor region (some figures give lines of constant percent moisture). The figure is intended for evaluating properties at superheated vapor states and for two-phase liquid–vapor mixtures. Liquid data are seldom shown. In the superheated vapor region, constant-temperature lines become nearly horizontal as pressure is reduced. These states are shown, approximately, as the shaded area on Fig. 6.3. This area corresponds to the shaded area on the temperature–entropy diagram of Fig. 6.2, where the ideal gas model provides a reasonable approximation.

FOR EXAMPLE... to illustrate the use of the Mollier diagram in SI units, consider two states of water. At state 1, $T_1 = 240°C$, $p_1 = 0.10$ MPa. The specific enthalpy and quality are required at state 2, where $p_2 = 0.01$ MPa and $s_2 = s_1$. Turning to Fig. A-8, state 1 is located in the superheated vapor region. Dropping a vertical line into the two-phase liquid–vapor region, state 2 is located. The quality and specific enthalpy at state 2 read from the figure agree closely with values obtained using Tables A-3 and A-4: $x_2 = 0.98$ and $h_2 = 2537$ kJ/kg.

6.3 Introducing the T dS Equations

Although the change in entropy between two states can be determined in principle by using Eq. 6.2a, such evaluations are generally conducted using the $T\,dS$ equations developed in this section. The $T\,dS$ equations allow entropy changes to be evaluated from other more readily determined property data. The use of the $T\,dS$ equations to evaluate entropy changes for an incompressible substance is illustrated in Sec. 6.4 and for ideal gases in Sec. 6.5. The importance of the *T dS* equations is greater than their role in assigning entropy values, however. In Chap. 11 they are used as a point of departure for deriving many important property relations for pure, simple compressible systems, including means for constructing the property tables giving u, h, and s.

The *T dS* equations are developed by considering a pure, simple compressible system undergoing an internally reversible process. In the absence of overall system motion and the effects of gravity, an energy balance in differential form is

$$(\delta Q)_{\substack{\text{int}\\\text{rev}}} = dU + (\delta W)_{\substack{\text{int}\\\text{rev}}} \tag{6.6}$$

By definition of simple compressible system (Sec. 3.1), the work is

$$(\delta W)_{\substack{\text{int}\\\text{rev}}} = p\,dV \tag{6.7a}$$

On rearrangement of Eq. 6.2b, the heat transfer is

$$(\delta Q)_{\substack{\text{int}\\\text{rev}}} = T\,dS \tag{6.7b}$$

Substituting Eq. 6.7 into Eq. 6.6, the ***first T dS equation*** results

$$T\,dS = dU + p\,dV \tag{6.8}$$

first T dS equation

The *second T dS* equation is obtained from Eq. 6.8 using $H = U + pV$. Forming the differential

$$dH = dU + d(pV) = dU + p\,dV + V\,dp$$

On rearrangement

$$dU + p\,dV = dH - V\,dp$$

Substituting this into Eq. 6.8 gives the ***second T dS equation***

$$T\,dS = dH - V\,dp \tag{6.9}$$

second T dS equation

The *T dS* equations can be written on a unit mass basis as

$$T\,ds = du + p\,dv \tag{6.10a}$$

$$T\,ds = dh - v\,dp \tag{6.10b}$$

or on a per mole basis as

$$T\,d\bar{s} = d\bar{u} + p\,d\bar{v} \tag{6.11a}$$

$$T\,d\bar{s} = d\bar{h} - \bar{v}\,dp \tag{6.11b}$$

Although the *T dS* equations are derived by considering an internally reversible process, an entropy change obtained by integrating these equations is the change for *any* process, reversible or irreversible, between two equilibrium states of a system. Because entropy is a property, the change in entropy between two states is independent of the details of the process linking the states.

To show the use of the *T dS* equations, consider a change in phase from saturated liquid to saturated vapor at constant temperature and pressure. Since pressure is constant, Eq. 6.10b reduces to give

$$ds = \frac{dh}{T}$$

Then, because temperature is also constant during the phase change

$$s_g - s_f = \frac{h_g - h_f}{T} \tag{6.12}$$

This relationship shows how $s_g - s_f$ is calculated for tabulation in property tables.

➡ **FOR EXAMPLE...** consider Refrigerant 134a at 0°C. From Table A-10, $h_g - h_f =$ 197.21 kJ/kg, so with Eq. 6.12

$$s_g - s_f = \frac{197.21\,\text{kJ/kg}}{273.15\,\text{K}} = 0.7220\frac{\text{kJ}}{\text{kg}\cdot\text{K}}$$

which is the value calculated using s_f and s_g from the table. To give a similar example in English units, consider Refrigerant 134a at 0°F. From Table A-10E, $h_g - h_f =$ 90.12 Btu/lb, so

$$s_g - s_f = \frac{90.12\,\text{Btu/lb}}{459.67°\text{R}} = 0.1961\frac{\text{Btu}}{\text{lb}\cdot°\text{R}}$$

which agrees with the value calculated using s_f and s_g from the table. ⬅

6.4 Entropy Change of an Incompressible Substance

In this section, Eq. 6.10a of Sec. 6.3 is used to evaluate the entropy change between two states of an incompressible substance. The incompressible substance model introduced in Sec. 3.10.2 assumes that the specific volume (density) is constant and the specific internal energy depends solely on temperature. Thus, $du = c(T)dT$, where c denotes the specific heat of the substance, and Eq. 6.10a reduces to give

$$ds = \frac{c(T)dT}{T} + \frac{p\cancel{dv}^{0}}{T} = \frac{c(T)dT}{T}$$

On integration, the change in specific entropy is

$$s_2 - s_1 = \int_{T_1}^{T_2} \frac{c(T)}{T}dT$$

When the specific heat is constant, this becomes

$$s_2 - s_1 = c\ln\frac{T_2}{T_1} \quad \text{(incompressible, constant } c\text{)} \tag{6.13}$$

Equation 6.13, along with Eqs. 3.20 giving Δu and Δh, respectively, are applicable to liquids and solids modeled as incompressible. Specific heats of some common liquids and solids are given in Table A-19.

➡ **FOR EXAMPLE...** consider a system consisting of liquid water initially at $T_1 = 300$ K, $p_1 = 2$ bar undergoing a process to a final state at $T_2 = 323$ K, $p_2 = 1$ bar. There are two ways to evaluate the change in specific entropy in this case. The first approach is to use Eq. 6.5 together with saturated liquid data from Table A-2. That is, $s_1 \approx s_f(T_1) = 0.3954$ kJ/kg · K and $s_2 \approx s_f(T_2) = 0.7038$ kJ/kg · K, giving $s_2 - s_1 = 0.308$ kJ/kg · K. The second approach is to use the incompressible model. That is, with Eq. 6.13 and $c = 4.18$ kJ/kg · K from Table A-19, we get

$$s_2 - s_1 = c \ln\frac{T_2}{T_1}$$

$$= \left(4.18 \frac{\text{kJ}}{\text{kg} \cdot \text{K}}\right) \ln\left(\frac{323 \text{ K}}{300 \text{ K}}\right) = 0.309 \text{ kJ/kg} \cdot \text{K}$$

Comparing the values obtained for the change in specific entropy using the two approaches considered here, we see they are in agreement. ◀

6.5 Entropy Change of an Ideal Gas

In this section, the $T\,dS$ equation of Sec. 6.3, Eq. 6.10, is used to evaluate the entropy change between two states of an ideal gas. For a quick review of ideal gas model relations, see Table 6.1.

It is convenient to begin with Eq. 6.10 expressed as

$$ds = \frac{du}{T} + \frac{p}{T} dv \tag{6.14}$$

$$ds = \frac{dh}{T} - \frac{v}{T} dp \tag{6.15}$$

For an ideal gas, $du = c_v(T)\,dT$, $dh = c_p(T)\,dT$, and $pv = RT$. With these relations, Eqs. 6.14 and 6.15 become, respectively

$$ds = c_v(T)\frac{dT}{T} + R\frac{dv}{v} \quad \text{and} \quad ds = c_p(T)\frac{dT}{T} - R\frac{dp}{p} \tag{6.16}$$

On integration, Eq. 6.16 gives

$$s(T_2, v_2) - s(T_1, v_1) = \int_{T_1}^{T_2} c_v(T)\frac{dT}{T} + R \ln\frac{v_2}{v_1} \tag{6.17}$$

$$s(T_2, p_2) - s(T_1, p_1) = \int_{T_1}^{T_2} c_p(T)\frac{dT}{T} - R \ln\frac{p_2}{p_1} \tag{6.18}$$

Since R is a constant, the last terms of Eq. 6.16 can be integrated directly. However, because c_v and c_p are functions of temperature for ideal gases, it is necessary to have information about the functional relationships before the integration of the first term in these equations can be performed. Since the two specific heats are related by

$$c_p(T) = c_v(T) + R \tag{3.44}$$

where R is the gas constant, knowledge of either specific function suffices.

6.5.1 Using Ideal Gas Tables

As for internal energy and enthalpy changes, the evaluation of entropy changes for ideal gases can be reduced to a convenient tabular approach. We begin by introducing a new variable $s°(T)$ as

$$s°(T) = \int_{T'}^{T} \frac{c_p(T)}{T} dT \tag{6.19}$$

where T' is an arbitrary reference temperature.

Table 6.1

Ideal Gas Model Review

Equations of State:

$$pv = RT \tag{3.32}$$
$$pV = mRT \tag{3.33}$$

Changes in u and h:

$$u(T_2) - u(T_1) = \int_{T_1}^{T_2} c_v(T)dT \tag{3.40}$$

$$h(T_2) - h(T_1) = \int_{T_1}^{T_2} c_p(T)dT \tag{3.43}$$

Constant Specific Heats	Variable Specific Heats
$u(T_2) - u(T_1) = c_v(T_2 - T_1)$ (3.50) $h(T_2) - h(T_1) = c_p(T_2 - T_1)$ (3.51) See Tables A-20, 21 for c_v and c_p data.	$u(T)$ and $h(T)$ are evaluated from Table A-22 for air (mass basis) and Table A-23 for several other gases (molar basis).

The integral of Eq. 6.18 can be expressed in terms of $s°$ as follows

$$\int_{T_1}^{T_2} c_p \frac{dT}{T} = \int_{T'}^{T_2} c_p \frac{dT}{T} - \int_{T'}^{T_1} c_p \frac{dT}{T}$$
$$= s°(T_2) - s°(T_1)$$

Thus, Eq. 6.18 can be written as

$$s(T_2, p_2) - s(T_1, p_1) = s°(T_2) - s°(T_1) - R \ln \frac{p_2}{p_1} \tag{6.20a}$$

or on a per mole basis as

$$\bar{s}(T_2, p_2) - \bar{s}(T_1, p_1) = \bar{s}°(T_2) - \bar{s}°(T_1) - \bar{R} \ln \frac{p_2}{p_1} \tag{6.20b}$$

Because $s°$ depends only on temperature, it can be tabulated versus temperature, like h and u. For air as an ideal gas, $s°$ with units of kJ/kmol · K is given in Table A-22. Values of $\bar{s}°$ for several other common gases are given in Table A-23 with units of kJ/kmol · K. When the reference temperature T' of Eq. 6.19 is selected as 0 K, the variable $s°$ is identified as the *absolute* entropy. See Sec. 13.5 for further discussion.

Using Eq. 6.20 and the tabulated values for $s°$ or $\bar{s}°$, as appropriate, entropy changes can be determined that account explicitly for the variation of specific heat with temperature.

➡ **FOR EXAMPLE...** let us evaluate the change in specific entropy, in kJ/kg · K, of air modeled as an ideal gas from a state where $T_1 = 300$ K and $p_1 = 1$ bar to a state where $T_2 = 1000$ K and $p_2 = 3$ bar. Using Eq. 6.20a and data from Table A-22

$$s_2 - s_1 = s°(T_2) - s°(T_1) - R \ln \frac{p_2}{p_1}$$
$$= (2.96770 - 1.70203) \frac{\text{kJ}}{\text{kg} \cdot \text{K}} - \frac{8.314}{28.97} \frac{\text{kJ}}{\text{kg} \cdot \text{K}} \ln \frac{3\,\text{bar}}{1\,\text{bar}}$$
$$= 0.9504 \text{ kJ/kg} \cdot \text{K}$$ ⬅

If a table giving $s°$ (or $\bar{s}°$) is not available for a particular gas of interest, the integrals of Eqs. 6.17 and 6.18 can be performed analytically or numerically using specific heat data such as provided in Tables A-20 and A-21.

6.5.2 Assuming Constant Specific Heats

When the specific heats c_v and c_p are taken as constants, Eqs. 6.17 and 6.18 reduce, respectively, to

$$s(T_2, v_2) - s(T_1, v_1) = c_v \ln\frac{T_2}{T_1} + R \ln\frac{v_2}{v_1} \tag{6.21}$$

$$s(T_2, p_2) - s(T_1, p_1) = c_p \ln\frac{T_2}{T_1} - R \ln\frac{p_2}{p_1} \tag{6.22}$$

These equations, along with Eqs. 3.50 and 3.51 giving Δu and Δh, respectively, are applicable when assuming the ideal gas model with constant specific heats.

➡ **FOR EXAMPLE...** let us determine the change in specific entropy, in kJ/kg · K, of air as an ideal gas undergoing a process from $T_1 = 300$ K, $p_1 = 1$ bar to $T_2 = 400$ K, $p_2 = 5$ bar. Because of the relatively small temperature range, we assume a constant value of c_p evaluated at 350 K. Using Eq. 6.22 and $c_p = 1.008$ kJ/kg · K from Table A-20

$$\Delta s = c_p \ln\frac{T_2}{T_1} - R \ln\frac{p_2}{p_1}$$

$$= \left(1.008 \frac{\text{kJ}}{\text{kg} \cdot \text{K}}\right) \ln\left(\frac{400\text{ K}}{300\text{ K}}\right) - \left(\frac{8.314}{28.97} \frac{\text{kJ}}{\text{kg} \cdot \text{K}}\right) \ln\left(\frac{5\text{ bar}}{1\text{ bar}}\right)$$

$$= -0.1719 \text{ kJ/kg} \cdot \text{K}$$ ⬅

6.5.3 Computer Retrieval

For gases modeled as ideal gases, *IT directly* returns $s(T, p)$ using the following special form of Eq. 6.18:

$$s(T, p) - s(T_{\text{ref}}, p_{\text{ref}}) = \int_{T_{\text{ref}}}^{T} \frac{c_p(T)}{T} dT - R \ln\frac{p}{p_{\text{ref}}}$$

and the following choice of reference state and reference value: $T_{\text{ref}} = 0$ K, $p_{\text{ref}} = 1$ atm, and $s(T_{\text{ref}}, p_{\text{ref}}) = 0$, giving

$$s(T, p) = \int_{0}^{T} \frac{c_p(T)}{T} dT - R \ln\frac{p}{p_{\text{ref}}}$$

Such reference state and reference value choices equip *IT* for use in combustion applications; see Sec. 13.5.1 for further discussion.

Changes in specific entropy evaluated using *IT* agree with entropy *changes* evaluated using ideal gas tables. ➡ **FOR EXAMPLE...** consider a process of air as an ideal gas from $T_1 = 300$ K, $p_1 = 1$ bar to $T_2 = 1000$ K, $p_2 = 3$ bar. The change in specific entropy, denoted as dels, is determined in SI units using *IT* as follows:

```
p1 = 1//bar
T1 = 300//K
p2 = 3
T2 = 1000
s1 = s_TP("Air",T1,p1)
s2 = s_TP("Air",T2,p2)
dels = s2 − s1
```

The software returns values of $s_1 = 1.706$, $s_2 = 2.656$, and dels = 0.9501, all in units of kJ/kg · K. This value for Δs agrees with the value obtained using Table A-22: 0.9504 kJ/kg · K, as shown in the concluding example of Sec. 6.5.1. ◀ Note that *IT* returns specific entropy directly and does not use the special function $s°$.

6.6 Entropy Change in Internally Reversible Processes of Closed Systems

In this section the relationship between entropy change and heat transfer for internally reversible processes is considered. The concepts introduced have important applications in subsequent sections of the book. The present discussion is limited to the case of closed systems. Similar considerations for control volumes are presented in Sec. 6.13.

As a closed system undergoes an internally reversible process, its entropy can increase, decrease, or remain constant. This can be brought out using

$$dS = \left(\frac{\delta Q}{T}\right)_{\substack{\text{int}\\\text{rev}}} \tag{6.2b}$$

entropy transfer

isentropic process

which indicates that when a closed system undergoing an internally reversible process receives energy by heat transfer, the system experiences an increase in entropy. Conversely, when energy is removed from the system by heat transfer, the entropy of the system decreases. This can be interpreted to mean that an ***entropy transfer** accompanies* heat transfer. The direction of the entropy transfer is the same as that of the heat transfer. In an *adiabatic* internally reversible process, entropy remains constant. A constant-entropy process is called an ***isentropic process.***

6.6.1 Area Representation of Heat Transfer

On rearrangement, Eq. 6.2b gives

$$(\delta Q)_{\substack{\text{int}\\\text{rev}}} = T\,dS$$

Integrating from an initial state 1 to a final state 2

$$Q_{\substack{\text{int}\\\text{rev}}} = \int_1^2 T\,dS \tag{6.23}$$

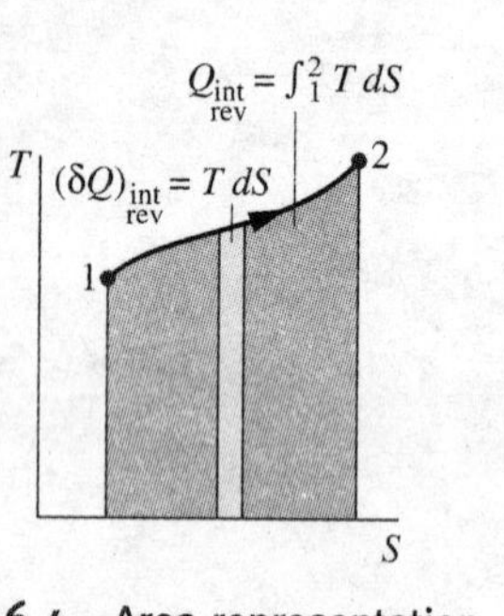

Fig. 6.4 Area representation of heat transfer for an internally reversible process of a closed system.

From Eq. 6.23 it can be concluded that an energy transfer by heat to a closed system during an internally reversible process can be represented as an area on a temperature–entropy diagram. Figure 6.4 illustrates the area representation of heat transfer for an arbitrary internally reversible process in which temperature varies. Carefully note that temperature must be in kelvins, and the area is the entire area under the curve (shown shaded). Also note that the area representation of heat transfer is not valid for irreversible processes, as will be demonstrated later.

6.6.2 Carnot Cycle Application

Carnot cycle

To provide an example illustrating both the entropy change that accompanies heat transfer and the area representation of heat transfer, consider Fig. 6.5*a*, which shows a ***Carnot power cycle*** (Sec. 5.10). The cycle consists of four internally reversible processes in series: two isothermal processes alternated with two adiabatic processes. In Process 2–3, heat transfer to the system occurs while the temperature of the system remains

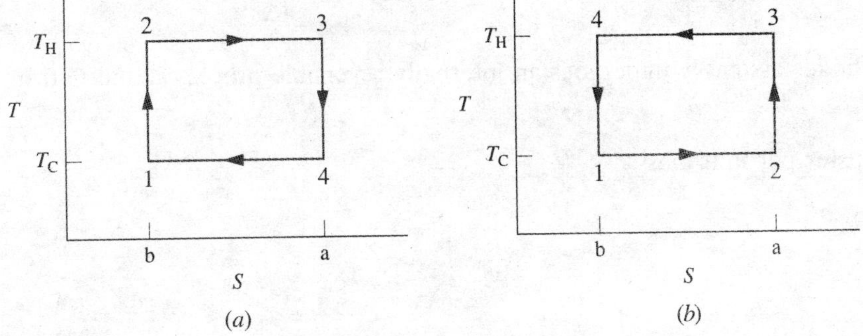

Fig. 6.5 Carnot cycles on the temperature–entropy diagram. (*a*) Power cycle. (*b*) Refrigeration or heat pump cycle.

constant at T_H. The system entropy increases due to the accompanying entropy transfer. For this process, Eq. 6.23 gives $Q_{23} = T_H(S_3 - S_2)$, so area 2–3–a–b–2 on Fig. 6.5*a* represents the heat transfer during the process. Process 3–4 is an adiabatic and internally reversible process and thus is an isentropic (constant-entropy) process. Process 4–1 is an isothermal process at T_C during which heat is transferred *from* the system. Since entropy transfer accompanies the heat transfer, system entropy decreases. For this process, Eq. 6.23 gives $Q_{41} = T_C(S_1 - S_4)$, which is negative in value. Area 4–1–b–a–4 on Fig. 6.5*a* represents the *magnitude* of the heat transfer Q_{41}. Process 1–2, which completes the cycle, is adiabatic and internally reversible (isentropic).

The net work of any cycle is equal to the net heat transfer, so *enclosed* area 1–2–3–4–1 represents the net work of the cycle. The thermal efficiency of the cycle also can be expressed in terms of areas:

$$\eta = \frac{W_{\text{cycle}}}{Q_{23}} = \frac{\text{area 1–2–3–4–1}}{\text{area 2–3–a–b–2}}$$

The numerator of this expression is $(T_H - T_C)(S_3 - S_2)$ and the denominator is $T_H(S_3 - S_2)$, so the thermal efficiency can be given in terms of temperatures only as $\eta = 1 - T_C/T_H$. If the cycle were executed as shown in Fig. 6.5*b*, the result would be a Carnot refrigeration or heat pump cycle. In such a cycle, heat is transferred to the system while its temperature remains at T_C, so entropy increases during Process 1–2. In Process 3–4 heat is transferred from the system while the temperature remains constant at T_H and entropy decreases.

6.6.3 Work and Heat Transfer in an Internally Reversible Process of Water

To further illustrate concepts introduced in this section, Example 6.1 considers water undergoing an internally reversible process while contained in a piston–cylinder assembly.

Example 6.1 INTERNALLY REVERSIBLE PROCESS OF WATER

Water, initially a saturated liquid at 150°C (423.15 K), is contained in a piston–cylinder assembly. The water undergoes a process to the corresponding saturated vapor state, during which the piston moves freely in the cylinder. If the change of state is brought about by heating the water as it undergoes an internally reversible process at constant pressure and temperature, determine the work and heat transfer per unit of mass, each in kJ/kg.

Solution

Known: Water contained in a piston–cylinder assembly undergoes an internally reversible process at 150°C from saturated liquid to saturated vapor.

Find: Determine the work and heat transfer per unit mass.

Schematic and Given Data:

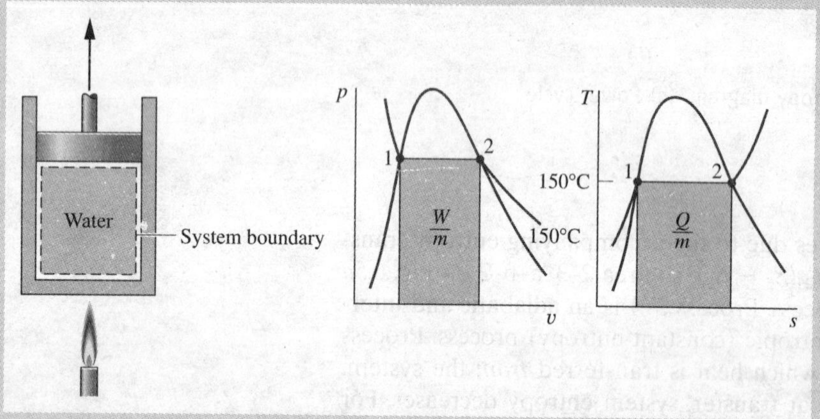

Fig. E6.1

Engineering Model:

1. The water in the piston–cylinder assembly is a closed system.
2. The process is internally reversible.
3. Temperature and pressure are constant during the process.
4. There is no change in kinetic or potential energy between the two end states.

Analysis: At constant pressure the work is

$$\frac{W}{m} = \int_1^2 p\, dv = p(v_2 - v_1)$$

With values from Table A-2 at 150°C

$$\frac{W}{m} = (4.758\,\text{bar})(0.3928 - 1.0905 \times 10^{-3})\left(\frac{\text{m}^3}{\text{kg}}\right)\left|\frac{10^5\,\text{N/m}^2}{1\,\text{bar}}\right|\left|\frac{1\,\text{kJ}}{10^3\,\text{N}\cdot\text{m}}\right|$$

$$= 186.38\,\text{kJ/kg}$$

Since the process is internally reversible and at constant temperature, Eq. 6.23 gives

$$Q = \int_1^2 T\, dS = m\int_1^2 T\, ds$$

or

$$\frac{Q}{m} = T(s_2 - s_1)$$

With values from Table A-2

❶ $$\frac{Q}{m} = (423.15\,\text{K})(6.8379 - 1.8418)\,\text{kJ/kg}\cdot\text{K} = 2114.1\,\text{kJ/kg}$$

As shown in the accompanying figure, the work and heat transfer can be represented as areas on p–v and T–s diagrams, respectively.

❶ The heat transfer can be evaluated alternatively from an energy balance written on a unit mass basis as

$$u_2 - u_1 = \frac{Q}{m} - \frac{W}{m}$$

Introducing $W/m = p(v_2 - v_1)$ and solving

$$\begin{aligned}\frac{Q}{m} &= (u_2 - u_1) + p(v_2 - v_1)\\ &= (u_2 + pv_2) - (u_1 + pv_1)\\ &= h_2 - h_1\end{aligned}$$

From Table A-2 at 150°C, $h_2 - h_1 = 2114.3$ kJ/kg, which agrees with the value for Q/m obtained in the solution.

Skills Developed: *Ability to...*
- ✓ evaluate work and heat transfer for an internally reversible process, and represent them as areas on p–v and T–s diagrams, respectively.
- ✓ retrieve entropy data for water.

Quick Quiz

If the initial and final states were saturation states at 100°C (373.15 K), determine the work and heat transfer per unit of mass, each in kJ/kg.
Ans. 170 kJ/kg, 2257 kJ/kg

6.7 Entropy Balance for Closed Systems

In this section, we begin our study of the ***entropy balance***. The entropy balance is an expression of the second law that is particularly effective for thermodynamic analysis. The current presentation is limited to closed systems. The entropy balance is extended to control volumes in Sec. 6.9.

entropy balance

Just as mass and energy are accounted for by mass and energy balances, respectively, entropy is accounted for by an entropy balance. In Eq. 5.2, the entropy balance is introduced in words as

$$\begin{bmatrix}\textit{change}\text{ in the amount of}\\ \text{entropy contained within}\\ \text{the system during some}\\ \text{time interval}\end{bmatrix} = \begin{bmatrix}\text{net amount of}\\ \text{entropy }\textit{transferred in}\\ \text{across the system}\\ \text{boundary during the}\\ \text{time interval}\end{bmatrix} + \begin{bmatrix}\text{amount of entropy }\textit{produced}\\ \textit{within}\text{ the system}\\ \text{during the time interval}\end{bmatrix}$$

In symbols, the ***closed system entropy balance*** takes the form

$$\underbrace{S_2 - S_1}_{\text{entropy change}} = \underbrace{\int_1^2 \left(\frac{\delta Q}{T}\right)_{\text{b}}}_{\text{entropy transfer}} + \underbrace{\sigma}_{\text{entropy production}} \tag{6.24}$$

closed system entropy balance

where subscript b signals that the integrand is evaluated at the system boundary. For the development of Eq. 6.24, see the box.

It is sometimes convenient to use the entropy balance expressed in differential form

$$dS = \left(\frac{\delta Q}{T}\right)_{\text{b}} + \delta\sigma \tag{6.25}$$

Note that the differentials of the nonproperties Q and σ are shown, respectively, as δQ and $\delta\sigma$. When there are no internal irreversibilities, $\delta\sigma$ vanishes and Eq. 6.25 reduces to Eq. 6.2b.

Developing the Closed System Entropy Balance

The entropy balance for closed systems can be developed using the *Clausius inequality* expressed by Eq. 5.13 (Sec. 5.11) and the defining equation for entropy change, Eq. 6.2a, as follows:

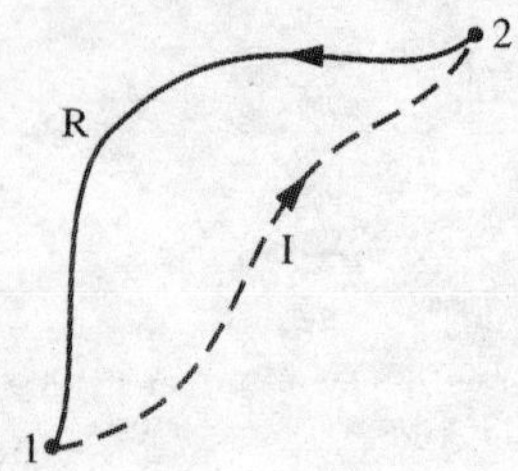

Fig. 6.6 Cycle used to develop the entropy balance.

Shown in Fig. 6.6 is a cycle executed by a closed system. The cycle consists of process I, during which internal irreversibilities are present, followed by internally reversible process R. For this cycle, Eq. 5.13 takes the form

$$\int_1^2 \left(\frac{\delta Q}{T}\right)_b + \int_2^1 \left(\frac{\delta Q}{T}\right)_{\substack{\text{int}\\ \text{rev}}} = -\sigma \tag{a}$$

where the first integral is for process I and the second is for process R. The subscript b in the first integral serves as a reminder that the integrand is evaluated at the system boundary. The subscript is not required in the second integral because temperature is uniform throughout the system at each intermediate state of an internally reversible process. Since no irreversibilities are associated with process R, the term σ_{cycle} of Eq. 5.13, which accounts for the effect of irreversibilities during the cycle, refers only to process I and is shown in Eq. (a) simply as σ.

Applying the definition of entropy change, Eq. 6.2a, we can express the second integral of Eq. (a) as

$$\int_2^1 \left(\frac{\delta Q}{T}\right)_{\substack{\text{int}\\ \text{rev}}} = S_1 - S_2 \tag{b}$$

With this, Eq. (a) becomes

$$\int_1^2 \left(\frac{\delta Q}{T}\right)_b + (S_1 - S_2) = -\sigma \tag{c}$$

On rearrangement, Eq. (c) gives Eq. 6.24, the closed system entropy balance.

In each of its alternative forms the entropy balance can be regarded as a statement of the second law of thermodynamics. For the analysis of engineering systems, the entropy balance is a more effective means for applying the second law than the Clausius and Kelvin-Planck statements given in Chap. 5.

6.7.1 Interpreting the Closed System Entropy Balance

If the end states are fixed, the entropy change on the left side of Eq. 6.24 can be evaluated independently of the details of the process. However, the two terms on the right side depend explicitly on the nature of the process and cannot be determined solely from knowledge of the end states. The first term on the right side of Eq. 6.24 is associated with heat transfer to or from the system during the process. This term can be interpreted as the ***entropy transfer accompanying heat transfer.*** The direction of entropy transfer is the same as the direction of the heat transfer, and the same sign convention applies as for heat transfer: A positive value means that entropy is

entropy transfer accompanying heat transfer

transferred into the system, and a negative value means that entropy is transferred out. When there is no heat transfer, there is no entropy transfer.

The entropy change of a system is not accounted for solely by the entropy transfer, but is due in part to the second term on the right side of Eq. 6.24 denoted by σ. The term σ is positive when internal irreversibilities are present during the process and vanishes when no internal irreversibilities are present. This can be described by saying that ***entropy is produced*** (or *generated*) within the system by the action of irreversibilities. The second law of thermodynamics can be interpreted as requiring that entropy is produced by irreversibilities and conserved only in the limit as irreversibilities are reduced to zero. Since σ measures the effect of irreversibilities present within the system during a process, its value depends on the nature of the process and not solely on the end states. Entropy production is *not* a property.

entropy production

When applying the entropy balance to a closed system, it is essential to remember the requirements imposed by the second law on entropy production: The second law requires that entropy *production* be positive, or zero, in value

$$\sigma: \begin{cases} > 0 & \text{irreversibilities present within the system} \\ = 0 & \text{no irreversibilities present within the system} \end{cases} \tag{6.26}$$

The value of the entropy production cannot be negative. In contrast, the *change* in entropy of the system may be positive, negative, or zero:

$$S_2 - S_1: \begin{cases} > 0 \\ = 0 \\ < 0 \end{cases} \tag{6.27}$$

Like other properties, entropy change for a process between two specified states can be determined without knowledge of the details of the process.

6.7.2 Evaluating Entropy Production and Transfer

The objective in many applications of the entropy balance is to evaluate the entropy production term. However, the value of the entropy production for a given process of a system often does not have much significance by itself. The significance is normally determined through comparison. For example, the entropy production within a given component might be compared to the entropy production values of the other components included in an overall system formed by these components. By comparing entropy production values, the components where appreciable irreversibilities occur can be identified and rank ordered. This allows attention to be focused on the components that contribute most to inefficient operation of the overall system.

To evaluate the entropy transfer term of the entropy balance requires information regarding both the heat transfer and the temperature on the boundary where the heat transfer occurs. The entropy transfer term is not always subject to direct evaluation, however, because the required information is either unknown or not defined, such as when the system passes through states sufficiently far from equilibrium. In such applications, it may be convenient, therefore, to enlarge the system to include enough of the immediate surroundings that the temperature on the boundary of the *enlarged system* corresponds to the temperature of the surroundings away from the immediate vicinity of the system, T_f. The entropy transfer term is then simply Q/T_f. However, as the irreversibilities present would not be just for the system of interest but for the enlarged system, the entropy production term would account for the effects of internal irreversibilities within the original system and external irreversibilities present within that portion of the surroundings included within the enlarged system.

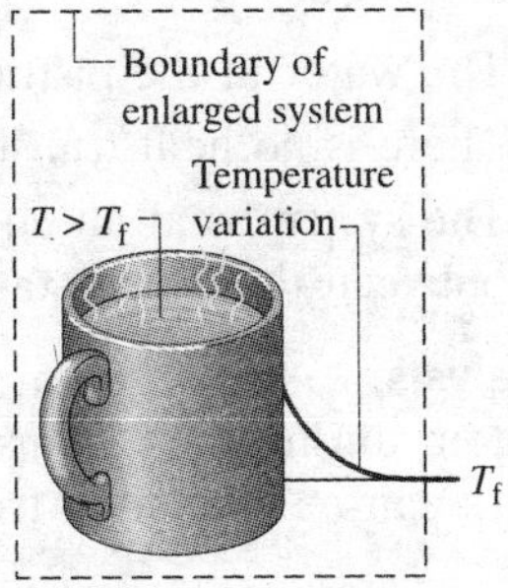

6.7.3 Applications of the Closed System Entropy Balance

The following examples illustrate the use of the energy and entropy balances for the analysis of closed systems. Property relations and property diagrams also contribute significantly in developing solutions. Example 6.2 reconsiders the system and end states of Example 6.1 to demonstrate that entropy is produced when internal irreversibilities are present and that the amount of entropy production is not a property. In Example 6.3, the entropy balance is used to determine the minimum theoretical compression work.

Example 6.2 IRREVERSIBLE PROCESS OF WATER

Water, initially a saturated liquid at 150°C, is contained within a piston–cylinder assembly. The water undergoes a process to the corresponding saturated vapor state, during which the piston moves freely in the cylinder. There is no heat transfer with the surroundings. If the change of state is brought about by the action of a paddle wheel, determine the net work per unit mass, in kJ/kg, and the amount of entropy produced per unit mass, in kJ/kg · K.

Solution

Known: Water contained in a piston–cylinder assembly undergoes an adiabatic process from saturated liquid to saturated vapor at 150°C. During the process, the piston moves freely, and the water is rapidly stirred by a paddle wheel.

Find: Determine the net work per unit mass and the entropy produced per unit mass.

Schematic and Given Data:

❶

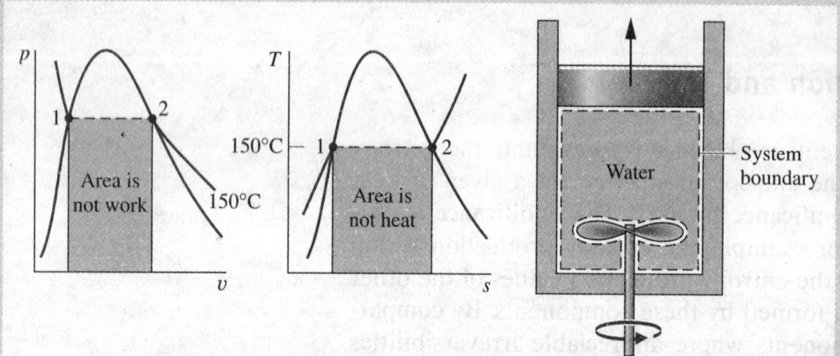

Fig. E6.2

Engineering Model:

1. The water in the piston–cylinder assembly is a closed system.
2. There is no heat transfer with the surroundings.
3. The system is at an equilibrium state initially and finally. There is no change in kinetic or potential energy between these two states.

Analysis: As the volume of the system increases during the process, there is an energy transfer by work from the system during the expansion, as well as an energy transfer by work to the system via the paddle wheel. The *net* work can be evaluated from an energy balance, which reduces with assumptions 2 and 3 to

$$\Delta U + \cancelto{0}{\Delta KE} + \cancelto{0}{\Delta PE} = \cancelto{0}{Q} - W$$

On a unit mass basis, the energy balance is then

$$\frac{W}{m} = -(u_2 - u_1)$$

With specific internal energy values from Table A-2 at 150°C, $u_1 = 631.68$ kJ/kg, $u_2 = 2559.5$ kJ/kg, we get

$$\frac{W}{m} = -1927.82 \frac{\text{kJ}}{\text{kg}}$$

The minus sign indicates that the work input by stirring is greater in magnitude than the work done by the water as it expands.

The amount of entropy produced is evaluated by applying the entropy balance Eq. 6.24. Since there is no heat transfer, the term accounting for entropy transfer vanishes

$$\Delta S = \cancel{\int_1^2 \left(\frac{\delta Q}{T}\right)_b}^{0} + \sigma$$

On a unit mass basis, this becomes on rearrangement

$$\frac{\sigma}{m} = s_2 - s_1$$

With specific entropy values from Table A-2 at 150°C, $s_1 = 1.8418$ kJ/kg · K, $s_2 = 6.8379$ kJ/kg · K, we get

❷ $$\frac{\sigma}{m} = 4.9961 \frac{\text{kJ}}{\text{kg} \cdot \text{K}}$$

---

❶ Although each end state is an equilibrium state at the same pressure and temperature, the pressure and temperature are not necessarily uniform throughout the system at *intervening* states, nor are they necessarily constant in value during the process. Accordingly, there is no well-defined "path" for the process. This is emphasized by the use of dashed lines to represent the process on these *p–v* and *T–s* diagrams. The dashed lines indicate only that a process has taken place, and no "area" should be associated with them. In particular, note that the process is adiabatic, so the "area" below the dashed line on the *T–s* diagram can have no significance as heat transfer. Similarly, the work cannot be associated with an area on the *p–v* diagram.

❷ The change of state is the same in the present example as in Example 6.1. However, in Example 6.1 the change of state is brought about by heat transfer while the system undergoes an internally reversible process. Accordingly, the value of entropy production for the process of Example 6.1 is zero. Here, fluid friction is present during the process and the entropy production is positive in value. Accordingly, different values of entropy production are obtained for two processes between the *same* end states. This demonstrates that entropy production is not a property.

Skills Developed: *Ability to...*
- ✓ apply the closed system energy and entropy balances.
- ✓ retrieve property data for water.

Quick Quiz

If the initial and final states were saturation states at 100°C, determine the net work, in kJ/kg, and the amount of entropy produced, in kJ/kg · K.

Ans. −2087.56 kJ/kg, 6.048 kJ/kg · K

As an illustration of second law reasoning, minimum theoretical compression work is evaluated in Example 6.3 using the fact that the entropy production term of the entropy balance cannot be negative.

Example 6.3 EVALUATING MINIMUM THEORETICAL COMPRESSION WORK

Refrigerant 134a is compressed adiabatically in a piston–cylinder assembly from saturated vapor at 0°C to a final pressure of 0.7 MPa. Determine the minimum theoretical work input required per unit mass of refrigerant, in kJ/kg.

Solution

Known: Refrigerant 134a is compressed without heat transfer from a specified initial state to a specified final pressure.

Find: Determine the minimum theoretical work input required per unit of mass.

Schematic and Given Data:

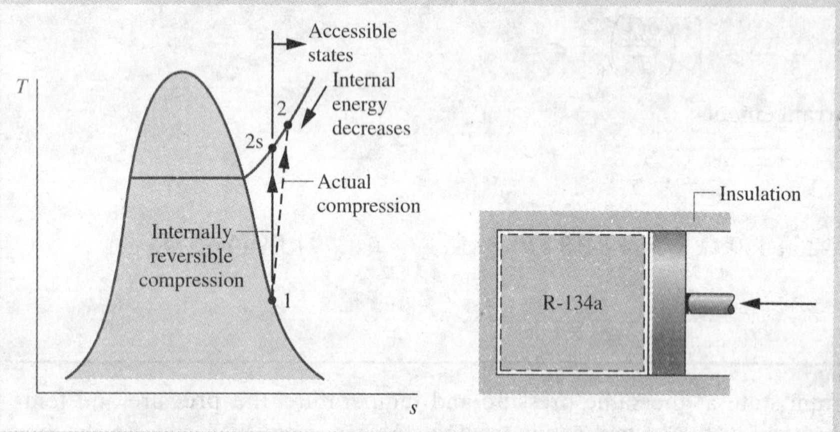

Fig. E6.3

Engineering Model:

1. The Refrigerant 134a is a closed system.
2. There is no heat transfer with the surroundings.
3. The initial and final states are equilibrium states. There is no change in kinetic or potential energy between these states.

Analysis: An expression for the work can be obtained from an energy balance. By applying assumptions 2 and 3, we get

$$\Delta U + \cancelto{0}{\Delta KE} + \cancelto{0}{\Delta PE} = \cancelto{0}{Q} - W$$

When written on a unit mass basis, the work *input* is then

$$\left(-\frac{W}{m}\right) = u_2 - u_1$$

The specific internal energy u_1 can be obtained from Table A-10E as $u_1 = 227.06$ kJ/kg. Since u_1 is known, the value for the work input depends on the specific internal energy u_2. The minimum work input corresponds to the smallest allowed value for u_2, determined using the second law as follows.

Applying the entropy balance, Eq. 6.24, we get

$$\Delta S = \cancelto{0}{\int_1^2 \left(\frac{\delta Q}{T}\right)_b} + \sigma$$

where the entropy transfer term is set equal to zero because the process is adiabatic. Thus, the *allowed* final states must satisfy

$$s_2 - s_1 = \frac{\sigma}{m} \geq 0$$

The restriction indicated by the foregoing equation can be interpreted using the accompanying *T–s* diagram. Since σ cannot be negative, states with $s_2 < s_1$ are not accessible adiabatically. When irreversibilities are present during the compression, entropy is produced, so $s_2 > s_1$. The state labeled 2s on the diagram would be attained in the limit as irreversibilities are reduced to zero. This state corresponds to an *isentropic* compression.

By inspection of Table A-12, we see that when pressure is fixed, the specific internal energy decreases as temperature decreases. Thus, the smallest allowed value for u_2 corresponds to state 2s. Interpolating in Table A-12 at 0.7 MPa, with $s_{2s} = s_1 = 0.9190$ kJ/kg · K, we find that $u_{2s} = 244.32$ kJ/kg, which corresponds to a temperature at state 2s of about 38.3°C. Finally, the *minimum* work input is

❶ $$\left(-\frac{W}{m}\right)_{\text{min}} = u_{2s} - u_1 = 244.32 - 227.06 = 17.26 \text{ kJ/kg}$$

❶ The effect of irreversibilities exacts a penalty on the work input required: A greater work input is needed for the actual adiabatic compression process than for an internally reversible adiabatic process between the same initial state and the same final pressure.

Skills Developed: Ability to...
- ✓ apply the closed system energy and entropy balances.
- ✓ retrieve property data for Refrigerant 134a.

Quick Quiz

If the refrigerant were compressed adiabatically to a final state where $p_2 = 0.7$ MPa, $T_2 = 49$°C, determine the work input, in kJ/kg, and the amount of entropy produced, in kJ/kg · K.

Ans. 39.9 kJ/kg, 36.43 J/kg · K

6.7.4 Closed System Entropy Rate Balance

If the temperature T_b is constant, Eq. 6.24 reads

$$S_2 - S_1 = \frac{Q}{T_b} + \sigma$$

where Q/T_b represents the *amount* of entropy transferred through the portion of the boundary at temperature T_b. Similarly, the quantity $\dot{Q}/T_j$ represents the *time rate* of entropy transfer through the portion of the boundary whose instantaneous temperature is T_j. This quantity appears in the closed system entropy rate balance considered next.

On a time rate basis, the ***closed system entropy rate balance*** is

$$\frac{dS}{dt} = \sum_j \frac{\dot{Q}_j}{T_j} + \dot{\sigma} \qquad (6.28)$$

closed system entropy rate balance

where dS/dt is the time rate of change of entropy of the system. The term $\dot{Q}_j/T_j$ represents the time rate of entropy transfer through the portion of the boundary whose instantaneous temperature is T_j. The term $\dot{\sigma}$ accounts for the time rate of entropy production due to irreversibilities within the system.

To pinpoint the relative significance of the internal and external irreversibilities, Example 6.4 illustrates the application of the entropy rate balance to a system and to an enlarged system consisting of the system and a portion of its immediate surroundings.

Example 6.4 PINPOINTING IRREVERSIBILITIES

Referring to Example 2.4, evaluate the rate of entropy production $\dot{\sigma}$, in kW/K, for (*a*) the gearbox as the system and (*b*) an enlarged system consisting of the gearbox and enough of its surroundings that heat transfer occurs at the temperature of the surroundings away from the immediate vicinity of the gearbox, $T_f = 293$ K (20°C).

Solution

Known: A gearbox operates at steady state with known values for the power input through the high-speed shaft, power output through the low-speed shaft, and heat transfer rate. The temperature on the outer surface of the gearbox and the temperature of the surroundings away from the gearbox are also known.

Find: Evaluate the entropy production rate $\dot{\sigma}$ for each of the two specified systems shown in the schematic.

Schematic and Given Data:

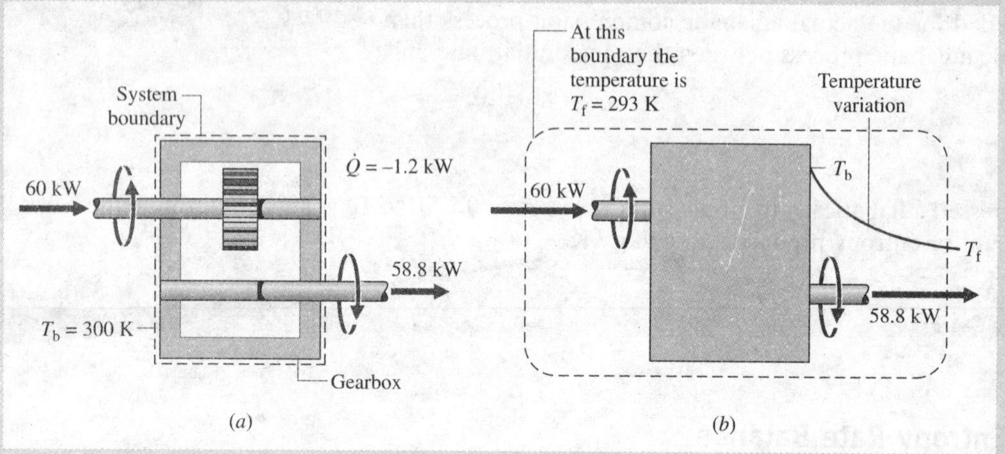

Fig. E6.4

Engineering Model:

1. In part (a), the gearbox is taken as a closed system operating at steady state, as shown on the accompanying sketch labeled with data from Example 2.4.
2. In part (b), the gearbox and a portion of its surroundings are taken as a closed system, as shown on the accompanying sketch labeled with data from Example 2.4.
3. The temperature of the outer surface of the gearbox and the temperature of the surroundings do not vary.

Analysis:

(a) To obtain an expression for the entropy production rate, begin with the entropy balance for a closed system on a time rate basis: Eq. 6.28. Since heat transfer takes place only at temperature T_b, the entropy rate balance reduces at steady state to

$$\cancelto{0}{\frac{dS}{dt}} = \frac{\dot{Q}}{T_b} + \dot{\sigma}$$

Solving

$$\dot{\sigma} = -\frac{\dot{Q}}{T_b}$$

Introducing the known values for the heat transfer rate $\dot{Q}$ and the surface temperature T_b

$$\dot{\sigma} = -\frac{(-1.2\ \text{kW})}{(300\ \text{K})} = 4 \times 10^{-3}\ \text{kW/K}$$

(b) Since heat transfer takes place at temperature T_f for the enlarged system, the entropy rate balance reduces at steady state to

$$\cancelto{0}{\frac{dS}{dt}} = \frac{\dot{Q}}{T_f} + \dot{\sigma}$$

Solving

$$\dot{\sigma} = -\frac{\dot{Q}}{T_f}$$

Introducing the known values for the heat transfer rate $\dot{Q}$ and the temperature T_f

❶ $$\dot{\sigma} = -\frac{(-1.2 \text{ kW})}{(293 \text{ K})} = 4.1 \times 10^{-3} \text{ kW/K}$$

---

❶ The value of the entropy production rate calculated in part (a) gauges the significance of irreversibilities associated with friction and heat transfer *within* the gearbox. In part (b), an additional source of irreversibility is included in the enlarged system, namely the irreversibility associated with the heat transfer from the outer surface of the gearbox at T_b to the surroundings at T_f. In this case, the irreversibilities within the gearbox are dominant, accounting for about 98% of the total rate of entropy production.

Skills Developed: *Ability to...*
- ✓ apply the closed system entropy rate balance.
- ✓ develop an engineering model.

Quick Quiz

If the power delivered were 59.32 kW, evaluate the outer surface temperature, in K, and the rate of entropy production, in kW/K, for the gearbox as the system, keeping input power, h, and A from Example 2.4 the same.
Ans. 297K, 2.3×10^{-3} kW/K

6.8 Directionality of Processes

Our study of the second law of thermodynamics began in Sec. 5.1 with a discussion of the *directionality* of processes. In this section we consider two related aspects for which there are significant applications: the increase in entropy principle and a statistical interpretation of entropy.

6.8.1 Increase of Entropy Principle

In the present discussion, we use the closed system energy and entropy balances to introduce the increase of entropy principle. Discussion centers on an enlarged system consisting of a system and that portion of the surroundings affected by the system as it undergoes a process. Since all energy and mass transfers taking place are included within the boundary of the enlarged system, the enlarged system is an *isolated* system.

An energy balance for the isolated system reduces to

$$\Delta E]_{\text{isol}} = 0 \tag{6.29a}$$

because no energy transfers take place across its boundary. Thus, the energy of the isolated system remains constant. Since energy is an extensive property, its value for the isolated system is the sum of its values for the system and surroundings, respectively, so Eq. 6.29a can be written as

$$\Delta E]_{\text{system}} + \Delta E]_{\text{surr}} = 0 \tag{6.29b}$$

In either of these forms, the conservation of energy principle places a constraint on the processes that can occur. For a process to take place, it is necessary for the energy of the system plus the surroundings to remain constant. However, not all processes for which this constraint is satisfied can actually occur. Processes also must satisfy the second law, as discussed next.

An entropy balance for the isolated system reduces to

$$\Delta S]_{\text{isol}} = \int_1^2 \left(\frac{\delta Q}{T}\right)_{\text{b}}^{0} + \sigma_{\text{isol}}$$

or

$$\Delta S]_{\text{isol}} = \sigma_{\text{isol}} \tag{6.30a}$$

where σ_{isol} is the total amount of entropy produced within the system and its surroundings. Since entropy is produced in all actual processes, the only processes that can occur are those for which the entropy of the isolated system increases. This is known as the ***increase of entropy principle.*** The increase of entropy principle is sometimes considered an alternative statement of the second law.

increase of entropy principle

Since entropy is an extensive property, its value for the isolated system is the sum of its values for the system and surroundings, respectively, so Eq. 6.30a can be written as

$$\Delta S]_{\text{system}} + \Delta S]_{\text{surr}} = \sigma_{\text{isol}} \tag{6.30b}$$

Notice that this equation does not require the entropy change to be positive for both the system and surroundings but only that the *sum* of the changes is positive. In either of these forms, the increase of entropy principle dictates the direction in which any process can proceed: the direction that causes the total entropy of the system *plus* surroundings to increase.

We observed previously the tendency of systems left to themselves to undergo processes until a condition of equilibrium is attained (Sec. 5.1). The increase of entropy principle suggests that the entropy of an isolated system increases as the state of equilibrium is approached, with the equilibrium state being attained when the entropy reaches a maximum. This interpretation is considered again in Sec. 14.1, which deals with equilibrium criteria.

Example 6.5 illustrates the increase of entropy principle.

Example 6.5 QUENCHING A HOT METAL BAR

A 0.3 kg metal bar initially at 1200 K is removed from an oven and quenched by immersing it in a closed tank containing 9 kg of water initially at 300 K. Each substance can be modeled as incompressible. An appropriate constant specific heat value for the water is $c_w = 4.2$ kJ/kg · K, and an appropriate value for the metal is $c_m = 0.42$ kJ/kg · K. Heat transfer from the tank contents can be neglected. Determine (a) the final equilibrium temperature of the metal bar and the water, in K, and (b) the amount of entropy produced, in kJ/K.

Solution

Known: A hot metal bar is quenched by immersing it in a tank containing water.

Find: Determine the final equilibrium temperature of the metal bar and the water, and the amount of entropy produced.

Schematic and Given Data:

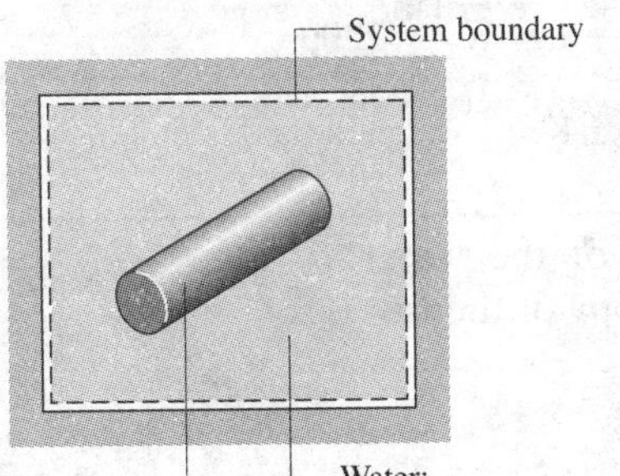

Engineering Model:

1. The metal bar and the water within the tank form a system, as shown on the accompanying sketch.
2. There is no energy transfer by heat or work: The system is isolated.
3. There is no change in kinetic or potential energy.
4. The water and metal bar are each modeled as incompressible with known specific heats.

Fig. E6.5

Analysis:

(a) The final equilibrium temperature can be evaluated from an energy balance for the isolated system

$$\Delta \overset{0}{\cancel{KE}} + \Delta \overset{0}{\cancel{PE}} + \Delta U = \overset{0}{\cancel{Q}} - \overset{0}{\cancel{W}}$$

where the indicated terms vanish by assumptions 2 and 3. Since internal energy is an extensive property, its value for the isolated system is the sum of the values for the water and metal, respectively. Thus, the energy balance becomes

$$\Delta U]_{\text{water}} + \Delta U]_{\text{metal}} = 0$$

Using Eq. 3.20a to evaluate the internal energy changes of the water and metal in terms of the constant specific heats

$$m_w c_w (T_f - T_{wi}) + m_m c_m (T_f - T_{mi}) = 0$$

where T_f is the final equilibrium temperature, and T_{wi} and T_{mi} are the initial temperatures of the water and metal, respectively. Solving for T_f and inserting values

$$T_f = \frac{m_w(c_w/c_m)T_{wi} + m_m T_{mi}}{m_w(c_w/c_m) + m_m}$$

$$= \frac{(9 \text{ kg})(10)(300 \text{ K}) + (0.3 \text{ kg})(1200 \text{ K})}{(9 \text{ kg})(10) + (0.3 \text{ kg})}$$

$$= 303 \text{ K}$$

(b) The amount of entropy production can be evaluated from an entropy balance. Since no heat transfer occurs between the isolated system and its surroundings, there is no accompanying entropy transfer, and an entropy balance for the isolated system reduces to

$$\Delta S = \overset{0}{\cancel{\int_1^2 \left(\frac{\delta Q}{T}\right)_b}} + \sigma$$

Entropy is an extensive property, so its value for the isolated system is the sum of its values for the water and the metal, respectively, and the entropy balance becomes

$$\Delta S]_{\text{water}} + \Delta S]_{\text{metal}} = \sigma$$

Evaluating the entropy changes using Eq. 6.13 for incompressible substances, the foregoing equation can be written as

$$\sigma = m_w c_w \ln \frac{T_f}{T_{wi}} + m_m c_m \ln \frac{T_f}{T_{mi}}$$

Inserting values

$$\sigma = (9\text{ kg})\left(4.2\frac{\text{kJ}}{\text{kg}\cdot\text{K}}\right)\ln\frac{303}{300} + (0.3\text{ kg})\left(0.42\frac{\text{kJ}}{\text{kg}\cdot\text{K}}\right)\ln\frac{303}{1200} \quad ❶$$

$$= \left(0.3761\frac{\text{kJ}}{\text{K}}\right) + \left(-0.1734\frac{\text{kJ}}{\text{K}}\right) = 0.2027\text{ kJ/K} \quad ❷$$

❶ The metal bar experiences a *decrease* in entropy. The entropy of the water *increases*. In accord with the increase of entropy principle, the entropy of the isolated system *increases*.

❷ The value of σ is sensitive to roundoff in the value of T_f.

Skills Developed: *Ability to...*
✓ apply the closed system energy and entropy balances.
✓ apply the incompressible substance model.

Quick Quiz

If the mass of the metal bar were 0.2 kg, determine the final equilibrium temperature, in K, and the amount of entropy produced, in kJ/K, keeping all other given data the same.
Ans. 296 K, 0.1049 kJ/K

6.8.2 Statistical Interpretation of Entropy

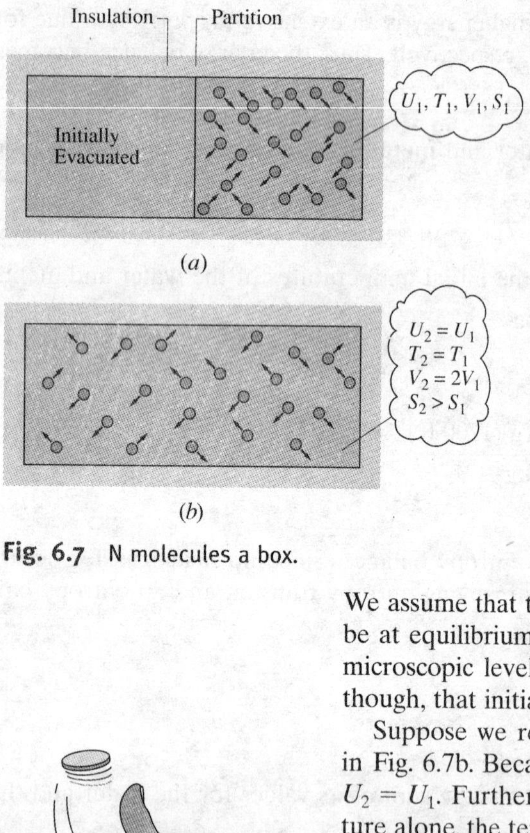

Fig. 6.7 N molecules a box.

Building on the increase of entropy principle, in this section we introduce an interpretation of entropy from a microscopic perspective based on *probability*.

In *statistical thermodynamics*, entropy is associated with the notion of microscopic *disorder*. From previous considerations we know that in a spontaneous process of an isolated system, the system moves toward equilibrium and the entropy increases. From the microscopic viewpoint, this is equivalent to saying that as an isolated system moves toward equilibrium our knowledge of the condition of individual particles making up the system decreases, which corresponds to an increase in microscopic disorder and a related increase in entropy.

We use an elementary thought experiment to bring out some basic ideas needed to understand this view of entropy. Actual microscopic analysis of systems is more complicated than the discussion given here, but the essential concepts are the same.

Consider N molecules initially contained in one half of the box shown in Fig. 6.7. The entire box is considered an isolated system. We assume that the ideal gas model applies. In the initial condition, the gas appears to be at equilibrium in terms of temperature, pressure, and other properties. But, on the microscopic level the molecules are moving about randomly. We do know *for sure*, though, that initially all molecules are on the right side of the vessel.

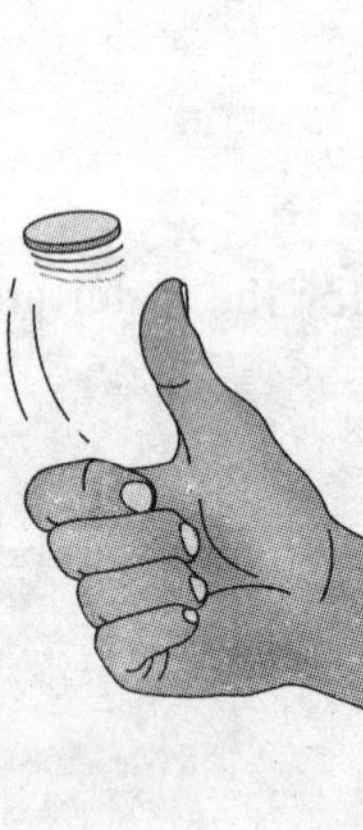

Suppose we remove the partition and wait until equilibrium is reached again, as in Fig. 6.7b. Because the system is isolated, the internal energy U does not change: $U_2 = U_1$. Further, because the internal energy of an ideal gas depends on temperature alone, the temperature is unchanged: $T_2 = T_1$. Still, at the final state a given molecule has twice the volume in which to move: $V_2 = 2V_1$. Just like a coin toss, the probability that the molecule is on one side or the other is now ½, which is the same as the volume ratio V_1/V_2. In the final condition, we have *less knowledge* about where each molecule is than we did originally.

We can evaluate the change in entropy for the process of Fig. 6.7 by applying Eq. 6.17, expressed in terms of volumes and on a molar basis. The entropy change for the constant-temperature process is

Horizons

A *New* Definition of Entropy?

Physicist Constantino Tsallis has proposed a new statistical definition of entropy some say will shake the foundations of science as we know it. The Boltzmann relation, providing the link between the microscopic interpretation of entropy and observations of macroscopic system behavior for over a century, is being called into question. Scientists are weighing in on whether the theory provides new fundamental understanding or simply amplifies our thinking.

Supporters of the new definition of entropy say it extends the scope of statistical thermodynamics to new and exciting classes of problems. They claim it fills a gap in theory first noted by Einstein and allows researchers to explain phenomena ranging from the movement of microorganisms to the motions of stars, and even swings in the stock market. Skeptics say it only provides an empirical "fudge" factor with no new fundamental understanding.

The new statistical definition reduces to the Boltzmann relation for systems at equilibrium. But for systems at states far from equilibrium the new definition shines, say its supporters. The theory uses the mathematical notion of *fractals* to account for nonequilibrium behavior and describes systems on the verge of *chaos* by new statistical descriptions. Time will tell whether the new definition is accepted as the basis of statistical mechanics, or takes its place on a long list of bright ideas that fell by the wayside of science.

$$(S_2 - S_1)/n = \bar{R}\ln(V_2/V_1) \tag{6.31}$$

where n is the amount of substance on a molar basis (Eq. 1.8). Next, we consider how the entropy change would be evaluated from a microscopic point of view.

Through more complete molecular modeling and statistical analysis, the total number of positions and velocities – ***microstates*** – available to a single molecule can be calculated. This total is called the ***thermodynamic probability,*** w. For a system of N molecules, the thermodynamic probability is w^N. In statistical thermodynamics, entropy is considered to be proportional to $\ln(w)^N$. That is, $S \propto N\ln(w)$. This gives the ***Boltzmann relation***

microstates

thermodynamic probability

$$S/N = k\ln w \tag{6.32}$$

boltzmann relation

where the proportionality factor, k, is called *Boltzmann's constant.*

Applying Eq. 6.32 to the process of Fig. 6.7, we get

$$\begin{aligned}(S_2 - S_1)/N &= k\ln(w_2) - k\ln(w_1)\\ &= k\ln(w_2/w_1)\end{aligned} \tag{6.33}$$

Comparing Eqs. 6.31 and 6.33, the expressions for entropy change coincide when $k = n\bar{R}/N$ and $w_2/w_1 = V_2/V_1$. The first of these expressions allows Boltzmann's constant to be evaluated, giving $k = 1.3806 \times 10^{-23}$ J/K. Also, since $V_2 > V_1$ and $w_2 > w_1$, Eqs. 6.31 and 6.33 each predict an increase of entropy owing to entropy production during the irreversible adiabatic expansion in this example.

From Eq. 6.33, we see that any process that increases the number of possible microstates of a system increases its entropy and conversely. Hence, for an *isolated* system, processes occur only in such a way that the number of microstates available to the system increases, resulting in our having less knowledge about the condition of individual particles. Because of this concept of decreased knowledge, entropy reflects the microscopic ***disorder*** of the system. We can then say that the only processes an isolated system can undergo are those that increase the disorder of the system. This interpretation is consistent with the idea of directionality of processes discussed previously.

disorder

The notion of entropy as a measure of disorder is sometimes used in fields other than thermodynamics. The concept is employed in information theory, statistics, biology, and even in some economic and social modeling. In these applications, the term entropy is used as a measure of disorder without the physical aspects of the thought experiment used here necessarily being implied.

Bio... connections

Do living things violate the second law of thermodynamics because they seem to create order from disorder? Living things are not isolated systems as considered in the previous discussion of entropy and disorder. Living things interact with their surroundings and are influenced by their surroundings. For instance, plants grow into highly ordered cellular structures synthesized from atoms and molecules originating in the Earth and its atmosphere. Through interactions with their surroundings, plants exist in highly organized states and are able to produce within themselves even more organized, lower entropy states. In keeping with the second law, states of lower entropy can be realized within a system as long as the *total* entropy of the system and its surroundings increases. The self-organizing tendency of living things as systems is widely observed and fully in accord with the second law.

6.9 Entropy Rate Balance for Control Volumes

Thus far the discussion of the entropy balance concept has been restricted to the case of closed systems. In the present section the entropy balance is extended to control volumes.

Like mass and energy, entropy is an extensive property, so it too can be transferred into or out of a control volume by streams of matter. Since this is the principal difference between the closed system and control volume forms, the ***control volume entropy rate balance*** can be obtained by modifying Eq. 6.28 to account for these entropy transfers. The result is

control volume entropy rate balance

$$\underbrace{\frac{dS_{cv}}{dt}}_{\substack{\text{rate of}\\\text{entropy}\\\text{change}}} = \underbrace{\sum_j \frac{\dot{Q}_j}{T_j} + \sum_i \dot{m}_i s_i - \sum_e \dot{m}_e s_e}_{\substack{\text{rates of}\\\text{entropy}\\\text{transfer}}} + \underbrace{\dot{\sigma}_{cv}}_{\substack{\text{rate of}\\\text{entropy}\\\text{production}}} \tag{6.34}$$

entropy transfer accompanying mass flow

where dS_{cv}/dt represents the time rate of change of entropy within the control volume. The terms $\dot{m}_i s_i$ and $\dot{m}_e s_e$ account, respectively, for rates of ***entropy transfer accompanying mass flow*** into and out of the control volume. The term $\dot{Q}_j$ represents the time rate of heat transfer at the location on the boundary where the instantaneous temperature is T_j. The ratio $\dot{Q}_j/T_j$ accounts for the accompanying rate of entropy *transfer*. The term $\dot{\sigma}_{cv}$ denotes the time rate of entropy *production* due to irreversibilities *within* the control volume.

Integral Form of the Entropy Rate Balance

As for the cases of the control volume mass and energy rate balances, the entropy rate balance can be expressed in terms of local properties to obtain forms that are more generally applicable. Thus, the term $S_{cv}(t)$, representing the total entropy associated with the control volume at time t, can be written as a volume integral

$$S_{cv}(t) = \int_V \rho s \, dV$$

where ρ and s denote, respectively, the local density and specific entropy. The rate of entropy transfer accompanying heat transfer can be expressed more generally as an integral over the surface of the control volume

$$\begin{bmatrix}\text{time rate of entropy}\\\text{transfer accompanying}\\\text{heat transfer}\end{bmatrix} = \int_A \left(\frac{\dot{q}}{T}\right)_b dA$$

where $\dot{q}$ is the *heat flux*, the time rate of heat transfer per unit of surface area, through the location on the boundary where the instantaneous temperature is T. The subscript "b" is added as a reminder that the integrand is evaluated on the boundary of the control volume. In addition, the terms accounting for entropy transfer accompanying mass flow can be expressed as integrals over the inlet and exit flow areas, resulting in the following form of the entropy rate balance

$$\frac{d}{dt}\int_V \rho s\, dV = \int_A \left(\frac{\dot{q}}{T}\right)_b dA + \sum_i \left(\int_A s\rho V_n\, dA\right)_i - \sum_e \left(\int_A s\rho V_n\, dA\right)_e + \dot{\sigma}_{cv} \quad (6.35)$$

where V_n denotes the normal component in the direction of flow of the velocity relative to the flow area. In some cases, it is also convenient to express the entropy production rate as a volume integral of the local volumetric rate of entropy production within the control volume. The study of Eq. 6.35 brings out the assumptions underlying Eq. 6.34. Finally, note that for a closed system the sums accounting for entropy transfer at inlets and exits drop out, and Eq. 6.35 reduces to give a more general form of Eq. 6.28.

6.10 Rate Balances for Control Volumes at Steady State

Since a great many engineering analyses involve control volumes at steady state, it is instructive to list steady-state forms of the balances developed for mass, energy, and entropy. At steady state, the conservation of mass principle takes the form

$$\sum_i \dot{m}_i = \sum_e \dot{m}_e \quad (4.6)$$

The energy rate balance at steady state is

$$0 = \dot{Q}_{cv} - \dot{W}_{cv} + \sum_i \dot{m}_i\left(h_i + \frac{V_i^2}{2} + gz_i\right) - \sum_e \dot{m}_e\left(h_e + \frac{V_e^2}{2} + gz_e\right) \quad (4.18)$$

Finally, the ***steady-state form of the entropy rate balance*** is obtained by reducing Eq. 6.34 to give

$$0 = \sum_j \frac{\dot{Q}_j}{T_j} + \sum_i \dot{m}_i s_i - \sum_e \dot{m}_e s_e + \dot{\sigma}_{cv} \quad (6.36)$$

steady-state entropy rate balance

These equations often must be solved simultaneously, together with appropriate property relations.

Mass and energy are conserved quantities, but entropy is not conserved. Equation 4.6 indicates that at steady state the total rate of mass flow into the control volume equals the total rate of mass flow out of the control volume. Similarly, Eq. 4.18 indicates that the total rate of energy transfer into the control volume equals the total rate of energy transfer out of the control volume. However, Eq. 6.36 requires that the rate at which entropy is transferred out must *exceed* the rate at which entropy enters, the difference being the rate of entropy production within the control volume owing to irreversibilities.

6.10.1 One-Inlet, One-Exit Control Volumes at Steady State

Since many applications involve one-inlet, one-exit control volumes at steady state, let us also list the form of the entropy rate balance for this important case. Thus, Eq. 6.36 reduces to read

$$0 = \sum_j \frac{\dot{Q}_j}{T_j} + \dot{m}(s_1 - s_2) + \dot{\sigma}_{cv} \quad (6.37)$$

Or, on dividing by the mass flow rate $\dot{m}$ and rearranging

$$s_2 - s_1 = \frac{1}{\dot{m}}\left(\sum_j \frac{\dot{Q}_j}{T_j}\right) + \frac{\dot{\sigma}_{cv}}{\dot{m}} \quad (6.38)$$

The two terms on the right side of Eq. 6.38 denote, respectively, the rate of entropy transfer accompanying heat transfer and the rate of entropy production within the control volume, each *per unit of mass flowing through the control volume*. From Eq. 6.38 it can be concluded that the entropy of a unit of mass passing from inlet to exit can increase, decrease, or remain the same. Furthermore, because the value of the second term on the right can never be negative, a decrease in the specific entropy from inlet to exit can be realized only when more entropy is transferred out of the control volume accompanying heat transfer than is produced by irreversibilities within the control volume. When the value of this entropy transfer term is positive, the specific entropy at the exit is greater than the specific entropy at the inlet whether internal irreversibilities are present or not. In the special case where there is no entropy transfer accompanying heat transfer, Eq. 6.38 reduces to

$$s_2 - s_1 = \frac{\dot{\sigma}_{cv}}{\dot{m}} \quad (6.39)$$

Accordingly, when irreversibilities are present within the control volume, the entropy of a unit of mass increases as it passes from inlet to exit. In the limiting case in which no irreversibilities are present, the unit mass passes through the control volume with no change in its entropy—that is, isentropically.

6.10.2 Applications of the Rate Balances to Control Volumes at Steady State

The following examples illustrate the use of the mass, energy, and entropy balances for the analysis of control volumes at steady state. Carefully note that property relations and property diagrams also play important roles in arriving at solutions.

In Example 6.6, we evaluate the rate of entropy production within a turbine operating at steady state when there is heat transfer from the turbine.

Example 6.6 ENTROPY PRODUCTION IN A STEAM TURBINE

Steam enters a turbine with a pressure of 30 bar, a temperature of 400°C, and a velocity of 160 m/s. Saturated vapor at 100°C exits with a velocity of 100 m/s. At steady state, the turbine develops work equal to 540 kJ per kg of steam flowing through the turbine. Heat transfer between the turbine and its surroundings occurs at an average outer surface temperature of 350 K. Determine the rate at which entropy is produced within the turbine per kg of steam flowing, in kJ/kg · K. Neglect the change in potential energy between inlet and exit.

Solution

Known: Steam expands through a turbine at steady state for which data are provided.

Find: Determine the rate of entropy production per kg of steam flowing.

Schematic and Given Data:

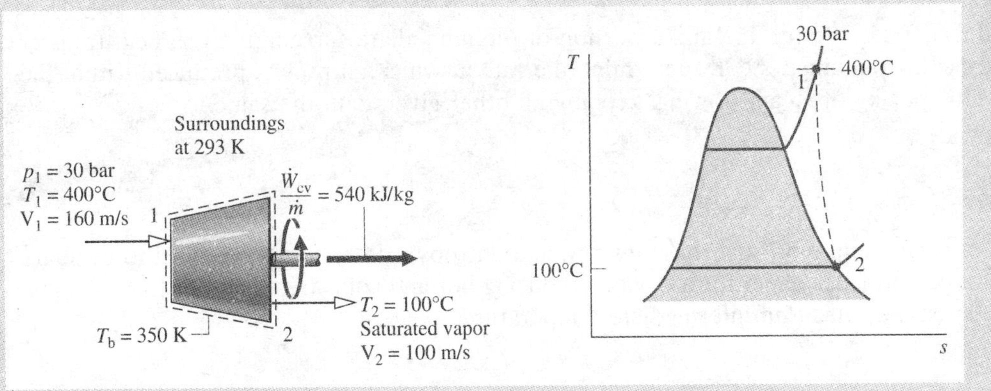

Fig. E6.6

Engineering Model:

1. The control volume shown on the accompanying sketch is at steady state.
2. Heat transfer from the turbine to the surroundings occurs at a specified average outer surface temperature.
3. The change in potential energy between inlet and exit can be neglected.

Analysis: To determine the entropy production per unit mass flowing through the turbine, begin with mass and entropy rate balances for the one-inlet, one-exit control volume at steady state:

$$\dot{m}_1 = \dot{m}_2$$

$$0 = \sum_j \frac{\dot{Q}_j}{T_j} + \dot{m}_1 s_1 - \dot{m}_2 s_2 + \dot{\sigma}_{cv}$$

Since heat transfer occurs only at $T_b = 350$ K, the first term on the right side of the entropy rate balance reduces to $\dot{Q}_{cv}/T_b$. Combining the mass and entropy rate balances

$$0 = \frac{\dot{Q}_{cv}}{T_b} + \dot{m}(s_1 - s_2) + \dot{\sigma}_{cv}$$

where $\dot{m}$ is the mass flow rate. Solving for $\dot{\sigma}_{cv}/\dot{m}$

$$\frac{\dot{\sigma}_{cv}}{\dot{m}} = -\frac{\dot{Q}_{cv}/\dot{m}}{T_b} + (s_2 - s_1)$$

The heat transfer rate, $\dot{Q}_{cv}/\dot{m}$, required by this expression is evaluated next.

Reduction of the mass and energy rate balances results in

$$\frac{\dot{Q}_{cv}}{\dot{m}} = \frac{\dot{W}_{cv}}{\dot{m}} + (h_2 - h_1) + \left(\frac{V_2^2 - V_1^2}{2}\right)$$

where the potential energy change from inlet to exit is dropped by assumption 3. From Table A-4 at 30 bar, 400°C, $h_1 = 3230.9$ kJ/kg, and from Table A-2, $h_2 = h_g(100°C) = 2676.1$ kJ/kg. Thus

$$\frac{\dot{Q}_{cv}}{\dot{m}} = 540\frac{\text{kJ}}{\text{kg}} + (2676.1 - 3230.9)\left(\frac{\text{kJ}}{\text{kg}}\right) + \left[\frac{(100)^2 - (160)^2}{2}\right]\left(\frac{\text{m}^2}{\text{s}^2}\right)\left|\frac{1\text{ N}}{1\text{ kg}\cdot\text{m/s}^2}\right|\left|\frac{1\text{ kJ}}{10^3\text{ N}\cdot\text{m}}\right|$$

$$= 540 - 554.8 - 7.8 = -22.6\text{ kJ/kg}$$

From Table A-2, $s_2 = 7.3549$ kJ/kg · K, and from Table A-4, $s_1 = 6.9212$ kJ/kg · K. Inserting values into the expression for entropy production

$$\frac{\dot{\sigma}_{cv}}{\dot{m}} = -\frac{(-22.6\text{ kJ/kg})}{350\text{ K}} + (7.3549 - 6.9212)\left(\frac{\text{kJ}}{\text{kg}\cdot\text{K}}\right)$$

$$= 0.0646 + 0.4337 = 0.498\text{ kJ/kg}\cdot\text{K}$$

Skills Developed: *Ability to...*

- ✓ apply the control volume mass, energy and entropy rate balances.
- ✓ retrieve property data for water.

Quick Quiz

If the boundary were located to include the turbine and a portion of the immediate surroundings so heat transfer occurs at the temperature of the surroundings, 293 K, determine the rate at which entropy is produced within the enlarged control volume, in kJ/K per kg of steam flowing, keeping all other given data the same.
Ans. 0.511 kJ/kg · K

In Example 6.7, the mass, energy, and entropy rate balances are used to evaluate a performance claim for a device producing hot and cold streams of air from a single stream of air at an intermediate temperature.

Example 6.7 EVALUATING A PERFORMANCE CLAIM

An inventor claims to have developed a device requiring no energy transfer by work or heat transfer, yet able to produce hot and cold streams of air from a single stream of air at an intermediate temperature. The inventor provides steady-state test data indicating that when air enters at a temperature of 21°C and a pressure of 5 bar, separate streams of air exit at temperatures of −18°C and 79°C, respectively, and each at a pressure of 1 bar. Sixty percent of the mass entering the device exits at the lower temperature. Evaluate the inventor's claim, employing the ideal gas model for air and ignoring changes in the kinetic and potential energies of the streams from inlet to exit.

Solution

Known: Data are provided for a device that at steady state produces hot and cold streams of air from a single stream of air at an intermediate temperature without energy transfers by work or heat.

Find: Evaluate whether the device can operate as claimed.

Schematic and Given Data:

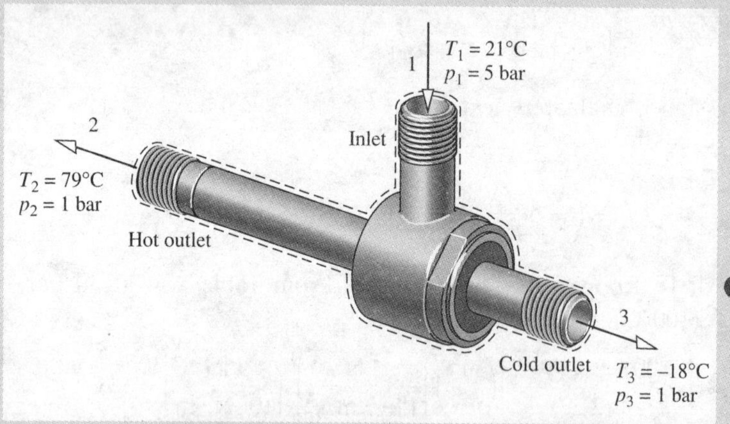

Engineering Model:

1. The control volume shown on the accompanying sketch is at steady state.
2. For the control volume, $\dot{W}_{cv} = 0$ and $\dot{Q}_{cv} = 0$.
3. Changes in the kinetic and potential energies from inlet to exit can be ignored.
4. The air is modeled as an ideal gas with constant $c_p = 1.0$ kJ/kg · K. ❶

Fig. E6.7

Analysis: For the device to operate as claimed, the conservation of mass and energy principles must be satisfied. The second law of thermodynamics also must be satisfied; and in particular the rate of entropy production cannot be negative. Accordingly, the mass, energy and entropy rate balances are considered in turn.

With assumptions 1–3, the mass and energy rate balances reduce, respectively, to

$$\dot{m}_1 = \dot{m}_2 + \dot{m}_3$$

$$0 = \dot{m}_1 h_1 - \dot{m}_2 h_2 - \dot{m}_3 h_3$$

Since $\dot{m}_3 = 0.6\dot{m}_1$, it follows from the mass rate balance that $\dot{m}_2 = 0.4\dot{m}_1$. By combining the mass and energy rate balances and evaluating changes in specific enthalpy using constant c_p, the energy rate balance is also satisfied. That is

$$\begin{aligned} 0 &= (\dot{m}_2 + \dot{m}_3)h_1 - \dot{m}_2 h_2 - \dot{m}_3 h_3 \\ &= \dot{m}_2(h_1 - h_2) + \dot{m}_3(h_1 - h_3) \\ &= 0.4\dot{m}_1[c_p(T_1 - T_2)] + 0.6\dot{m}_1[c_p(T_1 - T_3)] \\ &= 0.4(T_1 - T_2) + 0.6(T_1 - T_3) \\ &= 0.4(-105) + 0.6(70) \\ &= 0 \end{aligned}$$ ❷

Accordingly, with the given data the conservation of mass and energy principles are satisfied.

Since no significant heat transfer occurs, the entropy rate balance at steady state reads

$$0 = \sum_j \frac{\dot{Q}_j}{T_j}^{\,0} + \dot{m}_1 s_1 - \dot{m}_2 s_2 - \dot{m}_3 s_3 + \dot{\sigma}_{cv}$$

Combining the mass and entropy rate balances

$$\begin{aligned} 0 &= (\dot{m}_2 + \dot{m}_3)s_1 - \dot{m}_2 s_2 - \dot{m}_3 s_3 + \dot{\sigma}_{cv} \\ &= \dot{m}_2(s_1 - s_2) + \dot{m}_3(s_1 - s_3) + \dot{\sigma}_{cv} \\ &= 0.4\dot{m}_1(s_1 - s_2) + 0.6\dot{m}_1(s_1 - s_3) + \dot{\sigma}_{cv} \end{aligned}$$

Solving for $\dot{\sigma}_{cv}/\dot{m}_1$ and using Eq. 6.22 to evaluate changes in specific entropy

$$\frac{\dot{\sigma}_{cv}}{\dot{m}_1} = 0.4\left[c_p \ln\frac{T_2}{T_1} - R\ln\frac{p_2}{p_1}\right] + 0.6\left[c_p \ln\frac{T_3}{T_1} - R\ln\frac{p_3}{p_1}\right]$$

❸ $$= 0.4\left[\left(1.0\frac{\text{kJ}}{\text{kg}\cdot\text{K}}\right)\ln\frac{352}{294} - \left(\frac{8.314}{28.97}\frac{\text{kJ}}{\text{kg}\cdot\text{K}}\right)\ln\frac{1}{5.0}\right] + 0.6\left[\left(1.0\frac{\text{kJ}}{\text{kg}\cdot\text{K}}\right)\ln\frac{255}{294} - \left(\frac{8.314}{28.97}\frac{\text{kJ}}{\text{kg}\cdot\text{K}}\right)\ln\frac{1}{5.0}\right]$$

❹ $$= 0.454 \text{ kJ/kg}\cdot\text{K}$$

Thus, the second law of thermodynamics is also satisfied.

❺ On the basis of this evaluation, the inventor's claim does not violate principles of thermodynamics.

---

❶ Since the specific heat c_p of air varies little over the temperature interval from 0 to 79°C, c_p can be taken as constant. From Table A-20, $c_p = 1.0$ kJ/kg · K.

❷ Since temperature *differences* are involved in this calculation, the temperatures can be either in K or °C.

❸ In this calculation involving temperature *ratios,* the temperatures are in K. Temperatures in °C should not be used.

❹ If the value of the rate of entropy production had been negative or zero, the claim would be rejected. A negative value is impossible by the second law and a zero value would indicate operation without irreversibilities.

❺ Such devices *do* exist. They are known as *vortex tubes* and are used in industry for *spot cooling.*

Skills Developed: *Ability to...*
- ✓ apply the control volume mass, energy and entropy rate balances.
- ✓ apply the ideal gas model with constant c_p.

Quick Quiz

If the inventor would claim that the hot and cold streams exit the device at 5 bar, evaluate the revised claim, keeping all other given data the same.

Ans. Claim invalid.

In Example 6.8, we evaluate and compare the rates of entropy production for three components of a heat pump system. Heat pumps are considered in detail in Chap. 10.

Example 6.8 ENTROPY PRODUCTION IN HEAT PUMP COMPONENTS

Components of a heat pump for supplying heated air to a dwelling are shown in the schematic below. At steady state, Refrigerant 22 enters the compressor at −5°C, 3.5 bar and is compressed adiabatically to 75°C, 14 bar. From the compressor, the refrigerant passes through the condenser, where it condenses to liquid at 28°C, 14 bar. The refrigerant then expands through a throttling valve to 3.5 bar. The states of the refrigerant are shown on the accompanying *T–s* diagram. Return air from the dwelling enters the condenser at 20°C, 1 bar with a volumetric flow rate of 0.42 m^3/s and exits at 50°C with a negligible change in pressure. Using the ideal gas model for the air and neglecting kinetic and potential energy effects, **(a)** determine the rates of entropy production, in kW/K, for control volumes enclosing the condenser, compressor, and expansion valve, respectively. **(b)** Discuss the sources of irreversibility in the components considered in part (a).

Solution

Known: Refrigerant 22 is compressed adiabatically, condensed by heat transfer to air passing through a heat exchanger, and then expanded through a throttling valve. Steady-state operating data are known.

Find: Determine the entropy production rates for control volumes enclosing the condenser, compressor, and expansion valve, respectively, and discuss the sources of irreversibility in these components.

Schematic and Given Data:

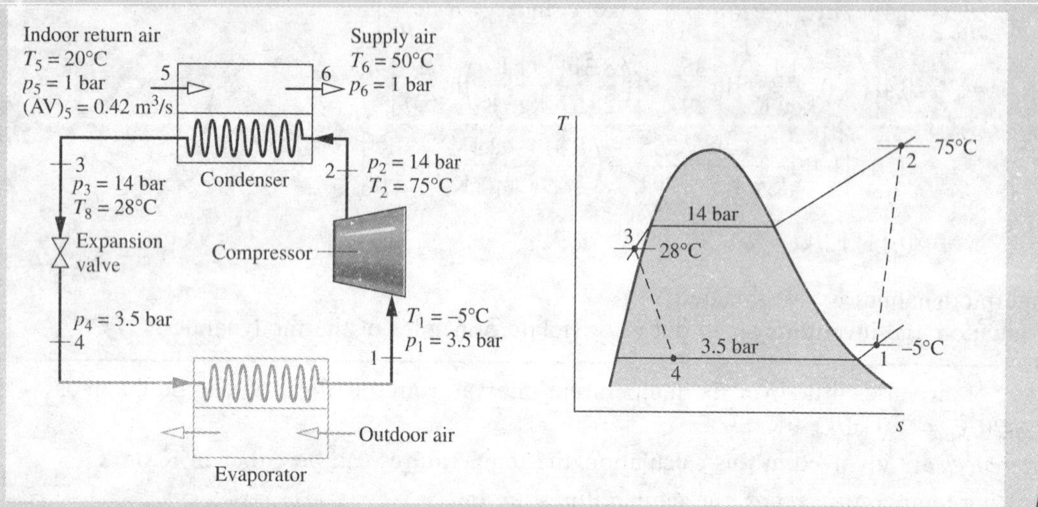

Fig. E6.8

Engineering Model:

1. Each component is analyzed as a control volume at steady state.
2. The compressor operates adiabatically, and the expansion across the valve is a *throttling process*.
3. For the control volume enclosing the condenser, $\dot{W}_{cv} = 0$ and $\dot{Q}_{cv} = 0$.
4. Kinetic and potential energy effects can be neglected.

❶ 5. The air is modeled as an ideal gas with constant $c_p = 1.005$ kJ/kg · K.

Analysis:

(a) Let us begin by obtaining property data at each of the principal refrigerant states located on the accompanying schematic and *T–s* diagram. At the inlet to the compressor, the refrigerant is a superheated vapor at −5°C, 3.5 bar, so from Table A-9, $s_1 = 0.9572$ kJ/kg · K. Similarly, at state 2, the refrigerant is a superheated vapor at 75°C, 14 bar, so interpolating in Table A-9 gives $s_2 = 0.98225$ kJ/kg · K and $h_2 = 294.17$ kJ/kg.

State 3 is compressed liquid at 28°C, 14 bar. From Table A-7, $s_3 \approx s_f(28°C) = 0.2936$ kJ/kg · K and $h_3 \approx h_f(28°C) = 79.05$ kJ/kg. The expansion through the valve is a *throttling process*, so $h_3 = h_4$. Using data from Table A-8, the quality at state 4 is

$$x_4 = \frac{(h_4 - h_{f4})}{(h_{fg})_4} = \frac{(79.05 - 33.09)}{(212.91)} = 0.216$$

and the specific entropy is

$$s_4 = s_{f4} + x_4(s_{g4} - s_{f4}) = 0.1328 + 0.216(0.9431 - 0.1328) = 0.3078 \text{ kJ/kg} \cdot \text{K}$$

Condenser

Consider the control volume enclosing the condenser. With assumptions 1 and 3, the entropy rate balance reduces to

$$0 = \dot{m}_{\text{ref}}(s_2 - s_3) + \dot{m}_{\text{air}}(s_5 - s_6) + \dot{\sigma}_{\text{cond}}$$

To evaluate $\dot{\sigma}_{\text{cond}}$ requires the two mass flow rates, $\dot{m}_{\text{air}}$ and $\dot{m}_{\text{ref}}$, and the change in specific entropy for the air. These are obtained next.

Evaluating the mass flow rate of air using the ideal gas model (assumption 5)

$$\dot{m}_{\text{air}} = \frac{(\text{AV})_5}{v_5} = (\text{AV})_5 \frac{p_5}{RT_5}$$

$$= \left(0.42 \frac{\text{m}^3}{\text{s}}\right) \frac{(1 \text{ bar})}{\left(\frac{8.314}{28.97} \frac{\text{kJ}}{\text{kg} \cdot \text{K}}\right)(293 \text{ K})} \left|\frac{10^5 \text{ N/m}^2}{1 \text{ bar}}\right| \left|\frac{1 \text{ kJ}}{10^3 \text{ N} \cdot \text{m}}\right| = 0.5 \text{ kg/s}$$

The refrigerant mass flow rate is determined using an energy balance for the control volume enclosing the condenser together with assumptions 1, 3, and 4 to obtain

$$\dot{m}_{\text{ref}} = \frac{\dot{m}_{\text{air}}(h_6 - h_5)}{(h_2 - h_3)}$$

With assumption 5, $h_6 - h_5 = c_p(T_6 - T_5)$. Inserting values

❷ $$\dot{m}_{\text{ref}} = \frac{\left(0.5 \frac{\text{kg}}{\text{s}}\right)\left(1.005 \frac{\text{kJ}}{\text{kg} \cdot \text{K}}\right)(323 - 293)\text{K}}{(294.17 - 79.05) \text{ kJ/kg}} = 0.07 \text{ kg/s}$$

Using Eq. 6.22, the change in specific entropy of the air is

$$s_6 - s_5 = c_p \ln \frac{T_6}{T_5} - R \ln \frac{p_6}{p_5}$$

$$= \left(1.005 \frac{\text{kJ}}{\text{kg} \cdot \text{K}}\right) \ln\left(\frac{323}{293}\right) - R \ln\left(\frac{1.0}{1.0}\right)^{0} = 0.098 \text{ kJ/kg} \cdot \text{K}$$

Finally, solving the entropy balance for $\dot{\sigma}_{\text{cond}}$ and inserting values

$$\dot{\sigma}_{\text{cond}} = \dot{m}_{\text{ref}}(s_3 - s_2) + \dot{m}_{\text{air}}(s_6 - s_5)$$

$$= \left[\left(0.07 \frac{\text{kg}}{\text{s}}\right)(0.2936 - 0.98225) \frac{\text{kJ}}{\text{kg} \cdot \text{K}} + (0.5)(0.098)\right] \left|\frac{1 \text{ kW}}{1 \text{ kJ/s}}\right|$$

$$= 7.95 \times 10^{-4} \frac{\text{kW}}{\text{K}}$$

Compressor

For the control volume enclosing the compressor, the entropy rate balance reduces with assumptions 1 and 3 to

$$0 = \dot{m}_{\text{ref}}(s_1 - s_2) + \dot{\sigma}_{\text{comp}}$$

or

$$\begin{aligned}\dot{\sigma}_{\text{comp}} &= \dot{m}_{\text{ref}}(s_2 - s_1)\\ &= \left(0.07\frac{\text{kg}}{\text{s}}\right)(0.98225 - 0.9572)\left(\frac{\text{kJ}}{\text{kg}\cdot\text{K}}\right)\left|\frac{1\text{ kW}}{1\text{ kJ/s}}\right|\\ &= 17.5 \times 10^{-4}\text{ kW/K}\end{aligned}$$

Valve

Finally, for the control volume enclosing the throttling valve, the entropy rate balance reduces to

$$0 = \dot{m}_{\text{ref}}(s_3 - s_4) + \dot{\sigma}_{\text{valve}}$$

Solving for $\dot{\sigma}_{\text{valve}}$ and inserting values

$$\begin{aligned}\dot{\sigma}_{\text{valve}} = \dot{m}_{\text{ref}}(s_4 - s_3) &= \left(0.07\frac{\text{kg}}{\text{s}}\right)(0.3078 - 0.2936)\left(\frac{\text{kJ}}{\text{kg}\cdot\text{K}}\right)\left|\frac{1\text{ kW}}{1\text{ kJ/s}}\right|\\ &= 9.94 \times 10^{-4}\text{ kW/K}\end{aligned}$$

(b) The following table summarizes, in rank order, the calculated entropy production rates:

❸

Component	$\dot{\sigma}_{\text{cv}}$ (kW/K)
compressor	17.5×10^{-4}
valve	9.94×10^{-4}
condenser	7.95×10^{-4}

Entropy production in the compressor is due to fluid friction, mechanical friction of the moving parts, and internal heat transfer. For the valve, the irreversibility is primarily due to fluid friction accompanying the expansion across the valve. The principal source of irreversibility in the condenser is the temperature difference between the air and refrigerant streams. In this example, there are no pressure drops for either stream passing through the condenser, but slight pressure drops due to fluid friction would normally contribute to the irreversibility of condensers. The evaporator lightly shown in Fig. E6.8 has not been analyzed.

❶ Due to the relatively small temperature change of the air, the specific heat c_p can be taken as constant at the average of the inlet and exit air temperatures.

❷ Temperatures in K are used to evaluate $\dot{m}_{\text{ref}}$, but since a temperature *difference* is involved the same result would be obtained if temperatures in °C were used. Temperatures in K, and not °C, are required when a temperature *ratio* is involved, as in Eq. 6.22 used to evaluate $s_6 - s_5$.

❸ By focusing attention on reducing irreversibilities at the sites with the highest entropy production rates, *thermodynamic* improvements may be possible. However, costs and other constraints must be considered, and can be overriding.

Skills Developed: *Ability to...*

- ✓ apply the control volume mass, energy and entropy rate balances.
- ✓ develop an engineering model.
- ✓ retrieve property data for Refrigerant 22.
- ✓ apply the ideal gas model with constant c_p.

Quick Quiz

If the compressor operated adiabatically *and* without internal irreversibilities, determine the temperature of the refrigerant at the compressor exit, in °C, keeping the compressor inlet state and exit pressure the same.

Ans. 65°C

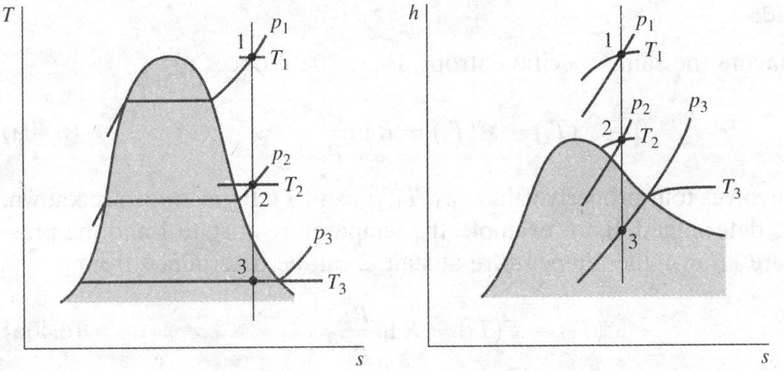

Fig. 6.8 *T–s* and *h–s* diagrams showing states having the same value of specific entropy.

6.11 *Isentropic Processes*

The term *isentropic* means constant entropy. Isentropic processes are encountered in many subsequent discussions. The object of the present section is to show how properties are related at any two states of a process in which there is no change in specific entropy.

6.11.1 General Considerations

The properties at states having the same specific entropy can be related using the graphical and tabular property data discussed in Sec. 6.2. For example, as illustrated by Fig. 6.8, temperature–entropy and enthalpy–entropy diagrams are particularly convenient for determining properties at states having the same value of specific entropy. All states on a vertical line passing through a given state have the same entropy. If state 1 on Fig. 6.8 is fixed by pressure p_1 and temperature T_1, states 2 and 3 are readily located once one additional property, such as pressure or temperature, is specified. The values of several other properties at states 2 and 3 can then be read directly from the figures.

Tabular data also can be used to relate two states having the same specific entropy. For the case shown in Fig. 6.8, the specific entropy at state 1 could be determined from the superheated vapor table. Then, with $s_2 = s_1$ and one other property value, such as p_2 or T_2, state 2 could be located in the superheated vapor table. The values of the properties v, u, and h at state 2 can then be read from the table. An illustration of this procedure is given in Sec. 6.2.1. Note that state 3 falls in the two-phase liquid–vapor regions of Fig. 6.8. Since $s_3 = s_1$, the quality at state 3 could be determined using Eq. 6.4. With the quality known, other properties such as v, u, and h could then be evaluated. Computer retrieval of entropy data provides an alternative to tabular data.

6.11.2 Using the Ideal Gas Model

Figure 6.9 shows two states of an ideal gas having the same value of specific entropy. Let us consider relations among pressure, specific volume, and temperature at these states, first using the ideal gas tables and then assuming specific heats are constant.

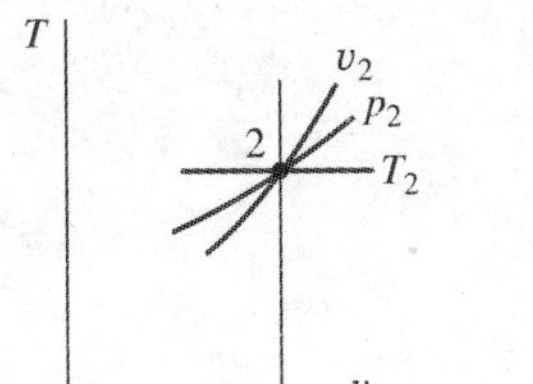

Fig. 6.9 Two states of an ideal gas where $s_2 = s_1$.

Ideal Gas Tables

For two states having the same specific entropy, Eq. 6.20a reduces to

$$0 = s°(T_2) - s°(T_1) - R \ln \frac{p_2}{p_1} \tag{6.40a}$$

Equation 6.40a involves four property values: p_1, T_1, p_2, and T_2. If any three are known, the fourth can be determined. If, for example, the temperature at state 1 and the pressure ratio p_2/p_1 are known, the temperature at state 2 can be determined from

$$s°(T_2) = s°(T_1) + R \ln \frac{p_2}{p_1} \tag{6.40b}$$

Since T_1 is known, $s°(T_1)$ would be obtained from the appropriate table, the value of $s°(T_2)$ would be calculated, and temperature T_2 would then be determined by interpolation. If p_1, T_1, and T_2 are specified and the pressure at state 2 is the unknown, Eq. 6.40a would be solved to obtain

$$p_2 = p_1 \exp\left[\frac{s°(T_2) - s°(T_1)}{R}\right] \tag{6.40c}$$

Equations 6.40 can be used when $s°$ (or $\bar{s}°$) data are known, as for the gases of Tables A-22 and A-23.

AIR. For the special case of *air* modeled as an ideal gas, Eq. 6.40c provides the basis for an alternative tabular approach for relating the temperatures and pressures at two states having the same specific entropy. To introduce this, rewrite the equation as

$$\frac{p_2}{p_1} = \frac{\exp[s°(T_2)/R]}{\exp[s°(T_1)/R]}$$

The quantity $\exp[s°(T)/R]$ appearing in this expression is solely a function of temperature, and is given the symbol $p_r(T)$. A tabulation of p_r versus temperature for *air* is provided in Tables A-22.[1] In terms of the function p_r, the last equation becomes

$$\frac{p_2}{p_1} = \frac{p_{r2}}{p_{r1}} \quad (s_1 = s_2, \text{ air only}) \tag{6.41}$$

where $p_{r1} = p_r(T_1)$ and $p_{r2} = p_r(T_2)$. The function p_r is sometimes called the *relative pressure*. Observe that p_r is not truly a pressure, so the name relative pressure is misleading. Also, be careful not to confuse p_r with the reduced pressure of the compressibility diagram.

A relation between specific volumes and temperatures for two states of air having the same specific entropy can also be developed. With the ideal gas equation of state, $v = RT/p$, the ratio of the specific volumes is

$$\frac{v_2}{v_1} = \left(\frac{RT_2}{p_2}\right)\left(\frac{p_1}{RT_1}\right)$$

Then, since the two states have the same specific entropy, Eq. 6.41 can be introduced to give

$$\frac{v_2}{v_1} = \left[\frac{RT_2}{p_r(T_2)}\right]\left[\frac{p_r(T_1)}{RT_1}\right]$$

[1] The values of p_r determined with this definition are inconveniently large, so they are divided by a scale factor before tabulating to give a convenient range of numbers.

The ratio $RT/p_r(T)$ appearing on the right side of the last equation is solely a function of temperature, and is given the symbol $v_r(T)$. Values of v_r for *air* are tabulated versus temperature in Tables A-22. In terms of the function v_r, the last equation becomes

$$\frac{v_2}{v_1} = \frac{v_{r2}}{v_{r1}} \quad (s_1 = s_2, \text{ air only}) \tag{6.42}$$

where $v_{r1} = v_r(T_1)$ and $v_{r2} = v_r(T_2)$. The function v_r is sometimes called the *relative volume*. Despite the name given to it, $v_r(T)$ is not truly a volume. Also, be careful not to confuse it with the pseudoreduced specific volume of the compressibility diagram.

Assuming Constant Specific Heats

Take Note...

When applying the software *IT* to relate two states of an ideal gas having the same value of specific entropy, *IT* returns specific entropy *directly* and does not employ the special functions $s°$, p_r, and v_r.

Let us consider next how properties are related for isentropic processes of an ideal gas when the specific heats are constants. For any such case, Eqs. 6.21 and 6.22 reduce to the equations

$$0 = c_p \ln\frac{T_2}{T_1} - R\ln\frac{p_2}{p_1}$$

$$0 = c_v \ln\frac{T_2}{T_1} + R\ln\frac{v_2}{v_1}$$

Introducing the ideal gas relations

$$c_p = \frac{kR}{k-1}, \qquad c_v = \frac{R}{k-1} \tag{3.47}$$

where k is the specific heat ratio and R is the gas constant, these equations can be solved, respectively, to give

$$\frac{T_2}{T_1} = \left(\frac{p_2}{p_1}\right)^{(k-1)/k} \quad (s_1 = s_2, \text{ constant } k) \tag{6.43}$$

$$\frac{T_2}{T_1} = \left(\frac{v_1}{v_2}\right)^{k-1} \quad (s_1 = s_2, \text{ constant } k) \tag{6.44}$$

The following relation can be obtained by eliminating the temperature ratio from Eqs. 6.43 and 6.44:

$$\frac{p_2}{p_1} = \left(\frac{v_1}{v_2}\right)^{k} \quad (s_1 = s_2, \text{ constant } k) \tag{6.45}$$

Previously, we have identified an internally reversible process described by $pv^n = constant$, where n is a constant, as a *polytropic process.* From the form of Eq. 6.45, it can be concluded that the polytropic process $pv^k = constant$ of an ideal gas with constant specific heat ratio k is an isentropic process. We observed in Sec. 3.15 that a polytropic process of an ideal gas for which $n = 1$ is an isothermal (constant-temperature) process. For *any* fluid, $n = 0$ corresponds to an isobaric (constant-pressure) process and $n = \pm\infty$ corresponds to an isometric (constant-volume) process. Polytropic processes corresponding to these values of n are shown in Fig. 6.10 on p–v and T–s diagrams.

6.11.3 Illustrations: Isentropic Processes of Air

Means for evaluating data for isentropic processes of air modeled as an ideal gas are illustrated in the next two examples. In Example 6.9, we consider three alternative methods.

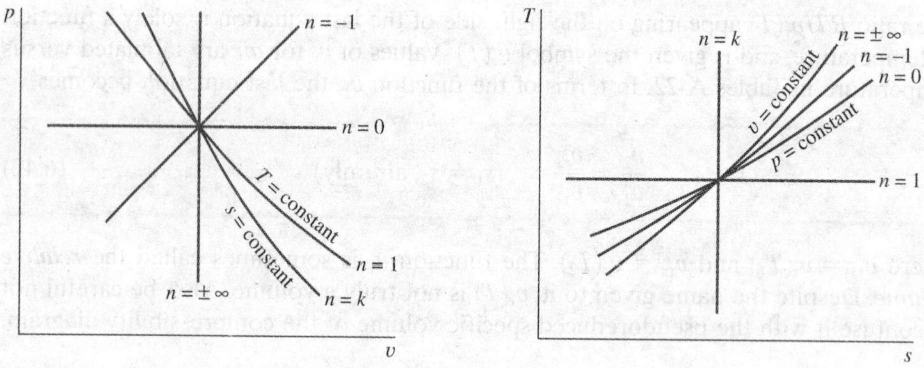

Fig. 6.10 Polytropic processes on *p–v* and *T–s* diagrams.

Example 6.9 ISENTROPIC PROCESS OF AIR

Air undergoes an isentropic process from $p_1 = 1$ bar, $T_1 = 300$ K to a final state where the temperature is $T_2 = 650$ K. Employing the ideal gas model, determine the final pressure p_2, in bar. Solve using (a) p_r data from Table A-22, (b) *Interactive Thermodynamics: IT,* and (c) a constant specific heat ratio k evaluated at the mean temperature, 475 K, from Table A-20.

Solution

Known: Air undergoes an isentropic process from a state where pressure and temperature are known to a state where the temperature is specified.

Find: Determine the final pressure using (a) p_r data, (b) *IT,* and (c) a constant value for the specific heat ratio k.

Schematic and Given Data:

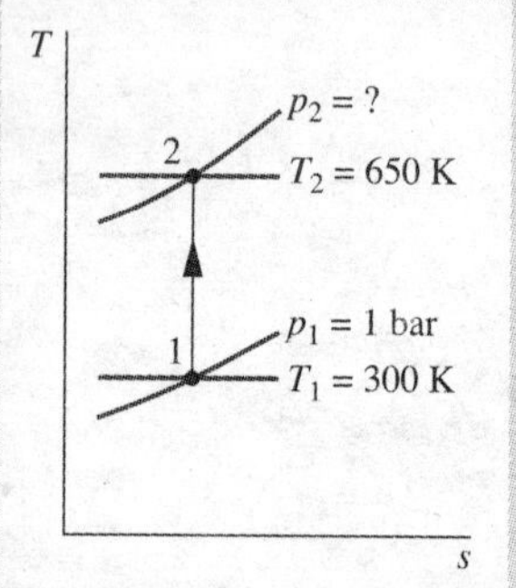

Engineering Model:

1. A quantity of air as the system undergoes an isentropic process.
2. The air can be modeled as an ideal gas.
3. In part (c), the specific heat ratio is constant.

Fig. E6.9

Analysis:

(a) The pressures and temperatures at two states of an ideal gas having the same specific entropy are related by Eq. 6.41

$$\frac{p_2}{p_1} = \frac{p_{r2}}{p_{r1}}$$

Solving

$$p_2 = p_1 \frac{p_{r2}}{p_{r1}}$$

With p_r values from Table A-22

$$p = (1\text{ bar})\frac{21.18}{1.3860} = 15.77\text{ bar}$$

(b) The *IT* solution follows:

```
T1 = 300 // K
p1 = 1 // bar
T2 = 650 // K
s_TP("Air", T1,p1) = s_TP("Air",T2,p2)

// Result: p2 = 15.77 bar
```

❶ (beside the s_TP line)

(c) When the specific heat ratio k is assumed constant, the temperatures and pressures at two states of an ideal gas having the same specific entropy are related by Eq. 6.43. Thus

$$p_2 = p_1\left(\frac{T_2}{T_1}\right)^{k/(k-1)}$$

From Table A-20 at the mean temperature, 202°C (475 K), $k = 1.39$. Inserting values into the above expression

❷ $$p_2 = (1\text{ bar})\left(\frac{650}{300}\right)^{1.39/0.39} = 15.81\text{ bar}$$

❶ *IT* returns a value for p_2 even though it is an implicit variable in the specific entropy function. Also note that *IT* returns values for specific entropy *directly* and does not employ special functions such as $s°$, p_r, and v_r.

❷ The close agreement between the answer obtained in part (c) and that of parts (a), (b) can be attributed to the use of an appropriate value for the specific heat ratio k.

Skills Developed: *Ability to...*
- ✓ analyze an isentropic process using Table A-22 data,
- ✓ *Interactive Thermodynamics*, and
- ✓ a constant specific heat ratio k.

Quick Quiz

Determine the final pressure, in bar, using a constant specific heat ratio k evaluated at $T_1 = 300$ K. Expressed as a percent, how much does this pressure value differ from that of part (c)?

Ans. 14.72 bar, −5%

Another illustration of an isentropic process of an ideal gas is provided in Example 6.10 dealing with air leaking from a tank.

Example 6.10 AIR LEAKING FROM A TANK

A rigid, well-insulated tank is filled initially with 5 kg of air at a pressure of 5 bar and a temperature of 500 K. A leak develops, and air slowly escapes until the pressure of the air remaining in the tank is 1 bar. Employing the ideal gas model, determine the amount of mass remaining in the tank and its temperature.

Solution

Known: A leak develops in a rigid, insulated tank initially containing air at a known state. Air slowly escapes until the pressure in the tank is reduced to a specified value.

Find: Determine the amount of mass remaining in the tank and its temperature.

Schematic and Given Data:

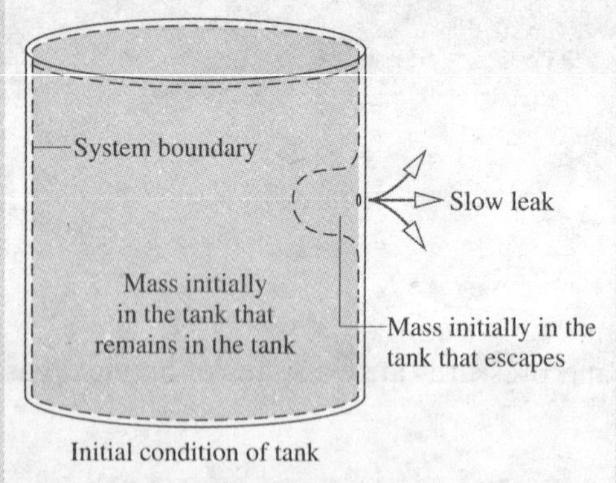

Engineering Model:

1. As shown on the accompanying sketch, the closed system is the mass initially in the tank that remains in the tank.
2. There is no significant heat transfer between the system and its surroundings.
3. Irreversibilities within the tank can be ignored as the air slowly escapes.
4. The air is modeled as an ideal gas.

Fig. E6.10

Analysis: With the ideal gas equation of state, the mass initially in the tank that *remains* in the tank at the end of the process is

$$m_2 = \frac{p_2 V}{(\bar{R}/M)T_2}$$

where p_2 and T_2 are the final pressure and temperature, respectively. Similarly, the initial amount of mass within the tank, m_1, is

$$m_1 = \frac{p_1 V}{(\bar{R}/M)T_1}$$

where p_1 and T_1 are the initial pressure and temperature, respectively. Eliminating volume between these two expressions, the mass of the system is

$$m_2 = \left(\frac{p_2}{p_1}\right)\left(\frac{T_1}{T_2}\right)m_1$$

Except for the final temperature of the air remaining in the tank, T_2, all required values are known. The remainder of the solution mainly concerns the evaluation of T_2.

❶ For the closed system under consideration, there are no significant irreversibilities (assumption 3), and no heat transfer occurs (assumption 2). Accordingly, the entropy balance reduces to

$$\Delta S = \int_1^2 \left(\frac{\delta Q}{T}\right)_b^{\,0} + \sigma^{0} = 0$$

Since the system mass remains constant, $\Delta S = m_2 \, \Delta s$, so

$$\Delta s = 0$$

That is, the initial and final states of the system have the same value of *specific* entropy.

Using Eq. 6.41

$$p_{r2} = \left(\frac{p_2}{p_1}\right)p_{r1}$$

where $p_1 = 5$ bar and $p_2 = 1$ bar. With $p_{r1} = 8.411$ from Table A-22 at 500 K, the previous equation gives $p_{r2} = 1.6822$. Using this to interpolate in Table A-22, $T_2 = 317$ K.

Finally, inserting values into the expression for system mass

$$m_2 = \left(\frac{1\text{ bar}}{5\text{ bar}}\right)\left(\frac{500\text{ K}}{317\text{ K}}\right)(5\text{ kg}) = 1.58\text{ kg}$$

❶ This problem also could be solved by considering a control volume enclosing the tank. The state of the control volume would change with time as air escapes. The details of the analysis are left as an exercise.

Skills Developed: *Ability to...*
- ✓ develop an engineering model.
- ✓ apply the closed system entropy balance.
- ✓ analyze an isentropic process.

Quick Quiz

Evaluate the tank volume, in m^3.

Ans. 1.43 m^3

6.12 Isentropic Efficiencies of Turbines, Nozzles, Compressors, and Pumps

Engineers make frequent use of efficiencies and many different efficiency definitions are employed. In the present section, *isentropic* efficiencies for turbines, nozzles, compressors, and pumps are introduced. Isentropic efficiencies involve a comparison between the actual performance of a device and the performance that would be achieved under idealized circumstances for the same inlet state and the same exit pressure. These efficiencies are frequently used in subsequent sections of the book.

6.12.1 Isentropic Turbine Efficiency

To introduce the isentropic turbine efficiency, refer to Fig. 6.11, which shows a turbine expansion on a Mollier diagram. The state of the matter entering the turbine and the exit pressure are fixed. Heat transfer between the turbine and its surroundings is ignored, as are kinetic and potential energy effects. With these assumptions, the mass and energy rate balances reduce, at steady state, to give the work developed per unit of mass flowing through the turbine

$$\frac{\dot{W}_{cv}}{\dot{m}} = h_1 - h_2$$

Since state 1 is fixed, the specific enthalpy h_1 is known. Accordingly, the value of the work depends on the specific enthalpy h_2 only, and increases as h_2 is reduced. The *maximum* value for the turbine work corresponds to the smallest *allowed* value for the specific enthalpy at the turbine exit. This can be determined using the second law as follows.

Since there is no heat transfer, the allowed exit states are constrained by Eq. 6.39

$$\frac{\dot{\sigma}_{cv}}{\dot{m}} = s_2 - s_1 \geq 0$$

Because the entropy production $\dot{\sigma}_{cv}/\dot{m}$ cannot be negative, states with $s_2 < s_1$ are not accessible in an adiabatic expansion. The only states that actually can be attained *adiabatically* are those with $s_2 > s_1$. The state labeled "2s" on Fig. 6.11 would be attained only in the limit of no internal irreversibilities. This corresponds to an isentropic expansion through the turbine. For fixed exit pressure, the specific enthalpy h_2

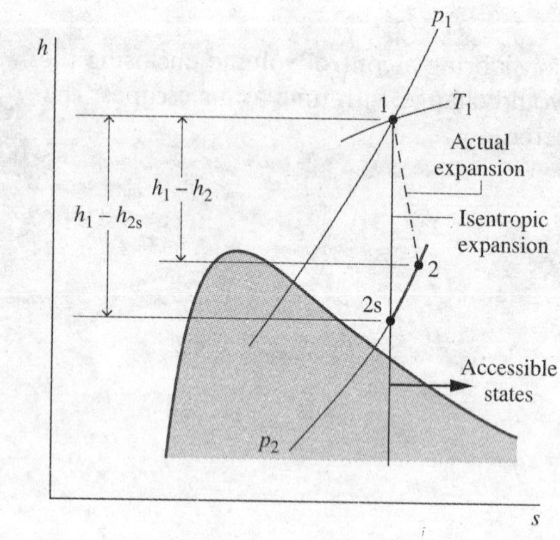

Fig. 6.11 Comparison of actual and isentropic expansions through a turbine.

decreases as the specific entropy s_2 decreases. Therefore, the *smallest allowed* value for h_2 corresponds to state 2s, and the *maximum* value for the turbine work is

$$\left(\frac{\dot{W}_{cv}}{\dot{m}}\right)_s = h_1 - h_{2s}$$

In an actual expansion through the turbine $h_2 > h_{2s}$, and thus less work than the maximum would be developed. This difference can be gauged by the ***isentropic turbine efficiency*** defined by

isentropic turbine efficiency

$$\eta_t = \frac{\dot{W}_{cv}/\dot{m}}{(\dot{W}_{cv}/\dot{m})_s} = \frac{h_1 - h_2}{h_1 - h_{2s}} \tag{6.46}$$

Both the numerator and denominator of this expression are evaluated for the same inlet state and the same exit pressure. The value of η_t is typically 0.7 to 0.9 (70–90%).

The two examples to follow illustrate the isentropic turbine efficiency concept. In Example 6.11 the isentropic efficiency of a steam turbine is known and the objective is to determine the turbine work.

Example 6.11 EVALUATING TURBINE WORK USING THE ISENTROPIC EFFICIENCY

A steam turbine operates at steady state with inlet conditions of $p_1 = 5$ bar, $T_1 = 320°C$. Steam leaves the turbine at a pressure of 1 bar. There is no significant heat transfer between the turbine and its surroundings, and kinetic and potential energy changes between inlet and exit are negligible. If the isentropic turbine efficiency is 75%, determine the work developed per unit mass of steam flowing through the turbine, in kJ/kg.

Solution

Known: Steam expands through a turbine operating at steady state from a specified inlet state to a specified exit pressure. The turbine efficiency is known.

Find: Determine the work developed per unit mass of steam flowing through the turbine.

Schematic and Given Data:

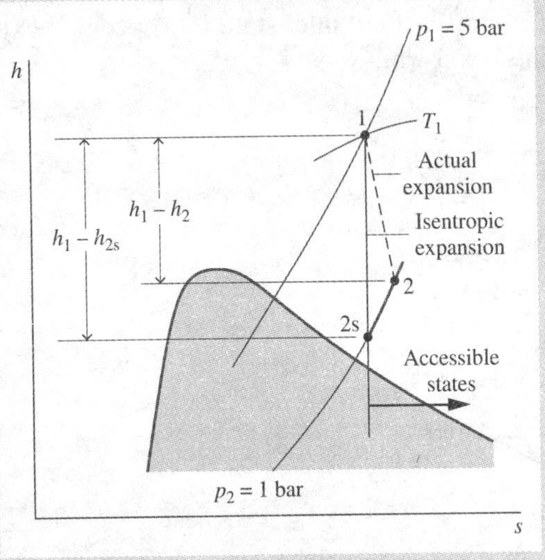

Engineering Model:

1. A control volume enclosing the turbine is at steady state.
2. The expansion is adiabatic and changes in kinetic and potential energy between the inlet and exit can be neglected.

Fig. E6.11

Analysis: The work developed can be determined using the isentropic turbine efficiency, Eq. 6.46, which on rearrangement gives

$$\frac{\dot{W}_{cv}}{\dot{m}} = \eta_t \left(\frac{\dot{W}_{cv}}{\dot{m}}\right)_s = \eta_t(h_1 - h_{2s})$$

From Table A-4, $h_1 = 3105.6$ kJ/kg and $s_1 = 7.5308$ kJ/kg · K. The exit state for an isentropic expansion is fixed
❶ by $p_2 = 1$ bar and $s_{2s} = s_1$. Interpolating with specific entropy in Table A-4 at 1 bar gives $h_{2s} = 2743.0$ kJ/kg. Substituting values

❷ $$\frac{\dot{W}_{cv}}{\dot{m}} = 0.75(3105.6 - 2743.0) = 271.95 \text{ kJ/kg}$$

---

❶ At 2s, the temperature is about 133°C.

❷ The effect of irreversibilities is to exact a penalty on the work output of the turbine. The work is only 75% of what it would be for an isentropic expansion between the given inlet state and the turbine exhaust pressure. This is clearly illustrated in terms of enthalpy differences on the accompanying h–s diagram.

Skills Developed: *Ability to...*
- ✓ apply the isentropic turbine efficiency, Eq. 6.46.
- ✓ retrieve steam table data.

Quick Quiz

Determine the temperature of the steam at the turbine exit, in °C.
Ans. 179°C

Example 6.12 is similar to Example 6.11, but here the working substance is air as an ideal gas. Moreover, in this case the turbine work is known and the objective is to determine the isentropic turbine efficiency.

Example 6.12 EVALUATING THE ISENTROPIC TURBINE EFFICIENCY

A turbine operating at steady state receives air at a pressure of $p_1 = 3.0$ bar and a temperature of $T_1 = 390$ K. Air exits the turbine at a pressure of $p_2 = 1.0$ bar. The work developed is measured as 74 kJ per kg of air flowing through the turbine. The turbine operates adiabatically, and changes in kinetic and potential energy between inlet and exit can be neglected. Using the ideal gas model for air, determine the isentropic turbine efficiency.

Solution

Known: Air expands adiabatically through a turbine at steady state from a specified inlet state to a specified exit pressure. The work developed per kg of air flowing through the turbine is known.

Find: Determine the turbine efficiency.

Schematic and Given Data:

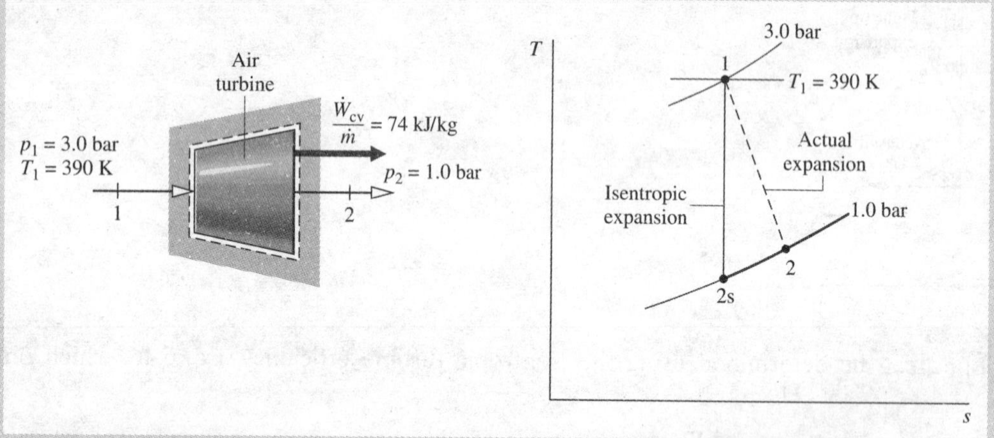

Fig. E6.12

Engineering Model:

1. The control volume shown on the accompanying sketch is at steady state.
2. The expansion is adiabatic and changes in kinetic and potential energy between inlet and exit can be neglected.
3. The air is modeled as an ideal gas.

Analysis: The numerator of the isentropic turbine efficiency, Eq. 6.46, is known. The denominator is evaluated as follows.

The work developed in an isentropic expansion from the given inlet state to the specified exit pressure is

$$\left(\frac{\dot{W}_{cv}}{\dot{m}}\right)_s = h_1 - h_{2s}$$

From Table A-22 at 390 K, $h_1 = 390.88$ kJ/kg. To determine h_{2s}, use Eq. 6.41

$$p_r(T_{2s}) = \left(\frac{p_2}{p_1}\right) p_r(T_1)$$

With $p_1 = 3.0$ bar, $p_2 = 1.0$ bar, and $p_{r1} = 3.481$ from Table A-22 at 390 K

$$p_r(T_{2s}) = \left(\frac{1.0}{3.0}\right)(3.481) = 1.1603$$

Interpolation in Table A-22 gives $h_{2s} = 285.27$ kJ/kg. Thus

$$\left(\frac{\dot{W}_{cv}}{\dot{m}}\right)_s = 390.88 - 285.27 = 105.6 \text{ kJ/kg}$$

Substituting values into Eq. 6.46

$$\eta_t = \frac{\dot{W}_{cv}/\dot{m}}{(\dot{W}_{cv}/\dot{m})_s} = \frac{74 \text{ kJ/kg}}{105.6 \text{ kJ/kg}} = 0.70\ (70\%)$$

Skills Developed: *Ability to...*
- ✔ apply the isentropic turbine efficiency, Eq. 6.46.
- ✔ retrieve data for air as an ideal gas.

Quick Quiz

Determine the rate of entropy production, in kJ/K per kg of air flowing through the turbine.

Ans. 0.105 kJ/kg · K

6.12.2 Isentropic Nozzle Efficiency

A similar approach to that for turbines can be used to introduce the isentropic efficiency of nozzles operating at steady state. The ***isentropic nozzle efficiency*** is defined as the ratio of the actual specific kinetic energy of the gas leaving the nozzle, $V_2^2/2$, to the kinetic energy at the exit that would be achieved in an isentropic expansion between the same inlet state and the same exhaust pressure, $(V_2^2/2)_s$

$$\eta_{\text{nozzle}} = \frac{V_2^2/2}{(V_2^2/2)_s} \tag{6.47}$$

isentropic nozzle efficiency

Nozzle efficiencies of 95% or more are common, indicating that well-designed nozzles are nearly free of internal irreversibilities.

In Example 6.13, the objective is to determine the isentropic efficiency of a steam nozzle.

Example 6.13 EVALUATING THE ISENTROPIC NOZZLE EFFICIENCY

Steam enters a nozzle operating at steady state at $p_1 = 9.7$ bar and $T_1 = 316°C$ with a velocity of 30 m/s. The pressure and temperature at the exit are $p_2 = 2.8$ bar and $T_2 = 177°C$. There is no significant heat transfer between the nozzle and its surroundings, and changes in potential energy between inlet and exit can be neglected. Determine the nozzle efficiency.

Solution

Known: Steam expands through a nozzle at steady state from a specified inlet state to a specified exit state. The velocity at the inlet is known.

Find: Determine the nozzle efficiency.

Schematic and Given Data:

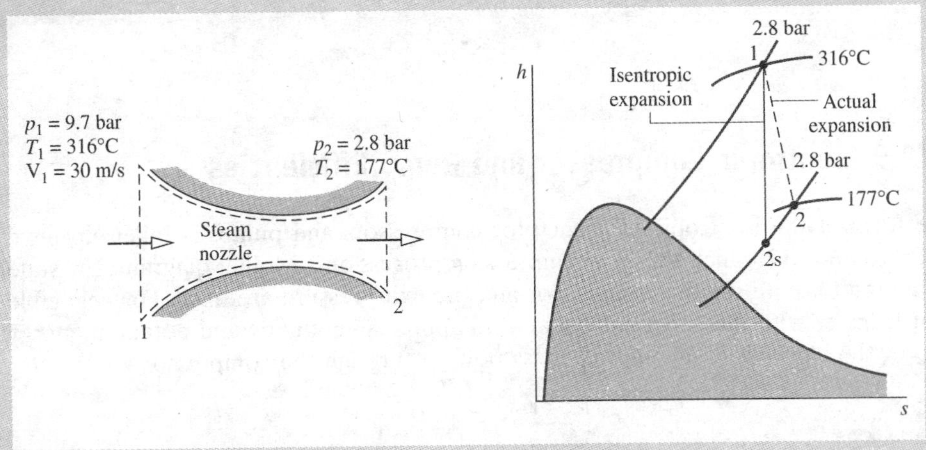

Fig. E6.13

Engineering Model:

1. The control volume shown on the accompanying sketch operates adiabatically at steady state.
2. For the control volume, $\dot{W}_{cv} = 0$ and the change in potential energy between inlet and exit can be neglected.

Analysis: The nozzle efficiency given by Eq. 6.47 requires the actual specific kinetic energy at the nozzle exit and the specific kinetic energy that would be achieved at the exit in an isentropic expansion from the given inlet state to the given exit pressure. The energy rate balance for a one-inlet, one-exit control volume at steady state (Eq. 4.20b) reduces to give

$$\frac{V_2^2}{2} = h_1 - h_2 + \frac{V_1^2}{2}$$

This equation applies for both the actual expansion and the isentropic expansion.

From Table A-4 at $T_1 = 320°C$ and $p_1 = 1$ MPa, $h_1 = 3093.9$ kJ/kg, $s_1 = 7.1962$ kJ/kg · K. Also, with $T_2 = 180°C$ and $p_2 = 0.3$ MPa, $h_2 = 2823.9$ kJ/kg. Thus, the actual specific kinetic energy at the exit in kJ/kg is

$$\frac{V_2^2}{2} = (3093.9 - 2823.9)\frac{\text{kJ}}{\text{kg}} + \frac{(30)^2}{2}\left(\frac{\text{m}}{\text{s}}\right)^2\left|\frac{1\text{N}}{1\text{ kg}\cdot\text{m/s}^2}\right|\left|\frac{1\text{ kJ}}{10^3\text{ N}\cdot\text{m}}\right|$$

$$= 270.45\frac{\text{kJ}}{\text{kg}}$$

Interpolating in Table A-4 at 0.3 MPa, with $s_{2s} = s_1 = 7.1962$ kJ/kg · K, results in $h_{2s} = 2813.3$ kJ/kg. Accordingly, the specific kinetic energy at the exit for an isentropic expansion is

$$\left(\frac{V_2^2}{2}\right)_s = 3093.9 - 2813.3 + \frac{(30)^2}{2(10^3)} = 281.05 \text{ kJ/kg}$$

Substituting values into Eq. 6.47

$$\eta_{\text{nozzle}} = \frac{(V_2^2/2)}{(V_2^2/2)_s} = \frac{270.45}{281.05} = 0.962\ (96.2\%)$$

❶

❶ The principal irreversibility in nozzles is friction between the flowing gas or liquid and the nozzle wall. The effect of friction is that a smaller exit kinetic energy, and thus a smaller exit velocity, is realized than would have been obtained in an isentropic expansion to the same pressure.

Skills Developed: *Ability to...*
- ✓ apply the control volume energy rate balance.
- ✓ apply the isentropic nozzle efficiency, Eq. 6.47.
- ✓ retrieve steam table data.

Quick Quiz

Determine the temperature, in °C, corresponding to state 2s in Fig. E6.13.
Ans. 166°C

6.12.3 Isentropic Compressor and Pump Efficiencies

The form of the isentropic efficiency for compressors and pumps is taken up next. Refer to Fig. 6.12, which shows a compression process on a Mollier diagram. The state of the matter entering the compressor and the exit pressure are fixed. For negligible heat transfer with the surroundings and no appreciable kinetic and potential energy effects, the work *input* per unit of mass flowing through the compressor is

$$\left(-\frac{\dot{W}_{cv}}{\dot{m}}\right) = h_2 - h_1$$

Since state 1 is fixed, the specific enthalpy h_1 is known. Accordingly, the value of the work input depends on the specific enthalpy at the exit, h_2. The above expression shows that the magnitude of the work input decreases as h_2 decreases. The *minimum* work

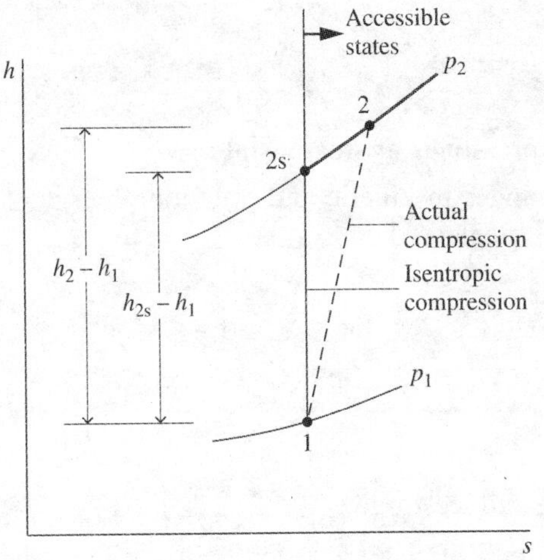

Fig. 6.12 Comparison of actual and isentropic compressions.

input corresponds to the smallest *allowed* value for the specific enthalpy at the compressor exit. With similar reasoning as for the turbine, the smallest allowed enthalpy at the exit state would be achieved in an isentropic compression from the specified inlet state to the specified exit pressure. The minimum work *input* is given, therefore, by

$$\left(-\frac{\dot{W}_{cv}}{\dot{m}}\right)_s = h_{2s} - h_1$$

In an actual compression, $h_2 > h_{2s}$, and thus more work than the minimum would be required. This difference can be gauged by the ***isentropic compressor efficiency*** defined by

$$\eta_c = \frac{(-\dot{W}_{cv}/\dot{m})_s}{(-\dot{W}_{cv}/\dot{m})} = \frac{h_{2s} - h_1}{h_2 - h_1} \tag{6.48}$$

isentropic compressor efficiency

Both the numerator and denominator of this expression are evaluated for the same inlet state and the same exit pressure. The value of η_c is typically 75 to 85% for compressors. An ***isentropic pump efficiency,*** η_p, is defined similarly.

isentropic pump efficiency

In Example 6.14, the isentropic efficiency of a refrigerant compressor is evaluated, first using data from property tables and then using *IT*.

Example 6.14 EVALUATING THE ISENTROPIC COMPRESSOR EFFICIENCY

For the compressor of the heat pump system in Example 6.8, determine the power, in kW, and the isentropic efficiency using **(a)** data from property tables, **(b)** *Interactive Thermodynamics: IT.*

Solution

Known: Refrigerant 22 is compressed adiabatically at steady state from a specified inlet state to a specified exit state. The mass flow rate is known.

Find: Determine the compressor power and the isentropic efficiency using (a) property tables, (b) *IT*.

Schematic and Given Data:

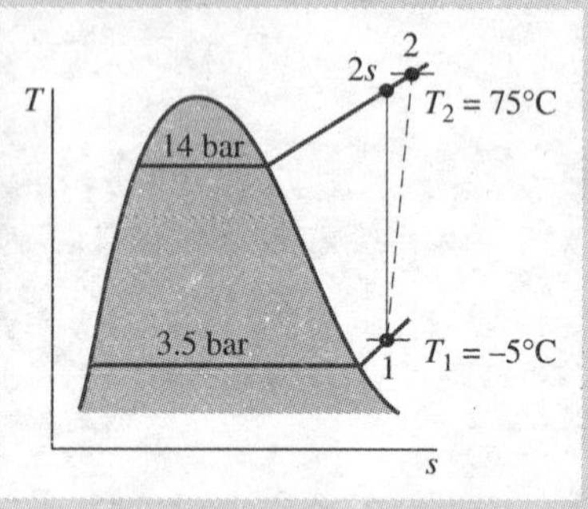

Engineering Model:

1. A control volume enclosing the compressor is at steady state.
2. The compression is adiabatic, and changes in kinetic and potential energy between the inlet and the exit can be neglected.

Fig. E6.14

Analysis: **(a)** By assumptions 1 and 2, the mass and energy rate balances reduce to give

$$\dot{W}_{cv} = \dot{m}(h_1 - h_2)$$

From Table A-9, $h_1 = 249.75$ kJ/kg and $h_2 = 294.17$ kJ/kg. Thus

$$\dot{W}_{cv} = (0.07\text{ kg/s})(249.75 - 294.17)\text{ kJ/kg}\left|\frac{1\text{ kW}}{1\text{ kJ/s}}\right| = -3.11\text{ kW}$$

The isentropic compressor efficiency is determined using Eq. 6.48

$$\eta_c = \frac{(-\dot{W}_{cv}/\dot{m})_s}{(-\dot{W}_{cv}/\dot{m})} = \frac{(h_{2s} - h_1)}{(h_2 - h_1)}$$

In this expression, the denominator represents the work input per unit mass of refrigerant flowing for the actual compression process, as considered above. The numerator is the work input for an isentropic compression between the initial state and the same exit pressure. The isentropic exit state is denoted as state 2s on the accompanying T–s diagram.

From Table A-9, $s_1 = 0.9572$ kJ/kg · K. With $s_{2s} = s_1$, interpolation in Table A-9 at 14 bar gives $h_{2s} = 285.58$ kJ/kg. Substituting values

$$\eta_c = \frac{(285.58 - 249.75)}{(294.17 - 249.75)} = 0.81(81\%)$$

(b) The *IT* program follows. In the program, $\dot{W}_{cv}$ is denoted as Wdot, $\dot{m}$ as mdot, and η_c as eta_c.

```
// Given Data:
T1 = -5 // °C
p1 = 3.5 // bar
T2 = 75 // °C
p2 = 14 // bar
mdot = 0.07 // kg/s

// Determine the specific enthalpies.
h1 = h_PT("R22",p1,T1)
h2 = h_PT("R22",p2,T2)

// Calculate the power.
Wdot = mdot * (h1 - h2)
// Find h2s:
❶ s1 = s_PT("R22",p1,T1)
s2s = s_Ph("R22",p2,h2s)
s2s = s1

// Determine the isentropic compressor efficiency.
eta_c = (h2s - h1) / (h2 - h1)
```

Use the **Solve** button to obtain: $\dot{W}_{cv} = -3.111$ kW and $\eta_c = 80.58\%$, which agree closely with the values obtained above.

❶ Note that *IT* solves for the value of h_{2s} even though it is an implicit variable in the specific entropy function.

Skills Developed: *Ability to...*
- ✓ apply the control volume energy rate balance.
- ✓ apply the isentropic compressor efficiency, Eq. 6.48.
- ✓ retrieve data for Refrigerant 22.

Quick Quiz

Determine the minimum theoretical work input, in kJ per kg flowing, for an adiabatic compression from state 1 to the exit pressure of 14 bar.
Ans. 35.83 kJ/kg

Energy & Environment

Government is partnering with business and industry to identify and promote energy-efficient products. ENERGY STAR® is a voluntary labeling program operated jointly by the US Environmental Protection Agency and the Department of Energy that saved enough energy in one recent year to power 10 million homes while reducing air pollution equivalent to removing 10 million cars from the road—all without sacrificing product features consumers want.

The government agencies sign agreements with program partners who then can use the ENERGY STAR® logo on products that meet specified energy efficiency and performance criteria. The logo now appears on many major appliances, office equipment, lighting products, and consumer electronics to encourage consumers to buy products that reduce energy consumption and greenhouse gas emissions. The program also covers new homes as well as commercial and industrial buildings.

Compared to the annual energy requirements of typical new equipment, ENERGY STAR® products save the following percentages:

- household appliances: 10–50%
- office equipment: 30–70%
- residential and commercial lighting: 70–90%
- consumer electronics: 20–40%
- residential heating and cooling equipment: 10–30%

By choosing ENERGY STAR® products, consumers have already saved billions on their utility bills and helped the environment, government sources say.

6.13 Heat Transfer and Work in Internally Reversible, Steady-State Flow Processes

This section concerns one-inlet, one-exit control volumes at steady state. The objective is to introduce expressions for the heat transfer and the work in the absence of internal irreversibilities. The resulting expressions have several important applications.

6.13.1 Heat Transfer

For a control volume at steady state in which the flow is both *isothermal* and *internally reversible*, the appropriate form of the entropy rate balance is

$$0 = \frac{\dot{Q}_{cv}}{T} + \dot{m}(s_1 - s_2) + \cancel{\dot{\sigma}_{cv}}^{\,0}$$

where 1 and 2 denote the inlet and exit, respectively, and $\dot{m}$ is the mass flow rate. Solving this equation, the heat transfer per unit of mass passing through the control volume is

$$\frac{\dot{Q}_{cv}}{\dot{m}} = T(s_2 - s_1)$$

More generally, the temperature would vary as the gas or liquid flows through the control volume. However, we can consider the temperature variation to consist of a series of infinitesimal steps. Then, the heat transfer per unit of mass would be given as

$$\left(\frac{\dot{Q}_{cv}}{\dot{m}}\right)_{\substack{\text{int}\\ \text{rev}}} = \int_1^2 T\,ds \tag{6.49}$$

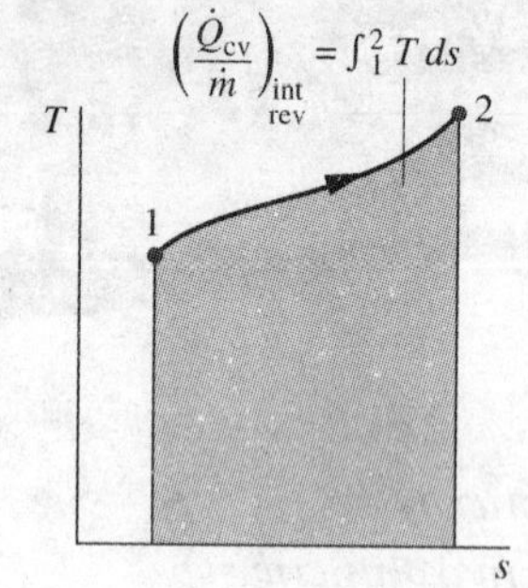

Fig. 6.13 Area representation of heat transfer for an internally reversible flow process.

The subscript "int rev" serves to remind us that the expression applies only to control volumes in which there are no internal irreversibilities. The integral of Eq. 6.49 is performed from inlet to exit. When the states visited by a unit mass as it passes reversibly from inlet to exit are described by a curve on a T–s diagram, the magnitude of the heat transfer per unit of mass flowing can be represented as the area *under* the curve, as shown in Fig. 6.13.

6.13.2 Work

The work per unit of mass passing through the control volume can be found from an energy rate balance, which reduces at steady state to give

$$\frac{\dot{W}_{cv}}{\dot{m}} = \frac{\dot{Q}_{cv}}{\dot{m}} + (h_1 - h_2) + \left(\frac{V_1^2 - V_2^2}{2}\right) + g(z_1 - z_2)$$

This equation is a statement of the conservation of energy principle that applies when irreversibilities are present within the control volume as well as when they are absent. However, if consideration is restricted to the internally reversible case, Eq. 6.49 can be introduced to obtain

$$\left(\frac{\dot{W}_{cv}}{\dot{m}}\right)_{\substack{\text{int}\\ \text{rev}}} = \int_1^2 T\,ds + (h_1 - h_2) + \left(\frac{V_1^2 - V_2^2}{2}\right) + g(z_1 - z_2) \tag{6.50}$$

where the subscript "int rev" has the same significance as before.

Since internal irreversibilities are absent, a unit of mass traverses a sequence of equilibrium states as it passes from inlet to exit. Entropy, enthalpy, and pressure changes are therefore related by Eq. 6.10b

$$T\,ds = dh - v\,dp$$

which on integration gives

$$\int_1^2 T\,ds = (h_2 - h_1) - \int_1^2 v\,dp$$

Introducing this relation, Eq. 6.50 becomes

$$\left(\frac{\dot{W}_{cv}}{\dot{m}}\right)_{\substack{\text{int}\\ \text{rev}}} = -\int_1^2 v\,dp + \left(\frac{V_1^2 - V_2^2}{2}\right) + g(z_1 - z_2) \tag{6.51a}$$

Fig. 6.14 Area representation of $\int_1^2 v\,dp$.

When the states visited by a unit of mass as it passes reversibly from inlet to exit are described by a curve on a p–v diagram as shown in Fig. 6.14, the magnitude of the integral $\int v\,dp$ is represented by the shaded area *behind* the curve.

Equation 6.51a is applicable to devices such as turbines, compressors, and pumps. In many of these cases, there is no significant change in kinetic or potential energy from inlet to exit, so

$$\left(\frac{\dot{W}_{cv}}{\dot{m}}\right)_{\substack{int\\rev}} = -\int_1^2 v\, dp \qquad (\Delta ke = \Delta pe = 0) \tag{6.51b}$$

This expression shows that the work is related to the magnitude of the specific volume of the gas or liquid as it flows from inlet to exit. **FOR EXAMPLE...** consider two devices: a pump through which liquid water passes and a compressor through which water vapor passes. For the *same pressure rise,* the pump would require a much smaller work *input* per unit of mass flowing than would the compressor because the liquid specific volume is much smaller than that of vapor. This conclusion is also qualitatively correct for actual pumps and compressors, where irreversibilities are present during operation.

If the specific volume remains approximately constant, as in many applications with liquids, Eq. 6.51b becomes

$$\left(\frac{\dot{W}_{cv}}{\dot{m}}\right)_{\substack{int\\rev}} = -v(p_2 - p_1) \qquad (v = \text{constant}, \Delta ke = \Delta pe = 0) \tag{6.51c}$$

Equation 6.51a also can be applied to study the performance of control volumes at steady state in which $\dot{W}_{cv}$ is zero, as in the case of nozzles and diffusers. For any such case, the equation becomes

$$\int_1^2 v\, dp + \left(\frac{V_2^2 - V_1^2}{2}\right) + g(z_2 - z_1) = 0 \tag{6.52}$$

Bernoulli equation

which is a form of the ***Bernoulli equation*** frequently used in fluid mechanics.

Energy & Environment

For coal-fired power plants, the cost of transporting coal from mine to plant is a major contributor to the coal cost per ton. For long distances, pumping coal-water mixtures through pipelines may be less costly with fewer safety and environmental issues than truck and train transport. In *coal-slurry* pipelines, a slurry of pulverized coal and water is pumped. In *coal-log* pipelines, cylinders (logs) of compacted coal suspended in water are pumped. Water requirements are significant with both approaches, ranging up to millions of gallons per day in some applications. In coal-slurry pipelines the mass ratio of water to coal is about one to one, whereas in coal-log pipelines it is approximately one to three, making coal-log pipelines more attractive in terms of water use. Coal-slurry pipelines are deployed in several countries including the United States. The newer coal-log concept has yet to be commercialized.

6.13.3 Work In Polytropic Processes

In the discussion of Fig. 6.10, we have identified an internally reversible process described by $pv^n = constant$, where n is a constant, as a *polytropic process.* If each unit of mass passing through a one-inlet, one-exit control volume undergoes a polytropic process, then introducing $pv^n = constant$ in Eq. 6.5lb, and performing the integration, gives the work per unit of mass in the absence of internal irreversibilities and significant changes in kinetic and potential energy. That is,

$$\left(\frac{\dot{W}_{cv}}{\dot{m}}\right)_{\substack{int\\rev}} = -\int_1^2 v\, dp = -(constant)^{1/n}\int_1^2 \frac{dp}{p^{1/n}}$$

$$= -\frac{n}{n-1}(p_2v_2 - p_1v_1) \qquad (\text{polytropic}, n \neq 1) \tag{6.53}$$

for any value of n except $n = 1$. When $n = 1$, $pv = constant$, and the work is

$$\left(\frac{\dot{W}_{cv}}{\dot{m}}\right)_{\substack{int\\rev}} = -\int_1^2 v\,dp = -constant\int_1^2 \frac{dp}{p}$$

$$= -(p_1 v_1)\ln(p_2/p_1) \qquad \text{(polytropic, } n = 1\text{)} \qquad (6.54)$$

Equations 6.53 and 6.54 apply generally to polytropic processes of *any* gas (or liquid).

IDEAL GAS CASE. For the special case of an ideal gas, Eq. 6.53 becomes

$$\left(\frac{\dot{W}_{cv}}{\dot{m}}\right)_{\substack{int\\rev}} = -\frac{nR}{n-1}(T_2 - T_1) \qquad \text{(ideal gas, } n \neq 1\text{)} \qquad (6.55a)$$

For a polytropic process of an ideal gas, Eq. 3.56 applies:

$$\frac{T_2}{T_1} = \left(\frac{p_2}{p_1}\right)^{(n-1)/n}$$

Thus, Eq. 6.55a can be expressed alternatively as

$$\left(\frac{\dot{W}_{cv}}{\dot{m}}\right)_{\substack{int\\rev}} = -\frac{nRT_1}{n-1}\left[\left(\frac{p_2}{p_1}\right)^{(n-1)/n} - 1\right] \qquad \text{(ideal gas, } n \neq 1\text{)} \qquad (6.55b)$$

For the case of an ideal gas, Eq. 6.54 becomes

$$\left(\frac{\dot{W}_{cv}}{\dot{m}}\right)_{\substack{int\\rev}} = -RT\ln(p_2/p_1) \qquad \text{(ideal gas, } n = 1\text{)} \qquad (6.56)$$

In Example 6.15, we consider air modeled as an ideal gas undergoing a polytropic compression process at steady state.

Example 6.15 POLYTROPIC COMPRESSION OF AIR

An air compressor operates at steady state with air entering at $p_1 = 1$ bar, $T_1 = 20°C$, and exiting at $p_2 = 5$ bar. Determine the work and heat transfer per unit of mass passing through the device, in kJ/kg, if the air undergoes a polytropic process with $n = 1.3$. Neglect changes in kinetic and potential energy between the inlet and the exit. Use the ideal gas model for air.

Solution

Known: Air is compressed in a polytropic process from a specified inlet state to a specified exit pressure.

Find: Determine the work and heat transfer per unit of mass passing through the device.

Schematic and Given Data:

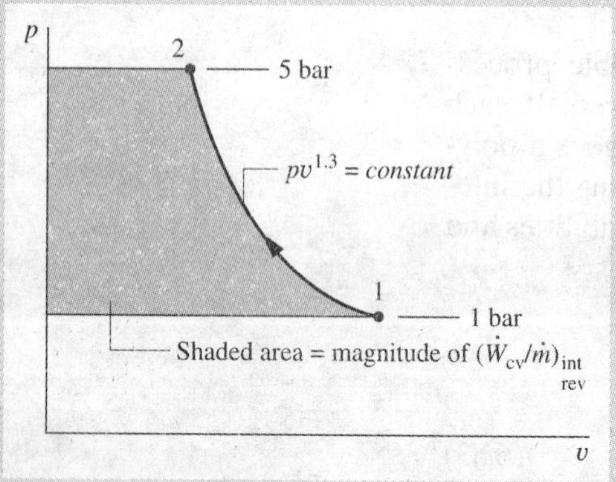

❶

Engineering Model:

1. A control volume enclosing the compressor is at steady state.
2. The air undergoes a polytropic process with $n = 1.3$.
3. The air behaves as an ideal gas.
4. Changes in kinetic and potential energy from inlet to exit can be neglected.

Fig. E6.15

Analysis: The work is obtained using Eq. 6.55a, which requires the temperature at the exit, T_2. The temperature T_2 can be found using Eq. 3.56

$$T_2 = T_1\left(\frac{p_2}{p_1}\right)^{(n-1)/n} = 293\left(\frac{5}{1}\right)^{(1.3-1)/1.3} = 425 \text{ K}$$

Substituting known values into Eq. 6.55a then gives

$$\frac{\dot{W}_{cv}}{\dot{m}} = -\frac{nR}{n-1}(T_2 - T_1) = -\frac{1.3}{1.3-1}\left(\frac{8.314}{28.97}\frac{\text{kJ}}{\text{kg}\cdot\text{K}}\right)(425-293) \text{ K}$$
$$= -164.2 \text{ kJ/kg}$$

The heat transfer is evaluated by reducing the mass and energy rate balances with the appropriate assumptions to obtain

$$\frac{\dot{Q}_{cv}}{\dot{m}} = \frac{\dot{W}_{cv}}{\dot{m}} + h_2 - h_1$$

Using the temperatures T_1 and T_2, the required specific enthalpy values are obtained from Table A-22 as $h_1 = 293.17$ kJ/kg and $h_2 = 426.35$ kJ/kg. Thus

$$\frac{\dot{Q}_{cv}}{\dot{m}} = -164.15 + (426.35 - 293.17) = -31 \text{ kJ/kg}$$

❶ The states visited in the polytropic compression process are shown by the curve on the accompanying p–v diagram. The magnitude of the work per unit of mass passing through the compressor is represented by the shaded area *behind* the curve.

Skills Developed: *Ability to...*

- ✓ analyze a polytropic process of an ideal gas.
- ✓ apply the control volume energy rate balance.

Quick Quiz

If the air were to undergo a polytropic process with $n = 1.0$, determine the work and heat transfer, each in kJ per kg of air flowing, keeping all other given data the same.

Ans. −135.3 kJ/kg

✓ Chapter Summary and Study Guide

In this chapter, we have introduced the property entropy and illustrated its use for thermodynamic analysis. Like mass and energy, entropy is an extensive property that can be transferred across system boundaries. Entropy transfer accompanies both heat transfer and mass flow. Unlike mass and energy, entropy is not conserved but is *produced* within systems whenever internal irreversibilities are present.

The use of entropy balances is featured in this chapter. Entropy balances are expressions of the second law that account for the entropy of systems in terms of entropy transfers and entropy production. For processes of closed systems, the entropy balance is Eq. 6.24, and a corresponding rate form is Eq. 6.28. For control volumes, rate forms include Eq. 6.34 and the companion steady-state expression given by Eq. 6.36.

The following checklist provides a study guide for this chapter. When your study of the text and end-of-chapter exercises has been completed you should be able to

- ✓ write out meanings of the terms listed in the margins throughout the chapter and understand each of the related concepts. The subset of key concepts listed below is particularly important in subsequent chapters.
- ✓ apply entropy balances in each of several alternative forms, appropriately modeling the case at hand, correctly observing sign conventions, and carefully applying SI and English units.
- ✓ use entropy data appropriately, to include
  - –retrieving data from Tables A-2 through A-18, using Eq. 6.4 to evaluate the specific entropy of two-phase liquid–vapor mixtures, sketching T–s and h–s diagrams and locating

states on such diagrams, and appropriately using Eqs. 6.5 and 6.13.

–determining Δs of ideal gases using Eq. 6.20 for variable specific heats together with Tables A-22 and A-23, and using Eqs. 6.21 and 6.22 for constant specific heats.

–evaluating isentropic efficiencies for turbines, nozzles, compressors, and pumps from Eqs. 6.46, 6.47, and 6.48, respectively, including for ideal gases the appropriate use of Eqs. 6.41–6.42 for variable specific heats and Eqs. 6.43–6.45 for constant specific heats.

✔ apply Eq. 6.23 for closed systems and Eqs. 6.49 and 6.51 for one-inlet, one-exit control volumes at steady state, correctly observing the restriction to internally reversible processes.

Key Engineering Concepts

entropy change p. 256
T–s diagram p. 259
Mollier diagram p. 260
entropy balance p. 269
entropy transfer p. 266, 270, 282
entropy production p. 271
entropy rate balance p. 275, 282, 283
isentropic efficiencies p. 298, 301, 303

Key Equations

$S_2 - S_1 = \int_1^2 \left(\frac{\delta Q}{T}\right)_b + \sigma$	(6.24)	Closed system entropy balance.
$\frac{dS}{dt} = \sum_j \frac{\dot{Q}_j}{T_j} + \dot{\sigma}$	(6.28)	Closed system entropy rate balance.
$\frac{dS_{cv}}{dt} = \sum_j \frac{\dot{Q}_j}{T_j} + \sum_i \dot{m}_i s_i - \sum_e \dot{m}_e s_e + \dot{\sigma}_{cv}$	(6.34)	Control volume entropy rate balance.
$0 = \sum_j \frac{\dot{Q}_j}{T_j} + \sum_i \dot{m}_i s_i - \sum_e \dot{m}_e s_e + \dot{\sigma}_{cv}$	(6.36)	Steady-state control volume entropy rate balance.
$\eta_t = \frac{\dot{W}_{cv}/\dot{m}}{(\dot{W}_{cv}/\dot{m})_s} = \frac{h_1 - h_2}{h_1 - h_{2s}}$	(6.46)	Isentropic turbine efficiency.
$\eta_{nozzle} = \frac{V_2^2/2}{(V_2^2/2)_s}$	(6.47)	Isentropic nozzle efficiency.
$\eta_c = \frac{(-\dot{W}_{cv}/\dot{m})_s}{(-\dot{W}_{cv}/\dot{m})} = \frac{h_{2s} - h_1}{h_2 - h_1}$	(6.48)	Isentropic compressor (and pump) efficiency.

Ideal Gas Model Relations

$s(T_2, v_2) - s(T_1, v_1) = \int_{T_1}^{T_2} c_v(T)\frac{dT}{T} + R\ln\frac{v_2}{v_1}$	(6.17)	Change in specific entropy; general form for T and v as independent properties.
$s(T_2, v_2) - s(T_1, v_1) = c_v \ln\frac{T_2}{T_1} + R\ln\frac{v_2}{v_1}$	(6.21)	Constant specific heat, c_v.
$s(T_2, p_2) - s(T_1, p_1) = \int_{T_1}^{T_2} c_p(T)\frac{dT}{T} - R\ln\frac{p_2}{p_1}$	(6.18)	Change in specific entropy; general form for T and p as independent properties.
$s(T_2, p_2) - s(T_1, p_1) = s^\circ(T_2) - s^\circ(T_1) - R\ln\frac{p_2}{p_1}$	(6.20a)	s° for air from Table A-22. ($\bar{s}^\circ$ for other gases from Table A-23).

$s(T_2, p_2) - s(T_1, p_1) = c_p \ln \frac{T_2}{T_1} - R \ln \frac{p_2}{p_1}$	(6.22)	Constant specific heat, c_p.
$\frac{p_2}{p_1} = \frac{p_{r2}}{p_{r1}}$	(6.41)	$s_1 = s_2$ (air only), p_r and v_r from Table A-22.
$\frac{v_2}{v_1} = \frac{v_{r2}}{v_{r1}}$	(6.42)	
$\frac{T_2}{T_1} = \left(\frac{p_2}{p_1}\right)^{(k-1)/k}$	(6.43)	$s_1 = s_2$, constant specific heat ratio k.
$\frac{T_2}{T_1} = \left(\frac{v_1}{v_2}\right)^{k-1}$	(6.44)	
$\frac{p_2}{p_1} = \left(\frac{v_1}{v_2}\right)^{k}$	(6.45)	

Exercises: Things Engineers Think About

1. Of the extensive properties mass, energy, and entropy, which are conserved? Which are produced?

2. Is it possible for entropy to be negative? For entropy *change* to be negative? For entropy *production* to be negative?

3. By what means can entropy be transferred across the boundary of a closed system? Across the boundary of a control volume?

4. Is it possible for the entropy of *both* a closed system and its surroundings to *decrease* during a process? Both to *increase* during a process?

5. What happens to the entropy produced within an insulated, one-inlet, one-exit control volume operating at steady state?

6. If a closed system would undergo an internally reversible process and an irreversible process between the same end states, how would the changes in entropy for the two processes compare? How would the amounts of entropy produced compare?

7. Referring to Example 3.1, the entropy of the system changes. Why? Repeat for Example 3.2.

8. Referring to Fig. 2.3, if systems A and B operate adiabatically, does the entropy of each increase, decrease, or remain the same?

9. The two power cycles shown to the same scale in the figure are composed of internally reversible processes. Compare the net work developed by these cycles. Which cycle has the greater thermal efficiency?

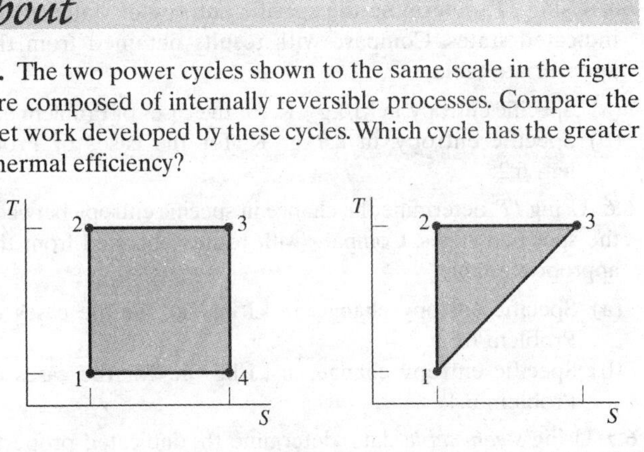

10. Sketch the T–s diagram for the Carnot vapor cycle of Fig. 5.15.

11. Sketch T–s and p–v diagrams of an ideal gas executing a power cycle consisting of four internally reversible processes in series: constant specific volume, constant pressure, isentropic, isothermal.

12. Could friction be less for a slurry of pulverized coal and water flowing through a pipeline than for water alone? Explain.

13. Reducing irreversibilities within a system can improve its *thermodynamic* performance, but steps taken in this direction are usually constrained by other considerations. What are some of these?

Problems: Developing Engineering Skills

Using Entropy Data and Concepts

6.1 Using the tables for water, determine the specific entropy at the indicated states, in kJ/kg · K. In each case, locate the state by hand on a sketch of the T–s diagram.

(a) $p = 5.0$ MPa, $T = 400°C$
(b) $p = 5.0$ MPa, $T = 100°C$
(c) $p = 5.0$ MPa, $u = 1872.5$ kJ/kg
(d) $p = 5.0$ MPa, saturated vapor

6.2 Using the tables for water, determine the specific entropy at the indicated states, in kJ/kg · K. In each case, locate the state by hand on a sketch of the T–s diagram.

(a) $p = 6.9$ MPa, $T = 400°C$
(b) $p = 6.9$ MPa, $T = 150°C$
(c) $p = 6.9$ MPa, $h = 2170$ kJ/kg
(d) $p = 6.9$ MPa, saturated vapor

6.3 Using the appropriate table, determine the change in specific entropy between the specified states, in kJ/kg · K.

(a) water, $p_1 = 10$ MPa, $T_1 = 400°C$, $p_2 = 10$ MPa, $T_2 = 100°C$.
(b) Refrigerant 134a, $h_1 = 111.44$ kJ/kg, $T_1 = -40°C$, saturated vapor at $p_2 = 5$ bar.
(c) air as an ideal gas, $T_1 = 7°C$, $p_1 = 2$ bar, $T_2 = 327°C$, $p_2 = 1$ bar.
(d) hydrogen (H_2) as an ideal gas, $T_1 = 727°C$, $p_1 = 1$ bar, $T_2 = 25°C$, $p_2 = 3$ bar.

6.4 Using the appropriate table, determine the change in specific entropy between the specified states, in kJ/kg · K.

(a) water, $p_1 = 6.9$ MPa, $T_1 = 427°C$, $p_2 = 6.9$ MPa, $T_2 = 38°C$.
(b) Refrigerant 134a, $h_1 = 111.43$ kJ/kg, $T_1 = -40°C$, saturated vapor at $p_2 = 276$ kPa.
(c) air as an ideal gas, $T_1 = 4°C$, $p_1 = 2$ bar, $T_2 = 216°C$, $p_2 = 1$ bar.
(d) carbon dioxide as an ideal gas, $T_1 = 438°C$, $p_1 = 1$ bar, $T_2 = 25°C$, $p_2 = 3.04$ bar.

6.5 Using *IT*, determine the specific entropy of water at the indicated states. Compare with results obtained from the appropriate table.

(a) Specific entropy, in kJ/kg · K, for the cases of Problem 6.1.
(b) Specific entropy, in kJ/kg · K, for the cases of Problem 6.2.

6.6 Using *IT*, determine the change in specific entropy between the specified states. Compare with results obtained from the appropriate table.

(a) Specific entropy change, in kJ/kg · K, for the cases of Problem 6.3.
(b) Specific entropy change, in kJ/kg · K, for the cases of Problem 6.4.

6.7 Using *steam table* data, determine the indicated property data for a process in which there is no change in specific entropy between state 1 and state 2. In each case, locate the states on a sketch of the *T–s* diagram.

(a) $T_1 = 40°C$, $x_1 = 100\%$, $p_2 = 150$ kPa. Find T_2, in °C, and Δh, in kJ/kg.
(b) $T_1 = 10°C$, $x_1 = 75\%$, $p_2 = 1$ MPa. Find T_2, in °C, and Δu, in kJ/kg.

6.8 Using the appropriate table, determine the indicated property for a process in which there is no change in specific entropy between state 1 and state 2.

(a) water, $p_1 = 101.3$ kPa, $T_1 = 260°C$, $p_2 = 0.7$ MPa Find T_2 in °C.
(b) water, $T_1 = 10°C$, $x_1 = 0.75$, saturated vapor at state 2. Find p_2 in bar.
(c) air as an ideal gas, $T_1 = 27°C$, $p_1 = 1.5$ bar, $T_2 = 127°C$. Find p_2 in bar.
(d) air as an ideal gas, $T_1 = 38°C$, $p_1 = 3$ bar, $p_2 = 2$ bar. Find T_2 in °C.
(e) Refrigerant 134a, $T_1 = 20°C$, $p_1 = 5$ bar, $p_2 = 1$ bar. Find v_2 in m^3/kg.

6.9 Using *IT*, obtain the property data requested in (a) Problem 6.7, (b) Problem 6.8, and compare with data obtained from the appropriate table.

6.10 Propane undergoes a process from state 1, where $p_1 = 1.4$ MPa, $T_1 = 60°C$, to state 2, where $p_2 = 1.0$ MPa, during which the change in specific entropy is $s_2 - s_1 = -0.35$ kJ/kg · K. At state 2, determine the temperature, in °C, and the specific enthalpy, in kJ/kg.

6.11 Air in a piston–cylinder assembly undergoes a process from state 1, where $T_1 = 300$ K, $p_1 = 100$ kPa, to state 2, where $T_2 = 500$ K, $p_2 = 650$ kPa. Using the ideal gas model for air, determine the change in specific entropy between these states, in kJ/kg · K, if the process occurs (a) without internal irreversibilities, (b) with internal irreversibilities.

6.12 Methane gas (CH_4) enters a compressor at 298 K, 1 bar and exits at 2 bar and temperature *T*. Employing the ideal gas model, determine *T*, in K, if there is no change in specific entropy from inlet to exit.

6.13 One-quarter kmol of nitrogen gas (N_2) undergoes a process from $p_1 = 138$ kPa, $T_1 = 278$ K to $p_2 = 1$ MPa. For the process $W = -528$ kJ and $Q = -132.8$ kJ. Employing the ideal gas model, determine

(a) T_2, in K.
(b) the change in entropy, in kJ/K.

Show the initial and final states on a *T–s* diagram.

6.14 One kilogram of water contained in a piston–cylinder assembly, initially at 160°C, 150 kPa, undergoes an isothermal compression process to saturated liquid. For the process, $W = -471.5$ kJ. Determine for the process,

(a) the heat transfer, in kJ.
(b) the change in entropy, in kJ/K.

Show the process on a sketch of the *T–s* diagram.

6.15 One-tenth kmol of carbon monoxide (CO) in a piston-cylinder assembly undergoes a process from $p_1 = 150$ kPa, $T_1 = 300$ K to $p_2 = 500$ kPa, $T_2 = 370$ K. For the process, $W = -300$ kJ. Employing the ideal gas model, determine

(a) the heat transfer, in kJ.
(b) the change in entropy, in kJ/K.

Show the process on a sketch of the *T–s* diagram.

6.16 Argon in a piston–cylinder assembly is compressed from state 1, where $T_1 = 300$ K, $V_1 = 1$ m^3, to state 2, where $T_2 = 200$ K. If the change in specific entropy is $s_2 - s_1 = -0.27$ kJ/kg · K, determine the final volume, in m^3. Assume the ideal gas model with $k = 1.67$.

6.17 Steam enters a turbine operating at steady state at 1 MPa, 200°C and exits at 40°C with a quality of 83%. Stray heat transfer and kinetic and potential energy effects are negligible. Determine (a) the power developed by the turbine, in kJ per kg of steam flowing, (b) the change in specific entropy from inlet to exit, in kJ/K per kg of steam flowing.

6.18 Answer the following true or false. Explain.

(a) The change of entropy of a closed system is the same for every process between two specified states.
(b) The entropy of a fixed amount of an ideal gas increases in every isothermal compression.
(c) The specific internal energy and enthalpy of an ideal gas are each functions of temperature alone but its specific entropy depends on two independent intensive properties.

(d) One of the $T\,ds$ equations has the form $T\,ds = du - p\,dv$.
(e) The entropy of a fixed amount of an incompressible substance increases in every process in which temperature decreases.

6.19 A closed system consists of an ideal gas with constant specific heat ratio k.

(a) The gas undergoes a process in which temperature increases from T_1 to T_2. Show that the entropy change for the process is greater if the change in state occurs at constant pressure than if it occurs at constant volume. Sketch the processes on p–v and T–s coordinates.
(b) Using the results of (a), show on T–s coordinates that a line of constant specific volume passing through a state has a greater slope than a line of constant pressure passing through that state.
(c) The gas undergoes a process in which pressure increases from p_1 to p_2. Show that the ratio of the entropy change for an isothermal process to the entropy change for a constant-volume process is $(1 - k)$. Sketch the processes on p–v and T–s coordinates.

Analyzing Internally Reversible Processes

6.20 One kilogram of water in a piston-cylinder assembly undergoes the two internally reversible processes in series shown in Fig. P6.20. For each process, determine, in kJ, the heat transfer and the work.

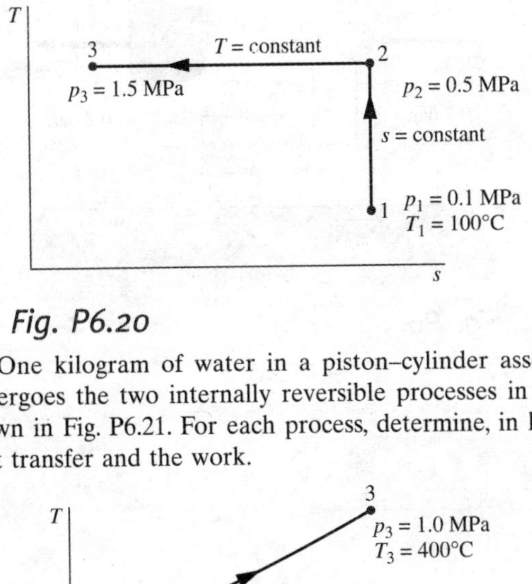

Fig. P6.20

6.21 One kilogram of water in a piston–cylinder assembly undergoes the two internally reversible processes in series shown in Fig. P6.21. For each process, determine, in kJ, the heat transfer and the work.

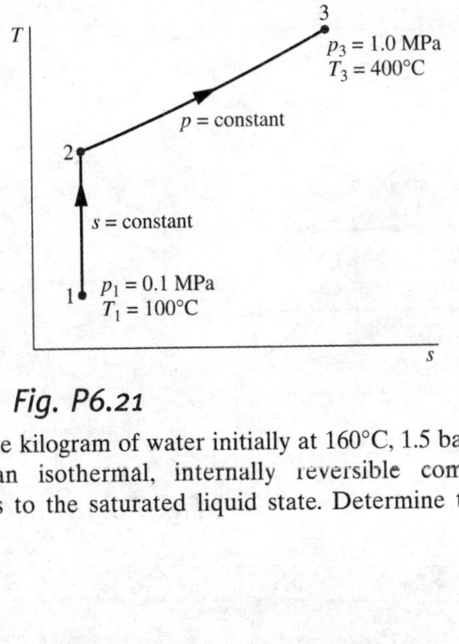

Fig. P6.21

6.22 One kilogram of water initially at 160°C, 1.5 bar, undergoes an isothermal, internally reversible compression process to the saturated liquid state. Determine the work and heat transfer, each in kJ. Sketch the process on p–v and T–s coordinates. Associate the work and heat transfer with areas on these diagrams.

6.23 One kg mass of water initially a saturated liquid at 1 bar undergoes a constant-pressure, internally reversible expansion to $x = 90\%$. Determine the work and heat transfer, each in kJ. Sketch the process on p–v and T–s coordinates. Associate the work and heat transfer with areas on these diagrams.

6.24 A gas initially at 14 bar and 60°C expands to a final pressure of 2.8 bar in an isothermal, internally reversible process. Determine the heat transfer and the work, each in kJ per kg of gas, if the gas is (a) Refrigerant 134a, (b) air as an ideal gas. Sketch the processes on p–v and T–s coordinates.

6.25 Reconsider the data of Problem 6.24, but now suppose the gas expands to 2.8 bar isentropically. Determine the work, in kJ per kg of gas, if the gas is (a) Refrigerant 134a, (b) air as an ideal gas. Sketch the processes on p–v and T–s coordinates.

6.26 Nitrogen (N_2) initially occupying 0.5 m^3 at 1.0 bar, 20°C, undergoes an internally reversible compression during which $pV^{1.30} = constant$ to a final state where the temperature is 200°C. Assuming the ideal gas model, determine

(a) the pressure at the final state, in bar.
(b) the work and heat transfer, each in kJ.
(c) the entropy change, in kJ/K.

6.27 Air in a piston–cylinder assembly and modeled as an ideal gas undergoes two internally reversible processes in series from state 1, where $T_1 = 290$ K, $p_1 = 1$ bar.

Process 1-2: Compression to $p_2 = 5$ bar during which $pV^{1.19} = constant$.

Process 2-3: Isentropic expansion to $p_3 = 1$ bar.

(a) Sketch the two processes in series on T–s coordinates.
(b) Determine the temperature at state 2, in K.
(c) Determine the net work, in kJ.

6.28 Air in a piston–cylinder assembly undergoes a thermodynamic cycle consisting of three internally reversible processes in series.

Process 1-2: Constant-volume heating from $T_1 = 288$ K, $p_1 = 0.1$ MPa to $p_2 = 0.42$ MPa.

Process 2-3: Constant-pressure cooling.

Process 3-1: Isothermal heating to the initial state.

Employing the ideal gas model with $c_p = 1$ kJ/kg · K, evaluate the change in specific entropy, in kJ/kg · K, for each process. Sketch the cycle on p–v and T–s coordinates. For each process, associate work and heat transfer with areas on these diagrams, respectively.

6.29 One-tenth kilogram of a gas in a piston–cylinder assembly undergoes a Carnot power cycle for which the isothermal expansion occurs at 800 K. The change in specific entropy of the gas during the isothermal compression, which occurs at 400 K, is -25 kJ/kg · K. Determine (a) the net work developed per cycle, in kJ, and (b) the thermal efficiency.

6.30 One kg mass of air as an ideal gas contained within a piston–cylinder assembly undergoes a Carnot power cycle. At the beginning of the isothermal expansion, the temperature is 890 K and the pressure is 8.27 MPa. The isothermal compression occurs at 278 K and the heat added per cycle is 42.2 kJ. Assuming the ideal gas model for the air, determine

(a) the pressures at the end of the isothermal expansion, the adiabatic expansion, and the isothermal compression, each in kPa.
(b) the net work developed per cycle, in kJ.
(c) the thermal efficiency.

6.31 Air in a piston–cylinder assembly undergoes a Carnot power cycle. The isothermal expansion and compression processes occur at 1400 K and 350 K, respectively. The pressures at the beginning and end of the isothermal compression are 100 kPa and 500 kPa, respectively. Assuming the ideal gas model with $c_p = 1.005$ kJ/kg · K, determine

(a) the pressures at the beginning and end of the isothermal expansion, each in kPa.
(b) the heat transfer and work, in kJ/kg, for each process.
(c) the thermal efficiency.

6.32 Water in a piston–cylinder assembly undergoes a Carnot power cycle. At the beginning of the isothermal expansion, the temperature is 250°C and the quality is 80%. The isothermal expansion continues until the pressure is 2 MPa. The adiabatic expansion then occurs to a final temperature of 175°C.

(a) Sketch the cycle on *T–s* coordinates.
(b) Determine the heat transfer and work, in kJ/kg, for each process.
(c) Evaluate the thermal efficiency.

6.33 A Carnot power cycle operates at steady state as shown in Fig. 5.15 with water as the working fluid. The boiler pressure is 1.4 MPa, with saturated liquid entering and saturated vapor exiting. The condenser pressure is 138 kPa.

(a) Sketch the cycle on *T–s* coordinates.
(b) Determine the heat transfer and work for each process, in kJ per kg of water flowing.
(c) Evaluate the thermal efficiency.

6.34 Figure P6.34 shows a Carnot heat pump cycle operating at steady state with ammonia as the working fluid. The condenser temperature is 49°C, with saturated vapor entering and saturated liquid exiting. The evaporator temperature is −12°C.

(a) Determine the heat transfer and work for each process, in kJ/kg of ammonia flowing.
(b) Evaluate the coefficient of performance for the heat pump.
(c) Evaluate the coefficient of performance for a Carnot refrigeration cycle operating as shown in the figure.

6.35 Figure P6.35 gives the schematic of a vapor power plant in which water steadily circulates through the four components shown. The water flows through the boiler and condenser at constant pressure, and flows through the turbine and pump adiabatically.

(a) Sketch the cycle on *T–s* coordinates.
(b) Determine the thermal efficiency and compare with the thermal efficiency of a Carnot cycle operating between the same maximum and minimum temperatures.

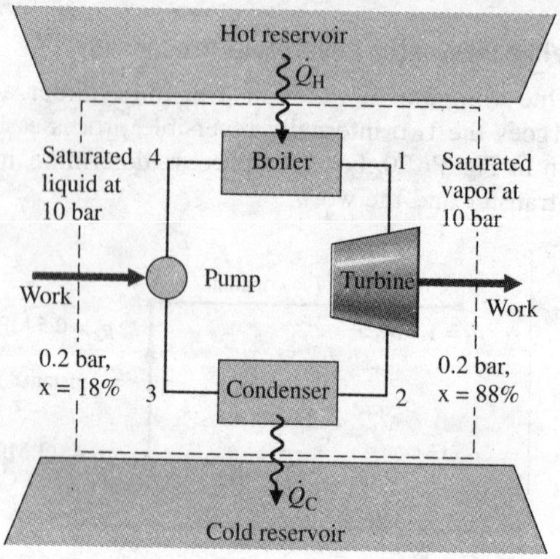

Fig. P6.35

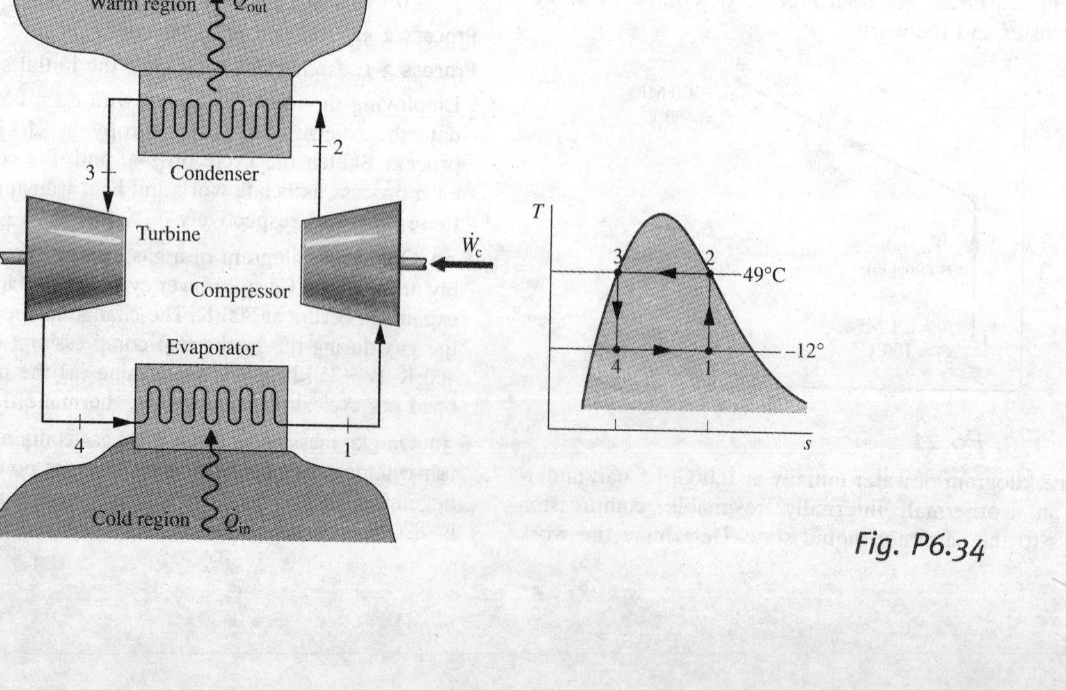

Fig. P6.34

Applying the Entropy Balance: Closed Systems

6.36 A closed system undergoes a process in which work is done on the system and the heat transfer Q occurs only at temperature T_b. For each case, determine whether the entropy change of the system is positive, negative, zero, or indeterminate.

(a) internally reversible process, $Q > 0$.
(b) internally reversible process, $Q = 0$.
(c) internally reversible process, $Q < 0$.
(d) internal irreversibilities present, $Q > 0$.
(e) internal irreversibilities present, $Q = 0$.
(f) internal irreversibilities present, $Q < 0$.

6.37 Answer the following true or false. Explain.

(a) A process that violates the second law of thermodynamics violates the first law of thermodynamics.
(b) When a net amount of work is done on a closed system undergoing an internally reversible process, a net heat transfer of energy from the system also occurs.
(c) One corollary of the second law of thermodynamics states that the change in entropy of a closed system must be greater than zero or equal to zero.
(d) A closed system can experience an increase in entropy only when irreversibilities are present within the system during the process.
(e) Entropy is produced in every internally reversible process of a closed system.
(f) In an adiabatic and internally reversible process of a closed system, the entropy remains constant.
(g) The energy of an isolated system must remain constant, but the entropy can only decrease.

6.38 A fixed mass of water m, initially a saturated liquid, is brought to a saturated vapor condition while its pressure and temperature remain constant. Volume change is the only work mode.

(a) Derive expressions for the work and heat transfer in terms of the mass m and properties that can be obtained directly from the *steam tables*.
(b) Demonstrate that this process is internally reversible.

6.39 Five kg of water contained in a piston-cylinder assembly expand from an initial state where $T_1 = 400°C$, $p_1 = 700$ kPa to a final state where $T_2 = 200°C$, $p_2 = 300$ kPa, with no significant effects of kinetic and potential energy. The accompanying table provides additional data at the two states. It is claimed that the water undergoes an adiabatic process between these states, while developing work. Evaluate this claim.

State	T(°C)	p(kPa)	v(m^3/kg)	u(kJ/kg)	h(kJ/kg)	s(kJ/kg · K)
1	400	700	0.4397	2960.9	3268.7	7.6350
2	200	300	0.7160	2650.7	2865.5	7.3115

6.40 Two m^3 of air in a rigid, insulated container fitted with a paddle wheel is initially at 293 K, 200 kPa. The air receives 710 kJ by work from the paddle wheel. Assuming the ideal gas model with $c_v = 0.72$ kJ/kg · K, determine for the air (a) the mass, in kg, (b) final temperature, in K, and (c) the amount of entropy produced, in kJ/K.

6.41 A rigid, insulated container fitted with a paddle wheel contains 2.27 kg of water, initially at 127°C and a quality of 60%. The water is stirred until the temperature is 177°C. For the water, determine (a) the work, in kJ, and (b) the amount of entropy produced, in kJ/K.

6.42 One kilogram of air contained in a piston–cylinder assembly is initially at 1 bar and 450 K. Can a final state at 2 bar and 350 K be attained in an adiabatic process?

6.43 One kg mass of Refrigerant 134a contained within a piston–cylinder assembly undergoes a process from a state where the pressure is 0.8 MPa and the quality is 40% to a state where the temperature is 10°C and the refrigerant is saturated liquid. Determine the change in specific entropy of the refrigerant, in kJ/kg · K. Can this process be accomplished adiabatically?

6.44 Refrigerant 134a contained in a piston–cylinder assembly rapidly expands from an initial state where $T_1 = 60°C$, $p_1 = 1.4$ MPa to a final state where $p_2 = 34.5$ kPa and the quality, x_2, is (a) 99%, (b) 95%. In each case, determine if the process can occur adiabatically. If yes, determine the work, in kJ/kg, for an adiabatic expansion between these states. If no, determine the direction of the heat transfer.

6.45 One kg of air contained in a piston–cylinder assembly undergoes a process from an initial state where $T_1 = 300$ K, $v_1 = 0.8$ m^3/kg to a final state where $T_2 = 420$ K, $v_2 = 0.2$ m^3/kg. Can this process occur adiabatically? If yes, determine the work, in kJ, for an adiabatic process between these states. If no, determine the direction of the heat transfer. Assume the ideal gas model for air.

6.46 Air as an ideal gas contained within a piston–cylinder assembly is compressed between two specified states. In each of the following cases, can the process occur adiabatically? If yes, determine the work in appropriate units for an adiabatic process between these states. If no, determine the direction of the heat transfer.

(a) State 1: $p_1 = 0.1$ MPa, $T_1 = 27°C$. State 2: $p_2 = 0.5$ MPa, $T_2 = 207°C$. Use Table A-22 data.
(b) State 1: $p_1 = 3$ bar, $T_1 = 27°C$. State 2: $p_2 = 10.1$ bar, $T_2 = 116°C$. Assume $c_p = 1$ kJ/kg · K.

6.47 A piston–cylinder assembly initially contains 0.04 m^3 of water at 1.0 MPa, 320°C. The water expands adiabatically to a final pressure of 0.1 MPa. Develop a plot of the work done by the water, in kJ, versus the amount of entropy produced, in kJ/K.

6.48 Two kg mass of steam contained in a piston-cylinder assembly expands adiabatically from state 1, where $p_1 = 0.7$ MPa, $T_1 = 260°C$, to a final pressure of 70 kPa. Develop a plot of the work done by the water, in kJ, versus the amount of entropy produced, in kJ/K.

6.49 One-tenth kilogram of water, initially at 300 kPa, 200°C, is compressed in a piston–cylinder assembly to 1500 kPa, 210°C. If heat transfer from the water occurs at an average temperature of 205°C, determine the minimum theoretical work of compression, in kJ.

6.50 One-half kilogram of propane initially at 4 bar, 30°C undergoes a process to 14 bar, 100°C while being rapidly compressed in a piston–cylinder assembly. Heat transfer with the surroundings at 20°C occurs through a *thin* wall.

The net work is measured as −72.5 kJ. Kinetic and potential energy effects can be ignored. Determine whether it is possible for the work measurement to be correct.

6.51 One kg mass of carbon dioxide (CO_2) in a piston-cylinder assembly, initially at 21°C and 1.24 MPa, expands isothermally to a final pressure of 0.1 MPa while receiving energy by heat transfer through a wall separating the carbon dioxide from a thermal energy reservoir at 49°C.

(a) For the carbon dioxide as the system, evaluate the work and heat transfer, each in kJ, and the amount of entropy produced, in kJ/K. Model the carbon dioxide as an ideal gas.

(b) Evaluate the entropy production for an enlarged system that includes the carbon dioxide and the wall, assuming the state of the wall remains unchanged. Compare with the entropy production of part (a) and comment on the difference.

6.52 Two kilograms of Refrigerant 134a initially at 1.4 bar, 60°C, are compressed to saturated vapor at 60°C. During this process, the temperature of the refrigerant departs by no more than 0.01°C from 60°C. Determine the minimum theoretical heat transfer from the refrigerant during the process, in kJ.

6.53 An inventor claims that the electricity-generating unit shown in Fig. P6.53 receives a heat transfer at the rate of 264 kJ/s at a temperature of 278 K, a second heat transfer at the rate of 360 kJ/s at 390 K, and a third at the rate of 528 kJ at 555 K. For operation at steady state, evaluate this claim.

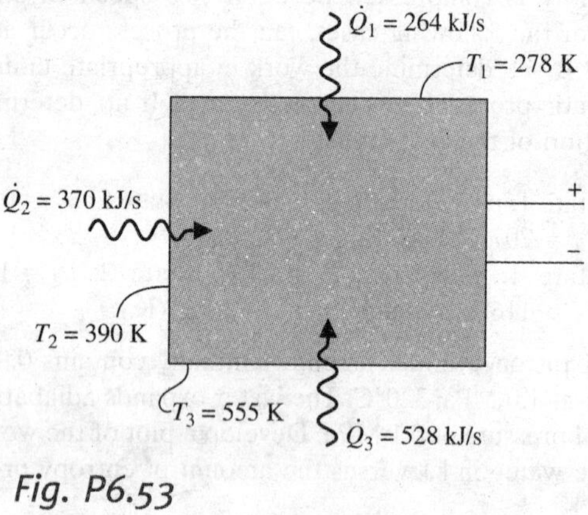

Fig. P6.53

6.54 For the silicon chip of Example 2.5, determine the rate of entropy production, in kW/K. What is the cause of entropy production in this case?

6.55 At steady state, the 20-W curling iron shown in Fig. P6.55 has an outer surface temperature of 82°C. For the curling iron, determine the rate of heat transfer, in kJ/h, and the rate of entropy production, in kJ/h · K.

6.56 A thermally insulated 30-ohm resistor receives a current of 6 amps. The mass of the resistor is 0.1 kg, its specific heat is 0.8 kJ/kg · K, and its initial temperature is 21°C. For the resistor, develop plots of the temperature, in K, and the amount of entropy produced, in kJ/K, versus time ranging from 0 to 3 s.

6.57 An electric water heater having a 200 liter capacity employs an electric resistor to heat water from 23 to 55°C. The outer surface of the resistor remains at an average temperature of 80°C. Heat transfer from the outside of the water heater is negligible and the states of the resistor and the tank holding the water do not change significantly. Modeling the water as incompressible, determine the amount of entropy produced, in kJ/K, for

(a) the water as the system.

(b) the overall water heater including the resistor.

Compare the results of parts (a) and (b), and discuss.

6.58 An electric motor at steady state draws a current of 10 amp with a voltage of 110 V. The output shaft develops a torque of 10.2 N · m and a rotational speed of 1000 RPM.

(a) If the outer surface of the motor is at 42°C, determine the rate of entropy production within the motor, in kW/K.

(b) Evaluate the rate of entropy production, in kW/K, for an enlarged system that includes the motor and enough of the nearby surroundings that heat transfer occurs at the ambient temperature, 22°C.

6.59 An electric motor operating at steady state draws a current of 10 amp with a voltage of 220 V. The output shaft rotates at 1000 RPM with a torque of 16 N · m applied to an external load. The rate of heat transfer *from* the motor *to* its surroundings is related to the surface temperature T_b and the ambient temperature T_0 by $hA(T_b - T_0)$, where h = 100 W/m² · K, A = 0.195 m², and T_0 = 293 K. Energy transfers are considered positive in the directions indicated by the arrows on Fig. P6.59.

(a) Determine the temperature T_b, in K.

(b) For the motor as the system, determine the rate of entropy production, in kW/K.

(c) If the system boundary is located to take in enough of the nearby surroundings for heat transfer to take place at temperature T_0, determine the rate of entropy production, in kW/K, for the enlarged system.

Fig. P6.59

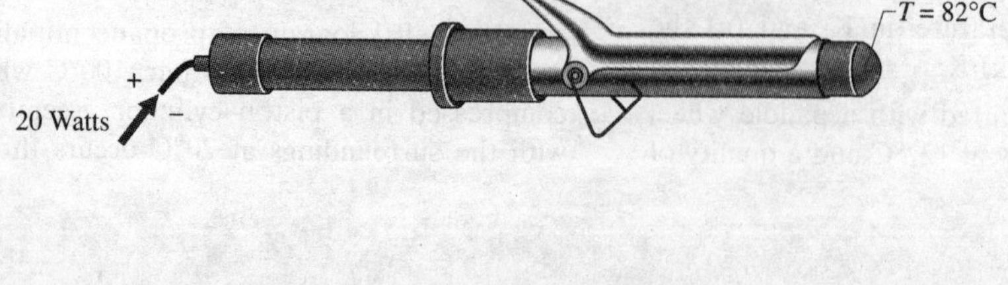

Fig. P6.55

6.60 At steady state, work at a rate of 25 kW is done by a paddle wheel on a *slurry* contained within a closed, rigid tank. Heat transfer from the tank occurs at a temperature of 250°C to surroundings that, away from the immediate vicinity of the tank, are at 27°C. Determine the rate of entropy production, in kW/K,

(a) for the tank and its contents as the system.
(b) for an enlarged system including the tank and enough of the nearby surroundings for the heat transfer to occur at 27°C.

6.61 An isolated system consists of a closed aluminum vessel of mass 0.1 kg containing 1 kg of used engine oil, each initially at 55°C, immersed in a 10-kg bath of liquid water, initially at 20°C. The system is allowed to come to equilibrium. Determine

(a) the final temperature, in degrees centigrade.
(b) the entropy changes, each in kJ/K, for the aluminum vessel, the oil, and the water.
(c) the amount of entropy produced, in kJ/K.

6.62 In a heat-treating process, a 1-kg metal part, initially at 1075 K, is quenched in a tank containing 100 kg of water, initially at 295 K. There is negligible heat transfer between the contents of the tank and their surroundings. Taking the specific heat of the metal part and water as constant at 0.5 kJ/kg · K and 4.2 kJ/kg · K, respectively, determine (a) the final equilibrium temperature after quenching, in K, and (b) the amount of entropy produced within the tank, in kJ/K.

6.63 A system initially contains 1 kg of liquid water at 27°C and 0.2 kg of ice at 0°C. The system attains an equilibrium state, while pressure remains constant at 1 atm. If heat transfer with the surroundings is negligible, determine

(a) the final temperature, in °C.
(b) the amount of entropy produced in the process, in kJ/K.

For water, the specific enthalpy change for a phase change from solid to liquid at 1 bar is 335 kJ/kg.

6.64 A system initially consists of a rivet at 982°C whose mass is 0.2 kg and a two-phase solid–liquid mixture of water at 0°C, 1 bar in which the mass of ice is 1.2 kg and the mass of liquid is 2.27 kg. The specific heat of the rivet is 0.5 kJ/kg · K. The system attains an equilibrium state while pressure remains constant. If heat transfer with the surroundings is negligible, determine

(a) the final temperature, in °C.
(b) the amount of entropy produced, in kJ/K.

For water, the specific enthalpy change for a phase change from solid to liquid at 1 bar is 335 kJ/kg.

6.65 Two insulated tanks are connected by a valve. One tank initially contains 0.45 kg of air at 93°C, 1 bar, and the other contains 0.9 kg of air at 38°C, 2 bar. The valve is opened and the two quantities of air are allowed to mix until equilibrium is attained. Employing the ideal gas model with $c_v = 0.7$ kJ/kg · K, determine

(a) the final temperature, in °C.
(b) the final pressure, in bar.
(c) the amount of entropy produced, in kJ/K.

6.66 An insulated cylinder is initially divided into halves by a frictionless, thermally conducting piston. On one side of the piston is 1 m^3 of a gas at 300 K, 2 bar. On the other side is 1 m^3 of the same gas at 300 K, 1 bar. The piston is released and equilibrium is attained, with the piston experiencing no change of state. Employing the ideal gas model for the gas, determine

(a) the final temperature, in K.
(b) the final pressure, in bar.
(c) the amount of entropy produced, in kJ/kg.

6.67 An insulated vessel is divided into two equal-sized compartments connected by a valve. Initially, one compartment contains steam at 0.34 MPa and 370°C, and the other is evacuated. The valve is opened and the steam is allowed to fill the entire volume. Determine

(a) the final temperature, in °C.
(b) the amount of entropy produced, in kJ/kg · K.

6.68 An insulated, rigid tank is divided into two compartments by a frictionless, thermally conducting piston. One compartment initially contains 1 m^3 of saturated water vapor at 4 MPa and the other compartment contains 1 m^3 of water vapor at 20 MPa, 800°C. The piston is released and equilibrium is attained, with the piston experiencing no change of state. For the water as the system, determine

(a) the final pressure, in MPa.
(b) the final temperature, in °C.
(c) the amount of entropy produced, in kJ/K.

6.69 A system consisting of air initially at 300 K and 1 bar experiences the two different types of interactions described below. In each case, the system is brought from the initial state to a state where the temperature is 500 K, while volume remains constant.

(a) The temperature rise is brought about adiabatically by stirring the air with a paddle wheel. Determine the amount of entropy produced, in kJ/kg · K.
(b) The temperature rise is brought about by heat transfer from a reservoir at temperature T. The temperature at the system boundary where heat transfer occurs is also T. Plot the amount of entropy produced, in kJ/kg · K, versus T for $T \geq 500$ K. Compare with the result of (a) and discuss.

6.70 A cylindrical copper rod of base area A and length L is insulated on its lateral surface. One end of the rod is in contact with a wall at temperature T_H. The other end is in contact with a wall at a lower temperature T_C. At steady state, the rate at which energy is conducted into the rod from the hot wall is

$$\dot{Q}_H = \frac{\kappa A(T_H - T_C)}{L}$$

where κ is the thermal conductivity of the copper rod.

(a) For the rod as the system, obtain an expression for the time rate of entropy production in terms of A, L, T_H, T_C, and κ.
(b) If $T_H = 327$°C, $T_C = 77$°C, $\kappa = 0.4$ kW/m · K, A = 0.1 m^2, plot the heat transfer rate $\dot{Q}_H$, in kW, and the time rate of entropy production, in kW/K, each versus L ranging from 0.01 to 1.0 m. Discuss.

6.71 Figure P6.71 shows a system consisting of air in a rigid container fitted with a paddle wheel and in contact with a thermal energy reservoir. By heating and/or stirring, the air can achieve a specified increase in temperature from T_1 to T_2 in alternative ways. Discuss how the temperature increase of the air might be achieved with (a) minimum entropy production, and (b) maximum entropy production. Assume that the temperature on the boundary where heat transfer to the air occurs, T_b, is the same as the reservoir temperature. Let $T_1 < T_b < T_2$. The ideal gas model applies to the air.

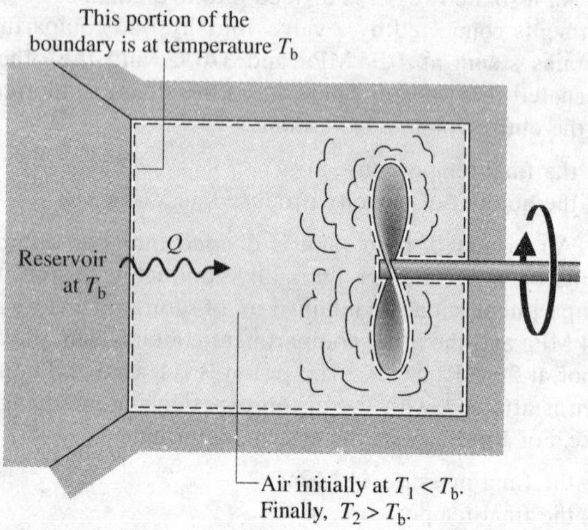

Fig. P6.71

6.72 An isolated system of total mass m is formed by mixing two equal masses of the same liquid initially at the temperatures T_1 and T_2. Eventually, the system attains an equilibrium state. Each mass is incompressible with constant specific heat c.

(a) Show that the amount of entropy produced is

$$\sigma = mc \ln \left[\frac{T_1 + T_2}{2(T_1 T_2)^{1/2}} \right]$$

(b) Demonstrate that σ must be positive.

6.73 A cylindrical rod of length L insulated on its lateral surface is initially in contact at one end with a wall at temperature T_H and at the other end with a wall at a lower temperature T_C. The temperature within the rod initially varies linearly with position z according to

$$T(z) = T_H - \left(\frac{T_H - T_C}{L} \right) z$$

The rod is then insulated on its ends and eventually comes to a final equilibrium state where the temperature is T_f. Evaluate T_f in terms of T_H and T_C and show that the amount of entropy produced is

$$\sigma = mc \left(1 + \ln T_f + \frac{T_C}{T_H - T_C} \ln T_C - \frac{T_H}{T_H - T_C} \ln T_H \right)$$

where c is the specific heat of the rod.

6.74 A system undergoing a thermodynamic cycle receives Q_H at temperature T'_H and discharges Q_C at temperature T'_C. There are no other heat transfers.

(a) Show that the net work developed per cycle is given by

$$W_{cycle} = Q_H \left(1 - \frac{T'_C}{T'_H} \right) - T'_C \sigma$$

where σ is the amount of entropy produced per cycle owing to irreversibilities within the system.

(b) If the heat transfers Q_H and Q_C are with hot and cold reservoirs, respectively, what is the relationship of T'_H to the temperature of the hot reservoir T_H and the relationship of T'_C to the temperature of the cold reservoir T_C?

(c) Obtain an expression for W_{cycle} if there are (i) no internal irreversibilities, (ii) no internal *or* external irreversibilities.

6.75 A thermodynamic power cycle receives energy by heat transfer from an incompressible body of mass m and specific heat c initially at temperature T_H. The cycle discharges energy by heat transfer to another incompressible body of mass m and specific heat c initially at a lower temperature T_C. There are no other heat transfers. Work is developed by the cycle until the temperature of each of the two bodies is the same. Develop an expression for the maximum theoretical amount of work that can be developed, W_{max}, in terms of m, c, T_H, and T_C.

6.76 At steady state, an insulated mixing chamber receives two liquid streams of the same substance at temperatures T_1 and T_2, and mass flow rates $\dot{m}_1$ and $\dot{m}_2$, respectively. A single stream exits at T_3 and $\dot{m}_3$. Using the incompressible substance model with constant specific heat c, obtain an expression for

(a) T_3 in terms of T_1, T_2, and the ratio of mass flow rates $\dot{m}_1/\dot{m}_3$.

(b) the rate of entropy production per unit of mass exiting the chamber in terms of c, T_1/T_2, and $\dot{m}_1/\dot{m}_3$.

(c) For fixed values of c and T_1/T_2, determine the value of $\dot{m}_1/\dot{m}_3$ for which the rate of entropy production is a maximum.

6.77 The temperature of an incompressible substance of mass m and specific heat c is reduced from T_0 to T ($< T_0$) by a refrigeration cycle. The cycle receives energy by heat transfer at T from the substance and discharges energy by heat transfer at T_0 to the surroundings. There are no other heat transfers. Plot (W_{min}/mcT_0) versus T/T_0 ranging from 0.8 to 1.0, where W_{min} is the minimum theoretical work *input* required.

6.78 The temperature of a 0.354-L can of soft drink is reduced from 20 to 5°C by a refrigeration cycle. The cycle receives energy by heat transfer from the soft drink and discharges energy by heat transfer at 20°C to the surroundings. There are no other heat transfers. Determine the minimum theoretical work input required, in kJ, assuming the soft drink is an incompressible liquid with the properties of liquid water. Ignore the aluminum can.

6.79 As shown in Fig. P6.79, a turbine is located between two tanks. Initially, the smaller tank contains steam at 3.0 MPa, 280°C and the larger tank is evacuated. Steam is allowed to flow from the smaller tank, through the turbine, and into the larger tank until equilibrium is attained. If heat transfer with the surroundings is negligible, determine the maximum theoretical work that can be developed, in kJ.

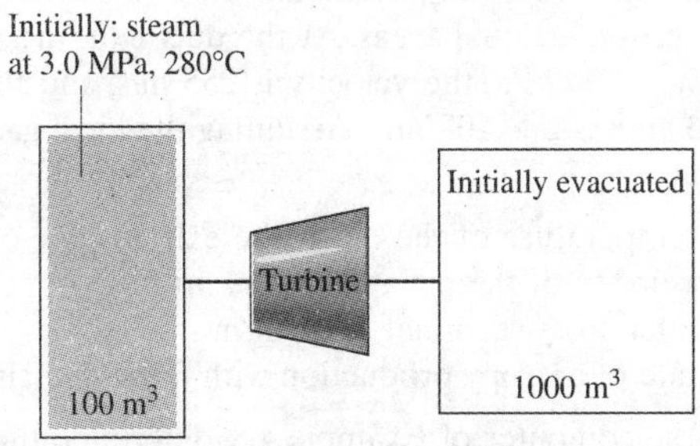

Fig. P6.79

Applying the Entropy Balance: Control Volumes

6.80 A gas flows through a one-inlet, one-exit control volume operating at steady state. Heat transfer at the rate $\dot{Q}_{cv}$ takes place only at a location on the boundary where the temperature is T_b. For each of the following cases, determine whether the specific entropy of the gas at the exit is greater than, equal to, or less than the specific entropy of the gas at the inlet:

(a) no internal irreversibilities, $\dot{Q}_{cv} = 0$.
(b) no internal irreversibilities, $\dot{Q}_{cv} < 0$.
(c) no internal irreversibilities, $\dot{Q}_{cv} > 0$.
(d) internal irreversibilities, $\dot{Q}_{cv} \geq 0$.

6.81 Steam at 10 bar, 600°C, 50 m/s enters an insulated turbine operating at steady state and exits at 0.35 bar, 100 m/s. The work developed per kg of steam flowing is claimed to be (a) 1000 kJ/kg, (b) 500 kJ/kg. Can either claim be correct? Explain.

6.82 Air enters an insulated turbine operating at steady state at 6.5 bar, 687°C and exits at 1 bar, 327°C. Neglecting kinetic and potential energy changes and assuming the ideal gas model, determine

(a) the work developed, in kJ per kg of air flowing through the turbine.
(b) whether the expansion is internally reversible, irreversible, or impossible.

6.83 Propane at 0.1 MPa, 20°C enters an insulated compressor operating at steady state and exits at 0.4 MPa, 90°C. Neglecting kinetic and potential energy effects, determine

(a) the power required by the compressor, in kJ per kg of propane flowing.
(b) the rate of entropy production within the compressor, in kJ/K per kg of propane flowing.

6.84 By injecting liquid water into superheated steam, the *desuperheater* shown in Fig. P6.84 has a saturated vapor stream at its exit. Steady-state operating data are provided in the accompanying table. Stray heat transfer and all kinetic and potential energy effects are negligible. (a) Locate states 1, 2, and 3 on a sketch of the *T–s* diagram. (b) Determine the rate of entropy production within the desuperheater, in kW/K.

State	p(MPa)	T(°C)	$v \times 10^3$(m³/kg)	u(kJ/kg)	h(kJ/kg)	s(kJ/kg·K)
1	2.7	40	1.0066	167.2	169.9	0.5714
2	2.7	300	91.01	2757.0	3002.8	6.6001
3	2.5	sat. vap.	79.98	2603.1	2803.1	6.2575

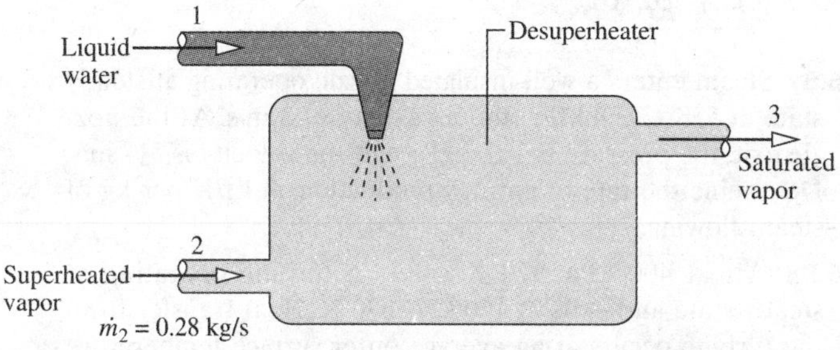

Fig. P6.84

6.85 An inventor claims that at steady state the device shown in Fig. P6.85 develops power from entering and exiting streams of water at a rate of 1174.9 kW. The accompanying table provides data for inlet 1 and exits 3 and 4. The pressure at inlet 2 is 1 bar. Stray heat transfer and kinetic and potential energy effects are negligible. Evaluate the inventor's claim.

State	$\dot{m}$(kg/s)	p (bar)	T(°C)	v (m³/kg)	u (kJ/kg)	h (kJ/kg)	s(kJ/kg·K)
1	4	1	450	3.334	3049.0	3382.4	8.6926
3	5	2	200	1.080	2654.4	2870.5	7.5066
4	3	4	400	0.773	2964.4	3273.4	7.8985

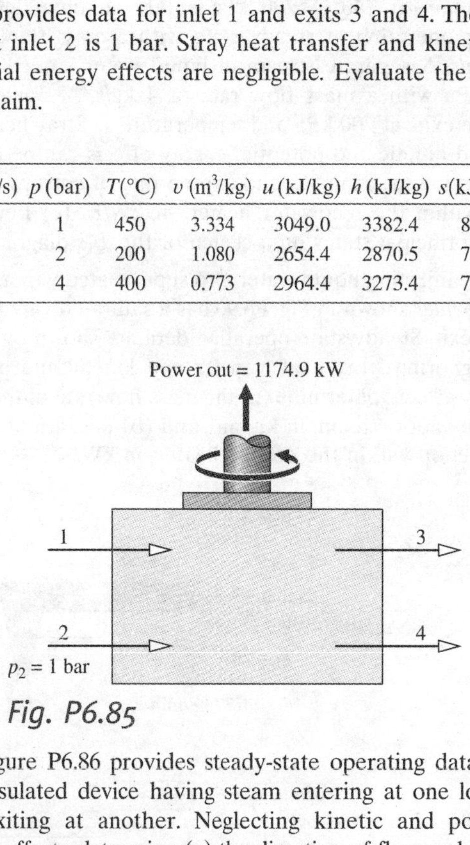

Fig. P6.85

6.86 Figure P6.86 provides steady-state operating data for a well-insulated device having steam entering at one location and exiting at another. Neglecting kinetic and potential energy effects, determine (a) the direction of flow and (b) the power output or input, as appropriate, in kJ per kg of steam flowing.

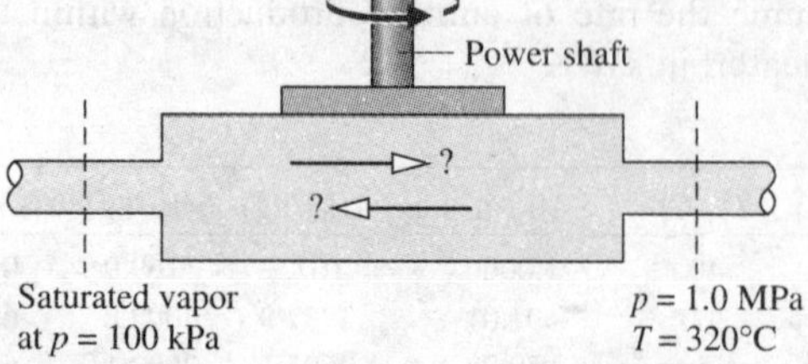

Fig. P6.86

6.87 Steam enters a well-insulated nozzle operating at steady state at 538°C, 3.4 MPa and a velocity of 3 m/s. At the nozzle exit, the pressure is 101.3 kPa and the velocity is 317 m/s. Determine the rate of entropy production, in kJ/K per kg of steam flowing.

6.88 Air at 400 kPa, 970 K enters a turbine operating at steady state and exits at 100 kPa, 670 K. Heat transfer from the turbine occurs at an average outer surface temperature of 315 K at the rate of 30 kJ per kg of air flowing. Kinetic and potential energy effects are negligible. For air as an ideal gas with c_p = 1.1 kJ/kg · K, determine (a) the rate power is developed, in kJ per kg of air flowing, and (b) the rate of entropy production within the turbine, in kJ/K per kg of air flowing.

6.89 Steam at 240°C, 700 kPa enters an open feedwater heater operating at steady state with a mass flow rate of 0.5 kg/s. A separate stream of liquid water enters at 45°C, 700 kPa with a mass flow rate of 4 kg/s. A single mixed stream exits at 700 kPa and temperature T. Stray heat transfer and kinetic and potential energy effects can be ignored. Determine (a) T, in °C, and (b) the rate of entropy production within the feedwater heater, in kW/K. (c) Locate the three principal states on a sketch of the T–s diagram.

6.90 By injecting liquid water into superheated vapor, the *desuperheater* shown in Fig. P6.90 has a saturated vapor stream at its exit. Steady-state operating data are shown on the figure. Ignoring stray heat transfer and kinetic and potential energy effects, determine (a) the mass flow rate of the superheated vapor stream, in kg/min, and (b) the rate of entropy production within the desuperheater, in kW/K.

6.91 Air at 600 kPa, 330 K enters a well-insulated, horizontal pipe having a diameter of 1.2 cm and exits at 120 kPa, 300 K. Applying the ideal gas model for air, determine at steady state (a) the inlet and exit velocities, each in m/s, (b) the mass flow rate, in kg/s, and (c) the rate of entropy production, in kW/K.

6.92 At steady state, air at 200 kPa, 52°C and a mass flow rate of 0.5 kg/s enters an insulated duct having differing inlet and exit cross-sectional areas. At the duct exit, the pressure of the air is 100 kPa, the velocity is 255 m/s, and the cross-sectional area is 2×10^{-3} m^2. Assuming the ideal gas model, determine

(a) the temperature of the air at the exit, in °C.
(b) the velocity of the air at the inlet, in m/s.
(c) the inlet cross-sectional area, in m^2.
(d) the rate of entropy production within the duct, in kW/K.

6.93 For the computer of Example 4.8, determine the rate of entropy production, in W/K, when air exits at 32°C. Ignore the change in pressure between the inlet and exit.

6.94 For the computer of Problem 4.86, determine the rate of entropy production, in kW/K, ignoring the change in pressure between the inlet and exit.

6.95 For the water-jacketed electronics housing of Problem 4.87, determine the rate of entropy production, in kW/K, when water exits at 26°C.

6.96 Electronic components are mounted on the inner surface of a horizontal cylindrical duct whose inner diameter is 0.2 m, as shown in Fig. P6.96. To prevent overheating of the electronics, the cylinder is cooled by a stream of air flowing through it and by convection from its outer surface. Air enters the duct at 25°C, 1 bar and a velocity of 0.3 m/s and exits at 40°C with negligible changes in kinetic energy and

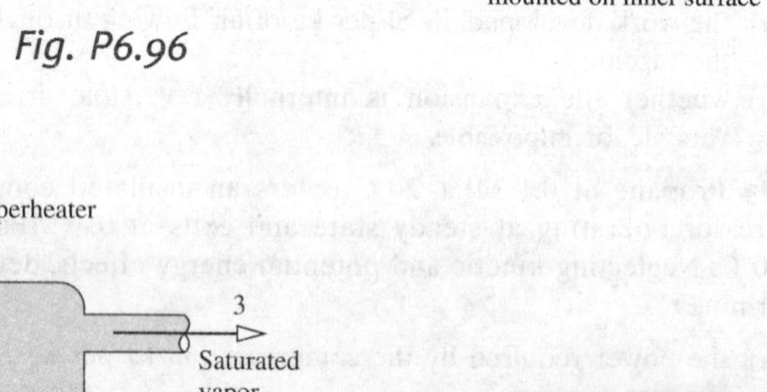

Fig. P6.96

Fig. P6.90

pressure. Convective cooling occurs on the outer surface to the surroundings, which are at 25°C, in accord with hA = 3.4 W/K, where h is the heat transfer coefficient and A is the surface area. The electronic components require 0.20 kW of electric power. For a control volume enclosing the cylinder, determine at steady state (a) the mass flow rate of the air, in kg/s, (b) the temperature on the outer surface of the duct, in °C, and (c) the rate of entropy production, in W/K. Assume the ideal gas model for air.

6.97 Air enters a turbine operating at steady state at 500 kPa, 860 K and exits at 100 kPa. A temperature sensor indicates that the exit air temperature is 460 K. Stray heat transfer and kinetic and potential energy effects are negligible, and the air can be modeled as an ideal gas. Determine if the exit temperature reading can be correct. It yes, determine the power developed by the turbine for an expansion between these states, in kJ per kg of air flowing. If no, provide an explanation with supporting calculations.

6.98 Figure P6.98 provides steady-state test data for a control volume in which two entering streams of air mixed to form a single exiting stream. Stray heat transfer and kinetic and potential energy effects are negligible. A hard-to-read photocopy of the data sheet indicates that the pressure of the exiting stream is either 1.0 MPa or 1.8 MPa. Assuming the ideal gas model for air with c_p = 1.02 kJ/kg · K, determine if either or both of these pressure values can be correct.

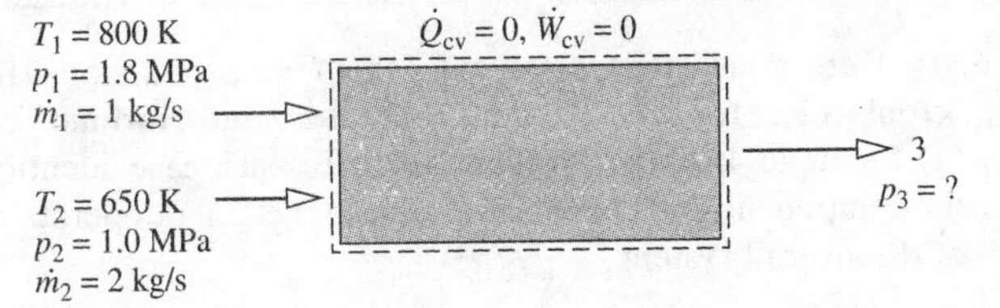

Fig. P6.98

6.99 Hydrogen gas (H_2) at 35°C and pressure p enters an insulated control volume operating at steady state for which $\dot{W}_{cv}$ = 0. Half of the hydrogen exits the device at 2 bar and 90°C and the other half exits at 2 bar and −20°C. The effects of kinetic and potential energy are negligible. Employing the ideal gas model with constant c_p = 14.3 kJ/kg · K, determine the minimum possible value for the inlet pressure p, in bar.

6.100 According to test data, a new type of engine takes in streams of water at 200°C, 3 bar and 100°C, 3 bar. The mass flow rate of the higher temperature stream is twice that of the other. A single stream exits at 3.0 bar with a mass flow rate of 5400 kg/h. There is no significant heat transfer between the engine and its surroundings, and kinetic and potential energy effects are negligible. For operation at steady state, determine the *maximum* theoretical rate that power can be developed, in kW.

6.101 A patent application describes a device for chilling water. At steady state, the device receives energy by heat transfer at a location on its surface where the temperature is 282°C and discharges energy by heat transfer to the surroundings at another location on its surface where the temperature is 38°C. A warm liquid water stream enters at 38°C, 1 bar and a cool stream exits at temperature T and 1 bar. The device requires no power input to operate, there are no significant effects of kinetic and potential energy, and the water can be modeled as incompressible. Plot the minimum theoretical heat addition required, in kJ/kg of cool water exiting the device, versus T ranging from 16° to 38°C.

6.102 Steam at 3.8 MPa, 370°C enters an insulated turbine operating at steady state with a mass flow rate of 0.45 kg/s. A two-phase liquid–vapor mixture exits the turbine at 101.3 kPa with quality x. Plot the power developed, in kJ/s, and the rate of entropy production, in kJ/K · s, each versus x.

6.103 Refrigerant 134a at 138 kPa, −18°C enters a compressor operating at steady state with a mass flow rate of 115 kg/h and exits at 1 MPa, 104°C. Heat transfer occurs from the compressor to its surroundings, which are at 21°C. Changes in kinetic and potential energy can be ignored. The power input is claimed to be 1.5 kW. Determine whether this claim can be correct.

6.104 Steam enters a horizontal 15-cm-diameter pipe as a saturated vapor at 5 bar with a velocity of 10 m/s and exits at 4.5 bar with a quality of 95%. Heat transfer from the pipe to the surroundings at 300 K takes place at an average outer surface temperature of 400 K. For operation at steady state, determine

(a) the velocity at the exit, in m/s.
(b) the rate of heat transfer from the pipe, in kW.
(c) the rate of entropy production, in kW/K, for a control volume comprising only the pipe and its contents.
(d) the rate of entropy production, in kW/K, for an enlarged control volume that includes the pipe and enough of its immediate surroundings so that heat transfer from the control volume occurs at 300 K.

Why do the answers of parts (c) and (d) differ?

6.105 Air at 500 kPa, 500 K and a mass flow of 600 kg/h enters a pipe passing overhead in a factory space. At the pipe exit, the pressure and temperature of the air are 475 kPa and 450 K, respectively. Air can be modeled as an ideal gas with k = 1.39. Kinetic and potential energy effects can be ignored. Determine at steady state, (a) the rate of heat transfer, in kW, for a control volume comprising the pipe and its contents, and (b) the rate of entropy production, in kW/K, for an enlarged control volume that includes the pipe and enough of its surroundings that heat transfer occurs at the ambient temperature, 300 K.

6.106 Steam enters a turbine operating at steady state at 6 MPa, 600°C with a mass flow rate of 125 kg/min and exits as saturated vapor at 20 kPa, producing power at a rate of 2 MW. Kinetic and potential energy effects can be ignored. Determine (a) the rate of heat transfer, in kW, for a control volume including the turbine and its contents, and (b) the rate of entropy production, in kW/K, for an enlarged control volume that includes the turbine and enough of its surroundings that heat transfer occurs at the ambient temperature, 27°C.

6.107 Air enters a compressor operating at steady state at 1 bar, 22°C with a volumetric flow rate of 1 m^3/min and is compressed to 4 bar, 177°C. The power input is 3.5 kW. Employing the ideal gas model and ignoring kinetic and potential energy effects, obtain the following results:

(a) For a control volume enclosing the compressor only, determine the heat transfer rate, in kW, and the change in specific entropy from inlet to exit, in kJ/kg · K. What additional information would be required to evaluate the rate of entropy production?
(b) Calculate the rate of entropy production, in kW/K, for an enlarged control volume enclosing the compressor and a portion of its immediate surroundings so that heat transfer occurs at the ambient temperature, 22°C.

6.108 Carbon dioxide (CO_2) enters a nozzle operating at steady state at 28 bar, 267°C, and 50 m/s. At the nozzle exit, the conditions are 1.2 bar, 67°C, 580 m/s, respectively.
(a) For a control volume enclosing the nozzle only, determine the heat transfer, in kJ, and the change in specific entropy, in kJ/K, each per kg of carbon dioxide flowing through the nozzle. What additional information would be required to evaluate the rate of entropy production?
(b) Evaluate the rate of entropy production, in kJ/K per kg of carbon dioxide flowing, for an enlarged control volume enclosing the nozzle and a portion of its immediate surroundings so that the heat transfer occurs at the ambient temperature, 25°C.

6.109 A counterflow heat exchanger operates at steady state with negligible kinetic and potential energy effects. In one stream, liquid water enters at 15°C and exits at 23°C with a negligible change in pressure. In the other stream, Refrigerant 22 enters at 12 bar, 90°C with a mass flow rate of 150 kg/h and exits at 12 bar, 28°C. Heat transfer from the outer surface of the heat exchanger can be ignored. Determine
(a) the mass flow rate of the liquid water stream, in kg/h.
(b) the rate of entropy production within the heat exchanger, in kW/K.

6.110 Saturated water vapor at 100 kPa enters a counterflow heat exchanger operating at steady state and exits at 20°C with a negligible change in pressure. Ambient air at 275 K, 1 bar enters in a separate stream and exits at 290 K, 1 bar. The air mass flow rate is 170 times that of the water. The air can be modeled as an ideal gas with $c_p = 1.005$ kJ/kg · K. Kinetic and potential energy effects can be ignored.
(a) For a control volume enclosing the heat exchanger, evaluate the rate of heat transfer, in kJ per kg of water flowing.
(b) For an enlarged control volume that includes the heat exchanger and enough of its immediate surroundings that heat transfer from the control volume occurs at the ambient temperature, 275 K, determine the rate of entropy production, in kJ/K per kg of water flowing.

6.111 Figure P6.111 shows data for a portion of the ducting in a ventilation system operating at steady state. The ducts are well insulated and the pressure is very nearly 1 bar throughout. Assuming the ideal gas model for air with $c_p = 1$ kJ/kg · K, and ignoring kinetic and potential energy effects, determine (a) the temperature of the air at the exit, in °C, (b) the exit diameter, in m, and (c) the rate of entropy production within the duct, in kJ/min.

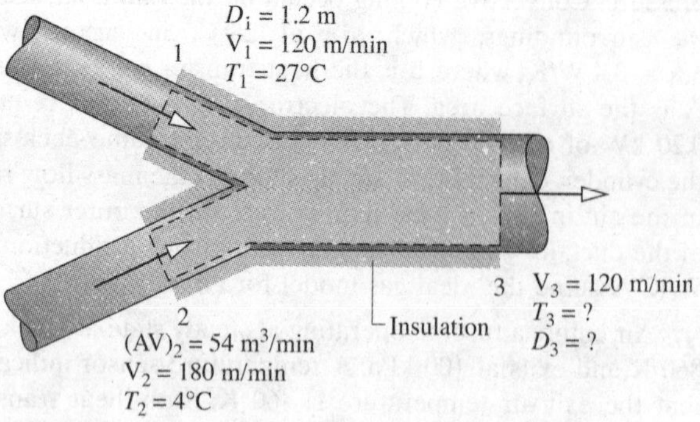

Fig. P6.111

6.112 Air flows through an insulated circular duct having a diameter of 2 cm. Steady-state pressure and temperature data obtained by measurements at two locations, denoted as 1 and 2, are given in the accompanying table. Modeling air as an ideal gas with $c_p = 1.005$ kJ/kg · K, determine (a) the direction of the flow, (b) the velocity of the air, in m/s, at each of the two locations, and (c) the mass flow rate of the air, in kg/s.

Measurement location	1	2
Pressure (kPa)	100	500
Temperature (°C)	20	50

6.113 Determine the rates of entropy production, in kJ/min · K, for the steam generator and turbine of (a) Example 4.10, (b) Problem 4.101. In each case, identify the component that contributes most to inefficient operation of the overall system.

6.114 Air as an ideal gas flows through the compressor and heat exchanger shown in Fig. P6.114. A separate liquid water stream also flows through the heat exchanger. The data given are for operation at steady state. Stray heat transfer to the surroundings can be neglected, as can all kinetic and potential energy changes. Determine
(a) the compressor power, in kW, and the mass flow rate of the cooling water, in kg/s.
(b) the rates of entropy production, each in kW/K, for the compressor and heat exchanger.

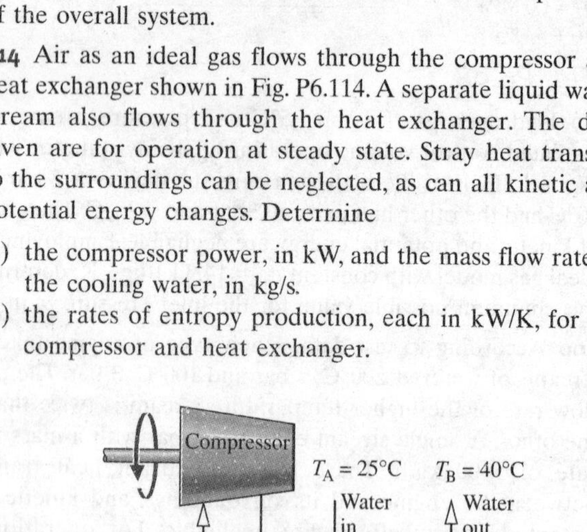

Fig. P6.114

6.115 Figure P6.115 shows several components in series, all operating at steady state. Liquid water enters the boiler at 60 bar. Steam exits the boiler at 60 bar, 540°C and undergoes a throttling process to 40 bar before entering the turbine. Steam expands adiabatically through the turbine to 5 bar, 240°C, and then undergoes a throttling process to 1 bar before entering the condenser. Kinetic and potential energy effects can be ignored.

(a) Locate each of the states 2–5 on a sketch of the T–s diagram.
(b) Determine the power developed by the turbine, in kJ per kg of steam flowing.
(c) For the valves and the turbine, evaluate the rate of entropy production, each in kJ/K per kg of steam flowing.
(d) Using the result of part (c), place the components in rank order, beginning with the component contributing the most to inefficient operation of the overall system. Comment.

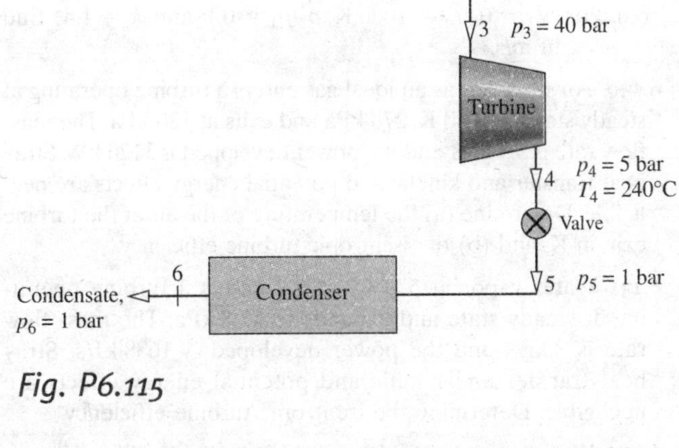

Fig. P6.115

6.116 Air as an ideal gas flows through the turbine and heat exchanger arrangement shown in Fig. P6.116. Steady-state data are given on the figure. Stray heat transfer and kinetic and potential energy effects can be ignored. Determine

(a) temperature T_3, in K.
(b) the power output of the second turbine, in kW.
(c) the rates of entropy production, each in kW/K, for the turbines and heat exchanger.
(d) Using the result of part (c), place the components in rank order, beginning with the component contributing most to inefficient operation of the overall system.

6.117 Figure P6.117 shows a simple vapor power plant operating at steady state with water as the working fluid. Data at key locations are given on the figure. The mass flow rate of the water circulating through the components is 109 kg/s. Stray heat transfer and kinetic and potential energy effects can be ignored. Determine

(a) the mass flow rate of the cooling water, in kg/s.
(b) the thermal efficiency.
(c) the rates of entropy production, each in kW/K, for the turbine, condenser, and pump.
(d) Using the results of part (c), place the components in rank order, beginning with the component contributing most to inefficient operation of the overall system.

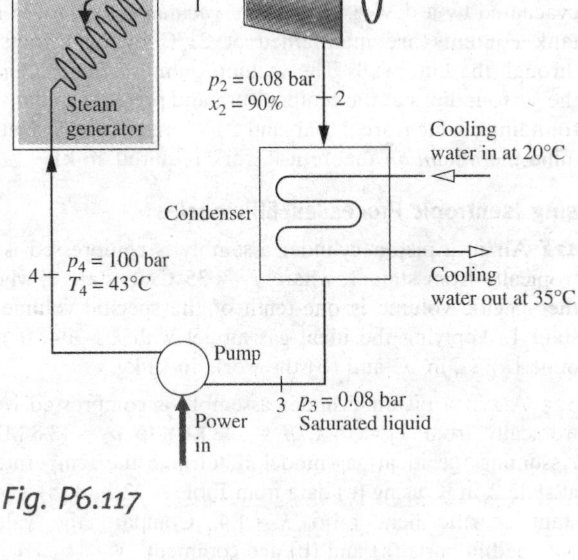

Fig. P6.117

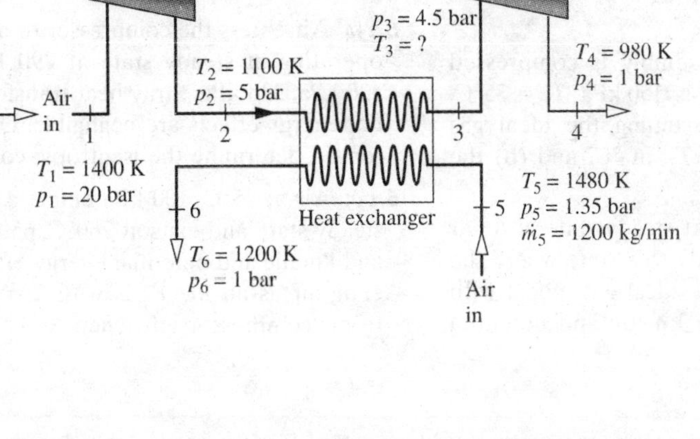

Fig. P6.116

6.118 For the control volume of Example 4.12, determine the amount of entropy produced during filling, in kJ/K. Repeat for the case where no work is developed by the turbine.

6.119 Steam is contained in a large vessel at 690 kPa, 232°C. Connected to the vessel by a valve is an initially evacuated tank having a volume of 0.027 m^3. The valve is opened until the tank is filled with steam at pressure p. The filling is adiabatic, kinetic and potential energy effects are negligible, and the state of the large vessel remains constant.

(a) If p = 690 kPa, determine the final temperature of the steam within the tank, in °C, and the amount of entropy produced within the tank, in kJ/K.

(b) Plot the quantities of part (a) versus pressure p ranging from 69 kPa to 690 kPa.

6.120 A well-insulated, rigid tank of volume 10 m^3 is connected by a valve to a large-diameter supply line carrying air at 227°C and 10 bar. The tank is initially evacuated. Air is allowed to flow into the tank until the tank pressure is p. Using the ideal gas model with constant specific heat ratio k, plot tank temperature, in K, the mass of air in the tank, in kg, and the amount of entropy produced, in kJ/K, versus p in bar.

6.121 A 5 m^3 tank initially filled with air at 1 bar and 21°C is evacuated by a device known as a *vacuum pump,* while the tank contents are maintained at 21°C by heat transfer through the tank walls. The vacuum pump discharges air to the surroundings at the temperature and pressure of the surroundings, which are 1 bar and 21°C, respectively. Determine the *minimum* theoretical work required, in kJ.

Using Isentropic Processes/Efficiencies

6.122 Air in a piston–cylinder assembly is compressed isentropically from state 1, where T_1 = 35°C, to state 2, where the specific volume is one-tenth of the specific volume at state 1. Applying the ideal gas model with k = 1.4, determine (a) T_2, in °C and (b) the work, in kJ/kg.

6.123 Air in a piston–cylinder assembly is compressed isentropically from T_1 = 16°C, p_1 = 138 kPa to p_2 = 13.8 MPa. Assuming the ideal gas model, determine the temperature at state 2, in K, using (a) data from Table A-22, and (b) a constant specific heat ratio, k = 1.4. Compare the values obtained in parts (a) and (b) and comment.

6.124 Propane undergoes an isentropic expansion from an initial state where T_1 = 40°C, p_1 = 1 MPa to a final state where the temperature and pressure are T_2, p_2, respectively. Determine

(a) p_2, in kPa, when T_2 = −40°C

(b) T_2, in °C, when p_2 = 0.8 MPa

6.125 Argon in a piston–cylinder assembly is compressed isentropically from state 1, where p_1 = 150 kPa, T_1 = 35°C, to state 2, where p_2 = 300 kPa. Assuming the ideal gas model with k = 1.67, determine (a) T_2, in °C, and (b) the work, in kJ per kg of argon.

6.126 Air enters a turbine operating at steady state at 6 bar and 1100 K and expands isentropically to a state where the temperature is 700 K. Employing the ideal gas model with data from Table A-22, and ignoring kinetic and potential energy changes, determine the pressure at the exit, in bar, and the work, in kJ per kg of air flowing.

6.127 A rigid, well-insulated tank having a volume of 0.2 m^3 is filled initially with Refrigerant 134a vapor at a pressure of 10 bar and a temperature of 40°C. A leak develops and refrigerant slowly escapes until the pressure within the tank becomes 1 bar. Determine

(a) the final temperature of the refrigerant within the tank, in °C.

(b) the amount of mass that exits the tank, in kg.

6.128 Air in a piston–cylinder assembly is compressed isentropically from an initial state where T_1 = 340 K to a final state where the pressure is 90% greater than at state 1. Assuming the ideal gas model, determine (a) T_2, in K, and (b) the work, in kJ/kg.

6.129 An ideal gas with constant specific heat ratio k enters a nozzle operating at steady state at pressure p_1, temperature T_1, and velocity V_1. The air expands isentropically to a pressure of p_2.

(a) Develop an expression for the velocity at the exit, V_2, in terms of k, R, V_1, T_1, p_1, and p_2, only.

(b) For V_1 = 0, T_1 = 1000 K, p_2/p_1 = 0.1 and k = 1.4, find V_2, in m/s.

6.130 Air modeled as an ideal gas enters a turbine operating at steady state at 1040 K, 278 kPa and exits at 120 kPa. The mass flow rate is 5.5 kg/s, and the power developed is 1120 kW. Stray heat transfer and kinetic and potential energy effects are negligible. Determine (a) the temperature of the air at the turbine exit, in K, and (b) the isentropic turbine efficiency.

6.131 Water vapor at 538°C, 1 MPa enters a turbine operating at steady state and expands to 13.8 kPa. The mass flow rate is 2 kg/s and the power developed is 1688 kJ/s. Stray heat transfer and kinetic and potential energy effects are negligible. Determine the isentropic turbine efficiency.

6.132 Water vapor at 6 MPa, 600°C enters a turbine operating at steady state and expands to 10 kPa. The mass flow rate is 2 kg/s, and the power developed is 2626 kW. Stray heat transfer and kinetic and potential energy effects are negligible. Determine (a) the isentropic turbine efficiency and (b) the rate of entropy production within the turbine, in kW/K.

6.133 Water vapor at 5.5 MPa, 538°C enters a turbine operating at steady state and expands to 13.8 kPa. The mass flow rate is 2.5 kg/s, and the isentropic turbine efficiency is 92%. Stray heat transfer and kinetic and potential energy effects are negligible. Determine the power developed by the turbine, in kW.

6.134 Air enters the compressor of a gas turbine power plant operating at steady state at 290 K, 100 kPa and exits at 420 kPa, 330 kPa. Stray heat transfer and kinetic and potential energy effects are negligible. Using the ideal gas model for air, determine the isentropic compressor efficiency.

6.135 Air at 25°C, 100 kPa enters a compressor operating at steady state and exits at 260°C, 650 kPa. Stray heat transfer and kinetic and potential energy effects are negligible. Modeling air as an ideal gas with k = 1.4, determine the isentropic compressor efficiency.

6.136 Air enters an insulated compressor operating at steady state at 1 bar, 350 K with a mass flow rate of 1 kg/s and exits at 4 bar. The isentropic compressor efficiency is 82%. Determine the power input, in kW, and the rate of entropy production, in kW/K, using the ideal gas model with data from Table A-22.

6.137 Refrigerant 134a enters a compressor operating at steady state as saturated vapor at −6.7°C and exits at a pressure of 0.8 MPa. There is no significant heat transfer with the surroundings, and kinetic and potential energy effects can be ignored.

(a) Determine the minimum theoretical work input required, in kJ per kg of refrigerant flowing through the compressor, and the corresponding exit temperature, in °C.
(b) If the refrigerant exits at a temperature of 49°C, determine the isentropic compressor efficiency.

6.138 Air enters an insulated compressor operating at steady state at 0.95 bar, 27°C with a mass flow rate of 4000 kg/h and exits at 8.7 bar. Kinetic and potential energy effects are negligible.

(a) Determine the minimum theoretical power input required, in kW, and the corresponding exit temperature, in °C.
(b) If the exit temperature is 347°C, determine the power input, in kW, and the isentropic compressor efficiency.

6.139 Water vapor enters an insulated nozzle operating at steady state at 0.7 MPa, 320°C, 35 m/s and expands to 0.15 MPa. If the isentropic nozzle efficiency is 94%, determine the velocity at the exit, in m/s.

6.140 Helium gas at 450 K, 0.3 MPa and a velocity of 3 m/s enters an insulated nozzle operating at steady state and exits at 370 K, 0.17 MPa. Modeling helium as an ideal gas with $k = 1.67$, determine (a) the velocity at the nozzle exit, in m/s, (b) the isentropic nozzle efficiency, and (c) the rate of entropy production within the nozzle, in kJ/K · kg of helium flowing.

6.141 Air modeled as an ideal gas enters a one-inlet, one-exit control volume operating at steady state at 690 kPa, 500 K and expands adiabatically to 0.17 MPa. Kinetic and potential energy effects are negligible. Determine the rate of entropy production, in kJ/K · kg of air flowing,

(a) if the control volume encloses a turbine having an isentropic turbine efficiency of 89.1%.
(b) if the control volume encloses a throttling valve.

6.142 Ammonia enters a valve as a saturated liquid at 7 bar with a mass flow rate of 0.06 kg/min and undergoes a throttling process to a pressure of 1 bar. Determine the rate of entropy production, in kW/K. If the valve were replaced by a power-recovery turbine operating at steady state, determine the maximum theoretical power that could be developed, in kW. Would you recommend using such a turbine? In each case, ignore heat transfer with the surroundings and changes in kinetic and potential energy.

6.143 Figure P6.143 provides the schematic of a heat pump using Refrigerant 134a as the working fluid, together with steady-state data at key points. The mass flow rate of the refrigerant is 7 kg/min, and the power input to the compressor is 5.17 kW. (a) Determine the coefficient of performance for the heat pump. (b) If the valve were replaced by a turbine, power could be produced, thereby reducing the power requirement of the heat pump system. Would you recommend this *power-saving* measure? Explain.

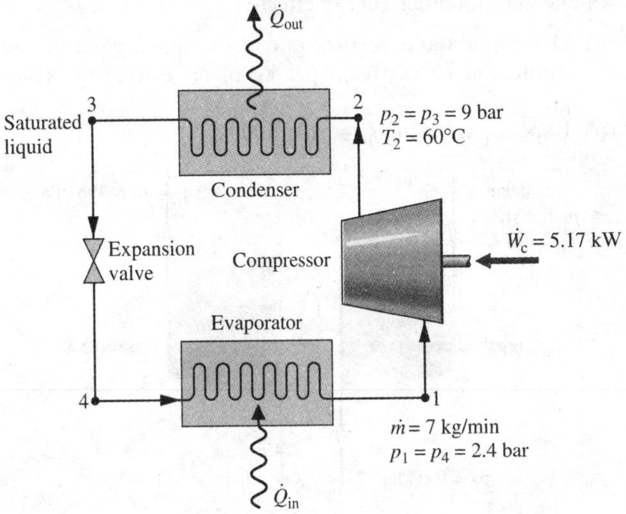

Fig. P6.143

6.144 Air enters an insulated diffuser operating at steady state at 1 bar, −3°C, and 260 m/s and exits with a velocity of 130 m/s. Employing the ideal gas model and ignoring potential energy, determine

(a) the temperature of the air at the exit, in °C.
(b) The maximum attainable exit pressure, in bar.

6.145 As shown in Fig. P6.145, air enters the diffuser of a jet engine at 18 kPa, 216 K with a velocity of 265 m/s, all data corresponding to high-altitude flight. The air flows adiabatically through the diffuser, decelerating to a velocity of 50 m/s at the diffuser exit. Assume steady-state operation, the ideal gas model for air, and negligible potential energy effects.

(a) Determine the temperature of the air at the exit of the diffuser, in K.
(b) If the air would undergo an isentropic process as it flows through the diffuser, determine the pressure of the air at the diffuser exit, in kPa.
(c) If friction were present, would the pressure of the air at the diffuser exit be greater than, less than, or equal to the value found in part (b)? Explain.

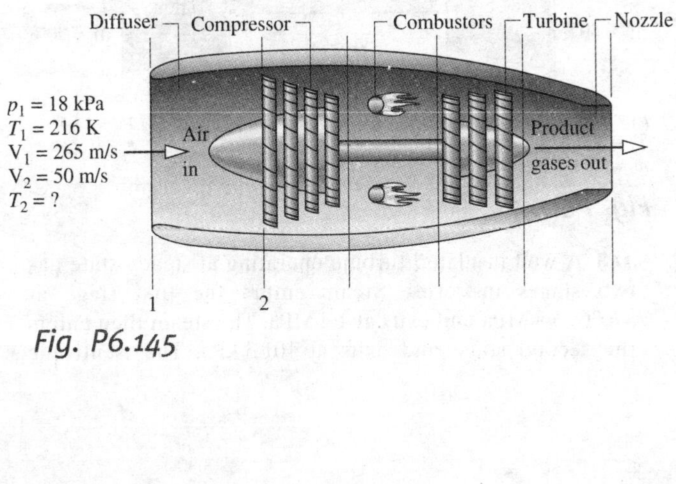

Fig. P6.145

6.146 As shown in Fig. P6.146, a steam turbine having an isentropic turbine efficiency of 90% drives an air compressor having an isentropic compressor efficiency of 85%. Steady-state operating data are provided on the figure. Assume the ideal gas model for air, and ignore stray heat transfer and kinetic and potential energy effects.

(a) Determine the mass flow rate of the steam entering the turbine, in kg of steam per kg of air exiting the compressor.

(b) Repeat part (a) if $\eta_t = \eta_c = 100\%$

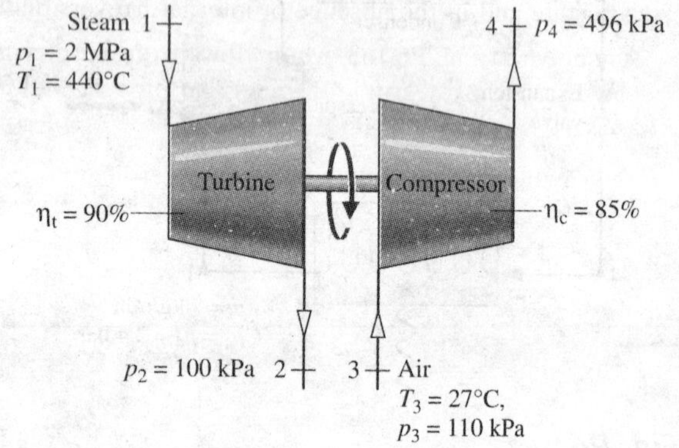

Fig. P6.146

6.147 Figure P6.147 shows a power system operating at steady state consisting of three components in series: an air compressor having an isentropic compressor efficiency of 80%, a heat exchanger, and a turbine having an isentropic turbine efficiency of 90%. Air enters the compressor at 1 bar, 300K with a mass flow rate of 5.8 kg/s and exits at a pressure of 10 bar. Air enters the turbine at 10 bar, 1400 K and exits at a pressure of 1 bar. Air can be modeled as an ideal gas. Stray heat transfer and kinetic and potential energy effects are negligible. Determine, in kW, (a) the power required by the compressor, (b) the power developed by the turbine, and (c) the *net* power output of the overall power system.

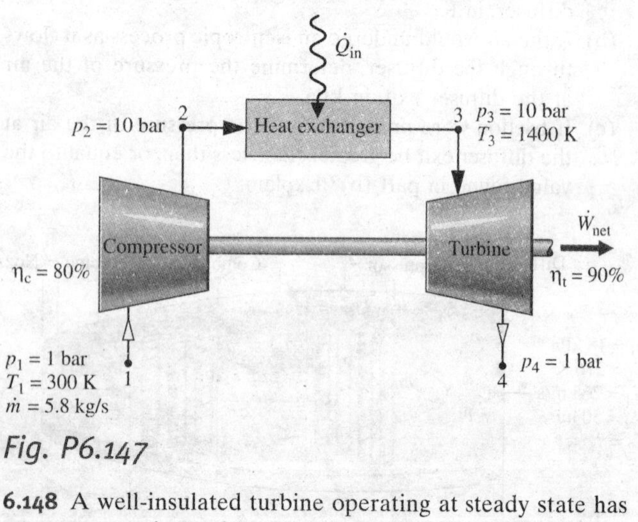

Fig. P6.147

6.148 A well-insulated turbine operating at steady state has two stages in series. Steam enters the first stage at 370°C, 3.8 MPa and exits at 1.4 MPa. The steam then enters the second stage and exits at 101.3 kPa. The isentropic efficiencies of the stages are 88% and 92%, respectively. Show the principal states on a *T–s* diagram. At the exit of the second stage, determine the temperature, in °C, if superheated vapor exits or the quality if a two-phase liquid–vapor mixture exits. Also determine the work developed by each stage, in kJ/kg of steam flowing.

6.149 A rigid tank is filled initially with 5.0 kg of air at a pressure of 0.5 MPa and a temperature of 500 K. The air is allowed to discharge through a turbine into the atmosphere, developing work until the pressure in the tank has fallen to the atmospheric level of 0.1 MPa. Employing the ideal gas model for the air, determine the *maximum* theoretical amount of work that could be developed, in kJ. Ignore heat transfer with the atmosphere and changes in kinetic and potential energy.

6.150 A tank initially containing air at 30.4 bar and 282°C is connected to a small turbine. Air discharges from the tank through the turbine, which produces work in the amount of 105 kJ. The pressure in the tank falls to 3.1 bar during the process and the turbine exhausts to the atmosphere at 1 bar. Employing the ideal gas model for the air and ignoring irreversibilities within the tank and the turbine, determine the volume of the tank, in m^3. Heat transfer with the atmosphere and changes in kinetic and potential energy are negligible.

6.151 Air enters a 3600-kW turbine operating at steady state with a mass flow rate of 18 kg/s at 800°C, 3 bar and a velocity of 100 m/s. The air expands adiabatically through the turbine and exits at a velocity of 150 m/s. The air then enters a diffuser where it is decelerated isentropically to a velocity of 10 m/s and a pressure of 1 bar. Employing the ideal gas model, determine

(a) the pressure and temperature of the air at the turbine exit, in bar and °C, respectively.

(b) the rate of entropy production in the turbine, in kW/K.

Show the processes on a *T–s* diagram.

Analyzing Internally Reversible Flow Processes

6.152 Air enters a compressor operating at steady state with a volumetric flow rate of 8 m^3/min at 23°C, 0.12 MPa. The air is compressed isothermally without internal irreversibilities, exiting at 1.5 MPa. Kinetic and potential energy effects can be ignored. Evaluate the work required and the heat transfer, each in kW.

6.153 Refrigerant 134a enters a compressor operating at steady state at 1.8 bar, −10°C with a volumetric flow rate of 2.4×10^{-2} m^3/s. The refrigerant is compressed to a pressure of 9 bar in an internally reversible process according to $pv^{1.04} = constant$. Neglecting kinetic and potential energy effects, determine

(a) the power required, in kW.

(b) the rate of heat transfer, in kW.

6.154 An air compressor operates at steady state with air entering at p_1 = 0.1 MPa, T_1 = 16°C. The air undergoes a polytropic process, and exits at p_2 = 0.5 MPa, T_2 = 146°C. (a) Evaluate the work and heat transfer, each in kJ/kg of air flowing. (b) Sketch the process on *p–v* and *T–s* diagrams and associate areas on the diagrams with work and heat transfer, respectively. Assume the ideal gas model for air and neglect changes in kinetic and potential energy.

6.155 An air compressor operates at steady state with air entering at $p_1 = 1$ bar, $T_1 = 17°C$ and exiting at $p_2 = 5$ bar. The air undergoes a polytropic process for which the compressor work input is 162.2 kJ per kg of air flowing. Determine (a) the temperature of the air at the compressor exit, in °C, and (b) the heat transfer, in kJ per kg of air flowing. (c) Sketch the process on p–v and T–s diagrams and associate areas on the diagrams with work and heat transfer, respectively. Assume the ideal gas model for air and neglect changes in kinetic and potential energy.

6.156 Compare the work required at steady state to compress *water vapor* isentropically to 3 MPa from the saturated vapor state at 0.1 MPa to the work required to pump *liquid water* isentropically to 3 MPa from the saturated liquid state at 0.1 MPa, each in kJ per kg of water flowing through the device. Kinetic and potential energy effects can be ignored.

6.157 A pump operating at steady state receives saturated liquid water at 50°C with a mass flow rate of 20 kg/s. The pressure of the water at the pump exit is 1 MPa. If the pump operates with negligible internal irreversibilities and negligible changes in kinetic and potential energy, determine the power required in kW.

6.158 A pump operating at steady state receives liquid water at 20 °C, 100 kPa with a mass flow rate of 53 kg/min. The pressure of the water at the pump exit is 5 MPa. The isentropic pump efficiency is 70%. Stray heat transfer and changes in kinetic and potential energy are negligible. Determine the power required by the pump, in kW.

6.159 A pump operating at steady state receives liquid water at 50°C, 1.5 MPa. The pressure of the water at the pump exit is 15 MPa. The magnitude of the work required by the pump is 18 kJ per kg of water flowing. Stray heat transfer and changes in kinetic and potential energy are negligible. Determine the isentropic pump efficiency.

6.160 A 5-kilowatt pump operating at steady state draws in liquid water at 1 bar, 15°C and delivers it at 5 bar at an elevation 6 m above the inlet. There is no significant change in velocity between the inlet and exit, and the local acceleration of gravity is 9.8 m/s^2. Would it be possible to pump 7.5 m^3 in 10 min or less? Explain.

6.161 An electrically driven pump operating at steady state draws water from a pond at a pressure of 1 bar and a rate of 40 kg/s and delivers the water at a pressure of 4 bar. There is no significant heat transfer with the surroundings, and changes in kinetic and potential energy can be neglected. The isentropic pump efficiency is 80%. Evaluating electricity at 8 cents per kW · h, estimate the hourly cost of running the pump.

6.162 Water behind a dam enters an intake pipe at a pressure of 0.1 MPa and velocity of 1.2 m/s, flows through a hydraulic turbine-generator, and exits at a point 90 m below the intake at 101.3 kPa, 10°C, 10.5 m/s. The diameter of the exit pipe is 1.2 m and the local acceleration of gravity is 9.81 m/s^2. Evaluating the electricity generated at 8 cents per kW · h, determine the value of the power produced, in $/day, for operation at steady state and in the absence of internal irreversibilities.

6.163 As shown in Fig. P6.163, water flows from an elevated reservoir through a hydraulic turbine. The pipe diameter is constant, and operation is at steady state. Estimate the minimum mass flow rate, in kg/s, that would be required for a turbine power output of 1 MW. The local acceleration of gravity is 9.8 m/s^2.

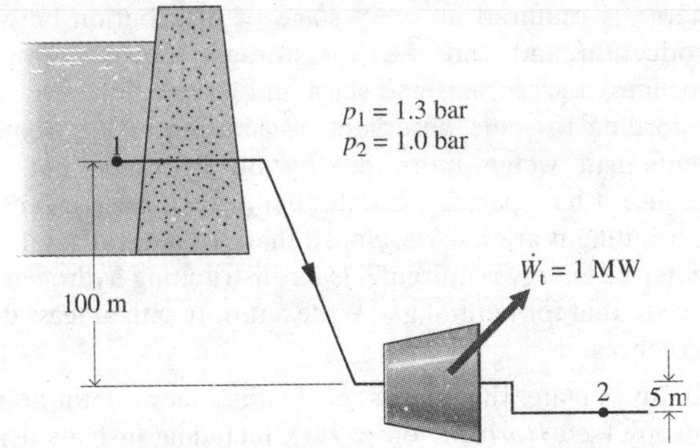

Fig. P6.163

6.164 Nitrogen (N_2) enters a nozzle operating at steady state at 0.2 MPa, 550K with a velocity of 1 m/s and undergoes a polytropic expansion with $n = 1.3$ to 0.15 MPa. Using the ideal gas model with $k = 1.4$, and ignoring potential energy effects, determine (a) the exit velocity, in m/s, and (b) the rate of heat transfer, in kJ per kg of gas flowing.

6.165 Carbon monoxide enters a nozzle operating at steady state at 5 bar, 200°C with a velocity of 1 m/s and undergoes a polytropic expansion to 1 bar and an exit velocity of 630 m/s. Using the ideal gas model and ignoring potential energy effects, determine

(a) the exit temperature, in °C.
(b) the rate of heat transfer, in kJ per kg of gas flowing.

Design & open ended problems: exploring engineering practice

6.1D Using the ENERGY STAR® *Home Improvement Toolbox,* obtain a rank-ordered list of the top three cost-effective improvements that would enhance the overall energy efficiency of your home. Develop a plan to implement the improvements. Write a report, including at least three references.

6.2D A thermodynamic diagram used in the refrigeration engineering field has the specific enthalpy and the natural logarithm of pressure as coordinates. Inspection of such a diagram suggests that in the vapor region constant-entropy lines are nearly linear, and thus the relation between s, ln p and h might be expressible in the vapor region as

$$h(s, p) = (\mathrm{A}\,s + \mathrm{B})\ln p + (\mathrm{C}\,s + \mathrm{D})$$

where A, B, C, and D are constants. Investigate the viability of this expression for pressure ranging up to 10 bar, using h, s, p, T, and v data for Refrigerant 134a, together with Eqs. 11.26 and 11.27 deduced from the second *Tds* equation, Eq. 6.10b. Summarize your conclusions in a memorandum.

6.3D Ocean thermal energy conversion (OTEC) power plants generate electricity on ships or platforms at sea by exploiting the naturally occurring decrease of the temperature of ocean water with depth. One proposal for the use of OTEC-generated electricity is to produce and commercialize ammonia in three steps: Hydrogen (H_2) would first be obtained by electrolysis of desalted sea water. The hydrogen would then be reacted with nitrogen (N_2) from the atmosphere to obtain ammonia (NH_3). Finally, liquid ammonia would be shipped to shore, where it would be reprocessed into hydrogen or used as a feedstock. Some say a major drawback with the proposal is whether *current technology* can be integrated to provide cost-competitive end products. Investigate this issue and summarize your findings in a report with at least three references.

6.4D Natural gas is currently playing a significant role in meeting our energy needs and hydrogen may be just as important in years ahead. For natural gas *and* hydrogen, energy is required at every stage of distribution between production and end use: for storage, transportation by pipelines, trucks, trains and ships, and liquefaction, if needed. According to some observers, distribution energy requirements will weigh more heavily on hydrogen not only because it has special attributes but also because means for distributing it are less developed than for natural gas. Investigate the energy requirements for distributing hydrogen relative to that for natural gas. Write a report with at least three references.

6.5D Investigate the merits of Tsallis's new definition of entropy (see *Horizons* on p. 281), including insights it may provide about physical systems verging on chaos. Explain what is meant by chaos in this context. Summarize your findings in a memorandum.

6.6D Classical economics was developed largely in analogy to the notion of mechanical equilibrium. Some observers are now saying that a macroeconomic system is more like a thermodynamic system than a mechanical one. Further, they say the failure of traditional economic theories to account for recent economic behavior may be partially due to not recognizing the role of entropy in controlling economic change and equilibrium, similar to the role of entropy in thermodynamics. Write a report, including at least three references, on how the second law and entropy are used in economics.

6.7D The *maximum entropy method* is widely used in the field of astronomical data analysis. Over the last three decades, considerable work has been done using the method for data filtering and removing features in an image that are caused by the telescope itself rather than from light coming from the sky (called deconvolution). To further such aims, refinements of the method have evolved over the years. Investigate the maximum entropy method as it is used today in astronomy, and summarize the state-of-the-art in a memorandum.

6.8D Elementary thermodynamic modeling, including the use of the temperature–entropy diagram for water and a form of the *Bernoulli equation* has been employed to study certain types of volcanic eruptions. (See L. G. Mastin, "Thermodynamics of Gas and Steam-Blast Eruptions," *Bull. Volcanol.*, 57, 85-98, 1995.) Write a report *critically evaluating* the underlying assumptions and application of thermodynamic principles, as reported in the article. Include at least three references.

6.9D The world's longest *coal-slurry pipeline* connects the Black Mesa Coal Mine in Arizona to the 1,600-MW Mojave Generating Station in Laughlin, Nevada, over 430 kilometers away. When operating, the 45 cm-diameter pipeline system uses vast amounts of water to transport millions of tons of coal from mine to power plant. Develop engineering and economic analyses for converting the existing pipeline to a more water-thrifty *coal-log pipeline.* Consider factors such as coal log preparation prior to transmission and drying after delivery to the power plant, pumping power requirements, and equipment costs. Write a report, including at least three references, in which the feasibility of such a conversion is critically evaluated.

6.10D In recent decades, many have written about the relationship between life in the biosphere and the second law of thermodynamics. Among these are Nobel Prize winners Erwin Schrodinger (Physics, 1933) and Ilya Prigogine (Chemistry, 1977). Contemporary observers such as Eric Schneider also have weighed in. Survey and *critically evaluate* such contributions to the literature. Summarize your conclusions in a report having at least three references.

7

Exergy Analysis

Engineering Context

The **objective** of this chapter is to introduce *exergy analysis,* which uses the conservation of mass and conservation of energy principles together with the second law of thermodynamics for the design and analysis of thermal systems.

The importance of developing thermal systems that make effective use of nonrenewable resources such as oil, natural gas, and coal is apparent. Exergy analysis is particularly suited for furthering the goal of more efficient resource use, since it enables the locations, types, and true magnitudes of waste and loss to be determined. This information can be used to design thermal systems, guide efforts to reduce sources of inefficiency in existing systems, and evaluate system economics.

Learning Objectives

When you complete your study of this chapter you will be able to...

- ✔ demonstrate understanding of key concepts related to exergy analysis... including the exergy reference environment, the dead state, exergy transfer, and exergy destruction.
- ✔ evaluate exergy at a state and exergy change between two states, using appropriate property data.
- ✔ apply exergy balances to closed systems and to control volumes at steady state.
- ✔ define and evaluate exergetic efficiencies.
- ✔ apply exergy costing to heat loss and simple cogeneration systems.

7.1 Introducing Exergy

Energy is conserved in every device or process. It cannot be destroyed. Energy entering a system with fuel, electricity, flowing streams of matter, and so on can be accounted for in the products and by-products. However, the energy conservation idea alone is inadequate for depicting some important aspects of resource utilization.

➡ FOR EXAMPLE... Figure 7.1*a* shows an *isolated system* consisting initially of a small container of fuel surrounded by air in abundance. Suppose the fuel burns (Fig. 7.1*b*) so that finally there is a slightly warm mixture of combustion products and air as shown in Fig. 7.1*c*. The total *quantity* of energy associated with the system is constant because no energy transfers take place across the boundary of an isolated system. Still, the initial fuel–air combination is intrinsically more useful than the final warm mixture. For instance, the fuel might be used in some device to generate electricity or produce superheated steam, whereas the uses of the final slightly warm mixture are far more limited in scope. We can say that the system has a greater *potential for use* initially than it has finally. Since nothing but a final warm mixture is achieved in the process, this potential is largely wasted. More precisely, the initial potential is largely *destroyed* because of the irreversible nature of the process. ⬅

Anticipating the main results of this chapter, *exergy* is the property that quantifies *potential for use.* The foregoing example illustrates that, unlike energy, exergy is not conserved but is destroyed by irreversibilities.

Subsequent discussion shows that exergy not only can be destroyed by irreversibilities but also can be transferred *to* and *from* systems Exergy transferred from a system to its surroundings without use typically represents a *loss.* Improved energy resource utilization can be realized by reducing exergy destruction within a system and/or reducing losses. An objective in exergy analysis is to identify sites where exergy destructions and losses occur and rank order them for significance. This allows attention to be centered on aspects of system operation that offer the greatest opportunities for cost-effective improvements.

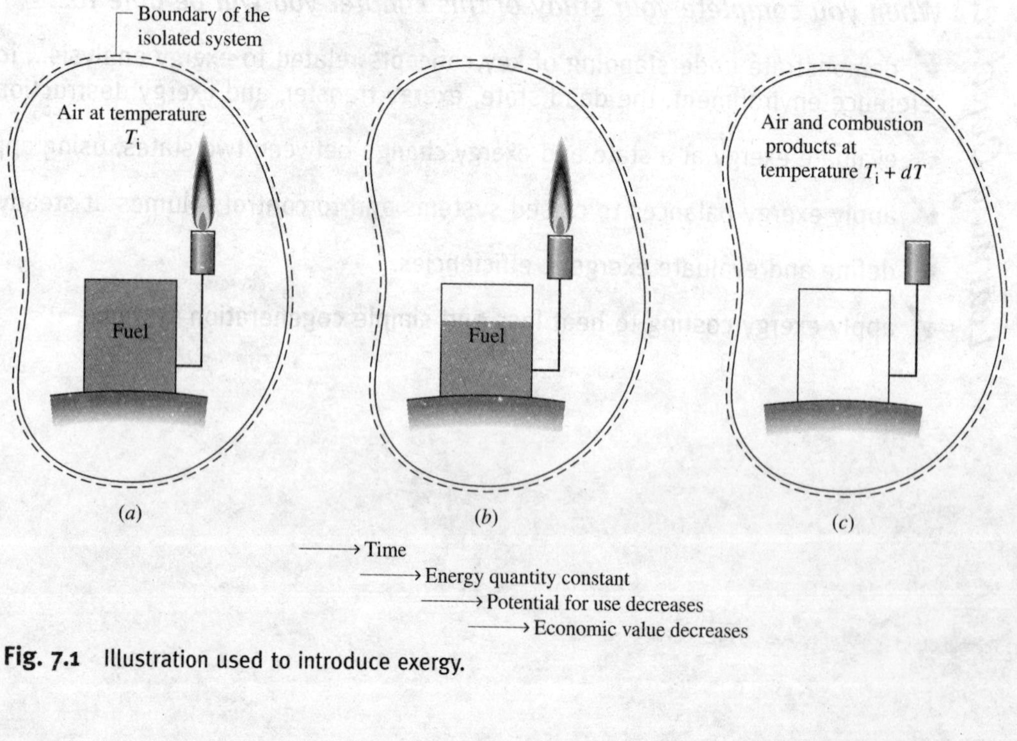

Fig. 7.1 Illustration used to introduce exergy.

Returning to Fig. 7.1, note that the fuel present initially has economic value while the final slightly warm mixture has little value. Accordingly, economic value decreases in this process. From such considerations we might infer there is a link between exergy and economic value, and this is the case as we will see in subsequent discussions.

7.2 Conceptualizing Exergy

The introduction to the second law in Chap. 5 also provides a basis for the exergy concept, as considered next.

Principal conclusions of the discussion of Fig. 5.1 given on p. 215 are that

- a potential for developing work exists whenever two systems at different states are brought into communication, and
- work can be developed as the two systems are allowed to come into equilibrium.

In Fig. 5. 1a, for example, a body initially at an elevated temperature T_i placed in contact with the atmosphere at temperature T_0 cools spontaneously. To conceptualize how work might be developed in this case, see Fig. 7.2. The figure shows an overall system with three elements: the body, a power cycle, and the atmosphere at T_0 and p_0. The atmosphere is presumed to be large enough that its temperature and pressure remain constant.

Instead of the body cooling spontaneously as considered in Fig. 5.1a, Fig. 7.2 shows that if the heat transfer Q during cooling is passed to the power cycle, work W_c can be developed, while Q_0 is discharged to the atmosphere. These are the only energy transfers. The work W_c is *fully available* for lifting a weight or, equivalently, as shaft work or electrical work. Ultimately the body cools to T_0, and no more work would be developed. At equilibrium, the body and atmosphere each possess energy, but there no longer is any potential for developing work from the two because no further interaction can occur between them.

Note that work W_c also could be developed by the system of Fig. 7.2 if the initial temperature of the body were *less* than that of the atmosphere: $T_i < T_0$. In such a case, the directions of the heat transfers Q and Q_0 shown on Fig. 7.2 would each reverse. Work could be developed as the body *warms* to equilibrium with the atmosphere.

Since there is no net change of state for the power cycle of Fig. 7.2, we conclude that the work W_c is realized solely because the initial state of the body differs from that of the atmosphere. *Exergy is the maximum theoretical value of such work.*

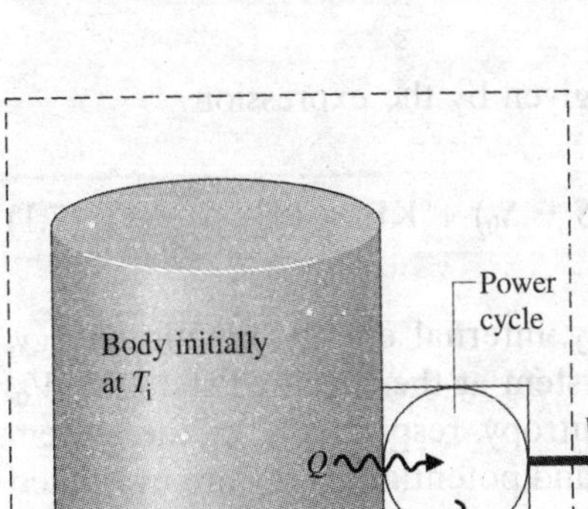

Fig. 7.2 Overall system of body, power cycle, and atmosphere used to conceptualize exergy.

7.2.1 Environment and Dead State

For thermodynamic analysis involving the exergy concept, it is necessary to model the atmosphere used in the foregoing discussion. The resulting model is called the ***exergy reference environment,*** or simply the ***environment.***

environment

In this book the environment is regarded to be a simple compressible system that is *large* in extent and *uniform* in temperature, T_0, and pressure, p_0. In keeping with the idea that the environment represents a portion of the physical world, the values for both p_0 and T_0 used throughout a particular analysis are normally taken as typical ambient conditions, such as 1 bar and 25°C. Additionally, the intensive properties of the environment do not change significantly as a result of any process under consideration, and the environment is free of irreversibilities.

dead state

When a system of interest is at T_0 and p_0 and *at rest* relative to the environment, we say the system is at the ***dead state.*** At the dead state there can be no interaction between system and environment, and thus no potential for developing work.

7.2.2 Defining Exergy

The discussion to this point of the current section can be summarized by the following ***definition of exergy:***

definition of exergy

> Exergy is the maximum theoretical work obtainable from an *overall* system consisting of a system and the environment as the system comes into equilibrium with the environment (passes to the dead state).

Interactions between the system and the environment may involve auxiliary devices, such as the power cycle of Fig. 7.2, that at least in principle allow the realization of the work. The work developed is fully available for lifting a weight or, equivalently, as shaft work or electrical work. We might expect that the maximum theoretical work would be obtained when there are no irreversibilities, and as considered in the next section this is the case.

7.3 Exergy of a System

exergy of a system

The ***exergy of a system,*** E, at a specified state is given by the expression

$$\mathsf{E} = (U - U_0) + p_0(V - V_0) - T_0(S - S_0) + \mathrm{KE} + \mathrm{PE} \tag{7.1}$$

where U, KE, PE, V, and S denote, respectively, internal energy, kinetic energy, potential energy, volume, and entropy of the system at the specified state. U_0, V_0, and S_0 denote internal energy, volume, and entropy, respectively, of the system when at the dead state. In this chapter kinetic and potential energy are evaluated relative to the environment. Thus, when the system is at the dead state, it is at rest relative to the environment and the values of its kinetic and potential energies are zero: $\mathrm{KE}_0 = \mathrm{PE}_0 = 0$. By inspection of Eq. 7.1, the units of exergy are seen to be the same as those of energy.

Take Note...

In this book, E and e are used for exergy and specific exergy, respectively, while E and e denote energy and specific energy, respectively. Such notation is in keeping with standard practice. The appropriate concept, exergy or energy, will be clear in context. Still, care is required to avoid mistaking the symbols for these concepts.

Equation 7.1 can be derived by applying energy and entropy balances to the overall system shown in Fig. 7.3 consisting of a system and an environment. See the box for the derivation of Eq. 7.1.

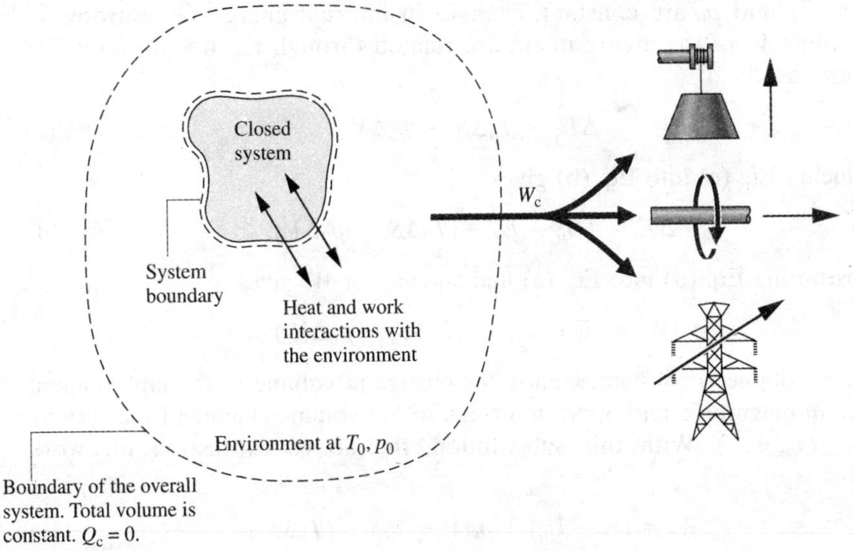

Fig. 7.3 Overall system of system and environment used to evaluate exergy.

Evaluating Exergy of a System

Referring to Fig. 7.3, exergy is the maximum theoretical value of the work W_c obtainable as the system comes into equilibrium with the environment—that is, as the system passes to the dead state.

In keeping with the discussion thus far, the system plus the environment is referred to as the *overall* system. The boundary of the overall system is located so there is no energy transfer across it by heat transfer: $Q_c = 0$. Moreover, the boundary of the overall system is located so that the total volume remains constant, even though the volumes of the system and environment can vary. Accordingly, the work W_c shown on the figure is the *only* energy transfer across the boundary of the overall system and is *fully available* for lifting a weight, turning a shaft, or producing electricity in the surroundings. Next, we apply the energy and entropy balances to determine the maximum theoretical value for W_c.

Energy Balance. Consider a process where the system and the environment come to equilibrium. The energy balance for the overall system is

$$\Delta E_c = \cancelto{0}{Q_c} - W_c \tag{a}$$

where W_c is the work developed by the overall system and ΔE_c is the change in energy of the overall system: the sum of the energy changes of the system and the environment. The energy of the system initially is denoted by E, which includes the kinetic, potential, and internal energies of the system. Since the kinetic and potential energies are evaluated relative to the environment, the energy of the system at the dead state is just its internal energy, U_0. Accordingly, ΔE_c can be expressed as

$$\Delta E_c = (U_0 - E) + \Delta U_e \tag{b}$$

where ΔU_e is the change in internal energy of the environment.

Since T_0 and p_0 are constant, changes in internal energy U_e, entropy S_e, and volume V_e of the environment are related through Eq. 6.8, the *first TdS* equation, as

$$\Delta U_e = T_0 \Delta S_e - p_0 \Delta V_e \tag{c}$$

Introducing Eq. (c) into Eq. (b) gives

$$\Delta E_c = (U_0 - E) + (T_0 \Delta S_e - p_0 \Delta V_e) \tag{d}$$

Substituting Eq. (d) into Eq. (a) and solving for W_c gives

$$W_c = (E - U_0) - (T_0 \Delta S_e - p_0 \Delta V_e)$$

The total volume is constant. Hence, the change in volume of the environment is equal in magnitude and opposite in sign to the volume change of the system: $\Delta V_e = -(V_0 - V)$. With this substitution, the above expression for work becomes

$$W_c = (E - U_0) + p_0(V - V_0) - (T_0 \Delta S_e) \tag{e}$$

This equation gives the work for the overall system as the system passes to the dead state. The maximum theoretical work is determined using the entropy balance as follows.

Entropy Balance. The entropy balance for the overall system reduces to

$$\Delta S_c = \sigma_c$$

where the entropy transfer term is omitted because no heat transfer takes place across the boundary of the overall system. The term σ_c accounts for entropy production due to irreversibilities as the system comes into equilibrium with the environment. The entropy change ΔS_c is the sum of the entropy changes for the system and environment, respectively. That is

$$\Delta S_c = (S_0 - S) + \Delta S_e$$

where S and S_0 denote the entropy of the system at the given state and the dead state, respectively. Combining the last two equations

$$(S_0 - S) + \Delta S_e = \sigma_c \tag{f}$$

Eliminating ΔS_e between Eqs. (e) and (f) results in

$$W_c = (E - U_0) + p_0(V - V_0) - T_0(S - S_0) - T_0\sigma_c \tag{g}$$

With $E = U + \text{KE} + \text{PE}$, Eq. (g) becomes

$$W_c = \underline{(U - U_0) + p_0(V - V_0) - T_0(S - S_0) + \text{KE} + \text{PE}} - T_0\sigma_c \tag{h}$$

The value of the underlined term in Eq. (h) is determined by the two end states of the system—the given state and the dead state—and is independent of the details of the process linking these states. However, the value of the term $T_0\sigma_c$ depends on the nature of the process as the system passes to the dead state. In accordance with the second law, $T_0\sigma_c$ is positive when irreversibilities are present and vanishes in the limiting case where there are no irreversibilities. The value of $T_0\sigma_c$ cannot be negative. Hence, the *maximum* theoretical value for the work of the overall system W_c is obtained by setting $T_0\sigma_c$ to zero in Eq. (h). By definition, this maximum value is the exergy, E. Accordingly, Eq. 7.1 is seen to be the appropriate expression for evaluating exergy.

7.3.1 Exergy Aspects

In this section, we list five important aspects of the exergy concept:

1. Exergy is a measure of the departure of the state of a system from that of the environment. It is therefore an attribute of the system and environment together. However, once the environment is specified, a value can be assigned to exergy in terms of property values for the system only, so exergy can be regarded as a property of the system. Exergy is an extensive property.
2. The value of exergy cannot be negative. If a system were at any state other than the dead state, the system would be able to change its condition *spontaneously* toward the dead state; this tendency would cease when the dead state was reached. No work must be done to effect such a spontaneous change. Accordingly, any change in state of the system to the dead state can be accomplished with *at least zero* work being developed, and thus the *maximum* work (exergy) cannot be negative.
3. Exergy is not conserved but is destroyed by irreversibilities. A limiting case is when exergy is completely destroyed, as would occur if a system were permitted to undergo a spontaneous change to the dead state with no provision to obtain work. The potential to develop work that existed originally would be completely wasted in such a spontaneous process.
4. Exergy has been viewed thus far as the *maximum* theoretical work obtainable from an overall system of system plus environment as the system passes *from* a given state *to* the dead state. Alternatively, exergy can be regarded as the magnitude of the *minimum* theoretical work *input* required to bring the system *from* the dead state *to* the given state. Using energy and entropy balances as above, we can readily develop Eq. 7.1 from this viewpoint. This is left as an exercise.
5. When a system is at the dead state, it is in *thermal* and *mechanical* equilibrium with the environment, and the value of exergy is zero. More precisely, the *thermomechanical* contribution to exergy is zero. This modifying term distinguishes the exergy concept of the present chapter from another contribution to exergy introduced in Sec. 13.6, where the contents of a system at the dead state are permitted to enter into chemical reaction with environmental components and in so doing develop additional work. This contribution to exergy is called *chemical exergy.* Although the chemical exergy concept is important in the second law analysis of systems involving reacting mixtures, including combustion, the thermomechanical exergy concept suffices for a wide range of thermodynamic evaluations.

Bio... connections

The U.S. poultry industry produces billions of kilograms of meat annually, with chicken production accounting for over 80% of the total. The annual amount of waste produced by these birds also reaches into the billions of kilograms. The waste may be more than can be managed by disposal over cropland as fertilizer. Some of the excess can be used to manufacture fertilizer pellets for commercial and domestic use. Despite its relatively low chemical exergy, poultry waste also can be used to produce methane, through *anaerobic* digestion. The methane can be burned in power plants to make electricity or process steam. Digester systems are available for use right on the farm. These are positive developments for an important sector of the U.S. agricultural economy that has received adverse publicity for concerns over arsenic content of poultry waste, run-off of waste into streams and rivers, and excessive odor and fly infestation in the vicinity of huge farming operations.

7.3.2 Specific Exergy

Although exergy is an extensive property, it is often convenient to work with it on a unit mass or molar basis. Expressing Eq. 7.1 on a unit mass basis, the ***specific exergy,*** e, is

specific exergy

$$\mathsf{e} = (u - u_0) + p_0(v - v_0) - T_0(s - s_0) + \mathrm{V}^2/2 + gz \tag{7.2}$$

Take Note...

When kinetic and potential energy are evaluated relative to the dead state, they are rightfully considered as *exergy*. But for simplicity of expression in the present chapter, we refer to these terms—whether viewed as energy or exergy—as *accounting for the effects of motion and gravity*. The meaning will be clear in context.

where u, v, s, $\mathrm{V}^2/2$, and gz are the specific internal energy, volume, entropy, kinetic energy, and potential energy, respectively, at the state of interest; u_0, v_0, and s_0 are specific properties at the dead state: at T_0, p_0. In Eq. 7.2, the kinetic and potential energies are measured relative to the environment and thus contribute their full values to the exergy magnitude because, in principle, each could be fully converted to work were the system brought to rest at zero elevation relative to the environment. Finally, by inspection of Eq. 7.2, note that the units of specific exergy are the same as for specific energy, kJ/kg.

The specific exergy at a specified state requires properties at that state and at the dead state. **FOR EXAMPLE...** let us use Eq. 7.2 to determine the specific exergy of saturated water vapor at 120°C, having a velocity of 30 m/s and an elevation of 6 m, each relative to an exergy reference environment where $T_0 = 298$ K (25°C), $p_0 = 1$ bar, and $g = 9.8$ m/s^2. For water as saturated vapor at 120°C, Table A-2 gives $v = 0.8919$ m^3/kg, $u = 2529.3$ kJ/kg, $s = 7.1296$ kJ/kg · K. At the dead state, where $T_0 = 298$ K (25°C) and $p_0 = 1$ bar, water is a liquid. Thus, with Eqs. 3.11, 3.12, and 6.5 and values from Table A-2, we get $v_0 = 1.0029 \times 10^{-3}$ m^3/kg, $u_0 = 104.88$ kJ/kg, $s_0 = 0.3674$ kJ/kg · K. Substituting values

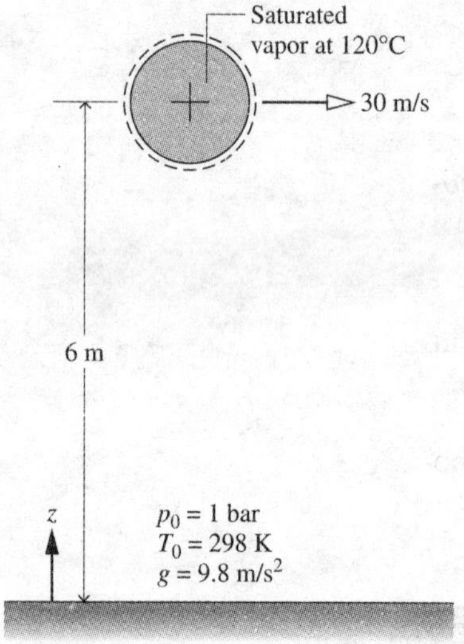

$$\begin{aligned}
\mathsf{e} &= (u - u_0) + p_0(v - v_0) - T_0(s - s_0) + \frac{\mathrm{V}^2}{2} + gz \\
&= \left[(2529.3 - 104.88)\frac{\text{kJ}}{\text{kg}}\right] \\
&\quad + \left[\left(1.01325 \times 10^5 \frac{\text{N}}{\text{m}^2}\right)(0.8919 - 1.0029 \times 10^{-3})\frac{\text{m}^3}{\text{kg}}\right]\left|\frac{1\text{ kJ}}{10^3\text{ N}\cdot\text{m}}\right| \\
&\quad - \left[(298\text{ K})(7.1296 - 0.3674)\frac{\text{kJ}}{\text{kg}\cdot\text{K}}\right] \\
&\quad + \left[\frac{(30\text{ m/s})^2}{2} + \left(9.8\frac{\text{m}}{\text{s}^2}\right)(6\text{ m})\right]\left|\frac{1\text{ N}}{1\text{ kg}\cdot\text{m/s}^2}\right|\left|\frac{1\text{ kJ}}{10^3\text{ N}\cdot\text{m}}\right| \\
&= (2424.42 + 90.27 - 2015.14 + 0.45 + 0.06)\frac{\text{kJ}}{\text{kg}} = 500\frac{\text{kJ}}{\text{kg}}
\end{aligned}$$

The following example illustrates the use of Eq. 7.2 together with ideal gas property data.

Example 7.1 EXERGY OF EXHAUST GAS

A cylinder of an internal combustion engine contains 2450 cm^3 of gaseous combustion products at a pressure of 7 bar and a temperature of 867°C just before the exhaust valve opens. Determine the specific exergy of the gas, in kJ/kg. Ignore the effects of motion and gravity, and model the combustion products as air as an ideal gas. Take $T_0 = 300$ K (27°C) and $p_0 = 1.013$ bar.

Solution

Known: Gaseous combustion products at a specified state are contained in the cylinder of an internal combustion engine.

Find: Determine the specific exergy.

Schematic and Given Data:

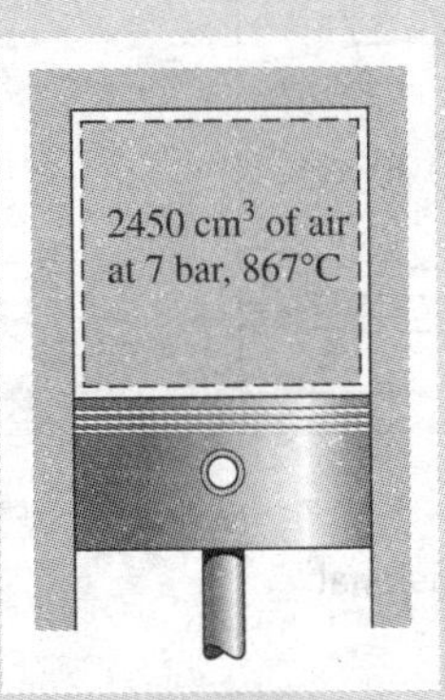

Engineering Model:

1. The gaseous combustion products are a closed system.
2. The combustion products are modeled as air as an ideal gas.
3. The effects of motion and gravity can be ignored.
4. $T_0 = 300$ K (27°C) and $p_0 = 1.013$ bar.

Fig. E7.1

Analysis: With assumption 3, Eq. 7.2 becomes

$$\text{e} = u - u_0 + p_0(v - v_0) - T_0(s - s_0)$$

The internal energy and entropy terms are evaluated using data from Table A-22, as follows:

$$\begin{aligned} u - u_0 &= (880.35 - 214.07)\ \text{kJ/kg} \\ &= 666.28\ \text{kJ/kg} \\ s - s_0 &= s°(T) - s°(T_0) - \frac{\overline{R}}{M}\ln\frac{p}{p_0} \\ &= \left(3.11883 - 1.70203 - \left(\frac{8.314}{28.97}\right)\ln\left(\frac{7}{1.013}\right)\right)\frac{\text{kJ}}{\text{kg}\cdot\text{K}} \\ &= 0.8621\ \text{kJ/kg}\cdot\text{K} \\ T_0(s - s_0) &= (300\ \text{K})(0.8621\ \text{kJ/kg}\cdot\text{K}) \\ &= 258.62\ \text{kJ/kg} \end{aligned}$$

The $p_0(v - v_0)$ term is evaluated using the ideal gas equation of state: $v = (\overline{R}/M)T/p$ and $v_0 = (\overline{R}/M)T_0/p_0$, so

$$\begin{aligned} p_0(v - v_0) &= \frac{\overline{R}}{M}\left(\frac{p_0 T}{p} - T_0\right) \\ &= \frac{8.314}{28.97}\left(\frac{(1.013)(1140)}{7} - 300\right)\frac{\text{kJ}}{\text{kg}} \\ &= -38.75\ \text{kJ/kg} \end{aligned}$$

Substituting values into the above expression for the specific exergy

$$\begin{aligned} \text{e} &= (666.28 + (-38.75) - 258.62)\ \text{kJ/kg} \\ &= 368.91\ \text{kJ/kg} \end{aligned}$$ ❶

❶ If the gases are discharged directly to the surroundings, the potential for developing work quantified by the exergy value determined in the solution is wasted. However, by venting the gases through a turbine, some work could be developed. This principle is utilized by the *turbochargers* added to some internal combustion engines.

Skills Developed: *Ability to...*
- ✓ evaluate specific exergy.
- ✓ apply the ideal gas model.

Quick Quiz

To what elevation, in m, would a 1-kg mass have to be raised from zero elevation with respect to the reference environment for its exergy to equal that of the gas in the cylinder? Assume $g = 9.81$ m/s^2.
Ans. 197 m.

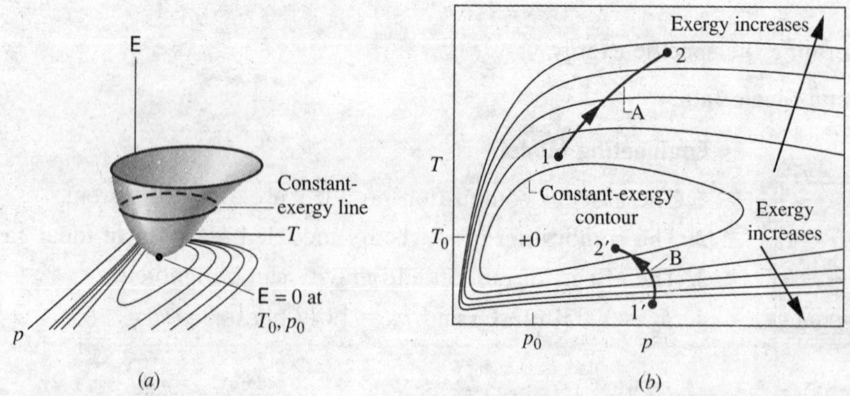

Fig. 7.4 Exergy-temperature-pressure surface for a gas. (a) Three-dimensional view. (b) Constant exergy contours on a *T-p* diagram.

7.3.3 Exergy Change

A closed system at a given state can attain new states by various means, including work and heat interactions with its surroundings. The exergy value associated with a new state generally differs from the exergy value at the initial state. Using Eq. 7.1, we can determine the change in exergy between the two states. At the initial state

$$E_1 = (U_1 - U_0) + p_0(V_1 - V_0) - T_0(S_1 - S_0) + KE_1 + PE_1$$

At the final state

$$E_2 = (U_2 - U_0) + p_0(V_2 - V_0) - T_0(S_2 - S_0) + KE_2 + PE_2$$

Subtracting these we get the ***exergy change***

exergy change

$$E_2 - E_1 = (U_2 - U_1) + p_0(V_2 - V_1) - T_0(S_2 - S_1) + (KE_2 - KE_1) + (PE_2 - PE_1) \quad (7.3)$$

Note that the dead state values U_0, V_0, S_0 cancel when we subtract the expressions for E_1 and E_2.

Exergy *change* can be illustrated using Fig. 7.4, which shows an exergy-temperature-pressure surface for a gas together with constant-exergy contours projected on temperature-pressure coordinates. For a system undergoing Process A, exergy increases as its state *moves away* from the dead state (from 1 to 2). In process B, exergy decreases as the state moves *toward* the dead state (from 1′ to 2′).

7.4 Closed System Exergy Balance

Like energy, exergy can be transferred across the boundary of a closed system. The change in exergy of a system during a process would not necessarily equal the net exergy transferred because exergy would be destroyed if irreversibilities were present within the system during the process. The concepts of exergy change, exergy transfer, and exergy destruction are related by the closed system exergy balance introduced in this section. The exergy balance concept is extended to control volumes in Sec. 7.5. Exergy balances are expressions of the second law of thermodynamics and provide the basis for exergy analysis.

7.4.1 Introducing the Closed System Exergy Balance

The ***closed system exergy balance*** is given by Eq. 7.4a. See the box for its development.

$$\underbrace{E_2 - E_1}_{\text{exergy change}} = \underbrace{\int_1^2 \left(1 - \frac{T_0}{T_b}\right)\delta Q - [W - p_0(V_2 - V_1)]}_{\text{exergy transfers}} - \underbrace{T_0\sigma}_{\text{exergy destruction}} \quad (7.4a)$$

closed system exergy balance

For specified end states and given values of p_0 and T_0, the exergy change $E_2 - E_1$ on the left side of Eq. 7.4a can be evaluated from Eq. 7.3. The underlined terms on the right depend explicitly on the nature of the process, however, and cannot be determined by knowing only the end states and the values of p_0 and T_0. These terms are interpreted in the discussions of Eqs. 7.5–7.7, respectively.

Developing the Exergy Balance

The exergy balance for a closed system is developed by combining the closed system energy and entropy balances. The forms of the energy and entropy balances used are, respectively

$$\Delta U + \Delta KE + \Delta PE = \left(\int_1^2 \delta Q\right) - W$$

$$\Delta S = \int_1^2 \left(\frac{\delta Q}{T}\right)_b + \sigma$$

where W and Q represent, respectively, work and heat transfer between the system and its surroundings. In the entropy balance, T_b denotes the temperature on the system boundary where δQ occurs. The term σ accounts for entropy produced within the system by internal irreversibilities.

As the first step in deriving the exergy balance, multiply the entropy balance by the temperature T_0 and subtract the resulting expression from the energy balance to obtain

$$(\Delta U + \Delta KE + \Delta PE) - T_0\Delta S = \left(\int_1^2 \delta Q\right) - T_0\int_1^2 \left(\frac{\delta Q}{T}\right)_b - W - T_0\sigma$$

Collecting the terms involving δQ on the right side and introducing Eq. 7.3 on the left side, we get

$$(E_2 - E_1) - p_0(V_2 - V_1) = \int_1^2 \left(1 - \frac{T_0}{T_b}\right)\delta Q - W - T_0\sigma$$

On rearrangement, this expression gives Eq. 7.4a, the closed system exergy balance.

Since Eq. 7.4a is obtained by deduction from the energy and entropy balances, it is not an independent result, but can be used in place of the entropy balance as an expression of the second law.

The first underlined term on the right side of Eq. 7.4a is associated with heat transfer to or from the system during the process. It is interpreted as the ***exergy transfer accompanying heat.*** That is

exergy transfer accompanying heat

$$\mathsf{E}_q = \begin{bmatrix} \textit{exergy} \text{ transfer} \\ \text{accompanying heat} \end{bmatrix} = \int_1^2 \left(1 - \frac{T_0}{T_b}\right)\delta Q \tag{7.5}$$

where T_b denotes the temperature on the boundary where heat transfer occurs.

The second underlined term on the right side of Eq. 7.4a is associated with work. It is interpreted as the ***exergy transfer accompanying work.*** That is

exergy transfer accompanying work

$$\mathsf{E}_w = \begin{bmatrix} \textit{exergy} \text{ transfer} \\ \text{accompanying work} \end{bmatrix} = [W - p_0(V_2 - V_1)] \tag{7.6}$$

The third underlined term on the right side of Eq. 7.4a accounts for the ***destruction of exergy*** due to irreversibilities within the system. It is symbolized by E_d. That is

exergy destruction

$$\mathsf{E}_d = T_0\sigma \tag{7.7}$$

With Eqs. 7.5, 7.6 and 7.7, Eq. 7.4a is expressed alternatively as

$$\mathsf{E}_2 - \mathsf{E}_1 = \mathsf{E}_q - \mathsf{E}_w - \mathsf{E}_d \tag{7.4b}$$

As for the concept of exergy itself (Sec. 7.2.2), exergy transfer is conceptualized in terms of work. See the box for a summary.

Conceptualizing Exergy Transfer

In exergy analysis, heat transfer and work are expressed in terms of a *common measure:* work *fully available* for lifting a weight or, equivalently, as shaft or electrical work. This is the significance of the exergy transfer expressions given by Eqs. 7.5 and 7.6, respectively.

Without regard for the nature of the surroundings with which the system is *actually* interacting, we interpret the *magnitudes* of these exergy transfers as the maximum theoretical work that *could* be developed *were* the system interacting with the environment, as follows:

- On recognizing the term $(1 - T_0/T_b)$ as the Carnot efficiency (Eq. 5.9), the quantity $(1 - T_0/T_b)\delta Q$ appearing in Eq. 7.5 is interpreted as the work developed by a reversible power cycle receiving energy δQ by heat transfer at temperature T_b and discharging energy by heat transfer to the environment at temperature $T_0 < T_b$. When T_b is below T_0, we think of the work developed by a reversible power cycle receiving a heat transfer from the environment at T_0 and discharging δQ at temperature $T_b < T_0$.
- The exergy transfer given by Eq. 7.6 is the work W of the system less the work required to displace the environment whose pressure is p_0, namely $p_0(V_2 - V_1)$.

See Example 7.2 for an illustration of these interpretations.

To summarize, in each of its forms Eq. 7.4 states that the change in exergy of a closed system can be accounted for in terms of exergy transfers and the destruction of exergy due to irreversibilities within the system.

When applying the exergy balance, it is essential to observe the requirements imposed by the second law on the exergy destruction: In accordance with the second law, the exergy destruction is positive when irreversibilities are present within the system during the process and vanishes in the limiting case where there are no irreversibilities. That is

$$\mathsf{E}_d : \begin{cases} >0 & \text{irreversibilities present within the system} \\ =0 & \text{no irreversibilities present within the system} \end{cases} \tag{7.8}$$

The value of the exergy destruction cannot be negative. Moreover, exergy destruction is *not* a property. On the other hand, exergy *is* a property, and like other properties, the *change* in exergy of a system can be positive, negative, or zero

$$\mathsf{E}_2 - \mathsf{E}_1 : \begin{cases} >0 \\ =0 \\ <0 \end{cases}$$

For an *isolated* system, no heat or work interactions with the surroundings occur, and thus there are no transfers of exergy between the system and its surroundings. Accordingly, the exergy balance reduces to give

$$\Delta \mathsf{E}\big]_{\text{isol}} = -\mathsf{E}_d\big]_{\text{isol}} \tag{7.9}$$

Since the exergy destruction must be positive in any actual process, the only processes of an isolated system that occur are those for which the exergy of the isolated system *decreases*. For exergy, this conclusion is the counterpart of the increase of entropy principle (Sec. 6.8.1) and, like the increase of entropy principle, can be regarded as an alternative statement of the second law.

In Example 7.2, we consider exergy change, exergy transfer, and exergy destruction for the process of water considered in Example 6.1. This example should be quickly reviewed before studying the current example.

Example 7.2 EXPLORING EXERGY CHANGE, TRANSFER, AND DESTRUCTION

Water, initially a saturated liquid at 150°C (423.15 K), is contained in a piston–cylinder assembly. The water is heated to the corresponding saturated vapor state in an internally reversible process at constant temperature and pressure. For $T_0 = 20°\text{C}$ (293.15 K), $p_0 = 1$ bar, and ignoring the effects of motion and gravity, determine per unit of mass, each in kJ/kg, (a) the change in exergy, (b) the exergy transfer accompanying heat, (c) the exergy transfer accompanying work, and (d) the exergy destruction.

Solution

Known: Water contained in a piston–cylinder assembly undergoes an internally reversible process at 150°C from saturated liquid to saturated vapor.

Find: Determine the change in exergy, the exergy transfers accompanying heat and work, and the exergy destruction.

Schematic and Given Data:

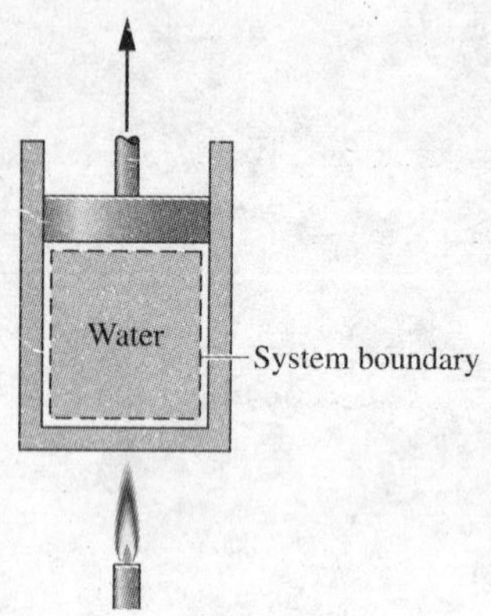

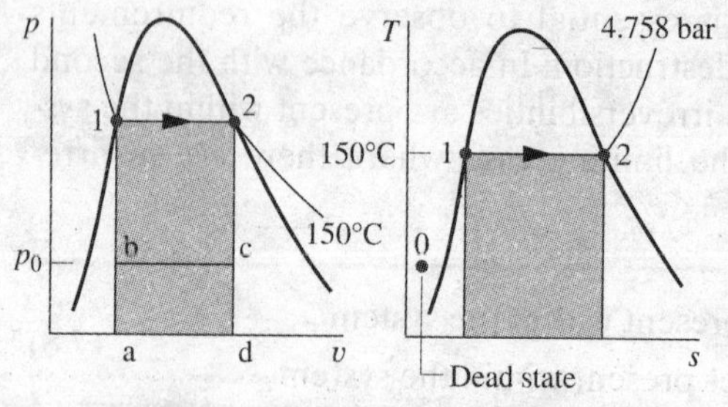

Data from Example 6.1:

W/m = 186.38 kJ/kg, Q/m = 2114.1 kJ/kg

State	v (m³/kg)	u (kJ/kg)	s (kJ/kg·K)
1	1.0905×10^{-3}	631.68	1.8418
2	0.3928	2559.5	6.8379

Engineering Model:

1. The water in the piston-cylinder assembly is a closed system.
2. The process is internally reversible.
3. Temperature and pressure are constant during the process.
4. Ignore the effects of motion and gravity.
5. $T_0 = 293.15$ K, $p_0 = 1$ bar.

Fig. E7.2

Analysis:

(a) Using Eq. 7.3 together with assumption 4, we have per unit of mass

$$e_2 - e_1 = u_2 - u_1 + p_0(v_2 - v_1) - T_0(s_2 - s_1) \tag{a}$$

With data from Fig. E7.2

$$e_2 - e_1 = (2559.5 - 631.68)\frac{\text{kJ}}{\text{kg}} + \left(1.0 \times 10^5 \frac{\text{N}}{\text{m}^2}\right)(0.3928 - (1.0905 \times 10^{-3}))\frac{\text{m}^3}{\text{kg}}\left|\frac{1\text{ kJ}}{10^3\text{N}\cdot\text{m}}\right|$$

$$-293.15\text{ K}(6.8379 - 1.8418)\frac{\text{kJ}}{\text{kg}\cdot\text{K}}$$

$$= (1927.82 + 39.17 - 1464.61)\frac{\text{kJ}}{\text{kg}} = 502.38\frac{\text{kJ}}{\text{kg}}$$

(b) Noting that temperature remains constant, Eq. 7.5, on a per unit of mass basis, reads

❶

$$\frac{E_q}{m} = \left(1 - \frac{T_0}{T}\right)\frac{Q}{m} \tag{b}$$

With $Q/m = 2114.1$ kJ/kg from Fig. E7.2

$$\frac{E_q}{m} = \left(1 - \frac{293.15\text{ K}}{423.15\text{ K}}\right)\left(2114.1\frac{\text{kJ}}{\text{kg}}\right) = 649.49\frac{\text{kJ}}{\text{kg}}$$

(c) With $W/m = 186.38$ kJ/kg from Fig. E7.2 and $p_0(v_2 - v_1) = 39.17$ kJ/kg from part (a), Eq. 7.6 gives, per unit of mass

❷

$$\frac{E_w}{m} = \frac{W}{m} - p_0(v_2 - v_1) \tag{c}$$

$$= (186.38 - 39.17)\frac{\text{kJ}}{\text{kg}} = 147.21\frac{\text{kJ}}{\text{kg}}$$

(d) Since the process is internally reversible, the exergy destruction is necessarily zero. This can be checked by inserting the results of parts (a)–(c) into an exergy balance. Thus, solving Eq. 7.4b for the exergy destruction per unit of mass, evaluating terms, and allowing for roundoff, we get

$$\frac{\mathsf{E}_d}{m} = -(\mathsf{e}_2 - \mathsf{e}_1) + \frac{\mathsf{E}_q}{m} - \frac{\mathsf{E}_w}{m}$$

$$= (-502.38 + 649.49 - 147.21)\frac{\text{kJ}}{\text{kg}} = 0$$

Alternatively, the exergy destruction can be evaluated using Eq. 7.7 together with the entropy production obtained from an entropy balance. This is left as an exercise.

---

❶ Recognizing the term $(1 - T_0/T)$ as the Carnot efficiency (Eq. 5.9), the right side of Eq. (b) can be interpreted as the work that *could be* developed by a reversible power cycle *were* it to receive energy Q/m at temperature T and discharge energy to the environment by heat transfer at T_0.

❷ The right side of Eq. (c) shows that *if* the system were interacting with the environment, all of the work W/m, represented by area 1-2-d-a-1 on the p–v diagram of Fig. E7.2, would not be fully available for lifting a weight. A portion would be spent in pushing aside the environment at pressure p_0. This portion is given by $p_0(v_2 - v_1)$, and is represented by area a-b-c-d-a on the p–v diagram of Fig. E7.2.

Skills Developed: *Ability to...*
- ✓ evaluate exergy change.
- ✓ evaluate exergy transfer accompanying heat and work.
- ✓ evaluate exergy destruction.

Quick Quiz

If heating from saturated liquid to saturated vapor would occur at 100°C (373.15 K), evaluate the exergy transfers accompanying heat and work, each in kJ/kg.

Ans. 484, 0.

7.4.2 Closed System Exergy Rate Balance

As in the case of the mass, energy, and entropy balances, the exergy balance can be expressed in various forms that may be more suitable for particular analyses. A convenient form is the *closed system exergy rate balance* given by

$$\frac{d\mathsf{E}}{dt} = \sum_j \left(1 - \frac{T_0}{T_j}\right)\dot{Q}_j - \left(\dot{W} - p_0\frac{dV}{dt}\right) - \dot{\mathsf{E}}_d \qquad (7.10)$$

where $d\mathsf{E}/dt$ is the time rate of change of exergy. The term $(1 - T_0/T_j)\dot{Q}_j$ represents the time rate of exergy transfer accompanying heat transfer at the rate $\dot{Q}_j$ occurring where the instantaneous temperature on the boundary is T_j. The term $\dot{W}$ represents the time rate of energy transfer by work. The accompanying rate of exergy transfer is given by $(\dot{W} - p_0 dV/dt)$, where dV/dt is the time rate of change of system volume. The term $\dot{\mathsf{E}}_d$ accounts for the time rate of exergy destruction due to irreversibilities within the system.

At steady state, $d\mathsf{E}/dt = dV/dt = 0$, and Eq. 7.10 reduces to give the ***steady-state form of the exergy rate balance.***

$$0 = \sum_j \left(1 - \frac{T_0}{T_j}\right)\dot{Q}_j - \dot{W} - \dot{\mathsf{E}}_d \qquad (7.11a)$$

steady-state form of the closed system exergy rate balance

Note that for a system at steady state, the rate of exergy transfer accompanying the power $\dot{W}$ is simply the power.

The rate of exergy transfer accompanying heat transfer at the rate $\dot{Q}_j$ occurring where the temperature is T_j is compactly expressed as

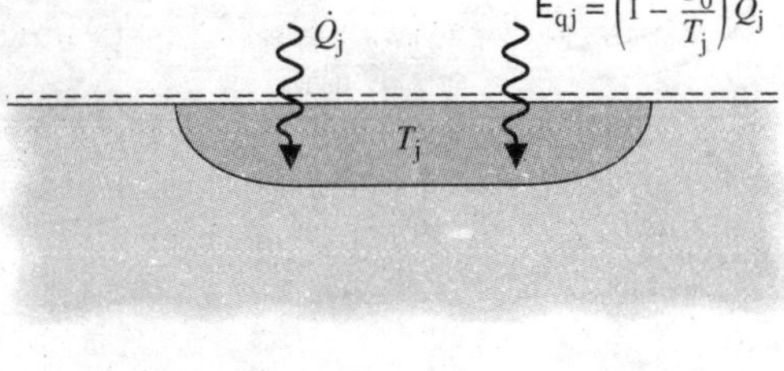

$$\dot{\mathsf{E}}_{qj} = \left(1 - \frac{T_0}{T_j}\right)\dot{Q}_j \qquad (7.12)$$

Using Eq. 7.12, Eq. 7.11a reads

$$0 = \sum_j \dot{E}_{qj} - \dot{W} - \dot{E}_d \qquad (7.11b)$$

In Eqs. 7.11, the rate of exergy destruction within the system, $\dot{E}_d$, is related to the rate of entropy production within the system by $\dot{E}_d = T_0\dot{\sigma}$.

7.4.3 Exergy Destruction and Loss

Most thermal systems are supplied with exergy inputs derived directly or indirectly from the consumption of fossil fuels. Accordingly, *avoidable* destructions and losses of exergy represent the waste of these resources. By devising ways to reduce such inefficiencies, better use can be made of fuels. The exergy balance can be applied to determine the locations, types, and true magnitudes of energy resource waste, and thus can play an important part in developing strategies for more effective fuel use.

In Example 7.3, the steady-state form of the closed system energy and exergy rate balances are applied to an oven wall to evaluate exergy destruction and exergy loss, which are interpreted in terms of fossil fuel use.

Example 7.3 EVALUATING EXERGY DESTRUCTION IN AN OVEN WALL

The wall of an industrial drying oven is constructed by sandwiching 0.066m-thick insulation, having a thermal conductivity $\kappa = 0.05 \times 10^{-3}$ kW/m · K, between thin metal sheets. At steady state, the inner metal sheet is at $T_1 = 575$ K and the outer sheet is at $T_2 = 310$ K. Temperature varies linearly through the wall. The temperature of the surroundings away from the oven is 293 K. Determine, in kW per m^2 of wall surface area, (a) the rate of heat transfer through the wall, (b) the rates of exergy transfer accompanying heat transfer at the inner and outer wall surfaces, and (c) the rate of exergy destruction within the wall. Let $T_0 = 293$ K.

Solution

Known: Temperature, thermal conductivity, and wall-thickness data are provided for a plane wall at steady state.

Find: For the wall, determine (a) the rate of heat transfer through the wall, (b) the rates of exergy transfer accompanying heat transfer at the inner and outer surfaces, and (c) the rate of exergy destruction, each per m^2 of wall surface area.

Schematic and Given Data:

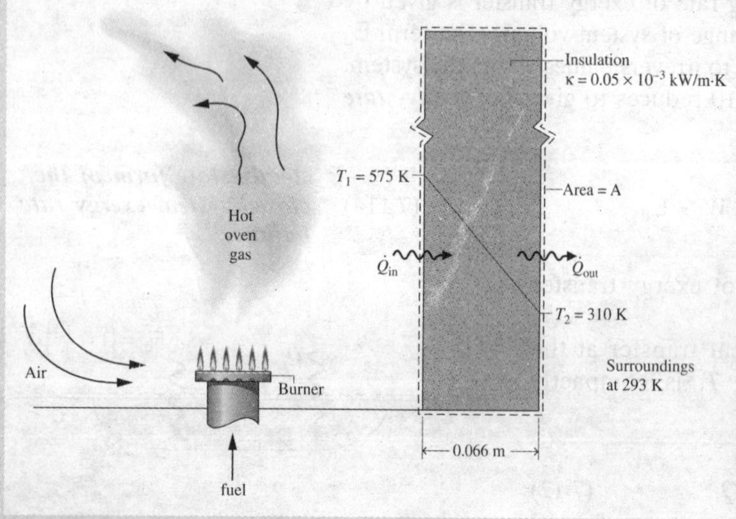

Engineering Model:

1. The closed system shown in the accompanying sketch is at steady state.
2. Temperature varies linearly through the wall.
3. $T_0 = 293$ K.

Fig. E7.3

Analysis: **(a)** At steady state, an energy rate balance for the system reduces to give $\dot{Q}_{in} = \dot{Q}_{out}$—namely, the rates of heat transfer into and out of the wall are equal. Let $\dot{Q}$ denote the common heat transfer rate. Using Eq. 2.3.1 with assumption 2, the heat transfer rate is given by

$$(\dot{Q}/A) = -\kappa\left[\frac{T_2 - T_1}{L}\right]$$

$$= -\left(0.05 \times 10^{-3}\frac{\text{kW}}{\text{m} \cdot \text{K}}\right)\left[\frac{(310 - 575)\text{ K}}{0.066\text{ m}}\right] = 0.2\frac{\text{kW}}{\text{m}^2}$$

(b) The rates of exergy transfer accompanying heat transfer are evaluated using Eq. 7.12. At the inner surface

$$(\dot{E}_{q1}/A) = \left[1 - \frac{T_0}{T_1}\right](\dot{Q}/A)$$

$$= \left[1 - \frac{293}{575}\right]\left(0.2\frac{\text{kW}}{\text{m}^2}\right) = 0.1\frac{\text{kW}}{\text{m}^2}$$

At the outer surface

$$(\dot{E}_{q2}/A) = \left[1 - \frac{T_0}{T_2}\right](\dot{Q}/A)$$

❶ $$= \left[1 - \frac{293}{310}\right]\left(0.2\frac{\text{kW}}{\text{m}^2}\right) = 0.01\frac{\text{kW}}{\text{m}^2}$$

(c) The rate of exergy destruction within the wall is evaluated using the exergy rate balance. Since $\dot{W} = 0$, Eq. 7.11b gives

❷ $$(\dot{E}_d/A) = (\dot{E}_{q1}/A) - (\dot{E}_{q2}/A)$$

❸ $$= (0.1 - 0.01)\frac{\text{kW}}{\text{m}^2} = 0.09\frac{\text{kW}}{\text{m}^2}$$

---

❶ The rates of heat transfer are the same at the inner and outer walls, but the rates of exergy transfer at these locations are much different. The rate of exergy transfer at the high-temperature inner wall is 10 times the rate of exergy transfer at the low-temperature outer wall. At each of these locations the exergy transfers provide a truer measure of thermodynamic value than the heat transfer rate. This is clearly seen at the outer wall, where the small exergy transfer indicates minimal potential for use, and thus minimal thermodynamic value.

❷ The exergy transferred into the wall at $T_1 = 575$ K is either destroyed within the wall owing to spontaneous heat transfer or transferred out of the wall at $T_2 = 310$ K where it is *lost* to the surroundings. Exergy transferred to the surroundings accompanying stray heat transfer, as in the present case, is ultimately destroyed in the surroundings. Thicker insulation and/or insulation having a lower thermal conductivity value would reduce the heat transfer rate and thus lower the exergy destruction and loss.

❸ In this example, the exergy destroyed and lost has its origin in the fuel supplied. Thus, cost-effective measures to reduce exergy destruction and loss have benefits in terms of saving fuel.

Skills Developed: *Ability to...*
- ✓ apply the energy and exergy rate balances.
- ✓ evaluate exergy transfer accompanying heat transfer.
- ✓ evaluate exergy destruction.

Quick Quiz

If the thermal conductivity were reduced to 0.04×10^{-3} kW/m · K, owing to a different selection of insulation material, while the insulation thickness were increased to 0.076 m, determine the rate of exergy destruction in the wall, in kW per m² of wall surface area, keeping the same temperatures of the walls and surroundings.

Ans. 0.06 kW/m².

Horizons

Oil from Shale and Sand Deposits—A Long Way Off?

Conventional oil supplies are now widely anticipated to decline in years ahead. But the impact could be lessened if cost-effective and environmentally benign technologies can be developed to recover oil-like substances from abundant oil shale and oil sand deposits in the United States and Canada. Production means available today are both costly and inefficient in terms of exergy demands for the blasting, digging, transporting, crushing, and heating of the materials rendered into oil. Current production means not only use increasingly scarce natural gas and large amounts of water, but also can cause wide-scale environmental damage. Although significant rewards await developers of improved technologies, the challenges also are significant. Some say that efforts are better directed to using conventional oil more efficiently and to developing alternatives to oil-based fuels.

7.4.4 Exergy Accounting

exergy accounting

In the next example, we reconsider the gearbox of Examples 2.4 and 6.4 from an exergy perspective to introduce ***exergy accounting,*** in which the various terms of an exergy balance for a system are systematically evaluated and compared.

Example 7.4 EXERGY ACCOUNTING FOR A GEARBOX

For the gearbox of Examples 2.4 and 6.4(a), develop a full exergy accounting of the power input. Let $T_0 = 293$ K.

Solution

Known: A gearbox operates at steady state with known values for the power input, power output, and heat transfer rate. The temperature on the outer surface of the gearbox is also known.

Find: Develop a full exergy accounting of the input power.

Schematic and Given Data:

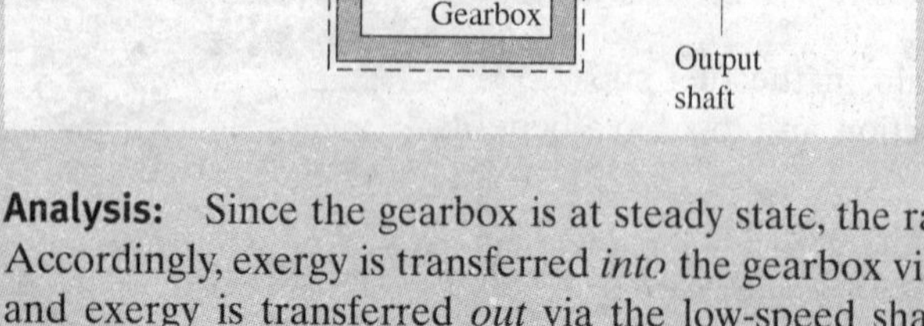

Engineering Model:

1. The gearbox is taken as a closed system operating at steady state.
2. The temperature at the outer surface does not vary.
3. $T_0 = 293$ K.

Fig. E7.4

Analysis: Since the gearbox is at steady state, the rate of exergy transfer accompanying power is simply the power. Accordingly, exergy is transferred *into* the gearbox via the high-speed shaft at a rate equal to the power *input,* 60 kW, and exergy is transferred *out* via the low-speed shaft at a rate equal to the power *output,* 58.8 kW. Additionally, exergy is transferred out accompanying stray heat transfer and destroyed by irreversibilities within the gearbox.

The rate of exergy transfer accompanying heat transfer is evaluated using Eq. 7.12. That is

$$\dot{E}_q = \left(1 - \frac{T_0}{T_b}\right)\dot{Q}$$

With $\dot{Q} = -1.2$ kW and $T_b = 300$ K from Fig. E7.4, we get

$$\dot{E}_q = \left(1 - \frac{293}{300}\right)(-1.2 \text{ kW})$$
$$= -0.03 \text{ kW}$$

where the minus sign denotes exergy transfer *from* the system.

The rate of exergy destruction is evaluated using the exergy rate balance. On rearrangement, and noting that $\dot{W} = \dot{W}_1 + \dot{W}_2 = -1.2$ kW, Eq. 7.11b gives

❶ $$\dot{E}_d = \dot{E}_q - \dot{W} = -0.03 \text{ kW} - (-1.2 \text{ kW}) = 1.17 \text{ kW}$$

The analysis is summarized by the following exergy *balance sheet* in terms of exergy magnitudes on a rate basis:

Rate of exergy in:	
high-speed shaft	60.00 kW (100%)
Disposition of the exergy:	
• Rate of exergy out	
low-speed shaft	58.80 kW (98%)
❷ stray heat transfer	0.03 kW (0.05%)
• Rate of exergy destruction	1.17 kW (1.95%)
	60.00 kW (100%)

❶ Alternatively, the rate of exergy destruction is calculated from $\dot{E}_d = T_0\dot{\sigma}$, where $\dot{\sigma}$ is the rate of entropy production. From the solution to Example 6.4(a), $\dot{\sigma} = 4 \times 10^{-3}$ kW/K. Then

$$\dot{E}_d = T_0\dot{\sigma}$$
$$= (293 \text{ K})(4 \times 10^{-3} \text{ kW/K})$$
$$= 1.17 \text{ kW}$$

❷ The difference between the input and output power is accounted for primarily by the rate of exergy destruction and only secondarily by the exergy transfer accompanying heat transfer, which is small by comparison. The exergy balance sheet provides a sharper picture of performance than the energy balance sheet of Example 2.4, which ignores the effect of irreversibilities within the system and overstates the significance of the heat transfer.

Skills Developed: *Ability to...*
✓ apply the exergy rate balance.
✓ develop an exergy accounting.

Quick Quiz

By inspection of the exergy balance sheet, specify an exergy-based efficiency for the gear box.
Ans. 98%.

7.5 Exergy Rate Balance for Control Volumes at Steady State

In this section, the exergy balance is extended to a form applicable to control volumes at steady state. The control volume form is generally the most useful for engineering analysis.

The exergy rate balance for a control volume can be derived using an approach like that employed in the box of Sec. 4.1, where the control volume form of the mass rate balance is obtained by transforming the closed system form. However, as in the developments of the energy and entropy rate balances for control volumes (Secs. 4.4 and 6.9, respectively), the present derivation is conducted less formally by modifying the closed system rate form, Eq. 7.10, to account for the exergy transfers at the inlets and exits. The result is

$$\frac{d\mathsf{E}_{cv}}{dt} = \sum_j \left(1 - \frac{T_0}{T_j}\right)\dot{Q}_j - \left(\dot{W}_{cv} - p_0 \frac{dV_{cv}}{dt}\right) + \underline{\sum_i \dot{m}_i \mathsf{e}_{fi}} - \underline{\sum_e \dot{m}_e \mathsf{e}_{fe}} - \dot{\mathsf{E}}_d$$

where the underlined terms account for exergy transfer where mass enters and exits the control volume, respectively.

At steady state, $d\mathsf{E}_{cv}/dt = dV_{cv}/dt = 0$, giving the ***steady-state exergy rate balance***

steady-state exergy rate balance: control volumes

$$0 = \sum_j \left(1 - \frac{T_0}{T_j}\right)\dot{Q}_j - \dot{W}_{cv} + \sum_i \dot{m}_i \mathsf{e}_{fi} - \sum_e \dot{m}_e \mathsf{e}_{fe} - \dot{\mathsf{E}}_d \qquad (7.13a)$$

where e_{fi} accounts for the exergy per unit of mass entering at inlet i and e_{fe} accounts for the exergy per unit of mass exiting at exit e. These terms, known as the ***specific flow exergy,*** are expressed as

specific flow exergy

$$\mathsf{e}_f = h - h_0 - T_0(s - s_0) + \frac{V^2}{2} + gz \qquad (7.14)$$

where h and s represent the specific enthalpy and entropy, respectively, at the inlet or exit under consideration; h_0 and s_0 represent the respective values of these properties when evaluated at T_0, p_0. See the box for the derivation of Eq. 7.14.

Evaluating the Specific Flow Exergy

The exergy associated with a flowing stream of matter is the maximum theoretical work that *could* be obtained *were* the stream brought to the dead state while heat transfer occurs only with the environment at T_0. The work developed is fully available for lifting a weight or, equivalently as shaft work or electrical work.

To evaluate the exergy associated with a stream at a state given by h, s, V, and z, let us think of the stream being fed to the control volume operating at steady state shown in Fig. 7.5. At the exit of the control volume, the respective properties are those corresponding to the dead state: h_0, s_0, $V_0 = 0$, $z_0 = 0$.

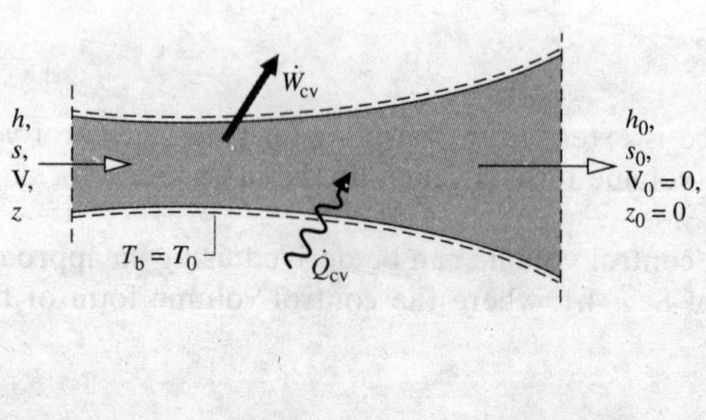

Fig. 7.5 Control volume used to evaluate the specific flow exergy of a stream.

For the control volume of Fig. 7.5, energy and entropy balances read, respectively,

$$0 = \dot{Q}_{cv} - \dot{W}_{cv} + \dot{m}\left[(h - h_0) + \frac{(V^2 - (0)^2)}{2} + g(z - 0)\right] \quad \text{(a)}$$

$$0 = \frac{\dot{Q}_{cv}}{T_0} + \dot{m}(s - s_0) + \dot{\sigma}_{cv} \quad \text{(b)}$$

Eliminating $\dot{Q}_{cv}$ between Eqs. (a) and (b), the work developed per unit of mass flowing is

$$\frac{\dot{W}_{cv}}{\dot{m}} = \left[\underline{(h - h_0) - T_0(s - s_0) + \frac{V^2}{2} + gz}\right] - T_0\left(\frac{\dot{\sigma}_{cv}}{\dot{m}}\right) \quad \text{(c)}$$

The value of the underlined term in Eq. (c) is determined by two states: the given state and the dead state. However, the value of the entropy production term, which cannot be negative, depends on the nature of the flow. Hence, the maximum theoretical work that could be developed, per unit of mass flowing, corresponds to a zero value for the entropy production—that is, when the flow through the control volume of Fig. 7.5 is internally reversible. The specific flow exergy, e_f, is this maximum work value, and thus Eq. 7.14 is seen to be the appropriate expression.

The steady-state exergy rate balance, Eq. 7.13a, can be expressed more compactly as Eq. 7.13b

$$0 = \sum_j \dot{E}_{qj} - \dot{W}_{cv} + \sum_i \dot{E}_{fi} - \sum_e \dot{E}_{fe} - \dot{E}_d \quad (7.13b)$$

where

$$\dot{E}_{qj} = \left(1 - \frac{T_0}{T_j}\right)\dot{Q}_j \quad (7.15)$$

$$\dot{E}_{fi} = \dot{m}_i e_{fi} \quad (7.16a)$$

$$\dot{E}_{fe} = \dot{m}_e e_{fe} \quad (7.16b)$$

are exergy transfer rates. Equation 7.15 has the same interpretation as given for Eq 7.5 in the box on p. 340, only on a time rate basis. Also note that at steady state the rate of exergy transfer accompanying the power $\dot{W}_{cv}$ is simply the power. Finally, the rate of exergy destruction within the control volume, $\dot{E}_d$, is related to the rate of entropy production by $T_0\dot{\sigma}_{cv}$.

If there is a single inlet and a single exit, denoted by 1 and 2, respectively, the steady-state exergy rate balance, Eq. 7.13a, reduces to

$$0 = \sum_j \left(1 - \frac{T_0}{T_j}\right)\dot{Q}_j - \dot{W}_{cv} + \dot{m}(e_{f1} - e_{f2}) - \dot{E}_d \quad (7.17)$$

where $\dot{m}$ is the mass flow rate. The term $(e_{f1} - e_{f2})$ is evaluated using Eq. 7.14 as

$$e_{f1} - e_{f2} = (h_1 - h_2) - T_0(s_1 - s_2) + \frac{V_1^2 - V_2^2}{2} + g(z_1 - z_2) \quad (7.18)$$

Take Note...

When the rate of exergy destruction $\dot{E}_d$ is the objective, it can be determined either from an exergy rate balance or from $\dot{E}_d = T_0\dot{\sigma}_{cv}$, where $\dot{\sigma}_{cv}$ is the rate of entropy production evaluated from an entropy rate balance. The second of these procedures normally requires fewer property evaluations and less computation.

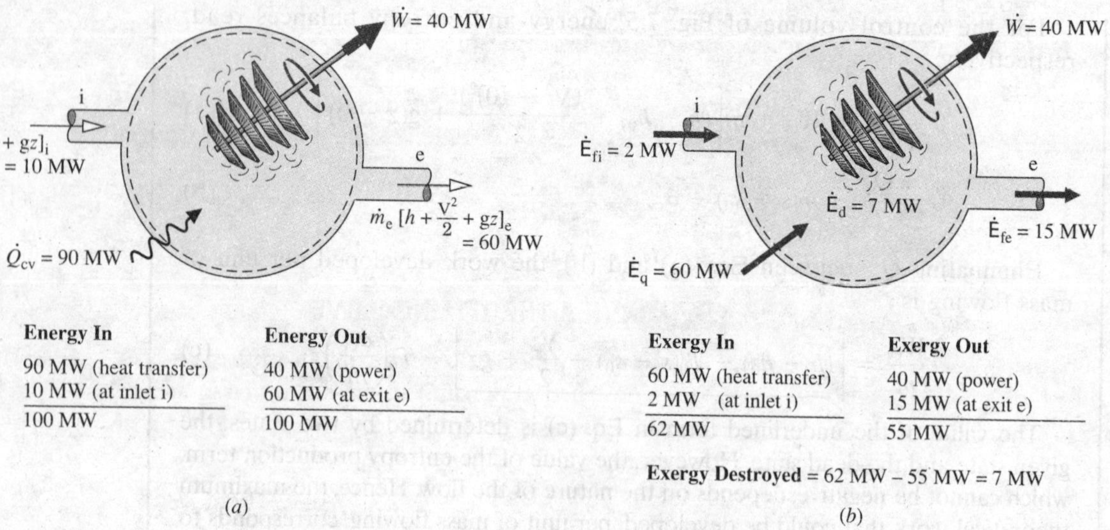

Fig. 7.6 Comparing energy and exergy for a control volume at steady state. (a) Energy analysis. (b) Exergy analysis.

7.5.1 Comparing Energy and Exergy for Control Volumes at Steady State

Although energy and exergy share common units, and exergy transfer accompanies energy transfer, energy and exergy are *fundamentally different* concepts. Energy and exergy relate, respectively, to the first and second laws of thermodynamics:

- Energy is *conserved*. Exergy is *destroyed* by irreversibilities.
- Exergy expresses energy transfer by work, heat, and mass flow in terms of a *common measure*—namely, work that is *fully available* for lifting a weight or, equivalently, as shaft or electrical work.

FOR EXAMPLE... Figure 7.6a shows energy transfer rates for a one-inlet, one-exit control volume at steady state. This includes energy transfers by work and heat and the energy transfers in and out where mass flows across the boundary. Figure 7.6b shows the same control volume, but now labeled with exergy transfer rates. Be sure to note that the *magnitudes* of the exergy transfers accompanying heat transfer and mass flow *differ* from the corresponding energy transfer magnitudes. These exergy transfer rates are calculated using Eqs. 7.15 and 7.16, respectively. At steady state, the rate of exergy transfer accompanying the power $\dot{W}_{cv}$ is simply the power. In accord with the conservation of energy principle, the total rate energy enters the control volume *equals* the total rate energy exits. However, the total rate exergy enters the control volume *exceeds* the total rate exergy exits. The difference between these exergy values is the rate at which exergy is destroyed by irreversibilities, in accord with the second law.

To summarize, exergy gives a sharper picture of performance than energy because exergy expresses all energy transfers on a common basis and accounts explicitly for the effect of irreversibilities through the exergy destruction concept.

7.5.2 Evaluating Exergy Destruction in Control Volumes at Steady State

The following examples illustrate the use of mass, energy, and exergy rate balances for the evaluation of exergy destruction in control volumes at steady state. Property data also play an important role in arriving at solutions. The first example involves the expansion of steam through a valve (a throttling process, Sec. 4.10). From an energy perspective, the expansion occurs without loss. Yet, as shown in Example 7.5, such a valve is a site of inefficiency quantified thermodynamically in terms of exergy destruction.

Example 7.5 EXERGY DESTRUCTION IN A THROTTLING VALVE

Superheated water vapor enters a valve at 3.0 MPa, 320°C and exits at a pressure of 0.5 MPa. The expansion is a throttling process. Determine the exergy destruction per unit of mass flowing, in kJ/kg. Let T_0 = 25°C, p_0 = 1 bar.

Solution

Known: Water vapor expands in a throttling process through a valve from a specified inlet state to a specified exit pressure.

Find: Determine the exergy destruction per unit of mass flowing.

Schematic and Given Data:

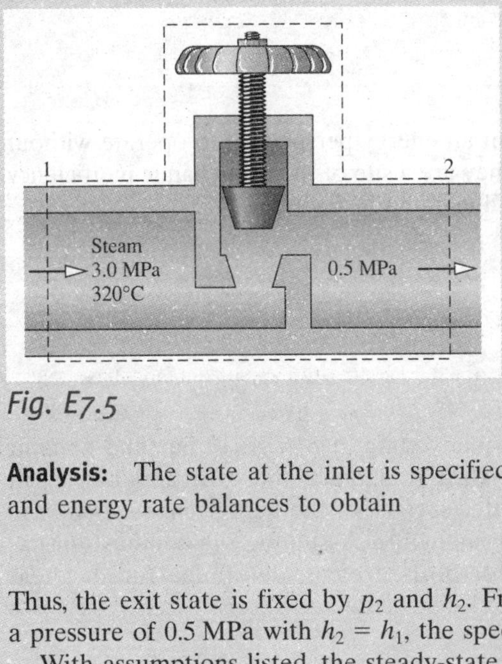

Fig. E7.5

Engineering Model:

1. The control volume shown in the accompanying figure is at steady state.
2. For the throttling process, $\dot{Q}_{cv} = \dot{W}_{cv} = 0$, and the effects of motion and gravity can be ignored.
3. T_0 = 25°C, p_0 = 1 bar.

Analysis: The state at the inlet is specified. The state at the exit can be fixed by reducing the steady-state mass and energy rate balances to obtain

$$h_2 = h_1 \tag{a}$$

Thus, the exit state is fixed by p_2 and h_2. From Table A-4, h_1 = 3043.4 kJ/s, s_1 = 6.6245 kJ/kg · K. Interpolating at a pressure of 0.5 MPa with $h_2 = h_1$, the specific entropy at the exit is s_2 = 7.4223 kJ/kg · K.

With assumptions listed, the steady-state form of the exergy rate balance, Eq. 7.17, reduces to

$$0 = \sum_j \left(1 - \frac{T_0}{T_j}\right)\cancelto{0}{\dot{Q}_j} - \cancelto{0}{\dot{W}_{cv}} + \dot{m}(\mathsf{e}_{f1} - \mathsf{e}_{f2}) - \dot{\mathsf{E}}_d$$

Dividing by the mass flow rate $\dot{m}$ and solving, the exergy destruction per unit of mass flowing is

$$\frac{\dot{\mathsf{E}}_d}{\dot{m}} = (\mathsf{e}_{f1} - \mathsf{e}_{f2}) \tag{b}$$

Introducing Eq. 7.18, using Eq. (a), and ignoring the effects of motion and gravity

$$\mathsf{e}_{f1} - \mathsf{e}_{f2} = \cancelto{0}{(h_1 - h_2)} - T_0(s_1 - s_2) + \cancelto{0}{\frac{V_1^2 - V_2^2}{2}} + \cancelto{0}{g(z_1 - z_2)}$$

Eq. (b) becomes

❶ $$\frac{\dot{E}_d}{\dot{m}} = T_0(s_2 - s_1) \quad \text{(c)}$$

Inserting values,

❷ $$\frac{\dot{E}_d}{\dot{m}} = 25°\text{C}\,(7.4223 - 6.6245)\frac{\text{kJ}}{\text{kg} \cdot \text{K}} = 237.7 \text{ kJ/kg}$$

❶ Equation (c) can be obtained alternatively beginning with the relationship $\dot{E}_d = T_0\dot{\sigma}_{cv}$ and then evaluating the rate of entropy production $\dot{\sigma}_{cv}$ from an entropy balance. The details are left as an exercise.

❷ Energy is conserved in the throttling process, but exergy is destroyed. The source of the exergy destruction is the uncontrolled expansion that occurs.

Skills Developed: *Ability to...*
✓ apply the energy and exergy rate balances.
✓ evaluate exergy destruction.

Quick Quiz

If air modeled as an ideal gas were to undergo a throttling process, evaluate the exergy destruction, in kJ per kg of air flowing, for the same inlet conditions and exit pressure as in this example.
Ans. 157.5 kJ/kg.

Although heat exchangers appear from an energy perspective to operate without loss when stray heat transfer is ignored, they are a site of thermodynamic inefficiency quantified by exergy destruction. This is illustrated in Example 7.6.

Example 7.6 EXERGY DESTRUCTION IN A HEAT EXCHANGER

Compressed air enters a counterflow heat exchanger operating at steady state at 610 K, 10 bar and exits at 860 K, 9.7 bar. Hot combustion gas enters as a separate stream at 1020 K, 1.1 bar and exits at 1 bar. Each ❶ stream has a mass flow rate of 90 kg/s. Heat transfer between the outer surface of the heat exchanger and the surroundings can be ignored. The effects of motion and gravity are negligible. Assuming the combustion gas stream has the properties of air, and using the ideal gas model for both streams, determine for the heat exchanger

(a) the exit temperature of the combustion gas, in K.

(b) the net change in the flow exergy rate from inlet to exit of each stream, in MW.

(c) the rate exergy is destroyed, in MW.

Let $T_0 = 300$ K, $p_0 = 1$ bar.

Solution

Known: Steady-state operating data are provided for a counterflow heat exchanger.

Find: For the heat exchanger, determine the exit temperature of the combustion gas, the change in the flow exergy rate from inlet to exit of each stream, and the rate exergy is destroyed.

Schematic and Given Data:

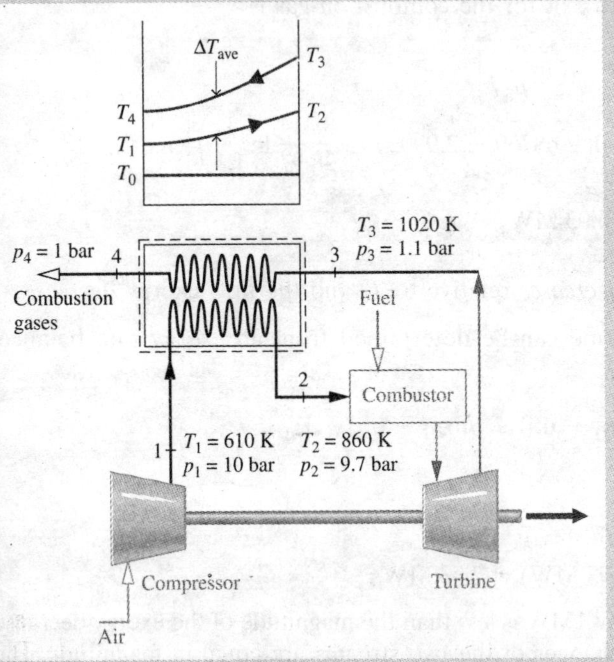

Fig. E7.6

Engineering Model:

1. The control volume shown in the accompanying figure is at steady state.
2. For the control volume, $\dot{Q}_{cv} = 0$, $\dot{W}_{cv} = 0$, and the effects of motion and gravity are negligible.
3. Each stream has the properties of air modeled as an ideal gas.
4. $T_0 = 300$ K, $p_0 = 1$ bar.

Analysis:

(a) The temperature T_4 of the exiting combustion gases can be found by reducing the mass and energy rate balances for the control volume at steady state to obtain

$$0 = \underline{\dot{Q}_{cv}} - \underline{\dot{W}_{cv}} + \dot{m}\left[(h_1 - h_2) + \underline{\left(\frac{V_1^2 - V_2^2}{2}\right)} + \underline{g(z_1 - z_2)}\right] + \dot{m}\left[(h_3 - h_4) + \underline{\left(\frac{V_3^2 - V_4^2}{2}\right)} + \underline{g(z_3 - z_4)}\right]$$

where $\dot{m}$ is the common mass flow rate of the two streams. The underlined terms drop out by listed assumptions, leaving

$$0 = \dot{m}(h_1 - h_2) + \dot{m}(h_3 - h_4)$$

Dividing by $\dot{m}$ and solving for h_4

$$h_4 = h_3 + h_1 - h_2$$

From Table A-22, $h_1 = 617.53$ kJ/kg, $h_2 = 888.27$ kJ/kg, $h_3 = 1068.89$ kJ/kg. Inserting values

$$h_4 = 1068.89 + 617.53 - 888.27 = 798.15 \text{ kJ/kg}$$

Interpolating in Table A-22 gives $T_4 = 778$ K (505°C).

(b) The net change in the flow exergy rate from inlet to exit for the air stream flowing from 1 to 2 can be evaluated using Eq. 7.18, neglecting the effects of motion and gravity. With Eq. 6.20a and data from Table A-22

$$\dot{m}(\mathsf{e}_{f2} - \mathsf{e}_{f1}) = \dot{m}[(h_2 - h_1) - T_0(s_2 - s_1)]$$

$$= \dot{m}\left[(h_2 - h_1) - T_0\left(s_2^\circ - s_1^\circ - R\ln\frac{p_2}{p_1}\right)\right]$$

$$= 90\frac{\text{kg}}{\text{s}}\left[(888.27 - 617.53)\frac{\text{kJ}}{\text{kg}} - 300\text{ K}\left(2.79783 - 2.42644 - \frac{8.314}{28.97}\ln\frac{9.7}{10}\right)\frac{\text{kJ}}{\text{kg}\cdot\text{K}}\right]$$

$$= 14{,}103\frac{\text{kJ}}{\text{s}}\left|\frac{1\text{ MW}}{10^3\text{ kJ/s}}\right| = 14.1\text{ MW}$$

❷ As the air flows from 1 to 2, its temperature *increases* relative to T_0 and the flow exergy *increases.* Similarly, the change in the flow exergy rate from inlet to exit for the combustion gas is

$$\dot{m}(e_{f4} - e_{f3}) = \dot{m}\left[(h_4 - h_3) - T_0\left(s_4^\circ - s_3^\circ - R\ln\frac{p_4}{p_3}\right)\right]$$

$$= 90\left[(798.15 - 1068.89) - 300\left(2.68769 - 2.99034 - \frac{8.314}{28.97}\ln\frac{1}{1.1}\right)\right]$$

$$= -16{,}934\frac{\text{kJ}}{\text{s}}\left|\frac{1\text{ MW}}{10^3\text{ kJ/s}}\right| = -16.93\text{ MW}$$

As the combustion gas flows from 3 to 4, its temperature *decreases* relative to T_0 and the flow exergy *decreases.*

❸ **(c)** The rate of exergy destruction within the control volume can be determined from an exergy rate balance, Eq. 7.13a,

$$0 = \sum_j \left(1 - \frac{T_0}{T_j}\right)\overset{0}{\dot{Q}_j} - \overset{0}{\dot{W}_{cv}} + \dot{m}(e_{f1} - e_{f2}) + \dot{m}(e_{f3} - e_{f4}) - \dot{E}_d$$

Solving for $\dot{E}_d$ and inserting known values

$$\dot{E}_d = \dot{m}(e_{f1} - e_{f2}) + \dot{m}(e_{f3} - e_{f4})$$

❹ $$= (-14.1\text{ MW}) + (16.93\text{ MW}) = 2.83\text{ MW}$$

Comparing results, the exergy increase of the compressed air: 14.1MW is less than the magnitude of the exergy decrease of the combustion gas: 16.93 MW, even though the energy changes of the two streams are equal in magnitude. The difference in these exergy values is the exergy destroyed: 2.83 MW. Thus, energy is conserved but exergy is not.

---

❶ Heat exchangers of this type are known as *regenerators* (see Sec. 9.7).

❷ The variation in temperature of each stream passing through the heat exchanger is sketched in the schematic. The dead state temperature T_0 also is shown on the schematic for reference.

❸ Alternatively, the rate of exergy destruction can be determined using $\dot{E}_d = T_0\dot{\sigma}_{cv}$, where $\dot{\sigma}_{cv}$ is the rate of entropy production evaluated from an entropy rate balance. This is left as an exercise.

❹ Exergy is destroyed by irreversibilities associated with fluid friction and stream-to-stream heat transfer. The pressure drops for the streams are indicators of frictional irreversibility. The temperature difference between the streams is an indicator of heat transfer irreversibility.

Skills Developed: **Ability to...**
- ✓ apply the energy and exergy rate balances.
- ✓ evaluate exergy destruction.

Quick Quiz

If the mass flow rate of each stream were 105 kg/s, what would be the rate of exergy destruction, in MW? **Ans.** 3.3 MW.

In previous discussions we have noted the effect of irreversibilities on *thermodynamic* performance. Some *economic* consequences of irreversibilities are considered in the next example.

Example 7.7 COST OF EXERGY DESTRUCTION

For the heat pump of Examples 6.8 and 6.14, determine the exergy destruction rates, each in kW, for the compressor, condenser, and throttling valve. If exergy is valued at $0.08 per kW · h, determine the daily cost of electricity to operate the compressor and the daily cost of exergy destruction in each component. Let T_0 = 273 K (0°C), which corresponds to the temperature of the outside air.

Solution

Known: Refrigerant 22 is compressed adiabatically, condensed by heat transfer to air passing through a heat exchanger, and then expanded through a throttling valve. Data for the refrigerant and air are known.

Find: Determine the daily cost to operate the compressor. Also determine the exergy destruction rates and associated daily costs for the compressor, condenser, and throttling valve.

Schematic and Given Data:

See Examples 6.8 and 6.14.

Engineering Model:

1. See Examples 6.8 and 6.14.
2. $T_0 = 273$ K (0°C).

Analysis: The rates of exergy destruction can be calculated using

$$\dot{E}_d = T_0\dot{\sigma}$$

together with data for the entropy production rates from Example 6.8. That is,

$$(\dot{E}_d)_{comp} = (273\ \text{K})(17.5 \times 10^{-4})\left(\frac{\text{kW}}{\text{K}}\right) = 0.478\ \text{kW}$$

$$(\dot{E}_d)_{valve} = (273)(9.94 \times 10^{-4}) = 0.271\ \text{kW}$$

$$(\dot{E}_d)_{cond} = (273)(7.95 \times 10^{-4}) = 0.217\ \text{kW}$$

The costs of exergy destruction are, respectively

$$\left(\begin{array}{c}\text{Daily cost of exergy destruction due}\\ \text{to compressor irreversibilities}\end{array}\right) = (0.478\ \text{kW})\left(\frac{\$0.08}{\text{kW}\cdot\text{h}}\right)\left|\frac{24\ \text{h}}{\text{day}}\right| = \$0.92$$

1 $$\left(\begin{array}{c}\text{Daily cost of exergy destruction due to}\\ \text{irreversibilities in the throttling valve}\end{array}\right) = (0.271)(0.08)|24| = \$0.52$$

$$\left(\begin{array}{c}\text{Daily cost of exergy destruction due to}\\ \text{irreversibilities in the condenser}\end{array}\right) = (0.217)(0.08)|24| = \$0.42$$

From the solution to Example 6.14, the magnitude of the compressor power is 3.11 kW. Thus, the daily cost is

$$\left(\begin{array}{c}\text{Daily cost of electricity}\\ \text{to operate compressor}\end{array}\right) = (3.11\ \text{kW})\left(\frac{\$0.08}{\text{kW}\cdot\text{h}}\right)\left|\frac{24\ \text{h}}{\text{day}}\right| = \$5.97$$

1 Associating exergy destruction with operating costs provides a rational basis for seeking cost-effective design improvements. Although it may be possible to select components that would destroy less exergy, the trade-off between any resulting reduction in operating cost and the potential increase in equipment cost must be carefully considered.

Skills Developed: *Ability to...*
- ✔ evaluate exergy destruction.
- ✔ conduct an elementary economic evaluation using exergy.

Quick Quiz

Expressed as a percent, how much of the cost of electricity to operate the compressor is attributable to exergy destruction in the three components?

Ans. 31%.

7.5.3 Exergy Accounting in Control Volumes at Steady State

For a control volume, the location, types, and true magnitudes of inefficiency and loss can be pinpointed by systematically evaluating and comparing the various terms of the exergy balance for the control volume. This is called ***exergy accounting.***

exergy accounting

The next two examples provide illustrations of exergy accounting in control volumes. The first involves the steam turbine with stray heat transfer considered previously in Example 6.6, which should be quickly reviewed before studying the current example.

Example 7.8 EXERGY ACCOUNTING OF A STEAM TURBINE

Steam enters a turbine with a pressure of 30 bar, a temperature of 400°C, and a velocity of 160 m/s. Steam exits as saturated vapor at 100°C with a velocity of 100 m/s. At steady state, the turbine develops work at a rate of 540 kJ per kg of steam flowing through the turbine. Heat transfer between the turbine and its surroundings occurs at an average outer surface temperature of 350 K. Develop a full accounting of the *net exergy carried in* by the steam, in kJ per unit mass of steam flowing. Let $T_0 = 25°C$, $p_0 = 1$ bar.

Solution

Known: Steam expands through a turbine for which steady-state data are provided.

Find: Develop a full exergy accounting of the *net exergy carried in* by the steam, in kJ per unit mass of steam flowing.

Schematic and Given Data: See Fig. E6.6. From Example 6.6, $\dot{W}_{cv}/\dot{m} = 540$ kJ/kg, $\dot{Q}_{cv}/\dot{m} = -22.6$ kJ/kg.

Engineering Model:

1. See the solution to Example 6.6.
2. $T_0 = 25°C$, $p_0 = 1$ bar.

Analysis: The *net exergy carried in* per unit mass of steam flowing is obtained using Eq. 7.18

$$\mathsf{e}_{f1} - \mathsf{e}_{f2} = (h_1 - h_2) - T_0(s_1 - s_2) + \left(\frac{V_1^2 - V_2^2}{2}\right) + \cancelto{0}{g(z_1 - z_2)}$$

From Table A-4, $h_1 = 3230.9$ kJ/kg, $s_1 = 6.9212$ kJ/kg · K. From Table A-2, $h_2 = 2676.1$ kJ/kg, $s_2 = 7.3549$ kJ/kg · K. Hence, the net rate exergy is carried in is

$$\mathsf{e}_{f1} - \mathsf{e}_{f2} = \left[(3230.9 - 2676.1)\frac{\text{kJ}}{\text{kg}} - 298\,(6.9212 - 7.3549)\frac{\text{kJ}}{\text{kg}} + \left[\frac{(160)^2 - (100)^2}{2}\right]\left(\frac{\text{m}}{\text{s}}\right)^2\left|\frac{1\text{ N}}{1\text{ kg}\cdot\text{m/s}^2}\right|\left|\frac{1\text{ kJ}}{10^3\text{ N}\cdot\text{m}}\right|\right]$$

$$= 691.84 \text{ kJ/kg}$$

The net exergy carried in can be accounted for in terms of the exergy transfers accompanying work and heat transfer and the exergy destruction within the control volume. At steady state, the exergy transfer accompanying work is simply the work, or $\dot{W}_{cv}/\dot{m} = 540$ kJ/kg. The quantity $\dot{Q}_{cv}/\dot{m}$ is evaluated in the solution to Example 6.6 using the steady-state forms of the mass and energy rate balances: $\dot{Q}_{cv}/\dot{m} = -22.6$ kJ/kg. The accompanying exergy transfer is

$$\frac{\dot{\mathsf{E}}_q}{\dot{m}} = \left(1 - \frac{T_0}{T_b}\right)\left(\frac{\dot{Q}_{cv}}{\dot{m}}\right)$$

$$= \left(1 - \frac{298}{350}\right)\left(-22.6\frac{\text{kJ}}{\text{kg}}\right)$$

$$= -3.36\frac{\text{kJ}}{\text{kg}}$$

where T_b denotes the temperature on the boundary where heat transfer occurs.

The exergy destruction can be determined by rearranging the steady-state form of the exergy rate balance, Eq. 7.17, to give

1 $$\frac{\dot{E}_d}{\dot{m}} = \left(1 - \frac{T_0}{T_b}\right)\left(\frac{\dot{Q}_{cv}}{\dot{m}}\right) - \frac{\dot{W}_{cv}}{\dot{m}} + (e_{f1} - e_{f2})$$

Substituting values

$$\frac{\dot{E}_d}{\dot{m}} = -3.36 - 540 + 691.84 = 148.48 \text{ kJ/kg}$$

The analysis is summarized by the following exergy *balance sheet* in terms of exergy magnitudes on a rate basis:

Net rate of exergy in:	691.84 kJ/kg (100%)
Disposition of the exergy:	
• Rate of exergy out	
work	540.00 kJ/kg (78.05%)
heat transfer	3.36 kJ/kg (0.49%)
• Rate of exergy destruction	148.48 kJ/kg (21.46%)
	691.84 kJ/kg (100%)

Note that the exergy transfer accompanying heat transfer is small relative to the other terms.

Skills Developed: *Ability to...*
- ✓ evaluate exergy quantities for an exergy accounting.
- ✓ develop an exergy accounting.

1 The exergy destruction can be determined alternatively using $\dot{E}_d = T_0\dot{\sigma}_{cv}$, where $\dot{\sigma}_{cv}$ is the rate of entropy production from an entropy balance. The solution to Example 6.6 provides $\dot{\sigma}_{cv}/\dot{m} = 0.4983$ kJ/kg · K.

Quick Quiz

By inspection of the exergy balance sheet, specify an exergy-based efficiency for the turbine. **Ans.** 78.05%.

The next example illustrates the use of exergy accounting to identify opportunities for improving thermodynamic performance of the waste heat recovery system considered in Example 4.10, which should be quickly reviewed before studying the current example.

Example 7.9 EXERGY ACCOUNTING OF A WASTE HEAT RECOVERY SYSTEM

Suppose the system of Example 4.10 is one option under consideration for utilizing the combustion products discharged from an industrial process.

(a) Develop a full accounting of the *net* exergy carried in by the combustion products.

(b) Discuss the design implications of the results.

Solution

Known: Steady-state operating data are provided for a heat-recovery steam generator and a turbine.

Find: Develop a full accounting of the *net* rate exergy is carried in by the combustion products and discuss the implications for design.

Schematic and Given Data:

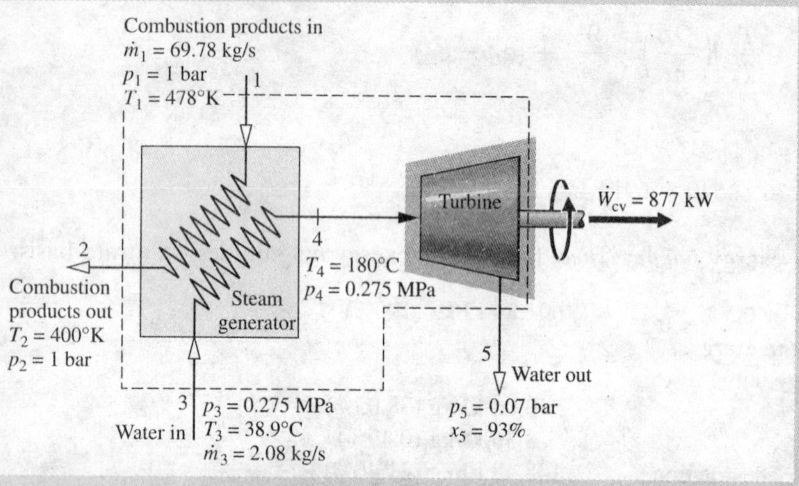

Engineering Model:

1. See solution to Example 4.10.

2. T_0 = 298°K.

Fig. E7.9

Analysis:

(a) We begin by determining the *net rate exergy is carried into* the control volume. Modeling the combustion products as an ideal gas, the net rate is determined using Eq. 7.18 together with Eq. 6.20a as

$$\dot{m}_1[\mathrm{e}_{f1} - \mathrm{e}_{f2}] = \dot{m}_1[h_1 - h_2 - T_0(s_1 - s_2)]$$

$$= \dot{m}_1\left[h_1 - h_2 - T_0\left(s_1^\circ - s_2^\circ - R\ln\frac{p_1}{p_2}\right)\right]$$

With data from Table A-22, h_1 = 480.35 kJ/kg, h_2 = 400.97 kJ/kg, s_1° = 2.173 kJ/kg · °K, s_2° = 1.992 kJ/kg · °K, and $p_2 = p_1$, we have

$$\dot{m}_1[\mathrm{e}_{f1} - \mathrm{e}_{f2}] = 69.8\ \text{kg/s}\left[(480.35 - 400.97)\frac{\text{kJ}}{\text{kg}} - 298°\text{K}\,(2.173 - 1.992)\frac{\text{kJ}}{\text{kg}\cdot°\text{C}}\right]\frac{\text{kJ}}{\text{s}}$$

$$= 1775.78\ \text{kJ/s}$$

Next, we determine the rate exergy is carried *out* of the control volume. Exergy is carried out of the control volume by work at a rate of 876.8 kJ/s, as shown on the schematic. Additionally, the *net* rate exergy is carried *out* by the water stream is

$$\dot{m}_3[\mathrm{e}_{f5} - \mathrm{e}_{f3}] = \dot{m}_3[h_5 - h_3 - T_0(s_5 - s_3)]$$

From Table A-2, $h_3 \approx h_f(39°\text{C})$ = 162.82 Kg/s, $s_3 \approx s_f(39°\text{C})$ = 0.5598 kJ/kg · °K. Using saturation data at 0.07 bar from Table A-3 with x_5 = 0.93 gives h_5 = 2403.27 kJ/kg and s_5 = 7.739 kJ/kg · °K. Substituting values

$$\dot{m}_3[\mathrm{e}_{f5} - \mathrm{e}_{f3}] = 2.08\frac{\text{kg}}{\text{s}}\left[(2403.27 - 162.82)\frac{\text{kJ}}{\text{kg}} - 298\,(7.739 - 0.5598)\frac{\text{kJ}}{\text{kg}\cdot°\text{K}}\right]$$

$$= 209.66\ \text{kJ/s}$$

Next, the rate exergy is destroyed in the heat-recovery steam generator can be obtained from an exergy rate balance applied to a control volume enclosing the steam generator. That is, Eq. 7.13a takes the form

$$0 = \sum_j \left(1 - \frac{T_0}{T_j}\right)\overset{0}{\dot{Q}_j} - \overset{0}{\dot{W}_{cv}} + \dot{m}_1(\mathrm{e}_{f1} - \mathrm{e}_{f2}) + \dot{m}_3(\mathrm{e}_{f3} - \mathrm{e}_{f4}) - \dot{\mathrm{E}}_d$$

Evaluating $(\mathrm{e}_{f3} - \mathrm{e}_{f4})$ with Eq. 7.18 and solving for $\dot{\mathrm{E}}_d$

$$\dot{\mathrm{E}}_d = \dot{m}_1(\mathrm{e}_{f1} - \mathrm{e}_{f2}) + \dot{m}_3[h_3 - h_4 - T_0(s_3 - s_4)]$$

The first term on the right is evaluated above. Then, with h_4 = 2825 kJ/kg, s_4 = 7.2196 kJ/kg · °K at 180°C, 0.275 MPa from Table A-4, and previously determined values for h_3 and s_3

$$\dot{E}_d = 1775.78\frac{kJ}{s} + 2.08\frac{kg}{s}\left[(162 - 2825)\frac{kJ}{kg} - 298\,(0.559 - 7.2196)\frac{kJ}{kg \cdot °K}\right]$$
$$= 366.1 \text{ kJ/s}$$

Finally, the rate exergy is destroyed in the turbine can be obtained from an exergy rate balance applied to a control volume enclosing the turbine. That is, Eq. 7.17 takes the form

$$0 = \sum_j \left(1 - \frac{T_0}{T_j}\right)\dot{Q}_j^{\,0} - \dot{W}_{cv} + \dot{m}_4(e_{f4} - e_{f5}) - \dot{E}_d$$

Solving for $\dot{E}_d$, evaluating $(e_{f4} - e_{f5})$ with Eq. 7.18, and using previously determined values

$$\dot{E}_d = -\dot{W}_{cv} + \dot{m}_4[h_4 - h_5 - T_0(s_4 - s_5)]$$

❶ $$= -876.8\frac{kJ}{s} + 2.08\frac{kg}{s}\left[(2885 - 2403)\frac{kJ}{kg} - 298°K\,(7.2196 - 7.739)\frac{kJ}{kg \cdot °C}\right]$$
$$= 320.2 \text{ kJ/s}$$

The analysis is summarized by the following exergy *balance sheet* in terms of exergy magnitudes on a rate basis:

Net rate of exergy in:	1772.8 kJ/s (100%)
Disposition of the exergy:	
• Rate of exergy out	
power developed	876.8 kJ/s (49.5%)
water stream	209.66 kJ/s (11.8%)
• Rate of exergy destruction	
heat-recovery steam generator	366.12 kJ/s (22.6%)
turbine	320.2 kJ/s (18%)
	1772.8 kJ/s (100%)

(b) The exergy balance sheet suggests an opportunity for improved *thermodynamic* performance because over 50% of the net exergy carried in is either destroyed by irreversibilities or carried out by the water stream. Better thermodynamic performance might be achieved by modifying the design. For example, we might reduce the heat transfer irreversibility by specifying a heat-recovery steam generator with a smaller stream-to-stream temperature difference, and/or reduce friction by specifying a turbine with a higher isentropic efficiency. Thermodynamic performance alone would not determine the *preferred* option, however, for other factors such as cost must be considered, and can be overriding. Further discussion of the use of exergy analysis in design is provided in Sec. 7.7.1.

Skills Developed: Ability to...
- ✓ evaluate exergy quantities for an exergy accounting.
- ✓ develop an exergy accounting.

❶ Alternatively, the rates of exergy destruction in control volumes enclosing the heat-recovery steam generator and turbine can be determined using $\dot{E}_d = T_0\dot{\sigma}_{cv}$, where $\dot{\sigma}_{cv}$ is the rate of entropy production for the respective control volume evaluated from an entropy rate balance. This is left as an exercise.

Quick Quiz

For the turbine of the waste heat recovery system, evaluate the isentropic turbine efficiency. Comment.
Ans. 74%.

7.6 Exergetic (Second Law) Efficiency

The objective of this section is to show the use of the exergy concept in assessing the effectiveness of energy resource utilization. As part of the presentation, the ***exergetic efficiency*** concept is introduced and illustrated. Such efficiencies are also known as *second law* efficiencies.

exergetic efficiency

7.6.1 Matching End Use to Source

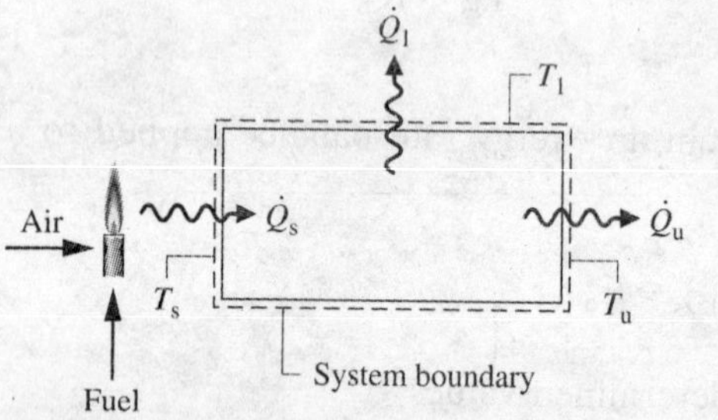

Fig. 7.7 Schematic used to discuss the efficient use of fuel.

Tasks such as space heating, heating in industrial furnaces, and process steam generation commonly involve the combustion of coal, oil, or natural gas. When the products of combustion are at a temperature significantly greater than required by a given task, the end use is not well matched to the source and the result is inefficient use of the fuel burned. To illustrate this simply, refer to Fig. 7.7, which shows a closed system receiving a heat transfer at the rate $\dot{Q}_s$ at a *source* temperature T_s and delivering $\dot{Q}_u$ at a *use* temperature T_u. Energy is lost to the surroundings by heat transfer at a rate $\dot{Q}_l$ across a portion of the surface at T_l. All energy transfers shown on the figure are in the directions indicated by the arrows.

Assuming that the system of Fig. 7.7 operates at steady state and there is no work, the closed system energy and exergy rate balances Eqs. 2.37 and 7.10 reduce, respectively, to

$$\frac{d\cancel{E}^{\,0}}{dt} = (\dot{Q}_s - \dot{Q}_u - \dot{Q}_l) - \cancel{\dot{W}}^{\,0}$$

$$\frac{d\cancel{\mathsf{E}}^{\,0}}{dt} = \left[\left(1 - \frac{T_0}{T_s}\right)\dot{Q}_s - \left(1 - \frac{T_0}{T_u}\right)\dot{Q}_u - \left(1 - \frac{T_0}{T_l}\right)\dot{Q}_l\right] - \left[\cancel{\dot{W}}^{\,0} - p_0\frac{d\cancel{V}^{\,0}}{dt}\right] - \dot{\mathsf{E}}_d$$

These equations can be rewritten as follows

$$\dot{Q}_s = \dot{Q}_u + \dot{Q}_l \tag{7.19a}$$

$$\left(1 - \frac{T_0}{T_s}\right)\dot{Q}_s = \left(1 - \frac{T_0}{T_u}\right)\dot{Q}_u + \left(1 - \frac{T_0}{T_l}\right)\dot{Q}_l + \dot{\mathsf{E}}_d \tag{7.19b}$$

Equation 7.19a indicates that the energy carried in by heat transfer, $\dot{Q}_s$, is either used, $\dot{Q}_u$, or lost to the surroundings, $\dot{Q}_l$. This can be described by an efficiency in terms of energy rates in the form product/input as

$$\eta = \frac{\dot{Q}_u}{\dot{Q}_s} \tag{7.20}$$

In principle, the value of η can be increased by applying insulation to reduce the loss. The limiting value, when $\dot{Q}_l = 0$, is $\eta = 1$ (100%).

Equation 7.19b shows that the exergy carried into the system accompanying the heat transfer $\dot{Q}_s$ is either transferred from the system accompanying the heat transfers $\dot{Q}_u$ and $\dot{Q}_l$ or destroyed by irreversibilities within the system. This can be described by an efficiency in terms of exergy rates in the form product/input as

$$\varepsilon = \frac{(1 - T_0/T_u)\dot{Q}_u}{(1 - T_0/T_s)\dot{Q}_s} \tag{7.21a}$$

Introducing Eq. 7.20 into Eq. 7.21a results in

$$\varepsilon = \eta\left(\frac{1 - T_0/T_u}{1 - T_0/T_s}\right) \tag{7.21b}$$

The parameter ε, defined with reference to the exergy concept, may be called an *exergetic* efficiency. Note that η and ε each gauge how effectively the input is converted to the product. The parameter η does this on an energy basis, whereas ε does it on an exergy basis. As discussed next, the value of ε is generally less than unity even when $\eta = 1$.

Equation 7.21b indicates that a value for η as close to unity as practical is important for proper utilization of the exergy transferred from the hot combustion gas to the system. However, this alone would not ensure effective utilization. The temperatures

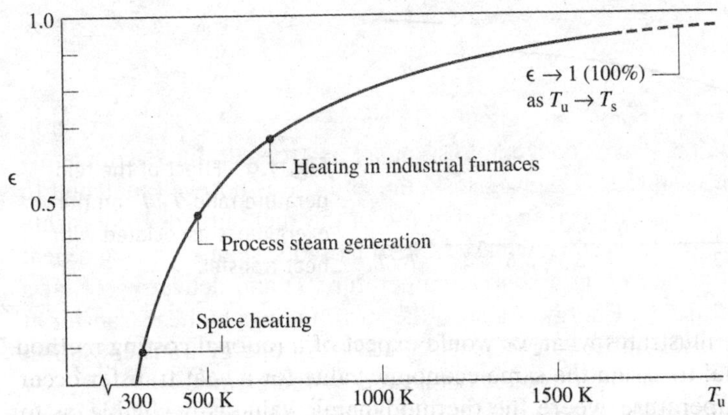

Fig. 7.8 Effect of use temperature T_u on the exergetic efficiency ε (T_s = 2200 K, η = 100%).

T_s and T_u are also important, with exergy utilization improving as the use temperature T_u approaches the source temperature T_s. For proper utilization of exergy, therefore, it is desirable to have a value for η as close to unity as practical and also a good *match* between the source and use temperatures.

To emphasize the central role of temperature in exergetic efficiency considerations, a graph of Eq. 7.21b is provided in Fig. 7.8. The figure gives the exergetic efficiency ε versus the use temperature T_u for an assumed source temperature T_s = 2200 K. Figure 7.8 shows that ε tends to unity (100%) as the use temperature approaches T_s. In most cases, however, the use temperature is substantially below T_s. Indicated on the graph are efficiencies for three applications: space heating at T_u = 320 K, process steam generation at T_u = 480 K, and heating in industrial furnaces at T_u = 700 K. These efficiency values suggest that fuel is used far more effectively in the higher use-temperature industrial applications than in the lower use-temperature space heating. The especially low exergetic efficiency for space heating reflects the fact that fuel is consumed to produce only slightly warm air, which from an exergy perspective has considerably less utility. The efficiencies given on Fig. 7.8 are actually on the *high* side, for in constructing the figure we have assumed η to be unity (100%). Moreover, as additional destruction and loss of exergy would be associated with combustion, the overall efficiency from fuel input to end use would be much less than indicated by the values shown on the figure.

Costing Heat Loss

For the system in Fig. 7.7, it is instructive to consider further the rate of exergy loss accompanying the heat loss $\dot{Q}_l$, that is $(1 - T_0/T_l)\dot{Q}_l$. This expression measures the *true* thermodynamic value of the heat loss and is graphed in Fig. 7.9. The figure shows that the thermodynamic value of the heat loss depends *significantly* on the temperature at which the heat loss occurs. Stray heat transfer, such as $\dot{Q}_l$, usually occurs at relatively low temperature, and thus has relatively low thermodynamic value. We might expect that the *economic* value of such a loss varies similarly with temperature, and this is the case. ➡ **FOR EXAMPLE...** since the source of the exergy loss by heat transfer is the fuel input (see Fig. 7.7), the economic value of the loss can be accounted for in terms of the *unit cost* of fuel based on exergy, c_F (in \$/kW · h, for example), as follows

$$\begin{bmatrix} \text{Cost rate of heat loss} \\ \dot{Q}_l \text{ at temperature } T_l \end{bmatrix} = c_F(1 - T_0/T_l)\dot{Q}_l \qquad (7.22)$$

Equation 7.22 shows that the cost of such a loss is less at lower temperatures than at higher temperatures. ⬅

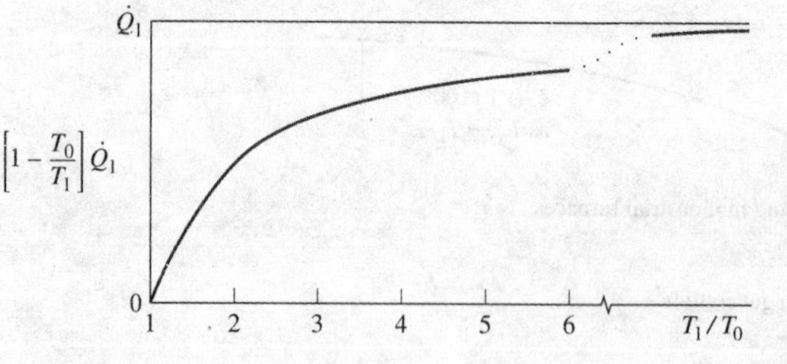

Fig. 7.9 Effect of the temperature ratio T_1/T_0 on the exergy loss associated with heat transfer.

The above example illustrates what we would expect of a rational costing method. It would not be rational to assign the same economic value for a heat transfer occurring near ambient temperature, where the thermodynamic value is negligible, as for an equal heat transfer occurring at a higher temperature, where the thermodynamic value is significant. Indeed, it would be incorrect to assign the *same cost* to heat loss independent of the temperature at which the loss is occurring. For further discussion of exergy costing, see Sec. 7.7.2.

7.6.2 Exergetic Efficiencies of Common Components

Exergetic efficiency expressions can take many different forms. Several examples are given in the current section for thermal system components of practical interest. In every instance, the efficiency is derived by the use of the exergy rate balance. The approach used here serves as a model for the development of exergetic efficiency expressions for other components. Each of the cases considered involves a control volume at steady state, and we assume no heat transfer between the control volume and its surroundings. The current presentation is not exhaustive. Many other exergetic efficiency expressions can be written.

Turbines

For a turbine operating at steady state with no heat transfer with its surroundings, the steady-state form of the exergy rate balance, Eq. 7.17, reduces as follows:

$$0 = \sum_j \left(1 - \frac{T_0}{T_j}\right)\dot{Q}_j - \dot{W}_{cv} + \dot{m}(e_{f1} - e_{f2}) - \dot{E}_d$$

This equation can be rearranged to read

$$e_{f1} - e_{f2} = \frac{\dot{W}_{cv}}{\dot{m}} + \frac{\dot{E}_d}{\dot{m}} \tag{7.23}$$

The term on the left of Eq. 7.23 is the decrease in flow exergy from turbine inlet to exit. The equation shows that the flow exergy decreases because the turbine develops work, $\dot{W}_{cv}/\dot{m}$, and exergy is destroyed, $\dot{E}_d/\dot{m}$. A parameter that gauges how effectively the flow exergy decrease is converted to the desired product is the *exergetic turbine efficiency*

$$\varepsilon = \frac{\dot{W}_{cv}/\dot{m}}{e_{f1} - e_{f2}} \tag{7.24}$$

This particular exergetic efficiency is sometimes referred to as the *turbine effectiveness.* Carefully note that the exergetic turbine efficiency is defined differently from the isentropic turbine efficiency introduced in Sec. 6.12.

FOR EXAMPLE... the exergetic efficiency of the turbine considered in Example 6.11 is 81.2% when $T_0 = 298$ K. It is left as an exercise to verify this value.

Compressors and Pumps

For a compressor or pump operating at steady state with no heat transfer with its surroundings, the exergy rate balance, Eq. 7.17, can be placed in the form

$$\left(-\frac{\dot{W}_{cv}}{\dot{m}}\right) = e_{f2} - e_{f1} + \frac{\dot{E}_d}{\dot{m}}$$

Thus, the exergy *input* to the device, $-\dot{W}_{cv}/\dot{m}$, is accounted for either as an increase in the flow exergy between inlet and exit or as exergy destroyed. The effectiveness of the conversion from work input to flow exergy increase is gauged by the *exergetic compressor* (or pump) *efficiency*

$$\varepsilon = \frac{e_{f2} - e_{f1}}{(-\dot{W}_{cv}/\dot{m})} \tag{7.25}$$

FOR EXAMPLE... the exergetic efficiency of the compressor considered in Example 6.14 is 84.6% when $T_0 = 273$ K. It is left as an exercise to verify this value.

Heat Exchanger without Mixing

The heat exchanger shown in Fig. 7.10 operates at steady state with no heat transfer with its surroundings and both streams at temperatures above T_0. The exergy rate balance, Eq. 7.13a, reduces to

$$0 = \sum_j \left(1 - \frac{T_0}{T_j}\right)\dot{Q}_j^{\,0} - \dot{W}_{cv}^{\,0} + (\dot{m}_h e_{f1} + \dot{m}_c e_{f3}) - (\dot{m}_h e_{f2} + \dot{m}_c e_{f4}) - \dot{E}_d$$

where $\dot{m}_h$ is the mass flow rate of the hot stream and $\dot{m}_c$ is the mass flow rate of the cold stream. This can be rearranged to read

$$\dot{m}_h(e_{f1} - e_{f2}) = \dot{m}_c(e_{f4} - e_{f3}) + \dot{E}_d \tag{7.26}$$

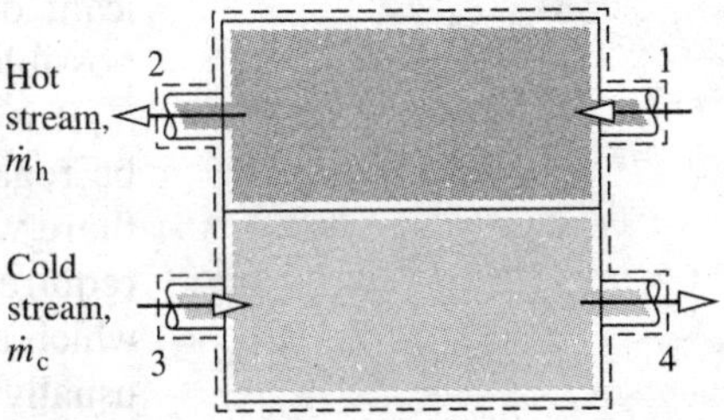

Fig. 7.10 Counterflow heat exchanger.

The term on the left of Eq. 7.26 accounts for the decrease in the exergy of the hot stream. The first term on the right accounts for the increase in exergy of the cold stream. Regarding the hot stream as supplying the exergy increase of the cold stream as well as the exergy destroyed, we can write an *exergetic heat exchanger efficiency* as

$$\varepsilon = \frac{\dot{m}_c(e_{f4} - e_{f3})}{\dot{m}_h(e_{f1} - e_{f2})} \tag{7.27}$$

FOR EXAMPLE... the exergetic efficiency of the heat exchanger of Example 7.6 is 83.3%. It is left as an exercise to verify this value.

Direct Contact Heat Exchanger

The direct contact heat exchanger shown in Fig. 7.11 operates at steady state with no heat transfer with its surroundings. The exergy rate balance, Eq. 7.13a, reduces to

$$0 = \sum_j \left(1 - \frac{T_0}{T_j}\right)\dot{Q}_j^{\,0} - \dot{W}_{cv}^{\,0} + \dot{m}_1 e_{f1} + \dot{m}_2 e_{f2} - \dot{m}_3 e_{f3} - \dot{E}_d$$

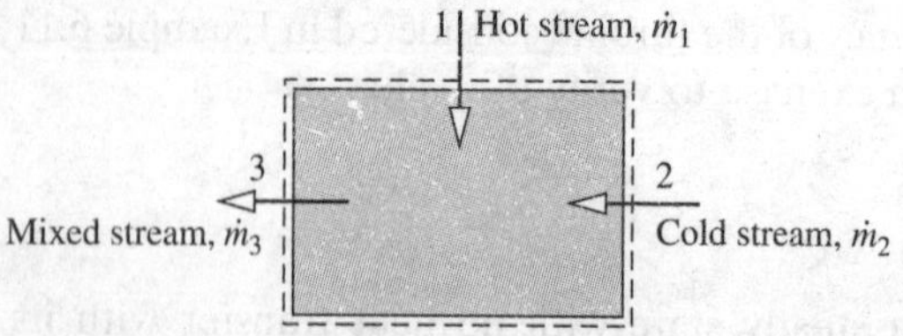

Fig. 7.11 Direct contact heat exchanger.

With $\dot{m}_3 = \dot{m}_1 + \dot{m}_2$ from a mass rate balance, this can be written as

$$\dot{m}_1(e_{f1} - e_{f3}) = \dot{m}_2(e_{f3} - e_{f2}) + \dot{E}_d \tag{7.28}$$

The term on the left of Eq. 7.28 accounts for the decrease in the exergy of the hot stream between inlet and exit. The first term on the right accounts for the increase in the exergy of the cold stream between inlet and exit. Regarding the hot stream as supplying the exergy increase of the cold stream as well as the exergy destroyed by irreversibilities, we can write an *exergetic efficiency* for a direct contact heat exchanger as

$$\varepsilon = \frac{\dot{m}_2(e_{f3} - e_{f2})}{\dot{m}_1(e_{f1} - e_{f3})} \tag{7.29}$$

7.6.3 Using Exergetic Efficiencies

Exergetic efficiencies are useful for distinguishing means for utilizing fossil fuels that are thermodynamically effective from those that are less so. Exergetic efficiencies also can be used to evaluate the effectiveness of engineering measures taken to improve the performance of a thermal system. This is done by comparing the efficiency values determined before and after modifications have been made to show how much improvement has been achieved. Moreover, exergetic efficiencies can be used to gauge the potential for improvement in the performance of a given thermal system by comparing the efficiency of the system to the efficiency of like systems. A significant difference between these values would suggest that improved performance is possible.

It is important to recognize that the limit of 100% exergetic efficiency should not be regarded as a practical objective. This theoretical limit could be attained only if there were no exergy destructions or losses. To achieve such idealized processes might require extremely long times to execute processes and/or complex devices, both of which are at odds with the objective of profitable operation. In practice, decisions are usually made on the basis of *total* costs. An increase in efficiency to reduce fuel consumption, or otherwise utilize fuels better, normally requires additional expenditures for facilities and operations. Accordingly, an improvement might not be implemented if an increase in total cost would result. The trade-off between fuel savings and additional investment invariably dictates a lower efficiency than might be achieved *theoretically* and may even result in a lower efficiency than could be achieved using the *best available* technology.

cogeneration

Various methods are used to improve energy resource utilization. All such methods must achieve their objectives cost-effectively. One method is ***cogeneration,*** which sequentially produces power and a heat transfer (or process steam) for some desired use. An aim of cogeneration is to develop the power and heat transfer using an *integrated* system with a total expenditure that is less than would be required to develop them individually. Further discussions of cogeneration are provided in Secs. 7.7.2 and 8.5. Two other methods employed to improve energy resource utilization are ***power recovery*** and ***waste heat recovery.*** Power recovery can be accomplished by inserting a turbine into a pressurized gas or liquid stream to capture some of the exergy that would otherwise be destroyed in a spontaneous expansion. Waste heat recovery contributes

power recovery
waste heat recovery

to overall efficiency by using some of the exergy that would otherwise be discarded to the surroundings, as in the exhaust gases of large internal combustion engines. An illustration of waste heat recovery is provided by Example 7.9.

Energy & Environment

A type of exergetic efficiency known as the *well-to-wheel efficiency* is used to compare different options for powering vehicles. The calculation of this efficiency begins at the well, where the oil or natural gas feedstock is extracted from the ground, and ends with the power delivered to a vehicle's wheels. The efficiency accounts separately for how effectively the vehicle's fuel is produced from feedstock, called the *well-to-fuel tank efficiency*, and how effectively the vehicle's power plant converts its fuel to power, called the *fuel tank-to-wheel efficiency*. The product of these gives the *overall* well-to-wheel efficiency.

The table below gives sample well-to-wheel efficiency values for three power plant options as reported by an automobile manufacturer:

	Well-to-tank (fuel production efficiency) (%)	×	Tank-to-wheel (vehicle efficiency) (%)	=	Well-to-wheel (overall efficiency) (%)
Conventional gasoline-fueled engine	88	×	16	=	14
Hydrogen-fueled fuel cell[a]	58	×	38	=	22
Gasoline-fueled hybrid electric	88	×	32	=	28

[a]Hydrogen produced from natural gas.

These data show that vehicles using conventional internal combustion engines do not fare well in terms of the well-to-wheel efficiency. The data also show that fuel-cell vehicles operating on hydrogen have the best tank-to-wheel efficiency of the three options, but lose out on an overall basis to hybrid vehicles, which enjoy a higher well-to-tank efficiency. Still, the well-to-wheel efficiency is just one consideration when making policy decisions concerning different options for powering vehicles.

7.7 Thermoeconomics

Thermal systems typically experience significant work and/or heat interactions with their surroundings, and they can exchange mass with their surroundings in the form of hot and cold streams, including chemically reactive mixtures. Thermal systems appear in almost every industry, and numerous examples are found in our everyday lives. Their design involves the application of principles from thermodynamics, fluid mechanics, and heat transfer, as well as such fields as materials, manufacturing, and mechanical design. The design of thermal systems also requires the explicit consideration of engineering economics, for cost is always a consideration. The term ***thermoeconomics*** may be applied to this general area of application, although it is often applied more narrowly to methodologies combining exergy and economics for optimizing the design and operation of thermal systems.

thermoeconomics

7.7.1 Using Exergy in Design

To illustrate the use of exergy in design, consider Fig. 7.12 showing a thermal system consisting of a power-generating unit and a *heat-recovery steam generator*. The power-generating unit develops an electric power output and combustion products that enter

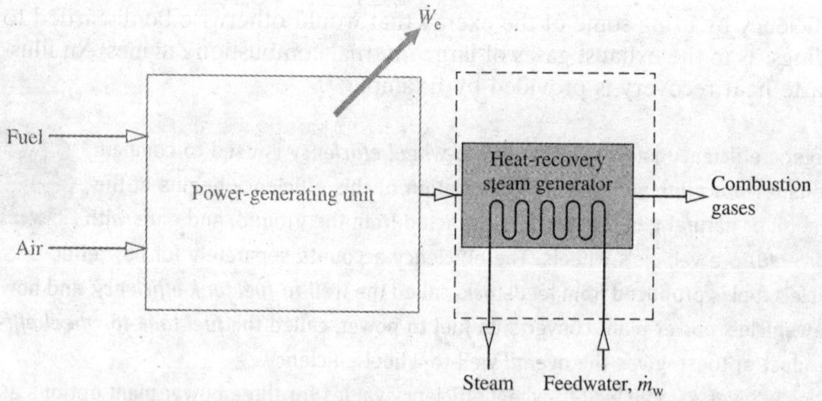

Fig. 7.12 Figure used to illustrate the use of exergy in design.

the heat recovery unit. Feedwater also enters the heat-recovery steam generator with a mass flow rate of $\dot{m}_w$, receives exergy by heat transfer from the combustion gases, and exits as steam at some desired condition for use in another process. The combustion products entering the heat-recovery steam generator can be regarded as having economic value. Since the source of the exergy of the combustion products is the fuel input (Fig. 7.12), the economic value can be accounted for in terms of the cost of fuel, as we have done in Sec. 7.6.1 when costing heat loss.

From our study of the second law of thermodynamics we know that the average temperature difference, ΔT_{ave}, between two streams passing through a heat exchanger is a measure of irreversibility and that the irreversibility of the heat transfer vanishes as the temperature difference approaches zero. For the heat-recovery steam generator in Fig. 7.12, this source of exergy destruction exacts an economic penalty in terms of fuel cost. Figure 7.13 shows the annual *fuel* cost attributed to the irreversibility of the heat exchanger as a function of ΔT_{ave}. The fuel cost increases with increasing ΔT_{ave}, because the irreversibility is directly related to the temperature difference.

From the study of heat transfer, we know that there is an inverse relation between ΔT_{ave} and the surface area required for a specified heat transfer rate. More heat transfer area means a larger, more costly heat exchanger—that is, a greater *capital* cost. Figure 7.13 also shows the annualized *capital cost* of the heat exchanger as a function of ΔT_{ave}. The capital cost decreases as ΔT_{ave} increases.

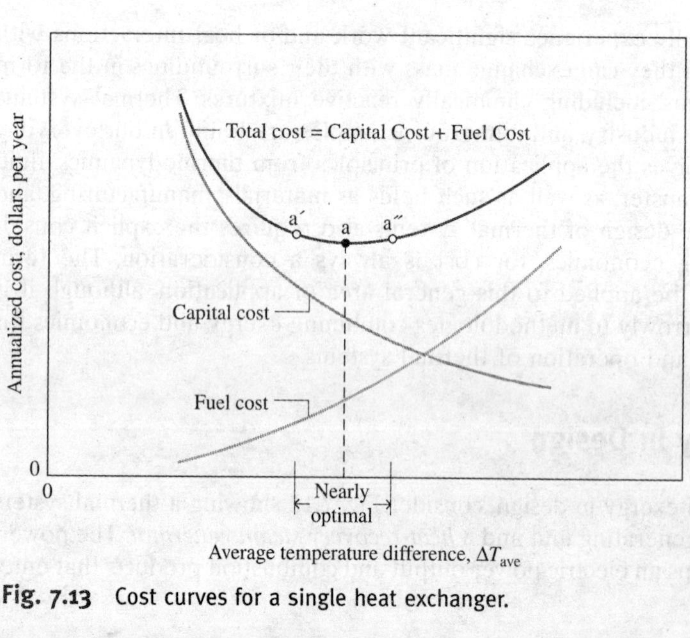

Fig. 7.13 Cost curves for a single heat exchanger.

The *total cost* is the sum of the capital cost and the fuel cost. The total cost curve shown in Fig. 7.13 exhibits a minimum at the point labeled a. Notice, however, that the curve is relatively flat in the neighborhood of the minimum, so there is a range of ΔT_{ave} values that could be considered *nearly optimal* from the standpoint of minimum total cost. If reducing the fuel cost were deemed more important than minimizing the capital cost, we might choose a design that would operate at point a′. Point a″ would be a more desirable operating point if capital cost were of greater concern. Such trade-offs are common in design situations.

The actual design process can differ significantly from the simple case considered here. For one thing, costs cannot be determined as precisely as implied by the curves in Fig. 7.13. Fuel prices may vary widely over time, and equipment costs may be difficult to predict as they often depend on a bidding procedure. Equipment is manufactured in discrete sizes, so the cost also would not vary continuously as shown in the figure. Furthermore, thermal systems usually consist of several components that interact with one another. Optimization of components individually, as considered for the heat exchanger, usually does not guarantee an optimum for the overall system. Finally, the example involves only ΔT_{ave} as a design variable. Often, several design variables must be considered and optimized simultaneously.

7.7.2 Exergy Costing of a Cogeneration System

Another important aspect of thermoeconomics is the use of exergy for *allocating* costs to the products of a thermal system. This involves assigning to each product the total cost to produce it, namely the cost of fuel and other inputs plus the cost of owning and operating the system (e.g., capital cost, operating and maintenance costs). Such costing is a common problem in plants where utilities such as electrical power, chilled water, compressed air, and steam are generated in one department and used in others. The plant operator needs to know the cost of generating each utility to ensure that the other departments are charged properly according to the type and amount of each utility used. Common to all such considerations are fundamentals from engineering economics, including procedures for annualizing costs, appropriate means for allocating costs, and reliable cost data.

To explore further the costing of thermal systems, consider the simple *cogeneration system* operating at steady state shown in Fig. 7.14. The system consists of a boiler and a turbine, with each having no significant heat transfer to its surroundings. The figure is labeled with exergy transfer rates associated with the flowing streams, where the subscripts F, a, P, and w denote fuel, combustion air, combustion products, and feedwater, respectively. The subscripts 1 and 2 denote high- and low-pressure steam, respectively. Means for evaluating the exergies of the fuel and combustion products are introduced in Chap. 13. The cogeneration system has two principal products: electricity, denoted by $\dot{W}_e$, and low-pressure steam for use in some process. The objective is to determine the cost at which each product is generated.

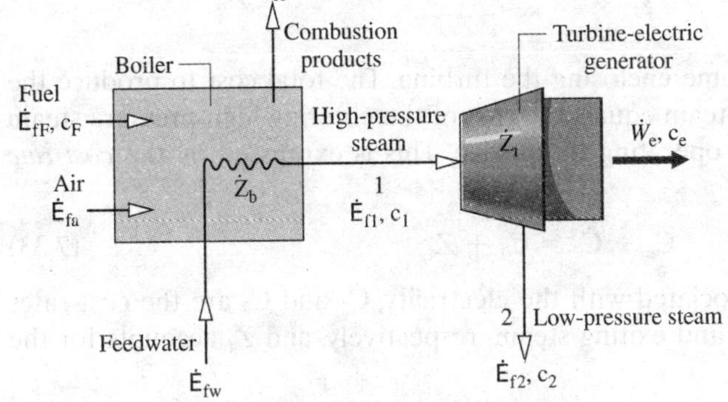

Fig. 7.14 Simple cogeneration system.

Boiler Analysis

Let us begin by evaluating the cost of the high-pressure steam produced by the boiler. For this, we consider a control volume enclosing the boiler. Fuel and air enter the boiler separately and combustion products exit. Feedwater enters and high-pressure steam exits. The total cost to produce the exiting streams equals the total cost of the entering streams plus the cost of owning and operating the boiler. This is expressed by the following ***cost rate balance*** for the boiler

cost rate balance

$$\dot{C}_1 + \dot{C}_P = \dot{C}_F + \dot{C}_a + \dot{C}_w + \dot{Z}_b \tag{7.30}$$

where $\dot{C}$ is the cost rate of the respective stream and $\dot{Z}_b$ accounts for the cost rate associated with owning and operating the boiler (each in \$ per hour, for example). In the present discussion, the cost rate $\dot{Z}_b$ is presumed known from a previous economic analysis.

Although the cost rates denoted by $\dot{C}$ in Eq. 7.30 are evaluated by various means in practice, the present discussion features the use of exergy for this purpose. Since exergy measures the true thermodynamic values of the work, heat, and other interactions between a system and its surroundings as well as the effect of irreversibilities within the system, exergy is a rational basis for assigning costs. With exergy costing, each of the cost rates is evaluated in terms of the associated rate of exergy transfer and a *unit cost.* Thus, for an entering or exiting stream, we write

$$\dot{C} = c\dot{E}_f \tag{7.31}$$

exergy unit cost

where c denotes the ***cost per unit of exergy*** (in cents per kW · h, for example) and $\dot{E}_f$ is the associated exergy transfer rate.

For simplicity, we assume the feedwater and combustion air enter the boiler with negligible exergy and cost, and the combustion products are discharged directly to the surroundings with negligible cost. Thus Eq. 7.30 reduces as follows

$$\dot{C}_1 + \cancelto{0}{\dot{C}_P} = \dot{C}_F + \cancelto{0}{\dot{C}_a} + \cancelto{0}{\dot{C}_w} + \dot{Z}_b$$

Then, with Eq. 7.31 we get

$$c_1\dot{E}_{f1} = c_F\dot{E}_{fF} + \dot{Z}_b \tag{7.32a}$$

Solving for c_1, the unit cost of the high-pressure steam is

$$c_1 = c_F\left(\frac{\dot{E}_{fF}}{\dot{E}_{f1}}\right) + \frac{\dot{Z}_b}{\dot{E}_{f1}} \tag{7.32b}$$

This equation shows that the unit cost of the high-pressure steam is determined by two contributions related, respectively, to the cost of the fuel and the cost of owning and operating the boiler. Due to exergy destruction and loss, less exergy exits the boiler with the high-pressure steam than enters with the fuel. Thus, $\dot{E}_{fF}/\dot{E}_{f1}$ is invariably greater than one, and the unit cost of the high-pressure steam is invariably greater than the unit cost of the fuel.

Turbine Analysis

Next, consider a control volume enclosing the turbine. The total cost to produce the electricity and low-pressure steam equals the cost of the entering high-pressure steam plus the cost of owning and operating the device. This is expressed by the *cost rate balance* for the turbine

$$\dot{C}_e + \dot{C}_2 = \dot{C}_1 + \dot{Z}_t \tag{7.33}$$

where $\dot{C}_e$ is the cost rate associated with the electricity, $\dot{C}_1$ and $\dot{C}_2$ are the cost rates associated with the entering and exiting steam, respectively, and $\dot{Z}_t$ accounts for the cost rate associated with owning and operating the turbine. With exergy costing, each of the cost rates $\dot{C}_e$, $\dot{C}_1$, and $\dot{C}_2$ is evaluated in terms of the associated rate of exergy transfer and a unit cost. Equation 7.33 then appears as

$$c_e \dot{W}_e + c_2 \dot{E}_{f2} = c_1 \dot{E}_{f1} + \dot{Z}_t \tag{7.34a}$$

The unit cost c_1 in Eq. 7.34a is given by Eq. 7.32b. In the present discussion, the same unit cost is assigned to the low-pressure steam; that is, $c_2 = c_1$. This is done on the basis that the purpose of the turbine is to generate electricity, and thus all costs associated with owning and operating the turbine should be charged to the power generated. We can regard this decision as a part of the *cost accounting* considerations that accompany the thermoeconomic analysis of thermal systems. With $c_2 = c_1$, Eq. 7.34a becomes

$$c_e \dot{W}_e = c_1 (\dot{E}_{f1} - \dot{E}_{f2}) + \dot{Z}_t \tag{7.34b}$$

The first term on the right side accounts for the cost of the exergy used and the second term accounts for the cost of the system itself.

Solving Eq. 7.34b for c_e, and introducing the exergetic turbine efficiency ε from Eq. 7.24

$$c_e = \frac{c_1}{\varepsilon} + \frac{\dot{Z}_t}{\dot{W}_e} \tag{7.34c}$$

This equation shows that the unit cost of the electricity is determined by the cost of the high-pressure steam and the cost of owning and operating the turbine. Because of exergy destruction within the turbine, the exergetic efficiency is invariably less than one, and therefore the unit cost of electricity is invariably greater than the unit cost of the high-pressure steam.

Summary

By applying cost rate balances to the boiler and the turbine, we are able to determine the cost of each product of the cogeneration system. The unit cost of the electricity is determined by Eq. 7.34c and the unit cost of the low-pressure steam is determined by the expression $c_2 = c_1$ together with Eq. 7.32b. The example to follow provides a detailed illustration. The same general approach is applicable for costing the products of a wide-ranging class of thermal systems.[1]

Example 7.10 EXERGY COSTING OF A COGENERATION SYSTEM

A cogeneration system consists of a natural gas-fueled boiler and a steam turbine that develops power and provides steam for an industrial process. At steady state, fuel enters the boiler with an exergy rate of 100 MW. Steam exits the boiler at 50 bar, 466°C with an exergy rate of 35 MW. Steam exits the turbine at 5 bar, 205°C and a mass flow rate of 26.15 kg/s. The unit cost of the fuel is 1.44 cents per kW · h of exergy. The costs of owning and operating the boiler and turbine are, respectively, \$1080/h and \$92/h. The feedwater and combustion air enter with negligible exergy and cost. The combustion products are discharged directly to the surroundings with negligible cost. Heat transfer with the surroundings and the effects of motion and gravity are negligible. Let $T_0 = 298$ K.

(a) For the turbine, determine the power and the rate exergy exits with the steam, each in MW.

(b) Determine the unit costs of the steam exiting the boiler, the steam exiting the turbine, and the power, each in cents per kW · h of exergy.

(c) Determine the cost rates of the steam exiting the turbine and the power, each in \$/h.

[1]See A. Bejan, G. Tsatsaronis, and M. J. Moran, *Thermal Design and Optimization*, John Wiley & Sons, New York, 1996.

Solution

Known: Steady-state operating data are known for a cogeneration system that produces both electricity and low-pressure steam for an industrial process.

Find: For the turbine, determine the power and the rate exergy exits with the steam. Determine the unit costs of the steam exiting the boiler, the steam exiting the turbine, and the power developed. Also determine the cost rates of the low-pressure steam and power.

Schematic and Given Data:

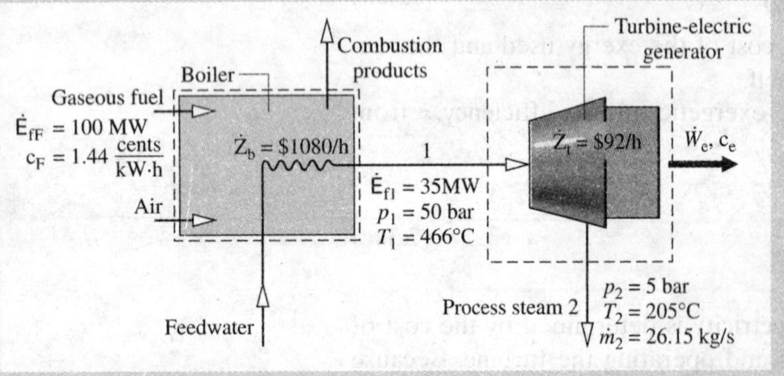

Fig. E7.10

Engineering Model:

1. Each control volume shown in the accompanying figure is at steady state.
2. For each control volume, $\dot{Q}_{cv} = 0$ and the effects of motion and gravity are negligible.
3. The feedwater and combustion air enter the boiler with negligible exergy and cost.
4. The combustion products are discharged directly to the surroundings with negligible cost.
5. For the environment, $T_0 = 298$ K.

Analysis:

(a) With assumption 2, the mass and energy rate balances for a control volume enclosing the turbine reduce at steady state to give

$$\dot{W}_e = \dot{m}(h_1 - h_2)$$

From Table A-4, $h_1 = 3353.54$ kJ/kg and $h_2 = 2865.96$ kJ/kg. Thus

$$\dot{W}_e = \left(26.15\frac{\text{kg}}{\text{s}}\right)(3353.54 - 2865.96)\left(\frac{\text{kJ}}{\text{kg}}\right)\left|\frac{1\text{ MW}}{10^3\text{ kJ/s}}\right|$$

$$= 12.75\text{ MW}$$

Using Eq. 7.18, the difference in the rates at which exergy enters and exits the turbine with the steam is

$$\dot{E}_{f2} - \dot{E}_{f1} = \dot{m}(e_{f2} - e_{f1})$$
$$= \dot{m}[h_2 - h_1 - T_0(s_2 - s_1)]$$

Solving for $\dot{E}_{f2}$

$$\dot{E}_{f2} = \dot{E}_{f1} + \dot{m}[h_2 - h_1 - T_0(s_2 - s_1)]$$

With known values for $\dot{E}_{f1}$ and $\dot{m}$, and data from Table A-4: $s_1 = 6.8773$ kJ/kg · K and $s_2 = 7.0806$ kJ/kg · K, the rate exergy exits with the steam is

$$\dot{E}_{f2} = 35\text{ MW} + \left(26.15\frac{\text{kg}}{\text{s}}\right)\left[(2865.96 - 3353.54)\frac{\text{kJ}}{\text{kg}} - 298\text{ K}\,(7.0806 - 6.8773)\frac{\text{kJ}}{\text{kg}\cdot\text{K}}\right]\left|\frac{1\text{ MW}}{10^3\text{ kJ/s}}\right|$$

$$= 20.67\text{ MW}$$

(b) For a control volume enclosing the boiler, the cost rate balance reduces with assumptions 3 and 4 to give

$$c_1\dot{E}_{f1} = c_F\dot{E}_{fF} + \dot{Z}_b$$

where $\dot{E}_{fF}$ is the exergy rate of the entering fuel, c_F and c_1 are the unit costs of the fuel and exiting steam, respectively, and $\dot{Z}_b$ is the cost rate associated with owning and operating the boiler. Solving for c_1 and inserting known values

$$c_1 = c_F\left(\frac{\dot{E}_{fF}}{\dot{E}_{f1}}\right) + \frac{\dot{Z}_b}{\dot{E}_{f1}}$$

$$= \left(1.44\frac{\text{cents}}{\text{kW}\cdot\text{h}}\right)\left(\frac{100\text{ MW}}{35\text{ MW}}\right) + \left(\frac{1080\text{ \$/h}}{35\text{ MW}}\right)\left|\frac{1\text{ MW}}{10^3\text{ MW}}\right|\left|\frac{100\text{ cents}}{1\$}\right|$$

$$= (4.11 + 3.09)\frac{\text{cents}}{\text{kW}\cdot\text{h}} = 7.2\frac{\text{cents}}{\text{kW}\cdot\text{h}}$$

The cost rate balance for the control volume enclosing the turbine is

$$c_e\dot{W}_e + c_2\dot{E}_{f2} = c_1\dot{E}_{f1} + \dot{Z}_t$$

where c_e and c_2 are the unit costs of the power and the exiting steam, respectively, and $\dot{Z}_t$ is the cost rate associated with owning and operating the turbine. Assigning the same unit cost to the steam entering and exiting the turbine, $c_2 = c_1 = 7.2$ cents/kW · h, and solving for c_e ❶

$$c_e = c_1\left[\frac{\dot{E}_{f1} - \dot{E}_{f2}}{\dot{W}_e}\right] + \frac{\dot{Z}_t}{\dot{W}_e}$$

Inserting known values

$$c_e = \left(7.2\frac{\text{cents}}{\text{kW}\cdot\text{h}}\right)\left[\frac{(35 - 20.67)\text{ MW}}{12.75\text{ MW}}\right] + \left(\frac{92\$/\text{h}}{12.75\text{ MW}}\right)\left|\frac{1\text{ MW}}{10^3\text{ kW}}\right|\left|\frac{100\text{ cents}}{1\$}\right|$$

❷
$$= (8.09 + 0.72)\frac{\text{cents}}{\text{kW}\cdot\text{h}} = 8.81\frac{\text{cents}}{\text{kW}\cdot\text{h}}$$

(c) For the low-pressure steam and power, the cost rates are, respectively,

$$\dot{C}_2 = c_2\dot{E}_{f2}$$

$$= \left(7.2\frac{\text{cents}}{\text{kW}\cdot\text{h}}\right)(20.67\text{ MW})\left|\frac{10^3\text{ kW}}{1\text{ MW}}\right|\left|\frac{1\$}{100\text{ cents}}\right|$$

$$= \$1488/\text{h}$$

❸
$$\dot{C}_e = c_e\dot{W}_e$$

$$= \left(8.81\frac{\text{cents}}{\text{kW}\cdot\text{h}}\right)(12.75\text{ MW})\left|\frac{10^3\text{ kW}}{1\text{ MW}}\right|\left|\frac{1\$}{100\text{ cents}}\right|$$

$$= \$1123/\text{h}$$

---

❶ The purpose of the turbine is to generate power, and thus all costs associated with owning and operating the turbine are charged to the power generated.

❷ Observe that the unit costs c_1 and c_e are significantly greater than the unit cost of the fuel.

❸ Although the unit cost of the steam is less than the unit cost of the power, the steam *cost* rate is greater because the associated exergy rate is much greater.

Skills Developed: Ability to...
- ✓ evaluate exergy quantities required for exergy costing.
- ✓ apply exergy costing.

Quick Quiz

If the unit cost of the fuel were to double to 2.88 cents/kW · h, what would be the change in the unit cost of power, expressed as a percent, keeping all other given data the same?

Ans. +53%

✓ Chapter Summary and Study Guide

In this chapter, we have introduced the property exergy and illustrated its use for thermodynamic analysis. Like mass, energy, and entropy, exergy is an extensive property that can be transferred across system boundaries. Exergy transfer accompanies heat transfer, work, and mass flow. Like entropy, exergy is not conserved. Exergy is destroyed within systems whenever internal irreversibilities are present. Entropy production corresponds to exergy destruction.

The use of exergy balances is featured in this chapter. Exergy balances are expressions of the second law that account for exergy in terms of exergy transfers and exergy destruction. For processes of closed systems, the exergy balance is given by Eqs. 7.4 and the companion steady-state forms are Eqs. 7.11. For control volumes, the steady-state expressions are given by Eqs. 7.13. Control volume analyses account for exergy transfer at inlets and exits in terms of flow exergy.

The following checklist provides a study guide for this chapter. When your study of the text and end-of-chapter exercises has been completed you should be able to

- ✓ write out meanings of the terms listed in the margins throughout the chapter and understand each of the related concepts. The subset of key concepts listed below is particularly important.
- ✓ evaluate specific exergy at a given state using Eq. 7.2 and exergy change between two states using Eq. 7.3, each relative to a specified reference environment.
- ✓ apply exergy balances in each of several alternative forms, appropriately modeling the case at hand, correctly observing sign conventions, and carefully applying SI units.
- ✓ evaluate the specific flow exergy relative to a specified reference environment using Eq. 7.14.
- ✓ define and evaluate exergetic efficiencies for thermal system components of practical interest.
- ✓ apply exergy costing to heat loss and simple cogeneration systems.

Key Engineering Concepts

exergy *p. 332*
exergy reference environment *p. 332*
dead state *p. 332*
specific exergy *p. 336*
exergy change *p. 338*
closed system exergy balance *p. 339*
exergy transfer *p. 340*
exergy destruction *p. 340*
flow exergy *p. 348*
exergy rate balance *p. 343, 348, 349*
exergy accounting *p. 346, 355*
exergetic efficiency *p. 359*

Key Equations

$\mathsf{E} = (U - U_0) + p_0(V - V_0) - T_0(S - S_0) + \text{KE} + \text{PE}$	(7.1)	Exergy of a system.
$\mathsf{e} = (u - u_0) + p_0(v - v_0) - T_0(s - s_0) + \text{V}^2/2 + gz$	(7.2)	Specific exergy.
$\mathsf{E}_2 - \mathsf{E}_1 = (U_2 - U_1) + p_0(V_2 - V_1) - T_0(S_2 - S_1) + (\text{KE}_2 - \text{KE}_1) + (\text{PE}_2 - \text{PE}_1)$	(7.3)	Exergy change.
$\mathsf{E}_2 - \mathsf{E}_1 = \mathsf{E}_q - \mathsf{E}_w - \mathsf{E}_d$	(7.4b)	Closed system exergy balance. See Eqs. 7.5–7.7 for E_q, E_w, E_d, respectively.
$0 = \sum_j \left(1 - \frac{T_0}{T_j}\right)\dot{Q}_j - \dot{W} - \dot{\mathsf{E}}_d$	(7.11a)	Steady-state closed system exergy rate balance.
$0 = \sum_j \left(1 - \frac{T_0}{T_j}\right)\dot{Q}_j - \dot{W}_{cv} + \sum_i \dot{m}_i \mathsf{e}_{fi} - \sum_e \dot{m}_e \mathsf{e}_{fe} - \dot{\mathsf{E}}_d$	(7.13a)	Steady-state control volume exergy rate balance.
$\mathsf{e}_f = h - h_0 - T_0(s - s_0) + \frac{\text{V}^2}{2} + gz$	(7.14)	Specific flow exergy.

Exercises: things engineers think about

1. When you hear the term "energy crisis" used by the news media, do the media really mean *exergy* crisis? Explain.

2. A convenience store sells gasoline and bottled drinking water at nearly the same price per liter. Comment.

3. For each case illustrated in Fig. 5.1 (Sec. 5.1), identify the relevant intensive property difference between the system and its surroundings that underlies the *potential for work.* For cases (a) and (b) discuss whether work could be developed if the particular intensive property value for the system were *less* than for the surroundings.

4. Is it possible for exergy to be negative? For exergy *change* to be negative? For exergy *destruction* to be negative?

5. Do those helium-filled blimps flying over stadiums on game days have exergy? Explain.

6. Does a system consisting of an evacuated space of volume V have exergy? Explain.

7. When an automobile brakes to rest, what happens to the exergy associated with its motion?

8. A block of ice melts when left in a sunny location. Does its exergy increase or decrease? Explain.

9. When evaluating exergy destruction, is it *necessary* to use an exergy balance? Explain.

10. Referring to the heat exchanger of Fig. 7.10, if both the hot and cold streams are at temperatures less than T_0, what form would the exergetic efficiency take?

11. Apart from the well-to-wheel efficiency, what other considerations are used to compare different options for powering vehicles?

12. A gasoline-fueled generator is claimed by its inventor to produce electricity at a lower unit cost than the unit cost of the fuel used, where each cost is based on exergy. Comment.

Problems: developing engineering skills

Exploring Energy Concepts

7.1 By inspection of Fig. P7.1 giving a T–v diagram for water, indicate whether exergy would increase, decrease, or remain the same in (a) Process 1-2, (b) Process 3-4, (c) Process 5-6. Explain.

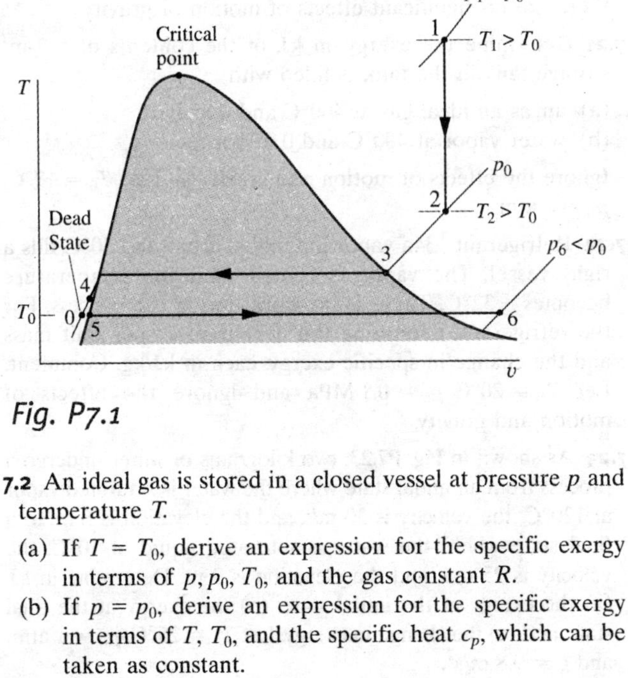

Fig. P7.1

7.2 An ideal gas is stored in a closed vessel at pressure p and temperature T.

(a) If $T = T_0$, derive an expression for the specific exergy in terms of p, p_0, T_0, and the gas constant R.

(b) If $p = p_0$, derive an expression for the specific exergy in terms of T, T_0, and the specific heat c_p, which can be taken as constant.

Ignore the effects of motion and gravity.

7.3 Consider an evacuated tank of volume V. For the space inside the tank as the system, show that the exergy given by $\mathsf{E} = p_0 V$. Discuss.

7.4 Equal molar amounts of carbon dioxide and helium are maintained at the same temperature and pressure. Which has the greater value for exergy relative to the same reference environment? Assume the ideal gas model with constant c_v for each gas. There are no significant effects of motion and gravity.

7.5 Two solid blocks, each having mass m and specific heat c, and initially at temperatures T_1 and T_2, respectively, are brought into contact, insulated on their outer surfaces, and allowed to come into thermal equilibrium.

(a) Derive an expression for the exergy destruction in terms of m, c, T_1, T_2, and the temperature of the environment, T_0.

(b) Demonstrate that the exergy destruction cannot be negative.

(c) What is the source of exergy destruction in this case?

7.6 A system undergoes a refrigeration cycle while receiving Q_C by heat transfer at temperature T_C and discharging energy Q_H by heat transfer at a higher temperature T_H. There are no other heat transfers.

(a) Using energy and exergy balances, show that the net work input to the cycle cannot be zero.

(b) Show that the coefficient of performance of the cycle can be expressed as

$$\beta = \left(\frac{T_C}{T_H - T_C}\right)\left(1 - \frac{T_H \mathsf{E}_d}{T_0(Q_H - Q_C)}\right)$$

where E_d is the exergy destruction and T_0 is the temperature of the exergy reference environment.

(c) Using the result of part (b), obtain an expression for the maximum theoretical value for the coefficient of performance.

7.7 When matter flows across the boundary of a control volume, an energy transfer by work, called *flow work*, occurs. The rate is $\dot{m}(pv)$ where $\dot{m}$, p, and v denote the mass flow rate, pressure, and specific volume, respectively, of the matter crossing the boundary (see Sec. 4.4.2). Show that the *exergy transfer accompanying flow work* is given by $\dot{m}(pv - p_0v)$, where p_0 is the pressure at the dead state.

7.8 When matter flows across the boundary of a control volume, an exergy transfer accompanying mass flow occurs, which is given by $\dot{m}e$, where e is the specific exergy (Eq. 7.2) and $\dot{m}$ is the mass flow rate. An exergy transfer accompanying flow work, which is given by the result of Problem 7.7, also occurs at the boundary. Show that the sum of these exergy transfers is given by $\dot{m}e_f$, where $\dot{m}e_f$ is the specific flow exergy (Eq. 7.14).

7.9 For an ideal gas with constant specific heat ratio k, show that in the absence of significant effects of motion and gravity the specific flow exergy can be expressed as

$$\frac{e_f}{c_p T_0} = \frac{T}{T_0} - 1 - \ln\frac{T}{T_0} + \ln\left(\frac{p}{p_0}\right)^{(k-1)/k}$$

(a) For $k = 1.2$ develop plots of e_f/c_pT_0 versus T/T_0 for $p/p_0 = 0.25, 0.5, 1, 2, 4$. Repeat for $k = 1.3$ and 1.4.
(b) The specific flow exergy can take on negative values when $p/p_0 < 1$. What does a negative value mean physically?

7.10 An ideal gas with constant specific heat ratio k enters a turbine operating at steady state at T_1 and p_1 and expands adiabatically to T_2 and p_2. When would the value of the exergetic turbine efficiency exceed the value of the isentropic turbine efficiency? Discuss. Ignore the effects of motion and gravity.

Evaluating Exergy

7.11 A system consists of 5 kg of water at 10°C and 1 bar. Determine the exergy, in kJ, if the system is at rest and zero elevation relative to an exergy reference environment for which $T_0 = 20°C$, $p_0 = 1$ bar.

7.12 Determine the exergy, in kJ, at 0.7 bar, 90°C for 1 kg of (a) water, (b) Refrigerant 134a, (c) air as an ideal gas with c_p constant. In each case, the mass is at rest and zero evaluation relative to an exergy reference environment for which $T_0 = 20°C$, $p_0 = 1$ bar.

7.13 Determine the specific exergy, in kJ/kg, at 0.01°C of water as a (a) saturated vapor, (b) saturated liquid, (c) saturated solid. In each case, consider a fixed mass at rest and zero elevation relative to an exergy reference environment for which $T_0 = 20°C$, $p_0 = 1$ bar.

7.14 Determine the specific exergy, in kJ/kg of

(a) saturated water vapor at 100°C.
(b) saturated liquid water at 4°C.
(c) ammonia at −40°C, 40 kPa.

In each case, consider a fixed mass at rest and zero elevation relative to an exergy reference environment for which $T_0 = 20°C$, $p_0 = 100$ kPa.

7.15 A balloon filled with helium at 20°C, 1 bar and a volume of 0.5 m³ is moving with a velocity of 15 m/s at an elevation of 0.5 km relative to an exergy reference environment for which $T_0 = 20°C$, $p_0 = 1$ bar. Using the ideal gas model, determine the specific exergy of the helium, in kJ.

7.16 A vessel contains carbon dioxide. Using the ideal gas model
(a) determine the specific exergy of the gas, in kJ/kg, at $p = 620$ kPa and $T = 90°C$.
(b) plot the specific exergy of the gas, in kJ/kg, versus pressure ranging from 100 to 620 kPa, for $T = 25°C$.
(c) plot the specific exergy of the gas, in kJ/kg, versus temperature ranging from 25°C to 90°C, for $p = 100$ kPa.

The gas is at rest and zero elevation relative to an exergy reference environment for which $T_0 = 25°C$, $p_0 = 100$ kPa.

7.17 Oxygen (O_2) at temperature T and 1 bar fills a balloon at rest on the surface of the earth at a location where the ambient temperature is 4°C and the ambient pressure is 1 bar. Using the ideal gas model with $c_p = 0.92$ kJ/kg·K, plot the specific exergy of the oxygen, in kJ/kg, relative to the earth and its atmosphere at this location versus T ranging from 250 to 330 K.

7.18 A vessel contains 1 kg of air at pressure p and 90°C. Using the ideal gas model, plot the specific exergy of the air, in kJ/kg, for p ranging from 0.5 to 2 bar. The air is at rest and negligible elevation relative to an exergy reference environment for which $T_0 = 15°C$, $p_0 = 1$ bar.

7.19 Determine the specific exergy, in kJ/kg, at 0.6 bar, −10°C of (a) ammonia, (b) Refrigerant 22, (c) Refrigerant 134a. Let $T_0 = 0°C$, $p_0 = 1$ bar and ignore the effects of motion and gravity.

7.20 Consider a two-phase solid–vapor mixture of water at −10°C. Each phase present has the same mass. Determine the specific exergy, in kJ/kg, if $T_0 = 20°C$, $p_0 = 1$ bar, and there are no significant effects of motion or gravity.

7.21 Determine the exergy, in kJ, of the contents of a 2-m³ storage tank, if the tank is filled with

(a) air as an ideal gas at 400°C and 0.35 bar.
(b) water vapor at 400°C and 0.35 bar.

Ignore the effects of motion and gravity and let $T_0 = 17°C$, $p_0 = 1$ atm.

7.22 Refrigerant 134a vapor initially at 1 bar and 20°C fills a rigid vessel. The vapor is cooled until the temperature becomes −32°C. There is no work during the process. For the refrigerant, determine the heat transfer per unit mass and the change in specific exergy, each in kJ/kg. Comment. Let $T_0 = 20°C$, $p_0 = 0.1$ MPa and ignore the effects of motion and gravity.

7.23 As shown in Fig. P7.23, two kilograms of water undergo a process from an initial state where the water is saturated vapor at 120°C, the velocity is 30 m/s, and the elevation is 6 m, to a final state where the water is saturated liquid at 10°C, the velocity is 25 m/s, and the elevation is 3 m. Determine in kJ, (a) the exergy at the initial state, (b) the exergy at the final state, and (c) the change in exergy. Let $T_0 = 25°C$, $p_0 = 1$ atm, and $g = 9.8$ m/s².

7.24 Two kilograms of air initially at 90°C and 350 kPa undergo two processes in series:

Process 1–2: Isothermal to $p_2 = 70$ kPa
Process 2–3: Constant pressure to $T_3 = -20°C$

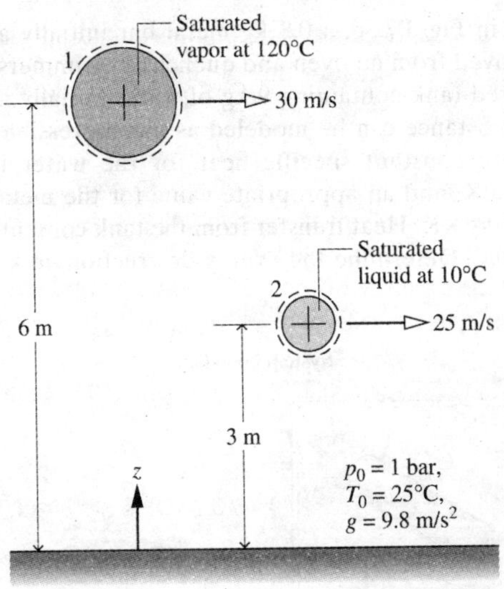

Fig. P7.23

Employing the ideal gas model

(a) represent each process on a p–v diagram and indicate the dead state.

(b) determine the change in exergy for each process, in kJ.

Let $T_0 = 25°C, p_0 = 100$ kPa and ignore the effects of motion and gravity.

7.25 One kilogram of air initially at 860 K, 3 bar fills a rigid tank. The air is cooled to 580 K, 2 bar. For the air modeled as an ideal gas

(a) indicate the initial state, final state, and dead state on a T–v diagram.
(b) determine the heat transfer, in kJ.
(c) determine the change in exergy, in kJ, and interpret the sign using the T–v diagram of part (a).

Let $T_0 = 290$ K, $p_0 = 1$ bar and ignore the effects of motion and gravity.

7.26 Consider 1 kg of steam initially at 20 bar and 240°C as the system. Determine the change in exergy, in kJ, for each of the following processes:

(a) The system is heated at constant pressure until its volume doubles.
(b) The system expands isothermally until its volume doubles.

Let $T_0 = 20°C, p_0 = 1$ bar and ignore the effects of motion and gravity.

Applying the Exergy Balance: Closed Systems

7.27 One kilogram of water in a piston-cylinder assembly, initially at 1.5 bar and 200°C, cools at constant pressure with no internal irreversibilities to a final state where the water is a saturated liquid. For the water as the system, determine the work, the heat transfer, and the amounts of exergy transfer accompanying work and heat transfer, each in kJ. Let $T_0 =$ 20°C, $p_0 = 1$ bar and ignore the effects of motion and gravity.

7.28 One kilogram of air in a piston-cylinder assembly, initially at 1 bar and 25°C, is heated at constant pressure with no internal irreversibilities to a final temperature of 177°C. Employing the ideal gas model, determine the work, the heat transfer, and the amounts of exergy transfer accompanying work and heat transfer, each in kJ. Let $T_0 = 298$ K, $p_0 = 1$ bar and ignore the effects of motion and gravity.

7.29 As shown in Fig. P7.29, 1.11 kg of Refrigerant 134a is contained in a rigid, insulated vessel. The refrigerant is initially saturated vapor at −28°C. The vessel is fitted with a paddle wheel from which a mass is suspended. As the mass descends a certain distance, the refrigerant is stirred until it attains a final equilibrium state at a pressure of 1.4 bar. The only significant changes in state are experienced by the refrigerant and the suspended mass. Determine, in kJ,

(a) the change in exergy of the refrigerant.
(b) the change in exergy of the suspended mass.
(c) the change in exergy of an isolated system of the vessel and pulley-mass assembly.
(d) the destruction of exergy within the isolated system.

Let $T_0 = 293$ K(20°C), $p_0 = 1$ bar.

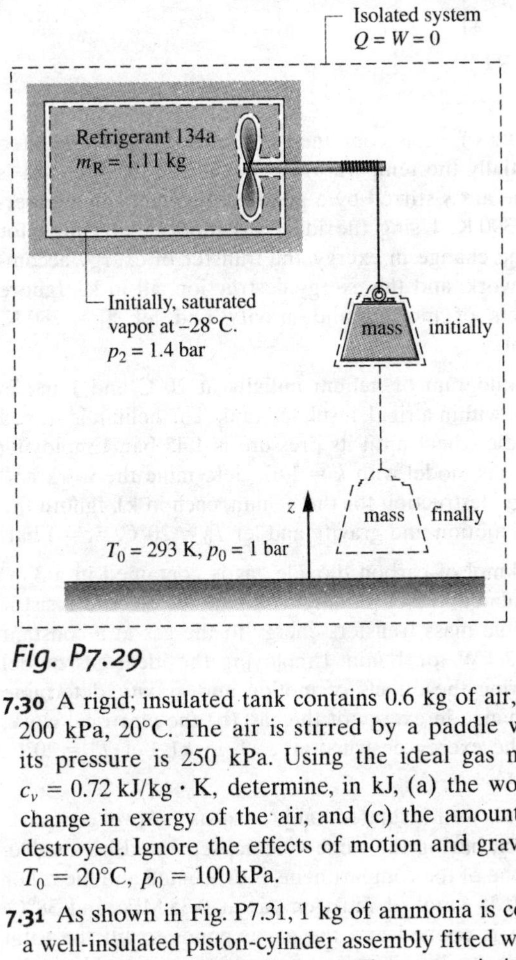

Fig. P7.29

7.30 A rigid, insulated tank contains 0.6 kg of air, initially at 200 kPa, 20°C. The air is stirred by a paddle wheel until its pressure is 250 kPa. Using the ideal gas model with $c_v = 0.72$ kJ/kg · K, determine, in kJ, (a) the work, (b) the change in exergy of the air, and (c) the amount of exergy destroyed. Ignore the effects of motion and gravity, and let $T_0 = 20°C, p_0 = 100$ kPa.

7.31 As shown in Fig. P7.31, 1 kg of ammonia is contained in a well-insulated piston-cylinder assembly fitted with an electrical resistor of negligible mass. The ammonia is initially at 140 kPa and a quality of 80%. The resistor is activated until the volume of the ammonia increases by 25%, while its pressure varies negligibly. Determine, in kJ,

(a) the amount of energy transfer by electrical work and the accompanying exergy transfer.
(b) the amount of energy transfer by work to the piston and the accompanying exergy transfer.

(c) the change in exergy of the ammonia.
(d) the amount of exergy destruction.

Ignore the effects of motion and gravity and let T_0 = 15°C, p_0 = 1 bar.

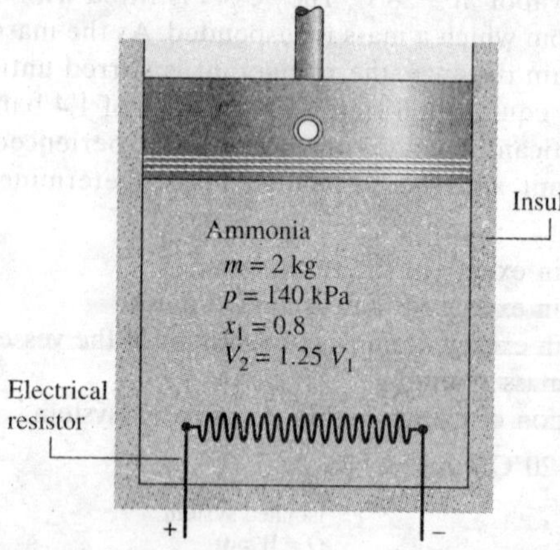

Fig. P7.31

7.32 One kg of air is contained in a closed, rigid, insulated tank. Initially the temperature is 280 K and the pressure is 1 bar. The air is stirred by a paddle wheel until its temperature is 390 K. Using the ideal gas model, determine for the air the change in exergy, the transfer of exergy accompanying work, and the exergy destruction, all in kJ. Ignore the effects of motion and gravity and let T_0 = 280 K, p_0 = 1 bar.

7.33 One kilogram of helium initially at 20°C and 1 bar is contained within a rigid, insulated tank. The helium is stirred by a paddle wheel until its pressure is 1.45 bar. Employing the ideal gas model with k = 1.67, determine the work and the exergy destruction for the helium, each in kJ. Ignore the effects of motion and gravity and let T_0 = 20°C, p_0 = 1 bar.

7.34 One kmol of carbon dioxide gas is contained in a 3 m^3 rigid, insulated vessel initially at 4 bar. An electric resistor of negligible mass transfers energy to the gas at a constant rate of 12 kW for 1 min. Employing the ideal gas model and ignoring the effects of motion and gravity, determine (a) the change in exergy of the gas, (b) the electrical work, and (c) the exergy destruction, each in kJ. Let T_0 = 20°C, p_0 = 1 bar.

7.35 A rigid, well-insulated tank consists of two compartments, each having the same volume, separated by a valve. Initially, one of the compartments is evacuated and the other contains 0.25 kmol of nitrogen gas at 0.35 MPa and 38°C. The valve is opened and the gas expands to fill the total volume, eventually achieving an equilibrium state. Using the ideal gas model for the nitrogen

(a) determine the final temperature, in °C, and final pressure, in MPa.
(b) evaluate the exergy destruction, in kJ.
(c) What is the cause of exergy destruction in this case?

Let T_0 = 21°C, p_0 = 1 bar.

7.36 As shown in Fig. P7.36, a 0.3 kg metal bar initially at 1200 K is removed from an oven and quenched by immersing it in a closed tank containing 9 kg of water initially at 300 K. Each substance can be modeled as incompressible. An appropriate constant specific heat for the water is c_w = 4.2 kJ/kg · K, and an appropriate value for the metal is c_m = 0.42 kJ/kg · K. Heat transfer from the tank contents can be neglected. Determine the exergy destruction, in kJ. Let T_0 = 25°C.

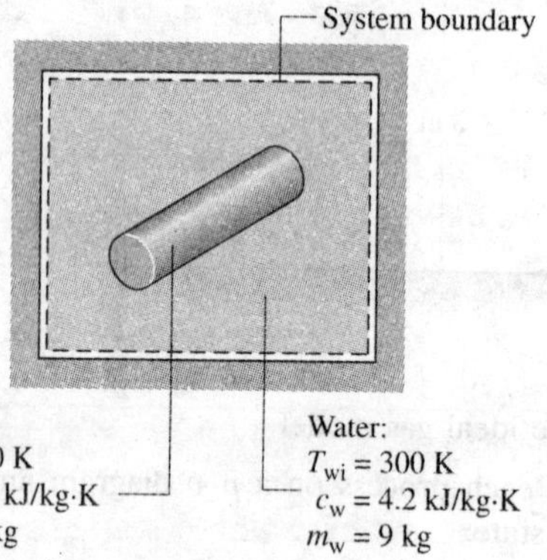

Fig. P7.36

7.37 Figure P7.37 provides steady-state data for a composite of a hot plate and two solid layers. Perform a full exergy accounting, in kW, of the electrical power provided to the composite, including the exergy transfer accompanying heat transfer from the composite and the destruction of exergy in the hot plate and each of the two layers. Let T_0 = 300 K.

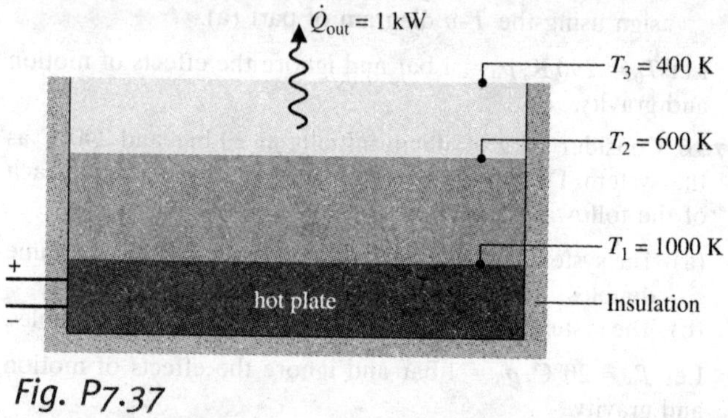

Fig. P7.37

7.38 As shown in Fig. P7.38, heat transfer at a rate of 528 kJ/h takes place through the inner surface of a wall. Measurements made during steady-state operation reveal temperatures of T_1 = 1500 K and T_2 = 600 K at the inner and outer surfaces, respectively. Determine, in kJ/h

(a) the rates of exergy transfer accompanying heat at the inner and outer surfaces of the wall.
(b) the rate of exergy destruction.
(c) What is the cause of exergy destruction in this case?

Let T_0 = 300 K.

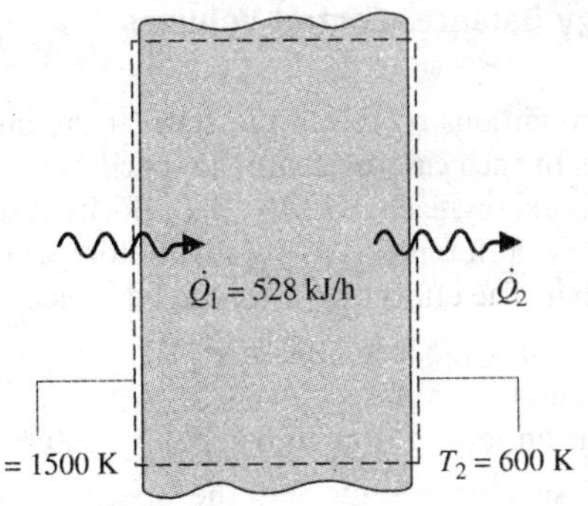

Fig. P7.38

7.39 Figure P7.39 provides steady-state data for the outer wall of a dwelling on a day when the indoor temperature is maintained at 20°C and the outdoor temperature is 0°C. The heat transfer rate through the wall is 1100 W. Determine, in W, the rate of exergy destruction (a) within the wall, and (b) within the enlarged system shown on the figure by the dashed line. Comment. Let T_0 = 0 °C.

7.40 A gearbox operating at steady state receives 2 kW along the input shaft and delivers 1.4 kW along the output shaft. The outer surface of the gearbox is at 40°C. For the gearbox, (a) determine, in kJ/s, the rate of heat transfer and (b) perform a full exergy accounting, in kJ/s, of the input power. Let T_0 = 20°C.

7.41 A gearbox operating at steady state receives 15 kW along its input shaft, delivers power along its output shaft, and is cooled on its outer surface according to $hA(T_b - T_0)$, where T_b = 45°C is the temperature of the outer surface and T_0 = 5°C is the temperature of the surroundings far from the gearbox. The product of the heat transfer coefficient h and outer surface area A is 0.018 kW/K. For the gearbox, determine, in hp, a full exergy accounting of the input power. Let T_0 = 5 °C.

7.42 At steady state, an electric motor develops power along its output shaft of 0.4 kW while drawing 4 amps at 120 V. The outer surface of the motor is at 50°C. For the motor, (a) determine, in kW, the rate of heat transfer and (b) perform a full exergy accounting, in kW, of the electrical power input. Let T_0 = 15°C.

7.43 For the curling iron of Problem 6.55, perform a full exergy accounting, in kW, of the electrical power supplied. Let T_0 = 20°C.

7.44 As shown in Fig. P7.44, a silicon chip measuring 5 mm on a side and 1 mm in thickness is embedded in a ceramic substrate. At steady state, the chip has an electrical power input of 0.225 W. The top surface of the chip is exposed to a coolant whose temperature is 20°C. The heat transfer coefficient for convection between the chip and the coolant is 150 W/m² · K. Heat transfer by conduction between the chip and the substrate is negligible. Determine (a) the surface temperature of the chip, in °C, and (b) the rate of exergy destruction within the chip, in W. What causes the exergy destruction in this case? Let T_0 = 293 K.

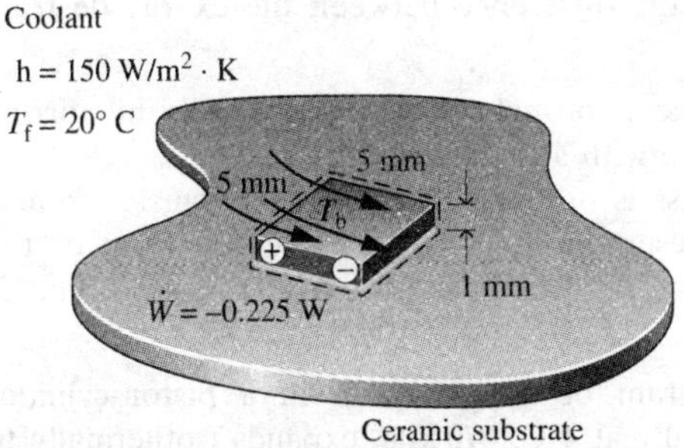

Fig. P7.44

7.45 An electric water heater having a 200 liter capacity heats water from 23 to 55°C. Heat transfer from the outside of the water heater is negligible, and the states of the electrical heating element and the tank holding the water do not change significantly. Perform a full exergy accounting, in kJ, of the electricity supplied to the water heater. Model the water as incompressible with a specific heat c = 4.18 kJ/kg · K. Let T_0 = 23°C.

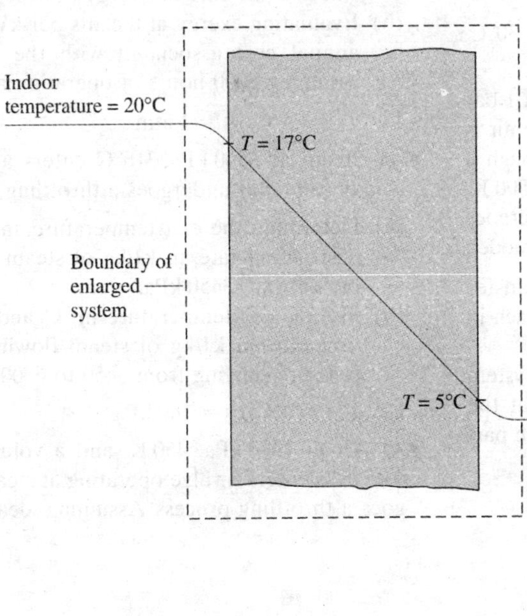

Fig. P7.39

7.46 A thermal reservoir at 1200 K is separated from another thermal reservoir at 300 K by a cylindrical rod insulated on its lateral surfaces. At steady state, energy transfer by conduction takes place through the rod. The rod diameter is 2 cm, the length is L, and the thermal conductivity is 0.4 kW/m · K. Plot the following quantities, each in kW, versus L ranging from 0.01 to 1 m: the rate of conduction through the rod, the rates of exergy transfer accompanying heat transfer into and out of the rod, and the rate of exergy destruction. Let $T_0 = 300$ K.

7.47 Two kilograms of a two-phase liquid–vapor mixture of water initially at 300°C and $x_1 = 0.5$ undergo the two different processes described below. In each case, the mixture is brought from the initial state to a saturated vapor state, while the volume remains constant. For each process, determine the change in exergy of the water, the net amounts of exergy transfer by work and heat, and the amount of exergy destruction, each in kJ. Let $T_0 = 300$ K, $p_0 = 1$ bar, and ignore the effects of motion and gravity. Comment on the difference between the exergy destruction values.

(a) The process is brought about adiabatically by stirring the mixture with a paddle wheel.
(b) The process is brought about by heat transfer from a thermal reservoir at 630 K. The temperature of the water at the location where the heat transfer occurs is 630 K.

7.48 One kilogram of oxygen (O_2), in a piston-cylinder assembly, initially at 4°C, 670 kPa, expands isothermally to a final pressure of 70 kPa while receiving energy by heat transfer through a thin wall separating the oxygen from a thermal reservoir at 30°C.

(a) For the oxygen as the system, evaluate the work, heat transfer, exergy transfers accompanying work and heat transfer, and amount of exergy destruction, each in kJ.
(b) Evaluate the amount of exergy destruction, in kJ, for an enlarged system that includes the oxygen and the wall, assuming the state of the wall remains unchanged. Compare with the exergy destruction of part (a) and comment.

Use the ideal gas model for the oxygen and let $T_0 = 30$°C, $p_0 = 100$ kPa.

7.49 Four and a half kilograms of mass of air initially at 1 bar and 330 K is contained within a closed, rigid tank. The air is slowly warmed, receiving 370 kJ by heat transfer through a wall separating the gas from a thermal reservoir at 500 K. This is the only energy transfer. Assuming the air undergoes an internally reversible process and using the ideal gas model,

(a) determine the change in exergy, the exergy transfer accompanying heat, and the exergy destruction, each in kJ, for the air as the system.
(b) evaluate the exergy destruction for an enlarged system that includes the air and the wall, assuming that the state of the wall remains unchanged. Compare with part (a) and comment.

Let $T_0 = 4$°C, $p_0 = 1$ bar.

Applying the Exergy Balance: Control Volumes at Steady State

7.50 The following conditions represent the state at the inlet to a control volume. In each case, evaluate the specific exergy and the specific flow exergy, each in kJ/kg. The velocity is relative to an exergy reference environment for which $T_0 = 20$°C, $p_0 = 1$ bar. The effect of gravity can be neglected.

(a) water vapor at 100 bar, 520°C, 100 m/s.
(b) ammonia at 3 bar, 0°C, 5 m/s.
(c) nitrogen (N_2) as an ideal gas at 50 bar, 527°C, 200 m/s.

7.51 Determine the specific exergy and the specific flow exergy, each in kJ/kg, for steam at 4.1 MPa, 480°C, with V = 45 m/s and $z = 15$ m. The velocity and elevation are relative to an exergy reference environment for which $T_0 = 20$°C, $p_0 = 1$ bar, and $g = 9.8$ m/s^2.

7.52 Water at 25°C, 1 bar is drawn from a mountain lake 1 km above a valley and allowed to flow through a hydraulic turbine-generator into a lake on the valley floor. For operation at steady state, determine the minimum theoretical mass flow rate, in kg/s, required to generate electricity at a rate of 1 MW. Let $T_0 = 25$°C, $p_0 = 1$ bar and ignore the effects of motion.

7.53 At steady state, hot gaseous products of combustion cool from 1540°C to 130°C as they flow through a pipe. Owing to negligible fluid friction, the flow occurs at nearly constant pressure. Applying the ideal gas model with $c_p = 1.04$ kJ/kg · K, determine the exergy transfer accompanying heat transfer from the gas, in kJ/kg of gas flowing. Let $T_0 = 15$°C and ignore the effects of motion and gravity.

7.54 For the simple vapor power plant of Problem 6.117, evaluate, in MW, (a) the *net* rate energy exits the plant with the cooling water and (b) the *net* rate exergy exits the plant with the cooling water. Comment. Let $T_0 = 20$°C, $p_0 = 1$ bar and ignore the effects of motion and gravity.

7.55 Water vapor enters a valve with a mass flow rate of 2.7 kg/s at a temperature of 280°C and a pressure of 30 bar and undergoes a throttling process to 20 bar.

(a) Determine the flow exergy rates at the valve inlet and exit and the rate of exergy destruction, each in kW.
(b) Evaluating exergy at 8 cents per kW · h, determine the annual cost associated with the exergy destruction, assuming 8000 hours of operation annually.

Let $T_0 = 25$°C, $p_0 = 1$ atm.

7.56 Steam at 6900 kPa, 316°C enters a valve operating at steady state and undergoes a throttling process.

(a) Determine the exit temperature, in °C, and the exergy destruction rate, in kJ/kg of steam flowing, for an exit pressure of 3450 kPa.
(b) Plot the exit temperature, in °C, and the exergy destruction rate, in kJ/kg of steam flowing, each versus exit pressure ranging from 3450 to 6900 kPa.

Let $T_0 = 20$°C, $p_0 = 100$ kPa.

7.57 Air at 1380 kPa, 450 K, and a volumetric flow rate of 0.05 m^3/s enters a valve operating at steady state and undergoes a throttling process. Assuming ideal gas behavior

(a) determine the rate of exergy destruction, in kJ/s, for an exit pressure of 100 kPa.

(b) plot the exergy destruction rate, in kJ/s, versus exit pressure ranging from 100 to 1380 kPa.

Let $T_0 = 295$ K, $p_0 = 100$ kPa.

7.58 Water vapor at 4.0 MPa and 400°C enters an insulated turbine operating at steady state and expands to saturated vapor at 0.1 MPa. The effects of motion and gravity can be neglected. Determine the work developed and the exergy destruction, each in kJ per kg of water vapor passing through the turbine. Let $T_0 = 27$°C, $p_0 = 0.1$ MPa.

7.59 Air enters an insulated turbine operating at steady state at 8 bar, 500 K, and 150 m/s. At the exit the conditions are 1 bar, 320 K, and 10 m/s. There is no significant change in elevation. Determine the work developed and the exergy destruction, each in kJ per kg of air flowing. Let $T_0 = 300$ K, $p_0 = 1$ bar.

7.60 Air enters a turbine operating at steady state with a pressure of 520 kPa, a temperature of 450 K, and a velocity of 120 m/s. At the turbine exit, the conditions are 100 kPa, 330 K, and 30 m/s. Heat transfer from the turbine to its surroundings takes place at an average surface temperature of 345 K. The rate of heat transfer is 4.65 kJ/kg of air passing through the turbine. For the turbine, determine the work developed and the exergy destruction, each in kJ/kg of air flowing. Let $T_0 = 5$°C, $p_0 = 100$ kPa.

7.61 Steam enters a turbine operating at steady state at 6 MPa, 500°C with a mass flow rate of 400 kg/s. Saturated vapor exits at 8 kPa. Heat transfer from the turbine to its surroundings takes place at a rate of 8 MW at an average surface temperature of 180°C. The effects of motion and gravity are negligible.

(a) For a control volume enclosing the turbine, determine the power developed and the rate of exergy destruction, each in MW.

(b) If the turbine is located in a facility where the ambient temperature is 27°C, determine the rate of exergy destruction for an enlarged control volume that includes the turbine and its immediate surroundings so the heat transfer takes place at the ambient temperature. Explain why the exergy destruction values of parts (a) and (b) differ.

Let $T_0 = 300$ K, $p_0 = 100$ kPa.

7.62 An insulated turbine operating at steady state receives steam at 2800 kPa, 315°C and exhausts at 7 kPa. Plot the exergy destruction rate, in kJ/kg of steam flowing, versus turbine isentropic efficiency ranging from 70 to 100%. The effects of motion and gravity are negligible and $T_0 = 15$°C, $p_0 = 1$ bar.

7.63 Air enters a compressor operating at steady state at $T_1 = 300$ K, $p_1 = 1$ bar with a velocity of 70 m/s. At the exit, $T_2 = 540$ K, $p_2 = 5$ bar and the velocity is 150 m/s. The air can be modeled as an ideal gas with $c_p = 1.01$ kJ/kg · K. Stray heat transfer can be ignored. Determine, in kJ per kg of air flowing, (a) the power required by the compressor and (b) the rate of exergy destruction within the compressor. Let $T_0 = 300$ K, $p_0 = 1$ bar. Ignore the effects of motion and gravity.

7.64 Air at 1 bar, 17°C, and a mass flow rate of 0.3 kg/s enters an insulated compressor operating at steady state and exits at 3 bar, 147°C. Determine the power required by the compressor and the rate of exergy destruction, each in kW. Ignore the effects of motion and gravity. Let $T_0 = 17$°C, $p_0 = 1$ bar.

7.65 Refrigerant 134a at −10°C, 1.4 bar, and a mass flow rate of 280 kg/h enters an insulated compressor operating at steady state and exits at 9 bar. The isentropic compressor efficiency is 82%. Determine

(a) the temperature of the refrigerant exiting the compressor, in °C.

(b) the power input to the compressor, in kW.

(c) the rate of exergy destruction, in kW.

Ignore the effects of motion and gravity and let $T_0 = 20$°C, $p_0 = 1$ bar.

7.66 Air is compressed in an axial-flow compressor operating at steady state from 27°C, 1 bar to a pressure of 2.1 bar. The work required is 94.6 kJ per kg of air flowing. Heat transfer from the compressor occurs at an average surface temperature of 40°C at the rate of 14 kJ per kg of air flowing. The effects of motion and gravity can be ignored. Let $T_0 = 20$°C, $p_0 = 1$ bar. Assuming ideal gas behavior, determine (a) the temperature of the air at the exit, in °C, and (b) the rate of exergy destruction within the compressor, in kJ per kg of air flowing.

7.67 Figure P7.67 shows a device to develop power using a heat transfer from a high-temperature industrial process together with a steam input. The figure provides data for steady-state operation. All surfaces are well insulated, except for the one at 527°C, across which heat transfer occurs at a rate of 4.21 kW. The device develops power at a rate of 6 kW. Determine, in kW,

(a) the rate exergy enters accompanying heat transfer.

(b) the *net* rate exergy is carried in by the steam, $(\dot{E}_{f1} - \dot{E}_{f2})$.

(c) the rate of exergy destruction within the device.

Ignore the effects of motion and gravity and let $T_0 = 293$ K, $p_0 = 1$ bar.

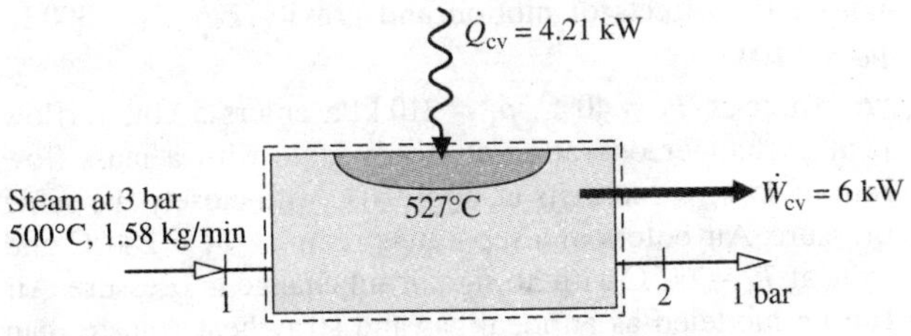

Fig. P7.67

7.68 Determine the rate of exergy destruction, in kJ/s, for the duct system of Problem 6.111. Let $T_0 = 280$ K, $p_0 = 1$ bar.

7.69 For the *vortex tube* of Example 6.7, determine the rate of exergy destruction, in kJ/kg of air entering. Referring to this value for exergy destruction, comment on the inventor's claim. Let $T_0 = 295$ K, $p_0 = 1$ bar.

7.70 Steam at 1.4 MPa and 350°C with a mass flow rate of 0.125 kg/s enters an insulated turbine operating at steady state and exhausts at 100 kPa. Plot the temperature of the exhaust steam, in °C, the power developed by the turbine, in kW, and the rate of exergy destruction within the turbine, in kW, each versus the isentropic turbine efficiency ranging from 0 to 100%. Ignore the effects of motion and gravity. Let T_0 = 20°C, p_0 = 0.1 MPa.

7.71 Steam enters an insulated turbine operating at steady state at 700 kPa, 260°C, with a mass flow rate of 38 kg/s and expands to a pressure of 1 bar. The isentropic turbine efficiency is 80%. If exergy is valued at 8 cents per kW · h, determine

(a) the value of the power produced, in $/h.
(b) the cost of the exergy destroyed, in $/h.
(c) Plot the values of the power produced and the exergy destroyed, each in $/h, versus isentropic efficiency ranging from 80 to 100%.

Ignore the effects of motion and gravity. Let T_0 = 20°C, p_0 = 1 bar.

7.72. Air at T_1 = 720 K, p_1 = 110 kPa enters a counterflow heat exchanger operating at steady state and exits at p_2 = 100 kPa. A separate stream of air enters at T_3 = 470 K, p_3 = 415 kPa and exits at T_4 = 555 K, p_4 = 345 kPa. The mass flow rates of the streams are equal. Stray heat transfer and the effects of motion and gravity can be ignored. Assuming the ideal gas model with c_p = 1 kJ/kg · K, determine (a) T_2, in K, and (b) the rate of exergy destruction within the heat exchanger, in kJ/kg of air flowing. Let T_0 = 290 K, p_0 = 1 bar.

7.73 A counterflow heat exchanger operating at steady state has water entering as saturated vapor at 1 bar with a mass flow rate of 2 kg/s and exiting as saturated liquid at 1 bar. Air enters in a separate stream at 300 K, 1 bar and exits at 335 K with a negligible change in pressure. Heat transfer between the heat exchanger and its surroundings is negligible. Determine

(a) the change in the flow exergy rate of each stream, in kW.
(b) the rate of exergy destruction in the heat exchanger, in kW.

Ignore the effects of motion and gravity. Let T_0 = 300 K, p_0 = 1 bar.

7.74 Water at T_1 = 40°C, p_1 = 210 kPa enters a counterflow heat exchanger operating at steady state with a mass flow rate of 45 kg/s and exits at T_2 = 94°C with closely the same pressure. Air enters in a separate stream at T_3 = 280°C and exits at T_4 = 60°C with no significant change in pressure. Air can be modeled as an ideal gas and stray heat transfer can be ignored. Determine (a) the mass flow rate of the air, in kg/s, and (b) the rate of exergy destruction within the heat exchanger, in kJ/s. Ignore the effects of motion and gravity and let T_0 = 15°C, p_0 = 1 bar.

7.75 Air enters a counterflow heat exchanger operating at steady state at 22°C, 0.1 MPa and exits at 7°C. Refrigerant 134a enters at 0.2 MPa, a quality of 0.2, and a mass flow rate of 30 kg/h. Refrigerant exits at 0°C. Stray heat transfer is negligible and there is no significant change in pressure for either stream.

(a) For the Refrigerant 134a stream, determine the rate of heat transfer, in kJ/h.
(b) For each of the streams, evaluate the change in flow exergy rate, in kJ/h. Compare the values.

Let T_0 = 22°C, p_0 = 0.1 MPa, and ignore the effects of motion and gravity.

7.76 Liquid water enters a heat exchanger operating at steady state at T_1 = 15°C, p_1 = 1 bar and exits at T_2 = 70°C with a negligible change in pressure. In a separate stream, steam enters at T_3 = 140 kPa, x_3 = 92% and exits at T_4 = 60°C, p_4 = 124 kPa. Stray heat transfer and the effects of motion and gravity are negligible. Let T_0 = 15°C, p_0 = 1 bar. Determine (a) the ratio of the mass flow rates of the two streams and (b) the rate of exergy destruction, in kJ/kg of steam entering the heat exchanger.

7.77 Argon enters a nozzle operating at steady state at 1300 K, 360 kPa with a velocity of 10 m/s and exits the nozzle at 900 K, 130 kPa. Stray heat transfer can be ignored. Modeling argon as an ideal gas with k = 1.67, determine (a) the velocity at the exit, in m/s, and (b) the rate of exergy destruction, in kJ per kg of argon flowing. Let T_0 = 293 K, p_0 = 1 bar.

7.78 Nitrogen (N_2) enters a well-insulated nozzle operating at steady state at 520 kPa, 670 K, 20 m/s. At the nozzle exit, the pressure is 140 kPa. The isentropic nozzle efficiency is 90%. For the nozzle, determine the exit velocity, in m/s, and the exergy destruction rate, in kJ/kg of nitrogen flowing. Let T_0 = 20°C, p_0 = 100 kPa.

7.79 Shown in Fig. P7.79 is an open feedwater heater operating at steady state with liquid entering at inlet 1 with T_1 = 40°C and p_1 = 7.0 bar. Water vapor at T_2 = 200°C and p_2 = 7.0 bar enters at inlet 2. Saturated liquid water exits with a pressure of p_3 = 7.0 bar. Ignoring heat transfer with the surroundings and the effects of motion and gravity, determine

(a) the ratio of mass flow rates, $\dot{m}_1/\dot{m}_2$.
(b) the rate of exergy destruction, in kJ per kg of liquid exiting.

Let T_0 = 25°C, p_0 = 1 atm.

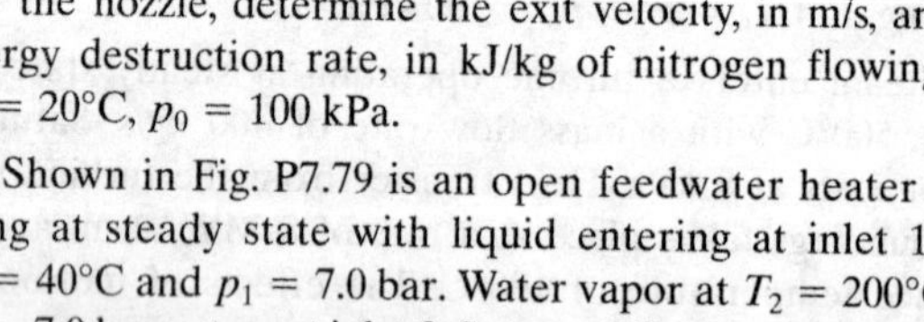

Fig. P7.79

7.80 Steam at 0.7 MPa, 355°C enters an open feedwater heater operating at steady state. A separate stream of liquid water enters at 0.7MPa, 35°C. A single mixed stream exits as saturated liquid at 0.7 MPa with a mass flow rate of 20 kg/s. Stray heat transfer and the effects of motion and gravity can be ignored. Let T_0 = 20°C, p_0 = 1 atm. Determine (a) the mass flow rates of the entering streams, each in kg/s,

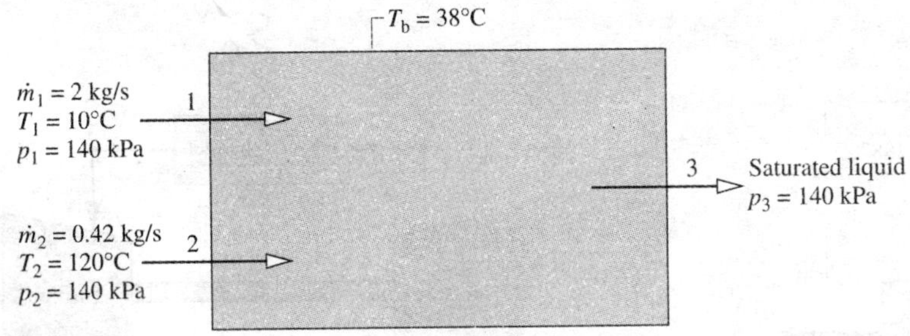

Fig. P7.81

(b) the rate of exergy destruction, in kW, and (c) the cost of the exergy destroyed, in \$/year, for 8000 hours of operation annually. Evaluate exergy at 8 cents per kW · h.

7.81 Figure P7.81 provides steady-state operating data for a mixing chamber in which entering liquid and vapor streams of water mix to form an exiting saturated liquid stream. Heat transfer from the mixing chamber to its surroundings occurs at an average surface temperature of 38°C. The effects of motion and gravity are negligible. Let $T_0 = 20°C$, $p_0 = 1$ bar. For the mixing chamber, determine, each in kJ/s, (a) the rate of heat transfer and the accompanying rate of exergy transfer and (b) the rate of exergy destruction.

7.82 Liquid water at 140 kPa, 10°C enters a mixing chamber operating at steady state with a mass flow rate of 2.3 kg/s and mixes with a separate stream of steam entering at 140 kPa, 120°C with a mass flow rate of 0.17 kg/s. A single mixed stream exits at 140 kPa, 55°C. Heat transfer from the mixing chamber occurs to its surroundings. Neglect the effects of motion and gravity and let $T_0 = 20°C$, $p_0 = 1$ bar. Determine the rate of exergy destruction, in kg/s, for a control volume including the mixing chamber and enough of its immediate surroundings that heat transfer occurs at 20°C.

7.83 At steady state, steam with a mass flow rate of 4.5 kg/s enters a turbine at 430°C and 4140 kPa and expands to 414 kPa. The power developed by the turbine is 2126 kW. The steam then passes through a counterflow heat exchanger with a negligible change in pressure, exiting at 430°C. Air enters the heat exchanger in a separate stream at 1.1 bar, 550°C and exits at 1 bar, 330°C. The effects of motion and gravity can be ignored and there is no significant heat transfer between either component and its surroundings. Determine

(a) the mass flow rate of air, in kg/s.
(b) the rates of exergy destruction in the turbine and heat exchanger, each in kJ/s.

Evaluating exergy at 8 cents per kW · h, determine the hourly cost of each of the exergy destructions found in part (b). Let $T_0 = 4°C$, $p_0 = 1$ bar.

7.84 A gas turbine operating at steady state is shown in Fig. P7.84. Air enters the compressor with a mass flow rate of 5 kg/s at 0.95 bar and 22°C and exits at 5.7 bar. The air then passes through a heat exchanger before entering the turbine at 1100 K, 5.7 bar. Air exits the turbine at 0.95 bar. The compressor and turbine operate adiabatically and the effects of motion and gravity can be ignored. The compressor and turbine isentropic efficiencies are 82 and 85%, respectively. Using the ideal gas model for air, determine, each in kW,

(a) the *net* power developed.
(b) the rates of exergy destruction for the compressor and turbine.
(c) the *net* rate exergy is carried out of the plant at the turbine exit, $(\dot{E}_{f4} - \dot{E}_{f1})$.

Let $T_0 = 22°C$, $p_0 = 0.95$ bar.

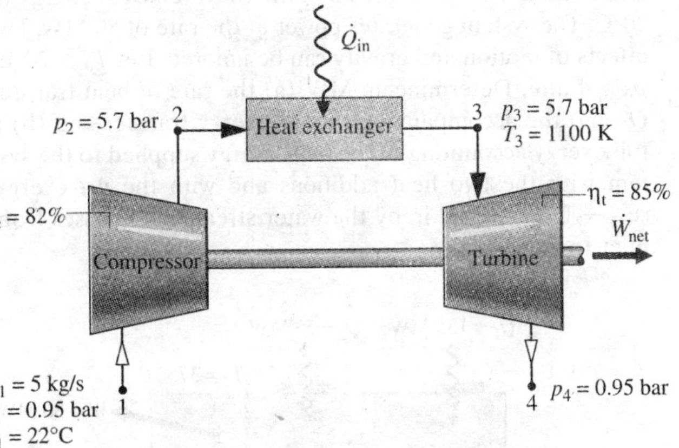

Fig. P7.84

7.85 Figure P7.85 shows a gas turbine power plant operating at steady state consisting of a compressor, a heat exchanger, and a turbine. Air enters the compressor with a mass flow rate of 3.9 kg/s at 0.95 bar, 22°C and exits the turbine at 0.95 bar, 421°C. Heat transfer to the air as it flows through the heat exchanger occurs at an average temperature of 488°C. The compressor and turbine operate adiabatically. Using the ideal gas model for the air, and neglecting the effects of motion and gravity, determine, in MW,

(a) the rate of exergy transfer accompanying heat transfer to the air flowing through the heat exchanger.
(b) the *net* rate exergy is carried out of the plant at the turbine exit, $(\dot{E}_{f2} - \dot{E}_{f1})$.
(c) the rate of exergy destruction within the power plant.

Compare the results of (a)−(c), and comment. Let $T_0 = 295$ K(22°C), $p_0 = 0.95$ bar.

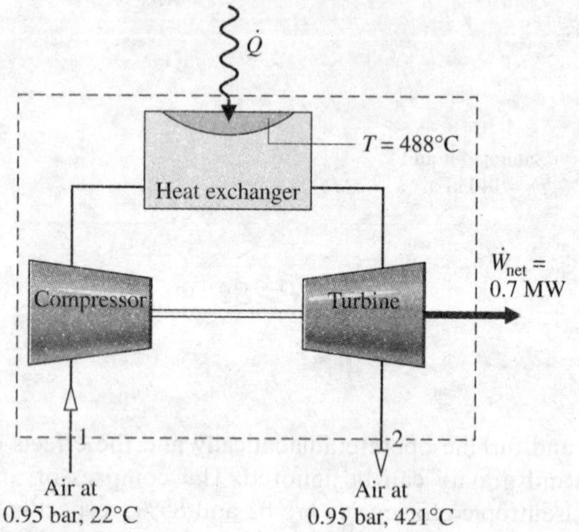

Fig. P7.85

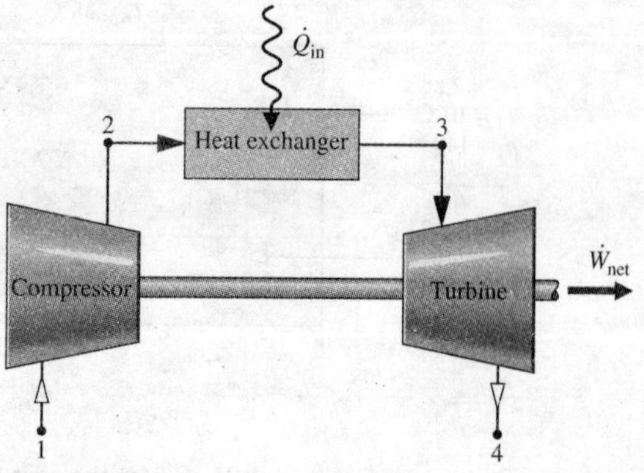

Fig. P7.87

State	*p* (kPa)	*T* (K)	*h* (kJ/kg)	s^0 (kJ/kg K)[a]
1	100	290	290.16	1.6680
2	500	505	508.17	2.2297
3	500	875	904.99	2.8170
4	100	635	643.93	2.4688

[a]s^0 is the variable appearing in Eq. 6.20a and Table A-22.

7.86 Figure P7.86 shows a power-generating system at steady state. Saturated liquid water enters at 80 bar with a mass flow rate of 94 kg/s. Saturated liquid exits at 0.08 bar with the same mass flow rate. As indicated by arrows, three heat transfers occur, each at a specified temperature in the direction of the arrow: The first adds 135 MW at 295°C, the second adds 55 MW at 375°C, and the third removes energy at 20°C. The system generates power at the rate of 80 MW. The effects of motion and gravity can be ignored. Let T_0 = 20°C, p_0 = 1 atm. Determine, in MW, (a) the rate of heat transfer $\dot{Q}_3$ and the accompanying rate of exergy transfer and (b) a full exergy accounting of the *total* exergy supplied to the system with the two heat additions and with the *net* exergy, $(\dot{E}_{f1} - \dot{E}_{f2})$, carried in by the water stream as it passes from inlet to exit.

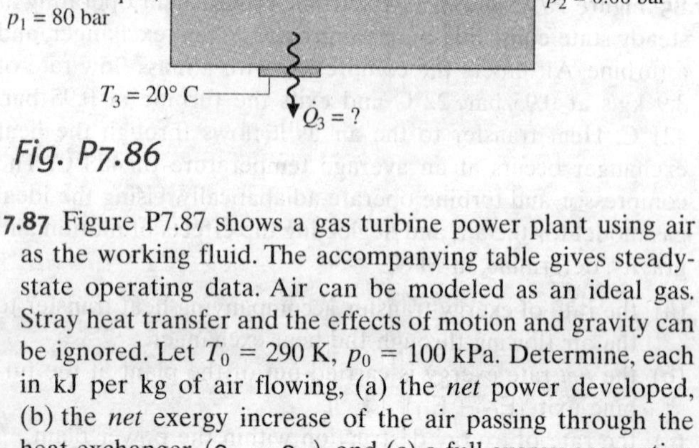

Fig. P7.86

7.87 Figure P7.87 shows a gas turbine power plant using air as the working fluid. The accompanying table gives steady-state operating data. Air can be modeled as an ideal gas. Stray heat transfer and the effects of motion and gravity can be ignored. Let T_0 = 290 K, p_0 = 100 kPa. Determine, each in kJ per kg of air flowing, (a) the *net* power developed, (b) the *net* exergy increase of the air passing through the heat exchanger, $(e_{f3} - e_{f2})$, and (c) a full exergy accounting based on the exergy supplied to the plant found in part (b). Comment.

7.88 Carbon dioxide (CO_2) gas enters a turbine operating at steady state at 50 bar, 500 K with a velocity of 50 m/s. The inlet area is 0.02 m^2. At the exit, the pressure is 20 bar, the temperature is 440 K, and the velocity is 10 m/s. The power developed by the turbine is 3 MW, and heat transfer occurs across a portion of the surface where the average temperature is 462 K. Assume ideal gas behavior for the carbon dioxide and neglect the effect of gravity. Let T_0 = 298 K, p_0 = 1 bar.

(a) Determine the rate of heat transfer, in kW.
(b) Perform a full exergy accounting, in kW, based on the *net* rate exergy is carried into the turbine by the carbon dioxide.

7.89 For the air compressor of Problem 7.66 perform a full exergy accounting, in kJ per kg, of air flowing, based on the work input.

7.90 A compressor fitted with a water jacket and operating at steady state takes in air with a volumetric flow rate of 900 m^3/h at 22°C, 0.95 bar and discharges air at 317°C, 8 bar. Cooling water enters the water jacket at 20°C, 100 kPa with a mass flow rate of 1400 kg/h and exits at 30°C and essentially the same pressure. There is no significant heat transfer from the outer surface of the water jacket to its surroundings, and the effects of motion and gravity can be ignored. For the water-jacketed compressor, perform a full exergy accounting of the power input. Let T_0 = 20°C, p_0 = 1 bar.

7.91 According to a laboratory data sheet, air enters a compressor operating at steady state at 1 bar, 20°C with a volumetric flow rate of 9 m^3/min and exits at 5 bar, 160°C. Cooling water is circulated through a water jacket enclosing the compressor at a rate of 8.6 kg/min, entering at 17°C, and exiting at 25°C with a negligible change in pressure. There is no significant heat transfer from the outer surface of the water jacket, and the effects of motion on gravity are negligible. Let T_0 = 17°C, p_0 = 1 bar. Perform a full exergy accounting in kW, based on the compressor power input and comment.

7.92 Refrigerant 134a enters a water-jacketed compressor operating at steady state at 10°C, 3.2 bar and exits at 70°C, 10 bar. Cooling water enters as a separate stream at 20°C and exits at 32°C with no significant change in pressure. The refrigerant mass flow rate is 1.63 kg/s, and the power *input* to the compressor is 55.2 kJ per kg of refrigerant flowing. Assuming no heat transfer between the outer surface of the water jacket and the surroundings, and neglecting the effects of motion and gravity

(a) determine the mass flow rate of cooling water, in kg/s, and

(b) perform a full exergy accounting, in kW, based on the compressor power input and comment.

Let $T_0 = 20°C, p_0 = 1$ bar.

7.93 For the compressor and heat exchanger of Problem 6.114, perform a full exergy accounting, in kW, based on the compressor power input. Let $T_0 = 300$ K, $p_0 = 96$ kPa.

7.94 Figure P7.94 shows liquid water at 550 kPa, 150°C entering a flash chamber though a valve at the rate of 10 kg/s. At the valve exit, the pressure is 290 kPa. Saturated liquid at 275 kPa exits from the bottom of the flash chamber and saturated vapor at 275 kPa exits from near the top. The vapor stream is fed to a steam turbine having an isentropic efficiency of 90% and an exit pressure of 14 kPa. For steady-state operation, negligible heat transfer with the surroundings, and no significant effects of motion and gravity, perform a full exergy accounting, in kJ/s, of the *net* rate at which exergy is supplied: $(\dot{E}_{f1} - \dot{E}_{f3} - \dot{E}_{f5})$. Let $T_0 = 280$ K, $p_0 = 1$ bar.

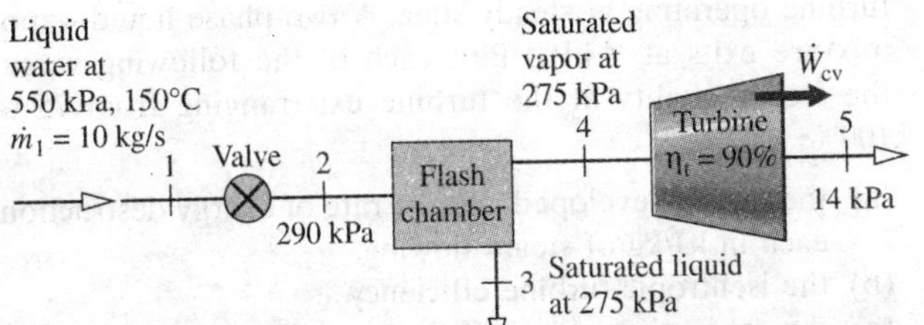

Fig. P7.94

7.95 Figure P7.95 provides steady-state operating data for a throttling valve in parallel with a steam turbine having an isentropic turbine efficiency of 90%. The streams exiting the valve and the turbine mix in a mixing chamber. Heat transfer with the surroundings and the effects of motion and gravity can be neglected. Determine

(a) the power developed by the turbine, in kJ/s.

(b) the mass flow rates through the turbine and valve, each in kg/s.

(c) a full exergy accounting, in kJ/s, of the *net* rate at which exergy is supplied: $(\dot{E}_{f1} - \dot{E}_{f4})$.

Let $T_0 = 280$ K, $p_0 = 1$ bar.

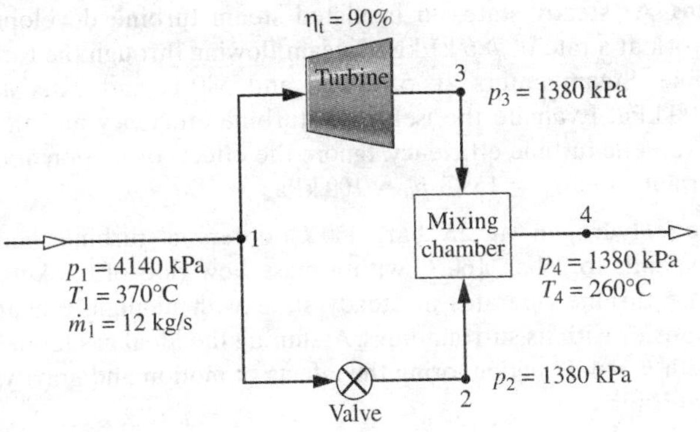

Fig. P7.95

Using Exergetic Efficiencies

7.96 For the water heater of Problem 7.45, devise and evaluate an exergetic efficiency.

7.97 Plot the exergetic efficiency given by Eq. 7.21b versus T_u/T_0 for $T_s/T_0 = 8.0$ and $\eta = 0.4, 0.6, 0.8, 1.0$. What can be learned from the plot when T_u/T_0 is fixed? When ε is fixed? Discuss.

7.98 From an input of electricity, an electric resistance furnace operating at steady state delivers energy by heat transfer to a process at the rate $\dot{Q}_u$ at a *use* temperature T_u. There are no other significant energy transfers.

(a) Devise an exergetic efficiency for the furnace.

(b) Plot the efficiency obtained in part (a) versus the use temperature ranging from 300 to 900 K. Let $T_0 = 20°C$.

7.99 Steam enters a turbine operating at steady state at $p_1 = 12$ MPa, $T_1 = 700°C$ and exits at $p_2 = 0.6$ MPa. The isentropic turbine efficiency is 88%. Property data are provided in the accompanying table. Stray heat transfer and the effects of motion and gravity are negligible. Let $T_0 = 300$ K, $p_0 = 100$ kPa. Determine (a) the power developed and the rate of exergy destruction, each in kJ per kg of steam flowing, and (b) the exergetic turbine efficiency.

State	p (MPa)	T(°C)	h (kJ/kg)	s (kJ/kg · K)
Turbine inlet	12	700	3858.4	7.0749
Turbine exit	0.6	(η_t = 88%)	3017.5	7.2938

7.100 Saturated liquid water at 0.01 MPa enters a power plant pump operating at a steady state. Liquid water exits the pump at 10 MPa. The isentropic pump efficiency is 90%. Property data are provided in the accompanying table. Stray heat transfer and the effects of motion and gravity are negligible. Let $T_0 = 300$ K, $p_0 = 100$ kPa. Determine (a) the power required by the pump and the rate of exergy destruction, each in kJ per kg of water flowing, and (b) the exergetic pump efficiency.

State	p (MPa)	h (kJ/kg)	s (kJ/kg·K)
Pump inlet	0.01	191.8	0.6493
Pump exit	10	204.5	0.6531

7.101 At steady state, an insulated steam turbine develops work at a rate of 786 kJ/kg of steam flowing through the turbine. Steam enters at 5515 kPa and 540°C and exits at 100 kPa. Evaluate the isentropic turbine efficiency and the exergetic turbine efficiency. Ignore the effects of motion and gravity. Let T_0 = 15°C, p_0 = 100 kPa.

7.102 Hydrogen at 25 bar, 450°C enters a turbine and expands to 2 bar, 160°C with a mass flow rate of 0.2 kg/s. The turbine operates at steady state with negligible heat transfer with its surroundings. Assuming the ideal gas model with k = 1.37 and ignoring the effects of motion and gravity, determine

(a) the isentropic turbine efficiency.
(b) the exergetic turbine efficiency.

Let T_0 = 25°C, p_0 = 1 atm.

7.103 Steam enters a turbine operating at steady state at 5 MPa, 600°C and exits at 50 kPa. Stray heat transfer and the effects of motion and gravity can be ignored. If the rate of exergy destruction is 95.4 kJ per kg of steam flowing, determine (a) the isentropic turbine efficiency and (b) the exergetic turbine efficiency. Let T_0 = 293 K, p_0 = 1 bar.

7.104 Air enters an insulated turbine operating at steady state with a pressure of 4 bar, a temperature of 450 K, and a volumetric flow rate of 5 m^3/s. At the exit, the pressure is 1 bar. The isentropic turbine efficiency is 84%. Assuming the ideal gas model and ignoring the effects of motion and gravity, determine

(a) the power developed and the exergy destruction rate, each in kW.
(b) the exergetic turbine efficiency.

Let T_0 = 20°C, p_0 = 1 bar.

7.105 Steam at 1.4 MPa, 350°C enters a turbine operating at steady state with a mass flow rate of 0.12 kg/s and exits at 100 kPa, 115°C. Stray heat transfer and the effects of motion and gravity can be ignored. Let T_0 = 20°C, p_0 = 100 kPa. Determine for the turbine (a) the power developed and the rate of exergy destruction, each in kJ/s, and (b) the isentropic and exergetic turbine efficiencies.

7.106 Figure P7.106 shows a turbine operating at steady state with steam entering at p_1 = 30 bar, T_1 = 350°C and a mass flow rate of 30 kg/s. Process steam is extracted at p_2 = 5 bar, T_2 = 200°C. The remaining steam exits at p_3 = 0.15 bar, x_3 = 90%, and a mass flow rate of 25 kg/s. Stray heat transfer and the effects of motion and gravity are negligible. Let T_0 = 25°C, p_0 = 1 bar. The accompanying table provides property data at key states. For the turbine, determine the power developed and rate of exergy destruction, each in MW. Also devise and evaluate an exergetic efficiency for the turbine.

State	p (bar)	T(°C)	h (kJ/kg)	s (kJ/kg · K)
1	30	350	3115.3	6.7428
2	5	200	2855.4	7.0592
3	0.15	(x = 90%)	2361.7	7.2831

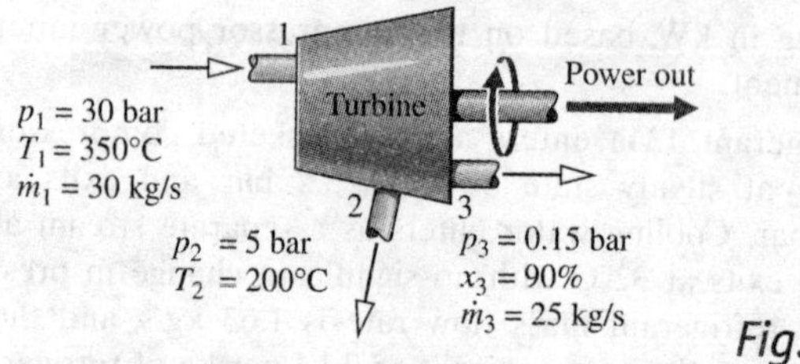

Fig. P7.106

7.107 For the turbine and heat exchanger arrangement of Problem 6.116, evaluate an exergetic efficiency for (a) each turbine, (b) the heat exchanger, and (c) an overall control volume enclosing the turbines and heat exchanger. Comment. Let T_0 = 300 K, p_0 = 1 bar.

7.108 Determine the exergetic efficiency for the turbine of

(a) Problem 6.130. Let T_0 = 300 K.
(b) Problem 6.131. Let T_0 = 280 K.
(c) Problem 6.132. Let T_0 = 295 K.
(d) Problem 6.133. Let T_0 = 300 K.

7.109 Steam at 2760 kPa, 315°C enters a well-insulated turbine operating at steady state and exits as saturated vapor at a pressure p.

(a) For p = 345 kPa, determine the exergy destruction rate, in kJ/kg of steam expanding through the turbine, and the turbine exergetic and isentropic efficiencies.
(b) Plot the exergy destruction rate, in kJ/kg of steam flowing, and the exergetic efficiency and isentropic efficiency, each versus pressure p ranging from 7 to 350 kPa.

Ignore the effects of motion and gravity and let T_0 = 15°C, p_0 = 1 bar.

7.110 Saturated water vapor at 2760 kPa enters an insulated turbine operating at steady state. A two-phase liquid–vapor mixture exits at 4 kPa. Plot each of the following versus the steam quality at the turbine exit ranging from 75 to 100%

(a) the power developed and the rate of exergy destruction, each in kJ/kg of steam flowing.
(b) the isentropic turbine efficiency.
(c) the exergetic turbine efficiency.

Let T_0 = 15°C, p_0 = 1 bar. Ignore the effects of motion and gravity.

7.111 Argon enters an insulated turbine operating at steady state at 1000°C and 2 MPa and exhausts at 350 kPa. The mass flow rate is 0.5 kg/s. Plot each of the following versus the turbine exit temperature, in °C

(a) the power developed, in kW.
(b) the rate of exergy destruction in the turbine, in kW.
(c) the exergetic turbine efficiency.

For argon, use the ideal gas model with k = 1.67. Ignore the effects of motion and gravity. Let T_0 = 20°C, p_0 = 1 bar.

7.112 An insulated steam turbine at steady state can be operated at part-load conditions by throttling the steam to a lower pressure before it enters the turbine. Before throttling, the steam is at 1380 kPa, 315°C. After throttling, the pressure is 1034 kPa. At the turbine exit, the steam is at 7 kPa and a quality x. For the turbine, plot the rate of exergy destruction, in kJ per kg of steam flowing, and exergetic efficiency, each versus x ranging from 90 to 100%. Ignore the effects of motion and gravity and let $T_0 = 15°C$, $p_0 = 1$ bar.

7.113 A pump operating at steady state takes in saturated liquid water at 34 kPa and discharges water at 10 MPa. The isentropic pump efficiency is 75%. Heat transfer with the surroundings and the effects of motion and gravity can be neglected. If $T_0 = 20°C$, determine for the pump

(a) exergy destruction, in kJ/kg of water flowing.
(b) the exergetic efficiency.

7.114 A compressor operating at steady state takes in 1 kg/s of air at 1 bar and 25°C and compresses it to 8 bar and 160°C. The power input to the compressor is 230 kW, and heat transfer occurs from the compressor to the surroundings at an average surface temperature of 50°C.

(a) Perform a full exergy accounting of the power input to the compressor.
(b) Devise and evaluate an exergetic efficiency for the compressor.
(c) Evaluating exergy at 8 cents per kW · h, determine the hourly costs of the power input, exergy loss associated with heat transfer, and exergy destruction.

Neglect the effects of motion and gravity. Let $T_0 = 25°C$, $p_0 = 1$ bar.

7.115 Refrigerant 134a as saturated vapor at −10°C enters a compressor operating at steady state with a mass flow rate of 0.3 kg/s. At the compressor exit the pressure of the refrigerant is 5 bar. Stray heat transfer and the effects of motion and gravity can be ignored. If the rate of exergy destruction within the compressor much be kept less than 2.4 kW, determine the allowed ranges for (a) the power required by the compressor, in kW, and (b) the isentropic and exergetic compressor efficiencies. Let $T_0 = 298$K, $p_0 = 1$ bar.

7.116 Determine the exergetic efficiency for the compressor of

(a) Problem 6.134 Let $T_0 = 290$K.
(b) Problem 6.135. Let $T_0 = 298$K.

7.117 For the turbine-compressor arrangement of Problem 6.146, determine the exergetic efficiency for (a) the turbine, (b) the compressor, (c) an overall control volume enclosing the turbine and compressor. Let $T_0 = 300$ K.

7.118 A counterflow heat exchanger operates at steady state. Air flows on both sides with a mass flow rate of 1 kg/s. On one side, air enters at 470 K, 414 kPa and exits at 555 K, 345 kPa. On the other side, air enters at 720 K, 110 kPa and exits at 640 K and 100 kPa. Heat transfer between the heat exchanger and its surroundings can be ignored, as can the effects of motion and gravity. Evaluate for the heat exchanger

(a) the rate of exergy destruction, in kJ/s.
(b) the exergetic efficiency given by Eq. 7.27.

Let $T_0 = 15°C, p_0 = 1$ bar.

7.119 A counterflow heat exchanger operating at steady state has oil and liquid water flowing in separate streams. The oil is cooled from 440 to 300 K, while the water temperature increases from 290 to 305 K. Neither stream experiences a significant pressure change. The mass flow rate of the oil is 0.4 kg/s. The oil and water can be regarded as incompressible with constant specific heats of 2 and 4 kJ/kg · K, respectively. Heat transfer between the heat exchanger and its surroundings can be ignored, as can the effects of motion and gravity. Determine

(a) the mass flow rate of the water, in kg/s.
(b) the exergetic efficiency given by Eq. 7.27.
(c) the hourly cost of exergy destruction if exergy is valued at 8 cents per kW · h.

Let $T_0 = 15°C, p_0 = 1$ bar.

7.120 In the boiler of a power plant are tubes through which water flows as it is brought from 0.8 MPa, 150°C to 240°C at essentially constant pressure. The total mass flow rate of the water is 100 kg/s. Combustion gases passing over the tubes cool from 1067 to 547°C at essentially constant pressure. The combustion gases can be modeled as air as an ideal gas. There is no significant heat transfer from the boiler to its surroundings. Assuming steady state and neglecting the effects of motion and gravity, determine

(a) the mass flow rate of the combustion gases, in kg/s.
(b) the rate of exergy destruction, in kJ/s.
(c) the exergetic efficiency given by Eq. 7.27.

Let $T_0 = 25°C, p_0 = 1$ bar.

7.121 In the boiler of a power plant are tubes through which water flows as it is brought from a saturated liquid condition at 8 MPa to 760°C at essentially constant pressure. Combustion gases passing over the tubes cool from 1260°C to temperature T at essentially constant pressure. The mass flow rates of the steam and combustion gases are 176 and 1420 kg/s, respectively. The combustion gases can be modeled as air as an ideal gas. There is no significant heat transfer from the boiler to its surroundings. Assuming steady state and neglecting the effects of motion and gravity, determine

(a) the exit temperature T of the combustion gases, in °C.
(b) the exergy destruction rate, in kJ/s.
(c) the exergetic efficiency given by Eq. 7.27.

Let $T_0 = 20°C, p_0 = 1$ bar.

7.122 Refrigerant 134a enters a counterflow heat exchanger operating at steady state at −20°C and a quality of 35% and exits as saturated vapor at −20°C. Air enters as a separate stream with a mass flow rate of 4 kg/s and is cooled at a constant pressure of 1 bar from 300 to 260 K. Heat transfer between the heat exchanger and its surroundings can be ignored, as can the effects of motion and gravity.

(a) As in Fig. E7.6, sketch the variation with position of the temperature of each stream. Locate T_0 on the sketch.
(b) Determine the rate of exergy destruction within the heat exchanger, in kW.
(c) Devise and evaluate an exergetic efficiency for the heat exchanger.

Let $T_0 = 300$ K, $p_0 = 1$ bar.

7.123 Ammonia enters a counterflow heat exchanger at −20°C, with a quality of 35%, and exits as saturated vapor at −20°C. Air at 300 K, 1 bar enters the heat exchanger in a separate stream with a mass flow rate of 240 kg/min and exits at 285 K, 0.98 bar. The heat exchanger is at steady state. There is no appreciable heat transfer between the heat exchanger and its surroundings. Neglect the effects of motion and gravity and let T_0 = 300 K, p_0 = 1 bar.

(a) As in Fig. E7.6, sketch the variation with position of the temperature of each stream. Locate T_0 on the sketch.

(b) Determine the mass flow rate of the ammonia, in kg/s.

(c) Devise and evaluate an exergetic efficiency for the heat exchanger.

7.124 Liquid water at 95°C, 1 bar enters a direct-contact heat exchanger operating at steady state and mixes with a stream of liquid water entering at 15°C, 1 bar. A single liquid stream exits at 1 bar. The entering streams have equal mass flow rates. Neglecting heat transfer with the surroundings and the effects of motion and gravity, determine for the heat exchanger

(a) the rate of exergy destruction, in kJ per kg of liquid exiting.

(b) the exergetic efficiency given by Eq. 7.29.

Let T_0 = 15°C, p_0 = 1 bar.

7.125 Figure P7.125 shows a *cogeneration* system producing two useful products: net power and process steam. The accompanying table provides steady-state mass flow rate, temperature, pressure, and flow exergy data at the 10 numbered states on the figure. Stray heat transfer and the effects of motion and gravity can be ignored. Let T_0 = 298.15 K, p_0 = 1.013 bar. Determine, in MW,

(a) the *net* rate exergy is carried out with the process steam, $(\dot{E}_{f9} - \dot{E}_{f8})$.

(b) the *net* rate exergy is carried out with the combustion products, $(\dot{E}_{f7} - \dot{E}_{f1})$.

(c) the rates of exergy destruction in the air preheater, heat-recovery steam generator, and the combustion chamber.

Devise and evaluate an exergetic efficiency for the overall cogeneration system.

State	Substance	Mass Flow Rate (kg/s)	Temperature (K)	Pressure (bar)	Flow Exergy Rate, $\dot{E}_f$ (MW)
1	Air	91.28	298.15	1.013	0.00
2	Air	91.28	603.74	10.130	27.54
3	Air	91.28	850.00	9.623	41.94
4	Combustion products	92.92	1520.00	9.142	101.45
5	Combustion products	92.92	1006.16	1.099	38.78
6	Combustion products	92.92	779.78	1.066	21.75
7	Combustion products	92.92	426.90	1.013	2.77
8	Water	14.00	298.15	20.000	0.06
9	Water	14.00	485.57	20.000	12.81
10	Methane	1.64	298.15	12.000	84.99

7.126 Figure P7.126 shows a *combined* gas turbine-vapor power plant operating at steady state. The gas turbine is numbered 1-5. The vapor power plant is numbered 6-9. The accompanying table gives data at these numbered states. The total *net* power output is 45 MW and the mass flow rate of the water flowing through the vapor power plant is 15.6 kg/s. The ideal gas model applies to the air. Stray heat transfer and the effects of motion and gravity can be ignored. Let T_0 = 300 K, p_0 = 100 kPa. Determine

(a) the mass flow rate of the air flowing through the gas turbine, in kg/s.

(b) the *net* rate exergy is carried out with the exhaust air stream, $(\dot{E}_{f5} - \dot{E}_{f1})$, in MW.

(c) the rate of exergy destruction in the compressor and pump, each in MW.

(d) the *net* rate of exergy increase of the air flowing through the combustor, $(\dot{E}_{f3} - \dot{E}_{f2})$, in MW.

Devise and evaluate an exergetic efficiency for the overall combined power plant.

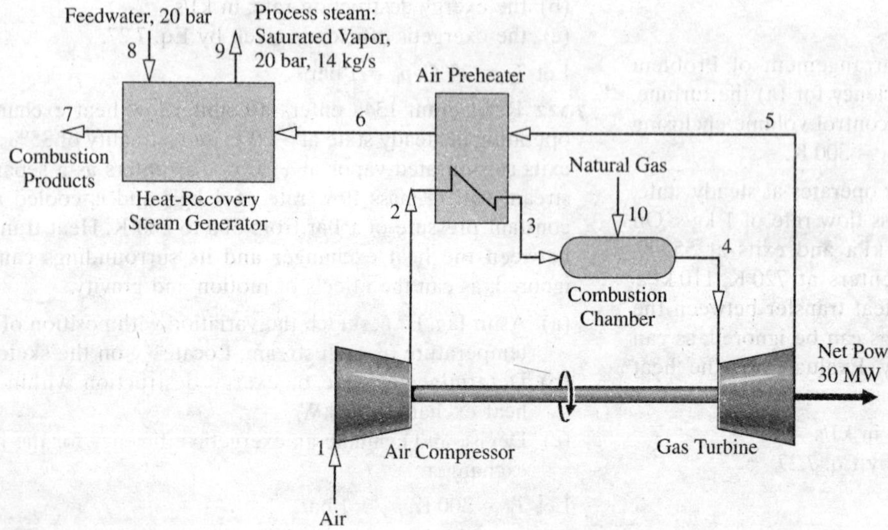

Fig. P7.125

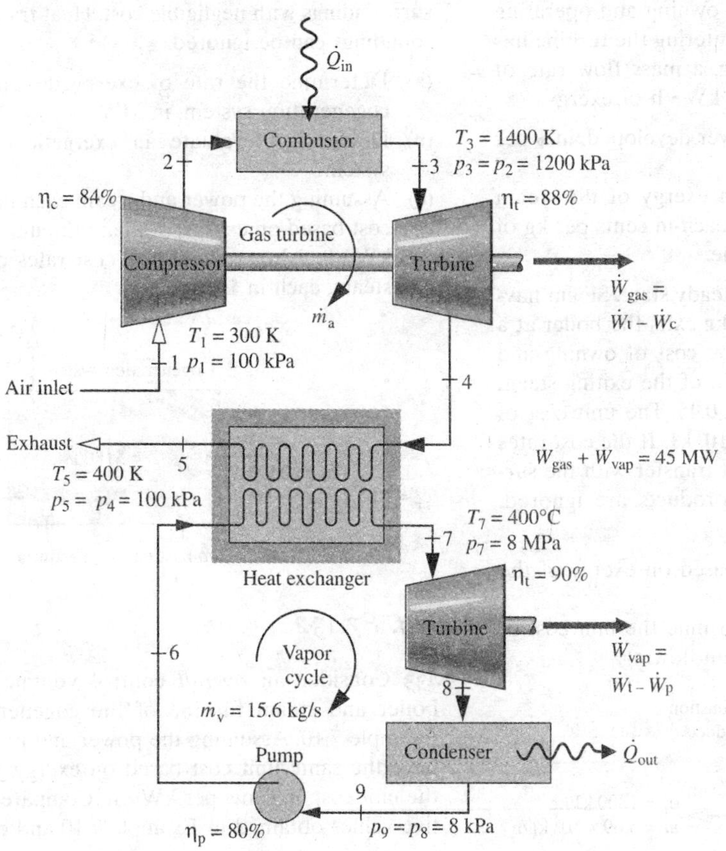

Fig. P7.126

Gas Turbine[a]			Vapor Cycle		
State	h (kJ/kg)	s^0 (kJ/kg · K)	State	h (kJ/kg)	s (kJ/kg · K)
1	300.19	1.7020	6	183.96	0.5975
2	669.79	2.5088	7	3138.30	6.3634
3	1515.42	3.3620	8	2104.74	6.7282
4	858.02	2.7620	9	173.88	0.5926
5	400.98	1.9919			

[a]s^0 is the variable appearing in Eq. 6.20a and Table A-22.

Considering Thermoeconomics

7.127 The total cost rate for a device varies with the pressure drop for flow through the device, $(p_1 - p_2)$, as follows:

$$\dot{C} = c_1(p_1 - p_2)^{-1/3} + c_2(p_1 - p_2)$$

where the c's are constants incorporating economic factors. The first term on the right side of this equation accounts for the capital cost and the second term on the right accounts for the operating cost (pumping power).

(a) Sketch a plot of $\dot{C}$ versus $(p_1 - p_2)$.

(b) At the point of minimum total cost rate, evaluate the contributions of the capital and operating cost rates to the total cost rate, each in percent. Discuss.

7.128 The rate of heat transfer from the outer surface of an electric water heater to the surroundings is given by $hA(T_b - T_f)$, where $hA = 8.9 \times 10^{-3}$ kW/K, T_b is the surface temperature, in K, and $T_f = 293$ K is the temperature of the surroundings at a distance. Evaluating electricity at 8 cents per kW · h

(a) determine the cost of the heat loss, in \$ per year, when $T_b = 297$ K.

(b) plot the cost of the heat loss, in \$ per year, versus T_b ranging from 297 K to 317 K.

Let $T_0 = 293$ K.

7.129 A system operating at steady state generates electricity at the rate $\dot{W}_e$. The cost rate of the fuel input is $\dot{C}_F = c_F\dot{E}_{fF}$, where c_F is the unit cost of fuel based on exergy. The cost of owning and operating the system is

$$\dot{Z} = c\left(\frac{\varepsilon}{1-\varepsilon}\right)\dot{W}_e$$

where $\varepsilon = \dot{W}_e/\dot{E}_{fF}$, and c is a constant incorporating economic factors. $\dot{C}_F$ and $\dot{Z}$ are the only significant cost rates for the system.

(a) Derive an expression for the unit cost of electricity, c_e, based on $\dot{W}_e$ in terms of ε and the ratios c_e/c_F and c/c_F only.

(b) For fixed c/c_F, derive an expression for the value of ε corresponding to the minimum value of c_e/c_F.

(c) Plot the ratio c_e/c_F versus ε for $c/c_F = 0.25$, 1.0, and 4.0. For each specified c/c_F, evaluate the minimum value of c_e/c_F and the corresponding value of ε.

7.130 At steady state, a turbine with an exergetic efficiency of 85% develops 18×10^7 kW · h of work annually (8000 operating hours). The annual cost of owning and operating the turbine is \$5.0 × 10^5. The steam entering the turbine has a specific flow exergy of 1500 kJ/kg, a mass flow rate of 40 kg/s, and is valued at \$0.0182 per kW · h of exergy.

(a) Evaluate the unit cost of the power developed, in \$ per kW · h.
(b) Evaluate the unit cost based on exergy of the steam entering and exiting the turbine, each in cents per kg of steam flowing through the turbine.

7.131 Figure P7.131 shows a boiler at steady state. Steam having a specific flow exergy of 1300 kJ/kg exits the boiler at a mass flow rate of 5.69 × 10^4 kg/h. The cost of owning and operating the boiler is \$91/h. The ratio of the exiting steam exergy to the entering fuel exergy is 0.45. The unit cost of the fuel based on exergy is \$1.50 per 10^6 kJ. If the cost rates of the combustion air, feedwater, heat transfer with the surroundings, and exiting combustion products are ignored, develop

(a) an expression for the unit cost based on exergy of the steam exiting the boiler.
(b) Using the result of part (a), determine the unit cost of the steam, in cents per kg of steam flowing.

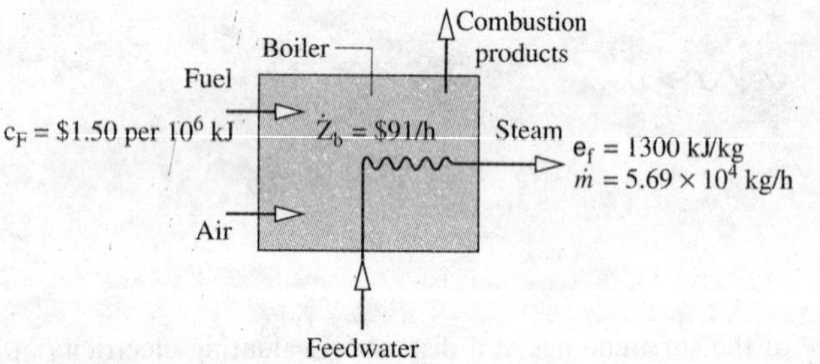

Fig. P7.131

7.132 A cogeneration system operating at steady state is shown schematically in Fig. P7.132. The exergy transfer rates of the entering and exiting streams are shown on the figure, in MW. The fuel, produced by reacting coal with steam, has a unit cost of 5.85 cents per kW · h of exergy. The cost of owning and operating the system is \$1800/h. The feedwater and combustion air enter with negligible exergy and cost. The combustion products are discharged directly to the surroundings with negligible cost. Heat transfer with the surroundings can be ignored.

(a) Determine the rate of exergy destruction within the cogeneration system, in MW.
(b) Devise and evaluate an exergetic efficiency for the system.
(c) Assuming the power and steam each have the same unit cost based on exergy, evaluate the unit cost, in cents per kW · h. Also evaluate the cost rates of the power and steam, each in \$/h.

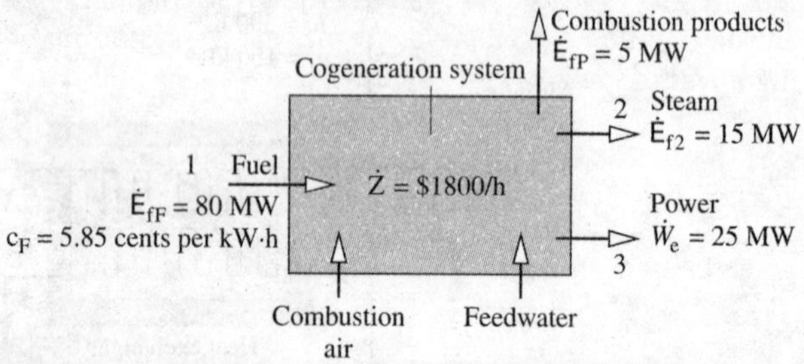

Fig. P7.132

7.133 Consider an *overall* control volume comprising the boiler and steam turbine of the cogeneration system of Example 7.10. Assuming the power and process steam each have the same unit cost based on exergy: $c_e = c_2$, evaluate the unit cost, in cents per kW · h. Compare with the respective values obtained in Example 7.10 and comment.

7.134 The table below gives alternative specifications for the state of the process steam exiting the turbine of Example 7.10. The cost of owning and operating the turbine, in \$/h, varies with the power $\dot{W}_e$, in MW, according to $\dot{Z}_t = 7.2\dot{W}_e$. All other data remain unchanged.

p_2 (bar)	40	30	20	9	5	2	1
T_2(°C)	436	398	349	262	205	128	sat

Plot versus p_2, in bar

(a) the power $\dot{W}_e$, in MW.
(b) the unit costs of the power and process steam, each in cents per kW · h of exergy.
(c) the unit cost of the process steam, in cents per kg of steam flowing.

Design & open ended problems: exploring engineering practice

7.1D A utility charges households the same per kW · h for space heating via steam radiators as it does for electricity. Critically evaluate this costing practice and prepare a memorandum summarizing your principal conclusions.

7.2D Proposals for water-fueled cars have appeared on the Internet. One inventor claims that the car's battery starts the engine, but once the engine is running the car's alternator powers an on-board unit that extracts hydrogen from water. It is further claimed that the hydrogen fuels the engine that powers the alternator. As long as there is water available on-board, the inventor says the entire system is self-sufficient. Investigate the feasibility of water-fueled cars. Write a report, including at least three references.

7.3D You have been invited to testify before a committee of your state legislature that is crafting regulations pertaining to the production of electricity using poultry waste as fuel. Develop a slide presentation providing a balanced assessment, including engineering, public health, and economic considerations.

7.4D *Anaerobic digestion* is a proven means of producing methane from livestock waste. To provide for the space heating, water heating, and cooking needs of a typical farm dwelling in your locale, determine the size of the anaerobic digester and the number of waste-producing animals required. Select animals from poultry, swine, and cattle, as appropriate. Place your findings in a report, including an economic evaluation and at least three references.

7.5D The manager of a construction company seeks your advice in deciding which type of tool is best to use for a range of construction tasks: air tools driven by compressed air from a storage tank or electric tools. Investigate this issue, including electricity requirements, cost, safety, and ease-of-use. Place your recommendations in a memorandum.

7.6D Tankless microwave water heating systems have been introduced that not only quickly provide hot water but also significantly reduce the exergy destruction inherent in domestic water heating with conventional electrical and gas-fueled water heaters. For a 230 m^2 dwelling in your locale, investigate the feasibility of using a microwave water heating system. Include a detailed economic evaluation accounting for equipment, installation, and operating costs. Place your findings in a memorandum.

7.7D In the 1840s, British engineers developed *atmospheric railways* that featured a large-diameter tube located between the tracks and stretching the entire length of the railroad. Pistons attached by struts to the rail cars moved inside the tube. As shown in Fig. P7.7D, piston motion was achieved by maintaining a vacuum ahead of the piston while the atmosphere was allowed to act behind it. Although several such railways came into use, limitations of the technology then available eventually ended this mode of transportation. Investigate the feasibility of combining the atmospheric railway concept with today's technology to develop rail service for commuting within urban areas. Write a report, including at least three references.

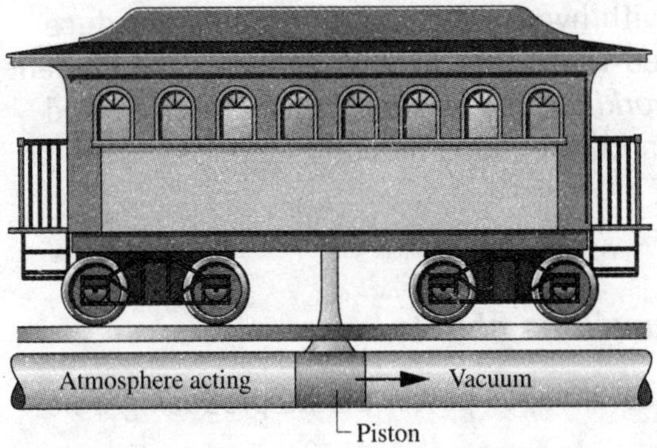

Fig. P7.7D

7.8D A government agency has solicited proposals for technology in the area of *exergy* harvesting. The aim is to develop small-scale devices to generate power for rugged-duty applications with power requirements ranging from hundreds of milliwatts to several watts. The power must be developed only from *ambient sources*, such as thermal and chemical gradients, naturally occurring fuels (tree sap, plants, waste matter, etc.), wind, solar, sound and vibration, and mechanical motion including human motion. The devices also must operate with little or no human intervention. Devise a system that would meet these requirements. Clearly identify its intended application and explain its operating principles. Estimate its size, weight, and expected power output.

7.9D Satisfactory operation of the electricity grid requires that the rate at which power is added to the grid balances the rate at which power is removed. This is challenging for the grid operators because power usage fluctuates with both the time of day and the day of the week. Various storage means are available for use when electricity supply exceeds demand. These include, but are not necessarily limited to, batteries, flywheels, compressed air storage, thermal storage, and the production of hydrogen by electrolysis. Such approaches can be used singly or in combination. Considering both thermodynamic and economic aspects, critically evaluate means for electricity storage. Write a report, including at least three references.

7.10D *Pinch analysis* (or *pinch technology*) is a popular methodology for optimizing the design of heat exchanger networks in complex thermal systems. Pinch analysis uses a primarily graphical approach to implement second-law reasoning. Write a paper, including at least three references, in which the role of pinch analysis within *thermoeconomics* is discussed.

8

Vapor Power Systems

Engineering Context

An important engineering goal is to devise systems that accomplish desired types of energy conversion. The present chapter and the next are concerned with several types of power-generating systems, each of which produces a net power output from a fossil fuel, nuclear, or solar input. In these chapters, we describe some of the practical arrangements employed for power production and illustrate how such power plants can be modeled thermodynamically. The discussion is organized into three main areas of application: vapor power plants, gas turbine power plants, and internal combustion engines. These power systems, together with hydroelectric power plants, produce virtually all of the electrical and mechanical power used worldwide. The **objective** of the present chapter is to study *vapor* power plants in which the *working fluid* is alternately vaporized and condensed. Chapter 9 is concerned with gas turbines and internal combustion engines in which the working fluid remains a gas.

Learning Objectives

When you complete your study of this chapter you will be able to...

- ✔ Demonstrate an understanding of the basic principles of vapor power plants producing a net power output from a fossil fuel, nuclear, or solar input.
- ✔ Develop and analyze thermodynamic models of vapor power plants based on the Rankine cycle and its modifications, including:
  - ✔ sketching schematic and accompanying *T–s* diagrams.
  - ✔ evaluating property data at principal states in the cycle.
  - ✔ applying mass, energy, entropy, and balances for the basic processes.
  - ✔ determining power cycle performance, thermal efficiency, net power output, and mass flow rates.
- ✔ Explain the effects on Rankine cycle performance of varying key parameters.
- ✔ Discuss the principal sources of exergy destruction and loss in vapor power plants.

8.1 Modeling Vapor Power Systems

The processes taking place in power-generating systems are sufficiently complicated that idealizations are required to develop thermodynamic models. Such modeling is an important initial step in engineering design. Although the study of simplified models generally leads only to qualitative conclusions about the performance of the corresponding actual devices, models often allow deductions about how changes in major operating parameters affect actual performance. They also provide relatively simple settings in which to discuss the functions and benefits of features intended to improve overall performance.

The vast majority of electrical generating plants are variations of vapor power plants in which water is the working fluid. The basic components of a simplified fossil-fuel vapor power plant are shown schematically in Fig. 8.1. To facilitate thermodynamic analysis, the overall plant can be broken down into the four major subsystems identified by the letters A through D on the diagram. The focus of our considerations in this chapter is subsystem A, where the important energy conversion from *heat to work* occurs. But first, let us briefly consider the other subsystems.

The function of subsystem B is to supply the energy required to vaporize the water passing through the boiler. In fossil-fuel plants, this is accomplished by heat transfer *to* the working fluid passing through tubes and drums in the boiler *from* the hot gases produced by the combustion of a fossil fuel. In nuclear plants, the origin of the energy is a controlled nuclear reaction taking place in an isolated reactor building. Pressurized water, a liquid metal, or a gas such as helium can be used to transfer energy released in the nuclear reaction to the working fluid in specially designed heat exchangers. Solar power plants have receivers for concentrating and collecting solar radiation to vaporize the working fluid. Regardless of the energy source, the vapor produced in the boiler passes through a turbine, where it expands to a lower pressure. The shaft of the turbine is connected to an electric generator (subsystem D). The vapor

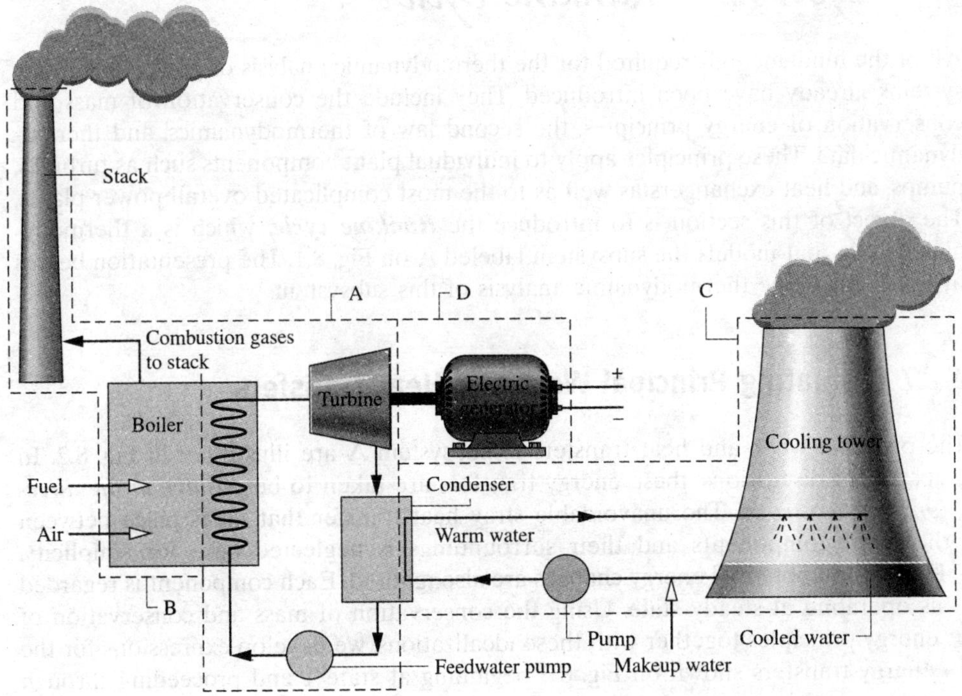

Fig. 8.1 Components of a simple vapor power plant.

leaving the turbine passes through the condenser, where it condenses on the outside of tubes carrying cooling water. The cooling water circuit comprises subsystem C. For the plant shown, the cooling water is sent to a cooling tower, where energy taken up in the condenser is rejected to the atmosphere. The cooling water is then recirculated through the condenser.

Concern for the environment and safety considerations govern what is allowable in the interactions between subsystems B and C and their surroundings. One of the major difficulties in finding a site for a vapor power plant is access to sufficient quantities of cooling water. For this reason and to minimize *thermal pollution* effects, most power plants now employ cooling towers. In addition to the question of cooling water, the safe processing and delivery of fuel, the control of pollutant discharges, and the disposal of wastes are issues that must be dealt with in both fossil-fueled and nuclear-fueled plants to ensure safety and operation with an acceptable level of environmental impact. Solar power plants are generally regarded as nonpolluting and safe but as yet are not widely used.

Returning now to subsystem A of Fig. 8.1, observe that each unit of mass periodically undergoes a thermodynamic cycle as the working fluid circulates through the series of four interconnected components. Accordingly, several concepts related to thermodynamic *power cycles* introduced in previous chapters are important for the present discussions. You will recall that the conservation of energy principle requires that the net work developed by a power cycle equals the net heat added. An important deduction from the second law is that the thermal efficiency, which indicates the extent to which the heat added is converted to a net work output, must be less than 100%. Previous discussions also have indicated that improved thermodynamic performance accompanies the reduction of irreversibilities. The extent to which irreversibilities can be reduced in power-generating systems depends on thermodynamic, economic, and other factors, however.

8.2 Analyzing Vapor Power Systems—Rankine Cycle

All of the fundamentals required for the thermodynamic analysis of power-generating systems already have been introduced. They include the conservation of mass and conservation of energy principles, the second law of thermodynamics, and thermodynamic data. These principles apply to individual plant components such as turbines, pumps, and heat exchangers as well as to the most complicated overall power plants. The object of this section is to introduce the ***Rankine cycle,*** which is a thermodynamic cycle that models the subsystem labeled A on Fig. 8.1. The presentation begins by considering the thermodynamic analysis of this subsystem.

Rankine cycle

8.2.1 Evaluating Principal Work and Heat Transfers

The principal work and heat transfers of subsystem A are illustrated in Fig. 8.2. In subsequent discussions, these energy transfers are taken to be *positive in the directions of the arrows*. The unavoidable stray heat transfer that takes place between the plant components and their surroundings is neglected here for simplicity. Kinetic and potential energy changes are also ignored. Each component is regarded as operating at steady state. Using the conservation of mass and conservation of energy principles together with these idealizations, we develop expressions for the energy transfers shown on Fig. 8.2 beginning at state 1 and proceeding through each component in turn.

Take Note...

When analyzing vapor power cycles, we take energy transfers as positive in the directions of arrows on system schematics and write energy balances accordingly.

Turbine

Vapor from the boiler at state 1, having an elevated temperature and pressure, expands through the turbine to produce work and then is discharged to the condenser at state 2 with relatively low pressure. Neglecting heat transfer with the surroundings, the mass and energy rate balances for a control volume around the turbine reduce at steady state to give

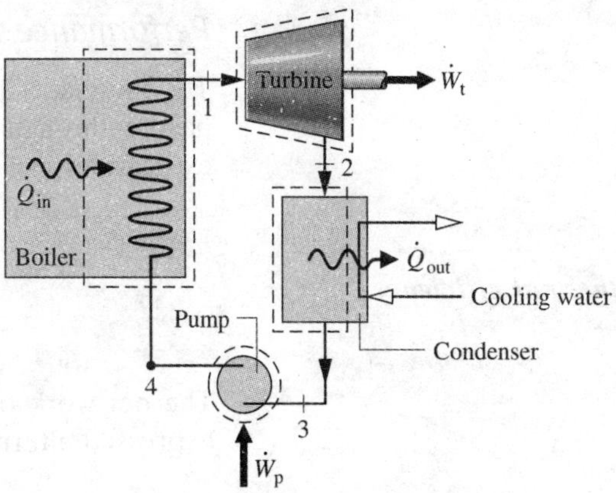

Fig. 8.2 Principal work and heat transfers of subsystem A.

$$0 = \cancelto{0}{\dot{Q}_{cv}} - \dot{W}_t + \dot{m}\left[h_1 - h_2 + \cancelto{0}{\frac{V_1^2 - V_2^2}{2}} + \cancelto{0}{g(z_1 - z_2)}\right]$$

or

$$\frac{\dot{W}_t}{\dot{m}} = h_1 - h_2 \tag{8.1}$$

where $\dot{m}$ denotes the mass flow rate of the working fluid, and $\dot{W}_t/\dot{m}$ is the rate at which work is developed per unit of mass of steam passing through the turbine. As noted above, kinetic and potential energy changes are ignored.

Condenser

In the condenser there is heat transfer from the vapor to cooling water flowing in a separate stream. The vapor condenses and the temperature of the cooling water increases. At steady state, mass and energy rate balances for a control volume enclosing the condensing side of the heat exchanger give

$$\frac{\dot{Q}_{out}}{\dot{m}} = h_2 - h_3 \tag{8.2}$$

where $\dot{Q}_{out}/\dot{m}$ is the rate at which energy is transferred by heat *from* the working fluid to the cooling water per unit mass of working fluid passing through the condenser. This energy transfer is positive in the direction of the arrow on Fig. 8.2.

Pump

The liquid condensate leaving the condenser at 3 is pumped from the condenser into the higher pressure boiler. Taking a control volume around the pump and assuming no heat transfer with the surroundings, mass and energy rate balances give

$$\frac{\dot{W}_p}{\dot{m}} = h_4 - h_3 \tag{8.3}$$

where $\dot{W}_p/\dot{m}$ is the rate of power *input* per unit of mass passing through the pump. This energy transfer is positive in the direction of the arrow on Fig. 8.2.

Boiler

The working fluid completes a cycle as the liquid leaving the pump at 4, called the boiler ***feedwater***, is heated to saturation and evaporated in the boiler. Taking a control volume enclosing the boiler tubes and drums carrying the feedwater from state 4 to state 1, mass and energy rate balances give *feedwater*

$$\frac{\dot{Q}_{in}}{\dot{m}} = h_1 - h_4 \tag{8.4}$$

where $\dot{Q}_{in}/\dot{m}$ is the rate of heat transfer from the energy source into the working fluid per unit mass passing through the boiler.

Performance Parameters

The thermal efficiency gauges the extent to which the energy input to the working fluid passing through the boiler is converted to the *net* work output. Using the quantities and expressions just introduced, the ***thermal efficiency*** of the power cycle of Fig. 8.2 is

thermal efficiency

$$\eta = \frac{\dot{W}_t/\dot{m} - \dot{W}_p/\dot{m}}{\dot{Q}_{in}/\dot{m}} = \frac{(h_1 - h_2) - (h_4 - h_3)}{h_1 - h_4} \tag{8.5a}$$

The net work output equals the net heat input. Thus, the thermal efficiency can be expressed alternatively as

$$\eta = \frac{\dot{Q}_{in}/\dot{m} - \dot{Q}_{out}/\dot{m}}{\dot{Q}_{in}/\dot{m}} = 1 - \frac{\dot{Q}_{out}/\dot{m}}{\dot{Q}_{in}/\dot{m}}$$

$$= 1 - \frac{(h_2 - h_3)}{(h_1 - h_4)} \tag{8.5b}$$

heat rate

The ***heat rate*** is the amount of energy added by heat transfer to the cycle, usually in kJ, to produce a unit of net work output, usually in kW · h. Accordingly, the heat rate, which is inversely proportional to the thermal efficiency, has units of $\frac{\text{kJ}}{\text{kW} \cdot \text{h}}$.

Another parameter used to describe power plant performance is the ***back work ratio,*** or bwr, defined as the ratio of the pump work input to the work developed by the turbine. With Eqs. 8.1 and 8.3, the back work ratio for the power cycle of Fig. 8.2 is

back work ratio

$$\text{bwr} = \frac{\dot{W}_p/\dot{m}}{\dot{W}_t/\dot{m}} = \frac{(h_4 - h_3)}{(h_1 - h_2)} \tag{8.6}$$

Examples to follow illustrate that the change in specific enthalpy for the expansion of vapor through the turbine is normally many times greater than the increase in enthalpy for the liquid passing through the pump. Hence, the back work ratio is characteristically quite low for vapor power plants.

Provided states 1 through 4 are fixed, Eqs. 8.1 through 8.6 can be applied to determine the thermodynamic performance of a simple vapor power plant. Since these equations have been developed from mass and energy rate balances, they apply equally for actual performance when irreversibilities are present and for idealized performance in the absence of such effects. It might be surmised that the irreversibilities of the various power plant components can affect overall performance, and this is the case. Even so, it is instructive to consider an idealized cycle in which irreversibilities are assumed absent, for such a cycle establishes an *upper limit* on the performance of the Rankine cycle. The ideal cycle also provides a simple setting in which to study various aspects of vapor power plant performance.

Energy & Environment

The United States relies heavily upon abundant coal reserves to generate electric power, but these systems require a great deal of clean-up, experts say. Awareness of the health and environmental impacts of coal have led to increasingly stringent regulations on coal-burning power plants. As a result, the search for new *clean-coal* technologies has intensified.

According to industry sources, controlling particulate emissions and safely disposing of millions of tons of coal waste once were the main concerns. Sulfur dioxide removal then became an issue due to concern over acid rain. More recently, nitric oxide (NO_X), mercury, and fine particle (less than 3 microns) emissions were recognized as especially harmful. Strides have been made in developing more effective sulfur dioxide scrubbers and particulate capture devices, but stricter environmental standards demand new approaches.

One promising technology is fluidized bed combustion, where a powdered coal-limestone mixture churns in air to enhance the burning. The limestone removes some sulfur *during* combustion rather than waiting to remove the sulfur *after* combustion, as in conventional boilers. Nitric oxide formation is also less because of the relatively low temperatures of fluidized bed combustion. Another innovation is the integrated gasification combined-cycle plant, or IGCC, that promises to be cleaner and have higher thermal efficiency than conventional plants. In an IGCC, coal is converted to a cleaner-burning combustible gas that is used in a power plant combining gas and steam turbines.

8.2.2 Ideal Rankine Cycle

If the working fluid passes through the various components of the simple vapor power cycle without irreversibilities, frictional pressure drops would be absent from the boiler and condenser, and the working fluid would flow through these components at constant pressure. Also, in the absence of irreversibilities and heat transfer with the surroundings, the processes through the turbine and pump would be isentropic. A cycle adhering to these idealizations is the ***ideal Rankine cycle*** shown in Fig. 8.3.

ideal Rankine cycle

Referring to Fig. 8.3, we see that the working fluid undergoes the following series of internally reversible processes:

Process 1–2: Isentropic expansion of the working fluid through the turbine from saturated vapor at state 1 to the condenser pressure.

Process 2–3: Heat transfer *from* the working fluid as it flows at constant pressure through the condenser with saturated liquid at state 3.

Process 3–4: Isentropic compression in the pump to state 4 in the compressed liquid region.

Process 4–1: Heat transfer *to* the working fluid as it flows at constant pressure through the boiler to complete the cycle.

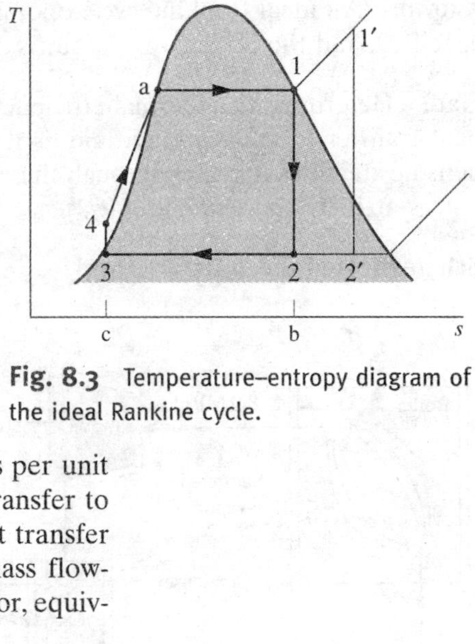

Fig. 8.3 Temperature–entropy diagram of the ideal Rankine cycle.

The ideal Rankine cycle also includes the possibility of superheating the vapor, as in cycle 1′–2′–3–4–1′. The importance of superheating is discussed in Sec. 8.3.

Since the ideal Rankine cycle consists of internally reversible processes, areas under the process lines of Fig. 8.3 can be interpreted as heat transfers per unit of mass flowing. Applying Eq. 6.49, area 1–b–c–4–a–1 represents the heat transfer to the working fluid passing through the boiler and area 2–b–c–3–2 is the heat transfer from the working fluid passing through the condenser, each per unit of mass flowing. The enclosed area 1–2–3–4–a–1 can be interpreted as the net heat input or, equivalently, the net work output, each per unit of mass flowing.

Because the pump is idealized as operating without irreversibilities, Eq. 6.51b can be invoked as an alternative to Eq. 8.3 for evaluating the pump work. That is

$$\left(\frac{\dot{W}_p}{\dot{m}}\right)_{\substack{\text{int}\\\text{rev}}} = \int_3^4 v\,dp \tag{8.7a}$$

where the minus sign has been dropped for consistency with the positive value for pump work in Eq. 8.3. The subscript "int rev" has been retained as a reminder that this expression is restricted to an internally reversible process through the pump. No such designation is required by Eq. 8.3, however, because it expresses the conservation of mass and energy principles and thus is not restricted to processes that are internally reversible.

Take Note...

For cycles, we modify the problem-solving methodology: The **Analysis** begins with a systematic evaluation of required property data at each numbered state. This reinforces what we know about the components, since given information and assumptions are required to fix the states.

Evaluation of the integral of Eq. 8.7a requires a relationship between the specific volume and pressure for the process. Because the specific volume of the liquid normally varies only slightly as the liquid flows from the inlet to the exit of the pump, a plausible approximation to the value of the integral can be had by taking the specific volume at the pump inlet, v_3, as constant for the process. Then

$$\left(\frac{\dot{W}_p}{\dot{m}}\right)_{\substack{\text{int}\\\text{rev}}} \approx v_3(p_4 - p_3) \tag{8.7b}$$

The next example illustrates the analysis of an ideal Rankine cycle.

Example 8.1 IDEAL RANKINE CYCLE

Steam is the working fluid in an ideal Rankine cycle. Saturated vapor enters the turbine at 8.0 MPa and saturated liquid exits the condenser at a pressure of 0.008 MPa. The *net* power output of the cycle is 100 MW. Determine for the cycle **(a)** the thermal efficiency, **(b)** the back work ratio, **(c)** the mass flow rate of the steam, in kg/h, **(d)** the rate of heat transfer, $\dot{Q}_{in}$, into the working fluid as it passes through the boiler, in MW, **(e)** the rate of heat transfer, $\dot{Q}_{out}$, from the condensing steam as it passes through the condenser, in MW, **(f)** the mass flow rate of the condenser cooling water, in kg/h, if cooling water enters the condenser at 15°C and exits at 35°C.

Solution

Known: An ideal Rankine cycle operates with steam as the working fluid. The boiler and condenser pressures are specified, and the net power output is given.

Find: Determine the thermal efficiency, the back work ratio, the mass flow rate of the steam, in kg/h, the rate of heat transfer to the working fluid as it passes through the boiler, in MW, the rate of heat transfer from the condensing steam as it passes through the condenser, in MW, the mass flow rate of the condenser cooling water, which enters at 15°C and exits at 35°C.

Schematic and Given Data:

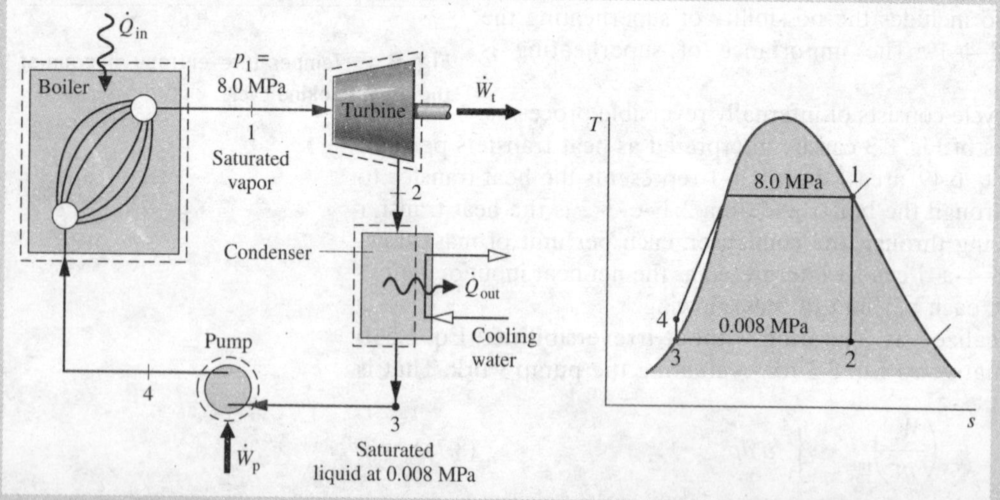

Fig. E8.1

Engineering Model:

1. Each component of the cycle is analyzed as a control volume at steady state. The control volumes are shown on the accompanying sketch by dashed lines.
2. All processes of the working fluid are internally reversible.

3. The turbine and pump operate adiabatically.

4. Kinetic and potential energy effects are negligible.

5. Saturated vapor enters the turbine. Condensate exits the condenser as saturated liquid.

❶ **Analysis:** To begin the analysis, we fix each of the principal states located on the accompanying schematic and T–s diagrams. Starting at the inlet to the turbine, the pressure is 8.0 MPa and the steam is a saturated vapor, so from Table A-3, $h_1 = 2758.0$ kJ/kg and $s_1 = 5.7432$ kJ/kg · K.

State 2 is fixed by $p_2 = 0.008$ MPa and the fact that the specific entropy is constant for the adiabatic, internally reversible expansion through the turbine. Using saturated liquid and saturated vapor data from Table A-3, we find that the quality at state 2 is

$$x_2 = \frac{s_2 - s_f}{s_g - s_f} = \frac{5.7432 - 0.5926}{7.6361} = 0.6745$$

The enthalpy is then

$$h_2 = h_f + x_2 h_{fg} = 173.88 + (0.6745)2403.1$$
$$= 1794.8 \text{ kJ/kg}$$

State 3 is saturated liquid at 0.008 MPa, so $h_3 = 173.88$ kJ/kg.

State 4 is fixed by the boiler pressure p_4 and the specific entropy $s_4 = s_3$. The specific enthalpy h_4 can be found by interpolation in the compressed liquid tables. However, because compressed liquid data are relatively sparse, it is more convenient to solve Eq. 8.3 for h_4, using Eq. 8.7b to approximate the pump work. With this approach

$$h_4 = h_3 + \dot{W}_p/\dot{m} = h_3 + v_3(p_4 - p_3)$$

By inserting property values from Table A-3

$$h_4 = 173.88 \text{ kJ/kg} + (1.0084 \times 10^{-3} \text{ m}^3/\text{kg})(8.0 - 0.008)\text{MPa} \left| \frac{10^6 \text{ N/m}^2}{1 \text{ MPa}} \right| \left| \frac{1 \text{ kJ}}{10^3 \text{ N} \cdot \text{m}} \right|$$
$$= 173.88 + 8.06 = 181.94 \text{ kJ/kg}$$

(a) The *net* power developed by the cycle is

$$\dot{W}_{\text{cycle}} = \dot{W}_t - \dot{W}_p$$

Mass and energy rate balances for control volumes around the turbine and pump give, respectively,

$$\frac{\dot{W}_t}{\dot{m}} = h_1 - h_2 \quad \text{and} \quad \frac{\dot{W}_p}{\dot{m}} = h_4 - h_3$$

where $\dot{m}$ is the mass flow rate of the steam. The rate of heat transfer to the working fluid as it passes through the boiler is determined using mass and energy rate balances as

$$\frac{\dot{Q}_{\text{in}}}{\dot{m}} = h_1 - h_4$$

The thermal efficiency is then

$$\eta = \frac{\dot{W}_t - \dot{W}_p}{\dot{Q}_{\text{in}}} = \frac{(h_1 - h_2) - (h_4 - h_3)}{h_1 - h_4}$$
$$= \frac{[(2758.0 - 1794.8) - (181.94 - 173.88)] \text{ kJ/kg}}{(2758.0 - 181.94) \text{ kJ/kg}}$$
$$= 0.371 \ (37.1\%)$$

(b) The back work ratio is

❷
$$\text{bwr} = \frac{\dot{W}_p}{\dot{W}_t} = \frac{h_4 - h_3}{h_1 - h_2} = \frac{(181.94 - 173.88)\text{ kJ/kg}}{(2758.0 - 1794.8)\text{ kJ/kg}}$$
$$= \frac{8.06}{963.2} = 8.37 \times 10^{-3}\ (0.84\%)$$

(c) The mass flow rate of the steam can be obtained from the expression for the net power given in part (a). Thus

$$\dot{m} = \frac{\dot{W}_{\text{cycle}}}{(h_1 - h_2) - (h_4 - h_3)}$$
$$= \frac{(100\text{ MW})|10^3\text{ kW/MW}|\,|3600\text{ s/h}|}{(963.2 - 8.06)\text{ kJ/kg}}$$
$$= 3.77 \times 10^5\text{ kg/h}$$

(d) With the expression for $\dot{Q}_{\text{in}}$ from part (a) and previously determined specific enthalpy values

$$\dot{Q}_{\text{in}} = \dot{m}(h_1 - h_4)$$
$$= \frac{(3.77 \times 10^5\text{ kg/h})(2758.0 - 181.94)\text{ kJ/kg}}{|3600\text{ s/h}|\,|10^3\text{ kW/MW}|}$$
$$= 269.77\text{ MW}$$

(e) Mass and energy rate balances applied to a control volume enclosing the steam side of the condenser give

$$\dot{Q}_{\text{out}} = \dot{m}(h_2 - h_3)$$
$$= \frac{(3.77 \times 10^5\text{ kg/h})(1794.8 - 173.88)\text{ kJ/kg}}{|3600\text{ s/h}|\,|10^3\text{ kW/MW}|}$$
$$= 169.75\text{ MW}$$

❸ Note that the ratio of $\dot{Q}_{\text{out}}$ to $\dot{Q}_{\text{in}}$ is 0.629 (62.9%).

Alternatively, $\dot{Q}_{\text{out}}$ can be determined from an energy rate balance on the *overall* vapor power plant. At steady state, the net power developed equals the net rate of heat transfer to the plant

$$\dot{W}_{\text{cycle}} = \dot{Q}_{\text{in}} - \dot{Q}_{\text{out}}$$

Rearranging this expression and inserting values

$$\dot{Q}_{\text{out}} = \dot{Q}_{\text{in}} - \dot{W}_{\text{cycle}} = 269.77\text{ MW} - 100\text{ MW} = 169.77\text{ MW}$$

The slight difference from the above value is due to round-off.

(f) Taking a control volume around the condenser, the mass and energy rate balances give at steady state

$$0 = \overset{0}{\cancel{\dot{Q}_{\text{cv}}}} - \overset{0}{\cancel{\dot{W}_{\text{cv}}}} + \dot{m}_{\text{cw}}(h_{\text{cw,in}} - h_{\text{cw,out}}) + \dot{m}(h_2 - h_3)$$

where $\dot{m}_{\text{cw}}$ is the mass flow rate of the cooling water. Solving for $\dot{m}_{\text{cw}}$

$$\dot{m}_{\text{cw}} = \frac{\dot{m}(h_2 - h_3)}{(h_{\text{cw,out}} - h_{\text{cw,in}})}$$

The numerator in this expression is evaluated in part (e). For the cooling water, $h \approx h_f(T)$, so with saturated liquid enthalpy values from Table A-2 at the entering and exiting temperatures of the cooling water

$$\dot{m}_{\text{cw}} = \frac{(169.75\text{ MW})|10^3\text{ kW/MW}|\,|3600\text{ s/h}|}{(146.68 - 62.99)\text{ kJ/kg}} = 7.3 \times 10^6\text{ kg/h}$$

❶ Note that a slightly revised problem-solving methodology is used in this example problem: We begin with a systematic evaluation of the specific enthalpy at each numbered state.

❷ Note that the back work ratio is relatively low for the Rankine cycle. In the present case, the work required to operate the pump is less than 1% of the turbine output.

❸ In this example, 62.9% of the energy added to the working fluid by heat transfer is subsequently discharged to the cooling water. Although considerable energy is carried away by the cooling water, its exergy is small because the water exits at a temperature only a few degrees greater than that of the surroundings. See Sec. 8.6 for further discussion.

Skills Developed: *Ability to...*

✓ sketch the *T-s* diagram of the basic Rankine cycle.

✓ fix each of the principal states and retrieve necessary property data.

✓ apply mass, energy, and entropy balances.

✓ calculate performance parameters for the cycle.

Quick Quiz

If the mass flow rate of steam were 150 kg/s, what would be the net power, in MW, and the thermal efficiency?

Ans. 143.2 MW, 37.1%.

8.2.3 Effects of Boiler and Condenser Pressures on the Rankine Cycle

In discussing Fig. 5.12 (Sec. 5.9.1), we observed that the thermal efficiency of power cycles tends to increase as the average temperature at which energy is added by heat transfer increases and/or the average temperature at which energy is rejected decreases. (For elaboration, see box.) Let us apply this idea to study the effects on performance of the ideal Rankine cycle of changes in the boiler and condenser pressures. Although these findings are obtained with reference to the ideal Rankine cycle, they also hold qualitatively for actual vapor power plants.

Considering the Effect of Temperature on Thermal Efficiency

Since the ideal Rankine cycle consists entirely of internally reversible processes, an expression for thermal efficiency can be obtained in terms of *average* temperatures during the heat interaction processes. Let us begin the development of this expression by recalling that areas under the process lines of Fig. 8.3 can be interpreted as the heat transfer per unit of mass flowing through the respective components. For example, the total area 1–b–c–4–a–1 represents the heat transfer into the working fluid per unit of mass passing through the boiler. In symbols,

$$\left(\frac{\dot{Q}_{in}}{\dot{m}}\right)_{\substack{int\\rev}} = \int_4^1 T\,ds = \text{area 1–b–c–4–a–1}$$

The integral can be written in terms of an average temperature of heat addition, $\overline{T}_{in}$, as follows:

$$\left(\frac{\dot{Q}_{in}}{\dot{m}}\right)_{\substack{int\\rev}} = \overline{T}_{in}(s_1 - s_4)$$

where the overbar denotes *average.* Similarly, area 2–b–c–3–2 represents the heat transfer from the condensing steam per unit of mass passing through the condenser

$$\left(\frac{\dot{Q}_{out}}{\dot{m}}\right)_{\substack{int\\rev}} = T_{out}(s_2 - s_3) = \text{area } 2\text{–b–c–3–2}$$
$$= T_{out}(s_1 - s_4)$$

where T_{out} denotes the temperature on the steam side of the condenser of the ideal Rankine cycle pictured in Fig. 8.3. The thermal efficiency of the ideal Rankine cycle can be expressed in terms of these heat transfers as

$$\eta_{ideal} = 1 - \frac{(\dot{Q}_{out}/\dot{m})_{\substack{int\\rev}}}{(\dot{Q}_{in}/\dot{m})_{\substack{int\\rev}}} = 1 - \frac{T_{out}}{\overline{T}_{in}} \tag{8.8}$$

By the study of Eq. 8.8, we conclude that the thermal efficiency of the ideal cycle tends to increase as the average temperature at which energy is added by heat transfer increases and/or the temperature at which energy is rejected decreases. With similar reasoning, these conclusions can be shown to apply to the other ideal cycles considered in this chapter and the next.

Figure 8.4*a* shows two ideal cycles having the same condenser pressure but different boiler pressures. By inspection, the average temperature of heat addition is seen to be greater for the higher-pressure cycle 1′–2′–3′–4′–1′ than for cycle 1–2–3–4–1. It follows that increasing the boiler pressure of the ideal Rankine cycle tends to increase the thermal efficiency.

Figure 8.4*b* shows two cycles with the same boiler pressure but two different condenser pressures. One condenser operates at atmospheric pressure and the other at *less than* atmospheric pressure. The temperature of heat rejection for cycle 1–2–3–4–1 condensing at atmospheric pressure is 100°C. The temperature of heat rejection for the lower-pressure cycle 1–2″–3″–4″–1 is correspondingly lower, so this cycle has the greater thermal efficiency. It follows that decreasing the condenser pressure tends to increase the thermal efficiency.

The lowest feasible condenser pressure is the saturation pressure corresponding to the ambient temperature, for this is the lowest possible temperature for heat rejection to the surroundings. The goal of maintaining the lowest practical turbine exhaust (condenser) pressure is a primary reason for including the condenser in a power plant. Liquid water at atmospheric pressure could be drawn into the boiler by a pump, and steam could be discharged directly to the atmosphere at the turbine exit. However, by including a condenser in which the steam side is operated at a pressure *below*

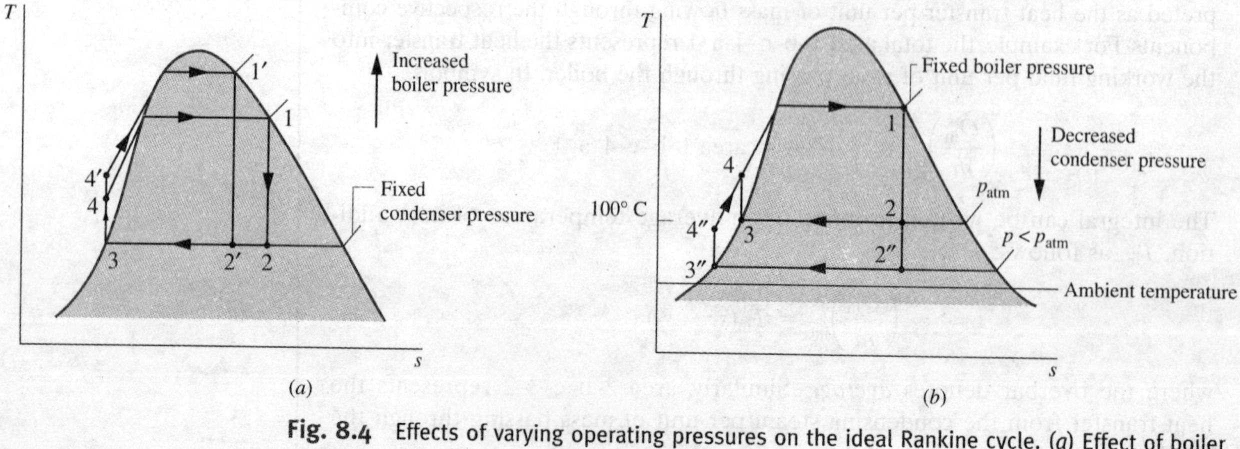

Fig. 8.4 Effects of varying operating pressures on the ideal Rankine cycle. (*a*) Effect of boiler pressure. (*b*) Effect of condenser pressure.

atmospheric, the turbine has a lower-pressure region in which to discharge, resulting in a significant increase in net work and thermal efficiency. The addition of a condenser also allows the working fluid to flow in a closed loop. This arrangement permits continual circulation of the working fluid, so purified water that is less corrosive than tap water can be used economically.

Comparison with Carnot Cycle

Referring to Fig. 8.5, the ideal Rankine cycle 1–2–3–4–4′–1 has a lower thermal efficiency than the Carnot cycle 1–2–3′–4′–1 having the same maximum temperature T_H and minimum temperature T_C because the average temperature between 4 and 4′ is less than T_H. Despite the greater thermal efficiency of the Carnot cycle, it has two shortcomings as a model for the simple vapor power cycle. First, the heat passing to the working fluid of a vapor power plant is usually obtained from hot products of combustion cooling at approximately constant pressure. To exploit fully the energy released on combustion, the hot products should be cooled as much as possible. The first portion of the heating process of the Rankine cycle shown in Fig. 8.5, Process 4–4′, is achieved by cooling the combustion products *below* the maximum temperature T_H. With the Carnot cycle, however, the combustion products would be cooled *at the most* to T_H. Thus, a smaller portion of the energy released on combustion would be used. The second shortcoming of the Carnot vapor power cycle involves the pumping process. Note that the state 3′ of Fig. 8.5 is a two-phase liquid–vapor mixture. Significant practical problems are encountered in developing pumps that handle two-phase mixtures, as would be required by Carnot cycle 1–2–3′–4′–1. It is far easier to condense the vapor completely and handle only liquid in the pump, as is done in the Rankine cycle. Pumping from 3 to 4 and constant-pressure heating without work from 4 to 4′ are processes that can be closely achieved in practice.

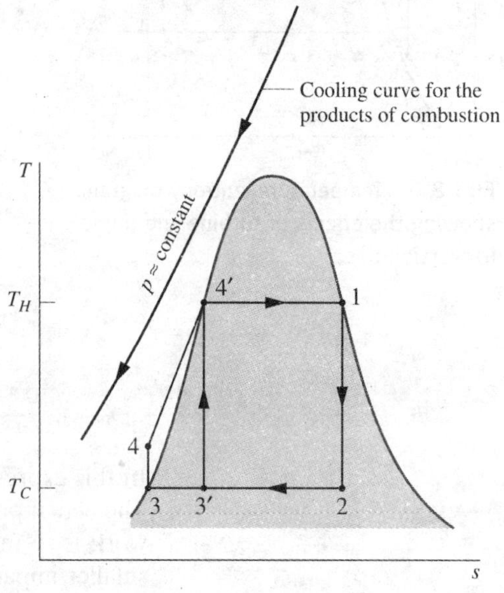

Fig. 8.5 Illustration used to compare the ideal Rankine cycle with the Carnot cycle.

8.2.4 Principal Irreversibilities and Losses

Irreversibilities and losses are associated with each of the four subsystems shown in Fig. 8.1. Some of these effects have a more pronounced influence on performance than others. Let us consider the irreversibilities and losses associated with the Rankine cycle.

Turbine

The principal irreversibility experienced by the working fluid is associated with the expansion through the turbine. Heat transfer from the turbine to the surroundings represents a loss, but since it is usually of secondary importance, this loss is ignored in subsequent discussions. As illustrated by Process 1–2 of Fig. 8.6, an actual adiabatic expansion through the turbine is accompanied by an increase in entropy. The work developed per unit of mass in this process is less than for the corresponding isentropic expansion 1–2s. The isentropic turbine efficiency η_t introduced in Sec. 6.12 allows the effect of irreversibilities within the turbine to be accounted for in terms of the actual and isentropic work amounts. Designating the states as in Fig. 8.6, the isentropic turbine efficiency is

$$\eta_t = \frac{(\dot{W}_t/\dot{m})}{(\dot{W}_t/\dot{m})_s} = \frac{h_1 - h_2}{h_1 - h_{2s}} \tag{8.9}$$

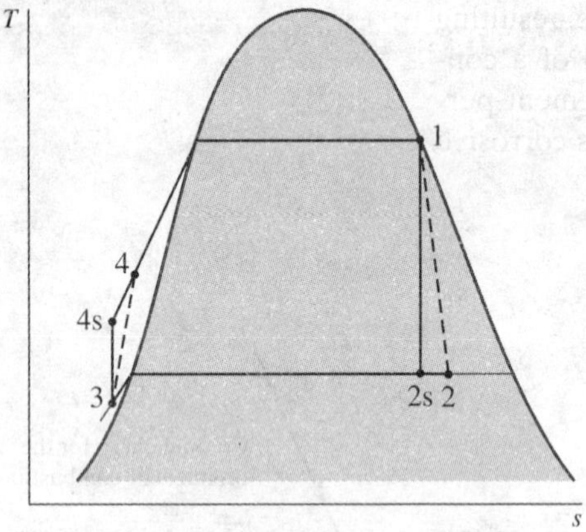

Fig. 8.6 Temperature–entropy diagram showing the effects of turbine and pump irreversibilities.

where the numerator is the actual work developed per unit of mass passing through the turbine and the denominator is the work for an isentropic expansion from the turbine inlet state to the turbine exhaust pressure. Irreversibilities within the turbine significantly reduce the net power output of the plant.

Pump

The work input to the pump required to overcome frictional effects also reduces the net power output of the plant. In the absence of heat transfer to the surroundings, there would be an increase in entropy across the pump. Process 3–4 of Fig. 8.6 illustrates the actual pumping process. The work input for this process is *greater* than for the corresponding isentropic process 3–4s. The isentropic pump efficiency η_p introduced in Sec. 6.12 allows the effect of irreversibilities within the pump to be accounted for in terms of the actual and isentropic work amounts. Designating the states as in Fig. 8.6, the isentropic pump efficiency is

$$\eta_p = \frac{(\dot{W}_p/\dot{m})_s}{(\dot{W}_p/\dot{m})} = \frac{h_{4s} - h_3}{h_4 - h_3} \tag{8.10}$$

In this expression, the pump work for the isentropic process appears in the numerator. The actual pump work, being the larger quantity, is the denominator. Because the pump work is so much less than the turbine work, irreversibilities in the pump have a much smaller impact on the net work of the cycle than do irreversibilities in the turbine.

Other Nonidealities

The turbine and pump irreversibilities mentioned above are *internal* irreversibilities experienced by the working fluid as it flows around the closed loop of the Rankine cycle. The most significant sources of irreversibility for a fossil-fueled vapor power plant, however, are associated with the combustion of the fuel and the subsequent heat transfer from the hot combustion products to the cycle working fluid. These effects occur in the surroundings of the subsystem labeled A on Fig. 8.1 and thus are *external* irreversibilities for the Rankine cycle. These irreversibilities are considered further in Sec. 8.6 and Chap. 13 using the exergy concept.

Another effect that occurs in the surroundings is the energy discharge to the cooling water as the working fluid condenses. Although considerable energy is carried away by the cooling water, its *utility* is extremely limited. For condensers in which steam condenses near the ambient temperature, the cooling water experiences a temperature rise of only a few degrees over the temperature of the surroundings in passing through the condenser and thus has limited usefulness. Accordingly, the significance of this loss is far less than suggested by the magnitude of the energy transferred to the cooling water. The utility of condenser cooling water is considered further in Sec. 8.6 using the exergy concept.

In addition to the foregoing, there are several other sources of nonideality. For example, stray heat transfers from the outside surfaces of the plant components have detrimental effects on performance, since such losses reduce the extent of conversion from heat input to work output. Frictional effects resulting in pressure drops are sources of internal irreversibility as the working fluid flows through the boiler, condenser, and piping connecting the various components. Detailed thermodynamic analyses would account for these effects. For simplicity, however, they are ignored in the subsequent discussions. Thus, Fig. 8.6 shows no pressure drops for flow through the boiler and condenser or between plant components. Another effect on performance is suggested by the placement of state 3 on Fig. 8.6. At this state, the temperature of the working fluid exiting the condenser would be lower than the saturation temperature corresponding to the condenser pressure. This is disadvantageous because a greater heat transfer would be required in the boiler to bring the water to saturation.

In the next example, the ideal Rankine cycle of Example 8.1 is modified to include the effects of irreversibilities in the turbine and pump.

Example 8.2 RANKINE CYCLE WITH IRREVERSIBILITIES

Reconsider the vapor power cycle of Example 8.1, but include in the analysis that the turbine and the pump each have an isentropic efficiency of 85%. Determine for the modified cycle **(a)** the thermal efficiency, **(b)** the mass flow rate of steam, in kg/h, for a net power output of 100 MW, **(c)** the rate of heat transfer $\dot{Q}_{in}$ into the working fluid as it passes through the boiler, in MW, **(d)** the rate of heat transfer $\dot{Q}_{out}$ from the condensing steam as it passes through the condenser, in MW, **(e)** the mass flow rate of the condenser cooling water, in kg/h, if cooling water enters the condenser at 15°C and exits as 35°C. Discuss the effects on the vapor cycle of irreversibilities within the turbine and pump.

Solution

Known: A vapor power cycle operates with steam as the working fluid. The turbine and pump both have efficiencies of 85%.

Find: Determine the thermal efficiency, the mass flow rate, in kg/h, the rate of heat transfer to the working fluid as it passes through the boiler, in MW, the heat transfer rate from the condensing steam as it passes through the condenser, in MW, and the mass flow rate of the condenser cooling water, in kg/h. Discuss.

Schematic and Given Data:

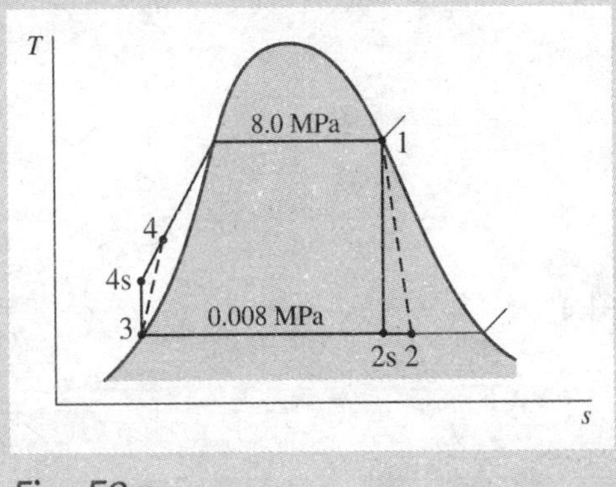

Fig. E8.2

Engineering Model:

1. Each component of the cycle is analyzed as a control volume at steady state.
2. The working fluid passes through the boiler and condenser at constant pressure. Saturated vapor enters the turbine. The condensate is saturated at the condenser exit.
3. The turbine and pump each operate adiabatically with an efficiency of 85%.
4. Kinetic and potential energy effects are negligible.

Analysis: Owing to the presence of irreversibilities during the expansion of the steam through the turbine, there is an increase in specific entropy from turbine inlet to exit, as shown on the accompanying T–s diagram. Similarly, there is an increase in specific entropy from pump inlet to exit. Let us begin the analysis by fixing each of the principal states. State 1 is the same as in Example 8.1, so $h_1 = 2758.0$ kJ/kg and $s_1 = 5.7432$ kJ/kg · K.

The specific enthalpy at the turbine exit, state 2, can be determined using the turbine efficiency

$$\eta_t = \frac{\dot{W}_t/\dot{m}}{(\dot{W}_t/\dot{m})_s} = \frac{h_1 - h_2}{h_1 - h_{2s}}$$

where h_{2s} is the specific enthalpy at state 2s on the accompanying T–s diagram. From the solution to Example 8.1, $h_{2s} = 1794.8$ kJ/kg. Solving for h_2 and inserting known values

$$h_2 = h_1 - \eta_t(h_1 - h_{2s})$$
$$= 2758 - 0.85(2758 - 1794.8) = 1939.3 \text{ kJ/kg}$$

State 3 is the same as in Example 8.1, so $h_3 = 173.88$ kJ/kg.

To determine the specific enthalpy at the pump exit, state 4, reduce mass and energy rate balances for a control volume around the pump to obtain $\dot{W}_p/\dot{m} = h_4 - h_3$. On rearrangement, the specific enthalpy at state 4 is

$$h_4 = h_3 + \dot{W}_p/\dot{m}$$

To determine h_4 from this expression requires the pump work, which can be evaluated using the pump efficiency η_p, as follows. By definition

$$\eta_p = \frac{(\dot{W}_p/\dot{m})_s}{(\dot{W}_p/\dot{m})}$$

The term $(\dot{W}_p/\dot{m})_s$ can be evaluated using Eq. 8.7b. Then solving for $\dot{W}_p/\dot{m}$ results in

$$\frac{\dot{W}_p}{\dot{m}} = \frac{v_3(p_4 - p_3)}{\eta_p}$$

The numerator of this expression was determined in the solution to Example 8.1. Accordingly,

$$\frac{\dot{W}_p}{\dot{m}} = \frac{8.06 \text{ kJ/kg}}{0.85} = 9.48 \text{ kJ/kg}$$

The specific enthalpy at the pump exit is then

$$h_4 = h_3 + \dot{W}_p/\dot{m} = 173.88 + 9.48 = 183.36 \text{ kJ/kg}$$

(a) The net power developed by the cycle is

$$\dot{W}_{\text{cycle}} = \dot{W}_t - \dot{W}_p = \dot{m}[(h_1 - h_2) - (h_4 - h_3)]$$

The rate of heat transfer to the working fluid as it passes through the boiler is

$$\dot{Q}_{\text{in}} = \dot{m}(h_1 - h_4)$$

Thus, the thermal efficiency is

$$\eta = \frac{(h_1 - h_2) - (h_4 - h_3)}{h_1 - h_4}$$

Inserting values

$$\eta = \frac{(2758 - 1939.3) - 9.48}{2758 - 183.36} = 0.314 \ (31.4\%)$$

(b) With the net power expression of part (a), the mass flow rate of the steam is

$$\dot{m} = \frac{\dot{W}_{\text{cycle}}}{(h_1 - h_2) - (h_4 - h_3)}$$

$$= \frac{(100 \text{ MW})|3600 \text{ s/h}||10^3 \text{ kW/MW}|}{(818.7 - 9.48) \text{ kJ/kg}} = 4.449 \times 10^5 \text{ kg/h}$$

(c) With the expression for $\dot{Q}_{\text{in}}$ from part (a) and previously determined specific enthalpy values

$$\dot{Q}_{\text{in}} = \dot{m}(h_1 - h_4)$$

$$= \frac{(4.449 \times 10^5 \text{ kg/h})(2758 - 183.36) \text{ kJ/kg}}{|3600 \text{ s/h}||10^3 \text{ kW/MW}|} = 318.2 \text{ MW}$$

(d) The rate of heat transfer from the condensing steam to the cooling water is

$$\dot{Q}_{\text{out}} = \dot{m}(h_2 - h_3)$$

$$= \frac{(4.449 \times 10^5 \text{ kg/h})(1939.3 - 173.88) \text{ kJ/kg}}{|3600 \text{ s/h}||10^3 \text{ kW/MW}|} = 218.2 \text{ MW}$$

(e) The mass flow rate of the cooling water can be determined from

$$\dot{m}_{cw} = \frac{\dot{m}(h_2 - h_3)}{(h_{cw,out} - h_{cw,in})}$$

$$= \frac{(218.2\,\text{MW})|10^3\,\text{kW/MW}||3600\,\text{s/h}|}{(146.68 - 62.99)\,\text{kJ/kg}} = 9.39 \times 10^6\,\text{kg/h}$$

Skills Developed: *Ability to...*

- ✓ sketch the *T-s* diagram of the Rankine cycle with turbine and pump irreversiblities.
- ✓ fix each of the principal states and retrieve necessary property data.
- ✓ apply mass, energy, and entropy balances.
- ✓ calculate performance parameters for the cycle.

Quick Quiz

If the mass flow rate of steam were 150 kg/s, what would be the pump power required, in kW, and the back work ratio?

Ans. 1422 kW, 0.0116.

The effect of irreversibilities within the turbine and pump can be gauged by comparing the present values with their counterparts in Example 8.1. In this example, the turbine work per unit of mass is less and the pump work per unit of mass is greater than in Example 8.1. The thermal efficiency in the present case is less than in the ideal case of the previous example. For a fixed net power output (100 MW), the smaller net work output per unit mass in the present case dictates a greater mass flow rate of steam. The magnitude of the heat transfer to the cooling water is greater in this example than in Example 8.1; consequently, a greater mass flow rate of cooling water would be required.

8.3 Improving Performance—Superheat and Reheat

The representations of the vapor power cycle considered thus far do not depict actual vapor power plants faithfully, for various modifications are usually incorporated to improve overall performance. In this section we consider two cycle modifications known as *superheat* and *reheat*. Both features are normally incorporated into vapor power plants.

Let us begin the discussion by noting that an increase in the boiler pressure or a decrease in the condenser pressure may result in a reduction of the steam quality at the exit of the turbine. This can be seen by comparing states 2′ and 2″ of Figs. 8.4*a* and 8.4*b* to the corresponding state 2 of each diagram. If the quality of the mixture passing through the turbine becomes too low, the impact of liquid droplets in the flowing liquid–vapor mixture can erode the turbine blades, causing a decrease in the turbine efficiency and an increased need for maintenance. Accordingly, common practice is to maintain at least 90% quality ($x \geqq 0.9$) at the turbine exit. The cycle modifications known as *superheat* and *reheat* permit advantageous operating pressures in the boiler and condenser and yet offset the problem of low quality of the turbine exhaust.

Superheat

First, let us consider ***superheat.*** As we are not limited to having saturated vapor at the turbine inlet, further energy can be added by heat transfer to the steam, ***superheat*** bringing it to a superheated vapor condition at the turbine inlet. This is accomplished in a separate heat exchanger called a superheater. The combination of boiler and superheater is referred to as a *steam generator*. Figure 8.3 shows an ideal Rankine cycle with superheated vapor at the turbine inlet: cycle 1′–2′–3–4–1′. The cycle with superheat has a higher average temperature of heat addition than the cycle without superheating (cycle 1–2–3–4–1), so the thermal efficiency is higher. Moreover, the quality at turbine exhaust state 2′ is greater than at state 2, which would be the turbine exhaust state without superheating. Accordingly, superheating also tends to alleviate the problem of low steam quality at the turbine exhaust. With sufficient superheating, the turbine exhaust state may even fall in the superheated vapor region.

Reheat

reheat

A further modification normally employed in vapor power plants is ***reheat.*** With reheat, a power plant can take advantage of the increased efficiency that results with higher boiler pressures and yet avoid low-quality steam at the turbine exhaust. In the ideal reheat cycle shown in Fig. 8.7, the steam does not expand to the condenser pressure in a single stage. The steam expands through a first-stage turbine (Process 1–2) to some pressure between the steam generator and condenser pressures. The steam is then reheated in the steam generator (Process 2–3). Ideally, there would be no pressure drop as the steam is reheated. After reheating, the steam expands in a second-stage turbine to the condenser pressure (Process 3–4). The principal advantage of reheat is to increase the quality of the steam at the turbine exhaust. This can be seen from the *T–s* diagram of Fig. 8.7 by comparing state 4 with state 4′, the turbine exhaust state without reheating. When computing the thermal efficiency of a reheat cycle, it is necessary to account for the work output of both turbine stages as well as the total heat addition occurring in the vaporization/superheating and reheating processes. This calculation is illustrated in Example 8.3.

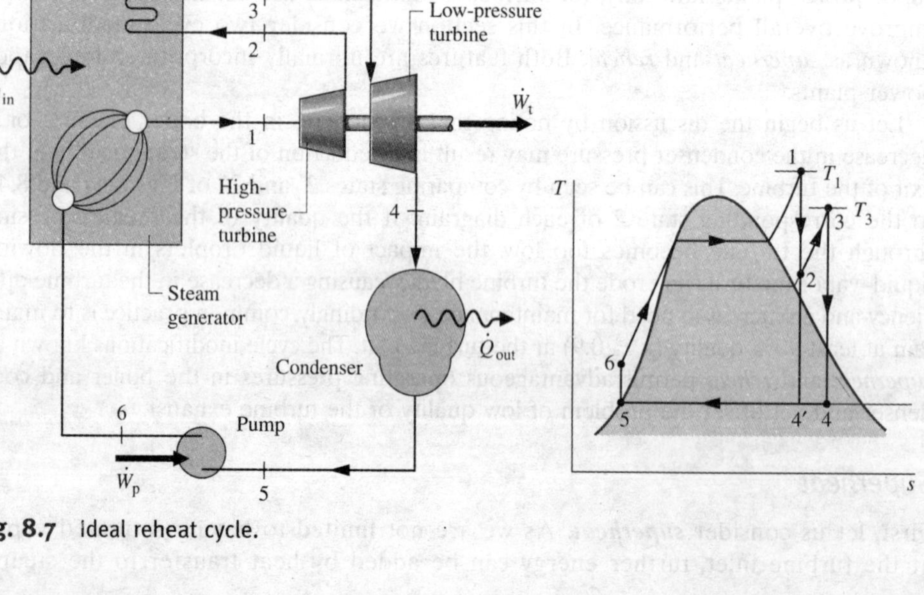

Fig. 8.7 Ideal reheat cycle.

Supercritical Cycle

The temperature of the steam entering the turbine is restricted by metallurgical limitations imposed by the materials used to fabricate the superheater, reheater, and turbine. High pressure in the steam generator also requires piping that can withstand great stresses at elevated temperatures. Although these factors limit the gains that can be realized through superheating and reheating, improved materials and methods of fabrication have permitted significant increases over the years in the maximum allowed cycle temperatures and steam generator pressures, with corresponding increases in thermal efficiency. This has progressed to the extent that vapor power plants can be designed to operate with steam generator pressures exceeding the critical pressure of water (22.1 MPa), and turbine inlet temperatures exceeding 600°C. Figure 8.8 shows an ideal reheat cycle with a supercritical steam generator pressure. Observe that no phase change occurs during the heat addition process from 6 to 1.

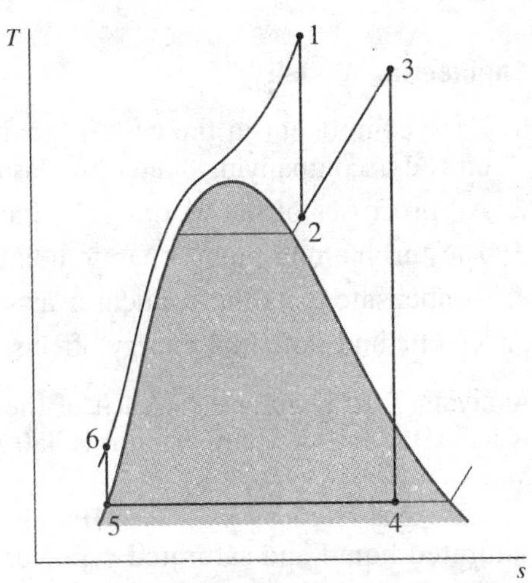

Fig. 8.8 Supercritical ideal reheat cycle.

In the next example, the ideal Rankine cycle of Example 8.1 is modified to include superheat and reheat.

Example 8.3 IDEAL REHEAT CYCLE

Steam is the working fluid in an ideal Rankine cycle with superheat and reheat. Steam enters the first-stage turbine at 8.0 MPa, 480°C, and expands to 0.7 MPa. It is then reheated to 440°C before entering the second-stage turbine, where it expands to the condenser pressure of 0.008 MPa. The *net* power output is 100 MW. Determine **(a)** the thermal efficiency of the cycle, **(b)** the mass flow rate of steam, in kg/h, **(c)** the rate of heat transfer $\dot{Q}_{out}$ from the condensing steam as it passes through the condenser, in MW. Discuss the effects of reheat on the vapor power cycle.

Solution

Known: An ideal reheat cycle operates with steam as the working fluid. Operating pressures and temperatures are specified, and the net power output is given.

Find: Determine the thermal efficiency, the mass flow rate of the steam, in kg/h, and the heat transfer rate from the condensing steam as it passes through the condenser, in MW. Discuss.

Schematic and Given Data:

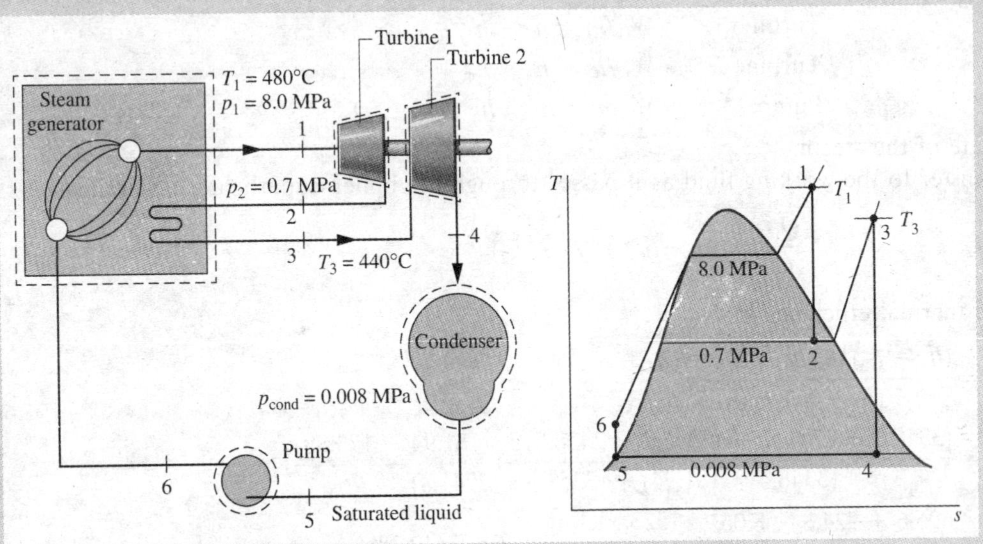

Fig. E8.3

Engineering Model:

1. Each component in the cycle is analyzed as a control volume at steady state. The control volumes are shown on the accompanying sketch by dashed lines.
2. All processes of the working fluid are internally reversible.
3. The turbine and pump operate adiabatically.
4. Condensate exits the condenser as saturated liquid.
5. Kinetic and potential energy effects are negligible.

Analysis: To begin, we fix each of the principal states. Starting at the inlet to the first turbine stage, the pressure is 8.0 MPa and the temperature is 480°C, so the steam is a superheated vapor. From Table A-4, $h_1 = 3348.4$ kJ/kg and $s_1 = 6.6586$ kJ/kg · K.

State 2 is fixed by $p_2 = 0.7$ MPa and $s_2 = s_1$ for the isentropic expansion through the first-stage turbine. Using saturated liquid and saturated vapor data from Table A-3, the quality at state 2 is

$$x_2 = \frac{s_2 - s_f}{s_g - s_f} = \frac{6.6586 - 1.9922}{6.708 - 1.9922} = 0.9895$$

The specific enthalpy is then

$$\begin{aligned} h_2 &= h_f + x_2 h_{fg} \\ &= 697.22 + (0.9895)2066.3 = 2741.8\,\text{kJ/kg} \end{aligned}$$

State 3 is superheated vapor with $p_3 = 0.7$ MPa and $T_3 = 440$°C, so from Table A-4, $h_3 = 3353.3$ kJ/kg and $s_3 = 7.7571$ kJ/kg · K.

To fix state 4, use $p_4 = 0.008$ MPa and $s_4 = s_3$ for the isentropic expansion through the second-stage turbine. With data from Table A-3, the quality at state 4 is

$$x_4 = \frac{s_4 - s_f}{s_g - s_f} = \frac{7.7571 - 0.5926}{8.2287 - 0.5926} = 0.9382$$

The specific enthalpy is

$$h_4 = 173.88 + (0.9382)2403.1 = 2428.5\text{ kJ/kg}$$

State 5 is saturated liquid at 0.008 MPa, so $h_5 = 173.88$ kJ/kg. Finally, the state at the pump exit is the same as in Example 8.1, so $h_6 = 181.94$ kJ/kg.

(a) The *net* power developed by the cycle is

$$\dot{W}_{\text{cycle}} = \dot{W}_{t1} + \dot{W}_{t2} - \dot{W}_p$$

Mass and energy rate balances for the two turbine stages and the pump reduce to give, respectively

Turbine 1:	$\dot{W}_{t1}/\dot{m} = h_1 - h_2$
Turbine 2:	$\dot{W}_{t2}/\dot{m} = h_3 - h_4$
Pump:	$\dot{W}_p/\dot{m} = h_6 - h_5$

where $\dot{m}$ is the mass flow rate of the steam.

The total rate of heat transfer to the working fluid as it passes through the boiler–superheater and reheater is

$$\frac{\dot{Q}_{\text{in}}}{\dot{m}} = (h_1 - h_6) + (h_3 - h_2)$$

Using these expressions, the thermal efficiency is

$$\begin{aligned} \eta &= \frac{(h_1 - h_2) + (h_3 - h_4) - (h_6 - h_5)}{(h_1 - h_6) + (h_3 - h_2)} \\ &= \frac{(3348.4 - 2741.8) + (3353.3 - 2428.5) - (181.94 - 173.88)}{(3348.4 - 181.94) + (3353.3 - 2741.8)} \\ &= \frac{606.6 + 924.8 - 8.06}{3166.5 + 611.5} = \frac{1523.3\text{ kJ/kg}}{3778\text{ kJ/kg}} = 0.403\,(40.3\%) \end{aligned}$$

(b) The mass flow rate of the steam can be obtained with the expression for net power given in part (a).

$$\dot{m} = \frac{\dot{W}_{\text{cycle}}}{(h_1 - h_2) + (h_3 - h_4) - (h_6 - h_5)}$$

$$= \frac{(100 \text{ MW})|3600 \text{ s/h}||10^3 \text{ kW/MW}|}{(606.6 + 924.8 - 8.06) \text{ kJ/kg}} = 2.363 \times 10^5 \text{ kg/h}$$

(c) The rate of heat transfer from the condensing steam to the cooling water is

$$\dot{Q}_{\text{out}} = \dot{m}(h_4 - h_5)$$

$$= \frac{2.363 \times 10^5 \text{ kg/h } (2428.5 - 173.88) \text{ kJ/kg}}{|3600 \text{ s/h}||10^3 \text{ kW/MW}|} = 148 \text{ MW}$$

To see the effects of reheat, we compare the present values with their counterparts in Example 8.1. With superheat and reheat, the thermal efficiency is increased over that of the cycle of Example 8.1. For a specified net power output (100 MW), a larger thermal efficiency means that a smaller mass flow rate of steam is required. Moreover, with a greater thermal efficiency the rate of heat transfer to the cooling water is also less, resulting in a reduced demand for cooling water. With reheating, the steam quality at the turbine exhaust is substantially increased over the value for the cycle of Example 8.1.

Skills Developed: *Ability to...*
- ✓ sketch the *T-s* diagram of the ideal Rankine cycle with reheat.
- ✓ fix each of the principal states and retrieve necessary property data.
- ✓ apply mass, energy, and entropy balances.
- ✓ calculate performance parameters for the cycle.

Quick Quiz

What is the rate of heat addition for the reheat process, in MW, and what percent is that value of the total heat addition to the cycle?

Ans. 40.1 MW, 16.2%.

The following example illustrates the effect of turbine irreversibilities on the ideal reheat cycle of Example 8.3.

Example 8.4 REHEAT CYCLE WITH TURBINE IRREVERSIBILITY

Reconsider the reheat cycle of Example 8.3, but include in the analysis that each turbine stage has the same isentropic efficiency. **(a)** If $\eta_t = 85\%$, determine the thermal efficiency. **(b)** Plot the thermal efficiency versus turbine stage efficiency ranging from 85 to 100%.

Solution

Known: A reheat cycle operates with steam as the working fluid. Operating pressures and temperatures are specified. Each turbine stage has the same isentropic efficiency.

Find: If $\eta_t = 85\%$, determine the thermal efficiency. Also, plot the thermal efficiency versus turbine stage efficiency ranging from 85 to 100%.

Schematic and Given Data:

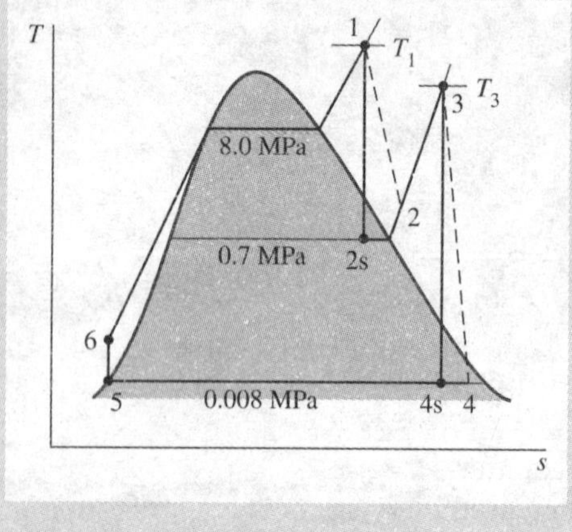

Engineering Model:

1. As in Example 8.3, each component is analyzed as a control volume at steady state.
2. Except for the two turbine stages, all processes are internally reversible.
3. The turbine and pump operate adiabatically.
4. The condensate exits the condenser as saturated liquid.
5. Kinetic and potential energy effects are negligible.

Fig. E8.4a

Analysis:

(a) From the solution to Example 8.3, the following specific enthalpy values are known, in kJ/kg: $h_1 = 3348.4$, $h_{2s} = 2741.8$, $h_3 = 3353.3$, $h_{4s} = 2428.5$, $h_5 = 173.88$, $h_6 = 181.94$.

The specific enthalpy at the exit of the first-stage turbine, h_2, can be determined by solving the expression for the turbine efficiency to obtain

$$\begin{aligned} h_2 &= h_1 - \eta_t(h_1 - h_{2s}) \\ &= 3348.4 - 0.85(3348.4 - 2741.8) = 2832.8 \text{ kJ/kg} \end{aligned}$$

The specific enthalpy at the exit of the second-stage turbine can be found similarly:

$$\begin{aligned} h_4 &= h_3 - \eta_t(h_3 - h_{4s}) \\ &= 3353.3 - 0.85(3353.3 - 2428.5) = 2567.2 \text{ kJ/kg} \end{aligned}$$

The thermal efficiency is then

1

$$\begin{aligned} \eta &= \frac{(h_1 - h_2) + (h_3 - h_4) - (h_6 - h_5)}{(h_1 - h_6) + (h_3 - h_2)} \\ &= \frac{(3348.4 - 2832.8) + (3353.3 - 2567.2) - (181.94 - 173.88)}{(3348.4 - 181.94) + (3353.3 - 2832.8)} \\ &= \frac{1293.6 \text{ kJ/kg}}{3687.0 \text{ kJ/kg}} = 0.351\ (35.1\%) \end{aligned}$$

(b) The *IT* code for the solution follows, where etat1 is η_{t1}, etat2 is η_{t2}, eta is η, Wnet $= \dot{W}_{net}/\dot{m}$, and Qin $= \dot{Q}_{in}/\dot{m}$.

```
// Fix the states
T1 = 480// °C
p1 = 80 // bar
h1 = h_PT ("Water/Steam", p1, T1)
s1 = s_PT ("Water/Steam", p1, T1)

p2 = 7 // bar
h2s = h_Ps ("Water/Steam" , p2, s1)
etat1 = 0.85
h2 = h1 – etat1 * (h1 – h2s)

T3 = 440 // °C
p3 = p2
h3 = h_PT ("Water/Steam", p3, T3)
s3 = s_PT ("Water/Steam", p3, T3)

p4 = 0.08//bar
h4s = h_Ps ("Water/Steam", p4, s3)
```

```
etat2 = etat1
h4 = h3 – etat2 * (h3 – h4s)

p5 = p4
h5 = hsat_Px ("Water/Steam", p5, 0) // kJ/kg
v5 = vsat_Px ("Water/Steam", p5, 0) // m³/kg

p6 = p1
h6 = h5 + v5 * (p6 – p5) * 100// The 100 in this expression is a unit conversion factor.

// Calculate thermal efficiency
Wnet = (h1 – h2) + (h3 – h4) – (h6 – h5)
Qin = (h1 – h6) + (h3 – h2)
eta = Wnet/Qin
```

Using the **Explore** button, sweep eta from 0.85 to 1.0 in steps of 0.01. Then, using the **Graph** button, obtain the following plot:

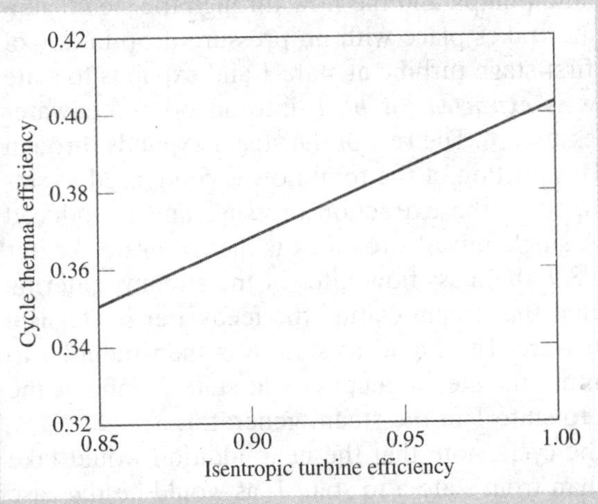

Fig. E8.4b

From Fig. E8.4b, we see that the cycle thermal efficiency increases from 0.351 to 0.403 as turbine stage efficiency increases from 0.85 to 1.00, as expected based on the results of Examples 8.3(a) and 8.4(a). Turbine isentropic efficiency is seen to have a significant effect on cycle thermal efficiency.

Skills Developed: *Ability to...*

- ✓ sketch the *T-s* diagram of the Rankine cycle with reheat, including turbine and pump irreversibilities.
- ✓ fix each of the principal states and retrieve necessary property data.
- ✓ apply mass, energy, and entropy balances.
- ✓ calculate performance parameters for the cycle.

❶ Owing to the irreversibilities present in the turbine stages, the net work per unit of mass developed in the present case is significantly less than in the case of Example 8.3. The thermal efficiency is also considerably less.

Quick Quiz

If the temperature T_3 were increased to 480°C, would you expect the thermal efficiency to increase, decrease, or stay the same?

Ans. Increase.

8.4 Improving Performance—Regenerative Vapor Power Cycle

Another commonly used method for increasing the thermal efficiency of vapor power plants is *regenerative feedwater heating,* or simply **regeneration.** This is the subject of the present section.

regeneration

To introduce the principle underlying regenerative feedwater heating, consider Fig. 8.3 once again. In cycle 1–2–3–4–a–1, the working fluid would enter the boiler as a compressed liquid at state 4 and be heated while in the liquid phase to state a. With regenerative feedwater heating, the working fluid would enter the boiler at a state *between* 4 and a. As a result, the average temperature of heat addition would be increased, thereby tending to increase the thermal efficiency.

8.4.1 Open Feedwater Heaters

open feedwater heater

Let us consider how regeneration can be accomplished using an ***open feedwater heater,*** a direct contact-type heat exchanger in which streams at different temperatures mix to form a stream at an intermediate temperature. Shown in Fig. 8.9 are the schematic diagram and the associated *T–s* diagram for a regenerative vapor power cycle having one open feedwater heater. For this cycle, the working fluid passes isentropically through the turbine stages and pumps, and the flow through the steam generator, condenser, and feedwater heater takes place with no pressure drop in any of these components. Steam enters the first-stage turbine at state 1 and expands to state 2, where a fraction of the total flow is *extracted,* or *bled,* into an open feedwater heater operating at the extraction pressure, p_2. The rest of the steam expands through the second-stage turbine to state 3. This portion of the total flow is condensed to saturated liquid, state 4, and then pumped to the extraction pressure and introduced into the feedwater heater at state 5. A single mixed stream exits the feedwater heater at state 6. For the case shown in Fig. 8.9, the mass flow rates of the streams entering the feedwater heater are chosen so that the stream exiting the feedwater heater is a saturated liquid at the extraction pressure. The liquid at state 6 is then pumped to the steam generator pressure and enters the steam generator at state 7. Finally, the working fluid is heated from state 7 to state 1 in the steam generator.

Referring to the *T–s* diagram of the cycle, note that the heat addition would take place from state 7 to state 1, rather than from state a to state 1, as would be the case without regeneration. Accordingly, the amount of energy that must be supplied from the combustion of a fossil fuel, or another source, to vaporize and superheat the steam would be reduced. This is the desired outcome. Only a portion of the total flow expands through the second-stage turbine (Process 2–3), however, so less work would

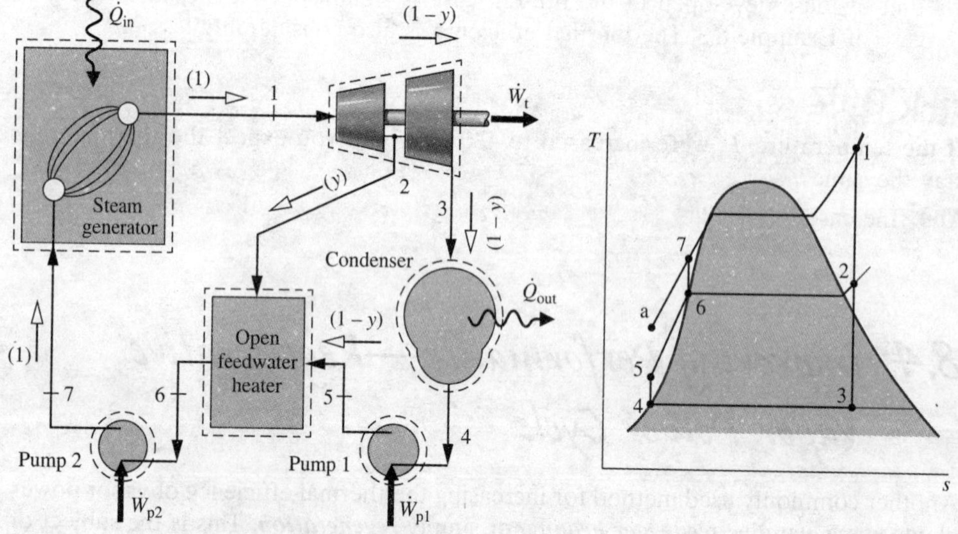

Fig. 8.9 Regenerative vapor power cycle with one open feedwater heater.

be developed as well. In practice, operating conditions are chosen so that the reduction in heat added more than offsets the decrease in net work developed, resulting in an increased thermal efficiency in regenerative power plants.

Cycle Analysis

Consider next the thermodynamic analysis of the regenerative cycle illustrated in Fig. 8.9. An important initial step in analyzing any regenerative vapor cycle is the evaluation of the mass flow rates through each of the components. Taking a single control volume enclosing both turbine stages, the mass rate balance reduces at steady state to

$$\dot{m}_2 + \dot{m}_3 = \dot{m}_1$$

where $\dot{m}_1$ is the rate at which mass enters the first-stage turbine at state 1, $\dot{m}_2$ is the rate at which mass is extracted and exits at state 2, and $\dot{m}_3$ is the rate at which mass exits the second-stage turbine at state 3. Dividing by $\dot{m}_1$ places this on the basis of a *unit of mass* passing through the first-stage turbine

$$\frac{\dot{m}_2}{\dot{m}_1} + \frac{\dot{m}_3}{\dot{m}_1} = 1$$

Denoting the fraction of the total flow extracted at state 2 by y ($y = \dot{m}_2/\dot{m}_1$), the fraction of the total flow passing through the second-stage turbine is

$$\frac{\dot{m}_3}{\dot{m}_1} = 1 - y \qquad (8.11)$$

The fractions of the total flow at various locations are indicated on Fig. 8.9.

The fraction y can be determined by applying the conservation of mass and conservation of energy principles to a control volume around the feedwater heater. Assuming no heat transfer between the feedwater heater and its surroundings and ignoring kinetic and potential energy effects, the mass and energy rate balances reduce at steady state to give

$$0 = yh_2 + (1 - y)h_5 - h_6$$

Solving for y

$$y = \frac{h_6 - h_5}{h_2 - h_5} \qquad (8.12)$$

Equation 8.12 allows the fraction y to be determined when states 2, 5, and 6 are fixed.

Expressions for the principal work and heat transfers of the regenerative cycle can be determined by applying mass and energy rate balances to control volumes around the individual components. Beginning with the turbine, the total work is the sum of the work developed by each turbine stage. Neglecting kinetic and potential energy effects and assuming no heat transfer with the surroundings, we can express the total turbine work on the basis of a unit of mass passing through the first-stage turbine as

$$\frac{\dot{W}_t}{\dot{m}_1} = (h_1 - h_2) + (1 - y)(h_2 - h_3) \qquad (8.13)$$

The total pump work is the sum of the work required to operate each pump individually. On the basis of a unit of mass passing through the first-stage turbine, the total pump work is

$$\frac{\dot{W}_p}{\dot{m}_1} = (h_7 - h_6) + (1 - y)(h_5 - h_4) \qquad (8.14)$$

The energy added by heat transfer to the working fluid passing through the steam generator, per unit of mass expanding through the first-stage turbine, is

$$\frac{\dot{Q}_{in}}{\dot{m}_1} = h_1 - h_7 \tag{8.15}$$

and the energy rejected by heat transfer to the cooling water is

$$\frac{\dot{Q}_{out}}{\dot{m}_1} = (1 - y)(h_3 - h_4) \tag{8.16}$$

The following example illustrates the analysis of a regenerative cycle with one open feedwater heater, including the evaluation of properties at state points around the cycle and the determination of the fractions of the total flow at various locations.

Example 8.5 REGENERATIVE CYCLE WITH OPEN FEEDWATER HEATER

Consider a regenerative vapor power cycle with one open feedwater heater. Steam enters the turbine at 8.0 MPa, 480°C and expands to 0.7 MPa, where some of the steam is extracted and diverted to the open feedwater heater operating at 0.7 MPa. The remaining steam expands through the second-stage turbine to the condenser pressure of 0.008 MPa. Saturated liquid exits the open feedwater heater at 0.7 MPa. The isentropic efficiency of each turbine stage is 85% and each pump operates isentropically. If the net power output of the cycle is 100 MW, determine **(a)** the thermal efficiency and **(b)** the mass flow rate of steam entering the first turbine stage, in kg/h.

Solution

Known: A regenerative vapor power cycle operates with steam as the working fluid. Operating pressures and temperatures are specified; the efficiency of each turbine stage and the net power output are also given.

Find: Determine the thermal efficiency and the mass flow rate into the turbine, in kg/h.

Schematic and Given Data:

❶

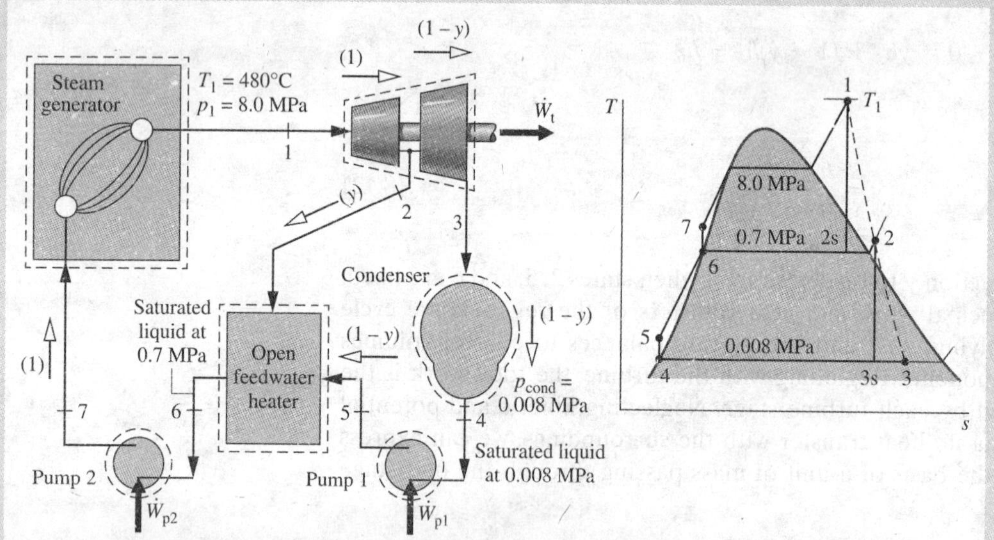

Fig. E8.5

Engineering Model:

1. Each component in the cycle is analyzed as a steady-state control volume. The control volumes are shown in the accompanying sketch by dashed lines.
2. All processes of the working fluid are internally reversible, except for the expansions through the two turbine stages and mixing in the open feedwater heater.
3. The turbines, pumps, and feedwater heater operate adiabatically.

4. Kinetic and potential energy effects are negligible.

5. Saturated liquid exits the open feedwater heater, and saturated liquid exits the condenser.

Analysis: The specific enthalpy at states 1 and 4 can be read from the steam tables. The specific enthalpy at state 2 is evaluated in the solution to Example 8.4. The specific entropy at state 2 can be obtained from the steam tables using the known values of enthalpy and pressure at this state. In summary, $h_1 = 3348.4$ kJ/kg, $h_2 = 2832.8$ kJ/kg, $s_2 = 6.8606$ kJ/kg · K, $h_4 = 173.88$ kJ/kg.

The specific enthalpy at state 3 can be determined using the efficiency of the second-stage turbine

$$h_3 = h_2 - \eta_t(h_2 - h_{3s})$$

With $s_{3s} = s_2$, the quality at state 3s is $x_{3s} = 0.8208$; using this, we get $h_{3s} = 2146.3$ kJ/kg. Hence

$$h_3 = 2832.8 - 0.85(2832.8 - 2146.3) = 2249.3 \text{ kJ/kg}$$

State 6 is saturated liquid at 0.7 MPa. Thus, $h_6 = 697.22$ kJ/kg.

Since the pumps are assumed to operate with no irreversibilities, the specific enthalpy values at states 5 and 7 can be determined as

$$h_5 = h_4 + v_4(p_5 - p_4)$$

$$= 173.88 + (1.0084 \times 10^{-3})(\text{m}^3/\text{kg})(0.7 - 0.008) \text{ MPa} \left|\frac{10^6 \text{ N/m}^2}{1 \text{ MPa}}\right| \left|\frac{1 \text{ kJ}}{10^3 \text{ N} \cdot \text{m}}\right| = 174.6 \text{ kJ/kg}$$

$$h_7 = h_6 + v_6(p_7 - p_6)$$

$$= 697.22 + (1.1080 \times 10^{-3})(8.0 - 0.7)|10^3| = 705.3 \text{ kJ/kg}$$

Applying mass and energy rate balances to a control volume enclosing the open heater, we find the fraction y of the flow extracted at state 2 from

$$y = \frac{h_6 - h_5}{h_2 - h_5} = \frac{697.22 - 174.6}{2832.8 - 174.6} = 0.1966$$

(a) On the basis of a unit of mass passing through the first-stage turbine, the total turbine work output is

$$\frac{\dot{W}_t}{\dot{m}_1} = (h_1 - h_2) + (1 - y)(h_2 - h_3)$$

$$= (3348.4 - 2832.8) + (0.8034)(2832.8 - 2249.3) = 984.4 \text{ kJ/kg}$$

The total pump work per unit of mass passing through the first-stage turbine is

$$\frac{\dot{W}_p}{\dot{m}_1} = (h_7 - h_6) + (1 - y)(h_5 - h_4)$$

$$= (705.3 - 697.22) + (0.8034)(174.6 - 173.88) = 8.7 \text{ kJ/kg}$$

The heat added in the steam generator per unit of mass passing through the first-stage turbine is

$$\frac{\dot{Q}_{in}}{\dot{m}_1} = h_1 - h_7 = 3348.4 - 705.3 = 2643.1 \text{ kJ/kg}$$

The thermal efficiency is then

$$\eta = \frac{\dot{W}_t/\dot{m}_1 - \dot{W}_p/\dot{m}_1}{\dot{Q}_{in}/\dot{m}_1} = \frac{984.4 - 8.7}{2643.1} = 0.369 \, (36.9\%)$$

(b) The mass flow rate of the steam entering the turbine, $\dot{m}_1$, can be determined using the given value for the net power output, 100 MW. Since

$$\dot{W}_{cycle} = \dot{W}_t - \dot{W}_p$$

and

$$\frac{\dot{W}_t}{\dot{m}_1} = 984.4 \text{ kJ/kg} \quad \text{and} \quad \frac{\dot{W}_p}{\dot{m}_1} = 8.7 \text{ kJ/kg}$$

it follows that

$$\dot{m}_1 = \frac{(100 \text{ MW})|3600 \text{ s/h}|}{(984.4 - 8.7) \text{ kJ/kg}} \left| \frac{10^3 \text{ kJ/s}}{1 \text{ MW}} \right| = 3.69 \times 10^5 \text{ kg/h}$$

❶ Note that the fractions of the total flow at various locations are labeled on the figure.

Skills Developed: *Ability to...*
- ✓ sketch the *T-s* diagram of the regenerative vapor power cycle with one open feedwater heater.
- ✓ fix each of the principal states and retrieve necessary property data.
- ✓ apply mass, energy, and entropy balances.
- ✓ calculate performance parameters for the cycle.

Quick Quiz

If the mass flow rate of steam entering the first-stage turbine were 150 kg/s, what would be the net power, in MW, and the fraction of steam extracted, *y*?

Ans. 146.4 MW, 0.1966.

8.4.2 Closed Feedwater Heaters

closed feedwater heater

Regenerative feedwater heating also can be accomplished with ***closed feedwater heaters.*** Closed heaters are shell-and-tube-type recuperators in which the feedwater temperature increases as the extracted steam condenses on the outside of the tubes carrying the feedwater. Since the two streams do not mix, they can be at different pressures. The diagrams of Fig. 8.10 show two different schemes for removing the condensate from closed feedwater heaters. In Fig. 8.10*a*, this is accomplished by means of a pump whose function is to pump the condensate forward to a higher-pressure point in the cycle. In Fig. 8.10*b*, the condensate is allowed to pass through a *trap* into a feedwater heater operating at a lower pressure or into the condenser. A trap is a type of valve that permits only liquid to pass through to a region of lower pressure.

A regenerative vapor power cycle having one closed feedwater heater with the condensate trapped into the condenser is shown schematically in Fig. 8.11. For this cycle, the working fluid passes isentropically through the turbine stages and pumps, and there are no pressure drops accompanying the flow through the other components. The *T–s* diagram shows the principal states of the cycle. The total steam flow expands through the first-stage turbine from state 1 to state 2. At this location, a fraction of the flow is

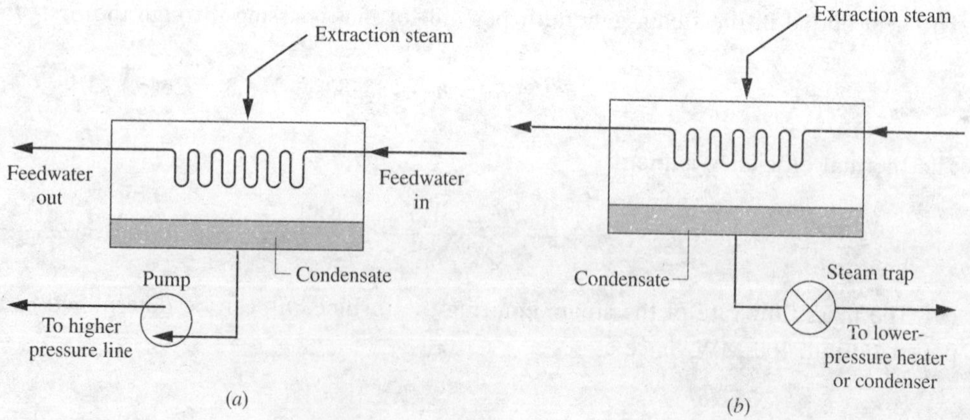

Fig. 8.10 Examples of closed feedwater heaters.

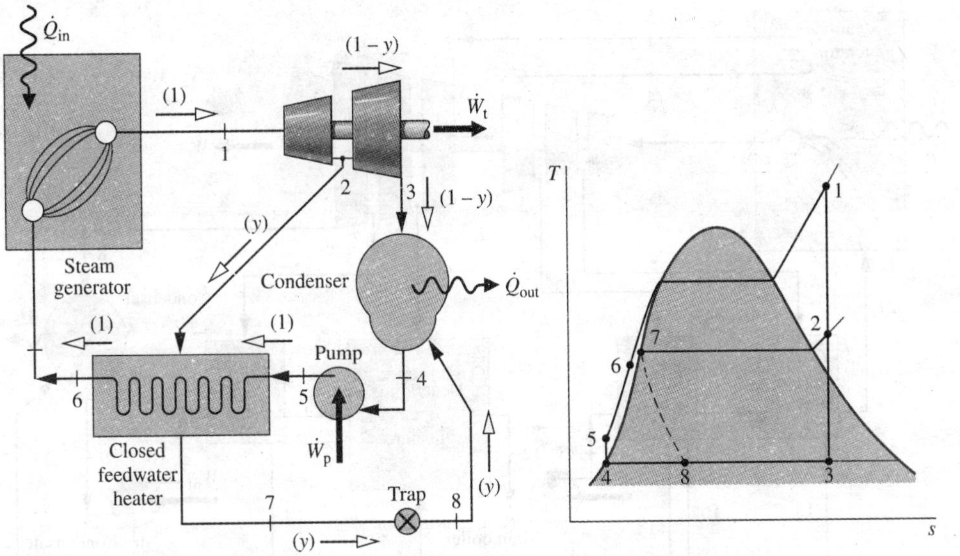

Fig. 8.11 Regenerative vapor power cycle with one closed feedwater heater.

bled into the closed feedwater heater, where it condenses. Saturated liquid at the extraction pressure exits the feedwater heater at state 7. The condensate is then trapped into the condenser, where it is reunited with the portion of the total flow passing through the second-stage turbine. The expansion from state 7 to state 8 through the trap is irreversible, so it is shown by a dashed line on the *T–s* diagram. The total flow exiting the condenser as saturated liquid at state 4 is pumped to the steam generator pressure and enters the feedwater heater at state 5. The temperature of the feedwater is increased in passing through the feedwater heater. The feedwater then exits at state 6. The cycle is completed as the working fluid is heated in the steam generator at constant pressure from state 6 to state 1. Although the closed heater shown on the figure operates with no pressure drop in either stream, there is a source of irreversibility due to the stream-to-stream temperature differences.

Cycle Analysis

The schematic diagram of the cycle shown in Fig. 8.11 is labeled with the fractions of the total flow at various locations. This is usually helpful in analyzing such cycles. The fraction of the total flow extracted, y, can be determined by applying the conservation of mass and conservation of energy principles to a control volume around the closed heater. Assuming no heat transfer between the feedwater heater and its surroundings and neglecting kinetic and potential energy effects, the mass and energy rate balances reduce at steady state to give

$$0 = y(h_2 - h_7) + (h_5 - h_6)$$

Solving for y

$$y = \frac{h_6 - h_5}{h_2 - h_7} \tag{8.17}$$

The principal work and heat transfers are evaluated as discussed previously.

8.4.3 Multiple Feedwater Heaters

The thermal efficiency of the regenerative cycle can be increased by incorporating several feedwater heaters at suitably chosen pressures. The number of feedwater heaters used is based on economic considerations, since incremental increases in ther-

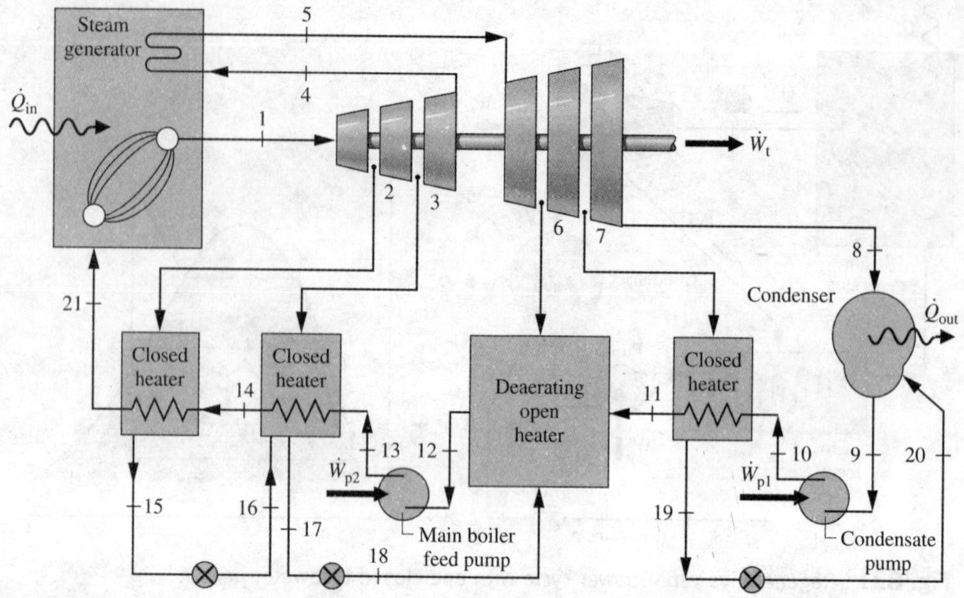

Fig. 8.12 Example of a power plant layout.

mal efficiency achieved with each additional heater must justify the added capital costs (heater, piping, pumps, etc.). Power plant designers use computer programs to simulate the thermodynamic and economic performance of different designs to help them decide on the number of heaters to use, the types of heaters, and the pressures at which they should operate.

Figure 8.12 shows the layout of a power plant with three closed feedwater heaters and one open heater. Power plants with multiple feedwater heaters ordinarily have at least one open feedwater heater operating at a pressure greater than atmospheric pressure so that oxygen and other dissolved gases can be vented from the cycle. This procedure, known as ***deaeration***, is needed to maintain the purity of the working fluid in order to minimize corrosion. Actual power plants have many of the same basic features as the one shown in the figure.

deaeration

In analyzing regenerative vapor power cycles with multiple feedwater heaters, it is good practice to base the analysis on a unit of mass entering the first-stage turbine. To clarify the quantities of matter flowing through the various plant components, the fractions of the total flow removed at each extraction point and the fraction of the total flow remaining at each state point in the cycle should be labeled on a schematic diagram of the cycle. The fractions extracted are determined from mass and energy rate balances for control volumes around each of the feedwater heaters, starting with the highest-pressure heater and proceeding to each lower-pressure heater in turn. This procedure is used in the next example that involves a reheat–regenerative vapor power cycle with two feedwater heaters, one open feedwater heater and one closed feedwater heater.

Example 8.6 REHEAT–REGENERATIVE CYCLE WITH TWO FEEDWATER HEATERS

Consider a reheat–regenerative vapor power cycle with two feedwater heaters, a closed feedwater heater and an open feedwater heater. Steam enters the first turbine at 8.0 MPa, 480°C and expands to 0.7 MPa. The steam is reheated to 440°C before entering the second turbine, where it expands to the condenser pressure of 0.008 MPa. Steam is extracted from the first turbine at 2 MPa and fed to the closed feedwater heater. Feedwater leaves the closed heater at 205°C and 8.0 MPa, and condensate exits as saturated liquid at 2 MPa. The condensate is trapped into the open feedwater heater. Steam extracted from the second turbine at 0.3 MPa is also fed into the open feedwater heater, which operates at 0.3 MPa. The stream exiting the open feedwater heater is saturated liquid at 0.3 MPa. The *net* power output of the cycle is 100 MW. There is no stray heat transfer from any component to its surroundings. If the working fluid experiences no irreversibilities as it passes through the turbines, pumps, steam generator, reheater, and condenser, determine **(a)** the thermal efficiency, **(b)** the mass flow rate of the steam entering the first turbine, in kg/h.

Solution

Known: A reheat–regenerative vapor power cycle operates with steam as the working fluid. Operating pressures and temperatures are specified, and the net power output is given.

Find: Determine the thermal efficiency and the mass flow rate entering the first turbine, in kg/h.

Schematic and Given Data:

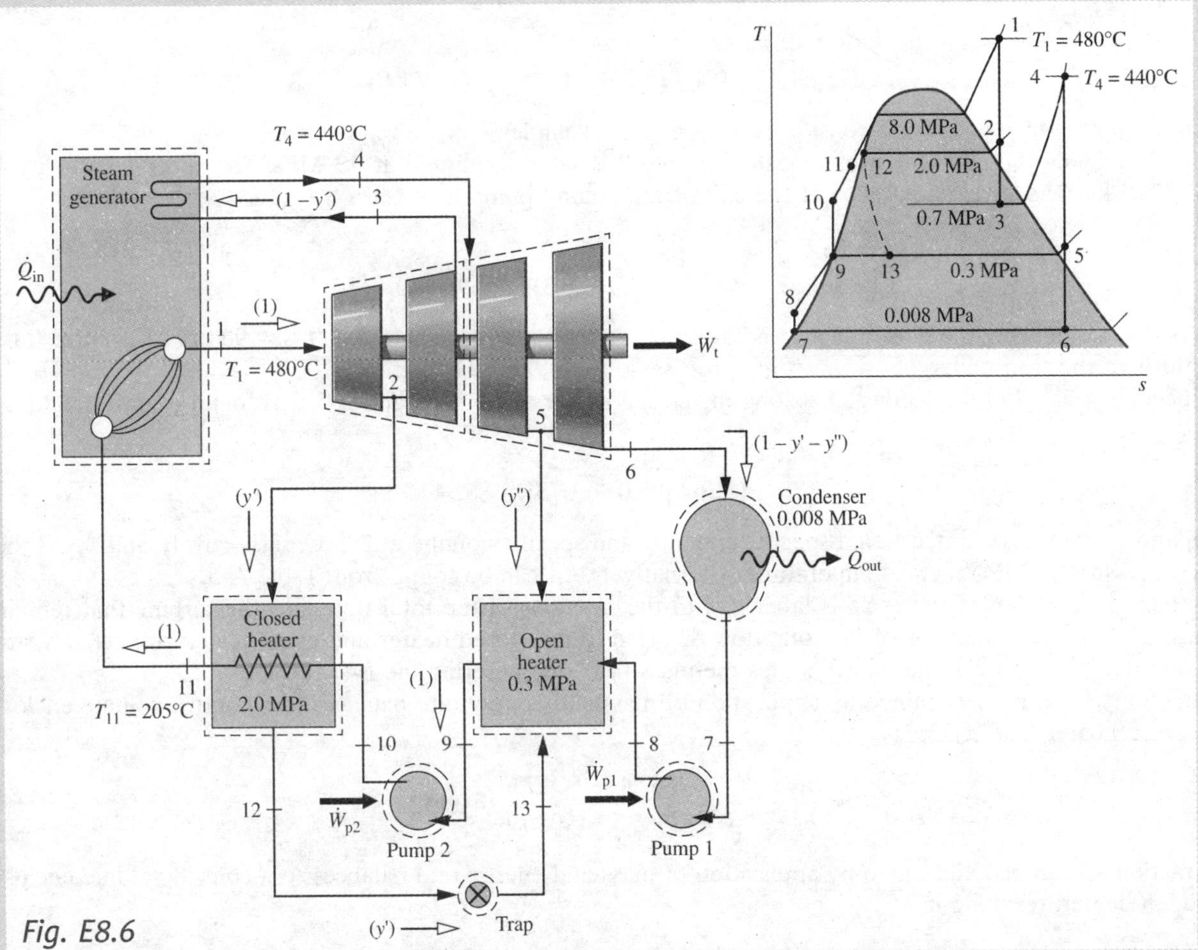

Fig. E8.6

Engineering Model:

1. Each component in the cycle is analyzed as a control volume at steady state. The control volumes are shown on the accompanying sketch by dashed lines.
2. There is no stray heat transfer from any component to its surroundings.
3. The working fluid undergoes internally reversible processes as it passes through the turbines, pumps, steam generator, reheater, and condenser.
4. The expansion through the trap is a *throttling* process.
5. Kinetic and potential energy effects are negligible.
6. Condensate exits the closed heater as a saturated liquid at 2 MPa. Feedwater exits the open heater as a saturated liquid at 0.3 MPa. Condensate exits the condenser as a saturated liquid.

Analysis: Let us determine the specific enthalpies at the principal states of the cycle. State 1 is the same as in Example 8.3, so $h_1 = 3348.4$ kJ/kg and $s_1 = 6.6586$ kJ/kg · K.

State 2 is fixed by $p_2 = 2.0$ MPa and the specific entropy s_2, which is the same as that of state 1. Interpolating in Table A-4, we get $h_2 = 2963.5$ kJ/kg. The state at the exit of the first turbine is the same as at the exit of the first turbine of Example 8.3, so $h_3 = 2741.8$ kJ/kg.

State 4 is superheated vapor at 0.7 MPa, 440°C. From Table A-4, $h_4 = 3353.3$ kJ/kg and $s_4 = 7.7571$ kJ/kg · K. Interpolating in Table A-4 at $p_5 = 0.3$ MPa and $s_5 = s_4 = 7.7571$ kJ/kg · K, the enthalpy at state 5 is $h_5 = 3101.5$ kJ/kg.

Using $s_6 = s_4$, the quality at state 6 is found to be $x_6 = 0.9382$. So

$$\begin{aligned} h_6 &= h_f + x_6 h_{fg} \\ &= 173.88 + (0.9382)2403.1 = 2428.5 \text{ kJ/kg} \end{aligned}$$

At the condenser exit, $h_7 = 173.88$ kJ/kg. The specific enthalpy at the exit of the first pump is

$$\begin{aligned} h_8 &= h_7 + v_7(p_8 - p_7) \\ &= 173.88 + (1.0084)(0.3 - 0.008) = 174.17 \text{ kJ/kg} \end{aligned}$$

The required unit conversions were considered in previous examples.

The liquid leaving the open feedwater heater at state 9 is saturated liquid at 0.3 MPa. The specific enthalpy is $h_9 = 561.47$ kJ/kg. The specific enthalpy at the exit of the second pump is

$$\begin{aligned} h_{10} &= h_9 + v_9(p_{10} - p_9) \\ &= 561.47 + (1.0732)(8.0 - 0.3) = 569.73 \text{ kJ/kg} \end{aligned}$$

The condensate leaving the closed heater is saturated at 2 MPa. From Table A-3, $h_{12} = 908.79$ kJ/kg. The fluid passing through the trap undergoes a throttling process, so $h_{13} = 908.79$ kJ/kg.

The specific enthalpy of the feedwater exiting the closed heater at 8.0 MPa and 205°C is found using Eq. 3.13 as

$$\begin{aligned} h_{11} &= h_f + v_f(p_{11} - p_{sat}) \\ &= 875.1 + (1.1646)(8.0 - 1.73) = 882.4 \text{ kJ/kg} \end{aligned}$$

where h_f and v_f are the saturated liquid specific enthalpy and specific volume at 205°C, respectively, and p_{sat} is the saturation pressure in MPa at this temperature. Alternatively, h_{11} can be found from Table A-5.

The schematic diagram of the cycle is labeled with the fractions of the total flow into the turbine that remain at various locations. The fractions of the total flow diverted to the closed heater and open heater, respectively, are $y' = \dot{m}_2/\dot{m}_1$ and $y'' = \dot{m}_5/\dot{m}_1$, where $\dot{m}_1$ denotes the mass flow rate entering the first turbine.

The fraction y' can be determined by application of mass and energy rate balances to a control volume enclosing the closed heater. The result is

$$y' = \frac{h_{11} - h_{10}}{h_2 - h_{12}} = \frac{882.4 - 569.73}{2963.5 - 908.79} = 0.1522$$

The fraction y'' can be determined by application of mass and energy rate balances to a control volume enclosing the open heater, resulting in

$$0 = y''h_5 + (1 - y' - y'')h_8 + y'h_{13} - h_9$$

Solving for y''

$$\begin{aligned} y'' &= \frac{(1 - y')h_8 + y'h_{13} - h_9}{h_8 - h_5} \\ &= \frac{(0.8478)174.17 + (0.1522)908.79 - 561.47}{174.17 - 3101.5} \\ &= 0.0941 \end{aligned}$$

(a) The following work and heat transfer values are expressed on the basis of a unit mass entering the first turbine. The work developed by the first turbine per unit of mass entering is the sum

$$\frac{\dot{W}_{t1}}{\dot{m}_1} = (h_1 - h_2) + (1 - y')(h_2 - h_3)$$
$$= (3348.4 - 2963.5) + (0.8478)(2963.5 - 2741.8)$$
$$= 572.9 \text{ kJ/kg}$$

Similarly, for the second turbine

$$\frac{\dot{W}_{t2}}{\dot{m}_1} = (1 - y')(h_4 - h_5) + (1 - y' - y'')(h_5 - h_6)$$
$$= (0.8478)(3353.3 - 3101.5) + (0.7537)(3101.5 - 2428.5)$$
$$= 720.7 \text{ kJ/kg}$$

For the first pump

$$\frac{\dot{W}_{p1}}{\dot{m}_1} = (1 - y' - y'')(h_8 - h_7)$$
$$= (0.7537)(174.17 - 173.88) = 0.22 \text{ kJ/kg}$$

and for the second pump

$$\frac{\dot{W}_{p2}}{\dot{m}_1} = (h_{10} - h_9)$$
$$= 569.73 - 561.47 = 8.26 \text{ kJ/kg}$$

The total heat added is the sum of the energy added by heat transfer during boiling/superheating and reheating. When expressed on the basis of a unit of mass entering the first turbine, this is

$$\frac{\dot{Q}_{in}}{\dot{m}_1} = (h_1 - h_{11}) + (1 - y')(h_4 - h_3)$$
$$= (3348.4 - 882.4) + (0.8478)(3353.3 - 2741.8)$$
$$= 2984.4 \text{ kJ/kg}$$

With the foregoing values, the thermal efficiency is

$$\eta = \frac{\dot{W}_{t1}/\dot{m}_1 + \dot{W}_{t2}/\dot{m}_1 - \dot{W}_{p1}/\dot{m}_1 - \dot{W}_{p2}/\dot{m}_1}{\dot{Q}_{in}/\dot{m}_1}$$
$$= \frac{572.9 + 720.7 - 0.22 - 8.26}{2984.4} = 0.431\ (43.1\%)$$

(b) The mass flow rate entering the first turbine can be determined using the given value of the net power output. Thus

$$\dot{m}_1 = \frac{\dot{W}_{cycle}}{\dot{W}_{t1}/\dot{m}_1 + \dot{W}_{t2}/\dot{m}_1 - \dot{W}_{p1}/\dot{m}_1 - \dot{W}_{p2}/\dot{m}_1}$$

❶ $$= \frac{(100\text{ MW})|3600\text{ s/h}||10^3\text{ kW/MW}|}{1285.1\text{ kJ/kg}} = 2.8 \times 10^5 \text{ kg/h}$$

❶ Compared to the corresponding values determined for the simple Rankine cycle of Example 8.1, the thermal efficiency of the present regenerative cycle is substantially greater and the mass flow rate is considerably less.

Skills Developed: *Ability to...*
- ✓ sketch the *T-s* diagram of the reheat–regenerative vapor power cycle with one closed and one open feedwater heater.
- ✓ fix each of the principal states and retrieve necessary property data.
- ✓ apply mass, energy, and entropy balances.
- ✓ calculate performance parameters for the cycle.

Quick Quiz

If each turbine stage had an isentropic efficiency of 85%, at which numbered states would the specific enthalpy values change?
Ans. 2, 3, 5, and 6.

8.5 Other Vapor Cycle Aspects

In this section we consider aspects of vapor cycles related to working fluid characteristics, binary vapor cycles, and cogeneration systems.

Working Fluid Characteristics

Water is used as the working fluid in the vast majority of vapor power systems because it is plentiful and low in cost, nontoxic, chemically stable, and relatively noncorrosive. In addition, water has a relatively large change in specific enthalpy when it vaporizes at ordinary steam generator pressures, which tends to limit the mass flow rate for a desired power plant output. The properties of liquid water and water vapor are also such that the back work ratios achieved are characteristically quite low, and the techniques of superheat, reheat, and regeneration can be effective for increasing power plant efficiencies.

Water is less satisfactory insofar as some other desirable working fluid characteristics are concerned. For example, the critical temperature of water is only 374.14°C, which is about 225°C below the maximum allowable turbine inlet temperatures. Accordingly, to achieve a high average temperature of heat addition and realize the attendant higher thermal efficiency, it may be necessary for the steam generator to operate at supercritical pressures. This requires costly piping and heat exchanger tubes capable of withstanding great stresses. Another undesirable characteristic of water is that its saturation pressure at ordinary condenser temperatures is well below atmospheric pressure. As a result, air can leak into the system, necessitating the use of special ejector pumps attached to the condenser or deaerating feedwater heaters to remove the air.

Although water has some shortcomings as a working fluid, no other single working fluid has been found that is more satisfactory overall for large electrical generating plants. Still, vapor power cycles intended for special uses may employ working fluids that are better matched to the application at hand than water. Cycles that operate at relatively low temperatures may perform best with a refrigerant such as ammonia as the working fluid. Power systems for high-temperature applications may employ substances having desirable performance characteristics at these temperatures. Moreover, water may be used together with some other substance in a *binary* vapor cycle to achieve better overall performance than could be realized with water alone.

Binary Vapor Cycle

binary cycle

In a ***binary vapor power cycle*** two working fluids are used, one with good high-temperature characteristics and another with good characteristics at the lower-temperature end of the operating range. ➥ **FOR EXAMPLE...** Fig. 8.13 shows a schematic diagram and an accompanying T–s diagram of a binary vapor cycle using water and a suitable liquid metal, with each substance in both the liquid and vapor phases. In this arrangement, two ideal Rankine cycles are combined, with the heat rejection from the high-temperature cycle (the topping cycle) being used as the energy input for the low-temperature cycle. This energy transfer is accomplished in an interconnecting heat exchanger, which serves as the condenser for the metal cycle and the boiler for the water cycle. Since the increase in the specific enthalpy of the water as it passes through the heat exchanger is typically several times the magnitude of the specific enthalpy decrease of the metal, several units of mass of metal must circulate in the topping cycle for each unit of mass of water in the other cycle. ⬅

Binary vapor power cycles can operate with higher average temperatures of heat addition than conventional cycles using water only and thus can attain higher thermal efficiencies. However, the higher efficiencies achieved in this manner must justify the increased costs related to the construction and operation of the more complex cycle arrangement.

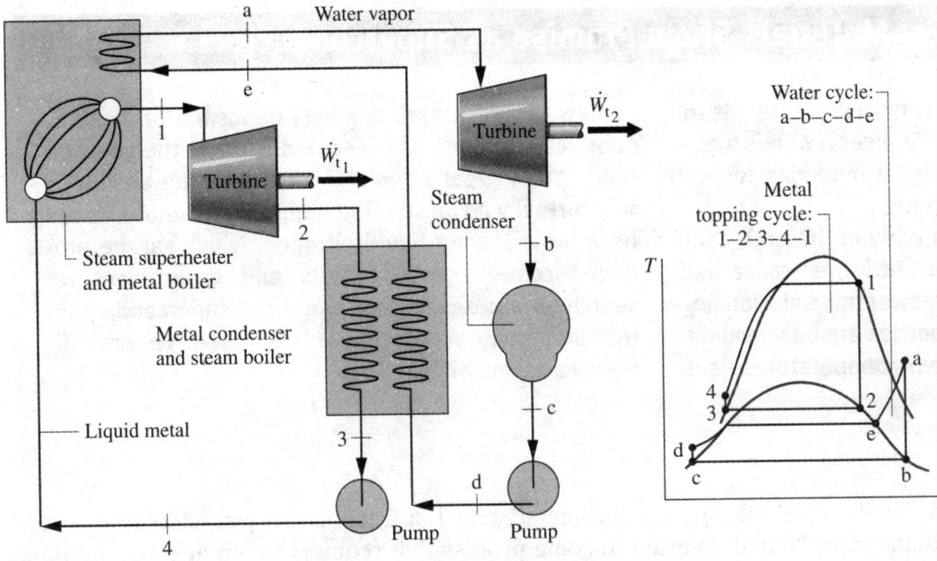

Fig. 8.13 Water–metal binary vapor cycle.

Cogeneration

The binary cycle considered above is just one example of how systems can be linked to obtain overall systems that utilize fuel more effectively. Other examples are discussed in Secs. 7.6 and 7.7, including the multiple-use strategy known as ***cogeneration.*** In each of the following two paragraphs we consider applications of cogeneration.

cogeneration

Direct combustion heating and the generation of process steam together account for a substantial portion of industrial energy resource use. But because the heat and steam are often required at a relatively low temperature, good use is not made of the relatively high-temperature products of combustion obtained by burning fuel. This source of inefficiency can be reduced with a cogeneration arrangement in which fuel is consumed to produce both electricity and steam (or heat), but with a total cost less than would be required to produce them individually. A schematic of such a cogeneration system is shown in Fig. 8.14. The figure is labeled with the fractions of the total flow entering the steam turbine that remain at various locations. On the basis

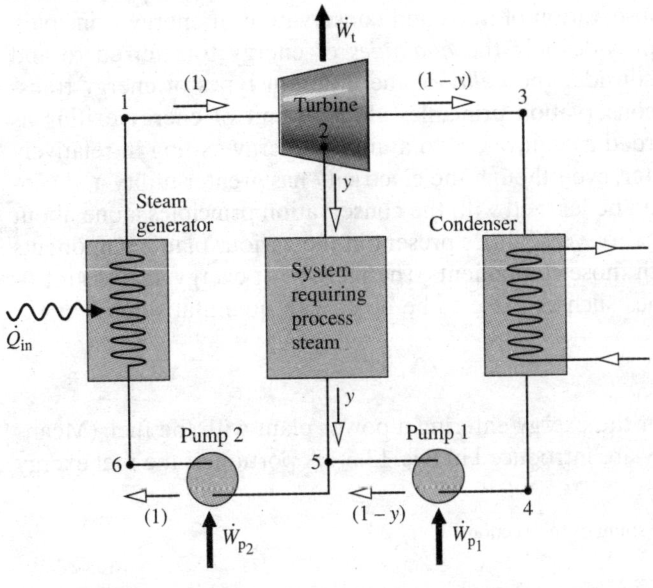

Fig. 8.14 Schematic of a cogeneration system in which process steam is bled from the turbine.

Horizons

New Materials Show Promise as Superconductors

Distributing electricity from power plants to consumers over long distances requires high-voltage transmission lines to minimize losses associated with electrical resistance.

Superconductors, discovered early in the 20th century, can conduct electricity with little resistance but have been impractical for use in power transmission because the superconducting properties are only evident when they are cooled to *cryogenic* temperatures near 0K, which is costly.

New ceramic-based coatings deposited on metallic substrates that act as superconductors in the 50–100K range show great promise for transmitting electricity with virtually no losses. The required temperatures can be achieved using liquid nitrogen, which can be produced cost-effectively. Utility and government researchers are developing prototype superconductors that some day may increase U.S. power system efficiency and reliability.

of a unit of mass entering the turbine at state 1, a fraction y is extracted at an intermediate point 20 and diverted to some process that requires steam at this condition. The remaining fraction, $1 - y$, expands to the turbine exit at state 3, producing power in addition to that developed by the first turbine stage. Eventually this fraction rejoins the amount extracted and the combined stream is returned to the steam generator. When no process steam is required, all of the steam generated by the steam generator is allowed to expand through the turbine.

district heating

Industries such as pulp and paper production and food processing, which require steam for various processes in addition to electricity for operating machines, lighting, etc., are particularly well suited for the use of cogeneration. Another cogeneration arrangement being used increasingly involves ***district heating.*** In this application a power plant is integrated into a community so that it provides electricity for industrial, commercial, and domestic use together with steam for process needs, space heating, and domestic water heating. District heating is commonly used in northern Europe and is being increasingly used in the United States.

8.6 Case Study: Exergy Accounting of a Vapor Power Plant[1]

The discussions to this point show that a useful picture of power plant performance can be obtained with the conservation of mass and conservation of energy principles. However, these principles provide only the *quantities* of energy transferred to and from the plant and do not consider the *utility* of the different types of energy transfer. For example, with the conservation principles alone, a unit of energy exiting as generated electricity is regarded as equivalent to a unit of energy exiting in relatively low-temperature cooling water, even though the electricity has greater utility and economic value. Also, nothing can be learned with the conservation principles alone about the relative significance of the irreversibilities present in the various plant components and the losses associated with those components. The method of exergy analysis introduced in Chap. 7 allows issues such as these to be dealt with quantitatively.

Exergy Accounting

In this section we account for the exergy entering a power plant with the fuel. (Means for evaluating the fuel exergy are introduced in Sec. 13.6.) A portion of the fuel exergy

[1]Chapter 7 is a prerequisite for the study of this section.

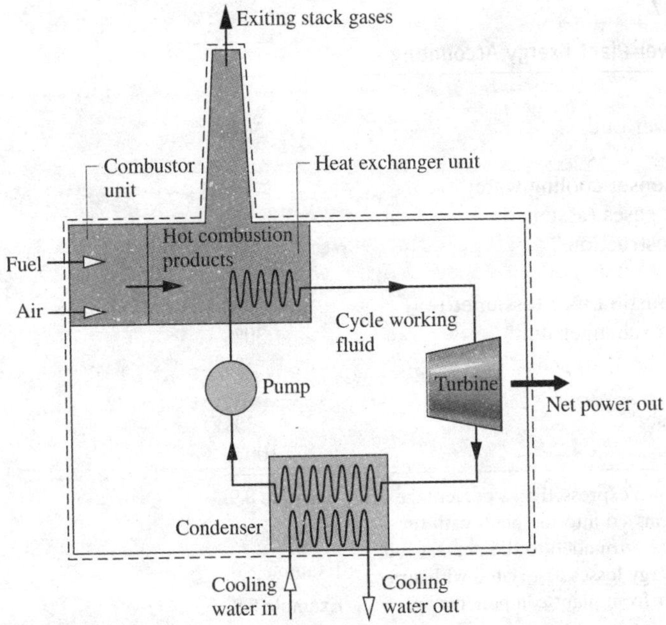

Fig. 8.15 Power plant schematic for the exergy analysis case study.

is ultimately returned to the plant surroundings as the net work developed. However, the largest part is either destroyed by irreversibilities within the various plant components or carried from the plant by the cooling water, stack gases, and unavoidable heat transfers with the surroundings. These considerations are illustrated in the present section by three solved examples, treating respectively the boiler, turbine and pump, and condenser of a simple vapor power plant.

The irreversibilities present in each power plant component exact a tariff on the exergy supplied to the plant, as measured by the exergy destroyed in that component. The component levying the greatest tariff is the boiler, for a significant portion of the exergy entering the plant with the fuel is destroyed by irreversibilities within it. There are two main sources of irreversibility in the boiler: (1) the irreversible heat transfer occurring between the hot combustion gases and the working fluid of the vapor power cycle flowing through the boiler tubes, and (2) the combustion process itself. To simplify the present discussion, the boiler is considered to consist of a combustor unit in which fuel and air are burned to produce hot combustion gases, followed by a heat exchanger unit where the cycle working fluid is vaporized as the hot gases cool. This idealization is illustrated in Fig. 8.15.

For the purposes of illustration, let us assume that 30% of the exergy entering the combustion unit with the fuel is destroyed by the combustion irreversibility and 1% of the fuel exergy exits the heat exchanger unit with the stack gases. The corresponding values for an actual power plant might differ from these nominal values. However, they provide characteristic values for discussion. (Means for evaluating the combustion exergy destruction and the exergy accompanying the exiting stack gases are introduced in Chap. 13.) Using the foregoing values for the combustion exergy destruction and stack gas loss, it follows that a *maximum* of 69% of the fuel exergy remains for transfer from the hot combustion gases to the cycle working fluid. It is from this portion of the fuel exergy that the net work developed by the plant is obtained. In Examples 8.7 through 8.9, we account for the exergy supplied by the hot combustion gases passing through the heat exchanger unit. The principal results of this series of examples are reported in Table 8.1. Carefully note that the values of Table 8.1 are keyed to the vapor power plant of Example 8.2 and thus have only qualitative significance for vapor power plants in general.

Table 8.1

Vapor Power Plant Exergy Accounting[a]

Outputs	
Net power out[b]	30%
Losses	
Condenser cooling water[c]	1%
Stack gases (assumed)	1%
Exergy Destruction	
Boiler	
Combustion unit (assumed)	30%
Heat exchanger unit[d]	30%
Turbine[e]	5%
Pump[f]	—
Condenser[g]	3%
Total	100%

[a]All values are expressed as a percentage of the exergy carried into the plant with the fuel. Values are rounded to the nearest full percent. Exergy losses associated with stray heat transfer from plant components are ignored.

[b]Example 8.8

[c]Example 8.9.

[d]Example 8.7.

[e]Example 8.8.

[f]Example 8.8.

[g]Example 8.9.

Case Study Conclusions

The entries of Table 8.1 suggest some general observations about vapor power plant performance. First, the table shows that the exergy destructions are more significant than the plant losses. The largest portion of the exergy entering the plant with the fuel is destroyed, with exergy destruction in the boiler overshadowing all others. By contrast, the loss associated with heat transfer to the cooling water is relatively unimportant. The cycle thermal efficiency (calculated in the solution to Example 8.2) is 31.4%, so over two-thirds (68.6%) of the *energy* supplied to the cycle working fluid is subsequently carried out by the condenser cooling water. By comparison, the amount of *exergy* carried out is virtually negligible because the temperature of the cooling water is raised only a few degrees over that of the surroundings and thus has limited utility. The loss amounts to only 1% of the exergy entering the plant with the fuel. Similarly, losses accompanying unavoidable heat transfer with the surroundings and the exiting stack gases typically amount only to a few percent of the exergy entering the plant with the fuel and are generally overstated when considered from the perspective of conservation of energy alone.

An exergy analysis allows the sites where destructions or losses occur to be identified and rank ordered for significance. This knowledge is useful in directing attention to aspects of plant performance that offer the greatest opportunities for improvement through the application of practical engineering measures. However, the decision to adopt any particular modification is governed by economic considerations that take into account both economies in fuel use and the costs incurred to achieve those economies.

The calculations presented in the following examples illustrate the application of exergy principles through the analysis of a simple vapor power plant. There are no fundamental difficulties, however, in applying the methodology to actual power plants, including consideration of the combustion process. The same procedures also can be used for exergy accounting of the gas turbine power plants considered in Chap. 9 and other types of thermal systems.

The following example illustrates the exergy analysis of the heat exchanger unit of the boiler of the case study vapor power plant.

Example 8.7 CYCLE EXERGY ANALYSIS—HEAT EXCHANGER UNIT

The heat exchanger unit of the boiler of Example 8.2 has a stream of water entering as a liquid at 8.0 MPa and exiting as a saturated vapor at 8.0 MPa. In a separate stream, gaseous products of combustion cool at a constant pressure of 1 bar from 1107 to 547°C. The gaseous stream can be modeled as air as an ideal gas. Let $T_0 = 22°C$, $p_0 = 1$ bar. Determine **(a)** the net rate at which exergy is carried into the heat exchanger unit by the gas stream, in MW, **(b)** the net rate at which exergy is carried from the heat exchanger by the water stream, in MW, **(c)** the rate of exergy destruction, in MW, **(d)** the exergetic efficiency given by Eq. 7.27.

Solution

Known: A heat exchanger at steady state has a water stream entering and exiting at known states and a separate gas stream entering and exiting at known states.

Find: Determine the net rate at which exergy is carried into the heat exchanger by the gas stream, in MW, the net rate at which exergy is carried from the heat exchanger by the water stream, in MW, the rate of exergy destruction, in MW, and the exergetic efficiency.

Schematic and Given Data:

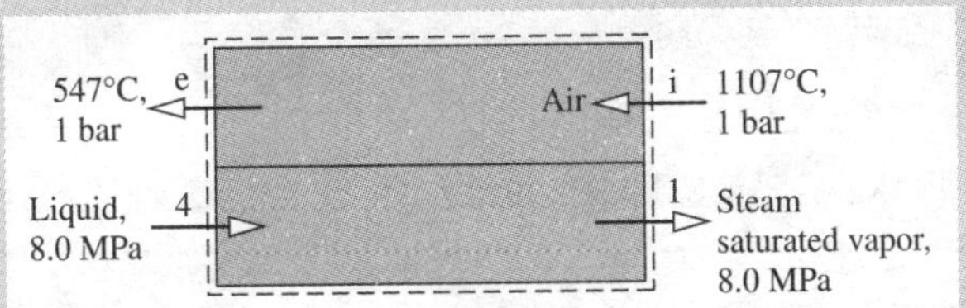

Fig. E8.7

Engineering Model:

1. The control volume shown in the accompanying figure operates at steady state with $\dot{Q}_{cv} = \dot{W}_{cv} = 0$.
2. Kinetic and potential energy effects can be ignored.
3. The gaseous combustion products are modeled as air as an ideal gas.
4. The air and the water each pass through the steam generator at constant pressure.
5. Only 69% of the exergy entering the plant with the fuel remains after accounting for the stack loss and combustion exergy destruction.
6. $T_0 = 22°C$, $p_0 = 1$ bar.

Analysis: The analysis begins by evaluating the mass flow rate of the air in terms of the mass flow rate of the water. The air and water pass through the boiler in separate streams. Hence, at steady state the conservation of mass principle requires

$$\dot{m}_i = \dot{m}_e \quad \text{(air)}$$
$$\dot{m}_4 = \dot{m}_1 \quad \text{(water)}$$

Using these relations, an energy rate balance for the overall control volume reduces a steady state to

$$0 = \overset{0}{\cancel{\dot{Q}_{cv}}} - \overset{0}{\cancel{\dot{W}_{cv}}} + \dot{m}_a(h_i - h_e) + \dot{m}(h_4 - h_1)$$

where $\dot{Q}_{cv} = \dot{W}_{cv} = 0$ by assumption 1, and the kinetic and potential energy terms are dropped by assumption 2. In this equation $\dot{m}_a$ and $\dot{m}$ denote, respectively, the mass flow rates of the air and water. On solving

$$\frac{\dot{m}_a}{\dot{m}} = \frac{h_1 - h_4}{h_i - h_e}$$

The solution to Example 8.2 gives $h_1 = 2758$ kJ/kg and $h_4 = 183.36$ kJ/kg. From Table A-22, $h_i = 1491.44$ kJ/kg and $h_e = 843.98$ kJ/kg. Hence

$$\frac{\dot{m}_a}{\dot{m}} = \frac{2758 - 183.36}{1491.44 - 843.98} = 3.977 \frac{\text{kg (air)}}{\text{kg (steam)}}$$

From Example 8.2, $\dot{m} = 4.449 \times 10^5$ kg/h. Thus, $\dot{m}_a = 17.694 \times 10^5$ kg/h.

(a) The net rate at which exergy is carried into the heat exchanger unit by the gaseous stream can be evaluated using Eq. 7.18

$$\begin{bmatrix}\text{net rate at which exergy}\\ \text{is carried in by the}\\ \text{gaseous stream}\end{bmatrix} = \dot{m}_a(e_{fi} - e_{fe})$$

$$= \dot{m}_a[h_i - h_e - T_0(s_i - s_e)]$$

Since the gas pressure remains constant, Eq. 6.20a giving the change in specific entropy of an ideal gas reduces to $s_i - s_e = s_i^\circ - s_e^\circ$. Thus, with h and s° values from Table A-22

$$\dot{m}_a(e_{fi} - e_{fe}) = (17.694 \times 10^5 \text{ kg/h})[(1491.44 - 843.98) \text{ kJ/kg} - (295 \text{ K})(3.34474 - 2.74504) \text{ kJ/kg} \cdot \text{K}]$$

$$= \frac{8.326 \times 10^8 \text{ kJ/h}}{|3600 \text{ s/h}|}\left|\frac{1 \text{ MW}}{10^3 \text{ kJ/s}}\right| = 231.28 \text{ MW}$$

(b) The net rate at which exergy is carried out of the boiler by the water stream is determined similarly

$$\begin{bmatrix}\text{net rate at which exergy}\\ \text{is carried out by the}\\ \text{water stream}\end{bmatrix} = \dot{m}(e_{f1} - e_{f4})$$

$$= \dot{m}[h_1 - h_4 - T_0(s_1 - s_4)]$$

From Table A-3, $s_1 = 5.7432$ kJ/kg · K. Double interpolation in Table A-5 at 8.0 MPa and $h_4 = 183.36$ kJ/kg gives $s_4 = 0.5957$ kJ/kg · K. Substituting known values

$$\dot{m}(e_{f1} - e_{f4}) = (4.449 \times 10^5)[(2758 - 183.36) - 295(5.7432 - 0.5957)]$$

❶

$$= \frac{4.699 \times 10^8 \text{ kJ/h}}{|3600 \text{ s/h}|}\left|\frac{1 \text{ MW}}{10^3 \text{ kJ/s}}\right| = 130.53 \text{ MW}$$

(c) The rate of exergy destruction can be evaluated by reducing the exergy rate balance to obtain

$$\dot{E}_d = \dot{m}_a(e_{fi} - e_{fe}) + \dot{m}(e_{f4} - e_{f1})$$

❷ With the results of parts (a) and (b)

❸

$$\dot{E}_d = 231.28 \text{ MW} - 130.53 \text{ MW} = 100.75 \text{ MW}$$

(d) The exergetic efficiency given by Eq. 7.27 is

$$\varepsilon = \frac{\dot{m}(e_{f1} - e_{f4})}{\dot{m}_a(e_{fi} - e_{fe})} = \frac{130.53 \text{ MW}}{231.28 \text{ MW}} = 0.564\ (56.4\%)$$

This calculation indicates that 43.6% of the exergy supplied to the heat exchanger unit by the cooling combustion products is destroyed. However, since only 69% of the exergy entering the plant with the fuel is assumed to remain after the stack loss and combustion exergy destruction are accounted for (assumption 5), it can be concluded that 0.69 × 43.6% = 30% of the exergy entering the plant with the fuel is destroyed within the heat exchanger. This is the value listed in Table 8.1.

---

❶ Since energy is conserved, the rate at which energy is transferred to the water as it flows through the heat exchanger *equals* the rate at which energy is transferred from the cooling gas passing through the heat exchanger. By contrast, the decrease in exergy of the gas as it passes through the heat exchanger *exceeds* the increase in the exergy of the water by the amount of exergy destroyed.

Skills Developed: Ability to...
✓ perform exergy analysis of a power plant steam generator.

❷ The rate of exergy destruction can be determined alternatively by evaluating the rate of entropy production, $\dot{\sigma}_{cv}$, from an entropy rate balance and multiplying by T_0 to obtain $\dot{E}_d = T_0\dot{\sigma}_{cv}$.

❸ Underlying the assumption that each stream passes through the heat exchanger at constant pressure is the neglect of friction as an irreversibility. Thus, the only contributor to exergy destruction in this case is heat transfer from the higher-temperature combustion products to the vaporizing water.

Quick Quiz

If the gaseous products of combustion are cooled to 517°C (h_e = 810.99 kJ/kg), what is the mass flow rate of the gaseous products, in kg/h?

Ans. 16.83×10^5 kg/h.

In the next example, we determine the exergy destruction rates in the turbine and pump of the case study vapor power plant.

Example 8.8 CYCLE EXERGY ANALYSIS—TURBINE AND PUMP

Reconsider the turbine and pump of Example 8.2. Determine for each of these components the rate at which exergy is destroyed, in MW. Express each result as a percentage of the exergy entering the plant with the fuel. Let T_0 = 22°C, p_0 = 1 bar.

Solution

Known: A vapor power cycle operates with steam as the working fluid. The turbine and pump each has an isentropic efficiency of 85%.

Find: For the turbine and the pump individually, determine the rate at which exergy is destroyed, in MW. Express each result as a percentage of the exergy entering the plant with the fuel.

Schematic and Given Data:

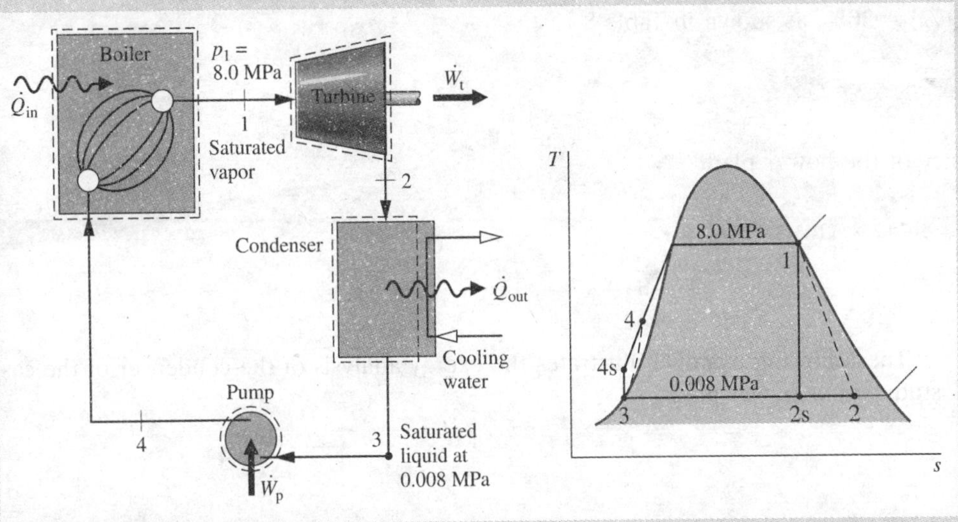

Fig. E8.8

Engineering Model:

1. The turbine and the pump can each be analyzed as a control volume at steady state.
2. The turbine and pump operate adiabatically and each has an efficiency of 85%.
3. Kinetic and potential energy effects are negligible.
4. Only 69% of the exergy entering the plant with the fuel remains after accounting for the stack loss and combustion exergy destruction.
5. T_0 = 22°C, p_0 = 1 bar.

Analysis: The rate of exergy destruction can be found by reducing the exergy rate balance or by use of the relationship $\dot{E}_d = T_0 \dot{\sigma}_{cv}$, where $\dot{\sigma}_{cv}$ is the rate of entropy production from an entropy rate balance. With either approach, the rate of exergy destruction for the turbine can be expressed as

$$\dot{E}_d = \dot{m}T_0(s_2 - s_1)$$

From Table A-3, $s_1 = 5.7432$ kJ/kg · K. Using $h_2 = 1939.3$ kJ/kg from the solution to Example 8.2, the value of s_2 can be determined from Table A-3 as $s_2 = 6.2021$ kJ/kg · K. Substituting values

$$\dot{E}_d = (4.449 \times 10^5 \text{ kg/h})(295 \text{ K})(6.2021 - 5.7432)(\text{kJ/kg} \cdot \text{K})$$

$$= \left(0.602 \times 10^8 \frac{\text{kJ}}{\text{h}}\right)\left|\frac{1 \text{ h}}{3600 \text{ s}}\right|\left|\frac{1 \text{ MW}}{10^3 \text{ kJ/s}}\right| = 16.72 \text{ MW}$$

From the solution to Example 8.7, the net rate at which exergy is supplied by the cooling combustion gases is 231.28 MW. The turbine rate of exergy destruction expressed as a percentage of this is (16.72/231.28)(100%) = 7.23%. However, since only 69% of the entering fuel exergy remains after the stack loss and combustion exergy destruction are accounted for, it can be concluded that 0.69 × 7.23% = 5% of the exergy entering the plant with the fuel is destroyed within the turbine. This is the value listed in Table 8.1.

Similarly, the exergy destruction rate for the pump is

$$\dot{E}_d = \dot{m}T_0(s_4 - s_3)$$

With s_3 from Table A-3 and s_4 from the solution to Example 8.7

$$\dot{E}_d = (4.449 \times 10^5 \text{ kg/h})(295 \text{ K})(0.5957 - 0.5926)(\text{kJ/kg} \cdot \text{K})$$

$$= \left(4.07 \times 10^5 \frac{\text{kJ}}{\text{h}}\right)\left|\frac{1 \text{ h}}{3600 \text{ s}}\right|\left|\frac{1 \text{ MW}}{10^3 \text{ kJ/s}}\right| = 0.11 \text{ MW}$$

Expressing this as a percentage of the exergy entering the plant as calculated above, we have (0.11/231.28)(69%) = 0.03%. This value is rounded to zero in Table 8.1.

The net power output of the vapor power plant of Example 8.2 is 100 MW. Expressing this as a percentage of the rate at which exergy is carried into the plant with the fuel, (100/231.28)(69%) = 30%, as shown in Table 8.1.

Skills Developed: *Ability to...*
✓ perform exergy analysis of a power plant turbine and pump.

Quick Quiz

What is the exergetic efficiency of the power plant?

Ans. 30%.

The following example illustrates the exergy analysis of the condenser of the case study vapor power plant.

Example 8.9 CYCLE EXERGY ANALYSIS—CONDENSER

The condenser of Example 8.2 involves two separate water streams. In one stream, a two-phase liquid–vapor mixture enters at 0.008 MPa and exits as a saturated liquid at 0.008 MPa. In the other stream, cooling water enters at 15°C and exits at 35°C. **(a)** Determine the net rate at which exergy is carried from the condenser by the cooling water, in MW. Express this result as a percentage of the exergy entering the plant with the fuel. **(b)** Determine for the condenser the rate of exergy destruction, in MW. Express this result as a percentage of the exergy entering the plant with the fuel. Let $T_0 = 22$°C and $p_0 = 1$ bar.

Solution

Known: A condenser at steady state has two streams: (1) a two-phase liquid–vapor mixture entering and condensate exiting at known states, and (2) a separate cooling water stream entering and exiting at known temperatures.

Find: Determine the net rate at which exergy is carried from the condenser by the cooling water stream and the rate of exergy destruction for the condenser. Express both quantities in MW and as percentages of the exergy entering the plant with the fuel.

Schematic and Given Data:

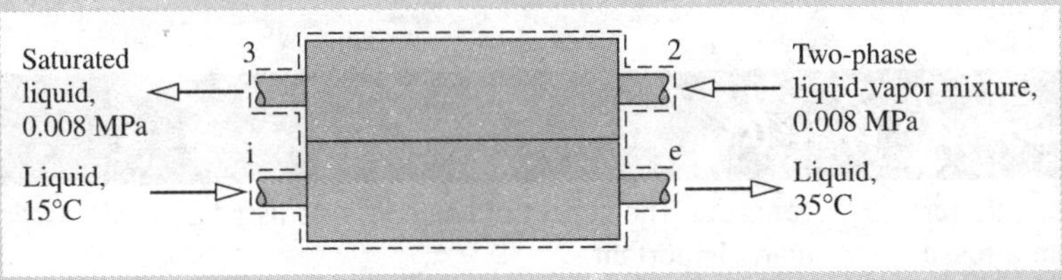

Fig. E8.9

Engineering Model:

1. The control volume shown on the accompanying figure operates at steady state with $\dot{Q}_{cv} = \dot{W}_{cv} = 0$.
2. Kinetic and potential energy effects can be ignored.
3. Only 69% of the fuel exergy remains after accounting for the stack loss and combustion exergy destruction.
4. $T_0 = 22°C$, $p_0 = 1$ bar.

Analysis:

(a) The net rate at which exergy is carried out of the condenser can be evaluated by using Eq. 7.18:

$$\begin{bmatrix} \text{net rate at which exergy} \\ \text{is carried out by the} \\ \text{cooling water} \end{bmatrix} = \dot{m}_{cw}(e_{fe} - e_{fi})$$

$$= \dot{m}_{cw}[h_e - h_i - T_0(s_e - s_i)]$$

where $\dot{m}_{cw}$ is the mass flow rate of the cooling water from the solution to Example 8.2. With saturated liquid values for specific enthalpy and entropy from Table A-2 at the specified inlet and exit temperatures of the cooling water

$$\dot{m}_{cw}(e_{fe} - e_{fi}) = (9.39 \times 10^6 \text{ kg/h})[(146.68 - 62.99) \text{ kJ/kg} - (295 \text{ K})(0.5053 - 0.2245) \text{ kJ/kg} \cdot \text{K}]$$

$$= \frac{8.019 \times 10^6 \text{ kJ/h}}{|3600 \text{ s/h}|}\left|\frac{1 \text{ MW}}{10^3 \text{ kJ/s}}\right| = 2.23 \text{ MW}$$

Expressing this as a percentage of the exergy entering the plant with the fuel, we get (2.23/231.28)(69%) = 1%. This is the value listed in Table 8.1.

(b) The rate of exergy destruction for the condenser can be evaluated by reducing the exergy rate balance. Alternatively, the relationship $\dot{E}_d = T_0\dot{\sigma}_{cv}$ can be employed, where $\dot{\sigma}_{cv}$ is the time rate of entropy production for the condenser determined from an entropy rate balance. With either approach, the rate of exergy destruction can be expressed as

$$\dot{E}_d = T_0[\dot{m}(s_3 - s_2) + \dot{m}_{cw}(s_e - s_i)]$$

Substituting values

$$\dot{E}_d = 295[(4.449 \times 10^5)(0.5926 - 6.2021) + (9.39 \times 10^6)(0.5053 - 0.2245)]$$

$$= \frac{416.1 \times 10^5 \text{ kJ/h}}{|3600 \text{ s/h}|}\left|\frac{1 \text{ MW}}{10^3 \text{ kJ/s}}\right| = 11.56 \text{ MW}$$

Expressing this as a percentage of the exergy entering the plant with the fuel, we get (11.56/231.28)(69%) = 3%. This is the value listed in Table 8.1.

Skills Developed: Ability to...
✓ perform exergy analysis of a power plant condenser.

Quick Quiz

Referring to data from Example 8.2, what percent of the *energy* supplied to the steam passing through the steam generator is carried out by the cooling water?
Ans. 68.6%.

✓ Chapter Summary and Study Guide

In this chapter we have considered practical arrangements for vapor power plants producing a net power output from a fossil fuel, nuclear, or solar input. We have illustrated how vapor power plants are modeled thermodynamically and considered the principal irreversibilities and losses associated with such plants. The main components of *simple* vapor power plants are modeled by the Rankine cycle. In this chapter, we also have introduced modifications to the simple vapor power cycle aimed at improving overall performance. These include superheat, reheat, regeneration, supercritical operation, cogeneration, and binary cycles. We have also included a case study to illustrate the application of exergy analysis to vapor power plants.

The following checklist provides a study guide for this chapter. When your study of the text and end-of-chapter exercises has been completed you should be able to

- ✓ write out the meanings of the terms listed in the margin throughout the chapter and understand each of the related concepts. The subset of key concepts listed below is particularly important.
- ✓ sketch schematic diagrams and accompanying *T–s* diagrams of Rankine, reheat, and regenerative vapor power cycles.
- ✓ apply conservation of mass and energy, the second law, and property data to determine power cycle performance, including thermal efficiency, net power output, and mass flow rates.
- ✓ discuss the effects on Rankine cycle performance of varying steam generator pressure, condenser pressure, and turbine inlet temperature.
- ✓ discuss the principal sources of exergy destruction and loss in vapor power plants.

Key Engineering Concepts

Rankine cycle p. 392
thermal efficiency p. 394
back work ratio p. 394
superheat p. 405
reheat p. 406
regeneration p. 411
cogeneration p. 423

Key Equations

$\eta = \dfrac{\dot{W}_t/\dot{m} - \dot{W}_p/\dot{m}}{\dot{Q}_{in}/\dot{m}} = \dfrac{(h_1 - h_2) - (h_4 - h_3)}{h_1 - h_4}$	(8.5a)	Thermal efficiency for the Rankine cycle of Fig. 8.2
$\text{bwr} = \dfrac{\dot{W}_p/\dot{m}}{\dot{W}_t/\dot{m}} = \dfrac{(h_4 - h_3)}{(h_1 - h_2)}$	(8.6)	Back work ratio for the Rankine cycle of Fig. 8.2
$\left(\dfrac{\dot{W}_p}{\dot{m}}\right)_{\substack{\text{int}\\ \text{rev}}} \approx v_3(p_4 - p_3)$	(8.7b)	Approximation for internally reversible pump work for the Rankine cycle of Fig. 8.2

Exercises: things engineers think about

1. Many utility companies offer special rates for "green power." What does that mean?

2. Referring to your monthly electric bills, make a plot of your monthly kW · h usage for the last year. Comment.

3. How much does the temperature of condenser cooling water vary in your area throughout the year? How would that affect the thermal efficiency of local power plants?

4. Power plants located in arid areas may accomplish condensation by air cooling instead of cooling with water flowing through the condenser. How might air cooling affect the thermal efficiency?

5. How does the Rankine cycle differ from the Carnot cycle of Fig. 5.15?

6. Keeping states 1 and 2 fixed, how would Fig. 8.6 be modified if the frictional pressure drops through the boiler and condenser were considered? How would the pressure drops affect the net power of the cycle?

7. Brainstorm some ways to use the cooling water exiting the condenser of a large power plant.

8. What effects on a river's ecology might result from a power plant's use of river water for condenser cooling?

9. Referring to Fig. 8.1, what environmental impacts might result from the two plumes shown on the figure?

10. Referring to Example 8.5, what are the principal sources of *internal* irreversibility?

12. Some prototype binary cycle power plants have been built in the last 50 years, but to date they have not found commercial application. Why?

13. There are many locations where solar energy is plentiful, yet to date there is no widespread use of solar energy for electric power generation. What issues have kept solar energy from being widely used?

Problems: developing engineering skills

Analyzing Rankine Cycles

8.1 Water is the working fluid in an ideal Rankine cycle. The condenser pressure is 6 kPa, and saturated vapor enters the turbine at 10 MPa. Determine the heat transfer rates, in kJ per kg of steam flowing, for the working fluid passing through the boiler and condenser and calculate the thermal efficiency.

8.2 Water is the working fluid in an ideal Rankine cycle. Superheated vapor enters the turbine at 10 MPa, 480°C, and the condenser pressure is 6 kPa. Determine for the cycle

(a) the rate of heat transfer to the working fluid passing through the steam generator, in kJ per kg of steam flowing.
(b) the thermal efficiency.
(c) the rate of heat transfer from the working fluid passing through the condenser to the cooling water, in kJ per kg of steam flowing.

8.3 Water is the working fluid in a Carnot vapor power cycle. Saturated liquid enters the boiler at a pressure of 10 MPa, and saturated vapor enters the turbine. The condenser pressure is 6 kPa. Determine

(a) the thermal efficiency.
(b) the back work ratio.
(c) the heat transfer to the working fluid per unit mass passing through the boiler, in kJ/kg.
(d) the heat transfer from the working fluid per unit mass passing through the condenser, in kJ/kg.

8.4 Plot each of the quantities calculated in Problem 8.2 versus condenser pressure ranging from 6 kPa to 0.1 MPa. Discuss.

8.5 Plot each of the quantities calculated in Problem 8.2 versus steam generator pressure ranging from 4 MPa to 20 MPa. Maintain the turbine inlet temperature at 480°C. Discuss.

8.6 Water is the working fluid in an ideal Rankine cycle. Saturated vapor enters the turbine at 16 MPa, and the condenser pressure is 8 kPa. The mass flow rate of steam entering the turbine is 120 kg/s. Determine

(a) the net power developed, in kW.
(b) the rate of heat transfer to the steam passing through the boiler, in kW.
(c) the thermal efficiency.
(d) the mass flow rate of condenser cooling water, in kg/s, if the cooling water undergoes a temperature increase of 18°C with negligible pressure change in passing through the condenser.

8.7 Water is the working fluid in a Carnot vapor power cycle. Saturated liquid enters the boiler at 16 MPa, and saturated vapor enters the turbine. The condenser pressure is 8 kPa. The mass flow rate of steam entering the turbine is 120 kg/s. Determine

(a) the thermal efficiency.
(b) the back work ratio.
(c) the net power developed, in kW.
(d) the rate of heat transfer from the working fluid passing through the condenser, in kW.

8.8 Plot each of the quantities calculated in Problem 8.6 versus turbine inlet temperature ranging from the saturation temperature at 16 MPa to 560°C. Discuss.

8.9 Water is the working fluid in an ideal Rankine cycle. The pressure and temperature at the turbine inlet are 11 MPa and 590°C, respectively, and the condenser pressure is 6.9 kPa. The mass flow rate of steam entering the turbine is 635×10^3 kg/h. The cooling water experiences a temperature increase from 16 to 27°C, with negligible pressure drop, as it passes through the condenser. Determine for the cycle

(a) the net power developed, in kJ/h.
(b) the thermal efficiency.
(c) the mass flow rate of cooling water, in kg/h.

8.10 Plot each of the quantities calculated in Problem 8.9 versus condenser pressure ranging from 2.8 kPa to 101.3 kPa. Maintain a constant mass flow rate of steam. Discuss.

8.11 Plot each of the quantities calculated in Problem 8.9 versus steam generator pressure ranging from 2 MPa to 28 MPa. Maintain the turbine inlet temperature at 590°C and a constant mass flow rate of steam. Discuss.

8.12 A power plant based on the Rankine cycle is under development to provide a net power output of 10 MW. Solar collectors are to be used to generate Refrigerant 22 vapor at 1.6 MPa, 50°C, for expansion through the turbine. Cooling water is available at 20°C. Specify the preliminary design of the cycle and estimate the thermal efficiency and the refrigerant and cooling water flow rates, in kg/h.

8.13 Refrigerant 134a is the working fluid in a solar power plant operating on a Rankine cycle. Saturated vapor at 60°C enters the turbine, and the condenser operates at a pressure of 6 bar. The rate of energy input to the collectors from solar radiation is 0.4 kW per m^2 of collector surface area. Determine the minimum possible solar collector surface area, in m^2, per kW of power developed by the plant.

8.14 On the south coast of the island of Hawaii, lava flows continuously into the ocean. It is proposed to anchor a floating power plant off-shore of the lava flow that uses ammonia as the working fluid. The plant would exploit the temperature variation between the warm water near the surface at 54°C and seawater at 10°C from a depth of 150 m to produce power. Figure P8.14 shows the configuration and provides additional data. Using the properties of pure water for the seawater and modeling the power plant as a Rankine cycle, determine

(a) the thermal efficiency.
(b) the mass flow rate of ammonia, in kg/min, for a net power output of 224 kW.

8.15 The cycle of Problem 8.3 is modified to include the effects of irreversibilities in the adiabatic expansion and compression processes. If the states at the turbine and pump inlets remain unchanged, repeat parts (a)–(d) of Problem 8.3 for the modified Carnot cycle with $\eta_t = 0.80$ and $\eta_p = 0.70$.

8.16 Steam enters the turbine of a simple vapor power plant with a pressure of 10 MPa and temperature T, and expands adiabatically to 6 kPa. The isentropic turbine efficiency is 85%. Saturated liquid exits the condenser at 6 kPa and the isentropic pump efficiency is 82%.

(a) For T = 580°C, determine the turbine exit quality and the cycle thermal efficiency.
(b) Plot the quantities of part (a) versus T ranging from 580 to 700°C.

8.17 Water is the working fluid in a Rankine cycle. Superheated vapor enters the turbine at 10 MPa, 480°C, and the condenser pressure is 6 kPa. The turbine and pump have isentropic efficiencies of 80 and 70%, respectively. Determine for the cycle

(a) the rate of heat transfer to the working fluid passing through the steam generator, in kJ per kg of steam flowing.
(b) the thermal efficiency.
(c) the rate of heat transfer from the working fluid passing through the condenser to the cooling water, in kJ per kg of steam flowing.

8.18 Steam enters the turbine of a Rankine cycle at 16 MPa, 560°C. The condenser pressure is 8 kPa. The turbine and

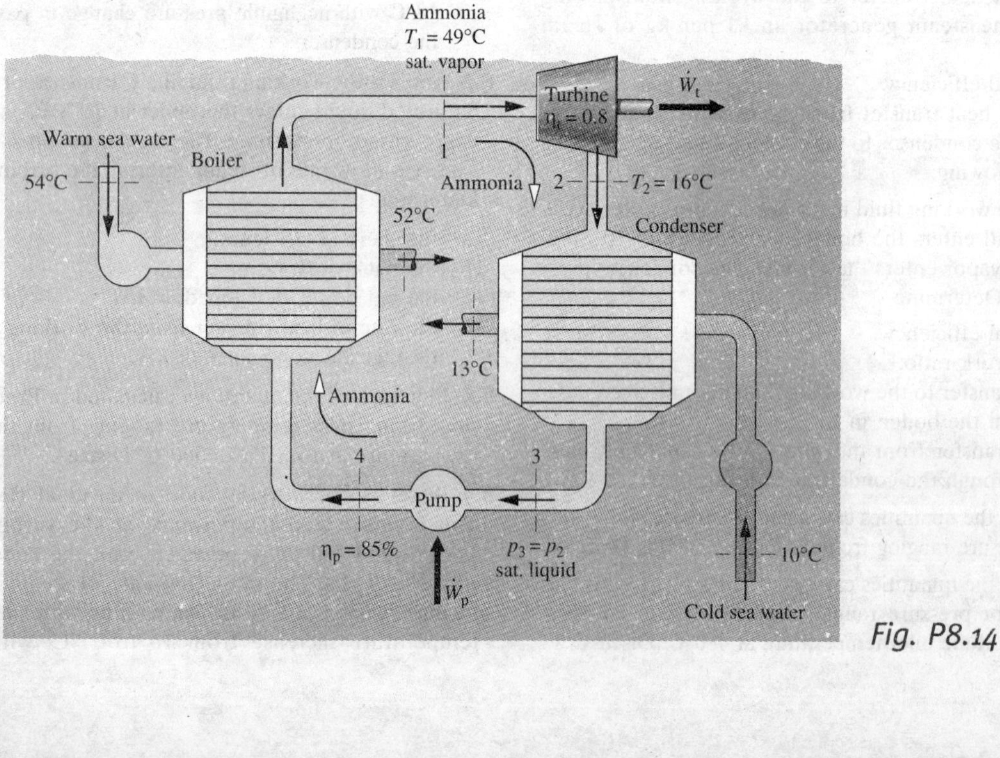

Fig. P8.14

pump each have isentropic efficiencies of 85%, and the mass flow rate of steam entering the turbine is 120 kg/s. Determine

(a) the net power developed, in kW.
(b) the rate of heat transfer to the steam passing through the boiler, in kW.
(c) the thermal efficiency.

Plot each of the quantities in parts (a)–(c) if the turbine and pump isentropic efficiencies remain equal to each other but vary from 80 to 100%.

8.19 Reconsider the cycle of Problem 8.9, but include in the analysis that the turbine and pump have isentropic efficiencies of 88%. The mass flow rate is unchanged. Determine for the modified cycle

(a) the net power developed, in kJ/h.
(b) the rate of heat transfer to the working fluid passing through the steam generator, in kJ/h.
(c) the thermal efficiency.
(d) the volumetric flow rate of cooling water entering the condenser, in m^3/min.

8.20 Water is the working fluid in a Rankine cycle. Superheated vapor enters the turbine at 8 MPa, 560°C with a mass flow rate of 7.8 kg/s and exits at 8 kPa. Saturated liquid enters the pump at 8 kPa. The isentropic turbine efficiency is 88%, and the isentropic pump efficiency is 82%. Cooling water enters the condenser at 18°C and exits at 36°C with no significant change in pressure. Determine

(a) the net power developed, in kW.
(b) the thermal efficiency.
(c) the mass flow rate of cooling water, in kg/s.

8.21 Figure P8.21 provides steady-state operating data for a vapor power plant using water as the working fluid. The mass flow rate of water is 12 kg/s. The turbine and pump operate adiabatically but not reversibly. Determine

(a) the thermal efficiency.
(b) the rates of heat transfer $\dot{Q}_{in}$ and $\dot{Q}_{out}$, each in kW.

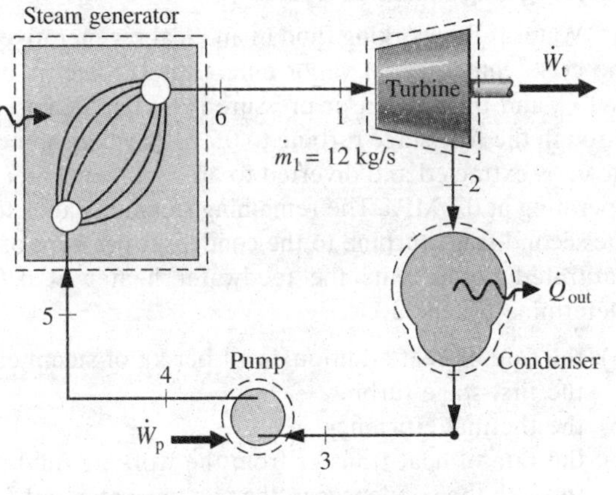

State	p	T (°C)	h (kJ/kg)
1	6 MPa	500	3422.2
2	10 kPa	- - -	1633.3
3	10 kPa	Sat.	191.83
4	7.5 MPa	- - -	199.4
5	7 MPa	40	167.57
6	6 MPa	550	3545.3

Fig. P8.21

8.22 Superheated steam at 8 MPa and 480°C leaves the steam generator of a vapor power plant. Heat transfer and frictional effects in the line connecting the steam generator and the turbine reduce the pressure and temperature at the turbine inlet to 7.6 MPa and 440°C, respectively. The pressure at the exit of the turbine is 10 kPa, and the turbine operates adiabatically. Liquid leaves the condenser at 8 kPa, 36°C. The pressure is increased to 8.6 MPa across the pump. The turbine and pump isentropic efficiencies are 88%. The mass flow rate of steam is 79.53 kg/s. Determine

(a) the net power output, in kW.
(b) the thermal efficiency.
(c) the rate of heat transfer from the line connecting the steam generator and the turbine, in kW.
(d) the mass flow rate of condenser cooling water, in kg/s, if the cooling water enters at 15°C and exits at 35°C with negligible pressure change.

8.23 Modify Problem 8.9 as follows. Steam leaves the steam generator at 11 MPa, 590°C, but due to heat transfer and frictional effects in the line connecting the steam generator and the turbine, the pressure and temperature at the turbine inlet are reduced to 10.3 MPa and 538°C, respectively. Also, condensate leaves the condenser at 5.5 kPa, 32°C and is pumped to 11.7 MPa before entering the steam generator. Determine for the cycle

(a) the net power developed, in kJ/h.
(b) the thermal efficiency.
(c) the heat rate, in kJ/kW · h.
(d) the mass flow rate of cooling water, in kg/h.

8.24 Steam enters the turbine of a vapor power plant at 4.1 MPa, 538°C and exits as a two-phase liquid–vapor mixture at temperature T. Condensate exits the condenser at a temperature 2.78°C lower than T and is pumped to 4.1 MPa. The turbine and pump isentropic efficiencies are 90 and 80%, respectively. The net power developed is 1 MW.

(a) For $T = 27$°C, determine the steam quality at the turbine exit, the steam mass flow rate, in kg/h, and the thermal efficiency.
(b) Plot the quantities of part (a) versus T ranging from 27°C to 41°C.

8.25 Superheated steam at 18 MPa, 560°C, enters the turbine of a vapor power plant. The pressure at the exit of the turbine is 0.06 bar, and liquid leaves the condenser at 0.045 bar, 26°C. The pressure is increased to 18.2 MPa across the pump. The turbine and pump have isentropic efficiencies of 82 and 77%, respectively. For the cycle, determine

(a) the net work per unit mass of steam flow, in kJ/kg.
(b) the heat transfer to steam passing through the boiler, in kJ per kg of steam flowing.
(c) the thermal efficiency.
(d) the heat transfer to cooling water passing through the condenser, in kJ per kg of steam condensed.

8.26 In the preliminary design of a power plant, water is chosen as the working fluid and it is determined that the turbine inlet temperature may not exceed 520°C. Based on expected cooling water temperatures, the condenser is to operate at a pressure of 0.06 bar. Determine the steam generator pressure required if the isentropic turbine efficiency is 80% and the quality of steam at the turbine exit must be at least 90%.

Considering Reheat and Supercritical Cycles

8.27 Steam at 10 MPa, 600°C enters the first-stage turbine of an ideal Rankine cycle with reheat. The steam leaving the reheat section of the steam generator is at 500°C, and the condenser pressure is 6 kPa. If the quality at the exit of the second-stage turbine is 90%, determine the cycle thermal efficiency.

8.28 The ideal Rankine cycle of Problem 8.9 is modified to include reheat. In the modified cycle, steam expands through the first-stage turbine to saturated vapor and then is reheated to 538°C. If the mass flow rate of steam in the modified cycle is the same as in Problem 8.9, determine for the modified cycle

(a) the net power developed, in kJ/h.
(b) the rate of heat transfer to the working fluid in the reheat process, in kJ/h.
(c) the thermal efficiency.

8.29 Water is the working fluid in an ideal Rankine cycle with reheat. Superheated vapor enters the turbine at 10 MPa, 480°C, and the condenser pressure is 6 kPa. Steam expands through the first-stage turbine to 0.7 MPa and then is reheated to 480°C. Determine for the cycle

(a) the rate of heat addition, in kJ per kg of steam entering the first-stage turbine.
(b) the thermal efficiency.
(c) the rate of heat transfer from the working fluid passing through the condenser to the cooling water, in kJ per kg of steam entering the first-stage turbine.

8.30 For the cycle of Problem 8.29, reconsider the analysis assuming the pump and each turbine stage has an isentropic efficiency of 80%. Answer the same questions as in Problem 8.29 for the modified cycle.

8.31 Investigate the effects on cycle performance as the reheat pressure and final reheat temperature take on other values. Construct suitable plots and discuss for the cycle of

(a) Problem 8.29.
(b) Problem 8.30.

8.32 An ideal Rankine cycle with reheat uses water as the working fluid. The conditions at the inlet to the first-stage turbine are $p_1 = 17.2$ MPa, $T_1 = 538°C$. The steam is reheated at constant pressure p between the turbine stages to 538°C. The condenser pressure is 6.9 kPa.

(a) If $p/p_1 = 0.2$, determine the cycle thermal efficiency and the steam quality at the exit of the second-stage turbine.
(b) Plot the quantities of part (a) versus the pressure ratio p/p_1 ranging from 0.05 to 1.0.

8.33 Steam at 32 MPa, 520°C enters the first stage of a supercritical reheat cycle including three turbine stages. Steam exiting the first-stage turbine at pressure p is reheated at constant pressure to 440°C, and steam exiting the second-stage turbine at 0.5 MPa is reheated at constant pressure to 360°C. Each turbine stage and the pump has an isentropic efficiency of 85%. The condenser pressure is 8 kPa.

(a) For $p = 4$ MPa, determine the net work per unit mass of steam flowing, in kJ/kg, and the thermal efficiency.
(b) Plot the quantities of part (a) versus p ranging from 0.5 to 10 MPa.

8.34 Steam at 33 MPa, 538°C enters the first stage of a supercritical reheat cycle including two turbine stages. The steam exiting the first-stage turbine at 4.1 MPa is reheated at constant pressure to 538°C. Each turbine stage and the pump has an isentropic efficiency of 85%. The condenser pressure is 6.9 kPa. If the net power output of the cycle is 100 MW, determine

(a) the rate of heat transfer to the working fluid passing through the steam generator, in MW.
(b) the rate of heat transfer from the working fluid passing through the condenser, in MW.
(c) the cycle thermal efficiency.

8.35 An ideal Rankine cycle with reheat uses water as the working fluid. The conditions at the inlet to the first-stage turbine are 11 MPa, 650°C and the steam is reheated between the turbine stages to 650°C. For a condenser pressure of 6.9 kPa, plot the cycle thermal efficiency versus reheat pressure for pressures ranging from 0.4 to 8.3 MPa.

8.36 An ideal Rankine cycle with reheat uses water as the working fluid. The conditions at the inlet to the first turbine stage are 11 MPa, 650°C and the steam is reheated between the turbine stages to 650°C. For a condenser pressure of 6.9 kPa, plot the cycle thermal efficiency versus reheat pressure for pressures ranging from 0.4 to 8.3 MPa.

Analyzing Regenerative Cycles

8.37 Water is the working fluid in an ideal regenerative Rankine cycle. Superheated vapor enters the turbine at 10 MPa, 480°C, and the condenser pressure is 6 kPa. Steam expands through the first-stage turbine to 0.7 MPa where some of the steam is extracted and diverted to an open feedwater heater operating at 0.7 MPa. The remaining steam expands through the second-stage turbine to the condenser pressure of 6 kPa. Saturated liquid exits the feedwater heater at 0.7 MPa. Determine for the cycle

(a) the rate of heat addition, in kJ per kg of steam entering the first-stage turbine.
(b) the thermal efficiency.
(c) the rate of heat transfer from the working fluid passing through the condenser to the cooling water, in kJ per kg of steam entering the first-stage turbine.

8.38 For the cycle of Problem 8.37, reconsider the analysis assuming the pump and each turbine stage has an isentropic efficiency of 80%. Answer the same questions as in Problem 8.37 for the modified cycle.

8.39 Investigate the effects on cycle performance as the feedwater heater pressure takes on other values. Construct suitable plots and discuss for the cycle of

(a) Problem 8.37.
(b) Problem 8.38.

8.40 A power plant operates on a regenerative vapor power cycle with one open feedwater heater. Steam enters the first turbine stage at 12 MPa, 520°C and expands to 1 MPa, where some of the steam is extracted and diverted to the open feedwater heater operating at 1 MPa. The remaining steam expands through the second turbine stage to the condenser pressure of 6 kPa. Saturated liquid exits the open feedwater heater at 1 MPa. For isentropic processes in the turbines and pumps, determine for the cycle (a) the thermal efficiency and (b) the mass flow rate into the first turbine stage, in kg/h, for a net power output of 330 MW.

8.41 Reconsider the cycle of Problem 8.40 as the feedwater heater pressure takes on other values. Plot the thermal efficiency and the rate of exergy destruction within the feedwater heater, in kW, versus the feedwater heater pressure ranging from 0.5 to 10 MPa. Let $T_0 = 293$ K.

8.42 Compare the results of Problem 8.40 with those for an ideal Rankine cycle having the same turbine inlet conditions and condenser pressure, but no regenerator.

8.43 For the cycle of Problem 8.40, investigate the effects on cycle performance as the feedwater heater pressure takes on other values. Construct suitable plots and discuss. Assume that each turbine stage and each pump has an isentropic efficiency of 80%.

8.44 Modify the ideal Rankine cycle of Problem 8.9 to include one open feedwater heater operating at 690 kPa. Saturated liquid exits the open feedwater heater at 690 kPa. The mass flow rate of steam into the first turbine stage is the same as the mass flow rate of steam in Problem 8.9. Answer the same questions about the modified cycle as in Problem 8.9 and discuss the results.

8.45 Reconsider the cycle of Problem 8.44, but include in the analysis that the isentropic efficiency of each turbine stage is 88% and of each pump is 80%.

8.46 Water is the working fluid in an ideal regenerative Rankine cycle with one open feedwater heater. Superheated vapor enters the first-stage turbine at 16 MPa, 560°C, and the condenser pressure is 8 kPa. The mass flow rate of steam entering the first-stage turbine is 120 kg/s. Steam expands through the first-stage turbine to 1 MPa where some of the steam is extracted and diverted to an open feedwater heater operating at 1 MPa. The remainder expands through the second-stage turbine to the condenser pressure of 8 kPa. Saturated liquid exits the feedwater heater at 1 MPa. Determine

(a) the net power developed, in kW.
(b) the rate of heat transfer to the steam passing through the boiler, in kW.
(c) the thermal efficiency.
(d) the mass flow rate of condenser cooling water, in kg/s, if the cooling water undergoes a temperature increase of 18°C with negligible pressure change in passing through the condenser.

8.47 Reconsider the cycle of Problem 8.46, but include in the analysis that each turbine stage and pump has an isentropic efficiency of 85%.

8.48 For the cycle of Problem 8.47, investigate the effects on cycle performance as the feedwater heater pressure takes on other values. Construct suitable plots and discuss.

8.49 Water is the working fluid in an ideal regenerative Rankine cycle with one closed feedwater heater. Superheated vapor enters the turbine at 10 MPa, 480°C, and the condenser pressure is 6 kPa. Steam expands through the first-stage turbine where some is extracted and diverted to a closed feedwater heater at 0.7 MPa. Condensate drains from the feedwater heater as saturated liquid at 0.7 MPa and is trapped into the condenser. The feedwater leaves the heater at 10 MPa and a temperature equal to the saturation temperature at 0.7 MPa. Determine for the cycle

(a) the rate of heat transfer to the working fluid passing through the steam generator, in kJ per kg of steam entering the first-stage turbine.
(b) the thermal efficiency.
(c) the rate of heat transfer from the working fluid passing through the condenser to the cooling water, in kJ per kg of steam entering the first-stage turbine.

8.50 For the cycle of Problem 8.49, reconsider the analysis assuming the pump and each turbine stage have isentropic efficiencies of 80%. Answer the same questions as in Problem 8.49 for the modified cycle.

8.51 For the cycle of Problem 8.50, investigate the effects on cycle performance as the extraction pressure takes on other values. Assume that condensate drains from the closed feedwater heater as saturated liquid at the extraction pressure. Also, feedwater leaves the heater at 10 MPa and a temperature equal to the saturation temperature at the extraction pressure. Construct suitable plots and discuss.

8.52 A power plant operates on a regenerative vapor power cycle with one closed feedwater heater. Steam enters the first turbine stage at 120 bar, 520°C and expands to 10 bar, where some of the steam is extracted and diverted to a closed feedwater heater. Condensate exiting the feedwater heater as saturated liquid at 10 bar passes through a trap into the condenser. The feedwater exits the heater at 120 bar with a temperature of 170°C. The condenser pressure is 0.06 bar. For isentropic processes in each turbine stage and the pump, determine for the cycle (a) the thermal efficiency and (b) the mass flow rate into the first-stage turbine, in kg/h, if the net power developed is 320 MW.

8.53 Reconsider the cycle of Problem 8.52, but include in the analysis that each turbine stage has an isentropic efficiency of 82%. The pump efficiency remains 100%.

8.54 Modify the cycle of Problem 8.49 such that the saturated liquid condensate from the feedwater heater at 0.7 MPa is pumped into the feedwater line rather than being trapped into the condenser. Answer the same questions about the modified cycle as in Problem 8.49. List advantages and disadvantages of each scheme for removing condensate from the closed feedwater heater.

8.55 Modify the ideal Rankine cycle of Problem 8.9 to include one closed feedwater heater using extracted steam at 690 kPa. Condensate exiting the heater as saturated liquid at 690 kPa passes through a trap into the condenser. The feedwater leaves the heater at 11 MPa and a temperature equal to the saturation temperature at 690 kPa. The mass flow rate of steam entering the first-stage turbine is the same as the steam flow rate in Problem 8.9. Answer the same questions about the modified cycle as in Problem 8.9 and discuss the results.

8.56 Reconsider the cycle of Problem 8.55, but include in the analysis that each turbine stage has an isentropic efficiency of 88% and the pump efficiency is 80%.

8.57 Water is the working fluid in an ideal regenerative Rankine cycle with one closed feedwater heater. Superheated vapor enters the turbine at 16 MPa, 560°C, and the condenser pressure is 8 kPa. The cycle has a closed feedwater heater using extracted steam at 1 MPa. Condensate drains from the feedwater heater as saturated liquid at 1 MPa and is trapped into the condenser. The feedwater leaves the heater at 16 MPa and a temperature equal to the saturation temperature at 1 MPa. The mass flow rate of steam entering the first-stage turbine is 120 kg/s. Determine

(a) the net power developed, in kW.
(b) the rate of heat transfer to the steam passing through the boiler, in kW.
(c) the thermal efficiency.
(d) the mass flow rate of condenser cooling water, in kg/s, if the cooling water undergoes a temperature increase of 18°C with negligible pressure change in passing through the condenser.

8.58 Reconsider the cycle of Problem 8.57, but include in the analysis that the isentropic efficiencies of the turbine stages and the pump are 85%.

8.59 Referring to Fig. 8.12, if the fractions of the total flow entering the first turbine stage (state 1) extracted at states 2, 3, 6, and 7 are y_2, y_3, y_6, and y_7, respectively, what are the fractions of the total flow at states 8, 11, and 17?

8.60 Consider a regenerative vapor power cycle with two feedwater heaters, a closed one and an open one, as shown in Figure P8.60. Steam enters the first turbine stage at 12 MPa, 480°C, and expands to 2 MPa. Some steam is extracted at 2 MPa and fed to the closed feedwater heater. The remainder expands through the second-stage turbine to 0.3 MPa, where an additional amount is extracted and fed into the open feedwater heater operating at 0.3 MPa. The steam expanding through the third-stage turbine exits at the condenser pressure of 6 kPa.

Feedwater leaves the closed heater at 210°C, 12 MPa, and condensate exiting as saturated liquid at 2 MPa is trapped into the open feedwater heater. Saturated liquid at 0.3 MPa leaves the open feedwater heater. Assume all pumps and turbine stages operate isentropically. Determine for the cycle

(a) the rate of heat transfer to the working fluid passing through the steam generator, in kJ per kg of steam entering the first-stage turbine.
(b) the thermal efficiency.
(c) the rate of heat transfer from the working fluid passing through the condenser to the cooling water, in kJ per kg of steam entering the first-stage turbine.

8.61 For the cycle of Problem 8.60, reconsider the analysis assuming the pump and each turbine stage has an isentropic efficiency of 80%. Answer the same questions as in Problem 8.60 for the modified cycle.

8.62 For the cycle of Problem 8.60, investigate the effects on cycle performance as the higher extraction pressure takes on other values. The operating conditions for the open feedwater heater are unchanged from those in Problem 8.60. Assume that condensate drains from the closed feedwater heater as saturated liquid at the higher extraction pressure. Also, feedwater leaves the heater at 12 MPa and a temperature equal to the saturation temperature at the extraction pressure. Construct suitable plots and discuss.

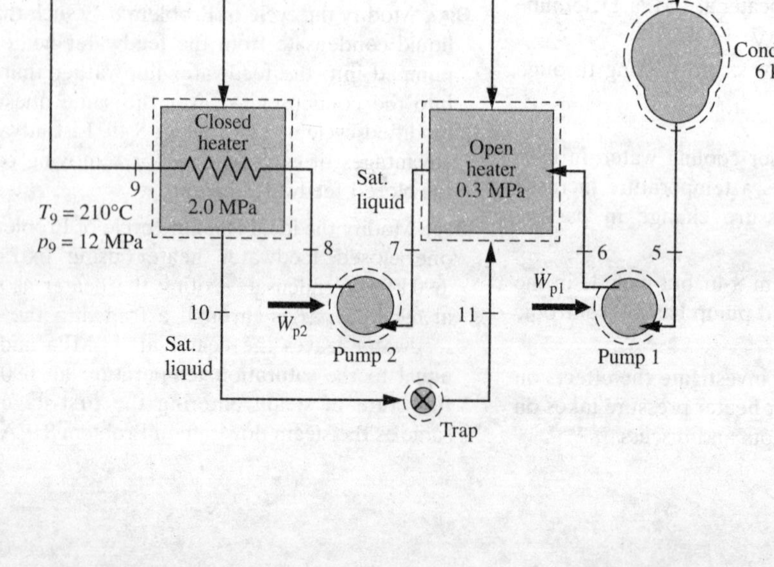

Fig. P8.60

8.63 A power plant operates on a regenerative vapor power cycle with two feedwater heaters. Steam enters the first turbine stage as 12 MPa, 520°C and expands in three stages to the condenser pressure of 6 kPa. Between the first and second stages, some steam is diverted to a closed feedwater heater at 1 MPa, with saturated liquid condensate being pumped ahead into the boiler feedwater line. The feedwater leaves the closed heater at 12 MPa, 170°C. Steam is extracted between the second and third turbine stages at 0.15 MPa and fed into an open feedwater heater operating at that pressure. Saturated liquid at 0.15 MPa leaves the open feedwater heater. For isentropic processes in the pumps and turbines, determine for the cycle (a) the thermal efficiency and (b) the mass flow rate into the first-stage turbine, in kg/h, if the net power developed is 320 MW.

8.64 Reconsider the cycle of Problem 8.63, but include in the analysis that each turbine stage has an isentropic efficiency of 82% and each pump an efficiency of 100%.

8.65 Modify the ideal Rankine cycle of Problem 8.9 to include one closed feedwater heater using extracted steam at 1.7 MPa and one open feedwater heater operating at 0.3 MPa. Saturated liquid condensate drains from the closed heater at 1.7 MPa and is trapped into the open heater. The feedwater leaves the closed heater at 11 MPa, 200°C, and saturated liquid leaves the open heater at 0.3 MPa. The mass flow rate of steam entering the first-stage turbine is the same as the steam flow rate in Problem 8.9. Answer the same questions about the modified cycle as in Problem 8.9 and discuss the results.

8.66 Reconsider the cycle of Problem 8.65, but include in the analysis that each turbine stage has an isentropic efficiency of 88% and each pump an efficiency of 80%.

8.67 Water is the working fluid in a Rankine cycle modified to include one closed feedwater heater and one open feedwater heater. Superheated vapor enters the turbine at 16 MPa, 560°C, and the condenser pressure is 8 kPa. The mass flow rate of steam entering the first-stage turbine is 120 kg/s. The closed feedwater heater uses extracted steam at 4 MPa, and the open feedwater heater uses extracted steam at 0.3 MPa. Saturated liquid condensate drains from the closed feedwater heater at 4 MPa and is trapped into the open feedwater heater. The feedwater leaves the closed heater at 16 MPa and a temperature equal to the saturation temperature at 4 MPa. Saturated liquid leaves the open heater at 0.3 MPa. Assume all turbine stages and pumps operate isentropically. Determine

(a) the net power developed, in kW.
(b) the rate of heat transfer to the steam passing through the steam generator, in kW.
(c) the thermal efficiency.
(d) the mass flow rate of condenser cooling water, in kg/s, if the cooling water undergoes a temperature increase of 18°C with negligible pressure change in passing through the condenser.

8.68 Reconsider the cycle of Problem 8.67, but include in the analysis that the isentropic efficiencies of the turbine stages and pumps are 85%.

8.69 Consider a regenerative vapor power cycle with two feedwater heaters, a closed one and an open one, and reheat. Steam enters the first turbine stage at 12 MPa, 480°C, and expands to 2 MPa. Some steam is extracted at 2 MPa and fed to the closed feedwater heater. The remainder is reheated at 2 MPa to 440°C and then expands through the second-stage turbine to 0.3 MPa, where an additional amount is extracted and fed into the open feedwater heater operating at 0.3 MPa. The steam expanding through the third-stage turbine exits at the condenser pressure of 6 kPa. Feedwater leaves the closed heater at 210°C, 12 MPa, and condensate exiting as saturated liquid at 2 MPa is trapped into the open feedwater heater. Saturated liquid at 0.3 MPa leaves the open feedwater heater. Assume all pumps and turbine stages operate isentropically. Determine for the cycle

(a) the rate of heat transfer to the working fluid passing through the steam generator, in kJ per kg of steam entering the first-stage turbine.
(b) the thermal efficiency.
(c) the rate of heat transfer from the working fluid passing through the condenser to the cooling water, in kJ per kg of steam entering the first-stage turbine.

8.70 Reconsider the cycle of Problem 8.69, but include in the analysis that the turbine stage and pumps all have isentropic efficiencies of 80%. Answer the same questions about the modified cycle as in Problem 8.69.

8.71 For the cycle of Problem 8.70, plot thermal efficiency versus turbine stage and pump isentropic efficiencies for values ranging from 80 to 100%. Discuss.

8.72 Modify the regenerative vapor power cycle in Problem 8.65 to include reheat at 1.7 MPa. The portion of the flow that passes through the second turbine stage is reheated to 482°C before it enters the turbine. Determine the thermal efficiency of the modified cycle.

8.73 Reconsider the cycle of Problem 8.72, but include in the analysis that each turbine stage has an isentropic efficiency of 88% and each pump an efficiency of 80%.

8.74 Steam enters the first turbine stage of a vapor power cycle with reheat and regeneration at 32 MPa, 600°C, and expands to 8 MPa. A portion of the flow is diverted to a closed feedwater heater at 8 MPa, and the remainder is reheated to 560°C before entering the second turbine stage. Expansion through the second turbine stage occurs to 1 MPa, where another portion of the flow is diverted to a second closed feedwater heater at 1 MPa. The remainder of the flow expands through the third turbine stage to 0.15 MPa, where a portion of the flow is diverted to an open feedwater heater operating at 0.15 MPa, and the rest expands through the fourth turbine stage to the condenser pressure of 6 kPa. Condensate leaves each closed feedwater heater as saturated liquid at the respective extraction pressure. The feedwater streams leave each closed feedwater heater at a temperature equal to the saturation temperature at the respective extraction pressure. The condensate streams from the closed heaters each pass through traps into the next lower-pressure feedwater heater. Saturated liquid exiting the open heater is pumped to the steam generator pressure. If each turbine stage has an isentropic efficiency of 85% and the pumps operate isentropically

(a) sketch the layout of the cycle and number the principal state points.
(b) determine the thermal efficiency of the cycle.
(c) calculate the mass flow rate into the first turbine stage, in kg/h, for a net power output of 500 MW.

8.75 Steam enters the first turbine stage of a vapor power plant with reheat and regeneration at 12.4 MPa, 590°C and expands in five stages to a condenser pressure of 6.9 kPa. Reheat is at 690 kPa to 538°C. The cycle includes three feedwater heaters. Closed heaters operate at 4.1 MPa and 1.1 MPa, with the drains from each trapped into the next lower-pressure feedwater heater. The feedwater leaving each closed heater is at the saturation temperature corresponding to the extraction pressure. An open feedwater heater operates at 0.14 MPa. The pumps operate isentropically, and each turbine stage has an isentropic efficiency of 88%.

(a) Sketch the layout of the cycle and number the principal state points.
(b) Determine the thermal efficiency of the cycle.
(c) Determine the heat rate, in kJ/kW · h.
(d) Calculate the mass flow rate into the first turbine stage, in kg/h, for a net power output of 3.2×10^9 kJ/h.

Considering Other Vapor Cycle Aspects

8.76 A binary vapor power cycle consists of two ideal Rankine cycles with steam and ammonia as the working fluids. In the steam cycle, superheated vapor enters the turbine at 6 MPa, 640°C, and saturated liquid exits the condenser at 60°C. The heat rejected from the steam cycle is provided to the ammonia cycle, producing saturated vapor at 50°C, which enters the ammonia turbine. Saturated liquid leaves the ammonia condenser at 1 MPa. For a net power output of 20 MW from the binary cycle, determine

(a) the power output of the steam and ammonia turbines, respectively, in MW.
(b) the rate of heat addition to the binary cycle, in MW.
(c) the thermal efficiency.

8.77 A binary vapor cycle consists of two Rankine cycles with steam and ammonia as the working fluids. In the steam cycle, superheated vapor enters the turbine at 6.2 MPa, 590°C, and saturated liquid exits the condenser at 60°C. The heat rejected from the steam cycle is provided to the ammonia cycle, producing saturated vapor at 49°C, which enters the ammonia turbine. Saturated liquid leaves the ammonia condenser at 24°C. Each turbine has an isentropic efficiency of 90% and the pumps operate isentropically. The net power output of the binary cycle is 7.4×10^7 kJ/h.

(a) Determine the quality at the exit of each turbine, the mass flow rate of each working fluid, in kg/h, and the overall thermal efficiency of the binary cycle.
(b) Compare the binary cycle performance to that of a single Rankine cycle using water as the working fluid and condensing at 24°C. The turbine inlet state, isentropic turbine efficiency, and net power output all remain the same.

8.78 Figure P8.78 shows a vapor power cycle with reheat and regeneration. The steam generator produces vapor at 6.9 MPa, 430°C. Some of this steam expands through the first turbine stage to 690 kPa and the remainder is directed to the heat exchanger. The steam exiting the first turbine stage enters the flash chamber. Saturated vapor and saturated liquid at 690 kPa exit the flash chamber as separate streams. The vapor is reheated in the heat exchanger to 277°C before entering the second turbine stage. The open feedwater heater operates at 690 kPa, and the condenser pressure is 6.9 kPa. Each turbine stage has an isentropic efficiency of 88% and the pumps operate isentropically. For a net power output of 5.27×10^9 kJ/h, determine

(a) the mass flow rate through the steam generator, in kg/h.
(b) the thermal efficiency of the cycle.
(c) the rate of heat transfer to the cooling water passing through the condenser, in kJ/h.

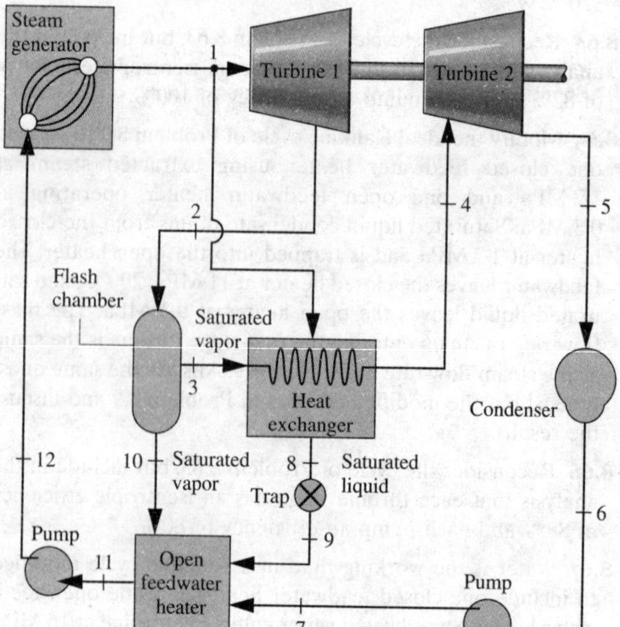

Fig. P8.78

8.79 Water is the working fluid in a cogeneration cycle that generates electricity and provides heat for campus buildings. Steam at 2 MPa, 320°C, enters a two-stage turbine with a mass flow rate of 0.82 kg/s. A fraction of the total flow, 0.141, is extracted between the two stages at 0.15 MPa to provide for building heating, and the remainder expands through the second stage to the condenser pressure of 0.06 bar. Condensate returns from the campus buildings at 0.1 MPa, 60°C and passes through a trap into the condenser, where it is reunited with the main feedwater flow. Saturated liquid leaves the condenser at 0.06 bar. Each turbine stage has an isentropic efficiency of 80%, and the pumping process can be considered isentropic. Determine

(a) the rate of heat transfer to the working fluid passing through the steam generator, in kJ/h.
(b) the net power developed, in kJ/h.
(c) the rate of heat transfer for building heating, in kJ/h.
(d) the rate of heat transfer to the cooling water passing through the condenser, in kJ/h.

8.80 Consider a cogeneration system operating as illustrated in Fig. 8.14. The steam generator provides 10^6 kg/h of steam at 8 MPa, 480°C, of which 4×10^5 kg/h is extracted between the first and second turbine stages at 1 MPa and diverted to a process heating load. Condensate returns from the process heating load at 0.95 MPa, 120°C and is mixed with liquid exiting the lower-pressure pump at 0.95 MPa. The entire flow is then pumped to the steam generator pressure. Saturated liquid at 8 kPa leaves the condenser. The turbine stages and the pumps operate with isentropic efficiencies of 86 and 80%, respectively. Determine

(a) the heating load, in kJ/h.
(b) the power developed by the turbine, in kW.
(c) the rate of heat transfer to the working fluid passing through the steam generator, in kJ/h.

8.81 Figure P8.81 shows a combined heat and power system (CHP) providing turbine power, process steam, and steam for a factory space heating load. Operating data are given on the figure at key states in the cycle. For the system, determine

(a) the rates that steam is extracted as process steam and for the heating load, each in kg/h.
(b) the rates of heat transfer for the process steam and the heating load, each in kJ/h.
(c) the net power developed, in kJ/h.

Devise and evaluate an overall energy-based efficiency for the combined heat and power system.

8.82 Figure P8.82 shows the schematic diagram of a cogeneration cycle. In the steam cycle, superheated vapor enters the turbine with a mass flow rate of 5 kg/s at 40 bar, 440°C and expands isentropically to 1.5 bar. Half of the flow is extracted at 1.5 bar and used for industrial process heating. The rest of the steam passes through a heat exchanger, which serves as the boiler of the Refrigerant 134a cycle and

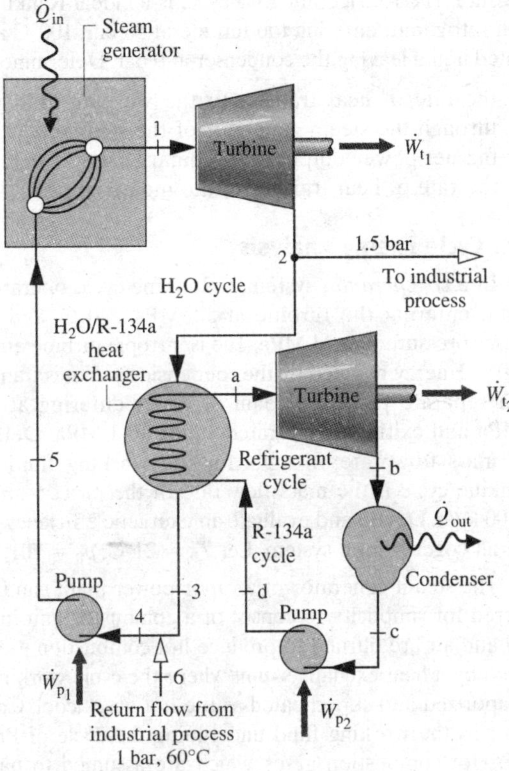

Fig. P8.82

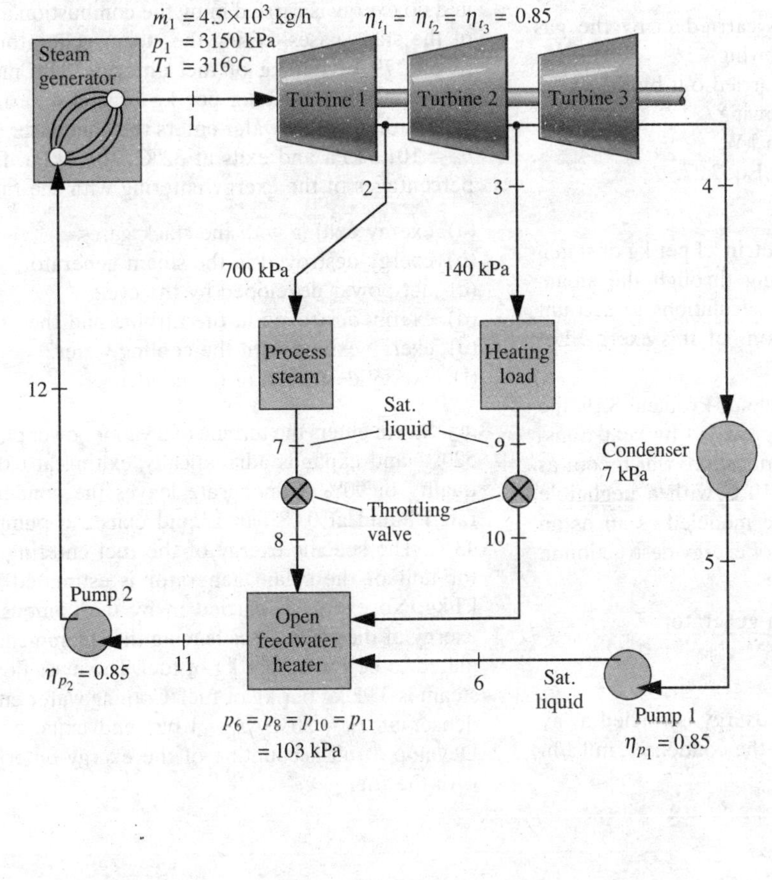

Fig. P8.81

the condenser of the steam cycle. The condensate leaves the heat exchanger as saturated liquid at 1 bar, where it is combined with the return flow from the process, at 60°C and 1 bar, before being pumped isentropically to the steam generator pressure. The Refrigerant 134a cycle is an ideal Rankine cycle with refrigerant entering the turbine at 16 bar, 100°C and saturated liquid leaving the condenser at 9 bar. Determine, in kW,

(a) the rate of heat transfer to the working fluid passing through the steam generator of the steam cycle.
(b) the net power output of the binary cycle.
(c) the rate of heat transfer to the industrial process.

Vapor Cycle Exergy Analysis

8.83 In a *cogeneration* system, a Rankine cycle operates with steam entering the turbine at 5.5 MPa, 370°C, and a condenser pressure of 1.24 MPa. The isentropic turbine efficiency is 80%. Energy rejected by the condensing steam is transferred to a separate process stream of water entering at 120°C, 1 MPa and exiting as saturated vapor at 1 MPa. Determine the mass flow rate, in kg/h, for the working fluid of the Rankine cycle if the mass flow rate of the process stream is 22700 kg/h. Devise and evaluate an exergetic efficiency for the overall cogeneration system. Let $T_0 = 21°C, p_0 = 101.3$ kPa.

8.84 The steam generator of a vapor power plant can be considered for simplicity to consist of a combustor unit in which fuel and air are burned to produce hot combustion gases, followed by a heat exchanger unit where the cycle working fluid is vaporized and superheated as the hot gases cool. Consider water as the working fluid undergoing the cycle of Problem 8.17. Hot combustion gases, which are assumed to have the properties of air, enter the heat exchanger portion of the steam generator at 1200 K and exit at 600 K with a negligible change in pressure. Determine for the heat exchanger unit

(a) the net rate at which exergy is carried in by the gas stream, in kJ per kg of steam flowing.
(b) the net rate at which exergy is carried out by the water stream, in kJ per kg of steam flowing.
(c) the rate of exergy destruction, in kW.
(d) the exergetic efficiency given by Eq. 7.27.

Let $T_0 = 15°C, p_0 = 0.1$ MPa

8.85 Determine the rate of exergy input, in kJ per kg of steam flowing, to the working fluid passing through the steam generator in Problem 8.17. Perform calculations to account for all outputs, losses, and destructions of this exergy. Let $T_0 = 15°C, p_0 = 0.1$ MPa.

8.86 In the steam generator of the cycle of Problem 8.19, the energy input to the working fluid is provided by heat transfer from hot gaseous products of combustion, which cool as a separate stream from 190°C to 810°C with a negligible pressure drop. The gas stream can be modeled as air as an ideal gas. Determine, in kJ/h, the rate of exergy destruction in the

(a) heat exchanger unit of the steam generator.
(b) turbine and pump.
(c) condenser.

Also, calculate the net rate at which exergy is carried away by the cooling water passing through the condenser, in kJ/h. Let $T_0 = 16°C, p_0 = 101.3$ kPa.

8.87 For the regenerative vapor power cycle of Problem 8.67, calculate the rates of exergy destruction in the feedwater heaters in kW. Express each as a fraction of the flow exergy increase of the working fluid passing through the steam generator. Let $T_0 = 16°C, p_0 = 1$ bar.

8.88 Determine the rate of exergy input, in MW, to the working fluid passing through the steam generator in Problem 8.74. Perform calculations to account for all outputs, losses, and destructions of this exergy. Let $T_0 = 15°C, p_0 = 1$ bar.

8.89 Determine the rate of exergy input, in kJ/h, to the working fluid passing through the steam generator in Problem 8.75. Perform calculations to account for all outputs, losses, and destructions of this exergy. Let $T_0 = 15°C$, $p_0 = 1$ bar.

8.90 Determine the rate of exergy transfer, in kJ/h, to the working fluid passing through the steam generator in Problem 8.78. Perform calculations to account for all outputs, losses, and destructions of this exergy. Let $T_0 = 15°C$, $p_0 = 1$ bar.

8.91 Determine the rate of exergy transfer, in kJ per kg of steam entering the first-stage turbine, to the working fluid passing through the steam generator in Problem 8.46. Perform calculations to account for all outputs, losses, and destructions of this exergy. Let $T_0 = 15°C$, $p_0 = 0.1$ MPa.

8.92 Steam enters the turbine of a vapor power plant at 482°C, 3.5 MPa and expands adiabatically, exiting at 6.9 kPa with a quality of 97%. Condensate leaves the condenser as saturated liquid at 6.9 kPa. The isentropic pump efficiency is 80%. The specific exergy of the fuel entering the combustor unit of the steam generator is estimated to be 51920 kJ/kg, and no exergy is carried in by the combustion air. The exergy of the stack gases exiting the steam generator is estimated to be 779 kJ per kg of fuel entering. The mass flow rate of the steam is 15.1 kg per kg of fuel entering the steam generator. Cooling water enters the condenser at $T_0 = 25°C$, $p_0 = 101.3$ kPa and exits at 32°C, 101.3 kPa. Determine, as percentages of the exergy entering with the fuel, the

(a) exergy exiting with the stack gases.
(b) exergy destroyed in the steam generator.
(c) net power developed by the cycle.
(d) exergy destroyed in the turbine and the pump.
(e) exergy exiting with the cooling water.
(f) exergy destroyed in the condenser.

8.93 Steam enters the turbine of a vapor power plant at 100 bar, 520°C and expands adiabatically, exiting at 0.08 bar with a quality of 90%. Condensate leaves the condenser as saturated liquid at 0.08 bar. Liquid exits the pump at 100 bar, 43°C. The specific exergy of the fuel entering the combustor unit of the steam generator is estimated to be 14,700 kJ/kg. No exergy is carried in by the combustion air. The exergy of the stack gases leaving the steam generator is estimated to be 150 kJ per kg of fuel. The mass flow rate of the steam is 3.92 kg per kg of fuel. Cooling water enters the condenser at $T_0 = 20°C$, $p_0 = 1$ bar and exits at 35°C, 1 bar. Develop a full accounting of the exergy entering the plant with the fuel.

Design & open ended problems: exploring engineering practice

8.1D Develop the preliminary design for a power plant to produce 1000 MW of electricity using water as the working fluid and a supercritical pressure of 24 MPa and with a maximum cycle temperature of 600°C. Include in your specifications whether reheat and regeneration should be included, and if so, at what pressures. Show a schematic of your cycle layout and include in your report appropriate analysis, including computer plots, to justify the configuration you recommend.

8.2D Vast quantities of water circulate through the condensers of large power plants, exiting at temperatures 10 to 15°C above the ambient temperature. For a 500-MW power plant, investigate possible uses of the condenser cooling water. Suggest a particular application. For that application, estimate how much exergy would be provided annually and the annual economic value of that exergy. Discuss the feasibility of your proposed use.

8.3D Early commercial vapor power plants operated with turbine inlet conditions of about 12 bar and 200°C. Plants are under development today that can operate at over 34 MPa, with turbine inlet temperatures of 650°C or higher. How have steam generator and turbine designs changed over the years to allow for such increases in pressure and temperature? Discuss.

8.4D Study the feasibility of installing a 500-MW field of wind turbines off the coast of Chicago in Lake Michigan. Determine the number and type of turbines needed, estimate the cost of owning and operating the system, and perform an economic analysis. Investigate the availability of federal and state tax credits as part of your economic analysis. Compare the cost per kilowatt-hour for your proposed system with the average cost of electricity in the Chicago area from conventional fossil-fueled and nuclear power plants. Write a report of your findings including at least three references.

8.5D Consider the feasibility of using *biomass* to fuel a 200-MW electric power plant in a rural area in your locale. Analyze the advantages and disadvantages of biomass in comparison with coal and natural gas. Include in your analysis material handling issues, plant operations, environmental considerations, and costs. Prepare a slide presentation of your recommendations.

8.6D A 450 m^2 research outpost is being designed for studying global change in Antarctica. The outpost will house five scientists along with their communication and research equipment. Develop the preliminary design of an array of *photovoltaic cells* to provide all necessary power from solar energy during months of continuous sunlight. Specify the number and type of cells needed and present schematics of the system you propose.

8.7D Owing to strong local winds and large elevation differences on the Hawaiian island of Maui, it may be a suitable place to combine a wind farm with *pumped hydro* energy storage. At times when the wind turbines produce excess power, water is pumped to reservoirs at higher elevations. The water is released during periods of high electric demand through hydraulic turbines to produce electricity. Develop a proposal to meet 30% of the island's power needs by the year 2020 using this renewable energy concept. In your report, list advantages and disadvantages of the proposed system. Include at least three references.

8.8D Many hospitals, industrial facilities, college campuses, and towns employ *cogeneration* plants that provide district heating/cooling and generate electricity. Sketch the main components and layout of such an existing cogeneration system in your locale. Determine the heating and cooling loads served by the system and its electrical generating capacity. Estimate the cost savings observed through the use of cogeneration compared to meeting these loads individually. Discuss.

8.9D A significant supply of *geothermal* hot water at 12 bar, 180°C exists at a particular location. Develop the preliminary design of a 10-MW electric power plant using *flash evaporation* to provide steam to the turbine-generator. Show system schematics and specify operating pressures and temperatures within the cycle. Propose a way to dispose of the excess water and spent brine from the flashing process.

8.10D Figure P8.10D shows a *thermosyphon* Rankine power plant having one end immersed in a *solar pond* and the other exposed to the ambient air. At the lower end, a working fluid is vaporized by heat transfer from the hot water of the solar pond. The vapor flows upward through an axial-flow turbine to the upper end, where condensation occurs by heat transfer to the ambient air. The condensate drains to the lower end by gravity. Evaluate the feasibility of producing power in the 0.1–0.25 kW range using such a device. What working fluids would be suitable? Write a report of your findings.

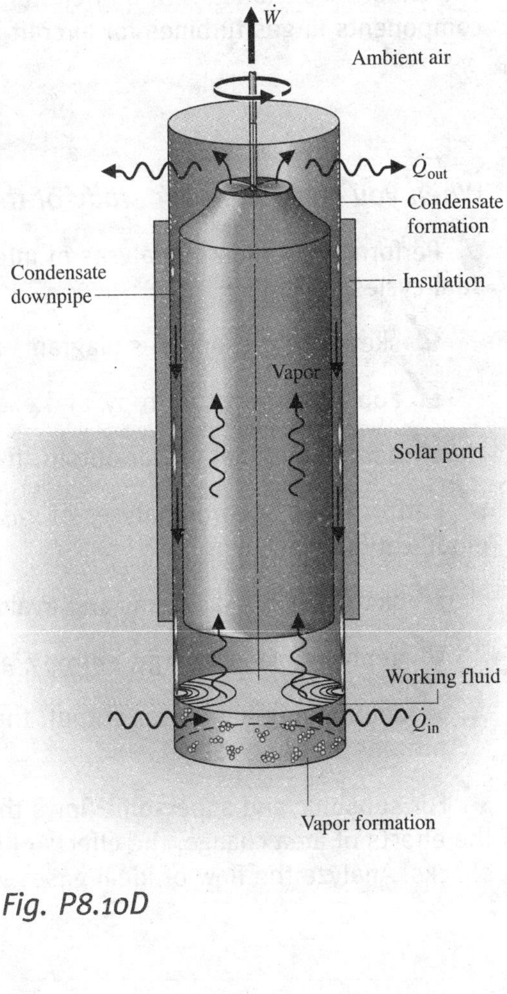

Fig. P8.10D

9

Gas Power Systems

Engineering Context

The vapor power systems studied in Chap. 8 use working fluids that are alternately vaporized and condensed. The **objective** of the present chapter is to study power systems utilizing working fluids that are always a gas. Included in this group are gas turbines and internal combustion engines of the spark-ignition and compression-ignition types. In the first part of the chapter, internal combustion engines are considered. Gas turbine power plants are discussed in the second part of the chapter. The chapter concludes with a brief study of compressible flow in nozzles and diffusers, which are components in gas turbines for aircraft propulsion and other devices of practical importance.

Learning Objectives

When you complete your study of this chapter, you will be able to...

- ✔ Perform air-standard analyses of internal combustion engines based on the Otto, Diesel, and dual cycles, including:
  - ✔ sketching *p–v* and *T–s* diagrams and evaluating property data at principal states.
  - ✔ applying energy, entropy, and exergy balances.
  - ✔ determining net power output, thermal efficiency, and mean effective pressure.
- ✔ Perform air-standard analyses of gas turbine power plants based on the Brayton cycle and its modifications, including:
  - ✔ sketching *T–s* diagrams and evaluating property data at principal states.
  - ✔ applying mass, energy, entropy, and exergy balances.
  - ✔ determining net power output, thermal efficiency, back work ratio, and the effects of compressor pressure ratio.
- ✔ For subsonic and supersonic flows through nozzles and diffusers, demonstrate understanding of the effects of area change, the effects of back pressure and the occurrence of choking and normal shocks. Analyze the flow of ideal gases with constant specific heats.

Internal Combustion Engines

This part of the chapter deals with *internal* combustion engines. Although most gas turbines are also internal combustion engines, the name is usually applied to *reciprocating* internal combustion engines of the type commonly used in automobiles, trucks, and buses. These engines differ from the power plants considered in Chap. 8 because the processes occur within reciprocating piston–cylinder arrangements and not in interconnected series of different components.

Two principal types of reciprocating internal combustion engines are the ***spark-ignition*** engine and the ***compression-ignition*** engine. In a spark-ignition engine, a mixture of fuel and air is ignited by a spark plug. In a compression-ignition engine, air is compressed to a high enough pressure and temperature that combustion occurs spontaneously when fuel is injected. Spark-ignition engines have advantages for applications requiring power up to about 225 kW. Because they are relatively light and lower in cost, spark-ignition engines are particularly suited for use in automobiles. Compression-ignition engines are normally preferred for applications when fuel economy and relatively large amounts of power are required (heavy trucks and buses, locomotives and ships, auxiliary power units). In the middle range, spark-ignition and compression-ignition engines are used.

spark-ignition
compression-ignition

9.1 Introducing Engine Terminology

Figure 9.1 is a sketch of a reciprocating internal combustion engine consisting of a piston that moves within a cylinder fitted with two valves. The sketch is labeled with some special terms. The *bore* of the cylinder is its diameter. The *stroke* is the distance the piston moves in one direction. The piston is said to be at *top dead center* when it has moved to a position where the cylinder volume is a minimum. This minimum volume is known as the *clearance* volume. When the piston has moved to the position of maximum cylinder volume, the piston is at *bottom dead center*. The volume swept out by the piston as it moves from the top dead center to the bottom dead center position is called the *displacement volume*. The ***compression ratio*** r is defined as the volume at bottom dead center divided by the volume at top dead center. The reciprocating motion of the piston is converted to rotary motion by a crank mechanism.

compression ratio

In a *four-stroke* internal combustion engine, the piston executes four distinct strokes within the cylinder for every two revolutions of the crankshaft. Fig. 9.2 gives a pressure–volume diagram such as might be displayed electronically.

1. With the intake valve open, the piston makes an *intake stroke* to draw a fresh charge into the cylinder. For spark-ignition engines, the charge is a combustible mixture of fuel and air. Air alone is the charge in compression-ignition engines.
2. With both valves closed, the piston undergoes a *compression stroke*, raising the temperature and pressure of the charge. This requires work input from the piston to the cylinder contents. A combustion process is then initiated, resulting in a high-pressure, high-temperature gas mixture. Combustion is induced near the end of the compression stroke in spark-ignition engines by the spark plug. In compression-ignition engines, combustion is initiated by injecting fuel into the hot compressed air, beginning near the

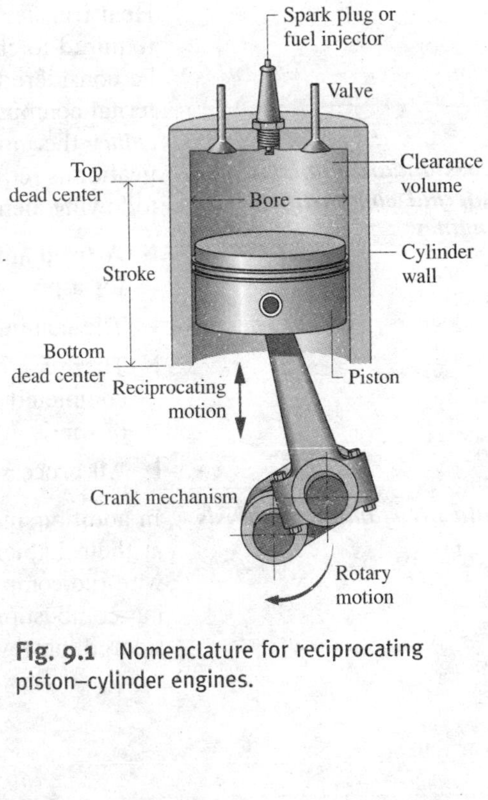

Fig. 9.1 Nomenclature for reciprocating piston–cylinder engines.

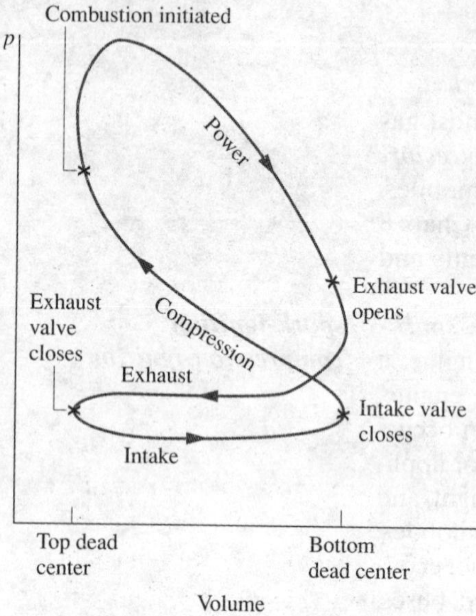

Fig. 9.2 Pressure–volume diagram for a reciprocating internal combustion engine.

end of the compression stroke and continuing through the first part of the expansion.

3. A *power* stroke follows the compression stroke, during which the gas mixture expands and work is done on the piston as it returns to bottom dead center.
4. The piston then executes an *exhaust stroke* in which the burned gases are purged from the cylinder through the open exhaust valve.

Smaller engines operate on *two-stroke* cycles. In two-stroke engines, the intake, compression, expansion, and exhaust operations are accomplished in one revolution of the crankshaft. Although internal combustion engines undergo *mechanical* cycles, the cylinder contents do not execute a *thermodynamic* cycle, for matter is introduced with one composition and is later discharged at a different composition.

A parameter used to describe the performance of reciprocating piston engines is the *mean effective pressure,* or mep. The ***mean effective pressure*** is the theoretical constant pressure that, if it acted on the piston during the power stroke, would produce the same *net* work as actually developed in one cycle. That is

mean effective pressure

$$\text{mep} = \frac{\text{net work for one cycle}}{\text{displacement volume}} \tag{9.1}$$

For two engines of equal displacement volume, the one with a higher mean effective pressure would produce the greater net work and, if the engines run at the same speed, greater power.

AIR-STANDARD ANALYSIS. A detailed study of the performance of a reciprocating internal combustion engine would take into account many features. These would include the combustion process occurring within the cylinder and the effects of irreversibilities associated with friction and with pressure and temperature gradients. Heat transfer between the gases in the cylinder and the cylinder walls and the work required to charge the cylinder and exhaust the products of combustion also would be considered. Owing to these complexities, accurate modeling of reciprocating internal combustion engines normally involves computer simulation. To conduct *elementary* thermodynamic analyses of internal combustion engines, considerable simplification is required. One procedure is to employ an ***air-standard analysis*** having the following elements:

air-standard analysis: internal combustion engines

- A fixed amount of air modeled as an ideal gas is the working fluid. See Table 9.1 for a review of ideal gas relations.
- The combustion process is replaced by a heat transfer from an external source.
- There are no exhaust and intake processes as in an actual engine. The cycle is completed by a constant-volume heat transfer process taking place while the piston is at the bottom dead center position.
- All processes are internally reversible.

cold air-standard analysis

In addition, in a ***cold air-standard analysis,*** the specific heats are assumed constant at their ambient temperature values. With an air-standard analysis, we avoid dealing with the complexities of the combustion process and the change of composition during combustion. A comprehensive analysis requires that such complexities be considered, however. For a discussion of combustion, see Chap. 13.

Although an air-standard analysis simplifies the study of internal combustion engines considerably, values for the mean effective pressure and operating temperatures and pressures calculated on this basis may depart significantly from those of actual engines. Accordingly, air-standard analysis allows internal combustion engines to be examined only qualitatively. Still, insights concerning actual performance can result with such an approach.

Table 9.1

Ideal Gas Model Review

Equations of State:

$$pv = RT \quad (3.32)$$

$$pV = mRT \quad (3.33)$$

Changes in u and h:

$$u(T_2) - u(T_1) = \int_{T_1}^{T_2} c_v(T)\,dT \quad (3.40)$$

$$h(T_2) - h(T_1) = \int_{T_1}^{T_2} c_p(T)\,dT \quad (3.43)$$

Constant Specific Heats	Variable Specific Heats
$u(T_2) - u(T_1) = c_v(T_2 - T_1)$ (3.50) $h(T_2) - h(T_1) = c_p(T_2 - T_1)$ (3.51) See Tables A-20, 21 for data.	$u(T)$ and $h(T)$ are evaluated from appropriate tables: Tables A-22 for air (mass basis) and A-23 for other gases (molar basis).

Changes in s:

$$s(T_2, v_2) - s(T_1, v_1) = \int_{T_1}^{T_2} c_v(T)\frac{dT}{T} + R\ln\frac{v_2}{v_1} \quad (6.17)$$

$$s(T_2, p_2) - s(T_1, p_1) = \int_{T_1}^{T_2} c_p(T)\frac{dT}{T} - R\ln\frac{p_2}{p_1} \quad (6.18)$$

Constant Specific Heats	Variable Specific Heats
$s(T_2, v_2) - s(T_1, v_1) = c_v \ln\frac{T_2}{T_1} + R\ln\frac{v_2}{v_1}$ (6.21) $s(T_2, p_2) - s(T_1, p_1) = c_p \ln\frac{T_2}{T_1} + R\ln\frac{p_2}{p_1}$ (6.22) See Tables A-20, 21 for data.	$s(T_2, p_2) - s(T_1, p_1) = s°(T_2) - s°(T_1) - R\ln\frac{p_2}{p_1}$ (6.20a) where $s°(T)$ is evaluated from appropriate tables: Tables A-22 for air (mass basis) and A-23 for other gases (molar basis).

Relating states of equal specific entropy: $\Delta s = 0$:

Constant Specific Heats	Variable Specific Heats − Air Only
$\frac{T_2}{T_1} = \left(\frac{p_2}{p_1}\right)^{(k-1)/k}$ (6.43) $\frac{T_2}{T_1} = \left(\frac{v_1}{v_2}\right)^{k-1}$ (6.44) $\frac{p_2}{p_1} = \left(\frac{v_1}{v_2}\right)^{k}$ (6.45) where $k = c_p/c_v$ is given in Tables A-20 for several gases.	$\frac{p_2}{p_1} = \frac{p_{r2}}{p_{r1}}$ (air only) (6.41) $\frac{v_2}{v_1} = \frac{v_{r2}}{v_{r1}}$ (air only) (6.42) where p_r and v_r are provided for air in Table A-22.

Energy & Environment

Nature needed 500 million years to create the world's stock of readily accessible oil, but some observers predict we will consume most of what's left within the next 50 years. Still, the important issue, they say, is not when the world runs out of oil, but when production will peak. After that, production must decline, and unless demand is reduced, oil prices will rise. This will end the era of cheap oil we enjoy and pose challenges for society.

The rate any well can produce oil typically rises to a maximum and then, when about half the oil has been pumped out, begins to fall as the remaining oil becomes increasingly difficult to extract. Using this model for the world oil supply as a whole, economists predict a peak in oil production by 2020, or even as soon as 2010. That will lead to shortages and higher prices at the pump, which will have global economic and political repercussions.

In the remainder of this part of the chapter, we consider three cycles that adhere to air-standard cycle idealizations: the Otto, Diesel, and dual cycles. These cycles differ from each other only in the way the heat addition process that replaces combustion in the actual cycle is modeled.

9.2 Air-Standard Otto Cycle

Otto cycle

The air-standard Otto cycle is an ideal cycle that assumes the heat addition occurs instantaneously while the piston is at top dead center. The ***Otto cycle*** is shown on the p–v and T–s diagrams of Fig. 9.3. The cycle consists of four internally reversible processes in series:

- ▶ Process 1–2 is an isentropic compression of the air as the piston moves from bottom dead center to top dead center.
- ▶ Process 2–3 is a constant-volume heat transfer to the air from an external source while the piston is at top dead center. This process is intended to represent the ignition of the fuel–air mixture and the subsequent rapid burning.
- ▶ Process 3–4 is an isentropic expansion (power stroke).
- ▶ Process 4–1 completes the cycle by a constant-volume process in which heat is rejected from the air while the piston is at bottom dead center.

Since the air-standard Otto cycle is composed of internally reversible processes, areas on the T–s and p–v diagrams of Fig. 9.3 can be interpreted as heat and work, respectively. On the T–s diagram, area 2–3–a–b–2 represents the heat added per unit

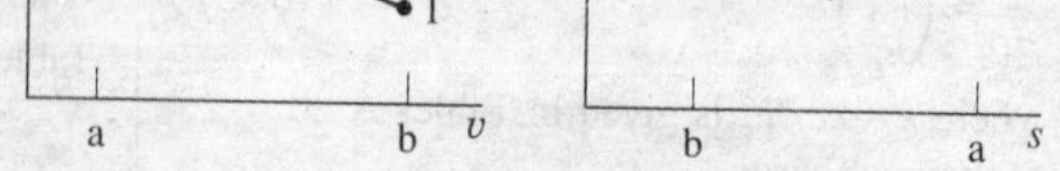

Fig. 9.3 p–v and T–s diagrams of the air-standard Otto cycle.

of mass and area 1–4–a–b–1 the heat rejected per unit of mass. On the p–v diagram, area 1–2–a–b–1 represents the work input per unit of mass during the compression process and area 3–4–b–a–3 is the work done per unit of mass in the expansion process. The enclosed area of each figure can be interpreted as the net work output or, equivalently, the net heat added.

CYCLE ANALYSIS. The air-standard Otto cycle consists of two processes in which there is work but no heat transfer, Processes 1–2 and 3–4, and two processes in which there is heat transfer but no work, Processes 2–3 and 4–1. Expressions for these energy transfers are obtained by reducing the closed system energy balance assuming that changes in kinetic and potential energy can be ignored. The results are

$$\frac{W_{12}}{m} = u_2 - u_1, \qquad \frac{W_{34}}{m} = u_3 - u_4$$
$$\frac{Q_{23}}{m} = u_3 - u_2, \qquad \frac{Q_{41}}{m} = u_4 - u_1 \tag{9.2}$$

Carefully note that in writing Eqs. 9.2, we have departed from our usual sign convention for heat and work. Thus, W_{12}/m is a positive number representing the work *input* during compression and Q_{41}/m is a positive number representing the heat *rejected* in Process 4–1. The *net work* of the cycle is expressed as

Take Note...

When analyzing air-standard cycles, it is frequently convenient to regard all work and heat transfers as positive quantities and write the energy balance accordingly.

$$\frac{W_{\text{cycle}}}{m} = \frac{W_{34}}{m} - \frac{W_{12}}{m} = (u_3 - u_4) - (u_2 - u_1)$$

Alternatively, the net work can be evaluated as the *net heat added*

$$\frac{W_{\text{cycle}}}{m} = \frac{Q_{23}}{m} - \frac{Q_{41}}{m} = (u_3 - u_2) - (u_4 - u_1)$$

which, on rearrangement, can be placed in the same form as the previous expression for net work.

The thermal efficiency is the ratio of the net work of the cycle to the heat added.

$$\eta = \frac{(u_3 - u_2) - (u_4 - u_1)}{u_3 - u_2} = 1 - \frac{u_4 - u_1}{u_3 - u_2} \tag{9.3}$$

When air table data are used to conduct an analysis involving an air-standard Otto cycle, the specific internal energy values required by Eq. 9.3 can be obtained from Table A-22 as appropriate. The following relationships based on Eq. 6.42 apply for the isentropic processes 1–2 and 3–4

$$v_{r2} = v_{r1}\left(\frac{V_2}{V_1}\right) = \frac{v_{r1}}{r} \tag{9.4}$$

$$v_{r4} = v_{r3}\left(\frac{V_4}{V_3}\right) = r v_{r3} \tag{9.5}$$

where r denotes the compression ratio. Note that since $V_3 = V_2$ and $V_4 = V_1$, $r = V_1/V_2 = V_4/V_3$. The parameter v_r is tabulated versus temperature for air in Tables A-22.

When the Otto cycle is analyzed on a cold air-standard basis, the following expressions based on Eq. 6.44 would be used for the isentropic processes in place of Eqs. 9.4 and 9.5, respectively

$$\frac{T_2}{T_1} = \left(\frac{V_1}{V_2}\right)^{k-1} = r^{k-1} \quad \text{(constant } k\text{)} \tag{9.6}$$

$$\frac{T_4}{T_3} = \left(\frac{V_3}{V_4}\right)^{k-1} = \frac{1}{r^{k-1}} \quad \text{(constant } k\text{)} \tag{9.7}$$

where k is the specific heat ratio, $k = c_p/c_v$.

EFFECT OF COMPRESSION RATIO ON PERFORMANCE. By referring to the T–s diagram of Fig. 9.3, we can conclude that the Otto cycle thermal efficiency increases as the compression ratio increases. An increase in the compression ratio changes the cycle from 1–2–3–4–1 to 1–2′–3′–4–1. Since the average temperature of heat addition is greater in the latter cycle and both cycles have the same heat rejection process, cycle 1–2′–3′–4–1 would have the greater thermal efficiency. The increase in thermal efficiency with compression ratio is also brought out simply by the following development on a cold air-standard basis. For constant c_v, Eq. 9.3 becomes

$$\eta = 1 - \frac{c_v(T_4 - T_1)}{c_v(T_3 - T_2)}$$

On rearrangement

$$\eta = 1 - \frac{T_1}{T_2}\left(\frac{T_4/T_1 - 1}{T_3/T_2 - 1}\right)$$

From Eqs. 9.6 and 9.7 above, $T_4/T_1 = T_3/T_2$, so

$$\eta = 1 - \frac{T_1}{T_2}$$

Finally, introducing Eq. 9.6

$$\eta = 1 - \frac{1}{r^{k-1}} \quad \text{(constant } k\text{)} \tag{9.8}$$

Equation 9.8 indicates that the cold air-standard Otto cycle thermal efficiency is a function of compression ratio and k. This relationship is shown in Fig. 9.4 for $k = 1.4$, representing ambient air.

The foregoing discussion suggests that it is advantageous for internal combustion engines to have high compression ratios, and this is the case. The possibility of autoignition, or "knock," places an upper limit on the compression ratio of spark-ignition engines, however. After the spark has ignited a portion of the fuel–air mixture, the rise in pressure accompanying combustion compresses the remaining charge. Autoignition can occur if the temperature of the unburned mixture becomes too high before the mixture is consumed by the flame front. Since the temperature attained by the air–fuel mixture during the compression stroke increases as the compression ratio increases, the likelihood of autoignition occurring increases with the compression ratio. Autoignition may result in high-pressure waves in the cylinder (manifested by a knocking or pinging sound) that can lead to loss of power as well as engine damage. Fuels formulated with tetraethyl lead are resistant to autoignition and thus allow relatively high compression ratios. The *unleaded* gasoline in common use today because of environmental concerns over air pollution limits the compression ratios of spark-ignition engines to approximately 9. Higher compression ratios can be achieved in compression-ignition engines because air alone is compressed. Compression ratios in the range of 12 to 20

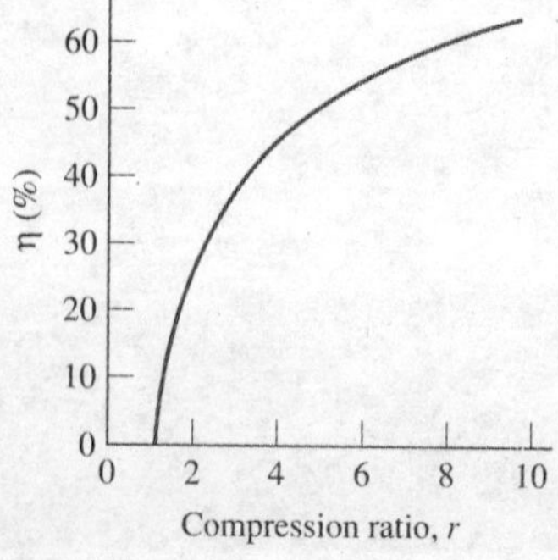

Fig. 9.4 Thermal efficiency of the cold air-standard Otto cycle, $k = 1.4$.

are typical. Compression-ignition engines also can use less refined fuels having higher ignition temperatures than the volatile fuels required by spark-ignition engines.

In the next example, we illustrate the analysis of the air-standard Otto cycle. Results are compared with those obtained on a cold air-standard basis.

Example 9.1 ANALYZING THE OTTO CYCLE

The temperature at the beginning of the compression process of an air-standard Otto cycle with a compression ratio of 8 is 300°K, the pressure is 1 bar, and the cylinder volume is 560 cm^3. The maximum temperature during the cycle is 2000°K. Determine **(a)** the temperature and pressure at the end of each process of the cycle, **(b)** the thermal efficiency, and **(c)** the mean effective pressure, in bars.

Solution

Known: An air-standard Otto cycle with a given value of compression ratio is executed with specified conditions at the beginning of the compression stroke and a specified maximum temperature during the cycle.

Find: Determine the temperature and pressure at the end of each process, the thermal efficiency, and mean effective pressure, in bars.

Schematic and Given Data:

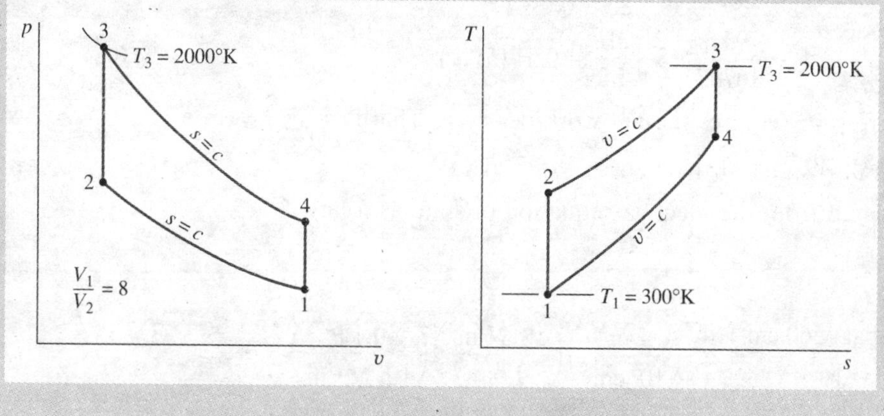

Fig. E9.1

Engineering Model:

1. The air in the piston–cylinder assembly is the closed system.
2. The compression and expansion processes are adiabatic.
3. All processes are internally reversible.
4. The air is modeled as an ideal gas.
5. Kinetic and potential energy effects are negligible.

Analysis:

(a) The analysis begins by determining the temperature, pressure, and specific internal energy at each principal state of the cycle. At T_1 = 300°K, Table A-22 gives u_1 = 214.07 kJ/kg and v_{r1} = 621.2.

For the isentropic compression Process 1–2

$$v_{r2} = \frac{V_2}{V_1} v_{r1} = \frac{v_{r1}}{r} = \frac{621.2}{8} = 77.65$$

Interpolating with v_{r2} in Table A-22, we get $T_2 = 673$ K and $u_2 = 491.2$ kJ/kg. With the ideal gas equation of state

$$p_2 = p_1 \frac{T_2}{T_1} \frac{V_1}{V_2} = (1 \text{ bar}) \left(\frac{673°\text{K}}{300°\text{K}} \right) 8 = 17.95 \text{ bars}$$

The pressure at state 2 can be evaluated alternatively by using the isentropic relationship, $p_2 = p_1(p_{r2}/p_{r1})$.

Since Process 2–3 occurs at constant volume, the ideal gas equation of state gives

$$p_3 = p_2 \frac{T_3}{T_2} = (17.95 \text{ bars}) \left(\frac{2000 \text{ K}}{673 \text{ K}} \right) = 53.3 \text{ bars}$$

At $T_3 = 2000$ K, Table A-22 gives $u_3 = 1678.7$ kJ/kg and $v_{r3} = 2.776$.

For the isentropic expansion process 3–4

$$v_{r4} = v_{r3} \frac{V_4}{V_3} = v_{r3} \frac{V_1}{V_2} = 2.776(8) = 22.21$$

Interpolating in Table A-22 with v_{r4} gives $T_4 = 1043$ K, $u_4 = 795.8$ kJ/kg. The pressure at state 4 can be found using the isentropic relationship $p_4 = p_3(p_{r4}/p_{r3})$ or the ideal gas equation of state applied at states 1 and 4. With $V_4 = V_1$, the ideal gas equation of state gives

$$p_4 = p_1 \frac{T_4}{T_1} = (1 \text{ bar}) \left(\frac{1043 \text{ K}}{300 \text{ K}} \right) = 3.48 \text{ bars}$$

(b) The thermal efficiency is

$$\eta = 1 - \frac{Q_{41}/m}{Q_{23}/m} = 1 - \frac{u_4 - u_1}{u_3 - u_2}$$

$$= 1 - \frac{795.8 - 214.07}{1678.7 - 491.2} = 0.51 \ (51\%)$$

(c) To evaluate the mean effective pressure requires the net work per cycle. That is

$$W_{\text{cycle}} = m[(u_3 - u_4) - (u_2 - u_1)]$$

where m is the mass of the air, evaluated from the ideal gas equation of state as follows:

$$m = \frac{p_1 V_1}{(\bar{R}/M)T_1}$$

$$= \frac{(1 \text{ bar})(560 \text{ cm}^3)}{\left(\frac{8.314}{28.97} \frac{\text{kJ}}{\text{kg} \cdot \text{K}} \right)(300 \text{ K})} \left(\frac{1 \text{ m}^3}{10^6 \text{ cm}^3} \right) \left(\frac{10^5 \text{ N/m}^2}{1 \text{ bar}} \right) \left(\frac{1 \text{ kJ}}{10^3 \text{ N} \cdot \text{m}} \right)$$

$$= 6.5 \times 10^{-4} \text{ kg}$$

Inserting values into the expression for W_{cycle}

$$W_{\text{cycle}} = (6.5 \times 10^{-4} \text{ kg}) [(1678.7 - 795.8) - (491.2 - 214.07)] \text{ kJ/kg}$$

$$= 0.394 \text{ kJ}$$

The displacement volume is $V_1 - V_2$, so the mean effective pressure is given by

$$\text{mep} = \frac{W_{\text{cycle}}}{V_1 - V_2} = \frac{W_{\text{cycle}}}{V_1(1 - V_2/V_1)}$$

❶ $$= \frac{0.394 \text{ kJ}}{(5601.3)/1 - \text{kg}} \left(\frac{10^6 \text{ cm}^3}{1 \text{ m}^3} \right) \left(\frac{10^3 \text{ N} \cdot \text{m}}{1 \text{ kJ}} \right) \left(\frac{1 \text{ bar}}{10^5 \text{ N} \cdot \text{m}} \right)$$

$$= 8.04 \text{ bar}$$

❶ This solution utilizes Table A-22 for air, which accounts explicitly for the variation of the specific heats with temperature. A solution also can be developed on a cold air-standard basis in which constant specific heats are assumed. This solution is left as an exercise, but for comparison the results are presented for the case $k = 1.4$ in the following table:

Parameter	Air-Standard Analysis	Cold Air-Standard Analysis, $k = 1.4$
T_2	673 K	689 K
T_3	2000 K	2000 K
T_4	1043 K	871 K
η	0.51 (51%)	0.565 (56.5%)
mep	8.03 bar	7.05 bar

Skills Developed: *Ability to...*
- ✓ sketch the Otto cycle p–v and T–s diagrams.
- ✓ evaluate temperatures and pressures at each principal state and retrieve necessary property data.
- ✓ calculate thermal efficiency and mean effective pressure.

Quick Quiz

Determine the heat addition and the heat rejection for the cycle, each in kJ.
Ans. $Q_{23} = 0.791$ kJ, $Q_{41} = 0.388$ kJ.

9.3 Air-Standard Diesel Cycle

The air-standard Diesel cycle is an ideal cycle that assumes the heat addition occurs during a constant-pressure process that starts with the piston at top dead center. The ***Diesel cycle*** is shown on p–v and T–s diagrams in Fig. 9.5. The cycle consists of four internally reversible processes in series. The first process from state 1 to state 2 is the same as in the Otto cycle: an isentropic compression. Heat is not transferred to the working fluid at constant volume as in the Otto cycle, however. In the Diesel cycle, heat is transferred to the working fluid at *constant pressure.* Process 2–3 also makes up the first part of the power stroke. The isentropic expansion from state 3 to state 4 is the remainder of the power stroke. As in the Otto cycle, the cycle is completed by constant-volume Process 4–1 in which heat is rejected from the air while the piston is at bottom dead center. This process replaces the exhaust and intake processes of the actual engine.

Diesel cycle

Since the air-standard Diesel cycle is composed of internally reversible processes, areas on the T–s and p–v diagrams of Fig. 9.5 can be interpreted as heat and work, respectively. On the T–s diagram, area 2–3–a–b–2 represents the heat added per unit

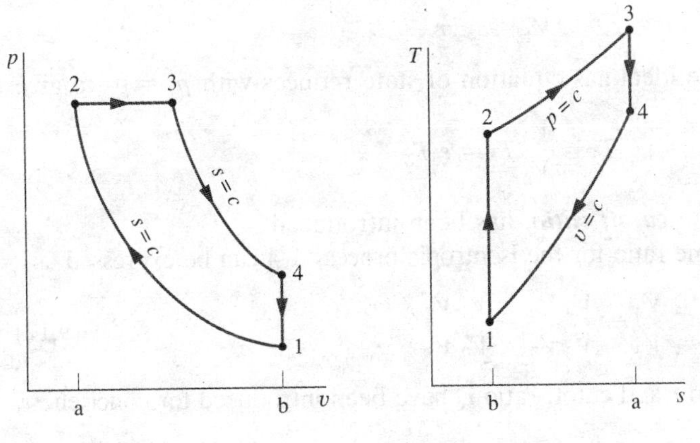

Fig. 9.5 p–v and T–s diagrams of the air-standard Diesel cycle.

of mass and area 1–4–a–b–1 is the heat rejected per unit of mass. On the p–v diagram, area 1–2–a–b–1 is the work input per unit of mass during the compression process. Area 2–3–4–b–a–2 is the work done per unit of mass as the piston moves from top dead center to bottom dead center. The enclosed area of each figure is the net work output, which equals the net heat added.

CYCLE ANALYSIS. In the Diesel cycle the heat addition takes place at constant pressure. Accordingly, Process 2–3 involves both work and heat. The work is given by

$$\frac{W_{23}}{m} = \int_2^3 p\,dv = p_2(v_3 - v_2) \tag{9.9}$$

The heat added in Process 2–3 can be found by applying the closed system energy balance

$$m(u_3 - u_2) = Q_{23} - W_{23}$$

Introducing Eq. 9.9 and solving for the heat transfer

$$\begin{aligned}\frac{Q_{23}}{m} &= (u_3 - u_2) + p(v_3 - v_2) = (u_3 + pv_3) - (u_2 + pv_2)\\ &= h_3 - h_2 \end{aligned} \tag{9.10}$$

where the specific enthalpy is introduced to simplify the expression. As in the Otto cycle, the heat rejected in Process 4–1 is given by

$$\frac{Q_{41}}{m} = u_4 - u_1$$

The thermal efficiency is the ratio of the net work of the cycle to the heat added

$$\eta = \frac{W_{\text{cycle}}/m}{Q_{23}/m} = 1 - \frac{Q_{41}/m}{Q_{23}/m} = 1 - \frac{u_4 - u_1}{h_3 - h_2} \tag{9.11}$$

As for the Otto cycle, the thermal efficiency of the Diesel cycle increases with the compression ratio.

To evaluate the thermal efficiency from Eq. 9.11 requires values for u_1, u_4, h_2, and h_3 or equivalently the temperatures at the principal states of the cycle. Let us consider next how these temperatures are evaluated. For a given initial temperature T_1 and compression ratio r, the temperature at state 2 can be found using the following isentropic relationship and v_r data

$$v_{r2} = \frac{V_2}{V_1} v_{r1} = \frac{1}{r} v_{r1}$$

To find T_3, note that the ideal gas equation of state reduces with $p_3 = p_2$ to give

$$T_3 = \frac{V_3}{V_2} T_2 = r_c T_2$$

cutoff ratio

where $r_c = V_3/V_2$, called the ***cutoff ratio,*** has been introduced.

Since $V_4 = V_1$, the volume ratio for the isentropic process 3–4 can be expressed as

$$\frac{V_4}{V_3} = \frac{V_4}{V_2}\frac{V_2}{V_3} = \frac{V_1}{V_2}\frac{V_2}{V_3} = \frac{r}{r_c} \tag{9.12}$$

where the compression ratio r and cutoff ratio r_c have been introduced for conciseness.

Using Eq. 9.12 together with v_{r3} at T_3, the temperature T_4 can be determined by interpolation once v_{r4} is found from the isentropic relationship

$$v_{r4} = \frac{V_4}{V_3} v_{r3} = \frac{r}{r_c} v_{r3}$$

In a *cold air-standard analysis,* the appropriate expression for evaluating T_2 is provided by

$$\frac{T_2}{T_1} = \left(\frac{V_1}{V_2}\right)^{k-1} = r^{k-1} \qquad \text{(constant } k\text{)}$$

The temperature T_4 is found similarly from

$$\frac{T_4}{T_3} = \left(\frac{V_3}{V_4}\right)^{k-1} = \left(\frac{r_c}{r}\right)^{k-1} \qquad \text{(constant } k\text{)}$$

where Eq. 9.12 has been used to replace the volume ratio.

EFFECT OF COMPRESSION RATIO ON PERFORMANCE. As for the Otto cycle, the thermal efficiency of the Diesel cycle increases with increasing compression ratio. This can be brought out simply using a *cold* air-standard analysis. On a cold air-standard basis, the thermal efficiency of the Diesel cycle can be expressed as

$$\eta = 1 - \frac{1}{r^{k-1}}\left[\frac{r_c^k - 1}{k(r_c - 1)}\right] \qquad \text{(constant } k\text{)} \qquad (9.13)$$

where r is the compression ratio and r_c the cutoff ratio. The derivation is left as an exercise. This relationship is shown in Fig. 9.6 for $k = 1.4$. Equation 9.13 for the Diesel cycle differs from Eq. 9.8 for the Otto cycle only by the term in brackets, which for $r_c > 1$ is greater than unity. Thus, when the compression ratio is the same, the thermal efficiency of the cold air-standard Diesel cycle would be less than that of the cold air-standard Otto cycle.

In the next example, we illustrate the analysis of the air-standard Diesel cycle.

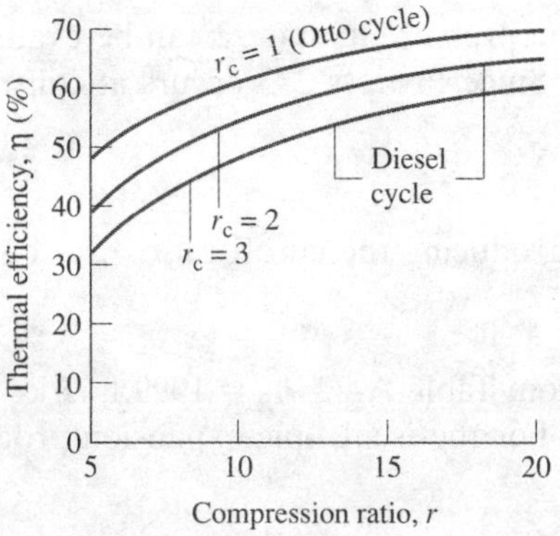

Fig. 9.6 Thermal efficiency of the cold air-standard Diesel cycle, $k = 1.4$.

Example 9.2 ANALYZING THE DIESEL CYCLE

At the beginning of the compression process of an air-standard Diesel cycle operating with a compression ratio of 18, the temperature is 300 K and the pressure is 0.1 MPa. The cutoff ratio for the cycle is 2. Determine **(a)** the temperature and pressure at the end of each process of the cycle, **(b)** the thermal efficiency, **(c)** the mean effective pressure, in MPa.

Solution

Known: An air-standard Diesel cycle is executed with specified conditions at the beginning of the compression stroke. The compression and cutoff ratios are given.

Find: Determine the temperature and pressure at the end of each process, the thermal efficiency, and mean effective pressure.

Schematic and Given Data:

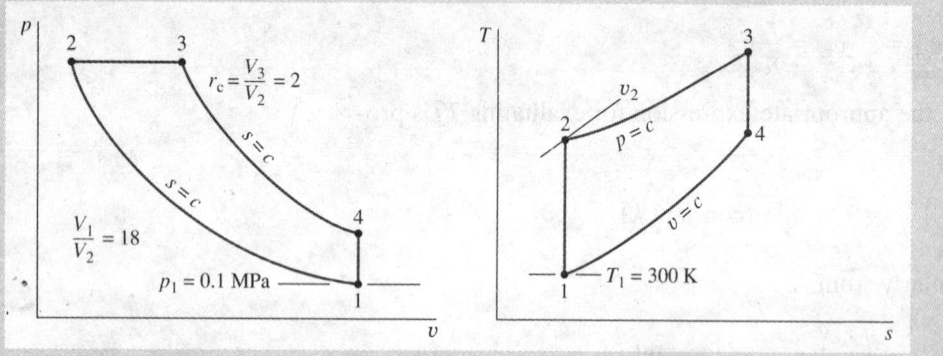

Fig. E9.2

Engineering Model:

1. The air in the piston–cylinder assembly is the closed system.
2. The compression and expansion processes are adiabatic.
3. All processes are internally reversible.
4. The air is modeled as an ideal gas.
5. Kinetic and potential energy effects are negligible.

Analysis:

(a) The analysis begins by determining properties at each principal state of the cycle. With $T_1 = 300$ K, Table A-22 gives $u_1 = 214.07$ kJ/kg and $v_{r1} = 621.2$. For the isentropic compression process 1–2

$$v_{r2} = \frac{V_2}{V_1} v_{r1} = \frac{v_{r1}}{r} = \frac{621.2}{18} = 34.51$$

Interpolating in Table A-22, we get $T_2 = 898.3$ K and $h_2 = 930.98$ kJ/kg. With the ideal gas equation of state

$$p_2 = p_1 \frac{T_2}{T_1} \frac{V_1}{V_2} = (0.1)\left(\frac{898.3}{300}\right)(18) = 5.39 \text{ MPa}$$

The pressure at state 2 can be evaluated alternatively using the isentropic relationship, $p_2 = p_1(p_{r2}/p_{r1})$.

Since Process 2–3 occurs at constant pressure, the ideal gas equation of state gives

$$T_3 = \frac{V_3}{V_2} T_2$$

Introducing the cutoff ratio, $r_c = V_3/V_2$

$$T_3 = r_c T_2 = 2(898.3) = 1796.6 \text{ K}$$

From Table A-22, $h_3 = 1999.1$ kJ/kg and $v_{r3} = 3.97$.

For the isentropic expansion process 3–4

$$v_{r4} = \frac{V_4}{V_3} v_{r3} = \frac{V_4}{V_2} \frac{V_2}{V_3} v_{r3}$$

Introducing $V_4 = V_1$, the compression ratio r, and the cutoff ratio r_c, we have

$$v_{r4} = \frac{r}{r_c} v_{r3} = \frac{18}{2}(3.97) = 35.73$$

Interpolating in Table A-22 with v_{r4}, we get $u_4 = 664.3$ kJ/kg and $T_4 = 887.7$ K. The pressure at state 4 can be found using the isentropic relationship $p_4 = p_3(p_{r4}/p_{r3})$ or the ideal gas equation of state applied at states 1 and 4. With $V_4 = V_1$, the ideal gas equation of state gives

$$p_4 = p_1 \frac{T_4}{T_1} = (0.1 \text{ MPa})\left(\frac{887.7 \text{ K}}{300 \text{ K}}\right) = 0.3 \text{ MPa}$$

(b) The thermal efficiency is found using

❶
$$\eta = 1 - \frac{Q_{41}/m}{Q_{23}/m} = 1 - \frac{u_4 - u_1}{h_3 - h_2}$$

$$= 1 - \frac{664.3 - 214.07}{1999.1 - 930.98} = 0.578 \ (57.8\%)$$

(c) The mean effective pressure written in terms of specific volumes is

$$\text{mep} = \frac{W_{\text{cycle}}/m}{v_1 - v_2} = \frac{W_{\text{cycle}}/m}{v_1(1 - 1/r)}$$

The net work of the cycle equals the net heat added

$$\frac{W_{\text{cycle}}}{m} = \frac{Q_{23}}{m} - \frac{Q_{41}}{m} = (h_3 - h_2) - (u_4 - u_1)$$

$$= (1999.1 - 930.98) - (664.3 - 214.07)$$

$$= 617.9 \text{ kJ/kg}$$

The specific volume at state 1 is

$$v_1 = \frac{(\bar{R}/M)T_1}{p_1} = \frac{\left(\dfrac{8314 \text{ N} \cdot \text{m}}{28.97 \text{ kg} \cdot \text{K}}\right)(300 \text{ K})}{10^5 \text{ N/m}^2} = 0.861 \text{ m}^3/\text{kg}$$

Inserting values

$$\text{mep} = \frac{617.9 \text{ kJ/kg}}{0.861(1 - 1/18) \text{ m}^3/\text{kg}} \left|\frac{10^3 \text{ N} \cdot \text{m}}{1 \text{ kJ}}\right| \left|\frac{1 \text{ MPa}}{10^6 \text{ N/m}^2}\right|$$

$$= 0.76 \text{ MPa}$$

❶ This solution uses the air tables, which account explicitly for the variation of the specific heats with temperature. Note that Eq. 9.13 based on the assumption of *constant* specific heats has not been used to determine the thermal efficiency. The cold air-standard solution of this example is left as an exercise.

Skills Developed: *Ability to...*
- ✓ sketch the Diesel cycle *p-v* and *T-s* diagrams.
- ✓ evaluate temperatures and pressures at each principal state and retrieve necessary property data.
- ✓ calculate the thermal efficiency and mean effective pressure.

Quick Quiz

If the mass of air is 0.0123 kg, what is the *displacement volume,* in L?
Ans. 10 L.

9.4 Air-Standard Dual Cycle

The pressure–volume diagrams of actual internal combustion engines are not described well by the Otto and Diesel cycles. An air-standard cycle that can be made to approximate the pressure variations more closely is the *air-standard dual cycle.* The ***dual cycle*** is shown in Fig. 9.7. As in the Otto and Diesel cycles, Process 1–2 is an isentropic compression. The heat addition occurs in two steps, however: Process 2–3 is a

dual cycle

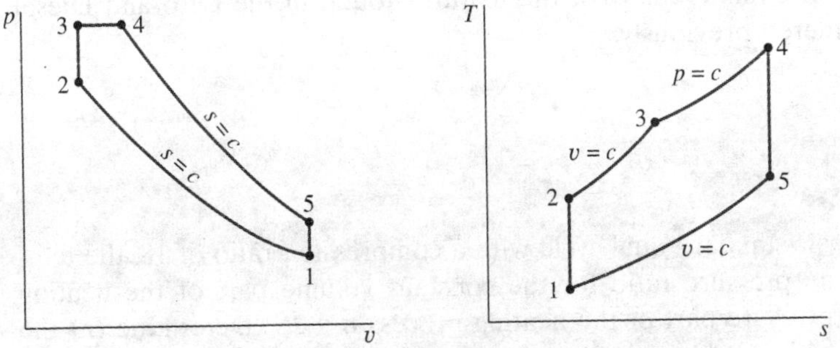

Fig. 9.7 *p–v* and *T–s* diagrams of the air-standard dual cycle.

constant-volume heat addition; Process 3–4 is a constant-pressure heat addition. Process 3–4 also makes up the first part of the power stroke. The isentropic expansion from state 4 to state 5 is the remainder of the power stroke. As in the Otto and Diesel cycles, the cycle is completed by a constant-volume heat rejection process, Process 5–1. Areas on the T–s and p–v diagrams can be interpreted as heat and work, respectively, as in the cases of the Otto and Diesel cycles.

Cycle Analysis

Since the dual cycle is composed of the same types of processes as the Otto and Diesel cycles, we can simply write down the appropriate work and heat transfer expressions by reference to the corresponding earlier developments. Thus, during the isentropic compression process 1–2 there is no heat transfer, and the work is

$$\frac{W_{12}}{m} = u_2 - u_1$$

As for the corresponding process of the Otto cycle, in the constant-volume portion of the heat addition process, Process 2–3, there is no work, and the heat transfer is

$$\frac{Q_{23}}{m} = u_3 - u_2$$

In the constant-pressure portion of the heat addition process, Process 3–4, there is both work and heat transfer, as for the corresponding process of the Diesel cycle

$$\frac{W_{34}}{m} = p(v_4 - v_3) \qquad \text{and} \qquad \frac{Q_{34}}{m} = h_4 - h_3$$

During the isentropic expansion process 4–5 there is no heat transfer, and the work is

$$\frac{W_{45}}{m} = u_4 - u_5$$

Finally, the constant-volume heat rejection process 5–1 that completes the cycle involves heat transfer but no work

$$\frac{Q_{51}}{m} = u_5 - u_1$$

The thermal efficiency is the ratio of the net work of the cycle to the *total* heat added

$$\eta = \frac{W_{\text{cycle}}/m}{(Q_{23}/m + Q_{34}/m)} = 1 - \frac{Q_{51}/m}{(Q_{23}/m + Q_{34}/m)}$$

$$= 1 - \frac{(u_5 - u_1)}{(u_3 - u_2) + (h_4 - h_3)} \tag{9.14}$$

The example to follow provides an illustration of the analysis of an air-standard dual cycle. The analysis exhibits many of the features found in the Otto and Diesel cycle examples considered previously.

Example 9.3 ANALYZING THE DUAL CYCLE

At the beginning of the compression process of an air-standard dual cycle with a compression ratio of 18, the temperature is 300 K and the pressure is 0.1 MPa. The pressure ratio for the constant volume part of the heating process is 1.5:1. The volume ratio for the constant pressure part of the heating process is 1.2:1. Determine **(a)** the thermal efficiency and **(b)** the mean effective pressure, in MPa.

Solution

Known: An air-standard dual cycle is executed in a piston–cylinder assembly. Conditions are known at the beginning of the compression process, and necessary volume and pressure ratios are specified.

Find: Determine the thermal efficiency and the mep, in MPa.

Schematic and Given Data:

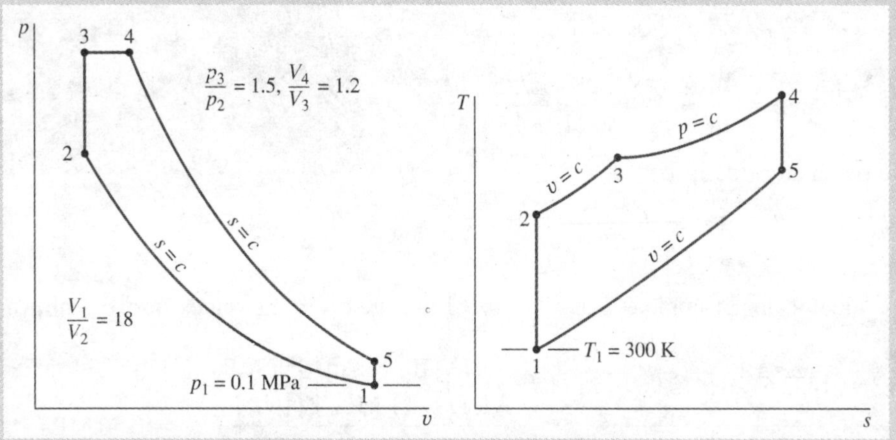

Fig. E9.3

Engineering Model:

1. The air in the piston–cylinder assembly is the closed system.
2. The compression and expansion processes are adiabatic.
3. All processes are internally reversible.
4. The air is modeled as an ideal gas.
5. Kinetic and potential energy effects are negligible.

Analysis: The analysis begins by determining properties at each principal state of the cycle. States 1 and 2 are the same as in Example 9.2, so $u_1 = 214.07$ kJ/kg, $T_2 = 898.3$ K, $u_2 = 673.2$ kJ/kg. Since Process 2–3 occurs at constant volume, the ideal gas equation of state reduces to give

$$T_3 = \frac{p_3}{p_2} T_2 = (1.5)(898.3) = 1347.5 \text{ K}$$

Interpolating in Table A-22, we get $h_3 = 1452.6$ kJ/kg and $u_3 = 1065.8$ kJ/kg.

Since Process 3–4 occurs at constant pressure, the ideal gas equation of state reduces to give

$$T_4 = \frac{V_4}{V_3} T_3 = (1.2)(1347.5) = 1617 \text{ K}$$

From Table A-22, $h_4 = 1778.3$ kJ/kg and $v_{r4} = 5.609$.

Process 4–5 is an isentropic expansion, so

$$v_{r5} = v_{r4} \frac{V_5}{V_4}$$

The volume ratio V_5/V_4 required by this equation can be expressed as

$$\frac{V_5}{V_4} = \frac{V_5}{V_3} \frac{V_3}{V_4}$$

With $V_5 = V_1$, $V_2 = V_3$, and given volume ratios

$$\frac{V_5}{V_4} = \frac{V_1}{V_2} \frac{V_3}{V_4} = 18\left(\frac{1}{1.2}\right) = 15$$

Inserting this in the above expression for v_{r5}

$$v_{r5} = (5.609)(15) = 84.135$$

Interpolating in Table A-22, we get $u_5 = 475.96$ kJ/kg.

(a) The thermal efficiency is

$$\eta = 1 - \frac{Q_{51}/m}{(Q_{23}/m + Q_{34}/m)} = 1 - \frac{(u_5 - u_1)}{(u_3 - u_2) + (h_4 - h_3)}$$

$$= 1 - \frac{(475.96 - 214.07)}{(1065.8 - 673.2) + (1778.3 - 1452.6)}$$

$$= 0.635\ (63.5\%)$$

(b) The mean effective pressure is

$$\text{mep} = \frac{W_{\text{cycle}}/m}{v_1 - v_2} = \frac{W_{\text{cycle}}/m}{v_1(1 - 1/r)}$$

The net work of the cycle equals the net heat added, so

$$\text{mep} = \frac{(u_3 - u_2) + (h_4 - h_3) - (u_5 - u_1)}{v_1(1 - 1/r)}$$

The specific volume at state 1 is evaluated in Example 9.2 as $v_1 = 0.861\ \text{m}^3/\text{kg}$. Inserting values into the above expression for mep

$$\text{mep} = \frac{[(1065.8 - 673.2) + (1778.3 - 1452.6) - (475.96 - 214.07)]\left(\frac{\text{kJ}}{\text{kg}}\right)\left|\frac{10^3\ \text{N}\cdot\text{m}}{1\ \text{kJ}}\right|\left|\frac{1\ \text{MPa}}{10^6\ \text{N/m}^2}\right|}{0.861(1 - 1/18)\ \text{m}^3/\text{kg}} = 0.56\ \text{MPa}$$

Skills Developed: *Ability to...*
- ✓ sketch the Dual cycle *p-v* and *T-s* diagrams.
- ✓ evaluate temperatures and pressures at each principal state and retrieve necessary property data.
- ✓ calculate the thermal efficiency and mean effective pressure.

Quick Quiz

Evaluate the total heat addition and the net work of the cycle, each in kJ per kg of air.
Ans. $Q_{\text{in}}/m = 718\ \text{kJ/kg}$, $W_{\text{cycle}}/m = 456\ \text{kJ/kg}$.

Energy & Environment

Diesel engines for small pickup trucks, sport utility vehicles, and passenger cars are likely to become more prevalent in the United States in coming years because of their fuel-efficient operation. Increased fuel efficiency is achieved in part through the use of advanced electronic engine controls and fuel-injection technology. Technologies that reduce emissions, noise, and odor are also making light-duty diesels a viable alternative to the gasoline engines that dominate the U.S. market. Although diesels are commonly used in Europe, strict emissions standards must be met for particulates and nitrogen oxides (NO_x) in order for small diesels to be widely used in the U.S. Meeting such standards adds cost and requires that challenging design trade-offs be made between efficiency, noise, and emissions.

One technology under development uses special filters to remove tiny soot particles from diesel exhaust and eliminate odor. Removing NO_x is proving to be more of a challenge because greater concentrations of NO_x tend to be produced in high-temperature oxygen-rich combustion typical of diesels. Another new technology is to use a chemical "trap" that removes the NO_x during short bursts of high-temperature combustion but requires complex air-fuel controls for the engine. Researchers say that catalysts also can be used to remove NO_x from the exhaust. If cost-effective solutions can be found, diesel-driven automobiles are likely to appear in many U.S. automobile showrooms.

Gas Turbine Power Plants

This part of the chapter deals with gas turbine power plants. Gas turbines tend to be lighter and more compact than the vapor power plants studied in Chap. 8. The favorable power-output-to-weight ratio of gas turbines makes them well suited for transportation applications (aircraft propulsion, marine power plants, and so on). Gas turbines are also commonly used for stationary power generation.

9.5 Modeling Gas Turbine Power Plants

Gas turbine power plants may operate on either an open or closed basis. The open mode pictured in Fig. 9.8*a* is more common. This is an engine in which atmospheric air is continuously drawn into the compressor, where it is compressed to a high pressure. The air then enters a combustion chamber, or combustor, where it is mixed with fuel and combustion occurs, resulting in combustion products at an elevated temperature. The combustion products expand through the turbine and are subsequently discharged to the surroundings. Part of the turbine work developed is used to drive the compressor; the remainder is available to generate electricity, to propel a vehicle, or for other purposes. In the system pictured in Fig. 9.8*b*, the working fluid receives an energy input by heat transfer from an external source, for example a gas-cooled nuclear reactor. The gas exiting the turbine is passed through a heat exchanger, where it is cooled prior to reentering the compressor.

An idealization often used in the study of open gas turbine power plants is that of an ***air-standard analysis.*** In an air-standard analysis two assumptions are always made:

air-standard analysis: gas turbines

- The working fluid is air, which behaves as an ideal gas.
- The temperature rise that would be brought about by combustion is accomplished by a heat transfer from an external source.

With an air-standard analysis, we avoid dealing with the complexities of the combustion process and the change of composition during combustion. An air-standard analysis simplifies the study of gas turbine power plants considerably. However, numerical values calculated on this basis may provide only qualitative indications of power plant

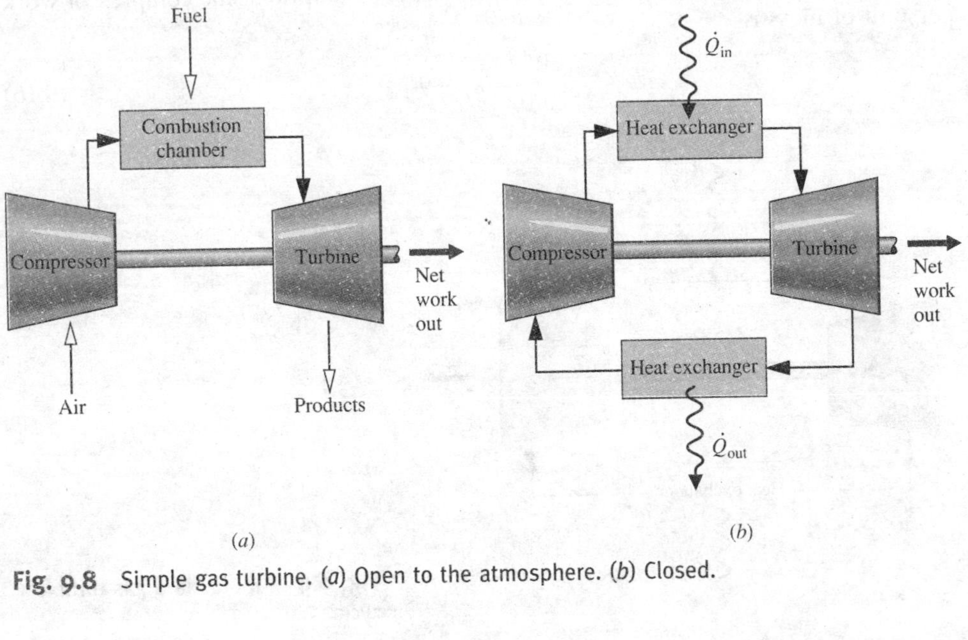

Fig. 9.8 Simple gas turbine. (*a*) Open to the atmosphere. (*b*) Closed.

performance. Sufficient information about combustion and the properties of products of combustion is known (Chap. 13) that the study of gas turbines can be conducted without the foregoing assumptions. Still, we can learn some important aspects of gas turbine operation by proceeding on the basis of an air-standard analysis, as follows.

9.6 Air-Standard Brayton Cycle

A schematic diagram of an air-standard gas turbine is shown in Fig. 9.9. The directions of the principal energy transfers are indicated on this figure by arrows. In accordance with the assumptions of an air-standard analysis, the temperature rise that would be achieved in the combustion process is brought about by a heat transfer to the working fluid from an external source and the working fluid is considered to be air as an ideal gas. With the air-standard idealizations, air would be drawn into the compressor at state 1 from the surroundings and later returned to the surroundings at state 4 with a temperature greater than the ambient temperature. After interacting with the surroundings, each unit mass of discharged air would eventually return to the same state as the air entering the compressor, so we may think of the air passing through the components of the gas turbine as undergoing a thermodynamic cycle. A simplified representation of the states visited by the air in such a cycle can be devised by regarding the turbine exhaust air as restored to the compressor inlet state by passing through a heat exchanger where heat rejection to the surroundings occurs. The cycle that results with this further idealization is called the air-standard ***Brayton cycle.***

Brayton cycle

9.6.1 Evaluating Principal Work and Heat Transfers

The following expressions for the work and heat transfers of energy that occur at steady state are readily derived by reduction of the control volume mass and energy rate balances. These energy transfers are positive in the directions of the arrows in Fig. 9.9. Assuming the turbine operates adiabatically and with negligible effects of kinetic and potential energy, the work developed per unit of mass is

$$\frac{\dot{W}_t}{\dot{m}} = h_3 - h_4 \tag{9.15}$$

where $\dot{m}$ denotes the mass flow rate. With the same assumptions, the compressor work per unit of mass is

$$\frac{\dot{W}_c}{\dot{m}} = h_2 - h_1 \tag{9.16}$$

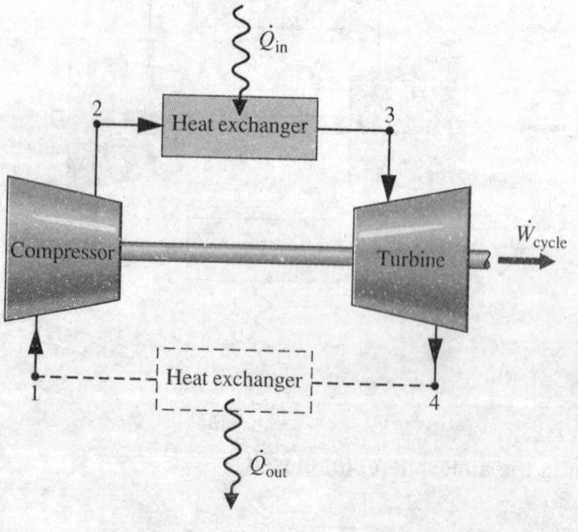

Fig. 9.9 Air-standard gas turbine cycle

The symbol $\dot{W}_c$ denotes work *input* and takes on a positive value. The heat added to the cycle per unit of mass is

$$\frac{\dot{Q}_{in}}{\dot{m}} = h_3 - h_2 \tag{9.17}$$

The heat rejected per unit of mass is

$$\frac{\dot{Q}_{out}}{\dot{m}} = h_4 - h_1 \tag{9.18}$$

where $\dot{Q}_{out}$ is positive in value.

The thermal efficiency of the cycle in Fig. 9.9 is

$$\eta = \frac{\dot{W}_t/\dot{m} - \dot{W}_c/\dot{m}}{\dot{Q}_{in}/\dot{m}} = \frac{(h_3 - h_4) - (h_2 - h_1)}{h_3 - h_2} \tag{9.19}$$

The back work ratio for the cycle is

$$\text{bwr} = \frac{\dot{W}_c/\dot{m}}{\dot{W}_t/\dot{m}} = \frac{h_2 - h_1}{h_3 - h_4} \tag{9.20}$$

For the same pressure rise, a gas turbine compressor would require a much greater work input per unit of mass flow than the pump of a vapor power plant because the average specific volume of the gas flowing through the compressor would be many times greater than that of the liquid passing through the pump (see discussion of Eq. 6.51b in Sec. 6.13). Hence, a relatively large portion of the work developed by the turbine is required to drive the compressor. Typical back work ratios of gas turbines range from 40 to 80%. In comparison, the back work ratios of vapor power plants are normally only 1 or 2%.

If the temperatures at the numbered states of the cycle are known, the specific enthalpies required by the foregoing equations are readily obtained from the ideal gas table for air, Table A-22. Alternatively, with the sacrifice of some accuracy, the variation of the specific heats with temperature can be ignored and the specific heats taken as constant. The air-standard analysis is then referred to as a *cold air-standard analysis*. As illustrated by the discussion of internal combustion engines given previously, the chief advantage of the assumption of constant specific heats is that simple expressions for quantities such as thermal efficiency can be derived, and these can be used to deduce qualitative indications of cycle performance without involving tabular data.

Since Eqs. 9.15 through 9.20 have been developed from mass and energy rate balances, they apply equally when irreversibilities are present and in the absence of irreversibilities. Although irreversibilities and losses associated with the various power plant components have a pronounced effect on overall performance, it is instructive to consider an idealized cycle in which they are assumed absent. Such a cycle establishes an upper limit on the performance of the air-standard Brayton cycle. This is considered next.

9.6.2 Ideal Air-Standard Brayton Cycle

Ignoring irreversibilities as the air circulates through the various components of the Brayton cycle, there are no frictional pressure drops, and the air flows at constant pressure through the heat exchangers. If stray heat transfers to the surroundings are also ignored, the processes through the turbine and compressor are isentropic. The ideal cycle shown on the p–v and T–s diagrams in Fig. 9.10 adheres to these idealizations.

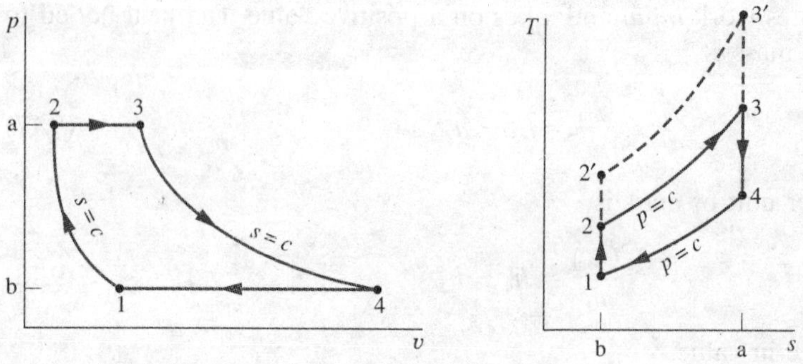

Fig. 9.10 Air-standard ideal Brayton cycle.

Areas on the T–s and p–v diagrams of Fig. 9.10 can be interpreted as heat and work, respectively, per unit of mass flowing. On the T–s diagram, area 2–3–a–b–2 represents the heat added per unit of mass and area 1–4–a–b–1 is the heat rejected per unit of mass. On the p–v diagram, area 1–2–a–b–1 represents the compressor work input per unit of mass and area 3–4–b–a–3 is the turbine work output per unit of mass (Sec. 6.13). The enclosed area on each figure can be interpreted as the net work output or, equivalently, the net heat added.

When air table data are used to conduct an analysis involving the ideal Brayton cycle, the following relationships, based on Eq. 6.41, apply for the isentropic processes 1–2 and 3–4

$$p_{r2} = p_{r1}\frac{p_2}{p_1} \tag{9.21}$$

$$p_{r4} = p_{r3}\frac{p_4}{p_3} = p_{r3}\frac{p_1}{p_2} \tag{9.22}$$

Recall that p_r is tabulated versus temperature in Tables A-22. Since the air flows through the heat exchangers of the ideal cycle at constant pressure, it follows that $p_4/p_3 = p_1/p_2$. This relationship has been used in writing Eq. 9.22.

When an ideal Brayton cycle is analyzed on a cold air-standard basis, the specific heats are taken as constant. Equations 9.21 and 9.22 are then replaced, respectively, by the following expressions, based on Eq. 6.43

$$T_2 = T_1\left(\frac{p_2}{p_1}\right)^{(k-1)/k} \tag{9.23}$$

$$T_4 = T_3\left(\frac{p_4}{p_3}\right)^{(k-1)/k} = T_3\left(\frac{p_1}{p_2}\right)^{(k-1)/k} \tag{9.24}$$

where k is the specific heat ratio, $k = c_p/c_v$.

In the next example, we illustrate the analysis of an ideal air-standard Brayton cycle and compare results with those obtained on a cold air-standard basis.

Example 9.4 ANALYZING THE IDEAL BRAYTON CYCLE

Air enters the compressor of an ideal air-standard Brayton cycle at 100 kPa, 300 K, with a volumetric flow rate of 5 m^3/s. The compressor pressure ratio is 10. The turbine inlet temperature is 1400 K. Determine **(a)** the thermal efficiency of the cycle, **(b)** the back work ratio, **(c)** the *net* power developed, in kW.

Solution

Known: An ideal air-standard Brayton cycle operates with given compressor inlet conditions, given turbine inlet temperature, and a known compressor pressure ratio.

Find: Determine the thermal efficiency, the back work ratio, and the *net* power developed, in kW.

Schematic and Given Data:

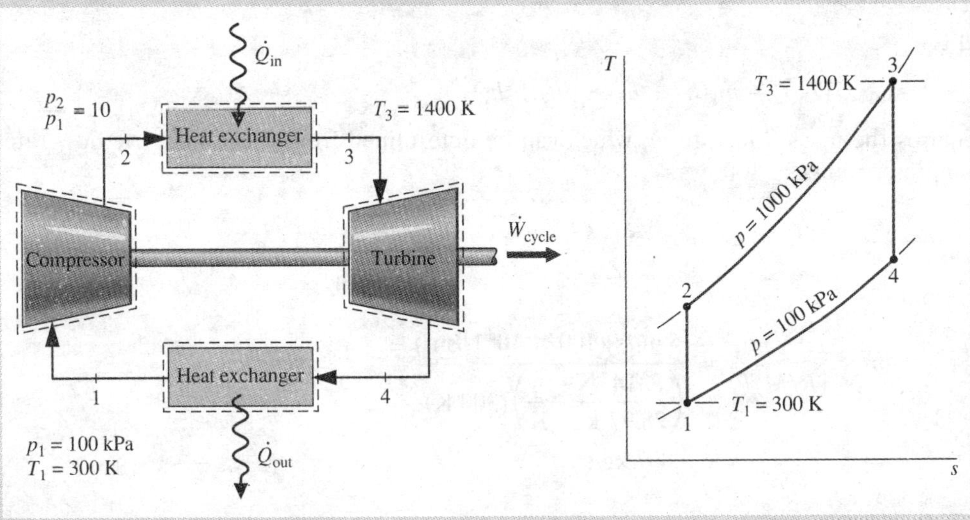

Fig. E9.4

Engineering Model:

1. Each component is analyzed as a control volume at steady state. The control volumes are shown on the accompanying sketch by dashed lines.
2. The turbine and compressor processes are isentropic.
3. There are no pressure drops for flow through the heat exchangers.
4. Kinetic and potential energy effects are negligible.
5. The working fluid is air modeled as an ideal gas.

❶ **Analysis:** The analysis begins by determining the specific enthalpy at each numbered state of the cycle. At state 1, the temperature is 300 K. From Table A-22, $h_1 = 300.19$ kJ/kg and $p_{r1} = 1.386$.

Since the compressor process is isentropic, the following relationship can be used to determine h_2

$$p_{r2} = \frac{p_2}{p_1} p_{r1} = (10)(1.386) = 13.86$$

Then, interpolating in Table A-22, we obtain $h_2 = 579.9$ kJ/kg.

The temperature at state 3 is given as $T_3 = 1400$ K. With this temperature, the specific enthalpy at state 3 from Table A-22 is $h_3 = 1515.4$ kJ/kg. Also, $p_{r3} = 450.5$.

The specific enthalpy at state 4 is found by using the isentropic relation

$$p_{r4} = p_{r3} \frac{p_4}{p_3} = (450.5)(1/10) = 45.05$$

Interpolating in Table A-22, we get $h_4 = 808.5$ kJ/kg.

(a) The thermal efficiency is

$$\eta = \frac{(\dot{W}_t/\dot{m}) - (\dot{W}_c/\dot{m})}{\dot{Q}_{in}/\dot{m}}$$

$$= \frac{(h_3 - h_4) - (h_2 - h_1)}{h_3 - h_2} = \frac{(1515.4 - 808.5) - (579.9 - 300.19)}{1515.4 - 579.9}$$

$$= \frac{706.9 - 279.7}{935.5} = 0.457 \ (45.7\%)$$

(b) The back work ratio is

❷ $$\text{bwr} = \frac{\dot{W}_c/\dot{m}}{\dot{W}_t/\dot{m}} = \frac{h_2 - h_1}{h_3 - h_4} = \frac{279.7}{706.9} = 0.396\ (39.6\%)$$

(c) The net power developed is

$$\dot{W}_{\text{cycle}} = \dot{m}[(h_3 - h_4) - (h_2 - h_1)]$$

To evaluate the net power requires the mass flow rate $\dot{m}$, which can be determined from the volumetric flow rate and specific volume at the compressor inlet as follows

$$\dot{m} = \frac{(\text{AV})_1}{v_1}$$

Since $v_1 = (\overline{R}/M)T_1/p_1$, this becomes

$$\dot{m} = \frac{(\text{AV})_1 p_1}{(\overline{R}/M)T_1} = \frac{(5\ \text{m}^3/\text{s})(100 \times 10^3\ \text{N/m}^2)}{\left(\frac{8314}{28.97}\frac{\text{N}\cdot\text{m}}{\text{kg}\cdot\text{K}}\right)(300\ \text{K})}$$
$$= 5.807\ \text{kg/s}$$

Finally,

$$\dot{W}_{\text{cycle}} = (5.807\ \text{kg/s})(706.9 - 279.7)\left(\frac{\text{kJ}}{\text{kg}}\right)\left|\frac{1\ \text{kW}}{1\ \text{kJ/s}}\right| = 2481\ \text{kW}$$

❶ The use of the ideal gas table for air is featured in this solution. A solution also can be developed on a cold air-standard basis in which constant specific heats are assumed. The details are left as an exercise, but for comparison the results are presented for the case $k = 1.4$ in the following table:

Parameter	Air-Standard Analysis	Cold Air-Standard Analysis, $k = 1.4$
T_2	574.1 K	579.2 K
T_4	787.7 K	725.1 K
η	0.457	0.482
bwr	0.396	0.414
$\dot{W}_{\text{cycle}}$	2481 kW	2308 kW

Skills Developed: *Ability to...*

- ✓ sketch the schematic of the basic air-standard gas turbine and the *T-s* diagram for the corresponding ideal Brayton cycle.
- ✓ evaluate temperatures and pressures at each principal state and retrieve necessary property data.
- ✓ calculate the thermal efficiency and back work ratio.

❷ The value of the back work ratio in the present gas turbine case is significantly greater than the back work ratio of the simple vapor power cycle of Example 8.1.

Quick Quiz

Determine the rate of heat transfer to the air passing through the combustor, in kW. **Ans.** 5432 kW.

EFFECT OF PRESSURE RATIO ON PERFORMANCE. Conclusions that are qualitatively correct for actual gas turbines can be drawn from a study of the ideal Brayton cycle. The first of these conclusions is that the thermal efficiency increases with increasing pressure ratio across the compressor. ➡ **FOR EXAMPLE...** referring again to the *T–s* diagram of Fig. 9.10, we see that an increase in the pressure ratio changes the cycle from 1–2–3–4–1 to 1–2′–3′–4–1. Since the average temperature of heat addition is greater in the latter cycle and both cycles have the same heat rejection process, cycle 1–2′–3′–4–1 would have the greater thermal efficiency.

The increase in thermal efficiency with the pressure ratio across the compressor is also brought out simply by the following development, in which the specific heat c_p, and thus the specific heat ratio k, is assumed constant. For constant c_p, Eq. 9.19 becomes

$$\eta = \frac{c_p(T_3 - T_4) - c_p(T_2 - T_1)}{c_p(T_3 - T_2)} = 1 - \frac{(T_4 - T_1)}{(T_3 - T_2)}$$

Or, on further rearrangement

$$\eta = 1 - \frac{T_1}{T_2}\left(\frac{T_4/T_1 - 1}{T_3/T_2 - 1}\right)$$

From Eqs. 9.23 and 9.24 above, $T_4/T_1 = T_3/T_2$, so

$$\eta = 1 - \frac{T_1}{T_2}$$

Finally, introducing Eq. 9.23

$$\eta = 1 - \frac{1}{(p_2/p_1)^{(k-1)/k}} \quad (k \text{ constant}) \qquad (9.25)$$

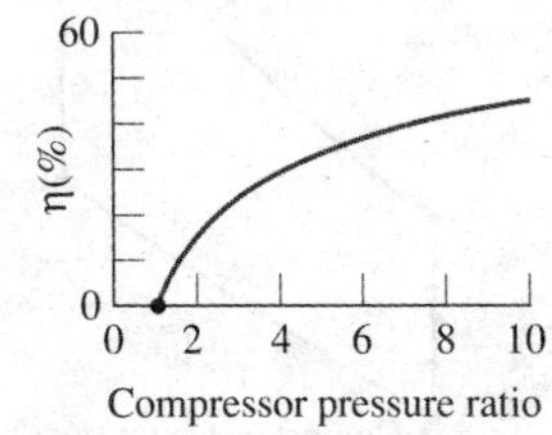

Fig. 9.11 Ideal Brayton cycle thermal efficiency versus compressor pressure ratio, $k = 1.4$.

By inspection of Eq. 9.25, it can be seen that the cold air-standard ideal Brayton cycle thermal efficiency is a function of the pressure ratio across the compressor. This relationship is shown in Fig. 9.11 for $k = 1.4$.

There is a limit of about 1700 K imposed by metallurgical considerations on the maximum allowed temperature at the turbine inlet. It is instructive therefore to consider the effect of compressor pressure ratio on thermal efficiency when the turbine inlet temperature is restricted to the maximum allowable temperature. The T–s diagrams of two ideal Brayton cycles having the same turbine inlet temperature but different compressor pressure ratios are shown in Fig. 9.12. Cycle A has a greater pressure ratio than cycle B and thus the greater thermal efficiency. However, cycle B has a larger enclosed area and thus the greater net work developed per unit of mass flow. Accordingly, for cycle A to develop the same net *power* output as cycle B, a larger mass flow rate would be required, and this might dictate a larger system. These considerations are important for gas turbines intended for use in vehicles where engine weight must be kept small. For such applications, it is desirable to operate near the compressor pressure ratio that yields the most work per unit of mass flow and not the pressure ratio for the greatest thermal efficiency.

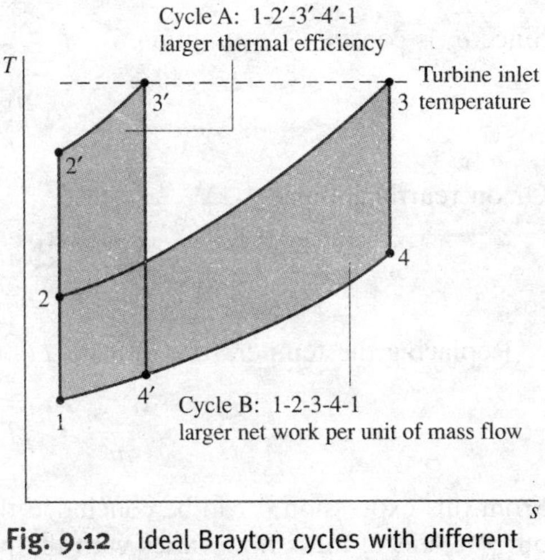

Fig. 9.12 Ideal Brayton cycles with different pressure ratios and the same turbine inlet temperature.

Example 9.5 provides an illustration of the determination of the compressor pressure ratio for maximum net work per unit of mass flow for the cold air-standard Brayton cycle.

Example 9.5 COMPRESSOR PRESSURE RATIO FOR MAXIMUM NET WORK

Determine the pressure ratio across the compressor of an ideal Brayton cycle for the maximum net work output per unit of mass flow if the state at the compressor inlet and the temperature at the turbine inlet are fixed. Use a cold air-standard analysis and ignore kinetic and potential energy effects. Discuss.

Solution

Known: An ideal Brayton cycle operates with a specified state at the inlet to the compressor and a specified turbine inlet temperature.

Find: Determine the pressure ratio across the compressor for the maximum net work output per unit of mass flow, and discuss the result.

Schematic and Given Data:

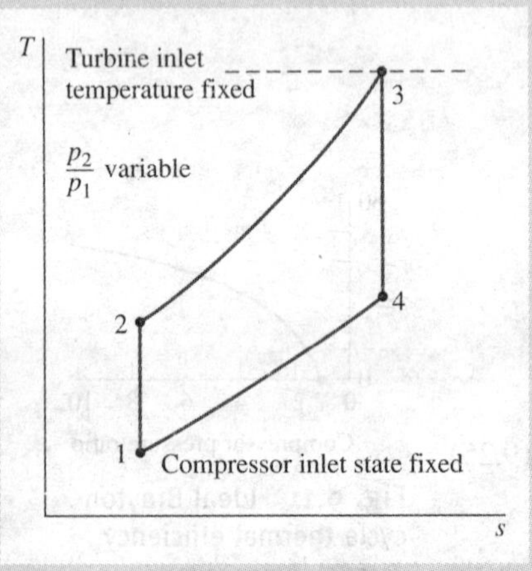

Engineering Model:

1. Each component is analyzed as a control volume at steady state.
2. The turbine and compressor processes are isentropic.
3. There are no pressure drops for flow through the heat exchangers.
4. Kinetic and potential energy effects are negligible.
5. The working fluid is air modeled as an ideal gas.
6. The specific heat c_p and the specific heat ratio k are constant.

Fig. E9.5

Analysis: The net work of the cycle per unit of mass flow is

$$\frac{\dot{W}_{\text{cycle}}}{\dot{m}} = (h_3 - h_4) - (h_2 - h_1)$$

Since c_p is constant (assumption 6)

$$\frac{\dot{W}_{\text{cycle}}}{\dot{m}} = c_p[(T_3 - T_4) - (T_2 - T_1)]$$

Or on rearrangement

$$\frac{\dot{W}_{\text{cycle}}}{\dot{m}} = c_p T_1\left(\frac{T_3}{T_1} - \frac{T_4}{T_3}\frac{T_3}{T_1} - \frac{T_2}{T_1} + 1\right)$$

Replacing the temperature ratios T_2/T_1 and T_4/T_3 by using Eqs. 9.23 and 9.24, respectively, gives

$$\frac{\dot{W}_{\text{cycle}}}{\dot{m}} = c_p T_1\left[\frac{T_3}{T_1} - \frac{T_3}{T_1}\left(\frac{p_1}{p_2}\right)^{(k-1)/k} - \left(\frac{p_2}{p_1}\right)^{(k-1)/k} + 1\right]$$

From this expression it can be concluded that for specified values of T_1, T_3, and c_p, the value of the net work output per unit of mass flow varies with the pressure ratio p_2/p_1 only.

To determine the pressure ratio that maximizes the net work output per unit of mass flow, first form the derivative

$$\begin{aligned}\frac{\partial(\dot{W}_{\text{cycle}}/\dot{m})}{\partial(p_2/p_1)} &= \frac{\partial}{\partial(p_2/p_1)}\left\{c_p T_1\left[\frac{T_3}{T_1} - \frac{T_3}{T_1}\left(\frac{p_1}{p_2}\right)^{(k-1)/k} - \left(\frac{p_2}{p_1}\right)^{(k-1)/k} + 1\right]\right\} \\ &= c_p T_1\left(\frac{k-1}{k}\right)\left[\left(\frac{T_3}{T_1}\right)\left(\frac{p_1}{p_2}\right)^{-1/k}\left(\frac{p_1}{p_2}\right)^2 - \left(\frac{p_2}{p_1}\right)^{-1/k}\right] \\ &= c_p T_1\left(\frac{k-1}{k}\right)\left[\left(\frac{T_3}{T_1}\right)\left(\frac{p_1}{p_2}\right)^{(2k-1)/k} - \left(\frac{p_2}{p_1}\right)^{-1/k}\right]\end{aligned}$$

When the partial derivative is set to zero, the following relationship is obtained

$$\frac{p_2}{p_1} = \left(\frac{T_3}{T_1}\right)^{k/[2(k-1)]}$$

By checking the sign of the second derivative, we can verify that the net work per unit of mass flow is a maximum when this relationship is satisfied.

For gas turbines intended for transportation, it is desirable to keep engine size small. Thus, the gas turbine should operate near the compressor pressure ratio that yields the most work per unit of mass flow. The present example provides an elementary illustration of how the compressor pressure ratio for maximum net work per unit of mass flow is affected by the constraint of a fixed turbine inlet temperature.

Skills Developed: Ability to...
- ✓ complete the detailed derivation of a thermodynamic expression.
- ✓ use calculus to maximize a function.

Quick Quiz

For an ideal cold air-standard Brayton cycle with a compressor inlet temperature of 300 K and a maximum cycle temperature of 1800K, what is the compressor pressure ratio that maximizes the net power output per unit mass flow? Assume $k = 1.4$.
Ans. 23.

9.6.3 Considering Gas Turbine Irreversibilities and Losses

The principal state points of an air-standard gas turbine might be shown more realistically as in Fig. 9.13*a*. Because of frictional effects within the compressor and turbine, the working fluid would experience increases in specific entropy across these components. Owing to friction, there also would be pressure drops as the working fluid passes through the heat exchangers. However, because frictional pressure drops are less significant sources of irreversibility, we ignore them in subsequent discussions and for simplicity show the flow through the heat exchangers as occurring at constant pressure. This is illustrated by Fig. 9.13*b*. Stray heat transfers from the power plant components to the surroundings represent losses, but these effects are usually of secondary importance and are also ignored in subsequent discussions.

As the effect of irreversibilities in the turbine and compressor becomes more pronounced, the work developed by the turbine decreases and the work input to the compressor increases, resulting in a marked decrease in the net work of the power plant. Accordingly, if an appreciable amount of net work is to be developed by the

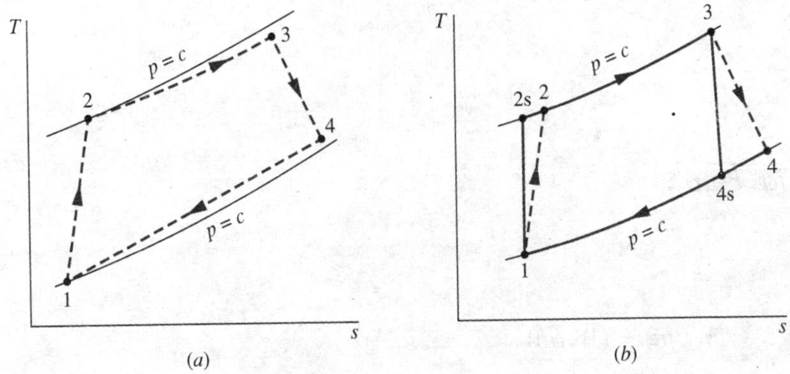

Fig. 9.13 Effects of irreversibilities on the air-standard gas turbine.

plant, relatively high turbine and compressor efficiencies are required. After decades of developmental effort, efficiencies of 80 to 90% can now be achieved for the turbines and compressors in gas turbine power plants. Designating the states as in Fig. 9.13*b*, the isentropic turbine and compressor efficiencies are given by

$$\eta_t = \frac{(\dot{W}_t/\dot{m})}{(\dot{W}_t/\dot{m})_s} = \frac{h_3 - h_4}{h_3 - h_{4s}}$$

$$\eta_c = \frac{(\dot{W}_c/\dot{m})_s}{(\dot{W}_c/\dot{m})} = \frac{h_{2s} - h_1}{h_2 - h_1}$$

The effects of irreversibilities in the turbine and compressor are important. Still, among the irreversibilities of actual gas turbine power plants, combustion irreversibility is the most significant by far. An air-standard analysis does not allow this irreversibility to be evaluated, however, and means introduced in Chap. 13 must be applied. Combustion irreversibility is also briefly considered in Sec. 9.10.

Example 9.6 brings out the effect of turbine and compressor irreversibilities on plant performance.

Example 9.6 BRAYTON CYCLE WITH IRREVERSIBILITIES

Reconsider Example 9.4, but include in the analysis that the turbine and compressor each have an isentropic efficiency of 80%. Determine for the modified cycle **(a)** the thermal efficiency of the cycle, **(b)** the back work ratio, and **(c)** the *net* power developed, in kW.

Solution

Known: An air-standard Brayton cycle operates with given compressor inlet conditions, given turbine inlet temperature, and known compressor pressure ratio. The compressor and turbine each have an isentropic efficiency of 80%.

Find: Determine the thermal efficiency, the back work ratio, and the net power developed, in kW.

Schematic and Given Data:

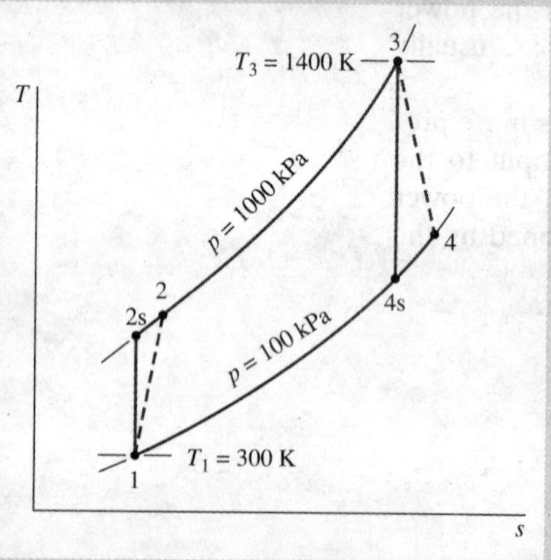

Fig. E9.6

Engineering Model:

1. Each component is analyzed as a control volume at steady state.
2. The compressor and turbine are adiabatic.
3. There are no pressure drops for flow through the heat exchangers.
4. Kinetic and potential energy effects are negligible.
5. The working fluid is air modeled as an ideal gas.

Analysis:

(a) The thermal efficiency is given by

$$\eta = \frac{(\dot{W}_t/\dot{m}) - (\dot{W}_c/\dot{m})}{\dot{Q}_{in}/\dot{m}}$$

The work terms in the numerator of this expression are evaluated using the given values of the compressor and turbine isentropic efficiencies as follows:

The turbine work per unit of mass is

$$\frac{\dot{W}_t}{\dot{m}} = \eta_t\left(\frac{\dot{W}_t}{\dot{m}}\right)_s$$

❶ where η_t is the turbine efficiency. The value of $(\dot{W}_t/\dot{m})_s$ is determined in the solution to Example 9.4 as 706.9 kJ/kg. Thus

$$\frac{\dot{W}_t}{\dot{m}} = 0.8(706.9) = 565.5 \text{ kJ/kg}$$

For the compressor, the work per unit of mass is

$$\frac{\dot{W}_c}{\dot{m}} = \frac{(\dot{W}_c/\dot{m})_s}{\eta_c}$$

where η_c is the compressor efficiency. The value of $(\dot{W}_c/\dot{m})_s$ is determined in the solution to Example 9.4 as 279.7 kJ/kg, so

$$\frac{\dot{W}_c}{\dot{m}} = \frac{279.7}{0.8} = 349.6 \text{ kJ/kg}$$

The specific enthalpy at the compressor exit, h_2, is required to evaluate the denominator of the thermal efficiency expression. This enthalpy can be determined by solving

$$\frac{\dot{W}_c}{\dot{m}} = h_2 - h_1$$

to obtain

$$h_2 = h_1 + \dot{W}_c/\dot{m}$$

Inserting known values

$$h_2 = 300.19 + 349.6 = 649.8 \text{ kJ/kg}$$

The heat transfer to the working fluid per unit of mass flow is then

$$\frac{\dot{Q}_{in}}{\dot{m}} = h_3 - h_2 = 1515.4 - 649.8 = 865.6 \text{ kJ/kg}$$

where h_3 is from the solution to Example 9.4.

Finally, the thermal efficiency is

$$\eta = \frac{565.5 - 349.6}{865.6} = 0.249\ (24.9\%)$$

(b) The back work ratio is

$$\text{bwr} = \frac{\dot{W}_c/\dot{m}}{\dot{W}_t/\dot{m}} = \frac{349.6}{565.5} = 0.618\ (61.8\%)$$

(c) The mass flow rate is the same as in Example 9.4. The net power developed by the cycle is then

❷
$$\dot{W}_{cycle} = \left(5.807\frac{\text{kg}}{\text{s}}\right)(565.5 - 349.6)\frac{\text{kJ}}{\text{kg}}\left|\frac{1\text{ kW}}{1\text{ kJ/s}}\right| = 1254 \text{ kW}$$

❶ The solution to this example on a cold air-standard basis is left as an exercise.

❷ Irreversibilities within the turbine and compressor have a significant impact on the performance of gas turbines. This is brought out by comparing the results of the present example with those of Example 9.4. Irreversibilities result in an increase in the work of compression and a reduction in work output of the turbine. The back work ratio is greatly increased and the thermal efficiency significantly decreased.

Skills Developed: *Ability to...*
- ✓ sketch the schematic of the basic air-standard gas turbine and the *T–s* diagram for the corresponding Brayton cycle with compressor and turbine irreversibilities.
- ✓ evaluate temperatures and pressures at each principal state and retrieve necessary property data.
- ✓ calculate the thermal efficiency and back work ratio.

Quick Quiz

What would be the thermal efficiency and back work ratio if the isentropic turbine efficiency were 70% keeping isentropic compressor efficiency and other given data the same?
Ans. $\eta = 16.8\%$, $\text{bwr} = 70.65\%$.

9.7 Regenerative Gas Turbines

The turbine exhaust temperature of a gas turbine is normally well above the ambient temperature. Accordingly, the hot turbine exhaust gas has a potential for use (exergy) that would be irrevocably lost were the gas discarded directly to the surroundings. One way of utilizing this potential is by means of a heat exchanger called a ***regenerator,*** which allows the air exiting the compressor to be *preheated* before entering the combustor, thereby reducing the amount of fuel that must be burned in the combustor. The combined cycle arrangement considered in Sec. 9.10 is another way to utilize the hot turbine exhaust gas.

regenerator

An air-standard Brayton cycle modified to include a regenerator is illustrated in Fig. 9.14. The regenerator shown is a counterflow heat exchanger through which the hot turbine exhaust gas and the cooler air leaving the compressor pass in opposite directions. Ideally, no frictional pressure drop occurs in either stream. The turbine exhaust gas is cooled from state 4 to state y, while the air exiting the compressor is heated from state 2 to state x. Hence, a heat transfer from a source external to the cycle is required only to increase the air temperature from state x to state 3, rather

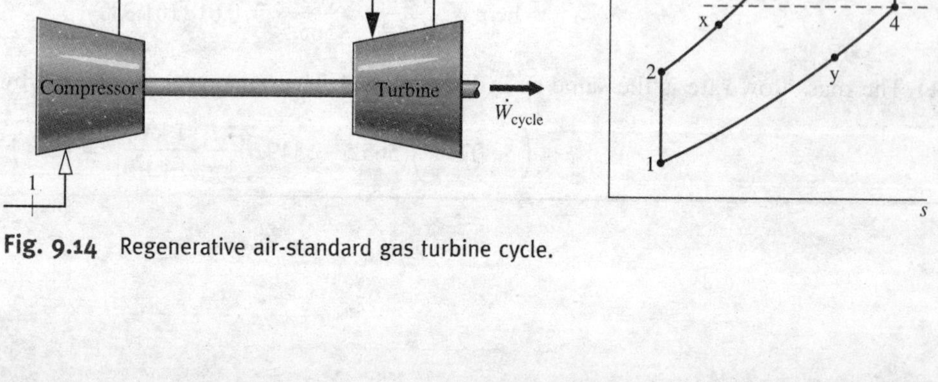

Fig. 9.14 Regenerative air-standard gas turbine cycle.

than from state 2 to state 3, as would be the case without regeneration. The heat added per unit of mass is then given by

$$\frac{\dot{Q}_{\text{in}}}{\dot{m}} = h_3 - h_x \tag{9.26}$$

The net work developed per unit of mass flow is not altered by the addition of a regenerator. Thus, since the heat added is reduced, the thermal efficiency increases.

REGENERATOR EFFECTIVENESS. From Eq. 9.26 it can be concluded that the external heat transfer required by a gas turbine power plant decreases as the specific enthalpy h_x increases and thus as the temperature T_x increases. Evidently, there is an incentive in terms of fuel saved for selecting a regenerator that provides the greatest practical value for this temperature. To consider the *maximum* theoretical value for T_x, refer to Fig. 9.15, which shows temperature variations of the hot and cold streams of a counterflow heat exchanger.

- ▶ First, refer to Fig. 9.15*a*. Since a finite temperature difference between the streams is required for heat transfer to occur, the temperature of the cold stream at each location, denoted by the coordinate z, is less than that of the hot stream. In particular, the temperature of the colder stream as it exits the heat exchanger is less than the temperature of the incoming hot stream. If the heat transfer area were increased, providing more opportunity for heat transfer between the two streams, there would be a smaller temperature difference at each location.
- ▶ In the limiting case of infinite heat transfer area, the temperature difference would approach zero at all locations, as illustrated in Fig. 9.15*b*, and the heat transfer would approach reversibility. In this limit, the exit temperature of the colder stream would approach the temperature of the incoming hot stream. Thus, the highest possible temperature that could be achieved by the colder stream is the temperature of the incoming hot gas.

Referring again to the regenerator of Fig. 9.14, we can conclude from the discussion of Fig. 9.15 that the maximum theoretical value for the temperature T_x is the turbine exhaust temperature T_4, obtained if the regenerator were operating reversibly. The *regenerator effectiveness*, η_{reg}, is a parameter that gauges the departure of an actual

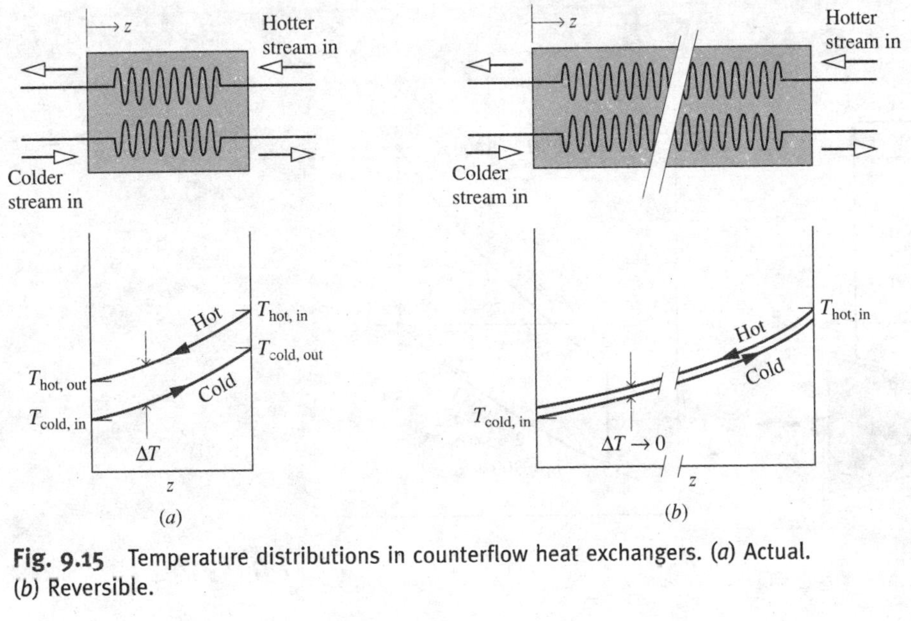

Fig. 9.15 Temperature distributions in counterflow heat exchangers. (*a*) Actual. (*b*) Reversible.

regenerator effectiveness

regenerator from such an ideal regenerator. The ***regenerator effectiveness*** is defined as the ratio of the actual enthalpy increase of the air flowing through the compressor side of the regenerator to the maximum theoretical enthalpy increase. That is,

$$\eta_{reg} = \frac{h_x - h_2}{h_4 - h_2} \tag{9.27}$$

As heat transfer approaches reversibility, h_x approaches h_4 and η_{reg} tends to unity (100%).

In practice, regenerator effectiveness values typically range from 60 to 80%, and thus the temperature T_x of the air exiting on the compressor side of the regenerator is normally well below the turbine exhaust temperature. To increase the effectiveness above this range would require greater heat transfer area, resulting in equipment costs that might cancel any advantage due to fuel savings. Moreover, the greater heat transfer area that would be required for a larger effectiveness can result in a significant frictional pressure drop for flow through the regenerator, thereby affecting overall performance. The decision to add a regenerator is influenced by considerations such as these, and the final decision is primarily an economic one.

In Example 9.7, we analyze an air-standard Brayton cycle with regeneration and explore the effect on thermal efficiency as the regenerator effectiveness varies.

Example 9.7 BRAYTON CYCLE WITH REGENERATION

A regenerator is incorporated in the cycle of Example 9.4. **(a)** Determine the thermal efficiency for a regenerator effectiveness of 80%. **(b)** Plot the thermal efficiency versus regenerator effectiveness ranging from 0 to 80%.

Solution

Known: A regenerative gas turbine operates with air as the working fluid. The compressor inlet state, turbine inlet temperature, and compressor pressure ratio are known.

Find: For a regenerator effectiveness of 80%, determine the thermal efficiency. Also plot the thermal efficiency versus the regenerator effectiveness ranging from 0 to 80%.

Schematic and Given Data:

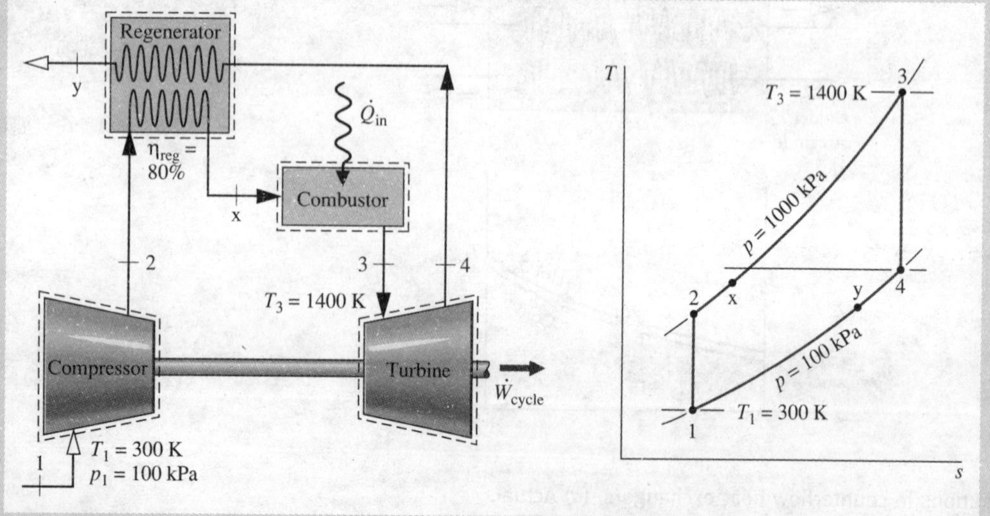

Fig. E9.7a

Engineering Model:

1. Each component is analyzed as a control volume at steady state. The control volumes are shown on the accompanying sketch by dashed lines.
2. The compressor and turbine processes are isentropic.
3. There are no pressure drops for flow through the heat exchangers.
4. The regenerator effectiveness is 80% in part (a).
5. Kinetic and potential energy effects are negligible.
6. The working fluid is air modeled as an ideal gas.

Analysis:

(a) The specific enthalpy values at the numbered states on the *T–s* diagram are the same as those in Example 9.4: $h_1 = 300.19$ kJ/kg, $h_2 = 579.9$ kJ/kg, $h_3 = 1515.4$ kJ/kg, $h_4 = 808.5$ kJ/kg.

To find the specific enthalpy h_x, the regenerator effectiveness is used as follows: By definition

$$\eta_{reg} = \frac{h_x - h_2}{h_4 - h_2}$$

Solving for h_x

$$h_x = \eta_{reg}(h_4 - h_2) + h_2$$
$$= (0.8)(808.5 - 579.9) + 579.9 = 762.8 \text{ kJ/kg}$$

With the specific enthalpy values determined above, the thermal efficiency is

❶ $$\eta = \frac{(\dot{W}_t/\dot{m}) - (\dot{W}_c/\dot{m})}{(\dot{Q}_{in}/\dot{m})} = \frac{(h_3 - h_4) - (h_2 - h_1)}{(h_3 - h_x)}$$

$$= \frac{(1515.4 - 808.5) - (579.9 - 300.19)}{(1515.4 - 762.8)}$$

❷ $$= 0.568\ (56.8\%)$$

(b) The *IT* code for the solution follows, where η_{reg} is denoted as etareg, η is eta, $\dot{W}_{comp}/\dot{m}$ is Wcomp, and so on.

```
// Fix the states
T1 = 300//K
p1 = 100//kPa
h1 = h_T("Air", T1)
s1 = s_TP("Air", T1, p1)

p2 = 1000//kPa
s2 = s_TP("Air", T2, p2)
s2 = s1
h2 = h_T("Air", T2)

T3 = 1400//K
p3 = p2
h3 = h_T("Air", T3)
s3 = s_TP("Air", T3, p3)

p4 = p1
s4 = s_TP("Air", T4, p4)
s4 = s3
h4 = h_T("Air", T4)
etareg = 0.8
hx = etareg*(h4 – h2) + h2

// Thermal efficiency
Wcomp = h2 – h1
```

```
Wturb = h3 – h4
Qin = h3 – hx
eta = (Wturb – Wcomp) / Qin
```

Using the **Explore** button, sweep etareg from 0 to 0.8 in steps of 0.01. Then, using the **Graph** button, obtain the following plot:

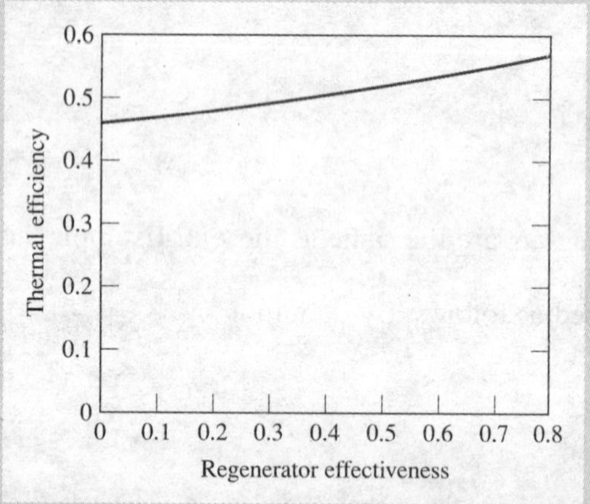

Fig. E9.7b

❸ From the computer data, we see that the cycle thermal efficiency increases from 0.456, which agrees closely with the result of Example 9.4 (no regenerator), to 0.567 for a regenerator effectiveness of 80%, which agrees closely with the result of part (a). This trend is also seen in the accompanying graph. Regenerator effectiveness is seen to have a significant effect on cycle thermal efficiency.

---

❶ The values for work per unit of mass flow of the compressor and turbine are unchanged by the addition of the regenerator. Thus, the back work ratio and net work output are not affected by this modification.

❷ Comparing the present thermal efficiency value with the one determined in Example 9.4, it should be evident that the thermal efficiency can be increased significantly by means of regeneration.

❸ The regenerator allows improved fuel utilization to be achieved by transferring a portion of the exergy in the hot turbine exhaust gas to the cooler air flowing on the other side.

Skills Developed: *Ability to...*

- ✓ sketch the schematic of the regenerative gas turbine and the *T–s* diagram for the corresponding air-standard cycle.
- ✓ evaluate temperatures and pressures at each principal state and retrieve necessary property data.
- ✓ calculate the thermal efficiency.

Quick Quiz

What would be the thermal efficiency if the regenerator effectiveness were 100%? **Ans.** 60.4%.

9.8 Regenerative Gas Turbines with Reheat and Intercooling

Two modifications of the basic gas turbine that increase the net work developed are multistage expansion with *reheat* and multistage compression with *intercooling*. When used in conjunction with regeneration, these modifications can result in substantial increases in thermal efficiency. The concepts of reheat and intercooling are introduced in this section.

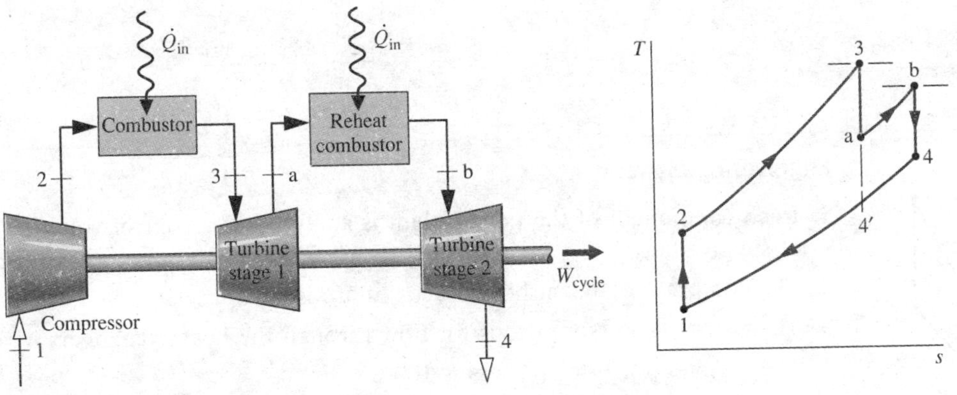

Fig. 9.16 Ideal gas turbine with reheat.

9.8.1 Gas Turbines with Reheat

For metallurgical reasons, the temperature of the gaseous combustion products entering the turbine must be limited. This temperature can be controlled by providing air in excess of the amount required to burn the fuel in the combustor (see Chap. 13). As a consequence, the gases exiting the combustor contain sufficient air to support the combustion of additional fuel. Some gas turbine power plants take advantage of the excess air by means of a multistage turbine with a ***reheat combustor*** between the stages. With this arrangement the net work per unit of mass flow can be increased. Let us consider reheat from the vantage point of an air-standard analysis. ***reheat***

The basic features of a two-stage gas turbine with reheat are brought out by considering an ideal air-standard Brayton cycle modified as shown in Fig. 9.16. After expansion from state 3 to state a in the first turbine, the gas is reheated at constant pressure from state a to state b. The expansion is then completed in the second turbine from state b to state 4. The ideal Brayton cycle without reheat, 1–2–3–4′–1, is shown on the same *T–s* diagram for comparison. Because lines of constant pressure on a *T–s* diagram diverge slightly with increasing entropy, the total work of the two-stage turbine is greater than that of a single expansion from state 3 to state 4′. Thus, the *net* work for the reheat cycle is greater than that of the cycle without reheat. Despite the increase in net work with reheat, the cycle thermal efficiency would not necessarily increase because a greater total heat addition would be required. However, the temperature at the exit of the turbine is higher with reheat than without reheat, so the potential for regeneration is enhanced.

When reheat and regeneration are used together, the thermal efficiency can increase significantly. The following example provides an illustration.

Example 9.8 BRAYTON CYCLE WITH REHEAT AND REGENERATION

Consider a modification of the cycle of Example 9.4 involving reheat and regeneration. Air enters the compressor at 100 kPa, 300 K and is compressed to 1000 kPa. The temperature at the inlet to the first turbine stage is 1400 K. The expansion takes place isentropically in two stages, with reheat to 1400 K between the stages at a constant pressure of 300 kPa. A regenerator having an effectiveness of 100% is also incorporated in the cycle. Determine the thermal efficiency.

Solution

Known: An ideal air-standard gas turbine cycle operates with reheat and regeneration. Temperatures and pressures at principal states are specified.

Find: Determine the thermal efficiency.

Schematic and Given Data:

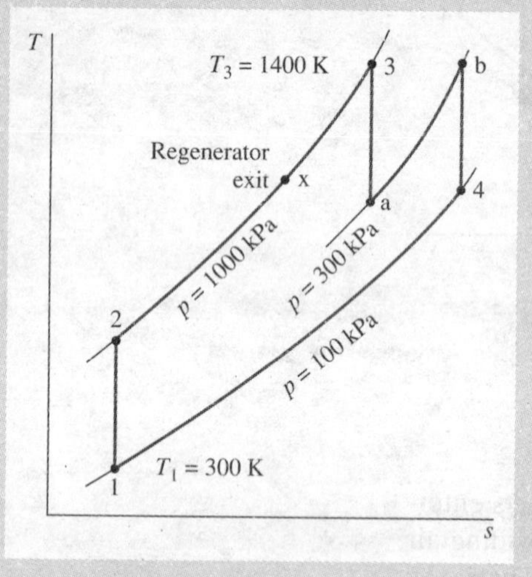

Engineering Model:

1. Each component of the power plant is analyzed as a control volume at steady state.
2. The compressor and turbine processes are isentropic.
3. There are no pressure drops for flow through the heat exchangers.
4. The regenerator effectiveness is 100%.
5. Kinetic and potential energy effects are negligible.
6. The working fluid is air modeled as an ideal gas.

Fig. E9.8

Analysis: We begin by determining the specific enthalpies at each principal state of the cycle. States 1, 2, and 3 are the same as in Example 9.4: $h_1 = 300.19$ kJ/kg, $h_2 = 579.9$ kJ/kg, $h_3 = 1515.4$ kJ/kg. The temperature at state b is the same as at state 3, so $h_b = h_3$.

Since the first turbine process is isentropic, the enthalpy at state a can be determined using p_r data from Table A-22 and the relationship

$$p_{ra} = p_{r3}\frac{p_a}{p_3} = (450.5)\frac{300}{1000} = 135.15$$

Interpolating in Table A-22, we get $h_a = 1095.9$ kJ/kg.

The second turbine process is also isentropic, so the enthalpy at state 4 can be determined similarly. Thus

$$p_{r4} = p_{rb}\frac{p_4}{p_b} = (450.5)\frac{100}{300} = 150.17$$

Interpolating in Table A-22, we obtain $h_4 = 1127.6$ kJ/kg. Since the regenerator effectiveness is 100%, $h_x = h_4 = 1127.6$ kJ/kg.

The thermal efficiency calculation must take into account the compressor work, the work of *each* turbine, and the *total* heat added. Thus, on a unit mass basis

$$\eta = \frac{(h_3 - h_a) + (h_b - h_4) - (h_2 - h_1)}{(h_3 - h_x) + (h_b - h_a)}$$

$$= \frac{(1515.4 - 1095.9) + (1515.4 - 1127.6) - (579.9 - 300.19)}{(1515.4 - 1127.6) + (1515.4 - 1095.9)}$$

❶ $= 0.654\ (65.4\%)$

❶ Comparing the present value with the thermal efficiency determined in part (a) of Example 9.4, we can conclude that the use of reheat coupled with regeneration can result in a substantial increase in thermal efficiency.

Skills Developed: *Ability to...*

- ✓ sketch the schematic of the regenerative gas turbine with reheat and the *T–s* diagram for the corresponding air-standard cycle.
- ✓ evaluate temperatures and pressures at each principal state and retrieve necessary property data.
- ✓ calculate the thermal efficiency.

Quick Quiz

What percentage of the total heat addition occurs in the reheat process? **Ans.** 52%.

9.8.2 Compression with Intercooling

The net work output of a gas turbine also can be increased by reducing the compressor work input. This can be accomplished by means of multistage compression with intercooling. The present discussion provides an introduction to this subject.

Let us first consider the work input to compressors at steady state, assuming that irreversibilities are absent and changes in kinetic and potential energy from inlet to exit are negligible. The p–v diagram of Fig. 9.17 shows two possible compression paths from a specified state 1 to a specified final pressure p_2. Path 1–2′ is for an adiabatic compression. Path 1–2 corresponds to a compression with heat transfer *from* the working fluid to the surroundings. The area to the left of each curve equals the *magnitude* of the work per unit mass of the respective process (see Sec. 6.13). The smaller area to the left of Process 1–2 indicates that the work of this process is less than for the adiabatic compression from 1 to 2′. This suggests that cooling a gas *during* compression is advantageous in terms of the work-input requirement.

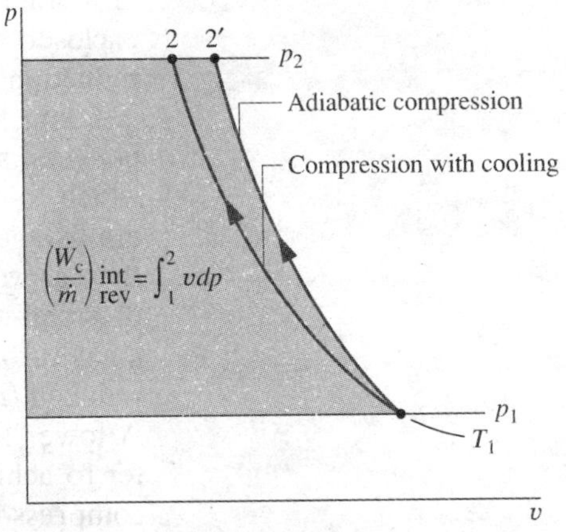

Fig. 9.17 Internally reversible compression processes between two fixed pressures.

Although cooling a gas *as it is compressed* would reduce the work, a heat transfer rate high enough to effect a significant reduction in work is difficult to achieve in practice. A practical alternative is to separate the work and heat interactions into separate processes by letting compression take place in stages with heat exchangers, called ***intercoolers,*** cooling the gas between stages. Figure 9.18 illustrates a two-stage compressor with an intercooler. The accompanying p–v and T–s diagrams show the states for internally reversible processes:

intercooler

- Process 1–c is an isentropic compression from state 1 to state c where the pressure is p_i.
- Process c–d is constant-pressure cooling from temperature T_c to T_d.
- Process d–2 is an isentropic compression to state 2.

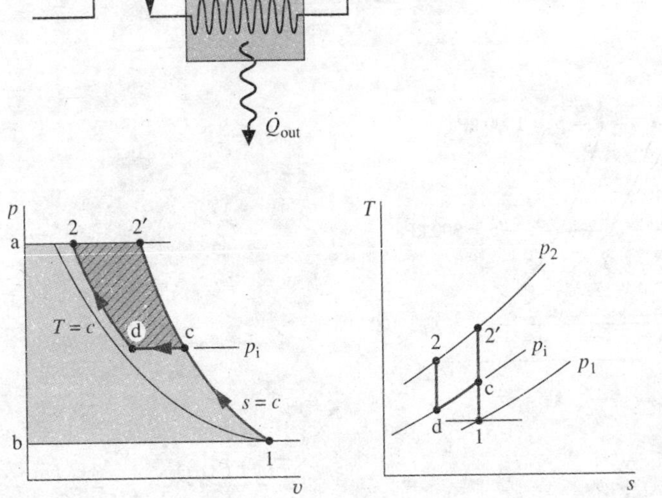

Fig. 9.18 Two-stage compression with intercooling.

The work input per unit of mass flow is represented on the p–v diagram by shaded area 1–c–d–2–a–b–1. Without intercooling the gas would be compressed isentropically in a single stage from state 1 to state 2′ and the work would be represented by enclosed area 1–2′–a–b–1. The crosshatched area on the p–v diagram represents the reduction in work that would be achieved with intercooling.

Some large compressors have several stages of compression with intercooling between stages. The determination of the number of stages and the conditions at which to operate the various intercoolers is a problem in optimization. The use of multistage compression with intercooling in a gas turbine power plant increases the net work developed by reducing the compression work. By itself, though, compression with intercooling would not necessarily increase the thermal efficiency of a gas turbine because the temperature of the air entering the combustor would be reduced (compare temperatures at states 2′ and 2 on the T–s diagram of Fig. 9.18). A lower temperature at the combustor inlet would require additional heat transfer to achieve the desired turbine inlet temperature. The lower temperature at the compressor exit enhances the potential for regeneration, however, so when intercooling is used in conjunction with regeneration, an appreciable increase in thermal efficiency can result.

In the next example, we analyze a two-stage compressor with intercooling between the stages. Results are compared with those for a single stage of compression.

Example 9.9 COMPRESSION WITH INTERCOOLING

Air is compressed from 100 kPa, 300 K to 1000 kPa in a two-stage compressor with intercooling between stages. The intercooler pressure is 300 kPa. The air is cooled back to 300 K in the intercooler before entering the second compressor stage. Each compressor stage is isentropic. For steady-state operation and negligible changes in kinetic and potential energy from inlet to exit, determine **(a)** the temperature at the exit of the second compressor stage and **(b)** the total compressor work input per unit of mass flow. **(c)** Repeat for a single stage of compression from the given inlet state to the final pressure.

Solution

Known: Air is compressed at steady state in a two-stage compressor with intercooling between stages. Operating pressures and temperatures are given.

Find: Determine the temperature at the exit of the second compressor stage and the total work input per unit of mass flow. Repeat for a single stage of compression.

Schematic and Given Data:

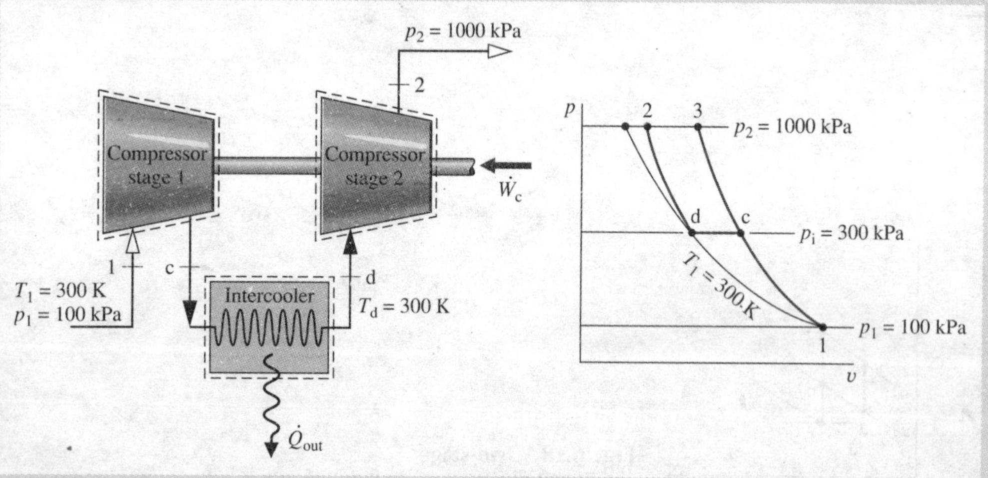

Fig. E9.9

Engineering Model:

1. The compressor stages and intercooler are analyzed as control volumes at steady state. The control volumes are shown on the accompanying sketch by dashed lines.
2. The compression processes are isentropic.
3. There is no pressure drop for flow through the intercooler.
4. Kinetic and potential energy effects are negligible.
5. The air is modeled as an ideal gas.

Analysis:

(a) The temperature at the exit of the second compressor stage, T_2, can be found using the following relationship for the isentropic process d–2

$$p_{r2} = p_{rd}\frac{p_2}{p_d}$$

With p_{rd} at $T_d = 300$ K from Table A-22, $p_2 = 1000$ kPa, and $p_d = 300$ kPa,

$$p_{r2} = (1.386)\frac{1000}{300} = 4.62$$

Interpolating in Table A-22, we get $T_2 = 422$ K and $h_2 = 423.8$ kJ/kg

(b) The total compressor work input per unit of mass is the sum of the work inputs for the two stages. That is

$$\frac{\dot{W}_c}{\dot{m}} = (h_c - h_1) + (h_2 - h_d)$$

From Table A-22 at $T_1 = 300$ K, $h_1 = 300.19$ kJ/kg. Since $T_d = T_1$, $h_d = 300.19$ kJ/kg. To find h_c, use p_r data from Table A-22 together with $p_1 = 100$ kPa and $p_c = 300$ kPa to write

$$p_{rc} = p_{r1}\frac{p_c}{p_1} = (1.386)\frac{300}{100} = 4.158$$

Interpolating in Table A-22, we obtain $h_c = 411.3$ kJ/kg. Hence, the total compressor work per unit of mass is

$$\frac{\dot{W}_c}{\dot{m}} = (411.3 - 300.19) + (423.8 - 300.19) = 234.7 \text{ kJ/kg}$$

(c) For a single isentropic stage of compression, the exit state would be state 3 located on the accompanying p–v diagram. The temperature at this state can be determined using

$$p_{r3} = p_{r1}\frac{p_3}{p_1} = (1.386)\frac{1000}{100} = 13.86$$

Interpolating in Table A-22, we get $T_3 = 574$ K and $h_3 = 579.9$ kJ/kg.

The work input for a single stage of compression is then

$$\frac{\dot{W}_c}{\dot{m}} = h_3 - h_1 = 579.9 - 300.19 = 279.7 \text{ kJ/kg}$$

This calculation confirms that a smaller work input is required with two-stage compression and intercooling than with a single stage of compression. With intercooling, however, a much lower gas temperature is achieved at the compressor exit.

Skills Developed: *Ability to...*

- ✓ sketch the schematic of a two-stage compressor with intercooling between the stages and the corresponding T–s diagram.
- ✓ evaluate temperatures and pressures at each principal state and retrieve necessary property data.
- ✓ apply energy and entropy balances.

Quick Quiz

In this case, what is the percentage reduction in compressor work with two-stage compression and intercooling compared to a single stage of compression?
Ans. 16.1%.

Referring again to Fig. 9.18, the size of the crosshatched area on the p–v diagram representing the reduction in work with intercooling depends on both the temperature T_d at the exit of the intercooler and the intercooler pressure p_i. By properly selecting T_d and p_i, the total work input to the compressor can be minimized. For example, if the pressure p_i is specified, the work input would decrease (crosshatched area would increase) as the temperature T_d approaches T_1, the temperature at the inlet to the compressor. For air entering the compressor from the surroundings, T_1 would be the limiting temperature that could be achieved at state d through heat transfer with the surroundings only. Also, for a specified value of the temperature T_d, the pressure p_i can be selected so that the total work input is a minimum (crosshatched area is a maximum).

Example 9.10 provides an illustration of the determination of the intercooler pressure for minimum total work using a cold air-standard analysis.

Example 9.10 INTERCOOLER PRESSURE FOR MINIMUM COMPRESSOR WORK

If the inlet state and the exit pressure are specified for a two-stage compressor operating at steady state, show that the minimum total work input is required when the pressure ratio is the same across each stage. Use a cold air-standard analysis assuming that each compression process is isentropic, there is no pressure drop through the intercooler, and the temperature at the inlet to each compressor stage is the same. Kinetic and potential energy effects can be ignored.

Solution

Known: A two-stage compressor with intercooling operates at steady state under specified conditions.

Find: Show that the minimum total work input is required when the pressure ratio is the same across each stage.

Schematic and Given Data:

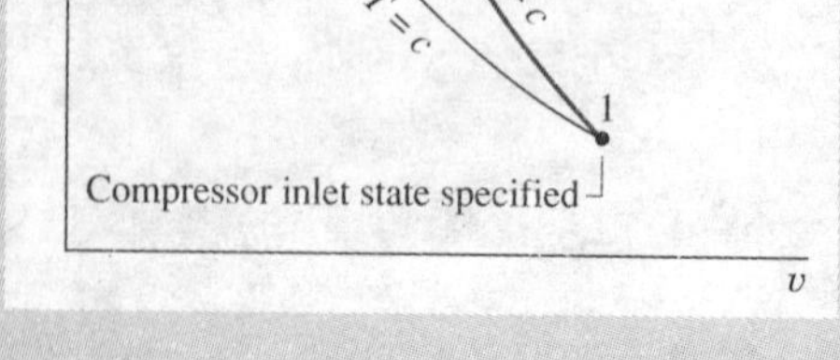

Engineering Model:

1. The compressor stages and intercooler are analyzed as control volumes at steady state.
2. The compression processes are isentropic.
3. There is no pressure drop for flow through the intercooler.
4. The temperature at the inlet to both compressor stages is the same.
5. Kinetic and potential energy effects are negligible.
6. The working fluid is air modeled as an ideal gas.
7. The specific heat c_p and the specific heat ratio k are constant.

Fig. E9.10

Analysis: The total compressor work input per unit of mass flow is

$$\frac{\dot{W}_c}{\dot{m}} = (h_c - h_1) + (h_2 - h_d)$$

Since c_p is constant

$$\frac{\dot{W}_c}{\dot{m}} = c_p(T_c - T_1) + c_p(T_2 - T_d)$$

With $T_d = T_1$ (assumption 4), this becomes on rearrangement

$$\frac{\dot{W}_c}{\dot{m}} = c_p T_1 \left(\frac{T_c}{T_1} + \frac{T_2}{T_1} - 2 \right)$$

Since the compression processes are isentropic and the specific heat ratio k is constant, the pressure and temperature ratios across the compressor stages are related, respectively, by

$$\frac{T_c}{T_1} = \left(\frac{p_i}{p_1} \right)^{(k-1)/k} \quad \text{and} \quad \frac{T_2}{T_d} = \left(\frac{p_2}{p_i} \right)^{(k-1)/k}$$

In the second of these equations, $T_d = T_1$ by assumption 4.

Collecting results

$$\frac{\dot{W}_c}{\dot{m}} = c_p T_1 \left[\left(\frac{p_i}{p_1} \right)^{(k-1)/k} + \left(\frac{p_2}{p_i} \right)^{(k-1)/k} - 2 \right]$$

Hence, for specified values of T_1, p_1, p_2, and c_p, the value of the total compressor work input varies with the intercooler pressure only. To determine the pressure p_i that minimizes the total work, form the derivative

$$\frac{\partial(\dot{W}_c/\dot{m})}{\partial p_i} = \frac{\partial}{\partial p_i} \left\{ c_p T_1 \left[\left(\frac{p_i}{p_1} \right)^{(k-1)/k} + \left(\frac{p_2}{p_i} \right)^{(k-1)/k} - 2 \right] \right\}$$

$$= c_p T_1 \left(\frac{k-1}{k} \right) \left[\left(\frac{p_i}{p_1} \right)^{-1/k} \left(\frac{1}{p_1} \right) + \left(\frac{p_2}{p_i} \right)^{-1/k} \left(-\frac{p_2}{p_i^2} \right) \right]$$

$$= c_p T_1 \left(\frac{k-1}{k} \right) \frac{1}{p_i} \left[\left(\frac{p_i}{p_1} \right)^{(k-1)/k} - \left(\frac{p_2}{p_i} \right)^{(k-1)/k} \right]$$

When the partial derivative is set to zero, the desired relationship is obtained

1 $$\frac{p_i}{p_1} = \frac{p_2}{p_i}$$

By checking the sign of the second derivative, it can be verified that the total compressor work is a minimum.

Skills Developed: *Ability to...*
- ✓ complete the detailed derivation of a thermodynamic expression.
- ✓ use calculus to maximize a function.

---

1 This relationship is for a two-stage compressor. Appropriate relations can be obtained similarly for multistage compressors.

Quick Quiz

Air enters an ideal two-stage compressor with intercooling at 100 kPa. The overall compressor pressure ratio is 12. What pressure, in bar, would minimize the total work input required?

Ans. 3.464 bar.

9.8.3 Reheat and Intercooling

Reheat between turbine stages and intercooling between compressor stages provide two important advantages: The net work output is increased, and the potential for regeneration is enhanced. Accordingly, when reheat and intercooling are used together with regeneration, a substantial improvement in performance can be realized. One arrangement incorporating reheat, intercooling, and regeneration is shown in Fig. 9.19. This gas turbine has two stages of compression and two turbine stages. The accompanying T–s diagram is drawn to indicate irreversibilities in the compressor

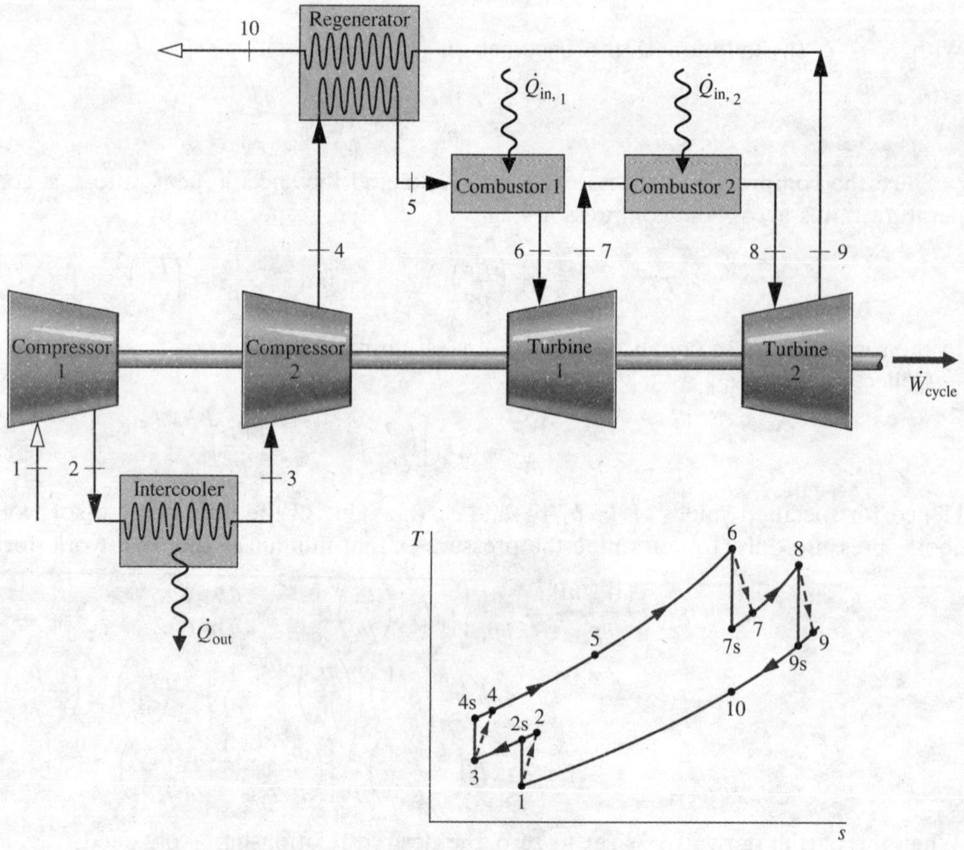

Fig. 9.19 Regenerative gas turbine with intercooling and reheat.

and turbine stages. The pressure drops that would occur as the working fluid passes through the intercooler, regenerator, and combustors are not shown.

Example 9.11 illustrates the analysis of a regenerative gas turbine with intercooling and reheat.

Example 9.11 REGENERATIVE GAS TURBINE WITH INTERCOOLING AND REHEAT

A regenerative gas turbine with intercooling and reheat operates at steady state. Air enters the compressor at 100 kPa, 300 K with a mass flow rate of 5.807 kg/s. The pressure ratio across the two-stage compressor is 10. The pressure ratio across the two-stage turbine is also 10. The intercooler and reheater each operate at 300 kPa. At the inlets to the turbine stages, the temperature is 1400 K. The temperature at the inlet to the second compressor stage is 300 K. The isentropic efficiency of each compressor and turbine stage is 80%. The regenerator effectiveness is 80%. Determine **(a)** the thermal efficiency, **(b)** the back work ratio, and **(c)** the net power developed, in kW.

Solution

Known: An air-standard regenerative gas turbine with intercooling and reheat operates at steady state. Operating pressures and temperatures are specified. Turbine and compressor isentropic efficiencies are given and the regenerator effectiveness is known.

Find: Determine the thermal efficiency, back work ratio, and net power developed, in kW.

Schematic and Given Data:

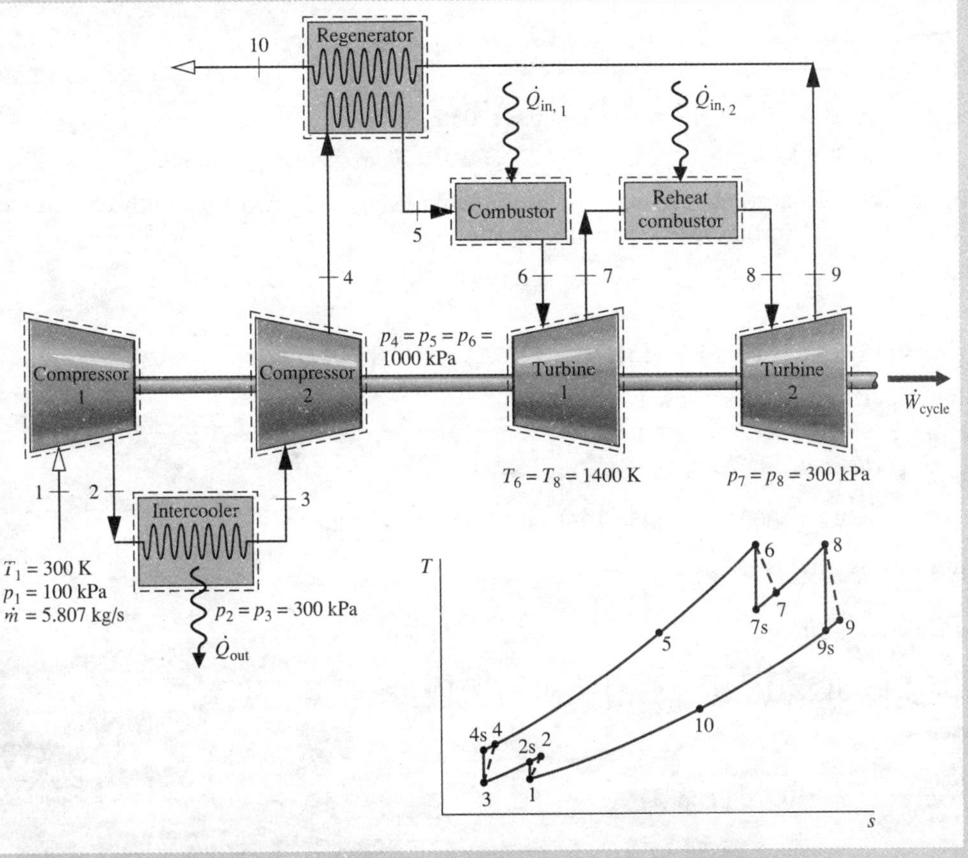

Engineering Model:

1. Each component is analyzed as a control volume at steady state. The control volumes are shown on the accompanying sketch by dashed lines.
2. There are no pressure drops for flow through the heat exchangers.
3. The compressor and turbine are adiabatic.
4. Kinetic and potential energy effects are negligible.
5. The working fluid is air modeled as an ideal gas.

Fig. E9.11

Analysis: We begin by listing the specific enthalpies at the principal states of this cycle. The enthalpies at states 1, 2s, 3, and 4s are obtained from the solution to Example 9.9 where these states are designated as 1, c, d, and 2, respectively. Thus, $h_1 = h_3 = 300.19$ kJ/kg, $h_{2s} = 411.3$ kJ/kg, $h_{4s} = 423.8$ kJ/kg.

The specific enthalpies at states 6, 7s, 8, and 9s are obtained from the solution to Example 9.8, where these states are designated as 3, a, b, and 4, respectively. Thus, $h_6 = h_8 = 1515.4$ kJ/kg, $h_{7s} = 1095.9$ kJ/kg, $h_{9s} = 1127.6$ kJ/kg.

The specific enthalpy at state 4 can be determined using the isentropic efficiency of the second compressor stage

$$\eta_c = \frac{h_{4s} - h_3}{h_4 - h_3}$$

Solving for h_4

$$h_4 = h_3 + \frac{h_{4s} - h_3}{\eta_c} = 300.19 + \left(\frac{423.8 - 300.19}{0.8}\right)$$
$$= 454.7 \text{ kJ/kg}$$

Similarly, the specific enthalpy at state 2 is $h_2 = 439.1$ kJ/kg.

The specific enthalpy at state 9 can be determined using the isentropic efficiency of the second turbine stage

$$\eta_t = \frac{h_8 - h_9}{h_8 - h_{9s}}$$

Solving for h_9

$$h_9 = h_8 - \eta_t(h_8 - h_{9s}) = 1515.4 - 0.8(1515.4 - 1127.6)$$
$$= 1205.2 \text{ kJ/kg}$$

Similarly, the specific enthalpy at state 7 is $h_7 = 1179.8$ kJ/kg.

The specific enthalpy at state 5 can be determined using the regenerator effectiveness

$$\eta_{reg} = \frac{h_5 - h_4}{h_9 - h_4}$$

Solving for h_5

$$h_5 = h_4 + \eta_{reg}(h_9 - h_4) = 454.7 + 0.8(1205.2 - 454.7)$$
$$= 1055.1 \text{ kJ/kg}$$

(a) The thermal efficiency must take into account the work of both turbine stages, the work of both compressor stages, and the total heat added. The total turbine work per unit of mass flow is

$$\frac{\dot{W}_t}{\dot{m}} = (h_6 - h_7) + (h_8 - h_9)$$

$$= (1515.4 - 1179.8) + (1515.4 - 1205.2) = 645.8 \text{ kJ/kg}$$

The total compressor work input per unit of mass flow is

$$\frac{\dot{W}_c}{\dot{m}} = (h_2 - h_1) + (h_4 - h_3)$$

$$= (439.1 - 300.19) + (454.7 - 300.19) = 293.4 \text{ kJ/kg}$$

The total heat added per unit of mass flow is

$$\frac{\dot{Q}_{in}}{\dot{m}} = (h_6 - h_5) + (h_8 - h_7)$$

$$= (1515.4 - 1055.1) + (1515.4 - 1179.8) = 795.9 \text{ kJ/kg}$$

Calculating the thermal efficiency

$$\eta = \frac{645.8 - 293.4}{795.9} = 0.443\ (44.3\%)$$

(b) The back work ratio is

$$\text{bwr} = \frac{\dot{W}_c/\dot{m}}{\dot{W}_t/\dot{m}} = \frac{293.4}{645.8} = 0.454\ (45.4\%)$$

(c) The net power developed is

$$\dot{W}_{cycle} = \dot{m}(\dot{W}_t/\dot{m} - \dot{W}_c/\dot{m})$$

❶ $$= \left(5.807\frac{\text{kg}}{\text{s}}\right)(645.8 - 293.4)\frac{\text{kJ}}{\text{kg}}\left|\frac{1 \text{ kJ/s}}{1 \text{ kW}}\right| = 2046 \text{ kW}$$

❶ Comparing the thermal efficiency, back work ratio, and net power values of the current example with the corresponding values of Example 9.6, it should be evident that gas turbine power plant performance can be increased significantly by coupling reheat and intercooling with regeneration.

Skills Developed: *Ability to...*
- ✓ sketch the schematic of the regenerative gas turbine with intercooling and reheat and the T–s diagram for the corresponding air-standard cycle.
- ✓ evaluate temperatures and pressures at each principal state and retrieve necessary property data.
- ✓ calculate the thermal efficiency, back work ratio, and net power developed.

Quick Quiz

Using the results of parts (a) and (c), determine the total rate of heat addition to the cycle, in kW.
Ans. 4619 kW.

9.9 Gas Turbines for Aircraft Propulsion

turbojet engine

Gas turbines are particularly suited for aircraft propulsion because of their favorable power-to-weight ratios. The ***turbojet engine*** is commonly used for this purpose. As illustrated in Fig. 9.20, this type of engine consists of three main sections: the diffuser,

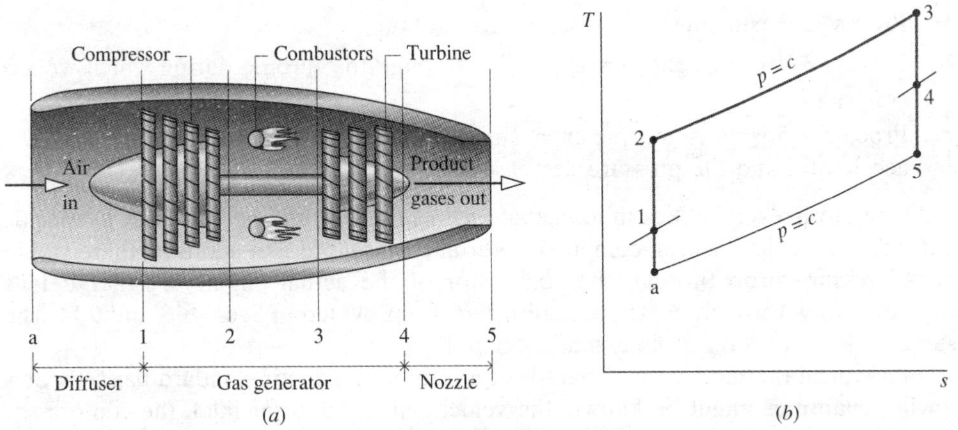

Fig. 9.20 Turbojet engine schematic and accompanying ideal *T*–*s* diagram.

the gas generator, and the nozzle. The diffuser placed before the compressor decelerates the incoming air relative to the engine. A pressure rise known as the ***ram effect*** is associated with this deceleration. The gas generator section consists of a compressor, combustor, and turbine, with the same functions as the corresponding components of a stationary gas turbine power plant. In a turbojet engine, the turbine power output need only be sufficient to drive the compressor and auxiliary equipment, however. The gases leave the turbine at a pressure significantly greater than atmospheric and expand through the nozzle to a high velocity before being discharged to the surroundings. The overall change in the velocity of the gases relative to the engine gives rise to the propulsive force, or thrust. Some turbojets are equipped with an ***afterburner***, as shown in Fig. 9.21. This is essentially a reheat device in which additional fuel is injected into the gas exiting the turbine and burned, producing a higher temperature at the nozzle inlet than would be achieved otherwise. As a consequence, a greater nozzle exit velocity is attained, resulting in increased thrust.

ram effect

afterburner

TURBOJET ANALYSIS. The *T*–*s* diagram of the processes in an ideal turbojet engine is shown in Fig. 9.20*b*. In accordance with the assumptions of an air-standard analysis, the working fluid is air modeled as an ideal gas. The diffuser, compressor, turbine, and nozzle processes are isentropic, and the combustor operates at constant pressure.

- ▶ Process a–1 shows the pressure rise that occurs in the diffuser as the air decelerates isentropically through this component.
- ▶ Process 1–2 is an isentropic compression.

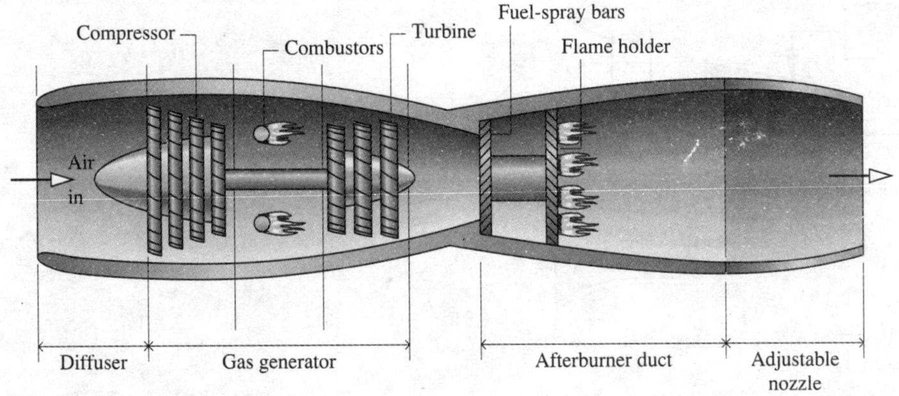

Fig. 9.21 Schematic of a turbojet engine with afterburner.

- Process 2–3 is a constant-pressure heat addition.
- Process 3–4 is an isentropic expansion through the turbine during which work is developed.
- Process 4–5 is an isentropic expansion through the nozzle in which the air accelerates and the pressure decreases.

Owing to irreversibilities in an actual engine, there would be increases in specific entropy across the diffuser, compressor, turbine, and nozzle. In addition, there would be a pressure drop through the combustor of the actual engine. Further details regarding flow through nozzles and diffusers are provided in Secs. 9.13 and 9.14. The subject of combustion is discussed in Chap. 13.

In a typical thermodynamic analysis of a turbojet on an air-standard basis, the following quantities might be known: the velocity at the diffuser inlet, the compressor pressure ratio, and the turbine inlet temperature. The objective of the analysis would be to determine the velocity at the nozzle exit. Once the nozzle exit velocity is determined, the thrust is determined by applying Newton's second law of motion in a form suitable for a control volume (Sec. 9.12). All principles required for the thermodynamic analysis of turbojet engines on an air-standard basis have been introduced. Example 9.12 provides an illustration.

Example 9.12 ANALYZING A TURBOJET ENGINE

Air enters a turbojet engine at 0.8 bar, 240 K, and an inlet velocity of 1000 km/h (278 m/s). The pressure ratio across the compressor is 8. The turbine inlet temperature is 1200 K and the pressure at the nozzle exit is 0.8 bar. The work developed by the turbine equals the compressor work input. The diffuser, compressor, turbine, and nozzle processes are isentropic, and there is no pressure drop for flow through the combustor. For operation at steady state, determine the velocity at the nozzle exit and the pressure at each principal state. Neglect kinetic energy at the exit of all components except the nozzle and neglect potential energy throughout.

Solution

Known: An ideal turbojet engine operates at steady state. Key operating conditions are specified.

Find: Determine the velocity at the nozzle exit, in m/s, and the pressure, in bar, at each principal state.

Schematic and Given Data:

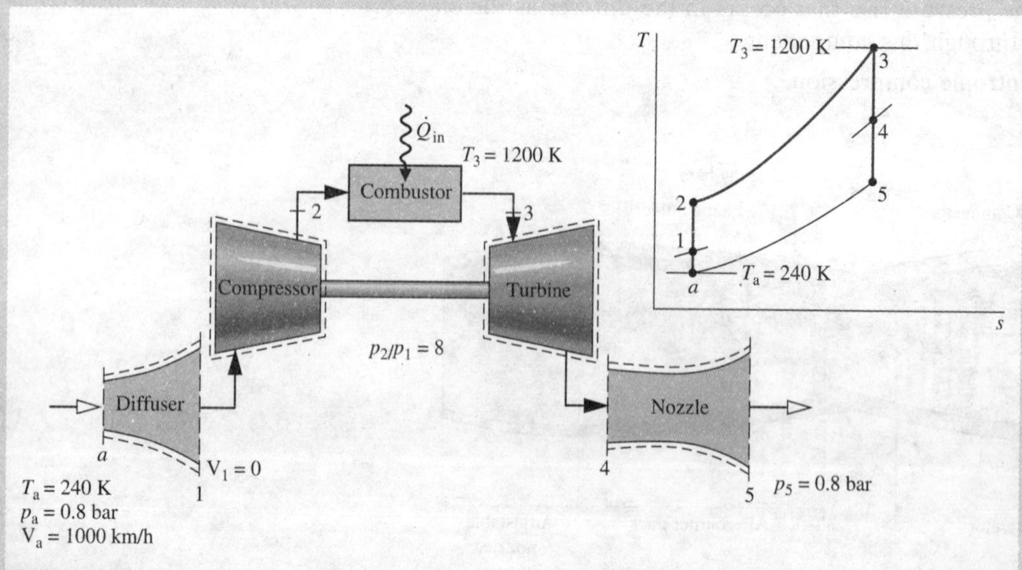

Fig. E9.12

Engineering Model:

1. Each component is analyzed as a control volume at steady state. The control volumes are shown on the accompanying sketch by dashed lines.
2. The diffuser, compressor, turbine, and nozzle processes are isentropic.
3. There is no pressure drop for flow through the combustor.
4. The turbine work output equals the work required to drive the compressor.
5. Except at the inlet and exit of the engine, kinetic energy effects can be ignored. Potential energy effects are negligible throughout.
6. The working fluid is air modeled as an ideal gas.

Analysis: To determine the velocity at the exit to the nozzle, the mass and energy rate balances for a control volume enclosing this component reduce at steady state to give

$$0 = \dot{Q}_{cv}^{\,0} - \dot{W}_{cv}^{\,0} + \dot{m}\left[(h_4 - h_5) + \left(\frac{V_4^{2\,0} - V_5^2}{2}\right) + g(z_4 - z_5)^{0}\right]$$

where $\dot{m}$ is the mass flow rate. The inlet kinetic energy is dropped by assumption 5. Solving for V_5

$$V_5 = \sqrt{2(h_4 - h_5)}$$

This expression requires values for the specific enthalpies h_4 and h_5 at the nozzle inlet and exit, respectively. With the operating parameters specified, the determination of these enthalpy values is accomplished by analyzing each component in turn, beginning with the diffuser. The pressure at each principal state can be evaluated as a part of the analyses required to find the enthalpies h_4 and h_5.

Mass and energy rate balances for a control volume enclosing the diffuser reduce to give

$$h_1 = h_a + \frac{V_a^2}{2}$$

With h_a from Table A-22 and the given value of V_a

1 $$h_1 = 240.02 \text{ kJ/kg} + \left[\frac{(278)^2}{2}\right]\left(\frac{\text{m}^2}{\text{s}^2}\right)\left|\frac{1\text{ N}}{1\text{ kg}\cdot\text{m/s}^2}\right|\left|\frac{1\text{ kJ}}{10^3\text{ N}\cdot\text{m}}\right|$$

$$= 278.7 \text{ kJ/kg}$$

Interpolating in Table A-22 gives $p_{r1} = 1.070$. The flow through the diffuser is isentropic, so pressure p_1 is

$$p_1 = \frac{p_{r1}}{p_{ra}} p_a$$

With p_r data from Table A-22 and the known value of p_a

$$p_1 = \frac{1.070}{0.6355}(0.8 \text{ bar}) = 1.347 \text{ bar}$$

Using the given compressor pressure ratio, the pressure at state 2 is $p_2 = 8(1.347 \text{ bar}) = 10.78$ bar.

The flow through the compressor is also isentropic. Thus

$$p_{r2} = p_{r1}\frac{p_2}{p_1} = 1.070(8) = 8.56$$

Interpolating in Table A-22, we get $h_2 = 505.5$ kJ/kg.

At state 3 the temperature is given as $T_3 = 1200$ K. From Table A-22, $h_3 = 1277.79$ kJ/kg. By assumption 3, $p_3 = p_2$. The work developed by the turbine is just sufficient to drive the compressor (assumption 4). That is

$$\frac{\dot{W}_t}{\dot{m}} = \frac{\dot{W}_c}{\dot{m}}$$

or

$$h_3 - h_4 = h_2 - h_1$$

Solving for h_4

$$h_4 = h_3 + h_1 - h_2 = 1277.79 + 278.7 - 505.5$$
$$= 1051 \text{ kJ/kg}$$

Interpolating in Table A-22 with h_4 gives $p_{r4} = 116.0$.

The expansion through the turbine is isentropic, so

$$p_4 = p_3 \frac{p_{r4}}{p_{r3}}$$

With $p_3 = p_2$ and p_r data from Table A-22

$$p_4 = (10.78 \text{ bar}) \frac{116.0}{238.0} = 5.25 \text{ bar}$$

The expansion through the nozzle is isentropic to $p_5 = 0.8$ bar. Thus

$$p_{r5} = p_{r4} \frac{p_5}{p_4} = (116.0) \frac{0.8}{5.25} = 17.68$$

From Table A-22, $h_5 = 621.3$ kJ/kg, which is the remaining specific enthalpy value required to determine the velocity at the nozzle exit.

Using the values for h_4 and h_5 determined above, the velocity at the nozzle exit is

$$V_5 = \sqrt{2(h_4 - h_5)}$$

$$= \sqrt{2\,(1051 - 621.3)\frac{\text{kJ}}{\text{kg}}\left(\frac{1\text{kg m/s}^2}{1\text{ N}}\right)\left(\frac{10^3\text{N}\cdot\text{m}}{1\text{ kJ}}\right)}$$

❷ $$= 927 \text{ m/s}$$

❶ Note the unit conversions required here and in the calculation of V_5.

❷ The increase in the velocity of the air as it passes through the engine gives rise to the thrust produced by the engine. A detailed analysis of the forces acting on the engine requires Newton's second law of motion in a form suitable for control volumes (see Sec. 9.12.1).

Skills Developed: *Ability to...*

- ✓ sketch the schematic of the turbojet engine and the *T-s* diagram for the corresponding air-standard cycle.
- ✓ evaluate temperatures and pressures at each principal state and retrieve necessary property data.
- ✓ apply mass, energy, and entropy balances.
- ✓ calculate the nozzle exit velocity.

Quick Quiz

Referring to Eq. 6.47, what would be the nozzle exit velocity, in m/s, if the nozzle isentropic efficiency were 90%? **Ans.** 880 m/s.

OTHER APPLICATIONS. Other related applications of the gas turbine include *turboprop* and *turbofan* engines. The turboprop engine shown in Fig. 9.22*a* consists of a gas turbine in which the gases are allowed to expand through the turbine to atmospheric pressure. The net power developed is directed to a propeller, which provides thrust to the aircraft. Turboprops are efficient propulsion devices for speeds of up to about 600 km/h. In the turbofan shown in Fig. 9.22*b*, the core of the engine is much like a turbojet, and some thrust is obtained from expansion through the nozzle. However, a set of large-diameter blades attached to the front of the engine accelerates air around the core. This *bypass flow* provides additional thrust for takeoff, whereas the core of

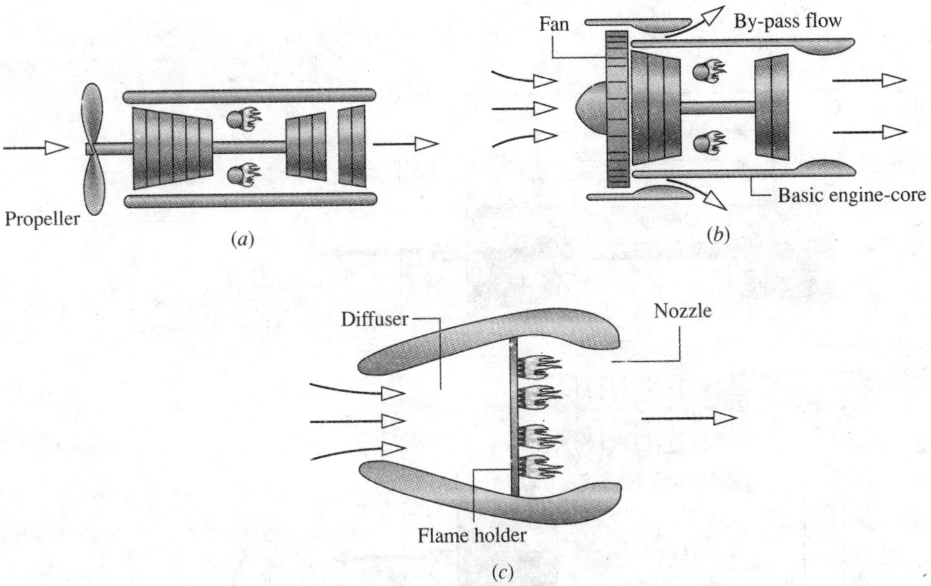

Fig. 9.22 Other examples of aircraft engines. (*a*) Turboprop. (*b*) Turbofan. (*c*) Ramjet.

the engine provides the primary thrust for cruising. Turbofan engines are commonly used for commercial aircraft with flight speeds of up to about 1000 km/h. A particularly simple type of engine known as a ramjet is shown in Fig. 9.22*c*. This engine requires neither a compressor nor a turbine. A sufficient pressure rise is obtained by decelerating the high-speed incoming air in the diffuser (ram effect). For the ramjet to operate, therefore, the aircraft must already be in flight at high speed. The combustion products exiting the combustor are expanded through the nozzle to produce the thrust.

In each of the engines mentioned thus far, combustion of the fuel is supported by air brought into the engines from the atmosphere. For very high-altitude flight and space travel, where this is no longer possible, *rockets* may be employed. In these applications, both fuel and an oxidizer (such as liquid oxygen) are carried on board the craft. Thrust is developed when the high-pressure gases obtained on combustion are expanded through a nozzle and discharged from the rocket.

9.10 Combined Gas Turbine–Vapor Power Cycle

A ***combined power cycle*** couples two power cycles such that the energy discharged by heat transfer from one cycle is used partly or wholly as the input for the other cycle. The binary vapor cycle introduced in Sec. 8.5 is an example of a *combined power cycle*. In the present section, a combined gas turbine–vapor power cycle is considered. ***combined cycle***

The stream exiting the turbine of a gas turbine is at a high temperature. One way the potential (exergy) of this high-temperature gas stream can be used, thereby improving overall fuel utilization, is by a regenerator that allows the turbine exhaust gas to preheat the air between the compressor and combustor (Sec. 9.7). Another method is provided by the combined cycle shown in Fig. 9.23, involving a gas turbine cycle and a vapor power cycle. The two power cycles are coupled so that the heat transfer to the vapor cycle is provided by the gas turbine cycle, which may be called the ***topping cycle.*** ***topping cycle***

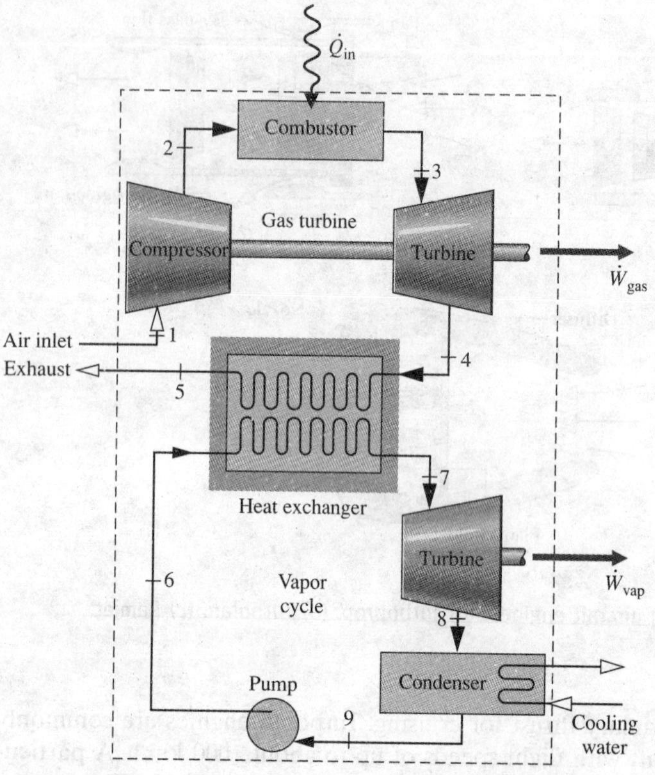

Fig. 9.23 Combined gas turbine–vapor power plant.

The combined cycle has the gas turbine's high average temperature of heat addition and the vapor cycle's low average temperature of heat rejection, and thus a thermal efficiency greater than either cycle would have individually. For many applications, combined cycles are economical, and they are increasingly being used worldwide for electric power generation.

With reference to Fig. 9.23, the thermal efficiency of the combined cycle is

$$\eta = \frac{\dot{W}_{gas} + \dot{W}_{vap}}{\dot{Q}_{in}} \tag{9.28}$$

where $\dot{W}_{gas}$ is the *net* power developed by the gas turbine and $\dot{W}_{vap}$ is the *net* power developed by the vapor cycle. $\dot{Q}_{in}$ denotes the *total* rate of heat transfer to the combined cycle, including additional heat transfer, if any, to superheat the vapor entering the vapor turbine. The evaluation of the quantities appearing in Eq. 9.28 follows the procedures described in the sections on vapor cycles and gas turbines.

The relation for the energy transferred from the gas cycle to the vapor cycle for the system of Fig. 9.23 is obtained by applying the mass and energy rate balances to a control volume enclosing the heat exchanger. For steady-state operation, negligible heat transfer with the surroundings, and no significant changes in kinetic and potential energy, the result is

$$\dot{m}_v(h_7 - h_6) = \dot{m}_g(h_4 - h_5) \tag{9.29}$$

where $\dot{m}_g$ and $\dot{m}_v$ are the mass flow rates of the gas and vapor, respectively.

As witnessed by relations such as Eqs. 9.28 and 9.29, combined cycle performance can be analyzed using mass and energy balances. To complete the analysis, however, the second law is required to assess the impact of irreversibilities and the true magnitudes of losses. Among the irreversibilities, the most significant is the exergy destroyed by combustion. About 30% of the exergy entering the combustor with the fuel is destroyed by combustion irreversibility. An analysis of the gas turbine on an air-standard basis does not allow this exergy destruction to be evaluated, however, and means introduced in Chap. 13 must be applied for this purpose.

Energy & Environment

A new combined-cycle power plant in Wales is the first to achieve 60% thermal efficiency, once the "four minute mile" of power plant technology. The 480-megawatt system provides power to adjacent chemical manufacturing plants and to the energy grid.

Development efforts began in the early 1990s, when the best available technology combining gas and steam turbines achieved thermal efficiencies of only 50%. Each percentage increase is estimated to save $15–20 million dollars of natural gas over the life of a typical system, according to industry sources. Electric rates with the new combined cycle in Wales are said to be 30% less than were previously available.

The efficiency gains are achieved because natural gas is burned at a temperature nearly 167°C higher than before. Advanced blade cooling methods, new alloys, and unique single crystal blade construction allow the turbines to withstand higher temperatures. The high thermal efficiency also results in the lowest carbon dioxide production per megawatt of electricity of any gas turbine in the world. Nitric oxide (NO_X) emission levels are half the average of other turbines now in use.

The next example illustrates the use of mass and energy balances, the second law, and property data to analyze combined cycle performance.

Example 9.13 COMBINED CYCLE EXERGY ACCOUNTING

A combined gas turbine–vapor power plant has a net power output of 45 MW. Air enters the compressor of the gas turbine at 100 kPa, 300 K, and is compressed to 1200 kPa. The isentropic efficiency of the compressor is 84%. The condition at the inlet to the turbine is 1200 kPa, 1400 K. Air expands through the turbine, which has an isentropic efficiency of 88%, to a pressure of 100 kPa. The air then passes through the interconnecting heat exchanger and is finally discharged at 400 K. Steam enters the turbine of the vapor power cycle at 8 MPa, 400°C, and expands to the condenser pressure of 8 kPa. Water enters the pump as saturated liquid at 8 kPa. The turbine and pump of the vapor cycle have isentropic efficiencies of 90 and 80%, respectively.

(a) Determine the mass flow rates of the air and the steam, each in kg/s, and the net power developed by the gas turbine and vapor power cycle, each in MW.

(b) Develop a full accounting of the *net* rate of exergy increase as the air passes through the gas turbine combustor. Discuss.

Let $T_0 = 300$ K, $p_0 = 100$ kPa.

Solution

Known: A combined gas turbine–vapor power plant operates at steady state with a known net power output. Operating pressures and temperatures are specified. Turbine, compressor, and pump efficiencies are also given.

Find: Determine the mass flow rate of each working fluid, in kg/s, and the net power developed by each cycle, in MW. Develop a full accounting of the exergy increase of the air passing through the gas turbine combustor and discuss the results.

Schematic and Given Data:

❶

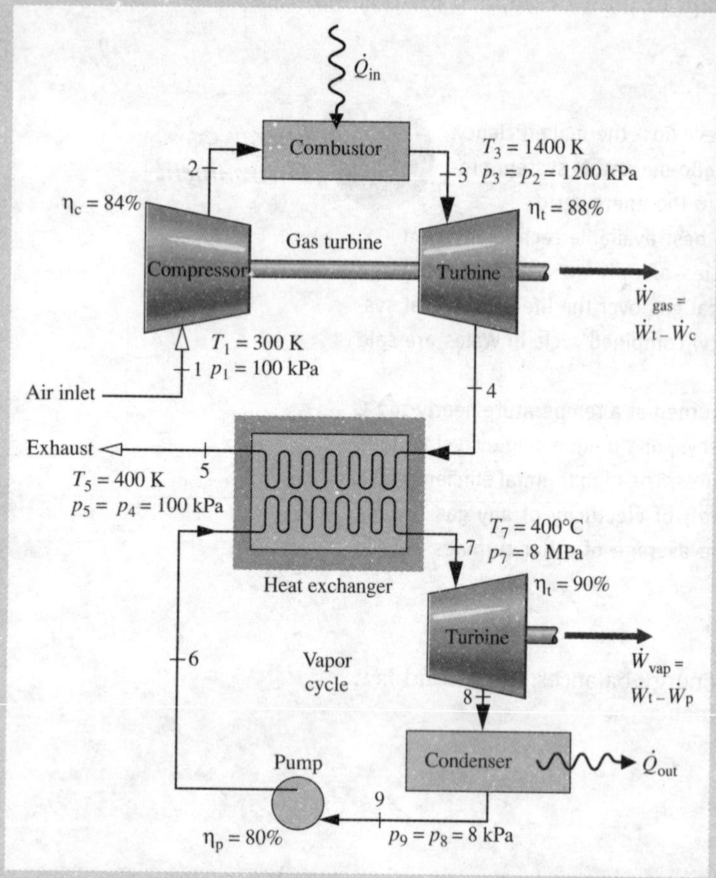

Fig. E9.13

Engineering Model:

1. Each component on the accompanying sketch is analyzed as a control volume at steady state.
2. The turbines, compressor, pump, and interconnecting heat exchanger operate adiabatically.
3. Kinetic and potential energy effects are negligible.
4. There are no pressure drops for flow through the combustor, interconnecting heat exchanger, and condenser.
5. An air-standard analysis is used for the gas turbine.
6. $T_0 = 300$ K, $p_0 = 100$ kPa.

Analysis: The property data given in the table below are determined using procedures illustrated in previous solved examples of Chaps. 8 and 9. The details are left as an exercise.

Gas Turbine			Vapor Cycle		
State	h (kJ/kg)	$s°$ (kJ/kg · K)	State	h (kJ/kg)	s (kJ/kg · K)
1	300.19	1.7020	6	183.96	0.5975
2	669.79	2.5088	7	3138.30	6.3634
3	1515.42	3.3620	8	2104.74	6.7282
4	858.02	2.7620	9	173.88	0.5926
5	400.98	1.9919			

(a) To determine the mass flow rates of the vapor, $\dot{m}_v$, and the air, $\dot{m}_g$, begin by applying mass and energy rate balances to the interconnecting heat exchanger to obtain

$$0 = \dot{m}_g(h_4 - h_5) + \dot{m}_v(h_6 - h_7)$$

or

$$\frac{\dot{m}_v}{\dot{m}_g} = \frac{h_4 - h_5}{h_7 - h_6} = \frac{858.02 - 400.98}{3138.3 - 183.96} = 0.1547$$

Mass and energy rate balances applied to the gas turbine and vapor power cycles give the net power developed by each, respectively

$$\dot{W}_{gas} = \dot{m}_g[(h_3 - h_4) - (h_2 - h_1)]$$
$$\dot{W}_{vap} = \dot{m}_v[(h_7 - h_8) - (h_6 - h_9)]$$

With $\dot{W}_{net} = \dot{W}_{gas} + \dot{W}_{vap}$

$$\dot{W}_{net} = \dot{m}_g\left\{[(h_3 - h_4) - (h_2 - h_1)] + \frac{\dot{m}_v}{\dot{m}_g}[(h_7 - h_8) - (h_6 - h_9)]\right\}$$

Solving for $\dot{m}_g$, and inserting $\dot{W}_{net} = 45$ MW = 45,000 kJ/s and $\dot{m}_v/\dot{m}_a = 0.1547$, we get

$$\dot{m}_g = \frac{45{,}000 \text{ kJ/s}}{\{[(1515.42 - 858.02) - (669.79 - 300.19)] + 0.1547[(3138.3 - 2104.74) - (183.96 - 173.88)]\} \text{ kJ/kg}}$$
$$= 100.87 \text{ kg/s}$$

and

$$\dot{m}_v = (0.1547)\dot{m}_g = 15.6 \text{ kg/s}$$

Using these mass flow rate values and specific enthalpies from the table above, the net power developed by the gas turbine and vapor power cycles, respectively, is

$$\dot{W}_{gas} = \left(100.87\frac{\text{kg}}{\text{s}}\right)\left(287.8\frac{\text{kJ}}{\text{kg}}\right)\left|\frac{1 \text{ MW}}{10^3 \text{ kJ/s}}\right| = 29.03 \text{ MW}$$

$$\dot{W}_{vap} = \left(15.6\frac{\text{kg}}{\text{s}}\right)\left(1023.5\frac{\text{kJ}}{\text{kg}}\right)\left|\frac{1 \text{ MW}}{10^3 \text{ kJ/s}}\right| = 15.97 \text{ MW}$$

(b) The *net* rate of exergy increase of the air passing through the combustor is (Eq. 7.18)

$$\dot{E}_{f3} - \dot{E}_{f2} = \dot{m}_g[h_3 - h_2 - T_0(s_3 - s_2)]$$
$$= \dot{m}_g[h_3 - h_2 - T_0(s_3^\circ - s_2^\circ - R \ln p_3/p_2)]$$

With assumption 4, we have

$$\dot{E}_{f3} - \dot{E}_{f2} = \dot{m}_g\left[h_3 - h_2 - T_0\left(s_3^\circ - s_2^\circ - R \overset{0}{\cancel{\ln\frac{p_3}{p_2}}}\right)\right]$$
$$= \left(100.87\frac{\text{kJ}}{\text{s}}\right)\left[(1515.42 - 669.79)\frac{\text{kJ}}{\text{kg}} - 300 \text{ K}(3.3620 - 2.5088)\frac{\text{kJ}}{\text{kg} \cdot \text{K}}\right]$$
$$= 59{,}480\frac{\text{kJ}}{\text{s}}\left|\frac{1 \text{ MW}}{10^3 \text{ kJ/s}}\right| = 59.48 \text{ MW}$$

The *net* rate exergy is carried out by the exhaust air stream at 5 is

$$\dot{E}_{f5} - \dot{E}_{f1} = \dot{m}_g\left[h_5 - h_1 - T_0\left(s_5^\circ - s_1^\circ - R \overset{0}{\cancel{\ln\frac{p_5}{p_1}}}\right)\right]$$
$$= \left(100.87\frac{\text{kg}}{\text{s}}\right)[(400.98 - 300.19) - 300(1.9919 - 1.7020)]\left(\frac{\text{kJ}}{\text{kg}}\right)\left|\frac{1 \text{ MW}}{10^3 \text{ kJ/s}}\right|$$
$$= 1.39 \text{ MW}$$

The *net* rate exergy is carried out as the water passes through the condenser is

$$\dot{E}_{f8} - \dot{E}_{f9} = \dot{m}_v[h_8 - h_9 - T_0(s_8 - s_9)]$$
$$= \left(15.6\frac{\text{kg}}{\text{s}}\right)\left[(2104.74 - 173.88)\frac{\text{kJ}}{\text{kg}} - 300 \text{ K}(6.7282 - 0.5926)\frac{\text{kJ}}{\text{kg} \cdot \text{K}}\right]\left|\frac{1 \text{ MW}}{10^3 \text{ kJ/s}}\right|$$
$$= 1.41 \text{ MW}$$

The rates of exergy destruction for the air turbine, compressor, steam turbine, pump, and interconnecting heat
❶ exchanger are evaluated using $\dot{E}_d = T_0\dot{\sigma}_{cv}$, respectively, as follows:
Air turbine:

$$\begin{aligned}\dot{E}_d &= \dot{m}_g T_0(s_4 - s_3)\\ &= \dot{m}_g T_0(s_4^\circ - s_3^\circ - R\ln p_4/p_3)\\ &= \left(100.87\frac{\text{kg}}{\text{s}}\right)(300\text{ K})\left[(2.7620 - 3.3620)\frac{\text{kJ}}{\text{kg}\cdot\text{K}} - \left(\frac{8.314}{28.97}\frac{\text{kJ}}{\text{kg}\cdot\text{K}}\right)\ln\left(\frac{100}{1200}\right)\right]\left|\frac{1\text{ MW}}{10^3\text{ kJ/s}}\right|\\ &= 3.42\text{ MW}\end{aligned}$$

Compressor:

$$\begin{aligned}\dot{E}_d &= \dot{m}_g T_0(s_2 - s_1)\\ &= \dot{m}_g T_0(s_2^\circ - s_1^\circ - R\ln p_2/p_1)\\ &= (100.87)(300)\left[(2.5088 - 1.7020) - \frac{8.314}{28.97}\ln\left(\frac{1200}{100}\right)\right]\left|\frac{1}{10^3}\right|\\ &= 2.83\text{ MW}\end{aligned}$$

Steam turbine:

$$\begin{aligned}\dot{E}_d &= \dot{m}_v T_0(s_8 - s_7)\\ &= (15.6)(300)(6.7282 - 6.3634)\left|\frac{1}{10^3}\right|\\ &= 1.71\text{ MW}\end{aligned}$$

Pump:

$$\begin{aligned}\dot{E}_d &= \dot{m}_v T_0(s_6 - s_9)\\ &= (15.6)(300)(0.5975 - 0.5926)\left|\frac{1}{10^3}\right|\\ &= 0.02\text{ MW}\end{aligned}$$

Heat exchanger:

$$\begin{aligned}\dot{E}_d &= T_0[\dot{m}_g(s_5 - s_4) + \dot{m}_v(s_7 - s_6)]\\ &= (300\text{ K})\left[\left(100.87\frac{\text{kg}}{\text{s}}\right)(1.9919 - 2.7620)\frac{\text{kJ}}{\text{kg}\cdot\text{K}} + \left(15.6\frac{\text{kg}}{\text{s}}\right)(6.3634 - 0.5975)\frac{\text{kJ}}{\text{kg}\cdot\text{K}}\right]\left|\frac{1\text{ MW}}{10^3\text{ kJ/s}}\right|\\ &= 3.68\text{ MW}\end{aligned}$$

❷ The results are summarized by the following exergy rate *balance sheet* in terms of exergy magnitudes on a rate basis:

Net exergy increase of the gas passing through the combustor:	59.48 MW	100%	(70%)*
Disposition of the exergy:			
• Net power developed			
gas turbine cycle	29.03 MW	48.8%	(34.2%)
vapor cycle	15.97 MW	26.8%	(18.8%)
Subtotal	45.00 MW	75.6%	(53.0%)
• Net exergy lost			
with exhaust gas at state 5	1.39 MW	2.3%	(1.6%)
from water passing through condenser	1.41 MW	2.4%	(1.7%)
• Exergy destruction			
air turbine	3.42 MW	5.7%	(4.0%)
compressor	2.83 MW	4.8%	(3.4%)
steam turbine	1.71 MW	2.9%	(2.0%)
pump	0.02 MW	—	—
heat exchanger	3.68 MW	6.2%	(4.3%)

*Estimation based on fuel exergy. For discussion, see note 2.

The subtotals given in the table under the *net power developed* heading indicate that the combined cycle is effective in generating power from the exergy supplied. The table also indicates the relative significance of the exergy destructions in the turbines, compressor, pump, and heat exchanger, as well as the relative significance of the exergy losses. Finally, the table indicates that the total of the exergy destructions overshadows the losses.

Skills Developed: *Ability to...*
- ✓ apply mass and energy balances.
- ✓ evaluate exergy quantities.
- ✓ develop an exergy accounting.

❶ The development of the appropriate expressions for the rates of entropy production in the turbines, compressor, pump, and heat exchanger is left as an exercise.

❷ In this exergy balance sheet, the percentages shown in parentheses are estimates based on the fuel exergy. Although combustion is the most significant source of irreversibility, the exergy destruction due to combustion cannot be evaluated using an air-standard analysis. Calculations of exergy destruction due to combustion (Chap. 13) reveal that approximately 30% of the exergy entering the combustor with the fuel would be destroyed, leaving about 70% of the fuel exergy for subsequent use. Accordingly, the value 59.48 MW for the net exergy increase of the air passing through the combustor is assumed to be 70% of the fuel exergy supplied. The other percentages in parentheses are obtained by multiplying the corresponding percentages, based on the exergy increase of the air passing through the combustor, by the factor 0.7. Since they account for combustion irreversibility, the table values in parentheses give the more accurate picture of combined cycle performance.

Quick Quiz

What is the overall exergetic efficiency of the combined cycle based on exergy entering with the fuel? **Ans.** 53.0%.

9.11 Ericsson and Stirling Cycles

Significant increases in the thermal efficiency of gas turbine power plants can be achieved through intercooling, reheat, and regeneration. There is an economic limit to the number of stages that can be employed, and normally there would be no more than two or three. Nonetheless, it is instructive to consider the situation where the number of stages of both intercooling and reheat becomes indefinitely large. Figure 9.24*a* shows an *ideal* closed regenerative gas turbine cycle with several stages of compression and expansion and a regenerator whose effectiveness is 100%. Each intercooler is assumed to return the working fluid to the temperature T_C at the inlet to the first compression stage and each reheater restores the working fluid to the temperature T_H at the inlet to the first turbine stage. The regenerator allows the heat input for Process 2–3 to be obtained from the heat rejected in Process 4–1. Accordingly, all the heat added *externally* would occur in the reheaters, and all the heat rejected to the surroundings would take place in the intercoolers.

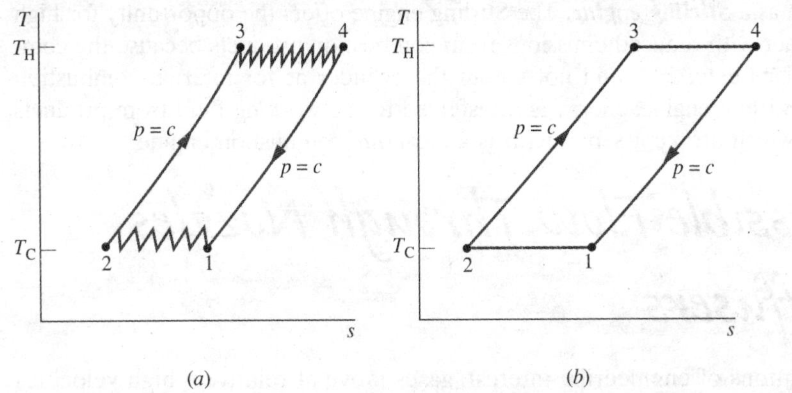

Fig. 9.24 Ericsson cycle as a limit of ideal gas turbine operation using multistage compression with intercooling, multistage expansion with reheating, and regeneration.

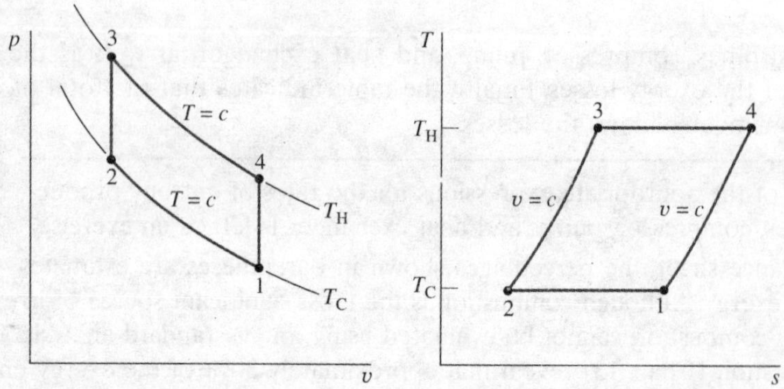

Fig. 9.25 p–v and T–s diagrams of the Stirling cycle.

In the limit, as an infinite number of reheat and intercooler stages is employed, all the heat added would occur when the working fluid is at its highest temperature, T_H, and all the heat rejected would take place when the working fluid is at its lowest temperature, T_C. The limiting cycle, shown in Fig. 9.24*b*, is called the ***Ericsson cycle.*** Since irreversibilities are presumed absent and all the heat is supplied and rejected isothermally, the thermal efficiency of the Ericsson cycle equals that of *any* reversible power cycle operating with heat addition at the temperature T_H and heat rejection at the temperature T_C: $\eta_{max} = 1 - T_C/T_H$. This expression was applied previously to evaluate the thermal efficiency of Carnot power cycles. Although the details of the Ericsson cycle differ from those of the Carnot cycle, both cycles have the same value of thermal efficiency when operating between the temperatures T_H and T_C.

Ericsson cycle

STIRLING CYCLE. Another cycle that employs a regenerator is the *Stirling* cycle, shown on the p–v and T–s diagrams of Fig. 9.25. The cycle consists of four internally reversible processes in series: isothermal compression from state 1 to state 2 at temperature T_C, constant-volume heating from state 2 to state 3, isothermal expansion from state 3 to state 4 at temperature T_H, and constant-volume cooling from state 4 to state 1 to complete the cycle. A regenerator whose effectiveness is 100% allows the heat rejected during Process 4–1 to be used as the heat input in Process 2–3. Accordingly, all the heat added to the working fluid externally would take place in the isothermal process 3–4 and all the heat rejected to the surroundings would occur in the isothermal process 1–2. It can be concluded, therefore, that the thermal efficiency of the Stirling cycle is given by the same expression as for the Carnot and Ericsson cycles.

The Ericsson and Stirling cycles are principally of theoretical interest as examples of cycles that exhibit the same thermal efficiency as the Carnot cycle. However, a practical engine of the piston–cylinder type that operates on a closed regenerative cycle having features in common with the Stirling cycle has been under study in recent years. This engine is known as a ***Stirling engine.*** The Stirling engine offers the opportunity for high efficiency together with reduced emissions from combustion products because the combustion takes place externally and not within the cylinder as for internal combustion engines. In the Stirling engine, energy is transferred to the working fluid from products of combustion, which are kept separate. It is an *external* combustion engine.

Stirling engine

Compressible Flow Through Nozzles and Diffusers

In many applications of engineering interest, gases move at relatively high velocities and exhibit appreciable changes in density. The flows through the nozzles and diffusers of jet engines discussed in Sec. 9.9 are important examples. Other examples are the flows through wind tunnels, shock tubes, and steam ejectors. These flows are known as ***compressible flows.*** In this part of the chapter, we introduce some of the principles involved in analyzing compressible flows.

compressible flow

9.12 Compressible Flow Preliminaries

Concepts introduced in this section play important roles in the study of compressible flows. The momentum equation is introduced in a form applicable to the analysis of control volumes at steady state. The velocity of sound is also defined, and the concepts of Mach number and stagnation state are discussed.

9.12.1 Momentum Equation for Steady One-Dimensional Flow

The analysis of compressible flows requires the principles of conservation of mass and energy, the second law of thermodynamics, and relations among the thermodynamic properties of the flowing gas. In addition, Newton's second law of motion is required. Application of Newton's second law of motion to systems of fixed mass (closed systems) involves the familiar form

$$\mathbf{F} = m\mathbf{a}$$

where $\mathbf{F}$ is the resultant force acting *on* a system of mass m and $\mathbf{a}$ is the acceleration. The object of the present discussion is to introduce Newton's second law of motion in a form appropriate for the study of the control volumes considered in subsequent discussions.

Consider the control volume shown in Fig. 9.26, which has a single inlet, designated by 1, and a single exit, designated by 2. The flow is assumed to be one-dimensional at these locations. The energy and entropy rate equations for such a control volume have terms that account for energy and entropy transfers, respectively, at the inlets and exits. Momentum also can be carried into or out of the control volume at the inlets and exits, and such transfers can be accounted for as

$$\begin{bmatrix} \text{time rate of momentum} \\ \text{transfer into or} \\ \text{out of a control volume} \\ \text{accompanying mass flow} \end{bmatrix} = \dot{m}\mathbf{V} \tag{9.30}$$

In this expression, the momentum per unit of mass flowing across the boundary of the control volume is given by the velocity vector $\mathbf{V}$. In accordance with the one-dimensional flow model, the vector is normal to the inlet or exit and oriented in the direction of flow

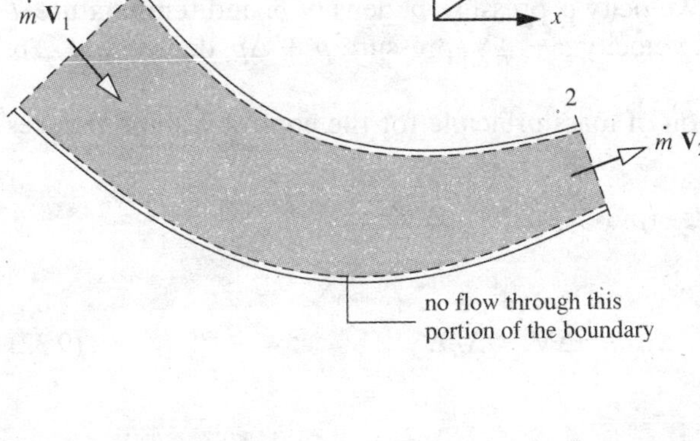

Fig. 9.26 One-inlet, one-exit control volume at steady state labeled with momentum transfers accompanying mass flow.

In words, Newton's second law of motion for control volumes is

$$\begin{bmatrix} \text{time rate of change} \\ \text{of momentum contained} \\ \text{within the control volume} \end{bmatrix} = \begin{bmatrix} \text{resultant force} \\ \text{acting } on \text{ the} \\ \text{control volume} \end{bmatrix} + \begin{bmatrix} \text{net rate at which momentum is} \\ \text{transferred into the control} \\ \text{volume accompanying mass flow} \end{bmatrix}$$

At steady state, the total amount of momentum contained in the control volume is constant with time. Accordingly, when applying Newton's second law of motion to control volumes at steady state, it is necessary to consider only the momentum accompanying the incoming and outgoing streams of matter and the forces acting on the control volume. Newton's law then states that the resultant force **F** acting *on* the control volume equals the difference between the rates of momentum exiting and entering the control volume accompanying mass flow. This is expressed by the following ***momentum equation***

momentum equation

$$\mathbf{F} = \dot{m}_2\mathbf{V}_2 - \dot{m}_1\mathbf{V}_1 = \dot{m}(\mathbf{V}_2 - \mathbf{V}_1) \tag{9.31}$$

Since $\dot{m}_1 = \dot{m}_2$ at steady state, the common mass flow is designated in this expression simply as $\dot{m}$. The resultant force includes the forces due to pressure acting at the inlet and exit, forces acting on the portion of the boundary through which there is no mass flow, and the force of gravity. The expression of Newton's second law of motion given by Eq. 9.31 suffices for subsequent discussions. More general control volume formulations are normally provided in fluid mechanics texts.

9.12.2 Velocity of Sound and Mach Number

A sound wave is a small pressure disturbance that propagates through a gas, liquid, or solid at a velocity c that depends on the properties of the medium. In this section we obtain an expression that relates the *velocity of sound,* or sonic velocity, to other properties. The velocity of sound is an important property in the study of compressible flows.

MODELING PRESSURE WAVES. Let us begin by referring to Fig. 9.27*a*, which shows a pressure wave moving to the right with a velocity of magnitude c. The wave is generated by a small displacement of the piston. As shown on the figure, the pressure, density, and temperature in the region to the left of the wave depart from the respective values of the undisturbed fluid to the right of the wave, which are designated simply p, ρ, and T. After the wave has passed, the fluid to its left is in steady motion with a velocity of magnitude ΔV.

Figure 9.27*a* shows the wave from the point of view of a stationary observer. It is easier to analyze this situation from the point of view of an observer at rest relative to the wave, as shown in Fig. 9.27*b*. By adopting this viewpoint, a steady-state analysis can be applied to the control volume identified on the figure. To an observer at rest relative to the wave, it appears as though the fluid is moving toward the stationary wave from the right with velocity c, pressure p, density ρ, and temperature T and moving away on the left with velocity $c - \Delta V$, pressure $p + \Delta p$, density $\rho + \Delta\rho$, and temperature $T + \Delta T$.

At steady state, the conservation of mass principle for the control volume reduces to $\dot{m}_1 = \dot{m}_2$, or

$$\rho A c = (\rho + \Delta\rho)A(c - \Delta V)$$

On rearrangement

$$0 = c\,\Delta\rho - \rho\Delta V - \overset{0}{\cancel{\Delta\rho\Delta V}} \tag{9.32}$$

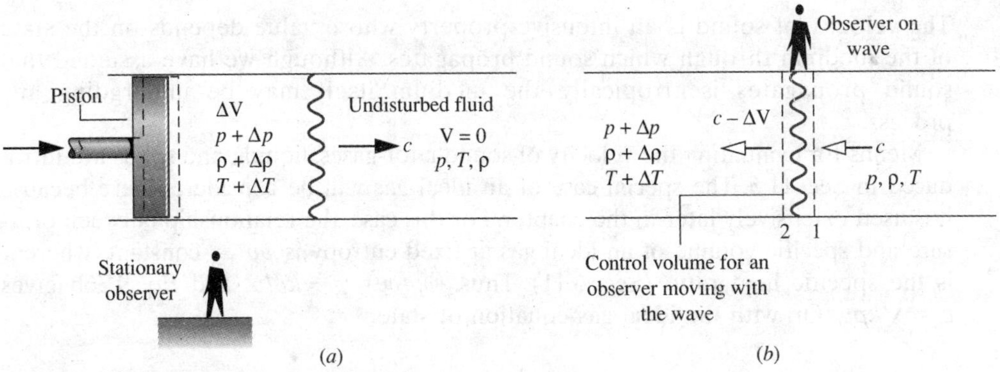

Fig. 9.27 Illustrations used to analyze the propagation of a sound wave. (*a*) Propagation of a pressure wave through a quiescent fluid, relative to a stationary observer. (*b*) Observer at rest relative to the wave.

If the disturbance is weak, the third term on the right of Eq. 9.32 can be neglected, leaving

$$\Delta V = (c/\rho)\Delta\rho \tag{9.33}$$

Next, the momentum equation, Eq. 9.31, is applied to the control volume under consideration. Since the thickness of the wave is small, shear forces at the wall are negligible. The effect of gravity is also ignored. Hence, the only significant forces acting on the control volume in the direction of flow are the forces due to pressure at the inlet and exit. With these idealizations, the component of the momentum equation in the direction of flow reduces to

$$\begin{aligned} pA - (p + \Delta p)A &= \dot{m}(c - \Delta V) - \dot{m}c \\ &= \dot{m}(c - \Delta V - c) \\ &= (\rho A c)(-\Delta V) \end{aligned}$$

or

$$\Delta p = \rho c \Delta V \tag{9.34}$$

Combining Eqs. 9.33 and 9.34 and solving for c

$$c = \sqrt{\frac{\Delta p}{\Delta \rho}} \tag{9.35}$$

SOUND WAVES. For all sound waves the differences in pressure, density, and temperature across the wave are quite small. In particular, $\Delta\rho << \rho$, justifying the neglect of the third term of Eq. 9.32. Furthermore, the ratio $\Delta p/\Delta\rho$ in Eq. 9.35 can be interpreted as the derivative of pressure with respect to density across the wave. Experiments also indicate that the relation between pressure and density across a sound wave is nearly *isentropic*. The expression for the ***velocity of sound*** then becomes

velocity of sound

$$c = \sqrt{\left(\frac{\partial p}{\partial \rho}\right)_s} \tag{9.36a}$$

or in terms of specific volume

$$c = \sqrt{-v^2\left(\frac{\partial p}{\partial v}\right)_s} \tag{9.36b}$$

The velocity of sound is an intensive property whose value depends on the state of the medium through which sound propagates. Although we have assumed that sound propagates isentropically, the medium itself may be undergoing any process.

Means for evaluating the velocity of sound c for gases, liquids, and solids are introduced in Sec. 11.5. The special case of an ideal gas will be considered here because it is used extensively later in the chapter. For this case, the relationship between pressure and specific volume of an ideal gas at fixed entropy is pv^k = constant, where k is the specific heat ratio (Sec. 6.11). Thus, $(\partial p/\partial v)_s = -kp/v$, and Eq. 9.36b gives $c = \sqrt{kpv}$. Or, with the ideal gas equation of state

$$c = \sqrt{kRT} \quad \text{(ideal gas)} \tag{9.37}$$

FOR EXAMPLE... to illustrate the use of Eq. 9.37, let us calculate the velocity of sound in air at 300 K and 650 K. From Table A-20 at 300 K, $k = 1.4$. Thus

$$c = \sqrt{1.4\left(\frac{8314}{28.97}\frac{\text{N}\cdot\text{m}}{\text{kg}\cdot\text{K}}\right)(300\ \text{K})\left|\frac{1\ \text{kg}\cdot\text{m/s}^2}{1\ \text{N}}\right|} = 347\frac{\text{m}}{\text{s}}$$

At 650 K, $k = 1.37$, and $c = 506$ m/s, as can be verified.

Mach number

Mach number

In subsequent discussions, the ratio of the velocity V at a state in a flowing fluid to the value of the sonic velocity c at the same state plays an important role. This ratio is called the ***Mach number M***

$$M = \frac{\text{V}}{c} \tag{9.38}$$

supersonic
subsonic

When $M > 1$, the flow is said to be ***supersonic;*** when $M < 1$, the flow is ***subsonic;*** and when $M = 1$, the flow is *sonic*. The term *hypersonic* is used for flows with Mach numbers much greater than one, and the term *transonic* refers to flows where the Mach number is close to unity.

Bio... connections

The propagation of elastic waves, such as sound waves, has important implications related to injury in living things. During impact such as a collision or falling to the ground, elastic waves are created that cause some bodily material to move relative to the rest of the body. The waves can propagate at supersonic, transonic, or subsonic speeds depending on the nature of the impact, and the resulting trauma can cause serious damage. The waves may be focused into a small area, causing localized damage, or they may be reflected at the boundary of organs and cause more widespread damage.

An example of the focusing of waves occurs in some head injuries. An impact to the skull causes flexural and compression waves to move along the curved surface and arrive at the far side of the skull simultaneously. Waves also propagate through the softer brain tissue. Consequently, concussions, skull fractures, and other injuries can appear at locations away from the site of the original impact.

Central to studying traumatic injury is to obtain data on speed of sound and other elastic characteristics of organs and tissues. For humans the speed of sound varies widely, from approximately 30–45 m/s in spongy lung tissue to about 1600 m/s in muscle and 3500 m/s in bone. Because the speed of sound in the lungs is relatively low, impacts such as in vehicle collisions or even air-bag deployment can set up waves that propagate supersonically. Medical personnel responding to traumas are trained to check for lung injuries.

The study of wave phenomena in the body constitutes an important area in the field of *biomechanics.*

9.12.3 Determining Stagnation State Properties

When dealing with compressible flows, it is often convenient to work with properties evaluated at a reference state known as the ***stagnation state.*** The stagnation state is the state a flowing fluid would attain if it were decelerated to zero velocity isentropically. We might imagine this as taking place in a diffuser operating at steady state. By reducing an energy balance for such a diffuser, it can be concluded that the enthalpy at the stagnation state associated with an actual state in the flow where the specific enthalpy is h and the velocity is V is given by

stagnation state

$$h_o = h + \frac{V^2}{2} \tag{9.39}$$

stagnation enthalpy

The enthalpy designated here as h_o is called the ***stagnation enthalpy.*** The pressure p_o and temperature T_o at a stagnation state are called the ***stagnation pressure*** and ***stagnation temperature,*** respectively.

stagnation pressure and temperature

9.13 Analyzing One-Dimensional Steady Flow in Nozzles and Diffusers

Although the subject of compressible flow arises in a great many important areas of engineering application, the remainder of this presentation is concerned only with flow through nozzles and diffusers. Texts dealing with compressible flow should be consulted for discussion of other areas of application.

In the present section we determine the shapes required by nozzles and diffusers for subsonic and supersonic flow. This is accomplished using mass, energy, entropy, and momentum principles, together with property relationships. In addition, we study how the flow through nozzles is affected as conditions at the nozzle exit are changed. The presentation concludes with an analysis of normal shocks, which can exist in supersonic flows.

9.13.1 Exploring the Effects of Area Change in Subsonic and Supersonic Flows

The objective of the present discussion is to establish criteria for determining whether a nozzle or diffuser should have a converging, diverging, or converging–diverging shape. This is accomplished using differential equations relating the principal variables that are obtained using mass and energy balances together with property relations, as considered next.

GOVERNING DIFFERENTIAL EQUATIONS. Let us begin by considering a control volume enclosing a nozzle or diffuser. At steady state, the mass flow rate is constant, so

$$\rho \mathrm{AV} = \text{constant}$$

In differential form

$$d(\rho \mathrm{AV}) = 0$$

$$\mathrm{AV}\, d\rho + \rho \mathrm{A}\, d\mathrm{V} + \rho \mathrm{V}\, d\mathrm{A} = 0$$

or on dividing each term by $\rho \mathrm{AV}$

$$\frac{d\rho}{\rho} + \frac{d\mathrm{V}}{\mathrm{V}} + \frac{d\mathrm{A}}{\mathrm{A}} = 0 \tag{9.40}$$

Assuming $\dot{Q}_{\mathrm{cv}} = \dot{W}_{\mathrm{cv}} = 0$ and negligible potential energy effects, an energy rate balance reduces to give

$$h_2 + \frac{\mathrm{V}_2^2}{2} = h_1 + \frac{\mathrm{V}_1^2}{2}$$

Introducing Eq. 9.39, it follows that the stagnation enthalpies at states 1 and 2 are equal: $h_{\mathrm{o}2} = h_{\mathrm{o}1}$. Since any state downstream of the inlet can be regarded as state 2, the following relationship between the specific enthalpy and kinetic energy must be satisfied at each state

$$h + \frac{\mathrm{V}^2}{2} = h_{\mathrm{o}1} \qquad \text{(constant)}$$

In differential form this becomes

$$dh = -\mathrm{V}\, d\mathrm{V} \tag{9.41}$$

This equation shows that if the velocity increases (decreases) in the direction of flow, the specific enthalpy must decrease (increase) in the direction of flow, and conversely.

In addition to Eqs. 9.40 and 9.41 expressing conservation of mass and energy, relationships among properties must be taken into consideration. Assuming the flow occurs isentropically, the property relation (Eq. 6.10b)

$$T\, ds = dh - \frac{dp}{\rho}$$

reduces to give

$$dh = \frac{1}{\rho} dp \tag{9.42}$$

This equation shows that when pressure increases or decreases in the direction of flow, the specific enthalpy changes in the same way.

Forming the differential of the property relation $p = p(\rho, s)$

$$dp = \left(\frac{\partial p}{\partial \rho}\right)_s d\rho + \left(\frac{\partial p}{\partial s}\right)_\rho ds$$

The second term vanishes in isentropic flow. Introducing Eq. 9.36a, we have

$$dp = c^2\, d\rho \tag{9.43}$$

which shows that when pressure increases or decreases in the direction of flow, density changes in the same way.

Additional conclusions can be drawn by combining the above differential equations. Combining Eqs. 9.41 and 9.42 results in

$$\frac{1}{\rho}dp = -V\,dV \tag{9.44}$$

which shows that if the velocity increases (decreases) in the direction of flow, the pressure must decrease (increase) in the direction of flow, and conversely.

Eliminating dp between Eqs. 9.43 and 9.44 and combining the result with Eq. 9.40 gives

$$\frac{dA}{A} = -\frac{dV}{V}\left[1 - \left(\frac{V}{c}\right)^2\right]$$

or with the *Mach number M*

$$\frac{dA}{A} = -\frac{dV}{V}(1 - M^2) \tag{9.45}$$

VARIATION OF AREA WITH VELOCITY. Equation 9.45 shows how area must vary with velocity. The following four cases can be identified:

Case 1: Subsonic nozzle. $dV > 0, M < 1 \Rightarrow dA < 0$: The duct *converges* in the direction of flow.

Case 2: Supersonic nozzle. $dV > 0, M > 1 \Rightarrow dA > 0$: The duct *diverges* in the direction of flow.

Case 3: Supersonic diffuser. $dV < 0, M > 1 \Rightarrow dA < 0$: The duct *converges* in the direction of flow.

Case 4: Subsonic diffuser. $dV < 0, M < 1 \Rightarrow dA > 0$: The duct *diverges* in the direction of flow.

The conclusions reached above concerning the nature of the flow in subsonic and supersonic nozzles and diffusers are summarized in Fig. 9.28. From Fig. 9.28*a*, we see that to accelerate a fluid flowing subsonically, a converging nozzle must be used, but once $M = 1$ is achieved, further acceleration can occur only in a diverging nozzle. From Fig. 9.28*b*, we see that a converging diffuser is required to decelerate a fluid flowing supersonically, but once $M = 1$ is achieved, further deceleration can occur only in a diverging diffuser. These findings suggest that a Mach number of unity can occur only at the location in a nozzle or diffuser where the cross-sectional area is a minimum. This location of minimum area is called the ***throat.***

throat

The developments of this section have not required the specification of an equation of state; thus, the conclusions hold for all gases. Moreover, although the conclusions

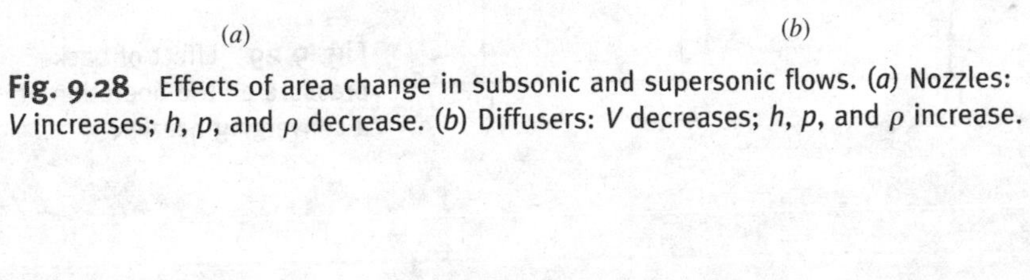

Fig. 9.28 Effects of area change in subsonic and supersonic flows. (*a*) Nozzles: *V* increases; *h*, *p*, and *ρ* decrease. (*b*) Diffusers: *V* decreases; *h*, *p*, and *ρ* increase.

have been drawn under the restriction of isentropic flow through nozzles and diffusers, they are at least qualitatively valid for actual flows because the flow through well-designed nozzles and diffusers is nearly isentropic. Isentropic nozzle efficiencies (Sec. 6.12) in excess of 95% can be attained in practice.

9.13.2 Effects of Back Pressure on Mass Flow Rate

back pressure

In the present discussion we consider the effect of varying the *back pressure* on the rate of mass flow through nozzles. The ***back pressure*** is the pressure in the exhaust region outside the nozzle. The case of converging nozzles is taken up first and then converging–diverging nozzles are considered.

CONVERGING NOZZLES. Figure 9.29 shows a converging duct with stagnation conditions at the inlet, discharging into a region in which the back pressure p_B can be varied. For the series of cases labeled a through e, let us consider how the mass flow rate $\dot{m}$ and nozzle exit pressure p_E vary as the back pressure is decreased while keeping the inlet conditions fixed.

When $p_B = p_E = p_o$, there is no flow, so $\dot{m} = 0$. This corresponds to case a of Fig. 9.29. If the back pressure p_B is decreased, as in cases b and c, there will be flow through the nozzle. As long as the flow is subsonic at the exit, information about changing conditions in the exhaust region can be transmitted upstream. Decreases in back pressure thus result in greater mass flow rates and new pressure variations within the nozzle. In each instance, the velocity is subsonic throughout the nozzle and the exit pressure equals the back pressure. The exit Mach number increases as p_B decreases, however, and eventually a Mach number of unity will be attained at the nozzle exit. The corresponding pressure is denoted by p^*, called the *critical pressure*. This case is represented by d on Fig. 9.29.

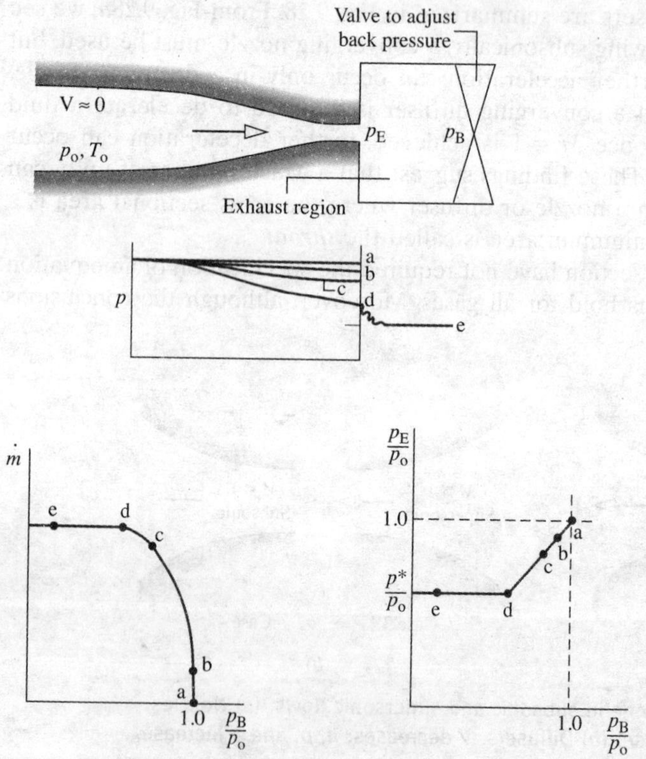

Fig. 9.29 Effect of back pressure on the operation of a converging nozzle.

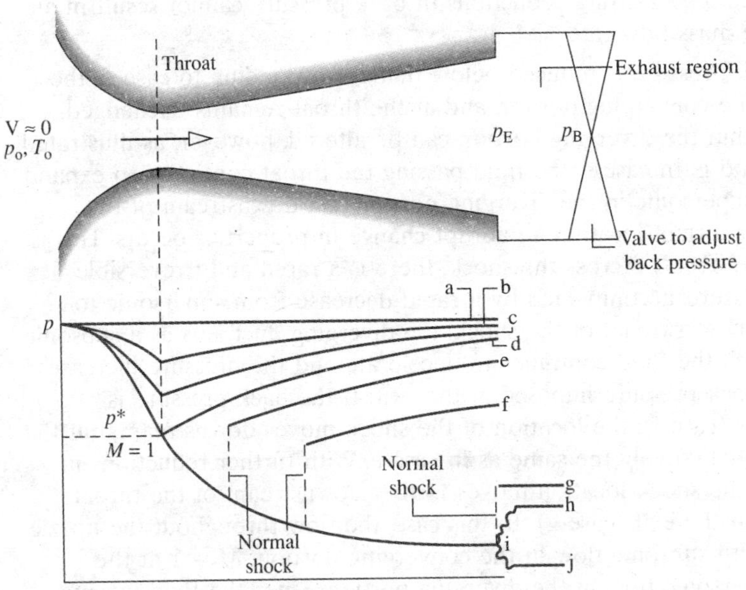

Fig. 9.30 Effect of back pressure on the operation of a converging–diverging nozzle.

Recalling that the Mach number cannot increase beyond unity in a converging section, let us consider next what happens when the back pressure is reduced further to a value less than p^*, such as represented by case e. Since the velocity at the exit equals the velocity of sound, information about changing conditions in the exhaust region no longer can be transmitted upstream past the exit plane. Accordingly, reductions in p_B below p^* have no effect on flow conditions in the nozzle. Neither the pressure variation within the nozzle nor the mass flow rate is affected. Under these conditions, the nozzle is said to be ***choked.*** When a nozzle is choked, the mass flow rate is the *maximum possible for the given stagnation conditions.* For p_B less than p^*, the flow expands outside the nozzle to match the lower back pressure, as shown by case e of Fig. 9.29. The pressure variation outside the nozzle cannot be predicted using the one-dimensional flow model.

choked flow: converging nozzle

CONVERGING–DIVERGING NOZZLES. Figure 9.30 illustrates the effects of varying back pressure on a *converging–diverging* nozzle. The series of cases labeled a through j is considered next.

▶ Let us first discuss the cases designated a, b, c, and d. Case a corresponds to $p_B = p_E = p_o$ for which there is no flow. When the back pressure is slightly less than p_o (case b), there is some flow, and the flow is subsonic throughout the nozzle. In accordance with the discussion of Fig. 9.28, the greatest velocity and lowest pressure occur at the throat, and the diverging portion acts as a diffuser in which pressure increases and velocity decreases in the direction of flow. If the back pressure is reduced further, corresponding to case c, the mass flow rate and velocity at the throat are greater than before. Still, the flow remains subsonic throughout and qualitatively the same as case b. As the back pressure is reduced, the Mach number at the throat increases, and eventually a Mach number of unity is attained there (case d). As before, the greatest velocity and lowest pressure occur at the throat, and the diverging portion remains a subsonic diffuser. However, because the throat velocity is sonic, the nozzle is now ***choked:*** The *maximum mass flow rate has been attained for the given*

choked flow: converging–diverging nozzle

stagnation conditions. Further reductions in back pressure cannot result in an increase in the mass flow rate.

- When the back pressure is reduced below that corresponding to case d, the flow through the converging portion and at the throat remains unchanged. Conditions within the diverging portion can be altered, however, as illustrated by cases e, f, and g. In case e, the fluid passing the throat continues to expand and becomes supersonic in the diverging portion just downstream of the throat; but at a certain location an abrupt change in properties occurs. This is called a ***normal shock.*** Across the shock, there is a rapid and irreversible increase in pressure, accompanied by a rapid decrease from supersonic to subsonic flow. Downstream of the shock, the diverging duct acts as a subsonic diffuser in which the fluid continues to decelerate and the pressure increases to match the back pressure imposed at the exit. If the back pressure is reduced further (case f), the location of the shock moves downstream, but the flow remains qualitatively the same as in case e. With further reductions in back pressure, the shock location moves farther downstream of the throat until it stands at the exit (case g). In this case, the flow throughout the nozzle is isentropic, with subsonic flow in the converging portion, $M = 1$ at the throat, and supersonic flow in the diverging portion. Since the fluid leaving the nozzle passes through a shock, it is subsonic just downstream of the exit plane.

normal shock

- Finally, let us consider cases h, i, and j where the back pressure is less than that corresponding to case g. In each of these cases, the flow through the nozzle is not affected. The adjustment to changing back pressure occurs outside the nozzle. In case h, the pressure decreases continuously as the fluid expands isentropically through the nozzle and then increases to the back pressure outside the nozzle. The compression that occurs outside the nozzle involves *oblique shock waves*. In case i, the fluid expands isentropically to the back pressure and no shocks occur within or outside the nozzle. In case j, the fluid expands isentropically through the nozzle and then expands outside the nozzle to the back pressure through *oblique expansion* waves. Once $M = 1$ is achieved at the throat, the mass flow rate is fixed at the maximum value for the given stagnation conditions, so the mass flow rate is the same for back pressures corresponding to cases d through j. The pressure variations outside the nozzle involving oblique waves cannot be predicted using the one-dimensional flow model.

9.13.3 Flow Across a Normal Shock

We have seen that under certain conditions a rapid and abrupt change of state called a shock takes place in the diverging portion of a supersonic nozzle. In a *normal* shock, this change of state occurs across a plane normal to the direction of flow. The object of the present discussion is to develop means for determining the change of state across a normal shock.

MODELING NORMAL SHOCKS. A control volume enclosing a normal shock is shown in Fig. 9.31. The control volume is assumed to be at steady state with $\dot{W}_{cv} = 0, \dot{Q}_{cv} = 0$ and negligible effects of potential energy. The thickness of the shock is very small (on the order of 10^{-5} cm). Thus, there is no significant change in flow area across the shock, even though it may occur in a diverging passage, and the forces acting at the wall can be neglected relative to the pressure forces acting at the upstream and downstream locations denoted by x and y, respectively.

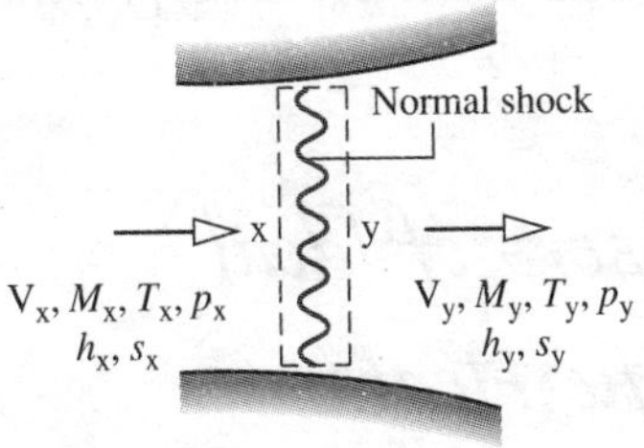

Fig. 9.31 Control volume enclosing a normal shock.

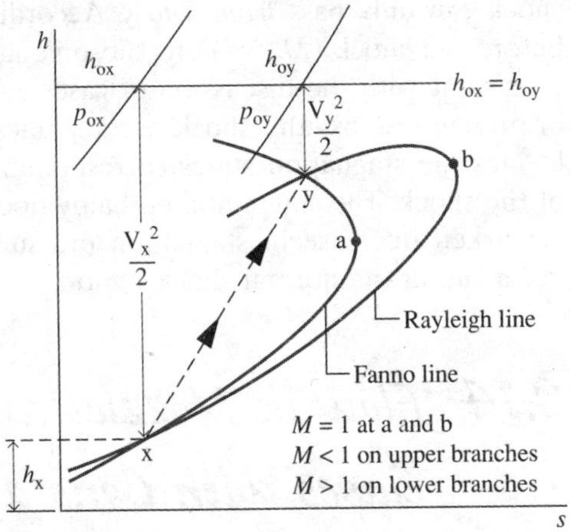

Fig. 9.32 Intersection of Fanno and Rayleigh lines as a solution to the normal shock equations.

The upstream and downstream states are related by the following equations:

Mass:

$$\rho_x V_x = \rho_y V_y \tag{9.46}$$

Energy:

$$h_x + \frac{V_x^2}{2} = h_y + \frac{V_y^2}{2} \tag{9.47a}$$

or

$$h_{ox} = h_{oy} \tag{9.47b}$$

Momentum:

$$p_x - p_y = \rho_y V_y^2 - \rho_x V_x^2 \tag{9.48}$$

Entropy:

$$s_y - s_x = \dot{\sigma}_{cv}/\dot{m} \tag{9.49}$$

When combined with property relations for the particular fluid under consideration, Eqs. 9.46, 9.47, and 9.48 allow the downstream conditions to be determined for specified upstream conditions. Equation 9.49, which corresponds to Eq. 6.39, leads to the important conclusion that the downstream state *must* have greater specific entropy than the upstream state, or $s_y > s_x$.

FANNO AND RAYLEIGH LINES. The mass and energy equations, Eqs. 9.46 and 9.47, can be combined with property relations for the particular fluid to give an equation that when plotted on an *h–s* diagram is called a ***Fanno line.*** Similarly, the mass and momentum equations, Eqs. 9.46 and 9.48, can be combined to give an equation that when plotted on an *h–s* diagram is called a ***Rayleigh line.*** Fanno and Rayleigh lines are sketched on *h–s* coordinates in Fig. 9.32. It can be shown that the point of maximum entropy on each line, points a and b, corresponds to $M = 1$. It also can be shown that the upper and lower branches of each line correspond, respectively, to subsonic and supersonic velocities.

Fanno line

Rayleigh line

The downstream state y must satisfy the mass, energy, and momentum equations simultaneously, so state y is fixed by the intersection of the Fanno and Rayleigh lines passing through state x. Since $s_y > s_x$, it can be concluded that the flow across the shock can only pass *from* x *to* y. Accordingly, the velocity changes from supersonic before the shock ($M_x > 1$) to subsonic after the shock ($M_y < 1$). This conclusion is consistent with the discussion of cases e, f, and g in Fig. 9.30. A significant increase in pressure across the shock accompanies the decrease in velocity. Figure 9.32 also locates the stagnation states corresponding to the states upstream and downstream of the shock. The stagnation enthalpy does not change across the shock, but there is a marked decrease in stagnation pressure associated with the irreversible process occurring in the normal shock region.

9.14 Flow in Nozzles and Diffusers of Ideal Gases with Constant Specific Heats

The discussion of flow in nozzles and diffusers presented in Sec. 9.13 requires no assumption regarding the equation of state, and therefore the results obtained hold generally. Attention is now restricted to ideal gases with constant specific heats. This case is appropriate for many practical problems involving flow through nozzles and diffusers. The assumption of constant specific heats also allows the derivation of relatively simple closed-form equations.

ISENTROPIC FLOW FUNCTIONS. Let us begin by developing equations relating a state in a compressible flow to the corresponding stagnation state. For the case of an ideal gas with constant c_p, Eq. 9.39 becomes

$$T_o = T + \frac{V^2}{2c_p}$$

where T_o is the stagnation temperature. Introducing $c_p = kR/(k-1)$, together with Eqs. 9.37 and 9.38, the relation between the temperature T and the Mach number M of the flowing gas and the corresponding stagnation temperature T_o is

$$\frac{T_o}{T} = 1 + \frac{k-1}{2}M^2 \tag{9.50}$$

With Eq. 6.43, a relationship between the temperature T and pressure p of the flowing gas and the corresponding stagnation temperature T_o and the stagnation pressure p_o is

$$\frac{p_o}{p} = \left(\frac{T_o}{T}\right)^{k/(k-1)}$$

Introducing Eq. 9.50 into this expression gives

$$\frac{p_o}{p} = \left(1 + \frac{k-1}{2}M^2\right)^{k/(k-1)} \tag{9.51}$$

Although sonic conditions may not actually be attained in a particular flow, it is convenient to have an expression relating the area A at a given section to the area A* that *would be* required for sonic flow ($M = 1$) at the same mass flow rate and stagnation state. These areas are related through

$$\rho AV = \rho^* A^* V^*$$

where ρ^* and V* are the density and velocity, respectively, when $M = 1$. Introducing the ideal gas equation of state, together with Eqs. 9.37 and 9.38, and solving for A/A*

$$\frac{A}{A^*} = \frac{1}{M}\frac{p^*}{p}\left(\frac{T}{T^*}\right)^{1/2} = \frac{1}{M}\frac{p^*/p_o}{p/p_o}\left(\frac{T/T_o}{T^*/T_o}\right)^{1/2}$$

where T^* and p^* are the temperature and pressure, respectively, when $M = 1$. Then with Eqs. 9.50 and 9.51

$$\frac{A}{A^*} = \frac{1}{M}\left[\left(\frac{2}{k+1}\right)\left(1 + \frac{k-1}{2}M^2\right)\right]^{(k+1)/2(k-1)} \tag{9.52}$$

The variation of A/A^* with M is given in Fig. 9.33 for $k = 1.4$. The figure shows that a unique value of A/A^* corresponds to any choice of M. However, for a given value of A/A^* other than unity, there are two possible values for the Mach number, one subsonic and one supersonic. This is consistent with the discussion of Fig. 9.28, where it was found that a converging–diverging passage with a section of minimum area is required to accelerate a flow from subsonic to supersonic velocity.

Equations 9.50, 9.51, and 9.52 allow the ratios T/T_o, p/p_o, and A/A^* to be computed and tabulated with the Mach number as the single independent variable for a specified value of k. Table 9.2 provides a tabulation of this kind for $k = 1.4$. Such a table facilitates the analysis of flow through nozzles and diffusers. Equations 9.50, 9.51, and 9.52 also can be readily evaluated using calculators and computer software such as *Interactive Thermodynamics: IT*.

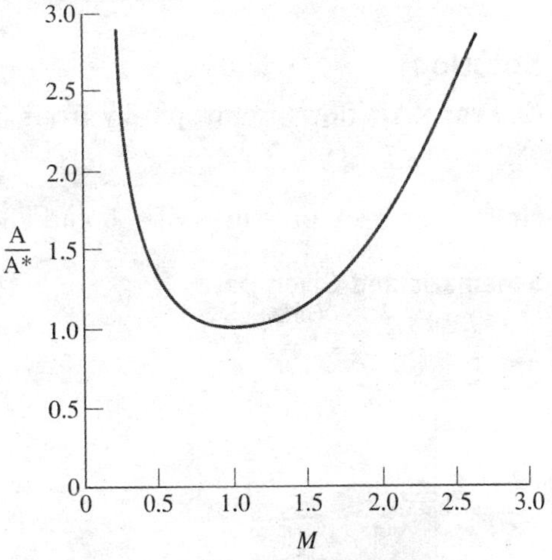

Fig. 9.33 Variation of A/A^* with Mach number in isentropic flow for $k = 1.4$.

Table 9.2

Isentropic Flow Functions for an Ideal Gas with $k = 1.4$

M	T/T_o	p/p_o	A/A^*
0	1.000 00	1.000 00	∞
0.10	0.998 00	0.993 03	5.8218
0.20	0.992 06	0.972 50	2.9635
0.30	0.982 32	0.939 47	2.0351
0.40	0.968 99	0.895 62	1.5901
0.50	0.952 38	0.843 02	1.3398
0.60	0.932 84	0.784 00	1.1882
0.70	0.910 75	0.720 92	1.094 37
0.80	0.886 52	0.656 02	1.038 23
0.90	0.860 58	0.591 26	1.008 86
1.00	0.833 33	0.528 28	1.000 00
1.10	0.805 15	0.468 35	1.007 93
1.20	0.776 40	0.412 38	1.030 44
1.30	0.747 38	0.360 92	1.066 31
1.40	0.718 39	0.314 24	1.1149
1.50	0.689 65	0.272 40	1.1762
1.60	0.661 38	0.235 27	1.2502
1.70	0.633 72	0.202 59	1.3376
1.80	0.606 80	0.174 04	1.4390
1.90	0.580 72	0.149 24	1.5552
2.00	0.555 56	0.127 80	1.6875
2.10	0.531 35	0.109 35	1.8369
2.20	0.508 13	0.093 52	2.0050
2.30	0.485 91	0.079 97	2.1931
2.40	0.464 68	0.068 40	2.4031

In Example 9.14, we consider the effect of back pressure on flow in a converging nozzle. The *first step* of the analysis is to check whether the flow is choked.

Example 9.14 EFFECT OF BACK PRESSURE: CONVERGING NOZZLE

A converging nozzle has an exit area of 0.001 m^2. Air enters the nozzle with negligible velocity at a pressure of 1.0 MPa and a temperature of 360 K. For isentropic flow of an ideal gas with $k = 1.4$, determine the mass flow rate, in kg/s, and the exit Mach number for back pressures of **(a)** 500 kPa and **(b)** 784 kPa.

Solution

Known: Air flows isentropically from specified stagnation conditions through a converging nozzle with a known exit area.

Find: For back pressures of 500 and 784 kPa, determine the mass flow rate, in kg/s, and the exit Mach number.

Schematic and Given Data:

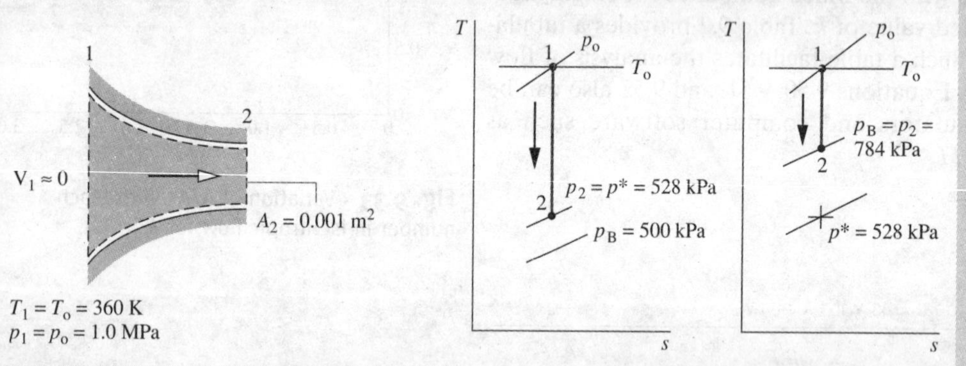

Fig. E9.14

Engineering Model:

1. The control volume shown in the accompanying sketch operates at steady state.
2. The air is modeled as an ideal gas with $k = 1.4$.
3. Flow through the nozzle is isentropic.

Analysis: The *first step* is to check whether the flow is choked. With $k = 1.4$ and $M = 1.0$, Eq. 9.51 gives
❶ $p^*/p_o = 0.528$. Since $p_o = 1.0$ MPa, the critical pressure is $p^* = 528$ kPa. Thus, for back pressures of 528 kPa or less, the Mach number is unity at the exit and the nozzle is choked.

(a) From the above discussion, it follows that for a back pressure of 500 kPa, the nozzle is choked. At the exit, $M_2 = 1.0$ and the exit pressure equals the critical pressure, $p_2 = 528$ kPa. The mass flow rate is the maximum value that can be attained for the given stagnation properties. With the ideal gas equation of state, the mass flow rate is

$$\dot{m} = \rho_2 A_2 V_2 = \frac{p_2}{RT_2} A_2 V_2$$

The exit area A_2 required by this expression is specified as 10^{-3} m^2. Since $M = 1$ at the exit, the exit temperature T_2 can be found from Eq. 9.50, which on rearrangement gives

$$T_2 = \frac{T_o}{1 + \dfrac{k-1}{2}M^2} = \frac{360 \text{ K}}{1 + \left(\dfrac{1.4 - 1}{2}\right)(1)^2} = 300 \text{ K}$$

Then, with Eq. 9.37, the exit velocity V_2 is

$$V_2 = \sqrt{kRT_2}$$

$$= \sqrt{1.4\left(\frac{8314\ \text{N}\cdot\text{m}}{28.97\ \text{kg}\cdot\text{K}}\right)(300\ \text{K})\left|\frac{1\ \text{kg}\cdot\text{m/s}^2}{1\ \text{N}}\right|} = 347.2\ \text{m/s}$$

Finally

$$\dot{m} = \frac{(528 \times 10^3\ \text{N/m}^2)(10^{-3}\ \text{m}^2)(347.2\ \text{m/s})}{\left(\frac{8314\ \text{N}\cdot\text{m}}{28.97\ \text{kg}\cdot\text{K}}\right)(300\ \text{K})} = 2.13\ \text{kg/s}$$

(b) Since the back pressure of 784 kPa is greater than the critical pressure determined above, the flow throughout the nozzle is subsonic and the exit pressure equals the back pressure, $p_2 = 784$ kPa. The exit Mach number can be found by solving Eq. 9.51 to obtain

$$M_2 = \left\{\frac{2}{k-1}\left[\left(\frac{p_o}{p_2}\right)^{(k-1)/k} - 1\right]\right\}^{1/2}$$

Inserting values

$$M_2 = \left\{\frac{2}{1.4-1}\left[\left(\frac{1 \times 10^6}{7.84 \times 10^5}\right)^{0.286} - 1\right]\right\}^{1/2} = 0.6$$

With the exit Mach number known, the exit temperature T_2 can be found from Eq. 9.50 as 336 K. The exit velocity is then

$$V_2 = M_2c_2 = M_2\sqrt{kRT_2} = 0.6\sqrt{1.4\left(\frac{8314}{28.97}\right)(336)}$$

$$= 220.5\ \text{m/s}$$

The mass flow rate is

$$\dot{m} = \rho_2 A_2 V_2 = \frac{p_2}{RT_2}A_2V_2 = \frac{(784 \times 10^3)(10^{-3})(220.5)}{(8314/28.97)(336)}$$

$$= 1.79\ \text{kg/s}$$

---

❶ The use of Table 9.2 reduces some of the computation required in the solution. It is left as an exercise to develop a solution using this table. Also, observe that the first step of the analysis is to check whether the flow is choked.

Skills Developed: *Ability to...*
- ✓ apply the ideal gas model with constant k in the analysis of isentropic flow through a converging nozzle.
- ✓ understand when choked flow occurs in a converging nozzle for different back pressures.
- ✓ determine conditions at the throat and the mass flow rate for different back pressures and a fixed stagnation state.

Quick Quiz

Using the isentropic flow functions in Table 9.2, determine the exit temperature and Mach number for a back pressure of 843 kPa.

Ans. 342.9 K, 0.5.

NORMAL SHOCK FUNCTIONS. Next, let us develop closed-form equations for normal shocks for the case of an ideal gas with constant specific heats. For this case, it follows from the energy equation, Eq. 9.47b, that there is no change in stagnation temperature across the shock, $T_{ox} = T_{oy}$. Then, with Eq. 9.50, the following expression for the ratio of temperatures across the shock is obtained

$$\frac{T_y}{T_x} = \frac{1 + \frac{k-1}{2}M_x^2}{1 + \frac{k-1}{2}M_y^2} \tag{9.53}$$

Rearranging Eq. 9.48

$$p_x + \rho_x V_x^2 = p_y + \rho_y V_y^2$$

Introducing the ideal gas equation of state, together with Eqs. 9.37 and 9.38, the ratio of the pressure downstream of the shock to the pressure upstream is

$$\frac{p_y}{p_x} = \frac{1 + kM_x^2}{1 + kM_y^2} \tag{9.54}$$

Similarly, Eq. 9.46 becomes

$$\frac{p_y}{p_x} = \sqrt{\frac{T_y}{T_x}}\frac{M_x}{M_y}$$

The following equation relating the Mach numbers M_x and M_y across the shock can be obtained when Eqs. 9.53 and 9.54 are introduced in this expression

$$M_y^2 = \frac{M_x^2 + \frac{2}{k-1}}{\frac{2k}{k-1}M_x^2 - 1} \tag{9.55}$$

The ratio of stagnation pressures across a shock p_{oy}/p_{ox} is often useful. It is left as an exercise to show that

$$\frac{p_{oy}}{p_{ox}} = \frac{M_x}{M_y}\left(\frac{1 + \frac{k-1}{2}M_y^2}{1 + \frac{k-1}{2}M_x^2}\right)^{(k+1)/2(k-1)} \tag{9.56}$$

Since there is no area change across a shock, Eqs. 9.52 and 9.56 combine to give

$$\frac{A_x^*}{A_y^*} = \frac{p_{oy}}{p_{ox}} \tag{9.57}$$

For specified values of M_x and specific heat ratio k, the Mach number downstream of a shock can be found from Eq. 9.55. Then, with M_x, M_y, and k known, the ratios T_y/T_x, p_y/p_x, and p_{oy}/p_{ox} can be determined from Eqs. 9.53, 9.54, and 9.56. Accordingly, tables can be set up giving M_y, T_y/T_x, p_y/p_x, and p_{oy}/p_{ox} versus the Mach number M_x as the single independent variable for a specified value of k. Table 9.3 is a tabulation of this kind for $k = 1.4$.

Table 9.3

Normal Shock Functions for an Ideal Gas with $k = 1.4$

M_x	M_y	p_y/p_x	T_y/T_x	p_{oy}/p_{ox}
1.00	1.000 00	1.0000	1.0000	1.000 00
1.10	0.911 77	1.2450	1.0649	0.998 92
1.20	0.842 17	1.5133	1.1280	0.992 80
1.30	0.785 96	1.8050	1.1909	0.979 35
1.40	0.739 71	2.1200	1.2547	0.958 19
1.50	0.701 09	2.4583	1.3202	0.929 78
1.60	0.668 44	2.8201	1.3880	0.895 20
1.70	0.640 55	3.2050	1.4583	0.855 73
1.80	0.616 50	3.6133	1.5316	0.812 68
1.90	0.595 62	4.0450	1.6079	0.767 35
2.00	0.577 35	4.5000	1.6875	0.720 88
2.10	0.561 28	4.9784	1.7704	0.674 22
2.20	0.547 06	5.4800	1.8569	0.628 12
2.30	0.534 41	6.0050	1.9468	0.583 31
2.40	0.523 12	6.5533	2.0403	0.540 15
2.50	0.512 99	7.1250	2.1375	0.499 02
2.60	0.503 87	7.7200	2.2383	0.460 12
2.70	0.495 63	8.3383	2.3429	0.423 59
2.80	0.488 17	8.9800	2.4512	0.389 46
2.90	0.481 38	9.6450	2.5632	0.357 73
3.00	0.475 19	10.333	2.6790	0.328 34
4.00	0.434 96	18.500	4.0469	0.138 76
5.00	0.415 23	29.000	5.8000	0.061 72
10.00	0.387 57	116.50	20.388	0.003 04
∞	0.377 96	∞	∞	0.0

In the next example, we consider the effect of back pressure on flow in a converging–diverging nozzle. Key elements of the analysis include determining whether the flow is choked and if a normal shock exists.

Example 9.15 EFFECT OF BACK PRESSURE: CONVERGING–DIVERGING NOZZLE

A converging–diverging nozzle operating at steady state has a throat area of 6.25 cm^2 and an exit area of 15 cm^2. Air enters the nozzle with a negligible velocity at a pressure of 6.8 bar and a temperature of 280 K. For air as an ideal gas with $k = 1.4$, determine the mass flow rate, in kg/s, the exit pressure, in bar, and exit Mach number for ❶ each of the five following cases. **(a)** Isentropic flow with $M = 0.7$ at the throat. **(b)** Isentropic flow with $M = 1$ at the throat and the diverging portion acting as a diffuser. **(c)** Isentropic flow with $M = 1$ at the throat and the diverging portion acting as a nozzle. **(d)** Isentropic flow through the nozzle with a normal shock standing at the exit. **(e)** A normal shock stands in the diverging section at a location where the area is 12.5 cm^2. Elsewhere in the nozzle, the flow is isentropic.

Solution

Known: Air flows from specified stagnation conditions through a converging–diverging nozzle having a known throat and exit area.

Find: The mass flow rate, exit pressure, and exit Mach number are to be determined for each of five cases.

Schematic and Given Data:

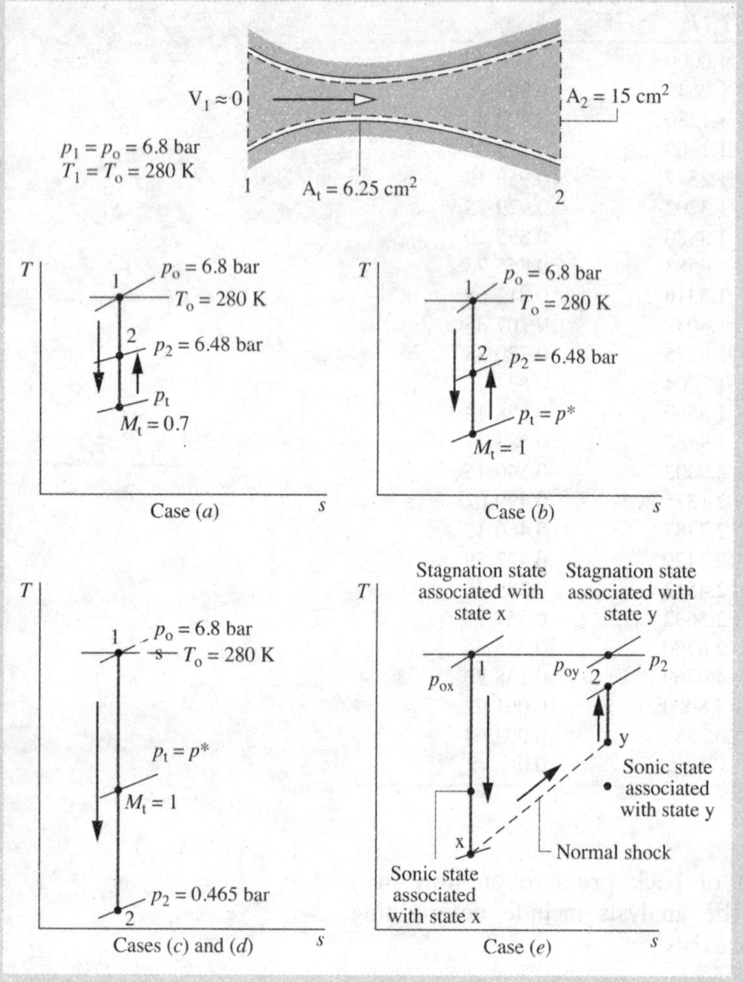

Engineering Model:

1. The control volume shown in the accompanying sketch operates at steady state. The *T–s* diagrams provided locate states within the nozzle.

2. The air is modeled as an ideal gas with $k = 1.4$.

3. Flow through the nozzle is isentropic throughout, except for case (*e*), where a shock stands in the diverging section.

Fig. E9.15

Analysis:

(a) The accompanying *T–s* diagram shows the states visited by the gas in this case. The following are known: the Mach number at the throat, $M_t = 0.7$, the throat area, $A_t = 6.25\ \text{cm}^2$, and the exit area, $A_2 = 15\ \text{cm}^2$. The exit Mach number M_2, exit temperature T_2, and exit pressure p_2 can be determined using the identity

$$\frac{A_2}{A^*} = \frac{A_2}{A_t}\frac{A_t}{A^*}$$

With $M_t = 0.7$, Table 9.2 gives $A_t/A^* = 1.09437$. Thus

$$\frac{A_2}{A^*} = \left(\frac{15\ \text{cm}^2}{6.25\ \text{cm}^2}\right)(1.09437) = 2.6265$$

The flow throughout the nozzle, including the exit, is subsonic. Accordingly, with this value for A_2/A^*, Table 9.2 gives $M_2 \approx 0.24$. For $M_2 = 0.24$, $T_2/T_o = 0.988$, and $p_2/p_o = 0.959$. Since the stagnation temperature and pressure are 280 K and 6.8 bar, respectively, it follows that $T_2 = 277$ K and $p_2 = 6.52$ bar.

The velocity at the exit is

$$V_2 = M_2 c_2 = M_2\sqrt{kRT_2}$$

$$= 0.24\sqrt{1.4\left(\frac{8.314}{28.97}\frac{\text{kJ}}{\text{kg}\cdot\text{K}}\right)(277\ \text{K})\left|\frac{1\ \text{kg}\cdot\text{m/s}^2}{1\ \text{N}}\right|\left|\frac{10^3\text{N}\cdot\text{m}}{1\ \text{kJ}}\right|}$$

$$= 80.1\ \text{m/s}$$

The mass flow rate is

$$\dot{m} = \rho_2 A_2 V_2 = \frac{p_2}{RT_2} A_2 V_2$$

$$= \frac{(6.52 \text{ bar})(15 \text{ cm}^2)(80.1 \text{ m/s})}{\left(\frac{8.314 \text{ kJ}}{28.97 \text{ kg} \cdot \text{K}}\right)(277°\text{K})} = 0.985 \text{ kg/s}$$

(b) The accompanying T–s diagram shows the states visited by the gas in this case. Since $M = 1$ at the throat, we have $A_t = A^*$, and thus, $A_2/A^* = 2.4$. Table 9.2 gives two Mach numbers for this ratio: $M \approx 0.26$ and $M \approx 2.4$. The diverging portion acts as a diffuser in the present part of the example; accordingly, the subsonic value is appropriate. The supersonic value is appropriate in part (c).

Thus, from Table 9.2 we have at $M_2 = 0.26$, $T_2/T_o = 0.986$, and $p_2/p_o = 0.953$. Since $T_o = 280$ K and $p_o = 6.8$ bar, it follows that $T_2 = 276$ K and $p_2 = 6.48$ bar.

The velocity at the exit is

$$V_2 = M_2 c_2 = M_2 \sqrt{kRT_2}$$

$$= 0.26 \sqrt{(1.4)\left(\frac{8.314}{28.97}\right)(276)(10^3)} = 86.58 \text{ m/s}$$

The mass flow rate is

$$\dot{m} = \frac{p_2}{RT_2} A_2 V_2 = \frac{(6.48)(15)(86.58)}{\left(\frac{8.314}{28.97}\right)(276)} = 1.062 \text{ kg/s}$$

This is the maximum mass flow rate for the specified geometry and stagnation conditions: the flow is choked.

(c) The accompanying T–s diagram shows the states visited by the gas in this case. As discussed in part (b), the exit Mach number in the present part of the example is $M_2 = 2.4$. Using this, Table 9.2 gives $p_2/p_o = 0.0684$. With $p_o = 6.8$ bar, the pressure at the exit is $p_2 = 0.465$ bar. Since the nozzle is choked, the mass flow rate is the same as found in part (b).

(d) Since a normal shock stands at the exit and the flow upstream of the shock is isentropic, the Mach number M_x and the pressure p_x correspond to the values found in part (c), $M_x = 2.4$, $p_x = 0.465$ bar. Then, from Table 9.3, $M_y \approx 0.52$ and $p_y/p_x = 6.5533$. The pressure downstream of the shock is thus 3.047 bar. This is the exit pressure. The mass flow is the same as found in part (b).

(e) The accompanying T–s diagram shows the states visited by the gas. It is known that a shock stands in the diverging portion where the area is $A_x = 12.5 \text{ cm}^2$. Since a shock occurs, the flow is sonic at the throat, so $A_x^* = A_t = 6.25 \text{ cm}^2$. The Mach number M_x can then be found from Table 9.2, by using $A_x/A_x^* = 2$, as $M_x = 2.2$.

The Mach number at the exit can be determined using the identity

$$\frac{A_2}{A_y^*} = \left(\frac{A_2}{A_x^*}\right)\left(\frac{A_x^*}{A_y^*}\right)$$

Introducing Eq. 9.57 to replace A_x^*/A_y^*, this becomes

$$\frac{A_2}{A_y^*} = \left(\frac{A_2}{A_x^*}\right)\left(\frac{p_{oy}}{p_{ox}}\right)$$

where p_{ox} and p_{oy} are the stagnation pressures before and after the shock, respectively. With $M_x = 2.2$, the ratio of stagnation pressures is obtained from Table 9.3 as $p_{oy}/p_{ox} = 0.62812$. Thus

$$\frac{A_2}{A_y^*} = \left(\frac{15 \text{ cm}^2}{6.25 \text{ cm}^2}\right)(0.62812) = 1.51$$

Using this ratio and noting that the flow is subsonic after the shock, Table 9.2 gives $M_2 \approx 0.43$, for which $p_2/p_{oy} = 0.88$.

The pressure at the exit can be determined using the identity

$$p_2 = \left(\frac{p_2}{p_{oy}}\right)\left(\frac{p_{oy}}{p_{ox}}\right) p_{ox} = (0.88)(0.628)\,(6.8 \text{ bar}) = 3.76 \text{ bar}$$

Since the flow is choked, the mass flow rate is the same as that found in part (b).

❶ Part (a) of the present example corresponds to the cases labeled b and c on Fig. 9.30. Part (c) corresponds to case d of Fig. 9.30. Part (d) corresponds to case g of Fig. 9.30 and part (e) corresponds to cases c and f.

Skills Developed: *Ability to...*

✓ analyze isentropic flow through a converging–diverging nozzle for an ideal gas with constant *k*.

✓ understand the occurrence of choked flow and normal shocks in a converging–diverging nozzle for different back pressures.

✓ analyze the flow through a converging–diverging nozzle when normal shocks are present for an ideal gas with constant *k*

Quick Quiz

What is the stagnation temperature, in °R, corresponding to the exit state for case (e)?
Ans. 280 K.

✓ Chapter Summary and Study Guide

In this chapter, we have studied the thermodynamic modeling of internal combustion engines, gas turbine power plants, and compressible flow in nozzles and diffusers. The modeling of cycles is based on the use of air-standard analysis, where the working fluid is considered to be air as an ideal gas.

The processes in internal combustion engines are described in terms of three air-standard cycles: the Otto, Diesel, and dual cycles, which differ from each other only in the way the heat addition process is modeled. For these cycles, we have evaluated the principal work and heat transfers along with two important performance parameters: the mean effective pressure and the thermal efficiency. The effect of varying compression ratio on cycle performance is also investigated.

The performance of simple gas turbine power plants is described in terms of the air-standard Brayton cycle. For this cycle, we evaluate the principal work and heat transfers along with two important performance parameters: the back-work ratio and the thermal efficiency. We also consider the effects on performance of irreversibilities and of varying compressor pressure ratio. Three modifications of the simple cycle to improve performance are introduced: regeneration, reheat, and compression with intercooling. Applications related to gas turbines are also considered, including aircraft propulsion systems and combined gas turbine–vapor power cycles. In addition, the Ericsson and Stirling cycles are introduced.

The chapter concludes with the study of compressible flow through nozzles and diffusers. We begin by introducing the momentum equation for steady, one-dimensional flow, the velocity of sound, and the stagnation state. We then consider the effects of area change and back pressure on performance in both subsonic and supersonic flows. Choked flow and the presence of normal shocks in such flows are investigated. Tables are introduced to facilitate analysis for the case of ideal gases with constant specific heat ratio, $k = 1.4$.

The following list provides a study guide for this chapter. When your study of the text and end-of-chapter exercises has been completed, you should be able to

- ✓ write out the meanings of the terms listed in the margin throughout the chapter and understand each of the related concepts. The subset of key concepts listed below is particularly important.
- ✓ sketch p–v and T–s diagrams of the Otto, Diesel, and dual cycles. Apply the closed system energy balance and the second law along with property data to determine the performance of these cycles, including mean effective pressure, thermal efficiency, and the effects of varying compression ratio.
- ✓ sketch schematic diagrams and accompanying T–s diagrams of the Brayton cycle and modifications involving regeneration, reheat, and compression with intercooling. In each case, be able to apply mass and energy balances, the second law, and property data to determine gas turbine power cycle performance, including thermal efficiency, back-work ratio, net power output, and the effects of varying compressor pressure ratio.
- ✓ analyze the performance of gas turbine–related applications involving aircraft propulsion and combined gas turbine–vapor power plants. You also should be able to apply the principles of this chapter to Ericsson and Stirling cycles.
- ✓ discuss for nozzles and diffusers the effects of area change in subsonic and supersonic flows, the effects of back pressure on mass flow rate, and the appearance and consequences of choking and normal shocks.
- ✓ analyze the flow in nozzles and diffusers of ideal gases with constant specific heats, as in Examples 9.14 and 9.15.

Key Engineering Concepts

- ***mean effective pressure*** p. 446
- ***air-standard analysis*** p. 446, 461
- ***Otto cycle*** p. 448
- ***Diesel cycle*** p. 453
- ***dual cycle*** p. 457
- ***Brayton cycle*** p. 462
- ***regenerator effectiveness*** p. 474
- ***turbojet engine*** p. 486
- ***combined cycle*** p. 491
- ***compressible flow*** p. 499
- ***momentum equation*** p. 500
- ***velocity of sound*** p. 501
- ***Mach number*** p. 502
- ***subsonic and supersonic flow*** p. 502
- ***stagnation state*** p. 503
- ***choked flow*** p. 507
- ***normal shock*** p. 508

Key Equations

$\text{mep} = \dfrac{\text{net work for one cycle}}{\text{displacement volume}}$	(9.1)	Mean effective pressure
Otto cycle		
$\eta = \dfrac{(u_3 - u_2) - (u_4 - u_1)}{u_3 - u_2} = 1 - \dfrac{u_4 - u_1}{u_3 - u_2}$	(9.3)	thermal efficiency
$\eta = 1 - \dfrac{1}{r^{k-1}}$	(9.8)	thermal efficiency (constant k)
Diesel cycle		
$\eta = \dfrac{W_{\text{cycle}}/m}{Q_{23}/m} = 1 - \dfrac{Q_{41}/m}{Q_{23}/m} = 1 - \dfrac{u_4 - u_1}{h_3 - h_2}$	(9.11)	thermal efficiency
$\eta = 1 - \dfrac{1}{r^{k-1}}\left[\dfrac{r_c^k - 1}{k(r_c - 1)}\right]$	(9.13)	thermal efficiency (constant k)
Brayton cycle		
$\eta = \dfrac{\dot{W}_t/\dot{m} - \dot{W}_c/\dot{m}}{\dot{Q}_{in}/\dot{m}} = \dfrac{(h_3 - h_4) - (h_2 - h_1)}{h_3 - h_2}$	(9.19)	thermal efficiency
$\eta = 1 - \dfrac{1}{(p_2/p_1)^{(k-1)/k}}$	(9.25)	thermal efficiency (constant k)
$\text{bwr} = \dfrac{\dot{W}_c/\dot{m}}{\dot{W}_t/\dot{m}} = \dfrac{h_2 - h_1}{h_3 - h_4}$	(9.20)	back work ratio
$\eta_{reg} = \dfrac{h_x - h_2}{h_4 - h_2}$	(9.27)	Regenerator effectiveness for the regenerative gas turbine cycle (Figure 9.14)
Compressible flow in nozzles and diffusers		
$\mathbf{F} = \dot{m}(\mathbf{V}_2 - \mathbf{V}_1)$	(9.31)	Momentum equation for steady-state one-dimensional flow
$c = \sqrt{kRT}$	(9.37)	Ideal gas velocity of sound
$M = \text{V}/c$	(9.38)	Mach number
$h_o = h + \text{V}^2/2$	(9.39)	Stagnation enthalpy
$\dfrac{T_o}{T} = 1 + \dfrac{k-1}{2}M^2$	(9.50)	Isentropic flow function relating temperature and stagnation temperature (constant k)
$\dfrac{p_o}{p} = \left(\dfrac{T_o}{T}\right)^{k/(k-1)} = \left(1 + \dfrac{k-1}{2}M^2\right)^{k/(k-1)}$	(9.51)	Isentropic flow function relating pressure and stagnation pressure (constant k)

Exercises: things engineers think about

1. Diesel engines are said to produce higher *torque* than gasoline engines. What does that mean?

2. Formula One race cars have 2.4 liter engines. What does that signify? How is your automobile's engine sized in liters?

3. A car magazine says that your car's engine has more power when the ambient temperature is low. Do you agree?

4. When operating at high elevations, cars can lose power. Why?

5. Why are the external surfaces of a lawn mower engine covered with fins?

6. The ideal Brayton and Rankine cycles are composed of the same four processes, yet look different when represented on a T–s diagram. Explain.

7. The term *regeneration* is used to describe the use of regenerative feedwater heaters in vapor power plants and regenerative heat exchangers in gas turbines. In what ways are the purposes of these devices similar? How do they differ?

8. What is the overall thermal efficiency of the combined cycle of Example 9.13?

9. You jump off a raft into the water in the middle of a lake. What direction does the raft move? Explain.

10. How would the T–s diagram of Fig. 9.20 appear if frictional effects for flow through the diffuser, compressor, turbine, and nozzle were considered?

11. How do *internal* and *external* combustion engines differ?

12. In which of the following media is the sonic velocity the greatest: air, steel, or water? Does sound propagate in a vacuum?

13. Can a shock stand *upstream* of the throat of a converging–diverging nozzle?

Problems: developing engineering skills

Otto, Diesel, and Dual Cycles

9.1 An air-standard Otto cycle has a compression ratio of 9. At the beginning of compression, $p_1 = 100$ kPa and $T_1 = 300$ K. The heat addition per unit mass of air is 1350 kJ/kg. Determine

(a) the net work, in kJ per kg of air.
(b) the thermal efficiency of the cycle.
(c) the mean effective pressure, in kPa.
(d) the maximum temperature in the cycle, in K.
(e) To investigate the effects of varying compression ratio, plot each of the quantities calculated in parts (a) through (d) for compression ratios ranging from 1 to 12.

9.2 Solve Problem 9.1 on a cold air-standard basis with specific heats evaluated at 300 K.

9.3 At the beginning of the compression process of an air-standard Otto cycle, $p_1 = 1$ bar, $T_1 = 290$ K, $V_1 = 400$ cm^3. The maximum temperature in the cycle is 2200 K and the compression ratio is 8. Determine

(a) the heat addition, in kJ.
(b) the net work, in kJ.
(c) the thermal efficiency.
(d) the mean effective pressure, in bar.
(e) Develop a full accounting of the exergy transferred to the air during the heat addition, in kJ.
(f) Devise and evaluate an exergetic efficiency for the cycle.

Let $T_0 = 290$ K, $p_0 = 1$ bar.

9.4 Plot each of the quantities specified in parts (a) through (d) of Problem 9.3 versus the compression ratio ranging from 2 to 12.

9.5 Solve Problem 9.3 on a cold air-standard basis with specific heats evaluated at 300 K.

9.6 A four-cylinder, four-stroke internal combustion engine operates at 2800 RPM. The processes within each cylinder are modeled as an air-standard Otto cycle with a pressure of 100 kPa, a temperature of 27°C, and a volume of 5.6×10^{-4} m^3 at the beginning of compression. The compression ratio is 10, and maximum pressure in the cycle is 2480 kPa. Determine, using a cold air-standard analysis with $k = 1.4$, the power developed by the engine, in horsepower, and the mean effective pressure, in kPa.

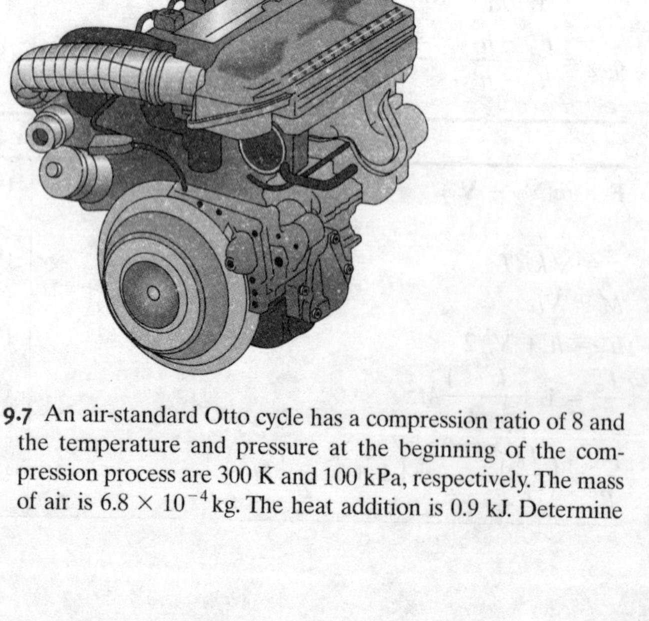

9.7 An air-standard Otto cycle has a compression ratio of 8 and the temperature and pressure at the beginning of the compression process are 300 K and 100 kPa, respectively. The mass of air is 6.8×10^{-4} kg. The heat addition is 0.9 kJ. Determine

(a) the maximum temperature, in K.
(b) the maximum pressure, in kPa.
(c) the thermal efficiency.
(d) To investigate the effects of varying compression ratio, plot each of the quantities calculated in parts (a) through (c) for compression ratios ranging from 2 to 12.

9.8 Solve Problem 9.7 on a cold air-standard basis with specific heats evaluated at 300 K.

9.9 At the beginning of the compression process in an air-standard Otto cycle, $p_1 = 100$ kPa and $T_1 = 294$ K. Plot the thermal efficiency and mean effective pressure, in kPa, for maximum cycle temperatures ranging from 1100 to 2800 K and compression ratios of 6, 8, and 10.

9.10 Solve Problem 9.9 on a cold air-standard basis using $k = 1.4$.

9.11 An air-standard Otto cycle has a compression ratio of 7.5. At the beginning of compression, $p_1 = 85$ kPa and $T_1 = 32°C$. The mass of air is 2 g, and the maximum temperature in the cycle is 960 K. Determine
(a) the heat rejection, in kJ.
(b) the net work, in kJ.
(c) the thermal efficiency.
(d) the mean effective pressure, in kPa.

9.12 The compression ratio of a cold air-standard Otto cycle is 9. At the end of the expansion process, the pressure is 655 kPa and the temperature is 550 K. The heat rejection from the cycle is 200 kJ/kg of air. Assuming $k = 1.4$, determine
(a) the net work, in kJ/kg of air.
(b) the thermal efficiency.
(c) the mean effective pressure, in kPa.

9.13 Consider a modification of the air-standard Otto cycle in which the isentropic compression and expansion processes are each replaced with polytropic processes having $n = 1.3$. The compression ratio is 9 for the modified cycle. At the beginning of compression, $p_1 = 1$ bar and $T_1 = 300$ K and $V_1 = 2270\ \text{cm}^3$. The maximum temperature during the cycle is 2000 K. Determine
(a) the heat transfer and work in kJ, for each process in the modified cycle.
(b) the thermal efficiency.
(c) the mean effective pressure, in bar.

9.14 A four-cylinder, four-stroke internal combustion engine has a bore of 65 mm and a stroke of 53 mm. The clearance volume is 12% of the cylinder volume at bottom dead center and the crankshaft rotates at 3600 RPM. The processes within each cylinder are modeled as an air-standard Otto cycle with a pressure of 100 kPa and a temperature of 38°C at the beginning of compression. The maximum temperature in the cycle is 2890 K. Based on this model, calculate the net work per cycle, in kJ, and the power developed by the engine, in kW.

9.15 At the beginning of the compression process in an air-standard Otto cycle, $p_1 = 1$ bar and $T_1 = 300$ K. The maximum cycle temperature is 2000 K. Plot the net work per unit of mass, in kJ/kg, the thermal efficiency, and the mean effective pressure, in bar, versus the compression ratio ranging from 2 to 14.

9.16 Investigate the effect of maximum cycle temperature on the net work per unit mass of air for air-standard Otto cycles with compression ratios of 5, 8, and 11. At the beginning of the compression process, $p_1 = 1$ bar and $T_1 = 295$ K. Let the maximum temperature in each case vary from 1000 to 2200 K.

9.17 The pressure-specific volume diagram of the air-standard *Lenoir* cycle is shown in Fig. P9.17. The cycle consists of constant volume heat addition, isentropic expansion, and constant pressure compression. For the cycle, $p_1 = 100$ kPa and $T_1 = 300$ K. The mass of air is 1.92×10^{-3} kg, and the maximum cycle temperature is 890 K. Assuming $c_v = 0.716$ kJ/kg · K determine for the cycle

(a) the net work, in kJ.
(b) the thermal efficiency.

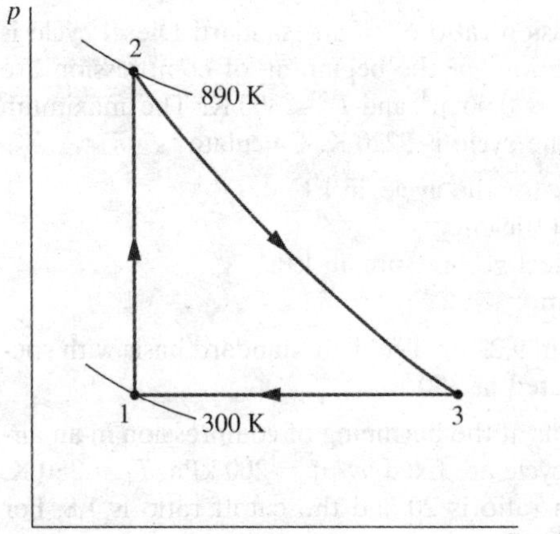

Fig. P9.17

9.18 The pressure-specific volume diagram of the air-standard *Atkinson* cycle is shown in Fig. P9.18. The cycle consists of isentropic compression, constant volume heat addition, isentropic expansion, and constant pressure compression. For a particular Atkinson cycle, the compression ratio during isentropic compression is 8.5. At the beginning of this compression process, $p_1 = 100$ kPa and $T_1 = 300$ K. The constant volume heat addition per unit mass of air is 1400 kJ/kg. (a) Sketch the cycle on T–s coordinates. Determine (b) the net work, in kJ per kg of air, (c) the thermal efficiency of the cycle, and (d) the mean effective pressure, in kPa.

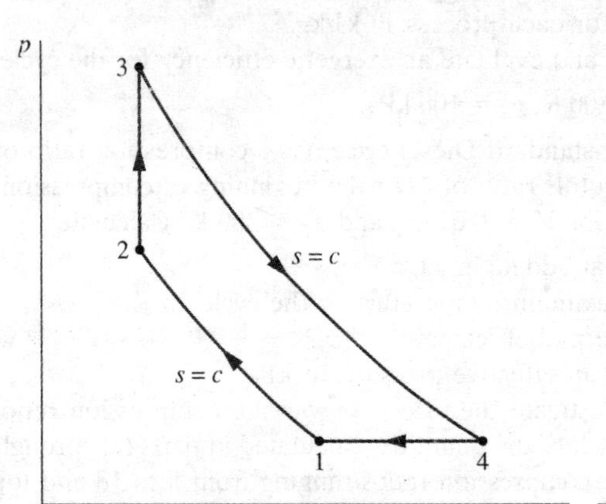

Fig. P9.18

9.19 On a cold air-standard basis, derive an expression for the thermal efficiency of the Atkinson cycle (see Fig. P9.18) in terms of the volume ratio during the isentropic compression, the pressure ratio for the constant volume process, and the specific heat ratio. Compare the thermal efficiencies of the cold air-standard Atkinson and Otto cycles, each having the same compression ratio and maximum temperature. Discuss.

9.20 The pressure and temperature at the beginning of compression of an air-standard Diesel cycle are 95 kPa and 300 K, respectively. At the end of the heat addition, the pressure is 7.2 MPa and the temperature is 2150 K. Determine

(a) the compression ratio.
(b) the cutoff ratio.
(c) the thermal efficiency of the cycle.
(d) the mean effective pressure, in kPa.

9.21 Solve Problem 9.20 on a cold air-standard basis with specific heats evaluated at 300 K.

9.22 The compression ratio of an air-standard Diesel cycle is 17 and the conditions at the beginning of compression are $p_1 = 100$ kPa, $V_1 = 0.06$ m^3, and $T_1 = 300$ K. The maximum temperature in the cycle is 2220 K. Calculate

(a) the net work for the cycle, in kJ.
(b) the thermal efficiency.
(c) the mean effective pressure, in kPa
(d) the cutoff ratio.

9.23 Solve Problem 9.22 on a cold air-standard basis with specific heats evaluated at 300 K.

9.24 The conditions at the beginning of compression in an air-standard Diesel cycle are fixed by $p_1 = 200$ kPa, $T_1 = 380$ K. The compression ratio is 20 and the cutoff ratio is 1.8. For $k = 1.4$, determine

(a) the maximum temperature, in K.
(b) the heat addition per unit mass, in kJ/kg.
(c) the net work per unit mass, in kJ/kg.
(d) the thermal efficiency.
(e) the mean effective pressure, in kPa.
(f) To investigate the effects of varying compression ratio, plot each of the quantities calculated in parts (a) through (e) for compression ratios ranging from 5 to 25.

9.25 For the Diesel cycle of Problem 9.24 with a compression ratio of 20 and a cutoff ratio of 1.8

(a) evaluate the exergy transfers accompanying heat and work for each process, in kJ/kg.
(b) devise and evaluate an exergetic efficiency for the cycle.

Let $T_0 = 300$ K, $p_0 = 100$ kPa.

9.26 An air-standard Diesel cycle has a compression ratio of 16 and a cutoff ratio of 2. At the beginning of compression, $p_1 = 100$ kPa, $V_1 = 0.01$ m^3, and $T_1 = 300$ K. Calculate

(a) the heat added, in kJ.
(b) the maximum temperature in the cycle, in K.
(c) the thermal efficiency.
(d) the mean effective pressure, in kPa
(e) To investigate the effects of varying compression ratio, plot each of the quantities calculated in parts (a) through (d) for compression ratios ranging from 5 to 18 and for cutoff ratios of 1.5, 2, and 2.5.

9.27 For the Diesel cycle of Problem 9.26 with a compression ratio of 16 and a cutoff ratio of 2

(a) evaluate the exergy transfers accompanying heat and work for each process, in kJ.
(b) devise and evaluate an exergetic efficiency for the cycle.

Let $T_0 = 300$ K, $p_0 = 100$ kPa.

9.28 The displacement volume of an internal combustion engine is 5.6 liters. The processes within each cylinder of the engine are modeled as an air-standard Diesel cycle with a cutoff ratio of 2.4. The state of the air at the beginning of compression is fixed by $p_1 = 95$ kPa, $T_1 = 27$°C, and $V_1 = 6.0$ liters. Determine the net work per cycle, in kJ, the power developed by the engine, in kW, and the thermal efficiency, if the cycle is executed 1500 times per min.

9.29 The state at the beginning of compression of an air-standard Diesel cycle is fixed by $p_1 = 100$ kPa and $T_1 = 310$ K. The compression ratio is 15. For cutoff ratios ranging from 1.5 to 2.5, plot

(a) the maximum temperature, in K.
(b) the pressure at the end of the expansion, in kPa.
(c) the net work per unit mass of air, in kJ/kg.
(d) the thermal efficiency.

9.30 An air-standard Diesel cycle has a maximum temperature of 1800 K. At the beginning of compression, $p_1 = 95$ kPa and $T_1 = 300$ K. The mass of air is 12 g. For compression ratios ranging from 15 to 25, plot

(a) the net work of the cycle, in kJ.
(b) the thermal efficiency.
(c) the mean effective pressure, in kPa.

9.31 At the beginning of compression in an air-standard Diesel cycle, $p_1 = 770$ kPa, $V_1 = 0.016$ m^3, and $T_1 = 315$ K. The compression ratio is 15 and the maximum cycle temperature is 1400 K. Determine

(a) the mass of air, in kg.
(b) the heat addition and heat rejection per cycle, each in kJ.
(c) the net work, in kJ, and the thermal efficiency.

9.32 At the beginning of the compression process in an air-standard Diesel cycle, $p_1 = 1$ bar and $T_1 = 300$ K. For maximum cycle temperatures of 1200, 1500, 1800, and 2100 K, plot the heat addition per unit of mass, in kJ/kg, the net work per unit of mass, in kJ/kg, the mean effective pressure, in bar, and the thermal efficiency, each versus compression ratio ranging from 5 to 20.

9.33 An air-standard dual cycle has a compression ratio of 9. At the beginning of compression, $p_1 = 100$ kPa, $T_1 = 300$ K, and $V_1 = 14$ L. The heat addition is 22.7 kJ, with one half added at constant volume and one half added at constant pressure. Determine

(a) the temperatures at the end of each heat addition process, in K.
(b) the net work of the cycle per unit mass of air, in kJ/kg.
(c) the thermal efficiency.
(d) the mean effective pressure, in kPa.

9.34 For the cycle in Problem 9.33, plot each of the quantities calculated in parts (a) through (d) versus the ratio of constant-volume heat addition to total heat addition varying from 0 to 1. Discuss.

9.35 Solve Problem 9.33 on a cold air-standard basis with specific heats evaluated at 300 K.

9.36 The thermal efficiency, η, of a cold air-standard dual cycle can be expressed as

$$\eta = 1 - \frac{1}{r^{k-1}}\left[\frac{r_p r_c^k - 1}{(r_p - 1) + k r_p (r_c - 1)}\right]$$

where r is compression ratio, r_c is cutoff ratio, and r_p is the pressure ratio for the constant volume heat addition. Derive this expression.

9.37 An air-standard dual cycle has a compression ratio of 16 and a cutoff ratio of 1.15. At the beginning of compression, $p_1 = 95$ kPa and $T_1 = 300$ K. The pressure increases by a factor of 2.2 during the constant volume heat addition process. If the mass of air is 0.04 kg, determine
(a) the heat addition at constant volume and at constant pressure, each in kJ.
(b) the net work of the cycle, in kJ.
(c) the heat rejection, in kJ.
(d) the thermal efficiency.

9.38 The pressure and temperature at the beginning of compression in an air-standard dual cycle are 95 kPa and 300 K, respectively. The compression ratio is 15 and the heat addition per unit mass of air is 1860 kJ/kg. At the end of the constant volume heat addition process, the pressure is 8.2 MPa. Determine
(a) the net work of the cycle per unit mass of air, in kJ/kg.
(b) the heat rejection for the cycle per unit mass of air, in kJ/kg.
(c) the thermal efficiency.
(d) the cutoff ratio.
(e) To investigate the effects of varying compression ratio, plot each of the quantities calculated in parts (a) through (d) for compression ratios ranging from 10 to 28.

9.39 An air-standard dual cycle has a compression ratio of 16. At the beginning of compression, $p_1 = 100$ kPa, $V_1 = 0.01$ m^3, and $T_1 = 10$°C. The pressure doubles during the constant volume heat addition process. For a maximum cycle temperature of 1670 K, determine
(a) the heat addition to the cycle, in kJ.
(b) the net work of the cycle, in kJ.
(c) the thermal efficiency.
(d) the mean effective pressure, in kPa.
(e) To investigate the effects of varying maximum cycle temperature, plot each of the quantities calculated in parts (a) through (d) for maximum cycle temperatures ranging from 1670 to 2220 K.

9.40 At the beginning of the compression process in an air-standard dual cycle, $p_1 = 1$ bar and $T_1 = 300$ K. The total heat addition is 1000 kJ/kg. Plot the net work per unit of mass, in kJ/kg, the mean effective pressure, in bar, and the thermal efficiency versus compression ratio for different fractions of constant volume and constant pressure heat addition. Consider compression ratio ranging from 10 to 20.

Brayton Cycle

9.41 Air enters the compressor of an ideal cold air-standard Brayton cycle at 100 kPa, 300 K, with a mass flow rate of 6 kg/s. The compressor pressure ratio is 10, and the turbine inlet temperature is 1400 K. For $k = 1.4$, calculate
(a) the thermal efficiency of the cycle.
(b) the back work ratio.
(c) the net power developed, in kW.

9.42 For the Brayton cycle of Problem 9.41, investigate the effects of varying compressor pressure ratio and turbine inlet temperature. Plot the same quantities calculated in Problem 9.41 for
(a) a compressor pressure ratio of 10 and turbine inlet temperatures ranging from 1000 to 1600 K.
(b) a turbine inlet temperature of 1400 K and compressor pressure ratios ranging from 2 to 20.

Discuss.

9.43 The rate of heat addition to an air-standard Brayton cycle is 9.9×10^5 kW. The pressure ratio for the cycle is 14 and the minimum and maximum temperatures are 300 K and 1670 K, respectively. Determine
(a) the thermal efficiency of the cycle.
(b) the mass flow rate of air, in kg/s.
(c) the net power developed by the cycle, in kJ/s.

9.44 Solve Problem 9.43 on a cold air-standard basis with specific heats evaluated at 300 K.

9.45 Consider an ideal air-standard Brayton cycle with minimum and maximum temperatures of 300 K and 1500 K, respectively. The pressure ratio is that which maximizes the net work developed by the cycle per unit mass of air flow. On a cold air-standard basis, calculate
(a) the compressor and turbine work per unit mass of air flow, each in kJ/kg.
(b) the thermal efficiency of the cycle.
(c) Plot the thermal efficiency versus the maximum cycle temperature ranging from 1200 to 1800 K.

9.46 On the basis of a cold air-standard analysis, show that the back work ratio of an ideal air-standard Brayton cycle equals the ratio of absolute temperatures at the compressor inlet and the turbine outlet.

9.47 The compressor inlet temperature of an ideal air-standard Brayton cycle is 300 K and the maximum allowable turbine inlet temperature is 1440 K. Plot the net work developed per unit mass of air flow, in kJ/kg, and the thermal efficiency versus compressor pressure ratio for pressure ratios ranging from 12 to 24. Using your plots, estimate the pressure ratio for maximum net work and the corresponding value of thermal efficiency. Compare the results to those obtained in analyzing the cycle on a cold air-standard basis.

9.48 The compressor inlet temperature for an ideal Brayton cycle is T_1 and the turbine inlet temperature is T_3. Using a cold air-standard analysis, show that the temperature T_2 at the compressor exit that maximizes the net work developed per unit mass of air flow is $T_2 = (T_1 T_3)^{1/2}$.

9.49 Air enters the compressor of a cold air-standard Brayton cycle at 100 kPa, 300 K, with a mass flow rate of 6 kg/s. The compressor pressure ratio is 10, and the turbine inlet temperature is 1400 K. The turbine and compressor each have isentropic efficiencies of 80%. For $k = 1.4$, calculate

(a) the thermal efficiency of the cycle.
(b) the back work ratio.
(c) the net power developed, in kW.
(d) the rates of exergy destruction in the compressor and turbine, respectively, each in kW, for $T_0 = 300$ K.

Plot the quantities calculated in parts (a) through (d) versus isentropic efficiency for equal compressor and turbine isentropic efficiencies ranging from 70 to 100%. Discuss.

9.50 Air enters the compressor of an air-standard Brayton cycle with a volumetric flow rate of 60 m^3/s at 0.8 bar, 280 K. The compressor pressure ratio is 20, and the maximum cycle temperature is 2100 K. For the compressor, the isentropic efficiency is 92% and for the turbine the isentropic efficiency is 95%. Determine

(a) the net power developed, in MW.
(b) the rate of heat addition in the combustor, in MW.
(c) the thermal efficiency of the cycle.

9.51 Air enters the compressor of a simple gas turbine at $p_1 = 95$ kPa, $T_1 = 300$ K. The isentropic efficiencies of the compressor and turbine are 83 and 87%, respectively. The compressor pressure ratio is 14 and the temperature at the turbine inlet is 1390 K. The net power developed is 1465 kW. On the basis of an air-standard analysis, calculate

(a) the volumetric flow rate of the air entering the compressor, in m^3/s.
(b) the temperatures at the compressor and turbine exits, each in K.
(c) the thermal efficiency of the cycle.

9.52 Solve Problem 9.51 on a cold air-standard basis with specific heats evaluated at 300 K.

9.53 Air enters the compressor of a simple gas turbine at 100 kPa, 300 K, with a volumetric flow rate of 5 m^3/s. The compressor pressure ratio is 10 and its isentropic efficiency is 85%. At the inlet to the turbine, the pressure is 950 kPa, and the temperature is 1400 K. The turbine has an isentropic efficiency of 88% and the exit pressure is 100 kPa. On the basis of an air-standard analysis,

(a) develop a full accounting of the *net* exergy increase of the air passing through the gas turbine combustor, in kW.
(b) devise and evaluate an exergetic efficiency for the gas turbine cycle.

Let $T_0 = 300$ K, $p_0 = 100$ kPa.

9.54 Air enters the compressor of a simple gas turbine at 100 kPa, 27°C, and exits at 600 kPa, 268°C. The air enters the turbine at 838°C, 600 kPa and expands to 492°C, 100 kPa. The compressor and turbine operate adiabatically, and kinetic and potential energy effects are negligible. On the basis of an air-standard analysis,

(a) develop a full accounting of the *net* exergy increase of the air passing through the gas turbine combustor, in kJ/kg.
(b) devise and evaluate an exergetic efficiency for the gas turbine cycle.

Let $T_0 = 27$°C, $p_0 = 100$ kPa.

Regeneration, Reheat, and Compression with Intercooling

9.55 Air enters the compressor of a cold air-standard Brayton cycle with regeneration at 100 kPa, 300 K, with a mass flow rate of 6 kg/s. The compressor pressure ratio is 10, and the turbine inlet temperature is 1400 K. The turbine and compressor each have isentropic efficiencies of 80% and the regenerator effectiveness is 80%. For $k = 1.4$, calculate

(a) the thermal efficiency of the cycle.
(b) the back work ratio.
(c) the net power developed, in kW.
(d) the rate of exergy destruction in the regenerator, in kW, for $T_0 = 300$ K.

9.56 Air enters the compressor of a regenerative air-standard Brayton cycle with a volumetric flow rate of 60 m^3/s at 0.8 bar, 280 K. The compressor pressure ratio is 20, and the maximum cycle temperature is 2100 K. For the compressor, the isentropic efficiency is 92% and for the turbine the isentropic efficiency is 95%. For a regenerator effectiveness of 85%, determine

(a) the net power developed, in MW.
(b) the rate of heat addition in the combustor, in MW.
(c) the thermal efficiency of the cycle.

Plot the quantities calculated in parts (a) through (c) for regenerator effectiveness values ranging from 0 to 100%. Discuss.

9.57 Reconsider Problem 9.51, but include a regenerator in the cycle. For regenerator effectiveness values ranging from 0 to 100%, plot

(a) the thermal efficiency.
(b) the percent decrease in heat addition to the air.

9.58 On the basis of a cold air-standard analysis, show that the thermal efficiency of an ideal regenerative gas turbine can be expressed as

$$\eta = 1 - \left(\frac{T_1}{T_3}\right)(r)^{(k-1)/k}$$

where r is the compressor pressure ratio, and T_1 and T_3 denote the temperatures at the compressor and turbine inlets, respectively.

9.59 An air-standard Brayton cycle has a compressor pressure ratio of 10. Air enters the compressor at $p_1 = 100$ kPa, $T_1 = 20$°C with a mass flow rate of 11 kg/s. The turbine inlet temperature is 1222 K. Calculate the thermal efficiency and the net power developed, in kW, if

(a) the turbine and compressor isentropic efficiencies are each 100%.
(b) the turbine and compressor isentropic efficiencies are 88 and 84%, respectively.
(c) the turbine and compressor isentropic efficiencies are 88 and 84%, respectively, and a regenerator with an effectiveness of 80% is incorporated.

9.60 Fig. P9.60 illustrates a gas turbine power plant that uses solar energy as the source of heat addition (see U.S. Patent no. 4,262,484). Operating data are given on the figure. Modeling the cycle as a Brayton cycle, and assuming no pressure drops in the heat exchanger or interconnecting piping, determine

(a) the thermal efficiency.
(b) the air mass flow rate, in kg/s, for a net power output of 500 kW.

9.61 Air enters the compressor of a regenerative gas turbine with a volumetric flow rate of 150 m^3/s at 100 kPa, 25°C, and is compressed to 410 kPa. The air then passes through the regenerator and exits at 620 K. The temperature at the turbine inlet is 950 K. The compressor and turbine each has an isentropic efficiency of 84%. Using an air-standard analysis, calculate

(a) the thermal efficiency of the cycle.
(b) the regenerator effectiveness.
(c) the net power output, in kJ/s.

9.62 Air enters the turbine of a gas turbine at 1200 kPa, 1200 K, and expands to 100 kPa in two stages. Between the stages, the air is reheated at a constant pressure of 350 kPa to 1200 K. The expansion through each turbine stage is isentropic. Determine, in kJ per kg of air flowing

(a) the work developed by each stage.
(b) the heat transfer for the reheat process.
(c) the increase in net work as compared to a single stage of expansion with no reheat.

9.63 Reconsider Problem 9.62 and include in the analysis that each turbine stage might have an isentropic efficiency less than 100%. Plot each of the quantities calculated in parts (a) through (c) of Problem 9.61 for values of the interstage pressure ranging from 100 to 1200 kPa and for isentropic efficiencies of 100%, 80%, and 60%.

9.64 Consider a two-stage turbine operating at steady state with reheat at constant pressure between the stages. Show that the maximum work is developed when the pressure ratio is the same across each stage. Use a cold air-standard analysis, assuming the inlet state and the exit pressure are specified, each expansion process is isentropic, and the temperature at the inlet to each turbine stage is the same. Kinetic and potential energy effects can be ignored.

9.65 Air enters the compressor of a cold air-standard Brayton cycle with regeneration and reheat at 100 kPa, 300 K, with a mass flow rate of 6 kg/s. The compressor pressure ratio is 10, and the inlet temperature for each turbine stage is 1400 K. The turbine stages and compressor each have isentropic efficiencies of 80% and the regenerator effectiveness is 80%. For $k = 1.4$, calculate

(a) the thermal efficiency of the cycle.
(b) the back work ratio.
(c) the net power developed, in kW.
(d) the rates of exergy destruction in the compressor and each turbine stage as well as the regenerator, in kW, for $T_0 = 300$ K.

9.66 If the inlet state and the exit pressure are specified for a two-stage turbine with reheat between the stages and operating at steady state, show that the maximum total work output is obtained when the pressure ratio is the same across each stage. Use a cold air-standard analysis assuming that each compression process is isentropic, there is no pressure drop through the reheater, and the temperature at the inlet

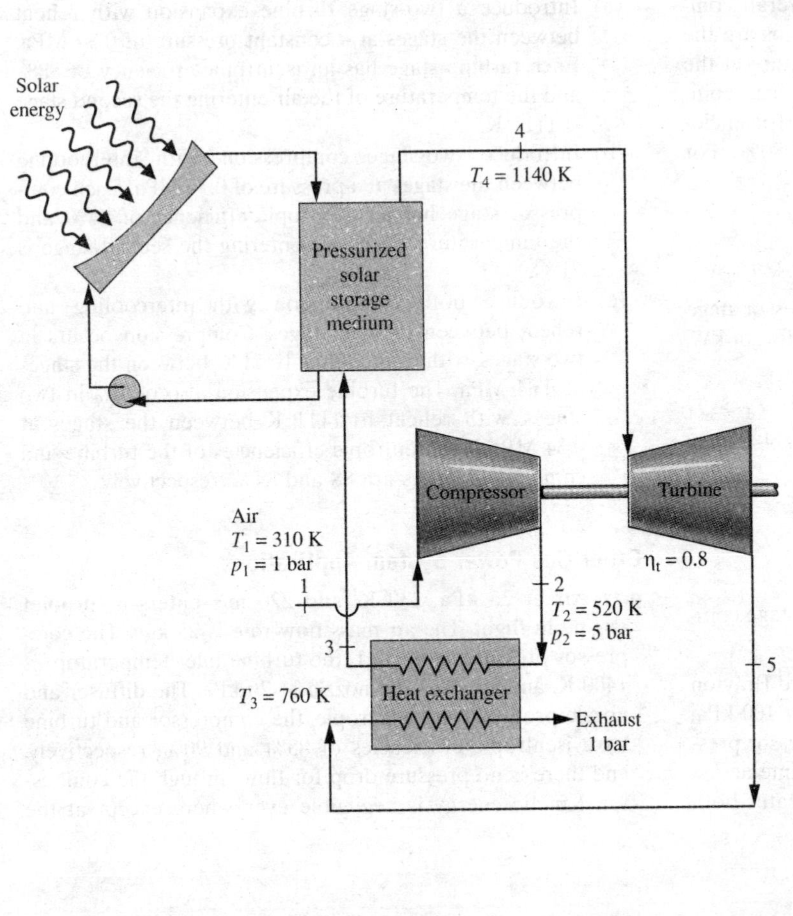

Fig. P9.60

to each turbine stage is the same. Kinetic and potential energy effects can be ignored.

9.67 A two-stage air compressor operates at steady state, compressing 10 m^3/min of air from 100 kPa, 300 K, to 1200 kPa. An intercooler between the two stages cools the air to 300 K at a constant pressure of 350 kPa. The compression processes are isentropic. Calculate the power required to run the compressor, in kW, and compare the result to the power required for isentropic compression from the same inlet state to the same final pressure.

9.68 Reconsider Problem 9.67 and include in the analysis that each compressor stage might have an isentropic efficiency less than 100%. Plot, in kW, (a) the power input to each stage, (b) the heat transfer rate for the intercooler, and (c) the decrease in power input as compared to a single stage of compression with no intercooling, for values of interstage pressure ranging from 100 to 1200 kPa and for isentropic efficiencies of 100%, 80%, and 60%.

9.69 Air enters a compressor operating at steady state at 96 kPa, 16°C, with a volumetric flow rate of 162 m^3/min. The compression occurs in two stages, with each stage being a polytropic process with $n = 1.27$. The air is cooled to 27°C between the stages by an intercooler operating at 310 kPa Air exits the compressor at 1 MPa. Determine, in kJ per min

(a) the power and heat transfer rate for each compressor stage.
(b) the heat transfer rate for the intercooler.

9.70 Air enters the first compressor stage of a cold air-standard Brayton cycle with regeneration and intercooling at 100 kPa, 300 K, with a mass flow rate of 6 kg/s. The overall compressor pressure ratio is 10, and the pressure ratios are the same across each compressor stage. The temperature at the inlet to the second compressor stage is 300 K. The compressor stages and turbine each have isentropic efficiencies of 80% and the regenerator effectiveness is 80%. For $k = 1.4$, calculate

(a) the thermal efficiency of the cycle.
(b) the back work ratio.
(c) the net power developed, in kW.
(d) the rates of exergy destruction in each compressor stage and the turbine stage as well as the regenerator, in kW, for $T_0 = 300$ K.

9.71 Referring to Example 9.10, show that if $T_d > T_1$ the pressure ratios across the two compressor stages are related by

$$\frac{p_i}{p_1} = \left(\frac{p_2}{p_i}\right)\left(\frac{T_d}{T_1}\right)^{k/(k-1)}$$

9.72 Rework Example 9.10 for the case of a three-stage compressor with intercooling between stages.

9.73 Air enters the compressor of a cold air-standard Brayton cycle with regeneration, intercooling, and reheat at 100 kPa, 300 K, with a mass flow rate of 6 kg/s. The compressor pressure ratio is 10, and the pressure ratios are the same across each compressor stage. The intercooler and reheater both operate at the same pressure. The temperature at the inlet to the second compressor stage is 300 K, and the inlet temperature for each turbine stage is 1400 K. The compressor and turbine stages each have isentropic efficiencies of 80% and the regenerator effectiveness is 80%. For $k = 1.4$, calculate

(a) the thermal efficiency of the cycle.
(b) the back work ratio.
(c) the net power developed, in kW.
(d) the rates of exergy destruction in the compressor and turbine stages as well as the regenerator, in kW, for $T_0 = 300$ K.

9.74 Air enters the compressor of a gas turbine at 100 kPa, 300 K. The air is compressed in two stages to 900 kPa, with intercooling to 300 K between the stages at a pressure of 300 kPa. The turbine inlet temperature is 1480 K and the expansion occurs in two stages, with reheat to 1420 K between the stages at a pressure of 300 kPa. The compressor and turbine stage efficiencies are 84 and 82%, respectively. The net power developed is 1.8 MW. Determine

(a) the volumetric flow rate, in m^3/s, at the inlet of each compressor stage.
(b) the thermal efficiency of the cycle.
(c) the back work ratio.

9.75 Reconsider Problem 9.74 but include a regenerator with an effectiveness of 75%.

9.76 For each of the following modifications of the cycle of part (c) of Problem 9.59, determine the thermal efficiency and net power developed, in kW.

(a) Introduce a two-stage turbine expansion with reheat between the stages at a constant pressure of 0.34 MPa. Each turbine stage has an isentropic efficiency of 88% and the temperature of the air entering the second stage is 1111 K.
(b) Introduce two-stage compression, with intercooling between the stages at a pressure of 0.34 MPa. Each compressor stage has an isentropic efficiency of 84% and the temperature of the air entering the second stage is 21°C.
(c) Introduce both compression with intercooling and reheat between turbine stages. Compression occurs in two stages, with intercooling to 21°C between the stages at 0.34 MPa. The turbine expansion also occurs in two stages, with reheat to 1111 K between the stages at 0.34 MPa. The isentropic efficiencies of the turbine and compressor stages are 88 and 84%, respectively.

Other Gas Power System Applications

9.77 Air at 26 kPa, 230 K, and 220 m/s enters a turbojet engine in flight. The air mass flow rate is 25 kg/s. The compressor pressure ratio is 11, the turbine inlet temperature is 1400 K, and air exits the nozzle at 26 kPa. The diffuser and nozzle processes are isentropic, the compressor and turbine have isentropic efficiencies of 85% and 90%, respectively, and there is no pressure drop for flow through the combustor. Kinetic energy is negligible everywhere except at the diffuser inlet and the nozzle exit. On the basis of air-standard analysis, determine

(a) the pressures, in kPa, and temperatures, in K, at each principal state.
(b) the rate of heat addition to the air passing through the combustor, in kJ/s.
(c) the velocity at the nozzle exit, in m/s.

9.78 For the turbojet in Problem 9.77, plot the velocity at the nozzle exit, in m/s, the pressure at the turbine exit, in kPa, and the rate of heat input to the combustor, in kW, each as a function of compressor pressure ratio in the range of 6 to 14. Repeat for turbine inlet temperatures of 1200 K and 1000 K.

9.79 Air enters the diffuser of a turbojet engine with a mass flow rate of 39 kg/s at 62 kPa, 233 K, and a velocity of 225 m/s. The pressure ratio for the compressor is 12, and its isentropic efficiency is 88%. Air enters the turbine at 1333 K with the same pressure as at the exit of the compressor. Air exits the nozzle at 62 kPa. The diffuser operates isentropically and the nozzle and turbine have isentropic efficiencies of 92% and 90%, respectively. On the basis of an air-standard analysis, calculate

(a) the rate of heat addition, in kJ/h.
(b) the pressure at the turbine exit, in kPa.
(c) the compressor power input, in kJ/h.
(d) the velocity at the nozzle exit, in m/s.

Neglect kinetic energy except at the diffuser inlet and the nozzle exit.

9.80 Consider the addition of an afterburner to the turbojet in Problem 9.77 that raises the temperature at the inlet of the nozzle to 1300 K. Determine the velocity at the nozzle exit, in m/s.

9.81 Consider the addition of an afterburner to the turbojet in Problem 9.79 that raises the temperature at the inlet of the nozzle to 1222 K. Determine the velocity at the nozzle exit, in m/s.

9.82 Air enters the diffuser of a ramjet engine at 40 kPa, 233 K, with a velocity of 480 m/s, and decelerates essentially to zero velocity. After combustion, the gases reach a temperature of 1111 K before being discharged through the nozzle at 40 kPa. On the basis of an air-standard analysis, determine

(a) the pressure at the diffuser exit, in kPa.
(b) the velocity at the nozzle exit, in m/s.

Neglect kinetic energy except at the diffuser inlet and the nozzle exit.

9.83 Air enters the diffuser of a ramjet engine at 40 kPa, 240 K, with a velocity of 2500 km/h and decelerates to negligible velocity. On the basis of an air-standard analysis, the heat addition is 1080 kJ per kg of air passing through the engine. Air exits the nozzle at 40 kPa. Determine

(a) the pressure at the diffuser exit, in kPa.
(b) the velocity at the nozzle exit, in m/s.

Neglect kinetic energy except at the diffuser inlet and the nozzle exit.

9.84 A turboprop engine consists of a diffuser, compressor, combustor, turbine, and nozzle. The turbine drives a propeller as well as the compressor. Air enters the diffuser with a volumetric flow rate of 83.7 m^3/s at 40 kPa, 240 K, and a velocity of 180 m/s, and decelerates essentially to zero velocity. The compressor pressure ratio is 10 and the compressor has an isentropic efficiency of 85%. The turbine inlet temperature is 1140 K, and its isentropic efficiency is 85%. The turbine exit pressure is 50 kPa. Flow through the diffuser and nozzle is isentropic. Using an air-standard analysis, determine

(a) the power delivered to the propeller, in MW.
(b) the velocity at the nozzle exit, in m/s.

Neglect kinetic energy except at the diffuser inlet and the nozzle exit.

9.85 A turboprop engine consists of a diffuser, compressor, combustor, turbine, and nozzle. The turbine drives a propeller as well as the compressor. Air enters the diffuser at 83 kPa, 256 K, with a volumetric flow rate of 850 m^3/min and a velocity of 156 m/s. In the diffuser, the air decelerates isentropically to negligible velocity. The compressor pressure ratio is 9, and the turbine inlet temperature is 1170 K. The turbine exit pressure is 172.5 kPa, and the air expands to 83 kPa through a nozzle. The compressor and turbine each has an isentropic efficiency of 87%, and the nozzle has an isentropic efficiency of 95%. Using an air-standard analysis, determine

(a) the power delivered to the propeller, in kW.
(b) the velocity at the nozzle exit, in m/s.

Neglect kinetic energy except at the diffuser inlet and the nozzle exit.

9.86 Helium is used in a combined cycle power plant as the working fluid in a simple closed gas turbine serving as the topping cycle for a vapor power cycle. A nuclear reactor is the source of energy input to the helium. Figure P9.86 provides steady-state operating data. Helium enters the compressor of the gas turbine at 1.4 MPa, 82°C with a mass flow rate of 4×10^5 kg/h and is compressed to 5.5 MPa. The isentropic efficiency of the compressor is 80%. The helium then passes through the reactor with a negligible decrease in pressure, exiting at 760°C. Next, the helium expands through the turbine, which has an isentropic efficiency of 80%, to a pressure of 1.4 MPa. The helium then passes through the interconnecting heat exchanger. A separate stream of liquid water enters the heat exchanger and exits as saturated vapor at 8 MPa. The vapor is superheated before entering the turbine at 425°C, 8 MPa. The steam expands through the turbine to 6.9 kPa and a quality of 0.9. Saturated liquid exits the condenser at 6.9 kPa. Cooling water passing through the condenser experiences a temperature rise from 16°C to 32°C. The isentropic pump efficiency is 100%. Stray heat transfer and kinetic and potential energy effects can be ignored. Determine

(a) the mass flow rates of the steam and the cooling water, each in kg/h.

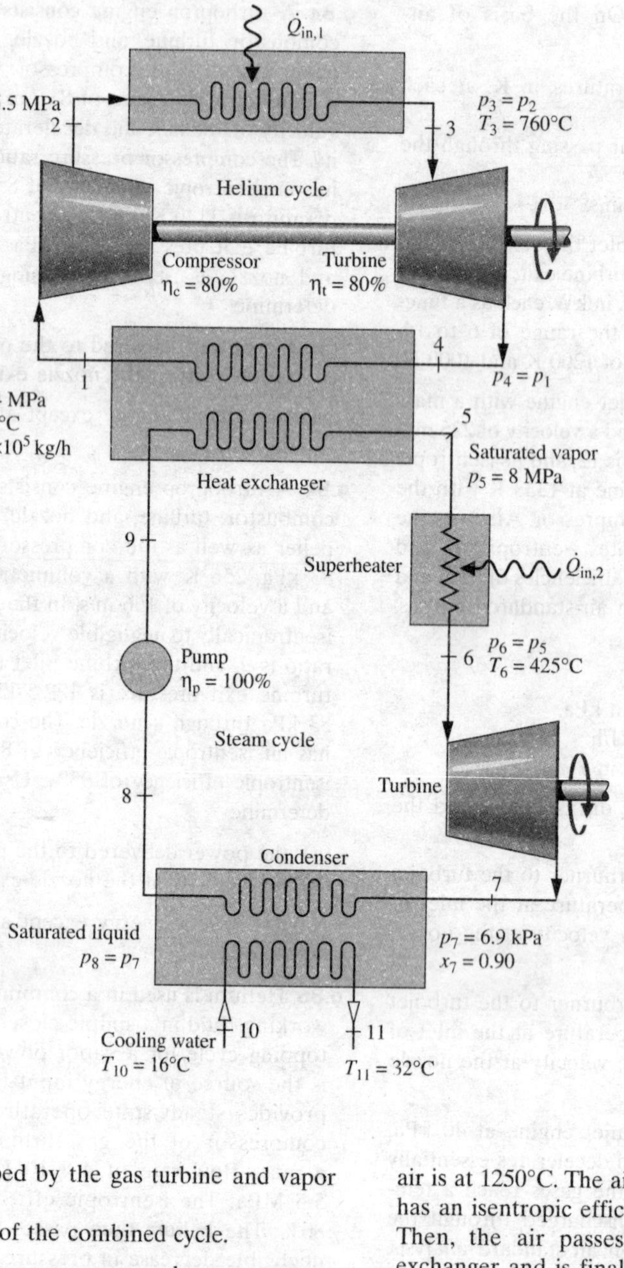

Fig. P9.86

(b) the net power developed by the gas turbine and vapor cycles, each in kJ/h.

(c) the thermal efficiency of the combined cycle.

9.87 A combined gas turbine-vapor power plant operates as shown in Fig. P9.87. Pressure and temperature data are given at principal states, and the net power developed by the gas turbine is 147 MW. Using air-standard analysis for the gas turbine, determine

(a) the net power, in MW, developed by the power plant.

(b) the overall thermal efficiency of the plant.

Develop a full accounting of the *net* rate of exergy increase of the air passing through the gas turbine combustor. Let $T_0 = 300$ K, $p_0 = 1$ bar.

9.88 Air enters the compressor of a combined gas turbine–vapor power plant (Fig. 9.23) at 1 bar, 25°C. The isentropic compressor efficiency is 85% and the compressor pressure ratio is 14. The air passing through the combustor receives energy by heat transfer at a rate of 50 MW with no significant decrease in pressure. At the inlet to the turbine the air is at 1250°C. The air expands through the turbine, which has an isentropic efficiency of 87%, to a pressure of 1 bar. Then, the air passes through the interconnecting heat exchanger and is finally discharged at 200°C, 1 bar. Steam enters the turbine of the vapor cycle at 12.5 MPa, 500°C, and expands to a condenser pressure of 0.1 bar. Water enters the pump as a saturated liquid at 0.1 bar. The turbine and pump have isentropic efficiencies of 90 and 100%, respectively. Cooling water enters the condenser at 20°C and exits at 35°C. Determine

(a) the mass flow rates of the air, steam, and cooling water, each in kg/s.

(b) the net power developed by the gas turbine cycle and the vapor cycle, respectively, each in MW.

(c) the thermal efficiency of the combined cycle.

(d) the *net* rate at which exergy is carried out with the exhaust air, $\dot{m}_{air}[e_{f5} - e_{f1}]$, in MW.

(e) the *net* rate at which exergy is carried out with the cooling water, in MW.

Let $T_0 = 20°C$, $p_0 = 1$ bar.

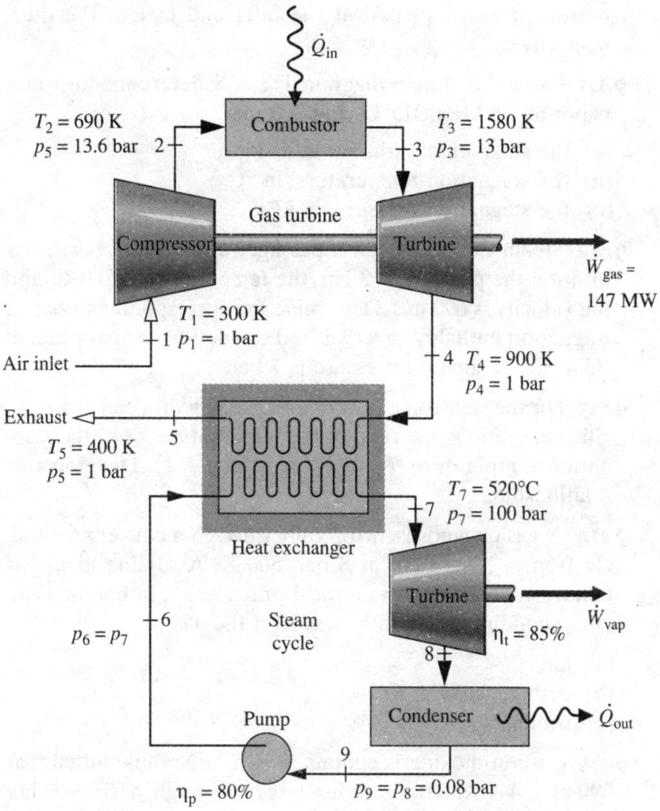

Fig. P9.87

9.89 A combined gas turbine–vapor power plant (Fig. 9.23) has a net power output of 100 MW. Air enters the compressor of the gas turbine at 100 kPa, 300 K, and is compressed to 1200 kPa. The isentropic efficiency of the compressor is 84%. The conditions at the inlet to the turbine are 1200 kPa and 1400 K. Air expands through the turbine, which has an isentropic efficiency of 88%, to a pressure of 100 kPa. The air then passes through the interconnecting heat exchanger, and is finally discharged at 480 K. Steam enters the turbine of the vapor power cycle at 8 MPa, 400°C, and expands to the condenser pressure of 8 kPa. Water enters the pump as saturated liquid at 8 kPa. The turbine and pump have isentropic efficiencies of 90 and 80%, respectively. Determine

(a) the mass flow rates of air and steam, each in kg/s.
(b) the thermal efficiency of the combined cycle.
(c) a full accounting of the *net* exergy increase of the air passing through the combustor of the gas turbine, $\dot{m}_{air}[e_{f3} - e_{f2}]$, in MW. Discuss.

Let $T_0 = 300$ K, $p_0 = 100$ kPa.

9.90 A simple gas turbine is the topping cycle for a simple vapor power cycle (Fig. 9.23). Air enters the compressor of the gas turbine at 15°C, 100 kPa, with a volumetric flow rate of 20 m^3/s. The compressor pressure ratio is 12 and the turbine inlet temperature is 1440 K. The compressor and turbine each have isentropic efficiencies of 88%. The air leaves the interconnecting heat exchanger at 460 K, 100 kPa. Steam enters the turbine of the vapor cycle at 7000 kPa, 480°C, and expands to the condenser pressure of 7 kPa. Water enters the pump as saturated liquid at 7 kPa. The turbine and pump efficiencies are 90 and 70%, respectively. Cooling water passing through the condenser experiences a temperature rise from 15 to 27°C with a negligible change in pressure. Determine

(a) the mass flow rates of the air, steam, and cooling water, each in kg/s.
(b) the net power developed by the gas turbine cycle and the vapor cycle, respectively, each in kJ/s.
(c) the thermal efficiency of the combined cycle.
(d) a full accounting of the *net* exergy increase of the air passing through the combustor of the gas turbine, $\dot{m}_{air}[e_{f3} - e_{f2}]$, in kJ/s. Discuss.

Let $T_0 = 300$ K, $p_0 = 100$ kPa.

9.91 Air enters the compressor of an Ericsson cycle at 300 K, 1 bar, with a mass flow rate of 5 kg/s. The pressure and temperature at the inlet to the turbine are 10 bar and 1400 K, respectively. Determine

(a) the net power developed, in kW.
(b) the thermal efficiency.
(c) the back work ratio.

9.92 For the cycle in Problem 9.91, plot the net power developed, in kW, for compressor pressure ratios ranging from 2 to 15. Repeat for turbine inlet temperatures of 1200 K and 1000 K.

9.93 Air is the working fluid in an Ericsson cycle. Expansion through the turbine takes place at a constant temperature of 1250 K. Heat transfer from the compressor occurs at 310 K. The compressor pressure ratio is 12. Determine

(a) the net work, in kJ/kg of air flowing.
(b) the thermal efficiency.

9.94 Nitrogen (N_2) is the working fluid of a Stirling cycle with a compression ratio of nine. At the beginning of the isothermal compression, the temperature, pressure, and volume are 310 K, 1 bar, and 0.008 m^3, respectively. The temperature during the isothermal expansion is 1000 K. Determine

(a) the net work, in kJ.
(b) the thermal efficiency.
(c) the mean effective pressure, in bar.

9.95 Helium is the working fluid in a Stirling cycle. In the isothermal compression, the helium is compressed from 100 kPa, 38°C, to 1000 kPa. The isothermal expansion occurs at 816°C. Determine

(a) the work and heat transfer, in kJ/kg of helium, for each process in the cycle.
(b) the thermal efficiency.

Compressible Flow

9.96 Calculate the thrust developed by the turbojet engine in Problem 9.77, in kN.

9.97 Calculate the thrust developed by the turbojet engine in Problem 9.79, in kN.

9.98 Calculate the thrust developed by the turbojet engine with afterburner in Problem 9.80, in kN.

9.99 Referring to the turbojet in Problem 9.79 and the modified turbojet in Problem 9.81, calculate the thrust developed by each engine, in N. Discuss.

9.100 Air enters the diffuser of a turbojet engine at 18 kPa, 216 K, with a volumetric flow rate of 230 m^3/s and a velocity of 265 m/s. The compressor pressure ratio is 15, and its isentropic efficiency is 87%. Air enters the turbine at 1360 K and the same pressure as at the exit of the compressor. The turbine isentropic efficiency is 89%, and the nozzle isentropic efficiency is 97%. The pressure at the nozzle exit is 18 kPa. On the basis of an air-standard analysis, calculate the thrust, in kN.

9.101 Calculate the ratio of the thrust developed to the mass flow rate of air, in N per kg/s, for the ramjet engine in Problem 9.83.

9.102 Air flows at steady state through a horizontal, well-insulated, constant-area duct of diameter 0.25 m. At the inlet, p_1 = 2.4 bar, T_1 = 430 K. The temperature of the air leaving the duct is 370 K. The mass flow rate is 600 kg/min. Determine the magnitude, in N, of the net horizontal force exerted by the duct wall on the air. In which direction does the force act?

9.103 Liquid water at 21°C flows at steady state through a 5 cm-diameter horizontal pipe. The mass flow rate is 12 kg/s. The pressure decreases by 14 kPa from inlet to exit of the pipe. Determine the magnitude, in N, and direction of the horizontal force required to hold the pipe in place.

9.104 Air enters a horizontal, well-insulated nozzle operating at steady state at 12 bar, 500K, with a velocity of 50 m/s and exits at 7 bar, 440 K. The mass flow rate is 1 kg/s. Determine the net force, in N, exerted by the air on the duct in the direction of flow.

9.105 Using the ideal gas model, determine the sonic velocity of

(a) air at 16°C.
(b) oxygen (O_2) at 500 K.
(c) argon at 300 K.

9.106 A flash of lightning is sighted and 3 seconds later thunder is heard. Approximately how far away was the lightning strike?

9.107 Using data from Table A-4, estimate the sonic velocity, in m/s, of steam of 60 bar, 360°C. Compare the result with the value predicted by the ideal gas model.

9.108 Plot the Mach number of carbon dioxide at 1 bar, 460 m/s, as a function of temperature in the range 250 to 1000 K.

9.109 An ideal gas flows through a duct. At a particular location, the temperature, pressure, and velocity are known. Determine the Mach number, stagnation temperature, in °K, and the stagnation pressure, in kPa for

(a) air at 154°C, 690 kPa, and a velocity of 420 m/s.
(b) helium at 290 K, 140 kPa, and a velocity of 270 m/s.
(c) nitrogen at 333 K, 345 kPa, and a velocity of 150 m/s.

9.110 For Problem 9.104, determine the values of the Mach number, the stagnation temperature, in K, and the stagnation pressure, in bar, at the inlet and exit of the duct, respectively.

9.111 Using the Mollier diagram, Fig. A-8, determine for water vapor at 3.5 MPa, 315°C, and 300 m/s.

(a) the stagnation enthalpy, in kJ/kg.
(b) the stagnation temperature, in °C.
(c) the stagnation pressure, in kPa.

9.112 Steam flows through a passageway, and at a particular location the pressure is 3 bar, the temperature is 280°C, and the velocity is 690 m/s. Determine the corresponding specific stagnation enthalpy, in kJ/kg, and stagnation temperature, in °C, if the stagnation pressure is 7 bar.

9.113 For the isentropic flow of an ideal gas with constant specific heat ratio k, the ratio of the temperature T^* to the stagnation temperature T_o is $T^*/T_o = 2/(k+1)$. Develop this relationship.

9.114 A gas expands isentropically through a converging nozzle from a large tank at 8 bar, 500 K. Assuming ideal gas behavior, determine the critical pressure p^*, in bar, and the corresponding temperature, in K, if the gas is

(a) air.
(b) carbon dioxide (CO_2).
(c) water vapor.

9.115 Carbon dioxide is contained in a large tank, initially at 690 kPa, 444 K. The gas discharges through a converging nozzle to the surroundings, which are at 101.3 kPa and the pressure in the tank drops. Estimate the pressure in the tank, in kPa, when the flow first ceases to be choked.

9.116 Steam expands isentropically through a converging nozzle operating at steady state from a large tank at 1.83 bar, 280°C. The mass flow rate is 2 kg/s, the flow is choked, and the exit plane pressure is 1 bar. Determine the diameter of the nozzle, in cm, at locations where the pressure is 1.5 bar, and 1 bar, respectively.

9.117 An ideal gas mixture with k = 1.31 and a molecular weight of 23 is supplied to a converging nozzle at p_o = 5 bar, T_o = 700 K, which discharges into a region where the pressure is 1 bar. The exit area is 30 cm^2. For steady isentropic flow through the nozzle, determine

(a) the exit temperature of the gas, in K.
(b) the exit velocity of the gas, in m/s.
(c) the mass flow rate, in kg/s.

9.118 An ideal gas expands isentropically through a converging nozzle from a large tank at 0.8 MPa, 333 K, and discharges into a region at 0.4 MPa. Determine the mass flow rate, in kg/s, for an exit flow area of 6.25 cm^2, if the gas is

(a) air, with k = 1.4.
(b) carbon dioxide, with k = 1.26.
(c) argon, with k = 1.667.

9.119 Air at p_o = 1.4 bar, T_o = 280 K expands isentropically through a converging nozzle and discharges to the atmosphere at 1 bar. The exit plane area is 0.0013 m^2.

(a) Determine the mass flow rate, in kg/s.
(b) If the supply region pressure, p_o, were increased to 2 bar, what would be the mass flow rate, in kg/s?

9.120 Air enters a nozzle operating at steady state at 310 kPa, 445 K, with a velocity of 146 m/s, and expands isentropically to an exit velocity of 457 m/s. Determine

(a) the exit pressure, in kPa.
(b) the ratio of the exit area to the inlet area.
(c) whether the nozzle is diverging only, converging only, or converging–diverging in cross section.

9.121 Air as an ideal gas with $k = 1.4$ enters a converging–diverging nozzle operating at steady state and expands isentropically as shown in Fig. P9.121. Using data from the figure and from Table 9.2 as needed, determine

(a) the stagnation pressure, in kPa, and the stagnation temperature, in K.
(b) the throat area, in mm^2.
(c) the exit area, in mm^2.

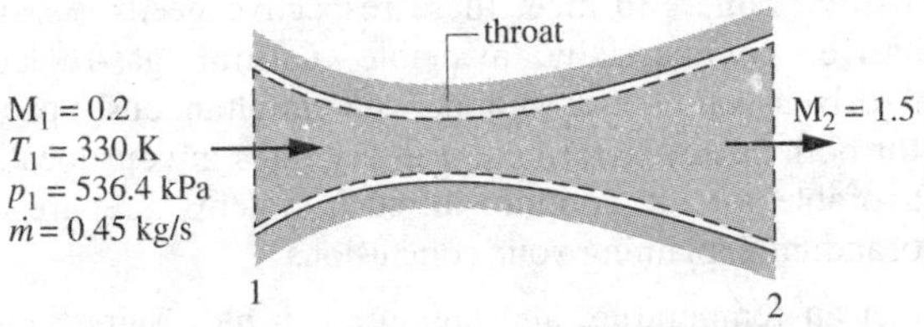

Fig. P9.121

9.122 Air as an ideal gas with $k = 1.4$ enters a diffuser operating at steady state at 4 bar, 290 K, with a velocity of 512 m/s. Assuming isentropic flow, plot the velocity, in m/s, the Mach number, and the area ratio A/A* for locations in the flow corresponding to pressures ranging from 4 to 14 bar.

9.123 A converging–diverging nozzle operating at steady state has a throat area of 3 cm^2 and an exit area of 6 cm^2. Air as an ideal gas with $k = 1.4$ enters the nozzle at 8 bar, 400 K, and a Mach number of 0.2, and flows isentropically throughout. If the nozzle is choked, and the diverging portion acts as a supersonic nozzle, determine the mass flow rate, in kg/s, and the Mach number, pressure, in bar, and temperature, in K, at the exit. Repeat if the diverging portion acts as a supersonic diffuser.

9.124 For the nozzle in Problem 9.123, determine the back pressure, in bar, for which a normal shock would stand at the exit plane.

9.125 For the nozzle in Problem 9.123, a normal shock stands in the diverging section at a location where the pressure is 2 bar. The flow is isentropic, except where the shock stands. Determine the back pressure, in bar.

9.126 Air as an ideal gas with $k = 1.4$ enters a converging–diverging duct with a Mach number of 2. At the inlet, the pressure is 180 kPa and the temperature is 250 K. A normal shock stands at a location in the converging section of the duct, with $M_x = 1.5$. At the exit of the duct, the pressure is 1030 kPa. The flow is isentropic everywhere except in the immediate vicinity of the shock. Determine temperature, in K, and the Mach number at the exit.

9.127 Air as an ideal gas with $k = 1.4$ undergoes a normal shock. The upstream conditions are $p_x = 0.5$ bar, $T_x = 280$ K, and $M_x = 1.8$. Determine

(a) the pressure p_y, in bar.
(b) the stagnation pressure p_{ox}, in bar.
(c) the stagnation temperature T_{ox}, in K.
(d) the change in specific entropy across the shock, in kJ/kg · K.
(e) Plot the quantities of parts (a)–(d) versus M_x ranging from 1.0 to 2.0. All other upstream conditions remain the same.

9.128 A converging–diverging nozzle operates at steady state with a mass flow rate of 0.3 kg/s. Air as an ideal gas with $k = 1.4$ flows through the nozzle, discharging to the atmosphere at 100 kPa and 300 K. A normal shock stands at the exit plane with $M_x = 2$. Up to the shock, the flow is isentropic. Determine

(a) the stagnation pressure p_{ox}, in kPa.
(b) the stagnation temperature T_{ox}, in K.
(c) the nozzle exit area, in mm^2.

9.129 For the nozzle in Problem 9.128, calculate the throat area, in mm^2, and the entropy produced, in kJ/K · kg of air flowing.

9.130 Air at 3.4 bar, 530 K, and a Mach number of 0.4 enters a converging–diverging nozzle operating at steady state. A normal shock stands in the diverging section at a location where the Mach number is $M_x = 1.8$. The flow is isentropic, except where the shock stands. If the air behaves as an ideal gas with $k = 1.4$, determine

(a) the stagnation temperature T_{ox}, in K.
(b) the stagnation pressure p_{ox}, in bar.
(c) the pressure p_x, in bar.
(d) the pressure p_y in bar.
(e) the stagnation pressure p_{oy}, in bar.
(f) the stagnation temperature T_{oy}, in K.

If the throat area is 7.6×10^{-4} m^2, and the exit plane pressure is 2.4 bar, determine the mass flow rate, in kg/s, and the exit area, in m^2.

9.131 Air as an ideal gas with $k = 1.4$ enters a converging–diverging channel at a Mach number of 1.6. A normal shock stands at the inlet to the channel. Downstream of the shock the flow is isentropic; the Mach number is unity at the throat; and the air exits at 138 kPa, 390 K, with negligible velocity. If the mass flow rate is 20 kg/s, determine the inlet and throat areas, in m^2.

9.132 Derive the following expressions: (a) Eq. 9.55, (b) Eq. 9.56, (c) Eq. 9.57.

9.133 Using *Interactive Thermodynamics: IT*, generate tables of the same isentropic flow functions as in Table 9.2 for specific heat ratios of 1.2, 1.3, 1.4, and 1.67 and Mach numbers ranging from 0 to 5.

9.134 Using *Interactive Thermodynamics: IT*, generate tables of the same normal shock functions as in Table 9.3 for specific heat ratios of 1.2, 1.3, 1.4, and 1.67 and Mach numbers ranging from 1 to 5.

Design & open ended problems: exploring engineering practice

9.1D With the development of high-strength metal–ceramic materials, internal combustion engines can now be built with no cylinder wall cooling. These *adiabatic engines* operate with cylinder wall temperatures as high as 927°C. What are the important considerations for adiabatic engine design? Are adiabatic engines likely to find widespread application? Discuss.

9.2D The Stirling engine was first patented in 1816 but has not been widely commercialized. Still, efforts continue to develop Stirling engine technology for practical uses such as vehicle propulsion. Prepare a memorandum summarizing the status of Stirling engine technology. Discuss the advantages and disadvantages of Stirling engines and assess the likelihood that they will be more widely used in the future.

9.3D An important factor leading to the increase in performance of gas turbines over the years has been the use of *turbine blade cooling* to allow higher turbine inlet temperatures. With the aid of diagrams, explain the various methods of turbine blade cooling that are commonly used. What are some of the most important research issues in gas turbine blade technology today? Discuss.

9.4D *Steam-injected* gas turbines use hot turbine exhaust gases to produce ste‹ m that is injected directly into the gas turbine system. Figure P9.4D illustrates two possible approaches to steam injection. In one approach, steam is injected directly into the combustor. In the other approach, steam is injected into the low-pressure turbine stages.

(a) What are the relative advantages and disadvantages of these steam injection approaches? How does steam injection lead to better power cycle performance?

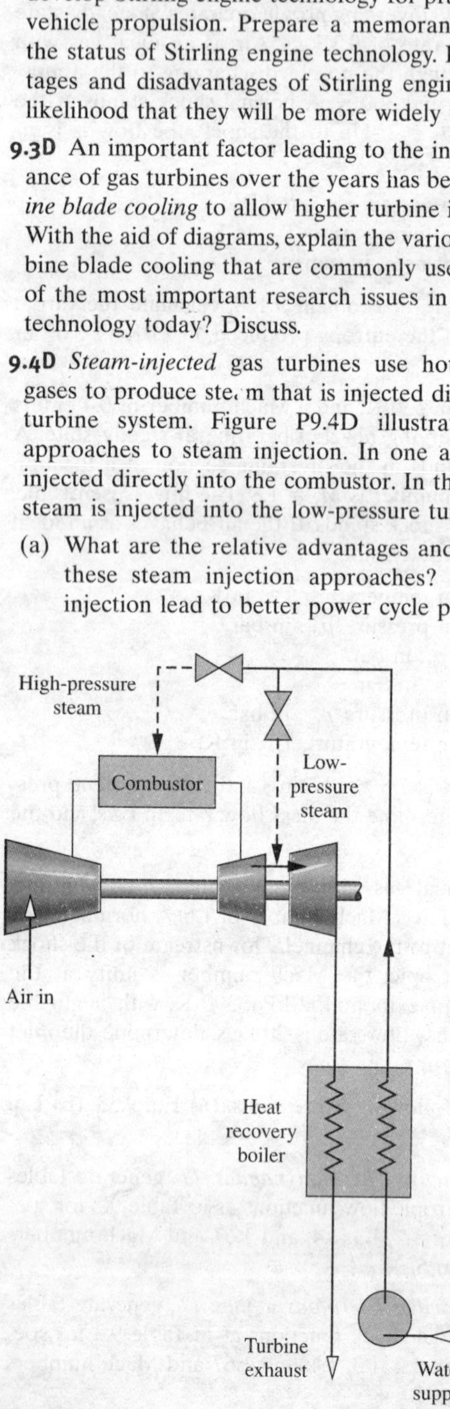

Fig. P9.4D

(b) Each of the steam injection approaches is inherently simpler than combined cycles like the one illustrated in Fig. 9.23. What are the other advantages and disadvantages of direct steam injection compared to combined cycles?

9.5D Closed Brayton power systems, receiving their energy input from radioisotopes, nuclear reactors, or solar collectors, have been suggested for meeting space vehicle power requirements. What are the advantages and disadvantages of closed Brayton cycles for spaceflight applications? What mission-specific design criteria might determine the selection of the system and the energy source?

9.6D Factories requiring compressed air and process heat commonly run electrically driven air compressors and natural gas-fired boilers to meet these respective needs. As an alternative, commercially available natural gas-fueled engine-driven compressor systems can simultaneously provide for both needs. Determine if utility rates in your locale are favorable for the adoption of such systems. Prepare a memorandum explaining your conclusions.

9.7D Inlet air temperature and pressure can have significant impact on gas turbine performance. Investigate the effect on the power developed by a General Electric LM 6000 gas turbine when the inlet air temperature and pressure vary about baseline values of 10°C and 1 bar, respectively. What are some ways that could be used to alter the inlet air temperature and pressure, and what are the economic implications? Prepare a memorandum summarizing your findings.

9.8D A manufacturing company currently purchases 2.4×10^5 MW · h of electricity annually from the local utility company. An aging boiler on the premises annually provides 4×10^8 kg of process steam at 20 bar. Consider the feasibility of acquiring a *cogeneration* system to meet these needs. The system would employ a natural gas–fueled gas turbine to produce the electricity and a heat-recovery steam generator to produce the steam. Using *thermoeconomic* principles, investigate the economic issues that should be considered in making a recommendation about the proposed cogeneration system. Write a report of your findings.

9.9D Develop the preliminary specifications for a 160-MW closed-cycle gas turbine power plant. Consider carbon dioxide and helium as possible gas turbine working fluids that circulate through a nuclear power unit where they absorb energy from the nuclear reaction. Sketch the schematic of your proposed cycle. For the two working fluids, compare the operating pressures and temperatures and estimate the expected performance of the cycle. Write a report that includes your analysis and design, and recommend a working fluid. Include at least three references.

9.10D As an engineer you must recommend whether or not to purchase 2 MW of electric power from the local utility company at a cost of \$0.06 per kW·h for a plant expansion or to purchase and operate a diesel engine generator set that runs on natural gas. Natural gas is priced at \$1.10 per thousand cubic meter, and each cubic meter has a *heating value* of 1055 kJ. Assume that the plant operates 7800 h per year. Specify diesel-generator equipment that would supply the required power and perform an economic analysis to determine which alternative is best. Present your recommendation in a memorandum with supporting analysis and calculations, including at least three references.

9.11D A financial services company has a computer server facility that requires very reliable electric power. The power demand is 3000 kW. The company has hired you as a consultant to study the feasibility of using 250-W *microturbines* for this application. Write a report discussing the pros and cons of such an arrangement compared to purchasing power from the local utility.

9.12D The inlet to a gas turbine aircraft engine must provide air to the compressor with as little a drop in stagnation pressure and with as small a drag force on the aircraft as possible. The inlet also serves as a diffuser to raise the pressure of the air entering the compressor while reducing the air velocity. Write a paper discussing the most common types of inlet designs for subsonic and supersonic applications. Include a discussion of how engine placement affects diffuser performance.

9.13D In what ways must the compressible flow analysis presented in this chapter be modified to model the flow through steam turbines? What special design considerations arise because of the possibility of condensation occurring within the blade rows and nozzles of steam turbines?

10

Refrigeration and Heat Pump Systems

Engineering Context

Refrigeration systems for food preservation and air conditioning play prominent roles in our everyday lives. Heat pumps also are used for heating buildings and for producing industrial process heat. There are many other examples of commercial and industrial uses of refrigeration, including air separation to obtain liquid oxygen and liquid nitrogen, liquefaction of natural gas, and production of ice.

The **objective** of this chapter is to describe some of the common types of refrigeration and heat pump systems presently in use and to illustrate how such systems can be modeled thermodynamically. The three principal types described are the vapor-compression, absorption, and reversed Brayton cycles. As for the power systems studied in Chaps. 8 and 9, both vapor and gas systems are considered. In vapor systems, the refrigerant is alternately vaporized and condensed. In gas refrigeration systems, the refrigerant remains a gas.

Learning Objectives

When you complete your study of this chapter, you will be able to...

- ✔ Demonstrate understanding of basic vapor-compression refrigeration and heat pump systems.
- ✔ Develop and analyze thermodynamic models of vapor-compression systems and their modifications, including
  - ✔ sketching schematic and accompanying *T–s* diagrams.
  - ✔ evaluating property data at principal states in the systems.
  - ✔ applying mass, energy, entropy, and exergy balances for the basic processes.
  - ✔ determining refrigeration and heat pump system performance, coefficient of performance and capacity.
- ✔ Explain the effects on vapor-compression system performance of varying key parameters.
- ✔ Demonstrate understanding of the operating principles of absorption and gas refrigeration systems, and perform thermodynamic analysis of gas systems.

10.1 Vapor Refrigeration Systems

The purpose of a refrigeration system is to maintain a *cold* region at a temperature below the temperature of its surroundings. This is commonly achieved using the vapor refrigeration systems that are the subject of the present section.

Carnot Refrigeration Cycle

To introduce some important aspects of vapor refrigeration, let us begin by considering a Carnot vapor refrigeration cycle. This cycle is obtained by reversing the Carnot vapor power cycle introduced in Sec. 5.10. Figure 10.1 shows the schematic and accompanying *T–s* diagram of a Carnot refrigeration cycle operating between a region at temperature T_C and another region at a higher temperature T_H. The cycle is executed by a refrigerant circulating steadily through a series of components. All processes are internally reversible. Also, since heat transfers between the refrigerant and each region occur with no temperature differences, there are no external irreversibilities. The energy transfers shown on the diagram are positive in the directions indicated by the arrows.

Let us follow the refrigerant as it passes steadily through each of the components in the cycle, beginning at the inlet to the evaporator. The refrigerant enters the evaporator as a two-phase liquid–vapor mixture at state 4. In the evaporator some of the refrigerant changes phase from liquid to vapor as a result of heat transfer from the region at temperature T_C to the refrigerant. The temperature and pressure of the refrigerant remain constant during the process from state 4 to state 1. The refrigerant is then compressed adiabatically from state 1, where it is a two-phase liquid–vapor mixture, to state 2, where it is a saturated vapor. During this process, the temperature of the refrigerant increases from T_C to T_H, and the pressure also increases. The refrigerant passes from the compressor into the condenser, where it changes phase from saturated vapor to saturated liquid as a result of heat transfer to the region at temperature T_H. The temperature and pressure remain constant in the process from state 2 to state 3. The refrigerant returns to the state at the inlet of the evaporator by expanding adiabatically through a turbine. In this process, from state 3 to state 4, the temperature decreases from T_H to T_C, and there is a decrease in pressure.

Since the Carnot vapor refrigeration cycle is made up of internally reversible processes, areas on the *T–s* diagram can be interpreted as heat transfers. Applying Eq. 6.49, area 1–a–b–4–1 is the heat added to the refrigerant from the cold region

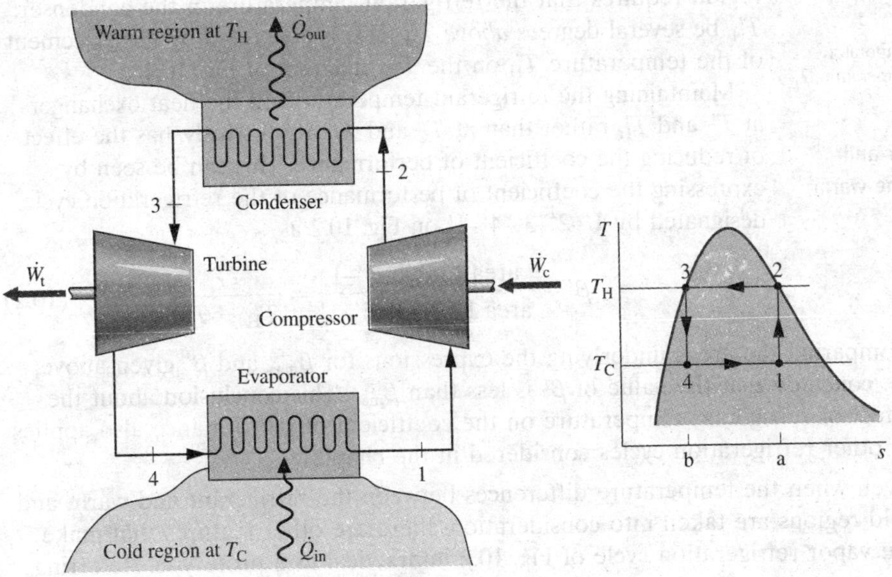

Fig. 10.1 Carnot vapor refrigeration cycle.

per unit mass of refrigerant flowing. Area 2–a–b–3–2 is the heat rejected from the refrigerant to the warm region per unit mass of refrigerant flowing. The enclosed area 1–2–3–4–1 is the *net* heat transfer *from* the refrigerant. The net heat transfer *from* the refrigerant equals the net work done *on* the refrigerant. The net work is the difference between the compressor work input and the turbine work output.

The *coefficient of performance* β of *any* refrigeration cycle is the ratio of the refrigeration effect to the net work input required to achieve that effect. For the Carnot vapor refrigeration cycle shown in Fig. 10.1, the coefficient of performance is

$$\beta_{max} = \frac{\dot{Q}_{in}/\dot{m}}{\dot{W}_c/\dot{m} - \dot{W}_t/\dot{m}} = \frac{\text{area 1–a–b–4–1}}{\text{area 1–2–3–4–1}} = \frac{T_C(s_a - s_b)}{(T_H - T_C)(s_a - s_b)}$$

which reduces to

$$\beta_{max} = \frac{T_C}{T_H - T_C} \qquad (10.1)$$

This equation, which corresponds to Eq. 5.10, represents the *maximum* theoretical coefficient of performance of any refrigeration cycle operating between regions at T_C and T_H.

Departures from the Carnot Cycle

Actual vapor refrigeration systems depart significantly from the Carnot cycle considered above and have coefficients of performance lower than would be calculated from Eq. 10.1. Three ways actual systems depart from the Carnot cycle are considered next.

▶ One of the most significant departures is related to the heat transfers between the refrigerant and the two regions. In actual systems, these heat transfers are not accomplished reversibly as presumed above. In particular, to achieve a rate of heat transfer sufficient to maintain the temperature of the cold region at T_C with a practical-sized evaporator requires the temperature of the refrigerant in the evaporator, T'_C, to be several degrees *below* T_C. This is illustrated by the placement of the temperature T'_C on the *T–s* diagram of Fig. 10.2. Similarly, to obtain a sufficient heat transfer rate from the refrigerant to the warm region requires that the refrigerant temperature in the condenser, T'_H, be several degrees *above* T_H. This is illustrated by the placement of the temperature T'_H on the *T–s* diagram of Fig. 10.2.

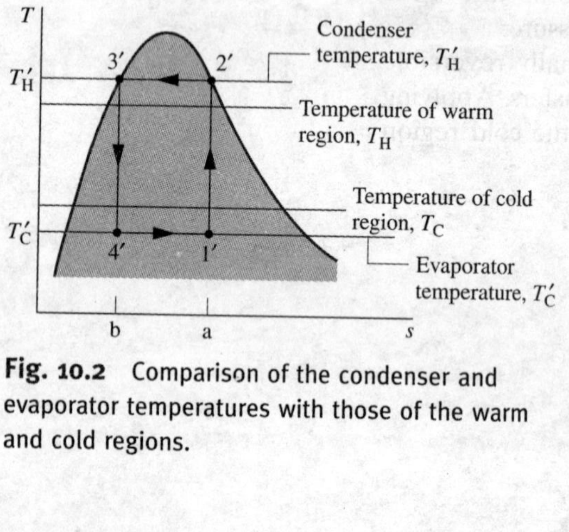

Fig. 10.2 Comparison of the condenser and evaporator temperatures with those of the warm and cold regions.

Maintaining the refrigerant temperatures in the heat exchangers at T'_C and T'_H rather than at T_C and T_H, respectively, has the effect of reducing the coefficient of performance. This can be seen by expressing the coefficient of performance of the refrigeration cycle designated by 1′–2′–3′–4′–1′ on Fig. 10.2 as

$$\beta' = \frac{\text{area 1′–a–b–4′–1}}{\text{area 1′–2′–3′–4′–1′}} = \frac{T'_C}{T'_H - T'_C} \qquad (10.2)$$

Comparing the areas underlying the expressions for β_{max} and β' given above, we conclude that the value of β' is less than β_{max}. This conclusion about the effect of refrigerant temperature on the coefficient of performance also applies to other refrigeration cycles considered in the chapter.

▶ Even when the temperature differences between the refrigerant and warm and cold regions are taken into consideration, there are other features that make the vapor refrigeration cycle of Fig. 10.2 impractical as a prototype. Referring again to the figure, note that the compression process from state 1′ to state 2′ occurs with the refrigerant as a two-phase liquid–vapor mixture. This is commonly referred to as *wet compression*. Wet compression is normally avoided because the presence of liquid droplets in the flowing liquid–vapor mixture can damage the compressor. In actual systems, the compressor handles vapor only. This is known as *dry compression*.

- Another feature that makes the cycle of Fig. 10.2 impractical is the expansion process from the saturated liquid state 3′ to the low-quality, two-phase liquid–vapor mixture state 4′. This expansion produces a relatively small amount of work compared to the work input in the compression process. The work output achieved by an actual turbine would be smaller yet because turbines operating under these conditions typically have low efficiencies. Accordingly, the work output of the turbine is normally sacrificed by substituting a simple throttling valve for the expansion turbine, with consequent savings in initial and maintenance costs. The components of the resulting cycle are illustrated in Fig. 10.3, where dry compression is presumed. This cycle, known as the *vapor-compression refrigeration cycle*, is the subject of the section to follow.

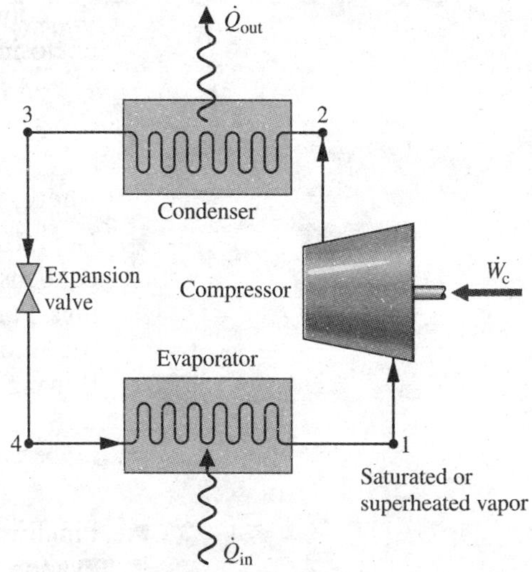

Fig. 10.3 Components of a vapor-compression refrigeration system.

10.2 Analyzing Vapor-Compression Refrigeration Systems

Vapor-compression refrigeration systems are the most common refrigeration systems in use today. The object of this section is to introduce some important features of systems of this type and to illustrate how they are modeled thermodynamically.

vapor-compression refrigeration

10.2.1 Evaluating Principal Work and Heat Transfers

Let us consider the steady-state operation of the vapor-compression system illustrated in Fig. 10.3. Shown on the figure are the principal work and heat transfers, which are positive in the directions of the arrows. Kinetic and potential energy changes are neglected in the following analyses of the components. We begin with the evaporator, where the desired refrigeration effect is achieved.

- As the refrigerant passes through the evaporator, heat transfer from the refrigerated space results in the vaporization of the refrigerant. For a control volume enclosing the refrigerant side of the evaporator, the mass and energy rate balances reduce to give the rate of heat transfer per unit mass of refrigerant flowing as

$$\frac{\dot{Q}_{in}}{\dot{m}} = h_1 - h_4 \tag{10.3}$$

where $\dot{m}$ is the mass flow rate of the refrigerant. The heat transfer rate $\dot{Q}_{in}$ is referred to as the ***refrigeration capacity.*** In the SI unit system, the capacity is normally expressed in kW. Another commonly used unit for the refrigeration capacity is the ***ton of refrigeration,*** which is equal to 211 kJ/min.

refrigeration capacity

ton of refrigeration

- The refrigerant leaving the evaporator is compressed to a relatively high pressure and temperature by the compressor. Assuming no heat transfer to or from the compressor, the mass and energy rate balances for a control volume enclosing the compressor give

$$\frac{\dot{W}_c}{\dot{m}} = h_2 - h_1 \tag{10.4}$$

where $\dot{W}_c/\dot{m}$ is the rate of power *input* per unit mass of refrigerant flowing.

- Next, the refrigerant passes through the condenser, where the refrigerant condenses and there is heat transfer from the refrigerant to the cooler surroundings. For a control volume enclosing the refrigerant side of the condenser, the rate of heat transfer from the refrigerant per unit mass of refrigerant flowing is

$$\frac{\dot{Q}_{out}}{\dot{m}} = h_2 - h_3 \tag{10.5}$$

- Finally, the refrigerant at state 3 enters the expansion valve and expands to the evaporator pressure. This process is usually modeled as a *throttling* process for which

$$h_4 = h_3 \tag{10.6}$$

The refrigerant pressure decreases in the irreversible adiabatic expansion, and there is an accompanying increase in specific entropy. The refrigerant exits the valve at state 4 as a two-phase liquid–vapor mixture.

In the vapor-compression system, the net power input is equal to the compressor power, since the expansion valve involves no power input or output. Using the quantities and expressions introduced above, the coefficient of performance of the vapor-compression refrigeration system of Fig. 10.3 is

$$\beta = \frac{\dot{Q}_{in}/\dot{m}}{\dot{W}_c/\dot{m}} = \frac{h_1 - h_4}{h_2 - h_1} \tag{10.7}$$

Provided states 1 through 4 are fixed, Eqs. 10.3 through 10.7 can be used to evaluate the principal work and heat transfers and the coefficient of performance of the vapor-compression system shown in Fig. 10.3. Since these equations have been developed by reducing mass and energy rate balances, they apply equally for actual performance when irreversibilities are present in the evaporator, compressor, and condenser and for idealized performance in the absence of such effects. Although irreversibilities in the evaporator, compressor, and condenser can have a pronounced effect on overall performance, it is instructive to consider an idealized cycle in which they are assumed absent. Such a cycle establishes an upper limit on the performance of the vapor-compression refrigeration cycle. It is considered next.

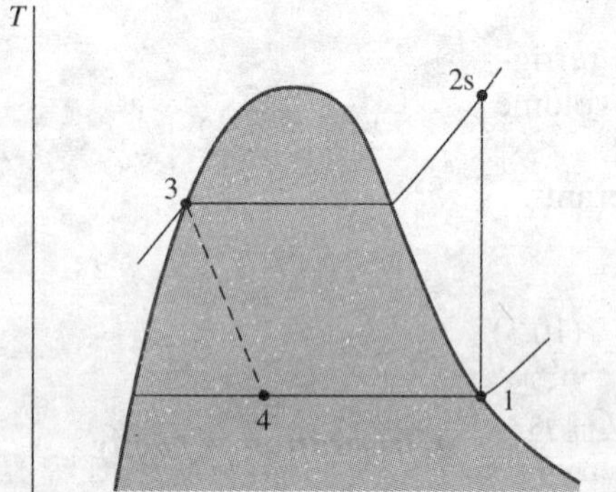

Fig. 10.4 *T–s* diagram of an ideal vapor-compression cycle.

10.2.2 Performance of Vapor-Compression Systems

IDEAL CYCLE. If irreversibilities within the evaporator and condenser are ignored, there are no frictional pressure drops, and the refrigerant flows at constant pressure through the two heat exchangers. If compression occurs without irreversibilities, and stray heat transfer to the surroundings is also ignored, the compression process is isentropic. With these considerations, the vapor-compression refrigeration cycle labeled 1–2s–3–4–1 on the *T–s* diagram of Fig. 10.4 results. The cycle consists of the following series of processes:

Process 1–2s: *Isentropic* compression of the refrigerant from state 1 to the condenser pressure at state 2s.

Process 2s–3: Heat transfer *from* the refrigerant as it flows at constant pressure through the condenser. The refrigerant exits as a liquid at state 3.

Process 3–4: *Throttling* process from state 3 to a two-phase liquid–vapor mixture at 4.

Process 4–1: Heat transfer *to* the refrigerant as it flows at constant pressure through the evaporator to complete the cycle.

All of the processes in the above cycle are internally reversible except for the throttling process. Despite the inclusion of this irreversible process, the cycle is commonly referred to as the ***ideal vapor-compression cycle.***

ideal vapor-compression cycle

The following example illustrates the application of the first and second laws of thermodynamics along with property data to analyze an ideal vapor-compression cycle.

Example 10.1 IDEAL VAPOR-COMPRESSION REFRIGERATION CYCLE

Refrigerant 134a is the working fluid in an ideal vapor-compression refrigeration cycle that communicates thermally with a cold region at 0°C and a warm region at 26°C. Saturated vapor enters the compressor at 0°C and saturated liquid leaves the condenser at 26°C. The mass flow rate of the refrigerant is 0.08 kg/s. Determine **(a)** the compressor power, in kW, **(b)** the refrigeration capacity, in tons, **(c)** the coefficient of performance, and **(d)** the coefficient of performance of a Carnot refrigeration cycle operating between warm and cold regions at 26 and 0°C, respectively.

Solution

Known: An ideal vapor-compression refrigeration cycle operates with Refrigerant 134a. The states of the refrigerant entering the compressor and leaving the condenser are specified, and the mass flow rate is given.

Find: Determine the compressor power, in kW, the refrigeration capacity, in tons, coefficient of performance, and the coefficient of performance of a Carnot vapor refrigeration cycle operating between warm and cold regions at the specified temperatures.

Schematic and Given Data:

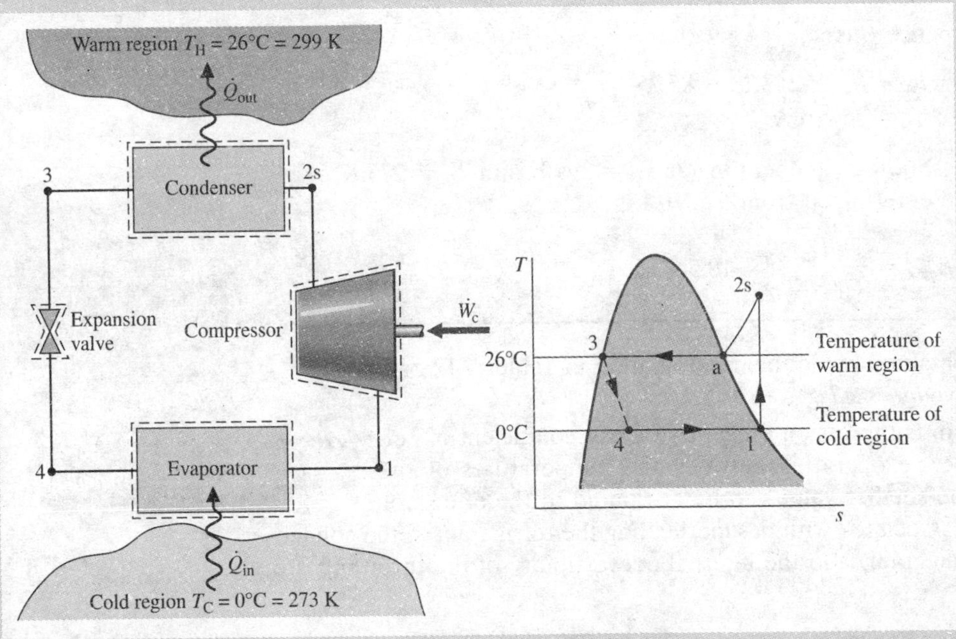

Fig. E10.1

Engineering Model:

1. Each component of the cycle is analyzed as a control volume at steady state. The control volumes are indicated by dashed lines on the accompanying sketch.

2. Except for the expansion through the valve, which is a throttling process, all processes of the refrigerant are internally reversible.
3. The compressor and expansion valve operate adiabatically.
4. Kinetic and potential energy effects are negligible.
5. Saturated vapor enters the compressor, and saturated liquid leaves the condenser.

Analysis: Let us begin by fixing each of the principal states located on the accompanying schematic and *T–s* diagrams. At the inlet to the compressor, the refrigerant is a saturated vapor at 0°C, so from Table A-10, $h_1 = 247.23$ kJ/kg and $s_1 = 0.9190$ kJ/kg · K.

The pressure at state 2s is the saturation pressure corresponding to 26°C, or $p_2 = 6.853$ bar. State 2s is fixed by p_2 and the fact that the specific entropy is constant for the adiabatic, internally reversible compression process. The ❶ refrigerant at state 2s is a superheated vapor with $h_{2s} = 264.7$ kJ/kg.

State 3 is saturated liquid at 26°C, so $h_3 = 85.75$ kJ/kg. The expansion through the valve is a throttling process (assumption 2), so $h_4 = h_3$.

(a) The compressor work input is

$$\dot{W}_c = \dot{m}(h_{2s} - h_1)$$

where $\dot{m}$ is the mass flow rate of refrigerant. Inserting values

$$\dot{W}_c = (0.08 \text{ kg/s})(264.7 - 247.23) \text{ kJ/kg} \left| \frac{1 \text{ kW}}{1 \text{ kJ/s}} \right|$$

$$= 1.4 \text{ kW}$$

(b) The refrigeration capacity is the heat transfer rate to the refrigerant passing through the evaporator. This is given by

$$\dot{Q}_{in} = \dot{m}(h_1 - h_4)$$

$$= (0.08 \text{ kg/s})|60 \text{ s/min}|(247.23 - 85.75) \text{ kJ/kg} \left| \frac{1 \text{ ton}}{211 \text{ kJ/min}} \right|$$

$$= 3.67 \text{ ton}$$

(c) The coefficient of performance β is

$$\beta = \frac{\dot{Q}_{in}}{\dot{W}_c} = \frac{h_1 - h_4}{h_{2s} - h_1} = \frac{247.23 - 85.75}{264.7 - 247.23} = 9.24$$

(d) For a Carnot vapor refrigeration cycle operating at $T_H = 299$ K and $T_C = 273$ K, the coefficient of performance determined from Eq. 10.1 is

❷
$$\beta_{max} = \frac{T_C}{T_H - T_C} = 10.5$$

❶ The value for h_{2s} can be obtained by double interpolation in Table A-12 or by using *Interactive Thermodynamics: IT*.

❷ As expected, the ideal vapor-compression cycle has a lower coefficient of performance than a Carnot cycle operating between the temperatures of the warm and cold regions. The smaller value can be attributed to the effects of the external irreversibility associated with desuperheating the refrigerant in the condenser (Process 2s–a on the *T–s* diagram) and the internal irreversibility of the throttling process.

Skills Developed: *Ability to...*
- ✓ sketch the *T–s* diagram of the ideal vapor compression refrigeration cycle.
- ✓ fix each of the principal states and retrieve necessary property data.
- ✓ calculate refrigeration capacity and coefficient of performance.
- ✓ compare with the corresponding Carnot refrigeration cycle.

Quick Quiz

Keeping all other given data the same, determine the mass flow rate of refrigerant, in kg/s, for a 10-ton refrigeration capacity.

Ans. 0.218 kg/s.

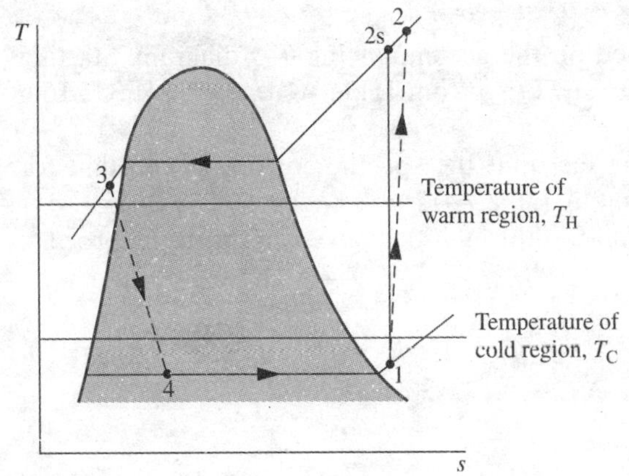

Fig. 10.5 *T–s* diagram of an actual vapor-compression cycle.

ACTUAL CYCLE. Figure 10.5 illustrates several features exhibited by *actual* vapor-compression systems. As shown in the figure, the heat transfers between the refrigerant and the warm and cold regions are not accomplished reversibly: the refrigerant temperature in the evaporator is less than the cold region temperature, T_C, and the refrigerant temperature in the condenser is greater than the warm region temperature, T_H. Such irreversible heat transfers have a significant effect on performance. In particular, the coefficient of performance decreases as the average temperature of the refrigerant in the evaporator decreases and as the average temperature of the refrigerant in the condenser increases. Example 10.2 provides an illustration.

Example 10.2 EFFECT OF IRREVERSIBLE HEAT TRANSFER ON PERFORMANCE

Modify Example 10.1 to allow for temperature differences between the refrigerant and the warm and cold regions as follows. Saturated vapor enters the compressor at −10°C. Saturated liquid leaves the condenser at a pressure of 9 bar. Determine for the modified vapor-compression refrigeration cycle **(a)** the compressor power, in kW, **(b)** the refrigeration capacity, in tons, **(c)** the coefficient of performance. Compare results with those of Example 10.1.

Solution

Known: An ideal vapor-compression refrigeration cycle operates with Refrigerant 134a as the working fluid. The evaporator temperature and condenser pressure are specified, and the mass flow rate is given.

Find: Determine the compressor power, in kW, the refrigeration capacity, in tons, and the coefficient of performance. Compare results with those of Example 10.1.

Schematic and Given Data:

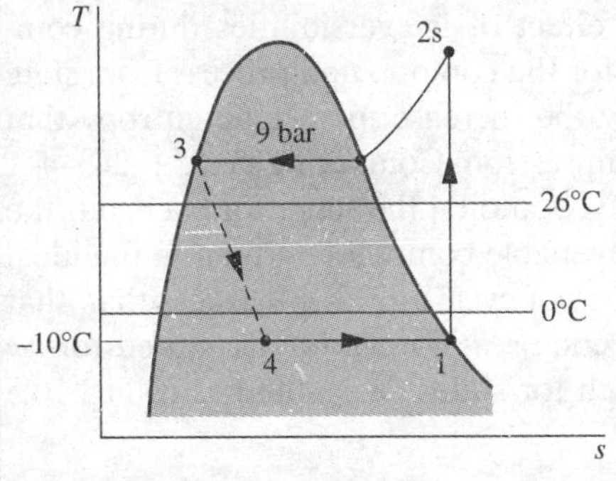

Fig. E10.2

Engineering Model:

1. Each component of the cycle is analyzed as a control volume at steady state. The control volumes are indicated by dashed lines on the sketch accompanying Example 10.1.
2. Except for the process through the expansion valve, which is a throttling process, all processes of the refrigerant are internally reversible.
3. The compressor and expansion valve operate adiabatically.
4. Kinetic and potential energy effects are negligible.
5. Saturated vapor enters the compressor, and saturated liquid exits the condenser.

Analysis: Let us begin by fixing each of the principal states located on the accompanying *T–s* diagram. Starting at the inlet to the compressor, the refrigerant is a saturated vapor at $-10°C$, so from Table A-10, $h_1 = 241.35$ kJ/kg and $s_1 = 0.9253$ kJ/kg · K.

The superheated vapor at state 2s is fixed by $p_2 = 9$ bar and the fact that the specific entropy is constant for the adiabatic, internally reversible compression process. Interpolating in Table A-12 gives $h_{2s} = 272.39$ kJ/kg.

State 3 is a saturated liquid at 9 bar, so $h_3 = 99.56$ kJ/kg. The expansion through the valve is a throttling process; thus, $h_4 = h_3$.

(a) The compressor power input is

$$\dot{W}_c = \dot{m}(h_{2s} - h_1)$$

where $\dot{m}$ is the mass flow rate of refrigerant. Inserting values

$$\dot{W}_c = (0.08 \text{ kg/s})(272.39 - 241.35) \text{ kJ/kg}\left|\frac{1 \text{ kW}}{1 \text{ kJ/s}}\right|$$
$$= 2.48 \text{ kW}$$

(b) The refrigeration capacity is

$$\dot{Q}_{in} = \dot{m}(h_1 - h_4)$$
$$= (0.08 \text{ kg/s})|60 \text{ s/min}|(241.35 - 99.56) \text{ kJ/kg}\left|\frac{1 \text{ ton}}{211 \text{ kJ/min}}\right|$$
$$= 3.23 \text{ ton}$$

(c) The coefficient or performance β is

$$\beta = \frac{\dot{Q}_{in}}{\dot{W}_c} = \frac{h_1 - h_4}{h_{2s} - h_1} = \frac{241.35 - 99.56}{272.39 - 241.35} = 4.57$$

Comparing the results of the present example with those of Example 10.1, we see that the power input required by the compressor is greater in the present case. Furthermore, the refrigeration capacity and coefficient of performance are smaller in this example than in Example 10.1. This illustrates the considerable influence on performance of irreversible heat transfer between the refrigerant and the cold and warm regions.

Skills Developed: *Ability to...*
- ✓ sketch the *T–s* diagram of the ideal vapor compression refrigeration cycle.
- ✓ fix each of the principal states and retrieve necessary property data.
- ✓ calculate refrigeration capacity and coefficient of performance.
- ✓ compare with the corresponding Carnot refrigeration cycle.

Quick Quiz

Determine the rate of heat transfer from the refrigerant passing through the condenser to the surroundings, in kW. **Ans.** 13.83 kW.

Referring again to Fig. 10.5, we can identify another key feature of actual vapor-compression system performance. This is the effect of irreversibilities during compression, suggested by the use of a dashed line for the compression process from state 1 to state 2. The dashed line is drawn to show the increase in specific entropy that would accompany an *adiabatic* irreversible compression. Comparing cycle 1–2–3–4–1 with cycle 1–2s–3–4–1, the refrigeration capacity would be the same for each, but the work input would be greater in the case of irreversible compression than in the ideal cycle. Accordingly, the coefficient of performance of cycle 1–2–3–4–1 is less than that of cycle 1–2s–3–4–1. The effect of irreversible compression can be accounted for by using the isentropic compressor efficiency, which for states designated as in Fig. 10.5 is given by

$$\eta_c = \frac{(\dot{W}_c/\dot{m})_s}{(\dot{W}_c/\dot{m})} = \frac{h_{2s} - h_1}{h_2 - h_1}$$

Additional departures from ideality stem from frictional effects that result in pressure drops as the refrigerant flows through the evaporator, condenser, and piping connecting the various components. These pressure drops are not shown on the T–s diagram of Fig. 10.5 and are ignored in subsequent discussions for simplicity.

Finally, two additional features exhibited by actual vapor-compression systems are shown in Fig. 10.5. One is the superheated vapor condition at the evaporator exit (state 1), which differs from the saturated vapor condition shown in Fig. 10.4. Another is the subcooling of the condenser exit state (state 3), which differs from the saturated liquid condition shown in Fig. 10.4.

Example 10.3 illustrates the effects of irreversible compression and condenser exit subcooling on the performance of the vapor-compression refrigeration system.

Example 10.3 ACTUAL VAPOR-COMPRESSION REFRIGERATION CYCLE

Reconsider the vapor-compression refrigeration cycle of Example 10.2, but include in the analysis that the compressor has an efficiency of 80%. Also, let the temperature of the liquid leaving the condenser be 30°C. Determine for the modified cycle **(a)** the compressor power, in kW, **(b)** the refrigeration capacity, in tons, **(c)** the coefficient of performance, and **(d)** the rates of exergy destruction within the compressor and expansion valve, in kW, for $T_0 = 299$ K (26°C).

Solution

Known: A vapor-compression refrigeration cycle has a compressor efficiency of 80%.

Find: Determine the compressor power, in kW, the refrigeration capacity, in tons, the coefficient of performance, and the rates of exergy destruction within the compressor and expansion valve, in kW.

Schematic and Given Data:

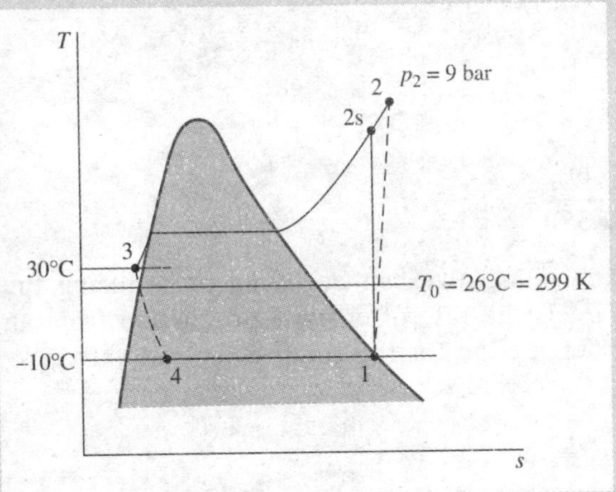

Fig. E10.3

Engineering Model:

1. Each component of the cycle is analyzed as a control volume at steady state.
2. There are no pressure drops through the evaporator and condenser.
3. The compressor operates adiabatically with an efficiency of 80%. The expansion through the valve is a throttling process.
4. Kinetic and potential energy effects are negligible.
5. Saturated vapor at −10°C enters the compressor, and liquid at 30°C leaves the condenser.
6. The environment temperature for calculating exergy is $T_0 = 299$ K (26°C).

Analysis: Let us begin by fixing the principal states. State 1 is the same as in Example 10.2, so $h_1 = 241.35$ kJ/kg and $s_1 = 0.9253$ kJ/kg · K.

Owing to the presence of irreversibilities during the adiabatic compression process, there is an increase in specific entropy from compressor inlet to exit. The state at the compressor exit, state 2, can be fixed using the compressor efficiency

$$\eta_c = \frac{(\dot{W}_c/\dot{m})_s}{\dot{W}_c/\dot{m}} = \frac{(h_{2s} - h_1)}{(h_2 - h_1)}$$

where h_{2s} is the specific enthalpy at state 2s, as indicated on the accompanying T–s diagram. From the solution to Example 10.2, $h_{2s} = 272.39$ kJ/kg. Solving for h_2 and inserting known values

$$h_2 = \frac{h_{2s} - h_1}{\eta_c} + h_1 = \frac{(272.39 - 241.35)}{(0.80)} + 241.35 = 280.15 \text{ kJ/kg}$$

State 2 is fixed by the value of specific enthalpy h_2 and the pressure, $p_2 = 9$ bar. Interpolating in Table A-12, the specific entropy is $s_2 = 0.9497$ kJ/kg · K.

The state at the condenser exit, state 3, is in the liquid region. The specific enthalpy is approximated using Eq. 3.14, together with saturated liquid data at 30°C, as follows: $h_3 \approx h_f = 91.49$ kJ/kg. Similarly, with Eq. 6.5, $s_3 \approx s_f = 0.3396$ kJ/kg · K.

The expansion through the valve is a throttling process; thus, $h_4 = h_3$. The quality and specific entropy at state 4 are, respectively

$$x_4 = \frac{h_4 - h_{f4}}{h_{g4} - h_{f4}} = \frac{91.49 - 36.97}{204.39} = 0.2667$$

and

$$\begin{aligned} s_4 &= s_{f4} + x_4(s_{g4} - s_{f4}) \\ &= 0.1486 + (0.2667)(0.9253 - 0.1486) = 0.3557 \text{ kJ/kg} \cdot \text{K} \end{aligned}$$

(a) The compressor power is

$$\begin{aligned} \dot{W}_c &= \dot{m}(h_2 - h_1) \\ &= (0.08 \text{ kg/s})(280.15 - 241.35) \text{ kJ/kg} \left| \frac{1 \text{ kW}}{1 \text{ kJ/s}} \right| = 3.1 \text{ kW} \end{aligned}$$

(b) The refrigeration capacity is

$$\begin{aligned} \dot{Q}_{in} &= \dot{m}(h_1 - h_4) \\ &= (0.08 \text{ kg/s})|60 \text{ s/min}|(241.35 - 91.49) \text{ kJ/kg} \left| \frac{1 \text{ ton}}{211 \text{ kJ/min}} \right| \\ &= 3.41 \text{ ton} \end{aligned}$$

(c) The coefficient of performance is

1 $$\beta = \frac{(h_1 - h_4)}{(h_2 - h_1)} = \frac{(241.35 - 91.49)}{(280.15 - 241.35)} = 3.86$$

(d) The rates of exergy destruction in the compressor and expansion valve can be found by reducing the exergy rate balance or using the relationship $\dot{E}_d = T_0\dot{\sigma}_{cv}$, where $\dot{\sigma}_{cv}$ is the rate of entropy production from an entropy rate balance. With either approach, the rates or exergy destruction for the compressor and valve are, respectively

$$(\dot{E}_d)_c = \dot{m}T_0(s_2 - s_1) \quad \text{and} \quad (\dot{E}_d)_{valve} = \dot{m}T_0(s_4 - s_3)$$

Substituting values

2 $$(\dot{E}_d)_c = \left(0.08\frac{\text{kg}}{\text{s}}\right)(299 \text{ K})(0.9497 - 0.9253)\frac{\text{kJ}}{\text{kg} \cdot \text{K}}\left|\frac{1 \text{ kW}}{1 \text{ kJ/s}}\right| = 0.58 \text{ kW}$$

and

$$(\dot{E}_d)_{valve} = (0.08)(299)(0.3557 - 0.3396) = 0.39 \text{ kW}$$

1 Irreversibilities in the compressor result in an increased compressor power requirement compared to the isentropic compression of Example 10.2. As a consequence, the coefficient of performance is lower.

Skills Developed: *Ability to...*

- ✓ sketch the T–s diagram of the vapor compression refrigeration cycle with irreversibilities in the compressor and subcooled liquid exiting the condenser.
- ✓ fix each of the principal states and retrieve necessary property data.
- ✓ calculate refrigeration capacity and coefficient of performance.

❷ The exergy destruction rates calculated in part (d) measure the effect of irreversibilities as the refrigerant flows through the compressor and valve. The percentages of the power input (exergy input) to the compressor destroyed in the compressor and valve are 18.7 and 12.6%, respectively.

Quick Quiz

What would be the coefficient of performance if the isentropic compressor efficiency were 100%?

Ans. 4.83.

10.3 Refrigerant Properties

From about 1940 to the early 1990s, the most common class of refrigerants used in vapor-compression refrigeration systems was the chlorine-containing CFCs (chlorofluorocarbons). Refrigerant 12 (CCl_2F_2) is one of these. Owing to concern about the effects of chlorine in refrigerants on the earth's protective ozone layer, international agreements have been implemented to phase out the use of CFCs. Classes of refrigerants containing various amounts of hydrogen in place of chlorine atoms have been developed that have less potential to deplete atmospheric ozone than do more fully chlorinated ones, such as Refrigerant 12. One such class, the HFCs, contain no chlorine. Refrigerant 134a (CF_3CH_2F) is the HFC considered by many to be an environmentally acceptable substitute for Refrigerant 12, and Refrigerant 134a has replaced Refrigerant 12 in many applications.

Refrigerant 22 ($CHClF_2$) is in the class called HCFCs that contains some hydrogen in place of the chlorine atoms. Although Refrigerant 22 and other HCFCs are widely used today, discussions are under way that will likely result in phasing out their use at some time in the future. Ammonia (NH_3), which was widely used in the early development of vapor-compression refrigeration, is again receiving some interest as an alternative to the CFCs because it contains no chlorine. Ammonia is also important in the absorption refrigeration systems discussed in Section 10.5. Hydrocarbons, such as propane (C_3H_8) and methane (CH_4), and gases, such as carbon dioxide (CO_2), are also under investigation for use as refrigerants.

Thermodynamic property data for ammonia, propane, and Refrigerants 22 and 134a are included in the appendix tables. These data allow us to study refrigeration and heat pump systems in common use and to investigate some of the effects on refrigeration cycles of using alternative working fluids.

A thermodynamic property diagram widely used in the refrigeration field is the pressure–enthalpy or ***p–h* diagram.** Figure 10.6 shows the main features of such a property diagram. The principal states of the vapor-compression cycles of Fig. 10.5 are located on this *p–h* diagram. It is left as an exercise to sketch the cycles of Examples 10.1, 10.2, and 10.3 on *p–h* diagrams. Property tables and *p–h* diagrams for many refrigerants are given in handbooks dealing with refrigeration.

p–h diagram

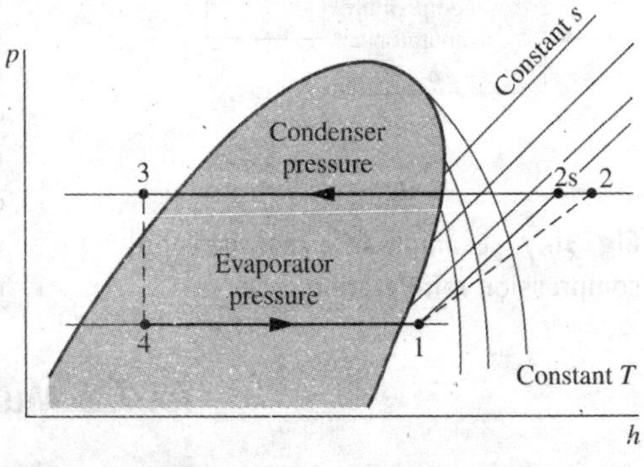

Fig. 10.6 Principal features of the pressure–enthalpy diagram for a typical refrigerant, with vapor-compression cycles superimposed.

SELECTING REFRIGERANTS. The temperatures of the refrigerant in the evaporator and condenser are governed by the temperatures of the cold and warm regions, respectively, with which the system interacts thermally. This, in turn, determines the operating pressures in the evaporator and condenser. Consequently, the selection of a refrigerant is based partly on the suitability of its pressure–temperature relationship in the range of the particular application. It is generally desirable to avoid excessively low pressures in the evaporator and excessively high pressures in the condenser. Other considerations in refrigerant selection include chemical stability, toxicity, corrosiveness, and cost.

The type of compressor also affects the choice of refrigerant. Centrifugal compressors are best suited for low evaporator pressures and refrigerants with large specific volumes at low pressure. Reciprocating compressors perform better over large pressure ranges and are better able to handle low specific volume refrigerants.

10.4 Cascade and Multistage Vapor-Compression Systems

Variations of the basic vapor-compression refrigeration cycle are used to improve performance or for special applications. Two variations are presented in this section. The first is a *combined cycle* arrangement in which refrigeration at relatively low temperature is achieved through a series of vapor-compression systems, with each normally employing a different refrigerant. In the second variation, the work of compression is reduced through *multistage compression with intercooling* between the stages. These variations are analogous to power cycle modifications considered in Chaps. 8 and 9.

10.4.1 Cascade Cycles

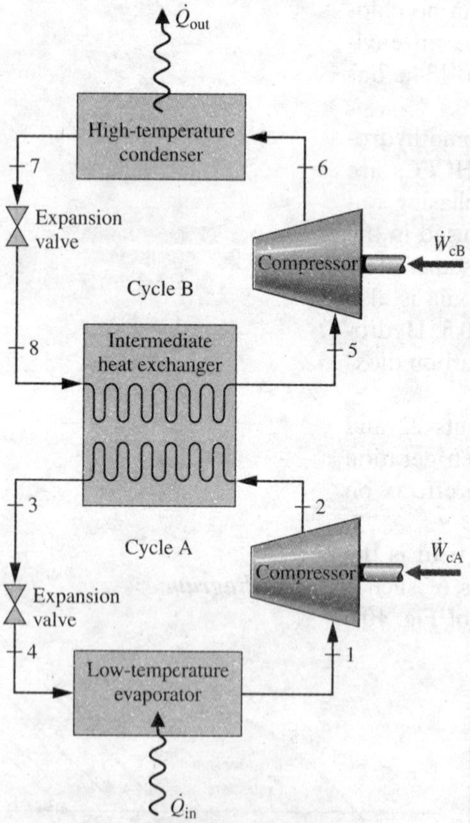

Fig. 10.7 Example of a cascade vapor-compression refrigeration cycle.

Combined cycle arrangements for refrigeration are called *cascade* cycles. In Fig. 10.7 a cascade cycle is shown in which *two* vapor-compression refrigeration cycles, labeled A and B, are arranged in series with a counterflow heat exchanger linking them. In the intermediate heat exchanger, the energy rejected during condensation of the refrigerant in the lower-temperature cycle A is used to evaporate the refrigerant in the higher-temperature cycle B. The desired refrigeration effect occurs in the low-temperature evaporator, and heat rejection from the overall cycle occurs in the high-temperature condenser. The coefficient of performance is the ratio of the refrigeration effect to the *total* work input

$$\beta = \frac{\dot{Q}_{in}}{\dot{W}_{cA} + \dot{W}_{cB}}$$

The mass flow rates in cycles A and B normally would be different. However, the mass flow rates are related by mass and energy rate balances on the interconnecting counterflow heat exchanger serving as the condenser for cycle A and the evaporator for cycle B. Although only two cycles are shown in Fig. 10.7, cascade cycles may employ three or more individual cycles.

A significant feature of the cascade system illustrated in Fig. 10.7 is that the refrigerants in the two or more stages can be selected to have reasonable evaporator and condenser pressures in the two or more temperature ranges. In a double cascade system, a refrigerant would be selected for cycle A that has a saturation pressure–temperature relationship that allows refrigeration at a relatively low temperature without excessively low evaporator pressures. The refrigerant for cycle B would have saturation characteristics that permit condensation at the required temperature without excessively high condenser pressures.

10.4.2 Multistage Compression with Intercooling

The advantages of multistage compression with intercooling between stages have been cited in Sec. 9.8, dealing with gas power systems. Intercooling is achieved in gas

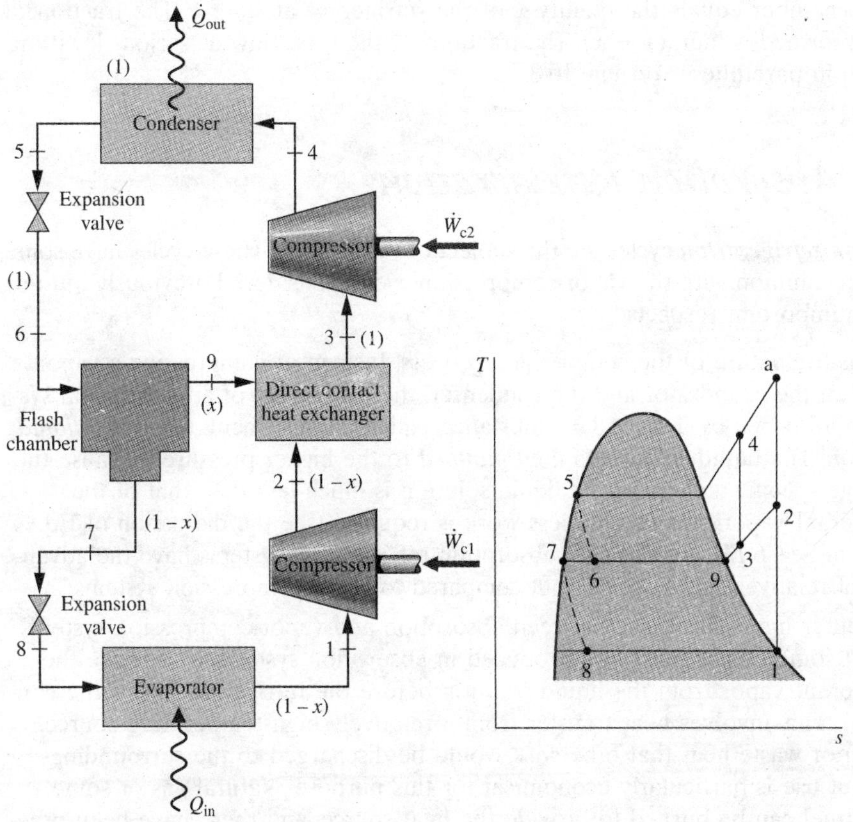

Fig. 10.8 Refrigeration cycle with two stages of compression and flash intercooling.

power systems by heat transfer to the lower-temperature surroundings. In refrigeration systems, the refrigerant temperature is below that of the surroundings for much of the cycle, so other means must be employed to accomplish intercooling and achieve the attendant savings in the required compressor work input. One arrangement for two-stage compression using the refrigerant itself for intercooling is shown in Fig. 10.8. The principal states of the refrigerant for an ideal cycle are shown on the accompanying T–s diagram.

Intercooling is accomplished in this cycle by means of a direct contact heat exchanger. Relatively low-temperature saturated vapor enters the heat exchanger at state 9, where it mixes with higher-temperature refrigerant leaving the first compression stage at state 2. A single mixed stream exits the heat exchanger at an intermediate temperature at state 3 and is compressed in the second compressor stage to the condenser pressure at state 4. Less work is required per unit of mass flow for compression from 1 to 2 followed by compression from 3 to 4 than for a single stage of compression 1–2–a. Since the refrigerant temperature entering the condenser at state 4 is lower than for a single stage of compression in which the refrigerant would enter the condenser at state a, the external irreversibility associated with heat transfer in the condenser is also reduced.

A central role is played in the cycle of Fig. 10.8 by a liquid–vapor separator, called a ***flash chamber***. Refrigerant exiting the condenser at state 5 expands through a valve and enters the flash chamber at state 6 as a two-phase liquid–vapor mixture with quality x. In the flash chamber, the liquid and vapor components separate into two streams. Saturated vapor exiting the flash chamber enters the heat exchanger at state 9, where intercooling is achieved as discussed above. Saturated liquid exiting the flash chamber at state 7 expands through a second valve into the evaporator. On the basis of a unit of mass flowing through the condenser, the fraction of the vapor formed in

flash chamber

the flash chamber equals the quality x of the refrigerant at state 6. The fraction of the liquid formed is then $(1 - x)$. The fractions of the total flow at various locations are shown in parentheses on Fig. 10.8.

10.5 Absorption Refrigeration

absorption refrigeration

Absorption refrigeration cycles are the subject of this section. These cycles have some features in common with the vapor-compression cycles considered previously but differ in two important respects:

- One is the nature of the compression process. Instead of compressing a vapor between the evaporator and the condenser, the refrigerant of an absorption system is absorbed by a secondary substance, called an absorbent, to form a *liquid solution*. The liquid solution is then *pumped* to the higher pressure. Because the average specific volume of the liquid solution is much less than that of the refrigerant vapor, significantly less work is required (see the discussion of Eq. 6.51b in Sec. 6.13). Accordingly, absorption refrigeration systems have the advantage of relatively small work input compared to vapor-compression systems.
- The other main difference between absorption and vapor-compression systems is that some means must be introduced in absorption systems to retrieve the refrigerant vapor from the liquid solution before the refrigerant enters the condenser. This involves heat transfer from a relatively high-temperature source. Steam or waste heat that otherwise would be discharged to the surroundings without use is particularly economical for this purpose. Natural gas or some other fuel can be burned to provide the heat source, and there have been practical applications of absorption refrigeration using alternative energy sources such as solar and geothermal energy.

The principal components of an absorption refrigeration system are shown schematically in Fig. 10.9. In this case, ammonia is the refrigerant and water is the

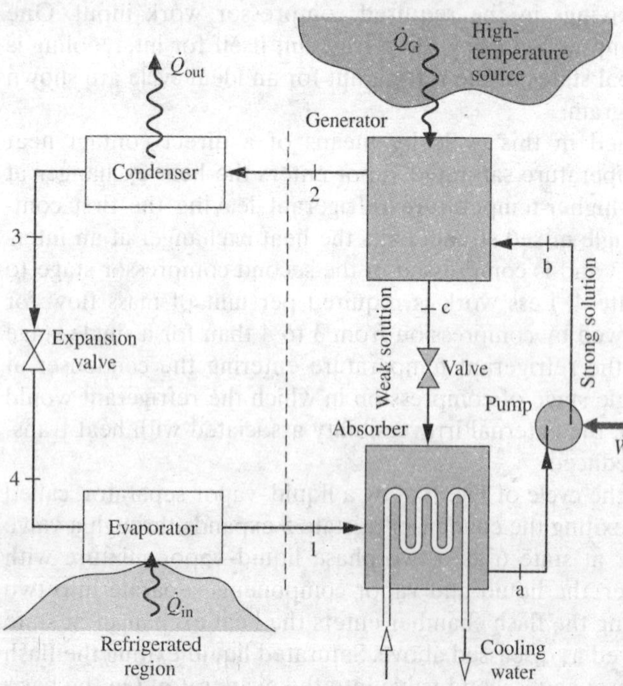

Fig. 10.9 Simple ammonia–water absorption refrigeration system.

Horizons

New Materials May Revive Thermoelectric Cooling

You can buy a thermoelectric cooler powered from the cigarette lighter outlet of your car. The same technology is used in space applications. These simple coolers have no moving parts and use no ozone-depleting refrigerants. Despite such advantages, thermoelectric cooling has found only specialized application because of low coefficients of performance that don't allow them to compete with vapor-compression systems. However, new materials and novel production methods involving engineering at a *nanometer* level may make thermoelectrics more competitive, material scientists say.

The basis of thermoelectric cooling is two dissimilar semiconductors coming together in specially designed electric circuits. Effective materials for thermoelectric cooling must have low thermal conductivity and high electrical conductivity, a rare combination in nature. One laboratory has produced better thermoelectric materials by alternately depositing 1- to 4-nanometer films of different materials on the same base. These *nano-engineered* materials hold promise for cooling electronic circuits, and perhaps even for improved thermoelectric coolers.

New materials are closing the performance gap, but much remains to be done before thermoelectric refrigerators are commonplace. Still, a vast market would be served if researchers succeed.

absorbent. Ammonia circulates through the condenser, expansion valve, and evaporator as in a vapor-compression system. However, the compressor is replaced by the absorber, pump, generator, and valve shown on the right side of the diagram.

- In the *absorber*, ammonia vapor coming from the evaporator at state 1 is absorbed by liquid water. The formation of this liquid solution is exothermic. Since the amount of ammonia that can be dissolved in water increases as the solution temperature decreases, cooling water is circulated around the absorber to remove the energy released as ammonia goes into solution and maintain the temperature in the absorber as low as possible. The strong ammonia–water solution leaves the absorber at point a and enters the *pump,* where its pressure is increased to that of the generator.
- In the *generator*, heat transfer from a high-temperature source drives ammonia vapor out of the solution (an endothermic process), leaving a weak ammonia–water solution in the generator. The vapor liberated passes to the condenser at state 2, and the remaining weak solution at c flows back to the absorber through a *valve*. The only work input is the power required to operate the pump, and this is small in comparison to the work that would be required to compress refrigerant vapor between the same pressure levels. However, costs associated with the heat source and extra equipment not required by vapor-compressor systems can cancel the advantage of a smaller work input.

Ammonia–water systems normally employ several modifications of the simple absorption cycle considered above. Two common modifications are illustrated in Fig. 10.10. In this cycle, a heat exchanger is included between the generator and the absorber that allows the strong water–ammonia solution entering the generator to be preheated by the weak solution returning from the generator to the absorber, thereby reducing the heat transfer to the generator, $\dot{Q}_G$. The other modification shown on the figure is the *rectifier* placed between the generator and the condenser. The function of the rectifier is to

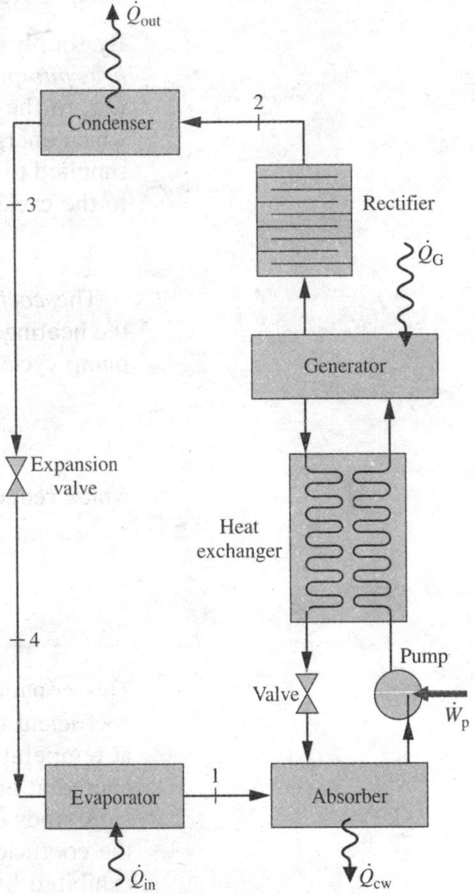

Fig. 10.10 Modified ammonia–water absorption system.

remove any traces of water from the refrigerant before it enters the condenser. This eliminates the possibility of ice formation in the expansion valve and the evaporator.

Another type of absorption system uses *lithium bromide* as the absorbent and *water* as the refrigerant. The basic principle of operation is the same as for ammonia–water systems. To achieve refrigeration at lower temperatures than are possible with water as the refrigerant, a lithium bromide–water absorption system may be combined with another cycle using a refrigerant with good low-temperature characteristics, such as ammonia, to form a cascade refrigeration system.

10.6 Heat Pump Systems

The objective of a heat pump is to maintain the temperature within a dwelling or other building above the temperature of the surroundings or to provide a heat transfer for certain industrial processes that occur at elevated temperatures. Heat pump systems have many features in common with the refrigeration systems considered thus far and may be of the vapor-compression or absorption type. Vapor-compression heat pumps are well suited for space heating applications and are commonly used for this purpose. Absorption heat pumps have been developed for industrial applications and are also increasingly being used for space heating. To introduce some aspects of heat pump operation, let us begin by considering the Carnot heat pump cycle.

Carnot Heat Pump Cycle

By simply changing our viewpoint, we can regard the cycle shown in Fig. 10.1 as a *heat pump*. The objective of the cycle now, however, is to deliver the heat transfer $\dot{Q}_{out}$ to the warm region, which is the space to be heated. At steady state, the rate at which energy is supplied to the warm region by heat transfer is the sum of the energy supplied to the working fluid from the cold region, $\dot{Q}_{in}$, and the net rate of work input to the cycle, $\dot{W}_{net}$. That is

$$\dot{Q}_{out} = \dot{Q}_{in} + \dot{W}_{net} \tag{10.8}$$

The *coefficient of performance* of *any* heat pump cycle is defined as the ratio of the heating effect to the net work required to achieve that effect. For the Carnot heat pump cycle of Fig. 10.1

$$\gamma_{max} = \frac{\dot{Q}_{out}/\dot{m}}{\dot{W}_c/\dot{m} - \dot{W}_t/\dot{m}} = \frac{\text{area 2–a–b–3–2}}{\text{area 1–2–3–4–1}}$$

which reduces to

$$\gamma_{max} = \frac{T_H(s_a - s_b)}{(T_H - T_C)(s_a - s_b)} = \frac{T_H}{T_H - T_C} \tag{10.9}$$

This equation, which corresponds to Eq. 5.11, represents the *maximum* theoretical coefficient of performance for any heat pump cycle operating between two regions at temperatures T_C and T_H. Actual heat pump systems have coefficients of performance that are lower than would be calculated from Eq. 10.9.

A study of Eq. 10.9 shows that as the temperature T_C of the cold region decreases, the coefficient of performance of the Carnot heat pump decreases. This trait is also exhibited by actual heat pump systems and suggests why heat pumps in which the role of the cold region is played by the local atmosphere (air-source heat pumps) normally require backup systems to provide heating on days when the ambient temperature becomes very low. If sources such as well water or the ground itself are used, relatively high coefficients of performance can be achieved despite low ambient air temperatures, and backup systems may not be required.

Vapor-Compression Heat Pumps

Actual heat pump systems depart significantly from the Carnot cycle model. Most systems in common use today are of the vapor-compression type. The method of analysis of *vapor-compression heat pumps* is the same as that of vapor-compression refrigeration cycles considered previously. Also, the previous discussions concerning the departure of actual systems from ideality apply for vapor-compression heat pump systems as for vapor-compression refrigeration cycles.

As illustrated by Fig. 10.11, a typical ***vapor-compression heat pump*** for space heating has the same basic components as the vapor-compression refrigeration system: compressor, condenser, expansion valve, and evaporator. The objective of the system is different, however. In a heat pump system, $\dot{Q}_{in}$ comes from the surroundings, and $\dot{Q}_{out}$ is directed to the dwelling as the desired effect. A net work input is required to accomplish this effect.

vapor-compression heat pump

The coefficient of performance of a simple vapor-compression heat pump with states as designated on Fig. 10.11 is

$$\gamma = \frac{\dot{Q}_{out}/\dot{m}}{\dot{W}_c/\dot{m}} = \frac{h_2 - h_3}{h_2 - h_1} \qquad (10.10)$$

The value of γ can never be less than unity.

Many possible sources are available for heat transfer to the refrigerant passing through the evaporator. These include the outside air, the ground, and lake, river, or well water. Liquid circulated through a solar collector and stored in an insulated tank also can be used as a source for a heat pump. Industrial heat pumps employ waste heat or warm liquid or gas streams as the low-temperature source and are capable of achieving relatively high condenser temperatures.

In the most common type of vapor-compression heat pump for space heating, the evaporator communicates thermally with the outside air. Such ***air-source heat pumps*** also can be used to provide cooling in the summer with the use of a reversing valve,

air-source heat pump

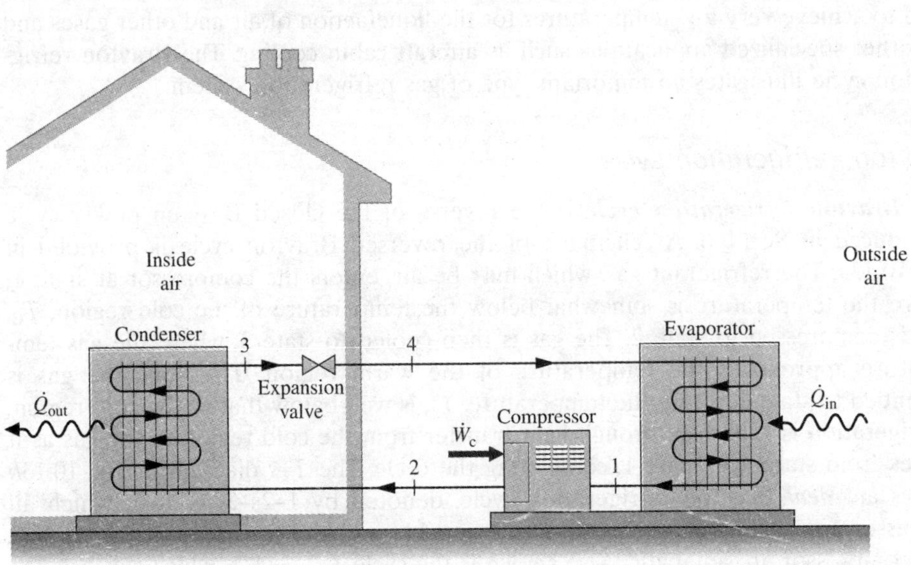

Fig. 10.11 Air-source vapor-compression heat pump system.

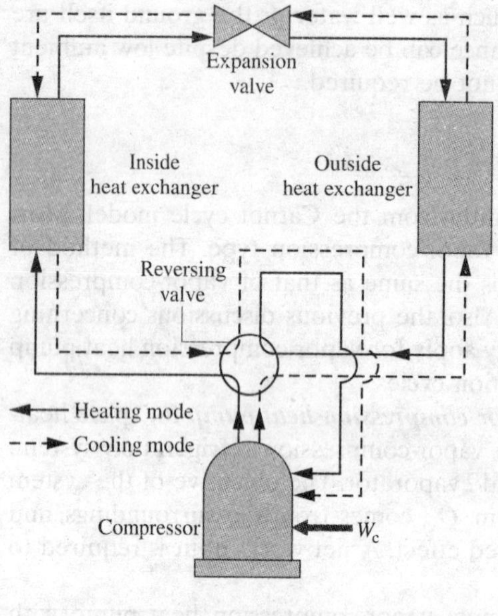

Fig. 10.12 Example of an air-to-air reversing heat pump.

as illustrated in Fig. 10.12. The solid lines show the flow path of the refrigerant in the heating mode, as described previously. To use the same components as an air conditioner, the valve is actuated, and the refrigerant follows the path indicated by the dashed line. In the cooling mode, the outside heat exchanger becomes the condenser, and the inside heat exchanger becomes the evaporator. Although heat pumps can be more costly to install and operate than other direct heating systems, they can be competitive when the potential for dual use is considered.

10.7 Gas Refrigeration Systems

gas refrigeration systems

All refrigeration systems considered thus far involve changes in phase. Let us now turn to ***gas refrigeration systems*** in which the working fluid remains a gas throughout. Gas refrigeration systems have a number of important applications. They are used to achieve very low temperatures for the liquefaction of air and other gases and for other specialized applications such as aircraft cabin cooling. The Brayton refrigeration cycle illustrates an important type of gas refrigeration system.

Brayton Refrigeration Cycle

Brayton refrigeration cycle

The ***Brayton refrigeration cycle*** is the reverse of the closed Brayton power cycle introduced in Sec. 9.6. A schematic of the reversed Brayton cycle is provided in Fig. 10.13*a*. The refrigerant gas, which may be air, enters the compressor at state 1, where the temperature is somewhat below the temperature of the cold region, T_C, and is compressed to state 2. The gas is then cooled to state 3, where the gas temperature approaches the temperature of the warm region, T_H. Next, the gas is expanded to state 4, where the temperature, T_4, is well below that of the cold region. Refrigeration is achieved through heat transfer from the cold region to the gas as it passes from state 4 to state 1, completing the cycle. The T–s diagram in Fig. 10.13*b* shows an *ideal* Brayton refrigeration cycle, denoted by 1–2s–3–4s–1, in which all processes are assumed to be internally reversible and the processes in the turbine and compressor are adiabatic. Also shown is the cycle 1–2–3–4–1, which suggests the

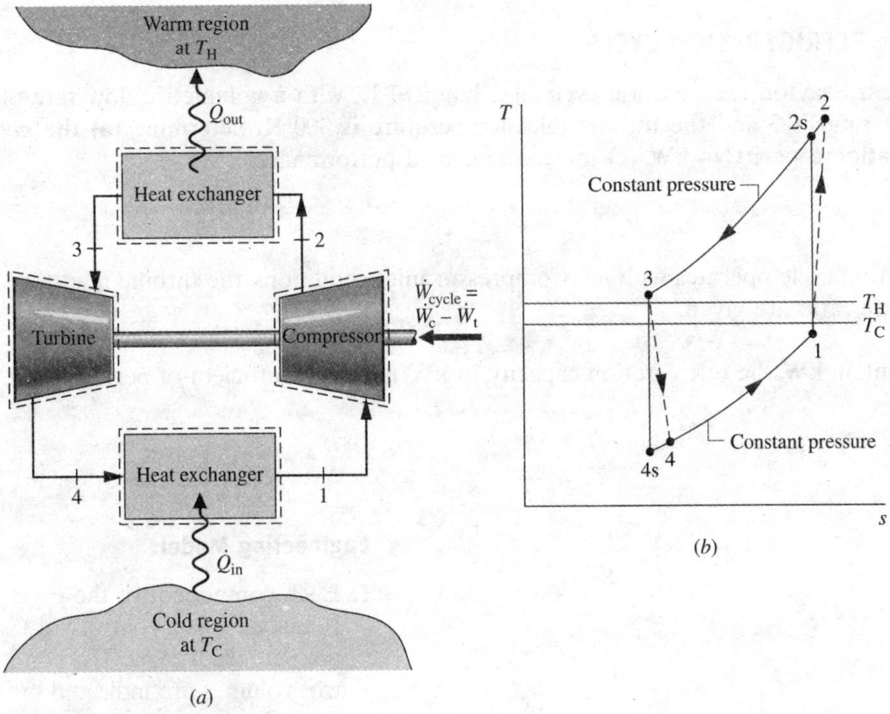

Fig. 10.13 Brayton refrigeration cycle.

effects of irreversibilities during adiabatic compression and expansion. Frictional pressure drops have been ignored.

CYCLE ANALYSIS. The method of analysis of the Brayton refrigeration cycle is similar to that of the Brayton power cycle. Thus, at steady state the work of the compressor and the turbine per unit of mass flow are, respectively

$$\frac{\dot{W}_c}{\dot{m}} = h_2 - h_1 \quad \text{and} \quad \frac{\dot{W}_t}{\dot{m}} = h_3 - h_4$$

In obtaining these expressions, heat transfer with the surroundings and changes in kinetic and potential energy have been ignored. In contrast to the vapor-compression cycle of Fig. 10.2, the work developed by the turbine of a Brayton refrigeration cycle is significant relative to the compressor work input.

The heat transfer from the cold region to the refrigerant gas circulating through the low-pressure heat exchanger, the refrigeration effect, is

$$\frac{\dot{Q}_{in}}{\dot{m}} = h_1 - h_4$$

The coefficient of performance is the ratio of the refrigeration effect to the net work input:

$$\beta = \frac{\dot{Q}_{in}/\dot{m}}{\dot{W}_c/\dot{m} - \dot{W}_t/\dot{m}} = \frac{(h_1 - h_4)}{(h_2 - h_1) - (h_3 - h_4)} \tag{10.11}$$

In the next example, we illustrate the analysis of an ideal Brayton refrigeration cycle.

Example 10.4 IDEAL BRAYTON REFRIGERATION CYCLE

Air enters the compressor of an ideal Brayton refrigeration cycle at 1 bar, 270°K, with a volumetric flow rate of 1.4 m^3/s. If the compressor pressure ratio is 3 and the turbine inlet temperature is 300°K, determine **(a)** the *net* power input, in kW, **(b)** the refrigeration capacity, in kW, **(c)** the coefficient of performance.

Solution

Known: An ideal Brayton refrigeration cycle operates with air. Compressor inlet conditions, the turbine inlet temperature, and the compressor pressure ratio are given.

Find: Determine the *net* power input, in kW, the refrigeration capacity, in kW, and the coefficient of performance.

Schematic and Given Data:

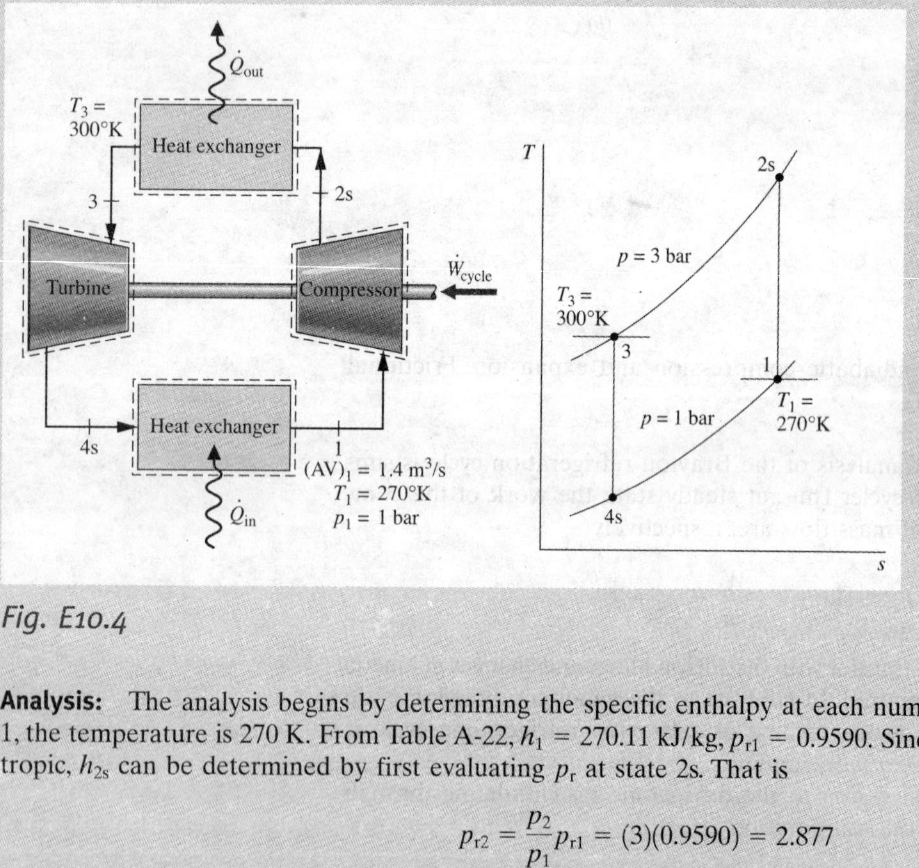

Fig. E10.4

Engineering Model:

1. Each component of the cycle is analyzed as a control volume at steady state. The control volumes are indicated by dashed lines on the accompanying sketch.
2. The turbine and compressor processes are isentropic.
3. There are no pressure drops through the heat exchangers.
4. Kinetic and potential energy effects are negligible.
5. The working fluid is air modeled as an ideal gas.

Analysis: The analysis begins by determining the specific enthalpy at each numbered state of the cycle. At state 1, the temperature is 270 K. From Table A-22, h_1 = 270.11 kJ/kg, p_{r1} = 0.9590. Since the compressor process is isentropic, h_{2s} can be determined by first evaluating p_r at state 2s. That is

$$p_{r2} = \frac{p_2}{p_1} p_{r1} = (3)(0.9590) = 2.877$$

Then, interpolating in Table A-22, we get h_{2s} = 370.1 kJ/kg.

The temperature at state 3 is given as T_3 = 300 K. From Table A-22, h_3 = 300.19 kJ/kg, p_{r3} = 1.3860. The specific enthalpy at state 4s is found by using the isentropic relation

$$p_{r4} = p_{r3} \frac{p_4}{p_3} = (1.3860)(1/3) = 0.462$$

Interpolating in Table A-22, we obtain h_{4s} = 219.0 kJ/kg.

(a) The net power input is

$$\dot{W}_{cycle} = \dot{m}[(h_{2s} - h_1) - (h_3 - h_{4s})]$$

This requires the mass flow rate $\dot{m}$, which can be determined from the volumetric flow rate and the specific volume at the compressor inlet:

$$\dot{m} = \frac{(\mathrm{AV})_1}{v_1}$$

Since $v_1 = (\overline{R}/M)T_1/p_1$

$$\dot{m} = \frac{(\mathrm{AV})_1 p_1}{(\overline{R}/M)T_1}$$

$$= \frac{(1.4\ \mathrm{m^3/s})(1\ \mathrm{bar})}{\left(\dfrac{8.314\ \mathrm{kJ}}{28.97}\right)(270\ \mathrm{K})}\left(\frac{10^5\ \mathrm{N/m^2}}{1\ \mathrm{bar}}\right)\left(\frac{1\ \mathrm{kJ}}{10^3\ \mathrm{N \cdot m}}\right)$$

$$= 1.807\ \mathrm{kg/s}$$

Finally

$$\dot{W}_{\mathrm{cycle}} = (1.807\ \mathrm{kg/s})[(370.1 - 270.11) - (300.19 - 219.0)]\ \mathrm{kJ/kg}\left(\frac{1\ \mathrm{kW}}{1\ \mathrm{kJ/s}}\right)$$

$$= 33.97\ \mathrm{kW}$$

(b) The refrigeration capacity is

$$\dot{Q}_{\mathrm{in}} = \dot{m}(h_1 - h_{4s})$$

$$= (1.807\ \mathrm{kg/s})[(270.11 - 219)]\ \mathrm{kJ/kg}\left(\frac{1\ \mathrm{kW}}{1\ \mathrm{kJ/s}}\right)$$

$$= 92.36\ \mathrm{kW}$$

(c) The coefficient of performance is

$$\beta = \frac{\dot{Q}_{\mathrm{in}}}{\dot{W}_{\mathrm{cycle}}} = \frac{92.36}{33.97} = 2.72$$

Irreversibilities within the compressor and turbine serve to decrease the coefficient of performance significantly from that of the corresponding ideal cycle because the compressor work requirement is increased and the turbine work output is decreased. This is illustrated in the example to follow.

Skills Developed: *Ability to...*

- ✓ sketch the *T–s* diagram of the ideal Brayton refrigeration cycle.
- ✓ fix each of the principal states and retrieve necessary property data.
- ✓ calculate refrigeration capacity and coefficient of performance.

Quick Quiz

Determine the refrigeration capacity in tons of refrigeration.

Ans. 25.53 ton.

Example 10.5 BRAYTON REFRIGERATION CYCLE WITH IRREVERSIBILITIES

Reconsider Example 10.4, but include in the analysis that the compressor and turbine each have an isentropic efficiency of 80%. Determine for the modified cycle **(a)** the *net* power input, in kW, **(b)** the refrigeration capacity, in kW, **(c)** the coefficient of performance, and interpret its value.

Solution

Known: A Brayton refrigeration cycle operates with air. Compressor inlet conditions, the turbine inlet temperature, and the compressor pressure ratio are given. The compressor and turbine each have an efficiency of 80%.

Find: Determine the *net* power input and the refrigeration capacity, each in kW. Also, determine the coefficient of performance and interpret its value.

Schematic and Given Data:

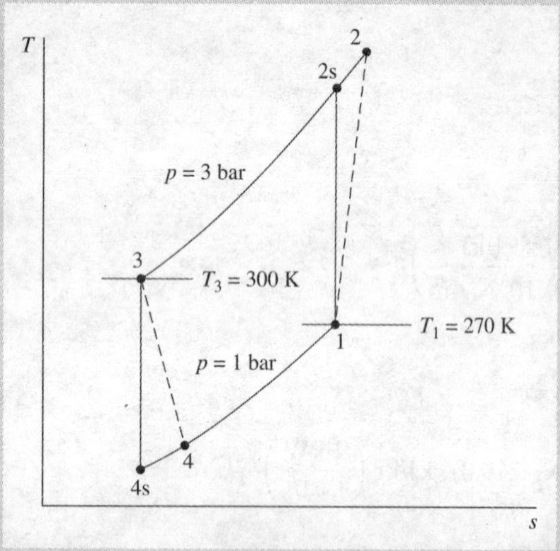

Engineering Model:

1. Each component of the cycle is analyzed as a control volume at steady state.
2. The compressor and turbine are adiabatic.
3. There are no pressure drops through the heat exchangers.
4. Kinetic and potential energy effects are negligible.
5. The working fluid is air modeled as an ideal gas.

Fig. E10.5

Analysis:

(a) The power input to the compressor is evaluated using the isentropic compressor efficiency, η_c. That is

$$\frac{\dot{W}_c}{\dot{m}} = \frac{(\dot{W}_c/\dot{m})_s}{\eta_c}$$

The value of the work per unit mass for the isentropic compression, $(\dot{W}_c/\dot{m})_s$, is determined with data from the solution in Example 10.4 as 99.99 kJ/kg. The actual power required is then

$$\dot{W}_c = \frac{\dot{m}(\dot{W}_c/\dot{m})_s}{\eta_c} = \frac{(1.807\text{ kg/s})(99.99\text{ kJ/kg})}{(0.8)} = 225.9\text{ kW}$$

The turbine power output is determined in a similar manner, using the turbine isentropic efficiency η_t. Thus, $\dot{W}_t/\dot{m} = \eta_t(\dot{W}_t/\dot{m})_s$. Using data from the solution to Example 10.4 gives $(\dot{W}_t/\dot{m})_s = 81.19$ kJ/kg. The actual turbine work is then

$$\dot{W}_t = \dot{m}\eta_t(\dot{W}_t/\dot{m})_s = (1.807\text{ kg/s})(0.8)(81.19\text{ kJ/kg}) = 117.4\text{ kW}$$

The *net* power input to the cycle is

$$\dot{W}_{cycle} = 225.9 - 117.4 = 108.5\text{ kW}$$

(b) The specific enthalpy at the turbine exit, h_4, is required to evaluate the refrigeration capacity. This enthalpy can be determined by solving $\dot{W}_t = \dot{m}(h_3 - h_4)$ to obtain $h_4 = h_3 - \dot{W}_t/\dot{m}$. Inserting known values

$$h_4 = 300.19 - \left(\frac{117.4}{1.807}\right) = 235.2\text{ kJ/kg}$$

The refrigeration capacity is then

$$\dot{Q}_{in} = \dot{m}(h_1 - h_4) = (1.807)(270.11 - 235.2) = 63.08\text{ kW}$$

(c) The coefficient of performance is

$$\beta = \frac{\dot{Q}_{in}}{\dot{W}_{cycle}} = \frac{63.08}{108.5} = 0.581$$

Skills Developed: *Ability to...*
- ✓ sketch the *T–s* diagram of the Brayton refrigeration cycle with irreversibilities in the turbine and compressor.
- ✓ fix each of the principal states and retrieve necessary property data.
- ✓ calculate refrigeration capacity and coefficient of performance.

The value of the coefficient of performance in this case is less than unity. This means that the refrigeration effect is smaller than the net work required to achieve it. Additionally, note that irreversibilities in the compressor and turbine have a significant effect on the performance of gas refrigeration systems. This is brought out by comparing the results of the present example with those of Example 10.4. Irreversibilities result in an increase in the work of compression and a reduction in the work output of the turbine. The refrigeration capacity is also reduced. The overall effect is that the coefficient of performance is decreased significantly.

Quick Quiz

Determine the coefficient of performance for a Carnot refrigeration cycle operating between reservoirs at 266.7 K and 300 K.

Ans. 8.

Additional Gas Refrigeration Applications

To obtain even moderate refrigeration capacities with the Brayton refrigeration cycle, equipment capable of achieving relatively high pressures and volumetric flow rates is needed. For most applications involving air conditioning and for ordinary refrigeration processes, vapor-compression systems can be built more cheaply and can operate with higher coefficients of performance than gas refrigeration systems. With suitable modifications, however, gas refrigeration systems can be used to achieve temperatures of about −150°C, which are well below the temperatures normally obtained with vapor systems.

Figure 10.14 shows the schematic and *T–s* diagram of an ideal Brayton cycle modified by the introduction of a regenerative heat exchanger. The heat exchanger allows the air entering the turbine at state 3 to be cooled *below* the warm region temperature T_H. In the subsequent expansion through the turbine, the air achieves a much lower temperature at state 4 than would have been possible without the regenerative heat exchanger. Accordingly, the refrigeration effect, achieved from state 4 to state b, occurs at a correspondingly lower average temperature.

An example of the application of gas refrigeration to cabin cooling in an aircraft is illustrated in Fig. 10.15. As shown in the figure, a small amount of high-pressure air is extracted from the main jet engine compressor and cooled by heat transfer to the ambient. The high-pressure air is then expanded through an auxiliary turbine to the pressure maintained in the cabin. The air temperature is reduced in the expansion and thus is able to fulfill its cabin cooling function. As an additional benefit, the turbine expansion can provide some of the auxiliary power needs of the aircraft. Size

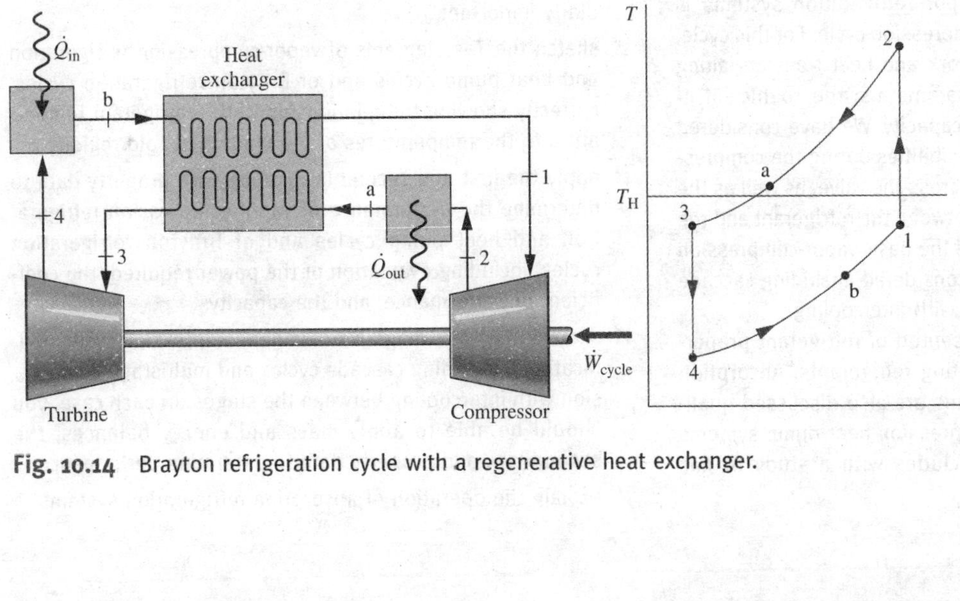

Fig. 10.14 Brayton refrigeration cycle with a regenerative heat exchanger.

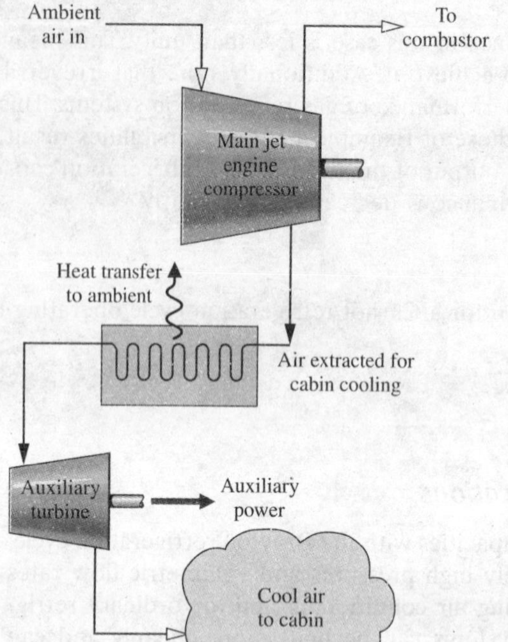

Fig. 10.15 An application of gas refrigeration to aircraft cabin cooling.

and weight are important considerations in the selection of equipment for use in aircraft. Open-cycle systems, like the example given here, utilize *compact* high-speed rotary turbines and compressors. Furthermore, since the air for cooling comes directly from the surroundings, there are fewer heat exchangers than would be needed if a separate refrigerant were circulated in a closed vapor-compression cycle.

✓ Chapter Summary and Study Guide

In this chapter, we have considered refrigeration and heat pump systems, including vapor systems where the refrigerant is alternately vaporized and condensed, and gas systems where the refrigerant remains a gas. The three principal types of refrigeration and heat pump systems discussed are the vapor-compression, absorption, and reversed Brayton cycles.

The performance of simple vapor refrigeration systems is described in terms of the vapor-compression cycle. For this cycle, we have evaluated the principal work and heat transfers along with two important performance parameters: the coefficient of performance and the refrigeration capacity. We have considered the effect on performance of irreversibilities during the compression process and in the expansion across the valve, as well as the effect of irreversible heat transfer between the refrigerant and the warm and cold regions. Variations of the basic vapor-compression refrigeration cycle also have been considered, including cascade cycles and multistage compression with intercooling.

Qualitative discussions are presented of refrigerant properties and of considerations in selecting refrigerants. Absorption refrigeration and heat pump systems are also discussed qualitatively. A discussion of vapor-compression heat pump systems is provided, and the chapter concludes with a study of gas refrigeration systems.

The following list provides a study guide for this chapter. When your study of the text and end-of-chapter exercises has been completed, you should be able to

- ✓ write out the meanings of the terms listed in the margin throughout the chapter and understand each of the related concepts. The subset of key concepts listed below is particularly important.
- ✓ sketch the *T–s* diagrams of vapor-compression refrigeration and heat pump cycles and of Brayton refrigeration cycles, correctly showing the relationship of the refrigerant temperature to the temperatures of the warm and cold regions.
- ✓ apply the first and second laws along with property data to determine the performance of vapor-compression refrigeration and heat pump cycles and of Brayton refrigeration cycles, including evaluation of the power required, the coefficient of performance, and the capacity.
- ✓ sketch schematic diagrams of vapor-compression cycle modifications, including cascade cycles and multistage compression with intercooling between the stages. In each case, you should be able to apply mass and energy balances, the second law, and property data to determine performance.
- ✓ explain the operation of absorption refrigeration systems.

Key Engineering Concepts

vapor-compression refrigeration *p. 537*
refrigeration capacity *p. 537*
ton of refrigeration *p. 537*
absorption refrigeration *p. 548*
vapor-compression heat pump *p. 551*
Brayton refrigeration cycle *p. 552*

Key Equations

$\beta_{max} = \dfrac{T_C}{T_H - T_C}$	(10.1)	Coefficient of performance of the Carnot refrigeration cycle (Fig. 10.1)
$\beta = \dfrac{\dot{Q}_{in}/\dot{m}}{\dot{W}_c/\dot{m}} = \dfrac{h_1 - h_4}{h_2 - h_1}$	(10.7)	Coefficient of performance of the vapor-compression refrigeration cycle (Fig. 10.3)
$\gamma_{max} = \dfrac{T_H}{T_H - T_C}$	(10.9)	Coefficient of performance of the Carnot heat pump cycle (Fig. 10.1)
$\gamma = \dfrac{\dot{Q}_{out}/\dot{m}}{\dot{W}_c/\dot{m}} = \dfrac{h_2 - h_3}{h_2 - h_1}$	(10.10)	Coefficient of performance of the vapor-compression heat pump cycle (Fig. 10.11)
$\beta = \dfrac{\dot{Q}_{in}/\dot{m}}{\dot{W}_c/\dot{m} - \dot{W}_t/\dot{m}} = \dfrac{(h_1 - h_4)}{(h_2 - h_1) - (h_3 - h_4)}$	(10.11)	Coefficient of performance of the Brayton refrigeration cycle (Fig. 10.13)

Exercises: things engineers think about

1. What are the temperatures inside the fresh food and freezer compartments of your refrigerator? Do you know what values are *recommended* for these temperatures?

2. You have a refrigerator located in your garage. Does it perform differently in the summer than in the winter? Explain.

3. Does it make sense to say that a room air conditioner refrigerates the room?

4. If it takes about 335 kJ to freeze 1 kg of water, how much ice could an ice maker having a 1-ton refrigeration capacity produce in 24 hours?

5. Abbe installs a dehumidifier to dry the walls of a small, closed basement room. When she enters the room later, it feels warm. Why?

6. Why does the indoor unit of a central air-conditioning system have a drain hose?

7. You recharge your automobile air conditioner with refrigerant from time to time yet seldom, if ever, your refrigerator. Why?

8. Would water be a suitable working fluid for use in a refrigerator?

9. Your new air conditioner has a label that lists an *EER* of 10 kJ/h per watt. What does that mean?

10. Would you recommend a domestic heat pump for use in Duluth, Minnesota? In San Diego, California?

11. Which has the bigger diameter, the tube feeding a refrigerator compressor, called the *suction line,* or the line at the compressor discharge? Explain.

12. Can the coefficient of performance of a heat pump have a value less than one?

13. You see an advertisement for a natural gas–fired absorption refrigeration system. How can *burning* natural gas play a role in achieving *cooling*?

Problems: developing engineering skills

Vapor Refrigeration Systems

10.1 Refrigerant 22 is the working fluid in a Carnot refrigeration cycle operating at steady state. The refrigerant enters the condenser as saturated vapor at 32°C and exits as saturated liquid. The evaporator operates at 0°C. What is the coefficient of performance of the cycle? Determine, in kJ per gram of refrigerant flowing

(a) the work input to the compressor.
(b) the work developed by the turbine.

(c) the heat transfer to the refrigerant passing through the evaporator.

10.2 Refrigerant 22 is the working fluid in a Carnot vapor refrigeration cycle for which the evaporator temperature is −30°C. Saturated vapor enters the condenser at 36°C, and saturated liquid exits at the same temperature. The mass flow rate of refrigerant is 10kg/min. Determine

(a) the rate of heat transfer to the refrigerant passing through the evaporator, in kW.
(b) the net power input to the cycle, in kW.
(c) the coefficient of performance.
(d) the refrigeration capacity, in tons.

10.3 A Carnot vapor refrigeration cycle operates between thermal reservoirs at 4°C and 32°C. For (a) Refrigerant 134a, (b) propane, (c) water, (d) Refrigerant 22, and (e) ammonia as the working fluid, determine the operating pressures in the condenser and evaporator, in kPa, and the coefficient of performance.

10.4 A Carnot vapor refrigeration cycle is used to maintain a cold region at 4°C when the ambient temperature is 32°C. Refrigerant 134a enters the condenser as saturated vapor at 965 kPa and leaves as saturated liquid at the same pressure. The evaporator pressure is 276 kPa. The mass flow rate of refrigerant is 0.05 kg/s. Calculate

(a) the compressor and turbine power, each in kW.
(b) the coefficient of performance.

10.5 For the cycle in Problem 10.4, determine

(a) the rates of heat transfer, in kW, for the refrigerant flowing through the evaporator and condenser, respectively.
(b) the rates and directions of exergy transfer accompanying each of these heat transfers, in kW. Let $T_0 = 32°C$.

10.6 An ideal vapor-compression refrigeration cycle operates at steady state with Refrigerant 134a as the working fluid. Saturated vapor enters the compressor at 2 bar, and saturated liquid exits the condenser at 8 bar. The mass flow rate of refrigerant is 7 kg/min. Determine

(a) the compressor power, in kW.
(b) the refrigerating capacity, in tons.
(c) the coefficient of performance.

10.7 Plot each of the quantities in Problem 10.6 versus evaporator temperature for evaporator pressures ranging from 0.6 to 4 bar, while the condenser pressure remains fixed at 8 bar.

10.8 Refrigerant 134a is the working fluid in an ideal vapor-compression refrigeration cycle operating at steady state. Refrigerant enters the compressor at 1.4 bar, −12°C, and the condenser pressure is 9 bar. Liquid exits the condenser at 32°C. The mass flow rate of refrigerant is 7 kg/min. Determine

(a) the compressor power, in kW.
(b) the refrigeration capacity, in tons.
(c) the coefficient of performance.

10.9 An ideal vapor-compression refrigeration system operates at steady state with Refrigerant 134a as the working fluid. Superheated vapor enters the compressor at 70 kPa, −18°C, and saturated liquid leaves the condenser at 1240 kPa. The refrigeration capacity is 8 tons. Determine

(a) the compressor power, in kW.
(b) the rate of heat transfer from the working fluid passing through the condenser, in kW.
(c) the coefficient of performance.

10.10 Refrigerant 22 enters the compressor of an ideal vapor-compression refrigeration system as saturated vapor at −40°C with a volumetric flow rate of 15 m^3/min. The refrigerant leaves the condenser at 32°C, 9 bar. Determine

(a) the compressor power, in kW.
(b) the refrigerating capacity, in tons.
(c) the coefficient of performance.

10.11 An ideal vapor-compression refrigeration cycle, with ammonia as the working fluid, has an evaporator temperature of −20°C and a condenser pressure of 12 bar. Saturated vapor enters the compressor, and saturated liquid exits the condenser. The mass flow rate of the refrigerant is 3 kg/min. Determine

(a) the coefficient of performance.
(b) the refrigerating capacity, in tons.

10.12 Refrigerant 134a enters the compressor of an ideal vapor-compression refrigeration cycle as saturated vapor at −23°C. The condenser pressure is 1100 kPa. The mass flow rate of refrigerant is 0.045 kg/s. Plot the coefficient of performance and the refrigerating capacity, in tons, versus the condenser exit temperature ranging from the saturation temperature at 1100 kPa to 32°C.

10.13 To determine the effect of changing the evaporator temperature on the performance of an ideal vapor-compression refrigeration cycle, plot the coefficient of performance and the refrigerating capacity, in tons, for the cycle in Problem 10.11 for saturated vapor entering the compressor at temperatures ranging from −40 to −10°C. All other conditions are the same as in Problem 10.11.

10.14 To determine the effect of changing condenser pressure on the performance of an ideal vapor-compression refrigeration cycle, plot the coefficient of performance and the refrigerating capacity, in tons, for the cycle in Problem 10.11 for condenser pressures ranging from 8 to 16 bar. All other conditions are the same as in Problem 10.11.

10.15 A vapor-compression refrigeration cycle operates at steady state with Refrigerant 134a as the working fluid. Saturated vapor enters the compressor at 2 bar, and saturated liquid exits the condenser at 8 bar. The isentropic compressor efficiency is 80%. The mass flow rate of refrigerant is 7 kg/min. Determine

(a) the compressor power, in kW.
(b) the refrigeration capacity, in tons.
(c) the coefficient of performance.

10.16 Modify the cycle in Problem 10.9 to have an isentropic compressor efficiency of 83% and let the temperature of the liquid leaving the condenser be 38°C. Determine, for the modified cycle,

(a) the compressor power, in kW.
(b) the rate of heat transfer from the working fluid passing through the condenser, in kW.
(c) the coefficient of performance.
(d) the rates of exergy destruction in the compressor and expansion valve, each in kW, for $T_0 = 38°C$.

10.17 A vapor-compression refrigeration system circulates Refrigerant 134a at a rate of 6 kg/min. The refrigerant enters the compressor at −10°C, 1.4 bar, and exits at 7 bar. The isentropic compressor efficiency is 67%. There are no appreciable pressure drops as the refrigerant flows through the condenser and evaporator. The refrigerant leaves the condenser at 7 bar, 24°C. Ignoring heat transfer between the compressor and its surroundings, determine

(a) the coefficient of performance.
(b) the refrigerating capacity, in tons.
(c) the rates of exergy destruction in the compressor and expansion valve, each in kW.
(d) the changes in specific flow exergy of the refrigerant passing through the evaporator and condenser, respectively, each in kJ/kg.

Let $T_0 = 21°C$, $p_0 = 1$ bar.

10.18 A vapor-compression refrigeration system, using ammonia as the working fluid, has evaporator and condenser pressures of 1380 kPa and 207 kPa, respectively. The refrigerant passes through each heat exchanger with a negligible pressure drop. At the inlet and exit of the compressor, the temperatures are −12°C and 150°C, respectively. The heat transfer rate from the working fluid passing through the condenser is 15 kW, and liquid exits at 1380 kPa, 32°C. If the compressor operates adiabatically, determine

(a) the compressor power input, in kW.
(b) the coefficient of performance.

10.19 If the minimum and maximum allowed refrigerant pressures are 1 and 10 bar, respectively, which of the following can be used as the working fluid in a vapor-compression refrigeration system that maintains a cold region at 0°C, while discharging energy by heat transfer to the surrounding air at 30°C: Refrigerant 22, Refrigerant 134a, ammonia, propane?

10.20 Consider the following vapor-compression refrigeration cycle used to maintain a cold region at temperature T_C when the ambient temperature is 27°C: Saturated vapor enters the compressor at −9°C below T_C, and the compressor operates adiabatically with an isentropic efficiency of 80%. Saturated liquid exits the condenser at 35°C. There are no pressure drops through the evaporator or condenser, and the refrigerating capacity is 1 ton. Plot refrigerant mass flow rate, in kg/s, coefficient of performance, and *refrigerating efficiency*, versus T_C ranging from 4°C to −32°C if the refrigerant is

(a) Refrigerant 134a.
(b) propane.
(c) Refrigerant 22.
(d) ammonia.

The refrigerating efficiency is defined as the ratio of the cycle coefficient of performance to the coefficient of performance of a Carnot refrigeration cycle operating between thermal reservoirs at the ambient temperature and the temperature of the cold region.

10.21 In a vapor-compression refrigeration cycle, ammonia exits the evaporator as saturated vapor at −22°C. The refrigerant enters the condenser at 16 bar and 160°C, and saturated liquid exits at 16 bar. There is no significant heat transfer between the compressor and its surroundings, and the refrigerant passes through the evaporator with a negligible change in pressure. If the refrigerating capacity is 150 kW, determine

(a) the mass flow rate of refrigerant, in kg/s.
(b) the power input to the compressor, in kW.
(c) the coefficient of performance.
(d) the isentropic compressor efficiency.

10.22 A vapor-compression refrigeration system with a capacity of 10 tons has superheated Refrigerant 134a vapor entering the compressor at 15°C, 4 bar, and exiting at 12 bar. The compression process can be modeled by $pv^{1.01} = constant$. At the condenser exit, the pressure is 11.6 bar, and the temperature is 44°C. The condenser is water-cooled, with water entering at 20°C and leaving at 30°C with a negligible change in pressure. Heat transfer from the outside of the condenser can be neglected. Determine

(a) the mass flow rate of the refrigerant, in kg/s.
(b) the power input and the heat transfer rate for the compressor, each in kW.
(c) the coefficient of performance.
(d) the mass flow rate of the cooling water, in kg/s.
(e) the rates of exergy destruction in the condenser and expansion valve, each expressed as a percentage of the power input. Let $T_0 = 20°C$.

10.23 The capacity of a propane vapor-compression refrigeration system is 5 tons. Saturated vapor at −18°C enters the compressor, and superheated vapor leaves at 50°C, 1240 kPa. Heat transfer from the compressor to its surroundings occurs at a rate of 8 kJ/kg of refrigerant passing through the compressor. Liquid refrigerant enters the expansion valve at 30°C, 1240 kPa. The condenser is water-cooled, with water entering at 18°C and leaving at 27°C with a negligible change in pressure. Determine

(a) the compressor power input, in kW.
(b) the mass flow rate of cooling water through the condenser, in kg/s.
(c) the coefficient of performance.

10.24 A window-mounted air conditioner supplies 19 m^3/min of air at 15°C, 1 bar to a room. Air returns from the room to the evaporator of the unit at 22°C. The air conditioner operates at steady state on a vapor-compression refrigeration cycle with Refrigerant-22 entering the compressor at 4 bar, 10°C. Saturated liquid refrigerant at 9 bar leaves the condenser. The compressor has an isentropic efficiency of 70%, and refrigerant exits the compressor at 9 bar. Determine the compressor power, in kW, the refrigeration capacity, in tons, and the coefficient of performance.

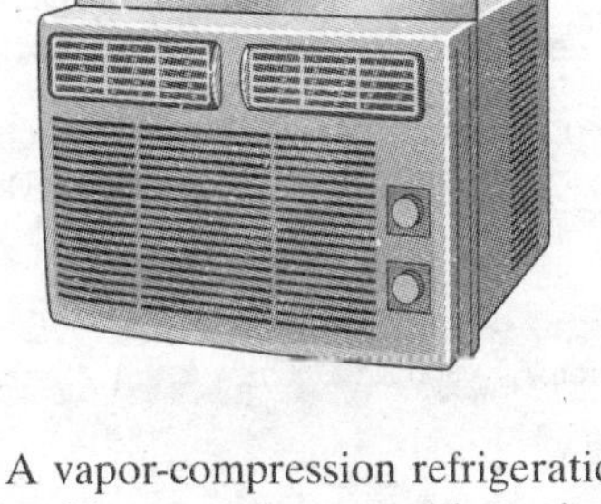

Fig. P10.24

10.25 A vapor-compression refrigeration system for a household refrigerator has a refrigerating capacity of 0.3 kW.

Refrigerant enters the evaporator at −23°C and exits at −18°C. The isentropic compressor efficiency is 80%. The refrigerant condenses at 35°C and exits the condenser subcooled at 32°C. There are no significant pressure drops in the flows through the evaporator and condenser. Determine the evaporator and condenser pressures, each in kPa, the mass flow rate of refrigerant, in kg/s, the compressor power input, in kW, and the coefficient of performance for (a) Refrigerant 134a and (b) propane as the working fluid.

10.26 A vapor-compression air-conditioning system operates at steady state as shown in Fig. P10.26. The system maintains a cool region at 15°C and discharges energy by heat transfer to the surroundings at 32°C. Refrigerant 134a enters the compressor as a saturated vapor at 4°C and is compressed adiabatically to 1103 kPa. The isentropic compressor efficiency is 80%. Refrigerant exits the condenser as a saturated liquid at 1103 kPa. The mass flow rate of the refrigerant is 0.07 kg/s. Kinetic and potential energy changes are negligible as are changes in pressure for flow through the evaporator and condenser. Determine

(a) the power required by the compressor, in kJ/s.
(b) the coefficient of performance.
(c) the rates of exergy destruction in the compressor and expansion valve, each in kJ/s.
(d) the rates of exergy destruction and exergy transfer accompanying heat transfer, each in kJ/s, for a control volume comprising the evaporator and a portion of the cool region such that heat transfer takes place at $T_C = 288$ K (15°C).
(e) the rates of exergy destruction and exergy transfer accompanying heat transfer, each in kJ/s, for a control volume enclosing the condenser and a portion of the surroundings such that heat transfer takes place at $T_H = 305$ K (32°C).

Let $T_0 = 305$ K.

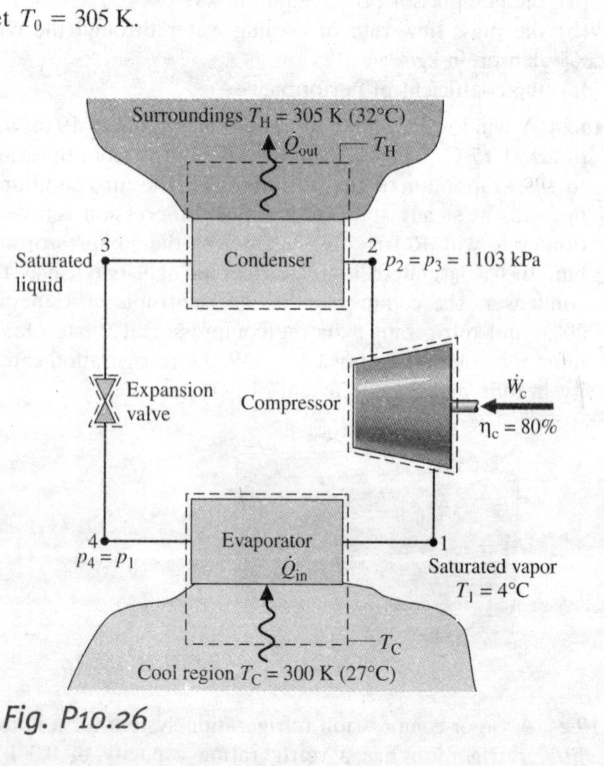

Fig. P10.26

10.27 A vapor-compression refrigeration cycle with Refrigerant 134a as the working fluid operates with an evaporator temperature of 10°C and a condenser pressure of 1240 kPa. Refrigerant enters the condenser at 60°C and exits as saturated liquid. The cycle has a refrigeration capacity of 5 tons. Determine

(a) the refrigerant mass flow rate, in kg/s.
(b) the compressor isentropic efficiency.
(c) the compressor power, in kW.
(d) the coefficient of performance.

Plot each of the quantities calculated in parts (b) through (d) for compressor exit temperatures varying from 54°C to 60°C.

Cascade and Multistage Systems

10.28 A vapor-compression refrigeration system operates with the cascade arrangement of Fig. 10.7. Refrigerant 22 is the working fluid in the high-temperature cycle and Refrigerant 134a is used in the low-temperature cycle. For the Refrigerant 134a cycle, the working fluid enters the compressor as saturated vapor at −34°C and is compressed isentropically to 345 kPa. Saturated liquid leaves the intermediate heat exchanger at 345 kPa and enters the expansion valve. For the Refrigerant 22 cycle, the working fluid enters the compressor as saturated vapor at a temperature −15°C below that of the condensing temperature of the Refrigerant 134a in the intermediate heat exchanger. The Refrigerant 22 is compressed isentropically to 1724 kPa. Saturated liquid then enters the expansion valve at 1724 kPa. The refrigerating capacity of the cascade system is 20 tons. Determine

(a) the power input to each compressor, in kJ/s.
(b) the overall coefficient of performance of the cascade cycle.
(c) the rate of exergy destruction in the intermediate heat exchanger, in kJ/s. Let $T_0 = 27$°C, $p_0 = 100$ kPa.

10.29 A vapor-compression refrigeration system uses the arrangement shown in Fig. 10.8 for two-stage compression with intercooling between the stages. Refrigerant 134a is the working fluid. Saturated vapor at −30°C enters the first compressor stage. The flash chamber and direct contact heat exchanger operate at 4 bar, and the condenser pressure is 12 bar. Saturated liquid streams at 12 and 4 bar enter the high- and low-pressure expansion valves, respectively. If each compressor operates isentropically and the refrigerating capacity of the system is 10 tons, determine

(a) the power input to each compressor, in kW.
(b) the coefficient of performance.

10.30 Figure P10.30 shows a two-stage vapor-compression refrigeration system with ammonia as the working fluid. The system uses a direct contact heat exchanger to achieve intercooling. The evaporator has a refrigerating capacity of 30 tons and produces −30°C saturated vapor at its exit. In the first compressor stage, the refrigerant is compressed adiabatically to 0.55 MPa, which is the pressure in the direct contact heat exchanger. Saturated vapor at 0.55 MPa enters the second compressor stage and is compressed adiabatically to 1.7 MPa. Each compressor stage has an isentropic efficiency of 85%. There are no significant pressure drops as the refrigerant passes through the heat exchangers. Saturated liquid enters each expansion valve. Determine

(a) the ratio of mass flow rates, $\dot{m}_3/\dot{m}_1$.
(b) the power input to each compressor stage, in kW.
(c) the coefficient of performance.
(d) Plot each of the quantities calculated in parts (a)–(c) versus the direct-contact heat exchanger pressure ranging from 0.14 to 1.4 MPa. Discuss.

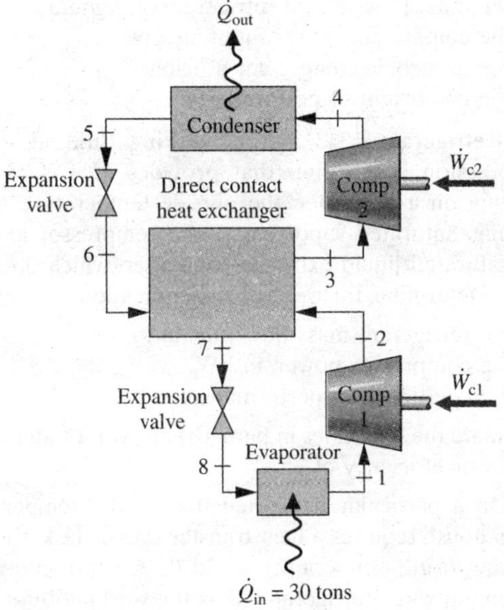

Fig. P10.30

10.31 Figure P10.31 shows a two-stage, vapor-compression refrigeration system with two evaporators and a direct contact heat exchanger. Saturated vapor from evaporator 1 enters compressor 1 at 0.12 MPa and exits at 0.5 MPa. Evaporator 2 operates at 0.5 MPa, with saturated vapor exiting at state 8. The condenser pressure is 1.4 MPa, and saturated liquid refrigerant exits the condenser. Each compressor stage has an isentropic efficiency of 80%. The refrigeration capacity of each evaporator is shown on the figure. Sketch the *T–s* diagram of the cycle and determine

(a) the temperatures, in °C, of the refrigerant in each evaporator.
(b) the power input to each compressor stage, in kW.
(c) the overall coefficient of performance.

10.32 Figure P10.32 shows the schematic diagram of a vapor-compression refrigeration system with two evaporators using Refrigerant 134a as the working fluid. This arrangement is used to achieve refrigeration at two different temperatures with a single compressor and a single condenser. The low-temperature evaporator operates at −18°C with saturated vapor at its exit and has a refrigerating capacity of 3 tons. The higher-temperature evaporator produces saturated vapor at 3.2 bar at its exit and has a refrigerating capacity of 2 tons. Compression is isentropic to the condenser pressure of 10 bar. There are no significant pressure drops in the flows through the condenser and the two evaporators, and the refrigerant leaves the condenser as saturated liquid at 10 bar. Calculate

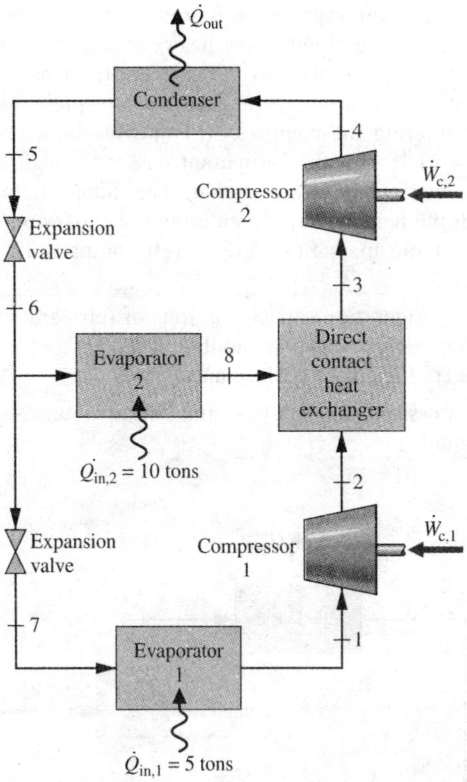

Fig. P10.31

(a) the mass flow rate of refrigerant through each evaporator, in kg/min.
(b) the compressor power input, in kW.
(c) the rate of heat transfer from the refrigerant passing through the condenser, in kW.

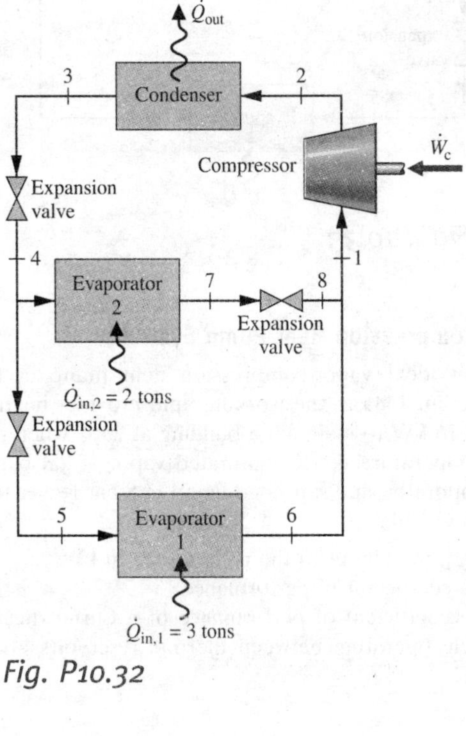

Fig. P10.32

10.33 An ideal vapor-compression refrigeration cycle is modified to include a counterflow heat exchanger, as shown in Fig. P10.33. Ammonia leaves the evaporator as saturated vapor at 1.0 bar and is heated at constant pressure to 5°C before entering the compressor. Following isentropic compression to 18 bar, the refrigerant passes through the condenser, exiting at 40°C, 18 bar. The liquid then passes through the heat exchanger, entering the expansion valve at 18 bar. If the mass flow rate of refrigerant is 12 kg/min, determine

(a) the refrigeration capacity, in tons of refrigeration.
(b) the compressor power input, in kW.
(c) the coefficient of performance.

Discuss possible advantages and disadvantages of this arrangement.

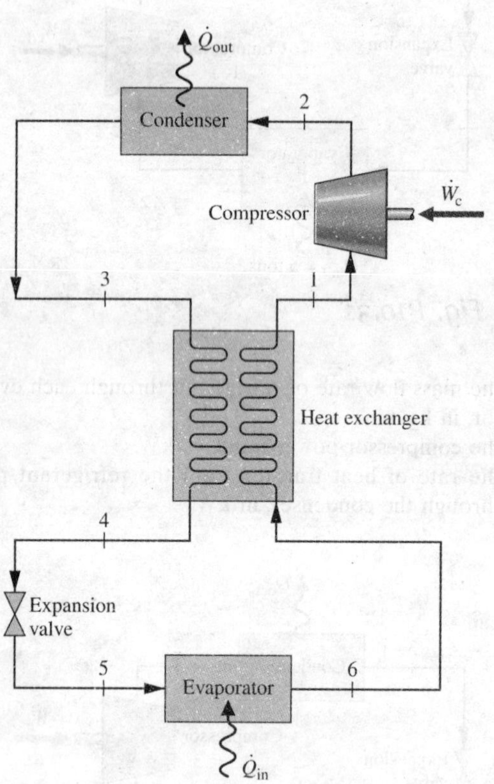

Fig. P10.33

Vapor-Compression Heat Pump Systems

10.34 An ideal vapor-compression heat pump cycle with Refrigerant 134a as the working fluid provides heating at a rate of 15 kW to maintain a building at 20°C when the outside temperature is 5°C. Saturated vapor at 2.4 bar leaves the evaporator, and saturated liquid at 8 bar leaves the condenser. Calculate

(a) the power input to the compressor, in kW.
(b) the coefficient of performance.
(c) the coefficient of performance of a Carnot heat pump cycle operating between thermal reservoirs at 20 and 5°C.

10.35 Refrigerant 134a is the working fluid in a vapor-compression heat pump system with a heating capacity of 63,300 kJ/h. The condenser operates at 1.4 MPa, and the evaporator temperature is −18°C. The refrigerant is a saturated vapor at the evaporator exit and a liquid at 43°C at the condenser exit. Pressure drops in the flows through the evaporator and condenser are negligible. The compression process is adiabatic, and the temperature at the compressor exit is 82°C. Determine

(a) the mass flow rate of refrigerant, in kg/min.
(b) the compressor power input, in kW.
(c) the isentropic compressor efficiency.
(d) the coefficient of performance.

10.36 Refrigerant 134a is the working fluid in a vapor-compression heat pump that provides 35 kW to heat a dwelling on a day when the outside temperature is below freezing. Saturated vapor enters the compressor at 1.6 bar, and saturated liquid exits the condenser, which operates at 8 bar. Determine, for isentropic compression

(a) the refrigerant mass flow rate, in kg/s.
(b) the compressor power, in kW.
(c) the coefficient of performance.

Recalculate the quantities in parts (b) and (c) for an isentropic compressor efficiency of 75%.

10.37 On a particular day when the outside temperature is 5°C, a house requires a heat transfer rate of 12 kW to maintain the inside temperature at 20°C. A vapor-compression heat pump with Refrigerant 22 as the working fluid is to be used to provide the necessary heating. Specify appropriate evaporator and condenser pressures of a cycle for this purpose. Let the refrigerant be saturated vapor at the evaporator exit and saturated liquid at the condenser exit. Calculate

(a) the mass flow rate of refrigerant, in kg/min.
(b) the compressor power, in kW.
(c) the coefficient of performance.

10.38 Repeat the calculations of Problem 10.37 for Refrigerant 134a as the working fluid. Compare the results with those of Problem 10.37 and discuss.

10.39 A process requires a heat transfer rate of 5.3×10^4 kJ/min at 77°C. It is proposed that a Refrigerant 134a vapor-compression heat pump be used to develop the process heat using a waste water stream at 52°C as the lower-temperature source. Specify appropriate evaporator and condenser pressures of a cycle for this purpose. Let the refrigerant be saturated vapor at the evaporator exit and saturated liquid at the condenser exit. Calculate

(a) the mass flow rate of refrigerant, in kg/h.
(b) the compressor power, in kW.
(c) the coefficient of performance.

10.40 A vapor-compression heat pump with a heating capacity of 500 kJ/min is driven by a power cycle with a thermal efficiency of 25%. For the heat pump, Refrigerant 134a is compressed from saturated vapor at −10°C to the condenser pressure of 10 bar. The isentropic compressor efficiency is 80%. Liquid enters the expansion valve at 9.6 bar, 34°C. For the power cycle, 80% of the heat rejected is transferred to the heated space.

(a) Determine the power input to the heat pump compressor, in kW.
(b) Evaluate the ratio of the total rate that heat is delivered to the heated space to the rate of heat input to the power cycle. Discuss.

10.41 Refrigerant 134a enters the compressor of a vapor-compression heat pump at 0.1 MPa, −18°C and is compressed adiabatically to 1.1 MPa, 70°C. Liquid enters the expansion valve at 1.1 MPa, 35°C. At the valve exit, the pressure is 0.1 MPa.

(a) Determine the isentropic compressor efficiency.
(b) Determine the coefficient of performance.
(c) Perform a full exergy accounting of the compressor power input, in kJ per kg of refrigerant flowing. Discuss.

Let $T_0 = 267$ K.

10.42 A *geothermal* heat pump operating at steady state with Refrigerant-22 as the working fluid is shown schematically in Fig. P10.42. The heat pump uses 13°C water from wells as the thermal source. Operating data are shown on the figure for a day in which the outside air temperature is −6.7°C. Assume adiabatic operation of the compressor. For the heat pump, determine

(a) the volumetric flow rate of heated air to the house, in m³/min.
(b) the isentropic compressor efficiency.
(c) the compressor power, in kW.
(d) the coefficient of performance.
(e) the volumetric flow rate of water from the geothermal wells, in L/min.

For $T_0 = -6.7$°C, perform a full exergy accounting of the compressor power input, and devise and evaluate at second law efficiency for the heat pump system.

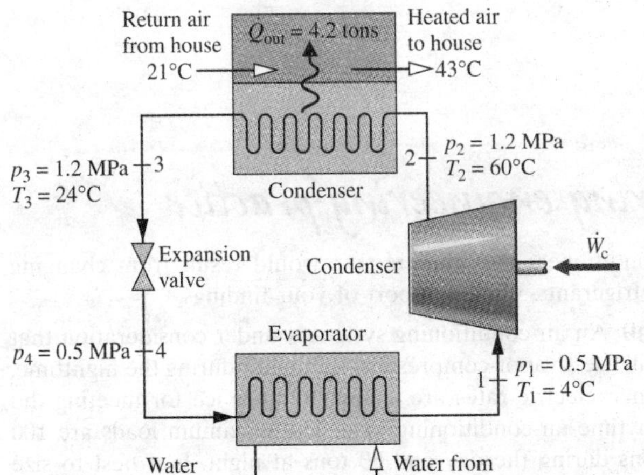

Fig. P10.42

Gas Refrigeration Systems

10.43 Air enters the compressor of an ideal Brayton refrigeration cycle at 100 kPa, 270 K. The compressor pressure ratio is 3, and the temperature at the turbine inlet is 310 K. Determine

(a) the net work input, per unit mass of air flow, in kJ/kg.
(b) the refrigeration capacity, per unit mass of air flow, in kJ/kg.
(c) the coefficient of performance.
(d) the coefficient of performance of a Carnot refrigeration cycle operating between thermal reservoirs at $T_C = 270$ K and $T_H = 310$ K, respectively.

10.44 Reconsider Problem 10.43, but include in the analysis that the compressor and turbine have isentropic efficiencies of 80 and 88%, respectively. For the modified cycle

(a) determine the coefficient of performance.
(b) develop an exergy accounting of the compressor power input, in kJ per kg of air flowing. Discuss.

Let $T_0 = 310$ K.

10.45 Plot the quantities calculated in parts (a) through (c) of Problem 10.43 versus the compressor pressure ratio ranging from 2 to 6. Repeat for compressor and turbine isentropic efficiencies of 95%, 90%, and 80%.

10.46 An ideal Brayton refrigeration cycle has a compressor pressure ratio of 6. At the compressor inlet, the pressure and temperature of the entering air are 0.14 MPa and 256 K. The temperature at the inlet of the turbine is 340 K. For a refrigerating capacity of 15 tons, determine

(a) the mass flow rate, in kg/min.
(b) the net power input, in kJ/min.
(c) the coefficient of performance.

10.47 Reconsider Problem 10.46, but include in the analysis that the compressor and turbine have isentropic efficiencies of 78% and 92%, respectively.

10.48 Air enters the compressor of an ideal Brayton refrigeration cycle at 140 kPa, 270 K, and is compressed to 420 kPa. At the turbine inlet, the temperature is 320 K and the volumetric flow rate is 0.4 m³/s. Determine

(a) the mass flow rate, in kg/s.
(b) the net power input, in kW.
(c) the refrigerating capacity, in kW.
(d) the coefficient of performance.

10.49 Air enters the compressor of a Brayton refrigeration cycle at 100 kPa, 260 K, and is compressed adiabatically to 300 kPa. Air enters the turbine at 300 kPa, 300 K, and expands adiabatically to 100 kPa. For the cycle

(a) determine the net work per unit mass of air flow, in kJ/kg, and the coefficient of performance if the compressor and turbine isentropic efficiencies are both 100%.
(b) plot the net work per unit mass of air flow, in kJ/kg, and the coefficient of performance for equal compressor and turbine isentropic efficiencies ranging from 80 to 100%.

10.50 The Brayton refrigeration cycle of Problem 10.43 is modified by the introduction of a regenerative heat exchanger. In the modified cycle, compressed air enters the regenerative heat exchanger at 310 K and is cooled to 280 K before entering the turbine. Determine, for the modified cycle,

(a) the lowest temperature, in K.
(b) the net work input per unit mass of air flow, in kJ/kg.

(c) the refrigeration capacity, per unit mass of air flow, in kJ/kg.
(d) the coefficient of performance.

10.51 Reconsider Problem 10.50, but include in the analysis that the compressor and turbine have isentropic efficiencies of 85 and 88% respectively. Answer the same questions as in Problem 10.50.

10.52 Plot the quantities calculated in parts (a) through (d) of Problem 10.50 versus the compressor pressure ratio ranging from 3 to 7. Repeat for equal compressor and turbine isentropic efficiencies of 95%, 90%, and 80%.

10.53 Consider a Brayton refrigeration cycle with a regenerative heat exchanger. Air enters the compressor at 267 K, 0.1 MPa and is compressed isentropically to 0.3 MPa. Compressed air enters the regenerative heat exchanger at 300 K and is cooled to 267 K before entering the turbine. The expansion through the turbine is isentropic. If the refrigeration capacity is 15 tons, calculate

(a) the volumetric flow rate at the compressor inlet, in m^3/min.
(b) the coefficient of performance.

10.54 Reconsider Problem 10.53, but include in the analysis that the compressor and turbine each have isentropic efficiencies of 88%. Answer the same questions for the modified cycle as in Problem 10.53.

10.55 Air at 2 bar, 380 K is extracted from a main jet engine compressor for cabin cooling. The extracted air enters a heat exchanger where it is cooled at constant pressure to 320 K through heat transfer with the ambient. It then expands adiabatically to 0.95 bar through a turbine and is discharged into the cabin. The turbine has an isentropic efficiency of 75%. If the mass flow rate of the air is 1.0 kg/s, determine

(a) the power developed by the turbine, in kW.
(b) the rate of heat transfer from the air to the ambient, in kW.

10.56 Air at 0.22 MPa, 378 K is extracted from a main jet engine compressor for cabin cooling. The extracted air enters a heat exchanger where it is cooled at constant pressure to 333 K through heat transfer with the ambient. It then expands adiabatically to 100 kPa through a turbine and is discharged into the cabin at 278 K with a mass flow rate of 100 kg/min. Determine

(a) the power developed by the turbine, in kW.
(b) the isentropic turbine efficiency.
(c) the rate of heat transfer from the air to the ambient, in kJ/min.

10.57 Air undergoes a *Stirling refrigeration cycle,* which is the reverse of the Stirling power cycle introduced in Sec. 9.11. At the beginning of the isothermal compression, the pressure and temperature are 100 kPa and 300 K, respectively. The compression ratio is 6, and the temperature during the isothermal expansion is 100 K. Determine

(a) the heat transfer during the isothermal expansion, in kJ per kg of air.
(b) the net work for the cycle, in kJ per kg of air.
(c) the coefficient of performance.

10.58 Air undergoes an *Ericsson refrigeration cycle,* which is the reverse of the Ericsson power cycle introduced in Sec. 9.11. At the beginning of the isothermal compression, the pressure and temperature are 100 kPa and 310 K, respectively. The pressure ratio during the isothermal compression is 3. During the isothermal expansion the temperature is 270 K. Determine

(a) the heat transfer for the isothermal expansion, per unit mass of air flow, in kJ/kg.
(b) the net work, per unit mass of air flow, in kJ/kg.
(c) the coefficient of performance.

Design & open ended problems: exploring engineering practice

10.1D A vapor-compression refrigeration system using propane is being designed for a household food freezer. The refrigeration system must maintain a temperature of −18°C within the freezer compartment when the temperature of the room is 32°C. Under these conditions, the steady-state heat transfer rate from the room into the freezer compartment is 1580 kJ/h. Specify operating pressures and temperatures at key points within the refrigeration system and estimate the refrigerant mass flow rate and compressor power required. Investigate and report on safety issues related to the use of propane as a refrigerant.

10.2D Refrigerant 22 is widely used as the working fluid in air conditioners and industrial chillers. However, its use is likely to be phased out in the future due to concerns about ozone depletion. Investigate which environmentally acceptable working fluids are under consideration to replace Refrigerant 22 for these uses. Determine the design issues for air conditioners and chillers that would result from changing refrigerants. Write a report of your findings.

10.3D An air-conditioning system is under consideration that will use a vapor-compression ice maker during the nighttime, when electric rates are lowest, to store ice for meeting the daytime air-conditioning load. The maximum loads are 100 tons during the day and 50 tons at night. Is it best to size the system to make enough ice at night to carry the entire daytime load or to use a smaller chiller that runs both day and night? Base your strategy on the day–night electric rate structure of your local electric utility company.

10.4D A vertical, closed-loop *geothermal* heat pump is under consideration for a new 4500 m^2 school building. The design capacity is 100 tons for both heating and cooling. The local water table is 45 m, and the ground water temperature is 13°C. Specify a ground source heat pump as well as the number and depth of wells for this application and develop a layout of vertical wells and piping required by the system.

10.5D Investigate the economic feasibility of using a waste heat-recovery heat pump for domestic water heating that employs ventilation air being discharged from a dwelling as the source. Assume typical hot water use of a family of four living in a 200 m^2 single-family dwelling in your locale. Write a report of your findings.

10.6D List the major design issues involved in using ammonia as the refrigerant in a system to provide chilled water at 4°C for air conditioning a college campus in your locale. Develop a layout of the equipment room and a schematic of the chilled water distribution system. Label the diagrams with key temperatures and make a list of capacities of each of the major pieces of equipment.

10.7D Carbon dioxide (CO_2) is an inexpensive, nontoxic, and non-flammable candidate for use as a working fluid in vapor-compression systems. Investigate the suitability of using CO_2 as the working fluid in a heat pump-water heater requiring a power input ranging from 50 to 100 kW and providing hot water ranging from 50 to 90°C. Assume that ground water is used as the source. Specify the typical operating pressures for such applications and estimate the variation in coefficient of performance for the range of hot water temperature. Prepare a memorandum summarizing your results.

10.8D In *induced hypothermia,* the temperature of a particular organ, such as the heart, is lowered to reduce the metabolic rate during surgery. Cooling is achieved by circulating blood through a heat exchanger outside the body. When the cooled blood is reintroduced through blood vessels in the organ, it cools the organ to the desired temperature. Develop the preliminary design of a vapor-compression refrigeration system to cool blood during heart surgery. Determine the necessary temperature requirements and specify a refrigerant, operating pressures and temperatures for the working fluid, and the refrigeration capacity.

10.9D Select a thermoelectric cooling system for a dormitory-sized refrigerator. Survey other students to determine their needs in order to size the unit. What type and number of thermoelectric modules are used in the unit selected, and what is its power requirement? Summarize your findings in a memorandum.

10.10D Determine the current status of *magnetic refrigeration* technology for use in the 80 to 300 K range. Does this technology hold promise as an economical alternative to vapor-compression systems? Discuss.

11

Thermodynamic Relations

Engineering Context

As seen in previous chapters, application of thermodynamic principles to engineering systems requires data for specific internal energy, enthalpy, entropy, and other properties. The **objective** of the present chapter is to introduce thermodynamic relations that allow *u, h, s,* and other thermodynamic properties of simple compressible systems to be evaluated from data that are more readily measured. Primary emphasis is on systems involving a single chemical species such as water or a mixture such as air. An introduction to general property relations for mixtures and solutions is also included.

Means are available for determining pressure, temperature, volume, and mass experimentally. In addition, the relationships between the specific heats c_v and c_p and temperature at relatively low pressure are accessible experimentally. Values for certain other thermodynamic properties also can be measured without great difficulty. However, specific internal energy, enthalpy, and entropy are among those properties that are not easily obtained experimentally, so we resort to computational procedures to determine values for these.

Learning Objectives

When you complete your study of this chapter you will be able to...

- ✓ calculate p–v–T data using equations of state involving two or more constants.
- ✓ demonstrate understanding of exact differentials involving properties and utilize the property relations developed from the exact differentials summarized in Table 11.1.
- ✓ evaluate Δu, Δh, and Δs, using the Clapeyron equation when considering phase change, and using equations of state and specific heat relations when considering single phases.
- ✓ demonstrate understanding of how tables of thermodynamic properties are constructed.
- ✓ evaluate Δh and Δs using generalized enthalpy and entropy departure charts.
- ✓ utilize mixture rules, such as Kay's Rule, to relate pressure, volume, and temperature of mixtures.
- ✓ apply thermodynamic relations to multicomponent systems.

11.1 Using Equations of State

An essential ingredient for the calculation of properties such as the specific internal energy, enthalpy, and entropy of a substance is an accurate representation of the relationship among pressure, specific volume, and temperature. The p–v–T relationship can be expressed alternatively: There are *tabular* representations, as exemplified by the steam tables. The relationship also can be expressed *graphically,* as in the p–v–T surface and compressibility factor charts. *Analytical* formulations, called ***equations of state,*** constitute a third general way of expressing the p–v–T relationship. Computer software such as *Interactive Thermodynamics: IT* also can be used to retrieve p–v–T data.

equations of state

The virial equation and the ideal gas equation are examples of analytical equations of state introduced in previous sections of the book. Analytical formulations of the p–v–T relationship are particularly convenient for performing the mathematical operations required to calculate u, h, s, and other thermodynamic properties. The object of the present section is to expand on the discussion of p–v–T relations for simple compressible substances presented in Chap. 3 by introducing some commonly used equations of state.

11.1.1 Getting Started

Recall from Sec. 3.11 that the ***virial equation of state*** can be derived from the principles of statistical mechanics to relate the p–v–T behavior of a gas to the forces between molecules. In one form, the compressibility factor Z is expanded in inverse powers of specific volume as

$$Z = 1 + \frac{B(T)}{\bar{v}} + \frac{C(T)}{\bar{v}^2} + \frac{D(T)}{\bar{v}^3} + \cdots \qquad (11.1)$$

virial equation

The coefficients B, C, D, etc. are called, respectively, the second, third, fourth, etc., virial coefficients. Each virial coefficient is a function of temperature alone. In principle, the virial coefficients are calculable if a suitable model for describing the forces of interaction between the molecules of the gas under consideration is known. Future advances in refining the theory of molecular interactions may allow the virial coefficients to be predicted with considerable accuracy from the fundamental properties of the molecules involved. However, at present, just the first two or three coefficients can be calculated and only for gases consisting of relatively simple molecules. Equation 11.1 also can be used in an empirical fashion in which the coefficients become parameters whose magnitudes are determined by fitting p–v–T data in particular realms of interest. Only the first few coefficients can be found this way, and the result is a *truncated* equation valid only for certain states.

In the limiting case where the gas molecules are assumed not to interact in any way, the second, third, and higher terms of Eq. 11.1 vanish and the equation reduces to $Z = 1$. Since $Z = p\bar{v}/\bar{R}T$, this gives the ideal gas equation of state $p\bar{v} = \bar{R}/T$. The ideal gas equation of state provides an acceptable approximation at many states, including but not limited to states where the pressure is low relative to the critical pressure and/or the temperature is high relative to the critical temperature of the substance under consideration. At many other states, however, the ideal gas equation of state provides a poor approximation.

Over 100 equations of state have been developed in an attempt to improve on the ideal gas equation of state and yet avoid the complexities inherent in a full virial series. In general, these equations exhibit little in the way of fundamental physical significance and are mainly empirical in character. Most are developed for gases, but some describe the p–v–T behavior of the liquid phase, at least qualitatively. Every equation of state is restricted to particular states. This realm of applicability is often indicated by giving an interval of pressure, or density, where the equation can be expected to represent the p–v–T behavior faithfully. When it is not stated, the realm of applicability of a given equation can be approximated by expressing the equation in terms of the compressibility factor Z and the reduced properties p_R, T_R, v'_R and superimposing the result on a generalized compressibility chart or comparing with tabulated compressibility data obtained from the literature.

11.1.2 Two-Constant Equations of State

Equations of state can be classified by the number of adjustable constants they include. Let us consider some of the more commonly used equations of state in order of increasing complexity, beginning with two-constant equations of state.

Van Der Waals Equation

An improvement over the ideal gas equation of state based on elementary molecular arguments was suggested in 1873 by van der Waals, who noted that gas molecules actually occupy more than the negligibly small volume presumed by the ideal gas model and also exert long-range attractive forces on one another. Thus, not all of the volume of a container would be available to the gas molecules, and the force they exert on the container wall would be reduced because of the attractive forces that exist between molecules. Based on these elementary molecular arguments, the ***van der Waals equation of state*** is

van der Waals equation

$$p = \frac{\overline{R}T}{\overline{v} - b} - \frac{a}{\overline{v}^2} \tag{11.2}$$

The constant b is intended to account for the finite volume occupied by the molecules, the term $a/\overline{v}^2$ accounts for the forces of attraction between molecules, and $\overline{R}$ is the universal gas constant. Note than when a and b are set to zero, the ideal gas equation of state results.

The van der Waals equation gives pressure as a function of temperature and specific volume and thus is *explicit* in pressure. Since the equation can be solved for temperature as a function of pressure and specific volume, it is also explicit in temperature. However, the equation is cubic in specific volume, so it cannot generally be solved for specific volume in terms of temperature and pressure. The van der Waals equation is *not* explicit in specific volume.

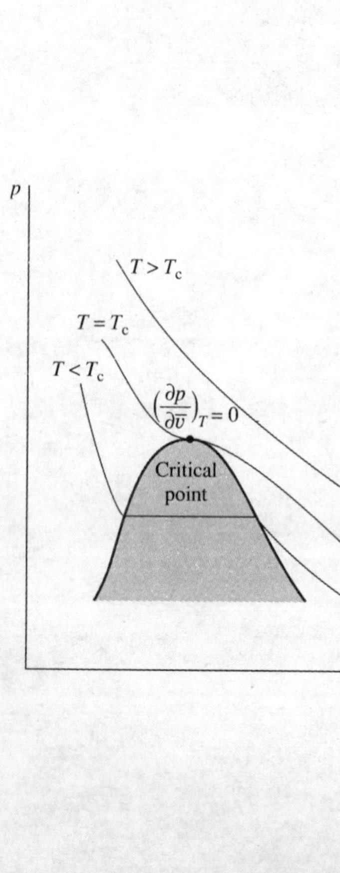

EVALUATING *a* AND *b*. The van der Waals equation is a *two-constant* equation of state. For a specified substance, values for the constants a and b can be found by fitting the equation to p–v–T data. With this approach several sets of constants might be required to cover all states of interest. Alternatively, a single set of constants for the van der Waals equation can be determined by noting that the critical isotherm passes through a point of inflection at the critical point, and the slope is zero there. Expressed mathematically, these conditions are, respectively

$$\left(\frac{\partial^2 p}{\partial \overline{v}^2}\right)_T = 0, \quad \left(\frac{\partial p}{\partial \overline{v}}\right)_T = 0 \quad \text{(critical point)} \tag{11.3}$$

Although less overall accuracy normally results when the constants a and b are determined using critical point behavior than when they are determined by fitting p–v–T data in a particular region of interest, the advantage of this approach is that the van der Waals constants can be expressed in terms of the critical pressure p_c and critical temperature T_c, as demonstrated next.

For the van der Waals equation at the critical point

$$p_c = \frac{\overline{R}T_c}{\overline{v}_c - b} - \frac{a}{\overline{v}_c^2}$$

Applying Eqs. 11.3 with the van der Waals equation gives

$$\left(\frac{\partial^2 p}{\partial \overline{v}^2}\right)_T = \frac{2\overline{R}T_c}{(\overline{v}_c - b)^3} - \frac{6a}{\overline{v}_c^4} = 0$$

$$\left(\frac{\partial p}{\partial \overline{v}}\right)_T = -\frac{\overline{R}T_c}{(\overline{v}_c - b)^2} + \frac{2a}{\overline{v}_c^3} = 0$$

Solving the foregoing three equations for a, b, and $\overline{v}_c$ in terms of the critical pressure and critical temperature

$$a = \frac{27}{64}\frac{\overline{R}^2T_c^2}{p_c} \tag{11.4a}$$

$$b = \frac{\overline{R}T_c}{8p_c} \tag{11.4b}$$

$$\overline{v}_c = \frac{3}{8}\frac{\overline{R}T_c}{p_c} \tag{11.4c}$$

Values of the van der Waals constants a and b determined from Eqs. 11.4a and 11.4b for several common substances are given in Table A-24 for pressure in bar, specific volume in m^3/kmol, and temperature in K.

GENERALIZED FORM. Introducing the compressibility factor $Z = p\overline{v}/\overline{R}T$, the reduced temperature $T_R = T/T_c$, the pseudoreduced specific volume $v'_R = p_c\overline{v}/\overline{R}T_c$, and the foregoing expressions for a and b, the van der Waals equation can be written in terms of Z, v'_R, and T_R as

$$Z = \frac{v'_R}{v'_R - 1/8} - \frac{27/64}{T_R v'_R} \tag{11.5}$$

or alternatively in terms of Z, T_R, and p_R as

$$Z^3 - \left(\frac{p_R}{8T_R} + 1\right)Z^2 + \left(\frac{27p_R}{64T_R^2}\right)Z - \frac{27p_R^2}{512T_R^3} = 0 \tag{11.6}$$

The details of these developments are left as exercises. Equation 11.5 can be evaluated for specified values of v'_R and T_R and the resultant Z values located on a generalized compressibility chart to show approximately where the equation performs satisfactorily. A similar approach can be taken with Eq. 11.6.

The compressibility factor at the critical point yielded by the van der Waals equation is determined from Eq. 11.4c as

$$Z_c = \frac{p_c\overline{v}_c}{\overline{R}T_c} = 0.375$$

Actually, Z_c varies from about 0.23 to 0.33 for most substances (see Tables A-1). Accordingly, with the set of constants given by Eqs. 11.4, the van der Waals equation is inaccurate in the vicinity of the critical point. Further study would show inaccuracy in other regions as well, so this equation is not suitable for many thermodynamic evaluations. The van der Waals equation is of interest to us primarily because it is the simplest model that accounts for the departure of actual gas behavior from the ideal gas equation of state.

Redlich–Kwong Equation

Three other two-constant equations of state that have been widely used are the Berthelot, Dieterici, and Redlich–Kwong equations. The ***Redlich–Kwong equation,*** considered by many to be the best of the two-constant equations of state, is

Redlich–Kwong equation

$$p = \frac{\overline{R}T}{\overline{v} - b} - \frac{a}{\overline{v}(\overline{v} + b)T^{1/2}} \tag{11.7}$$

This equation, proposed in 1949, is mainly empirical in nature, with no rigorous justification in terms of molecular arguments. The Redlich–Kwong equation is explicit in pressure but not in specific volume or temperature. Like the van der Waals equation, the Redlich–Kwong equation is cubic in specific volume.

Although the Redlich–Kwong equation is somewhat more difficult to manipulate mathematically than the van der Waals equation, it is more accurate, particularly at higher pressures. In recent years, several modified forms of this equation have been proposed to achieve improved accuracy. The two-constant Redlich–Kwong equation performs better than some equations of state having several adjustable constants; still, two-constant equations of state tend to be limited in accuracy as pressure (or density) increases. Increased accuracy at such states normally requires equations with a greater number of adjustable constants.

EVALUATING *a* AND *b*. As for the van der Waals equation, the constants a and b in Eq. 11.7 can be determined for a specified substance by fitting the equation to p–v–T data, with several sets of constants required to represent accurately all states of interest. Alternatively, a single set of constants in terms of the critical pressure and critical temperature can be evaluated using Eqs. 11.3, as for the van der Waals equation. The result is

$$a = a'\frac{\overline{R}^2 T_c^{5/2}}{p_c} \quad \text{and} \quad b = b'\frac{\overline{R}T_c}{p_c} \tag{11.8}$$

where $a' = 0.42748$ and $b' = 0.08664$. Evaluation of these constants is left as an exercise. Values of the Redlich–Kwong constants a and b determined from Eqs. 11.8 for several common substances are given in Table A-24 for pressure in bar, specific volume in m^3/kmol, and temperature in K.

GENERALIZED FORM. Introducing the compressibility factor Z, the reduced temperature T_R, the pseudoreduced specific volume v'_R, and the foregoing expressions for a and b, the Redlich–Kwong equation can be written as

$$Z = \frac{v'_R}{v'_R - b'} - \frac{a'}{(v'_R + b')T_R^{3/2}} \tag{11.9}$$

Equation 11.9 can be evaluated at specified values of v'_R and T_R and the resultant Z values located on a generalized compressibility chart to show the regions where the equation performs satisfactorily. With the constants given by Eqs. 11.8, the compressibility factor at the critical point yielded by the Redlich–Kwong equation is $Z_c = 0.333$, which is at the high end of the range of values for most substances, indicating that inaccuracy in the vicinity of the critical point should be expected.

In Example 11.1, the pressure of a gas is determined using three equations of state and the generalized compressibility chart. The results are compared.

Example 11.1 COMPARING EQUATIONS OF STATE

A cylindrical tank containing 4.0 kg of carbon monoxide gas at −50°C has an inner diameter of 0.2 m and a length of 1 m. Determine the pressure, in bar, exerted by the gas using **(a)** the generalized compressibility chart, **(b)** the ideal gas equation of state, **(c)** the van der Waals equation of state, **(d)** the Redlich–Kwong equation of state. Compare the results obtained.

Solution

Known: A cylindrical tank of known dimensions contains 4.0 kg of CO gas at −50°C.

Find: Determine the pressure exerted by the gas using four alternative methods.

Schematic and Given Data:

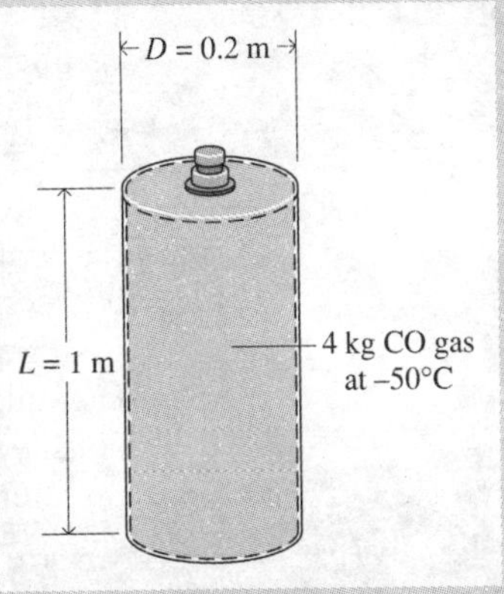

Engineering Model:

1. As shown in the accompanying figure, the closed system is taken as the gas.
2. The system is at equilibrium.

Fig. E11.1

Analysis: The molar specific volume of the gas is required in each part of the solution. Let us begin by evaluating it. The volume occupied by the gas is

$$V = \left(\frac{\pi D^2}{4}\right)L = \frac{\pi(0.2\text{ m})^2(1.0\text{ m})}{4} = 0.0314\text{ m}^3$$

The molar specific volume is then

$$\bar{v} = Mv = M\left(\frac{V}{m}\right) = \left(28\frac{\text{kg}}{\text{kmol}}\right)\left(\frac{0.0314\text{ m}^3}{4.0\text{ kg}}\right) = 0.2198\frac{\text{m}^3}{\text{kmol}}$$

(a) From Table A-1 for CO, $T_c = 133$ K, $p_c = 35$ bar. Thus, the reduced temperature T_R and pseudoreduced specific volume v'_R are, respectively

$$T_R = \frac{223\text{ K}}{133\text{ K}} = 1.68$$

$$v'_R = \frac{\bar{v}p_c}{\bar{R}T_c} = \frac{(0.2198\text{ m}^3/\text{kmol})(35 \times 10^5\text{ N/m}^2)}{(8314\text{ N}\cdot\text{m/kmol}\cdot\text{K})(133\text{ K})} = 0.696$$

Turning to Fig. A-2, $Z \approx 0.9$. Solving $Z = p\bar{v}/\bar{R}T$ for pressure and inserting known values

$$p = \frac{Z\bar{R}T}{\bar{v}} = \frac{0.9(8314\text{ N}\cdot\text{m/kmol}\cdot\text{K})(223\text{ K})}{(0.2198\text{ m}^3/\text{kmol})}\left|\frac{1\text{ bar}}{10^5\text{ N/m}^2}\right| = 75.9\text{ bar}$$

(b) The ideal gas equation of state gives

$$p = \frac{\bar{R}T}{\bar{v}} = \frac{(8314\text{ N}\cdot\text{m/kmol}\cdot\text{K})(223\text{ K})}{(0.2198\text{ m}^3/\text{kmol})}\left|\frac{1\text{ bar}}{10^5\text{ N/m}^2}\right| = 84.4\text{ bar}$$

(c) For carbon monoxide, the van der Waals constants a and b given by Eqs. 11.4 can be read directly from Table A-24. Thus

$$a = 1.474\text{ bar}\left(\frac{\text{m}^3}{\text{kmol}}\right)^2 \quad\text{and}\quad b = 0.0395\frac{\text{m}^3}{\text{kmol}}$$

Substituting into Eq. 11.2

$$p = \frac{\overline{R}T}{\overline{v} - b} - \frac{a}{\overline{v}^2}$$

$$= \frac{(8314\ \text{N}\cdot\text{m/kmol}\cdot\text{K})(223\ \text{K})}{(0.2198 - 0.0395)(\text{m}^3/\text{kmol})}\left|\frac{1\ \text{bar}}{10^5\ \text{N/m}^2}\right| - \frac{1.474\ \text{bar}\ (\text{m}^3/\text{kmol})^2}{(0.2198\ \text{m}^3/\text{kmol})^2}$$

$$= 72.3\ \text{bar}$$

Alternatively, the values for v_R' and T_R obtained in the solution of part (a) can be substituted into Eq. 11.5, giving $Z = 0.86$. Then, with $p = Z\overline{R}T/\overline{v}$, $p = 72.5$ bar. The slight difference is attributed to roundoff.

(d) For carbon monoxide, the Redlich–Kwong constants given by Eqs. 11.8 can be read directly from Table A-24. Thus

$$a = \frac{17.22\ \text{bar}\ (\text{m}^6)(\text{K})^{1/2}}{(\text{kmol})^2} \quad \text{and} \quad b = 0.02737\ \text{m}^3/\text{kmol}$$

Substituting into Eq. 11.7

$$p = \frac{\overline{R}T}{\overline{v} - b} - \frac{a}{\overline{v}(\overline{v} + b)T^{1/2}}$$

$$= \frac{(8314\ \text{N}\cdot\text{m/kmol}\cdot\text{K})(223\ \text{K})}{(0.2198 - 0.02737)\ \text{m}^3/\text{kmol}}\left|\frac{1\ \text{bar}}{10^5\ \text{N/m}^2}\right| - \frac{17.22\ \text{bar}}{(0.2198)(0.24717)(223)^{1/2}}$$

$$= 75.1\ \text{bar}$$

Alternatively, the values for v_R' and T_R obtained in the solution of part (a) can be substituted into Eq. 11.9, giving $Z = 0.89$. Then, with $p = Z\overline{R}T/\overline{v}$, $p = 75.1$ bar. In comparison to the value of part (a), the ideal gas equation of state predicts a pressure that is 11% higher and the van der Waals equation gives a value that is 5% lower. The Redlich–Kwong value is about 1% less than the value obtained using the compressibility chart.

Skills Developed: *Ability to...*
- ✓ calculate pressure using the compressibility chart, ideal gas model, and the van der Waals and Redlich-Kwong equations of state.
- ✓ perform unit conversions correctly.

Quick Quiz

Using the pressure and T_R values of part (a), check the value of Z using Fig. A-2. **Ans.** $Z \approx 0.9$.

11.1.3 Multiconstant Equations of State

To fit the p–v–T data of gases over a wide range of states, Beattie and Bridgeman proposed in 1928 a pressure-explicit equation involving five constants in addition to the gas constant. The ***Beattie–Bridgeman equation*** can be expressed in a truncated virial form as

Beattie–Bridgeman equation

$$p = \frac{\overline{R}T}{\overline{v}} + \frac{\beta}{\overline{v}^2} + \frac{\gamma}{\overline{v}^3} + \frac{\delta}{\overline{v}^4} \tag{11.10}$$

where

$$\begin{aligned} \beta &= B\overline{R}T - A - c\overline{R}/T^2 \\ \gamma &= -Bb\overline{R}T + Aa - Bc\overline{R}/T^2 \\ \delta &= Bbc\overline{R}/T^2 \end{aligned} \tag{11.11}$$

The five constants a, b, c, A, and B appearing in these equations are determined by curve fitting to experimental data.

Benedict, Webb, and Rubin extended the Beattie–Bridgeman equation of state to cover a broader range of states. The resulting equation, involving eight constants in addition to the gas constant, has been particularly successful in predicting the p–v–T behavior of *light hydrocarbons.* The ***Benedict–Webb–Rubin equation*** is

$$p = \frac{\bar{R}T}{\bar{v}} + \left(B\bar{R}T - A - \frac{C}{T^2}\right)\frac{1}{\bar{v}^2} + \frac{(b\bar{R}T - a)}{\bar{v}^3} + \frac{a\alpha}{\bar{v}^6} + \frac{c}{\bar{v}^3 T^2}\left(1 + \frac{\gamma}{\bar{v}^2}\right)\exp\left(-\frac{\gamma}{\bar{v}^2}\right) \tag{11.12}$$

Benedict–Webb–Rubin equation

Values of the constants appearing in Eq. 11.12 for five common substances are given in Table A-24 for pressure in bar, specific volume in m^3/kmol, and temperature in K. Because Eq. 11.12 has been so successful, its realm of applicability has been extended by introducing additional constants.

Equations 11.10 and 11.12 are merely representative of multiconstant equations of state. Many other multiconstant equations have been proposed. With high-speed computers, equations having 50 or more constants have been developed for representing the p–v–T behavior of different substances.

11.2 Important Mathematical Relations

Values of two independent intensive properties are sufficient to fix the state of a simple compressible system of specified mass and composition (Sec. 3.1). All other intensive properties can be determined as functions of the two independent properties: $p = p(T, v)$, $u = u(T, v)$, $h = h(T, v)$, and so on. These are all functions of two independent variables of the form $z = z(x, y)$, with x and y being the independent variables. It might also be recalled that the differential of every property is *exact* (Sec. 2.2.1). The differentials of nonproperties such as work and heat are inexact. Let us review briefly some concepts from calculus about functions of two independent variables and their differentials.

The ***exact differential*** of a function z, continuous in the variables x and y, is

exact differential

$$dz = \left(\frac{\partial z}{\partial x}\right)_y dx + \left(\frac{\partial z}{\partial y}\right)_x dy \tag{11.13a}$$

This can be expressed alternatively as

$$dz = M\,dx + N\,dy \tag{11.13b}$$

where $M = (\partial z/\partial x)_y$ and $N = (\partial z/\partial y)_x$. The coefficient M is the partial derivative of z with respect to x (the variable y being held constant). Similarly, N is the partial derivative of z with respect to y (the variable x being held constant).

If the coefficients M and N have continuous first partial derivatives, the order in which a second partial derivative of the function z is taken is immaterial. That is

$$\frac{\partial}{\partial y}\left[\left(\frac{\partial z}{\partial x}\right)_y\right]_x = \frac{\partial}{\partial x}\left[\left(\frac{\partial z}{\partial y}\right)_x\right]_y \tag{11.14a}$$

or

$$\left(\frac{\partial M}{\partial y}\right)_x = \left(\frac{\partial N}{\partial x}\right)_y \tag{11.14b}$$

which can be called the ***test for exactness,*** as discussed next.

test for exactness

In words, Eqs. 11.14 indicate that the mixed second partial derivatives of the function z are equal. The relationship in Eqs. 11.14 is both a necessary and sufficient condition for the *exactness* of a differential expression, and it may therefore be used as a test for exactness. When an expression such as $M\,dx + N\,dy$ does not meet this test, no function z exists whose differential is equal to this expression. In thermodynamics, Eq. 11.14 is not generally used to test exactness but rather to develop additional property relations. This is illustrated in Sec. 11.3 to follow.

Two other relations among partial derivatives are listed next for which applications are found in subsequent sections of this chapter. These are

$$\left(\frac{\partial x}{\partial y}\right)_z \left(\frac{\partial y}{\partial x}\right)_z = 1 \tag{11.15}$$

and

$$\left(\frac{\partial y}{\partial z}\right)_x \left(\frac{\partial z}{\partial x}\right)_y \left(\frac{\partial x}{\partial y}\right)_z = -1 \tag{11.16}$$

FOR EXAMPLE... consider the three quantities x, y, and z, any two of which may be selected as the independent variables. Thus, we can write $x = x(y, z)$ and $y = y(x, z)$. The differentials of these functions are, respectively

$$dx = \left(\frac{\partial x}{\partial y}\right)_z dy + \left(\frac{\partial x}{\partial z}\right)_y dz \quad \text{and} \quad dy = \left(\frac{\partial y}{\partial x}\right)_z dx + \left(\frac{\partial y}{\partial z}\right)_x dz$$

Eliminating dy between these two equations results in

$$\left[1 - \left(\frac{\partial x}{\partial y}\right)_z \left(\frac{\partial y}{\partial x}\right)_z\right] dx = \left[\left(\frac{\partial x}{\partial y}\right)_z \left(\frac{\partial y}{\partial z}\right)_x + \left(\frac{\partial x}{\partial z}\right)_y\right] dz \tag{11.17}$$

Since x and z can be varied independently, let us hold z constant and vary x. That is, let $dz = 0$ and $dx \neq 0$. It then follows from Eq. 11.17 that the coefficient of dx must vanish, so Eq. 11.15 must be satisfied. Similarly, when $dx = 0$ and $dz \neq 0$, the coefficient of dz in Eq. 11.17 must vanish. Introducing Eq. 11.15 into the resulting expression and rearranging gives Eq. 11.16. The details are left as an exercise.

APPLICATION. An equation of state $p = p(T, v)$ provides a specific example of a function of two independent variables. The partial derivatives $(\partial p/\partial T)_v$ and $(\partial p/\partial v)_T$ of $p(T, v)$ are important for subsequent discussions. The quantity $(\partial p/\partial T)_v$ is the partial derivative of p with respect to T (the variable v being held constant). This partial derivative represents the slope at a point on a line of constant specific volume (isometric) projected onto the p–T plane. Similarly, the partial derivative $(\partial p/\partial v)_T$ is the partial derivative of p with respect to v (the variable T being held constant). This partial derivative represents the slope at a point on a line of constant temperature (isotherm) projected on the p–v plane. The partial derivatives $(\partial p/\partial T)_v$ and $(\partial p/\partial v)_T$ are themselves intensive properties because they have unique values at each state.

The p–v–T surfaces given in Figs. 3.1 and 3.2 are graphical representations of functions of the form $p = p(v, T)$. Figure 11.1 shows the liquid, vapor, and two-phase regions of a p–v–T surface projected onto the p–v and p–T planes. Referring first to Fig. 11.1*a*, note that several isotherms are sketched. In the single-phase regions, the partial derivative $(\partial p/\partial v)_T$ giving the slope is negative at each state along an isotherm except at the critical point, where the partial derivative vanishes. Since the isotherms are horizontal in the two-phase liquid–vapor region, the partial derivative $(\partial p/\partial v)_T$ vanishes there as well. For these states, pressure is independent of specific volume and is a function of temperature only: $p = p_{sat}(T)$.

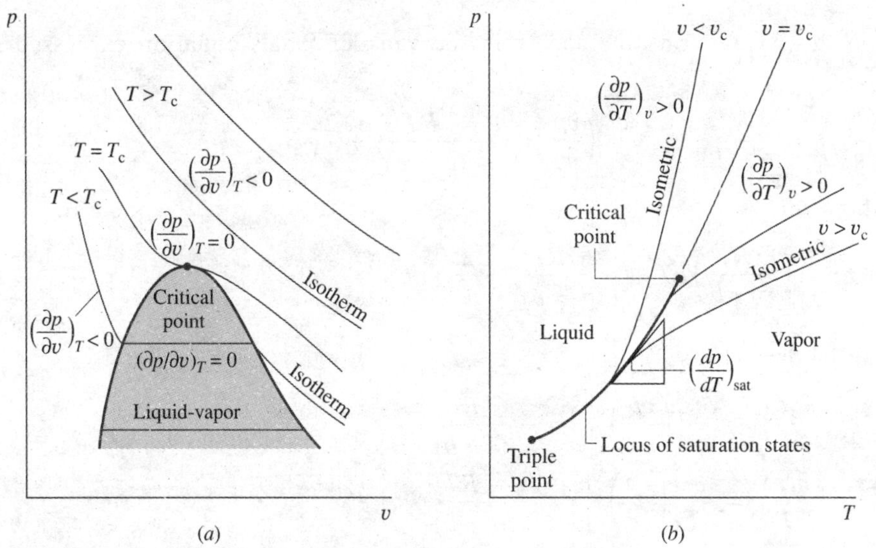

Fig. 11.1 Diagrams used to discuss $(\partial p/\partial v)_T$ and $(\partial p/\partial T)_v$. *(a)* p–v diagram. *(b)* Phase diagram.

Figure 11.1*b* shows the liquid and vapor regions with several isometrics (constant specific volume lines) superimposed. In the single-phase regions, the isometrics are nearly straight or are slightly curved and the partial derivative $(\partial p/\partial T)_v$ is positive at each state along the curves. For the two-phase liquid–vapor states corresponding to a specified value of temperature, the pressure is independent of specific volume and is determined by the temperature only. Hence, the slopes of the isometrics passing through the two-phase states corresponding to a specified temperature are all equal, being given by the slope of the saturation curve at that temperature, denoted simply as $(dp/dT)_{sat}$. For these two-phase states, $(\partial p/\partial T)_v = (dp/dT)_{sat}$.

In this section, important aspects of functions of two variables have been introduced. The following example illustrates some of these ideas using the van der Waals equation of state.

Example 11.2 APPLYING MATHEMATICAL RELATIONS

For the van der Waals equation of state, **(a)** determine an expression for the exact differential dp, **(b)** show that the mixed second partial derivatives of the result obtained in part (a) are equal, and **(c)** develop an expression for the partial derivative $(\partial v/\partial T)_p$.

Solution

Known: The equation of state is the van der Waals equation.

Find: Determine the differential dp, show that the mixed second partial derivatives of dp are equal, and develop an expression for $(\partial v/\partial T)_p$.

Analysis:

(a) By definition, the differential of a function $p = p(T, v)$ is

$$dp = \left(\frac{\partial p}{\partial T}\right)_v dT + \left(\frac{\partial p}{\partial v}\right)_T dv$$

The partial derivatives appearing in this expression obtained from the van der Waals equation expressed as $p = RT/(v - b) - a/v^2$ are

$$M = \left(\frac{\partial p}{\partial T}\right)_v = \frac{R}{v - b}, \qquad N = \left(\frac{\partial p}{\partial v}\right)_T = -\frac{RT}{(v - b)^2} + \frac{2a}{v^3}$$

Accordingly, the differential takes the form

$$dp = \left(\frac{R}{v - b}\right) dT + \left[\frac{-RT}{(v - b)^2} + \frac{2a}{v^3}\right] dv$$

(b) Calculating the mixed second partial derivatives

$$\left(\frac{\partial M}{\partial v}\right)_T = \frac{\partial}{\partial v}\left[\left(\frac{\partial p}{\partial T}\right)_v\right]_T = -\frac{R}{(v - b)^2}$$

$$\left(\frac{\partial N}{\partial T}\right)_v = \frac{\partial}{\partial T}\left[\left(\frac{\partial p}{\partial v}\right)_T\right]_v = -\frac{R}{(v - b)^2}$$

Thus, the mixed second partial derivatives are equal, as expected.

(c) An expression for $(\partial v/\partial T)_p$ can be derived using Eqs. 11.15 and 11.16. Thus, with $x = p$, $y = v$, and $z = T$, Eq. 11.16 gives

$$\left(\frac{\partial v}{\partial T}\right)_p \left(\frac{\partial p}{\partial v}\right)_T \left(\frac{\partial T}{\partial p}\right)_v = -1$$

or

$$\left(\frac{\partial v}{\partial T}\right)_p = -\frac{1}{(\partial p/\partial v)_T(\partial T/\partial p)_v}$$

Then, with $x = T$, $y = p$, and $z = v$, Eq. 11.15 gives

$$\left(\frac{\partial T}{\partial p}\right)_v = \frac{1}{(\partial p/\partial T)_v}$$

Combining these results

$$\left(\frac{\partial v}{\partial T}\right)_p = -\frac{(\partial p/\partial T)_v}{(\partial p/\partial v)_T}$$

The numerator and denominator of this expression have been evaluated in part (a), so

❶ $$\left(\frac{\partial v}{\partial T}\right)_p = -\frac{R/(v - b)}{[-RT/(v - b)^2 + 2a/v^3]}$$

which is the desired result.

Skills Developed: *Ability to...*

✓ use Eqs. 11.15 and 11.16 together with the van der Waals equation of state to develop a thermodynamic property relation.

❶ Since the van der Waals equation is cubic in specific volume, it can be solved for $v(T, p)$ at only certain states. Part (c) shows how the partial derivative $(\partial v/\partial T)_p$ can be evaluated at states where it exists.

Quick Quiz

Using the results obtained, develop an expression for $\left(\frac{\partial v}{\partial T}\right)_p$ of an ideal gas.

Ans. v/T.

11.3 Developing Property Relations

In this section, several important property relations are developed, including the expressions known as the *Maxwell* relations. The concept of a *fundamental thermodynamic function* is also introduced. These results, which are important for subsequent discussions, are obtained for simple compressible systems of fixed chemical composition using the concept of an exact differential.

11.3.1 Principal Exact Differentials

The principal results of this section are obtained using Eqs. 11.18, 11.19, 11.22, and 11.23. The first two of these equations are derived in Sec. 6.3, where they are referred to as the *T ds equations.* For present purposes, it is convenient to express them as

$$du = T\,ds - p\,dv \tag{11.18}$$

$$dh = T\,ds + v\,dp \tag{11.19}$$

The other two equations used to obtain the results of this section involve, respectively, the specific ***Helmholtz function*** ψ defined by

$$\psi = u - Ts \tag{11.20}$$

Helmholtz function

and the specific ***Gibbs function*** g defined by

$$g = h - Ts \tag{11.21}$$

Gibbs function

The Helmholtz and Gibbs functions are properties because each is defined in terms of properties. From inspection of Eqs. 11.20 and 11.21, the units of ψ and g are the same as those of u and h. These two new properties are introduced solely because they contribute to the present discussion, and no physical significance need be attached to them at this point.

Forming the differential $d\psi$

$$d\psi = du - d(Ts) = du - T\,ds - s\,dT$$

Substituting Eq. 11.18 into this gives

$$d\psi = -p\,dv - s\,dT \tag{11.22}$$

Similarly, forming the differential dg

$$dg = dh - d(Ts) = dh - T\,ds - s\,dT$$

Substituting Eq. 11.19 into this gives

$$dg = v\,dp - s\,dT \tag{11.23}$$

11.3.2 Property Relations from Exact Differentials

The four differential equations introduced above, Eqs. 11.18, 11.19, 11.22, and 11.23, provide the basis for several important property relations. Since only properties are involved, each is an exact differential exhibiting the general form $dz = M\,dx + N\,dy$ considered in Sec. 11.2. Underlying these exact differentials are, respectively, functions of the form $u(s, v)$, $h(s, p)$, $\psi(v, T)$, and $g(T, p)$. Let us consider these functions in the order given.

The differential of the function $u = u(s, v)$ is

$$du = \left(\frac{\partial u}{\partial s}\right)_v ds + \left(\frac{\partial u}{\partial v}\right)_s dv$$

By comparison with Eq. 11.18, we conclude that

$$T = \left(\frac{\partial u}{\partial s}\right)_v \tag{11.24}$$

$$-p = \left(\frac{\partial u}{\partial v}\right)_s \tag{11.25}$$

The differential of the function $h = h(s, p)$ is

$$dh = \left(\frac{\partial h}{\partial s}\right)_p ds + \left(\frac{\partial h}{\partial p}\right)_s dp$$

By comparison with Eq. 11.19, we conclude that

$$T = \left(\frac{\partial h}{\partial s}\right)_p \tag{11.26}$$

$$v = \left(\frac{\partial h}{\partial p}\right)_s \tag{11.27}$$

Similarly, the coefficients $-p$ and $-s$ of Eq. 11.22 are partial derivatives of $\psi(v, T)$

$$-p = \left(\frac{\partial \psi}{\partial v}\right)_T \tag{11.28}$$

$$-s = \left(\frac{\partial \psi}{\partial T}\right)_v \tag{11.29}$$

and the coefficients v and $-s$ of Eq. 11.23 are partial derivatives of $g(T, p)$

$$v = \left(\frac{\partial g}{\partial p}\right)_T \tag{11.30}$$

$$-s = \left(\frac{\partial g}{\partial T}\right)_p \tag{11.31}$$

As each of the four differentials introduced above is exact, the second mixed partial derivatives are equal. Thus, in Eq. 11.18, T plays the role of M in Eq. 11.14b and $-p$ plays the role of N in Eq. 11.14b, so

$$\left(\frac{\partial T}{\partial v}\right)_s = -\left(\frac{\partial p}{\partial s}\right)_v \tag{11.32}$$

In Eq. 11.19, T and v play the roles of M and N in Eq. 11.14b, respectively. Thus

$$\left(\frac{\partial T}{\partial p}\right)_s = \left(\frac{\partial v}{\partial s}\right)_p \tag{11.33}$$

Similarly, from Eqs. 11.22 and 11.23 follow

$$\left(\frac{\partial p}{\partial T}\right)_v = \left(\frac{\partial s}{\partial v}\right)_T \tag{11.34}$$

$$\left(\frac{\partial v}{\partial T}\right)_p = -\left(\frac{\partial s}{\partial p}\right)_T \tag{11.35}$$

Equations 11.32 through 11.35 are known as the ***Maxwell relations.*** ***Maxwell relations***

Since each of the properties T, p, v, s appears on the left side of two of the eight equations, Eqs. 11.24 through 11.31, four additional property relations can be obtained by equating such expressions. They are

$$\left(\frac{\partial u}{\partial s}\right)_v = \left(\frac{\partial h}{\partial s}\right)_p, \qquad \left(\frac{\partial u}{\partial v}\right)_s = \left(\frac{\partial \psi}{\partial v}\right)_T$$
$$\left(\frac{\partial h}{\partial p}\right)_s = \left(\frac{\partial g}{\partial p}\right)_T, \qquad \left(\frac{\partial \psi}{\partial T}\right)_v = \left(\frac{\partial g}{\partial T}\right)_p \tag{11.36}$$

Equations 11.24 through 11.36, which are listed in Table 11.1 for ease of reference, are 16 property relations obtained from Eqs. 11.18, 11.19, 11.22, and 11.23, using the concept of an exact differential. Since Eqs. 11.19, 11.22, and 11.23 can themselves be derived from Eq. 11.18, the important role of the first $T\,dS$ equation in developing property relations is apparent.

Table 11.1

Summary of Property Relations from Exact Differentials

Basic relations:

from $u = u(s, v)$

$$T = \left(\frac{\partial u}{\partial s}\right)_v \tag{11.24}$$

$$-p = \left(\frac{\partial u}{\partial v}\right)_s \tag{11.25}$$

from $h = h(s, p)$

$$T = \left(\frac{\partial h}{\partial s}\right)_p \tag{11.26}$$

$$v = \left(\frac{\partial h}{\partial p}\right)_s \tag{11.27}$$

from $\psi = \psi(v, T)$

$$-p = \left(\frac{\partial \psi}{\partial v}\right)_T \tag{11.28}$$

$$-s = \left(\frac{\partial \psi}{\partial T}\right)_v \tag{11.29}$$

from $g = g(T, p)$

$$v = \left(\frac{\partial g}{\partial p}\right)_T \tag{11.30}$$

$$-s = \left(\frac{\partial g}{\partial T}\right)_p \tag{11.31}$$

Maxwell relations:

$$\left(\frac{\partial T}{\partial v}\right)_s = -\left(\frac{\partial p}{\partial s}\right)_v \tag{11.32}$$

$$\left(\frac{\partial T}{\partial p}\right)_s = \left(\frac{\partial v}{\partial s}\right)_p \tag{11.33}$$

$$\left(\frac{\partial p}{\partial T}\right)_v = \left(\frac{\partial s}{\partial v}\right)_T \tag{11.34}$$

$$\left(\frac{\partial v}{\partial T}\right)_p = -\left(\frac{\partial s}{\partial p}\right)_T \tag{11.35}$$

Additional relations:

$$\left(\frac{\partial u}{\partial s}\right)_v = \left(\frac{\partial h}{\partial s}\right)_p \qquad \left(\frac{\partial u}{\partial v}\right)_s = \left(\frac{\partial \psi}{\partial v}\right)_T \tag{11.36}$$

$$\left(\frac{\partial h}{\partial p}\right)_s = \left(\frac{\partial g}{\partial p}\right)_T \qquad \left(\frac{\partial \psi}{\partial T}\right)_v = \left(\frac{\partial g}{\partial T}\right)_p$$

The utility of these 16 property relations is demonstrated in subsequent sections of this chapter. However, to give a specific illustration at this point, suppose the partial derivative $(\partial s/\partial v)_T$ involving entropy is required for a certain purpose. The Maxwell relation Eq. 11.34 would allow the derivative to be determined by evaluating the partial derivative $(\partial p/\partial T)_v$, which can be obtained using p–v–T data only. Further elaboration is provided in Example 11.3.

Example 11.3 APPLYING THE MAXWELL RELATIONS

Evaluate the partial derivative $(\partial s/\partial v)_T$ for water vapor at a state fixed by a temperature of 240°C and a specific volume of 0.4646 m^3/kg. **(a)** Use the Redlich–Kwong equation of state and an appropriate Maxwell relation. **(b)** Check the value obtained using steam table data.

Solution

Known: The system consists of a fixed amount of water vapor at 240°C and 0.4646 m^3/kg.

Find: Determine the partial derivative $(\partial s/\partial v)_T$ employing the Redlich–Kwong equation of state, together with a Maxwell relation. Check the value obtained using steam table data.

Engineering Model:

1. The system consists of a fixed amount of water at a known equilibrium state.
2. Accurate values for $(\partial s/\partial T)_v$ in the neighborhood of the given state can be determined from the Redlich–Kwong equation of state.

Analysis:

(a) The Maxwell relation given by Eq. 11.34 allows $(\partial s/\partial v)_T$ to be determined from the p–v–T relationship. That is

$$\left(\frac{\partial s}{\partial v}\right)_T = \left(\frac{\partial p}{\partial T}\right)_v$$

The partial derivative $(\partial p/\partial T)_v$ obtained from the Redlich–Kwong equation, Eq. 11.7, is

$$\left(\frac{\partial p}{\partial T}\right)_v = \frac{\overline{R}}{\overline{v} - b} + \frac{a}{2\overline{v}(\overline{v} + b)T^{3/2}}$$

At the specified state, the temperature is 513 K and the specific volume on a molar basis is

$$\overline{v} = 0.4646\frac{\text{m}^3}{\text{kg}}\left(\frac{18.02\ \text{kg}}{\text{kmol}}\right) = 8.372\frac{\text{m}^3}{\text{kmol}}$$

From Table A-24

$$a = 142.59\ \text{bar}\left(\frac{\text{m}^3}{\text{kmol}}\right)^2(\text{K})^{1/2}, \qquad b = 0.0211\frac{\text{m}^3}{\text{kmol}}$$

Substituting values into the expression for $(\partial p/\partial T)_v$

$$\left(\frac{\partial p}{\partial T}\right)_v = \frac{\left(8314\dfrac{\text{N}\cdot\text{m}}{\text{kmol}\cdot\text{K}}\right)}{(8.372 - 0.0211)\dfrac{\text{m}^3}{\text{kmol}}} + \frac{142.59\ \text{bar}\left(\dfrac{\text{m}^3}{\text{kmol}}\right)^2(\text{K})^{1/2}}{2\left(8.372\dfrac{\text{m}^2}{\text{kmol}}\right)\left(8.3931\dfrac{\text{m}^3}{\text{kmol}}\right)(513\ \text{K})^{3/2}}\left|\frac{10^5\ \text{N/m}^2}{1\ \text{bar}}\right|$$

$$= \left(1004.3\frac{\text{N}\cdot\text{m}}{\text{m}^3\cdot\text{K}}\right)\left|\frac{1\ \text{kJ}}{10^3\ \text{N}\cdot\text{m}}\right|$$

$$= 1.0043\frac{\text{kJ}}{\text{m}^3\cdot\text{K}}$$

Accordingly

$$\left(\frac{\partial s}{\partial v}\right)_T = 1.0043 \frac{\text{kJ}}{\text{m}^3 \cdot \text{K}}$$

(b) A value for $(\partial s/\partial v)_T$ can be estimated using a graphical approach with steam table data, as follows: At 240°C, Table A-4 provides the values for specific entropy s and specific volume v tabulated below

T = 240°C		
p (bar)	s (kJ/kg · K)	v (m^3/kg)
1.0	7.9949	2.359
1.5	7.8052	1.570
3.0	7.4774	0.781
5.0	7.2307	0.4646
7.0	7.0641	0.3292
10.0	6.8817	0.2275

With the values for s and v listed in the table, the plot in Fig. E11.3*a* giving s versus v can be prepared. Note that a line representing the tangent to the curve at the given state is shown on the plot. The pressure at this state is 5 bar. The slope of the tangent is $(\partial s/\partial v)_T \approx 1.0 \text{ kJ/m}^3 \cdot \text{K}$. Thus, the value of $(\partial s/\partial v)_T$ obtained using the Redlich–Kwong equation agrees closely with the result determined graphically using steam table data.

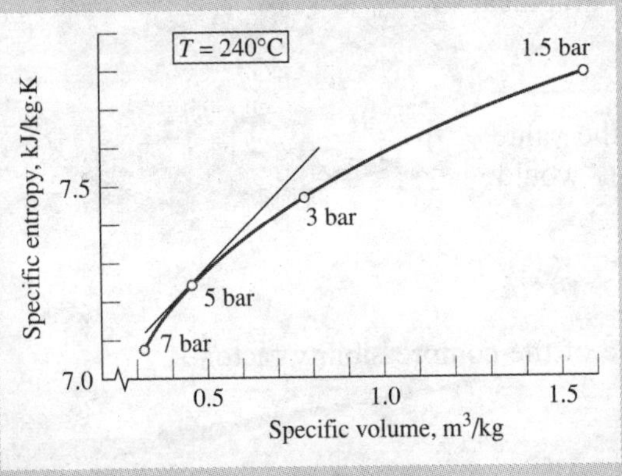

Fig. E11.3a

Alternative Solution: Alternatively, the partial derivative $(\partial s/\partial v)_T$ can be estimated using numerical methods and computer-generated data. The following *IT* code illustrates *one way* the partial derivative, denoted dsdv, can be estimated:

```
v = 0.4646 // m³/kg
T = 240 // °C
v2 = v + dv
v1 = v – dv
dv = 0.2
v2 = v_PT ("Water/Steam", p2, T)
v1 = v_PT ("Water/Steam", p1, T)
s2 = s_PT ("Water/Steam", p2, T)
s1 = s_PT ("Water/Steam", p1, T)
dsdv = (s2 – s1)/(v2 – v1)
```

Using the **Explore** button, sweep dv from 0.0001 to 0.2 in steps of 0.001. Then, using the **Graph** button, the following graph can be constructed:

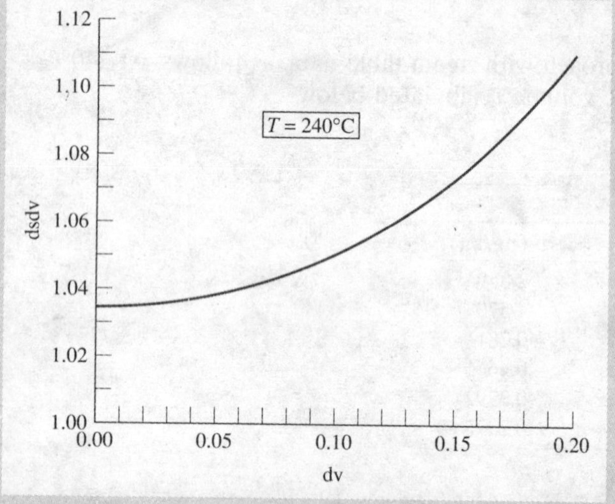

Fig. E11.3b

From the computer data, the *y*-intercept of the graph is

1 $$\left(\frac{\partial s}{\partial v}\right)_T = \lim_{\Delta v \to 0}\left(\frac{\Delta s}{\Delta v}\right)_T \approx 1.033 \frac{\text{kJ}}{\text{m}^3 \cdot \text{K}}$$

This answer is an estimate because it relies on a numerical approximation of the partial derivative based on the equation of state that underlies the *steam tables*. The values obtained using the Redlich–Kwong equation of state and the graphical method agree with this result.

1 It is left as an exercise to show that, in accordance with Eq. 11.34, the value of $(\partial p/\partial T)_v$ estimated by a procedure like the one used for $(\partial s/\partial v)_T$ would agree with the value obtained here.

Skills Developed: *Ability to...*
- ✓ apply a Maxwell relation to evaluate a thermodynamic expression.
- ✓ apply the Redlich–Kwong equation.
- ✓ perform a comparison with data from the steam table using graphical and computer-based methods.

Quick Quiz

For steam at $T = 240°\text{C}$, $v = 0.4646\ \text{m}^3/\text{kg}$, $p = 5$ bar, calculate the value of the compressibility factor Z.
Ans. 0.981.

11.3.3 Fundamental Thermodynamic Functions

A *fundamental thermodynamic function* provides a complete description of the thermodynamic state. In the case of a pure substance with two independent properties, the ***fundamental thermodynamic function*** can take one of the following four forms:

fundamental function

$$\begin{aligned} u &= u(s, v) \\ h &= h(s, p) \\ \psi &= \psi(T, v) \\ g &= g(T, p) \end{aligned} \tag{11.37}$$

Of the four fundamental functions listed in Eqs. 11.37, the Helmholtz function ψ and the Gibbs function g have the greatest importance for subsequent discussions (see

Sec. 11.6.2). Accordingly, let us discuss the fundamental function concept with reference to ψ and g.

In principle, all properties of interest can be determined from a fundamental thermodynamic function by differentiation and combination. ➡ **FOR EXAMPLE...** consider a fundamental function of the form $\psi(T, v)$. The properties v and T, being the independent variables, are specified to fix the state. The pressure p at this state can be determined from Eq. 11.28 by differentiation of $\psi(T, v)$. Similarly, the specific entropy s at the state can be found from Eq. 11.29 by differentiation. By definition, $\psi = u - Ts$, so the specific internal energy is obtained as

$$u = \psi + Ts$$

With u, p, and v known, the specific enthalpy can be found from the definition $h = u + pv$. Similarly, the specific Gibbs function is found from the definition, $g = h - Ts$. The specific heat c_v can be determined by further differentiation, $c_v = (\partial u/\partial T)_v$. Other properties can be calculated with similar operations. ⬅

➡ **FOR EXAMPLE...** consider a fundamental function of the form $g(T, p)$. The properties T and p are specified to fix the state. The specific volume and specific entropy at this state can be determined by differentiation from Eqs. 11.30 and 11.31, respectively. By definition, $g = h - Ts$, so the specific enthalpy is obtained as

$$h = g + Ts$$

With h, p, and v known, the specific internal energy can be found from $u = h - pv$. The specific heat c_p can be determined by further differentiation, $c_p = (\partial h/\partial T)_p$. Other properties can be calculated with similar operations. ⬅

Like considerations apply for functions of the form $u(s, v)$ and $h(s, p)$, as can readily be verified. Note that a Mollier diagram provides a graphical representation of the fundamental function $h(s, p)$.

11.4 Evaluating Changes in Entropy, Internal Energy, and Enthalpy

With the introduction of the Maxwell relations, we are in a position to develop thermodynamic relations that allow changes in entropy, internal energy, and enthalpy to be evaluated from measured property data. The presentation begins by considering relations applicable to phase changes and then turns to relations for use in single-phase regions.

11.4.1 Considering Phase Change

The object of this section is to develop relations for evaluating the changes in specific entropy, internal energy, and enthalpy accompanying a change of phase at fixed temperature and pressure. A principal role is played by the *Clapeyron equation,* which allows the change in enthalpy during vaporization, sublimation, or melting at a constant temperature to be evaluated from pressure-specific volume–temperature data pertaining to the phase change. Thus, the present discussion provides important examples of how p–v–T measurements can lead to the determination of other property changes, namely Δs, Δu, and Δh for a change of phase.

Consider a change in phase from saturated liquid to saturated vapor at fixed temperature. For an isothermal phase change, pressure also remains constant, so Eq. 11.19 reduces to

$$dh = T\, ds$$

Integration of this expression gives

$$s_g - s_f = \frac{h_g - h_f}{T} \tag{11.38}$$

Hence, the change in specific entropy accompanying a phase change from saturated liquid to saturated vapor at temperature T can be determined from the temperature and the change in specific enthalpy.

The change in specific internal energy during the phase change can be determined using the definition $h = u + pv$.

$$u_g - u_f = h_g - h_f - p(v_g - v_f) \tag{11.39}$$

Thus, the change in specific internal energy accompanying a phase change at temperature T can be determined from the temperature and the changes in specific volume and enthalpy.

CLAPEYRON EQUATION. The change in specific enthalpy required by Eqs. 11.38 and 11.39 can be obtained using the Clapeyron equation. To derive the Clapeyron equation, begin with the Maxwell relation

$$\left(\frac{\partial s}{\partial v}\right)_T = \left(\frac{\partial p}{\partial T}\right)_v \tag{11.34}$$

During a phase change at fixed temperature, the pressure is independent of specific volume and is determined by temperature alone. Thus, the quantity $(\partial p/\partial T)_v$ is determined by the temperature and can be represented as

$$\left(\frac{\partial p}{\partial T}\right)_v = \left(\frac{dp}{dT}\right)_{\text{sat}}$$

where "sat" indicates that the derivative is the slope of the saturation pressure–temperature curve at the point determined by the temperature held constant during the phase change (Sec. 11.2). Combining the last two equations gives

$$\left(\frac{\partial s}{\partial v}\right)_T = \left(\frac{dp}{dT}\right)_{\text{sat}}$$

Since the right side of this equation is fixed when the temperature is specified, the equation can be integrated to give

$$s_g - s_f = \left(\frac{dp}{dT}\right)_{\text{sat}} (v_g - v_f)$$

Clapeyron equation

Introducing Eq. 11.38 into this expression results in the ***Clapeyron equation***

$$\left(\frac{dp}{dT}\right)_{\text{sat}} = \frac{h_g - h_f}{T(v_g - v_f)} \tag{11.40}$$

Equation 11.40 allows $(h_g - h_f)$ to be evaluated using only p–v–T data pertaining to the phase change. In instances when the enthalpy change is also measured, the Clapeyron equation can be used to check the consistency of the data. Once the specific enthalpy change is determined, the corresponding changes in specific entropy and specific internal energy can be found from Eqs. 11.38 and 11.39, respectively.

Equations 11.38, 11.39, and 11.40 also can be written for sublimation or melting occurring at constant temperature and pressure. In particular, the Clapeyron equation would take the form

$$\left(\frac{dp}{dT}\right)_{sat} = \frac{h'' - h'}{T(v'' - v')} \tag{11.41}$$

where $''$ and $'$ denote the respective phases, and $(dp/dT)_{sat}$ is the slope of the relevant saturation pressure–temperature curve.

The Clapeyron equation shows that the slope of a saturation line on a phase diagram depends on the signs of the specific volume and enthalpy changes accompanying the phase change. In most cases, when a phase change takes place with an increase in specific enthalpy, the specific volume also increases, and $(dp/dT)_{sat}$ is positive. However, in the case of the melting of ice and a few other substances, the specific volume decreases on melting. The slope of the saturated solid–liquid curve for these few substances is negative, as was pointed out in Sec. 3.2.2 in the discussion of phase diagrams.

An approximate form of Eq. 11.40 can be derived when the following two idealizations are justified: (1) v_f is negligible in comparison to v_g, and (2) the pressure is low enough that v_g can be evaluated from the ideal gas equation of state as $v_g = RT/p$. With these, Eq. 11.40 becomes

$$\left(\frac{dp}{dT}\right)_{sat} = \frac{h_g - h_f}{RT^2/p}$$

which can be rearranged to read

$$\left(\frac{d \ln p}{dT}\right)_{sat} = \frac{h_g - h_f}{RT^2} \tag{11.42}$$

Equation 11.42 is called the ***Clausius–Clapeyron equation.*** A similar expression applies for the case of sublimation.

Clausius–Clapeyron equation

The use of the Clapeyron equation in any of the foregoing forms requires an accurate representation for the relevant saturation pressure–temperature curve. This must not only depict the pressure–temperature variation accurately but also enable accurate values of the derivative $(dp/dT)_{sat}$ to be determined. Analytical representations in the form of equations are commonly used. Different equations for different portions of the pressure–temperature curves may be required. These equations can involve several constants. One form that is used for the vapor-pressure curves is the four-constant equation

$$\ln p_{sat} = A + \frac{B}{T} + C \ln T + DT$$

in which the constants A, B, C, D are determined empirically.

The use of the Clapeyron equation for evaluating changes in specific entropy, internal energy, and enthalpy accompanying a phase change at fixed T and p is illustrated in the next example.

Example 11.4 APPLYING THE CLAPEYRON EQUATION

Using p–v–T data for saturated water, calculate at 100°C **(a)** $h_g - h_f$, **(b)** $u_g - u_f$, **(c)** $s_g - s_f$. Compare with the respective steam table value.

Solution

Known: The system consists of a unit mass of saturated water at 100°C.

Find: Using saturation data, determine at 100°C the change on vaporization of the specific enthalpy, specific internal energy, and specific entropy, and compare with the respective steam table value.

Analysis: For comparison, Table A-2 gives at 100°C, $h_g - h_f = 2257.0$ kJ/kg, $u_g - u_f = 2087.6$ kJ/kg, $s_g - s_f = 6.048$ kJ/kg · K.

(a) The value of $h_g - h_f$ can be determined from the Clapeyron equation, Eq. 11.40, expressed as

$$h_g - h_f = T(v_g - v_f)\left(\frac{dp}{dT}\right)_{sat}$$

This equation requires a value for the slope $(dp/dT)_{sat}$ of the saturation pressure–temperature curve at the specified temperature.

The required value for $(dp/dT)_{sat}$ at 100°C can be estimated graphically as follows. Using saturation pressure–temperature data from the steam tables, the accompanying plot can be prepared. Note that a line drawn tangent to the curve at 100°C is shown on the plot. The slope of this tangent line is about 3570 N/m² · K. Accordingly, at 100°C

$$\left(\frac{dp}{dT}\right)_{sat} \approx 3570 \frac{\text{N}}{\text{m}^2 \cdot \text{K}}$$

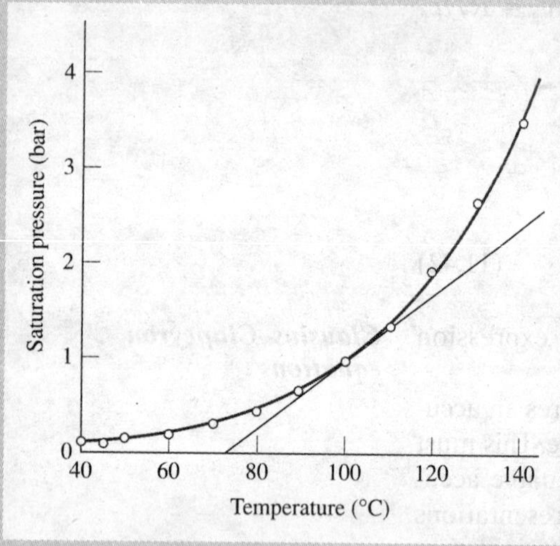

Fig. E11.4

Inserting values into the above equation for $h_g - h_f$ gives

$$h_g - h_f = (373.15\ \text{K})(1.673 - 1.0435 \times 10^{-3})\left(\frac{\text{m}^3}{\text{kg}}\right)\left(3570 \frac{\text{N}}{\text{m}^2 \cdot \text{K}}\right)\left|\frac{1\ \text{kJ}}{10^3\ \text{N} \cdot \text{m}}\right|$$

$$= 2227\ \text{kJ/kg}$$

This value is about 1% less than the value read from the steam tables.

❶ Alternatively, the derivative $(dp/dT)_{sat}$ can be estimated using numerical methods and computer-generated data. The following *IT* code illustrates *one way* the derivative, denoted dpdT, can be estimated:

```
T = 100 // °C
dT = 0.001
T1 = T – dT
T2 = T + dT
p1 = Psat ("Water/Steam", T1) // bar
p2 = Psat ("Water/Steam", T2) // bar
dpdT = ((p2 – p1) / (T2 – T1)) * 100000
```

Using the **Explore** button, sweep dT from 0.001 to 0.01 in steps of 0.001. Then, reading the limiting value from the computer data

$$\left(\frac{dp}{dT}\right)_{sat} \approx 3616 \frac{\text{N}}{\text{m}^2 \cdot \text{K}}$$

When this value is used in the above expression for $h_g - h_f$, the result is $h_g - h_f = 2256$ kJ/kg, which agrees very closely with the value read from the steam tables.

(b) With Eq. 11.39

$$u_g - u_f = h_g - h_f - p_{sat}(v_g - v_f)$$

Inserting the *IT* result for $(h_g - h_f)$ from part (a) together with saturation data at 100°C

$$u_g - u_f = 2256\frac{\text{kJ}}{\text{kg}} - \left(1.014 \times 10^5\frac{\text{N}}{\text{m}^2}\right)\left(1.672\frac{\text{m}^3}{\text{kg}}\right)\left|\frac{1\text{ kJ}}{10^3\text{ N}\cdot\text{m}}\right|$$
$$= 2086.5\frac{\text{kJ}}{\text{kg}}$$

which also agrees closely with the value from the steam tables.

(c) With Eq. 11.38 and the *IT* result for $(h_g - h_f)$ from part (a)

$$s_g - s_f = \frac{h_g - h_f}{T} = \frac{2256\text{ kJ/kg}}{373.15\text{ K}} = 6.046\frac{\text{kJ}}{\text{kg}\cdot\text{K}}$$

which again agrees very closely with the steam table value.

Skills Developed: *Ability to...*

- ✓ use the Clapeyron Equation with *p-v-T* data for saturated water to evaluate, u_{fg}, h_{fg}, and s_{fg}.
- ✓ use graphical and computer-based methods to evaluate thermodynamic property data and relations.

❶ Also, $(dp/dT)_{sat}$ might be obtained by differentiating an analytical expression for the vapor pressure curve, as discussed on page 587 above.

Quick Quiz

Use the *IT* result $(dp/dT)_{sat} = 3616$ N/m^2 · K to extrapolate the saturation pressure, in bar, at 105°C.

Ans. 1.195 bar.

11.4.2 Considering Single-Phase Regions

The objective of the present section is to derive expressions for evaluating Δs, Δu, and Δh between states in single-phase regions. These expressions require both p–v–T data and appropriate specific heat data. Since single-phase regions are under present consideration, any two of the properties pressure, specific volume, and temperature can be regarded as the independent properties that fix the state. Two convenient choices are T, v and T, p.

T AND v AS INDEPENDENT PROPERTIES. With temperature and specific volume as the independent properties that fix the state, the specific entropy can be regarded as a function of the form $s = s(T, v)$. The differential of this function is

$$ds = \left(\frac{\partial s}{\partial T}\right)_v dT + \left(\frac{\partial s}{\partial v}\right)_T dv$$

The partial derivative $(\partial s/\partial v)_T$ appearing in this expression can be replaced using the Maxwell relation, Eq. 11.34, giving

$$ds = \left(\frac{\partial s}{\partial T}\right)_v dT + \left(\frac{\partial p}{\partial T}\right)_v dv \tag{11.43}$$

The specific internal energy also can be regarded as a function of T and v: $u = u(T, v)$. The differential of this function is

$$du = \left(\frac{\partial u}{\partial T}\right)_v dT + \left(\frac{\partial u}{\partial v}\right)_T dv$$

With $c_v = (\partial u/\partial T)_v$

$$du = c_v\, dT + \left(\frac{\partial u}{\partial v}\right)_T dv \tag{11.44}$$

Substituting Eqs. 11.43 and 11.44 into $du = T\,ds - p\,dv$ and collecting terms results in

$$\left[\left(\frac{\partial u}{\partial v}\right)_T + p - T\left(\frac{\partial p}{\partial T}\right)_v\right] dv = \left[T\left(\frac{\partial s}{\partial T}\right)_v - c_v\right] dT \tag{11.45}$$

Since specific volume and temperature can be varied independently, let us hold specific volume constant and vary temperature. That is, let $dv = 0$ and $dT \neq 0$. It then follows from Eq. 11.45 that

$$\left(\frac{\partial s}{\partial T}\right)_v = \frac{c_v}{T} \tag{11.46}$$

Similarly, suppose that $dT = 0$ and $dv \neq 0$. It then follows that

$$\left(\frac{\partial u}{\partial v}\right)_T = T\left(\frac{\partial p}{\partial T}\right)_v - p \tag{11.47}$$

Equations 11.46 and 11.47 are additional examples of useful thermodynamic property relations. ➡ **FOR EXAMPLE...** Equation 11.47, which expresses the dependence of the specific internal energy on specific volume at fixed temperature, allows us to demonstrate that the internal energy of a gas whose equation of state is $pv = RT$ depends on temperature alone, a result first discussed in Sec. 3.12. Equation 11.47 requires the partial derivative $(\partial p/\partial T)_v$. If $p = RT/v$, the derivative is $(\partial p/\partial T)_v = R/v$. Introducing this, Eq. 11.47 gives

$$\left(\frac{\partial u}{\partial v}\right)_T = T\left(\frac{\partial p}{\partial T}\right)_v - p = T\left(\frac{R}{v}\right) - p = p - p = 0$$

This demonstrates that when $pv = RT$, the specific internal energy is independent of specific volume and depends on temperature alone. ⬅

Continuing the discussion, when Eq. 11.46 is inserted in Eq. 11.43, the following expression results.

$$ds = \frac{c_v}{T} dT + \left(\frac{\partial p}{\partial T}\right)_v dv \tag{11.48}$$

Inserting Eq. 11.47 into Eq. 11.44 gives

$$du = c_v\, dT + \left[T\left(\frac{\partial p}{\partial T}\right)_v - p\right] dv \tag{11.49}$$

Observe that the right sides of Eqs. 11.48 and 11.49 are expressed solely in terms of p, v, T, and c_v.

Changes in specific entropy and internal energy between two states are determined by integration of Eqs. 11.48 and 11.49, respectively.

$$s_2 - s_1 = \int_1^2 \frac{c_v}{T} dT + \int_1^2 \left(\frac{\partial p}{\partial T}\right)_v dv \tag{11.50}$$

$$u_2 - u_1 = \int_1^2 c_v\, dT + \int_1^2 \left[T\left(\frac{\partial p}{\partial T}\right)_v - p\right] dv \tag{11.51}$$

To integrate the first term on the right of each of these expressions, the variation of c_v with temperature at one fixed specific volume (isometric) is required. Integration of the second term requires knowledge of the p–v–T relation at the states of interest. An equation of state explicit in pressure would be particularly convenient for evaluating the integrals involving $(\partial p/\partial T)_v$. The accuracy of the resulting specific entropy and internal energy changes would depend on the accuracy of this derivative. In cases where the integrands of Eqs. 11.50 and 11.51 are too complicated to be integrated in closed form they may be evaluated numerically. Whether closed-form or numerical integration is used, attention must be given to the path of integration.

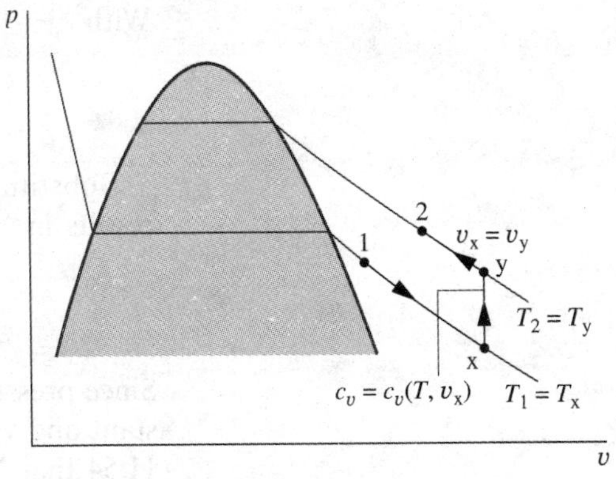

Fig. 11.2 Integration path between two vapor states.

FOR EXAMPLE... let us consider the evaluation of Eq. 11.51. Referring to Fig. 11.2, if the specific heat c_v is known as a function of temperature along the isometric (constant specific volume) passing through the states x and y, one possible path of integration for determining the change in specific internal energy between states 1 and 2 is 1–x–y–2. The integration would be performed in three steps. Since the temperature is constant from state 1 to state x, the first integral of Eq. 11.51 would vanish, so

$$u_x - u_1 = \int_{v_1}^{v_x} \left[T\left(\frac{\partial p}{\partial T}\right)_v - p \right] dv$$

From state x to y, the specific volume is constant and c_v is known as a function of temperature only, so

$$u_y - u_x = \int_{T_x}^{T_y} c_v \, dT$$

where $T_x = T_1$ and $T_y = T_2$. From state y to state 2, the temperature is constant once again, and

$$u_2 - u_y = \int_{v_y = v_x}^{v_2} \left[T\left(\frac{\partial p}{\partial T}\right)_v - p \right] dv$$

When these are added, the result is the change in specific internal energy between states 1 and 2.

***T* AND *p* AS INDEPENDENT PROPERTIES.** In this section a presentation parallel to that considered above is provided for the choice of temperature and pressure as the independent properties. With this choice for the independent properties, the specific entropy can be regarded as a function of the form $s = s(T, p)$. The differential of this function is

$$ds = \left(\frac{\partial s}{\partial T}\right)_p dT + \left(\frac{\partial s}{\partial p}\right)_T dp$$

The partial derivative $(\partial s/\partial p)_T$ appearing in this expression can be replaced using the Maxwell relation, Eq. 11.35, giving

$$ds = \left(\frac{\partial s}{\partial T}\right)_p dT - \left(\frac{\partial v}{\partial T}\right)_p dp \tag{11.52}$$

The specific enthalpy also can be regarded as a function of T and p: $h = h(T, p)$. The differential of this function is

$$dh = \left(\frac{\partial h}{\partial T}\right)_p dT + \left(\frac{\partial h}{\partial p}\right)_T dp$$

With $c_p = (\partial h/\partial T)_p$

$$dh = c_p\, dT + \left(\frac{\partial h}{\partial p}\right)_T dp \tag{11.53}$$

Substituting Eqs. 11.52 and 11.53 into $dh = T\,ds + v\,dp$ and collecting terms results in

$$\left[\left(\frac{\partial h}{\partial p}\right)_T + T\left(\frac{\partial v}{\partial T}\right)_p - v\right] dp = \left[T\left(\frac{\partial s}{\partial T}\right)_p - c_p\right] dT \tag{11.54}$$

Since pressure and temperature can be varied independently, let us hold pressure constant and vary temperature. That is, let $dp = 0$ and $dT \neq 0$. It then follows from Eq. 11.54 that

$$\left(\frac{\partial s}{\partial T}\right)_p = \frac{c_p}{T} \tag{11.55}$$

Similarly, when $dT = 0$ and $dp \neq 0$, Eq. 11.54 gives

$$\left(\frac{\partial h}{\partial p}\right)_T = v - T\left(\frac{\partial v}{\partial T}\right)_p \tag{11.56}$$

Equations 11.55 and 11.56, like Eqs. 11.46 and 11.47, are useful thermodynamic property relations.

When Eq. 11.55 is inserted in Eq. 11.52, the following equation results:

$$ds = \frac{c_p}{T} dT - \left(\frac{\partial v}{\partial T}\right)_p dp \tag{11.57}$$

Introducing Eq. 11.56 into Eq. 11.53 gives

$$dh = c_p\, dT + \left[v - T\left(\frac{\partial v}{\partial T}\right)_p\right] dp \tag{11.58}$$

Observe that the right sides of Eqs. 11.57 and 11.58 are expressed solely in terms of p, v, T, and c_p.

Changes in specific entropy and enthalpy between two states are found by integrating Eqs. 11.57 and 11.58, respectively

$$s_2 - s_1 = \int_1^2 \frac{c_p}{T} dT - \int_1^2 \left(\frac{\partial v}{\partial T}\right)_p dp \tag{11.59}$$

$$h_2 - h_1 = \int_1^2 c_p\, dT + \int_1^2 \left[v - T\left(\frac{\partial v}{\partial T}\right)_p\right] dp \tag{11.60}$$

To integrate the first term on the right of each of these expressions, the variation of c_p with temperature at one fixed pressure (isobar) is required. Integration of the second term requires knowledge of the p–v–T behavior at the states of interest. An equation of state explicit in v would be particularly convenient for evaluating the integrals involving $(\partial v/\partial T)_p$. The accuracy of the resulting specific entropy and enthalpy changes would depend on the accuracy of this derivative.

Changes in specific enthalpy and internal energy are related through $h = u + pv$ by

$$h_2 - h_1 = (u_2 - u_1) + (p_2v_2 - p_1v_1) \tag{11.61}$$

Hence, only one of Δh and Δu need be found by integration. Then, the other can be evaluated from Eq. 11.61. Which of the two property changes is found by integration depends on the information available. Δh would be found using Eq. 11.60 when an equation of state explicit in v and c_p as a function of temperature at some fixed pressure are known. Δu would be found from Eq. 11.51 when an equation of state explicit in p and c_v as a function of temperature at some specific volume are known. Such issues are considered in Example 11.5.

Example 11.5 EVALUATING Δs, Δu, AND Δh OF A GAS

Using the Redlich–Kwong equation of state, develop expressions for the changes in specific entropy, internal energy, and enthalpy of a gas between two states where the temperature is the same, $T_1 = T_2$, and the pressures are p_1 and p_2, respectively.

Solution

Known: Two states of a unit mass of a gas as the system are fixed by p_1 and T_1 at state 1 and p_2, T_2 $(= T_1)$ at state 2.

Find: Determine the changes in specific entropy, internal energy, and enthalpy between these two states.

Schematic and Given Data:

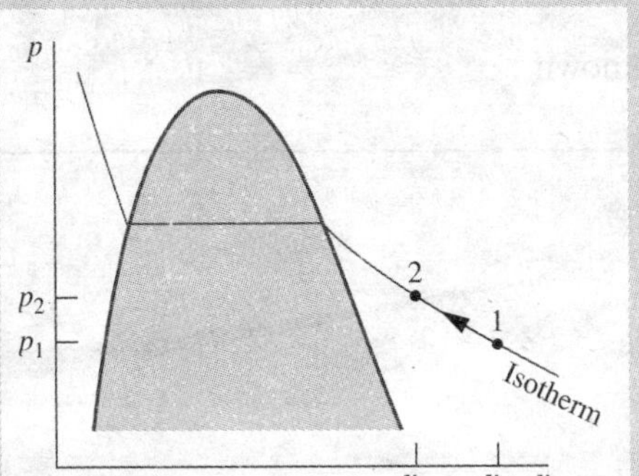

Engineering Model: The Redlich–Kwong equation of state represents the p–v–T behavior at these states and yields accurate values for $(\partial p/\partial T)_v$.

Fig. E11.5

Analysis: The Redlich–Kwong equation of state is explicit in pressure, so Eqs. 11.50 and 11.51 are selected for determining $s_2 - s_1$ and $u_2 - u_1$. Since $T_1 = T_2$, an isothermal path of integration between the two states is convenient. Thus, these equations reduce to give

$$s_2 - s_1 = \int_1^2 \left(\frac{\partial p}{\partial T}\right)_v dv$$

$$u_2 - u_1 = \int_1^2 \left[T\left(\frac{\partial p}{\partial T}\right)_v - p\right] dv$$

The limits for each of the foregoing integrals are the specific volumes v_1 and v_2 at the two states under consideration. Using p_1, p_2, and the known temperature, these specific volumes would be determined from the Redlich–Kwong equation of state. Since this equation is not explicit in specific volume, the use of an equation solver such as *Interactive Thermodynamics: IT* is recommended.

The above integrals involve the partial derivative $(\partial p/\partial T)_v$, which can be determined from the Redlich–Kwong equation of state as

$$\left(\frac{\partial p}{\partial T}\right)_v = \frac{R}{v-b} + \frac{a}{2v(v+b)T^{3/2}}$$

Inserting this into the expression for $(s_2 - s_1)$ gives

$$s_2 - s_1 = \int_{v_1}^{v_2} \left[\frac{R}{v-b} + \frac{a}{2v(v+b)T^{3/2}}\right] dv$$
$$= \int_{v_1}^{v_2} \left[\frac{R}{v-b} + \frac{a}{2bT^{3/2}}\left(\frac{1}{v} - \frac{1}{v+b}\right)\right] dv$$
$$= R \ln\left(\frac{v_2-b}{v_1-b}\right) + \frac{a}{2bT^{3/2}}\left[\ln\left(\frac{v_2}{v_1}\right) - \ln\left(\frac{v_2+b}{v_1+b}\right)\right]$$
$$= R \ln\left(\frac{v_2-b}{v_1-b}\right) + \frac{a}{2bT^{3/2}} \ln\left[\frac{v_2(v_1+b)}{v_1(v_2+b)}\right]$$

With the Redlich–Kwong equation, the integrand of the expression for $(u_2 - u_1)$ becomes

$$\left[T\left(\frac{\partial p}{\partial T}\right)_v - p\right] = T\left[\frac{R}{v-b} + \frac{a}{2v(v+b)T^{3/2}}\right] - \left[\frac{RT}{v-b} - \frac{a}{v(v+b)T^{1/2}}\right]$$
$$= \frac{3a}{2v(v+b)T^{1/2}}$$

Accordingly

$$u_2 - u_1 = \int_{v_1}^{v_2} \frac{3a}{2v(v+b)T^{1/2}} dv = \frac{3a}{2bT^{1/2}} \int_{v_1}^{v_2} \left(\frac{1}{v} - \frac{1}{v+b}\right) dv$$
$$= \frac{3a}{2bT^{1/2}}\left[\ln\frac{v_2}{v_1} - \ln\left(\frac{v_2+b}{v_1+b}\right)\right] = \frac{3a}{2bT^{1/2}} \ln\left[\frac{v_2(v_1+b)}{v_1(v_2+b)}\right]$$

Finally, $(h_2 - h_1)$ would be determined using Eq. 11.61 together with the known values of $(u_2 - u_1)$, p_1, v_1, p_2, and v_2.

Skills Developed: *Ability to...*
✓ perform differentiations and integrations required to evaluate Δu and Δs using the two-constant Redlich–Kwong equation of state.

Quick Quiz

Using results obtained, develop expressions for Δu and Δs of an ideal gas.
Ans. $\Delta u = 0$, $\Delta s = R\ln(v_2/v_1)$.

11.5 Other Thermodynamic Relations

The presentation to this point has been directed mainly at developing thermodynamic relations that allow changes in u, h, and s to be evaluated from measured property data. The object of the present section is to introduce several other thermodynamic relations that are useful for thermodynamic analysis. Each of the properties considered has a common attribute: it is defined in terms of a partial derivative of some other property. The specific heats c_v and c_p are examples of this type of property.

11.5.1 Volume Expansivity, Isothermal and Isentropic Compressibility

In single-phase regions, pressure and temperature are independent, and we can think of the specific volume as being a function of these two, $v = v(T, p)$. The differential of such a function is

$$dv = \left(\frac{\partial v}{\partial T}\right)_p dT + \left(\frac{\partial v}{\partial p}\right)_T dp$$

Two thermodynamic properties related to the partial derivatives appearing in this differential are the ***volume expansivity*** β, also called the *coefficient of volume expansion*

volume expansivity

$$\beta = \frac{1}{v}\left(\frac{\partial v}{\partial T}\right)_p \qquad (11.62)$$

and the ***isothermal compressibility*** κ

isothermal compressibility

$$\kappa = -\frac{1}{v}\left(\frac{\partial v}{\partial p}\right)_T \qquad (11.63)$$

By inspection, the unit for β is seen to be the reciprocal of that for temperature and the unit for κ is the reciprocal of that for pressure. The volume expansivity is an indication of the change in volume that occurs when temperature changes while pressure remains constant. The isothermal compressibility is an indication of the change in volume that takes place when pressure changes while temperature remains constant. The value of κ is positive for all substances in all phases.

The volume expansivity and isothermal compressibility are thermodynamic properties, and like specific volume are functions of T and p. Values for β and κ are provided in handbooks of engineering data. Table 11.2 gives values of these properties for liquid water at a pressure of 1 bar versus temperature. For a pressure of 1 bar, water has a *state of maximum density* at about 4°C. At this state, the value of β is zero.

The ***isentropic compressibility*** α is an indication of the change in volume that occurs when pressure changes while entropy remains constant

isentropic compressibility

$$\alpha = -\frac{1}{v}\left(\frac{\partial v}{\partial p}\right)_s \qquad (11.64)$$

The unit for α is the reciprocal of that for pressure.

The isentropic compressibility is related to the speed at which sound travels in the substance, and such speed measurements can be used to determine α. In Sec. 9.12, the ***velocity of sound,*** or *sonic velocity,* is introduced as

velocity of sound

$$c = \sqrt{-v^2\left(\frac{\partial p}{\partial v}\right)_s} \qquad (9.36b)$$

Table 11.2

Volume Expansivity β and Isothermal Compressibility κ of Liquid Water at 1 bar versus Temperature

T (°C)	Density (kg/m³)	$\beta \times 10^6$ $(K)^{-1}$	$\kappa \times 10^6$ $(bar)^{-1}$
0	999.84	−68.14	50.89
10	999.70	87.90	47.81
20	998.21	206.6	45.90
30	995.65	303.1	44.77
40	992.22	385.4	44.24
50	988.04	457.8	44.18

The relationship of the isentropic compressibility and the velocity of sound can be obtained using the relation between partial derivatives expressed by Eq. 11.15. Identifying p with x, v with y, and s with z, we have

$$\left(\frac{\partial p}{\partial v}\right)_s = \frac{1}{(\partial v/\partial p)_s}$$

With this, the previous two equations can be combined to give

$$c = \sqrt{v/\alpha} \tag{11.65}$$

The details are left as an exercise.

11.5.2 Relations Involving Specific Heats

In this section, general relations are obtained for the difference between specific heats $(c_p - c_v)$ and the ratio of specific heats c_p/c_v.

EVALUATING $(c_p - c_v)$. An expression for the difference between c_p and c_v can be obtained by equating the two differentials for entropy given by Eqs. 11.48 and 11.57 and rearranging to obtain

$$(c_p - c_v)\,dT = T\left(\frac{\partial p}{\partial T}\right)_v dv + T\left(\frac{\partial v}{\partial T}\right)_p dp$$

Considering the equation of state $p = p(T, v)$, the differential dp can be expressed as

$$dp = \left(\frac{\partial p}{\partial T}\right)_v dT + \left(\frac{\partial p}{\partial v}\right)_T dv$$

Eliminating dp between the last two equations and collecting terms gives

$$\left[(c_p - c_v) - T\left(\frac{\partial v}{\partial T}\right)_p\left(\frac{\partial p}{\partial T}\right)_v\right] dT = T\left[\left(\frac{\partial v}{\partial T}\right)_p\left(\frac{\partial p}{\partial v}\right)_T + \left(\frac{\partial p}{\partial T}\right)_v\right] dv$$

Since temperature and specific volume can be varied independently, the coefficients of the differentials in this expression must vanish, so

$$c_p - c_v = T\left(\frac{\partial v}{\partial T}\right)_p\left(\frac{\partial p}{\partial T}\right)_v \tag{11.66}$$

$$\left(\frac{\partial p}{\partial T}\right)_v = -\left(\frac{\partial v}{\partial T}\right)_p\left(\frac{\partial p}{\partial v}\right)_T \tag{11.67}$$

Introducing Eq. 11.67 into Eq. 11.66 gives

$$c_p - c_v = -T\left(\frac{\partial v}{\partial T}\right)_p^2\left(\frac{\partial p}{\partial v}\right)_T \tag{11.68}$$

This equation allows c_v to be calculated from observed values of c_p, or conversely, knowing only p–v–T data. ➡ **FOR EXAMPLE...** for the special case of an ideal gas, Eq. 11.68 reduces to Eq. 3.44, as can readily be shown. ⬅

The right side of Eq. 11.68 can be expressed in terms of the volume expansivity β and the isothermal compressibility κ. Introducing Eqs. 11.62 and 11.63, we get

$$c_p - c_v = v\frac{T\beta^2}{\kappa} \tag{11.69}$$

In developing this result, the relationship between partial derivatives expressed by Eq. 11.15 has been used.

Several important conclusions about the specific heats c_p and c_v can be drawn from Eq. 11.69. ➡ **FOR EXAMPLE...** since the factor β^2 cannot be negative and κ is positive for all substances in all phases, the value of c_p is always greater than, or equal to, c_v. The specific heats would be equal when $\beta = 0$, as occurs in the case of water at 1 atmosphere and 4°C, where water is at its state of maximum density. The two specific heats also become equal as the temperature approaches absolute zero. For some liquids and solids at certain states, c_p and c_v differ only slightly. For this reason, tables often give the specific heat of a liquid or solid without specifying whether it is c_p or c_v. The data reported are normally c_p values, since these are more easily determined for liquids and solids. ⬅

EVALUATING c_p/c_v. Next, let us obtain expressions for the ratio of specific heats. Employing Eq. 11.16, we can rewrite Eqs. 11.46 and 11.55, respectively, as

$$\frac{c_v}{T} = \left(\frac{\partial s}{\partial T}\right)_v = \frac{-1}{(\partial v/\partial s)_T(\partial T/\partial v)_s}$$

$$\frac{c_p}{T} = \left(\frac{\partial s}{\partial T}\right)_p = \frac{-1}{(\partial p/\partial s)_T(\partial T/\partial p)_s}$$

Forming the ratio of these equations gives

$$\frac{c_p}{c_v} = \frac{(\partial v/\partial s)_T(\partial T/\partial v)_s}{(\partial p/\partial s)_T(\partial T/\partial p)_s} \tag{11.70}$$

Since $(\partial s/\partial p)_T = 1/(\partial p/\partial s)_T$ and $(\partial p/\partial T)_s = 1/(\partial T/\partial p)_s$, Eq. 11.70 can be expressed as

$$\frac{c_p}{c_v} = \left[\left(\frac{\partial v}{\partial s}\right)_T\left(\frac{\partial s}{\partial p}\right)_T\right]\left[\left(\frac{\partial p}{\partial T}\right)_s\left(\frac{\partial T}{\partial v}\right)_s\right] \tag{11.71}$$

Finally, the chain rule from calculus allows us to write $(\partial v/\partial p)_T = (\partial v/\partial s)_T(\partial s/\partial p)_T$ and $(\partial p/\partial v)_s = (\partial p/\partial T)_s(\partial T/\partial v)_s$, so Eq. 11.71 becomes

$$k = \frac{c_p}{c_v} = \left(\frac{\partial v}{\partial p}\right)_T\left(\frac{\partial p}{\partial v}\right)_s \tag{11.72}$$

This can be expressed alternatively in terms of the isothermal and isentropic compressibilities as

$$k = \frac{\kappa}{\alpha} \tag{11.73}$$

Solving Eq. 11.72 for $(\partial p/\partial v)_s$ and substituting the resulting expression into Eq. 9.36b gives the following relationship involving the velocity of sound c and the specific heat ratio k

$$c = \sqrt{-kv^2(\partial p/\partial v)_T} \tag{11.74}$$

Equation 11.74 can be used to determine c knowing the specific heat ratio and p–v–T data, or to evaluate k knowing c and $(\partial p/\partial v)_T$. ➡ **FOR EXAMPLE...** in the special case of an ideal gas, Eq. 11.74 reduces to

$$c = \sqrt{kRT} \quad \text{(ideal gas)} \tag{9.37}$$

as can easily be verified. ⬅

In the next example we illustrate the use of specific heat relations introduced above.

Example 11.6 USING SPECIFIC HEAT RELATIONS

For liquid water at 1 bar and 20°C, estimate **(a)** the percent error in c_v that would result if it were assumed that $c_p = c_v$, **(b)** the velocity of sound, in m/s.

Solution

Known: The system consists of a fixed amount of liquid water at 1 bar and 20°C.

Find: Estimate the percent error in c_v that would result if c_v were approximated by c_p, and the velocity of sound, in m/s.

Analysis:

(a) Equation 11.69 gives the difference between c_p and c_v. Table 11.2 provides the required values for the volume expansivity β, the isothermal compressibility κ, and the specific volume. Thus

$$c_p - c_v = v\frac{T\beta^2}{\kappa}$$

$$= \left(\frac{1}{998.21\ \text{kg/m}^3}\right)(293\ \text{K})\left(\frac{206.6 \times 10^{-6}}{\text{K}}\right)^2\left(\frac{\text{bar}}{45.90 \times 10^{-6}}\right)$$

$$= \left(272.96 \times 10^{-6}\frac{\text{bar}\cdot\text{m}^3}{\text{kg}\cdot\text{K}}\right)\left|\frac{10^5\ \text{N/m}^2}{1\ \text{bar}}\right|\left|\frac{1\ \text{kJ}}{10^3\ \text{N}\cdot\text{m}}\right|$$

$$= 0.027\frac{\text{kJ}}{\text{kg}\cdot\text{K}}$$

❶ Interpolating in Table A-19 at 20°C gives $c_p = 4.188$ kJ/kg · K Thus, the value of c_v is

$$c_v = 4.188 - 0.027 = 4.161\ \text{kJ/kg}\cdot\text{K}$$

Using these values, the percent error in approximating c_v by c_p is

❷ $$\left(\frac{c_p - c_v}{c_v}\right)(100) = \left(\frac{0.027}{4.161}\right)(100) = 0.6\%$$

(b) The velocity of sound at this state can be determined using Eq. 11.65. The required value for the isentropic compressibility α is calculable in terms of the specific heat ratio k and the isothermal compressibility κ. With Eq. 11.73, $\alpha = \kappa/k$. Inserting this into Eq. 11.65 results in the following expression for the velocity of sound

$$c = \sqrt{\frac{kv}{\kappa}}$$

The values of v and κ required by this expression are the same as used in part (a). Also, with the values of c_p and c_v from part (a), the specific heat ratio is $k = 1.006$. Accordingly

❸ $$c = \sqrt{\frac{(1.006)(10^6)\ \text{bar}}{(998.21\ \text{kg/m}^3)(45.90)}\left|\frac{10^5\ \text{N/m}^2}{1\ \text{bar}}\right|\left|\frac{1\ \text{kg}\cdot\text{m/s}^2}{1\ \text{N}}\right|} = 1482\ \text{m/s}$$

❶ Consistent with the discussion of Sec. 3.10.1, we take c_p at 1 bar and 20°C as the saturated liquid value at 20°C.

❷ The result of part (a) shows that for liquid water at the given state, c_p and c_v are closely equal.

❸ For comparison, the velocity of sound in air at 1 bar, 20°C is about 343 m/s, which can be checked using Eq. 9.37.

Skills Developed: *Ability to...*
- ✓ apply specific heat relations to liquid water.
- ✓ evaluate velocity of sound for liquid water.

Quick Quiz

A submarine moves at a speed of 37 km/h. Using the sonic velocity calculated in part (b), estimate the Mach number of the vessel relative to the water.
Ans. 0.0069.

11.5.3 Joule–Thomson Coefficient

The value of the specific heat c_p can be determined from p–v–T data and the Joule–Thomson coefficient. The ***Joule–Thomson coefficient*** μ_J is defined as

Joule–Thomson coefficient

$$\mu_J = \left(\frac{\partial T}{\partial p}\right)_h \tag{11.75}$$

Like other partial differential coefficients introduced in this section, the Joule–Thomson coefficient is defined in terms of thermodynamic properties only and thus is itself a property. The units of μ_J are those of temperature divided by pressure.

A relationship between the specific heat c_p and the Joule–Thomson coefficient μ_J can be established by using Eq. 11.16 to write

$$\left(\frac{\partial T}{\partial p}\right)_h \left(\frac{\partial p}{\partial h}\right)_T \left(\frac{\partial h}{\partial T}\right)_p = -1$$

The first factor in this expression is the Joule–Thomson coefficient and the third is c_p. Thus

$$c_p = \frac{-1}{\mu_J(\partial p/\partial h)_T}$$

With $(\partial h/\partial p)_T = 1/(\partial p/\partial h)_T$ from Eq. 11.15, this can be written as

$$c_p = -\frac{1}{\mu_J}\left(\frac{\partial h}{\partial p}\right)_T \tag{11.76}$$

The partial derivative $(\partial h/\partial p)_T$, called the *constant-temperature coefficient,* can be eliminated from Eq. 11.76 by use of Eq. 11.56. The following expression results:

$$c_p = \frac{1}{\mu_J}\left[T\left(\frac{\partial v}{\partial T}\right)_p - v\right] \tag{11.77}$$

Equation 11.77 allows the value of c_p at a state to be determined using p–v–T data and the value of the Joule–Thomson coefficient at that state. Let us consider next how the Joule–Thomson coefficient can be found experimentally.

EXPERIMENTAL EVALUATION. The Joule–Thomson coefficient can be evaluated experimentally using an apparatus like that pictured in Fig. 11.3. Consider first Fig. 11.3*a*, which shows a porous plug through which a gas (or liquid) may pass. During operation at steady state, the gas enters the apparatus at a specified temperature T_1 and pressure p_1 and expands through the plug to a lower pressure p_2, which is controlled by an outlet valve. The temperature T_2 at the exit is measured. The apparatus is designed so that the gas undergoes a *throttling* process (Sec. 4.10) as it expands from 1 to 2. Accordingly, the exit state fixed by p_2 and T_2 has the same value for the specific enthalpy as at the inlet, $h_2 = h_1$. By progressively lowering the outlet pressure, a finite sequence of such exit states can be visited, as indicated on Fig. 11.3*b*. A curve may be drawn through the set of data points. Such a curve is called an isenthalpic (constant enthalpy) curve. An isenthalpic curve is the locus of all points representing equilibrium states of the same specific enthalpy.

The slope of an isenthalpic curve at any state is the Joule–Thomson coefficient at that state. The slope may be positive, negative, or zero in value. States where the coefficient has a zero value are called ***inversion states.*** Notice that not all lines of constant h have an inversion state. The uppermost curve of Fig. 11.3*b*, for example,

inversion states

Horizons

Small Power Plants Pack Punch

An innovation in power systems moving from concept to reality promises to help keep computer networks humming, hospital operating rooms lit, and shopping centers thriving. Called *distributed power systems,* compact power plants provide electricity for small loads or are linked for larger applications. With distributed power, consumers hope to avoid unpredictable price swings and brownouts that have plagued some regions of the country.

At the heart of many distributed power systems are cleverly engineered gas turbine units, called *microturbines,* less than one-fifth the size of utility gas turbines with fewer moving parts. They also emit less nitric oxide (NO_X) pollution.

Although the cost per kilowatt hour may be higher with microturbines, some customers are willing to pay more to gain control over their electric supply. Computer networks and hospitals need high reliability, since even short disruptions can be disastrous. Businesses such as shopping centers also must avoid costly service interruptions. With microturbine-based systems, the needed reliability is provided by modular units that can run virtually unattended for years.

always has a negative slope. Throttling a gas from an initial state on this curve would result in an increase in temperature. However, for isenthalpic curves having an inversion state, the temperature at the exit of the apparatus may be greater than, equal to, or less than the initial temperature, depending on the exit pressure specified. For states to the right of an inversion state, the value of the Joule–Thomson coefficient would be negative. For these states, the temperature would increase as the pressure at the exit of the apparatus is reduced. At states to the left of an inversion state, the value of the Joule–Thomson coefficient would be positive. For these states, the temperature would decrease as the pressure at the exit of the device is reduced. This can be used to advantage in systems designed to *liquefy gases.*

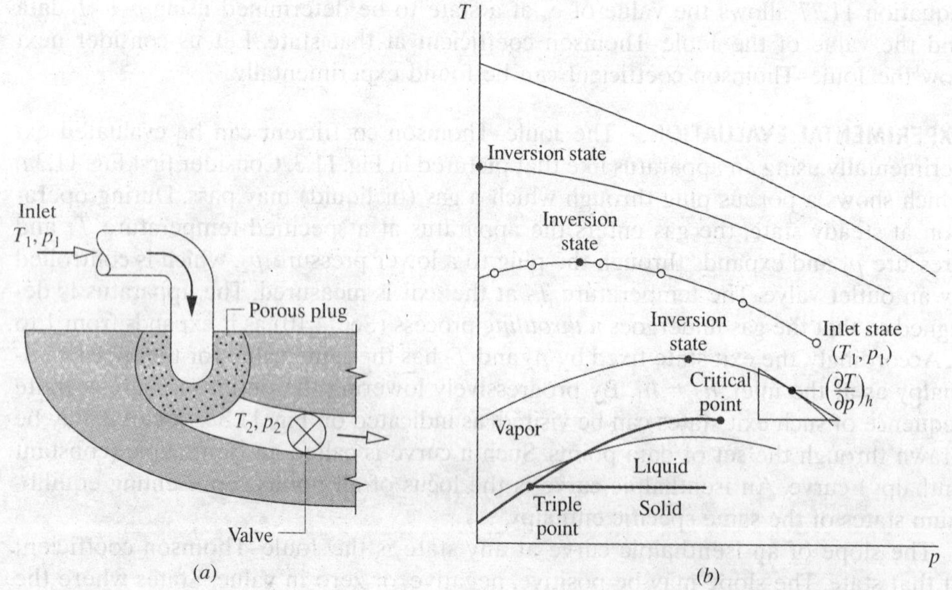

Fig. 11.3 Joule–Thomson expansion. (*a*) Apparatus. (*b*) Isenthalpics on a *T–p* diagram.

11.6 Constructing Tables of Thermodynamic Properties

The objective of this section is to utilize the thermodynamic relations introduced thus far to describe how tables of thermodynamic properties can be constructed. The characteristics of the tables under consideration are embodied in the tables for water and the refrigerants presented in the Appendix. The methods introduced in this section are extended in Chap. 13 for the analysis of reactive systems, such as gas turbine and vapor power systems involving combustion. The methods of this section also provide the basis for computer retrieval of thermodynamic property data.

Two different approaches for constructing property tables are considered:

- The presentation of Sec. 11.6.1 employs the methods introduced in Sec. 11.4 for assigning specific enthalpy, specific internal energy, and specific entropy to states of pure, simple compressible substances using p–v–T data, together with a limited amount of specific heat data. The principal mathematical operation of this approach is *integration.*
- The approach of Sec. 11.6.2 utilizes the fundamental thermodynamic function concept introduced in Sec. 11.3.3. Once such a function has been constructed, the principal mathematical operation required to determine all other properties is *differentiation.*

11.6.1 Developing Tables by Integration Using p–v–T and Specific Heat Data

In principle, all properties of present interest can be determined using

$$c_p = c_{p0}(T) \qquad (11.78)$$
$$p = p(v, T), \qquad v = v(p, T)$$

In Eqs. 11.78, $c_{p0}(T)$ is the specific heat c_p for the substance under consideration extrapolated to zero pressure. This function might be determined from data obtained calorimetrically or from spectroscopic data, using equations supplied by statistical mechanics. Specific heat expressions for several gases are given in Tables A-21. The expressions $p(v, T)$ and $v(p, T)$ represent functions that describe the saturation pressure–temperature curves, as well as the p–v–T relations for the single-phase regions. These functions may be tabular, graphical, or analytical in character. Whatever their forms, however, the functions must not only represent the p–v–T data accurately but also yield accurate values for derivatives such as $(\partial v/\partial T)_p$ and $(dp/dT)_{sat}$.

Figure 11.4 shows eight states of a substance. Let us consider how values can be assigned to specific enthalpy and specific entropy at these states. The same procedures can be used to assign property values at other states of interest. Note that when h has been assigned to a state, the specific internal energy at that state can be found from $u = h - pv$.

- Let the state denoted by 1 on Fig. 11.4 be selected as the datum state for enthalpy and entropy. Any value can be assigned to h and s at this state, but a value of zero would be usual. It should be noted that the use of an arbitrary datum state and arbitrary reference values for specific enthalpy and specific entropy suffices only for evaluations involving differences in property values between states of the same composition, for then datums cancel.

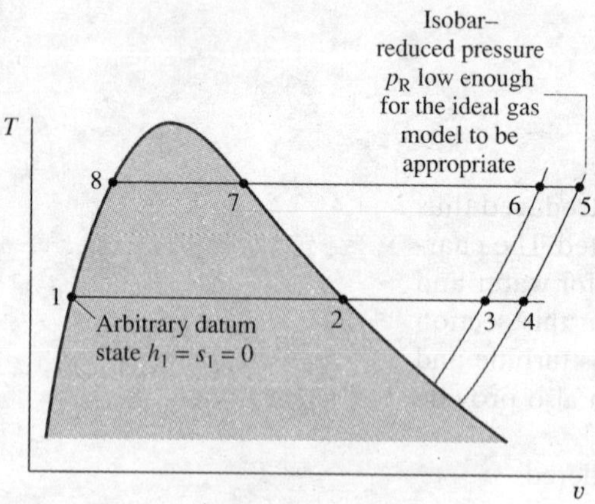

Fig. 11.4 T–v diagram used to discuss how h and s can be assigned to liquid and vapor states.

- Once a value is assigned to enthalpy at state 1, the enthalpy at the saturated vapor state, state 2, can be determined using the Clapeyron equation, Eq. 11.40

$$h_2 - h_1 = T_1(v_2 - v_1)\left(\frac{dp}{dT}\right)_{\text{sat}}$$

where the derivative $(dp/dT)_{\text{sat}}$ and the specific volumes v_1 and v_2 are obtained from appropriate representations of the p–v–T data for the substance under consideration. The specific entropy at state 2 is found using Eq. 11.38 in the form

$$s_2 - s_1 = \frac{h_2 - h_1}{T_1}$$

- Proceeding at constant temperature from state 2 to state 3, the entropy and enthalpy are found by means of Eqs. 11.59 and 11.60, respectively. Since temperature is fixed, these equations reduce to give

$$s_3 - s_2 = -\int_{p_2}^{p_3}\left(\frac{\partial v}{\partial T}\right)_p dp \quad \text{and} \quad h_3 - h_2 = \int_{p_2}^{p_3}\left[v - T\left(\frac{\partial v}{\partial T}\right)_p\right] dp$$

With the same procedure, s_4 and h_4 can be determined.

- The isobar (constant-pressure line) passing through state 4 is assumed to be at a low enough pressure for the ideal gas model to be appropriate. Accordingly, to evaluate s and h at states such as 5 on this isobar, the only required information would be $c_{p0}(T)$ and the temperatures at these states. Thus, since pressure is fixed, Eqs. 11.59 and 11.60 give, respectively

$$s_5 - s_4 = \int_{T_4}^{T_5} c_{p0}\frac{dT}{T} \quad \text{and} \quad h_5 - h_4 = \int_{T_4}^{T_5} c_{p0}\, dT$$

- Specific entropy and enthalpy values at states 6 and 7 are found from those at state 5 by the same procedure used in assigning values at states 3 and 4 from those at state 2. Finally, s_8 and h_8 are obtained from the values at state 7 using the Clapeyron equation.

11.6.2 Developing Tables by Differentiating a Fundamental Thermodynamic Function

Property tables also can be developed using a fundamental thermodynamic function. It is convenient for this purpose to select the independent variables of the fundamental function from among pressure, specific volume (density), and temperature. This indicates the use of the Helmholtz function $\psi(T, v)$ or the Gibbs function $g(T, p)$. The properties of water tabulated in Tables A-2 through A-6 have been calculated using the Helmholtz function. Fundamental functions also have been employed successfully to evaluate the properties of other substances.

The development of a fundamental function requires considerable mathematical manipulation and numerical evaluation. Prior to the advent of high-speed computers, the evaluation of properties by this means was not feasible, and the approach described in Sec. 11.6.1 was used exclusively. The fundamental function approach involves three steps:

1. The first step is the selection of a functional form in terms of the appropriate pair of independent properties and a set of adjustable coefficients, which may number 50 or more. The functional form is specified on the basis of both theoretical and practical considerations.

2. Next, the coefficients in the fundamental function are determined by requiring that a set of carefully selected property values and/or observed conditions be satisfied in a least-squares sense. This generally involves the use of property data requiring the assumed functional form to be differentiated one or more times, such as p–v–T and specific heat data.
3. When all coefficients have been evaluated, the function is carefully tested for accuracy by using it to evaluate properties for which accepted values are known. These may include properties requiring differentiation of the fundamental function two or more times. For example, velocity of sound and Joule–Thomson data might be used.

This procedure for developing a fundamental function is not routine and can be accomplished only with a computer. However, once a suitable fundamental function is established, extreme accuracy in and consistency among the thermodynamic properties is possible.

The form of the Helmholtz function used in constructing the steam tables from which Tables A-2 through A-6 have been extracted is

$$\psi(\rho, T) = \psi_0(T) + RT[\ln \rho + \rho Q(\rho, \tau)] \tag{11.79}$$

where ψ_0 and Q are given as the sums listed in Table 11.3. The independent variables are density and temperature. The variable τ denotes $1000/T$. Values for pressure, specific internal energy, and specific entropy can be determined by differentiation of Eq. 11.79. Values for the specific enthalpy and Gibbs function are found from $h = u + pv$ and $g = \psi + pv$, respectively. The specific heat c_v is evaluated by further differentiation, $c_v = (\partial u/\partial T)_v$. With similar operations, other properties can be evaluated. Property values for water calculated from Eq. 11.79 are in excellent agreement with experimental data over a wide range of conditions.

Table 11.3

Fundamental Equation Used to Construct the Steam Tables[a,b]

$$\psi = \psi_0(T) + RT[\ln \rho + \rho Q(\rho, \tau)] \tag{1}$$

where

$$\psi_0 = \sum_{i=1}^{6} C_i/\tau^{i-1} + C_7 \ln T + C_8 \ln T/\tau \tag{2}$$

and

$$Q = (\tau - \tau_c)\sum_{j=1}^{7}(\tau - \tau_{aj})^{j-2}\left[\sum_{i=1}^{8} A_{ij}(\rho - \rho_{aj})^{i-1} + e^{-E\rho}\sum_{i=9}^{10} A_{ij}\rho^{i-9}\right] \tag{3}$$

In (1), (2), and (3), T denotes temperature on the Kelvin scale, τ denotes $1000/T$, ρ denotes density in g/cm^3, $R = 4.6151$ bar · cm^3/g · K or 0.46151 J/g · K, $\tau_c = 1000/T_c = 1.544912$, $E = 4.8$, and

$$\tau_{aj} = \tau_c (j = 1) \qquad \rho_{aj} = 0.634 (j = 1)$$
$$= 2.5 (j > 1) \qquad = 1.0 (j > 1)$$

The coefficients for ψ_0 in J/g are given as follows:

$C_1 = 1857.065$	$C_4 = 36.6649$	$C_7 = 46.0$
$C_2 = 3229.12$	$C_5 = -20.5516$	$C_8 = -1011.249$
$C_3 = -419.465$	$C_6 = 4.85233$	

Values for the coefficients A_{ij} are listed in the original source.[a]

[a] J. H. Keenan, F. G. Keyes, P. G. Hill, and J. G. Moore, *Steam Tables*, Wiley, New York, 1969.
[b] Also see L. Haar, J. S. Gallagher, and G. S. Kell, *NBS/NRC Steam Tables*, Hemisphere, Washington, D.C., 1984. The properties of water are determined in this reference using a different functional form for the Helmholtz function than given by Eqs. (1)–(3).

Energy & Environment

Due to the phase-out of chlorine-containing CFC refrigerants because of ozone layer and global warming concerns, new substances and mixtures with no chlorine have been developed in recent years as possible alternatives. This has led to significant research efforts to provide the necessary thermodynamic property data for analysis and design.

The National Institute of Standards and Technology (NIST) has led governmental efforts to provide accurate data. Specifically, data have been developed for high-accuracy p–v–T equations of state from which fundamental functions can be obtained. The equations are carefully validated using data for velocity of sound, Joule–Thomson coefficient, saturation pressure–temperature relations, and specific heats. Such data were used to calculate the property values in Tables A-7 to A-18 in the Appendix. NIST has developed REFPROP, a computer data base that is the current standard for refrigerant and refrigerant mixture properties.

Example 11.7 illustrates the use of a fundamental function to determine thermodynamic properties for computer evaluation and to develop tables.

Example 11.7 DETERMINING PROPERTIES USING A FUNDAMENTAL FUNCTION

The following expression for the Helmholtz function has been used to determine the properties of water:

$$\psi(\rho, T) = \psi_0(T) + RT[\ln \rho + \rho Q(\rho, \tau)]$$

where ρ denotes density and τ denotes $1000/T$. The functions ψ_0 and Q are sums involving the indicated independent variables and a number of adjustable constants (see Table 11.3). Obtain expressions for **(a)** pressure, **(b)** specific entropy, and **(c)** specific internal energy resulting from this fundamental function.

Solution

Known: An expression for the Helmholtz function ψ is given.

Find: Determine the expressions for pressure, specific entropy, and specific internal energy that result from this fundamental function.

Analysis: The expressions developed below for p, s, and u require only the functions $\psi_0(T)$ an $Q(\rho, \tau)$. Once these functions are determined, p, s, and u can each be determined as a function of density and temperature using elementary mathematical operations.

(a) When expressed in terms of density instead of specific volume, Eq. 11.28 becomes

$$p = \rho^2\left(\frac{\partial \psi}{\partial \rho}\right)_T$$

as can easily be verified. When T is held constant τ is also constant. Accordingly, the following is obtained on differentiation of the given function:

$$\left(\frac{\partial \psi}{\partial \rho}\right)_T = RT\left[\frac{1}{\rho} + Q(\rho, \tau) + \rho\left(\frac{\partial Q}{\partial \rho}\right)_\tau\right]$$

Combining these equations gives an expression for pressure

$$p = \rho RT\left[1 + \rho Q + \rho^2\left(\frac{\partial Q}{\partial \rho}\right)_\tau\right] \tag{a}$$

(b) From Eq. 11.29

$$s = -\left(\frac{\partial \psi}{\partial T}\right)_\rho$$

Differentiation of the given expression for ψ yields

$$\left(\frac{\partial\psi}{\partial T}\right)_\rho = \frac{d\psi_0}{dT} + \left[R(\ln\rho + \rho Q) + RT\rho\left(\frac{\partial Q}{\partial\tau}\right)_\rho\frac{d\tau}{dT}\right]$$

$$= \frac{d\psi_0}{dT} + \left[R(\ln\rho + \rho Q) + RT\rho\left(\frac{\partial Q}{\partial\tau}\right)_\rho\left(-\frac{1000}{T^2}\right)\right]$$

$$= \frac{d\psi_0}{dT} + R\left[\ln\rho + \rho Q - \rho\tau\left(\frac{\partial Q}{\partial\tau}\right)_\rho\right]$$

Combining results gives

$$s = -\frac{d\psi_0}{dT} - R\left[\ln\rho + \rho Q - \rho\tau\left(\frac{\partial Q}{\partial\tau}\right)_\rho\right] \tag{b}$$

(c) By definition, $\psi = u - Ts$. Thus, $u = \psi + Ts$. Introducing the given expression for ψ together with the expression for s from part (b) results in

$$u = [\psi_0 + RT(\ln\rho + \rho Q)] + T\left\{-\frac{d\psi_0}{dT} - R\left[\ln\rho + \rho Q - \rho\tau\left(\frac{\partial Q}{\partial\tau}\right)_\rho\right]\right\}$$

$$= \psi_0 - T\frac{d\psi_0}{dT} + RT\rho\tau\left(\frac{\partial Q}{\partial\tau}\right)_\rho$$

This can be written more compactly by noting that

$$T\frac{d\psi_0}{dT} = T\frac{d\psi_0}{d\tau}\frac{d\tau}{dT} = T\frac{d\psi_0}{d\tau}\left(-\frac{1000}{T^2}\right) = -\tau\frac{d\psi_0}{d\tau}$$

Thus,

$$\psi_0 - T\frac{d\psi_0}{dT} = \psi_0 + \tau\frac{d\psi_0}{d\tau} = \frac{d(\psi_0\tau)}{d\tau}$$

Finally, the expression for u becomes

$$u = \frac{d(\psi_0\tau)}{d\tau} + RT\rho\tau\left(\frac{\partial Q}{\partial\tau}\right)_\rho \tag{c}$$

Skills Developed: *Ability to...*
- ✓ derive expressions for pressure, specific entropy, and specific internal energy based on a fundamental function.

Quick Quiz

Using results obtained, how can an expression be developed for h?

Ans. $h = u + p/\rho$. Substitute Eq. (c) for u and Eq. (a) for p and collect terms.

11.7 Generalized Charts for Enthalpy and Entropy

Generalized charts giving the compressibility factor Z in terms of the reduced properties p_R, T_R, and v'_R are introduced in Sec. 3.11. With such charts, estimates of p–v–T data can be obtained rapidly knowing only the critical pressure and critical temperature for the substance of interest. The objective of the present section is to introduce generalized charts that allow changes in enthalpy and entropy to be estimated.

Generalized Enthalpy Departure Chart

The change in specific enthalpy of a gas (or liquid) between two states fixed by temperature and pressure can be evaluated using the identity

$$h(T_2, p_2) - h(T_1, p_1) = [h^*(T_2) - h^*(T_1)] + \underline{\{[h(T_2, p_2) - h^*(T_2)] - [h(T_1, p_1) - h^*(T_1)]\}} \quad (11.80)$$

The term $[h(T, p) - h^*(T)]$ denotes the specific enthalpy of the substance relative to that of its ideal gas model when both are at the same temperature. The superscript * is used in this section to identify ideal gas property values. Thus, Eq. 11.80 indicates that the change in specific enthalpy between the two states equals the enthalpy change determined using the ideal gas model plus a correction that accounts for the departure from ideal gas behavior. The correction is shown underlined in Eq. 11.80. The ideal gas term can be evaluated using methods introduced in Chap. 3. Next, we show how the correction term is evaluated in terms of the *enthalpy departure.*

DEVELOPING THE ENTHALPY DEPARTURE. The variation of enthalpy with pressure at fixed temperature is given by Eq. 11.56 as

$$\left(\frac{\partial h}{\partial p}\right)_T = v - T\left(\frac{\partial v}{\partial T}\right)_p$$

Integrating from pressure p' to pressure p at fixed temperature T

$$h(T, p) - h(T, p') = \int_{p'}^{p} \left[v - T\left(\frac{\partial v}{\partial T}\right)_p\right] dp$$

This equation is not altered fundamentally by adding and subtracting $h^*(T)$ on the left side. That is

$$[h(T, p) - h^*(T)] - [h(T, p') - h^*(T)] = \int_{p'}^{p} \left[v - T\left(\frac{\partial v}{\partial T}\right)_p\right] dp \quad (11.81)$$

As pressure tends to zero at fixed temperature, the enthalpy of the substance approaches that of its ideal gas model. Accordingly, as p' tends to zero

$$\lim_{p' \to 0} [h(T, p') - h^*(T)] = 0$$

In this limit, the following expression is obtained from Eq. 11.81 for the specific enthalpy of a substance relative to that of its ideal gas model when both are at the same temperature:

$$h(T, p) - h^*(T) = \int_{0}^{p} \left[v - T\left(\frac{\partial v}{\partial T}\right)_p\right] dp \quad (11.82)$$

This also can be thought of as the change in enthalpy as the pressure is increased from zero to the given pressure while temperature is held constant. Using p–v–T data only, Eq. 11.82 can be evaluated at states 1 and 2 and thus the correction term of Eq. 11.80 evaluated. Let us consider next how this procedure can be conducted in terms of compressibility factor data and the reduced properties T_R and p_R.

The integral of Eq. 11.82 can be expressed in terms of the compressibility factor Z and the reduced properties T_R and p_R as follows. Solving $Z = pv/RT$ gives

$$v = \frac{ZRT}{p}$$

On differentiation

$$\left(\frac{\partial v}{\partial T}\right)_p = \frac{RZ}{p} + \frac{RT}{p}\left(\frac{\partial Z}{\partial T}\right)_p$$

With the previous two expressions, the integrand of Eq. 11.82 becomes

$$v - T\left(\frac{\partial v}{\partial T}\right)_p = \frac{ZRT}{p} - T\left[\frac{RZ}{p} + \frac{RT}{p}\left(\frac{\partial Z}{\partial T}\right)_p\right] = -\frac{RT^2}{p}\left(\frac{\partial Z}{\partial T}\right)_p \tag{11.83}$$

Equation 11.83 can be written in terms of reduced properties as

$$v - T\left(\frac{\partial v}{\partial T}\right)_p = -\frac{RT_c}{p_c} \cdot \frac{T_R^2}{p_R}\left(\frac{\partial Z}{\partial T_R}\right)_{p_R}$$

Introducing this into Eq. 11.82 gives on rearrangement

$$\frac{h^*(T) - h(T, p)}{RT_c} = T_R^2 \int_0^{p_R} \left(\frac{\partial Z}{\partial T_R}\right)_{p_R} \frac{dp_R}{p_R}$$

Or, on a per mole basis, the ***enthalpy departure*** is

enthalpy departure

$$\frac{\bar{h}^*(T) - \bar{h}(T, p)}{\bar{R}T_c} = T_R^2 \int_0^{p_R} \left(\frac{\partial Z}{\partial T_R}\right)_{p_R} \frac{dp_R}{p_R} \tag{11.84}$$

The right side of Eq. 11.84 depends only on the reduced temperature T_R and reduced pressure p_R. Accordingly, the quantity $(\bar{h}^* - \bar{h})/\bar{R}T_c$, the enthalpy departure, is a function only of these two reduced properties. Using a generalized equation of state giving Z as a function of T_R and p_R, the enthalpy departure can readily be evaluated with a computer. Tabular representations are also found in the literature. Alternatively, the graphical representation provided in Fig. A-4 can be employed.

EVALUATING ENTHALPY CHANGE. The change in specific enthalpy between two states can be evaluated by expressing Eq. 11.80 in terms of the enthalpy departure as

$$\bar{h}_2 - \bar{h}_1 = \underline{\bar{h}_2^* - \bar{h}_1^*} - \underline{\bar{R}T_c\left[\left(\frac{\bar{h}^* - \bar{h}}{\bar{R}T_c}\right)_2 - \left(\frac{\bar{h}^* - \bar{h}}{\bar{R}T_c}\right)_1\right]} \tag{11.85}$$

The first underlined term in Eq. 11.85 represents the change in specific enthalpy between the two states assuming ideal gas behavior. The second underlined term is the correction that must be applied to the ideal gas value for the enthalpy change to obtain the actual value for the enthalpy change. The quantity $(\bar{h}^* - \bar{h})/\bar{R}T_c$ at state 1 would be calculated with an equation giving $Z(T_R, p_R)$ or obtained from tables or the generalized enthalpy departure chart, Fig. A-4, using the reduced temperature T_{R1} and reduced pressure p_{R1} corresponding to the temperature T_1 and pressure p_1 at the initial state, respectively. Similarly, $(\bar{h}^* - \bar{h})/\bar{R}T_c$ at state 2 would be evaluated using T_{R2} and p_{R2}. The use of Eq. 11.85 is illustrated in the next example.

Example 11.8 USING THE GENERALIZED ENTHALPY DEPARTURE CHART

Nitrogen enters a turbine operating at steady state at 100 bar and 300 K and exits at 40 bar and 245 K. Using the enthalpy departure chart, determine the work developed, in kJ per kg of nitrogen flowing, if heat transfer with the surroundings can be ignored. Changes in kinetic and potential energy from inlet to exit also can be neglected.

Solution

Known: A turbine operating at steady state has nitrogen entering at 100 bar and 300 K and exiting at 40 bar and 245 K.

Find: Using the enthalpy departure chart, determine the work developed.

Schematic and Given Data:

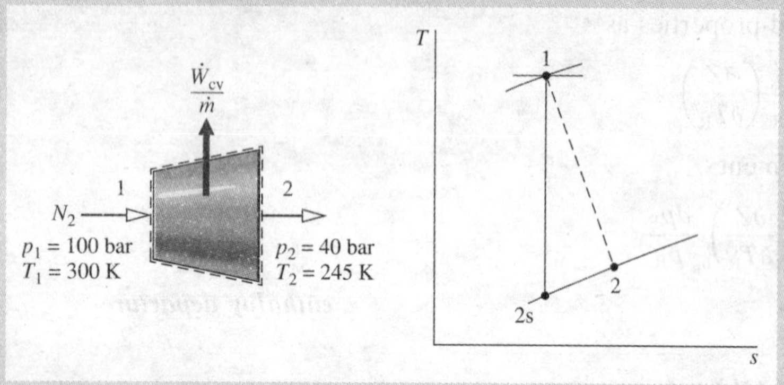

Fig. E11.8

Engineering Model:

1. The control volume shown on the accompanying figure operates at steady state.
2. There is no significant heat transfer between the control volume and its surroundings.
3. Changes in kinetic and potential energy between inlet and exit can be neglected.
4. Equilibrium property relations apply at the inlet and exit.

Analysis: The mass and energy rate balances reduce at steady state to give

$$0 = \frac{\dot{Q}_{cv}}{\dot{m}} - \frac{\dot{W}_{cv}}{\dot{m}} + \left[h_1 - h_2 + \frac{V_1^2 - V_2^2}{2} + g(z_1 - z_2)\right]$$

where $\dot{m}$ is the mass flow rate. Dropping the heat transfer term by assumption 2 and the kinetic and potential energy terms by assumption 3 gives on rearrangement

$$\frac{\dot{W}_{cv}}{\dot{m}} = h_1 - h_2$$

The term $h_1 - h_2$ can be evaluated as follows:

$$h_1 - h_2 = \frac{1}{M}\left\{\bar{h}_1^* - \bar{h}_2^* - \bar{R}T_c\left[\left(\frac{\bar{h}^* - \bar{h}}{\bar{R}T_c}\right)_1 - \left(\frac{\bar{h}^* - \bar{h}}{\bar{R}T_c}\right)_2\right]\right\}$$

In this expression, M is the molecular weight of nitrogen and the other terms have the same significance as in Eq. 11.85. With specific enthalpy values from Table A-23 at $T_1 = 300$ K and $T_2 = 245$ K, respectively

$$\bar{h}_1^* - \bar{h}_2^* = 8723 - 7121 = 1602 \text{ kJ/kmol}$$

The terms $(\bar{h}^* - \bar{h})/\bar{R}T_c$ at states 1 and 2 required by the above expression for $h_1 - h_2$ can be determined from Fig. A-4. First, the reduced temperature and reduced pressure at the inlet and exit must be determined. From Tables A-1, $T_c = 126$ K, $p_c = 33.9$ bar. Thus, at the inlet

$$T_{R1} = \frac{300}{126} = 2.38, \qquad p_{R1} = \frac{100}{33.9} = 2.95$$

At the exit

$$T_{R2} = \frac{245}{126} = 1.94, \qquad p_{R2} = \frac{40}{33.9} = 1.18$$

By inspection of Fig. A-4

❶ $$\left(\frac{\bar{h}^* - \bar{h}}{\bar{R}T_c}\right)_1 \approx 0.5, \qquad \left(\frac{\bar{h}^* - \bar{h}}{\bar{R}T_c}\right)_2 \approx 0.31$$

Substituting values

$$\frac{\dot{W}_{cv}}{\dot{m}} = \frac{1}{28\frac{\text{kg}}{\text{kmol}}}\left[1602\frac{\text{kJ}}{\text{kmol}} - \left(8.314\frac{\text{kJ}}{\text{kmol}\cdot\text{K}}\right)(126\text{ K})(0.5 - 0.31)\right] = 50.1\text{ kJ/kg}$$

❶ Due to inaccuracy in reading values from a graph such as Fig. A-4, we cannot expect extreme accuracy in the final calculated result.

Skills Developed: *Ability to...*

✓ use data from the generalized enthalpy departure chart to calculate the change in enthalpy of nitrogen.

Quick Quiz

Determine the work developed, in kJ per kg of nitrogen flowing, assuming the ideal gas model.

Ans. 57.2 kJ/kg.

Generalized Entropy Departure Chart

A generalized chart that allows changes in specific entropy to be evaluated can be developed in a similar manner to the generalized enthalpy departure chart introduced above. The difference in specific entropy between states 1 and 2 of a gas (or liquid) can be expressed as the identity

$$\begin{aligned} s(T_2, p_2) - s(T_1, p_1) &= s^*(T_2, p_2) - s^*(T_1, p_1) \\ &\quad + \underline{\{[s(T_2, p_2) - s^*(T_2, p_2)] - [s(T_1, p_1) - s^*(T_1, p_1)]\}} \end{aligned} \tag{11.86}$$

where $[s(T, p) - s^*(T, p)]$ denotes the specific entropy of the substance relative to that of its ideal gas model when both are at the same temperature and pressure. Equation 11.86 indicates that the change in specific entropy between the two states equals the entropy change determined using the ideal gas model plus a correction (shown underlined) that accounts for the departure from ideal gas behavior. The ideal gas term can be evaluated using methods introduced in Sec. 6.5. Let us consider next how the correction term is evaluated in terms of the *entropy departure.*

DEVELOPING THE ENTROPY DEPARTURE. The following Maxwell relation gives the variation of entropy with pressure at fixed temperature:

$$\left(\frac{\partial s}{\partial p}\right)_T = -\left(\frac{\partial v}{\partial T}\right)_p \tag{11.35}$$

Integrating from pressure p' to pressure p at fixed temperature T gives

$$s(T, p) - s(T, p') = -\int_{p'}^{p}\left(\frac{\partial v}{\partial T}\right)_p dp \tag{11.87}$$

For an ideal gas, $v = RT/p$, so $(\partial v/\partial T)_p = R/p$. Using this in Eq. 11.87, the change in specific entropy assuming ideal gas behavior is

$$s^*(T, p) - s^*(T, p') = -\int_{p'}^{p}\frac{R}{p}dp \tag{11.88}$$

Subtracting Eq. 11.88 from Eq. 11.87 gives

$$[s(T, p) - s^*(T, p)] - [s(T, p') - s^*(T, p')] = \int_{p'}^{p}\left[\frac{R}{p} - \left(\frac{\partial v}{\partial T}\right)_p\right] dp \tag{11.89}$$

Since the properties of a substance tend to merge into those of its ideal gas model as pressure tends to zero at fixed temperature, we have

$$\lim_{p' \to 0} [s(T, p') - s^*(T, p')] = 0$$

Thus, in the limit as p' tends to zero, Eq. 11.89 becomes

$$s(T, p) - s^*(T, p) = \int_0^p \left[\frac{R}{p} - \left(\frac{\partial v}{\partial T} \right)_p \right] dp \tag{11.90}$$

Using p–v–T data only, Eq. 11.90 can be evaluated at states 1 and 2 and thus the correction term of Eq. 11.86 evaluated.

entropy departure

Equation 11.90 can be expressed in terms of the compressibility factor Z and the reduced properties T_R and p_R. The result, on a per mole basis, is the ***entropy departure***

$$\frac{\bar{s}^*(T, p) - \bar{s}(T, p)}{\bar{R}} = \frac{\bar{h}^*(T) - \bar{h}(T, p)}{\bar{R} T_R T_c} + \int_0^{p_R} (Z - 1) \frac{dp_R}{p_R} \tag{11.91}$$

The right side of Eq. 11.91 depends only on the reduced temperature T_R and reduced pressure p_R. Accordingly, the quantity $(\bar{s}^* - \bar{s})/\bar{R}$, the entropy departure, is a function only of these two reduced properties. As for the enthalpy departure, the entropy departure can be evaluated with a computer using a generalized equation of state giving Z as a function of T_R and p_R. Alternatively, tabular data from the literature or the graphical representation provided in Fig. A-5 can be employed.

EVALUATING ENTROPY CHANGE. The change in specific entropy between two states can be evaluated by expressing Eq. 11.86 in terms of the entropy departure as

$$\bar{s}_2 - \bar{s}_1 = \underline{\bar{s}_2^* - \bar{s}_1^*} - \underline{\bar{R}\left[\left(\frac{\bar{s}^* - \bar{s}}{\bar{R}} \right)_2 - \left(\frac{\bar{s}^* - \bar{s}}{\bar{R}} \right)_1 \right]} \tag{11.92}$$

The first underlined term in Eq. 11.92 represents the change in specific entropy between the two states assuming ideal gas behavior. The second underlined term is the correction that must be applied to the ideal gas value for entropy change to obtain the actual value for the entropy change. The quantity $(\bar{s}^* - \bar{s})_1/\bar{R}$ appearing in Eq. 11.92 would be calculated with an equation giving $Z(T_R, p_R)$ or obtained from the generalized entropy departure chart, Fig. A-5, using the reduced temperature T_{R1} and reduced pressure p_{R1} corresponding to the temperature T_1 and pressure p_1 at the initial state, respectively. Similarly, $(\bar{s}^* - \bar{s})_2/\bar{R}$ would be evaluated using T_{R2} and p_{R2}. The use of Eq. 11.92 is illustrated in the next example.

Example 11.9 USING THE GENERALIZED ENTROPY DEPARTURE CHART

For the case of Example 11.8, determine **(a)** the rate of entropy production, in kJ/kg · K, and **(b)** the isentropic turbine efficiency.

Solution

Known: A turbine operating at steady state has nitrogen entering at 100 bar and 300 K and exiting at 40 bar and 245 K.

Find: Determine the rate of entropy production, in kJ/kg · K, and the isentropic turbine efficiency.

Schematic and Given Data: See Fig. E11.8.

Engineering Model: See Example 11.8.

Analysis:

(a) At steady state, the control volume form of the entropy rate equation reduces to give

$$\frac{\dot{\sigma}_{cv}}{\dot{m}} = s_2 - s_1$$

The change in specific entropy required by this expression can be written as

$$s_2 - s_1 = \frac{1}{M}\left\{\bar{s}_2^* - \bar{s}_1^* - \bar{R}\left[\left(\frac{\bar{s}^* - \bar{s}}{\bar{R}}\right)_2 - \left(\frac{\bar{s}^* - \bar{s}}{\bar{R}}\right)_1\right]\right\}$$

where M is the molecular weight of nitrogen and the other terms have the same significance as in Eq. 11.92.

The change in specific entropy $s_2^* - s_1^*$ can be evaluated using

$$\bar{s}_2^* - \bar{s}_1^* = \bar{s}°(T_2) - \bar{s}°(T_1) - \bar{R}\ln\frac{p_2}{p_1}$$

With values from Table A-23

$$\bar{s}_2^* - \bar{s}_1^* = 185.775 - 191.682 - 8.314\ln\frac{40}{100} = 1.711\frac{\text{kJ}}{\text{kmol}\cdot\text{K}}$$

The terms $(\bar{s}^* - \bar{s})/\bar{R}$ at the inlet and exit can be determined from Fig. A-5. Using the reduced temperature and reduced pressure values calculated in the solution to Example 11.8, inspection of Fig. A-5 gives

$$\left(\frac{\bar{s}^* - \bar{s}}{\bar{R}}\right)_1 \approx 0.21, \qquad \left(\frac{\bar{s}^* - \bar{s}}{\bar{R}}\right)_2 \approx 0.14$$

Substituting values

$$\frac{\dot{\sigma}_{cv}}{\dot{m}} = \frac{1}{(28\text{ kg/kmol})}\left[1.711\frac{\text{kJ}}{\text{kmol}\cdot\text{K}} - 8.314\frac{\text{kJ}}{\text{kmol}\cdot\text{K}}(0.14 - 0.21)\right]$$

$$= 0.082\frac{\text{kJ}}{\text{kg}\cdot\text{K}}$$

(b) The isentropic turbine efficiency is defined in Sec. 6.12 as

$$\eta_t = \frac{(\dot{W}_{cv}/\dot{m})}{(\dot{W}_{cv}/\dot{m})_s}$$

where the denominator is the work that would be developed by the turbine if the nitrogen expanded isentropically from the specified inlet state to the specified exit pressure. Thus, it is necessary to fix the state, call it 2s, at the turbine exit for an expansion in which there is no change in specific entropy from inlet to exit. With $(\bar{s}_{2s} - \bar{s}_1) = 0$ and procedures similar to those used in part (a)

$$0 = \bar{s}_{2s}^* - \bar{s}_1^* - \bar{R}\left[\left(\frac{\bar{s}^* - \bar{s}}{\bar{R}}\right)_{2s} - \left(\frac{\bar{s}^* - \bar{s}}{\bar{R}}\right)_1\right]$$

$$0 = \left[\bar{s}°(T_{2s}) - \bar{s}°(T_1) - \bar{R}\ln\left(\frac{p_2}{p_1}\right)\right] - \bar{R}\left[\left(\frac{\bar{s}^* - \bar{s}}{\bar{R}}\right)_{2s} - \left(\frac{\bar{s}^* - \bar{s}}{\bar{R}}\right)_1\right]$$

Using values from part (a), the last equation becomes

$$0 = \bar{s}°(T_{2s}) - 191.682 - 8.314\ln\frac{40}{100} - \bar{R}\left(\frac{\bar{s}^* - \bar{s}}{\bar{R}}\right)_{2s} + 1.746$$

or

$$\bar{s}°(T_{2s}) - \bar{R}\left(\frac{\bar{s}^* - \bar{s}}{\bar{R}}\right)_{2s} = 182.3$$

The temperature T_{2s} can be determined in an iterative procedure using $\bar{s}°$ data from Table A-23 and $(\bar{s}^* - \bar{s})/\bar{R}$ from Fig. A-5 as follows: First, a value for the temperature T_{2s} is assumed. The corresponding value of $\bar{s}°$ can then be obtained from Table A-23. The reduced temperature $(T_R)_{2s} = T_{2s}/T_c$, together with $p_{R2} = 1.18$, allows a value for $(\bar{s}^* - \bar{s})/\bar{R}$ to be obtained from Fig. A-5. The procedure continues until agreement with the value on the right side of the above equation is obtained. Using this procedure, T_{2s} is found to be closely 228 K.

With the temperature T_{2s} known, the work that would be developed by the turbine if the nitrogen expanded isentropically from the specified inlet state to the specified exit pressure can be evaluated from

$$\left(\frac{\dot{W}_{cv}}{\dot{m}}\right)_s = h_1 - h_{2s}$$

$$= \frac{1}{M}\left\{(\bar{h}_1^* - \bar{h}_{2s}^*) - \bar{R}T_c\left[\left(\frac{\bar{h}^* - \bar{h}}{\bar{R}T_c}\right)_1 - \left(\frac{\bar{h}^* - \bar{h}}{\bar{R}T_c}\right)_{2s}\right]\right\}$$

From Table A-23, $\bar{h}_{2s}^* = 6654$ kJ/kmol. From Fig. A-4 at $P_{R2} = 1.18$ and $(T_R)_{2s} = 228/126 = 1.81$

$$\left(\frac{\bar{h}^* - \bar{h}}{\bar{R}T_c}\right)_{2s} \approx 0.36$$

Values for the other terms in the expression for $(\dot{W}_{cv}/\dot{m})_s$ are obtained in the solution to Example 11.8. Finally

$$\left(\frac{\dot{W}_{cv}}{\dot{m}}\right)_s = \frac{1}{28}[8723 - 6654 - (8.314)(126)(0.5 - 0.36)] = 68.66 \text{ kJ/kg}$$

With the work value from Example 11.8, the turbine efficiency is

1 $$\eta_t = \frac{(\dot{W}_{cv}/\dot{m})}{(\dot{W}_{cv}/\dot{m})_s} = \frac{50.1}{68.66} = 0.73(73\%)$$

1 We cannot expect extreme accuracy when reading data from a generalized chart such as Fig. A-5, which affects the final calculated result.

Skills Developed: *Ability to...*
- ✓ use data from the generalized entropy departure chart to calculate the entropy production.
- ✓ use data from the generalized enthalpy and entropy departure charts to calculate isentropic turbine efficiency.
- ✓ use an iterative procedure to calculate the temperature at the end of an isentropic process using data from the generalized entropy departure chart.

Quick Quiz

Determine the rate of entropy production, in kJ/K per kg of nitrogen flowing, assuming the ideal gas model.

Ans. 0.061 kJ/kg · K.

11.8 p–v–T Relations for Gas Mixtures

Many systems of interest involve mixtures of two or more components. The principles of thermodynamics introduced thus far are applicable to systems involving mixtures, but to apply them requires that the properties of mixtures be evaluated. Since an unlimited variety of mixtures can be formed from a given set of pure components by varying the relative amounts present, the properties of mixtures are available in tabular, graphical, or equation forms only in particular cases such as air. Generally, special means are required for determining the properties of mixtures. In this section, methods for evaluating the p–v–T relations for pure components introduced in previous sections of the book are adapted to obtain plausible estimates for gas mixtures. In Sec. 11.9 some general aspects of property evaluation for multicomponent systems are introduced. The case of ideal gas mixtures is taken up in Chap. 12.

To evaluate the properties of a mixture requires knowledge of the composition. The composition can be described by giving the *number of moles* (kmol) of each component present. The total number of moles, n, is the sum of the number of moles of each of the components

$$n = n_1 + n_2 + \cdots + n_j = \sum_{i=1}^{j} n_i \tag{11.93}$$

The *relative* amounts of the components present can be described in terms of *mole fractions.* The mole fraction y_i of component i is defined as

$$y_i = \frac{n_i}{n} \tag{11.94}$$

Dividing each term of Eq. 11.93 by the total number of moles and using Eq. 11.94

$$1 = \sum_{i=1}^{j} y_i \tag{11.95}$$

That is, the sum of the mole fractions of all components present is equal to unity.

Most techniques for estimating mixture properties are empirical in character and are not derived from fundamental principles. The realm of validity of any particular technique can be established only by comparing predicted property values with empirical data. The brief discussion to follow is intended only to show how certain of the procedures for evaluating the p–v–T relations of pure components introduced previously can be extended to gas mixtures.

MIXTURE EQUATION OF STATE. One way the p–v–T relation for a gas mixture can be estimated is by applying to the overall mixture an equation of state such as introduced in Sec. 11.1. The constants appearing in the equation selected would be *mixture values* determined with empirical combining rules developed for the equation. For example, mixture values of the constants a and b for use in the van der Waals and Redlich–Kwong equations would be obtained using relations of the form

$$a = \left(\sum_{i=1}^{j} y_i a_i^{1/2} \right)^2, \qquad b = \left(\sum_{i=1}^{j} y_i b_i \right) \tag{11.96}$$

where a_i and b_i are the values of the constants for component i and y_i is the mole fraction. Combination rules for obtaining mixture values for the constants in other equations of state also have been suggested.

KAY'S RULE. The principle of corresponding states method for single components introduced in Sec. 3.4 can be extended to mixtures by regarding the mixture as if it were a single pure component having critical properties calculated by one of several mixture rules. Perhaps the simplest of these, requiring only the determination of a mole fraction averaged critical temperature T_c and critical pressure p_c, is ***Kay's rule***

Kay's rule

$$T_c = \sum_{i=1}^{j} y_i T_{c,i}, \qquad p_c = \sum_{i=1}^{j} y_i p_{c,i} \tag{11.97}$$

where $T_{c,i}$, $p_{c,i}$, and y_i are the critical temperature, critical pressure, and mole fraction of component i, respectively. Using T_c and p_c, the mixture compressibility factor Z is obtained as for a single pure component. The unknown quantity from among the pressure p, volume V, temperature T, and total number of moles n of the gas mixture can then be obtained by solving

$$Z = \frac{pV}{n\bar{R}T} \tag{11.98}$$

Mixture values for T_c and p_c also can be used to enter the generalized enthalpy departure and entropy departure charts introduced in Sec. 11.7.

additive pressure rule

ADDITIVE PRESSURE RULE. Additional means for estimating p–v–T relations for mixtures are provided by empirical mixture rules, of which several are found in the engineering literature. Among these are the *additive pressure* and *additive volume* rules. According to the ***additive pressure rule,*** the pressure of a gas mixture occupying volume V at temperature T is expressible as a sum of pressures exerted by the individual components

$$p = p_1 + p_2 + p_3 + \cdots \big]_{T,V} \tag{11.99a}$$

where the pressures p_1, p_2, etc., are evaluated by considering the respective components to be at the volume and temperature of the mixture. These pressures would be determined using tabular or graphical p–v–T data or a suitable equation of state.

An alternative expression of the additive pressure rule in terms of compressibility factors can be obtained. Since component i is considered to be at the volume and temperature of the mixture, the compressibility factor Z_i for this component is $Z_i = p_iV/n_i\bar{R}T$, so the pressure p_i is

$$p_i = \frac{Z_i n_i \bar{R}T}{V}$$

Similarly, for the mixture

$$p = \frac{Zn\bar{R}T}{V}$$

Substituting these expressions into Eq. 11.99a and reducing gives the following relationship between the compressibility factors for the mixture Z and the mixture components Z_i

$$Z = \sum_{i=1}^{j} y_i Z_i \big]_{T,V} \tag{11.99b}$$

The compressibility factors Z_i are determined assuming that component i occupies the entire volume of the mixture at the temperature T.

additive volume rule

ADDITIVE VOLUME RULE. The underlying assumption of the ***additive volume rule*** is that the volume V of a gas mixture at temperature T and pressure p is expressible as the sum of volumes occupied by the individual components

$$V = V_1 + V_2 + V_3 + \cdots \big]_{p,T} \tag{11.100a}$$

where the volumes V_1, V_2, etc., are evaluated by considering the respective components to be at the pressure and temperature of the mixture. These volumes would be determined from tabular or graphical p–v–T data or a suitable equation of state.

An alternative expression of the additive volume rule in terms of compressibility factors can be obtained. Since component i is considered to be at the pressure and temperature of the mixture, the compressibility factor Z_i for this component is $Z_i = pV_i/n_i\bar{R}T$, so the volume V_i is

$$V_i = \frac{Z_i n_i \bar{R} T}{p}$$

Similarly, for the mixture

$$V = \frac{Z n \bar{R} T}{p}$$

Substituting these expressions into Eq. 11.100a and reducing gives

$$Z = \sum_{i=1}^{j} y_i Z_i]_{p,T} \tag{11.100b}$$

The compressibility factors Z_i are determined assuming that component i exists at the temperature T and pressure p of the mixture.

The next example illustrates alternative means for estimating the pressure of a gas mixture.

Example 11.10 ESTIMATING MIXTURE PRESSURE BY ALTERNATIVE MEANS

A mixture consisting of 0.18 kmol of methane (CH_4) and 0.274 kmol of butane (C_4H_{10}) occupies a volume of 0.241 m^3 at a temperature of 238°C. The experimental value for the pressure is 68.9 bar. Calculate the pressure, in bar, exerted by the mixture by using **(a)** the ideal gas equation of state, **(b)** Kay's rule together with the generalized compressibility chart, **(c)** the van der Waals equation, and **(d)** the rule of additive pressures employing the generalized compressibility chart. Compare the calculated values with the known experimental value.

Solution

Known: A mixture of two specified hydrocarbons with known molar amounts occupies a known volume at a specified temperature.

Find: Determine the pressure, in bar, using four alternative methods, and compare the results with the experimental value.

Schematic and Given Data:

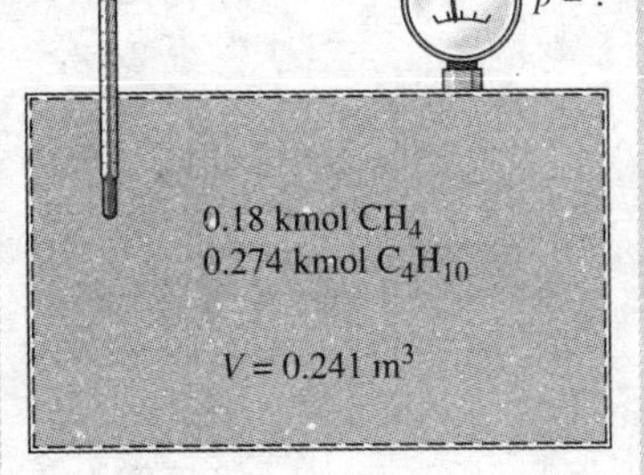

Engineering Model: As shown in the accompanying figure, the system is the mixture.

Fig. E11.10

Analysis: The total number of moles of mixture n is

$$n = 0.18 + 0.274 = 0.454 \text{ kmol}$$

Thus, the mole fractions of the methane and butane are, respectively

$$y_1 = 0.396 \quad \text{and} \quad y_2 = 0.604$$

The specific volume of the mixture on a molar basis is

$$\bar{v} = \frac{0.241 \text{ m}^3}{(0.18 + 0.274) \text{ kmol}} = 0.531 \frac{\text{m}^3}{\text{kmol}}$$

(a) Substituting values into the ideal gas equation of state

$$p = \frac{\bar{R}T}{\bar{v}} = \frac{(8314 \text{ N} \cdot \text{m/kmol} \cdot \text{K})(511 \text{ K})}{(0.531 \text{ m}^3\text{/kmol})} \left| \frac{1 \text{ bar}}{10^5 \text{ N/m}^2} \right|$$
$$= 80.01 \text{ bar}$$

(b) To apply Kay's rule, the critical temperature and pressure for each component are required. From Table A-1, for methane

$$T_{c1} = 191 \text{ K}, \qquad p_{c1} = 46.4 \text{ bar}$$

and for butane

$$T_{c2} = 425 \text{ K}, \qquad p_{c2} = 38.0 \text{ bar}$$

Thus, with Eqs. 11.97

$$T_c = y_1 T_{c1} + y_2 T_{c2} = (0.396)(191) + (0.604)(425) = 332.3 \text{ K}$$
$$p_c = y_1 p_{c1} + y_2 p_{c2} = (0.396)(46.4) + (0.604)(38.0) = 41.33 \text{ bar}$$

Treating the mixture as a pure component having the above values for the critical temperature and pressure, the following reduced properties are determined for the mixture:

$$T_R = \frac{T}{T_c} = \frac{511}{332.3} = 1.54$$

$$v'_R = \frac{\bar{v} p_c}{\bar{R} T_c} = \frac{(0.531)(41.33)|10^5|}{(8314)(332.3)}$$
$$= 0.794$$

Turning to Fig. A-2, $Z \approx 0.88$. The mixture pressure is then found from

$$p = \frac{Zn\bar{R}T}{V} = Z\frac{\bar{R}T}{\bar{v}} = 0.88 \frac{(8314)(511)}{(0.531)|10^5|}$$
$$= 70.4 \text{ bar}$$

(c) Mixture values for the van der Waals constants can be obtained using Eqs. 11.96. This requires values of the van der Waals constants for each of the two mixture components. Table A-24 gives the following values for methane:

$$a_1 = 2.293 \text{ bar}\left(\frac{\text{m}^3}{\text{kmol}}\right)^2, \qquad b_1 = 0.0428 \frac{\text{m}^3}{\text{kmol}}$$

Similarly, from Table A-24 for butane

$$a_2 = 13.86 \text{ bar}\left(\frac{\text{m}^3}{\text{kmol}}\right)^2, \qquad b_2 = 0.1162 \frac{\text{m}^3}{\text{kmol}}$$

Then, the first of Eqs. 11.96 gives a mixture value for the constant a as

$$a = (y_1 a_1^{1/2} + y_2 a_2^{1/2})^2 = [0.396(2.293)^{1/2} + 0.604(13.86)^{1/2}]^2$$
$$= 8.113 \text{ bar}\left(\frac{\text{m}^3}{\text{kmol}}\right)^2$$

Substituting into the second of Eqs. 11.96 gives a mixture value for the constant b

$$b = y_1 b_1 + y_2 b_2 = (0.396)(0.0428) + (0.604)(0.1162)$$
$$= 0.087 \frac{\text{m}^3}{\text{kmol}}$$

Inserting the mixture values for a and b into the van der Waals equation together with known data

$$p = \frac{RT}{\bar{v} - b} - \frac{a}{\bar{v}^2}$$
$$= \frac{(8314 \text{ N}\cdot\text{m/kmol}\cdot\text{K})(511 \text{ K})}{(0.531 - 0.087)(\text{m}^3/\text{kmol})}\left|\frac{1 \text{ bar}}{10^5 \text{ N/m}^2}\right| - \frac{8.113 \text{ bar}(\text{m}^3/\text{kmol})^2}{(0.531 \text{ m}^3/\text{kmol})^2}$$
$$= 66.91 \text{ bar}$$

(d) To apply the additive pressure rule with the generalized compressibility chart requires that the compressibility factor for each component be determined assuming that the component occupies the entire volume at the mixture temperature. With this assumption, the following reduced properties are obtained for methane

$$T_{R1} = \frac{T}{T_{c1}} = \frac{511}{191} = 2.69$$
$$v'_{R1} = \frac{\bar{v}_1 p_{c1}}{RT_{c1}} = \frac{(0.241 \text{ m}^3/0.18 \text{ kmol})(46.4 \text{ bar})}{(8314 \text{ N}\cdot\text{m/kmol}\cdot\text{K})(191 \text{ K})}\left|\frac{10^5 \text{ N/m}^2}{1 \text{ bar}}\right| = 3.91$$

With these reduced properties, Fig. A-2 gives $Z_1 \approx 1.0$.

Similarly, for butane

$$T_{R2} = \frac{T}{T_{c2}} = \frac{511}{425} = 1.2$$
$$v'_{R2} = \frac{\bar{v}_2 p_{c2}}{RT_{c2}} = \frac{(0.88)(38)|10^5|}{(8314)(425)} = 0.95$$

From Fig. A-2, $Z_2 \approx 0.8$.

The compressibility factor for the mixture determined from Eq. 11.99b is

$$Z = y_1 Z_1 + y_2 Z_2 = (0.396)(1.0) + (0.604)(0.8) = 0.88.$$

Accordingly, the same value for pressure as determined in part (b) using Kay's rule results: $p = 70.4$ bar.

In this particular example, the ideal gas equation of state gives a value for pressure that exceeds the experimental value by nearly 16%. Kay's rule and the rule of additive pressures give pressure values about 3% greater than the experimental value. The van der Waals equation with mixture values for the constants gives a pressure value about 3% less than the experimental value.

Skills Developed: *Ability to...*
✓ calculate the pressure of a gas mixture using four alternative methods.

Quick Quiz

Convert the mixture analysis from a molar basis to a mass fraction basis.

Ans. Methane: 0.153, Butane: 0.847.

11.9 Analyzing Multicomponent Systems[1]

In the preceding section we considered means for evaluating the p–v–T relation of gas mixtures by extending methods developed for pure components. The current section is devoted to the development of some general aspects of the properties of systems with two or more components. Primary emphasis is on the case of *gas mixtures*, but the methods developed also apply to *solutions*. When liquids and solids are under consideration, the term ***solution*** is sometimes used in place of mixture. The present discussion is limited to nonreacting mixtures or solutions in a single phase. The effects of chemical reactions and equilibrium between different phases are taken up in Chaps. 13 and 14.

solution

To describe multicomponent systems, composition must be included in our thermodynamic relations. This leads to the definition and development of several new concepts, including the *partial molal property*, the *chemical potential*, and the *fugacity*.

11.9.1 Partial Molal Properties

In the present discussion we introduce the concept of a *partial molal* property and illustrate its use. This concept plays an important role in subsequent discussions of multicomponent systems.

DEFINING PARTIAL MOLAL PROPERTIES. Any extensive thermodynamic property X of a single-phase, single-component system is a function of two independent intensive properties and the size of the system. Selecting temperature and pressure as the independent properties and the number of moles n as the measure of size, we have $X = X(T, p, n)$. For a single-phase, *multicomponent* system, the extensive property X must then be a function of temperature, pressure, and the number of moles of each component present, $X = X(T, p, n_1, n_2, \ldots, n_j)$.

If each mole number is increased by a factor α, the size of the system increases by the same factor, and so does the value of the extensive property X. That is

$$\alpha X(T, p, n_1, n_2, \ldots, n_j) = X(T, p, \alpha n_1, \alpha n_2, \ldots, \alpha n_j)$$

Differentiating with respect to α while holding temperature, pressure, and the mole numbers fixed and using the chain rule on the right side gives

$$X = \frac{\partial X}{\partial(\alpha n_1)} n_1 + \frac{\partial X}{\partial(\alpha n_2)} n_2 + \cdots + \frac{\partial X}{\partial(\alpha n_j)} n_j$$

This equation holds for all values of α. In particular, it holds for $\alpha = 1$. Setting $\alpha = 1$

$$X = \sum_{i=1}^{j} n_i \frac{\partial X}{\partial n_i}\bigg)_{T,p,n_l} \qquad (11.101)$$

where the subscript n_l denotes that all n's except n_i are held fixed during differentiation.

partial molal property

The ***partial molal property*** $\overline{X}_i$ is by definition

$$\overline{X}_i = \frac{\partial X}{\partial n_i}\bigg)_{T,p,n_l} \qquad (11.102)$$

The partial molal property $\overline{X}_i$ is a property of the mixture and not simply a property of component i, for $\overline{X}_i$ depends in general on temperature, pressure, *and* mixture com-

[1]This section may be deferred until Secs. 12.1–12.4 have been studied.

position: $\overline{X}_i(T, p, n_1, n_2, \ldots, n_j)$. Partial molal properties are intensive properties of the mixture.

Introducing Eq. 11.102, Eq. 11.101 becomes

$$X = \sum_{i=1}^{j} n_i \overline{X}_i \tag{11.103}$$

This equation shows that the extensive property X can be expressed as a weighted sum of the partial molal properties $\overline{X}_i$.

Selecting the extensive property X in Eq. 11.103 to be volume, internal energy, enthalpy, and entropy, respectively, gives

$$V = \sum_{i=1}^{j} n_i \overline{V}_i, \quad U = \sum_{i=1}^{j} n_i \overline{U}_i, \quad H = \sum_{i=1}^{j} n_i \overline{H}_i, \quad S = \sum_{i=1}^{j} n_i \overline{S}_i \tag{11.104}$$

where $\overline{V}_i, \overline{U}_i, \overline{H}_i, \overline{S}_i$ denote the partial molal volume, internal energy, enthalpy, and entropy. Similar expressions can be written for the Gibbs function G and the Helmholtz function Ψ. Moreover, the relations between these extensive properties: $H = U + pV$, $G = H - TS$, $\Psi = U - TS$ can be differentiated with respect to n_i while holding temperature, pressure, and the remaining n's constant to produce corresponding relations among partial molal properties: $\overline{H}_i = \overline{U}_i + p\overline{V}_i$, $\overline{G}_i = \overline{H}_i - T\overline{S}_i$, $\overline{\Psi}_i = \overline{U}_i - T\overline{S}_i$, where $\overline{G}_i$ and $\overline{\Psi}_i$ are the partial molal Gibbs function and Helmholtz function, respectively. Several additional relations involving partial molal properties are developed later in this section.

EVALUATING PARTIAL MOLAL PROPERTIES. Partial molal properties can be evaluated by several methods, including the following:

- If the property X can be measured, $\overline{X}_i$ can be found by extrapolating a plot giving $(\Delta X/\Delta n_i)_{T,p,n_l}$ versus Δn_i. That is

$$\overline{X}_i = \left(\frac{\partial X}{\partial n_i}\right)_{T,p,n_l} = \lim_{\Delta n_i \to 0} \left(\frac{\Delta X}{\Delta n_i}\right)_{T,p,n_l}$$

- If an expression for X in terms of its independent variables is known, $\overline{X}_i$ can be evaluated by differentiation. The derivative can be determined analytically if the function is expressed analytically or found numerically if the function is in tabular form.
- When suitable data are available, a simple graphical procedure known as the ***method of intercepts*** can be used to evaluate partial molal properties. In principle, the method can be applied for any extensive property. To introduce this method, let us consider the volume of a system consisting of two components, A and B. For this system, Eq. 11.103 takes the form

method of intercepts

$$V = n_A \overline{V}_A + n_B \overline{V}_B$$

where $\overline{V}_A$ and $\overline{V}_B$ are the partial molal volumes of A and B, respectively. Dividing by the number of moles of mixture n

$$\frac{V}{n} = y_A \overline{V}_A + y_B \overline{V}_B$$

where y_A and y_B denote the mole fractions of A and B, respectively. Since $y_A + y_B = 1$, this becomes

$$\frac{V}{n} = (1 - y_B)\overline{V}_A + y_B \overline{V}_B = \overline{V}_A + y_B(\overline{V}_B - \overline{V}_A)$$

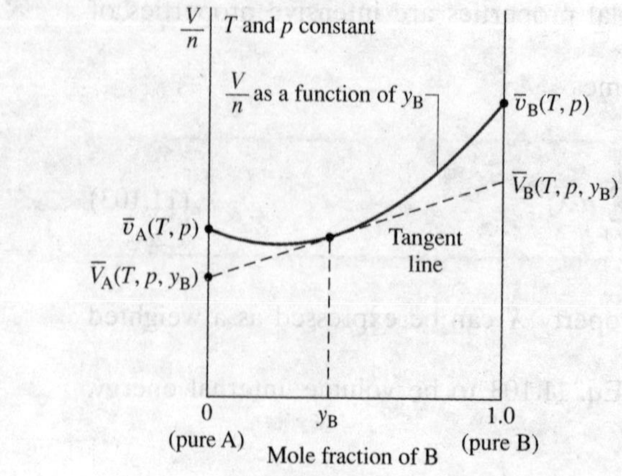

Fig. 11.5 Illustration of the evaluation of partial molal volumes by the method of intercepts.

This equation provides the basis for the method of intercepts. For example, refer to Fig. 11.5, in which V/n is plotted as a function of y_B at constant T and p. At a specified value for y_B, a tangent to the curve is shown on the figure. When extrapolated, the tangent line intersects the axis on the left at $\overline{V}_A$ and the axis on the right at $\overline{V}_B$. These values for the partial molal volumes correspond to the particular specifications for T, p, and y_B. At fixed temperature and pressure, $\overline{V}_A$ and $\overline{V}_B$ vary with y_B and are not equal to the molar specific volumes of *pure* A and *pure* B, denoted on the figure as $\bar{v}_A$ and $\bar{v}_B$, respectively. The values of $\bar{v}_A$ and $\bar{v}_B$ are fixed by temperature and pressure only.

EXTENSIVE PROPERTY CHANGES ON MIXING. Let us conclude the present discussion by evaluating the change in volume on mixing of pure components at the same temperature and pressure, a result for which an application is given later in the section. The total volume of the pure components before mixing is

$$V_{\text{components}} = \sum_{i=1}^{j} n_i \bar{v}_i$$

where $\bar{v}_i$ is the molar specific volume of pure component i. The volume of the mixture is

$$V_{\text{mixture}} = \sum_{i=1}^{j} n_i \overline{V}_i$$

where $\overline{V}_i$ is the partial molal volume of component i in the mixture. The volume change on mixing is

$$\Delta V_{\text{mixing}} = V_{\text{mixture}} - V_{\text{components}} = \sum_{i=1}^{j} n_i \overline{V}_i - \sum_{i=1}^{j} n_i \bar{v}_i$$

or

$$\Delta V_{\text{mixing}} = \sum_{i=1}^{j} n_i(\overline{V}_i - \bar{v}_i) \tag{11.105}$$

Similar results can be obtained for other extensive properties, for example,

$$\Delta U_{\text{mixing}} = \sum_{i=1}^{j} n_i(\overline{U}_i - \bar{u}_i)$$

$$\Delta H_{\text{mixing}} = \sum_{i=1}^{j} n_i(\overline{H}_i - \bar{h}_i) \tag{11.106}$$

$$\Delta S_{\text{mixing}} = \sum_{i=1}^{j} n_i(\overline{S}_i - \bar{s}_i)$$

In Eqs. 11.106, $\bar{u}_i$, $\bar{h}_i$, and $\bar{s}_i$ denote the molar internal energy, enthalpy, and entropy of pure component i. The symbols $\bar{U}_i$, $\bar{H}_i$, and $\bar{S}_i$ denote the respective partial molal properties.

11.9.2 Chemical Potential

Of the partial molal properties, the partial molal Gibbs function is particularly useful in describing the behavior of mixtures and solutions. This quantity plays a central role in the criteria for both chemical and phase equilibrium (Chap. 14). Because of its importance in the study of multicomponent systems, the partial molal Gibbs function of component i is given a special name and symbol. It is called the ***chemical potential*** of component i and symbolized by μ_i

chemical potential

$$\mu_i = \bar{G}_i = \left(\frac{\partial G}{\partial n_i}\right)_{T,p,n_l} \tag{11.107}$$

Like temperature and pressure, the chemical potential μ_i is an *intensive* property.

Applying Eq. 11.103 together with Eq. 11.107, the following expression can be written:

$$G = \sum_{i=1}^{j} n_i \mu_i \tag{11.108}$$

Expressions for the internal energy, enthalpy, and Helmholtz function can be obtained from Eq. 11.108, using the definitions $H = U + pV$, $G = H - TS$, and $\Psi = U - TS$. They are

$$U = TS - pV + \sum_{i=1}^{j} n_i \mu_i$$

$$H = TS + \sum_{i=1}^{j} n_i \mu_i \tag{11.109}$$

$$\Psi = -pV + \sum_{i=1}^{j} n_i \mu_i$$

Other useful relations can be obtained as well. Forming the differential of $G(T, p, n_1, n_2, \ldots, n_j)$

$$dG = \left(\frac{\partial G}{\partial p}\right)_{T,n} dp + \left(\frac{\partial G}{\partial T}\right)_{p,n} dT + \sum_{i=1}^{j} \left(\frac{\partial G}{\partial n_i}\right)_{T,p,n_l} dn_i \tag{11.110}$$

The subscripts n in the first two terms indicate that all n's are held fixed during differentiation. Since this implies fixed composition, it follows from Eqs. 11.30 and 11.31 that

$$V = \left(\frac{\partial G}{\partial p}\right)_{T,n} \quad \text{and} \quad -S = \left(\frac{\partial G}{\partial T}\right)_{p,n} \tag{11.111}$$

With Eqs. 11.107 and 11.111, Eq. 11.110 becomes

$$dG = V\,dp - S\,dT + \sum_{i=1}^{j} \mu_i\,dn_i \tag{11.112}$$

which for a multicomponent system is the counterpart of Eq. 11.23.

Another expression for dG is obtained by forming the differential of Eq. 11.108. That is

$$dG = \sum_{i=1}^{j} n_i \, d\mu_i + \sum_{i=1}^{j} \mu_i \, dn_i$$

Gibbs–Duhem equation

Combining this equation with Eq. 11.112 gives the ***Gibbs–Duhem equation***

$$\sum_{i=1}^{j} n_i \, d\mu_i = V\,dp - S\,dT \tag{11.113}$$

11.9.3 Fundamental Thermodynamic Functions for Multicomponent Systems

A *fundamental thermodynamic function* provides a complete description of the thermodynamic state of a system. In principle, all properties of interest can be determined from such a function by differentiation and/or combination. Reviewing the developments of Sec. 11.9.2, we see that a function $G(T, p, n_1, n_2, \ldots, n_j)$ is a fundamental thermodynamic function for a multicomponent system.

Functions of the form $U(S, V, n_1, n_2, \ldots, n_j)$, $H(S, p, n_1, n_2, \ldots, n_j)$, and $\Psi(T, V, n_1, n_2, \ldots, n_j)$ also can serve as fundamental thermodynamic functions for multicomponent systems. To demonstrate this, first form the differential of each of Eqs. 11.109 and use the Gibbs–Duhem equation, Eq. 11.113, to reduce the resultant expressions to obtain

$$dU = T\,dS - p\,dV + \sum_{i=1}^{j} \mu_i \, dn_i \tag{11.114a}$$

$$dH = T\,dS + V\,dp + \sum_{i=1}^{j} \mu_i \, dn_i \tag{11.114b}$$

$$d\Psi = -p\,dV - S\,dT + \sum_{i=1}^{j} \mu_i \, dn_i \tag{11.114c}$$

For multicomponent systems, these are the counterparts of Eqs. 11.18, 11.19, and 11.22, respectively.

The differential of $U(S, V, n_1, n_2, \ldots, n_j)$ is

$$dU = \left(\frac{\partial U}{\partial S}\right)_{V,n} dS + \left(\frac{\partial U}{\partial V}\right)_{S,n} dV + \sum_{i=1}^{j} \left(\frac{\partial U}{\partial n_i}\right)_{S,V,n_l} dn_i$$

Comparing this expression term by term with Eq. 11.114a, we have

$$T = \left(\frac{\partial U}{\partial S}\right)_{V,n}, \quad -p = \left(\frac{\partial U}{\partial V}\right)_{S,n}, \quad \mu_i = \left(\frac{\partial U}{\partial n_i}\right)_{S,V,n_l} \tag{11.115a}$$

That is, the temperature, pressure, and chemical potentials can be obtained by differentiation of $U(S, V, n_1, n_2, \ldots, n_j)$. The first two of Eqs. 11.115a are the counterparts of Eqs. 11.24 and 11.25.

A similar procedure using a function of the form $H(S, p, n_1, n_2, \ldots, n_j)$ together with Eq. 11.114b gives

$$T = \left(\frac{\partial H}{\partial S}\right)_{p,n}, \quad V = \left(\frac{\partial H}{\partial p}\right)_{S,n}, \quad \mu_i = \left(\frac{\partial H}{\partial n_i}\right)_{S,p,n_l} \tag{11.115b}$$

where the first two of these are the counterparts of Eqs. 11.26 and 11.27. Finally, with $\Psi(S, V, n_1, n_2, \ldots, n_j)$ and Eq. 11.114c

$$-p = \left(\frac{\partial \Psi}{\partial V}\right)_{T,n}, \quad -S = \left(\frac{\partial \Psi}{\partial T}\right)_{V,n}, \quad \mu_i = \left(\frac{\partial \Psi}{\partial n_i}\right)_{T,V,n_l} \tag{11.115c}$$

The first two of these are the counterparts of Eqs. 11.28 and 11.29. With each choice of fundamental function, the remaining extensive properties can be found by combination using the definitions $H = U + pV$, $G = H - TS$, $\Psi = U - TS$.

The foregoing discussion of fundamental thermodynamic functions has led to several property relations for multicomponent systems that correspond to relations obtained previously. In addition, counterparts of the Maxwell relations can be obtained by equating mixed second partial derivatives. For example, the first two terms on the right of Eq. 11.112 give

$$\left.\frac{\partial V}{\partial T}\right)_{p,n} = -\left.\frac{\partial S}{\partial p}\right)_{T,n} \tag{11.116}$$

which corresponds to Eq. 11.35. Numerous relationships involving chemical potentials can be derived similarly by equating mixed second partial derivatives. An important example from Eq. 11.112 is

$$\left.\frac{\partial \mu_i}{\partial p}\right)_{T,n} = \left.\frac{\partial V}{\partial n_i}\right)_{T,p,n_l}$$

Recognizing the right side of this equation as the partial molal volume, we have

$$\left.\frac{\partial \mu_i}{\partial p}\right)_{T,n} = \overline{V}_i \tag{11.117}$$

This relationship is applied later in the section.

The present discussion concludes by listing four different expressions derived above for the chemical potential in terms of other properties. In the order obtained, they are

$$\mu_i = \left.\frac{\partial G}{\partial n_i}\right)_{T,p,n_l} = \left.\frac{\partial U}{\partial n_i}\right)_{S,V,n_l} = \left.\frac{\partial H}{\partial n_i}\right)_{S,p,n_l} = \left.\frac{\partial \Psi}{\partial n_i}\right)_{T,V,n_l} \tag{11.118}$$

Only the first of these partial derivatives is a partial molal property, however, for the term *partial molal* applies only to partial derivatives where the independent variables are temperature, pressure, and the number of moles of each component present.

11.9.4 Fugacity

The chemical potential plays an important role in describing multicomponent systems. In some instances, however, it is more convenient to work in terms of a related property, the fugacity. The fugacity is introduced in the present discussion.

Single-Component Systems

Let us begin by taking up the case of a system consisting of a single component. For this case, Eq. 11.108 reduces to give

$$G = n\mu \qquad \text{or} \qquad \mu = \frac{G}{n} = \overline{g}$$

That is, for a pure component the chemical potential equals the Gibbs function per mole. With this equation, Eq. 11.30 written on a per mole basis becomes

$$\left.\frac{\partial \mu}{\partial p}\right)_T = \overline{v} \tag{11.119}$$

For the special case of an ideal gas, $\bar{v} = \bar{R}T/p$, and Eq. 11.119 assumes the form

$$\left(\frac{\partial \mu^*}{\partial p}\right)_T = \frac{\bar{R}T}{p}$$

where the asterisk denotes ideal gas. Integrating at constant temperature

$$\mu^* = \bar{R}T \ln p + C(T) \tag{11.120}$$

where $C(T)$ is a function of integration. Since the pressure p can take on values from zero to plus infinity, the ln p term of this expression, and thus the chemical potential, has an inconvenient range of values from minus infinity to plus infinity. Equation 11.120 also shows that the chemical potential can be determined only to within an arbitrary constant.

INTRODUCING FUGACITY. Because of the above considerations, it is advantageous for many types of thermodynamic analyses to use fugacity in place of the chemical potential, for it is a well-behaved function that can be more conveniently evaluated. We introduce the ***fugacity*** f by the expression

fugacity

$$\mu = \bar{R}T \ln f + C(T) \tag{11.121}$$

Comparing Eq. 11.121 with Eq. 11.120, the fugacity is seen to play the same role in the general case as pressure plays in the ideal gas case. Fugacity has the same units as pressure.

Substituting Eq. 11.121 into Eq. 11.119 gives

$$\bar{R}T\left(\frac{\partial \ln f}{\partial p}\right)_T = \bar{v} \tag{11.122}$$

Integration of Eq. 11.122 while holding temperature constant can determine the fugacity only to within a constant term. However, since ideal gas behavior is approached as pressure tends to zero, the constant term can be fixed by requiring that the fugacity of a pure component equals the pressure in the limit of zero pressure. That is

$$\lim_{p \to 0} \frac{f}{p} = 1 \tag{11.123}$$

Equations 11.122 and 11.123 then *completely determine* the fugacity function.

EVALUATING FUGACITY. Let us consider next how the fugacity can be evaluated. With $Z = p\bar{v}/\bar{R}T$, Eq. 11.122 becomes

$$\bar{R}T\left(\frac{\partial \ln f}{\partial p}\right)_T = \frac{\bar{R}TZ}{p}$$

or

$$\left(\frac{\partial \ln f}{\partial p}\right)_T = \frac{Z}{p}$$

Subtracting $1/p$ from both sides and integrating from pressure p' to pressure p at fixed temperature T

$$[\ln f - \ln p]_{p'}^{p} = \int_{p'}^{p} (Z - 1)\, d \ln p$$

or

$$\left[\ln \frac{f}{p}\right]_{p'}^{p} = \int_{p'}^{p} (Z - 1)\, d \ln p$$

Taking the limit as p' tends to zero and applying Eq. 11.123 results in

$$\ln \frac{f}{p} = \int_0^p (Z - 1)\, d \ln p$$

When expressed in terms of the reduced pressure, $p_R = p/p_c$, the above equation is

$$\ln \frac{f}{p} = \int_0^{p_R} (Z - 1)\, d \ln p_R \qquad (11.124)$$

Since the compressibility factor Z depends on the reduced temperature T_R and reduced pressure p_R, it follows that the right side of Eq. 11.124 depends on these properties only. Accordingly, the quantity $\ln f/p$ is a function only of these two reduced properties. Using a generalized equation of state giving Z as a function of T_R and p_R, $\ln f/p$ can readily be evaluated with a computer. Tabular representations are also found in the literature. Alternatively, the graphical representation presented in Fig. A-6 can be employed.

FOR EXAMPLE... to illustrate the use of Fig. A-6, consider two states of water vapor at the same temperature, 400°C. At state 1 the pressure is 200 bar, and at state 2 the pressure is 240 bar. The change in the chemical potential between these states can be determined using Eq. 11.121 as

$$\mu_2 - \mu_1 = \overline{R}T \ln \frac{f_2}{f_1} = \overline{R}T \ln\left(\frac{f_2}{p_2}\frac{p_2}{p_1}\frac{p_1}{f_1}\right)$$

Using the critical temperature and pressure of water from Table A-1, at state 1 $p_{R1} = 0.91$, $T_{R1} = 1.04$, and at state 2 $p_{R2} = 1.09$, $T_{R2} = 1.04$. By inspection of Fig. A-6, $f_1/p_1 = 0.755$ and $f_2/p_2 = 0.7$. Inserting values in the above equation

$$\mu_2 - \mu_1 = (8.314)(673.15) \ln\left[(0.7)\left(\frac{240}{200}\right)\left(\frac{1}{0.755}\right)\right] = 597 \text{ kJ/kmol}$$

For a pure component, the chemical potential equals the Gibbs function per mole, $\bar{g} = \bar{h} - T\bar{s}$. Since the temperature is the same at states 1 and 2, the change in the chemical potential can be expressed as $\mu_2 - \mu_1 = \bar{h}_2 - \bar{h}_1 - T(\bar{s}_2 - \bar{s}_1)$. Using steam table data, the value obtained with this expression is 597 kJ/kmol, which agrees with the value determined from the generalized fugacity coefficient chart.

Multicomponent Systems

The fugacity of a component i in a mixture can be defined by a procedure that parallels the definition for a pure component. For a pure component, the development begins with Eq. 11.119, and the fugacity is introduced by Eq. 11.121. These are then used to write the pair of equations, Eqs. 11.122 and 11.123, from which the fugacity can be evaluated. For a mixture, the development begins with Eq. 11.117, the counterpart of Eq. 11.119, and the fugacity $\bar{f}_i$ of component i is introduced by

$$\mu_i = \overline{R}T \ln \bar{f}_i + C_i(T) \qquad (11.125)$$

which parallels Eq. 11.121. The pair of equations that allow the ***fugacity of a mixture component***, $\bar{f}_i$, to be evaluated are

fugacity of a mixture component

$$\overline{R}T\left(\frac{\partial \ln \bar{f}_i}{\partial p}\right)_{T,n} = \overline{V}_i \qquad (11.126a)$$

$$\lim_{p\to 0}\left(\frac{\bar{f}_i}{y_i p}\right) = 1 \qquad (11.126b)$$

The symbol $\bar{f}_i$ denotes the fugacity of component i in the *mixture* and should be carefully distinguished in the presentation to follow from f_i, which denotes the fugacity of *pure i*.

DISCUSSION. Referring to Eq. 11.126b, note that in the ideal gas limit the fugacity $\bar{f}_i$ is not required to equal the pressure p as for the case of a pure component, but to equal the quantity $y_i p$. To see that this is the appropriate limiting quantity, consider a system consisting of a mixture of gases occupying a volume V at pressure p and temperature T. If the overall mixture behaves as an ideal gas, we can write

$$p = \frac{n\bar{R}T}{V} \tag{11.127}$$

where n is the total number of moles of mixture. Recalling from Sec. 3.12 that an ideal gas can be regarded as composed of molecules that exert negligible forces on one another and whose volume is negligible relative to the total volume, we can think of each component i as behaving as if it were an ideal gas alone at the temperature T and volume V. Thus, the pressure exerted by component i would not be the mixture pressure p but the pressure p_i given by

$$p_i = \frac{n_i\bar{R}T}{V} \tag{11.128}$$

where n_i is the number of moles of component i. Dividing Eq. 11.128 by Eq. 11.127

$$\frac{p_i}{p} = \frac{n_i\bar{R}T/V}{n\bar{R}T/V} = \frac{n_i}{n} = y_i$$

On rearrangement

$$p_i = y_i p \tag{11.129}$$

Accordingly, the quantity $y_i p$ appearing in Eq. 11.126b corresponds to the pressure p_i.

Summing both sides of Eq. 11.129, we obtain

$$\sum_{i=1}^{j} p_i = \sum_{i=1}^{j} y_i p = p \sum_{i=1}^{j} y_i$$

Or, since the sum of the mole fractions equals unity

$$p = \sum_{i=1}^{j} p_i \tag{11.130}$$

In words, Eq. 11.130 states that the sum of the pressures p_i equals the mixture pressure. This gives rise to the designation *partial pressure* for p_i. With this background, we now see that Eq. 11.126b requires the fugacity of component i to approach the partial pressure of component i as pressure p tends to zero. Comparing Eqs. 11.130 and 11.99a, we also see that the *additive pressure rule* is exact for ideal gas mixtures. This special case is considered further in Sec. 12.2 under the heading *Dalton model.*

EVALUATING FUGACITY IN A MIXTURE. Let us consider next how the fugacity of component i in a mixture can be expressed in terms of quantities that can be evaluated. For a pure component i, Eq. 11.122 gives

$$\bar{R}T\left(\frac{\partial \ln f_i}{\partial p}\right)_T = \bar{v}_i \tag{11.131}$$

where $\bar{v}_i$ is the molar specific volume of pure i. Subtracting Eq. 11.131 from Eq. 11.126a

$$\bar{R}T\left[\frac{\partial \ln (\bar{f}_i/f_i)}{\partial p}\right]_{T,n} = \bar{V}_i - \bar{v}_i \tag{11.132}$$

Integrating from pressure p' to pressure p at fixed temperature and mixture composition

$$\overline{R}T\left[\ln\left(\frac{\bar{f}_i}{f_i}\right)\right]_{p'}^{p} = \int_{p'}^{p} (\overline{V}_i - \bar{v}_i)\,dp$$

In the limit as p' tends to zero, this becomes

$$\overline{R}T\left[\ln\left(\frac{\bar{f}_i}{f_i}\right) - \lim_{p'\to 0}\ln\left(\frac{\bar{f}_i}{f_i}\right)\right] = \int_{0}^{p} (\overline{V}_i - \bar{v}_i)\,dp$$

Since $f_i \to p'$ and $\bar{f}_i \to y_i p'$ as pressure p' tends to zero

$$\lim_{p'\to 0}\ln\left(\frac{\bar{f}_i}{f_i}\right) \to \ln\left(\frac{y_i p'}{p'}\right) = \ln y_i$$

Therefore, we can write

$$\overline{R}T\left[\ln\left(\frac{\bar{f}_i}{f_i}\right) - \ln y_i\right] = \int_{0}^{p} (\overline{V}_i - \bar{v}_i)\,dp$$

or

$$\overline{R}T\ln\left(\frac{\bar{f}_i}{y_i f_i}\right) = \int_{0}^{p} (\overline{V}_i - \bar{v}_i)\,dp \qquad (11.133)$$

in which $\bar{f}_i$ is the fugacity of component i at pressure p in a mixture of given composition at a given temperature, and f_i is the fugacity of pure i at the same temperature and pressure. Equation 11.133 expresses the relation between $\bar{f}_i$ and f_i in terms of the difference between $\overline{V}_i$ and $\bar{v}_i$, a measurable quantity.

11.9.5 Ideal Solution

The task of evaluating the fugacities of the components in a mixture is considerably simplified when the mixture can be modeled as an ideal solution. An ***ideal solution*** is a mixture for which

ideal solution

$$\bar{f}_i = y_i f_i \quad \text{(ideal solution)} \qquad (11.134)$$

Equation 11.134, known as the ***Lewis–Randall rule,*** states that the fugacity of each component in an ideal solution is equal to the product of its mole fraction and the fugacity of the pure component at the same temperature, pressure, and state of aggregation (gas, liquid, or solid) as the mixture. Many gaseous mixtures at low to moderate pressures are adequately modeled by the Lewis–Randall rule. The ideal gas mixtures considered in Chap. 12 are an important special class of such mixtures. Some liquid solutions also can be modeled with the Lewis–Randall rule.

Lewis–Randall rule

As consequences of the definition of an ideal solution, the following characteristics are exhibited:

- Introducing Eq. 11.134 into Eq. 11.132, the left side vanishes, giving $\overline{V}_i - \bar{v}_i = 0$, or

$$\overline{V}_i = \bar{v}_i \qquad (11.135)$$

Thus, the partial molal volume of each component in an ideal solution is equal to the molar specific volume of the corresponding pure component at the same temperature and pressure. When Eq. 11.135 is introduced in Eq. 11.105, it can be concluded that there is no volume change on mixing pure components to form an ideal solution.

With Eq. 11.135, the volume of an ideal solution is

$$V = \sum_{i=1}^{j} n_i \overline{V}_i = \sum_{i=1}^{j} n_i \overline{v}_i = \sum_{i=1}^{j} V_i \quad \text{(ideal solution)} \tag{11.136}$$

where V_i is the volume that pure component i would occupy when at the temperature and pressure of the mixture. Comparing Eqs. 11.136 and 11.100a, the *additive volume rule* is seen to be exact for ideal solutions.

► It also can be shown that the partial molal internal energy of each component in an ideal solution is equal to the molar internal energy of the corresponding pure component at the same temperature and pressure. A similar result applies for enthalpy. In symbols

$$\overline{U}_i = \overline{u}_i, \qquad \overline{H}_i = \overline{h}_i \tag{11.137}$$

With these expressions, it can be concluded from Eqs. 11.106 that there is no change in internal energy or enthalpy on mixing pure components to form an ideal solution.

With Eqs. 11.137, the internal energy and enthalpy of an ideal solution are

$$U = \sum_{i=1}^{j} n_i \overline{u}_i \quad \text{and} \quad H = \sum_{i=1}^{j} n_i \overline{h}_i \quad \text{(ideal solution)} \tag{11.138}$$

where $\overline{u}_i$ and $\overline{h}_i$ denote, respectively, the molar internal energy and enthalpy of pure component i at the temperature and pressure of the mixture.

Although there is no change in V, U, or H on mixing pure components to form an ideal solution, we expect an entropy increase to result from the *adiabatic* mixing of different pure components because such a process is irreversible: The separation of the mixture into the pure components would never occur spontaneously. The entropy change on adiabatic mixing is considered further for the special case of ideal gas mixtures in Sec. 12.4.

The Lewis–Randall rule requires that the fugacity of mixture component i be evaluated in terms of the fugacity of pure component i at the same temperature and pressure as the mixture and in the *same state of aggregation.* For example, if the mixture were a gas at T, p, then f_i would be determined for pure i at T, p and as a gas. However, at certain temperatures and pressures of interest a component of a gaseous mixture may, as a pure substance, be a liquid or solid. An example is an air–water vapor mixture at 20°C and 1 bar. At this temperature and pressure, water exists not as a vapor but as a liquid. Although not considered here, means have been developed that allow the ideal solution model to be useful in such cases.

11.9.6 Chemical Potential for Ideal Solutions

The discussion of multicomponent systems concludes with the introduction of expressions for evaluating the chemical potential for ideal solutions used in subsequent sections of the book.

Consider a reference state where component i of a multicomponent system is pure at the temperature T of the system and a reference-state pressure p_{ref}. The difference in the chemical potential of i between a specified state of the multicomponent system and the reference state is obtained with Eq. 11.125 as

$$\mu_i - \mu_i^\circ = \overline{R}T \ln \frac{\bar{f}_i}{f_i^\circ} \tag{11.139}$$

where the superscript ° denotes property values at the reference state. The fugacity ratio appearing in the logarithmic term is known as the ***activity***, a_i, of component i in the mixture. That is **activity**

$$a_i = \frac{\bar{f}_i}{f_i^\circ} \tag{11.140}$$

For subsequent applications, it suffices to consider the case of gaseous mixtures. For gaseous mixtures, p_{ref} is specified as 1 bar, so μ_i° and f_i° in Eq. 11.140 are, respectively, the chemical potential and fugacity of pure i at temperature T and 1 bar.

Since the chemical potential of a pure component equals the Gibbs function per mole, Eq. 11.139 can be written as

$$\mu_i = \bar{g}_i^\circ + \bar{R}T \ln a_i \tag{11.141}$$

where $\bar{g}_i^\circ$ is the Gibbs function per mole of pure component i evaluated at temperature T and 1 bar: $\bar{g}_i^\circ = \bar{g}_i\,(T, 1\text{ bar})$.

For an ideal solution, the Lewis–Randall rule applies and the activity is

$$a_i = \frac{y_i f_i}{f_i^\circ} \tag{11.142}$$

where f_i is the fugacity of pure component i at temperature T and pressure p. Introducing Eq. 11.142 into Eq. 11.141

$$\mu_i = \bar{g}_i^\circ + \bar{R}T \ln \frac{y_i f_i}{f_i^\circ}$$

or

$$\mu_i = \bar{g}_i^\circ + \bar{R}T \ln \left[\underline{\left(\frac{f_i}{p}\right)\left(\frac{p_{ref}}{f_i^\circ}\right)} \frac{y_i p}{p_{ref}} \right] \quad \text{(ideal solution)} \tag{11.143}$$

In principle, the ratios of fugacity to pressure shown underlined in this equation can be evaluated from Eq. 11.124 or the generalized fugacity chart, Fig. A-6, developed from it. If component i behaves as an ideal gas at both T, p and T, p_{ref}, $f_i/p = f_i^\circ/p_{ref} = 1$, and Eq. 11.143 reduces to

$$\mu_i = \bar{g}_i^\circ + \bar{R}T \ln \frac{y_i p}{p_{ref}} \quad \text{(ideal gas)} \tag{11.144}$$

✓ Chapter Summary and Study Guide

In this chapter, we introduce thermodynamic relations that allow u, h, and s as well as other properties of simple compressible systems to be evaluated using property data that are more readily measured. The emphasis is on systems involving a single chemical species such as water or a mixture such as air. An introduction to general property relations for mixtures and solutions is also included.

Equations of state relating p, v, and T are considered, including the virial equation and examples of two-constant and multiconstant equations. Several important property relations based on the mathematical characteristics of exact differentials are developed, including the Maxwell relations. The concept of a fundamental thermodynamic function is discussed. Means for evaluating changes in specific internal energy, enthalpy, and entropy are developed and applied to phase change and to single-phase processes. Property relations are introduced involving the volume expansivity, isothermal and isentropic compressibilities, velocity of sound, specific heats and specific heat ratio, and the Joule–Thomson coefficient.

Additionally, we describe how tables of thermodynamic properties are constructed using the property relations and methods developed in this chapter. Such procedures also provide the basis for data retrieval by computer software. Also described are means for using the generalized enthalpy and entropy departure charts and the generalized fugacity coefficient chart to evaluate enthalpy, entropy, and fugacity, respectively.

We also consider p–v–T relations for gas mixtures of known composition, including Kay's rule. The chapter concludes with a discussion of property relations for multicomponent systems, including partial molal properties, chemical potential, fugacity, and activity. Ideal solutions and the Lewis–Randall rule are introduced as a part of that presentation.

The following checklist provides a study guide for this chapter. When your study of the text and end-of-chapter exercises has been completed you should be able to write out the meanings of the terms listed in the margins throughout the chapter and understand each of the related concepts. The subset of key concepts listed below is particularly important. Additionally, for systems involving a single species you should be able to

- ✔ calculate p–v–T data using equations of state such as the Redlich–Kwong and Benedict–Webb–Rubin equations.
- ✔ use the 16 property relations summarized in Table 11.1 and explain how the relations are obtained.
- ✔ evaluate Δs, Δu, and Δh, using the Clapeyron equation when considering phase change, and using equations of state and specific heat relations when considering single phases.
- ✔ use the property relations introduced in Sec. 11.5 such as those involving the specific heats, the volume expansivity, and the Joule–Thomson coefficient.
- ✔ explain how tables of thermodynamic properties, such as Tables A-2 through A-18, are constructed.
- ✔ use the generalized enthalpy and entropy departure charts, Figs. A-4 and A-5, to evaluate Δh and Δs.

For a *gas mixture* of known composition, you should be able to

- ✔ apply the methods introduced in Sec. 11.8 for relating pressure, specific volume, and temperature—Kay's rule, for example.

For *multicomponent systems*, you should be able to

- ✔ evaluate extensive properties in terms of the respective partial molal properties.
- ✔ evaluate partial molal volumes using the *method of intercepts*.
- ✔ evaluate fugacity using data from the generalized fugacity coefficient chart, Fig. A-6.
- ✔ apply the ideal solution model.

Key Engineering Concepts

equation of state p. 569
exact differential p. 575
test for exactness p. 575
Helmholtz function p. 579
Gibbs function p. 579
Maxwell relations p. 581
fundamental function p. 584
Clapeyron equation p. 586
Joule–Thomson coefficient p. 599
enthalpy and entropy departures p. 607, 610
Kay's rule p. 613
method of intercepts p. 619
chemical potential p. 621
fugacity p. 624
Lewis–Randall rule p. 627

Key Equations

Equations of State		
$Z = 1 + \dfrac{B(T)}{\bar{v}} + \dfrac{C(T)}{\bar{v}^2} + \dfrac{D(T)}{\bar{v}^3} + \ldots$	(11.1)	Virial equation of state
$p = \dfrac{\bar{R}T}{\bar{v} - b} - \dfrac{a}{\bar{v}^2}$	(11.2)	van der Waals equation of state
$p = \dfrac{\bar{R}T}{\bar{v} - b} - \dfrac{a}{\bar{v}(\bar{v} + b)T^{1/2}}$	(11.7)	Redlich–Kwong equation of state

Mathematical Relations for Properties		
$\frac{\partial}{\partial y}\left[\left(\frac{\partial z}{\partial x}\right)_y\right]_x = \frac{\partial}{\partial x}\left[\left(\frac{\partial z}{\partial y}\right)_x\right]_y$	(11.14a)	Test for exactness
$\left(\frac{\partial M}{\partial y}\right)_x = \left(\frac{\partial N}{\partial x}\right)_y$	(11.14b)	
$\left(\frac{\partial x}{\partial y}\right)_z\left(\frac{\partial y}{\partial x}\right)_z = 1$	(11.15)	Important relations among partial derivatives of properties
$\left(\frac{\partial y}{\partial z}\right)_x\left(\frac{\partial z}{\partial x}\right)_y\left(\frac{\partial x}{\partial y}\right)_z = -1$	(11.16)	
Table 11.1	(11.24–11.36)	Summary of property relations from exact differentials
Expressions for Δu, Δh, and Δs		
$\left(\frac{dp}{dT}\right)_{sat} = \frac{h_g - h_f}{T(v_g - v_f)}$	(11.40)	Clapeyron equation
$s_2 - s_1 = \int_1^2 \frac{c_v}{T}dT + \int_1^2\left(\frac{\partial p}{\partial T}\right)_v dv$	(11.50)	Expressions for changes in s and u with T and v as independent variables
$u_2 - u_1 = \int_1^2 c_v dT + \int_1^2\left[T\left(\frac{\partial p}{\partial T}\right)_v - p\right]dv$	(11.51)	
$s_2 - s_1 = \int_1^2 \frac{c_p}{T}dT - \int_1^2\left(\frac{\partial v}{\partial T}\right)_p dp$	(11.59)	Expressions for changes in s and h with T and p as independent variables
$h_2 - h_1 = \int_1^2 c_p dT + \int_1^2\left[v - T\left(\frac{\partial v}{\partial T}\right)_p\right]dp$	(11.60)	
$\bar{h}_2 - \bar{h}_1 = \bar{h}_2^* - \bar{h}_1^* - \bar{R}T_c\left[\left(\frac{\bar{h}^* - \bar{h}}{\bar{R}T_c}\right)_2 - \left(\frac{\bar{h}^* - \bar{h}}{\bar{R}T_c}\right)_1\right]$	(11.85)	Evaluating enthalpy and entropy changes in terms of generalized enthalpy and entropy departures and data from Figs. A-4 and A-5, respectively
$\bar{s}_2 - \bar{s}_1 = \bar{s}_2^* - \bar{s}_1^* - \bar{R}\left[\left(\frac{\bar{s}^* - \bar{s}}{\bar{R}}\right)_2 - \left(\frac{\bar{s}^* - \bar{s}}{\bar{R}}\right)_1\right]$	(11.92)	
Additional Thermodynamic Relations		
$\psi = u - Ts$	(11.20)	Helmholtz function
$g = h - Ts$	(11.21)	Gibbs function
$c = \sqrt{-v^2\left(\frac{\partial p}{\partial v}\right)_s} = \sqrt{-kv^2\left(\frac{\partial p}{\partial v}\right)_T}$	(9.36b and 11.74)	Expressions for velocity of sound
$\mu_J = \left(\frac{\partial T}{\partial p}\right)_h$	(11.75)	Joule–Thomson coefficient
Properties of Multicomponent Mixtures		
$T_c = \sum_{i=1}^{j} y_i T_{c,i}, \quad p_c = \sum_{i=1}^{j} y_i p_{c,i}$	(11.97)	Kay's rule for critical temperature and pressure of mixtures
$\bar{X}_i = \left(\frac{\partial X}{\partial n_i}\right)_{T,p,n_l}$	(11.102)	Partial molal property $\bar{X}_i$ and its relation to extensive property X
$X = \sum_{i=1}^{j} n_i \bar{X}_i$	(11.103)	X as a weighted sum of partial molal properties

$\mu_i = \overline{G}_i = \left.\frac{\partial G}{\partial n_i}\right)_{T,p,n_l}$	(11.107)	Chemical potential of species *i* in a mixture
$\overline{R}T\left(\frac{\partial \ln f}{\partial p}\right)_T = \overline{v}$	(11.122)	Expressions for evaluating fugacity of a single-component system
$\lim\limits_{p\to 0} \frac{f}{p} = 1$	(11.123)	
$\overline{R}T\left(\frac{\partial \ln \overline{f}_i}{\partial p}\right)_{T,n} = \overline{V}_i$	(11.126a)	Expressions for evaluating fugacity of mixture component *i*
$\lim\limits_{p\to 0}\left(\frac{\overline{f}_i}{y_i p}\right) = 1$	(11.126b)	
$\overline{f}_i = y_i f_i$	(11.134)	Lewis–Randall rule for ideal solutions
$\mu_i = \overline{g}_i^\circ + \overline{R}T \ln \frac{y_i p}{p_{ref}}$	(11.144)	Chemical potential of component *i* in an ideal gas mixture

Exercises: things engineers think about

1. What is an advantage of using the Redlich–Kwong equation of state in the generalized form given by Eq. 11.9 instead of Eq. 11.7? A disadvantage?

2. To determine the specific volume of superheated water vapor at a known pressure and temperature, when would you use each of the following: the *steam tables,* the generalized compressibility chart, an equation of state, the ideal gas model?

3. If the function $p = p(T, v)$ is an equation of state, is $(\partial p/\partial T)_v$ a property? What are the independent variables of $(\partial p/\partial T)_v$?

4. In the expression $(\partial u/\partial T)_v$, what does the subscript v signify?

5. Explain how a Mollier diagram provides a graphical representation of the fundamental function $h(s, p)$.

6. How is the Clapeyron equation used?

7. For a gas whose equation of state is $p\overline{v} = \overline{R}T$, are the specific heats $\overline{c}_p$ and $\overline{c}_v$ *necessarily* functions of temperature alone?

8. Referring to the p–T diagram for water, explain why ice melts under the blade of an ice skate.

9. Can you devise a way to determine the specific heat $\overline{c}_p$ of a gas by *direct* measurement? *Indirectly,* using other measured data?

10. For an ideal gas, what is the value of the Joule–Thomson coefficient?

11. At what states is the entropy departure negligible? The fugacity coefficient, f/p, closely equal to unity?

12. In Eq. 11.107, what do the subscripts T, p, and n_l signify? What does i denote?

13. How does Eq. 11.108 reduce for a system consisting of a pure substance? Repeat for an ideal gas mixture.

14. If two different liquids of known volumes are mixed, is the final volume *necessarily* equal to the sum of the original volumes?

15. For a binary solution at temperature T and pressure p, how would you determine the specific heat $\overline{c}_p$? Repeat for an ideal solution and for an ideal gas mixture.

Problems: developing engineering skills

Using Equations of State

11.1 Owing to safety requirements, the pressure within a 0.5 m³ cylinder should not exceed 52 bar. Check the pressure within the cylinder if filled with 45 kg of CO_2 maintained at 100°C using the

(a) van der Waals equation.
(b) compressibility chart.
(c) ideal gas equation of state.

11.2 Five kilograms mass of propane have a volume of 0.06 m³ and a pressure of 4 MPa. Determine the temperature, in K, using the

(a) van der Waals equation.
(b) compressibility chart.
(c) ideal gas equation of state.
(d) propane tables.

11.3 The pressure within a 23.3-m^3 tank should not exceed 105 bar. Check the pressure within the tank if filled with 1000 kg of water vapor maintained at 360°C using the

(a) ideal gas equation of state.
(b) van der Waals equation.
(c) Redlich–Kwong equation.
(d) compressibility chart.
(e) steam tables.

11.4 Estimate the pressure of water vapor at a temperature of 500°C and a density of 24 kg/m^3 using the

(a) steam tables.
(b) compressibility chart.
(c) Redlich–Kwong equation.
(d) van der Waals equation.
(e) ideal gas equation of state.

11.5 Methane gas flows through a pipeline with a volumetric flow rate of 0.3 m^3/s at a pressure of 183 bar and a temperature of 13°C. Determine the mass flow rate, in kg/s, using the

(a) ideal gas equation of state.
(b) van der Waals equation.
(c) compressibility chart.

11.6 Determine the specific volume of water vapor at 20 MPa and 400°C, in m^3/kg, using the

(a) steam tables.
(b) compressibility chart.
(c) Redlich–Kwong equation.
(d) van der Waals equation.
(e) ideal gas equation of state.

11.7 A vessel whose volume is 1 m^3 contains 4 kmol of methane at 100°C. Owing to safety requirements, the pressure of the methane should not exceed 12 MPa. Check the pressure using the

(a) ideal gas equation of state.
(b) Redlich–Kwong equation.
(c) Benedict–Webb–Rubin equation.

11.8 Methane gas at 100 bar and −18°C is stored in a 10-m^3 tank. Determine the mass of methane contained in the tank, in kg, using the

(a) ideal gas equation of state.
(b) van der Waals equation.
(c) Benedict–Webb–Rubin equation.

11.9 Using the Benedict–Webb–Rubin equation of state, determine the volume, in m^3, occupied by 165 kg of methane at a pressure of 200 bar and temperature of 400 K. Compare with the results obtained using the ideal gas equation of state and the generalized compressibility chart.

11.10 A rigid tank contains 1 kg of oxygen (O_2) at $p_1 = 40$ bar, $T_1 = 180$ K. The gas is cooled until the temperature drops to 150 K. Determine the volume of the tank, in m^3, and the final pressure, in bar, using the

(a) ideal gas equation of state.
(b) Redlich–Kwong equation.
(c) compressibility chart.

11.11 One kilogram mass of air initially occupying a volume of 0.01 m^3 at a pressure of 7 MPa expands isothermally and without irreversibilities until the volume is 0.06 m^3. Using the Redlich–Kwong equation of state, determine the

(a) temperature, in K.
(b) final pressure, in MPa.
(c) work developed in the process, in kJ.

11.12 Water vapor initially at 240°C, 1 MPa expands in a piston–cylinder assembly isothermally and without internal irreversibilities to a final pressure of 0.1 MPa. Evaluate the work done, in kJ/kg. Use a truncated virial equation of state with the form

$$Z = 1 + \frac{B}{v} + \frac{C}{v^2}$$

where B and C are evaluated from steam table data at 240°C and pressures ranging from 0 to 1 MPa.

11.13 Referring to the virial series, Eqs. 3.30 and 3.31, show that $\hat{B} = B/\bar{R}T$, $\hat{C} = (C - B^2)/\bar{R}^2T^2$.

11.14 Express Eq. 11.5, the van der Waals equation, in terms of the compressibility factor Z

(a) as a virial series in v'_R. [*Hint:* Expand the $(v'_R - 1/8)^{-1}$ term of Eq. 11.5 in a series.]
(b) as a virial series in p_R.
(c) Dropping terms involving $(p_R)^2$ and higher in the virial series of part (b), obtain the following approximate form:

$$Z = 1 + \left(\frac{1}{8} - \frac{27/64}{T_R}\right)\frac{p_R}{T_R}$$

(d) Compare the compressibility factors determined from the equation of part (c) with tabulated compressibility factors from the literature for $0 < p_R < 0.6$ and each of $T_R = 1.0, 1.2, 1.4, 1.6, 1.8, 2.0$. Comment on the realm of validity of the approximate form.

11.15 The Berthelot equation of state has the form

$$p = \frac{\bar{R}T}{\bar{v} - b} - \frac{a}{T\bar{v}^2}$$

(a) Using Eqs. 11.3, show that

$$a = \frac{27}{64}\frac{\bar{R}^2T_c^3}{p_c}, \qquad b = \frac{1}{8}\frac{\bar{R}T_c}{p_c}$$

(b) Express the equation in terms of the compressibility factor Z, the reduced temperature T_R, and the pseudoreduced specific volume, v'_R.

11.16 The Beattie–Bridgeman equation of state can be expressed as

$$p = \frac{RT(1 - \varepsilon)(v + B)}{v^2} - \frac{A}{v^2}$$

where

$$A = A_0\left(1 - \frac{a}{v}\right), \qquad B = B_0\left(1 - \frac{b}{v}\right)$$

$$\varepsilon = \frac{c}{vT^3}$$

and A_0, B_0, a, b, and c are constants. Express this equation of state in terms of the reduced pressure, p_R, reduced temperature, T_R, pseudoreduced specific volume, v'_R, and appropriate dimensionless constants.

11.17 The Dieterici equation of state is

$$p = \left(\frac{RT}{v-b}\right)\exp\left(\frac{-a}{RTv}\right)$$

(a) Using Eqs. 11.3, show that

$$a = \frac{4R^2T_c^2}{p_c e^2}, \qquad b = \frac{RT_c}{p_c e^2}$$

(b) Show that the equation of state can be expressed in terms of compressibility chart variables as

$$Z = \left(\frac{v'_R}{v'_R - 1/e^2}\right)\exp\left(\frac{-4}{T_R v'_R e^2}\right)$$

(c) Convert the result of part (b) to a virial series in v'_R. (*Hint:* Expand the $(v'_R - 1/e^2)^{-1}$ term in a series. Also expand the exponential term in a series.)

11.18 The Peng–Robinson equation of state has the form

$$p = \frac{RT}{v-b} - \frac{a}{v^2 - c^2}$$

Using Eqs. 11.3, evaluate the constants a, b, c in terms of the critical pressure p_c, critical temperature T_c, and critical compressibility factor Z_c.

11.19 The p–v–T relation for chlorofluorinated hydrocarbons can be described by the Carnahan–Starling–DeSantis equation of state

$$\frac{p\bar{v}}{\bar{R}T} = \frac{1+\beta+\beta^2-\beta^3}{(1+\beta)^3} - \frac{a}{\bar{R}T(\bar{v}+b)}$$

where $\beta = b/4\bar{v}$, $a = a_0 \exp(a_1T + a_2T^2)$, and $b = b_0 + b_1T + b_2T^2$. For Refrigerants 12 and 13, the required coefficients for T in K, a in J · L/(mol)^2, and b in L/mol are given in Table P11.19. Specify which of the two refrigerants would allow the smaller amount of mass to be stored in a 10-m^3 vessel at 0.2 MPa, 80°C.

Using Relations from Exact Differentials

11.20 The differential of pressure obtained from a certain equation of state is given by *one* of the following expressions. Determine the equation of state.

$$dp = \frac{2(v-b)}{RT}dv + \frac{(v-b)^2}{RT^2}dT$$

$$dp = -\frac{RT}{(v-b)^2}dv + \frac{R}{v-b}dT$$

11.21 Introducing $\delta Q_{\text{int rev}} = T\,dS$ into Eq. 6.8 gives

$$\delta Q_{\text{int rev}} = dU + p\,dV$$

Using this expression together with the test for exactness, demonstrate that $Q_{\text{int rev}}$ is not a property.

11.22 Show that Eq. 11.16 is satisfied by an equation of state with the form $p = [RT/(v-b)] + a$.

11.23 For the functions $x = x(y, w)$, $y = y(z, w)$, $z = z(x, w)$, demonstrate that

$$\left(\frac{\partial x}{\partial y}\right)_w\left(\frac{\partial y}{\partial z}\right)_w\left(\frac{\partial z}{\partial x}\right)_w = 1$$

11.24 Using Eq. 11.35, check the consistency of

(a) the steam tables at 2 MPa, 400°C.
(b) the Refrigerant 134a tables at 2 bar, 50°C.

11.25 Using Eq. 11.35, check the consistency of

(a) the steam tables at 0.7 MPa, 316°C.
(b) the Refrigerant 134a tables at 0.3 MPa, 38°C.

11.26 At a pressure of 1 bar, liquid water has a state of *maximum* density at about 4°C. What can be concluded about $(\partial s/\partial p)_T$ at

(a) 3°C?
(b) 4°C?
(c) 5°C?

11.27 A gas enters a compressor operating at steady state and is compressed isentropically. Does the specific enthalpy increase or decrease as the gas passes from the inlet to the exit?

11.28 Show that T, p, h, ψ, and g can each be determined from a fundamental thermodynamic function of the form $u = u(s, v)$.

11.29 Evaluate p, s, u, h, c_v, and c_p for a substance for which the Helmholtz function has the form

$$\psi = -RT\ln\frac{v}{v'} - cT'\left[1 - \frac{T}{T'} + \frac{T}{T'}\ln\frac{T}{T'}\right]$$

where v' and T' denote specific volume and temperature, respectively, at a reference state, and c is a constant.

11.30 The Mollier diagram provides a graphical representation of the fundamental thermodynamic function $h = h(s, p)$. Show that at any state fixed by s and p the properties T, v, u, ψ, and g can be evaluated using data obtained from the diagram.

11.31 Derive the relation $c_p = -T(\partial^2 g/\partial T^2)_p$.

Evaluating Δs, Δu, and Δh

11.32 Using p–v–T data for saturated ammonia from Table A-13, calculate at −7°C.

(a) $h_g - h_f$.
(b) $u_g - u_f$.
(c) $s_g - s_f$.

Compare with the values obtained using table data.

	$a_0 \times 10^{-3}$	$a_1 \times 10^3$	$a_2 \times 10^6$	b_0	$b_1 \times 10^4$	$b_2 \times 10^8$
R-12	3.52412	−2.77230	−0.67318	0.15376	−1.84195	−5.03644
R-13	2.29813	−3.41828	−1.52430	0.12814	−1.84474	−10.7951

Table P11.19

11.33 Using p–v–T data for saturated water from the steam tables, calculate at 50°C

(a) $h_g - h_f$.
(b) $u_g - u_f$.
(c) $s_g - s_f$.

Compare with the values obtained using steam table data.

11.34 Using h_{fg}, v_{fg}, and p_{sat} at −12°C from the Refrigerant 134a tables, estimate the saturation pressure at −7°C. Comment on the accuracy of your estimate.

11.35 Using h_{fg}, v_{fg}, and p_{sat} at 26°C from the ammonia tables, estimate the saturation pressure at 30°C. Comment on the accuracy of your estimate.

11.36 Using triple-point data for water from Table A-6, estimate the saturation pressure at −40°C. Compare with the value listed in Table A-6.

11.37 At 0°C, the specific volumes of saturated solid water (ice) and saturated liquid water are, respectively, $v_i = 1.0911 \times 10^{-3}\ \text{m}^3/\text{kg}$ and $v_f = 1.0002 \times 10^{-3}\ \text{m}^3/\text{kg}$, and the change in specific enthalpy on melting is $h_{if} = 333.4$ kJ/kg. Calculate the melting temperature of ice at (a) 250 bar, (b) 500 bar. Locate your answers on a sketch of the p–T diagram for water.

11.38 The line representing the two-phase solid–liquid region on the phase diagram slopes to the left for substances that expand on freezing and to the right for substances that contract on freezing (Sec. 3.2.2). Verify this for the cases of lead that contracts on freezing and bismuth that expands on freezing.

11.39 Consider a four-legged chair at rest on an ice rink. The total mass of the chair and a person sitting on it is 80 kg. If the ice temperature is −2°C, determine the minimum total area, in cm^2, the tips of the chair legs can have before the ice in contact with the legs would melt. Use data from Problem 11.37 and let the local acceleration of gravity be 9.8 m/s^2.

11.40 Over a certain temperature interval, the saturation pressure–temperature curve of a substance is represented by an equation of the form $\ln p_{sat} = A - B/T$, where A and B are empirically determined constants.

(a) Obtain expressions for $h_g - h_f$ and $s_g - s_f$ in terms of p–v–T data and the constant B.
(b) Using the results of part (a), calculate $h_g - h_f$ and $s_g - s_f$ for water vapor at 25°C and compare with steam table data.

11.41 Using data for water from Table A-2, determine the constants A and B to give the best fit in a least-squares sense to the saturation pressure in the interval from 20 to 30°C by the equation $\ln p_{sat} = A - B/T$. Using this equation, determine dp_{sat}/dT at 25°C. Calculate $h_g - h_f$ at 25°C and compare with the steam table value.

11.42 Over limited intervals of temperature, the saturation pressure–temperature curve for two-phase liquid–vapor states can be represented by an equation of the form $\ln p_{sat} = A - B/T$, where A and B are constants. Derive the following expression relating any three states on such a portion of the curve:

$$\frac{p_{sat,3}}{p_{sat,1}} = \left(\frac{p_{sat,2}}{p_{sat,1}}\right)^{\tau}$$

where $\tau = T_2(T_3 - T_1)/T_3(T_2 - T_1)$.

11.43 Use the result of Problem 11.42 to determine

(a) the saturation pressure at 30°C using saturation pressure–temperature data at 20 and 40°C from Table A-2. Compare with the table value for saturation pressure at 30°C.
(b) the saturation temperature at 0.006 MPa using saturation pressure–temperature data at 20 to 40°C from Table A-2. Compare with the saturation temperature at 0.006 MPa given in Table A-3.

11.44 Complete the following exercises dealing with slopes:

(a) At the triple point of water, evaluate the ratio of the slope of the vaporization line to the slope of the sublimation line. Use steam table data to obtain a numerical value for the ratio.
(b) Consider the superheated vapor region of a temperature–entropy diagram. Show that the slope of a constant specific volume line is greater than the slope of a constant pressure line through the same state.
(c) An enthalpy–entropy diagram (Mollier diagram) is often used in analyzing steam turbines. Obtain an expression for the slope of a constant-pressure line on such a diagram in terms of p–v–T data only.
(d) A pressure–enthalpy diagram is often used in the refrigeration industry. Obtain an expression for the slope of an isentropic line on such a diagram in terms of p–v–T data only.

11.45 Using only p–v–T data from the ammonia tables, evaluate the changes in specific enthalpy and entropy for a process from 0.5 MPa, 4°C to 0.1 MPa, 4°C. Compare with the table values.

11.46 One kmol of argon at 300 K is initially confined to one side of a rigid, insulated container divided into equal volumes of 0.2 m^3 by a partition. The other side is initially evacuated. The partition is removed and the argon expands to fill the entire container. Using the van der Waals equation of state, determine the final temperature of the argon, in K. Repeat using the ideal gas equation of state.

11.47 Obtain the relationship between c_p and c_v for a gas that obeys the equation of state $p(v - b) = RT$.

11.48 The p–v–T relation for a certain gas is represented closely by $v = RT/p + B - A/RT$, where R is the gas constant and A and B are constants. Determine expressions for the changes in specific enthalpy, internal energy, and entropy, $[h(p_2, T) - h(p_1, T)]$, $[u(p_2, T) - u(p_1, T)]$, and $[s(p_2, T) - s(p_1, T)]$, respectively.

11.49 Develop expressions for the specific enthalpy, internal energy, and entropy changes $[h(v_2, T) - h(v_1, T)]$, $[u(v_2, T) - u(v_1, T)]$, $[s(v_2, T) - s(v_1, T)]$, using the

(a) van der Waals equation of state.
(b) Redlich-Kwong equation of state.

11.50 At certain states, the p–v–T data of a gas can be expressed as $Z = 1 - Ap/T^4$, where Z is the compressibility factor and A is a constant.

(a) Obtain an expression for $(\partial p/\partial T)_v$ in terms of p, T, A, and the gas constant R.
(b) Obtain an expression for the change in specific entropy, $[s(p_2, T) - s(p_1, T)]$.

(c) Obtain an expression for the change in specific enthalpy, $[h(p_2, T) - h(p_1, T)]$.

11.51 For a gas whose p–v–T behavior is described by $Z = 1 + Bp/RT$, where B is a function of temperature, derive expressions for the specific enthalpy, internal energy, and entropy changes, $[h(p_2, T) - h(p_1, T)]$, $[u(p_2, T) - u(p_1, T)]$, and $[s(p_2, T) - s(p_1, T)]$.

11.52 For a gas whose p–v–T behavior is described by $Z = 1 + B/v + C/v^2$, where B and C are functions of temperature, derive an expression for the specific entropy change, $[s(v_2, T) - s(v_1, T)]$.

Using Other Thermodynamic Relations

11.53 The volume of a 1-kg copper sphere is not allowed to vary by more than 0.1%. If the pressure exerted on the sphere is increased from 10 bar while the temperature remains constant at 300 K, determine the maximum allowed pressure, in bar. Average values of ρ, β, and κ are 8888 kg/m^3, 49.2×10^{-6} (K)$^{-1}$, and 0.776×10^{-11} m^2/N respectively.

11.54 The volume of a 0.5 kg copper sphere is not allowed to vary by more than 0.1%. If the pressure exerted on the sphere is increased from 1 bar while the temperature remains constant at 27°C, determine the maximum allowed pressure, in bar. Average values of ρ, β, and κ are 9300 kg/m^3, 4.91×10^{-5} (K)$^{-1}$, and 7.5×10^{-12} m^2/N, respectively.

11.55 Develop expressions for the volume expansivity β and the isothermal compressibility κ for

(a) an ideal gas.
(b) a gas whose equation of state is $p(v - b) = RT$.
(c) a gas obeying the van der Waals equation.

11.56 Derive expressions for the volume expansivity β and the isothermal compressibility κ in terms of T, p, Z, and the first partial derivatives of Z. For gas states with $p_R < 3$, $T_R < 2$, determine the sign of κ. Discuss.

11.57 Show that the isothermal compressibility κ is always greater than or equal to the isentropic compressibility α.

11.58 Prove that $(\partial\beta/\partial p)_T = -(\partial\kappa/\partial T)_p$.

11.59 For aluminum at 0°C, ρ = 2700 kg/m^3, $\beta = 71.4 \times 10^{-8}$ (K)$^{-1}$, $\kappa = 1.34 \times 10^{-13}$ m^2/N, and c_p = 0.9211 kJ/kg · K. Determine the percent error in c_v that would result if it were assumed that $c_p = c_v$.

11.60 Estimate the temperature rise, in °C, of mercury, initially at 0°C and 1 bar if its pressure were raised to 1000 bar isentropically. For mercury at 0°C, c_p = 28.0 kJ/kmol · K, $\bar{v}$ = 0.0147 m^3/kmol, and $\beta = 17.8 \times 10^{-5}$ (K)$^{-1}$.

11.61 At certain states, the p–v–T data for a particular gas can be represented as $Z = 1 - Ap/T^4$, where Z is the compressibility factor and A is a constant. Obtain an expression for the specific heat c_p in terms of the gas constant R, specific heat ratio k, and Z. Verify that your expression reduces to Eq. 3.47a when $Z = 1$.

11.62 For a gas obeying the van der Waals equation of state,

(a) show that $(\partial c_v/\partial v)_T = 0$.
(b) develop an expression for $c_p - c_v$.
(c) develop expressions for $[u(T_2, v_2) - u(T_1, v_1)]$ and $[s(T_2, v_2) - s(T_1, v_1)]$.
(d) complete the Δu and Δs evaluations if $c_v = a + bT$, where a and b are constants.

11.63 If the value of the specific heat c_v of air is 823 J/kg · K at T_1 = 538°C, v_1 = 2.19 m^3/kg, determine the value of c_v at T_2 = 538°C, v_2 = 0.0033 m^3/kg. Assume that air obeys the Berthelot equation of state

$$p = \frac{RT}{v - b} - \frac{a}{Tv^2}$$

where

$$a = \frac{27}{64}\frac{R^2T_c^3}{p_c} \qquad b = \frac{1}{8}\frac{RT_c}{p_c}$$

11.64 Show that the specific heat ratio k can be expressed as $k = c_p\kappa/(c_p\kappa - Tv\beta^2)$. Using this expression together with data from the steam tables, evaluate k for water vapor at 1.4 MPa, 260°C.

11.65 For liquid water at 40°C, 1 bar estimate

(a) c_v, in kJ/kg · K.
(b) the velocity of sound, in m/s.

Use data from Table 11.2, as required.

11.66 Using steam table data, estimate the velocity of sound in liquid water at (a) 20°C, 50 bar, (b) 10°C, 10.3 MPa.

11.67 At a certain location in a *wind tunnel*, a stream of air is at 260°C, 1 bar and has a velocity of 635 m/s. Determine the Mach number at this location.

11.68 For a gas obeying the equation of state $p(v - b) = RT$, where b is a positive constant, can the temperature be reduced in a Joule–Thomson expansion?

11.69 A gas is described by $v = RT/p - A/T + B$, where A and B are constants. For the gas

(a) obtain an expression for the temperatures at the Joule–Thomson inversion states.
(b) obtain an expression for $c_p - c_v$.

11.70 Determine the *maximum* Joule–Thomson inversion temperature in terms of the critical temperature T_c predicted by the

(a) van der Waals equation.
(b) Redlich–Kwong equation.
(c) Dieterici equation given in Problem 11.17.

11.71 Derive an equation for the Joule–Thomson coefficient as a function of T and v for a gas that obeys the van der Waals equation of state and whose specific heat c_v is given by $c_v = A + BT + CT^2$, where A, B, C are constants. Evaluate the temperatures at the *inversion* states in terms of R, v, and the van der Waals constants a and b.

11.72 Show that Eq. 11.77 can be written as

$$\mu_J = \frac{T^2}{c_p}\left(\frac{\partial(v/T)}{\partial T}\right)_p$$

(a) Using this result, obtain an expression for the Joule–Thomson coefficient for a gas obeying the equation of state

$$v = \frac{RT}{p} - \frac{Ap}{T^2}$$

where A is a constant.

(b) Using the result of part (a), determine c_p, in kJ/kg · K, for CO_2 at 400 K, 1 bar, where $\mu_J = 0.57$ K/bar. For CO_2, $A = 2.78 \times 10^{-3}$ m^5 · K^2/kg · N.

Developing Property Data

11.73 If the specific heat c_v of a gas obeying the van der Waals equation is given at a particular pressure, p', by $c_v = A + BT$, where A and B are constants, develop an expression for the change in specific entropy between any two states 1 and 2: $[s(T_2, p_2) - s(T_1, p_1)]$.

11.74 For air, write a computer program that evaluates the change in specific enthalpy from a state where the temperature is 25°C and the pressure is 1 bar to a state where the temperature is T and the pressure is p. Use the van der Waals equation of state and account for the variation of the ideal gas specific heat as in Table A-21.

11.75 Using the Redlich–Kwong equation of state, determine the changes in specific enthalpy, in kJ/kmol, and entropy, in kJ/kmol · K, for ethylene between 400 K, 1 bar and 400 K, 100 bar.

11.76 Using the Benedict–Webb–Rubin equation of state together with a specific heat relation from Table A-21, determine the change in specific enthalpy, in kJ/kmol, for methane between 300 K, 1 bar and 400 K, 200 bar.

11.77 Using the Redlich–Kwong equation of state together with an appropriate specific heat relation, determine the final temperature for an isentropic expansion of nitrogen from 400 K, 250 bar to 5 bar.

11.78 A certain pure, simple compressible substance has the following property relations. The p–v–T relationship in the vapor phase is

$$v = \frac{RT}{p} - \frac{Bp}{T^2}$$

where v is in m^3/kg, T is in K, p is in N/m^2, $R = 267$ N · m/kg · K, and $B = 0.041$ m^5 · (K)^2/kg · N. The saturation pressure, in kPa is described by

$$\ln p_{sat} = 12 - \frac{2400}{T}$$

The Joule–Thomson coefficient at 69 kPa, 93°C is 4.7×10^{-5} K · m^2/N. The ideal gas specific heat c_{p0} is constant over the temperature range −18°C to 150°C.

(a) Complete the accompanying table of property values

T	p	v_f	v_g	h_f	h_g	s_f	s_g
−18°C		0.03		0		0.000	
38°C		0.03					

for p in kPa, v in m^3/kg, h in kJ/kg, and s in kJ/kg · K.

(b) Evaluate v, h, s at the state fixed by 103 kPa, 150°C.

Using Enthalpy and Entropy Departures

11.79 Beginning with Eq. 11.90, derive Eq. 11.91.

11.80 Derive an expression giving

(a) the internal energy of a substance relative to that of its ideal gas model at the same temperature: $[u(T, v) - u^*(T)]$.

(b) the entropy of a substance relative to that of its ideal gas model at the same temperature and specific volume: $[s(T, v) - s^*(T, v)]$.

11.81 Derive expressions for the enthalpy and entropy departures using an equation of state with the form $Z = 1 + Bp_R$, where B is a function of the reduced temperature, T_R.

11.82 The following expression for the enthalpy departure is convenient for use with equations of state that are explicit in pressure:

$$\frac{\bar{h}^*(T) - \bar{h}(T, \bar{v})}{\bar{R}T_c} = T_R\left[1 - Z - \frac{1}{\bar{R}T}\int_{\infty}^{\bar{v}}\left[T\left(\frac{\partial p}{\partial T}\right)_{\bar{v}} - p\right]d\bar{v}\right]$$

(a) Derive this expression.

(b) Using the given expression, evaluate the enthalpy departure for a gas obeying the Redlich–Kwong equation of state.

(c) Using the result of part (b), determine the change in specific enthalpy, in kJ/kmol, for CO_2 undergoing an isothermal process at 300 K from 50 to 20 bar.

11.83 Using the equation of state of Problem 11.14 (c), evaluate v and c_p for water vapor at 550°C, 20 MPa and compare with data from Table A-4 and Fig. 3.9, respectively. Discuss.

11.84 Ethylene at 67°C, 10 bar enters a compressor operating at steady state and is compressed isothermally without internal irreversibilities to 100 bar. Kinetic and potential energy changes are negligible. Evaluate in kJ per kg of ethylene flowing through the compressor

(a) the work required.

(b) the heat transfer.

11.85 Methane at 27°C, 10 MPa enters a turbine operating at steady state, expands adiabatically through a 5 : 1 pressure ratio, and exits at −48°C. Kinetic and potential energy effects are negligible. If $\bar{c}_{po} = 35$ kJ/kmol · K, determine the work developed per kg of methane flowing through the turbine. Compare with the value obtained using the ideal gas model.

11.86 Nitrogen (N_2) enters a compressor operating at steady state at 1.5 MPa, 300 K and exits at 8 MPa, 500 K. If the work input is 240 kJ per kg of nitrogen flowing, determine the heat transfer, in kJ per kg of nitrogen flowing. Ignore kinetic and potential energy effects.

11.87 Oxygen (O_2) enters a control volume operating at steady state with a mass flow rate of 9 kg/min at 100 bar, 287 K and is compressed adiabatically to 150 bar, 400 K. Determine the power required, in kW, and the rate of entropy production, in kW/K. Ignore kinetic and potential energy effects.

11.88 Argon gas enters a turbine operating at steady state at 100 bar, 325 K and expands adiabatically to 40 bar, 235 K with no significant changes in kinetic or potential energy. Determine

(a) the work developed, in kJ per kg of argon flowing through the turbine.

(b) the amount of entropy produced, in kJ/K per kg of argon flowing.

11.89 Oxygen (O_2) undergoes a throttling process from 100 bar, 300 K to 20 bar. Determine the temperature after throttling, in K, and compare with the value obtained using the ideal gas model.

11.90 Water vapor enters a turbine operating at steady state at 30 MPa, 600°C and expands adiabatically to 6 MPa with no significant change in kinetic or potential energy. If the isentropic turbine efficiency is 80%, determine the work developed, in kJ per kg of steam flowing, using the generalized property charts. Compare with the result obtained using steam table data. Discuss.

11.91 Oxygen (O_2) enters a nozzle operating at steady state at 60 bar, 300 K, 1 m/s and expands isentropically to 30 bar. Determine the velocity at the nozzle exit, in m/s.

11.92 A quantity of nitrogen gas undergoes a process at a constant pressure of 80 bar from 220 to 300 K. Determine the work and heat transfer for the process, each in kJ per kmol of nitrogen.

11.93 A closed, rigid, insulated vessel having a volume of 0.142 m^3 contains oxygen (O_2) initially at 100 bar, 7°C. The oxygen is stirred by a paddle wheel until the pressure becomes 150 bar. Determine the

(a) final temperature, in °C.
(b) work, in kJ.
(c) amount of exergy destroyed in the process, in kJ.

Let T_0 = 7°C.

Evaluating *p–v–T* for Gas Mixtures

11.94 A preliminary design calls for a 1 kmol mixture of CO_2 and C_2H_6 (ethane) to occupy a volume of 0.15 m^3 at a temperature of 400 K. The mole fraction of CO_2 is 0.3. Owing to safety requirements, the pressure should not exceed 180 bar. Check the pressure using

(a) the ideal gas equation of state.
(b) Kay's rule together with the generalized compressibility chart.
(c) the additive pressure rule together with the generalized compressibility chart.

Compare and discuss these results.

11.95 A gaseous mixture with a molar composition of 60% CO and 40% H_2 enters a turbine operating at steady state at 150°C, 14 MPa and exits at 100°C, 1 bar with a volumetric flow rate of 540 m^3/min. Estimate the volumetric flow rate at the turbine inlet, in m^3/min, using Kay's rule. What value would result from using the ideal gas model? Discuss.

11.96 A 0.1-m^3 cylinder contains a gaseous mixture with a molar composition of 97% CO and 3% CO_2 initially at 138 bar. Due to a leak, the pressure of the mixture drops to 129 bar while the temperature remains constant at 30°C. Using Kay's rule, estimate the amount of mixture, in kmol, that leaks from the cylinder.

11.97 A gaseous mixture consisting of 0.75 kmol of hydrogen (H_2) and 0.25 kmol of nitrogen (N_2) occupies 0.085 m^3 at 25°C. Estimate the pressure, in bar, using

(a) the ideal gas equation of state.
(b) Kay's rule together with the generalized compressibility chart.
(c) the van der Waals equation together with mixture values for the constants *a* and *b*.
(d) the rule of additive pressure together with the generalized compressibility chart.

11.98 A gaseous mixture of 0.25 kmol of methane and 0.25 kmol of propane occupies a volume of 0.21 m^3 at a temperature of 90°C. Estimate the pressure using the following procedures and compare each estimate with the measured value of pressure, 50.7 bar:

(a) the ideal gas equation of state.
(b) Kay's rule together with the generalized compressibility chart.
(c) the van der Waals equation together with mixture values for the constants *a* and *b*.
(d) the rule of additive pressures together with the van der Waals equation.
(e) the rule of additive pressures together with the generalized compressibility chart.
(f) the rule of additive volumes together with the van der Waals equation.

11.99 One kmol of a gaseous mixture occupies a volume of 0.048 m^3 at 100°C. The mixture consists of 69.5% carbon dioxide and 30.5% ethylene (C_2H_4) (molar basis). Estimate the mixture pressure, in bar, using

(a) the ideal gas equation of state.
(b) Kay's rule together with the generalized compressibility chart.
(c) the additive pressure rule together with the generalized compressibility chart.
(d) the van der Waals equation together with mixture values for the constants *a* and *b*.

11.100 Air having an approximate molar composition of 79% N_2 and 21% O_2 fills a 0.36-m^3 vessel. The mass of mixture is 100 kg. The measured pressure and temperature are 101 bar and 180 K, respectively. Compare the measured pressure with the pressure predicted using

(a) the ideal gas equation of state.
(b) Kay's rule.
(c) the additive pressure rule with the Redlich–Kwong equation.
(d) the additive volume rule with the Redlich–Kwong equation.

11.101 A gaseous mixture consisting of 50% argon and 50% nitrogen (molar basis) is contained in a closed tank at 20 bar, −96°C. Estimate the specific volume, in m^3/kg, using

(a) the ideal gas equation of state.
(b) Kay's rule together with the generalized compressibility chart.
(c) the Redlich–Kwong equation with mixture values for *a* and *b*.
(d) the additive volume rule together with the generalized compressibility chart.

11.102 Using the Carnahan–Starling–DeSantis equation of state introduced in Problem 11.19, together with the following expressions for the mixture values of *a* and *b*:

$$a = y_1^2 a_1 + 2y_1y_2(1 - f_{12})(a_1a_2)^{1/2} + y_2^2 a_2$$
$$b = y_1b_1 + y_2b_2$$

where f_{12} is an empirical *interaction* parameter, determine the pressure, in kPa, at v = 0.005 m^3/kg, T = 180°C for a mixture of Refrigerants 12 and 13, in which Refrigerant 12 is 40% by mass. For a mixture of Refrigerants 12 and 13, f_{12} = 0.035.

11.103 A rigid vessel initially contains carbon dioxide gas at 32°C and pressure p. Ethylene gas is allowed to flow into the tank until a mixture consisting of 20% carbon dioxide and 80% ethylene (molar basis) exists within the tank at a temperature of 43°C and a pressure of 110 bar. Determine the pressure p, in bar, using Kay's rule together with the generalized compressibility chart.

11.104 Two tanks having equal volumes are connected by a valve. One tank contains carbon dioxide gas at 38°C and pressure p. The other tank contains ethylene gas at 38°C and 10.2 MPa. The valve is opened and the gases mix, eventually attaining equilibrium at 38°C and pressure p' with a composition of 20% carbon dioxide and 80% ethylene (molar basis). Using Kay's rule and the generalized compressibility chart, determine in kPa,

(a) the initial pressure of the carbon dioxide, p.
(b) the final pressure of the mixture, p'.

Analyzing Multicomponent Systems

11.105 A binary solution at 25°C consists of 59 kg of ethyl alcohol (C_2H_5OH) and 41 kg of water. The respective partial molal volumes are 0.0573 and 0.0172 m^3/kmol. Determine the total volume, in m^3. Compare with the volume calculated using the molar specific volumes of the pure components, each a liquid at 25°C, in the place of the partial molal volumes.

11.106 The following data are for a binary solution of ethane (C_2H_6) and pentane (C_5H_{12}) at a certain temperature and pressure:

mole fraction of ethane	0.2	0.3	0.4	0.5	0.6	0.7	0.8
volume (in m^3) per kmol of solution	0.119	0.116	0.112	0.109	0.107	0.107	0.11

Estimate

(a) the specific volumes of pure ethane and pure pentane, each in m^3/kmol.
(b) the partial molal volumes of ethane and pentane for an equimolar solution, each in m^3/kmol.

11.107 The following data are for a binary mixture of carbon dioxide and methane at a certain temperature and pressure:

mole fraction of methane	0.000	0.204	0.406	0.606	0.847	1.000
volume (in m^3) per kmol of mixture	0.090	0.180	0.210	0.230	0.250	0.255

Estimate

(a) the specific volumes of pure carbon dioxide and pure methane, each in m^3/kmol.
(b) the partial molal volumes of carbon dioxide and methane for an equimolar mixture, each in m^3/kmol.

11.108 Using p–v–T data from the steam tables, determine the fugacity of water as a saturated vapor at (a) 280°C, (b) 260°C. Compare with the values obtained from the generalized fugacity chart.

11.109 Determine the fugacity, in bar, for

(a) butane at 555 K, 150 bar.
(b) methane at 49°C, 5.5 MPa.
(c) benzene at 495 K, 137 bar.

11.110 Using the equation of state of Problem 11.14 (c), evaluate the fugacity of ammonia at 750 K, 100 bar and compare with the value obtained from Fig. A-6.

11.111 Using tabulated compressibility data from the literature, evaluate f/p at T_R = 1.40 and p_R = 2.0. Compare with the value obtained from Fig. A-6.

11.112 Consider the truncated virial expansion

$$Z = 1 + \hat{B}(T_R)p_R + \hat{C}(T_R)p_R^2 + \hat{D}(T_R)p_R^3$$

(a) Using tabulated compressibility data from the literature, evaluate the coefficients $\hat{B}$, $\hat{C}$, and $\hat{D}$ for $0 < p_R < 1.0$ and each of T_R = 1.0, 1.2, 1.4, 1.6, 1.8, 2.0.
(b) Obtain an expression for ln (f/p) in terms of T_R and p_R. Using the coefficients of part (a), evaluate f/p at selected states and compare with tabulated values from the literature.

11.113 Derive the following approximation for the fugacity of a liquid at temperature T and pressure p:

$$f(T, p) \approx f_{\text{sat}}^L(T)\exp\left\{\frac{v_f(T)}{RT}[p - p_{\text{sat}}(T)]\right\}$$

where $f_{\text{sat}}^L(T)$ is the fugacity of the saturated liquid at temperature T. For what range of pressures might the approximation $f(T, p) \approx f_{\text{sat}}^L(T)$ apply?

11.114 Beginning with Eq. 11.122,

(a) evaluate ln f for a gas obeying the Redlich–Kwong equation of state.
(b) Using the result of part (a), evaluate the fugacity, in bar, for Refrigerant 134a at 90°C, 10 bar. Compare with the fugacity value obtained from the generalized fugacity chart.

11.115 Consider a one-inlet, one-exit control volume at steady state through which the flow is internally reversible and isothermal. Show that the work per unit of mass flowing can be expressed in terms of the fugacity f as

$$\left(\frac{\dot{W}_{\text{cv}}}{\dot{m}}\right)_{\substack{\text{int}\\\text{rev}}} = -RT\ln\left(\frac{f_2}{f_1}\right) + \frac{V_1^2 - V_2^2}{2} + g(z_1 - z_2)$$

11.116 Methane expands isothermally and without irreversibilities through a turbine operating at steady state, entering at 61 bar, 25°C and exiting at 1 bar. Using data from the generalized fugacity chart, determine the work developed, in kJ per kg of methane flowing. Ignore kinetic and potential energy effects.

11.117 Propane (C_3H_8) enters a turbine operating at steady state at 100 bar, 400 K and expands isothermally without irreversibilities to 10 bar. There are no significant changes in kinetic or potential energy. Using data from the generalized fugacity chart, determine the power developed, in kW, for a mass flow rate of 50 kg/min.

11.118 Ethane (C_2H_6) is compressed isothermally without irreversibilities at a temperature of 320 K from 5 to 40 bar. Using data from the generalized fugacity and enthalpy departure charts, determine the work of compression and the heat transfer, each in kJ per kg of ethane flowing. Assume steady-state operation and neglect kinetic and potential energy effects.

11.119 Methane enters a turbine operating at steady state at 100 bar, 275 K and expands isothermally without irreversibilities to 15 bar. There are no significant changes in kinetic or potential energy. Using data from the generalized fugacity and enthalpy departure charts, determine the power developed and heat transfer, each in kW, for a mass flow rate of 0.5 kg/s.

11.120 Methane flows isothermally and without irreversibilities through a horizontal pipe operating at steady state, entering at 50 bar, 300 K, 10 m/s and exiting at 40 bar. Using data from the generalized fugacity chart, determine the velocity at the exit, in m/s.

11.121 Determine the fugacity, in bar, for pure ethane at 310 K, 20.4 bar and as a component with a mole fraction of 0.35 in an ideal solution at the same temperature and pressure.

11.122 Denoting the *solvent* and *solute* in a dilute binary liquid solution at temperature T and pressure p by the subscripts 1 and 2, respectively, show that if the fugacity of the solute is proportional to its mole fraction in the solution: $\bar{f}_2 = \kappa y_2$, where k is a constant (*Henry's rule*), then the fugacity of the solvent is $\bar{f}_1 = y_1 f_1$, where y_1 is the solvent mole fraction and f_1 is the fugacity of pure 1 at T, p.

11.123 A tank contains 310 kg of a gaseous mixture of 70% ethane and 30% nitrogen (molar basis) at 311 K and 170 bar. Determine the volume of the tank, in m^3, using data from the generalized compressibility chart together with (a) Kay's rule, (b) the ideal solution model. Compare with the measured tank volume of 1 m^3.

11.124 A tank contains a mixture of 75% argon and 25% ethylene on a molar basis at 25°C, 82.5 bar. For 71 kg of mixture, estimate the tank volume, in m^3, using

(a) the ideal gas equation of state.
(b) Kay's rule together with data from the generalized compressibility chart.
(c) the ideal solution model together with data from the generalized compressibility chart.

11.125 A tank contains a mixture of 70% ethane and 30% nitrogen (N_2) on a molar basis at 400 K, 200 bar. For 2130 kg of mixture, estimate the tank volume, in m^3, using

(a) the ideal gas equation of state.
(b) Kay's rule together with data from the generalized compressibility chart.
(c) the ideal solution model together with data from the generalized compressibility chart.

11.126 An equimolar mixture of O_2 and N_2 enters a compressor operating at steady state at 10 bar, 220 K with a mass flow rate of 1 kg/s. The mixture exits at 60 bar, 400 K with no significant change in kinetic or potential energy. Stray heat transfer from the compressor can be ignored. Determine for the compressor

(a) the power required, in kW.
(b) the rate of entropy production, in kW/K.

Assume the mixture is modeled as an ideal solution. For the pure components:

	10 bar, 220 K		60 bar, 400 K	
	h (kJ/kg)	s (kJ/kg · K)	h (kJ/kg)	s (kJ/kg · K)
Oxygen	195.6	5.521	358.2	5.601
Nitrogen	224.1	5.826	409.8	5.911

11.127 A gaseous mixture with a molar analysis of 70% CH_4 and 30% N_2, enters a compressor operating at steady state at 10 bar, 250 K and a molar flow rate of 6 kmol/h. The mixture exits the compressor at 100 bar. During compression, the temperature of the mixture departs from 250 K by no more than 0.1 K. The power required by the compressor is reported to be 6 kW. Can this value be correct? Explain. Ignore kinetic and potential energy effects. Assume the mixture is modeled as an ideal solution. For the pure components at 250 K:

	h (kJ/kg)		s (kJ/kg · K)	
	10 bar	100 bar	10 bar	100 bar
Methane	506.0	358.6	10.003	8.3716
Nitrogen	256.18	229.68	5.962	5.188

11.128 The departure of a binary solution from ideal solution behavior is gauged by the *activity coefficient*, $\gamma_i = a_i/y_i$, where a_i is the activity of component i and y_i is its mole fraction in the solution (i = 1, 2). Introducing Eq. 11.140, the activity coefficient can be expressed alternatively as $\gamma_i = \bar{f}_i/y_i f_i^\circ$. Using this expression together with the *Gibbs–Duhem* equation, derive the following relation among the activity coefficients and the mole fractions for a solution at temperature T and pressure p:

$$\left(y_1 \frac{\partial \ln \gamma_1}{\partial y_1}\right)_{p,T} = \left(y_2 \frac{\partial \ln \gamma_2}{\partial y_2}\right)_{p,T}$$

How might this expression be used?

Design & open ended problems: exploring engineering practice

11.1D Compressed natural gas (CNG) is being used as a fuel to replace gasoline for automobile engines. Aluminum cylinders wrapped in a fibrous composite can provide lightweight, economical, and safe on-board storage. The storage vessels should hold enough CNG for 160 km to 200 km of urban travel, at storage pressures up to 21 MPa, and with a maximum total mass of 68 kg. Adhering to applicable U.S. Department of Transportation standards, specify both the size and number of cylinders that would meet the above design constraints.

11.2D Develop the preliminary design of a thermal storage system that would recover automobile engine *waste heat* for later use in improving the engine cold-start performance. Among the specifications are: Reliable operation down to an ambient temperature of −30°C, a storage duration of 16 hours, and no more than 15 minutes of urban driving to return the storage medium to its maximum temperature of 200°C. Specify the storage medium and determine whether the medium should be charged by the engine exhaust gases, the engine coolant, or some combination. Explain how the system would be configured and where it would be located in the automobile.

11.3D Figure P11.3D shows the schematic of a hydraulic accumulator in the form of a cylindrical pressure vessel with a piston separating a hydraulic fluid from a charge of nitrogen gas. The device has been proposed as a means for storing some of the exergy of a decelerating vehicle as it comes to rest. The exergy is stored by compressing the nitrogen. When the vehicle accelerates again, the gas expands and returns some exergy to the hydraulic fluid which is in communication with the vehicle's drive train, thereby assisting the vehicle to accelerate. In a proposal for one such device, the nitrogen operates in the range 50–150 bar and 200–350 K. Develop a thermodynamic model of the accumulator and use the model to assess its suitability for vehicle deceleration/acceleration.

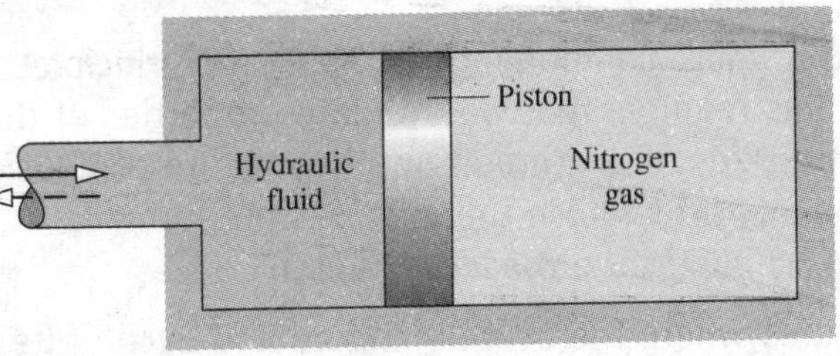

Fig. P11.3D

11.4D To investigate liquid–vapor phase transition behavior, construct a p–v diagram for water showing isotherms in the range $0.7 < T_R < 1.2$ by solving the van der Waals equation of state for pressure and specific volume at constant temperature. Superimpose the vapor dome on the diagram using saturated liquid and saturated vapor data from the steam tables. Interpret the behavior of the various isotherms in the liquid and vapor regions on the diagram. Referring to the literature as necessary, explain the behavior of the isotherms in the two-phase, liquid–vapor region, carefully distinguishing among *stable, metastable,* and *unstable* states. Write a paper discussing the plot and summarizing your findings.

11.5D In the experiment for the *regelation* of ice, a small-diameter wire weighted at each end is draped over a block of ice. The loaded wire is observed to cut slowly through the ice without leaving a trace. In one such set of experiments, a weighted 1.00-mm-diameter wire is reported to have passed through 0°C ice at a rate of 54 mm/h. Perform the regelation experiment and propose a plausible explanation for this phenomenon.

11.6D A portable refrigeration machine requiring no external power supply and using carbon dioxide at its *triple point* is described in U.S. Patent No. 4,096,707. Estimate the cost of the initial carbon dioxide charge required by such a machine to maintain a 1.8 m by 2.4 m by 4.5 m cargo container at 2°C for up to 24 hours, if the container is fabricated from sheet metal covered with a 2.5 cm layer of polystyrene. Would you recommend the use of such a refrigeration machine? Discuss.

11.7D During a phase change from liquid to vapor at fixed pressure, the temperature of a binary *nonazeotropic* solution such as an ammonia–water solution increases rather than remains constant as for a pure substance. This attribute is exploited in both the *Kalina* power cycle and in the *Lorenz* refrigeration cycle. Write a report assessing the status of technologies based on these cycles. Discuss the principal advantages of using binary nonazeotropic solutions. What are some of the main design issues related to their use in power and refrigeration systems?

11.8D The following data are known for a 100-ton ammonia–water absorption system like the one shown in Fig. 10.10. The pump is to handle 259 kg of strong solution per minute. The generator conditions are 1.2 MPa, 104°C. The absorber is at 0.2 MPa with strong solution exiting at 27°C. For the evaporator, the pressure is 0.21 MPa and the exit temperature is −12°C. Specify the type and size, in horsepower, of the pump required. Justify your choices.

11.9D The *Servel* refrigerator works on an absorption principle and requires no moving parts. An energy input by heat transfer is used to drive the cycle, and the refrigerant circulates due to its natural buoyancy. This type of refrigerator is commonly employed in mobile applications, such as recreational vehicles. Liquid propane is burned to provide the required energy input during mobile operation, and electric power is used when the vehicle is parked and can be connected to an electrical outlet. Investigate the principles of operation of commercially available Servel-type systems, and study their feasibility for solar-activated operation. Consider applications in remote locations where electricity or gas is not available. Write a report summarizing your findings.

11.10D An inventor has proposed a new type of marine engine *fueled* by fresh water stored on board and seawater drawn in from the surrounding ocean. At steady state, the engine would develop power from freshwater and seawater streams, each entering the engine at the ambient temperature and pressure. A single mixed stream would be discharged at the ambient temperature and pressure. Critically evaluate this proposal.

12

Ideal Gas Mixture and Psychrometric Applications

Engineering Context

Many systems of interest involve gas mixtures of two or more components. To apply the principles of thermodynamics introduced thus far to these systems requires that we evaluate properties of the mixtures. Means are available for determining the properties of mixtures from the mixture composition and the properties of the individual pure components from which the mixtures are formed. Methods for this purpose are discussed both in Chap. 11 and in the present chapter.

The **objective** of the present chapter is to study mixtures where the overall mixture and each of its components can be modeled as ideal gases. General ideal gas mixture considerations are provided in the first part of the chapter. Understanding the behavior of ideal gas mixtures of air and water vapor is prerequisite to considering air-conditioning processes in the second part of the chapter. In those processes, we sometimes must consider the presence of liquid water as well. We will also need to know how to handle ideal gas mixtures when we study the subjects of combustion and chemical equilibrium in Chapters 13 and 14, respectively.

Learning Objectives

When you complete your study of this chapter you will be able to...

- ✔ describe ideal gas mixture composition in terms of mass fractions or mole fractions.
- ✔ use the *Dalton model* to relate pressure, volume, and temperature and to calculate changes in *U*, *H*, and *S* for ideal gas mixtures.
- ✔ apply mass, energy, and entropy balances to systems involving ideal gas mixtures, including mixing processes.
- ✔ demonstrate understanding of psychrometric terminology, including humidity ratio, relative humidity, mixture enthalpy, and dew point temperature.
- ✔ use the psychrometric chart to represent common air-conditioning processes and to retrieve data.
- ✔ apply mass, energy, and entropy balances to analyze air-conditioning processes and cooling towers.

Ideal Gas Mixtures: General Considerations

12.1 Describing Mixture Composition

To specify the state of a mixture requires the composition and the values of two independent intensive properties such as temperature and pressure. The object of the present section is to consider ways for describing mixture composition. In subsequent sections, we show how mixture properties other than composition can be evaluated.

Consider a closed system consisting of a gaseous mixture of two or more components. The composition of the mixture can be described by giving the *mass* or the *number of moles* of each component present. With Eq. 1.8, the mass, the number of moles, and the molecular weight of a component i are related by

$$n_i = \frac{m_i}{M_i} \tag{12.1}$$

where m_i is the mass, n_i is the number of moles, and M_i is the molecular weight of component i, respectively. When m_i is expressed in terms of the kilogram, n_i is in kmol. However, any unit of mass can be used in this relationship.

The total mass of the mixture, m, is the sum of the masses of its components

$$m = m_1 + m_2 + \cdots + m_j = \sum_{i=1}^{j} m_i \tag{12.2}$$

The *relative* amounts of the components present in the mixture can be specified in terms of ***mass fractions.*** The mass fraction mf_i of component i is defined as

mass fractions

$$mf_i = \frac{m_i}{m} \tag{12.3}$$

A listing of the mass fractions of the components of a mixture is sometimes referred to as a ***gravimetric analysis.***

gravimetric analysis

Dividing each term of Eq. 12.2 by the total mass of mixture m and using Eq. 12.3

$$1 = \sum_{i=1}^{j} mf_i \tag{12.4}$$

That is, the sum of the mass fractions of all the components in a mixture is equal to unity.

The total number of moles in a mixture, n, is the sum of the number of moles of each of its components

$$n = n_1 + n_2 + \cdots + n_j = \sum_{i=1}^{j} n_i \tag{12.5}$$

The *relative* amounts of the components present in the mixture also can be described in terms of ***mole fractions.*** The mole fraction y_i of component i is defined as

mole fractions

$$y_i = \frac{n_i}{n} \tag{12.6}$$

molar analysis

volumetric analysis

A listing of the mole fractions of the components of a mixture may be called a ***molar analysis.*** An analysis of a mixture in terms of mole fractions is also called a ***volumetric analysis.***

Dividing each term of Eq. 12.5 by the total number of moles of mixture n and using Eq. 12.6

$$1 = \sum_{i=1}^{j} y_i \tag{12.7}$$

That is, the sum of the mole fractions of all the components in a mixture is equal to unity.

apparent molecular weight

The *apparent (or average) molecular weight* of the mixture, M, is defined as the ratio of the total mass of the mixture, m, to the total number of moles of mixture, n

$$M = \frac{m}{n} \tag{12.8}$$

Equation 12.8 can be expressed in a convenient alternative form. With Eq. 12.2, it becomes

$$M = \frac{m_1 + m_2 + \cdots + m_j}{n}$$

Introducing $m_i = n_i M_i$ from Eq. 12.1

$$M = \frac{n_1 M_1 + n_2 M_2 + \cdots + n_j M_j}{n}$$

Finally, with Eq. 12.6, the apparent molecular weight of the mixture can be calculated as a mole-fraction average of the component molecular weights

$$M = \sum_{i=1}^{j} y_i M_i \tag{12.9}$$

dry air

FOR EXAMPLE... consider the case of air. A sample of *atmospheric air* contains several gaseous components including water vapor and contaminants such as dust, pollen, and pollutants. The term ***dry air*** refers only to the gaseous components when all water vapor and contaminants have been removed. The molar analysis of a typical sample of dry air is given in Table 12.1. Selecting molecular weights for nitrogen, oxygen, argon, and carbon dioxide from Table A-1, and neglecting the trace substances neon, helium, etc., the apparent molecular weight of dry air obtained from Eq. 12.9 is

$$\begin{aligned} M &\approx 0.7808(28.02) + 0.2095(32.00) + 0.0093(39.94) + 0.0003(44.01) \\ &= 28.97 \text{ kg/kmol} \end{aligned}$$

This value, which is the entry for air in Tables A-1, would not be altered significantly if the trace substances were also included in the calculation.

Table 12.1

Approximate Composition of Dry Air

Component	Mole Fraction (%)
Nitrogen	78.08
Oxygen	20.95
Argon	0.93
Carbon dioxide	0.03
Neon, helium, methane, and others	0.01

Next, we consider two examples illustrating, respectively, the conversion from an analysis in terms of mole fractions to an analysis in terms of mass fractions, and conversely.

Example 12.1 CONVERTING MOLE FRACTIONS TO MASS FRACTIONS

The molar analysis of the gaseous products of combustion of a certain hydrocarbon fuel is CO_2, 0.08; H_2O, 0.11; O_2, 0.07; N_2, 0.74. **(a)** Determine the apparent molecular weight of the mixture. **(b)** Determine the composition in terms of mass fractions (gravimetric analysis).

Solution

Known: The molar analysis of the gaseous products of combustion of a hydrocarbon fuel is given.

Find: Determine (a) the apparent molecular weight of the mixture, (b) the composition in terms of mass fractions.

Analysis:

(a) Using Eq. 12.9 and approximate molecular weights from Table A-1

$$M = 0.08(44) + 0.11(18) + 0.07(32) + 0.74(28)$$
$$= 28.46 \text{ kg/kmol}$$

(b) Equations 12.1, 12.3, and 12.6 are the key relations required to determine the composition in terms of mass fractions.

1 Although the actual amount of mixture is not known, the calculations can be based on any convenient amount. Let us base the solution on 1 kmol of mixture. Then, with Eq. 12.6 the amount n_i of each component present in kmol is numerically equal to the mole fraction, as listed in column (ii) of the accompanying table. Column (iii) of the table gives the respective molecular weights of the components.

Column (iv) of the table gives the mass m_i of each component, in kg per kmol of mixture, obtained with Eq. 12.1 in the form $m_i = M_i n_i$. The values of column (iv) are obtained by multiplying each value of column (ii) by the corresponding value of column (iii). The sum of the values in column (iv) is the mass of the mixture: kg of mixture per kmol of mixture. Note that this sum is just the apparent mixture molecular weight determined in part (a). Finally, using Eq. 12.3, column (v) gives the mass fractions as a percentage. The values of column (v) are obtained by dividing the values of column (iv) by the column (iv) total and multiplying by 100.

(i) Component	(ii)[a] n_i	×	(iii) M_i	=	(iv)[b] m_i	(v) mf_i %
CO_2	0.08	×	44	=	3.52	12.37
H_2O	0.11	×	18	=	1.98	6.96
O_2	0.07	×	32	=	2.24	7.87
N_2	0.74	×	28	=	20.72	72.80
	1.00				28.46	100.00

[a]Entries in this column have units of kmol per kmol of mixture. For example, the first entry is 0.08 kmol of CO_2 per kmol of mixture.

[b]Entries in this column have units of kg per kmol of mixture. For example, the first entry is 3.52 kg of CO_2 per kmol of mixture. The column sum, 28.46, has units of kg of mixture per kmol of mixture.

Skills Developed: *Ability to...*
- ✓ calculate the apparent molecular weight with known mole fractions.
- ✓ determine the gravimetric analysis given the molar analysis.

1 If the solution to part (b) were conducted on the basis of some other assumed amount of mixture—for example, 100 kmol—the same result for the mass fractions would be obtained, as can be verified.

Quick Quiz

Determine the mass, in kg, of CO_2 in 0.5 kmol of mixture.
Ans. 1.76 kg.

Example 12.2 CONVERTING MASS FRACTIONS TO MOLE FRACTIONS

A gas mixture has the following composition in terms of mass fractions: H_2, 0.10; N_2, 0.60; CO_2, 0.30. Determine **(a)** the composition in terms of mole fractions and **(b)** the apparent molecular weight of the mixture.

Solution

Known: The gravimetric analysis of a gas mixture is known.

Find: Determine the analysis of the mixture in terms of mole fractions (molar analysis) and the apparent molecular weight of the mixture.

Analysis:

(a) Equations 12.1, 12.3, and 12.6 are the key relations required to determine the composition in terms of mole fractions.

Although the actual amount of mixture is not known, the calculation can be based on any convenient amount. ❶ Let us base the solution on 100 kg. Then, with Eq. 12.3, the amount m_i of each component present, in kg, is equal to the mass fraction multiplied by 100 kg. The values are listed in column (ii) of the accompanying table. Column (iii) of the table gives the respective molecular weights of the components.

Column (iv) of the table gives the amount n_i of each component, in kmol per 100 kg of mixture, obtained using Eq. 12.1. The values of column (iv) are obtained by dividing each value of column (ii) by the corresponding value of column (iii). The sum of the values of column (iv) is the total amount of mixture, in kmol per 100 kg of mixture. Finally, using Eq. 12.6, column (v) gives the mole fractions as a percentage. The values of column (v) are obtained by dividing the values of column (iv) by the column (iv) total and multiplying by 100.

❷

(i) Component	(ii)[a] m_i	$\div$	(iii) M_i	$=$	(iv)[b] n_i	(v) y_i %
H_2	10	$\div$	2	$=$	5.00	63.9
N_2	60	$\div$	28	$=$	2.14	27.4
CO_2	30	$\div$	44	$=$	0.68	8.7
	100				7.82	100.0

[a]Entries in this column have units of kg per 100 kg of mixture. For example, the first entry is 10 kg of H_2 per 100 kg of mixture.

[b]Entries in this column have units of kmol per 100 kg of mixture. For example, the first entry is 5.00 kmol of H_2 per 100 kg of mixture. The column sum, 7.82, has units of kmol of mixture per 100 kg of mixture.

(b) The apparent molecular weight of the mixture can be found by using Eq. 12.9 and the calculated mole fractions. The value can be determined alternatively by using the column (iv) total giving the total amount of mixture in kmol per 100 kg of mixture. Thus, with Eq. 12.8

$$M = \frac{m}{n} = \frac{100 \text{ kg}}{7.82 \text{ kmol}} = 12.79 \frac{\text{kg}}{\text{kmol}}$$

❶ If the solution to part (a) were conducted on the basis of some other assumed amount of mixture, the same result for the mass fractions would be obtained, as can be verified.

❷ Although H_2 has the smallest mass fraction, its mole fraction is the largest.

Skills Developed: *Ability to...*
- ✓ determine the molar analysis given the gravimetric analysis.

Quick Quiz

How many kmol of H_2 would be present in 200 kg of mixture?
Ans. 10 kmol.

12.2 Relating p, V, and T for Ideal Gas Mixtures

The definitions given in Sec. 12.1 apply generally to mixtures. In the present section we are concerned only with *ideal gas* mixtures and introduce two models used in conjunction with this idealization: the *Dalton model* and the *Amagat model.*

Consider a system consisting of a number of gases contained within a closed vessel of volume V as shown in Fig. 12.1. The temperature of the gas mixture is T and the pressure is p. The overall mixture is considered an ideal gas, so p, V, T, and the total number of moles of mixture n are related by the ideal gas equation of state

$$p = n\frac{\overline{R}T}{V} \tag{12.10}$$

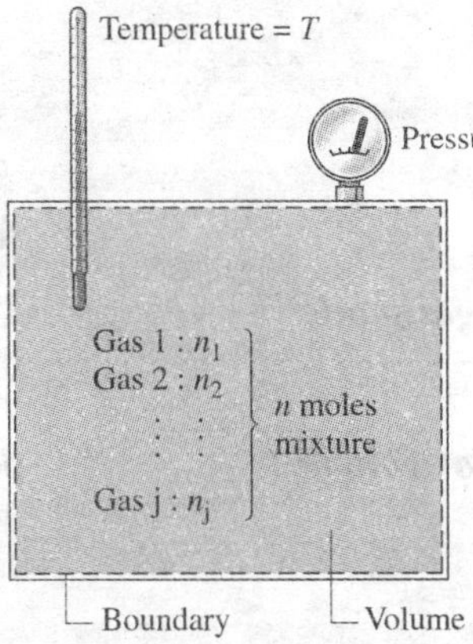

Fig. 12.1 Mixture of several gases.

With reference to this system, let us consider in turn the Dalton and Amagat models.

Dalton Model

The Dalton model is consistent with the concept of an ideal gas as being made up of molecules that exert negligible forces on one another and whose volume is negligible relative to the volume occupied by the gas (Sec. 3.12). In the absence of significant intermolecular forces, the behavior of each component would be unaffected by the presence of the other components. Moreover, if the volume occupied by the molecules is a very small fraction of the total volume, the molecules of each gas present may be regarded as free to roam throughout the full volume. In keeping with this simple picture, the ***Dalton model*** assumes that each mixture component behaves as an ideal gas as if it were *alone at the temperature T and volume V of the mixture.*

Dalton model

It follows from the Dalton model that the individual components would not exert the mixture pressure p but rather a *partial pressure*. As shown below, the sum of the partial pressures equals the mixture pressure. The ***partial pressure*** of component i, p_i, is the pressure that n_i moles of component i would exert if the component were alone in the volume V at the mixture temperature T. The partial pressure can be evaluated using the ideal gas equation of state

partial pressure

$$p_i = \frac{n_i\overline{R}T}{V} \tag{12.11}$$

Dividing Eq. 12.11 by Eq. 12.10

$$\frac{p_i}{p} = \frac{n_i\overline{R}T/V}{n\overline{R}T/V} = \frac{n_i}{n} = y_i$$

Thus, the partial pressure of component i can be evaluated in terms of its mole fraction y_i and the mixture pressure p

$$p_i = y_i p \tag{12.12}$$

To show that the sum of partial pressures equals the mixture pressure, sum both sides of Eq. 12.12 to obtain

$$\sum_{i=1}^{j} p_i = \sum_{i=1}^{j} y_i p = p\sum_{i=1}^{j} y_i$$

Since the sum of the mole fractions is unity (Eq. 12.7), this becomes

$$p = \sum_{i=1}^{j} p_i \tag{12.13}$$

Amagat Model

Amagat model

partial volume

The underlying assumption of the ***Amagat model*** is that each mixture component behaves as an ideal gas as if it existed separately at the pressure p and temperature T of the mixture. The volume that n_i moles of component i would occupy if the component existed at p and T is called the ***partial volume,*** V_i, of component i. As shown below, the sum of the partial volumes equals the total volume. The partial volume can be evaluated using the ideal gas equation of state

$$V_i = \frac{n_i \overline{R} T}{p} \tag{12.14}$$

Dividing Eq. 12.14 by the total volume V

$$\frac{V_i}{V} = \frac{n_i \overline{R}T/p}{n\overline{R}T/p} = \frac{n_i}{n} = y_i$$

Thus, the partial volume of component i also can be evaluated in terms of its mole fraction y_i and the total volume

$$V_i = y_i V \tag{12.15}$$

This relationship between volume fraction and mole fraction underlies the use of the term *volumetric analysis* as signifying an analysis of a mixture in terms of mole fractions.

To show that the sum of partial volumes equals the total volume, sum both sides of Eq. 12.15 to obtain

$$\sum_{i=1}^{j} V_i = \sum_{i=1}^{j} y_i V = V \sum_{i=1}^{j} y_i$$

Since the sum of the mole fractions equals unity, this becomes

$$V = \sum_{i=1}^{j} V_i \tag{12.16}$$

Closing Comment

The Dalton and Amagat models are special cases, respectively, of the additive pressure and additive volume rules introduced in Sec. 11.8, which do not require the assumption of the ideal gas model. The concept of an ideal gas mixture is a special case of the *ideal solution* concept introduced in Sec. 11.9.

12.3 Evaluating U, H, S, and Specific Heats

To apply the conservation of energy principle to a system involving an ideal gas mixture requires evaluation of the internal energy, enthalpy, or specific heats of the mixture at various states. Similarly, to conduct an analysis using the second law normally requires the entropy of the mixture. The objective of the present section is to develop means to evaluate these properties for ideal gas mixtures.

Evaluating U and H

Consider a closed system consisting of an ideal gas mixture. Extensive properties of the mixture, such as U, H, or S, can be found by adding the contribution of each component *at the condition at which the component exists in the mixture*. Let us apply this model to internal energy and enthalpy.

Since the internal energy and enthalpy of ideal gases are functions of temperature only, the values of these properties for each component present in the mixture are determined by the mixture temperature alone. Accordingly

$$U = U_1 + U_2 + \cdots + U_j = \sum_{i=1}^{j} U_i \tag{12.17}$$

$$H = H_1 + H_2 + \cdots + H_j = \sum_{i=1}^{j} H_i \tag{12.18}$$

where U_i and H_i are the internal energy and enthalpy, respectively, of component i evaluated at the mixture temperature.

Equations 12.17 and 12.18 can be rewritten on a molar basis as

$$n\bar{u} = n_1\bar{u}_1 + n_2\bar{u}_2 + \cdots + n_j\bar{u}_j = \sum_{i=1}^{j} n_i\bar{u}_i \tag{12.19}$$

and

$$n\bar{h} = n_1\bar{h}_1 + n_2\bar{h}_2 + \cdots + n_j\bar{h}_j = \sum_{i=1}^{j} n_i\bar{h}_i \tag{12.20}$$

where $\bar{u}$ and $\bar{h}$ are the specific internal energy and enthalpy of the *mixture* per mole of mixture, and $\bar{u}_i$ and $\bar{h}_i$ are the specific internal energy and enthalpy of *component i* per mole of i. Dividing by the total number of moles of mixture n gives expressions for the specific internal energy and enthalpy of the mixture per mole of mixture, respectively

$$\bar{u} = \sum_{i=1}^{j} y_i\bar{u}_i \tag{12.21}$$

$$\bar{h} = \sum_{i=1}^{j} y_i\bar{h}_i \tag{12.22}$$

Each of the molar internal energy and enthalpy terms appearing in Eqs. 12.19 through 12.22 is evaluated at the mixture temperature only.

Evaluating c_v and c_p

Differentiation of Eqs. 12.21 and 12.22 with respect to temperature results, respectively, in the following expressions for the specific heats $\bar{c}_v$ and $\bar{c}_p$ of the mixture on a molar basis

$$\bar{c}_v = \sum_{i=1}^{j} y_i\bar{c}_{v,i} \tag{12.23}$$

$$\bar{c}_p = \sum_{i=1}^{j} y_i\bar{c}_{p,i} \tag{12.24}$$

That is, the mixture specific heats $\bar{c}_p$ and $\bar{c}_v$ are mole-fraction averages of the respective component specific heats. The specific heat ratio for the mixture is $k = \bar{c}_p/\bar{c}_v$.

Evaluating S

The entropy of a mixture can be found, as for U and H, by adding the contribution of each component at the condition at which the component exists in the mixture. The entropy of an ideal gas depends on two properties, not on temperature alone as for internal energy and enthalpy. Accordingly, for the mixture

$$S = S_1 + S_2 + \cdots + S_j = \sum_{i=1}^{j} S_i \tag{12.25}$$

where S_i is the entropy of component i evaluated at the mixture temperature T and partial pressure p_i (or at temperature T and total volume V).

Equation 12.25 can be written on a molar basis as

$$n\bar{s} = n_1\bar{s}_1 + n_2\bar{s}_2 + \cdots + n_j\bar{s}_j = \sum_{i=1}^{j} n_i\bar{s}_i \tag{12.26}$$

where $\bar{s}$ is the entropy of the *mixture* per mole of mixture and $\bar{s}_i$ is the entropy of *component* i per mole of i. Dividing by the total number of moles of mixture gives an expression for the entropy of the mixture per mole of mixture

$$\bar{s} = \sum_{i=1}^{j} y_i\bar{s}_i \tag{12.27}$$

The specific entropies $\bar{s}_i$ of Eqs. 12.26 and 12.27 are evaluated at the mixture temperature T and the partial pressure p_i.

Working on a Mass Basis

In cases where it is convenient to work on a mass basis, the foregoing expressions would be written with the mass of the mixture, m, and the mass of component i in the mixture, m_i, replacing, respectively, the number of moles of mixture, n, and the number of moles of component i, n_i. Similarly, the mass fraction of component i, mf_i, would replace the mole fraction, y_i. All specific internal energies, enthalpies, and entropies would be evaluated on a unit mass basis rather than on a per mole basis as above. The development of the appropriate expressions is left as an exercise. By using the molecular weight of the mixture or of component i, as appropriate, data can be converted from a mass basis to a molar basis, or conversely, with relations of the form

$$\bar{u} = Mu, \quad \bar{h} = Mh, \quad \bar{c}_p = Mc_p, \quad \bar{c}_v = Mc_v, \quad \bar{s} = Ms \tag{12.28}$$

for the mixture, and

$$\bar{u}_i = M_i u_i, \quad \bar{h} = M_i h_i, \quad \bar{c}_{p,i} = M_i c_{p,i}, \quad \bar{c}_{v,i} = M_i c_{v,i}, \quad \bar{s}_i = M_i s_i \tag{12.29}$$

for component i.

12.4 Analyzing Systems Involving Mixtures

To perform thermodynamic analyses of systems involving *nonreacting* ideal gas mixtures requires no new fundamental principles. The conservation of mass and energy principles and the second law of thermodynamics are applicable in the forms previously introduced. The only new aspect is the proper evaluation of the required property data for the mixtures involved. This is illustrated in the present section, which deals with two classes of problems involving mixtures. In Sec. 12.4.1 the mixture is already formed, and we study processes in which there is no change in composition. Section 12.4.2 considers the formation of mixtures from individual components that are initially separate.

12.4.1 Mixture Processes at Constant Composition

In the present section we are concerned with the case of ideal gas mixtures undergoing processes during which the composition remains constant. The number of moles of each component present, and thus the total number of moles of mixture, remain the same throughout the process. This case is shown schematically in Fig. 12.2, which is labeled with expressions for U, H, and S of a mixture at the initial and final states of a process undergone by the mixture. In accordance with the discussion of Sec. 12.3, the specific internal energies and enthalpies of the components are evaluated at the temperature of the mixture. The specific entropy of each component is evaluated at the mixture temperature and the partial pressure of the component in the mixture.

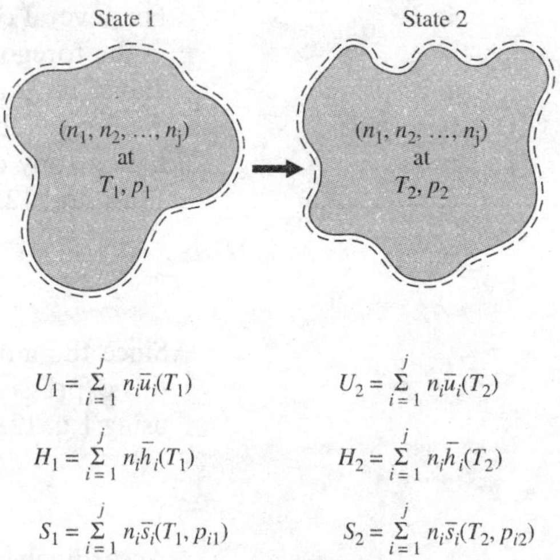

Fig. 12.2 Process of an ideal gas mixture.

The changes in the internal energy and enthalpy of the mixture during the process are given, respectively, by

$$U_2 - U_1 = \sum_{i=1}^{j} n_i[\bar{u}_i(T_2) - \bar{u}_i(T_1)] \tag{12.30}$$

$$H_2 - H_1 = \sum_{i=1}^{j} n_i[\bar{h}_i(T_2) - \bar{h}_i(T_1)] \tag{12.31}$$

where T_1 and T_2 denote the temperature at the initial and final states. Dividing by the number of moles of mixture, n, expressions for the change in internal energy and enthalpy of the mixture per mole of mixture result

$$\Delta\bar{u} = \sum_{i=1}^{j} y_i[\bar{u}_i(T_2) - \bar{u}_i(T_1)] \tag{12.32}$$

$$\Delta\bar{h} = \sum_{i=1}^{j} y_i[\bar{h}_i(T_2) - \bar{h}_i(T_1)] \tag{12.33}$$

Similarly, the change in entropy for the mixture is

$$S_2 - S_1 = \sum_{i=1}^{j} n_i[\bar{s}_i(T_2, p_{i2}) - \bar{s}_i(T_1, p_{i1})] \tag{12.34}$$

where p_{i1} and p_{i2} denote, respectively, the initial and final partial pressures of component i. Dividing by the total moles of mixture, Eq. 12.34 becomes

$$\Delta\bar{s} = \sum_{i=1}^{j} y_i[\bar{s}_i(T_2, p_{i2}) - \bar{s}_i(T_1, p_{i1})] \tag{12.35}$$

Companion expressions for Eqs. 12.30 through 12.35 on a mass basis also can be written. This is left as an exercise.

The foregoing expressions giving the changes in internal energy, enthalpy, and entropy of the mixture are written in terms of the respective property changes of the components. Accordingly, different datums might be used to assign specific enthalpy values to the various components because the datums would cancel when the component enthalpy changes are calculated. Similar remarks apply to the cases of internal energy and entropy.

Using Ideal Gas Tables

For several common gases modeled as ideal gases, the quantities $\overline{u}_i$ and $\overline{h}_i$ appearing in the foregoing expressions can be evaluated as functions of temperature only from Tables A-22 and A-23. Table A-22 for air gives these quantities on a *mass* basis. Table A-23 gives them on a *molar* basis. The ideal gas tables also can be used to evaluate the entropy change. The change in specific entropy of component i required by Eqs. 12.34 and 12.35 can be determined with Eq. 6.20b as

$$\Delta \overline{s}_i = \overline{s}_i^\circ(T_2) - \overline{s}_i^\circ(T_1) - \overline{R} \ln \frac{p_{i2}}{p_{i1}}$$

Since the mixture composition remains constant, the ratio of the partial pressures in this expression is the same as the ratio of the mixture pressures, as can be shown by using Eq. 12.12 to write

$$\frac{p_{i2}}{p_{i1}} = \frac{y_i p_2}{y_i p_1} = \frac{p_2}{p_1}$$

Accordingly, when the composition is constant, the change in the specific entropy of component i is simply

$$\Delta \overline{s}_i = \overline{s}_i^\circ(T_2) - \overline{s}_i^\circ(T_1) - \overline{R} \ln \frac{p_2}{p_1} \tag{12.36}$$

where p_1 and p_2 denote, respectively, the initial and final *mixture* pressures. The terms $\overline{s}_i^\circ$ of Eq. 12.36 can be obtained as functions of temperature for several common gases from Table A-23. Table A-22 for air gives s° versus temperature.

Assuming Constant Specific Heats

When the component specific heats $\overline{c}_{v,i}$ and $\overline{c}_{p,i}$ are taken as constants, the specific internal energy, enthalpy, and entropy changes of the mixture and the components of the mixture are given by

$$\Delta \overline{u} = \overline{c}_v (T_2 - T_1), \qquad \Delta \overline{u}_i = \overline{c}_{v,i}(T_2 - T_1) \tag{12.37}$$

$$\Delta \overline{h} = \overline{c}_p (T_2 - T_1), \qquad \Delta \overline{h}_i = \overline{c}_{p,i}(T_2 - T_1) \tag{12.38}$$

$$\Delta \overline{s} = \overline{c}_p \ln \frac{T_2}{T_1} - \overline{R} \ln \frac{p_2}{p_1}, \qquad \Delta \overline{s}_i = \overline{c}_{p,i} \ln \frac{T_2}{T_1} - \overline{R} \ln \frac{p_2}{p_1} \tag{12.39}$$

where the mixture specific heats $\overline{c}_v$ and $\overline{c}_p$ are evaluated from Eqs. 12.23 and 12.24, respectively, using data from Tables A-20 or the literature, as required. The expression for $\Delta \overline{u}$ can be obtained formally by substituting the above expression for $\Delta \overline{u}_i$ into Eq. 12.32 and using Eq. 12.23 to simplify the result. Similarly, the expressions for $\Delta \overline{h}$ and $\Delta \overline{s}$ can be obtained by inserting $\Delta \overline{h}_i$ and $\Delta \overline{s}_i$ into Eqs. 12.33 and 12.35, respectively, and using Eq. 12.24 to simplify. In the equations for entropy change, the ratio of mixture pressures replaces the ratio of partial pressures as discussed above. Similar expressions can be written for the mixture specific internal energy, enthalpy, and entropy changes on a mass basis. This is left as an exercise.

Using Computer Software

The changes in internal energy, enthalpy, and entropy required in Eqs. 12.32, 12.33, and 12.35, respectively, also can be evaluated using computer software. *Interactive Thermodynamics: IT* provides data for a large number of gases modeled as ideal gases, and its use is illustrated in Example 12.4 below.

The next example illustrates the use of ideal gas mixture relations for analyzing a compression process.

Example 12.3 COMPRESSING AN IDEAL GAS MIXTURE

A mixture of 0.3 kg of carbon dioxide and 0.2 kg of nitrogen is compressed from $p_1 = 1$ bar, $T_1 = 300$ K to $p_2 = 3$ bar in a polytropic process for which $n = 1.25$. Determine **(a)** the final temperature, in K, **(b)** the work, in kJ, **(c)** the heat transfer, in kJ, and **(d)** the change in entropy of the mixture, in kJ/K.

Solution

Known: A mixture of 0.3 kg of CO_2 and 0.2 kg of N_2 is compressed in a polytropic process for which $n = 1.25$. At the initial state, $p_1 = 1$ bar, $T_1 = 300$°K. At the final state, $p_2 = 3$ bar.

Find: Determine the final temperature, in K, the work, in kJ, the heat transfer, in kJ, and the entropy change of the mixture in, kJ/K.

Schematic and Given Data:

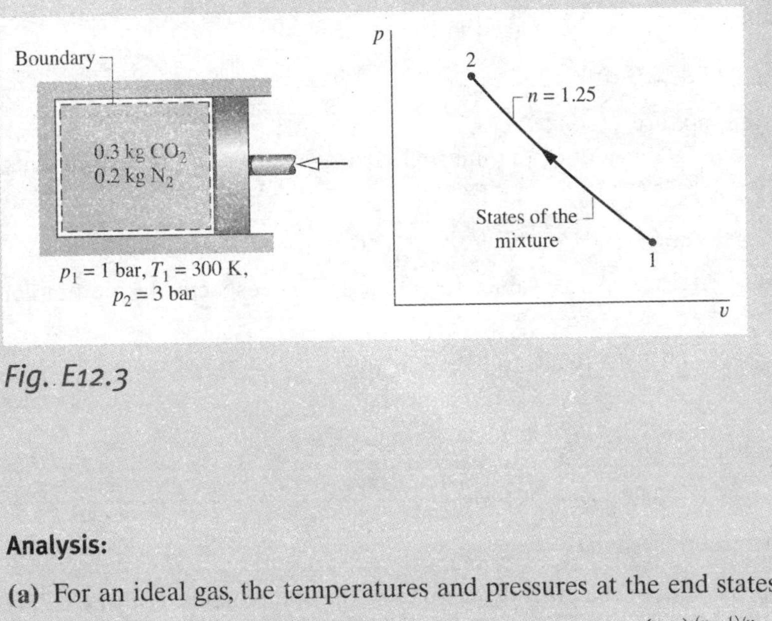

Fig. E12.3

Engineering Model:

1. As shown in the accompanying figure, the system is the mixture of CO and N_2. The mixture composition remains constant during the compression.
2. Each mixture component behaves as if it were an ideal gas occupying the entire system volume at the mixture temperature. The overall mixture acts as an ideal gas.
3. The compression process is a polytropic process for which $n = 1.25$.
4. The changes in kinetic and potential energy between the initial and final states can be ignored.

Analysis:

(a) For an ideal gas, the temperatures and pressures at the end states of a polytropic process are related by Eq. 3.56

$$T_2 = T_1\left(\frac{p_2}{p_1}\right)^{(n-1)/n}$$

Inserting values

$$T_2 = 300\left(\frac{3}{1}\right)^{0.2} = 374 \text{ K}$$

(b) The work for the compression process is given by

$$W = \int_1^2 p\,dV$$

Introducing $pV^n = constant$ and performing the integration

$$W = \frac{p_2V_2 - p_1V_1}{1 - n}$$

With the ideal gas equation of state, this reduces to

$$W = \frac{m(\overline{R}/M)(T_2 - T_1)}{1 - n}$$

The mass of the mixture is $m = 0.3 + 0.2 = 0.5$ kg. The apparent molecular weight of the mixture can be calculated using $M = m/n$, where n is the total number of moles of mixture. With Eq. 12.1, the numbers of moles of CO_2 and N_2 are, respectively

$$n_{CO_2} = \frac{0.3}{44} = 0.0068\,\text{kmol}, \qquad n_{N_2} = \frac{0.2}{28} = 0.0071\,\text{kmol}$$

The total number of moles of mixture is then $n = 0.0139$ kmol. The apparent molecular weight of the mixture is $M = 0.5/0.0139 = 35.97$.

Calculating the work

$$W = \frac{(0.5\,\text{kg})\left(\dfrac{8.314\,\text{kJ}}{35.97\,\text{kg}\cdot{}^\circ\text{K}}\right)(374\,\text{K} - 300\,\text{K})}{1 - 1.25}$$
$$= -34.21\,\text{kJ}$$

where the minus sign indicates that work is done on the mixture, as expected.

(c) With assumption 4, the closed system energy balance can be placed in the form

$$Q = \Delta U + W$$

where ΔU is the change in internal energy of the mixture.

The change in internal energy of the mixture equals the sum of the internal energy changes of the components. With Eq. 12.30

❶ $$\Delta U = n_{CO_2}[\overline{u}_{CO_2}(T_2) - \overline{u}_{CO_2}(T_1)] + n_{N_2}[\overline{u}_{N_2}(T_2) - \overline{u}_{N_2}(T_1)]$$

This form is convenient because Table A-23 gives internal energy values for N_2 and CO_2, respectively, on a molar basis. With values from this table

$$\Delta U = (0.0068)(9198 - 6939) + (0.0071)(7770 - 6229)$$
$$= 26.3\,\text{kJ}$$

Inserting values for ΔU and W into the expression for Q

$$Q = +26.3 - 34.21 = -7.91\,\text{kJ}$$

where the minus sign signifies a heat transfer from the system.

(d) The change in entropy of the mixture equals the sum of the entropy changes of the components. With Eq. 12.34

$$\Delta S = n_{CO_2}\,\Delta\overline{s}_{CO_2} + n_{N_2}\,\Delta\overline{s}_{N_2}$$

where $\Delta\overline{s}_{N_2}$ and $\Delta\overline{s}_{CO_2}$ are evaluated using Eq. 12.36 and values of $\overline{s}^\circ$ for N_2 and CO_2 from Table A-23. That is

❷ $$\Delta S = 0.0068\left(222.475 - 213.915 - 8.314\frac{3}{1}\right) + 0.0071\left(198.105 - 191.682 - 8.314\frac{3}{1}\right)$$
$$= -0.0231\,\text{kJ/K}$$

Entropy decreases in the process because entropy is transferred from the system accompanying heat transfer.

Skills Developed: *Ability to...*
- ✓ analyze a polytropic process of a closed system for a mixture of ideal gases.
- ✓ apply ideal gas mixture principles.
- ✓ determine changes in internal energy and entropy for ideal gas mixtures using tabular data.

---

❶ In view of the relatively small temperature change, the changes in the internal energy and entropy of the mixture can be evaluated alternatively using the constant specific heat relations, Eqs. 12.37 and 12.39, respectively. In these equations, $\bar{c}_v$ and $\bar{c}_p$ are specific heats for the mixture determined using Eqs. 12.23 and 12.24 together with appropriate specific heat values for the components chosen from Table A-20.

❷ Since the composition remains constant, the ratio of mixture pressures equals the ratio of partial pressures, so Eq. 12.36 can be used to evaluate the component specific entropy changes required here.

Quick Quiz

Recalling that polytropic processes are internally reversible, determine for the system the amount of entropy transfer accompanying heat transfer, in kJ/K.
Ans. −0.0105 kJ/K.

The next example illustrates the application of ideal gas mixture principles for the analysis of a mixture expanding isentropically through a nozzle. The solution features the use of table data and *IT* as an alternative.

Example 12.4 GAS MIXTURE EXPANDING ISENTROPICALLY THROUGH A NOZZLE

A gas mixture consisting of CO_2 and O_2 with mole fractions 0.8 and 0.2, respectively, expands isentropically and at steady state through a nozzle from 700 K, 5 atm, 3 m/s to an exit pressure of 1 atm. Determine **(a)** the temperature at the nozzle exit, in K, **(b)** the entropy changes of the CO_2 and O_2 from inlet to exit, in kJ/kmol · K, and **(c)** the exit velocity, in m/s.

Solution

Known: A gas mixture consisting of CO_2 and O_2 in specified proportions expands isentropically through a nozzle from specified inlet conditions to a given exit pressure.

Find: Determine the temperature at the nozzle exit, in K, the entropy changes of the CO_2 and O_2 from inlet to exit, in kJ/kmol · K, and the exit velocity, in m/s.

Schematic and Given Data:

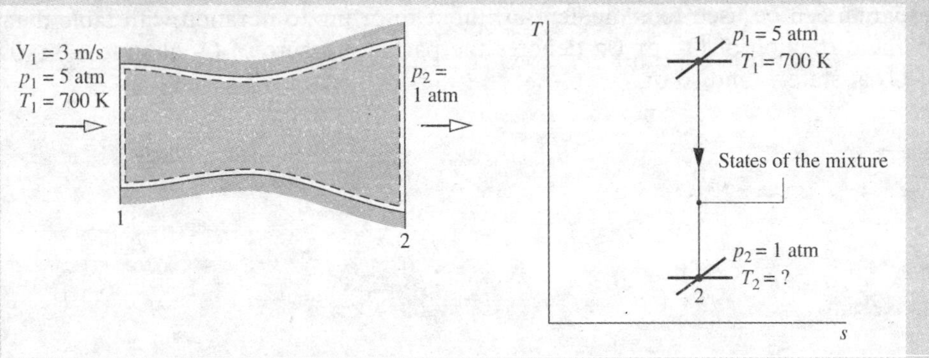

Fig. E12.4

Engineering Model:

1. The control volume shown by the dashed line on the accompanying figure operates at steady state.
2. The mixture composition remains constant as the mixture expands isentropically through the nozzle. The overall mixture and each mixture component act as ideal gases. The state of each component is defined by the temperature and the partial pressure of the component.
3. The change in potential energy between inlet and exit can be ignored.

Analysis:

(a) The temperature at the exit can be determined using the fact that the expansion occurs isentropically: $\bar{s}_2 - \bar{s}_1 = 0$. As there is no change in the specific entropy of the *mixture* between inlet and exit, Eq. 12.35 can be used to write

$$\bar{s}_2 - \bar{s}_1 = y_{O_2}\,\Delta\bar{s}_{O_2} + y_{CO_2}\,\Delta\bar{s}_{CO_2} = 0 \tag{a}$$

The change in specific entropy of each component can be determined using Eq. 12.36. Thus, Eq. (a) becomes

$$y_{O_2}\left[\bar{s}^\circ_{O_2}(T_2) - \bar{s}^\circ_{O_2}(T_1) - \bar{R}\ln\frac{p_2}{p_1}\right] + y_{CO_2}\left[\bar{s}^\circ_{CO_2}(T_2) - \bar{s}^\circ_{CO_2}(T_1) - \bar{R}\ln\frac{p_2}{p_1}\right] = 0$$

On rearrangement

$$y_{O_2}\bar{s}^\circ_{O_2}(T_2) + y_{CO_2}\bar{s}^\circ_{CO_2}(T_2) = y_{O_2}\bar{s}^\circ_{O_2}(T_1) + y_{CO_2}\bar{s}^\circ_{CO_2}(T_1) + (y_{O_2} + y_{CO_2})\bar{R}\ln\frac{p_2}{p_1}$$

The sum of mole fractions equals unity, so the coefficient of the last term on the right side is $(y_{O_2} + y_{CO_2}) = 1$.
Introducing given data, and values of $\bar{s}^\circ$ for O_2 and CO_2 at $T_1 = 700$ K from Table A-23

$$0.2\bar{s}^\circ_{O_2}(T_2) + 0.8\bar{s}^\circ_{CO_2}(T_2) = 0.2(231.358) + 0.8(250.663) + 8.314\ln\frac{1}{5}$$

or

$$0.2\bar{s}^\circ_{O_2}(T_2) + 0.8\bar{s}^\circ_{CO_2}(T_2) = 233.42 \text{ kJ/kmol}\cdot\text{K}$$

To determine the temperature T_2 requires an iterative approach with the above equation: A final temperature T_2 is assumed, and the $\bar{s}^\circ$ values for O_2 and CO_2 are found from Table A-23. If these two values do not satisfy the equation, another temperature is assumed. The procedure continues until the desired agreement is attained. In the present case

$$\text{at } T = 510 \text{ K:} \quad 0.2(221.206) + 0.8(235.700) = 232.80$$
$$\text{at } T = 520 \text{ K:} \quad 0.2(221.812) + 0.8(236.575) = 233.62$$

Linear interpolation between these values gives $T_2 = 517.6$ K.

Alternative Solution:

Alternatively, the following *IT* program can be used to evaluate T_2 without resorting to iteration with table data. In the program, yO2 denotes the mole fraction of O_2, p1_O2 denotes the partial pressure of O_2 at state 1, s1_O2 denotes the entropy per mole of O_2 at state 1, and so on.

```
T1 = 700 // K
p1 = 5 // bar
p2 = 1 // bar
yO2 = 0.2
yCO2 = 0.8
p1_O2 = yO2 * p1
p1_CO2 = yCO2 * p1
p2_O2 = yO2 * p2
p2_CO2 = yCO2 * p2
s1_O2 = s_TP ("O2",T1,p1_O2)
s1_CO2 = s_TP ("CO2",T1,p1_CO2)
s2_O2 = s_TP ("O2",T2,p2_O2)
s2_CO2 = s_TP ("CO2",T2,p2_CO2)
// When expressed in terms of these quantities, Eq. (a) takes the form
yO2 * (s2_O2 – s1_O2) + yCO2 * (s2_CO2 – s1_CO2) = 0
```

Using the **Solve** button, the result is $T_2 = 517.6$ K, which agrees with the value obtained using table data. Note that *IT* provides the value of specific entropy for each component directly and does not return $\bar{s}°$ of the ideal gas tables.

❶ **(b)** The change in the specific entropy for each of the components can be determined using Eq. 12.36. For O_2

$$\Delta \bar{s}_{O_2} = \bar{s}^\circ_{O_2}(T_2) - \bar{s}^\circ_{O_2}(T_1) - \bar{R} \ln \frac{p_2}{p_1}$$

Inserting $\bar{s}°$ values for O_2 from Table A-23

$$\Delta \bar{s}_{O_2} = 221.667 - 231.358 - 8.314 \ln(0.2) = 3.69 \text{ kJ/kmol} \cdot \text{K}$$

Similarly, with CO_2 data from Table A-23

$$\Delta \bar{s}_{CO_2} = \bar{s}^\circ_{CO_2}(T_2) - \bar{s}^\circ_{CO_2}(T_1) - \bar{R} \ln \frac{p_2}{p_1}$$
$$= 236.365 - 250.663 - 8.314 \ln(0.2)$$

❷ $$= -0.92 \text{ kJ/kmol} \cdot \text{K}$$

(c) Reducing the energy rate balance for the one-inlet, one-exit control volume at steady state

$$0 = h_1 - h_2 + \frac{V_1^2 - V_2^2}{2}$$

where h_1 and h_2 are the enthalpy *of the mixture,* per unit mass of mixture, at the inlet and exit, respectively. Solving for V_2

$$V_2 = \sqrt{V_1^2 + 2(h_1 - h_2)}$$

The term $(h_1 - h_2)$ in the expression for V_2 can be evaluated as

$$h_1 - h_2 = \frac{\bar{h}_1 - \bar{h}_2}{M} = \frac{1}{M}[y_{O_2}(\bar{h}_1 - \bar{h}_2)_{O_2} + y_{CO_2}(\bar{h}_1 - \bar{h}_2)_{CO_2}]$$

where M is the apparent molecular weight of mixture, and the molar specific enthalpies of O_2 and CO_2 are from Table A-23. With Eq. 12.9, the apparent molecular weight of the mixture is

$$M = 0.8(44) + 0.2(32) = 41.6 \text{ kg/kmol}$$

Then, with enthalpy values at $T_1 = 700$ K and $T_2 = 517.6$ K from Table A-23

$$h_1 - h_2 = \frac{1}{41.6}[0.2(21{,}184 - 15{,}320) + 0.8(27{,}125 - 18{,}468)]$$
$$= 194.7 \text{ kJ/kg}$$

Finally,

❸ $$V_2 = \sqrt{\left(3\frac{\text{m}}{\text{s}}\right)^2 + 2\left(194.7\frac{\text{kJ}}{\text{kg}}\right)\left|\frac{1 \text{ kg} \cdot \text{m/s}^2}{1 \text{ N}}\right|\left|\frac{10^3 \text{ N} \cdot \text{m}}{1 \text{ kJ}}\right|} = 624 \text{ m/s}$$

❶ Parts (b) and (c) can be solved alternatively using *IT*. These parts also can be solved using a constant c_p together with Eqs. 12.38 and 12.39. Inspection of Table A-20 shows that the specific heats of CO_2 and O_2 increase only slightly with temperature over the interval from 518 to 700 K, and so suitable constant values of c_p for the components and the overall mixture can be readily determined. These alternative solutions are left as exercises.

Skills Developed: *Ability to...*

- ✓ analyze the isentropic expansion of an ideal gas mixture flowing through a nozzle.
- ✓ apply ideal gas mixture principles together with mass and energy balances to calculate the exit velocity of a nozzle.
- ✓ determine the exit temperature for a given inlet state and a given exit pressure using tabular data and alternatively using *IT*.

❷ Each component experiences an entropy change as it passes from inlet to exit. The increase in entropy of the oxygen and the decrease in entropy of the carbon dioxide are due to an entropy transfer accompanying heat transfer from the CO_2 to the O_2 as they expand through the nozzle. However, as indicated by Eq. (a), there is no change in the entropy of the *mixture* as it expands through the nozzle.

❸ Note the use of unit conversion factors in the calculation of V_2.

Quick Quiz

What would be the exit velocity, in m/s, if the isentropic nozzle efficiency were 90%?

Ans. 592 m/s.

12.4.2 Mixing of Ideal Gases

Thus far, we have considered only mixtures that have already been formed. Now let us take up cases where ideal gas mixtures are formed by mixing gases that are initially separate. Such mixing is irreversible because the mixture forms spontaneously, and a work input from the surroundings would be required to separate the gases and return them to their respective initial states. In this section, the irreversibility of mixing is demonstrated through calculations of the entropy production.

Three factors contribute to the production of entropy in mixing processes:

1. The gases are initially at different temperatures.
2. The gases are initially at different pressures.
3. The gases are distinguishable from one another.

Entropy is produced when any of these factors is present during a mixing process. This is illustrated in the next example, where different gases, initially at different temperatures and pressures, are mixed.

Example 12.5 ADIABATIC MIXING AT CONSTANT TOTAL VOLUME

Two rigid, insulated tanks are interconnected by a valve. Initially 0.79 kmol of nitrogen at 2 bar and 250 K fills one tank. The other tank contains 0.21 kmol of oxygen at 1 bar and 300 K. The valve is opened and the gases are allowed to mix until a final equilibrium state is attained. During this process, there are no heat or work interactions between the tank contents and the surroundings. Determine **(a)** the final temperature of the mixture, in K, **(b)** the final pressure of the mixture, in bar, and **(c)** the amount of entropy produced in the mixing process, in kJ/K.

Solution

Known: Nitrogen and oxygen, initially separate at different temperatures and pressures, are allowed to mix without heat or work interactions with the surroundings until a final equilibrium state is attained.

Find: Determine the final temperature of the mixture, in K, the final pressure of the mixture, in bar, and the amount of entropy produced in the mixing process, in kJ/K.

Schematic and Given Data:

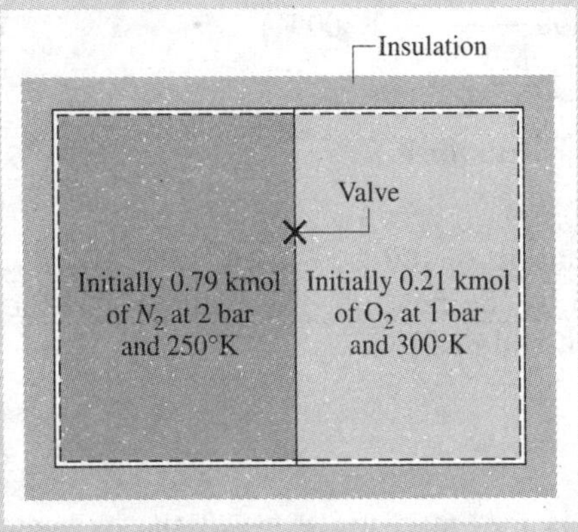

Engineering Model:

1. The system is taken to be the nitrogen and the oxygen together.
2. When separate, each of the gases behaves as an ideal gas. The final mixture also acts as an ideal gas. Each mixture component occupies the total volume and exhibits the mixture temperature.
3. No heat or work interactions occur with the surroundings, and there are no changes in kinetic and potential energy.

Fig. E12.5

Analysis:

(a) The final temperature of the mixture can be determined from an energy balance. With assumption 3, the closed system energy balance reduces to

$$\Delta U = \overset{0}{\cancel{Q}} - \overset{0}{\cancel{W}} \quad \text{or} \quad U_2 - U_1 = 0$$

The initial internal energy of the system, U_1, equals the sum of the internal energies of the two gases when separate

$$U_1 = n_{N_2}\bar{u}_{N_2}(T_{N_2}) + n_{O_2}\bar{u}_{O_2}(T_{O_2})$$

where $T_{N_2} = 250°K$ is the initial temperature of the nitrogen and $T_{O_2} = 300°K$ is the initial temperature of the oxygen. The final internal energy of the system, U_2, equals the sum of the internal energies of the two gases evaluated at the final mixture temperature T_2

$$U_2 = n_{N_2}\bar{u}_{N_2}(T_2) + n_{O_2}\bar{u}_{O_2}(T_2)$$

Collecting the last three equations

$$n_{N_2}[\bar{u}_{N_2}(T_2) - \bar{u}_{N_2}(T_{N_2})] + n_{O_2}[\bar{u}_{O_2}(T_2) - \bar{u}_{O_2}(T_{O_2})] = 0$$

The temperature T_2 can be determined using specific internal energy data from Table A-23 and an iterative procedure like that employed in part (a) of Example 12.4. However, since the specific heats of N_2 and O_2 vary little over the temperature interval from 250 to 300°K, the solution can be conducted accurately on the basis of constant specific heats. Hence, the foregoing equation becomes

$$n_{N_2}\bar{c}_{v,N_2}(T_2 - T_{N_2}) + n_{O_2}\bar{c}_{v,O_2}(T_2 - T_{O_2}) = 0$$

Solving for T_2

$$T_2 = \frac{n_{N_2}\bar{c}_{v,N_2}T_{N_2} + n_{O_2}\bar{c}_{v,O_2}T_{O_2}}{n_{N_2}\bar{c}_{v,N_2} + n_{O_2}\bar{c}_{v,O_2}}$$

Selecting c_v values for N_2 and O_2 from Table A-20 at the average of the initial temperatures of the gases, 275 K, and using the respective molecular weights to convert to a molar basis

$$\bar{c}_{v,N_2} = \left(28.01\frac{\text{kg}}{\text{mol}}\right)\left(0.743\frac{\text{kJ}}{\text{kg}\cdot\text{K}}\right) = 20.82\frac{\text{kJ}}{\text{kmol}\cdot°\text{K}}$$

$$\bar{c}_{v,O_2} = \left(32.0\frac{\text{kg}}{\text{mol}}\right)\left(0.656\frac{\text{kJ}}{\text{kg}\cdot\text{K}}\right) = 20.99\frac{\text{kJ}}{\text{kmol}\cdot°\text{K}}$$

Substituting values into the expression for T_2

$$T_2 = \frac{(0.79\,\text{kmol})\left(20.82\dfrac{\text{kJ}}{\text{kmol}\cdot\text{K}}\right)(250\text{ K}) + (0.21\,\text{kmol})\left(20.99\dfrac{\text{kJ}}{\text{kmol}\cdot\text{K}}\right)(300\text{ K})}{(0.79\,\text{kmol})\left(20.82\dfrac{\text{kJ}}{\text{kmol}\cdot\text{K}}\right) + (0.21\,\text{kmol})\left(20.99\dfrac{\text{kJ}}{\text{kmol}\cdot\text{K}}\right)}$$

$$= 261°\text{K}$$

(b) The final mixture pressure p_2 can be determined using the ideal gas equation of state, $p_2 = n\overline{R}T_2/V$, where n is the total number of moles of mixture and V is the total volume occupied by the mixture. The volume V is the sum of the volumes of the two tanks, obtained with the ideal gas equation of state as follows

$$V = \frac{n_{N_2}\overline{R}T_{N_2}}{p_{N_2}} + \frac{n_{O_2}\overline{R}T_{O_2}}{p_{O_2}}$$

where $p_{N_2} = 2$ atm is the initial pressure of the nitrogen and $p_{O_2} = 1$ atm is the initial pressure of the oxygen. Combining results and reducing

$$p_2 = \frac{(n_{N_2} + n_{O_2})T_2}{\left(\dfrac{n_{N_2}T_{N_2}}{p_{N_2}} + \dfrac{n_{O_2}T_{O_2}}{p_{O_2}}\right)}$$

Substituting values

$$p_2 = \frac{(1.0\,\text{kmol})(261°\text{K})}{\left[\dfrac{(0.79\,\text{kmol})(250°\text{K})}{2\,\text{bar}} + \dfrac{(0.21\,\text{kmol})(300°\text{K})}{1\,\text{bar}}\right]}$$

$$= 1.62\text{ bar}$$

(c) Reducing the closed system form of the entropy balance

$$S_2 - S_1 = \int_1^2 \left(\frac{\delta Q}{T}\right)_b^{\!\!0} + \sigma$$

where the entropy transfer term drops out for the adiabatic mixing process. The initial entropy of the system, S_1, is the sum of the entropies of the gases at the respective initial states

$$S_1 = n_{N_2}\bar{s}_{N_2}(T_{N_2}, p_{N_2}) + n_{O_2}\bar{s}_{O_2}(T_{O_2}, p_{O_2})$$

The final entropy of the system, S_2, is the sum of the entropies of the individual components, each evaluated at the final mixture temperature and the partial pressure of the component in the mixture

$$S_2 = n_{N_2}\bar{s}_{N_2}(T_2, y_{N_2}p_2) + n_{O_2}\bar{s}_{O_2}(T_2, y_{O_2}p_2)$$

Collecting the last three equations

$$\sigma = n_{N_2}[\bar{s}_{N_2}(T_2, y_{N_2}p_2) - \bar{s}_{N_2}(T_{N_2}, p_{N_2})] + n_{O_2}[\bar{s}_{O_2}(T_2, y_{O_2}p_2) - \bar{s}_{O_2}(T_{O_2}, p_{O_2})]$$

Evaluating the change in specific entropy of each gas in terms of a constant specific heat $\bar{c}_p$, this becomes

$$\sigma = n_{N_2}\left(\bar{c}_{p,N_2}\ln\frac{T_2}{T_{N_2}} - \overline{R}\ln\frac{y_{N_2}p_2}{p_{N_2}}\right) + n_{O_2}\left(\bar{c}_{p,O_2}\ln\frac{T_2}{T_{O_2}} - \overline{R}\ln\frac{y_{O_2}p_2}{p_{O_2}}\right)$$

The required values for $\bar{c}_p$ can be found by adding $\overline{R}$ to the $\bar{c}_v$ values found previously (Eq. 3.45)

$$\bar{c}_{p,N_2} = 29.13\frac{\text{kJ}}{\text{kmol}\cdot\text{K}} \qquad \bar{c}_{p,O_2} = 29.30\frac{\text{kJ}}{\text{kmol}\cdot°\text{K}}$$

Since the total number of moles of mixture $n = 0.79 + 0.21 = 1.0$, the mole fractions of the two gases are $y_{N_2} = 0.79$ and $y_{O_2} = 0.21$.

Substituting values into the expression for σ gives

$$\sigma = 0.79\,\text{kmol}\left[29.13\frac{\text{kJ}}{\text{kmol}\cdot\text{K}}\ln\left(\frac{261°\text{K}}{250°\text{K}}\right) - 8.314\frac{\text{kJ}}{\text{kmol}\cdot\text{K}}\ln\left(\frac{(0.79)(1.62\text{ bar})}{2\text{ bar}}\right)\right]$$

$$+\,0.21\,\text{kmol}\left[29.30\frac{\text{kJ}}{\text{kmol}\cdot\text{K}}\ln\left(\frac{261°\text{K}}{300°\text{K}}\right) - 8.314\frac{\text{kJ}}{\text{kmol}\cdot\text{K}}\ln\left(\frac{(0.21)(1.62\text{ bar})}{1\text{ bar}}\right)\right]$$

❶ $$= 5.0\text{ kJ/°K}$$

❶ Entropy is produced when different gases, initially at different temperatures and pressures, are followed to mix.

Skills Developed: Ability to...
- ✓ analyze the adiabatic mixing of two ideal gases at constant total volume.
- ✓ apply energy and entropy balances to the mixing of two gases.
- ✓ apply ideal gas mixture principles assuming constant specific heats.

Quick Quiz

Determine the total volume of the final mixture, in m^3.

Ans. 6 m^3.

In the next example, we consider a control volume at steady state where two incoming streams form a mixture. A single stream exits.

Example 12.6 ADIABATIC MIXING OF TWO STREAMS

At steady state, 100 m^3/min of dry air at 32°C and 1 bar is mixed adiabatically with a stream of oxygen (O_2) at 127°C and 1 bar to form a mixed stream at 47°C and 1 bar. Kinetic and potential energy effects can be ignored. Determine **(a)** the mass flow rates of the dry air and oxygen, in kg/min, **(b)** the mole fractions of the dry air and oxygen in the exiting mixture, and **(c)** the time rate of entropy production, in kJ/K · min.

Solution

Known: At steady state, 100 m^3/min of dry air at 32°C and 1 bar is mixed adiabatically with an oxygen stream at 127°C and 1 bar to form a mixed stream at 47°C and 1 bar.

Find: Determine the mass flow rates of the dry air and oxygen, in kg/min, the mole fractions of the dry air and oxygen in the exiting mixture, and the time rate of entropy production, in kJ/K · min.

Schematic and Given Data:

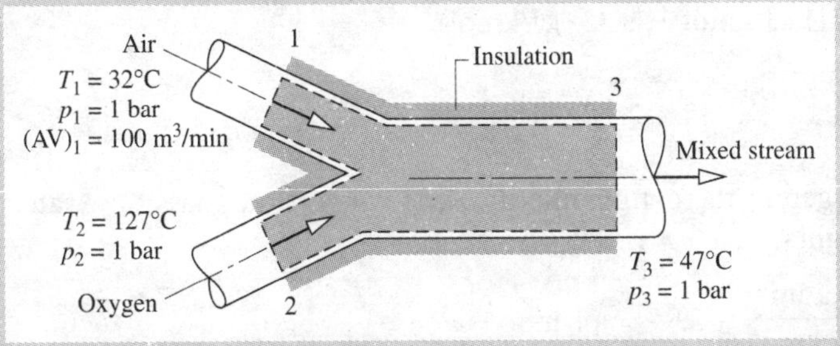

Fig. E12.6

Engineering Model:

1. The control volume identified by the dashed line on the accompanying figure operates at steady state.
2. No heat transfer occurs with the surroundings.

3. Kinetic and potential energy effects can be ignored, and $\dot{W}_{cv} = 0$.

4. The entering gases can be regarded as ideal gases. The exiting mixture can be regarded as an ideal gas mixture.

5. The dry air is treated as a pure component.

Analysis:

(a) The mass flow rate of the dry air entering the control volume can be determined from the given volumetric flow rate $(AV)_1$

$$\dot{m}_{a1} = \frac{(AV)_1}{v_{a1}}$$

where v_{a1} is the specific volume of the air at 1. Using the ideal gas equation of state

$$v_{a1} = \frac{(\overline{R}/M_a)T_1}{p_1} = \frac{\left(\dfrac{8314\ \text{N}\cdot\text{m}}{28.97\ \text{kg}\cdot\text{K}}\right)(305\ \text{K})}{10^5\ \text{N/m}^2} = 0.875\ \frac{\text{m}^3}{\text{kg}}$$

The mass flow rate of the dry air is then

$$\dot{m}_{a1} = \frac{100\ \text{m}^3/\text{min}}{0.875\ \text{m}^3/\text{kg}} = 114.29\ \frac{\text{kg}}{\text{min}}$$

The mass flow rate of the oxygen can be determined using mass and energy rate balances. At steady state, the amounts of dry air and oxygen contained within the control volume do not vary. Thus, for each component individually it is necessary for the incoming and outgoing mass flow rates to be equal. That is

$$\dot{m}_{a1} = \dot{m}_{a3} \quad \text{(dry air)}$$

$$\dot{m}_{o2} = \dot{m}_{o3} \quad \text{(oxygen)}$$

Using assumptions 1–3 together with the foregoing mass flow rate relations, the energy rate balance reduces to

$$0 = \dot{m}_a h_a(T_1) + \dot{m}_o h_o(T_2) - [\dot{m}_a h_a(T_3) + \dot{m}_o h_o(T_3)]$$

where $\dot{m}_a$ and $\dot{m}_o$ denote the mass flow rates of the dry air and oxygen, respectively. The enthalpy of the mixture at the exit is evaluated by summing the contributions of the air and oxygen, each at the mixture temperature. Solving for $\dot{m}_o$

$$\dot{m}_o = \dot{m}_a \left[\frac{h_a(T_3) - h_a(T_1)}{h_o(T_2) - h_o(T_3)}\right]$$

The specific enthalpies can be obtained from Tables A-22 and A-23. Since Table A-23 gives enthalpy values on a molar basis, the molecular weight of oxygen is introduced into the denominator to convert the molar enthalpy values to a mass basis

$$\dot{m}_o = \frac{(114.29\ \text{kg/min})(320.29\ \text{kJ/kg} - 305.22\ \text{kJ/kg})}{\left(\dfrac{1}{32\ \text{kg/kmol}}\right)(11{,}711\ \text{kJ/kmol} - 9{,}325\ \text{kJ/kmol})}$$

$$= 23.1\ \frac{\text{kg}}{\text{min}}$$

(b) To obtain the mole fractions of the dry air and oxygen in the exiting mixture, first convert the mass flow rates to molar flow rates using the respective molecular weights

$$\dot{n}_a = \frac{\dot{m}_a}{M_a} = \frac{114.29\ \text{kg/min}}{28.97\ \text{kg/kmol}} = 3.95\ \text{kmol/min}$$

$$\dot{n}_o = \frac{\dot{m}_o}{M_o} = \frac{23.1\ \text{kg/min}}{32\ \text{kg/kmol}} = 0.72\ \text{kmol/min}$$

where $\dot{n}$ denotes molar flow rate. The molar flow rate of the mixture $\dot{n}$ is the sum

$$\dot{n} = \dot{n}_a + \dot{n}_o = 3.95 + 0.72 = 4.67 \text{ kmol/min}$$

The mole fractions of the air and oxygen are, respectively

❶ $$y_a = \frac{\dot{n}_a}{\dot{n}} = \frac{3.95}{4.67} = 0.846 \quad \text{and} \quad y_o = \frac{\dot{n}_o}{\dot{n}} = \frac{0.72}{4.67} = 0.154$$

(c) For the control volume at steady state, the entropy rate balance reduces to

$$0 = \dot{m}_a s_a(T_1, p_1) + \dot{m}_o s_o(T_2, p_2) - [\dot{m}_a s_a(T_3, y_a p_3) + \dot{m}_o s_o(T_3, y_o p_3)] + \dot{\sigma}$$

The specific entropy of each component in the exiting ideal gas mixture is evaluated at its partial pressure in the mixture and at the mixture temperature. Solving for $\dot{\sigma}$

$$\dot{\sigma} = \dot{m}_a[s_a(T_3, y_a p_3) - s_a(T_1, p_1)] + \dot{m}_o[s_o(T_3, y_o p_3) - s_o(T_2, p_2)]$$

Since $p_1 = p_3$, the specific entropy change of the dry air is

$$\begin{aligned} s_a(T_3, y_a p_3) - s_a(T_1, p_1) &= s_a^\circ(T_3) - s_a^\circ(T_1) - \frac{\overline{R}}{M_a} \ln \frac{y_a p_3}{p_1} \\ &= s_a^\circ(T_3) - s_a^\circ(T_1) - \frac{\overline{R}}{M_a} \ln y_a \end{aligned}$$

The s_a° terms are evaluated from Table A-22. Similarly, since $p_2 = p_3$, the specific entropy change of the oxygen is

$$s_o(T_3, y_o p_3) - s_o(T_2, p_2) = \frac{1}{M_o}[\bar{s}_o^\circ(T_3) - \bar{s}_o^\circ(T_2) - \overline{R} \ln y_o]$$

The $\bar{s}_o^\circ$ terms are evaluated from Table A-23. Note the use of the molecular weights M_a and M_o in the last two equations to obtain the respective entropy changes on a mass basis.

The expression for the rate of entropy production becomes

$$\dot{\sigma} = \dot{m}_a\left[s_a^\circ(T_3) - s_a^\circ(T_1) - \frac{\overline{R}}{M_a}\ln y_a\right] + \frac{\dot{m}_o}{M_o}[\bar{s}_o^\circ(T_3) - \bar{s}_o^\circ(T_2) - \overline{R}\ln y_o]$$

Substituting values

$$\begin{aligned} \dot{\sigma} &= \left(114.29 \frac{\text{kg}}{\text{min}}\right)\left[1.7669 \frac{\text{kJ}}{\text{kg} \cdot \text{K}} - 1.71865 \frac{\text{kJ}}{\text{kg} \cdot \text{K}} - \left(\frac{8.314}{28.97} \frac{\text{kJ}}{\text{kg} \cdot \text{K}}\right) \ln 0.846\right] \\ &+ \left(\frac{23.1 \text{ kg/min}}{32 \text{ kg/kmol}}\right)\left[207.112 \frac{\text{kJ}}{\text{kmol} \cdot \text{K}} - 213.765 \frac{\text{kJ}}{\text{kmol} \cdot \text{K}} - \left(8.314 \frac{\text{kJ}}{\text{kmol} \cdot \text{K}}\right) \ln 0.154\right] \end{aligned}$$

❷ $$= 17.42 \frac{\text{kJ}}{\text{K} \cdot \text{min}}$$

❶ This calculation is based on dry air modeled as a pure component (assumption 5). However, since O_2 is a component of dry air (Table 12.1), the *actual* mole fraction of O_2 in the exiting mixture is greater than given here.

❷ Entropy is produced when different gases, initially at different temperatures, are allowed to mix.

Skills Developed: *Ability to...*

- ✓ analyze the adiabatic mixing of two ideal gas streams at steady state.
- ✓ apply ideal mixture principles together with mass, energy, and entropy rate balances.

Quick Quiz

What are the mass fractions of air and oxygen in the exiting mixture?

Ans. $mf_{air} = 0.832$, $mf_{O_2} = 0.168$

Bio... Connections

In the 1970s, concern for reducing energy use prompted ventilation standards for buildings to be decreased to as low as 0.14 m^3/min per occupant, a 60–70% reduction over previous practice. With less outside air being brought into buildings, less energy was needed to condition the air to desired temperature levels. However, with lower rates of ventilation to dilute contaminants, occupants experienced increased instances of headaches, eye and throat irritation, nausea, and other illness attributed to indoor contaminants. The term *sick building syndrome* was coined to describe indoor air quality conditions that can lead to acute health problems for the occupants.

The most common contaminants arise from chemical or biological sources. Chemical contaminants include volatile organic compounds (VOCs) such as found in adhesives, formaldehyde that can leach from building materials, and tobacco smoke. Improperly vented heaters and furnaces can cause dangerous inorganic compounds such as carbon monoxide and nitrogen oxides to build up in indoor spaces. Biological sources include bacteria, molds, pollen, and viruses that may breed in water accumulating in drain pans or duct work. Insects and bird droppings can also be sources of indoor air contamination. The bacterium *Legionella* was named for acute respiratory irritation and congestion experienced by American Legion convention-goers in a hotel in 1976. The fatal outbreak was attributed to bacteria in stagnant water within the air handling system.

More recently, recommended ventilation rates per occupant have been increased to the 0.4–0.6 m^3/min range, depending on the application, to ensure acceptable indoor air quality. Such rates are specified for particular occupied spaces within buildings, and careful design is needed to ensure that air distribution systems deliver required ventilation for each space. Inadequate system installation and improper maintenance can give rise to indoor air quality problems, even when appropriate standards have been applied in the design. Indoor air quality continues to be a significant concern.

Psychrometric Applications

psychrometrics

The remainder of this chapter is concerned with the study of systems involving mixtures of dry air and water vapor. A condensed water phase also may be present. Knowledge of the behavior of such systems is essential for the analysis and design of air-conditioning devices, cooling towers, and industrial processes requiring close control of the vapor content in air. The study of systems involving dry air and water is known as ***psychrometrics.***

12.5 Introducing Psychrometric Principles

The object of the present section is to introduce some important definitions and principles used in the study of systems involving of dry air and water.

12.5.1 Moist Air

moist air

The term ***moist air*** refers to a mixture of dry air and water vapor in which the dry air is treated as if it were a pure component. As can be verified by reference to appropriate property data, the overall mixture and each mixture component behave as ideal gases at the states under present consideration. Accordingly, for the applications to be considered, the ideal gas mixture concepts introduced previously apply directly.

Shown in Fig. 12.3 is a closed system consisting of moist air occupying a volume V at mixture pressure p and mixture temperature T. The overall mixture is assumed to obey the ideal gas equation of state. Thus

$$p = \frac{n\bar{R}T}{V} = \frac{m(\bar{R}/M)T}{V} \tag{12.40}$$

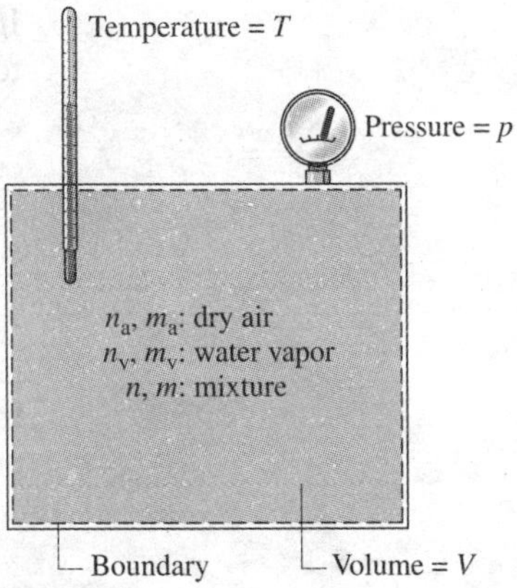

Fig. 12.3 Mixture of dry air and water vapor.

where n, m, and M denote the moles, mass, and molecular weight of the mixture, respectively, and $n = m/M$. Each mixture component is considered to act as if it existed alone in the volume V at the mixture temperature T while exerting a part of the pressure. The mixture pressure is the sum of the partial pressures of the dry air and the water vapor: $p = p_a + p_v$.

Using the ideal gas equation of state, the partial pressures p_a and p_v of the dry air and water vapor are, respectively

$$p_a = \frac{n_a\bar{R}T}{V} = \frac{m_a(\bar{R}/M_a)T}{V}, \qquad p_v = \frac{n_v\bar{R}T}{V} = \frac{m_v(\bar{R}/M_v)T}{V} \tag{12.41a}$$

where n_a and n_v denote the moles of dry air and water vapor, respectively; m_a, m_v, M_a, and M_v are the respective masses and molecular weights. The amount of water vapor present is normally much less than the amount of dry air. Accordingly, the values of n_v, m_v, and p_v are small relative to the corresponding values of n_a, m_a, and p_a.

Forming ratios with Eqs. 12.40 and 12.41a, we get the following alternative expressions for p_a and p_v

$$p_a = y_a\,p, \qquad p_v = y_v\,p \tag{12.41b}$$

where y_a and y_v are the mole fractions of the dry air and water vapor, respectively.

A typical state of water vapor in moist air is shown in Fig. 12.4. At this state, fixed by the partial pressure p_v and the mixture temperature T, the vapor is superheated. When the partial pressure of the water vapor corresponds to the saturation pressure of water at the mixture temperature, p_g of Fig. 12.4, the mixture is said to be *saturated*. ***Saturated air*** is a mixture of dry air and saturated water vapor. The amount of water vapor in moist air varies from zero in dry air to a maximum, depending on the pressure and temperature, when the mixture is saturated.

saturated air

12.5.2 Humidity Ratio, Relative Humidity, and Mixture Enthalpy

A given moist air sample can be described in a number of ways. The mixture can be described in terms of the moles of dry air and water vapor present or in terms of the respective mole fractions. Alternatively, the mass of dry air and water vapor, or the respective mass fractions, can be specified. The composition also can be indicated by means of the ***humidity ratio*** ω, defined as the ratio of the mass of the water vapor to the mass of dry air

humidity ratio

$$\omega = \frac{m_v}{m_a} \tag{12.42}$$

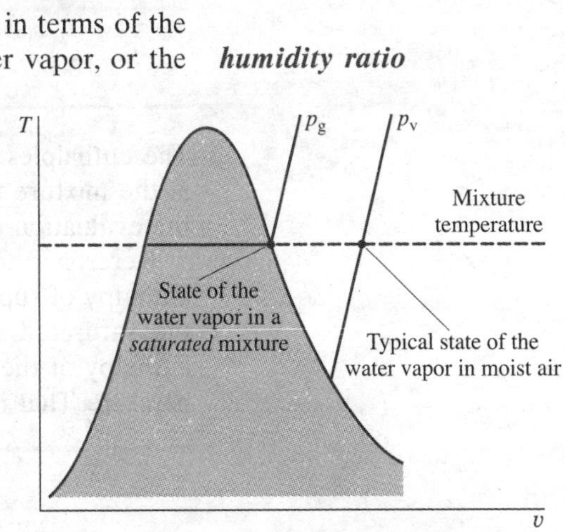

Fig. 12.4 T–v diagram for water vapor in an air–water mixture.

The humidity ratio is sometimes referred to as the *specific humidity*.

The humidity ratio can be expressed in terms of partial pressures and molecular weights by solving Eqs. 12.41a for m_a and m_v, respectively, and substituting the resulting expressions into Eq. 12.42 to obtain

$$\omega = \frac{m_v}{m_a} = \frac{M_v p_v V/\bar{R}T}{M_a p_a V/\bar{R}T} = \frac{M_v p_v}{M_a p_a}$$

Introducing $p_a = p - p_v$ and noting that the ratio of the molecular weight of water to that of dry air is approximately 0.622, this expression can be written as

$$\omega = 0.622 \frac{p_v}{p - p_v} \tag{12.43}$$

Moist air also can be described in terms of the *relative humidity* ϕ, defined as the ratio of the mole fraction of water vapor y_v in a given moist air sample to the mole fraction $y_{v,sat}$ in a saturated moist air sample at the same mixture temperature and pressure

$$\phi = \left.\frac{y_v}{y_{v,sat}}\right)_{T,p}$$

relative humidity

Since $p_v = y_v p$ and $p_g = y_{v,sat} p$, the ***relative humidity*** can be expressed as

$$\phi = \left.\frac{p_v}{p_g}\right)_{T,p} \tag{12.44}$$

The pressures in this expression for the relative humidity are labeled on Fig. 12.4.

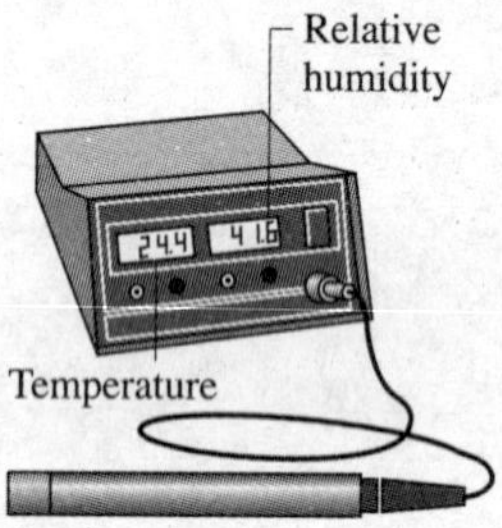

The humidity ratio and relative humidity can be measured. For laboratory measurements of humidity ratio, a *hygrometer* can be used in which a moist air sample is exposed to suitable chemicals until the moisture present is absorbed. The amount of water vapor is determined by weighing the chemicals. Continuous recording of the relative humidity can be accomplished by means of transducers consisting of resistance- or capacitance-type sensors whose electrical characteristics change with relative humidity.

Evaluating H, U, and S.

The values of H, U, and S for moist air modeled as an ideal gas mixture can be found by adding the contribution of each component at the condition at which the component exists in the mixture. For example, the enthalpy H of a given moist air sample is

$$H = H_a + H_v = m_a h_a + m_v h_v \tag{12.45}$$

mixture enthalpy

Dividing by m_a and introducing the humidity ratio gives the ***mixture enthalpy*** *per unit mass of dry air*

$$\frac{H}{m_a} = h_a + \frac{m_v}{m_a} h_v = h_a + \omega h_v \tag{12.46}$$

The enthalpies of the dry air and water vapor appearing in Eq. 12.46 are evaluated at the mixture temperature. An approach similar to that for enthalpy also applies to the evaluation of the internal energy of moist air.

Reference to steam table data or a Mollier diagram for water shows that the enthalpy of superheated water vapor at *low vapor pressures* is very closely given by the saturated vapor value corresponding to the given temperature. Hence, the enthalpy of the water vapor h_v in Eq. 12.46 can be taken as h_g at the mixture temperature. That is

$$h_v \approx h_g(T) \tag{12.47}$$

This approach is used in the remainder of the chapter. Enthalpy data for water vapor as an ideal gas from Table A-23 are *not used* for h_v because the enthalpy datum of the ideal gas tables differs from that of the steam tables. These different datums can lead to error when studying systems that contain both water vapor and a liquid or solid phase of water. The enthalpy of dry air, h_a, can be obtained from the appropriate ideal gas table, Table A-22, however, because air is a gas at all states under present consideration and is closely modeled as an ideal gas at these states.

When evaluating the entropy of moist air, the contribution of each component is determined at the mixture temperature and the partial pressure of the component in the mixture. Using Eq. 6.18, it can be shown that the specific entropy of the water vapor is given by $s_v(T, p_v) = s_g(T) - R \ln \phi$, where s_g is the specific entropy of saturated vapor at temperature T from the steam tables and ϕ is the relative humidity.

Using Computer Software

Property functions for moist air are listed under the **Properties** menu of *Interactive Thermodynamics: IT*. Functions are included for humidity ratio, relative humidity, specific enthalpy and entropy as well as other psychrometric properties introduced later. The methods used for evaluating these functions correspond to the methods discussed in this chapter, and the values returned by the computer software agree closely with those obtained by hand calculations with table data. The use of *IT* for psychrometric evaluations is illustrated in examples later in the chapter.

12.5.3 Modeling Moist Air in Equilibrium with Liquid Water

Thus far, our study of psychrometrics has been conducted as an application of the ideal gas mixture principles introduced in the first part of this chapter. However, many systems of interest are composed of a mixture of dry air and water vapor in contact with a liquid (or solid) water phase. To study these systems requires additional considerations.

Shown in Fig. 12.5 is a vessel containing liquid water, above which is a mixture of water vapor and dry air. If no interactions with the surroundings are allowed, liquid will evaporate until eventually the gas phase becomes saturated and the system attains an equilibrium state. For many engineering applications, systems consisting of moist air in *equilibrium* with a liquid water phase can be described simply and accurately with the following idealizations:

- ▶ The dry air and water vapor behave as independent ideal gases.
- ▶ The equilibrium between the liquid phase and the water vapor is not significantly disturbed by the presence of the air.
- ▶ The partial pressure of the water vapor equals the saturation pressure of water corresponding to the temperature of the mixture: $p_v = p_g(T)$.

Similar considerations apply for systems consisting of moist air in equilibrium with a solid water phase. The presence of the air actually alters the partial pressure of the vapor from the saturation pressure by a small amount whose magnitude is calculated in Sec. 14.6.

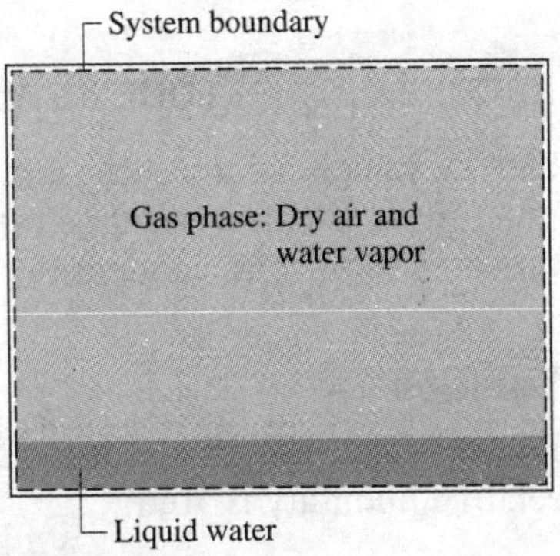

Fig. 12.5 System consisting of moist air in contact with liquid water.

12.5.4 Evaluating the Dew Point Temperature

A significant aspect of the behavior of moist air is that partial condensation of the water vapor can occur when the temperature is reduced. This type of phenomenon is commonly encountered in the condensation of vapor on windowpanes and on pipes carrying cold water. The formation of dew on grass is another familiar example.

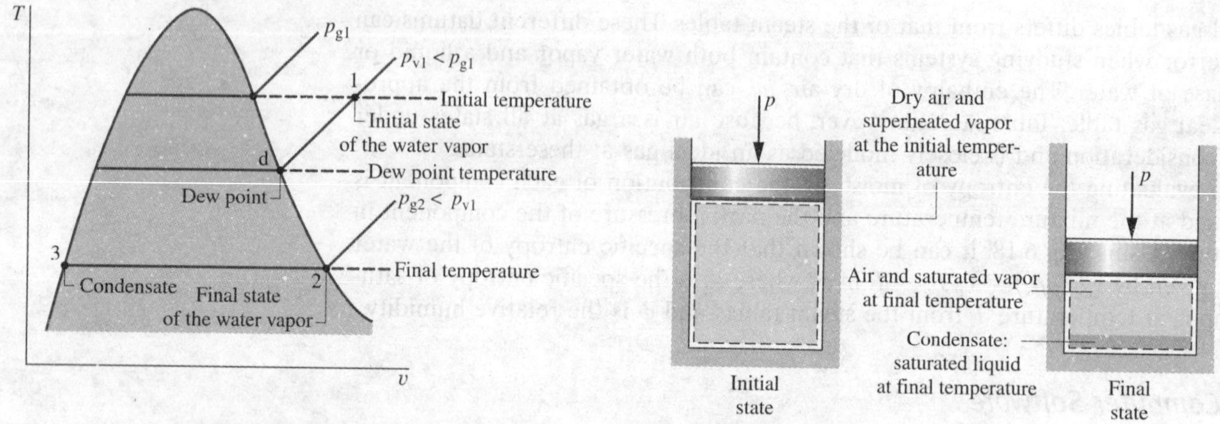

Fig. 12.6 States of water for moist air cooled at constant mixture pressure.

To study such condensation, consider a closed system consisting of a sample of moist air that is cooled at *constant* pressure, as shown in Fig. 12.6. The property diagram given on this figure locates states of the water vapor. Initially, the water vapor is superheated at state 1. In the first part of the cooling process, both the system pressure *and* the composition of the moist air would remain constant. Accordingly, since $p_v = y_v p$, the *partial pressure* of the water vapor would remain constant, and the water vapor would cool at constant p_v from state 1 to state d, called the *dew point*. The saturation temperature corresponding to p_v is called the ***dew point temperature.*** This temperature is labeled on Fig. 12.6.

dew point temperature

In the next part of the cooling process, the system would be cooled *below* the dew point temperature and some of the water vapor initially present would condense. At the final state, the system would consist of a gas phase of dry air and water vapor in equilibrium with a liquid water phase. The vapor that remains can be regarded as saturated at the final temperature, state 2 of Fig. 12.6, with a partial pressure equal to the saturation pressure p_{g2} corresponding to this temperature. The condensate would be a saturated liquid at the final temperature, state 3 of Fig. 12.6. Note that the partial pressure of the water vapor at the final state, p_{g2}, is less than the initial value, p_{v1}. Owing to condensation, the partial pressure decreases because the amount of water vapor present at the final state is less than at the initial state. Since the amount of dry air is unchanged, the mole fraction of water vapor in the moist air also decreases.

In the next two examples, we illustrate the use of psychrometric properties introduced thus far. The examples consider, respectively, cooling moist air at constant pressure and at constant volume.

Example 12.7 COOLING MOIST AIR AT CONSTANT PRESSURE

A 1 kg sample of moist air initially at 21°C, 1 bar, and 70% relative humidity is cooled to 5°C while keeping the pressure constant. Determine **(a)** the initial humidity ratio, **(b)** the dew point temperature, in °C, and **(c)** the amount of water vapor that condenses, in kg.

Solution

Known: A 1 kg sample of moist air is cooled at a constant mixture pressure of 1 bar from 21 to 5°C. The initial relative humidity is 70%.

Find: Determine the initial humidity ratio, the dew point temperature, in °C, and the amount of water vapor that condenses, in kg.

Schematic and Given Data:

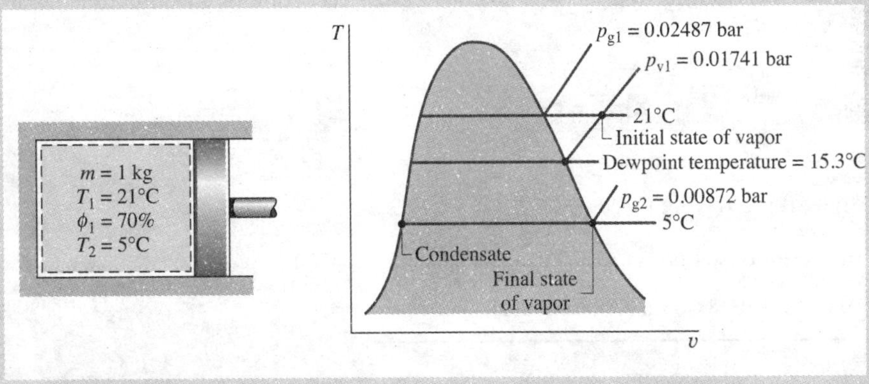

Fig. E12.7

Engineering Model:

1. The 1 kg sample of moist air is taken as the closed system. The system pressure remains constant at 1 bar.
2. The gas phase can be treated as an ideal gas mixture. Each mixture component acts as an ideal gas existing alone in the volume occupied by the gas phase at the mixture temperature.
3. When a liquid water phase is present, the water vapor exists as a saturated vapor at the system temperature. The liquid present is a saturated liquid at the system temperature.

Analysis:

(a) The initial humidity ratio can be evaluated from Eq. 12.43. This requires the partial pressure of the water vapor, p_{v1}, which can be found from the given relative humidity and p_g from Table A-2 at 21°C as follows

$$p_{v1} = \phi p_g = (0.7)(0.02487)\text{ bar} = 0.01741\text{ bar}$$

Inserting values in Eq. 12.43

$$\omega_1 = 0.622\left(\frac{0.01741}{1 - 0.01741}\right) = 0.011\frac{\text{kg (vapor)}}{\text{kg (dry air)}}$$

(b) The dew point temperature is the saturation temperature corresponding to the partial pressure, p_{v1}. Interpolation in Table A-2 gives $T = 15.3°C$. The dew point temperature is labeled on the accompanying property diagram.

(c) The amount of condensate, m_w, equals the difference between the initial amount of water vapor in the sample, m_{v1}, and the final amount of water vapor, m_{v2}. That is

$$m_w = m_{v1} - m_{v2}$$

To evaluate m_{v1}, note that the system initially consists of 1 kg of dry air and water vapor, so 1 kg $= m_a + m_{v1}$, where m_a is the mass of dry air present in the sample. Since $\omega_1 = m_{v1}/m_a$, $m_a = m_{v1}/\omega_1$. With this we get

$$1\text{ kg} = \frac{m_{v1}}{\omega_1} + m_{v1} = m_{v1}\left(\frac{1}{\omega_1} + 1\right)$$

Solving for m_{v1}

$$m_{v1} = \frac{1\text{ kg}}{(1/\omega_1) + 1}$$

Inserting the value of ω_1 determined in part (a)

$$m_{v1} = \frac{1\text{ kg}}{(1/0.011) + 1} = 0.0109\text{ kg (vapor)}$$

❶ The mass of dry air present is then $m_a = 1 - 0.0109 = 0.9891$ kg (dry air).

Next, let us evaluate m_{v2}. With assumption 3, the partial pressure of the water vapor remaining in the system at the final state is the saturation pressure corresponding to 5°C: $p_g = 0.00872$ bar. Accordingly, the humidity ratio after cooling is found from Eq. 12.43 as

$$\omega_2 = 0.622\left(\frac{0.00872}{1 - 0.00872}\right) = 0.0054 \frac{\text{kg (vapor)}}{\text{kg (dry air)}}$$

The mass of the water vapor present at the final state is then

$$m_{v2} = \omega_2 m_a = (0.0054)(0.9891) = 0.0053 \text{ kg (vapor)}$$

Finally, the amount of water vapor that condenses is

$$m_w = m_{v1} - m_{v2} = 0.0109 - 0.0053 = 0.0056 \text{ kg (condensate)}$$

❶ The amount of water vapor present in a typical moist air mixture is considerably less than the amount of dry air present.

Skills Developed: *Ability to...*

- ✓ apply psychrometric terminology and principles.
- ✓ demonstrate understanding of the dew point temperature and the formation of liquid condensate when pressure is constant.
- ✓ retrieve property data for water.

Quick Quiz

Determine the quality of the two-phase, liquid–vapor mixture and the relative humidity of the gas phase at the final state.

Ans. 47%, 100%.

Example 12.8 COOLING MOIST AIR AT CONSTANT VOLUME

An air–water vapor mixture is contained in a rigid, closed vessel with a volume of 35 m^3 at 1.5 bar, 120°C, and $\phi = 10\%$. The mixture is cooled at constant volume until its temperature is reduced to 22°C. Determine **(a)** the dew point temperature corresponding to the initial state, in °C, **(b)** the temperature at which condensation actually begins, in °C, and **(c)** the amount of water condensed, in kg.

Solution

Known: A rigid, closed tank with a volume of 35 m^3 containing moist air initially at 1.5 bar, 120°C, and $\phi = 10\%$ is cooled to 22°C.

Find: Determine the dew point temperature at the initial state, in °C, the temperature at which condensation actually begins, in °C, and the amount of water condensed, in kg.

Schematic and Given Data:

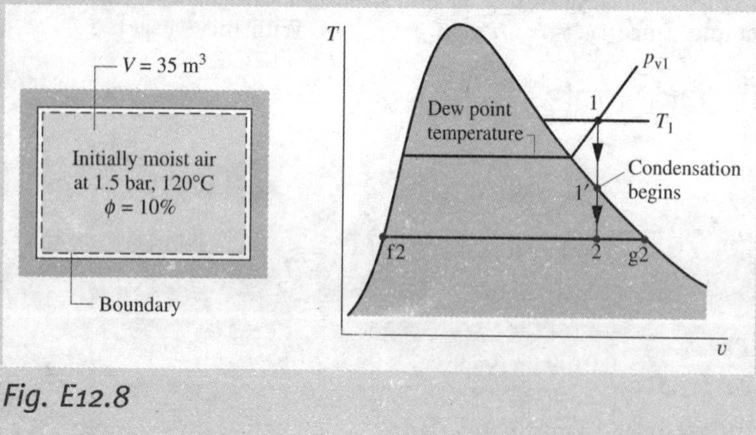

Fig. E12.8

Engineering Model:

1. The contents of the tank are taken as a closed system. The system volume remains constant.
2. The gas phase can be treated as an ideal gas mixture. Each mixture component acts as an ideal gas existing alone in the volume occupied by the gas phase at the mixture temperature.
3. When a liquid water phase is present, the water vapor exists as a saturated vapor at the system temperature. The liquid is a saturated liquid at the system temperature.

Analysis:

(a) The dew point temperature at the initial state is the saturation temperature corresponding to the partial pressure p_{v1}. With the given relative humidity and the saturation pressure at 120°C from Table A-2

$$p_{v1} = \phi_1 p_{g1} = (0.10)(1.985) = 0.1985 \text{ bar}$$

Interpolating in Table A-2 gives the dew point temperature as 60°C, which is the temperature condensation would begin *if* the moist air were cooled at *constant pressure.*

(b) Whether the water exists as a vapor only, or as liquid *and* vapor, it occupies the full volume, which remains constant. Accordingly, since the total mass of the water present is also constant, the water undergoes the constant specific volume process illustrated on the accompanying T–v diagram. In the process from state 1 to state 1′, the water exists as a vapor only. For the process from state 1′ to state 2, the water exists as a two-phase liquid–vapor mixture. Note that pressure does not remain constant during the cooling process from state 1 to state 2.

State 1′ on the T–v diagram denotes the state where the water vapor first becomes saturated. The saturation temperature at this state is denoted as T'. Cooling to a temperature less than T' would result in condensation of some of the water vapor present. Since state 1′ is a saturated vapor state, the temperature T' can be found by interpolating in Table A-2 with the specific volume of the water at this state. The specific volume of the vapor at state 1′ equals the specific volume of the vapor at state 1, which can be evaluated from the ideal gas equation

$$v_{v1} = \frac{(\overline{R}/M_v)T_1}{p_{v1}} = \left(\frac{8314}{18}\frac{\text{N}\cdot\text{m}}{\text{kg}\cdot\text{K}}\right)\left(\frac{393 \text{ K}}{0.1985 \times 10^5 \text{ N/m}^2}\right)$$

$$= 9.145 \frac{\text{m}^3}{\text{kg}}$$

❶ Interpolation in Table A-2 with $v_{v1} = v_g$ gives $T' = 56°\text{C}$. This is the temperature at which condensation begins.

(c) The amount of condensate equals the difference between the initial and final amounts of water vapor present. The mass of the water vapor present initially is

$$m_{v1} = \frac{V}{v_{v1}} = \frac{35 \text{ m}^3}{9.145 \text{ m}^3/\text{kg}} = 3.827 \text{ kg}$$

The mass of water vapor present finally can be determined from the *quality*. At the final state, the water forms a two-phase liquid–vapor mixture having a specific volume of 9.145 m³/kg. Using this specific volume value, the quality x_2 of the liquid–vapor mixture can be found as

$$x_2 = \frac{v_{v2} - v_{f2}}{v_{g2} - v_{f2}} = \frac{9.145 - 1.0022 \times 10^{-3}}{51.447 - 1.0022 \times 10^{-3}} = 0.178$$

where v_{f2} and v_{g2} are the saturated liquid and saturated vapor specific volumes at $T_2 = 22°\text{C}$, respectively.

Using the quality together with the known total amount of water present, 3.827 kg, the mass of the water vapor contained in the system at the final state is

$$m_{v2} = (0.178)(3.827) = 0.681 \text{ kg}$$

The mass of the condensate, m_{w2}, is then

$$m_{w2} = m_{v1} - m_{v2} = 3.827 - 0.681 = 3.146 \text{ kg}$$

❶ When a moist air mixture is cooled at constant mixture volume, the temperature at which condensation begins is not the dew point temperature corresponding to the initial state. In this case, condensation begins at 56°C, but the dew point temperature at the initial state, determined in part (a), is 60°C.

Skills Developed: *Ability to...*
- ✓ apply psychrometric terminology and principles.
- ✓ demonstrate understanding of the onset of condensation when cooling moist air at constant volume.
- ✓ retrieve property data for water.

Quick Quiz

Determine the specific humidity at the initial state and the amount of dry air present, in kg.

Ans. 0.0949, 40.33 kg.

No additional fundamental concepts are required for the study of closed systems involving mixtures of dry air and water vapor. Example 12.9, which builds on Example 12.8, brings out some special features of the use of conservation of mass and conservation of energy in analyzing this kind of system. Similar considerations can be used to study other closed systems involving moist air.

Example 12.9 EVALUATING HEAT TRANSFER FOR MOIST AIR COOLING AT CONSTANT VOLUME

An air–water vapor mixture is contained in a rigid, closed vessel with a volume of 35 m^3 at 1.5 bar, 120°C, and $\phi = 10\%$. The mixture is cooled until its temperature is reduced to 22°C. Determine the heat transfer during the process, in kJ.

Solution

Known: A rigid, closed tank with a volume of 35 m^3 containing moist air initially at 1.5 bar, 120°C, and $\phi = 10\%$ is cooled to 22°C.

Find: Determine the heat transfer for the process, in kJ.

Schematic and Given Data: See the figure for Example 12.8.

Engineering Model:

1. The contents of the tank are taken as a closed system. The system volume remains constant.
2. The gas phase can be treated as an ideal gas mixture. Each component acts as an ideal gas existing alone in the volume occupied by the gas phase at the mixture temperature.
3. When a liquid water phase is present, the water vapor exists as a saturated vapor and the liquid is a saturated liquid, each at the system temperature.
4. There is no work during the cooling process and no change in kinetic or potential energy.

Analysis: Reduction of the closed system energy balance using assumption 4 results in

$$\Delta U = Q - \cancelto{0}{W}$$

or

$$Q = U_2 - U_1$$

where

$$U_1 = m_a u_{a1} + m_{v1} u_{v1} = m_a u_{a1} + m_{v1} u_{g1}$$

and

$$U_2 = m_a u_{a2} + m_{v2} u_{v2} + m_{w2} u_{w2} = m_a u_{a2} + m_{v2} u_{g2} + m_{w2} u_{f2}$$

In these equations, the subscripts a, v, and w denote, respectively, dry air, water vapor, and liquid water. The specific internal energy of the water vapor at the initial state can be approximated as the saturated vapor value at T_1. At the final state, the water vapor is assumed to exist as a saturated vapor, so its specific internal energy is u_g at T_2. The liquid water at the final state is saturated, so its specific internal energy is u_f at T_2.

Collecting the last three equations

❶ $$Q = \underline{m_a(u_{a2} - u_{a1})} + \underline{m_{v2} u_{g2} + m_{w2} u_{f2} - m_{v1} u_{g1}}$$

The mass of dry air, m_a, can be found using the ideal gas equation of state together with the partial pressure of the dry air at the initial state obtained using $p_{v1} = 0.1985$ bar from the solution to Example 12.8

$$m_a = \frac{p_{a1}V}{(\bar{R}/M_a)T_1} = \frac{[(1.5 - 0.1985) \times 10^5 \text{ N/m}^2](35 \text{ m}^3)}{(8314/28.97 \text{ N} \cdot \text{m/kg} \cdot \text{K})(393 \text{ K})}$$
$$= 40.389 \text{ kg}$$

Then, evaluating internal energies of dry air and water from Tables A-22 and A-2, respectively

$$Q = 40.389(210.49 - 281.1) + 0.681(2405.7) + 3.146(92.32) - 3.827(2529.3)$$
$$= -2851.87 + 1638.28 + 290.44 - 9679.63 = -10{,}603 \text{ kJ}$$

The values for m_{v1}, m_{v2}, and m_{w2} are from the solution to Example 12.8.

❶ The first underlined term in this equation for Q is evaluated with specific internal energies from the ideal gas table for air, Table A-22. Steam table data are used to evaluate the second underlined term. The different datums for internal energy underlying these tables cancel because each of these two terms involves internal energy *differences*. Since the specific heat c_{va} for dry air varies only slightly over the interval from 120 to 22°C (Table A-20), the specific internal energy change of the dry air could be evaluated alternatively using a constant c_{va} value.

Skills Developed: *Ability to...*
- ✓ apply psychrometric terminology and principles.
- ✓ apply the energy balance to the cooling of moist air at constant volume.
- ✓ retrieve property data for water.

Quick Quiz

Calculate the change in internal energy of the *dry air,* in kJ, assuming a constant specific heat c_{va} interpolated from Table A-20 at the average of the initial and final temperatures.
Ans. −2854 kJ.

12.5.5 Evaluating Humidity Ratio Using the Adiabatic-Saturation Temperature

The humidity ratio ω of an air–water vapor mixture can be determined, in principle, knowing the values of three mixture properties: the pressure p, the temperature T, and the ***adiabatic-saturation temperature*** T_{as} introduced in this section. The relationship among these quantities is obtained by applying conservation of mass and conservation of energy to an *adiabatic saturator* (see box).

adiabatic-saturation temperature

Equations 12.48 and 12.49 give the humidity ratio ω in terms of the adiabatic-saturation temperature and other quantities:

$$\omega = \frac{h_a(T_{as}) - h_a(T) + \omega'[h_g(T_{as}) - h_f(T_{as})]}{h_g(T) - h_f(T_{as})} \tag{12.48}$$

where h_f and h_g denote the enthalpies of saturated liquid water and saturated water vapor, respectively, obtained from the steam tables at the indicated temperatures. The enthalpies of the dry air h_a can be obtained from the ideal gas table for air. Alternatively, $h_a(T_{as}) - h_a(T) = c_{pa}(T_{as} - T)$, where c_{pa} is an appropriate constant value for the specific heat of dry air. The humidity ratio ω' appearing in Eq. 12.48 is

$$\omega' = 0.622\frac{p_g(T_{as})}{p - p_g(T_{as})} \tag{12.49}$$

where $p_g(T_{as})$ is the saturation pressure at the adiabatic-saturation temperature and p is the mixture pressure.

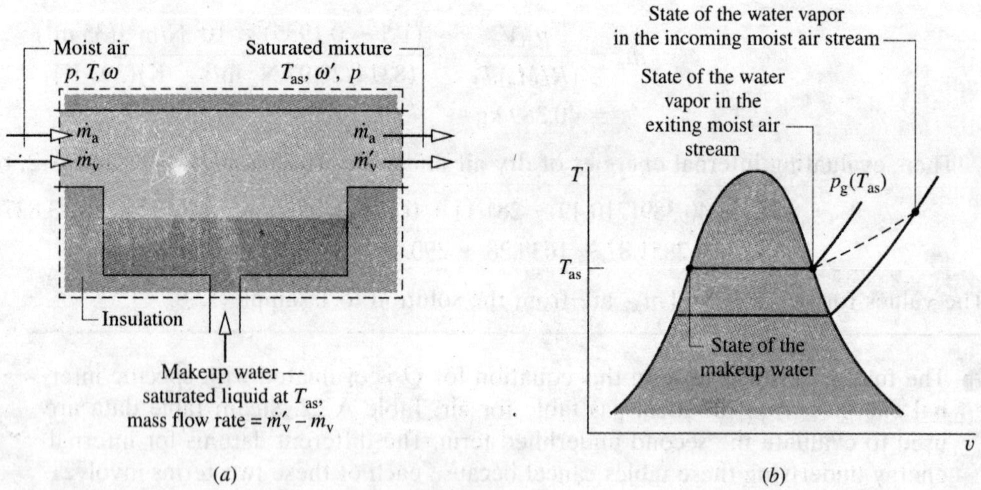

Fig. 12.7 Adiabatic saturator. (*a*) Schematic. (*b*) Process representation.

Modeling an Adiabatic Saturator

Figure 12.7 shows the schematic and process representations of an adiabatic saturator, which is a two-inlet, single-exit device through which moist air passes. The device is assumed to operate at steady state and without significant heat transfer with its surroundings. An air–water vapor mixture of *unknown* humidity ratio ω enters the adiabatic saturator at a known pressure p and temperature T. As the mixture passes through the device, it comes into contact with a pool of water. If the entering mixture is not saturated ($\phi < 100\%$), some of the water would evaporate. The energy required to evaporate the water would come from the moist air, so the mixture temperature would decrease as the air passes through the duct. For a sufficiently long duct, the mixture would be saturated as it exits ($\phi = 100\%$). Since a saturated mixture would be achieved without heat transfer with the surroundings, the temperature of the exiting mixture is the *adiabatic-saturation temperature*. As indicated on Fig. 12.7, a steady flow of makeup water at temperature T_{as} is added at the same rate at which water is evaporated. The pressure of the mixture is assumed to remain constant as it passes through the device. Equation 12.48 giving the humidity ratio ω of the entering moist air in terms of p, T, and T_{as} can be obtained by applying conservation of mass and conservation of energy to the adiabatic saturator, as follows.

At steady state, the mass flow rate of the dry air entering the device, $\dot{m}_a$, must equal the mass flow rate of the dry air exiting. The mass flow rate of the makeup water is the difference between the exiting and entering vapor flow rates denoted by $\dot{m}_v$ and $\dot{m}'_v$, respectively. These flow rates are labeled on Fig. 12.7*a*. At steady state, the energy rate balance reduces to

$$(\dot{m}_a h_a + \dot{m}_v h_v)_{\substack{\text{moist air}\\\text{entering}}} + [(\dot{m}'_v - \dot{m}_v)h_w]_{\substack{\text{makeup}\\\text{water}}} = (\dot{m}_a h_a + \dot{m}'_v h_v)_{\substack{\text{moist air}\\\text{exiting}}}$$

Several assumptions underlie this expression: Each of the two moist air streams is modeled as an ideal gas mixture of dry air and water vapor. Heat transfer with the surroundings is assumed to be negligible. There is no work $\dot{W}_{cv}$, and changes in kinetic and potential energy are ignored.

Dividing by the mass flow rate of the dry air, $\dot{m}_a$, the energy rate balance can be written on the basis of a unit mass of dry air passing through the device as

$$(h_a + \omega h_g)_{\substack{\text{moist air}\\\text{entering}}} + [(\omega' - \omega)h_f]_{\substack{\text{makeup}\\\text{water}}} = (h_a + \omega' h_g)_{\substack{\text{moist air}\\\text{exiting}}} \qquad (12.50)$$

where $\omega = \dot{m}_v/\dot{m}_a$ and $\omega' = \dot{m}'_v/\dot{m}_a$.

For the exiting saturated mixture, the partial pressure of the water vapor is the saturation pressure corresponding to the adiabatic-saturation temperature, $p_g(T_{as})$. Accordingly, the humidity ratio ω' can be evaluated knowing T_{as} and the mixture pressure p, as indicated by Eq. 12.49. In writing Eq. 12.50, the specific enthalpy of the entering water vapor has been evaluated as that of saturated water vapor at the temperature of the incoming mixture, in accordance with Eq. 12.47. Since the exiting mixture is saturated, the enthalpy of the water vapor at the exit is given by the saturated vapor value at T_{as}. The enthalpy of the makeup water is evaluated as that of saturated liquid at T_{as}.

When Eq. 12.50 is solved for ω, Eq. 12.48 results. The details of the solution are left as an exercise. Although derived with reference to an adiabatic saturator, the relationship provided by Eq. 12.48 applies generally to moist air mixtures and is not restricted to this type of system or even to control volumes. The relationship allows the humidity ratio ω to be determined for *any* moist air mixture for which the pressure p, temperature T, and adiabatic-saturation temperature T_{as} are known.

12.6 Psychrometers: Measuring the Wet-Bulb and Dry-Bulb Temperatures

For moist air mixtures in the normal pressure and temperature ranges of psychrometrics, the readily measured wet-bulb temperature is an important parameter.

The ***wet-bulb temperature*** is read from a wet-bulb thermometer, which is an ordinary liquid-in-glass thermometer whose bulb is enclosed by a wick moistened with water. The term ***dry-bulb temperature*** refers simply to the temperature that would be measured by a thermometer placed in the mixture. Often a wet-bulb thermometer is mounted together with a dry-bulb thermometer to form an instrument called a ***psychrometer.***

wet-bulb temperature

dry-bulb temperature

psychrometer

The psychrometer of Fig. 12.8*a* is whirled in the air whose wet- and dry-bulb temperatures are to be determined. This induces air flow over the two thermometers. For the psychrometer of Fig. 12.8*b*, the air flow is induced by a battery-operated fan. In

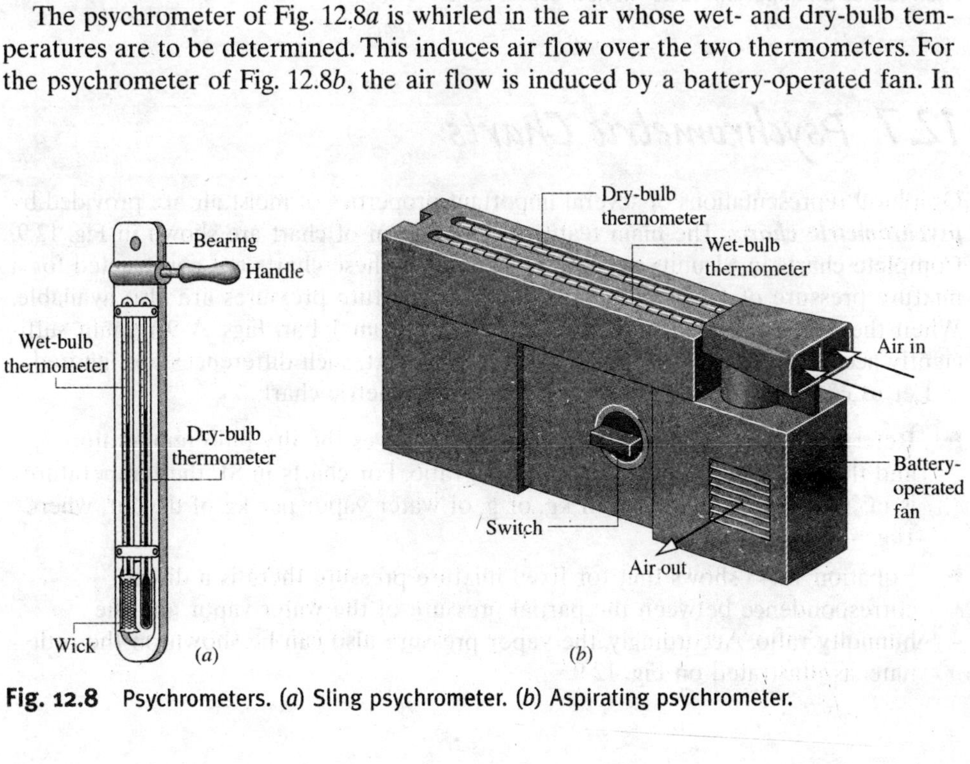

Fig. 12.8 Psychrometers. (*a*) Sling psychrometer. (*b*) Aspirating psychrometer.

each type of psychrometer, if the surrounding air is not saturated, water in the wick of the wet-bulb thermometer evaporates and the temperature of the remaining water falls below the dry-bulb temperature. Eventually a steady-state condition is attained by the wet-bulb thermometer. The wet- and dry-bulb temperatures are then read from the respective thermometers. The wet-bulb temperature depends on the rates of heat and mass transfer between the moistened wick and the air. Since these depend in turn on the geometry of the thermometer, air velocity, supply water temperature, and other factors, the wet-bulb temperature is not a mixture property.

For moist air mixtures in the normal pressure and temperature ranges of psychrometric applications, the adiabatic saturation temperature introduced in Sec. 12.5.5 is closely approximated by the wet-bulb temperature. Accordingly, the humidity ratio for any such mixture can be calculated by using the wet-bulb temperature in Eqs. 12.48 and 12.49 in place of the adiabatic-saturation temperature. Close agreement between the adiabatic-saturation and wet-bulb temperatures is not generally found for moist air departing from normal psychrometric conditions.

Bio... Connections

The *National Weather Service* is finding better ways to help measure our misery during cold snaps so we can avoid weather dangers. The wind chill index, for many years based on a single 1945 study, was recently upgraded using new physiological data and computer modeling to better reflect the perils of cold winds and freezing temperatures.

The new wind chill index is a standardized "temperature" that accounts for both the actual air temperature and the wind speed. The formula on which it is based uses measurements of skin tissue thermal resistance and computer models of the wind patterns over the human face, together with principles of heat transfer. Using the new index, an air temperature of −15°C and a wind speed of 40 km per hour correspond to a wind chill temperature of −40°C . The old index assigned a wind chill of only −29°C to the same conditions. With the new information, people are better armed to avoid exposure that can lead to such serious medical problems as frostbite.

The improved measure was developed by universities, international scientific societies, and government in an effort that led to the new standard being adopted in the United States. Further upgrades are in the works to include the amount of cloud cover in the formula, since solar radiation is also an important factor in how cold it feels.

12.7 Psychrometric Charts

psychrometric chart

Graphical representations of several important properties of moist air are provided by ***psychrometric charts.*** The main features of one form of chart are shown in Fig. 12.9. Complete charts in SI units are given in Fig. A-9. These charts are constructed for a mixture pressure of 1 bar, but charts for other mixture pressures are also available. When the mixture pressure differs only slightly from 1 bar, Figs. A-9 remain sufficiently accurate for engineering analyses. In this text, such differences are ignored.

Let us consider several features of the psychrometric chart:

- Referring to Fig. 12.9, note that the abscissa gives the dry-bulb temperature and the ordinate provides the humidity ratio. For charts in SI, the temperature is in °C and ω is expressed in kg, or g, of water vapor per kg of dry air, where 1kg = 15,432 grains.
- Equation 12.43 shows that for fixed mixture pressure there is a direct correspondence between the partial pressure of the water vapor and the humidity ratio. Accordingly, the vapor pressure also can be shown on the ordinate, as illustrated on Fig. 12.9.

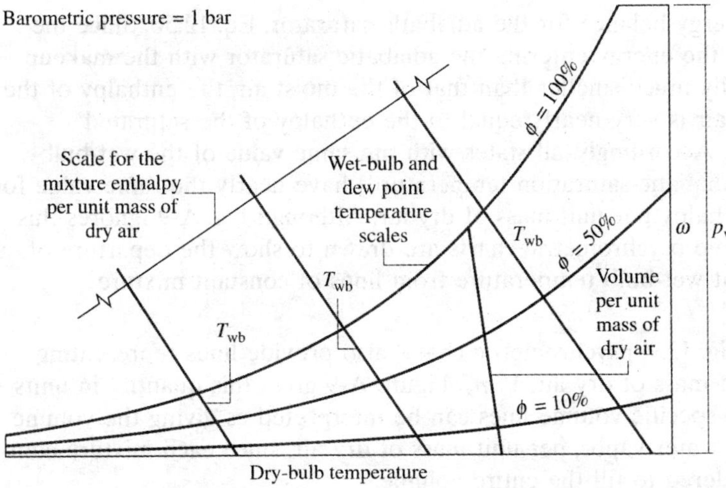

Fig. 12.9 Psychrometric chart.

- Curves of constant relative humidity are shown on psychrometric charts. On Fig. 12.9, curves labeled ϕ = 100, 50, and 10% are indicated. Since the dew point is the state where the mixture becomes saturated when cooled at constant vapor pressure, the dew point temperature corresponding to a given moist air state can be determined by following a line of constant ω (constant p_v) to the saturation line, ϕ = 100%. The dew point temperature and dry-bulb temperature are identical for states on the saturation curve.
- Psychrometric charts also give values of the mixture enthalpy per unit mass of dry air in the mixture: $h_a + \omega h_v$. In Fig. A-9, the mixture enthalpy has units of kJ per kg of dry air. The numerical values provided on these charts are determined relative to the following *special* reference states and reference values. In Fig. A-9, the enthalpy of the dry air h_a is determined relative to a zero value at 0°C, and not 0 K as in Table A-22. Accordingly, in place of Eq. 3.49 used to develop the enthalpy data of Tables A-22, the following expression is employed to evaluate the enthalpy of the dry air for use on the psychrometric chart:

$$h_a = \int_{273.15\text{ K}}^{T} c_{pa}\, dT = c_{pa} T(°\text{C}) \tag{12.51}$$

where c_{pa} is a constant value for the specific heat c_p of dry air and T(°C) denotes the temperature in °C. In the temperature ranges of Fig. A-9, c_{pa} can be taken as 1.005 kJ/kg · K. On Fig. A-9 the enthalpy of the water vapor h_v is evaluated as h_g at the dry-bulb temperature of the mixture from Table A-2, as appropriate.

- Another important parameter on psychrometer charts is the wet-bulb temperature. As illustrated by Fig. A-9, constant T_{wb} lines run from the upper left to the lower right of the chart. The relationship between the wet-bulb temperature and other chart quantities is provided by Eq. 12.48. The wet-bulb temperature can be used in this equation in place of the adiabatic-saturation temperature for the states of moist air located on Fig. A-9.
- Lines of constant wet-bulb temperature are approximately lines of constant mixture enthalpy per unit mass of dry air. This feature can be brought out by

study of the energy balance for the adiabatic saturator, Eq. 12.50. Since the contribution of the energy entering the adiabatic saturator with the makeup water is normally much smaller than that of the moist air, the enthalpy of the entering moist air is very nearly equal to the enthalpy of the saturated mixture exiting. Accordingly, all states with the same value of the wet-bulb temperature (adiabatic-saturation temperature) have nearly the same value for the mixture enthalpy per unit mass of dry air. Although Fig. A-9 ignores this slight effect, some psychrometric charts are drawn to show the departure of lines of constant wet-bulb temperature from lines of constant mixture enthalpy.

- As shown on Fig. 12.9, psychrometric charts also provide lines representing volume per unit mass of dry air, V/m_a. Figure A-9 gives this quantity in units of m^3/kg. These specific volume lines can be interpreted as giving the volume of dry air or of water vapor, per unit mass of dry air, since each mixture component is considered to fill the entire volume.

The psychrometric chart is easily used. ➡ **FOR EXAMPLE...** a psychrometer indicates that in a classroom the dry-bulb temperature is 20°C and the wet-bulb temperature is 15°C. Locating the mixture state on Fig. A-9 corresponding to the intersection of these temperatures, we read $\omega = 0.0092$ kg(vapor)/kg(dry air) and $\phi = 63\%$. ⬅

12.8 Analyzing Air-Conditioning Processes

The purpose of the present section is to study typical air-conditioning processes using the psychrometric principles developed in this chapter. Specific illustrations are provided in the form of solved examples involving control volumes at steady state. In each case, the methodology introduced in Sec. 12.8.1 is employed to arrive at the solution. To reinforce the psychrometric principles developed in the present chapter, the required psychrometric parameters are determined in most cases using tabular data provided in the appendix. It is left as an exercise to check these values by means of a psychrometric chart.

12.8.1 Applying Mass and Energy Balances to Air-Conditioning Systems

The object of this section is to illustrate the use of the conservation of mass and conservation of energy principles in analyzing systems involving mixtures of dry air and water vapor in which a condensed water phase may be present. The same basic solution approach that has been used in thermodynamic analyses considered thus far is applicable. The only new aspect is the use of the special vocabulary and parameters of psychrometrics.

Systems that accomplish air-conditioning processes such as heating, cooling, humidification, or dehumidification are normally analyzed on a control volume basis. To consider a typical analysis, refer to Fig. 12.10, which shows a two-inlet, single-exit control volume at steady state. A moist air stream enters at 1, a moist air stream exits at 2, and a water-only stream enters at 3. The water-only stream may be a liquid or a vapor. Heat transfer at the rate $\dot{Q}_{cv}$ can occur between the control volume and its surroundings. Depending on the application, the value of $\dot{Q}_{cv}$ might be positive, negative, or zero.

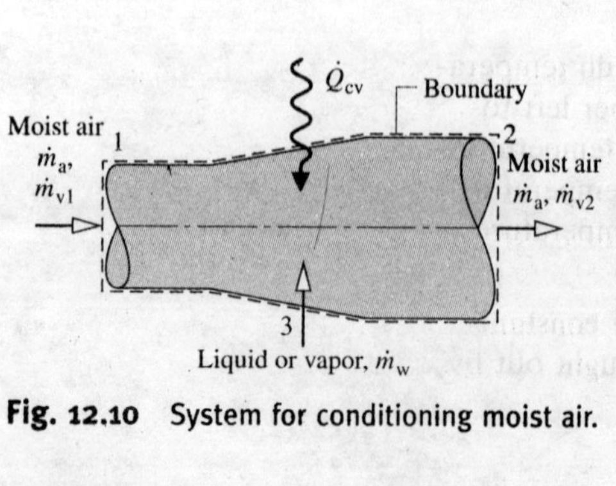

Fig. 12.10 System for conditioning moist air.

Mass Balance

At steady state, the amounts of dry air and water vapor contained within the control volume cannot vary. Thus, for each component individually it is necessary for the total incoming and outgoing mass flow rates to be equal. That is

$$\dot{m}_{a1} = \dot{m}_{a2} \quad \text{(dry air)}$$
$$\dot{m}_{v1} + \dot{m}_{w} = \dot{m}_{v2} \quad \text{(water)}$$

For simplicity, the constant mass flow rate of the dry air is denoted by $\dot{m}_a$. The mass flow rates of the water vapor can be expressed conveniently in terms of humidity ratios as $\dot{m}_{v1} = \omega_1 \dot{m}_a$ and $\dot{m}_{v2} = \omega_2 \dot{m}_a$. With these expressions, the mass balance for water becomes

$$\dot{m}_w = \dot{m}_a(\omega_2 - \omega_1) \quad \text{(water)} \tag{12.52}$$

When water is added at 3, ω_2 is greater than ω_1.

Energy Balance

Assuming $\dot{W}_{cv} = 0$ and ignoring all kinetic and potential energy effects, the energy rate balance reduces at steady state to

$$0 = \dot{Q}_{cv} + (\dot{m}_a h_{a1} + \dot{m}_{v1} h_{v1}) + \dot{m}_w h_w - (\dot{m}_a h_{a2} + \dot{m}_{v2} h_{v2}) \tag{12.53}$$

In this equation, the entering and exiting moist air streams are regarded as ideal gas mixtures of dry air and water vapor.

Equation 12.53 can be cast into a form that is particularly convenient for the analysis of air-conditioning systems. First, with Eq. 12.47 the enthalpies of the entering and exiting water vapor can be evaluated as the saturated vapor enthalpies corresponding to the temperatures T_1 and T_2, respectively, giving

$$0 = \dot{Q}_{cv} + (\dot{m}_a h_{a1} + \dot{m}_{v1} h_{g1}) + \dot{m}_w h_w - (\dot{m}_a h_{a2} + \dot{m}_{v2} h_{g2})$$

Then, with $\dot{m}_{v1} = \omega_1 \dot{m}_a$ and $\dot{m}_{v2} = \omega_2 \dot{m}_a$, the equation can be expressed as

$$0 = \dot{Q}_{cv} + \dot{m}_a(h_{a1} + \omega_1 h_{g1}) + \dot{m}_w h_w - \dot{m}_a(h_{a2} + \omega_2 h_{g2}) \tag{12.54}$$

Finally, introducing Eq. 12.52, the energy rate balance becomes

$$0 = \dot{Q}_{cv} + \dot{m}_a[\underline{(h_{a1} - h_{a2})} + \underline{\omega_1 h_{g1} + (\omega_2 - \omega_1)h_w - \omega_2 h_{g2}}] \tag{12.55}$$

The first underlined term of Eq. 12.55 can be evaluated from Tables A-22 giving the ideal gas properties of air. Alternatively, since relatively small temperature differences are normally encountered in the class of systems under present consideration, this term can be evaluated as $h_{a1} - h_{a2} = c_{pa}(T_1 - T_2)$, where c_{pa} is a constant value for the specific heat of dry air. The second underlined term of Eq. 12.55 can be evaluated using steam table data together with known values for ω_1 and ω_2.

Modeling Summary

As suggested by the foregoing development, several simplifying assumptions are commonly used when analyzing the air-conditioning systems under present consideration. In addition to the assumption of steady-state operation, one-dimensional flow is assumed to apply at locations where matter crosses the boundary of the control volume, and the effects of kinetic and potential energy at these locations are neglected. In most cases there is no work, except for flow work where matter crosses the boundary. Further simplifications also may be required in particular cases.

12.8.2 Conditioning Moist Air at Constant Composition

Building air-conditioning systems frequently heat or cool a moist air stream with no change in the amount of water vapor present. In such cases the humidity ratio ω remains constant, while relative humidity and other moist air parameters vary. Example 12.10 gives an elementary illustration using the methodology of Sec. 12.8.1.

Example 12.10 HEATING MOIST AIR IN A DUCT

Moist air enters a duct at 10°C, 80% relative humidity, and a volumetric flow rate of 150 m³/min. The mixture is heated as it flows through the duct and exits at 30°C. No moisture is added or removed, and the mixture pressure remains approximately constant at 1 bar. For steady-state operation, determine **(a)** the rate of heat transfer, in kJ/min, and **(b)** the relative humidity at the exit. Changes in kinetic and potential energy can be ignored.

Solution

Known: Moist air that enters a duct at 10°C and ϕ = 80% with a volumetric flow rate of 150 m³/min is heated at constant pressure and exits at 30°C. No moisture is added or removed.

Find: Determine the rate of heat transfer, in kJ/min, and the relative humidity at the exit.

Schematic and Given Data:

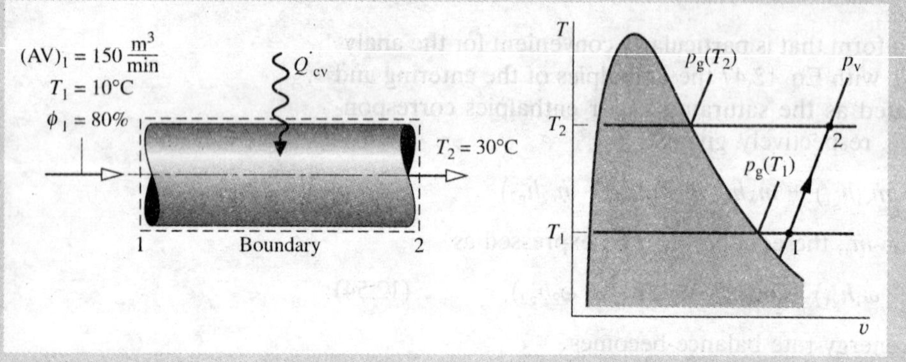

Fig. E12.10a

Engineering Model:

1. The control volume shown in the accompanying figure operates at steady state.
2. The changes in kinetic and potential energy between inlet and exit can be ignored and $\dot{W}_{cv} = 0$.
3. The entering and exiting moist air streams can be regarded as ideal gas mixtures.

Analysis:

(a) The heat transfer rate $\dot{Q}_{cv}$ can be determined from the mass and energy rate balances. At steady state, the amounts of dry air and water vapor contained within the control volume cannot vary. Thus, for each component individually it is necessary for the incoming and outgoing mass flow rates to be equal. That is

$$\dot{m}_{a1} = \dot{m}_{a2} \quad \text{(dry air)}$$
$$\dot{m}_{v1} = \dot{m}_{v2} \quad \text{(water)}$$

For simplicity, the constant mass flow rates of the dry air and water vapor are denoted, respectively, by $\dot{m}_a$ and $\dot{m}_v$. From these considerations, it can be concluded that the humidity ratio is the same at the inlet and exit: $\omega_1 = \omega_2$.

The steady-state form of the energy rate balance reduces with assumption 2 to

$$0 = \dot{Q}_{cv} - \overset{0}{\dot{W}_{cv}} + (\dot{m}_a h_{a1} + \dot{m}_v h_{v1}) - (\dot{m}_a h_{a2} + \dot{m}_v h_{v2})$$

In writing this equation, the incoming and outgoing moist air streams are regarded as ideal gas mixtures of dry air and water vapor.

Solving for $\dot{Q}_{cv}$

$$\dot{Q}_{cv} = \dot{m}_a(h_{a2} - h_{a1}) + \dot{m}_v(h_{v2} - h_{v1})$$

Noting that $\dot{m}_v = \omega\dot{m}_a$, where ω is the humidity ratio, the expression for $\dot{Q}_{cv}$ can be written in the form

❶
$$\dot{Q}_{cv} = \dot{m}_a[(h_{a2} - h_{a1}) + \omega(h_{v2} - h_{v1})] \tag{1}$$

To evaluate $\dot{Q}_{cv}$ from this expression requires the specific enthalpies of the dry air and water vapor at the inlet and exit, the mass flow rate of the dry air, and the humidity ratio.

The specific enthalpies of the dry air are obtained from Table A-22 at the inlet and exit temperatures T_1 and T_2, respectively: h_{a1} = 283.1 kJ/kg, h_{a2} = 303.2 kJ/kg. The specific enthalpies of the water vapor are found using $h_v \approx h_g$ and data from Table A-2 at T_1 and T_2, respectively: h_{g1} = 2519.8 kJ/kg, h_{g2} = 2556.3 kJ/kg.

The mass flow rate of the dry air can be determined from the volumetric flow rate at the inlet $(AV)_1$

$$\dot{m}_a = \frac{(AV)_1}{v_{a1}}$$

In this equation, v_{a1} is the specific volume of the dry air evaluated at T_1 and the partial pressure of the dry air p_{a1}. Using the ideal gas equation of state

$$v_{a1} = \frac{(\bar{R}/M)T_1}{p_{a1}}$$

The partial pressure p_{a1} can be determined from the mixture pressure p and the partial pressure of the water vapor p_{v1}: $p_{a1} = p - p_{v1}$. To find p_{v1}, use the given inlet relative humidity and the saturation pressure at 10°C from Table A-2

$$p_{v1} = \phi_1 p_{g1} = (0.8)(0.01228 \text{ bar}) = 0.0098 \text{ bar}$$

Since the mixture pressure is 1 bar, it follows that p_{a1} = 0.9902 bar. The specific volume of the dry air is then

$$v_{a1} = \frac{\left(\dfrac{8314 \text{ N}\cdot\text{m}}{28.97 \text{ kg}\cdot\text{K}}\right)(283 \text{ K})}{(0.9902 \times 10^5 \text{ N/m}^2)} = 0.82 \text{ m}^3/\text{kg}$$

Using this value, the mass flow rate of the dry air is

$$\dot{m}_a = \frac{150 \text{ m}^3/\text{min}}{0.82 \text{ m}^3/\text{kg}} = 182.9 \text{ kg/min}$$

The humidity ratio v can be found from

$$\omega = 0.622\left(\frac{p_{v1}}{p - p_{v1}}\right) = 0.622\left(\frac{0.0098}{1 - 0.0098}\right)$$
$$= 0.00616 \frac{\text{kg (vapor)}}{\text{kg (dry air)}}$$

Finally, substituting values into Eq. (1) we get

$$\dot{Q}_{cv} = 182.9[(303.2 - 283.1) + (0.00616)(2556.3 - 2519.8)]$$
$$= 3717 \text{ kJ/min}$$

(b) The states of the water vapor at the duct inlet and exit are located on the accompanying T–v diagram. Both the composition of the moist air and the mixture pressure remain constant, so the partial pressure of the water vapor at the exit equals the partial pressure of the water vapor at the inlet: $p_{v2} = p_{v1}$ = 0.0098 bar. The relative humidity at the exit is then

❷
$$\phi_2 = \frac{p_{v2}}{p_{g2}} = \frac{0.0098}{0.04246} = 0.231\,(23.1\%)$$

where p_{g2} is from Table A-2 at 30°C.

Alternative Psychrometric Chart Solution: Let us consider an alternative solution using the psychrometric chart. As shown on the sketch of the psychrometric chart, Fig. E12.10*b*, the state of the moist air at the inlet is defined by ϕ_1 = 80% and a dry-bulb temperature of 10°C. From the solution to part (a), we know that the humidity ratio has the same value at the exit as at the inlet. Accordingly, the state of the moist air at the exit is fixed by $\omega_2 = \omega_1$

❸ and a dry-bulb temperature of 30°C. By inspection of Fig. A-9, the relative humidity at the duct exit is about 23%, and thus in agreement with the result of part (b).

The rate of heat transfer can be evaluated from the psychrometric chart using the following expression obtained by rearranging Eq. (1) of part (a) to read

$$\dot{Q}_{cv} = \dot{m}_a[(h_a + \omega h_v)_2 - (h_a + \omega h_v)_1] \tag{2}$$

To evaluate $\dot{Q}_{cv}$ from this expression requires values for the mixture enthalpy per unit mass of dry air $(h_a + \omega h_v)$ at the inlet and exit. These can be determined by inspection of the psychrometric chart, Fig. A-9, as $(h_a + \omega h_v)_1 = 25.7$ kJ/kg (dry air), $(h_a + \omega h_v)_2 = 45.9$ kJ/kg (dry air).

Using the specific volume value v_{a1} from the chart at the inlet state together with the given volumetric flow rate at the inlet, the mass flow rate of the dry air is found as

$$\dot{m}_a = \frac{150 \text{ m}^3/\text{min}}{0.81 \text{ m}^3/\text{kg(dry air)}} = 185 \frac{\text{kg(dry air)}}{\text{min}}$$

Substituting values into the energy rate balance, Eq. (2), we get

$$\dot{Q}_{cv} = 185 \frac{\text{kg(dry air)}}{\text{min}}(45.9 - 25.7)\frac{\text{kJ}}{\text{kg(dry air)}}$$

$$= 3737 \frac{\text{kJ}}{\text{min}}$$

which agrees closely with the result obtained in part (a), as expected.

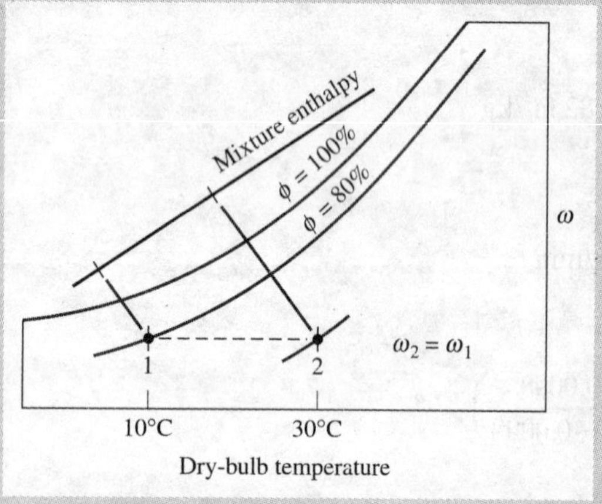

Fig. E12.10b

❶ The first underlined term in this equation for $\dot{Q}_{cv}$ is evaluated with specific enthalpies from the ideal gas table for air, Table A-22. Steam table data are used to evaluate the second underlined term. Note that the different datums for enthalpy underlying these tables cancel because each of the two terms involves enthalpy *differences* only. Since the specific heat c_{pa} for dry air varies only slightly over the interval from 10 to 30°C (Table A-20), the specific enthalpy change of the dry air could be evaluated alternatively with $c_{pa} = 1.005$ kJ/kg · K.

❷ No water is added or removed as the moist air passes through the duct at constant pressure; accordingly, the humidity ratio ω and the partial pressures p_v and p_a remain constant. However, because the saturation pressure increases as the temperature increases from inlet to exit, the *relative humidity* decreases: $\phi_2 < \phi_1$.

❸ The mixture pressure, 1 bar, differs slightly from the pressure used to construct the psychrometric chart, 1 atm. This difference is ignored.

Skills Developed: *Ability to...*
- ✓ apply psychrometric terminology and principles.
- ✓ apply mass and energy balances for heating at constant composition in a control volume at steady state.
- ✓ retrieve necessary property data.

Quick Quiz

Using the psychrometric chart, what is the dew point temperature, in °C, for the moist air entering? At the exit? **Ans.** ≈ 7°C, same.

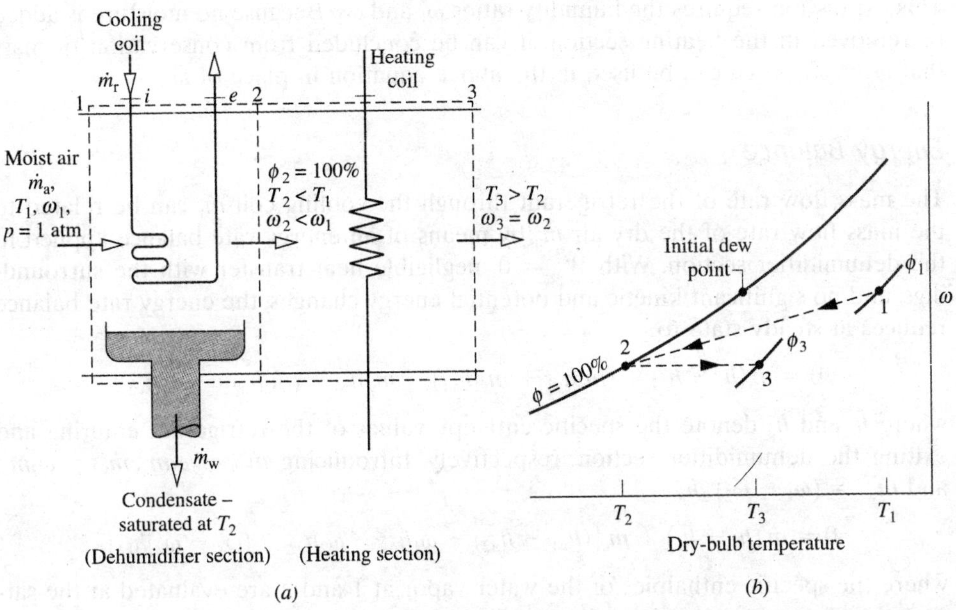

Fig. 12.11 Dehumidification. *(a)* Equipment schematic. *(b)* Psychrometric chart representation.

12.8.3 Dehumidification

When a moist air stream is cooled at constant mixture pressure to a temperature below its dew point temperature, some condensation of the water vapor initially present would occur. Figure 12.11 shows the schematic of a dehumidifier using this principle. Moist air enters at state 1 and flows across a cooling coil through which a refrigerant or chilled water circulates. Some of the water vapor initially present in the moist air condenses, and a saturated moist air mixture exits the dehumidifier section at state 2. Although water would condense at various temperatures, the condensed water is assumed to be cooled to T_2 before it exits the dehumidifier. Since the moist air leaving the humidifier is saturated at a temperature lower than the temperature of the moist air entering, the moist air stream might be unsuitable for direct use in occupied spaces. However, by passing the stream through a following heating section, it can be brought to a condition most occupants would regard as comfortable. Let us sketch the procedure for evaluating the rates at which condensate exits and refrigerant circulates.

Mass Balance

The mass flow rate of the condensate $\dot{m}_w$ can be related to the mass flow rate of the dry air $\dot{m}_a$ by applying conservation of mass separately for the dry air and water passing through the dehumidifier section. At steady state

$$\dot{m}_{a1} = \dot{m}_{a2} \quad \text{(dry air)}$$

$$\dot{m}_{v1} = \dot{m}_w + \dot{m}_{v2} \quad \text{(water)}$$

The common mass flow rate of the dry air is denoted as $\dot{m}_a$. Solving for the mass flow rate of the condensate

$$\dot{m}_w = \dot{m}_{v1} - \dot{m}_{v2}$$

Introducing $\dot{m}_{v1} = \omega_1 \dot{m}_a$ and $\dot{m}_{v2} = \omega_2 \dot{m}_a$, the amount of water condensed per unit mass of dry air passing through the device is

$$\frac{\dot{m}_w}{\dot{m}_a} = \omega_1 - \omega_2$$

This expression requires the humidity ratios ω_1 and ω_2. Because no moisture is added or removed in the heating section, it can be concluded from conservation of mass that $\omega_2 = \omega_3$, so ω_3 can be used in the above equation in place of ω_2.

Energy Balance

The mass flow rate of the refrigerant through the cooling coil $\dot{m}_r$ can be related to the mass flow rate of the dry air $\dot{m}_a$ by means of an energy rate balance applied to the dehumidifier section. With $\dot{W}_{cv} = 0$, negligible heat transfer with the surroundings, and no significant kinetic and potential energy changes, the energy rate balance reduces at steady state to

$$0 = \dot{m}_r(h_i - h_e) + (\dot{m}_a h_{a1} + \dot{m}_{v1} h_{v1}) - \dot{m}_w h_w - (\dot{m}_a h_{a2} + \dot{m}_{v2} h_{v2})$$

where h_i and h_e denote the specific enthalpy values of the refrigerant entering and exiting the dehumidifier section, respectively. Introducing $\dot{m}_{v1} = \omega_1 \dot{m}_a$, $\dot{m}_{v2} = \omega_2 \dot{m}_a$, and $\dot{m}_w = (\omega_1 - \omega_2)\, \dot{m}_a$

$$0 = \dot{m}_r(h_i - h_e) + \dot{m}_a[(h_{a1} - h_{a2}) + \omega_1 h_{g1} - \omega_2 h_{g2} - (\omega_1 - \omega_2) h_{f2}]$$

where the specific enthalpies of the water vapor at 1 and 2 are evaluated at the saturated vapor values corresponding to T_1 and T_2, respectively. Since the condensate is assumed to exit as a saturated liquid at T_2, $h_w = h_{f2}$. Solving for the refrigerant mass flow rate per unit mass of dry air flowing through the device

$$\frac{\dot{m}_r}{\dot{m}_a} = \frac{(h_{a1} - h_{a2}) + \omega_1 h_{g1} - \omega_2 h_{g2} - (\omega_1 - \omega_2) h_{f2}}{h_e - h_i}$$

The accompanying psychrometric chart, Fig. 12.11*b*, illustrates important features of the processes involved. As indicated by the chart, the moist air first cools from state 1, where the temperature is T_1 and the humidity ratio is ω_1, to state 2, where the mixture is saturated ($\phi_2 = 100\%$), the temperature is $T_2 < T_1$, and the humidity ratio is $\omega_2 < \omega_1$. During the subsequent heating process, the humidity ratio would remain constant, $\omega_2 = \omega_3$, and the temperature would increase to T_3. Since the states visited would not be equilibrium states, these processes are indicated on the psychrometric chart by dashed lines. The example to follow provides a specific illustration.

Example 12.11 DEHUMIDIFIER

Moist air at 30°C and 50% relative humidity enters a dehumidifier operating at steady state with a volumetric flow rate of 280 m^3/min. The moist air passes over a cooling coil and water vapor condenses. Condensate exits the dehumidifier saturated at 10°C. Saturated moist air exits in a separate stream at the same temperature. There is no significant loss of energy by heat transfer to the surroundings and pressure remains constant at 1.013 bar. Determine **(a)** the mass flow rate of the dry air, in kg/min, **(b)** the rate at which water is condensed, in kg per kg of dry air flowing through the control volume, and **(c)** the required refrigerating capacity, in tons.

Solution

Known: Moist air enters a dehumidifier at 30°C and 50% relative humidity with a volumetric flow rate of 280 m^3/min. Condensate and moist air exit in separate streams at 10°C.

Determine: Find the mass flow rate of the dry air, in kg/min, the rate at which water is condensed, in kg per kg of dry air, and the required refrigerating capacity, in tons.

Schematic and Given Data:

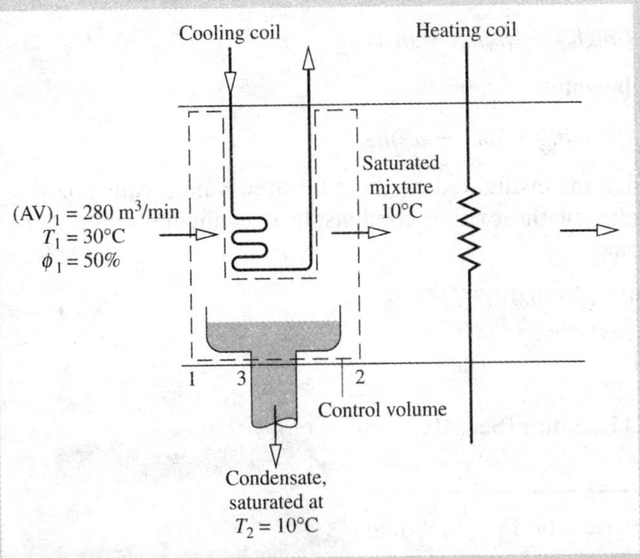

Engineering Model:

1. The control volume shown in the accompanying figure operates at steady state. Changes in kinetic and potential energy can be neglected, and $\dot{W}_{cv} = 0$.
2. There is no significant heat transfer to the surroundings.
3. The pressure remains constant throughout at 1.013 bar.
4. At location 2, the moist air is saturated. The condensate exits at location 3 as a saturated liquid at temperature T_2.

Fig. E12.11

Analysis:

(a) At steady state, the mass flow rates of the dry air entering and exiting are equal. The common mass flow rate of the dry air can be determined from the volumetric flow rate at the inlet

$$\dot{m}_a = \frac{(AV)_1}{v_{a1}}$$

The specific volume of the dry air at inlet 1, v_{a1}, can be evaluated using the ideal gas equation of state, so

$$\dot{m}_a = \frac{(AV)_1}{(\overline{R}/M_a)(T_1/p_{a1})}$$

The partial pressure of the dry air p_{a1} can be determined from $p_{a1} = p_1 - p_{v1}$. Using the relative humidity at the inlet ϕ_1 and the saturation pressure at 30°C from Table A-2

$$p_{v1} = \phi_1 p_{g1} = (0.5)(0.04246) = 0.02123 \text{ bar}$$

Thus, $p_{a1} = 1.013 - 0.02123 = 0.99177$ bar. Inserting values into the expression for $\dot{m}_a$ gives

$$\dot{m}_a = \frac{(280 \text{ m}^3/\text{min})(0.99177 \times 10^5 \text{ N/m}^2)}{(8314/28.97 \text{ N}\cdot\text{m/kg}\cdot\text{K})(303 \text{ K})} = 319.35 \text{ kg/min}$$

(b) Conservation of mass for the water requires $\dot{m}_{v1} = \dot{m}_{v2} + \dot{m}_w$. With $\dot{m}_{v1} = \omega_1 \dot{m}_a$ and $\dot{m}_{v2} = \omega_2 \dot{m}_a$, the rate at which water is condensed per unit mass of dry air is

$$\frac{\dot{m}_w}{\dot{m}_a} = \omega_1 - \omega_2$$

The humidity ratios ω_1 and ω_2 can be evaluated using Eq. 12.43. Thus, ω_1 is

$$\omega_1 = 0.622\left(\frac{p_{v1}}{p_1 - p_{v1}}\right) = 0.622\left(\frac{0.02123}{0.99177}\right) = 0.0133 \frac{\text{kg(vapor)}}{\text{kg(dry air)}}$$

Since the moist air is saturated at 10°C, p_{v2} equals the saturation pressure at 10°C: $p_g = 0.01228$ bar from Table A-2. Equation 12.43 then gives $\omega_2 = 0.0076$ kg (vapor)/kg(dry air). With these values for ω_1 and ω_2

$$\frac{\dot{m}_w}{\dot{m}_a} = 0.0133 - 0.0076 = 0.0057 \frac{\text{kg(condensate)}}{\text{kg(dry air)}}$$

(c) The rate of heat transfer $\dot{Q}_{cv}$ between the moist air stream and the refrigerant coil can be determined using an energy rate balance. With assumptions 1 and 2, the steady-state form of the energy rate balance reduces to

❶ $$0 = \dot{Q}_{cv} + (\dot{m}_a h_{a1} + \dot{m}_{v1} h_{v1}) - \dot{m}_w h_w - (\dot{m}_a h_{a2} + \dot{m}_{v2} h_{v2})$$

With $\dot{m}_{v1} = \omega_1 \dot{m}_a$, $\dot{m}_{v2} = \omega_2 \dot{m}_a$, and $\dot{m}_w = (\omega_1 - \omega_2)\dot{m}_a$, this becomes

$$\dot{Q}_{cv} = \dot{m}_a[(h_{a2} - h_{a1}) - \omega_1 h_{g1} + \omega_2 h_{g2} + (\omega_1 - \omega_2) h_{f2}]$$

where the specific enthalpies of the water vapor at 1 and 2 are evaluated at the saturated vapor values corresponding to T_1 and T_2, respectively, and the specific enthalpy of the exiting condensate is evaluated as h_f at T_2. Selecting enthalpies from Tables A-2 and A-22, as appropriate

$$\dot{Q}_{cv} = (319.35)[(283.1 - 303.2) - 0.0133(2556.3) + 0.0076(2519.8) + 0.0057(42.01)] = -11{,}084 \text{ kJ/min}$$

Since 1 ton of refrigeration equals a heat transfer rate of 211 kJ/min (Sec. 10.2), the required refrigerating capacity is 52.5 tons.

❶ If a psychrometric chart were used to obtain data, this expression for $\dot{Q}_{cv}$ would be rearranged to read

$$\dot{Q}_{cv} = \dot{m}_a[\underline{(h_a + \omega h_v)_2} - \underline{(h_a + \omega h_v)_1} + (\omega_1 - \omega_2) h_w]$$

The underlined terms and humidity ratios ω_1 and ω_2 would be read directly from the chart; the specific enthalpy h_w would be obtained from Table A-2 as h_f at T_2.

Skills Developed: *Ability to...*
- ✓ apply psychrometric terminology and principles.
- ✓ apply mass and energy balances for a dehumidification process in a control volume at steady state.
- ✓ retrieve property data for water.

Quick Quiz

Using the psychrometric chart, determine the wet-bulb temperature of the moist air entering the dehumidifier, in °C. **Ans.** ≈22°C.

12.8.4 Humidification

It is often necessary to increase the moisture content of the air circulated through occupied spaces. One way to accomplish this is to inject steam. Alternatively, liquid water can be sprayed into the air. Both cases are shown schematically in Fig. 12.12*a*. The temperature of the moist air as it exits the humidifier depends on the condition of the water introduced. When relatively high-temperature steam is injected, both the humidity ratio and the dry-bulb temperature would be increased. This is illustrated

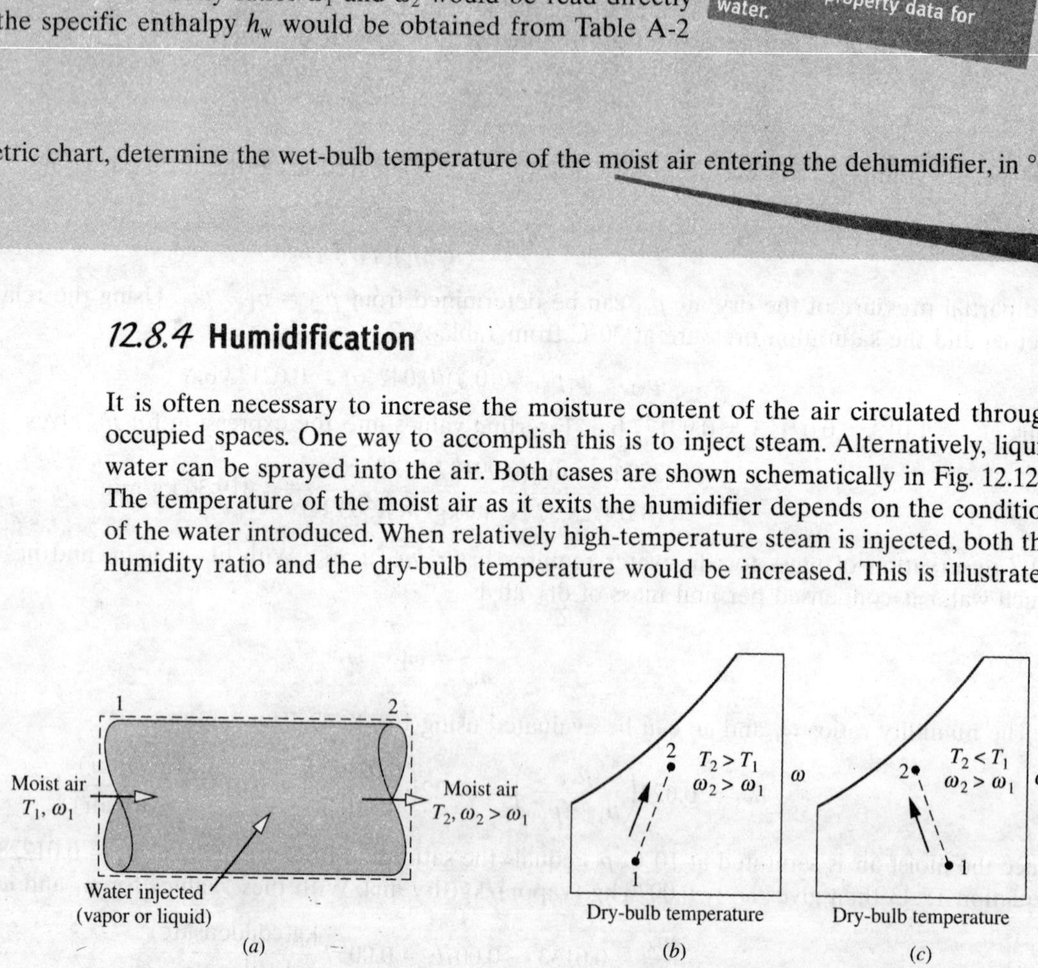

Fig. 12.12 Humidification. (*a*) Control volume. (*b*) Steam injected. (*c*) Liquid injected.

by the accompanying psychrometric chart of Fig. 12.12*b*. If liquid water was injected instead of steam, the moist air may exit the humidifier with a *lower* temperature than at the inlet. This is illustrated in Fig. 12.12*c*. The example to follow illustrates the case of steam injection. The case of liquid water injection is considered further in the next section.

Example 12.12 STEAM-SPRAY HUMIDIFIER

Moist air with a temperature of 22°C and a wet-bulb temperature of 9°C enters a steam-spray humidifier. The mass flow rate of the dry air is 90 kg/min. Saturated water vapor at 110°C is injected into the mixture at a rate of 52 kg/h. There is no heat transfer with the surroundings, and the pressure is constant throughout at 1 bar. Using the psychrometric chart, determine at the exit **(a)** the humidity ratio and **(b)** the temperature, in °C.

Solution

Known: Moist air enters a humidifier at a temperature of 22°C and a wet-bulb temperature of 9°C. The mass flow rate of the dry air is 90 kg/min. Saturated water vapor at 110°C is injected into the mixture at a rate of 52 kg/h.

Find: Using the psychrometric chart, determine at the exit the humidity ratio and the temperature, in °C.

Schematic and Given Data:

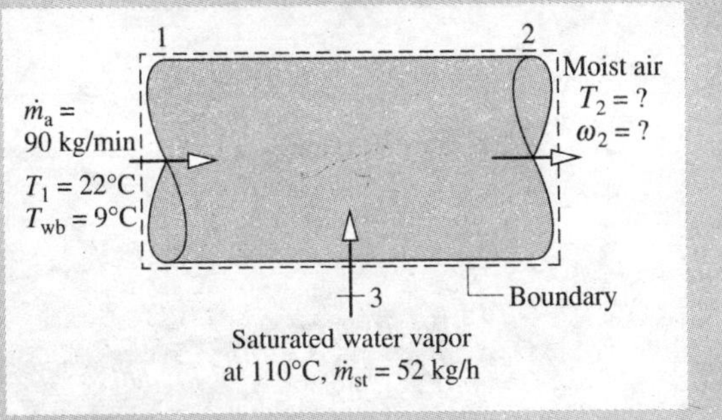

Engineering Model:

1. The control volume shown in the accompanying figure operates at steady state. Changes in kinetic and potential energy can be neglected and $\dot{W}_{cv} = 0$.
2. There is no heat transfer with the surroundings.
3. The pressure remains constant throughout at 1 bar. Figure A-9 remains valid at this pressure.

Fig. E12.12

Analysis:

(a) The humidity ratio at the exit ω_2 can be found from mass rate balances on the dry air and water individually. Thus

$$\dot{m}_{a1} = \dot{m}_{a2} \quad \text{(dry air)}$$
$$\dot{m}_{v1} + \dot{m}_{st} = \dot{m}_{v2} \quad \text{(water)}$$

With $\dot{m}_{v1} = \omega_1 \dot{m}_a$, and $\dot{m}_{v2} = \omega_2 \dot{m}_a$, where $\dot{m}_a$ is the mass flow rate of the air, the second of these becomes

$$\omega_2 = \omega_1 + \frac{\dot{m}_{st}}{\dot{m}_a}$$

Using the inlet dry-bulb temperature, 22°C, and the inlet wet-bulb temperature, 9°C, the value of the humidity ratio ω_1 can be found by inspection of the psychrometric chart, Fig. A-9. The result is $\omega_1 = 0.002$ kg(vapor)/kg(dry air). This value should be verified as an exercise. Inserting values into the expression for ω_2

$$\omega_2 = 0.002 + \frac{(52\ \text{kg/h})|1\ \text{h}/60\ \text{min}|}{90\ \text{kg/min}} = 0.0116 \frac{\text{kg(vapor)}}{\text{kg(dry air)}}$$

(b) The temperature at the exit can be determined using an energy rate balance. With assumptions 1 and 2, the steady-state form of the energy rate balance reduces to a special case of Eq. 12.55. Namely

$$0 = h_{a1} - h_{a2} + \omega_1 h_{g1} + (\omega_2 - \omega_1) h_{g3} - \omega_2 h_{g2} \quad (1)$$

In writing this, the specific enthalpies of the water vapor at 1 and 2 are evaluated as the respective saturated vapor values, and h_{g3} denotes the enthalpy of the saturated vapor injected into the moist air.

Equation (1) can be rearranged in the following form suitable for use with the psychrometric chart.

❶
$$(h_a + \omega h_g)_2 = (h_a + \omega h_g)_1 + (\omega_2 - \omega_1)h_{g3} \tag{2}$$

The first term on the right can be obtained from Fig. A-9 at the state defined by the intersection of the inlet dry-bulb temperature, 22°C, and the inlet wet-bulb temperature, 9°C: 27.2 kJ/kg(dry air). The second term on the right can be evaluated with the known humidity ratios ω_1 and ω_2 and the value of h_{g3} from Table A-2: 2691.5 kJ/kg(vapor). The state at the exit is fixed by ω_2 and $(h_a + \omega h_g)_2 = 53$ kJ/kg(dry air). The temperature at the exit can then be read directly from the chart. The result is $T_2 \approx 23.5$°C.

Alternative Solution:

❷ The following program allows T_2 to be determined using *IT*, where $\dot{m}_a$ is denoted as mdota, $\dot{m}_{st}$ is denoted as mdotst, w1 and w2 denote ω_1 and ω_2, respectively, and so on.

```
// Given data
T1 = 22 // °C
Twb1 = 9 // °C
mdota = 90 // kg/min
p = 1 // bar
Tst = 110 // °C
mdotst = (52 / 60) // converting kg/h to kg/min

// Evaluate humidity ratios
w1 = w_TTwb (T1,Twb1,p)
w2 = w1 + (mdotst / mdota)

// Denoting the enthalpy of moist air at state 1 by
// h1, etc., the energy balance, Eq. (1), becomes
0 = h1 – h2 + (w2 – w1)*hst

// Evaluate enthalpies
h1 = ha_Tw(T1,w1)
h2 = ha_Tw(T2,w2)
hst = hsat_Px("Water/Steam",psat,1)
psat = Psat_T("Water/Steam ",Tst)
```

Using the **Solve** button, the result is $T_2 = 23.4$°C, which agrees closely with the values obtained above, as expected.

❶ A solution of Eq. (2) using data from Tables A-2 and A-22 requires an iterative (trial) procedure. The result is $T_2 = 24$°C, as can be verified.

❷ Note the use of special *Moist Air* functions listed in the **Properties** menu of *IT*.

Skills Developed: *Ability to...*
- ✓ apply psychrometric terminology and principles.
- ✓ apply mass and energy balances for a spray humidification process in a control volume at steady state.
- ✓ retrieve necessary property data using the psychrometric chart.

Quick Quiz

Using the psychrometric chart, what is the relative humidity at the exit?
Ans. ≈63%.

12.8.5 Evaporative Cooling

Cooling in hot, relatively dry climates can be accomplished by *evaporative cooling*. This involves either spraying liquid water into air or forcing air through a soaked pad that is kept replenished with water, as shown in Fig. 12.13. Owing to the low humidity of

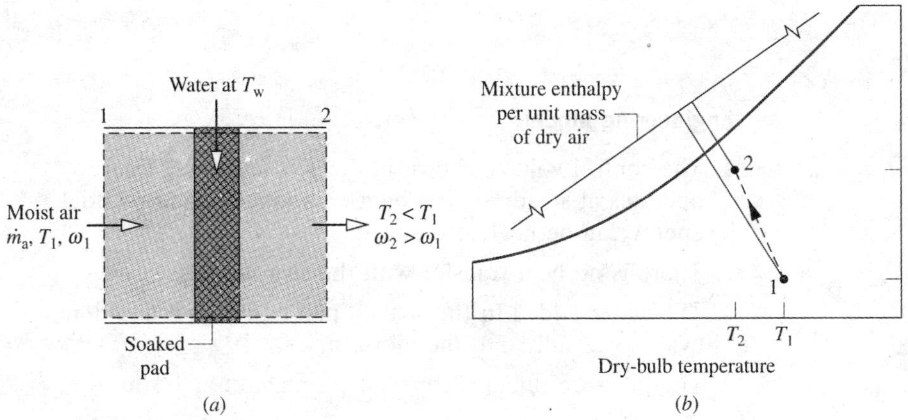

Fig. 12.13 Evaporative cooling. (*a*) Equipment schematic. (*b*) Psychrometric chart representation.

the moist air entering at state 1, part of the injected water evaporates. The energy for evaporation is provided by the air stream, which is reduced in temperature and exits at state 2 with a lower temperature than the entering stream. Because the incoming air is relatively dry, the additional moisture carried by the exiting moist air stream is normally beneficial.

For negligible heat transfer with the surroundings, no work $\dot{W}_{cv}$, and no significant changes in kinetic and potential energy, the steady-state forms of the mass and energy rate balances reduce for the control volume of Fig. 12.13*a* to

$$(h_{a2} + \omega_2 h_{g2}) = \underline{(\omega_2 - \omega_1)h_f} + (h_{a1} + \omega_1 h_{g1})$$

where h_f denotes the specific enthalpy of the liquid stream entering the control volume. All the injected water is assumed to evaporate into the moist air stream. The underlined term accounts for the energy carried in with the injected liquid water. This term is normally much smaller in magnitude than either of the two moist air enthalpy terms. Accordingly, the enthalpy of the moist air varies only slightly, as illustrated on the psychrometric chart of Fig. 12.13*b*. Recalling that lines of constant mixture enthalpy are closely lines of constant wet-bulb temperature (Sec. 12.7), it follows that evaporative cooling takes place at a nearly constant wet-bulb temperature.

In the next example, we consider the analysis of an evaporative cooler.

Example 12.13 EVAPORATIVE COOLER

Air at 38°C and 10% relative humidity enters an evaporative cooler with a volumetric flow rate of 140 m^3/min. Moist air exits the cooler at 21°C. Water is added to the soaked pad of the cooler as a liquid at 21°C and evaporates fully into the moist air. There is no heat transfer with the surroundings and the pressure is constant throughout at 1 bar. Determine **(a)** the mass flow rate of the water to the soaked pad, in kg/h, and **(b)** the relative humidity of the moist air at the exit to the evaporative cooler.

Solution

Known: Air at 38°C and ϕ = 10% enters an evaporative cooler with a volumetric flow rate of 140 m^3/min. Moist air exits the cooler at 21°C. Water is added to the soaked pad of the cooler at 21°C.

Find: Determine the mass flow rate of the water to the soaked pad, in kg/h, and the relative humidity of the moist air at the exit of the cooler.

Schematic and Given Data:

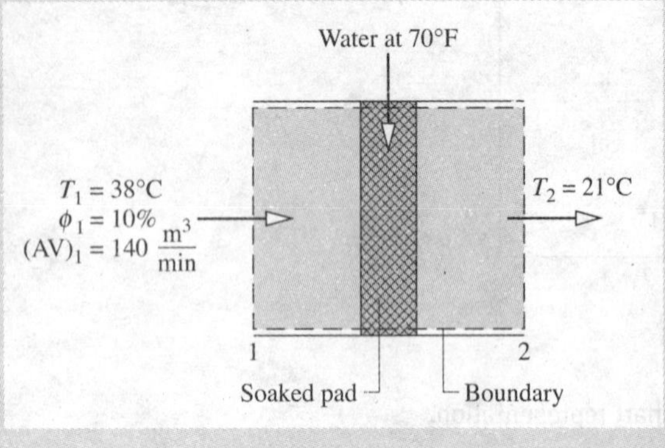

Engineering Model:

1. The control volume shown in the accompanying figure operates at steady state. Changes in kinetic and potential energy can be neglected and $\dot{W}_{cv} = 0$.
2. There is no heat transfer with the surroundings.
3. The water added to the soaked pad enters as a liquid and evaporates fully into the moist air.
4. The pressure remains constant throughout at 1 atm.

Fig. E12.13

Analysis:

(a) Applying conservation of mass to the dry air and water individually as in previous examples gives

$$\dot{m}_w = \dot{m}_a(\omega_2 - \omega_1)$$

where $\dot{m}_w$ is the mass flow rate of the water to the soaked pad. To find $\dot{m}_w$ requires ω_1, $\dot{m}_a$, and ω_2. These will now be evaluated in turn.

The humidity ratio ω_1 can be found from Eq. 12.43, which requires p_{v1}, the partial pressure of the moist air entering the control volume. Using the given relative humidity ϕ_1 and p_g at T_1 from Table A-2, we have $p_{v1} = \phi_1 p_{g1} = 0.0066$ bar. With this, $\omega_1 = 0.00408$ kg(vapor)/kg(dry air).

The mass flow rate of the dry air $\dot{m}_a$ can be found as in previous examples using the volumetric flow rate and specific volume of the dry air. Thus

$$\dot{m}_a = \frac{(AV)_1}{v_{a1}}$$

The specific volume of the dry air can be evaluated from the ideal gas equation of state. The result is $v_{a1} = 0.887\ m^3/kg$ (dry air). Inserting values, the mass flow rate of the dry air is

$$\dot{m}_a = \frac{140\ m^3/min}{0.887\ m^3/kg(\text{dry air})} = 157.8 \frac{kg(\text{dry air})}{min}$$

To find the humidity ratio ω_2, reduce the steady-state forms of the mass and energy rate balances using assumption 1 to obtain

$$0 = (\dot{m}_a h_{a1} + \dot{m}_{v1} h_{v1}) + \dot{m}_w h_w - (\dot{m}_a h_{a2} + \dot{m}_{v2} h_{v2})$$

With the same reasoning as in previous examples, this can be expressed as

❶ $$0 = (h_a + \omega h_g)_1 + \underline{(\omega_2 - \omega_1)h_f} - (h_a + \omega h_g)_2$$

where h_f denotes the specific enthalpy of the water entering the control volume at 21°C. Solving for ω_2

$$\omega_2 = \frac{h_{a1} - h_{a2} + \omega_1(h_{g1} - h_f)}{h_{g2} - h_f} = \frac{c_{pa}(T_1 - T_2) + \omega_1(h_{g1} - h_f)}{h_{g2} - h_f}$$

❷ where $c_{pa} = 1.005$ kJ/kg · K. With h_f, h_{g1}, and h_{g2} from Table A-2

$$\omega_2 = \frac{(1.005)(38 - 21) + 0.00408(2570.7 - 88.14)}{2451.8}$$

$$= 0.011 \frac{kg(\text{vapor})}{kg(\text{dry air})}$$

Substituting values for $\dot{m}_a$, ω_1, and ω_2 into the expression for $\dot{m}_w$

$$\dot{m}_w = \left[157.8 \frac{\text{kg(dry air)}}{\text{min}} \left| \frac{60\,\text{min}}{\text{h}} \right| \right](0.011 - 0.00408)\frac{\text{kg(water)}}{\text{kg(dry air)}}$$
$$= 65.5 \frac{\text{kg(water)}}{\text{h}}$$

(b) The relative humidity of the moist air at the exit can be determined using Eq. 12.44. The partial pressure of the water vapor required by this expression can be found by solving Eq. 12.43 to obtain

$$p_{v2} = \frac{\omega_2 p}{\omega_2 + 0.622}$$

Inserting values

$$p_{v2} = \frac{(0.011)(1.01325\text{ bar})}{(0.011 + 0.622)} = 0.0176\,\text{bar}$$

At 21°C, the saturation pressure is 0.02487 bar. Thus, the relative humidity at the exit is

$$\phi_2 = \frac{0.0176}{0.02487} = 0.708(70.8\%)$$

---

❶ Since the underlined term in this equation is much smaller than either of the moist air enthalpies, the enthalpy of the moist air remains nearly constant, and thus evaporative cooling takes place at nearly constant wet-bulb temperature. This can be verified by locating the incoming and outgoing moist air states on the psychrometric chart.

❷ A constant value of the specific heat c_{pa} has been used here to evaluate the term $(h_{a1} - h_{a2})$. As shown in previous examples, this term can be evaluated alternatively using the ideal gas table for air.

Skills Developed: *Ability to...*
- ✓ apply psychrometric terminology and principles.
- ✓ apply mass and energy balances for an evaporative cooling process in a control volume at steady state.
- ✓ retrieve property data for dry air and water.

Quick Quiz

Using steam table data, what is the dew point temperature at the exit, in °C?
Ans. 15.3°C.

12.8.6 Adiabatic Mixing of Two Moist Air Streams

A common process in air-conditioning systems is the mixing of moist air streams, as shown in Fig. 12.14. The objective of the thermodynamic analysis of such a process is normally to fix the flow rate and state of the exiting stream for specified flow rates and states of each of the two inlet streams. The case of adiabatic mixing is governed by Eqs. 12.56 to follow.

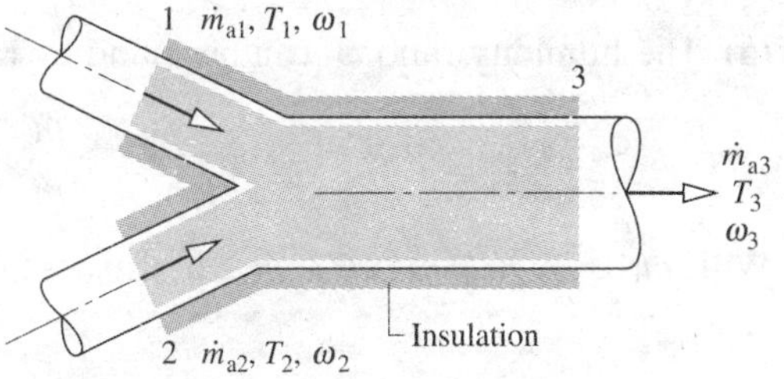

Fig. 12.14 Adiabatic mixing of two moist air streams.

The mass rate balances for the dry air and water vapor at steady state are, respectively,

$$\dot{m}_{a1} + \dot{m}_{a2} = \dot{m}_{a3} \quad \text{(dry air)} \qquad (12.56a)$$
$$\dot{m}_{v1} + \dot{m}_{v2} = \dot{m}_{v3} \quad \text{(water vapor)}$$

With $\dot{m}_v = \omega \dot{m}_a$, the water vapor mass balance becomes

$$\omega_1 \dot{m}_{a1} + \omega_2 \dot{m}_{a2} = \omega_3 \dot{m}_{a3} \quad \text{(water vapor)} \qquad (12.56b)$$

Assuming $\dot{Q}_{cv} = \dot{W}_{cv} = 0$ and ignoring the effects of kinetic and potential energy, the energy rate balance reduces at steady state to

$$\dot{m}_{a1}(h_{a1} + \omega_1 h_{g1}) + \dot{m}_{a2}(h_{a2} + \omega_2 h_{g2}) = \dot{m}_{a3}(h_{a3} + \omega_3 h_{g3}) \qquad (12.56c)$$

where the enthalpies of the entering and exiting water vapor are evaluated as the saturated vapor values at the respective dry-bulb temperatures.

If the inlet flow rates and states are known, Eqs. 12.56 are three equations in three unknowns: $\dot{m}_{a3}$, ω_3, and $(h_{a3} + \omega_3 h_{g3})$. The solution of these equations is illustrated by the next example.

Example 12.14 ADIABATIC MIXING OF MOIST STREAMS

A stream consisting of 142 m³/min of moist air at a temperature of 5°C and a humidity ratio of 0.002 kg(vapor)/kg(dry air) is mixed adiabatically with a second stream consisting of 425 m³/min of moist air at 24°C and 50% relative humidity. The pressure is constant throughout at 1 bar. Using the psychrometric chart, determine **(a)** the humidity ratio and **(b)** the temperature of the exiting mixed stream, in °C.

Solution

Known: A moist air stream at 5°C, $\omega = 0.002$ kg(vapor)/kg(dry air), and a volumetric flow rate of 142 m³/min is mixed adiabatically with a stream consisting of 425 m³/min of moist air at 24°C and $\phi = 50\%$.

Find: Determine the humidity ratio and the temperature, in °C, of the mixed stream exiting the control volume.

Schematic and Given Data:

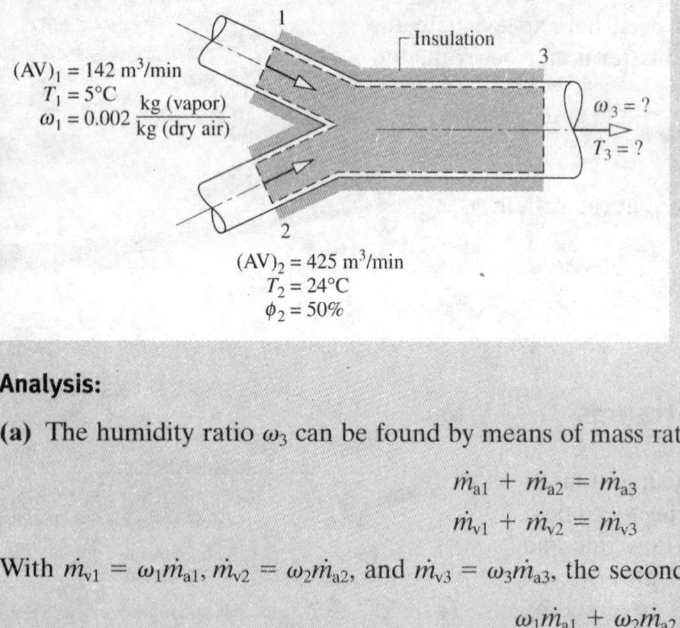

Engineering Model:

1. The control volume shown in the accompanying figure operates at steady state. Changes in kinetic and potential energy can be neglected and $\dot{W}_{cv} = 0$.
2. There is no heat transfer with the surroundings.
3. The pressure remains constant throughout at 1 bar.

Fig. E12.14

Analysis:

(a) The humidity ratio ω_3 can be found by means of mass rate balances for the dry air and water vapor, respectively

$$\dot{m}_{a1} + \dot{m}_{a2} = \dot{m}_{a3} \quad \text{(dry air)}$$
$$\dot{m}_{v1} + \dot{m}_{v2} = \dot{m}_{v3} \quad \text{(water vapor)}$$

With $\dot{m}_{v1} = \omega_1 \dot{m}_{a1}$, $\dot{m}_{v2} = \omega_2 \dot{m}_{a2}$, and $\dot{m}_{v3} = \omega_3 \dot{m}_{a3}$, the second of these balances becomes

$$\omega_1 \dot{m}_{a1} + \omega_2 \dot{m}_{a2} = \omega_3 \dot{m}_{a3}$$

Solving

$$\omega_3 = \frac{\omega_1 \dot{m}_{a1} + \omega_2 \dot{m}_{a2}}{\dot{m}_{a3}}$$

Since $\dot{m}_{a3} = \dot{m}_{a1} + \dot{m}_{a2}$, this can be expressed as

$$\omega_3 = \frac{\omega_1 \dot{m}_{a1} + \omega_2 \dot{m}_{a2}}{\dot{m}_{a1} + \dot{m}_{a2}}$$

To determine ω_3 requires values for ω_2, $\dot{m}_{a1}$, and $\dot{m}_{a2}$. The mass flow rates of the dry air, $\dot{m}_{a1}$ and $\dot{m}_{a2}$, can be found as in previous examples using the given volumetric flow rates

$$\dot{m}_{a1} = \frac{(AV)_1}{v_{a1}}, \qquad \dot{m}_{a2} = \frac{(AV)_2}{v_{a2}}$$

The values of v_{a1}, v_{a2}, and ω_2 are readily found from the psychrometric chart, Fig. A-9. Thus, at $\omega_1 = 0.002$ and $T_1 = 5°C$, $v_{a1} = 0.79\ m^3/kg(\text{dry air})$. At $\phi_2 = 50\%$ and $T_2 = 24°C$, $v_{a2} = 0.855\ m^3/kg(\text{dry air})$ and $\omega_2 = 0.0094$. The mass flow rates of the dry air are then $\dot{m}_{a1} = 180$ kg(dry air)/min and $\dot{m}_{a2} = 497$ kg(dry air)/min. Inserting values into the expression for ω_3

$$\omega_3 = \frac{(0.002)(180) + (0.0094)(497)}{180 + 497} = 0.0074 \frac{\text{kg(vapor)}}{\text{kg(dry air)}}$$

(b) The temperature T_3 of the exiting mixed stream can be found from an energy rate balance. Reduction of the energy rate balance using assumptions 1 and 2 gives

$$\dot{m}_{a1}(h_a + \omega h_v)_1 + \dot{m}_{a2}(h_a + \omega h_v)_2 = \dot{m}_{a3}(h_a + \omega h_v)_3 \tag{1}$$

Solving

$$(h_a + \omega h_v)_3 = \frac{\dot{m}_{a1}(h_a + \omega h_v)_1 + \dot{m}_{a2}(h_a + \omega h_v)_2}{\dot{m}_{a1} + \dot{m}_{a2}} \tag{2}$$

With $(h_a + \omega h_v)_1 = 10$ kJ/kg(dry air) and $(h_a + \omega h_v)_2 = 47.8$ kJ/kg(dry air) from Fig. A-9 and other known values

$$(h_a + \omega h_v)_3 = \frac{180(10) + 497(47.8)}{180 + 497} = 37.7 \frac{\text{kJ}}{\text{kg(dry air)}}$$

This value for the enthalpy of the moist air at the exit, together with the previously determined value for ω_3, fixes the state of the exiting moist air. From inspection of Fig. A-9, $T_3 = 19°C$.

Alternative Solutions:

The use of the psychrometric chart facilitates the solution for T_3. Without the chart, an iterative solution of Eq. (2) using table data could be used as noted in the solution to Example 12.12. Alternatively, T_3 can be determined using the following *IT* program, where ϕ_2 is denoted as phi2, the volumetric flow rates at 1 and 2 are denoted as AV1 and AV2, respectively, and so on.

```
// Given data
T1 = 5 // °C
w1 = 0.002 // kg(vapor) / kg(dry air)
AV1 = 142 // m³/min
T2 = 24 // °C
phi2 = 0.5
AV2 = 425 // m³/min
p = 1 // bar

// Mass balances for water vapor and dry air:
w1 * mdota1 + w2 * mdota2 = w3 * mdota3
mdota1 + mdota2 = mdota3
// Evaluate mass flow rates of dry air
mdota1 = AV1 / va1
va1 = va_Tw(T1, w1, p)
mdota2 = AV2 / va2
va2 = va_Tphi(T2, phi2, p)

// Determine w2
w2 = w_Tphi(T2, phi2, p)
// The energy balance, Eq. (1), reads
```

❶

```
mdota1 * h1 + mdota2 * h2 = mdota3 * h3
h1 = ha_Tw(T1, w1)
h2 = ha_Tphi(T2, phi2, p)
h3 = ha_Tw(T3, w3)
```

Using the **Solve** button, the result is $T_3 = 19.01°C$ and $\omega_3 = 0.00745$ kg (vapor)/kg (dry air), which agree with the psychrometric chart solution.

❶ Note the use here of special *Moist Air* functions listed in the **Properties** menu of *IT*.

Skills Developed: *Ability to...*
- ✓ apply psychrometric terminology and principles.
- ✓ apply mass and energy balances for an adiabatic mixing process of two moist air streams in a control volume at steady state.
- ✓ retrieve property data for moist air using the psychrometric chart.

Quick Quiz

Using the psychrometric chart, what is the relative humidity at the exit? **Ans.** ≈53%.

12.9 Cooling Towers

Power plants invariably discharge considerable energy to their surroundings by heat transfer (Chap. 8). Although water drawn from a nearby river or lake can be employed to carry away this energy, cooling towers provide an alternative in locations where sufficient cooling water cannot be obtained from natural sources or where concerns for the environment place a limit on the temperature at which cooling water can be returned to the surroundings. Cooling towers also are frequently employed to provide chilled water for applications other than those involving power plants.

Cooling towers can operate by *natural* or *forced* convection. Also they may be *counterflow, cross-flow,* or a combination of these. A schematic diagram of a forced-convection, counterflow cooling tower is shown in Fig. 12.15. The warm water to be cooled enters at 1 and is sprayed from the top of the tower. The falling water usually passes through a series of baffles intended to keep it broken up into fine drops

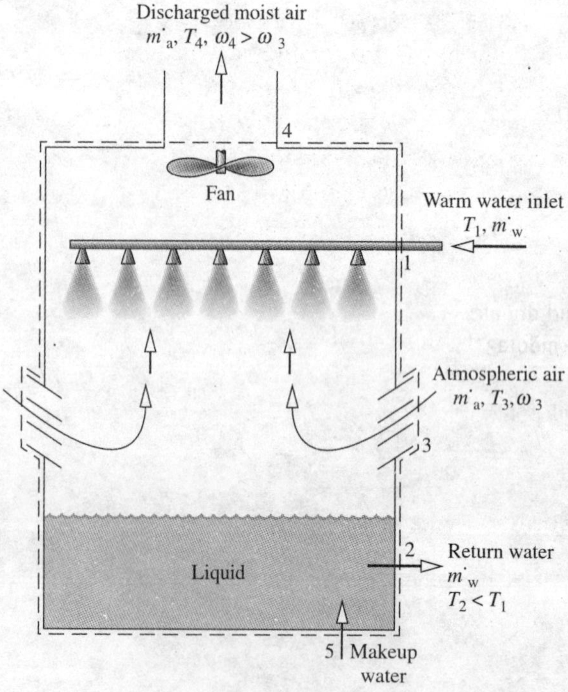

Fig. 12.15 Schematic of a cooling tower.

to promote evaporation. Atmospheric air drawn in at 3 by the fan flows upward, counter to the direction of the falling water droplets. As the two streams interact, a small fraction of the water stream evaporates into the moist air, which exits at 4 with a greater humidity ratio than the incoming moist air at 3. The energy required for evaporation is provided mainly by the portion of the incoming water stream that does not evaporate, with the result that the water exiting at 2 is at a lower temperature than the water entering at 1. Since some of the incoming water is evaporated into the moist air stream, an equivalent amount of makeup water is added at 5 so that the return mass flow rate of the cool water equals the mass flow rate of the warm water entering at 1.

For operation at steady state, mass balances for the dry air and water and an energy balance on the overall cooling tower provide information about cooling tower performance. In applying the energy balance, heat transfer with the surroundings is usually neglected. The power input to the fan of forced-convection towers also may be negligible relative to other energy rates involved. The example to follow illustrates the analysis of a cooling tower using conservation of mass and energy together with property data for the dry air and water.

Example 12.15 POWER PLANT COOLING TOWER

Water exiting the condenser of a power plant at 38°C enters a cooling tower with a mass flow rate of 4.5×10^7 kg/h. A stream of cooled water is returned to the condenser from a cooling tower with a temperature of 30°C and the same flow rate. Makeup water is added in a separate stream at 20°C. Atmospheric air enters the cooling tower at 25°C and 35% relative humidity. Moist air exits the tower at 35°C and 90% relative humidity. Determine the mass flow rates of the dry air and the makeup water, in kg/h. The cooling tower operates at steady state. Heat transfer with the surroundings and the fan power can each be neglected, as can changes in kinetic and potential energy. The pressure remains constant throughout at 1 atm.

Solution

Known: A liquid water stream enters a cooling tower from a condenser at 38°C with a known mass flow rate. A stream of cooled water is returned to the condenser at 30°C and the same flow rate. Makeup water is added at 20°C. Atmospheric air enters the tower at 25°C and $\phi = 35\%$. Moist air exits the tower at 35°C and $\phi = 90\%$.

Find: Determine the mass flow rates of the dry air and the makeup water, in kg/h.

Schematic and Given Data:

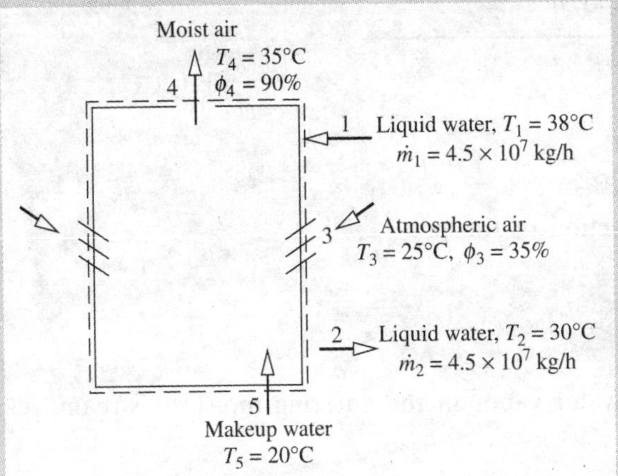

Engineering Model:

1. The control volume shown in the accompanying figure operates at steady state. Heat transfer with the surroundings can be neglected, as can changes in kinetic and potential energy; also $\dot{W}_{cv} = 0$.
2. To evaluate specific enthalpies, each liquid stream is regarded as a saturated liquid at the corresponding specified temperature.
3. The pressure is constant throughout at 1 atm.

Fig. E12.15

Analysis: The required mass flow rates can be found from mass and energy rate balances. Mass balances for the dry air and water individually reduce at steady state to

$$\dot{m}_{a3} = \dot{m}_{a4} \qquad \text{(dry air)}$$

$$\dot{m}_1 + \dot{m}_5 + \dot{m}_{v3} = \dot{m}_2 + \dot{m}_{v4} \qquad \text{(water)}$$

The common mass flow rate of the dry air is denoted as $\dot{m}_a$. Since $\dot{m}_1 = \dot{m}_2$, the second of these equations becomes

$$\dot{m}_5 = \dot{m}_{v4} - \dot{m}_{v3}$$

With $\dot{m}_{v3} = \omega_3 \dot{m}_a$ and $\dot{m}_{v4} = \omega_4 \dot{m}_a$

$$\dot{m}_5 = \dot{m}_a(\omega_4 - \omega_3)$$

Accordingly, the two required mass flow rates, $\dot{m}_a$ and $\dot{m}_5$, are related by this equation. Another equation relating the flow rates is provided by the energy rate balance.

Reducing the energy rate balance with assumption 1 results in

$$0 = \dot{m}_1 h_{w1} + (\dot{m}_a h_{a3} + \dot{m}_{v3} h_{v3}) + \dot{m}_5 h_{w5} - \dot{m}_2 h_{w2} - (\dot{m}_a h_{a4} + \dot{m}_{v4} h_{v4})$$

Evaluating the enthalpies of the water vapor as the saturated vapor values at the respective temperatures and the enthalpy of each liquid stream as the saturated liquid enthalpy at the respective temperature, the energy rate equation becomes

$$0 = \dot{m}_1 h_{f1} + (\dot{m}_a h_{a3} + \dot{m}_{v3} h_{g3}) + \dot{m}_5 h_{f5} - \dot{m}_2 h_{f2} - (\dot{m}_a h_{a4} + \dot{m}_{v4} h_{g4})$$

Introducing $\dot{m}_1 = \dot{m}_2$, $\dot{m}_5 = \dot{m}_a(\omega_4 - \omega_3)$, $\dot{m}_{v3} = \omega_3 \dot{m}_a$, and $\dot{m}_{v4} = \omega_4 \dot{m}_a$ and solving for $\dot{m}_a$

❶ $$\dot{m}_a = \frac{\dot{m}_1(h_{f1} - h_{f2})}{h_{a4} - h_{a3} + \omega_4 h_{g4} - \omega_3 h_{g3} - (\omega_4 - \omega_3)h_{f5}}$$

The humidity ratios ω_3 and ω_4 required by this expression can be determined from Eq. 12.43, using the partial pressure of the water vapor obtained with the respective relative humidity. Thus, $\omega_3 = 0.00688$ kg(vapor)/kg(dry air) and $\omega_4 = 0.0327$ kg(vapor)/kg(dry air).

With enthalpies from Tables A-2 and A-22, as appropriate, and the known values for ω_3, ω_4, and $\dot{m}_1$, the expression for $\dot{m}_a$ becomes

$$\dot{m}_a = \frac{(4.5 \times 10^7)(159.21 - 125.79)}{(308.2 - 298.2) + (0.0327)(2565.3) - (0.00688)(2547.2) - (0.0258)(83.96)}$$
$$= 2.03 \times 10^7 \text{ kg/h}$$

Finally, inserting known values into the expression for $\dot{m}_5$ results in

$$\dot{m}_5 = (2.03 \times 10^7)(0.0327 - 0.00688) = 5.24 \times 10^5 \text{ kg/h}$$

❶ This expression for $\dot{m}_a$ can be rearranged to read

$$\dot{m}_a = \frac{\dot{m}_1(h_{f1} - h_{f2})}{\underline{(h_{a4} + \omega_4 h_{g4})} - \underline{(h_{a3} + \omega_3 h_{g3})} - (\omega_4 - \omega_3)h_{f5}}$$

The underlined terms and ω_3 and ω_4 can be obtained by inspection of the psychrometric chart.

Skills Developed: *Ability to...*
- ✓ apply psychrometric terminology and principles.
- ✓ apply mass and energy balances for a cooling tower process in a control volume at steady state.
- ✓ retrieve property data for dry air and water.

Quick Quiz

Using steam table data, determine the partial pressure of the water vapor in the entering moist air stream, p_{v3}, in bar.

Ans. 0.0111 bar.

✓ Chapter Summary and Study Guide

In this chapter we have applied the principles of thermodynamics to systems involving ideal gas mixtures, including the special case of *psychrometric* applications involving air–water vapor mixtures, possibly in the presence of liquid water. Both closed system and control volume applications are presented.

The first part of the chapter deals with general ideal gas mixture considerations and begins by describing mixture composition in terms of the mass fractions or mole fractions. Two models are then introduced for the p–v–T relation of ideal gas mixtures: the Dalton model, which includes the partial pressure concept, and the Amagat model. Means are also introduced for evaluating the enthalpy, internal energy, and entropy of a mixture by adding the contribution of each component at its condition in the mixture. Applications are considered where ideal gas mixtures undergo processes at constant composition as well as where ideal gas mixtures are formed from their component gases.

In the second part of the chapter, we study *psychrometrics*. Special terms commonly used in psychrometrics are introduced, including moist air, humidity ratio, relative humidity, mixture enthalpy, and the dew point, dry-bulb, and wet-bulb temperatures. The *psychrometric chart,* which gives a graphical representation of important moist air properties, is introduced. The principles of conservation of mass and energy are formulated in terms of psychrometric quantities, and typical air-conditioning applications are considered, including dehumidification and humidification, evaporative cooling, and mixing of moist air streams. A discussion of cooling towers is also provided.

The following list provides a study guide for this chapter. When your study of the text and end-of-chapter exercises has been completed, you should be able to

- ✓ write out the meanings of the terms listed in the margin throughout the chapter and understand each of the related concepts. The subset of key concepts listed below is particularly important.
- ✓ describe mixture composition in terms of mass fractions or mole fractions.
- ✓ relate pressure, volume, and temperature of ideal gas mixtures using the Dalton model, and evaluate U, H, c_v, c_p, and S of ideal gas mixtures in terms of the mixture composition and the respective contribution of each component.
- ✓ apply the conservation of mass and energy principles and the second law of thermodynamics to systems involving ideal gas mixtures.

For psychrometric applications, you should be able to

- ✓ evaluate the humidity ratio, relative humidity, mixture enthalpy, and dew point temperature.
- ✓ use the psychrometric chart.
- ✓ apply the conservation of mass and energy principles and the second law of thermodynamics to analyze air-conditioning processes and cooling towers.

Key Engineering Concepts

mass fraction p. 643
gravimetric analysis p. 643
mole fraction p. 643
molar (volumetric) analysis p. 644
apparent molecular weight p. 644
Dalton model p. 647
partial pressure p. 647
psychrometrics p. 664
moist air p. 664
humidity ratio p. 665
relative humidity p. 666
mixture enthalpy p. 666
dew point temperature p. 668
dry-bulb temperature p. 675
wet-bulb temperature p. 675
psychrometric chart p. 676

Key Equations

Ideal Gas Mixtures: General Considerations

$mf_i = m_i/m$	(12.3)	Analysis in terms of mass fractions
$1 = \sum_{i=1}^{j} mf_i$	(12.4)	
$y_i = n_i/n$	(12.6)	Analysis in terms of mole fractions
$1 = \sum_{i=1}^{j} y_i$	(12.7)	

$M = \sum_{i=1}^{j} y_i M_i$	(12.9)	Apparent molecular weight
$p_i = y_i p$ $p = \sum_{i=1}^{j} p_i$	(12.12) (12.13)	Partial pressure of component i and relation to mixture pressure p
$\bar{u} = \sum_{i=1}^{j} y_i \bar{u}_i$ $\bar{h} = \sum_{i=1}^{j} y_i \bar{h}_i$ $\bar{s} = \sum_{i=1}^{j} y_i \bar{s}_i$	(12.21) (12.22) (12.27)	Internal energy, enthalpy, and entropy per mole of mixture. $\bar{u}_i$ and $\bar{h}_i$ evaluated at mixture temperature T. $\bar{s}_i$ evaluated at T and partial pressure p_i.
$\bar{c}_v = \sum_{i=1}^{j} y_i \bar{c}_{v,i}$ $\bar{c}_p = \sum_{i=1}^{j} y_i \bar{c}_{p,i}$	(12.23) (12.24)	Mixture specific heats on a molar basis
Psychrometric Applications		
$\omega = \frac{m_v}{m_a} = 0.622 \frac{p_v}{p - p_v}$	(12.42, 12.43)	Humidity ratio
$\phi = \left.\frac{p_v}{p_g}\right)_{T,p}$	(12.44)	Relative humidity
$\frac{H}{m_a} = h_a + \omega h_v$	(12.46)	Mixture enthalpy per unit mass of dry air

Exercises: things engineers think about

1. In an *equimolar* mixture of O_2 and N_2, are the mass fractions equal ?

2. The molar analysis of an ideal gas mixture is $\{y_{CO_2} = 0.4, y_{N_2} = 0.25, y_{O_2}\}$. How many kmol of oxygen are present in 5 kmol of mixture?

3. How would you calculate the specific heat ratio, k, at 300 K of a mixture of N_2, O_2, and CO_2 if the molar analysis were known?

4. If two different ideal gases mix spontaneously, is the process irreversible? Discuss

5. A rigid, insulated container has two compartments, each containing air at the same temperature and pressure on both sides of the partition. If the partition is removed and the air on both sides mix, is any entropy produced?

6. Which component of the fuel–air mixture in a cylinder of an automobile engine would have the greater mass fraction?

7. Why does a dog pant?

8. Which do you think is most closely related to human comfort, the humidity ratio or the relative humidity?

9. How do you explain the different rates of evaporation from a dish of water in winter and summer?

10. Why does your bathroom mirror often fog up when you shower?

11. Although water vapor in air is typically a superheated vapor, why can we use the saturated vapor value, $h_g(T)$, to represent its enthalpy?

12. Can the dry-bulb and wet-bulb temperatures be equal?

13. How do you explain the water dripping from the tailpipe of an automobile on a cold morning?

14. How does an automobile's windshield defroster achieve its purpose?

15. Would you recommend an evaporative cooling system for use in Florida? In Arizona?

Problems: developing engineering skills

Determining Mixture Composition

12.1 Answer the following questions involving a mixture of two gases:

(a) When would the analysis of the mixture in terms of mass fractions be *identical* to the analysis in terms of mole fractions?

(b) When would the apparent molecular weight of the mixture equal the average of the molecular weights of the two gases?

12.2 The molar analysis of a gas mixture at 30°C, 2 bar is 40% N_2, 50% CO_2, 10% CH_4. Determine

(a) the analysis in terms of mass fractions.

(b) the partial pressure of each component, in bar.

(c) the volume occupied by 10 kg of mixture, in m^3.

12.3 The analysis on a molar basis of a gas mixture at 5°C, 1 bar is 20% Ar, 35% CO_2, 45% O_2. Determine

(a) the analysis in terms of mass fractions.

(b) the partial pressure of each component, in kPa.

(c) the volume occupied by 5 kg of mixture, in m^3.

12.4 The molar analysis of a gas mixture at 25°C, 0.1 MPa is 60% N_2, 30% CO_2, 10% O_2. Determine

(a) the analysis in terms of mass fractions.

(b) the partial pressure of each component, in MPa.

(c) the volume occupied by 50 kg of the mixture, in m^3.

12.5 The analysis on a mass basis of a gas mixture at 4°C, 140 kPa is 60% CO_2, 25% CO, 15% O_2. Determine

(a) the analysis in terms of mole fractions.

(b) the partial pressure of each component, in kPa.

(c) the volume occupied by 10 kg of the mixture, in m^3.

12.6 Natural gas at 23°C, 1 bar enters a furnace with the following molar analysis: 40% propane (C_3H_8), 40% ethane (C_2H_6), 20% methane (CH_4). Determine

(a) the analysis in terms of mass fractions.

(b) the partial pressure of each component, in bar.

(c) the mass flow rate, in kg/s, for a volumetric flow rate of 20 m^3/s.

12.7 A rigid vessel having a volume of 3 m^3 initially contains a mixture at 21°C, 1 bar consisting of 79% N_2 and 21% O_2 on a molar basis. Helium is allowed to flow into the vessel until the pressure is 2 bar. If the final temperature of the mixture within the vessel is 27°C, determine the mass, in kg, of each component present.

12.8 Nitrogen (N_2) at 150 kPa, 40°C occupies a closed, rigid container having a volume of 1 m^3. If 2 kg of oxygen (O_2) is added to the container, what is the molar analysis of the resulting mixture? If the temperature remains constant, what is the pressure of the mixture, in kPa? Assume ideal gas behavior.

12.9 A flue gas in which the mole fraction of H_2S is 0.002 enters a *scrubber* operating at steady state at 93°C, 1 bar and a volumetric flow rate of 540 m^3/h. If the scrubber removes 92% (molar basis) of the entering H_2S, determine the rate at which H_2S is removed, in kg/h. Comment on why H_2S should be removed from the gas stream.

12.10 A control volume operating at steady state has two entering streams and a single exiting stream. A mixture with a mass flow rate of 11.67 kg/min and a molar analysis 9% CH_4, 91% air enters at one location and is diluted by a separate stream of air entering at another location. The molar analysis of the air is 21% O_2, 79% N_2. If the mole fraction of CH_4 in the exiting stream is required to be 5%, determine

(a) the molar flow rate of the entering air, in kmol/min.

(b) the mass flow rate of oxygen in the exiting stream, in kg/min.

Considering Constant-Composition Processes

12.11 A gas mixture consists of 2 kg of N_2 and 3 kg of He. Determine

(a) the composition in terms of mass fractions.

(b) the composition in terms of mole fractions.

(c) the heat transfer, in kJ, required to increase the mixture temperature from 21 to 66°C, while keeping the pressure constant.

(d) the change in entropy of the mixture for the process of part (c), in kJ/K.

For parts (c) and (d), use the ideal gas model with constant specific heats.

12.12 Two kg of a mixture having an analysis on a mass basis of 30% N_2, 40% CO_2, 30% O_2 is compressed adiabatically from 1 bar, 300 K to 4 bar, 500 K. Determine

(a) the work, in kJ.

(b) the amount of entropy produced, in kJ/K.

12.13 As illustrated in Fig. P12.13, an ideal gas mixture in a piston-cylinder assembly has a molar analysis of 30% carbon dioxide (CO_2) and 70% nitrogen (N_2). The mixture is cooled at constant pressure from 425 to 325 K. Assuming constant specific heats evaluated at 375 K, determine the heat transfer and the work, each in kJ per kg of mixture.

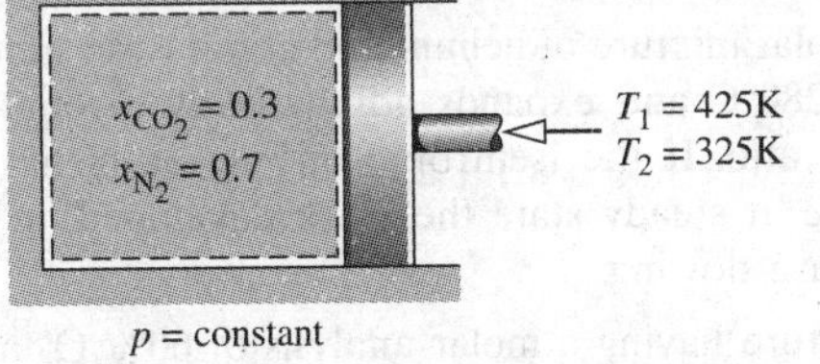

Fig. P12.13

12.14 A mixture consisting of 0.3 kmol of N_2 and 0.2 kmol of O_2 is compressed isothermally at 556 K from 1 to 3 bar. During the process, there is energy transfer by heat from the mixture to the surroundings, which are at 4°C. For the mixture, determine

(a) the work, in kJ.

(b) the heat transfer, in kJ.

(c) the amount of entropy produced, in kJ/K.

For an enlarged system that includes the mixture and enough of its immediate surroundings that heat transfer occurs at 4°C, determine the amount of entropy produced, in kJ/K. Discuss.

12.15 A mixture of 2.5 kg of oxygen and 1 kg of argon is compressed from 138 kPa, 27°C to 550 kPa, 120°C. Can this process be accomplished adiabatically? Explain.

12.16 A mixture having a molar analysis of 50% CO_2, 33.3% CO, and 16.7% O_2 enters a compressor operating at steady state at 37°C, 1 bar, 40 m/s with a mass flow rate of 1 kg/s and exits at 237°C, 30 m/s. The rate of heat transfer *from* the compressor to its surroundings is 5% of the power *input*.

(a) Neglecting potential energy effects, determine the power input to the compressor, in kW.
(b) If the compression is polytropic, evaluate the polytropic exponent n and the exit pressure, in bar.

12.17 A mixture of 2 kg of H_2 and 4 kg of N_2 is compressed in a piston–cylinder assembly in a polytropic process for which $n = 1.2$. The temperature increases from 22 to 150°C. Using constant values for the specific heats, determine

(a) the heat transfer, in kJ.
(b) the entropy change, in kJ/K.

12.18 A gas turbine receives a mixture having the following molar analysis: 10% CO_2, 19% H_2O, 71% N_2 at 720 K, 0.35 MPa and a volumetric flow rate of 3.2 m^3/s. Products exit the turbine at 380 K, 0.11 MPa. For adiabatic operation with negligible kinetic and potential energy effects, determine the power developed at steady state, in kW.

12.19 A gas mixture at 1500 K with the molar analysis 10% CO_2, 20% H_2O, 70% N_2 enters a waste-heat boiler operating at steady state, and exits the boiler at 600 K. A separate stream of saturated liquid water enters at 25 bar and exits as saturated vapor with a negligible pressure drop. Ignoring stray heat transfer and kinetic and potential energy changes, determine the mass flow rate of the exiting saturated vapor, in kg per kmol of gas mixture.

12.20 An equimolar mixture of helium and carbon dioxide enters an insulated nozzle at 127°C, 5 bar, 30 m/s and expands isentropically to a pressure of 3.24 bar. Determine the temperature, in °C, and the velocity, in m/s, at the nozzle exit. Neglect potential energy effects.

12.21 An equimolar mixture of helium and argon gases enters a turbine at 1280°C and expands adiabatically through a 7.5:1 pressure ratio. If the isentropic turbine efficiency is 80%, determine at steady state the work developed, in kJ per kg of mixture flowing.

12.22 A gas mixture having a molar analysis of 60% O_2 and 40% N_2 enters an insulated compressor operating at steady state at 1 bar, 20°C with a mass flow rate of 0.5 kg/s and is compressed to 5.4 bar. Kinetic and potential energy effects are negligible. For an isentropic compressor efficiency of 78%, determine

(a) the temperature at the exit, in °C.
(b) the power required, in kW.
(c) the rate of entropy production, in kW/K.

12.23 A mixture having an analysis on a mass basis of 80% N_2, 20% CO_2 enters a nozzle operating at steady state at 1000 K with a velocity of 5 m/s and expands adiabatically through a 7.5:1 pressure ratio, exiting with a velocity of 900 m/s. Determine the isentropic nozzle efficiency.

12.24 A mixture having a molar analysis of 60% N_2 and 40% CO_2 enters an insulated compressor operating at steady state at 1 bar, 30°C with a mass flow rate of 1 kg/s and is compressed to 3 bar, 147°C. Neglecting kinetic and potential energy effects, determine

(a) the power required, in kW.
(b) the isentropic compressor efficiency.
(c) the rate of exergy destruction, in kW, for $T_0 = 300$ K.

12.25 An equimolar mixture of N_2 and CO_2 enters a heat exchanger at −40°C, 3.5 MPa and exits at 260°C, 3.5 MPa. The heat exchanger operates at steady state, and kinetic and potential energy effects are negligible.

(a) Using the ideal gas mixture concepts of the present chapter, determine the rate of heat transfer to the mixture, in kJ per kmol of mixture flowing.
(b) Compare with the value of the heat transfer determined using the generalized enthalpy chart (Fig. A-4), together with Kay's rule (see Sec. 11.8).

12.26 Natural gas having a molar analysis of 60% methane (CH_4) and 40% ethane (C_2H_6) enters a compressor at 340 K, 6 bar and is compressed isothermally without internal irreversibilities to 20 bar. The compressor operates at steady state, and kinetic and potential energy effects are negligible.

(a) Assuming ideal gas behavior, determine for the compressor the work and heat transfer, each in kJ per kmol of mixture flowing.
(b) Compare with the values for work and heat transfer, respectively, determined assuming ideal solution behavior (Sec. 11.9.5). For the pure components at 340 K:

	h (kJ/kg)		s (kJ/Kg · K)	
	6 bar	20 bar	6 bar	20 bar
Methane	715.33	704.40	10.9763	10.3275
Ethane	462.39	439.13	7.3493	6.9680

Forming Mixtures

12.27 One kilogram of argon at 27°C, 1 bar is contained in a rigid tank connected by a valve to another rigid tank containing 0.8 kg of O_2 at 127°C, 5 bar. The valve is opened, and the gases are allowed to mix, achieving an equilibrium state at 87°C. Determine

(a) the volume of each tank, in m^3.
(b) the final pressure, in bar.
(c) the heat transfer to or from the gases during the process, in kJ.
(d) the entropy change of each gas, in kJ/K.

12.28 Using the ideal gas model with constant specific heats, determine the mixture temperature, in K, for each of two cases:

(a) Initially, 0.6 kmol of O_2 at 500 K is separated by a partition from 0.4 kmol of H_2 at 300 K in a rigid insulated vessel. The partition is removed and the gases mix to obtain a final equilibrium state.
(b) Oxygen (O_2) at 500 K and a molar flow rate of 0.6 kmol/s enters an insulated control volume operating at steady

state and mixes with H_2 entering as a separate stream at 300 K and a molar flow rate of 0.4 kmol/s. A single mixed stream exits. Kinetic and potential energy effects can be ignored.

12.29 A system consists initially of n_A moles of gas A at pressure p and temperature T and n_B moles of gas B separate from gas A but at the same pressure and temperature. The gases are allowed to mix with no heat or work interactions with the surroundings. The final equilibrium pressure and temperature are p and T, respectively, and the mixing occurs with no change in total volume.

(a) Assuming ideal gas behavior, obtain an expression for the entropy produced in terms of $\bar{R}$, n_A, and n_B.
(b) Using the result of part (a), demonstrate that the entropy produced has a positive value.
(c) Would entropy be produced when samples of the *same* gas at the *same* temperature and pressure mix? Explain.

12.30 Determine the amount of entropy produced, in kJ/K when 0.5 kg of H_2 at 21°C, 1 bar are allowed to mix adiabatically to a final equilibrium state with 10 kg of **(a)** CO_2 and **(b)** H_2, initially at the same temperature and pressure.

12.31 An insulated tank has two compartments connected by a valve. Initially, one compartment contains 0.7 kg of CO_2 at 500 K, 6.0 bar and the other contains 0.3 kg of N_2 at 300 K, 6.0 bar. The valve is opened and the gases are allowed to mix until equilibrium is achieved. Determine

(a) the final temperature, in K.
(b) the final pressure, in bar.
(c) the amount of entropy produced, in kJ/K.

12.32 An insulated tank having a total volume of 1.62 m^3 is divided into two compartments. Initially one compartment having a volume of 0.54 m^3 contains 2 kg of carbon monoxide (CO) at 260°C and the other contains 0.4 kg of helium (He) at 15°C. The gases are allowed to mix until an equilibrium state is attained. Determine

(a) the final temperature, in °C.
(b) the final pressure, in kPa.
(c) the exergy destruction, in kJ, for T_0 = 15°C.

12.33 A rigid insulated tank has two compartments. Initially one compartment is filled with 1 kmol of argon at 65°C, 350 kPa and the other is filled with 0.3 kmol of helium at – 18°C, 103.5 kPa. The gases are allowed to mix until an equilibrium state is attained. Determine

(a) the final temperature, in °C.
(b) the final pressure, in bars.
(c) the amount of entropy produced, in kJ/K.

12.34 A rigid insulated tank has two compartments. Initially one contains 0.5 kmol of carbon dioxide (CO_2) at 27°C, 2 bar and the other contains 1 kmol of oxygen (O_2) at 152°C, 5 bar. The gases are allowed to mix while 500 kJ of energy are added by electrical work. Determine

(a) the final temperature, in °C.
(b) the final pressure, in bar.
(c) the change in exergy, in kJ, for T_0 = 20°C.
(d) the exergy destruction, in kJ.

12.35 Air at 40°C, 1 atm and a volumetric flow rate of 50 m^3/min enters an insulated control volume operating at steady state and mixes with helium entering as a separate stream at 100°C, 1 atm and a volumetric flow rate of 20 m^3/min. A single mixed stream exits at 1 atm. Ignoring kinetic and potential energy effects, determine for the control volume

(a) the temperature of the exiting mixture, in °C.
(b) the rate of entropy production, in kW/K.

12.36 Argon (Ar), at 300 K, 1 bar with a mass flow rate of 1 kg/s enters the insulated mixing chamber shown in Fig. P12.36 and mixes with carbon dioxide (CO_2) entering as a separate stream at 575 K, 1 bar with a mass flow rate of 0.5 kg/s. The mixture exits at 1 bar. Assume ideal gas behavior with k = 1.67 for Ar and k = 1.25 for CO_2. For steady-state operation, determine

(a) the molar analysis of the exiting mixture.
(b) the temperature of the exiting mixture, in K.
(c) the rate of entropy production, in kW/K.

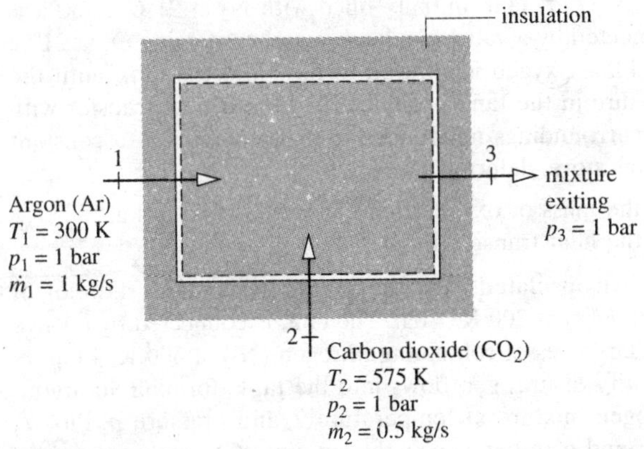

Fig. P12.36

12.37 Nitrogen (N_2) at 50°C, 138 kPa and a volumetric flow rate of 90 m^3/min enters an insulated control volume operating at steady state and mixes with oxygen (O_2) entering as a separate stream at 93°C, 138 kPa and a mass flow rate of 25 kg/min. A single mixed stream exits at 120 kPa. Kinetic and potential energy effects can be ignored. Using the ideal gas model with constant specific heats, determine for the control volume

(a) the temperature of the exiting mixture, in °C.
(b) the rate of exergy destruction, in kJ/s, for T_0 = 4°C.

12.38 Air at 77°C, 1 bar, and a molar flow rate of 0.1 kmol/s enters an insulated mixing chamber operating at steady state and mixes with water vapor entering at 277°C, 1 bar, and a molar flow rate of 0.3 kmol/s. The mixture exits at 1 bar. Kinetic and potential energy effects can be ignored. For the chamber, determine

(a) the temperature of the exiting mixture, in °C.
(b) the rate of entropy production, in kW/K.

12.39 A gas mixture required in an industrial process is prepared by first allowing carbon monoxide (CO) at 27°C, 125 kPa to enter an insulated mixing chamber operating at steady state and mix with argon (Ar) entering at 190°C, 125 kPa. The mixture exits the chamber at 60°C, 110 kPa and is then allowed to expand in a throttling process through a valve to 101.3 kPa. Determine

(a) the mass and molar analyses of the mixture.
(b) the temperature of the mixture at the exit of the valve, in °C.
(c) the rates of exergy destruction for the mixing chamber and the valve, each in kJ per kg of mixture, for $T_0 = 4°C$.

Kinetic and potential energy effects can be ignored.

12.40 Helium at 400 K, 1 bar enters an insulated mixing chamber operating at steady state, where it mixes with argon entering at 300 K, 1 bar. The mixture exits at a pressure of 1 bar. If the argon mass flow rate is x times that of helium, plot versus x

(a) the exit temperature, in K.
(b) the rate of exergy destruction within the chamber, in kJ per kg of helium entering.

Kinetic and potential energy effects can be ignored. Let $T_0 = 300$ K.

12.41 A 27 m^3 tank initially filled with N_2 at 21°C, 35 kPa is connected by a valve to a large vessel containing O_2 at 21°C, 138 kPa. Oxygen is allowed to flow into the tank until the pressure in the tank becomes 103 kPa. If heat transfer with the surroundings maintains the tank contents at a constant temperature, determine

(a) the mass of oxygen that enters the tank, in kg.
(b) the heat transfer, in kJ.

12.42 An insulated, rigid tank initially contains 1 kmol of argon (Ar) at 300 K, 1 bar. The tank is connected by a valve to a large vessel containing nitrogen (N_2) at 500 K, 4 bar. A quantity of nitrogen flows into the tank, forming an argon-nitrogen mixture at temperature T and pressure p. Plot T, in K, and p, in bar, versus the amount of N_2 within the tank, in kmol.

12.43 A stream of air (21% O_2 and 79% N_2 on a molar basis) at 300 K and 0.1 MPa is to be *separated* into pure oxygen and nitrogen streams, each at 300 K and 0.1 MPa. A device to achieve the separation is claimed to require a work input at steady state of 1200 kJ per kmol of air. Heat transfer between the device and its surroundings occurs at 300 K. Ignoring kinetic and potential energy effects, evaluate whether the work value can be as claimed.

12.44 A device is being designed to *separate* into components a natural gas consisting of CH_4 and C_2H_6 in which the mole fraction of C_2H_6, denoted by y, may vary from 0.05 to 0.50. The device will receive natural gas at 20°C, 1 atm with a volumetric flow rate of 100 m^3/s. Separate streams of CH_4 and C_2H_6 will exit, each at 20°C, 1 atm. Heat transfer between the device and its surroundings occurs at 20°C. Ignoring kinetic and potential energy effects, plot versus y the minimum theoretical work input required at steady state, in kW.

Exploring Psychrometric Principles

12.45 A water pipe at 5°C runs above ground between two buildings. The surrounding air is at 35°C. What is the maximum relative humidity the air can have before condensation occurs on the pipe?

12.46 A can of soft drink at a temperature of 4°C is taken from a refrigerator into a room where the temperature is 21°C and the relative humidity is 70%. Explain why beads of moisture form on the can's outer surface. Provide supporting calculations.

12.47 On entering a dwelling maintained at 20°C from the outdoors where the temperature is 10°C, a person's eyeglasses are observed *not* to become fogged. A humidity gauge indicates that the relative humidity in the dwelling is 55%. Can this reading be correct? Provide supporting calculations.

12.48 A large room contains moist air at 30°C, 102 kPa. The partial pressure of water vapor is 1.5 kPa. Determine

(a) the relative humidity.
(b) the humidity ratio.
(c) the dew point temperature, in °C.
(d) the mass of dry air, in kg, if the mass of water vapor is 10 kg.

12.49 To what temperature, in °C, must moist air with a humidity ratio of 5×10^{-3} be cooled at a constant pressure of 2 bar to become saturated moist air?

12.50 A fixed amount of moist air initially at 1 bar and a relative humidity of 60% is compressed isothermally until condensation of water begins. Determine the pressure of the mixture at the onset of condensation, in bar. Repeat if the initial relative humidity is 90%.

12.51 Figure P12.51 shows a dryer operating at steady state. Damp fabric containing 60% moisture by mass enters on a conveyor and exits with a moisture content of 6% by mass at a rate of 215 kg/h. Dry air at 66°C, 1 bar enters, and moist air at 55°C, 1 bar, and 50% relative humidity exits. Determine the required volumetric flow rate of the entering dry air, in m^3/h.

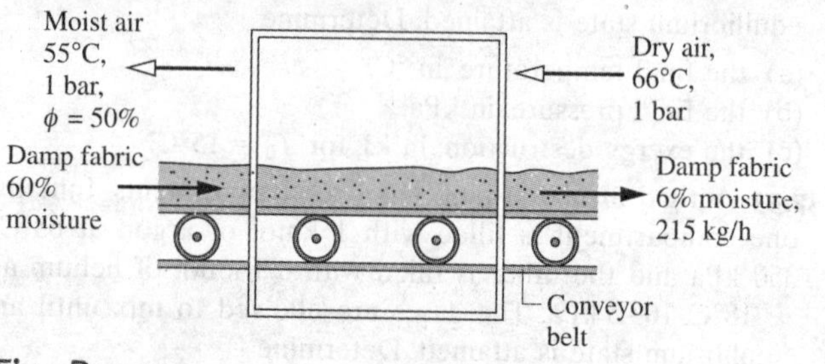

Fig. P12.51

12.52 One kg of moist air initially at 27°C, 1 bar, 50% relative humidity is compressed isothermally to 3 bar. If condensation occurs, determine the amount of water condensed, in kg. If there is no condensation, determine the final relative humidity.

12.53 A vessel whose volume is 0.5 m^3 initially contains dry air at 0.2 MPa and 20°C. Water is added to the vessel until the air is saturated at 20°C. Determine the

(a) mass of water added, in kg.
(b) final pressure in the vessel, in bar.

12.54 Wet grain at 20°C containing 40% moisture by mass enters a dryer operating at steady state. Dry air enters the dryer at 90°C, 1 atm at a rate of 15 kg per kg of wet grain entering. Moist air exits the dryer at 38°C, 1 atm, and 52% relative humidity. For the grain exiting the dryer, determine the percent moisture by mass.

12.55 Figure P12.55 shows a spray dryer operating at steady state. The mixture of liquid water and suspended solid particles entering through the spray head contains 30% solid matter by mass. Dry air enters at 177°C, 1 atm, and moist air exits at 85°C, 1 atm, and 21% relative humidity with a volumetric flow rate of 310 m^3/min. The dried particles exit separately. Determine

(a) the volumetric flow rate of the entering dry air, in m^3/min.
(b) the rate that dried particles exit, in kg/min.

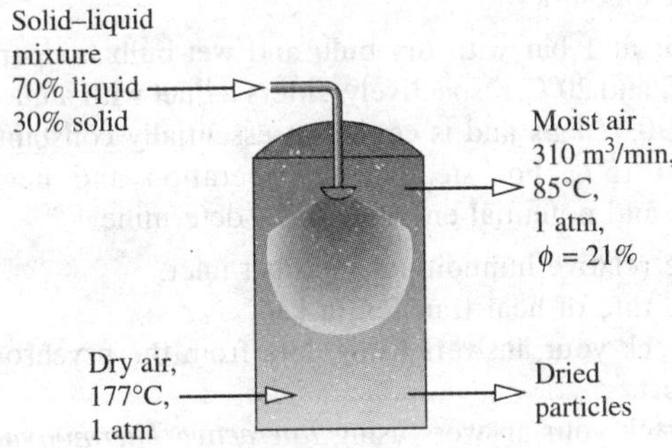

Fig. P12.55

12.56 A mixture of nitrogen and water vapor at 93°C, 1 bar has the molar analysis 80% N_2, 20% water vapor. If the mixture is cooled at constant pressure, determine the temperature, in °C, at which water vapor begins to condense.

12.57 A system consisting initially of 0.5 m^3 of air at 35°C, 1 bar, and 70% relative humidity is cooled at constant pressure to 29°C. Determine the work and heat transfer for the process, each in kJ.

12.58 A closed, rigid tank having a volume of 3 m^3 initially contains air at 100°C, 4.4 bar, and 40% relative humidity. Determine the heat transfer, in kJ, if the tank contents are cooled to **(a)** 80°C, **(b)** 20°C.

12.59 A closed, rigid tank initially contains 0.5 m^3 of moist air in equilibrium with 0.1 m^3 of liquid water at 80°C and 0.1 MPa. If the tank contents are heated to 200°C, determine

(a) the final pressure, in MPa.
(b) the heat transfer, in kJ.

12.60 A closed, rigid cylindrical tank having a height of 6 m and a diameter of 2 m initially contains air at 150°C, 550 kPa, and 10% relative humidity. If the tank contents are cooled to 77°C, determine

(a) if condensation occurs.
(b) the final pressure, in kPa.
(c) the heat transfer, in kJ.
(d) the change in entropy, in kJ/K.

12.61 Combustion products with the molar analysis of 10% CO_2, 20% H_2O, 70% N_2 enter an engine's exhaust pipe at 538°C, 1 bar and are cooled as they pass through the pipe, exiting at 38°C, 1 bar. Determine the heat transfer at steady state, in kJ/kg of entering mixture.

12.62 Air at 16°C, 101.3 kPa, and 75% relative humidity enters an insulated compressor operating at steady state and is compressed to 690 kPa. The isentropic compressor efficiency is η_c.

(a) For $\eta_c = 0.8$, determine the temperature, in K, of the exiting air, and the work input required and the exergy destruction, each in kJ/kg of dry air flowing. Let $T_0 = 290$ K.
(b) Plot each of the quantities determined in part (a) versus η_c ranging from 0.7 to 1.0.

12.63 Air at 35°C, 3 bar, 30% relative humidity, and a velocity of 50 m/s expands isentropically through a nozzle. Determine the lowest exit pressure, in bar, that can be attained without condensation. For this exit pressure, determine the exit velocity, in m/s. The nozzle operates at steady state and without significant potential energy effects.

12.64 A closed, rigid tank having a volume of 1 m^3 contains a mixture of carbon dioxide (CO_2) and water vapor at 75°C. The respective masses are 12.3 kg of carbon dioxide and 0.05 kg of water vapor. If the tank contents are cooled to 20°C, determine the heat transfer, in kJ, assuming ideal gas behavior.

12.65 Gaseous combustion products at 427°C, 1 bar and a volumetric flow rate of 0.135 m^3/s enter a counterflow heat exchanger operating at steady state and exit at 93°C. The molar analysis of the products is 7.1% CO_2, 4.3% O_2, 14.3% H_2O, 74.3% N_2. A separate moist air stream enters the heat exchanger at 15°C, 1 bar, and 30% relative humidity and exits at 38°C. Determine the mass flow rate of the entering moist air stream, in kg/s. The pressure drops of the two streams can be ignored, as can stray heat transfer and kinetic and potential energy effects.

12.66 Moist air at 20°C, 1.05 bar, 85% relative humidity and a volumetric flow rate of 0.3 m^3/s enters a well-insulated compressor operating at steady state. If moist air exits at 100°C, 2.0 bar, determine

(a) the relative humidity at the exit.
(b) the power input, in kW.
(c) the rate of entropy production, in kW/K.

12.67 Moist air at 32°C, 1 bar, 60% relative humidity and a volumetric flow rate of 54 m^3/min enters a control volume at steady state and flows along a surface maintained at 4°C, through which heat transfer occurs. Saturated moist air and condensate, each at 12°C, exit the control volume. For the control volume, $\dot{W}_{cv} = 0$, and kinetic and potential energy effects are negligible. Determine

(a) the rate of heat transfer, in kJ/min.
(b) the rate of entropy production, in kJ/K · min.

12.68 Moist air at 20°C, 1 atm, 43% relative humidity and a volumetric flow rate of 900 m^3/h enters a control volume at steady state and flows along a surface maintained at 65°C, through which heat transfer occurs. Liquid water at 20°C is injected at a rate of 5 kg/h and evaporates into the flowing stream. For the control volume, $\dot{W}_{cv} = 0$, and kinetic and potential energy effects are negligible. Moist air exits at 32°C, 1 atm. Determine

(a) the rate of heat transfer, in kW.
(b) the rate of entropy production, in kW/K.

12.69 Using Eq. 12.48, determine the humidity ratio and relative humidity for each case below.

(a) The dry-bulb and wet-bulb temperatures in a conference room at 1 bar are 24 and 16°C, respectively.
(b) The dry-bulb and wet-bulb temperatures in a factory space at 1 bar are 25°C and 15°C, respectively.
(c) Repeat parts (a) and (b) using the psychrometric chart.
(d) Repeat parts (a) and (b) using *Interactive Thermodynamics: IT*.

12.70 Using the psychrometric chart, Fig. A-9, determine

(a) the relative humidity, the humidity ratio, and the specific enthalpy of the mixture, in kJ per kg of dry air, corresponding to dry-bulb and wet-bulb temperatures of 30 and 25°C, respectively.
(b) the humidity ratio, mixture specific enthalpy, and wet-bulb temperature corresponding to a dry-bulb temperature of 30°C and 60% relative humidity.
(c) the dew point temperature corresponding to dry-bulb and wet-bulb temperatures of 30 and 20°C, respectively.
(d) Repeat parts (a)–(c) using *Interactive Thermodynamics: IT*.

12.71 Using the psychrometric chart, Fig. A-9, determine

(a) the dew point temperature corresponding to dry-bulb and wet-bulb temperatures of 30°C and 25°C, respectively.
(b) the humidity ratio, the specific enthalpy of the mixture, in kJ per kg of dry air, and the wet-bulb temperature corresponding to a dry-bulb temperature of 30°C and 60% relative humidity.
(c) the relative humidity, humidity ratio, and mixture specific enthalpy corresponding to dry-bulb and wet-bulb temperatures of 30°C and 20°C, respectively.
(d) Repeat parts (a)–(c) using *Interactive Thermodynamics: IT*.

12.72 A fixed amount of air initially at 52°C, 1 atm, and 10% relative humidity is cooled at constant pressure to 15°C. Using the psychrometric chart, determine whether condensation occurs. If so, evaluate the amount of water condensed, in kg per kg of dry air. If there is no condensation, determine the relative humidity at the final state.

12.73 A fan within an insulated duct delivers moist air at the duct exit at 22°C, 60% relative humidity, and a volumetric flow rate of 0.5 m^3/s. At steady state, the power input to the fan is 1.3 kW. Using the psychrometric chart, determine the temperature and relative humidity at the duct inlet.

12.74 The mixture enthalpy per unit mass of dry air, in kJ/kg(a), represented on Fig. A-9 can be approximated closely from the expression

$$\frac{H}{m_a} = 1.005\ T(°C) + \omega[2501.7 + 1.82\ T(°C)]$$

Noting all significant assumptions, develop the above expressions.

Considering Air-Conditioning Applications

12.75 Each case listed gives the dry-bulb temperature and relative humidity of the moist air stream entering an air-conditioning system: **(a)** 30°C, 40%, **(b)** 17°C, 60%, **(c)** 25°C, 70%, **(d)** 15°C, 40%, **(e)** 27°C, 30%. The condition of the moist air stream exiting the system must satisfy these *constraints:* $22 \le T_{db} \le 27°C$, $40 \le \phi \le 60\%$. In each case, develop a schematic of equipment and processes from Sec. 12.9 that would achieve the desired result. Sketch the processes on a psychrometric chart.

12.76 Moist air enters an air-conditioning system as shown in Fig. 12.11 at 26°C, $\phi = 80\%$ and a volumetric flow rate of 0.47 m^3/s. At the exit of the heating section the moist air is at 26°C, $\phi = 50\%$. For operation at steady state, and neglecting kinetic and potential energy effects, determine

(a) the rate energy is removed by heat transfer in the dehumidifier section, in tons.
(b) the rate energy is added by heat transfer in the heating section, in kW.

12.77 Air at 1 bar with dry-bulb and wet-bulb temperatures of 30°C and 20°C, respectively, enters a duct with a mass flow rate of 0.07 kg/s and is cooled at essentially constant pressure to 15°C. For steady-state operation and negligible kinetic and potential energy effects, determine

(a) the relative humidity at the duct inlet.
(b) the rate of heat transfer, in kJ/s.
(c) Check your answers using data from the psychrometric chart.
(d) Check your answers using *Interactive Thermodynamics: IT*.

12.78 Air at 35°C, 1 bar, and 50% relative humidity enters a dehumidifier operating at steady state. Saturated moist air and condensate exit in separate streams, each at 15°C. Neglecting kinetic and potential energy effects, determine

(a) the heat transfer from the moist air, in kJ per kg of dry air.
(b) the amount of water condensed, in kg per kg of dry air.
(c) Check your answers using data from the psychrometric chart.
(d) Check your answers using *Interactive Thermodynamics: IT*.

12.79 Air at 30°C, 1 bar, and 70% relative humidity enters a dehumidifier operating at steady state with a mass flow rate of 1 kg/s. Saturated moist air and condensate exit in separate streams, each at 10°C. Neglecting kinetic and potential energy effects, determine

(a) the rate of heat transfer from the moist air, in tons.
(b) the rate water is condensed, in kg/s.
(c) Check your answers using data from the psychrometric chart.
(d) Check your answers using *Interactive Thermodynamics: IT*.

12.80 Moist air at 28°C, 1 bar, and 50% relative humidity flows through a duct operating at steady state. The air is cooled at essentially constant pressure and exits at 20°C. Determine the heat transfer rate, in kJ per kg of dry air flowing, and the relative humidity at the exit.

12.81 An air conditioner operating at steady state takes in moist air at 28°C, 1 bar, and 70% relative humidity. The moist air first passes over a cooling coil in the dehumidifier unit and some water vapor is condensed. The rate of heat transfer between the moist air and the cooling coil is 11 tons. Saturated moist air and condensate streams exit the dehumidifier unit at the same temperature. The moist air then passes through a heating unit, exiting at 24°C, 1 bar, and 40% relative humidity. Neglecting kinetic and potential energy effects, determine

(a) the temperature of the moist air exiting the dehumidifer unit, in °C.
(b) the volumetric flow rate of the air entering the air conditioner, in m^3/min.
(c) the rate water is condensed, in kg/min.
(d) the rate of heat transfer to the air passing through the heating unit, in kW.

12.82 Figure P12.82 shows a compressor followed by an aftercooler. Atmospheric air at 100 kPa, 30°C, and 75% relative humidity enters the compressor with a volumetric flow rate of 0.05 m^3/s. The compressor power input is 10 kW. The moist air exiting the compressor at 670 kPa, 200°C flows through the aftercooler, where it is cooled at constant pressure, exiting saturated at 40°C. Condensate also exits the aftercooler at 40°C. For steady-state operation and negligible kinetic and potential energy effects, determine

(a) the rate of heat transfer from the compressor to its surroundings, in kJ/s.
(b) the mass flow rate of the condensate, in kg/s.
(c) the rate of heat transfer from the moist air to the refrigerant circulating in the cooling coil, in tons of refrigeration.

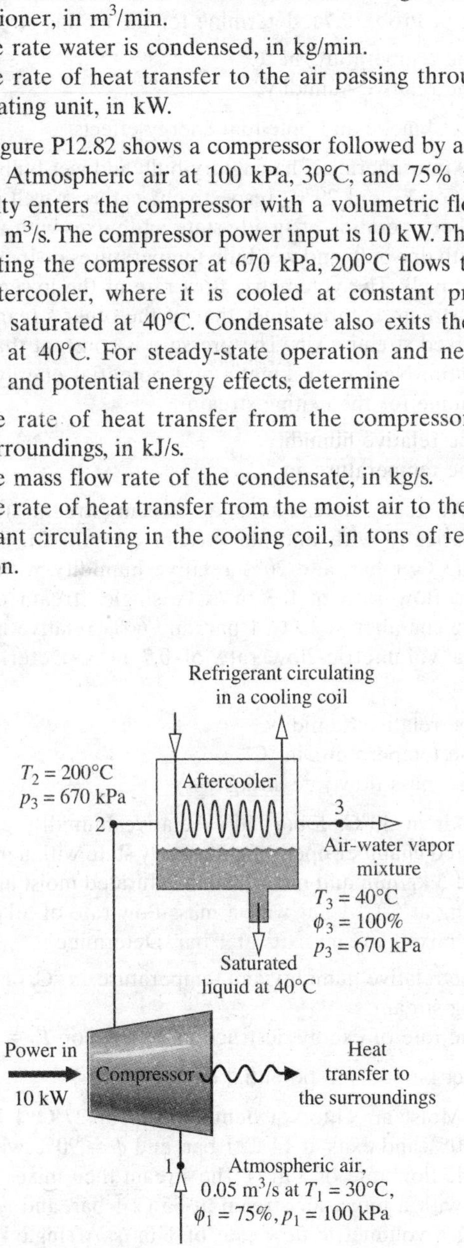

Fig. P12.82

12.83 Outside air at 10°C, 1 bar, and 40% relative humidity enters an air-conditioning device operating at steady state. Liquid water is injected at 5°C and a moist air stream exits with a volumetric flow rate of 0.5 m^3/s at 30°C, 1 bar and a relative humidity of 40%. Neglecting kinetic and potential energy effects, determine

(a) the rate water is injected, in kg/s.
(b) the rate of heat transfer to the moist air, in kJ/s.

12.84 An air-conditioning system consists of a spray section followed by a reheater. Moist air at 32°C and $\phi = 77\%$ enters the system and passes through a water spray, leaving the spray section cooled and saturated with water. The moist air is then heated to 25°C and $\phi = 45\%$ with no change in the amount of water vapor present. For operation at steady state, determine

(a) the temperature of the moist air leaving the spray section, in °C.
(b) the change the amount of water vapor contained in the moist air passing through the system, in kg per kg of dry air.

Locate the principal states on a psychrometric chart.

12.85 Moist air at 35°C, 1 bar, and a relative humidity of 30% enters a steam-spray humidification device operating at steady state with a volumetric flow rate of 3 m^3/s. Saturated water vapor at 110°C is sprayed into the moist air, which then exits the device at a relative humidity of 50%. Heat transfer between the device and its surroundings can be ignored, as can kinetic and potential energy effects. Determine

(a) the temperature of the exiting moist air stream, in °C.
(b) the rate at which steam is injected, in kg/s.

12.86 For the steam-spray humidifier in Prob. 12.85, determine the exergy destruction rate, in kJ/s. Let $T_0 = 35°C$.

12.87 Outside air at 10°C, 1 bar, and 40% relative humidity enters an air conditioner operating at steady state with a mass flow rate of 1.5 kg/s. The air is first heated at essentially constant pressure to 30°C. Liquid water at 15°C is then injected, bringing the air to 25°C, 1 bar. Determine

(a) the rate of heat transfer to the air passing through the heating section, in kJ/s.
(b) the rate water is injected, in kg/s.
(c) the relative humidity at the exit of the humidification section.

Kinetic and potential energy effects can be ignored.

12.88 Moist air at 27°C, 1 bar, and 50% relative humidity enters an evaporative cooling unit operating at steady state consisting of a heating section followed by a soaked pad evaporative cooler operating adiabatically. The air passing through the heating section is heated to 45°C. Next, the air passes through a soaked pad exiting with 50% relative humidity. Using data from the psychrometric chart, determine

(a) the humidity ratio of the entering moist air mixture.
(b) the rate of heat transfer to the moist air passing through the heating section, in kJ per kg of mixture.
(c) the humidity ratio and temperature, in °C, at the exit of the evaporative cooling section.

12.89 In an industrial dryer operating at steady state, atmospheric air at 30°C, 1 bar, and 65% relative humidity is first heated to 140°C at constant pressure. The heated air is then allowed to pass over the materials being dried, exiting the dryer at 65°C, 1 bar, and 30% relative humidity. If moisture is to be removed from the materials at a rate of 0.3 kg/s, determine

(a) the mass flow rate of dry air required, in kg/s.
(b) the rate of heat transfer to the air as it passes through the heating section, in kJ/s.

Neglect kinetic and potential energy effects.

12.90 At steady state, moist air is to be supplied to a classroom at a specified volumetric flow rate and temperature T. Air is removed from the classroom in a separate stream at a temperature of 27°C and 50% relative humidity. Moisture is added to the air in the room from the occupants at a rate of 4.5 kg/h. The moisture can be regarded as saturated vapor at 33°C. Heat transfer into the occupied space from all sources is estimated to occur at a rate of 34,000 kJ/h. The pressure remains uniform at 1 atm.

(a) For a supply air volumetric flow rate of 40 m^3/min, determine the supply air temperature T, in °C, and the relative humidity.
(b) Plot the supply air temperature, in °C, and relative humidity, each versus the supply air volumetric flow rate ranging from 35 to 90 m^3/min.

12.91 At steady state, a device for heating and humidifying air has 0.12m^3/s of air at 4°C, 1 bar, and 80% relative humidity entering at one location, 0.5 m^3/s of air at 15°C, 1 bar, and 80% relative humidity entering at another location, and liquid water injected at 12°C. A single moist air stream exits at 30°C, 1 bar, and 35% relative humidity. Determine

(a) the rate of heat transfer to the device, in kJ/s.
(b) the rate at which liquid water is injected, in kg/s.

Neglect kinetic and potential energy effects.

12.92 Air at 35°C, 1 bar, and 10% relative humidity enters an evaporative cooler operating at steady state. The volumetric flow rate of the incoming air is 50 m^3/min. Liquid water at 20°C enters the cooler and fully evaporates. Moist air exits the cooler at 25°C, 1 bar. If there is no significant heat transfer between the device and its surroundings, determine

(a) the rate at which liquid enters, in kg/min.
(b) the relative humidity at the exit.
(c) the rate of exergy destruction, in kJ/min, for T_0 = 20°C.

Neglect kinetic and potential energy effects.

12.93 Using Eqs. 12.56b and 12.56c, show that

$$\frac{\dot{m}_{a1}}{\dot{m}_{a2}} = \frac{\omega_3 - \omega_2}{\omega_1 - \omega_3} = \frac{(h_{a3} + \omega_3 h_{g3}) - (h_{a2} + \omega_2 h_{g2})}{(h_{a1} + \omega_1 h_{g1}) - (h_{a3} + \omega_3 h_{g3})}$$

Employ this relation to demonstrate on a psychrometric chart that state 3 of the mixture lies on a straight line connecting the initial states of the two streams before mixing.

12.94 For the adiabatic mixing process in Example 12.14, plot the exit temperature, in °C, versus the volumetric flow rate of stream 2 ranging from 0 to 1400 m^3/min. Discuss the plot as $(AV)_2$ goes to zero and as $(AV)_2$ becomes large.

12.95 A stream consisting of 35 m^3/min of moist air at 14°C, 1 atm, 80% relative humidity mixes adiabatically with a stream consisting of 80 m^3/min of moist air at 40°C, 1 atm, 40% relative humidity, giving a single mixed stream at 1 atm. Using the psychrometric chart together with the procedure of Prob. 12.93, determine the relative humidity and temperature, in °C, of the exiting stream.

12.96 At steady state, a stream of air at 15°C, 1 atm, 30% relative humidity is mixed adiabatically with a stream of air at 30°C, 1 bar, 70% relative humidity. The mass flow rate of the higher-temperature stream is twice that of the other stream. A single mixed stream exits at 1 bar. Using the result of Prob. 12.74, determine for the exiting stream

(a) the temperature, in °C.
(b) the relative humidity.

Neglect kinetic and potential energy effects.

12.97 Atmospheric air having dry-bulb and wet-bulb temperatures of 33 and 29°C, respectively, enters a well-insulated chamber operating at steady state and mixes with air entering with dry-bulb and wet-bulb temperatures of 16 and 12°C, respectively. The volumetric flow rate of the lower temperature stream is three times that of the other stream. A single mixed stream exits. The pressure is constant throughout at 1 atm. Neglecting kinetic and potential energy effects, determine for the exiting stream

(a) the relative humidity.
(b) the temperature, in °C.

12.98 At steady state, a moist air stream (stream 1) is mixed adiabatically with another stream (stream 2). Stream 1 is at 13°C, 1 bar, and 20% relative humidity, with a volumetric flow rate of 0.3 m^3/s. A single stream exits the mixing chamber at 19°C, 1 bar, and 60% relative humidity, with a volumetric flow rate of 0.7 m^3/s. Determine for stream 2

(a) the relative humidity.
(b) the temperature, in °C.
(c) the mass flow rate, in kg/s.

12.99 Air at 30°C, 1 bar, 50% relative humidity enters an insulated chamber operating at steady state with a mass flow rate of 3 kg/min and mixes with a saturated moist air stream entering at 5°C, 1 bar with a mass flow rate of 5 kg/min. A single mixed stream exits at 1 bar. Determine

(a) the relative humidity and temperature, in °C, of the exiting stream.
(b) the rate of exergy destruction, in kW, for T_0 = 20°C.

Neglect kinetic and potential energy effects.

12.100 Moist air enters a dehumidifier at 27°C, 1 bar, and ϕ = 60% and exits at 14°C, 1 bar, and ϕ = 90% with a volumetric flow rate of 5 m^3/s. The stream then mixes adiabatically with a moist air stream at 35°C, 1 bar, and ϕ = 47% having a volumetric flow rate of 1 m^3/s. A single moist air stream exits the mixing chamber. Stray heat transfer and kinetic and potential energy effects can be ignored. Determine at steady state

(a) the rate water is removed from the moist air passing through the dehumidifier, in kg/s.
(b) the temperature, in °C, and the relative humidity of the moist air at the mixing chamber exit.

12.101 A stream of air (stream 1) at 15°C, 1 bar, 30% relative humidity is mixed adiabatically with a stream of air (stream 2) at 32°C, 1 bar, 80% relative humidity. A single stream (stream 3) exits the mixing chamber at temperature T_3 and 1 bar. Assume steady state and ignore kinetic and potential energy effects. Letting r denote the ratio of dry air mass flow rates $\dot{m}_{a1}/\dot{m}_{a2}$

(a) determine T_3, in °C, for $r = 2$.
(b) plot T_3, in °C, versus r ranging from 0 to 10.

12.102 Figure P12.102 shows the adiabatic mixing of two moist air streams at steady state. Kinetic and potential energy effects are negligible. Determine the rate of exergy destruction, in kJ/s, for $T_0 = 35$°C.

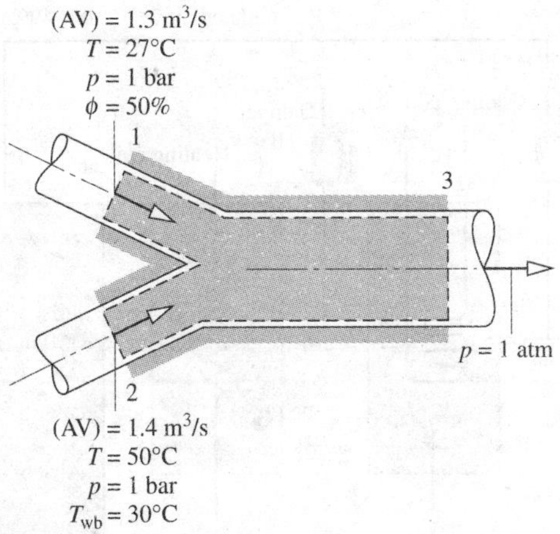

Fig. P12.102

Analyzing Cooling Towers

12.103 In the condenser of a power plant, energy is discharged by heat transfer at a rate of 836 MW to cooling water that exits the condenser at 40°C into a cooling tower. Cooled water at 20°C is returned to the condenser. Atmospheric air enters the tower at 25°C, 1 bar, 35% relative humidity. Moist air exits at 35°C, 1 bar, 90% relative humidity. Makeup water is supplied at 20°C. For operation at steady state, determine the mass flow rate, in kg/s, of

(a) the entering atmospheric air.
(b) the makeup water.

Ignore kinetic and potential energy effects.

12.104 Liquid water at 50°C enters a cooling tower operating at steady state with a mass flow rate of 64 kg/s. Atmospheric air enters at 27°C, 1 bar, 30% relative humidity. Saturated air exits at 38°C, 1 bar. No makeup water is provided. Plot the mass flow rate of dry air required, in kg/s, versus the temperature at which cooled water exits the tower. Consider temperatures ranging from 15 to 32°C. Ignore kinetic and potential energy effects.

12.105 Liquid water at 54°C and a mass flow rate of 4.5×10^4 kg/s enters a cooling tower operating at steady state. Liquid water exits the tower at 20°C. No makeup water is provided. Atmospheric air enters at 1 bar with a dry-bulb temperature of 10°C and a wet-bulb temperature of 2°C. Saturated air exits at 50°C, 1 bar. Ignoring kinetic and potential energy effects, determine the mass flow rate of the cooled water stream exiting the tower, in kg/s.

12.106 Liquid water at 38°C and a volumetric flow rate of 760 L/min enters a cooling tower operating at steady state. Atmospheric air enters at 1 bar with a dry-bulb temperature of 27°C and a wet-bulb temperature of 15°C. Moist air exits the cooling tower at 32°C and 90% relative humidity. Makeup water is provided at 27°C. Plot the mass flow rates of the dry air and makeup water, each in kg/s, versus return water temperature ranging from 27 to 38°C. Ignore kinetic and potential energy effects.

12.107 Liquid water at 50°C enters a forced draft cooling tower operating at steady state. Cooled water exits the tower with a mass flow rate of 80 kg/min. No makeup water is provided. A fan located within the tower draws in atmospheric air at 17°C, 0.098 MPa, 60% relative humidity with a volumetric flow rate of 110 m³/min. Saturated air exits the tower at 30°C, 0.098 MPa. The power input to the fan is 8 kW. Ignoring kinetic and potential energy effects, determine

(a) the mass flow rate of the liquid stream entering, in kg/min.
(b) the temperature of the cooled liquid stream exiting, in °C.

12.108 Liquid water at 43°C and a volumetric flow rate of 0.1 m³/s enters a cooling tower operating at steady state. Cooled water exits the cooling tower at 30°C. Atmospheric air enters the tower at 27°C, 1 bar, 40% relative humidity, and saturated moist air at 40°C, 1 bar exits the cooling tower. Determine

(a) the mass flow rates of the dry air and the cooled water, each in kg/s.
(b) the rate of exergy destruction within the cooling tower, in kJ/s, for $T_0 = 25$°C.

Ignore kinetic and potential energy effects.

Design & open ended problems: exploring engineering practice

12.1D Using methods in the most current *ASHRAE Handbook of HVAC Applications*, estimate the rate of evaporation from the surface of an Olympic-size swimming pool. The swimming hall is maintained at 30°C with a dew point temperature of 21°C. Based on your estimate, determine the load in kW, that the added moisture places on the air-conditioning system. Write a memorandum that summarizes your analysis and include at least three references.

12.2D Figure P12.2D shows a spray cooler included on the schematic of a plant layout. Arguing that it can only be the *addition* of makeup water at location 3 that is intended, an

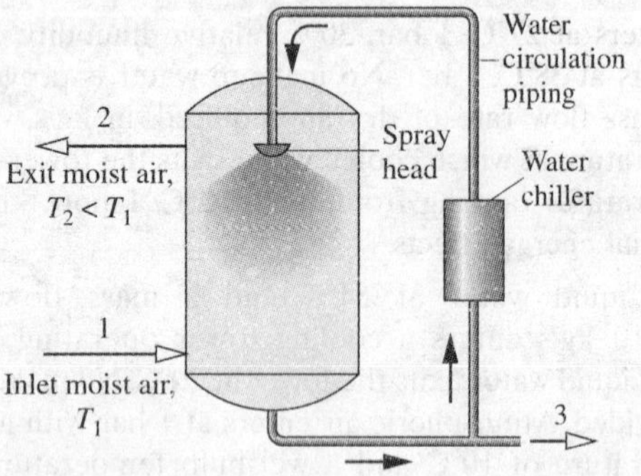

Fig. P12.2D

engineer orders the direction of the arrow at this location on the schematic to be reversed. Evaluate the engineer's order and write a memorandum explaining your evaluation.

12.3D Control of microbial contaminants is crucial in hospital operating rooms. Figure P12.3D illustrates the proposed design of a hospital air-handling system aimed at effective infection control. Write a critique of this design, taking into consideration issues including, but not necessarily limited to the following: Why is 100% outside air specified rather than a recirculated air-filtration arrangement requiring minimum outside air? Why is steam humidification used rather than the more common water-spray method? Why are three filters with increasing efficiency specified, and why are they located as shown? What additional features might be incorporated to promote air quality?

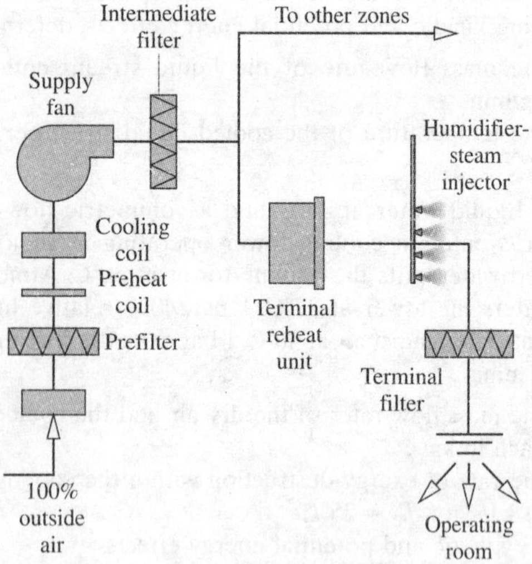

Fig. P12.3D

12.4D Write a report explaining the way the human body regulates its temperature, both under cold and hot climate conditions. Based on these *thermoregulation* mechanisms, discuss designs of *personal protective clothing* for fire fighters and other first responders. How are such clothing systems designed to provide protection against chemical or biological agents while maintaining sufficient thermal comfort to allow for strenuous physical activity? Include in your report at least three references.

12.5D Figure P12.5D shows a system for supplying a space with 2100 m³/min of conditioned air at a dry-bulb temperature of 22°C and a relative humidity of 60% when the outside air is at a dry-bulb temperature of 35°C and a relative humidity of 55%. Dampers A and B can be set to give three alternative operating modes: (1) Both dampers closed (no use of recirculated air). (2) Damper A open and damper B closed. One-third of the conditioned air comes from outside air. (3) Both dampers open. One-third of the conditioned air comes from outside air. One-third of the recirculated air bypasses the dehumidifier via open damper B, and the rest flows through the damper A. Which of the three operating modes should be used? Discuss.

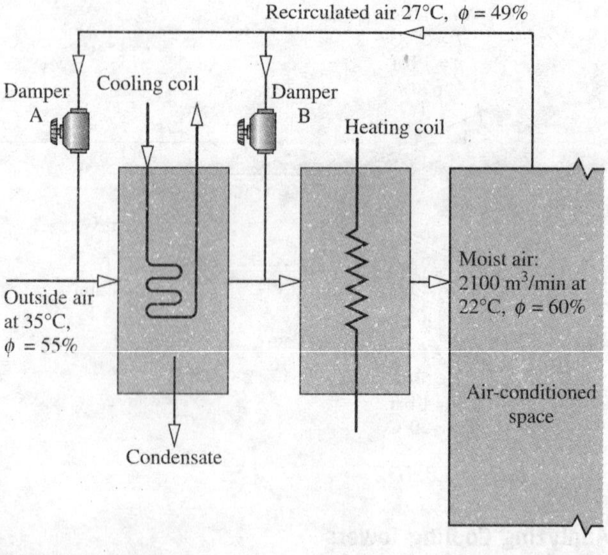

Fig. P12.5D

12.6D Tunnel-type spray coolers are used to cool vegetables. In one application, tomatoes on a 1.2 m wide conveyor belt pass under a 1°C water spray and cool from 20 to 4°C. The water is collected in a reservoir below and recirculated through the evaporator of an ammonia refrigeration unit. Recommend the length of the tunnel and the conveyor speed. Estimate the number of tomatoes that can be cooled in an hour with your design. Write a report including calculations and at least three references.

12.7D A laboratory test facility requires 1.4 m³/s of conditioned air at 38°C and relative humidities ranging from 20 to 50%. As shown in Fig. P12.7D, the supply system takes in ambient air at temperatures ranging from 21 to 32°C and relative humidities ranging from 50 to 100%. As some dehumidification is needed to achieve the temperature and humidity control required by the test facility, two strategies are under consideration: (1) Use a conventional refrigeration system to cool the incoming air below its dew point temperature. (2) Use a *rotary desiccant dehumidification* unit. Develop schematics and design specifications, based on each of these strategies, that would allow the supply system to provide the required temperature and humidity control for this application.

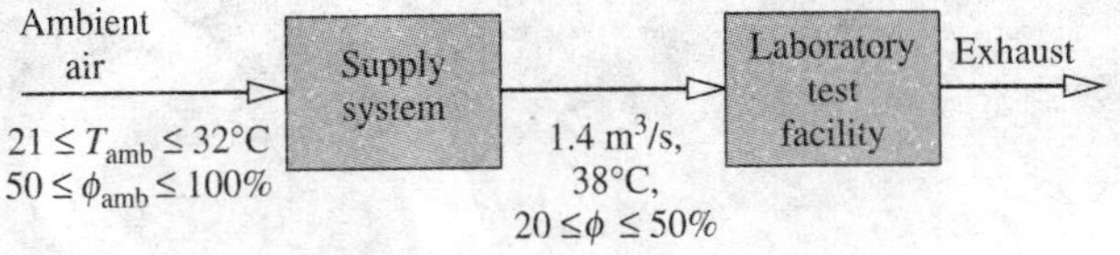

Fig. P12.7D

12.8D An air-handling system is being designed for a 12 m × 12 m × 2.5 m biological research facility that houses 3000 laboratory mice. The indoor conditions must be maintained at 24°C, 60% relative humidity when the outdoor air conditions are 32°C, 70% relative humidity. Develop a preliminary design of an air-conditioning and distribution system that satisfies National Institute of Health (NIH) standards for animal facilities. Assume a *biological safety level of one* (BSL-1), and that two thirds of the floor space is devoted to animal care. Since an interruption in ventilation or air conditioning could place the laboratory animals under stress and compromise the research under way in the facility, account for *redundancy* in your design.

12.9D Figure P12.9D shows a steam-injected gas-turbine cogeneration system that produces power and process steam: a simplified *Cheng* cycle. The steam is generated by a heat-recovery steam generator (HRSG) in which the hot gas exiting the turbine at state 4 is cooled and discharged at state 5. A separate stream of return water enters the HRSG at state 6 and steam exits at states 7 and 8. The superheated steam exiting at 7 is injected into the combustor. In what ways does the steam-injected system shown in the figure differ from the humid air turbine (HAT) cycle? Draw the schematic of a HAT cycle. Discuss appropriate applications of each cycle.

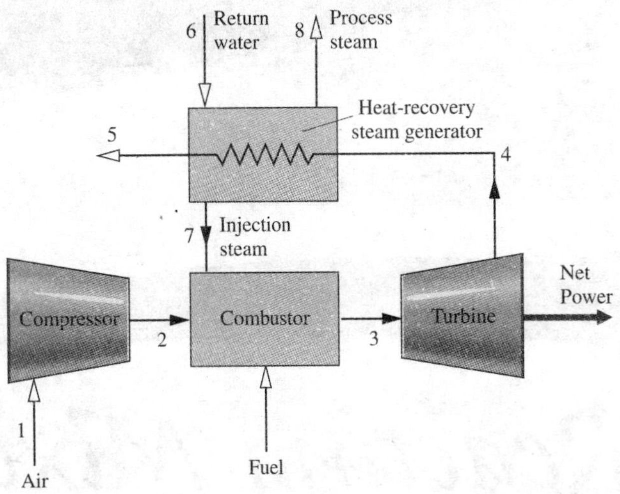

Fig. P12.9D

12.10D Most supermarkets use air-conditioning systems designed primarily to control temperature, and these typically produce a relative humidity of about 55%. At this relative humidity a substantial amount of energy must be used to keep frost and condensation from forming inside refrigerated display cases. Investigate technologies available for reducing the overall humidity levels to 40–45% and thereby reducing problems associated with frost and condensation. Estimate the potential cost savings associated with such a strategy for a supermarket in your locale.

12.11D About half the air we breathe on some airplanes is fresh air, and the rest is recirculated. Investigate the typical equipment schematics for providing a blend of fresh and recirculated filtered air to the passenger cabins of commercial airplanes. What types of filters are used and how do they work? Write a report including at least three references.

13

Reacting Mixtures and Combustion

Engineering Context

The **objective** of this chapter is to study systems involving chemical reactions. Since the *combustion* of hydrocarbon fuels occurs in most power-producing devices (Chaps. 8 and 9), combustion is emphasized. The thermodynamic analysis of reacting systems is primarily an extension of principles introduced thus far. The concepts applied in the *first part* of the chapter dealing with combustion fundamentals remain the same: conservation of mass, conservation of energy, and the second law. It is necessary, though, to modify the methods used to evaluate specific enthalpy, internal energy, and entropy, by accounting for changing chemical composition. Only the manner in which these properties are evaluated represents a departure from previous practice, for once appropriate values are determined they are used as in earlier chapters in the energy and entropy balances for the system under consideration. In the *second part* of the chapter, the exergy concept of Chap. 7 is extended by introducing chemical exergy.

The principles developed in this chapter allow the equilibrium composition of a mixture of chemical substances to be determined. This topic is studied in the next chapter. The subject of *dissociation* is also deferred until then. Prediction of *reaction rates* is not within the scope of classical thermodynamics, so the topic of chemical kinetics, which deals with reaction rates, is not discussed in this text.

Learning Objectives

When you complete your study of this chapter you will be able to...

✓ demonstrate understanding of key concepts, including complete combustion, theoretical air, enthalpy of formation, and adiabatic flame temperature.

✓ determine balanced reaction equations for combustion of hydrocarbon fuels.

✓ apply mass, energy, and entropy balances to closed systems and control volumes involving chemical reactions.

✓ perform exergy analyses, including chemical exergy and the evaluation of exergetic efficiencies.

Combustion Fundamentals

13.1 Introducing Combustion

When a chemical reaction occurs, the bonds within molecules of the ***reactants*** are broken, and atoms and electrons rearrange to form ***products.*** In combustion reactions, rapid oxidation of combustible elements of the fuel results in energy release as combustion products are formed. The three major combustible chemical elements in most common fuels are carbon, hydrogen, and sulfur. Sulfur is usually a relatively unimportant contributor to the energy released, but it can be a significant cause of pollution and corrosion problems. Combustion is ***complete*** when all the carbon present in the fuel is burned to carbon dioxide, all the hydrogen is burned to water, all the sulfur is burned to sulfur dioxide, and all other combustible elements are fully oxidized. When these conditions are not fulfilled, combustion is *incomplete*.

reactants
products
complete combustion

In this chapter, we deal with combustion reactions expressed by chemical equations of the form

$$\text{reactants} \rightarrow \text{products}$$

or

$$\text{fuel} + \text{oxidizer} \rightarrow \text{products}$$

When dealing with chemical reactions, it is necessary to remember that mass is conserved, so the mass of the products equals the mass of the reactants. The total mass of each chemical *element* must be the same on both sides of the equation, even though the elements exist in different chemical compounds in the reactants and products. However, the number of moles of products may differ from the number of moles of reactants.

FOR EXAMPLE... consider the complete combustion of hydrogen with oxygen

$$1H_2 + \tfrac{1}{2}O_2 \rightarrow 1H_2O \quad (13.1)$$

In this case, the reactants are hydrogen and oxygen. Hydrogen is the fuel and oxygen is the oxidizer. Water is the only product of the reaction. The numerical coefficients in the equation, which precede the chemical symbols to give equal amounts of each chemical element on both sides of the equation, are called ***stoichiometric coefficients.*** In words, Eq. 13.1 states

stoichiometric coefficients

$$1 \text{ kmol } H_2 + \tfrac{1}{2} \text{ kmol } O_2 \rightarrow 1 \text{ kmol } H_2O$$

Note that the total numbers of moles on the left and right sides of Eq. 13.1 are not equal. However, because mass is conserved, the total mass of reactants must equal the total mass of products. Since 1 kmol of H_2 equals 2 kg, $\tfrac{1}{2}$ kmol of O_2 equals 16 kg, and 1 kmol of H_2O equals 18 kg, Eq. 13.1 can be interpreted as stating

$$2 \text{ kg } H_2 + 16 \text{ kg } O_2 \rightarrow 18 \text{ kg } H_2O$$

In the remainder of this section, consideration is given to the makeup of the fuel, oxidizer, and combustion products typically involved in engineering combustion applications.

Fuels

A *fuel* is simply a combustible substance. In this chapter emphasis is on hydrocarbon fuels, which contain hydrogen and carbon. Sulfur and other chemical substances also may be present. Hydrocarbon fuels can exist as liquids, gases, and solids.

Liquid hydrocarbon fuels are commonly derived from crude oil through distillation and cracking processes. Examples are gasoline, diesel fuel, kerosene, and other types of fuel oils. Most liquid fuels are mixtures of hydrocarbons for which compositions are usually given in terms of mass fractions. For simplicity in combustion calculations, gasoline is often modeled as octane, C_8H_{18}, and diesel fuel as dodecane, $C_{12}H_{26}$.

Gaseous hydrocarbon fuels are obtained from natural gas wells or are produced in certain chemical processes. Natural gas normally consists of several different hydrocarbons, with the major constituent being methane, CH_4. The compositions of gaseous fuels are usually given in terms of mole fractions. Both gaseous and liquid hydrocarbon fuels can be synthesized from coal, oil shale, and tar sands.

ultimate analysis

Coal is a familiar solid fuel. Its composition varies considerably with the location from which it is mined. For combustion calculations, the composition of coal is usually expressed as an ultimate analysis. The ***ultimate analysis*** gives the composition on a *mass basis* in terms of the relative amounts of chemical elements (carbon, sulfur, hydrogen, nitrogen, oxygen) and ash.

Energy & Environment

Environmental concerns and increased reliance on scarce oil supplies, coupled with plentiful coal reserves in the United States, have brought renewed interest in processes that convert coal into gaseous and liquid fuels suitable for transportation and power plants. As oil prices rise, the costly venture of coal conversion becomes more feasible, and government and coal industry sources expect that millions of barrels of fuel will be produced from coal by the year 2030.

Some of the most promising technologies under development are based on the *Fischer–Tropsch* process, in which coal is gasified at high temperature and converted to a liquid. The coal liquefaction process allows mercury, sulfur, and heavy metals to be removed, reducing emissions when the fuel is burned. Critics point out, though, that the process doesn't eliminate carbon dioxide emissions thought by many to cause global warming. In addition, they say, 40% of the coal's exergy value is lost in the conversion process. Still, interest in reducing dependence on imported oil is prompting coal gasification and liquefaction to be seriously considered.

Modeling Combustion Air

Oxygen is required in every combustion reaction. Pure oxygen is used only in special applications such as cutting and welding. In most combustion applications, air provides the needed oxygen. The composition of a typical sample of dry air is given in Table 12.1. For the combustion calculations of this book, however, the following model is used for simplicity:

- All components of air other than oxygen are lumped together with nitrogen. Accordingly, air is considered to be 21% oxygen and 79% nitrogen on a molar basis. With this idealization the molar ratio of the nitrogen to the oxygen is $0.79/0.21 = 3.76$. When air supplies the oxygen in a combustion reaction, therefore, every mole of oxygen is accompanied by 3.76 moles of nitrogen.
- We also assume that the nitrogen present in the combustion air does *not* undergo chemical reaction. That is, nitrogen is regarded as inert. The nitrogen in the products is at the same temperature as the other products, however, so the nitrogen undergoes a change of state if the products are at a temperature other than the air temperature before combustion. If high enough temperatures

Take Note...

In this model, air is assumed to contain no water vapor. When *moist* air is involved in combustion, the water vapor present must be considered in writing the combustion equation.

are attained, nitrogen can form compounds such as nitric oxide and nitrogen dioxide. Even trace amounts of oxides of nitrogen appearing in the exhaust of internal combustion engines can be a source of air pollution.

AIR–FUEL RATIO. Two parameters that are frequently used to quantify the amounts of fuel and air in a particular combustion process are the air–fuel ratio and its reciprocal, the fuel–air ratio. The ***air–fuel ratio*** is simply the ratio of the amount of air in a reaction to the amount of fuel. The ratio can be written on a molar basis (moles of air divided by moles of fuel) or on a mass basis (mass of air divided by mass of fuel). Conversion between these values is accomplished using the molecular weights of the air, M_{air}, and fuel, M_{fuel},

air–fuel ratio

$$\frac{\text{mass of air}}{\text{mass of fuel}} = \frac{\text{moles of air} \times M_{air}}{\text{moles of fuel} \times M_{fuel}} = \frac{\text{moles of air}}{\text{moles of fuel}}\left(\frac{M_{air}}{M_{fuel}}\right)$$

or

$$AF = \overline{AF}\left(\frac{M_{air}}{M_{fuel}}\right) \tag{13.2}$$

where $\overline{AF}$ is the air–fuel ratio on a molar basis and AF is the ratio on a mass basis. For the combustion calculations of this book the molecular weight of air is taken as 28.97. Tables A-1 provide the molecular weights of several important hydrocarbons. Since AF is a ratio, it has the same value whether the quantities of air and fuel are expressed in SI units.

THEORETICAL AIR. The minimum amount of air that supplies sufficient oxygen for the complete combustion of all the carbon, hydrogen, and sulfur present in the fuel is called the ***theoretical amount of air.*** For complete combustion with the theoretical amount of air, the products would consist of carbon dioxide, water, sulfur dioxide, the nitrogen accompanying the oxygen in the air, and any nitrogen contained in the fuel. No free oxygen would appear in the products.

theoretical air

➡ **FOR EXAMPLE...** let us determine the theoretical amount of air for the complete combustion of methane. For this reaction, the products contain only carbon dioxide, water, and nitrogen. The reaction is

$$CH_4 + a(O_2 + 3.76N_2) \rightarrow bCO_2 + cH_2O + dN_2 \tag{13.3}$$

where a, b, c, and d represent the numbers of moles of oxygen, carbon dioxide, water, and nitrogen. In writing the left side of Eq. 13.3, 3.76 moles of nitrogen are considered to accompany each mole of oxygen. Applying the conservation of mass principle to the carbon, hydrogen, oxygen, and nitrogen, respectively, results in four equations among the four unknowns

$$\begin{aligned} &\text{C:} && b = 1 \\ &\text{H:} && 2c = 4 \\ &\text{O:} && 2b + c = 2a \\ &\text{N:} && d = 3.76a \end{aligned}$$

Solving these equations, the *balanced* chemical equation is

$$CH_4 + 2(O_2 + 3.76N_2) \rightarrow CO_2 + 2H_2O + 7.52N_2 \tag{13.4}$$

The coefficient 2 before the term $(O_2 + 3.76N_2)$ in Eq. 13.4 is the number of moles of *oxygen* in the combustion air, per mole of fuel, and *not* the amount of air. The amount of combustion air is 2 moles of oxygen *plus* 2×3.76 moles of nitrogen, giving a total of 9.52 moles of air per mole of fuel. Thus, for the reaction given by Eq. 13.4 the air–fuel ratio on a molar basis is 9.52. To calculate the air–fuel ratio on a mass basis, use Eq. 13.2 to write

$$AF = \overline{AF}\left(\frac{M_{\text{air}}}{M_{\text{fuel}}}\right) = 9.52\left(\frac{28.97}{16.04}\right) = 17.19$$

Normally the amount of air supplied is either greater or less than the theoretical amount. The amount of air actually supplied is commonly expressed in terms of the *percent of theoretical air*. For example, 150% of theoretical air means that the air actually supplied is 1.5 times the theoretical amount of air. The amount of air supplied can be expressed alternatively as a ***percent excess air*** or a *percent deficiency* of air. Thus, 150% of theoretical air is equivalent to 50% excess air, and 80% of theoretical air is the same as a 20% deficiency of air.

percent of theoretical air

percent excess air

FOR EXAMPLE... consider the *complete* combustion of methane with 150% theoretical air (50% excess air). The balanced chemical reaction equation is

$$CH_4 + (1.5)(2)(O_2 + 3.76N_2) \rightarrow CO_2 + 2H_2O + O_2 + 11.28N_2 \qquad (13.5)$$

In this equation, the amount of air per mole of fuel is 1.5 times the theoretical amount determined by Eq. 13.4. Accordingly, the air–fuel ratio is 1.5 times the air–fuel ratio determined for Eq. 13.4. Since complete combustion is assumed, the products contain only carbon dioxide, water, nitrogen, and oxygen. The excess air supplied appears in the products as uncombined oxygen and a greater amount of nitrogen than in Eq. 13.4, based on the theoretical amount of air.

equivalence ratio

The ***equivalence ratio*** is the ratio of the actual fuel–air ratio to the fuel–air ratio for complete combustion with the theoretical amount of air. The reactants are said to form a *lean* mixture when the equivalence ratio is less than unity. When the ratio is greater than unity, the reactants are said to form a *rich* mixture.

In Example 13.1, we use conservation of mass to obtain balanced chemical reactions. The air–fuel ratio for each of the reactions is also calculated.

Example 13.1 DETERMINING THE AIR–FUEL RATIO

Determine the air–fuel ratio on both a molar and mass basis for the complete combustion of octane, C_8H_{18}, with **(a)** the theoretical amount of air, **(b)** 150% theoretical air (50% excess air).

Solution

Known: Octane, C_8H_{18}, is burned completely with (a) the theoretical amount of air, (b) 150% theoretical air.

Find: Determine the air–fuel ratio on a molar and a mass basis.

Engineering Model:

1. Each mole of oxygen in the combustion air is accompanied by 3.76 moles of nitrogen.
2. The nitrogen is inert.
3. Combustion is complete.

Analysis:

(a) For complete combustion of C_8H_{18} with the theoretical amount of air, the products contain carbon dioxide, water, and nitrogen only. That is

$$C_8H_{18} + a(O_2 + 3.76N_2) \rightarrow bCO_2 + cH_2O + dN_2$$

Applying the conservation of mass principle to the carbon, hydrogen, oxygen, and nitrogen, respectively, gives

$$\begin{aligned} &\text{C:} && b = 8 \\ &\text{H:} && 2c = 18 \\ &\text{O:} && 2b + c = 2a \\ &\text{N:} && d = 3.76a \end{aligned}$$

Solving these equations, $a = 12.5$, $b = 8$, $c = 9$, $d = 47$. The balanced chemical equation is

$$C_8H_{18} + 12.5(O_2 + 3.76N_2) \rightarrow 8CO_2 + 9H_2O + 47N_2$$

The air–fuel ratio on a molar basis is

$$\overline{AF} = \frac{12.5 + 12.5(3.76)}{1} = \frac{12.5(4.76)}{1} = 59.5\frac{\text{kmol (air)}}{\text{kmol (fuel)}}$$

The air–fuel ratio expressed on a mass basis is

$$AF = \left[59.5\frac{\text{kmol (air)}}{\text{kmol (fuel)}}\right]\left[\frac{28.97\frac{\text{kg (air)}}{\text{kmol (air)}}}{114.22\frac{\text{kg (fuel)}}{\text{kmol (fuel)}}}\right] = 15.1\frac{\text{kg (air)}}{\text{kg (fuel)}}$$

(b) For 150% theoretical air, the chemical equation for complete combustion takes the form

❶ $$C_8H_{18} + 1.5(12.5)(O_2 + 3.76N_2) \rightarrow bCO_2 + cH_2O + dN_2 + eO_2$$

Applying conservation of mass

$$\begin{aligned} &\text{C:} && b = 8 \\ &\text{H:} && 2c = 18 \\ &\text{O:} && 2b + c + 2e = (1.5)(12.5)(2) \\ &\text{N:} && d = (1.5)(12.5)(3.76) \end{aligned}$$

Solving this set of equations, $b = 8$, $c = 9$, $d = 70.5$, $e = 6.25$, giving a balanced chemical equation

$$C_8H_{18} + 18.75(O_2 + 3.76N_2) \rightarrow 8CO_2 + 9H_2O + 70.5N_2 + 6.25O_2$$

The air–fuel ratio on a molar basis is

$$\overline{AF} = \frac{18.75(4.76)}{1} = 89.25\frac{\text{kmol (air)}}{\text{kmol (fuel)}}$$

On a mass basis, the air–fuel ratio is 22.6 kg (air)/kg (fuel), as can be verified.

❶ When complete combustion occurs with *excess air,* oxygen appears in the products, in addition to carbon dioxide, water, and nitrogen.

Skills Developed: *Ability to...*

✓ balance a chemical reaction equation for complete combustion with theoretical air and with excess air.

✓ apply definitions of air–fuel ratio on mass and molar bases.

Quick Quiz

For the condition in part (b), determine the *equivalence ratio.*

Ans. 0.67.

Determining Products of Combustion

In each of the illustrations given above, complete combustion is assumed. For a hydrocarbon fuel, this means that the only allowed products are CO_2, H_2O, and N_2, with O_2 also present when excess air is supplied. If the fuel is specified and combustion is complete, the respective amounts of the products can be determined by applying the conservation of mass principle to the chemical equation. The procedure for obtaining the balanced reaction equation of an *actual* reaction where combustion is incomplete is not always so straightforward.

Combustion is the result of a series of very complicated and rapid chemical reactions, and the products formed depend on many factors. When fuel is burned in the cylinder of an internal combustion engine, the products of the reaction vary with the temperature and pressure in the cylinder. In combustion equipment of all kinds, the degree of mixing of the fuel and air is a controlling factor in the reactions that occur once the fuel and air mixture is ignited. Although the amount of air supplied in an actual combustion process may exceed the theoretical amount, it is not uncommon for some carbon monoxide and unburned oxygen to appear in the products. This can be due to incomplete mixing, insufficient time for complete combustion, and other factors. When the amount of air supplied is less than the theoretical amount of air, the products may include both CO_2 and CO, and there also may be unburned fuel in the products. Unlike the complete combustion cases considered above, the products of combustion of an actual combustion process and their relative amounts can be determined only by *measurement*.

Among several devices for measuring the composition of products of combustion are the *Orsat analyzer, gas chromatograph, infrared analyzer,* and *flame ionization detector*. Data from these devices can be used to determine the mole fractions of the gaseous products of combustion. The analyses are often reported on a "dry" basis. In a ***dry product analysis,*** the mole fractions are given for all gaseous products *except* the water vapor. In Examples 13.2 and 13.3, we show how analyses of the products of combustion on a dry basis can be used to determine the balanced chemical reaction equations.

dry product analysis

Since water is formed when hydrocarbon fuels are burned, the mole fraction of water vapor in the gaseous products of combustion can be significant. If the gaseous products of combustion are cooled at constant mixture pressure, the *dew point temperature* is reached when water vapor begins to condense. Since water deposited on duct work, mufflers, and other metal parts can cause corrosion, knowledge of the dew point temperature is important. Determination of the dew point temperature is illustrated in Example 13.2, which also features a dry product analysis.

Example 13.2 USING A DRY PRODUCT ANALYSIS

Methane, CH_4, is burned with dry air. The molar analysis of the products on a dry basis is CO_2, 9.7%; CO, 0.5%; O_2, 2.95%; and N_2, 86.85%. Determine **(a)** the air–fuel ratio on both a molar and a mass basis, **(b)** the percent theoretical air, **(c)** the dew point temperature of the products, in °C, if the mixture were cooled at 1 atm.

Solution

Known: Methane is burned with dry air. The molar analysis of the products on a dry basis is provided.

Find: Determine (a) the air–fuel ratio on both a molar and a mass basis, (b) the percent theoretical air, and (c) the dew point temperature of the products, in °C, if cooled at 1 atm.

Engineering Model:

1. Each mole of oxygen in the combustion air is accompanied by 3.76 moles of nitrogen, which is inert.

2. The products form an ideal gas mixture.

Analysis:

❶ **(a)** The solution is conveniently conducted on the basis of 100 kmol of dry products. The chemical equation then reads

$$aCH_4 + b(O_2 + 3.76N_2) \rightarrow 9.7CO_2 + 0.5CO + 2.95O_2 + 86.85N_2 + cH_2O$$

In addition to the assumed 100 kmol of dry products, water must be included as a product.

Applying conservation of mass to carbon, hydrogen, and oxygen, respectively

$$\text{C:} \quad 9.7 + 0.5 = a$$

$$\text{H:} \quad 2c = 4a$$

$$\text{O:} \quad (9.7)(2) + 0.5 + 2(2.95) + c = 2b$$

❷ Solving this set of equations gives $a = 10.2$, $b = 23.1$, $c = 20.4$. The balanced chemical equation is

$$10.2CH_4 + 23.1(O_2 + 3.76N_2) \rightarrow 9.7CO_2 + 0.5CO + 2.95O_2 + 86.85N_2 + 20.4H_2O$$

On a molar basis, the air–fuel ratio is

$$\overline{AF} = \frac{23.1(4.76)}{10.2} = 10.78 \frac{\text{kmol (air)}}{\text{kmol (fuel)}}$$

On a mass basis

$$AF = (10.78)\left(\frac{28.97}{16.04}\right) = 19.47 \frac{\text{kg (air)}}{\text{kg (fuel)}}$$

(b) The balanced chemical equation for the *complete combustion* of methane with the *theoretical amount* of air is

$$CH_4 + 2(O_2 + 3.76N_2) \rightarrow CO_2 + 2H_2O + 7.52N_2$$

The theoretical air–fuel ratio on a molar basis is

$$(\overline{AF})_{\text{theo}} = \frac{2(4.76)}{1} = 9.52 \frac{\text{kmol (air)}}{\text{kmol (fuel)}}$$

The percent theoretical air is then found from

$$\%\text{ theoretical air} = \frac{(\overline{AF})}{(\overline{AF})_{\text{theo}}}$$

$$= \frac{10.78 \text{ kmol (air)/kmol (fuel)}}{9.52 \text{ kmol (air)/kmol (fuel)}} = 1.13\ (113\%)$$

(c) To determine the dew point temperature requires the partial pressure of the water vapor p_v. The partial pressure p_v is found from $p_v = y_v p$, where y_v is the mole fraction of the water vapor in the combustion products and p is 1 atm.

Referring to the balanced chemical equation of part (a), the mole fraction of the water vapor is

$$y_v = \frac{20.4}{100 + 20.4} = 0.169$$

❸ Thus, $p_v = 0.169$ atm $= 0.1712$ bar. Interpolating in Table A-2, $T = 57°$C.

Skills Developed: *Ability to...*

- ✓ balance a chemical reaction equation for incomplete combustion given the analysis of dry products of combustion.
- ✓ apply definitions of air–fuel ratio on mass and molar bases as well as percent theoretical air.
- ✓ determine the dew point temperature of combustion products.

---

❶ The solution could be obtained on the basis of any assumed amount of dry products—for example, 1 kmol. With some other assumed amount, the values of the coefficients of the balanced chemical equation would differ from those obtained in the solution, but the air–fuel ratio, the value for the percent of theoretical air, and the dew point temperature would be unchanged.

❷ The three unknown coefficients, a, b, and c, are evaluated here by application of conservation of mass to carbon, hydrogen, and oxygen. As a check, note that the nitrogen also balances

$$\text{N:} \quad b(3.76) = 86.85$$

This confirms the accuracy of both the given product analysis and the calculations conducted to determine the unknown coefficients.

❸ If the products of combustion were cooled at constant pressure below the dew point temperature of 57°C, some condensation of the water vapor would occur.

Quick Quiz

Recalculate the dew point temperature as in part (c) if the air supply were *moist* air, including 3.53 kmol of additional water vapor.

Ans. 59°C.

In Example 13.3, a fuel mixture having a known molar analysis is burned with air, giving products with a known dry analysis.

Example 13.3 BURNING NATURAL GAS WITH EXCESS AIR

A natural gas has the following molar analysis: CH_4, 80.62%; C_2H_6, 5.41%; C_3H_8, 1.87%; C_4H_{10}, 1.60%; N_2, 10.50%. The gas is burned with dry air, giving products having a molar analysis on a dry basis: CO_2, 7.8%; CO, 0.2%; O_2, 7%; N_2, 85%. **(a)** Determine the air–fuel ratio on a molar basis. **(b)** Assuming ideal gas behavior for the fuel mixture, determine the amount of products in kmol that would be formed from 100 m^3 of fuel mixture at 300 K and 1 bar. **(c)** Determine the percent of theoretical air.

Solution

Known: A natural gas with a specified molar analysis burns with dry air, giving products having a known molar analysis on a dry basis.

Find: Determine the air–fuel ratio on a molar basis, the amount of products in kmol that would be formed from 100 m^3 of natural gas at 300 K and 1 bar, and the percent of theoretical air.

Engineering Model:

1. Each mole of oxygen in the combustion air is accompanied by 3.76 moles of nitrogen, which is inert.
2. The fuel mixture can be modeled as an ideal gas.

Analysis:

(a) The solution can be conducted on the basis of an assumed amount of fuel mixture or on the basis of an assumed amount of dry products. Let us illustrate the first procedure, basing the solution on 1 kmol of fuel mixture. The chemical equation then takes the form

$$(0.8062CH_4 + 0.0541C_2H_6 + 0.0187C_3H_8 + 0.0160C_4H_{10} + 0.1050N_2)$$
$$+a(O_2 + 3.76N_2) \rightarrow b(0.078CO_2 + 0.002CO + 0.07O_2 + 0.85N_2) + cH_2O$$

The products consist of b kmol of dry products and c kmol of water vapor, each per kmol of fuel mixture.

Applying conservation of mass to carbon

$$b(0.078 + 0.002) = 0.8062 + 2(0.0541) + 3(0.0187) + 4(0.0160)$$

Solving gives $b = 12.931$. Conservation of mass for hydrogen results in

$$2c = 4(0.8062) + 6(0.0541) + 8(0.0187) + 10(0.0160)$$

which gives $c = 1.93$. The unknown coefficient a can be found from either an oxygen balance or a nitrogen balance. Applying conservation of mass to oxygen

$$12.931[2(0.078) + 0.002 + 2(0.07)] + 1.93 = 2a$$

❶ giving $a = 2.892$.

The balanced chemical equation is then

$$(0.8062CH_4 + 0.0541C_2H_6 + 0.0187C_3H_8 + 0.0160C_4H_{10} + 0.1050N_2) + 2.892(O_2 + 3.76N_2) \rightarrow 12.931(0.078CO_2 + 0.002CO + 0.07O_2 + 0.85N_2) + 1.93H_2O$$

The air–fuel ratio on a molar basis is

$$\overline{AF} = \frac{(2.892)(4.76)}{1} = 13.77 \frac{\text{kmol (air)}}{\text{kmol (fuel)}}$$

(b) By inspection of the chemical reaction equation, the total amount of products is $b + c = 12.931 + 1.93 = 14.861$ kmol of products per kmol of fuel. The amount of fuel in kmol, n_F, present in 100 m^3 of fuel mixture at 300 K and 1 bar can be determined from the ideal gas equation of state as

$$n_F = \frac{pV}{\bar{R}T} = \frac{(10^5\ \text{N/m}^2)(100\ \text{m}^3)}{(8314\ \text{N} \cdot \text{m/kmol} \cdot \text{K})(300\ \text{K})} = 4.01\ \text{kmol (fuel)}$$

Accordingly, the amount of product mixture that would be formed from 100 m^3 of fuel mixture is (14.861)(4.01) = 59.59 kmol of product gas.

(c) The balanced chemical equation for the *complete combustion* of the fuel mixture with the *theoretical amount* of air is

$$(0.8062CH_4 + 0.0541C_2H_6 + 0.0187C_3H_8 + 0.0160C_4H_{10} + 0.1050N_2) + 2(O_2 + 3.76N_2) \rightarrow 1.0345CO_2 + 1.93H_2O + 7.625N_2$$

The theoretical air–fuel ratio on a molar basis is

$$(\overline{AF})_{\text{theo}} = \frac{2(4.76)}{1} = 9.52 \frac{\text{kmol (air)}}{\text{kmol (fuel)}}$$

The percent theoretical air is then

$$\%\ \text{theoretical air} = \frac{13.77\ \text{kmol (air)/kmol (fuel)}}{9.52\ \text{kmol (air)/kmol (fuel)}} = 1.45\ (145\%)$$

---

❶ A check on both the accuracy of the given molar analyses and the calculations conducted to determine the unknown coefficients is obtained by applying conservation of mass to nitrogen. The amount of nitrogen in the reactants is

$$0.105 + (3.76)(2.892) = 10.98\ \text{kmol/kmol of fuel}$$

The amount of nitrogen in the products is (0.85)(12.931) = 10.99 kmol/kmol of fuel. The difference can be attributed to round-off.

Skills Developed: Ability to...

✓ balance a chemical reaction equation for incomplete combustion of a fuel mixture given the analysis of dry products of combustion.

✓ apply the definition of air–fuel ratio on a molar basis as well as percent theoretical air.

Quick Quiz

Determine the mole fractions of the products of combustion.

Ans. $y_{CO_2} = 0.0679$, $y_{CO} = 0.0017$, $y_{O_2} = 0.0609$, $y_{N_2} = 0.7396$, $y_{H_2O} = 0.1299$.

13.2 Conservation of Energy—Reacting Systems

The objective of the present section is to illustrate the application of the conservation of energy principle to reacting systems. The forms of the conservation of energy principle introduced previously remain valid whether or not a chemical reaction occurs within the system. However, the methods used for evaluating the properties of reacting systems differ somewhat from the practices used to this point.

13.2.1 Evaluating Enthalpy for Reacting Systems

In each of the tables of thermodynamic properties used thus far, values for the specific internal energy, enthalpy, and entropy are given relative to some arbitrary datum state where the enthalpy (or alternatively the internal energy) and entropy are set to zero. This approach is satisfactory for evaluations involving *differences* in property values between states of the same composition, for then arbitrary datums cancel. However, when a chemical reaction occurs, reactants disappear and products are formed, so differences cannot be calculated for all substances involved. For reacting systems, it is necessary to evaluate h, u, and s in such a way that there are no subsequent ambiguities or inconsistencies in evaluating properties. In this section, we will consider how this is accomplished for h and u. The case of entropy is handled differently and is taken up in Sec. 13.5.

standard reference state

An enthalpy datum for the study of reacting systems can be established by assigning arbitrarily a value of zero to the enthalpy of the *stable elements* at a state called the ***standard reference state*** and defined by $T_{ref} = 298.15$ K (25°C) and $p_{ref} = 1$ atm. Note that only *stable* elements are assigned a value of zero enthalpy at the standard state. The term stable simply means that the particular element is in a chemically stable form. For example, at the standard state the stable forms of hydrogen, oxygen, and nitrogen are H_2, O_2, and N_2 and not the monatomic H, O, and N. No ambiguities or conflicts result with this choice of datum.

enthalpy of formation

ENTHALPY OF FORMATION. Using the datum introduced above, enthalpy values can be assigned to *compounds* for use in the study of reacting systems. The enthalpy of a compound at the standard state equals its *enthalpy of formation,* symbolized $\bar{h}_f^\circ$. The ***enthalpy of formation*** is the energy released or absorbed when the compound is formed from its elements, the compound and elements all being at T_{ref} and p_{ref}. The enthalpy of formation is usually determined by application of procedures from statistical thermodynamics using observed spectroscopic data.

The enthalpy of formation also can be found in principle by measuring the heat transfer in a reaction in which the compound is formed from the elements. **FOR EXAMPLE...** consider the simple reactor shown in Fig. 13.1, in which carbon and oxygen each enter at T_{ref} and p_{ref} and react completely at steady state to form carbon dioxide at the same temperature and pressure. Carbon dioxide is *formed* from carbon and oxygen according to

$$C + O_2 \rightarrow CO_2 \tag{13.6}$$

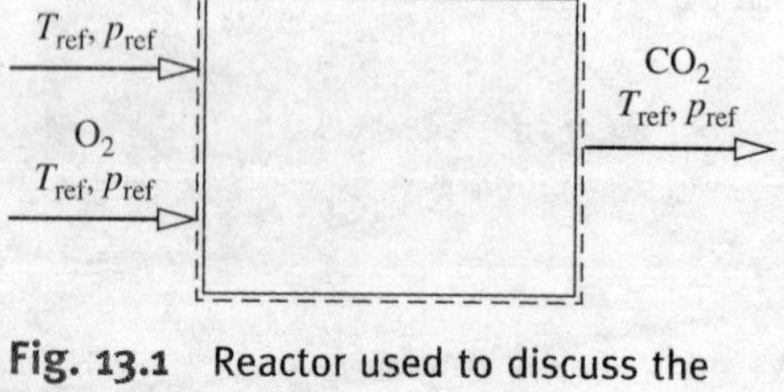

Fig. 13.1 Reactor used to discuss the enthalpy of formation concept.

This reaction would be *exothermic,* so for the carbon dioxide to exit at the same temperature as the entering elements, there would be a heat transfer from the reactor to its surroundings. The rate of heat transfer and the enthalpies of the incoming and exiting streams are related by the energy rate balance

$$0 = \dot{Q}_{cv} + \dot{m}_C h_C + \dot{m}_{O_2} h_{O_2} - \dot{m}_{CO_2} h_{CO_2}$$

where $\dot{m}$ and h denote, respectively, mass flow rate and specific enthalpy. In writing this equation, we have assumed no work $\dot{W}_{cv}$ and negligible effects of kinetic and potential energy. For enthalpies on a molar basis, the energy rate balance appears as

$$0 = \dot{Q}_{cv} + \dot{n}_C \bar{h}_C + \dot{n}_{O_2} \bar{h}_{O_2} - \dot{n}_{CO_2} \bar{h}_{CO_2}$$

where $\dot{n}$ and $\bar{h}$ denote, respectively, the molar flow rate and enthalpy per mole. Solving for the specific enthalpy of carbon dioxide and noting from Eq. 13.6 that all molar flow rates are equal

$$\bar{h}_{CO_2} = \frac{\dot{Q}_{cv}}{\dot{n}_{CO_2}} + \frac{\dot{n}_C}{\dot{n}_{CO_2}} \bar{h}_C + \frac{\dot{n}_{O_2}}{\dot{n}_{CO_2}} \bar{h}_{O_2} = \frac{\dot{Q}_{cv}}{\dot{n}_{CO_2}} + \bar{h}_C + \bar{h}_{O_2} \tag{13.7}$$

Since carbon and oxygen are stable elements at the standard state, $\bar{h}_C = \bar{h}_{O_2} = 0$, and Eq. 13.7 becomes

$$\bar{h}_{CO_2} = \frac{\dot{Q}_{cv}}{\dot{n}_{CO_2}} \tag{13.8}$$

Accordingly, the value assigned to the specific enthalpy of carbon dioxide at the standard state, the enthalpy of formation, equals the heat transfer, per mole of CO_2, between the reactor and its surroundings. If the heat transfer could be measured accurately, it would be found to equal −393,520 kJ per kmol of carbon dioxide formed.

Table A-25 gives values of the enthalpy of formation for several compounds in units of kJ/kmol. In this text, the superscript ° is used to denote properties at 1 bar. For the case of the enthalpy of formation, the reference temperature T_{ref} is also intended by this symbol. The values of $\bar{h}_f^\circ$ listed in Table A-25 for CO_2 correspond to those given in the previous example.

The sign associated with the enthalpy of formation values appearing in Table A-25 corresponds to the sign convention for heat transfer. If there is heat transfer *from* a reactor in which a compound is formed from its elements (an *exothermic* reaction as in the previous example), the enthalpy of formation has a negative sign. If a heat transfer *to* the reactor is required (an *endothermic* reaction), the enthalpy of formation is positive.

Evaluating Enthalpy

The specific enthalpy of a compound at a state other than the standard state is found by adding the specific enthalpy change $\Delta\bar{h}$ between the standard state and the state of interest to the enthalpy of formation

$$\bar{h}(T, p) = \bar{h}_f^\circ + [\bar{h}(T, p) - \bar{h}(T_{ref}, p_{ref})] = \bar{h}_f^\circ + \Delta\bar{h} \tag{13.9}$$

That is, the enthalpy of a compound is composed of $\bar{h}_f^\circ$, associated with the formation of the compound from its elements, and $\Delta\bar{h}$, associated with a change of state at constant composition. An arbitrary choice of datum can be used to determine $\Delta\bar{h}$, since it is a *difference* at constant composition. Accordingly, $\Delta\bar{h}$ can be evaluated from tabular sources such as the steam tables, the ideal gas tables when appropriate, and so on. Note that as a consequence of the enthalpy datum adopted for the stable elements, the specific enthalpy determined from Eq. 13.9 is often negative.

Table A-25 provides two values of the enthalpy of formation of water. One is for liquid water and the other is for water vapor. Under equilibrium conditions, water exists only as a liquid at 25°C and 1 bar. The vapor value listed is

Take Note...

When applying Eq. 13.9 to water vapor, we use the *vapor* value of the enthalpy of formation of water, $\bar{h}_f^\circ(g)$, from Table A-25 together with $\Delta\bar{h}$ for water vapor from the ideal gas table, Table A-23.

for a *hypothetical* ideal gas state in which water is a vapor at 25°C and 1 bar. The difference between the two enthalpy of formation values is given closely by the enthalpy of vaporization $\bar{h}_{fg}$ at T_{ref}

$$\bar{h}_f^\circ(g) - \bar{h}_f^\circ(l) \approx \bar{h}_{fg} \tag{13.10}$$

Similar considerations apply to other substances for which liquid and vapor values for $\bar{h}_f^\circ$ are listed in Tables A-25.

13.2.2 Energy Balances for Reacting Systems

Several considerations enter when writing energy balances for systems involving combustion. Some of these apply generally, without regard for whether combustion takes place. For example, it is necessary to consider if significant work and heat transfers take place and if the respective values are known or unknown. Also, the effects of kinetic and potential energy must be assessed. Other considerations are related directly to the occurrence of combustion. For example, it is important to know the state of the fuel before combustion occurs. Whether the fuel is a liquid, a gas, or a solid is important. It is necessary to consider whether the fuel is premixed with the combustion air or the fuel and air enter a reactor separately. The state of the combustion products also must be assessed. It is important to know whether the products of combustion are a gaseous mixture or whether some of the water formed on combustion has condensed.

Control Volumes at Steady State

To illustrate the many considerations involved when writing energy balances for reacting systems, we consider special cases of broad interest, highlighting the underlying assumptions. Let us begin by considering the steady-state reactor shown in Fig. 13.2, in which a hydrocarbon fuel C_aH_b burns completely with the theoretical amount of air according to

$$C_aH_b + \left(a + \frac{b}{4}\right)(O_2 + 3.76N_2) \rightarrow aCO_2 + \frac{b}{2}H_2O + \left(a + \frac{b}{4}\right)3.76N_2 \tag{13.11}$$

The fuel enters the reactor in a stream separate from the combustion air, which is regarded as an ideal gas mixture. The products of combustion also are assumed to form an ideal gas mixture. Kinetic and potential energy effects are ignored.

With the foregoing idealizations, the mass and energy rate balances for the two-inlet, single-exit reactor can be used to obtain the following equation on a *per mole of fuel basis:*

$$\frac{\dot{Q}_{cv}}{\dot{n}_F} - \frac{\dot{W}_{cv}}{\dot{n}_F} = \left[a\bar{h}_{CO_2} + \frac{b}{2}\bar{h}_{H_2O} + \left(a + \frac{b}{4}\right)3.76\bar{h}_{N_2}\right] - \bar{h}_F - \left[\left(a + \frac{b}{4}\right)\bar{h}_{O_2} + \left(a + \frac{b}{4}\right)3.76\bar{h}_{N_2}\right] \tag{13.12a}$$

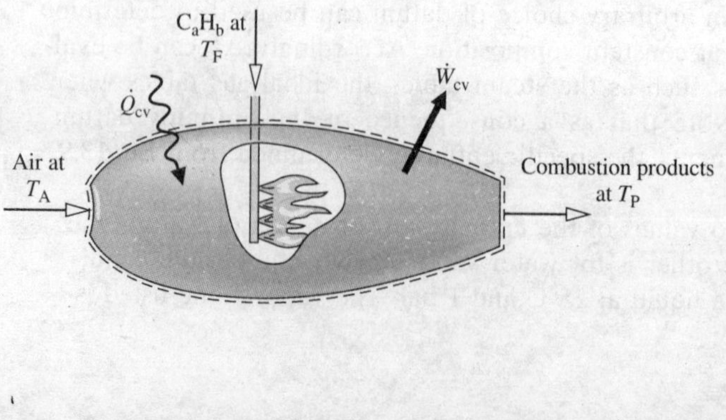

Fig. 13.2 Reactor at steady state.

where $\dot{n}_F$ denotes the molar flow rate of the fuel. Note that each coefficient on the right side of this equation is the same as the coefficient of the corresponding substance in the reaction equation.

The first underlined term on the right side of Eq. 13.12a is the enthalpy of the exiting gaseous products of combustion *per mole of fuel.* The second underlined term on the right side is the enthalpy of the combustion air *per mole of fuel.* Note that the enthalpies of the combustion products and the air have been evaluated by adding the contribution of each component present in the respective ideal gas mixtures. The symbol $\bar{h}_F$ denotes the molar enthalpy of the fuel. Equation 13.12a can be expressed more concisely as

$$\frac{\dot{Q}_{cv}}{\dot{n}_F} - \frac{\dot{W}_{cv}}{\dot{n}_F} = \bar{h}_P - \bar{h}_R \tag{13.12b}$$

where $\bar{h}_P$ and $\bar{h}_R$ denote, respectively, the enthalpies of the products and reactants per mole of fuel.

EVALUATING ENTHALPY TERMS. Once the energy balance has been written, the next step is to evaluate the individual enthalpy terms. Since each component of the combustion products is assumed to behave as an ideal gas, its contribution to the enthalpy of the products depends solely on the temperature of the products, T_P. Accordingly, for each component of the products, Eq. 13.9 takes the form

$$\bar{h} = \bar{h}_f^\circ + [\bar{h}(T_P) - \bar{h}(T_{ref})] \tag{13.13}$$

In Eq. 13.13, $\bar{h}_f^\circ$ is the enthalpy of formation from Table A-25, as appropriate. The second term accounts for the change in enthalpy from the temperature T_{ref} to the temperature T_P. For several common gases, this term can be evaluated from tabulated values of enthalpy versus temperature in Table A-23, as appropriate. Alternatively, the term can be obtained by integration of the ideal gas specific heat $\bar{c}_p$ obtained from Table A-21 or some other source of data. A similar approach would be employed to evaluate the enthalpies of the oxygen and nitrogen in the combustion air. For these

$$\bar{h} = \cancelto{0}{\bar{h}_f^\circ} + [\bar{h}(T_A) - \bar{h}(T_{ref})] \tag{13.14}$$

where T_A is the temperature of the air entering the reactor. Note that the enthalpy of formation for oxygen and nitrogen is zero by definition and thus drops out of Eq. 13.14 as indicated. The evaluation of the enthalpy of the fuel is also based on Eq. 13.9. If the fuel can be modeled as an ideal gas, the fuel enthalpy is obtained using an expression of the same form as Eq. 13.13, with the temperature of the incoming fuel replacing T_P.

With the foregoing considerations, Eq. 13.12a takes the form

$$\begin{aligned}\frac{\dot{Q}_{cv}}{\dot{n}_F} - \frac{\dot{W}_{cv}}{\dot{n}_F} &= a(\bar{h}_f^\circ + \Delta\bar{h})_{CO_2} + \frac{b}{2}(\bar{h}_f^\circ + \Delta\bar{h})_{H_2O} + \left(a + \frac{b}{4}\right)3.76(\cancelto{0}{\bar{h}_f^\circ} + \Delta\bar{h})_{N_2} \\ &- (\bar{h}_f^\circ + \Delta\bar{h})_F - \left(a + \frac{b}{4}\right)(\cancelto{0}{\bar{h}_f^\circ} + \Delta\bar{h})_{O_2} - \left(a + \frac{b}{4}\right)3.76(\cancelto{0}{\bar{h}_f^\circ} + \Delta\bar{h})_{N_2}\end{aligned} \tag{13.15a}$$

The terms set to zero in this expression are the enthalpies of formation of oxygen and nitrogen.

Equation 13.15a can be written more concisely as

$$\frac{\dot{Q}_{cv}}{\dot{n}_F} - \frac{\dot{W}_{cv}}{\dot{n}_F} = \sum_P n_e(\bar{h}_f^\circ + \Delta\bar{h})_e - \sum_R n_i(\bar{h}_f^\circ + \Delta\bar{h})_i \tag{13.15b}$$

where i denotes the incoming fuel and air streams, and e the exiting combustion products. The coefficients n_i and n_e correspond to the respective coefficients of the reaction equation giving the moles of reactants and products *per mole of fuel,* respectively. Although Eqs. 13.15 have been developed with reference to the reaction of Eq. 13.11, equations having the same general forms would be obtained for other combustion reactions.

In Examples 13.4 and 13.5, the energy balance is applied together with tabular property data to analyze control volumes at steady state involving combustion.

Example 13.4 ANALYZING AN INTERNAL COMBUSTION ENGINE

Liquid octane enters an internal combustion engine operating at steady state with a mass flow rate of 1.8×10^{-3} kg and is mixed with the theoretical amount of air. The fuel and air enter the engine at 25°C and 1 atm. The mixture burns completely and combustion products leave the engine at 890 K. The engine develops a power output of 37 kW. Determine the rate of heat transfer from the engine, in kW, neglecting kinetic and potential energy effects.

Solution

Known: Liquid octane and the theoretical amount of air enter an internal combustion engine operating at steady state in separate streams at 25°C, 1 atm. Combustion is complete and the products exit at 890 K. The power developed by the engine and fuel mass flow rate are specified.

Find: Determine the rate of heat transfer from the engine, in kW.

Schematic and Given Data:

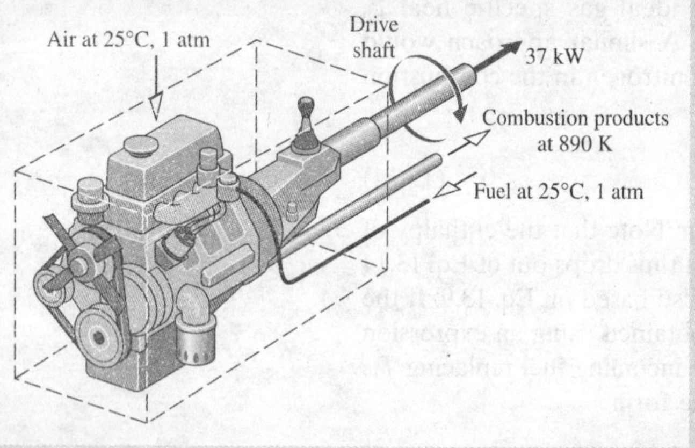

Fig. E13.4

Engineering Model:

1. The control volume identified by a dashed line on the accompanying figure operates at steady state.
2. Kinetic and potential energy effects can be ignored.
3. The combustion air and the products of combustion each form ideal gas mixtures.
4. Each mole of oxygen in the combustion air is accompanied by 3.76 moles of nitrogen. The nitrogen is inert and combustion is complete.

Analysis: The balanced chemical equation for complete combustion with the theoretical amount of air is obtained from the solution to Example 13.1 as

$$C_8H_{18} + 12.5O_2 + 47N_2 \rightarrow 8CO_2 + 9H_2O + 47N_2$$

The energy rate balance reduces, with assumptions 1–3, to give

$$\frac{\dot{Q}_{cv}}{\dot{n}_F} = \frac{\dot{W}_{cv}}{\dot{n}_F} + \bar{h}_P - \bar{h}_R$$

$$= \frac{\dot{W}_{cv}}{\dot{n}_F} + \{8[\bar{h}_f^\circ + \Delta\bar{h}]_{CO_2} + 9[\bar{h}_f^\circ + \Delta\bar{h}]_{H_2O(g)} + 47[\cancelto{0}{\bar{h}_f^\circ} + \Delta\bar{h}]_{N_2}\}$$

$$- \{[\bar{h}_f^\circ + \cancelto{0}{\Delta\bar{h}}]_{C_8H_{18}(l)} + 12.5[\cancelto{0}{\bar{h}_f^\circ} + \cancelto{0}{\Delta\bar{h}}]_{O_2} + 47[\cancelto{0}{\bar{h}_f^\circ} + \cancelto{0}{\Delta\bar{h}}]_{N_2}\}$$

where each coefficient is the same as the corresponding term of the balanced chemical equation and Eq. 13.9 has been used to evaluate enthalpy terms. The enthalpy of formation terms for oxygen and nitrogen are zero, and $\Delta\bar{h} = 0$ for each of the reactants, because the fuel and combustion air enter at 25°C.

With the enthalpy of formation for $C_8H_{18}(l)$ from Table A-25

$$\bar{h}_R = (\bar{h}_f^\circ)_{C_8H_{18}(l)} = -249{,}910 \text{ kJ/kmol (fuel)}$$

With enthalpy of formation values for CO_2 and $H_2O(g)$ from Table A-25, and enthalpy values for N_2, H_2O, and CO_2 from Table A-23

$$\begin{aligned}\bar{h}_P &= 8[-393{,}520 + (36{,}876 - 9364)] + 9[-241{,}820 + (31{,}429 - 9{,}904)]\\ &\quad +47[26{,}568 - 8{,}669]\\ &= -4{,}069{,}466 \text{ kJ/kmol}\end{aligned}$$

Using the molecular weight of the fuel from Table A-1, the molar flow rate of the fuel is

$$\dot{n}_F = \frac{1.8 \times 10^{-3} \text{ kg(fuel)/s}}{114.22 \text{ kg(fuel)/kmol (fuel)}} = 1.58 \times 10^{-5} \text{ kmol(fuel)/s}$$

Inserting values into the expression for the rate of heat transfer

$$\begin{aligned}\dot{Q}_{cv} &= \dot{W}_{cv} + \dot{n}_F(\bar{h}_P - \bar{h}_R)\\ &= 37 \text{ kW} + \left[1.58 \times 10^{-5} \frac{\text{kmol (fuel)}}{\text{s}}\right]\\ &\quad [-4{,}069{,}466 - (-249{,}910)]\left(\frac{\text{kJ}}{\text{kmol (fuel)}}\right)\left(\frac{1 \text{ kW}}{1 \text{ kg/s}}\right)\\ &= -23.3 \text{ kW}\end{aligned}$$

Skills Developed: *Ability to...*

- ✓ balance a chemical reaction equation for complete combustion of octane with theoretical air.
- ✓ apply the control volume energy balance to a reacting system.
- ✓ evaluate enthalpy values appropriately.

Quick Quiz

If the density of octane is 0.7 kg/L, how many liters of fuel would be used in 2 h of continuous operation of the engine?

Ans. 18.5 L.

Example 13.5 ANALYZING A COMBUSTION CHAMBER

Methane gas at 400 K and 1 atm enters a combustion chamber, where it is mixed with air entering at 500 K and 1 atm. The products of combustion exit at 1800 K and 1 atm with the product analysis given in Example 13.2. For operation at steady state, determine the rate of heat transfer from the combustion chamber in kJ per kmol of fuel. Neglect kinetic and potential energy effects. The average value for the specific heat $\bar{c}_p$ of methane between 298 and 400 K is 38 kJ/kmol · K.

Solution

Known: Methane gas, air, and combustion products enter and exit a combustion chamber at steady state in separate streams with specified temperatures and a pressure of 1 atm. An analysis of the dry products of combustion is provided.

Find: Determine the rate of heat transfer, in kJ per kmol of fuel.

Schematic and Given Data:

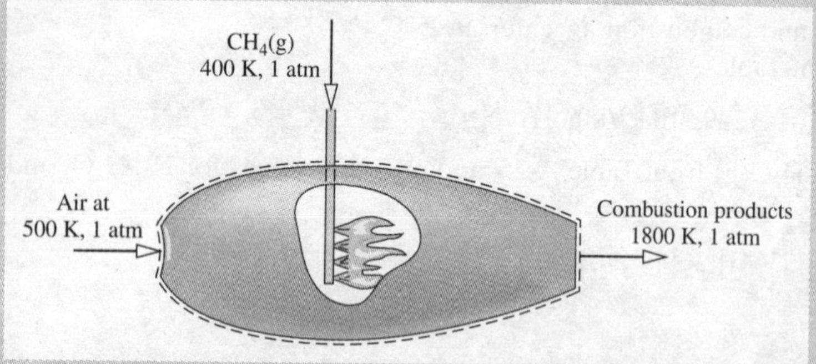

Fig. E13.5

Engineering Model:

1. The control volume identified by a dashed line on the accompanying figure operates at steady state with $\dot{W}_{cv} = 0$.
2. Kinetic and potential energy effects can be ignored.
3. The fuel is modeled as an ideal gas with $\bar{c}_p = 38$ kJ/kmol · K. The combustion air and the products of combustion each form ideal gas mixtures.
4. Each mole of oxygen in the combustion air is accompanied by 3.76 moles of nitrogen, which is inert.

Analysis: When expressed on a per mole of fuel basis, the balanced chemical equation obtained in the solution to Example 13.2 takes the form

$$CH_4 + 2.265O_2 + 8.515N_2 \rightarrow 0.951CO_2 + 0.049CO + 0.289O_2 + 8.515N_2 + 2H_2O$$

The energy rate balance reduces with assumptions 1–3 to give

$$\frac{\dot{Q}_{cv}}{\dot{n}_F} = \bar{h}_P - \bar{h}_R$$

$$= [0.951(\bar{h}_f^\circ + \Delta\bar{h})_{CO_2} + 0.049(\bar{h}_f^\circ + \Delta\bar{h})_{CO} + 0.289(\overset{0}{\cancel{\bar{h}_f^\circ}} + \Delta\bar{h})_{O_2}$$

$$+ 8.515(\overset{0}{\cancel{\bar{h}_f^\circ}} + \Delta\bar{h})_{N_2} + 2(\bar{h}_f^\circ + \Delta\bar{h})_{H_2O(g)}] - [(\bar{h}_f^\circ + \Delta\bar{h})_{CH_4}$$

$$+ 2.265(\overset{0}{\cancel{\bar{h}_f^\circ}} + \Delta\bar{h})_{O_2} + 8.515(\overset{0}{\cancel{\bar{h}_f^\circ}} + \Delta\bar{h})_{N_2}]$$

where each coefficient in the equation is the same as the coefficient of the corresponding term of the balanced chemical equation and Eq. 13.9 has been used to evaluate the enthalpy terms. The enthalpies of formation for oxygen and nitrogen are set to zero.

Consider first the reactants. With the enthalpy of formation for methane from Table A-25, the given $\bar{c}_p$ value for methane, and enthalpy values for nitrogen and oxygen from Table A-23

$$\begin{aligned}\bar{h}_R &= (\bar{h}_f^\circ + \bar{c}_p\Delta T)_{CH_4} + 2.265(\Delta\bar{h})_{O_2} + 8.515(\Delta\bar{h})_{N_2} \\ &= [-74{,}850 + 38(400 - 298)] + 2.265[14{,}770 - 8682] + 8.515[14{,}581 - 8669] \\ &= -6844 \text{ kJ/kmol } (CH_4)\end{aligned}$$

Next consider the products. With enthalpy of formation values for CO_2, CO, and $H_2O(g)$ from Table A-25 and enthalpy values from Table A-23

$$\begin{aligned}\bar{h}_P &= 0.951[-393{,}520 + (88{,}806 - 9364)] \\ &\quad + 0.049[-110{,}530 + (58{,}191 - 8669)] \\ &\quad + 0.289(60{,}371 - 8682) + 8.515(57{,}651 - 8669) \\ &\quad + 2[-241{,}820 + (72{,}513 - 9904)] \\ &= -228{,}079 \text{ kJ/kmol } (CH_4)\end{aligned}$$

Inserting values into the expression for the rate of heat transfer

$$\frac{\dot{Q}_{cv}}{\dot{n}_F} = \bar{h}_P - \bar{h}_R = -228{,}079 - (-6844) = -221{,}235 \text{ kJ/kmol } (CH_4)$$

Skills Developed: *Ability to...*

- ✓ apply the control volume energy balance to a reacting system.
- ✓ evaluate enthalpy values appropriately.

Quick Quiz

Determine the heat transfer rate, in kW, for a fuel mass flow rate of 0.036 kg/s.
Ans. −496.5 kW.

CONDENSATION OF COMBUSTION PRODUCTS. When the combustion products of a hydrocarbon fuel are cooled sufficiently, condensation of a portion of the water formed will occur. The energy balance must then be written to account for the presence of water in the products as both a liquid and a vapor. For cooling of combustion products at constant pressure, the dew point temperature method of Example 13.2 is used to determine the temperature at the onset of condensation.

Closed Systems

Let us consider next a closed system involving a combustion process. In the absence of kinetic and potential energy effects, the appropriate form of the energy balance is

$$U_P - U_R = Q - W$$

where U_R denotes the internal energy of the reactants and U_P denotes the internal energy of the products. If the reactants and products form ideal gas mixtures, the energy balance can be expressed as

$$\sum_{P} n\bar{u} - \sum_{R} n\bar{u} = Q - W \qquad (13.16)$$

where the coefficients n on the left side are the coefficients of the reaction equation giving the moles of each reactant or product.

Since each component of the reactants and products behaves as an ideal gas, the respective specific internal energies of Eq. 13.16 can be evaluated as $\bar{u} = \bar{h} - \bar{R}T$, so the equation becomes

$$Q - W = \sum_{P} n(\bar{h} - \bar{R}T_P) - \sum_{R} n(\bar{h} - \bar{R}T_R) \qquad (13.17a)$$

where T_P and T_R denote the temperature of the products and reactants, respectively. With expressions of the form of Eq. 13.13 for each of the reactants and products, Eq. 13.17a can be written alternatively as

$$\begin{aligned} Q - W &= \sum_{P} n(\bar{h}_f^\circ + \Delta\bar{h} - \bar{R}T_P) - \sum_{R} n(\bar{h}_f^\circ + \Delta\bar{h} - \bar{R}T_R) \\ &= \sum_{P} n(\bar{h}_f^\circ + \Delta\bar{h}) - \sum_{R} n(\bar{h}_f^\circ + \Delta\bar{h}) - \bar{R}T_P \sum_{P} n + \bar{R}T_R \sum_{R} n \end{aligned} \qquad (13.17b)$$

The enthalpy of formation terms are obtained from Table A-25. The $\Delta\bar{h}$ terms are evaluated as discussed above.

The foregoing concepts are illustrated in Example 13.6, where a gaseous mixture burns in a closed, rigid container.

Example 13.6 ANALYZING COMBUSTION AT CONSTANT VOLUME

A mixture of 1 kmol of gaseous methane and 2 kmol of oxygen initially at 25°C and 1 atm burns completely in a closed, rigid container. Heat transfer occurs until the products are cooled to 900 K. If the reactants and products each form ideal gas mixtures, determine (a) the amount of heat transfer, in kJ, and (b) the final pressure, in atm.

Solution

Known: A mixture of gaseous methane and oxygen, initially at 25°C and 1 atm, burns completely within a closed rigid container. The products are cooled to 900 K.

Find: Determine the amount of heat transfer, in kJ, and the final pressure of the combustion products, in atm.

Schematic and Given Data:

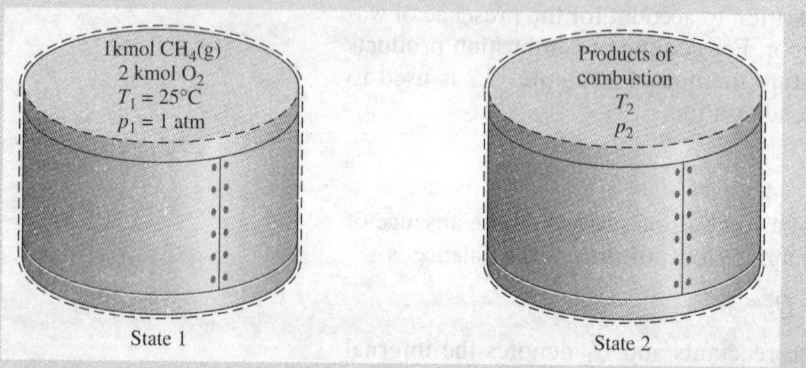

Fig. E13.6

Engineering Model:

1. The contents of the closed, rigid container are taken as the system.
2. Kinetic and potential energy effects are absent, and $W = 0$.
3. Combustion is complete.
4. The initial mixture and the products of combustion each form ideal gas mixtures.
5. The initial and final states are equilibrium states.

Analysis: The chemical reaction equation for the complete combustion of methane with oxygen is

$$CH_4 + 2O_2 \rightarrow CO_2 + 2H_2O$$

(a) With assumptions 2 and 3, the closed system energy balance takes the form

$$U_P - U_R = Q - \overset{0}{\cancel{W}}$$

or

$$Q = U_P - U_R = (1\bar{u}_{CO_2} + 2\bar{u}_{H_2O(g)}) - (1\bar{u}_{CH_4(g)} + 2\bar{u}_{O_2})$$

Each coefficient in this equation is the same as the corresponding term of the balanced chemical equation.

Since each reactant and product behaves as an ideal gas, the respective specific internal energies can be evaluated as $\bar{u} = \bar{h} - \bar{R}T$. The energy balance then becomes

$$Q = [1(\bar{h}_{CO_2} - \bar{R}T_2) + 2(\bar{h}_{H_2O(g)} - \bar{R}T_2)] - [1(\bar{h}_{CH_4(g)} - \bar{R}T_1) + 2(\bar{h}_{O_2} - \bar{R}T_1)]$$

where T_1 and T_2 denote, respectively, the initial and final temperatures. Collecting like terms

$$Q = (\bar{h}_{CO_2} + 2\bar{h}_{H_2O(g)} - \bar{h}_{CH_4(g)} - 2\bar{h}_{O_2}) + 3\bar{R}(T_1 - T_2)$$

The specific enthalpies are evaluated in terms of the respective enthalpies of formation to give

$$Q = [(\bar{h}_f^\circ + \Delta\bar{h})_{CO_2} + 2(\bar{h}_f^\circ + \Delta\bar{h})_{H_2O(g)}$$
$$- (\bar{h}_f^\circ + \overset{0}{\cancel{\Delta\bar{h}}})_{CH_4(g)} - 2(\overset{0}{\cancel{\bar{h}_f^\circ}} + \overset{0}{\cancel{\Delta\bar{h}}})_{O_2}] + 3\bar{R}(T_1 - T_2)$$

Since the methane and oxygen are initially at 25°C, $\Delta\bar{h} = 0$ for each of these reactants. Also, $\bar{h}_f^\circ = 0$ for oxygen.

With enthalpy of formation values for CO_2, $H_2O(g)$, and $CH_4(g)$ from Table A-25 and enthalpy values for H_2O and CO_2 from Table A-23

$$\begin{aligned} Q &= [-393{,}520 + (37{,}405 - 9364)] + 2[-241{,}820 + (31{,}828 - 9904)] \\ &\quad -(-74{,}850) + 3(8.314)(298 - 900) \\ &= -745{,}436 \text{ kJ} \end{aligned}$$

(b) By assumption 3, the initial mixture and the products of combustion each form ideal gas mixtures. Thus, for the reactants

$$p_1 V = n_R \overline{R} T_1$$

where n_R is the total number of moles of reactants and p_1 is the initial pressure. Similarly, for the products

$$p_2 V = n_P \overline{R} T_2$$

where n_P is the total number of moles of products and p_2 is the final pressure.

Since $n_R = n_P = 3$ and volume is constant, these equations combine to give

$$p_2 = \frac{T_2}{T_1} p_1 = \left(\frac{900\ \text{K}}{298\ \text{K}}\right)(1\ \text{atm}) = 3.02\ \text{atm}$$

Skills Developed: *Ability to...*

- ✓ apply the closed system energy balance to a reacting system.
- ✓ evaluate property data appropriately.
- ✓ apply the ideal gas equation of state.

Quick Quiz

Calculate the volume of the system, in m^3.
Ans. 73.36 m^3.

13.2.3 Enthalpy of Combustion and Heating Values

Although the enthalpy of formation concept underlies the formulations of the energy balances for reactive systems presented thus far, the enthalpy of formation of fuels is not always tabulated. ➡ **FOR EXAMPLE...** fuel oil and coal are normally composed of several individual chemical substances, the relative amounts of which may vary considerably, depending on the source. Owing to the wide variation in composition that these fuels can exhibit, we do not find their enthalpies of formation listed in Table A-25 or similar compilations of thermophysical data. ⬅ In many cases of practical interest, however, the *enthalpy of combustion,* which is accessible experimentally, can be used to conduct an energy analysis when enthalpy of formation data are lacking.

enthalpy of combustion

The ***enthalpy of combustion*** $\overline{h}_{RP}$ is defined as the difference between the enthalpy of the products and the enthalpy of the reactants when *complete* combustion occurs at a *given temperature and pressure*. That is

$$\overline{h}_{RP} = \sum_P n_e \overline{h}_e - \sum_R n_i \overline{h}_i \qquad (13.18)$$

where the n's correspond to the respective coefficients of the reaction equation giving the moles of reactants and products per mole of fuel. When the enthalpy of combustion is expressed on a unit mass of fuel basis, it is designated h_{RP}. Tabulated values are usually given at the standard temperature T_{ref} and pressure p_{ref} introduced in Sec. 13.2.1. The symbol $\overline{h}^\circ_{RP}$ or h°_{RP} is used for data at this temperature and pressure.

higher and lower heating values

The *heating value* of a fuel is a positive number equal to the magnitude of the enthalpy of combustion. Two heating values are recognized by name: the ***higher heating value*** (HHV) and the ***lower heating value*** (LHV). The higher heating value is obtained when all the water formed by combustion is a liquid; the lower heating value is obtained when all the water formed by combustion is a vapor. The higher heating value exceeds the lower heating value by the energy that would be required to vaporize the liquid formed. Values for the HHV and LHV also depend on whether the fuel is a liquid or a gas. Heating value data for several hydrocarbons are provided in Table A-25.

The calculation of the enthalpy of combustion, and the associated heating value, using table data is illustrated in the next example.

Example 13.7 CALCULATING ENTHALPY OF COMBUSTION

Calculate the enthalpy of combustion of gaseous methane, in kJ per kg of fuel, (a) at 25°C, 1 atm with liquid water in the products, (b) at 25°C, 1 atm with water vapor in the products. (c) Repeat part (b) at 1000 K, 1 atm.

Solution

Known: The fuel is gaseous methane.

Find: Determine the enthalpy of combustion, in kJ per kg of fuel, (a) at 25°C, 1 atm with liquid water in the products, (b) at 25°C, 1 atm with water vapor in the products, (c) at 1000 K, 1 atm with water vapor in the products.

Engineering Model:

1. Each mole of oxygen in the combustion air is accompanied by 3.76 moles of nitrogen, which is inert.
2. Combustion is complete, and both reactants and products are at the same temperature and pressure.
3. The ideal gas model applies for methane, the combustion air, and the gaseous products of combustion.

Analysis: The combustion equation is

$$CH_4 + 2O_2 + 7.52N_2 \rightarrow CO_2 + 2H_2O + 7.52N_2$$

The enthalpy of combustion is, from Eq. 13.18

$$\bar{h}_{RP} = \sum_P n_e(\bar{h}_f^\circ + \Delta\bar{h})_e - \sum_R n_i(\bar{h}_f^\circ + \Delta\bar{h})_i$$

Introducing the coefficients of the combustion equation and evaluating the specific enthalpies in terms of the respective enthalpies of formation

$$\bar{h}_{RP} = \bar{h}_{CO_2} + 2\bar{h}_{H_2O} - \bar{h}_{CH_4(g)} - 2\bar{h}_{O_2}$$

$$= (\bar{h}_f^\circ + \Delta\bar{h})_{CO_2} + 2(\bar{h}_f^\circ + \Delta\bar{h})_{H_2O} - (\bar{h}_f^\circ + \Delta\bar{h})_{CH_4(g)} - 2(\overset{0}{\cancel{\bar{h}_f^\circ}} + \Delta\bar{h})_{O_2}$$

For nitrogen, the enthalpy terms of the reactants and products cancel. Also, the enthalpy of formation of oxygen is zero by definition. On rearrangement, the enthalpy of combustion expression becomes

$$\bar{h}_{RP} = (\bar{h}_f^\circ)_{CO_2} + 2(\bar{h}_f^\circ)_{H_2O} - (\bar{h}_f^\circ)_{CH_4(g)} + [(\Delta\bar{h})_{CO_2} + 2(\Delta\bar{h})_{H_2O} - (\Delta\bar{h})_{CH_4(g)} - 2(\Delta\bar{h})_{O_2}]$$

$$= \bar{h}_{RP}^\circ + [(\Delta\bar{h})_{CO_2} + 2(\Delta\bar{h})_{H_2O} - (\Delta\bar{h})_{CH_4(g)} - 2(\Delta\bar{h})_{O_2}]$$

The values for $\bar{h}_{RP}^\circ$ and $(\Delta\bar{h})_{H_2O}$ depend on whether the water in the products is a liquid or a vapor.

(a) Since the reactants and products are at 25°C, 1 atm in this case, the $\Delta\bar{h}$ terms drop out of the above expression for $\bar{h}_{RP}$. Thus, for liquid water in the products, the enthalpy of combustion is

$$\bar{h}_{RP}^\circ = (\bar{h}_f^\circ)_{CO_2} + 2(\bar{h}_f^\circ)_{H_2O(l)} - (\bar{h}_f^\circ)_{CH_4(g)}$$

With enthalpy of formation values from Table A-25

$$\bar{h}_{RP}^\circ = -393{,}520 + 2(-285{,}830) - (-74{,}850) = -890{,}330 \text{ kJ/kmol (fuel)}$$

Dividing by the molecular weight of methane places this result on a unit mass of fuel basis

$$h_{RP}^\circ = \frac{-890{,}330 \text{ kJ/kmol (fuel)}}{16.04 \text{ kg (fuel)/kmol (fuel)}} = -55{,}507 \text{ kJ/kg (fuel)}$$

which agrees with the higher heating value of methane given in Table A-25.

(b) As in part (a), the $\Delta\bar{h}$ terms drop out of the above expression for $\bar{h}_{RP}$, which for water vapor in the products reduces to $\bar{h}_{RP}^\circ$, where

$$\bar{h}_{RP}^\circ = (\bar{h}_f^\circ)_{CO_2} + 2(\bar{h}_f^\circ)_{H_2O(g)} - (\bar{h}_f^\circ)_{CH_4(g)}$$

With enthalpy of formation values from Table A-25

$$\bar{h}_{RP}^\circ = -393{,}520 + 2(-241{,}820) - (-74{,}850) = -802{,}310 \text{ kJ/kmol (fuel)}$$

On a unit of mass of fuel basis, the enthalpy of combustion for this case is

$$h_{RP}^{\circ} = \frac{-802{,}310}{16.04} = -50{,}019 \text{ kJ/kg (fuel)}$$

which agrees with the lower heating value of methane given in Table A-25.

(c) For the case where the reactants and products are at 1000 K, 1 atm, the term $\bar{h}_{RP}^{\circ}$ in the above expression for $\bar{h}_{RP}$ has the value determined in part (b): $\bar{h}_{RP}^{\circ} = -802{,}310$ kJ/kmol (fuel), and the $\Delta\bar{h}$ terms for O_2, $H_2O(g)$, and CO_2 can be evaluated using specific enthalpies at 298 and 1000 K from Table A-23. The results are

$$(\Delta\bar{h})_{O_2} = 31{,}389 - 8682 = 22{,}707 \text{ kJ/kmol}$$

$$(\Delta\bar{h})_{H_2O(g)} = 35{,}882 - 9904 = 25{,}978 \text{ kJ/kmol}$$

$$(\Delta\bar{h})_{CO_2} = 42{,}769 - 9364 = 33{,}405 \text{ kJ/kmol}$$

For methane, the $\bar{c}_p$ expression of Table A-21 can be used to obtain

$$(\Delta\bar{h})_{CH_4(g)} = \int_{298}^{1000} \bar{c}_p \, dT$$

$$= \bar{R}\left(3.826T - \frac{3.979}{10^3}\frac{T^2}{2} + \frac{24.558}{10^6}\frac{T^3}{3} - \frac{22.733}{10^9}\frac{T^4}{4} + \frac{6.963}{10^{12}}\frac{T^5}{5}\right)_{298}^{1000}$$

❶ $$= 38{,}189 \text{ kJ/kmol (fuel)}$$

Substituting values into the expression for the enthalpy of combustion

$$\bar{h}_{RP} = -802{,}310 + [33{,}405 + 2(25{,}978) - 38{,}189 - 2(22{,}707)]$$

$$= -800{,}522 \text{ kJ/kmol (fuel)}$$

On a unit mass basis

❷ $$h_{RP} = \frac{-800{,}552}{16.04} = -49{,}910 \text{ kJ/kg (fuel)}$$

❶ Using *Interactive Thermodynamics: IT*, we get 38,180 kJ/kmol (fuel).

❷ Comparing the values of parts (b) and (c), the enthalpy of combustion of methane is seen to vary little with temperature. The same is true for many hydrocarbon fuels. This fact is sometimes used to simplify combustion calculations.

Skills Developed: *Ability to...*
- ✓ calculate the enthalpy of combustion at standard temperature and pressure.
- ✓ calculate the enthalpy of combustion at an elevated temperature and standard pressure.

Quick Quiz

Calculate the lower heating value of methane, in kJ/kg (fuel) at 25°C, 1 atm.
Ans. 50,019 kJ/kg.

CALORIMETRY. When enthalpy of formation data are available for *all* the reactants and products, the enthalpy of combustion can be calculated directly from Eq. 13.18, as illustrated in Example 13.7. Otherwise, it must be obtained experimentally using devices known as *calorimeters*. Both constant-volume (bomb calorimeters) and flow-through devices are employed for this purpose. Consider as an illustration a reactor operating at steady state in which the fuel is burned completely with air. For the products to be returned to the same temperature as the reactants, a heat transfer from the reactor would be required. From an energy rate balance, the required heat transfer is

$$\frac{\dot{Q}_{cv}}{\dot{n}_F} = \sum_P n_e\bar{h}_e - \sum_R n_i\bar{h}_i \qquad (13.19)$$

where the symbols have the same significance as in previous discussions. The heat transfer per mole of fuel, $\dot{Q}_{cv}/\dot{n}_F$, would be determined from measured data. Comparing Eq. 13.19 with the defining equation, Eq. 13.18, we have $\bar{h}_{RP} = \dot{Q}_{cv}/\dot{n}_F$. In accord with the usual sign convention for heat transfer, the enthalpy of combustion would be negative.

As noted previously, the enthalpy of combustion can be used for energy analyses of reacting systems. ➡ **FOR EXAMPLE...** consider a control volume at steady state for which the energy rate balance takes the form

$$\frac{\dot{Q}_{cv}}{\dot{n}_F} - \frac{\dot{W}_{cv}}{\dot{n}_F} = \sum_P n_e(\bar{h}_f^\circ + \Delta\bar{h})_e - \sum_R n_i(\bar{h}_f^\circ + \Delta\bar{h})_i$$

All symbols have the same significance as in previous discussions. This equation can be rearranged to read

$$\frac{\dot{Q}_{cv}}{\dot{n}_F} - \frac{\dot{W}_{cv}}{\dot{n}_F} = \underline{\sum_P n_e(\bar{h}_f^\circ)_e - \sum_R n_i(\bar{h}_f^\circ)_i} + \sum_P n_e(\Delta\bar{h})_e - \sum_R n_i(\Delta\bar{h})_i$$

For a complete reaction, the underlined term is just the enthalpy of combustion $\bar{h}_{RP}^\circ$, at T_{ref} and p_{ref}. Thus, the equation becomes

$$\frac{\dot{Q}_{cv}}{\dot{n}_F} - \frac{\dot{W}_{cv}}{\dot{n}_F} = \bar{h}_{RP}^\circ + \sum_P n_e(\Delta\bar{h})_e - \sum_R n_i(\Delta\bar{h})_i \qquad (13.20)$$

The right side of Eq. 13.20 can be evaluated with an experimentally determined value for $\bar{h}_{RP}^\circ$ and $\Delta\bar{h}$ values for the reactants and products determined as discussed previously. ⬅

13.3 Determining the Adiabatic Flame Temperature

Let us reconsider the reactor at steady state pictured in Fig. 13.2. In the absence of work $\dot{W}_{cv}$ and appreciable kinetic and potential energy effects, the energy liberated on combustion is transferred from the reactor in two ways only: by energy accompanying the exiting combustion products and by heat transfer to the surroundings. The smaller the heat transfer, the greater the energy carried out with the combustion products and thus the greater the temperature of the products. The temperature that would be achieved by the products in the limit of adiabatic operation of the reactor is called the ***adiabatic flame temperature*** or ***adiabatic combustion*** temperature.

adiabatic flame temperature

The adiabatic flame temperature can be determined by use of the conservation of mass and conservation of energy principles. To illustrate the procedure, let us suppose that the combustion air and the combustion products each form ideal gas mixtures. Then, with the other assumptions stated above, the energy rate balance on a per mole of fuel basis, Eq. 13.12b, reduces to the form $\bar{h}_P = \bar{h}_R$—that is

$$\sum_P n_e\bar{h}_e = \sum_R n_i\bar{h}_i \qquad (13.21a)$$

where i denotes the incoming fuel and air streams and e the exiting combustion products. With this expression, the adiabatic flame temperature can be determined using table data or computer software, as follows.

USING TABLE DATA. When using Eq. 13.9 with table data to evaluate enthalpy terms, Eq. 13.21a takes the form

$$\sum_P n_e(\bar{h}_f^\circ + \Delta\bar{h})_e = \sum_R n_i(\bar{h}_f^\circ + \Delta\bar{h})_i$$

or

$$\sum_{\text{P}} n_e(\Delta\bar{h})_e = \sum_{\text{R}} n_i(\Delta\bar{h})_i + \sum_{\text{R}} n_i\bar{h}^\circ_{\text{fi}} - \sum_{\text{P}} n_e\bar{h}^\circ_{\text{fe}} \qquad (13.21b)$$

The n's are obtained on a per mole of fuel basis from the balanced chemical reaction equation. The enthalpies of formation of the reactants and products are obtained from Table A-25. Enthalpy of combustion data might be employed in situations where the enthalpy of formation for the fuel is not available. Knowing the states of the reactants as they enter the reactor, the $\Delta\bar{h}$ terms for the reactants can be evaluated as discussed previously. Thus, all terms on the right side of Eq. 13.21b can be evaluated. The terms $(\Delta\bar{h})_e$ on the left side account for the changes in enthalpy of the products from T_{ref} to the unknown adiabatic flame temperature. Since the unknown temperature appears in each term of the sum on the left side of the equation, determination of the adiabatic flame temperature requires *iteration:* A temperature for the products is assumed and used to evaluate the left side of Eq. 13.21b. The value obtained is compared with the previously determined value for the right side of the equation. The procedure continues until satisfactory agreement is attained. Example 13.8 gives an illustration.

USING COMPUTER SOFTWARE. Thus far we have emphasized the use of Eq. 13.9 together with table data when evaluating the specific enthalpies required by energy balances for reacting systems. Such enthalpy values also can be retrieved using *Interactive Thermodynamics: IT*. With *IT*, the quantities on the right side of Eq. 13.9 are evaluated by software, and $\bar{h}$ data are returned *directly*. ➡ **FOR EXAMPLE...** consider CO_2 at 500 K modeled as an ideal gas. The specific enthalpy is obtained from *IT* as follows:

```
T = 500 // K
h = h_T("CO2", T)
```

Choosing K for the temperature unit and moles for the amount under the **Units** menu, *IT* returns $h = -3.852 \times 10^5$ kJ/kmol.

This value agrees with the value calculated from Eq. 13.9 using enthalpy data for CO_2 from Table A-23, as follows

$$\begin{aligned}\bar{h} &= \bar{h}^\circ_{\text{f}} + [\bar{h}(500\text{ K}) - \bar{h}(298\text{ K})] \\ &= -393{,}520 + [17{,}678 - 9364] \\ &= -3.852 \times 10^5\text{ kJ/kmol}\end{aligned}$$ ⬅

As suggested by this discussion, *IT* is also useful for analyzing reacting systems. In particular, the equation solver and property retrieval features of *IT* allow the adiabatic flame temperature to be determined without the iteration required when using table data. This is illustrated in Example 13.8.

Example 13.8 DETERMINING THE ADIABATIC FLAME TEMPERATURE

Liquid octane at 25°C, 1 atm enters a well-insulated reactor and reacts with air entering at the same temperature and pressure. For steady-state operation and negligible effects of kinetic and potential energy, determine the temperature of the combustion products for complete combustion with **(a)** the theoretical amount of air, **(b)** 400% theoretical air.

Solution

Known: Liquid octane and air, each at 25°C and 1 atm, burn completely within a well-insulated reactor operating at steady state.

Find: Determine the temperature of the combustion products for (a) the theoretical amount of air and (b) 400% theoretical air.

Schematic and Given Data:

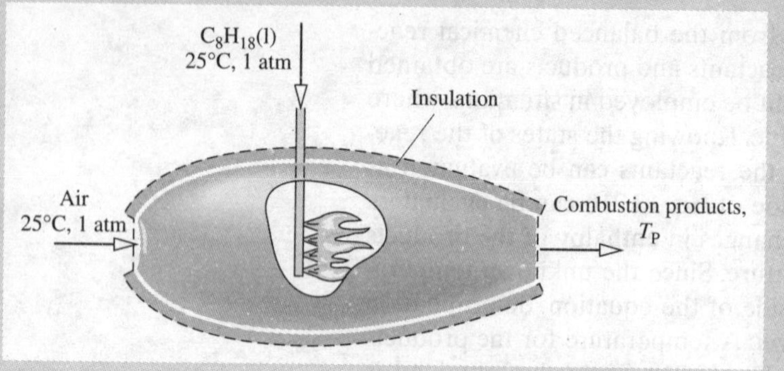

Fig. E13.8

Engineering Model:

1. The control volume indicated on the accompanying figure by a dashed line operates at steady state.
2. For the control volume, $\dot{Q}_{cv} = 0$, $\dot{W}_{cv} = 0$, and kinetic and potential effects are negligible.
3. The combustion air and the products of combustion each form ideal gas mixtures.
4. Combustion is complete.
5. Each mole of oxygen in the combustion air is accompanied by 3.76 moles of nitrogen, which is inert.

Analysis: At steady state, the control volume energy rate balance Eq. 13.12b reduces with assumptions 2 and 3 to give Eq. 13.21a

$$\sum_P n_e \bar{h}_e = \sum_R n_i \bar{h}_i \quad (1)$$

When Eq. 13.9 and table data are used to evaluate the enthalpy terms, Eq. (1) is written as

$$\sum_P n_e(\bar{h}_f^\circ + \Delta\bar{h})_e = \sum_R n_i(\bar{h}_f^\circ + \Delta\bar{h})_i$$

On rearrangement, this becomes

$$\sum_P n_e(\Delta\bar{h})_e = \sum_R n_i(\Delta\bar{h})_i + \sum_R n_i\bar{h}_{fi}^\circ - \sum_P n_e\bar{h}_{fe}^\circ$$

which corresponds to Eq. 13.21b. Since the reactants enter at 25°C, the $(\Delta\bar{h})_i$ terms on the right side vanish, and the energy rate equation becomes

$$\sum_P n_e(\Delta\bar{h})_e = \sum_R n_i\bar{h}_{fi}^\circ - \sum_P n_e\bar{h}_{fe}^\circ \quad (2)$$

(a) For combustion of liquid octane with the theoretical amount of air, the chemical equation is

$$C_8H_{18}(l) + 12.5O_2 + 47N_2 \rightarrow 8CO_2 + 9H_2O(g) + 47N_2$$

Introducing the coefficients of this equation, Eq. (2) takes the form

$$8(\Delta\bar{h})_{CO_2} + 9(\Delta\bar{h})_{H_2O(g)} + 47(\Delta\bar{h})_{N_2}$$
$$= [(\bar{h}_f^\circ)_{C_8H_{18}(l)} + 12.5\cancelto{0}{(\bar{h}_f^\circ)_{O_2}} + 47\cancelto{0}{(\bar{h}_f^\circ)_{N_2}}]$$
$$- [8(\bar{h}_f^\circ)_{CO_2} + 9(\bar{h}_f^\circ)_{H_2O(g)} + 47\cancelto{0}{(\bar{h}_f^\circ)_{N_2}}]$$

The right side of the above equation can be evaluated with enthalpy of formation data from Table A-25, giving

$$8(\Delta\bar{h})_{CO_2} + 9(\Delta\bar{h})_{H_2O(g)} + 47(\Delta\bar{h})_{N_2} = 5{,}074{,}630 \text{ kJ/kmol (fuel)}$$

Each $\Delta\bar{h}$ term on the left side of this equation depends on the temperature of the products, T_P. This temperature can be determined by an iterative procedure.

The following table gives a summary of the iterative procedure for three trial values of T_P. Since the summation of the enthalpies of the products equals 5,074,630 kJ/kmol, the actual value of T_P is in the interval from 2350 to 2400 K. Interpolation between these temperatures gives $T_P = 2395$ K.

	2500 K	2400 K	2350 K
$8(\Delta\bar{h})_{CO_2}$	975,408	926,304	901,816
$9(\Delta\bar{h})_{H_2O(g)}$	890,676	842,436	818,478
$47(\Delta\bar{h})_{N_2}$	3,492,664	3,320,597	3,234,869
$\sum_{P} n_e(\Delta\bar{h})_e$	5,358,748	5,089,337	4,955,163

Alternative Solution:

The following *IT* code can be used as an alternative to iteration with table data, where hN2_R and hN2_P denote the enthalpy of N_2 in the reactants and products, respectively, and so on. In the **Units** menu, select temperature in K and amount of substance in moles.

```
TR = 25 + 273.15 // K
// Evaluate reactant and product enthalpies, hR and hP, respectively
hR = hC8H18 + 12.5 * hO2_R + 47 * hN2_R
hP = 8 * hCO2_P + 9 * hH2O_P + 47 * hN2_P
hC8H18 = −249910 // kJ/kmol (Value from Table A-25)
hO2_R = h_T("O2",TR)
hN2_R = h_T("N2" TR)
hCO2_P = h_T("CO2",TP)
hH2O_P = h_T("H2O",TP)
hN2_P = h_T("N2",TP)
// Energy balance
hP = hR
```

Using the **Solve** button, the result is TP = 2394 K, which agrees closely with the result obtained above.

(b) For complete combustion of liquid octane with 400% theoretical air, the chemical equation is

$$C_8H_{18}(l) + 50O_2 + 188N_2 \rightarrow 8CO_2 + 9H_2O + 37.5O_2 + 188N_2$$

Equation (2), the energy rate balance, reduces for this case to

$$8(\Delta\bar{h})_{CO_2} + 9(\Delta\bar{h})_{H_2O(g)} + 37.5(\Delta\bar{h})_{O_2} + 188(\Delta\bar{h})_{N_2} = 5{,}074{,}630 \text{ kJ/kmol (fuel)}$$

❶ Observe that the right side has the same value as in part (a). Proceeding iteratively as above, the temperature of the products is $T_P = 962$ K. The use of *IT* to solve part (b) is left as an exercise.

Skills Developed: *Ability to...*
- ✓ apply the control volume energy balance to calculate the adiabatic flame temperature.
- ✓ evaluate enthalpy values appropriately.

❶ The temperature determined in part (b) is considerably lower than the value found in part (a). This shows that once enough oxygen has been provided for complete combustion, bringing in more air dilutes the combustion products, lowering their temperature.

Quick Quiz

If octane gas entered instead of liquid octane, would the adiabatic flame temperature increase, decrease, or stay the same?
Ans. Increase.

CLOSING COMMENTS. For a specified fuel and specified temperature and pressure of the reactants, the *maximum* adiabatic flame temperature is for complete combustion with the theoretical amount of air. The measured value of the temperature of the combustion products may be several hundred degrees below the calculated maximum adiabatic flame temperature, however, for several reasons:

- ▶ Once adequate oxygen has been provided to permit complete combustion, bringing in more air dilutes the combustion products, lowering their temperature.

- Incomplete combustion also tends to reduce the temperature of the products, and combustion is seldom complete (see Sec. 14.4).
- Heat losses can be reduced but not altogether eliminated.
- As a result of the high temperatures achieved, some of the combustion products may dissociate. Endothermic dissociation reactions lower the product temperature. The effect of dissociation on the adiabatic flame temperature is considered in Sec. 14.4.

13.4 Fuel Cells

fuel cell

A ***fuel cell*** is an *electrochemical* device in which fuel and an oxidizer (normally oxygen from air) undergo a chemical reaction, providing electrical current to an external circuit and producing products. The fuel and oxidizer react catalytically in stages on separate electrodes: the anode and the cathode. An electrolyte separating the two electrodes allows passage of ions formed by reaction. Depending on the type of fuel cell, the ions may be positively or negatively charged. The reaction is *not* a combustion process.

Rates of reaction in fuel cells are limited by the time it takes for diffusion of chemical species through the electrodes and the electrolyte and by the speed of the chemical reactions themselves. These features, together with other aspects of fuel cell operation, result in internal irreversibilities that are inherently less significant than encountered in power producing devices relying on combustion.

By avoiding highly irreversible combustion, fuel cells have the *potential* of providing more power from a given supply of fuel and oxidizer, while forming fewer undesirable products, than internal combustion engines and gas turbines. In contrast to power plants studied in previous chapters, fuel cells can produce electric power without moving parts or utilizing intermediate heat exchangers. Fuel cells do not operate as thermodynamic power cycles, and thus the notion of a limiting thermal efficiency imposed by the second law is not applicable. Despite these thermodynamic advantages, widespread use of fuel cells has not occurred thus far owing primarily to cost.

For scale, Fig. 13.3*a* shows a *solid oxide* fuel cell module. Figure 13.3*b* gives the schematic of a *proton exchange membrane* fuel cell, which is discussed next as a representative case.

PROTON EXCHANGE MEMBRANE FUEL CELL. The fuel cell shown in Fig. 13.3*b* operates with hydrogen (H_2) as the fuel and oxygen (O_2) as the oxidizer. This type of cell

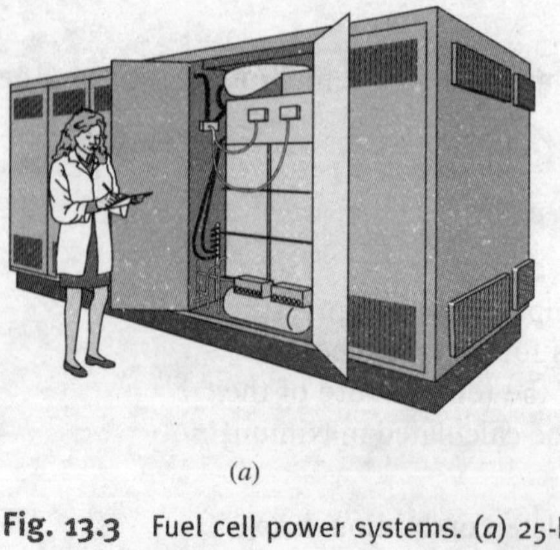
(*a*)

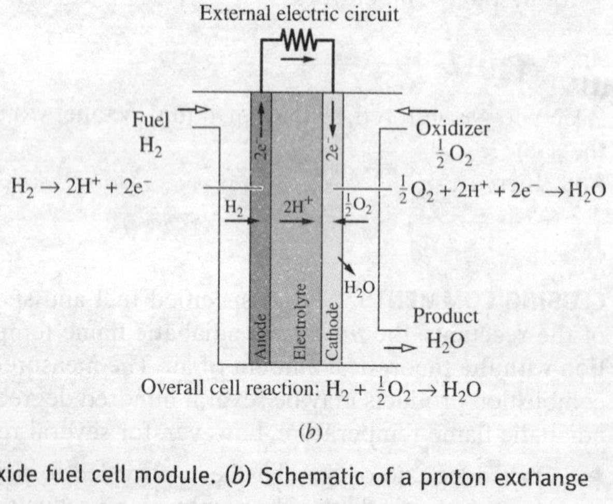

(*b*)

Fig. 13.3 Fuel cell power systems. (*a*) 25-kW solid oxide fuel cell module. (*b*) Schematic of a proton exchange membrane fuel cell.

Horizons

Goodbye Batteries, Hello Fuel Cells?

Power needs of cellular phones, laptops, and other portable electronic devices are increasing so rapidly that the battery industry is struggling to keep up. Some observers say that today's batteries won't be able to provide enough power, are too heavy, and don't last long enough to meet the needs of quickly evolving electronics. Pocket-size fuel cells might prove to be a viable alternative.

To meet consumer needs, companies are rushing to develop small fuel cells that promise to provide power up to 10 times longer on a single charge than conventional batteries. These cells can be charged instantly just by adding more fuel. Battery companies are fighting back with a new generation of batteries, known as lithium ion batteries, already used in watches, flash cameras, and rechargeable power packs. Lithium ion batteries provide several times the output of similar-size alkaline batteries and can be recharged numerous times. To compete, fuel cells must prove themselves as reliable and versatile as batteries, and stakes are high in the consumer electronics market.

Like the fuel cells being developed to power cars and generate electricity for homes and offices, the pocket-size versions face stiff challenges on the way to market. They rely on costly precious-metal catalysts to operate and are difficult to manufacture. They run on combustible fuels, typically hydrogen or methanol, and this brings concerns over safety. Still, many think fuel cells for portable electronics will be the first fuel cells most of us will see because of strong consumer demand for cost-competitive, longer lasting, instantly rechargeable power.

is known as a *proton exchange membrane fuel cell* (PEMFC). At the anode, hydrogen ions (H^+) and electrons are produced. At the cathode, ions and electrons are consumed. The reactions at these electrodes and the *overall* cell reaction are labeled on the figure. The only products of this fuel cell are water and the power generated.

Charge-carrying hydrogen ions are conducted through the electrolytic membrane. For acceptable ion conductivity, a high membrane water content is required. This requirement restricts the fuel cell to operating in the range 60–100°C, which is below the boiling point of water. Cooling is generally needed to maintain the fuel cell at the operating temperature. Due to such relatively low temperature operation, costly platinum catalysts are required at both the anode and cathode to increase ionization reaction rates. Catalytic activity is more important in lower temperature fuel cells because reaction rates tend to decrease with decreasing temperature.

Major automakers are beginning to introduce vehicles powered by proton exchange membrane fuel cells. Fuel cell *stacks* formed from several hundred individual fuel cells are required to meet automotive needs. These fuel cell stacks are integrated with several components necessary to support fuel cell operation, including components that provide fuel, oxidizer, and coolant. Power requirements, irreversibilities, and losses associated with the support components conspire to give a much lower net power output for a given fuel input than obtainable from the stand-alone fuel cell.

Proton exchange membrane fuel cells also have potential to replace batteries in portable devices such as cellular phones, laptop computers, and video players.

FUELS. With today's technology, hydrogen is preferred for fuel cell applications because of its exceptional ability to produce electrons when suitable catalysts are used, while in principle producing no harmful emissions from the fuel cell itself. Methanol (CH_3OH) and carbon monoxide can be used directly as fuels in some applications, but often with performance penalties.

Since hydrogen is not naturally occurring, it must be produced chemically from hydrocarbons, including natural gas, by electrolysis of water, or by other means. Hydrogen production technologies involving chemical processes can generate undesirable emissions. Irreversibilities and losses inherent in hydrogen production also

Table 13.1

Characteristics of Major Fuel Cell Types

	Proton Exchange Membrane Fuel Cell (PEMFC)	*Phosphoric Acid Fuel Cell (PAFC)*	*Molten Carbonate Fuel Cell (MCFC)*	*Solid Oxide Fuel Cell (SOFC)*
Transportation application	- automotive power	- large vehicle power		- vehicle auxiliary power - heavy vehicle propulsion
Other applications	- portable power - small-scale stationary power	- on-site cogeneration - electric power generation	- on-site cogeneration - electric power generation	- on-site cogeneration - electric power generation
Electrolyte	ion exchange membrane	liquid phosphoric acid	liquid molten carbonate	solid oxide ceramic
Charge carrier	H^+	H^+	CO_3^{--}	O^{--}
Operating temperature	60–100°C	150–220°C	600–700°C	800–1000°C
Fuel reforming	external	external	internal or external	internal or external

Source: Appleby, A. J., and Foulkes, F. R., 1993, *Fuel Cell Handbook,* Krieger Publishing Company, Malabar, Florida. Hirschenhofer, J. H., Stauffer, D. B., Engleman, R. R., and Klett, M. G., 1998, *Fuel Cell Handbook,* Fourth Edition, DOE/FETC-99/1076.

result in a lower *overall* conversion of *primary* input to fuel cell-generated power than reported for fuel cells directly fueled with hydrogen.

In some applications, hydrogen is provided directly to the fuel cell from storage as a compressed gas, a cryogenic liquid, or atoms absorbed within metallic structures. Hydrogen also can be produced at the point of use. *Internal reforming* refers to applications where hydrogen production is integrated with the fuel cell. When hydrogen is produced separately from the fuel cell itself, this is known as *external reforming.* Owing to thermal limitations of current technology, internal reforming is feasible only in higher-temperature *molten carbonate* and *solid oxide* fuel cells.

Table 13.1 summarizes the most promising fuel cell technologies currently under investigation. Included are potential applications and other characteristics. For more detailed discussions, see the sources listed in Table 13.1.

13.5 Absolute Entropy and the Third Law of Thermodynamics

Thus far our analyses of reacting systems have been conducted using the conservation of mass and conservation of energy principles. In the present section some of the implications of the second law of thermodynamics for reacting systems are considered. The discussion continues in the second part of this chapter dealing with the exergy concept, and in the next chapter where the subject of chemical equilibrium is taken up.

13.5.1 Evaluating Entropy for Reacting Systems

The property entropy plays an important part in quantitative evaluations using the second law of thermodynamics. When reacting systems are under consideration, the same problem arises for entropy as for enthalpy and internal energy: A common datum must be used to assign entropy values for each substance involved in the reaction. This is accomplished using the *third law* of thermodynamics and the *absolute entropy* concept.

The ***third law*** deals with the entropy of substances at the absolute zero of temperature. Based on empirical evidence, this law states that the entropy of a pure crystalline substance is zero at the absolute zero of temperature, 0 K. Substances not having a pure crystalline structure at absolute zero have a nonzero value of entropy at absolute zero. The experimental evidence on which the third law is based is obtained primarily from studies of chemical reactions at low temperatures and specific heat measurements at temperatures approaching absolute zero.

third law of thermodynamics

ABSOLUTE ENTROPY. For present considerations, the importance of the third law is that it provides a datum relative to which the entropy of each substance participating in a reaction can be evaluated so that no ambiguities or conflicts arise. The entropy relative to this datum is called the ***absolute entropy.*** The change in entropy of a substance between absolute zero and any given state can be determined from precise measurements of energy transfers and specific heat data or from procedures based on statistical thermodynamics and observed molecular data.

absolute entropy

Table A-25 gives the value of the absolute entropy for selected substances at the standard reference state, $T_{ref} = 298.15$ K, $p_{ref} = 1$ atm. Two values of absolute entropy for water are provided. One is for liquid water and the other is for water vapor. As for the case of the enthalpy of formation of water considered previously, the vapor value listed is for a hypothetical ideal gas state in which water is a vapor at 25°C and $p_{ref} = 1$ atm. Table A-23 gives tabulations of absolute entropy versus temperature at a pressure of 1 atm for selected gases. The absolute entropy at 1 atm and temperature T is designated as $s°(T)$ or $\bar{s}°(T)$, depending on whether the value is on a unit mass or per mole basis. In these tables, ideal gas behavior is assumed for the gases.

USING ABSOLUTE ENTROPY. When the absolute entropy is known at the standard state, the specific entropy at any other state can be found by adding the specific entropy change between the two states to the absolute entropy at the standard state. Similarly, when the absolute entropy is known at the pressure p_{ref} and temperature T, the absolute entropy at the same temperature and any pressure p can be found from

$$\bar{s}(T, p) = \bar{s}(T, p_{ref}) + [\bar{s}(T, p) - \bar{s}(T, p_{ref})]$$

The second term on the right side of this equation can be evaluated for an *ideal gas* by using Eq. 6.20b, giving

$$\bar{s}(T, p) = \bar{s}°(T) - \bar{R} \ln \frac{p}{p_{ref}} \quad \text{(ideal gas)} \qquad (13.22)$$

where $\bar{s}°(T)$ is the absolute entropy at temperature T and pressure $p_{ref} = 1$ atm.

The entropy of the ith component of an ideal gas mixture is evaluated at the mixture temperature T and the *partial* pressure p_i: $\bar{s}_i(T, p_i)$. The partial pressure is given by $p_i = y_i p$, where y_i is the mole fraction of component i and p is the mixture pressure. Thus, Eq. 13.22 takes the form

$$\bar{s}_i(T, p_i) = \bar{s}_i°(T) - \bar{R} \ln \frac{p_i}{p_{ref}}$$

or

$$\bar{s}_i(T, p_i) = \bar{s}_i^\circ(T) - \bar{R} \ln \frac{y_i p}{p_{\text{ref}}} \quad \left(\begin{array}{c}\text{component } i \text{ of an}\\ \text{ideal gas mixture}\end{array}\right) \tag{13.23}$$

where $\bar{s}_i^\circ(T)$ is the absolute entropy of component i at temperature T and p_{ref}.

Finally, note that *Interactive Thermodynamics* (*IT*) returns absolute entropy directly and does not use the special function s°.

13.5.2 Entropy Balances for Reacting Systems

Many of the considerations that enter when energy balances are written for reacting systems also apply to entropy balances. The writing of entropy balances for reacting systems will be illustrated by referring to special cases of broad interest.

CONTROL VOLUMES AT STEADY STATE. Let us begin by reconsidering the steady-state reactor of Fig. 13.2, for which the combustion reaction is given by Eq. 13.11. The combustion air and the products of combustion are each assumed to form ideal gas mixtures. The entropy rate balance for the two-inlet, single-exit reactor can be expressed on a *per mole of fuel* basis as

$$0 = \sum_j \frac{\dot{Q}_j/T_j}{\dot{n}_F} + \bar{s}_F + \underline{\left[\left(a + \frac{b}{4}\right)\bar{s}_{O_2} + \left(a + \frac{b}{4}\right)3.76\bar{s}_{N_2}\right]}$$
$$- \underline{\left[a\bar{s}_{CO_2} + \frac{b}{2}\bar{s}_{H_2O} + \left(a + \frac{b}{4}\right)3.76\bar{s}_{N_2}\right]} + \frac{\dot{\sigma}_{cv}}{\dot{n}_F} \tag{13.24}$$

where $\dot{n}_F$ is the molar flow rate of the fuel and the coefficients appearing in the underlined terms are the same as those for the corresponding substances in the reaction equation.

The specific entropies of Eq. 13.24 are absolute entropies. Let us consider how the entropies are evaluated for the combustion products and the combustion air. The entropies of the combustion products would be evaluated from Eq. 13.23, using the temperature, pressure, and composition of the products. The entropies of the entering oxygen and nitrogen would be evaluated similarly, using the temperature, pressure, and composition of the combustion air. If the fuel and air entered the reactor as an ideal gas mixture, the entropies of the mixture components would be evaluated from Eq. 13.23 using the appropriate partial pressures. Such considerations are illustrated in Example 13.9.

Example 13.9 EVALUATING ENTROPY PRODUCTION FOR A REACTOR

Liquid octane at 25°C, 1 atm enters a well-insulated reactor and reacts with air entering at the same temperature and pressure. The products of combustion exit at 1 atm pressure. For steady-state operation and negligible effects of kinetic and potential energy, determine the rate of entropy production, in kJ/K per kmol of fuel, for complete combustion with **(a)** the theoretical amount of air, **(b)** 400% theoretical air.

Solution

Known: Liquid octane and air, each at 25°C and 1 atm, burn completely within a well-insulated reactor operating at steady state. The products of combustion exit at 1 atm pressure.

Find: Determine the rate of entropy production, in kJ/K per kmol of fuel, for combustion with (a) the theoretical amount of air, (b) 400% theoretical air.

Schematic and Given Data:

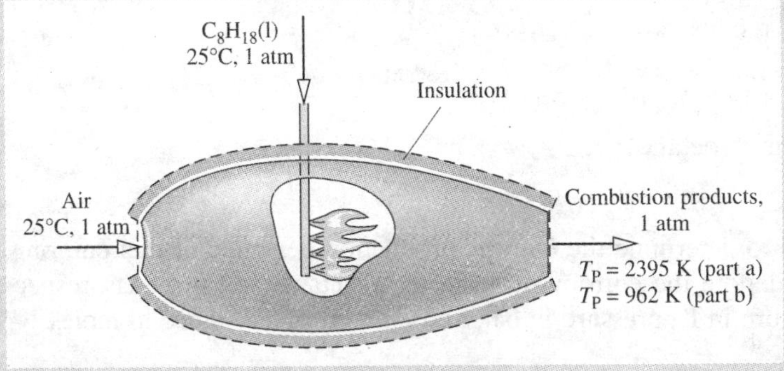

Fig. E13.9

Engineering Model:

1. The control volume shown on the accompanying figure by a dashed line operates at steady state and without heat transfer with its surroundings.
2. Combustion is complete. Each mole of oxygen in the combustion air is accompanied by 3.76 moles of nitrogen, which is inert.
3. The combustion air can be modeled as an ideal gas mixture, as can the products of combustion.
4. The reactants enter at 25°C, 1 atm. The products exit at a pressure of 1 atm.

Analysis: The temperature of the exiting products of combustion T_P was evaluated in Example 13.8 for each of the two cases. For combustion with the theoretical amount of air, T_P = 2395 K. For complete combustion with 400% theoretical air, T_P = 962 K.

(a) For combustion of liquid octane with the theoretical amount of air, the chemical equation is

$$C_8H_{18}(l) + 12.5O_2 + 47N_2 \rightarrow 8CO_2 + 9H_2O(g) + 47N_2$$

With assumptions 1 and 3, the entropy rate balance on a per mole of fuel basis, Eq. 13.24, takes the form

$$0 = \sum_j \frac{\dot{Q}_j/T_j}{\dot{n}_F}^{\,0} + \bar{s}_F + (12.5\bar{s}_{O_2} + 47\bar{s}_{N_2}) - (8\bar{s}_{CO_2} + 9\bar{s}_{H_2O(g)} + 47\bar{s}_{N_2}) + \frac{\dot{\sigma}_{cv}}{\dot{n}_F}$$

or on rearrangement

$$\frac{\dot{\sigma}_{cv}}{\dot{n}_F} = (8\bar{s}_{CO_2} + 9\bar{s}_{H_2O(g)} + 47\bar{s}_{N_2}) - \bar{s}_F - (12.5\bar{s}_{O_2} + 47\bar{s}_{N_2}) \qquad (1)$$

Each coefficient of this equation is the same as for the corresponding term of the balanced chemical equation.

The fuel enters the reactor separately at T_{ref}, p_{ref}. The absolute entropy of liquid octane required by the entropy balance is obtained from Table A-25 as 360.79 kJ/kmol · K.

The oxygen and nitrogen in the combustion air enter the reactor as components of an ideal gas mixture at T_{ref}, p_{ref}. With Eq. 13.23 and absolute entropy data from Table A-23

$$\bar{s}_{O_2} = \bar{s}^{\circ}_{O_2}(T_{ref}) - \bar{R}\ln\frac{y_{O_2}p_{ref}}{p_{ref}}$$
$$= 205.03 - 8.314 \ln 0.21 = 218.01 \text{ kJ/kmol} \cdot \text{K}$$

$$\bar{s}_{N_2} = \bar{s}^{\circ}_{N_2}(T_{ref}) - \bar{R}\ln\frac{y_{N_2}p_{ref}}{p_{ref}}$$
$$= 191.5 - 8.314 \ln 0.79 = 193.46 \text{ kJ/kmol} \cdot \text{K}$$

The product gas exits as an ideal gas mixture at 1 atm, 2395 K with the following composition: $y_{CO_2} = 8/64 = 0.125$, $y_{H_2O(g)} = 9/64 = 0.1406$, $y_{N_2} = 47/64 = 0.7344$. With Eq. 13.23 and absolute entropy data at 2395 K from Table A-23

$$\bar{s}_{CO_2} = \bar{s}^{\circ}_{CO_2} - \bar{R}\ln y_{CO_2}$$
$$= 320.173 - 8.314 \ln 0.125 = 337.46 \text{ kJ/kmol} \cdot \text{K}$$
$$\bar{s}_{H_2O} = 273.986 - 8.314 \ln 0.1406 = 290.30 \text{ kJ/kmol} \cdot \text{K}$$
$$\bar{s}_{N_2} = 258.503 - 8.314 \ln 0.7344 = 261.07 \text{ kJ/kmol} \cdot \text{K}$$

Inserting values into Eq. 1, the expression for the rate of entropy production

$$\frac{\dot{\sigma}_{cv}}{\dot{n}_F} = 8(337.46) + 9(290.30) + 47(261.07)$$

$$- 360.79 - 12.5(218.01) - 47(193.46)$$

$$= 5404 \text{ kJ/kmol (octane)} \cdot \text{K}$$

Alternative Solution:

As an alternative, the following *IT* code can be used to determine the entropy production per mole of fuel entering, where sigma denotes $\dot{\sigma}_{cv}/\dot{n}_F$, and sN2_R and sN2_P denote the entropy of N_2 in the reactants and products, respectively, and so on. In the **Units** menu, select temperature in K, pressure in bar, and amount of substance in moles.

```
TR = 25 + 273.15 // K
p = 1.01325 // bar
TP = 2394 // K (Value from the IT alternative solution of Example 13.8)

// Determine the partial pressures
pO2_R = 0.21 * p
pN2_R = 0.79 * p
pCO2_P = (8/64) * p
pH2O_P = (9/64) * p
pN2_P = (47/64) * p

// Evaluate the absolute entropies
sC8H18 = 360.79 // kJ/kmol K (from Table A-25)
sO2_R = s_TP("O2", TR, pO2_R)
sN2_R = s_TP("N2", TR, pN2_R)
❶ sCO2_P = s_TP("CO2", TP, pCO2_P)
sH2O_P = s_TP("H2O", TP, pH2O_P)
sN2_P = s_TP("N2", TP, pN2_P)

// Evaluate the reactant and product entropies, sR and sP, respectively
sR = sC8H18 + 12.5 * sO2_R + 47 * sN2_R
sR = 8 * sCO2_P + 9 * sH2O_P + 47 * sN2_P

// Entropy balance, Eq. (1)
sigma = sP – sR
```

Using the **Solve** button, the result is sigma = 5404 kJ/kmol (octane) · K, which agrees with the result obtained above.

(b) The complete combustion of liquid octane with 400% theoretical air is described by the following chemical equation:

$$C_8H_{18}(l) + 50O_2 + 188N_2 \rightarrow 8CO_2 + 9H_2O(g) + 37.5O_2 + 188N_2$$

The entropy rate balance on a per mole of fuel basis takes the form

$$\frac{\dot{\sigma}_{cv}}{\dot{n}_F} = (8\bar{s}_{CO_2} + 9\bar{s}_{H_2O(g)} + 37.5\bar{s}_{O_2} + 188\bar{s}_{N_2}) - \bar{s}_F - (50\bar{s}_{O_2} + 188\bar{s}_{N_2})$$

The specific entropies of the reactants have the same values as in part (a). The product gas exits as an ideal gas mixture at 1 atm, 962 K with the following composition: $y_{CO_2} = 8/242.5 = 0.033$, $y_{H_2O(g)} = 9/242.5 = 0.0371$, $y_{O_2} = 37.5/242.5 = 0.1546$, $y_{N_2} = 0.7753$. With the same approach as in part (a)

$$\bar{s}_{CO_2} = 267.12 - 8.314 \ln 0.033 = 295.481 \text{ kJ/kmol} \cdot \text{K}$$

$$\bar{s}_{H_2O} = 231.01 - 8.314 \ln 0.0371 = 258.397 \text{ kJ/kmol} \cdot \text{K}$$

$$\bar{s}_{O_2} = 242.12 - 8.314 \ln 0.1546 = 257.642 \text{ kJ/kmol} \cdot \text{K}$$

$$\bar{s}_{N_2} = 226.795 - 8.314 \ln 0.7753 = 228.911 \text{ kJ/kmol} \cdot \text{K}$$

Inserting values into the expression for the rate of entropy production

$$\frac{\dot{\sigma}_{cv}}{\dot{n}_F} = 8(295.481) + 9(258.397) + 37.5(257.642) + 188(228.911)$$

$$- 360.79 - 50(218.01) - 188(193.46)$$

$$= 9754 \text{ kJ/kmol (octane)} \cdot \text{K}$$

❷ The use of *IT* to solve part (b) is left as an exercise.

❶ For several gases modeled as ideal gases, *IT* directly returns the absolute entropies required by entropy balances for reacting systems. The entropy data obtained from *IT* agree with values calculated from Eq. 13.23 using table data.

❷ Although the rates of entropy production calculated in this example are positive, as required by the second law, this does not mean that the proposed reactions necessarily would occur, for the results are based on the assumption of *complete* combustion. The possibility of achieving complete combustion with specified reactants at a given temperature and pressure can be investigated with the methods of Chap. 14, dealing with chemical equilibrium. For further discussion, see Sec. 14.4.1.

Skills Developed: *Ability to...*

✓ apply the control volume entropy balance to a reacting system.

✓ evaluate entropy values appropriately based on absolute entropies.

Quick Quiz

How do combustion product temperature and rate of entropy production vary, respectively, as percent excess air increases? Assume complete combustion.

Ans. Decrease, increase.

CLOSED SYSTEMS. Next consider an entropy balance for a process of a closed system during which a chemical reaction occurs

$$S_P - S_R = \int \left(\frac{\delta Q}{T}\right)_b + \sigma \tag{13.25}$$

S_R and S_P denote, respectively, the entropy of the reactants and the entropy of the products. When the reactants and products form ideal gas mixtures, the entropy balance can be expressed on a *per mole of fuel* basis as

$$\sum_P n\bar{s} - \sum_R n\bar{s} = \frac{1}{n_F}\int \left(\frac{\delta Q}{T}\right)_b + \frac{\sigma}{n_F} \tag{13.26}$$

where the coefficients n on the left are the coefficients of the reaction equation giving the moles of each reactant or product *per mole of fuel*. The entropy terms would be evaluated from Eq. 13.23 using the temperature, pressure, and composition of the reactants or products, as appropriate. The fuel would be mixed with the oxidizer, so this must be taken into account when determining the partial pressures of the reactants. Example 13.10 provides an illustration of the evaluation of entropy change for combustion at constant volume.

Example 13.10 ENTROPY CHANGE FOR COMBUSTION AT CONSTANT VOLUME

Determine the change in entropy of the system of Example 13.6 in kJ/K.

Solution

Known: A mixture of gaseous methane and oxygen, initially at 25°C and 1 atm, burns completely within a closed rigid container. The products are cooled to 900 K, 3.02 atm.

Find: Determine the change in entropy for the process in kJ/K.

Schematic and Given Data: See Fig. E13.6.

Engineering Model:

1. The contents of the container are taken as the system.
2. The initial mixture can be modeled as an ideal gas mixture, as can the products of combustion.
3. Combustion is complete.

Analysis: The chemical equation for the complete combustion of methane with oxygen is

$$CH_4 + 2O_2 \rightarrow CO_2 + 2H_2O$$

The change in entropy for the process of the closed system is $\Delta S = S_P - S_R$, where S_R and S_P denote, respectively, the initial and final entropies of the system. Since the initial mixture forms an ideal gas mixture (assumption 2), the entropy of the reactants can be expressed as the sum of the contributions of the components, each evaluated at the mixture temperature and the partial pressure of the component. That is

$$S_R = 1\bar{s}_{CH_4}(T_1, y_{CH_4}p_1) + 2\bar{s}_{O_2}(T_1, y_{O_2}p_1)$$

where $y_{CH_4} = 1/3$ and $y_{O_2} = 2/3$ denote, respectively, the mole fractions of the methane and oxygen in the initial mixture. Similarly, since the products of combustion form an ideal gas mixture (assumption 2)

$$S_P = 1\bar{s}_{CO_2}(T_2, y_{CO_2}p_2) + 2\bar{s}_{H_2O}(T_2, y_{H_2O}p_2)$$

where $y_{CO_2} = 1/3$ and $y_{H_2O} = 2/3$ denote, respectively, the mole fractions of the carbon dioxide and water vapor in the products of combustion. In these equations, p_1 and p_2 denote the pressure at the initial and final states, respectively.

The specific entropies required to determine S_R can be calculated from Eq. 13.23. Since $T_1 = T_{ref}$ and $p_1 = p_{ref}$, absolute entropy data from Table A-25 can be used as follows

$$\bar{s}_{CH_4}(T_1, y_{CH_4}p_1) = \bar{s}^\circ_{CH_4}(T_{ref}) - \bar{R}\ln\frac{y_{CH_4}p_{ref}}{p_{ref}}$$

$$= 186.16 - 8.314\ln\frac{1}{3} = 195.294 \text{ kJ/kmol} \cdot \text{K}$$

Similarly

$$\bar{s}_{O_2}(T_1, y_{O_2}p_1) = \bar{s}^\circ_{O_2}(T_{ref}) - \bar{R}\ln\frac{y_{O_2}p_{ref}}{p_{ref}}$$

$$= 205.03 - 8.314\ln\frac{2}{3} = 208.401 \text{ kJ/kmol} \cdot \text{K}$$

At the final state, the products are at $T_2 = 900$ K and $p_2 = 3.02$ atm. With Eq. 13.23 and absolute entropy data from Tables A-23

$$\bar{s}_{CO_2}(T_2, y_{CO_2}p_2) = \bar{s}^\circ_{CO_2}(T_2) - \bar{R}\ln\frac{y_{CO_2}p_2}{p_{ref}}$$

$$= 263.559 - 8.314\ln\frac{(1/3)(3.02)}{1} = 263.504 \text{ kJ/kmol} \cdot \text{K}$$

$$\bar{s}_{H_2O}(T_2, y_{H_2O}p_2) = \bar{s}^\circ_{H_2O}(T_2) - \bar{R}\ln\frac{y_{H_2O}p_2}{p_{ref}}$$

$$= 228.321 - 8.314\ln\frac{(2/3)(3.02)}{1} = 222.503 \text{ kJ/kmol} \cdot \text{K}$$

Finally, the entropy change for the process is

$$\Delta S = S_P - S_R$$

$$= [263.504 + 2(222.503)] - [195.294 + 2(208.401)]$$

$$= 96.414 \text{ kJ/K}$$

Skills Developed: *Ability to...*

✓ apply the closed system entropy balance to a reacting system.

✓ evaluate entropy values appropriately based on absolute entropies.

Quick Quiz

Applying the entropy balance, Eq. 13.25, is σ greater than, less than, or equal to ΔS?

Ans. Greater than.

13.5.3 Evaluating Gibbs Function for Reacting Systems

The thermodynamic property known as the Gibbs function plays a role in the second part of this chapter dealing with exergy analysis. The *specific Gibbs function* $\bar{g}$, introduced in Sec. 11.3, is

$$\bar{g} = \bar{h} - T\bar{s} \qquad (13.27)$$

The procedure followed in setting a datum for the Gibbs function closely parallels that used in defining the enthalpy of formation: To each stable element at the standard state is assigned a zero value of the Gibbs function. The ***Gibbs function of formation*** of a compound equals the change in the Gibbs function for the reaction in which the compound is formed from its elements, the compound and the elements all being at T_{ref} and p_{ref}. Table A-25 gives the Gibbs function of formation, $\bar{g}^\circ_f$, at 25°C and 1 atm for selected substances.

Gibbs function of formation

The Gibbs function at a state other than the standard state is found by adding to the Gibbs function of formation the change in the specific Gibbs function $\Delta\bar{g}$ between the standard state and the state of interest

$$\bar{g}(T, p) = \bar{g}^\circ_f + [\bar{g}(T, p) - \bar{g}(T_{ref}, p_{ref})] = \bar{g}^\circ_f + \Delta\bar{g} \qquad (13.28a)$$

With Eq. 13.27, $\Delta\bar{g}$ can be written as

$$\Delta\bar{g} = [\bar{h}(T, p) - \bar{h}(T_{ref}, p_{ref})] - [T\bar{s}(T, p) - T_{ref}\bar{s}(T_{ref}, p_{ref})] \qquad (13.28b)$$

The Gibbs function of component i in an ideal gas mixture is evaluated at the *partial pressure* of component i and the mixture temperature.

The procedure for determining the Gibbs function of formation is illustrated in the next example.

Example 13.11 DETERMINING THE GIBBS FUNCTION OF FORMATION

Determine the Gibbs function of formation of methane at the standard state, 25°C and 1 atm, in kJ/kmol, and compare with the value given in Table A-25.

Solution

Known: The compound is methane.

Find: Determine the Gibbs function of formation at the standard state, in kJ/kmol, and compare with the Table A-25 value.

Assumptions: In the formation of methane from carbon and hydrogen (H_2), the carbon and hydrogen are each initially at 25°C and 1 atm. The methane formed is also at 25°C and 1 atm.

Analysis: Methane is formed from carbon and hydrogen according to $C + 2H_2 \rightarrow CH_4$. The change in the Gibbs function for this reaction is

$$\begin{aligned}\bar{g}_P - \bar{g}_R &= (\bar{h} - T\bar{s})_{CH_4} - (\bar{h} - T\bar{s})_C - 2(\bar{h} - T\bar{s})_{H_2} \\ &= (\bar{h}_{CH_4} - \bar{h}_C - 2\bar{h}_{H_2}) - T(\bar{s}_{CH_4} - \bar{s}_C - 2\bar{s}_{H_2})\end{aligned}$$

where $\bar{g}_P$ and $\bar{g}_R$ denote, respectively, the Gibbs functions of the reactants and products, each per kmol of methane.

In the present case, all substances are at the same temperature and pressure, 25°C and 1 atm, which correspond to the standard reference state values. At the standard reference state, the enthalpies and Gibbs functions for carbon and hydrogen are zero by definition. Thus, in the above equation $\bar{g}_R = \bar{h}_C = \bar{h}_{H_2} = 0$. Also, $\bar{g}_P = (\bar{g}_f^\circ)_{CH_4}$, giving

$$(\bar{g}_f^\circ)_{CH_4} = (\bar{h}_f^\circ)_{CH_4} - T_{ref}(\bar{s}^\circ_{CH_4} - \bar{s}^\circ_C - 2\bar{s}^\circ_{H_2})$$

where the superscript ° denotes properties at T_{ref}, p_{ref}. With enthalpy of formation and absolute entropy data from Table A-25

$$(\bar{g}_f^\circ)_{CH_4} = -74{,}850 - 298.15[186.16 - 5.74 - 2(130.57)] = -50{,}783 \text{ kJ/kmol}$$

The slight difference between the calculated value for the Gibbs function of formation of methane and the value from Table A-25 can be attributed to round-off.

Skills Developed: *Ability to...*
- ✓ apply the definition of Gibbs function of formation to calculate $\bar{g}_f^\circ$.

Quick Quiz

Using the method applied in the example, calculate $\bar{g}_f^\circ$ for monatomic oxygen at the standard state, in kJ/kmol. Begin by writing $\frac{1}{2}O_2 \rightarrow O$.

Ans. 231,750 kJ/kmol.

Chemical Exergy[1]

The objective of this part of the chapter is to extend the exergy concept introduced in Chap. 7 by considering the role of chemical composition. To distinguish the current considerations from those introduced previously, let the system be a specified amount of a hydrocarbon fuel C_aH_b at temperature T_0 and pressure p_0 and let the exergy reference environment consist of a gas phase at T_0, p_0 involving nitrogen, oxygen, water vapor, and carbon dioxide. Since the system is in thermal and mechanical equilibrium with the environment, the value of exergy would be zero, as discussed in Sec. 7.3.1. More precisely, we should say that the *thermomechanical* contribution to the

[1]Study of Chap. 7 is a prerequisite for this part of the chapter.

exergy magnitude has a value of zero, for a *chemical* contribution related to composition can be defined that has a nonzero value. This aspect of the exergy concept is the subject of the present section.

13.6 Introducing Chemical Exergy

Exergy is introduced in Chap. 7 through study of an overall system consisting of a system of interest and an exergy reference environment. The object of the development of Sec. 7.3 is an expression for the maximum theoretical work obtainable from the overall system as the system comes into thermal and mechanical equilibrium with the environment. The ***thermomechanical exergy*** is this value for work. We begin the present section by studying an overall system formed by an environment and an amount of a hydrocarbon fuel at T_0, p_0. The object is to evaluate the work obtainable by allowing the fuel to react with oxygen from the environment to produce the environmental components carbon dioxide and water, each at its respective state in the environment. The ***chemical exergy*** is, by definition, the maximum theoretical work that could be developed by the overall system. The sum of the thermomechanical and chemical exergies is the *total exergy* associated with a given system at a specified state, relative to a specified exergy reference environment.

thermomechanical exergy

chemical exergy

Chemical Exergy of a Hydrocarbon: C_aH_b

Consider an overall system formed by an environment and an amount of a hydrocarbon fuel, C_aH_b. To help us visualize how work might be obtained through the reaction of the fuel with environmental components, a fuel cell operating at steady state is included, as shown in Fig. 13.4. Referring to this figure, fuel enters the cell at temperature T_0 and pressure p_0. At another location, oxygen enters from the environment. Assuming the environment consists of an ideal gas mixture, the oxygen would enter at its condition within the environment: temperature T_0 and partial pressure $y^e_{O_2}p_0$, where $y^e_{O_2}$ is the mole fraction of the oxygen in the exergy reference environment. The fuel and oxygen react *completely* within the cell to produce carbon dioxide and *water vapor,* which exit in separate streams at their conditions within the environment: temperature T_0 and the partial pressures $y^e_{CO_2}p_0$ and $y^e_{H_2O}p_0$, respectively. The reaction is given by

$$C_aH_b + \left(a + \frac{b}{4}\right)O_2 \rightarrow aCO_2 + \frac{b}{2}H_2O \qquad (13.29)$$

For steady-state operation, the energy rate balance for a control volume enclosing the fuel cell reduces to give

$$\frac{\dot{W}_{cv}}{\dot{n}_F} = \frac{\dot{Q}_{cv}}{\dot{n}_F} + \bar{h}_F + \left(a + \frac{b}{4}\right)\bar{h}_{O_2} - a\bar{h}_{CO_2} - \frac{b}{2}\bar{h}_{H_2O} \qquad (13.30)$$

Kinetic and potential energy effects are regarded as negligible. Since the fuel cell is at steady state, its volume does not change with time, so no portion of $\dot{W}_{cv}$ is required to displace the environment. Thus, Eq. 13.30 gives the work developed by the overall system. Heat transfer is assumed to occur with the environment only at temperature T_0. An entropy rate balance for the control volume enclosing the fuel cell takes the form

$$0 = \frac{\dot{Q}_{cv}/\dot{n}_F}{T_0} + \bar{s}_F + \left(a + \frac{b}{4}\right)\bar{s}_{O_2} - a\bar{s}_{CO_2} - \frac{b}{2}\bar{s}_{H_2O} + \frac{\dot{\sigma}_{cv}}{\dot{n}_F} \qquad (13.31)$$

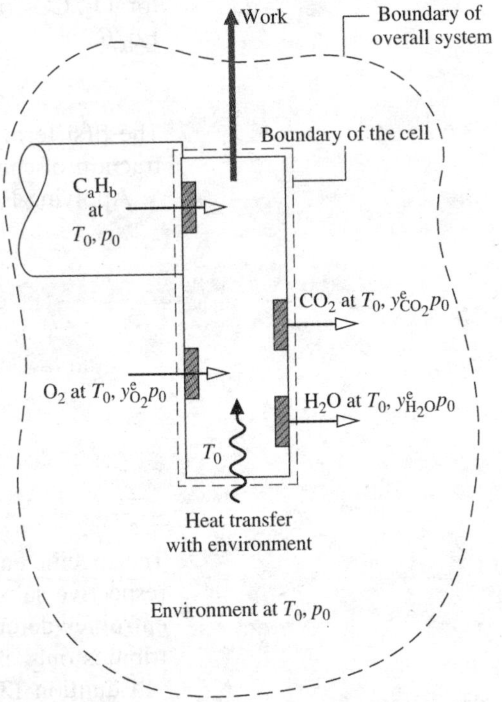

Fig. 13.4 Illustration used to introduce the fuel chemical exergy concept.

Eliminating the heat transfer rate between Eqs. 13.30 and 13.31 results in

$$\frac{\dot{W}_{cv}}{\dot{n}_F} = \underline{\left[\bar{h}_F + \left(a + \frac{b}{4}\right)\bar{h}_{O_2} - a\bar{h}_{CO_2} - \frac{b}{2}\bar{h}_{H_2O}\right]}$$
$$- T_0\underline{\left[\bar{s}_F + \left(a + \frac{b}{4}\right)\bar{s}_{O_2} - a\bar{s}_{CO_2} - \frac{b}{2}\bar{s}_{H_2O}\right]} - T_0\frac{\dot{\sigma}_{cv}}{\dot{n}_F} \tag{13.32}$$

In Eq. 13.32, the specific enthalpy and entropy of the fuel are evaluated at T_0 and p_0. The values of the specific enthalpies in the first underlined term can be determined knowing only the temperature T_0. The specific entropies in the second underlined term can be determined knowing the temperature, pressure, and composition of the environment. Accordingly, once the environment is specified, all enthalpy and entropy terms of Eq. 13.32 can be regarded as known and independent of the nature of the processes occurring within the control volume. The term $T_0\dot{\sigma}_{cv}$ depends explicity on the nature of such processes, however. In accordance with the second law, $T_0\dot{\sigma}_{cv}$ is positive whenever internal irreversibilities are present, vanishes in the limiting case of no irreversibilities, and is never negative. The *maximum theoretical value* for the work developed is obtained when no irreversibilities are present. Setting $T_0\dot{\sigma}_{cv}$ to zero in Eq. 13.32, the following expression for the *chemical exergy* results

$$\bar{e}^{ch} = \left[\bar{h}_F + \left(a + \frac{b}{4}\right)\bar{h}_{O_2} - a\bar{h}_{CO_2} - \frac{b}{2}\bar{h}_{H_2O}\right]$$
$$- T_0\left[\bar{s}_F + \left(a + \frac{b}{4}\right)\bar{s}_{O_2} - a\bar{s}_{CO_2} - \frac{b}{2}\bar{s}_{H_2O}\right] \tag{13.33}$$

The superscript ch is used to distinguish this contribution to the exergy magnitude from the thermomechanical exergy introduced in Chap. 7.

WORKING EQUATIONS. For computational convenience, the chemical exergy given by Eq. 13.33 can be written as Eq. 13.35 by replacing the specific entropy terms for O_2, CO_2, and H_2O using the following expression obtained by reduction of Eq. 6.20b

$$\bar{s}_i(T_0, y_i^e p_0) = \bar{s}_i(T_0, p_0) - \bar{R}\ln y_i^e \tag{13.34}$$

The first term on the right is the absolute entropy at T_0 and p_0, and y_i^e is the mole fraction of component i in the environment.

Applying Eq. 13.34, Eq. 13.33 becomes

$$\bar{e}^{ch} = \left[\bar{h}_F + \left(a + \frac{b}{4}\right)\bar{h}_{O_2} - a\bar{h}_{CO_2} - \frac{b}{2}\bar{h}_{H_2O(g)}\right](T_0, p_0)$$
$$- T_0\left[\bar{s}_F + \left(a + \frac{b}{4}\right)\bar{s}_{O_2} - a\bar{s}_{CO_2} - \frac{b}{2}\bar{s}_{H_2O(g)}\right](T_0, p_0) \tag{13.35}$$
$$+ \bar{R}T_0 \ln\left[\frac{(y^e_{O_2})^{a+b/4}}{(y^e_{CO_2})^a (y^e_{H_2O})^{b/2}}\right]$$

The specific enthalpy terms are determined using the enthalpies of formation for the respective substances. The specific entropies appearing in the equation are absolute entropies determined as described in Sec. 13.5. The logarithmic term normally contributes only a few percent to the chemical exergy magnitude.

Equation 13.35 can be expressed alternatively in terms of the Gibbs functions of the respective substances as

$$\bar{e}^{ch} = \underline{\left[\bar{g}_F + \left(a + \frac{b}{4}\right)\bar{g}_{O_2} - a\bar{g}_{CO_2} - \frac{b}{2}\bar{g}_{H_2O(g)}\right](T_0, p_0)} + \bar{R}T_0 \ln\left[\frac{(y^e_{O_2})^{a+b/4}}{(y^e_{CO_2})^a (y^e_{H_2O})^{b/2}}\right] \tag{13.36}$$

The specific Gibbs functions are evaluated at the temperature T_0 and pressure p_0 of the environment. These terms can be determined with Eq. 13.28 as

$$\bar{g}(T_0, p_0) = \bar{g}^\circ_f + [\bar{g}(T_0, p_0) - \bar{g}(T_{ref}, p_{ref})] \tag{13.37}$$

where $\bar{g}^\circ_f$ is the Gibbs function of formation. For the *special case* where T_0 and p_0 are the same as T_{ref} and p_{ref}, respectively, the second term on the right of Eq. 13.37 vanishes and the specific Gibbs function is just the Gibbs function of formation. Finally, note that the underlined term of Eq. 13.36 can be written more compactly as $-\Delta G$: the negative of the change in Gibbs function for the reaction, Eq. 13.29, regarding each substance as separate at temperature T_0 and pressure p_0.

Chemical Exergy of Other Substances

The method introduced above for evaluating the chemical exergy of pure hydrocarbons also can be used in principle for substances other than hydrocarbons: The chemical exergy is the maximum theoretical work that could be developed by a fuel cell into which a substance of interest enters at T_0, p_0 and reacts completely with environmental components to produce environmental components. All environmental components involved enter and exit the cell at their conditions within the environment. By describing the environment appropriately, this method can be applied to all substances of practical interest.[2] In the following discussion, we limit consideration to the compounds CO, H_2O, N_2, O_2, and CO_2, for these participate in the elementary combustion reactions serving as the focus of the present chapter. Of these five compounds, only carbon monoxide is not among the substances present in the environment we have been considering. Let us take up the compounds in the order listed.

▶ Paralleling the development of Eq. 13.36 for the case of pure carbon monoxide, at T_0, p_0 the reaction within the cell is $CO + \frac{1}{2}O_2 \rightarrow CO_2$, and the chemical exergy is given by

$$\bar{e}^{ch}_{CO} = [\bar{g}_{CO} + \tfrac{1}{2}\bar{g}_{O_2} - \bar{g}_{CO_2}](T_0, p_0) + \bar{R}T_0 \ln\left[\frac{(y^e_{O_2})^{1/2}}{y^e_{CO_2}}\right] \tag{13.38}$$

If carbon monoxide is not pure but a component of an ideal gas mixture at T_0, p_0, each component i of the mixture would enter the cell at temperature T_0 and the respective partial pressure $y_i p_0$. The contribution of carbon monoxide to the chemical exergy of the mixture, per mole of CO, is then given by Eq. 13.38 with the mole fraction of carbon monoxide in the mixture, y_{CO}, appearing in the numerator of the logarithmic term that then reads $\ln[y_{CO}(y^e_{O_2})^{1/2}/y^e_{CO_2}]$. This consideration is important when evaluating the exergy of combustion products involving carbon monoxide.

▶ Next, consider the case of pure water at T_0, p_0. Water is present as a vapor within the environment under present consideration but normally is a liquid

[2]For further discussion see M. J. Moran, *Availability Analysis: A Guide to Efficient Energy Use*, ASME Press, New York, 1989, pp. 169–170.

when at T_0, p_0. Thus, water would enter the cell as a liquid at T_0, p_0 and exit as a vapor at T_0, $y^e_{H_2O}p_0$, with *no cell reaction occurring*. The chemical exergy is

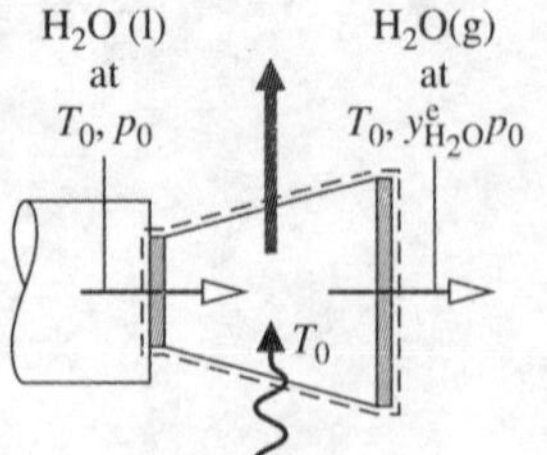

$$\bar{e}^{ch}_{H_2O} = [\bar{g}_{H_2O(l)} - \bar{g}_{H_2O(g)}](T_0, p_0) + \bar{R}T_0 \ln\left(\frac{1}{y^e_{H_2O}}\right) \tag{13.39}$$

► Next, consider N_2, O_2, CO_2, each pure at T_0, p_0. Nitrogen, oxygen, and carbon dioxide are present within the environment, and normally are gases when at T_0, p_0. For each case, the gas would enter the cell at T_0, p_0 and exit at $T_0, y^e p_0$, where y^e is the mole fraction of N_2, O_2, or CO_2 in the environment, as appropriate. *No cell reaction would occur.* The chemical exergy is given simply by a logarithmic term of the form

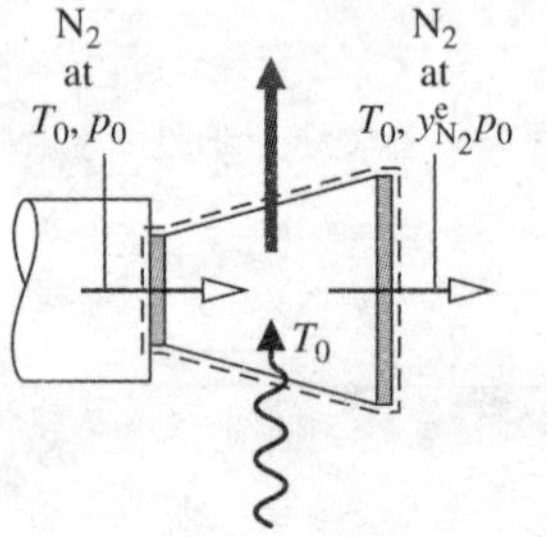

$$\bar{e}^{ch} = \bar{R}T_0 \ln\left(\frac{1}{y^e}\right) \tag{13.40}$$

► Finally, for an ideal gas mixture at T_0, p_0 consisting *only* of substances present as gases in the environment, the chemical exergy is obtained by summing the contributions of each of the components. The result, per mole of mixture, is

$$\bar{e}^{ch} = \bar{R}T_0 \sum_i y_i \ln\left(\frac{y_i}{y^e_i}\right) \tag{13.41a}$$

where y_i and y^e_i denote, respectively, the mole fraction of component i in the mixture at T_0, p_0 and in the environment. Expressing the logarithmic term as $(\ln(1/y^e_i) + \ln y_i)$ and introducing Eq. 13.40 for each gas, Eq. 13.41a can be written alternatively as

$$\bar{e}^{ch} = \sum_i y_i \bar{e}^{ch}_i + \bar{R}T_0 \sum_i y_i \ln y_i \tag{13.41b}$$

The development of Eqs. 13.38 through 13.41 is left as an exercise. Subsequent examples and end-of-chapter problems illustrate their use.

13.7 Standard Chemical Exergy

The exergy reference environments used thus far to calculate exergy values are adequate for a wide range of practical applications, including combustion. However, for many cases of interest the environment must be extended to include other substances. In applications involving coal, for example, sulfur dioxide or some other sulfur-bearing compound would generally appear among the environmental components. Furthermore, once the environment is determined, a series of calculations would be required to obtain exergy values for the substances of interest. These complexities can be sidestepped by using a table of *standard chemical exergies*.

standard chemical exergy

Standard chemical exergy values are based on a standard exergy reference environment exhibiting standard values of the environmental temperature T_0 and pressure p_0 such as 298.15 K and 1 atm, respectively. The exergy reference environment also consists of a set of reference substances with standard concentrations reflecting as closely as possible the chemical makeup of the natural environment. To exclude the possibility of developing work from interactions among parts of the environment, these reference substances must be in equilibrium mutually.

The reference substances generally fall into three groups: gaseous components of the atmosphere, solid substances from the Earth's crust, and ionic and nonionic substances from the oceans. A common feature of standard exergy reference environments is a gas phase, intended to represent air, that includes N_2, O_2, CO_2, $H_2O(g)$, and other gases. The ith gas present in this gas phase is assumed to be at temperature T_0 and the partial pressure $p_i^e = y_i^e p_0$.

Two alternative standard exergy reference environments are commonly used, called here *Model I* and *Model II*. For each of these models, Table A-26 gives values of the standard chemical exergy for several substances, in units of kJ/kmol, together with a brief description of the underlying rationale. The methods employed to determine the tabulated standard chemical exergy values are detailed in the references accompanying the tables. Only one of the two models should be used in a particular analysis.

The use of a table of standard chemical exergies often simplifies the application of exergy principles. However, the term "standard" is somewhat misleading, for there is no one specification of the environment that suffices for *all* applications. Still, chemical exergies calculated relative to alternative specifications of the environment are generally in good agreement. For a broad range of engineering applications, the convenience of using standard values generally outweighs the slight lack of accuracy that might result. In particular, the effect of slight variations in the values of T_0 and p_0 about their standard values can be neglected.

Standard Chemical Exergy of a Hydrocarbon: C_aH_b

In principle, the standard chemical exergy of a substance *not* present in the environment can be evaluated by considering an idealized reaction of the substance involving other substances for which the chemical exergies *are known*. To illustrate this for the case of a pure hydrocarbon fuel C_aH_b at T_0, p_0, refer to the control volume at steady state shown in Fig. 13.5 where the fuel reacts with oxygen to form carbon dioxide and *liquid water* according to Eq. 13.29. All substances are assumed to enter and exit at T_0, p_0 and heat transfer occurs only at temperature T_0.

Assuming no irreversibilities, an exergy rate balance reads

$$0 = \sum_j \left[1 - \frac{T_0}{T_j}\right]^{0}\left(\frac{\dot{Q}_j}{\dot{n}_F}\right) - \left(\frac{\dot{W}_{cv}}{\dot{n}_F}\right)_{\substack{int\\rev}} + \bar{e}_F^{ch} + \left(a + \frac{b}{4}\right)\bar{e}_{O_2}^{ch} - a\bar{e}_{CO_2}^{ch} - \left(\frac{b}{2}\right)\bar{e}_{H_2O(l)}^{ch} - \dot{E}_d^{\,0}$$

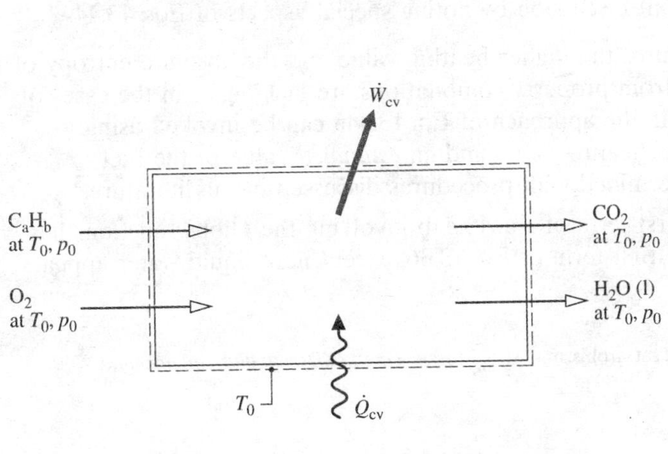

Fig. 13.5 Reactor used to introduce the standard chemical exergy of C_aH_b.

where, as before, the subscript F denotes the fuel. Solving for the fuel chemical exergy, we obtain

$$\bar{e}_F^{ch} = \left(\frac{\dot{W}_{cv}}{\dot{n}_F}\right)_{\substack{int\\rev}} + a\bar{e}_{CO_2}^{ch} + \left(\frac{b}{2}\right)\bar{e}_{H_2O(l)}^{ch} - \left(a + \frac{b}{4}\right)\bar{e}_{O_2}^{ch} \tag{13.42}$$

Paralleling the development of Eq. 13.32 for the case of Fig. 13.5, we also have

$$\left(\frac{\dot{W}_{cv}}{\dot{n}_F}\right)_{\substack{int\\rev}} = \underline{\left[\bar{h}_F + \left(a + \frac{b}{4}\right)\bar{h}_{O_2} - a\bar{h}_{CO_2} - \frac{b}{2}\bar{h}_{H_2O(l)}\right](T_0, p_0)}$$
$$- T_0\left[\bar{s}_F + \left(a + \frac{b}{4}\right)\bar{s}_{O_2} - a\bar{s}_{CO_2} - \frac{b}{2}\bar{s}_{H_2O(l)}\right](T_0, p_0) \tag{13.43}$$

The underlined term in Eq. 13.43 is recognized from Sec. 13.2 as the molar higher heating value $\overline{HHV}(T_0, p_0)$. Introducing Eq. 13.43 into Eq. 13.42, we have

$$\bar{e}_F^{ch} = \overline{HHV}(T_0, p_0) - T_0\left[\bar{s}_F + \left(a + \frac{b}{4}\right)\bar{s}_{O_2} - a\bar{s}_{CO_2} - \frac{b}{2}\bar{s}_{H_2O(l)}\right](T_0, p_0)$$
$$+ a\bar{e}_{CO_2}^{ch} + \left(\frac{b}{2}\right)\bar{e}_{H_2O(l)}^{ch} - \left(a + \frac{b}{4}\right)\bar{e}_{O_2}^{ch} \tag{13.44a}$$

Equations 13.42 and 13.43 can be expressed alternatively in terms of molar Gibbs functions as follows

$$\bar{e}_F^{ch} = \left[\bar{g}_F + \left(a + \frac{b}{4}\right)\bar{g}_{O_2} - a\bar{g}_{CO_2} - \frac{b}{2}\bar{g}_{H_2O(l)}\right](T_0, p_0)$$
$$+ a\bar{e}_{CO_2}^{ch} + \left(\frac{b}{2}\right)\bar{e}_{H_2O(l)}^{ch} - \left(a + \frac{b}{4}\right)\bar{e}_{O_2}^{ch} \tag{13.44b}$$

With Eqs. 13.44, the standard chemical exergy of the hydrocarbon C_aH_b can be calculated using the standard chemical exergies of O_2, CO_2, and $H_2O(l)$, together with selected property data: the higher heating value and absolute entropies, or Gibbs functions. ➡ **FOR EXAMPLE...** consider the case of methane, CH_4, and $T_0 = 298.15$ K (25°C), $p_0 = 1$ atm. For this application we can use Gibbs function data directly from Table A-25, and standard chemical exergies for CO_2, $H_2O(l)$, and O_2 from Table A-26 (Model II), since each source corresponds to 298 K, 1 atm. With a = 1, b = 4, Eq. 13.44b gives 831,680 kJ/kmol. This agrees with the value listed for methane in Table A-26 for Model II. ⬅

We conclude the present discussion by noting special aspects of Eqs. 13.44:

- First, Eq. 13.44a requires the higher heating value and the absolute entropy of the fuel. When data from property compilations are lacking, as in the cases of coal, char, and fuel oil, the approach of Eq. 13.44a can be invoked using a *measured* or *estimated* heating value and an *estimated* value of the fuel absolute entropy determined with procedures discussed in the literature.[3]
- Next, note that the first term of Eq. 13.44b involving the Gibbs functions has the same form as the first term of Eq. 13.36, except here liquid water appears.

[3]See, for example, A. Bejan, G. Tsatsaronis, and M. J. Moran, *Thermal Design and Optimization*, Wiley, New York, 1996, Sec. 3.4.3.

Also, the first term of Eq. 13.44b can be written more compactly as $-\Delta G$: the negative of the change in Gibbs function for the reaction.

▶ Finally, note that only the underlined terms of Eqs. 13.44 require chemical exergy data relative to the model selected for the exergy reference environment.

In Example 13.12 we compare the use of Eq. 13.36 and Eq. 13.44b for evaluating the chemical exergy of a pure hydrocarbon fuel.

Example 13.12 EVALUATING THE CHEMICAL EXERGY OF OCTANE

Determine the chemical exergy of liquid octane at 25°C, 1 atm, in kJ/kg. (a) Using Eq. 13.36, evaluate the chemical exergy for an environment consisting of a gas phase at 25°C, 1 atm obeying the ideal gas model with the following composition on a molar basis: N_2, 75.67%; O_2, 20.35%; H_2O, 3.12%; CO_2, 0.03%; other, 0.83%. (b) Evaluate the chemical exergy using Eq. 13.44b and standard chemical exergies from Table A-26 (Model II).

Solution

Known: The fuel is liquid octane.

Find: Determine the chemical exergy (a) using Eq. 13.36 relative to an environment consisting of a gas phase at 25°C, 1 atm with a specified composition, (b) using Eq. 13.44b and standard chemical exergies.

Schematic and Given Data:

❶

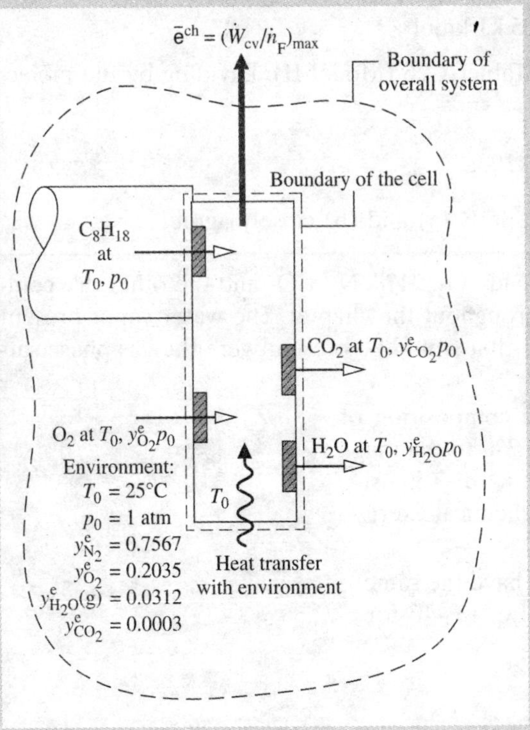

Fig. E13.12

Engineering Model: As shown in Fig. E13.12, the environment for part (a) consists of an ideal gas mixture with the molar analysis: N_2, 75.67%; O_2, 20.35%; H_2O, 3.12%; CO_2, 0.03%; other, 0.83%. For part (b), Model II of Table A-26 applies.

Analysis: Complete combustion of liquid octane with O_2 is described by

$$C_8H_{18}(l) + 12.5O_2 \rightarrow 8CO_2 + 9H_2O$$

(a) For this reaction equation, Eq. 13.36 takes the form

$$\bar{e}^{ch} = [\bar{g}_{C_8H_{18}(l)} + 12.5\bar{g}_{O_2} - 8\bar{g}_{CO_2} - 9\bar{g}_{H_2O(g)}](T_0, p_0) + \bar{R}T_0 \ln\left[\frac{(y^e_{O_2})^{12.5}}{(y^e_{CO_2})^8 (y^e_{H_2O(g)})^9}\right]$$

Since $T_0 = T_{ref}$ and $p_0 = p_{ref}$, the required specific Gibbs functions are just the Gibbs functions of formation from Table A-25. With the given composition of the environment and data from Table A-25, the above equation gives

❷
$$\bar{e}^{ch} = [6610 + 12.5(0) - 8(-394{,}380) - 9(-228{,}590)] + 8.314(298.15)\ln\left[\frac{(0.2035)^{12.5}}{(0.0003)^8(0.0312)^9}\right] = 5{,}218{,}960 + 188{,}883 = 5{,}407{,}843 \text{ kJ/kmol}$$

Dividing by the molecular weight, the chemical exergy is obtained on a unit mass basis

$$e^{ch} = \frac{5{,}407{,}843}{114.22} = 47{,}346 \text{ kJ/kg}$$

(b) Using coefficients from the reaction equation above, Eq. 13.44b reads

❸
$$\bar{e}^{ch} = [\bar{g}_{C_8H_{18}(l)} + 12.5\bar{g}_{O_2} - 8\bar{g}_{CO_2} - 9\bar{g}_{H_2O(l)}](T_0, p_0) + 8\bar{e}^{ch}_{CO_2} + 9\bar{e}^{ch}_{H_2O(l)} - 12.5\bar{e}^{ch}_{O_2}$$

With data from Table A-25 and Model II of Table A-26, the above equation gives

$$\bar{e}^{ch} = [6610 + 12.5(0) - 8(-394{,}380) - 9(-237{,}180)] + 8(19{,}870) + 9(900) - 12.5(3970) = 5{,}296{,}270 + 117{,}435 = 5{,}413{,}705 \text{ kJ/kmol}$$

As expected, this agrees closely with the value listed for octane in Table A-26 (Model II). Dividing by the molecular weight, the chemical exergy is obtained on a unit mass basis

$$e^{ch} = \frac{5{,}413{,}705}{114.22} = 47{,}397 \text{ kJ/kg}$$

The chemical exergies determined with the two approaches used in parts (a) and (b) closely agree.

❶ A molar analysis of the environment of part (a) on a *dry* basis reads, O_2: 21%, N_2, CO_2 and the other dry components: 79%. This is consistent with the dry air analysis used throughout the chapter. The water vapor present in the assumed environment corresponds to the amount of vapor that would be present were the gas phase saturated with water at the specified temperature and pressure.

❷ The value of the logarithmic term of Eq. 13.36 depends on the composition of the environment. In the present case, this term contributes about 3% to the magnitude of the chemical exergy. The contribution of the logarithmic term is usually small. In such instances, a satisfactory approximation to the chemical exergy can be obtained by omitting the term.

❸ The terms of Eqs. 13.36 and 13.44b involving the Gibbs functions have the same form, except $\bar{g}_{H_2O}$ appears in Eq. 13.36 for water vapor and in Eq. 13.44b for liquid water.

Skills Developed: *Ability to...*
- ✓ calculate the chemical exergy of a hydrocarbon fuel relative to a specified reference environment.
- ✓ calculate the chemical exergy of a hydrocarbon fuel based on standard chemical exergies.

Quick Quiz

Would the higher heating value (HHV) provide a plausible estimate of the chemical exergy in this case?
Ans. Yes, it is approximately 1% greater than the value calculated in part (a).

Standard Chemical Exergy of Other Substances

By paralleling the development given above for hydrocarbon fuels leading to Eq. 13.44b, we can in principle determine the standard chemical exergy of any substance not present in the environment. With such a substance playing the role of the fuel in the previous development, we consider a reaction of the substance involving other substances for which the standard exergies *are known,* and write

$$\bar{e}^{ch} = -\Delta G + \underline{\sum_P n\bar{e}^{ch} - \sum_R n\bar{e}^{ch}} \qquad (13.45)$$

where ΔG is the change in Gibbs function for the reaction, regarding each substance as separate at temperature T_0 and pressure p_0. The underlined term corresponds to the underlined term of Eq. 13.44b and is evaluated using the *known* standard chemical exergies, together with the n's giving the moles of these reactants and products per mole of the substance whose chemical exergy is being evaluated.

FOR EXAMPLE... consider the case of ammonia, NH_3, and $T_0 = 298.15$ K (25°C), $p_0 = 1$ atm. Letting NH_3 play the role of the hydrocarbon in the development leading to Eq. 13.44b, we can consider any reaction of NH_3 involving other substances for which the standard chemical exergies are known. For the reaction

$$NH_3 + \tfrac{3}{4}O_2 \rightarrow \tfrac{1}{2}N_2 + \tfrac{3}{2}H_2O$$

Eq. 13.45 takes the form

$$\begin{aligned}\bar{e}^{ch}_{NH_3} &= [\bar{g}_{NH_3} + \tfrac{3}{4}\bar{g}_{O_2} - \tfrac{1}{2}\bar{g}_{N_2} - \tfrac{3}{2}\bar{g}_{H_2O(l)}](T_0, p_0) \\ &\quad + \tfrac{1}{2}\bar{e}^{ch}_{N_2} + \tfrac{3}{2}\bar{e}^{ch}_{H_2O(l)} - \tfrac{3}{4}\bar{e}^{ch}_{O_2}\end{aligned}$$

Using Gibbs function data from Table A-25, and standard chemical exergies for O_2, N_2, and $H_2O(l)$ from Table A-26 (Model II), $\bar{e}^{ch}_{NH_3} = 337{,}910$ kJ/kmol. This agrees closely with the value for ammonia listed in Table A-26 for Model II.

Finally, note that Eq. 13.41b is valid for mixtures containing gases other than those present in the reference environment, for example, gaseous fuels. This equation also can be applied to mixtures (and solutions) that do not adhere to the ideal gas model. In all such applications, the terms $\bar{e}^{ch}_i$ are selected from a table of standard chemical exergies.

13.8 Exergy Summary

The exergy associated with a specified state of a system is the sum of two contributions: the thermomechanical contribution introduced in Chap. 7 and the chemical contribution introduced in this chapter. On a unit mass basis, the ***total exergy*** is

total exergy

$$e = \underline{(u - u_0) + p_0(v - v_0) - T_0(s - s_0) + \frac{V^2}{2} + gz} + e^{ch} \qquad (13.46)$$

where the underlined term is the thermomechanical contribution (Eq. 7.2) and e^{ch} is the chemical contribution evaluated as in Sec. 13.6 or 13.7. Similarly, the specific flow exergy associated with a specified state is the sum

$$e_f = \underline{h - h_0 - T_0(s - s_0) + \frac{V^2}{2} + gz} + e^{ch} \qquad (13.47)$$

where the underlined term is the thermomechanical contribution (Eq. 7.14) and e^{ch} is the chemical contribution.

When evaluating the thermomechanical contributions, we can think of bringing the system without change in composition from the specified state to T_0, p_0, the condition where the system is in thermal and mechanical equilibrium with the environment. Depending on the nature of the system, this may be a *hypothetical* condition.

When a difference in exergy or flow exergy between states of the same composition is evaluated, the chemical contribution cancels, leaving just the difference in the thermomechanical contributions. For such a calculation, it is unnecessary to evaluate the chemical exergy explicitly. However, for many evaluations it is necessary to account explicitly for the chemical exergy contribution. This is brought out in subsequent examples.

Illustrations

The following two examples illustrate the exergy principles considered above. In Example 13.13, Eq. 13.47 is applied to evaluate the specific flow exergy of a steam leak.

Example 13.13 EVALUATING THE FLOW EXERGY OF A STEAM LEAK

Steam at 5 bar, 240°C leaks from a line in a vapor power plant. Evaluate the flow exergy of the steam, in kJ/kg, relative to an environment at 25°C, 1 atm in which the mole fraction of water vapor is $y^e_{H_2O} = 0.0303$.

Solution

Known: Water vapor at a known state is specified. The environment is also described.

Find: Determine the flow exergy of the water vapor, in kJ/kg.

Engineering Model:

1. The environment consists of a gas phase that obeys the ideal gas model. The mole fraction of water vapor in the environment is 0.0303.
2. Neglect the effects of motion and gravity.

Analysis: With assumption 2, the specific flow exergy is given by Eq. 13.47 as

$$e_f = \underline{(h - h_0) - T_0(s - s_0)} + e^{ch}$$

The underlined term is the thermomechanical contribution to the flow exergy, evaluated as in Chap. 7. With data from the steam tables, and noting that water is a liquid at T_0, p_0

$$\begin{aligned} h - h_0 - T_0(s - s_0) &= (2939.9 - 104.9) - 298(7.2307 - 0.3674) \\ &= 789.7 \text{ kJ/kg} \end{aligned}$$

where h_0 and s_0 are approximated as the saturated liquid values at T_0.

The chemical exergy contribution to the flow exergy relative to the specified environment is evaluated using Eq. 13.39. With data from Table A-25

$$\begin{aligned} e^{ch} &= \frac{1}{M}\left\{[\bar{g}_{H_2O(l)} - \bar{g}_{H_2O(g)}](T_0, p_0) + \bar{R}T_0 \ln\left(\frac{1}{y^e_{H_2O}}\right)\right\} \\ &= \frac{1}{18}\left\{[-237{,}180 - (-228{,}590)] + (8.314)(298)\ln\left(\frac{1}{0.0303}\right)\right\} \\ &= \frac{73.1 \text{ kJ/kmol}}{18 \text{ kg/kmol}} = 4.1 \text{ kJ/kg} \end{aligned}$$

Adding the thermomechanical and chemical contributions, the flow exergy of steam at the specified state is

$$e_f = 789.7 + 4.1 = 793.8 \text{ kJ/kg}$$

In this case, the chemical exergy contributes less than 1% to the total flow exergy magnitude.

Skills Developed: *Ability to...*
✓ determine the flow exergy, including the chemical exergy contribution, of pure substances present in the reference environment.

Quick Quiz

If the mass flow rate of the steam leak is 0.06 kg/s, and the flow exergy is valued at \$0.08/kW · h, what is the value of one day's steam loss?
Ans. \$91/day.

In Example 13.14, we evaluate the flow exergy of combustion products.

Example 13.14 EVALUATING THE FLOW EXERGY OF COMBUSTION PRODUCTS

Methane gas enters a reactor and burns completely with 140% theoretical air. Combustion products exit as a mixture at temperature T and a pressure of 1 atm. For $T = 480$ and 1560 K, evaluate the flow exergy of the combustion products, in kJ per kmol of fuel. Perform calculations relative to an environment consisting of an ideal gas mixture at 25°C, 1 atm with the molar analysis, $y^e_{N_2} = 0.7567$, $y^e_{O_2} = 0.2035$, $y^e_{H_2O} = 0.0303$, $y^e_{CO_2} = 0.0003$.

Solution

Known: Methane gas reacts completely with 140% of the theoretical amount of air. Combustion products exit the reactor at 1 atm and a specified temperature. The environment is also specified.

Find: Determine the flow exergy of the combustion products, in kJ per kmol of fuel, for each of two given temperatures.

Engineering Model:

1. The combustion products are modeled as an ideal gas mixture at all states considered.
2. Neglect the effects of motion and gravity.

Analysis: For 140% theoretical air, the reaction equation for complete combustion of methane is

$$CH_4 + 2.8(O_2 + 3.76N_2) \rightarrow CO_2 + 2H_2O + 10.53N_2 + 0.8O_2$$

The flow exergy is given by Eq. 13.47, which involves chemical and thermomechanical contributions. Since the ❶ combustion products form an ideal gas mixture when at T_0, p_0 (assumption 1) and each component is present within the environment, the chemical exergy contribution, per mole of fuel, is obtained from the following expression patterned after Eq. 13.41a

$$\bar{e}^{ch} = \bar{R}T_0\left[1\ln\left(\frac{y_{CO_2}}{y^e_{CO_2}}\right) + 2\ln\left(\frac{y_{H_2O}}{y^e_{H_2O}}\right) + 10.53\ln\left(\frac{y_{N_2}}{y^e_{N_2}}\right) + 0.8\ln\left(\frac{y_{O_2}}{y^e_{O_2}}\right)\right]$$

From the reaction equation, the mole fractions of the components of the products are $y_{CO_2} = 0.0698$, $y_{H_2O} = 0.1396$, $y_{N_2} = 0.7348$, $y_{O_2} = 0.0558$. Substituting these values together with the respective environmental mole fractions, we obtain $\bar{e}^{ch} = 17{,}750$ kJ per kmol of fuel.

Applying ideal gas mixture principles, the thermomechanical contribution to the flow exergy, per mole of fuel, is

$$\begin{aligned}\bar{h} - \bar{h}_0 - T_0(\bar{s} - \bar{s}_0) = {} & [\bar{h}(T) - \bar{h}(T_0) - T_0(\bar{s}^\circ(T) - \bar{s}^\circ(T_0) - \bar{R}\ln(y_{CO_2}p/y_{CO_2}p_0))]_{CO_2} \\ & + 2[\bar{h}(T) - \bar{h}(T_0) - T_0(\bar{s}^\circ(T) - \bar{s}^\circ(T_0) - \bar{R}\ln(y_{H_2O}p/y_{H_2O}p_0))]_{H_2O} \\ & + 10.53[\bar{h}(T) - \bar{h}(T_0) - T_0(\bar{s}^\circ(T) - \bar{s}^\circ(T_0) - \bar{R}\ln(y_{N_2}p/y_{N_2}p_0))]_{N_2} \\ & + 0.8[\bar{h}(T) - \bar{h}(T_0) - T_0(\bar{s}^\circ(T) - \bar{s}^\circ(T_0) - \bar{R}\ln(y_{O_2}p/y_{O_2}p_0))]_{O_2}\end{aligned}$$

Since $p = p_0$, each of the logarithm terms drop out, and with $\bar{h}$ and $\bar{s}^\circ$ data at T_0 from Table A-23, the thermomechanical contribution reads

$$\begin{aligned}\bar{h} - \bar{h}_0 - T_0(\bar{s} - \bar{s}_0) = {} & [\bar{h}(T) - 9364 - 298(\bar{s}^\circ(T) - 213.685)]_{CO_2} \\ & + 2[\bar{h}(T) - 9904 - 298(\bar{s}^\circ(T) - 188.72)]_{H_2O} \\ & + 10.53[\bar{h}(T) - 8669 - 298(\bar{s}^\circ(T) - 191.502)]_{N_2} \\ & + 0.8[\bar{h}(T) - 8682 - 298(\bar{s}^\circ(T) - 205.033)]_{O_2}\end{aligned}$$

Then, with $\bar{h}$ and $\bar{s}^\circ$ from Table A-23 at $T = 480$ and 1560 K, respectively, the following results are obtained

$$T = 480\text{ K}: \bar{h} - \bar{h}_0 - T_0(\bar{s} - \bar{s}_0) = 17{,}712\text{ kJ per kmol of fuel}$$
$$T = 1560\text{ K}: \bar{h} - \bar{h}_0 - T_0(\bar{s} - \bar{s}_0) = 390{,}853\text{ kJ per kmol of fuel}$$

Adding the two contributions, the flow exergy of the combustion products at each of the specified states is

❷
$$T = 480\text{ K}: \bar{e}_f = 35{,}462\text{ kJ per mol of fuel}$$
$$T = 1560\text{ K}: \bar{e}_f = 408{,}603\text{ kJ per mol of fuel}$$

---

❶ This is a *hypothetical* state for the combustion products because condensation of some of the water vapor present would occur were the products brought to T_0, p_0. An exergy evaluation explicitly taking such condensation into account is considered in Bejan, Tsatsaronis, and Moran, *Thermal Design and Optimization,* p. 129, p. 138.

❷ The chemical contribution to the flow exergy is relatively unimportant in the higher-temperature case, amounting only to about 4% of the flow exergy. Chemical exergy accounts for about half of the exergy in the lower-temperature case, however.

Skills Developed: *Ability to...*
✓ determine the flow exergy, including the chemical exergy contribution, of gaseous products of combustion.

Quick Quiz

If fuel enters with a mass flow rate of 3.5×10^{-3} kg/s, determine the flow exergy rate of the exiting combustion gases for part (b), in kJ/s.

Ans. 90.8 kJ/s.

13.9 Exergetic (Second Law) Efficiencies of Reacting Systems

Devices designed to do work by utilization of a combustion process, such as vapor and gas power plants and internal combustion engines, invariably have irreversibilities and losses associated with their operation. Accordingly, actual devices produce work equal to only a fraction of the maximum theoretical value that might be obtained in idealized circumstances. The vapor power plant exergy analysis of Sec. 8.6 and the combined cycle exergy analysis of Example 9.13 provide illustrations.

The performance of devices intended to do work can be evaluated as the ratio of the actual work developed to the maximum theoretical work. This ratio is a type of *exergetic (second law) efficiency*. The relatively low exergetic efficiency exhibited by many common power-producing devices suggests that thermodynamically more thrifty ways of utilizing the fuel to develop power might be possible. However, efforts in this direction must be tempered by the economic imperatives that govern the practical application of all devices. The trade-off between fuel savings and the additional costs required to achieve those savings must be carefully weighed.

The fuel cell provides an illustration of a relatively fuel-efficient device. We noted previously (Sec. 13.4) that the chemical reactions in fuel cells are more controlled than the rapidly occurring, highly irreversible combustion reactions taking place in conventional power-producing devices. In principle, fuel cells can achieve greater exergetic efficiencies than many such devices.

The example to follow illustrates the evaluation of an exergetic efficiency for an internal combustion engine.

Example 13.15 EXERGETIC EFFICIENCY OF AN INTERNAL COMBUSTION ENGINE

Devise and evaluate an exergetic efficiency for the internal combustion engine of Example 13.4. For the fuel, use the chemical exergy value determined in Example 13.12(a).

Solution

Known: Liquid octane and the theoretical amount of air enter an internal combustion engine operating at steady state in separate streams at 25°C, 1 atm, and burn completely. The combustion products exit at 890°C. The power developed by the engine is 37 kW, and the fuel mass flow rate is 1.8×10^{-3} kg/s.

Find: Devise and evaluate an exergetic efficiency for the engine using the fuel chemical exergy value determined in Example 13.12(a).

Schematic and Given Data: See Fig. E13.4.

Engineering Model:

1. See the assumptions listed in the solution to Example 13.4.
2. The environment is the same as used in Example 13.12(a).
3. The combustion air enters at the condition of the environment.
4. Kinetic and potential energy effects are negligible.

Analysis: An exergy balance can be used in formulating an exergetic efficiency for the engine: At steady state, the rate at which exergy enters the engine equals the rate at which exergy exits plus the rate at which exergy is destroyed within the engine. As the combustion air enters at the condition of the environment, and thus with zero exergy, exergy enters the engine only with the fuel. Exergy exits the engine accompanying heat and work, and with the products of combustion.

If the power developed is taken to be the *product* of the engine, and the heat transfer and exiting product gas are regarded as *losses*, an exergetic efficiency expression that gauges the extent to which the exergy entering the engine with the fuel is converted to the product is

❶
$$\varepsilon = \frac{\dot{W}_{cv}}{\dot{E}_F}$$

where $\dot{E}_F$ denotes the rate at which exergy enters with the fuel.

Since the fuel enters the engine at 25°C and 1 atm, which correspond to the values of T_0 and p_0 of the environment, and kinetic and potential energy effects are negligible, the exergy of the fuel is just the chemical exergy evaluated in Example 13.12(a). There is no thermomechanical contribution. Thus

$$\dot{E}_F = \dot{m}_F e^{ch} = \left(1.8 \times 10^{-3} \frac{kg}{s}\right)\left(47{,}346 \frac{kJ}{kg}\right)\left|\frac{1\ kW}{1\ kJ/s}\right| = 85.2\ kW$$

The exergetic efficiency is then

$$\varepsilon = \frac{37\ kW}{85.2\ kW} = 0.434\ (43.4\%)$$

❶ The "waste heat" from large engines may be utilizable by some other device—for example, an absorption heat pump. In such cases, some of the exergy accompanying heat transfer and the exiting product gas might be included in the numerator of the exergetic efficiency expression. Since a greater portion of the entering fuel exergy would be utilized in such arrangements, the value of ε would be greater than that evaluated in the solution.

Skills Developed: *Ability to...*
✓ devise and evaluate an exergetic efficiency for an internal combustion engine.

Quick Quiz

For the internal combustion engine, list sources of exergy loss.

Ans. Flow exergy of the exiting combustion products, exergy loss associated with heat transfer to the surroundings.

In the next example, we evaluate an exergetic efficiency for a reactor.

Example 13.16 EXERGETIC EFFICIENCY OF A REACTOR

For the reactor of Example 13.9, determine the exergy destruction, in kJ per kmol of fuel, and devise and evaluate an exergetic efficiency. Consider **(a)** complete combustion with the theoretical amount of air **(b)** complete combustion with 400% theoretical air. For the fuel, use the chemical exergy value determined in Example 13.12(a).

Solution

Known: Liquid octane and air, each at 25°C and 1 atm, burn completely in a well-insulated reactor operating at steady state. The products of combustion exit at 1 atm pressure.

Find: Determine the exergy destruction, in kJ per kmol of fuel, and evaluate an exergetic efficiency for complete combustion with (a) the theoretical amount of air, (b) 400% theoretical air.

Schematic and Given Data: See Fig. E13.9.

Engineering Model:

1. See assumptions listed in Example 13.9.
2. The environment is the same as that in Example 13.12(a).
3. The combustion air enters at the condition of the environment.
4. Kinetic and potential energy effects are negligible.

Analysis: An exergy balance can be used in formulating an exergetic efficiency: At steady state, the rate at which exergy enters the reactor equals the rate at which exergy exits plus the rate at which exergy is destroyed within the reactor. Since the combustion air enters at the condition of the environment, and thus with zero exergy, exergy enters the reactor only with the fuel. The reactor is well insulated, so there is no exergy transfer accompanying heat transfer. There is also no work $\dot{W}_{cv}$. Accordingly, exergy exits only with the combustion products. An exergetic efficiency can be written as

$$\varepsilon = \frac{\dot{E}_{products}}{\dot{E}_F}$$

where $\dot{E}_F$ is the rate at which exergy enters with the fuel and $\dot{E}_{products}$ is the rate at which exergy exits with the combustion products. Using the exergy balance for the reactor, given above in words, the exergetic efficiency expression can be written alternatively as

$$\varepsilon = \frac{\dot{E}_F - \dot{E}_d}{\dot{E}_F} = 1 - \frac{\dot{E}_d}{\dot{E}_F}$$

The exergy destruction term appearing in the last expression can be found from the relation

$$\frac{\dot{E}_d}{\dot{n}_F} = T_0 \frac{\dot{\sigma}_{cv}}{\dot{n}_F}$$

where T_0 is the temperature of the environment and $\dot{\sigma}_{cv}$ is the rate of entropy production. The rate of entropy production is evaluated in the solution to Example 13.9 for each of the two cases. For the case of complete combustion with the theoretical amount of air

$$\frac{\dot{E}_d}{\dot{n}_F} = (298\ \text{K})\left(5404 \frac{\text{kJ}}{\text{kmol} \cdot \text{K}}\right) = 1{,}610{,}392 \frac{\text{kJ}}{\text{kmol}}$$

Similarly, for the case of complete combustion with 400% of the theoretical amount of air

$$\frac{\dot{E}_d}{\dot{n}_F} = (298)(9754) = 2{,}906{,}692 \frac{\text{kJ}}{\text{kmol}}$$

Since the fuel enters the reactor at 25°C, 1 atm, which correspond to the values of T_0 and p_0 of the environment, and kinetic and potential effects are negligible, the exergy of the fuel is just the chemical exergy evaluated in Example 13.12(a). There is no thermomechanical contribution. Thus, for the case of complete combustion with the theoretical amount of air

$$\varepsilon = 1 - \frac{1{,}610{,}392}{5{,}407{,}843} = 0.702\ (70.2\%)$$

❶ Similarly, for the case of complete combustion with 400% of the theoretical amount of air

$$\varepsilon = 1 - \frac{2{,}906{,}692}{5{,}407{,}843} = 0.463\ (46.3\%)$$

---

❶ The calculated efficiency values show that a substantial portion of the fuel exergy is destroyed in the combustion process. In the case of combustion with the theoretical amount of air, about 30% of the fuel exergy is destroyed. In the excess air case, over 50% of the fuel exergy is destroyed. Further exergy destructions would take place as the hot gases are utilized. It might be evident, therefore, that the overall conversion from fuel input to end use would have a relatively low exergetic efficiency. The vapor power plant exergy analysis of Sec. 8.6 illustrates this point.

Skills Developed: *Ability to...*
- ✓ determine exergy destruction for a chemical reactor.
- ✓ devise and evaluate an appropriate exergetic efficiency.

Quick Quiz

List factors that contribute to exergy destruction within the control volume.
Ans. Spontaneous mixing of fuel and air, spontaneous heat transfer within the control volume, spontaneous chemical reaction.

✓ Chapter Summary and Study Guide

In this chapter we have applied the principles of thermodynamics to systems involving chemical reactions, with emphasis on systems involving the combustion of hydrocarbon fuels. We also have extended the notion of exergy to include chemical exergy.

The first part of the chapter begins with a discussion of concepts and terminology related to fuels, combustion air, and products of combustion. The application of energy balances to reacting systems is then considered, including control volumes at steady state and closed systems. To evaluate the specific enthalpies required in such applications, the enthalpy of formation concept is introduced and illustrated. The determination of the adiabatic flame temperature is considered as an application.

The use of the second law of thermodynamics is also discussed. The absolute entropy concept is developed to provide the specific entropies required by entropy balances for systems involving chemical reactions. The related Gibbs function of formation concept is introduced. The first part of the chapter also includes a discussion of fuel cells.

In the second part of the chapter, we extend the exergy concept of Chap. 7 by introducing chemical exergy. The *standard* chemical exergy concept is also discussed. Means are developed and illustrated for evaluating the chemical exergies of hydrocarbon fuels and other substances. The presentation concludes with a discussion of exergetic efficiencies of reacting systems.

The following list provides a study guide for this chapter. When your study of the text and end-of-chapter exercises has been completed, you should be able to

- ✓ write out the meaning of the terms listed in the margin throughout the chapter and understand each of the related concepts. The subset of key concepts listed below is particularly important.
- ✓ determine balanced reaction equations for the combustion of hydrocarbon fuels, including complete and incomplete combustion with various percentages of theoretical air.
- ✓ apply energy balances to systems involving chemical reactions, including the evaluation of enthalpy using Eq. 13.9 and the evaluation of the adiabatic flame temperature.
- ✓ apply entropy balances to systems involving chemical reactions, including the evaluation of the entropy produced.
- ✓ evaluate the chemical exergy of hydrocarbon fuels and other substances using Eqs. 13.35 and 13.36, as well as the standard chemical exergy using Eqs. 13.44 and 13.45.
- ✓ apply exergy analysis, including chemical exergy and the evaluation of exergetic efficiencies.

Key Engineering Concepts

complete combustion p. 711
air–fuel ratio p. 713
theoretical air p. 713
percent of theoretical air p. 714
dry product analysis p. 716
enthalpy of formation p. 720
heating values p. 729
adiabatic flame temperature p. 732
fuel cell p. 736
absolute entropy p. 739
chemical exergy p. 747
standard chemical exergy p. 750

Key Equations

$AF = \overline{AF}\left(\dfrac{M_{air}}{M_{fuel}}\right)$	(13.2)	Relation between air–fuel ratios on mass and molar bases
$\bar{h}(T,p) = \bar{h}_f^\circ + [\bar{h}(T,p) - \bar{h}(T_{ref}, p_{ref})] = \bar{h}_f^\circ + \Delta\bar{h}$	(13.9)	Evaluating enthalpy at T, p in terms of enthalpy of formation
$\dfrac{\dot{Q}_{cv}}{\dot{n}_F} - \dfrac{\dot{W}_{cv}}{\dot{n}_F} = \sum_P n_e(\bar{h}_f^\circ + \Delta\bar{h})_e - \sum_R n_i(\bar{h}_f^\circ + \Delta\bar{h})_i$	(13.15b)	Energy rate balance for a control volume at steady state per mole of fuel entering
$Q - W = \sum_P n(\bar{h}_f^\circ + \Delta\bar{h}) - \sum_R n(\bar{h}_f^\circ + \Delta\bar{h}) - \bar{R}T_P\sum_P n + \bar{R}T_R\sum_R n$	(13.17b)	Closed system energy balance
$\bar{s}(T,p) = \bar{s}^\circ(T) - \bar{R}\ln\dfrac{p}{p_{ref}}$	(13.22)	Absolute entropy of an ideal gas (molar basis) at T, p
$\bar{s}_i(T,p_i) = \bar{s}_i^\circ(T) - \bar{R}\ln\dfrac{y_i p}{p_{ref}}$	(13.23)	Absolute entropy for component i of an ideal gas mixture (molar basis) at T, p

$\bar{g}(T, p) = \bar{g}_f^\circ + [\bar{g}(T, p) - \bar{g}(T_{ref}, p_{ref})] = \bar{g}_f^\circ + \Delta\bar{g}$	(13.28a)	Evaluating Gibbs function at T, p in terms of Gibbs function of formation
where $\Delta\bar{g} = [\bar{h}(T, p) - \bar{h}(T_{ref}, p_{ref})] - [T\bar{s}(T, p) - T_{ref}\bar{s}(T_{ref}, p_{ref})]$	(13.28b)	(see Tables A-25 for $\bar{g}_f^\circ$ values)
$\text{e}_f = h - h_0 - T_0(s - s_0) + \frac{V^2}{2} + gz + \text{e}^{ch}$	(13.47)	Specific flow exergy including thermomechanical and chemical contributions (see Secs. 13.6 and 13.7 for chemical exergy expressions)

Exercises: Things Engineers Think About

1. If 1 kmol of carbon (C) burns completely with 1 kmol of oxygen (O_2), how many kmol of products are formed? Is mass conserved?

2. If a hydrocarbon is burned with the theoretical amount of air, can the combustion be complete? If it is burned with less than theoretical air, can the combustion be complete?

3. If an engine burns *rich,* is the percent of theoretical air greater than 100% or less than 100%?

4. When I burn wood in my fireplace, do I contribute to *global warming*? Explain.

5. If the enthalpy of formation of carbon dioxide (CO_2) were assigned a value of zero, would the enthalpy of formation of pure carbon be positive or negative in value?

6. If I knew the higher heating value of a hydrocarbon, C_aH_b, how would I determine its lower heating value?

7. Why are some *high-efficiency,* residential natural gas furnaces equipped with drain tubes?

8. How might I define an adiabatic flame temperature for constant volume combustion?

9. What irreversibilities are present in the internal combustion engine of Example 13.4? The combustion chamber of Example 13.5?

10. Why can't I assign *absolute* enthalpy values that are independent of the choice of reference state like I can for entropy?

11. What is an advantage of using standard chemical exergies? A disadvantage?

12. How might I define an exergetic efficiency for the hydrogen–oxygen fuel cell of Fig. 13.3*b*?

13. How might I define an exergetic efficiency for the reactor of Example 13.5?

Problems: Developing Engineering Skills

Working with Reaction Equations

13.1 A vessel contains a mixture of 60% O_2 and 40% CO on a mass basis. Determine the percent excess or percent deficiency of oxygen, as appropriate.

13.2 Ten grams of propane (C_3H_8) burns with just enough oxygen (O_2) for complete combustion. Determine the amount of oxygen required and the amount of combustion products formed, each in grams.

13.3 Ethane (C_2H_6) burns completely with the theoretical amount of air. Determine the air–fuel ratio on a (a) molar basis, (b) mass basis.

13.4 A gas turbine burns octane (C_8H_{18}) completely with 400% of theoretical air. Determine the amount of N_2 in the products, in kmol per kmol of fuel.

13.5 One hundred kmol of butane (C_4H_{10}) together with 4000 kmol of air enter a furnace per unit of time. Carbon dioxide, carbon monoxide, and unburned fuel appear in the products of combustion exiting the furnace. Determine the percent excess or percent deficiency of air, whichever is appropriate.

13.6 Propane (C_3H_8) is burned with air. For each case, obtain the balanced reaction equation for complete combustion

(a) with the theoretical amount of air.
(b) with 20% excess air.
(c) with 20% excess air, but only 90% of the propane being consumed in the reaction.

13.7 Butane (C_4H_{10}) burns completely with air. The equivalence ratio is 0.9. Determine

(a) the balanced reaction equation.
(b) the percent excess air.

13.8 Methane (CH_4) burns completely with the *stoichiometric* amount of hydrogen peroxide (H_2O_2). Determine the balanced reaction equation.

13.9 A fuel mixture with the molar analysis 40% CH_3OH, 50% C_2H_5OH, and 10% N_2 burns completely with 33% excess air. Determine

(a) the balanced reaction equation.
(b) the air–fuel ratio, both on a molar and mass basis.

13.10 A gas mixture with the molar analysis 25% H_2, 25% CO, 50% O_2 reacts to form products consisting of CO_2, H_2O, and O_2 only. Determine the amount of each product, in kg per kg of mixture.

13.11 A natural gas with the molar analysis 78% CH_4, 13% C_2H_6, 6% C_3H_8, 1.7% C_4H_{10}, 1.3% N_2 burns completely with 40% excess air in a reactor operating at steady state. If the molar flow rate of the fuel is 0.5 kmol/h, determine the molar flow rate of the air, in kmol/h.

13.12 A fuel mixture with the molar analysis of 20% CH_4, 40% H_2, 40% NH_3 burns completely with 150% of theoretical oxygen. Determine the mole fraction of each product.

13.13 Coal with the mass analysis 77.54% C, 4.28% H, 1.46% S, 7.72% O, 1.34% N, 7.66% noncombustible ash burns completely with 120% of theoretical air. Determine

(a) the balanced reaction equation.
(b) the amount of SO_2 produced, in kg per kg of coal.

13.14 A coal sample has a mass analysis of 80.4% carbon, 3.9% hydrogen (H_2), 5.0% oxygen (O_2), 1.1% nitrogen (N_2), 1.1% sulfur, and the rest is noncombustible ash. For complete combustion with 120% of the theoretical amount of air, determine the air–fuel ratio on a mass basis.

13.15 A sample of dried feedlot manure is being tested for use as a fuel. The mass analysis of the sample is 42.7% carbon, 5.5% hydrogen (H), 31.3% oxygen (O), 2.4% nitrogen (N), 0.3% sulfur, and 17.8% noncombustible ash. The sample is burned completely with 120% of theoretical air. Determine

(a) the balanced reaction equation.
(b) the air–fuel ratio on a mass basis.

13.16 A sample of dried Appanoose County coal has a mass analysis of 71.1% carbon, 5.1% hydrogen (H), 9.0% oxygen (O), 1.4% nitrogen (N_2), 5.8% sulfur, and the rest noncombustible ash. For complete combustion with the theoretical amount of air, determine

(a) the amount of SO_2 produced, in kg per kg of coal.
(b) the air–fuel ratio on a mass basis.

13.17 Octane (C_8H_{18}) burns completely with 120% of theoretical air. Determine

(a) the air–fuel ratio on a molar and mass basis.
(b) the dew point temperature of the combustion products, in °C, when cooled at 1 atm.

13.18 Butane (C_4H_{10}) burns completely with 150% of theoretical air. If the combustion products are cooled at 1 atm to temperature T, plot the amount of water vapor condensed, in kmol per kmol of fuel, versus T ranging from 20 to 60°C.

13.19 Ethylene (C_2H_4) burns completely with air and the combustion products are cooled to temperature T at 1 atm. The air–fuel ratio on a mass basis is AF.

(a) Determine for $AF = 15$ and $T = 21$°C, the percent excess air and the amount of water vapor condensed, in kg per kmol of fuel.
(b) Plot the amount of water vapor condensed, in kg per kmol of fuel, versus T ranging from 21 to 38°C, for $AF = 15, 20, 25, 30$.

13.20 A gaseous fuel mixture with a molar analysis of 72% CH_4, 9% H_2, 14% N_2, 2% O_2, and 3% CO_2 burns completely with moist air to form *gaseous* products at 1 atm consisting of CO_2, H_2O, and N_2 only. If the dew point temperature of the products is 60°C, determine the amount of water vapor present in the combustion air, in kmol per kmol of fuel mixture.

13.21 The gas driven off when low-grade coal is burned with insufficient air for complete combustion is known as *producer gas*. A particular producer gas has the following volumetric analysis: 3.8% CH_4, 0.1% C_2H_6, 4.8% CO_2, 11.7% H_2, 0.6% O_2, 23.2% CO, and the remainder N_2. Determine, for complete combustion with the theoretical amount of air

(a) the molar analysis of the dry products of combustion.
(b) the amount of water vapor condensed, in $\frac{\text{kmol}}{\text{kmol}}$ of producer gas, if the products are cooled to 21°C at a constant pressure of 1 atm.

13.22 Acetylene (C_2H_2) enters a torch and burns completely with 110% of theoretical air entering at 25°C, 1 atm, 50% relative humidity. Obtain the balanced reaction equation, and determine the dew point temperature of the products, in °C, at 1 atm.

13.23 Butane (C_4H_{10}) burns completely with 160% of theoretical air at 20°C, 1 atm, and 90% relative humidity. Determine

(a) the balanced reaction equation.
(b) the dew point temperature, in °C, of the products, when cooled at 1 atm.

13.24 Ethane (C_2H_6) enters a furnace and burns completely with 130% of theoretical air entering at 25°C, 85 kPa, 50% relative humidity. Determine

(a) the balanced reaction equation.
(b) the dew point temperature of the combustion products, in °C, at 85 kPa.

13.25 Pentane (C_5H_{12}) burns completely with the theoretical amount of air at 25°C, 1 atm, 75% relative humidity. Determine

(a) the balanced reaction equation.
(b) the dew point temperature of the combustion products at 1 atm.
(c) the amount of water condensed, in kmol/kmol of fuel, if the combustion products are cooled to 25°C, at 1 atm.

13.26 A liquid fuel mixture that is 40% octane (C_8H_{18}) and 60% decane ($C_{10}H_{22}$) by mass is burned completely with 10% excess air at 25°C, 1 atm, 80% relative humidity.

(a) Determine the equivalent hydrocarbon composition, C_aH_b, of a fuel that would have the same carbon–hydrogen ratio on a mass basis as the fuel mixture.
(b) If the combustion products are cooled to 25°C at a pressure of 1 atm, determine the amount of water vapor that condenses, in kg per kg of fuel mixture.

13.27 Hydrogen (H_2) enters a combustion chamber with a mass flow rate of 6×10^{-4} kg/s and burns with air entering at 30°C, 1 atm with a volumetric flow rate of 3.5×10^{-2} m^3/s. Determine the percent of theoretical air used.

13.28 Carbon burns with 80% theoretical air yielding CO_2, CO, and N_2 only. Determine

(a) the balanced reaction equation.
(b) the air–fuel ratio on a mass basis.
(c) the analysis of the products on a molar basis.

13.29 Propane (C_3H_8) reacts with 80% of theoretical air to form products including CO_2, CO, H_2O, and N_2 only. Determine

(a) the balanced reaction equation.
(b) the air–fuel ratio on a mass basis.
(c) the analysis of the products on a dry molar basis.

13.30 Hexane (C_6H_{14}) burns with dry air to give products with the dry molar analysis 8.5% CO_2, 5.2% CO, 3% O_2, 83.3% N_2. Determine

(a) the balanced reaction equation.
(b) the percent of theoretical air.
(c) the dew point temperature, in °C, of the products at 1 atm.

13.31 The components of the exhaust gas of a spark-ignition engine using a fuel mixture represented as C_8H_{17} have a dry molar analysis of 8.7% CO_2, 8.9% CO, 0.3% O_2, 3.7% H_2, 0.3% CH_4, and 78.1% N_2. Determine the equivalence ratio.

13.32 The combustion of a hydrocarbon fuel, represented as C_aH_b, results in products with the dry molar analysis 11% CO_2, 0.5% CO, 2% CH_4, 1.5% H_2, 6% O_2, and 79% N_2. Determine the air–fuel ratio on (a) a molar basis, (b) a mass basis.

13.33 Decane ($C_{10}H_{22}$) burns with 95% of theoretical air, producing a gaseous mixture of CO_2, CO, H_2O, and N_2. Determine

(a) the air–fuel ratio on a molar basis.
(b) the analysis of the products on a dry molar basis.

13.34 Butane (C_4H_{10}) burns with air, giving products having the dry molar analysis 11.0% CO_2, 1.0% CO, 3.5% O_2, 84.5% N_2. Determine

(a) the percent theoretical air.
(b) the dew point temperature of the combustion products, in °C, at 1 bar.

13.35 A natural gas with the volumetric analysis 97.3% CH_4, 2.3% CO_2, 0.4% N_2 is burned with air in a furnace to give products having a dry molar analysis of 9.20% CO_2, 3.84% O_2, 0.64% CO, and the remainder N_2. Determine

(a) the percent theoretical air.
(b) the dew point temperature, in °C, of the combustion products at 1 atm.

13.36 A fuel oil having an analysis on a mass basis of 85.7% C, 14.2% H, 0.1% inert matter burns with air to give products with a dry molar analysis of 12.29% CO_2; 3.76% O_2; 83.95% N_2. Determine the air–fuel ratio on a mass basis.

13.37 Ethyl alcohol (C_2H_5OH) burns with air. The product gas is analyzed and the laboratory report gives only the following percentages on a dry molar basis: 6.9% CO_2, 1.4% CO, 0.5% C_2H_5OH. Assuming the remaining components consist of O_2 and N_2, determine

(a) the percentages of O_2 and N_2 in the dry molar analysis.
(b) the percent excess air.

13.38 A fuel oil with the mass analysis 87% C, 11% H, 1.4% S, 0.6% inert matter burns with 120% of theoretical air. The hydrogen and sulfur are completely oxidized, but 95% of the carbon is oxidized to CO_2 and the remainder to CO.

(a) Determine the balanced reaction equation.
(b) For the CO and SO_2, determine the amount, in kmol per 10^6 kmol of combustion products (that is, the amount in *parts per million*).

13.39 Pentane (C_5H_{12}) burns with air so that a fraction x of the carbon is converted to CO_2. The remaining carbon appears as CO. There is no free O_2 in the products. Develop plots of the air–fuel ratio and the percent of theoretical air versus x, for x ranging from zero to unity.

13.40 For each of the following mixtures, determine the equivalence ratio and indicate if the mixture is lean or rich:

(a) 1 kmol of butane (C_4H_{10}) and 32 kmol of air.
(b) 1 kg of propane (C_3H_8) and 6.5 kg of air.

13.41 Octane (C_8H_{18}) enters an engine and burns with air to give products with the dry molar analysis of CO_2, 10.5%; CO, 5.8%; CH_4, 0.9%; H_2, 2.6%; O_2, 0.3%; N_2, 79.9%. Determine the equivalence ratio.

13.42 Methane (CH_4) burns with air to form products consisting of CO_2, CO, H_2O, and N_2 only. If the equivalence ratio is 1.25, determine the balanced reaction equation.

Applying the First Law to Reacting Systems

13.43 Butane (C_4H_{10}) at 25°C, 1 atm enters a combustion chamber operating at steady state and burns completely with the theoretical amount of air entering at the same conditions. If the products exit at 93°C, 1 atm, determine the rate of heat transfer from the combustion chamber, in kJ per kmol of fuel. Kinetic and potential energy effects are negligible.

13.44 Propane (C_3H_8) at 298 K, 1 atm, enters a combustion chamber operating at steady state with a molar flow rate of 0.7 kmol/s and burns completely with 200% of theoretical air entering at 298 K, 1 atm. Kinetic and potential energy effects are negligible. If the combustion products exit at 560 K, 1 atm, determine the rate of heat transfer for the combustion chamber, in kW. Repeat for an exit temperature of 298 K.

13.45 Methane (CH_4) at 25°C, 1 atm enters a furnace operating at steady state and burns completely with 140% of theoretical air entering at 400 K, 1 atm. The products of combustion exit at 700 K, 1 atm. Kinetic and potential energy effects are negligible. If the rate of heat transfer *from* the furnace to the surroundings is 400 kW, determine the mass flow rate of methane, in kg/s.

13.46 Methane gas (CH_4) at 25°C, 1 atm enters a steam generator operating at steady state. The methane burns completely with 140% of theoretical air entering at 127°C, 1 atm. Products of combustion exit at 427°C, 1 atm. In a separate stream, saturated liquid water enters at 8 MPa and exits as superheated vapor at 480°C with a negligible pressure drop. If the vapor mass flow rate is 3.7×10^5 kg/h, determine the volumetric flow rate of the methane, in m^3/h.

13.47 Liquid ethanol (C_2H_5OH) at 25°C, 1 atm enters a combustion chamber operating at steady state and burns with air entering at 227°C, 1 atm. The fuel flow rate is 25 kg/s and the equivalence ratio is 1.2. Heat transfer from the combustion chamber to the surroundings is at a rate of 3.75×10^5 kJ/s.

Products of combustion, consisting of CO_2, CO, $H_2O(g)$, and N_2, exit. Ignoring kinetic and potential energy effects, determine

(a) the exit temperature, in K.
(b) the air–fuel ratio on a mass basis.

13.48 Benzene gas (C_6H_6) at 25°C, 1 atm enters a combustion chamber operating at steady state and burns with 95% theoretical air entering at 25°C, 1 atm. The combustion products exit at 1000 K and include only CO_2, CO, H_2O, and N_2. Determine the mass flow rate of the fuel, in kg/s, to provide heat transfer at a rate of 1000 kW.

13.49 Liquid propane (C_3H_8) at 25°C, 1 atm, enters a well-insulated reactor operating at steady state. Air enters at the same temperature and pressure. For liquid propane, $\bar{h}_f^\circ = -118{,}900$ kJ/kmol. Determine the temperature of the combustion products, in K, for complete combustion with

(a) the theoretical amount of air.
(b) 300% of theoretical air.

13.50 The energy required to vaporize the working fluid passing through the boiler of a simple vapor power plant is provided by the complete combustion of methane with 110% of theoretical air. The fuel and air enter in separate streams at 25°C, 1 atm. Products of combustion exit the stack at 150°C, 1 atm. Plot the mass flow rate of fuel required, in kg/h per MW of power developed by the plant versus the plant thermal efficiency, η. Consider η in the range 30–40%. Kinetic and potential energy effects are negligible.

13.51 Methane (CH_4) at 25°C, enters the combustor of a simple open gas turbine power plant and burns completely with 400% of theoretical air entering the compressor at 25°C, 1 atm. Products of combustion exit the turbine at 577°C, 1 atm. The rate of heat transfer from the gas turbine is estimated as 10% of the net power developed. Determine the net power output, in MW, if the fuel mass flow rate is 1200 kg/h. Kinetic and potential energy effects are negligible.

13.52 Octane gas C_8H_{18} at 25°C enters a jet engine and burns completely with 300% of theoretical air entering at 25°C, 1 atm with a volumetric flow rate of 42 m^3/s. Products of combustion exit at 990 K, 1 atm. If the fuel and air enter with negligible velocities, determine the *thrust* produced by the engine in kN.

13.53 Figure P13.53 provides data for a boiler and air preheater operating at steady state. Methane (CH_4) entering the boiler at 25°C, 1 atm is burned completely with 170% of theoretical air. Ignoring stray heat transfer and kinetic and potential energy effects, determine the temperature, in °C, of the combustion air entering the boiler from the preheater.

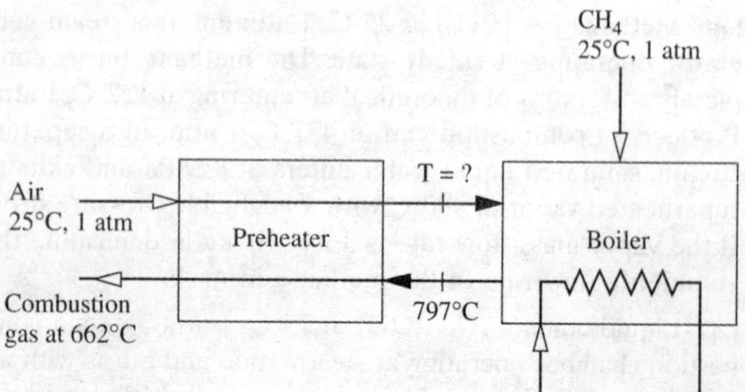

Fig. P13.53

13.54 A rigid tank initially contains 7.27 kg of CH_4 and 44 kg of O_2 at 25°C, 1 atm. After complete combustion, the pressure in the tank is 3.352 atm. Determine the heat transfer, in kJ.

13.55 A closed rigid vessel initially contains a gaseous mixture at 25°C, 1 atm with the molar analysis of 25% ethylene (C_2H_4), 75% oxygen (O_2). The mixture burns completely and the products are cooled to 500 K. Determine the heat transfer between the vessel and its surroundings, in kJ per kmol of fuel present initially, and the final pressure, in atm.

13.56 A closed, rigid vessel initially contains a gaseous mixture of 1 kmol of pentane (C_5H_{12}) and 150% of theoretical air at 25°C, 1 atm. If the mixture burns completely, determine the heat transfer from the vessel, in kJ, and the final pressure, in atm, for a final temperature of 800 K.

13.57 Determine the enthalpy of combustion for gaseous butane (C_4H_{10}), in kJ per kmol of fuel and kJ per kg of fuel, at 25°C, 1 atm, assuming

(a) water vapor in the products.
(b) liquid water in the products.

13.58 Plot the enthalpy of combustion for gaseous propane (C_3H_8), in kJ per kmol of fuel, at 1 atm versus temperature in the interval 25 to 260°C. Assume water vapor in the products. For propane, let $c_p = 1.72$ kJ/kg · K.

13.59 Plot the enthalpy of combustion for gaseous methane (CH_4), in kJ per kmol of fuel, at 1 atm versus temperature in the interval from 298 to 1000 K. Assume water vapor in the products. For methane, let $\bar{c}_p = 19.07 + 31.1(T/1000)$ kJ/kmol · K, where T is in K.

13.60 For the *producer gas* of Prob. 13.21, determine the enthalpy of combustion, in kJ per kmol of mixture, at 25°C, 1 atm, assuming water vapor in the products.

13.61 Determine the higher heating value, in kJ per kmol of fuel and in kJ per kg of fuel, at 25°C, 1 atm for

(a) liquid octane (C_8H_{18}).
(b) gaseous hydrogen (H_2)
(c) liquid methanol (CH_3OH).
(d) gaseous butane (C_4H_{10}).

Compare with the values listed in Table A-25.

13.62 For a natural gas with a molar analysis of 86.5% CH_4, 8% C_2H_6, 2% C_3H_8, 3.5% N_2, determine the lower heating value, in kJ per kmol of fuel and in kJ per kg of fuel, at 25°C, 1 atm.

13.63 Liquid octane (C_8H_{18}) at 25°C, 1 atm enters an insulated reactor operating at steady state and burns with 90% of theoretical air at 25°C, 1 atm to form products consisting of CO_2, CO, H_2O, and N_2 only. Determine the temperature of the exiting products, in K. Compare with the results of Example 13.8 and comment.

13.64 For each of the following fuels, plot the adiabatic flame temperature, in K, versus percent excess air for complete combustion in a combustor operating at steady state. The reactants enter at 25°C, 1 atm.

(a) carbon.
(b) hydrogen (H_2).
(c) liquid octane (C_8H_{18}).

13.65 Propane gas (C_3H_8) at 25°C, 1 atm enters an insulated reactor operating at steady state and burns completely with air entering at 25°C, 1 atm. Plot the adiabatic flame temperature versus percent of theoretical air ranging from 100 to 400%. Why does the adiabatic flame temperature vary with increasing combustion air?

13.66 Hydrogen (H_2) at 25°C, 1 atm enters an insulated reactor operating at steady state and burns completely with x% of theoretical air entering at 25°C 1 atm. Plot the adiabatic flame temperature for x ranging from 100 to 400%.

13.67 Methane gas (CH_4) at 25°C, 1 atm enters an insulated reactor operating at steady state and burns completely with x% of theoretical air entering at 25°C, 1 atm. Plot the adiabatic flame temperature for x ranging from 100 to 400%.

13.68 Methane (CH_4) at 25°C, 1 atm enters an insulated reactor operating at steady state and burns with the theoretical amount of air entering at 25°C, 1 atm. The products contain CO_2, CO, H_2O, O_2, and N_2, and exit at 2260 K. Determine the fractions of the entering carbon in the fuel that burn to CO_2 and CO, respectively.

13.69 Propane gas (C_3H_8) at 25°C, 1 atm enters an insulated reactor operating at steady state and burns completely with air entering at 25°C, 1 atm. Determine the percent of theoretical air if the combustion products exit at

(a) 616°C.
(b) 1227°C.

Neglect kinetic and potential energy effects.

13.70 Liquid methanol (CH_3OH) at 25°C, 1 atm enters an insulated reactor operating at steady state and burns completely with air entering at 100°C, 1 atm. If the combustion products exit at 1256°C, determine the percent excess air used. Neglect kinetic and potential energy effects.

13.71 Methane (CH_4) at 25°C enters the combustor of a gas turbine power plant operating at steady state and burns completely with air entering at 204°C. The temperature of the products of combustion flowing from the combustor to the turbine depends on the percent excess air for combustion. Plot the percent excess air versus combustion product temperatures ranging from 760°C to 980°C. There is no significant heat transfer between the combustor and its surroundings, and kinetic and potential energy effects can be ignored.

13.72 Air enters the compressor of a simple gas turbine power plant at 21°C, 1 atm, is compressed adiabatically to 275 kPa, and then enters the combustion chamber where it burns completely with propane gas (C_3H_8) entering at 25°C, 275 kPa and a molar flow rate of 0.77 kmol/h. The combustion products at 727°C, 275 kPa enter the turbine and expand adiabatically to a pressure of 1 atm. The isentropic compressor efficiency is 83.3% and the isentropic turbine efficiency is 90%. Determine at steady state

(a) the percent of theoretical air required.
(b) the net power developed, in kW.

13.73 A mixture of gaseous octane (C_8H_{18}) and 200% of theoretical air, initially at 25°C, 1 atm, reacts completely in a rigid vessel.

(a) If the vessel were well-insulated, determine the temperature, in °C, and the pressure, in atm, of the combustion products.
(b) If the combustion products were cooled at constant volume to 25°C, determine the final pressure, in atm, and the heat transfer, in kJ per kmol of fuel.

13.74 A mixture of methane (CH_4) and 200% of theoretical air, initially at 25°C, 1 atm, reacts completely in an insulated vessel. Determine the temperature, in °C, of the combustion products if the reaction occurs

(a) at constant volume.
(b) at constant pressure in a piston–cylinder assembly.

13.75 A 5×10^{-3} kg sample of liquid benzene (C_6H_6) together with 20% excess air, initially at 25°C and 1 atm, reacts completely in a rigid, insulated vessel. Determine the temperature, in °C, and the pressure, in atm, of the combustion products.

Applying the Second Law to Reacting Systems

13.76 Carbon monoxide (CO) at 25°C, 1 atm enters an insulated reactor operating at steady state and reacts completely with the theoretical amount of air entering in a separate stream at 25°C, 1 atm. The products of combustion exit as a mixture at 1 atm. For the reactor, determine the rate of entropy production, in kJ/K per kmol of CO entering. Neglect kinetic and potential energy effects.

13.77 Methane (CH_4) at 25°C, 1 atm enters an insulated reactor operating at steady state and burns completely with air entering in a separate stream at 25°C, 1 atm. The products of combustion exit as a mixture at 1 atm. For the reactor, determine the rate of entropy production, in kJ/K per kmol of methane entering, for combustion with

(a) the theoretical amount of air.
(b) 200% of theoretical air.

Neglect kinetic and potential energy effects.

13.78 Carbon monoxide (CO) reacts with water vapor in an insulated reactor operating at steady state to form hydrogen (H_2) and carbon dioxide (CO_2). The products exit as a mixture at 1 atm. For the reactor, determine the rate of entropy production, in kJ/K per kmol of carbon monoxide entering. Neglect kinetic and potential energy effects. Consider two cases:

(a) the carbon monoxide and water vapor enter the reactor in separate streams, each at 400 K, 1 atm.
(b) the carbon monoxide and water vapor enter the reactor as a mixture at 400 K, 1 atm.

Explain why the answers in parts (a) and (b) differ.

13.79 A gaseous mixture of butane (C_4H_{10}) and 80% excess air at 25°C, 3 atm enters a reactor. Complete combustion occurs, and the products exit as a mixture at 1200 K, 3 atm. Coolant enters an outer jacket as a saturated liquid and saturated vapor exits at essentially the same pressure. No significant heat transfer occurs from the outer surface of the jacket, and kinetic and potential energy effects are negligible. Determine for the jacketed reactor

(a) the mass flow rate of the coolant, in kg per kmol of fuel.
(b) the rate of entropy production, in kJ/K per kmol of fuel.
(c) the rate of exergy destruction, in kJ per kmol of fuel, for $T_0 = 25°C$.

Consider each of two coolants: water at 1 bar and ammonia at 10 bar.

13.80 Liquid ethanol (C_2H_5OH) at 25°C, 1 atm enters a reactor operating at steady state and burns completely with 130% of theoretical air entering in a separate stream at 25°C, 1 atm. Combustion products exit at 227°C, 1 atm. Heat transfer from the reactor takes place at an average surface temperature T_b. For T_b ranging from 25 to 200°C, determine the rate of exergy destruction within the reactor, in kJ per kmol of fuel. Kinetic and potential energy effects are negligible. Let $T_0 = 25°C$.

13.81 A gaseous mixture of ethane (C_2H_6) and the theoretical amount of air at 25°C, 1 atm enters a reactor operating at steady state and burns completely. Combustion products exit at 627°C, 1 atm. Heat transfer from the reactor takes place at an average surface temperature T_b. For T_b ranging from 25 to 600°C, determine the rate of exergy destruction within the reactor, in kJ per kmol of fuel. Kinetic and potential energy effects are negligible. Let $T_0 = 25°C$.

13.82 Determine the change in the Gibbs function, in kJ per kmol of methane, at 25°C, 1 atm for $CH_4 + 2O_2 \rightarrow CO_2 + 2H_2O$, using

(a) Gibbs function of formation data.
(b) enthalpy of formation data, together with absolute entropy data.

13.83 Determine the change in the Gibbs function, in kJ per kmol of hydrogen, at 25°C, 1 atm for $H_2 + \frac{1}{2}O_2 \rightarrow H_2O(g)$, using

(a) Gibbs function of formation data.
(b) enthalpy of formation data, together with absolute entropy data.

13.84 Separate streams of hydrogen (H_2) and oxygen (O_2) at 25°C, 1 atm enter a fuel cell operating at steady state, and liquid water exits at 25°C, 1 atm. The hydrogen flow rate is 2×10^{-4} kmol/s. If the fuel cell operates isothermally at 25°C, determine the maximum theoretical power it can develop and the accompanying rate of heat transfer, each in kW. Kinetic and potentially energy effects are negligible.

13.85 Streams of methane (CH_4) and oxygen (O_2), each at 25°C, 1 atm, enter a fuel cell operating at steady state. Streams of carbon dioxide and water exit separately at 25°C, 1 atm. If the fuel cell operates isothermally at 25°C, 1 atm, determine the maximum theoretical work that it can develop, in kJ per kmol of methane. Ignore kinetic and potential energy effects.

13.86 Streams of hydrogen (H_2) and oxygen (O_2), each at 1 atm, enter a fuel cell operating at steady state and water vapor exits at 1 atm. If the cell operates isothermally at **(a)** 300 K, **(b)** 400 K, and **(c)** 500 K, determine the maximum theoretical work that can be developed by the cell in each case, in kJ per kmol of hydrogen flowing, and comment. Heat transfer with the surroundings takes place at the cell temperature, and kinetic and potential energy effects can be ignored.

13.87 An inventor has developed a device that at steady state takes in liquid water at 25°C, 1 atm with a mass flow rate of 4 kg/h and produces separate streams of hydrogen (H_2) and oxygen (O_2), each at 25°C, 1 atm. The inventor claims that the device requires an electrical power input of 14.6 kW when operating isothermally at 25°C. Heat transfer with the surroundings occurs, but kinetic and potential energy effects can be ignored. Evaluate the inventor's claim.

13.88 Coal with a mass analysis of 88% C, 6% H, 4% O, 1% N, and 1% S burns completely with the theoretical amount of air. Determine

(a) the amount of SO_2 produced, in kg per kg of coal.
(b) the air–fuel ratio on a mass basis.
(c) For environmental reasons, it is desired to separate the SO_2 from the combustion products by supplying the products at 340 K, 1 atm to a device operating isothermally at 340 K. At steady state, a stream of SO_2 and a stream of the remaining gases exit, each at 340 K, 1 atm. If the coal is burned at a rate of 10 kg/s, determine the minimum theoretical power input required by the device, in kW. Heat transfer with the surroundings occurs, but kinetic and potential energy effects can be ignored.

Using Chemical Exergy

13.89 For **(a)** carbon, **(b)** hydrogen (H_2), **(c)** methane (CH_4), **(d)** carbon monoxide, **(e)** liquid methanol (CH_3OH), **(f)** nitrogen (N_2), **(g)** oxygen (O_2), **(h)** carbon dioxide, and **(i)** water, determine the chemical exergy, in kJ/kg, relative to the following environment in which the gas phase obeys the ideal gas model:

Environment $T_0 = 298.15$ K (25°C), $p_0 = 1$ atm		
Gas Phase:	Component	y^e (%)
	N_2	75.67
	O_2	20.35
	$H_2O(g)$	3.12
	CO_2	0.03
	Other	0.83

13.90 The accompanying table shows an environment consisting of a gas phase and a condensed water phase. The gas phase forms an ideal gas mixture.

Environment $T_0 = 298.15$ K (25°C), $p_0 = 1$ atm		
Condensed Phase: $H_2O(l)$ at T_0, p_0		
Gas Phase:	Component	y^e (%)
	N_2	75.67
	O_2	20.35
	$H_2O(g)$	3.12
	CO_2	0.03
	Other	0.83

(a) Show that the chemical exergy of the hydrocarbon C_aH_b can be determined as

$$\bar{e}^{ch} = \left[\bar{g}_F + \left(a + \frac{b}{4}\right)\bar{g}_{O_2} - a\bar{g}_{CO_2} - \frac{b}{2}\bar{g}_{H_2O(l)}\right] + \bar{R}T_0 \ln\left[\frac{(y^e_{O_2})^{a+b/4}}{(y^e_{CO_2})^a}\right]$$

(b) Using the result of part (a), repeat parts (a) through (c) of Problem 13.89.

13.91 The compounds listed in the table can be represented as $C_aH_bO_c$. Following the approach of Sec. 13.6, develop expressions for the chemical exergy of this group of compounds. Begin by writing Eq. 13.29 as

$$C_aH_bO_c + [a + b/4 - c/2]O_2 \rightarrow aCO_2 + b/2\, H_2O$$

	C	H_2	C_aH_b	CO	CO_2	H_2O(liq.)
a	1	0	a	1	1	0
b	0	2	b	0	0	2
c	0	0	0	1	2	1

13.92 Showing all important steps, derive **(a)** Eq. 13.38, **(b)** Eq. 13.39, **(c)** Eq. 13.40, **(d)** Eqs. 13.41a, b, and **(e)** Eqs. 13.44 a, b.

13.93 Using data from Tables A-25 and A-26, together with Eq. 13.45, determine the standard molar chemical exergy, in kJ/kmol, of

(a) ammonia (NH_3).
(b) propane (C_3H_8).

13.94 The chemical exergies of common hydrocarbons C_aH_b can be represented in terms of their respective lower heating value, LHV, by an expression of the form

$$\frac{\bar{e}^{ch}}{(\text{LHV})} = c_1 + c_2(b/a) + c_3/a$$

where c_1, c_2, and c_3 are constants. Evaluate the constants relative to the environment of Problem 13.89 to obtain an expression valid for several selected **(a)** gaseous hydrocarbons, **(b)** liquid hydrocarbons.

13.95 Evaluate the specific flow exergy of nitrogen (N_2), in kJ/kg, at 93°C, 4 atm. Neglect the effects of motion and gravity. Perform calculations

(a) relative to the environment of Problem 13.89.
(b) using data from Table A-26.

13.96 Evaluate the specific flow exergy of water vapor, in kJ/kg, at 200°C, 1 bar. Neglect the effects of motion and gravity. Perform calculations

(a) relative to the environment of Problem 13.89.
(b) using data from Table A-26.

13.97 Evaluate the specific flow exergy of an equimolar mixture of oxygen (O_2) and nitrogen (N_2), in kJ/kg, at 20°C, 1 atm. Neglect the effects of motion and gravity. Perform calculations

(a) relative to the environment of Problem 13.89.
(b) using data from Table A-26.

13.98 A mixture of methane gas (CH_4) and 150% of theoretical air enters a combustion chamber at 25°C, 1 atm. Determine the specific flow exergy of the entering mixture, in kJ per kmol of methane. Ignore the effects of motion and gravity. Perform calculations

(a) relative to the environment of Problem 13.89.
(b) using data from Table A-26.

13.99 A mixture having an analysis on a molar basis of 85% dry air, 15% CO enters a device at 125°C, 2.1 atm, and a velocity of 250 m/s. If the mass flow rate is 1.0 kg/s, determine the rate exergy enters, in kW. Neglect the effect of gravity. Perform calculations

(a) relative to the environment of Problem 13.89.
(b) using data from Table A-26.

13.100 The following flow rates in kg/h are reported for the exiting SNG (substitute natural gas) stream in a certain process for producing SNG from bituminous coal:

CH_4	194,905 kg/h
CO_2	4,125 kg/h
N_2	1,697 kg/h
H_2	261 kg/h
CO	92 kg/h
H_2O	27 kg/h

If the SNG stream is at 25°C, 1 atm, determine the rate at which exergy exits. Perform calculations relative to the environment of Problem 13.89. Neglect the effects of motion and gravity.

Exergy Analysis of Reacting and Psychrometric Systems

13.101 Carbon at 25°C, 1 atm enters an insulated reactor operating at steady state and reacts completely with the theoretical amount of air entering separately at 25°C, 1 atm. For the reactor, **(a)** determine the rate of exergy destruction, in kJ per kmol of carbon, and **(b)** evaluate an exergetic efficiency. Perform calculations relative to the environment of Problem 13.89. Neglect the effects of motion and gravity.

13.102 Propane gas (C_3H_8) at 25°C, 1 atm and a volumetric flow rate of 0.03 m³/min enters a furnace operating at steady state and burns completely with 200% of theoretical air entering at 25°C, 1 atm. Combustion products exit at 227°C, 1 atm. The furnace provides energy by heat transfer at 227°C for an industrial process. For the furnace, compare the rate of exergy transfer accompanying heat transfer with the rate of exergy destruction, each in kJ/min. Let T_0 = 25°C and ignore kinetic and potential energy effects.

13.103 Figure P13.103 shows a coal gasification reactor making use of the *carbon–steam* process. The energy required for the endothermic reaction is supplied by an electrical resistor. The reactor operates at steady state, with no stray heat transfers and negligible kinetic and potential energy effects. Evaluate in kJ per kmol of carbon entering

(a) the required electrical input.
(b) the exergy entering with the carbon.
(c) the exergy entering with the steam.

(d) the exergy exiting with the product gas.
(e) the exergy destruction within the reactor.

Perform calculations relative to the environment of Problem 13.89. Ignore kinetic and potential energy effects.

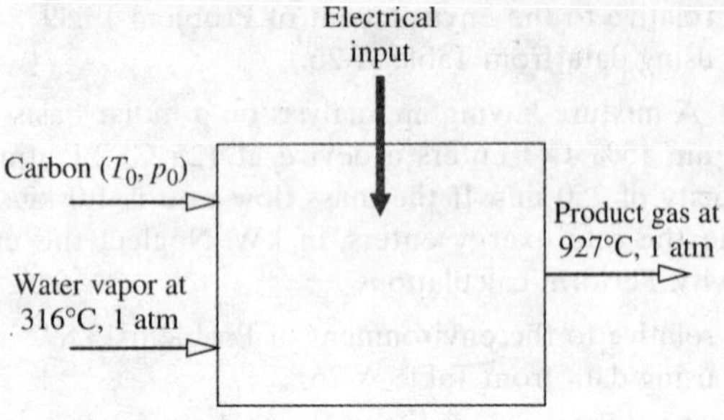

$C + 1.25H_2O(g) \rightarrow CO + H_2 + 0.25H_2O(g)$

Fig. P13.103

13.104 Carbon monoxide at 25°C, 1 atm enters an insulated reactor operating at steady state and reacts completely with the theoretical amount of air entering in a separate stream at 25°C, 1 atm. The products exit as a mixture at 1 atm. Determine in kJ per kmol of CO

(a) the exergy entering with the carbon monoxide.
(b) the exergy exiting with the products.
(c) the rate of exergy destruction.

Also, evaluate an exergetic efficiency for the reactor. Perform calculations relative to the environment of Problem 13.89. Ignore kinetic and potential energy effects.

13.105 Acetylene gas (C_2H_2) at 25°C, 1 atm enters an insulated reactor operating at steady state and burns completely with 180% of theoretical air, entering in a separate stream at 25°C, 1 atm. The products exit as a mixture at 1 atm. Determine in kJ per kmol of fuel

(a) the exergy of the fuel entering the reactor.
(b) the exergy exiting with the products.
(c) the rate of exergy destruction.

Also, evaluate an exergetic efficiency for the reactor. Perform calculations relative to the environment of Problem 13.89. Ignore kinetic and potential energy effects.

13.106 Liquid octane (C_8H_{18}) at 25°C, 1 atm and a mass flow rate of 0.57 kg/h enters a small internal combustion engine operating at steady state. The fuel burns with air entering the engine in a separate stream at 25°C, 1 atm. Combustion products exit at 670 K, 1 atm with a dry molar analysis of 11.4% CO_2, 2.9% CO, 1.6% O_2, and 84.1% N_2. If the engine develops power at the rate of 1 kW, determine

(a) the rate of heat transfer from the engine, in kW.
(b) an exergetic efficiency for the engine.

Use the environment of Problem 13.89 and neglect kinetic and potential energy effects.

13.107 Evaluate an exergetic efficiency for the gas turbine power plant of Problem 13.51. Base exergy values on the environment of Problem 13.89.

13.108 Figure P13.108 shows a simple vapor power plant. The fuel is methane that enters at 25°C, 1 atm and burns completely with 200% theoretical air entering at 25°C, 1 atm. Steam exits the steam generator at 480°C, 3.45 MPa. The vapor expands through the turbine and exits at 6.9 kPa, and a quality of 97%. At the condenser exit, the pressure is 6.9 kPa and the water is a saturated liquid. The plant operates at steady state with no stray heat transfers from any plant component. Pump work and all kinetic and potential energy effects are negligible. Determine

(a) the balanced reaction equation.
(b) the vapor mass flow rate, in kg per kmol of fuel.
(c) the cooling water mass flow rate, in kg per kmol of fuel.
(d) each of the following, expressed as a percent of the exergy entering the steam generator with the fuel, (i) the exergy exiting with the stack gases, (ii) the exergy destroyed in the steam generator, (iii) the power developed by the turbine, (iv) the exergy destroyed in the turbine, (v) the exergy exiting with the cooling water, (vi) the exergy destroyed in the condenser.

Base exergy values on the environment of Problem 13.89.

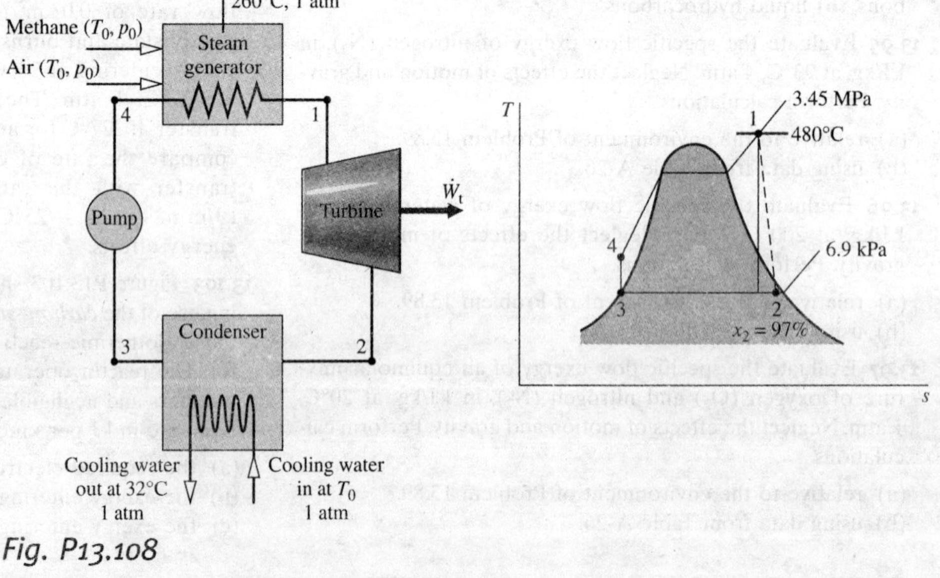

Fig. P13.108

13.109 Consider a furnace operating at steady state idealized as shown in Fig. P13.109. The fuel is methane, which enters at 25°C, 1 atm and burns completely with 200% theoretical air entering at the same temperature and pressure. The furnace delivers energy by heat transfer at an average temperature of 60°C. Combustion products at 600 K, 1 atm are discharged to the surroundings. There are no stray heat transfers, and kinetic and potential energy effects can be ignored. Determine in kJ per kmol of fuel

(a) the exergy entering the furnace with the fuel.
(b) the exergy exiting with the products.
(c) the rate of exergy destruction.

Also, evaluate an exergetic efficiency for the furnace and comment. Perform calculations relative to the environment of Problem 13.89.

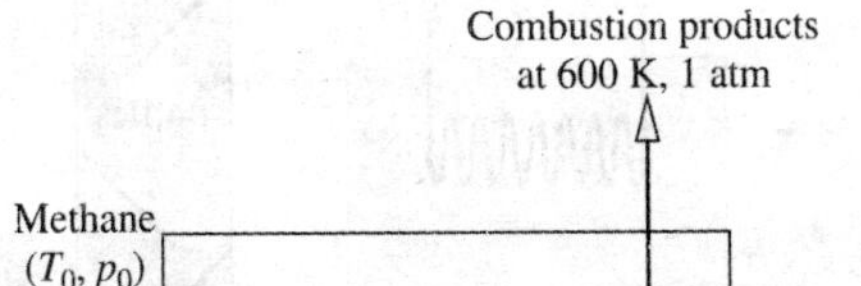

Fig. P13.109

13.110 Coal enters the combustor of a power plant with a mass analysis of 49.8% C, 3.5% H, 6.8% O, 6.4% S, 14.1% H_2O, and 19.4% noncombustible ash. The higher heating value of the coal is measured as 21,220 kJ/kg, and the lower heating value on a dry basis, $(LHV)_d$, is 20,450 kJ/kg. The following expression can be used to estimate the chemical exergy of the coal, in kJ/kg:

$$e^{ch} = (LHV)_d\left(1.0438 + 0.0013\frac{h}{c} + 0.1083\frac{o}{c} + 0.0549\frac{n}{c}\right) + 6740s$$

where h/c, o/c, and n/c denote, respectively, the mass ratio of hydrogen to carbon, oxygen to carbon, and nitrogen to carbon, and s is the mass fraction of sulfur in kg per kg of fuel.[4] The environment is closely the same as in Problem 13.90, but extended appropriately to account for the presence of sulfur in the coal.

(a) Using the above expression, calculate the chemical exergy of the coal, in kJ/kg.
(b) Compare the answer of part (a) with the values that would result by approximating the chemical exergy with each of the measured heating values.
(c) What data would be required to determine the chemical exergy in this case using the methodology of Sec. 13.6? Discuss.

[4]Moran, *Availability Analysis,* pp. 192–193.

13.111 For psychrometric applications such as those considered in Chap. 12, the environment often can be modeled simply as an ideal gas mixture of water vapor and dry air at temperature T_0 and pressure p_0. The composition of the environment is defined by the dry air and water vapor mole fractions y_a^e, y_v^e, respectively. Show that relative to such an environment the flow exergy of a moist air stream at temperature T and pressure p with dry air and water vapor mole fractions y_a and y_v, respectively, can be expressed on a molar basis as

$$\bar{e}_f = T_0\left\{(y_a\bar{c}_{pa} + y_v\bar{c}_{pv})\left[\left(\frac{T}{T_0}\right) - 1 - \ln\left(\frac{T}{T_0}\right)\right] + \bar{R}\ln\left(\frac{p}{p_0}\right)\right\} + \bar{R}T_0\left[y_a\ln\left(\frac{y_a}{y_a^e}\right) + y_v\ln\left(\frac{y_v}{y_v^e}\right)\right]$$

where $\bar{c}_{pa}$ and $\bar{c}_{pv}$ denote the molar specific heats of dry air and water vapor, respectively. Neglect the effects of motion and gravity.

13.112 For each of the following, use the result of Problem 13.111 to determine the specific flow exergy, in kJ/kg, relative to an environment consisting of moist air at 20°C, 1 atm, $\phi = 100\%$

(a) moist air at 20°C, 1 atm, $\phi = 100\%$.
(b) moist air at 20°C, 1 atm, $\phi = 50\%$.
(c) dry air at 20°C, 1 atm.

13.113 Using the result of Problem 13.111, determine the flow exergy at locations 1, 2, and 3 and the rate of exergy destruction, each in kJ/min, for the device of Problem 12.102. Let the environment be a mixture of dry air and water vapor at 35°C, 1 atm with $y_v^e = 0.022$, $y_a^e = 0.978$. Also, let $c_{pa} = 1004$ J/kJ · K and $c_{pv} = 1840$ J/kg · K.

13.114 Figure P13.114 provides data for a cooling tower at steady state. Determine the rate of exergy destruction, in kJ/min. Let the environment be a mixture of dry air and water vapor at 32°C, 1 atm with $y_v^e = 0.024$, $y_a^e = 0.976$. Also, let $c_{pa} = 1004$ J/kg · K and $c_{pv} = 1842$ J/kg · K.

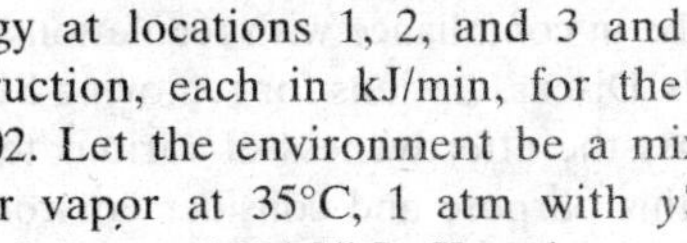

Fig. P13.114

Design & Open Ended Problems: Exploring Engineering Practice

13.1D The term *acid rain* is frequently used today. Define what is meant by the term. Write a report in which you discuss the origin and consequences of acid rain as well as options for its control. Include at least three references.

13.2D Many observers have expressed concern that the release of CO_2 into the atmosphere due to the combustion of fossil fuels is contributing to *global warming*. Write a paper reviewing the scientific evidence regarding the contribution of fossil fuel combustion to global warming. Compare and contrast this evidence with comparable data for the combustion of *biofuels* derived from plant matter. Include at least three references.

13.3D About 234 million tires are discarded in the U.S. annually, adding to a stockpile of two to three billion scrap tires from previous years. Write a memorandum discussing the potential advantages and disadvantages of using scrap tires as a fuel. How does the heating value of scrap tires compare with the heating values of gasoline and commonly used coals? How does the ultimate analysis of scrap tires compare with the ultimate analysis of commonly used coals?

13.4D A coal with the following ultimate analysis:

$$\begin{Bmatrix} 75.8\%\ \text{C}, 5.1\%\ \text{H}, 8.2\%\ \text{O}, 1.5\%\ \text{N}, \\ 1.6\%\ \text{S}, 7.8\%\ \text{noncombustible ash} \end{Bmatrix}$$

is burned in a power plant boiler. If all of the sulfur present in the combustible portion of the fuel forms SO_2, determine if the plant would be in compliance with regulations regarding sulfur emissions. Discuss options for removing SO_2 from the stack gas. Assess the effectiveness of various technologies available for this purpose and consider environmental problems associated with them.

13.5D Figure P13.5D shows a natural gas–fired boiler for steam generation integrated with a *direct-contact condensing* heat exchanger that discharges warm water for tasks such as space heating, water heating, and combustion air preheating. For 7200 h of operation annually, estimate the annual savings for fuel, in dollars, by integrating these functions. What other cost considerations would enter into the decision to install such a condensing heat exchanger? Discuss.

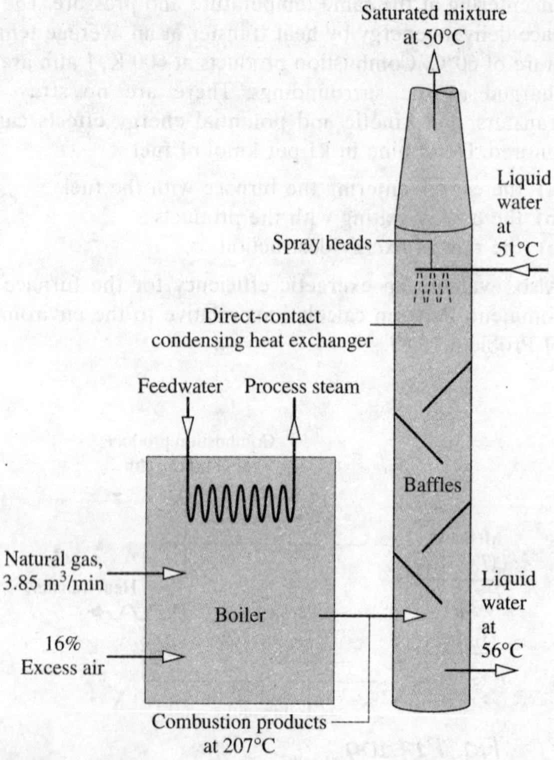

Fig. P13.5D

13.6D A factory requires 3750 kW of electric power and high-quality steam at 107°C with a mass flow rate of 2.2 kg/s. Two options are under consideration:

Option 1: A single boiler generates steam at 2.0 MPa, 320°C, supplying a turbine that exhausts to a condenser at 0.007 MPa. Steam is extracted from the turbine at 107°C, returning as a liquid to the boiler after use.

Option 2: A boiler generates steam at 2.0 MPa, 320°C supplying a turbine that exhausts to a condenser at 0.007 MPa. A separate process steam boiler generates the required steam at 107°C, which is returned as a liquid to the boiler after use.

The boilers are fired with natural gas and 20% excess air. Perform a complete thermoeconomic analysis based on 7200 hours of operation annually.

13.7D Figure P13.7D illustrates the schematic of an air preheater for the boiler of a coal-fired power plant. An alternative design would eliminate the air preheater, supplying air to the furnace directly at 30°C and discharging flue gas at 260°C. All other features would remain unchanged. On the basis of thermodynamic and economic analyses, recommend one of the two options for further consideration. The analysis of the coal on a mass basis is

$$\{76\%\ \text{C}, 5\%\ \text{H}, 8\%\ \text{O}, 11\%\ \text{noncombustible ash}\}$$

The coal higher heating value is 26,000 kJ/kg. The molar analysis of the flue gas on a dry basis is

$$\{7.8\%\ CO_2, 1.2\%\ CO, 11.4\%\ O_2, 79.6\%\ N_2\}$$

13.8D Fuel or chemical leaks and spills can have catastrophic ramifications; thus the hazards associated with such events must be well understood. Prepare a memorandum for one of the following:

(a) Experience with interstate pipelines shows that propane leaks are usually much more hazardous than leaks of natural gas or liquids such as gasoline. Why is this so?

(b) The most important parameter in determining the accidental rate of release from a fuel or chemical storage vessel is generally the size of the opening. Roughly how much faster would such a substance be released from a 1-cm hole than from a 1-mm hole? What are the implications of this?

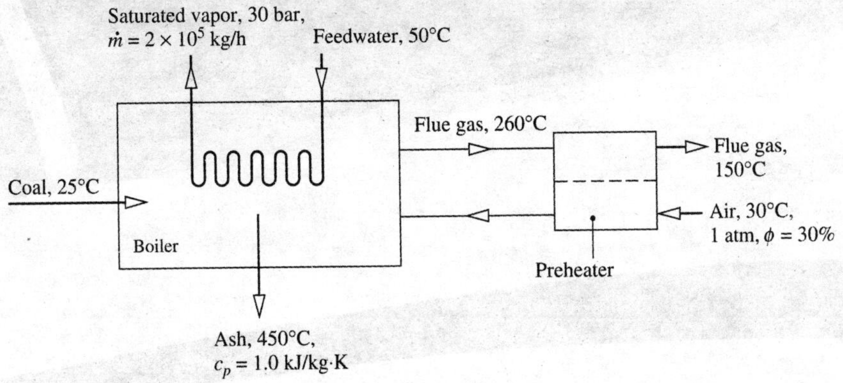

Fig. P13.7D

13.9D By the year 2010, as much as 60,000 MW of electric power could be generated worldwide by fuel cells. A step in this direction has been taken at Santa Clara, California, where a 2-MW fuel cell demonstration power plant was installed under the auspices of a consortium of electric and gas utilities. Report on the Santa Clara demonstration project in a brief memorandum. Include a schematic of the system. For further study of fuel cells:

(a) What are the principal features that make fuel cell technology attractive for power generation? Is fuel cell performance limited by the Carnot efficiency? Discuss.

(b) The types of fuel cells listed in Table 13.1 are considered to hold commercial promise. Investigate the specific realms of application of these fuel cell types. Write a report including at least three references.

(c) *Stack life* and *installed cost* are two parameters considered critical for fuel cell development. What is meant by stack life and why is it important? What is the projected installed cost, in \$ per kW, of current fuel cell technology for large-scale power generation? To be competitive with conventional power systems such as gas turbines, what should be the target installed cost for fuel cells? Discuss.

(d) What principal issues must be resolved for fuel cells to be widely used to power automobiles and trucks? Write a report including at least three references.

13.10D Mounting evidence suggests that more frequent severe weather events, melting of glaciers and polar icecaps, and other observed changes in nature are linked to emissions from fossil fuel combustion. Researchers have developed detailed computer simulations to predict global weather patterns and investigate the effects of CO_2 and other gaseous emissions on the weather. Investigate the underlying modeling assumptions and characteristics of computer simulations used to predict global weather patterns. Write a report, including at least three references, of your findings.

13.11D Investigate the advantages and disadvantages of hydrogen and methanol as fuels for fuel cell-powered portable electronic devices. Write a report including at least three references.

14

Chemical and Phase Equilibrium

Engineering Context

The **objective** of the present chapter is to consider the concept of equilibrium in greater depth than has been done thus far. In the first part of the chapter, we develop the fundamental concepts used to study chemical and phase equilibrium. In the second part of the chapter, the study of reacting systems initiated in Chap. 13 is continued with a discussion of *chemical* equilibrium in a single phase. Particular emphasis is placed on the case of reacting ideal gas mixtures. The third part of the chapter concerns *phase* equilibrium. The equilibrium of multicomponent, multiphase, nonreacting systems is considered and the *phase rule* is introduced.

Learning Objectives

When you complete your study of this chapter you will be able to...

✓ demonstrate understanding of key concepts related to chemical and phase equilibrium, including criteria for equilibrium, the equilibrium constant, and the Gibbs phase rule.

✓ apply the equilibrium constant relationship, Eq. 14.35, to relate pressure, temperature, and equilibrium constant for ideal gas mixtures involving individual and multiple reactions.

✓ use chemical equilibrium concepts with the energy balance.

✓ determine the equilibrium flame temperature.

✓ apply the Gibbs phase rule, Eq. 14.68.

Equilibrium Fundamentals

In this part of the chapter, fundamental concepts are developed that are useful in the study of chemical and phase equilibrium. Among these are equilibrium criteria and the chemical potential concept.

14.1 Introducing Equilibrium Criteria

thermodynamic equilibrium

A system is said to be in ***thermodynamic equilibrium*** if, when it is isolated from its surroundings, there would be no macroscopically observable changes. An important requirement for equilibrium is that the temperature be uniform throughout the system or each part of the system in thermal contact. If this condition were not met, spontaneous heat transfer from one location to another could occur when the system was isolated. There must also be no unbalanced forces between parts of the system. These conditions ensure that the system is in thermal and mechanical equilibrium, but there is still the possibility that complete equilibrium does not exist. A process might occur involving a chemical reaction, a transfer of mass between phases, or both. The objective of this section is to introduce criteria that can be applied to decide whether a system in a particular state is in equilibrium. These criteria are developed using the conservation of energy principle and the second law of thermodynamics as discussed next.

Consider the case of a simple compressible system of fixed mass for which temperature and pressure are uniform with position throughout. In the absence of overall system motion and ignoring the influence of gravity, the energy balance in differential form (Eq. 2.36) is

$$dU = \delta Q - \delta W$$

If volume change is the only work mode and pressure is uniform with position throughout the system, $\delta W = p\,dV$. Introducing this in the energy balance and solving for δQ gives

$$\delta Q = dU + p\,dV$$

Since temperature is uniform with position throughout the system, the entropy balance in differential form (Eq. 6.25) is

$$dS = \frac{\delta Q}{T} + \delta\sigma$$

Eliminating δQ between the last two equations

$$T\,dS - dU - p\,dV = T\,\delta\sigma \tag{14.1}$$

Entropy is produced in all actual processes and conserved only in the absence of irreversibilities. Hence, Eq. 14.1 provides a constraint on the direction of processes. The only processes allowed are those for which $\delta\sigma \geq 0$. Thus

$$T\,dS - dU - p\,dV \geq 0 \tag{14.2}$$

Equation 14.2 can be used to study equilibrium under various conditions. **FOR EXAMPLE...** a process taking place in an insulated, constant-volume vessel, where $dU = 0$ and $dV = 0$, must be such that

$$dS]_{U,V} \geq 0 \tag{14.3}$$

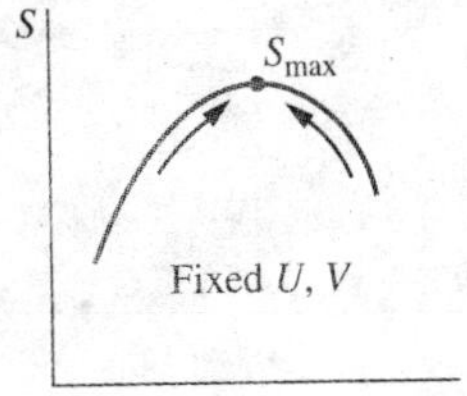

Equation 14.3 suggests that changes of state of a closed system at constant internal energy and volume can occur only in the direction of *increasing entropy*. The expression also implies that entropy approaches a *maximum* as a state of equilibrium is approached. This is a special case of the increase of entropy principle introduced in Sec. 6.8.1.

Gibbs function

An important case for the study of chemical and phase equilibria is one in which temperature and pressure are fixed. For this, it is convenient to employ the ***Gibbs function*** in extensive form

$$G = H - TS = U + pV - TS$$

Forming the differential

$$dG = dU + p\,dV + V\,dp - T\,dS - S\,dT$$

or on rearrangement

$$dG - V\,dp + S\,dT = -(T\,dS - dU - p\,dV)$$

Except for the minus sign, the right side of this equation is the same as the expression appearing in Eq. 14.2. Accordingly, Eq. 14.2 can be written as

$$dG - V\,dp + S\,dT \le 0 \qquad (14.4)$$

where the inequality reverses direction because of the minus sign noted above.

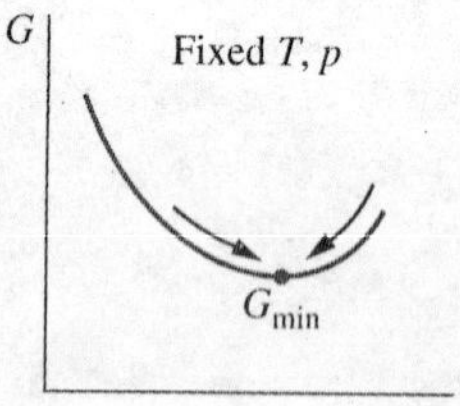

It can be concluded from Eq. 14.4 that any process taking place at a specified temperature and pressure ($dT = 0$ and $dp = 0$) must be such that

$$dG]_{T,p} \le 0 \qquad (14.5)$$

This inequality indicates that the Gibbs function of a system at fixed T and p *decreases* during an irreversible process. Each step of such a process results in a decrease in the Gibbs function of the system and brings the system closer to equilibrium. The equilibrium state is the one having the *minimum* value of the Gibbs function. Therefore, when

equilibrium criterion

$$dG]_{T,p} = 0 \qquad (14.6)$$

we have equilibrium. In subsequent discussions, we refer to Eq. 14.6 as the ***criterion for equilibrium.***

Equation 14.6 provides a relationship among the properties of a system when it is *at* an equilibrium state. The manner in which the equilibrium state is reached is unimportant, however, for once an equilibrium state is obtained, a system exists at a particular T and p and no further spontaneous changes can take place. When applying Eq. 14.6, therefore, we may specify the temperature T and pressure p, but it is unnecessary to require additionally that the system actually achieves equilibrium at fixed T and fixed p.

14.1.1 Chemical Potential and Equilibrium

In the present discussion, the Gibbs function is considered further as a prerequisite for application of the equilibrium criterion $dG]_{T,p} = 0$ introduced above. Let us begin by noting that any *extensive* property of a single-phase, single-component system is a function of two independent intensive properties and the size of the system. Selecting temperature and pressure as the independent properties and the number of moles n as the measure of size, the Gibbs function can be expressed in the form

$G = G(T, p, n)$. For a single-phase, ***multicomponent system,*** G may then be considered a function of temperature, pressure, and the number of moles of each component present, written $G = G(T, p, n_1, n_2, \ldots, n_j)$.

multicomponent system

If each mole number is multiplied by α, the size of the system changes by the same factor and so does the value of every extensive property. Thus, for the Gibbs function we may write

$$\alpha G(T, p, n_1, n_2, \ldots, n_j) = G(T, p, \alpha n_1, \alpha n_2, \ldots, \alpha n_j)$$

Differentiating with respect to α while holding temperature, pressure, and the mole numbers fixed and using the chain rule on the right side gives

$$G = \frac{\partial G}{\partial(\alpha n_1)} n_1 + \frac{\partial G}{\partial(\alpha n_2)} n_2 + \cdots + \frac{\partial G}{\partial(\alpha n_j)} n_j$$

This equation holds for all values of α. In particular, it holds for $\alpha = 1$. Setting $\alpha = 1$, the following expression results:

$$G = \sum_{i=1}^{j} n_i \left(\frac{\partial G}{\partial n_i}\right)_{T,p,n_l} \tag{14.7}$$

where the subscript n_l denotes that all n's except n_i are held fixed during differentiation.

The partial derivatives appearing in Eq. 14.7 have such importance for our study of chemical and phase equilibrium that they are given a special name and symbol. The ***chemical potential*** of component i, symbolized by μ_i, is defined as

$$\mu_i = \left(\frac{\partial G}{\partial n_i}\right)_{T,p,n_l} \tag{14.8}$$

chemical potential

The chemical potential is an *intensive property*. With Eq. 14.8, Eq. 14.7 becomes

$$G = \sum_{i=1}^{j} n_i \mu_i \tag{14.9}$$

The equilibrium criterion given by Eq. 14.6 can be written in terms of chemical potentials, providing an expression of fundamental importance for subsequent discussions of equilibrium. Forming the differential of $G(T, p, n_1, \ldots, n_j)$ while holding temperature and pressure fixed results in

$$dG]_{T,p} = \sum_{i=1}^{j} \left(\frac{\partial G}{\partial n_i}\right)_{T,p,n_l} dn_i$$

The partial derivatives are recognized from Eq. 14.8 as the chemical potentials, so

$$dG]_{T,p} = \sum_{i=1}^{j} \mu_i \, dn_i \tag{14.10}$$

With Eq. 14.10, the equilibrium criterion $dG]_{T,p} = 0$ can be placed in the form

$$\sum_{i=1}^{j} \mu_i \, dn_i = 0 \tag{14.11}$$

Like Eq. 14.6, from which it is obtained, this equation provides a relationship among properties of a system when the system is *at* an equilibrium state where the temperature is T and the pressure is p. Like Eq. 14.6, this equation applies to a particular state, and the manner in which that state is attained is not important.

14.1.2 Evaluating Chemical Potentials

Means for evaluating the chemical potentials for two cases of interest are introduced in this section: a single phase of a pure substance and an ideal gas mixture.

SINGLE PHASE OF A PURE SUBSTANCE. An elementary case considered later in this chapter is that of equilibrium between two phases involving a pure substance. For a single phase of a pure substance, Eq. 14.9 becomes simply

$$G = n\mu$$

or

$$\mu = \frac{G}{n} = \bar{g} \tag{14.12}$$

That is, the chemical potential is just the Gibbs function per mole.

IDEAL GAS MIXTURE. An important case for the study of chemical equilibrium is that of an ideal gas mixture. The enthalpy and entropy of an ideal gas mixture are given by

$$H = \sum_{i=1}^{j} n_i \bar{h}_i(T) \quad \text{and} \quad S = \sum_{i=1}^{j} n_i \bar{s}_i(T, p_i)$$

where $p_i = y_i p$ is the partial pressure of component i. Accordingly, the Gibbs function takes the form

$$G = H - TS = \sum_{i=1}^{j} n_i \bar{h}_i(T) - T \sum_{i=1}^{j} n_i \bar{s}_i(T, p_i)$$

$$= \sum_{i=1}^{j} n_i [\bar{h}_i(T) - T\bar{s}_i(T, p_i)] \qquad \text{(ideal gas)} \tag{14.13}$$

Introducing the molar Gibbs function of component i

$$\bar{g}_i(T, p_i) = \bar{h}_i(T) - T\bar{s}_i(T, p_i) \qquad \text{(ideal gas)} \tag{14.14}$$

Equation 14.13 can be expressed as

$$G = \sum_{i=1}^{j} n_i \bar{g}_i(T, p_i) \qquad \text{(ideal gas)} \tag{14.15}$$

Comparing Eq. 14.15 to Eq. 14.9 suggests that

$$\mu_i = \bar{g}_i(T, p_i) \qquad \text{(ideal gas)} \tag{14.16}$$

That is, the chemical potential of component i in an ideal gas mixture is equal to its Gibbs function per mole of i, evaluated at the mixture temperature and the partial pressure of i in the mixture. Equation 14.16 can be obtained formally by taking the partial derivative of Eq. 14.15 with respect to n_i, holding temperature, pressure, and the remaining n's constant, and then applying the definition of chemical potential, Eq. 14.8.

The chemical potential of component i in an ideal gas mixture can be expressed in an alternative form that is somewhat more convenient for subsequent applications. Using Eq. 13.23, Eq. 14.14 becomes

$$\begin{aligned}\mu_i &= \bar{h}_i(T) - T\bar{s}_i(T, p_i)\\ &= \bar{h}_i(T) - T\left(\bar{s}_i^\circ(T) - \bar{R}\ln\frac{y_i p}{p_{\text{ref}}}\right)\\ &= \bar{h}_i(T) - T\bar{s}_i^\circ(T) + \bar{R}T\ln\frac{y_i p}{p_{\text{ref}}}\end{aligned}$$

where p_{ref} is 1 atm and y_i is the mole fraction of component i in a mixture at temperature T and pressure p. The last equation can be written compactly as

$$\mu_i = \bar{g}_i^\circ + \bar{R}T\ln\frac{y_i p}{p_{\text{ref}}} \qquad \text{(ideal gas)} \tag{14.17}$$

where $\bar{g}_i^\circ$ is the Gibbs function of component i evaluated at temperature T and a pressure of 1 atm. Additional details concerning the chemical potential concept are provided in Sec. 11.9. Equation 14.17 is the same as Eq. 11.144 developed there.

Bio... connections

The human body, as well as all living things, exist in a type of equilibrium that is considered in biology to be a *dynamic* equilibrium called *homeostasis.* The term refers to the ability of the body to regulate its internal state, such as body temperature, within specific limits necessary for life despite the conditions of the external environment, or the level of exertion. Within the body, feedback mechanisms regulate numerous variables that must be kept under control. If this dynamic equilibrium cannot be maintained because of the severe influence of the surroundings, living things can become sick and die.

The numerous control mechanisms in the body are varied, but they share some common elements. Generally, they involve a feedback loop that includes a way to sense a fluctuation of a variable from its desired condition and generate a corrective action to bring the variable back into the desired range. Feedback loops exist within the body at the levels of cells, body fluid systems (such as the circulatory system), tissues, and organs. These feedback loops also interact to maintain homeostasis much as a home thermostat regulates the heating and cooling system in a house. The mechanisms are highly stable as long as the external conditions are not too severe.

An example of the complexity of homeostasis is the way in which the body maintains the levels of glucose in the blood within desired limits. Glucose is an essential "fuel" for the processes within cells. Glucose absorbed by the digestive system from food, or stored as glycogen in the liver, is distributed by the blood stream to the cells. The body senses the level of glucose in the blood and various glands produce hormones to *stimulate* the conversion of glucose from stored glycogen, if needed to supplement food intake, or to *inhibit* the release of glucose, as necessary to maintain the desired levels. Several different hormones and numerous organs within the body are involved. The result is a balance that is highly stable within the limits needed by the body for homeostatic equilibrium.

Chemical Equilibrium

In this part of the chapter, the equilibrium criterion $dG]_{T,p} = 0$ introduced in Sec. 14.1 is used to study the equilibrium of reacting mixtures. The objective is to establish the composition present at equilibrium for a specified temperature and pressure. An important parameter for determining the equilibrium composition is the *equilibrium constant.* The equilibrium constant is introduced and its use illustrated by several solved examples. The discussion is concerned only with equilibrium states of reacting systems, and no information can be deduced about the *rates of reaction.* Whether an equilibrium mixture would form quickly or slowly can be determined only by considering the *chemical kinetics,* a topic that is not treated in this text.

14.2 Equation of Reaction Equilibrium

In Chap. 13 the conservation of mass and conservation of energy principles are applied to reacting systems by assuming that the reactions can occur as written. However, the extent to which a chemical reaction proceeds is limited by many factors. In general, the composition of the products actually formed from a given set of reactants, and the relative amounts of the products, can be determined only from experiment. Knowledge of the composition that would be present were a reaction to proceed to equilibrium is frequently useful, however. The *equation of reaction equilibrium* introduced in the present section provides the basis for determining the equilibrium composition of a reacting mixture.

INTRODUCTORY CASE. Consider a closed system consisting initially of a gaseous mixture of hydrogen and oxygen. A number of reactions might take place, including

$$1H_2 + \tfrac{1}{2}O_2 \rightleftarrows 1H_2O \tag{14.18}$$

$$1H_2 \rightleftarrows 2H \tag{14.19}$$

$$1O_2 \rightleftarrows 2O \tag{14.20}$$

Let us consider for illustration purposes only the first of the reactions given above, in which hydrogen and oxygen combine to form water. At equilibrium, the system will consist in general of three components: H_2, O_2, and H_2O, for not all of the hydrogen and oxygen initially present need be reacted. *Changes* in the amounts of these components during each differential step of the reaction leading to the formation of an equilibrium mixture are governed by Eq. 14.18. That is

$$dn_{H_2} = -dn_{H_2O}, \qquad dn_{O_2} = -\tfrac{1}{2}dn_{H_2O} \tag{14.21a}$$

where dn denotes a differential change in the respective component. The minus signs signal that the amounts of hydrogen and oxygen present decrease when the reaction proceeds toward the right. Equations 14.21a can be expressed alternatively as

$$\frac{-dn_{H_2}}{1} = \frac{-dn_{O_2}}{\frac{1}{2}} = \frac{dn_{H_2O}}{1} \tag{14.21b}$$

which emphasizes that increases and decreases in the components are proportional to the stoichiometric coefficients of Eq. 14.18.

Equilibrium is a condition of *balance.* Accordingly, as suggested by the direction of the arrows in Eq. 14.18, when the system is at equilibrium, the tendency of the hydrogen and oxygen to form water is just balanced by the tendency of water to dissociate into oxygen and hydrogen. The equilibrium criterion $dG]_{T,p} = 0$ can be used to determine the composition at an equilibrium state where the temperature is T and the pressure is p. This requires evaluation of the differential $dG]_{T,p}$ in terms of system properties.

For the present case, Eq. 14.10 giving the difference in the Gibbs function of the mixture between two states having the same temperature and pressure, but compositions that differ infinitesimally, takes the following form:

$$dG]_{T,p} = \mu_{H_2}\, dn_{H_2} + \mu_{O_2}\, dn_{O_2} + \mu_{H_2O}\, dn_{H_2O} \tag{14.22}$$

The changes in the mole numbers are related by Eqs. 14.21. Hence

$$dG]_{T,p} = (-1\mu_{H_2} - \tfrac{1}{2}\mu_{O_2} + 1\mu_{H_2O})\, dn_{H_2O}$$

At equilibrium, $dG]_{T,p} = 0$, so the term in parentheses must be zero. That is

$$-1\mu_{H_2} - \tfrac{1}{2}\mu_{O_2} + 1\mu_{H_2O} = 0$$

When expressed in a form that resembles Eq. 14.18, this becomes

$$1\mu_{H_2} + \tfrac{1}{2}\mu_{O_2} = 1\mu_{H_2O} \tag{14.23}$$

Equation 14.23 is the *equation of reaction equilibrium* for the case under consideration. The chemical potentials are functions of temperature, pressure, and composition. Thus, the composition that would be present at equilibrium for a given temperature and pressure can be determined, in principle, by solving this equation. The solution procedure is described in the next section.

GENERAL CASE. The foregoing development can be repeated for reactions involving any number of components. Consider a closed system containing *five* components, A, B, C, D, and E, at a given temperature and pressure, subject to a single chemical reaction of the form

$$\nu_A A + \nu_B B \rightleftarrows \nu_C C + \nu_D D \tag{14.24}$$

where the ν's are stoichiometric coefficients. Component E is assumed to be inert and thus does not appear in the reaction equation. As we will see, component E does influence the equilibrium composition even though it is not involved in the chemical reaction. The form of Eq. 14.24 suggests that at equilibrium the tendency of A and B to form C and D is just balanced by the tendency of C and D to form A and B.

The stoichiometric coefficients ν_A, ν_B, ν_C, and ν_D do not correspond to the respective number of moles of the components present. The amounts of the components present are designated n_A, n_B, n_C, n_D, and n_E. However, *changes* in the amounts of the components present do bear a definite relationship to the values of the stoichiometric coefficients. That is

$$\frac{-dn_A}{\nu_A} = \frac{-dn_B}{\nu_B} = \frac{dn_C}{\nu_C} = \frac{dn_D}{\nu_D} \tag{14.25a}$$

where the minus signs indicate that A and B would be consumed when C and D are produced. Since E is inert, the amount of this component remains constant, so $dn_E = 0$.

Introducing a proportionality factor $d\varepsilon$, Eqs. 14.25a take the form

$$\frac{-dn_A}{\nu_A} = \frac{-dn_B}{\nu_B} = \frac{dn_C}{\nu_C} = \frac{dn_D}{\nu_D} = d\varepsilon$$

from which the following expressions are obtained:

$$\begin{aligned} dn_A &= -\nu_A\, d\varepsilon, \qquad & dn_B &= -\nu_B\, d\varepsilon \\ dn_C &= \nu_C\, d\varepsilon, \qquad & dn_D &= \nu_D\, d\varepsilon \end{aligned} \tag{14.25b}$$

The parameter ε is sometimes referred to as the ***extent of reaction.*** *extent of reaction*

For the system under present consideration, Eq. 14.10 takes the form

$$dG]_{T,p} = \mu_A\, dn_A + \mu_B\, dn_B + \mu_C\, dn_C + \mu_D\, dn_D + \mu_E\, dn_E$$

Introducing Eqs. 14.25b and noting that $dn_E = 0$, this becomes

$$dG]_{T,p} = (-\nu_A\mu_A - \nu_B\mu_B + \nu_C\mu_C + \nu_D\mu_D)\,d\varepsilon$$

At equilibrium, $dG]_{T,p} = 0$, so the term in parentheses must be zero. That is

$$-\nu_A\mu_A - \nu_B\mu_B + \nu_C\mu_C + \nu_D\mu_D = 0$$

or when written in a form resembling Eq. 14.24

equation of reaction equilibrium

$$\nu_A\mu_A + \nu_B\mu_B = \nu_C\mu_C + \nu_D\mu_D \tag{14.26}$$

For the present case, Eq. 14.26 is the ***equation of reaction equilibrium.*** In principle, the composition that would be present at equilibrium for a given temperature and pressure can be determined by solving this equation. The solution procedure is simplified through the *equilibrium constant* concept introduced in the next section.

14.3 Calculating Equilibrium Compositions

The objective of the present section is to show how the equilibrium composition of a system at a specified temperature and pressure can be determined by solving the equation of reaction equilibrium. An important part is played in this by the *equilibrium constant.*

14.3.1 Equilibrium Constant for Ideal Gas Mixtures

The first step in solving the equation of reaction equilibrium, Eq. 14.26, for the equilibrium composition is to introduce expressions for the chemical potentials in terms of temperature, pressure, and composition. For an ideal gas mixture, Eq. 14.17 can be used for this purpose. When this expression is introduced for each of the components A, B, C, and D, Eq. 14.26 becomes

$$\nu_A\left(\bar{g}_A^\circ + \bar{R}T\ln\frac{y_A p}{p_{ref}}\right) + \nu_B\left(\bar{g}_B^\circ + \bar{R}T\ln\frac{y_B p}{p_{ref}}\right)$$
$$= \nu_C\left(\bar{g}_C^\circ + \bar{R}T\ln\frac{y_C p}{p_{ref}}\right) + \nu_D\left(\bar{g}_D^\circ + \bar{R}T\ln\frac{y_D p}{p_{ref}}\right) \tag{14.27}$$

where $\bar{g}_i^\circ$ is the Gibbs function of component i evaluated at temperature T and the pressure $p_{ref} = 1$ atm. Equation 14.27 is the basic working relation for chemical equilibrium in a mixture of ideal gases. However, subsequent calculations are facilitated if it is written in an alternative form, as follows.

Collect like terms and rearrange Eq. 14.27 as

$$(\nu_C\bar{g}_C^\circ + \nu_D\bar{g}_D^\circ - \nu_A\bar{g}_A^\circ - \nu_B\bar{g}_B^\circ)$$
$$= -\bar{R}T\left(\nu_C\ln\frac{y_C p}{p_{ref}} + \nu_D\ln\frac{y_D p}{p_{ref}} - \nu_A\ln\frac{y_A p}{p_{ref}} - \nu_B\ln\frac{y_B p}{p_{ref}}\right) \tag{14.28}$$

The term on the left side of Eq. 14.28 can be expressed concisely as ΔG°. That is

$$\Delta G^\circ = \nu_C\bar{g}_C^\circ + \nu_D\bar{g}_D^\circ - \nu_A\bar{g}_A^\circ - \nu_B\bar{g}_B^\circ \tag{14.29a}$$

which is the change in the Gibbs function for the reaction given by Eq. 14.24 if each reactant and product were separate at temperature T and a pressure of 1 atm.

This expression can be written alternatively in terms of specific enthalpies and entropies as

$$\Delta G^\circ = \nu_C(\bar{h}_C - T\bar{s}_C^\circ) + \nu_D(\bar{h}_D - T\bar{s}_D^\circ) - \nu_A(\bar{h}_A - T\bar{s}_A^\circ) - \nu_B(\bar{h}_B - T\bar{s}_B^\circ)$$

$$= (\nu_C\bar{h}_C + \nu_D\bar{h}_D - \nu_A\bar{h}_A - \nu_B\bar{h}_B) - T(\nu_C\bar{s}_C^\circ + \nu_D\bar{s}_D^\circ - \nu_A\bar{s}_A^\circ - \nu_B\bar{s}_B^\circ) \quad (14.29b)$$

Since the enthalpy of an ideal gas depends on temperature only, the $\bar{h}$'s of Eq. 14.29b are evaluated at temperature T. As indicated by the superscript °, each of the entropies is evaluated at temperature T and a pressure of 1 atm.

Introducing Eq. 14.29a into Eq. 14.28 and combining the terms involving logarithms into a single expression gives

$$-\frac{\Delta G^\circ}{\bar{R}T} = \ln\left[\frac{y_C^{\nu_C} y_D^{\nu_D}}{y_A^{\nu_A} y_B^{\nu_B}}\left(\frac{p}{p_{\text{ref}}}\right)^{\nu_C+\nu_D-\nu_A-\nu_B}\right] \quad (14.30)$$

Equation 14.30 is simply the form taken by the equation of reaction equilibrium, Eq. 14.26, for an ideal gas mixture subject to the reaction Eq. 14.24. As illustrated by subsequent examples, similar expressions can be written for other reactions.

Equation 14.30 can be expressed concisely as

$$-\frac{\Delta G^\circ}{\bar{R}T} = \ln K(T) \quad (14.31)$$

where K is the ***equilibrium constant*** defined by

$$K(T) = \frac{y_C^{\nu_C} y_D^{\nu_D}}{y_A^{\nu_A} y_B^{\nu_B}}\left(\frac{p}{p_{\text{ref}}}\right)^{\nu_C+\nu_D-\nu_A-\nu_B} \quad (14.32)$$

equilibrium constant

Given the values of the stoichiometric coefficients, ν_A, ν_B, ν_C, and ν_D and the temperature T, the left side of Eq. 14.31 can be evaluated using either of Eqs. 14.29 together with the appropriate property data. The equation can then be solved for the value of the equilibrium constant K. Accordingly, for selected reactions K can be evaluated and tabulated against temperature. It is common to tabulate $\log_{10}K$ or $\ln K$ versus temperature, however. A tabulation of $\log_{10}K$ values over a range of temperatures for several reactions is provided in Table A-27, which is extracted from a more extensive compilation.

The terms in the numerator and denominator of Eq. 14.32 correspond, respectively, to the products and reactants of the reaction given by Eq. 14.24 as it proceeds from left to right as written. For the *inverse* reaction $\nu_C C + \nu_D D \rightleftarrows \nu_A A + \nu_B B$, the equilibrium constant takes the form

$$K^* = \frac{y_A^{\nu_A} y_B^{\nu_B}}{y_C^{\nu_C} y_D^{\nu_D}}\left(\frac{p}{p_{\text{ref}}}\right)^{\nu_A+\nu_B-\nu_C-\nu_D} \quad (14.33)$$

Comparing Eqs. 14.32 and 14.33, it follows that the value of K^* is just the reciprocal of K: $K^* = 1/K$. Accordingly,

$$\log_{10}K^* = -\log_{10}K \quad (14.34)$$

Hence, Table A-27 can be used both to evaluate K for the reactions listed proceeding in the direction left to right and to evaluate K^* for the inverse reactions proceeding in the direction right to left.

Example 14.1 illustrates how the $\log_{10}K$ values of Table A-27 are determined. Subsequent examples show how the $\log_{10}K$ values can be used to evaluate equilibrium compositions.

Example 14.1 EVALUATING THE EQUILIBRIUM CONSTANT AT A SPECIFIED TEMPERATURE

Evaluate the equilibrium constant, expressed as $\log_{10}K$, for the reaction $CO + \frac{1}{2}O_2 \leftrightarrows CO_2$ at **(a)** 298 K and **(b)** 2000 K. Compare with the value obtained from Table A-27.

Solution

Known: The reaction is $CO + \frac{1}{2}O_2 \rightleftarrows CO_2$.

Find: Determine the equilibrium constant for $T = 298$ K (25°C) and $T = 2000$ K.

Engineering Model: The ideal gas model applies.

Analysis: The equilibrium constant requires the evaluation of $\Delta G°$ for the reaction. Invoking Eq. 14.29b for this purpose, we have

$$\Delta G° = (\bar{h}_{CO_2} - \bar{h}_{CO} - \tfrac{1}{2}\bar{h}_{O_2}) - T(\bar{s}°_{CO_2} - \bar{s}°_{CO} - \tfrac{1}{2}\bar{s}°_{O_2})$$

where the enthalpies are evaluated at temperature T and the absolute entropies are evaluated at temperature T and a pressure of 1 atm. Using Eq. 13.13, the enthalpies are evaluated in terms of the respective enthalpies of formation, giving

$$\Delta G° = [(\bar{h}°_f)_{CO_2} - (\bar{h}°_f)_{CO} - \tfrac{1}{2}\cancelto{0}{(\bar{h}°_f)_{O_2}}] + [(\Delta\bar{h})_{CO_2} - (\Delta\bar{h})_{CO} - \tfrac{1}{2}(\Delta\bar{h})_{O_2}] - T(\bar{s}°_{CO_2} - \bar{s}°_{CO} - \tfrac{1}{2}\bar{s}°_{O_2})$$

where the $\Delta\bar{h}$ terms account for the change in specific enthalpy from $T_{ref} = 298$ K to the specified temperature T. The enthalpy of formation of oxygen is zero by definition.

(a) When $T = 298$ K, the $\Delta\bar{h}$ terms of the above expression for $\Delta G°$ vanish. The required enthalpy of formation and absolute entropy values can be read from Table A-25, giving

$$\Delta G° = [(-393{,}520) - (-110{,}530) - \tfrac{1}{2}(0)] - 298[213.69 - 197.54 - \tfrac{1}{2}(205.03)]$$
$$= -257{,}253 \text{ kJ/kmol}$$

With this value for $\Delta G°$, Eq. 14.31 gives

$$\ln K = -\frac{(-257{,}253 \text{ kJ/kmol})}{(8.314 \text{ kJ/kmol}\cdot\text{K})(298 \text{ K})} = 103.83$$

which corresponds to $\log_{10}K = 45.093$.

Table A-27 gives the logarithm to the base 10 of the equilibrium constant for the inverse reaction: $CO_2 \rightleftarrows CO + \frac{1}{2}O_2$. That is, $\log_{10}K^* = -45.066$. Thus, with Eq. 14.34, $\log_{10}K = 45.066$, which agrees closely with the calculated value.

(b) When $T = 2000$ K, the $\Delta\bar{h}$ and $\bar{s}°$ terms for O_2, CO, and CO_2 required by the above expression for $\Delta G°$ are evaluated from Tables A-23. The enthalpy of formation values are the same as in part (a). Thus

$$\Delta G° = [(-393{,}520) - (-110{,}530) - \tfrac{1}{2}(0)] + [(100{,}804 - 9364) - (65408 - 8669)$$
$$- \tfrac{1}{2}(67{,}881 - 8682)] - 2000[309.210 - 258.600 - \tfrac{1}{2}(268.655)]$$
$$= -282{,}990 + 5102 + 167{,}435 = -110{,}453 \text{ kJ/kmol}$$

With this value, Eq. 14.31 gives

$$\ln K = -\frac{(-110{,}453)}{(8.314)(2000)} = 6.643$$

which corresponds to $\log_{10}K = 2.885$.

At 2000 K, Table A-27 gives $\log_{10}K^* = -2.884$. With Eq. 14.34, $\log_{10}K = 2.884$, which is in agreement with the calculated value.

Using the procedures described above, it is straightforward to determine $\log_{10}K$ versus temperature for each of several specified reactions and tabulate the results as in Table A-27.

Skills Developed: *Ability to...*
- ✓ evaluate $\log_{10}K$ based on Eq. 14.31 and data from Tables A-23 and A-25.
- ✓ use the relation of Eq. 14.34 for inverse reactions.

Quick Quiz

If ln K = 23.535 for the given reaction, use Table A-27 to determine T, in K.
Ans. 1000 K.

14.3.2 Illustrations of the Calculation of Equilibrium Compositions for Reacting Ideal Gas Mixtures

It is often convenient to express Eq. 14.32 explicitly in terms of the number of moles that would be present at equilibrium. Each mole fraction appearing in the equation has the form $y_i = n_i/n$, where n_i is the amount of component i in the equilibrium mixture and n is the total number of moles of mixture. Hence, Eq. 14.32 can be rewritten as

$$K = \frac{n_C^{\nu_C} n_D^{\nu_D}}{n_A^{\nu_A} n_B^{\nu_B}} \left(\frac{p/p_{ref}}{n} \right)^{\nu_C + \nu_D - \nu_A - \nu_B} \tag{14.35}$$

The value of n must include not only the reacting components A, B, C, and D but also all inert components present. Since inert component E has been assumed present, we would write $n = n_A + n_B + n_C + n_D + n_E$.

Equation 14.35 provides a relationship among the temperature, pressure, and composition of an ideal gas mixture at equilibrium. Accordingly, if any two of temperature, pressure, and composition are known, the third can be found by solving this equation. ➡ **FOR EXAMPLE...** suppose that the temperature T and pressure p are known and the object is the equilibrium composition. With temperature known, the value of K can be obtained from Table A-27. The n's of the reacting components A, B, C, and D can be expressed in terms of a single unknown variable through application of the conservation of mass principle to the various chemical species present. Then, since the pressure is known, Eq. 14.35 constitutes a single equation in a single unknown, which can be solved using an *equation solver* or iteratively with a hand calculator. ⬅

In Example 14.2, we apply Eq. 14.35 to study the effect of pressure on the equilibrium composition of a mixture of CO_2, CO, and O_2.

Example 14.2 DETERMINING EQUILIBRIUM COMPOSITION GIVEN TEMPERATURE AND PRESSURE

One kilomole of carbon monoxide, CO, reacts with $\frac{1}{2}$ kmol of oxygen, O_2, to form an equilibrium mixture of CO_2 CO, and O_2 at 2500 K and **(a)** 1 atm, **(b)** 10 atm. Determine the equilibrium composition in terms of mole fractions.

Solution

Known: A system initially consisting of 1 kmol of CO and $\frac{1}{2}$ kmol of O_2 reacts to form an equilibrium mixture of CO_2 CO, and O_2. The temperature of the mixture is 2500 K and the pressure is (a) 1 atm, (b) 10 atm.

Find: Determine the equilibrium composition in terms of mole fractions.

Engineering Model: The equilibrium mixture is modeled as an ideal gas mixture.

Analysis: Equation 14.35 relates temperature, pressure, and composition for an ideal gas mixture at equilibrium. If any two are known, the third can be determined using this equation. In the present case, T and p are known, and the composition is unknown.

Applying conservation of mass, the overall balanced chemical reaction equation is

$$1CO + \frac{1}{2}O_2 \rightarrow zCO + \frac{z}{2}O_2 + (1 - z)CO_2$$

where z is the amount of CO, in kmol, present in the equilibrium mixture. Note that $0 \le z \le 1$.

The total number of moles n in the equilibrium mixture is

$$n = z + \frac{z}{2} + (1 - z) = \frac{2 + z}{2}$$

Accordingly, the molar analysis of the equilibrium mixture is

$$y_{CO} = \frac{2z}{2 + z}, \quad y_{O_2} = \frac{z}{2 + z}, \quad y_{CO_2} = \frac{2(1 - z)}{2 + z}$$

At equilibrium, the tendency of CO and O_2 to form CO_2 is just balanced by the tendency of CO_2 to form CO and O_2, so we have $CO_2 \rightleftarrows CO + \frac{1}{2}O_2$. Accordingly, Eq. 14.35 takes the form

$$K = \frac{z(z/2)^{1/2}}{(1 - z)}\left[\frac{p/p_{ref}}{(2 + z)/2}\right]^{1+1/2-1} = \frac{z}{1 - z}\left(\frac{z}{2 + z}\right)^{1/2}\left(\frac{p}{p_{ref}}\right)^{1/2}$$

At 2500 K, Table A-27 gives $\log_{10}K = -1.44$. Thus, $K = 0.0363$. Inserting this value into the last expression

$$0.0363 = \frac{z}{1 - z}\left(\frac{z}{2 + z}\right)^{1/2}\left(\frac{p}{p_{ref}}\right)^{1/2} \quad \text{(a)}$$

(a) When $p = 1$ atm, Eq. (a) becomes

$$0.0363 = \frac{z}{1 - z}\left(\frac{z}{2 + z}\right)^{1/2}$$

Using an equation solver or iteration with a hand calculator, $z = 0.129$. The equilibrium composition in terms of mole fractions is then

$$y_{CO} = \frac{2(0.129)}{2.129} = 0.121, \quad y_{O_2} = \frac{0.129}{2.129} = 0.061, \quad y_{CO_2} = \frac{2(1 - 0.129)}{2.129} = 0.818$$

(b) When $p = 10$ atm, Eq. (a) becomes

$$0.0363 = \frac{z}{1 - z}\left(\frac{z}{2 + z}\right)^{1/2}(10)^{1/2}$$

❶ Solving, $z = 0.062$. The corresponding equilibrium composition in terms of mole fractions is $y_{CO} = 0.06$, $y_{O_2} = 0.03$, $y_{CO_2} = 0.91$.

---

❶ Comparing the results of parts (a) and (b) we conclude that the extent to which the reaction proceeds toward completion (the extent to which CO_2 is formed) is increased by increasing the pressure.

Skills Developed: *Ability to...*
- ✓ apply Eq. 14.35 to determine equilibrium composition given temperature and pressure.
- ✓ retrieve and use data from Table A-27.

Quick Quiz

If $z = 0.0478$ (corresponding to $p = 22.4$ atm, $T = 2500$ K), what would be the mole fraction of each constituent of the equilibrium mixture?

Ans. $y_{CO} = 0.0467$, $y_{O_2} = 0.0233$, $y_{CO_2} = 0.9300$.

In Example 14.3, we determine the temperature of an equilibrium mixture when the pressure and composition are known.

Example 14.3 DETERMINING EQUILIBRIUM TEMPERATURE GIVEN PRESSURE AND COMPOSITION

Measurements show that at a temperature T and a pressure of 1 atm, the equilibrium mixture for the system of Example 14.2 has the composition $y_{CO} = 0.298$, $y_{O_2} = 0.149$, $y_{CO_2} = 0.553$. Determine the temperature T of the mixture, in K.

Solution

Known: The pressure and composition of an equilibrium mixture of CO, O_2, and CO_2 are specified.

Find: Determine the temperature of the mixture, in K.

Engineering Model: The mixture can be modeled as an ideal gas mixture.

Analysis: Equation 14.35 relates temperature, pressure, and composition for an ideal gas mixture at equilibrium. If any two are known, the third can be found using this equation. In the present case, composition and pressure are known, and the temperature is the unknown.

Equation 14.35 takes the same form here as in Example 14.2. Thus, when $p = 1$ atm, we have

$$K(T) = \frac{z}{1-z}\left(\frac{z}{2+z}\right)^{1/2}$$

where z is the amount of CO, in kmol, present in the equilibrium mixture and T is the temperature of the mixture.

The solution to Example 14.2 gives the following expression for the mole fraction of the CO in the mixture: $y_{CO} = 2z/(2+z)$. Since $y_{CO} = 0.298$, $z = 0.35$.

Inserting this value for z into the expression for the equilibrium constant gives $K = 0.2078$. Thus, $\log_{10} K = -0.6824$.

❶ Interpolation in Table A-27 then gives $T = 2881$ K.

---

❶ Comparing this example with part (a) of Example 14.2, we conclude that the extent to which the reaction proceeds to completion (the extent to which CO_2 is formed) is decreased by increasing the temperature.

Skills Developed: *Ability to...*

- ✓ apply Eq. 14.35 to determine temperature given pressure and equilibrium composition.
- ✓ retrieve and use data from Table A-27.

Quick Quiz

Determine the temperature, in K, for a pressure of 2 atm if the equilibrium composition were unchanged.

Ans. ≈2970 K.

In Example 14.4, we consider the effect of an inert component on the equilibrium composition.

Example 14.4 CONSIDERING THE EFFECT OF AN INERT COMPONENT

One kilomole of carbon monoxide reacts with the theoretical amount of air to form an equilibrium mixture of CO_2, CO, O_2, and N_2 at 2500 K and 1 atm. Determine the equilibrium composition in terms of mole fractions, and compare with the result of Example 14.2.

Solution

Known: A system initially consisting of 1 kmol of CO and the theoretical amount of air reacts to form an equilibrium mixture of CO_2, CO, O_2, and N_2. The temperature and pressure of the mixture are 2500 K and 1 atm.

Find: Determine the equilibrium composition in terms of mole fractions, and compare with the result of Example 14.2.

Engineering Model: The equilibrium mixture can be modeled as an ideal gas mixture.

Analysis: For a *complete reaction* of CO with the theoretical amount of air

$$CO + \tfrac{1}{2}O_2 + 1.88N_2 \rightarrow CO_2 + 1.88N_2$$

Accordingly, the reaction of CO with the theoretical amount of air to form CO_2, CO, O_2, and N_2 is

$$CO + \tfrac{1}{2}O_2 + 1.88N_2 \rightarrow zCO + \frac{z}{2}O_2 + (1-z)CO_2 + 1.88N_2$$

where z is the amount of CO, in kmol, present in the equilibrium mixture.

The total number of moles n in the equilibrium mixture is

$$n = z + \frac{z}{2} + (1-z) + 1.88 = \frac{5.76 + z}{2}$$

The composition of the equilibrium mixture in terms of mole fractions is

$$y_{CO} = \frac{2z}{5.76+z}, \quad y_{O_2} = \frac{z}{5.76+z}, \quad y_{CO_2} = \frac{2(1-z)}{5.76+z}, \quad y_{N_2} = \frac{3.76}{5.76+z}$$

At equilibrium we have $CO_2 \rightleftarrows CO + \tfrac{1}{2}O_2$. So Eq. 14.35 takes the form

$$K = \frac{z(z/2)^{1/2}}{(1-z)}\left[\frac{p/p_{ref}}{(5.76+z)/2}\right]^{1/2}$$

The value of K is the same as in the solution to Example 14.2, $K = 0.0363$. Thus, since $p = 1$ atm, we have

$$0.0363 = \frac{z}{1-z}\left(\frac{z}{5.76+z}\right)^{1/2}$$

Solving, $z = 0.175$. The corresponding equilibrium composition is $y_{CO} = 0.059$, $y_{CO_2} = 0.278$, $y_{O_2} = 0.029$, $y_{N_2} = 0.634$.

Comparing this example with Example 14.2, we conclude that the presence of the inert component nitrogen reduces the extent to which the reaction proceeds toward completion at the specified temperature and pressure (reduces the extent to which CO_2 is formed).

Skills Developed: *Ability to...*

- ✓ apply Eq. 14.35 to determine equilibrium composition for given temperature and pressure in the presence of an inert component.
- ✓ retrieve and use data from Table A-27.

Quick Quiz

Determine the amounts, in kmol, of each component of the equilibrium mixture.

Ans. $n_{CO} = 0.175$, $n_{O_2} = 0.0875$, $n_{CO_2} = 0.8250$, $n_{N_2} = 1.88$.

In the next example, the equilibrium concepts of this chapter are applied together with the energy balance for reacting systems developed in Chap. 13.

Example 14.5 USING EQUILIBRIUM CONCEPTS AND THE ENERGY BALANCE

Carbon dioxide at 25°C, 1 atm enters a reactor operating at steady state and dissociates, giving an equilibrium mixture of CO_2, CO, and O_2 that exits at 3200 K, 1 atm. Determine the heat transfer to the reactor, in kJ per kmol of CO_2 entering. The effects of kinetic and potential energy can be ignored and $\dot{W}_{cv} = 0$.

Solution

Known: Carbon dioxide at 25°C, 1 atm enters a reactor at steady state. An equilibrium mixture of CO_2, CO, and O_2 exits at 3200 K, 1 atm.

Find: Determine the heat transfer to the reactor, in kJ per kmol of CO_2 entering.

Schematic and Given Data:

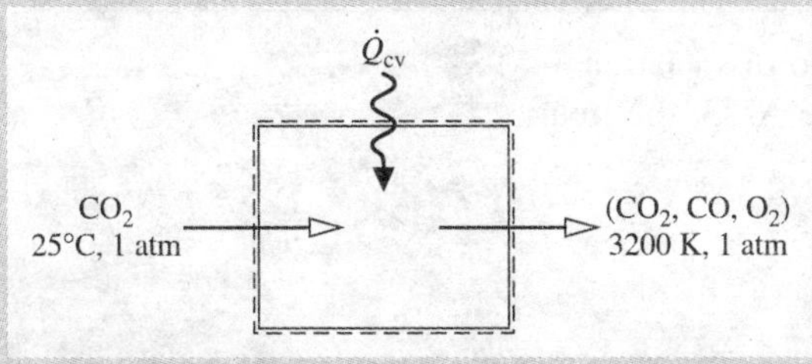

Fig. E14.5

Engineering Model:

1. The control volume shown on the accompanying sketch by a dashed line operates at steady state with $\dot{W}_{cv} = 0$. Kinetic energy and potential energy effects can be ignored.
2. The entering CO_2 is modeled as an ideal gas.
3. The exiting mixture of CO_2, CO, and O_2 is an equilibrium ideal gas mixture.

Analysis: The required heat transfer can be determined from an energy rate balance for the control volume, but first the composition of the exiting equilibrium mixture must be determined.

Applying the conservation of mass principle, the overall dissociation reaction is described by

$$CO_2 \rightarrow zCO_2 + (1-z)CO + \left(\frac{1-z}{2}\right)O_2$$

where z is the amount of CO_2, in kmol, present in the mixture exiting the control volume, per kmol of CO_2 entering. The total number of moles n in the mixture is then

$$n = z + (1-z) + \left(\frac{1-z}{2}\right) = \frac{3-z}{2}$$

The exiting mixture is assumed to be an equilibrium mixture (assumption 3). Thus, for the mixture we have $CO_2 \rightleftarrows CO + \frac{1}{2}O_2$. Equation 14.35 takes the form

$$K = \frac{(1-z)[(1-z)/2]^{1/2}}{z}\left[\frac{p/p_{ref}}{(3-z)/2}\right]^{1+1/2-1}$$

Rearranging and noting that $p = 1$ atm

$$K = \left(\frac{1-z}{z}\right)\left(\frac{1-z}{3-z}\right)^{1/2}$$

At 3200 K, Table A-27 gives $\log_{10}K = -0.189$. Thus, $K = 0.647$, and the equilibrium constant expression becomes

$$0.647 = \left(\frac{1-z}{z}\right)\left(\frac{1-z}{3-z}\right)^{1/2}$$

Solving, $z = 0.422$. The composition of the exiting equilibrium mixture, in kmol per kmol of CO_2 entering, is then $(0.422CO_2, 0.578CO, 0.289O_2)$.

When expressed per kmol of CO_2 entering the control volume, the energy rate balance reduces by assumption 1 to

$$0 = \frac{\dot{Q}_{cv}}{\dot{n}_{CO_2}} - \frac{\cancelto{0}{\dot{W}_{cv}}}{\dot{n}_{CO_2}} + \bar{h}_{CO_2} - (0.422\bar{h}_{CO_2} + 0.578\bar{h}_{CO} + 0.289\bar{h}_{O_2})$$

Solving for the heat transfer per kmol of CO_2 entering, and evaluating each enthalpy in terms of the respective enthalpy of formation

$$\frac{\dot{Q}_{cv}}{\dot{n}_{CO_2}} = 0.422(\bar{h}_f^\circ + \Delta\bar{h})_{CO_2} + 0.578(\bar{h}_f^\circ + \Delta\bar{h})_{CO} + 0.289(\cancelto{0}{\bar{h}_f^\circ} + \Delta\bar{h})_{O_2} - (\bar{h}_f^\circ + \cancelto{0}{\Delta\bar{h}})_{CO_2}$$

The enthalpy of formation of O_2 is zero by definition; $\Delta\bar{h}$ for the CO_2 at the inlet vanishes because CO_2 enters at 25°C. With enthalpy of formation values from Tables A-25 and $\Delta\bar{h}$ values for O_2, CO, and CO_2, from Table A-23

$$\frac{\dot{Q}_{cv}}{\dot{n}_{CO_2}} = 0.422[-393{,}520 + (174{,}695 - 9364)] + 0.578[-110{,}530 + (109{,}667 - 8669)]$$
$$+ 0.289(114{,}809 - 8682) - (-393{,}520)$$
$$= 322{,}385 \text{ kJ/kmol } (CO_2)$$

❶

❶ For comparison, let us determine the heat transfer if we assume no dissociation — namely, when CO_2 alone exits the reactor. With data from Table A-23, the heat transfer is

$$\frac{\dot{Q}_{cv}}{\dot{n}_{CO_2}} = \bar{h}_{CO_2}(3200 \text{ K}) - \bar{h}_{CO_2}(298 \text{ K})$$
$$= 174{,}695 - 9364 = 165{,}331 \text{ kJ/kmol}(CO_2)$$

The value is much less than the value obtained in the solution above because the dissociation of CO_2 requires more energy input (an endothermic reaction).

Skills Developed: *Ability to...*
✓ apply Eq. 14.35 together with the energy balance for reacting systems to determine heat transfer for a reactor.
✓ retrieve and use data from Tables A-23, A-25, and A-27.

Quick Quiz

Determine the heat transfer rate, in kW, and the molar flow rate of mixture exiting, in kmol/s, for a flow rate of 3.1×10^{-5} kmol/s of CO_2 entering.

Ans. 10 kW, 4×10^{-5} kmol/s.

14.3.3 Equilibrium Constant for Mixtures and Solutions[1]

The procedures that led to the equilibrium constant for reacting ideal gas mixtures can be followed for the general case of reacting mixtures by using the fugacity and activity concepts introduced in Sec. 11.9. In principle, equilibrium compositions of such mixtures can be determined with an approach paralleling the one for ideal gas mixtures.

Equation 11.141 can be used to evaluate the chemical potentials appearing in the equation of reaction equilibrium (Eq. 14.26). The result is

$$\nu_A(\bar{g}_A^\circ + \bar{R}T \ln a_A) + \nu_B(\bar{g}_B^\circ + \bar{R}T \ln a_B) = \nu_C(\bar{g}_C^\circ + \bar{R}T \ln a_C) + \nu_D(\bar{g}_D^\circ + \bar{R}T \ln a_D) \tag{14.36}$$

where $\bar{g}_i^\circ$ is the Gibbs function of pure component i at temperature T and the pressure $p_{ref} = 1$ atm, and a_i is the *activity* of that component.

Collecting terms and employing Eq. 14.29a, Eq. 14.36 becomes

$$-\frac{\Delta G^\circ}{\bar{R}T} = \ln\left(\frac{a_C^{\nu_C} a_D^{\nu_D}}{a_A^{\nu_A} a_B^{\nu_B}}\right) \tag{14.37}$$

This equation can be expressed in the same form as Eq. 14.31 by defining the equilibrium constant as

$$K = \frac{a_C^{\nu_C} a_D^{\nu_D}}{a_A^{\nu_A} a_B^{\nu_B}} \tag{14.38}$$

[1]This section requires study of Sec. 11.9.

Since Table A-27 and similar compilations are constructed simply by evaluating $-\Delta G^\circ/\bar{R}T$ for specified reactions at several temperatures, such tables can be employed to evaluate the more general equilibrium constant given by Eq. 14.38. However, before Eq. 14.38 can be used to determine the equilibrium composition for a known value of K, it is necessary to evaluate the activity of the various mixture components. Let us illustrate this for the case of mixtures that can be modeled as *ideal solutions*.

IDEAL SOLUTIONS. For an ideal solution, the activity of component i is given by

$$a_i = \frac{y_i f_i}{f_i^\circ} \tag{11.142}$$

where f_i is the fugacity of pure i at the temperature T and pressure p of the mixture, and f_i° is the fugacity of pure i at temperature T and the pressure p_{ref}. Using this expression to evaluate a_A, a_B, a_C, and a_D, Eq. 14.38 becomes

$$K = \frac{(y_C f_C/f_C^\circ)^{\nu_C}(y_D f_D/f_D^\circ)^{\nu_D}}{(y_A f_A/f_A^\circ)^{\nu_A}(y_B f_B/f_B^\circ)^{\nu_B}} \tag{14.39a}$$

which can be expressed alternatively as

$$K = \left[\frac{(f_C/p)^{\nu_C}(f_D/p)^{\nu_D}}{(f_A/p)^{\nu_A}(f_B/p)^{\nu_B}}\right]\left[\frac{(f_A^\circ/p_{\text{ref}})^{\nu_A}(f_B^\circ/p_{\text{ref}})^{\nu_B}}{(f_C^\circ/p_{\text{ref}})^{\nu_C}(f_D^\circ/p_{\text{ref}})^{\nu_D}}\right]\left[\underline{\frac{y_C^{\nu_C} y_D^{\nu_D}}{y_A^{\nu_A} y_B^{\nu_B}}\left(\frac{p}{p_{\text{ref}}}\right)^{\nu_C+\nu_D-\nu_A-\nu_B}}\right] \tag{14.39b}$$

The ratios of fugacity to pressure in this equation can be evaluated, in principle, from Eq. 11.124 or the generalized fugacity chart, Fig. A-6, developed from it. In the special case when each component behaves as an ideal gas at both T, p and T, p_{ref}, these ratios equal unity and Eq. 14.39b reduces to the underlined term, which is just Eq. 14.32.

14.4 Further Examples of the Use of the Equilibrium Constant

Some additional aspects of the use of the equilibrium constant are introduced in this section: the equilibrium flame temperature, the van't Hoff equation, and chemical equilibrium for ionization reactions and simultaneous reactions. To keep the presentation at an introductory level, only the case of ideal gas mixtures is considered.

14.4.1 Determining Equilibrium Flame Temperature

In this section, the effect of incomplete combustion on the adiabatic flame temperature, introduced in Sec. 13.3, is considered using concepts developed in the present chapter. We begin with a review of some ideas related to the adiabatic flame temperature by considering a reactor operating at steady state for which no significant heat transfer with the surroundings takes place.

Let carbon monoxide gas entering at one location react *completely* with the theoretical amount of air entering at another location as follows:

$$CO + \tfrac{1}{2}O_2 + 1.88N_2 \rightarrow CO_2 + 1.88N_2$$

As discussed in Sec. 13.3, the products would exit the reactor at a temperature we have designated the *maximum* adiabatic flame temperature. This temperature can be determined by solving a *single* equation, the energy equation. At such an elevated temperature, however, there would be a tendency for CO_2 to dissociate

$$CO_2 \rightarrow CO + \tfrac{1}{2}O_2$$

Since dissociation requires energy (an endothermic reaction), the temperature of the products would be *less than* the maximum adiabatic temperature found under the assumption of complete combustion.

When dissociation takes place, the gaseous products exiting the reactor would not be CO_2 and N_2, but a mixture of CO_2, CO, O_2, and N_2. The balanced chemical reaction equation would read

$$CO + \tfrac{1}{2}O_2 + 1.88N_2 \rightarrow z\,CO + (1 - z)CO_2 + \frac{z}{2}O_2 + 1.88N_2 \qquad (14.40)$$

where z is the amount of CO, in kmol, present in the exiting mixture for each kmol of CO entering the reactor.

Accordingly, there are *two* unknowns: z and the temperature of the exiting stream. To solve a problem with two unknowns requires two equations. One is provided by an energy equation. If the exiting gas mixture is in equilibrium, the other equation is provided by the equilibrium constant, Eq. 14.35. The temperature of the products may then be called the ***equilibrium flame temperature.*** The equilibrium constant used to evaluate the equilibrium flame temperature would be determined with respect to $CO_2 \rightleftarrows CO + \tfrac{1}{2}O_2$.

equilibrium flame temperature

Although only the dissociation of CO_2 has been discussed, other products of combustion may dissociate, for example

$$H_2O \rightleftarrows H_2 + \tfrac{1}{2}O_2$$
$$H_2O \rightleftarrows OH + \tfrac{1}{2}H_2$$
$$O_2 \rightleftarrows 2O$$
$$N_2 \rightleftarrows 2H$$
$$N_2 \rightleftarrows 2N$$

When there are many dissociation reactions, the study of chemical equilibrium is facilitated by the use of computers to solve the *simultaneous* equations that result. Simultaneous reactions are considered in Sec. 14.4.4. The following example illustrates how the equilibrium flame temperature is determined when one dissociation reaction occurs.

Example 14.6 DETERMINING THE EQUILIBRIUM FLAME TEMPERATURE

Carbon monoxide at 25°C, 1 atm enters a well-insulated reactor and reacts with the theoretical amount of air entering at the same temperature and pressure. An equilibrium mixture of CO_2, CO, O_2, and N_2 exits the reactor at a pressure of 1 atm. For steady-state operation and negligible effects of kinetic and potential energy, determine the composition and temperature of the exiting mixture in K.

Solution

Known: Carbon monoxide at 25°C, 1 atm reacts with the theoretical amount of air at 25°C, 1 atm to form an equilibrium mixture of CO_2, CO, O_2, and N_2 at temperature T and a pressure of 1 atm.

Find: Determine the composition and temperature of the exiting mixture.

Schematic and Given Data:

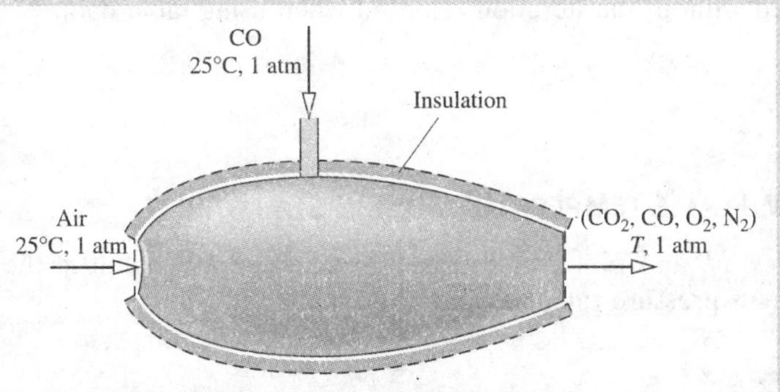

Engineering Model:

1. The control volume shown on the accompanying sketch by a dashed line operates at steady state with $\dot{Q}_{cv} = 0$, $\dot{W}_{cv} = 0$, and negligible effects of kinetic and potential energy.
2. The entering gases are modeled as ideal gases.
3. The exiting mixture is an ideal gas mixture at equilibrium.

Fig. E14.6

Analysis: The overall reaction is the same as in the solution to Example 14.4

$$\text{CO} + \tfrac{1}{2}\text{O}_2 + 1.88\text{N}_2 \rightarrow z\,\text{CO} + \frac{z}{2}\text{O}_2 + (1 - z)\text{CO}_2 + 1.88\text{N}_2$$

By assumption 3, the exiting mixture is an equilibrium mixture. The equilibrium constant expression developed in the solution to Example 14.4 is

$$K(T) = \frac{z(z/2)^{1/2}}{(1 - z)}\left(\frac{p/p_{\text{ref}}}{(5.76 + z)/2}\right)^{1/2} \quad \text{(a)}$$

Since $p = 1$ atm, Eq. (a) reduces to

$$K(T) = \frac{z}{(1 - z)}\left(\frac{z}{5.76 + z}\right)^{1/2} \quad \text{(b)}$$

This equation involves two unknowns: z and the temperature T of the exiting equilibrium mixture.

Another equation involving the two unknowns is obtained from the energy rate balance, Eq. 13.12b, which reduces with assumption 1 to

$$\bar{h}_R = \bar{h}_P \quad \text{(c)}$$

where

$$\bar{h}_R = (\bar{h}_f^\circ + \overset{0}{\cancel{\Delta\bar{h}}})_{CO} + \frac{1}{2}(\overset{0}{\cancel{\bar{h}_f^\circ}} + \overset{0}{\cancel{\Delta\bar{h}}})_{O_2} + 1.88(\overset{0}{\cancel{\bar{h}_f^\circ}} + \overset{0}{\cancel{\Delta\bar{h}}})_{N_2}$$

and

$$\bar{h}_P = z(\bar{h}_f^\circ + \Delta\bar{h})_{CO} + \frac{z}{2}(\overset{0}{\cancel{\bar{h}_f^\circ}} + \Delta\bar{h})_{O_2} + (1 - z)(\bar{h}_f^\circ + \Delta\bar{h})_{CO_2} + 1.88(\overset{0}{\cancel{\bar{h}_f^\circ}} + \Delta\bar{h})_{N_2}$$

The enthalpy of formation terms set to zero are those for oxygen and nitrogen. Since the reactants enter at 25°C, the corresponding $\Delta\bar{h}$ terms also vanish. Collecting and rearranging, we get

$$z(\Delta\bar{h})_{CO} + \frac{z}{2}(\Delta\bar{h})_{O_2} + (1 - z)(\Delta\bar{h})_{CO_2} + 1.88(\Delta\bar{h})_{N_2} + (1 - z)[(\bar{h}_f^\circ)_{CO_2} - (\bar{h}_f^\circ)_{CO}] = 0 \quad \text{(d)}$$

Equations (b) and (d) are simultaneous equations involving the unknowns z and T. When solved *iteratively* using *tabular data*, the results are $z = 0.125$ and $T = 2399$ K, as can be verified. The composition of the equilibrium mixture, in kmol per kmol of CO entering the reactor, is then 0.125CO, 0.0625O_2, 0.875CO_2, 1.88N_2.

Skills Developed: *Ability to...*
- ✓ apply Eq. 14.35 together with the energy balance for reacting systems to determine equilibrium flame temperature.
- ✓ retrieve and use data from Tables A-23, A-25, and A-27.

Quick Quiz

If the CO and air each entered at 500°C, would the equilibrium flame temperature increase, decrease, or stay constant?

Ans. Increase.

As illustrated by Example 14.7, the equation solver and property retrieval features of *Interactive Thermodynamics: IT* allow the equilibrium flame temperature and composition to be determined without the iteration required when using table data.

Example 14.7 DETERMINING THE EQUILIBRIUM FLAME TEMPERATURE USING SOFTWARE

Solve Example 14.6 using *Interactive Thermodynamics: IT* and plot equilibrium flame temperature and z, the amount of CO present in the exiting mixture, each versus pressure ranging from 1 to 10 atm.

Solution

Known: See Example 14.6.

Find: Using *IT*, plot the equilibrium flame temperature and the amount of CO present in the exiting mixture of Example 14.6, each versus pressure ranging from 1 to 10 atm.

Engineering Model: See Example 14.6.

Analysis: Equation (a) of Example 14.6 provides the point of departure for the *IT* solution

$$K(T) = \frac{z(z/2)^{1/2}}{(1 - z)}\left[\frac{p/p_{\text{ref}}}{(5.76 + z)/2}\right]^{1/2} \tag{a}$$

For a given pressure, this expression involves two unknowns: z and T.

Also, from Example 14.6, we use the energy balance, Eq. (c)

$$\bar{h}_R = \bar{h}_P \tag{c}$$

where

$$\bar{h}_R = (\bar{h}_{CO})_R + \tfrac{1}{2}(\bar{h}_{O_2})_R + 1.88(\bar{h}_{N_2})_R$$

and

$$\bar{h}_P = z(\bar{h}_{CO})_P + (z/2)(\bar{h}_{O_2})_P + (1 - z)(\bar{h}_{CO_2})_P + 1.88(\bar{h}_{N_2})_P$$

where the subscripts R and P denote reactants and products, respectively, and z denotes the amount of CO in the products, in kmol per kmol of CO entering.

With pressure known, Eqs. (a) and (c) can be solved for T and z using the following *IT* code. Choosing SI from the **Units** menu and amount of substance in moles, and letting **hCO_R** denote the specific enthalpy of CO in the reactants, and so on, we have

```
// Given data
TR = 25 + 273.15 // K
p = 1 // atm
pref = 1 // atm

// Evaluating the equilibrium constant using Eq. (a)
K = ((z * (z/2)^0.5) / (1 – z)) * ((p / pref) / ((5.76 + z) / 2))^0.5

// Energy balance: Eq. (c)
hR = hP
hR = hCO_R + (1/2) * hO2_R + 1.88 * hN2_R
hP = z * hCO_P + (z /2) * hO2_P + (1 – z) * hCO2_P + 1.88 * hN2_P

hCO_R = h_T("CO",TR)
hO2_R = h_T("O2",TR)
hN2_R = h_T("N2",TR)
```

```
hCO_P = h_T("CO",T)
hO2_P = h_T("O2",T)
hCO2_P = h_T ("CO2",T)
hN2_P = h_T ("N2",T)

/* To obtain data for the equilibrium constant use the Look-up Table
option under the Edit menu. Load the file "eqco2.lut". Data for
CO2 ⇄ CO + 1/2 O2 from Table A-27 are stored in the look-up table
as T in column 1 and log10(K) in column 2. To retrieve the data use */

log(K) = lookupvall(eqco2, 1, T,2)
```

❶ (marker beside the comment above)

Obtain a solution for $p = 1$ using the **Solve** button. To ensure rapid convergence, restrict T and K to positive values, and set a lower limit of 0.001 and an upper limit of 0.999 for z. The results are $T = 2399$ K and $z = 0.1249$, which agree with the values obtained in Example 14.6.

Now, use the **Explore** button and sweep p from 1 to 10 atm in steps of 0.01. Using the **Graph** button, construct the following plots:

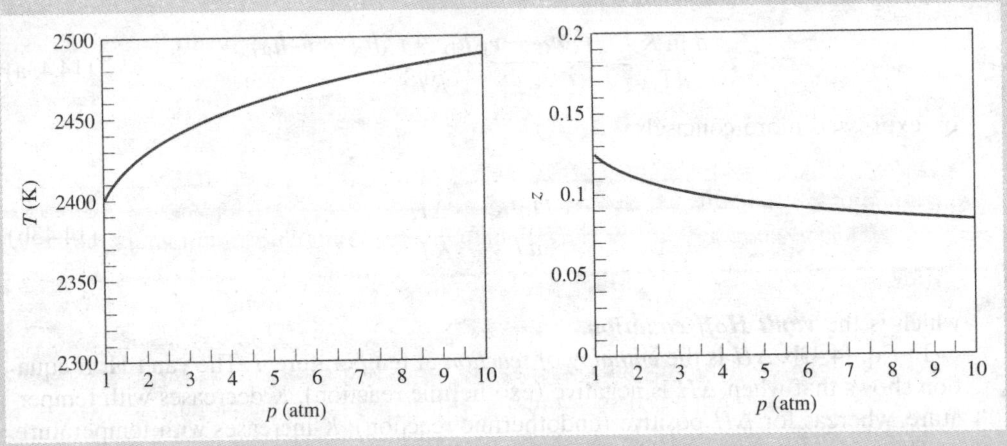

Fig. E14.7

From Fig. E14.7, we see that as pressure increases more CO is oxidized to CO_2 (z decreases) and temperature increases.

❶ Similar files are included in *IT* for each of the reactions in Table A-27.

Skills Developed: *Ability to...*

✓ apply Eq. 14.35 together with the energy balance for reacting systems to determine equilibrium flame temperature.

✓ perform equilibrium calculations using *Interactive Thermodynamics: IT.*

Quick Quiz

If the CO and air each entered at 500°C, determine the equilibrium flame temperature in K using *Interactive Thermodynamics: IT.*

Ans. 2575.

14.4.2 Van't Hoff Equation

The dependence of the equilibrium constant on temperature exhibited by the values of Table A-27 follows from Eq. 14.31. An alternative way to express this dependence is given by the van't Hoff equation, Eq. 14.43b.

The development of this equation begins by introducing Eq. 14.29b into Eq. 14.31 to obtain on rearrangement

$$\bar{R}T \ln K = -[(\nu_C \bar{h}_C + \nu_D \bar{h}_D - \nu_A \bar{h}_A - \nu_B \bar{h}_B) - T(\nu_C \bar{s}^\circ_C + \nu_D \bar{s}^\circ_D - \nu_A \bar{s}^\circ_A - \nu_B \bar{s}^\circ_B)] \tag{14.41}$$

Each of the specific enthalpies and entropies in this equation depends on temperature alone. Differentiating with respect to temperature

$$\bar{R}T\frac{d\ln K}{dT} + \bar{R}\ln K = -\left[\underline{\nu_C\left(\frac{d\bar{h}_C}{dT} - T\frac{d\bar{s}^\circ_C}{dT}\right)} + \underline{\nu_D\left(\frac{d\bar{h}_D}{dT} - T\frac{d\bar{s}^\circ_D}{dT}\right)}\right.$$
$$\left.\underline{-\nu_A\left(\frac{d\bar{h}_A}{dT} - T\frac{d\bar{s}^\circ_A}{dT}\right)} - \underline{\nu_B\left(\frac{d\bar{h}_B}{dT} - T\frac{d\bar{s}^\circ_B}{dT}\right)}\right]$$
$$+ (\nu_C\bar{s}^\circ_C + \nu_D\bar{s}^\circ_D - \nu_A\bar{s}^\circ_A - \nu_B\bar{s}^\circ_B)$$

From the definition of $\bar{s}^\circ(T)$ (Eq. 6.19), we have $d\bar{s}^\circ/dT = \bar{c}_p/T$. Moreover, $d\bar{h}/dT = \bar{c}_p$. Accordingly, each of the underlined terms in the above equation vanishes identically, leaving

$$\bar{R}T\frac{d\ln K}{dT} + \bar{R}\ln K = (\nu_C\bar{s}^\circ_C + \nu_D\bar{s}^\circ_D - \nu_A\bar{s}^\circ_A - \nu_B\bar{s}^\circ_B) \tag{14.42}$$

Using Eq. 14.41 to evaluate the second term on the left and simplifying the resulting expression, Eq. 14.42 becomes

$$\frac{d\ln K}{dT} = \frac{(\nu_C\bar{h}_C + \nu_D\bar{h}_D - \nu_A\bar{h}_A - \nu_B\bar{h}_B)}{\bar{R}T^2} \tag{14.43a}$$

or, expressed more concisely

van't Hoff equation

$$\frac{d\ln K}{dT} = \frac{\Delta H}{\bar{R}T^2} \tag{14.43b}$$

which is the ***van't Hoff equation.***

In Eq. 14.43b, ΔH is the *enthalpy of reaction* at temperature T. The van't Hoff equation shows that when ΔH is negative (exothermic reaction), K decreases with temperature, whereas for ΔH positive (endothermic reaction), K increases with temperature.

The enthalpy of reaction ΔH is often very nearly constant over a rather wide interval of temperature. In such cases, Eq. 14.43b can be integrated to yield

$$\ln\frac{K_2}{K_1} = -\frac{\Delta H}{\bar{R}}\left(\frac{1}{T_2} - \frac{1}{T_1}\right) \tag{14.44}$$

where K_1 and K_2 denote the equilibrium constants at temperatures T_1 and T_2, respectively. This equation shows that $\ln K$ is linear in $1/T$. Accordingly, plots of $\ln K$ versus $1/T$ can be used to determine ΔH from experimental equilibrium composition data. Alternatively, the equilibrium constant can be determined using enthalpy data.

14.4.3 Ionization

The methods developed for determining the equilibrium composition of a reactive ideal gas mixture can be applied to systems involving ionized gases, also known as *plasmas*. In previous sections we considered the chemical equilibrium of systems where dissociation is a factor. For example, the dissociation reaction of diatomic nitrogen

$$N_2 \rightleftarrows 2N$$

can occur at elevated temperatures. At still higher temperatures, ionization may take place according to

$$N \rightleftarrows N^+ + e^- \tag{14.45}$$

That is, a nitrogen atom loses an electron, yielding a singly ionized nitrogen atom N^+ and a free electron e^-. Further heating can result in the loss of additional electrons until all electrons have been removed from the atom.

For some cases of practical interest, it is reasonable to think of the neutral atoms, positive ions, and electrons as forming an ideal gas mixture. With this idealization, ionization equilibrium can be treated in the same manner as the chemical equilibrium of reacting ideal gas mixtures. The change in the Gibbs function for the equilibrium ionization reaction required to evaluate the ionization-equilibrium constant can be calculated as a function of temperature by using the procedures of statistical thermodynamics. In general, the extent of ionization increases as the temperature is raised and the pressure is lowered.

Example 14.8 illustrates the analysis of ionization equilibrium.

Example 14.8 CONSIDERING IONIZATION EQUILIBRIUM

Consider an equilibrium mixture at 2000 K consisting of Cs, Cs^+, and e^-, where Cs denotes neutral cesium atoms, Cs^+ singly ionized cesium ions, and e^- free electrons. The ionization-equilibrium constant at this temperature for

$$Cs \rightleftarrows Cs^+ + e^-$$

is $K = 15.63$. Determine the pressure, in atmospheres, if the ionization of Cs is 95% complete, and plot percent completion of ionization versus pressure ranging from 0 to 10 atm.

Solution

Known: An equilibrium mixture of Cs, Cs^+, e^- is at 2000 K. The value of the equilibrium constant at this temperature is known.

Find: Determine the pressure of the mixture if the ionization of Cs is 95% complete. Plot percent completion versus pressure.

Engineering Model: Equilibrium can be treated in this case using ideal gas mixture equilibrium considerations.

Analysis: The ionization of cesium to form a mixture of Cs, Cs^+, and e^- is described by

$$Cs \rightarrow (1 - z)Cs + zCs^+ + ze^-$$

where z denotes the extent of ionization, ranging from 0 to 1. The total number of moles of mixture n is

$$n = (1 - z) + z + z = 1 + z$$

At equilibrium, we have $Cs \rightleftarrows Cs^+ + e^-$, so Eq. 14.35 takes the form

$$K = \frac{(z)(z)}{(1 - z)}\left[\frac{p/p_{ref}}{(1 + z)}\right]^{1+1-1} = \left(\frac{z^2}{1 - z^2}\right)\left(\frac{p}{p_{ref}}\right) \qquad (1)$$

Solving for the ratio p/p_{ref} and introducing the known value of K

$$\frac{p}{p_{ref}} = (15.63)\left(\frac{1 - z^2}{z^2}\right)$$

For p_{ref} = 1 atm and z = 0.95 (95%), p = 1.69 atm. Using an equation-solver and plotting package, the following plot can be constructed:

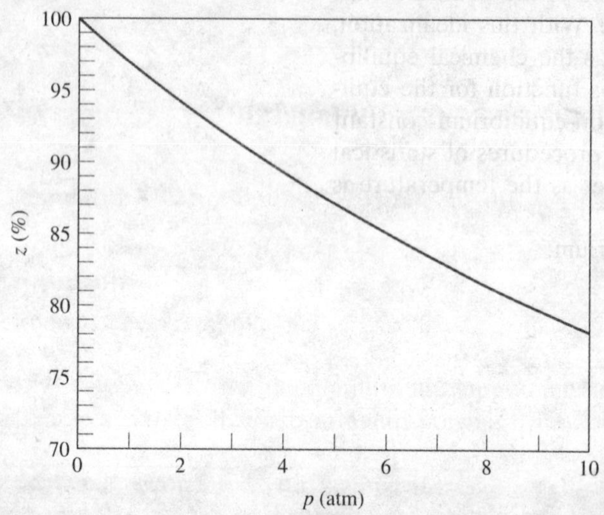

Fig. E14.8

Skills Developed: Ability to...
✓ apply Eq. 14.35 to determine the extent of ionization of cesium given temperature and pressure.

Figure E14.8 shows that ionization tends to occur to a lesser extent as pressure is raised. Ionization also tends to occur to a greater extent as temperature is raised at fixed pressure.

Quick Quiz

Solving Eq. (1) for z, determine the percent of ionization of Cs at T = 1600 K (K = 0.78) and p = 1 atm.
Ans. 66.2%.

14.4.4 Simultaneous Reactions

Let us return to the discussion of Sec. 14.2 and consider the possibility of more than one reaction among the substances present within a system. For the present application, the closed system is assumed to contain a mixture of *eight* components A, B, C, D, E, L, M, and N, subject to *two* independent reactions

$$\text{(1)} \quad \nu_A A + \nu_B B \rightleftarrows \nu_C C + \nu_D D \tag{14.24}$$

$$\text{(2)} \quad \nu_{A'} A + \nu_L L \rightleftarrows \nu_M M + \nu_N N \tag{14.46}$$

As in Sec. 14.2, component E is inert. Also, note that component A has been taken as common to both reactions but with a possibly different stoichiometric coefficient ($\nu_{A'}$ is not necessarily equal to ν_A).

The stoichiometric coefficients of the above equations do not correspond to the numbers of moles of the respective components present within the system, but *changes* in the amounts of the components are related to the stoichiometric coefficients by

$$\frac{-dn_A}{\nu_A} = \frac{-dn_B}{\nu_B} = \frac{dn_C}{\nu_C} = \frac{dn_D}{\nu_D} \tag{14.25a}$$

following from Eq. 14.24, and

$$\frac{-dn_A}{\nu_{A'}} = \frac{-dn_L}{\nu_L} = \frac{dn_M}{\nu_M} = \frac{dn_N}{\nu_N} \tag{14.47a}$$

following from Eq. 14.46. Introducing a proportionality factor $d\varepsilon_1$, Eqs. 14.25a may be represented by

$$dn_A = -\nu_A\, d\varepsilon_1, \qquad dn_B = -\nu_B\, d\varepsilon_1$$
$$dn_C = \nu_C\, d\varepsilon_1, \qquad dn_D = \nu_D\, d\varepsilon_1 \tag{14.25b}$$

Similarly, with the proportionality factor $d\varepsilon_2$, Eqs. 14.47a may be represented by

$$dn_A = -\nu_{A'}\, d\varepsilon_2, \qquad dn_L = -\nu_L\, d\varepsilon_2$$
$$dn_M = \nu_M\, d\varepsilon_2, \qquad dn_N = \nu_N\, d\varepsilon_2 \tag{14.47b}$$

Component A is involved in both reactions, so the total change in A is given by

$$dn_A = -\nu_A\, d\varepsilon_1 - \nu_{A'}\, d\varepsilon_2 \tag{14.48}$$

Also, we have $dn_E = 0$ because component E is inert.

For the system under present consideration, Eq. 14.10 is

$$\begin{aligned} dG]_{T,p} &= \mu_A\, dn_A + \mu_B\, dn_B + \mu_C\, dn_C + \mu_D\, dn_D \\ &\quad + \mu_E\, dn_E + \mu_L\, dn_L + \mu_M\, dn_M + \mu_N\, dn_N \end{aligned} \tag{14.49}$$

Introducing the above expressions giving the changes in the n's, this becomes

$$\begin{aligned} dG]_{T,p} &= (-\nu_A\mu_A - \nu_B\mu_B + \nu_C\mu_C + \nu_D\mu_D)\, d\varepsilon_1 \\ &\quad + (-\nu_{A'}\mu_A - \nu_L\mu_L + \nu_M\mu_M + \nu_N\mu_N)\, d\varepsilon_2 \end{aligned} \tag{14.50}$$

Since the two reactions are independent, $d\varepsilon_1$ and $d\varepsilon_2$ can be independently varied. Accordingly, when $dG]_{T,p} = 0$, the terms in parentheses must be zero and *two* equations of reaction equilibrium result, one corresponding to each of the foregoing reactions

$$\nu_A\mu_A + \nu_B\mu_B = \nu_C\mu_C + \nu_D\mu_D \tag{14.26b}$$

$$\nu_{A'}\mu_A + \nu_L\mu_L = \nu_M\mu_M + \nu_N\mu_N \tag{14.51}$$

The first of these equations is exactly the same as that obtained in Sec. 14.2. For the case of reacting ideal gas mixtures, this equation can be expressed as

$$-\left(\frac{\Delta G^\circ}{RT}\right)_1 = \ln\left[\frac{y_C^{\nu_C} y_D^{\nu_D}}{y_A^{\nu_A} y_B^{\nu_B}}\left(\frac{p}{p_{\text{ref}}}\right)^{\nu_C+\nu_D-\nu_A-\nu_B}\right] \tag{14.52}$$

Similarly, Eq. 14.51 can be expressed as

$$-\left(\frac{\Delta G^\circ}{RT}\right)_2 = \ln\left[\frac{y_M^{\nu_M} y_N^{\nu_N}}{y_A^{\nu_{A'}} y_L^{\nu_L}}\left(\frac{p}{p_{\text{ref}}}\right)^{\nu_M+\nu_N-\nu_{A'}-\nu_L}\right] \tag{14.53}$$

In each of these equations, the ΔG° term is evaluated as the change in Gibbs function for the respective reaction, regarding each reactant and product as separate at temperature T and a pressure of 1 atm.

From Eq. 14.52 follows the equilibrium constant

$$K_1 = \frac{y_C^{\nu_C} y_D^{\nu_D}}{y_A^{\nu_A} y_B^{\nu_B}}\left(\frac{p}{p_{\text{ref}}}\right)^{\nu_C+\nu_D-\nu_A-\nu_B} \tag{14.54}$$

and from Eq. 14.53 follows

$$K_2 = \frac{y_M^{\nu_M} y_N^{\nu_N}}{y_A^{\nu_{A'}} y_L^{\nu_L}}\left(\frac{p}{p_{\text{ref}}}\right)^{\nu_M+\nu_N-\nu_{A'}-\nu_L} \tag{14.55}$$

The equilibrium constants K_1 and K_2 can be determined from Table A-27 or a similar compilation. The mole fractions appearing in these expressions must be evaluated by considering *all* the substances present within the system, including the inert substance E. Each mole fraction has the form $y_i = n_i/n$, where n_i is the amount of component i in the equilibrium mixture and

$$n = n_A + n_B + n_C + n_D + n_E + n_L + n_M + n_N \tag{14.56}$$

The n's appearing in Eq. 14.56 can be expressed in terms of *two* unknown variables through application of the conservation of mass principle to the various chemical species present. Accordingly, for a specified temperature and pressure, Eqs. 14.54 and 14.55 give *two* equations in *two* unknowns. The composition of the system at equilibrium can be determined by solving these equations simultaneously. This procedure is illustrated by Example 14.9.

The procedure discussed in this section can be extended to systems involving several simultaneous independent reactions. The number of simultaneous equilibrium constant expressions that results equals the number of independent reactions. As these equations are nonlinear and require simultaneous solution, the use of a computer is usually required.

Example 14.9 CONSIDERING EQUILIBRIUM WITH SIMULTANEOUS REACTIONS

As a result of heating, a system consisting initially of 1 kmol of CO_2, $\frac{1}{2}$ kmol of O_2, and $\frac{1}{2}$ kmol of N_2 forms an equilibrium mixture of CO_2, CO, O_2, N_2, and NO at 3000 K, 1 atm. Determine the composition of the equilibrium mixture.

Solution

Known: A system consisting of specified amounts of CO_2, O_2, and N_2 is heated to 3000 K, 1 atm, forming an equilibrium mixture of CO_2, CO, O_2, N_2, and NO.

Find: Determine the equilibrium composition.

Engineering Model: The final mixture is an equilibrium mixture of ideal gases.

Analysis: The overall reaction has the form

1 $$1\,CO_2 + \tfrac{1}{2}O_2 + \tfrac{1}{2}N_2 \rightarrow a\,CO + b\,NO + c\,CO_2 + d\,O_2 + e\,N_2$$

Applying conservation of mass to carbon, oxygen, and nitrogen, the five unknown coefficients can be expressed in terms of any two of the coefficients. Selecting a and b as the unknowns, the following balanced equation results:

$$1\,CO_2 + \tfrac{1}{2}O_2 + \tfrac{1}{2}N_2 \rightarrow a\,CO + b\,NO + (1-a)CO_2 + \tfrac{1}{2}(1+a-b)O_2 + \tfrac{1}{2}(1-b)N_2$$

The total number of moles n in the mixture formed by the products is

$$n = a + b + (1-a) + \tfrac{1}{2}(1+a-b) + \tfrac{1}{2}(1-b) = \frac{4+a}{2}$$

At equilibrium, two independent reactions relate the components of the product mixture:

1. $CO_2 \rightleftarrows CO + \frac{1}{2}O_2$
2. $\frac{1}{2}O_2 + \frac{1}{2}N_2 \rightleftarrows NO$

For the first of these reactions, the form taken by the equilibrium constant when $p = 1$ atm is

$$K_1 = \frac{a[\frac{1}{2}(1+a-b)]^{1/2}}{(1-a)}\left[\frac{1}{(4+a)/2}\right]^{1+1/2-1} = \frac{a}{1-a}\left(\frac{1+a-b}{4+a}\right)^{1/2}$$

Similarly, the equilibrium constant for the second of the reactions is

$$K_2 = \frac{b}{[\frac{1}{2}(1+a-b)]^{1/2}[\frac{1}{2}(1-b)]^{1/2}}\left[\frac{1}{(4+a)/2}\right]^{1-1/2-1/2} = \frac{2b}{[(1+a-b)(1-b)]^{1/2}}$$

At 3000 K, Table A-27 provides $\log_{10}K_1 = -0.485$ and $\log_{10}K_2 = -0.913$, giving $K_1 = 0.3273$ and $K_2 = 0.1222$. Accordingly, the two equations that must be solved simultaneously for the two unknowns a and b are

$$0.3273 = \frac{a}{1-a}\left(\frac{1+a-b}{4+a}\right)^{1/2}, \qquad 0.1222 = \frac{2b}{[(1+a-b)(1-b)]^{1/2}}$$

The solution is $a = 0.3745$, $b = 0.0675$, as can be verified. The composition of the equilibrium mixture, in kmol per kmol of CO_2 present initially, is then $0.3745CO$, $0.0675NO$, $0.6255CO_2$, $0.6535O_2$, $0.4663N_2$.

Skills Developed: Ability to...
- apply Eq. 14.35 to determine equilibrium composition given temperature and pressure for two simultaneous equilibrium reactions.
- retrieve and use data from Table A-27.

❶ If high enough temperatures are attained, nitrogen can combine with oxygen to form components such as nitric oxide. Even trace amounts of oxides of nitrogen in products of combustion can be a source of air pollution.

Quick Quiz

Determine the mole fractions of the components of the equilibrium mixture.
Ans. $y_{CO} = 0.171$, $y_{NO} = 0.031$, $y_{CO_2} = 0.286$, $y_{O_2} = 0.299$, $y_{N_2} = 0.213$.

Phase Equilibrium

In this part of the chapter the equilibrium condition $dG]_{T,p} = 0$ introduced in Sec. 14.1 is used to study the equilibrium of multicomponent, multiphase, nonreacting systems. The discussion begins with the elementary case of equilibrium between two phases of a pure substance and then turns to the general case of several components present in several phases.

14.5 Equilibrium between Two Phases of a Pure Substance

Consider the case of a system consisting of two phases of a pure substance at equilibrium. Since the system is at equilibrium, each phase is at the same temperature and pressure. The Gibbs function for the system is

$$G = n'\bar{g}'(T, p) + n''\bar{g}''(T, p) \tag{14.57}$$

where the primes ′ and ″ denote phases 1 and 2, respectively.

Forming the differential of G at fixed T and p

$$dG]_{T,p} = \bar{g}'\, dn' + \bar{g}''\, dn'' \tag{14.58}$$

Since the total amount of the pure substance remains constant, an increase in the amount present in one of the phases must be compensated by an equivalent decrease in the amount present in the other phase. Thus, we have $dn'' = -dn'$, and Eq. 14.58 becomes

$$dG]_{T,p} = (\bar{g}' - \bar{g}'')\, dn'$$

At equilibrium, $dG]_{T,p} = 0$, so

$$\bar{g}' = \bar{g}'' \tag{14.59}$$

At equilibrium, the molar Gibbs functions of the phases are equal.

CLAPEYRON EQUATION. Equation 14.59 can be used to derive the *Clapeyron* equation, obtained by other means in Sec. 11.4. For two phases at equilibrium, variations in pressure are uniquely related to variations in temperature: $p = p_{\text{sat}}(T)$; thus, differentiation of Eq. 14.59 with respect to temperature gives

$$\left.\frac{\partial \bar{g}'}{\partial T}\right)_p + \left.\frac{\partial \bar{g}'}{\partial p}\right)_T \frac{dp_{\text{sat}}}{dT} = \left.\frac{\partial \bar{g}''}{\partial T}\right)_p + \left.\frac{\partial \bar{g}''}{\partial p}\right)_T \frac{dp_{\text{sat}}}{dT}$$

With Eqs. 11.30 and 11.31, this becomes

$$-\bar{s}' + \bar{v}' \frac{dp_{\text{sat}}}{dT} = -\bar{s}'' + \bar{v}'' \frac{dp_{\text{sat}}}{dT}$$

Or on rearrangement

$$\frac{dp_{\text{sat}}}{dT} = \frac{\bar{s}'' - \bar{s}'}{\bar{v}'' - \bar{v}'}$$

This can be expressed alternatively by noting that, with $\bar{g} = \bar{h} - T\bar{s}$, Eq. 14.59 becomes

$$\bar{h}' - T\bar{s}' = \bar{h}'' - T\bar{s}''$$

or

$$\bar{s}'' - \bar{s}' = \frac{\bar{h}'' - \bar{h}'}{T} \tag{14.60}$$

Clapeyron equation

Combining results, the ***Clapeyron equation*** is obtained

$$\frac{dp_{\text{sat}}}{dT} = \frac{1}{T}\left(\frac{\bar{h}'' - \bar{h}'}{\bar{v}'' - \bar{v}'}\right) \tag{14.61}$$

Applications of the Clapeyron equation are provided in Chap. 11.

A special form of Eq. 14.61 for a system at equilibrium consisting of a liquid or solid phase and a vapor phase can be obtained simply. If the specific volume of the liquid or solid, $\bar{v}'$, is negligible compared with the specific volume of the vapor, $\bar{v}''$, and the vapor can be treated as an ideal gas, $\bar{v}'' = \bar{R}T/p_{\text{sat}}$, Eq. 14.61 becomes

$$\frac{dp_{\text{sat}}}{dT} = \frac{\bar{h}'' - \bar{h}'}{\bar{R}T^2/p_{\text{sat}}}$$

or

Clausius–Clapeyron equation

$$\frac{d \ln p_{\text{sat}}}{dT} = \frac{\bar{h}'' - \bar{h}'}{\bar{R}T^2} \tag{14.62}$$

which is the ***Clausius–Clapeyron equation.*** The similarity in form of Eq. 14.62 and the van't Hoff equation, Eq. 14.43b, may be noted. The van't Hoff equation for chemical equilibrium is the counterpart of the Clausius–Clapeyron equation for phase equilibrium.

14.6 Equilibrium of Multicomponent, Multiphase Systems

The equilibrium of systems that may involve several phases, each having a number of components present, is considered in this section. The principal result is the Gibbs phase rule, which summarizes important limitations on multicomponent, multiphase systems at equilibrium.

14.6.1 Chemical Potential and Phase Equilibrium

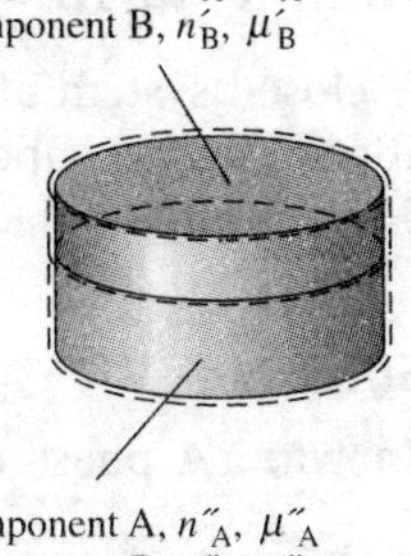

Fig. 14.1 System consisting of two components in two phases.

Figure 14.1 shows a system consisting of *two* components A and B in *two* phases 1 and 2 that are at the same temperature and pressure. Applying Eq. 14.10 to each of the phases

$$\begin{aligned} dG']_{T,p} &= \mu'_A\, dn'_A + \mu'_B\, dn'_B \\ dG'']_{T,p} &= \mu''_A\, dn''_A + \mu''_B\, dn''_B \end{aligned} \tag{14.63}$$

where as before the primes identify the two phases.

When matter is transferred between the two phases in the absence of chemical reaction, the total amounts of A and B must remain constant. Thus, the increase in the amount present in one of the phases must be compensated by an equivalent decrease in the amount present in the other phase. That is

$$dn''_A = -dn'_A, \qquad dn''_B = -dn'_B \tag{14.64}$$

With Eqs. 14.63 and 14.64, the change in the Gibbs function for the system is

$$\begin{aligned} dG]_{T,p} &= dG']_{T,p} + dG'']_{T,p} \\ &= (\mu'_A - \mu''_A)\, dn'_A + (\mu'_B - \mu''_B)\, dn'_B \end{aligned} \tag{14.65}$$

Since n'_A and n'_B can be varied independently, it follows that when $dG]_{T,p} = 0$, the terms in parentheses are zero, resulting in

$$\mu'_A = \mu''_A \quad \text{and} \quad \mu'_B = \mu''_B \tag{14.66}$$

At equilibrium, the chemical potential of each component is the same in each phase.

The significance of the chemical potential for phase equilibrium can be brought out simply by reconsidering the system of Fig. 14.1 in the special case when the chemical potential of component B is the same in both phases: $\mu'_B = \mu''_B$. With this constraint, Eq. 14.65 reduces to

$$dG]_{T,p} = (\mu'_A - \mu''_A)\, dn'_A$$

Any spontaneous process of the system taking place at a fixed temperature and pressure must be such that the Gibbs function decreases: $dG]_{T,p} < 0$. Thus, with the above expression we have

$$(\mu'_A - \mu''_A)\, dn'_A < 0$$

Accordingly,

- when the chemical potential of A is greater in phase 1 than in phase 2 ($\mu'_A > \mu''_A$), it follows that $dn'_A < 0$. That is, substance A passes from phase 1 to phase 2.
- when the chemical potential of A is greater in phase 2 than in phase 1 ($\mu''_A > \mu'_A$), it follows that $dn'_A > 0$. That is, substance A passes from phase 2 to phase 1.

At equilibrium, the chemical potentials are equal ($\mu'_A = \mu''_A$), and there is no net transfer of A between the phases.

With this reasoning, we see that the chemical potential can be regarded as a measure of the *escaping tendency* of a component. If the chemical potential of a component is not the same in each phase, there will be a tendency for that component to pass from the phase having the higher chemical potential for that component to the phase having the lower chemical potential. When the chemical potential is the same in both phases, there is no tendency for a net transfer to occur from one phase to the other.

In Example 14.10, we apply phase equilibrium principles to provide a rationale for the model introduced in Sec. 12.5.3 for moist air in contact with liquid water.

Example 14.10 EQUILIBRIUM OF MOIST AIR IN CONTACT WITH LIQUID WATER

A closed system at a temperature of 20°C and a pressure of 1 atm consists of a pure liquid water phase in equilibrium with a vapor phase composed of water vapor and dry air. Determine the departure, in percent, of the partial pressure of the water vapor from the saturation pressure of water at 20°C.

Solution

Known: A phase of liquid water only is in equilibrium with *moist air* at 20°C and 1 bar.

Find: Determine the percentage departure of the partial pressure of the water vapor in the moist air from the saturation pressure of water at 20°C.

Schematic and Given Data:

1 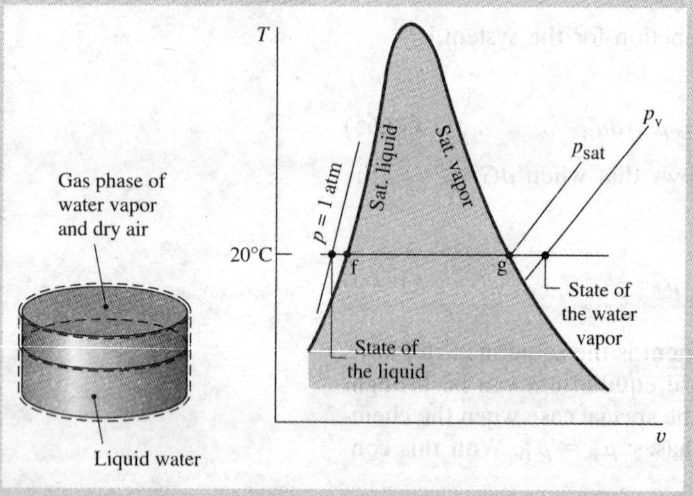

Fig. E14.10

Engineering Model:

1. The gas phase can be modeled as an ideal gas mixture.
2. The liquid phase is pure water only.

Analysis: For phase equilibrium, the chemical potential of the water must have the same value in both phases: $\mu_{\text{l}} = \mu_{\text{v}}$, where μ_{l} and μ_{v} denote, respectively, the chemical potentials of the pure liquid water in the liquid phase and the water vapor in the vapor phase.

The chemical potential μ_{l} is the Gibbs function per mole of pure liquid water (Eq. 14.12)

$$\mu_{\text{l}} = \bar{g}(T, p)$$

Since the vapor phase is assumed to form an ideal gas mixture, the chemical potential μ_{v} equals the Gibbs function per mole evaluated at temperature T and the partial pressure p_{v} of the water vapor (Eq. 14.16)

$$\mu_{\text{v}} = \bar{g}(T, p_{\text{v}})$$

For phase equilibrium, $\mu_{\text{l}} = \mu_{\text{v}}$, or

$$\bar{g}(T, p_{\text{v}}) = \bar{g}(T, p)$$

With $\bar{g} = \bar{h} - T\bar{s}$, this can be expressed alternatively as

$$\bar{h}(T, p_{\text{v}}) - T\bar{s}(T, p_{\text{v}}) = \bar{h}(T, p) - T\bar{s}(T, p)$$

The water vapor is modeled as an ideal gas. Thus, the enthalpy is given closely by the saturated vapor value at temperature T

$$\bar{h}(T, p_{\text{v}}) \approx \bar{h}_{\text{g}}$$

Furthermore, with Eq. 6.20b, the difference between the specific entropy of the water vapor and the specific entropy at the corresponding saturated vapor state is

$$\bar{s}(T, p_{\text{v}}) - \bar{s}_{\text{g}}(T) = -\bar{R} \ln \frac{p_{\text{v}}}{p_{\text{sat}}}$$

or

$$\bar{s}(T, p_v) = \bar{s}_g(T) - \bar{R} \ln \frac{p_v}{p_{sat}}.$$

where p_{sat} is the saturation pressure at temperature T.

With Eq. 3.13, the enthalpy of the liquid is closely

$$\bar{h}(T, p) \approx \bar{h}_f + \bar{v}_f(p - p_{sat})$$

where $\bar{v}_f$ and $\bar{h}_f$ are the saturated liquid specific volume and enthalpy at temperature T. Furthermore, with Eq. 6.5

$$\bar{s}(T, p) \approx \bar{s}_f(T)$$

where $\bar{s}_f$ is the saturated liquid specific entropy at temperature T.

Collecting the foregoing expressions, we have

$$\bar{h}_g - T\left(\bar{s}_g - \bar{R} \ln \frac{p_v}{p_{sat}}\right) = \bar{h}_f + \bar{v}_f(p - p_{sat}) - T\bar{s}_f$$

or

$$\bar{R}T \ln \frac{p_v}{p_{sat}} = \bar{v}_f(p - p_{sat}) - \underline{[(\bar{h}_g - \bar{h}_f) - T(\bar{s}_g - \bar{s}_f)]}$$

The underlined term vanishes by Eq. 14.60, leaving

$$\ln \frac{p_v}{p_{sat}} = \frac{\bar{v}_f(p - p_{sat})}{\bar{R}T} \quad \text{or} \quad \frac{p_v}{p_{sat}} = \exp \frac{\bar{v}_f(p - p_{sat})}{\bar{R}T}$$

With data from Table A-2 at 20°C, $v_f = 1.0018 \times 10^{-3}\ \text{m}^3/\text{kg}$ and $p_{sat} = 0.0239$ bar, we have

$$\frac{v_f(p - p_{sat})}{RT} = \frac{1.0018 \times 10^{-3}\ \text{m}^3/\text{kg}\,(1 - 0.0239) \times 10^5\ \text{N/m}^2}{\left(\dfrac{8314\ \text{N} \cdot \text{m}}{18.02\ \text{kg} \cdot \text{K}}\right)(293.15\ \text{K})}$$

$$= 7.23 \times 10^{-4}$$

Finally

$$\frac{p_v}{p_{sat}} = \exp(7.23 \times 10^{-4}) = 1.00072$$

When expressed as a percentage, the departure of p_v from p_{sat} is

❷ $$\left(\frac{p_v - p_{sat}}{p_{sat}}\right)(100) = (1.00072 - 1)(100) = 0.072\%$$

---

❶ For phase equilibrium, there would be a small, but finite, concentration of air in the liquid water phase. However, this small amount of dissolved air is ignored in the present development.

❷ The departure of p_v from p_{sat} is negligible at the specified conditions. This suggests that at normal temperatures and pressures the equilibrium between the liquid water phase and the water vapor is not significantly disturbed by the presence of the dry air. Accordingly, the partial pressure of the water vapor can be taken as equal to the saturation pressure of the water at the system temperature. This model, introduced in Sec. 12.5.3, is used extensively in Chap. 12.

Skills Developed: *Ability to...*

✓ apply the concept of phase equilibrium expressed in Eq. 14.66 to an air–water–vapor mixture in equilibrium with liquid water.

Quick Quiz

Using the methods of Sec. 12.5.2, determine the specific humidity, ω, of the air–water–vapor mixture.

Ans. 0.01577.

14.6.2 Gibbs Phase Rule

The requirement for equilibrium of a system consisting of two components and two phases, given by Eqs. 14.66, can be extended with similar reasoning to nonreacting multicomponent, multiphase systems. At equilibrium, the chemical potential of each component must be the same in all phases. For the case of N components that are present in P phases we have, therefore, the following set of $N(P-1)$ equations:

$$N\left\{\begin{array}{l} \overbrace{\mu_1^1 = \mu_1^2 = \mu_1^3 = \cdots = \mu_1^P}^{P-1} \\ \mu_2^1 = \mu_2^2 = \mu_2^3 = \cdots = \mu_2^P \\ \vdots \\ \mu_N^1 = \mu_N^2 = \mu_N^3 = \cdots = \mu_N^P \end{array}\right. \tag{14.67}$$

where μ_i^j denotes the chemical potential of the ith component in the jth phase. This set of equations provides the basis for the *Gibbs phase rule,* which allows the determination of the number of *independent intensive* properties that may be arbitrarily specified in order to fix the *intensive* state of the system. The number of independent intensive properties is called the ***degrees of freedom*** (or the *variance*).

degrees of freedom

Since the chemical potential is an intensive property, its value depends on the relative proportions of the components present and not on the amounts of the components. In other words, in a given phase involving N components at temperature T and pressure p, the chemical potential is determined by the *mole fractions* of the components present and not the respective n's. However, as the mole fractions add to unity, at most $N-1$ of the mole fractions can be independent. Thus, for a system involving N components, there are at most $N-1$ independently variable mole fractions for each phase. For P phases, therefore, there are at most $P(N-1)$ independently variable mole fractions. In addition, the temperature and pressure, which are the same in each phase, are two further intensive properties, giving a maximum of $P(N-1)+2$ independently variable intensive properties for the system. But because of the $N(P-1)$ equilibrium conditions represented by Eqs. 14.67 among these properties, the number of intensive properties that are freely variable, the degrees of freedom F, is

Gibbs phase rule

$$F = [P(N-1)+2] - N(P-1) = 2 + N - P \tag{14.68}$$

which is the ***Gibbs phase rule.***

In Eq. 14.68, F is the number of intensive properties that may be arbitrarily specified and that must be specified to fix the intensive state of a nonreacting system at equilibrium. ➡ **FOR EXAMPLE...** let us apply the Gibbs phase rule to a liquid solution consisting of water and ammonia such as considered in the discussion of absorption refrigeration (Sec. 10.5). This solution involves two components and a single phase: $N = 2$ and $P = 1$. Equation 14.68 then gives $F = 3$, so the intensive state is fixed by giving the values of *three* intensive properties, such as temperature, pressure, and the ammonia (or water) mole fraction. ⬅

The phase rule summarizes important limitations on various types of systems. For example, for a system involving a single component such as water, $N = 1$ and Eq. 14.68 becomes

$$F = 3 - P \tag{14.69}$$

▶ The minimum number of phases is one, corresponding to $P = 1$. For this case, Eq. 14.69 gives $F = 2$. That is, *two* intensive properties must be specified to fix the intensive state of the system. This requirement is familiar from our use of the steam tables and similar property tables. To obtain properties of superheated vapor, say, from such tables requires that we give values for *any two* of the tabulated properties, for example, T and p.

- When two phases are present in a system involving a single component, $N = 1$ and $P = 2$. Equation 14.69 then gives $F = 1$. That is, the intensive state is determined by a single intensive property value. For example, the intensive states of the separate phases of an equilibrium mixture of liquid water and water vapor are completely determined by specifying the temperature.
- The minimum allowable value for the degrees of freedom is zero: $F = 0$. For a single-component system, Eq. 14.69 shows that this corresponds to $P = 3$, a three-phase system. Thus, *three* is the maximum number of different phases of a pure component that can coexist in equilibrium. Since there are no degrees of freedom, both temperature and pressure are fixed at equilibrium. For example, there is only a single temperature 0.01°C and a single pressure 0.6113 kPa (0.006 atm) for which ice, liquid water, and water vapor are in equilibrium.

The phase rule given here must be modified for application to systems in which chemical reactions occur. Furthermore, the system of equations, Eqs. 14.67, giving the requirements for phase equilibrium at a specified temperature and pressure can be expressed alternatively in terms of partial molal Gibbs functions, fugacities, and activities, all of which are introduced in Sec. 11.9. To use any such expression to determine the equilibrium composition of the different phases present within a system at equilibrium requires a model for each phase that allows the relevant quantities—the chemical potentials, fugacities, and so on—to be evaluated for the components present in terms of system properties that can be determined. For example, a gas phase might be modeled as an ideal gas mixture or, at higher pressures, as an ideal solution.

✓ Chapter Summary and Study Guide

In this chapter, we have studied chemical equilibrium and phase equilibrium. The chapter opens by developing criteria for equilibrium and introducing the chemical potential. In the second part of the chapter, we study the chemical equilibrium of ideal gas mixtures using the equilibrium constant concept. We also utilize the energy balance and determine the equilibrium flame temperature as an application. The final part of the chapter concerns phase equilibrium, including multicomponent, multiphase systems and the Gibbs phase rule.

The following list provides a study guide for this chapter. When your study of the text and end-of-chapter exercises has been completed, you should be able to

- ✓ write out the meaning of the terms listed in the margin throughout the chapter and understand each of the related concepts. The subset of key concepts listed below is particularly important.
- ✓ apply the equilibrium constant relationship, Eq. 14.35, to determine the third quantity when *any two* of temperature, pressure, and equilibrium composition of an ideal gas mixture are known. Special cases include applications with simultaneous reactions and systems involving ionized gases.
- ✓ use chemical equilibrium concepts with the energy balance, including determination of the equilibrium flame temperature.
- ✓ apply Eq. 14.43b, the van't Hoff equation, to determine the enthalpy of reaction when the equilibrium constant is known, and conversely.
- ✓ apply the Gibbs phase rule, Eq. 14.68.

Key Engineering Concepts

Gibbs function p. 776
equilibrium criterion p. 776
chemical potential p. 777
equation of reaction equilibrium p. 782
equilibrium constant p. 783
Gibbs phase rule p. 806

Key Equations

$dG]_{T,p} = 0$	(14.6)	Equilibrium criterion
$\mu_i = \left(\frac{\partial G}{\partial n_i}\right)_{T,p,n_l}$	(14.8)	Chemical potential of component *i* in a mixture.
$G = \sum_{i=1}^{j} n_i \bar{g}_i(T, p_i)$	(14.15)	Gibbs function and chemical potential relations for ideal gas mixtures.
$\mu_i = \bar{g}_i(T, p_i)$	(14.16)	
$\mu_i = \bar{g}_i^\circ + \bar{R}T \ln \frac{y_i p}{p_{ref}}$	(14.17)	
$K(T) = \frac{y_C^{\nu_C} y_D^{\nu_D}}{y_A^{\nu_A} y_B^{\nu_B}}\left(\frac{p}{p_{ref}}\right)^{\nu_C+\nu_D-\nu_A-\nu_B}$	(14.32)	Equilibrium constant expressions for an equilibrium mixture of ideal gases.
$K = \frac{n_C^{\nu_C} n_D^{\nu_D}}{n_A^{\nu_A} n_B^{\nu_B}}\left(\frac{p/p_{ref}}{n}\right)^{\nu_C+\nu_D-\nu_A-\nu_B}$	(14.35)	
$\frac{d \ln K}{dT} = \frac{\Delta H}{\bar{R}T^2}$	(14.43b)	van't Hoff equation.
$\bar{g}' = \bar{g}''$	(14.59)	Phase equilibrium criterion for a pure substance
$\mu_A' = \mu_A'' \quad \mu_B' = \mu_B''$	(14.66)	Phase equilibrium criteria for two-component, two-phase systems.
$F = 2 + N - P$	(14.68)	Gibbs phase rule.

Exercises: things engineers think about

1. Why is using the Gibbs function advantageous when studying chemical and phase equilibrium?

2. For Eq. 14.6 to apply at equilibrium, must a system attain equilibrium at fixed T and p?

3. Show that $(dA)_{T,V} = 0$ is a valid equilibrium criterion, where $A = U - TS$ is the *Helmholtz function*.

4. A mixture of 1 kmol of CO and $\frac{1}{2}$ kmol of O_2 is held at ambient temperature and pressure. After 100 hours only an insignificant amount of CO_2 has formed. Why?

5. Why might oxygen contained in an iron tank be treated as *inert* in a thermodynamic analysis even though iron *oxidizes* in the presence of oxygen?

6. For $CO_2 + H_2 \rightleftarrows CO + H_2O$, how does pressure affect the equilibrium composition?

7. For each of the reactions listed in Table A-27, the value of $\log_{10}K$ increases with increasing temperature. What does this imply?

8. For each of the reactions listed in Table A-27, the value of the equilibrium constant K at 298 K is relatively small. What does this imply?

9. If a system initially containing CO_2 and H_2O were held at fixed T, p, list chemical species that *might* be present at equilibrium.

10. Using Eq. 14.12 together with phase equilibrium considerations, suggest how the chemical potential of a mixture component could be evaluated.

11. Note 1 of Example 14.10 refers to the small amount of air that would be dissolved in the liquid phase. For equilibrium, what must be true of the chemical potentials of air in the liquid and gas phases?

12. Water can exist in a number of different solid phases. Can liquid water, water vapor, and two phases of ice exist in equilibrium?

Problems: developing engineering skills

Working with the Equilibrium Constant

14.1 Determine the change in the Gibbs function ΔG° at 25°C in kJ/kmol, for the reaction

$$CH_4(g) + 2O_2 \rightleftarrows CO_2 + 2H_2O(g)$$

using

(a) Gibbs function of formation data.
(b) enthalpy of formation and absolute entropy data.

14.2 Calculate the equilibrium constant, expressed as $\log_{10}K$, for $CO_2 \rightleftarrows CO + \frac{1}{2}O_2$ at (a) 500 K, (b) 1000 K. Compare with values from Table A-27.

14.3 Calculate the equilibrium constant, expressed as $\log_{10}K$, for the *water-gas* reaction $CO + H_2O(g) \rightleftarrows CO_2 + H_2$ at (a) 298 K, (b) 1000 K. Compare with values from Table A-27.

14.4 Calculate the equilibrium constant, expressed as $\log_{10}K$, for $H_2O \rightleftarrows H_2 + \frac{1}{2}O_2$ at (a) 298 K, (b) 2000 K. Compare with values from Table A-27.

14.5 Using data from Table A-27, determine $\log_{10}K$ at 2500 K for

(a) $H_2O \rightleftarrows H_2 + \frac{1}{2}O_2$.
(b) $H_2 + \frac{1}{2}O_2 \rightleftarrows H_2O$.
(c) $2H_2O \rightleftarrows 2H_2 + O_2$.

14.6 In Table A-27, $\log_{10}K$ is nearly linear in $1/T$: $\log_{10}K = C_1 + C_2/T$, where C_1 and C_2 are constants. For selected reactions listed in the table

(a) verify this by plotting $\log_{10}K$ versus $1/T$ for temperature ranging from 2100 to 2500 K.
(b) evaluate C_1 and C_2 for any pair of adjacent table entries in the temperature range of part (a).

14.7 Determine the relationship between the ideal gas equilibrium constants K_1 and K_2 for the following two alternative ways of expressing the ammonia synthesis reaction:

1. $\frac{1}{2}N_2 + \frac{3}{2}H_2 \rightleftarrows NH_3$
2. $N_2 + 3H_2 \rightleftarrows 2NH_3$

14.8 Consider the reactions

1. $CO + H_2O \rightleftarrows H_2 + CO_2$
2. $2CO_2 \rightleftarrows 2CO + O_2$
3. $2H_2O \rightleftarrows 2H_2 + O_2$

Show that $K_1 = (K_3/K_2)^{1/2}$.

14.9 Consider the reactions

1. $CO_2 + H_2 \rightleftarrows CO + H_2O$
2. $CO_2 \rightleftarrows CO + \frac{1}{2}O_2$
3. $H_2O \rightleftarrows H_2 + \frac{1}{2}O_2$

(a) Show that $K_1 = K_2/K_3$.
(b) Evaluate $\log_{10}K_1$ at 298 K, 1 atm using the expression from part (a), together with $\log_{10}K$ data from Table A-27.
(c) Check the value for $\log_{10}K_1$ obtained in part (b) by applying Eq. 14.31 to reaction 1.

14.10 Evaluate the equilibrium constant at 2000 K for $CH_4 + H_2O \rightleftarrows 3H_2 + CO$. At 2000 K, $\log_{10}K = 7.469$ for $C + \frac{1}{2}O_2 \rightleftarrows CO$, and $\log_{10}K = -3.408$ for $C + 2H_2 \rightleftarrows CH_4$.

14.11 For each of the following dissociation reactions, determine the equilibrium compositions:

(a) One kmol of N_2O_4 dissociates to form an equilibrium ideal gas mixture of N_2O_4 and NO_2 at 25°C, 2 atm. For $N_2O_4 \rightleftarrows 2NO_2$, $\Delta G° = 5400$ kJ/kmol at 25°C.
(b) One kmol of CH_4 dissociates to form an equilibrium ideal gas mixture of CH_4, H_2, and C at 1000 K, 5 atm. For $C + 2H_2 \rightleftarrows CH_4$, $\log_{10}K = 1.011$ at 1000 K.

14.12 Determine the extent to which dissociation occurs in the following cases: One kmol of H_2O dissociates to form an equilibrium mixture of H_2O, H_2, and O_2 at 2616°C, 1.25 atm. One kmol of CO_2 dissociates to form an equilibrium mixture of CO_2, CO, and O_2 at the same temperature and pressure.

14.13 One kmol of carbon reacts with 2 kmol of oxygen (O_2) to form an equilibrium mixture of CO_2, CO, and O_2 at 2727°C, 1 atm. Determine the equilibrium composition.

14.14 The following exercises involve oxides of nitrogen:

(a) One kmol of N_2O_4 dissociates at 25°C, 1 atm to form an equilibrium ideal gas mixture of N_2O_4 and NO_2 in which the amount of N_2O_4 present is 0.8154 kmol. Determine the amount of N_2O_4 that would be present in an equilibrium mixture at 25°C, 0.5 atm.
(b) A gaseous mixture consisting of 1 kmol of NO, 10 kmol of O_2, and 40 kmol of N_2 reacts to form an equilibrium ideal gas mixture of NO_2, NO, and O_2 at 500 K, 0.1 atm. Determine the composition of the equilibrium mixture. For $NO + \frac{1}{2}O_2 \rightleftarrows NO_2$, $K = 120$ at 500 K.
(c) An equimolar mixture of O_2 and N_2 reacts to form an equilibrium ideal gas mixture of O_2, N_2, and NO. Plot the mole fraction of NO in the equilibrium mixture versus equilibrium temperature ranging from 1200 to 2000 K.

Why are oxides of nitrogen of concern?

14.15 One kmol of CO_2 dissociates to form an equilibrium ideal gas mixture of CO_2, CO, and O_2 at temperature T and pressure p.

(a) For $T = 3000$ K, plot the amount of CO present, in kmol, versus pressure for $1 \le p \le 10$ atm.
(b) For $p = 1$ atm plot the amount of CO present, in kmol, versus temperature for $2000 \le T \le 3500$ K.

14.16 One kmol of H_2O dissociates to form an equilibrium ideal gas mixture of H_2O, H_2, and O_2 at temperature T and pressure p.

(a) For $T = 3000$°K, plot the amount of H_2 present, in kmol, versus pressure ranging from 1 to 10 atm.
(b) For $p = 1$ atm, plot the amount of H_2 present, in kmol, versus temperature ranging from 2000 to 3500 K.

14.17 One kmol of H_2O together with x kmol of N_2 (inert) forms an equilibrium mixture at 3000 K, 1 atm consisting of H_2O, H_2, O_2, and N_2. Plot the amount of H_2 present in the equilibrium mixture, in kmol, versus x ranging from 0 to 2.

14.18 An equimolar mixture of CO and O_2 reacts to form an equilibrium mixture of CO_2, CO, and O_2 at 3000 K. Determine the effect of pressure on the composition of the equilibrium mixture. Will lowering the pressure while keeping the temperature fixed increase or decrease the amount of CO_2 present? Explain.

14.19 An equimolar mixture of CO and $H_2O(g)$ reacts to form an equilibrium mixture of CO_2, CO, H_2O, and H_2 at 1727°C, 1 atm.

(a) Will lowering the temperature increase or decrease the amount of H_2 present? Explain.
(b) Will decreasing the pressure while keeping the temperature constant increase or decrease the amount of H_2 present? Explain.

14.20 Determine the temperature, in K, at which 9% of diatomic hydrogen (H_2) dissociates into monatomic hydrogen (H) at a pressure of 10 atm. For a greater percentage of

H_2 at the same pressure, would the temperature be *higher* or *lower*? Explain.

14.21 Two kmol of CO_2 dissociate to form an equilibrium mixture of CO_2, CO, and O_2 in which 1.8 kmol of CO_2 is present. Plot the temperature of the equilibrium mixture, in K, versus the pressure p for $0.5 \le p \le 10$ atm.

14.22 One kmol of $H_2O(g)$ dissociates to form an equilibrium mixture of $H_2O(g)$, H_2, and O_2 in which the amount of water vapor present is 0.95 kmol. Plot the temperature of the equilibrium mixture, in K, versus the pressure p for $1 \le p \le 10$ atm.

14.23 A vessel initially containing 1 kmol of $H_2O(g)$ and x kmol of N_2 forms an equilibrium mixture at 1 atm consisting of $H_2O(g)$, H_2, O_2, and N_2 in which 0.5 kmol of $H_2O(g)$ is present. Plot x versus the temperature T for $3000 \le T \le 3600$ K.

14.24 A vessel initially containing 2 kmol of N_2 and 1 kmol of O_2 forms an equilibrium mixture at 1 atm consisting of N_2, O_2, and NO. Plot the amount of NO formed versus temperature T for $2000 \le T \le 3500$ K.

14.25 A vessel initially containing 1 kmol of CO and 4.76 kmol of dry air forms an equilibrium mixture of CO_2, CO, O_2, and N_2 at 3000 K, 1 atm. Determine the equilibrium composition.

14.26 A vessel initially containing 1 kmol of O_2, 2 kmol of N_2, and 1 kmol of Ar forms an equilibrium mixture of O_2, N_2, NO, and Ar at 2727°C, 1 atm. Determine the equilibrium composition.

14.27 One kmol of CO and 0.5 kmol of O_2 react to form a mixture at temperature T and pressure p consisting of CO_2, CO, and O_2. If 0.35 kmol of CO is present in an equilibrium mixture when the pressure is 1 atm, determine the amount of CO present in an equilibrium mixture at the same temperature if the pressure were 10 atm.

14.28 A vessel initially contains 1 kmol of H_2 and 4 kmol of N_2. An equilibrium mixture of H_2, H, and N_2 forms at 3000 K, 1 atm. Determine the equilibrium composition. If the pressure were increased while keeping the temperature fixed, would the amount of monatomic hydrogen in the equilibrium mixture increase or decrease? Explain.

14.29 Dry air enters a heat exchanger. An equilibrium mixture of N_2, O_2, and NO exits at 2139°C, 1 atm. Determine the mole fraction of NO in the exiting mixture. Will the amount of NO increase or decrease as temperature decreases at fixed pressure? Explain.

14.30 A gaseous mixture with a molar analysis of 20% CO_2, 40% CO, and 40% O_2 enters a heat exchanger and is heated at constant pressure. An equilibrium mixture of CO_2, CO, and O_2 exits at 3000 K, 1.5 bar. Determine the molar analysis of the exiting mixture.

14.31 An ideal gas mixture with the molar analysis 30% CO, 10% CO_2, 40% H_2O, 20% inert gas enters a reactor operating at steady state. An equilibrium mixture of CO, CO_2, H_2O, H_2, and the inert gas exits at 1 atm.

(a) If the equilibrium mixture exits at 1200 K, determine on a molar basis the ratio of the H_2 in the equilibrium mixture to the H_2O in the entering mixture.

(b) If the mole fraction of CO present in the equilibrium mixture is 7.5%, determine the temperature of the equilibrium mixture, in K.

14.32 A mixture of 1 kmol CO and 0.5 kmol O_2 in a closed vessel, initially at 1 atm and 300 K, reacts to form an equilibrium mixture of CO_2, CO, and O_2 at 2500 K. Determine the final pressure, in atm.

14.33 Methane burns with 90% of theoretical air to form an equilibrium mixture of CO_2, CO, $H_2O(g)$, H_2, and N_2 at 1000 K, 1 atm. Determine the composition of the equilibrium mixture, per kmol of mixture.

14.34 Octane (C_8H_{18}) burns with air to form an equilibrium mixture of CO_2, H_2, CO, $H_2O(g)$, and N_2 at 1700 K, 1 atm. Determine the composition of the products, in kmol per kmol of fuel, for an equivalence ratio of 1.2.

14.35 Acetylene gas (C_2H_2) at 25°C, 1 atm enters a reactor operating at steady state and burns with 40% excess air entering at 25°C, 1 atm, 80% relative humidity. An equilibrium mixture of CO_2, H_2O, O_2, NO, and N_2 exits at 2200 K, 0.9 atm. Determine, per kmol of C_2H_2 entering, the composition of exiting mixture.

Chemical Equilibrium and the Energy Balance

14.36 Carbon dioxide gas at 25°C, 5.1 atm enters a heat exchanger operating at steady state. An equilibrium mixture of CO_2, CO, and O_2 exits at 2527°C, 5 atm. Determine, per kmol of CO_2 entering,

(a) the composition of the exiting mixture.
(b) the heat transfer to the gas stream, in kJ.

Neglect kinetic and potential energy effects.

14.37 Saturated water vapor at 103 kPa enters a heat exchanger operating at steady state. An equilibrium mixture of $H_2O(g)$, H_2, and O_2 exits at 2627°C, 1 atm. Determine, per kmol of steam entering,

(a) the composition of the exiting mixture.
(b) the heat transfer to the flowing stream, in kJ.

Neglect kinetic and potential energy effects.

14.38 Carbon at 25°C, 1 atm enters a reactor operating at steady state and burns with oxygen entering at 127°C, 1 atm. The entering streams have equal molar flow rates. An equilibrium mixture of CO_2, CO, and O_2 exits at 2727°C, 1 atm. Determine, per kmol of carbon,

(a) the composition of the exiting mixture.
(b) the heat transfer between the reactor and its surroundings, in kJ.

Neglect kinetic and potential energy effects.

14.39 An equimolar mixture of carbon monoxide and water vapor at 93°C, 1 atm enters a reactor operating at steady state. An equilibrium mixture of CO_2, CO, $H_2O(g)$, and H_2 exits at 1227°C, 1 atm. Determine the heat transfer between the reactor and its surroundings, in kJ per kmol of CO entering. Neglect kinetic and potential energy effects.

14.40 Carbon dioxide (CO_2) and oxygen (O_2) in a 1:2 molar ratio enter a reactor operating at steady state in separate streams at 1 atm, 127°C and 1 atm, 277°C, respectively. An equilibrium mixture of CO_2, CO, and O_2 exits at 1 atm.

If the mole fraction of CO in the exiting mixture is 0.1, determine the rate of heat transfer from the reactor, in kJ per kmol of CO_2 entering. Ignore kinetic and potential energy effects.

14.41 Methane gas at 25°C, 1 atm enters a reactor operating at steady state and burns with 80% of theoretical air entering at 227°C, 1 atm. An equilibrium mixture of CO_2, CO, $H_2O(g)$, H_2, and N_2 exits at 1427°C, 1 atm. Determine, per kmol of methane entering,

(a) the composition of the exiting mixture.
(b) the heat transfer between the reactor and its surroundings, in kJ.

Neglect kinetic and potential energy effects.

14.42 Gaseous propane (C_3H_8) at 25°C, 1 atm enters a reactor operating at steady state and burns with 80% of theoretical air entering separately at 25°C, 1 atm. An equilibrium mixture of CO_2, CO, $H_2O(g)$, H_2, and N_2 exits at 1227°C, 1 atm. Determine the heat transfer between the reactor and its surroundings, in kJ per kmol of propane entering. Neglect kinetic and potential energy effects.

14.43 Gaseous propane (C_3H_8) at 25°C, 1 atm enters a reactor operating at steady state and burns with the theoretical amount of air entering separately at 115°C, 1 atm. An equilibrium mixture of CO_2, CO, $H_2O(g)$ O_2, and N_2 exits at 1730°C, 1 atm. Determine the heat transfer between the reactor and its surroundings, in kJ per kmol of propane entering. Neglect kinetic and potential energy effects.

14.44 One kmol of CO_2 initially at temperature T and 1 atm is heated at constant pressure until a final state is attained consisting of an equilibrium mixture of CO_2, CO, and O_2 in which the amount of CO_2 present is 0.422 kmol. Determine the heat transfer and the work, each in kJ, if T is (a) 298 K, (b) 400 K.

14.45 Hydrogen gas (H_2) at 25°C, 1 atm enters an insulated reactor operating at steady state and reacts with 250% excess oxygen entering at 227°C, 1 atm. The products of combustion exit at 1 atm. Determine the temperature of the products, in K, if

(a) combustion is complete.
(b) an equilibrium mixture of H_2O, H_2, and O_2 exits.

Kinetic and potential energy effects are negligible.

14.46 For each case of Problem 14.45, determine the rate of entropy production, in kJ/K per kmol of H_2 entering. What can be concluded about the possibility of achieving complete combustion?

14.47 Hydrogen (H_2) at 25°C, 1 atm enters an insulated reactor operating at steady state and reacts with 100% of theoretical air entering at 25°C, 1 atm. The products of combustion exit at temperature T and 1 atm. Determine T, in K, if

(a) combustion is complete.
(b) an equilibrium mixture of H_2O, H_2, O_2, and N_2 exits.

14.48 Methane at 25°C, 1 atm enters an insulated reactor operating at steady state and burns with 90% of theoretical air entering separately at 25°C, 1 atm. The products exit at 1 atm as an equilibrium mixture of CO_2, CO, $H_2O(g)$ H_2, and N_2. Determine the temperature of the exiting products, in K. Kinetic and potential energy effects are negligible.

14.49 Carbon monoxide at 25°C, 1 atm enters an insulated reactor operating at steady state and burns with air entering at 25°C, 1 atm. The products exit at 1 atm as an equilibrium mixture of CO_2, CO, O_2, and N_2. Determine the temperature of the equilibrium mixture, in K, if the combustion occurs with

(a) 80% of theoretical air.
(b) 100% of theoretical air.

Kinetic and potential energy effects are negligible.

14.50 For each case of Problem 14.49, determine the rate of exergy destruction, in kJ per kmol of CO entering the reactor. Let T_0 = 298 K.

14.51 Carbon monoxide at 25°C, 1 atm enters an insulated reactor operating at steady state and burns with excess oxygen (O_2) entering at 25°C, 1 atm. The products exit at 2950 K, 1 atm as an equilibrium mixture of CO_2, CO, and O_2. Determine the percent excess oxygen. Kinetic and potential energy effects are negligible.

14.52 A gaseous mixture of carbon monoxide and the theoretical amount of air at 130°C, 1.5 atm enters an insulated reactor operating at steady state. An equilibrium mixture of CO_2, CO, O_2, and N_2 exits at 1.5 atm. Determine the temperature of the exiting mixture, in K. Kinetic and potential energy effects are negligible.

14.53 Methane at 25°C, 1 atm enters an insulated reactor operating at steady state and burns with oxygen entering at 127°C, 1 atm. An equilibrium mixture of CO_2, CO, O_2, and $H_2O(g)$ exits at 3250 K, 1 atm. Determine the rate at which oxygen enters the reactor, in kmol per kmol of methane. Kinetic and potential energy effects are negligible.

14.54 Methane at 25°C, 1 atm enters an insulated reactor operating at steady state and burns with the theoretical amount of air entering at 25°C, 1 atm. An equilibrium mixture of CO_2, CO, O_2, $H_2O(g)$, and N_2 exits at 1 atm.

(a) Determine the temperature of the exiting products, in K.
(b) Determine the rate of exergy destruction, in kJ per kmol of methane entering, for T_0 = 298 K.

Kinetic and potential energy effects are negligible.

14.55 Methane gas at 25°C, 1 atm enters an insulated reactor operating at steady state, where it burns with x times the theoretical amount of air entering at 25°C, 1 atm. An equilibrium mixture of CO_2, CO, O_2, H_2O, and N_2 exits at 1 atm. For selected values of x ranging from 1 to 4, determine the temperature of the exiting equilibrium mixture, in K. Kinetic and potential energy effects are negligible.

14.56 A mixture consisting of 1 kmol of carbon monoxide (CO), 0.5 kmol of oxygen (O_2), and 1.88 kmol of nitrogen (N_2), initially at 227°C, 1 atm, reacts in a closed, rigid, insulated vessel, forming an equilibrium mixture of CO_2, CO, O_2, and N_2. Determine the final equilibrium pressure, in atm.

14.57 A mixture consisting of 1 kmol of CO and the theoretical amount of air, initially at 60°C, 1 atm, reacts in a closed, rigid, insulated vessel to form an equilibrium mixture. An analysis of the products shows that there are 0.808 kmol of CO_2, 0.192 kmol of CO, and 0.096 kmol of O_2 present. The temperature of the final mixture is measured as 2465°C. Check the consistency of these data.

Using the van't Hoff Equation, Ionization

14.58 Estimate the enthalpy of reaction at 2000 K, in kJ/kmol, for $CO_2 \rightleftarrows CO + \frac{1}{2}O_2$ using the van't Hoff equation and equilibrium constant data. Compare with the value obtained for the enthalpy of reaction using enthalpy data.

14.59 Estimate the enthalpy of reaction at 2000 K, in kJ/kmol, for $H_2O \rightleftarrows H_2 + \frac{1}{2}O_2$, using the van't Hoff equation and equilibrium constant data. Compare with the value obtained for the enthalpy of reaction using enthalpy data.

14.60 Estimate the equilibrium constant at 2800 K for $CO_2 \rightleftarrows CO + \frac{1}{2}O_2$ using the equilibrium constant at 2000 K from Table A-27, together with the van't Hoff equation and enthalpy data. Compare with the value for the equilibrium constant obtained from Table A-27.

14.61 Estimate the equilibrium constant at 2800 K for the reaction $H_2O \rightleftarrows H_2 + \frac{1}{2}O_2$ using the equilibrium constant at 2500 K from Table A-27, together with the van't Hoff equation and enthalpy data. Compare with the value for the equilibrium constant obtained from Table A-27.

14.62 At 25°C, $\log_{10}K = 8.9$ for $C + 2H_2 \rightleftarrows CH_4$. Assuming that the enthalpy of reaction does not vary much with temperature, estimate the value of $\log_{10}K$ at 500°C.

14.63 If the ionization-equilibrium constants for $Cs \rightleftarrows Cs^+ + e^-$ at 1600 and 2000 K are $K = 0.78$ and $K = 15.63$, respectively, estimate the enthalpy of ionization, in kJ/kmol, at 1800 K using the van't Hoff equation.

14.64 An equilibrium mixture at 2000 K, 1 atm consists of Cs, Cs^+, and e^-. Based on 1 kmol of Cs present initially, determine the percent ionization of cesium. At 2000 K, the ionization-equilibrium constant for $Cs \rightleftarrows Cs^+ + e^-$ is $K = 15.63$.

14.65 An equilibrium mixture at 10,000 K and pressure p consists of Ar, Ar^+, and e^-. Based on 1 kmol of neutral argon present initially, plot the percent ionization of argon versus pressure for $0.01 \leq p \leq 0.05$ atm. At 10,000 K, the ionization-equilibrium constant for $Ar \rightleftarrows Ar^+ + e^-$ is $K = 4.2 \times 10^{-4}$.

14.66 At 2000 K and pressure p, 1 kmol of Na ionizes to form an equilibrium mixture of Na, Na^+, and e^- in which the amount of Na present is x kmol. Plot the pressure, in atm, versus x for $0.2 \leq x \leq 0.3$ kmol. At 2000 K, the ionization-equilibrium constant for $Na \rightleftarrows Na^+ + e^-$ is $K = 0.668$.

14.67 At 12,000 K and 6 atm, 1 kmol of N ionizes to form an equilibrium mixture of N, N^+, and e^- in which the amount of N present is 0.95 kmol. Determine the ionization-equilibrium constant at this temperature for $N \rightleftarrows N^+ + e^-$.

Considering Simultaneous Reactions

14.68 Carbon dioxide (CO_2), oxygen (O_2), and nitrogen (N_2) enter a reactor operating at steady state with equal molar flow rates. An equilibrium mixture of CO_2, O_2, N_2, CO, and NO exits at 3000 K, 5 atm. Determine the molar analysis of the equilibrium mixture.

14.69 An equimolar mixture of carbon monoxide and water vapor enters a heat exchanger operating at steady state. An equilibrium mixture of CO, CO_2, O_2, H_2O(g), and H_2 exits at 2227°C, 1 atm. Determine the molar analysis of the exiting equilibrium mixture.

14.70 A closed vessel initially contains a gaseous mixture consisting of 1.5 kmol of CO_2, 3 kmol of CO, and 0.5 kmol of H_2. An equilibrium mixture at 2325°C, 1 atm is formed containing CO_2, CO, H_2O, H_2, and O_2. Determine the composition of the equilibrium mixture.

14.71 Butane (C_4H_{10}) burns with 100% excess air to form an equilibrium mixture at 1400 K, 20 atm consisting of CO_2, O_2, H_2O(g), N_2, NO, and NO_2. Determine the balanced reaction equation. For $N_2 + 2O_2 \rightleftarrows 2NO_2$ at 1400 K, $K = 8.4 \times 10^{-10}$.

14.72 One kmol of H_2O(g) dissociates to form an equilibrium mixture at 2778 K, 1 atm consisting of H_2O(g), H_2, O_2, and OH. Determine the equilibrium composition.

14.73 Steam enters a heat exchanger operating at steady state. An equilibrium mixture of H_2O, H_2, O_2, H, and OH exits at temperature T, 1 atm. Determine the molar analysis of the exiting equilibrium mixture for
(a) $T = 2800$ K.
(b) $T = 3000$ K.

Considering Phase Equilibrium

14.74 For a two-phase liquid–vapor mixture of water at 100°C, use tabulated property data to show that the specific Gibbs functions of the saturated liquid and saturated vapor are equal. Repeat for a two-phase liquid–vapor mixture of Refrigerant 134a at 20°C.

14.75 Using the Clapeyron equation, solve the following problems from Chap. 11: (a) 11.32, (b) 11.33, (c) 11.34, (d) 11.35, (e) 11.40.

14.76 A closed system at 20°C, 1 bar consists of a pure liquid water phase in equilibrium with a vapor phase composed of water vapor and dry air. Determine the departure, in percent, of the partial pressure of the water vapor from the saturation pressure of pure water at 20°C.

14.77 Derive an expression for estimating the pressure at which graphite and diamond exist in equilibrium at 25°C in terms of the specific volume, specific Gibbs function, and isothermal compressibility of each phase at 25°C, 1 atm. Discuss.

14.78 An isolated system has two phases, denoted by A and B, each of which consists of the same two substances, denoted by 1 and 2. Show that necessary conditions for equilibrium are

1. the temperature of each phase is the same, $T_A = T_B$.
2. the pressure of each phase is the same, $p_A = p_B$.
3. the chemical potential of each component has the same value in each phase, $\mu_1^A = \mu_1^B$, $\mu_2^A = \mu_2^B$.

14.79 An isolated system has two phases, denoted by A and B, each of which consists of the same two substances, denoted by 1 and 2. The phases are separated by a freely moving *thin* wall permeable only by substance 2. Determine the necessary conditions for equilibrium.

14.80 Referring to Problem 14.79, let each phase be a binary mixture of argon and helium and the wall be permeable only to argon. If the phases initially are at the conditions tabulated below, determine the final equilibrium temperature, pressure, and composition in the two phases.

	T(K)	p(MPa)	n(kmol)	y_{Ar}	y_{He}
Phase A	300	0.2	6	0.5	0.5
Phase B	400	0.1	5	0.8	0.2

14.81 Figure P14.81 shows an ideal gas mixture at temperature T and pressure p containing substance k, separated from a gas phase of pure k at temperature T and pressure p' by a semipermeable membrane that allows only k to pass through. Assuming the ideal gas model also applies to the pure gas phase, determine the relationship between p and p' for there to be no net transfer of k through the membrane.

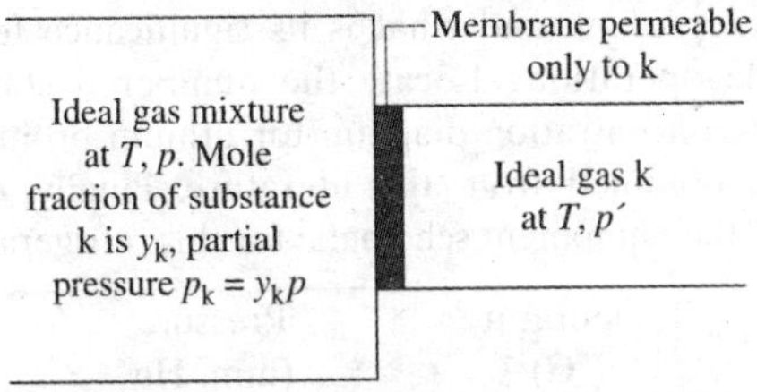

Fig. P14.81

14.82 What is the maximum number of homogeneous phases that can exist at equilibrium for a system involving

(a) one component?
(b) two components?
(c) three components?

14.83 Determine the number of degrees of freedom for systems composed of

(a) ice and liquid water.
(b) ice, liquid water, and water vapor.
(c) liquid water and water vapor.
(d) water vapor only.
(e) water vapor and dry air.
(f) liquid water, water vapor, and dry air.
(g) ice, water vapor, and dry air.
(h) N_2 and O_2 at 20°C, 1 atm.
(i) a liquid phase and a vapor phase, each of which contains ammonia and water.
(j) liquid mercury, liquid water, and a vapor phase of mercury and water.
(k) liquid acetone and a vapor phase of acetone and N_2.

14.84 Develop the *phase rule* for chemically reacting systems.

14.85 Apply the result of Problem 14.84 to determine the number of degrees of freedom for the gas phase reaction:

$$CH_4 + H_2O \rightleftarrows CO + 3H_2$$

14.86 For a gas–liquid system in equilibrium at temperature T and pressure p, *Raoult's law* models the relation between the partial pressure of substance i in the gas phase, p_i, and the mole fraction of substance i in the liquid phase, y_i, as follows:

$$p_i = y_i p_{sat,i}(T)$$

where $p_{sat,i}(T)$ is the saturation pressure of pure i at temperature T. The gas phase is assumed to form an ideal gas mixture; thus, $p_i = x_i p$ where x_i is the mole fraction of i in the gas phase. Apply Raoult's law to the following cases, which are representative of conditions that might be encountered in ammonia–water absorption systems (Sec. 10.5):

(a) Consider a two-phase, liquid–vapor ammonia–water system in equilibrium at 20°C. The mole fraction of ammonia in the liquid phase is 80%. Determine the pressure, in bar, and the mole fraction of ammonia in the vapor phase.
(b) Determine the mole fractions of ammonia in the liquid and vapor phases of a two-phase ammonia–water system in equilibrium at 40°C, 12 bar.

Design & open ended problems: exploring engineering practice

14.1D Using appropriate software, develop plots giving the variation with equivalence ratio of the equilibrium products of octane–air mixtures at 30 atm and selected temperatures ranging from 1700 to 2800 K. Consider equivalence ratios in the interval from 0.2 to 1.4 and equilibrium products including, but not necessarily limited to, CO_2, CO, H_2O, O_2, O, H_2, N_2, NO, OH. Under what conditions is the formation of nitric oxide (NO) and carbon monoxide (CO) most significant? Discuss.

14.2D Spark-ignition engine exhaust gases contain several air pollutants including the oxides of nitrogen, NO and NO_2, collectively known as NO_x. Additionally, the exhaust gases may contain carbon monoxide (CO) and unburned or partially burned hydrocarbons (HC).

(a) The pollutant amounts actually present depend on engine design and operating conditions, and typically differ significantly from values calculated on the basis of chemical equilibrium. Discuss both the reasons for these discrepancies and possible mechanisms by which such pollutants are formed in an actual engine.
(b) For spark-ignition engines, the average production of pollutants upstream of the catalyst, in g per mile of vehicle travel, are nitric oxides, 1.5; hydrocarbons, 2; and carbon monoxide, 20. For a city in your locale having a population of 100,000 or more, estimate the annual amount, in kg, of each pollutant that would be discharged if automobiles had no emission control devices. Repeat if the vehicles adhere to current U.S. government emissions standards.

14.3D The Federal Clean Air Act of 1970 and succeeding Clean Air Act Amendments target the oxides of nitrogen NO and NO_2, collectively known as NO_x, as significant air pollutants. NO_x is formed in combustion via three primary mechanisms: *thermal* NO_x formation, *prompt* NO_x formation, and *fuel* NO_x formation. Discuss these formation mechanisms, including a discussion of thermal NO_x formation by the *Zeldovich mechanism*. What is the role of NO_x in the formation of ozone? What are some NO_x reduction strategies?

14.4D The amount of sulfur dioxide (SO_2) present in *off gases* from industrial processes can be reduced by oxidizing the

SO_2 to SO_3 at an elevated temperature in a catalytic reactor. The SO_3 can be reacted in turn with water to form sulfuric acid that has economic value. For an off gas at 1 atm having the molar analysis of 12% SO_2, 8% O_2, 80% N_2, estimate the range of temperatures at which a *substantial* conversion of SO_2 to SO_3 might be realized.

14.5D A gaseous mixture of hydrogen (H_2) and carbon monoxide (CO) enters a catalytic reactor and a gaseous mixture of methanol (CH_3OH), hydrogen, and carbon monoxide exits. At the preliminary process design stage, a plausible estimate is required of the inlet hydrogen mole fraction, y_{H_2}, the temperature of the exiting mixture, T_e, and the pressure of the exiting mixture, p_e, subject to the following four constraints: (1) $0.5 \le y_{H_2} \le 0.75$, (2) $300 \le T_e \le 400$ K, (3) $1 \le p_e \le 10$ atm, and (4) the exiting mixture contains at least 75% methanol on a molar basis. Obtain this estimate.

14.6D When systems in thermal, mechanical, and chemical equilibrium are *perturbed,* changes within the systems can occur, leading to a new equilibrium state. The effects of perturbing the system considered in developing Eqs. 14.32 and 14.33 can be determined by study of these equations. For example, at fixed pressure and temperature it can be concluded that an increase in the amount of the inert component E would lead to increases in n_C and n_D when $\Delta\nu = (\nu_C + \nu_D - \nu_A - \nu_B)$ is positive, to decreases in n_C and n_D when $\Delta\nu$ is negative, and no change when $\Delta\nu = 0$.

(a) For a system consisting of NH_3, N_2, and H_2 at fixed pressure and temperature, subject to the reaction

$$2NH_3(g) = N_2(g) + 3H_2(g)$$

investigate the effects, in turn, of additions in the amounts present of NH_3, H_2, and N_2.

(b) For the *general case* of Eqs. 14.32 and 14.33, investigate the effects, in turn, of additions of A, B, C, and D.

14.7D With reference to the equilibrium constant data of Table A-27:

(a) For each of the tabulated reactions plot $\log_{10}K$ versus $1/T$ and determine the slope of the line of best fit. What is the thermodynamic significance of the slope? Check your conclusion about the slope using data from the *JANAF* tables.[2]

(b) A text states that the magnitude of the equilibrium constant often signals the importance of a reaction, and offers this *rule of thumb:* When $K < 10^{-3}$, the extent of the reaction is usually not significant, whereas when $K > 10^{3}$ the reaction generally proceeds closely to equilibrium. Confirm or deny this rule.

14.8D (a) For an equilibrium ideal gas mixture of N_2, H_2, and NH_3, evaluate the equilibrium constant from an expression you derive from the *van't Hoff equation* that requires only standard state enthalpy of formation and Gibbs function of formation data together with suitable analytical expressions in terms of temperature for the ideal gas specific heats of N_2, H_2, NH_3.

(b) For the synthesis of ammonia by

$\frac{1}{2}N_2 + \frac{3}{2}H_2 \rightarrow NH_3$

provide a recommendation for the ranges of temperature and pressure for which the mole fraction of ammonia in the mixture is at least 0.5.

14.9D Figure P14.9D gives a table of data for a lithium bromide–water absorption refrigeration cycle together with the sketch of a property diagram showing the cycle. The property diagram plots the vapor pressure versus the lithium bromide concentration. Apply the *phase rule* to verify that the numbered states are fixed by the property values provided. What does the *crystallization* line on the equilibrium diagram represent, and what is its significance for adsorption cycle operation? Locate the numbered states on an enthalpy–concentration diagram for lithium bromide–water solutions obtained from the literature. Finally, develop a sketch of the equipment schematic for this refrigeration cycle.

State	Temperature (°C)	Pressure (mm. Hg)	$(mf)_{LiBr}$ (%)
1	46	6.75	63.3
2	40	6.75	59.5
3	75	41.25	59.5
4	89	75.00	59.5
5	102	75.00	64.0
6	57	11.25	64.0
7	49	8.00	63.3

14.10D U.S. Patent 5,298,233 describes a means for converting industrial wastes to carbon dioxide and water vapor. Hydrogen- and carbon-containing feed, such as organic or inorganic sludge, low-grade fuel oil, or municipal garbage, is introduced into a molten bath consisting of two immiscible molten metal phases. The carbon and hydrogen of the feed are converted, respectively, to dissolved carbon and dissolved hydrogen. The dissolved carbon is oxidized in the first molten metal phase to carbon dioxide, which is released to the atmosphere. The dissolved hydrogen migrates to the second molten metal phase, where it is oxidized to form water vapor, which is also released from the bath. Critically evaluate this technology for waste disposal. Is the technology promising commercially? Compare with alternative waste management practices such as pyrolysis and incineration.

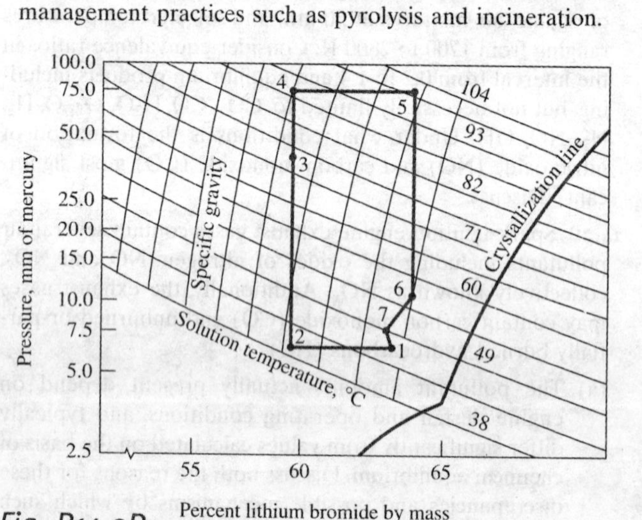

Fig. P14.9D

[2]Stull, D. R., and H. Prophet, *JANAF Thermochemical Tables,* 2nd ed., NSRDS-NBS 37, National Bureau of Standards, Washington, DC, June 1971.

Index to Tables in SI Units

Table A-1 **Atomic or Molecular Weights and Critical Properties of Selected Elements and Compounds**

Substance	Chemical Formula	M (kg/kmol)	T_c (K)	p_c (bar)	$Z_c = \frac{p_c v_c}{RT_c}$
Acetylene	C_2H_2	26.04	309	62.8	0.274
Air (equivalent)	—	28.97	133	37.7	0.284
Ammonia	NH_3	17.03	406	112.8	0.242
Argon	Ar	39.94	151	48.6	0.290
Benzene	C_6H_6	78.11	563	49.3	0.274
Butane	C_4H_{10}	58.12	425	38.0	0.274
Carbon	C	12.01	—	—	—
Carbon dioxide	CO_2	44.01	304	73.9	0.276
Carbon monoxide	CO	28.01	133	35.0	0.294
Copper	Cu	63.54	—	—	—
Ethane	C_2H_6	30.07	305	48.8	0.285
Ethyl alcohol	C_2H_5OH	46.07	516	63.8	0.249
Ethylene	C_2H_4	28.05	283	51.2	0.270
Helium	He	4.003	5.2	2.3	0.300
Hydrogen	H_2	2.016	33.2	13.0	0.304
Methane	CH_4	16.04	191	46.4	0.290
Methyl alcohol	CH_3OH	32.04	513	79.5	0.220
Nitrogen	N_2	28.01	126	33.9	0.291
Octane	C_8H_{18}	114.22	569	24.9	0.258
Oxygen	O_2	32.00	154	50.5	0.290
Propane	C_3H_8	44.09	370	42.7	0.276
Propylene	C_3H_6	42.08	365	46.2	0.276
Refrigerant 12	CCl_2F_2	120.92	385	41.2	0.278
Refrigerant 22	$CHClF_2$	86.48	369	49.8	0.267
Refrigerant 134a	CF_3CH_2F	102.03	374	40.7	0.260
Sulfur dioxide	SO_2	64.06	431	78.7	0.268
Water	H_2O	18.02	647.3	220.9	0.233

Sources: Adapted from *International Critical Tables* and L. C. Nelson and E. F. Obert, Generalized Compressibility Charts, *Chem. Eng., 61:* 203 (1954).

H_2O

Table A-2 **Properties of Saturated Water (Liquid–Vapor): Temperature Table**

Pressure Conversions:
1 bar = 0.1 MPa
= 10^2 kPa

Temp. °C	Press. bar	Specific Volume m^3/kg Sat. Liquid $v_f \times 10^3$	Sat. Vapor v_g	Internal Energy kJ/kg Sat. Liquid u_f	Sat. Vapor u_g	Enthalpy kJ/kg Sat. Liquid h_f	Evap. h_{fg}	Sat. Vapor h_g	Entropy kJ/kg · K Sat. Liquid s_f	Sat. Vapor s_g	Temp. °C
.01	0.00611	1.0002	206.136	0.00	2375.3	0.01	2501.3	2501.4	0.0000	9.1562	.01
4	0.00813	1.0001	157.232	16.77	2380.9	16.78	2491.9	2508.7	0.0610	9.0514	4
5	0.00872	1.0001	147.120	20.97	2382.3	20.98	2489.6	2510.6	0.0761	9.0257	5
6	0.00935	1.0001	137.734	25.19	2383.6	25.20	2487.2	2512.4	0.0912	9.0003	6
8	0.01072	1.0002	120.917	33.59	2386.4	33.60	2482.5	2516.1	0.1212	8.9501	8
10	0.01228	1.0004	106.379	42.00	2389.2	42.01	2477.7	2519.8	0.1510	8.9008	10
11	0.01312	1.0004	99.857	46.20	2390.5	46.20	2475.4	2521.6	0.1658	8.8765	11
12	0.01402	1.0005	93.784	50.41	2391.9	50.41	2473.0	2523.4	0.1806	8.8524	12
13	0.01497	1.0007	88.124	54.60	2393.3	54.60	2470.7	2525.3	0.1953	8.8285	13
14	0.01598	1.0008	82.848	58.79	2394.7	58.80	2468.3	2527.1	0.2099	8.8048	14
15	0.01705	1.0009	77.926	62.99	2396.1	62.99	2465.9	2528.9	0.2245	8.7814	15
16	0.01818	1.0011	73.333	67.18	2397.4	67.19	2463.6	2530.8	0.2390	8.7582	16
17	0.01938	1.0012	69.044	71.38	2398.8	71.38	2461.2	2532.6	0.2535	8.7351	17
18	0.02064	1.0014	65.038	75.57	2400.2	75.58	2458.8	2534.4	0.2679	8.7123	18
19	0.02198	1.0016	61.293	79.76	2401.6	79.77	2456.5	2536.2	0.2823	8.6897	19
20	0.02339	1.0018	57.791	83.95	2402.9	83.96	2454.1	2538.1	0.2966	8.6672	20
21	0.02487	1.0020	54.514	88.14	2404.3	88.14	2451.8	2539.9	0.3109	8.6450	21
22	0.02645	1.0022	51.447	92.32	2405.7	92.33	2449.4	2541.7	0.3251	8.6229	22
23	0.02810	1.0024	48.574	96.51	2407.0	96.52	2447.0	2543.5	0.3393	8.6011	23
24	0.02985	1.0027	45.883	100.70	2408.4	100.70	2444.7	2545.4	0.3534	8.5794	24
25	0.03169	1.0029	43.360	104.88	2409.8	104.89	2442.3	2547.2	0.3674	8.5580	25
26	0.03363	1.0032	40.994	109.06	2411.1	109.07	2439.9	2549.0	0.3814	8.5367	26
27	0.03567	1.0035	38.774	113.25	2412.5	113.25	2437.6	2550.8	0.3954	8.5156	27
28	0.03782	1.0037	36.690	117.42	2413.9	117.43	2435.2	2552.6	0.4093	8.4946	28
29	0.04008	1.0040	34.733	121.60	2415.2	121.61	2432.8	2554.5	0.4231	8.4739	29
30	0.04246	1.0043	32.894	125.78	2416.6	125.79	2430.5	2556.3	0.4369	8.4533	30
31	0.04496	1.0046	31,165	129.96	2418.0	129.97	2428.1	2558.1	0.4507	8.4329	31
32	0.04759	1.0050	29.540	134.14	2419.3	134.15	2425.7	2559.9	0.4644	8.4127	32
33	0.05034	1.0053	28.011	138.32	2420.7	138.33	2423.4	2561.7	0.4781	8.3927	33
34	0.05324	1.0056	26.571	142.50	2422.0	142.50	2421.0	2563.5	0.4917	8.3728	34
35	0.05628	1.0060	25.216	146.67	2423.4	146.68	2418.6	2565.3	0.5053	8.3531	35
36	0.05947	1.0063	23.940	150.85	2424.7	150.86	2416.2	2567.1	0.5188	8.3336	36
38	0.06632	1.0071	21.602	159.20	2427.4	159.21	2411.5	2570.7	0.5458	8.2950	38
40	0.07384	1.0078	19.523	167.56	2430.1	167.57	2406.7	2574.3	0.5725	8.2570	40
45	0.09593	1.0099	15.258	188.44	2436.8	188.45	2394.8	2583.2	0.6387	8.1648	45

H_2O

Table A-2 ***(Continued)***

		Specific Volume m^3/kg		Internal Energy kJ/kg		Enthalpy kJ/kg			Entropy kJ/kg · K		
Temp. °C	Press. bar	Sat. Liquid $v_f \times 10^3$	Sat. Vapor v_g	Sat. Liquid u_f	Sat. Vapor u_g	Sat. Liquid h_f	Evap. h_{fg}	Sat. Vapor h_g	Sat. Liquid s_f	Sat. Vapor s_g	Temp. °C
50	.1235	1.0121	12.032	209.32	2443.5	209.33	2382.7	2592.1	.7038	8.0763	50
55	.1576	1.0146	9.568	230.21	2450.1	230.23	2370.7	2600.9	.7679	7.9913	55
60	.1994	1.0172	7.671	251.11	2456.6	251.13	2358.5	2609.6	.8312	7.9096	60
65	.2503	1.0199	6.197	272.02	2463.1	272.06	2346.2	2618.3	.8935	7.8310	65
70	.3119	1.0228	5.042	292.95	2469.6	292.98	2333.8	2626.8	.9549	7.7553	70
75	.3858	1.0259	4.131	313.90	2475.9	313.93	2321.4	2635.3	1.0155	7.6824	75
80	.4739	1.0291	3.407	334.86	2482.2	334.91	2308.8	2643.7	1.0753	7.6122	80
85	.5783	1.0325	2.828	355.84	2488.4	355.90	2296.0	2651.9	1.1343	7.5445	85
90	.7014	1.0360	2.361	376.85	2494.5	376.92	2283.2	2660.1	1.1925	7.4791	90
95	.8455	1.0397	1.982	397.88	2500.6	397.96	2270.2	2668.1	1.2500	7.4159	95
100	1.014	1.0435	1.673	418.94	2506.5	419.04	2257.0	2676.1	1.3069	7.3549	100
110	1.433	1.0516	1.210	461.14	2518.1	461.30	2230.2	2691.5	1.4185	7.2387	110
120	1.985	1.0603	0.8919	503.50	2529.3	503.71	2202.6	2706.3	1.5276	7.1296	120
130	2.701	1.0697	0.6685	546.02	2539.9	546.31	2174.2	2720.5	1.6344	7.0269	130
140	3.613	1.0797	0.5089	588.74	2550.0	589.13	2144.7	2733.9	1.7391	6.9299	140
150	4.758	1.0905	0.3928	631.68	2559.5	632.20	2114.3	2746.5	1.8418	6.8379	150
160	6.178	1.1020	0.3071	674.86	2568.4	675.55	2082.6	2758.1	1.9427	6.7502	160
170	7.917	1.1143	0.2428	718.33	2576.5	719.21	2049.5	2768.7	2.0419	6.6663	170
180	10.02	1.1274	0.1941	762.09	2583.7	763.22	2015.0	2778.2	2.1396	6.5857	180
190	12.54	1.1414	0.1565	806.19	2590.0	807.62	1978.8	2786.4	2.2359	6.5079	190
200	15.54	1.1565	0.1274	850.65	2595.3	852.45	1940.7	2793.2	2.3309	6.4323	200
210	19.06	1.1726	0.1044	895.53	2599.5	897.76	1900.7	2798.5	2.4248	6.3585	210
220	23.18	1.1900	0.08619	940.87	2602.4	943.62	1858.5	2802.1	2.5178	6.2861	220
230	27.95	1.2088	0.07158	986.74	2603.9	990.12	1813.8	2804.0	2.6099	6.2146	230
240	33.44	1.2291	0.05976	1033.2	2604.0	1037.3	1766.5	2803.8	2.7015	6.1437	240
250	39.73	1.2512	0.05013	1080.4	2602.4	1085.4	1716.2	2801.5	2.7927	6.0730	250
260	46.88	1.2755	0.04221	1128.4	2599.0	1134.4	1662.5	2796.6	2.8838	6.0019	260
270	54.99	1.3023	0.03564	1177.4	2593.7	1184.5	1605.2	2789.7	2.9751	5.9301	270
280	64.12	1.3321	0.03017	1227.5	2586.1	1236.0	1543.6	2779.6	3.0668	5.8571	280
290	74.36	1.3656	0.02557	1278.9	2576.0	1289.1	1477.1	2766.2	3.1594	5.7821	290
300	85.81	1.4036	0.02167	1332.0	2563.0	1344.0	1404.9	2749.0	3.2534	5.7045	300
320	112.7	1.4988	0.01549	1444.6	2525.5	1461.5	1238.6	2700.1	3.4480	5.5362	320
340	145.9	1.6379	0.01080	1570.3	2464.6	1594.2	1027.9	2622.0	3.6594	5.3357	340
360	186.5	1.8925	0.006945	1725.2	2351.5	1760.5	720.5	2481.0	3.9147	5.0526	360
374.14	220.9	3.155	0.003155	2029.6	2029.6	2099.3	0	2099.3	4.4298	4.4298	374.14

Source: Tables A-2 through A-5 are extracted from J. H. Keenan, F. G. Keyes, P. G. Hill, and J. G. Moore, *Steam Tables,* Wiley, New York, 1969.

Table A-3 **Properties of Saturated Water (Liquid–Vapor): Pressure Table**

Pressure Conversions:
1 bar = 0.1 MPa
= 10^2 kPa

		Specific Volume m^3/kg		Internal Energy kJ/kg		Enthalpy kJ/kg			Entropy kJ/kg · K		
Press. bar	Temp. °C	Sat. Liquid $v_f \times 10^3$	Sat. Vapor v_g	Sat. Liquid u_f	Sat. Vapor u_g	Sat. Liquid h_f	Evap. h_{fg}	Sat. Vapor h_g	Sat. Liquid s_f	Sat. Vapor s_g	Press. bar
0.04	28.96	1.0040	34.800	121.45	2415.2	121.46	2432.9	2554.4	0.4226	8.4746	0.04
0.06	36.16	1.0064	23.739	151.53	2425.0	151.53	2415.9	2567.4	0.5210	8.3304	0.06
0.08	41.51	1.0084	18.103	173.87	2432.2	173.88	2403.1	2577.0	0.5926	8.2287	0.08
0.10	45.81	1.0102	14.674	191.82	2437.9	191.83	2392.8	2584.7	0.6493	8.1502	0.10
0.20	60.06	1.0172	7.649	251.38	2456.7	251.40	2358.3	2609.7	0.8320	7.9085	0.20
0.30	69.10	1.0223	5.229	289.20	2468.4	289.23	2336.1	2625.3	0.9439	7.7686	0.30
0.40	75.87	1.0265	3.993	317.53	2477.0	317.58	2319.2	2636.8	1.0259	7.6700	0.40
0.50	81.33	1.0300	3.240	340.44	2483.9	340.49	2305.4	2645.9	1.0910	7.5939	0.50
0.60	85.94	1.0331	2.732	359.79	2489.6	359.86	2293.6	2653.5	1.1453	7.5320	0.60
0.70	89.95	1.0360	2.365	376.63	2494.5	376.70	2283.3	2660.0	1.1919	7.4797	0.70
0.80	93.50	1.0380	2.087	391.58	2498.8	391.66	2274.1	2665.8	1.2329	7.4346	0.80
0.90	96.71	1.0410	1.869	405.06	2502.6	405.15	2265.7	2670.9	1.2695	7.3949	0.90
1.00	99.63	1.0432	1.694	417.36	2506.1	417.46	2258.0	2675.5	1.3026	7.3594	1.00
1.50	111.4	1.0528	1.159	466.94	2519.7	467.11	2226.5	2693.6	1.4336	7.2233	1.50
2.00	120.2	1.0605	0.8857	504.49	2529.5	504.70	2201.9	2706.7	1.5301	7.1271	2.00
2.50	127.4	1.0672	0.7187	535.10	2537.2	535.37	2181.5	2716.9	1.6072	7.0527	2.50
3.00	133.6	1.0732	0.6058	561.15	2543.6	561.47	2163.8	2725.3	1.6718	6.9919	3.00
3.50	138.9	1.0786	0.5243	583.95	2546.9	584.33	2148.1	2732.4	1.7275	6.9405	3.50
4.00	143.6	1.0836	0.4625	604.31	2553.6	604.74	2133.8	2738.6	1.7766	6.8959	4.00
4.50	147.9	1.0882	0.4140	622.25	2557.6	623.25	2120.7	2743.9	1.8207	6.8565	4.50
5.00	151.9	1.0926	0.3749	639.68	2561.2	640.23	2108.5	2748.7	1.8607	6.8212	5.00
6.00	158.9	1.1006	0.3157	669.90	2567.4	670.56	2086.3	2756.8	1.9312	6.7600	6.00
7.00	165.0	1.1080	0.2729	696.44	2572.5	697.22	2066.3	2763.5	1.9922	6.7080	7.00
8.00	170.4	1.1148	0.2404	720.22	2576.8	721.11	2048.0	2769.1	2.0462	6.6628	8.00
9.00	175.4	1.1212	0.2150	741.83	2580.5	742.83	2031.1	2773.9	2.0946	6.6226	9.00
10.0	179.9	1.1273	0.1944	761.68	2583.6	762.81	2015.3	2778.1	2.1387	6.5863	10.0
15.0	198.3	1.1539	0.1318	843.16	2594.5	844.84	1947.3	2792.2	2.3150	6.4448	15.0
20.0	212.4	1.1767	0.09963	906.44	2600.3	908.79	1890.7	2799.5	2.4474	6.3409	20.0
25.0	224.0	1.1973	0.07998	959.11	2603.1	962.11	1841.0	2803.1	2.5547	6.2575	25.0
30.0	233.9	1.2165	0.06668	1004.8	2604.1	1008.4	1795.7	2804.2	2.6457	6.1869	30.0
35.0	242.6	1.2347	0.05707	1045.4	2603.7	1049.8	1753.7	2803.4	2.7253	6.1253	35.0
40.0	250.4	1.2522	0.04978	1082.3	2602.3	1087.3	1714.1	2801.4	2.7964	6.0701	40.0
45.0	257.5	1.2692	0.04406	1116.2	2600.1	1121.9	1676.4	2798.3	2.8610	6.0199	45.0
50.0	264.0	1.2859	0.03944	1147.8	2597.1	1154.2	1640.1	2794.3	2.9202	5.9734	50.0
60.0	275.6	1.3187	0.03244	1205.4	2589.7	1213.4	1571.0	2784.3	3.0267	5.8892	60.0
70.0	285.9	1.3513	0.02737	1257.6	2580.5	1267.0	1505.1	2772.1	3.1211	5.8133	70.0
80.0	295.1	1.3842	0.02352	1305.6	2569.8	1316.6	1441.3	2758.0	3.2068	5.7432	80.0
90.0	303.4	1.4178	0.02048	1350.5	2557.8	1363.3	1378.9	2742.1	3.2858	5.6772	90.0
100.	311.1	1.4524	0.01803	1393.0	2544.4	1407.6	1317.1	2724.7	3.3596	5.6141	100.
110.	318.2	1.4886	0.01599	1433.7	2529.8	1450.1	1255.5	2705.6	3.4295	5.5527	110.

H_2O

Table A-3 *(Continued)*

Press. bar	Temp. °C	Specific Volume m³/kg Sat. Liquid $v_f \times 10^3$	Sat. Vapor v_g	Internal Energy kJ/kg Sat. Liquid u_f	Sat. Vapor u_g	Enthalpy kJ/kg Sat. Liquid h_f	Evap. h_{fg}	Sat. Vapor h_g	Entropy kJ/kg · K Sat. Liquid s_f	Sat. Vapor s_g	Press. bar
120.	324.8	1.5267	0.01426	1473.0	2513.7	1491.3	1193.6	2684.9	3.4962	5.4924	120.
130.	330.9	1.5671	0.01278	1511.1	2496.1	1531.5	1130.7	2662.2	3.5606	5.4323	130.
140.	336.8	1.6107	0.01149	1548.6	2476.8	1571.1	1066.5	2637.6	3.6232	5.3717	140.
150.	342.2	1.6581	0.01034	1585.6	2455.5	1610.5	1000.0	2610.5	3.6848	5.3098	150.
160.	347.4	1.7107	0.009306	1622.7	2431.7	1650.1	930.6	2580.6	3.7461	5.2455	160.
170.	352.4	1.7702	0.008364	1660.2	2405.0	1690.3	856.9	2547.2	3.8079	5.1777	170.
180.	357.1	1.8397	0.007489	1698.9	2374.3	1732.0	777.1	2509.1	3.8715	5.1044	180.
190.	361.5	1.9243	0.006657	1739.9	2338.1	1776.5	688.0	2464.5	3.9388	5.0228	190.
200.	365.8	2.036	0.005834	1785.6	2293.0	1826.3	583.4	2409.7	4.0139	4.9269	200.
220.9	374.1	3.155	0.003155	2029.6	2029.6	2099.3	0	2099.3	4.4298	4.4298	220.9

Table A-4 Properties of Superheated Water Vapor

Pressure Conversions:
1 bar = 0.1 MPa
= 10^2 kPa

H_2O

T °C	v m³/kg	u kJ/kg	h kJ/kg	s kJ/kg · K	v m³/kg	u kJ/kg	h kJ/kg	s kJ/kg · K
	p = 0.06 bar = 0.006 MPa (T_{sat} = 36.16°C)				p = 0.35 bar = 0.035 MPa (T_{sat} = 72.69°C)			
Sat.	23.739	2425.0	2567.4	8.3304	4.526	2473.0	2631.4	7.7158
80	27.132	2487.3	2650.1	8.5804	4.625	2483.7	2645.6	7.7564
120	30.219	2544.7	2726.0	8.7840	5.163	2542.4	2723.1	7.9644
160	33.302	2602.7	2802.5	8.9693	5.696	2601.2	2800.6	8.1519
200	36.383	2661.4	2879.7	9.1398	6.228	2660.4	2878.4	8.3237
240	39.462	2721.0	2957.8	9.2982	6.758	2720.3	2956.8	8.4828
280	42.540	2781.5	3036.8	9.4464	7.287	2780.9	3036.0	8.6314
320	45.618	2843.0	3116.7	9.5859	7.815	2842.5	3116.1	8.7712
360	48.696	2905.5	3197.7	9.7180	8.344	2905.1	3197.1	8.9034
400	51.774	2969.0	3279.6	9.8435	8.872	2968.6	3279.2	9.0291
440	54.851	3033.5	3362.6	9.9633	9.400	3033.2	3362.2	9.1490
500	59.467	3132.3	3489.1	10.1336	10.192	3132.1	3488.8	9.3194

T °C	v m³/kg	u kJ/kg	h kJ/kg	s kJ/kg · K	v m³/kg	u kJ/kg	h kJ/kg	s kJ/kg · K
	p = 0.70 bar = 0.07 MPa (T_{sat} = 89.95°C)				p = 1.0 bar = 0.10 MPa (T_{sat} = 99.63°C)			
Sat.	2.365	2494.5	2660.0	7.4797	1.694	2506.1	2675.5	7.3594
100	2.434	2509.7	2680.0	7.5341	1.696	2506.7	2676.2	7.3614
120	2.571	2539.7	2719.6	7.6375	1.793	2537.3	2716.6	7.4668
160	2.841	2599.4	2798.2	7.8279	1.984	2597.8	2796.2	7.6597
200	3.108	2659.1	2876.7	8.0012	2.172	2658.1	2875.3	7.8343
240	3.374	2719.3	2955.5	8.1611	2.359	2718.5	2954.5	7.9949
280	3.640	2780.2	3035.0	8.3162	2.546	2779.6	3034.2	8.1445
320	3.905	2842.0	3115.3	8.4504	2.732	2841.5	3114.6	8.2849
360	4.170	2904.6	3196.5	8.5828	2.917	2904.2	3195.9	8.4175
400	4.434	2968.2	3278.6	8.7086	3.103	2967.9	3278.2	8.5435
440	4.698	3032.9	3361.8	8.8286	3.288	3032.6	3361.4	8.6636
500	5.095	3131.8	3488.5	8.9991	3.565	3131.6	3488.1	8.8342

T °C	v m³/kg	u kJ/kg	h kJ/kg	s kJ/kg · K	v m³/kg	u kJ/kg	h kJ/kg	s kJ/kg · K
	p = 1.5 bar = 0.15 MPa (T_{sat} = 111.37°C)				p = 3.0 bar = 0.30 MPa (T_{sat} = 133.55°C)			
Sat.	1.159	2519.7	2693.6	7.2233	0.606	2543.6	2725.3	6.9919
120	1.188	2533.3	2711.4	7.2693				
160	1.317	2595.2	2792.8	7.4665	0.651	2587.1	2782.3	7.1276
200	1.444	2656.2	2872.9	7.6433	0.716	2650.7	2865.5	7.3115
240	1.570	2717.2	2952.7	7.8052	0.781	2713.1	2947.3	7.4774
280	1.695	2778.6	3032.8	7.9555	0.844	2775.4	3028.6	7.6299
320	1.819	2840.6	3113.5	8.0964	0.907	2838.1	3110.1	7.7722
360	1.943	2903.5	3195.0	8.2293	0.969	2901.4	3192.2	7.9061
400	2.067	2967.3	3277.4	8.3555	1.032	2965.6	3275.0	8.0330
440	2.191	3032.1	3360.7	8.4757	1.094	3030.6	3358.7	8.1538
500	2.376	3131.2	3487.6	8.6466	1.187	3130.0	3486.0	8.3251
600	2.685	3301.7	3704.3	8.9101	1.341	3300.8	3703.2	8.5892

H_2O

Table A-4 (Continued)

T °C	v m³/kg	u kJ/kg	h kJ/kg	s kJ/kg · K	v m³/kg	u kJ/kg	h kJ/kg	s kJ/kg · K
	p = 5.0 bar = 0.50 MPa (T_{sat} = 151.86°C)				p = 7.0 bar = 0.70 MPa (T_{sat} = 164.97°C)			
Sat.	0.3749	2561.2	2748.7	6.8213	0.2729	2572.5	2763.5	6.7080
180	0.4045	2609.7	2812.0	6.9656	0.2847	2599.8	2799.1	6.7880
200	0.4249	2642.9	2855.4	7.0592	0.2999	2634.8	2844.8	6.8865
240	0.4646	2707.6	2939.9	7.2307	0.3292	2701.8	2932.2	7.0641
280	0.5034	2771.2	3022.9	7.3865	0.3574	2766.9	3017.1	7.2233
320	0.5416	2834.7	3105.6	7.5308	0.3852	2831.3	3100.9	7.3697
360	0.5796	2898.7	3188.4	7.6660	0.4126	2895.8	3184.7	7.5063
400	0.6173	2963.2	3271.9	7.7938	0.4397	2960.9	3268.7	7.6350
440	0.6548	3028.6	3356.0	7.9152	0.4667	3026.6	3353.3	7.7571
500	0.7109	3128.4	3483.9	8.0873	0.5070	3126.8	3481.7	7.9299
600	0.8041	3299.6	3701.7	8.3522	0.5738	3298.5	3700.2	8.1956
700	0.8969	3477.5	3925.9	8.5952	0.6403	3476.6	3924.8	8.4391

T °C	v m³/kg	u kJ/kg	h kJ/kg	s kJ/kg · K	v m³/kg	u kJ/kg	h kJ/kg	s kJ/kg · K
	p = 10.0 bar = 1.0 MPa (T_{sat} = 179.91°C)				p = 15.0 bar = 1.5 MPa (T_{sat} = 198.32°C)			
Sat.	0.1944	2583.6	2778.1	6.5865	0.1318	2594.5	2792.2	6.4448
200	0.2060	2621.9	2827.9	6.6940	0.1325	2598.1	2796.8	6.4546
240	0.2275	2692.9	2920.4	6.8817	0.1483	2676.9	2899.3	6.6628
280	0.2480	2760.2	3008.2	7.0465	0.1627	2748.6	2992.7	6.8381
320	0.2678	2826.1	3093.9	7.1962	0.1765	2817.1	3081.9	6.9938
360	0.2873	2891.6	3178.9	7.3349	0.1899	2884.4	3169.2	7.1363
400	0.3066	2957.3	3263.9	7.4651	0.2030	2951.3	3255.8	7.2690
440	0.3257	3023.6	3349.3	7.5883	0.2160	3018.5	3342.5	7.3940
500	0.3541	3124.4	3478.5	7.7622	0.2352	3120.3	3473.1	7.5698
540	0.3729	3192.6	3565.6	7.8720	0.2478	3189.1	3560.9	7.6805
600	0.4011	3296.8	3697.9	8.0290	0.2668	3293.9	3694.0	7.8385
640	0.4198	3367.4	3787.2	8.1290	0.2793	3364.8	3783.8	7.9391

T °C	v m³/kg	u kJ/kg	h kJ/kg	s kJ/kg · K	v m³/kg	u kJ/kg	h kJ/kg	s kJ/kg · K
	p = 20.0 bar = 2.0 MPa (T_{sat} = 212.42°C)				p = 30.0 bar = 3.0 MPa (T_{sat} = 233.90°C)			
Sat.	0.0996	2600.3	2799.5	6.3409	0.0667	2604.1	2804.2	6.1869
240	0.1085	2659.6	2876.5	6.4952	0.0682	2619.7	2824.3	6.2265
280	0.1200	2736.4	2976.4	6.6828	0.0771	2709.9	2941.3	6.4462
320	0.1308	2807.9	3069.5	6.8452	0.0850	2788.4	3043.4	6.6245
360	0.1411	2877.0	3159.3	6.9917	0.0923	2861.7	3138.7	6.7801
400	0.1512	2945.2	3247.6	7.1271	0.0994	2932.8	3230.9	6.9212
440	0.1611	3013.4	3335.5	7.2540	0.1062	3002.9	3321.5	7.0520
500	0.1757	3116.2	3467.6	7.4317	0.1162	3108.0	3456.5	7.2338
540	0.1853	3185.6	3556.1	7.5434	0.1227	3178.4	3546.6	7.3474
600	0.1996	3290.9	3690.1	7.7024	0.1324	3285.0	3682.3	7.5085
640	0.2091	3362.2	3780.4	7.8035	0.1388	3357.0	3773.5	7.6106
700	0.2232	3470.9	3917.4	7.9487	0.1484	3466.5	3911.7	7.7571

H₂O

Table A-4 **(Continued)**

T °C	v m³/kg	u kJ/kg	h kJ/kg	s kJ/kg · K	v m³/kg	u kJ/kg	h kJ/kg	s kJ/kg · K
	p = 40 bar = 4.0 MPa (T_{sat} = 250.4°C)				p = 60 bar = 6.0 MPa (T_{sat} = 275.64°C)			
Sat.	0.04978	2602.3	2801.4	6.0701	0.03244	2589.7	2784.3	5.8892
280	0.05546	2680.0	2901.8	6.2568	0.03317	2605.2	2804.2	5.9252
320	0.06199	2767.4	3015.4	6.4553	0.03876	2720.0	2952.6	6.1846
360	0.06788	2845.7	3117.2	6.6215	0.04331	2811.2	3071.1	6.3782
400	0.07341	2919.9	3213.6	6.7690	0.04739	2892.9	3177.2	6.5408
440	0.07872	2992.2	3307.1	6.9041	0.05122	2970.0	3277.3	6.6853
500	0.08643	3099.5	3445.3	7.0901	0.05665	3082.2	3422.2	6.8803
540	0.09145	3171.1	3536.9	7.2056	0.06015	3156.1	3517.0	6.9999
600	0.09885	3279.1	3674.4	7.3688	0.06525	3266.9	3658.4	7.1677
640	0.1037	3351.8	3766.6	7.4720	0.06859	3341.0	3752.6	7.2731
700	0.1110	3462.1	3905.9	7.6198	0.07352	3453.1	3894.1	7.4234
740	0.1157	3536.6	3999.6	7.7141	0.07677	3528.3	3989.2	7.5190
	p = 80 bar = 8.0 MPa (T_{sat} = 295.06°C)				p = 100 bar = 10.0 MPa (T_{sat} = 311.06°C)			
Sat.	0.02352	2569.8	2758.0	5.7432	0.01803	2544.4	2724.7	5.6141
320	0.02682	2662.7	2877.2	5.9489	0.01925	2588.8	2781.3	5.7103
360	0.03089	2772.7	3019.8	6.1819	0.02331	2729.1	2962.1	6.0060
400	0.03432	2863.8	3138.3	6.3634	0.02641	2832.4	3096.5	6.2120
440	0.03742	2946.7	3246.1	6.5190	0.02911	2922.1	3213.2	6.3805
480	0.04034	3025.7	3348.4	6.6586	0.03160	3005.4	3321.4	6.5282
520	0.04313	3102.7	3447.7	6.7871	0.03394	3085.6	3425.1	6.6622
560	0.04582	3178.7	3545.3	6.9072	0.03619	3164.1	3526.0	6.7864
600	0.04845	3254.4	3642.0	7.0206	0.03837	3241.7	3625.3	6.9029
640	0.05102	3330.1	3738.3	7.1283	0.04048	3318.9	3723.7	7.0131
700	0.05481	3443.9	3882.4	7.2812	0.04358	3434.7	3870.5	7.1687
740	0.05729	3520.4	3978.7	7.3782	0.04560	3512.1	3968.1	7.2670
	p = 120 bar = 12.0 MPa (T_{sat} = 324.75°C)				p = 140 bar = 14.0 MPa (T_{sat} = 336.75°C)			
Sat.	0.01426	2513.7	2684.9	5.4924	0.01149	2476.8	2637.6	5.3717
360	0.01811	2678.4	2895.7	5.8361	0.01422	2617.4	2816.5	5.6602
400	0.02108	2798.3	3051.3	6.0747	0.01722	2760.9	3001.9	5.9448
440	0.02355	2896.1	3178.7	6.2586	0.01954	2868.6	3142.2	6.1474
480	0.02576	2984.4	3293.5	6.4154	0.02157	2962.5	3264.5	6.3143
520	0.02781	3068.0	3401.8	6.5555	0.02343	3049.8	3377.8	6.4610
560	0.02977	3149.0	3506.2	6.6840	0.02517	3133.6	3486.0	6.5941
600	0.03164	3228.7	3608.3	6.8037	0.02683	3215.4	3591.1	6.7172
640	0.03345	3307.5	3709.0	6.9164	0.02843	3296.0	3694.1	6.8326
700	0.03610	3425.2	3858.4	7.0749	0.03075	3415.7	3846.2	6.9939
740	0.03781	3503.7	3957.4	7.1746	0.03225	3495.2	3946.7	7.0952

Table A-4 **(Continued)**

p = 160 bar = 16.0 MPa (T_{sat} = 347.44°C)

T °C	v m³/kg	u kJ/kg	h kJ/kg	s kJ/kg · K
Sat.	0.00931	2431.7	2580.6	5.2455
360	0.01105	2539.0	2715.8	5.4614
400	0.01426	2719.4	2947.6	5.8175
440	0.01652	2839.4	3103.7	6.0429
480	0.01842	2939.7	3234.4	6.2215
520	0.02013	3031.1	3353.3	6.3752
560	0.02172	3117.8	3465.4	6.5132
600	0.02323	3201.8	3573.5	6.6399
640	0.02467	3284.2	3678.9	6.7580
700	0.02674	3406.0	3833.9	6.9224
740	0.02808	3486.7	3935.9	7.0251

p = 180 bar = 18.0 MPa (T_{sat} = 357.06°C)

T °C	v m³/kg	u kJ/kg	h kJ/kg	s kJ/kg · K
Sat.	0.00749	2374.3	2509.1	5.1044
360	0.00809	2418.9	2564.5	5.1922
400	0.01190	2672.8	2887.0	5.6887
440	0.01414	2808.2	3062.8	5.9428
480	0.01596	2915.9	3203.2	6.1345
520	0.01757	3011.8	3378.0	6.2960
560	0.01904	3101.7	3444.4	6.4392
600	0.02042	3188.0	3555.6	6.5696
640	0.02174	3272.3	3663.6	6.6905
700	0.02362	3396.3	3821.5	6.8580
740	0.02483	3478.0	3925.0	6.9623

p = 200 bar = 20.0 MPa (T_{sat} = 365.81°C)

T °C	v m³/kg	u kJ/kg	h kJ/kg	s kJ/kg · K
Sat.	0.00583	2293.0	2409.7	4.9269
400	0.00994	2619.3	2818.1	5.5540
440	0.01222	2774.9	3019.4	5.8450
480	0.01399	2891.2	3170.8	6.0518
520	0.01551	2992.0	3302.2	6.2218
560	0.01689	3085.2	3423.0	6.3705
600	0.01818	3174.0	3537.6	6.5048
640	0.01940	3260.2	3648.1	6.6286
700	0.02113	3386.4	3809.0	6.7993
740	0.02224	3469.3	3914.1	6.9052
800	0.02385	3592.7	4069.7	7.0544

p = 240 bar = 24.0 MPa

T °C	v m³/kg	u kJ/kg	h kJ/kg	s kJ/kg · K
400	0.00673	2477.8	2639.4	5.2393
440	0.00929	2700.6	2923.4	5.6506
480	0.01100	2838.3	3102.3	5.8950
520	0.01241	2950.5	3248.5	6.0842
560	0.01366	3051.1	3379.0	6.2448
600	0.01481	3145.2	3500.7	6.3875
640	0.01588	3235.5	3616.7	6.5174
700	0.01739	3366.4	3783.8	6.6947
740	0.01835	3451.7	3892.1	6.8038
800	0.01974	3578.0	4051.6	6.9567

p = 280 bar = 28.0 MPa

T °C	v m³/kg	u kJ/kg	h kJ/kg	s kJ/kg · K
400	0.00383	2223.5	2330.7	4.7494
440	0.00712	2613.2	2812.6	5.4494
480	0.00885	2780.8	3028.5	5.7446
520	0.01020	2906.8	3192.3	5.9566
560	0.01136	3015.7	3333.7	6.1307
600	0.01241	3115.6	3463.0	6.2823
640	0.01338	3210.3	3584.8	6.4187
700	0.01473	3346.1	3758.4	6.6029
740	0.01558	3433.9	3870.0	6.7153
800	0.01680	3563.1	4033.4	6.8720
900	0.01873	3774.3	4298.8	7.1084

p = 320 bar = 32.0 MPa

T °C	v m³/kg	u kJ/kg	h kJ/kg	s kJ/kg · K
400	0.00236	1980.4	2055.9	4.3239
440	0.00544	2509.0	2683.0	5.2327
480	0.00722	2718.1	2949.2	5.5968
520	0.00853	2860.7	3133.7	5.8357
560	0.00963	2979.0	3287.2	6.0246
600	0.01061	3085.3	3424.6	6.1858
640	0.01150	3184.5	3552.5	6.3290
700	0.01273	3325.4	3732.8	6.5203
740	0.01350	3415.9	3847.8	6.6361
800	0.01460	3548.0	4015.1	6.7966
900	0.01633	3762.7	4285.1	7.0372

Table A-5 Properties of Compressed Liquid Water

T °C	$v \times 10^3$ m^3/kg	u kJ/kg	h kJ/kg	s kJ/kg · K	$v \times 10^3$ m^3/kg	u kJ/kg	h kJ/kg	s kJ/kg · K
	p = 25 bar = 2.5 MPa (T_{sat} = 223.99°C)				p = 50 bar = 5.0 MPa (T_{sat} = 263.99°C)			
20	1.0006	83.80	86.30	.2961	.9995	83.65	88.65	.2956
40	1.0067	167.25	169.77	.5715	1.0056	166.95	171.97	.5705
80	1.0280	334.29	336.86	1.0737	1.0268	333.72	338.85	1.0720
100	1.0423	418.24	420.85	1.3050	1.0410	417.52	422.72	1.3030
140	1.0784	587.82	590.52	1.7369	1.0768	586.76	592.15	1.7343
180	1.1261	761.16	763.97	2.1375	1.1240	759.63	765.25	2.1341
200	1.1555	849.9	852.8	2.3294	1.1530	848.1	853.9	2.3255
220	1.1898	940.7	943.7	2.5174	1.1866	938.4	944.4	2.5128
Sat.	1.1973	959.1	962.1	2.5546	1.2859	1147.8	1154.2	2.9202
	p = 75 bar = 7.5 MPa (T_{sat} = 290.59°C)				p = 100 bar = 10.0 MPa (T_{sat} = 311.06°C)			
20	.9984	83.50	90.99	.2950	.9972	83.36	93.33	.2945
40	1.0045	166.64	174.18	.5696	1.0034	166.35	176.38	.5686
80	1.0256	333.15	340.84	1.0704	1.0245	332.59	342.83	1.0688
100	1.0397	416.81	424.62	1.3011	1.0385	416.12	426.50	1.2992
140	1.0752	585.72	593.78	1.7317	1.0737	584.68	595.42	1.7292
180	1.1219	758.13	766.55	2.1308	1.1199	756.65	767.84	2.1275
220	1.1835	936.2	945.1	2.5083	1.1805	934.1	945.9	2.5039
260	1.2696	1124.4	1134.0	2.8763	1.2645	1121.1	1133.7	2.8699
Sat.	1.3677	1282.0	1292.2	3.1649	1.4524	1393.0	1407.6	3.3596
	p = 150 bar = 15.0 MPa (T_{sat} = 342.24°C)				p = 200 bar = 20.0 MPa (T_{sat} = 365.81°C)			
20	.9950	83.06	97.99	.2934	.9928	82.77	102.62	.2923
40	1.0013	165.76	180.78	.5666	.9992	165.17	185.16	.5646
80	1.0222	331.48	346.81	1.0656	1.0199	330.40	350.80	1.0624
100	1.0361	414.74	430.28	1.2955	1.0337	413.39	434.06	1.2917
140	1.0707	582.66	598.72	1.7242	1.0678	580.69	602.04	1.7193
180	1.1159	753.76	770.50	2.1210	1.1120	750.95	773.20	2.1147
220	1.1748	929.9	947.5	2.4953	1.1693	925.9	949.3	2.4870
260	1.2550	1114.6	1133.4	2.8576	1.2462	1108.6	1133.5	2.8459
300	1.3770	1316.6	1337.3	3.2260	1.3596	1306.1	1333.3	3.2071
Sat.	1.6581	1585.6	1610.5	3.6848	2.036	1785.6	1826.3	4.0139
	p = 250 bar = 25 MPa				p = 300 bar = 30.0 MPa			
20	.9907	82.47	107.24	.2911	.9886	82.17	111.84	.2899
40	.9971	164.60	189.52	.5626	.9951	164.04	193.89	.5607
100	1.0313	412.08	437.85	1.2881	1.0290	410.78	441.66	1.2844
200	1.1344	834.5	862.8	2.2961	1.1302	831.4	865.3	2.2893
300	1.3442	1296.6	1330.2	3.1900	1.3304	1287.9	1327.8	3.1741

Pressure Conversions:
1 bar = 0.1 MPa
= 10^2 kPa

H_2O

Table A-6 **Properties of Saturated Water (Solid–Vapor): Temperature Table**

Pressure Conversions:
1 bar = 0.1 MPa
= 10^2 kPa

Temp. °C	Pressure kPa	Specific Volume m³/kg Sat. Solid $v_i \times 10^3$	Sat. Vapor v_g	Internal Energy kJ/kg Sat. Solid u_i	Subl. u_{ig}	Sat. Vapor u_g	Enthalpy kJ/kg Sat. Solid h_i	Subl. h_{ig}	Sat. Vapor h_g	Entropy kJ/kg · K Sat. Solid s_i	Subl. s_{ig}	Sat. Vapor s_g
.01	.6113	1.0908	206.1	−333.40	2708.7	2375.3	−333.40	2834.8	2501.4	−1.221	10.378	9.156
0	.6108	1.0908	206.3	−333.43	2708.8	2375.3	−333.43	2834.8	2501.3	−1.221	10.378	9.157
−2	.5176	1.0904	241.7	−337.62	2710.2	2372.6	−337.62	2835.3	2497.7	−1.237	10.456	9.219
−4	.4375	1.0901	283.8	−341.78	2711.6	2369.8	−341.78	2835.7	2494.0	−1.253	10.536	9.283
−6	.3689	1.0898	334.2	−345.91	2712.9	2367.0	−345.91	2836.2	2490.3	−1.268	10.616	9.348
−8	.3102	1.0894	394.4	−350.02	2714.2	2364.2	−350.02	2836.6	2486.6	−1.284	10.698	9.414
−10	.2602	1.0891	466.7	−354.09	2715.5	2361.4	−354.09	2837.0	2482.9	−1.299	10.781	9.481
−12	.2176	1.0888	553.7	−358.14	2716.8	2358.7	−358.14	2837.3	2479.2	−1.315	10.865	9.550
−14	.1815	1.0884	658.8	−362.15	2718.0	2355.9	−362.15	2837.6	2475.5	−1.331	10.950	9.619
−16	.1510	1.0881	786.0	−366.14	2719.2	2353.1	−366.14	2837.9	2471.8	−1.346	11.036	9.690
−18	.1252	1.0878	940.5	−370.10	2720.4	2350.3	−370.10	2838.2	2468.1	−1.362	11.123	9.762
−20	.1035	1.0874	1128.6	−374.03	2721.6	2347.5	−374.03	2838.4	2464.3	−1.377	11.212	9.835
−22	.0853	1.0871	1358.4	−377.93	2722.7	2344.7	−377.93	2838.6	2460.6	−1.393	11.302	9.909
−24	.0701	1.0868	1640.1	−381.80	2723.7	2342.0	−381.80	2838.7	2456.9	−1.408	11.394	9.985
−26	.0574	1.0864	1986.4	−385.64	2724.8	2339.2	−385.64	2838.9	2453.2	−1.424	11.486	10.062
−28	.0469	1.0861	2413.7	−389.45	2725.8	2336.4	−389.45	2839.0	2449.5	−1.439	11.580	10.141
−30	.0381	1.0858	2943	−393.23	2726.8	2333.6	−393.23	2839.0	2445.8	−1.455	11.676	10.221
−32	.0309	1.0854	3600	−396.98	2727.8	2330.8	−396.98	2839.1	2442.1	−1.471	11.773	10.303
−34	.0250	1.0851	4419	−400.71	2728.7	2328.0	−400.71	2839.1	2438.4	−1.486	11.872	10.386
−36	.0201	1.0848	5444	−404.40	2729.6	2325.2	−404.40	2839.1	2434.7	−1.501	11.972	10.470
−38	.0161	1.0844	6731	−408.06	2730.5	2322.4	−408.06	2839.0	2430.9	−1.517	12.073	10.556
−40	.0129	1.0841	8354	−411.70	2731.3	2319.6	−411.70	2838.9	2427.2	−1.532	12.176	10.644

Source: J. H. Keenan, F. G. Keyes, P. G. Hill, and J. G. Moore, *Steam Tables,* Wiley, New York, 1978.

R-22

Table A-7 **Properties of Saturated Refrigerant 22 (Liquid–Vapor): Temperature Table**

Pressure Conversions:
1 bar = 0.1 MPa
= 10^2 kPa

Temp. °C	Press. bar	Specific Volume m³/kg: Sat. Liquid $v_f \times 10^3$	Sat. Vapor v_g	Internal Energy kJ/kg: Sat. Liquid u_f	Sat. Vapor u_g	Enthalpy kJ/kg: Sat. Liquid h_f	Evap. h_{fg}	Sat. Vapor h_g	Entropy kJ/kg · K: Sat. Liquid s_f	Sat. Vapor s_g	Temp. °C
−60	0.3749	0.6833	0.5370	−21.57	203.67	−21.55	245.35	223.81	−0.0964	1.0547	−60
−50	0.6451	0.6966	0.3239	−10.89	207.70	−10.85	239.44	228.60	−0.0474	1.0256	−50
−45	0.8290	0.7037	0.2564	−5.50	209.70	−5.44	236.39	230.95	−0.0235	1.0126	−45
−40	1.0522	0.7109	0.2052	−0.07	211.68	0.00	233.27	233.27	0.0000	1.0005	−40
−36	1.2627	0.7169	0.1730	4.29	213.25	4.38	230.71	235.09	0.0186	0.9914	−36
−32	1.5049	0.7231	0.1468	8.68	214.80	8.79	228.10	236.89	0.0369	0.9828	−32
−30	1.6389	0.7262	0.1355	10.88	215.58	11.00	226.77	237.78	0.0460	0.9787	−30
−28	1.7819	0.7294	0.1252	13.09	216.34	13.22	225.43	238.66	0.0551	0.9746	−28
−26	1.9345	0.7327	0.1159	15.31	217.11	15.45	224.08	239.53	0.0641	0.9707	−26
−22	2.2698	0.7393	0.0997	19.76	218.62	19.92	221.32	241.24	0.0819	0.9631	−22
−20	2.4534	0.7427	0.0926	21.99	219.37	22.17	219.91	242.09	0.0908	0.9595	−20
−18	2.6482	0.7462	0.0861	24.23	220.11	24.43	218.49	242.92	0.0996	0.9559	−18
−16	2.8547	0.7497	0.0802	26.48	220.85	26.69	217.05	243.74	0.1084	0.9525	−16
−14	3.0733	0.7533	0.0748	28.73	221.58	28.97	215.59	244.56	0.1171	0.9490	−14
−12	3.3044	0.7569	0.0698	31.00	222.30	31.25	214.11	245.36	0.1258	0.9457	−12
−10	3.5485	0.7606	0.0652	33.27	223.02	33.54	212.62	246.15	0.1345	0.9424	−10
−8	3.8062	0.7644	0.0610	35.54	223.73	35.83	211.10	246.93	0.1431	0.9392	−8
−6	4.0777	0.7683	0.0571	37.83	224.43	38.14	209.56	247.70	0.1517	0.9361	−6
−4	4.3638	0.7722	0.0535	40.12	225.13	40.46	208.00	248.45	0.1602	0.9330	−4
−2	4.6647	0.7762	0.0501	42.42	225.82	42.78	206.41	249.20	0.1688	0.9300	−2
0	4.9811	0.7803	0.0470	44.73	226.50	45.12	204.81	249.92	0.1773	0.9271	0
2	5.3133	0.7844	0.0442	47.04	227.17	47.46	203.18	250.64	0.1857	0.9241	2
4	5.6619	0.7887	0.0415	49.37	227.83	49.82	201.52	251.34	0.1941	0.9213	4
6	6.0275	0.7930	0.0391	51.71	228.48	52.18	199.84	252.03	0.2025	0.9184	6
8	6.4105	0.7974	0.0368	54.05	229.13	54.56	198.14	252.70	0.2109	0.9157	8
10	6.8113	0.8020	0.0346	56.40	229.76	56.95	196.40	253.35	0.2193	0.9129	10
12	7.2307	0.8066	0.0326	58.77	230.38	59.35	194.64	253.99	0.2276	0.9102	12
16	8.1268	0.8162	0.0291	63.53	231.59	64.19	191.02	255.21	0.2442	0.9048	16
20	9.1030	0.8263	0.0259	68.33	232.76	69.09	187.28	256.37	0.2607	0.8996	20
24	10.164	0.8369	0.0232	73.19	233.87	74.04	183.40	257.44	0.2772	0.8944	24
28	11.313	0.8480	0.0208	78.09	234.92	79.05	179.37	258.43	0.2936	0.8893	28
32	12.556	0.8599	0.0186	83.06	235.91	84.14	175.18	259.32	0.3101	0.8842	32
36	13.897	0.8724	0.0168	88.08	236.83	89.29	170.82	260.11	0.3265	0.8790	36
40	15.341	0.8858	0.0151	93.18	237.66	94.53	166.25	260.79	0.3429	0.8738	40
45	17.298	0.9039	0.0132	99.65	238.59	101.21	160.24	261.46	0.3635	0.8672	45
50	19.433	0.9238	0.0116	106.26	239.34	108.06	153.84	261.90	0.3842	0.8603	50
60	24.281	0.9705	0.0089	120.00	240.24	122.35	139.61	261.96	0.4264	0.8455	60

Source: Tables A-7 through A-9 are calculated based on equations from A. Kamei and S. W. Beyerlein, "A Fundamental Equation for Chlorodifluoromethane (R-22)," *Fluid Phase Equilibria,* Vol. 80, No. 11, 1992, pp. 71–86.

Table A-8 **Properties of Saturated Refrigerant 22 (Liquid–Vapor): Pressure Table**

Pressure Conversions:
1 bar = 0.1 MPa
= 10^2 kPa

Press. bar	Temp. °C	Specific Volume m^3/kg Sat. Liquid $v_f \times 10^3$	Sat. Vapor v_g	Internal Energy kJ/kg Sat. Liquid u_f	Sat. Vapor u_g	Enthalpy kJ/kg Sat. Liquid h_f	Evap. h_{fg}	Sat. Vapor h_g	Entropy kJ/kg · K Sat. Liquid s_f	Sat. Vapor s_g	Press. bar
0.40	−58.86	0.6847	0.5056	−20.36	204.13	−20.34	244.69	224.36	−0.0907	1.0512	0.40
0.50	−54.83	0.6901	0.4107	−16.07	205.76	−16.03	242.33	226.30	−0.0709	1.0391	0.50
0.60	−51.40	0.6947	0.3466	−12.39	207.14	−12.35	240.28	227.93	−0.0542	1.0294	0.60
0.70	−48.40	0.6989	0.3002	−9.17	208.34	−9.12	238.47	229.35	−0.0397	1.0213	0.70
0.80	−45.73	0.7026	0.2650	−6.28	209.41	−6.23	236.84	230.61	−0.0270	1.0144	0.80
0.90	−43.30	0.7061	0.2374	−3.66	210.37	−3.60	235.34	231.74	−0.0155	1.0084	0.90
1.00	−41.09	0.7093	0.2152	−1.26	211.25	−1.19	233.95	232.77	−0.0051	1.0031	1.00
1.25	−36.23	0.7166	0.1746	4.04	213.16	4.13	230.86	234.99	0.0175	0.9919	1.25
1.50	−32.08	0.7230	0.1472	8.60	214.77	8.70	228.15	236.86	0.0366	0.9830	1.50
1.75	−28.44	0.7287	0.1274	12.61	216.18	12.74	225.73	238.47	0.0531	0.9755	1.75
2.00	−25.18	0.7340	0.1123	16.22	217.42	16.37	223.52	239.88	0.0678	0.9691	2.00
2.25	−22.22	0.7389	0.1005	19.51	218.53	19.67	221.47	241.15	0.0809	0.9636	2.25
2.50	−19.51	0.7436	0.0910	22.54	219.55	22.72	219.57	242.29	0.0930	0.9586	2.50
2.75	−17.00	0.7479	0.0831	25.36	220.48	25.56	217.77	243.33	0.1040	0.9542	2.75
3.00	−14.66	0.7521	0.0765	27.99	221.34	28.22	216.07	244.29	0.1143	0.9502	3.00
3.25	−12.46	0.7561	0.0709	30.47	222.13	30.72	214.46	245.18	0.1238	0.9465	3.25
3.50	−10.39	0.7599	0.0661	32.82	222.88	33.09	212.91	246.00	0.1328	0.9431	3.50
3.75	−8.43	0.7636	0.0618	35.06	223.58	35.34	211.42	246.77	0.1413	0.9399	3.75
4.00	−6.56	0.7672	0.0581	37.18	224.24	37.49	209.99	247.48	0.1493	0.9370	4.00
4.25	−4.78	0.7706	0.0548	39.22	224.86	39.55	208.61	248.16	0.1569	0.9342	4.25
4.50	−3.08	0.7740	0.0519	41.17	225.45	41.52	207.27	248.80	0.1642	0.9316	4.50
4.75	−1.45	0.7773	0.0492	43.05	226.00	43.42	205.98	249.40	0.1711	0.9292	4.75
5.00	0.12	0.7805	0.0469	44.86	226.54	45.25	204.71	249.97	0.1777	0.9269	5.00
5.25	1.63	0.7836	0.0447	46.61	227.04	47.02	203.48	250.51	0.1841	0.9247	5.25
5.50	3.08	0.7867	0.0427	48.30	227.53	48.74	202.28	251.02	0.1903	0.9226	5.50
5.75	4.49	0.7897	0.0409	49.94	227.99	50.40	201.11	251.51	0.1962	0.9206	5.75
6.00	5.85	0.7927	0.0392	51.53	228.44	52.01	199.97	251.98	0.2019	0.9186	6.00
7.00	10.91	0.8041	0.0337	57.48	230.04	58.04	195.60	253.64	0.2231	0.9117	7.00
8.00	15.45	0.8149	0.0295	62.88	231.43	63.53	191.52	255.05	0.2419	0.9056	8.00
9.00	19.59	0.8252	0.0262	67.84	232.64	68.59	187.67	256.25	0.2591	0.9001	9.00
10.00	23.40	0.8352	0.0236	72.46	233.71	73.30	183.99	257.28	0.2748	0.8952	10.00
12.00	30.25	0.8546	0.0195	80.87	235.48	81.90	177.04	258.94	0.3029	0.8864	12.00
14.00	36.29	0.8734	0.0166	88.45	236.89	89.68	170.49	260.16	0.3277	0.8786	14.00
16.00	41.73	0.8919	0.0144	95.41	238.00	96.83	164.21	261.04	0.3500	0.8715	16.00
18.00	46.69	0.9104	0.0127	101.87	238.86	103.51	158.13	261.64	0.3705	0.8649	18.00
20.00	51.26	0.9291	0.0112	107.95	239.51	109.81	152.17	261.98	0.3895	0.8586	20.00
24.00	59.46	0.9677	0.0091	119.24	240.22	121.56	140.43	261.99	0.4241	0.8463	24.00

Table A-9 Properties of Superheated Refrigerant 22 Vapor

T °C	v m³/kg	u kJ/kg	h kJ/kg	s kJ/kg · K	v m³/kg	u kJ/kg	h kJ/kg	s kJ/kg · K
	p = 0.4 bar = 0.04 MPa (T_{sat} = −58.86°C)				p = 0.6 bar = 0.06 MPa (T_{sat} = −51.40°C)			
Sat.	0.50559	204.13	224.36	1.0512	0.34656	207.14	227.93	1.0294
−55	0.51532	205.92	226.53	1.0612				
−50	0.52787	208.26	229.38	1.0741	0.34895	207.80	228.74	1.0330
−45	0.54037	210.63	232.24	1.0868	0.35747	210.20	231.65	1.0459
−40	0.55284	213.02	235.13	1.0993	0.36594	212.62	234.58	1.0586
−35	0.56526	215.43	238.05	1.1117	0.37437	215.06	237.52	1.0711
−30	0.57766	217.88	240.99	1.1239	0.38277	217.53	240.49	1.0835
−25	0.59002	220.35	243.95	1.1360	0.39114	220.02	243.49	1.0956
−20	0.60236	222.85	246.95	1.1479	0.39948	222.54	246.51	1.1077
−15	0.61468	225.38	249.97	1.1597	0.40779	225.08	249.55	1.1196
−10	0.62697	227.93	253.01	1.1714	0.41608	227.65	252.62	1.1314
−5	0.63925	230.52	256.09	1.1830	0.42436	230.25	255.71	1.1430
0	0.65151	233.13	259.19	1.1944	0.43261	232.88	258.83	1.1545
	p = 0.8 bar = 0.08 MPa (T_{sat} = −45.73°C)				p = 1.0 bar = 0.10 MPa (T_{sat} = −41.09°C)			
Sat.	0.26503	209.41	230.61	1.0144	0.21518	211.25	232.77	1.0031
−45	0.26597	209.76	231.04	1.0163				
−40	0.27245	212.21	234.01	1.0292	0.21633	211.79	233.42	1.0059
−35	0.27890	214.68	236.99	1.0418	0.22158	214.29	236.44	1.0187
−30	0.28530	217.17	239.99	1.0543	0.22679	216.80	239.48	1.0313
−25	0.29167	219.68	243.02	1.0666	0.23197	219.34	242.54	1.0438
−20	0.29801	222.22	246.06	1.0788	0.23712	221.90	245.61	1.0560
−15	0.30433	224.78	249.13	1.0908	0.24224	224.48	248.70	1.0681
−10	0.31062	227.37	252.22	1.1026	0.24734	227.08	251.82	1.0801
−5	0.31690	229.98	255.34	1.1143	0.25241	229.71	254.95	1.0919
0	0.32315	232.62	258.47	1.1259	0.25747	232.36	258.11	1.1035
5	0.32939	235.29	261.64	1.1374	0.26251	235.04	261.29	1.1151
10	0.33561	237.98	264.83	1.1488	0.26753	237.74	264.50	1.1265
	p = 1.5 bar = 0.15 MPa (T_{sat} = −32.08°C)				p = 2.0 bar = 0.20 MPa (T_{sat} = −25.18°C)			
Sat.	0.14721	214.77	236.86	0.9830	0.11232	217.42	239.88	0.9691
−30	0.14872	215.85	238.16	0.9883				
−25	0.15232	218.45	241.30	1.0011	0.11242	217.51	240.00	0.9696
−20	0.15588	221.07	244.45	1.0137	0.11520	220.19	243.23	0.9825
−15	0.15941	223.70	247.61	1.0260	0.11795	222.88	246.47	0.9952
−10	0.16292	226.35	250.78	1.0382	0.12067	225.58	249.72	1.0076
−5	0.16640	229.02	253.98	1.0502	0.12336	228.30	252.97	1.0199
0	0.16987	231.70	257.18	1.0621	0.12603	231.03	256.23	1.0310
5	0.17331	234.42	260.41	1.0738	0.12868	233.78	259.51	1.0438
10	0.17674	237.15	263.66	1.0854	0.13132	236.54	262.81	1.0555
15	0.18015	239.91	266.93	1.0968	0.13393	239.33	266.12	1.0671
20	0.18355	242.69	270.22	1.1081	0.13653	242.14	269.44	1.0786
25	0.18693	245.49	273.53	1.1193	0.13912	244.97	272.79	1.0899

Pressure Conversions:
1 bar = 0.1 MPa
= 10^2 kPa

R-22

R-22

Table A-9 **(Continued)**

T °C	v m³/kg	u kJ/kg	h kJ/kg	s kJ/kg · K	v m³/kg	u kJ/kg	h kJ/kg	s kJ/kg · K
	p = 2.5 bar = 0.25 MPa (T_{sat} = −19.51°C)				p = 3.0 bar = 0.30 MPa (T_{sat} = −14.66°C)			
Sat.	0.09097	219.55	242.29	0.9586	0.07651	221.34	244.29	0.9502
−15	0.09303	222.03	245.29	0.9703				
−10	0.09528	224.79	248.61	0.9831	0.07833	223.96	247.46	0.9623
−5	0.09751	227.55	251.93	0.9956	0.08025	226.78	250.86	0.9751
0	0.09971	230.33	255.26	1.0078	0.08214	229.61	254.25	0.9876
5	0.10189	233.12	258.59	1.0199	0.08400	232.44	257.64	0.9999
10	0.10405	235.92	261.93	1.0318	0.08585	235.28	261.04	1.0120
15	0.10619	238.74	265.29	1.0436	0.08767	238.14	264.44	1.0239
20	0.10831	241.58	268.66	1.0552	0.08949	241.01	267.85	1.0357
25	0.11043	244.44	272.04	1.0666	0.09128	243.89	271.28	1.0472
30	0.11253	247.31	275.44	1.0779	0.09307	246.80	274.72	1.0587
35	0.11461	250.21	278.86	1.0891	0.09484	249.72	278.17	1.0700
40	0.11669	253.13	282.30	1.1002	0.09660	252.66	281.64	1.0811
	p = 3.5 bar = 0.35 MPa (T_{sat} = −10.39°C)				p = 4.0 bar = 0.40 MPa (T_{sat} = −6.56°C)			
Sat.	0.06605	222.88	246.00	0.9431	0.05812	224.24	247.48	0.9370
−10	0.06619	223.10	246.27	0.9441				
−5	0.06789	225.99	249.75	0.9572	0.05860	225.16	248.60	0.9411
0	0.06956	228.86	253.21	0.9700	0.06011	228.09	252.14	0.9542
5	0.07121	231.74	256.67	0.9825	0.06160	231.02	225.66	0.9670
10	0.07284	234.63	260.12	0.9948	0.06306	233.95	259.18	0.9795
15	0.07444	237.52	263.57	1.0069	0.06450	236.89	262.69	0.9918
20	0.07603	240.42	267.03	1.0188	0.06592	239.83	266.19	1.0039
25	0.07760	243.34	270.50	1.0305	0.06733	242.77	269.71	1.0158
30	0.07916	246.27	273.97	1.0421	0.06872	245.73	273.22	1.0274
35	0.08070	249.22	227.46	1.0535	0.07010	248.71	276.75	1.0390
40	0.08224	252.18	280.97	1.0648	0.07146	251.70	280.28	1.0504
45	0.08376	255.17	284.48	1.0759	0.07282	254.70	283.83	1.0616
	p = 4.5 bar = 0.45 MPa (T_{sat} = −3.08°C)				p = 5.0 bar = 0.50 MPa (T_{sat} = 0.12°C)			
Sat.	0.05189	225.45	248.80	0.9316	0.04686	226.54	249.97	0.9269
0	0.05275	227.29	251.03	0.9399				
5	0.05411	230.28	254.63	0.9529	0.04810	229.52	253.57	0.9399
10	0.05545	233.26	258.21	0.9657	0.04934	232.55	257.22	0.9530
15	0.05676	236.24	261.78	0.9782	0.05056	235.57	260.85	0.9657
20	0.05805	239.22	265.34	0.9904	0.05175	238.59	264.47	0.9781
25	0.05933	242.20	268.90	1.0025	0.05293	241.61	268.07	0.9903
30	0.06059	245.19	272.46	1.0143	0.05409	244.63	271.68	1.0023
35	0.06184	248.19	276.02	1.0259	0.05523	247.66	275.28	1.0141
40	0.06308	251.20	279.59	1.0374	0.05636	250.70	278.89	1.0257
45	0.06430	254.23	283.17	1.0488	0.05748	253.76	282.50	1.0371
50	0.06552	257.28	286.76	1.0600	0.05859	256.82	286.12	1.0484
55	0.06672	260.34	290.36	1.0710	0.05969	259.90	289.75	1.0595

Table A-9 **(Continued)**

T °C	v m³/kg	u kJ/kg	h kJ/kg	s kJ/kg · K	v m³/kg	u kJ/kg	h kJ/kg	s kJ/kg · K
	p = 5.5 bar = 0.55 MPa (T_{sat} = 3.08°C)				p = 6.0 bar = 0.60 MPa (T_{sat} = 5.85°C)			
Sat.	0.04271	227.53	251.02	0.9226	0.03923	228.44	251.98	0.9186
5	0.04317	228.72	252.46	0.9278				
10	0.04433	231.81	256.20	0.9411	0.04015	231.05	255.14	0.9299
15	0.04547	234.89	259.90	0.9540	0.04122	234.18	258.91	0.9431
20	0.04658	237.95	263.57	0.9667	0.04227	237.29	262.65	0.9560
25	0.04768	241.01	267.23	0.9790	0.04330	240.39	266.37	0.9685
30	0.04875	244.07	270.88	0.9912	0.04431	243.49	270.07	0.9808
35	0.04982	247.13	274.53	1.0031	0.04530	246.58	273.76	0.9929
40	0.05086	250.20	278.17	1.0148	0.04628	249.68	277.45	1.0048
45	0.05190	253.27	281.82	1.0264	0.04724	252.78	281.13	1.0164
50	0.05293	256.36	285.47	1.0378	0.04820	255.90	284.82	1.0279
55	0.05394	259.46	289.13	1.0490	0.04914	259.02	288.51	1.0393
60	0.05495	262.58	292.80	1.0601	0.05008	262.15	292.20	1.0504

T °C	v m³/kg	u kJ/kg	h kJ/kg	s kJ/kg · K	v m³/kg	u kJ/kg	h kJ/kg	s kJ/kg · K
	p = 7.0 bar = 0.70 MPa (T_{sat} = 10.91°C)				p = 8.0 bar = 0.80 MPa (T_{sat} = 15.45°C)			
Sat.	0.03371	230.04	253.64	0.9117	0.02953	231.43	255.05	0.9056
15	0.03451	232.70	256.86	0.9229				
20	0.03547	235.92	260.75	0.9363	0.03033	234.47	258.74	0.9182
25	0.03639	239.12	264.59	0.9493	0.03118	237.76	262.70	0.9315
30	0.03730	242.29	268.40	0.9619	0.03202	241.04	266.66	0.9448
35	0.03819	245.46	272.19	0.9743	0.03283	244.28	270.54	0.9574
40	0.03906	248.62	275.96	0.9865	0.03363	247.52	274.42	0.9700
45	0.03992	251.78	279.72	0.9984	0.03440	250.74	278.26	0.9821
50	0.04076	254.94	283.48	1.0101	0.03517	253.96	282.10	0.9941
55	0.04160	258.11	287.23	1.0216	0.03592	257.18	285.92	1.0058
60	0.04242	261.29	290.99	1.0330	0.03667	260.40	289.74	1.0174
65	0.04324	264.48	294.75	1.0442	0.03741	263.64	293.56	1.0287
70	0.04405	267.68	298.51	1.0552	0.03814	266.87	297.38	1.0400

T °C	v m³/kg	u kJ/kg	h kJ/kg	s kJ/kg · K	v m³/kg	u kJ/kg	h kJ/kg	s kJ/kg · K
	p = 9.0 bar = 0.90 MPa (T_{sat} = 19.59°C)				p = 10.0 bar = 1.00 MPa (T_{sat} = 23.40°C)			
Sat.	0.02623	232.64	256.25	0.9001	0.02358	233.71	257.28	0.8952
20	0.02630	232.92	256.59	0.9013				
30	0.02789	239.73	264.83	0.9289	0.02457	238.34	262.91	0.9139
40	0.02939	246.37	272.82	0.9549	0.02598	245.18	271.17	0.9407
50	0.03082	252.95	280.68	0.9795	0.02732	251.90	279.22	0.9660
60	0.03219	259.49	288.46	1.0033	0.02860	258.56	287.15	0.9902
70	0.03353	266.04	296.21	1.0262	0.02984	265.19	295.03	1.0135
80	0.03483	272.62	303.96	1.0484	0.03104	271.84	302.88	1.0361
90	0.03611	279.23	311.73	1.0701	0.03221	278.52	310.74	1.0580
100	0.03736	285.90	319.53	1.0913	0.03337	285.24	318.61	1.0794
110	0.03860	292.63	327.37	1.1120	0.03450	292.02	326.52	1.1003
120	0.03982	299.42	335.26	1.1323	0.03562	298.85	334.46	1.1207
130	0.04103	306.28	343.21	1.1523	0.03672	305.74	342.46	1.1408
140	0.04223	313.21	351.22	1.1719	0.03781	312.70	350.51	1.1605
150	0.04342	320.21	359.29	1.1912	0.03889	319.74	358.63	1.1790

R-22

Table A-9 (*Continued*)

T °C	v m³/kg	u kJ/kg	h kJ/kg	s kJ/kg · K	v m³/kg	u kJ/kg	h kJ/kg	s kJ/kg · K
	p = 12.0 bar = 1.20 MPa (T_{sat} = 30.25°C)				p = 14.0 bar = 1.40 MPa (T_{sat} = 36.29°C)			
Sat.	0.01955	235.48	258.94	0.8864	0.01662	236.89	260.16	0.8786
40	0.02083	242.63	267.62	0.9146	0.01708	239.78	263.70	0.8900
50	0.02204	249.69	276.14	0.9413	0.01823	247.29	272.81	0.9186
60	0.02319	256.60	284.43	0.9666	0.01929	254.52	281.53	0.9452
70	0.02428	263.44	292.58	0.9907	0.02029	261.60	290.01	0.9703
80	0.02534	270.25	300.66	1.0139	0.02125	268.60	298.34	0.9942
90	0.02636	277.07	308.70	1.0363	0.02217	275.56	306.60	1.0172
100	0.02736	283.90	316.73	1.0582	0.02306	282.52	314.80	1.0395
110	0.02834	290.77	324.78	1.0794	0.02393	289.49	323.00	1.0612
120	0.02930	297.69	332.85	1.1002	0.02478	296.50	331.19	1.0823
130	0.03024	304.65	340.95	1.1205	0.02562	303.55	339.41	1.1029
140	0.03118	311.68	349.09	1.1405	0.02644	310.64	347.65	1.1231
150	0.03210	318.77	357.29	1.1601	0.02725	317.79	355.94	1.1429
160	0.03301	325.92	365.54	1.1793	0.02805	324.99	364.26	1.1624
170	0.03392	333.14	373.84	1.1983	0.02884	332.26	372.64	1.1815
	p = 16.0 bar = 1.60 MPa (T_{sat} = 41.73°C)				p = 18.0 bar = 1.80 MPa (T_{sat} = 46.69°C)			
Sat.	0.01440	238.00	261.04	0.8715	0.01265	238.86	261.64	0.8649
50	0.01533	244.66	269.18	0.8971	0.01301	241.72	265.14	0.8758
60	0.01634	252.29	278.43	0.9252	0.01401	249.86	275.09	0.9061
70	0.01728	259.65	287.30	0.9515	0.01492	257.57	284.43	0.9337
80	0.01817	266.86	295.93	0.9762	0.01576	265.04	293.40	0.9595
90	0.01901	274.00	304.42	0.9999	0.01655	272.37	302.16	0.9839
100	0.01983	281.09	312.82	1.0228	0.01731	279.62	310.77	1.0073
110	0.02062	288.18	321.17	1.0448	0.01804	286.83	319.30	1.0299
120	0.02139	295.28	329.51	1.0663	0.01874	294.04	327.78	1.0517
130	0.02214	302.41	337.84	1.0872	0.01943	301.26	336.24	1.0730
140	0.02288	309.58	346.19	1.1077	0.02011	308.50	344.70	1.0937
150	0.02361	316.79	354.56	1.1277	0.02077	315.78	353.17	1.1139
160	0.02432	324.05	362.97	1.1473	0.02142	323.10	361.66	1.1338
170	0.02503	331.37	371.42	1.1666	0.02207	330.47	370.19	1.1532
	p = 20.0 bar = 2.00 MPa (T_{sat} = 51.26°C)				p = 24.0 bar = 2.4 MPa (T_{sat} = 59.46°C)			
Sat.	0.01124	239.51	261.98	0.8586	0.00907	240.22	261.99	0.8463
60	0.01212	247.20	271.43	0.8873	0.00913	240.78	262.68	0.8484
70	0.01300	255.35	281.36	0.9167	0.01006	250.30	274.43	0.8831
80	0.01381	263.12	290.74	0.9436	0.01085	258.89	284.93	0.9133
90	0.01457	270.67	299.80	0.9689	0.01156	267.01	294.75	0.9407
100	0.01528	278.09	308.65	0.9929	0.01222	274.85	304.18	0.9663
110	0.01596	285.44	317.37	1.0160	0.01284	282.53	313.35	0.9906
120	0.01663	292.76	326.01	1.0383	0.01343	290.11	322.35	1.0137
130	0.01727	300.08	334.61	1.0598	0.01400	297.64	331.25	1.0361
140	0.01789	307.40	343.19	1.0808	0.01456	305.14	340.08	1.0577
150	0.01850	314.75	351.76	1.1013	0.01509	312.64	348.87	1.0787
160	0.01910	322.14	360.34	1.1214	0.01562	320.16	357.64	1.0992
170	0.01969	329.56	368.95	1.1410	0.01613	327.70	366.41	1.1192
180	0.02027	337.03	377.58	1.1603	0.01663	335.27	375.20	1.1388

Table A-10 Properties of Saturated Refrigerant 134a (Liquid–Vapor): Temperature Table

Pressure Conversions: 1 bar = 0.1 MPa = 10^2 kPa

Temp. °C	Press. bar	Specific Volume m³/kg		Internal Energy kJ/kg		Enthalpy kJ/kg			Entropy kJ/kg · K		Temp. °C
		Sat. Liquid $v_f \times 10^3$	Sat. Vapor v_g	Sat. Liquid u_f	Sat. Vapor u_g	Sat. Liquid h_f	Evap. h_{fg}	Sat. Vapor h_g	Sat. Liquid s_f	Sat. Vapor s_g	
−40	0.5164	0.7055	0.3569	−0.04	204.45	0.00	222.88	222.88	0.0000	0.9560	−40
−36	0.6332	0.7113	0.2947	4.68	206.73	4.73	220.67	225.40	0.0201	0.9506	−36
−32	0.7704	0.7172	0.2451	9.47	209.01	9.52	218.37	227.90	0.0401	0.9456	−32
−28	0.9305	0.7233	0.2052	14.31	211.29	14.37	216.01	230.38	0.0600	0.9411	−28
−26	1.0199	0.7265	0.1882	16.75	212.43	16.82	214.80	231.62	0.0699	0.9390	−26
−24	1.1160	0.7296	0.1728	19.21	213.57	19.29	213.57	232.85	0.0798	0.9370	−24
−22	1.2192	0.7328	0.1590	21.68	214.70	21.77	212.32	234.08	0.0897	0.9351	−22
−20	1.3299	0.7361	0.1464	24.17	215.84	24.26	211.05	235.31	0.0996	0.9332	−20
−18	1.4483	0.7395	0.1350	26.67	216.97	26.77	209.76	236.53	0.1094	0.9315	−18
−16	1.5748	0.7428	0.1247	29.18	218.10	29.30	208.45	237.74	0.1192	0.9298	−16
−12	1.8540	0.7498	0.1068	34.25	220.36	34.39	205.77	240.15	0.1388	0.9267	−12
−8	2.1704	0.7569	0.0919	39.38	222.60	39.54	203.00	242.54	0.1583	0.9239	−8
−4	2.5274	0.7644	0.0794	44.56	224.84	44.75	200.15	244.90	0.1777	0.9213	−4
0	2.9282	0.7721	0.0689	49.79	227.06	50.02	197.21	247.23	0.1970	0.9190	0
4	3.3765	0.7801	0.0600	55.08	229.27	55.35	194.19	249.53	0.2162	0.9169	4
8	3.8756	0.7884	0.0525	60.43	231.46	60.73	191.07	251.80	0.2354	0.9150	8
12	4.4294	0.7971	0.0460	65.83	233.63	66.18	187.85	254.03	0.2545	0.9132	12
16	5.0416	0.8062	0.0405	71.29	235.78	71.69	184.52	256.22	0.2735	0.9116	16
20	5.7160	0.8157	0.0358	76.80	237.91	77.26	181.09	258.36	0.2924	0.9102	20
24	6.4566	0.8257	0.0317	82.37	240.01	82.90	177.55	260.45	0.3113	0.9089	24
26	6.8530	0.8309	0.0298	85.18	241.05	85.75	175.73	261.48	0.3208	0.9082	26
28	7.2675	0.8362	0.0281	88.00	242.08	88.61	173.89	262.50	0.3302	0.9076	28
30	7.7006	0.8417	0.0265	90.84	243.10	91.49	172.00	263.50	0.3396	0.9070	30
32	8.1528	0.8473	0.0250	93.70	244.12	94.39	170.09	264.48	0.3490	0.9064	32
34	8.6247	0.8530	0.0236	96.58	245.12	97.31	168.14	265.45	0.3584	0.9058	34
36	9.1168	0.8590	0.0223	99.47	246.11	100.25	166.15	266.40	0.3678	0.9053	36
38	9.6298	0.8651	0.0210	102.38	247.09	103.21	164.12	267.33	0.3772	0.9047	38
40	10.164	0.8714	0.0199	105.30	248.06	106.19	162.05	268.24	0.3866	0.9041	40
42	10.720	0.8780	0.0188	108.25	249.02	109.19	159.94	269.14	0.3960	0.9035	42
44	11.299	0.8847	0.0177	111.22	249.96	112.22	157.79	270.01	0.4054	0.9030	44
48	12.526	0.8989	0.0159	117.22	251.79	118.35	153.33	271.68	0.4243	0.9017	48
52	13.851	0.9142	0.0142	123.31	253.55	124.58	148.66	273.24	0.4432	0.9004	52
56	15.278	0.9308	0.0127	129.51	255.23	130.93	143.75	274.68	0.4622	0.8990	56
60	16.813	0.9488	0.0114	135.82	256.81	137.42	138.57	275.99	0.4814	0.8973	60
70	21.162	1.0027	0.0086	152.22	260.15	154.34	124.08	278.43	0.5302	0.8918	70
80	26.324	1.0766	0.0064	169.88	262.14	172.71	106.41	279.12	0.5814	0.8827	80
90	32.435	1.1949	0.0046	189.82	261.34	193.69	82.63	276.32	0.6380	0.8655	90
100	39.742	1.5443	0.0027	218.60	248.49	224.74	34.40	259.13	0.7196	0.8117	100

Source: Tables A-10 through A-12 are calculated based on equations from D. P. Wilson and R. S. Basu, "Thermodynamic Properties of a New Stratospherically Safe Working Fluid—Refrigerant 134a," *ASHRAE Trans.*, Vol. 94, Pt. 2, 1988, pp. 2095–2118.

Table A-11 **Properties of Saturated Refrigerant 134a (Liquid–Vapor): Pressure Table**

Pressure Conversions:
1 bar = 0.1 MPa
= 10^2 kPa

Press. bar	Temp. °C	Specific Volume m³/kg: Sat. Liquid $v_f \times 10^3$	Sat. Vapor v_g	Internal Energy kJ/kg: Sat. Liquid u_f	Sat. Vapor u_g	Enthalpy kJ/kg: Sat. Liquid h_f	Evap. h_{fg}	Sat. Vapor h_g	Entropy kJ/kg · K: Sat. Liquid s_f	Sat. Vapor s_g	Press. bar
0.6	−37.07	0.7097	0.3100	3.41	206.12	3.46	221.27	224.72	0.0147	0.9520	0.6
0.8	−31.21	0.7184	0.2366	10.41	209.46	10.47	217.92	228.39	0.0440	0.9447	0.8
1.0	−26.43	0.7258	0.1917	16.22	212.18	16.29	215.06	231.35	0.0678	0.9395	1.0
1.2	−22.36	0.7323	0.1614	21.23	214.50	21.32	212.54	233.86	0.0879	0.9354	1.2
1.4	−18.80	0.7381	0.1395	25.66	216.52	25.77	210.27	236.04	0.1055	0.9322	1.4
1.6	−15.62	0.7435	0.1229	29.66	218.32	29.78	208.19	237.97	0.1211	0.9295	1.6
1.8	−12.73	0.7485	0.1098	33.31	219.94	33.45	206.26	239.71	0.1352	0.9273	1.8
2.0	−10.09	0.7532	0.0993	36.69	221.43	36.84	204.46	241.30	0.1481	0.9253	2.0
2.4	−5.37	0.7618	0.0834	42.77	224.07	42.95	201.14	244.09	0.1710	0.9222	2.4
2.8	−1.23	0.7697	0.0719	48.18	226.38	48.39	198.13	246.52	0.1911	0.9197	2.8
3.2	2.48	0.7770	0.0632	53.06	228.43	53.31	195.35	248.66	0.2089	0.9177	3.2
3.6	5.84	0.7839	0.0564	57.54	230.28	57.82	192.76	250.58	0.2251	0.9160	3.6
4.0	8.93	0.7904	0.0509	61.69	231.97	62.00	190.32	252.32	0.2399	0.9145	4.0
5.0	15.74	0.8056	0.0409	70.93	235.64	71.33	184.74	256.07	0.2723	0.9117	5.0
6.0	21.58	0.8196	0.0341	78.99	238.74	79.48	179.71	259.19	0.2999	0.9097	6.0
7.0	26.72	0.8328	0.0292	86.19	241.42	86.78	175.07	261.85	0.3242	0.9080	7.0
8.0	31.33	0.8454	0.0255	92.75	243.78	93.42	170.73	264.15	0.3459	0.9066	8.0
9.0	35.53	0.8576	0.0226	98.79	245.88	99.56	166.62	266.18	0.3656	0.9054	9.0
10.0	39.39	0.8695	0.0202	104.42	247.77	105.29	162.68	267.97	0.3838	0.9043	10.0
12.0	46.32	0.8928	0.0166	114.69	251.03	115.76	155.23	270.99	0.4164	0.9023	12.0
14.0	52.43	0.9159	0.0140	123.98	253.74	125.26	148.14	273.40	0.4453	0.9003	14.0
16.0	57.92	0.9392	0.0121	132.52	256.00	134.02	141.31	275.33	0.4714	0.8982	16.0
18.0	62.91	0.9631	0.0105	140.49	257.88	142.22	134.60	276.83	0.4954	0.8959	18.0
20.0	67.49	0.9878	0.0093	148.02	259.41	149.99	127.95	277.94	0.5178	0.8934	20.0
25.0	77.59	1.0562	0.0069	165.48	261.84	168.12	111.06	279.17	0.5687	0.8854	25.0
30.0	86.22	1.1416	0.0053	181.88	262.16	185.30	92.71	278.01	0.6156	0.8735	30.0

R-134a

Table A-12 **Properties of Superheated Refrigerant 134a Vapor**

Pressure Conversions:
1 bar = 0.1 MPa
= 10^2 kPa

R-134a

T °C	*v* m³/kg	*u* kJ/kg	*h* kJ/kg	*s* kJ/kg · K	*v* m³/kg	*u* kJ/kg	*h* kJ/kg	*s* kJ/kg · K
	p = 0.6 bar = 0.06 MPa (T_{sat} = −37.07°C)				p = 1.0 bar = 0.10 MPa (T_{sat} = −26.43°C)			
Sat.	0.31003	206.12	224.72	0.9520	0.19170	212.18	231.35	0.9395
−20	0.33536	217.86	237.98	1.0062	0.19770	216.77	236.54	0.9602
−10	0.34992	224.97	245.96	1.0371	0.20686	224.01	244.70	0.9918
0	0.36433	232.24	254.10	1.0675	0.21587	231.41	252.99	1.0227
10	0.37861	239.69	262.41	1.0973	0.22473	238.96	261.43	1.0531
20	0.39279	247.32	270.89	1.1267	0.23349	246.67	270.02	1.0829
30	0.40688	255.12	279.53	1.1557	0.24216	254.54	278.76	1.1122
40	0.42091	263.10	288.35	1.1844	0.25076	262.58	287.66	1.1411
50	0.43487	271.25	297.34	1.2126	0.25930	270.79	296.72	1.1696
60	0.44879	279.58	306.51	1.2405	0.26779	279.16	305.94	1.1977
70	0.46266	288.08	315.84	1.2681	0.27623	287.70	315.32	1.2254
80	0.47650	296.75	325.34	1.2954	0.28464	296.40	324.87	1.2528
90	0.49031	305.58	335.00	1.3224	0.29302	305.27	334.57	1.2799

T °C	*v* m³/kg	*u* kJ/kg	*h* kJ/kg	*s* kJ/kg · K	*v* m³/kg	*u* kJ/kg	*h* kJ/kg	*s* kJ/kg · K
	p = 1.4 bar = 0.14 MPa (T_{sat} = −18.80°C)				p = 1.8 bar = 0.18 MPa (T_{sat} = −12.73°C)			
Sat.	0.13945	216.52	236.04	0.9322	0.10983	219.94	239.71	0.9273
−10	0.14549	223.03	243.40	0.9606	0.11135	222.02	242.06	0.9362
0	0.15219	230.55	251.86	0.9922	0.11678	229.67	250.69	0.9684
10	0.15875	238.21	260.43	1.0230	0.12207	237.44	259.41	0.9998
20	0.16520	246.01	269.13	1.0532	0.12723	245.33	268.23	1.0304
30	0.17155	253.96	277.97	1.0828	0.13230	253.36	277.17	1.0604
40	0.17783	262.06	286.96	1.1120	0.13730	261.53	286.24	1.0898
50	0.18404	270.32	296.09	1.1407	0.14222	269.85	295.45	1.1187
60	0.19020	278.74	305.37	1.1690	0.14710	278.31	304.79	1.1472
70	0.19633	287.32	314.80	1.1969	0.15193	286.93	314.28	1.1753
80	0.20241	296.06	324.39	1.2244	0.15672	295.71	323.92	1.2030
90	0.20846	304.95	334.14	1.2516	0.16148	304.63	333.70	1.2303
100	0.21449	314.01	344.04	1.2785	0.16622	313.72	343.63	1.2573

T °C	*v* m³/kg	*u* kJ/kg	*h* kJ/kg	*s* kJ/kg · K	*v* m³/kg	*u* kJ/kg	*h* kJ/kg	*s* kJ/kg · K
	p = 2.0 bar = 0.20 MPa (T_{sat} = −10.09°C)				p = 2.4 bar = 0.24 MPa (T_{sat} = −5.37°C)			
Sat.	0.09933	221.43	241.30	0.9253	0.08343	224.07	244.09	0.9222
−10	0.09938	221.50	241.38	0.9256				
0	0.10438	229.23	250.10	0.9582	0.08574	228.31	248.89	0.9399
10	0.10922	237.05	258.89	0.9898	0.08993	236.26	257.84	0.9721
20	0.11394	244.99	267.78	1.0206	0.09399	244.30	266.85	1.0034
30	0.11856	253.06	276.77	1.0508	0.09794	252.45	275.95	1.0339
40	0.12311	261.26	285.88	1.0804	0.10181	260.72	285.16	1.0637
50	0.12758	269.61	295.12	1.1094	0.10562	269.12	294.47	1.0930
60	0.13201	278.10	304.50	1.1380	0.10937	277.67	303.91	1.1218
70	0.13639	286.74	314.02	1.1661	0.11307	286.35	313.49	1.1501
80	0.14073	295.53	323.68	1.1939	0.11674	295.18	323.19	1.1780
90	0.14504	304.47	333.48	1.2212	0.12037	304.15	333.04	1.2055
100	0.14932	313.57	343.43	1.2483	0.12398	313.27	343.03	1.2326

Table A-12 **(Continued)**

T °C	v m³/kg	u kJ/kg	h kJ/kg	s kJ/kg · K	v m³/kg	u kJ/kg	h kJ/kg	s kJ/kg · K
	p = 2.8 bar = 0.28 MPa (T_{sat} = −1.23°C)				p = 3.2 bar = 0.32 MPa (T_{sat} = 2.48°C)			
Sat.	0.07193	226.38	246.52	0.9197	0.06322	228.43	248.66	0.9177
0	0.07240	227.37	247.64	0.9238				
10	0.07613	235.44	256.76	0.9566	0.06576	234.61	255.65	0.9427
20	0.07972	243.59	265.91	0.9883	0.06901	242.87	264.95	0.9749
30	0.08320	251.83	275.12	1.0192	0.07214	251.19	274.28	1.0062
40	0.08660	260.17	284.42	1.0494	0.07518	259.61	283.67	1.0367
50	0.08992	268.64	293.81	1.0789	0.07815	268.14	293.15	1.0665
60	0.09319	277.23	303.32	1.1079	0.08106	276.79	302.72	1.0957
70	0.09641	285.96	312.95	1.1364	0.08392	285.56	312.41	1.1243
80	0.09960	294.82	322.71	1.1644	0.08674	294.46	322.22	1.1525
90	0.10275	303.83	332.60	1.1920	0.08953	303.50	332.15	1.1802
100	0.10587	312.98	342.62	1.2193	0.09229	312.68	342.21	1.2076
110	0.10897	322.27	352.78	1.2461	0.09503	322.00	352.40	1.2345
120	0.11205	331.71	363.08	1.2727	0.09774	331.45	362.73	1.2611
	p = 4.0 bar = 0.40 MPa (T_{sat} = 8.93°C)				p = 5.0 bar = 0.50 MPa (T_{sat} = 15.74°C)			
Sat.	0.05089	231.97	252.32	0.9145	0.04086	235.64	256.07	0.9117
10	0.05119	232.87	253.35	0.9182				
20	0.05397	241.37	262.96	0.9515	0.04188	239.40	260.34	0.9264
30	0.05662	249.89	272.54	0.9837	0.04416	248.20	270.28	0.9597
40	0.05917	258.47	282.14	1.0148	0.04633	256.99	280.16	0.9918
50	0.06164	267.13	291.79	1.0452	0.04842	265.83	290.04	1.0229
60	0.06405	275.89	301.51	1.0748	0.05043	274.73	299.95	1.0531
70	0.06641	284.75	311.32	1.1038	0.05240	283.72	309.92	1.0825
80	0.06873	293.73	321.23	1.1322	0.05432	292.80	319.96	1.1114
90	0.07102	302.84	331.25	1.1602	0.05620	302.00	330.10	1.1397
100	0.07327	312.07	341.38	1.1878	0.05805	311.31	340.33	1.1675
110	0.07550	321.44	351.64	1.2149	0.05988	320.74	350.68	1.1949
120	0.07771	330.94	362.03	1.2417	0.06168	330.30	361.14	1.2218
130	0.07991	340.58	372.54	1.2681	0.06347	339.98	371.72	1.2484
140	0.08208	350.35	383.18	1.2941	0.06524	349.79	382.42	1.2746
	p = 6.0 bar = 0.60 MPa (T_{sat} = 21.58°C)				p = 7.0 bar = 0.70 MPa (T_{sat} = 26.72°C)			
Sat.	0.03408	238.74	259.19	0.9097	0.02918	241.42	261.85	0.9080
30	0.03581	246.41	267.89	0.9388	0.02979	244.51	265.37	0.9197
40	0.03774	255.45	278.09	0.9719	0.03157	253.83	275.93	0.9539
50	0.03958	264.48	288.23	1.0037	0.03324	263.08	286.35	0.9867
60	0.04134	273.54	298.35	1.0346	0.03482	272.31	296.69	1.0182
70	0.04304	282.66	308.48	1.0645	0.03634	281.57	307.01	1.0487
80	0.04469	291.86	318.67	1.0938	0.03781	290.88	317.35	1.0784
90	0.04631	301.14	328.93	1.1225	0.03924	300.27	327.74	1.1074
100	0.04790	310.53	339.27	1.1505	0.04064	309.74	338.19	1.1358
110	0.04946	320.03	349.70	1.1781	0.04201	319.31	348.71	1.1637
120	0.05099	329.64	360.24	1.2053	0.04335	328.98	359.33	1.1910
130	0.05251	339.38	370.88	1.2320	0.04468	338.76	370.04	1.2179
140	0.05402	349.23	381.64	1.2584	0.04599	348.66	380.86	1.2444
150	0.05550	359.21	392.52	1.2844	0.04729	358.68	391.79	1.2706
160	0.05698	369.32	403.51	1.3100	0.04857	368.82	402.82	1.2963

R-134a

Table A-12 **(Continued)**

T °C	v m³/kg	u kJ/kg	h kJ/kg	s kJ/kg · K	v m³/kg	u kJ/kg	h kJ/kg	s kJ/kg · K
	p = 8.0 bar = 0.80 MPa (T_{sat} = 31.33°C)				p = 9.0 bar = 0.90 MPa (T_{sat} = 35.53°C)			
Sat.	0.02547	243.78	264.15	0.9066	0.02255	245.88	266.18	0.9054
40	0.02691	252.13	273.66	0.9374	0.02325	250.32	271.25	0.9217
50	0.02846	261.62	284.39	0.9711	0.02472	260.09	282.34	0.9566
60	0.02992	271.04	294.98	1.0034	0.02609	269.72	293.21	0.9897
70	0.03131	280.45	305.50	1.0345	0.02738	279.30	303.94	1.0214
80	0.03264	289.89	316.00	1.0647	0.02861	288.87	314.62	1.0521
90	0.03393	299.37	326.52	1.0940	0.02980	298.46	325.28	1.0819
100	0.03519	308.93	337.08	1.1227	0.03095	308.11	335.96	1.1109
110	0.03642	318.57	347.71	1.1508	0.03207	317.82	346.68	1.1392
120	0.03762	328.31	358.40	1.1784	0.03316	327.62	357.47	1.1670
130	0.03881	338.14	369.19	1.2055	0.03423	337.52	368.33	1.1943
140	0.03997	348.09	380.07	1.2321	0.03529	347.51	379.27	1.2211
150	0.04113	358.15	391.05	1.2584	0.03633	357.61	390.31	1.2475
160	0.04227	368.32	402.14	1.2843	0.03736	367.82	401.44	1.2735
170	0.04340	378.61	413.33	1.3098	0.03838	378.14	412.68	1.2992
180	0.04452	389.02	424.63	1.3351	0.03939	388.57	424.02	1.3245

T °C	v m³/kg	u kJ/kg	h kJ/kg	s kJ/kg · K	v m³/kg	u kJ/kg	h kJ/kg	s kJ/kg · K
	p = 10.0 bar = 1.00 MPa (T_{sat} = 39.39°C)				p = 12.0 bar = 1.20 MPa (T_{sat} = 46.32°C)			
Sat.	0.02020	247.77	267.97	0.9043	0.01663	251.03	270.99	0.9023
40	0.02029	248.39	268.68	0.9066				
50	0.02171	258.48	280.19	0.9428	0.01712	254.98	275.52	0.9164
60	0.02301	268.35	291.36	0.9768	0.01835	265.42	287.44	0.9527
70	0.02423	278.11	302.34	1.0093	0.01947	275.59	298.96	0.9868
80	0.02538	287.82	313.20	1.0405	0.02051	285.62	310.24	1.0192
90	0.02649	297.53	324.01	1.0707	0.02150	295.59	321.39	1.0503
100	0.02755	307.27	334.82	1.1000	0.02244	305.54	332.47	1.0804
110	0.02858	317.06	345.65	1.1286	0.02335	315.50	343.52	1.1096
120	0.02959	326.93	356.52	1.1567	0.02423	325.51	354.58	1.1381
130	0.03058	336.88	367.46	1.1841	0.02508	335.58	365.68	1.1660
140	0.03154	346.92	378.46	1.2111	0.02592	345.73	376.83	1.1933
150	0.03250	357.06	389.56	1.2376	0.02674	355.95	388.04	1.2201
160	0.03344	367.31	400.74	1.2638	0.02754	366.27	399.33	1.2465
170	0.03436	377.66	412.02	1.2895	0.02834	376.69	410.70	1.2724
180	0.03528	388.12	423.40	1.3149	0.02912	387.21	422.16	1.2980

T °C	v m³/kg	u kJ/kg	h kJ/kg	s kJ/kg · K	v m³/kg	u kJ/kg	h kJ/kg	s kJ/kg · K
	p = 14.0 bar = 1.40 MPa (T_{sat} = 52.43°C)				p = 16.0 bar = 1.60 MPa (T_{sat} = 57.92°C)			
Sat.	0.01405	253.74	273.40	0.9003	0.01208	256.00	275.33	0.8982
60	0.01495	262.17	283.10	0.9297	0.01233	258.48	278.20	0.9069
70	0.01603	272.87	295.31	0.9658	0.01340	269.89	291.33	0.9457
80	0.01701	283.29	307.10	0.9997	0.01435	280.78	303.74	0.9813
90	0.01792	293.55	318.63	1.0319	0.01521	291.39	315.72	1.0148
100	0.01878	303.73	330.02	1.0628	0.01601	301.84	327.46	1.0467
110	0.01960	313.88	341.32	1.0927	0.01677	312.20	339.04	1.0773
120	0.02039	324.05	352.59	1.1218	0.01750	322.53	350.53	1.1069
130	0.02115	334.25	363.86	1.1501	0.01820	332.87	361.99	1.1357
140	0.02189	344.50	375.15	1.1777	0.01887	343.24	373.44	1.1638
150	0.02262	354.82	386.49	1.2048	0.01953	353.66	384.91	1.1912
160	0.02333	365.22	397.89	1.2315	0.02017	364.15	396.43	1.2181
170	0.02403	375.71	409.36	1.2576	0.02080	374.71	407.99	1.2445
180	0.02472	386.29	420.90	1.2834	0.02142	385.35	419.62	1.2704
190	0.02541	396.96	432.53	1.3088	0.02203	396.08	431.33	1.2960
200	0.02608	407.73	444.24	1.3338	0.02263	406.90	443.11	1.3212

R-134a

Ammonia

Table A-13 Properties of Saturated Ammonia (Liquid–Vapor): Temperature Table

Pressure Conversions:
1 bar = 0.1 MPa
= 10^2 kPa

		Specific Volume m^3/kg		Internal Energy kJ/kg		Enthalpy kJ/kg			Entropy kJ/kg · K		
Temp. °C	Press. bar	Sat. Liquid $v_f \times 10^3$	Sat. Vapor v_g	Sat. Liquid u_f	Sat. Vapor u_g	Sat. Liquid h_f	Evap. h_{fg}	Sat. Vapor h_g	Sat. Liquid s_f	Sat. Vapor s_g	Temp. °C
−50	0.4086	1.4245	2.6265	−43.94	1264.99	−43.88	1416.20	1372.32	−0.1922	6.1543	−50
−45	0.5453	1.4367	2.0060	−22.03	1271.19	−21.95	1402.52	1380.57	−0.0951	6.0523	−45
−40	0.7174	1.4493	1.5524	−0.10	1277.20	0.00	1388.56	1388.56	0.0000	5.9557	−40
−36	0.8850	1.4597	1.2757	17.47	1281.87	17.60	1377.17	1394.77	0.0747	5.8819	−36
−32	1.0832	1.4703	1.0561	35.09	1286.41	35.25	1365.55	1400.81	0.1484	5.8111	−32
−30	1.1950	1.4757	0.9634	43.93	1288.63	44.10	1359.65	1403.75	0.1849	5.7767	−30
−28	1.3159	1.4812	0.8803	52.78	1290.82	52.97	1353.68	1406.66	0.2212	5.7430	−28
−26	1.4465	1.4867	0.8056	61.65	1292.97	61.86	1347.65	1409.51	0.2572	5.7100	−26
−22	1.7390	1.4980	0.6780	79.46	1297.18	79.72	1335.36	1415.08	0.3287	5.6457	−22
−20	1.9019	1.5038	0.6233	88.40	1299.23	88.68	1329.10	1417.79	0.3642	5.6144	−20
−18	2.0769	1.5096	0.5739	97.36	1301.25	97.68	1322.77	1420.45	0.3994	5.5837	−18
−16	2.2644	1.5155	0.5291	106.36	1303.23	106.70	1316.35	1423.05	0.4346	5.5536	−16
−14	2.4652	1.5215	0.4885	115.37	1305.17	115.75	1309.86	1425.61	0.4695	5.5239	−14
−12	2.6798	1.5276	0.4516	124.42	1307.08	124.83	1303.28	1428.11	0.5043	5.4948	−12
−10	2.9089	1.5338	0.4180	133.50	1308.95	133.94	1296.61	1430.55	0.5389	5.4662	−10
−8	3.1532	1.5400	0.3874	142.60	1310.78	143.09	1289.86	1432.95	0.5734	5.4380	−8
−6	3.4134	1.5464	0.3595	151.74	1312.57	152.26	1283.02	1435.28	0.6077	5.4103	−6
−4	3.6901	1.5528	0.3340	160.88	1314.32	161.46	1276.10	1437.56	0.6418	5.3831	−4
−2	3.9842	1.5594	0.3106	170.07	1316.04	170.69	1269.08	1439.78	0.6759	5.3562	−2
0	4.2962	1.5660	0.2892	179.29	1317.71	179.96	1261.97	1441.94	0.7097	5.3298	0
2	4.6270	1.5727	0.2695	188.53	1319.34	189.26	1254.77	1444.03	0.7435	5.3038	2
4	4.9773	1.5796	0.2514	197.80	1320.92	198.59	1247.48	1446.07	0.7770	5.2781	4
6	5.3479	1.5866	0.2348	207.10	1322.47	207.95	1240.09	1448.04	0.8105	5.2529	6
8	5.7395	1.5936	0.2195	216.42	1323.96	217.34	1232.61	1449.94	0.8438	5.2279	8
10	6.1529	1.6008	0.2054	225.77	1325.42	226.75	1225.03	1451.78	0.8769	5.2033	10
12	6.5890	1.6081	0.1923	235.14	1326.82	236.20	1217.35	1453.55	0.9099	5.1791	12
16	7.5324	1.6231	0.1691	253.95	1329.48	255.18	1201.70	1456.87	0.9755	5.1314	16
20	8.5762	1.6386	0.1492	272.86	1331.94	274.26	1185.64	1459.90	1.0404	5.0849	20
24	9.7274	1.6547	0.1320	291.84	1334.19	293.45	1169.16	1462.61	1.1048	5.0394	24
28	10.993	1.6714	0.1172	310.92	1336.20	312.75	1152.24	1465.00	1.1686	4.9948	28
32	12.380	1.6887	0.1043	330.07	1337.97	332.17	1134.87	1467.03	1.2319	4.9509	32
36	13.896	1.7068	0.0930	349.32	1339.47	351.69	1117.00	1468.70	1.2946	4.9078	36
40	15.549	1.7256	0.0831	368.67	1340.70	371.35	1098.62	1469.97	1.3569	4.8652	40
45	17.819	1.7503	0.0725	393.01	1341.81	396.13	1074.84	1470.96	1.4341	4.8125	45
50	20.331	1.7765	0.0634	417.56	1342.42	421.17	1050.09	1471.26	1.5109	4.7604	50

Source: Tables A-13 through A-15 are calculated based on equations from L. Haar and J. S. Gallagher, "Thermodynamic Properties of Ammonia," *J. Phys. Chem. Reference Data*, Vol. 7, 1978, pp. 635–792.

Table A-14 **Properties of Saturated Ammonia (Liquid–Vapor): Pressure Table**

Pressure Conversions:
1 bar = 0.1 MPa
= 10^2 kPa

		Specific Volume m^3/kg		Internal Energy kJ/kg		Enthalpy kJ/kg			Entropy kJ/kg · K		
Press. bar	Temp. °C	Sat. Liquid $v_f \times 10^3$	Sat. Vapor v_g	Sat. Liquid u_f	Sat. Vapor u_g	Sat. Liquid h_f	Evap. h_{fg}	Sat. Vapor h_g	Sat. Liquid s_f	Sat. Vapor s_g	Press. bar
0.40	−50.36	1.4236	2.6795	−45.52	1264.54	−45.46	1417.18	1371.72	−0.1992	6.1618	0.40
0.50	−46.53	1.4330	2.1752	−28.73	1269.31	−28.66	1406.73	1378.07	−0.1245	6.0829	0.50
0.60	−43.28	1.4410	1.8345	−14.51	1273.27	−14.42	1397.76	1383.34	−0.0622	6.0186	0.60
0.70	−40.46	1.4482	1.5884	−2.11	1276.66	−2.01	1389.85	1387.84	−0.0086	5.9643	0.70
0.80	−37.94	1.4546	1.4020	8.93	1279.61	9.04	1382.73	1391.78	0.0386	5.9174	0.80
0.90	−35.67	1.4605	1.2559	18.91	1282.24	19.04	1376.23	1395.27	0.0808	5.8760	0.90
1.00	−33.60	1.4660	1.1381	28.03	1284.61	28.18	1370.23	1398.41	0.1191	5.8391	1.00
1.25	−29.07	1.4782	0.9237	48.03	1289.65	48.22	1356.89	1405.11	0.2018	5.7610	1.25
1.50	−25.22	1.4889	0.7787	65.10	1293.80	65.32	1345.28	1410.61	0.2712	5.6973	1.50
1.75	−21.86	1.4984	0.6740	80.08	1297.33	80.35	1334.92	1415.27	0.3312	5.6435	1.75
2.00	−18.86	1.5071	0.5946	93.50	1300.39	93.80	1325.51	1419.31	0.3843	5.5969	2.00
2.25	−16.15	1.5151	0.5323	105.68	1303.08	106.03	1316.83	1422.86	0.4319	5.5558	2.25
2.50	−13.67	1.5225	0.4821	116.88	1305.49	117.26	1308.76	1426.03	0.4753	5.5190	2.50
2.75	−11.37	1.5295	0.4408	127.26	1307.67	127.68	1301.20	1428.88	0.5152	5.4858	2.75
3.00	−9.24	1.5361	0.4061	136.96	1309.65	137.42	1294.05	1431.47	0.5520	5.4554	3.00
3.25	−7.24	1.5424	0.3765	146.06	1311.46	146.57	1287.27	1433.84	0.5864	5.4275	3.25
3.50	−5.36	1.5484	0.3511	154.66	1313.14	155.20	1280.81	1436.01	0.6186	5.4016	3.50
3.75	−3.58	1.5542	0.3289	162.80	1314.68	163.38	1274.64	1438.03	0.6489	5.3774	3.75
4.00	−1.90	1.5597	0.3094	170.55	1316.12	171.18	1268.71	1439.89	0.6776	5.3548	4.00
4.25	−0.29	1.5650	0.2921	177.96	1317.47	178.62	1263.01	1441.63	0.7048	5.3336	4.25
4.50	1.25	1.5702	0.2767	185.04	1318.73	185.75	1257.50	1443.25	0.7308	5.3135	4.50
4.75	2.72	1.5752	0.2629	191.84	1319.91	192.59	1252.18	1444.77	0.7555	5.2946	4.75
5.00	4.13	1.5800	0.2503	198.39	1321.02	199.18	1247.02	1446.19	0.7791	5.2765	5.00
5.25	5.48	1.5847	0.2390	204.69	1322.07	205.52	1242.01	1447.53	0.8018	5.2594	5.25
5.50	6.79	1.5893	0.2286	210.78	1323.06	211.65	1237.15	1448.80	0.8236	5.2430	5.50
5.75	8.05	1.5938	0.2191	216.66	1324.00	217.58	1232.41	1449.99	0.8446	5.2273	5.75
6.00	9.27	1.5982	0.2104	222.37	1324.89	223.32	1227.79	1451.12	0.8649	5.2122	6.00
7.00	13.79	1.6148	0.1815	243.56	1328.04	244.69	1210.38	1455.07	0.9394	5.1576	7.00
8.00	17.84	1.6302	0.1596	262.64	1330.64	263.95	1194.36	1458.30	1.0054	5.1099	8.00
9.00	21.52	1.6446	0.1424	280.05	1332.82	281.53	1179.44	1460.97	1.0649	5.0675	9.00
10.00	24.89	1.6584	0.1285	296.10	1334.66	297.76	1165.42	1463.18	1.1191	5.0294	10.00
12.00	30.94	1.6841	0.1075	324.99	1337.52	327.01	1139.52	1466.53	1.2152	4.9625	12.00
14.00	36.26	1.7080	0.0923	350.58	1339.56	352.97	1115.82	1468.79	1.2987	4.9050	14.00
16.00	41.03	1.7306	0.0808	373.69	1340.97	376.46	1093.77	1470.23	1.3729	4.8542	16.00
18.00	45.38	1.7522	0.0717	394.85	1341.88	398.00	1073.01	1471.01	1.4399	4.8086	18.00
20.00	49.37	1.7731	0.0644	414.44	1342.37	417.99	1053.27	1471.26	1.5012	4.7670	20.00

Pressure Conversions:
1 bar = 0.1 MPa
= 10^2 kPa

Ammonia

Table A-15 **Properties of Superheated Ammonia Vapor**

T °C	v m³/kg	u kJ/kg	h kJ/kg	s kJ/kg · K	v m³/kg	u kJ/kg	h kJ/kg	s kJ/kg · K
	p = 0.4 bar = 0.04 MPa (T_{sat} = −50.36°C)				p = 0.6 bar = 0.06 MPa (T_{sat} = −43.28°C)			
Sat.	2.6795	1264.54	1371.72	6.1618	1.8345	1273.27	1383.34	6.0186
−50	2.6841	1265.11	1372.48	6.1652				
−45	2.7481	1273.05	1382.98	6.2118				
−40	2.8118	1281.01	1393.48	6.2573	1.8630	1278.62	1390.40	6.0490
−35	2.8753	1288.96	1403.98	6.3018	1.9061	1286.75	1401.12	6.0946
−30	2.9385	1296.93	1414.47	6.3455	1.9491	1294.88	1411.83	6.1390
−25	3.0015	1304.90	1424.96	6.3882	1.9918	1303.01	1422.52	6.1826
−20	3.0644	1312.88	1435.46	6.4300	2.0343	1311.13	1433.19	6.2251
−15	3.1271	1320.87	1445.95	6.4711	2.0766	1319.25	1443.85	6.2668
−10	3.1896	1328.87	1456.45	6.5114	2.1188	1327.37	1454.50	6.3077
−5	3.2520	1336.88	1466.95	6.5509	2.1609	1335.49	1465.14	6.3478
0	3.3142	1344.90	1477.47	6.5898	2.2028	1343.61	1475.78	6.3871
5	3.3764	1352.95	1488.00	6.6280	2.2446	1351.75	1486.43	6.4257
	p = 0.8 bar = 0.08 MPa (T_{sat} = −37.94°C)				p = 1.0 bar = 0.10 MPa (T_{sat} = −33.60°C)			
Sat.	1.4021	1279.61	1391.78	5.9174	1.1381	1284.61	1398.41	5.8391
−35	1.4215	1284.51	1398.23	5.9446				
−30	1.4543	1292.81	1409.15	5.9900	1.1573	1290.71	1406.44	5.8723
−25	1.4868	1301.09	1420.04	6.0343	1.1838	1299.15	1417.53	5.9175
−20	1.5192	1309.36	1430.90	6.0777	1.2101	1307.57	1428.58	5.9616
−15	1.5514	1317.61	1441.72	6.1200	1.2362	1315.96	1439.58	6.0046
−10	1.5834	1325.85	1452.53	6.1615	1.2621	1324.33	1450.54	6.0467
−5	1.6153	1334.09	1463.31	6.2021	1.2880	1332.67	1461.47	6.0878
0	1.6471	1342.31	1474.08	6.2419	1.3136	1341.00	1472.37	6.1281
5	1.6788	1350.54	1484.84	6.2809	1.3392	1349.33	1483.25	6.1676
10	1.7103	1358.77	1495.60	6.3192	1.3647	1357.64	1494.11	6.2063
15	1.7418	1367.01	1506.35	6.3568	1.3900	1365.95	1504.96	6.2442
20	1.7732	1375.25	1517.10	6.3939	1.4153	1374.27	1515.80	6.2816
	p = 1.5 bar = 0.15 MPa (T_{sat} = −25.22°C)				p = 2.0 bar = 0.20 MPa (T_{sat} = −18.86°C)			
Sat.	0.7787	1293.80	1410.61	5.6973	0.59460	1300.39	1419.31	5.5969
−25	0.7795	1294.20	1411.13	5.6994				
−20	0.7978	1303.00	1422.67	5.7454				
−15	0.8158	1311.75	1434.12	5.7902	0.60542	1307.43	1428.51	5.6328
−10	0.8336	1320.44	1445.49	5.8338	0.61926	1316.46	1440.31	5.6781
−5	0.8514	1329.08	1456.79	5.8764	0.63294	1325.41	1452.00	5.7221
0	0.8689	1337.68	1468.02	5.9179	0.64648	1334.29	1463.59	5.7649
5	0.8864	1346.25	1479.20	5.9585	0.65989	1343.11	1475.09	5.8066
10	0.9037	1354.78	1490.34	5.9981	0.67320	1351.87	1486.51	5.8473
15	0.9210	1363.29	1501.44	6.0370	0.68640	1360.59	1497.87	5.8871
20	0.9382	1371.79	1512.51	6.0751	0.69952	1369.28	1509.18	5.9260
25	0.9553	1380.28	1523.56	6.1125	0.71256	1377.93	1520.44	5.9641
30	0.9723	1388.76	1534.60	6.1492	0.72553	1386.56	1531.67	6.0014

Table A-15 **(Continued)**

T °C	v m³/kg	u kJ/kg	h kJ/kg	s kJ/kg · K	v m³/kg	u kJ/kg	h kJ/kg	s kJ/kg · K
	p = 2.5 bar = 0.25 MPa (T_{sat} = −13.67°C)				p = 3.0 bar = 0.30 MPa (T_{sat} = −9.24°C)			
Sat.	0.48213	1305.49	1426.03	5.5190	0.40607	1309.65	1431.47	5.4554
−10	0.49051	1312.37	1435.00	5.5534				
−5	0.50180	1321.65	1447.10	5.5989	0.41428	1317.80	1442.08	5.4953
0	0.51293	1330.83	1459.06	5.6431	0.42382	1327.28	1454.43	5.5409
5	0.52393	1339.91	1470.89	5.6860	0.43323	1336.64	1466.61	5.5851
10	0.53482	1348.91	1482.61	5.7278	0.44251	1345.89	1478.65	5.6280
15	0.54560	1357.84	1494.25	5.7685	0.45169	1355.05	1490.56	5.6697
20	0.55630	1366.72	1505.80	5.8083	0.46078	1364.13	1502.36	5.7103
25	0.56691	1375.55	1517.28	5.8471	0.46978	1373.14	1514.07	5.7499
30	0.57745	1384.34	1528.70	5.8851	0.47870	1382.09	1525.70	5.7886
35	0.58793	1393.10	1540.08	5.9223	0.48756	1391.00	1537.26	5.8264
40	0.59835	1401.84	1551.42	5.9589	0.49637	1399.86	1548.77	5.8635
45	0.60872	1410.56	1562.74	5.9947	0.50512	1408.70	1560.24	5.8998
	p = 3.5 bar = 0.35 MPa (T_{sat} = −5.36°C)				p = 4.0 bar = 0.40 MPa (T_{sat} = −1.90°C)			
Sat.	0.35108	1313.14	1436.01	5.4016	0.30942	1316.12	1439.89	5.3548
0	0.36011	1323.66	1449.70	5.4522	0.31227	1319.95	1444.86	5.3731
10	0.37654	1342.82	1474.61	5.5417	0.32701	1339.68	1470.49	5.4652
20	0.39251	1361.49	1498.87	5.6259	0.34129	1358.81	1495.33	5.5515
30	0.40814	1379.81	1522.66	5.7057	0.35520	1377.49	1519.57	5.6328
40	0.42350	1397.87	1546.09	5.7818	0.36884	1395.85	1543.38	5.7101
60	0.45363	1433.55	1592.32	5.9249	0.39550	1431.97	1590.17	5.8549
80	0.48320	1469.06	1638.18	6.0586	0.42160	1467.77	1636.41	5.9897
100	0.51240	1504.73	1684.07	6.1850	0.44733	1503.64	1682.58	6.1169
120	0.54136	1540.79	1730.26	6.3056	0.47280	1539.85	1728.97	6.2380
140	0.57013	1577.38	1776.92	6.4213	0.49808	1576.55	1775.79	6.3541
160	0.59876	1614.60	1824.16	6.5330	0.52323	1613.86	1823.16	6.4661
180	0.62728	1652.51	1872.06	6.6411	0.54827	1651.85	1871.16	6.5744
200	0.65572	1691.15	1920.65	6.7460	0.57322	1690.56	1919.85	6.6796
	p = 4.5 bar = 0.45 MPa (T_{sat} = 1.25°C)				p = 5.0 bar = 0.50 MPa (T_{sat} = 4.13°C)			
Sat.	0.27671	1318.73	1443.25	5.3135	0.25034	1321.02	1446.19	5.2765
10	0.28846	1336.48	1466.29	5.3962	0.25757	1333.22	1462.00	5.3330
20	0.30142	1356.09	1491.72	5.4845	0.26949	1353.32	1488.06	5.4234
30	0.31401	1375.15	1516.45	5.5674	0.28103	1372.76	1513.28	5.5080
40	0.32631	1393.80	1540.64	5.6460	0.29227	1391.74	1537.87	5.5878
60	0.35029	1430.37	1588.00	5.7926	0.31410	1428.76	1585.81	5.7362
80	0.37369	1466.47	1634.63	5.9285	0.33535	1465.16	1632.84	5.8733
100	0.39671	1502.55	1681.07	6.0564	0.35621	1501.46	1679.56	6.0020
120	0.41947	1538.91	1727.67	6.1781	0.37681	1537.97	1726.37	6.1242
140	0.44205	1575.73	1774.65	6.2946	0.39722	1574.90	1773.51	6.2412
160	0.46448	1613.13	1822.15	6.4069	0.41749	1612.40	1821.14	6.3537
180	0.48681	1651.20	1870.26	6.5155	0.43765	1650.54	1869.36	6.4626
200	0.50905	1689.97	1919.04	6.6208	0.45771	1689.38	1918.24	6.5681

Table A-15 (Continued)

T °C	v m³/kg	u kJ/kg	h kJ/kg	s kJ/kg · K	v m³/kg	u kJ/kg	h kJ/kg	s kJ/kg · K
	p = 5.5 bar = 0.55 MPa (T_{sat} = 6.79°C)				p = 6.0 bar = 0.60 MPa (T_{sat} = 9.27°C)			
Sat.	0.22861	1323.06	1448.80	5.2430	0.21038	1324.89	1451.12	5.2122
10	0.23227	1329.88	1457.63	5.2743	0.21115	1326.47	1453.16	5.2195
20	0.24335	1350.50	1484.34	5.3671	0.22155	1347.62	1480.55	5.3145
30	0.25403	1370.35	1510.07	5.4534	0.23152	1367.90	1506.81	5.4026
40	0.26441	1389.64	1535.07	5.5345	0.24118	1387.52	1532.23	5.4851
50	0.27454	1408.53	1559.53	5.6114	0.25059	1406.67	1557.03	5.5631
60	0.28449	1427.13	1583.60	5.6848	0.25981	1425.49	1581.38	5.6373
80	0.30398	1463.85	1631.04	5.8230	0.27783	1462.52	1629.22	5.7768
100	0.32307	1500.36	1678.05	5.9525	0.29546	1499.25	1676.52	5.9071
120	0.34190	1537.02	1725.07	6.0753	0.31281	1536.07	1723.76	6.0304
140	0.36054	1574.07	1772.37	6.1926	0.32997	1573.24	1771.22	6.1481
160	0.37903	1611.66	1820.13	6.3055	0.34699	1610.92	1819.12	6.2613
180	0.39742	1649.88	1868.46	6.4146	0.36390	1649.22	1867.56	6.3707
200	0.41571	1688.79	1917.43	6.5203	0.38071	1688.20	1916.63	6.4766
	p = 7.0 bar = 0.70 MPa (T_{sat} = 13.79°C)				p = 8.0 bar = 0.80 MPa (T_{sat} = 17.84°C)			
Sat.	0.18148	1328.04	1455.07	5.1576	0.15958	1330.64	1458.30	5.1099
20	0.18721	1341.72	1472.77	5.2186	0.16138	1335.59	1464.70	5.1318
30	0.19610	1362.88	1500.15	5.3104	0.16948	1357.71	1493.29	5.2277
40	0.20464	1383.20	1526.45	5.3958	0.17720	1378.77	1520.53	5.3161
50	0.21293	1402.90	1551.95	5.4760	0.18465	1399.05	1546.77	5.3986
60	0.22101	1422.16	1576.87	5.5519	0.19189	1418.77	1572.28	5.4763
80	0.23674	1459.85	1625.56	5.6939	0.20590	1457.14	1621.86	5.6209
100	0.25205	1497.02	1673.46	5.8258	0.21949	1494.77	1670.37	5.7545
120	0.26709	1534.16	1721.12	5.9502	0.23280	1532.24	1718.48	5.8801
140	0.28193	1571.57	1768.92	6.0688	0.24590	1569.89	1766.61	5.9995
160	0.29663	1609.44	1817.08	6.1826	0.25886	1607.96	1815.04	6.1140
180	0.31121	1647.90	1865.75	6.2925	0.27170	1646.57	1863.94	6.2243
200	0.32571	1687.02	1915.01	6.3988	0.28445	1685.83	1913.39	6.3311
	p = 9.0 bar = 0.90 MPa (T_{sat} = 21.52°C)				p = 10.0 bar = 1.00 MPa (T_{sat} = 24.89°C)			
Sat.	0.14239	1332.82	1460.97	5.0675	0.12852	1334.66	1463.18	5.0294
30	0.14872	1352.36	1486.20	5.1520	0.13206	1346.82	1478.88	5.0816
40	0.15582	1374.21	1514.45	5.2436	0.13868	1369.52	1508.20	5.1768
50	0.16263	1395.11	1541.47	5.3286	0.14499	1391.07	1536.06	5.2644
60	0.16922	1415.32	1567.61	5.4083	0.15106	1411.79	1562.86	5.3460
80	0.18191	1454.39	1618.11	5.5555	0.16270	1451.60	1614.31	5.4960
100	0.19416	1492.50	1667.24	5.6908	0.17389	1490.20	1664.10	5.6332
120	0.20612	1530.30	1715.81	5.8176	0.18478	1528.35	1713.13	5.7612
140	0.21788	1568.20	1764.29	5.9379	0.19545	1566.51	1761.96	5.8823
160	0.22948	1606.46	1813.00	6.0530	0.20598	1604.97	1810.94	5.9981
180	0.24097	1645.24	1862.12	6.1639	0.21638	1643.91	1860.29	6.1095
200	0.25237	1684.64	1911.77	6.2711	0.22670	1683.44	1910.14	6.2171

Table A-15 (Continued)

T °C	v m³/kg	u kJ/kg	h kJ/kg	s kJ/kg · K	v m³/kg	u kJ/kg	h kJ/kg	s kJ/kg · K
	p = 12.0 bar = 1.20 MPa (T_{sat} = 30.94°C)				p = 14.0 bar = 1.40 MPa (T_{sat} = 36.26°C)			
Sat.	0.10751	1337.52	1466.53	4.9625	0.09231	1339.56	1468.79	4.9050
40	0.11287	1359.73	1495.18	5.0553	0.09432	1349.29	1481.33	4.9453
60	0.12378	1404.54	1553.07	5.2347	0.10423	1396.97	1542.89	5.1360
80	0.13387	1445.91	1606.56	5.3906	0.11324	1440.06	1598.59	5.2984
100	0.14347	1485.55	1657.71	5.5315	0.12172	1480.79	1651.20	5.4433
120	0.15275	1524.41	1707.71	5.6620	0.12986	1520.41	1702.21	5.5765
140	0.16181	1563.09	1757.26	5.7850	0.13777	1559.63	1752.52	5.7013
160	0.17072	1601.95	1806.81	5.9021	0.14552	1598.92	1802.65	5.8198
180	0.17950	1641.23	1856.63	6.0145	0.15315	1638.53	1852.94	5.9333
200	0.18819	1681.05	1906.87	6.1230	0.16068	1678.64	1903.59	6.0427
220	0.19680	1721.50	1957.66	6.2282	0.16813	1719.35	1954.73	6.1485
240	0.20534	1762.63	2009.04	6.3303	0.17551	1760.72	2006.43	6.2513
260	0.21382	1804.48	2061.06	6.4297	0.18283	1802.78	2058.75	6.3513
280	0.22225	1847.04	2113.74	6.5267	0.19010	1845.55	2111.69	6.4488
	p = 16.0 bar = 1.60 MPa (T_{sat} = 41.03°C)				p = 18.0 bar = 1.80 MPa (T_{sat} = 45.38°C)			
Sat.	0.08079	1340.97	1470.23	4.8542	0.07174	1341.88	1471.01	4.8086
60	0.08951	1389.06	1532.28	5.0461	0.07801	1380.77	1521.19	4.9627
80	0.09774	1434.02	1590.40	5.2156	0.08565	1427.79	1581.97	5.1399
100	0.10539	1475.93	1644.56	5.3648	0.09267	1470.97	1637.78	5.2937
120	0.11268	1516.34	1696.64	5.5008	0.09931	1512.22	1690.98	5.4326
140	0.11974	1556.14	1747.72	5.6276	0.10570	1552.61	1742.88	5.5614
160	0.12663	1595.85	1798.45	5.7475	0.11192	1592.76	1794.23	5.6828
180	0.13339	1635.81	1849.23	5.8621	0.11801	1633.08	1845.50	5.7985
200	0.14005	1676.21	1900.29	5.9723	0.12400	1673.78	1896.98	5.9096
220	0.14663	1717.18	1951.79	6.0789	0.12991	1715.00	1948.83	6.0170
240	0.15314	1758.79	2003.81	6.1823	0.13574	1756.85	2001.18	6.1210
260	0.15959	1801.07	2056.42	6.2829	0.14152	1799.35	2054.08	6.2222
280	0.16599	1844.05	2109.64	6.3809	0.14724	1842.55	2107.58	6.3207
	p = 20.0 bar = 2.00 MPa (T_{sat} = 49.37°C)							
Sat.	0.06445	1342.37	1471.26	4.7670				
60	0.06875	1372.05	1509.54	4.8838				
80	0.07596	1421.36	1573.27	5.0696				
100	0.08248	1465.89	1630.86	5.2283				
120	0.08861	1508.03	1685.24	5.3703				
140	0.09447	1549.03	1737.98	5.5012				
160	0.10016	1589.65	1789.97	5.6241				
180	0.10571	1630.32	1841.74	5.7409				
200	0.11116	1671.33	1893.64	5.8530				
220	0.11652	1712.82	1945.87	5.9611				
240	0.12182	1754.90	1998.54	6.0658				
260	0.12706	1797.63	2051.74	6.1675				
280	0.13224	1841.03	2105.50	6.2665				

Table A-16 **Properties of Saturated Propane (Liquid–Vapor): Temperature Table**

Pressure Conversions:
1 bar = 0.1 MPa
= 10^2 kPa

Temp. °C	Press. bar	Specific Volume m³/kg Sat. Liquid $v_f \times 10^3$	Sat. Vapor v_g	Internal Energy kJ/kg Sat. Liquid u_f	Sat. Vapor u_g	Enthalpy kJ/kg Sat. Liquid h_f	Evap. h_{fg}	Sat. Vapor h_g	Entropy kJ/kg · K Sat. Liquid s_f	Sat. Vapor s_g	Temp. °C
−100	0.02888	1.553	11.27	−128.4	319.5	−128.4	480.4	352.0	−0.634	2.140	−100
−90	0.06426	1.578	5.345	−107.8	329.3	−107.8	471.4	363.6	−0.519	2.055	−90
−80	0.1301	1.605	2.774	−87.0	339.3	−87.0	462.4	375.4	−0.408	1.986	−80
−70	0.2434	1.633	1.551	−65.8	349.5	−65.8	453.1	387.3	−0.301	1.929	−70
−60	0.4261	1.663	0.9234	−44.4	359.9	−44.3	443.5	399.2	−0.198	1.883	−60
−50	0.7046	1.694	0.5793	−22.5	370.4	−22.4	433.6	411.2	−0.098	1.845	−50
−40	1.110	1.728	0.3798	−0.2	381.0	0.0	423.2	423.2	0.000	1.815	−40
−30	1.677	1.763	0.2585	22.6	391.6	22.9	412.1	435.0	0.096	1.791	−30
−20	2.444	1.802	0.1815	45.9	402.4	46.3	400.5	446.8	0.190	1.772	−20
−10	3.451	1.844	0.1309	69.8	413.2	70.4	388.0	458.4	0.282	1.757	−10
0	4.743	1.890	0.09653	94.2	423.8	95.1	374.5	469.6	0.374	1.745	0
4	5.349	1.910	0.08591	104.2	428.1	105.3	368.8	474.1	0.410	1.741	4
8	6.011	1.931	0.07666	114.3	432.3	115.5	362.9	478.4	0.446	1.737	8
12	6.732	1.952	0.06858	124.6	436.5	125.9	356.8	482.7	0.482	1.734	12
16	7.515	1.975	0.06149	135.0	440.7	136.4	350.5	486.9	0.519	1.731	16
20	8.362	1.999	0.05525	145.4	444.8	147.1	343.9	491.0	0.555	1.728	20
24	9.278	2.024	0.04973	156.1	448.9	158.0	337.0	495.0	0.591	1.725	24
28	10.27	2.050	0.04483	166.9	452.9	169.0	329.9	498.9	0.627	1.722	28
32	11.33	2.078	0.04048	177.8	456.7	180.2	322.4	502.6	0.663	1.720	32
36	12.47	2.108	0.03659	188.9	460.6	191.6	314.6	506.2	0.699	1.717	36
40	13.69	2.140	0.03310	200.2	464.3	203.1	306.5	509.6	0.736	1.715	40
44	15.00	2.174	0.02997	211.7	467.9	214.9	298.0	512.9	0.772	1.712	44
48	16.40	2.211	0.02714	223.4	471.4	227.0	288.9	515.9	0.809	1.709	48
52	17.89	2.250	0.02459	235.3	474.6	239.3	279.3	518.6	0.846	1.705	52
56	19.47	2.293	0.02227	247.4	477.7	251.9	269.2	521.1	0.884	1.701	56
60	21.16	2.340	0.02015	259.8	480.6	264.8	258.4	523.2	0.921	1.697	60
65	23.42	2.406	0.01776	275.7	483.6	281.4	243.8	525.2	0.969	1.690	65
70	25.86	2.483	0.01560	292.3	486.1	298.7	227.7	526.4	1.018	1.682	70
75	28.49	2.573	0.01363	309.5	487.8	316.8	209.8	526.6	1.069	1.671	75
80	31.31	2.683	0.01182	327.6	488.2	336.0	189.2	525.2	1.122	1.657	80
85	34.36	2.827	0.01011	347.2	486.9	356.9	164.7	521.6	1.178	1.638	85
90	37.64	3.038	0.008415	369.4	482.2	380.8	133.1	513.9	1.242	1.608	90
95	41.19	3.488	0.006395	399.8	467.4	414.2	79.5	493.7	1.330	1.546	95
96.7	42.48	4.535	0.004535	434.9	434.9	454.2	0.0	457.2	1.437	1.437	96.7

Source: Tables A-16 through A-18 are calculated based on B. A. Younglove and J. F. Ely, "Thermophysical Properties of Fluids. II. Methane, Ethane, Propane, Isobutane and Normal Butane," *J. Phys. Chem. Ref. Data,* Vol. 16, No. 4, 1987, pp. 577–598.

Propane

Pressure Conversions:
1 bar = 0.1 MPa
= 10^2 kPa

Table A-17 **Properties of Saturated Propane (Liquid–Vapor): Pressure Table**

		Specific Volume m^3/kg		Internal Energy kJ/kg		Enthalpy kJ/kg			Entropy kJ/kg · K		
Press. bar	Temp. °C	Sat. Liquid $v_f \times 10^3$	Sat. Vapor v_g	Sat. Liquid u_f	Sat. Vapor u_g	Sat. Liquid h_f	Evap. h_{fg}	Sat. Vapor h_g	Sat. Liquid s_f	Sat. Vapor s_g	Press. bar
0.05	−93.28	1.570	6.752	−114.6	326.0	−114.6	474.4	359.8	−0.556	2.081	0.05
0.10	−83.87	1.594	3.542	−95.1	335.4	−95.1	465.9	370.8	−0.450	2.011	0.10
0.25	−69.55	1.634	1.513	−64.9	350.0	−64.9	452.7	387.8	−0.297	1.927	0.25
0.50	−56.93	1.672	0.7962	−37.7	363.1	−37.6	440.5	402.9	−0.167	1.871	0.50
0.75	−48.68	1.698	0.5467	−19.6	371.8	−19.5	432.3	412.8	−0.085	1.841	0.75
1.00	−42.38	1.719	0.4185	−5.6	378.5	−5.4	425.7	420.3	−0.023	1.822	1.00
2.00	−25.43	1.781	0.2192	33.1	396.6	33.5	406.9	440.4	0.139	1.782	2.00
3.00	−14.16	1.826	0.1496	59.8	408.7	60.3	393.3	453.6	0.244	1.762	3.00
4.00	−5.46	1.865	0.1137	80.8	418.0	81.5	382.0	463.5	0.324	1.751	4.00
5.00	1.74	1.899	0.09172	98.6	425.7	99.5	372.1	471.6	0.389	1.743	5.00
6.00	7.93	1.931	0.07680	114.2	432.2	115.3	363.0	478.3	0.446	1.737	6.00
7.00	13.41	1.960	0.06598	128.2	438.0	129.6	354.6	484.2	0.495	1.733	7.00
8.00	18.33	1.989	0.05776	141.0	443.1	142.6	346.7	489.3	0.540	1.729	8.00
9.00	22.82	2.016	0.05129	152.9	447.6	154.7	339.1	493.8	0.580	1.726	9.00
10.00	26.95	2.043	0.04606	164.0	451.8	166.1	331.8	497.9	0.618	1.723	10.00
11.00	30.80	2.070	0.04174	174.5	455.6	176.8	324.7	501.5	0.652	1.721	11.00
12.00	34.39	2.096	0.03810	184.4	459.1	187.0	317.8	504.8	0.685	1.718	12.00
13.00	37.77	2.122	0.03499	193.9	462.2	196.7	311.0	507.7	0.716	1.716	13.00
14.00	40.97	2.148	0.03231	203.0	465.2	206.0	304.4	510.4	0.745	1.714	14.00
15.00	44.01	2.174	0.02997	211.7	467.9	215.0	297.9	512.9	0.772	1.712	15.00
16.00	46.89	2.200	0.02790	220.1	470.4	223.6	291.4	515.0	0.799	1.710	16.00
17.00	49.65	2.227	0.02606	228.3	472.7	232.0	285.0	517.0	0.824	1.707	17.00
18.00	52.30	2.253	0.02441	236.2	474.9	240.2	278.6	518.8	0.849	1.705	18.00
19.00	54.83	2.280	0.02292	243.8	476.9	248.2	272.2	520.4	0.873	1.703	19.00
20.00	57.27	2.308	0.02157	251.3	478.7	255.9	265.9	521.8	0.896	1.700	20.00
22.00	61.90	2.364	0.01921	265.8	481.7	271.0	253.0	524.0	0.939	1.695	22.00
24.00	66.21	2.424	0.01721	279.7	484.3	285.5	240.1	525.6	0.981	1.688	24.00
26.00	70.27	2.487	0.01549	293.1	486.2	299.6	226.9	526.5	1.021	1.681	26.00
28.00	74.10	2.555	0.01398	306.2	487.5	313.4	213.2	526.6	1.060	1.673	28.00
30.00	77.72	2.630	0.01263	319.2	488.1	327.1	198.9	526.0	1.097	1.664	30.00
35.00	86.01	2.862	0.009771	351.4	486.3	361.4	159.1	520.5	1.190	1.633	35.00
40.00	93.38	3.279	0.007151	387.9	474.7	401.0	102.3	503.3	1.295	1.574	40.00
42.48	96.70	4.535	0.004535	434.9	434.9	454.2	0.0	454.2	1.437	1.437	42.48

Table A-18 Properties of Superheated Propane

Pressure Conversions:
1 bar = 0.1 MPa
= 10^2 kPa

T °C	v m³/kg	u kJ/kg	h kJ/kg	s kJ/kg · K	v m³/kg	u kJ/kg	h kJ/kg	s kJ/kg · K
	p = 0.05 bar = 0.005 MPa (T_{sat} = −93.28°C)				p = 0.1 bar = 0.01 MPa (T_{sat} = −83.87°C)			
Sat.	6.752	326.0	359.8	2.081	3.542	367.3	370.8	2.011
−90	6.877	329.4	363.8	2.103				
−80	7.258	339.8	376.1	2.169	3.617	339.5	375.7	2.037
−70	7.639	350.6	388.8	2.233	3.808	350.3	388.4	2.101
−60	8.018	361.8	401.9	2.296	3.999	361.5	401.5	2.164
−50	8.397	373.3	415.3	2.357	4.190	373.1	415.0	2.226
−40	8.776	385.1	429.0	2.418	4.380	385.0	428.8	2.286
−30	9.155	397.4	443.2	2.477	4.570	397.3	443.0	2.346
−20	9.533	410.1	457.8	2.536	4.760	410.0	457.6	2.405
−10	9.911	423.2	472.8	2.594	4.950	423.1	472.6	2.463
0	10.29	436.8	488.2	2.652	5.139	436.7	488.1	2.520
10	10.67	450.8	504.1	2.709	5.329	450.6	503.9	2.578
20	11.05	270.6	520.4	2.765	5.518	465.1	520.3	2.634
	p = 0.5 bar = 0.05 MPa (T_{sat} = −56.93°C)				p = 1.0 bar = 0.1 MPa (T_{sat} = −42.38°C)			
Sat.	0.796	363.1	402.9	1.871	0.4185	378.5	420.3	1.822
−50	0.824	371.3	412.5	1.914				
−40	0.863	383.4	426.6	1.976	0.4234	381.5	423.8	1.837
−30	0.903	396.0	441.1	2.037	0.4439	394.2	438.6	1.899
−20	0.942	408.8	455.9	2.096	0.4641	407.3	453.7	1.960
−10	0.981	422.1	471.1	2.155	0.4842	420.7	469.1	2.019
0	1.019	435.8	486.7	2.213	0.5040	434.4	484.8	2.078
10	1.058	449.8	502.7	2.271	0.5238	448.6	501.0	2.136
20	1.096	464.3	519.1	2.328	0.5434	463.3	517.6	2.194
30	1.135	479.2	535.9	2.384	0.5629	478.2	534.5	2.251
40	1.173	494.6	553.2	2.440	0.5824	493.7	551.9	2.307
50	1.211	510.4	570.9	2.496	0.6018	509.5	569.7	2.363
60	1.249	526.7	589.1	2.551	0.6211	525.8	587.9	2.419
	p = 2.0 bar = 0.2 MPa (T_{sat} = −25.43°C)				p = 3.0 bar = 0.3 MPa (T_{sat} = −14.16°C)			
Sat.	0.2192	396.6	440.4	1.782	0.1496	408.7	453.6	1.762
−20	0.2251	404.0	449.0	1.816				
−10	0.2358	417.7	464.9	1.877	0.1527	414.7	460.5	1.789
0	0.2463	431.8	481.1	1.938	0.1602	429.0	477.1	1.851
10	0.2566	446.3	497.6	1.997	0.1674	443.8	494.0	1.912
20	0.2669	461.1	514.5	2.056	0.1746	458.8	511.2	1.971
30	0.2770	476.3	531.7	2.113	0.1816	474.2	528.7	2.030
40	0.2871	491.9	549.3	2.170	0.1885	490.1	546.6	2.088
50	0.2970	507.9	567.3	2.227	0.1954	506.2	564.8	2.145
60	0.3070	524.3	585.7	2.283	0.2022	522.7	583.4	2.202
70	0.3169	541.1	604.5	2.339	0.2090	539.6	602.3	2.258
80	0.3267	558.4	623.7	2.394	0.2157	557.0	621.7	2.314
90	0.3365	576.1	643.4	2.449	0.2223	574.8	641.5	2.369

Table A-18 **(Continued)**

T °C	v m³/kg	u kJ/kg	h kJ/kg	s kJ/kg · K	v m³/kg	u kJ/kg	h kJ/kg	s kJ/kg · K
	p = 4.0 bar = 0.4 MPa (T_{sat} = −5.46°C)				p = 5.0 bar = 0.5 MPa (T_{sat} = 1.74°C)			
Sat.	0.1137	418.0	463.5	1.751	0.09172	425.7	471.6	1.743
0	0.1169	426.1	472.9	1.786				
10	0.1227	441.2	490.3	1.848	0.09577	438.4	486.3	1.796
20	0.1283	456.6	507.9	1.909	0.1005	454.1	504.3	1.858
30	0.1338	472.2	525.7	1.969	0.1051	470.0	522.5	1.919
40	0.1392	488.1	543.8	2.027	0.1096	486.1	540.9	1.979
50	0.1445	504.4	562.2	2.085	0.1140	502.5	559.5	2.038
60	0.1498	521.1	581.0	2.143	0.1183	519.4	578.5	2.095
70	0.1550	538.1	600.1	2.199	0.1226	536.6	597.9	2.153
80	0.1601	555.7	619.7	2.255	0.1268	554.1	617.5	2.209
90	0.1652	573.5	639.6	2.311	0.1310	572.1	637.6	2.265
100	0.1703	591.8	659.9	2.366	0.1351	590.5	658.0	2.321
110	0.1754	610.4	680.6	2.421	0.1392	609.3	678.9	2.376
	p = 6.0 bar = 0.6 MPa (T_{sat} = 7.93°C)				p = 7.0 bar = 0.7 MPa (T_{sat} = 13.41°C)			
Sat.	0.07680	432.2	478.3	1.737	0.06598	438.0	484.2	1.733
10	0.07769	435.6	482.2	1.751				
20	0.08187	451.5	500.6	1.815	0.06847	448.8	496.7	1.776
30	0.08588	467.7	519.2	1.877	0.07210	465.2	515.7	1.840
40	0.08978	484.0	537.9	1.938	0.07558	481.9	534.8	1.901
50	0.09357	500.7	556.8	1.997	0.07896	498.7	554.0	1.962
60	0.09729	517.6	576.0	2.056	0.08225	515.9	573.5	2.021
70	0.1009	535.0	595.5	2.113	0.08547	533.4	593.2	2.079
80	0.1045	552.7	615.4	2.170	0.08863	551.2	613.2	2.137
90	0.1081	570.7	635.6	2.227	0.09175	569.4	633.6	2.194
100	0.1116	589.2	656.2	2.283	0.09482	587.9	654.3	2.250
110	0.1151	608.0	677.1	2.338	0.09786	606.8	675.3	2.306
120	0.1185	627.3	698.4	2.393	0.1009	626.2	696.8	2.361
	p = 8.0 bar = 0.8 MPa (T_{sat} = 18.33°C)				p = 9.0 bar = 0.9 MPa (T_{sat} = 22.82°C)			
Sat.	0.05776	443.1	489.3	1.729	0.05129	447.2	493.8	1.726
20	0.05834	445.9	492.6	1.740				
30	0.06170	462.7	512.1	1.806	0.05355	460.0	508.2	1.774
40	0.06489	479.6	531.5	1.869	0.05653	477.2	528.1	1.839
50	0.06796	496.7	551.1	1.930	0.05938	494.7	548.1	1.901
60	0.07094	514.0	570.8	1.990	0.06213	512.2	568.1	1.962
70	0.07385	531.6	590.7	2.049	0.06479	530.0	588.3	2.022
80	0.07669	549.6	611.0	2.107	0.06738	548.1	608.7	2.081
90	0.07948	567.9	631.5	2.165	0.06992	566.5	629.4	2.138
100	0.08222	586.5	652.3	2.221	0.07241	585.2	650.4	2.195
110	0.08493	605.6	673.5	2.277	0.07487	604.3	671.7	2.252
120	0.08761	625.0	695.1	2.333	0.07729	623.7	693.3	2.307
130	0.09026	644.8	717.0	2.388	0.07969	643.6	715.3	2.363
140	0.09289	665.0	739.3	2.442	0.08206	663.8	737.7	2.418

Table A-18 **(Continued)**

T °C	v m³/kg	u kJ/kg	h kJ/kg	s kJ/kg · K	v m³/kg	u kJ/kg	h kJ/kg	s kJ/kg · K
	p = 10.0 bar = 1.0 MPa (T_{sat} = 26.95°C)				p = 12.0 bar = 1.2 MPa (T_{sat} = 34.39°C)			
Sat.	0.04606	451.8	497.9	1.723	0.03810	459.1	504.8	1.718
30	0.04696	457.1	504.1	1.744				
40	0.04980	474.8	524.6	1.810	0.03957	469.4	516.9	1.757
50	0.05248	492.4	544.9	1.874	0.04204	487.8	538.2	1.824
60	0.05505	510.2	565.2	1.936	0.04436	506.1	559.3	1.889
70	0.05752	528.2	585.7	1.997	0.04657	524.4	580.3	1.951
80	0.05992	546.4	606.3	2.056	0.04869	543.1	601.5	2.012
90	0.06226	564.9	627.2	2.114	0.05075	561.8	622.7	2.071
100	0.06456	583.7	648.3	2.172	0.05275	580.9	644.2	2.129
110	0.06681	603.0	669.8	2.228	0.05470	600.4	666.0	2.187
120	0.06903	622.6	691.6	2.284	0.05662	620.1	688.0	2.244
130	0.07122	642.5	713.7	2.340	0.05851	640.1	710.3	2.300
140	0.07338	662.8	736.2	2.395	0.06037	660.6	733.0	2.355
	p = 14.0 bar = 1.4 MPa (T_{sat} = 40.97°C)				p = 16.0 bar = 1.6 MPa (T_{sat} = 46.89°C)			
Sat.	0.03231	465.2	510.4	1.714	0.02790	470.4	515.0	1.710
50	0.03446	482.6	530.8	1.778	0.02861	476.7	522.5	1.733
60	0.03664	501.6	552.9	1.845	0.03075	496.6	545.8	1.804
70	0.03869	520.4	574.6	1.909	0.03270	516.2	568.5	1.871
80	0.04063	539.4	596.3	1.972	0.03453	535.7	590.9	1.935
90	0.04249	558.6	618.1	2.033	0.03626	555.2	613.2	1.997
100	0.04429	577.9	639.9	2.092	0.03792	574.8	635.5	2.058
110	0.04604	597.5	662.0	2.150	0.03952	594.7	657.9	2.117
120	0.04774	617.5	684.3	2.208	0.04107	614.8	680.5	2.176
130	0.04942	637.7	706.9	2.265	0.04259	635.3	703.4	2.233
140	0.05106	658.3	729.8	2.321	0.04407	656.0	726.5	2.290
150	0.05268	679.2	753.0	2.376	0.04553	677.1	749.9	2.346
160	0.05428	700.5	776.5	2.431	0.04696	698.5	773.6	2.401
	p = 18.0 bar = 1.8 MPa (T_{sat} = 52.30°C)				p = 20.0 bar = 2.0 MPa (T_{sat} = 57.27°C)			
Sat.	0.02441	474.9	518.8	1.705	0.02157	478.7	521.8	1.700
60	0.02606	491.1	538.0	1.763	0.02216	484.8	529.1	1.722
70	0.02798	511.4	561.8	1.834	0.02412	506.3	554.5	1.797
80	0.02974	531.6	585.1	1.901	0.02585	527.1	578.8	1.867
90	0.03138	551.5	608.0	1.965	0.02744	547.6	602.5	1.933
100	0.03293	571.5	630.8	2.027	0.02892	568.1	625.9	1.997
110	0.03443	591.7	653.7	2.087	0.03033	588.5	649.2	2.059
120	0.03586	612.1	676.6	2.146	0.03169	609.2	672.6	2.119
130	0.03726	632.7	699.8	2.204	0.03299	630.0	696.0	2.178
140	0.03863	653.6	723.1	2.262	0.03426	651.2	719.7	2.236
150	0.03996	674.8	746.7	2.318	0.03550	672.5	743.5	2.293
160	0.04127	696.3	770.6	2.374	0.03671	694.2	767.6	2.349
170	0.04256	718.2	794.8	2.429	0.03790	716.2	792.0	2.404
180	0.04383	740.4	819.3	2.484	0.03907	738.5	816.6	2.459

Table A-18 **(Continued)**

T °C	v m³/kg	u kJ/kg	h kJ/kg	s kJ/kg · K	v m³/kg	u kJ/kg	h kJ/kg	s kJ/kg · K
	p = 22.0 bar = 2.2 MPa (T_{sat} = 61.90°C)				p = 24.0 bar = 2.4 MPa (T_{sat} = 66.21°C)			
Sat.	0.01921	481.8	524.0	1.695	0.01721	484.3	525.6	1.688
70	0.02086	500.5	546.4	1.761	0.01802	493.7	536.9	1.722
80	0.02261	522.4	572.1	1.834	0.01984	517.0	564.6	1.801
90	0.02417	543.5	596.7	1.903	0.02141	539.0	590.4	1.873
100	0.02561	564.5	620.8	1.969	0.02283	560.6	615.4	1.941
110	0.02697	585.3	644.6	2.032	0.02414	581.9	639.8	2.006
120	0.02826	606.2	668.4	2.093	0.02538	603.2	664.1	2.068
130	0.02949	627.3	692.2	2.153	0.02656	624.6	688.3	2.129
140	0.03069	648.6	716.1	2.211	0.02770	646.0	712.5	2.188
150	0.03185	670.1	740.2	2.269	0.02880	667.8	736.9	2.247
160	0.03298	691.9	764.5	2.326	0.02986	689.7	761.4	2.304
170	0.03409	714.1	789.1	2.382	0.03091	711.9	786.1	2.360
180	0.03517	736.5	813.9	2.437	0.03193	734.5	811.1	2.416
	p = 26.0 bar = 2.6 MPa (T_{sat} = 70.27°C)				p = 30.0 bar = 3.0 MPa (T_{sat} = 77.72°C)			
Sat.	0.01549	486.2	526.5	1.681	0.01263	488.2	526.0	1.664
80	0.01742	511.0	556.3	1.767	0.01318	495.4	534.9	1.689
90	0.01903	534.2	583.7	1.844	0.01506	522.8	568.0	1.782
100	0.02045	556.4	609.6	1.914	0.01654	547.2	596.8	1.860
110	0.02174	578.3	634.8	1.981	0.01783	570.4	623.9	1.932
120	0.02294	600.0	659.6	2.045	0.01899	593.0	650.0	1.999
130	0.02408	621.6	684.2	2.106	0.02007	615.4	675.6	2.063
140	0.02516	643.4	708.8	2.167	0.02109	637.7	701.0	2.126
150	0.02621	665.3	733.4	2.226	0.02206	660.1	726.3	2.186
160	0.02723	687.4	758.2	2.283	0.02300	682.6	751.6	2.245
170	0.02821	709.9	783.2	2.340	0.02390	705.4	777.1	2.303
180	0.02918	732.5	808.4	2.397	0.02478	728.3	802.6	2.360
190	0.03012	755.5	833.8	2.452	0.02563	751.5	828.4	2.417
	p = 35.0 bar = 3.5 MPa (T_{sat} = 86.01°C)				p = 40.0 bar = 4.0 MPa (T_{sat} = 93.38°C)			
Sat.	0.00977	486.3	520.5	1.633	0.00715	474.7	503.3	1.574
90	0.01086	502.4	540.5	1.688				
100	0.01270	532.9	577.3	1.788	0.00940	512.1	549.7	1.700
110	0.01408	558.9	608.2	1.870	0.01110	544.7	589.1	1.804
120	0.01526	583.4	636.8	1.944	0.01237	572.1	621.6	1.887
130	0.01631	607.0	664.1	2.012	0.01344	597.4	651.2	1.962
140	0.01728	630.2	690.7	2.077	0.01439	621.9	679.5	2.031
150	0.01819	653.3	717.0	2.140	0.01527	645.9	707.0	2.097
160	0.01906	676.4	743.1	2.201	0.01609	669.7	734.1	2.160
170	0.01989	699.6	769.2	2.261	0.01687	693.4	760.9	2.222
180	0.02068	722.9	795.3	2.319	0.01761	717.3	787.7	2.281
190	0.02146	746.5	821.6	2.376	0.01833	741.2	814.5	2.340
200	0.02221	770.3	848.0	2.433	0.01902	765.3	841.4	2.397

Table A-19 **Properties of Selected Solids and Liquids: c_p, ρ, and κ**

Substance	Specific Heat, c_p (kJ/kg · K)	Density, ρ (kg/m^3)	Thermal Conductivity, κ (W/m · K)
Selected Solids, 300K			
Aluminium	0.903	2700	237
Coal, anthracite	1.260	1350	0.26
Copper	0.385	8930	401
Granite	0.775	2630	2.79
Iron	0.447	7870	80.2
Lead	0.129	11300	35.3
Sand	0.800	1520	0.27
Silver	0.235	10500	429
Soil	1.840	2050	0.52
Steel (AISI 302)	0.480	8060	15.1
Tin	0.227	7310	66.6
Building Materials, 300K			
Brick, common	0.835	1920	0.72
Concrete (stone mix)	0.880	2300	1.4
Glass, plate	0.750	2500	1.4
Hardboard, siding	1.170	640	0.094
Limestone	0.810	2320	2.15
Plywood	1.220	545	0.12
Softwoods (fir, pine)	1.380	510	0.12
Insulating Materials, 300K			
Blanket (glass fiber)	—	16	0.046
Cork	1.800	120	0.039
Duct liner (glass fiber, coated)	0.835	32	0.038
Polystyrene (extruded)	1.210	55	0.027
Vermiculite fill (flakes)	0.835	80	0.068
Saturated Liquids			
Ammonia, 300K	4.818	599.8	0.465
Mercury, 300K	0.139	13529	8.540
Refrigerant 22, 300K	1.267	1183.1	0.085
Refrigerant 134a, 300K	1.434	1199.7	0.081
Unused Engine Oil, 300K	1.909	884.1	0.145
Water, 275K	4.211	999.9	0.574
300K	4.179	996.5	0.613
325K	4.182	987.1	0.645
350K	4.195	973.5	0.668
375K	4.220	956.8	0.681
400K	4.256	937.4	0.688

Source: Drawn from several sources, these data are only representative. Values can vary depending on temperature, purity, moisture content, and other factors.

Table A-20 **Ideal Gas Specific Heats of Some Common Gases (kJ/kg · K)**

Temp. K	Air c_p	Air c_v	Air k	Nitrogen, N_2 c_p	Nitrogen, N_2 c_v	Nitrogen, N_2 k	Oxygen, O_2 c_p	Oxygen, O_2 c_v	Oxygen, O_2 k	Temp. K
250	1.003	0.716	1.401	1.039	0.742	1.400	0.913	0.653	1.398	250
300	1.005	0.718	1.400	1.039	0.743	1.400	0.918	0.658	1.395	300
350	1.008	0.721	1.398	1.041	0.744	1.399	0.928	0.668	1.389	350
400	1.013	0.726	1.395	1.044	0.747	1.397	0.941	0.681	1.382	400
450	1.020	0.733	1.391	1.049	0.752	1.395	0.956	0.696	1.373	450
500	1.029	0.742	1.387	1.056	0.759	1.391	0.972	0.712	1.365	500
550	1.040	0.753	1.381	1.065	0.768	1.387	0.988	0.728	1.358	550
600	1.051	0.764	1.376	1.075	0.778	1.382	1.003	0.743	1.350	600
650	1.063	0.776	1.370	1.086	0.789	1.376	1.017	0.758	1.343	650
700	1.075	0.788	1.364	1.098	0.801	1.371	1.031	0.771	1.337	700
750	1.087	0.800	1.359	1.110	0.813	1.365	1.043	0.783	1.332	750
800	1.099	0.812	1.354	1.121	0.825	1.360	1.054	0.794	1.327	800
900	1.121	0.834	1.344	1.145	0.849	1.349	1.074	0.814	1.319	900
1000	1.142	0.855	1.336	1.167	0.870	1.341	1.090	0.830	1.313	1000

Temp. K	Carbon Dioxide, CO_2 c_p	Carbon Dioxide, CO_2 c_v	Carbon Dioxide, CO_2 k	Carbon Monoxide, CO c_p	Carbon Monoxide, CO c_v	Carbon Monoxide, CO k	Hydrogen, H_2 c_p	Hydrogen, H_2 c_v	Hydrogen, H_2 k	Temp. K
250	0.791	0.602	1.314	1.039	0.743	1.400	14.051	9.927	1.416	250
300	0.846	0.657	1.288	1.040	0.744	1.399	14.307	10.183	1.405	300
350	0.895	0.706	1.268	1.043	0.746	1.398	14.427	10.302	1.400	350
400	0.939	0.750	1.252	1.047	0.751	1.395	14.476	10.352	1.398	400
450	0.978	0.790	1.239	1.054	0.757	1.392	14.501	10.377	1.398	450
500	1.014	0.825	1.229	1.063	0.767	1.387	14.513	10.389	1.397	500
550	1.046	0.857	1.220	1.075	0.778	1.382	14.530	10.405	1.396	550
600	1.075	0.886	1.213	1.087	0.790	1.376	14.546	10.422	1.396	600
650	1.102	0.913	1.207	1.100	0.803	1.370	14.571	10.447	1.395	650
700	1.126	0.937	1.202	1.113	0.816	1.364	14.604	10.480	1.394	700
750	1.148	0.959	1.197	1.126	0.829	1.358	14.645	10.521	1.392	750
800	1.169	0.980	1.193	1.139	0.842	1.353	14.695	10.570	1.390	800
900	1.204	1.015	1.186	1.163	0.866	1.343	14.822	10.698	1.385	900
1000	1.234	1.045	1.181	1.185	0.888	1.335	14.983	10.859	1.380	1000

Source: Adapted from K. Wark, *Thermodynamics,* 4th ed., McGraw-Hill, New York, 1983, as based on "Tables of Thermal Properties of Gases," NBS Circular 564, 1955.

Table A-21 **Variation of $\bar{c}_p$ with Temperature for Selected Ideal Gases**

$$\frac{\bar{c}_p}{\bar{R}} = \alpha + \beta T + \gamma T^2 + \delta T^3 + \varepsilon T^4$$

T is in K, equations valid from 300 to 1000 K

Gas	α	$\beta \times 10^3$	$\gamma \times 10^6$	$\delta \times 10^9$	$\varepsilon \times 10^{12}$
CO	3.710	−1.619	3.692	−2.032	0.240
CO_2	2.401	8.735	−6.607	2.002	0
H_2	3.057	2.677	−5.810	5.521	−1.812
H_2O	4.070	−1.108	4.152	−2.964	0.807
O_2	3.626	−1.878	7.055	−6.764	2.156
N_2	3.675	−1.208	2.324	−0.632	−0.226
Air	3.653	−1.337	3.294	−1.913	0.2763
SO_2	3.267	5.324	0.684	−5.281	2.559
CH_4	3.826	−3.979	24.558	−22.733	6.963
C_2H_2	1.410	19.057	−24.501	16.391	−4.135
C_2H_4	1.426	11.383	7.989	−16.254	6.749
Monatomic gases[a]	2.5	0	0	0	0

[a] For monatomic gases, such as He, Ne, and Ar, $\bar{c}_p$ is constant over a wide temperature range and is very nearly equal to $5/2\,\bar{R}$.

Source: Adapted from K. Wark, *Thermodynamics,* 4th ed., McGraw-Hill, New York, 1983, as based on NASA SP-273, U.S. Government Printing Office, Washington, DC, 1971.

Table A-22 **Ideal Gas Properties of Air**

T(K), h and u(kJ/kg), s° (kJ/kg · K)

				when $\Delta s = 0$[1]						when $\Delta s = 0$	
T	h	u	s°	p_r	v_r	T	h	u	s°	p_r	v_r
200	199.97	142.56	1.29559	0.3363	1707.	450	451.80	322.62	2.11161	5.775	223.6
210	209.97	149.69	1.34444	0.3987	1512.	460	462.02	329.97	2.13407	6.245	211.4
220	219.97	156.82	1.39105	0.4690	1346.	470	472.24	337.32	2.15604	6.742	200.1
230	230.02	164.00	1.43557	0.5477	1205.	480	482.49	344.70	2.17760	7.268	189.5
240	240.02	171.13	1.47824	0.6355	1084.	490	492.74	352.08	2.19876	7.824	179.7
250	250.05	178.28	1.51917	0.7329	979.	500	503.02	359.49	2.21952	8.411	170.6
260	260.09	185.45	1.55848	0.8405	887.8	510	513.32	366.92	2.23993	9.031	162.1
270	270.11	192.60	1.59634	0.9590	808.0	520	523.63	374.36	2.25997	9.684	154.1
280	280.13	199.75	1.63279	1.0889	738.0	530	533.98	381.84	2.27967	10.37	146.7
285	285.14	203.33	1.65055	1.1584	706.1	540	544.35	389.34	2.29906	11.10	139.7
290	290.16	206.91	1.66802	1.2311	676.1	550	554.74	396.86	2.31809	11.86	133.1
295	295.17	210.49	1.68515	1.3068	647.9	560	565.17	404.42	2.33685	12.66	127.0
300	300.19	214.07	1.70203	1.3860	621.2	570	575.59	411.97	2.35531	13.50	121.2
305	305.22	217.67	1.71865	1.4686	596.0	580	586.04	419.55	2.37348	14.38	115.7
310	310.24	221.25	1.73498	1.5546	572.3	590	596.52	427.15	2.39140	15.31	110.6
315	315.27	224.85	1.75106	1.6442	549.8	600	607.02	434.78	2.40902	16.28	105.8
320	320.29	228.42	1.76690	1.7375	528.6	610	617.53	442.42	2.42644	17.30	101.2
325	325.31	232.02	1.78249	1.8345	508.4	620	628.07	450.09	2.44356	18.36	96.92
330	330.34	235.61	1.79783	1.9352	489.4	630	638.63	457.78	2.46048	19.84	92.84
340	340.42	242.82	1.82790	2.149	454.1	640	649.22	465.50	2.47716	20.64	88.99
350	350.49	250.02	1.85708	2.379	422.2	650	659.84	473.25	2.49364	21.86	85.34
360	360.58	257.24	1.88543	2.626	393.4	660	670.47	481.01	2.50985	23.13	81.89
370	370.67	264.46	1.91313	2.892	367.2	670	681.14	488.81	2.52589	24.46	78.61
380	380.77	271.69	1.94001	3.176	343.4	680	691.82	496.62	2.54175	25.85	75.50
390	390.88	278.93	1.96633	3.481	321.5	690	702.52	504.45	2.55731	27.29	72.56
400	400.98	286.16	1.99194	3.806	301.6	700	713.27	512.33	2.57277	28.80	69.76
410	411.12	293.43	2.01699	4.153	283.3	710	724.04	520.23	2.58810	30.38	67.07
420	421.26	300.69	2.04142	4.522	266.6	720	734.82	528.14	2.60319	32.02	64.53
430	431.43	307.99	2.06533	4.915	251.1	730	745.62	536.07	2.61803	33.72	62.13
440	441.61	315.30	2.08870	5.332	236.8	740	756.44	544.02	2.63280	35.50	59.82

1. p_r and v_r data for use with Eqs. 6.41 and 6.42, respectively.

Table A-22 (Continued)

T	h	u	s°	p_r (when $\Delta s = 0$[1])	v_r	T	h	u	s°	p_r (when $\Delta s = 0$)	v_r
750	767.29	551.99	2.64737	37.35	57.63	1300	1395.97	1022.82	3.27345	330.9	11.275
760	778.18	560.01	2.66176	39.27	55.54	1320	1419.76	1040.88	3.29160	352.5	10.747
770	789.11	568.07	2.67595	41.31	53.39	1340	1443.60	1058.94	3.30959	375.3	10.247
780	800.03	576.12	2.69013	43.35	51.64	1360	1467.49	1077.10	3.32724	399.1	9.780
790	810.99	584.21	2.70400	45.55	49.86	1380	1491.44	1095.26	3.34474	424.2	9.337
800	821.95	592.30	2.71787	47.75	48.08	1400	1515.42	1113.52	3.36200	450.5	8.919
820	843.98	608.59	2.74504	52.59	44.84	1420	1539.44	1131.77	3.37901	478.0	8.526
840	866.08	624.95	2.77170	57.60	41.85	1440	1563.51	1150.13	3.39586	506.9	8.153
860	888.27	641.40	2.79783	63.09	39.12	1460	1587.63	1168.49	3.41247	537.1	7.801
880	910.56	657.95	2.82344	68.98	36.61	1480	1611.79	1186.95	3.42892	568.8	7.468
900	932.93	674.58	2.84856	75.29	34.31	1500	1635.97	1205.41	3.44516	601.9	7.152
920	955.38	691.28	2.87324	82.05	32.18	1520	1660.23	1223.87	3.46120	636.5	6.854
940	977.92	708.08	2.89748	89.28	30.22	1540	1684.51	1242.43	3.47712	672.8	6.569
960	1000.55	725.02	2.92128	97.00	28.40	1560	1708.82	1260.99	3.49276	710.5	6.301
980	1023.25	741.98	2.94468	105.2	26.73	1580	1733.17	1279.65	3.50829	750.0	6.046
1000	1046.04	758.94	2.96770	114.0	25.17	1600	1757.57	1298.30	3.52364	791.2	5.804
1020	1068.89	776.10	2.99034	123.4	23.72	1620	1782.00	1316.96	3.53879	834.1	5.574
1040	1091.85	793.36	3.01260	133.3	22.39	1640	1806.46	1335.72	3.55381	878.9	5.355
1060	1114.86	810.62	3.03449	143.9	21.14	1660	1830.96	1354.48	3.56867	925.6	5.147
1080	1137.89	827.88	3.05608	155.2	19.98	1680	1855.50	1373.24	3.58335	974.2	4.949
1100	1161.07	845.33	3.07732	167.1	18.896	1700	1880.1	1392.7	3.5979	1025	4.761
1120	1184.28	862.79	3.09825	179.7	17.886	1750	1941.6	1439.8	3.6336	1161	4.328
1140	1207.57	880.35	3.11883	193.1	16.946	1800	2003.3	1487.2	3.6684	1310	3.944
1160	1230.92	897.91	3.13916	207.2	16.064	1850	2065.3	1534.9	3.7023	1475	3.601
1180	1254.34	915.57	3.15916	222.2	15.241	1900	2127.4	1582.6	3.7354	1655	3.295
1200	1277.79	933.33	3.17888	238.0	14.470	1950	2189.7	1630.6	3.7677	1852	3.022
1220	1301.31	951.09	3.19834	254.7	13.747	2000	2252.1	1678.7	3.7994	2068	2.776
1240	1324.93	968.95	3.21751	272.3	13.069	2050	2314.6	1726.8	3.8303	2303	2.555
1260	1348.55	986.90	3.23638	290.8	12.435	2100	2377.4	1775.3	3.8605	2559	2.356
1280	1372.24	1004.76	3.25510	310.4	11.835	2150	2440.3	1823.8	3.8901	2837	2.175
						2200	2503.2	1872.4	3.9191	3138	2.012
						2250	2566.4	1921.3	3.9474	3464	1.864

Units: T(K), h and u(kJ/kg), s° (kJ/kg · K)

Source: Table A-22 is based on J. H. Keenan and J. Kaye, *Gas Tables,* Wiley, New York, 1945.

Table A-23 Ideal Gas Properties of Selected Gases

Enthalpy $\bar{h}(T)$ and internal energy $\bar{u}(T)$, in kJ/kmol. Absolute entropy at 1 atm $\bar{s}^\circ(T)$, in kJ/kmol · K.

	Carbon Dioxide, CO_2 ($\bar{h}_f^\circ = -393{,}520$ kJ/kmol)			Carbon Monoxide, CO ($\bar{h}_f^\circ = -110{,}530$ kJ/kmol)			Water Vapor, H_2O ($\bar{h}_f^\circ = -241{,}820$ kJ/kmol)			Oxygen, O_2 ($\bar{h}_f^\circ = 0$ kJ/kmol)			Nitrogen, N_2 ($\bar{h}_f^\circ = 0$ kJ/kmol)			
T(K)	$\bar{h}$	$\bar{u}$	$\bar{s}^\circ$	$\bar{h}$	$\bar{u}$	$\bar{s}^\circ$	$\bar{h}$	$\bar{u}$	$\bar{s}^\circ$	$\bar{h}$	$\bar{u}$	$\bar{s}^\circ$	$\bar{h}$	$\bar{u}$	$\bar{s}^\circ$	T(K)
0	0	0	0	0	0	0	0	0	0	0	0	0	0	0	0	0
220	6,601	4,772	202.966	6,391	4,562	188.683	7,295	5,466	178.576	6,404	4,575	196.171	6,391	4,562	182.638	220
230	6,938	5,026	204.464	6,683	4,771	189.980	7,628	5,715	180.054	6,694	4,782	197.461	6,683	4,770	183.938	230
240	7,280	5,285	205.920	6,975	4,979	191.221	7,961	5,965	181.471	6,984	4,989	198.696	6,975	4,979	185.180	240
250	7,627	5,548	207.337	7,266	5,188	192.411	8,294	6,215	182.831	7,275	5,197	199.885	7,266	5,188	186.370	250
260	7,979	5,817	208.717	7,558	5,396	193.554	8,627	6,466	184.139	7,566	5,405	201.027	7,558	5,396	187.514	260
270	8,335	6,091	210.062	7,849	5,604	194.654	8,961	6,716	185.399	7,858	5,613	202.128	7,849	5,604	188.614	270
280	8,697	6,369	211.376	8,140	5,812	195.173	9,296	6,968	186.616	8,150	5,822	203.191	8,141	5,813	189.673	280
290	9,063	6,651	212.660	8,432	6,020	196.735	9,631	7,219	187.791	8,443	6,032	204.218	8,432	6,021	190.695	290
298	9,364	6,885	213.685	8,669	6,190	197.543	9,904	7,425	188.720	8,682	6,203	205.033	8,669	6,190	191.502	298
300	9,431	6,939	213.915	8,723	6,229	197.723	9,966	7,472	188.928	8,736	6,242	205.213	8,723	6,229	191.682	300
310	9,807	7,230	215.146	9,014	6,437	198.678	10,302	7,725	190.030	9,030	6,453	206.177	9,014	6,437	192.638	310
320	10,186	7,526	216.351	9,306	6,645	199.603	10,639	7,978	191.098	9,325	6,664	207.112	9,306	6,645	193.562	320
330	10,570	7,826	217.534	9,597	6,854	200.500	10,976	8,232	192.136	9,620	6,877	208.020	9,597	6,853	194.459	330
340	10,959	8,131	218.694	9,889	7,062	201.371	11,314	8,487	193.144	9,916	7,090	208.904	9,888	7,061	195.328	340
350	11,351	8,439	219.831	10,181	7,271	202.217	11,652	8,742	194.125	10,213	7,303	209.765	10,180	7,270	196.173	350
360	11,748	8,752	220.948	10,473	7,480	203.040	11,992	8,998	195.081	10,511	7,518	210.604	10,471	7,478	196.995	360
370	12,148	9,068	222.044	10,765	7,689	203.842	12,331	9,255	196.012	10,809	7,733	211.423	10,763	7,687	197.794	370
380	12,552	9,392	223.122	11,058	7,899	204.622	12,672	9,513	196.920	11,109	7,949	212.222	11,055	7,895	198.572	380
390	12,960	9,718	224.182	11,351	8,108	205.383	13,014	9,771	197.807	11,409	8,166	213.002	11,347	8,104	199.331	390
400	13,372	10,046	225.225	11,644	8,319	206.125	13,356	10,030	198.673	11,711	8,384	213.765	11,640	8,314	200.071	400
410	13,787	10,378	226.250	11,938	8,529	206.850	13,699	10,290	199.521	12,012	8,603	214.510	11,932	8,523	200,794	410
420	14,206	10,714	227.258	12,232	8,740	207.549	14,043	10,551	200.350	12,314	8,822	215.241	12,225	8,733	201.499	420
430	14,628	11,053	228.252	12,526	8,951	208.252	14,388	10,813	201.160	12,618	9,043	215.955	12,518	8,943	202.189	430
440	15,054	11,393	229.230	12,821	9,163	208.929	14,734	11,075	201.955	12,923	9,264	216.656	12,811	9,153	202.863	440
450	15,483	11,742	230.194	13,116	9,375	209.593	15,080	11,339	202.734	13,228	9,487	217.342	13,105	9,363	203.523	450
460	15,916	12,091	231.144	13,412	9,587	210.243	15,428	11,603	203.497	13,535	9,710	218.016	13,399	9,574	204.170	460
470	16,351	12,444	232.080	13,708	9,800	210.880	15,777	11,869	204.247	13,842	9,935	218.676	13,693	9,786	204.803	470
480	16,791	12,800	233.004	14,005	10,014	211.504	16,126	12,135	204.982	14,151	10,160	219.326	13,988	9,997	205.424	480
490	17,232	13,158	233.916	14,302	10,228	212.117	16,477	12,403	205.705	14,460	10,386	219.963	14,285	10,210	206.033	490
500	17,678	13,521	234.814	14,600	10,443	212.719	16,828	12,671	206.413	14,770	10,614	220.589	14,581	10,423	206.630	500
510	18,126	13,885	235.700	14,898	10,658	213.310	17,181	12,940	207.112	15,082	10,842	221.206	14,876	10,635	207.216	510
520	18,576	14,253	236.575	15,197	10,874	213.890	17,534	13,211	207.799	15,395	11,071	221.812	15,172	10,848	207.792	520
530	19,029	14,622	237.439	15,497	11,090	214.460	17,889	13,482	208.475	15,708	11,301	222.409	15,469	11,062	208.358	530
540	19,485	14,996	238.292	15,797	11,307	215.020	18,245	13,755	209.139	16,022	11,533	222.997	15,766	11,277	208.914	540
550	19,945	15,372	239.135	16,097	11,524	215.572	18,601	14,028	209.795	16,338	11,765	223.576	16,064	11,492	209.461	550
560	20,407	15,751	239.962	16,399	11,743	216.115	18,959	14,303	210.440	16,654	11,998	224.146	16,363	11,707	209.999	560
570	20,870	16,131	240.789	16,701	11,961	216.649	19,318	14,579	211.075	16,971	12,232	224.708	16,662	11,923	210.528	570
580	21,337	16,515	241.602	17,003	12,181	217.175	19,678	14,856	211.702	17,290	12,467	225.262	16,962	12,139	211.049	580
590	21,807	16,902	242.405	17,307	12,401	217.693	20,039	15,134	212.320	17,609	12,703	225.808	17,262	12,356	211.562	590

$\bar{h}$ and $\bar{u}$ in kJ/kmol, $\bar{s}^\circ$ in kJ/kmol · K

T(K)	Carbon Dioxide, CO_2 ($\bar{h}_f^\circ = -393{,}520$ kJ/kmol)			Carbon Monoxide, CO ($\bar{h}_f^\circ = -110{,}530$ kJ/kmol)			Water Vapor, H_2O ($\bar{h}_f^\circ = -241{,}820$ kJ/kmol)			Oxygen, O_2 ($\bar{h}_f^\circ = 0$ kJ/kmol)			Nitrogen, N_2 ($\bar{h}_f^\circ = 0$ kJ/kmol)			T(K)
	$\bar{h}$	$\bar{u}$	$\bar{s}^\circ$	$\bar{h}$	$\bar{u}$	$\bar{s}^\circ$	$\bar{h}$	$\bar{u}$	$\bar{s}^\circ$	$\bar{h}$	$\bar{u}$	$\bar{s}^\circ$	$\bar{h}$	$\bar{u}$	$\bar{s}^\circ$	
600	22,280	17,291	243.199	17,611	12,622	218.204	20,402	15,413	212.920	17,929	12,940	226.346	17,563	12,574	212.066	600
610	22,754	17,683	243.983	17,915	12,843	218.708	20,765	15,693	213.529	18,250	13,178	226.877	17,864	12,792	212.564	610
620	23,231	18,076	244.758	18,221	13,066	219.205	21,130	15,975	214.122	18,572	13,417	227.400	18,166	13,011	213.055	620
630	23,709	18,471	245.524	18,527	13,289	219.695	21,495	16,257	214.707	18,895	13,657	227.918	18,468	13,230	213.541	630
640	24,190	18,869	246.282	18,833	13,512	220.179	21,862	16,541	215.285	19,219	13,898	228.429	18,772	13,450	214.018	640
650	24,674	19,270	247.032	19,141	13,736	220.656	22,230	16,826	215.856	19,544	14,140	228.932	19,075	13,671	214.489	650
660	25,160	19,672	247,773	19,449	13,962	221.127	22,600	17,112	216.419	19,870	14,383	229.430	19,380	13,892	214.954	660
670	25,648	20,078	248.507	19,758	14,187	221.592	22,970	17,399	216.976	20,197	14,626	229.920	19,685	14,114	215.413	670
680	26,138	20,484	249.233	20,068	14,414	222.052	23,342	17,688	217.527	20,524	14,871	230.405	19,991	14,337	215.866	680
690	26,631	20,894	249.952	20,378	14,641	222.505	23,714	17,978	218.071	20,854	15,116	230.885	20,297	14,560	216.314	690
700	27,125	21,305	250.663	20,690	14,870	222.953	24,088	18,268	218.610	21,184	15,364	231.358	20,604	14,784	216.756	700
710	27,622	21,719	251.368	21,002	15,099	223.396	24,464	18,561	219.142	21,514	15,611	231.827	20,912	15,008	217.192	710
720	28,121	22,134	252.065	21,315	15,328	223.833	24,840	18,854	219.668	21,845	15,859	232.291	21,220	15,234	217.624	720
730	28,622	22,552	252.755	21,628	15,558	224.265	25,218	19,148	220.189	22,177	16,107	232.748	21,529	15,460	218.059	730
740	29,124	22,972	253.439	21,943	15,789	224.692	25,597	19,444	220.707	22,510	16,357	233.201	21,839	15,686	218.472	740
750	29,629	23,393	254.117	22,258	16,022	225.115	25,977	19,741	221.215	22,844	16,607	233.649	22,149	15,913	218.889	750
760	30,135	23,817	254.787	22,573	16,255	225.533	26,358	20,039	221.720	23,178	16,859	234.091	22,460	16,141	219.301	760
770	30,644	24,242	255,452	22,890	16,488	225.947	26,741	20,339	222.221	23,513	17,111	234.528	22,772	16,370	219.709	770
780	31,154	24,669	256.110	23,208	16,723	226.357	27,125	20,639	222.717	23,850	17,364	234.960	23,085	16,599	220.113	780
790	31,665	25,097	256.762	23,526	16,957	226.762	27,510	20,941	223.207	24,186	17,618	235.387	23,398	16,830	220.512	790
800	32,179	25,527	257.408	23,844	17,193	227.162	27,896	21,245	223.693	24,523	17,872	235.810	23,714	17,061	220.907	800
810	32,694	25,959	258.048	24,164	17,429	227.559	28,284	21,549	224.174	24,861	18,126	236.230	24,027	17,292	221.298	810
820	33,212	26,394	258.682	24,483	17,665	227.952	28,672	21,855	224.651	25,199	18,382	236.644	24,342	17,524	221.684	820
830	33,730	26,829	259.311	24,803	17,902	228.339	29,062	22,162	225.123	25,537	18,637	237.055	24,658	17,757	222.067	830
840	34,251	27,267	259.934	25,124	18,140	228.724	29,454	22,470	225.592	25,877	18,893	237.462	24,974	17,990	222.447	840
850	34,773	27,706	260.551	25,446	18,379	229,106	29,846	22,779	226.057	26,218	19,150	237.864	25,292	18,224	222.822	850
860	35,296	28,125	261.164	25,768	18,617	229,482	30,240	23,090	226.517	26,559	19,408	238.264	25,610	18,459	223.194	860
870	35,821	28,588	261.770	26.091	18,858	229.856	30,635	23,402	226.973	26,899	19,666	238.660	25,928	18,695	223.562	870
880	36,347	29,031	262.371	26,415	19,099	230.227	31,032	23,715	227.426	27,242	19,925	239.051	26,248	18,931	223.927	880
890	36,876	29,476	262.968	26,740	19,341	230.593	31,429	24,029	227.875	27,584	20,185	239.439	26,568	19,168	224.288	890
900	37,405	29,922	263.559	27,066	19,583	230.957	31,828	24,345	228.321	27,928	20,445	239.823	26,890	19,407	224.647	900
910	37,935	30,369	264.146	27,392	19,826	231.317	32,228	24,662	228.763	28,272	20,706	240.203	27,210	19,644	225.002	910
920	38,467	30,818	264.728	27,719	20,070	231.674	32,629	24,980	229.202	28,616	20,967	240.580	27,532	19,883	225.353	920
930	39,000	31,268	265.304	28,046	20,314	232.028	33,032	25,300	229.637	28,960	21,228	240.953	27,854	20,122	225.701	930
940	39,535	31,719	265.877	28,375	20,559	232.379	33,436	25,621	230.070	29,306	21,491	241.323	28,178	20,362	226.047	940
950	40,070	32,171	266.444	28,703	20,805	232.727	33,841	25,943	230.499	29,652	21,754	241.689	28,501	20,603	226.389	950
960	40,607	32,625	267.007	29,033	21,051	233.072	34,247	26,265	230.924	29,999	22,017	242.052	28,826	20,844	226.728	960
970	41,145	33,081	267.566	29,362	21,298	233.413	34,653	26,588	231.347	30,345	22,280	242.411	29,151	21,086	227.064	970
980	41,685	33,537	268.119	29,693	21,545	233.752	35,061	26,913	231.767	30,692	22,544	242.768	29,476	21,328	227.398	980
990	42,226	33,995	268.670	30,024	21,793	234.088	35,472	27,240	232.184	31.041	22,809	243.120	29,803	21,571	227.728	990

Table A-23 **(Continued)**

	Carbon Dioxide, CO_2 ($\bar{h}_f^\circ = -393{,}520$ kJ/kmol)			Carbon Monoxide, CO ($\bar{h}_f^\circ = -110{,}530$ kJ/kmol)			Water Vapor, H_2O ($\bar{h}_f^\circ = -241{,}820$ kJ/kmol)			Oxygen, O_2 ($\bar{h}_f^\circ = 0$ kJ/kmol)			Nitrogen, N_2 ($\bar{h}_f^\circ = 0$ kJ/kmol)			
T(K)	$\bar{h}$	$\bar{u}$	$\bar{s}^\circ$	$\bar{h}$	$\bar{u}$	$\bar{s}^\circ$	$\bar{h}$	$\bar{u}$	$\bar{s}^\circ$	$\bar{h}$	$\bar{u}$	$\bar{s}^\circ$	$\bar{h}$	$\bar{u}$	$\bar{s}^\circ$	T(K)
1000	42,769	34,455	269.215	30,355	22,041	234.421	35,882	27,568	232.597	31,389	23,075	243.471	30,129	21,815	228.057	1000
1020	43,859	35,378	270.293	31,020	22,540	235.079	36,709	28,228	233.415	32,088	23,607	244.164	30,784	22,304	228.706	1020
1040	44,953	36,306	271.354	31,688	23,041	235.728	37,542	28,895	234.223	32,789	24,142	244.844	31,442	22,795	229.344	1040
1060	46,051	37,238	272.400	32,357	23,544	236.364	38,380	29,567	235.020	33,490	24,677	245.513	32,101	23,288	229.973	1060
1080	47,153	38,174	273.430	33,029	24,049	236.992	39,223	30,243	235.806	34,194	25,214	246.171	32,762	23,782	230.591	1080
1100	48,258	39,112	274.445	33,702	24,557	237.609	40,071	30,925	236.584	34,899	25,753	246.818	33,426	24,280	231.199	1100
1120	49,369	40,057	275.444	34,377	25,065	238.217	40,923	31,611	237.352	35,606	26,294	247.454	34,092	24,780	231.799	1120
1140	50,484	41,006	276.430	35,054	25,575	238.817	41,780	32,301	238.110	36,314	26,836	248.081	34,760	25,282	232.391	1140
1160	51,602	41,957	277.403	35,733	26,088	239.407	42,642	32,997	238.859	37,023	27,379	248.698	35,430	25,786	232.973	1160
1180	52,724	42,913	278.362	36,406	26,602	239.989	43,509	33,698	239.600	37,734	27,923	249.307	36,104	26,291	233.549	1180
1200	53,848	43,871	279.307	37,095	27,118	240.663	44,380	34,403	240.333	38,447	28,469	249.906	36,777	26,799	234.115	1200
1220	54,977	44,834	280.238	37,780	27,637	241.128	45,256	35,112	241.057	39,162	29,018	250.497	37,452	27,308	234.673	1220
1240	56,108	45,799	281.158	38,466	28,426	241.686	46,137	35,827	241.773	39,877	29,568	251.079	38,129	27,819	235.223	1240
1260	57,244	46,768	282.066	39,154	28,678	242.236	47,022	36,546	242.482	40,594	30,118	251.653	38,807	28,331	235.766	1260
1280	58,381	47,739	282.962	39,884	29,201	242.780	47,912	37,270	243.183	41,312	30,670	252.219	39,488	28,845	236.302	1280
1300	59,522	48,713	283.847	40,534	29,725	243.316	48,807	38,000	243.877	42,033	31,224	252.776	40,170	29,361	236.831	1300
1320	60,666	49,691	284.722	41,266	30,251	243.844	49,707	38,732	244.564	42,753	31,778	253.325	40,853	29,878	237.353	1320
1340	61,813	50,672	285.586	41,919	30,778	244.366	50,612	39,470	245.243	43,475	32,334	253.868	41,539	30,398	237.867	1340
1360	62,963	51,656	286.439	42,613	31,306	244.880	51,521	40,213	245.915	44,198	32,891	254.404	42,227	30,919	238.376	1360
1380	64,116	52,643	287.283	43,309	31,836	245.388	52,434	40,960	246.582	44,923	33,449	254.932	42,915	31,441	238.878	1380
1400	65,271	53,631	288.106	44,007	32,367	245.889	53,351	41,711	247.241	45,648	34,008	255.454	43,605	31,964	239.375	1400
1420	66,427	54,621	288.934	44,707	32,900	246.385	54,273	42,466	247.895	46,374	34,567	255.968	44,295	32,489	239.865	1420
1440	67,586	55,614	289.743	45,408	33,434	246.876	55,198	43,226	248.543	47,102	35,129	256.475	44,988	33,014	240.350	1440
1460	68,748	56,609	290.542	46,110	33,971	247.360	56,128	43,989	249.185	47,831	35,692	256.978	45,682	33,543	240.827	1460
1480	69,911	57,606	291.333	46,813	34,508	247.839	57,062	44,756	249.820	48,561	36,256	257.474	46,377	34,071	241.301	1480
1500	71,078	58,606	292.114	47,517	35,046	248.312	57,999	45,528	250.450	49,292	36,821	257.965	47,073	34,601	241.768	1500
1520	72,246	59,609	292.888	48,222	35,584	248.778	58,942	46,304	251.074	50,024	37,387	258.450	47,771	35,133	242.228	1520
1540	73,417	60,613	292.654	48,928	36,124	249.240	59,888	47,084	251.693	50,756	37,952	258.928	48,470	35,665	242.685	1540
1560	74,590	61,620	294.411	49,635	36,665	249.695	60,838	47,868	252.305	51,490	38,520	259.402	49,168	36,197	243.137	1560
1580	76,767	62,630	295.161	50,344	37,207	250.147	61,792	48,655	252.912	52,224	39,088	259.870	49,869	36,732	243.585	1580
1600	76,944	63,741	295.901	51,053	37,750	250.592	62,748	49,445	253.513	52,961	39,658	260.333	50,571	37,268	244.028	1600
1620	78,123	64,653	296.632	51,763	38,293	251.033	63,709	52,240	254.111	53,696	40,227	260.791	51,275	37,806	244.464	1620
1640	79,303	65,668	297.356	52,472	38,837	251.470	64,675	51,039	254.703	54,434	40,799	261.242	51,980	38,344	244.896	1640
1660	80,486	66,592	298.072	53,184	39,382	251.901	65,643	51,841	255.290	55,172	41,370	261.690	52,686	38,884	245.324	1660
1680	81,670	67,702	298.781	53,895	39,927	252.329	66,614	52,646	255.873	55,912	41,944	262.132	53,393	39,424	245.747	1680
1700	82,856	68,721	299.482	54,609	40,474	252.751	67,589	53,455	256.450	56,652	42,517	262.571	54,099	39,965	246.166	1700
1720	84,043	69,742	300.177	55,323	41,023	253.169	68,567	54,267	257.022	57,394	43,093	263.005	54,807	40,507	246.580	1720
1740	85,231	70,764	300.863	56,039	41,572	253.582	69,550	55,083	257.589	58,136	43,669	263.435	55,516	41,049	246.990	1740

$\bar{h}$ and $\bar{u}$ in kJ/kmol, $\bar{s}^\circ$ in kJ/kmol · K

Table A-23 ***Continued)***

$\bar{h}$ and $\bar{u}$ in kJ/kmol, $\bar{s}^\circ$ in kJ/kmol · K

T(K)	Carbon Dioxide, CO_2 ($\bar{h}_f^\circ = -393{,}520$ kJ/kmol)			Carbon Monoxide, CO ($\bar{h}_f^\circ = -110{,}530$ kJ/kmol)			Water Vapor, H_2O ($\bar{h}_f^\circ = -241{,}820$ kJ/kmol)			Oxygen, O_2 ($\bar{h}_f^\circ = 0$ kJ/kmol)			Nitrogen, N_2 ($\bar{h}_f^\circ = 0$ kJ/kmol)			T(K)
	$\bar{h}$	$\bar{u}$	$\bar{s}^\circ$	$\bar{h}$	$\bar{u}$	$\bar{s}^\circ$	$\bar{h}$	$\bar{u}$	$\bar{s}^\circ$	$\bar{h}$	$\bar{u}$	$\bar{s}^\circ$	$\bar{h}$	$\bar{u}$	$\bar{s}^\circ$	
1760	86,420	71,787	301.543	56,756	42,123	253.991	70,535	55,902	258.151	58,800	44,247	263.861	56,227	41,594	247.396	1760
1780	87,612	72,812	302.271	57,473	42,673	254.398	71,523	56,723	258.708	59,624	44,825	264.283	56,938	42,139	247.798	1780
1800	88,806	73,840	302.884	58,191	43,225	254.797	72,513	57,547	259.262	60,371	45,405	264.701	57,651	42,685	248.195	1800
1820	90,000	74,868	303.544	58,910	43,778	255.194	73,507	58,375	259.811	61,118	45,986	265.113	58,363	43,231	248.589	1820
1840	91,196	75,897	304.198	59,629	44,331	255.587	74,506	59,207	260.357	61,866	46,568	265.521	59,075	43,777	248.979	1840
1860	92,394	76,929	304.845	60,351	44,886	255.976	75,506	60,042	260.898	62,616	47,151	265.925	59,790	44,324	249.365	1860
1880	93,593	77,962	305.487	61,072	45,441	256.361	76,511	60,880	261.436	63,365	47,734	266.326	60,504	44,873	249.748	1880
1900	94,793	78,996	306.122	61,794	45,997	256.743	77,517	61,720	261.969	64,116	48,319	266.722	61,220	45,423	250.128	1900
1920	95,995	80,031	306.751	62,516	46,552	257.122	78,527	62,564	262.497	64,868	48,904	267.115	61,936	45,973	250.502	1920
1940	97,197	81,067	307.374	63,238	47,108	257.497	79,540	63,411	263.022	65,620	49,490	267.505	62,654	46,524	250.874	1940
1960	98,401	82,105	307.992	63,961	47,665	257.868	80,555	64,259	263.542	66,374	50,078	267.891	63,381	47,075	251.242	1960
1980	99,606	83,144	308.604	64,684	48,221	258.236	81,573	65,111	264.059	67,127	50,665	268.275	64,090	47,627	251.607	1980
2000	100,804	84,185	309.210	65,408	48,780	258.600	82,593	65,965	264.571	67,881	51,253	268.655	64,810	48,181	251.969	2000
2050	103,835	86,791	310.701	67,224	50,179	259.494	85,156	68,111	265.838	69,772	52,727	269.588	66,612	49,567	252.858	2050
2100	106,864	89,404	312.160	69,044	51,584	260.370	87,735	70,275	267.081	71,668	54,208	270.504	68,417	50,957	253.726	2100
2150	109,898	92,023	313.589	70,864	52,988	261.226	90,330	72,454	268.301	73,573	55,697	271.399	70,226	52,351	254.578	2150
2200	112,939	94,648	314.988	72,688	54,396	262.065	92,940	74,649	269.500	75,484	57,192	272.278	72,040	53,749	255.412	2200
2250	115,984	97,277	316.356	74,516	55,809	262.887	95,562	76,855	270.679	77,397	58,690	273.136	73,856	55,149	256.227	2250
2300	119,035	99,912	317.695	76,345	57,222	263.692	98,199	79,076	271.839	79,316	60,193	273.981	75,676	56,553	257.027	2300
2350	122,091	102,552	319.011	78,178	58,640	264.480	100,846	81,308	272.978	81,243	61,704	274.809	77,496	57,958	257.810	2350
2400	125,152	105,197	320.302	80,015	60,060	265.253	103,508	83,553	274.098	83,174	63,219	275.625	79,320	59,366	258.580	2400
2450	128,219	107,849	321.566	81,852	61,482	266.012	106,183	85,811	275.201	85,112	64,742	276.424	81,149	60,779	259.332	2450
2500	131,290	110,504	322.808	83,692	62,906	266.755	108,868	88,082	276.286	87,057	66,271	277.207	82,981	62,195	260.073	2500
2550	134,368	113,166	324.026	85,537	64,335	267.485	111,565	90,364	277.354	89,004	67,802	277.979	84,814	63,613	260.799	2550
2600	137,449	115,832	325.222	87,383	65,766	268.202	114,273	92,656	278.407	90,956	69,339	278.738	86,650	65,033	261.512	2600
2650	140,533	118,500	326.396	89,230	67,197	268.905	116,991	94,958	279,441	92,916	70,883	279.485	88,488	66,455	262.213	2650
2700	143,620	121,172	327.549	91,077	68,628	269.596	119,717	97,269	280.462	94,881	72,433	280.219	90,328	67,880	262.902	2700
2750	146,713	123,849	328.684	92,930	70,066	270.285	122,453	99,588	281.464	96,852	73,987	280.942	92,171	69,306	263.577	2750
2800	149,808	126,528	329.800	94,784	71,504	270.943	125,198	101,917	282.453	98,826	75,546	281.654	94,014	70,734	264.241	2800
2850	152,908	129,212	330.896	96,639	72,945	271.602	127,952	104,256	283.429	100,808	77,112	282.357	95,859	72,163	264.895	2850
2900	156,009	131,898	331.975	98,495	74,383	272.249	130,717	106,605	284.390	102,793	78,682	283.048	97,705	73,593	265.538	2900
2950	159,117	134,589	333.037	100,352	75,825	272.884	133,486	108,959	285.338	104,785	80,258	283.728	99,556	75,028	266.170	2950
3000	162,226	137,283	334.084	102,210	77,267	273.508	136,264	111,321	286.273	106,780	81,837	284.399	101,407	76,464	266.793	3000
3050	165,341	139,982	335.114	104,073	78,715	274.123	139,051	113,692	287.194	108,778	83,419	285.060	103,260	77,902	267.404	3050
3100	168,456	142,681	336.126	105,939	80,164	274.730	141,846	116,072	288.102	110,784	85,009	285.713	105,115	79,341	268.007	3100
3150	171,576	145,385	337.124	107,802	81,612	275.326	144,648	118,458	288.999	112,795	86,601	286.355	106,972	80,782	268.601	3150
3200	174,695	148,089	338.109	109,667	83,061	275.914	147,457	120,851	289.884	114,809	88,203	286.989	108,830	82,224	269.186	3200
3250	177,822	150,801	339.069	111,534	84,513	276.494	150,272	123,250	290.756	116,827	89,804	287.614	110,690	83,668	269.763	3250

Source: Table A-23 is based on the JANAF Thermochemical Tables, NSRDS-NBS-37, 1971.

Table A-24 **Constants for the van der Waals, Redlich–Kwong, and Benedict–Webb–Rubin Equations of State**

1. van der Waals and Redlich–Kwong: Constants for pressure in bar, specific volume in m^3/kmol, and temperature in K

Substance	van der Waals		Redlich–Kwong	
	a bar $\left(\frac{m^3}{kmol}\right)^2$	b $\frac{m^3}{kmol}$	a bar $\left(\frac{m^3}{kmol}\right)^2 K^{1/2}$	b $\frac{m^3}{kmol}$
Air	1.368	0.0367	15.989	0.02541
Butane (C_4H_{10})	13.86	0.1162	289.55	0.08060
Carbon dioxide (CO_2)	3.647	0.0428	64.43	0.02963
Carbon monoxide (CO)	1.474	0.0395	17.22	0.02737
Methane (CH_4)	2.293	0.0428	32.11	0.02965
Nitrogen (N_2)	1.366	0.0386	15.53	0.02677
Oxygen (O_2)	1.369	0.0317	17.22	0.02197
Propane (C_3H_8)	9.349	0.0901	182.23	0.06242
Refrigerant 12	10.49	0.0971	208.59	0.06731
Sulfur dioxide (SO_2)	6.883	0.0569	144.80	0.03945
Water (H_2O)	5.531	0.0305	142.59	0.02111

Source: Calculated from critical data.

2. Benedict–Webb–Rubin: Constants for pressure in bar, specific volume in m^3/kmol, and temperature in K

Substance	a	A	b	B	c	C	α	γ
C_4H_{10}	1.9073	10.218	0.039998	0.12436	3.206×10^5	1.006×10^6	1.101×10^{-3}	0.0340
CO_2	0.1386	2.7737	0.007210	0.04991	1.512×10^4	1.404×10^5	8.47×10^{-5}	0.00539
CO	0.0371	1.3590	0.002632	0.05454	1.054×10^3	8.676×10^3	1.350×10^{-4}	0.0060
CH_4	0.0501	1.8796	0.003380	0.04260	2.579×10^3	2.287×10^4	1.244×10^{-4}	0.0060
N_2	0.0254	1.0676	0.002328	0.04074	7.381×10^2	8.166×10^3	1.272×10^{-4}	0.0053

Source: H. W. Cooper and J. C. Goldfrank, *Hydrocarbon Processing, 46* (12): 141 (1967).

Table A-25 **Thermochemical Properties of Selected Substances at 298K and 1 atm**

Substance	Formula	Molar Mass, M (kg/kmol)	Enthalpy of Formation, $\bar{h}_f^\circ$ (kJ/kmol)	Gibbs Function of Formation, $\bar{g}_f^\circ$ (kJ/kmol)	Absolute Entropy, $\bar{s}^\circ$ (kJ/kmol · K)	Heating Values: Higher, HHV (kJ/kg)	Heating Values: Lower, LHV (kJ/kg)
Carbon	$C(s)$	12.01	0	0	5.74	32,770	32,770
Hydrogen	$H_2(g)$	2.016	0	0	130.57	141,780	119,950
Nitrogen	$N_2(g)$	28.01	0	0	191.50	—	—
Oxygen	$O_2(g)$	32.00	0	0	205.03	—	—
Carbon monoxide	$CO(g)$	28.01	−110,530	−137,150	197.54	—	—
Carbon dioxide	$CO_2(g)$	44.01	−393,520	−394,380	213.69	—	—
Water	$H_2O(g)$	18.02	−241,820	−228,590	188.72	—	—
Water	$H_2O(l)$	18.02	−285,830	−237,180	69.95	—	—
Hydrogen peroxide	$H_2O_2(g)$	34.02	−136,310	−105,600	232.63	—	—
Ammonia	$NH_3(g)$	17.03	−46,190	−16,590	192.33	—	—
Oxygen	$O(g)$	16.00	249,170	231,770	160.95	—	—
Hydrogen	$H(g)$	1.008	218,000	203,290	114.61	—	—
Nitrogen	$N(g)$	14.01	472,680	455,510	153.19	—	—
Hydroxyl	$OH(g)$	17.01	39,460	34,280	183.75	—	—
Methane	$CH_4(g)$	16.04	−74,850	−50,790	186.16	55,510	50,020
Acetylene	$C_2H_2(g)$	26.04	226,730	209,170	200.85	49,910	48,220
Ethylene	$C_2H_4(g)$	28.05	52,280	68,120	219.83	50,300	47,160
Ethane	$C_2H_6(g)$	30.07	−84,680	−32,890	229.49	51,870	47,480
Propylene	$C_3H_6(g)$	42.08	20,410	62,720	266.94	48,920	45,780
Propane	$C_3H_8(g)$	44.09	−103,850	−23,490	269.91	50,350	46,360
Butane	$C_4H_{10}(g)$	58.12	−126,150	−15,710	310.03	49,500	45,720
Pentane	$C_5H_{12}(g)$	72.15	−146,440	−8,200	348.40	49,010	45,350
Octane	$C_8H_{18}(g)$	114.22	−208,450	17,320	463.67	48,260	44,790
Octane	$C_8H_{18}(l)$	114.22	−249,910	6,610	360.79	47,900	44,430
Benzene	$C_6H_6(g)$	78.11	82,930	129,660	269.20	42,270	40,580
Methyl alcohol	$CH_3OH(g)$	32.04	−200,890	−162,140	239.70	23,850	21,110
Methyl alcohol	$CH_3OH(l)$	32.04	−238,810	−166,290	126.80	22,670	19,920
Ethyl alcohol	$C_2H_5OH(g)$	46.07	−235,310	−168,570	282.59	30,590	27,720
Ethyl alcohol	$C_2H_5OH(l)$	46.07	−277,690	174,890	160.70	29,670	26,800

Source: Based on JANAF Thermochemical Tables, NSRDS-NBS-37, 1971; *Selected Values of Chemical Thermodynamic Properties,* NBS Tech. Note 270-3, 1968; and *API Research Project 44,* Carnegie Press, 1953. Heating values calculated.

Table A-26 **Standard Molar Chemical Exergy, $\bar{e}^{ch}$ (kJ/kmol), of Selected Substances at 298 K and p_0**

Substance	Formula	Model I[a]	Model II[b]
Nitrogen	$N_2(g)$	640	720
Oxygen	$O_2(g)$	3,950	3,970
Carbon dioxide	$CO_2(g)$	14,175	19,870
Water	$H_2O(g)$	8,635	9,500
Water	$H_2O(l)$	45	900
Carbon (graphite)	C(s)	404,590	410,260
Hydrogen	$H_2(g)$	235,250	236,100
Sulfur	S(s)	598,160	609,600
Carbon monoxide	CO(g)	269,410	275,100
Sulfur dioxide	$SO_2(g)$	301,940	313,400
Nitrogen monoxide	NO(g)	88,850	88,900
Nitrogen dioxide	$NO_2(g)$	55,565	55,600
Hydrogen sulfide	$H_2S(g)$	799,890	812,000
Ammonia	$NH_3(g)$	336,685	337,900
Methane	$CH_4(g)$	824,350	831,650
Ethane	$C_2H_6(g)$	1,482,035	1,495,840
Methyl alcohol	$CH_3OH(g)$	715,070	722,300
Methyl alcohol	$CH_3OH(l)$	710,745	718,000
Ethyl alcohol	$C_2H_5OH(g)$	1,348,330	1,363,900
Ethyl alcohol	$C_2H_5OH(l)$	1,342,085	1,357,700

[a] J. Ahrendts, "Die Exergie Chemisch Reaktionsfähiger Systeme," *VDI-Forschungsheft,* VDI-Verlag, Dusseldorf, 579, 1977. Also see "Reference States," *Energy—The International Journal, 5:* 667–677, 1980. In Model I, p_0 = 1.019 atm. This model attempts to impose a criterion that the reference environment be in equilibrium. The reference substances are determined assuming restricted chemical equilibrium for nitric acid and nitrates and unrestricted thermodynamic equilibrium for all other chemical components of the atmosphere, the oceans, and a portion of the Earth's crust. The chemical composition of the gas phase of this model approximates the composition of the natural atmosphere.

[b] J. Szargut, D. R. Morris, and F. R. Steward, *Exergy Analysis of Thermal, Chemical, and Metallurgical Processes,* Hemisphere, New York, 1988. In Model II, p_0 = 1.0 atm. In developing this model a reference substance is selected for each chemical element from among substances that contain the element being considered and that are abundantly present in the natural environment, even though the substances are not in completely mutual stable equilibrium. An underlying rationale for this approach is that substances found abundantly in nature have little economic value. On an overall basis, the chemical composition of the exergy reference environment of Model II is closer than Model I to the composition of the natural environment, but the equilibrium criterion is not always satisfied.

Table A-27 **Logarithms to the Base 10 of the Equilibrium Constant *K***

Temp. K	$\log_{10} K$								Temp. °R
	$H_2 \leftrightharpoons 2H$	$O_2 \leftrightharpoons 2O$	$N_2 \leftrightharpoons 2N$	$\frac{1}{2}O_2 + \frac{1}{2}N_2 \leftrightharpoons NO$	$H_2O \leftrightharpoons H_2 + \frac{1}{2}O_2$	$H_2O \leftrightharpoons OH + \frac{1}{2}H_2$	$CO_2 \leftrightharpoons CO + \frac{1}{2}O_2$	$CO_2 + H_2 \leftrightharpoons CO + H_2O$	
298	−71.224	−81.208	−159.600	−15.171	−40.048	−46.054	−45.066	−5.018	537
500	−40.316	−45.880	−92.672	−8.783	−22.886	−26.130	−25.025	−2.139	900
1000	−17.292	−19.614	−43.056	−4.062	−10.062	−11.280	−10.221	−0.159	1800
1200	−13.414	−15.208	−34.754	−3.275	−7.899	−8.811	−7.764	+0.135	2160
1400	−10.630	−12.054	−28.812	−2.712	−6.347	−7.021	−6.014	+0.333	2520
1600	−8.532	−9.684	−24.350	−2.290	−5.180	−5.677	−4.706	+0.474	2880
1700	−7.666	−8.706	−22.512	−2.116	−4.699	−5.124	−4.169	+0.530	3060
1800	−6.896	−7.836	−20.874	−1.962	−4.270	−4.613	−3.693	+0.577	3240
1900	−6.204	−7.058	−19.410	−1.823	−3.886	−4.190	−3.267	+0.619	3420
2000	−5.580	−6.356	−18.092	−1.699	−3.540	−3.776	−2.884	+0.656	3600
2100	−5.016	−5.720	−16.898	−1.586	−3.227	−3.434	−2.539	+0.688	3780
2200	−4.502	−5.142	−15.810	−1.484	−2.942	−3.091	−2.226	+0.716	3960
2300	−4.032	−4.614	−14.818	−1.391	−2.682	−2.809	−1.940	+0.742	4140
2400	−3.600	−4.130	−13.908	−1.305	−2.443	−2.520	−1.679	+0.764	4320
2500	−3.202	−3.684	−13.070	−1.227	−2.224	−2.270	−1.440	+0.784	4500
2600	−2.836	−3.272	−12.298	−1.154	−2.021	−2.038	−1.219	+0.802	4680
2700	−2.494	−2.892	−11.580	−1.087	−1.833	−1.823	−1.015	+0.818	4860
2800	−2.178	−2.536	−10.914	−1.025	−1.658	−1.624	−0.825	+0.833	5040
2900	−1.882	−2.206	−10.294	−0.967	−1.495	−1.438	−0.649	+0.846	5220
3000	−1.606	−1.898	−9.716	−0.913	−1.343	−1.265	−0.485	+0.858	5400
3100	−1.348	−1.610	−9.174	−0.863	−1.201	−1.103	−0.332	+0.869	5580
3200	−1.106	−1.340	−8.664	−0.815	−1.067	−0.951	−0.189	+0.878	5760
3300	−0.878	−1.086	−8.186	−0.771	−0.942	−0.809	−0.054	+0.888	5940
3400	−0.664	−0.846	−7.736	−0.729	−0.824	−0.674	+0.071	+0.895	6120
3500	−0.462	−0.620	−7.312	−0.690	−0.712	−0.547	+0.190	+0.902	6300

Source: Based on data from the JANAF Thermochemical Tables, NSRDS-NBS-37, 1971.

Index to Tables in English Units

Table A-1E **Atomic or Molecular Weights and Critical Properties of Some Selected Elements and Compounds**

Substance	Chemical Formula	M (lb/lbmol)	T_c (°R)	p_c (atm)	$Z_c = \frac{p_c v_c}{RT_c}$
Acetylene	C_2H_2	26.04	556	62	0.274
Air (equivalent)	—	28.97	239	37.2	0.284
Ammonia	NH_3	17.03	730	111.3	0.242
Argon	Ar	39.94	272	47.97	0.290
Benzene	C_6H_6	78.11	1013	48.7	0.274
Butane	C_4H_{10}	58.12	765	37.5	0.274
Carbon	C	12.01	—	—	—
Carbon dioxide	CO_2	44.01	548	72.9	0.276
Carbon monoxide	CO	28.01	239	34.5	0.294
Copper	Cu	63.54	—	—	—
Ethane	C_2H_6	30.07	549	48.2	0.285
Ethyl alcohol	C_2H_5OH	46.07	929	63.0	0.249
Ethylene	C_2H_4	28.05	510	50.5	0.270
Helium	He	4.003	9.33	2.26	0.300
Hydrogen	H_2	2.016	59.8	12.8	0.304
Methane	CH_4	16.04	344	45.8	0.290
Methyl alcohol	CH_3OH	32.04	924	78.5	0.220
Nitrogen	N_2	28.01	227	33.5	0.291
Octane	C_8H_{18}	114.22	1025	24.6	0.258
Oxygen	O_2	32.00	278	49.8	0.290
Propane	C_3H_8	44.09	666	42.1	0.276
Propylene	C_3H_6	42.08	657	45.6	0.276
Refrigerant 12	CCl_2F_2	120.92	693	40.6	0.278
Refrigerant 22	$CHClF_2$	86.48	665	49.1	0.267
Refrigerant 134a	CF_3CH_2F	102.03	673	40.2	0.260
Sulfur dioxide	SO_2	64.06	775	77.7	0.268
Water	H_2O	18.02	1165	218.0	0.233

Sources: Adapted from *International Critical Tables* and L. C. Nelson and E. F. Obert, Generalized Compressibility Charts, *Chem. Eng., 617:* 203 (1954).

Table A-2E **Properties of Saturated Water (Liquid–Vapor): Temperature Table**

		Specific Volume ft³/lb		Internal Energy Btu/lb		Enthalpy Btu/lb			Entropy Btu/lb · °R		
Temp. °F	Press. lbf/in.²	Sat. Liquid v_f	Sat. Vapor v_g	Sat. Liquid u_f	Sat. Vapor u_g	Sat. Liquid h_f	Evap. h_{fg}	Sat. Vapor h_g	Sat. Liquid s_f	Sat. Vapor s_g	Temp. °F
32	0.0886	0.01602	3305	−.01	1021.2	−.01	1075.4	1075.4	−.00003	2.1870	32
35	0.0999	0.01602	2948	2.99	1022.2	3.00	1073.7	1076.7	0.00607	2.1764	35
40	0.1217	0.01602	2445	8.02	1023.9	8.02	1070.9	1078.9	0.01617	2.1592	40
45	0.1475	0.01602	2037	13.04	1025.5	13.04	1068.1	1081.1	0.02618	2.1423	45
50	0.1780	0.01602	1704	18.06	1027.2	18.06	1065.2	1083.3	0.03607	2.1259	50
52	0.1917	0.01603	1589	20.06	1027.8	20.07	1064.1	1084.2	0.04000	2.1195	52
54	0.2064	0.01603	1482	22.07	1028.5	22.07	1063.0	1085.1	0.04391	2.1131	54
56	0.2219	0.01603	1383	24.08	1029.1	24.08	1061.9	1085.9	0.04781	2.1068	56
58	0.2386	0.01603	1292	26.08	1029.8	26.08	1060.7	1086.8	0.05159	2.1005	58
60	0.2563	0.01604	1207	28.08	1030.4	28.08	1059.6	1087.7	0.05555	2.0943	60
62	0.2751	0.01604	1129	30.09	1031.1	30.09	1058.5	1088.6	0.05940	2.0882	62
64	0.2952	0.01604	1056	32.09	1031.8	32.09	1057.3	1089.4	0.06323	2.0821	64
66	0.3165	0.01604	988.4	34.09	1032.4	34.09	1056.2	1090.3	0.06704	2.0761	66
68	0.3391	0.01605	925.8	36.09	1033.1	36.09	1055.1	1091.2	0.07084	2.0701	68
70	0.3632	0.01605	867.7	38.09	1033.7	38.09	1054.0	1092.0	0.07463	2.0642	70
72	0.3887	0.01606	813.7	40.09	1034.4	40.09	1052.8	1092.9	0.07839	2.0584	72
74	0.4158	0.01606	763.5	42.09	1035.0	42.09	1051.7	1093.8	0.08215	2.0526	74
76	0.4446	0.01606	716.8	44.09	1035.7	44.09	1050.6	1094.7	0.08589	2.0469	76
78	0.4750	0.01607	673.3	46.09	1036.3	46.09	1049.4	1095.5	0.08961	2.0412	78
80	0.5073	0.01607	632.8	48.08	1037.0	48.09	1048.3	1096.4	0.09332	2.0356	80
82	0.5414	0.01608	595.0	50.08	1037.6	50.08	1047.2	1097.3	0.09701	2.0300	82
84	0.5776	0.01608	559.8	52.08	1038.3	52.08	1046.0	1098.1	0.1007	2.0245	84
86	0.6158	0.01609	527.0	54.08	1038.9	54.08	1044.9	1099.0	0.1044	2.0190	86
88	0.6562	0.01609	496.3	56.07	1039.6	56.07	1043.8	1099.9	0.1080	2.0136	88
90	0.6988	0.01610	467.7	58.07	1040.2	58.07	1042.7	1100.7	0.1117	2.0083	90
92	0.7439	0.01611	440.9	60.06	1040.9	60.06	1041.5	1101.6	0.1153	2.0030	92
94	0.7914	0.01611	415.9	62.06	1041.5	62.06	1040.4	1102.4	0.1189	1.9977	94
96	0.8416	0.01612	392.4	64.05	1041.2	64.06	1039.2	1103.3	0.1225	1.9925	96
98	0.8945	0.01612	370.5	66.05	1042.8	66.05	1038.1	1104.2	0.1261	1.9874	98
100	0.9503	0.01613	350.0	68.04	1043.5	68.05	1037.0	1105.0	0.1296	1.9822	100
110	1.276	0.01617	265.1	78.02	1046.7	78.02	1031.3	1109.3	0.1473	1.9574	110
120	1.695	0.01621	203.0	87.99	1049.9	88.00	1025.5	1113.5	0.1647	1.9336	120
130	2.225	0.01625	157.2	97.97	1053.0	97.98	1019.8	1117.8	0.1817	1.9109	130
140	2.892	0.01629	122.9	107.95	1056.2	107.96	1014.0	1121.9	0.1985	1.8892	140
150	3.722	0.01634	97.0	117.95	1059.3	117.96	1008.1	1126.1	0.2150	1.8684	150
160	4.745	0.01640	77.2	127.94	1062.3	127.96	1002.2	1130.1	0.2313	1.8484	160
170	5.996	0.01645	62.0	137.95	1065.4	137.97	996.2	1134.2	0.2473	1.8293	170
180	7.515	0.01651	50.2	147.97	1068.3	147.99	990.2	1138.2	0.2631	1.8109	180
190	9.343	0.01657	41.0	158.00	1071.3	158.03	984.1	1142.1	0.2787	1.7932	190
200	11.529	0.01663	33.6	168.04	1074.2	168.07	977.9	1145.9	0.2940	1.7762	200

Table A-2E **(Continued)**

Temp. °F	Press. lbf/in.^{2}	Specific Volume ft^3/lb Sat. Liquid v_f	Sat. Vapor v_g	Internal Energy Btu/lb Sat. Liquid u_f	Sat. Vapor u_g	Enthalpy Btu/lb Sat. Liquid h_f	Evap. h_{fg}	Sat. Vapor h_g	Entropy Btu/lb · °R Sat. Liquid s_f	Sat. Vapor s_g	Temp. °F
210	14.13	0.01670	27.82	178.1	1077.0	178.1	971.6	1149.7	0.3091	1.7599	210
212	14.70	0.01672	26.80	180.1	1077.6	180.2	970.3	1150.5	0.3121	1.7567	212
220	17.19	0.01677	23.15	188.2	1079.8	188.2	965.3	1153.5	0.3241	1.7441	220
230	20.78	0.01685	19.39	198.3	1082.6	198.3	958.8	1157.1	0.3388	1.7289	230
240	24.97	0.01692	16.33	208.4	1085.3	208.4	952.3	1160.7	0.3534	1.7143	240
250	29.82	0.01700	13.83	218.5	1087.9	218.6	945.6	1164.2	0.3677	1.7001	250
260	35.42	0.01708	11.77	228.6	1090.5	228.8	938.8	1167.6	0.3819	1.6864	260
270	41.85	0.01717	10.07	238.8	1093.0	239.0	932.0	1170.9	0.3960	1.6731	270
280	49.18	0.01726	8.65	249.0	1095.4	249.2	924.9	1174.1	0.4099	1.6602	280
290	57.53	0.01735	7.47	259.3	1097.7	259.4	917.8	1177.2	0.4236	1.6477	290
300	66.98	0.01745	6.472	269.5	1100.0	269.7	910.4	1180.2	0.4372	1.6356	300
310	77.64	0.01755	5.632	279.8	1102.1	280.1	903.0	1183.0	0.4507	1.6238	310
320	89.60	0.01765	4.919	290.1	1104.2	290.4	895.3	1185.8	0.4640	1.6123	320
330	103.00	0.01776	4.312	300.5	1106.2	300.8	887.5	1188.4	0.4772	1.6010	330
340	117.93	0.01787	3.792	310.9	1108.0	311.3	879.5	1190.8	0.4903	1.5901	340
350	134.53	0.01799	3.346	321.4	1109.8	321.8	871.3	1193.1	0.5033	1.5793	350
360	152.92	0.01811	2.961	331.8	1111.4	332.4	862.9	1195.2	0.5162	1.5688	360
370	173.23	0.01823	2.628	342.4	1112.9	343.0	854.2	1197.2	0.5289	1.5585	370
380	195.60	0.01836	2.339	353.0	1114.3	353.6	845.4	1199.0	0.5416	1.5483	380
390	220.2	0.01850	2.087	363.6	1115.6	364.3	836.2	1200.6	0.5542	1.5383	390
400	247.1	0.01864	1.866	374.3	1116.6	375.1	826.8	1202.0	0.5667	1.5284	400
410	276.5	0.01878	1.673	385.0	1117.6	386.0	817.2	1203.1	0.5792	1.5187	410
420	308.5	0.01894	1.502	395.8	1118.3	396.9	807.2	1204.1	0.5915	1.5091	420
430	343.3	0.01909	1.352	406.7	1118.9	407.9	796.9	1204.8	0.6038	1.4995	430
440	381.2	0.01926	1.219	417.6	1119.3	419.0	786.3	1205.3	0.6161	1.4900	440
450	422.1	0.01943	1.1011	428.6	1119.5	430.2	775.4	1205.6	0.6282	1.4806	450
460	466.3	0.01961	0.9961	439.7	1119.6	441.4	764.1	1205.5	0.6404	1.4712	460
470	514.1	0.01980	0.9025	450.9	1119.4	452.8	752.4	1205.2	0.6525	1.4618	470
480	565.5	0.02000	0.8187	462.2	1118.9	464.3	740.3	1204.6	0.6646	1.4524	480
490	620.7	0.02021	0.7436	473.6	1118.3	475.9	727.8	1203.7	0.6767	1.4430	490
500	680.0	0.02043	0.6761	485.1	1117.4	487.7	714.8	1202.5	0.6888	1.4335	500
520	811.4	0.02091	0.5605	508.5	1114.8	511.7	687.3	1198.9	0.7130	1.4145	520
540	961.5	0.02145	0.4658	532.6	1111.0	536.4	657.5	1193.8	0.7374	1.3950	540
560	1131.8	0.02207	0.3877	548.4	1105.8	562.0	625.0	1187.0	0.7620	1.3749	560
580	1324.3	0.02278	0.3225	583.1	1098.9	588.6	589.3	1178.0	0.7872	1.3540	580
600	1541.0	0.02363	0.2677	609.9	1090.0	616.7	549.7	1166.4	0.8130	1.3317	600
620	1784.4	0.02465	0.2209	638.3	1078.5	646.4	505.0	1151.4	0.8398	1.3075	620
640	2057.1	0.02593	0.1805	668.7	1063.2	678.6	453.4	1131.9	0.8681	1.2803	640
660	2362	0.02767	0.1446	702.3	1042.3	714.4	391.1	1105.5	0.8990	1.2483	660
680	2705	0.03032	0.1113	741.7	1011.0	756.9	309.8	1066.7	0.9350	1.2068	680
700	3090	0.03666	0.0744	801.7	947.7	822.7	167.5	990.2	0.9902	1.1346	700
705.4	3204	0.05053	0.05053	872.6	872.6	902.5	0	902.5	1.0580	1.0580	705.4

Source: Tables A-2E through A-6E are extracted from J. H. Keenan, F. G. Keyes, P. G. Hill, and J. G. Moore, *Steam Tables,* Wiley, New York, 1969.

H$_2$O

Table A-3E **Properties of Saturated Water (Liquid–Vapor): Pressure Table**

		Specific Volume ft³/lb		Internal Energy Btu/lb		Enthalpy Btu/lb			Entropy Btu/lb · °R			
Press. lbf/in.²	Temp. °F	Sat. Liquid v_f	Sat. Vapor v_g	Sat. Liquid u_f	Sat. Vapor u_g	Sat. Liquid h_f	Evap. h_{fg}	Sat. Vapor h_g	Sat. Liquid s_f	Evap. s_{fg}	Sat. Vapor s_g	Press. lbf/in.²
0.4	72.84	0.01606	792.0	40.94	1034.7	40.94	1052.3	1093.3	0.0800	1.9760	2.0559	0.4
0.6	85.19	0.01609	540.0	53.26	1038.7	53.27	1045.4	1098.6	0.1029	1.9184	2.0213	0.6
0.8	94.35	0.01611	411.7	62.41	1041.7	62.41	1040.2	1102.6	0.1195	1.8773	1.9968	0.8
1.0	101.70	0.01614	333.6	69.74	1044.0	69.74	1036.0	1105.8	0.1327	1.8453	1.9779	1.0
1.2	107.88	0.01616	280.9	75.90	1046.0	75.90	1032.5	1108.4	0.1436	1.8190	1.9626	1.2
1.5	115.65	0.01619	227.7	83.65	1048.5	83.65	1028.0	1111.7	0.1571	1.7867	1.9438	1.5
2.0	126.04	0.01623	173.75	94.02	1051.8	94.02	1022.1	1116.1	0.1750	1.7448	1.9198	2.0
3.0	141.43	0.01630	118.72	109.38	1056.6	109.39	1013.1	1122.5	0.2009	1.6852	1.8861	3.0
4.0	152.93	0.01636	90.64	120.88	1060.2	120.89	1006.4	1127.3	0.2198	1.6426	1.8624	4.0
5.0	162.21	0.01641	73.53	130.15	1063.0	130.17	1000.9	1131.0	0.2349	1.6093	1.8441	5.0
6.0	170.03	0.01645	61.98	137.98	1065.4	138.00	996.2	1134.2	0.2474	1.5819	1.8292	6.0
7.0	176.82	0.01649	53.65	144.78	1067.4	144.80	992.1	1136.9	0.2581	1.5585	1.8167	7.0
8.0	182.84	0.01653	47.35	150.81	1069.2	150.84	988.4	1139.3	0.2675	1.5383	1.8058	8.0
9.0	188.26	0.01656	42.41	156.25	1070.8	156.27	985.1	1141.4	0.2760	1.5203	1.7963	9.0
10	193.19	0.01659	38.42	161.20	1072.2	161.23	982.1	1143.3	0.2836	1.5041	1.7877	10
14.696	211.99	0.01672	26.80	180.10	1077.6	180.15	970.4	1150.5	0.3121	1.4446	1.7567	14.696
15	213.03	0.01672	26.29	181.14	1077.9	181.19	969.7	1150.9	0.3137	1.4414	1.7551	15
20	227.96	0.01683	20.09	196.19	1082.0	196.26	960.1	1156.4	0.3358	1.3962	1.7320	20
25	240.08	0.01692	16.31	208.44	1085.3	208.52	952.2	1160.7	0.3535	1.3607	1.7142	25
30	250.34	0.01700	13.75	218.84	1088.0	218.93	945.4	1164.3	0.3682	1.3314	1.6996	30
35	259.30	0.01708	11.90	227.93	1090.3	228.04	939.3	1167.4	0.3809	1.3064	1.6873	35
40	267.26	0.01715	10.50	236.03	1092.3	236.16	933.8	1170.0	0.3921	1.2845	1.6767	40
45	274.46	0.01721	9.40	243.37	1094.0	243.51	928.8	1172.3	0.4022	1.2651	1.6673	45
50	281.03	0.01727	8.52	250.08	1095.6	250.24	924.2	1174.4	0.4113	1.2476	1.6589	50
55	287.10	0.01733	7.79	256.28	1097.0	256.46	919.9	1176.3	0.4196	1.2317	1.6513	55
60	292.73	0.01738	7.177	262.1	1098.3	262.2	915.8	1178.0	0.4273	1.2170	1.6443	60
65	298.00	0.01743	6.647	267.5	1099.5	267.7	911.9	1179.6	0.4345	1.2035	1.6380	65
70	302.96	0.01748	6.209	272.6	1100.6	272.8	908.3	1181.0	0.4412	1.1909	1.6321	70
75	307.63	0.01752	5.818	277.4	1101.6	277.6	904.8	1182.4	0.4475	1.1790	1.6265	75
80	312.07	0.01757	5.474	282.0	1102.6	282.2	901.4	1183.6	0.4534	1.1679	1.6213	80
85	316.29	0.01761	5.170	286.3	1103.5	286.6	898.2	1184.8	0.4591	1.1574	1.6165	85
90	320.31	0.01766	4.898	290.5	1104.3	290.8	895.1	1185.9	0.4644	1.1475	1.6119	90
95	324.16	0.01770	4.654	294.5	1105.0	294.8	892.1	1186.9	0.4695	1.1380	1.6075	95
100	327.86	0.01774	4.434	298.3	1105.8	298.6	889.2	1187.8	0.4744	1.1290	1.6034	100
110	334.82	0.01781	4.051	305.5	1107.1	305.9	883.7	1189.6	0.4836	1.1122	1.5958	110
120	341.30	0.01789	3.730	312.3	1108.3	312.7	878.5	1191.1	0.4920	1.0966	1.5886	120
130	347.37	0.01796	3.457	318.6	1109.4	319.0	873.5	1192.5	0.4999	1.0822	1.5821	130
140	353.08	0.01802	3.221	324.6	1110.3	325.1	868.7	1193.8	0.5073	1.0688	1.5761	140
150	358.48	0.01809	3.016	330.2	1111.2	330.8	864.2	1194.9	0.5142	1.0562	1.5704	150
160	363.60	0.01815	2.836	335.6	1112.0	336.2	859.8	1196.0	0.5208	1.0443	1.5651	160

H_2O

Table A-3E ***(Continued)***

		Specific Volume ft³/lb		Internal Energy Btu/lb		Enthalpy Btu/lb			Entropy Btu/lb · °R			
Press. lbf/in.²	Temp. °F	Sat. Liquid v_f	Sat. Vapor v_g	Sat. Liquid u_f	Sat. Vapor u_g	Sat. Liquid h_f	Evap. h_{fg}	Sat. Vapor h_g	Sat. Liquid s_f	Evap. s_{fg}	Sat. Vapor s_g	Press. lbf/in.²
170	368.47	0.01821	2.676	340.8	1112.7	341.3	855.6	1196.9	0.5270	1.0330	1.5600	170
180	373.13	0.01827	2.553	345.7	1113.4	346.3	851.5	1197.8	0.5329	1.0223	1.5552	180
190	377.59	0.01833	2.405	350.4	1114.0	351.0	847.5	1198.6	0.5386	1.0122	1.5508	190
200	381.86	0.01839	2.289	354.9	1114.6	355.6	843.7	1199.3	0.5440	1.0025	1.5465	200
250	401.04	0.01865	1.845	375.4	1116.7	376.2	825.8	1202.1	0.5680	0.9594	1.5274	250
300	417.43	0.01890	1.544	393.0	1118.2	394.1	809.8	1203.9	0.5883	0.9232	1.5115	300
350	431.82	0.01912	1.327	408.7	1119.0	409.9	795.0	1204.9	0.6060	0.8917	1.4977	350
400	444.70	0.01934	1.162	422.8	1119.5	424.2	781.2	1205.5	0.6218	0.8638	1.4856	400
450	456.39	0.01955	1.033	435.7	1119.6	437.4	768.2	1205.6	0.6360	0.8385	1.4745	450
500	467.13	0.01975	0.928	447.7	1119.4	449.5	755.8	1205.3	0.6490	0.8154	1.4644	500
550	477.07	0.01994	0.842	458.9	1119.1	460.9	743.9	1204.8	0.6611	0.7941	1.4451	550
600	486.33	0.02013	0.770	469.4	1118.6	471.7	732.4	1204.1	0.6723	0.7742	1.4464	600
700	503.23	0.02051	0.656	488.9	1117.0	491.5	710.5	1202.0	0.6927	0.7378	1.4305	700
800	518.36	0.02087	0.569	506.6	1115.0	509.7	689.6	1199.3	0.7110	0.7050	1.4160	800
900	532.12	0.02123	0.501	523.0	1112.6	526.6	669.5	1196.0	0.7277	0.6750	1.4027	900
1000	544.75	0.02159	0.446	538.4	1109.9	542.4	650.0	1192.4	0.7432	0.6471	1.3903	1000
1100	556.45	0.02195	0.401	552.9	1106.8	557.4	631.0	1188.3	0.7576	0.6209	1.3786	1100
1200	567.37	0.02232	0.362	566.7	1103.5	571.7	612.3	1183.9	0.7712	0.5961	1.3673	1200
1300	577.60	0.02269	0.330	579.9	1099.8	585.4	593.8	1179.2	0.7841	0.5724	1.3565	1300
1400	587.25	0.02307	0.302	592.7	1096.0	598.6	575.5	1174.1	0.7964	0.5497	1.3461	1400
1500	596.39	0.02346	0.277	605.0	1091.8	611.5	557.2	1168.7	0.8082	0.5276	1.3359	1500
1600	605.06	0.02386	0.255	616.9	1087.4	624.0	538.9	1162.9	0.8196	0.5062	1.3258	1600
1700	613.32	0.02428	0.236	628.6	1082.7	636.2	520.6	1156.9	0.8307	0.4852	1.3159	1700
1800	621.21	0.02472	0.218	640.0	1077.7	648.3	502.1	1150.4	0.8414	0.4645	1.3060	1800
1900	628.76	0.02517	0.203	651.3	1072.3	660.1	483.4	1143.5	0.8519	0.4441	1.2961	1900
2000	636.00	0.02565	0.188	662.4	1066.6	671.9	464.4	1136.3	0.8623	0.4238	1.2861	2000
2250	652.90	0.02698	0.157	689.9	1050.6	701.1	414.8	1115.9	0.8876	0.3728	1.2604	2250
2500	668.31	0.02860	0.131	717.7	1031.0	730.9	360.5	1091.4	0.9131	0.3196	1.2327	2500
2750	682.46	0.03077	0.107	747.3	1005.9	763.0	297.4	1060.4	0.9401	0.2604	1.2005	2750
3000	695.52	0.03431	0.084	783.4	968.8	802.5	213.0	1015.5	0.9732	0.1843	1.1575	3000
3203.6	705.44	0.05053	0.0505	872.6	872.6	902.5	0	902.5	1.0580	0	1.0580	3203.6

Table A-4E **Properties of Superheated Water Vapor**

T °F	v ft³/lb	u Btu/lb	h Btu/lb	s Btu/lb · °R	v ft³/lb	u Btu/lb	h Btu/lb	s Btu/lb · °R
	p = 1 lbf/in.² (T_{sat} = 101.7°F)				p = 5 lbf/in.² (T_{sat} = 162.2°F)			
Sat.	333.6	1044.0	1105.8	1.9779	73.53	1063.0	1131.0	1.8441
150	362.6	1060.4	1127.5	2.0151				
200	392.5	1077.5	1150.1	2.0508	78.15	1076.0	1148.6	1.8715
250	422.4	1094.7	1172.8	2.0839	84.21	1093.8	1171.7	1.9052
300	452.3	1112.0	1195.7	2.1150	90.24	1111.3	1194.8	1.9367
400	511.9	1147.0	1241.8	2.1720	102.24	1146.6	1241.2	1.9941
500	571.5	1182.8	1288.5	2.2235	114.20	1182.5	1288.2	2.0458
600	631.1	1219.3	1336.1	2.2706	126.15	1219.1	1335.8	2.0930
700	690.7	1256.7	1384.5	2.3142	138.08	1256.5	1384.3	2.1367
800	750.3	1294.4	1433.7	2.3550	150.01	1294.7	1433.5	2.1775
900	809.9	1333.9	1483.8	2.3932	161.94	1333.8	1483.7	2.2158
1000	869.5	1373.9	1534.8	2.4294	173.86	1373.9	1534.7	2.2520
	p = 10 lbf/in.² (T_{sat} = 193.2°F)				p = 14.7 lbf/in.² (T_{sat} = 212.0°F)			
Sat.	38.42	1072.2	1143.3	1.7877	26.80	1077.6	1150.5	1.7567
200	38.85	1074.7	1146.6	1.7927				
250	41.95	1092.6	1170.2	1.8272	28.42	1091.5	1168.8	1.7832
300	44.99	1110.4	1193.7	1.8592	30.52	1109.6	1192.6	1.8157
400	51.03	1146.1	1240.5	1.9171	34.67	1145.6	1239.9	1.8741
500	57.04	1182.2	1287.7	1.9690	38.77	1181.8	1287.3	1.9263
600	63.03	1218.9	1335.5	2.0164	42.86	1218.6	1335.2	1.9737
700	69.01	1256.3	1384.0	2.0601	46.93	1256.1	1383.8	2.0175
800	74.98	1294.6	1433.3	2.1009	51.00	1294.4	1433.1	2.0584
900	80.95	1333.7	1483.5	2.1393	55.07	1333.6	1483.4	2.0967
1000	86.91	1373.8	1534.6	2.1755	59.13	1373.7	1534.5	2.1330
1100	92.88	1414.7	1586.6	2.2099	63.19	1414.6	1586.4	2.1674
	p = 20 lbf/in.² (T_{sat} = 228.0°F)				p = 40 lbf/in.² (T_{sat} = 267.3°F)			
Sat.	20.09	1082.0	1156.4	1.7320	10.50	1093.3	1170.0	1.6767
250	20.79	1090.3	1167.2	1.7475				
300	22.36	1108.7	1191.5	1.7805	11.04	1105.1	1186.8	1.6993
350	23.90	1126.9	1215.4	1.8110	11.84	1124.2	1211.8	1.7312
400	25.43	1145.1	1239.2	1.8395	12.62	1143.0	1236.4	1.7606
500	28.46	1181.5	1286.8	1.8919	14.16	1180.1	1284.9	1.8140
600	31.47	1218.4	1334.8	1.9395	15.69	1217.3	1333.4	1.8621
700	34.47	1255.9	1383.5	1.9834	17.20	1255.1	1382.4	1.9063
800	37.46	1294.3	1432.9	2.0243	18.70	1293.7	1432.1	1.9474
900	40.45	1333.5	1483.2	2.0627	20.20	1333.0	1482.5	1.9859
1000	43.44	1373.5	1534.3	2.0989	21.70	1373.1	1533.8	2.0223
1100	46.42	1414.5	1586.3	2.1334	23.20	1414.2	1585.9	2.0568

Table A-4E (*Continued*)

T °F	v ft³/lb	u Btu/lb	h Btu/lb	s Btu/lb·°R	v ft³/lb	u Btu/lb	h Btu/lb	s Btu/lb·°R
	p = 60 lbf/in.² (T_{sat} = 292.7°F)				p = 80 lbf/in.² (T_{sat} = 312.1°F)			
Sat.	7.17	1098.3	1178.0	1.6444	5.47	1102.6	1183.6	1.6214
300	7.26	1101.3	1181.9	1.6496				
350	7.82	1121.4	1208.2	1.6830	5.80	1118.5	1204.3	1.6476
400	8.35	1140.8	1233.5	1.7134	6.22	1138.5	1230.6	1.6790
500	9.40	1178.6	1283.0	1.7678	7.02	1177.2	1281.1	1.7346
600	10.43	1216.3	1332.1	1.8165	7.79	1215.3	1330.7	1.7838
700	11.44	1254.4	1381.4	1.8609	8.56	1253.6	1380.3	1.8285
800	12.45	1293.0	1431.2	1.9022	9.32	1292.4	1430.4	1.8700
900	13.45	1332.5	1481.8	1.9408	10.08	1332.0	1481.2	1.9087
1000	14.45	1372.7	1533.2	1.9773	10.83	1372.3	1532.6	1.9453
1100	15.45	1413.8	1585.4	2.0119	11.58	1413.5	1584.9	1.9799
1200	16.45	1455.8	1638.5	2.0448	12.33	1455.5	1638.1	2.0130
	p = 100 lbf/in.² (T_{sat} = 327.8°F)				p = 120 lbf/in.² (T_{sat} = 341.3°F)			
Sat.	4.434	1105.8	1187.8	1.6034	3.730	1108.3	1191.1	1.5886
350	4.592	1115.4	1200.4	1.6191	3.783	1112.2	1196.2	1.5950
400	4.934	1136.2	1227.5	1.6517	4.079	1133.8	1224.4	1.6288
450	5.265	1156.2	1253.6	1.6812	4.360	1154.3	1251.2	1.6590
500	5.587	1175.7	1279.1	1.7085	4.633	1174.2	1277.1	1.6868
600	6.216	1214.2	1329.3	1.7582	5.164	1213.2	1327.8	1.7371
700	6.834	1252.8	1379.2	1.8033	5.682	1252.0	1378.2	1.7825
800	7.445	1291.8	1429.6	1.8449	6.195	1291.2	1428.7	1.8243
900	8.053	1331.5	1480.5	1.8838	6.703	1330.9	1479.8	1.8633
1000	8.657	1371.9	1532.1	1.9204	7.208	1371.5	1531.5	1.9000
1100	9.260	1413.1	1584.5	1.9551	7.711	1412.8	1584.0	1.9348
1200	9.861	1455.2	1637.7	1.9882	8.213	1454.9	1637.3	1.9679
	p = 140 lbf/in.² (T_{sat} = 353.1°F)				p = 160 lbf/in.² (T_{sat} = 363.6°F)			
Sat.	3.221	1110.3	1193.8	1.5761	2.836	1112.0	1196.0	1.5651
400	3.466	1131.4	1221.2	1.6088	3.007	1128.8	1217.8	1.5911
450	3.713	1152.4	1248.6	1.6399	3.228	1150.5	1246.1	1.6230
500	3.952	1172.7	1275.1	1.6682	3.440	1171.2	1273.0	1.6518
550	4.184	1192.5	1300.9	1.6945	3.646	1191.3	1299.2	1.6785
600	4.412	1212.1	1326.4	1.7191	3.848	1211.1	1325.0	1.7034
700	4.860	1251.2	1377.1	1.7648	4.243	1250.4	1376.0	1.7494
800	5.301	1290.5	1427.9	1.8068	4.631	1289.9	1427.0	1.7916
900	5.739	1330.4	1479.1	1.8459	5.015	1329.9	1478.4	1.8308
1000	6.173	1371.0	1531.0	1.8827	5.397	1370.6	1530.4	1.8677
1100	6.605	1412.4	1583.6	1.9176	5.776	1412.1	1583.1	1.9026
1200	7.036	1454.6	1636.9	1.9507	6.154	1454.3	1636.5	1.9358

H_2O

Table A-4E ***(Continued)***

T °F	v ft³/lb	u Btu/lb	h Btu/lb	s Btu/lb · °R	v ft³/lb	u Btu/lb	h Btu/lb	s Btu/lb · °R
	p = 180 lbf/in.² (T_{sat} = 373.1°F)				p = 200 lbf/in.² (T_{sat} = 381.8°F)			
Sat.	2.533	1113.4	1197.8	1.5553	2.289	1114.6	1199.3	1.5464
400	2.648	1126.2	1214.4	1.5749	2.361	1123.5	1210.8	1.5600
450	2.850	1148.5	1243.4	1.6078	2.548	1146.4	1240.7	1.5938
500	3.042	1169.6	1270.9	1.6372	2.724	1168.0	1268.8	1.6239
550	3.228	1190.0	1297.5	1.6642	2.893	1188.7	1295.7	1.6512
600	3.409	1210.0	1323.5	1.6893	3.058	1208.9	1322.1	1.6767
700	3.763	1249.6	1374.9	1.7357	3.379	1248.8	1373.8	1.7234
800	4.110	1289.3	1426.2	1.7781	3.693	1288.6	1425.3	1.7660
900	4.453	1329.4	1477.7	1.8174	4.003	1328.9	1477.1	1.8055
1000	4.793	1370.2	1529.8	1.8545	4.310	1369.8	1529.3	1.8425
1100	5.131	1411.7	1582.6	1.8894	4.615	1411.4	1582.2	1.8776
1200	5.467	1454.0	1636.1	1.9227	4.918	1453.7	1635.7	1.9109
	p = 250 lbf/in.² (T_{sat} = 401.0°F)				p = 300 lbf/in.² (T_{sat} = 417.4°F)			
Sat.	1.845	1116.7	1202.1	1.5274	1.544	1118.2	1203.9	1.5115
450	2.002	1141.1	1233.7	1.5632	1.636	1135.4	1226.2	1.5365
500	2.150	1163.8	1263.3	1.5948	1.766	1159.5	1257.5	1.5701
550	2.290	1185.3	1291.3	1.6233	1.888	1181.9	1286.7	1.5997
600	2.426	1206.1	1318.3	1.6494	2.004	1203.2	1314.5	1.6266
700	2.688	1246.7	1371.1	1.6970	2.227	1244.0	1368.3	1.6751
800	2.943	1287.0	1423.2	1.7301	2.442	1285.4	1421.0	1.7187
900	3.193	1327.6	1475.3	1.7799	2.653	1326.3	1473.6	1.7589
1000	3.440	1368.7	1527.9	1.8172	2.860	1367.7	1526.5	1.7964
1100	3.685	1410.5	1581.0	1.8524	3.066	1409.6	1579.8	1.8317
1200	3.929	1453.0	1634.8	1.8858	3.270	1452.2	1633.8	1.8653
1300	4.172	1496.3	1689.3	1.9177	3.473	1495.6	1688.4	1.8973
	p = 350 lbf/in.² (T_{sat} = 431.8°F)				p = 400 lbf/in.² (T_{sat} = 444.7°F)			
Sat.	1.327	1119.0	1204.9	1.4978	1.162	1119.5	1205.5	1.4856
450	1.373	1129.2	1218.2	1.5125	1.175	1122.6	1209.5	1.4901
500	1.491	1154.9	1251.5	1.5482	1.284	1150.1	1245.2	1.5282
550	1.600	1178.3	1281.9	1.5790	1.383	1174.6	1277.0	1.5605
600	1.703	1200.3	1310.6	1.6068	1.476	1197.3	1306.6	1.5892
700	1.898	1242.5	1365.4	1.6562	1.650	1240.4	1362.5	1.6397
800	2.085	1283.8	1418.8	1.7004	1.816	1282.1	1416.6	1.6844
900	2.267	1325.0	1471.8	1.7409	1.978	1323.7	1470.1	1.7252
1000	2.446	1366.6	1525.0	1.7787	2.136	1365.5	1523.6	1.7632
1100	2.624	1408.7	1578.6	1.8142	2.292	1407.8	1577.4	1.7989
1200	2.799	1451.5	1632.8	1.8478	2.446	1450.7	1621.8	1.8327
1300	2.974	1495.0	1687.6	1.8799	2.599	1494.3	1686.8	1.8648

Table A-4E **(Continued)**

T °F	v ft³/lb	u Btu/lb	h Btu/lb	s Btu/lb · °R	v ft³/lb	u Btu/lb	h Btu/lb	s Btu/lb · °R
	p = 450 lbf/in.² (T_{sat} = 456.4°F)				p = 500 lbf/in.² (T_{sat} = 467.1°F)			
Sat.	1.033	1119.6	1205.6	1.4746	0.928	1119.4	1205.3	1.4645
500	1.123	1145.1	1238.5	1.5097	0.992	1139.7	1231.5	1.4923
550	1.215	1170.7	1271.9	1.5436	1.079	1166.7	1266.6	1.5279
600	1.300	1194.3	1302.5	1.5732	1.158	1191.1	1298.3	1.5585
700	1.458	1238.2	1359.6	1.6248	1.304	1236.0	1356.7	1.6112
800	1.608	1280.5	1414.4	1.6701	1.441	1278.8	1412.1	1.6571
900	1.752	1322.4	1468.3	1.7113	1.572	1321.0	1466.5	1.6987
1000	1.894	1364.4	1522.2	1.7495	1.701	1363.3	1520.7	1.7371
1100	2.034	1406.9	1576.3	1.7853	1.827	1406.0	1575.1	1.7731
1200	2.172	1450.0	1630.8	1.8192	1.952	1449.2	1629.8	1.8072
1300	2.308	1493.7	1685.9	1.8515	2.075	1493.1	1685.1	1.8395
1400	2.444	1538.1	1741.7	1.8823	2.198	1537.6	1741.0	1.8704
	p = 600 lbf/in.² (T_{sat} = 486.3°F)				p = 700 lbf/in.² (T_{sat} = 503.2°F)			
Sat.	0.770	1118.6	1204.1	1.4464	0.656	1117.0	1202.0	1.4305
500	0.795	1128.0	1216.2	1.4592				
550	0.875	1158.2	1255.4	1.4990	0.728	1149.0	1243.2	1.4723
600	0.946	1184.5	1289.5	1.5320	0.793	1177.5	1280.2	1.5081
700	1.073	1231.5	1350.6	1.5872	0.907	1226.9	1344.4	1.5661
800	1.190	1275.4	1407.6	1.6343	1.011	1272.0	1402.9	1.6145
900	1.302	1318.4	1462.9	1.6766	1.109	1315.6	1459.3	1.6576
1000	1.411	1361.2	1517.8	1.7155	1.204	1358.9	1514.9	1.6970
1100	1.517	1404.2	1572.7	1.7519	1.296	1402.4	1570.2	1.7337
1200	1.622	1447.7	1627.8	1.7861	1.387	1446.2	1625.8	1.7682
1300	1.726	1491.7	1683.4	1.8186	1.476	1490.4	1681.7	1.8009
1400	1.829	1536.5	1739.5	1.8497	1.565	1535.3	1738.1	1.8321
	p = 800 lbf/in.² (T_{sat} = 518.3°F)				p = 900 lbf/in.² (T_{sat} = 532.1°F)			
Sat.	0.569	1115.0	1199.3	1.4160	0.501	1112.6	1196.0	1.4027
550	0.615	1138.8	1229.9	1.4469	0.527	1127.5	1215.2	1.4219
600	0.677	1170.1	1270.4	1.4861	0.587	1162.2	1260.0	1.4652
650	0.732	1197.2	1305.6	1.5186	0.639	1191.1	1297.5	1.4999
700	0.783	1222.1	1338.0	1.5471	0.686	1217.1	1331.4	1.5297
800	0.876	1268.5	1398.2	1.5969	0.772	1264.9	1393.4	1.5810
900	0.964	1312.9	1455.6	1.6408	0.851	1310.1	1451.9	1.6257
1000	1.048	1356.7	1511.9	1.6807	0.927	1354.5	1508.9	1.6662
1100	1.130	1400.5	1567.8	1.7178	1.001	1398.7	1565.4	1.7036
1200	1.210	1444.6	1623.8	1.7526	1.073	1443.0	1621.7	1.7386
1300	1.289	1489.1	1680.0	1.7854	1.144	1487.8	1687.3	1.7717
1400	1.367	1534.2	1736.6	1.8167	1.214	1533.0	1735.1	1.8031

H_2O

Table A-4E **(Continued)**

T °F	v ft³/lb	u Btu/lb	h Btu/lb	s Btu/lb · °R	v ft³/lb	u Btu/lb	h Btu/lb	s Btu/lb · °R
	p = 1000 lbf/in.² (T_{sat} = 544.7°F)				p = 1200 lbf/in.² (T_{sat} = 567.4°F)			
Sat.	0.446	1109.0	1192.4	1.3903	0.362	1103.5	1183.9	1.3673
600	0.514	1153.7	1248.8	1.4450	0.402	1134.4	1223.6	1.4054
650	0.564	1184.7	1289.1	1.4822	0.450	1170.9	1270.8	1.4490
700	0.608	1212.0	1324.6	1.5135	0.491	1201.3	1310.2	1.4837
800	0.688	1261.2	1388.5	1.5665	0.562	1253.7	1378.4	1.5402
900	0.761	1307.3	1448.1	1.6120	0.626	1301.5	1440.4	1.5876
1000	0.831	1352.2	1505.9	1.6530	0.685	1347.5	1499.7	1.6297
1100	0.898	1396.8	1562.9	1.6908	0.743	1393.0	1557.9	1.6682
1200	0.963	1441.5	1619.7	1.7261	0.798	1438.3	1615.5	1.7040
1300	1.027	1486.5	1676.5	1.7593	0.853	1483.8	1673.1	1.7377
1400	1.091	1531.9	1733.7	1.7909	0.906	1529.6	1730.7	1.7696
1600	1.215	1624.4	1849.3	1.8499	1.011	1622.6	1847.1	1.8290

T °F	v ft³/lb	u Btu/lb	h Btu/lb	s Btu/lb · °R	v ft³/lb	u Btu/lb	h Btu/lb	s Btu/lb · °R
	p = 1400 lbf/in.² (T_{sat} = 587.2°F)				p = 1600 lbf/in.² (T_{sat} = 605.1°F)			
Sat.	0.302	1096.0	1174.1	1.3461	0.255	1087.4	1162.9	1.3258
600	0.318	1110.9	1193.1	1.3641				
650	0.367	1155.5	1250.5	1.4171	0.303	1137.8	1227.4	1.3852
700	0.406	1189.6	1294.8	1.4562	0.342	1177.0	1278.1	1.4299
800	0.471	1245.8	1367.9	1.5168	0.403	1237.7	1357.0	1.4953
900	0.529	1295.6	1432.5	1.5661	0.466	1289.5	1424.4	1.5468
1000	0.582	1342.8	1493.5	1.6094	0.504	1338.0	1487.1	1.5913
1100	0.632	1389.1	1552.8	1.6487	0.549	1385.2	1547.7	1.6315
1200	0.681	1435.1	1611.4	1.6851	0.592	1431.8	1607.1	1.6684
1300	0.728	1481.1	1669.6	1.7192	0.634	1478.3	1666.1	1.7029
1400	0.774	1527.2	1727.8	1.7513	0.675	1524.9	1724.8	1.7354
1600	0.865	1620.8	1844.8	1.8111	0.755	1619.0	1842.6	1.7955

T °F	v ft³/lb	u Btu/lb	h Btu/lb	s Btu/lb · °R	v ft³/lb	u Btu/lb	h Btu/lb	s Btu/lb · °R
	p = 1800 lbf/in.² (T_{sat} = 621.2°F)				p = 2000 lbf/in.² (T_{sat} = 636.0°F)			
Sat.	0.218	1077.7	1150.4	1.3060	0.188	1066.6	1136.3	1.2861
650	0.251	1117.0	1200.4	1.3517	0.206	1091.1	1167.2	1.3141
700	0.291	1163.1	1259.9	1.4042	0.249	1147.7	1239.8	1.3782
750	0.322	1198.6	1305.9	1.4430	0.280	1187.3	1291.1	1.4216
800	0.350	1229.1	1345.7	1.4753	0.307	1220.1	1333.8	1.4562
900	0.399	1283.2	1416.1	1.5291	0.353	1276.8	1407.6	1.5126
1000	0.443	1333.1	1480.7	1.5749	0.395	1328.1	1474.1	1.5598
1100	0.484	1381.2	1542.5	1.6159	0.433	1377.2	1537.2	1.6017
1200	0.524	1428.5	1602.9	1.6534	0.469	1425.2	1598.6	1.6398
1300	0.561	1475.5	1662.5	1.6883	0.503	1472.7	1659.0	1.6751
1400	0.598	1522.5	1721.8	1.7211	0.537	1520.2	1718.8	1.7082
1600	0.670	1617.2	1840.4	1.7817	0.602	1615.4	1838.2	1.7692

H_2O

Table A-4E **(Continued)**

T °F	*v* ft³/lb	*u* Btu/lb	*h* Btu/lb	*s* Btu/lb · °R	*v* ft³/lb	*u* Btu/lb	*h* Btu/lb	*s* Btu/lb · °R
	p = 2500 lbf/in.² (T_{sat} = 668.3°F)				p = 3000 lbf/in.² (T_{sat} = 695.5°F)			
Sat.	0.1306	1031.0	1091.4	1.2327	0.0840	968.8	1015.5	1.1575
700	0.1684	1098.7	1176.6	1.3073	0.0977	1003.9	1058.1	1.1944
750	0.2030	1155.2	1249.1	1.3686	0.1483	1114.7	1197.1	1.3122
800	0.2291	1195.7	1301.7	1.4112	0.1757	1167.6	1265.2	1.3675
900	0.2712	1259.9	1385.4	1.4752	0.2160	1241.8	1361.7	1.4414
1000	0.3069	1315.2	1457.2	1.5262	0.2485	1301.7	1439.6	1.4967
1100	0.3393	1366.8	1523.8	1.5704	0.2772	1356.2	1510.1	1.5434
1200	0.3696	1416.7	1587.7	1.6101	0.3086	1408.0	1576.6	1.5848
1300	0.3984	1465.7	1650.0	1.6465	0.3285	1458.5	1640.9	1.6224
1400	0.4261	1514.2	1711.3	1.6804	0.3524	1508.1	1703.7	1.6571
1500	0.4531	1562.5	1772.1	1.7123	0.3754	1557.3	1765.7	1.6896
1600	0.4795	1610.8	1832.6	1.7424	0.3978	1606.3	1827.1	1.7201
	p = 3500 lbf/in.²				p = 4000 lbf/in.²			
650	0.0249	663.5	679.7	0.8630	0.0245	657.7	675.8	0.8574
700	0.0306	759.5	779.3	0.9506	0.0287	742.1	763.4	0.9345
750	0.1046	1058.4	1126.1	1.2440	0.0633	960.7	1007.5	1.1395
800	0.1363	1134.7	1223.0	1.3226	0.1052	1095.0	1172.9	1.2740
900	0.1763	1222.4	1336.5	1.4096	0.1462	1201.5	1309.7	1.3789
1000	0.2066	1287.6	1421.4	1.4699	0.1752	1272.9	1402.6	1.4449
1100	0.2328	1345.2	1496.0	1.5193	0.1995	1333.9	1481.6	1.4973
1200	0.2566	1399.2	1565.3	1.5624	0.2213	1390.1	1553.9	1.5423
1300	0.2787	1451.1	1631.7	1.6012	0.2414	1443.7	1622.4	1.5823
1400	0.2997	1501.9	1696.1	1.6368	0.2603	1495.7	1688.4	1.6188
1500	0.3199	1552.0	1759.2	1.6699	0.2784	1546.7	1752.8	1.6526
1600	0.3395	1601.7	1831.6	1.7010	0.2959	1597.1	1816.1	1.6841
	p = 4400 lbf/in.²				p = 4800 lbf/in.²			
650	0.0242	653.6	673.3	0.8535	0.0237	649.8	671.0	0.8499
700	0.0278	732.7	755.3	0.9257	0.0271	725.1	749.1	0.9187
750	0.0415	870.8	904.6	1.0513	0.0352	832.6	863.9	1.0154
800	0.0844	1056.5	1125.3	1.2306	0.06[illegible]	1011.2	1070.5	1.1827
900	0.1270	1183.7	1287.1	1.3548	0.1109	1164.8	1263.4	1.3310
1000	0.1552	1260.8	1387.2	1.4260	0.1385	1248.3	1317.4	1.4078
1100	0.1784	1324.7	1469.9	1.4809	0.1608	1315.3	1458.1	1.4653
1200	0.1989	1382.8	1544.7	1.5274	0.1802	1375.4	1535.4	1.5133
1300	0.2176	1437.7	1614.9	1.5685	0.1979	1431.7	1607.4	1.5555
1400	0.2352	1490.7	1682.3	1.6057	0.2143	1485.7	1676.1	1.5934
1500	0.2520	1542.7	1747.6	1.6399	0.2300	1538.2	1742.5	1.6282
1600	0.2681	1593.4	1811.7	1.6718	0.2450	1589.8	1807.4	1.6605

H_2O

Table A-5E Properties of Compressed Liquid Water

T °F	v ft³/lb	u Btu/lb	h Btu/lb	s Btu/lb · °R	v ft³/lb	u Btu/lb	h Btu/lb	s Btu/lb · °R
	p = 500 lbf/in.² (T_{sat} = 467.1°F)				p = 1000 lbf/in.² (T_{sat} = 544.7°F)			
32	0.015994	0.00	1.49	0.00000	0.015967	0.03	2.99	0.00005
50	0.015998	18.02	19.50	0.03599	0.015972	17.99	20.94	0.03592
100	0.016106	67.87	69.36	0.12932	0.016082	67.70	70.68	0.12901
150	0.016318	117.66	119.17	0.21457	0.016293	117.38	120.40	0.21410
200	0.016608	167.65	169.19	0.29341	0.016580	167.26	170.32	0.29281
300	0.017416	268.92	270.53	0.43641	0.017379	268.24	271.46	0.43552
400	0.018608	373.68	375.40	0.56604	0.018550	372.55	375.98	0.56472
Sat.	0.019748	447.70	449.53	0.64904	0.021591	538.39	542.38	0.74320
	p = 1500 lbf/in.² (T_{sat} = 596.4°F)				p = 2000 lbf/in.² (T_{sat} = 636.0°F)			
32	0.015939	0.05	4.47	0.00007	0.015912	0.06	5.95	0.00008
50	0.015946	17.95	22.38	0.03584	0.015920	17.91	23.81	0.03575
100	0.016058	67.53	71.99	0.12870	0.016034	67.37	73.30	0.12839
150	0.016268	117.10	121.62	0.21364	0.016244	116.83	122.84	0.21318
200	0.016554	166.87	171.46	0.29221	0.016527	166.49	172.60	0.29162
300	0.017343	267.58	272.39	0.43463	0.017308	266.93	273.33	0.43376
400	0.018493	371.45	376.59	0.56343	0.018439	370.38	377.21	0.56216
500	0.02024	481.8	487.4	0.6853	0.02014	479.8	487.3	0.6832
Sat.	0.02346	605.0	611.5	0.8082	0.02565	662.4	671.9	0.8623
	p = 3000 lbf/in.² (T_{sat} = 695.5°F)				p = 4000 lbf/in.²			
32	0.015859	0.09	8.90	0.00009	0.015807	0.10	11.80	0.00005
50	0.015870	17.84	26.65	0.03555	0.015821	17.76	29.47	0.03534
100	0.015987	67.04	75.91	0.12777	0.015942	66.72	78.52	0.12714
150	0.016196	116.30	125.29	0.21226	0.016150	115.77	127.73	0.21136
200	0.016476	165.74	174.89	0.29046	0.016425	165.02	177.18	0.28931
300	0.017240	265.66	275.23	0.43205	0.017174	264.43	277.15	0.43038
400	0.018334	368.32	378.50	0.55970	0.018235	366.35	379.85	0.55734
500	0.019944	476.2	487.3	0.6794	0.019766	472.9	487.5	0.6758
Sat.	0.034310	783.5	802.5	0.9732				

Table A-6E **Properties of Saturated Water (Solid–Vapor): Temperature Table**

		Specific Volume ft³/lb		Internal Energy Btu/lb			Enthalpy Btu/lb			Entropy Btu/lb · °R		
Temp. °F	Press. lbf/in.²	Sat. Solid v_i	Sat. Vapor $v_g \times 10^{-3}$	Sat. Solid u_i	Subl. u_{ig}	Sat. Vapor u_g	Sat. Solid h_i	Subl. h_{ig}	Sat. Vapor h_g	Sat. Solid s_i	Subl. s_{ig}	Sat. Vapor s_g
32.018	.0887	.01747	3.302	−143.34	1164.6	1021.2	−143.34	1218.7	1075.4	−.292	2.479	2.187
32	.0886	.01747	3.305	−143.35	1164.6	1021.2	−143.35	1218.7	1075.4	−.292	2.479	2.187
30	.0808	.01747	3.607	−144.35	1164.9	1020.5	−144.35	1218.9	1074.5	−.294	2.489	2.195
25	.0641	.01746	4.506	−146.84	1165.7	1018.9	−146.84	1219.1	1072.3	−.299	2.515	2.216
20	.0505	.01745	5.655	−149.31	1166.5	1017.2	−149.31	1219.4	1070.1	−.304	2.542	2.238
15	.0396	.01745	7.13	−151.75	1167.3	1015.5	−151.75	1219.7	1067.9	−.309	2.569	2.260
10	.0309	.01744	9.04	−154.17	1168.1	1013.9	−154.17	1219.9	1065.7	−.314	2.597	2.283
5	.0240	.01743	11.52	−156.56	1168.8	1012.2	−156.56	1220.1	1063.5	−.320	2.626	2.306
0	.0185	.01743	14.77	−158.93	1169.5	1010.6	−158.93	1220.2	1061.2	−.325	2.655	2.330
−5	.0142	.01742	19.03	−161.27	1170.2	1008.9	−161.27	1220.3	1059.0	−.330	2.684	2.354
−10	.0109	.01741	24.66	−163.59	1170.9	1007.3	−163.59	1220.4	1056.8	−.335	2.714	2.379
−15	.0082	.01740	32.2	−165.89	1171.5	1005.6	−165.89	1220.5	1054.6	−.340	2.745	2.405
−20	.0062	.01740	42.2	−168.16	1172.1	1003.9	−168.16	1220.6	1052.4	−.345	2.776	2.431
−25	.0046	.01739	55.7	−170.40	1172.7	1002.3	−170.40	1220.6	1050.2	−.351	2.808	2.457
−30	.0035	.01738	74.1	−172.63	1173.2	1000.6	−172.63	1220.6	1048.0	−.356	2.841	2.485
−35	.0026	.01737	99.2	−174.82	1173.8	998.9	−174.82	1220.6	1045.8	−.361	2.874	2.513
−40	.0019	.01737	133.8	−177.00	1174.3	997.3	−177.00	1220.6	1043.6	−.366	2.908	2.542

Table A-7E **Properties of Saturated Refrigerant 22 (Liquid–Vapor): Temperature Table**

		Specific Volume ft³/lb		Internal Energy Btu/lb		Enthalpy Btu/lb			Entropy Btu/lb · °R		
Temp. °F	Press. lbf/in.²	Sat. Liquid v_f	Sat. Vapor v_g	Sat. Liquid u_f	Sat. Vapor u_g	Sat. Liquid h_f	Evap. h_{fg}	Sat. Vapor h_g	Sat. Liquid s_f	Sat. Vapor s_g	Temp. °F
−80	4.781	0.01090	9.6984	−10.30	87.24	−10.29	106.11	95.82	−0.0257	0.2538	−80
−60	8.834	0.01113	5.4744	−5.20	89.16	−5.18	103.30	98.12	−0.0126	0.2458	−60
−55	10.187	0.01120	4.7933	−3.91	89.64	−3.89	102.58	98.68	−0.0094	0.2441	−55
−50	11.701	0.01126	4.2123	−2.62	90.12	−2.60	101.84	99.24	−0.0063	0.2424	−50
−45	13.387	0.01132	3.7147	−1.33	90.59	−1.30	101.10	99.80	−0.0031	0.2407	−45
−40	15.261	0.01139	3.2869	−0.03	91.07	0.00	100.35	100.35	0.0000	0.2391	−40
−35	17.335	0.01145	2.9176	1.27	91.54	1.31	99.59	100.90	0.0031	0.2376	−35
−30	19.624	0.01152	2.5976	2.58	92.00	2.62	98.82	101.44	0.0061	0.2361	−30
−25	22.142	0.01159	2.3195	3.89	92.47	3.94	98.04	101.98	0.0092	0.2347	−25
−20	24.906	0.01166	2.0768	5.21	92.93	5.26	97.24	102.50	0.0122	0.2334	−20
−15	27.931	0.01173	1.8644	6.53	93.38	6.59	96.43	103.03	0.0152	0.2321	−15
−10	31.233	0.01181	1.6780	7.86	93.84	7.93	95.61	103.54	0.0182	0.2308	−10
−5	34.829	0.01188	1.5138	9.19	94.28	9.27	94.78	104.05	0.0211	0.2296	−5
0	38.734	0.01196	1.3688	10.53	94.73	10.62	93.93	104.55	0.0240	0.2284	0
5	42.967	0.01204	1.2404	11.88	95.17	11.97	93.06	105.04	0.0270	0.2272	5
10	47.545	0.01212	1.1264	13.23	95.60	13.33	92.18	105.52	0.0298	0.2261	10
15	52.486	0.01220	1.0248	14.58	96.03	14.70	91.29	105.99	0.0327	0.2250	15
20	57.808	0.01229	0.9342	15.95	96.45	16.08	90.38	106.45	0.0356	0.2240	20
25	63.529	0.01237	0.8531	17.31	96.87	17.46	89.45	106.90	0.0384	0.2230	25
30	69.668	0.01246	0.7804	18.69	97.28	18.85	88.50	107.35	0.0412	0.2220	30
35	76.245	0.01255	0.7150	20.07	97.68	20.25	87.53	107.78	0.0441	0.2210	35
40	83.278	0.01265	0.6561	21.46	98.08	21.66	86.54	108.20	0.0468	0.2200	40
45	90.787	0.01275	0.6029	22.86	98.47	23.07	85.53	108.60	0.0496	0.2191	45
50	98.792	0.01285	0.5548	24.27	98.84	24.50	84.49	108.99	0.0524	0.2182	50
55	107.31	0.01295	0.5112	25.68	99.22	25.94	83.44	109.37	0.0552	0.2173	55
60	116.37	0.01306	0.4716	27.10	99.58	27.38	82.36	109.74	0.0579	0.2164	60
65	125.98	0.01317	0.4355	28.53	99.93	28.84	81.25	110.09	0.0607	0.2155	65
70	136.18	0.01328	0.4027	29.98	100.27	30.31	80.11	110.42	0.0634	0.2147	70
75	146.97	0.01340	0.3726	31.43	100.60	31.79	78.95	110.74	0.0661	0.2138	75
80	158.38	0.01352	0.3452	32.89	100.92	33.29	77.75	111.04	0.0689	0.2130	80
85	170.44	0.01365	0.3200	34.36	101.22	34.80	76.53	111.32	0.0716	0.2121	85
90	183.16	0.01378	0.2969	35.85	101.51	36.32	75.26	111.58	0.0743	0.2113	90
95	196.57	0.01392	0.2756	37.35	101.79	37.86	73.96	111.82	0.0771	0.2104	95
100	210.69	0.01407	0.2560	38.86	102.05	39.41	72.63	112.04	0.0798	0.2095	100
105	225.54	0.01422	0.2379	40.39	102.29	40.99	71.24	112.23	0.0825	0.2087	105
110	241.15	0.01438	0.2212	41.94	102.52	42.58	69.82	112.40	0.0852	0.2078	110
115	257.55	0.01455	0.2058	43.50	102.72	44.19	68.34	112.53	0.0880	0.2069	115
120	274.75	0.01472	0.1914	45.08	102.90	45.83	66.81	112.64	0.0907	0.2060	120
140	352.17	0.01555	0.1433	51.62	103.36	52.64	60.06	112.70	0.1019	0.2021	140

Source: Tables A-7E through A-9E are calculated based on equations from A. Kamei and S. W. Beyerlein, "A Fundamental Equation for Chlorodifluoromethane (R-22)," *Fluid Phase Equilibria,* Vol. 80, No. 11, 1992, pp. 71–86.

Table A-8E **Properties of Saturated Refrigerant 22 (Liquid–Vapor): Pressure Table**

		Specific Volume ft³/lb		Internal Energy Btu/lb		Enthalpy Btu/lb			Entropy Btu/lb · °R		
Press. lbf/in.²	Temp. °F	Sat. Liquid v_f	Sat. Vapor v_g	Sat. Liquid u_f	Sat. Vapor u_g	Sat. Liquid h_f	Evap. h_{fg}	Sat. Vapor h_g	Sat. Liquid s_f	Sat. Vapor s_g	Press. lbf/in.²
5	−78.62	0.01091	9.3014	−9.95	87.37	−9.93	105.92	95.98	−0.0248	0.2532	5
10	−55.66	0.01119	4.8769	−4.08	89.58	−4.06	102.67	98.61	−0.0098	0.2443	10
15	−40.67	0.01138	3.3402	−0.21	91.00	−0.17	100.45	100.28	−0.0004	0.2393	15
20	−29.22	0.01153	2.5518	2.78	92.07	2.83	98.70	101.52	0.0066	0.2359	20
25	−19.84	0.01166	2.0695	5.25	92.94	5.31	97.22	102.52	0.0123	0.2333	25
30	−11.82	0.01178	1.7430	7.38	93.67	7.44	95.91	103.35	0.0171	0.2313	30
35	−4.77	0.01189	1.5068	9.25	94.30	9.33	94.74	104.07	0.0212	0.2295	35
40	1.54	0.01198	1.3277	10.94	94.86	11.03	93.66	104.70	0.0249	0.2280	40
45	7.27	0.01207	1.1870	12.49	95.37	12.59	92.67	105.26	0.0283	0.2267	45
50	12.53	0.01216	1.0735	13.91	95.82	14.03	91.73	105.76	0.0313	0.2256	50
55	17.41	0.01224	0.9799	15.24	96.23	15.36	90.85	106.21	0.0341	0.2245	55
60	21.96	0.01232	0.9014	16.48	96.62	16.62	90.01	106.63	0.0367	0.2236	60
65	26.23	0.01239	0.8345	17.65	96.97	17.80	89.21	107.01	0.0391	0.2227	65
70	30.26	0.01247	0.7768	18.76	97.30	18.92	88.45	107.37	0.0414	0.2219	70
75	34.08	0.01254	0.7265	19.82	97.61	19.99	87.71	107.70	0.0435	0.2212	75
80	37.71	0.01260	0.6823	20.83	97.90	21.01	86.99	108.00	0.0456	0.2205	80
85	41.18	0.01267	0.6431	21.79	98.17	21.99	86.30	108.29	0.0475	0.2198	85
90	44.49	0.01274	0.6081	22.72	98.43	22.93	85.63	108.56	0.0494	0.2192	90
95	47.67	0.01280	0.5766	23.61	98.67	23.84	84.98	108.81	0.0511	0.2186	95
100	50.73	0.01286	0.5482	24.47	98.90	24.71	84.34	109.05	0.0528	0.2181	100
110	56.52	0.01298	0.4988	26.11	99.33	26.37	83.11	109.49	0.0560	0.2170	110
120	61.92	0.01310	0.4573	27.65	99.71	27.94	81.93	109.88	0.0590	0.2161	120
130	67.00	0.01321	0.4220	29.11	100.07	29.43	80.80	110.22	0.0618	0.2152	130
140	71.80	0.01332	0.3915	30.50	100.39	30.84	79.70	110.54	0.0644	0.2144	140
150	76.36	0.01343	0.3649	31.82	100.69	32.20	78.63	110.82	0.0669	0.2136	150
160	80.69	0.01354	0.3416	33.09	100.96	33.49	77.59	111.08	0.0693	0.2128	160
170	84.82	0.01365	0.3208	34.31	101.21	34.74	76.57	111.31	0.0715	0.2121	170
180	88.78	0.01375	0.3023	35.49	101.44	35.95	75.57	111.52	0.0737	0.2115	180
190	92.58	0.01386	0.2857	36.62	101.66	37.11	74.60	111.71	0.0757	0.2108	190
200	96.24	0.01396	0.2706	37.72	101.86	38.24	73.64	111.88	0.0777	0.2102	200
225	104.82	0.01422	0.2386	40.34	102.28	40.93	71.29	112.22	0.0824	0.2087	225
250	112.73	0.01447	0.2126	42.79	102.63	43.46	69.02	112.47	0.0867	0.2073	250
275	120.07	0.01473	0.1912	45.10	102.91	45.85	66.79	112.64	0.0908	0.2060	275
300	126.94	0.01499	0.1732	47.30	103.11	48.14	64.60	112.73	0.0946	0.2047	300
325	133.39	0.01525	0.1577	49.42	103.26	50.33	62.42	112.75	0.0982	0.2034	325
350	139.49	0.01552	0.1444	51.45	103.35	52.46	60.25	112.71	0.1016	0.2022	350

Table A-9E Properties of Superheated Refrigerant 22 Vapor

T °F	v ft³/lb	u Btu/lb	h Btu/lb	s Btu/lb · °R	v ft³/lb	u Btu/lb	h Btu/lb	s Btu/lb · °R
	$p = 5$ lbf/in.² ($T_{sat} = -78.62$°F)				$p = 10$ lbf/in.² ($T_{sat} = -55.66$°F)			
Sat.	9.3014	87.37	95.98	0.2532	4.8769	89.58	98.61	0.2443
−70	9.5244	88.31	97.13	0.2562				
−60	9.7823	89.43	98.48	0.2596				
−50	10.0391	90.55	99.84	0.2630	4.9522	90.23	99.40	0.2462
−40	10.2952	91.69	101.22	0.2663	5.0846	91.39	100.81	0.2496
−30	10.5506	92.84	102.61	0.2696	5.2163	92.57	102.23	0.2530
−20	10.8054	94.01	104.01	0.2728	5.3472	93.75	103.65	0.2563
−10	11.0596	95.19	105.43	0.2760	5.4775	94.95	105.09	0.2595
0	11.3133	96.39	106.87	0.2791	5.6073	96.16	106.55	0.2627
10	11.5666	97.60	108.31	0.2822	5.7366	97.39	108.01	0.2658
20	11.8195	98.83	109.77	0.2853	5.8655	98.63	109.49	0.2690
30	12.0720	100.07	111.25	0.2884	5.9941	99.88	110.98	0.2720
40	12.3242	101.33	112.74	0.2914	6.1223	101.15	112.49	0.2751

T °F	v ft³/lb	u Btu/lb	h Btu/lb	s Btu/lb · °R	v ft³/lb	u Btu/lb	h Btu/lb	s Btu/lb · °R
	$p = 15$ lbf/in.² ($T_{sat} = -40.67$°F)				$p = 20$ lbf/in.² ($T_{sat} = -29.22$°F)			
Sat.	3.3402	91.00	100.28	0.2393	2.5518	92.07	101.52	0.2359
−40	3.3463	91.08	100.38	0.2396				
−30	3.4370	92.28	101.83	0.2430				
−20	3.5268	93.49	103.28	0.2463	2.6158	93.21	102.90	0.2391
−10	3.6160	94.70	104.75	0.2496	2.6846	94.45	104.39	0.2424
0	3.7046	95.93	106.22	0.2529	2.7528	95.69	105.89	0.2457
10	3.7927	97.17	107.71	0.2561	2.8204	96.95	107.39	0.2490
20	3.8804	98.43	109.20	0.2592	2.8875	98.22	108.91	0.2522
30	3.9677	99.69	110.71	0.2623	2.9542	99.49	110.43	0.2553
40	4.0546	100.97	112.23	0.2654	3.0205	100.78	111.97	0.2584
50	4.1412	102.26	113.76	0.2684	3.0865	102.09	113.52	0.2615
60	4.2275	103.57	115.31	0.2714	3.1522	103.40	115.08	0.2645
70	4.3136	104.89	116.87	0.2744	3.2176	104.73	116.65	0.2675

T °F	v ft³/lb	u Btu/lb	h Btu/lb	s Btu/lb · °R	v ft³/lb	u Btu/lb	h Btu/lb	s Btu/lb · °R
	$p = 25$ lbf/in.² ($T_{sat} = -19.84$°F)				$p = 30$ lbf/in.² ($T_{sat} = -11.82$°F)			
Sat.	2.0695	92.94	102.52	0.2333	1.7430	93.67	103.35	0.2313
−10	2.1252	94.18	104.02	0.2367	1.7518	93.91	103.64	0.2319
0	2.1812	95.45	105.54	0.2400	1.7997	95.19	105.19	0.2353
10	2.2365	96.72	107.07	0.2433	1.8470	96.48	106.74	0.2386
20	2.2914	98.00	108.61	0.2466	1.8937	97.78	108.30	0.2419
30	2.3458	99.29	110.15	0.2498	1.9400	99.09	109.86	0.2451
40	2.3998	100.59	111.70	0.2529	1.9858	100.40	111.43	0.2483
50	2.4535	101.91	113.27	0.2560	2.0313	101.73	113.01	0.2514
60	2.5068	103.23	114.84	0.2590	2.0764	103.06	114.60	0.2545
70	2.5599	104.57	116.42	0.2621	2.1213	104.41	116.19	0.2576
80	2.6127	105.92	118.01	0.2650	2.1659	105.77	117.80	0.2606
90	2.6654	107.28	119.62	0.2680	2.2103	107.13	119.41	0.2635
100	2.7178	108.65	121.24	0.2709	2.2545	108.52	121.04	0.2665

R-22

Table A-9E **(Continued)**

T °F	v ft³/lb	u Btu/lb	h Btu/lb	s Btu/lb · °R	v ft³/lb	u Btu/lb	h Btu/lb	s Btu/lb · °R
	$p = 40$ lbf/in.² ($T_{sat} = 1.54$°F)				$p = 50$ lbf/in.² ($T_{sat} = 12.53$°F)			
Sat.	1.3277	94.86	104.70	0.2280	1.0735	95.82	105.76	0.2256
10	1.3593	95.99	106.06	0.2310				
20	1.3960	97.33	107.67	0.2343	1.0965	96.85	107.00	0.2282
30	1.4321	98.66	109.27	0.2376	1.1268	98.22	108.65	0.2316
40	1.4678	100.01	110.88	0.2409	1.1565	99.59	110.30	0.2349
50	1.5032	101.35	112.49	0.2441	1.1858	100.97	111.95	0.2382
60	1.5381	102.71	114.10	0.2472	1.2147	102.35	113.60	0.2414
70	1.5728	104.08	115.73	0.2503	1.2433	103.74	115.25	0.2445
80	1.6071	105.45	117.36	0.2534	1.2716	105.13	116.90	0.2476
90	1.6413	106.84	118.99	0.2564	1.2996	106.53	118.57	0.2507
100	1.6752	108.23	120.64	0.2593	1.3274	107.95	120.24	0.2537
110	1.7089	109.64	122.30	0.2623	1.3549	109.37	121.91	0.2567
120	1.7424	111.06	123.97	0.2652	1.3823	110.80	123.60	0.2596
	$p = 60$ lbf/in.² ($T_{sat} = 21.96$°F)				$p = 70$ lbf/in.² ($T_{sat} = 30.26$°F)			
Sat.	0.9014	96.62	106.63	0.2236	0.7768	97.30	107.37	0.2219
30	0.9226	97.75	108.00	0.2264				
40	0.9485	99.16	109.70	0.2298	0.7994	98.71	109.07	0.2254
50	0.9739	100.57	111.39	0.2332	0.8221	100.15	110.81	0.2288
60	0.9988	101.98	113.07	0.2365	0.8443	101.59	112.53	0.2321
70	1.0234	103.39	114.76	0.2397	0.8660	103.03	114.25	0.2354
80	1.0476	104.80	116.44	0.2428	0.8874	104.46	115.97	0.2386
90	1.0716	106.22	118.13	0.2459	0.9086	105.90	117.68	0.2418
100	1.0953	107.65	119.82	0.2490	0.9294	107.35	119.40	0.2449
110	1.1188	109.09	121.52	0.2520	0.9500	108.80	121.12	0.2479
120	1.1421	110.53	123.22	0.2549	0.9704	110.26	122.84	0.2509
130	1.1653	111.99	124.93	0.2579	0.9907	111.73	124.57	0.2539
140	1.1883	113.45	126.65	0.2608	1.0107	113.21	126.31	0.2568
	$p = 80$ lbf/in.² ($T_{sat} = 37.71$°F)				$p = 90$ lbf/in.² ($T_{sat} = 44.49$°F)			
Sat.	0.6823	97.90	108.00	0.2205	0.6081	98.43	108.56	0.2192
40	0.6871	98.24	108.42	0.2213				
50	0.7079	99.72	110.20	0.2248	0.6186	99.26	109.57	0.2212
60	0.7280	101.19	111.97	0.2283	0.6373	100.77	111.39	0.2247
70	0.7478	102.65	113.73	0.2316	0.6555	102.27	113.19	0.2282
80	0.7671	104.11	115.48	0.2349	0.6733	103.76	114.98	0.2315
90	0.7861	105.58	117.22	0.2381	0.6907	105.24	116.75	0.2348
100	0.8048	107.04	118.97	0.2412	0.7078	106.73	118.52	0.2380
110	0.8233	108.51	120.71	0.2443	0.7246	108.22	120.29	0.2411
120	0.8416	109.99	122.45	0.2474	0.7412	109.71	122.06	0.2442
130	0.8596	111.47	124.20	0.2504	0.7576	111.20	123.83	0.2472
140	0.8775	112.96	125.96	0.2533	0.7739	112.71	125.60	0.2502
150	0.8953	114.46	127.72	0.2562	0.7899	114.22	127.38	0.2531

Table A-9E **(Continued)**

T °F	v ft³/lb	u Btu/lb	h Btu/lb	s Btu/lb · °R	v ft³/lb	u Btu/lb	h Btu/lb	s Btu/lb · °R
	p = 100 lbf/in.² (T_{sat} = 50.73°F)				p = 120 lbf/in.² (T_{sat} = 61.92°F)			
Sat.	0.5482	98.90	109.05	0.2181	0.4573	99.71	109.88	0.2161
60	0.5645	100.33	110.79	0.2214				
80	0.5980	103.38	114.46	0.2284	0.4846	102.60	113.37	0.2227
100	0.6300	106.40	118.07	0.2349	0.5130	105.73	117.13	0.2295
120	0.6609	109.42	121.66	0.2412	0.5400	108.83	120.83	0.2360
140	0.6908	112.45	125.24	0.2473	0.5661	111.92	124.50	0.2422
160	0.7201	115.50	128.83	0.2532	0.5914	115.02	128.16	0.2482
180	0.7489	118.58	132.45	0.2589	0.6161	118.15	131.84	0.2541
200	0.7771	121.69	136.08	0.2645	0.6404	121.30	135.53	0.2597
220	0.8051	124.84	139.75	0.2700	0.6642	124.48	139.24	0.2653
240	0.8327	128.04	143.45	0.2754	0.6878	127.69	142.98	0.2707
260	0.8600	131.27	147.19	0.2806	0.7110	130.95	146.75	0.2760
280	0.8871	134.54	150.97	0.2858	0.7340	134.24	150.55	0.2812
300	0.9140	137.85	154.78	0.2909	0.7568	137.57	154.39	0.2863
	p = 140 lbf/in.² (T_{sat} = 71.80°F)				p = 160 lbf/in.² (T_{sat} = 80.69°F)			
Sat.	0.3915	100.39	110.54	0.2144	0.3416	100.96	111.08	0.2128
80	0.4028	101.76	112.20	0.2175				
100	0.4289	105.02	116.14	0.2246	0.3653	104.26	115.08	0.2201
120	0.4534	108.21	119.96	0.2313	0.3881	107.56	119.06	0.2271
140	0.4768	111.37	123.73	0.2377	0.4095	110.81	122.94	0.2337
160	0.4993	114.53	127.48	0.2439	0.4301	114.03	126.77	0.2400
180	0.5212	117.70	131.21	0.2498	0.4499	117.25	130.57	0.2460
200	0.5426	120.89	134.96	0.2556	0.4692	120.47	134.37	0.2518
220	0.5636	124.10	138.71	0.2612	0.4880	123.72	138.18	0.2575
240	0.5842	127.35	142.49	0.2666	0.5065	126.99	142.00	0.2631
260	0.6045	130.62	146.30	0.2720	0.5246	130.30	145.84	0.2685
280	0.6246	133.94	150.13	0.2773	0.5425	133.63	149.70	0.2738
300	0.6445	137.29	154.00	0.2824	0.5602	137.00	153.60	0.2790
320	0.6642	140.68	157.89	0.2875	0.5777	140.41	157.62	0.2841
	p = 180 lbf/in.² (T_{sat} = 88.78°F)				p = 200 lbf/in.² (T_{sat} = 96.24°F)			
Sat.	0.3023	101.44	111.52	0.2115	0.2706	101.86	111.88	0.2102
100	0.3154	103.44	113.95	0.2159	0.2748	102.56	112.73	0.2117
120	0.3369	106.88	118.11	0.2231	0.2957	106.15	117.10	0.2194
140	0.3570	110.21	122.11	0.2299	0.3148	109.59	121.25	0.2264
160	0.3761	113.50	126.04	0.2364	0.3327	112.96	125.28	0.2330
180	0.3943	116.78	129.92	0.2425	0.3497	116.29	129.25	0.2393
200	0.4120	120.05	133.78	0.2485	0.3661	119.61	133.17	0.2454
220	0.4292	123.33	137.64	0.2542	0.3820	122.94	137.08	0.2512
240	0.4459	126.64	141.50	0.2598	0.3975	126.27	140.99	0.2569
260	0.4624	129.96	145.38	0.2653	0.4126	129.63	144.91	0.2624
280	0.4786	133.32	149.28	0.2706	0.4275	133.01	148.84	0.2678
300	0.4946	136.71	153.20	0.2759	0.4422	136.42	152.79	0.2731
320	0.5104	140.13	157.15	0.2810	0.4566	139.86	156.77	0.2782
340	0.5260	143.59	161.12	0.2860	0.4709	143.33	160.77	0.2833

R-22

Table A-9E **(Continued)**

T °F	v ft³/lb	u Btu/lb	h Btu/lb	s Btu/lb · °R	v ft³/lb	u Btu/lb	h Btu/lb	s Btu/lb · °R
	p = 225 lbf/in.² (T_{sat} = 104.82°F)				p = 250 lbf/in.² (T_{sat} = 112.73°F)			
Sat.	0.2386	102.28	112.22	0.2087	0.2126	102.63	112.47	0.2073
120	0.2539	105.17	115.75	0.2149	0.2198	104.10	114.27	0.2104
140	0.2722	108.78	120.12	0.2223	0.2378	107.90	118.91	0.2183
160	0.2891	112.26	124.30	0.2291	0.2540	111.51	123.27	0.2255
180	0.3050	115.67	128.38	0.2356	0.2690	115.02	127.48	0.2321
200	0.3202	119.06	132.40	0.2418	0.2833	118.48	131.59	0.2385
220	0.3348	122.43	136.38	0.2477	0.2969	121.91	135.66	0.2445
240	0.3490	125.81	140.35	0.2535	0.3101	125.33	139.69	0.2504
260	0.3628	129.20	144.32	0.2591	0.3229	128.76	143.71	0.2560
280	0.3764	132.61	148.29	0.2645	0.3354	132.21	147.73	0.2616
300	0.3896	136.05	152.28	0.2699	0.3476	135.67	151.76	0.2669
320	0.4027	139.51	156.29	0.2751	0.3596	139.16	155.81	0.2722
340	0.4156	143.00	160.32	0.2802	0.3715	142.67	159.87	0.2773
360	0.4284	146.33	164.38	0.2852	0.3831	146.22	163.95	0.2824
	p = 275 lbf/in.² (T_{sat} = 120.07°F)				p = 300 lbf/in.² (T_{sat} = 126.94°F)			
Sat.	0.1912	102.91	112.64	0.2060	0.1732	103.11	112.73	0.2047
140	0.2092	106.96	117.61	0.2144	0.1849	105.93	116.20	0.2105
160	0.2250	110.73	122.19	0.2219	0.2006	109.89	121.04	0.2185
180	0.2395	144.35	126.54	0.2288	0.2146	133.64	125.56	0.2257
200	0.2530	117.88	130.77	0.2353	0.2276	117.26	129.91	0.2324
220	0.2659	121.38	134.91	0.2415	0.2399	120.83	134.15	0.2387
240	0.2782	124.85	139.02	0.2475	0.2516	124.35	138.33	0.2447
260	0.2902	128.32	143.10	0.2532	0.2629	127.87	142.47	0.2506
280	0.3018	131.80	147.17	0.2588	0.2739	131.38	146.59	0.2562
300	0.3132	135.29	151.24	0.2642	0.2845	134.90	150.71	0.2617
320	0.3243	138.80	155.32	0.2695	0.2949	138.44	154.83	0.2671
340	0.3353	142.34	159.41	0.2747	0.3051	142.00	158.95	0.2723
360	0.3461	145.90	163.53	0.2798	0.3152	145.58	163.09	0.2774
	p = 325 lbf/in.² (T_{sat} = 133.39°F)				p = 350 lbf/in.² (T_{sat} = 139.49°F)			
Sat.	0.1577	103.26	112.75	0.2034	0.1444	103.35	112.71	0.2022
140	0.1637	104.78	114.63	0.2066	0.1448	103.48	112.86	0.2024
160	0.1796	109.00	119.81	0.2151	0.1605	107.90	118.30	0.2113
180	0.1934	112.89	124.53	0.2226	0.1747	112.06	123.38	0.2194
200	0.2061	116.62	129.02	0.2295	0.1874	115.95	128.10	0.2267
220	0.2179	120.26	133.37	0.2360	0.1987	119.65	132.53	0.2333
240	0.2291	123.84	137.63	0.2422	0.2095	123.31	136.89	0.2396
260	0.2398	127.40	141.83	0.2481	0.2199	126.93	141.18	0.2457
280	0.2501	130.96	146.01	0.2538	0.2297	130.52	145.41	0.2514
300	0.2602	134.51	150.17	0.2593	0.2393	134.12	149.62	0.2571
320	0.2700	138.08	154.33	0.2647	0.2486	137.71	153.82	0.2626
340	0.2796	141.66	158.49	0.2700	0.2577	141.32	158.02	0.2679
360	0.2891	145.26	162.66	0.2752	0.2666	144.95	162.23	0.2730
380	0.2983	148.89	166.85	0.2802	0.2754	148.59	166.43	0.2781

Table A-10E **Properties of Saturated Refrigerant 134a (Liquid–Vapor): Temperature Table**

		Specific Volume ft³/lb		Internal Energy Btu/lb		Enthalpy Btu/lb			Entropy Btu/lb · °R		
Temp. °F	Press. lbf/in.²	Sat. Liquid v_f	Sat. Vapor v_g	Sat. Liquid u_f	Sat. Vapor u_g	Sat. Liquid h_f	Evap. h_{fg}	Sat. Vapor h_g	Sat. Liquid s_f	Sat. Vapor s_g	Temp. °F
−40	7.490	0.01130	5.7173	−0.02	87.90	0.00	95.82	95.82	0.0000	0.2283	−40
−30	9.920	0.01143	4.3911	2.81	89.26	2.83	94.49	97.32	0.0067	0.2266	−30
−20	12.949	0.01156	3.4173	5.69	90.62	5.71	93.10	98.81	0.0133	0.2250	−20
−15	14.718	0.01163	3.0286	7.14	91.30	7.17	92.38	99.55	0.0166	0.2243	−15
−10	16.674	0.01170	2.6918	8.61	91.98	8.65	91.64	100.29	0.0199	0.2236	−10
−5	18.831	0.01178	2.3992	10.09	92.66	10.13	90.89	101.02	0.0231	0.2230	−5
0	21.203	0.01185	2.1440	11.58	93.33	11.63	90.12	101.75	0.0264	0.2224	0
5	23.805	0.01193	1.9208	13.09	94.01	13.14	89.33	102.47	0.0296	0.2219	5
10	26.651	0.01200	1.7251	14.60	94.68	14.66	88.53	103.19	0.0329	0.2214	10
15	29.756	0.01208	1.5529	16.13	95.35	16.20	87.71	103.90	0.0361	0.2209	15
20	33.137	0.01216	1.4009	17.67	96.02	17.74	86.87	104.61	0.0393	0.2205	20
25	36.809	0.01225	1.2666	19.22	96.69	19.30	86.02	105.32	0.0426	0.2200	25
30	40.788	0.01233	1.1474	20.78	97.35	20.87	85.14	106.01	0.0458	0.2196	30
40	49.738	0.01251	0.9470	23.94	98.67	24.05	83.34	107.39	0.0522	0.2189	40
50	60.125	0.01270	0.7871	27.14	99.98	27.28	81.46	108.74	0.0585	0.2183	50
60	72.092	0.01290	0.6584	30.39	101.27	30.56	79.49	110.05	0.0648	0.2178	60
70	85.788	0.01311	0.5538	33.68	102.54	33.89	77.44	111.33	0.0711	0.2173	70
80	101.37	0.01334	0.4682	37.02	103.78	37.27	75.29	112.56	0.0774	0.2169	80
85	109.92	0.01346	0.4312	38.72	104.39	38.99	74.17	113.16	0.0805	0.2167	85
90	118.99	0.01358	0.3975	40.42	105.00	40.72	73.03	113.75	0.0836	0.2165	90
95	128.62	0.01371	0.3668	42.14	105.60	42.47	71.86	114.33	0.0867	0.2163	95
100	138.83	0.01385	0.3388	43.87	106.18	44.23	70.66	114.89	0.0898	0.2161	100
105	149.63	0.01399	0.3131	45.62	106.76	46.01	69.42	115.43	0.0930	0.2159	105
110	161.04	0.01414	0.2896	47.39	107.33	47.81	68.15	115.96	0.0961	0.2157	110
115	173.10	0.01429	0.2680	49.17	107.88	49.63	66.84	116.47	0.0992	0.2155	115
120	185.82	0.01445	0.2481	50.97	108.42	51.47	65.48	116.95	0.1023	0.2153	120
140	243.86	0.01520	0.1827	58.39	110.41	59.08	59.57	118.65	0.1150	0.2143	140
160	314.63	0.01617	0.1341	66.26	111.97	67.20	52.58	119.78	0.1280	0.2128	160
180	400.22	0.01758	0.0964	74.83	112.77	76.13	43.78	119.91	0.1417	0.2101	180
200	503.52	0.02014	0.0647	84.90	111.66	86.77	30.92	117.69	0.1575	0.2044	200
210	563.51	0.02329	0.0476	91.84	108.48	94.27	19.18	113.45	0.1684	0.1971	210

Source: Tables A-10E through A-12E are calculated based on equations from D. P. Wilson and R. S. Basu, "Thermodynamic Properties of a New Stratospherically Safe Working Fluid—Refrigerant 134a," *ASHRAE Trans.*, Vol. 94, Pt. 2, 1988, pp. 2095–2118.

Table A-11E **Properties of Saturated Refrigerant 134a (Liquid–Vapor): Pressure Table**

		Specific Volume ft³/lb		Internal Energy Btu/lb		Enthalpy Btu/lb			Entropy Btu/lb · °R		
Press. lbf/in.²	Temp. °F	Sat. Liquid v_f	Sat. Vapor v_g	Sat. Liquid u_f	Sat. Vapor u_g	Sat. Liquid h_f	Evap. h_{fg}	Sat. Vapor h_g	Sat. Liquid s_f	Sat. Vapor s_g	Press. lbf/in.²
5	−53.48	0.01113	8.3508	−3.74	86.07	−3.73	97.53	93.79	−0.0090	0.2311	5
10	−29.71	0.01143	4.3581	2.89	89.30	2.91	94.45	97.37	0.0068	0.2265	10
15	−14.25	0.01164	2.9747	7.36	91.40	7.40	92.27	99.66	0.0171	0.2242	15
20	−2.48	0.01181	2.2661	10.84	93.00	10.89	90.50	101.39	0.0248	0.2227	20
30	15.38	0.01209	1.5408	16.24	95.40	16.31	87.65	103.96	0.0364	0.2209	30
40	29.04	0.01232	1.1692	20.48	97.23	20.57	85.31	105.88	0.0452	0.2197	40
50	40.27	0.01252	0.9422	24.02	98.71	24.14	83.29	107.43	0.0523	0.2189	50
60	49.89	0.01270	0.7887	27.10	99.96	27.24	81.48	108.72	0.0584	0.2183	60
70	58.35	0.01286	0.6778	29.85	101.05	30.01	79.82	109.83	0.0638	0.2179	70
80	65.93	0.01302	0.5938	32.33	102.02	32.53	78.28	110.81	0.0686	0.2175	80
90	72.83	0.01317	0.5278	34.62	102.89	34.84	76.84	111.68	0.0729	0.2172	90
100	79.17	0.01332	0.4747	36.75	103.68	36.99	75.47	112.46	0.0768	0.2169	100
120	90.54	0.01360	0.3941	40.61	105.06	40.91	72.91	113.82	0.0839	0.2165	120
140	100.56	0.01386	0.3358	44.07	106.25	44.43	70.52	114.95	0.0902	0.2161	140
160	109.56	0.01412	0.2916	47.23	107.28	47.65	68.26	115.91	0.0958	0.2157	160
180	117.74	0.01438	0.2569	50.16	108.18	50.64	66.10	116.74	0.1009	0.2154	180
200	125.28	0.01463	0.2288	52.90	108.98	53.44	64.01	117.44	0.1057	0.2151	200
220	132.27	0.01489	0.2056	55.48	109.68	56.09	61.96	118.05	0.1101	0.2147	220
240	138.79	0.01515	0.1861	57.93	110.30	58.61	59.96	118.56	0.1142	0.2144	240
260	144.92	0.01541	0.1695	60.28	110.84	61.02	57.97	118.99	0.1181	0.2140	260
280	150.70	0.01568	0.1550	62.53	111.31	63.34	56.00	119.35	0.1219	0.2136	280
300	156.17	0.01596	0.1424	64.71	111.72	65.59	54.03	119.62	0.1254	0.2132	300
350	168.72	0.01671	0.1166	69.88	112.45	70.97	49.03	120.00	0.1338	0.2118	350
400	179.95	0.01758	0.0965	74.81	112.77	76.11	43.80	119.91	0.1417	0.2102	400
450	190.12	0.01863	0.0800	79.63	112.60	81.18	38.08	119.26	0.1493	0.2079	450
500	199.38	0.02002	0.0657	84.54	111.76	86.39	31.44	117.83	0.1570	0.2047	500

Table A-12E Properties of Superheated Refrigerant 134a Vapor

T °F	v ft³/lb	u Btu/lb	h Btu/lb	s Btu/lb · °R	v ft³/lb	u Btu/lb	h Btu/lb	s Btu/lb · °R
	p = 10 lbf/in.² (T_{sat} = −29.71°F)				p = 15 lbf/in.² (T_{sat} = −14.25°F)			
Sat.	4.3581	89.30	97.37	0.2265	2.9747	91.40	99.66	0.2242
−20	4.4718	90.89	99.17	0.2307				
0	4.7026	94.24	102.94	0.2391	3.0893	93.84	102.42	0.2303
20	4.9297	97.67	106.79	0.2472	3.2468	97.33	106.34	0.2386
40	5.1539	101.19	110.72	0.2553	3.4012	100.89	110.33	0.2468
60	5.3758	104.80	114.74	0.2632	3.5533	104.54	114.40	0.2548
80	5.5959	108.50	118.85	0.2709	3.7034	108.28	118.56	0.2626
100	5.8145	112.29	123.05	0.2786	3.8520	112.10	122.79	0.2703
120	6.0318	116.18	127.34	0.2861	3.9993	116.01	127.11	0.2779
140	6.2482	120.16	131.72	0.2935	4.1456	120.00	131.51	0.2854
160	6.4638	124.23	136.19	0.3009	4.2911	124.09	136.00	0.2927
180	6.6786	128.38	140.74	0.3081	4.4359	128.26	140.57	0.3000
200	6.8929	132.63	145.39	0.3152	4.5801	132.52	145.23	0.3072
	p = 20 lbf/in.² (T_{sat} = −2.48°F)				p = 30 lbf/in.² (T_{sat} = 15.38°F)			
Sat.	2.2661	93.00	101.39	0.2227	1.5408	95.40	103.96	0.2209
0	2.2816	93.43	101.88	0.2238				
20	2.4046	96.98	105.88	0.2323	1.5611	96.26	104.92	0.2229
40	2.5244	100.59	109.94	0.2406	1.6465	99.98	109.12	0.2315
60	2.6416	104.28	114.06	0.2487	1.7293	103.75	113.35	0.2398
80	2.7569	108.05	118.25	0.2566	1.8098	107.59	117.63	0.2478
100	2.8705	111.90	122.52	0.2644	1.8887	111.49	121.98	0.2558
120	2.9829	115.83	126.87	0.2720	1.9662	115.47	126.39	0.2635
140	3.0942	119.85	131.30	0.2795	2.0426	119.53	130.87	0.2711
160	3.2047	123.95	135.81	0.2869	2.1181	123.66	135.42	0.2786
180	3.3144	128.13	140.40	0.2922	2.1929	127.88	140.05	0.2859
200	3.4236	132.40	145.07	0.3014	2.2671	132.17	144.76	0.2932
220	3.5323	136.76	149.83	0.3085	2.3407	136.55	149.54	0.3003
	p = 40 lbf/in.² (T_{sat} = 29.04°F)				p = 50 lbf/in.² (T_{sat} = 40.27°F)			
Sat.	1.1692	97.23	105.88	0.2197	0.9422	98.71	107.43	0.2189
40	1.2065	99.33	108.26	0.2245				
60	1.2723	103.20	112.62	0.2331	0.9974	102.62	111.85	0.2276
80	1.3357	107.11	117.00	0.2414	1.0508	106.62	116.34	0.2361
100	1.3973	111.08	121.42	0.2494	1.1022	110.65	120.85	0.2443
120	1.4575	115.11	125.90	0.2573	1.1520	114.74	125.39	0.2523
140	1.5165	119.21	130.43	0.2650	1.2007	118.88	129.99	0.2601
160	1.5746	123.38	135.03	0.2725	1.2484	123.08	134.64	0.2677
180	1.6319	127.62	139.70	0.2799	1.2953	127.36	139.34	0.2752
200	1.6887	131.94	144.44	0.2872	1.3415	131.71	144.12	0.2825
220	1.7449	136.34	149.25	0.2944	1.3873	136.12	148.96	0.2897
240	1.8006	140.81	154.14	0.3015	1.4326	140.61	153.87	0.2969
260	1.8561	145.36	159.10	0.3085	1.4775	145.18	158.85	0.3039
280	1.9112	149.98	164.13	0.3154	1.5221	149.82	163.90	0.3108

Table A-12E (Continued)

T °F	v ft³/lb	u Btu/lb	h Btu/lb	s Btu/lb · °R	v ft³/lb	u Btu/lb	h Btu/lb	s Btu/lb · °R
	p = 60 lbf/in.² (T_{sat} = 49.89°F)				p = 70 lbf/in.² (T_{sat} = 58.35°F)			
Sat.	0.7887	99.96	108.72	0.2183	0.6778	101.05	109.83	0.2179
60	0.8135	102.03	111.06	0.2229	0.6814	101.40	110.23	0.2186
80	0.8604	106.11	115.66	0.2316	0.7239	105.58	114.96	0.2276
100	0.9051	110.21	120.26	0.2399	0.7640	109.76	119.66	0.2361
120	0.9482	114.35	124.88	0.2480	0.8023	113.96	124.36	0.2444
140	0.9900	118.54	129.53	0.2559	0.8393	118.20	129.07	0.2524
160	1.0308	122.79	134.23	0.2636	0.8752	122.49	133.82	0.2601
180	1.0707	127.10	138.98	0.2712	0.9103	126.83	138.62	0.2678
200	1.1100	131.47	143.79	0.2786	0.9446	131.23	143.46	0.2752
220	1.1488	135.91	148.66	0.2859	0.9784	135.69	148.36	0.2825
240	1.1871	140.42	153.60	0.2930	1.0118	140.22	153.33	0.2897
260	1.2251	145.00	158.60	0.3001	1.0448	144.82	158.35	0.2968
280	1.2627	149.65	163.67	0.3070	1.0774	149.48	163.44	0.3038
300	1.3001	154.38	168.81	0.3139	1.1098	154.22	168.60	0.3107
	p = 80 lbf/in.² (T_{sat} = 65.93°F)				p = 90 lbf/in.² (T_{sat} = 72.83°F)			
Sat.	0.5938	102.02	110.81	0.2175	0.5278	102.89	111.68	0.2172
80	0.6211	105.03	114.23	0.2239	0.5408	104.46	113.47	0.2205
100	0.6579	109.30	119.04	0.2327	0.5751	108.82	118.39	0.2295
120	0.6927	113.56	123.82	0.2411	0.6073	113.15	123.27	0.2380
140	0.7261	117.85	128.60	0.2492	0.6380	117.50	128.12	0.2463
160	0.7584	122.18	133.41	0.2570	0.6675	121.87	132.98	0.2542
180	0.7898	126.55	138.25	0.2647	0.6961	126.28	137.87	0.2620
200	0.8205	130.98	143.13	0.2722	0.7239	130.73	142.79	0.2696
220	0.8506	135.47	148.06	0.2796	0.7512	135.25	147.76	0.2770
240	0.8803	140.02	153.05	0.2868	0.7779	139.82	152.77	0.2843
260	0.9095	144.63	158.10	0.2940	0.8043	144.45	157.84	0.2914
280	0.9384	149.32	163.21	0.3010	0.8303	149.15	162.97	0.2984
300	0.9671	154.06	168.38	0.3079	0.8561	153.91	168.16	0.3054
320	0.9955	158.88	173.62	0.3147	0.8816	158.73	173.42	0.3122
	p = 100 lbf/in.² (T_{sat} = 79.17°F)				p = 120 lbf/in.² (T_{sat} = 90.54°F)			
Sat.	0.4747	103.68	112.46	0.2169	0.3941	105.06	113.82	0.2165
80	0.4761	103.87	112.68	0.2173				
100	0.5086	108.32	117.73	0.2265	0.4080	107.26	116.32	0.2210
120	0.5388	112.73	122.70	0.2352	0.4355	111.84	121.52	0.2301
140	0.5674	117.13	127.63	0.2436	0.4610	116.37	126.61	0.2387
160	0.5947	121.55	132.55	0.2517	0.4852	120.89	131.66	0.2470
180	0.6210	125.99	137.49	0.2595	0.5082	125.42	136.70	0.2550
200	0.6466	130.48	142.45	0.2671	0.5305	129.97	141.75	0.2628
220	0.6716	135.02	147.45	0.2746	0.5520	134.56	146.82	0.2704
240	0.6960	139.61	152.49	0.2819	0.5731	139.20	151.92	0.2778
260	0.7201	144.26	157.59	0.2891	0.5937	143.89	157.07	0.2850
280	0.7438	148.98	162.74	0.2962	0.6140	148.63	162.26	0.2921
300	0.7672	153.75	167.95	0.3031	0.6339	153.43	167.51	0.2991
320	0.7904	158.59	173.21	0.3099	0.6537	158.29	172.81	0.3060

Table A-12E (Continued)

T °F	v ft³/lb	u Btu/lb	h Btu/lb	s Btu/lb · °R	v ft³/lb	u Btu/lb	h Btu/lb	s Btu/lb · °R
	p = 140 lbf/in.² (T_{sat} = 100.56°F)				p = 160 lbf/in.² (T_{sat} = 109.55°F)			
Sat.	0.3358	106.25	114.95	0.2161	0.2916	107.28	115.91	0.2157
120	0.3610	110.90	120.25	0.2254	0.3044	109.88	118.89	0.2209
140	0.3846	115.58	125.54	0.2344	0.3269	114.73	124.41	0.2303
160	0.4066	120.21	130.74	0.2429	0.3474	119.49	129.78	0.2391
180	0.4274	124.82	135.89	0.2511	0.3666	124.20	135.06	0.2475
200	0.4474	129.44	141.03	0.2590	0.3849	128.90	140.29	0.2555
220	0.4666	134.09	146.18	0.2667	0.4023	133.61	145.52	0.2633
240	0.4852	138.77	151.34	0.2742	0.4192	138.34	150.75	0.2709
260	0.5034	143.50	156.54	0.2815	0.4356	143.11	156.00	0.2783
280	0.5212	148.28	161.78	0.2887	0.4516	147.92	161.29	0.2856
300	0.5387	153.11	167.06	0.2957	0.4672	152.78	166.61	0.2927
320	0.5559	157.99	172.39	0.3026	0.4826	157.69	171.98	0.2996
340	0.5730	162.93	177.78	0.3094	0.4978	162.65	177.39	0.3065
360	0.5898	167.93	183.21	0.3162	0.5128	167.67	182.85	0.3132
	p = 180 lbf/in.² (T_{sat} = 117.74°F)				p = 200 lbf/in.² (T_{sat} = 125.28°F)			
Sat.	0.2569	108.18	116.74	0.2154	0.2288	108.98	117.44	0.2151
120	0.2595	108.77	117.41	0.2166				
140	0.2814	113.83	123.21	0.2264	0.2446	112.87	121.92	0.2226
160	0.3011	118.74	128.77	0.2355	0.2636	117.94	127.70	0.2321
180	0.3191	123.56	134.19	0.2441	0.2809	122.88	133.28	0.2410
200	0.3361	128.34	139.53	0.2524	0.2970	127.76	138.75	0.2494
220	0.3523	133.11	144.84	0.2603	0.3121	132.60	144.15	0.2575
240	0.3678	137.90	150.15	0.2680	0.3266	137.44	149.53	0.2653
260	0.3828	142.71	155.46	0.2755	0.3405	142.30	154.90	0.2728
280	0.3974	147.55	160.79	0.2828	0.3540	147.18	160.28	0.2802
300	0.4116	152.44	166.15	0.2899	0.3671	152.10	165.69	0.2874
320	0.4256	157.38	171.55	0.2969	0.3799	157.07	171.13	0.2945
340	0.4393	162.36	177.00	0.3038	0.3926	162.07	176.60	0.3014
360	0.4529	167.40	182.49	0.3106	0.4050	167.13	182.12	0.3082
	p = 300 lbf/in.² (T_{sat} = 156.17°F)				p = 400 lbf/in.² (T_{sat} = 179.95°F)			
Sat.	0.1424	111.72	119.62	0.2132	0.0965	112.77	119.91	0.2102
160	0.1462	112.95	121.07	0.2155				
180	0.1633	118.93	128.00	0.2265	0.0965	112.79	119.93	0.2102
200	0.1777	124.47	134.34	0.2363	0.1143	120.14	128.60	0.2235
220	0.1905	129.79	140.36	0.2453	0.1275	126.35	135.79	0.2343
240	0.2021	134.99	146.21	0.2537	0.1386	132.12	142.38	0.2438
260	0.2130	140.12	151.95	0.2618	0.1484	137.65	148.64	0.2527
280	0.2234	145.23	157.63	0.2696	0.1575	143.06	154.72	0.2610
300	0.2333	150.33	163.28	0.2772	0.1660	148.39	160.67	0.2689
320	0.2428	155.44	168.92	0.2845	0.1740	153.69	166.57	0.2766
340	0.2521	160.57	174.56	0.2916	0.1816	158.97	172.42	0.2840
360	0.2611	165.74	180.23	0.2986	0.1890	164.26	178.26	0.2912
380	0.2699	170.94	185.92	0.3055	0.1962	169.57	184.09	0.2983
400	0.2786	176.18	191.64	0.3122	0.2032	174.90	189.94	0.3051

Table A-13E **Properties of Saturated Ammonia (Liquid–Vapor): Temperature Table**

		Specific Volume ft³/lb		Internal Energy Btu/lb		Enthalpy Btu/lb			Entropy Btu/lb · °R		
Temp. °F	Press. lbf/in.²	Sat. Liquid v_f	Sat. Vapor v_g	Sat. Liquid u_f	Sat. Vapor u_g	Sat. Liquid h_f	Evap. h_{fg}	Sat. Vapor h_g	Sat. Liquid s_f	Sat. Vapor s_g	Temp. °F
−60	5.548	0.02278	44.7537	−21.005	543.61	−20.97	610.56	589.58	−0.0512	1.4765	−60
−55	6.536	0.02288	38.3991	−15.765	545.11	−15.73	607.31	591.58	−0.0381	1.4627	−55
−50	7.664	0.02299	33.0880	−10.525	546.59	−10.49	604.04	593.54	−0.0253	1.4492	−50
−45	8.949	0.02310	28.6284	−5.295	548.04	−5.25	600.72	595.48	−0.0126	1.4361	−45
−40	10.405	0.02322	24.8672	−0.045	549.46	0.00	597.37	597.37	0.0000	1.4235	−40
−35	12.049	0.02333	21.6812	5.20	550.86	5.26	593.98	599.24	0.0124	1.4111	−35
−30	13.899	0.02345	18.9715	10.46	552.24	10.52	590.54	601.06	0.0247	1.3992	−30
−25	15.972	0.02357	16.6577	15.73	553.59	15.80	587.05	602.85	0.0369	1.3875	−25
−20	18.290	0.02369	14.6744	21.01	554.91	21.09	583.51	604.61	0.0490	1.3762	−20
−15	20.871	0.02381	12.9682	26.31	556.20	26.40	579.92	606.32	0.0610	1.3652	−15
−10	23.738	0.02393	11.4951	31.63	557.46	31.73	576.26	607.99	0.0729	1.3544	−10
−5	26.912	0.02406	10.2190	36.96	558.70	37.08	572.54	609.62	0.0847	1.3440	−5
0	30.416	0.02419	9.1100	42.32	559.91	42.45	568.76	611.22	0.0964	1.3338	0
5	34.275	0.02432	8.1430	47.69	561.08	47.85	564.92	612.76	0.1080	1.3238	5
10	38.512	0.02446	7.2974	53.09	562.23	53.27	561.00	614.27	0.1196	1.3141	10
15	43.153	0.02460	6.5556	58.52	563.34	58.72	557.01	615.73	0.1311	1.3046	15
20	48.224	0.02474	5.9032	63.97	564.43	64.19	552.95	617.14	0.1425	1.2953	20
25	53.752	0.02488	5.3278	69.43	565.48	69.68	548.82	618.51	0.1539	1.2862	25
30	59.765	0.02503	4.8188	74.93	566.49	75.20	544.62	619.82	0.1651	1.2774	30
35	66.291	0.02517	4.3675	80.44	567.48	80.75	540.34	621.09	0.1764	1.2687	35
40	73.359	0.02533	3.9664	85.98	568.42	86.33	535.97	622.30	0.1875	1.2602	40
45	81.000	0.02548	3.6090	91.55	569.33	91.93	531.54	623.46	0.1986	1.2518	45
50	89.242	0.02564	3.2897	97.13	570.21	97.55	527.02	624.57	0.2096	1.2436	50
55	98.118	0.02581	3.0040	102.73	571.04	103.20	522.42	625.62	0.2205	1.2356	55
60	107.66	0.02597	2.7476	108.35	571.83	108.87	517.74	626.61	0.2314	1.2277	60
65	117.90	0.02614	2.5171	113.99	572.59	114.56	512.97	627.54	0.2422	1.2199	65
70	128.87	0.02632	2.3095	119.65	573.29	120.28	508.12	628.40	0.2530	1.2123	70
75	140.60	0.02650	2.1220	125.33	573.95	126.02	503.18	629.20	0.2636	1.2048	75
80	153.13	0.02668	1.9524	131.02	574.57	131.78	498.15	629.93	0.2742	1.1973	80
85	166.50	0.02687	1.7988	136.73	575.13	137.56	493.03	630.59	0.2848	1.1900	85
90	180.73	0.02707	1.6593	142.46	575.65	143.37	487.81	631.18	0.2953	1.1827	90
95	195.87	0.02727	1.5324	148.21	576.10	149.20	482.49	631.68	0.3057	1.1756	95
100	211.96	0.02747	1.4168	153.98	576.51	155.05	477.06	632.11	0.3161	1.1685	100
105	229.02	0.02768	1.3113	159.76	576.85	160.94	471.52	632.46	0.3264	1.1614	105
110	247.10	0.02790	1.2149	165.58	577.13	166.85	465.86	632.71	0.3366	1.1544	110
115	266.24	0.02813	1.1266	171.41	577.34	172.80	460.08	632.88	0.3469	1.1475	115
120	286.47	0.02836	1.0456	177.28	577.48	178.79	454.16	632.95	0.3570	1.1405	120

Source: Tables A-13E through A-15E are calculated based on equations from L. Haar and J. S. Gallagher, "Thermodynamic Properties of Ammonia," *J. Phys. Chem. Reference Data*, Vol. 7, 1978, pp. 635–792.

Table A-14E **Properties of Saturated Ammonia (Liquid–Vapor): Pressure Table**

		Specific Volume ft³/lb		Internal Energy Btu/lb		Enthalpy Btu/lb			Entropy Btu/lb · °R		
Press. lbf/in.²	Temp. °F	Sat. Liquid v_f	Sat. Vapor v_g	Sat. Liquid u_f	Sat. Vapor u_g	Sat. Liquid h_f	Evap. h_{fg}	Sat. Vapor h_g	Sat. Liquid s_f	Sat. Vapor s_g	Press. lbf/in.²
5	−63.10	0.02271	49.320	−24.24	542.67	−24.22	612.56	588.33	−0.0593	1.4853	5
6	−57.63	0.02283	41.594	−18.51	544.32	−18.49	609.02	590.54	−0.0450	1.4699	6
7	−52.86	0.02293	36.014	−13.52	545.74	−13.49	605.92	592.42	−0.0326	1.4569	7
8	−48.63	0.02302	31.790	−9.09	546.98	−9.06	603.13	594.08	−0.0218	1.4456	8
9	−44.81	0.02311	28.477	−5.09	548.09	−5.05	600.60	595.55	−0.0121	1.4357	9
10	−41.33	0.02319	25.807	−1.44	549.09	−1.40	598.27	596.87	−0.0033	1.4268	10
12	−35.14	0.02333	21.764	5.06	550.82	5.11	594.08	599.18	0.0121	1.4115	12
14	−29.74	0.02345	18.843	10.73	552.31	10.79	590.36	601.16	0.0254	1.3986	14
16	−24.94	0.02357	16.631	15.80	553.60	15.87	587.01	602.88	0.0371	1.3874	16
18	−20.60	0.02367	14.896	20.38	554.75	20.46	583.94	604.40	0.0476	1.3775	18
20	−16.63	0.02377	13.497	24.58	555.78	24.67	581.10	605.76	0.0571	1.3687	20
25	−7.95	0.02399	10.950	33.81	557.97	33.92	574.75	608.67	0.0777	1.3501	25
30	−0.57	0.02418	9.229	41.71	559.77	41.84	569.20	611.04	0.0951	1.3349	30
35	5.89	0.02435	7.984	48.65	561.29	48.81	564.22	613.03	0.1101	1.3221	35
40	11.65	0.02450	7.041	54.89	562.60	55.07	559.69	614.76	0.1234	1.3109	40
45	16.87	0.02465	6.302	60.56	563.75	60.76	555.50	616.26	0.1354	1.3011	45
50	21.65	0.02478	5.705	65.77	564.78	66.00	551.59	617.60	0.1463	1.2923	50
55	26.07	0.02491	5.213	70.61	565.70	70.86	547.93	618.79	0.1563	1.2843	55
60	30.19	0.02503	4.801	75.13	566.53	75.41	544.46	619.87	0.1656	1.2770	60
65	34.04	0.02515	4.450	79.39	567.29	79.69	541.16	620.85	0.1742	1.2703	65
70	37.67	0.02526	4.1473	83.40	567.99	83.73	538.01	621.74	0.1823	1.2641	70
75	41.11	0.02536	3.8837	87.21	568.63	87.57	535.00	622.56	0.1900	1.2583	75
80	44.37	0.02546	3.6520	90.84	569.22	91.22	532.10	623.32	0.1972	1.2529	80
85	47.47	0.02556	3.4466	94.30	569.77	94.71	529.31	624.02	0.2040	1.2478	85
90	50.44	0.02566	3.2632	97.62	570.28	98.05	526.62	624.66	0.2106	1.2429	90
100	56.01	0.02584	2.9497	103.87	571.21	104.35	521.48	625.82	0.2227	1.2340	100
110	61.17	0.02601	2.6913	109.68	572.01	110.20	516.63	626.83	0.2340	1.2259	110
120	65.98	0.02618	2.4745	115.11	572.73	115.69	512.02	627.71	0.2443	1.2184	120
130	70.50	0.02634	2.2899	120.21	573.36	120.85	507.64	628.48	0.2540	1.2115	130
140	74.75	0.02649	2.1309	125.04	573.92	125.73	503.43	629.16	0.2631	1.2051	140
150	78.78	0.02664	1.9923	129.63	574.42	130.37	499.39	629.76	0.2717	1.1991	150
175	88.02	0.02699	1.7128	140.19	575.45	141.07	489.89	630.95	0.2911	1.1856	175
200	96.31	0.02732	1.5010	149.72	576.21	150.73	481.07	631.80	0.3084	1.1737	200
225	103.85	0.02764	1.3348	158.43	576.77	159.58	472.80	632.38	0.3240	1.1630	225
250	110.78	0.02794	1.2007	166.48	577.16	167.77	464.97	632.74	0.3382	1.1533	250
275	117.20	0.02823	1.0901	173.99	577.41	175.43	457.49	632.92	0.3513	1.1444	275
300	123.20	0.02851	0.9974	181.05	577.54	182.63	450.31	632.94	0.3635	1.1361	300

Table A-15E **Properties of Superheated Ammonia Vapor**

T °F	v ft³/lb	u Btu/lb	h Btu/lb	s Btu/lb · °R	v ft³/lb	u Btu/lb	h Btu/lb	s Btu/lb · °R
	p = 6 lbf/in.² (T_{sat} = −57.63°F)				p = 8 lbf/in.² (T_{sat} = −48.63°F)			
Sat.	41.594	544.32	590.54	1.4699	31.790	546.98	594.08	1.4456
−50	42.435	547.22	594.37	1.4793				
−40	43.533	551.03	599.40	1.4915	32.511	550.32	598.49	1.4562
−30	44.627	554.84	604.42	1.5033	33.342	554.19	603.58	1.4682
−20	45.715	558.66	609.45	1.5149	34.169	558.06	608.68	1.4799
−10	46.800	562.47	614.47	1.5261	34.992	561.93	613.76	1.4914
0	47.882	566.29	619.49	1.5372	35.811	565.79	618.84	1.5025
10	48.960	570.12	624.51	1.5480	36.627	569.66	623.91	1.5135
20	50.035	573.95	629.54	1.5586	37.440	573.52	628.99	1.5241
30	51.108	577.78	634.57	1.5690	38.250	577.40	634.06	1.5346
40	52.179	581.63	639.60	1.5791	39.058	581.27	639.13	1.5449
50	53.247	585.49	644.64	1.5891	39.865	585.16	644.21	1.5549
60	54.314	589.35	649.70	1.5990	40.669	589.05	649.29	1.5648
	p = 10 lbf/in.² (T_{sat} = −41.33°F)				p = 12 lbf/in.² (T_{sat} = −35.14°F)			
Sat.	25.807	549.09	596.87	1.4268	21.764	550.82	599.18	1.4115
−40	25.897	549.61	597.56	1.4284				
−30	26.571	553.54	602.74	1.4406	22.056	552.87	601.88	1.4178
−20	27.241	557.46	607.90	1.4525	22.621	556.85	607.12	1.4298
−10	27.906	561.37	613.05	1.4641	23.182	560.82	612.33	1.4416
0	28.568	565.29	618.19	1.4754	23.739	564.78	617.53	1.4530
10	29.227	569.19	623.31	1.4864	24.293	568.73	622.71	1.4642
20	29.882	573.10	628.43	1.4972	24.843	572.67	627.88	1.4750
30	30.535	577.01	633.55	1.5078	25.392	576.61	633.03	1.4857
40	31.186	580.91	638.66	1.5181	25.937	580.55	638.19	1.4961
50	31.835	584.82	643.77	1.5282	26.481	584.49	643.33	1.5063
60	32.482	588.74	648.89	1.5382	27.023	588.43	648.48	1.5163
70	33.127	592.66	654.01	1.5479	27.564	592.38	653.63	1.5261
	p = 14 lbf/in.² (T_{sat} = −29.74°F)				p = 16 lbf/in.² (T_{sat} = −24.94°F)			
Sat.	18.843	552.31	601.16	1.3986	16.631	553.60	602.88	1.3874
−20	19.321	556.24	606.33	1.4105	16.845	555.62	605.53	1.3935
−10	19.807	560.26	611.61	1.4223	17.275	559.69	610.88	1.4055
0	20.289	564.27	616.86	1.4339	17.701	563.75	616.19	1.4172
10	20.768	568.26	622.10	1.4452	18.124	567.79	621.48	1.4286
20	21.244	572.24	627.31	1.4562	18.544	571.81	626.75	1.4397
30	21.717	576.22	632.52	1.4669	18.961	575.82	632.00	1.4505
40	22.188	580.19	637.71	1.4774	19.376	579.82	637.23	1.4611
50	22.657	584.16	642.89	1.4877	19.789	583.82	642.45	1.4714
60	23.124	588.12	648.07	1.4977	20.200	587.81	647.66	1.4815
70	23.590	592.09	653.25	1.5076	20.609	591.80	652.86	1.4915
80	24.054	596.07	658.42	1.5173	21.017	595.80	658.07	1.5012
90	24.517	600.04	663.60	1.5268	21.424	599.80	663.27	1.5107

Table A-15E **(Continued)**

T °F	v ft³/lb	u Btu/lb	h Btu/lb	s Btu/lb · °R	v ft³/lb	u Btu/lb	h Btu/lb	s Btu/lb · °R
	$p = 18$ lbf/in.² ($T_{sat} = -20.60$°F)				$p = 20$ lbf/in.² ($T_{sat} = -16.63$°F)			
Sat.	14.896	554.75	604.40	1.3775	13.497	555.78	605.76	1.3687
−20	14.919	555.00	604.72	1.3783				
−10	15.306	559.13	610.14	1.3905	13.730	558.55	609.40	1.3769
0	15.688	563.23	615.52	1.4023	14.078	562.70	614.84	1.3888
10	16.068	567.31	620.87	1.4138	14.422	566.83	620.24	1.4005
20	16.444	571.37	626.18	1.4250	14.764	570.94	625.61	1.4118
30	16.818	575.42	631.47	1.4359	15.103	575.02	630.95	1.4228
40	17.189	579.46	636.75	1.4466	15.439	579.09	636.26	1.4335
50	17.558	583.48	642.00	1.4570	15.773	583.14	641.55	1.4440
60	17.925	587.50	647.25	1.4672	16.105	587.19	646.83	1.4543
70	18.291	591.52	652.48	1.4772	16.436	591.23	652.10	1.4643
80	18.655	595.53	657.71	1.4869	16.765	595.26	657.35	1.4741
90	19.018	599.55	662.94	1.4965	17.094	599.30	662.60	1.4838
	$p = 30$ lbf/in.² ($T_{sat} = -0.57$°F)				$p = 40$ lbf/in.² ($T_{sat} = 11.65$°F)			
Sat.	9.2286	559.77	611.04	1.3349	7.0414	562.60	614.76	1.3109
0	9.2425	560.02	611.36	1.3356				
10	9.4834	564.38	617.07	1.3479				
20	9.7209	568.70	622.70	1.3598	7.1965	566.39	619.69	1.3213
30	9.9554	572.97	628.28	1.3713	7.3795	570.86	625.52	1.3333
40	10.187	577.21	633.80	1.3824	7.5597	575.28	631.28	1.3450
50	10.417	581.42	639.28	1.3933	7.7376	579.65	636.96	1.3562
60	10.645	585.60	644.73	1.4039	7.9134	583.97	642.58	1.3672
70	10.871	589.76	650.15	1.4142	8.0874	588.26	648.16	1.3778
80	11.096	593.90	655.54	1.4243	8.2598	592.52	653.69	1.3881
90	11.319	598.04	660.91	1.4342	8.4308	596.75	659.20	1.3982
100	11.541	602.16	666.27	1.4438	8.6006	600.97	664.67	1.4081
110	11.762	606.28	671.62	1.4533	8.7694	605.17	670.12	1.4178
	$p = 50$ lbf/in.² ($T_{sat} = 21.65$°F)				$p = 60$ lbf/in.² ($T_{sat} = 30.19$°F)			
Sat.	5.7049	564.78	617.60	1.2923	4.8009	566.53	619.87	1.2770
40	5.9815	573.30	628.68	1.3149	4.9278	571.25	626.00	1.2894
60	6.2733	582.31	640.39	1.3379	5.1788	580.60	638.14	1.3133
80	6.5574	591.10	651.82	1.3595	5.4218	589.66	649.90	1.3355
100	6.8358	599.75	663.04	1.3799	5.6587	598.52	661.39	1.3564
120	7.1097	608.30	674.13	1.3993	5.8910	607.23	672.68	1.3762
140	7.3802	616.80	685.13	1.4180	6.1198	615.86	683.85	1.3951
160	7.6480	625.28	696.09	1.4360	6.3458	624.44	694.95	1.4133
200	8.1776	642.27	717.99	1.4702	6.7916	641.59	717.05	1.4479
240	8.7016	659.44	740.00	1.5026	7.2318	658.87	739.21	1.4805
280	9.2218	676.88	762.26	1.5336	7.6679	676.38	761.58	1.5116
320	9.7391	694.65	784.82	1.5633	8.1013	694.21	784.22	1.5414
360	10.254	712.79	807.73	1.5919	8.5325	712.40	807.20	1.5702

Table A-15E (Continued)

T °F	v ft³/lb	u Btu/lb	h Btu/lb	s Btu/lb · °R	v ft³/lb	u Btu/lb	h Btu/lb	s Btu/lb · °R
	p = 70 lbf/in.² (T_{sat} = 37.67°F)				p = 80 lbf/in.² (T_{sat} = 44.37°F)			
Sat.	4.1473	567.99	621.74	1.2641	3.6520	569.22	623.32	1.2529
40	4.1739	569.15	623.25	1.2671				
60	4.3962	578.85	635.84	1.2918	3.8084	577.06	633.48	1.2727
80	4.6100	588.19	647.95	1.3147	4.0006	586.69	645.95	1.2963
100	4.8175	597.26	659.70	1.3361	4.1862	595.98	657.99	1.3182
120	5.0202	606.14	671.22	1.3563	4.3668	605.04	669.73	1.3388
140	5.2193	614.91	682.56	1.3756	4.5436	613.94	681.25	1.3583
160	5.4154	623.60	693.79	1.3940	4.7175	622.74	692.63	1.3770
200	5.8015	640.91	716.11	1.4289	5.0589	640.22	715.16	1.4122
240	6.1818	658.29	738.42	1.4617	5.3942	657.71	737.62	1.4453
280	6.5580	675.89	760.89	1.4929	5.7256	675.39	760.20	1.4767
320	6.9314	693.78	783.62	1.5229	6.0540	693.34	783.02	1.5067
360	7.3026	712.02	806.67	1.5517	6.3802	711.63	806.15	1.5357
400	7.6721	730.63	830.08	1.5796	6.7047	730.29	829.61	1.5636
	p = 90 lbf/in.² (T_{sat} = 50.44°F)				p = 100 lbf/in.² (T_{sat} = 56.01°F)			
Sat.	3.2632	570.28	624.66	1.2429	2.9497	571.21	625.82	1.2340
60	3.3504	575.22	631.05	1.2553	2.9832	573.32	628.56	1.2393
80	3.5261	585.15	643.91	1.2796	3.1460	583.58	641.83	1.2644
100	3.6948	594.68	656.26	1.3021	3.3014	593.35	654.49	1.2874
120	3.8584	603.92	668.22	1.3231	3.4513	602.79	666.70	1.3088
140	4.0180	612.97	679.93	1.3430	3.5972	611.98	678.59	1.3290
160	4.1746	621.88	691.45	1.3619	3.7401	621.01	690.27	1.3481
200	4.4812	639.52	714.20	1.3974	4.0189	638.82	713.24	1.3841
240	4.7817	657.13	736.82	1.4307	4.2916	656.54	736.01	1.4176
280	5.0781	674.89	759.52	1.4623	4.5600	674.39	758.82	1.4493
320	5.3715	692.90	782.42	1.4924	4.8255	692.47	781.82	1.4796
360	5.6628	711.24	805.62	1.5214	5.0888	710.86	805.09	1.5087
400	5.9522	729.95	829.14	1.5495	5.3503	729.60	828.68	1.5368
	p = 110 lbf/in.² (T_{sat} = 61.17°F)				p = 120 lbf/in.² (T_{sat} = 65.98°F)			
Sat.	2.6913	572.01	626.83	1.2259	2.4745	572.73	627.71	1.2184
80	2.8344	581.97	639.71	1.2502	2.5744	580.33	637.53	1.2369
100	2.9791	592.00	652.69	1.2738	2.7102	590.63	650.85	1.2611
120	3.1181	601.63	665.14	1.2957	2.8401	600.46	663.57	1.2834
140	3.2528	610.98	677.24	1.3162	2.9657	609.97	675.86	1.3043
160	3.3844	620.13	689.07	1.3356	3.0879	619.24	687.86	1.3240
200	3.6406	638.11	712.27	1.3719	3.3254	637.40	711.29	1.3606
240	3.8905	655.96	735.20	1.4056	3.5563	655.36	734.39	1.3946
280	4.1362	673.88	758.13	1.4375	3.7829	673.37	757.43	1.4266
320	4.3788	692.02	781.22	1.4679	4.0065	691.58	780.61	1.4572
360	4.6192	710.47	804.56	1.4971	4.2278	710.08	804.02	1.4864
400	4.8578	729.26	828.21	1.5252	4.4473	728.92	827.74	1.5147

Table A-15E (Continued)

T °F	v ft³/lb	u Btu/lb	h Btu/lb	s Btu/lb · °R	v ft³/lb	u Btu/lb	h Btu/lb	s Btu/lb · °R
	p = 130 lbf/in.² (T_{sat} = 70.50°F)				p = 140 lbf/in.² (T_{sat} = 74.75°F)			
Sat.	2.2899	573.36	628.48	1.2115	2.1309	573.92	629.16	1.2051
80	2.3539	578.64	635.30	1.2243	2.1633	576.80	632.89	1.2119
100	2.4824	589.23	648.98	1.2492	2.2868	587.79	647.08	1.2379
120	2.6048	599.27	661.97	1.2720	2.4004	597.85	660.08	1.2604
140	2.7226	608.94	674.48	1.2932	2.5140	607.90	673.07	1.2828
160	2.8370	618.34	686.64	1.3132	2.6204	617.34	685.27	1.3025
180	2.9488	627.57	698.55	1.3321	2.7268	626.77	697.46	1.3222
200	3.0585	636.69	710.31	1.3502	2.8289	635.93	709.27	1.3401
240	3.2734	654.77	733.57	1.3844	3.0304	654.17	732.73	1.3747
280	3.4840	672.87	756.73	1.4166	3.2274	672.38	756.04	1.4071
320	3.6915	691.14	780.00	1.4472	3.4212	690.73	779.42	1.4379
360	3.8966	709.69	803.49	1.4766	3.6126	709.34	802.99	1.4674
400	4.1000	728.57	827.27	1.5049	3.8022	728.27	826.84	1.4958
	p = 150 lbf/in.² (T_{sat} = 78.78°F)				p = 200 lbf/in.² (T_{sat} = 96.31°F)			
Sat.	1.9923	574.42	629.76	1.1991	1.5010	576.21	631.80	1.1737
100	2.1170	586.33	645.13	1.2271	1.5190	578.52	634.77	1.1790
140	2.3332	606.84	671.65	1.2729	1.6984	601.34	664.24	1.2299
180	2.5343	625.95	696.35	1.3128	1.8599	621.77	690.65	1.2726
220	2.7268	644.43	720.17	1.3489	2.0114	641.07	715.57	1.3104
260	2.9137	662.70	743.63	1.3825	2.1569	659.90	739.78	1.3450
300	3.0968	681.02	767.04	1.4141	2.2984	678.62	763.74	1.3774
340	3.2773	699.54	790.57	1.4443	2.4371	697.44	787.70	1.4081
380	3.4558	718.35	814.34	1.4733	2.5736	716.50	811.81	1.4375
420	3.6325	737.50	838.39	1.5013	2.7085	735.86	836.17	1.4659
460	3.8079	757.01	862.78	1.5284	2.8420	755.57	860.82	1.4933
500	3.9821	776.91	887.51	1.5548	2.9742	775.65	885.80	1.5199
540	4.1553	797.19	912.60	1.5804	3.1054	796.10	911.11	1.5457
580	4.3275	817.85	938.05	1.6053	3.2357	816.94	936.77	1.5709
	p = 250 lbf/in.² (T_{sat} = 110.78°F)				p = 300 lbf/in.² (T_{sat} = 123.20°F)			
Sat.	1.2007	577.16	632.74	1.1533	0.9974	577.54	632.94	1.1361
140	1.3150	595.40	656.28	1.1936	1.0568	588.94	647.65	1.1610
180	1.4539	617.38	684.69	1.2395	1.1822	612.75	678.42	1.2107
220	1.5816	637.61	710.82	1.2791	1.2944	634.01	705.91	1.2524
260	1.7025	657.03	735.85	1.3149	1.3992	654.09	731.82	1.2895
300	1.8191	676.17	760.39	1.3481	1.4994	673.69	756.98	1.3235
340	1.9328	695.32	784.79	1.3794	1.5965	693.16	781.85	1.3554
380	2.0443	714.63	809.27	1.4093	1.6913	712.74	806.70	1.3857
420	2.1540	734.22	833.93	1.4380	1.7843	732.55	831.67	1.4148
460	2.2624	754.12	858.85	1.4657	1.8759	752.66	856.87	1.4428
500	2.3695	774.38	884.07	1.4925	1.9663	773.10	882.33	1.4699
540	2.4755	795.01	909.61	1.5186	2.0556	793.90	908.09	1.4962
580	2.5807	816.01	935.47	1.5440	2.1440	815.07	934.17	1.5218

Table A-16E **Properties of Saturated Propane (Liquid–Vapor): Temperature Table**

		Specific Volume ft³/lb		Internal Energy Btu/lb		Enthalpy Btu/lb			Entropy Btu/lb · °R		
Temp. °F	Press. lbf/in.²	Sat. Liquid v_f	Sat. Vapor v_g	Sat. Liquid u_f	Sat. Vapor u_g	Sat. Liquid h_f	Evap. h_{fg}	Sat. Vapor h_g	Sat. Liquid s_f	Sat. Vapor s_g	Temp. °F
−140	0.6053	0.02505	128.00	−51.33	139.22	−51.33	204.9	153.6	−0.139	0.501	−140
−120	1.394	0.02551	58.88	−41.44	143.95	−41.43	200.6	159.1	−0.109	0.481	−120
−100	2.888	0.02601	29.93	−31.34	148.80	−31.33	196.1	164.8	−0.080	0.465	−100
−80	5.485	0.02653	16.52	−21.16	153.73	−21.13	191.6	170.5	−0.053	0.452	−80
−60	9.688	0.02708	9.75	−10.73	158.74	−10.68	186.9	176.2	−0.026	0.441	−60
−40	16.1	0.02767	6.08	−0.08	163.80	0.00	181.9	181.9	0.000	0.433	−40
−20	25.4	0.02831	3.98	10.81	168.88	10.94	176.6	187.6	0.025	0.427	−20
0	38.4	0.02901	2.70	21.98	174.01	22.19	171.0	193.2	0.050	0.422	0
10	46.5	0.02939	2.25	27.69	176.61	27.94	168.0	196.0	0.063	0.420	10
20	55.8	0.02978	1.89	33.47	179.15	33.78	164.9	198.7	0.074	0.418	20
30	66.5	0.03020	1.598	39.34	181.71	39.71	161.7	201.4	0.087	0.417	30
40	78.6	0.03063	1.359	45.30	184.30	45.75	158.3	204.1	0.099	0.415	40
50	92.3	0.03110	1.161	51.36	186.74	51.89	154.7	206.6	0.111	0.414	50
60	107.7	0.03160	0.9969	57.53	189.30	58.16	151.0	209.2	0.123	0.413	60
70	124.9	0.03213	0.8593	63.81	191.71	64.55	147.0	211.6	0.135	0.412	70
80	144.0	0.03270	0.7433	70.20	194.16	71.07	142.9	214.0	0.147	0.411	80
90	165.2	0.03332	0.6447	76.72	196.46	77.74	138.4	216.2	0.159	0.410	90
100	188.6	0.03399	0.5605	83.38	198.71	84.56	133.7	218.3	0.171	0.410	100
110	214.3	0.03473	0.4881	90.19	200.91	91.56	128.7	220.3	0.183	0.409	110
120	242.5	0.03555	0.4254	97.16	202.98	98.76	123.3	222.1	0.195	0.408	120
130	273.3	0.03646	0.3707	104.33	204.92	106.17	117.5	223.7	0.207	0.406	130
140	306.9	0.03749	0.3228	111.70	206.64	113.83	111.1	225.0	0.220	0.405	140
150	343.5	0.03867	0.2804	119.33	208.05	121.79	104.1	225.9	0.233	0.403	150
160	383.3	0.04006	0.2426	127.27	209.16	130.11	96.3	226.4	0.246	0.401	160
170	426.5	0.04176	0.2085	135.60	209.81	138.90	87.4	226.3	0.259	0.398	170
180	473.4	0.04392	0.1771	144.50	209.76	148.35	76.9	225.3	0.273	0.394	180
190	524.3	0.04696	0.1470	154.38	208.51	158.94	63.8	222.8	0.289	0.387	190
200	579.7	0.05246	0.1148	166.65	204.16	172.28	44.2	216.5	0.309	0.376	200
206.1	616.1	0.07265	0.07265	186.99	186.99	195.27	0.0	195.27	0.343	0.343	206.1

Table A-17E **Properties of Saturated Propane (Liquid–Vapor): Pressure Table**

		Specific Volume ft³/lb		Internal Energy Btu/lb		Enthalpy Btu/lb			Entropy Btu/lb · °R		
Press. lbf/in.²	Temp. °F	Sat. Liquid v_f	Sat. Vapor v_g	Sat. Liquid u_f	Sat. Vapor u_g	Sat. Liquid h_f	Evap. h_{fg}	Sat. Vapor h_g	Sat. Liquid s_f	Sat. Vapor s_g	Press. lbf/in.²
0.75	−135.1	0.02516	104.8	−48.93	140.36	−48.93	203.8	154.9	−0.132	0.496	0.75
1.5	−118.1	0.02556	54.99	−40.44	144.40	−40.43	200.1	159.7	−0.106	0.479	1.5
3	−98.9	0.02603	28.9	−30.84	149.06	−30.83	196.0	165.1	−0.079	0.464	3
5	−83.0	0.02644	18.00	−22.75	152.96	−22.73	192.4	169.6	−0.057	0.454	5
7.5	−69.3	0.02682	12.36	−15.60	156.40	−15.56	189.1	173.6	−0.038	0.446	7.5
10	−58.8	0.02711	9.468	−10.10	159.04	−10.05	186.6	176.6	−0.024	0.441	10
20	−30.7	0.02796	4.971	4.93	166.18	5.03	179.5	184.6	0.012	0.430	20
30	−12.1	0.02858	3.402	15.15	170.93	15.31	174.5	189.8	0.035	0.425	30
40	2.1	0.02909	2.594	23.19	174.60	23.41	170.4	193.8	0.053	0.422	40
50	13.9	0.02954	2.099	29.96	177.63	30.23	166.8	197.1	0.067	0.419	50
60	24.1	0.02995	1.764	35.86	180.23	36.19	163.6	199.8	0.079	0.418	60
70	33.0	0.03033	1.520	41.14	182.50	41.53	160.6	202.2	0.090	0.416	70
80	41.1	0.03068	1.336	45.95	184.57	46.40	157.9	204.3	0.100	0.415	80
90	48.4	0.03102	1.190	50.38	186.36	50.90	155.3	206.2	0.109	0.414	90
100	55.1	0.03135	1.073	54.52	188.07	55.10	152.8	207.9	0.117	0.414	100
120	67.2	0.03198	0.8945	62.08	191.07	62.79	148.1	210.9	0.131	0.412	120
140	78.0	0.03258	0.7650	68.91	193.68	69.75	143.7	213.5	0.144	0.412	140
160	87.6	0.03317	0.6665	75.17	195.97	76.15	139.5	215.7	0.156	0.411	160
180	96.5	0.03375	0.5890	80.99	197.97	82.12	135.5	217.6	0.166	0.410	180
200	104.6	0.03432	0.5261	86.46	199.77	87.73	131.4	219.2	0.176	0.409	200
220	112.1	0.03489	0.4741	91.64	201.37	93.06	127.6	220.7	0.185	0.408	220
240	119.2	0.03547	0.4303	96.56	202.76	98.14	123.7	221.9	0.194	0.408	240
260	125.8	0.03606	0.3928	101.29	204.07	103.0	120.0	223.0	0.202	0.407	260
280	132.1	0.03666	0.3604	105.83	205.27	107.7	116.1	223.9	0.210	0.406	280
300	138.0	0.03727	0.3319	110.21	206.27	112.3	112.4	224.7	0.217	0.405	300
320	143.7	0.03790	0.3067	114.47	207.17	116.7	108.6	225.3	0.224	0.404	320
340	149.1	0.03855	0.2842	118.60	207.96	121.0	104.7	225.8	0.231	0.403	340
360	154.2	0.03923	0.2639	122.66	208.58	125.3	100.9	226.2	0.238	0.402	360
380	159.2	0.03994	0.2455	126.61	209.07	129.4	97.0	226.4	0.245	0.401	380
400	164.0	0.04069	0.2287	130.51	209.47	133.5	93.0	226.5	0.251	0.400	400
450	175.1	0.04278	0.1921	140.07	209.87	143.6	82.2	225.9	0.266	0.396	450
500	185.3	0.04538	0.1610	149.61	209.27	153.8	70.4	224.2	0.282	0.391	500
600	203.4	0.05659	0.1003	172.85	200.27	179.1	32.2	211.4	0.319	0.367	600
616.1	206.1	0.07265	0.07265	186.99	186.99	195.3	0.0	195.3	0.343	0.343	616.1

Table A-18E Properties of Superheated Propane

T °F	v ft³/lb	u Btu/lb	h Btu/lb	s Btu/lb · °R	v ft³/lb	u Btu/lb	h Btu/lb	s Btu/lb · °R
	p = 0.75 lbf/in.² (T_{sat} = −135.1°F)				p = 1.5 lbf/in.² (T_{sat} = −118.1°F)			
Sat.	104.8	140.4	154.9	0.496	54.99	144.4	159.7	0.479
−130	106.5	141.6	156.4	0.501				
−110	113.1	146.6	162.3	0.518	56.33	146.5	162.1	0.486
−90	119.6	151.8	168.4	0.535	59.63	151.7	168.2	0.503
−70	126.1	157.2	174.7	0.551	62.92	157.1	174.5	0.520
−50	132.7	162.7	181.2	0.568	66.20	162.6	181.0	0.536
−30	139.2	168.6	187.9	0.584	69.47	168.4	187.7	0.552
−10	145.7	174.4	194.7	0.599	72.74	174.4	194.6	0.568
10	152.2	180.7	201.9	0.615	76.01	180.7	201.8	0.583
30	158.7	187.1	209.2	0.630	79.27	187.1	209.1	0.599
50	165.2	193.8	216.8	0.645	82.53	193.8	216.7	0.614
70	171.7	200.7	224.6	0.660	85.79	200.7	224.5	0.629
90	178.2	207.8	232.6	0.675	89.04	207.8	232.5	0.644
	p = 5.0 lbf/in.² (T_{sat} = −83.0°F)				p = 10.0 lbf/in.² (T_{sat} = −58.8°F)			
Sat.	18.00	153.0	169.6	0.454	9.468	159.0	176.6	0.441
−80	18.15	153.8	170.6	0.456				
−60	19.17	159.4	177.1	0.473				
−40	20.17	165.1	183.8	0.489	9.957	80.9	99.3	1.388
−20	21.17	171.1	190.7	0.505	10.47	86.9	106.3	1.405
0	22.17	177.2	197.7	0.521	10.98	93.1	113.4	1.421
20	23.16	183.5	205.0	0.536	11.49	99.5	120.8	1.436
40	24.15	190.1	212.5	0.552	11.99	106.1	128.3	1.452
60	25.14	196.9	220.2	0.567	12.49	113.0	136.1	1.467
80	26.13	204.0	228.2	0.582	12.99	120.0	144.1	1.482
100	27.11	211.3	236.4	0.597	13.49	127.3	152.3	1.497
120	28.09	218.8	244.8	0.611	13.99	134.9	160.7	1.512
140	29.07	226.5	253.4	0.626	14.48	142.6	169.4	1.526
	p = 20.0 lbf/in.² (T_{sat} = −30.7°F)				p = 40.0 lbf/in.² (T_{sat} = 2.1°F)			
Sat.	4.971	166.2	184.6	0.430	2.594	174.6	193.8	0.422
−20	5.117	169.5	188.5	0.439				
0	5.385	175.8	195.8	0.455				
20	5.648	182.4	203.3	0.471	2.723	180.6	200.8	0.436
40	5.909	189.1	211.0	0.487	2.864	187.6	208.8	0.453
60	6.167	195.9	218.8	0.502	3.002	194.6	216.9	0.469
80	6.424	203.1	226.9	0.518	3.137	201.8	225.1	0.484
100	6.678	210.5	235.2	0.533	3.271	209.4	233.6	0.500
120	6.932	218.0	243.7	0.548	3.403	217.0	242.2	0.515
140	7.184	225.8	252.4	0.562	3.534	224.9	251.1	0.530
160	7.435	233.9	261.4	0.577	3.664	232.9	260.1	0.545
180	7.685	242.1	270.6	0.592	3.793	241.3	269.4	0.559
200	7.935	250.6	280.0	0.606	3.921	249.8	278.9	0.574

Table A-18E (Continued)

T °F	v ft³/lb	u Btu/lb	h Btu/lb	s Btu/lb · °R	v ft³/lb	u Btu/lb	h Btu/lb	s Btu/lb · °R
	p = 60.0 lbf/in.² (T_{sat} = 24.1°F)				p = 80.0 lbf/in.² (T_{sat} = 41.1°F)			
Sat.	1.764	180.2	199.8	0.418	1.336	184.6	204.3	0.415
30	1.794	182.4	202.3	0.384				
50	1.894	189.5	210.6	0.400	1.372	187.9	208.2	0.423
70	1.992	196.9	219.0	0.417	1.450	195.4	216.9	0.440
90	2.087	204.4	227.6	0.432	1.526	203.1	225.7	0.456
110	2.179	212.1	236.3	0.448	1.599	210.9	234.6	0.472
130	2.271	220.0	245.2	0.463	1.671	218.8	243.6	0.487
150	2.361	228.0	254.2	0.478	1.741	227.0	252.8	0.503
170	2.450	236.3	263.5	0.493	1.810	235.4	262.2	0.518
190	2.539	244.8	273.0	0.508	1.879	244.0	271.8	0.533
210	2.626	253.5	282.7	0.523	1.946	252.7	281.5	0.548
230	2.713	262.3	292.5	0.537	2.013	261.7	291.5	0.562
250	2.800	271.6	302.7	0.552	2.079	270.9	301.7	0.577
	p = 100 lbf/in.² (T_{sat} = 55.1°F)				p = 120 lbf/in.² (T_{sat} = 67.2°F)			
Sat.	1.073	188.1	207.9	0.414	0.8945	191.1	210.9	0.412
60	1.090	189.9	210.1	0.418				
80	1.156	197.8	219.2	0.435	0.9323	196.2	216.9	0.424
100	1.219	205.7	228.3	0.452	0.9887	204.3	226.3	0.441
120	1.280	213.7	237.4	0.468	1.043	212.5	235.7	0.457
140	1.340	221.9	246.7	0.483	1.094	220.8	245.1	0.473
160	1.398	230.2	256.1	0.499	1.145	229.2	254.7	0.489
180	1.454	238.8	265.7	0.514	1.194	237.9	264.4	0.504
200	1.510	247.5	275.5	0.529	1.242	246.7	274.3	0.520
220	1.566	256.4	285.4	0.544	1.289	255.6	284.3	0.534
240	1.620	265.6	295.6	0.559	1.336	264.8	294.5	0.549
260	1.674	274.9	305.9	0.573	1.382	274.2	304.9	0.564
280	1.728	284.4	316.4	0.588	1.427	283.8	315.5	0.579
	p = 140 lbf/in.² (T_{sat} = 78.0°F)				p = 160 lbf/in.² (T_{sat} = 87.6°F)			
Sat.	0.7650	193.7	213.5	0.412	0.6665	196.0	215.7	0.411
80	0.7705	213.3	214.5	0.413				
100	0.8227	222.9	224.2	0.431	0.6968	201.2	221.9	0.422
120	0.8718	232.4	233.8	0.448	0.7427	209.9	231.9	0.439
140	0.9185	242.1	243.5	0.464	0.7859	218.4	241.7	0.456
160	0.9635	251.7	253.2	0.480	0.8272	227.2	251.7	0.472
180	1.007	261.4	263.0	0.496	0.8669	235.9	261.6	0.488
200	1.050	271.4	273.0	0.511	0.9054	244.9	271.7	0.504
220	1.091	281.5	283.2	0.526	0.9430	254.0	282.0	0.519
240	1.132	291.7	293.5	0.541	0.9797	263.4	292.4	0.534
260	1.173	302.1	303.9	0.556	1.016	272.8	302.9	0.549
280	1.213	312.7	314.6	0.571	1.051	282.6	313.7	0.564
300	1.252	323.6	325.5	0.585	1.087	292.4	324.6	0.578

Table A-18E **(Continued)**

T °F	v ft³/lb	u Btu/lb	h Btu/lb	s Btu/lb · °R	v ft³/lb	u Btu/lb	h Btu/lb	s Btu/lb · °R
	p = 180 lbf/in.² (T_{sat} = 96.5°F)				p = 200 lbf/in.² (T_{sat} = 104.6°F)			
Sat.	0.5890	198.0	217.6	0.410	0.5261	199.8	219.2	0.409
100	0.5972	199.6	219.5	0.413				
120	0.6413	208.4	229.8	0.431	0.5591	206.8	227.5	0.424
140	0.6821	217.1	239.9	0.449	0.5983	215.8	238.0	0.441
160	0.7206	226.1	250.1	0.465	0.6349	224.9	248.4	0.458
180	0.7574	234.9	260.2	0.481	0.6694	233.9	258.7	0.475
200	0.7928	244.0	270.4	0.497	0.7025	243.1	269.1	0.491
220	0.8273	253.2	280.8	0.513	0.7345	252.4	279.6	0.506
240	0.8609	262.6	291.3	0.528	0.7656	261.7	290.1	0.522
260	0.8938	272.1	301.9	0.543	0.7960	271.4	300.9	0.537
280	0.9261	281.8	312.7	0.558	0.8257	281.1	311.7	0.552
300	0.9579	291.8	323.7	0.572	0.8549	291.1	322.8	0.567
320	0.9894	301.9	334.9	0.587	0.8837	301.3	334.0	0.581
	p = 220 lbf/in.² (T_{sat} = 112.1°F)				p = 240 lbf/in.² (T_{sat} = 119.2°F)			
Sat.	0.4741	201.4	220.7	0.408	0.4303	202.8	221.9	0.408
120	0.4906	205.1	225.1	0.416	0.4321	203.2	222.4	0.409
140	0.5290	214.4	236.0	0.435	0.4704	212.9	233.8	0.428
160	0.5642	223.6	246.6	0.452	0.5048	222.4	244.8	0.446
180	0.5971	232.9	257.2	0.469	0.5365	231.6	255.5	0.463
200	0.6284	242.1	267.7	0.485	0.5664	241.1	266.3	0.480
220	0.6585	251.5	278.3	0.501	0.5949	250.5	277.0	0.496
240	0.6875	261.0	289.0	0.516	0.6223	260.1	287.8	0.511
260	0.7158	270.6	299.8	0.532	0.6490	269.8	298.7	0.527
280	0.7435	280.5	310.8	0.547	0.6749	279.8	309.8	0.542
300	0.7706	290.5	321.9	0.561	0.7002	289.8	320.9	0.557
320	0.7972	300.6	333.1	0.576	0.7251	300.1	332.3	0.571
340	0.8235	311.0	344.6	0.591	0.7496	310.5	343.8	0.586
	p = 260 lbf/in.² (T_{sat} = 125.8°F)				p = 280 lbf/in.² (T_{sat} = 132.1°F)			
Sat.	0.3928	204.1	223.0	0.407	0.3604	205.3	223.9	0.406
130	0.4012	206.3	225.6	0.411				
150	0.4374	216.1	237.2	0.431	0.3932	214.5	234.9	0.424
170	0.4697	225.8	248.4	0.449	0.4253	224.4	246.5	0.443
190	0.4995	235.2	259.3	0.466	0.4544	234.1	257.7	0.461
210	0.5275	244.8	270.2	0.482	0.4815	243.8	268.8	0.477
230	0.5541	254.4	281.1	0.498	0.5072	253.5	279.8	0.494
250	0.5798	264.2	292.1	0.514	0.5317	263.3	290.9	0.510
270	0.6046	274.1	303.2	0.530	0.5553	273.3	302.1	0.525
290	0.6288	284.0	314.3	0.545	0.5783	283.4	313.4	0.540
310	0.6524	294.3	325.7	0.560	0.6007	293.5	324.7	0.555
330	0.6756	304.7	337.2	0.574	0.6226	304.0	336.3	0.570
350	0.6984	315.2	348.8	0.589	0.6441	314.6	348.0	0.585

Table A-18E **(Continued)**

T °F	v ft³/lb	u Btu/lb	h Btu/lb	s Btu/lb · °R	v ft³/lb	u Btu/lb	h Btu/lb	s Btu/lb · °R
	p = 320 lbf/in.² (T_{sat} = 143.7°F)				p = 360 lbf/in.² (T_{sat} = 154.2°F)			
Sat.	0.3067	207.2	225.3	0.404	0.2639	208.6	226.2	0.402
150	0.3187	210.7	229.6	0.412				
170	0.3517	221.4	242.3	0.432	0.2920	217.9	237.4	0.420
190	0.3803	231.7	254.2	0.450	0.3213	228.8	250.2	0.440
210	0.4063	241.6	265.7	0.468	0.3469	239.3	262.4	0.459
230	0.4304	251.6	277.1	0.485	0.3702	249.5	274.2	0.476
250	0.4533	261.6	288.5	0.501	0.3919	259.8	285.9	0.493
270	0.4751	271.7	299.9	0.517	0.4124	270.1	297.6	0.509
290	0.4961	281.9	311.3	0.532	0.4320	280.4	309.2	0.525
310	0.5165	292.3	322.9	0.548	0.4510	290.8	320.9	0.540
330	0.5364	302.7	334.5	0.563	0.4693	301.4	332.7	0.556
350	0.5559	313.4	346.3	0.577	0.4872	312.2	344.7	0.570
370	0.5750	324.2	358.3	0.592	0.5047	323.0	356.7	0.585
	p = 400 lbf/in.² (T_{sat} = 164.0°F)				p = 450 lbf/in.² (T_{sat} = 175.1°F)			
Sat.	0.2287	209.5	226.5	0.400	0.1921	209.9	225.9	0.396
170	0.2406	213.6	231.4	0.408				
190	0.2725	225.6	245.8	0.430	0.2205	220.7	239.1	0.416
210	0.2985	236.7	258.8	0.450	0.2486	233.0	253.7	0.439
230	0.3215	247.4	271.2	0.468	0.2719	244.3	267.0	0.458
250	0.3424	257.8	283.2	0.485	0.2925	255.2	279.6	0.476
270	0.3620	268.3	295.1	0.502	0.3113	266.0	292.0	0.493
290	0.3806	278.8	307.0	0.518	0.3290	276.8	304.2	0.510
310	0.3984	289.4	318.9	0.534	0.3457	287.6	316.4	0.526
330	0.4156	300.1	330.9	0.549	0.3617	298.4	328.5	0.542
350	0.4322	311.0	343.0	0.564	0.3772	309.4	340.8	0.557
370	0.4484	321.9	355.1	0.579	0.3922	320.4	353.1	0.572
390	0.4643	333.1	367.5	0.594	0.4068	331.7	365.6	0.587
	p = 500 lbf/in.² (T_{sat} = 185.3°F)				p = 600 lbf/in.² (T_{sat} = 203.4°F)			
Sat.	0.1610	209.3	224.2	0.391	0.1003	200.3	211.4	0.367
190	0.1727	213.8	229.8	0.399				
210	0.2066	228.6	247.7	0.426	0.1307	214.3	228.8	0.394
230	0.2312	240.9	262.3	0.448	0.1661	232.2	250.7	0.426
250	0.2519	252.4	275.7	0.467	0.1892	245.8	266.8	0.449
270	0.2704	263.6	288.6	0.485	0.2080	258.1	281.2	0.469
290	0.2874	274.6	301.2	0.502	0.2245	269.8	294.8	0.487
310	0.3034	285.6	313.7	0.519	0.2396	281.4	308.0	0.505
330	0.3186	296.6	326.1	0.534	0.2536	292.8	321.0	0.521
350	0.3331	307.7	338.6	0.550	0.2669	304.2	333.9	0.538
370	0.3471	318.9	351.0	0.565	0.2796	315.7	346.8	0.553
390	0.3607	330.2	363.6	0.580	0.2917	327.3	359.7	0.569
410	0.3740	341.7	376.3	0.595	0.3035	338.9	372.6	0.584

Table A-19E Properties of Selected Solids and Liquids: c_p, ρ, and κ

Substance	Specific Heat, c_p (Btu/lb · °R)	Density, ρ (lb/ft^3)	Thermal Conductivity, κ (Btu/h · ft · °R)
Selected Solids, 540°R			
Aluminum	0.216	169	137
Coal, anthracite	0.301	84.3	0.15
Copper	0.092	557	232
Granite	0.185	164	1.61
Iron	0.107	491	46.4
Lead	0.031	705	20.4
Sand	0.191	94.9	0.16
Silver	0.056	656	248
Soil	0.439	128	0.30
Steel (AISI 302)	0.115	503	8.7
Tin	0.054	456	38.5
Building Materials, 540°R			
Brick, common	0.199	120	0.42
Concrete (stone mix)	0.210	144	0.81
Glass, plate	0.179	156	0.81
Hardboard, siding	0.279	40	0.054
Limestone	0.193	145	1.24
Plywood	0.291	34	0.069
Softwoods (fir, pine)	0.330	31.8	0.069
Insulating Materials, 540°R			
Blanket (glass fiber)	—	1.0	0.027
Cork	0.43	7.5	0.023
Duct liner (glass fiber, coated)	0.199	2.0	0.022
Polystyrene (extruded)	0.289	3.4	0.016
Vermiculite fill (flakes)	0.199	5.0	0.039
Saturated Liquids			
Ammonia, 540°R	1.151	37.5	0.269
Mercury, 540°R	0.033	845	4.94
Refrigerant 22, 540°R	0.303	74.0	0.049
Refrigerant 134a, 540°R	0.343	75.0	0.047
Unused Engine Oil, 540°R	0.456	55.2	0.084
Water, 495°R	1.006	62.42	0.332
540°R	0.998	62.23	0.354
585°R	0.999	61.61	0.373
630°R	1.002	60.79	0.386
675°R	1.008	59.76	0.394
720°R	1.017	58.55	0.398

Source: Drawn from several sources, these data are only representative. Values can vary depending on temperature, purity, moisture content, and other factors.

Table A-19E

Table A-20E **Ideal Gas Specific Heats of Some Common Gases (Btu/lb · °R)**

Temp. °F	Air c_p	Air c_v	Air k	Nitrogen, N_2 c_p	Nitrogen, N_2 c_v	Nitrogen, N_2 k	Oxygen, O_2 c_p	Oxygen, O_2 c_v	Oxygen, O_2 k	Temp. °F
40	0.240	0.171	1.401	0.248	0.177	1.400	0.219	0.156	1.397	40
100	0.240	0.172	1.400	0.248	0.178	1.399	0.220	0.158	1.394	100
200	0.241	0.173	1.397	0.249	0.178	1.398	0.223	0.161	1.387	200
300	0.243	0.174	1.394	0.250	0.179	1.396	0.226	0.164	1.378	300
400	0.245	0.176	1.389	0.251	0.180	1.393	0.230	0.168	1.368	400
500	0.248	0.179	1.383	0.254	0.183	1.388	0.235	0.173	1.360	500
600	0.250	0.182	1.377	0.256	0.185	1.383	0.239	0.177	1.352	600
700	0.254	0.185	1.371	0.260	0.189	1.377	0.242	0.181	1.344	700
800	0.257	0.188	1.365	0.262	0.191	1.371	0.246	0.184	1.337	800
900	0.259	0.191	1.358	0.265	0.194	1.364	0.249	0.187	1.331	900
1000	0.263	0.195	1.353	0.269	0.198	1.359	0.252	0.190	1.326	1000
1500	0.276	0.208	1.330	0.283	0.212	1.334	0.263	0.201	1.309	1500
2000	0.286	0.217	1.312	0.293	0.222	1.319	0.270	0.208	1.298	2000
Temp. °F	**Carbon Dioxide, CO_2**			**Carbon Monoxide, CO**			**Hydrogen, H_2**			**Temp. °F**
40	0.195	0.150	1.300	0.248	0.177	1.400	3.397	2.412	1.409	40
100	0.205	0.160	1.283	0.249	0.178	1.399	3.426	2.441	1.404	100
200	0.217	0.172	1.262	0.249	0.179	1.397	3.451	2.466	1.399	200
300	0.229	0.184	1.246	0.251	0.180	1.394	3.461	2.476	1.398	300
400	0.239	0.193	1.233	0.253	0.182	1.389	3.466	2.480	1.397	400
500	0.247	0.202	1.223	0.256	0.185	1.384	3.469	2.484	1.397	500
600	0.255	0.210	1.215	0.259	0.188	1.377	3.473	2.488	1.396	600
700	0.262	0.217	1.208	0.262	0.191	1.371	3.477	2.492	1.395	700
800	0.269	0.224	1.202	0.266	0.195	1.364	3.494	2.509	1.393	800
900	0.275	0.230	1.197	0.269	0.198	1.357	3.502	2.519	1.392	900
1000	0.280	0.235	1.192	0.273	0.202	1.351	3.513	2.528	1.390	1000
1500	0.298	0.253	1.178	0.287	0.216	1.328	3.618	2.633	1.374	1500
2000	0.312	0.267	1.169	0.297	0.226	1.314	3.758	2.773	1.355	2000

Source: Adapted from K. Wark, *Thermodynamics,* 4th ed., McGraw-Hill, New York, 1983, as based on "Tables of Thermal Properties of Gases," NBS Circular 564, 1955.

Table A-21E Variation of $\bar{c}_p$ with Temperature for Selected Ideal Gases

$$\frac{\bar{c}_p}{\bar{R}} = \alpha + \beta T + \gamma T^2 + \delta T^3 + \varepsilon T^4$$

T is in °R, equations valid from 540 to 1800 °R

Gas	α	$\beta \times 10^3$	$\gamma \times 10^6$	$\delta \times 10^9$	$\varepsilon \times 10^{12}$
CO	3.710	−0.899	1.140	−0.348	0.0228
CO_2	2.401	4.853	−2.039	0.343	0
H_2	3.057	1.487	−1.793	0.947	−0.1726
H_2O	4.070	−0.616	1.281	−0.508	0.0769
O_2	3.626	−1.043	2.178	−1.160	0.2053
N_2	3.675	−0.671	0.717	−0.108	−0.0215
Air	3.653	−0.7428	1.017	−0.328	0.02632
NH_3	3.591	0.274	2.576	−1.437	0.2601
NO	4.046	−1.899	2.464	−1.048	0.1517
NO_2	3.459	1.147	2.064	−1.639	0.3448
SO_2	3.267	2.958	0.211	−0.906	0.2438
SO_3	2.578	8.087	−2.832	−0.136	0.1878
CH_4	3.826	−2.211	7.580	−3.898	0.6633
C_2H_2	1.410	10.587	−7.562	2.811	−0.3939
C_2H_4	1.426	6.324	2.466	−2.787	0.6429
Monatomic gases[a]	2.5	0	0	0	0

[a]For monatomic gases, such as He, Ne, and Ar, $\bar{c}_p$ is constant over a wide temperature range and is very nearly equal to $5/2\ \bar{R}$.

Source: Adapted from K. Wark, *Thermodynamics,* 4th ed., McGraw-Hill, New York, 1983, as based on NASA SP-273, U.S. Government Printing Office, Washington, DC, 1971.

Table A-21E

Table A-22E **Ideal Gas Properties of Air**

T(°R), h and u(Btu/lb), $s°$(Btu/lb · °R)

				when $\Delta s = 0$[1]						when $\Delta s = 0$	
T	h	u	$s°$	p_r	v_r	T	h	u	$s°$	p_r	v_r
360	85.97	61.29	0.50369	0.3363	396.6	940	226.11	161.68	0.73509	9.834	35.41
380	90.75	64.70	0.51663	0.4061	346.6	960	231.06	165.26	0.74030	10.61	33.52
400	95.53	68.11	0.52890	0.4858	305.0	980	236.02	168.83	0.74540	11.43	31.76
420	100.32	71.52	0.54058	0.5760	270.1	1000	240.98	172.43	0.75042	12.30	30.12
440	105.11	74.93	0.55172	0.6776	240.6	1040	250.95	179.66	0.76019	14.18	27.17
460	109.90	78.36	0.56235	0.7913	215.33	1080	260.97	186.93	0.76964	16.28	24.58
480	114.69	81.77	0.57255	0.9182	193.65	1120	271.03	194.25	0.77880	18.60	22.30
500	119.48	85.20	0.58233	1.0590	174.90	1160	281.14	201.63	0.78767	21.18	20.29
520	124.27	88.62	0.59172	1.2147	158.58	1200	291.30	209.05	0.79628	24.01	18.51
537	128.34	91.53	0.59945	1.3593	146.34	1240	301.52	216.53	0.80466	27.13	16.93
540	129.06	92.04	0.60078	1.3860	144.32	1280	311.79	224.05	0.81280	30.55	15.52
560	133.86	95.47	0.60950	1.5742	131.78	1320	322.11	231.63	0.82075	34.31	14.25
580	138.66	98.90	0.61793	1.7800	120.70	1360	332.48	239.25	0.82848	38.41	13.12
600	143.47	102.34	0.62607	2.005	110.88	1400	342.90	246.93	0.83604	42.88	12.10
620	148.28	105.78	0.63395	2.249	102.12	1440	353.37	254.66	0.84341	47.75	11.17
640	153.09	109.21	0.64159	2.514	94.30	1480	363.89	262.44	0.85062	53.04	10.34
660	157.92	112.67	0.64902	2.801	87.27	1520	374.47	270.26	0.85767	58.78	9.578
680	162.73	116.12	0.65621	3.111	80.96	1560	385.08	278.13	0.86456	65.00	8.890
700	167.56	119.58	0.66321	3.446	75.25	1600	395.74	286.06	0.87130	71.73	8.263
720	172.39	123.04	0.67002	3.806	70.07	1650	409.13	296.03	0.87954	80.89	7.556
740	177.23	126.51	0.67665	4.193	65.38	1700	422.59	306.06	0.88758	90.95	6.924
760	182.08	129.99	0.68312	4.607	61.10	1750	436.12	316.16	0.89542	101.98	6.357
780	186.94	133.47	0.68942	5.051	57.20	1800	449.71	326.32	0.90308	114.0	5.847
800	191.81	136.97	0.69558	5.526	53.63	1850	463.37	336.55	0.91056	127.2	5.388
820	196.69	140.47	0.70160	6.033	50.35	1900	477.09	346.85	0.91788	141.5	4.974
840	201.56	143.98	0.70747	6.573	47.34	1950	490.88	357.20	0.92504	157.1	4.598
860	206.46	147.50	0.71323	7.149	44.57	2000	504.71	367.61	0.93205	174.0	4.258
880	211.35	151.02	0.71886	7.761	42.01	2050	518.61	378.08	0.93891	192.3	3.949
900	216.26	154.57	0.72438	8.411	39.64	2100	532.55	388.60	0.94564	212.1	3.667
920	221.18	158.12	0.72979	9.102	37.44	2150	546.54	399.17	0.95222	233.5	3.410

1. p_r and v_r data for use with Eqs. 6.41 and 6.42, respectively.

Table A-22E **(Continued)**

T(°R), h and u(Btu/lb), $s°$(Btu/lb · °R)											
				when $\Delta s = 0$[1]						when $\Delta s = 0$	
T	h	u	$s°$	p_r	v_r	T	h	u	$s°$	p_r	v_r
2200	560.59	409.78	0.95868	256.6	3.176	3700	998.11	744.48	1.10991	2330	.5882
2250	574.69	420.46	0.96501	281.4	2.961	3750	1013.1	756.04	1.11393	2471	.5621
2300	588.82	431.16	0.97123	308.1	2.765	3800	1028.1	767.60	1.11791	2618	.5376
2350	603.00	441.91	0.97732	336.8	2.585	3850	1043.1	779.19	1.12183	2773	.5143
2400	617.22	452.70	0.98331	367.6	2.419	3900	1058.1	790.80	1.12571	2934	.4923
2450	631.48	463.54	0.98919	400.5	2.266	3950	1073.2	802.43	1.12955	3103	.4715
2500	645.78	474.40	0.99497	435.7	2.125	4000	1088.3	814.06	1.13334	3280	.4518
2550	660.12	485.31	1.00064	473.3	1.996	4050	1103.4	825.72	1.13709	3464	.4331
2600	674.49	496.26	1.00623	513.5	1.876	4100	1118.5	837.40	1.14079	3656	.4154
2650	688.90	507.25	1.01172	556.3	1.765	4150	1133.6	849.09	1.14446	3858	.3985
2700	703.35	518.26	1.01712	601.9	1.662	4200	1148.7	860.81	1.14809	4067	.3826
2750	717.83	529.31	1.02244	650.4	1.566	4300	1179.0	884.28	1.15522	4513	.3529
2800	732.33	540.40	1.02767	702.0	1.478	4400	1209.4	907.81	1.16221	4997	.3262
2850	746.88	551.52	1.03282	756.7	1.395	4500	1239.9	931.39	1.16905	5521	.3019
2900	761.45	562.66	1.03788	814.8	1.318	4600	1270.4	955.04	1.17575	6089	.2799
2950	776.05	573.84	1.04288	876.4	1.247	4700	1300.9	978.73	1.18232	6701	.2598
3000	790.68	585.04	1.04779	941.4	1.180	4800	1331.5	1002.5	1.18876	7362	.2415
3050	805.34	596.28	1.05264	1011	1.118	4900	1362.2	1026.3	1.19508	8073	.2248
3100	820.03	607.53	1.05741	1083	1.060	5000	1392.9	1050.1	1.20129	8837	.2096
3150	834.75	618.82	1.06212	1161	1.006	5100	1423.6	1074.0	1.20738	9658	.1956
3200	849.48	630.12	1.06676	1242	0.9546	5200	1454.4	1098.0	1.21336	10539	.1828
3250	864.24	641.46	1.07134	1328	0.9069	5300	1485.3	1122.0	1.21923	11481	.1710
3300	879.02	652.81	1.07585	1418	0.8621						
3350	893.83	664.20	1.08031	1513	0.8202						
3400	908.66	675.60	1.08470	1613	0.7807						
3450	923.52	687.04	1.08904	1719	0.7436						
3500	938.40	698.48	1.09332	1829	0.7087						
3550	953.30	709.95	1.09755	1946	0.6759						
3600	968.21	721.44	1.10172	2068	0.6449						
3650	983.15	732.95	1.10584	2196	0.6157						

Table A-23E Ideal Gas Properties of Selected Gases

Enthalpy $\bar{h}(T)$ and internal energy $\bar{u}(T)$ in Btu/lbmol. Absolute entropy at 1 atm $\bar{s}^\circ(T)$, in Btu/lbmol · °R.

	Carbon Dioxide, CO_2 ($\bar{h}_f^\circ = -169{,}300$ Btu/lbmol)			Carbon Monoxide, CO ($\bar{h}_f^\circ = -47{,}540$ Btu/lbmol)			Water Vapor, H_2O ($\bar{h}_f^\circ = -104{,}040$ Btu/lbmol)			Oxygen, O_2 ($\bar{h}_f^\circ = 0$ Btu/lbmol)			Nitrogen, N_2 ($\bar{h}_f^\circ = 0$ Btu/lbmol)			
T(°R)	$\bar{h}$	$\bar{u}$	$\bar{s}^\circ$	$\bar{h}$	$\bar{u}$	$\bar{s}^\circ$	$\bar{h}$	$\bar{u}$	$\bar{s}^\circ$	$\bar{h}$	$\bar{u}$	$\bar{s}^\circ$	$\bar{h}$	$\bar{u}$	$\bar{s}^\circ$	T(°R)
300	2108.2	1512.4	46.353	2081.9	1486.1	43.223	2367.6	1771.8	40.439	2073.5	1477.8	44.927	2082.0	1486.2	41.695	300
320	2256.6	1621.1	46.832	2220.9	1585.4	43.672	2526.8	1891.3	40.952	2212.6	1577.1	45.375	2221.0	1585.5	42.143	320
340	2407.3	1732.1	47.289	2359.9	1684.7	44.093	2686.0	2010.8	41.435	2351.7	1676.5	45.797	2360.0	1684.4	42.564	340
360	2560.5	1845.6	47.728	2498.8	1783.9	44.490	2845.1	2130.2	41.889	2490.8	1775.9	46.195	2498.9	1784.0	42.962	360
380	2716.4	1961.8	48.148	2637.9	1883.3	44.866	3004.4	2249.8	42.320	2630.0	1875.3	46.571	2638.0	1883.4	43.337	380
400	2874.7	2080.4	48.555	2776.9	1982.6	45.223	3163.8	2369.4	42.728	2769.1	1974.8	46.927	2777.0	1982.6	43.694	400
420	3035.7	2201.7	48.947	2916.0	2081.9	45.563	3323.2	2489.1	43.117	2908.3	2074.3	47.267	2916.1	2082.0	44.034	420
440	3199.4	2325.6	49.329	3055.0	2181.2	45.886	3482.7	2608.9	43.487	3047.5	2173.8	47.591	3055.1	2181.3	44.357	440
460	3365.7	2452.2	49.698	3194.0	2280.5	46.194	3642.3	2728.8	43.841	3186.9	2273.4	47.900	3194.1	2280.6	44.665	460
480	3534.7	2581.5	50.058	3333.0	2379.8	46.491	3802.0	2848.8	44.182	3326.5	2373.3	48.198	3333.1	2379.9	44.962	480
500	3706.2	2713.3	50.408	3472.1	2479.2	46.775	3962.0	2969.1	44.508	3466.2	2473.2	48.483	3472.2	2479.3	45.246	500
520	3880.3	2847.7	50.750	3611.2	2578.6	47.048	4122.0	3089.4	44.821	3606.1	2573.4	48.757	3611.3	2578.6	45.519	520
537	4027.5	2963.8	51.032	3725.1	2663.1	47.272	4258.0	3191.9	45.079	3725.1	2658.7	48.982	3729.5	2663.1	45.743	537
540	4056.8	2984.4	51.082	3750.3	2677.9	47.310	4282.4	3210.0	45.124	3746.2	2673.8	49.021	3750.3	2678.0	45.781	540
560	4235.8	3123.7	51.408	3889.5	2777.4	47.563	4442.8	3330.7	45.415	3886.6	2774.5	49.276	3889.5	2777.4	46.034	560
580	4417.2	3265.4	51.726	4028.7	2876.9	47.807	4603.7	3451.9	45.696	4027.3	2875.5	49.522	4028.7	2876.9	46.278	580
600	4600.9	3409.4	52.038	4168.0	2976.5	48.044	4764.7	3573.2	45.970	4168.3	2976.8	49.762	4167.9	2976.4	46.514	600
620	4786.6	3555.6	52.343	4307.4	3076.2	48.272	4926.1	3694.9	46.235	4309.7	3078.4	49.993	4307.1	3075.9	46.742	620
640	4974.9	3704.0	52.641	4446.9	3175.9	48.494	5087.8	3816.8	46.492	4451.4	3180.4	50.218	4446.4	3175.5	46.964	640
660	5165.2	3854.6	52.934	4586.6	3275.8	48.709	5250.0	3939.3	46.741	4593.5	3282.9	50.437	4585.8	3275.2	47.178	660
680	5357.6	4007.2	53.225	4726.2	3375.8	48.917	5412.5	4062.1	46.984	4736.2	3385.8	50.650	4725.3	3374.9	47.386	680
700	5552.0	4161.9	53.503	4866.0	3475.9	49.120	5575.4	4185.3	47.219	4879.3	3489.2	50.858	4864.9	3474.8	47.588	700
720	5748.4	4318.6	53.780	5006.1	3576.3	49.317	5738.8	4309.0	47.450	5022.9	3593.1	51.059	5004.5	3574.7	47.785	720
740	5946.8	4477.3	54.051	5146.4	3676.9	49.509	5902.6	4433.1	47.673	5167.0	3697.4	51.257	5144.3	3674.7	47.977	740
760	6147.0	4637.9	54.319	5286.8	3777.5	49.697	6066.9	4557.6	47.893	5311.4	3802.2	51.450	5284.1	3774.9	48.164	760
780	6349.1	4800.1	54.582	5427.4	3878.4	49.880	6231.7	4682.7	48.106	5456.4	3907.5	51.638	5424.2	3875.2	48.345	780
800	6552.9	4964.2	54.839	5568.2	3979.5	50.058	6396.9	4808.2	48.316	5602.0	4013.3	51.821	5564.4	3975.7	48.522	800
820	6758.3	5129.9	55.093	5709.4	4081.0	50.232	6562.6	4934.2	48.520	5748.1	4119.7	52.002	5704.7	4076.3	48.696	820
840	6965.7	5297.6	55.343	5850.7	4182.6	50.402	6728.9	5060.8	48.721	5894.8	4226.6	52.179	5845.3	4177.1	48.865	840
860	7174.7	5466.9	55.589	5992.3	4284.5	50.569	6895.6	5187.8	48.916	6041.9	4334.1	52.352	5985.9	4278.1	49.031	860
880	7385.3	5637.7	55.831	6134.2	4386.6	50.732	7062.9	5315.3	49.109	6189.6	4442.0	52.522	6126.9	4379.4	49.193	880
900	7597.6	5810.3	56.070	6276.4	4489.1	50.892	7230.9	5443.6	49.298	6337.9	4550.6	52.688	6268.1	4480.8	49.352	900
920	7811.4	5984.4	56.305	6419.0	4592.0	51.048	7399.4	5572.4	49.483	6486.7	4659.7	52.852	6409.6	4582.6	49.507	920
940	8026.8	6160.1	56.536	6561.7	4695.0	51.202	7568.4	5701.7	49.665	6636.1	4769.4	53.012	6551.2	4684.5	49.659	940
960	8243.8	6337.4	56.765	6704.9	4798.5	51.353	7738.0	5831.6	49.843	6786.0	4879.5	53.170	6693.1	4786.7	49.808	960
980	8462.2	6516.1	56.990	6848.4	4902.3	51.501	7908.2	5962.0	50.019	6936.4	4990.3	53.326	6835.4	4889.3	49.955	980
1000	8682.1	6696.2	57.212	6992.2	5006.3	51.646	8078.9	6093.0	50.191	7087.5	5101.6	53.477	6977.9	4992.0	50.099	1000
1020	8903.4	6877.8	57.432	7136.4	5110.8	51.788	8250.4	6224.8	50.360	7238.9	5213.3	53.628	7120.7	5095.1	50.241	1020
1040	9126.2	7060.9	57.647	7281.0	5215.7	51.929	8422.4	6357.1	50.528	7391.0	5325.7	53.775	7263.8	5198.5	50.380	1040
1060	9350.3	7245.3	57.861	7425.9	5320.9	52.067	8595.0	6490.0	50.693	7543.6	5438.6	53.921	7407.2	5302.2	50.516	1060

Table A-23E **(Continued)**

$\bar{h}$ and $\bar{u}$ in Btu/lbmol, $\bar{s}^\circ$ in Btu/lbmol · °R

	Carbon Dioxide, CO_2 ($\bar{h}_f^\circ = -169{,}300$ Btu/lbmol)			Carbon Monoxide, CO ($\bar{h}_f^\circ = -47{,}540$ Btu/lbmol)			Water Vapor, H_2O ($\bar{h}_f^\circ = -104{,}040$ Btu/lbmol)			Oxygen, O_2 ($\bar{h}_f^\circ = 0$ Btu/lbmol)			Nitrogen, N_2 ($\bar{h}_f^\circ = 0$ Btu/lbmol)			
T(°R)	$\bar{h}$	$\bar{u}$	$\bar{s}^\circ$	$\bar{h}$	$\bar{u}$	$\bar{s}^\circ$	$\bar{h}$	$\bar{u}$	$\bar{s}^\circ$	$\bar{h}$	$\bar{u}$	$\bar{s}^\circ$	$\bar{h}$	$\bar{u}$	$\bar{s}^\circ$	T(°R)
1080	9575.8	7431.1	58.072	7571.1	5426.4	52.203	8768.2	6623.5	50.854	7696.8	5552.1	54.064	7551.0	5406.2	50.651	1080
1100	9802.6	7618.1	58.281	7716.8	5532.3	52.337	8942.0	6757.5	51.013	7850.4	5665.9	54.204	7695.0	5510.5	50.783	1100
1120	10030.6	7806.4	58.485	7862.9	5638.7	52.468	9116.4	6892.2	51.171	8004.5	5780.3	54.343	7839.3	5615.2	50.912	1120
1140	10260.1	7996.2	58.689	8009.2	5745.4	52.598	9291.4	7027.5	51.325	8159.1	5895.2	54.480	7984.0	5720.1	51.040	1140
1160	10490.6	8187.0	58.889	8156.1	5851.5	52.726	9467.1	7163.5	51.478	8314.2	6010.6	54.614	8129.0	5825.4	51.167	1160
1180	10722.3	8379.0	59.088	8303.3	5960.0	52.852	9643.4	7300.1	51.630	8469.8	6126.5	54.748	8274.4	5931.0	51.291	1180
1200	10955.3	8572.3	59.283	8450.8	6067.8	52.976	9820.4	7437.4	51.777	8625.8	6242.8	54.879	8420.0	6037.0	51.413	1200
1220	11189.4	8766.6	59.477	8598.8	6176.0	53.098	9998.0	7575.2	51.925	8782.4	6359.6	55.008	8566.1	6143.4	51.534	1220
1240	11424.6	8962.1	59.668	8747.2	6284.7	53.218	10176.1	7713.6	52.070	8939.4	6476.9	55.136	8712.6	6250.1	51.653	1240
1260	11661.0	9158.8	59.858	8896.0	6393.8	53.337	10354.9	7852.7	52.212	9096.7	6594.5	55.262	8859.3	6357.2	51.771	1260
1280	11898.4	9356.5	60.044	9045.0	6503.1	53.455	10534.4	7992.5	52.354	9254.6	6712.7	55.386	9006.4	6464.5	51.887	1280
1300	12136.9	9555.3	60.229	9194.6	6613.0	53.571	10714.5	8132.9	52.494	9412.9	6831.3	55.508	9153.9	6572.3	51.001	1300
1320	12376.4	9755.0	60.412	9344.6	6723.2	53.685	10895.3	8274.0	52.631	9571.6	6950.2	55.630	9301.8	6680.4	52.114	1320
1340	12617.0	9955.9	60.593	9494.8	6833.7	53.799	11076.6	8415.5	52.768	9730.7	7069.6	55.750	9450.0	6788.9	52.225	1340
1360	12858.5	10157.7	60.772	9645.5	6944.7	53.910	11258.7	8557.9	52.903	9890.2	7189.4	55.867	9598.6	6897.8	52.335	1360
1380	13101.0	10360.5	60.949	9796.6	7056.1	54.021	11441.4	8700.9	53.037	10050.1	7309.6	55.984	9747.5	7007.0	52.444	1380
1400	13344.7	10564.5	61.124	9948.1	7167.9	54.129	11624.8	8844.6	53.168	10210.4	7430.1	56.099	9896.9	7116.7	52.551	1400
1420	13589.1	10769.2	61.298	10100.0	7280.1	54.237	11808.8	8988.9	53.299	10371.0	7551.1	56.213	10046.6	7226.7	52.658	1420
1440	13834.5	10974.8	61.469	10252.2	7392.6	54.344	11993.4	9133.8	53.428	10532.0	7672.4	56.326	10196.6	7337.0	52.763	1440
1460	14080.8	11181.4	61.639	10404.8	7505.4	54.448	12178.8	9279.4	53.556	10693.3	7793.9	56.437	10347.0	7447.6	52.867	1460
1480	14328.0	11388.9	61.800	10557.8	7618.7	54.522	12364.8	9425.7	53.682	10855.1	7916.0	56.547	10497.8	7558.7	52.969	1480
1500	14576.0	11597.2	61.974	10711.1	7732.3	54.665	12551.4	9572.7	53.808	11017.1	8038.3	56.656	10648.0	7670.1	53.071	1500
1520	14824.9	11806.4	62.138	10864.9	7846.4	54.757	12738.8	9720.3	53.932	11179.6	8161.1	56.763	10800.4	7781.9	53.171	1520
1540	15074.7	12016.5	62.302	11019.0	7960.8	54.858	12926.8	9868.6	54.055	11342.4	8284.2	56.869	10952.2	7893.9	53.271	1540
1560	15325.3	12227.3	62.464	11173.4	8075.4	54.958	13115.6	10017.6	54.117	11505.4	8407.4	56.975	11104.3	8006.4	53.369	1560
1580	15576.7	12439.0	62.624	11328.2	8190.5	55.056	13305.0	10167.3	54.298	11668.8	8531.1	57.079	11256.9	8119.2	53.465	1580
1600	15829.0	12651.6	62.783	11483.4	8306.0	55.154	13494.4	10317.6	54.418	11832.5	8655.1	57.182	11409.7	8232.3	53.561	1600
1620	16081.9	12864.8	62.939	11638.9	8421.8	55.251	13685.7	10468.6	54.535	11996.6	8779.5	57.284	11562.8	8345.7	53.656	1620
1640	16335.7	13078.9	63.095	11794.7	8537.9	55.347	13877.0	10620.2	54.653	12160.9	8904.1	57.385	11716.4	8459.6	53.751	1640
1660	16590.2	13293.7	63.250	11950.9	8654.4	55.411	14069.2	10772.7	54.770	12325.5	9029.0	57.484	11870.2	8573.6	53.844	1660
1680	16845.5	13509.2	63.403	12107.5	8771.2	55.535	14261.9	10925.6	54.886	12490.4	9154.1	57.582	12024.3	8688.1	53.936	1680
1700	17101.4	13725.4	63.555	12264.3	8888.3	55.628	14455.4	11079.4	54.999	12655.6	9279.6	57.680	12178.9	8802.9	54.028	1700
1720	17358.1	13942.4	63.704	12421.4	9005.7	55.720	14649.5	11233.8	55.113	12821.1	9405.4	57.777	12333.7	8918.0	54.118	1720
1740	17615.5	14160.1	63.853	12579.0	9123.6	55.811	14844.3	11388.9	55.226	12986.9	9531.5	57.873	12488.8	9033.4	54.208	1740
1760	17873.5	14378.4	64.001	12736.7	9241.6	55.900	15039.8	11544.7	55.339	13153.0	9657.9	57.968	12644.3	9149.2	54.297	1760
1780	18132.2	14597.4	64.147	12894.9	9360.0	55.990	15236.1	11701.2	55.449	13319.2	9784.4	58.062	12800.2	9265.3	54.385	1780
1800	18391.5	14816.9	64.292	13053.2	9478.6	56.078	15433.0	11858.4	55.559	13485.8	9911.2	58.155	12956.3	9381.7	54.472	1800
1820	18651.5	15037.2	64.435	13212.0	9597.7	56.166	15630.6	12016.3	55.668	13652.5	10038.2	58.247	13112.7	9498.4	54.559	1820
1840	18912.2	15258.2	64.578	13371.0	9717.0	56.253	15828.7	12174.7	55.777	13819.6	10165.6	58.339	13269.5	9615.5	54.645	1840
1860	19173.4	15479.7	64.719	13530.2	9836.5	56.339	16027.6	12333.9	55.884	13986.8	10293.1	58.428	13426.5	9732.8	54.729	1860

Table A-23E **(Continued)**

$\bar{h}$ and $\bar{u}$ in Btu/lbmol, $\bar{s}^\circ$ in Btu/lbmol · °R

	Carbon Dioxide, CO_2 ($\bar{h}_f^\circ = -169,300$ Btu/lbmol)			Carbon Monoxide, CO ($\bar{h}_f^\circ = -47,540$ Btu/lbmol)			Water Vapor, H_2O ($\bar{h}_f^\circ = -104,040$ Btu/lbmol)			Oxygen, O_2 ($\bar{h}_f^\circ = 0$ Btu/lbmol)			Nitrogen, N_2 ($\bar{h}_f^\circ = 0$ Btu/lbmol)			
T(°R)	$\bar{h}$	$\bar{u}$	$\bar{s}^\circ$	$\bar{h}$	$\bar{u}$	$\bar{s}^\circ$	$\bar{h}$	$\bar{u}$	$\bar{s}^\circ$	$\bar{h}$	$\bar{u}$	$\bar{s}^\circ$	$\bar{h}$	$\bar{u}$	$\bar{s}^\circ$	T(°R)
1900	19,698	15,925	64.999	13,850	10,077	56.509	16,428	12,654	56.097	14,322	10,549	58.607	13,742	9,968	54.896	1900
1940	20,224	16,372	65.272	14,170	10,318	56.677	16,830	12,977	56.307	14,658	10,806	58.782	14,058	10,205	55.061	1940
1980	20,753	16,821	65.543	14,492	10,560	56.841	17,235	13,303	56.514	14,995	11,063	58.954	14,375	10,443	55.223	1980
2020	21,284	17,273	65.809	14,815	10,803	57.007	17,643	13,632	56.719	15,333	11,321	59.123	14,694	10,682	55.383	2020
2060	21,818	17,727	66.069	15,139	11,048	57.161	18,054	13,963	56.920	15,672	11,581	59.289	15,013	10,923	55.540	2060
2100	22,353	18,182	66.327	15,463	11,293	57.317	18,467	14,297	57.119	16,011	11,841	59.451	15,334	11,164	55.694	2100
2140	22,890	18,640	66.581	15,789	11,539	57.470	18,883	14,633	57.315	16,351	12,101	59.612	15,656	11,406	55.846	2140
2180	23,429	19,101	66.830	16,116	11,787	57.621	19,301	14,972	57.509	16,692	12,363	59.770	15,978	11,649	55.995	2180
2220	23,970	19,561	67.076	16,443	12,035	57.770	19,722	15,313	57.701	17,036	12,625	59.926	16,302	11,893	56.141	2220
2260	24,512	20,024	67.319	16,722	12,284	57.917	20,145	15,657	57.889	17,376	12,888	60.077	16,626	12,138	56.286	2260
2300	25,056	20,489	67.557	17,101	12,534	58.062	20,571	16,003	58.077	17,719	13,151	60.228	16,951	12,384	56.429	2300
2340	25,602	20,955	67.792	17,431	12,784	58.204	20,999	16,352	58.261	18,062	13,416	60.376	17,277	12,630	56.570	2340
2380	26,150	21,423	68.025	17,762	13,035	58.344	21,429	16,703	58.445	18,407	13,680	60.522	17,604	12,878	56.708	2380
2420	26,699	21,893	68.253	18,093	13,287	58.482	21,862	17,057	58.625	18,572	13,946	60.666	17,932	13,126	56.845	2420
2460	27,249	22,364	68.479	18,426	13,541	58.619	22,298	17,413	58.803	19,097	14,212	60.808	18,260	13,375	56.980	2460
2500	27,801	22,837	68.702	18,759	13,794	58.754	22,735	17,771	58.980	19,443	14,479	60.946	18,590	13,625	57.112	2500
2540	28,355	23,310	68.921	19,093	14,048	58.885	23,175	18,131	59.155	19,790	14,746	61.084	18,919	13,875	57.243	2540
2580	28,910	23,786	69.138	19,427	14,303	59.016	23,618	18,494	59.328	20,138	15,014	61.220	19,250	14,127	57.372	2580
2620	29,465	24,262	69.352	19,762	14,559	59.145	24,062	18,859	59.500	20,485	15,282	61.354	19,582	14,379	57.499	2620
2660	30,023	24,740	69.563	20,098	14,815	59.272	24,508	19,226	59.669	20,834	15,551	61.486	19,914	14,631	57.625	2660
2700	30,581	25,220	69.771	20,434	15,072	59.398	24,957	19,595	59.837	21,183	15,821	61.616	20,246	14,885	57.750	2700
2740	31,141	25,701	69.977	20,771	15,330	59.521	25,408	19,967	60.003	21,533	16,091	61.744	20,580	15,139	57.872	2740
2780	31,702	26,181	70.181	21,108	15,588	59.644	25,861	20,340	60.167	21,883	16,362	61.871	20,914	15,393	57.993	2780
2820	32,264	26,664	70.382	21,446	15,846	59.765	26,316	20,715	60.330	22,232	16,633	61.996	21,248	15,648	58.113	2820
2860	32,827	27,148	70.580	21,785	16,105	59.884	26,773	21,093	60.490	22,584	16,905	62.120	21,584	15,905	58.231	2860
2900	33,392	27,633	70.776	22,124	16,365	60.002	27,231	21,472	60.650	22,936	17,177	62.242	21,920	16,161	58.348	2900
2940	33,957	28,118	70.970	22,463	16,625	60.118	27,692	21,853	60.809	23,288	17,450	62.363	22,256	16,417	58.463	2940
2980	34,523	28,605	71.160	22,803	16,885	60.232	28,154	22,237	60.965	23,641	17,723	62.483	22,593	16,675	58.576	2980
3020	35,090	29,093	71.350	23,144	17,146	60.346	28,619	22,621	61.120	23,994	17,997	62.599	22,930	16,933	58.688	3020
3060	35,659	29,582	71.537	23,485	17,408	60.458	29,085	23,085	61.274	24,348	18,271	62.716	23,268	17,192	58.800	3060
3100	36,228	30,072	71.722	23,826	17,670	60.569	29,553	23,397	61.426	24,703	18,546	62.831	23,607	17,451	58.910	3100
3140	36,798	30,562	71.904	24,168	17,932	60.679	30,023	23,787	61.577	25,057	18,822	62.945	23,946	17,710	59.019	3140
3180	37,369	31,054	72.085	24,510	18,195	60.787	30,494	24,179	61.727	25,413	19,098	63.057	24,285	17,970	59.126	3180
3220	37,941	31,546	72.264	24,853	18,458	60.894	30,967	24,572	61.874	25,769	19,374	63.169	24,625	18,231	59.232	3220
3260	38,513	32,039	72.441	25,196	18,722	61.000	31,442	24,968	62.022	26,175	19,651	63.279	24,965	18,491	59.338	3260
3300	39,087	32,533	72.616	25,539	18,986	61.105	31,918	25,365	62.167	26,412	19,928	63.386	25,306	18,753	59.442	3300
3340	39,661	33,028	72.788	25,883	19,250	61.209	32,396	25,763	62.312	26,839	20,206	63.494	25,647	19,014	59.544	3340
3380	40,236	33,524	72.960	26,227	19,515	61.311	[illegible]	26,164	62.454	27,197	20,485	63.601	25,989	19,277	59.646	3380
3420	40,812	34,020	73.129	26,572	19,780	61.412	[illegible]	[illegible]	62.597	27,555	20,763	63.706	26,331	19,539	59.747	3420
3460	41,338	34,517	73.297	26,917	20,045	61.513	33,839	26,968	62.738	27,914	21,043	63.811	26,673	19,802	59.846	3460

Table A-23E **(Continued)**

	$\bar{h}$ and $\bar{u}$ in Btu/lbmol, $\bar{s}^\circ$ in Btu/lbmol · °R															
	Carbon Dioxide, CO_2 ($\bar{h}_f^\circ = -169{,}300$ Btu/lbmol)			Carbon Monoxide, CO ($\bar{h}_f^\circ = -47{,}540$ Btu/lbmol)			Water Vapor, H_2O ($\bar{h}_f^\circ = -104{,}040$ Btu/lbmol)			Oxygen, O_2 ($\bar{h}_f^\circ = 0$ Btu/lbmol)			Nitrogen, N_2 ($\bar{h}_f^\circ = 0$ Btu/lbmol)			
T(°R)	$\bar{h}$	$\bar{u}$	$\bar{s}^\circ$	$\bar{h}$	$\bar{u}$	$\bar{s}^\circ$	$\bar{h}$	$\bar{u}$	$\bar{s}^\circ$	$\bar{h}$	$\bar{u}$	$\bar{s}^\circ$	$\bar{h}$	$\bar{u}$	$\bar{s}^\circ$	T(°R)
3500	41,965	35,015	73.462	27,262	20,311	61.612	34,324	27,373	62.875	28,273	21,323	63.914	27,016	20,065	59.944	3500
3540	42,543	35,513	73.627	27,608	20,576	61.710	34,809	27,779	63.015	28,633	21,603	64.016	27,359	20,329	60.041	3540
3580	43,121	36,012	73.789	27,954	20,844	61.807	35,296	28,187	63.153	28,994	21,884	64.114	27,703	20,593	60.138	3580
3620	43,701	36,512	73.951	28,300	21,111	61.903	35,785	28,596	63.288	29,354	22,165	64.217	28,046	20,858	60.234	3620
3660	44,280	37,012	74.110	28,647	21,378	61.998	36,274	29,006	63.423	29,716	22,447	64.316	28,391	21,122	60.328	3660
3700	44,861	37,513	74.267	28,994	21,646	62.093	36,765	29,418	63.557	30,078	22,730	64.415	28,735	21,387	60.422	3700
3740	45,442	38,014	74.423	29,341	21,914	62.186	37,258	29,831	63.690	30,440	23,013	64.512	29,080	21,653	60.515	3740
3780	46,023	38,517	74.578	29,688	22,182	62.279	37,752	30,245	63.821	30,803	23,296	64.609	29,425	21,919	60.607	3780
3820	46,605	39,019	74.732	30,036	22,450	62.370	38,247	30,661	63.952	31,166	23,580	64.704	29,771	22,185	60.698	3820
3860	47,188	39,522	74.884	30,384	22,719	62.461	38,743	31,077	64.082	31,529	23,864	64.800	30,117	22,451	60.788	3860
3900	47,771	40,026	75.033	30,733	22,988	61.511	39,240	31,495	64.210	31,894	24,149	64.893	30,463	22,718	60.877	3900
3940	48,355	40,531	75.182	31,082	23,257	62.640	39,739	31,915	64.338	32,258	24,434	64.986	30,809	22,985	60.966	3940
3980	48,939	41,035	75.330	31,431	23,527	62.728	40,239	32,335	64.465	32,623	24,720	65.078	31,156	23,252	61.053	3980
4020	49,524	41,541	75.477	31,780	23,797	62.816	40,740	32,757	64.591	32,989	25,006	65.169	31,503	23,520	61.139	4020
4060	50,109	42,047	75.622	32,129	24,067	62.902	41,242	33,179	64.715	33,355	25,292	65.260	31,850	23,788	61.225	4060
4100	50,695	42,553	75.765	32,479	24,337	62.988	41,745	33,603	64.839	33,722	25,580	65.350	32,198	24,056	61.310	4100
4140	51,282	43,060	75.907	32,829	24,608	63.072	42,250	34,028	64.962	34,089	25,867	64.439	32,546	24,324	61.395	4140
4180	51,868	43,568	76.048	33,179	24,878	63.156	42,755	34,454	65.084	34,456	26,155	65.527	32,894	24,593	61.479	4180
4220	52,456	44,075	76.188	33,530	25,149	63.240	43,267	34,881	65.204	34,824	26,444	65.615	33,242	24,862	61.562	4220
4260	53,044	44,584	76.327	33,880	25,421	63.323	43,769	35,310	65.325	35,192	26,733	65.702	33,591	25,131	61.644	4260
4300	53,632	45,093	76.464	34,231	25,692	63.405	44,278	35,739	65.444	35,561	27,022	65.788	33,940	25,401	61.726	4300
4340	54,221	45,602	76.601	34,582	25,934	63.486	44,788	36,169	65.563	35,930	27,312	65.873	34,289	25,670	61.806	4340
4380	54,810	46,112	76.736	34,934	26,235	63.567	45,298	36,600	65.680	36,300	27,602	65.958	34,638	25,940	61.887	4380
4420	55,400	46,622	76.870	35,285	26,508	63.647	45,810	37,032	65.797	36,670	27,823	66.042	34,988	26,210	61.966	4420
4460	55,990	47,133	77.003	35,637	26,780	63.726	46,322	37,465	65.913	37,041	28,184	66.125	35,338	26,481	62.045	4460
4500	56,581	47,645	77.135	35,989	27,052	63.805	46,836	37,900	66.028	37,412	28,475	66.208	35,688	26,751	62.123	4500
4540	57,172	48,156	77.266	36,341	27,325	63.883	47,350	38,334	66.142	37,783	28,768	66.290	36,038	27,022	62.201	4540
4580	57,764	48,668	77.395	36,693	27,598	63.960	47,866	38,770	66.255	38,155	29,060	66.372	36,389	27,293	62.278	4580
4620	58,356	49,181	77.581	37,046	27,871	64.036	48,382	39,207	66.368	38,528	29,353	66.453	36,739	27,565	62.354	4620
4660	58,948	49,694	77.652	37,398	28,144	64.113	48,899	39,645	66.480	38,900	29,646	66.533	37,090	27,836	62.429	4660
4700	59,541	50,208	77.779	37,751	28,417	64.188	49,417	40,083	66.591	39,274	29,940	66.613	37,441	28,108	62.504	4700
4740	60,134	50,721	77.905	38,104	28,691	64.263	49,936	40,523	66.701	39,647	30,234	66.691	37,792	28,379	62.578	4740
4780	60,728	51,236	78.029	38,457	28,965	64.337	50,455	40,963	66.811	40,021	30,529	66.770	38,144	28,651	62.652	4780
4820	61,322	51,750	78.153	38,811	29,239	64.411	50,976	41,404	66.920	40,396	30,824	66.848	38,495	28,924	62.725	4820
4860	61,916	52,265	78.276	39,164	29,513	64.484	51,497	41,856	67.028	40,771	31,120	66.925	38,847	29,196	62.798	4860
4900	62,511	52,781	78.398	39,518	29,787	64.556	52,019	42,288	67.135	41,146	31,415	67.003	39,199	29,468	62.870	4900
5000	64,000	54,071	78.698	40,403	30,473	64.735	53,327	43,398	67.401	42,086	32,157	67.193	40,080	30,151	63.049	5000
5100	65,491	55,363	78.994	41,289	31,161	64.910	54,640	44,512	67.662	43,021	32,901	67.380	40,962	30,834	63.223	5100
5200	66,984	56,658	79.284	42,176	31,849	65.082	55,957	45,631	67.918	43,974	33,648	67.562	41,844	31,518	63.395	5200
5300	68,471	57,954	79.569	43,063	32,538	65.252	57,279	46,754	68.172	44,922	34,397	67.743	42,728	32,203	63.563	5300

Table A-24E **Constants for the van der Waals, Redlich–Kwong, and Benedict–Webb–Rubin Equations of State**

1. van der Waals and Redlich–Kwong: Constants for pressure in atm, specific volume in ft^3/lbmol, and temperature in °R

Substance	van der Waals a atm $\left(\frac{ft^3}{lbmol}\right)^2$	van der Waals b $\frac{ft^3}{lbmol}$	Redlich–Kwong a atm $\left(\frac{ft^3}{lbmol}\right)^2 (°R)^{1/2}$	Redlich–Kwong b $\frac{ft^3}{lbmol}$
Air	345	0.586	5,409	0.4064
Butane (C_4H_{10})	3,509	1.862	98,349	1.2903
Carbon dioxide (CO_2)	926	0.686	21,972	0.4755
Carbon monoxide (CO)	372	0.632	5,832	0.4382
Methane (CH_4)	581	0.685	10,919	0.4751
Nitrogen (N_2)	346	0.618	5,280	0.4286
Oxygen (O_2)	349	0.509	5,896	0.3531
Propane (C_3H_8)	2,369	1.444	61,952	1.0006
Refrigerant 12	2,660	1.558	70,951	1.0796
Sulfur dioxide (SO_2)	1,738	0.910	49,032	0.6309
Water (H_2O)	1,400	0.488	48,418	0.3380

Source: Calculated from critical data.

2. Benedict–Webb–Rubin: Constants for pressure in atm, specific volume in ft^3/lbmol, and temperature in °R

Substance	a	A	b	B	c	C	α	γ
C_4H_{10}	7736.7	2587.6	10.26	1.9921	4.214×10^9	8.254×10^8	4.527	8.724
CO_2	562.3	702.4	1.850	0.7995	1.987×10^8	1.152×10^8	0.348	1.384
CO	150.6	344.1	0.675	0.8737	1.385×10^7	7.118×10^6	0.555	1.540
CH_4	203.0	476.0	0.867	0.6824	3.389×10^7	1.876×10^7	0.511	1.540
N_2	103.2	270.4	0.597	0.6526	9.700×10^6	6.700×10^6	0.523	1.360

Source: H. W. Cooper and J. C. Goldfrank, *Hydrocarbon Processing, 46* (12): 141 (1967).

Table A-25E **Thermochemical Properties of Selected Substances at 537°R and 1 atm**

Substance	Formula	Molar Mass, M (lb/lbmol)	Enthalpy of Formation, $\bar{h}_f^\circ$ (Btu/lbmol)	Gibbs Function of Formation, $\bar{g}_f^\circ$ (Btu/lbmol)	Absolute Entropy, $\bar{s}^\circ$ (Btu/lbmol · °R)	Heating Values: Higher, HHV (Btu/lb)	Heating Values: Lower, LHV (Btu/lb)
Carbon	$C(s)$	12.01	0	0	1.36	14,100	14,100
Hydrogen	$H_2(g)$	2.016	0	0	31.19	61,000	51,610
Nitrogen	$N_2(g)$	28.01	0	0	45.74	—	—
Oxygen	$O_2(g)$	32.00	0	0	48.98	—	—
Carbon monoxide	$CO(g)$	28.01	−47,540	−59,010	47.27	—	—
Carbon dioxide	$CO_2(g)$	44.01	−169,300	−169,680	51.03	—	—
Water	$H_2O(g)$	18.02	−104,040	−98,350	45.08	—	—
Water	$H_2O(l)$	18.02	−122,970	−102,040	16.71	—	—
Hydrogen peroxide	$H_2O_2(g)$	34.02	−58,640	−45,430	55.60	—	—
Ammonia	$NH_3(g)$	17.03	−19,750	−7,140	45.97	—	—
Oxygen	$O(g)$	16.00	107,210	99,710	38.47	—	—
Hydrogen	$H(g)$	1.008	93,780	87,460	27.39	—	—
Nitrogen	$N(g)$	14.01	203,340	195,970	36.61	—	—
Hydroxyl	$OH(g)$	17.01	16,790	14,750	43.92	—	—
Methane	$CH_4(g)$	16.04	−32,210	−21,860	44.49	23,880	21,520
Acetylene	$C_2H_2(g)$	26.04	97,540	87,990	48.00	21,470	20,740
Ethylene	$C_2H_4(g)$	28.05	22,490	29,306	52.54	21,640	20,290
Ethane	$C_2H_6(g)$	30.07	−36,420	−14,150	54.85	22,320	20,430
Propylene	$C_3H_6(g)$	42.08	8,790	26,980	63.80	21,050	19,700
Propane	$C_3H_8(g)$	44.09	−44,680	−10,105	64.51	21,660	19,950
Butane	$C_4H_{10}(g)$	58.12	−54,270	−6,760	74.11	21,300	19,670
Pentane	$C_5H_{12}(g)$	72.15	−62,960	−3,530	83.21	21,090	19,510
Octane	$C_8H_{18}(g)$	114.22	−89,680	7,110	111.55	20,760	19,270
Octane	$C_8H_{18}(l)$	114.22	−107,530	2,840	86.23	20,610	19,110
Benzene	$C_6H_6(g)$	78.11	35,680	55,780	64.34	18,180	17,460
Methyl alcohol	$CH_3OH(g)$	32.04	−86,540	−69,700	57.29	10,260	9,080
Methyl alcohol	$CH_3OH(l)$	32.04	−102,670	−71,570	30.30	9,760	8,570
Ethyl alcohol	$C_2H_5OH(g)$	46.07	−101,230	−72,520	67.54	13,160	11,930
Ethyl alcohol	$C_2H_5OH(l)$	46.07	−119,470	75,240	38.40	12,760	11,530

Source: Based on JANAF Thermochemical Tables, NSRDS-NBS-37, 1971; *Selected Values of Chemical Thermodynamic Properties,* NBS Tech. Note 270-3, 1968; and *API Research Project 44,* Carnegie Press, 1953. Heating values calculated.

Index to Figures and Charts

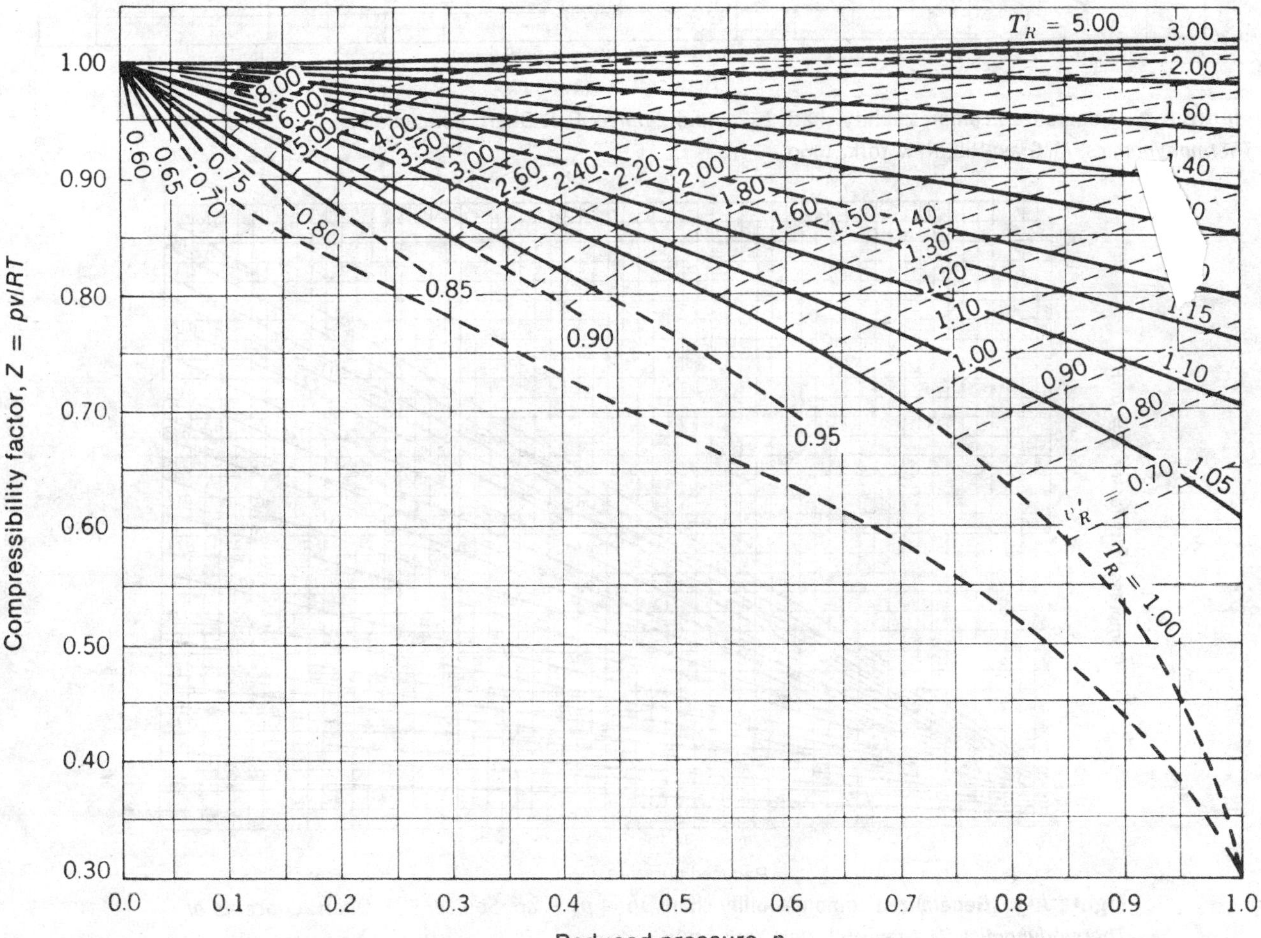

Figure A-1 Generalized compressibility chart, $p_R \leq 1.0$. *Source:* E. F. Obert, *Concepts of Thermodynamics,* McGraw-Hill, New York, 1960.

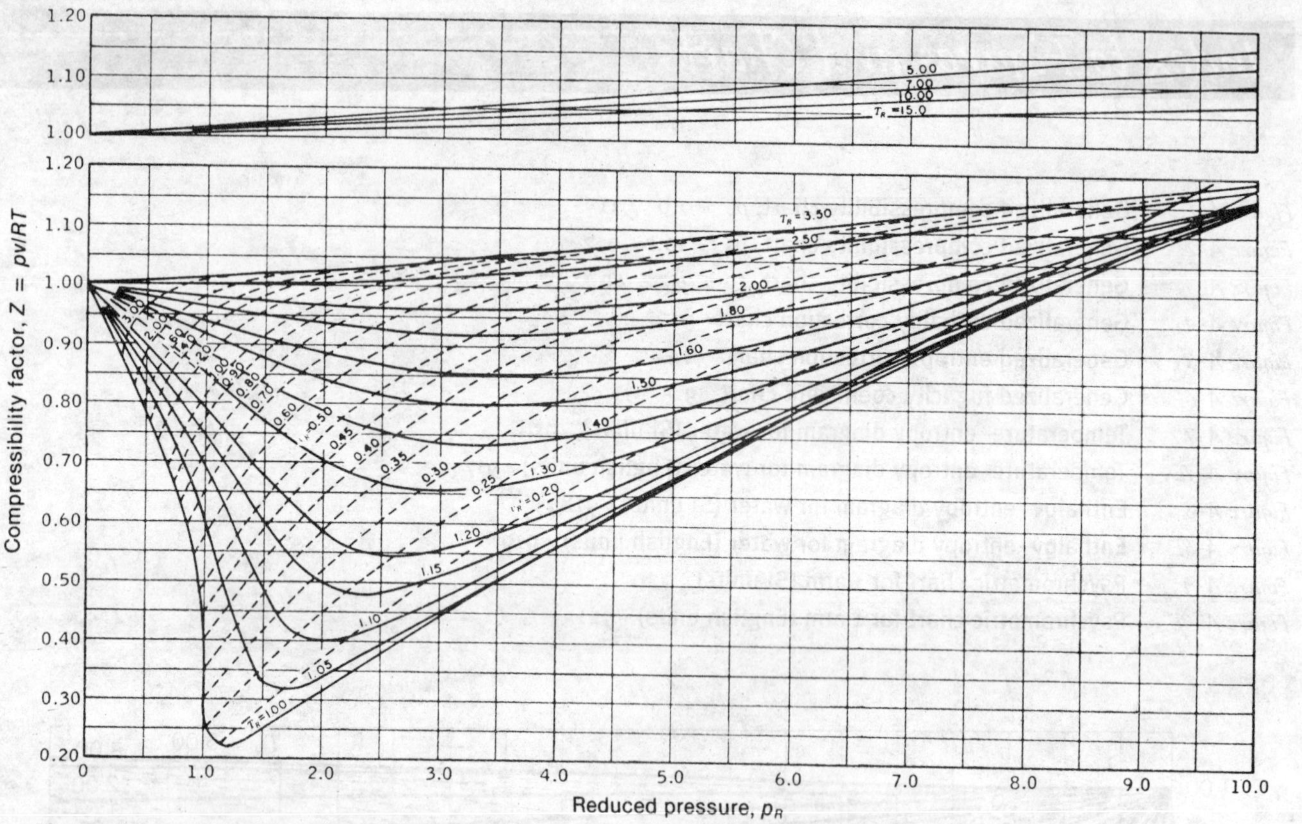

Figure A-2 Generalized compressibility chart, $p_R \leq 10.0$. *Source:* E. F. Obert, *Concepts of Thermodynamics,* McGraw-Hill, New York, 1960.

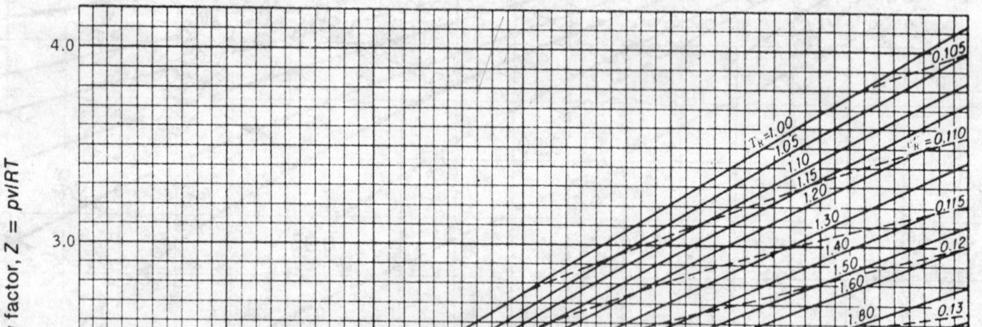

Figure A-3 Generalized compressibility chart, $10 \leq p_R \leq 40$. *Source:* E. F. Obert. *Concepts of Thermodynamics,* McGraw-Hill, New York, 1960.

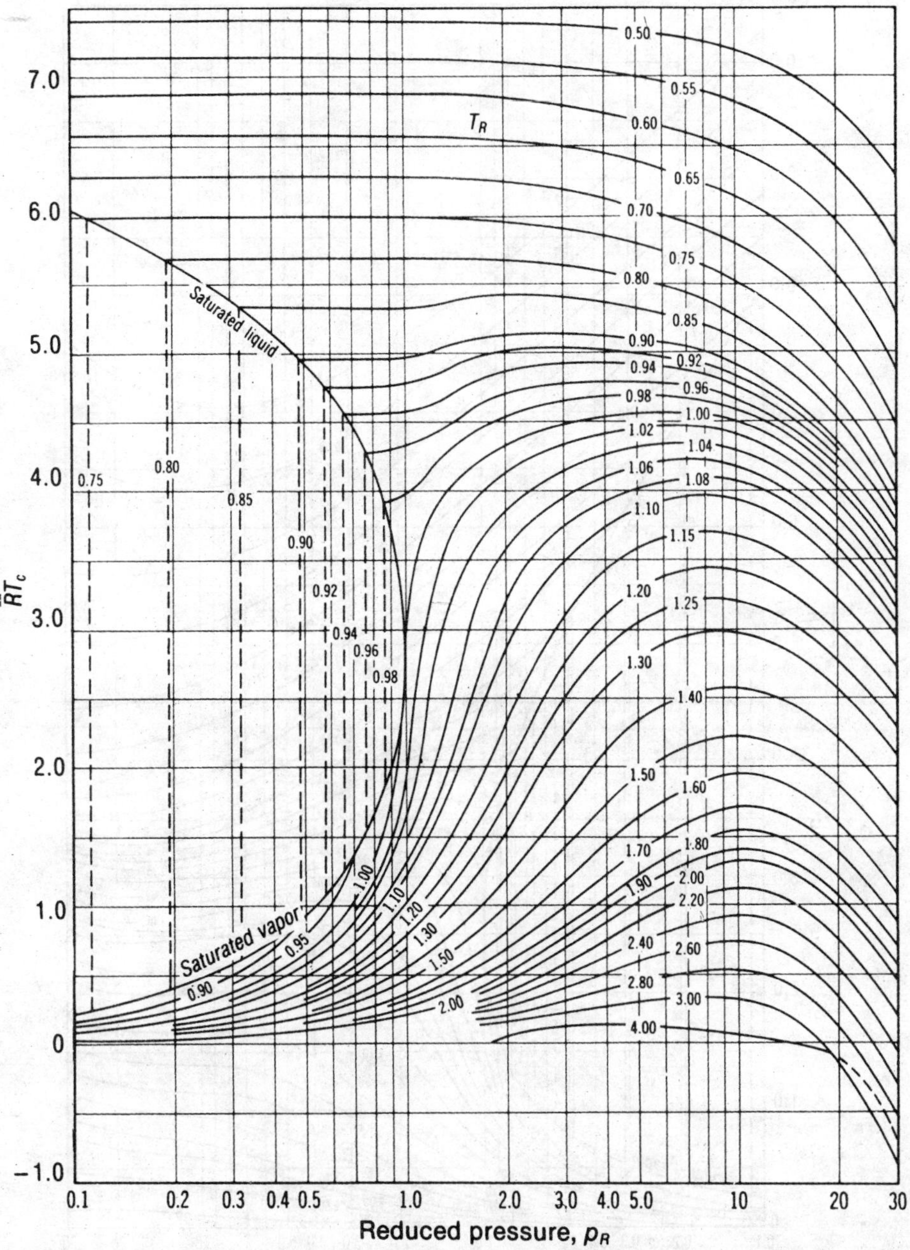

Figure A-4 Generalized enthalpy correction chart. *Source:* Adapted from G. J. Van Wylen and R. E. Sonntag, *Fundamentals of Classical Thermodynamics,* 3rd. ed., English/SI, Wiley, New York, 1986.

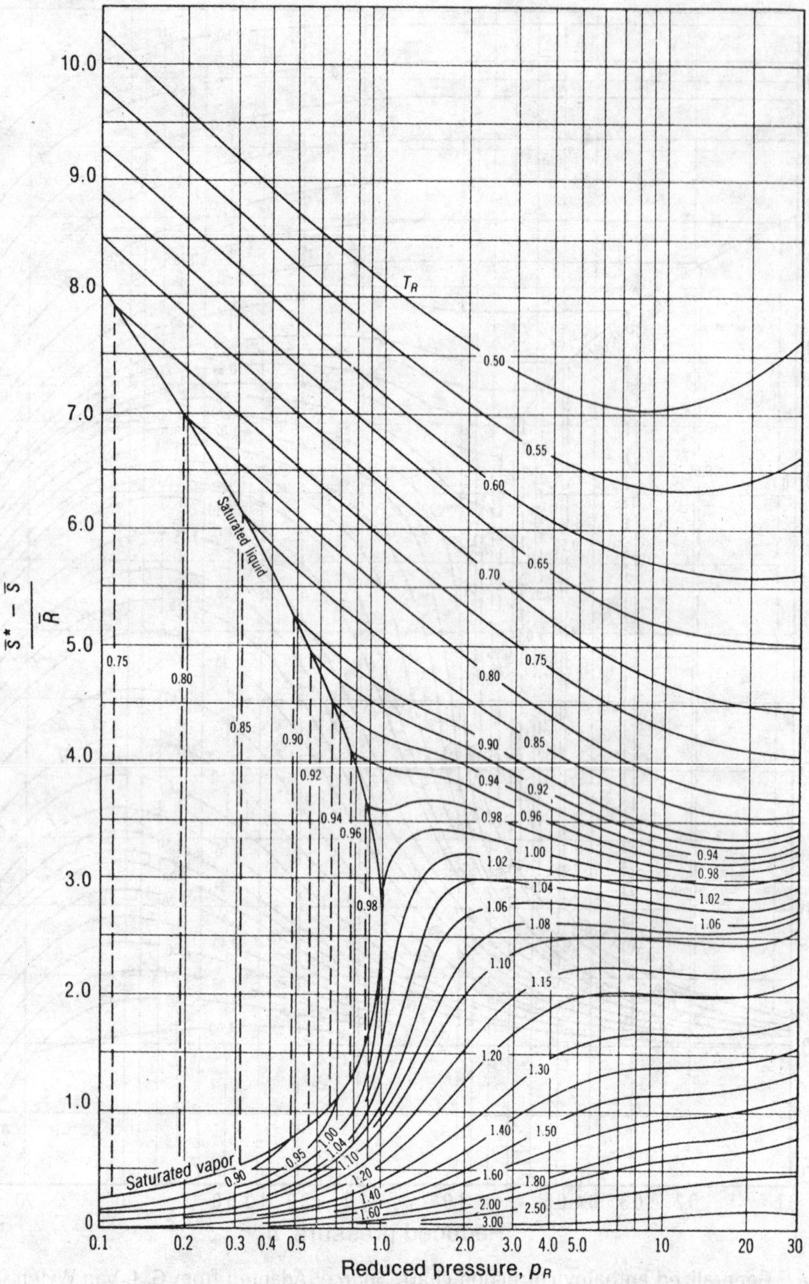

Figure A-5 Generalized entropy correction chart. *Source:* Adapted from G. J. Van Wylen and R. E. Sonntag, *Fundamentals of Classical Thermodynamics*, 3rd. ed., English/SI, Wiley, New York, 1986.

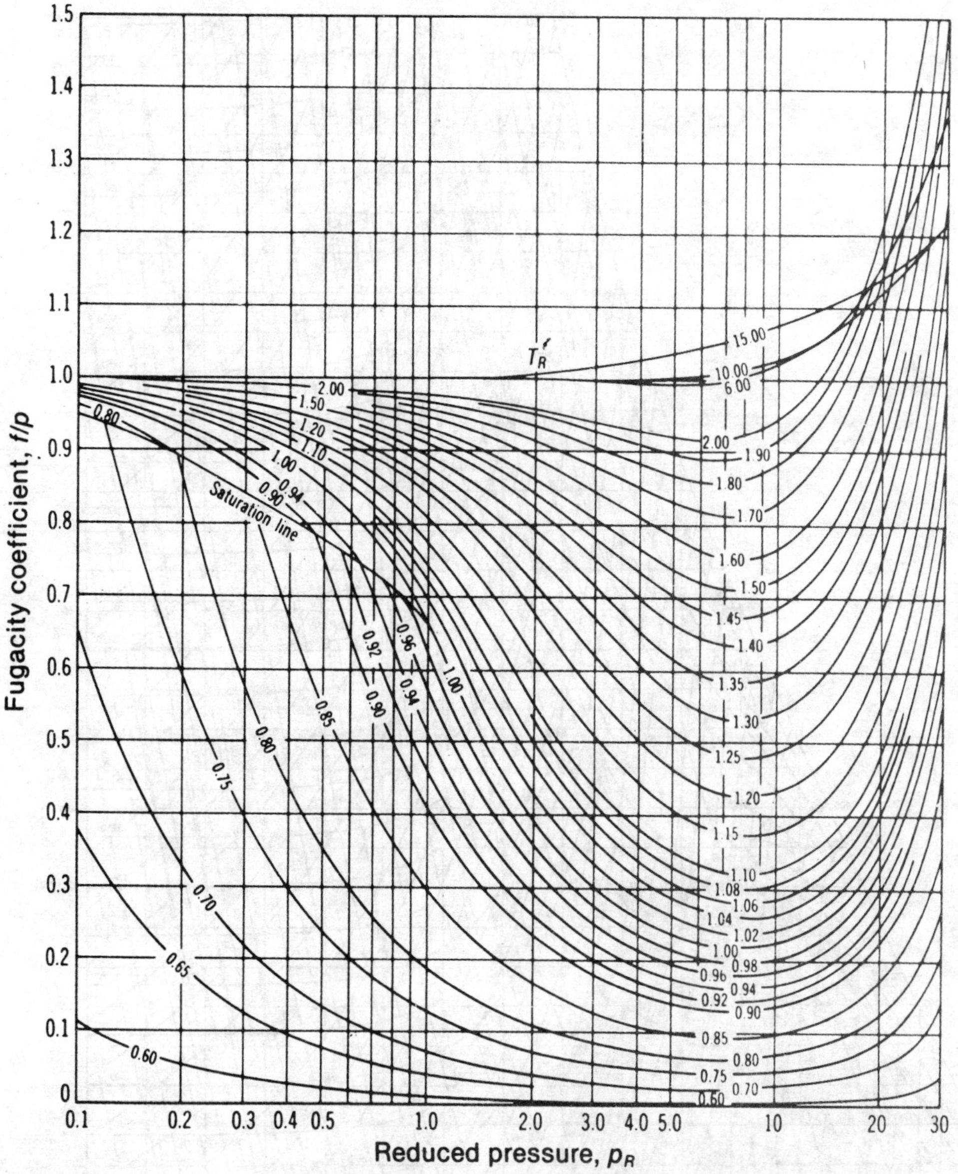

Figure A-6 Generalized fugacity coefficient chart. *Source:* G. J. Van Wylen and R. E. Sonntag, *Fundamentals of Classical Thermodynamics*, 3rd. ed., English/SI, Wiley, New York, 1986.

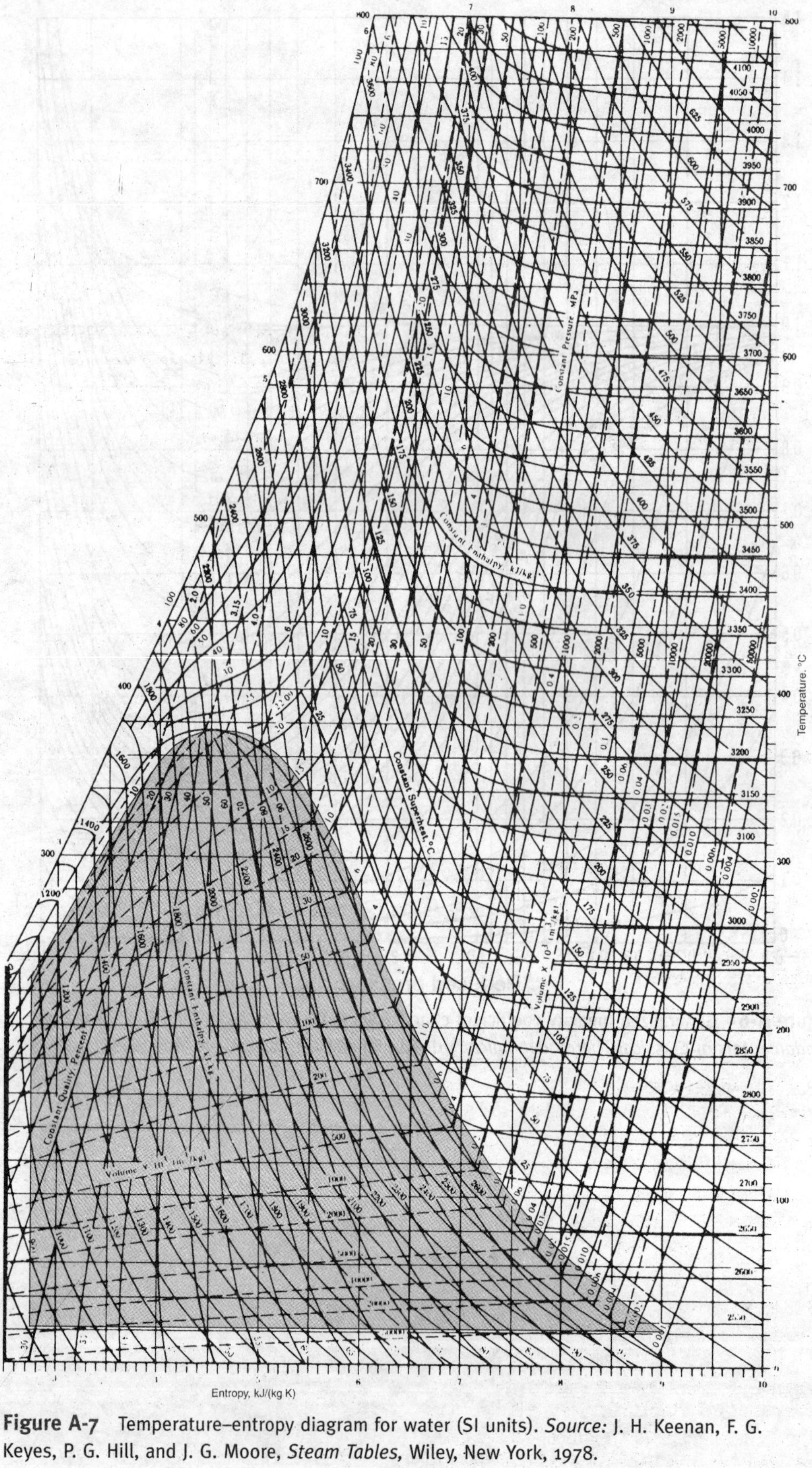

Figure A-7 Temperature–entropy diagram for water (SI units). *Source:* J. H. Keenan, F. G. Keyes, P. G. Hill, and J. G. Moore, *Steam Tables*, Wiley, New York, 1978.

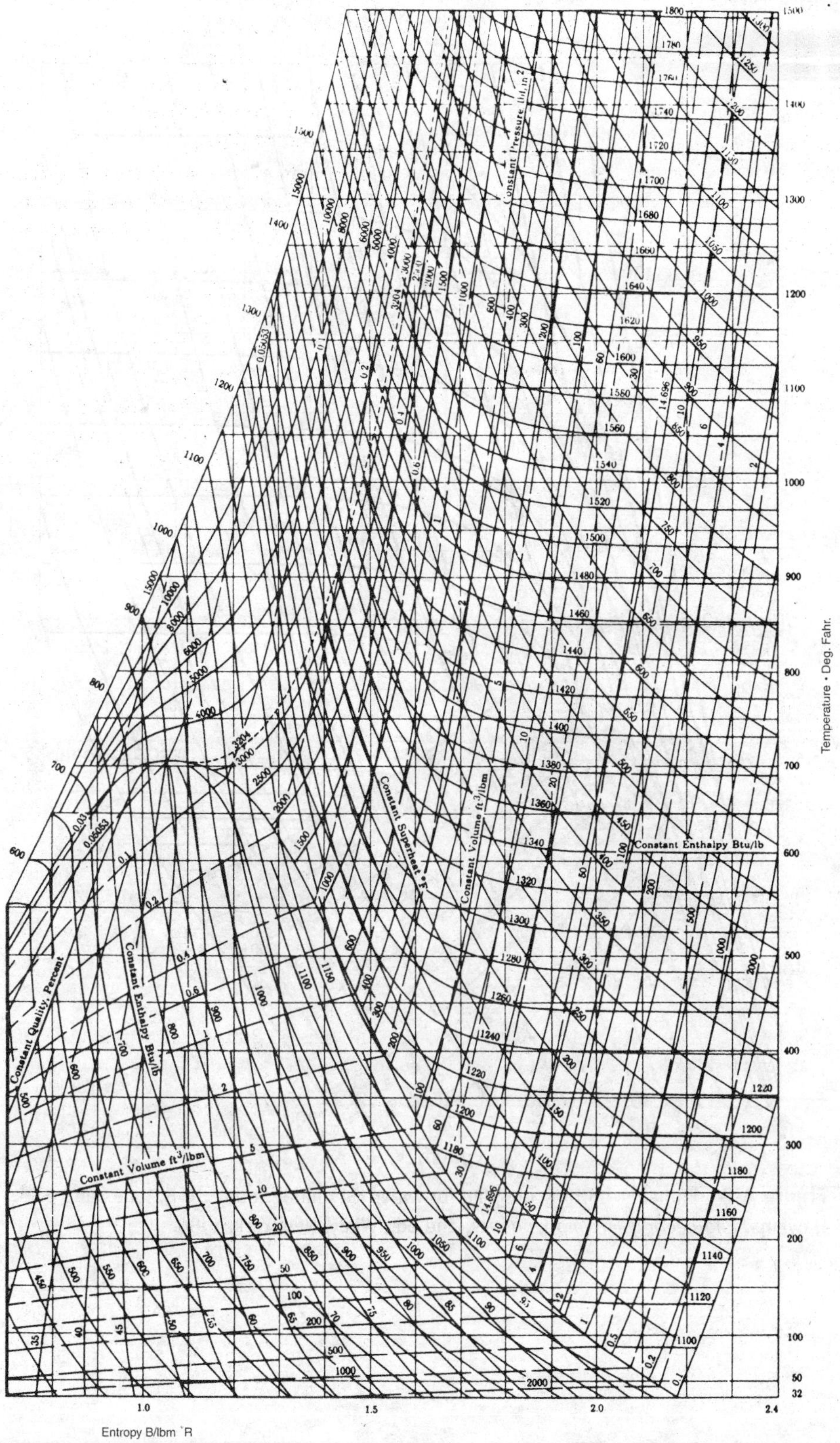

Figure A-7E Temperature–entropy diagram for water (English units). *Source:* J. H. Keenan, F. G. Keyes, P. G. Hill, and J. G. Moore, *Steam Tables,* Wiley, New York, 1969.

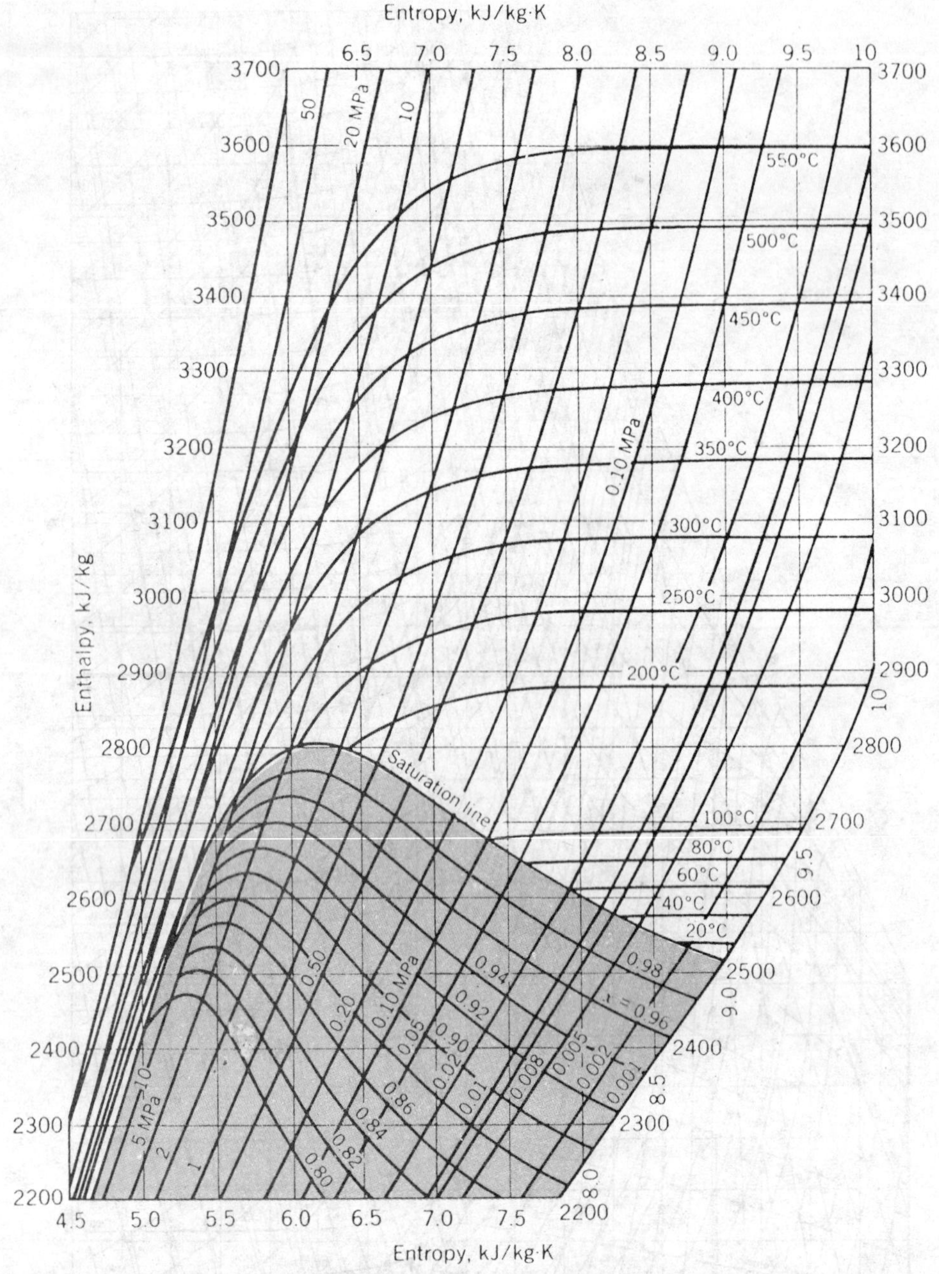

Figure A-8 Enthalpy–entropy diagram for water (SI units). *Source:* J. B. Jones and G. A. Hawkins, *Engineering Thermodynamics*, 2nd ed., Wiley, New York, 1986.

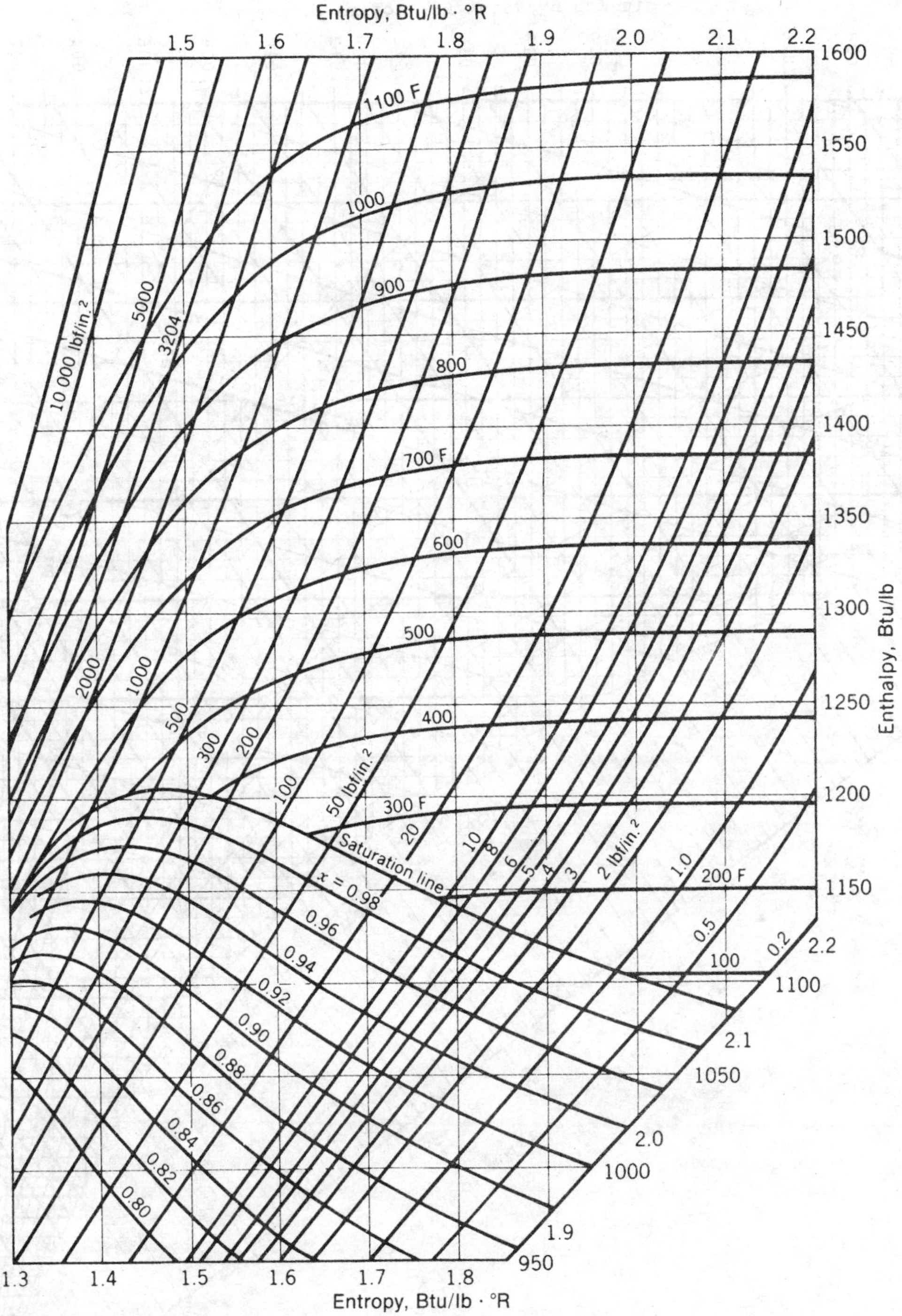

Figure A-8E Enthalpy–entropy diagram for water (English units). *Source:* J. B. Jones and G. A. Hawkins, *Engineering Thermodynamics,* 2nd ed., Wiley, New York, 1986.

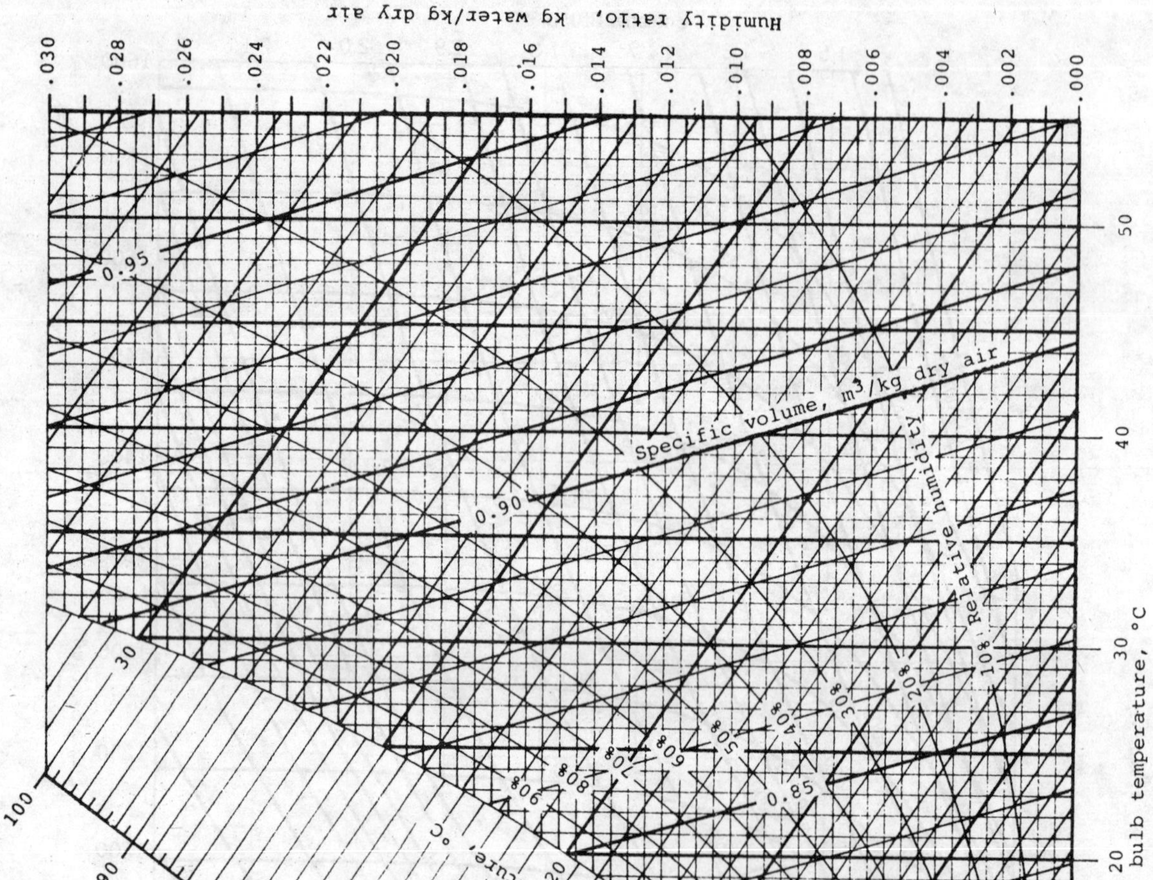

Figure A-9 Psychrometric chart for 1 atm (SI units). *Source:* Z. Zhang and M. B. Pate, "A Methodology for Implementing a Psychrometric Chart in a Computer Graphics System," *ASHRAE Transactions,* Vol. 94, Pt. 1, 1988.

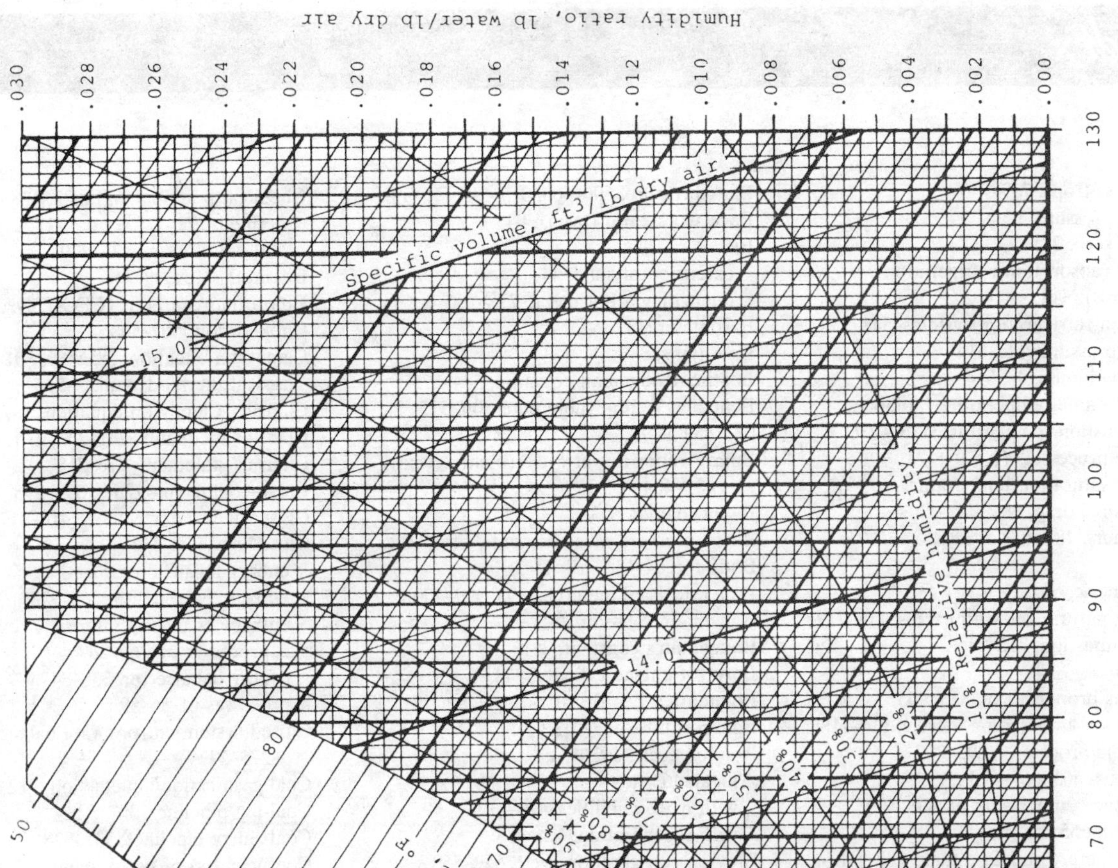

Figure A-9E Psychrometric chart for 1 atm (English units). *Source:* Z. Zhang and M. B. Pate, "A Methodology for Implementing a Psychrometric Chart in a Computer Graphics System," *ASHRAE Transactions*, Vol. 94, Pt. 1, 1988.

Index

Symbols

a	acceleration, activity
A	area
AF	air–fuel ratio
bwr	back work ratio
c	specific heat of an incompressible substance, velocity of sound
c	unit cost
$\dot{C}$	cost rate
C_aH_b	hydrocarbon fuel
c_p	specific heat at constant pressure, $\partial h/\partial T)_p$
c_v	specific heat at constant volume, $\partial u/\partial T)_v$
c_{p0}	specific heat c_p at zero pressure
e, E	energy per unit of mass, energy
e, E	exergy per unit of mass, exergy
e_f, $\dot{E}_f$	specific flow exergy, flow exergy rate
E_d, $\dot{E}_d$	exergy destruction, exergy destruction rate
$\dot{E}_q$	rate of exergy transfer accompanying heat transfer
E	electric field strength
$\mathscr{E}$	electrical potential, electromotive force (emf)
f	fugacity
$\bar{f}_i$	fugacity of component i in a mixture
F	degrees of freedom in the phase rule
F, F	force
FA	fuel–air ratio
g	acceleration of gravity
g, G	Gibbs function per unit of mass, Gibbs function
$\bar{g}_f^\circ$	Gibbs function of formation per mole at standard state
h, H	enthalpy per unit of mass, enthalpy
h	heat transfer coefficient
H	magnetic field strength
$\bar{h}_f^\circ$	enthalpy of formation per mole at standard state
$\bar{h}_{RP}$	enthalpy of combustion per mole
HHV	higher heating value
i	electric current
k	specific heat ratio: c_p/c_v
k	Boltzmann constant
K	equilibrium constant
ke, KE	kinetic energy per unit of mass, kinetic energy
l, L	length
LHV	lower heating value
m	mass
$\dot{m}$	mass flow rate
M	molecular weight, Mach number
M	magnetic dipole moment per unit volume
mep	mean effective pressure
mf	mass fraction
n	number of moles, polytropic exponent
N	number of components in the phase rule
p	pressure
p_{atm}	atmospheric pressure
p_i	pressure associated with mixture component i, partial pressure of i
p_r	relative pressure as used in Tables A-22
p_R	reduced pressure: p/p_c
P	number of phases in the phase rule
P	electric dipole moment per unit volume
pe, PE	potential energy per unit of mass, potential energy
$\dot{q}$	heat flux
Q	heat transfer
$\dot{Q}$	heat transfer rate
$\dot{Q}_x$	conduction rate
$\dot{Q}_c$, $\dot{Q}_e$	convection rate, thermal radiation rate
r	compression ratio
r_c	cutoff ratio
R	gas constant: $\bar{R}/M$, resultant force, electric resistance
$\bar{R}$	universal gas constant
s, S	entropy per unit of mass, entropy
s°	entropy function as used in Tables A-22, absolute entropy at the standard reference pressure as used in Table A-23
t	time
T	temperature
T_R	reduced temperature: T/T_c
$\mathscr{T}$	torque
u, U	internal energy per unit of mass, internal energy
v, V	specific volume, volume
V, V	Velocity
v_r	relative volume as used in Tables A-22
v_R'	pseudoreduced specific volume: $\bar{v}/(\bar{R}T_c/p_c)$
V_i	volume associated with mixture component i, partial volume of i
W	work
$\dot{W}$	rate of work, or power
x	quality, position
X	extensive property
y	mole fraction, mass flow rate ratio
z	elevation, position
Z	compressibility factor, electric charge
$\dot{Z}$	cost rate of owning/operating

Greek Letters

α	isentropic compressibility
β	coefficient of performance for a refrigerator, volume expansivity
γ	coefficient of performance for a heat pump, activity coefficient
Δ	change = final minus initial
ε	exergetic (second law) efficiency, emissivity, extent of reaction
η	efficiency, effectiveness
θ	temperature
κ	thermal conductivity, isothermal compressibility
μ	chemical potential
μ_J	Joule–Thomson coefficient
ν	stoichiometric coefficient
ρ	density
$\sigma, \dot{\sigma}$	entropy production, rate of entropy production
σ	normal stress, Stefan–Boltzmann constant
Σ	summation
τ	surface tension
ϕ	relative humidity
ψ, Ψ	Helmholtz function per unit of mass, Helmholtz function
ω	humidity ratio (specific humidity), angular velocity

Subscripts

a	dry air
ad	adiabatic
as	adiabatic saturation
ave	average
b	boundary
c	property at the critical point, compressor, overall system
cv	control volume
cw	cooling water
C	cold reservoir, low temperature
db	dry bulb
e	state of a substance exiting a control volume
e	exergy reference environment
f	property of saturated liquid, temperature of surroundings, final value
F	fuel
fg	difference in property for saturated vapor and saturated liquid
g	property of saturated vapor
H	hot reservoir, high temperature
i	state of a substance entering a control volume, mixture component
i	initial value, property of saturated solid
I	irreversible
ig, if	difference in property for saturated vapor (saturated liquid) and saturated solid
isol	isolated
int rev	internally reversible
j	portion of the boundary, number of components present in a mixture
n	normal component
p	pump
ref	reference value or state
reg	regenerator
res	reservoir
P	products
R	reversible, reactants
s	isentropic
sat	saturated
surr	surroundings
t	turbine
tp	triple point
0	property at the dead state, property of the surroundings
o	stagnation property
v	vapor
w	water
wb	wet bulb
x	upstream of a normal shock
y	downstream of a normal shock
1,2,3	different states of a system, different locations in space

Superscripts

ch	chemical exergy
e	component of the exergy reference environment
–	bar over symbol denotes property on a molar basis (over X, V, H, S, U, Ψ, G, the bar denotes partial molal property)
.	dot over symbol denotes time rate
°	property at standard state or standard pressure
*	ideal gas, quantity corresponding to sonic velocity